MODERN
DICTIONARY
of
ELECTRONICS

SEVENTH EDITION

REVISED AND UPDATED

Rudolf F. Graf is an author whose name is well-known to engineers, technicians, and hobbyists around the world. He graduated as an electronics engineer from Brooklyn Polytechnic Institute and did his graduate work New York University. Mr. Graf has been active in the electronics industry for more than fifty years in capacities ranging from design and consulting engineer, chief engineer, chief instructor at electronics and television schools, and consulting editor. He also held various sales and marketing positions. Mr. Graf is the author or co-author of more than 150 technical articles published by major magazines. He has written about 50 books on electricity and electronics, with more than 2 million copies in print, including the best-selling *Video Scrambling & Descrambling for Satellite & Cable TV, Second Edition* and the *Circuits* series of books, both published by Newnes. A number of his books have been translated into several European languages as well as Chinese, Japanese, and Russian.

MODERN
DICTIONARY
of
ELECTRONICS

SEVENTH EDITION

REVISED AND UPDATED

Rudolf F. Graf

Newnes

Boston   Oxford   Auckland   Johannesburg   Melbourne   New Delhi

Newnes is an imprint of Butterworth-Heinemann.

Copyright © 1999 by Rudolf F. Graf

 A member of the Reed Elsevier Group.

 Recognizing the importance of preserving what has been written, Butterworth-Heinemann prints its books on acid-free paper whenever possible.

 Butterworth-Heinemann supports the efforts of American Forests and the Global ReLeaf program in its campaign for the betterment of trees, forests, and our environment.

Library of Congress Cataloging-in-Publication Data

Graf, Rudolf F.
 Modern dictionary of electronics / Rudolf F. Graf. — 7th ed.,
revised and updated.
  p. cm.
 ISBN 0-7506-9866-7 (alk. paper)
 1. Electronics — Dictionaries. I. Title
TK7804.G67 1999
621.381'03 — dc21   99-17889
                                     CIP

British Library Cataloguing-in-Publication Data
A catalogue record for this book is available from the British Library.

The publisher offers special discounts on bulk orders of this book.
For information, please contact:
Manager of Special Sales
Butterworth-Heinemann
225 Wildwood Avenue
Woburn, MA 01801-2041
Tel: 781-904-2500
Fax: 781-904-2620

For information on all Butterworth-Heinemann publications available, contact
our World Wide Web home page at: http://www.bh.com

10 9 8 7 6 5 4 3 2 1

Typeset by Laser Words, Madras, India
Printed in the United States of America

It gives me great pleasure to dedicate this edition to Allison, Sheryl, Daniel, David, Russell and Scott, the loveliest children this side of heaven.

Preface

When the first edition of this dictionary was published in 1961, today's everyday items like color TVs, VCRs, CD players, computers, FAX machines, ATMs, cordless and cell phones, pagers, tape recorders, digital watches, pocket calculators, lasers, and many others too numerous to mention, were non-existent or mere laboratory curiosities. Since then, electronics has undergone significant changes based primarily on the meteoric expansion of integrated circuits and their apparently limitless applications. Vacuum tubes were replaced by semiconductors, and numerous technologies like ferrite core or bubble memories were relegated to the electronics graveyard. No other industry has ever grown so much and matured so fast, paced by technological advances that occur at a feverish pace. The first edition of this dictionary contained about 10,000 definitions of then current terms. And now, a scant 38 years later, this seventh edition contains approximately 25,000 terms—a clear indication of the phenomenal growth of our industry.

As technologies evolve and fresh products and concepts are introduced, suitable terminology must be developed to be able to communicate. The originators of the new words give them their initial meaning, but their exact definitions change with technological advances and through actual use by others. The contents of this dictionary is thus an analysis of words and their meanings as determined by common usage, written in a modern and popular style to provide clear and concise explanations of each entry. Continual updating of a work such as this is vital, so that those involved in the world of electronics have the power to communicate with those about them and to grasp new concepts as they emerge.

All entries are allowed as much space as is necessary for complete and meaningful definitions. Terms are explained clearly and precisely without excessive technical jargon. Original entries from the previous edition have been reviewed and many were revised to keep pace with current usage. Where more than one definition exists for a term, they are arranged numerically. This method, however, does not necessarily imply a preferred order of meanings. Important words from formative technologies that are no longer in use are retained in this edition for their historical interest.

My thanks go out to Ms. Tara Troxler Thomas and to Charles Thomas whose dedication to this project and skill at the word processor made it possible to deliver the manuscript for this work to the publisher in a timely fashion.

Industry and technical sources—notably the IEEE and the ASA—generously aided in defining many terms during the preparation of earlier editions of this work.

While this volume is as up-to-date as possible at the time of writing, the field of electronics is expanding so rapidly that new terms are constantly being developed and older terms take on broader or more specialized meanings. It is the intention of the publisher to periodically issue revised editions of this dictionary; thus suggestions for new terms and definitions are always welcome.

Rudolf F. Graf
February 1999

A

A — 1. Abbreviation for angstrom unit, used in expressing wavelength of light. Its length is 10^{-8} centimeter. 2. Chemical symbol for argon, an inert gas used in some electron tubes. 3. Letter symbol for area of a plane surface. 4. Letter symbol for ampere.

a — Letter symbol for atto- (10^{-18}).

A0 — The Federal Communications Commission (FCC) designation for radio emission consisting solely of an unmodulated carrier.

A1 — The FCC designation for radio emission consisting of a continuous-wave carrier keyed by telegraphy.

A-1 or **A.1** — The atomic time scale maintained by the U.S. Naval Observatory; presently it is based on weighted averages of frequencies from cesium-beam devices operated at a number of laboratories.

A2 — The FCC designation for radio emission consisting of a tone-modulated continuous wave.

A3 — The FCC designation for radio emission consisting of amplitude-modulated speech transmission.

A4 — The FCC designation for radio emission consisting of amplitude-modulated facsimile signals.

A5 — The FCC designation for radio emission consisting of amplitude-modulated television video signals.

A− **(A-minus** or **A-negative)** — Sometimes called F−. Negative terminal of an A battery or negative polarity of other sources of filament voltage. Denotes the terminal to which the negative side of the filament-voltage source should be connected.

A+ **(A-plus** or **A-positive)** — Sometimes called F+. Positive terminal of an A battery or positive polarity of other sources of filament voltage. The terminal to which the positive side of the filament voltage source should be connected.

ab- — The prefix attached to names of practical electric units to indicate the corresponding unit in the cgs (centimeter-gram-second) electromagnetic system, e.g., abampere, abvolt, abcoulomb.

abac — See alignment chart.

abampere — Centimeter-gram-second electromagnetic unit of current. The current that, when flowing through a wire 1 centimeter long bent into an arc with a radius of 1 centimeter, produces a magnetic field intensity of 1 oersted. One abampere is equal to 10 amperes.

A battery — Source of energy that heats the filaments of vacuum tubes in battery-operated equipment.

abbreviated dialing — 1. A system using special-grade circuits that require fewer than the usual number of dial pulses to connect two or more subscribers. 2. Ability of a phone system to require only two to four digits, while the network dials the balance of the seven to fourteen digits required.

abc — Also ABC. See automatic bass compensation.

abcoulomb — Centimeter-gram-second electromagnetic unit of electrical quantity. The quantity of electricity passing any point in an electrical circuit in 1 second when the current is 1 abampere. One abcoulomb is equal to 10 coulombs.

aberration — 1. In lenses, a defect that produces inexact focusing. Aberration may also occur in electron optical systems, causing a halo around the light spot. 2. In a cathode-ray tube, a defect in which the electron "lens" does not bring the electron beam to the same point of sharp focus at all points on the screen. 3. Failure of an optical lens to produce exact point-to-point correspondence between an object and its image. 4. Blurred focusing of light rays due to the difference in bending (refraction) imparted on different light frequencies (colors) as they pass through a lens.

abfarad — Centimeter-gram-second electromagnetic unit of capacitance. The capacitance of a capacitor when a charge of 1 abcoulomb produces a difference of potential of 1 abvolt between its plates. One abfarad is equal to 10^9 farads.

abhenry — Centimeter-gram-second electromagnetic unit of inductance. The inductance in a circuit in which an electromotive force of 1 abvolt is induced by a current changing at the rate of 1 abampere per second. One abhenry is equal to 10^{-9} henry.

abmho — Centimeter-gram-second electromagnetic unit of conductance. A conductor or circuit has a conductance of 1 abmho when a difference of potential of 1 abvolt between its terminals will cause a current of 1 abampere to flow through the conductor. One abmho is equal to 10^9 mho. Preferred term: absiemens.

abnormal glow — In a glow tube, a current discharge of such magnitude that the cathode area is entirely surrounded by a glow. A further increase in current results in a rise in its density and a drop in voltage.

abnormal propagation — The phenomenon of unstable or changing atmospheric and/or ionospheric conditions acting on transmitted radio waves. Such waves are prevented from following their normal path through space, causing difficulties and disruptions of communications.

abnormal reflections — See sporadic reflections.

abnormal termination — The shutdown of a computer program run or other process by the detection of an error by the associated hardware, indicating that some ongoing series of actions cannot be executed correctly.

abohm — Centimeter-gram-second electromagnetic unit of resistance. The resistance of a conductor when, with an unvarying current of 1 abampere flowing through it, the potential difference between the ends of the conductor is 1 abvolt. One abohm is equal to 10^{-9} ohm.

abort — 1. To cut short or break off (an action, operation, or procedure) with an aircraft, guided missile, or the like, especially because of equipment failure. An abort may occur at any point from start of countdown or takeoff to the destination. An abort can be caused by human technical or meteorological errors, miscalculation, or malfunctions. 2. The process of halting a computer program in an orderly fashion and returning control to the operator or operating system. 3. Abnormal termination of a computer program, caused by hardware or software malfunction or operator cancellation.

AB power pack — Assembly in a single unit of the A and B batteries of a battery-operated circuit. Also, a unit that supplies the necessary A and B voltages from an ac source of power.

abrasion machine — A laboratory device for determining the abrasive resistance of wire or cable. The two standard types of machines are the squirrel cage with square steel bars and the abrasive grit types.

abrasion resistance — A measure of the ability of a wire or wire covering to resist damage due to mechanical causes. Usually expressed as inches of abrasive tape travel.

abrasion soldering — Soldering difficult metals by abrading the surface oxide film beneath a pool of molten solder.

abrasive trimming — Trimming a ceramic capacitor or a film resistor to its nominal value by notching the surface with a finely adjusted stream of abrasive material such as aluminum oxide.

abscissa — Horizontal, or x, axis on a chart or graph.

absence-of-ground searching selector — In dial telephone systems, an automatic switch that rotates, or rises vertically and rotates, in search of an ungrounded contact.

absolute accuracy — 1. The tolerance of the full-scale set point referred to as the absolute voltage standard. 2. Parameter for a d/a converter. It is the overall accuracy of the converter, in which all levels are compared with an absolute standard. Absolute accuracy includes the combination of all nonlinearity and end-point errors.

absolute address — 1. An address used to specify the location in storage of a word in a computer program, not its position in the program. 2. A binary number assigned permanently as the address of a storage location in a computer. 3. A fixed location in the memory of the CPU, as opposed to a relative address, which is specified according to its distance from another location.

absolute altimeter — 1. Electronic instrument that furnishes altitude data with regard to the surface of the earth or any other surface immediately below the instrument, as distinguished from an aneroid altimeter, the readings of which depend on air pressure. 2. An altimeter that employs transmitted and reflected radio waves for its operation and thus does not depend on barometric pressure for its altitude indication.

absolute code — A code using absolute addresses and absolute operation codes; that is, a code that indicates the exact location where the reference operand is to be found or stored.

absolute coding — Coding written in machine language. It can be understood by the computer without processing.

absolute delay — The time interval between the transmission of two synchronized radio, loran, or radar signals from the same or different stations.

absolute digital position transducer — A digital position transducer, the output signal of which is indicative of absolute position. Also called encoder.

absolute efficiency — Ratio of the actual output of a transducer to that of a corresponding ideal transducer under similar conditions.

absolute error — 1. The amounts of error expressed in the same units as the quantity containing the error.

2. Loosely, the absolute value of the error, that is, the magnitude of the error without regard to its algebraic sign.

absolute gain of an antenna — The gain in a given direction when the reference antenna is an isotropic antenna isolated in space.

absolute instruction — A computer instruction that explicitly states, and causes the execution of, a specific operation.

absolute language — The language in which instructions must be given to the computer. The absolute language is determined when the computer is designed. Synonyms: machine language, machine code.

absolute loader — Program to load a computer program at specified numerical addresses.

absolute maximum rating — Limiting values of operating and environmental conditions, applicable to any electron device of a specified type as defined by its published data and not to be exceeded under the worst probable conditions. Those ratings beyond which the life and reliability of a device can be expected to decline.

absolute maximum supply voltage — The maximum supply voltage that may be applied without the danger of causing a permanent change in the characteristics of a circuit.

absolute minimum resistance — The resistance between the wiper and the termination of a potentiometer, when the wiper is adjusted to minimize that resistance.

absolute Peltier coefficient — The product of the absolute temperature and the absolute Seebeck coefficient of a material.

absolute power — Power level expressed in absolute units (e.g., watts or dBm).

absolute pressure transducer — 1. A pressure transducer that accepts two independent pressure sources simultaneously, and the output of which is proportional to the pressure difference between the sources. 2. A transducer that senses a range of pressures, which are referenced to a fixed pressure. The fixed pressure is normally total vacuum.

absolute scale — See Kelvin scale.

absolute Seebeck coefficient — The integral from absolute zero to the given temperature of the quotient of the Thomson coefficient of a material divided by its absolute temperature.

absolute spectral response — Output or response of a device, in terms of absolute power levels, as a function of wavelength.

absolute system of units — Also called coherent system of units. A system of units in which a small number of units is chosen as fundamental, e.g., units of mass, length, time, and charge. Such units are termed absolute units. All other units are derived from them by taking a definite proportional factor in each of those laws chosen as the basic laws for expressing the relationships between the physical quantities. The proportional factor is generally taken as unity.

absolute temperature — Temperature measured from absolute zero, a theoretical temperature level variously defined as $-273.2°C$, $-459.7°F$, or 0 K.

absolute temperature scale — Thermodynamic temperature scale, named for Lord Kelvin (1848), in which temperatures are given in kelvins (K). (In the SI system the degree sign and the word *degree* are not used for Kelvin temperatures.) The absolute zero of temperature is 0 K, $-273.2°C$, or $-459.7°F$. The kelvin is the same size as the Celsius degree.

absolute tolerance — Also called accuracy. The maximum deviation from the nominal resistance (or capacitance) value, usually given as a percentage of the nominal value.

absolute units — A system of units based on physical principles, in which a small number of units are chosen as fundamental and all other units are derived from them; e.g., abohm, abcoulomb, abhenry, etc.

absolute value — The numerical value of a number or symbol without reference to its algebraic sign. Thus, 3 is the absolute value of $|3|$ or $|-3|$. An absolute value is signified by placing vertical lines around the number or symbol.

absolute value device — A computing element that produces an output equal to the magnitude of the input signal, but always of one polarity.

absolute zero — Lowest possible point on the scale of absolute temperature; the point at which all molecular activity ceases. Absolute zero is variously defined as $-273.2°C$, $-459.7°F$, or $0 K$.

absorbed wave — A radio wave that becomes lost in the ionosphere due to molecular agitation and the accompanying energy loss it undergoes there. Absorption is most pronounced at low frequencies.

absorber — 1. In a nuclear reactor, a substance that absorbs neutrons without reproducing them. Such a substance may be useful in control of a reactor or, if unavoidably present, may impair the neutron economy. 2. Any material or device that absorbs and dissipates radiated energy. 3. In microwave terminology, a material or device that takes up and dissipates radiated energy. It may be used for shielding, to prevent reflection, or to transmit one or more radiation components selectively.

absorption — 1. Dissipation of the energy of a radio or sound wave into other forms as a result of its interaction with matter. 2. The process by which the number of particles or photons entering a body of matter is reduced by interaction of the particle or radiation with matter. Similarly, the reduction of the energy of a particle while traversing a body of matter. This term is sometimes erroneously used for capture. 3. Penetration of a substance into the body of another. 4. Conversion of radiant energy into other forms by passage through, or reflection from, matter. 5. The adhesion of a fluid in extremely thin layers to the surfaces of a solid. 6. Reduction in strength of an electromagnetic wave propagating through a medium, determined by dielectric properties of the material.

absorption attenuation — Loss in an optical fiber due primarily to impurities, including metals, such as cobalt, chromium, and iron, as well as OH ions.

absorption circuit — A tuned circuit that dissipates energy taken from another circuit or from a signal source. This effect is especially evident in a resonant circuit such as a wavemeter or wave trap.

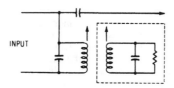

Absorption circuit.

absorption coefficient — 1. Measure of sound-absorbing characteristics of a unit area of a given material compared with the sound-absorbing characteristics of an open space (total absorption) having the same area. 2. Ratio of loss of intensity caused by absorption to the total original intensity of radiation.

absorption current — The current flowing into a capacitor following its initial charge, due to a gradual

penetration of the electric stress into the dielectric. Also, the current that flows out of a capacitor following its initial discharge.

absorption dynamometer — An instrument for measuring power, in which the energy of a revolving wheel or shaft is absorbed by the friction of a brake.

absorption fading — A slow type of fading, primarily caused by variations in the absorption rate along the radio path.

absorption frequency meter — *See* absorption wavemeter.

absorption loss — 1. That part of transmission loss due to dissipation or conversion of electrical energy into other forms (e.g., heat), either within the medium or attendant upon a reflection. 2. The loss of optical flux or energy caused by impurities in the transmission medium as well as intrinsic material absorption. Expressed in decibels per kilometer.

absorption marker — 1. A sharp dip on a frequency-response curve due to the absorption of energy by a circuit sharply tuned to the frequency at which the dip occurs. 2. A small pip or blank introduced on an oscilloscope trace to indicate a frequency point. It is so called because it is produced by a frequency-calibrated tuned trap similar to an absorption wavemeter.

absorption modulation — Also called loss modulation. A system for amplitude-modulating the output of a radio transmitter by means of a variable-impedance device (such as a microphone semiconductor or vacuum-tube circuit) inserted into or coupled to the output circuit.

absorption trap — A parallel-tuned circuit coupled either magnetically or capacitively to absorb and attenuate interfering signals.

absorption wavemeter — Also called absorption frequency meter. An instrument for measuring frequency. Its operation depends on the use of a tuned electrical circuit or cavity loosely coupled inductively to the source. Maximum energy will be absorbed at the resonant frequency, as indicated by a meter or other device. Frequency can then be determined by reference to a calibrated dial or chart.

absorptivity — A measure of the portion of incident radiation or sound energy absorbed by a material.

abstraction — A simplified description or specification of a system that emphasizes some of the system's details or properties while suppressing others. A high level of abstraction or a highly abstract machine is one in which very few machine details are apparent to a programmer, who sees only a broad set of machine concepts. Abstract machines are created by surrounding a primitive machine with layers of operating systems. To converse with higher levels of abstract machines, the user needs higher levels of languages.

A/B switch — A switch that selects one of two inputs (A or B) for routing to a common output while providing adequate isolating between the two signals.

A-B test — 1. Direct comparison of two sounds by playing first one and then the other. May be done with two tape recorders playing identical tapes (or the same tape), two speakers playing alternately from the same tape recorder, or two amplifiers playing alternately through one speaker, etc. 2. An audio comparison test for evaluating the relative performance of two or more components or systems by quickly changing from one to the other. The left- and right-hand channels or the record and replay sound signals are often designated A and B. A and B test facilities are installed at most high-fidelity dealers.

abvolt — Centimeter-gram-second electromagnetic unit of potential difference. The potential difference between two points when 1 erg of work is required to transfer 1 abcoulomb of positive electricity from a lower to a higher potential. An abvolt is equal to 10^{-8} volt.

ac — Abbreviation for alternating current.

ac bias — The alternating current, usually of a frequency several times higher than the highest signal frequency, that is fed to a record head in addition to the signal current. The ac bias serves to linearize the recording process.

accelerated aging — A test in which certain parameters, such as voltage and temperature, are increased above normal operating values to obtain observable deterioration in a relatively short period. The plotted results give expected service life under normal conditions. Also called accelerated life test.

accelerated graphics port — Abbreviated AGP. A slot inside PCs for high speed video to be used instead of the standard slot type, called PCI.

accelerated life test — Test conditions used to bring about, in a short time, the deteriorating effect obtained under normal service conditions.

accelerated service test — A service or bench test in which some service condition, such as speed, temperature, or continuity of operation, is exaggerated to obtain a result in a shorter time than that which elapses in normal service.

accelerating conductor or relay — A conductor or relay that causes the operation of a succeeding device to begin in the starting sequence after the proper conditions have been established.

accelerating electrode — An electrode in a cathode-ray or other electronic tube to which a positive potential is applied to increase the velocity of electrons or ions toward the anode. A klystron tube does not have an anode but does have accelerating electrodes.

accelerating time — The time required for a motor to reach full speed from a standstill (zero speed) position.

accelerating voltage — A high positive voltage applied to the accelerating electrode of a cathode-ray tube to increase the velocity of electrons in the beam.

acceleration — 1. The rate of change in velocity. Often expressed as a multiple of the acceleration of gravity ($g = 32.2$ ft/s^2). 2. The rate of change in velocity of a stepping motor measured in rad/s; it is the result of rotor torque divided by rotor and load inertia. 3. A vector quantity that specifies rate of change of velocity.

acceleration at stall — The value of servomotor angular acceleration calculated from the stall torque of the motor and the moment of inertia of the rotor. Also called torque-to-inertia ratio.

acceleration time — In a computer, the elapsed time between the interpretation of instructions to read or write on tape and the possibility of information transfer from the tape to the internal storage, or vice versa.

acceleration torque — Numerical difference between motor torque produced and load torque demanded at any given speed during the acceleration period. It is this net torque that is available to change the speed of the driven load.

acceleration voltage — Potential between a cathode and anode or other accelerating element in a vacuum tube. Its value determines the average velocity of the electrons.

accelerator — A device for imparting a very high velocity to charged particles such as electrons or protons. Fast-moving particles of this type are used in research or in studying the structure of the atom itself. 2. A circuit that speeds up a computer or monitor. Typically a circuit card with an extra processing chip and/or additional RAM.

accelerator board — An adapter with a microprocessor that makes a computer run faster.

accelerator dynamic test — A test performed on an accelerometer by means of which information is gathered pertaining to the overall behavior frequency response and/or natural frequency of the device.

accelerometer — 1. An instrument or device, often mounted in an aircraft, guided missile, or the like, used to sense accelerative forces and convert them into corresponding electrical quantities, usually for measuring, indicating, or recording purposes. It does not measure velocity or distance, only changes in velocity. 2. A transducer that measures acceleration and/or gravitational forces capable of imparting acceleration. 3. A sensor whose electrical output is proportional to acceleration.

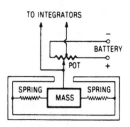

Accelerometer.

accentuation — Also called preemphasis. The emphasizing of any certain band of frequencies, to the exclusion of all others, in an amplifier or electronic device. Applied particularly to the higher audio frequencies in frequency-modulated (FM) transmitters.

accentuator — 1. Network or circuit used for preemphasis, that is, an increase in amplitude of a given band of usually audible frequencies. 2. A circuit or device, such as a filter, tone control, or equalizer, used to emphasize a band of frequencies, usually in the audio-frequency spectrum.

acceptable-environmental-range test — A test to determine the range of environmental conditions for which an apparatus maintains at least the minimum required reliability.

acceptable quality level — Abbreviated AQL. The maximum percentage of defective components considered to be acceptable as an average for a process or the lowest quality a supplier is permitted to present continually for acceptance. *Also see* AQL.

acceptance angle — 1. The solid angle within which all incident light rays will enter the core of an optical fiber. Expressed in degrees. 2. In fiber optics, a measure of the maximum angle within which light may be coupled from a source or emitter. It is measured relative to the fiber's axis. 3. The critical angle, measured from the core centerline, above which light will not enter an optical fiber. It is equal to the half-angle of the acceptance cone. 4. The maximum angle within which light will be accepted by an element, such as a detector.

acceptance cone — 1. A parameter that defines acceptable light-launching angles. Only light launched at angles within this cone will be waveguided (fiber optics). 2. A cone with an included angle twice that of the acceptance angle.

acceptance pattern — In fiber optics, a curve of total transmitted power plotted against the launch angle.

acceptance sampling plan — A plan for the inspection of samples as a basis for acceptance or rejection of a lot.

acceptor — Also called acceptor impurity. An impurity lacking sufficient valence electrons to complete the bonding arrangement in the crystal structure. When added to a semiconductor crystal, it accepts an electron from a neighboring atom and thus creates a hole in the lattice structure of the crystal, making a p-type semiconductor. 2. An impurity from column III of the periodic table, which adds a mobile hole to silicon, thereby making it more p-type and accepting of electrons. Boron is the primary acceptor used to dope silicon (compare with *donor*).

acceptor circuit — 1. A circuit that offers minimum opposition to a given signal. 2. A circuit tuned to respond to a single frequency.

acceptor impurity — *See* acceptor.

acceptor-type semiconductor — A p-type semiconductor.

access — 1. To gain access to a computer's memory location in which binary information is already stored or can be stored. 2. To open up a set of connections to allow reading from or writing into this location.

access arm — In a computer storage unit, a mechanical device that positions the reading and writing mechanism.

access code — 1. The preliminary digit or digits that a telephone user must dial to be connected to a particular outgoing trunk group. 2. A group of characters or numbers that identifies a user to a computer or any other secure system. 3. One or more numbers and/or symbols that are keyed into the repeater with a telephone tone pad to activate a repeater function, such as an autopatch.

access control — 1. The control of pedestrian and vehicular traffic through entrances and exits of a protected area or premises. 2. The process of limiting access to resources of a system to only authorized users, programs, processes, or other systems.

access grant — Multiprocessor system response that satisfies a previous service request.

access hole — A hole drilled through successive layers of a multilayer board to gain access to a land or pad location on one of the inside layers.

access method — 1. A data-management technique available for use in transferring data between the main storage and an input/output device. 2. A software component of a computer operating system that controls the flow of data between application programs and either local or remote peripheral devices.

access mode — 1. A technique used in COBOL to obtain a specific logic record from, or to place it into, a file assigned to a mass storage device. 2. The operation of an alarm system such that no alarm signal is given when the protected area is entered; however, a signal may be given if the sensor, annunciator, or control unit is tampered with or opened.

accessory card — An additional circuit card that can be mounted inside a personal computer and connected to the system bus.

access protocol — A defined set of procedures that function as an interface between a computer user and a network, enabling the user to employ the services of that network.

access provider — Telecommunications company that links businesses and individuals to the Internet using modem devices, high-speed ISDN lines, or dedicated links.

access time — Also called waiting time. 1. The time interval (called read time) between the instant of calling for data from a storage device and the instant of completion of delivery. 2. In a memory system, the time delay, at specified thresholds, from the presentation of an enable or address input pulse until the arrival of the memory data output. 3. The time required for a computer to move data between its memory section and its CPU. 4. A time interval that is characteristic of a storage device. Essentially, it is a measure of the time required to communicate with that device, or, more specifically, it is the time between the application of a specified input pulse (assuming that other necessary inputs are also present) and the availability of valid data signals at an output. The access time can be defined only with reference to an output signal. 5. The time required by a computer to begin delivering information after the memory or storage has been interrogated. 6. The time it takes a computer to retrieve a piece of information. With hard disks or compact discs, maximum access time is measured as the time it takes to move from one end of the disk to the other, find a piece of information, and transfer that information to RAM.

accidental jamming — Jamming caused by transmission from friendly equipment.

ac circuit breaker — A device that is used to close and interrupt an ac power circuit under normal conditions or to interrupt this circuit under faulty or emergency conditions.

accompanying audio (sound) channel — Also known as co-channel sound frequency. The rf carrier frequency that supplies the sound to accompany a television picture.

ac component — In a complex wave (i.e., one containing both ac and dc), the alternating, fluctuating, or pulsating member of the combination.

accordion — A type of contact used in some printed-circuit connectors. The contact spring is given a *z* shape to permit high deflection without excessive stress.

ac-coupled flip-flop — A flip-flop that changes state when triggered by the rise or fall of a clock pulse. There is a maximum allowable rise or fall time for proper triggering.

ac coupling — Coupling of one circuit to another circuit through a capacitor or other device that passes the varying portion but not the static (dc) characteristics of an electrical signal.

accumulation key — In a calculator, it automatically accumulates products and totals of successive calculations.

accumulator — 1. In an electronic computer, a device which stores a number and which, on receipt of another number, adds the two and stores the sum. An accumulator may have properties such as shifting, sensing signals, clearing, complementing, etc. 2. A chemical cell able to store electrical energy (British). Also called secondary cell. 3. The "scratch pad" section of the computer, in which arithmetic operations are carried out. 4. A register and related circuitry that hold an operand for arithmetic and logic operations. 5. A register or latch internal to the MPU where data is stored temporarily before being sent to another location internal or external to the MPU chip.

accuracy — 1. The maximum error in the measurement of a physical quantity in terms of the output of an instrument when referred to the individual instrument calibration. Usually given as a percentage of full scale. 2. The quality of freedom from mistake or error in an electronic computer, that is, of conformity to truth or to a rule. 3. The closeness with which a measured quantity approaches the true value of that quantity. (*See* true value.) 4. The degree to which a measured or calculated value conforms to the accepted standard or rule. 5. The measure of a meter's ability to indicate a value corresponding to the absolute value of electrical energy applied. Accuracy is expressed as a percentage of the meter's rated full-scale value. To be meaningful, accuracy specifications must always consider the effects of time, temperature, and humidity. 6. Confidence in the correlation between measurements in one location and another, or between a measurement and a recognized standard. 7. The correctness or certainty of position when the rotor of a stepping motor comes to rest. It is usually expressed as a percentage of the step angle, but can also be specified in degrees or minutes of arc. In steppers, the error is not cumulative, but occurs only at the completion of the last step. 8. The degree of freedom from error, that is, the degree of conformity to some standard. Accuracy is contrasted with *precision*. For example, four-place numbers are less precise than six-place numbers; however, a properly computed four-place number might be more accurate than an improperly computed six-place number. 9. As applied to an adc, the term describes the difference between the actual input voltage and the full-scale weighted equivalent of the binary code.

accuracy rating of an instrument — The limit, usually expressed as a percentage of full-scale value, not exceeded by errors when the instrument is used under reference conditions.

ACD — Abbreviation for automatic call distributor. A switching system that automatically distributes incoming calls to a centralized group of receivers in the sequence in which the calls are received. It holds calls until a receiver is available.

ac/dc — Electronic equipment capable of operation from either an ac or dc primary power source. Abbreviation for alternating current/direct current.

ac/dc receiver — A radio receiver designed to operate directly from either an ac or a dc source.

ac/dc ringing — A method of telephone ringing in which alternating current is used to operate a ringing device, and direct current is used to aid the action of a relay that stops the ringing when the called party answers.

ac directional overcurrent relay — A device that functions on a desired value of ac overcurrent flowing in a predetermined direction.

ac dump — The intentional, accidental, or conditional removal of all alternating-current power from a system or component. An ac dump usually results in the removal of all power, since direct current is usually supplied through a rectifier or converter.

ac erasing head — In magnetic recording, a device using alternating current to produce the magnetic field necessary for removal of previously recorded information.

acetate — A basic chemical compound in the mixture used to coat recording discs.

acetate base — The transparent plastic film that forms the tough backing for acetate magnetic recording tape.

acetate disc — A mechanical recording disc, either solid or laminated, made mostly from cellulose nitrate lacquer plus a lubricant.

acetate tape — A sound-recording tape with a smooth, transparent acetate backing. One side is coated with an oxide capable of being magnetized.

ac generator — 1. A rotating electrical machine that converts mechanical power into alternating current. Also known as an alternator. 2. A device, usually an oscillator, designed for the purpose of producing alternating current.

A channel — One of two stereo channels, usually the left.

achieved reliability — Reliability determined on the basis of actual performance of nominally identical items under equivalent environmental conditions. Also called operational reliability.

achromatic — 1. In color television, a term meaning a shade of gray from black to white, or the absence of color (without color). 2. Black-and-white television, as distinguished from color television. 3. Literally, color free. In an optical system, the term is used when chromatic aberration is corrected for at least two wavelengths. A

color that is defined as being achromatic is often referred to as gray. 4. Having no color; being a neutral such as black, white, or gray.

achromatic lens — A lens that has been corrected for chromatic aberration. Such a lens is capable of bringing all colors of light rays to approximately the same point of focus by combining a concave lens of flint glass with a convex lens of crown glass. A lens that transmits light without separating it into its constituent colors.

achromatic locus — Also called achromatic region. On a chromaticity diagram, an area that contains all points representing acceptable reference white standards.

achromatic region — *See* achromatic locus.

acicular — Needle-shaped; descriptive of the shape of the magnetizable particles composing the coating of a recording tape. Modern tapes are premagnetized during the coating process to line the "needles" up with the direction of the tape, thus providing maximum sensitivity from the oxide.

acid — A chemical compound that dissociates and forms hydrogen ions when in aqueous solution.

acid depolarizer — An acid, such as nitric acid, sometimes introduced into a primary cell to prevent polarization.

acid fluxes — Fluxes consisting of inorganic acids and salts, which are used when a surface to be soldered is below the ideal for rapid wetting. Also called corrosive fluxes.

acknowledge — A control signal used to complete a handshaking sequence in telecommunications. The acknowledge signal indicates that the information has been accepted by the receiving computer.

ac line — A power line delivering alternating current only.

ac line filter — A filter designed to dissipate or bypass to ground any extraneous signals or electrical noise on an ac power line, while causing virtually no reduction of the power-line voltage or power. Used to keep unwanted signals and noise out of sensitive equipment.

aclinic line — Also called isoclinic line. On a magnetic map, an imaginary line that connects points of equal magnetic inclination or dip.

ac magnetic biasing — In magnetic recording, the method used to remove random noise and/or previously recorded material from the wire or tape. This is done by introducing an alternating magnetic field at a substantially higher frequency than the highest frequency to be recorded.

ac noise — Noise that displays a rate of change that is fast relative to the response capability of the device.

ac noise immunity — A measure of a logic circuit's ability to maintain the prescribed logic state in the presence of such noise. It is defined in terms of the amplitude and pulse width of an input noise signal to which the element will not respond.

acorn tube — A button- or acorn-shaped vacuum tube with no base, designed for UHF applications. Electrodes are brought out through the glass envelope on the side, top, and bottom.

Acorn tube.

acoustic— Also acoustical. Pertaining to sound or the science of sound.

acoustic absorption loss— The energy lost by conversion into heat or other forms when sound passes through or is reflected by a medium.

acoustic absorptivity— The ratio of sound energy absorbed by a surface to the sound energy arriving at the surface. Equal to 1 minus the reflectivity of the surface.

acoustical— *See* acoustic.

acoustical attenuation constant— The real part of the acoustical propagation constant. The commonly used unit is the neper per section or per unit distance.

acoustical coupler— A device for connecting a telephone handset to a computer input port.

acoustical-electrical transducer— A device designed to transform sound energy into electrical energy and vice versa.

acoustical material— Any material considered in terms of its acoustical properties; especially, a material designed to absorb sound.

acoustical mode— A mode of crystal-lattice vibration that does not produce an oscillating dipole.

acoustical ohm— A measure of acoustic resistance, reactance, or impedance. One acoustical ohm is equal to a volume velocity of 1 cubic centimeter per second when produced by a sound pressure of 1 microbar.

acoustical phase constant— The imaginary part of the acoustical propagation constant. The commonly used unit is the radian per section or per unit distance.

acoustical reflectivity— *See* sound-reflection coefficient.

acoustical transmittivity— *See* sound-transmission coefficient.

acoustic burglar alarm— Also called acoustic intrusion detector. A burglar alarm that is responsive to sounds produced by an intruder. Concealed microphones connected to an audio amplifier trip an alarm when sounds within a predetermined range of frequencies exceed a predetermined normal level.

acoustic capacitance— In a sound medium, a measure of volume displacement per dyne per square centimeter. The unit is centimeter to the fifth power per dyne.

acoustic clarifier— A system of cones loosely attached to the baffle of a speaker and designed to vibrate and absorb energy during sudden loud sounds, thereby suppressing them.

acoustic compliance— 1. The measure of volume displacement of a sound medium when subjected to sound waves. 2. That type of acoustic reactance which corresponds to capacitive reactance in an electrical circuit.

acoustic coupler— 1. A device that converts digital signals into audio signals, enabling data to be transmitted over the telephone lines via a conventional telephone. 2. A modem device that connects a terminal or computer to the handset of a telephone.

acoustic coupling— Coupling resonator elements by mechanical means through the use of wires, rods, or nonelectroded sections of quartz or ceramic. The terms *acoustic* and *mechanical* can be used interchangeably.

acoustic delay line— A device that retards one or more signal vibrations by causing them to pass through a solid or liquid.

acoustic dispersion— The change of the speed of sound with frequency.

acoustic elasticity— 1. The compressibility of the air in a speaker enclosure as the cone moves backward. 2. The compressibility of any material through which sound is passed.

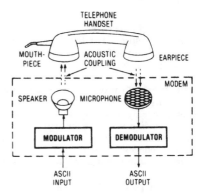

Acoustic coupler.

acoustic feedback— 1. Also called acoustic regeneration. The mechanical coupling of a portion of the sound waves from the output of an audio-amplifying system to a preceding part or input circuit (such as the microphone) of the system. When excessive, acoustic feedback will produce a howling sound in the speaker. 2. The pickup, by a turntable, of vibrations from the speaker. If these vibrations reach the cartridge, they will be reamplified, causing noise (usually a rumble, but in extreme cases a howl) and/or distortion. Also, feedback resulting from such sound waves setting some part of an amplifier circuit into vibration and thus modulating the currents in the circuit. Acoustic feedback usually causes howling or whistling.

acoustic filter— 1. A sound-absorbing device that selectively suppresses certain audio frequencies while allowing others to pass. 2. Any sound-absorbing or sound-transmitting arrangement, or combination of the two, that passes sound waves of desired frequency while attenuating or eliminating others.

acoustic frequency response— The voltage-attenuation frequency measured into a resistive load, producing a bandwidth approaching sufficiently close to the maximum.

acoustic generator— A transducer, such as a speaker, headphones, or a bell, that converts electrical, mechanical, or other forms of energy into sound.

acoustic homing system— 1. A system that uses a sound signal for guidance purposes. 2. A guidance method in which a missile homes in on noise generated by a target.

acoustic horn— Also called horn. 1. A tube of varying cross section having different terminal areas that change the acoustic impedance to control the directivity of the sound pattern. 2. A tapered tube (round or rectangular, but generally funnel shaped) that directs sound and, to some extent, amplifies it.

acoustic impedance— 1. Total opposition of a medium to sound waves. Equal to the force per unit area on the surface of the medium divided by the flux (volume velocity or linear velocity multiplied by area) through that surface. Expressed in ohms and equal to the mechanical impedance divided by the square of the surface area. One unit of acoustic impedance is equal to a volume velocity of 1 cubic centimeter per second produced by a pressure of 1 microbar. Acoustic impedance contains both acoustic resistance and acoustic reactance. 2. The degree of resistance to transmitting sound imparted by the characteristic elasticity of a given substance.

acoustic inertance— A type of acoustic reactance that corresponds to inductive reactance in an electrical circuit. (The resistance to movement or reactance offered by the sound medium because of the inertia of the effective mass of the medium.) Measured in acoustic ohms.

acoustic intensity — The limit approached by the quotient of acoustical power being transmitted at a given time through a given area divided by the area as the area approaches zero.

acoustic interferometer — An instrument for measuring the velocity or frequency of sound waves in a liquid or gas. This is done by observing the variations of sound pressure in a standing wave, established in the medium between a sound source and a reflector, as the reflector is moved or the frequency is varied.

acoustic intrusion detector — See acoustic burglar alarm.

acoustic labyrinth — A loudspeaker enclosure in which the rear of the loudspeaker is coupled to a tube which, at the resonant frequency of the loudspeaker, is one quarter of a wavelength long. The tube, folded upon itself in order to save space, gives the appearance of a labyrinth.

acoustic lens — 1. An array of obstacles that refract sound waves in the same way that an optical lens refracts light waves. The dimensions of these obstacles are small compared with the wavelengths of the sounds being focused. 2. A device that produces convergence or divergence of moving sound waves. When used with a loudspeaker, the acoustic lens widens the beam of the higher-frequency sound waves.

acoustic line — Mechanical equivalent of an electrical transmission line. Baffles, labyrinths, or resonators are placed at the rear of a speaker to help reproduce the very low audio frequencies.

acoustic memory — A computer memory using an acoustic delay line. The line employs a train of pulses in a medium such as mercury or quartz.

acoustic mine — Also called sonic mine. An underwater mine that is detonated by sound waves, such as those from a ship's propeller or engines.

acoustic mirage — The distortion of a sound wavefront by a large temperature gradient in air or water. This creates the illusion of two sound sources.

acoustic ohm — The unit of acoustic resistance, reactance, or impedance. One acoustic ohm is present when a sound pressure of 1 dyne per square centimeter produces a volume velocity of 1 cubic centimeter per second.

acoustic phase constant — The imaginary part of the acoustic propagation constant. The commonly used unit is the radian per section or per unit distance.

acoustic phonograph — A mechanical record player (now obsolete) in which the needle sets a thin diaphragm into vibration. The diaphragm in turn causes the air in a horn to vibrate, thus reproducing the recorded sound.

acoustic pickup — 1. In nonelectrical phonographs, the method of reproducing the material on a record by linking the needle directly to a flexible diaphragm. 2. In an acoustic phonograph, a pickup consisting of a needle, needle holder, and vibrating diaphragm.

acoustic radiator — In an electroacoustic transducer, the part that initiates the radiation of sound vibration. A speaker cone or headphone diaphragm is an example.

acoustic radiometer — An instrument for measuring sound intensity by determining the unidirectional steady-state pressure caused by the reflection or absorption of a sound wave at a boundary.

acoustic reactance — That part of acoustic impedance due to the effective mass of the medium, that is, to the inertia and elasticity of the medium through which the sound travels. The imaginary component of acoustic impedance; expressed in acoustic ohms.

acoustic reflectivity — The ratio of the rate of flow of sound energy reflected from the surface on the side of incidence to the incident rate of flow.

acoustic refraction — A bending of sound waves when passing obliquely from one medium to another in which the velocity of sound is different.

acoustic regeneration — See acoustic feedback.

acoustic resistance — That component of acoustic impedance responsible for the dissipation of energy due to friction between molecules of the air or other medium through which sound travels. Measured in acoustic ohms and analogous to electrical resistance.

acoustic resonance — An increase in sound intensity as reflected waves and direct waves that are in phase combine. May also be due to the natural vibration of air columns or solid bodies at a particular sound frequency.

acoustic resonator — An enclosure that intensifies those audio frequencies at which the enclosed air is set into natural vibration.

acoustics — 1. Science of production, transmission, reception, and effects of sound. 2. In a room or other location, those characteristics that control reflections of sound waves and thus the sound reception in it.

acoustic scattering — The irregular reflection, refraction, or diffraction of a sound wave in many directions.

acoustic shock — Physical pain, dizziness, and sometimes nausea brought on by hearing a loud, sudden sound. (The threshold of pain is about 120 dBm.)

acoustic surface-wave component — A passive electroacoustic device that has metallized interdigital transducer elements deposited on the surface of a piezoelectric substrate. The device allows acoustic energy to be generated, manipulated, and detected on the substrate surface. Most of the acoustic energy is confined to a region within one wavelength of the surface of the substrate. When the metallization is subjected to an alternating voltage, a strain develops between the interdigital fingers and also at the frequency of excitation. This alternating strain on the crystal surface launches a Rayleigh surface-wave front that travels in both directions and that originates from the center of the transducer. The wave exists as an electroacoustic vibration.

acoustic suspension — 1. A loudspeaker system in which the moving cone is held by an overcompliant suspension, the stiffness required for proper operation being supplied by air that is trapped behind the cone in a sealed enclosure. While relatively inefficient, such a system permits good bass reproduction in a unit of moderate size. 2. A speaker enclosure design in which the speaker cone is suspended in an airtight box. This enables the acoustic pressure of the air enclosed therein to provide the principal restoring force for the diaphragm of the speaker. It needs somewhat more power from the amplifier than a free speaker but has better low-frequency performance.

acoustic system — Arrangement of components in devices designed to reproduce audio frequencies in a specified manner.

acoustic telemetry — The utilization of sound energy for the transmission of information. It differs from other telemetry methods in that information derived from the received signal is encoded by the transmitting source.

acoustic transformer — A device that transmits power along a glass or ceramic rod and isolates the power supply from the signal input.

acoustic transmission — Direct transmission of sound energy without the intermediary of electric currents.

acoustic transmission system— An assembly of elements adapted for the transmission of sound.

acoustic treatment— Use of certain sound-absorbing materials to control the amount of reverberation in a room, hall, or other enclosure; that is, to make the room less live.

acoustic wave— A traveling vibration by which sound energy is transmitted in air, water, or the earth. The characteristics of these waves may be described in terms of change of pressure, particle displacement, or density.

acoustic wave filter— A device designed to separate sound waves of different frequencies. (Through electroacoustic transducers, such a filter may be associated with electric circuits.)

acoustoelectric effect— Generation of an electric current in a crystal by a traveling longitudinal sound wave.

acousto-optic Bragg cell— A modulation device that impresses analog information on light beams. This transducer is composed of two sets of interleaved electrodes of alternating polarities deposited on an optical waveguide. An electrical signal applied to each pair of adjacent electrodes buckles the film between them. This distortion changes the refractive index of the waveguide and creates physical waves in the film, commonly called surface waves. These waves are generated at a rate equal to that of the applied electrical signal.

acousto-optics— The study of the interactions between sound waves and light in a solid medium. Sound waves can be made to modulate, deflect, and focus light waves— an important characteristic in laser and holographic applications.

ac plate resistance— Also called dynamic plate resistance. Internal resistance of a vacuum tube to the flow of alternating current. Expressed in ohms, the ratio of a small change in plate voltage to the resultant change in plate current, other voltages being held constant.

ac power supply— A power supply that provides one or more ac output voltages, e.g., ac generator, dynamotor, inverter, or transformer.

acquisition— 1. The process of pointing an antenna or telescope so that it is properly oriented to allow gathering of tracking or telemetry data from a satellite or space probe. 2. In radar, the process between the initial location of a target and the final alignment of the tracking equipment on the target. 3. The gathering of data from transducers or a computer.

acquisition and tracking radar— A radar set that locks onto a strong signal and tracks the object emitting or reflecting the signal. May be airborne or on the ground. Tracking radars use a dish-type antenna reflector to produce a searchlight-type beam.

acquisition radar— A radar set that detects an approaching target and feeds approximate position data to a fire-control or missile-guidance radar, which then takes over the function of tracking the target.

acquisition range— Also called capture range. The range of input frequency about f_0 under which a phase-locked loop, which is initially unlocked, will become locked. This range is narrower than the normal tracking range and is a function of the loop-filter characteristics and the input amplitude.

acquisition time— 1. Time delay between request for data conversion and the holding of the analog value by a sample-and-hold amplifier. 2. In a sample-and-hold circuit, how long it takes after the sample command is given for the hold capacitor to be charged to a full-scale voltage change and to remain within a specified error band around its final value. 3. The time it takes for the output of a sample-and-hold circuit to change from its previous value to a new value when the circuit is switched from the hold mode to the sample mode. It includes the slew

time and settling time to within a certain error band of the final value and is usually specified for a full-scale change.

ac receiver— A radio receiver designed to operate from an ac source only.

ac reclosing relay— A device that controls the automatic reclosing and locking out of an ac circuit interrupter.

ac relay— A relay designed to operate from an alternating-current source.

ac resistance— Total resistance of a device in an ac circuit. *See also* high-frequency resistance.

acronym— A word formed from the first letter or letters of the words describing some item, e.g., FORTRAN from *fo*rmula *tran*slation.

across-the-line starting— Connection of a motor directly to the supply line for starting. Also called full-voltage starting.

ac signaling— Using ac signals or tones to transmit data and/or control signals.

ACTCRBS— Abbreviation for air traffic control radar beacon system. A control system in use worldwide. Air separation information exchanged between plane and air traffic controller must be sent by radio.

ac time overcurrent relay— A device that has either a definite or an inverse time characteristic and functions when the current in an ac circuit exceeds a predetermined value.

actinic— In radiation, the property of producing a chemical change, such as the photographic action of light.

actinium— A radioactive element discovered in pitchblende by the French chemist Debierne in 1889. Its atomic number is 89, its atomic weight 227, and its symbol Ac.

actinodielectric— A photoconductive dielectric.

actinoelectric— Exhibiting a temporary rise in electrical conductivity during exposure to light.

actinoelectric effect— 1. The property of some special materials whereby when an electric current is impressed on them, their resistance changes with light. 2. The property of certain materials (such as selenium, cadmium sulfide, germanium, and silicon) that causes them to change their electrical resistance or generate a voltage on exposure to light.

actinoelectricity— Electricity produced by the action of radiant energy on crystals.

actinometer— An instrument that measures the intensity of radiation by determining the amount of fluorescence produced by that radiation.

action area— In the rectifying junction of a metallic rectifier, that portion which carries the forward current.

action current— A brief and very small electric current that flows in a nerve during a nervous impulse.

action potential— 1. The instantaneous value of the voltage between excited and resting portions of an excitable living structure. 2. The voltage variations in a nerve or muscle cell when it is excited or fired by an appropriate stimulus. After a short time, the cell recovers its normal resting potential, typically about 80 millivolts. The interior of the cell is negative relative to the outside.

activate— To start an operation, usually by application of an appropriate enabling signal.

activating— 1. Chemically treating a basic metal to remove oxides and other passive films to make it more receptive to electroplating. 2. A treatment that renders nonconductive material receptive to electroless deposition. (Nonpreferred synonyms: seeding, catalyzing, and sensitizing.)

activation— 1. Making a substance artificially radioactive by placing it in an accelerator such as a cyclotron or by bombarding it with neutrons. 2. To treat the cathode or target of an electron tube in order to create or increase its emission. 3. The process of adding electrolytes to a cell to make it ready for operation. 4. Causing the acceleration of a chemical reaction.

activation time — In a cell or battery, the time interval from the moment activation is initiated to the moment the desired operating voltage is obtained.

activator — An additive that improves the action of an accelerator.

active — 1. Controlling power from a separate supply. 2. Requiring a power supply separate from the controls. 3. Containing, or connected to and using, a source of energy.

active area — The portion of the rectifying junction of a metallic rectifier that carries forward current.

active balance — In operation of a telephone repeater, the summation of all return currents at a terminal network balanced against the local circuit or drop impedance.

active circuit — A circuit that contains active elements such as transistors, diodes, or integrated circuits.

active communications satellite — A communications satellite in which on-board receivers and transmitters receive signals beamed at them from a ground terminal, amplify them greatly, and retransmit them to another ground terminal. Less sensitive receivers and less powerful transmitters can be used on the ground than are needed for passive satellites. Also called active comsat.

active component — 1. Those components in a circuit that have gain, or direct current flow, such as SCRs, transistors, thyristors, or tunnel diodes. They change the basic character of an applied electrical signal by rectification, amplification, switching, and so forth. (Passive elements like inductors, capacitors, and resistors have no gain characteristics.) 2. A device, the output of which is dependent on a source of power other than the main input signal. 3. A device capable of some dynamic function (such as amplification, oscillation, signal control) and which usually requires an external power supply for its operation. 4. Broadly, any device (including electromechanical relays) that can switch (or amplify) by application of low-level signals.

active computer — The one of two or more computers in an installation that is online and processing data.

active comsat — See active communications satellite.

active current — In an alternating current, a component in phase with the voltage. The working component as distinguished from the idle or wattless component.

active decoder — A device that is associated with a ground station and automatically indicates the radar beacon reply code that is received in terms of its number or letter designation.

active delay line — A digital delay module that incorporates a passive delay line and a series of logical gate circuits. These modules are used specifically with digital or logic signals. Also called digital delay line, digital delay unit, digital delay module, and digital programmable delay line.

active device — See active component.

active display — A display, such as a cathode-ray tube, electroluminescent display, or plasma panel, that presents information by emitting light.

active ECM — See jamming.

active electric network — An electric network containing one or more sources of energy.

active element — 1. The driven or self-excited element in a multielement antenna or antenna array. 2. Also known as the responsive element. That part of a detector on which the infrared energy is projected and which, when radiation falls on it, undergoes a physical change that results in an electrical signal. See active component.

active equalizer — An equalizer designed to correct deficiencies in a speaker system's response. Such equalizers, which are designed to precisely match specific speaker systems, usually connect between the amplifier and preamplifier, or in one of the amplifier's tape-monitor circuits.

active filter — 1. A device employing passive network elements and amplifiers. It is used for transmitting or rejecting signals in certain frequency ranges or for controlling the relative output of signals as a function of frequency. 2. A high-pass, low-pass, bandpass, or band-elimination filter that uses an active element, such as an operational amplifier, and relatively small capacitors, rather than the larger inductors and capacitors that would be required in a conventional passive filter. 3. A circuit whose gain depends on the frequency of the input signal. 4. A filter, consisting of an amplifier and suitable tuning elements, usually inserted in a feedback path. 5. A filter that uses active devices such as operational amplifiers to synthesize the filter response function. This technique has an advantage at high speeds because the need for inductors (with their poor high-frequency characteristics) is eliminated.

active guidance — See active homing.

active homing — A system whereby a missile homes in on a target by means of a radar aboard the missile. Also called active guidance.

active infrared detection — An infrared detection system in which a beam of infrared rays is transmitted toward one or more possible targets, and the rays reflected from the target are detected.

active infrared system — A system in which the object is irradiated by a source of infrared energy, which, in turn, is reflected by the object onto a detector. A snooperscope is an active infrared system.

active intrusion sensor — An active sensor that detects the presence of an intruder within the range of the sensor. Examples are an ultrasonic motion detector, a radio-frequency motion detector, and a photoelectric alarm system. See also passive intrusion sensor.

active jamming — 1. Intentional radiation or reradiation of electromagnetic waves to impair the use of a specific portion of the electromagnetic-wave spectrum. 2. Transmission or retransmission of signals for the express purpose of disrupting communications.

active junction — In a semiconductor, a change in n-type to p-type doping, or vice versa, by a diffusion step. On discrete transistors there are two active junctions, the collector-base junction and the emitter-base junction.

active leg — Within a transducer, an electrical element that changes its electrical characteristics as a function of the applied stimulus.

active line — In a U.S. television picture, one of the lines (approximately 488) that make up the picture. The remaining 37 of the 525 available lines are blanked; they are called inactive lines.

active maintenance downtime — The time during which work is actually being done on an item, from the recognition of an occurrence of failure to the time of restoration to normal operation. This includes both preventive and corrective maintenance.

active material — 1. In the plates of a storage battery, lead oxide or some other active substance that reacts chemically to produce electrical energy. 2. The fluorescent material, such as calcium tungstate, used on the screen of a cathode-ray tube.

active matrix — A display matrix with a transistor at each pixel location to individually store its state (on or off). Pixels in active matrix panels only need to be addressed when they are being turned on or off.

active mixer and modulator — A device requiring a source of electrical power and using nonlinear network elements to heterodyne or combine two or more electrical signals.

active network — 1. A network containing passive and active (gain) elements. 2. An electrical network that includes a source of energy.

active pressure — In an ac circuit, the pressure that produces a current, as distinguished from the voltage impressed on the circuit.

active probe — A test probe, generally used with an oscilloscope, that is so named because of the active components used within probe circuitry. These components consist of one or all of the following: transistors, diodes, integrated circuits, or FETs. If FETs are used, these probes are often referred to as FET probes.

active pull-up — An arrangement in which a transistor is used to replace the pull-up resistor in an integrated circuit in order to provide low output impedance without high power consumption.

active *RC* network — A network formed by resistors, capacitors, and active elements.

active redundancy — That redundancy wherein all redundant items are operating simultaneously rather than being switched on when needed.

active repair time — That portion of corrective maintenance downtime during which repair work is being done on the item, including preparation, fault location, part replacement, adjustment and recalibration, and final test time. It may also include part procurement time under shipboard or field conditions.

active satellite — A satellite that receives, regenerates, and retransmits signals between stations. *See also* communications satellite.

active sensor — A sensor that detects the disturbance of a radiation field that is generated by the sensor. *See also* passive sensor.

active sonar — *See* sonar.

active splitter — *See* line splitter.

active substrate — 1. A substrate in which active elements are formed to provide discrete or integrated devices. Examples of active substrates are single crystals of semiconductor materials within which are transistors, resistors, and diodes, or combinations of these elements. Another example is ferrite substrates within which electromagnetic fields are used to perform logical, gating, or memory functions. 2. A substrate for an integrated component in which parts display transistance. 3. A working part of the electronic circuit, which it supports physically. 4. In an integrated circuit, a substrate consisting of single-crystal semiconductor material into which the various IC components are formed; it acts as some or all of the components. This is in contrast to a substrate consisting of a dielectric, on whose surface the various components are deposited.

active swept-frequency interferometer radar — A dual radar system for air surveillance. It provides angle and range information of high precision for pinpointing target locations by trigonometric techniques.

active systems — In radio and radar, systems that require transmitting equipment, such as a beacon or transponder, to be carried in the vehicle.

active tracking system — Usually, a system that requires the addition of a transponder or responder on board the vehicle to repeat or retransmit information to the tracking equipment; e.g., dovap, secor, azusa.

active transducer — 1. A type of transducer in which its output waves depend on one or more sources of power, apart from the actuating waves. 2. A transducer that requires energy from local sources in addition to that which is received.

active trim — Trimming of a circuit element (usually resistors) in a circuit that is electrically activated and operating to obtain a specified functional output for the circuit. *See* functional trimming.

active wire — The wire of an armature winding that produces useful voltage. That portion of the winding in which induction takes place.

activity — 1. In a piezoelectric crystal, the magnitude of oscillation relative to the exciting voltage. 2. The intensity of a radioactive source. 3. Operations that result in the use or modification of the information in a computer file.

activity curve — A graph showing how the activity of a radioactive source varies with time.

activity ratio — The ratio of the number of records in a computer file that have activity to the total number of records in the file.

ac transducer — A transducer that, for proper operation, must be excited with alternating currents only. Also a device, the output of which appears in the form of an alternating current.

actual height — The highest altitude at which refraction of radio waves actually occurs.

actual power — The average of values of instantaneous power taken over one cycle.

actuating device — A mechanical or electrical device, either manual or automatic, that operates electrical contacts to bring about signal transmission.

actuating system — 1. In a device or vehicle, a system that supplies and transmits energy for the operation of a mechanism or other device. 2. A manually or automatically operated mechanical or electrical device that operates electrical contacts to effect signal transmission.

actuating time — The time at which a specified contact functions.

actuator — 1. In a servo system, the device that moves the load. 2. The part of a relay that converts electrical energy into mechanical motion. 3. Switch part to which an external force is applied to operate the switch. 4. A manual or automatic switch or sensor, such as a holdup button, magnetic switch, or thermostat, that causes a system to transmit an alarm signal when manually activated or when the device automatically senses an intruder or other unwanted condition. 5. A motorized arm that moves a satellite dish into position under the control of a receiver.

ACU — Abbreviation for automatic calling unit.

ac voltage — *See* alternating voltage.

acyclic machine — A direct-current machine in which the voltage generated in the active conductors maintains the same direction with respect to those conductors at all times.

a/d — Abbreviation for analog-to-digital. Also a-d, A-D, or A/D.

adapter — 1. A fitting designed to change the terminal arrangement of a jack, plug, socket, or other receptacle, so that other than the original electrical connections are possible. 2. An intermediate device that permits attachment of special accessories or provides special means for mounting. 3. A device for connecting two parts of an apparatus that would not be directly connectable because of incompatible dimensions, terminations, currents, voltages, frequencies, etc.

adaptive communication — A method in which automatic changes in the communications system allow for changing inputs or changing characteristics of the device or process being controlled. Also called self-adjusting communication or self-optimizing communication.

adaptive control — 1. A control method that uses sensors for real-time measurement of process variables with calculation and adjustment of control parameters as a method of achieving near-optimum process performance. 2. A method of control in which actions are continuously adjusted in response to feedback.

adaptive control system — A device whose parameters are automatically adjusted to compensate for changes in the dynamics of the process to be controlled. An AFC circuit utilizing temperature-compensating capacitors to correct for temperature changes is an example.

adaptive telemetry — Telemetry having the ability to select certain vital information or any change in a given signal.

adaptor — A device that locates and supports products to be tested. Generally, it is made of an insulating material with locator pins mounted to precisely position the product to a spring contact probe test pattern. Also, an adaptor serves as an intermediate between the circuit verifier and the interchangeable test head that contains the test pattern.

adc — Abbreviation for analog-to-digital converter. Also ADC.

Adcock antenna — A pair of vertical antennas separated by one-half wavelength or less and connected in phase opposition to produce a figure-8 directional pattern.

Adcock direction finder — A radio direction finder using one or more pairs of Adcock antennas for directional reception of vertically polarized radio waves.

Adcock radio range — A type of radio range utilizing four vertical antennas (Adcock antennas) placed at the corners of a square, with a fifth antenna in the center.

a/d converter — Abbreviation for analog-to-digital converter. Also a-d, A-D, or A/D converter. 1. A unit or device that converts an analog signal, that is, a signal in the form of a continuously variable voltage or current, to a digital signal. 2. A circuit that accepts information in a continuously varying ac or dc current or voltage and whose output is the same information in digital form. 3. A circuit or device for producing a set of digital output signals representing the magnitude of a voltage applied to its input.

add-and-subtract relay — A stepping relay capable of being operated so as to rotate the movable contact arm in either direction.

addend — A quantity that, when added to another quantity (called the augend), produces a result called the sum.

adder — 1. A device that forms the sum of two or more numbers or quantities impressed on it. 2. In a color TV receiver, a circuit that amplifies the receiver primary signal coming from the matrix. Usually there is one adder circuit for each receiver primary channel. 3. An arrangement of logic gates that adds two binary digits and produces sum and carry outputs.

add-in — Components (expansion boards, cartridges, or chips) that can increase a computer's capabilities, such as memory, graphics, and communications. Add-ins usually refer to an entire circuit board. *See also* add-on.

add-in memory — Additional computer memory that is added to a computer system within the computer's physical housing. Typically the add-in memory is inserted, in board form, into an available card slot on the assembly, the connections for which have already been placed on the existing computer. Additional memory may be in the form of semiconductor RAM, CCD, bubble memory, disk, or tape.

additional station — Any amateur radio station licensed to an amateur radio operator, normally for a specific land location other than the primary station.

addition record — A new record created during the processing of a file in a computer.

additive — Sometimes referred to as the key. A number, series of numbers, or alphabetical intervals added to a code to put it in a cipher.

additive color — A system that combines two colored lights to form a third.

additive primaries — Primary colors that can be mixed to form other colors, but which cannot themselves be produced by mixing other primaries. Red, green, and blue are the primaries in television because, when added in various proportions, they produce a wide range of other colors.

additive process — A printed-circuit manufacturing process in which a conductive pattern is formed on an insulating base by electrolytic chemical deposition.

additive synthesis — A technique for creating musical notes whereby sine waves are added together to create new waveforms. Frequently used in electric organs rather than in synthesizers.

additron — An electrostatically focused, beam-switching tube used as a binary adder in high-speed digital computers. (No longer used.)

add mode — Allows entry of numbers in a calculator to two decimal places without the need to enter the decimal point.

add-on — Circuitry or system that can be attached to a computer to increase memory or performance. *See also* add-in.

add-on component or **add-on device** — A discrete or integrated prepackaged or chip component that is attached to a film circuit to complete the circuit functions.

add-on memory — Additional computer memory that is added externally and is plug compatible with the computer system. The add-on memory is connected with an external connector cable to the computer, where provision has been made for memory expansion. Such a memory device is also available in its own housing, in which case it is physically placed beside the computer's main cabinet. Additional memory may be in the form of semiconductor RAM, CCD, bubble memory, disk, or tape.

address — 1. An expression, usually numerical, that designates a specific location in a storage or memory device or other source or destination of information in a computer. 2. An identification, as represented by a name, label, or number, for a register, location in storage, or any other data source or destination, such as the location of a station in a communications network. 3. Loosely, any part of an instruction that specifies the location of an operand for the instruction. 4. To select the location of a stored information set for access. 5. In computer technology, a number used by the central processing unit (CPU) to specify a location in memory. 6. Element(s) of a packet frame that identifies the source and/or destination stations by means of an agreed bit pattern. 7. A unique sequence of letters or numbers for the location of data or the identity of an intelligent device.

address bus — 1. A unidirectional bus over which digital information appears to identify either a particular memory location or a particular device. 2. The set of output pins from a microprocessor chip and the associated circuitry linking them to other devices for the purpose of addressing those chips or parts of them. *See also* bus system.

address characters — Blocks of alphanumeric characters that identify users or stations uniquely.

address comparator — In a computer, a device that ensures that an address being read is the right one.

address computation — The process by which the address part of an instruction in a digital computer is produced or modified.

address constant — *See* base address.

addressed memory — In a computer, memory sections containing each individual register.

address field — The portion of an instruction that specifies the location of a particular piece of information in a computer memory.

addressing mode — An addressing method. One of several different addressing methods possible in microprocessors.

address modification — In a computer, a change in the address portion of an instruction or command such that, if the routine which contains that instruction or command is repeated, the computer will go to a new address or location for data or instructions.

address part — In an electronic computer instruction, a portion of an expression designating location. *See also* instruction code.

address-routing indicator — Group of characters contained in a message heading that designates the destination of the message.

add-subtract time — The time required by a digital computer to perform addition or subtraction. It does not include the time required to obtain the quantities from storage and put the result back into storage.

add time — The time required in a digital computer to perform addition. It does not include the time required to obtain the quantities from storage and put the result back into storage.

a/d encoder — Analog-to-digital encoder. A device that changes an analog quantity into an equivalent digital representation. Also referred to as an a-d, A-D, or A/D encoder.

adf — *See* automatic direction finder. Also referred to as ADF.

adiabatic damping — A reduction in the size of an accelerator beam as the energy of the beam is increased.

adiabatic demagnetization — A technique used to obtain temperatures within thousandths of a degree of absolute zero. It consists of applying a magnetic field to a substance at a low temperature and in good thermal contact with its surroundings, insulating the substance thermally, and then removing the magnetic field.

A-display — Also called A-scan. A radarscope presentation in which time (distance or range) is one coordinate (usually horizontal) and the target appears displaced perpendicular to the time base.

adjacency — In character recognition, a condition in which the character-spacing reference lines of two characters printed consecutively on the same line are less than a specified distance apart.

adjacent- and alternate-channel selectivity — A measure of the ability of a receiver to differentiate between a desired signal and signals that differ in frequency from the desired signal by the width of one channel or two channels, respectively.

adjacent audio (sound) channel — The rf carrier frequency that contains the sound modulation associated with the next-lower-frequency television channel.

adjacent channel — That frequency band immediately above or below the one being considered.

adjacent-channel attenuation — *See* selectance.

adjacent-channel interference — Undesired signals received on one communication channel from a transmitter operating on a channel immediately above or below.

adjacent-channel selectivity — The ability of a receiver to reject signals on channels adjacent to the channel of the desired station.

adjacent conductor — Any conductor next to another conductor, either in the same multiconductor cable layer or in adjacent layers.

adjacent sound channel — In television, the rf channel containing the sound signal modulation of the next lower channel.

adjacent video carrier — The rf carrier that carries the picture modulation for the television channel immediately above the channel to which the viewer is tuned.

adjustable component — Any circuit component whose electrical value may be varied at will, e.g., adjustable capacitor, inductor, resistor, or load.

adjustable resistor — 1. A resistor that has the resistance wire partly exposed to enable the amount of resistance in use to be adjusted occasionally by the user. Adjustment requires the loosening of a screw, the subsequent moving of the lug, and retightening of the screw. 2. A fixed resistor with a movable contact (or tap) that can be positioned along the length of the resistive path.

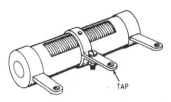

Adjustable resistor.

adjustable voltage divider — A wirewound resistor with one or more movable terminals that can be slid along the length of the exposed resistance wire until the desired voltage values are obtained.

adjusted circuit — Also called bolted-fault level. In a circuit, the current measured under short-circuit conditions with the leads that are normally connected to the circuit breaker bolted together.

adjusted decibels — An expression of the ratio of the noise level to a reference noise at any point in a transmission system, when the noise meter has been adjusted to allow for the interfering effect under specified conditions.

admittance — 1. The ease with which an alternating current flows in a circuit. The reciprocal of impedance; usually expressed in siemens. Symbol is Y or y. 2. The (sinusoidal) current in a circuit divided by the terminal voltage. 3. The vector sum of a resistive component of conductance and a reactive component of susceptance.

ADP — Abbreviation for automatic data processing.

ADSL — Abbreviation for Asymmetric Digital Subscriber Line. A video display terminal distribution video system delivering video over existing (i.e., copper) telephone lines.

adsorption — The deposition of a thin layer of gas or vapor particles onto the surface of a solid. The process is known as chemisorption if the deposited material is bound to the surface by a simple chemical bond.

ADU — Abbreviation for automatic dialing unit.

advance ball — In mechanical recording, a rounded support (often sapphire) that is attached to a cutter and rides on the surface of the recording medium. Its purpose is to maintain a uniform mean depth of cut and to correct for small irregularities on the surface of the disc.

advance calling — A telecommunications feature in which voice messages can be spoken into the telephone for automatic delivery at a prearranged time to any other telephone or telephones.

advanced license — A license issued by the FCC to amateur radio operators who are capable of sending and receiving Morse code at the rate of 13 words per minute and are familiar with general and intermediate radio theory and practice. Its privileges include exclusive use of certain frequencies.

advance wire — An alloy of copper and nickel, used in the manufacture of electric heating units and some wirewound resistors.

aeolight — A glow lamp that uses a cold cathode and a mixture of inert gases. Its illumination can be regulated with an applied signal voltage and it is often used as a modulation indicator for motion picture sound recording.

aerial — See antenna.

aerial cable — A cable installed on a pole line or similar overhead structure.

aerodiscone antenna — An aircraft antenna that is aerodynamically shaped and is physically small compared with other antennas having similar electrical characteristics. Its radiation pattern is omnidirectional and linearly polarized.

aerodrome control radio station — A radio station providing communications between an airport control tower and aircraft or mobile aeronautical radio stations.

aerodynamics — The science of the motion of air and other gases. Also, the forces acting on bodies such as aircraft when they move through such gases, or when such gases move against or around the bodies.

aeromagnetic — Pertaining to the magnetic field of the earth as surveyed from the air.

aeronautical advisory station — A station used for civil defense and advisory communications with private aircraft stations.

aeronautical broadcasting service — The broadcasting service intended for the transmission of information related to air navigation.

aeronautical broadcast station — A station that broadcasts information regarding air navigation and meteorological data pertinent to aircraft operation.

aeronautical fixed service — A fixed service intended for the transmission of information relating to air navigation and preparation for and safety of flight.

aeronautical fixed station — A station operating in the aeronautical fixed service.

aeronautical ground station — A radio station operated for the purpose of providing air-to-ground communications in connection with the operation of aircraft.

aeronautical marker-beacon station — A land station operating in the aeronautical radionavigation service and providing a signal to designate a small area above the station.

aeronautical mobile service — A radio service between aircraft and land stations or between aircraft stations.

aeronautical radio-beacon station — An aeronautical radionavigation land station transmitting signals that are used by aircraft and other vehicles to determine their position bearing or position in relation to the aeronautical radio beacon station.

aeronautical radionavigation service — A radionavigation service intended for use in the operation of aircraft.

aeronautical radio service — 1. Service carried on between aircraft stations and/or land stations. 2. Special radio for air navigation. 3. Service that includes aircraft-to-aircraft, aircraft-to-ground, and ground-to-aircraft communications important to the operation of aircraft.

aeronautical station — A land station (or in certain instances a shipboard station) in the aeronautical mobile service that carries on communications with aircraft stations.

aeronautical telecommunication — Electronic and nonelectronic communications used in the aeronautical service.

aeronautical telecommunication agency — An agency to which is assigned the responsibility for operating a station or stations in the aeronautical telecommunication service.

aeronautical telecommunication log — A record of the activities of an aeronautical telecommunication station.

aeronautical telecommunications — Any telegraph or telephone communications of signals, writing, images, and sounds of any nature by wire, radio, or other system or process of signaling, used in the aeronautical service.

aeronautical telecommunication service — Telecommunication service provided for aeronautical purposes.

aeronautical telecommunication station — A station in the aeronautical telecommunication service.

aeronautical utility land station — A land station located at an airport control tower and used for communications connected with the control of ground vehicles and aircraft on the ground.

aeronautical utility mobile station — A mobile station used at an airport for communications with aeronautical utility land stations, ground vehicles, and aircraft on the ground.

aerophare — See radio beacon.

AES — Abbreviation for Audio Engineering Society. A professional group; the official association of technical personnel, scientists, engineers, and executives in the audio field.

AF — See audio frequency.

AFC — See automatic frequency control.

afocal — An optical system with one set of object and image points at infinity. Literally, "without a focal length." An afocal system receives its input image from infinity and projects its output image to infinity.

AFSK — Abbreviation for audio-frequency shift keying. With this method of modulation, two tones (mark = 2125 Hz, space = 2295 Hz) are fed directly into the microphone jack of the transmitter.

afterglow — Also called phosphorescence. 1. The light that remains in a gas-discharge tube after the voltage has been removed, or on the phosphorescent screen of a cathode-ray tube after the exciting electron beam has been removed. 2. The luminosity that remains in a rarefied gas after an electrodeless discharge has traversed the gas.

afterpulse — In a photomultiplier, a spurious pulse induced by a preceding pulse.

AGC — See automatic gain control.

age — To maintain an electrical component in a specified environment, as with respect to pressure, temperature, applied voltage, etc., until its characteristics stabilize.

aggregate function — A command that performs calculations based on a set of values rather than on a single value.

agile receiver — A satellite receiver that can be tuned to any desired channel.

aging — 1. Storing a permanent magnet, capacitor, semiconductor, meter, or other device, sometimes with voltage applied, until its desired characteristics become essentially constant. 2. The change of a component or a material with time under defined environmental conditions, leading to improvement or deterioration of properties.

agonic line — An imaginary line on the earth's surface, all points of which have zero magnetic declination.

AGP — Abbreviation for accelerated graphics port.

AGREE — Advisory Group on Reliability of Electronics Equipment.

AI — *See* artificial intelligence.

aided tracking — A system of tracking a target signal in bearing, elevation, or range (or any combination of these variables) in which manual correction of the tracking error automatically corrects the rate at which the tracking mechanism moves.

AIEE — Abbreviation for American Institute of Electrical Engineers. Now merged with IRE to form IEEE.

air bearing — A means of supporting magnetic tape on an air film rather than by means of a sliding or rolling contact. Usually, an air bearing is a perforated cylinder; pressurized air flows through the perforations and forms a film that prevents the tape from contacting the cylinder.

airborne intercept radar — Short-range airborne radar employed by fighter and interceptor planes to track down their targets.

airborne long-range input — Airborne equipment designed to extend air-surveillance coverage seaward so that long-range interceptors may be used.

airborne moving-target indicator — A type of airborne-radar display that does not present essentially stationary objects.

airborne noise — Undesired sound in the form of fluctuations of air pressure about the atmospheric pressure as a mean.

airborne radar platform — Airborne surveillance and height-finding radar for early warning and control.

air capacitor — A capacitor in which air is the only dielectric material between its plates.

aircarrier aircraft station — A radio station aboard an aircraft that is engaged in or essential to the transportation of passengers or cargo for hire.

air cell — A primary cell in which depolarization at the positive electrode is accomplished chemically by reduction of the oxygen in the air.

air column — The air space within a horn of an acoustic chamber.

air condenser — *See* air capacitor.

air-cooled tube — An electron tube in which the generated heat is dissipated to the surrounding air directly, through metal heat-radiating fins, or with the aid of channels or chimneys that increase the air flow.

air-core cable — A telephone cable in which the interstices in the cable core are not filled with a moisture barrier.

air-core coil — A number of turns of spiral wire in which no metal is used in the center.

air-core transformer — A transformer (usually rf) having two or more coils wound around a nonmetallic core. Transformers wound around a solid insulating substance or on an insulating coil form are included in this category.

aircraft bonding — Electrically connecting together all the metal structure of the aircraft, including the engine and metal covering of the wiring.

aircraft flutter — Flickering (repetitive fading and intensifying) in a TV picture as the signal is reflected from flying aircraft. The reflected signal arrives in or out of phase with the normal signal and thus strengthens or weakens the latter.

aircraft station — A radio station installed on aircraft and continuously subject to human control.

air defense control center — Principal information, communications, and operations center from which all aircraft, antiaircraft operations, air-defense artillery, guided missiles, and air-raid warning functions of a specific area of air defense responsibility are supervised and coordinated.

air defense identification zone — Airspace of defined dimensions within which the ready identification, location, and control of aircraft is required.

air dielectric capacitor — A capacitor with a dielectric consisting of air.

airdrome control station — A station used for communication between an airport control tower and aircraft.

air environment — In communications electronics, all airborne equipment that is part of the communications-electronics system, as distinguished from the equipment on the ground, which belongs to the ground environment.

air gap — 1. A nonmagnetic discontinuity in a ferromagnetic circuit. For example, the space between the poles of a magnet — although filled with brass, wood, or any other nonmagnetic material — is nevertheless called an air gap. This gap increases magnetic reluctance and prevents saturation of the core. 2. The air space between two magnetically or electrically related objects.

air/ground control radio station — An aeronautical telecommunication station with the primary responsibility of handling communications related to the operation and control of aircraft in a given area.

air/ground liaison code — Set of symbols for a limited number of words, phrases, and sentences used for communications between air and ground forces.

air-motion transformer — A type of speaker in which the air is not pushed into vibration by a piston, but rather squeezed by the contractions of a folded diaphragm.

air navigation radio aids — Aeronautical ground stations, radio beacons, direction finders, and similar facilities.

airplane flutter rejection — The measure of a receiver's immunity to the effects of wavering signals produced by aircraft in the reception path.

airport beacon — A beacon (light or radio) to indicate the location of an airport.

airport control station — A station that furnishes communications between an airport control tower and aircraft in the immediate vicinity; messages are limited to those related to actual aviation needs.

airport radar control — The surveillance-radar portion of radar approach control.

airport runway beacon — A radio-range beacon that defines one or more approaches to an airport.

airport surface detection equipment — Abbreviated ASDE. 1. Radar that shows the movement of aircraft and other vehicles on the ground at an airport. Valuable tool at night and during low visibility. 2. A digital radar system used to track planes and vehicles on airport runways and up to 200 feet in altitude. Unlike previous surface radar systems, ASDE provides clear images in bad weather.

airport surveillance radar — 1. Abbreviated ASR. A short-range radar system that maintains constant surveillance over aircraft at the lower levels of flight. Distinct from air route surveillance radar (ARSR), which is long-range radar — 150-mile (241-km) radius — to control traffic between terminals. 2. An air-traffic-control radar that scans the airspace 30 to 60 miles (48 to 98 km) around an airport and displays the location of all aircraft below a certain altitude and all obstructions near the control tower.

air-position indicator — Airborne computing system that presents a continuous indication of aircraft position on the basis of aircraft heading, air speed, and elapsed time.

air-spaced coax — A coaxial cable in which air is basically the dielectric material. The conductor may be centered by means of a spirally wound synthetic filament, by beads, or by braided filaments. This construction is also referred to as an air dielectric.

air surveillance — Systematic observation of airspace by electronic, visual, or other means, primarily to identify and determine the movements of aircraft and missiles, friendly and enemy, in the airspace under observation.

airtime — Time spent on a cellular phone, which is usually billed to the subscriber on a per-minute basis.

air-to-ground communication — Transmission of radio signals from an aircraft to stations or other locations on the earth's surface, as differentiated from ground-to-air, air-to-air, or ground-to-ground communications.

air-to-ground radio frequency — The frequency or band of frequencies agreed upon for transmission from an aircraft to an aeronautical ground station.

air-to-surface missile — A missile designed to be dropped from an aircraft. An internal homing device or the aircraft's radio guides it to a surface target.

airwaves — Slang expression for radio waves used in radio and television broadcasting.

alacritized switch — 1. A mercury switch treated to yield a low adhesional force between the rolling surface and mercury pool, resulting in a decreased differential angle. 2. A mercury switch in which the tendency of the mercury to stick to the mating parts has been reduced.

alarm — A device that signals the existence of an abnormal condition by means of an audible or visible discrete change, or both, intended to attract attention. An alarm circuit produces or transmits an alarm signal.

alarm condition — A threatening condition, such as an intrusion, fire, or holdup, sensed by a detector.

alarm device — A device that signals a warning in response to an alarm condition, such as a bell, siren, or annunciator.

alarm discrimination — The ability of an alarm system to distinguish between those stimuli caused by an intrusion and those which are a part of the environment.

alarm hold — A means of holding an alarm once sensed. The typical magnetic trap does not hold or latch, and thus the reclosing of a trapped door resets the typical magnetic trap. A hold circuit applied to such a device indicates that the door has been opened and continues to so indicate until reset.

alarm line — A wired electrical circuit used for the transmission of alarm signals from the protected premises to a monitoring station.

alarm relay — A relay, other than an annunciator, used to operate, or to operate in connection with, a visual or audible alarm.

alarm state — The condition of a detector that causes a control unit in the secure mode to transmit an alarm signal.

alarm system — An assembly of equipment and devices designated and arranged to signal the presence of an alarm condition requiring urgent attention, such as unauthorized entry, fire, temperature rise, etc. The system may sound a local warning or alert the police, a central station, or a proprietary service.

albedo — The reflecting ability of an object. It is the ratio of the amount of light reflected compared with the amount received.

ALC — Abbreviation for automatic level (volume) control. 1. A special compressor circuit included in some tape recorders for automatically maintaining the recording volume within the required limits regardless of changes in the volume of the sound. 2. A circuit that automatically maintains recording levels within permissible limits, so that, no matter how loud or soft the sound being recorded,

the signal on the tape will not get strong enough to overmodulate and distort or soft enough to be lost in noise. Also known as automatic volume control (AVC).

Alexanderson alternator — An early mechanical generator used as a source of low-frequency power for transmission or induction heating. It is capable of generating frequencies as high as 200,000 hertz.

Alexanderson antenna — A vlf antenna consisting of a horizontal wire connected to ground at equally spaced points by vertical wires with base-loading coils; the transmitter is coupled to an end coil.

Alford antenna — A square loop antenna comprising four linear sides with their ends bent inward so that capacitive loading is provided to equalize the current around the loop.

algebraic adder — In a computer, an adder that provides the algebraic rather than arithmetic sum of the entered quantities.

algebraic logic — A calculator mode that permits all calculations to be done in the order in which they are written.

algebraic sum — The sum of two or more quantities combined according to their signs. (Compare with *arithmetic sum.*)

ALGOL — 1. An international problem language designed for the concise, efficient expression of arithmetic and logical processes and the control (iterative, etc.) of these processes. From *algo*rithmic *l*anguage. 2. A high-level language that has a context-free structure.

algorithm — 1. A set of rules or processes for solving a problem in a finite number of steps (for example, a full statement of an arithmetic procedure for finding the value of sin x with a stated precision). *See also* procedure. 2. A series of equations, some of which may state inequalities, that cause decisions to be made and the computational process to be altered based on these decisions. 3. A set of rules or directions for getting a specific output from a specific input. The distinguishing feature of an algorithm is that all vagueness must be eliminated; the rules must describe operations that are so simple and well defined they can be executed by a machine. Furthermore, an algorithm must always terminate after a finite number of steps. 4. An ordered sequence of mathematical steps that always produces the correct answer to a problem, though the solution may be more lengthy than necessary. 5. A set of well-defined procedures for the solution of a problem in a limited number of steps. Algorithms are implemented in a computer by a programmed sequence of instructions.

algorithmically generated pattern — An array of digital data automatically generated by a predetermined software routine or program.

algorithmic language — An arithmetic language by which a numerical procedure may be presented to a computer precisely and in a standard form.

algorithmic pattern generation — Real-time generation of input test patterns during test execution according to specified procedures, formulas, or algorithms. Also refers to procedures or algorithms used in automatic-test-generation software for specific fault sets.

alias — An alternate label. For instance, a label and one or more aliases may be used to identify the same data element or point in a computer program.

aliasing — 1. The introduction of error into the Fourier analysis of a discrete sampling of continuous data when components with frequencies too great to be analyzed with the sampling interval being used contribute to the amplitudes of lower-frequency components. 2. A phenomenon arising as a result of the sampling process in which high-frequency components of the original analog signal (whether information or noise) appear as lower frequencies in the sampled signal. Aliasing occurs when the sampling rate is less than twice the highest frequency existing in the original analog signal. 3. Undesirable distortion component that can arise in digital audio equipment when the input signal's frequency exceeds one-half of the digital circuitry's sampling rate. 4. The mistaking of some object or situation for another, especially because of the way data is examined. Examples: movies of propellers and wagon wheels that seem to turn backward, musical notes that are wrongly analyzed by sequential measurement (for instance, thought to be an octave too low), and, especially, "jaggies." 5. Undesirable stairstep distortions in computer-generated images caused by improper sampling techniques. The most common effect is a jagged edge along object boundaries.

aliasing noise—A distortion component that will be created if a sampled signal bandwidth is effectively greater than one-half the sampling rate.

align—1. To adjust the tuned circuits of a receiver or transmitter for maximum signal response. 2. To put into proper relative position, agreement, or coordination when placing parts of a photomask together or placing a photomask over an etched pattern in the oxide on a semiconductor wafer. 3. To adjust the tuning of a multistage device so that all stages are adjusted to the same frequency or so that they work together properly.

aligned bundle—*See* coherent bundle.

aligned-grid tube—A multigrid vacuum tube in which at least two of the grids are aligned one behind the other to give such effects as beam formation and noise suppression.

alignment—1. The process of adjusting components of a system for proper interrelationship. The term is applied especially to (*a*) the adjustment of tuned circuits in a receiver to obtain the desired frequency response and (*b*) the synchronization of components in a system. 2. In a tape recorder, the physical positioning of a tape head relative to the tape itself. Alignment in all respects must conform to rigid requirements in order for a recorder to function properly. 3. The accuracy or proper relative position of an image on a photomask with respect to an existing image on a substrate, as in a photoresist coating, or etched in the oxide of an oxidized silicon wafer. 4. A technique in the fabrication process of semiconductors by which a series of six to eight masks are successively registered to build up the various layers of a monolithic device. Each mask pattern must be accurately referenced to or aligned to all preceding mask patterns. 5. The accuracy of coordination or relative position of images on a semiconductor oxide coating and on the photomask, or any other images placed in relation to those.

alignment chart—Also called nomograph, nomogram, or abac. Chart or diagram consisting of two or more lines on which equations can be solved graphically. This is done by laying a straightedge on the two known values and reading the answer at the point where the straightedge intersects the scale for the value sought.

alignment pin—1. A pin in the center of the base of a tube. A projecting rib on the pin ensures that the tube is correctly inserted into its socket. 2. Any pin or device that will ensure the correct mating of two components designed to be connected.

alignment protractor—An instrument that indicates error in a pickup's lateral alignment. It fits on the center spindle of the turntable, and the pickup stylus fits into a small hole on the device. The correct indication is shown when the angle of lateral movement of the pickup head is at 90° to the tangent of the groove at any point, although minimal tracking error is expected with most pickup arms.

alignment tool—A special screwdriver or socket wrench used for adjusting trimmer or padder capacitors

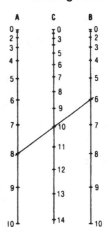

Alignment chart.

or cores in tuning inductances. It is usually constructed partly or entirely of nonmagnetic material. *See also* neutralizing tool.

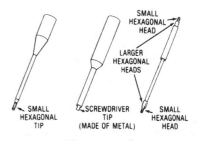

Alignment tools.

alive—1. Electrically connected to a source of potential difference, or electrically charged to have a potential different from that of the earth. 2. Energized. 3. Reverberant, as a room in which sound reflects and echoes.

alive circuit—A circuit that is energized.

alkali—A compound that forms hydroxyl ions when in aqueous solution. Also called a base.

alkaline cell—1. A primary cell, similar to the zinc-carbon cell, in which the negative electrode is granular zinc mixed with a potassium hydroxide (alkaline) electrolyte; the positive electrode is a polarizer in electrical contact with the outer metal can of the cell. A porous separator divides the electrodes. This type of cell delivers a terminal potential of 1.5 volts and has a 50 percent to 100 percent higher capacity than does a 1.5-volt zinc-carbon cell. Also called an alkaline-manganese cell. 2. A primary dry cell that has a very low internal resistance and high

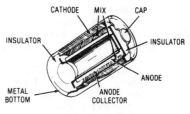

Alkaline cell.

service capacity. It is characterized by a relatively flat discharge curve under load.

all-channel tuning — Ability of a television set to receive all assigned channels, VHF and UHF, channels 2 through 83, as well as cable channels.

all-diffused monolithic integrated circuit — Also called compatible monolithic integrated circuit. A microcircuit consisting of a silicon substrate into which all the circuit parts (both active and passive elements) are fabricated by diffusion and related processes.

Allen screw — A screw having a hexagonal hole or socket in its head. Often used as a setscrew.

Allen wrench — A straight or bent hexagonal rod used to turn an Allen screw.

alligator clip — A spring-loaded metal clip with long, narrow meshing jaws similar to the jaws of an alligator; it is used for making temporary electrical connections, generally at the end of a test lead on interconnection wire.

SCREW

Alligator clip.

allocate — In a computer, to assign storage locations to main routines and subroutines, thus fixing the absolute values of symbolic addresses.

allocated channel — A channel assigned to a specific user.

allocated frequency band — A segment of the radio-frequency spectrum established by a competent authority that designates the use that may be made of the frequencies contained therein.

allocated-use circuit — 1. A circuit in which one or more channels have been allocated for the exclusive use of one or more services by a proprietary service; may be a unilateral or joint circuit. 2. Communication link specifically assigned to user(s) warranting such facilities.

allocation technique — The method of providing a process access to a shared resource.

allochromatic — Exhibiting photoelectric effects due to the inclusion of microscopic impurities or as a result of exposure to various types of radiation.

allophone — A variation in the pronunciation of a phoneme. An allophone can be regarded as the sound that results when a phoneme is placed in its environment.

allotter — In a telephone system, a distributor, associated with the finder control group relay assembly, that allots an idle linefinder in preparation for an additional call.

allotter relay — In a telephone system, a relay of the linefinder circuit, the functions of which are to preallot an idle linefinder to the next incoming call from the line and to guard relays.

allowable deviation — The permissible difference between any range of conditions and a reference condition.

alloy — 1. A composition of two or more elements, of which at least one is a metal. It may be a solid solution, a heterogeneous mixture, or a combination of both. 2. Method of making pn junctions by melting a metallic dopant so that it dissolves some of the semiconductor material and then hardens to produce a doped alloy.

alloy deposition — The process of depositing an alloy on a substrate during manufacturing.

alloy-diffused transistor — A transistor in which the base is diffused and the emitter is alloyed. The

collector is the semiconductor substrate into which alloying and diffusion are effected.

alloyed contact — An ohmic contact formed by an alloy process.

alloy junction — Also called fused junction. A junction produced by alloying one or more impurity metals to a semiconductor. A small button of impurity metal is placed at each desired location on the semiconductor wafer, heated to its melting point, and cooled rapidly. The impurity metal alloys with the semiconductor material to form a p or n region, depending on the impurity used.

alloy-junction photocell — A photodiode in which an alloy junction is produced by alloying (mixing) an indium disc with a thin wafer of n-type germanium.

alloy-junction transistor — Also called fused-junction transistor. A semiconductor wafer of p- or n-type impurities fused, or alloyed, into opposite sides of the wafer to provide emitter and base junctions. The base region comprises the original semiconductor wafer.

alloy process — A fabrication technique in which a small part of the semiconductor material is melted together with the desired metal and allowed to recrystallize. The alloy developed is usually intended to form a pn junction or an ohmic contact.

alloy transistor — A transistor in which the emitter and collector junctions are both alloy junctions.

all-pass filter — A network designed to produce a delay (phase shift) and an attenuation that is the same at all frequencies; a lumped-parameter delay line. Also called all-pass network.

all-pass network — A network designed to introduce phase shift or delay but not appreciable attenuation at any frequency.

all-relay central office — An automatic central-office dial switchboard in which relay circuits are used to make the line interconnections.

all-wave antenna — A receiving antenna suitable for use over a wide range of frequencies.

all-wave receiver — A receiver capable of receiving stations on all the commonly used wavelengths in shortwave bands as well as in the broadcast band.

alnico — An alloy consisting mainly of *al*uminum, *ni*ckel, and *co*balt plus iron. Various subscripts and combinations of letters are available. Material can be found both in cast and sintered form, including isotropic and anisotropic alloys. Capable of very high flux density and magnetic retentivity, the alloy is used in permanent magnets for speakers, magnetrons, etc.

alpha — 1. Emitter-to-collector current gain of a transistor connected as a common-base amplifier. For a junction transistor, alpha is less than unity, or 1. Alpha is usually defined as the ratio of a small change in collector current to the corresponding change in emitter current, when the collector-base voltage is kept constant. 2. Brain wave signals whose frequency is approximately 8 to 12 Hz. The associated mental state is relaxation, heightened awareness, elation, and in some cases, dreamlike.

alphabet — An ordered set of all the letters and associated marks used in a language, for example, the Morse code alphabet, the 128 characters of the U.S. ASCII alphabet.

alphabetic coding — A system of abbreviation used in preparing information for input into a computer. Information may then be reported in the form of letters and words as well as in numbers.

alphabetic-numeric — Having to do with the alphabetic letters, numerical digits, and special characters used in electronic data processing work.

alphabetic string — A character string containing only letters and special characters.

alpha cutoff frequency — The frequency at which the current gain of a common-base transistor stage has decreased to 0.707 of its low-frequency value. Gives a rough indication of the useful frequency range of the device.

alphameric (alphanumeric) — Generic term for alphabetic letters, numerical digits, and special characters that are machine processable.

alphameric characters — 1. A character set that mixes alphabetic characters, numeric characters, and usually punctuation characters. The alphabetic characters may be uppercase and/or lowercase or even in Japanese or Arabic script. 2. Consisting of letters and numbers. Also called alphameric or alphanumeric.

alphanumeric — 1. A generic term for alphabetic letters, numerical digits, and special ASCII characters that can be processed by a computer. A character set containing any combination of the above. 2. Consisting of letters and numbers. 3. All letters in the alphabet, the numbers 0 through 9, and special characters — such as -, /, *, $, (), +, and # — that are machine processable.

alphanumeric code — In computer practice or in communications, a code in which the letters of the alphabet are represented by numbers.

alphanumeric display — Device consisting of a typewriter-style keyboard and a display (CRT) screen on which text is viewed.

alphanumeric keys — Keys on a data entry device that resemble those on a standard keyboard. Usually they are used to manually input or edit text for the display system, although they can also be used in a function key mode.

alphanumeric reader — An instrument that reads alphabetic, numeric, and special characters by means of a photosensor that measures the varying intensity of the characters reflected from a light source.

alphanumeric readout — A type of digital readout that displays both letters and numerals.

alpha particle — A small, electrically charged particle thrown off at very high velocity by many radioactive materials, including uranium and radium. Identical to the nucleus of a helium atom, it is made up of two neutrons and two protons. Its electrical charge is positive and is equal in magnitude to twice that of an electron.

alpha ray — A stream of fast-moving alpha particles that produce intense ionization in gases through which they pass, are easily absorbed by matter, and produce a glow on a fluorescent screen. The lowest-frequency radioactive emissions.

alpha system — A signaling system in which the signaling code to be used is designated by alphabetic characters.

alpha-wave detector — A device that detects and displays alpha-wave segments of brain wave output. Used in biofeedback. Also called alpha-wave meter or sensor.

alpha-wave meter — See alpha-wave detector.

ALS — Abbreviation for advanced low-power Schottky (Texas Instruments). A low-power, high-speed transistor-transistor logic (TTL) family.

alterable memory — A storage medium that may be written into.

alteration switch — A manual switch on a computer console or a program-simulated switch that can be set on or off to control coded machine instructions.

alternate channel — A channel located two channels above or below the reference channel.

alternate-channel interference — Interference caused in one communication channel by a transmitter operating in the channel after an adjacent channel. See also second-channel interference.

alternate facility — A communications-electronics facility that is established for the purpose of replacing or supplementing another facility or facilities under real or simulated emergency conditions.

alternate frequency — The frequency assigned for use at a certain time, or for a certain purpose, to replace or supplement the frequency normally used.

alternate mode — A means of displaying on an oscilloscope the output signals of two or more channels by switching the channels, in sequence, after each sweep.

alternate route or routing — A secondary or backup communications path to be used if the normal (primary) routing is not possible.

alternate voice/data operation — Modem operations coordinated by voice over the same line that accommodates transmission. The modem is patched out of the circuit to allow this. A special switch, called an exclusion key, converts the line from voice to data.

alternating-charge characteristic — The function relating, under steady-state conditions, the instantaneous values of the alternating component of transferred charge to the corresponding instantaneous values of a specified periodic voltage applied to a nonlinear capacitor.

alternating current — Abbreviated ac. 1. A flow of electricity that reaches maximum in one direction, decreases to zero, then reverses itself and reaches maximum in the opposite direction. The cycle is repeated continuously. The number of such cycles per second is the frequency. The average value of voltage during any cycle is zero. 2. Any signal that varies with time. It usually means that the current actually changes polarity with time. The plot of current versus time usually is a sine wave that comprises a succession of instantaneous values, the greatest of which is the amplitude or peak value. The time taken by one complete cyclic repetition is the period, and the number of periods in one second is the frequency.

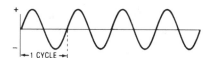

Alternating current.

alternating current/direct current — A term applied to electronic equipment indicating it is capable of operation from either an alternating-current or direct-current primary power source.

alternating-current erasing head — An erasing head used in magnetic recording, in which alternating current produces the magnetic field necessary for erasing. Alternating-current erasing is achieved by subjecting the medium to a number of cycles of a magnetic field of a decreasing magnitude. The medium is, therefore, essentially magnetically neutralized.

alternating-current generator — A rotary machine that generates alternating current when its rotor, which may be either the armature or the field, is rotated by an engine or a motor. Also called an alternator.

alternating-current pulse — An alternating-current wave of brief duration.

alternating-current transmission — In television, that form of transmission in which a fixed setting of the controls makes any instantaneous value of signal correspond to the same value of brightness only for a short time.

alternating flasher — A control that provides voltage first to one load and then to another load. This cycle repeats normally at a fixed rate per minute.

alternating quantity — A periodic quantity that has alternately positive and negative values, the average value of which is zero over a complete cycle.

alternating voltage — Also called ac voltage. Voltage that is continually varying in value and reverses its direction at regular intervals, such as that generated by an alternator or developed across a resistance or impedance through which alternating current is flowing.

alternation — One-half of a cycle — either when an alternating current goes positive and returns to zero, or when it goes negative and returns to zero. Two alternations make one cycle. The complete rise and fall of a current traveling in one direction, or one-half of an alternating-current cycle.

alternator — A device for converting mechanical energy into electrical energy in the form of an alternating current.

alternator transmitter — A radio transmitter that generates power by means of a radio-frequency alternator.

altimeter — An instrument that indicates the altitude of an aircraft above a specific reference level, usually sea level or the ground below the aircraft. It may be similar to an aneroid barometer, which utilizes the change of atmospheric pressure with altitude, or it may be electronic.

altimeter station — An airborne transmitter, the emissions from which are used to determine the altitude of an aircraft above the surface of the earth.

altitude delay — The synchronization delay introduced between the time of transmission of the radar pulse and the start of the trace on the indicator. This is done to eliminate the altitude circle on the plan-position-indicator display.

ALU — Abbreviation for arithmetic and logic unit. 1. A device that performs the basic mathematical operations such as addition, subtraction, multiplication, and division of numbers (usually binary) presented to its inputs and provides an output that is an appropriate function of the inputs. 2. The arithmetic and logic unit internal to the microprocessor chip. This register handles all arithmetic and logical operations carried out as part of a microprocessor instruction. 3. The part of a CPU that executes adds, subtracts, shifts, AND logic operations, OR logic operations, etc. 4. A complex array of gates that can be used to perform binary arithmetic, logic operations, shifts and rotates, and complementing. 5. One of the three essential components of a microprocessor, the other two being data registers and control. The ALU performs addition and subtraction, logic operations, masking, and shifting (multiplication and division).

alumina — 1. A ceramic used for insulators in electron tubes or substrates in thin-film circuits. It can withstand continuously high temperatures and has a low dielectric loss over a wide frequency range. Aluminum oxide (AL_2O_3). 2. The substrate material on which are deposited thin conductive and resistive layers for thin-film microwave integrated circuits.

aluminized-screen picture tube — A cathode-ray picture tube that has a thin layer of aluminum deposited on the back of its fluorescent surface to improve the brilliance of the image and also prevent ion-spot formation.

aluminizing — The process of applying a film of aluminum to a surface, usually by evaporation in a vacuum.

aluminum-electrolytic capacitor — A capacitor with two aluminum electrodes (the anode has the oxide film) separated by layers of absorbent paper saturated with the operating electrolyte. The aluminum-oxide film

or dielectric is repairable in the presence of an operating electrolyte.

aluminum-steel conductor — A composite conductor made up of a combination of aluminum and steel wires. In the usual construction, the aluminum wires surround the steel.

alumoweld — A thin coating of aluminum fused to a steel core. Used in line wire and cable messengers.

AM — *See* amplitude modulation.

amateur — Also called a ham. 1. A person licensed to operate radio transmitters as a hobby. Any amateur radio operator. 2. A nonprofessional, usually noncommercial, devotee of any technology (as a hobby).

amateur band — Any one of several radio frequency bands assigned for noncommercial use by licensed radio amateurs. In the United States, there are twelve such bands between 1.80 MHz and 1.3 GHz. Assignments are made by the Federal Communications Commission.

amateur call letters — Call letters and numbers assigned to amateur stations by the licensing authority. Call-letter combinations consist of a letter prefix denoting the country in which the station is situated, plus a number designating the location within the country, and two or more letters identifying the particular station. Example: K2ABC (K or W = United States, 2 = New York, and ABC = identification of individual licensee issued alphabetically except for special circumstances).

amateur extra license — A license issued by the FCC to amateur radio operators who are able to send and receive Morse code at the rate of 20 words per minute and who are familiar with general, intermediate, and advanced radio theory and practice. Its privileges include all authorized amateur rights and the exclusive rights to operate on certain frequencies.

amateur radio — The practice of operating electronic communications equipment as a hobby in the amateur service. Also refers to the equipment used for this purpose.

amateur radiocommunication — Noncommercial radiocommunication by or among radio stations solely with a personal aim and without pecuniary or business interest.

amateur radio license — The instrument of authorization issued by the Federal Communications Commission, comprised of a station license and, in the case of the primary station, incorporating an operator license.

amateur radio operation — Radiocommunication conducted by an amateur radio operator from an amateur radio station.

amateur radio operator — 1. A person interested in radio technique solely with a personal aim and without pecuniary interest, and holding a valid Federal Communications Commission license to operate amateur radio stations. 2. A private citizen who operates electronic communications equipment as a hobby.

amateur radio service — A radiocommunication service of self-training, intercommunication, and technical investigation carried on by amateur radio operators.

amateur service — A radiocommunication service that licensed operators with no pecuniary interest use for self-training, communication, and technical investigations.

amateur station — A radio transmitting station operated by one or more licensed amateur operators.

amateur-station call letters — *See* amateur call letters.

ambience — 1. Reverberant or reflected sound that reaches a listener's ear from all directions as sound waves "bounce" successively off the various surfaces of a listening area — the walls, ceiling, etc. The term is usually reserved for large areas such as auditoriums and concert halls, though home listening-rooms have their own ambience effects. 2. The indirect sounds heard in a concert hall or other large listening area that contribute to the overall auditory effect obtained when listening to live performances.

ambient — Surrounding. The surrounding environment coming into contact with the system or component in question. *See also* ambient noise; ambient temperature.

ambient level — The level of interference emanating from sources other than the test sample, such as inherent noise of the measuring device and extraneous radiated fields.

ambient light — Normal room light. Light existing in a room or other location that is characteristic of the environment.

ambient-light filter — A filter used in front of a television picture-tube screen to reduce the amount of ambient light reaching the screen and to minimize the reflections of light from the glass face of the tube.

ambient lighting — Lighting designed to provide a substantially uniform level of illumination throughout an area, exclusive of any provision for special local requirements.

ambient noise — 1. Acoustic noise in a room or other location. Usually measured with a sound-level meter. The term *room noise* commonly designates ambient noise at a telephone station. 2. Unwanted background noise picked up by a microphone, that is, any extraneous clatter in a room. Also any acoustic coloration that influences sounds, brought about by the acoustic properties of a room in which a recording is being made or replayed. 3. Interference present (in a communication line) at all times. 4. Background electrical noise in electrical measurements and operation.

ambient operating temperature — The temperature of the air surrounding an object, neglecting small localized variations.

ambient pressure — The general surrounding atmospheric pressure.

ambient temperature — 1. Temperature of air or liquid surrounding any electrical part or device. Usually refers to the effect of such temperature in aiding or retarding removal of heat by radiation and convection from the part or device in question. 2. The prevailing temperature in the immediate vicinity of an object; the temperature of its environment. 3. A temperature within a given volume, e.g., a room or building.

ambient temperature range — The range of environmental temperatures in the vicinity of a component or device over which it may be operated safely and within specifications. For forced-air cooled operation, the ambient temperature is measured at the air intake.

ambiguity — 1. An undesirable tendency of a synchro or servo system to seek a false null position in addition to the proper null position. 2. Inherent error resulting from multiple-bit changes in a polystropic code. (Proper logic design prevents such errors.)

ambiguous count — A count on an electronic scaler that is obviously impossible.

ambisonic reproduction — The recreation of the ambience of an original recording situation with associated directionality. Sound from every direction is picked up by a tetrahedral microphone array and is then encoded onto two channels, which, upon decoding, produce sound through several speakers in a continuous range of directions around the listener, thus approximating the original. It can be subdivided into periphonic and pantophonic systems, the former concerning a complete sphere of information, the latter relating to a horizontal circle. Pantophonic reproduction does not distinguish vertical directionality, but still achieves remarkable realism.

AM broadcast channel — Any of the 10-kHz wide bands of radio frequencies, which extend from 530 to 1710 kHz and are used for standard amplitude-modulated radio broadcasts.

American Institute of Electrical Engineers (AIEE) — Now merged with IRE to form the IEEE.

American Morse code — A system of dot-and-dash signals originated by Samuel F. B. Morse and still used to a limited extent for wire telegraphy in North America. It differs from the international Morse code used in radiotelegraph transmission.

American National Standards Institute, Inc. — Abbreviated ANSI. An independent, industry-wide association that establishes standards for the purpose of promoting consistency and interchangeability among the products of different manufacturers. Formerly United States of America Standards Institute (USASI) and American Standards Association (ASA).

American Radio Relay League (ARRL) — An organization of amateur radio operators.

American Standards Association — Abbreviated ASA. *See* American National Standards Institute, Inc.

American wire gage (AWG) — The standard system used for designating wire diameter. Gage sizes range from No. 40, the smallest diameter wire, to No. 4/0, the largest. AWG sizes are used for specifying both solid and stranded wire. Gage numbers have an inverse relationship to size, i.e., larger numbers have smaller diameter.

American wire gage (AWG)

AWG	Diameter (mm)	AWG	Diameter (mm)
1	7.35	21	0.723
2	6.54	22	0.644
3	5.83	23	0.573
4	5.19	24	0.511
5	4.62	25	0.455
6	4.12	26	0.405
7	3.67	27	0.361
8	3.26	28	0.321
9	2.91	29	0.286
10	2.59	30	0.255
11	2.31	31	0.227
12	2.05	32	0.202
13	1.83	33	0.180
14	1.63	34	0.160
15	1.45	35	0.143
16	1.29	36	0.127
17	1.15	37	0.113
18	1.02	38	0.101
19	0.912	39	0.090
20	0.812	40	0.080

AM/FM receiver — A device capable of converting either amplitude- or frequency-modulated signals into audio frequencies.

AM/FM tuner — A device capable of converting either amplitude- or frequency-modulated signals into low-level audio frequencies.

AML — Abbreviation for automatic modulation limiting. A circuit that uses an agc (automatic gain control) effect to prevent overmodulation. As a stronger voice signal is applied, this stage reduces the gain of the audio amplifier(s), keeping the modulation level below 100 percent.

ammeter— An instrument for measuring either direct or alternating electric current. Its scale is usually graduated in amperes, milliamperes, microamperes, or kiloamperes.

ammeter shunt— A low-resistance conductor placed in parallel with the meter movement so that most of the current flows through this conductor and only a small part passes through the movement itself. This arrangement extends the usable range of the meter.

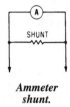

*Ammeter
shunt.*

amorphous— A characteristic, particularly of a crystal, determining that it has no regular structure.

amorphous silicon— A noncrystalline form of silicon used to fabricate transistors on large-area flat displays. Although it is not as good a semiconductor as crystalline silicon, amorphous silicon is much easier to lay down.

amorphous silicon cell— A photovoltaic cell made of silicon and hydrogen atoms deposited in an irregular atomic structure on substrate.

amortisseur winding—*See* damper winding.

amp— Abbreviation for ampere.

ampacity— The maximum current an insulated wire or cable can safely carry without exceeding either the insulation or jacket material limitations. Expressed in amperes. *See also* current-carrying capacity.

amperage— The number of amperes flowing in an electrical conductor or circuit.

ampere— Letter symbol: A. 1. A unit of electrical current or rate of flow of electrons. One volt across 1 ohm of resistance causes a current flow of 1 ampere. A flow of 1 coulomb per second equals 1 ampere. An unvarying current is passed through a solution of silver nitrate of standard concentration at a fixed temperature. A current that deposits silver at the rate of 0.001118 gram per second is equal to 1 ampere, or 6.25×10^{18} electrons per second passing a given point in a circuit. 2. The constant current which, if maintained in two straight parallel conductors of infinite length, of negligible circular sections, and placed 1 meter apart in a vacuum will produce between these conductors a force equal to 2×10^{-7} newtons per meter of length.

ampere-hour— A current of 1 ampere flowing for 1 hour. Multiplying the current in amperes by the time of flow in hours gives the total number of ampere-hours. Used mostly to indicate the amount of energy a storage battery can deliver before it needs recharging, or the energy a primary battery can deliver before it needs replacing. One ampere-hour equals 3600 coulombs.

ampere-hour capacity— The amount of current a battery can deliver in a specified length of time under specified conditions. For example, a 100-ampere-hour battery can supply 20 amperes for 5 hours.

ampere-hour efficiency— The number of ampere-hours obtained from a storage battery divided by the number of ampere-hours required to recharge the storage battery to its original condition.

ampere-hour meter— An electrical meter that measures and registers the amount or the integral, with respect to time, of the current that passes through it and is consumed in the circuit.

Ampère's rule— Current in a certain direction is equivalent to the motion of positive charges in that direction. The magnetic flux generated by a current in a wire encircles the current in the counterclockwise direction when the current is approaching the observer.

ampere-turn— A measure of magnetomotive force, especially as developed by an electric current, defined as the magnetomotive force developed by a coil of one turn through which a current of 1 ampere flows; that is, 1.26 gilberts.

amp-hr— Abbreviation for ampere-hour or ampere-hours.

amplidyne— A special direct-current generator used extensively in servo systems as a power amplifier. The response of its output voltage to changes in field excitation is very rapid, and its amplification factor is high.

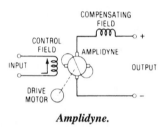

Amplidyne.

amplification— 1. Increase in size of a medium in its transmission from one point to another. May be expressed as a ratio or, by extension of the term, in decibels. 2. An increase in the magnitude of a signal brought about by passing through an amplifier.

amplification factor (λ)— 1. In a vacuum tube, the ratio of a small change in plate voltage to a small change in grid voltage required to produce the same change in plate current (all other electrode voltages and currents being held constant). 2. In any device, the ratio of output magnitude to input magnitude.

amplified AGC— An automatic gain-control (AGC) circuit in which the control voltage is amplified before being applied to the tube or transistor, the gain of which is to be controlled in accordance with the strength of the incoming signal.

amplified back bias— Degenerative voltage developed across a fast time-constant circuit within a stage of an amplifier and fed back into a preceding stage.

amplifier— 1. A device that draws power from a source other than the input signal and that produces as an output an enlarged reproduction of the essential features of its input. The amplifying element may be an electron tube, transistor, magnetic circuit, or any of various devices. 2. A device for increasing the magnitude of a signal by means of a varying control voltage, maintaining the signal's characteristic form as closely as possible to the original. 3. An electronic device for magnifying (and usually controlling) electrical signals. High-fidelity amplifiers consist of a preamplifier equalizer section, plus a power or basic amplifier section. In an integrated amplifier, both sections are built on one chassis and made available as a single unit. Alternately, the two sections are available as separate units. 4. Device for increasing power associated with a signal (voltage or current). Basic types include dc, ac, audio, linear, radio, video, differential, pulse, logarithmic.

amplifier noise — All spurious or unwanted signals, random or otherwise, that can be observed in a completely isolated amplifier in the absence of a genuine input signal.

amplifier nonlinearity — 1. The inability of an amplifier to produce an output at all times proportionate to its input. 2. Gain deviation from a straight line on a plot of amplifier output versus input (the transfer curve).

amplify — To increase in magnitude or strength, usually said of a current or voltage.

amplifying delay line — A delay line used in pulse-compression systems to amplify delayed superhigh-frequency signals.

amplistat — A self-saturating type of magnetic amplifier.

Amplitron — (Raytheon) A broadband crossed-field amplifier with a reentrant electron stream. The electron stream interacts with the backward wave of a nonreentrant rf structure.

amplitude — 1. The magnitude of variation in a changing quantity from its zero value. The word must be modified with an adjective such as peak, rms, maximum, etc., which designates the specific amplitude in question. 2. The level of an audio or other signal in voltage or current terms. 3. The extent to which an alternating or pulsating current or voltage swings from zero or from a mean value.

amplitude-controlled rectifier — A rectifier circuit in which a thyratron is the rectifying element.

amplitude density distribution — A function that gives the fraction of time that a voltage is within a narrow range.

amplitude distortion — Distortion that is present in an amplifier when the amplitude of the output signal fails to follow exactly any increase or decrease in the amplitude of the input signal. It results from nonlinearity of the transfer function and gives rise to harmonic and intermodulation distortion. No amplifier is completely free from the effect because its transfer function is slightly curved. The nature of the curvature determines the order of the distortion produced, but negative feedback and other circuit configurations help minimize the curvature within the dynamic range and hence keep the distortion at a very low level.

amplitude distribution function — A function that gives the fraction of time that a time-varying voltage is below a given level.

amplitude fading — Fading in which the amplitudes of all frequency components of a modulated carrier wave are uniformly attenuated.

amplitude-frequency distortion — The distortion that occurs when the various frequency components of a complex wave are not amplified, attenuated, or transmitted equally well.

amplitude-frequency response — The variation of gain, loss, amplification, or attenuation of a device or system as a function of frequency. Usually measured in the region where the transfer characteristic is essentially linear.

amplitude gate — *See* slicer.

amplitude-level selection — The choice of the voltage level at which an oscilloscope sweep is triggered.

amplitude limiter — A circuit or stage that automatically reduces the amplification to prevent signal peaks from exceeding a predetermined level.

amplitude-modulated transmitter — A transmitter in which the amplitude of its radio-frequency wave is varied at a low frequency rate — usually in the audio or video range. This low frequency is the intelligence (information) to be conveyed.

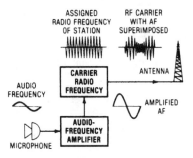

Amplitude-modulated transmitter.

amplitude-modulated wave — A constant-frequency waveform in which the amplitude varies in step with the frequency of an impressed signal.

amplitude modulation — Abbreviated AM. 1. Modulation in which the amplitude of a wave is the characteristic subject to variation. Those systems of modulation in which each component frequency (f) of the transmitted intelligence produces a pair of sideband frequencies at carrier frequency plus f and carrier frequency minus f. In special cases, the carrier may be suppressed; either the lower or upper sets of sideband frequencies may be suppressed; the lower set of sideband frequencies may be produced by one or more channels of information. The carrier may be transmitted without intelligence-carrying sideband frequencies. The resulting emission bandwidth is proportional to the highest frequency component of the intelligence transmitted. 2. A process in which the program information is imposed on a carrier signal of constant frequency by varying its amplitude in proportion to program level. Used on the standard broadcast band (530 to 1710 kHz) and on long-wave and shortwave bands.

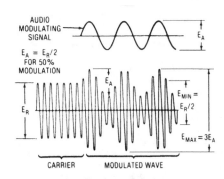

Amplitude modulation.

amplitude-modulation noise level — Undesired amplitude variations of a constant radio-frequency signal, especially in the absence of any intended modulation.

amplitude noise — The effect on radar accuracy of the fluctuations in amplitude of the signal returned by the target. These fluctuations are caused by any change in aspect if the target is not a point source.

amplitude of noise — When impulse-type noise is of random occurrence and so closely spaced that the individual waveshapes are not separated by the receiving equipment, then the noise has the waveshape and characteristics of random noise. Random-noise amplitude is proportional to the square root of the bandwidth. If the impulses are separated, the noise no longer has the waveshape of random noise and its amplitude is directly proportional to the bandwidth of the transmission system.

amplitude permeability — The relative permeability at a stated value of field strength and understated conditions, the field strength varying periodically with time and no direct magnetic-field component being present.

amplitude range — The ratio, usually expressed in decibels, between the upper and lower limits of program amplitudes that contain all significant energy contributions.

amplitude resonance — The condition that exists when any change in the period or frequency of the periodic agency (but not its amplitude) decreases the amplitude of the oscillation or vibration of the system.

amplitude response — The maximum output amplitude that can be obtained at various points over the frequency range of an instrument operated under rated conditions.

amplitude selection — The process of selecting that portion of a waveform which lies above or below a given value or between two given values.

amplitude separator — A television-receiver circuit that separates the control impulses from the video signal.

amplitude-shift keying — Abbreviated ask. The modulation of digital information on a carrier by changing the amplitude of the carrier.

amplitude-suppression ratio — In frequency modulation, the ratio of the magnitude of the undesired output to the magnitude of the desired output of an FM receiver when the applied signal is simultaneously amplitude and frequency modulated. Generally measured with an applied signal that is amplitude modulated 30 percent at a 400-hertz rate and is frequency modulated 30 percent of the maximum system deviation at a 1000-hertz rate.

amplitude versus frequency distortion — Distortion caused by the nonuniform attenuation or gain of the system, with respect to frequency under specified terminal conditions.

AM rejection ratio — The ratio of the recovered audio output produced by a desired FM signal with specified modulation, amplitude, and frequency to that produced by an AM signal, on the same carrier, with specified modulation index.

AM suppression — The ability of an FM tuner to reject AM signals. Expressed in decibels, it is the ratio between the tuner output with a 100-percent modulation FM signal to its output with a 30-percent modulated AM signal.

AM tuner — A device capable of converting amplitude-modulated signals into low-level audio frequencies.

amu — Abbreviation for atomic mass unit.

analog — 1. In electronic computers, a physical system in which the performance of measurements yields information concerning a class of mathematical problems. 2. Of or pertaining to the general class of devices or circuits in which the output varies as a continuous function of the input. 3. The representation of numerical quantities by means of physical variables, e.g., translation, rotation, voltage, resistance; contrasted with *digital*. 4. A continuous representation of phenomena in terms of points along a scale, each point merging imperceptibly into the next. An analog voltage, for example, may take any value. Real-world phenomena, such as heat and pressure, are analog (compare with *digital*).

analog adder — An analog circuit or device that receives two or more inputs and delivers an output that is equal to their sum.

analog amplifier — A device whose output is continuously proportional to the input stimulus.

analog channel — A computer channel in which the transmitted information can have any value between the defined limits of the channel.

analog circuit — A circuit in which the output varies as a continuous function of the input, as contrasted with digital circuits.

analog communications — A system of telecommunications employing a nominally continuous electrical signal that varies in frequency, amplitude, etc., in some direct correlation to nonelectrical information (sound, light, etc.) impressed on a transducer.

analog computer — 1. A computer operating on the principle of creating a physical (often electrical) analogy of the mathematical problem to be solved. Variables such as temperature, light, pressure, distance, angle, shaft speed, or flow are represented by the magnitude of a physical phenomenon such as voltage or current. The computer manipulates these variables in accordance with the mathematical formulas "analogued" on it. 2. A computer system in which both the input and output are continuously varying signals. 3. A computing machine that works on the principle of measuring, as distinguished from counting. 4. A computer that solves problems by setting up equivalent electric circuits and making measurements as the variables are changed in accordance with the corresponding physical phenomena. An analog computer gives approximate solutions, whereas a digital computer gives exact solutions. 5. A nondigital computer that manipulates linear (continuous) data to measure the effect of a change in one variable on all other variables in a particular problem. (Compare: *digital computer*.)

analog computing — Computing system in which continuous signals represent mechanical (or other) parameters.

analog data — 1. A physical representation of information such that the representation bears an exact relationship to the original information. The electrical signals on a telephone channel are an analog data representation of the original voice. 2. Data represented in a continuous form, as contrasted with digital data represented in a discrete (discontinuous) form. Analog data is usually represented by physical variables, such as voltage, resistance, rotation, etc.

analog input module — An I/O rack module that converts an analog signal from a user device to a digital signal that may be processed by the processor.

analog meter — An indicating instrument that employs a movable coil and pointer arrangement (or equivalent) to display values along a graduated scale.

analog multiplexer — 1. Circuit used for time-sharing of analog-to-digital converters between a number of different analog information channels. Consists of a group of analog switches arranged with inputs connected to the individual analog channels and outputs connected in common. 2. Two or more analog switches with separate inputs and a common output, with each gate separately controllable. Multiplexing is performed by sequentially turning on each switch one at a time, switching each individual input to a common output. 3. A device that selects one of several analog signals according to a digital code. Analog multiplexers (amux) are available in many forms; their chief application is as a front end in data-acquisition systems, enabling a single analog-to-digital converter to monitor more than one information channel.

analog network — A circuit or circuits that represent physical variables in such a manner as to permit the expression and solution of mathematical relationships between the variables, or to permit the solution directly by electric or electronic means.

analog output — 1. A signal (voltage) whose amplitude is continuously proportionate to the stimulus, the proportionality being limited by the resolution of the device. 2. An output quantity that varies smoothly over a continuous range of values rather than in discrete steps.

analog panel meter — *See* APM.

analog recording — A method of recording in which some characteristic of the record current, such as amplitude or frequency, is continuously varied in a manner analogous to the time variations of the original signal.

analog representation — A representation that does not have discrete values, but is continuously variable.

analog signal — 1. An electrical signal that varies continuously in both time and amplitude, as obtained from temperature or pressure, or speed transducers. A voltage level that changes in proportion to the change in a physical variable. 2. A signal representing a variable that may be continuously observed and continuously represented.

analog switch — 1. A device that either transmits an analog signal without distortion or completely blocks it. 2. Any solid-state device, with or without a driver, capable of bilaterally switching voltages or current. It has an input terminal, output terminal, and, ideally, no offset voltage, low *on* resistance, and extreme isolation between the signal being gated and control signals. 3. A means to interconnect two or more circuits whose information is represented in analog form using a network that may or may not be time divided and may or may not consist of linear elements.

analog-to-digital conversion — 1. The process of converting a continuously variable (analog) signal to a digital signal (binary code) that is a close approximation of the original signal. 2. The process of quantizing a continuous function.

analog-to-digital converter — Abbreviated a-d converter, adc, or ADC. 1. A circuit that changes a continuously varying voltage or current (analog) into a digital output. The input may be ac or dc, and the output may be serial or parallel, binary or decimal. 2. Device that translates analog signals (voltages, pressures, etc.) from sensors into numerical digital form (binary, decimal, etc.).

Analytical Engine — An early form of general-purpose digital computer invented in 1833 by Charles Babbage.

analyzer — 1. An instrument or other device designed to examine the functions of components, circuits, or systems and their relations to each other, as contrasted with an instrument designed to measure some specific parameter of such a system or circuit. 2. Of computers, a routine the purpose of which is to analyze a program written for the same or a different computer. This analysis may consist of summarizing instruction references to storage and tracing sequences of jumps. 3. An instrument that evaluates and/or measures one or more specific parameters (e.g., voltage, current, frequency, logic level, bit time, distortion). 4. A test assembly that checks the performance of, or locates trouble in, electronic equipment. Also called test set and tester.

anastigmat — A lens system designed so as to be free from the aberration called astigmatism.

anchor — An object, such as a metal rod, set into the ground to hold the end of a guy wire.

ancillary equipment — Equipment not directly employed in the operation of a system but necessary for logistic support, preparation for flight, or assessment of target damage; e.g., test equipment, vehicle transport.

AND circuit — Synonym for AND gate.

AND device — A device that has its output in the logic 1 state if and only if all the control signals are in the logic 1 state.

Anderson bridge — A bridge normally used for the comparison of self-inductance with capacitance. It is a six-branch network in which an outer loop of four arms is formed by four nonreactive resistors and the unknown inductor. An inner loop of three arms is formed by a capacitor and a fifth resistor in series with each other and in parallel with the arm opposite the unknown inductor. The detector is connected between the junction of the capacitor and the fifth resistor and at that end of the unknown inductor separated from a terminal of the capacitor by only one resistor. The source is connected to the other end of the unknown inductor and to the junction of the capacitor with two resistors of the outer loop. The balance is independent of frequency.

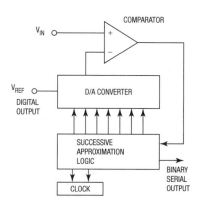

Analog-to-digital converter.

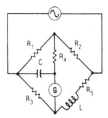

Anderson bridge.

AND gate — 1. In an electronic computer, a gate circuit with more than one control (input) terminal. No output signal will be produced unless a pulse is applied to all inputs simultaneously. 2. A binary circuit, with two or more inputs and a single output, in which the output is logic 1 only when all inputs are logic 1, and the output is logic 0 if any one of the inputs is logic 0.

AND/NOR gate — A single logic element that performs the operation of two AND gates with outputs feeding a NOR gate. No access to the internal logic elements is provided (i.e., no connection is available at the outputs of the AND gates).

analog transmission — Transmission of a continuously variable signal as opposed to a discretely variable one.

analog value — A continuously variable value, such as a current or voltage.

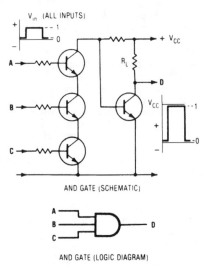

AND gate with three inputs.

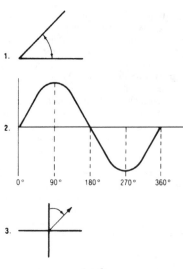

Angle.

AND/OR circuit — A gating circuit that produces a prescribed output condition when several possible combinations of input signals are applied. It exhibits the characteristics of the AND gate and the OR gate.

android — 1. A mobile mechanism possessing the ability to manipulate objects external to itself under the constant control of its own resident intelligence, operating within guidelines initially established and occasionally updated by a human being, a computer, or some other external intelligence. 2. Automaton of manlike form.

anechoic — Nonreflective, producing no echoes.

anechoic chamber — 1. A room or chamber specially designed to absorb all sound within, thus preventing sound reflections or reverberation. Such rooms are used for evaluation of microphones and speakers. 2. A room lined with material that traps sound waves so the sound is perfectly absorbed and the room is acoustically dead. Such a chamber is used for testing microphones and speakers. 3. A derived term for a room or enclosure that is designed to be echo free over a specified frequency range. Any sound reflections within this frequency range must be less than 10 percent of the source sound pressure.

anechoic enclosure — A special echo-free enclosure used for testing audio transducers, in which all wall surfaces have been covered with acoustically absorbent materials so that reflections of the sound waves are eliminated. Also known as a dead room or an anechoic room.

anechoic room — A room whose walls have been treated so as to make them absorb a particular kind of radiation almost completely; used for testing components of sound systems, radar systems, etc., in an environment free of reflections.

anelectronic — *See* anelectrotonus.

anelectrotonus — The reduced sensitivity produced in a nerve or muscle in the region of contact with the anode when an electric current is passed through it.

anemometer — An instrument used for measuring the force or speed of wind.

angels — Short-duration radar reflections in the lower atmosphere. Most often caused by birds, insects, organic particles, tropospheric layers, or water vapor.

angle — 1. A fundamental mathematical concept formed when two straight lines meet at a point. The lines are the sides of the angle, and the point of intersection is the vertex. 2. A measure of the distance along a wave or part of a cycle, measured in degrees. 3. The distance through which a rotating vector has progressed.

angle jamming — An electronic countermeasures technique in which azimuth and elevation information present in the modulation components of the returning echo pulse of a scanning fire-control radar is jammed by transmitting a pulse similar to the radar pulse but with angle information of erroneous phase.

angle modulation — Modulation in which the angle of a sine-wave carrier is the characteristic varied from its normal value by modulation. Phase and frequency modulation are particular forms of angle modulation.

angle noise — Tracking error introduced into radar by variations in the apparent angle of arrival of the echo from a target due to finite target size. (This effect is caused by variations in the phase front of the radiation from a multiple-point target as the target changes its aspect with respect to the observer.)

angle of arrival — Angle made between the line of propagation of a radio wave and the earth's surface at the receiving antenna.

angle of azimuth — The angle measured clockwise in a horizontal plane, usually from the north. The north used may be true north, Y-north, or magnetic north.

angle of beam — The angle that encloses most of the transmitted energy from a directional-antenna system.

angle of convergence — Angle formed by the lines of sight of both eyes when focusing on an object.

angle of deflection — The angle formed between the new position of the electron beam in a cathode-ray tube and the normal position before deflection.

angle of departure — The angle of the line of propagation of a radio wave with respect to a horizontal plane at the transmitting antenna.

angle of divergence — In cathode-ray tubes, a measure of its spread as the electron beam travels from cathode to the screen. The angle formed by an imaginary center line and the border line of the electron beam. In good tubes, this angle is less than 2°.

angle of elevation — The angle between the horizontal plane and the line ascending to the object.

angle of incidence — The angle between a wave or beam striking a surface and a line perpendicular to that surface.

angle of lag — The angular phase difference between one sinusoidal function and a second having the same frequency. Expressed in degrees, the amount the second function must be retarded to coincide with the first.

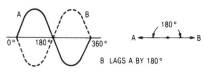

Angle of lag.

angle of lead — 1. The time or angle by which one alternating electrical quantity leads another of the same cyclic period. 2. The angle through which the commutator brushes of a generator or motor must be moved from the normal position to prevent sparking.

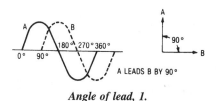

Angle of lead, 1.

angle of radiation — The angle between the surface of the earth and the center of the beam of energy radiated upward into the sky from a transmitting antenna.

angle of reflection — The angle between a wave or beam reflected from a surface and a line perpendicular to that surface. This angle lies in a common plane with the angle of incidence and is equal to it.

angle of refraction — The angle between a wave or beam as it passes through a medium and a line perpendicular to the surface of that medium. This angle lies in a common plane with the angle of incidence.

angle tracking noise — Any deviation of the tracking axis from the center of reflectivity of a target. The resultant of servo noise, receiver noise, angle noise, and amplitude noise.

angstrom unit — A unit of measurement of wavelength of light and other radiation. Equal to one ten-thousandth (10^{-4}) of a micrometer or one hundred-millionth of a centimeter (10^{-8} cm). The visible spectrum extends from about 4000 to 8000 angstrom units. Blue light has a wavelength in the region of 4700 angstroms; yellow, 5800; and red, 6500. A measure of wavelength equal to 10^{-10} meter, or 0.1 nanometer, the preferred term.

angular acceleration — The rate at which angular velocity changes with respect to time, generally expressed in radians per second.

angular accelerometer — A device capable of measuring the magnitude of, and/or variations in, angular acceleration.

angular aperture — The largest angular extent of wave surface that an objective can transmit.

angular deviation loss — The ratio of the response of a microphone or speaker on its principal axis to the response at a specified angle from the principal axis (expressed in decibels).

angular distance — The angle subtended by two bodies at the point of observation. It is equal to the distance of wavelengths multiplied by 2π radians or by 360°.

angular frequency — Frequency expressed in radians per second. It is equal to the number of hertz (cycles per second) multiplied by 2π.

angular length — Length expressed in radians or equivalent angular measure equal to 2π radians, or 360°, multiplied by the length in wavelengths.

angular momentum — The momentum that a body has by virtue of its rotational movement.

angular phase difference — Phase difference between two sinusoidal functions expressed as an angle.

angular rate — The rate of change of bearing.

angular resolution — The ability of a radar to distinguish between two targets solely on the basis of angular separation.

angular velocity — The rate at which an angle changes. Expressed in radians per second, the angular velocity of a periodic quantity is the frequency multiplied by 2π. If the periodic quantity results from uniform rotation of a vector, the angular velocity is the number of radians per second passed over by the rotating vector. Generally designated by the Greek letter omega (ω).

anharmonic oscillator — An oscillating system in which the restoring force is a nonlinear function of the displacement from equilibrium.

anhysteresis — The process whereby a material is magnetized by applying a unidirectional field upon which is superimposed an alternating field of gradually decreasing amplitude.

ANIK — The Canadian domestic satellite system used to transmit the network television feeds of the Canadian Broadcasting Corporation. All ANIK satellites are operated by TeleSat Canada of Ottawa. ANIK satellites have both 4-GHz C-band and 12-GHz Ku-band transponders. ANIK means *brother* in Inuit (Eskimo).

animation — A moving on-screen representation of the activities taking place in a simulation.

anion — 1. A negatively charged ion which, during electrolysis, is attracted toward the anode. A corresponding positive ion is called a cation. 2. A negative ion that moves toward the anode in a discharge tube, electrolytic cell, or similar device.

anisotropic — 1. Describing a substance that exhibits different magnetic, electrical, optical, and other physical properties when measured along axes in different directions. 2. A material that has characteristics such as wave propagation constant, magnetic permeability, conductivity, etc., that vary with direction; that is, not isotropic.

anisotropic body — A body in which the value of any given property depends on the direction of measurement, as opposed to a body that is isotropic.

anisotropic magnet — A magnetic material having a better magnetic characteristic along the preferred axis than along any other.

anisotropic material — A material having preferred orientation so that the magnetic characteristics are superior along a particular axis. This may be as a result of rolling, heat treatment in a magnetic field, or, in the case of some of the sintered magnets, the direction of press.

anisotropy — Directional dependence of magnetic properties, leading to the existence of easy or preferred directions of magnetization. Anisotropy of a particle may be related to its shape, to its crystalline structure, or to the existence of strains within it.

anneal — 1. To heat a metal to a predetermined temperature and then let it cool slowly. This process prevents brittleness and often stabilizes electrical characteristics.

2. To heat and then gradually cool in order to relieve mechanical stresses. Annealing copper makes it softer and less brittle.

annealed laminations — Laminations that have been annealed for transformers or choke coils.

annealed wire — Wire that has been softened by heating and gradual cooling to remove mechanical stresses.

annotation — An added descriptive comment or explanatory note.

annular — Ringed; ring-shaped.

annular conductor — A conductor consisting of a number of wires stranded in three reversed concentric layers surrounding a saturated hemp core. The core is usually made wholly or mostly of nonconducting material. This construction has the advantage of lower total ac resistance for a given cross-sectional area of conducting material by eliminating the greater skin effect at the center.

annular transistor — A mesa transistor in which the semiconductor regions are arranged in concentric circles about the emitter.

annulling network — An arrangement of impedance elements connected in parallel with filters to annul or cancel capacitive or inductive impedance at the extremes of the passband of a filter.

annunciation relay — 1. An electromagnetically operated signaling apparatus that indicates whether a current is flowing or has flowed in one or more circuits. 2. A nonautomatic reset device that gives a number of separate visual indications upon the functioning of protective devices, and which may also be arranged to perform a lockout function.

annunciator — 1. A visual device consisting of a number of pilot lights or drops. Each light or drop indicates the condition that exists or has existed in an associated circuit and is labeled accordingly. 2. A device for sounding an alarm or attracting attention. The indication is usually aural, but occasionally may be visual or both aural and visual. 3. An alarm-monitoring device that consists of a number of visible signals, such as flags or lamps indicating the status of the detectors in an alarm system or systems. Each circuit in the device is usually labeled to identify the location and condition being monitored. In addition to the visible signal, an audible signal is usually associated with the device. When an alarm condition is reported, a signal is indicated — visible, audible, or both. The visible signal is generally maintained until reset either manually or automatically.

anode — 1. The positive electrode, such as the plate of a vacuum tube; the element to which the principal stream of electrons flows. 2. In a cathode-ray tube, the electrodes connected to a source of positive potential. These anodes are used to concentrate and accelerate the electron beam for focusing. 3. The less noble and/or higher-potential electrode of an electrolytic cell, at which corrosion occurs. This may be an area on the surface of a metal or alloy, the more active metal in a cell composed of two dissimilar metals, or the positive electrode of an impressed-current system.

ANODE

TRIODE   DIODE

Anode.

anode-balancing coil — A set of mutually coupled windings used to maintain approximately equal currents in anodes operating in parallel from the same transformer terminal.

anode breakdown voltage — The potential required to cause conduction across the main gap of a gas tube when the starter gap is not conducting and all other tube elements are held at cathode potential.

anode-bypass capacitor — Also called plate-bypass capacitor. A capacitor connected between the anode and ground in an electron-tube circuit. Its purpose is to bypass high-frequency currents and keep them out of the load.

anode characteristic curve — A graph that shows how the anode current of an electron tube is affected by changes in the anode voltage.

anode circuit breaker — A device used in the anode circuits of a power rectifier for the primary purpose of interrupting the rectifier circuit if an arcback should occur.

anode current — The electron flow in the element designated as the anode. Usually signifies plate current.

anode dark space — In a gas tube, a narrow, dark zone next to the surface of the anode.

anode dissipation — The power dissipated as heat in the anode of an electron tube because of the bombardment by electrons and ions.

anode efficiency — *See* plate efficiency.

anode-load impedance — *See* plate-load impedance.

anode modulation — *See* plate modulation.

anode neutralization — Also called plate neutralization. A method of neutralization in which a portion of the anode-cathode ac voltage is shifted 180° and applied to the grid-cathode circuit through a neutralizing capacitor.

anode power input — *See* plate power input.

anode power supply — The means for supplying power to the plate of an electron tube at a more positive voltage than that of the cathode. Also called plate power supply.

anode pulse modulation — *See* plate pulse modulation.

anode rays — Positive ions coming from the anode of an electron tube; these ions are generally due to impurities in the metal of the anode.

anode saturation — *See* plate saturation.

anode sheath — A layer of electrons surrounding the anode in mercury-pool arc tubes.

anode strap — A metallic connector between selected anode segments of a multicavity magnetron, used principally for mode separation.

anode supply — Also called plate supply. The dc voltage source used in an electron-tube circuit to place the anode at a high positive potential with respect to the cathode.

anode terminal — 1. In a diode (semiconductor or tube), that terminal to which a positive dc voltage must be applied to forward-bias the diode. Compare with *cathode terminal*. 2. In a diode (semiconductor or tube), that terminal at which a negative dc voltage appears when the diode is employed as an ac rectifier (blocking). 3. That terminal which is internally connected to the anodic element of any device.

anode voltage — The potential difference existing between the anode and cathode.

anode voltage drop (of a glow-discharge, cold cathode tube) — Difference in potential between cathode and anode during conduction, caused by the electron flow through the tube resistance (*IR* drop)

anodic protection — Corrosion inhibition based on the electrolytic formation of a protective passive film on metals by applying to them a positive (anodic) potential; e.g., aluminum is anodized (oxidized) by a positive charge in a sulfuric acid solution.

anodic silver — A precious metal used in plating; fine silver in different configurations, such as shot, cones, bars, etc., is sacrificed during the silver-plating process.

anodization — The formation of an insulating oxide over certain elements, usually metals, by electrolytic action. The most commonly anodized materials are tantalum, aluminum, titanium, and niobium. Anodization is particularly useful where protection of a conductor is required. The base metal can form the conductor and the anodized surface layer can form the insulator.

anodize — To deposit a protective coating of oxide on a metal by means of an electrolytic process in which it is used as the anode.

anodizing — An electrochemical oxidation process used to improve the corrosion resistance or to enhance the appearance of a metal surface. Aluminum and magnesium parts are frequently anodized.

anomalous displacement current — Also called dielectric absorption. The current in addition to the normal leakage current in a circuit containing a capacitor with an imperfect dielectric after the normal charging or discharging current has become negligibly small.

anomalous photoconductivity — A spectral phenomenon in which the degree of the photoresponse of an illuminated semiconductor is determined by the wavelength composition of the incident light.

anomalous propagation — 1. Propagation that is unusual or abnormal. 2. The conduction of UHF signals through atmospheric ducts or layers in a manner similar to that of a waveguide. These atmospheric ducts carry the signals with less than normal attenuation over distances far beyond the optical path taken by UHF signals. Also called superrefraction. 3. In sonar, pronounced and rapid variations in the strength of the echo due to large, rapid focal fluctuations in propagation conditions.

anonymous FTP (File Transfer Protocol) — The procedure of connecting to a remote computer as an anonymous or guest user in order to transfer public files back to a local computer. *See also* FTP; protocol.

A-N radio range — A navigational aid that provides four equisignal zones for aircraft guidance. Deviation from the assigned course is indicated aurally by the Morse code letters A (·—) or N (—·). On-course position is indicated by an audible merging of the A and N code signals into a continuous tone.

ANSI — American National Standards Institute. The U.S. government organization with responsibility for the development and promulgation of (among others) data processing standards.

A-N signal — A radio-range, quadrant-designation signal that indicates to the pilot whether he or she is on course or to the right or left.

ANSI keyboard — Abbreviation for American National Standards Institute keyboard. A typewriter standard unit that offers a choice of uppercase characters only or uppercase and lowercase combined.

ANSI standards — A series of standards recommended by the American National Standards Institute.

answerback — 1. The response of a terminal to remote-control signals. 2. A signal sent by a data receiver to a data transmitter indicating that it is ready to receive data or is acknowledging the receipt of data. *See also* handshaking. 3. A reply message from a terminal, manually or automatically initiated, to verify that the right terminal has been accessed and is in operation.

answerback (W-R-U) system — A system capable of being remotely controlled by another station. When tripped by a unique access code, a short predetermined message is broadcast.

answerback unit — An electromechanical device used with a teletypewriter set to transmit a predetermined message of not more than 21 characters in response to a request signal. It can transmit either a five-level 7.42 unit code or an eight-level 11.0 unit code at speeds of up to 100 words per minute.

answering cord — The cord nearest the face of a telephone switchboard. It is used for answering subscriber's calls and calls on incoming trunks.

answering service — A business that contracts with subscribers to answer incoming telephone calls after a specified delay or when scheduled to do so. It may also provide other services, such as relaying fire or intrusion alarm signals to proper authorities.

answer lamp — In a telephone switchboard, a lamp that lights when an answer cord is plugged into a line jack; it extinguishes when the telephone answers and lights when the call is complete.

answer tone — Tone signal, with a frequency between 2025 and 2225 Hz and a duration of at least 1.5 s, used by an answering modem to indicate its ready condition to an originating modem.

antenna — Also called aerial. 1. That portion, usually wires or rods, of a radio transmitter or receiver station used for radiating waves into or receiving them from space. It changes electrical currents into electromagnetic radio waves, and vice versa. 2. A section of wire or a metallic device designed to intercept radio waves in the air and convert them to an electrical signal for feeding to a receiver. Under relatively difficult reception conditions, such as created by location, terrain, obstructions, etc., an antenna becomes fairly critical and should be one especially designed for its intended purposes. 3. A device for transmitting and receiving radio waves. Depending on their use and operating frequency, antennas can take the form of a single piece of wire, a dipole, a grid such as a yagi array, a horn, a helix, a sophisticated parabolic-shaped dish, or a phase array of active electronic elements of virtually any flat or convoluted surface. 4. A device that collects and focuses electromagnetic energy, i.e., contributes an energy gain. Gain is proportional to surface area for a microwave dish.

antenna array — 1. A combination of antennas assembled to obtain a desired pickup or rejection pattern. 2. An arrangement of two or more directional antennas, spaced and connected so that they are in phase and their effects are electrically additive.

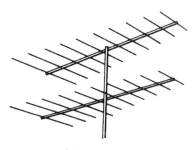

Antenna array.

antenna bandwidth — 1. The range of frequencies over which the impedance characteristics of the antenna are sufficiently uniform that the quality of the radiated signal is not significantly impaired. 2. The frequency range over which a certain antenna characteristic falls within acceptable limits. For instance, an antenna may have a bandwidth of 1 MHz over which the standing-wave ratio is 2 : 1 or less. 3. The frequency range throughout which an antenna will operate at a specified efficiency without the need for alteration or adjustment.

antenna beam width — The angle, in degrees, between two opposite half-power points of an antenna beam.

antenna coil — In a radio receiver or transmitter, the inductance through which antenna current flows.

antenna coincidence — That instance when two rotating, highly directional antennas are pointed toward each other.

antenna-conducted interference — Any signal that is generated within a transmitter or receiver and appears as an undesired signal at the antenna terminals of the device, e.g., harmonics of a transmitter signal, or the local-oscillator signal of the receiver.

antenna cores — Ferrite cores of various cross sections for use in radio antennas.

Antenna core.

antenna coupler — 1. A radio-frequency transformer used to connect an antenna to a transmission line or to connect a transmission line to a radio receiver. 2. A radio-frequency transformer, link circuit, or tuned line used to transfer radio-frequency energy from the final plate-tank circuit of a transmitter to the transmission line feeding the antenna.

antenna crosstalk — A measure of undesired power transfer through space from one antenna to another. Usually expressed in decibels, the ratio of power received by one antenna to the power transmitted by the other.

antenna current — The radio-frequency current that flows in an antenna.

antenna detector — A device consisting of an antenna and electronic equipment to warn aircraft crew members of their being observed by radar sets. The device is usually located in the nose or tail of the aircraft and illuminates a light on one or more panels when radar signals are detected.

antenna diplexer — A coupling device that permits several transmitters to share one antenna without troublesome interaction.

antenna-directivity diagram — A curve representing, in polar or Cartesian coordinates, a quantity proportional to the gain of an antenna in the various directions in a particular plane or cone.

antenna disconnect switch — A safety switch or interlock plug used to remove driving power from the antenna to prevent rotation while work is being performed.

antenna duplexer — A circuit or device that permits one antenna to be shared by two transmitters without undesirable interaction.

antenna effect — 1. Cause of error in a loop antenna due to the capacitance to ground. 2. In a navigational system, any undesirable output signal that results when a directional antenna acts as a nondirectional antenna. 3. The tendency of wires or metallic bodies to act as antennas, i.e., to radiate or pick up radio signals.

antenna effective area — In any specified direction, the square of the wavelength multiplied by the power gain (or directional gain) in that direction, and divided by 4π. (When power gain is used, the effective area is that for power reception; when directive gain is used, the effective area is that for directivity.)

antenna efficiency — The relative ability of an antenna to convert rf energy from a transmitter into electromagnetic waves. If the gain rating of a directional antenna is 10 dB, for example, it is often assumed that the effective radiated power will be 10 times greater than the rf power fed to it. However, if the antenna efficiency is, say, 50 percent, a loss of 3 dB, the true gain will be only 7 dB ($10 - 3 = 7$ dB).

antenna elevation — The physical height of an antenna above the earth.

antenna factor — The value of decibel that must be added to a two-terminal voltmeter reading to obtain the actual induced antenna open-circuit voltage or the electric-field strength.

antenna farm — A large plot of ground (5 to 2000 acres) surrounding a radio transmitting or receiving station that provides space and adequate clearance for the installation of several large antennas, such as rhombic antennas.

antenna field — 1. The region defined by a group of antennas. 2. A group of antennas placed in a geometric configuration that is specific for a particular trajectory measuring system. 3. The effective free-space energy distribution produced by an antenna or group of antennas.

antennafier — An integrated low-profile antenna and amplifier for use with compact, portable communications systems.

antenna front-to-back ratio — The ratio of field strength in front of a directional antenna (i.e., directly forward in the line of maximum directivity) to the field strength in back of the antenna (i.e., 180° from the front). Measured at a fixed distance from the radiator.

antenna gain — 1. The effectiveness of a directional antenna in a particular direction, compared against a standard (usually an isotropic antenna). The ratio of standard antenna power to the directional antenna power that will produce the same field strength in the desired direction. 2. The increase in signal level at the antenna terminals with reference to the level at the terminals of a half-wave dipole antenna, expressed in decibels. 3. For a given antenna, the ratio of signal strength (received or transmitted) to that obtained with a simple dipole antenna.

antenna ground system — That portion of an antenna closely associated with the earth and including an extensive conducting surface, which may be the earth itself.

antenna height — The average height above the terrain from 2 to 10 miles (3.2 to 16 km) from the antenna. In general, the antenna height will be different in each direction from the antenna. The average of these various heights is considered the antenna height above average terrain.

antenna height above average terrain — The height of the center of radiation of an antenna above an averaged value of the elevation above sea level for the surrounding terrain.

antenna illumination — Describes how a feedhorn "sees" the surface of a dish as well as the surrounding terrain.

antenna impedance — The impedance an antenna presents to a transmitter or receiver at the attachment point of the transmission line or feeder. It varies from about 50 to 600 ohms, depending on antenna type and installation.

antenna induced microvolts — The voltage that exists across the open-circuited antenna terminals, as calculated from a measurement.

antenna lens — An arrangement of metal vanes or dielectric material used to focus a microwave beam in a manner similar to an optical lens.

antenna lobe — *See* lobe.

antenna matching — Selection of components to make the impedance of an antenna equal to the characteristic impedance of its transmission line.

antennamitter — An integrated low-profile antenna and oscillator for use with compact, portable communications systems.

antenna pair — Two antennas located on a base line of accurately surveyed length. The signals received by these antennas are used to determine quantities related to a target position.

antenna pattern — Also called antenna polar diagram. A plot of angle versus free-space field intensity at a fixed distance in the horizontal plane passing through the center of the antenna.

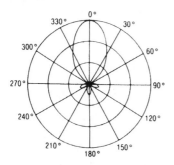

Antenna pattern.

antenna-pattern measuring equipment — Devices used to measure the relative field strength or intensity existing at any point or points in the space immediately surrounding an antenna.

antenna pedestal — A structure that supports an antenna assembly (motors, gears, synchros, rotating joints, etc.).

antenna polar diagram — *See* antenna pattern.

antenna polarization — The position of an antenna, with respect to the surface of the earth, that determines the wave polarization for which the antenna is most efficient. A vertical antenna radiates and receives vertically polarized waves; a horizontal antenna radiates and receives horizontally polarized waves broadside to itself and vertically polarized waves at high angles off its ends.

antenna power — The square of the antenna current of a transmitter, multiplied by the antenna resistance at the point where the current is measured.

antenna power gain — The power gain of an antenna in a given direction is four times the ratio of the radiation intensity in that direction to the total power delivered to the antenna. (The term is also applied to receiving antennas.)

antenna preamplifier — A low-noise rf amplifier, usually mast-mounted near the terminals of the receiving antennas, used to compensate for transmission-line loss and thereby improve the overall noise figure.

antenna reflector — In a directional-antenna array, an element that modifies the field pattern in order to reduce the field intensity behind the array and increase it in front. In a receiving antenna, the reflector reduces interference from stations behind the antenna.

antenna relay — A relay used in radio stations to automatically switch the antenna to the receiver or transmitter and thus protect the receiver circuits from the rf power of the transmitter.

antenna resistance — The total resistance of a transmitting antenna system at the operating frequency. The power supplied to the entire antenna circuit, divided by the square of the effective antenna current referred to the feed point. Antenna resistance is made up of such components as radiation resistance, ground resistance, radio-frequency resistance of conductors in the antenna circuit, and equivalent resistance due to corona, eddy currents, insulator leakage, and dielectric power loss.

antenna resonant frequency — The frequency (or frequencies) at which an antenna appears to be a pure resistance.

antenna stabilization — A system for holding a radar beam steady despite the roll and pitch of a ship or airplane.

antenna structure — A structure that includes the radiating system, its supporting structures, and appurtenances mounted thereon.

antenna switch — Switch used for connecting an antenna to or disconnecting it from a circuit.

antenna system — An assembly consisting of the antenna and the necessary electrical and mechanical devices for insulating, supporting, and/or rotating it.

antenna terminals — On an antenna, the points to which the lead-in (transmission line) is attached.

antenna tilt error — The angular difference between the antenna tilt angle shown on the mechanical indicator and the electrical center of the radar beam.

antennaverter — A receiving antenna and converter combined in a single unit that feeds directly into the receiver IF amplifier.

antenna wire — A wire, usually of high tensile strength, such as copperweld, bronze, etc., with or without insulation, used as an antenna for radio and electronic equipment.

antiaircraft missile — A guided missile launched from the surface against an airborne target.

anti-aliasing — The smoothing or removal of diagonal lines in digitized images at low resolutions that appear as stair-steps in order to recreate smoother diagonal lines.

anti-aliasing filter — A filter (normally low pass) that band-limits the input signal before sampling to less than half the sampling rate to prevent aliasing noise.

anticapacitance switch — A switch with widely separated legs, designed to keep capacitance at a minimum in the circuits being switched.

anticathode — Also called target. The target of an X-ray tube on which the stream of electrons from the cathode is focused and from which the X-rays are radiated.

anticlutter circuit — In a radar receiver, an auxiliary circuit that reduces undesired reflection in order to permit the detection of targets that otherwise would be obscured by such reflections.

anticlutter gain control — A device that automatically and gradually increases the gain of a radar receiver from low to maximum within a specified period after each transmitter pulse. In this way, short-range echoes producing clutter are amplified less than long-range echoes.

anticoincidence — A nonsimultaneous occurrence of two or more events (usually, ionizing events).

anticoincidence circuit — 1. A counter circuit that produces an output pulse when either of two input circuits receives a pulse, but not when the two inputs receive pulses simultaneously. 2. A circuit that provides an output only when all inputs are absent; a NAND circuit.

anticollision radar — A radar system used in an aircraft or ship to warn of possible collision.

antiferroelectricity — The property of a class of crystals that also undergo phase transitions from a higher to a lower symmetry. They differ from the ferroelectrics in having no electric dipole moment.

antiferroelectric materials — Those materials in which spontaneous electric polarization occurs in lines of ions; adjacent lines are polarized in an antiparallel arrangement.

antiferromagnetic materials — Those materials in which spontaneous magnetic polarization occurs in equivalent sublattices; the polarization in one sublattice is aligned antiparallel to the other.

antiferromagnetic resonance — The absorption of energy from an oscillating electromagnetic field by a system of processing spins located on two sublattices, with the spins on one sublattice going in one direction and the spins on the other sublattice in the opposite direction.

antiferromagnetism — A phenomenon of magnetism characterized by the elimination of magnetic moments and decrease in magnetic susceptibility with a decrease in temperature due to the equal power of atomic magnets.

antihunt — A stabilizing signal or equalizing circuit used in a closed-loop feedback system of a servomechanism to prevent the system from hunting, or oscillating. Special types of antihunt circuits are the anticipator, derivative, velocity feedback, and damper.

antihunt circuit — A circuit used to prevent excessive correction in a control system.

antihunt device — A device used in positioning systems to prevent hunting, or oscillation, of the load around an ordered position. The device may be mechanical or electrical. It usually involves some from of feedback.

antijamming — 1. Minimizing the effect of enemy electronic countermeasures to permit echoes from targets detected by radar to be visible on the indicator. 2. Controls or circuit features incorporated to minimize jamming.

antijamming radar data processing — Use of data from one or more radar sources to determine target range in the presence of jamming.

antilogarithm — The number from which a given logarithm is derived. For example, the logarithm of 4261 is 3.6295. Therefore the antilogarithm of 3.6295 is 4261.

antimagnetic — Made of alloys that will not remain in a magnetized state.

antimicrophonic — Specifically designed to prevent microphonics. Possessing the characteristic of not introducing undesirable noise or howling into a system.

antimissile missile — A missile that is launched to intercept and destroy another missile in flight.

antinode — The point on a transmission line at which the current is maximum and the voltage is minimum.

antinodes — Also called loops. The points of maximum displacement in a series of standing waves. Two similar and equal wave trains traveling at the same velocity in opposite directions along a straight line result in alternate antinodes and nodes along the line. Antinodes are separated from their adjacent nodes by half the wavelength of the wave motion.

antinoise carrier-operated device — A device commonly used to mute the audio output of a receiver during standby or no-carrier periods. Usually the automatic volume control voltage is used to control a squelch tube which, in turn, controls the bias applied to the first audio tube so that it is permitted to operate only when a carrier is present at the receiver input. Thus, the receiver output is heard when a signal is received, and is muted when no signal is present.

antinoise microphone — A microphone that discriminates against acoustic noise. A lip or throat microphone is an example.

antiphase — Two identical signals disposed in 180° phase opposition. When superimposed, they tend to cancel each other because their waveform patterns are of equal magnitude but opposite polarity.

antiproton — An elementary atomic particle that has the same mass as a proton but is negatively charged.

antirad — A material that inhibits damage caused by radiation.

antiresonance — A type of resonance in which a system offers maximum impedance at its resonant frequency.

antiresonant circuit — A parallel resonant circuit offering maximum impedance to the series passage of the resonant frequency.

antiresonant frequency — 1. The frequency at which the impedance of a system is very high. 2. Of a crystal unit, the frequency for a particular mode of vibration at which, neglecting dissipation, the effective impedance of the crystal unit is infinite.

antisidetone — 1. In a telephone circuit, special circuits and equipment that are so arranged that only a negligible amount of the power generated in the transmitter reaches the associated receiver. 2. Pertaining to the reduction or elimination of interference in telephone circuits between the microphone and earphone of the same telephone.

antisidetone circuit — A telephone circuit that prevents sound, introduced in the local transmitter, from being reproduced in the local receiver. (Reduces sidetones.)

antisidetone induction coil — An induction coil designed for use in an antisidetone telephone set.

antisidetone telephone set — A telephone set with an antisidetone circuit.

antiskating bias — A bias force applied to a pivoted pickup arm to counteract the inward force (toward the center of the record) resulting from the drag of the stylus in the groove and the offset angle of the head.

antiskating device — A mechanism found on modern phonograph pickups that provides a small outward force on a pickup arm. This counteracts the arm's tendency to move toward the turntable center (inward) due to offset geometry, and reduces stylus/groove friction.

antistatic agents — Methods employed to minimize static electricity in plastic materials. Such agents are of two basic types. Metallic devices that come into contact with the plastics and conduct the static to earth give complete neutralization initially, but because it is not modified, the surface of the material can become prone to further static accumulation during subsequent handling. Chemical additives, which are mixed with the compound during processing, give a reasonable degree of protection to the finished products.

antistatic cleaner — Substance used on phonograph records that helps to prevent the buildup of a static charge that attracts dust.

antistatic coating — 1. An electrically conductive layer for carrying off static charges that could accumulate on a surface. 2. A conductive coating applied to a TV or monitor screen (or on a glass panel immediately in front of the screen) that conducts away any static charge and prevents dust from adhering to the surface of the television.

antistatic sprays — Chemical agents which, when applied to circuits and plastic surfaces, leave a conductive coating that acts to repel dust and dirt and changes surface characteristics. Good antistatic sprays will leave a resistivity reading of 20 to 100 megohms per square inch ($3-15$ MΩ/cm^2) on plastics and 100 megohms or more per square inch on glass surfaces. *See* static eliminators.

antistickoff voltage — A small voltage, usually applied to the rotor winding of the coarse synchro control transformer in a two-speed system. The antistickoff voltage acts to eliminate the possibility of ambiguous behavior in the system.

antitransmit-receive box — A second transmit-receive switch used in a radar antenna system to minimize absorption of the echo signal in the transmitter circuit during the interval between transmitted pulses.

antitransmit-receive switch — Abbreviated atr switch. An automatic device employed in a radar system to prevent received energy from being absorbed in the transmitter.

antitransmit-receive tube — *See* atr tube.

antivoice-operated transmission — A method of radiocommunication in which a voice-activated circuit prevents the operation of the transmitter during reception of messages on an associated receiver.

aperiodic — 1. Having no fixed resonant frequency or repetitive characteristics or no tendency to vibrate. A circuit that will not resonate within its tuning range is often called aperiodic. 2. Not characterized by predictable periods or steps.

aperiodic antenna — An antenna designed to have a constant impedance over a wide frequency range (for example, a terminated rhombic antenna) due to the suppression of reflections within the antenna system.

aperiodic damping — Also called overdamping. The condition of a system when the amount of damping is so large that when the system is subjected to a single disturbance, either constant or instantaneous, the system comes to a position of rest without passing through that position. Although an aperiodically damped system is not strictly an oscillating system, it has such properties that it would become an oscillating system if the damping were sufficiently reduced.

aperiodic function — A function having no repetitive characteristics and not repeatable within a specified period.

aperiodic waveform — A nonrepeating, random, one-shot waveform.

aperture — 1. In a unidirectional antenna, that portion of the plane surface which is perpendicular to the direction of maximum radiation and through which the major part of the radiation passes. 2. In an opaque disc, the hole or window placed on either side of a lens to control the amount of light passing through. 3. Also called aperture time. The amount of certainty about the exact time when the encoder input was at the value represented by a given output code. In general, the aperture is equal to the conversion time; it may be reduced by the use of sample-and-hold circuits. 4. In an electron gun, the opening that determines the size of, and has an effect on, the shape of the electron beam. In television optics, it is the effective diameter of the lens that controls the amount of light reaching the photoconductive or photoemitting image pickup tube. 5. An opening that will pass light, electrons, or other forms of radiation.

aperture antenna — A type of antenna whose beam width is determined by the dimensions of a horn, lens, or reflector.

aperture compensation — Reduction of aperture distortion by boosting the high-frequency response of a television-camera video amplifier.

aperture correction — Compensation for the loss in sharpness of detail because of the finite dimensions of a scanning beam in the horizontal dimension.

aperture delay time — The time elapsed from the application of the hold (or encode) command until the sampling switch in a sample-and-hold circuit opens fully and the device actually takes the sample. Aperture delay time is a fixed delay time and is normally not an error source since the hold clock edge can be advanced to compensate for it.

aperture distortion — In a television signal, the distortion due to the finite dimension of the camera-tube scanning beam. The beam covers several mosaic globules simultaneously, resulting in a loss of picture detail.

aperture illumination — The field distribution in amplitude and phase through the aperture.

aperture jitter — Also called aperture uncertainty time. 1. In a sample-and-hold circuit, the time variation or uncertainty with which the switch opens, or the time variation in aperture delay. 2. A source of error in a sampling system, which determines the maximum slew rate limitation of the sampled analog input signal for a given system resolution.

aperture mask — Also called shadow mask. A thin sheet of perforated material placed directly behind the viewing screen in a three-gun color picture tube to prevent the excitation of any one color phosphor by either of the two electron beams not associated with that color.

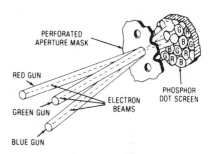

Aperture mask.

aperture plate — A ferrite memory plate containing a large number of uniformly spaced holes arranged in parallel rows and interconnected by plated conductors to provide a magnetic memory plate.

aperture time — 1. In a sample-and-hold circuit, the averaging time of a sample-hold during the sample-to-hold transition. 2. The time required by a sample-and-hold device to go from the sample mode into the hold mode, once the hold command has been received. The aperture time is generally a few nanoseconds, measured from the 50-percent point of the mode-control transition to the time when the output stops tracking the input. *See* aperture, 3.

aperture-time uncertainty — The possible variation in aperture time from one sample-to-hold transition to the next.

APL — Abbreviation for average picture level. The average luminance level of the part of a television line between blanking pulses.

APM — Abbreviation for analog panel meter. A scale-and-pointer meter capable of indicating a continuous, rather than incremental, range of values from zero to the rated full-scale value.

apogee — The point in an elliptical satellite orbit that is farthest from the surface of the earth. Geosynchronous satellites that maintain circular orbits around the earth are first launched into highly elliptical orbits with apogees of 22,237 miles. When the communication satellite reaches the appropriate apogee, a rocket motor is fired to place the satellite into its permanent circular orbit of 22,237 miles. *Also see* perigee.

A positive (A+ or A plus) — 1. Positive terminal of a battery or positive polarity of any other sources of voltage. 2. The terminal to which the positive side of the filament-voltage source of a vacuum tube should be connected.

A power supply — A power supply used as a source of heating current for the cathode or filament of a vacuum tube.

apparatus — 1. Any complex device. 2. Equipment or instruments used for a specific purpose.

apparatus wire and cable — Insulated wire and cable used in connecting electrical apparatus to a power source, also including wire and cable used in the apparatus itself.

apparent bearing — The direction from which the signal arrives with respect to some reference direction.

apparent power — In an ac circuit, the power value obtained by simple multiplication of current by voltage with no consideration of the effect of phase angle. (Compare with *true power*.)

apparent power loss — For voltage-measuring instruments, the product of nominal end-scale voltage and the resulting current. For current-measuring instruments, the product of the nominal end-scale current and the resulting voltage. For other types of instruments (for example, wattmeters), the apparent power loss is expressed for a stated value of current or voltage. Also called volt-ampere loss.

apparent source — *See* effective acoustic center.

Applegate diagram — A graphical representation of electron bunching in a velocity-modulated tube, showing their positions along the drift space. This bunching is plotted on the vertical coordinate, against time along the horizontal axis.

applet — A small computer program that performs a simple task.

AppleTalk — A networking protocol developed by Apple Computer for communication between Apple Computer products and other computers. This protocol is independent of what network it is layered on.

Appleton layer — In the ionosphere, a region of highly ionized air capable of reflecting or refracting radio waves back to earth. It is made up of the F_1 and F_2 layers.

apple tube — A color-television picture tube in which the three colors of phosphors are laid in fine vertical strips along the screen. The intensity of the electron beam is modulated as its sweeps over them so that each color is produced with appropriate brightness.

appliance — Any electrical equipment used in the home and capable of being operated by a nontechnical person. Included are units that perform some task that could be accomplished by other, more difficult means, but usually not those used for entertainment (radios, TVs, hi-fi sets, etc.).

appliance wire and cable — A classification of Underwriters' Laboratories, Inc., covering insulated wire and cable intended for internal wiring of appliances and equipment. Each construction satisfies the requirements for use in particular applications.

application — 1. The use of a computer for a specific purpose, e.g., designing a brochure or writing a letter. 2. System or problem to which a computer is applied. An application may be of the computational type, in which arithmetic computations predominate, or of the data-processing type, in which data-handling operations predominate. *See also* application program.

application factor — A modifier of the failure rate. It is based on deviations from rated operating stress (usually temperature and one electrical parameter).

application-oriented language — 1. A programming language that is primarily useful in some specialized area. 2. A problem-oriented programming language whose statements resemble or contain the terminology of the computer user.

application program — 1. A computer program intended to solve a problem or do a job, as distinct from systems programs, which control the operations of the computer system. 2. A computer program that performs a data-processing function rather than a control operation. 3. A program used to perform some logical or computational task that is important to the user rather than some internal computer function. 4. Software designed for a specific purpose, such as accounts payable, inventory, payroll, and word processing. 5. A computer program that accomplishes specific tasks, such as word processing.

application schematic diagram — Pictorial representation using symbols and lines to illustrate the interrelation of a number of circuits.

application-specific integrated circuit — *See* ASIC.

applications software — 1. A program that depends on the specific end application and is used to do the real work or apparent work that is visible to the user. Generally this is the software that is used for dedicated computer-based systems (systems designed to perform a single or specific set of functions). Typical applications include food and chemical processing, production control, automotive electronics, computer-controlled sewing machines, photographic equipment (both for computer-controlled cameras and for darkroom computerized processing), energy distribution systems, word processing, mailing lists, payrolls, and inventory. 2. Computer programs that perform specific tasks, such as word processing or database management.

applicators (applicator electrodes) — 1. In dielectric heating, the electrodes between which the dielectric item is placed and the electrostatic field developed. 2. Appropriately shaped conducting surfaces between which an alternating electric field is established for the purpose of producing dielectric heating. 3. In medical electronics, the electrodes applied to a patient undergoing diathermy or ultrasonic therapy.

applied voltage — 1. The potential between a terminal and a reference point in any circuit or device. 2. The voltage obtained when measuring between two given points in a circuit with voltage applied to the complete circuit. 3. The voltage presented to a circuit point or system input, as opposed to the voltage drop resulting from current through an element that results from the applied voltage.

applique circuit — A special circuit provided to modify existing equipment in order to allow for some special usage.

approach-control radar — Any radar set or system used in a ground-controlled approach system, e.g., an airport-surveillance radar, precision approach radar, etc.

approach path — In radio aircraft navigation, that portion of the flight path in the immediate vicinity of a landing area where such a flight path terminates at the touchdown point.

approved circuit — *See* protected wireline distribution system.

APT — Abbreviation for automatically programmed tool. A high-level or simplified programming language.

AQL — Abbreviation for acceptable quality level. A statistically defined quality level, in terms of percent defective accepted on an average of 95 percent of the time. In other words, a sampling plan with 1 percent AQL passes (accepts) lots 1 percent defective 95 percent of the time.

aquadag layer — Trademark of Acheson Industries, Inc. A conductive graphite coating on the inner side walls of some cathode-ray tubes. It serves as an electrostatic shield or as a postdeflection and an accelerating anode. Also applied to outer walls and grounded; here it serves, with the inner coating, as a capacitor to filter the applied high voltage.

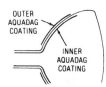

Aquadag coating.

arbiter — The section of a multiprocessor network's operating system that resolves simultaneous system-resource requests.

arbitrary function fitter — A circuit having an output voltage or current that is a presettable, adjustable, usually nonlinear function of the input voltage(s) or current(s) fed to it.

arbitrary waveform generator — *See* waveform generator.

arbor — *See* mandrel.

arc — 1. A luminous discharge of electricity through a gas. Characterized by a change in space potential in the immediate vicinity of the cathode; this change is approximately equal to the ionization potential of the gas. 2. A prolonged electrical discharge, or series of prolonged discharges, between two electrodes. (Both produce a bright-colored flame, as contrasted with a dim corona-glow discharge). 3. To form an arc. 4. The electric current in a flamelike stream of incandescent gas particles. 5. An electric current through air or across the surface of an insulator associated with high voltage; usually occurs when a contact is opened, or when deenergizing an inductive load.

arcback — Also called backfire. Failure of the rectifying action in a tube, resulting from the flow of a principal electron stream in the reverse direction due to the formation of a cathode spot on the anode. This action limits the peak inverse voltage that may be applied to a particular rectifier tube.

arc converter — A form of oscillator utilizing an electric arc to generate an alternating or pulsating current.

arc discharge — 1. A discharge between electrodes in gas or vapor. Characterized by a relatively low voltage drop and a high current density. 2. The sustained, luminous thermionic discharge between anode and cathode in a gas-filled tube.

arc-discharge tube — A gas-filled or mercury-vapor tube that utilizes ionic phenomena for switching, voltage regulation, or rectification.

arc drop — The voltage drop between the anode and cathode of a gas rectifier tube during conduction.

arc-drop loss — In a gas tube, the product of the instantaneous values of arc-drop voltage and current averaged over a complete cycle of operation.

arc-drop voltage — The voltage drop between the anode and cathode of a gas rectifier tube during conduction.

arc failure — 1. A flashover in the air near an insulation surface. 2. An electrical failure in the surface heated by a flashover arc. 3. An electrical failure in the surface damaged by the flashover arc.

arc function — An inverse trigonometric function.

arc furnace — An electric furnace heated by arcs between two or more electrodes.

architecture — 1. Organizational structure of a computing system, mainly referring to the CPU or microprocessor. 2. The manner in which the basic computer functions are organized and partitioned on the silicon chips. 3. The manner in which a system (such as a network or a computer) or program is structured.

archival — Pertaining to long-term storage of data.

archival backup — Backing up only files that have been changed since the last backup.

archive — 1. A procedure for transferring information from an online storage diskette or memory area to an offline storage medium. 2. To copy computer programs and data onto an auxiliary storage medium, such as a disk or tape, for long-term retention.

arcing — The production of an arc, e.g., at the brushes of a motor or at the contact of a switch.

arcing contacts — Special contacts on which the arc is drawn after the main contacts of a switch or circuit breaker have opened.

arcing time — 1. The interval between the parting, in a switch or circuit breaker, of the arcing contacts and the extension of the arc. 2. The time elapsing, in a fuse, from the severance of the fuse link to the final interruption of the circuit under the specified condition.

arc lamp — Source of brilliant artificial light obtained by an electric arc passing between two carbon rods. The arc is struck by bringing the two rods together and then rapidly separating them. As the arc burns, the carbon rods are vaporized away. A mechanism is employed to keep the space between the two rods constant. This type of lamp is used extensively in motion picture projectors and spotlights. The illumination of the arc lamp is derived from the incandescence of the positive electrode and from the heated, luminous, ionized gases or vapor that surround the arc.

arc oscillator — A negative-resistance oscillator comprising a sustained dc arc and a resonant circuit.

arcover — The (usually abrupt) creation of an arc between electrodes, contacts, or plates of a capacitor.

arcover resistance — The resistance of a material to the effects of a high-voltage, low-current arc (under prescribed conditions) passing across the surface of the material. The resistance is stated as a measure of total elapsed time required to form a conductive path on the surface (material carbonized by the arc).

arcover voltage — Under specified conditions, the minimum voltage required to create an arc between electrodes separated by a gas or liquid insulation.

arc percussive welding — A type of welding in which the materials to be welded are separated by a gap, across which an arc is struck; the arc melts the surfaces of the materials, and the materials are simultaneously brought together. *See also* pulse arc welding.

arc resistance — The length of time that a material can resist the formation of a conductive path by an arc adjacent to the surface of the material. Also called tracking resistance.

arc suppressor — A device, or combination of devices, used for arc suppression. *See* spark suppressor.

arc-through — In a gas tube, a loss of control with the result that a principal electron stream flows in the normal direction during what should be a nonconducting period.

area code — A three-digit number code identifying one of the geographic areas of the United States, Canada, and Mexico to permit direct distance dialing on the telephone system. The area code precedes the central office code in the complete 10-digit telephone number, and must be used when the called telephone is in a numbering plan area different from that of the calling telephone. The first digit of the area code is never a 1 or 0. *See also* direct distance dialing.

area protection — Protection of the inner space or volume of a secured area by means of a volumetric sensor.

area redistribution — A method of measuring the duration of irregularly shaped pulses. A rectangle is drawn having the same peak amplitude and the same area as the original pulse under consideration. Because the same time units are used in measuring the original and the new pulse, the width of the rectangle is considered the duration of the pulse.

area sensor — A sensor with a detection zone that approximates an area, such as a wall surface or the exterior of a safe.

A register — The accumulator for all arithmetical operations in a computer. Also called A accumulator.

argon — An inert gas used in discharge tubes and some electric lamps. It gives off a purple glow when ionized; its symbol is Ar.

argon glow lamp — A glow lamp containing argon gas that produces a pale blue violet light.

argument — 1. A variable upon which the value of a function depends. The arguments of a function are listed in parentheses after the function name. The computations specified by the function definition are made with the variables specified as arguments. 2. The number that a function works on to produce its results. 3. The independent variable of a function. Arguments can be passed as part of a subroutine call where they would be used in that subroutine.

arithmetic and logic unit — Computer element that can perform the basic data manipulations in the central processor. Usually it can add, subtract, complement, negate, rotate, AND, and OR. Abbreviated ALU.

arithmetic capability — The ability to do addition, subtraction, and in some cases multiplication and division.

arithmetic check — A check of a computation making use of the arithmetical properties of the computation.

arithmetic element — Synonym for arithmetic unit.

arithmetic mean — 1. Usually, the same as *average*. It is obtained by first adding quantities together and then dividing by the number of quantities involved. 2. A figure midway between two extremes and is found by adding the minimum and maximum together and dividing by two.

arithmetic operation — 1. In an electronic computer, the operations in which numerical quantities form the elements of the calculation, including the fundamental operations of arithmetic (addition, subtraction, multiplication, comparison, and division). 2. Adding, subtracting, incrementing, or decrementing data in registers or memory.

arithmetic organ — *See* arithmetic unit.

arithmetic shift — In a digital computer, the multiplication or division of a quantity by a power of the base used in the notation.

arithmetic statement — 1. An expression and a variable separated by an equals sign. The expression is evaluated and the resulting value is assigned to the variable. 2. Instruction specifying an arithmetic operation.

arithmetic sum — The sum of two or more quantities regardless of their signs. Compare with *algebraic sum*.

arithmetic symmetry — Filter response showing mirror-image symmetry about the center frequency when frequency is displayed on an arithmetic scale. Constant envelope delay in bandpass filters usually is accompanied by arithmetic symmetry in the phase and amplitude responses and generally requires a computer design. *See also* geometric symmetry.

arithmetic unit — Also called arithmetic element or arithmetic organ. In an automatic digital computer, that portion in which arithmetical and logical operations are performed on elements of information.

armature — The moving element in an electromechanical device, such as the rotating part of a generator or motor, the movable part of a relay, or the spring-mounted, iron portion of a bell or buzzer.

armature contacts — 1. Contacts mounted directly on the armature. 2. Sometimes used for movable contacts.

armature control of speed — The varying of voltage applied to the armature of a shunt-wound motor to control the motor's speed over the basic speed range.

armature core — An assembly of laminations forming the magnetic circuit of an armature.

armature gap — The space between the armature and pole face.

armature hesitation — A delay or momentary reversal of the motion of the armature.

armature-hesitation contact chatter — Chatter caused by delay or momentary reversal in direction of the armature motion of a relay during either the operate or the release stroke.

armature-impact contact chatter — Chatter caused by impact of the armature of a relay on the pole piece in operation, or on the backstop in release.

armature overtravel — That portion of the available stroke occurring after the contacts of a relay have touched.

armature reaction — In an armature, the reaction of the magnetic field produced by the current on the magnetic lines of force produced by the field coil of an electric motor or generator.

armature rebound — Return motion of a relay armature after striking the backstop.

armature-rebound contact chatter — Chatter caused by the partial return of the armature of a relay to its operated position as a result of rebound from the backstop in release.

armature relay — A relay operated by an electromagnet that, when energized, causes an armature to be attracted to a fixed pole or poles.

armature slot — In the core of an armature, a slot or groove into which the coils or windings are placed.

armature stud — In a relay, an insulating member that transmits the motion of the armature to an adjacent contact member.

armature travel — The distance traveled during operation by a specified point on the armature of a relay.

armature voltage control — A means of controlling the speed of a motor by changing the voltage applied to its armature windings.

armature wire — Stranded annealed copper wire, straight lay, soft loose white cotton braid. It is used for low-voltage, high-current rotor winding motors and generators. Straight lay permits forming in armature slots, and compressibility.

armchair copy — Amateur term for clear, static-free signals.

armed sweep — *See* single sweep.

arming the oscilloscope sweep — Closing a switch that enables the oscilloscope to trigger on the next pulse.

armor — A braid or wrapping of metal, usually steel, placed over the insulation of wire or cable to protect it from abrasion or crushing.

armor clamp — A fitting for gripping the armor of a cable at the point where the armor terminates or where the cable enters a junction box.

armored cable — Two or more insulated wires collectively provided with a metallic covering, primarily to protect the insulated wires from damage.

Armstrong frequency-modulation system — A phase-shift modulation system originally proposed by E. H. Armstrong.

Armstrong oscillator — An inductive feedback oscillator that consists of a tuned gate circuit and an untuned tickler coil in the drain circuit. Feedback is controlled by varying the coupling between the tickler and the gate circuit.

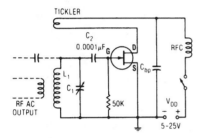

Armstrong oscillator.

arr — *See* automatic repeat request.

array — 1. In an antenna, a group of elements arranged to provide the desired directional characteristics. These elements may be antennas, reflectors, directors, etc. 2. A series of items, not necessarily arranged in a meaningful pattern. 3. The group of patterns on a wafer or in the artwork or photomask for semiconductor processing. *See* random-access memory.

array antenna — An antenna comprising a number of radiating elements, generally similar, arranged and excited to obtain directional effects.

array device — A group of many similar, basic, complex, or integrated devices without separate enclosures. Each has at least one of its electrodes connected to a common conductor, or all are connected in series.

array noise — Unwanted disturbance in a memory integrated circuit generated by the normal movement of data within the array.

array processor — 1. A computer optimized in architecture and instruction set to handle programs involving computations on large batches of data, such as fast Fourier transforms and large matrix computations. An array processor takes blocks of data and instructions from a host mini or large computer and performs the computations at speeds many times as high as those that are possible through the host computer alone. The host may be considered the data-organizing front end; the array processor is the processing unit. 2. A computer dedicated by its design to performing repetitive arithmetical calculations on large arrays of data with high precision, wide dynamic range, and high throughput. Usually most input/output operations and file management chores are left to the host computer in order to free the peripheral array processor to concentrate on its calculations. 3. A single computer that operates on one piece of data at a time. 4. A processor in a computer that performs matrix arithmetic much faster than is done in a standard computer. Capable of performing operations on all the elements in large matrices at one time. Also called a vector processor.

arrester — Also called a lightning arrester. 1. A protective device used to provide a bypass path directly to ground for lightning discharges that strike an antenna or other conductor. 2. A power-line device capable of reducing the voltage of a surge applied to its terminals, interrupting current, if present, and restoring itself to original operating conditions. 3. Device that diverts high voltages to ground and away from the equipment it protects.

ARRL — Abbreviation for American Radio Relay League.

arrowhead — A linearly polarized, frequency-independent, log-periodic antenna.

ARSR — Abbreviation for air route surveillance radar.

ARTCC — Abbreviation for air route traffic control center. A complex data-handling facility designed by Burroughs, IBM, and Raytheon to computerize as much in-route air traffic control as possible.

articulation — Sometimes called intelligibility. 1. In a communications system, the percentage of speech units understood by a listener. The word *articulation* is customarily used when the contextual relationships among the units of speech material are thought to play an unimportant role; the word *intelligibility* is used when the context is thought to play an important role in determining the listener's perception. 2. A quantitative measurement of the intelligibility of human speech, where 100 percent is completely understandable. For the typical sound reinforcement or other communications system, no more than a 15-percent articulation loss is acceptable. 3. The ability of a mechanism to pivot, grasp, or extend.

articulation equivalent — The articulation of speech reproduced over a complete telephone connection, expressed numerically in terms of the trunk loss of a working reference system that is adjusted to give equal articulation.

artificial antenna — Also called dummy antenna. A device that simulates a real antenna in its essential impedance characteristics and has the necessary power-handling capabilities, but which does not radiate or receive radio waves. Used mainly for testing and adjusting transmitters.

artificial ear — A microphone-equipped device for measuring the sound pressures developed by an earphone. To the earphone it presents an acoustic impedance equivalent to the impedance presented by the human ear.

artificial echo — 1. Received reflections of a transmitted pulse from an artificial target, such as an echo box, corner reflector, or other metallic reflecting surface. 2. A delayed signal from a pulsed radio-frequency signal generator.

artificial horizon — A gyroscopically operated instrument that shows, within limited degrees, the pitching and banking of an aircraft with respect to the horizon. Lines or marks on the face of the instrument represent the aircraft and the horizon. The relative positions of the two are then easily discernible.

artificial intelligence — Abbreviated AI 1. The design of computer and other data-processing machinery to perform increasingly higher-level cybernetic functions. 2. The capability of a device to perform functions that are normally associated with human intelligence, such as reasoning, learning, and self-improvement. Related to machine learning. 3. The imitation by artificial systems of characteristics described as intelligent when observed in humans. Artificial intelligence embraces concepts and theories from many different disciplines, including mathematics, cybernetics, computer science, psychology, biology, and others. 4. Overlapping subsets called expert systems, knowledge representations, inference schemes, program synthesis, scene analysis, and robotics. 5. The ability of a machine to perform certain complex functions normally associated with human intelligence, such as judgment, pattern recognition, understanding, learning, planning, and problem solving. 6. Computer programs developed to mimic human intelligence, such as reasoning, learning, problem solving, and making decisions. Artificial intelligence programs enable computers to perform tasks such as playing chess, proving mathematical theorems, etc. 7. An area of computer science dedicated to the development of machines that can learn, understand, interpret, and arrive at conclusions in a manner that would be considered intelligent if a person were doing it.

artificial ionization — Introduction of an artificial reflecting or scattering layer into the atmosphere to permit beyond-the-horizon communications.

artificial language — In computer terminology, a language designed for ease of communication in a particular area of activity, but one that is not yet natural to that area (as contrasted with a natural language evolved through long usage).

artificial line — A lumped-constant network designed to simulate some or all the characteristics of a transmission line over a desired frequency range.

artificial line duct — A balancing network simulating the impedance of the real line and distant terminal apparatus. It is employed in a duplex circuit to make the receiving device unresponsive to outgoing signal currents.

artificial load — Also called dummy load. A dissipative but essentially nonradiating device having the impedance characteristics of an antenna, transmission line, or other practical utilization circuit. Energy is dissipated in the form of heat. Used to test radio transmitters, engine generators, etc. Permits testing under load conditions without the creation of any standing waves or radiating a signal.

artificial radioactivity — Radioactivity induced in stable elements under controlled conditions by bombarding them with neutrons or high-energy, charged particles. Artificially radioactive elements emit beta and/or gamma rays.

artificial voice — A small speaker mounted in a specially shaped baffle that is proportioned to simulate the acoustical constants of the human head. It is used for calibrating and testing close-talking microphones.

Artos stripper — A machine that, when properly adjusted, will automatically measure to a predetermined length, cut, strip, count, and tie wire in bundles.

ARTS — Abbreviation for Automated Radar Terminal System. A multiprocessor computing system used at terminal radar approach controls (tracons) and airport towers, based on Sperry Univac 1140 computers. Various configurations, differing in size of memory and number of processors, exist at different facilities; the largest, ARTS IIIE, is installed at the New York tracon.

artwork — 1. A topological pattern of an integrated circuit, made with accurate dimensions so that it can be used in mask making. Generally, it is a large multiple of the final mask size, and final reduction is accomplished through the use of a step-and-repeat camera. 2. Detailed, original drawing (often developed with the aid of a computer) showing layout of an integrated circuit. 3. The images formed by drawing, scribing, or by cutting and stripping on a film or glass support, which are reduced, contact-printed, or stepped and repeated to make a photomask or intermediate. 4. Layouts and photographic films created to produce thick-film screens and thin-film masks.

ASA — Abbreviation for American Standards Association. *See* American National Standards Institute.

ASA code — A code that was recommended by the American Standards Association for industry-wide use in the transmission of information. Now ANSI code.

asbestos — A nonflammable material generally used for heat insulation, such as in a line-cord resistor. No longer used on new equipment.

A-scan — Also called A-display. On a cathode-ray indicator, a presentation in which time (range or distance) is one coordinate (horizontal) and signals appear as perpendicular deflections to the time scale (vertical).

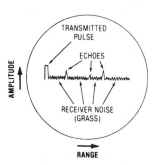

A-scan on a CRT.

ASCII — Acronym for American Standard Code for Information Interchange (pronounced "ask-ee"). 1. A standard code used extensively in data transmission, in which 128 numerals, letters, symbols, and special control codes are each represented by a seven-bit binary number 0000000 through 1111111. For example, numeral 5 is represented by 011 0101, letter K by 100 1011, percent symbol (%) by 010 0101, and start of text (STX) control code by 000 0010. 2. A standard code used by many computers, video-display terminals, teleprinters, and computer peripherals. A full eight-bit ASCII word may be transmitted in parallel or serial form, with the eighth bit often providing parity information. Keyboard encoders convert a single switch closure into an ASCII data word; character generators convert stored ASCII data words (and some timing commands) into groups of dots suitable for raster-scan display.

A-scope — An oscilloscope that uses an A-scan to present the range of a target as the distance along a horizontal line from the transmitted pulse pip to the target, or echo pip. Signals appear as vertical excursions of the horizontal line, or trace.

ASDE — *See* airport surface detection equipment.

as-fired — Values of thick-film resistors or smoothness of ceramic substrates as they come out of the firing furnace prior to trimming and polishing, respectively (if required).

ASI — An abbreviation for standards published by the American Standards Institute. Now American National Standards Institute.

ASIC — Abbreviation for application-specific integrated circuit. 1. Semiconductor circuits specifically designed to suit a customer's particular requirement, as opposed to general-purpose parts that can be used in many different systems or applications. 2. An integrated circuit designed to fill the specific requirement of a unique application.

ask — *See* amplitude-shift keying.

aspect ratio — 1. Ratio of frame width to frame height. 2. The ratio of an object's height to its width. In graphics this ratio usually pertains to the face of a rectangular CRT or to the characters or symbols drawn by the character generator. 3. The ratio between the length of a film resistor and its width: equal to the number of squares of the resistor. 4. The ratio of the width of a television picture tube to its height. In the United States, the television standard is 4 : 3.

asperities — Local microscopic points on an electrode surface at which there is considerable field enhancement. They lead to a dependence of electric strength on electrode area (area effect).

aspheric — 1. Not spherical; an optical element having one or more surfaces that are not spherical. 2. A mirror or lens surface that varies slightly from a true spherical surface. This is done to reduce lens aberrations.

ASR — Abbreviation for automatic send and receive and airport surveillance radar. 1. A terminal equipped with recording devices, usually a paper-tape reader and punch, which is capable of answering a call, recording a message, or sending data loaded in its tape reader without the need for an operator in attendance at the time of the call. Also used to specify terminals that have paper-type equipment used by the operator. 2. A teletypewriter that contains a keyboard, page printer, paper-tape transmitter, and paper-tape punch. Paper tape can be prepared offline, which can take place while hard copy is being received from the line or while other paper tape is being transmitted.

ASRA — Abbreviation for automatic stereophonic recording amplifier. An instrument developed by Columbia Broadcasting System for stereo recording. Compression of the vertical component of the stereo recording signal is automatically decreased or increased as required by the recording conditions.

assemble — 1. To collect, interpret, and coordinate the data required for a computer program, translate the data into computer language, and project it into the master routine for the computer to follow. 2. To translate from a symbolic program to a binary program by substituting binary operation codes for symbolic operation codes and replacing symbolic addresses with absolute or relocatable addresses.

assembler — 1. A program that prepares a program in machine language from a program in symbolic language by substituting absolute operation codes for symbolic operation codes and absolute or relocatable addresses for symbolic addresses. 2. A unit that converts the assembly language of a computer program into the machine language of the computer, accepting mnemonics and symbolic addresses instead of actual binary values for addresses, instructions, and data. 3. A program that accepts instructions, addresses, and data in symbolic form (character strings that represent machine instructions, addresses, data, among others). Then it automatically translates symbols into their corresponding numerical values. It permits symbolic addressing by assigning values to labels used to indicate program-jump locations. 4. A simple programming language that allows the programmer to define labels and fixed values and to then use these labels with a mnemonic instruction set to produce a machine code program. 5. Program that converts source-code (mnemonic) input into op-code (binary) machine language instructions. If such a program were not available, the programmer would have to enter all instructions in ones and zeros, a much more tedious and error-prone procedure. 6. A computer program that converts a higher-level (Englishlike) programming language into machine-readable instructions.

assembler program — Software, usually supplied by the computer manufacturer, to convert an assembly language application program into machine language.

assembly — 1. A complete operating unit, such as a radio receiver, made up of subassemblies, such as an amplifier and various components. 2. Process in which instructions written in symbolic form by the programmer are changed to machine language by the computer.

assembly language — 1. A computer language that has one-to-one correspondence with an assembly program. The assembly program directs a computer to operate on a program in symbolic language to produce a program in machine language. *See also* high-order language; machine language, 3; and source language. 2. Grouped alphabet characters, called mnemonics, that replace the numeric instructions of machine language. These mnemonics are easier to remember than machine instructions and hence easier to develop into a working program. 3. A machine-oriented language based primarily on a one-to-one relationship between machine instructions and user-supplied source code. 4. Microprocessor commands written in mnemonic form. Typically, three-letter abbreviations, called mnemonics, are used to represent each instruction, and each mnemonic can usually be equated to one machine-code instruction. 5. A human-oriented symbolic-mnemonic source language that is used by the programmer to encode programs and associated databases. Assembly language programs are read by the assembler and converted to executable machine language programs during the assembly process. Assembly language is easier to remember and manipulate than machine language. 6. Human-oriented varieties of machine languages. Precisely the same final program code can be produced from an assembler as by hand-coding machine language. However, assembly languages prove more convenient for people than the numeric-only machine languages.

assembly-language programming — *See* symbolic-language programming.

assembly-output language — An optional symbolic assembly language listing of the object-code output from a high-level language compiler. Can be quite helpful as a debugging tool because it shows exact machine code in a readable format.

assembly program — A program that enables a computer to assemble mnemonic language into machine language; for example, a FORTRAN assembly program. Also called assembly routine.

assembly robot — A computerized robot, probably a sensory model, designed specifically for assembly-line jobs. For light, batch-manufacturing applications, the arm's design may be fairly anthropomorphic.

assembly routine — *See* assembly program.

assertion checking — Evaluating a program by embedding statements that should always hold true.

assignable cause — A definitely identified factor contributing to a quality variation.

assigned frequency — The center of the frequency band assigned to a station.

assigned frequency band — The frequency band, the center of which coincides with the frequency assigned to the station, and the width of which equals the necessary bandwidth plus twice the absolute value of the frequency tolerance.

associative memory — A computer memory in which the data are stored and indexed by content, as in a dictionary, in contrast with the storage of a random-access memory. 2. A memory in which the storage locations are identified by their contents rather than by their addresses. Enables faster interrogation to retrieve a particular data element.

associative storage — Computer storage in which locations may be identified by specification of part or all of their contents. Also called parallel-search storage or content-addressed storage.

astable — 1. Pertaining to a device that has two temporary states: the device alternates between these states with a period and duty cycle determined by circuit time constants. *See also* bistable. 2. Refers to a device that has two temporary states. The device oscillates between the two states with a period and duty cycle predetermined by time constants.

astable circuit — A circuit that continuously alternates between its two unstable states at a frequency determined by the circuit constants. It can be readily synchronized by applying a repetitive input signal of slightly higher frequency. A blocking oscillator is an example of an astable circuit.

astable multivibrator (free-running) — A circuit having two momentarily stable states, between which it continuously alternates, remaining in each for a period controlled by the circuit parameters and switching rapidly from one to the other.

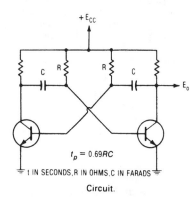

Circuit.

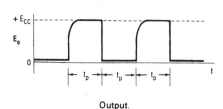

Output.

Astable multivibrator.

astatic — 1. Having no particular orientation or directional characteristics; such as a vertical antenna. 2. Being in neutral equilibrium; having no tendency toward any change of position.

astatic galvanometer — A sensitive galvanometer used for detecting small currents. Consists of two small magnetized needles of equal size and strength arranged in parallel and with their north and south poles adjacent, suspended inside the galvanometer coil. Since the resultant magnetic moment is zero, the earth's magnetic field does not affect the system.

A station — One of a pair of transmitting stations in a loran system. The A-station signal always occurs less than half a repetition period after the immediately preceding signal of the other station of the pair and more than half a repetition period before the next succeeding signal of the other station.

astigmatism — A type of spherical aberration in which the rays from a single point of an object do not converge on the image, thereby causing a blurred image. Astigmatism in an electron-beam tube is a focus defect in which electrons in different axial planes come to focus at different points.

astrionics — Electronics as involved with astronautics.

astrocompass — An instrument for determining direction relative to the stars. It is unaffected by the errors to which magnetic or gyrocompasses are subject.

astrodome — A rigid hemispherical structure used to cover large tracking instruments to protect them from the elements. It is usually constructed so that the dome rotates with the instrument.

astronautics — The science and art of operating space vehicles.

astrotracker — A device for tracking stars.

A-supply — The A battery, transformer filament winding, or other voltage source that supplies power for heating the filaments of vacuum tubes.

asynchronous motor — An ac motor whose speed is not proportional to the frequency of the supply voltage.

asymmetrical cell — A cell, such as a photoelectric cell, in which the impedance to the flow of current is greater in one direction than in the other direction.

asymmetrical distortion — Distortion affecting a two-condition or binary modulation or restitution, in which all the significant intervals corresponding to one of the two significant conditions have longer or shorter durations than the corresponding theoretical durations of the excitation. If this particular requirement is not met, distortion is present.

asymmetrical SCR — A fast silicon-controlled rectifier (SCR) with low reverse blocking, which is voltage that causes conduction without an input trigger at the gate input of the SCR.

asymmetric sideband transmission — *See* vestigial sideband transmission.

asymmetry control — In pH meters, an adjustment sometimes provided to compensate for differences in the electrodes.

asymptote — A line that comes nearer and nearer to a given curve but never touches it.

asymptotic breakdown voltage — A voltage that will break down insulation if applied over a long period.

asynchronous — 1. A communication method in which data is sent when it is ready without being referenced to a timing clock, rather than waiting until the receiver signals that it is ready to receive. 2. Transmission in which each data byte is preceded by a start bit and followed by one or more stop bits. Data transmission is intermittent, with an irregular time interval between data bytes. 3. Lacking a regular time relationship; not related through repeating time patterns. Hence, as applied to computer program execution, unexpected or unpredictable with respect to the instruction sequence. 4. Modems, terminals, and transmissions in which each character of information is individually framed (synchronized), usually by start and stop elements. The interval between characters is not fixed. 5. An external interface that can be started and stopped by a microprocessor or other equipment. The opposite is *synchronous*, which means that the data is randomly available. 6. Having no set pattern, cycle, or speed of transmission. 7. Not synchronized by a clocking signal; in code sets, character codes containing start and stop bits. 8. A mode of data transmission in which time intervals between transmitted characters may be of unequal length. Transmission is independently controlled by start and stop elements at the beginning and end of each character. 9. Pertaining to a mode of data communications that provides a variable time interval between characters during transmission.

asynchronous communication — 1. A method of transferring data in which the timing of character placement on connecting communication lines is not critical. Each transferred character is preceded by a start bit and followed by a stop bit, permitting the interval between characters to vary. 2. A relatively simple and cheap system for moving data between machines at speeds up to 1200 or 2400 baud. 3. Data communication of the start/stop type. Each character is sent individually without regular or predictable time relationships with other characters. 4. Method of communications in which data is sent as soon as it is ready.

asynchronous computer — 1. An automatic digital computer in which an operation is started by a signal denoting that the previous operation has been completed. 2. A computer in which each operation starts as a result of a signal generated by the completion of the previous operation or by the availability of the equipment required for the next operation.

asynchronous device — A device in which the speed of operation is not related to any frequency in the system to which it is connected.

asynchronous input/output — The ability to accept input data while simultaneously delivering output data.

asynchronous inputs — The terminals that affect the output state of a flip-flop independently of the clock terminals. Called set, preset, reset, or clear; sometimes referred to as dc inputs.

asynchronous logic — Logic networks whose operational speed depends only on the signal propagation through the network, rather than on clock pulses as in synchronous logic.

asynchronous machine — 1. Any machine in which its speed of operation is not proportionate to the frequency of the system to which the machine is connected. 2. A multiprocessor system whose processes occur as needed by, and whose operations follow, input data instead of an autonomous clock.

asynchronous operation — 1. Generally, an operation that is started by a signal at the completion of a previous operation. It proceeds at the maximum speed of the circuits until it is finished and then generates its own completion signal. 2. A mode in which entry of data into a flip-flop does not require a gating or clock pulse. 3. Operation of a switching network by a free-running signal that triggers successive instructions. The completion of one instruction triggers the next. 4. A computer operation that does not proceed in step with some external timing.

asynchronous shift register — A shift register that does not require a clock. Register segments are loaded and shifted only at data entry.

asynchronous transmission — Transmission in which each character of the information is synchronized individually (usually by the use of start and stop elements).

ATARS — Acronym for Automated Traffic Advisory and Resolution Service. A ground-based automatic collision-avoidance system being developed for use at air terminals.

ATC — 1. Abbreviation for automatic temperature control (General Motors). A means of automatically maintaining desired passenger compartment temperature in a vehicle. Temperature is sensed by a thermistor. Control valves are actuated by a vacuum motor to adjust the proportion of heated or cooled air. 2. Also abbreviation for automated technical control. A computer system used to maintain control of a data-communication network.

AT-cut crystal — A quartz-crystal slab cut at a 35° angle with respect to the optical, or Z-axis, of the crystal. It has practically a zero temperature coefficient and is used at frequencies of about 0.8 to 250 MHz.

ATE — Abbreviation for automatic test equipment.

ATG — See automatic test generation.

atmosphere — 1. The body of air surrounding the earth. 2. A unit of pressure defined as the pressure of 760 mm of mercury at 0°C. Approximately 14.7 pounds per square inch.

atmospheric absorption — The energy lost in the transmission of radio waves due to dissipation in the atmosphere.

atmospheric absorption noise — The dominant noise factor, at frequencies above 1000 MHz, caused by the absorption of energy from radio waves by oxygen and water vapor in the atmosphere.

atmospheric duct — Within the troposphere, a condition in which the variation of refractive index is such that the propagation of an abnormally large proportion of any radiation of sufficiently high frequency is confined within the limits of a stratum. This effect is most noticeable above 3000 MHz.

atmospheric electricity — Static electricity between clouds, or between clouds and the earth.

atmospheric noise — Also called atmospherics. 1. The noise heard during radio reception because of atmospheric interference. 2. A product of the discharging of lightning and other phenomena in the atmosphere.

atmospheric pressure — 1. The barometric pressure of air at a particular location on the earth's surface. The nominal, or standard, value of atmospheric pressure is 760 mm of mercury (14.7 pounds per square inch) at sea level. Atmospheric pressure decreases at higher altitudes. 2. Pressure exerted by the atmosphere on all things exposed to it. Although it varies constantly, it is considered as a standard to be normal when it is 14.7 lb/in^2 (1.033×10^4 kg/m^2) at sea level.

atmospheric radio wave — A radio wave that is propagated by reflections in the atmosphere. May include the ionospheric wave, the tropospheric wave, or both.

atmospheric radio window — That portion of the frequency spectrum that will allow radio-frequency waves to pass through the earth's atmosphere (approximately 10 to 10,000 MHz).

atmospheric refraction — The bending of the path of electromagnetic radiation from a distant point as the radiation passes obliquely through varying air densities.

atmospherics — Also referred to as static, atmospheric noise, and strays. In a radio tuner or receiver, noise due to natural weather phenomena and electrical charges existing in the atmosphere.

atom — 1. The smallest portion of an element that exhibits all properties of the element. It is pictured as composed of a positively charged nucleus containing almost all the mass of the atom, surrounded by one or more electrons. In the neutral atom, the number of electrons is such that their total charge (negative) exactly equals the positive charge in the nucleus. 2. The basic unit of a chemical element, consisting of a positively charged nucleus surrounded by a number of electrons sufficient to counterbalance the charge of the nucleus. The identity of an element, in a chemical sense, depends on the number of positive charges in the nucleus of its atom. The nucleus also contains particles that contribute mass but no charge. The stability of a nucleus depends on its ratio of charge to mass.

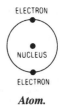

Atom.

atomic battery— *See* nuclear battery.

atomic charge— The electronic charge of an ion, equal to the number of ionization multiplied by the charge on one electron.

atomic fission— *See* fission.

atomic frequency— The natural vibration frequency of an atom.

atomic fuel— A fissionable material, i.e., one in which the atomic nucleus may be split to release energy.

atomic mass unit (unified)— One-twelfth of the mass of an atom of the ^{12}C nuclide. Use of the old atomic mass unit (amu), defined by reference oxygen, is deprecated.

atomic migration— The progressive transfer of a valence electron from one atom to another within the same molecule.

atomic number— The number of protons (positively charged particles) in the nucleus of an atom. All elements have different atomic numbers, which determine their positions in the periodic table. For example, the atomic number of hydrogen is 1, that of oxygen is 8, iron 26, lead 82, and uranium 92.

atomic ratio— The ratio of quantities of different substances to the number of atoms of each.

atomic theory— A generally accepted theory concerning the structure and composition of substances and compounds. It states that everything is composed of various combinations of ultimate particles called atoms.

atomic time— Time scales based on molecular or atomic resonance effects, which are apparently constant and equivalent (or nearly equivalent) to ephemeris time.

atomic weight— The approximate weight of the number of protons and neutrons in the nucleus of an atom. The atomic weight of oxygen, for example, is approximately 16 (actually it is 16.0044)— it contains 8 neutrons and 8 protons. Aluminum is 27 and contains 14 neutrons and 13 protons. If expressed in grams, these weights are called gram atomic weights.

atr switch— *See* antitransmit-receive switch.

atr tube— Abbreviation for antitransmit-receive tube. A gas-filled, radio-frequency switching tube used to isolate the transmitter while a pulse is being received.

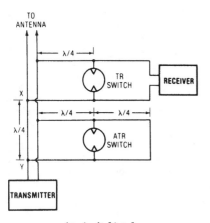

Atr (switch) tube.

ATS— Abbreviation for automotive anti-theft system (General Motors). A security system that causes the horn to blow, headlights and parking lights to flash, and the dome light to turn on if the hood, trunk, or doors are opened without use of a key.

attached foreign material— In a semiconductor, a foreign substance that cannot be removed when subjected to a nominal gas flow. Lint, silicon dust, etc., are not considered attached since they can be removed after die mounts.

attachment cap— *See* attachment plug.

attachment cord— *See* patch cord.

attachment plug— An assembly consisting of two or more blades projecting from a small insulating base, with provision for connecting the plug to a cord. Also called attachment cap.

attack— 1. The length of time it takes for a tone in an organ to reach full intensity after a key is depressed. On most organs this effect is adjustable by either a switch or potentiometer. 2. The action of a control system in response to a sudden error condition. 3. The responsiveness of an amplifier to signals with a fast rise-time, such as produced by percussive sounds of a transient nature. 4. The beginning of a sound or the initial transient of a musical note.

attack time— The interval required for an input signal, after suddenly increasing in amplitude, to attain a specified percentage (usually 63 percent) of the ultimate change in amplification or attenuation due to this increase.

attendant's switchboard— A switchboard, of one or more positions, that permits an operator in the central office to receive, transmit, or cut in on a call to or from one of the lines serviced by the office.

attended operation— Data-set applications in which individuals are required at both stations to establish the connection and transfer the data sets from talk (voice) mode to data mode. Compare with *unattended operation*.

attention display— A computer-generated tabular or vector message placed on the display tubes of a control facility to draw attention to a particular situation.

attenuate— To obtain a fractional part of or reduce in amplitude an action or signal.

attenuating— Decreasing electrical current, voltage, or power in a communicating channel. Refers to audio, radio, or carrier frequencies.

attenuation— 1. The decrease in amplitude of a signal during its transmission from one point to another. It may be expressed as a ratio or, by extension of the term, in decibels. 2. *See* insertion loss. 3. The decrease in amplitude of a signal at a specified frequency due to its transmission through a filter. This is expressed as a function of the amplitude ratio V_1/V_2, which is the reciprocal of the magnitude of the transmission function. 4. Optical power loss per unit length, usually expressed in decibels per kilometer. 5. Loss of electrical power in a length of cable. Amount of power leaving a length of cable as compared with the amount introduced. Attenuation is measured in decibels per 100 feet. 6. Applied to coaxial cables, the power drop or signal loss in a circuit, expressed in decibels. It is also the decrease in amplitude of a wave with distance in the direction of wave propagation when the amplitude at any given place is constant in time, or the decrease in amplitude with time at a given place. Attenuation is generally expressed in decibels per unit, usually 100 feet, and is indicative of the power loss.

attenuation constant— 1. The real component of the propagation constant. 2. For a traveling plane wave at a given frequency, the rate at which the amplitude of a field component (or the voltage or current) decreases exponentially in the direction of propagation, in nepers or decibels per unit length.

attenuation distortion— 1. In a circuit or system, its departure from uniform amplification or attenuation over the frequency range required for transmission. 2. Distortion that causes a decrease in the amplitude of a field component (voltage or current) in the direction of propagation.

attenuation equalizer — A corrective network designed to make the absolute value of the transfer impedance of two chosen pairs of terminals substantially constant for all frequencies within a desired range.

attenuation-frequency distortion — A form of wave distortion in which the relative magnitudes of the different frequency components of the wave are changed.

attenuation network — 1. A network providing relatively slight phase shift and substantially constant attenuation over a range of frequencies. 2. The arrangement of circuit elements, usually impedance elements, inserted in circuitry to introduce a known loss or to reduce the impedance level without reflections.

attenuation ratio — The magnitude of the propagation ratio, which indicates the relative decrease in energy.

attenuator — 1. A resistive network that provides reduction of the amplitude of an electrical signal without introducing appreciable phase or frequency distortion. 2. A distributed network that absorbs part of a signal and transmits the remainder with a minimum of distortion or delay. 3. Network for reducing signal level. Sometimes necessary in the input circuit of a tuner to avoid overloading by strong local signals. 4. An electronic transducer, either fixed or adjustable, that reduces the amplitude of a wave without causing significant distortion.

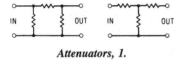

Attenuators, 1.

attenuator tube — A gas-filled, radio-frequency switching tube in which a gas discharge, initiated and regulated independently from the radio-frequency power, is used to control this rf power by reflection or absorption.

atto- — Prefix meaning 10^{-18}. Letter symbol is a.

attraction — The force that exists between two unlike magnetic poles (N and S), or between two unlike static charges (+ and −), or between two masses.

attribute — The manner in which a variable is handled by a computer.

audibility — 1. The ability to be heard, usually construed as being heard by the human ear. 2. The ratio of the strength of a specific sound to the strength of a sound that can barely be heard. Usually expressed in decibels.

audibility threshold — The minimum sound intensity that the average human ear can hear, normally considered to be 0 dB sound pressure level at 1000 Hz.

audible — Capable of being heard; in most contexts, by the average human ear.

audible ringing tone — That tone received by the calling telephone, indicating that the called telephone is being rung.

audible tones — Sounds composed of frequencies that the average human can detect.

audio — Pertaining to frequencies corresponding to a normally audible sound wave. These frequencies range roughly from 15 to 20,000 hertz.

audio amplifier — *See* audio-frequency amplifier.

audio band — The range of audio frequencies passed by an amplifier, receiver, transmitter, etc. *See also* audio frequency.

audio carrier — In the NTSC system, a carrier located 4.5 MHz above the video carrier; it extends to within 0.25 MHz of the assigned channel frequency. For channel 2, its active frequency is 59.75 MHz.

audio-channel wire — A small-diameter wire, shielded and jacketed, used primarily in radio and television for wiring consoles, panels, etc.

audio component — That portion of any wave or signal which contains frequencies in the audible range (between 15 and 20,000 hertz).

audio frequency — Abbreviated AF. Any frequency corresponding to a normally audible sound wave. Audio frequencies range roughly from 15 to 20,000 hertz.

audio-frequency amplification — An increase in voltage, current, or power of a signal at an audio frequency.

audio-frequency amplifier — Also called audio amplifier. A device that contains one or more electron tubes or transistors (or both) and is designed to amplify signals within a frequency range of about 15 to 20,000 hertz.

audio-frequency choke — An inductance used to impede the flow of audio-frequency currents.

audio-frequency noise — In the audio-frequency range, any electrical disturbance introduced from a source extraneous to the signal.

audio-frequency oscillator — An oscillator circuit using an electron tube, transistor, or other nonrotating device capable of producing audio signals.

audio-frequency peak limiter — A circuit generally used in the audio system of a radio transmitter to prevent overmodulation. It keeps the signal amplitude from exceeding a predetermined value.

audio-frequency shift keying — *See* AFSK.

audio-frequency shift modulator — A system of facsimile transmission over radio, in which the frequency shift required is applied through an 800-hertz shift of an audio signal rather than by shifting the transmitter frequency. The radio signal is modulated by the shifting audio signal, usually at 1500 to 2300 hertz.

audio-frequency spectrum — The full range of sounds we can hear. In a person with good hearing, the range of sounds in the audio spectrum is usually between 15 and 20,000 Hz. In older people it is usually 50 to 10,000 Hz.

audio-frequency transformer — Also called audio transformer. An iron-core transformer for use with audio-frequency currents to transfer signals from one circuit to another. Used for impedance matching or to permit maximum transfer of power.

audiogram — Also called threshold audiogram. A graph showing hearing loss, percentage of hearing loss, or percentage of hearing as a function of frequency.

audio-level meter — An instrument that measures audio-frequency power with reference to a predetermined level. Usually calibrated in decibels.

audiometer — An electronic instrument for measuring hearing acuity. In simple units, the listener is provided (usually through earphones) with an audio signal (commonly a pure tone) of known intensity and frequency. More complex instruments provide a variety of signals (pure tones, white noise, and speech) through a variety of output transducers (such as earphones, bone vibrators, or loudspeakers).

audio mixer — An amplifier circuit used for blending two or more audio signals, such as those delivered by microphones and record players.

audion — A three-electrode vacuum tube introduced by Dr. Lee de Forest.

audio oscillator — *See* audio-frequency oscillator.

audio output — The output signal from any audio equipment. It is generally measured in volts or watts, rms.

audio patch bay — Specific patch panels provided for termination of all audio circuits and equipment used in the channel and technical-control facility. This equipment can also be found in transmitting and receiving stations.

audio peak limiter — See audio-frequency peak limiter.

audiophile — A person who is interested in good musical reproduction for his or her own personal listening and who uses the latest audio equipment and techniques.

audio rectification — The phenomenon of rf signals being picked up, rectified, and amplified by audio circuits — notably by high-gain preamplifiers.

audio response — 1. A computer output technique that is formatted from stored words previously recorded in the computer. Programmed instructions often communicate with a student through this technique. 2. The fidelity with which audio-frequency equipment reproduces its input signal.

audio signal — An electrical signal whose frequency is within the audio range.

audio spectrum — 1. The continuous range of audio frequencies, extending from the lowest to the highest (from about 15 to 20,000 hertz). 2. The range of frequencies that can be detected as sound by the human ear.

audio subcarrier — Subcarriers of satellite video signals that are modulated by audio signals. The frequency range can be from 5 to 8 MHz but is usually 6.2 or 6.8 MHz.

audio taper — Semilogarithmic change of resistance. Used on tone controls in audio amplifiers to compensate for the lower sensitivity of the human ear when listening to low-volume sounds.

audio transformer — See audio-frequency transformer.

audiovisual — Involving both sight and sound (e.g., audiovisual education uses films, slides, phonograph records, and the like to supplement instruction).

audiovisual system — A system of communications that simultaneously transmits pictorial and audio signals.

auditing — Examination of software for consistency and traceability.

augend — In arithmetic addition, the number increased by having another number (called the addend) added to it.

augmented operation code — In a computer, an operation code that is further defined by information contained in another portion of an instruction.

aural — Pertaining to the ear or to the sense of hearing. This term is often used to distinguish between sound that is actually heard and sound represented by audio-frequency currents. See also audio.

aural radio range — A radio range whose courses are normally followed by interpretation of an aural signal.

aural signal — The signal corresponding to the sound portion of a television program. In general, the audible component of a signal.

aural transmitter — The equipment used to transmit the aural (sound) signals from a television broadcast station.

aurora — Sheets, streamers, or streaks of pale light often seen in the skies of the northern and southern hemispheres. The aurora borealis and aurora australis.

auroral absorption — Absorption of radio waves due to auroral activity. See also aurora.

auroral absorption index — A factor that relates the average auroral absorption with the geographic location of the points of reflection from the ionosphere.

authentication — Security measure designed to protect a communications system against fraudulent transmissions and establish the authenticity of a message by an authenticator within the transmission derived from certain predetermined elements of the message itself.

authenticator — A group of letters or numerals, or both, inserted at a predetermined point in a transmission or message for the purpose of attesting to the authenticity of the message or transmission.

authoring — The process of planning, designing, and producing multimedia applications.

authorized access switch — A device used to make an alarm system or some portion or zone of a system inoperative in order to permit authorized access through a protected port. A shunt is an example of such a device.

authorized carrier frequency — A specific carrier frequency authorized for use, from which the actual carrier frequency is permitted to deviate, solely because of frequency instability, by an amount not to exceed the frequency tolerance.

autoalarm — Also called automatic alarm receiver. 1. A device that is tuned to the international distress frequency of 500 kHz and that automatically actuates an alarm if any signal is received. 2. A circuit or device operated from a radio receiver to alert a radio operator that an incoming message is addressed to him or her. 3. Complete receiving, selecting, and warning device capable of being actuated automatically by intercepted radio-frequency signals forming the international automatic-alarm signal.

auto balance — A system for detecting errors in color balance in the white and black areas of the picture and automatically adjusting the white and black levels of both the red and blue signals as needed for correction.

auto call — An alerting device that sounds a preset code of signals in a building to page those persons whose code is being sounded.

autocondensation — A method of introducing high-frequency alternating current into living tissue for therapeutic purposes. The patient is connected as one plate of a capacitor to which the current is applied.

autoconduction — A method of introducing high-frequency alternating currents into living tissues for therapeutic purposes. The patient is placed inside a coil and acts essentially as the secondary of a transformer.

autocorrelation — 1. The correlation of a waveform with itself. It gives the Fourier transform of the power spectrum of the waveform (the power-density spectrum in the case of random signals). 2. A mathematical technique to measure the degree of rhythmic activity in physical phenomena that vary in a complex manner as a function of time.

autocorrelation function — A measure of the similarity between time-delayed and undelayed versions of the same signal, expressed as a function of delay.

autocorrelator — A circuit that distinguishes coherent programs (music or speech) from random noise (hiss) and operates filters that attenuate noise without an audible loss of program frequencies.

autodyne circuit — A vacuum-tube circuit that serves simultaneously as an oscillator and as a heterodyne detector.

autodyne reception — A type of radio reception employed in regenerative receivers for the reception of cw code signals. In this system the incoming signal beats with the signal from an oscillating detector to produce an audible beat frequency.

AUTOEXEC.BAT — Abbreviation for automatic execute batch file. The main file that tells a computer what to do when it is turned on. It literally automatically executes a batch of files.

auto iris control — An accessory unit that measures the video level of a TV camera and opens and closes the iris of the lens to compensate for light changes.

auto light range — The range of light (e.g., sunlight to moonlight) over which a TV camera is capable of automatically operating at specified output.

autoluminescence — The luminescence of a substance that is produced by energy within it (e.g., radioactive material).

auto-man — A locking switch that controls the method of operation (i.e., automatic or manual).

automata — A plural form of automaton.

automated communications — Combination of techniques and facilities by which intelligence is conveyed from one point to another without human effort.

automatic — 1. Self-regulating or self-acting; capable of producing a desired response to certain predetermined conditions. 2. Self-acting and self-regulating; operating without human intervention; often implying the presence of a feedback control system. 3. Pertaining to a process or device that, under specific conditions, performs its functions without intervention by a human operator.

automatic-alarm receiver — Complete receiving, selecting, and warning device capable of being actuated automatically by intercepted radio-frequency signals forming the international automatic-alarm signal.

automatic-alarm-signal keying device — A device that automatically keys the radio-telegraph transmitter on board a vessel to transmit the international automatic-alarm signal.

automatic answer — A feature by which a communicating word processor, twx, etc., may receive text without an operator in attendance.

automatic back bias — A radar-receiver technique that consists of one or more automatic gain-control loops to prevent large signals from overloading a receiver, whether by jamming or by actual echoes.

automatic bass compensation — Abbreviated abc. 1. A circuit used in a receiver or audio amplifier to make the bass notes sound more natural at low-volume settings. The circuit, which usually consists of resistors and capacitors connected to taps on the volume control, automatically compensates for the poor response of the human ear to weak sounds. 2. A circuit used in some audio equipment to increase the amplitude of the bass notes to make them appear more natural at low volume settings.

automatic bias — See self-bias.

automatic brightness control — A circuit used in television receivers to keep the average brightness of the reproduced image essentially constant. Its action is similar to that of an automatic volume-control circuit.

automatic call — A communications feature that allows a transmission control unit to automatically establish a connection with one or more message recipients.

automatic call distributor — See ACD.

automatic calling unit — Abbreviated ACU. A dialing device, supplied by the communication common carrier, that permits a business machine to dial calls automatically over the communication networks.

automatic carriage — A control mechanism for a typewriter or other listing device that can automatically control the feeding, spacing, skipping, and ejecting of paper or preprinted forms.

automatic check — An operation performed by equipment built into an electronic computer to automatically verify proper operation.

automatic chrominance control — A color-television circuit that automatically controls the gain of the chrominance bandpass amplifier by varying the bias.

automatic circuit breaker — A device that automatically opens a circuit, usually by electromagnetic means, when the current exceeds a safe value. Unlike a fuse, which must be replaced once it blows, the circuit breaker can be reset manually when the current is again within safe limits.

automatic coding — A technique by which a digital computer is programmed to perform a significant portion of the coding of a problem.

automatic color purifier — See automatic degausser.

automatic computer — A computer capable of processing a specified volume of work without a need for human intervention other than program changes.

automatic connections — Connections between users made by electronic switching equipment without human intervention.

automatic constant — In a calculator, a provision that allows the user to multiply or divide a series of numbers by the same divider or multiplier without reentering each time.

automatic contrast control — A television circuit that automatically changes the gain of the video intermediate-frequency and radio-frequency stages to maintain proper contrast in the television picture.

automatic controller — A device or instrument for measuring and regulating that operates by receiving a signal from a sensing device, comparing this signal with a desired value, and issuing signals for corrective action.

automatic crossover — 1. A type of current-limiting circuit on a power supply provided with an adjustment for setting the short-circuit current to an adjustable maximum value. 2. A term applied to bimodal power supplies (constant voltage/constant current) that describes the transferral from one operating mode to the other at a predetermined value of load resistance. Usually the crossover point is preset by means of front panel controls.

automatic current limiting — An overload-protection mechanism designed to limit the maximum output current of a power supply to a preset value. Usually it automatically restores the output when the overload is removed.

automatic cutout — A device, operated by electromagnetism or centrifugal force, to automatically disconnect some parts of an equipment after a predetermined operating limit has been reached.

automatic data processing — Abbreviated ADP. The processing of digital information by automatic computers and other machines. Also called integrated data processing.

automatic data-processing system — A system that includes electronic data-processing equipment together with auxiliary and connecting communications equipment.

automatic degausser — Also called automatic color purifier and degausser. An arrangement of degaussing coils mounted around a color-television picture tube. These coils are energized only for a short while after the set is turned on. They serve to demagnetize any parts of the picture tube that may have been affected by the earth's magnetic field or the magnetic field of any nearby home appliance.

automatic dial — A communications feature whereby a calling unit has the ability to automatically establish a connection with one or more message recipients.

automatic dialer — 1. Device that will automatically dial any of a group of preselected telephone numbers. 2. A device that automatically dials telephone numbers on a network.

automatic dialing unit — Abbreviated ADU. A device capable of generating dialing digits automatically.

automatic digital network — Automatic communications network for end-to-end message-switched digital data communication.

automatic direction finder — Abbreviated ADF. Also called an automatic radio compass. An electronic device, usually for marine or aviation application, which provides a radio bearing to any transmitter whose frequency is known but whose direction and location are not.

automatic electronic data-switching center — Communications center designed specifically for the automatic electronic transmission, reception, relay, and switching of digitalized data.

automatic error correction — A technique, usually requiring the use of special codes and/or automatic retransmission, that detects and corrects errors occurring in transmission. The degree of correction depends on coding and equipment configuration.

automatic exchange — A telephone exchange in which connections are made between subscribers by means of devices set in operation by the originating subscriber's instrument without the intervention of an operator.

automatic fine tuning — A circuit in a receiver that automatically maintains the correct tuner oscillator frequency and compensates for drift and for moderate amounts of inaccurate tuning. Similar to automatic frequency control.

automatic focusing — A method of electrostatically focusing a television picture tube; the focusing anode is internally connected through a resistor to the cathode and thus requires no external focusing voltage.

automatic frequency control — Abbreviated AFC. 1. A system that produces an error voltage in proportion to the amount by which an oscillator drifts away from its correct frequency, the error voltage acting to reverse the drift. 2. A control circuit in a receiver or tuner that compensates for small variations in the carrier signal frequency to provide a stable audio output. A circuit function in a tuner or receiver that keeps the unit accurately tuned to the desired station, eliminating any tendency to drift.

automatic frequency correction — *See* automatic frequency control.

automatic function key correction — When the wrong function key is depressed in a calculator, pressing the correct function key automatically replaces it.

automatic gain control — 1. A means of controlling gain of a receiver through feedback to suit the strength of the incoming signal. Ideally, the output level at the speaker should remain constant over a wide range of input signals. For low-level inputs the feedback signal is small and gain is high. For stronger signals the feedback loop cuts gain to prevent overload. 2. A self-acting compensation device that maintains the output of a transmission system constant within narrow limits in the face of wide variations in the attenuation of the system. 3. A radar circuit that prevents saturation of the radar receiver by long blocks of received signals or by a carrier modulated at low frequency. 4. A process by which gain is automatically adjusted as a function of input or other specified parameter.

automatic gain stabilization — A circuit, used in certain identification friend-or-foe equipment and radar beacon systems, which serves to maintain optimum sensitivity in a superregenerative stage by keeping the noise-pulse load constant. The system prevents random noises from triggering the automatic transmitter associated with the receiver.

automatic grid bias — Grid-bias voltage provided by the difference in potential across a resistance (or resistances) in the grid or cathode circuit due to grid or cathode current or both.

automatic intercept — Automatic recording of messages a caller may wish to leave when the called party is away from his telephone.

automatic level compensation — A system that automatically compensates for variations in the circuit. *See also* automatic volume control.

automatic level control — *See* ALC.

automatic light control — The process by which the illumination incident on the face of a television pickup device is automatically adjusted as a function of scene brightness.

automatic message accounting — *See* automatic toll ticketing.

automatic message-switching center — A center in which messages are routed automatically according to information they contain.

automatic modulation control — A transmitter circuit that reduces the gain for excessively strong audio input signals without affecting the strength of normal signals, thus permitting higher average modulation without overmodulation; this is equivalent to an increase in the carrier-frequency power output.

automatic noise-limiter — A circuit that automatically clips off all noise peaks above the highest peak of the desired signal being received. This circuit prevents strong atmospheric or human-made interference from being troublesome.

automatic numbering equipment — Equipment used in association with tape transmitters to transmit a channel number.

automatic pedestal control — The process by which the pedestal height of a television signal is automatically adjusted as a function of the input or other specified parameter.

automatic phase control — A circuit used in color television receivers to synchronize the burst signal with the 3.58-MHz color oscillator.

automatic pilot — *See* autopilot.

automatic programming — Any technique designed to simplify the writing and execution of programs in a computer. Examples are assembly programs that translate from the programmer's symbolic language to the machine language, those which assign absolute addresses to instruction and data words, and those which integrate subroutines into the main routine.

automatic quality control — A technique in which the quality of a product being processed is evaluated in terms of a predetermined standard, and proper corrective action is taken automatically if the quality falls below the standard.

automatic radio compass — A radio direction finder that automatically rotates the loop antenna to the correct position. A bearing can then be secured from the indicator dials without mechanical adjustment or calculation. *See also* automatic direction finder.

automatic record changer — An electrically operated mechanism that automatically feeds, plays, and rejects a number of records in a preset sequence. It consists of a motor, turntable, pickup arm, and changer. Some changers are designed to play automatically $16\frac{2}{3}$-, $33\frac{1}{3}$-, 45-, and 78-rpm records.

automatic regulation — In a power supply, the automatic holding of the output voltage or current to a constant value in spite of variations in the input voltage of load resistance.

automatic relay — A means of selective switching that causes automatic equipment to record and retransmit communications.

automatic repeater station — A station that receives signals and simultaneously retransmits them, but usually on a different frequency.

automatic repeat request — Abbreviated arr. A system of error checking in which an error-detecting code is included with the transmitted data. When the receiver is unable to verify the message, it initiates a retransmission.

automatic reset — The automatic reversion of a timer to its ready state after it has completed a timing cycle or after the input circuit has been interrupted.

automatic reset relay — Also called automatic reset. 1. A stepping relay that returns to its home position either when it reaches a predetermined contact position or when a pulsing circuit fails to energize the driving coil within a given time. It may either pulse forward or be spring-reset to the home position. 2. An overload relay that restores the circuit as soon as an overcurrent situation is corrected.

automatic retransmission — Retransmission of signals by a radio station whereby the retransmitting station is actuated solely by the presence of a received signal through electrical or electromechanical means, i.e., without any direct, positive action by the control operator.

automatic reverse — Ability of some four-track stereo tape recorders to play the second pair of stereo tracks automatically, in the reverse direction, without need to interchange the empty and full reels after the first pair of stereo tracks has been played.

automatic scanning — A variable-speed sweep of the entire frequency range of a radio-frequency interference meter. It may also include the scanning of a portion of this frequency range over a predetermined sector.

automatic scanning receiver — A receiver that can automatically and continuously sweep across a preselected frequency, either to stop when a signal is found or to plot signal occupancy within the frequency spectrum being swept.

automatic secure voice communications — A network that provides cryptographically secure voice communications through the use of a combination of wideband and narrow-band voice-digitizing techniques.

automatic selective control (or **transfer**) **relay** — A device that operates to select automatically between certain sources or conditions in an equipment, or performs a transfer operation automatically.

automatic send — A communications capability whereby a communicating system has the ability to automatically send out a message in an unattended mode.

automatic send/receive — A teletype-writer unit that includes a keyboard, printer, paper tape, reader/transmitter, and paper-tape punch. This combination of facilities may be used online or offline, and, in some cases, online and offline simultaneously.

automatic sensitivity control — 1. A circuit used for automatically maintaining receiver sensitivity at a predetermined level. It is similar to an automatic gain control, but it affects the receiver constantly rather than during the brief interval selected by the range gate. 2. The self-acting mechanism that varies the system sensitivity as a function of the specified control parameters. This may include automatic target control, automatic light control, or any combination thereof.

automatic sequencing — The ability of a computer to perform successive operations without additional instructions from a human being.

automatic short-circuiter — A device used in some forms of single-phase commutator motors to short-circuit the commutator bars automatically.

automatic short-circuit protection — An automatic current-limiting system that enables a power supply to continue operating at a limited current, and without damage, into any output overload, including a short circuit. The output voltage is restored to normal when the overload is removed, as distinguished from a fuse or circuit-breaker system that opens with overload and must be replaced or reclosed manually to restore power.

automatic shutoff — In a tape recorder, a switching arrangement that automatically shuts the recorder off when the tape breaks or runs out. Also, a switching arrangement that stops a record changer after the last record.

automatic starter — 1. A device which, after being given the initial impulse by means of a push button or similar device, starts a system or motor automatically in the proper sequence. 2. A self-acting starter that is completely controlled by master or pilot switches or some other sensing device.

automatic switchboard — Telephone switchboard in which the connections are made by the operation of remotely controlled switches.

automatic switch center — A switch center in which messages originating at any subscriber terminal are relayed automatically through one or more switching centers to their destinations.

automatic target control — The self-acting mechanism that controls the vidicon target potential as a function of the scene brightness.

automatic telegraph transmission — A form of telegraphy in which signals are transmitted mechanically from a perforated tape.

automatic telegraphy — A form of telegraphy in which signals are transmitted and/or received automatically.

automatic telephone dialer — A device that, when activated, automatically dials one or more preprogrammed telephone numbers (e.g., police, fire department) and relays a recorded voice or coded message giving the location and nature of the alarm; used with intrusion alarms and security systems.

automatic test generation — Abbreviated ATG. Automatic test-pattern generation (ATPG). Calculation of a specific set of input test patterns with a computer program providing algorithmic and heuristic routines.

automatic threshold variation — A constant false-alarm rate scheme that is an open-loop type of automatic gain control in which the decision threshold is varied continuously in proportion to the incoming intermediate frequency and video noise level.

automatic time switch — A combination of a switch with an electric or spring-wound clock arranged to turn an apparatus on and off at predetermined times.

automatic toll ticketing — System whereby toll calls are automatically recorded, timed, and toll tickets printed, under control of the calling telephone's dial pulses and without the intervention of an operator. Also called automatic message accounting.

automatic tracking — In radar, the process whereby a mechanism, actuated by the echo, automatically keeps the radar beam locked on the target; may also determine the range simultaneously.

automatic transfer equipment — Equipment that automatically transfers a load so that a source of power may be selected from one of several incoming lines.

automatic tuning — An electrical, mechanical, or electrical/mechanical system that automatically tunes a circuit to a predetermined frequency when a button or other control is operated.

automatic turntable — A record player whose tone arm is positioned for playing records when a control is operated, and which shuts off automatically at the end of play. *See* record changer.

automatic video-noise leveling — A constant false-alarm rate scheme in which the video-noise level at the output of the receiver is sampled at the end of each range sweep and the receiver gain is readjusted accordingly to maintain a constant video-noise level at the output.

automatic voltage regulator — A device or circuit that maintains a constant voltage, regardless of any variation in input voltage or load.

automatic volume compression — *See* volume compression.

automatic volume control — Abbreviated AVC. 1. A self-acting compensation device that maintains the output of a transmission system constant within narrow limits in the face of wide variations in attenuation in that system. 2. A self-acting device that maintains the output of a radio receiver or amplifier substantially constant within relatively narrow limits while the input voltage varies over a wide range. 3. *See* automatic level compensation.

automatic volume expansion — Also called volume expansion. An audio-frequency circuit that automatically increases the volume range by making loud portions louder and weak ones weaker. This is done to make radio reception sound more like the actual program, because the volume range of programs is generally compressed at the point of broadcast.

automatic zero and full-scale calibration correction — A system of zero and sensitivity stabilization in which electronic servos are used that compare demodulated "zero" and "full-scale" signals with reference voltages.

automation — 1. The method or act of making a manufacturing or processing system partially or fully automatic. 2. The entire field of investigation, design, development, application, and methods of rendering or making processes or machines self-acting or self-moving; rendering automatic. 3. Automatically controlled operation of an apparatus process or system by mechanical or electronic devices that take the place of observation, effort, and decision by a human operator.

automaton — 1. A device that automatically follows predetermined operations or responds to encoded instructions. 2. Any communication-linked set of elements. 3. A machine that exhibits living properties. 4. A mechanism, fixed or mobile, possessing the ability to manipulate objects external to itself under the constant control of a programming routine previously supplied by an external intelligence. 5. A machine that is designed to simulate the operations of living things.

automonitor — 1. To instruct an automatic digital computer to produce a record of its information-handling operations. 2. A program or routine for this purpose.

automotive analyzer — An instrument containing numerous automotive test features combined into one portable unit.

automotive electronics — The branch of engineering science that deals with the generation, control, conversion, and application of electricity in self-propelled vehicles.

automotive primary wire — Low-voltage single or multiconductor wire for automotive applications. Resistant to oil and weather.

autopatch — 1. A repeater-to-telephone connection that makes it possible to place a telephone call from a mobile transmitter that has DTMF capability. An access code is frequently required to establish the connection to the phone line. 2. A remotely controllable device frequently used at a repeater location that patches a radiocommunications system into a telephone land-line network. 3. A component that allows telephone calls to be placed through a repeater.

autopilot — Also called automatic pilot, gyropilot, or robot pilot. A device containing amplifiers, gyroscopes, and servomotors that automatically control and guide the flight of an aircraft or guided missile. The autopilot detects any deviation from the planned flight and automatically applies the necessary corrections to keep the aircraft or missile on course.

autopolarity — A feature of a digital voltmeter or digital multimeter wherein the correct polarity (either negative or positive) for a measured quantity is automatically indicated on the display.

auto radio — A radio receiver designed to be installed in an automobile and powered by the storage battery of the automobile.

autoregulation induction heater — An induction heater in which a desired control is effected by the change in characteristics of a magnetic charge as it is heated at or near its Curie point.

auto reverse (cassette deck) — Automatically reverses tape direction at the end of each side of an audio-cassette, for extended unattended listening.

auto single play — A turntable in which records are played one at a time and the arm performs a playing cycle, lifting itself from the record at the end (and in some cases shutting off the motor).

autostarter — *See* autotransformer starter.

Autosyn — A trade name of the Bendix Corp. for a remote-indicating instrument or system based on the synchronous-motor principle, in which the angular position of the rotor of a motor at the measuring source is duplicated by the rotor of the indicator motor.

auto tracking — Also called automatic tracking operation. A master/slave connection of two or more power supplies, each of which has one of its output terminals in common with one of the output terminals of all of the other supplies, such connections being characterized by one-knob control and proportional output voltage from all supplies. Useful where simultaneous turn-up, turn-down, or proportional control of all power supplies in a system is required.

autotransformer — 1. A transformer with a single winding (electrically) in which the whole winding acts as the primary winding, and only part of the winding acts as the secondary (step-down); or part of the winding acts as the primary, and the whole winding acts as the secondary (step-up). 2. A voltage, current, or impedance transforming device in which parts of one winding are common to both the primary and secondary circuits.

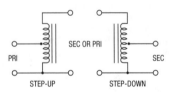

Autotransformers, 1.

autotransformer starter — Also called a compensator or autostarter. A motor starter having an autotransformer to furnish reduced voltage for starting. Includes the necessary switching mechanism.

autozero — Displaying all zeros in a digital multimeter or digital voltmeter when no measurement is being made.

aux — Abbreviation for auxiliary. Often applied to amplifier inputs, where it refers to an extra input facility as distinct from "mic," "tuner," "pickup," etc.

auxiliary actuator — A mechanism that may be attached to a switch to modify its characteristics.

auxiliary bass radiator — A parasitic (nonelectrically driven) unit resembling a bass speaker unit located in a loudspeaker enclosure as if it were an ordinary unit, to increase the movement of air and hence enhance the bass performance at a given enclosure volume.

auxiliary circuit — Any circuit other than the main circuit.

auxiliary contacts — In a switching device, contacts, in addition to main circuit contacts, that function with the movement of the latter.

auxiliary electrodes — Metallic electrodes pushed or driven into the earth to provide electrical contact for the purpose of performing measurements on grounding electrodes of grounded grid systems.

auxiliary equipment — Equipment not directly controlled by the central processing unit of a computer.

auxiliary function — In automatic control of machine tools, a machine function other than the control of the motion of a workpiece or cutter, e.g., control of machine lubrication and cooling.

auxiliary-link station — A station, other than a repeater station, at a specific land location, licensed only for the purpose of automatically relaying radio signals from that location to another specific land location.

auxiliary memory — *See* auxiliary storage.

auxiliary operation — An operation performed by equipment not under continuous control of the computer central processing unit.

auxiliary relay — 1. A relay that responds to the opening or closing of its operating circuit to assist another relay or device in the performance of a function. 2. A relay actuated by another relay and used to control secondary circuit functions such as signals, lights, or other devices. Also called slave relay.

auxiliary-station line filter — A line filter for use at repeater points to separate frequencies of different carrier systems using the same line pair. For example, such a filter might be used at a high-frequency carrier-system repeater to bypass the low-frequency carrier system and voice frequencies around the repeater.

auxiliary storage — 1. In a computer, storage that supplements the main storage, such as hard disks, floppy disks, magnetic tapes, or optical discs. 2. Storage capacity, such as magnetic tape, disk, or drum, in addition to the main memory of a computer. Also called auxiliary memory.

auxiliary switch — A switch actuated by some device such as a circuit breaker, for signaling, interlocking, or other purpose.

auxiliary transmitter — A transmitter held in readiness in case the main transmitter of a broadcasting station fails.

availability — 1. The ratio, expressed as a percent, of the time during a given period that an equipment is correctly operating to the total time in that period. Also called operating ratio. 2. The probability that a system is operating satisfactorily at any point in time when used under stated conditions, where the total time considered includes operating time, active repair time, administrative time, and logistic time.

available conversion gain — Ratio of available output-frequency power from the output terminals of a transducer to the available input-frequency power from the driving generator, with terminating conditions specified for all frequencies that may affect the result. Applies to outputs of such magnitude that the conversion transducer is operating in a substantially linear condition.

available gain — The ratio of the available power at the output terminals of the network to the available power at the input terminals of the network.

available line — In a facsimile system, that portion of a scanning line that can be used for picture signals.

Expressed as a percentage of the length of the scanning line.

available machine time — Time after the application of power during which a computer is operating correctly.

available power — 1. The mean square of the open-circuit terminal voltage of a linear source, divided by four times the resistive component of the source impedance. 2. Of a network, the power that would be delivered to a conjugately matched load. It is the maximum power that a network can deliver. Available power, though defined in terms of an output load impedance, is independent of that impedance.

available power gain — Sometime called completely matched power gain. Ratio of the available power from the output terminals of a linear transducer, under specified input-termination conditions, to the available power from the driving generator. The available power gain of an electrical transducer is maximum when the input-termination admittance is the conjugate of the driving-point admittance at the input terminals of the transducer.

available signal-to-noise ratio — Ratio of the available signal power at a point in a circuit to the available random-noise power.

avalanche — 1. Rapid generation of a current flow with reverse-bias conditions as electrons sweep across a semiconductor junction with enough energy to ionize other bonds and create electron-hole pairs, making the action regenerative. 2. One of the mechanisms responsible for voltage breakdown of semiconductor junctions and devices. When avalanche occurs, carriers moving through the crystal lattice have achieved sufficient kinetic energy to knock further carriers from the lattice, producing a snowballing increase in current level. Provided the current increase is limited externally, avalanche breakdown causes no permanent damage to the device.

avalanche breakdown — 1. In a semiconductor diode, a nondestructive breakdown caused by the cumulative multiplication of carriers through field-induced impact ionization. 2. In a reverse-biased semiconductor, the sudden, marked increase of reverse current at the bias voltage at which avalanche begins. The action resembles a breakdown but it is nondestructive when the current is limited by external means. 3. A breakdown that is caused by the action of a strong electric field which causes some free carriers to gain enough energy to liberate new hole-electron pairs by ionization.

avalanche conduction — A form of conduction in a semiconductor in which charged-particle collisions create additional hole-electron pairs.

avalanche current — The high current through a semiconductor junction in response to an avalanche voltage.

avalanche diode — 1. Also called breakdown diode. A silicon diode that has a high ratio of reverse-to-forward resistance until avalanche breakdown occurs. After breakdown the voltage drop across the diode is essentially constant and is independent of the current. Used for voltage regulating and voltage limiting. Originally called zener diode, before it was found that the zener effect had no significant role in the operation of diodes of this type. 2. A silicon diode in which avalanche breakdown occurs across the diode's pn junction. *See* IMPATT diode; TRAPATT diode.

avalanche impedance — *See* breakdown impedance.

avalanche noise — 1. A phenomenon in a semiconductor junction in which carriers in a high-voltage gradient develop sufficient energy to dislodge additional carriers through physical impact. 2. Electrical noise generated in a junction diode operated at the point at which avalanche just begins.

avalanche photodiode — 1. A photodiode that takes advantage of the avalanche multiplication of photocurrent. It is particularly suited to low-noise and/or high-speed applications. 2. A device that utilizes avalanche multiplication of photocurrent by means of hole electrons created by absorbed photons. When the device's reverse-bias voltage nears breakdown level, the hole-electron pairs collide with substrate atoms to produce multiple hole-electron pairs.

avalanche transistor — A transistor that, when operated at a high reverse-bias voltage, supplies a chain generation of electron-hole pairs.

avalanche voltage — The applied voltage at which avalanche breakdown occurs.

avalanching — A process resulting from high fields in a semiconductor device, in which an electron is accelerated by the field, hits an atom, and releases more electrons, which continue the sequence.

AVC — *See* automatic volume control.

average — *See* arithmetic mean.

average absolute pulse amplitude — The average of the absolute value of instantaneous amplitude taken over the pulse duration. Absolute value means the arithmetic value regardless of algebraic sign.

average brightness — The average illumination in a television picture.

average calculation operation — A typical computer calculating operation longer than an addition and shorter than a multiplication, often taken as the mean of nine additions and one multiplication.

average current — The arithmetic mean of the instantaneous currents of a complex wave, averaged over one half cycle.

average electrode current — The value obtained by integrating the instantaneous electrode current over an averaging time and dividing by the average time.

average life — *See* mean life, 1.

average noise factor — *See* average noise figure.

average noise figure — Also called average noise factor. In a transducer, the ratio of total output noise power to the portion attributable to thermal noise in the input termination, with the total noise being summed over frequencies from zero to infinity and the noise temperature of the input termination being standard (290 K).

average outgoing quality — The ultimate average quality of products shipped to the customer that results from composite sampling and screening techniques.

average power output of an amplitude-modulated transmitter — The radio-frequency power delivered to the transmitter output terminals, averaged over a modulation cycle.

average pulse amplitude — The average of the instantaneous amplitudes taken over the pulse duration.

average rate of transmission — Effective speed of transmission.

average value — 1. The value obtained by dividing the sum of a number of quantities by the number of quantities. The average value of a sine wave is 0.637 times the peak value. 2. The dc voltage of current amplitude that will transfer the same electrical charge to a capacitor as the ac waveform during a half period. Mathematically, it is the average of the absolute value of all the instantaneous amplitudes.

average voltage — The sum of the instantaneous voltages in a half-cycle waveshape, divided by the number of instantaneous voltages. In a sine wave, the average voltage is equal to 0.637 times the peak voltage.

aviation channels — A band of frequencies, below and above the standard broadcast band, assigned exclusively for aircraft and aviation applications.

aviation services — The aeronautical mobile and radionavigational services.

avionics — 1. An acronym designating the field of *avi*ation electr*onics*. 2. The branch of electronics that is concerned with aviation applications. 3. The design, production, and application of electronic devices and systems for use in aviation and astronautics.

Avogadro's number — The actual number of molecules in one gram-molecule, or of atoms in one gram-atom, of an element or any pure substance (6.023×10^{23} molecules/mole).

AWG — 1. American wire gage. A means of specifying wire diameter. The higher the number, the smaller the diameter. 2. A scale of gage sizes which, with the exception of the largest sizes, 4/0 through 1/0, increases with the descending values of wire diameter. For example, a 1 AWG wire has a diameter of 0.289 inch (7.34 mm) and a wire of 40 AWG has a diameter of 0.0031 inch (78.74 μm). AWG is applied to stranded as well as solid conductors.

axial leads — Leads coming out the ends and along the axes of a resistor, capacitor, or other axial part, rather than out the side.

Axial leads.

axial ratio — Ratio of the major axis to the minor axis of the polarization ellipse of a waveguide. This term is preferred over *ellipticity* because, mathematically, ellipticity is 1 minus the reciprocal of the axial ratio.

axis — The straight line, either real or imaginary, passing through a body around which the body revolves or around which parts of a body are symmetrically arranged.

Ayrton-Perry winding — 1. Two conductors connected in parallel so that the current flows in opposite directions in each conductor and thus neutralizes the inductance between the two. 2. A noninductive winding with two inductors that conduct current in opposite directions, with the opposing currents canceling the magnetic field.

Ayrton shunt — Also called universal shunt. A high-resistance parallel connection used to increase the range of a galvanometer without changing the damping.

azel display — A modified type of plan-position indicator presentation showing two separate radar displays on one cathode-ray screen. One display presents bearing information, and the other shows elevation.

az-el mount — An antenna mount that tracks satellites by moving in two directions: the azimuth in the horizontal plain, and elevation up from the horizon.

azimuth — 1. The angular measurement in a horizontal plane and in a clockwise direction. 2. In a tape recorder, the angle that recording and playback head gaps make with the line along which the tape moves. The head is oriented until this angle is 90°. 3. The vertical setting (alignment) of the head in a tape recorder. 4. Compass direction from due north measured in degrees clockwise. (True north can be found by sighting the star Polaris at

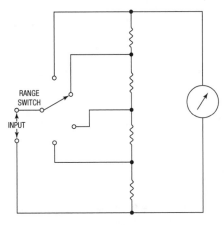

Ayrton shunt.

night, or by applying a local correction for magnetic deviation to a compass reading.)

azimuth alignment—Alignment of the recording and reproducing gaps so that their center lines lie parallel with one another. Misalignment of the gaps causes a loss in output at short wavelengths. *See* head alignment.

azimuth blanking—Blanking of the CRT screen in a radar receiver as the antenna scans a selected azimuth region.

azimuth-elevation mount—A movable dish-antenna mount and aiming system in which one pivot allows rotation in the horizontal plane about the azimuth angle from due north. The other pivot is the elevation above the horizon. (This mount can be more difficult to aim than a polar mount.)

azimuth gain reduction—A technique that allows control of the radar receiver system throughout any azimuth sectors.

azimuth loss—High-frequency losses that are caused by recording head misalignment.

azimuth rate—The rate of change of true bearing.

azimuth resolution—The angle or distance by which two targets must be separated in azimuth to be distinguished by a radar set when the targets are at the same range.

azimuth stabilization—The presentation of indications on a radar display so that north, or any specific reference line of direction, is always at the top of the screen.

azimuth-stabilized plan-position indicator—A plan-position indictator scope on which the reference bearing (usually true or magnetic north) remains fixed with respect to the indicator, regardless of the vehicle orientation.

azimuth versus amplitude—An electronic counter-countermeasures receiver with plan-position indicator-type display attached to the main antenna, used to display strobes due to jamming aircraft. It is useful in making passive position fixes when two or more radar sites can operate together.

azusa—A short base-line continuous-wave phase-comparison electronic tracking system operating on the C-band, in which a single station provides two direction cosines and slant range.

B

B—1. Symbol for the base of a transistor. 2. Symbol for magnetic flux. 3. Abbreviation for photometric brightness. 4. B or b. Abbreviation for susceptance.

B— (B-minus or **B-negative)**—Negative terminal of a B battery or the negative polarity of other sources of anode voltage. Denotes the terminal to which the negative side of the anode-voltage source should be connected.

B+ (B-plus or **B-positive)**—Positive terminal of a B battery or the positive polarity of other sources of anode voltage. The terminal to which the positive side of the anode voltage source should be connected.

babble—1. The aggregate crosstalk from a large number of disturbing channels. 2. In a carrier or other multiple-channel system, the unwanted disturbing sounds that result from the aggregate crosstalk or mutual interference from other channels. 3. Crosstalk from a large number of channels in a system.

babble signal—A type of electronic deception signal used to confuse enemy receivers. Generally, it has characteristics of energy transmission signals. It can be composed by superimposing incoming signals on previously recorded intercepted signals; this composite signal can then be radiated as a jamming signal.

BABS—Abbreviation for blind approach beacon system. A pulse-type ground-based navigation beacon used for runway approach. The BABS ground beacon is installed beyond the far end of the runway on the extended center line. When interrogated by an aircraft, it retransmits two diverging beams, one of short- and the other of long-duration pulses. The beams are transmitted alternately, but because of the first switching, the aircraft receives what appears to be a continuous transmission of both beams. The cathode-ray tube in the aircraft displays both long and short pulses superimposed on each other. When the aircraft is properly aligned with the runway, the pulses will be of equal amplitude.

back bias—1. A degenerative or regenerative voltage that is fed back to circuits before its originating point. Usually applied to a control anode of a tube. 2. A voltage applied to a grid of a tube (or tubes) to restore a condition that has been upset by some external cause. 3. *See also* reverse bias.

back bonding—Bonding active chips to a substrate using the back of the chip, leaving the face with its circuitry uppermost. The opposite of back bonding is face-down bonding.

backbone—A high-voltage, high-capacity transmission line or group of lines having a limited number of large-capacity connections between loads and points of generation.

backbone network—1. A network that links several smaller networks. 2. A transmission facility designed to interconnect low-speed distribution channels or clusters of dispersed user devices.

back contact—Relay, key, jack, or other contact designed to close a circuit and permit current to flow when, in the case of a relay, the armature has released or fallen back, or, in other cases, when the equipment is inoperative.

back current—Also called reverse current. The current that flows when reverse bias is applied to a semiconductor junction.

back diode—A tunnel diode that is usually chosen for its reverse-conduction characteristics.

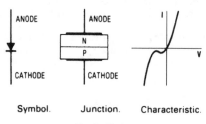

Symbol. Junction. Characteristic.

Back diode.

back echo—An echo due to the back lobe of an antenna.

back emission—Also called reverse emission. Emission from an electrode occurring only when the electrode has the opposite polarity from that required for normal conduction. A form of primary emission common to rectifiers during the inverse portion of their cycles.

back end—In semiconductor manufacturing, the package assembly and test stages of production. Includes burn-in and environmental test functions.

backfill—Filling an evacuated (hybrid) circuit package with dry inert gas prior to hermetic sealing of the package.

backfire—*See* arcback.

back focal length—Distance from the center of a lens to its principal focus on the side of the lens away from the object.

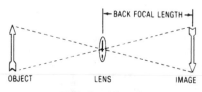

Back focal length.

back focus—The distance from the rear vertex of a lens to the focal plane, with the subject at infinite distance.

background—1. The picture white of the copy being scanned when the picture is black and white only. Also undesired printing in the recorded copy of the picture being transmitted, resulting in shading of the background area. 2. Noise heard during a radio program; this noise is caused by atmospheric interference or operation of the receiver at such high gain that inherent tube and circuit noises become noticeable. 3. A low-priority job that a computer works on when it isn't occupied by more pressing matters.

background control—In color television, a potentiometer used as a means of controlling the dc level of a color signal at one input of a tricolor picture tube. The setting of this control determines the average (or background) illumination produced by the associated color phosphor.

background count—Count caused by radiation from sources other than the one being measured.

background noise—1. The total system noise independent of the presence or absence of a signal. The signal is not included as part of the noise. 2. In a receiver, the noise in the absence of signal modulation in the carrier. 3. Any unwanted sound that intrudes upon program material, such as sounds produced as a result of surface imperfections of a disc record.

background processing—The automatic execution of lower-priority programs in a computer when the system resources are not being used for higher-priority (foreground) programs. For example, background program processing is temporarily suspended to service interrupt requests from I/O devices that require foreground processing.

background program—1. Low-priority program in a multiprogramming environment, which operates when the processor is not doing anything else. 2. That portion of the resident computer program that is run when no immediate pressing needs exist in the system.

background radiation—Radiation due to the presence of radioactive material in the vicinity of the measuring instrument.

background response—In radiation detectors, response caused by ionizing radiation from sources other than that to be measured.

backhaul—A terrestrial communications channel linking an earth station to a local switching network or population center.

back-haul—Use of excess circuit mileage by routing through switching centers that are not in a direct facility path between an originating office and a terminating office.

backing—Flexible material (usually cellulose acetate or polyester) on which is deposited the magnetic-oxide coat that "records" the signal on magnetic recording tape. Also known as base.

backing up—Making copies of files to prevent loss of their contents if the originals are damaged.

backlash—In a potentiometer, the maximum difference that occurs in shaft position when the shaft is moved to the same actual output-ratio point from opposite directions. Resolution and contact-width effects must be excluded from this measure.

backloaded horn—A speaker enclosure arranged so the sound from the front of the cone feeds directly into the room, and the sound from the rear feeds into the room via a folded horn.

back loading—A form of horn loading particularly applicable to low-frequency speakers; the rear radiating surface of the speaker feeds the horn, and the front part of the speaker is directly exposed to the room.

back lobe—1. In the radiation pattern of a directional antenna, that part which extends backward from the main lobe. 2. The three-dimensional petal-shaped pattern representing antenna directional response that is pointing away from the intended direction.

back metallization—Metal applied to the side of the transistor wafer opposite the active areas. Provides the collector contact in bipolar transistors and permits the transistor chips to be bonded to the package or thin-film circuit substrate.

back pitch—The winding pitch of the back end of the armature, that is, the end opposite the commutator.

backplane—1. Area of a computer or other equipment where various logic and control elements are interconnected. Often takes the form of a rat's nest of wires interconnecting printed-circuit cards in the back of computer racks or cabinets. 2. The physical area where printed circuit boards in a system plug in. Usually contains the buses of the system either in printed-circuit or wire-wrap form. Also called a motherboard. 3. A printed circuit card located in the back of a chassis. It has sockets into which specific modules fit for interconnection. 4. A printed circuit motherboard with connectors placed at intervals to allow connection and communication between daughterboards.

backplate—In a camera tube, the electrode to which the stored-charge image is capacitively coupled.

back porch—In a composite picture signal, that portion which lies between the trailing edge of a horizontal-sync pulse and the trailing edge of the corresponding blanking pulse. A color burst, if present, is not considered part of the back porch.

back-porch effect—The continuation of collector current in a transistor for a short time after the input signal has dropped to zero. The effect is due to storage of minority carriers in the base region. It also occurs in junction diodes.

back-porch tilt—The slope of the back porch from its normal horizontal position. Positive and negative refer, respectively, to upward and downward tilt to the right.

backscattering—1. Radiation of unwanted energy to the rear of an antenna. 2. The reflected radiation of energy from a target toward the illuminating radar.

back-shunt keying—A method of keying a transmitter, in which the radio-frequency energy is fed to the antenna when the telegraph key is closed and to an artificial load when the key is open.

backside illumination—A charge-coupled device fabrication technique employing thinned silicon, in which the image is impressed on the side opposite the MOS electrodes.

backspace—In a computer, an operation whose function is to move backward in a sequential file one record at a time.

backstop—That part of the relay that limits the movement of the armature away from the pole face or core. In some relays, a normally closed contact may serve as a backstop.

backswing—The amplitude of the first maximum excursion in the negative direction after the trailing edge of a pulse, expressed as a percentage of the 100-percent amplitude.

backtalk—Transfer of information to the active computer from a standby computer.

back-to-back circuit—Two tubes or semiconductor devices connected in parallel but in opposite directions so that they can be used to control current without introducing rectification. Also called inverse-parallel connection.

backup—1. A duplicate copy of a file or program. 2. A system, device, file or facility that can be used as an alternative in case of a malfunction or loss of data. 3. An item kept available to replace an item that fails to perform satisfactorily. 4. An item under development intended to perform the same general function performed by another item also under development. 5. The hardware and software resources available to recover after a degradation or failure of one or more system components. 6. A system or element, such as a circuit component, that is used to replace a similar system or component in case of failure of the latter.

backup control — *See* redundancy.

backup copy — A copy of a file or data set to be kept for reference in case the original file or data set is destroyed or lost.

backup facility — A communications-electronics facility that is established for the purpose of replacing or supplementing another facility or facilities under real or simulated emergency conditions.

backup item — An additional item to perform the general functions of another item. It may be secondary to an identified primary item or a parallel development to improve the probability of success in performing the general function.

backwall — The plate in a pot core that connects the center post to the sleeve.

backwall photovoltaic cell — A cell in which light must pass through the front electrode and a semiconductor layer before reaching the barrier layer.

backward-acting regulator — A transmission regulator in which the adjustment made by the regulator affects the quantity that caused the adjustment.

backward diode — A highly doped, alloyed germanium junction that operates on the principle of quantum-mechanical tunneling. The diode is "backward" because its easy current direction is in the negative-voltage rather than the positive-voltage region of the I/V curve. The backward diode has a negative-resistance region, but the resultant valley of its I/V curve is much less pronounced than in tunnel diodes.

backward wave — In a traveling-wave tube, a wave having a group velocity opposite the direction of electron-stream motion.

backward-wave oscillator — An oscillator employing a special vacuum tube in which oscillatory currents are produced by using an oscillatory electromagnetic field to bunch the electrons as they flow from cathode to anode.

backward-wave tube — A traveling-wave tube in which the electrons travel in a direction opposite to that in which the wave is propagated.

back wave — *See* spacing wave.

baffle — 1. In acoustics, a shielding structure or partition used to increase the effective length of the external transmission path between two points (e.g., between the front and back of an electroacoustic transducer). A baffle is often used in a speaker to increase the acoustic loading of the diaphragm. (Although this term sometimes is used to designate the entire cabinet, or enclosure, that houses a loudspeaker, strictly speaking it refers to the panel on which the speaker is mounted, usually the front panel of such an enclosure. The term derives from its original use in preventing, or baffling, the speaker's rear sound waves from interfering with its front waves.) 2. In a gas tube, an auxiliary member placed in the arc path and having no separate external connection. 3. A device for deflecting oil or gas in a circuit breaker. 4. A single shielding device designed to reduce the effect of ambient light on the operation of an optical transmission link. 5. An opaque shielding device designed to reduce the effect of stray light on an optical system.

baffle plate — A metal plate inserted into a waveguide to reduce the cross-sectional area for wave-conversion purposes.

bail — A loop of wire used to prevent permanent separation of two or more parts assembled together (e.g., the bail holding dust caps on round connectors).

Bakelite — A trademark of the Bakelite Corp. for its line of plastic and resins. Formerly, the term applied only to its phenolic compound used as an insulating material in the construction of radio parts.

bake-out — Subjecting an unsealed (hybrid) circuit package to an elevated temperature to bake out moisture and unwanted gases prior to final sealing.

balance — 1. The effect of blending the volume of various sounds coming over different microphones in order to present them in correct proportion. 2. The maintenance of equal average volume from both speaker systems of a stereo installation. 3. Relative volume, as between different voices or instruments, bass and treble, or left and right stereo channels. 4. Either a condition of symmetry in an electrical circuit, such as a Wheatstone bridge, or the condition of zero output from a device when properly energized. In the latter sense, depending on the nature of the excitation, two general categories of balance may be encountered: for dc excitation, resistive balance; for ac excitation, resistive and/or reactive balance.

balance control — 1. On a stereo amplifier, a differential gain control used to vary the volume of one speaker system relative to the other without affecting the overall volume level. As the volume of one speaker increases and the other decreases, the sound appears to shift from left to center to right, or vice versa. 2. A variable resistor used to compensate for any slight loss of signal in the right or left channel of a stereo amplifier. To some extent, this control can compensate for unbalanced speakers and be used for adjustment when the listener is not in an equidistant position between the two loudspeakers. 3. A variable component, such as a potentiometer or variable capacitor, used to balance bridges, null circuits, or phase speakers.

balanced — 1. Electrically alike and symmetrical with respect to ground. 2. Arranged to provide balance between certain sets of terminals. 3. A type of line in which both wires are electrically equal.

balanced amplifier — 1. An amplifier circuit with two identical signal branches connected to operate in phase opposition and with their input and output connections each balanced to ground; for example, a push-pull amplifier. 2. A transistor amplifier stage in which two identical transistors are used and the input signal and output power are equally divided between them.

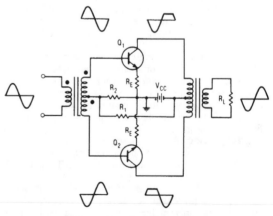

Balanced push-pull amplifier.

This technique produces approximately twice the output power of a single transistor stage, with generally improved dynamic range and reduced voltage standing-wave ratio.

balanced armature — An armature that is approximately in equilibrium with respect to both static and dynamic forces.

balanced-armature unit — The driving unit used in magnetic speakers, consisting of an iron armature pivoted between the poles of a permanent magnet and surrounded by coils carrying the audio-frequency current. Variations in the audio-frequency current cause corresponding changes in the armature magnetism and corresponding movements of the armature with respect to the poles of the permanent magnet.

balanced bridge — A bridge circuit with its components adjusted so that it has an output voltage of zero.

balanced circuit — 1. A circuit with two sides electrically alike and symmetrical to a common reference point, usually ground. 2. A circuit terminated by a network that has infinite impedance losses. 3. A circuit terminated by a network whose impedance balances that of the line, resulting in negligible return losses. 4. A circuit whose electrical midpoint is grounded, as opposed to the single-ended circuit, which has one side grounded. 5. A nulled bridge circuit. 6. Telephone circuit in which the two conductors are electrically balanced to each other and to ground. 7. A circuit so arranged that the impressed voltages on each conductor of the pair are equal in magnitude but opposite in polarity with respect to ground.

balanced converter — *See* balun.

balanced currents — Also called push-pull currents. In the two conductors of a balanced line, currents that are equal in value and opposite in direction at every point along the line.

balanced detector — A demodulator for frequency-modulation systems. In one form, the output consists of the rectified difference of the two voltages produced across two resonant circuits, one circuit being tuned slightly above the carrier frequency and the other slightly below.

balanced line — A line or circuit utilizing two identical conductors. Each conductor is operated so that the voltages on them at any transverse plane are equal in magnitude and opposite in polarity with respect to ground. Thus, the currents on the line are equal in magnitude and opposite in direction. A balanced line is preferred where minimum noise and crosstalk are desired.

balanced-line system — A system consisting of a generator, balanced line, and load adjusted so that the voltages of the two conductors at all transverse planes are equal in magnitude and opposite in polarity with respect to ground.

balanced low-pass filter — A low-pass filter designed to be used with a balanced line.

balanced magnetic switch — A magnetic switch that is operated by a balanced magnetic field in such a manner as to resist defeat with an external magnet. It signals an alarm when it detects either an increase or decrease in magnetic field strength.

balanced method — A method of measurement in which the reading is taken at zero. It may be a visual or audible reading; in the latter case, the null is the no-sound setting.

balanced modulator — An amplitude modulator in which the control grids of two tubes are connected for parallel operation, and the screen grids and plates for push-pull operation. After modulation, the output contains the two sidebands without the carrier.

balanced network — 1. A hybrid network in which the impedances of the opposite branches are equal. 2. A network in which the corresponding series impedance

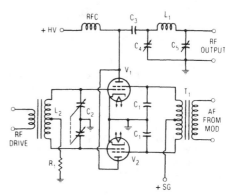

Balanced modulator.

elements are identical and symmetrical with respect to ground.

balanced oscillator — Any oscillator in which (*a*) the impedance centers of the tank circuits are at ground potential and (*b*) the voltages between either end and the centers are equal in magnitude and opposite in phase.

balanced output — A three-conductor output (as from a microphone) in which the signal voltage alternates above and below a third neutral circuit. This symmetrical arrangement tends to cancel any hum picked up by long lengths of interconnecting cable.

balanced output transformer — 1. A push-pull output transformer with a center-tapped primary winding. 2. An output transformer with a grounded center tap on its secondary winding.

balanced probe — A probe, used with an electronic voltmeter or oscilloscope, that has a balanced input and (usually) a single-ended output.

balanced telephone line — A telephone line that is floated with respect to ground so that the impedance measured from either side of the line to ground is equal to that of the other side to ground.

balanced termination — For a system or network having two output terminals, a load presenting the same impedance to ground for each output terminal.

balanced transmission line — A transmission line having equal conductor resistances per unit length and equal impedances from each conductor to earth and to other electrical circuits.

balanced voltages — Also called push-pull voltages. On the two conductors of a balanced line, voltages (relative to ground) that are equal in magnitude and opposite in polarity at every point along the line.

balanced-wire circuit — A circuit with two sides electrically alike and symmetrical to ground and other conductors. Commonly refers to a circuit the two sides of which differ only by chance.

balancer — In a direction finder, that portion used for improving the sharpness of the direction indication. It balances out the capacitance effect between the loop and ground.

balance stripe — A magnetic sound stripe placed on the edge of a motion-picture film opposite the main stripe; it provides mechanical balance for the film.

balance-to-unbalance transformer — A device for matching a pair of lines, balanced with respect to earth, to a pair of lines not balanced with respect to earth. *See also* balun.

balancing network — 1. An electrical network designed for use in a circuit in such a way that two branches of the circuit are made substantially conjugate (i.e., such that an electromotive force inserted into one branch produces no current in the other). 2. Electronic circuitry used to match two-wire to four-wire facilities, sometimes called a hybrid. Balancing is necessary to maximize power transfer and minimize echo. 3. Another name for a hybrid, a circuit that connects a two-wire line to a four-wire line and maximizes power transfer while minimizing echo.

balancing unit — An antenna-matching device used to permit efficient coupling of a transmitter or receiver having an unbalanced output circuit to an antenna having a balanced transmission line. 2. A device for converting balanced to unbalanced transmission lines, and vice versa, by placing suitable discontinuities at the junction between the lines instead of using lumped components.

ball — In face bonding, a method of providing chips with contact.

ballast — A device used with an electronic-discharge lamp to obtain the necessary circuit conditions (voltage, current, and waveform) for starting and operating.

ballasting — An integrated circuit design technique that prevents current hogging.

ballast lamp — A lamp that maintains a nearly constant current by increasing its resistance as the current increases.

ballast resistor — A special type of resistor used to compensate for fluctuations in alternating-current power-line voltage. It is usually connected in series with the power supply to a receiver or amplifier. The resistance of a ballast resistor increases rapidly with increases in current through it, thereby tending to maintain an essentially constant current despite variations in the line voltage.

ballast tube — A current-controlling resistance device designed to maintain a substantially constant current over a specified range of variations in the applied voltage to a series circuit.

ball bond — 1. A type of thermocompression bond in which a gold wire is flame-cut to produce a ball-shaped end that is then bonded to a metal pad by pressure and heat. 2. A bond formed when a ball-shaped end interconnecting wire is deformed by thermocompression against a metallized pad. The bond is also called a nail-head bond from the appearance of the flattened ball.

ball bonding — A bonding technique that uses a capillary tube to feed the bonding wire. The end of the wire is heated and melts, thus forming a large ball. The capillary and ball are then positioned on the contact area and the capillary is lowered. This forms a large bond. The capillary is then removed and a flame is applied, severing the wire and forming a new ball.

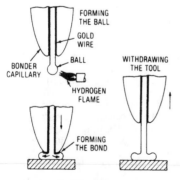

Ball bonding.

ballistic galvanometer — An instrument that indicates the effect of a sudden rush of electrical energy, such as the discharge current of a capacitor.

ballistic-missile early-warning system — An electronic system for providing detection and early warning of attack by enemy intercontinental ballistic missiles. Abbreviated BMEWS.

ballistics — A general term used to describe the dynamic characteristics of a meter movement — most notably, response time, damping, and overshoot.

ballistic trajectory — In the trajectory of a missile, the curve traced after the propulsive force is cut off and the body of the missile is acted upon only by gravity, aerodynamic drag, and wind.

balop — Contraction of balopticon, an apparatus for the projection of opaque images in conjuction with a television camera.

balopticon — See balop.

balun (*bal*anced *un*balanced) — 1. Also called balanced converter or bazooka. An acronym from *bal*anced to *un*balanced. A device used for matching an unbalanced coaxial transmission line to a balanced two-wire system. 2. Usually, a transformer designed to accept 75-ohm unbalanced input (coaxial cable) and deliver the signal at 300-ohm balanced (twin lead). Usable in the converse sense, and sometimes necessary for matching a tuner with 300-ohm balanced antenna terminals to a 75-ohm coaxial line. 3. An impedance-matching transformer device used to connect balanced twisted-pair cabling with unbalanced coaxial or other cabling systems.

banana jack — A jack that accepts a banana plug. Generally designed for panel mounting.

banana plug — A plug with a banana-shaped spring-metal tip and with elongated springs to provide a low-resistance compression fit.

band — 1. Any range of frequencies lying between two defined limits that is used for a specified purpose. 2. A group of radio channels assigned by the FCC to a particular type of radio service.

very low freq. (VLF)	10–30 kHz
low freq. (LF)	30–300 kHz
medium freq. (MF)	300–3000 kHz
high freq. (HF)	3–30 MHz
very high freq. (VHF)	30–300 MHz
ultrahigh freq. (UHF)	300–3000 MHz
superhigh freq. (SHF)	3000–30,000 MHz
extremely high freq. (EHF)	30–300 GHz

3. A group of tracks or channels on a magnetic drum in an electronic computer. *See also* track, 2. 4. In instrumentation, a range of values that represents the scope of operation of an instrument.

bandage — Rubber ribbon, about 4 inches wide, used as a temporary moisture protection for a splice in telephone or coaxial cable.

band center — The geometric mean between the limits of a band of frequencies.

band compression — Reduction of the frequency band needed to transmit a message while maintaining acceptable quality of the message.

banded cable — Two or more cables banded together by stainless-steel strapping.

band-elimination filter — 1. Also called band-stop filter. A wave filter with a single attenuation band, neither of the cutoff frequencies being zero or infinite. The filter passes frequencies on either side of this band. 2. A filter that attenuates frequencies within its rejection band, but passes frequencies above and below this band.

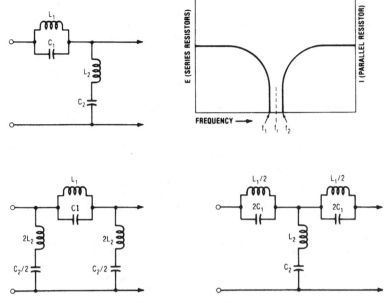

Band-elimination filters.

bandgap — 1. The energy difference between the conduction band and the valence band in a material. 2. The difference in the energy of an electron in a stable state and an electron in an excited state after absorbing energy. 3. The minimum energy required for a valence electron in a semiconductor to make a transition into the conduction band where the electron can move more freely throughout the crystal.

bandgap energy — The difference in energy between the conduction band and the valence band.

band marking — A continuous circumferential band applied to a conductor at regular intervals for identification.

band opening — A condition that results in greater-than-normal communication range on VHF and UHF amateur bands.

bandpass — 1. A specific range of frequencies that will be passed through a device. 2. The number of hertz expressing the difference between the limiting frequencies at which the desired fraction (usually half-power) of the maximum output is obtained.

bandpass amplifier — An amplifier that is tuned to pass only selected frequencies between preset limits.

bandpass amplifier circuit — A stage designed to uniformly amplify signals of certain frequencies only.

bandpass coupling — A coupling circuit with an essentially flat-topped frequency response so that a band of frequencies rather than a single frequency is coupled to a following circuit.

bandpass filter — A wave filter with a single transmission band, neither of the cutoff frequencies being zero

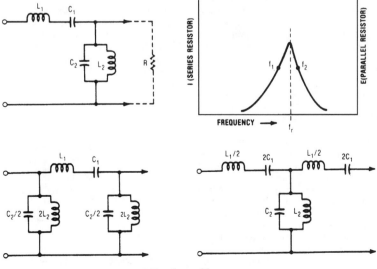

Bandpass filters.

or infinite. The filter attenuates frequencies on either side of this band.

bandpass flatness — The variations in gain in the bandpass of a filter or tuned circuit.

bandpass response — Also called flat-top response. The response characteristic in which a definite band of frequencies is transmitted uniformly.

band plan — A voluntary system of frequency allocations for each amateur radio band.

band-reject filter — A filter that does not pass a band of frequencies but passes both higher and lower frequencies. Sometimes called a notch filter.

band selector — Also called bandswitch. A switch used to select any one of the bands in which an electronic apparatus such as a receiver, signal generator, or transmitter is designed to operate.

B & S gage — Brown and Sharpe wire gage, in which the conductor sizes rise in geometrical progression. Adopted as the American wire gage (AWG) standard.

bandsplitting — Also called split-streaming. A technique for combining several data channels onto one transmission facility by interleaving the data on a bit-by-bit basis. *See also* multiplexing.

bandspreading — 1. The spreading of tuning indications over a wider range to facilitate tuning in a crowded band of frequencies. 2. The method of double-sideband transmission in which the frequency band of the modulating wave is shifted upward so that the sidebands produced by modulation are separated from the carrier by a frequency at least equal to the bandwidth of the original modulating wave. In this way, second-order distortion products may be filtered out of the demodulator output.

bandspread tuning control — A separate tuning control provided on some shortwave receivers to spread the stations in a single band of frequencies over an entire tuning dial.

band-stop filter — *See* band-elimination filter.

bandswitch — *See* band selector.

bandswitching — In a receiver, transmitter, or test instrument, the process of selecting one of two or more self-contained tuned circuits to change from one frequency spectrum to another within the frequency range of the device's intended operation.

bandwidth — 1. The range within the limits of a band. The width of a bandpass filter is generally taken as the limits between which its attenuation is not more than 3.0 decibels greater than its average attenuation throughout its passband. Also used in connection with receiver selectivity, transmitted frequency spectrum occupancy, etc. 2. In a given facsimile system, the difference in hertz between the highest and lowest frequency components required for the adequate transmission of the facsimile signals. 3. The least frequency interval of a wave, outside of which the power spectrum of a time-varying quantity is everywhere less than some specified fraction of its value at a reference frequency. 4. The range of frequencies of a device, within which its performance, with respect to some characteristic, conforms to a specified standard. 5. The range of audio frequencies over which an amplifier or receiver will respond and provide a useful output. 6. The complete range of frequencies over which a particular information system can function. Because it varies with length in optical fibers, bandwidth is typically expressed as a frequency-distance (megahertz-kilometer) product. 7. A range of frequencies available for signaling. In data transmission, the greater the bandwidth, the greater the capacity to transmit data bits. *See* hertz. 8. The number of hertz expressing the difference between the lower and upper limiting frequencies of a frequency band; also, the width of a band of frequencies. 9. The information-carrying capability of a communication line or channel.

10. In data communications, the difference between the highest and lowest frequencies of a band. A measure of the capacity of a communication channel, expressed in bits per second or bauds in digital communications channels, and in cycles per second or hertz in an analog system. 11. The difference between an analog signal's lowest frequency component and its highest signal component as measured in hertz (Hz). 12. The speed of a digital communications circuit in bits per second.

bandwidth limited gain control — A control that adjusts the gain of an amplifier while varying the bandwidth. An increase in gain reduces the bandwidth.

bang-bang controller — A discontinuity-type nonlinear system that contains time delay, dead space, and hysteresis.

bank — An aggregation of similar devices (e.g., transformers, lamps) connected and used together. In automatic switching, a bank is an assemblage of fixed contacts over which one or more wipers or brushes move to establish electric connections.

bank-and-wiper switch — A switch in which the electromagnetic ratchets or other mechanisms are used, first, to move the wipers to a desired group of terminals, and second, to move the wipers over the terminals of this group to the desired bank contacts.

bank winding — Also called banked winding. A compact multilayer form of coil winding used for reducing distributed capacitance. Single turns are wound successively in two or more layers, the entire winding proceeding from one end of the coil to the other without being returned.

bantam tube — A compact tube having a standard octal base but a considerably smaller glass envelope than the standard glass octal tube has.

bar — *See* microbar. 1. A subdivision of a crystal slab. 2. A vertical or horizontal line on a television screen, used for testing. 3. A symbol, placed over a letter, used to indicate the inverse, or complement, of a function. For example, inversion of A is $\bar{A}$, read "A bar" or "not A".

bar code — 1. A self-contained message with information encoded in the physical widths of bars and spaces in a printed pattern. 2. Coding of consumer products that uses combinations of bars of varying thicknesses. Designed to be read by an optical wand or bar code reader. 3. A printed bar-and-space representation of digital data configured to represent numeric or alphanumeric information. In a more restricted sense, a bar code is a sequence of binary ones and zeros that represents a character, or a set of sequences that represents a set of characters. The machine-readable pattern of bars and spaces is called the symbol. Therefore the term *bar code*, comprising the symbol and the code, represents messages with data encoded in the widths of printed bars and (in some cases) the spaces between the bars. 4. A code made up of a series of variable-width vertical lines that can be read by an optical bar reader.

Bar code.

bar code reader — A photoelectric scanner that reads bar codes by means of reflected light.

bar code scanner — An optical scanning device designed to read information printed in the form of bars of different size by detection and processing of the varying reflectivity of light in the bar code.

bare conductor — A conductor not covered with any insulating material.

bar generator — A generator of pulses or repeating waves that are equally separated in time. These pulses are synchronized by the synchronizing pulses of a television system so that they produce a stationary bar pattern on a television screen.

bar-graph display — A display presenting an illuminated line or bar whose length varies in proportion to some parameter being measured.

bar-graph monitoring oscilloscope — An oscilloscope for observation of commutated signals appearing as a series of bars with lengths proportional to channel modulation. The same oscilloscope is commonly used for setup and troubleshooting observations.

barium — An element, the oxide of which is used in the cathode coating of vacuum tubes.

barium titanate — A ceramic that has electric properties and is capable of withstanding much higher temperatures than Rochelle salt crystals. Used in crystal pickups, sonar transducers, and capacitors.

Barkhausen effect — A succession of abrupt changes that occur when the magnetizing force acting on a piece of iron or other magnetic material is varied.

Barkhausen interference — Interference caused by Barkhausen oscillations.

Barkhausen-Kurz oscillator — Circuit for generating ultrahigh frequencies. Its operation depends on the variation in the electrical field around the positive grid and less positive plate of a triode; the variation is caused by oscillatory electrons in the interelectrode spaces.

Barkhausen oscillation — A form of parasitic oscillation in the horizontal-output tube of a television receiver; it results in one or more narrow, dark, ragged vertical lines near the left side of the picture or raster.

Barkhausen oscillator — *See* Barkhausen-Kurz oscillator.

Barkhausen tube — *See* positive-grid oscillator tube.

bar magnet — A bar of metal that has been so strongly magnetized that it holds its magnetism and thereby serves as a permanent magnet.

barn — A unit of measure of nuclear cross sections. Equal to 10^{-24} square centimeter.

Barnett effect — The magnetization resulting from the rotation of a magnetic specimen. The rotation of a ferromagnet produces the same effect as placing the ferromagnet in a magnetic field directed along the axis of rotation. On the macroscopic model, the domains of a ferromagnet can be considered a group of electron systems, each acting as an independent gyroscope or gyrostat.

barometer — An instrument for measuring atmospheric pressure. There are two types of barometers commonly used in meteorology: the mercury barometer and the aneroid barometer.

barometric pressure — The weight of the atmosphere per unit of surface. The standard barometer reading at sea level and 59°F (15°C) is 29.92 inches (760 mm) of mercury absolute.

bar pattern — A pattern of repeating lines or bars on a television screen. When such a pattern is produced by pulses that are equally separated in time, the spacing between the bars on the television screen can be used to measure the linearity of the horizontal or vertical scanning systems.

bar quad — *See* B-quad.

barrage jamming — Simultaneous jamming of a number of adjacent channels or frequencies.

barrel — 1. The cylindrical portion of a solderless terminal, splice, or contact, in which the conductor is accommodated. 2. The portion(s) of a terminal or contact that is (are) crimped. When designed to receive the conductor, it is called the wire barrel. When designed to support or grip the insulation, it is called the insulation barrel.

bar relay — A relay in which a bar actuates several contacts simultaneously.

barrel distortion — 1. Negative distortion that causes a grid pattern to be imaged as barrel-shaped. 2. In television, distortion that makes the televised image appear to bulge outward on all sides like a barrel. 3. In camera or image tubes, the distortion that results in a monotonic decrease in radial magnification in the reproduced image away from the axis of symmetry of the electron optical system. 4. A distortion of a video image that results in the image resembling a barrel rather than a perfect rectangle. 5. Image distortion in an optical or video system, characterized by bending of the edges of a displayed grid into a barrel shape with convex sides and compressed corners. Can be corrected by image processing techniques.

barrel effect — The boomy or hollow voice quality obtained when the voice is transmitted from a reverberant environment; usually accompanied by a loss of intelligibility and a sense of loss of privacy for far-end users.

barretter — 1. A voltage-regulator tube consisting of an iron-wire filament in a hydrogen-filled envelope. The filament is connected in a series with the circuit to be regulated and maintains a constant current over a given voltage variation. 2. A positive coefficient resistor whose resistance increases as temperature increases.

barretter mount — A waveguide mount in which a barretter can be inserted to measure electromagnetic power.

barricade shield — A type of movable shield for protection from radiation.

barrier — 1. A partition for the insulation or isolation of electric circuits or electric arcs. 2. In a semiconductor, the electric field between the acceptor ions and the donor ions at a junction. *See* depletion layer.

barrier capacitance — *See* depletion-layer capacitance.

barrier-film rectifier — A rectifier in which a film having unilateral (single-direction) conductivity is in contact with metal or other normally conducting plates.

barrier grid — A grid close to, or in contact with, a storage surface of a charge storage tube. This grid establishes an equilibrium voltage for secondary-emission charging, and it serves to minimize redistribution.

barrier height — In a semiconductor, the difference in potential from one side of a barrier to the other.

barrier layer — *See* depletion layer.

barrier-layer cell — A type of photovoltaic cell in which light acting on the surface of the contact between layers of copper and cuprous oxide causes an electromotive force to be produced. *See* photovoltaic cell.

barrier-layer rectification — *See* depletion-layer rectification.

barrier plate — A layer of slow-diffusing metal (usually palladium or nickel) placed between two fast-diffusing materials to slow or prevent their interdiffusion.

barrier region — *See* depletion region.

barrier shield — A wall or enclosure shielding the operator from an area where radioactive material is being used or processed by remote-control equipment.

barrier strip — 1. A terminal strip with protective barriers between adjacent terminals. 2. A continuous section of dielectric material that insulates electrical circuits from each other or from ground.

barrier voltage — The voltage necessary to cause electrical conduction in a junction of two dissimilar materials, such as a pn junction diode.

bar test pattern — Special test pattern for adjusting color TV receivers or color encoders. The upper portion consists of vertical bars of saturated colors and white. The lower horizontal bars have black-and-white areas and I and Q signals.

base — 1. The region between the emitter and collector of a transistor which receives minority carriers injected from the emitter. It is the element that corresponds to the control grid of an electron tube. 2. In a vacuum tube, the insulated portion through which the electrodes are connected to the pins. 3. On a printed circuit board, the portion that supports the printed pattern. 4. A thin, strong, and flexible material, usually a polyester or acetate film, on which is deposited a magnetic formulation to make recording tape. *See also* alkali; backing; positional notation; radix. 5. One of the three regions that form a bipolar transistor. It physically separates the emitter and collector regions. Minority carriers are injected from the emitter into the base, where they subsequently either recombine or diffuse into the collector.

base address — A given address from which an absolute address is obtained by combination with a relative address. Also called address constant.

baseband — 1. The frequency band occupied by the aggregate of the transmitted signals used to modulate a carrier, before they combine with a carrier in the modulation process. 2. In CD-4 records, the left- or right-channel's band containing musical information for the front and back channels, recorded at listening frequencies in the standard sound spectrum (from 30 to 15,000 hertz). 3. The output signal of a video camera, videotape recorder, or satellite TV receiver before remodulation so that it can be viewed on an ordinary TV set. A signal in a satellite TV receiver goes from 4 GHz through the down-converter to become intermediate frequency and then through an FM modulator to become baseband. The American NTSC TV bandwidth is 4.2 MHz at baseband. 4. The basic direct output signal from a television camera, satellite television receiver, or videotape recorder. Baseband signals can be viewed only on monitors. To display the baseband signal on a conventional television set, a "modulator" is required to convert the baseband signal to one of the VHF or UHF television channels that the television set can be tuned to receive.

baseband frequency response — Response characteristics over the frequency band occupied by all of the signals that modulate a transmitted carrier.

baseband signal — A signal that is not modulated onto a carrier.

baseband signaling — A form of transmission that uses discrete pulses, without modulation. Also called baseband transmission.

baseband transmission — A method of using low-frequency transmission of signals across coaxial cables for short distances. *See also* baseband signaling.

baseband video — Same as composite video (CVS or CVBS). A composite video signal contains video picture information for color, brightness, and synchronization (horizontal and vertical).

base-coupled logic — Abbreviated bcl. A circuit configuration designed for subnanosecond propagation delays and rise and fall times; it can be used for a bit rate of more than 1G bit/s. Base-coupled logic circuits consist of a current-mode switch and emitter followers. Switching is done by means of a base-coupled current-mode switch.

base electrode — An ohmic or majority-carrier contact to the base region of a transistor.

base film — The plastic substrate that supports the coating of magnetic recording tape. The base film of most instrumentation and computer tapes is made of polyester. For less critical uses, cellulose acetate and polyvinyl chloride are employed.

basegroup — 1. Designation for a number of carrier channels in a particular frequency range that forms a basic unit (channel bank) for further modulation to a final frequency band. 2. Twelve communication-set paths capable of carrying the human voice on a telephone set; a unit of a frequency-division-multiplexing system's bandwidth allocation.

base insulator — Heavy-duty insulator used to support the weight of an antenna mast and to insulate the mast from the ground or some other surface.

base line — 1. In radar displays, the visual line representing the track of the radar scanning beam. 2. In graphical presentations, the horizontal scale, often representing time, bias, or some other variable.

base-line break — In radar, a technique that uses the characteristic break in the base line on an A-scope display due to a pulse signal of significant strength in noise jamming.

base load — In a dc converter, the current that must be taken from the base to maintain a saturated state.

base-loaded antenna — A vertical antenna the electrical height of which is increased by adding inductance in series at the base.

base material — 1. An insulating material (usually a copper-clad laminate) used to support a conductive pattern. 2. Rigid or flexible insulating substrate that supports conductive patterns or interconnections. *See* substrate.

base metal — Metal from which connectors, contacts, or other metal accessories are made and on which one or more metals or coatings may be deposited. Sometimes called basis metal.

base number — The radix of a number system (10 is the base number, or radix, for the decimal system; 2 for the binary system).

base-One peak voltage — The peak voltage measured across a resistor in series with base-One when a unijunction transistor is operated as a relaxation oscillator in a specified circuit.

base pin — One of the metal prongs on the base of an electron tube, which makes contact with springs in a tube socket.

base point — *See* radix point.

base region — In a transistor, the interelectrode region, between the emitter and collector junctions, into which minority carriers are injected.

base register — A high-speed computer storage location that contains the address of the first data word or instruction of a segment. All addresses within the segment are referenced to the start of the segment; during execution, the physical address of each memory reference is calculated by simply adding the contents of the base register to the relative addresses contained within the segment. Relocation is easily accomplished by simply moving a segment in memory and placing the new address in the base register.

base resistance — Resistance in series with the base lead in the common-T equivalent circuit of a transistor.

base resistor — The (external) resistor connected to the base of a bipolar transistor. In a common-emitter circuit, the base resistor is analogous to the grid resistor of a vacuum-tube amplifier or the gate resistor of a field-effect transistor amplifier.

base ring — Ohmic contact to the base region of power transistors; so called because it is ring-shaped.

base spreading resistance — In a transistor, the resistance of the base region caused by the resistance of the bulk material of the base region.

base station — 1. A fixed station used by a dispatcher to communicate with mobile units. 2. The central transmitter in a communications system that acts as the cell hub for communicating with handsets and/or mobile units. 3. A land station, in the land mobile service, carrying on a service with land mobile stations. (A base station may secondarily communicate with other base stations incident to communications with land mobile stations.) Sometimes defined as a station in a land mobile system that remains in a fixed location and communicates with mobile stations.

base-timing sequencing — Sharing of a transponder on a time basis between several ground transmitters through the use of coded timing signals.

base voltage — The voltage between the base terminals of a transistor and the reference point.

BASIC — 1. A simplified computer language intended for use in engineering applications. 2. Acronym for beginners all-purpose symbolic introduction code. An interpreter language that is one of the easier languages to learn and use. 3. An easy-to-learn and easy-to-use programming language developed at Dartmouth College. BASIC can be used to solve business problems as well as in scientific applications.

basic access method — A method of computer access in which each input/output statement results in a corresponding machine input/output operation.

basic frequency — In any wave, the frequency that is considered the most important. In a driven system, it would in general be the driving frequency, whereas in most periodic waves it would correspond to the fundamental frequency.

basic linkage — In a computer, a linkage that is used repeatedly in one routine, program, or system and that follows the same set of rules each time it is used.

basic processing unit — The principal section for control and data processing within a communications system.

basic protection — Fundamental lightning protection measures and/or devices, such as the use of gas tubes or carbon-block protectors, which are applied directly to transmission media at apparatus locations to provide initial voltage limitation.

basic Q — See nonloaded Q.

basic radio pager — A simple one-way communication device that emits audio sounds when it detects the proper sequence of selective tone signals sent by a dispatcher. The recipient must then telephone the signal originator to get a message. A radio pager is capable of receiving and decoding signals radiated from a call-service transmitter but has no transmitting facilities of its own.

basic rectifier — A metallic rectifier in which each rectifying element consists of a single metallic rectifier cell.

basic speed range — The range over which a motor and control are capable of delivering full load torque without overheating or clogging. It is obtained by armature voltage control.

basic television service — In any of the television-delivery services besides broadcasting (which is free), the monthly fee to subscribe to the lowest tier or set of services and programs, as distinguished from the premium services.

basis metal — See base metal.

basket winding — A coil winding in which adjacent turns are separated except at the points of crossing.

bass — Sounds in the low audio-frequency range. On the standard piano keyboard, all notes below middle C (261.63 hertz).

bass boost — A deliberate adjustment of the amplitude-frequency response of a system or component to accentuate the lower audio frequencies.

bass-boosting circuit — A circuit that attenuates the higher audio frequencies in order that low or bass frequencies will be emphasized by comparison.

bass compensation — Emphasizing the low-frequency response of an audio amplifier at low volume levels to compensate for the lowered sensitivity of the human ear to weak low frequencies.

bass control — A manual tone control that has the effect of changing the level of bass frequencies reproduced by an audio amplifier.

bass half-loudness points — The low frequency, in hertz, at which the sound power of a speaker rolls off to become half as loud as the rest of the tonal spectrum; a relative measure of a speaker's ability to reproduce low bass sounds.

bass reflex — A ported loudspeaker configuration using an acoustically tuned vent, through which low-frequency sound from inside the enclosure passes to strengthen and extend bass response.

bass-reflex enclosure — A type of speaker enclosure in which the rear wave from the speaker emerges through an auxiliary opening or port of critical dimensions to reinforce the bass tones. See also vented baffle.

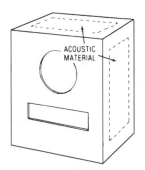

Bass-reflex enclosure.

bass response — 1. The extent to which a speaker or audio-frequency amplifier handles low audio frequencies. 2. The ability of any device to pick up or reproduce low audio frequencies.

bassy — Term applied to sound reproduction in which the low-frequency notes are overemphasized.

BAT — Abbreviation for battery.

batch — 1. A group of documents to be processed; an arbitrary subdivision of a job by the supervisor into smaller, more manageable parts. A batch is the smallest group of such documents accessible by name (number) for data entry, data verify, peripheral device transfers, etc. 2. A type of computer operating system in which jobs are processed one at a time and a job must complete its execution before the next is begun. 3. Computer processing mode in which a program is submitted and the result is delivered back. No interactive communication between program and user is possible. 4. The group of programs considered as a single unit for processing on a computer.

batch control sample — A representative batch extracted either at random or at specific intervals from a process or product for quality-control purposes. Results of equivalent tests of the batches are averaged to interpolate the quality of the total process.

batch environment — A situation in which a computer receives instructions and program(s) from a terminal or other peripheral device, then executes the requested operations at its own convenience.

batch file — A text file that contains operating system and program commands, used for standard procedures performed repetitiously.

batching — *See* batch processing.

batch mode — Operational mode in a computer in which input to or output from a process is transmitted as a single set of successive messages. Contrast with *interactive mode*.

batch process — A method of fabricating monolithic resistors, capacitors, and diodes with the same process at the same time.

batch processing — 1. In a computer, a method of processing in which a number of similar input items are grouped for processing during the same machine run. 2. Pertaining to the technique of executing a set of computer programs such that each is completed before the next program of the set is started. Loosely, the execution of computer programs serially. 3. A technique by which items to be processed must be coded and collected into groups prior to processing. 4. A data-processing technique in which input data is accumulated offline and processed in batches.

bat handle — Standard form of a toggle-switch lever, having a shape similar to that of a baseball bat.

bathtub capacitor — A type of capacitor enclosed in a metal housing having broadly rounded corners like those on a bathtub.

bathyconductorgraph — A device used from a moving ship to measure the electrical conductivity of seawater at various depths.

bathythermograph — A device that automatically plots a graph showing temperature as a function of depth when lowered into the sea.

Batten system — A method developed by W. E. Batten for coordinating single words in a computer to identify a document. Sometimes called peek-a-boo system.

battery — Abbreviated BAT. 1. A dc voltage source consisting of two or more cells that converts chemical, nuclear, solar, or thermal energy into electrical energy. 2. In communications, a source (not necessarily a storage device) of direct current or the current itself. 3. Two or more cells coupled together in series or parallel. In the former configuration the arrangement gives a greater voltage (two cells give twice the voltage, three cells give three times the voltage, and *n* cells give *n* times the voltage); the latter arrangement gives the same voltage as the individual cell but a greater current.

battery acid — A solution that serves as the electrolyte in a storage battery. In the common lead-acid storage battery, the electrolyte is diluted sulfuric acid.

battery cable — A single conductor cable, either insulated or uninsulated, used for carrying current from batteries to the point where power is needed. May also be used for grounding.

battery capacity — The amount of energy obtainable from a storage battery, usually expressed in ampere-hours.

battery charger — Device used to convert alternating current into a pulsating direct current that can be used for charging a storage battery.

battery clip — A metal clip with a terminal to which a connecting wire can be attached, and with spring jaws that can be quickly snapped onto a battery terminal or other point to which a temporary connection is desired.

battery life — The number of times that a battery can be charged and discharged. One complete charge and one complete discharge is called a cycle. The number of complete cycles a battery will give depends on construction of the battery, charging procedure, maintenance, and operation.

battery post adapter — A device connected to a battery post and used for connecting ammeter leads and simulating a charged battery.

battery pulses — Negative potential pulses from the central office battery that are applied to a telephone circuit.

battery receiver — A radio receiver that obtains its operating power from one or more batteries.

battery separator — An insulator that separates the positive and negative plates in a storage battery.

battle short — A switch for short-circuiting safety interlocks and lighting a red warning light.

bat wing — An element on an FM or TV transmitting or receiving antenna, so called because of its shape.

baud — 1. A unit of signaling speed derived from the duration of the shortest code element. 2. A unit of signaling speed equal to the number of discrete conditions or signal events per second. For example, in Morse code 1 baud equals one-half dot cycle per second; in a train of binary signals, 1 baud is 1 bit per second; and in a train of signals, each of which can assume one of eight different states, 1 baud is one 3-bit value per second. 3. A measurement of communication channel capacity as a function of time. For example, a 110-baud line is divided into 110 equal parts. Within each of these parts a certain amount of data can be placed, typically, one bit. This means that a speed of 110 baud is 110 bits per second. 4. The unit of modulation rate (or signaling speed). The reciprocal of the duration of the minimum signaling element (pulse width). 5. The number of times per second the line condition changes. If the line condition represents the presence or absence of a single bit (as in two-state signaling), then the signaling speed in bauds is the same as bits per second. However, if the signaling is not two-state, then bauds are not equal to bits per second. The latter condition exists, for instance, in "dibit" or four-state signaling, in which the baud rate is equivalent to the number of bits per second times two. 6. In an equal length code, 1 baud corresponds to a rate of one signal element per second. Thus, with a signal element duration of 20 ms, the modulation rate is 50 baud (per second). The term is both singular and plural. 7. A data-communication-rate unit used similarly to bits per second (bps) for low-speed data; the number of signal-level changes per second (regardless of the information the signals contain). 8. A measure of serial data flow between a computer and/or communication devices. One baud is equal to 1 bit per second (1 bps). 9. Unit of signaling speed. The speed in bauds is the number of line changes (in frequency, amplitude, etc.) or events per second. At low speeds, each event represents only one bit condition, and baud rate equals bps. As speed increases, each event represents more than one bit, and baud rate does not truly equal bps. But in common usage, baud rate and bps are often used interchangeably. 10. A variable unit of data transmission speed (as one bit per second). Often confused with bits per second (bps). 11. Signaling rate unit for analog communications. One baud is equal to one change of state per second. Where there are two possible states (e.g., two tone frequencies), 1 baud equals 1 bps. Where there are 2^n possible states (e.g., four possible phase shifts in a sinusoidal wave), 1 baud equals *n* bps. In low-speed transmission, in which modems with frequency-shift keying are used, the terms *baud* and *bps* may be used interchangeably. Otherwise, the two terms are never interchangeable.

Baudot code — 1. A data-transmission code in which one character is represented by five equal-length bits. This code is used in most dc teletypewriter machines, in which one start element and 1.42 stop elements are added. 2. Five-level code (plus one start and one stop bit) used primarily by amateurs in rtty communication. Only code allowed by the FCC without special waiver. 3. A code of 32 numbers used for alphabetic and symbolic communication. Invented in 1880 by the Frenchman J. M. E. Baudot, the code bears his name; the term *baud* is also derived from it. 4. Data-transmission code in which five bits represent one character. Use of shift letters/figures enables 64 alphanumeric characters to be represented. Baudot is used in many teleprinter systems with one start bit and 1.5 stop bits added.

baud rate — 1. The number of signal events per second occurring on a communications channel. Although not technically accurate, baud rate is commonly used to mean bit rate. *See* bps. 2. The speed at which data is transmitted, measured in symbols per second. This is not the same as bits per second, since each symbol can carry several bits of information. 3. The number of code elements transferred in one second based on the length of the shortest element. If each element consists of one bit, the baud rate would equal the number of bits per second. 4. Binary speed through a serial interface, traditionally defined as the number of signal elements per second. When each element is one bit, the baud rate equals the number of bits per second. 5. The number of bits transmitted per second in a serial data transmission system. The number of bits per second may also include control bits as well as data bits.

baud-rate generator — Oscillator, usually adjustable, that provides clock signals for connection of a peripheral. Typical rates are 110, 300, 9600 bauds and higher.

bay — 1. A portion of an antenna array. 2. A vertical compartment in which a radio transmitter or other equipment is housed.

bayonet base — A base having two projecting pins on opposite sides of a smooth cylindrical base; the pins engage corresponding slots in a bayonet socket and hold the base firmly in the socket.

bayonet coupling — 1. A quick-coupling device. Connection is accomplished by rotating two parts under pressure. Pins on the side of the male connector engage slots on the side of the female connector. 2. A quick-coupling device for plug and receptacle connectors, accomplished by rotation of a cam operating device designed over a short cylindrical stud or studs to bring the connector halves together.

bayonet socket — A socket for bayonet-base tubes or lamps; it has slots on opposite sides and one or more contact buttons at the bottom.

bazooka — *See* balun.

B battery — The battery that furnishes the required dc voltages to the plate and screen-grid electrodes of the vacuum tubes in a battery-operated circuit.

BBS — Abbreviation for bulletin board system. A system by which a group of users with common interests (such as a business, club, or professional society) can share information by posting it to an electronic bulletin board. Some subscriber-based BBSs offer limited Internet services. The system allows people to carry on discussions, upload and download files, and make announcements without the people being connected to the computer at the same time. *See also* bulletin board.

BC — Abbreviation for bare copper or bell cord.

BCAS — Abbreviation for beacon collision avoidance system. An airborne automatic collision-avoidance system.

BCD — Abbreviation for binary-coded decimal. 1. A system of representing numerical, alphabetic, and special characters in which individual decimal digits are represented by some binary code. For example, in an 8-4-2-1 BCD notation, 16 might be represented as 001 (for 1) and 0110 (for 6). In pure binary notation, 16 is 10000; the number 23 is represented by 0010 0011 in the BCD notation. 2. A code defined with 6 bits per character and capable, therefore, of representing up to 64 unique values, each representing one of the 26 letters of the English alphabet (uppercase only), a digit (0 to 9), or a special or punctuation character. BCD is no longer used. *See* EBCDIC.

BCD code — An eight-bit code (of which one bit is used for odd parity check) used by IBM. It provides 88 characters, plus 16 functions.

BCD counter — A counter in which each section consists of four flip-flops or stages, each section of which counts to nine (binary 1001) and then resets to zero (binary 0000). The outputs are in BCD form.

B channel — One of two stereo channels, usually the right, together with the microphone, speakers, or other equipment associated with this channel.

bci — Abbreviation for broadcast interference, a term denoting interference by transmitters with reception of broadcast signals on standard broadcast receivers.

B-display — On a radarscope, a type of presentation in which the target appears as a bright spot. Its bearing is indicated by the horizontal coordinate and its range by the vertical coordinate.

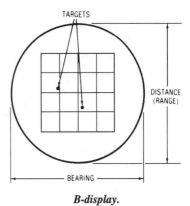

B-display.

beacon — 1. A device that emits a signal for use as a guidance or warning aid. Radar beacons aid the radar set in locating and identifying special targets that may otherwise be difficult or impossible to sense. 2. Low-power carrier transmitted by a satellite that supplies the controlling engineers on the ground with a means of monitoring telemetry data, tracking the satellite, or conducting propagation experiments. This tracking beacon is usually a horn or omni antenna.

beacon delay — The amount of inherent delay within the beacon, i.e., the time between the arrival of a signal and the response of the beacon. In a pulse beacon, delay ordinarily is measured between the leading 3-dB points of the triggering pulse and the reply pulse.

beacon receiver — A radio receiver for converting into perceptible signals the waves emanating from a radio beacon.

beacon skipping — A term used to describe a condition in which beacon return pulses are missing at the interrogating radar. Beacon skipping can be caused by interference, overinterrogation of the beacon, antenna nulls, or pattern minima.

beacon station — 1. A station broadcasting beacon signals for direction finding or navigation. 2. A radar transmitting station.

beacon stealing — The loss of beacon tracking by a (desired) radar due to (interfering) interrogation signals from another radar.

beacon time-sharing — A technique by which two or more radars may interrogate and track a long-recovery type of beacon without exceeding the duty cycle of the beacon. This technique is accomplished by the proper sequencing of the various radar interrogations. It is necessary to ensure that the total of all interrogations does not exceed the beacon duty cycle and that enough time is allowed for the modulator section of the beacon to recover before it receives the next interrogation.

beacon transmitter — A transmitter specially adapted for the transmission of beacon signals.

beaded coax — A coaxial cable in which the dielectric consists of beads made of various insulating materials.

beaded support — Ceramic and plastic beads used to support the inner conductor in coaxial transmission lines.

beaded transmission line — A line using beads to support the inner conductor in coaxial transmission lines.

bead thermistor — A thermistor consisting of a small bead of semiconducting material, such as germanium, placed between two wire leads. Used for microwave power measurement, temperature measurement, and as a protective device. The resistance decreases as the temperature increases.

beam — 1. A flow of electromagnetic radiation concentrated in a parallel, converging, or diverging pattern. 2. The unidirectional or approximately unidirectional flow of radiated energy or particles. 3. A shaft or column of light, a bundle of rays that may or may not consist of parallel, converging, or diverging rays.

beam-addressable technology — The applications of reversible writing with a laser beam on particular storage materials. In one method, an amorphous film is heated and then crystallized for writing. Bubble writing involves the formation of bubbles on the film-glass interface of a glass substrate by the intense light beam.

beam alignment — The adjustment of the electron beam in a camera tube (on tubes employing low-velocity scanning) to cause the beam to be perpendicular to the target surface.

beam angle — The angle between the directions, on either side of the axis, at which the intensity of the radio-frequency field drops to one-half the value it has on the axis.

beam antenna — 1. An antenna that concentrates its radiation into a narrow beam in a definite direction. 2. A directional antenna that radiates or intercepts more energy in one direction than in others.

beam bender — *See* ion trap.

beam bending — Deflection of the scanning beam of a camera tube by the electrostatic field of the charges stored on the target.

beam blanking — Interruption of the electron beam in a cathode-ray tube by the application of a pulse to the control grid or cathode.

beam breaker — A level-measuring device using a light or electrical beam or pneumatic jet between a source and a detector or reflector.

beam candlepower — The candlepower of a bare source that, if located at the same distance as the beam, would produce the same illumination as the beam.

beam convergence — The converging of the three electron beams of a three-gun color picture tube at a shadow-mask opening.

beam-coupling coefficient — In a microwave tube, the ratio of the amplitude, expressed in volts, of the velocity modulation produced by a gap to the radio-frequency gap voltage.

beam crossover — The point of overlap of a beam from an antenna that is nutated or rotated about the center line of the antenna radiation direction. The crossover point is normally at the half-power point. The received energy, when commutated into four quadrants, provides the necessary information for the servoamplifier error signal used to align the antenna to a target.

beam current — The current carried by the electron stream that forms the beam in a cathode-ray tube.

beam cutoff — In a television picture tube or cathode-ray tube, the condition in which the control-grid potential is so negative with respect to the cathode that electrons cannot flow and thereby form the beam.

beam-deflection tube — An electron-beam tube in which current to an output electrode is controlled by the transverse movement of an electron beam.

beam droop — A form of distortion of the normal rectilinear fan-shaped radiation pattern of a detection radar in which a portion of the fan is at a lower elevation than the rest of the fan.

beam-forming electrode — Electron-beam focusing elements in power tetrodes and cathode-ray tubes.

beam hole — An opening through a reactor shield and, generally, through the reactor reflector that permits a beam of radioactive particles or radiation to be used for experiments outside the reactor.

beam-index color tube — A color picture tube in which the signal generated by an electron beam after deflection is fed back to a control device or element in such a way that an image in color is provided.

beam lead — 1. A metal beam deposited directly onto the surface of the die as part of the wafer processing cycle in the fabrication of an integrated circuit. Upon separation of the individual die (normally by chemical etching instead of the conventional scribe-and-break technique), the cantilevered beam is left protruding from the edge of the chip and can be bonded directly to interconnecting pads on the circuit substrate without the need for individual wire interconnections. 2. A long structural member not supported everywhere along its length and subject to the forces of flexure, one end of which is permanently attached to a chip device and the other end intended to be bonded to another material, providing an electrical interconnection or mechanical support or both.

beam-lead bonding — 1. A free-bonding technique in which thick gold extensions of the thin-film terminals of semiconductor devices and circuits are electroformed so they extend beyond the edges of the chips. 2. A method of interconnecting ICs in a circuit by bonding beam leads located on the IC chip's back surface to the circuit's conducting paths. 3. A hybrid bonding technique that provides for multiple bonding simultaneously.

beam-lead device — An active or passive chip component possessing beam leads as its primary interconnection and means of mechanical attachment to a substrate.

beam-lead isolation — The method in which electrical isolation between IC elements is produced by interconnecting the elements with thick gold leads and selectively etching the silicon from between elements without affecting the gold leads. This process leaves the elements as separate units supported by the gold leads.

beam leads — 1. A generic term describing a system in which flat metallic leads extend from the edges of a chip component much as wooden beams extend from a roof overhang. These are then used to interconnect the component to film circuitry. 2. Techniques for attachment of lead frames to silicon chips, including vacuum and chemical deposition, diffusion thermal-compression techniques, welding, etc.

beam-lobe switching — A method of determining the direction of a remote object by comparison of the signals corresponding to two or more successive beam angles at directions slightly different from that of the object.

beam modulation — *See* z-axis modulation.

beam optics — A discipline within the broad study of optics that is specifically oriented toward the investigation of waves with small angular divergence.

beam parametric amplifier — A parametric amplifier in which a modulated electron beam provides a variable reactance.

beam-positioning magnet — A magnet used with a tricolor picture tube to influence the direction of one of the electron beams so that it will have the proper spatial relationship with the other two beams.

beam-power tube — An electron-beam tube in which directed electron beams are used to contribute substantially to its power-handling capability, and in which the control and screen grids essentially are aligned.

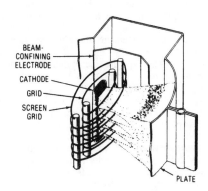

Beam-power tube.

beam relaxor — A type of sawtooth scanning-oscillator circuit that generates but does not amplify the current wave required for magnetic deflection in a single-beam-power pentode.

beam-rider control system — A system whereby the control station sends a beam to the target, and the missile follows this beam until it collides with the target.

beam-rider guidance — A form of missile guidance wherein a missile, through a self-contained mechanism, automatically guides itself along a beam transmitted by a radar.

beams — CB radio term for any type of directional antenna.

beam splitter — A device used for dividing a light beam (as by a transparent mirror) into two components,

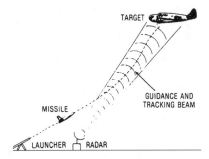

Beam-rider guidance.

one transmitted and the other reflected. 2. An optical device used to divide the beam of light into two beams. Either prisms or partial mirrors.

beam splitting — A process for increasing the accuracy of locating targets by radar. By noting the azimuths at which one radar scan first discloses a target and at which radar data from it ceases, beam splitting calculates the mean azimuth for the target.

beam-splitting mirror — In an oscilloscope camera system, a tilted, transparent mirror that allows rays to pass horizontally from the oscilloscope screen to the camera and also to be reflected vertically to the viewer's eye.

beam spreader — An optical element the purpose of which is to impart a small angular divergence to a collimated incident beam.

beam switching — A method of obtaining the bearing and/or elevation of an object more accurately by comparing the signals received when the beam is in a direction differing slightly in bearing and/or elevation. When these signals are equal, the object lies midway between the beam axes.

beam-switching tube — A multiposition, high-vacuum, constant-current distributor. The beam-switching tube consists of many identical arrays around a central cathode. Each array comprises a spade that automatically forms and locks the electron beam, a target-output electrode that gives the beam current its constant characteristics, and a high-impedance switching grid that switches the beam from target to target. A small cylindrical magnet, permanently attached to the glass envelope, provides a magnetic field. This field, in conjunction with an applied electric field, comprises the crossed fields necessary for operation of this tube. It is used in electronic switching and in distributing, such as counting, timing, sampling, frequency dividing, coding, matrixing, telemetering, and controlling.

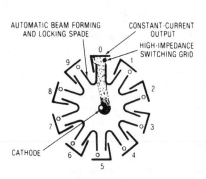

Beam-switching tube.

beam width — 1. The angular width of a radio, radar, or other beam measured between two reference lines. 2. The width of a radar beam measured between lines of half-power intensity. 3. Of a dish antenna, the angle of sky that can be illuminated (picked up or sent out) by the dish. Large dishes have narrow beam widths, which reduce noise from their sides. Small dishes have wider beam widths and are noisier but easier to aim.

bearing — 1. The horizontal direction of an object, or point, usually measured clockwise from a reference line or direction through 360°. 2. Support for a rotating shaft. 3. The horizontal angle at a given point, measured from a specific reference datum, to a second point relative to another as measured from a specific reference datum.

bearing cursor — A mechanical bearing line of a plan-position indicator type of display for reading the target bearing.

bearing loss — The loss of power through friction in the bearings of an electric motor (brushes removed and no current in the windings).

bearing resolution — The minimum angular separation in a horizontal plane between two targets at the same range that will allow an operator to obtain data on either individual target.

beat — Periodic variations that result from the superimposition of waves having different frequencies. The term is applied both to the linear addition of two waves, resulting in a periodic variation of amplitude, and to the nonlinear addition of two waves, resulting in new frequencies, of which the most important usually are the sum and difference of the original frequencies.

beat frequency — One of the two additional frequencies produced when two different frequencies are combined. One beat frequency is the sum of the two original frequencies; the other is the difference between them.

beat-frequency oscillator — Abbreviated BFO. An oscillator that produces a signal which mixes with another signal to provide frequencies equal to the sum and difference of the combined frequencies.

beating — The combining of two or more frequencies to produce sum and difference frequencies called beats.

beating-in — Interconnecting two transmitter oscillators and adjusting one until no beat frequency is heard in a connected receiver. The oscillators are then at the same frequency.

beating oscillator — See local oscillator.

beat marker — A marker pip resulting from the beat not between the sweep-generator signal and the signal from a marker oscillator; a marker is visible on an oscilloscope during visual alignment of a tuned circuit.

beat note — The difference frequency produced when two sinusoidal waves of different frequencies are applied to a nonlinear device.

beat reception — See heterodyne reception.

beats — 1. Beat notes that are generally at a sufficiently low audio frequency that they can be heard or counted. 2. The signal formed when two signals of different frequencies are simultaneously present in a nonlinear device. The frequency of the beat is equal to the difference in frequency of the two primary signals. For example, beats are produced in superheterodyne receivers, where the beat is between the incoming signal and the local oscillator in the receiver. 3. Periodic variations of amplitude that result when two periodic waves having different frequencies are superimposed.

beat tone — Musical tone due to beats, produced by the heterodyning of two high-frequency wave trains.

beaver tail — A fan-shaped radar beam, wide in the horizontal plane and narrow in the vertical plane. The beaver tail is swept up and down for height finding.

bed-of-nail tester — See in-circuit-tester.

bedspring — A broadside antenna array with a flat reflector.

before start — The interval before the starting circuit to a timer has been operated. The timer is fully reset and all contacts are in the precycle position.

bel — The fundamental unit in a logarithmic scale for expressing the ratio of two amounts of power. The number of bels is equal to the $\log_{10} P_1/P_2$, where P_1 is the power level being considered and P_2 is an arbitrary reference level. The decibel, equal to 1/10 bel, is a more commonly used unit.

B eliminator — A power pack that changes the ac power-line voltage to the dc source required by the vacuum tubes. In this way, batteries can be eliminated.

bell — 1. An electrical device consisting of a hammer vibrated by an electromagnet. The hammer strikes the sides of the bell and emits a ringing noise. The electromagnet attracts an armature or piece of soft iron forming part of the hammer lever. A contact breaker then opens the circuit and cuts off the attraction. A spring draws the hammer back to its original position, closing the circuit and repeating the action. 2. An electromechanical device in which an electrically vibrated clapper repeatedly strikes one or two gongs, which give out a musical tone.

Bellini-Tosi antenna — A direction-finding antenna comprising two vertical orthogonal triangular loops installed with their bases over ground and used with a goniometer.

Bellini-Tosi direction finder — An early radio direction-finder system consisting of two loop antennas at right angles to each other and connected to a goniometer.

bellows — 1. A pressure-sensing element consisting of a ridged metal cylinder closed at one end. A pressure difference between its outside and inside will cause the cylinder to expand or contract along its axis. 2. A mechanical pressure-sensing element consisting of a metallic bellows with a plate on one end. Pressure applied to the open end causes the plate to move. The amount of the movement becomes a measure of the applied pressure.

bellows contact — A contact in which a multileaf spring is folded. This type provides a more uniform spring rate over the full tolerance range of the mating unit.

bell-shaped curve — A statistical curve (so called from its characteristic shape) that exhibits a normal distribution of data. The curve typically describes the distribution of errors of measurement around the real value.

bell transformer — A small iron-core transformer; its primary coil is connected to an ac primary line, and its secondary coil delivers 10 to 20 volts for operation of a doorbell, buzzer, or chimes.

bell wire — Cotton-covered copper wire, usually No. 18, used for doorbell and thermostat connections in homes and for similar low-voltage work.

belt drive — A drive system used to rotate a turntable, in which the motor pulley drives the platter with a belt.

benchmark — 1. In connection with microprocessors, a frequently used routine or program selected for the purpose of comparing different makes of microprocessors. A flowchart in assembly language is written out for each microprocessor, and the execution of the benchmark by each unit is evaluated on paper. (It is not necessary to use hardware to measure capability by benchmark.) 2. A point of reference from which measurements can be made. 3. A test standard for measuring product performance. 4. A test point for facilitating measurement of a product run on several computers for the purposes of comparing speed, throughput, and ease of conversion. 5. The act of determining a benchmark. 6. Standard measure used to test performance.

benchmark problem — A problem used in the evaluation of the performance of computers relative to each other.

benchmark program — 1. A sample program used to evaluate and compare computers. In general, two computers will not use the same number of instructions, memory words, or cycles to solve the same problems. 2. A specific program written to measure the speed of a computer in a well-defined situation, such as serial transfer and 8-bit by 8-bit multiplication. 3. A set of standards used in testing a software or hardware product or system, from which a measurement can be made. Benchmarks are often run on a system to verify that it performs according to specifications.

bench test — A test in which service conditions are approximated, but the equipment is conventional laboratory equipment and not necessarily identical with that in which the product will be employed in normal service.

bend — A change in the direction of the longitudinal axis of a waveguide.

bend loss — 1. A form of attenuation that is caused by bending an optical fiber at a restrictive radius. The term also applies to losses due to minute distortions in the fiber itself caused by bending it. 2. A form of increased attenuation caused by allowing high-order modes to radiate from the side of an optical fiber. The two common types of bend losses are (*a*) those occurring when the fiber is curved around a restrictive radius of curvature and (*b*) microbend caused by small distortions of the fiber imposed by externally induced perturbations, such as poor cabling techniques.

bend radius — The minimum radius an optical fiber can bend before breaking.

bend waveguide — A section of waveguide in which the direction of the longitudinal axis is changed.

Benito — A cw navigational system in which the distance to an aircraft is determined on the ground by measuring the phase difference of an audio signal transmitted from the ground and retransmitted by the aircraft. Bearing information is obtained by ground direction finding of the aircraft signals.

bent gun — A TV picture-tube neck arrangement with an electron gun that is slanted to direct the undesired ion beam toward a positive electrode and still allow the electron beam to pass to the screen.

beryllia — 1. Beryllium-oxide (BeO) ceramics that have high thermal conductivity characteristics. Used as substrates in hybrid circuit manufacturing and for thick-film substrates in high-power applications. 2. Beryllium oxide is used in various forms as an insulator and structural element (as in resistor cores).

beryllium — An elemental metal whose atomic number is 4. Beryllium is present in various dielectrics and alloys used in electronics.

beryllium oxide (BeO) — A ceramic material having very high heat conductivity, good thermal shock resistance, and high strength. Used in metal/ceramic packages for higher-power microwave transistors and as substrates in some MIC power amplifiers.

Bessel function — A mathematical function used in the design of a filter for maximally constant time delay with little consideration for amplitude response. This function is very close to a Gaussian function.

best fit — An algorithm for computer memory allocation that searches the memory-free list for the unused memory block that is closest in size to that needed by the requesting task.

beta — Symbolized by the Greek letter *beta* (β). Also called current-transfer ratio. 1. The current gain of a transistor connected as a grounded-emitter amplifier; it is the ratio of a small change in collector current to the corresponding change in base current, with the collector voltage constant. 2. A parameter used to express the current gain of a bipolar transistor. There are many versions of beta, but all relate a change in collector current to the corresponding change in base current, with the collector-emitter voltage kept constant. 3. A symbol used to denote B quartz. 4. Brainwave signals whose frequency is approximately 13 to 28 Hz. The associated mental state is irritation, anger, jitteriness, frustration, worry, tension, etc. 5. A prerelease version of software, distributed to a selected group of users to test. By the end of a beta test, all major bugs should have been discovered and repaired.

beta circuit — In a feedback amplifier, the circuit that transmits a portion of the amplifier output back to the input.

beta cutoff frequency — The frequency at which the beta of a transistor is 3 decibels below the low-frequency value.

beta particle — A small electrically charged particle thrown off by many radioactive materials. It is identical with the electron and possesses the smallest negative electrical charge found in nature. Beta particles emerge from radioactive material at high speeds, sometimes close to the speed of light.

beta ray — 1. A stream of beta particles. 2. Electrons or positrons given off by a radioactive nucleus in the process of decay.

beta test — Testing computer software before it is released commercially.

betatron — A large doughnut-shaped accelerator that produces artificial beta radiation. Electrons (beta particles) are whirled through a changing magnetic field. They gain speed with each trip and emerge with high energies (on the order of 100 million electron volts in some instances).

bev — A billion electron volts. An electron possessing this much energy travels at a speed close to that of light — 186,000 miles a second (3×10^8 m/s).

bevatron — A very large circular accelerator in which protons are whirled between the poles of a huge magnet to produce energies in excess of one billion electron volts.

Beverage antenna — *See* wave antenna.

beyond-the-horizon propagation — *See* scatter propagation.

bezel — 1. A holder designed to receive and position the edges of a lens, meter, window, or dial glass. 2. The flange or cover used for holding an external graticule or CRT cover in front of the CRT in an oscilloscope. May also be used for mounting a trace-recording camera or other accessory item.

BFO — Abbreviation for beat-frequency oscillator.

B-H curve — Curve plotted on a graph to show successive states during magnetization of a ferromagnetic material. A normal magnetization curve is a portion of a symmetrical hysteresis loop. A virgin magnetization curve shows what happens the first time the material is magnetized.

B-H meter — A device for measuring the intrinsic hysteresis loop of a sample of magnetic material.

biamplification — The technique of splitting the audio-frequency spectrum into two sections and using individual power amplifiers to drive a separate woofer and tweeter. Crossover frequencies for the amplifiers usually vary between 500 and 1600 Hz. Biamplification has the advantages of allowing smaller-power amps to produce a given sound pressure level and reducing distortion effects produced by overdrive in one part of the frequency spectrum affecting the other part.

bias — 1. The electrical, mechanical, or magnetic force applied to a relay, semiconductor, vacuum tube, or other device for the purpose of establishing an electrical or mechanical reference level for the operation of the device. 2. Direct-current potential applied to the control grid of a vacuum tube. 3. Bias derived from a direct current, used on signaling or telegraph relays or electromagnets to secure the desired time spacing of transitions from marking to spacing. 4. A method of restraining a relay armature, by means of spring tension, to secure a desired time spacing of transitions from marking to spacing. 5. The average direct-current voltage between the control grid and cathode of a vacuum tube. 6. The effect on teletypewriter signals produced by the electrical characteristics of the line and the equipment. 7. Energy applied to a relay to hold it in a given position. 8. A high-frequency signal applied to the audio signal at the tape recording head to minimize distortion and noise and increase frequency response and efficiency. Although sometimes dc (fixed magnetic polarity) is used, the bias signal is usually above 40 kHz to avoid audible intermodulation distortion. Every tape formulation has slightly different bias requirements. 9. The sideways thrust of a pickup arm. 10. Communication-signal distortion related to bit timing. 11. The departure from a reference value of the average of a set of values.

bias cell — A dry cell used in the grid circuit of a vacuum type to provide the necessary C-bias voltage.

bias compensator — A device that counteracts the inward bias of a pickup arm as it tracks the record. The compensator exerts an outward force on the arm and generally can be adjusted to have a definite relationship to the playing weight of the pickup.

bias current — The current through the base-emitter junction of a transistor. It is adjusted to set the operating point of the transistor.

bias distortion — 1. Distortion resulting from operation on a nonlinear portion of the characteristic curve of a vacuum tube, semiconductor, or other device, due to improper biasing. 2. In teletype circuits, the uniform shifting of mark pulses from their proper position in relationship to the start pulses.

biased induction — Symbolized by B_b. The biased induction at a point in a magnetic material that is subjected simultaneously to a periodically varying magnetizing force and biasing magnetizing force is the algebraic mean of the maximum and minimum values of the magnetic induction at the point.

biased ringer — A polarized telephone bell whose armature is held at one end of its travel by a small biasing spring so it responds only to pulsating current of one polarity. Current pulses of the opposite polarity will merely attract the armature more strongly to the pole piece of the electromagnet but the bell will not ring.

bias-induced noise — The difference between bulk-erased and zero-modulation noise.

biasing magnetizing force — Symbolized by H_b. A biasing magnetizing force at a point in a magnetic material that is subjected simultaneously to a periodically varying magnetizing force and a constant magnetizing force is the algebraic mean of the maximum and minimum values of the combined magnetizing forces.

bias meter — A meter used in teletypewriter work for determining signal bias directly in percent. A positive reading indicates a marking signal bias; a negative reading indicates a spacing signal bias.

bias oscillator — An oscillator used in magnetic recorders to generate an ac signal in the range of 40 to 80 kHz for the purpose of magnetic biasing to obtain a linear recording characteristic. Usually the bias oscillator also serves as the erase oscillator.

bias port — In a fluidic device, the port at which a biasing signal is applied.

bias resistor — A resistance connected into a self-biasing vacuum-tube or semiconductor circuit to produce the voltage drop necessary to provide a desired biasing voltage.

bias-set frequency — In direct magnetic tape recording, a specified recording frequency employed during the adjustment of bias level for optimum record performance (not the frequency of the bias).

bias telegraph distortion — Distortion in which all mark pulses are lengthened (positive bias) or shortened (negative bias). It can be measured with a steady stream of unbiased reversals (square waves having equal-length mark and space pulses). The average lengthening or shortening does not give true bias distortion unless other types of distortion are negligible.

bias voltage — 1. A voltage, usually dc, used to set the operating point of a circuit above or below a reference voltage. 2. The base or grid voltage that establishes a semiconductor's or vacuum tube's desired dc operating voltage. 3. A steady voltage that presets the operating threshold or operating point of a circuit or device, such as a transistor or vacuum tube.

bias windings — Control windings of a saturable reactor, by means of which the operating condition is translated by an arbitrary amount.

biax — Two-hole, orthogonal, cubical ferrite computer memory elements.

BiCMOS — Abbreviation for bipolar complementary metal oxide semiconductor. An IC technology combining the linearity and speed advantages of bipolar and the low-power advantages of CMOS on a single IC. BiCMOS can operate at either ECL (emitter-coupled-logic) or TTL (transistor-transistor-logic) levels and is ideal for mixed-signal devices. (BiCMOS may eclipse CMOS, just as CMOS edged out MOS and bipolar circuits.)

biconical antenna — An antenna that is formed by two conical conductors, having a common axis and vertex, and excited at the vertex. When the vertex angle of one of the cones is 180°, the antenna is called a discone.

Biconical antenna.

bidirectional — 1. Responsive in opposite directions. An ordinary loop antenna is bidirectional because it has maximum response from the opposite directions in the plane of the loop. 2. Refers to a type of computer bus structure in which a single conductor is used to transmit data or signals in either direction between a peripheral device and a central processor or memory. 3. In open-reel or cassette recorders, the ability to play (and, in some cases, record) both stereo track pairs on a tape by reversing the tape's direction of motion without removing and replacing the tape reels or cassette. 4. In microphones, a figure-8 pickup pattern.

bidirectional antenna — An antenna having two directions of maximum response.

bidirectional bus — 1. In computers, a data path over which both input and output signals are routed. 2. In a computer, a bus that carries signals in either direction. The bus also carries special signals that tell the devices connected to it which way data is passing. 3. A bus used by any individual device for two-way transmission of messages, that is, both input and output.

bidirectional bus driver — A signal-driving device in a microcomputer that permits direct connection of a buffer-to-buffer arrangement on one end (the interface to I/O memories, etc.) and data inputs and outputs on the other. This permits bidirectional signals to pass and provides drive capability in both directions.

bidirectional current — A current that is both positive and negative.

bidirectional data bus — A data bus in which digital information can be transferred in either direction.

bidirectional diode thyristor — A two-terminal thyristor having substantially the same switching behavior in the first and third quadrants of the principal voltage-current characteristic.

bidirectional lines — Links between devices in a system that may carry information in either direction, but not both simultaneously.

bidirectional loudspeaker — A speaker that delivers sound waves to the front and rear.

bidirectional microphone — 1. A microphone in which the response predominates for sound incidences of 0° and 180°. 2. A microphone that is equally sensitive to sounds arriving from in front or in back, but discriminates against sounds arriving from the sides.

bidirectional printer — A printer that prints from left to right as well as from right to left, avoiding delays caused by carriage returns.

bidirectional pulses — Pulses, some of which rise in one direction and the remainder in the other direction.

bidirectional pulse train — A pulse train in which some pulses rise in one direction and the remainder in the other direction.

bidirectional thyristor — A thyristor that can be made conductive at any instant when the voltage between the main terminals is either positive or negative.

bidirectional transducer — See bilateral transducer.

bidirectional transistor — A transistor that is specified with parameter limits in both the normal and inverted configuration and has substantially the same electrical characteristics when the terminals normally designed as emitter and collector are interchanged. (Bidirectional transistors are sometimes called symmetrical transistors. The term, however, is deprecated because it might give the incorrect impression of an ideally symmetrical transistor.)

bidirectional triode thyristor — A three-terminal thyristor having substantially the same switching behavior in the first and third quadrants of the principal voltage-current characteristic.

bifet — Linear circuit that combines bipolar transistors with junction field-effect transistors on the same silicon chip and provides broader bandwidth, faster slewing, and higher impedance than standard bipolar devices when incorporated into monolithic operational amplifiers.

bifilar — A winding made noninductive by winding two wires carrying current in opposite directions together, side by side, as one wire.

bifilar resistor — A resistor wound with a wire doubled back on itself to reduce the inductance.

bifilar suspension — A type of galvanometer movement that is highly resistant to overloads, in which a D'Arsonval moving coil is supported at each end by two taut wires. The elimination of the pivot, with its attendant friction, results in superior sensitivity and precision.

bifilar transformer — A transformer in which the turns of the primary and secondary windings are wound

together side by side and in the same direction. This type of winding results in near-unity coupling, so that there is a very efficient transfer of energy from primary to secondary.

bifilar winding — 1. A method of winding noninductive resistors in which the wire is folded back on itself and then wound double, with the winding starting from the point at which the wire is folded. 2. A winding consisting of two insulated conductors side by side to produce (a) two balanced windings, (b) a resistor with minimum inductance, and (c) maximum coupling between two windings.

bifurcate — Describes lengthwise slotting of a flat spring contact, as used in a printed-circuit connector, to provide additional independently operating points of contact. Example: bifurcated contact.

bifurcated — Usually fork-shaped. Refers to physical construction of a contact whereby two mating portions make physical contact. Yet, if one tip section of the contact fails, the remaining section maintains the physical and electrical connection.

bifurcated connector — A hermaphroditic connector containing fork-shaped mating contacts.

bifurcated contact — 1. A movable contact that is forked (divided) to provide two contact-mating surfaces in parallel for a more reliable contact. 2. A connector contact (usually a flat spring) that is slotted lengthwise to provide additional, independently operating points of contact.

bilateral — Having a voltage-current characteristic curve that is symmetrical with respect to the origin, that is, being such that if a positive voltage produces a positive current magnitude, an equal negative voltage produces a negative current of the same magnitude.

bilateral amplifier — An amplifier capable of receiving as well as transmitting signals; it is used primarily in transceivers.

bilateral antenna — An antenna, such as a loop, having maximum response in exactly opposite directions (180° apart).

bilateral bearing — A bearing that indicates two possible directions of wave arrival. One of these is the true bearing, and the other is a bearing displaced 180° from the true bearing.

bilateral circuit — A circuit wherein equipment at opposite ends is managed, operated, and maintained by different services.

bilateral element — A two-terminal element, the voltage-current characteristic of which has odd symmetry around the origin.

bilateral network — 1. A network in which a given current flow in either direction results in the same voltage drop. 2. A network that passes current and signals equally well in both directions.

bilateral transducer — 1. Also called bidirectional transducer. A transducer capable of transmission simultaneously in both directions between at least two terminations. 2. A device capable of measuring stimuli in both a positive and a negative direction from a reference zero or rest position.

billboard antenna — An antenna array consisting of several bays of staked dipoles spaced 1/4 to 3/4 wavelength apart, with a large reflector placed behind the entire assembly. The required spacing of the dipoles tends to make the array inconveniently large at frequencies below the VHF range.

bimag — See tape-wound core.

bimetal — A union of two dissimilar metals (especially those having a different temperature coefficient of expansion), usually welded together over their entire surface.

bimetal cold-junction compensation — Automatic mechanical correction for ambient temperature change at the cold junction of a thermocouple, which would normally cause erroneous readings.

bimetallic strip — A strip formed of two dissimilar metals welded together. Because the metals have different temperature coefficients of expansion, the strip bends or curls when the temperature changes.

bimetallic switch — A temperature-sensitive switch that uses a bimetallic element.

bimetallic thermometer — 1. A device containing a bimetallic strip that expands or contracts as the temperature changes. A calibrated scale indicates the amount of change in temperature. 2. A strip of two metals having different coefficients of expansion bonded together in the form of a spiral or helix. Movement caused by a temperature change becomes a measure of temperature.

bimetallic wire — Any wire formed of two different metals joined together (not alloyed). It can include wire with a steel core for high-strength clad wire, or plated or coated wire.

bimetal mask — A mask formed by chemically etching openings in a metal film or plate where it is not protected by photoresist or other chemically resistant material.

bimorph cell — Two crystal elements (usually Rochelle salt) in rigid combination, arranged to act as a mechanical transforcer in headphones, microphones, pickups, and speakers.

BiMOS — Abbreviation for bipolar metal oxide semiconductor. A general term to refer to bipolar and metal oxide semiconductors on one chip. Sometimes used interchangeably with *BiCMOS*.

binary — 1. A numbering system using a base number, or radix, of 2. There are two digits (1 and 0) in the binary system instead of 10 as in the decimal. 2. Pertaining to a characteristic or property involving a selection, choice, or condition in which there are two possibilities or alternatives. 3. A bistable multivibrator. 4. Two-valued logic using only the values true and false. Represented in a computer circuit by the presence of current (equivalent to 1) or its absence (equivalent to 0). All computer programs are executed in binary form. 5. A counting system in which the value of any digit can only be 1 or 0. As with the decimal system, the right-hand digit denotes the number of units of the next value, etc. In decimal the units can be 0 to 9; in binary, the units are 0 and 1.

binary arithmetic — Mathematical operations performed with only the digits 0 and 1.

binary card — A card that contains data in column binary or row binary form.

binary cell — In an electronic computer, an elementary unit of storage that can be placed in either of two stable states.

binary chain — A series of binary circuits, each of which can exist in either one of two states, arranged so each circuit can affect or modify the condition of the next circuit.

binary channel — A transmission facility limited to the use of two symbols.

binary code — A method of representing numbers in a scale of 2 (on or off, high level or low level, one or zero, presence or absence of a signal) rather than the more familiar scale of 10 used in normal arithmetic. Electronic circuits designed to work in two defined states are much simpler and more reliable than those working in ten such states.

binary-coded — Expressed by a series of binary codes (0s and 1s).

binary-coded character — A decimal digit, alphabetic letter, punctuation mark, etc., represented by a fixed number of consecutive binary digits.

binary-coded decimal — Abbreviated BCD. A coding system in which each decimal digit from 0 to 9 is represented by four binary digits:

Decimal Digit	Binary Code
0	0000
1	0001
2	0010
3	0011
4	0100
5	0101
6	0110
7	0111
8	1000
9	1001

2. The representation of decimal numbers in binary form. It is useful in adc systems intended to drive decimal displays. Its advantage over decimal is that only four lines are needed to represent 10 digits. The disadvantage of coding digital-to-analog converters or analog-to-digital converters in BCD is that a full 4 bits could represent 16 digits, whereas only 10 digits are represented in BCD. 3. A numbering system used in many computers, in which the basic binary system is used to represent decimal numbers.

binary-coded digit — One element of a notation system for representing a decimal digit by a fixed number of binary positions.

binary-coded octal system — An octal numbering system in which each octal digit is represented by a three-place binary number.

binary counter — *See* binary scaler.

binary digit — 1. A character that represents one of the two digits in the number system that has a radix of 2. Also called bit. 2. Either of the digits, 0 or 1, that may be used to represent the binary conditions on or off. 3. A whole number in the binary scale of notation; this digit may be only 0 (zero) or 1 (one). It may be equivalent to an on or off condition, yes or no, etc.

binary file — A file stored in a binary format containing data or program instructions in a computer-readable format. Special software is required to display such a file.

binary incremental representation — Incremental representation in which the value of an increment is rounded to plus or minus one quantum and is represented by one binary digit.

binary magnetic core — A ring-shaped magnetic material that can be made to take either of two stable states of magnetic polarization.

binary notation — *See* binary number system.

binary number system — A number system employed in computers and digital systems, in which successive digits are coefficients of powers of the base 2, rather than the base 10. For example, the decimal number 13 is represented by the binary number $1101(1 \times 2 + 1 \times 2 + 0 \times 2 + 1 \times 2)$. Since the only values in the binary system are 0 and 1, quantities, or *bits* (binary digits), are represented electronically with either of two conditions, typically a high voltage representing a 1 and a low voltage representing a 0. Also called binary notation.

binary numeral — The binary representation of a number; for example, 101 is the binary numeral and V is the Roman numeral of the number of fingers on one hand.

binary phase-shift keying—Abbreviated BPSK. A modulation scheme that uses two phases to represent data; one phase represents a mark, and the other phase represents a space.

binary point—1. The point that marks the place between integral powers of 2 and fractional powers of 2 in a binary number. 2. The radix point in a mixed binary numeral, such as 110.011, separating the fractional part from the integer part. In the binary numeral 110.011, the binary point is between the two 0s.

binary pulse-code modulation—A form of pulse-code modulation in which the code for each element of information consists of one of two distinct kinds, e.g., pulses and spaces.

binary raster data—Computer data in an on/off format (ones and zeros) that is formed into scan lines that can be used to control a light source, such as a laser, for exposure of light-sensitive material.

binary scaler—Also called binary counter. 1. A counter that produces one output pulse for every two input pulses. 2. A counting circuit, each stage of which has two distinguishable states. 3. A flip-flop having a single input (called a T flip-flop). Each time a pulse appears at the input, the flip-flop changes state.

binary search—Also called dichotomizing search. 1. A search in which a set of items is divided into two parts; one part is rejected, and the process is repeated on the accepted part until those items with the desired property are found. 2. A search that starts in the middle of a database, first determining if the desired record is above or below the midpoint, then proceeding to the middle of the remaining records, and so on.

binary signal—A voltage or current that carries information in the form of changes between two possible values.

binary signaling—A communications mode in which information is passed by the presence and absence, or plus and minus variations, of one parameter of the signaling medium.

binary system—A system of mathematical computation based on powers of 2.

binary-to-decimal conversion—The process of converting a number written in binary notation to the equivalent number written in the ordinary decimal notation.

binary-to-hexadecimal conversion—The process of converting a numeral written in base 2 to the equivalent numeral written in base 16.

binary-to-octal conversion—The process of converting a numeral written in base 2 to the equivalent numeral written in base 8.

binary word—A related grouping of ones and zeros that has a meaning assigned by definition, or that has a weighted numerical value in the natural binary number system.

binaural—Two-channel sound in which each channel recorded is heard only through one ear. In recording, microphones are spaced to approximate the distance between a person's own ears. To hear the recording binaurally, the listener must use headphones. Compare with *stereo*.

binaural disc—A stereo record with two separate signals recorded in its grooves. Stereophonic sound is obtained by feeding each signal into its own speaker or headphone.

binaural effect—The effect that makes it possible for a person to distinguish the difference in arrival time or intensity of sound at his or her ears and thereby determine the direction from which a sound is arriving.

binaural recorder—A tape recorder that employs two separate recording channels, or systems, each with its own microphone, amplifier, recording and playback heads, and earphones. Recordings using both channels are made simultaneously on a single magnetic tape having two parallel tracks. During playback, the original sound is reproduced with depth and realism. For a true binaural effect, headphones are necessary.

binaural sound—Sound recorded or transmitted by pairs of equipment so as to give the listener the effect of having heard the original sound.

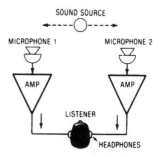

Binaural sound-reproducing system.

binder—1. A substance, such as cement, used to hold particles together and thus provide mechanical strength in, for example, carbon resistors and phonograph records. 2. Materials or substances added to thick-film compositions and unfired substrates to give sufficient strength for prefire handling.

binding—The act of assigning absolute addresses to a program.

binding energy—The minimum energy required to dissociate a nucleus into its component neutrons and protons. Neutron or proton binding energies are those energies required to remove a neutron or a proton, respectively, from a nucleus. Electron binding energy is that energy required to remove an electron from an atom or a molecule.

binding post—1. A bolt-and-nut terminal for making temporary electrical connections. 2. A device for clamping or holding electrical conductors in a rigid position.

Binding post.

binistor—A four-terminal controlled rectifier semiconductor that provides bistable negative-resistance characteristics.

binomial array—A directional-antenna array used for reducing minor lobes and providing maximum response in opposite directions.

biochemical fuel cell—An electrochemical generator of electrical power in which bi-organic matter is used as the fuel source. In the usual electrochemical reaction, air serves as the oxidant at the cathode and microorganisms are used to catalyze the oxidation of the bi-organic matter at the anode.

bioelectricity — Electric currents and potential differences that occur in living tissues. Muscle and nerve tissue, for example, are generators of bioelectricity, although the potential registered may be less than 1 millivolt in some cases.

bioelectric potential — *See* bioelectricity.

bioelectrogenesis — The practical application of electricity drawn directly from the bodies of animals, including humans, to power electronic devices and appliances.

bioelectronics — 1. The application of electronic theories and techniques to the problems of biology. 2. The integrated, long-term electronic control of various impaired physiologic systems by means of small, low-power electrical and electromechanical devices. (The pacemaker is therefore a bioelectronic instrument.)

bioengineering — *See* bionics.

biogalvanic battery — A device that makes use of reactions between metals and the oxygen and fluids in the body to generate electricity.

biologic energy — Energy that is produced by bodily processes and that can be used to supply electrical energy for implanted devices such as electronic cardiac pacemakers, bladder stimulators, etc. The biologic energy can result from muscle movement (such as that of the diaphragm), temperature differences, pressure differences, expansion of the aorta, oxidation of materials within the gastrointestinal tract, or other processes.

biomedical oscilloscope — An oscilloscope designed or modified to be used in medical applications. Such oscilloscopes have slow sweep rates and long-persistence screens because of the low frequencies of many biological signals.

bionics — Also called bioengineering. 1. The study of living systems so that their characteristics and functions can be related to the development of mechanical and electronic hardware. 2. The reduction of various life processes to mathematical terms to make possible duplication or simulation with systems hardware. 3. The art that treats electronic simulation of biological phenomena. 4. The emulation of biological components, "body parts," with electromechanical ones. 5. The application of observed operational processes of sophisticated living organisms to mechanical and electrical systems in order to analogize capabilities or efficiency.

BIOS — Abbreviation for basic input-output system. The software that translates operating system instructions into commands to and from the hardware components in a computer system.

biotelemetry — The process of remote measurement or recording of such biological variables as pulse rate, temperature, etc. Typically, the information is transmitted between the patient and the receiving equipment by a radio link.

biotelescanner — A device that can analyze and radio data on life forms during space exploration.

biphase suppressed carrier — A digital radio modulation scheme that uses one phase of a carrier to signify a 0 and the opposite phase to denote a 1. If the two phases are less than 180° apart, some residual carrier will be transmitted.

bipolar — 1. Having two poles. 2. Having to do with a device in which both majority and minority carriers are present. In connection with integrated circuits, the term describes a specific type of construction; bipolar and MOS are the two most common types of IC construction. 3. The semiconductor technology employing two-junction transistors. 4. A transistor structure whose electrical properties are determined within the silicon material. Memories using this technology are characteristically high-speed devices. 5. General name for npn

and pnp transistors, since working current passes through semiconductor material of both polarities (p and n). Current in the collector is controlled by a current between base and collector. Also applied to integrated circuits that use bipolar transistors. 6. In bipolar transistors, the working current consists of both positive and negative electrical charges. The first transistor and the first ICs were bipolar types. (Most present-day discrete transistors are also bipolar.) Bipolar IC transistors operate faster than unipolar (MOS) transistors and consume more power. They take up more space on a chip and cost more to manufacture. 7. In meters, the capability of measuring voltages or currents of either polarity — positive or negative — with respect to a reference point or ground. 8. One of several fundamental processes for fabricating ICs. A bipolar IC is made up of layers of silicon with differing electrical characteristics. Current flows between the layers when a voltage is applied to the junction or boundary between the layers. 9. A technology of IC fabrication that uses transistor switching elements based on majority carriers for switching and amplification. 10. Technology using transistors with both negative and positive charge carriers allowing current flow in only one direction (unlike MOS circuits). Bipolar memory is characterized by very fast access times and very high power consumption.

bipolar complementary metal oxide semiconductor — *See* BiCMOS.

bipolar device — 1. A semiconductor device in which there are both majority and minority carriers. (This is the case in all npn and pnp transistors.) 2. A current-driven electronic device with two poles. Operation relies on the flow of both electrons and holes.

bipolar electrode — An electrode without metallic connection with the current supply, one face of which acts as an anode surface and the opposite face as a cathode surface when an electric current is passed through a cell.

bipolar electrolytic capacitors — An electrolytic capacitor designed to withstand an alternating voltage and/or a reversal of the applied direct voltage. Also called nonpolar capacitor.

bipolar ferreed — A device comprising two sealed reed switches, a permanent magnet, and a winding with a semipermanent magnet. When the winding is pulsed with current in one direction, the reed contacts close; when the winding is pulsed in the opposite direction, the reed contacts open.

bipolar magnetic driving unit — A headphone or speaker unit having two magnetic poles acting directly on a flexible iron diaphragm.

bipolar memory cell — In a computer, a system comprising a storage latch, a pair of control gates, and an output gate. The control and output gates need not be part of the storage cell, but they usually are included because each latch requires both control and output gating.

bipolar pulse — 1. A pulse that has appreciable amplitude in both directions from the reference axis. 2. A current or voltage pulse that may be either positive or negative.

bipolar transistor — 1. A transistor that uses both negative and positive charge carriers. 2. An active semiconductor device formed by two pn junctions whose function is amplification of an electric current. Bipolar transistors are of two types: npn and pnp, depending on the manner in which the two pn junctions are combined. Bipolar transistors have three sections: emitter, base, and collector. Operation of a bipolar transistor depends on the migration of both electrons and holes, in contrast to field-effect transistors, in which only one polarity carrier predominates.

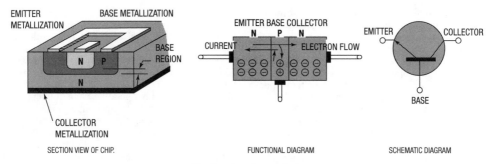

EMITTER METALLIZATION · BASE METALLIZATION · BASE REGION · COLLECTOR METALLIZATION · SECTION VIEW OF CHIP.

EMITTER BASE COLLECTOR · N P N · CURRENT · ELECTRON FLOW · FUNCTIONAL DIAGRAM

EMITTER · COLLECTOR · BASE · SCHEMATIC DIAGRAM

Bipolar (npn) transistor.

biquadratic filter — A filter transfer function that contains complete quadratic equations in both the numerator and denominator and provides the basis for implementing high-pass, low-pass, and single-frequency notch characteristics as well as band-reject realizations.

biquinary code — A mixed-radix notation in which each decimal digit to be represented is considered to be the sum of two digits, the first of which is 0 or 1 with significance 5 and the second of which is 0, 1, 2, 3, or 4 with significance 1.

biradial — Having an elliptical cross section. A term used with reference to phonograph styli. *See* elliptical stylus.

biradial stylus — Also called elliptical stylus. A stylus tip that has a small radius where it touches the walls of the record grooves, as distinguished from a conventional stylus, which has a hemispherical top used with lightweight pickup arms to reduce tracking distortion.

bird — Jargon or nickname for communication satellites.

birdcage — A defect in stranded wire in which the strand in the stripped portion between the covering of an insulated wire and a soldered connection (or an end-time lead) has separated from the normal lay of the strands.

bird-dogging — *See* hunting, 1.

birdnesting — Clumping together of chaff dipoles after they have been dropped from an aircraft.

Birmingham wire gage — Abbreviated BWG. The Birmingham wire gage was used extensively in Great Britain and the United States for many years, but is now obsolete. Its uses have persisted, however, for certain purposes, including galvanized steel wire for cable armor.

biscuit — *See* preform.

B-ISDN — Abbreviation for Broadband Integrated Services Digital Network. An ISDN service requiring a broadband channel operating at speeds greater than a single primary rate interface — 34 to as much as 6000 Mbps.

bislope triggering — Allows positive- and negative-going signal polarities to initiate waveform storage in both analog and digital oscilloscopes. A triggering threshold or level control on the front panel determines the trigger characteristics.

bistable — 1. A circuit element with two stable operating states, e.g., a flip-flop in which one transistor is saturated while the other is turned off. It changes state for each input pulse or trigger. 2. A device capable of assuming either one of two stable states. 3. Of or pertaining to the general class of devices that operate in either of two possible states in the presence or absence of the setting input. 4. Element that has two output possibilities and that will hold a given condition until switched.

bistable contacts — A contact combination in which the movable contact remains in its last operated position until the magnetic polarity of the coil is reversed.

bistable device — 1. Any device, such as a flip-flop, that has two stable states and may be readily switched from one state to the other. 2. A device with only two stable states, such as on and off.

bistable display — A matrix-controlled display that has information storage at the display surface and that requires that an element be addressed only once to ensure that it is on or off.

bistable latch — A rudimentary flip-flop that can be enabled to store a logical 1 or a logical 0. One bistable latch device is commonly used in memory and register circuits for the storage of each bit.

bistable multivibrator (flip-flop) — A circuit having two stable states; it will stay in either one indefinitely until appropriately triggered, after which it immediately switches to the other state.

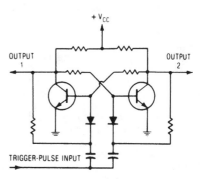

+ V_cc · OUTPUT 1 · OUTPUT 2 · TRIGGER-PULSE INPUT

Bistable multivibrator.

bistable relay — A relay that requires two pulses to complete one cycle composed of two conditions of operation. Also called locked, interlocked, and latching relay.

bistatic radar — Radar system in which the receiver and transmitter have separate antennas and are some distance apart.

biswitch — A two-terminal integrated device that basically performs the function of two pnpn switches interconnected so as to provide bilateral switching.

bisynchronous — 1. Data transmission in which synchronization of characters is controlled by timing signals generated at both the sending and receiving stations. 2. Method of transmitting computerized data that allows for multiple error detection.

bit — Abbreviation for binary digit. 1. A unit of information equal to one binary decision, or the designation of one of two possible and equally like values or states (such as 1 or 0) of anything used to store or convey information. It may also mean "yes" or "no." 2. The smallest part of information in a binary notation system. A bit is either a one (1) or a zero (0). In a BCD system, four bits represent one decimal digit. 3. The smallest unit of information, representing either a mark or a space (1 or 0). In data transmission, the common unit of speed is bits per second (bps). 4. In a computer, one bit is enough to tell the difference between yes or no, up or down, on or off, 1 or 0, in short, any two opposites. Computers represent information in the form of bits because their circuits can have only two states: on or off. 5. The bit can take the form of a magnetized spot, an electronic impulse, a positively charged magnetic core, etc. A number of bits together are used to represent a character in a computer (byte, word).

bit copier — A program that copies individual data bits on a disk without regard to their information content; it is not affected by improperly formatted data.

bit density — The number of bits of information contained in a given area, such as the number of bits written along an inch of magnetic tape.

bit diddling — A method of increasing storage efficiency by packing extra information into unused parts of a computer word.

bit interleave — A technique in time-division multiplexing in which bits of data are transmitted in one frame.

bit map — 1. A graphics image consisting of rows and columns of dots. 2. A screen display in which each pixel location corresponds to a unique main memory location accessible by the CPU. Also refers to images intended for display on this type of digital system.

bit-mapped display — A method of CRT display that uses a separate area of computer memory to specify the locations of individual pixels, resulting in high-quality images.

bit-mapped font — A font with characters formed by a pattern of dots.

bit parallel — 1. A method of simultaneously moving or transferring all bits in a contiguous set of bits over separate wires, one wire for each bit in the set. 2. Refers to a set of concurrent data bits present on a like number of signal lines used to carry information. Bit-parallel data bits may be acted on concurrently as a group (byte) or independently as individual data bits.

bit plane — The circuitry required to make a bit-map display, especially when combinable or stackable to allow more than one bit per pixel. Thus, two bit planes give each pixel four possible colors to select from; five planes yield 32.

bit rate — 1. The number of binary bits transmitted per unit time; for example, a bit rate of 80 means that 80 binary bits are transmitted per second. 2. The rate at which binary digits, or pulses representing them, pass a given point in a communication line. 3. The rate at which data bits (digital information) are transmitted over a communication path, normally expressed in bits per second (bps). Not to be confused with the data signaling rate (baud), which measures the rate of signal changes transmitted. 4. Informal term for data rate, when it is measured in bits per second. 5. The rate of transfer of information necessary to ensure acceptable reproduction of the information at the receiver.

bit-rate generators — Devices that provide the reference frequencies required by serial interfaces (TTY, UART, cassette, modem) and also furnish adjustment-free crystal stability with easily changed, multiple frequencies.

bit-rate-length product — The product of the bit rate that a fiber or cable is able to handle and the length for tolerable dispersion at the bit rate, with the product usually stated in units of megabits per kilometer/second. Typical bit-rate-length products for graded index fibers with a numerical aperture of 0.2 is 1000 MB per km/s for research fibers and 200 MB per km/s for production fibers. The product is a good measure of fiber performance in terms of transmission capability.

bit serial — A method of sequentially moving or transferring a contiguous set of bits one at a time over a single wire, according to a fixed sequence.

bit slice — 1. As in a 4-bit slice and a 2-bit slice. A multichip microprocessor in which the control section is contained on one chip, and one or more identical arithmetic and logic unit (ALU) sections and register sections are contained on separate chips called slices. For example, three 4-bit slices connected in parallel with the control section produce a 12-bit word microprocessor. 2. An enhanced subsection of a microprocessor's ALU, embodying an architecture that permits the cascading or stacking of devices to increase word bit size. (Support chips are required to construct a functional MPU.) 3. The arithmetic and logic unit (ALU) within the central processor unit (CPU) of a microcomputer. Circuits capable of handling a few bits — normally four but sometimes less — are stacked in parallel to provide an ALU and registers capable of handling some specific word length. For example, to get the 16-bit word length normally used for numerical control, 4-bit slices can be combined, with each slice on a separate chip. The slice excludes the control section, which must be implemented by external devices. 4. A microprocessor in which the CPU is partitioned into two or more silicon chips, each containing 2- or 4-bit register logic. 5. A microcomputer chip that is a quarter or an eighth of an entire processor. When a CPU is too complex or would dissipate too much heat to be put on a single chip it is sliced into 2- or 4-bit chunks that are then wired together on circuit cards.

bit-slice processor — A microprocessor whose word (or byte) capacity is achieved though the use of interrelated smaller-capacity processors, e.g., a 16-bit unit derived from eight 2-bit slices.

bits per second — Abbreviated bps. 1. The unit of information rate. It expresses the number of binary digits passed though a channel per second. 2. The speed at which modems send and receive data. For example, a 14.4 modem moves 14,400 bits per second; a 28.8 modem, also known as a V.34, is twice as fast. An earlier term, *baud*, is considered archaic.

bit stream — 1. A binary signal without regard to grouping according to character. 2. A continuous series of bits transmitted on a communications link.

bit string — A string of binary digits in which each bit position is considered an independent unit.

bit time — 1. In a serial binary computer, the time during which each bit appears. 2. The amount of time that one bit of information in a digital pattern remains in its 1 or 0 state.

black — A signal produced at any point in a facsimile system by the scanning of a selected area of subject copy having maximum density.

black and white — *See* monochrome.

black-and-white transmission — *See* monochrome transmission.

black area — An area with only encrypted signal present.

blackbody — 1. An idealized emitter for which total radiated energy and the spectral distribution of the energy are accurately known functions of temperature. 2. A solid that radiates or absorbs energy with no internal reflection of the energy at any wavelength. Physically it may be a hollow sphere coated on the side with lampblack and with an opening through which energy may enter or leave. 3. An ideal body that absorbs all incident light and therefore appears perfectly black at all wavelengths. The radiation emitted from such a body when it is hot is called blackbody radiation. The spectral energy density of blackbody radiation is the theoretical maximum for a body in thermal equilibrium. 4. An ideal body that would absorb all radiation incident on it. When heated by external means, the spectral energy distribution of radiated energy would follow curves shown on optical spectrum charts. The ideal blackbody is a perfectly absorbing body. It reflects none of the energy that may be incident on it. It radiates (perfectly) at a rate expressed by the Stefan-Boltzmann law, and the spectral distribution of radiation is expressed by Planck's radiation formula. When in thermal equilibrium, an ideal blackbody absorbs perfectly and radiates perfectly at the same rate. The radiation will be just equal to absorption if thermal equilibrium is to be maintained.

blackbody luminous efficiency — The efficiency of an incandescent blackbody as a source of visible light. It is a function of temperature.

blackbody radiation — *See* blackbody.

black box — 1. A term used loosely to refer to any subcomponent that is equipped with connects and disconnects so that it can be readily inserted into or removed from a specified place in a larger system (e.g., the complete missile or some major subdivision) without benefit of knowledge of its detailed internal structure. 2. A term pertaining to either the functional transformation that acts upon a specified input to give a particular output or to the apparatus for accomplishing this transformation (without regard to the detailed circuitry used). 3. A useful mathematical approach to an electronic circuit which concerns itself only with the input and output and ignores the interior elements, discrete or integrated. 4. An equipment specified only in terms of its performance.

black compression — Also called black saturation. The reduction in the gain of a television picture signal at those levels corresponding to dark areas in the picture with respect to the gain at that level corresponding to the midrange light value in the picture. The overall effect of black compression is to reduce contrast in the low lights of the picture.

blacker-than-black — 1. The amplitude region of the composite video signal below the reference black level in the direction of the synchronizing pulses. 2. That portion of the standard television signal devoted to the synchronizing signal.

blacker-than-black level — A voltage value used in an electronic television system for control impulses. It is greater than the value representing the black portions of the image.

black level — That level of the picture signal corresponding to the maximum limit of black peaks.

black light — 1. Invisible light radiation. May be either ultraviolet or infrared radiation, both of which are invisible. 2. A lamp that produces a principal portion of its radiation in the ultraviolet region.

black-light emitter — A source of electromagnetic radiation in the ultraviolet or infrared region, just outside the visible spectrum.

black matrix — Picture tube in which the color phosphors are surrounded by black for increased contrast.

black negative — The television picture signal in which the polarity of the voltage corresponding to black is negative with respect to that which corresponds to the white area of the picture signal.

black noise — In a spectrum of electromagnetic wave frequencies, a frequency spectrum of predominantly zero power level at all frequencies except for a few narrow bands or spikes, such as might be obtained when scannning a black area in facsimile transmission systems on which there are a few white spots or speckles on the surface.

blackout — 1. Interruption of radiocommunication due to excess absorption caused by solar flares. During severe blackouts, all frequencies above approximately 1500 kHz are absorbed excessively in the daylight zone. 2. Passive defense that consists of interrupting all forms of communication or identification. 3. A sudden, unexpected loss of all electrical power, typically lasting for many minutes or even hours.

black peak — A peak excursion of the picture signal in the black direction.

black-peak clipping — Limiting the amplitude of the picture signal to a preselected maximum black level, usually at blanking level.

black reference — The blanking level of pulses in a TV signal beyond which the sync pulse is in the blacker-than-black region.

black saturation — *See* black compression.

black scope — Cathode-ray tube being operated at the threshold of luminescence with no video signals applied.

black signal — Also called picture black. A signal produced at any point in a facsimile system by the scanning of a maximum density area of the subject copy.

black transmission — 1. In an amplitude-modulation facsimile system, a form of transmission in which the maximum transmitted power corresponds to the maximum density of the copy. 2. In a frequency-modulation system, a form of transmission in which the lowest transmitted frequency corresponds to the maximum density of the copy.

blade contact — A flat male contact designed to mate with a tuning fork or a flat-formed female contact. It is used in multiple-contact connectors.

blank — 1. The result of the final operation on a crystal. 2. To cut off the electron beam of a cathode-ray tube. 3. A code character to denote the presence of no information rather than the absence of information. In the Baudot code, it is composed of all spacing pulses. In paper tape, it is represented by a feed hole without intelligence holes. 4. A no-information condition in a data-recording medium or storage location. This vacancy can be represented by all spaces or all zeros, depending on the medium.

blank coil — Tape for perforation in which only the feed holes have been punched.

blank deleter — A device used to eliminate the receiving of blanks in perforated paper tape.

blanked picture signal — The signal resulting from adding blanking to a picture signal. Adding the sync signal to the blanked picture signal forms the composite picture signal.

blanketing — The overriding of a signal by a more powerful one or by interference, so that a receiver is unable to receive the desired signal.

blank groove — *See* unmodulated groove.

blanking — 1. The process of making a channel or device noneffective for the desired interval. In television, blanking is the substitution, during prescribed intervals, of the picture signal by a signal whose instantaneous amplitude is such as to make the return trace invisible. *See also* gating. 2. The process whereby the beam in an image pickup tube or CRT is cut off during the retrace period.

blanking level — Also called pedestal level. 1. In a composite picture signal, the level that separates the range of the composite picture signal containing picture information from the range containing synchronizing information. 2. Usually referred to as the front porch or back porch. At 0 IRE units, it is the level that will shut off the picture tube, resulting in the blackest possible picture.

blanking pulse — A square wave (positive or negative) used to switch off electronically a part of a television or radar set for a predetermined length of time.

blanking signal — A wave made up of recurrent pulses related in time to the scanning process and used to effect blanking. In television, this signal is composed of pulses at line and field frequencies, which usually originate in a central sync generator and are combined with the picture signal at the pickup equipment in order to form the blanked picture signal. The addition of a sync signal completes the composite picture signal.

blanking time — The length of time the electron beam of a cathode-ray tube is cut off.

blanking zone — *See* blanking pulse.

blank instruction — *See* no-operation instruction.

blank record — A recording disk on which no material has been recorded.

blank tape — Also called raw tape or virgin tape. 1. Tape on which nothing has been recorded. 2. Magnetic tape that has never been subjected to the recording process and is therefore substantially free from noise.

blast filter — Also called a pop filter. A dense mesh screen on a microphone, which minimizes overload caused by loud, close sounds.

blasting — 1. Overloading of an amplifier or speaker, resulting in severe distortion of loud sounds. 2. Severe audible distortion due to overloading of sound-reproducing equipment.

bleeder — A resistor connected across a power source to improve voltage regulation, provide a current path under no-load conditions, or dissipate stored energy on shut-off.

bleeder current — The current drawn continuously from a power supply by a resistor. Used to improve the voltage regulation of the power supply. (A technology no longer in use.)

bleeder resistor — 1. A resistor used to draw a fixed current. Also used, as a safety measure, to discharge a filter capacitor after the circuit is deenergized. 2. A resistor placed in the power supply of a radio receiver or other electronic device to stabilize the voltage supply.

bleeding — 1. Migration of plasticizers, waxes, or similar materials to the surface to form a film or bead. 2. In photomasking, poor edge definition or acuity caused by spread of image onto adjacent areas. 3. A condition in which a plated hole discharges process material or solution from crevices or voids. 5. During hybrid circuit manufacturing, the lateral spreading or diffusion of a printed film into adjacent areas beyond the geometric dimensions of the printing screen. This may occur during drying or firing.

bleeding whites — An overloading condition in which white areas in a television picture appear to flow into the black areas.

bleedout — The tendency of absorbed electrolytes, impurities, base materials, and preplates to diffuse to the surface of gold plating.

blemish — 1. On the storage surface of a charge-storage tube, an imperfection that produces a spurious output. 2. An area in a fiber or fiber bundle that has a reduced light transmission capability, i.e., increased attenuation, due to defective or broken fibers, foreign substances, or other spoilage.

blended data — Q-point that results from the combination of scanning data and tracking data to form a vector.

blending — A means of obtaining intermediate viscosities from materials of the same type but different viscosities. This term is also applied to resistive inks that can be blended with each other to achieve intermediate resistivities.

blind approach — An aircraft landing approach when visibility is poor, usually made with the aid of instruments and radiocommunication.

blind approach beacon system — *See* BABS

blind landing — Landing an aircraft entirely by means of instruments and electronic communications.

blind zone — An area from which echoes cannot be received; generally, an area shielded from the transmitter by some natural obstruction and therefore from which there can be no return.

blinking — 1. An ECM technique by which two aircraft separated a short distance and within the same azimuth resolution appear as one target to a tracking radar. The two aircraft alternately spot jam, causing the radar system to oscillate from one target to the other and making it impossible to obtain an accurate solution of the fire-control problem. 2. In pulse systems, a method of providing information in which the signal is modified at its source so that the presentation on the display scope alternately appears and disappears. In loran, this indicates that a station is malfunctioning.

blip — Sometimes referred to as pip. 1. On a cathode-ray display, a spot of light or a base-line irregularity representing the radar reflection from an object. 2. A discontinuity in the insulation of a wire.

blip-frame ratio — The ratio of the number of computer frames during which radar data was obtained to the total number of computer frames.

blip-scan ratio — The ratio between a single recognizable blip on a radarscope and the number of scans necessary to produce it. The blip-scan ratio of any given radar set varies with the range, antenna tilt, level of operator and set performance, target aspect, wind, etc.

blister — 1. The enclosure housing an airborne radar antenna. 2. A lump or raised section of a conductor or resistor caused by outgassing of the binder or vehicle during firing.

blistering — The development, during firing, of enclosed or broken macroscopic vesicles or bubbles in a body or in a glaze or other coating.

blivet — An excess of coating material, such as a lump around a dust particle on a wire or a surface. *See also* land, 2.

Bloch wall — The transition layer separating adjacent ferromagnetic domains.

block — 1. A group of computer words considered as a unit because they are in successive storage locations. 2. The set of locations or tape positions in which a block of words is stored. 3. A circuit assemblage that functions as a unit, such as a circuit building block of standard design or the logic block in a sequential circuit. 4. A set of contiguous bits and/or bytes that make up a definable quantity of information. 5. A section of information recorded on magnetic tape or disk. One block may consist of several records, that is, collections of information consisting of one or more related items; or a record may extend over several blocks, depending on the characteristics of the device and the needs of the programmer. 6. A group of consecutive words, characters, or bits that is handled as a single unit, particularly with respect to input/output operations. 7. A group of characters that is written or read as a physical unit as distinct from a logical unit (*see* record). A block may contain one or more complete records, or part of a record. 8. A string of data elements that is recorded or transmitted as a unit. 9. In word processing, a selected section of characters. In data management, a group of records. In communications, a fixed batch of data that is transferred together.

block address — A method of identifying words through use of an address that specifies the format and meaning of the words in the block of information.

block cancel character — A character used to signify that the preceding portion of the block is to be disregarded. Also called block ignore character.

block code — A special code or character used to separate blocks of data. A block code is used typically on paper tape and generally occurs at both the beginning and end of a block. Thus, the information on a paper tape containing a number of blocks would be started by a block code, there would be a block code between adjacent blocks, and the data would be ended by a block code.

block diagram — 1. A diagram in which the essential systems units are drawn as blocks, and their relationship to each other is indicated by appropriately connected lines. The path of the signal or energy may be indicated by lines or arrows. 2. In computer programming, a graphical representation of the data-processing procedures within the system. It is used by programmers as an aid to program development. 3. A diagram in which a system or computer program is represented by annotated boxes and interconnecting lines. Synonym: flowchart. 4. A chart that graphically depicts the functional relationships of hardware making up a system. The block diagram serves to indicate the various data and control signal paths between functional units of the system hardware. 5. A drawing in which circuit functions are represented as blocks of various geometries.

block downconversion — The process of lowering an entire band of frequencies in one step to some intermediate range to be processed by a receiver. Multiple block downconversion receivers are capable of independently selecting channels because each can process the entire block of signals.

block downconverter — 1. A device that converts an entire band (e.g., the 3.7–4.2 GHz C-band) down to a lower band of frequencies. 2. A type of downconverter that changes the microwave signal into an IF frequency that contains all the transponder frequencies (channels) of the satellite. The block downconverter allows inexpensive multiple receivers to tune all the channels simultaneously using one central downconverter — an advantage when using multiple receivers with a single antenna.

blocked impedance — The input impedance of a transducer when its output is connected to a load of infinite impedance.

blocked resistance — Resistance of an audio-frequency transducer when its moving elements are restrained so they cannot move; it represents the resistance due only to electrical loss.

blockette — In digital computer programming, a sub-group, or subdivision, of a group of consecutive machine words transferred as a unit.

block gap — 1. An area used to indicate the end of a block or record on a data medium. 2. An absence of data along a specified length of magnetic tape between adjacent blocks of data.

block-grid keying — A method of keying a continuous-wave transmitter by operating the amplifier stage as an electronic switch. During the spacing interval when the key is open, the bias on the control grid becomes highly negative and prevents the flow of plate current so that the tube has no output. During the marking interval when the key is closed, this bias is removed and full plate current flows.

block ignore character — See block cancel character.

blocking — 1. Application of an extremely high-bias voltage to a transistor, vacuum tube, or metallic rectifier to prevent current from flowing in the forward direction. 2. Combining two or more records into one block. 3. A condition in a switching system in which no paths or circuits are available to complete a call, resulting in a busy tone returned to the calling party. A denial or busy condition.

blocking capacitor — 1. A capacitor that introduces a comparatively high series impedance for limiting the flow of low-frequency alternating and direct current without materially affecting the flow of high-frequency alternating current. 2. A capacitor used to block direct current while allowing an alternating current of certain frequencies to pass.

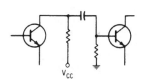

Blocking capacitor.

blocking layer — See depletion layer.

blocking oscillator — Also called squegging oscillator. 1. An electron-tube oscillator that operates intermittently as its grid bias increases during oscillation to a point

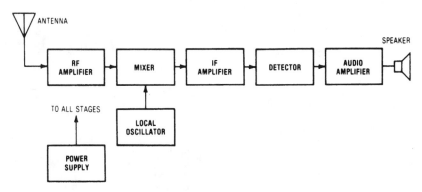

Block diagram, 1.

at which the oscillations stop, and then decreases until oscillation resumes. 2. A relaxation oscillator consisting of an amplifier (usually a single stage) whose output is coupled back to the input by means that include capacitance, resistance, and mutual inductance. 3. A relaxation-type oscillator that conducts for a short period of time and is cut off for a relatively long period.

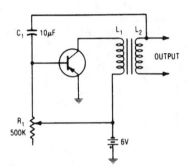

Blocking oscillator.

blocking-oscillator driver — A circuit that develops a square pulse used to drive the modulator tubes, and which usually contains a line-controlled blocking oscillator that shapes the pulse into the square wave.

blocking relay — A device that initiates a pilot signal to block tripping on external faults in a transmission line or in other apparatus under predetermined conditions, or which cooperates with other devices to block tripping or to block reclosing on an out-of-step condition or on power swings.

block length — In a computer, the total number of records, words, or characters contained in one block.

block loading — In a computer, a form of fetch in which the control sections of a load module are brought into continuous positions of main storage.

block mark — A method of indicating the end of one block of data and the start of another on tape or in data transmission. On magnetic tape, the block mark is a block gap; on paper tape it is a block code; in data transmission it is typically a pause or a code.

block-multiplexer channel — A computer-peripheral multiplexer channel that interleaves blocks of data.

block protector — A rectangular piece of carbon, Bakelite with a metal insert, or porcelain with a carbon insert that, in combination with each other, make one element of a protector. They form a gap that will break down and provide a path to ground for excessive voltages.

block size — The number of data elements in a block.

block sort — A computer sorting technique in which the file is first divided according to the most significant character of the key, and the separate portions are then sorted one at a time. It is used particularly for large files.

block transfer — In a computer, the process of transmitting one or more blocks of data.

blooming — 1. An increase in the size of the scanning spot on a cathode-ray tube, caused by defocusing when the brightness control is set too high. The result is expansion and consequent distortion of the image. May also be caused by insufficient high voltage. 2. The defocusing of regions of the picture where the brightness is at an excessive level, due to enlargement of spot size and halation of the fluorescent screen of the cathode-ray picture tube.

blooper — A radio receiver that is oscillating and radiating an undesired signal.

blow — The opening of a circuit because of excessive current, particularly when the current is heavy and a melting or breakdown point is reached and a fuse blows.

blower — 1. An electric fan used to supply moving air for cooling purposes. *See* fan. 2. A high-pressure device in which a rotating impeller moves an air mass in a spiral direction around the shaft. There are single-stage, dual-stage, and vacuum types.

blown jacket — The common term given to an outer covering of insulation of a cable that was applied by the controlled inflation of the cured jacket tube and the pulling of the cable through it.

blowout coil — An electromagnetic device used to establish a magnetic field in the space where an electrical circuit is broken and thus displace and extinguish the arc.

blowout magnet — A strong permanent magnet or electromagnet used for reducing or deflecting the arc between electrodes or contacts.

blue-beam magnet — A small permanent magnet used to adjust the static convergence of the electron beam for blue phosphor dots in a three-gun color picture tube.

blue glow — The glow normally seen in vacuum tubes containing mercury vapor; it is due to ionization of the molecules of mercury vapor.

blue gun — In a three-gun color picture tube, the electron gun whose beam strikes the phosphor dots emitting the blue primary color.

blue noise — In a spectrum of electromagnetic wave frequencies, a region in which the spectral density is proportional to the frequency (sloped) rather than independent of frequency (flat), as in white noise that is more of a uniformly distributed constant-amplitude frequency spectrum.

blue restorer — The dc restorer in the blue channel of a three-gun color-television picture-tube circuit.

blue video voltage — The signal voltage that controls the grid of the blue gun in a three-gun picture tube. This signal is a reproduction of the blue output signal of the color-television camera.

B-MAC — A method of transmitting and scrambling television signals. In such transmissions, MAC (multiplexed analog component) signals are time-multiplexed with a digital burst containing digitized sound, video synchronizing, authorization, and information.

BMEWS — *See* ballistic-missile early-warning system.

BNC — A bayonet-locking connector for miniature coax; BNC is said to be short for bayonet-neil-connector. Contrast with *TNC*.

BNC coax connector — A twist-lock connector for various types of RG-type coaxial cables.

BNC coax connector.

board — 1. A telephone or audio mixer panel containing patch jacks. 2. A circuit board.

board computer — A computer in which all electronic components are on a single circuit board.

board-test simulation — A testing technique in which the circuit to be tested is modeled, component by component and node by node, in the test system computer. From this model the system can calculate the correct response to any input pattern, plus predict failure modes and their responses. This allows only those patterns that identify faults to be used as the test pattern stimulus.

boat — 1. A container for materials to be evaporated or fired. 2. A wafer holder used in a diffusion furnace.

BOB — Abbreviation for break-out box.

bobbin — 1. A small insulated spool that serves as a support for a coil or wirewound resistor. 2. Spool used for taking up drawn wire and subsequently used for pay-out packages in cabling and stranding equipment.

bobbin core — *See* tape-wound core.

body capacitance — The capacitance to ground that is introduced into an electric circuit by the proximity of, or contact with, the human body.

body effect — Characteristic shift in threshold voltage resulting from bias applied to a semiconductor device's substrate.

body electrodes — Electrodes placed on or in the body to couple electrical impulses from the body to an external measuring or recording device.

boffle — A speaker enclosure, developed by H. A. Hartley, containing a group of stretched, resilient and sound-absorbing screens.

bogey — 1. The average, or published, value for a tube characteristic. A bogey tube would be one having all characteristics of a bogey value. 2. An average, published, or nominal value for some characteristic of a device.

bogey electron device — An electron device whose characteristics have the published nominal values for the type. A bogey electronic device for any particular application can be obtained by considering only those characteristics that are directly related to the application.

boilerplate — 1. A full-size model that simulates the weight, size, and shape, but not all of the functional features, of the actual item. 2. That part of the specifications of a component, piece of equipment, system, or the like that defines and describes the set of conditions of the sale.

boiling point — The temperature at which a liquid vaporizes when heated. The exact point depends on the absolute pressure at the liquid-vapor surface.

bolometer — 1. A radiation detector that converts incident radiation into heat, which, in turn, causes a temperature change in the material used in the detector. This change is then measured to give an indication of the amount of incident radiant energy. 2. A very sensitive thermometric instrument used for the detection and measurement of radiant energy. Its essential component is a short, narrow strip that is covered with a dead-black absorbing coating. It is mounted at the lower end of a long, cylindrical tube having a stop across it to exclude unwanted radiation. The electrical resistance of the strip changes with changes in temperature that arise from absorbing varying amounts of radiant energy.

bolted fault level — *See* adjusted circuit.

Boltzmann's constant — 1.38×10^{-16} J/K. Relates the average energy of a molecule to its absolute temperature.

bomb — 1. A computer program that fails spectacularly. 2. A programmer can bomb a computer system by deliberately writing a program that will disrupt the system. 3. In a computer, to fail or to crash.

bombardment — 1. The directing of high-speed electrons at an electrode, causing secondary emission of electrons, fluorescence, disintegration, or the production of X-rays. 2. The process of directing high-speed particles at atoms to cause ionization or transmutation.

bond — 1. Electrical interconnection made with a low-resistance material between a chassis, metal shield cans, or cable shielding braid, in order to eliminate undesirable interaction and interference resulting from high-impedance paths between them. 2. To make an electrical bond, an interconnection that performs a permanent electrical and/or mechanical function. 3. A low-resistance electrical connection between two ground connections or between similar parts of two circuits. *See* valence bond.

bondability — Those surface characteristics and conditions of cleanliness of a bonding area that must exist in order to provide a capability for successfully bonding an interconnection material by one of several methods, such as ultrasonic or thermocompression wire bonding.

bondable wire — An insulated wire whose surface has been treated to facilitate adherence to other materials, such as potting compounds. The term also could be applied to magnetic wires used in making coils where bonding the turns together is desirable.

bond deformation — The change in the form of the lead produced by the bonding tool, causing plastic flow, in making the bond.

bonded assembly — An assembly whose supporting frame and metallic noncircuit elements are connected so as to be electrically shorted together.

bonded-barrier transistor — A transistor made by alloying the base with the alloying material on the end of a wire.

bonded cables — Cables consisting of preinsulated conductors or multiconductor components that are laid in parallel and bonded into a flat cable.

bonded nr diode — An n-junction semiconductor device in which the negative resistance arises from a combination of avalanche breakdown and conductivity modulation due to the current through the junction.

bonded pickup — *See* bonded transducer.

bonded strain gage — A pressure transducer that uses a pressure-sensing system consisting of strain-gage elements firmly bonded to a pressure-responsive member. Thermal stability and insensitivity to shock and vibration are improved by means of this bonded construction.

bonded transducer — Also called bonded pickup. A transducer that employs the bonded strain-gage principle of transduction.

bonding — 1. Soldering or welding together various elements, shields, or housings of a device to prevent potential differences and possible interference. 2. A method used to produce good electrical contact between metallic parts of any device. Used extensively in automobiles and aircraft to prevent static buildup. Also refers to the connectors and straps used to bond equipment. 3. The means employed to obtain an electromagnetically homogenous mass having an equipotential surface. 4. The attachment of wire to a circuit. 5. The permanent joining of metallic parts to form an electrically conductive part. *See* ball bonding; die bonding; stitch bond; thermal compression bonding; wedge bonding; wire bond; wobble bond. 6. The process of connecting wires from the package leads to the chip (or die) bonding pads. Part of the assembly process. Alternately, the process of securing a semiconductor die to a lead frame or package.

bonding area — The area, defined by the extent of a metallization land or the top surface of the terminal, to which a lead is or is to be bonded. Also called bond site.

bonding conductor — 1. A conductor that serves to connect exposed metal surfaces together. 2. Device used to connect exposed metal to ground. It normally carries no current but is used to eliminate shock or spark hazards and ensures the operation of circuit protective devices in cases of insulation breakdown.

bonding island — *See* bonding pad.

bonding pad — Also called bonding island. 1. A relatively large metallic area at the edge of an integrated circuit chip; this area is connected through a thin metallic strip to some specific circuit point to which an external connection is to be made. 2. A metallized area at the end of a thin metallic strip to which a connection is to be made. *See also* beam lead.

bonding wire — Fine gold or aluminum wire for making electrical connections to a semiconductor die or a hybrid circuit between various bonding pads on the semiconductor device substrate and device terminals or substrate lands.

bond liftoff — The failure mode whereby the bonded lead separates from the surface to which it was bonded.

bond schedule — The values of the bonding machine parameters used when adjusting for bonding. For example, in ultrasonic bonding, the values of the bonding force, time, and ultrasonic power.

bond site — *See* bonding area.

bond strength — A measure of the amount of stress required to separate a layer of material from the base to which it is bonded. Peel strength, measured in pounds per inch of width, is obtained by peeling the layer; pull strength, measured in pounds per square inch, is obtained by a perpendicular pull applied to a surface of the layer.

bond-to-bond distance — The distance, measured from the bonding site on the die to the bond impression on the post, substrate land, or fingers, that must be bridged by a bonding wire or ribbon.

bond-to-chip distance — In beam-lead bonding, the distance from the heel of the bond to the component.

bone conduction — The process by which sound is conducted to the inner ear through the cranial bones.

book capacitor — A two-plate trimmer capacitor that has its plates hinged together like the pages of a book. The capacitance is varied by changing the angle between the plates.

Boolean algebra — 1. A system of mathematical logic dealing with classes, propositions, on-off circuit elements, etc., associated by operators such as AND, OR, NOT, EXCEPT, IF ... THEN, etc., thereby permitting computations and demonstration as in any mathematical system. Named after George Boole, famous English mathematician and logician, who introduced it in 1847. 2. Algebraic rules for manipulating logic equations. 3. Shorthand notation for expressing logic functions.

Boolean calculus — Boolean algebra modified to include time.

Boolean equation — Expression of relations between logic functions.

Boolean function — A mathematical function in Boolean algebra.

Boolean operator — The symbol or word used to specify the inclusion or exclusion of criteria such as AND, OR, NOT, etc. Same as logical operator.

boom — A mechanical support for a microphone, used in a television studio to suspend the microphone within range of the actors' voices but out of camera range.

boost capacitor — A capacitor used in the damper circuit of a television receiver to supply a boosted B voltage.

boost charge — The partial charge of a storage battery, usually at a high current rate for a short period.

boosted B voltage — In television receivers, the voltage resulting from the combination of the B-plus voltage from the power supply and the average value of voltage pulses coming through the damper tube from the horizontal deflection-coil circuit. The pulses are partially or wholly smoothed by filtering. This boosted voltage may be several hundred volts higher than the B-plus voltage.

booster — 1. A carrier-frequency amplifier, usually a self-contained unit, connected between the antenna or transmission line and a television or radio receiver. 2. An intermediate radio or TV station that retransmits signals from one fixed station to another. 3. A small, self-contained transformer designed to be connected to a cathode-ray tube socket to increase the filament voltage and thereby extend the life of the tube. 4. A device inserted into a line (or cable) to increase the voltage. Boosting generators are also used to raise the level of a dc line. Transformers are usually employed to boost ac voltages. The term booster also is applied to antenna preamplifiers.

booster amplifier — A circuit used to increase the output current of the voltage capabilities of an operational amplifier circuit without loss of accuracy (ideally) or inversion of polarity. Usually employed inside the loop for accuracy.

booster voltage — The additional voltage supplied by the damper tube of a television receiver to the horizontal-output, horizontal-oscillator, and vertical-output tubes, resulting in a greater sawtooth sweep output.

boot — 1. A form placed around the wire termination of a multiple-contact connector for the purpose of containing the liquid potting compound until it hardens. 2. A housing, usually made from a resilient material, used to protect connector or other terminals from moisture. 3. An accessory, usually of a flexible material, designed to be placed around the terminals of a component as a protective housing. 4. To make a computer operate by loading in a program. 5. The initial starting-up of a PC. The operating system is brought into main memory and takes over control. 6. The automatic routine that clears the memory, loads the operating system, and prepares a computer for use. A *cold boot* occurs when the power is first switched on. A *warm boot* refers to restarting a computer that is already turned on (clearing the memory and reloading the operating system) without first switching it off. Also called boot up.

booting — Loading a computer's memory with the necessary information so it can function. A cold boot occurs when a computer's power is turned on and there is nothing in memory.

boot loader — Also called bootstrap. A program in a minicomputer that usually works on a simple data format called core image. The data format, in this context, is the organization of the data as it appears on the input device from which the program is being loaded. Core image data is binary, bit-for-bit identical to what will appear in memory after loadings. The boot leader is used to bring simple programs into memory and run them immediately.

bootstrap — 1. A technique or device that brings itself into a desired state through its own action; for example, a routine whose first few instructions are sufficient to cause the rest of the routine to be brought into the computer from an input device. 2. A feedback technique that tends to improve linearity and input impedance of circuits operating over a wide range of input signals. *See* boot loader.

bootstrap circuit — A single-stage amplifier in which the output load is connected between the negative end of the plate supply and the cathode, the signal voltage being applied between the grid and cathode. The name *bootstrap* arises from the fact that the change in grid voltage also changes the potential of the input source (with respect to ground) by an amount equal to the output signal.

bootstrap driver — An electronic circuit used to generate a square pulse to drive a modulator tube.

bootstrap loader — Device for loading first instructions (usually only a few words) of a routine into memory, then using these instructions to bring in the rest of the routine.

bootstrapping — A feedback technique that tends to improve the linearity of circuits that operate over a wide input-signal range.

boresight — The direction along the principle axis of either a transmitting or a receiving antenna.

boresight error — The angular deviation of the electrical boresight of an antenna from its reference boresight.

boresighting — The initial alignment of a directional microwave or radar antenna system through use of an optical procedure or a fixed, known target.

boresight point — The area of maximum signal strength of a down-link signal from a satellite. The center of the transponder footprint.

BORSCHT — Abbreviation for Battery, Overvoltage, Ringing, Supervision, Coding, Hybrid, and Test. 1. An acronym for the function that must be performed in the central office of a telephone system when digital voice transmission occurs. 2. The seven functions (battery feed, overvoltage protection, ringing, supervision (or signaling), coding, hybrid, and testing) that are used in telecommunications and performed by a subscriber line interface circuit (SLIC).

boss — *See* land, 2.

BOT — Abbreviation for beginning of tape. *See* load point.

BOT/EOT markers — The reflective markers on the back (nonoxide) side of a magnetic tape, used to locate the beginning of the data (BOT) and provide an early warning of the end of the tape (EOT). Tape-drive sensors optically detect these markers.

bottom — To reach a point on an operating or characteristic curve at which a negative change in the independent variable, as, for example, in input, no longer produces a constant change in the dependent variable, as, for example, output.

bottoming — 1. A condition in which a stylus reaches the bottom of a record groove because its tip radius is smaller than optimum for the groove. Also the opposite of the pinch effect. 2. Excessive movement of the diaphragm of a headphone or the cone of a speaker so that the magnet or supporting structure is struck by the moving coil-piston assembly.

bottom metallization — The metallization that may be provided over the back portion of an uncased IC chip, facilitating its face-up attachment.

bounce — 1. An unnatural, sudden variation in the brightness of a television picture. 2. Sudden variations in a video picture presentation (brightness, size, etc.) independent of scene illumination.

bounce buffer — The electronic circuitry used to eliminate the effects of bounce of a mechanical switch.

boundary — An interface between p and n material at which donor and acceptor concentrations are equal.

boundary defect — In a crystal, the boundary area between two adjacent perfect crystal regions that are tilted slightly with respect to each other.

boundary-element method — A procedure for solving electromagnetic field problems by breaking the domains into smaller segments and by using integral equations in which the solution variables are confined within the boundaries.

boundary-layer photocell — *See* photovoltaic cell.

boundary marker — A transmitting device installed near the approach end of an airport runway and approximately on the localizer course line.

bound charge — On a conductor, the charge which, owing to the inductive action of a neighboring charge, will not escape to ground; residual charge.

bound circuit — A circuit designed to limit the excursion of a signal. The limit value it establishes may be nominal when used for protection, or highly precise when used operationally.

bound electron — An electron bound to the nucleus of an atom by electrostatic attraction.

bow-tie antenna — An antenna generally used for UHF reception. It consists of two triangles in the same plane, usually with a reflector behind them. The transmission line is connected to the points, which form a gap.

Bow-tie antenna.

boxcar — One of a series of pulses having long duration in comparison with the spaces between them.

boxcar circuit — A circuit used in radar to sample voltage waveforms and store the latest value sampled.

boxcar detector — A signal recovery instrument that is used either to retrieve the waveform of a repetitive signal from noise or to measure the amplitude of a repetitive pulse buried in noise. The detector has two modes of operation: scan and single-point. The former is used for waveform retrieval and the latter for pulse measurement.

boxcar integrator — A signal processor that uses a narrow filter to reduce the noise with little or no effect on the signal bandwidth. A simple integrator consists of a gated switch and a low-pass filter. During the time when the gate is closed, the repetitive input signal is applied to the low-pass filter, which acts as an integrator.

boxcar lengthener — A circuit that lengthens a series of pulses without changing their heights.

box pattern — A pin arrangement for plug-in packages in which the pins form four rows in either a square or rectangle.

bpi/cpi — Bytes per inch (bits per inch per track), the number of bytes per inch written on a magnetic tape, and cpi (characters per inch) are used interchangeably.

B-plus — 1. Symbol, B+. The positive dc voltage required for certain electrodes of tubes, transistors, etc. 2. The positive terminal of a B power supply.

B-plus boost — The positive voltage that is added to the low dc B+ voltage in a TV receiver by the action of the damper tube.

B power supply — A power supply that provides the plate and screen voltages applied to a vacuum tube.

bps — Abbreviation for bits per second. 1. In serial data transmission, the instantaneous bit speed within one character as transmitted. 2. A measurement of the speed at which data is transmitted and received by a modem. The larger the number, the faster the data is sent and received. Typical rates are 2400, 14,400, 28,800, and 56,000 bps. Often confused with *baud*, although the terms are not interchangeable.

BPSK — *See* binary phase-shift keying.

B-quad — A quad arrangement similar to the S-quad except for a short between the junction of the two sets of series elements. Also called bridge quad or bar quad.

Bragg's law — An expression of the conditions under which a system of parallel atomic layers in a crystal will reflect an X-ray beam with maximum intensity.

braid — 1. A weave of organic or inorganic fiber used as a covering for a conductor or group of conductors. 2. A woven metal tube used as shielding around a conductor or group of conductors. When flattened, it is used as a grounding strap.

braided wire — 1. A flexible wire made up of small strands woven together. 2. Woven bare or tinned copper wire used as shielding for wires and cables and as ground wire for batteries or heavy industrial equipment. There are many different types of construction.

brain waves — The patterns of lines produced on the moving chart of an electroencephalograph as the result of electrical potentials produced by the brain, picked up by electrodes, and amplified in the machine.

brake — An electromechanical friction device to stop and hold a load.

brake wire — Wires used in automotive and truck trailers to supply current to the electrical braking system.

braking magnet — *See* retarding magnet.

branch — 1. In an electronic network, a section between two adjacent branch points. 2. A portion of a network consisting of one or more two-terminal elements in series. 3. An instruction to a computer to follow one of several courses of action, depending on the nature of control events that occur later.

branch circuit — 1. That portion of the wiring system between the final overcurrent device protecting the circuit and the outlet. 2. (As applied to appliances) A circuit designed for the sole purpose of supplying an appliance or appliances; nothing else can be connected to this circuit, including lighting. 3. (General purpose) A circuit that supplies lighting and appliances. 4. (Individual) A circuit that supplies just one piece of equipment, such as a motor, an air conditioner, or a furnace. 5. (Multiwire) A circuit consisting of two or more underground conductors with a potential difference between them, and a grounded conductor with an equal potential between it and any one ungrounded conductor. (This may be a 120/240-volt three-wire circuit or a wye-connected circuit with two or more phase wires and a neutral.) It is not a circuit using two or more wires, connected to the same phase, and the neutral.

branch current — The current in the branch of a network.

branch impedance — In a passive branch, the impedance obtained by assuming a driving force across and a corresponding response in the branch, no other branch being electrically connected to the one under consideration.

branching — 1. In a computer, a method of selecting, on the basis of the computer results, the next operation to execute while the program is in progress. 2. The function of a computer that alters the logic path depending on some detected condition or data status. For example, the program would branch to a recorder routine when the projected available balance goes negative.

branching instructions — In a computer, conditional and unconditional jumping, calling subroutines, and return-looping operations.

branch order — An instruction used to link subroutines into the main program of a computer.

branch point — 1. In an electric network, the junction of more than two conductors. *See also* node, 1. 2. In a computer, a point in the routine where one of two or more choices is selected under control of a routine.

branch voltage — The voltage across a branch of a network.

branch windings — Forked polyphase transformer windings.

brass pounder — Amateur term for a Morse code operator, especially one who spends long hours handling traffic.

braze bonding — The joining of similar or dissimilar metals by introducing a braze filler metal at the joint and establishing a conventional brazed joint, and then diffusing the braze filler into the base metals by subsequent heat treating. Melting occurs in the braze filler independent of the base metals. Base-metal fusion is not required.

brazing — 1. A group of welding processes in which the filler is a nonferrous metal or alloy with a melting point greater than 1000°F (537°C) but lower than that of the metals or alloy to be joined. Brazing is sometimes referred to as hard soldering. 2. Joining two metal parts (usually made of iron or steel) together with a suitable melted copper-alloy metal.

breadboard — 1. Developmental or prototype version of a circuit. Solderless sockets and bus strips in modular form are often used to create expandable matrices for placement of ICs, capacitors, resistors, and so forth. 2. Perforated substrates that facilitate trial positioning of circuit components and wiring arrangements leading to final circuit construction and packaging. They are used in design, construction, and assembly.

breadboard circuit — A circuit simulation using discrete components or partially integrated components to prove the feasibility of a circuit.

breadboard construction — An arrangement in which electronic components are fastened temporarily to a board for experimental work.

breadboard model — 1. An assembly in rough form to prove the feasibility of a circuit, device, system, or principle. 2. An experimental model of a circuit in which the components are fastened temporarily to a chassis or board and electrically tested.

break — 1. In a communication circuit, the taking control of the circuit by a receiving operator or listening operator. The term is used in connection with half-duplex telegraph circuits and two-way telephone circuits equipped with voice-operated devices. 2. In a circuit-opening device, the minimum distance between the stationary and movable contacts when these contacts are open. 3. An open circuit or on-hook condition as determined by the dial of a telephone set. 4. A signal sent over a backward (secondary) channel by a receiving start-stop terminal on a half-duplex circuit (or, in some cases, over the transmit channel on a full-duplex circuit), usually to indicate a requirement to transmit. 5. The word used to interrupt a conversation on a repeater to indicate there is an emergency.

break alarm — 1. An alarm condition signaled by the opening or breaking of an electrical circuit. 2. The signal produced by a break alarm condition (sometimes referred to as an open-circuit alarm or trouble signal), designed to indicate possible system failure.

breakaway panels — Hardboards held together in a grouping by using breakaway tabs. Breakaway panels make handling easier for automatic insertion and wave soldering. When necessary, boards can be separated by snapping them apart much like a soda cracker.

break-before-make — 1. The action of opening a switching circuit before closing another associated circuit. 2. Movable contact that breaks one circuit before making the next circuit.

break-before-make contacts — Contacts that interrupt one circuit before establishing another.

break contact — 1. In a switching device, the contact that opens a circuit upon operation of the device (normally closed contact). 2. Contacts that open when a key or relay is operated.

break distance — The effective open-gap distance between the stationary and movable contacts.

breakdown — 1. An electric discharge through an insulator, insulation on wire, or other circuit separator, or between electrodes in a vacuum or gas-filled tube. 2. Phenomenon occurring in a reverse-biased semiconductor diode, the initiation of which is observed as a transition from a region of high dynamic resistance to a region of substantially lower dynamic resistance for increasing magnitude of reverse voltage. Also collector-emitter breakdown, punch-through, secondary breakdown, etc. 3. Initiation of a spark discharge between two electrodes. 4. A disruptive current (discharge) through insulation. 5. Failure of insulation for any reason. 6. A state of a circuit device in which the electric field exceeds a maximum allowed value. A sharp current increase results, which may destroy the device.

breakdown diode — *See* avalanche diode.

breakdown impedance — Also called avalanche impedance. The small-signal impedance at a specified direct current in the breakdown region of a semiconductor diode.

breakdown region — The entire region of the volt-ampere characteristic beyond the initiation of breakdown due to reverse current in a semiconductor-diode characteristic curve.

breakdown strength — *See* dielectric strength.

breakdown torque — The maximum torque a motor will develop, without an abrupt drop in speed, as the rated voltage is applied at the rated frequency.

breakdown voltage — 1. That voltage at which an insulator or dielectric ruptures, or at which ionization and conduction take place in a gas or a vapor. 2. The voltage measured at a specified current in the breakdown region of a semiconductor diode at which there is a substantial change in characteristics from those at lower voltages. Also called zener voltage. 3. The voltage required to jump an air gap. 4. The reverse-bias voltage applied to a pn junction for which large currents are drawn for relatively small increases in voltage. 5. The voltage at which a disruptive discharge takes place, either through or over the surface of insulation. 6. The voltage at which the insulation between two conductors will break down.

break elongation — The relative elongation of a specimen of recording tape or base film at the instant it breaks after having been stretched at a given rate.

breakerless ignition — A semiconductor electronic ignition system that does not use mechanical breaker contacts for timing or triggering purposes, but retains the distributor mechanism for distribution of the secondary voltage.

breaker points — 1. The low-voltage contacts that interrupt the current in the primary circuit of the ignition system of a gasoline engine. 2. A pair of movable points that are opened and closed to break and make the primary circuit.

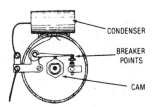

Breaker points.

break frequency — In a plot of log gain (attenuation) versus log frequency, the frequency at which the asymptotes of two adjacent linear slope segments meet.

break-in keying — In the operation of a radiotelegraph communication system, a method by which the receiver is capable of receiving signals during transmission spacing intervals.

break-in operation — A method of radiocommunication involving break-in keying that allows the receiving operator to interrupt the transmission.

break-in relay — A relay used for break-in operation.

break-make contact — Also called transfer contact. A contact form in which one contact opens its connection to another contact and then closes its connection to a third contact.

breakout — The exit point of a conductor or number of conductors along various points of a main cable of which the conductors are a part. This point is usually harnessed or sealed with some synthetic-rubber compound.

breakout box — Abbreviated BOB. 1. A testing device that permits the user to monitor the status of the various signals between two communicating devices, and to cross and tie interface leads, using jumper wires. 2. A device for monitoring and manipulating the interface signals between a DTE (data terminal equipment) and a DCE (data circuit-terminating equipment) or a pair of DTEs. 3. A device used to separately test each electrical line in a communications cable.

breakover — In a silicon-controlled rectifier or related device, a transition into forward conduction caused by the application of an excessively high anode voltage. In some cases this is destructive to the device.

breakover voltage — The value of positive anode voltage at which a silicon-controlled rectifier with the gate circuit open switches into the conductive state.

break period — The time interval during which the circuit contacts of a telephone dial are open.

breakpoint — 1. A place in a computer routine, specified by an instruction, instruction digit, or other condition, where the routine may be interrupted by external intervention or by a monitor routine. 2. A special instruction that may be inserted in a program to break off the normal program control and return control to a debug-type program. When a breakpoint is executed, the debug program indicates what the computer was doing at that point. 3. A hardware or software condition (bit pattern) that stops program execution; e.g., specific addresses or control signals. The user can then examine registers and memory locations, alter data generated by the program, and even request that execution resume from that point. 4. A point in a program where the program flow is interrupted for the purpose of testing the logic of the program up to that point. Usually a breakpoint will transfer control to a routine that will display test data.

breakpoint instruction — In the programming of a digital computer, an instruction that, together with a manual control, causes the computer to stop.

breakpoint switch — A manually operated switch that controls conditioned operation at breakpoints; it is used primarily in debugging.

breakup — *See* color breakup.

breathing — 1. Amplitude variations similar to bounce, but at a slow, regular rate. 2. Slow, rhythmic pulsations of a quantity, such as current, voltage, brightness, beat note, etc., readily noticeable on a CRT display.

breezeway — In the NTSC color system, that portion of the back porch between the trailing edge of the sync pulse and the start of the color burst.

B register — A computer register that stores a word which will change an instruction before the computer carries out that instruction.

bremsstrahlung—Electromagnetic radiation emitted by a fast-moving charged particle (usually an electron) when it is slowed down (or accelerated) and deflected by the electric field surrounding a positively charged atomic nucleus. X-rays produced in ordinary X-ray machines are bremsstrahlung. (In German, the term means braking radiation.)

brevity code—A code that has as its sole purpose the shortening of messages rather than concealment of content.

Brewster angle—The angle of incidence at which the reflection of parallel-polarized electromagnetic radiation at the interface between two dielectric media equals zero.

bridge—1. In a measuring system, an instrument in which part or all of a bridge circuit is used to measure one or more electrical quantities. 2. In a fully electronic stringed instrument, the part that converts the mechanical vibrations produced by the strings into electrical signals. 3. Equipment and techniques used to match circuits to each other, ensuring minimum transmission impairment. Bridging is normally required on multipoint data channels where several local loops or channels interconnect. 4. A device used singly, or with other like devices, to connect two or more similar local area networks (LANs) in order to allow workstations on one network to communicate with resources on the other network or networks as though they resided on the same network. 5. Equipment that connects two or more LANs that use the same protocol, allowing communication between devices on separate LANs.

bridge circuit—A network arranged so that, when an electromotive force is present in one branch, the response of a suitable detecting device in another branch may be zeroed by suitable adjustment of the electrical constants of still other branches.

bridged connection—A connection of a circuit across, or in parallel with, another circuit.

bridged T-network—A T-network in which the two series impedances of the T are bridged by a fourth impedance.

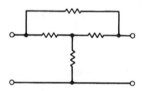

Bridged T-network.

bridge duplex system—A duplex system based on the Wheatstone-bridge principle in which a substantial neutrality of the receiving apparatus to the transmitted currents is obtained by an impedance balance. Received currents pass through the receiving relay, which is bridged between the points that are equipotential for the transmitted currents.

bridge hybrid—*See* hybrid junction, 2.

bridge quad—*See* B-quad.

bridge rectifier—A full-wave rectifier with four elements connected in the form of a bridge circuit so that dc voltage is obtained from one pair of opposite junctions when an alternating voltage is applied to the other pair of junctions.

bridge tap—An unterminated length of line attached (bridged) at some point between the extremities of a communication line; bridge taps are undesirable.

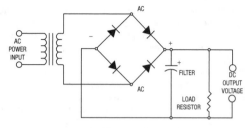

Bridge rectifier.

bridge transformer—*See* bridging transformer.

bridging—1. The shunting of one electrical circuit by another. 2. Connecting two electrical circuits in parallel. Usually the input impedances are large enough so as not to affect the signal level.

bridging amplifier—An amplifier with an input impedance sufficiently high (at least 10 times) that its input may be bridged across a circuit without substantially affecting the signal level of the circuit.

bridging connection—A parallel connection by means of which some of the signal energy in a circuit may be withdrawn with imperceptible effect on the normal operation of the circuit.

bridging contacts—1. A set of contacts in which some of the signal energy in a circuit may be withdrawn with imperceptible effect on the normal operation of the circuit. 2. A set of contacts in which the moving contact touches two stationary contacts simultaneously during transfer.

bridging gain—Ratio between the power a transducer delivers to a specified load impedance under specified operating conditions and the power dissipated in the reference impedance across which the transducer input is bridged. Usually expressed in decibels.

bridging loss—1. Ratio between the power dissipated in the reference impedance across which the input of a transducer is bridged and the power the transducer delivers to a specified load impedance under specified operating conditions. Usually expressed in decibels. 2. The loss resulting from bridging an impedance across a transmission circuit. Unless the impedance is a pure resistance, the loss will vary with frequency.

bridging transformer—Also called bridge transformer and hybrid coil. 1. A transformer designed to couple two circuits having at least nominal ohmic isolation and operating at different impedance levels, without introducing significant frequency or phase distortion or significant phase shift. 2. A transformer whose primary impedance is at least 10 times the source impedance of the device that feeds it. The secondary can be lower or higher than the primary impedance.

brightness—1. The attribute of visual perception in accordance with which an area appears to emit more or less light. Used with cathode-ray tubes. (*Luminance* is the recommended name for the photometric quantity that has also been called brightness.) 2. *See* luminance. 3. A surface measurement of light intensity per unit projected area. Usually expressed in footlamberts.

brightness control—1. In a television receiver, the control that varies the average brightness of the reproduced image. 2. The manual bias control of a cathode-ray tube that determines both the average brightness and the contrast of a picture.

brightness of image—The apparent luminance of an image observed by the eye.

brightness scale—A graduated range of stimuli perceived as having equivalent differences of brightness.

brightness signal — *See* luminance signal.

brilliance — 1. The degree of brightness and clarity in a reproduced cathode-ray tube. 2. The degree to which the higher audio frequencies sound like the original when reproduced by a receiver or public-address amplifier.

brilliance control — A potentiometer used in a three-way speaker system to adjust the output level of the tweeter for proper relative volume between the treble and the lower audio frequencies produced by the complete speaker system.

British Standard Wire Gage — A modification of the Birmingham wire gage; the legal standard of Great Britain for all wires. It is variously known as Standard Wire Gage (SWG), New British Standard (NBS), English Legal Standard, and Imperial Wire Gage.

British thermal unit — The energy required to raise the temperature of 1 pound of water 1 degree Fahrenheit.

broadband — 1. As applied to data transmission, the term denotes transmission facilities capable of handling frequencies greater than those required for high-grade voice communications (higher than 3 to 4 kilohertz). 2. Having an essentially uniform response over a wide range of frequencies. To design or adjust (an amplifier) for bandwidth. 3. A communications channel in which bandwidth is greater than 64 kb/s and that can provide higher-speed data communications than a standard telephone circuit. Also called wideband.

broadband amplifier — 1. An amplifier that has an essentially flat response over a wide frequency range. 2. An amplifier capable of amplifying a wide band of frequencies with minimal distortion.

broadband antenna — 1. An antenna that is capable of receiving a wide range of frequencies. 2. An antenna that is not sharply resonant. It provides adequate gain across a relatively wide band of frequencies. Examples: television receiving and log-periodic antennas.

broadband electrical noise — Also called random noise. A signal that contains a wide range of frequencies and has a randomly varying instantaneous amplitude.

broadband interference — Interference occupying a frequency range that is much greater than the bandwidth of the equipment being used to measure it.

broadband klystron — A klystron in which three or more externally loaded, stagger-tuned resonant cavities are used to broaden the bandwidth.

broadband random vibration — Single-frequency component vibrations that are random in both phase and amplitude at minute increments of frequency throughout a specified bandwidth. Typically it occurs when intense, high-power-level noise impinges on structures.

broadband tube (tr and pre-tr tubes) — A gas-filled, fixed-tuned tube incorporating a bandpass filter suitable for radio-frequency switching.

broadcast — 1. Radio or television transmission intended for public reception, for which receiving stations make no receipt. 2. To send messages or communicate simultaneously with many or all points in a circuit.

broadcast band — Radio frequency range between 530 and 1710 kHz in which all commercial AM broadcasting stations are assigned.

broadcasting — The transmitting of speech, music, or visual programs for commercial or public-service motives to a relatively large audience (as opposed to two-way radio, for example, which is utilitarian and is directed toward a limited audience).

broadcasting-satellite service — Abbreviated BSS. 1. International designation for direct-to-home satellite transmissions. In the Western Hemisphere, BSS operates in the 12.2- to 12.7-GHz frequency band. 2. A radio-communication service in which signals transmitted or retransmitted by satellites are used for direct reception by the general public.

broadcasting service — A radiocommunication service in which the transmissions are intended for direct reception by the general public. This service may include sound transmissions, television transmissions, or other types of transmissions.

broadcasting station — A station in the broadcasting service.

broadcast receiver — A receiver intended primarily for picking up standard broadcast stations.

broadcast teletext — A one-way system in which textual and graphic information in digital form is placed on unused portions of the television signal and retrieved by the user via a special terminal built into or attached to a TV set. The information is broadcast in a continuous cycle or loop and is retrieved by the user as it passes.

broadside array — 1. An antenna array whose direction of maximum radiation is perpendicular to the line or plane of the array (depending on whether the elements lie on a line or a plane). A uniform broadside array is a linear array whose elements contribute fields of equal amplitude and phase. 2. A unidirectional antenna array whose individual radiating elements are all in the same plane.

broad tuning — A tuned circuit or circuits that respond to frequencies within one band or channel, as well as to a considerable range of frequencies on each side.

Brownian motion noise — Also identified as thermal or Johnson noise. A random movement of microscopic particles, in organic or inorganic fluid suspension, caused by collision with surrounding molecules.

brownout — 1. Refers to low line voltage, which can cause misoperation and possible damage of equipment. For example, a motor trying to start at low voltage can actually be in a lock-rotor condition and overheat. 2. Deliberate lowering of the line voltage by a power company to reduce load demands. 3. A period of low-voltage electrical power due to unusually heavy power demands. Brownouts can cause computers to operate erratically or to crash. 4. A sudden, unexpected reduction in electrical power, usually lasting just a few seconds, but long enough to cause computer equipment to fail from insufficient power levels.

browse — To view data.

browser — 1. A software application that permits viewing and possibly searching of content in an information database, typically text, static images, or graphics, in a random or leisurely fashion at the user's discretion. 2. An application used to view information from the Internet. Browsers provide a user-friendly interface for navigating through and accessing the vast amount of information on the Internet.

browsing — 1. Exploring an online area, usually on the World Wide Web. 2. To look at files or computer listings in search of something interesting.

Bruce antenna — Original name of the rhombic end-fire antenna, which consists of a diamond-shaped arrangement of four wires with the feed line at one end and a resistive termination at the opposite end of the longer diagonal. Bruce's name also is given to a series-fed array of vertically polarized resonant rectangular loops (or half-loops over ground) with one-quarter-wave width and one-half-wave spacing.

brush — 1. A piece of conductive material, usually carbon or graphite, that rides on the commutator of a motor and forms the electrical connection between it and the power source. 2. A carbon block or metal leaf spring used to make sliding contact with a rotating contact, such as the commutator of a generator.

brush discharge — An intermittent discharge of electricity that starts from a conductor when its potential exceeds a certain value but is too low for the formation of an actual spark. It is generally accompanied by a whistling or crackling noise.

brush-discharge resistance — See corona resistance.

brush fluxing — A specialized wave-solder technique. A 360° bristled brush rotates in a foaming flux head to transfer the flux to the board. See also spray fluxing; wave fluxing; foam fluxing.

brush rocker — A movable rocker, or yoke, on which the brush holders of a dynamo or motor are fixed so that the position of the brushes on the commutator can be adjusted.

brush station — In a computer, a position where the holes of a punched card are sensed, particularly when this is done by a row of brushes sweeping electrical contacts.

brute force — The use of seemingly inefficient design in order to achieve a desired result. Sometimes this is done in order to avoid involved design procedures, critical adjustments, or the like, but often it is the only possible approach. For example, the miniaturization of low-frequency speakers requires brute force in the form of greatly increased amplifier power.

brute-force filter — A type of power-pack filter that depends on large values of capacitance and inductance to smooth out pulsations, rather than on the resonant effects of a tuned filter.

brute supply — A power supply that is completely unregulated. It employs no circuitry to maintain the output voltage constant with a changing input line or load variations. The output voltage varies in the same percentage and in the same direction as the input line voltage. If the load current changes, the output voltage changes inversely — an increase in load current drops the output voltage.

BSC — Abbreviation for binary synchronous communications. A line-control procedure for communicating, expressed in eight-bit EBCDIC, seven-bit U.S. ASCII, or six-bit transcode. The data code selected must include the required line-control characters, used according to specified rules.

B-scope — A type of radarscope that presents the range of an object by a vertical deflection of the signal on the screen, and the bearing by a horizontal deflection.

"B" service — FAA service pertaining to transmission and reception by teletype or radio of messages containing requests for and approval to conduct an aircraft flight, flight plans, in-flight progress reports, and arrival reports.

B-spline — A mathematical representation of a smooth curve.

B-spline surface — The mathematical description of a three-dimensional surface that passes through a set of B-splines.

BSS — See broadcasting-satellite service.

B supply — A source for supplying a positive voltage to the anodes and other positive electrodes of electron tubes. Sometimes called B+ supply.

BT-cut crystal — A piezoelectric plate that is cut from a quartz crystal at an angle of rotation (about the x-axis) of −49°. At approximately 25°C it has a zero temperature coefficient of frequency.

bubble — See dot, 2.

bubble logic — An obsolete form of magnetic logic.

bubble memory — 1. A bubble-memory chip surrounded by the two orthogonal coils, sandwiched between two permanent-bias magnets, which provide nonvolatility and stable domains. The coils create a rotating magnetic field that moves bubbles or cylindrical magnetic domains through a film on the chip. Films may consist of magnetic metals or synthetic ferrite or garnet crystals. In addition, control functions are located in aluminum-copper elements on the magnetic film. This entire grouping is then enclosed in a magnetic shield to prevent external magnetic fields (up to about 20 oersteds) from affecting the data. 2. A device that stores data as the absence or presence of tiny right-cylindrical domains of magnetization in a magnetic medium. Bubble memories operate as shift registers with complex control signals for the various magnetic fields required. 3. A nonvolatile storage technique that uses magnetic fields to create regions of magnetization. A pulsed field breaks the regions into isolated bubbles, free to move along the surface of the crystal sheet that contains the regions. The presence or absence of a bubble represents digital (bit, not bit) information. External electromagnetic fields manipulate the bubbles (information) past read/write locations within the memory. 4. Serial-access nonvolatile memory storing data in magnetic domains (bubbles) of opposite polarity from the rest of the cell. It is characterized by very low access times and high cost, but is able to store data after power is removed. This nonvolatility makes it well suited for portable applications in which disk storage is impractical. (Note: Bubble memories are no longer used.)

bubble sort — A sorting algorithm for putting the elements of a file in order.

buck — To oppose, as one voltage bucking another, or the magnetic fields of two coils bucking each other.

bucket — A general term for a specific reference in storage in a computer, such as a section of storage, the location of a word, a storage cell, etc.

bucket brigade — 1. A shift register that transfers information from stage to stage in response to timing signals. 2. A technique of building a photosensing array using field-effect transistors (FETs). Two FET amplifiers are at each sensing area and are interconnected as an element of a shift register. The charges sensed by the FETs are successively clocked from element to element. The electrical charges representing bits of stored information are transferred from one element to the next by means of clock pulses that raise the level of each element in the correct sequence, in the same way as a firefighting brigade passes a bucket of water down the line.

bucket curve — A graphical presentation of noise within a microwave system, displayed in a form that permits separation and identification of the individual idle and intermodulation (IM or cross-modulation) impairments.

bucking — Counteracting one quantity (such as a current or voltage) by opposing it with a like quantity of equal magnitude but opposite polarity.

bucking coil — A coil connected and positioned in such a way that its magnetic field opposes the magnetic field of another coil. The hum-bucking coil of an electrodynamic speaker is an example.

bucking voltage — A voltage that is opposite in polarity to another voltage in the circuit and hence bucks, or opposes, the latter voltage.

buckling — The warping of the plates of a battery due to an excessively high rate of charge or discharge.

buffer — 1. A circuit or component that isolates one electrical circuit from another. Usually refers to electron-tube amplifiers used for this purpose. 2. A vacuum-tube stage used chiefly to prevent undesirable interaction between two other stages. In a transmitter, it generally follows the master-oscillator stage. 3. An isolating circuit used in an electronic computer to avoid reaction of a driven circuit. 4. A storage device used to compensate for a difference in the rate of flow of information or the time or occurrence of events when transmitting information from one device to another. 5. A computer circuit having an output and a multiplicity of inputs and designed so that the output is energized. Thus, a buffer performs the circuit function equivalent to the logical OR. 6. A noninverting digital circuit element that may be used to handle a large fanout or to invert input and output levels. Normally, a buffer is an emitter follower. 7. A storage device in which data is temporarily stored during data transfers between peripherals and computers. A means of temporarily storing data must be used when a peripheral device inputs data faster than the computer can process it, or when the computer sends data faster than a peripheral can receive it. 8. A soft fiber-optic material that mechanically isolates individual fibers in a fiber-optic cable or bundle from small geometrical irregularities, distortions, or roughness of adjacent surfaces. The buffer has no optical function. 9. A high-speed area of storage in a computer that is temporarily reserved for use in performing the input/output operation, into which data is read or from which data is written. 10. A storage area in a computer for data that is used to compensate for a speed difference when transferring data from one device to another. Usually refers to an area reserved for I/O operations, into which data is read, or from which data is written. 11. A temporary-storage device used to compensate for a difference in data rate and data flow between two devices (typically a computer and a printer). *See also* spooler.

buffer amplifier — 1. An amplifier designed to isolate a preceding circuit from the effects of a following circuit. 2. An amplifier stage that is used to isolate a frequency-sensitive circuit from variations in the load presented by following stages. 3. A unity gain amplifier used to isolate the loading effect of one circuit from another.

buffer capacitor — A capacitor connected across the secondary of a vibrator transformer, or between the anode and cathode of a cold-cathode rectifier tube, to suppress voltage surges that might otherwise damage other parts in the circuit.

buffer circuitry — Circuitry necessary to adapt signals between two systems, e.g., between a test system and a board under test.

buffer computer — A computing system provided with a storage device so that input and output data may be stored temporarily to match the slower speed of input/output devices with the higher speed of the computer.

buffer/driver — A device used to increase the fanout or drive capability of a digital circuit.

buffered terminal — A terminal that contains storage equipment so that the rate at which it sends or receives data over its line does not need to agree exactly with the rate at which the data is entered or printed.

buffer element — A low-impedance inverting driver circuit. Because of its very low source impedance the element can supply substantially more output current than the basic circuit. As a consequence, the buffer element is valuable in driving heavily loaded circuits or minimizing rise-time deterioration due to capacitive loading.

buffer gate — A logic gate with a high output-drive capability, or fanout; a buffer gate is used when it is necessary to drive a large number of gate inputs from one gate function.

buffering — 1. Storing data between transfer operations. Data read from a disk is buffered before transfer to a system memory, and data to be written is buffered after transfer from a system memory. 2. The process of temporarily storing data in a software program or in RAM to allow transmission devices to accommodate differences in data transmission rates.

buffer storage — 1. A synchronizing element used between two different forms of storage (usually internal and external). 2. An input device in which information from external or secondary storage is assembled and stored ready for transfer to internal storage. 3. An output device into which information from internal storage is copied and held for transfer to secondary or external storage. Computation continues during transfers from buffer storage to secondary or internal storage or vice versa. 4. A device for the purpose of storing information temporarily during data transfers. 5. *See* cache.

buffer storage unit — A computer unit used for temporary storage of data before transmission to another destination. The unit often is able to accept and give back data at widely varying rates so that data transfer takes place efficiently for each device connected to it; that is, a high-speed device is not slowed by the presence of a slow-speed device in the same system unless the data transfer occurs directly between the two devices.

buffer stripper — A device that removes flat cable insulation from conductors; a unit of motorized buffing wheels that scrapes the insulation and brushes it away. Also called an abrasion stripper.

bug — 1. A semiautomatic telegraph sending key consisting of a lever that is moved to one side to produce a series of correctly spaced dots, and to the other side to produce a dash. 2. A circuit fault due to improper design or construction. 3. An electronic listening device, generally small and concealed, used for commercial or military espionage. 4. To plant a microphone or other sound sensor or to tap a communication line for the purpose of surreptitious listening or audio monitoring; loosely, to install a sensor in a specified location. 5. The microphone or other sensor used for the purpose of surreptitious listening. 6. A device or system used for surreptitiously sensing and removing audio information from a target area. In the sense in which it is usually used, a bug transmits information from the target area to the listening post via radio; however, the information may be transmitted over dedicated wire by varying dc, as in an ordinary telephone circuit, or over power wiring by carrier current, as in a wireless intercom. The sensor in this system (transducer that changes sound into electrical impulses) is normally a microphone, but might be something that has another primary purpose, such as a telephone ringer. 7. A term used to denote a mistake in a computer program or a malfunction in a computer hardware component. 8. A defect in hardware or software that causes a program to malfunction.

bugging height — The distance between the hybrid substrate and the lower surface of the beam-lead device that occurs because of deformation of beam leads during beam-lead bonding.

build — The increase of diameter due to insulation.

building-out — Addition to an electric structure of an element or elements electrically similar to an element or elements of the structure to bring a certain property or characteristic to a desired value.

building-out circuit — A short section of transmission line, or a network that is shunted across a transmission line, for the purpose of impedance matching.

building-out network — A network designed to be connected to a basic network so that the combination will simulate the sending-end impedance, neglecting dissipation, of a line having a termination other than that for which the basic network was designed.

building-out section — A short section of transmission line, either open or short-circuited at the far end, shunted across another transmission line for use on an impedance-matching transformer.

building wire — Insulated wires used in buildings for light and power, 60 volts or less. Usually not exposed to outdoor environment.

bulb — The glass envelope that encloses an incandescent lamp or an electronic tube.

bulb-temperature pickup — A temperature transducer in which the sensing element is enclosed in a metal tube or sheath to protect it against corrosive liquids or other contaminants.

bulge — The difference between the actual characteristic and a linear characteristic of the attenuation-frequency characteristic of a transmission line.

bulk degausser — *See* bulk eraser.

bulk effect — An effect, such as current, resistance, or resistivity, observed in the overall body of a sample of material, as opposed to a region within the material or on its surface.

bulk-erased noise — The noise arising when a bulk-erased tape is reproduced with the erase and record heads completely deenergized. Ideally, this noise is governed by the number of magnetic particles that pass by the head in unit time.

bulk eraser — Also called bulk degausser or degausser. 1. Equipment for erasing a roll of tape. The roll is usually rotated while a 60-hertz ac erasing field is decreased, either by withdrawing the roll from an electromagnet or by reducing the ac supply to an electromagnet. 2. A device used to erase an entire tape at one time. Bulk erasers are usually more effective than a recorder's erase heads.

bulk noise — *See* excess noise.

bulk resistance — The portion of the contact resistance that is due to the length, cross section, and material.

bulk resistivity — Resistance measured between opposite faces of a cube of homogeneous material.

bulk resistor — A resistor made by providing ohmic contacts between two points of a homogeneous, uniformly doped crystal of silicon material.

bulk storage — An external device, such as a tape recorder, on which a computer's contents can be recorded to be reloaded into the computer when needed.

bulletin board — A computer system operated so that people can access it via telephone lines and leave messages for others to see.

bump — A means of providing connections to terminal areas of a device. A small mound is formed on the device (or substrate) pads and is utilized as a contact for face-down bonding.

bump chip — A chip that has on its termination pads a bump of solder or other bonding material that is used to bond the chip to external contacts. It allows for simultaneous bonding of all leads, rather than one at a time as in wire bonding.

bump contacts — Small amounts of material formed on the chip substrate to register with terminal pads, as when the chip is employed in flip-chip circuits.

bunched pair — A group of pairs tied together or otherwise associated for identification.

buncher — 1. The input resonant cavity in a conventional klystron oscillator. 2. In a velocity-modulated tube, the electrode that concentrates the electrons in the constant-current electron beam into bunches.

buncher gap — *See* input gap.

buncher resonator — The input cavity resonator in a velocity-modulated tube. It serves to modify the velocity of the electrons in the beam.

bunching — 1. Grouping pairs together for identification and testing. 2. In a velocity-modulated electron stream, the action that produces an alternating convection-current component as a direct result of the differences of electron transit time produced by the velocity modulation.

bunching parameter — One-half of the product of the bunching angle in the absence of velocity modulation and the depth of velocity modulation.

bunching time — The time in the armature motion of a relay during which all three contacts of a bridging-contact combination are electrically connected.

bunching voltage — The radio-frequency voltage between the grids of the buncher resonator in a velocity-modulated tube such as a klystron. Generally, the term implies the peak value of this oscillating voltage.

bunch stranding — 1. A method in which a number of wires are twisted together in a common direction and with a uniform pitch to form a finished, stranded wire. 2. A group of wires of the same diameter twisted together without a predetermined pattern.

bundle — 1. A number of optical fibers grouped together, usually carrying a common signal. 2. A collection of glass or plastic fibers that transmit data in the form of optical energy. 3. A group of optical fibers contained within a single jacket.

bundle connector — Fiber-optic connector that joins fiber bundle to fiber bundle. It is used when fiber-optic cables penetrate bulkheads, when new cables are spliced into transmission networks, and for repair of breaks. *See* source connector; detector connector.

bundled cable — Individual insulated wires laced together to form a bundle to facilitate handling.

bundled software — Software that comes with a computer.

bundling — *See* lacing and harnessing.

buried cable — Also called direct burial cable. 1. A cable installed underground and not removable except by disturbing the soil. 2. A cable installed directly in the earth without use of underground conduit.

buried channel — Because charge trapping can occur at the surface of the $Si-SiO_2$ interface, a thin doped layer can be introduced in the silicon just below the oxide (typically by ion implantation) to prevent trapping of charges. (MIS technology term.)

buried layer — 1. A layer of very low resistivity, usually of n+ material, between the high-resistivity n-type collector region and p-type substrate of an integrated-circuit transistor. The buried layer tends to reduce the series collector resistance of the transistor without having an adverse effect on the breakdown voltage. 2. An underlying layer of a silicon IC formed by introducing impurities into the silicon, then covering it with additionally grown silicon.

buried resistors — Terminating film resistors deposited on inner layers of multilayer boards in order to reduce conductor lengths.

burn — *See* burned-in image.

burned-in image — Also called burn. An image that remains in a fixed position in the output signal of a camera tube after the camera has been turned to a different scene.

burn-in — 1. Operation of a device to stabilize its failure rate. 2. The operation of items prior to their ultimate application, intended to stabilize their characteristics and to identify early failures. 3. Phase of component testing in which infant mortality or early failures are screened out by running the circuit for a specified length of time (such as 168 hours). 4. Subjecting components to a high temperature, normally 125°C, for a specified period under an electrical power stress (normally 80 percent of rated power). The process is designed to accelerate the aging of a device beyond the infant mortality life stage. After an appropriate burn-in, the devices should have a very low failure rate, normally defined by military standards. 5. Operation of a component, module, or system under some increased stress, typically elevated temperature, so as to cause failure at the vendor's rather than at the user's plant. 6. A test of computer components that runs the system for a day or two to detect defective chips. 7. A test in which a product, such as a chip, is operated prior to its use to detect failures that would otherwise occur in about the first six months of the product's life.

burning — Programming a read-only memory chip.

burn-in period — The time during which components are operated at predetermined stress conditions prior to their installation in the user's equipment. *See* early-failure period.

burst — 1. A sudden increase in the strength of a signal. 2. The cosmic-ray effect on matter, causing a sudden intense ionization that gives rise to great numbers of ion pairs at once. 3. A group of events occurring together in time. 4. A number of events that occur within a short period of time. 5. One of a series of successions of bits, frames, or other elements of data, occurring at regular intervals. 6. A pulsed rf envelope, generally 2.3 μs (8 cycles) in duration, that appears on the back porch of the horizontal synch pulse. This 3.579545-MHz frequency is used to synchronize the television receiver's regenerated subcarrier color oscillator. 7. *See* color burst.

burst error — A series of consecutive errors in data transmission.

burst noise — An unwanted signal that is characterized by an excessively large interfering effect that is extended over a relatively short but finite time interval.

burst pedestal — A rectangular pulselike television signal that may be part of the color burst. The amplitude of the color-burst pedestal is measured from the alternating-current axis of the sine-wave portion of the horizontal pedestal.

burst pressure — The maximum pressure to which a device can be subjected without rupturing.

burst rate — The bit rate during the time of transmission from a terminal that transmits TDMA bursts.

burst separator output — The amplitude of the chroma reference burst at the output of the gated burst amplifier.

burst sequence — An arrangement of color-burst signals in which the polarity of the burst signal is the same at the start of each field so that the stability of color synchronization is improved.

burst transmission — Radio transmission in which messages are stored and then released at 10 to 100 or more times the normal speed. The received signals are recorded and returned to the normal rate for the user.

bus — 1. The term used to specify an uninsulated conductor (a bar or wire); may be solid, hollow, square, or round. 2. Sometimes used to specify a bus bar. 3. The communications path between two switching points. 4. Wire used to connect two terminals inside an electrical unit. A common point for electrical circuits to return. Can be bare, tinned, or insulated. 5. A power line that provides power to a large number of circuits. In computing, a bus is a group of wires that conveys information to a large number of devices. The information may be data, commands, or addresses, or all three in sequence. All the devices in the system are connected to the bus. Each device continually listens for a command addressed to it. Only one device is allowed to transmit over the bus at once. Bus-oriented systems are flexible and easy to expand. 6. A circuit over which data or power is transmitted or received. 7. A power or signal line used in common by several parts of a computer. 8. A signal path to which a number of inputs may be connected for feed to one or more outputs. 9. A group of conductors considered as a single entity that interconnects various parts of a system, e.g., data bus or address bus. 10. Usually a group of wires over which digital information is transmitted from one of several sources to one of several destinations. Commonly found in digital systems. 11. A group of wires that allows memory, CPU, and I/O devices to exchange words. 12. A connective link between multiple processing sites (colocated only), where any of the processing sites can transmit to any other, but only one way at a time. 13. A mechanical, logical, and electrical interconnection scheme for modular microcomputer circuit elements; provides a common path for data, address, and control information between those modular elements. 14. A group of conductors that provide time-shared communication paths for the transmission of information between equipment units. 15. A common pathway or channel between hardware devices. Can be serial (information travels one bit at a time) or parallel (information travels in groups of bits moving simultaneously along multiple parallel paths). 16. A data path shared by many devices (e.g., multipoint line) with one or more conductors for transmitting signals, data, or power. 17. The physical channel over which electric signals are transferred between the components of a system, along with the protocol rules governing the transfer.

bus analyzer — 1. A tool that captures the bus signals at a given instant for later analysis. 2. An instrument that analyzes digital data on the bus lines of a microprocessor.

bus architecture — A data-communications structure that consists of a common connection among a number of printer/plotter modules.

busback — The connection, by a common carrier, of the output portion of a circuit back to the input portion of a circuit. *See also* loopback test.

bus bar — 1. A heavy copper strap or bar used to carry high currents or to make a common connection between several circuits. 2. A heavy copper (or other metal such as aluminum) strip or bar used on switchboards and in power plants to carry heavy currents. 3. Interconnection device that distributes power from remote power supplies to cabinets or drawers, across connector backpanels, and on PC boards. These flat, rectangular components frequently consist of two or more conductor layers electrically insulated from one another and from other components by thin dielectric layers. Usually, conductor layers are bonded or laminated together to provide a rigid, mechanically and electrically stable package that is mounted with screws, clamps, or foam adhesive. Bus bars mounted on PC boards are typically soldered in place along with other components that are to be mounted on the boards.

bus driver — 1. An integrated circuit that is added to the data bus system in a computer to facilitate proper drive to the CPU when several memories are tied to the data bus line. Drivers are necessary because of capacitive loading, which slows down the data rate and prevents proper time sequencing of microprocessor operation. 2. A buffering device that increases the driving capability of a microprocessor that itself may be capable of driving no more than a single TTL load. 3. A circuit that amplifies a bus data or control signal sufficiently to ensure valid reception of that signal at the destination.

bushing — A mechanical device used as a lining for an opening to prevent abrasion to wire and cable. Also used as a low-cost method of insulating, anchoring, cushioning, and positioning. Usually a nonmetallic material is preferred.

business data processing — 1. Automatic data processing used in accounting or management. 2. Data processing for business purposes, such as recording and summarizing the financial transactions of a business.

business machine — Customer-provided equipment that is connected to the communications services of a common carrier for the purpose of data movement.

busing — The joining of two or more circuits together.

bus master — In a bus structure, in which control of data transfers on the bus is shared between the central processor and associated peripheral devices, the term bus master refers to the device controlling the current bus transaction.

bus organization — The manner in which many circuits are connected to common input and output lines (buses).

bus reactor — A current-limiting reactor connected between two different buses or two sections of the same bus for the purpose of limiting and localizing the disturbance due to a fault on either bus.

bus slave — In a bus structure, in which control of data transfers on the bus is shared between the central processor and associated peripheral devices, the device currently receiving or transmitting data from or to the bus master is referred to as the bus slave.

bus system — A network of paths inside a microprocessor that facilitates data flow. The important buses in a microprocessor are identified as data bus, address bus, and control bus.

bust this — Phrase used instead of a normal message ending to indicate that the entire message, including heading, is to be disregarded.

busy test — In telephony, a test to find out whether certain facilities that may be desired, such as a subscriber line or a trunk, are available for use.

busy tone — Interrupted low tone returned to the calling party to indicate that the called line is busy.

Butler antenna — An array antenna in which hybrid junctions are incorporated into the feed system to obtain a plurality of independent beams.

Butler oscillator — A two-tube (or transistor) crystal-controlled oscillator in which the crystal forms the positive feedback path when excited in its series-resonant mode.

butt connector — *See* butt contact.

butt contact — A hemispherically shaped contact designed to mate end to end without overlap, with axes aligned, against a similarly shaped contact. When properly aligned, the two convex surfaces form a reasonably good surface-to-surface contact, usually under spring pressure, with the ends designed to provide optimum surface contact.

butterfly capacitor — *See* butterfly resonator.

butterfly circuit — Frequency-determining element having no sliding contacts and providing simultaneous change of both inductance and capacitance. It is used to replace conventional tuning capacitors and coils in ultrahigh-frequency oscillator circuits. The rotor of the device resembles the opened wings of a butterfly.

butterfly resonator — Also called butterfly capacitor. A tuning device that combines both inductance and capacitance in such a manner that it exhibits resonant properties at very high and ultrahigh frequencies (characterized by a high tuning ratio and Q). So called because the shape of the rotor resembles the opened wings of a butterfly.

Butterworth filter — A filter network that exhibits the flattest possible response in the passband. The response is monotonic, rolling off smoothly into the stopband, where it approaches a constant slope of 6 dB/octave.

Butterworth function — A mathematical function used in designing a filter for maximally constant amplitude response with little consideration for time delay or phase response.

butt joint — 1. A splice or other connection formed by placing the ends of two conductors together and joining them by welding, brazing, or soldering. 2. A connection between two waveguides that maintains electrical continuity by providing physical contact between the ends.

button — 1. The metal container in which the carbon granules of a carbon microphone are held. 2. Also called dot. A piece of metal used for alloying onto the base wafer in making alloy transistors.

button capacitor — A fixed button-shaped ceramic or silvered-mica capacitor that, because of its disk shape and mode of terminal connection, offers very low internal inductance.

button-hook contact — A contact with a curved, hooklike termination, often located at the rear of hermetic headers to facilitate soldering or desoldering of leads.

buttonhook feed — A rod shaped like a question mark supporting the feedhorn and LNA. A buttonhook feed for use with commercial-grade antennas is often a hollow waveguide that directs signals from a feedhorn to an LNA behind the antenna.

buttons — In a computer, objects that, when clicked once, cause something to happen.

button silver-mica capacitor — A stack of silvered-mica sheets encased in a silver-plated brass housing. The high-potential terminal is connected through the center of the stack. The other capacitor terminal is formed by the metal shell, which connects at all points around the outer edge of the electrodes. This design permits the current to fan out in a 360° pattern from the center terminal, providing the shortest possible electrical path between the center terminal and chassis. The internal series inductance is thus kept small.

button stem — *See also* pressed stem. In a vacuum tube, the glass base onto which the mount structure is assembled. The pins may be sealed into the glass; if so, no base is needed. In some large tubes, the stiff wires are passed directly into the base pins to give added strength.

button up — To close or completely seal any operating device.

butt splice — A device for joining two conductors placed end to end with their axes in line (that is, conductors not overlapping).

buzzer — A signaling device in which an armature vibrates to produce a raucous, nonresonant sound.

BV_{EBO} — The reverse-breakdown voltage of the emitter-to-base junction of a transistor with the collector open-circuited.

BWG — Abbreviation for Birmingham wire gage.

BX cable — Insulated wires enclosed in flexible metal tubing or flexible spiral metal armor used in electrical wiring.

BX cable.

bypass — 1. A shunt (parallel) path around one or more elements of a circuit. 2. A secondary channel that permits routing of data in a computer sample around the data compressor, regardless of the value of the sample, at intervals determined by the operator.

bypass capacitor — A capacitor used for providing a comparatively low-impedance ac path around a circuit element.

bypass filter — A filter providing a low-attenuation path around some other circuit or equipment.

bypassing — Reducing high-frequency current in a high-impedance path by shunting that path with a bypass element (usually a capacitor).

B-Y signal — One of the three color-difference signals in color television. The B-Y signal forms a blue primary signal for the picture tube when combined — either inside or outside the picture tube — with a luminance, or Y, signal.

byte — 1. A single group of bits processed together (in parallel), usually 8 bits. 2. A sequence of adjacent binary digits, usually shorter than a word, operated on as a unit. 3. The number of bits that a computer processes as a unit. This may be equal to or less than the number of bits in a word. For example, both an 8-bit and a 16-bit length computer may process data in 8-bit bytes. 4. The smallest addressable unit of main storage in a computer system. The byte consists of 8 data bits. 5. A byte is universally used to represent a character. Microcomputer instructions require one, two, or three bytes. A word may be one byte long. One byte has two nibbles. Computers and microprocessors work with words of 4, 8, 12, 16, 24, or 32 bits. 6. A set of bits that represents a single character. Usually there are 8 or 10 bits in a byte, depending on how the measurement is being made. 7. A unit of data equal to 8 bits, and hence capable of storing any one of $2^8 = 256$ distinct values.

byte-multiplexer channel — A channel that interleaves bytes of data from different sources. Contrast with *selector channel*.

byte-serial — A sequence of bit-parallel data bytes used to carry information over a common bus.

C

C — 1. Symbol for capacitor, capacitance, carbon, coulomb, centigrade or Celsius, transistor collector, and (when lowercase) velocity of light. 2. A general-purpose programming language developed in the 1970s by Dennis Ritchie of AT&T, Bell Labs. Its generality, machine independence, and efficiency have made C popular for many application areas. The UNIX operating system is written in C, and the close linking of UNIX and C has made C the de facto standard language in engineering software development.

C− (C minus) — The negative terminal of a C battery, or the negative polarity of other sources of grid-bias voltage. Used to denote the terminal to which the negative side of a grid-bias voltage source should be connected.

C+ (C plus) — Positive terminal of a C battery, or the positive polarity of other sources of grid-bias voltage. The terminal to which the positive side of the grid-bias voltage source should be connected.

C² system — Abbreviation for command and control system.

C³ — Abbreviation for command, control, and communications.

C³I — Abbreviation for command, control, communications, and intelligence. Includes computers, displays, communications equipment, and other supporting devices that provide the means for communications as well as intelligence-gathering capability at all levels of military command.

C³L — Abbreviation for complementary contact-current logic.

C⁴ — Abbreviation for command, control, communications, and computer.

cabinet — 1. A protective housing for electrical or electronic equipment. 2. An enclosure for mounting equipment. If fabricated of steel, the cabinet helps attenuate electrostatic and electromagnetic radiation noise interference.

cable — 1. An assembly of one or more conductors, usually within a protective sheath, so arranged that the conductors can be used separately or in groups. 2. A stranded conductor (single-conductor cable) or a combination of conductors insulated from one another (multiple-conductor cable). 3. Jacketed combination of fiber bundles with cladding and strength-reinforcing components. 4. An assembly of insulated conductors into a compact form that is covered by a flexible, waterproof protective covering.

cable armor — In cable construction, a layer of steel wire or tape, or other extra-strength material, used to reinforce the lead wall.

cable assembly — A cable with plugs or connectors on each end for a specific purpose. It may be formed in various configurations.

cable attenuation — Reduction of signal intensity along a cable, usually expressed in decibels per foot, hundred feet, kilometer, mile, etc.

cable clamp — A device used to give mechanical support to the wire bundle or cable at the rear of a plug or receptacle.

cable complement — A group of cable pairs that have some common distinguishing characteristic.

cable core — That portion of an insulated cable lying under the protective covering or jacket.

cable coupler — A device used to join lengths of similar or dissimilar cable having the same electrical characteristics.

cable fill — The ratio of the number of pairs in use to the total number of pairs in a cable.

cable filler — Material used in multiple-conductor cables to occupy the spaces between the insulated conductors.

cable messenger — A stranded cable supported at intervals by poles or other structures and employed to furnish frequency points of support for conductors or cables.

cable modem — A device that enables a user to connect a computer to existing cable TV networks at Ethernet speeds and access Internet and/or online services. This device can be much faster than a telephone modem, operating at 14.4 kbs, 28.8 kbs, or higher.

cable Morse code — A three-element code used mainly in submarine-cable telegraphy. Dots and dashes are represented by positive and negative current impulses of equal length, and a space by the absence of current.

cable-ready television — A television receiver that can receive unscrambled cable television without the use of a converter.

cable run — The path occupied by a cable on cable racks or other support from one termination to another.

cable sheath — A protective covering of rubber, neoprene, resin, or lead over a wire or cable core.

cable splice — A connection between two or more separate lengths of cable. The conductors in one length are individually connected to conductors in the other length, and the protecting sheaths are so connected that protection is extended over the joint.

cable terminal — A means of electrically connecting a predetermined number of cable conductors in such a way that they can be individually selected and extended by conductors outside the cable.

cable TV — 1. Television system in which programs are received by a local central antenna and distributed by cable to individual homes. The term does not apply to a system serving fewer than 50 subscribers or serving only subscribers in one or more multiple-unit buildings under common ownership, control, or management. 2. Linking a TV set via cable to a system operator that, for a monthly fee, provides the viewer with typically 30 to 75 or more channels of programming. 3. A system for distributing television programming by a cable network rather than by electromagnetic radiation.

cable vault — A vault in which the outside plant cables are spliced to the tipping cables.

cabling — 1. The assembly of wire bundles extended from one physical structure to another to interconnect the circuits within structures. Cabling differs from wire jumpers in that it is understood to be external to the physical structures and may include tubing sheaths, zipper tubing, or rubber jackets. 2. Twisting together two or more insulated conductors by machine to form a cable. This also is a term loosely applied to bundling wires together, such as in the forming of wire harnesses. 3. In fiber optics, a method by which a group of fibers or bundle of fibers is mechanically assembled.

cache — 1. A small, fast memory built into a processor to give faster access to the data and instructions that a program uses repeatedly (also called cache memory, buffer storage). 2. A small, fast storage area interposed between the main memory and CPU to improve memory-transfer rates and processor efficiency. The name is derived from the fact that the memory is hidden from or transparent to the programmer. 3. A storage area for frequently accessed information. Retrieval of the information is faster from the cache than the original source. There are many types of cache, including RAM cache, secondary cache, disk cache, and cache memory, to name a few. 4. In a computer system, a small but very high-speed memory buffer situated between the processor and main memory. It operates on the principle that certain memory locations tend to be accessed very often (normally for reads). Thus, when a main memory location is read, it is stored in the cache at the same time.

cache memory — 1. A high-speed, low-capacity computer memory similar to a scratch-pad memory except that it has a larger capacity. 2. The fastest portion of the overall memory, which stores only the data that the computer may need in the immediate future. 3. A high-speed buffer memory used between the central processor and main memory. The cache is filled at medium speed from the main memory. Instructions and programs can operate at higher speed if found in the cache. If not found, a new segment is loaded. The cache contains the instruction or sequence of instructions most likely to be executed next. 4. A special, extra-fast part of RAM in which frequently accessed information is stored. Same as memory cache and RAM cache. 5. *See* cache. 6. A fast random-access memory system designed to store the most frequently accessed data in RAM.

cactus needle — A phonograph needle made from the thorn of a cactus plant.

CAD — Abbreviation for computer-aided design. 1. Use of a computer to aid in the design of complex MSI or LSI circuit arrays. CAD is especially useful for custom IC fabrication. 2. Man/computer interactions for the design and testing of customer MSI/LSI arrays and other complex engineering designs in a reasonable time frame. 3. Any system or process using a computer to aid in the creation or modification of a design. 4. High-performance design workstations that enable designers to manipulate parts diagrams and simulate operations, among other things. Can be linked to computer-aided manufacturing systems.

CADAM — *C*omputer-*a*ugmented *d*esign *a*nd *m*anufacturing system for design and analysis functions, developed by Lockheed for the aerospace industry and licensed by IBM.

CAD/CAM — Computer-aided design/computer-aided manufacturing. Two highly specialized technical applications of a computer to improve the productivity of the engineer.

cadmium — A metallic element widely used for plating steel hardware or chassis to improve its appearance and solderability and to prevent corrosion. It is also used in the manufacture of photocells.

cadmium cell — A standard cell used as a voltage reference; at 20°C its voltage is 1.0186 volts.

cadmium selenide photoconductive cell — A photoconductive cell that uses cadmium selenide as the semiconductor material. It has a fast response time and high sensitivity to longer light wavelengths, such as those emitted by incandescent lamps and some infrared light sources.

cadmium sulfide cell — *See* cadmium sulfide photoconductive cell.

cadmium sulfide photoconductive cell — A photoconductive cell in which a small wafer of cadmium sulfide is used to provide an extremely high dark-light resistance ratio. Some of the cells can be used directly as a light-controlled switch operated from the 120-volt ac power line.

CAE — Abbreviation for computer-aided engineering. The use of general-purpose computers and special application software to automate routine, iterative hardware engineering tasks.

cage — 1. A completely shielded enclosure. 2. A screened room that is covered with a grounded fine-mesh conductive screen on all sides to allow measurements within the cage without any influence or interference from extraneous signals.

cage antenna — An antenna comprising a number of wires connected in parallel and arranged in the form of a cage. This arrangement reduces the copper losses and increases the effective capacity.

Cage antenna.

calculating — Computing a result by multiplication, division, addition, or subtraction or by a combination of these operations. A data-processing function.

calculating punch — A punched-card machine that reads data from a group of cards and punches new data in the same or other cards.

calculator — 1. A device capable of performing arithmetic. 2. A calculator as in definition 1 that requires frequent manual intervention. 3. Generally and historically, a device for carrying out logical and arithmetical digital operations of any kind.

calculator mode — Also called fast answerback. An interactive computer system that has a mode which allows the terminal to be used like a desk calculator. The user types an expression, and the computer evaluates it and returns the answer immediately.

calendar age — Age of an item or object measured in terms of time elapsed since it was manufactured.

calendar life — That period of time, expressed in days, months, or years, during which an item may remain installed and in operation, but at the end of which the item should be removed and returned for repair, overhaul, or other maintenance.

calibrate — To ascertain, by measurement or by comparison with a standard, any variations in the readings of another instrument, or to correct the readings.

calibrated triggered sweep — In a cathode-ray oscilloscope, a sweep that occurs only when initiated by a pulse and that moves horizontally at a known rate.

calibration — 1. The process of comparing an instrument or device with a standard to determine its accuracy or to devise a corrected scale. 2. Taking measurements of various parts of electronic equipment to determine the performance level of the equipment and whether it conforms to technical order specifications. 3. Comparison of the performance of an item of test and measuring equipment with a reference standard traceable to the National Bureau of Standards or some other authoritative source or specification. 4. A test during which known values of measure are applied to a transducer, and its corresponding output readings are recorded.

calibration accuracy — Finite degree to which a device can be calibrated (influenced by sensitivity, resolution, and reproducibility of the device itself and the calibrating equipment). Expressed as a percentage of full scale.

calibration curve — A smooth curve connecting a series of calibration points.

calibration marker — On the screen of a radar indicator, the markings that divide the range scale into accurate intervals for range determination or checking against mechanical indicating dials, scales, or counters.

call — 1. A transmission made for the purpose of identifying the transmitting station and the station for which the transmission is intended. 2. To transfer control to a specified closed subroutine. 3. In communication, the action performed by the calling party, or the operations necessary in making a call, or the effective use made of a connection between two stations.

call accounting — Daily and monthly accounting reports for individual callers and for hierarchical groupings of callers in a communication system.

call announcer — Device for accepting pulses from an automatic telephone office and reproducing the corresponding number with speechlike sounds.

callback device — A device on the receiving end of a communications network that assures the authenticity of the sender by calling the sender back.

call circuit — A communication circuit between switching points used by traffic forces for transmitting switching instructions.

call diverter — A device that intercepts a call directed to a telephone subscriber and diverts it to another number.

caller ID — A telephone company service that provides the name and number of the caller from the phone company's information network. A caller ID device translates that data into usable form by displaying it on a screen. It works on standard telephones only.

call forwarding — A service available in some dial offices whereby a subscriber can have calls to his or her number forwarded to another phone by dialing the forwarding number to the equipment. All incoming calls are then forwarded automatically. The forwarding can

be discontinued at will, by command from the base telephone.

calligraphic display — Also known as a vector stroke or vector refresh. A refresh graphic system that draws the picture on the screen vector by vector. The monitor is driven by X and Y signals that control the beam's position for drawing. A vector list is continually processed and converted to the analog X and Y signals. The rate at which the picture is refreshed need not be synchronized to the ac line. Calligraphic displays draw smooth and very high-resolution lines. Since the display is refreshed from a vector list, any changes in this list immediately appear on the screen, creating dynamic displays. The main disadvantage is that increasingly complex pictures take longer to refresh, causing flicker.

call in — To transfer control of a digital computer temporarily from a main routine to a subroutine, which is inserted in the sequence of calculating operations to fulfill a subsidiary purpose.

call indicator — Device for accepting pulses from an automatic switching system and displaying the corresponding called number before an operator.

calling device — Apparatus that generates the pulses used to control the establishment of connections in an automatic telephone switching system.

calling mode — The ability to originate and/or answer on the dial-up network. A central-site modem normally has auto-call capability, meaning it can automatically dial a specific number through an automatic calling unit.

calling party control — *See* CPC.

calling sequence — In a computer, a sequence of instructions required to enter a subroutine. It may contain information required by the subroutine.

call letters — A series of government-assigned letters, or letters and numbers, that identify a transmitting station.

call number — In computer operations, a set of characters that identifies a subroutine and contains information with respect to parameters to be inserted in the subroutine or information related to the operands.

call-setup time — The overall length of time required to establish a switched call between pieces of data-terminal equipment.

call sign — 1. Any combination of characters or pronounceable words that identifies a communication facility, a command, an authority, an activity, or a unit; used primarily for establishing and maintaining communications. 2. The station identification assigned to a licensee by the FCC.

call word — A call number that is exactly the size of one machine word.

calomel electrode — An electrode consisting of mercury in contact with a solution of potassium chloride saturated with mercurous chloride (calomel). *See also* glass electrode.

calorimeter — An apparatus for measuring quantities of heat. Used to measure microwave power in terms of heat generated.

calorimeter system — A precision rf watt-meter as well as an efficient dummy load able to absorb energy at any frequency band. It can absorb and measure any level of microwave energy, and functions by circulating a known amount of liquid through a suitably designed low VSWR load. The load is located in a waveguide or coaxial section that mates to the terminal of the rf energy source being measured or absorbed.

CAM — Abbreviation for content addressable memory. 1. A computer memory in which information is retrieved by addressing the content (the data actually stored in the memory) rather than by selecting a physical location. *See also* associative storage. 2. Abbreviation for computer-aided manufacturing. The use of computer technology in manufacturing processes. It includes numerical and programmable controls, factory information management, robot controls, materials handling, and storage-and-retrieval systems.

cam actuator — An electromechanical device in which a switch is closed when the high spot of a rotating cam, or eccentric, is in a certain position.

camera — *See* television camera.

camera cable — A cable or group of wires that carry the picture signal from the television camera to the control room.

camera chain — A television camera, associated control units, power supplies, monitor, and connecting cables necessary to deliver a picture for broadcasting.

camera signal — The video-output signal of a television camera.

camera tube — 1. An electron-beam tube in which an electron current or charge-density image is formed from an optical image and scanned in a predetermined sequence to provide an electric signal. 2. The electron-beam tube of a television camera that converts an optical image into a pattern of electrostatic charges and then scans the pattern to produce a corresponding electric signal for transmission.

camera tube target — The storage surface of an electron-beam tube that is scanned by an electron beam to generate an output signal current corresponding to the charge-density pattern stored.

camp-on — Also called clamp-on. A method of holding a call for a line that is already in use and of signaling when the line becomes free.

can — 1. A metal shield placed around a tube, coil, or transformer to prevent electromagnetic or electrostatic interaction. 2. A metal package for enclosing a device, as opposed to a plastic or ceramic package.

Canadian Standards Association — Abbreviated CSA. In Canada, a body that issues standards and specifications prepared by various voluntary committees of government and industry.

canal ray — Also called positive ray. Streams of positive ions that flow from the anode to the cathode in an evacuated tube.

candela — Abbreviated cd. 1. Formerly candle. The unit of luminous intensity. The luminous intensity of 1/60th of 1 square centimeter of projected area of a blackbody radiator operating at the temperature of solidification of platinum (2046 K). Values for standards having other spectral distributions are derived by the use of accepted spectral luminous efficiency data for photopic vision. 2. International unit of luminous intensity, also called new candle. Prior to 1948, the standard was a specific type of candle and was termed candle, candle power, or international candle.

candela/cm² — Luminance unit called a stilb.

candle — The unit of luminous intensity. One candle is defined as the luminous intensity of 1/60th square centimeter of a blackbody radiator operating at the solidification temperature of platinum.

candlepower — 1. Luminous intensity expressed in terms of standard candles. 2. A measure of the intensity of light produced by a source. This standard of measurement is used in France, Britain, and the United States. One candlepower corresponds approximately to the light produced in the horizontal direction by an ordinary sperm candle weighing six to the pound and burning at the rate of 120 gr/hr. 3. The luminous intensity of a source of light expressed in candelas.

candoluminescence — 1. A phenomenon that produces white light without need for very high temperatures. 2. The luminescence of an incandescent material.

canned cycle — The use of preparatory functions on a punched tape to initiate a complete machining sequence; the need for much repetitive information in the program is thereby eliminated.

cannibalization — A method of maintenance or modification in which the required parts are removed from one system or assembly for installation on a similar system or assembly.

cantilever — The rod, or tube, that supports the stylus of a phonograph cartridge at its free end, is pivoted at or near its other end, and that transfers the stylus motion to the generating elements of the cartridge. Usually made of aluminum, but beryllium is used in some recent cartridges. Also known as the shank.

cantilevered contact — A spring contact in which the contact force is provided by one or more cantilevered springs. It permits more uniform contact pressure and is used almost exclusively in printed circuit board connectors.

capacitance — Abbreviated C. 1. Also called capacity. In a capacitor or a system of conductors and dielectrics, that property which permits the storage of electrically separated charges when potential differences exist between the conductors. The capacitance of a capacitor is defined as the ratio between the electric charge that has been transferred from one electrode to the other and the resultant difference in potential between the electrodes. The value of this ratio is dependent on the magnitude of the transferred charge. 2. Capacitance opposes any change in circuit voltage. A voltage change is delayed until the stored charges can be altered through current. The unit of capacitance is the farad. 3. The property of an electric system — comprised of conductors and associated dielectrics — that determines (a) the displacement currents in the system for a given rate of potential difference change between the conductors and (b) how much electrical charge will be stored in the dielectric for a given potential difference between the conductors.

capacitance alarm system — An alarm system in which a protected object is electrically connected as a capacitance sensor. The approach of an intruder causes sufficient change in capacitance to upset the balance of the system and initiate an alarm signal. Also called a proximity alarm system.

capacitance between two conductors — The ratio between the charge transferred from one conductor to the other and the resultant difference in the potentials of the two conductors when insulated from each other and from all other conductors.

capacitance bridge — A four-arm ac bridge for measuring capacitance by comparison against a standard capacitor.

capacitance detector — *See* capacitance sensor.

capacitance divider — A circuit made up of capacitors and used for measuring the value of a high-voltage pulse by making available only a small, known fraction of the total pulse voltage for measurement.

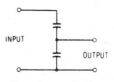

Capacitance divider.

capacitance level detector — A device with single or multiple probes that operates based on the fact that a change in capacitance level causes a change in probe capacitance.

capacitance meter — An instrument for measuring capacitance. If the scale is graduated in microfarads, the instrument is usually designated a microfaradmeter.

capacitance-operated intrusion detector — A boundary alarm system in which the approach of an intruder to an antenna wire encircling the protected area (a few feet above ground) changes the antenna-to-ground capacitance and thereby sets off the alarm.

capacitance ratio — The ratio of maximum to minimum capacitance, as determined from a capacitance characteristic, over a specified voltage range.

capacitance relay — An electronic circuit incorporating a relay that responds to a small change in capacitance, such as that created by bringing the hand or body near a pickup wire or plate.

capacitance sensor — A sensor that responds to a change in capacitance in a field containing a protected object or in a field within a protected area. Also called capacitance detector.

capacitance switch — A keyboard switch in which two pads on the circuit board under each keyswitch serve as capacitor plates connected to the drive and sense circuits. Depression of the key causes an increase in the series capacitance, coupling the two elements and creating an analog signal in the sense circuit.

capacitance tolerance — The maximum percentage deviation from the specified nominal value (at standard or stated environmental conditions) specified by the manufacturer.

capacitive coupling — Also called electrostatic coupling. The association of two or more circuits with one another by means of mutual capacitance between them. For example, between stages of an amplifier, that type of interconnection that employs a capacitor in the circuit, between the plate of one tube and the grid of the following tube or the collector of one transistor and the base of the following transistor.

capacitive diaphragm — A resonant window placed in a waveguide to provide the effect of capacitive reactance at the frequency being transmitted.

capacitive-discharge ignition — Also called capacitor-discharge ignition. An electronic ignition system used on internal combustion engines to provide nearly constant high voltage regardless of engine speed. A dc-to-dc step-up converter charges a capacitor when the distributor breaker points are closed; when they are open, the capacitor discharges through the ignition coil, thereby generating the ignition voltage

capacitive divider — Two or more capacitors placed in series across a source, making available a portion of the source voltage across each capacitor. The voltage across each capacitor will be inversely proportional to its capacitance.

capacitive feedback — The process of returning part of the energy in the plate or output circuit of a vacuum tube to the grid, or input, circuit by means of a capacitance common to both circuits.

capacitive load — A predominantly capacitive load, that is, one in which the current leads the voltage.

capacitive post — A metal post or screw extending at right angles to the E field in a waveguide. It provides capacitive susceptance in parallel with the waveguide for purposes of tuning or matching.

capacitive reactance — Symbolized by X_C. The impedance a capacitor offers to ac or pulsating dc. Measured in ohms and equal to $1/2\pi f C$, where f is in hertz and C is in farads.

capacitive speaker — See electrostatic speaker.

capacitive storage welding — A particular type of resistance welding whereby the energy is stored in banks of capacitors, which are then discharged through the primary of the welding transformer. The secondary current generates enough heat to produce the weld.

capacitive transduction — Conversion of the measurand into a change in capacitance.

capacitive tuning — Tuning by means of a variable capacitor.

capacitive voltage divider — A combination of capacitors connected in series to form a capacitive voltage dividing device for application with ac voltages.

capacitive welding — An electronic welding system in which energy stored in a capacitor is discharged through the joint to be welded. The resulting current develops the heat necessary for the operation.

capacitive window — A conductive diaphragm extended into a waveguide from one or both sidewalls to introduce the effect of capacitive susceptance in parallel with the waveguide.

capacitivity — See dielectric constant.

capacitor — 1. A device consisting essentially of two conducting surfaces separated by an insulating material or dielectric such as air, paper, mica, glass, plastic film, or oil. A capacitor stores electrical energy, blocks the flow of direct current, and permits the flow of alternating current to a degree dependent essentially on the capacitance and the frequency. 2. An electrical energy storage device used in the electronics industry for varied applications, notably as elements of resonant circuits, in coupling and bypass application, blockage of dc current, as frequency determining and timing elements, as filters and delay-line components, and in voltage transient suppression.

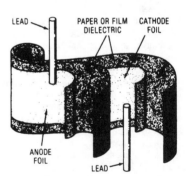

Capacitor (internal construction).

capacitor antenna — Also called condenser antenna. An antenna that consists of two conductors or systems of conductors and the essential characteristic of which is its capacitance.

capacitor bank — A number of capacitors connected together in series, parallel, or in series-parallel.

capacitor braking — A means of stopping an induction motor. The capacitor or capacitors can be applied to the winding after shut-off.

capacitor color code — Color dots or bands placed on capacitors to indicate one or more of the following: capacitance, capacitance tolerance, voltage rating, temperature coefficient, and the outside foil (on paper or film capacitors).

capacitor-discharge ignition — See capacitive-discharge ignition.

capacitor discharge system — An ignition system that stores its primary energy in a capacitor.

capacitor electrolyte — A current-conducting material (nonsolid or solid) serving as the cathode in an electrolytic capacitor.

capacitor filtering — A method for improving the form factor of a direct current by means of a parallel capacitor. Also, a means for increasing the magnitude of a rectified voltage.

capacitor-input filter — A power-supply filter in which a capacitor is connected directly across, or in parallel with, the rectifier output.

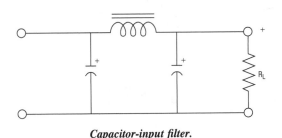

Capacitor-input filter.

capacitor losses — The active power dissipated by a capacitor.

capacitor microphone — *See* electrostatic microphone.

capacitor motor — A single-phase induction motor with the mean winding arranged for direct connection to the power source, together with an auxiliary winding connected in series with a capacitor.

capacitor pickup — A phonograph pickup that depends for its operation on the variation of its electrical capacitance.

capacitor-run motor — A single-phase induction motor using a capacitor in series with a second primary winding displaced 90° from the first. This winding remains in the circuit full time, converting the machine to polyphase operation.

capacitor series resistance — An equivalent resistance in series with a pure capacitance that gives the same resultant losses as the actual capacitor. The equivalent circuit does not represent the variation in capacitor losses with frequency.

capacitor speaker — *See* electrostatic speaker.

capacitor-start motor — 1. An ac split-phase induction motor in which a capacitor is connected in series with an auxiliary winding to provide a means of starting. The auxiliary circuit opens when the motor reaches a predetermined speed. 2. A type of motor that provides greater starting torque for a split-phase induction motor by connecting a capacitor in series with the auxiliary starting winding to provide a leading current vector of 90°, which momentarily converts the motor to a polyphase induction motor.

capacitor voltage — The voltage across the terminals of a.

capacity — 1. The current-output capability of a cell or battery a period of time. Usually expressed in ampere-hours (amp-hr). 2. Capacitance. 3. The limits, both upper and lower, of the items or numbers that may be processed in a computer register, in the accumulator. When quantities exceed the capacity, a computer interrupt develops and requires special handling. 4. The total quantity of data that a part of a computer can hold or handle, usually expressed as words per unit of time. 5. The capability of a

specific system to store data, accept transactions, process data, and generate reports. 6. In a calculator, the maximum number of digits that can be entered as one factor or obtained in a result. In most machines, the capacity is equivalent to the number of digits in the display. In a few machines, it is larger than the number of digits in the display and the flip-flop key is used to show the full result. 7. The maximum number of bits that a system can process, or transmit, per second; e.g., $C = H/T$, where T is time required for the processing or transmission and H is information content in bits.

capillary tool — A tool used in bonding; the wire is fed to the bonding surface of the tool through a bore located along the long axis of the tool.

capstan — 1. The driven spindle or shaft in a tape recorder — sometimes the motor shaft itself — that rotates against the tape (which is backed up by a rubber pressure or pinch roller), pulling it through the machine at a constant speed during recording and playback modes of operation. The rotational speed and diameter of the capstan thus determine the tape speed. 2. A revolving shaft or flangeless pulley that drives the tape by squeezing it against a pinch roller, and that controls the rate at which tape passes over the heads of the tape recorder or deck. 3. A rotating shaft that is connected to the motor in a tape recorder. It moves the tape at constant speed across the heads. The tape is pressed against the capstan by a pinch roller.

capstan idler — *See* pressure roller.

captive screw — Screw-type fastener that is retained when unscrewed and cannot easily be separated from the part it secured.

capture area — The area of the antenna elements that intercept radio signals.

capture bandwidth — The frequency range over which an unlocked, free-running oscillator can be brought into lock by either phase- or injection-locking techniques.

capture effect — 1. The selection of the stronger of two frequency-modulated signals of the same frequency, with the complete rejection of the weaker signal. If both signals are of equal strength, both may be accepted and no intelligible signal will result. 2. An effect occurring in a transducer (usually a demodulator) whereby the input wave having the largest magnitude controls the output.

capture range — The range of frequencies over which a phase-locked loop can detect a signal on the input and respond to it. This is sometimes called the lock-in range. *See* acquisition range.

capture ratio — 1. The ability of a tuner to reject unwanted FM stations and interference on the same frequency as a desired one, measured in dB. The lower the figure, the better the performance of the tuner. 2. The power ratio of two signals in the same channel required to keep the signal/interference ratio to a value of 30 dB referred to 100-percent modulation and 1-mV input signal level. The ratio of the powers of the two input signals is expressed in decibels; the smaller the dB number the better the capture ratio. Topflight tuners have a value as low as 1 dB, but 4.5 dB is usually sufficient. 3. A measure of an FM tuner's ability to discriminate against weaker signals arriving on the same frequency as the desired one.

carbon — One of the elements, consisting of a nonmetallic conductive material occurring as graphite, lampblack, diamond, etc. Its resistance is fifty to several hundred times that of copper and decreases as the temperature increases.

carbon arc — An electric discharge between two carbon rods that are touched together to start the arc and then separated slightly. The light comes from the heated carbon vapor. High-intensity arcs use cored carbons, the core being filled with the oxides of thorium and cerium, which radiate brightly when heated.

carbon brush—A current-carrying brush made of carbon, carbon and graphite, or carbon and copper.

carbon-composition resistor—Hot- or cold-molded fixed resistor made from mixtures of granulated carbon and ceramic binder. In some versions the composition forms a monolithic structure; in others the composition is thickly applied to a ceramic core. Hot-molded carbon-composition resistors are specified where low-cost, reliable resistors with tolerances of ± 5 and ± 10 percent are acceptable.

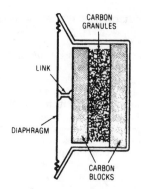

Carbon microphone.

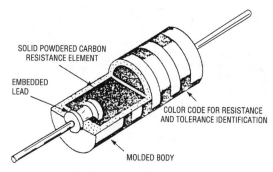

Carbon-composition resistor.

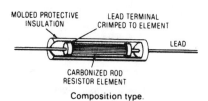

Composition type.

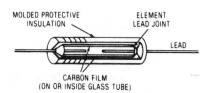

Deposited-film type.

Carbon resistor.

carbon-contact pickup—A phonograph pickup that depends for its operation on the variation in resistance of carbon contacts.

carbon-film resistor—1. A resistor formed by vacuum-depositing a thin carbon film on a ceramic form. 2. Carbon-film resistors are general-purpose, low-cost types with axial leads. Specification characteristics generally match those of carbon-composition resistors but at a lower cost.

carbonize—To coat with carbon.

carbonized filament—A thoriated-tungsten filament treated with carbon. A layer of tungsten carbide formed on the surface slows down the evaporation of the active emitting thorium and thus permits higher operating temperatures and much greater electron emission.

carbonized plate—An electron-tube anode that has been blackened with carbon to increase its heat dissipation.

carbon microphone—A microphone that depends for its operation on the variation in resistance of carbon contacts.

carbon-pile regulator—An arrangement of carbon discs whose series resistance decreases as more pressure or compression is applied.

carbon resistor—Also called composition resistor. A resistor consisting of carbon particles that are mixed with a binder molded into a cylindrical shape, and then baked. Terminal leads are attached to opposite ends. The resistance of a carbon resistor decreases as the temperature increases.

carbon transfer recording—A type of facsimile recording in which carbon particles are deposited on the record sheet in response to the received signal.

Carborundum—A compound of carbon and silicon used in crystals to rectify or detect radio waves.

carcinotron—A voltage-tuned, backward-wave oscillator tube used to generate frequencies ranging from UHF up to 100 GHz or more.

card—Nonpreferred term for printed circuit board. *See* printed circuit board.

card bed—A mechanical device for holding punch cards to be transported past the punching and reading stations.

card code—An arbitrary code in which holes punched in a card are assigned numeric or alphabetic values.

card column—One of 20 to 90 single-digit columns in a tabulating card. When punched, a column contains only one digit, one letter, or one special code.

card-edge connector—Also called edgeboard connector. A connector that mates with printed wiring leads running to the edge of a printed circuit board.

Cardew voltmeter—The earliest type of hot-wire instrument. It consisted of a small-diameter platinum-silver wire sufficiently long to give a resistance high enough to be connected directly across the circuit being measured. The wire was looped over pulleys and it expanded as current flowed, causing the pointer to rotate.

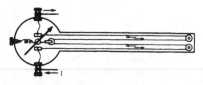

Cardew voltmeter.

card face — The printed side of a punched card if only one side is printed.

card feed — A mechanism that moves punch cards, one at a time, into a machine.

card field — On a punch card, the fixed columns in which the same type of information is routinely entered.

card hopper — *See* card stacker.

cardiac monitor — An instrument that usually has an oscilloscope display of the heart wave, and combines the features of several cardiac instruments, such as an electrocardiograph, cardiotachometer, etc. May also allow upper and lower limits to be set and trigger audible and/or visual alarms when these limits are exceeded.

cardiac pacemaker — 1. A device that controls the frequency of cardiac contractions. 2. A device that stimulates the heart and controls its rhythm by means of electrodes placed on the chest wall or implanted under the skin.

cardiac stimulator — Also called a pacemaker. An electronic device (sometimes implanted in the patient) that supplies electric pulses to stimulate regular heart action.

card image — 1. A representation in storage of the holes punched in a card, in such a manner that the holes are represented by one binary digit and the unpunched spaces are represented by the other binary digit. 2. In machine language, a duplication of the data contained in a punch card.

cardiograph — An instrument (or recording of instruments) for measuring the form or force of heart motion.

cardioid — 1. A heart-shaped polar response pattern, with strong rejection to signals arriving from the rear. 2. The quasi-heart-shaped sensitivity pattern of most unidirectional microphones. Hypercardioid and supercardioid microphones have basically similar patterns, but with longer, narrower areas of sensitivity at the front and slightly increased rear sensitivity. 3. A heart-shaped pattern that is typical of some directional antennas or microphones.

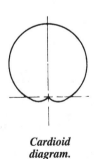

Cardioid diagram.

cardioid microphone — A microphone that has a heart-shaped response pattern that gives nearly uniform response for a range of about 180° in one direction and a minimum response in the opposite direction.

cardiostimulator — A device used to stimulate the heart and/or regulate its beat. *See also* pacemaker; defibrillator.

cardiotachometer — A measuring instrument that provides a meter reading proportional to the rate at which the heart beats.

card jam — A pile-up of cards in a machine.

card machine — A machine used to transfer information from or to punched cards.

card programmed — The capability of being programmed by punched cards.

card punch — 1. A device to record information in cards by punching holes in the cards to represent letters, digits, and special characters. 2. Device used in data handling systems to enter data on cards according to a desired code.

card reader — 1. A device designed to read information from punched cards and convert each hole into an electrical impulse for use in a computer system. 2. Device used in data handling systems to sense data on punched cards via a mechanical or photoelectric technique. 3. Equipment that takes a stack of punched cards and reads the information on them into the computer's memory or onto magnetic tape or disk for future reference. 4. A system that generally uses a photodetector to read punched cards for information by sensing the light transmitted through the punched holes.

card row — On a punched card, one of the horizontal lines of punching positions.

card sensing — The process of sensing or reading the information in punched cards and converting this information, usually into electrical pulses.

card stacker — A mechanism that stacks cards in a pocket or bin after they have passed through a computer. Also called card hopper.

card-to-tape — Having to do with equipment that transfers information directly from punched cards to punched tape or magnetic tape.

carillon — A bell tower designed to play from a keyboard. In an organ, this may be achieved by tube synthesis of bell-like tones struck with felt hammers, or completely electronically.

Carnot theorem — A thermodynamic principle that states that a cycle continuously operating between a low temperature and a high temperature can be no more efficient than a reversible cycle operating between the same temperatures.

carriage tape — *See* control tape.

carrier — 1. A wave of constant amplitude, frequency, and phase that can be modulated by changing amplitude, frequency, or pulse. 2. An entity capable of carrying an electric charge through a solid (e.g., holes and conduction electrons in semiconductors). 3. A wave that has at least one characteristic that can be varied from a known reference value by a modulation process. 4. That part of the modulated wave that corresponds to the unmodulated wave in a specified way. 5. A frequency on which a second, information-carrying signal is impressed. 6. Holder for electronic parts and devices that facilitates handling during processing, production, imprinting, or testing operations and protects such parts under transport. 7. An analog signal at a fixed amplitude and frequency that combines with an information-bearing signal in the modulation process to produce an output signal suitable for transmission. 8. A continuous signal modulated with a second, information-carrying signal.

carrier-amplitude regulation — The change in amplitude of the carrier wave in an amplitude-modulated transmitter when symmetrical modulation is applied.

carrier band — In CD-4. discs, the left- or right-channel's band that contains musical information recorded at very high frequencies (in the 20–45 kilohertz range). In playback, the demodulator recovers those frequencies, which have been frequency modulated.

carrier beat — In facsimile transmission, an undesirable heterodyne of signals, each synchronous with a different stable oscillator, causing a pattern in received copy. When one or more of the oscillators is fork controlled, this is called fork beat.

carrier chrominance signal — In color television, sidebands of a modulated chrominance subcarrier, plus any unsuppressed subcarrier, added to the monochrome signal to convey color information.

carrier color signal — In color television, sidebands of a modulated chrominance subcarrier, plus the chrominance subcarrier, if not suppressed, that are added to the monochrome signal to convey color information.

carrier concentration — The number of carriers in a cubic centimeter of semiconductor material.

carrier control — A control by the presence or absence of an rf carrier.

carrier current — 1. The current associated with a carrier wave. 2. High-frequency alternating current superimposed on ordinary telephone, telegraph, or power-line frequencies. The carrier may be tone modulated to operate switching relays or transmit data.

carrier-current communication — The superimposing of a high-frequency alternating current on ordinary telephone, telegraph, and power-line frequencies for telephone communication and control.

carrier-current control — Remote control in which the receiver and transmitter are coupled together through power lines.

carrier-current transmitter — A device that transmits signals via the standard ac power lines.

carrier frequency — 1. The frequency (hertz) of the wave modulated by the intelligence wave; usually a radio frequency (rf). 2. The reciprocal of the period of a periodic carrier. 3. The frequency of the unmodulated fundamental output from a radio transmitter.

carrier-frequency interconnection — In the formation of carrier networks, the transfer of groups of channels between terminals of wire-line cable or radio carrier systems at carrier frequencies.

carrier-frequency peak-pulse power — The power averaged over that carrier-frequency cycle that occurs at the maximum pulse of power (usually half the maximum instantaneous power).

carrier-frequency pulse — A carrier that is amplitude modulated by a pulse. The amplitude of the modulated carrier is zero before and after the pulse.

carrier-frequency range — The continuous range of frequencies within which a transmitter may normally operate. A transmitter may have more than one carrier-frequency range.

carrier-frequency stereo disc — A stereo disc with two laterally cut channels. One channel is cut in the usual manner. The second channel is employed to frequency modulate a supersonic carrier frequency. The playback cartridge delivers the signal for one channel plus the carrier frequency containing the other channel. The latter must then be demodulated to obtain the second channel.

carrier injection — The process whereby light is emitted at the junction of n- and p-type semiconductors when an external electric source is applied to drive the electrons and the holes into the junction.

carrier-isolating choke coil — An inductor inserted in series with a line on which carrier energy is applied to impede the flow of carrier energy beyond that point.

carrier leak — The carrier-frequency signal remaining after suppression in a suppressed carrier system.

carrier level — The strength, expressed in decibels, of an unmodulated carrier signal at a particular point in a system.

carrier lifetime — The time required for excess carriers doped into a semiconductor to recombine with other carriers of the opposite sign.

carrier line — A transmission line used for multiple-channel carrier communications.

carrier loading — The insertion of additional lump inductance in a cable section of a transmission line utilized for carrier transmission up to about 35 kHz. Loading serves to minimize impedance mismatch between cable and open wire and to reduce the cable attenuation.

carrier mobility — The average drift velocity of carriers per unit electric field in a homogeneous semiconductor. The mobility of electrons is usually different from that of holes.

carrier noise — Undesired variation of a radio-frequency carrier signal in the absence of intended modulation. Also called residual modulation.

carrier noise level — Also called residual modulation. The noise level produced by undesired variations of a radio-frequency signal in the absence of any intended modulation.

carrier on microwave — A means of transmitting many voice messages on one microwave radio channel. Transmission is point to point by microwave antennas mounted on towers or tall buildings.

carrier on wire — A means widely used by the telephone companies to transmit many voice messages on a single pair of wires. Circuits involving one or more carrier links never evidence dc continuity.

carrier-operated anti-noise device — A device whose purpose is to mute the audio output of a receiver during standby or intervals of no carrier.

carrier power — The rf power output of an AM transmitter when not modulated.

carrier power output rating — The power available at the output terminals of a transmitter when the output terminals are connected to the normal load circuit or to a circuit equivalent thereto.

carrier repeater — An assembly, including an amplifier (or amplifiers), filters, equalizers, level controls, etc., used to raise the carrier signal level to a value suitable for traversing a succeeding line section while maintaining an adequate signal-to-noise ratio.

carriers — Entities that carry an electrical charge and are also able to move relatively freely through a crystal lattice. The two most commonly encountered carriers are conduction band electrons, which are negatively charged, and valence band holes, which are positively charged.

carrier shift — 1. The transmission of radio teletypewriter messages by shifting the carrier frequency in one direction for a marking signal and in the opposite direction for a spacing signal. 2. The condition resulting from imperfect modulation, whereby the positive and negative excursion of the envelope pattern are unequal, thus effecting a change in the power associated with the carrier.

carrier signaling — In a telephone system, the method by which ringing, busy signals, or dial-signaling relays are operated by the transmission of a carrier-frequency tone.

carrier storage time (of a switching transistor) — The time interval between the beginning of the fall of the pulse applied to the input terminals and the beginning of the fall of the pulse generated by charge carriers at the output terminals. (The time is generally measured between the 90-percent values of the two pulse amplitudes.)

carrier suppression — The method of operation in which the carrier wave is not transmitted.

carrier swing — The total deviation of a frequency- or phase-modulated wave from the lowest to the highest instantaneous frequency.

carrier system — A means of obtaining a number of channels over a single wideband communication path by modulating each channel on a different carrier frequency at the originating end and demodulating at the receiving point to restore the signals to their original form.

carrier tap choke coil — A carrier-isolating choke coil inserted in series with a line tap.

carrier tap transmission choke coil — An inductor inserted in series with a line tap to control the amount of carrier energy flowing into the tap.

carrier telegraphy — The form of telegraphy in which the transmitted signal is formed by modulating the alternating current, under control of the transmitting apparatus, before supplying it to the line.

carrier telephony — Ordinarily applied only to wire telephony. That form of telephony in which carrier transmission is used, the modulating wave being at an audio frequency.

carrier terminal — Apparatus at one end of a carrier transmission system, whereby the processes of modulation, demodulation, filtering, amplification, and associated functions are effected.

carrier-to-noise ratio (C/N) — 1. Ratio of the magnitude of the carrier to the magnitude of the noise after selection and before any nonlinear process such as amplitude limiting and detection. This ratio is expressed in many different ways — for example, in terms of peak values in the case of impulse noise, and in terms of root-mean-square values in the case of random noise. 2. The ratio of the received carrier power to the noise power in a given bandwidth. The C/N is an indicator of how well an earth receiving station will perform in a particular location and is calculated from satellite power levels, antenna gain, and the combined antenna and LNA noise temperature.

carrier-transfer filter — A group of filters arranged to form a carrier-frequency crossover or bridge between two transmission circuits.

carrier transmission — 1. That form of electrical transmission in which a single-frequency wave is modulated by another wave containing the information. 2. A system for transmitting many voice channels over a common telephone circuit.

carrier-type dc amplifier — An amplifier system that converts a dc input to modulated ac, amplifies, and synchronously detects to provide amplifier dc output.

carrier wave — 1. The single-frequency transmitted wave that is modulated by another wave containing the information. 2. The basic frequency or pulse repetition rate of a signal, bearing no intrinsic intelligence until it is modulated by another signal that does bear intelligence. A carrier may be amplitude, phase, or frequency modulated.

carry — 1. A signal or expression produced in an electronic computer by an arithmetic operation on a one-digit place of two or more numbers expressed in positional notation and transferred to the next higher place for processing there. 2. A signal or expression — as defined in (1) — that arises when the sum of two digits in the same digit place equals or exceeds the base of the number system in use. If a carry into a digit place will result in a carry out of the same digit place and the normal adding circuit is bypassed when this new carry is generated, the result is called a high-speed or stand-on-nines carry. If the normal adding circuit is used, the result is called a cascade carry. If a carry resulting from the addition of carries is not allowed to propagate, the process is called a partial carry; if it is allowed to propagate, it is called a complete carry. A carry generated in the most significant digit place and sent directly to the least significant place is called an end-around carry. 3. In direct subtraction, a signal or expression — as defined in (1) — that arises when the difference between the digits is less than zero. Such a carry is frequently called a borrow. 4. The action of forwarding a carry. 5. The command directing a carry to be forwarded.

carry look-ahead — A circuit that predicts the final carry from propagate and generate signals supplied by partial adders. Used to speed up binary addition

significantly by eliminating the carry propagation (or ripple) delay.

carry time — The time required for a computer to transfer a carry digit to the next higher column and add it there.

cartridge — 1. The electromechanical transducer of a phonograph pickup head that converts stylus vibrations to an electrical signal. It is generally detachable and fits into the head shell of a pickup. Most cartridges are either magnetic or piezoelectric types. 2. Generally, any enclosed package containing a length of magnetic tape and its basic winding receptacles, designed to eliminate the need for handling or threading the tape. Specifically, the word *cartridge* is used to describe that variety of package that contains a continuous (endless) loop of tape on a single reel. 3. A film or tape magazine containing only one spool. 4. A plastic container that holds recording tape for easy loading into a matching recorder or player, especially the eight-track cartridge.

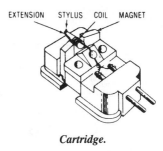

Cartridge.

cartridge fuse — 1. A tubular fuse whose end caps are enclosed in a glass or composition insulating tube to confine the arc or vapor when the fuse blows. 2. A short tube of fiber containing a fusible link or wire that is connected to metallic ferrules at the ends of the tube. Serves to interrupt excessive currents by melting of the fusible link.

cascadable counter — A logic counting block that has available the necessary connections to permit more than one counter to be operated in series, thus increasing the modulus of the counter subsystem.

cascade — Also called tandem. An arrangement of two or more similar circuits or amplifying stages in which the output of one circuit provides the input of the next.

cascade amplification — In a series of amplifiers, amplification by each of the preceding output.

cascade amplifier — 1. A multiple-stage amplifier in which the output of each stage is connected to the input of the next stage. 2. A multistage amplifier whose stages are forward coupled in succession.

cascade-amplifier klystron — A klystron that has three resonant cavities to provide increased power amplification and output. The extra resonator is located between the input and output resonators and is excited by the bunched beam energizing from the first resonator gap, thereby producing further bunching of the beam.

cascade connection — Two or more similar component devices arranged in tandem, with the output of one connected to the input of the next.

cascade control — Also called piggyback control. An automatic control system in which the control units, linked in chain fashion, feed into one another in succession. Each unit thus regulates the operation of the next in line.

cascaded carry — In a computer a system of executing the carry process in which carry information cannot

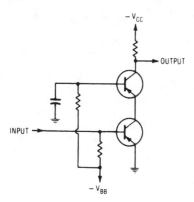

Cascade amplifier.

be passed on to place $(N + 1)$ unless the Nth place has received carry information or produced a carry.

cascaded feedback canceler — Also called velocity-shaped canceler. A sophisticated moving-target indicator canceler that provides clutter and chaff rejection.

cascaded systems — Multistorage operations; the input to each stage is the output of a preceding stage, thereby causing interdependence among the stages.

cascaded thermoelectric device — A thermoelectric device having two or more stages that are arranged thermally in series.

cascade image tube — An image tube that functions in low-light-level conditions by virtue of its series of stacked sections wherein the output of one section becomes the input for the next.

cascading — The connecting of two or more circuits in series so that the output from one provides the input to the next.

cascode amplifier — An amplifier using a neutralized grounded-cathode input stage followed by a grounded-grid output stage. The circuit has high gain, high input impedance, and low noise.

CASE — Abbreviation for computer-aided software engineering. A working environment consisting of programs and development tools that help automate the design and implementation of programs and procedures for business, engineering, and scientific computer systems.

case — In a computer, a set of data for use in a particular program.

case pressure — The total differential pressure in the internal cavity of a transducer and the ambient pressure. The term is commonly used to summarize the limiting combined differential and/or line-pressure capabilities of differential transducers.

case temperature — The temperature on the surface of the case at a designated point.

Cassegrain antenna — 1. An antenna whose feed is positioned near the vertex of the reflector, with a small subreflector placed near the focal point. The feed illuminates the subreflector, and the subreflector redirects the waves toward the main reflector, which then forms the radiated beam (called the secondary beam or pattern). The shapes of the subreflector and main reflector are so chosen that the secondary rays will emerge parallel to the main-reflector axis. 2. An antenna that utilizes a subreflector at the focal point that reflects energy to or from a feed located at the apex (center) of the main reflector.

Cassegrain feed — 1. A method of feeding a reflector antenna in which a waveguide located in the center of the main reflector feeds energy to a small reflector, which reflects it in turn to the main reflector. 2. An antenna feed design that includes a primary reflector, the dish, and a secondary reflector that redirects microwaves via a waveguide to a low-noise amplifier.

cassette — 1. A thin, flat, rectangular enclosure that contains a length of narrow magnetic recording tape permanently affixed to two flangeless floating reels that wind and unwind the tape while it passes an external recording and/or playback head. 2. A flat enclosure that contains two flangeless reels that link a narrow magnetic tape. 3. A sealed instant-load cartridge containing a length of tape and separate supply and takeup reels or hubs. Cassettes, unlike continuous-loop cartridges, can be rewound as well as fast-forwarded. 4. A film or tape magazine containing two spools. 5. Most commonly applied to the compact cassette developed by Philips, but also to a variety of micro and mini cassette systems that are not mutually compatible. 6. Recently applied to mass storage requirements of microcomputers and minicomputers. A digital cassette is certified for digital recording, which differs from the audio recording requirements. 7. A miniature reel-to-reel tape system retaining the flexibility and freedom of back-and-forth tape movement provided by a reel system; it eliminates the inconvenience of tape threading. 8. An open metal or plastic carrier used on IC production lines for moving groups of wafers.

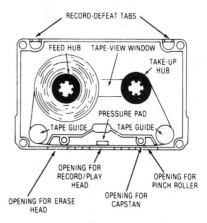

Cassette.

cassette recorder — A magnetic tape recording and playback device for entering or storing programs.

casting-out-nines check — A partial verification of an arithmetical operation on two or more numbers. It involves casting out nines from the number and from the results.

CAT — Abbreviation for computer-aided tomography, computer-assisted tomography, and computerized axial tomography. Literally, section graphics — the graphic display of a cross section of a piece of material or of the human body. Unlike ordinary X-ray photographs, which are produced directly by X-ray radiation on a photosensitive film (resulting in shadows on the film) computer-aided tomography displays are created by a computer after it processes the sensed X-ray signals. The big advantage of CAT is that a given display is not influenced by material in front of or behind the plane of interest, as is the case with ordinary X-ray photos. In fact, a three-dimensional picture can be obtained from multiple CAT displays, each displaced a small, finite distance from the adjacent one.

catadioptric—Optical process using both reflection and refraction of light.

catalog—An ordered compilation of descriptions of items, including sufficient information to afford access to the items. *See also* directory.

catalog search—A computerized data retrieval technique involving the search of keywords stored as key nouns or characteristics of the product. A display shows all items from the database with the specified keywords.

catalytic converter—A device that enhances certain chemical reactions that help to reduce the levels of undesirable exhaust gases.

cataphoresis—The migration toward the cathode of particles suspended in a liquid; movement of the particles is caused by the influence of an electrostatic field.

catastrophic failure—1. A sudden failure without warning, as opposed to degradation failure. 2. A failure the occurrence of which can prevent the satisfactory performance of an entire assembly or system.

catcher—In a velocity-modulated vacuum tube, an electrode on which the spaced electron groups induce a signal. The output of the tube is taken from this element.

catching diode—A diode connected to act as a short circuit when its anode becomes more positive than its cathode; the diode then tends to prevent the voltage of a circuit terminal from rising above the voltage at the cathode.

categorization—The process by which multiple addressed messages are separated to form individual messages for single addresses.

catena—A chain or connected series.

catenate—*See* concatenate.

cathamplifier—A push-pull vacuum-tube amplifier in which the push-pull transformer is in the cathode circuit.

cathode—1. In an electron tube, the electrode through which a primary source of electrons enters the interelectrode space. 2. General name for any negative electrode. 3. The lower-potential electrode of a corrosion cell, in which the action of the corrosion current may reduce or eliminate corrosion, or the negatively charged metallic parts of an impressed current system. 4. When a semiconductor diode is biased in the forward direction, that terminal of the diode that is negative with respect to the other terminal. 5. The negative electrode of a polar capacitor. 6. The negative pole of a plating apparatus at which positively charged ions leave the plating solution. A metal is deposited on the cathode. *See* anode. 7. In electrolytic plating, the workpiece being plated.

cathode activity—Measure of the efficiency of an emitter. The mathematical relationship between two values of emission current measured under two conditions of cathode temperature.

cathode bias—A method of biasing a vacuum tube by placing the biasing resistor in the common cathode-return circuit, thereby making the cathode more positive—rather than the grid more negative—with respect to ground.

cathode-coupled amplifier—A cascade amplifier in which the coupling between two stages is accomplished by a common cathode resistor.

cathode coupling—The use of an input or output element in the cathode circuit for coupling energy to another stage.

cathode current—*See* electrode current.

cathode-current density—The current per square centimeter of cathode area, expressed as amperes or milliamperes per centimeter squared.

cathode dark space—Also called Crookes' dark space. The relatively nonluminous region between the cathode and negative glow in a glow-discharge cold-cathode tube.

cathode emission—The process whereby electrons are emitted from the cathode structure.

cathode follower—Also called grounded-plate amplifier. A vacuum-tube circuit in which the input signal is applied to the control grid, and the output is taken from the cathode. Electrically, such a circuit possesses a high input impedance and a low output impedance characteristic and a gain of less than unity. The equivalent circuit using a transistor is called an emitter follower.

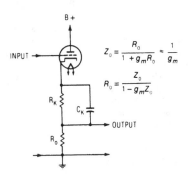

Cathode follower.

cathode glow—The apparent luminosity or glow that immediately envelops the cathode in a gas-discharge tube when operating at low pressures. The glow increases as the pressure decreases.

cathode guide—The element of a glow tube used in switching the neon glow from one indicated number to the next.

cathode heating time—The time required for the cathode to attain a specified condition, for example, a specified value of emission or a specified rate of change in emission.

cathode interface—A resistive and capacitive layer formed between the nickel sleeve and oxide coating of an indirectly heated cathode. Raising the cathode temperature will largely nullify the layer.

cathode keying—A method of keying a radiotelegraph transmitter by opening the plate return lead to the cathode or filament center tap.

cathode luminous sensitivity (of a multiplier phototube)—The photocathode current divided by the incident luminous flux.

cathode modulation—A form of amplitude modulation in which the modulating voltage is applied to the cathode circuit.

cathode pulse modulation—Modulation produced in an amplifier or oscillator by applying externally generated pulses to the cathode circuit.

cathode radiant sensitivity—The current leaving the photocathode divided by the incident radiant power of a given wavelength.

cathode ray—A stream of electrons emitted, under the influence of an electric field, from the cathode of an evacuated tube or from the ionized region nearby.

cathode-ray charge storage tube—A charge storage tube in which the desired information is written by means of a cathode-ray beam.

cathode-ray instrument—*See* electron-beam instrument.

cathode-ray oscillograph — An apparatus capable of producing, from a cathode-ray tube, a permanent record of the value of an electrical quantity as a function of time.

cathode-ray oscilloscope — Abbreviated CRO. 1. A test instrument that, when properly adjusted, makes possible the visual inspection of alternating-current signals. It consists of an amplifier, time-base generating circuits, and a cathode-ray tube for transformation of electrical energy into light energy. 2. A system wherein a supplied signal causes deflection of a CRT electron beam, thus forming a visible trace on the phosphor CRT screen that allows examination of electrical quantities.

cathode-ray output — A term used in data processing to describe a cathode-ray tube that displays graphic or character data.

cathode-ray storage — An electrostatic data storage device in which a cathode-ray beam provides access to the data.

cathode-ray tube — Abbreviated CRT. 1. A vacuum tube in which its electron beam can be focused to a small cross section on a luminescent screen and can be varied in position and intensity to produce a visible pattern. 2. A vacuum tube with an electron gun at one end and a fluorescent screen at the other. Electrons emitted from a heated cathode are accelerated by a series of annular anodes at progressively higher positive voltages. The electron beam is then deflected by two pairs of electrostatically charged plates between the gun and the screen. Electromagnets are often used in place of the deflector plates.

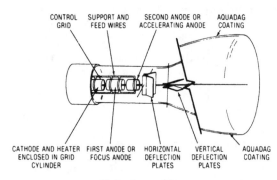

Cathode-ray tube.

cathode-ray-tube display — Abbreviated CRT display. 1. A device in which controlled electron beams are used to present data in visual form. 2. The data presentation produced by such a device. 3. In a calculator, a type of display resembling a small television tube. Usually, two to four rows of digits can be displayed simultaneously. 4. A high-speed device, similar to a television picture tube, that provides a visual nonpermanent display of system input/output data, such as instructions as they are being developed and data in storage. 5. A device, used to present data (alphanumeric, graphical, or a combination of the two), that incorporates a cathode-ray tube as the presenting element.

cathode-ray tuning indicator — Commonly called magic eye. A small-diameter cathode-ray tube that visually indicates whether an apparatus such as a radio receiver is tuned precisely to a station.

cathode resistor — A resistance connected in the cathode circuit of a tube so that the voltage drop across it will supply the proper cathode-biasing voltage.

cathode spot — On the cathode of an arc, the area from which electrons are emitted at a current density of thousands of amperes per square centimeter and where the temperature of the electrode is too low to account for such currents by thermionic emission.

cathode sputtering — The method of disintegrating the substance of the cathode by bombarding it with ions and then depositing it on another electrode or electron-tube envelope. *See also* sputtering, 1.

cathodic protection — Corrosion inhibition employed in protecting underground (or underwater) metal from electrochemical corrosion; it involves making the metal cathodic (negative) to a buried (or immersed) anode. Two methods are available for the required polarization: a dc power supply applying a positive polarity to a stainless steel or platinum plated titanium anode, or a dc self-generating sacrificial anode of zinc, magnesium, or aluminum alloy.

cathodofluorescence — Fluorescence that results from a material's exposure to cathode rays.

cathodoluminescence — Luminescence produced by the bombardment with high-energy electrons of a metal in a vacuum. Small amounts of the metal are vaporized in an excited state by the bombardment and emit radiation characteristic of the metal.

cation — 1. A positive ion that moves toward the cathode in a discharge tube, electrolytic cell, or similar equipment. The corresponding negative ion is called an anion. 2. An atom with a deficiency of electrons and therefore having a positive charge.

CATV — Abbreviation for community antenna television or cable television. A system of distributing TV signals to homes by cable.

catwhisker — A small, pointed wire used to make contact with a sensitive area on the surface of a crystal or semiconductor.

Cauer-elliptic filter — Also known as an elliptic or elliptical filter. A transfer function having an extremely high rate of attenuation (a very fast rolloff) near the corner frequency. Exhibits passband ripple and a renewal of gain beyond the stopband.

cavitation — The production of gas-filled cavities in a liquid when the pressure is reduced below a certain critical value with no change in the temperature. Ordinarily this is a destructive effect because the high pressures produced when these cavities collapse often damage mechanical components of hydraulic systems, but the effect is turned to advantage in ultrasonic cleaning.

cavitation noise — The noise produced in a liquid by the collapse of the bubbles created by cavitation.

cavity — A metallic enclosure that can be made to resonate at a desired microwave frequency. Primarily used to describe a cavity filter, which is a highly selective tuning element at microwave frequencies that may be used as the frequency-determining element of an oscillator or as a low-pass, bandpass, or highpass filter. Generally of fixed frequency, or may be mechanically tunable over a very limited frequency range.

cavity filter — A selective tuned device having the proper coupling means for insertion into a transmission line to produce attenuation of unwanted off-frequency signals.

cavity impedance — The impedance that appears across the gap of the cavity of a microwave tube.

cavity magnetron — A magnetron having a number of resonant cavities forming the anode; used as a microwave-transmitting oscillator.

cavity oscillator — Abbreviated CO. An oscillator in which the primary frequency-determining element is either a waveguide or coaxial cavity. Oscillator frequency can be mechanically tuned and voltage tuned (via a tuning varactor) over a relatively narrow band.

cavity radiation — The radiation (heat) emerging from a small hole leading to a constant-temperature enclosure. Such radiation is identical with blackbody radiation at the same temperature, no matter what the nature of the inner surface of the enclosure.

cavity resonator — 1. A space that is normally bounded by an electrically conducting surface and in which oscillating electromagnetic energy is stored; the resonant frequency is determined by the geometry of the enclosure. 2. A section of coaxial line or waveguide completely enclosed by conducting walls; it is often made variable for use as a wavemeter.

cavity resonator frequency meter — A cavity resonator used for determining the frequency of an electromagnetic wave.

cavity-tuned, absorption-type frequency meter — A device used for measuring frequency. Its operation depends on the use of an enclosure with a conductive inner wall; the resonant frequency of the wall is determined by its internal dimensions.

cavity-tuned, heterodyne-type frequency meter — A device for measuring frequency. Its operation depends on the use of an enclosure, the resonant frequency of which is determined by its internal dimensions.

cavity-tuned, transmission-type frequency meter — A device for measuring frequency. Its operation depends on the use of an enclosure with a conductive inner wall; the resonant frequency of the wall is determined by its internal dimensions.

C-band — 1. Microwave band in which the wavelengths are at or near 5.6 cm. It includes the top two sidebands of the S-band and the bottom three sidebands of the X-band. 2. The band of frequencies between 4 and 8 GHz, with the 6- and 4-GHz band being used for satellite communications. Specifically, the 3.7 to 4.2 GHz satellite communication band is used as the downlink frequencies in tandem with the 5.925 to 6.425 GHz band that serves as the uplink.

CBASIC — A version of BASIC that runs on the CP/M operating system. It is a structured language often preferred by programmers working in BASIC.

C battery — Also called grid battery. The energy source that supplies the voltage for biasing the grid of a vacuum tube.

CBC — Canadian Broadcasting Corporation. The government radio and television organization of Canada.

C-bias — *See* grid bias.

CCD — Abbreviation for charged-coupled device. 1. A semiconductor storage device in which an electrical charge is moved across the surface of a semiconductor by electrical control signals. Zeros or ones are represented by the absence or presence of a charge. A charge transfer system in which charges created by either an input diode or by an impinging photon are contained in MOS (metal-oxide semiconductor) or MIS (metal-insulated semiconductor) capacitors fabricated on a single crystal wafer. By varying electrode voltages successively, charge packets are moved from capacitor to capacitor to a single output amplifier. 2. Large buffer memory for minicomputer systems, interfacing between the magnetic storage disks or tape drives and the RAM, or as plug-compatible disk replacements. 3. A shift register formed by a string of closely spaced MOS capacitors. A CCD can store and transfer analog-charge signals — either electrons or holes — that may be introduced electrically or optically. The storing and transferring of charge occurs between potential wells at or near a silicon-silicon dioxide interface. The MOS capacitors, pulsed by a multiphase clock voltage, form these wells. For a three-phase, n-channel

CCD, the charges transferred between potential wells are electrons. 4. Functionally, a shift-register memory for either analog or digital information in which the data is represented as stored charges. The charges in a CCD are stored in a linear array of potential wells, with the potential of each well controlled by a voltage applied to an isolated metal-oxide semiconductor (MOS) capacitor above the well. By applying a traveling voltage wave to this linear array of capacitors, any charges within the wells are pushed along from well to well. For a digital memory, the potential wells are either uncharged or fully charged to represent 0 and 1. To implement an analog memory, the charge is varied linearly in proportion to the sampled input voltage. Thus, analog delay lines as well as digital memories are implementable with CCD technology. Also known as a bucket brigade device because of the way charge is transferred from one cell to another in a recirculating fashion. 5. For imaging devices, a self-scanning semiconductor array that utilizes MOS technology, surface storage, and information transfer by digital shift register techniques.

CCIF — Abbreviation for International Telephone Consultative Committee.

CCIR — Abbreviation for International Radio Consultative Committee.

CCITT — 1. Abbreviation for Comité Consulatif International Télégraphique et Téléphohique. The original French name of the committee that published international communications standards. Replaced by ITU. 2. Abbreviation for Consultative Committee for International Telephone and Telegraph. An international standards group that is a part of the International Telecommunications Union (ITU).

CCS — Abbreviation for continuous commercial service. Refers to the power rating of transformers, tubes, resistors, etc. Used for rating components in broadcasting stations and some industrial applications.

CCTV — Abbreviation for closed-circuit television.

CCTV camera — That part of a closed-circuit TV system that captures the picture and produces the video (picture) signal.

CCTV monitor — That part of a closed-circuit TV system that receives the picture from the CCTV camera and displays it on the picture tube.

ccw — Abbreviation for counterclockwise.

CD — Abbreviation for compact disc.

CD-4 — 1. A record-playback system for discrete discs. Invented by the Victor Co. of Japan (JVC) and developed by JVC and RCA Records, the system needs a demodulator and special cartridge with a special stylus for discrete four-channel playback. The system is not compatible with matrix quad discs and is not used for FM broadcasting. Also called quadradisc. 2. A recording and playback system similar in some respect to FM multiplex stereo broadcast and reception. Each wall of the record groove carries a single channel of information — left front plus left rear on the inner wall and right front plus right rear on the outer wall of the groove. In addition, each groove wall carries a 30-kHz FM subcarrier that is modulated by the front-minus-back difference signals that are needed to decode or demodulate the quadraphonic signal into four discrete channels. 3. A discrete four-channel disc recording and playback system, using a frequency-modulated 30-kHz carrier to convey additional information that can be combined with the audio output of the cartridge to produce four essentially independent program channels. Requires a cartridge frequency response to at least 45 kHz and a special demodulator.

CD-4 capability — The ability of a cartridge to reproduce the ultrasonic signals necessary for discrete four-channel disc reproduction using a CD-4 demodulator.

cdi — Abbreviation for collector-diffusion isolation.

CD-I — Abbreviation for compact disc-interactive. A home entertainment system based on a player that connects to any TV and stereo system, with information stored digitally on compact disc. It was introduced by Philips Consumer Electronics in 1991.

C-display — A type of radar display in which the signal is a bright spot, with the bearing as the horizontal coordinate and the elevation angle as the vertical coordinate.

CDMA — Abbreviation for code-division multiple access. Digital cellular system multiple-access standard that allows for higher capacity and greater security. Stations use spread-spectrum modulations and orthogonal codes to avoid interfering with one another. *See also* code-division multiple access.

CDPD — Abbreviation for Cellular Digital Packet Data. A standards-based technology for wireless communication of data.

CD-R — Abbreviation for compact disc-recordable (recordable CD). Same as CD-WO. CD format is compatible with CD-ROM and can be written to once and read many times.

CD-ROM — Abbreviation for compact disc-read-only memory. 1. A compact disc with a format for storing different types of information digitally, which can be played on a CD-ROM drive connected to a personal computer. 2. A compact 5-1/4-inch optical disc, typically used to store text, images, audio, video, and programs that can run on suitably equipped computers. CD-ROM drives come in single, double, and quad speeds (150 kbs, 300 Kbs, and 600 kbs, respectively). 3. A read-only optical disc capable of storing large amounts (up to 250,000 pages) of data.

CD-ROM drive — A peripheral device attached to a computer that allows it to read/play a CD-ROM disc. All CD-ROM players can also play back audio CDs, but require external headphones or speakers to hear them.

CD-WO — Compact disc-write once. Recordable compact disc. Same as CD-R.

ceiling — The maximum voltage that may be attained by an exciter under specified conditions.

celestial guidance — A system of guidance in which star sightings that are automatically taken during the flight of a missile provide position information used by the guidance equipment.

cell — 1. A single unit that produces a direct voltage by converting chemical energy into electrical energy. 2. A single unit that produces a direct voltage by converting radiant energy into electrical energy; for example, a solar or photovoltaic cell. 3. A single unit that produces a varying voltage drop because its resistance varies with illumination. 4. Elementary unit of storage. 5. In corrosion processes, a source of electric potential that is responsible for corrosion. It consists of an anode and a cathode immersed in an electrolyte and electrically bonded together. The anode and cathode may be separate metals or dissimilar areas on the same metal. The different metals will develop a difference in potential that is accompanied by corrosion of the anode. When this cell involves an electrolyte, as it does in corrosion processes, it is referred to as an electrolytic cell. 6. The geographic area served by a single low-power transmitter/receiver. A cellular system's service area is divided into multiple cells.

cell counter — An electronic instrument used to count white or red blood cells or other very small particles.

cell-type enclosure — A prefabricated basic shielded enclosure of double-walled copper-mesh construction. The original screen-room design.

cell-type tube (tr, atr, and pre-tr tubes) — A gas-filled radio-frequency switching tube that operates in an external resonant circuit. A tuning mechanism may be incorporated into the external resonant circuit or the tube.

cellular — Type of mobile telephone service in which the geographic serving area is divided into subregions (cells), each with its own antenna and switching node.

cellular radio — The terrestrial mobile telephone technology that increases the number of available channels by dividing an area into cells; each cell may use the same frequencies as other cells, except that adjacent cells may not use the same frequencies.

cellular system — 1. A mobile telephone system that divides large service areas into small cells, each with its own low-power transmitter. A telephone call is switched by computers from one transmitter to the next without interrupting the signal as a vehicle moves from cell to cell. Calls can be divided and frequencies reused over shorter intervals. 2. Method of mobile telephone service that divides radio communication service areas into small cells, or districts. The cellular approach

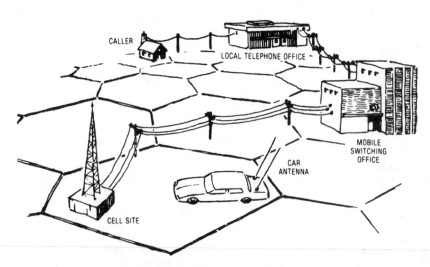

Cellular telephone system.

utilizes low-power transmitters that allow uninterrupted communications through sophisticated switching equipment linking the cells. This system, by reusing frequencies and standardizing service and equipment, has substantially improved mobile telephone service.

cellulose acetate — An inexpensive transparent plastic film used as the backing material for many recording tapes.

cellulose-nitrate disc — *See* lacquer disc.

Celsius temperature scale — Also called centrigrade temperature scale. A temperature scale based on the freezing point of water defined as 0°C and the boiling point defined as 100°C, both under conditions of normal atmospheric pressure (760 mm of mercury).

cent — A measure of frequency, defined as equal to 100th of a semitone.

center-fed antenna — An antenna in which the feeder wires are connected to the center of the radiator.

center feed — 1. Attaching feeder wires or a transmission line to the center of the radiator of an antenna. 2. Connection of signal input terminals to the center of a coil.

center frequency — Also called resting frequency. 1. The average frequency of the emitted wave when modulated by a symmetrical signal. 2. The frequency at the center of a spectrum display (for linear frequency scanning). It is usually tunable. 3. Also called free-running frequency. The frequency at which a phase-locked loop operates when not locked onto an incoming (input) signal.

centering control — One of two controls used to shift the position of the entire image on the screen of a cathode-ray tube. The horizontal-centering control moves the image to the right or left, and the vertical-centering control moves it up or down. *See also* framing control.

centering diode — A clamping circuit used in some types of plan-position indicators.

centering magnet — A magnet that centers the televised picture on the face of the tube. Also called framing magnet.

center of gravity — A point inside or outside a body and around which all parts of the body balance each other.

center of mass — On a line between two bodies, the point around which the two bodies would revolve freely as a system.

center poise — Scale of viscosity for insulating varnishes.

center ring — The part that supports the stator in an induction-motor housing. The motor end shields are attached to the ends of the center ring.

center tap — A connection at the electrical center of a winding, or midway between the electrical ends of a resistor or other portion at a circuit.

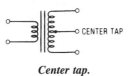

Center tap.

center-tapped inductor — An inductor that has a tap at half the total inductance.

center wire — A fine loop of wire used in proportional counters as an anode. A high voltage is applied to it to set the conditions for radiation measurement.

center-zero meter — A dc meter that has its zero point at the center of the scale, e.g., a dc galvanometer.

centi- — One hundredth (10^{-2}) of a specific quantity or dimension.

centigrade temperature scale — The older name for a Celsius temperature scale in the English-speaking countries. Officially abandoned by international agreement in 1948, but still in common usage.

centimeter waves — Microwave frequencies between 3 and 30 GHz, corresponding to wavelengths of 10 to 1 centimeters.

central battery exchange — Manual telephone exchange in which a battery situated at the exchange is the source of current for operating supervisory signals, for subscribers' calling signals, and for the current required to enable a subscriber to speak over his or her line.

central office — The facility at which a communications common carrier terminates customer lines and locates the equipment for interconnecting those lines.

central-office equipment — Apparatus used in a telephone central office to furnish communication services.

central-office line — *See* subscriber line.

central processing unit — Also called central processor; abbreviated CPU. 1. The part of a computer system that contains the main storage, arithmetic unit, and special register groups. Performs arithmetic operations, controls instruction processing, and provides timing signals and other housekeeping operations. 2. A group of registers and logic that form the arithmetic/logic unit plus another group of registers with associated decoding logic that form the control unit. Most MOS devices are single-chip CPUs, in that the registers hold as many bits as the word length of the unit. With bit-slice devices, however, central processing units of any bit width can be assembled essentially by connecting the bit-slice parts in parallel. Externally, a bit-slice device will appear to be a coherent single CPU capable of handling words of the desired bit length. 3. That part of a computer system that controls the interpretation and execution of instructions. In general, the CPU contains the following elements: arithmetic and logic unit (ALU), timing and control, accumulator, scratch-pad memory, program counter and address stack, instruction register, and I/O control logic. 4. That unit of a computing system that fetches, decodes, and executes programmed instructions and maintains the status of results as the program is executed.

central processor — *See* central processing unit; CPU.

central station — A control center to which alarm systems in a subscriber's premises are connected, where circuits are supervised and where personnel are maintained continuously to record and investigate alarm or trouble signals. Facilities are provided for the reporting of alarms to police and fire departments or to other outside agencies.

central station alarm system — An alarm system, or group of systems, the activities of which are transmitted to, recorded in, maintained by, and supervised from a central station. This differs from proprietary alarm systems in that the central station is owned and operated independently of the subscriber.

centrex — 1. A service offered by telephone companies. It uses central-office equipment to provide features comparable with those provided by a PBX. 2. An improved PBX system that also provides direct inward dialing (DID) and automatic number identification (ANI) of the calling PBX station.

centrifugal force — The force that acts on a rotating body and that tends to throw the body farther from the axis of its rotation.

centripetal force — The force that compels a rotating body to move inward toward the center of rotation.

Ceracircuits — A trademark of the Sprague Electric Company for hybrid thick-thin film integrated circuits that consist of discrete passive and semiconductor active elements attached to precision resistor substrates to form functional electronic modules.

ceramic — 1. A claylike material, consisting primarily of magnesium and aluminum oxides, that after molding and firing is used as an insulating material. It withstands high temperatures and is less fragile than glass. When glazed, it is called porcelain. 2. Pertaining to or made of clay or other silicates. 3. Piezoelectric part of a pickup, speaker, or microphone that acts as a transducer. It has characteristics that are similar to a crystal transducer, but it is more robust. 4. Nonmetallic and inorganic material, e.g., alumina, beryllia, or steatite, formed through heat processing, used in microelectronic substrates and component parts.

ceramic amplifier — An amplifier that makes use of the piezoelectric properties of ceramics such as barium titanate and the piezoresistive properties of semiconductors such as silicon. An ac signal applied through electrodes to a barium titanate bar produces deformation of the bar and the attached silicon strip, thereby producing a corresponding variation in resistance. This resistance change causes the load current to vary. The device is essentially a current amplifier with extremely high input impedance and low output impedance.

ceramic-based microcircuit — A microminiature circuit printed on a ceramic substrate. Usually consists of combinations of resistive, capacitive, or conductive elements fired on a waferlike piece of ceramic.

ceramic capacitor — A capacitor whose dielectric is a ceramic material such as steatite or barium titanate, the composition of which can be varied to give a wide range of temperature coefficients. The electrodes are usually silver coatings, fired on opposite sides of the ceramic disc or slab, or fired on the inside and outside of a ceramic tube. After connecting leads are soldered to the electrodes, the unit is usually given a protective insulating coating.

ceramic dielectric — 1. One of a great variety of ceramic materials used as a dielectric in capacitors: some typical materials are barium titanate, barium strontium titanate, and titanium dioxide. Different ceramic dielectrics provide the desired temperature coefficient of capacitance and medium-to-dielectric constants. 2. A ceramic such as isolantite, porcelain, or steatite, used as an insulator.

ceramic filter — Electrically coupled, two-terminal piezoelectric ceramic resonators in ladder and lattice configurations. Monolithic filters with ceramic substrates are also called ceramic filters.

ceramic microphone — 1. A microphone with a ceramic cartridge. 2. A transducer that uses a piezoelectric ceramic pickup element (barium titanate) to convert sound to electrical energy.

ceramic permanent magnet — A permanent, nonmetallic magnet made from pressed and sintered mixtures of metallic-oxide powders, usually oxides of barium and iron.

ceramic pickup — 1. A phonograph pickup with a ceramic cartridge. 2. A pickup whose generator system is based on piezoelectricity produced by the stressing of natural and man-polarized crystals.

ceramic transducer — See piezoelectric transducer.

CERDIP — 1. Abbreviation for *ceramic dual-inline package*. A package assembled with the leadframe sandwiched between two ceramic layers and sealed by firing a glass frit. 2. A ceramic dual-inline package for ICs.

Cerenkov counter — An instrument that detects high-energy charged particles by an analysis of the Cerenkov radiation emitted.

Cerenkov radiation — 1. Light emitted when charged particles pass through a transparent material at a velocity greater than that of light in that material. It can be seen, for example, as a blue glow in the water around the fuel elements of pool reactors. P. A. Cerenkov was the Russian scientist who first explained the origin of this light. 2. The radiation produced when a charged particle traverses a medium that has a refraction index considerably greater than unity.

Cerenkov rebatron radiator — A device in which a tightly bunched velocity-modulated electron beam is passed through a hole in a dielectric. The reaction between the higher velocity of the electrons passing through the hole and the slower velocity of the electromagnetic energy passing through the dielectric results in radiation at some frequency higher than the frequency of modulation of the electron beam.

cermet — 1. A metal-dielectric mixture used in making thick-film resistive elements. The first half of the term is derived from *cer*amic and the second half from *met*al. 2. A solid homogeneous material usually consisting of a finely divided admixture of a metal and ceramic in intimate contact. Cermet thick films are normally combinations of dielectric materials and metals.

cermet potentiometer — A potentiometer in which the resistive element is made by combining very fine particles of ceramic or glass with precious metals.

CERT — Abbreviation for character error-rate testing. Checking a data line with test characters.

certification — Verification that specified testing has been performed, and required parameter values have been attained.

certified magnetic tape — Magnetic tape that has been tested and is certified to be free from error over its entire recording surface.

cesium — A chemical element having a low work function. Used as a getter in vacuum tubes and in cesium-oxygen-silver photocell cathodes.

cesium-oxide cell — A photoemissive detector sensitive to wavelengths up to 1 micrometer. It has one sharp maximum of sensitivity at 350 nanometers and a broad maximum at 800 nanometers.

cesium-vapor lamp — A low-voltage arc lamp for producing infrared radiation.

cev — See corona extinction voltage.

CGA — Abbreviation for color graphics adapter. A color graphics system for IBM PCs and compatibles. Provides less resolution than EGA or VGA monitors.

C_{gk} — Symbol for grid-cathode capacitance in a vacuum tube.

CGM — Abbreviation for computer graphics metafile. A standard file format that stores object-oriented graphics in device-independent form, enabling them to work in different systems and programs.

C_{gp} — Symbol for grid-plate capacitance in a vacuum tube.

cgs — Abbreviation for centimeter-gram-second. These quantities of space, mass, and time are the basis of absolute units.

cgs electromagnetic system of units — A coherent system of units for expressing the magnitude of electrical and magnetic quantities. The most common fundamental units of these quantities are the centimeter, gram, and second. Their unit of current (abampere) is of such a magnitude that if maintained constant in two straight parallel conductors having an infinite length and negligible circular sections and placed 1 centimeter apart in a vacuum, a force equal to 2 dynes per centimeter of length will be produced.

cgs electrostatic system of units — A coherent system of units for expressing the magnitude of electrical and magnetic quantities. The most common fundamental units of these quantities are the centimeter, gram, and second. Their unit of electrical charge (stacoulomb) is of such a magnitude that two equal unit point charges 1 centimeter apart in a vacuum will repel each other with a force of 1 dyne.

chad — The piece of material removed when a hole or notch is formed in a storage medium such as punched tape or punched cards.

chadless — Pertaining to tape in which the data holes are deliberately not punched through and a flap of material remains attached to the tape.

chadless tape — A type of punched paper tape in which each chad is left fastened by about a quarter of the circumference of the hole. Chadless punched paper tape must be sensed by mechanical fingers, because chad interferes with reliable electrical or photoelectrical reading.

chad tape — Tape used in printing telegraph or teletypewriter operation, in which the perforations are severed from the tape to form holes that represent characters. Normally, the characters are not printed on chad tape.

chaff — A general name applied to radar-confusion reflectors that consist of thin, narrow, metallic strips of various length and frequency responses used to reflect radar echoes.

chain — A series of processing locations through which information must pass on a store-and-forward basis to reach a subsequent location.

chain calculations — In a calculator, series of continued operations in a single mode. Example: $118 \times 94 \times 116 \times 395$.

chained list — A list in which the items may be dispersed but in which each item contains an identifier for locating the next item to be considered.

chaining — 1. In a computer, a system of storing records such that each record belongs to a list or group of records and has a linking field for tracing the chain. 2. If a computer program is too big to fit into the memory, it can be written in a series of segments. The computer works on one segment at a time and continues operating this way until the program is finished. 3. The ability of an executing program to call another program that resides on a disk.

chaining search — A search technique in which each item contains an identifier for locating the next item to be considered.

chain printer — In a computer, a high-speed printer having type slugs carried on the links of a revolving chain.

chain radar beacon — A radar beacon with a very fast recovery time, so that simultaneous interrogation and tracking of the beacon by a number of radars is possible.

chain radar system — A radar system comprising a number of radars or radar stations located at various sites along a missile flight path. These radar stations are linked together by data and communication lines for target acquisition, target positioning, and/or data-recording purposes. The target-acquisition link makes it possible for any radar to position any other radar on target.

challenger — *See* interrogator.

chance failure — *See* random failure.

changer — A device that plays several phonograph records in sequence automatically. It consists of a turntable, an arm, and a record stacking and dropping mechanism.

channel — 1. A portion of the spectrum assigned for the operation of a specific carrier and the minimum number of sidebands necessary to convey intelligence. 2. A single path for transmitting electric signals. (Note: The word *path* includes separation by frequency division or time division. *Channel* may signify either a one-way or two-way path, providing communication in either one direction only or in two directions.) *See also* alternate channel. 3. In electronic computers, that portion of a storage medium which is accessible to a given reading station. 4. The path along which information, particularly a series of digits or characters, may flow. 5. In computer circulating storage, one recirculating path containing a fixed number of words stored serially by word. 6. An area, under the silicon dioxide of a planar surface, that has been changed from one type of conductivity to the opposite type. A channel is the conductive path between the source and drain in an IGFET. Generally, channels are undesirable in other instances. Thick oxides or heavily doped regions called channel stoppers are used to prevent channels. 7. A complete sound path. A single-channel, or monophonic, system has one channel. A stereophonic system has at least two full channels designated as left (A) and right (B). Monophonic material may be played through a stereo system; both channels will carry the same signal. Stereo material, if played on a monophonic system, will mix and emerge as monophonic sound. 8. The conducting charge layer between source and drain induced by the applied gate voltage. The charge layer is holes in a p-type device, and electrons in n-types. 9. An independent signal path. Stereo recorders have two such channels, quadraphonic ones have four. 10. A thin semiconductor layer between the source region and the drain region, whose conductance is controlled by the gate voltage. 11. The band of frequencies (including the assigned carrier frequency) within which a radio system must operate in order to prevent interference with stations on adjacent channels. 12. The part of a communications system that connects a message source to a message link; a path for electrical transmission between two or more points.

channel balance — Equal response on both left and right channels.

channel bank — 1. The part of a carrier multiplex terminal in which are performed the first step of modulation of the voice frequencies into a higher-frequency band and the final step in the demodulation of the received higher-frequency band into voice frequencies. 2. Communication equipment that multiplexes, typically used for multiplexing voice-grade channels.

channel capacity — 1. The maximum number of elementary digits that can be handled per unit time in a particular channel. 2. The maximum possible rate of information transmission through a channel at a specified error rate. It may be measured in bits per second or bauds. 3. The total number of individual channels in a system.

channel designator — A number assigned for reference purposes to a channel, tributary, or trunk. Also called channel sequence number.

channel diffusion stops — A narrow doped region beside each sensing channel in a CCD that prevents excess charges generated within a particular light-sensing site from spreading sideways.

channel effect — Current leakage over a surface path between the collector and emitter of some types of transistors.

channel frequency — The band of frequencies that must be handled by a carrier system to transmit a specific quantity of information.

channeling — The utilization of a modulation-frequency band for the simultaneous transmission from two or more communication channels in which the channel separation is accomplished by the use of carriers or subcarriers, each in a different discrete frequency band forming a subdivision of the main band. This term covers a special case of multiplex transmission.

channel interval — The time allocated to a channel, including on and off time.

channelization — The assignment of circuits to channels, and the arrangement of those channels into separate groups and supergroups.

channelizing — The process of subdividing wideband transmission facilities for the purpose of putting many different circuits requiring comparatively narrow bandwidths on a single wideband facility.

channel pulse — A telemetering pulse that, by its time or modulation characteristics, represents intelligence on a channel.

channel pulse synchronization — Synchronization of a local-channel rate oscillator by comparison and phase lock with separate channel-synchronizing pulses.

channel reliability — The percentage of time that the channel meets the arbitrary standards established by the user.

channel reversal — Shifting the outputs of a stereo system so the channel formerly heard from the left speaker now comes from the right and vice versa.

channel-reversing switch — A switch that reverses the connections of two speakers in a stereo system with respect to the channels, so that the channel heard previously from the right speaker is heard from the left and vice versa.

channel sampling rate — The number of times per second that individual channels are sampled. This is different from comutation rate, since it is possible for more than one channel to be applied to a given commutator input (with subcommutation).

channel selector — A switch or dial used for selecting a desired channel.

channel separation — 1. In stereo, the electrical or acoustical difference between the left and right channels. Inadequate separation can lessen the stereo effect; excessive separation can exaggerate it beyond natural proportions. 2. The degree to which the two signals in a stereo system are electrically isolated. Usually expressed as a ratio in decibels. 3. The amount of stereo program material from one channel appearing in the cartridge output for the other channel. Expressed in decibels relative to the desired channel output, with values of 20 to 30 dB (the higher figure being preferable) through most of the audible frequency range being typical of good cartridges. 4. The degree to which the signal in one amplifier is kept separate from an adjacent undriven amplifier. Channel separation for FM stereo decoders is typically 40 dB, whereas phono cartridge channel separation is typically between 20 and 30 dB.

channel sequence number — *See* channel designator.

channel shift — The interchange of communications channels; for example, the shift from a calling frequency to a working frequency.

channel shifter — A radiotelephone carrier circuit by means of which one or two voice-frequency channels are shifted from normal channels to higher voice-frequency channels as a means of reducing crosstalk between channels. At the receiving end, the channels are shifted back by a similar circuit.

channel stop — A barrier put in place during the manufacture of a semiconductor to prevent small leakage paths going to the outside of the chip.

channel strip — An amplifier or other device having a sufficiently wide bandpass to amplify one television channel.

channel subcarrier — The channel required to convey telemetric data involving a subcarrier band.

channel synchronizing pulse separator — A device for separating channel synchronizing pulses from commutated signals.

channel-to-channel connection — A device for rapid data transfer between two computers. A channel adapter is available that permits the connections between any two channels on any two systems. Data is transferred at the rate of the slower channel.

channel-utilization index — In a computer, the ratio between the information rate (per second) through a channel and the channel capacity (per second).

channel wave — Any elastic wave propagated in a sound channel because of a low-velocity layer in the solid earth, the sea, or the atmosphere.

character — 1. In electronic computers, one of a set of elementary symbols that may be collectively arranged in order to express information. These symbols may include the decimal digits 0 through 9, the letters A through Z, punctuation and typewriter symbols, and any other single symbol that a computer may read, store, or write. 2. One of a set of symbols used to present information on a display tube. 3. Part of a computer word that has a meaning in itself. For example, six bits recorded across a magnetic tape make up a character and signify a number or letter symbol. 4. A combination of holes punched in a line. 5. A letter, digit, or other symbol that is used as part of the organization, control, or representation of data. A character is often in the form of a spatial arrangement of adjacent or connected strokes. 6. A symbol, mark, or event that a data-processing machine can read, write, or store. It is used to represent data to a machine. 7. An alphabetic, numeric, or special graphic symbol. Each character is represented in its set by a unique binary code. 8. A letter, number, or sign made up of a specific number of bits. Nonprinting characters may be used for control functions. 9. A language unit consisting of bits. 10. A letter, digit, or other symbol that is the representation of data. A connected sequence of characters is called a character string.

character boundary — In character recognition, the largest rectangle having a side parallel to the document reference edge, each of the sides of which is tangent to a given character outline.

character check — Verification of the observance of rules for character formation.

character code — A special way of using a group of bits to represent a character. Different codes may be used in different equipment according to the internal design.

character crowding — Also called packing. The effect of reducing the time interval between subsequent characters read from tape. It is caused by a combination of mechanical skew, gap scatter, jitter, amplitude variation, etc.

character density — A measure of the number of recorded characters per unit of length or area.

character display tube — A form of cathode-ray tube in which the cathode-ray beam can be shaped, either by electrostatic or electromagnetic deflection or by passing the beam through a mask, into symbols or letters.

character emitter — In a computer, an electromechanical device that puts out coded pulses.

character generator — 1. A unit that accepts input in the form of one of the alphanumeric codes and prepares the electrical signals necessary for its display in the proper position on a dot matrix, TV system, or CRT. 2. That part of the display controller that draws alphanumeric characters and special symbols for the screen. A character is automatically drawn and spaced every time a character code is interpreted. 3. A hardware or software device that provides the means for formulating a character font and that also may provide some controlling function during printing. 4. A circuit that generates the letters or numbers on a display or printer. 5. A device that superimposes text over a video image, such as program credits at the end of a program.

character-generator cathode-ray tube — A cathode-ray tube that generates symbols for use in other displays. Basically, the tube operates by scanning specific characters on the target and generating them as video signals to other cathode-ray systems.

character-graphics color system — A system that can address color only to a block of pixels (character cell); all pixels within this block have the same color.

character interleave — Also byte interleave. A technique in time-division multiplexing in which bytes of data are transmitted in one frame.

characteristic — 1. An inherent and measurable property of a device. Such a property may be electrical, mechanical, thermal, hydraulic, electromagnetic, or nuclear; it can be expressed as a value for stated or recognized conditions. A characteristic may also be a set of related values (usually in graphical form). 2. The integral part of a logarithm to the base 10; also, the power of 10 by which the significant digits of a floating-point number are multiplied.

characteristic curve — 1. A graph plotted to show the relationship between changing values. 2. A plot of how any value changes with respect to another in a component or circuit: for example, a plot of how current changes as voltage is changed on a component or circuit.

characteristic distortion — 1. Displacement of signal transitions due to the persistence of transients caused by preceding transitions. 2. Repetitive displacement or disruption peculiar to specific parts of a teletypewriter signal. The two types of characteristic distortion are line and equipment.

characteristic frequency — The frequency that can be easily identified and measured in a given emission.

characteristic impedance — Also called surge impedance. 1. The driving-point impedance of a line if it were of infinite length. 2. In a delay line, the value of terminating resistance that provides minimum reflection to the network input and output. 3. The ratio of voltage to current at every point along a transmission line on which there are no standing waves. 4. The square root of the product of the open-and short-circuit impedance of the line. 5. The ratio of the applied voltage to the steady-state current that flows when a transmission line is terminated in a pure resistance that is equal to the characteristic impedance value of the cable. 6. That value of pure resistance which, when connected to the output terminals of a transmission line, makes the cable appear infinitely long; i.e., no signal is reflected back up the cable toward the source. 7. The ratio of voltage to current in a propagating wave, i.e., the impedance that is offered to this wave at any point of the line. In printed wiring, its value depends on the width of the conductor, the distance from the conductor to the ground plane(s) and the dielectric constant of the media between them. 8. A property of antenna transmission lines that is determined primarily by the diameter of the conductors and the spacing between them.

characteristic impedance of free space — The relationship between the electric and magnetic intensities of space due to the expansion of the impedance concept to electromagnetic fields.

characteristic spread — The range between the minimum and maximum values for a given characteristic that is considered normal in any large group of tubes or other devices.

characteristic telegraph distortion — Distortion that does not affect all signal pulses alike. Rather, the effect on each transition depends on the signal previously sent, because remnants of previous transitions or transients persist for one or more pulse lengths.

characteristic wave impedance — The ratio of the transverse electric vector to the transverse magnetic vector at the point it is crossed by an electromagnetic wave.

character parity — Adding an overhead bit to a character code to provide error-checking capability.

character reader — A computer input device that can directly recognize printed or written characters; they need not first be converted into punched holes in cards or paper or into polarized magnetic spots.

character read-out systems — Photoelectrically controlled, alphanumeric reading devices that convert characters to audible or sorting signals that can be fed to a computer, electric typewriter, tape punch, or other machine.

character recognition — The automatic identification of graphic, phonic, or other characters. *See also* magnetic-ink character recognition; optical character recognition.

character sensing — To detect the presence of characters optically, magnetically, electrostatically, etc.

character set — 1. An ordered group of unique representations called characters, such as the 26 letters of the English alphabet, 0 and 1 of the Boolean alphabet, the signals in the Morse code alphabet, the 128 characters of the U.S. ASCII alphabet, etc. 2. All the letters, numbers, and symbols used by a device or language.

character string — Two or more alphanumeric characters or special symbols (math, Greek, etc.) aligned in a textual format on the screen.

character subset — A selection from a character set of all characters having a specified common feature; for example, in the definition of a character set, the digits 0 through 9 are a character subset.

Charactron — Trade name of General Dynamics/Electronics for a specially constructed cathode-ray tube used to display alphanumeric characters and other special symbols directly on its screen.

charge — 1. The electrical energy stored in a capacitor or battery or held in an insulated object. 2. The quantity of electrical energy in (1) above. 3. In electrostatics, the amount of electricity present on any substance that has accumulated electric energy.

charge amplifier — 1. An operational amplifier with capacitive feedback. The output voltage that results from a charge signal input is returned to the input circuit through the feedback capacitor in the direction necessary to maintain the input circuit voltage at or near zero. Thus, the net charge input is stored in the feedback capacitor, producing a potential difference across it equal to the value of charge divided by the value of capacitance. This potential difference determines the relationship of the output voltage signal magnitude to the input charge signal magnitude. The transfer characteristic of the amplifier depends on the value of the feedback capacitor. 2. An amplifier whose output voltage is proportional to the input charge from a piezoelectric transducer.

charge carrier — A mobile hole or conduction electron in a semiconductor.

charge-coupled device — Abbreviated CCD. Semiconductor device arrayed so that the electric charge at the output end produces a stimulus to the next device.

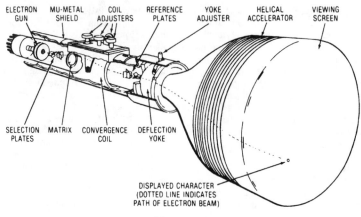

ELECTRON GUN   MU-METAL SHIELD   COIL ADJUSTERS   REFERENCE PLATES   YOKE ADJUSTER   HELICAL ACCELERATOR   VIEWING SCREEN

SELECTION PLATES   MATRIX   CONVERGENCE COIL   DEFLECTION YOKE

DISPLAYED CHARACTER
(DOTTED LINE INDICATES
PATH OF ELECTRON BEAM)

Charactron.

charge density—The charge per unit area on a surface, or charge per unit volume in space.

charged particle—An ion, an elementary particle that carries a positive or negative electric charge.

charge injection device—*See* CID.

charge injection imaging device—*See* CID.

charge offset—During the sample-to-hold transition of a sample/hold circuit, the charge transferred to the holding capacitor because of the switching process. It is usually expressed in millivolts.

charger—A device used to convert alternating current into a pulsating direct current that can be used for charging a storage battery.

charge retention—The ability of a battery to hold its energy once it has been charged.

charge-storage tube—A storage tube that retains information on its surface in the form of electric charges.

charge transfer—The process in which an ion takes an electron from a neutral atom of the same type, with a resultant transfer of electronic charge.

charging—1. The process of converting electrical energy to stored chemical energy. 2. The process of storing electrical energy in a capacitor.

charging current—The current produced when a dc voltage is first applied to a capacitor. This current decreases exponentially with time.

charging rate—1. The rate of current flow used in charging a battery. 2. The rate at which charging current flows into a capacitor or capacitance-resistance circuit. Expressed in amperes, milliamperes, or microamperes.

chart recorder—A data recorder that provides a record of the values of a physical parameter, in the form of a graph on a piece of chart paper, either with respect to time or to some other variable. The recording system consists of essentially three elements: a transducer to convert the variable to be measured (temperature, pressure, rpm, etc.) into an electrical signal, a signal conditioner to process the signal into a form such that it may be recorded, and the recording device.

chaser—1. An array of elements similar to a ring except that as each successive element is switched to the "on" condition, the others remain on as well; when all stages are on, the next pulse turns them all off and the process generally repeats. 2. A repeat-cycle flasher with three or more outputs each operating in sequence to the other. Normally used on signs or displays to create a moving effect.

chassis—1. A sheet-metal box, frame, or simple plate on which electronic components and their associated

circuitry can be mounted. 2. The entire equipment (less cabinet) when so assembled. *See also* printed circuit board.

chassis ground—A connection to the metal structure that supports the electrical components that make up the unit or system.

chat rooms—Areas on an online service, BBS, or the Internet that allow real-time, typed-in communication with other people.

chatter—1. A sustained rapid opening and closing of contacts due to variations in the coil current. 2. The vibration of a cutting stylus in a direction other than the direction in which it is driven.

chattering—Rapid audible cyclic action within an electromechanical device.

chatter time—The interval of time from initial actuation of a contact to the end of chatter.

cheater cord—An extension cord used to conduct power to a piece of equipment (especially a TV) by temporarily bypassing the safety interlock connector.

check—The partial or complete verification of the correctness of equipment operations, the existence of certain prescribed conditions, and/or the correctness of results.

check bit—1. A binary check digit. 2. The bit that is automatically added by the computer to an item of data when it is necessary to make it either even or odd parity. Synonym: parity bit.

check character—A character used to perform a check.

check digit—A digit added to each number in a coding system that allows for detection of errors in the recording of the code numbers. Through the use of the check digit and a predetermined mathematical formula, recording errors, such as digit reversal, can be noted.

checkerboard—*See* worst-case noise pattern.

checking code—Machine instructions that read part of a diskette to determine whether it has been copied.

checkout—A series of operations and calibration tests used to determine the condition and status of a system or element of the system.

checkpoint—1. In a computer routine, a point at which it is possible to store sufficient information to permit restarting the computation from that point. 2. The status of a long-running program is often recorded at frequent intervals called checkpoints. If something goes wrong, the program can be restarted at its last checkpoint instead of from the beginning.

checkpoint routine—A computer routine in which information for a checkpoint is stored.

check problem — A problem that, when incorrectly solved, indicates an error in the programming or operation of a computer.

check register — A special register provided in some computers to temporarily store transferred information for comparison with a second transfer of the same information in order to verify that the information transferred each time agrees precisely.

check routine — A program whose purpose is to determine whether a computer or a program is operating correctly.

checksum — 1. In a computer, a summation of digits or bits summed according to an arbitrary set of rules and primarily used for checking purposes. 2. A value that is the arithmetic sum of all the bytes in a program or program segment. As the program is loaded, the loader computes the sum of all bytes and compares the result with the checksum. If the two values are equal, it is assumed that the program segment was loaded without error. 3. A character added after a block of n words that contains the truncated binary sum of preceding nibbles. Used to verify the integrity of data in a ROM or on a tape. 4. Error detection data in a diskette sector based on the sum of all data bits in a block; if the sums as written and as calculated do not agree, an error is reported. 5. A simple error-checking technique used in data communications.

cheese antenna — An antenna with a cylindrical parabolic reflector enclosed by two plates perpendicular to the cylinder and so spaced that more than one mode can be propagated in the desired direction of polarization. It is fed on the focal line.

chelate — A molecule in which a central inorganic ion is covalently bonded to one or more organic molecules, with at least two bonds to each molecule; used as a laser dopant.

chemical deposition — The process of depositing a substance on a surface by means of the chemical reduction of a solution.

chemically deposited printed circuit — A printed circuit formed on a base by the reaction of chemicals alone. Dielectric, magnetic, and conductive circuits can be applied.

chemically reduced printed circuit — 1. A printed circuit formed by chemically reducing a metallic compound. 2. A printed circuit formed by the chemical erosion (etching) of portions of the metallic surface of a metal-clad insulative material.

chemical vapor deposition — Abbreviated CVD. A gaseous process that deposits insulating films or metal onto a wafer at elevated temperature. Often, reduced pressure is used to promote the chemical reaction.

chemisorption — *See* adsorption.

CHIL — Abbreviation for current-hogging injection logic.

Child's law — Also known as the three-halves power equation. It states that the current in a thermionic diode varies directly with the three-halves power of the anode voltage and inversely with the square of the distance between electrodes.

chip — 1. The uncased and normally leadless form of an electronic component part, either passive or active, discrete or integrated. 2. A single substrate on which all the active and passive circuit elements have been fabricated using one or all of the semiconductor techniques of diffusion, passivation, masking, photoresist, and epitaxial growth. A chip is not ready for use until packaged and provided with external connectors. The term is also applied to discrete capacitors and resistors that are small enough to be bonded to substrates by hybrid techniques. 3. A tiny piece of semiconductor material, scribed or etched from a semiconductor slice, on which one or more electronic components are formed. The total number of usable chips obtained from a wafer is the yield. 4. An unpackaged semiconductor device; a die incorporating an integrated circuit cut from a silicon wafer. By extension, every LSI package is commonly called a chip. 5. An electronic circuit element prior to having terminal connections added and prior to being encased for physical protection. 6. A single piece of semiconductor material (silicon, sapphire, germanium, etc.) containing one or more circuits, usually packaged as a unit. 7. Also called thread. In mechanical recording, the material removed from the recording medium by the recording stylus as it cuts the groove. 8. In punched cards, a piece of cardboard removed in the punching process.

chip and wire — A hybrid technology employing face-up bonded chip devices interconnected to the substrate conventionally, i.e., by flying wires.

chip architecture — The design or structure of an IC chip, incorporating arithmetic and logic unit, registers, and control-bus pathway configuration.

chip capacitor — 1. A capacitor of which the small dimensions or the nature and configuration of its terminations render it suitable for use mainly in hybrid circuits. 2. Discrete device that introduces capacitance into an electronic circuit, made in tiny wedge or rectangular shapes to be soldered onto hybrid circuits. 3. A subminiature capacitor (usually ceramic or solid tantalum) in chip form.

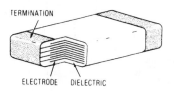

Chip ceramic (multilayer) capacitor.

chip carrier — 1. A low-profile component package, usually square, whose active chip cavity or mounting area is a large fraction of the package size and whose external connections are usually on the four sides of the package. 2. A physical package into which a die is mounted.

chip component — An unpackaged circuit element (active or passive) for use in hybrid microelectronics. Besides ICs, the term includes diodes, transistors, resistors, inductors, and capacitors.

chip-in-tape — An automated hybrid bonding technique that provides multiple, simultaneous bonding of leads by means of thermal and mechanical energy transmission through a deformable prepunched tape.

chip-level integration — The combination of two or more integrated-circuit functions and/or technologies on one IC to achieve miniaturization, reduce systems cost, and make new applications possible. Particularly important for signal processing and power control solutions.

chip-outs — Semiconductor die defects where fragments of silicon on the face have been chipped off in processing, leaving an active junction exposed.

chip resistor — A subminiature resistor formed on a small insulating substrate.

chip set — A set of integrated circuits that supplies all or most of the circuitry needed to build a functional item of electronic equipment. Most modems and computers are built from chip sets.

chip sets — The microprocessor chip in addition to RAMs, ROMs, and interface I/O devices. The chip sets mounted on a board are also referred to as the CPU portion of the microcomputer. Also called microcontroller.

Chireix antenna — Also called Chireix-Mesny antenna. Resonant series-fed array of square loops with half-wave sides. The loops feed each other in cascade, corner to corner, and the antenna resembles a double zigzag.

chirp — 1. An all-encompassing term for the various techniques of pulse expansion and pulse compression applied to pulse radar. A technique to expand narrow pulses to wide pulses for transmission, and to compress wide received pulses to the original narrow pulse width and waveshape. This improves the signal-to-noise ratio without degradation to the range resolution and range discrimination. 2. A colloquial expression for a coded pulse. In coding the pulse, the carrier frequency is increased in a linear manner for the duration of the pulse, and when the pulse is translated to an audio frequency, it sounds like a chirp. 3. A change in the pitch of code signals, generally due to poor regulation of the transmitter power supply. 4. A pulsed frequency-modulation scheme in which a carrier is swept over a wide frequency band during a given pulse interval.

chirp modulation — Swept-frequency modulation used in some radar and sonar equipment to increase the on-target energy and improve range resolution by making full use of the average power capability of the transmitter.

chirp radar — Radar in which a swept-frequency signal is transmitted, received after being returned from a target, and compressed in time to give a final narrow pulse called the chirp signal. This type of radar has high immunity to jamming and provides inherent rejection of random noise signals.

choke — 1. An inductance used to impede the flow of pulsating direct current or alternating current by means of its self-inductance. 2. An inductance used in a circuit to present a high impedance to frequencies appreciably limiting the flow of direct current. Also called choke coil. 3. A groove or other discontinuity in a waveguide surface so shaped and dimensioned as to impede the passage of guided waves within a limited frequency range.

Chokes.

choke coil — Also called impedance coil. An inductor (reactor) used to limit or suppress the flow of alternating current without appreciable effect on the flow of direct current.

choke flange — A waveguide flange with a grooved surface; the groove is so dimensioned that the flange forms part of a choke joint.

choke-input filter — A power-supply filter in which a choke is the first element in series with the input current from the rectifier.

choke joint — 1. A connector between two sections of transmission line in which the gap between sections to be connected is built out to form a series-branching transmission line carrying a standing wave, in which actual contact falls at or near a current minimum. 2. A joint for connecting two sections of waveguide together. Permits efficient energy transfer without the necessity of an electrical contact at the inside surface of the guide.

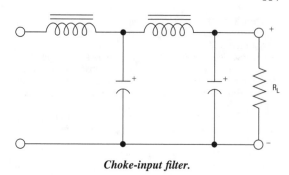

Choke-input filter.

cholesteric phase — An arrangement of liquid crystal molecules that occurs only in optically active substances and is considered to be a twisted nematic phase with a helical structure. It consists of layers resembling the smectic phase, but each layer has an order characteristic of a nematic phase.

chopped mode — A time-sharing method of displaying with a single CRT gun the output signals of two or more channels in sequence at a rate not referenced to the sweep.

chopper — 1. A device for interrupting a current or a light beam at regular intervals. Choppers are frequently used to facilitate amplification. 2. An electromechanical switch for the production of modified square waves. The waves are of the same frequency as a driving sine wave and bear a definite relationship to it. 3. An electromechanical or electronic device used to interrupt a dc or low-frequency ac signal at regular intervals to permit amplification of the signal by an ac amplifier. It may also be used as a demodulator to convert an ac signal to dc. 4. A rotating shutter for interrupting an otherwise continuous stream of particles. Choppers can release short bursts of neutrons with known energies. Used to measure nuclear cross sections.

chopper amplifier — A circuit that amplifies a low-level signal after it has gone through a chopper.

chopper stabilization — 1. The addition of a chopper amplifier to the regulator input circuitry of a regulated power supply in order to reduce output drift. 2. A method of improving the dc drift of an amplifier by utilization of chopper circuits.

chopper-stabilized amplifier — 1. An amplifier configuration utilizing a carrier-type dc amplifier to reduce the effect of input offset and drift of a direct-coupled amplifier. 2. An amplifier in which the dc input is chopped, simulating ac in order to overcome amplifier dc drift.

chopping — Removal by electronic means of one or both extremities of a wave at a predetermined level.

chopping frequency — The frequency at which a chopper interrupts a signal.

chord — A harmonious combination of tones sounded together through the use of one or more fingers on either or both hands. On chord organs, a full chord is selected by depressing a single chord button.

chord organ — An organ with provision for playing a variety of chords, each produced by means of a single button or key.

chorus — A natural electromagnetic phenomenon in the VLF range. Probably originates in the exosphere. Also called dawn chorus because it sounds like birds at dawn. It generally consists of a multitude of rising tones, each tone rising from 1–2 kHz to 3–4 kHz and usually lasting 0.1 to 0.5 second.

Christiansen antenna—A radiotelescope composed of two interferometer arrays placed at right angles. It resembles a Mills cross antenna.

Christmas-tree pattern—1. *See* optical pattern. 2. A pattern resembling a Christmas tree, sometimes produced on the screen of a television receiver when the horizontal oscillator falls out of sync.

chroma—1. That quality which characterizes a color without reference to its brightness; that quality which embraces hue and saturation. White, black, and gray have no chroma. 2. The quality of light perception that includes color and its purity. The purity of a color varies inversely with the amount of noise, that is, white light, mixed with it. Thus pink, red, and deep-red describe chromas for the visible wavelength in the vicinity of 650 nm. In parlance more relevant to optics, chroma pertains to the response of the eye to the combined effects of hue and saturation.

chroma-clear raster—Also called white raster. Looks like a clear raster, but each of the three guns in the CRT is operating under the influence of a color level determined by a white video signal. In this case, all TV set chroma circuits are working as though the TV set were receiving a color transmission of a completely white scene.

chroma control—A variable resistor that controls saturation by varying the level of chrominance signal fed to the demodulators of a color television receiver.

chroma detector—A circuit that detects the absence of chrominance information in a color encoder input. The chroma detector automatically deletes the color burst from the color encoder output when the absence of chrominance is detected.

chromatic aberration—1. An effect that causes refracted white light to produce an image with colored fringes due to the various colors being bent at different angles. 2. An optical lens defect that causes light-color separation because the optical material focuses different light colors at different points. A lens without this defect is said to be achromatic.

chromaticity—1. The combination of the hue and saturation attributes of color. 2. A term quantitatively descriptive of a color, and dependent on both hue and saturation, but without reference to brilliance.

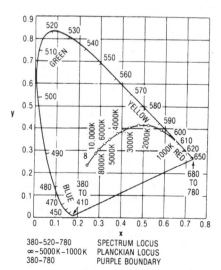

380–520–780	SPECTRUM LOCUS
∞ –5000 K –1000 K	PLANCKIAN LOCUS
380–780	PURPLE BOUNDARY

Chromaticity diagram.

chromaticity coordinates—Proportions of standard primaries (tristimulus values) required for a color match; ratios of each tristimulus value of a color to their sum. In the CIE calorimetric system, designated X, Y, and Z.

chromaticity diagram—A plane diagram formed when any one of the three chromaticity coordinates is plotted against another.

chromaticity flicker—The flicker that results from fluctuation of the chromaticity only.

chromatron—A color kinescope that has a single electron gun and whose color phosphors are laid out in parallel lines on its screen. The electron beam is directed to the correct phosphor by a deflection grid or wire grill near the face of the tube.

chrominance—1. Calorimetric difference between any color and a reference color having a specified chromaticity. In standard color-television transmission, the specified chromaticity is that of the zero subcarrier. 2. Pertaining to chroma, that is, to the mix of color and white light. Thus, pale pink may be either more or less bright than its pure constituent, red. An example of this is seen in a TV receiver, wherein the color signal is processed in the chrominance channel, but the brightness information is handled separately in the luminance channel. 3. A color term defining the hue and saturation of a color. Does not refer to brightness. 4. The hue and saturation of a color. The chrominance signal is modulated onto a 4.43 MHz carrier in the PAL television system and a 3.58 MHz carrier in the NTSC television system.

chrominance amplifier—The amplifier that separates the chrominance signal from the total video signal.

chrominance cancellation—A cancellation of the brightness variations produced by the chrominance signal on the screen of a monochrome picture tube.

chrominance-carrier reference—A continuous signal having the same frequency as the chrominance subcarrier, and a fixed phase with the color burst. The phase reference of carrier chrominance signals for modulation or demodulation.

chrominance channel—In color television, a combination of circuits designed to pass only those signals having to do with the reproduction of color.

chrominance component—Either of the I and Q signals that add to produce the complete chrominance signal in NTSC systems.

chrominance contrast—The color contrast between two adjacent surfaces of identical area, shape, texture, and luminance. The human eye is more sensitive to differences in color than it is to differences in brightness.

chrominance demodulator—A demodulator used in color-television reception for deriving video-frequency chrominance components from the chrominance signal and a sine wave of the chrominance subcarrier frequency.

chrominance gain control—In red, green, and blue matrix channels, variable resistors that individually adjust the primary-signal levels. Used in color television.

chrominance modulator—A modulator used in color-television transmission for generating the chrominance signal from the video-frequency chrominance components and the chrominance subcarrier.

chrominance primary—One of two transmission primaries, the amounts of which determine the chrominance of a color. Chrominance primaries have zero luminance and are not physical.

chrominance signal—1. The chrominance-subcarrier sidebands added to a monochrome television signal to convey color information. The components of the chrominance signal represent hue and saturation but do not include luminance or brightness. 2. That portion of

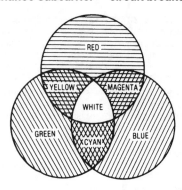

Chrominance primaries.

the total video signal that contains the color information. Without the chrominance signals, the received TV picture would be in black and white.

chrominance subcarrier — Also called color carrier. An rf signal that has a specific frequency of 3.579545 MHz and that is used as a carrier for the I and Q signals.

chrominance-subcarrier oscillator — In a color TV receiver, a crystal-controlled oscillator that generates the subcarrier signal for use in the chrominance demodulators.

chrominance video signals — Output voltages from the red, green, and blue sections of a color-television camera or receiver matrix.

chromium dioxide — A type of recording-tape coating that produces very good quality at low recording speeds. Because of its magnetic properties, it requires a higher value of bias current in the recorder. The high performance inherent in chromium dioxide tape can only be realized in a tape machine having provision for a CrO_2 bias setting. On a standard recorder, the chromium dioxide tape will appear to have high-frequency emphasis and may likely be difficult to erase. Chromium dioxide has a good dynamic range and a low noise level. Used in the cassette format with a suitable machine equipped with the Dolby system, it can make recordings that meet the best high-fidelity standards.

chronistor — A subminiature elapsed-time indicator that uses electroplating principles to totalize the operating time of equipment up to several thousand hours.

chronograph — An instrument for producing a graphical record of time as shown by a clock or other device.

chronoscope — An instrument for measuring very small intervals of time.

CID — Abbreviation for charge injection device or charge injection imaging device. 1. A memory in which the charge is stored in an X-Y addressable array of potential cells. For image arrays, the charge for each of the cells is generated by an associated photodioide. 2. A solid-state imaging device utilizing an image sensor composed of a two-dimensional array of coupled MOS charge-storage capacitors and designed to convert near infrared energy to electrical signals, providing broad gray shade or tonal rendition. The sensor collects minority carrier charge, generated by photon energy in the substrate near the charge-storage capacitors, and stores it in the surface inversion region. By injecting the stored charge into the substrate, and by monitoring the current, signal readout is achieved.

CIE — Initials of the Commission Internationale de l'Éclairage, or International Commission on Illumination.

CIE source — Standard light source representative of the quality of specified natural or artificial illumination.

CIE standard chromaticity diagram — A chromaticity diagram in which the X and Y chromaticity coordinates are plotted in rectangular coordinates.

CIM — Abbreviation for computer-integrated manufacturing. Applying information technology to production processes and organizational structure to streamline operations. Often focused on integrating systems and processes distributed across a company, such as order entry, scheduling, and production.

cinching — Longitudinal slippage between the layers in a tape pack as a result of acceleration or deceleration of the roll.

cipher — Cryptographic system in which arbitrary symbols or groups of symbols represent units of plain text of regular length, usually single letters, or in which units of plain text are rearranged, or both, according to certain predetermined rules.

cipher machine — Mechanical and/or electrical apparatus for enciphering and deciphering.

cipher telephony — A technique by which mechanical and/or electrical equipment is used for scrambling or unscrambling, or enciphering or decoding, radio or voice messages.

ciphertext or cryptogram — A secret form of a message.

ciphony — *See* cipher telephony.

circle cutter — A tool consisting of a center drill with an adjustable extension-arm cutter, used to cut holes in panels and chassis.

circle-dot mode — A method of storage of binary digits in a cathode-ray tube in which one kind of digit is represented by a small circle on the screen, and the other kind is represented by a similar circle with a concentric dot.

circle of confusion — The circular image of a point source due to the inherent aberrations in an optical system.

circlotron amplifier — A one-port, nonlinear cross-field high-power microwave amplifier that uses a magnetron as a negative-resistance element, much as a maser uses an active material.

circuit — 1. Path through which electrical signals flow. 2. An electronic path between two or more points capable of providing a number of channels. 3. A number of conductors connected together for the purpose of carrying an electrical current. 4. The interconnection of a number of devices in one or more closed paths to perform a desired electrical or electronic function. Examples of simple circuits are high- or low-pass filters, multivibrators, oscillators, and amplifiers. 5. A complete path of electron flow from a negative terminal of voltage source through a conductor and back to the positive terminal. 6. An electrical system using two or more wires in which the current flows from the source to one or more electrical devices and back again to the source of supply. 7. A complete, closed path. Confusion between circuit and network is common. *Circuit* refers to a closed path within a network. 8. An array of elements interconnected to perform functions beyond the range of single-element capability. *See* channel, 2.

circuit analysis — Careful determination of the nature and behavior of a circuit and its various parts. The analysis may be theoretical, practical, or both.

circuit analyzer — Also called multimeter. Several instruments or instrument circuits combined in a single enclosure and used in measuring two or more electrical quantities in a circuit.

circuit bonding jumper — The connection between portions of a conductor in a circuit to maintain required ampacity of the circuit.

circuit breaker — 1. An automatic device that, under abnormal conditions, will open a current-carrying circuit

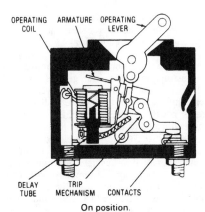

OPERATING COIL   ARMATURE   OPERATING LEVER

DELAY TUBE   TRIP MECHANISM   CONTACTS

On position.

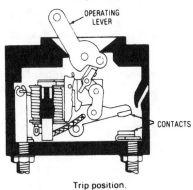

OPERATING LEVER

CONTACTS

Trip position.

Circuit breaker.

without damaging itself (unlike a fuse, which must be replaced when it blows). 2. A device for interrupting a circuit under normal or abnormal conditions by means of separable contacts. 3. An electromagnetic device that opens a circuit automatically when the current exceeds a predetermined value. It can be reset by operating a lever or by other means.

circuit-breaker cascade system — A system wherein the protective devices are arranged in order of ratings such that those in series will coordinate and provide the required protection.

circuit capacity — The number of communication channels that can be handled by a given circuit at the same time.

circuit card — A printed circuit board containing electronic components.

circuit commutated turn-off time — The time interval between the instant when the principal current has decreased to zero after external switching of the principal voltage circuit and the instant when a thyristor is capable of supporting a specified rate of rise of on-state voltage without turning on.

circuit component — An element of a circuit, such as a resistor, capacitor, diode, inductor transformer, integrated circuit, or transistor.

circuit density — The number of equivalent transistors per unit area of a IC chip.

circuit diagram — A drawing in which symbols and lines represent the components and wiring of an electronic circuit. Also called circuit schematic. *See* schematic diagram.

circuit dropout — A momentary interruption of a transmission because of the complete failure of a circuit.

circuit efficiency (of the output circuit of electron tubes) — Ratio of the power, at the desired frequency, delivered to a load at the output-circuit terminals of an oscillator or amplifier to the power, at the desired frequency, delivered to the output circuit by the electron stream.

circuit element — 1. Any basic constituent of a circuit except the interconnections. 2. A discrete unit of resistance, inductance, or capacitance that, when two or more are interconnected, forms an electric circuit.

circuit hole — On a printed circuit board, a hole that lies partially or completely within the conductive area.

circuit noise — The noise brought to the receiver electrically from a telephone system, but not the noise picked up acoustically by the telephone transmitters.

circuit noise level — At any point in a transmission system, the ratio of the circuit noise at that point to some arbitrary amount of circuit noise chosen as a reference. This ratio is usually expressed in decibels above reference noise, abbreviated dBrn, signifying the reading of a circuit-noise meter; or in adjusted decibels, abbreviated dBa, signifying the circuit-noise meter reading adjusted to represent interfering effect under specified conditions.

circuit-noise meter — Also called noise-measuring set. An instrument for measuring the circuit noise level. Through the use of a suitable frequency-weighting network and other characteristics, the instrument gives equal readings for noises of approximately equal interference. The readings are expressed in decibels above the reference noise.

circuit parameters — The values of the physical quantities associated with circuit elements; for example, the resistance (parameter) of a resistor (element), or the inductance per unit length (parameter) of a transmission line (element).

circuit protection — Automatic protection of a consequence-limiting nature used to minimize the danger of fire or smoke, as well as the disturbance to the rest of the system, that may result from electrical faults or prolonged electrical overloads.

circuit Q — The quality factor of a circuit, equal to the ratio of the circuit reactance to the circuit resistance. In a tuned resonant circuit, it determines the selectivity of the circuit; the higher the Q, the more selective the circuit.

circuit reentrancy — *See* reentrancy, 1.

circuit reliability — The percentage of time the circuit meets arbitrary standards set by the user.

Circuitron — A combination of active and passive components mounted in a single tube-type envelope and functioning as one or more complete operating stages.

circuit switching — A communication method in which an electrical connection between calling and called stations is established on demand for exclusive circuit use until the connection is released.

circuit synthesis — The development of a circuit by the use of theoretical or practical knowledge of basic electronics principles and component parameters. Compare *circuit analysis*.

circular antenna — A horizontally polarized antenna derived essentially from a half-wave antenna but having its elements bent into a circle.

circularly polarized loop vee — An airborne communications antenna that provides an omnidirectional radiation pattern for use in obtaining optimum near-horizon communications coverage.

circularly polarized wave — Applied usually to transverse waves. An electromagnetic wave for which the electric and/or magnetic field vector at a point describes a circle.

circular magnetic wave — A wave with circular magnetic lines of force.

circular mil — A unit of area equal to the area of a circle whose diameter is 1 mil (0.001 in or 25.4 μm); equal to square mil × 0.78540. Used chiefly in specifying cross-sectional areas of round conductors.

circular mil area — The square of the diameter of a round conductor measured in thousandths of an inch. The circular mil area of a braid is the sum of the circular mil area of each of the wires that make up the braid.

circular polarization — 1. Polarization such that the vector representing the wave has a constant magnitude and rotates continuously about a point. 2. Simultaneous transmission of vertically and horizontally polarized radio waves. 3. A method of transmitting signals in a rotating corkscrew-like pattern. Both right-hand rotating and left-hand rotating signals can be transmitted simultaneously on the same frequency, thereby doubling the capacity of a satellite to carry communications channels. 4. Polarization of electromagnetic waves whose electric field rotates uniformly along the signal path. Broadcasts used by Intelsat and other international satellites use circular polarization, not horizontally or vertically polarized waves as are common in North American and European transmissions.

circular scanning — Scanning in which the direction of maximum radiation generates a plane or a right circular cone with a vertex angle close to 180°.

circular trace — A CRO time base produced by applying sine waves of the same frequency and amplitude, but 90° out of phase, to the horizontal- and vertical-deflection plates of a cathode-ray tube. This results in a circular trace, and signals then give inward or outward radial deflections from the circle.

circular waveguide — A waveguide having a circular cross-sectional area.

circulating memory — 1. *See* circulating register. 2. A type of memory in which a data stream circulates in a loop. One example is a string of shift-register stages with the last output connected to the first input. At every clock pulse, a particular bit would be accessed as it passed a certain point in the circuit. Circulating memories also use other delay techniques, including electrical and acoustical delay lines.

circulating register — Also called circulating memory. A register (or memory) consisting of a means for delaying the information and a means for regenerating and reinserting it into the delaying means. This is accomplished as the information moves around a loop and returns to its starting place after a fixed delay.

circulating storage — A device using a delay line to store information in a train or pattern of pulses. The pulses at the output end are sensed, amplified, reshaped, and reinserted into the input end of the delay line.

circulator — 1. A microwave coupling device having a number of terminals so arranged that energy entering one terminal is transmitted to the next adjacent terminal in a particular direction. 2. An arrangement of phase shifters and waveguide or coax that distributes incoming signals among selected outputs. For example, a four-port circulator will transfer a signal entering from port 1 to port 2. In turn, a signal entering port 2 will leave only by port 3. A signal entering port 3 will leave by port 4, and a port-4 entering signal leaves by port 1. Can be used to isolate a transmitter and receiver when both are connected to the same antenna.

circumferential crimp — The type of crimp in which symmetrical indentations are formed in a barrel by crimping dies that completely surround the barrel.

circumnaural — A headphone in which the earpiece completely surrounds the wearer's ear and is sealed to the head to provide tight bass coupling.

CISC — Abbreviation for complex instruction set computer. A type of computer architecture that has long/complex instructions that are general purpose and powerful, but considered to be slower than RISC computers. Intel × 86 chips are CISC technology.

Citizens band — Abbreviated CB. A band of radio frequencies allocated to the Citizens Radio Service.

Citizens Radio Service — A radiocommunications service of fixed, land, and mobile stations intended for short-distance personal or business radiocommunications, radio signaling, and control of remote objects or devices by radio; all to the extent that these uses are not specifically prohibited by the FCC rules and regulations.

cladding — 1. A covering for the core of an optical fiber that provides optical insulation and protection. Generally fused to the fiber, it has a low index of refraction. 2. In fiber optics, a sheathing of a lower

Citizens band channels (Class D)

Channel No.	Frequency (MHz)
1	26.965
2	26.975
3	26.985
4	27.005
5	27.015
6	27.025
7	27.035
8	27.055
9	27.065
10	27.075
11	27.085
12	27.105
13	27.115
14	27.125
15	27.135
16	27.155
17	27.165
18	27.175
19	27.185
20	27.205
21	27.215
22	27.225
23*	27.255
24	27.235
25	27.245
26	27.265
27	27.275
28	27.285
29	27.295
30	27.305
31	27.315
32	27.325
33	27.335
34	27.345
35	27.355
36	27.365
37	27.375
38	27.385
39	27.395
40	27.405

*Shared with Class C radio control

refractive index material, intimately in contact with the core of a higher refractive index material, that serves to provide optical insulation and protection to the reflection interface. 3. A method of applying a layer of metal over another metal whereby the junction of the two metals is continuously welded. 4. A relatively thin layer or sheet of metal foil that is bonded to a laminate core to form the base material for printed circuits.

clamp — *See* clamping circuit.

clamper — *See* clamping circuit.

clamping — The process that establishes a fixed level for the picture signal at the beginning of each scanning line.

clamping circuit — A circuit that adds a fixed bias to a wave at each occurrence of some predetermined feature of the wave. This is done to hold the voltage or current of the feature at (clamp it to) a specified fixed or variable level. *See also* dc restorer.

clamping diode — A diode used to fix a voltage level at some point in a circuit.

clamp-on — *See* camp-on.

clamp-on ammeter — An ac ammeter with a built-in current transformer whose core can be clamped around the conductor in which current is to be measured.

clapper — 1. A hinged or pivoted armature. 2. In a bell, the ball or hammer that strikes the bell; in an electric bell, it is attached to the vibrating armature.

Clapp oscillator — A Colpitts-type oscillator using a series-resonant tank circuit for improved stability.

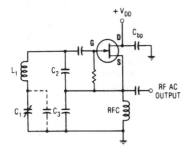

Clapp oscillator.

clarifier — A control on an SSB transceiver that enables adjustment of frequency so that the frequencies of the recovered audio signal will be essentially the same as the frequencies of the modulating signal fed to a distant transmitter. It is carefully adjusted so the received signal sounds natural. Its effective range is usually ±600 to ±1500 Hz.

Clark cell — An early standard cell that used an anode of mercury, a cathode of zinc amalgam, and an electrolyte containing zinc sulfate and mercurous sulfate. Its voltage is 1.433 at a temperature of 15°C.

Clarke belt — The circular orbital belt at 22,238 miles (35,863 kilometers) above the equator, named after the writer Arthur C. Clarke, in which satellites travel at the same speed as the earth's rotation. Also called the geosynchronous or geostationary orbit

Clarke orbit — That circular orbit in space 22,238 miles from the surface of the earth at which geosynchronous satellites are placed. This orbit was first postulated by the science fiction writer Arthur C. Clarke in *Wireless World* magazine in 1945. Satellites placed in these orbits, although traveling around the earth at thousands of miles an hour, appear to be stationary when

viewed from a point on earth because the earth is rotating upon its axis at the same angular rate that the satellite is traveling around the earth. *See also* geostationary.

class A amplifier — 1. A class of service of a power-amplifer stage in which both bias and drive signal (to be amplified) are adjusted to allow continuous output current at all times. Current through the load resistor generates the output voltage. Efficiency is low — theoretically reaching 50 percent for a pure sine wave, but in practice generally hovering between 20 and 30 percent. This class of service is generally limited to selected audio applications, amateur single-sideband linear rf service, and some CB. 2. An amplifier in which the grid bias and alternating grid voltage are such that plate current flows at all times. To denote that no grid current flows during any part of the input cycle, the suffix "1" is sometimes added to the letter or letters of the class identification. The suffix "2" denotes that grid current flows during part of the input cycle. 3. An amplifier in which the output transistors or tubes are operating permanently on linear portions of their transfer characteristics. Efficiency is low, but a constant current is drawn from the power supply whatever the signal level. Usually recognized by the use of a single transistor or tube driving the loudspeaker. 4. Operation that implies biasing the tubes or transistors to the middle parts of their transfer characteristics so that the device is driven upward on one half-cycle and downward on the other half-cycle. In a push-pull power amplifier, one of the device pair is driven upward on negative half-cycles while its partner is driven downward, the mode reversing on the positive half-cycles. The stage thus draws a constant current at all drive levels within the dynamic range of the amplifier. A class A power amplifier has an efficiency of almost 50 percent. 5. For a transistor amplifier, an amplifier with a single output device that has collector current for the full 360° of the input cycle.

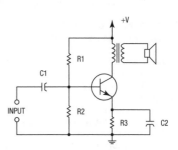

Class A amplifier.

class AB amplifier — 1. An amplifier in which the grid bias and alternating grid voltage are such that plate current flows for more than half but less than the entire electrical cycle. To denote that no grid current flows during any part of the input cycle, the suffix "1" is sometimes added to the letter or letters of the class indentification. The suffix "2" denotes that grid current flows during part of the cycle. 2. One type of power amplifier engineered so that at low drive level the stage operates at class A, while at increasing drive level the mode changes to class B.

class A computing device — A computing device for use in a commercial, industrial, or business environment, exclusive of any device marketed for use by the general public or intended to be used in the home.

class A0 emission — The incidental radiation of an unmodulated carrier wave from a station.

class A1 emission — A carrier wave (unmodulated by an audio frequency) keyed normally for telegraphy to transmit intelligence in the International Morse code at a speed not exceeding 40 words per minute (the average word is composed of five letters).

class A2 emission — A carrier wave that is amplitude modulated at audio frequencies not exceeding 1250 hertz. The modulated carrier wave is keyed normally for telegraphy to transmit intelligence in the International Morse code at a speed not exceeding 40 words per minute, the average word being composed of five letters.

class A3 emission — A carrier wave that is amplitude modulated at audio frequencies corresponding to those necessary for intelligible speed transmitted at the speed of conversation.

class A GFCI — A ground fault circuit interrupter that will trip when a fault current to ground is 5 milliamperes or more.

class A insulating material — A material or combination of materials, such as cotton, silk, and paper, suitably impregnated, coated, or immersed in a dielectric liquid such as oil. Other materials or combinations of materials may be included if shown to be capable of satisfactory operation at 105°C.

class A modulator — A class A amplifier used for supplying the signal power needed to modulate the carrier.

class A operation — Operation of a vacuum tube with grid bias such that plate current flows throughout 360° of the input cycle.

Class A signal area — A strong TV signal area, defined by the FCC as receiving a signal strength equal to or greater than approximately 2500 microvolts per meter for channels 2 through 6, 3500 microvolts per meter for channels 7 through 13, and 5000 microvolts per meter for channels 14 through 69.

Class A station — A station in the Citizens Radio Service licensed to be operated on an assigned frequency in the 460- to 470-MHz band with input power of 60 watts or less.

class A transistor amplifier — 1. An amplifier in which the input electrode and alternating input signal are biased so that output current flows at all times. 2. An amplifier with a single output device that has collector current for the full 360° of the input cycle.

class B amplifier — 1. A power-amplifier stage that offers improved efficiency with linear performance. Class B most commonly is configured in the push-pull arrangement popular in both highly linear audio and rf amplifiers. Adjustments in bias and drive effect a 50-percent output-current duty cycle for each active element. When the collector, drain, or plate voltage is at its maximum value, the current is zero. The fact that only one active element is on at any time creates the equivalent of a continuous current source for the load. Output power is the same as for class A, but input power is proportional to average load current. Efficiency can

reach 78 percent, though usually limited to 50 or 60 percent. Amplitude of the output is independent of the supply voltage (provided the amplifier is not saturated). 2. An amplifier in which the grid bias is approximately equal to the cutoff value so that, when no exciting grid voltage is applied, the plate current will be approximately zero and will flow for approximately half of each cycle when an alternating grid voltage is applied. To denote that no grid current flows during any part of the input cycle, the suffix "1" is sometimes added to the letter or letters of the class identification. The suffix "2" denotes that grid current flows during part of the cycle. 3. An amplifier in which two transistors or tubes operate on positive and negative half-cycles of the signal waveform. Each operates from a low initial current, but this rises as the signal level increases. Usually recognized by the use of two transistors or tubes operating in antiphase to drive the loudspeaker. 4. Transistor or tube (power) amplifier whose biasing is adjusted such that the push-pull transistors (or tubes) operate at a low no-drive current (called quiescent current). When drive is applied, the current in one of the pair rises while the partner is pushed into cutoff on one half-cycle, the mode reversing on the other half-cycle.

class B computing device — A computing device marketed for use in a residential environment, notwithstanding use to commercial, business, and industrial environment. Examples of such devices include electronic games, personal computers, calculators, and similar electronic devices marketed for use by the general public.

class B GFCI — A ground-fault circuit interrupter that will trip when a fault current to ground is 20 milliamperes or more.

class B insulating material — A material or combination of materials such as mica, glass fiber, asbestos, etc., suitably bonded. Other materials or combinations, not necessarily inorganic, may be included if shown to be capable of satisfactory operation at 130°C.

class B modulator — A class B amplifier used specifically for supplying the signal power needed to modulate a carrier.

class B operation — Operation of a vacuum tube with the triad bias set at or very near cutoff, so that plate current flows for approximately the positive half of each cycle of the input signal.

Class B station — A station in the Citizens Radio Service licensed to be operated on an authorized frequency in the 460- to 470-MHz band with input power of 5 watts or less.

class B transistor amplifier — 1. An amplifier in which the input electrode is biased so that when no alternating input signal is applied, the output current is approximately zero, and when an alternating input signal is applied, the output current flows for approximately half a cycle. 2. The most common type of audio amplifier, which basically consists of two output devices each of which conducts for 180° of the input cycle.

class C amplifier — 1. A power-amplifier stage in which bias is adjusted so that drive-signal voltge produces output current for less than half of the total cycle. Class C amplifiers normally operate in a saturated condition and are insensitive to drive variations. They can be modulated by variations of the power-supply voltage because in saturation they perform as voltage sources. Efficiency can theoretically approach 100 percent, but 60 to 80 percent is typical. Because this class is highly nonlinear, it is restricted to rf applications and enjoys wide popularity in broadcast amplifier applications. 2. An amplifier in which the grid bias is appreciably beyond the cutoff point, so that plate current is zero when no alternating grid voltage is applied, and plate current flows for appreciably less

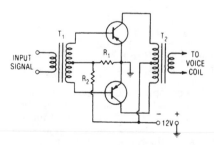

Class B audio amplifier.

than half of each cycle when an alternating grid voltage is applied. To denote that no grid current flows during any part of the input cycle, the suffix "1" is sometimes added to the letter or letters of the class identification. The suffix "2" denotes that grid current flows during part of the cycle.

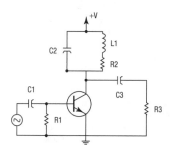

Class C amplifier.

class C insulating material — Insulation consisting entirely of mica, porcelain, glass, quartz, or similar inorganic materials. Other materials or combinations of materials may be included if shown to be capable of satisfactory operation at temperatures over 220°C.

class C operation — Operation of a vacuum tube with grid bias considerably greater than cutoff. The plate current is zero with no input signal to the grid and flows for appreciably less than one-half of each cycle of the input signal.

Class C station — A station in the Citizens Radio Service licensed to be operated on an authorized frequency in the 26.96- to 27.23-MHz band, or on the frequency 27.255 MHz, for the control of remote objects or devices by radio or for the remote actuation of devices that are used solely as a means of attracting attention, or on an authorized frequency in the 72- to 76-MHz band for the control of model aircraft only.

class C transistor amplifier — An amplifier in which the collector current flows for less than 180° of each input cycle. Although highly efficient, high distortion results and the load is frequently tuned to minimize this distortion (primarily used in rf power amplifiers).

class D amplifier — 1. A switching or sampling amplifier with extremely high efficiency (approaching 100 percent). The output devices are used as switches, voltage appearing across them only while they are off and current flowing only when they are saturated. 2. Also called pulse width modulator amplifier. In this amplifier a very high-frequency series of pulses are modulated in their width by the audio signal. The output stages need to conduct for a short interval only to amplify the tips of these pulses; when they do conduct, they are highly efficient — conducting as much as 90 to 95 percent. The high-frequency pulses associated with class D amplifiers (500 kHz or more) present special problems in the transistor switching speed and suitable transistor availability. 3. A power-amplifier stage in which push-pull configuration is driven as a two-position switch alternately connecting the output between V_{DD} and ground. The class D power amplifier delivers a square wave to a tuned circuit that passes only the fundamental (switching) frequency to the load. Extreme linearity and very high efficiency (approaching 100 percent) are possible. Applications consist mainly of audio and low-frequency broadcast equipment.

class D auxiliary power — An uninterruptible (no-break) power unit that makes use of stored energy to provide continuous power within specified tolerances for voltage and frequency.

Class D station — A station in the Citizens Radio Service licensed to be operated on an authorized frequency in the 26.965- to 27.405-MHz band, with input power of 5 watts or less, and to be used for radio telephony only.

class D telephone — A telephone restricted to use in special classes of service, such as fire alarm, guard alarm, and watchman services.

class E amplifier — Another switch-mode design amplifier (*see* class D amplifier) that uses only one active element. It combines the switching action of that element with the transient response of the tuned-load network to achieve efficiencies approaching 100 percent in practice. Use has been generally limited to medium-frequency rf designs.

class F amplifier — A power-amplifier stage very similar to both class D and class E. Class F amplifiers differ in that the tuned output circuit introduces a third-harmonic component properly phased to improve output-power capability. This configuration achieves very high efficiencies, approaching 90 percent in rf applications.

class F insulating material — A material or combination of materials such as mica, glass fiber, asbestos, etc., suitably bonded. Other materials or combinations of materials, not necessarily inorganic, may be included if shown to be capable of satisfactory operation at 155°C.

class G amplifier — 1. A power-amplifier stage in which two class B amplifiers with different supply voltages are combined. Small-amplitude signals are boosted by the one with the lower supply voltage, resulting in much higher average efficiency for speech and music. 2. An amplifier that uses a minimum of two pairs of output transistors. One pair is powered by a lower voltage supply than the other. When signal levels are relatively low, only the low-powered pair of transistors does the amplifying. When signals exceed the low-voltage supply amplitude, the other transistor pair, which operates from the higher voltage supply, takes over, while the first pair is simultaneously turned off. In this way each pair of transistors is always operating over its most efficient range, and overall amplifier efficiency is greater than with a class B design. Thus, less massive heat sinks are needed for the output transistors, and the complete amplifier or receiver is lighter in weight.

class H amplifier — 1. An amplifier somewhat similar to class G operation that uses only one set of output transistors, but these transistors are connected to two different power-supply voltages. The lower voltage powers the output devices for low-level signals, while the higher voltage takes over when the input signal amplitudes exceed the limits of the low-voltage supply. As in class G, this approach results in a more efficient use of the output transistors, and the audio signals themselves do not have to be switched from one device to another during the process. 2. A power-amplifier stage in which the supply voltage of the class B amplifier is varied by an efficient class S amplifier so it remains just above the minimum value required to prevent saturation. This configuration also achieves much higher average efficiency for speech and music signals.

class H insulating material — A material or combination of materials, such as silicone elastomer, mica, glass fiber, asbestos, etc., suitably bonded. Other materials or combinations of materials may be included if shown to be capable of satisfactory operation at 180°C.

class J oscilloscope — *See* J-scope.

class O insulating material — An unimpregnated material or combination of materials, such as cotton, silk, or paper. Other materials or combinations of materials may be included if shown to be capable of satisfactory operation at 90°C.

class S amplifier — A pulse-width-modulated audio amplifier in which the active elements are switched by a control frequency several times higher than the signal frequency being amplified. Class S offers an ideal efficiency of 90 percent.

clavier — Any keyboard, either hand or foot operated.

clean room — A confined area in which the humidity, temperature, and particulate matter are precisely controlled within specified units. The class designation of the clean room defines the maximum number of particles of 0.3-micron size or larger that may exist in one cubic foot of space anywhere in the designated area. For example, in a Class 1 clean room, only one particle of any kind may exist in one cubic foot of space. Newer clean rooms are typically Class 1 to 10, and are needed for manufacturing ICs with feature size close to 1 micron.

clear — 1. Also called reset. To restore a storage or memory device to a prescribed or nonprogrammed state, usually to zero or off (empty). 2. Remove all components of a calculation in a calculator. 3. In a calculator, to erase the contents of a display, memory, or storage register. 4. As used in security work, the term *clear* is synonymous with *reset*, meaning that a latched circuit is restored to normal state. 5. Signal to reset or set all signals to an initial known state (usually zero). 6. The process of setting the contents of a register, flag, or memory location to zero. 7. To erase the contents of a display or a memory or storage register.

clearance — The shortest distance through space between two live parts, between live parts and supports or other objects, or between any live part and grounded part.

clear channel — In the standard broadcast band, a channel such that the station assigned to it is free of objectionable interference through all of its primary service area and most of its secondary service area.

clear entry — Remove only the last number, not the entire calculation, in a calculator.

clear entry/clear all — In a calculator, a key used to clear the last entry or to clear the machine completely.

clearing — 1. Removal of a flaw or weak spot in the dielectric of a metallized capacitor by the electrical vaporization of the metallized electrode at the flaw. 2. The ability of a lightning protector to interrupt follow current before the operation of circuit fuses or breakers. In the case of a simple gap, clearing frequently requires some external assistance.

clearing ends — The operation of removing the sheath from the end of a cable, eliminating all moisture, and checking for crosses, shorts, and grounds in preparation for testing.

clearing-out drop — A drop signal associated with a cord or trunk circuit and operated by ringing current to attract the operator's attention.

clear input — An asynchronous input to a flip-flop used to set the Q output to logic zero.

clear memory key — Removes what is stored in a memory register of a calculator.

clear raster — A raster free of snow such as would be obtained in the absence of a video signal on either the cathodes or the grids of the three guns in the color CRT (mostly a function of bias conditions).

clear terminal — *See* reset terminal.

clear to send — *See* CTS.

click — To point a mouse pointer at a word or icon on a monitor, press a mouse button, and then release it quickly. Clicking is usually performed to select or deselect an item or to activate a program or a program feature.

click and pop suppressor — An audio-signal-processing accessory. It removes or greatly reduces the audible transient sounds resulting from scratches and blemishes on the surface of a phonograph record.

click filter — A capacitor and resistor connected across the contacts of a switch or relay to prevent a surge from being introduced into an adjacent circuit. *See also* key-click filter.

click-noise modulation — A clipping action performed to increase the bandwidth of a jamming signal. Results in more energy in the sidebands, correspondingly less energy in the carrier, and an increase in the ratio of average power to peak power.

client — A software program or computer that requests information from another computer.

client-server network — A network that uses a central computer (server) to store data that is accessed from other computers on the network (clients).

clipboard — A temporary storage place in a computer where text or graphics are stored.

clipper — A device whose output is zero or a fixed value for instantaneous input amplitudes up to a certain value, but is a function of the input for amplitudes exceeding the critical value.

clipper amplifier — An amplifier designed to limit the instantaneous value of its output to a predetermined maximum.

clipper-limiter — Also called slicer. A device whose output is a function of the instantaneous input amplitude for a range of values lying between two predetermined limits, but is approximately constant at another level for input values above the range.

clipping — 1. The loss of initial or final parts of words or syllables due to less than ideal operation of voice-operated devices. 2. Term used to express the clipping of the peaks of a waveform when an amplifier is driven beyond its power capacity. The flattening of the tips of the sine wave due to clipping. 3. Severe distortion caused by overloading the input of an amplifier. A sine-wave signal waveform has a flat top and bottom at the peaks when clipping occurs. 4. The deforming and distortion of speech signals due to limiting the maximum amplitude of the signals. 5. The shearing off of the peaks of a signal. For a picture signal, this may affect either the positive (white) or negative (black) peaks. For a composite video signal, the sync signal may be affected. 6. Removing parts of display elements that lie outside defined bounds. Also called scissoring. 7. The loss of one or more bits at the beginning of a transmission, typically caused by a delay in line turnaround or echo suppression. (May also occur in voice communication, with the loss of the beginning of an initial syllable.)

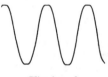

Clipping, 3.

clipping level — The signal level at which clipping (distortion) just begins to occur.

clock — 1. A pulse generator or signal waveform used to achieve synchronization of the timing of switching circuits and the memory in a digital computer system. It determines the speed of the CPU. 2. A timing device in a system: usually it provides a continuous series of timing pulses. 3. An electronic circuit that generates timing pulses to synchronize the operation of a computer as well as keep time. 4. A strobe signal that activates a certain sequence of operations. 5. An electronic circuit or device for producing precisely timed, repetitive voltage pulses of fixed frequency and amplitude.

clock cable — Cable of specific impedance and electrical characteristics used to distribute the clock (master) frequency where needed in digital computers.

clocked — Pertaining to the type of operation in which gating is added to a basic flip-flop to permit the flip-flop to change state only when there is a change in the clocking input or an enabling level of the clocking input is present.

clocked flip-flop — A flip-flop circuit designed so that it is triggered only if trigger and clock pulses are present at the same time.

clocked R-S flip-flop — A flip-flop in which two conditioning inputs control the state the flip-flop assumes upon arrival of the clock pulse. If the S (set) input is enabled, the flip-flop assumes the logic 1 condition when clocked; if the R (reset) input is enabled, the flip-flop assumes the logic 0 condition when clocked. A clock pulse must be applied to change the state of the flip-flop.

clock frequency — In digital computers, the master frequency of periodic pulses that are used to schedule the operation of the computer.

clock generator — A test-signal generator that supplies a chain of pulses identical with those supplied by the clock circuit of a digital computer.

clocking — Time-synchronizing communication information.

clock input — That flip-flop terminal whose condition or change of conditions controls the admission of data through the synchronous inputs and thereby controls the output state of the flip-flop. The clock signal permits data signals to enter the flip-flop and, after entry, directs the flip-flop to change state accordingly.

clock pulse — 1. The synchronization signal produced by a clock. 2. A pulse used to gate information into a flip-flop operated in the synchronous mode. (In JK flip-flops, the clock pulse causes counting if the data inputs are both held in logic 1.)

clock rate — 1. The rate at which a word or characters of a word (bits) are transferred from one internal computer element to another. Clock rate is expressed in cycles (in a parallel-operation machine, in words; in a serial-operation machine, in bits) per second. 2. The speed (frequency) at which a processor operates, as determined by the rate at which words or bits are transferred through internal logic sequences. 3. The minimum or maximum pulse rate at which adc counters may be driven. There is a fixed relationship between the minimum conversion rate and the clock rate, depending on the converter accuracy and type. All factors that affect conversion rate of an adc limit the clock rate. 4. The time rate at which pulses are emitted from a clock.

clock skew — 1. Phase shift in a single clock distribution system in a digital circuit. It results from different delays in clock driving elements and/or different distribution paths. 2. Unintentional time difference between clock edges; can exist between clock phases or between clock signals in different parts of a circuit.

clock slips — The relative shift of a system clock with respect to data in synchronous systems. Clock slips can cause modems to lose synchronization.

clock stagger — 1. Time separation of clock pulses in a multiphase clock system. 2. Voltage separation between the clock thresholds in a flip-flop.

clockwise-polarized wave — *See* right-handed polarized wave.

clone — 1. A PC designed to duplicate the behavior and performance of another personal computer, usually an IBM PC. 2. A copy that performs the same as the hardware, software, cellular phone, or computer on which it was based.

close-captioned TV — A text service for the hard-of-hearing TV audience that decodes a text subcarrier and displays it at the bottom of the TV frame on the accompanying video picture. It does not interfere with the standard audio FM subcarrier.

close coupling — 1. Coupling between two circuits so that (*a*) most of the power flowing in one is transferred to the other and (*b*) impedance changes in one circuit greatly affect the other. 2. Also called tight coupling. Any degree of coupling greater than critical coupling.

closed architecture — 1. A system whose characteristics are proprietary and therefore cannot be readily connected with other systems (compare with *open architecture*). 2. Equipment designed to work only with peripherals and accessories made by the same company.

closed array — An array that cannot be extended at either end.

closed circuit — 1. A complete electric circuit through which current may flow when a voltage is applied. 2. A program source, audio or video, that is not broadcast for general consumption, but is fed to remote monitored units by wire.

closed-circuit communication systems — Certain communication systems that are entirely self-contained and do not exchange intelligence with other facilities and systems.

closed-circuit jack — A jack that has its through circuits normally closed. Circuits are opened by inserting a mating plug.

closed-circuit signaling — Signaling in which current flows in the idle conditions and a signal is initiated by increasing or decreasing the current.

closed-circuit system — An intrusion alarm system in which the sensors of each zone are connected in series so that the same current exists in each sensor. When an activated sensor breaks the circuit or the connecting wire is cut, an alarm is transmitted for that zone.

closed-circuit television — Abbreviated CCTV. 1. A television system in which the television signals are not broadcast, but are transmitted over a closed circuit and received only by interconnected receivers. 2. Transmission and reception of video signals via wire carriers.

closed entry — A design that places a limit on the size of a mating part.

closed-entry contact — A female contact designed to prevent the entry of a device that has a cross-sectional dimension greater than that of a mating pin.

close-differential relay — A relay whose dropout value is specified close to its pickup value.

closed loop — 1. A circuit in which the output is continuously fed back to the input for constant comparison. 2. In a computer, a group of indefinitely repeated instructions. 3. A system with feedback control in which the output is used to control the input. 4. An automatic control system in which feedback is used to link a controlled process back to the original command signal. The feedback mechanism compares the actual controlled value with the desired value; if there is any difference, an error signal is created that helps correct the variation. In automation, feedback closes the loop. 5. A control arrangement in which data from the process or device being controlled is fed to the computer to affect the control operation, i.e., the computer can perform all control functions without intervention of an operator. *See also* open loop.

closed-loop bandwidth — The frequency at which the closed-loop gain drops 3 dB from its midband or dc value.

closed-loop drive — 1. A tape transport mechanism in which the tape's speed and tension are controlled by contact with a capstan at each end of the head assembly. 2. A tape transport system that drives both incoming and outgoing tape in order to control the portion of the tape contacting the heads and isolate it from the reels or cassette hubs. There are several closed-loop geometries regularly used with open-reel recorders, but dual-capstan drive is the most popular for both open-reel and cassette tapes.

closed-loop feedback — An automatic means of sensing speed variations and correcting to maintain close speed regulation.

closed-loop gain — 1. The response of a feedback circuit to a voltage inserted in series with the amplifier input. 2. The overall gain of an amplifier with an external negative-feedback loop.

closed-loop input impedance — The impedance looking into the input port of an amplifier with feedback.

closed-loop output impedance — The impedance looking into the output port of an amplifier with feedback.

closed-loop system — Automatic control equipment in which the system output is fed back for comparison with the input for the purpose of reducing any difference between input command and output response.

closed-loop voltage gain — The voltage gain of an amplifier with feedback.

closed magnetic circuit — A circuit in which the magnetic flux is conducted continually around a closed path through low-reluctance ferromagnetic materials, for example, a steel ring or a toroid core.

closed repeater — A repeater whose access is limited to a select group.

closed routine — In a computer, a routine that is entered by basic linkage from the main routine other than being inserted as a block of instructions within a main routine.

closed subroutine — In a computer, a subroutine not stored in the normal program sequence. Transfer is made from the program to the storage location of the subroutine, and then, following execution of the subroutine, control is returned to the main program.

close memory — Part of a directly addressable computer memory that provides fast cycle time and is usually employed for frequently used accesses.

close-talking microphone — Also called noise-canceling microphone. A microphone designed to be held close to the mouth of the speaker so that ambient noise will not degrade the speech.

closing rating — In a relay, conditions under which the contact must close, with a prescribed duty cycle and contact life.

cloud absorption — Absorption of electromagnetic radiation as a result of water drops and water vapor in a cloud.

cloud attenuation — Reduction in microwave radiation intensity due largely to scattering, rather than absorption, by clouds.

cloverleaf antenna — A nondirectional VHF transmitting antenna that consists of a number of horizontal four-element radiators arranged much like a four-leaf clover, stacked a half-wave apart vertically. These horizontal units are energized to give maximum radiation in the horizontal plane.

club station — A separate amateur radio station for use by the members of a bona fide amateur radio society and licensed to an amateur radio operator acting as the station trustee for the society.

cluster — 1. A group of user terminals co-located and connected to one controller, through which each terminal accesses a communication line. 2. A unit of storage that includes one or more sectors of a floppy or hard disk.

clustering — In a computer, the process of grouping things with similar characteristics. A properly programmed computer can take a list of items and group them into clusters.

clutter — Confusing, unwanted echoes that interfere with the observation of desired signals on a radar display.

clutter gating — The technique that provides switching between moving-target indicators and normal video. This results in normal video being displayed in regions with no clutter, and moving-target indicator video being switched in only for the clutter areas.

cm — 1. Abbreviation for circular mil. A system for specifying wire size by conductor area. Circular mils are obtained by multiplying the conductor diameter in inches by 1000 and squaring the result. 2. Letter symbol for centimeter.

CML — Abbreviation for current mode logic.

CMOS — *See* complementary metal-oxide semiconductor.

CMR — Abbreviation for common-mode rejection.

CMRR — Abbreviation for common-mode rejection ratio.

CMV — *See* common-mode voltage.

CNC — Abbreviation for computer numerical control. A means of operating production machines (most commonly punch presses and machining equipment) by numerical control instructions generated by CAD/CAM or from a programmable logic controller (PLC).

C network — A network composed of three impedance branches in series. The free ends are connected to one pair of terminals, and the junction points to another pair.

CO — 1. Abbreviation for central office. The switching equipment that provides local-exchange telephone service for a given geographical area; designated by the first three digits of the telephone number. 2. *See* cavity oscillator.

coarse-chrominance primary — Also called the Q signal. A zero-luminance transmission primary associated with the minimum bandwidth of chrominance transmission and chosen for its relatively small importance in contributing to the subjective sharpness of the color picture.

coast — On a radar, a memory feature that, when activated, causes the range and/or angle systems to continue to move in the same direction and at the same speed as an original target was moving. Used to prevent lock-up to a stronger target if approached by the target being tracked.

coastal refraction — Bending of the path of a direct radio wave as it crosses the coast at or near the surface. It is caused by differences in electrostatic conditions between soil and water.

coast station — A land-based radio station in the maritime mobile service. It carries on communication with shipboard stations.

coated cathode — In a vacuum tube, a cathode that has been coated with compounds so as to increase its electron emission (e.g., an oxide-coated cathode).

coated filament — A vacuum-tube filament that has been coated with metal oxides to increase electron emission.

coated tape — *See* magnetic-powder-coated tape.

coating — 1. The magnetic layer, consisting of oxide particles held in a binder, that is applied to the base film used for magnetic recordings. 2. The magnetizable material on one surface of a recording tape that stores the (audio) signals when recording.

coating thickness — The thickness of the magnetic coating applied to the base film. Modern tape coatings range in thickness from 170 to 650 μin (4.3–16.5 μm), with a preponderance of coatings being approximately 400 μin (10.1 μm) thick. In general, thin coatings give good resolution, at the expense of reduced output at long wavelengths; thick coatings give a high output at long wavelengths, at the expense of degraded resolution.

coax — Abbreviation for coaxial cable.

coaxial — Having a common axis.

coaxial antenna — An antenna comprised of a quarter-wavelength extension to the inner conductor of a coaxial line, and a radiating sleeve that in effect is formed by folding back the outer conductor of the coaxial line for approximately one-quarter wavelength.

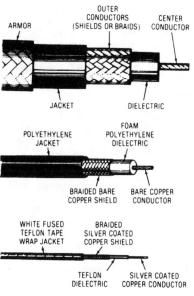

Coaxial cable.

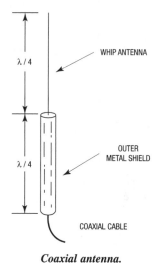

Coaxial antenna.

coaxial cable — Abbreviated coax. 1. A cylindrical transmission line made up of a central conductor (solid or stranded) inside a metallic tube or shield conductor separated by dielectric material in the form of spacers or a solid continuous extrusion. An insulating jacket is optional. 2. A cable consisting of two cylindrical conductors with a common axis. The two conductors are separated by a dielectric. The outer conductor, normally at ground potential, acts as a return path for current through the center conductor and prevents energy radiation from the cable. The outer conductor, or shield, is also commonly used to prevent external radiation from affecting the current in the inner conductor. The outer shield or conductor consists of woven strands of wire or is a metal sheath. 3. Two-conductor wire whose longitudinal axes are coincident; cable with a noise shield around a signal-carrying conductor. *See also* coaxial line.

coaxial cavity — A cylindrical resonating cavity that has a central conductor in contact with its movable pistons or other reflecting devices. The conductor serves to pick up a desired wave in the microwave region.

coaxial conductor — An electric conductor comprising outgoing and returning current paths with a common axis, with one of the paths surrounding the other throughout its length.

coaxial diode — A diode that has the same outer diameter and terminations as a coaxial cable, into which the diode is designed to be inserted.

coaxial-fed linear array — A beacon antenna having a uniform azimuth pattern.

coaxial filter — A passive, linear, essentially nondissipative network that transmits certain frequencies and rejects others.

coaxial line — Also called coaxial cable, coaxial transmission line, and concentric line. A transmission line in which one conductor completely surrounds the other, the two being coaxial and separated by a continuous solid dielectric or by dielectric spacers. Such a line has no external field and is not susceptible to external fields from other sources.

coaxial-line connector — A device used to provide a connection between two coaxial lines or between a coaxial line and the equipment.

coaxial-line frequency meter — A shorted section of coaxial line that acts as a resonant circuit and is calibrated in terms of frequency or wavelength.

coaxial relay — A type of relay used for switching high-frequency circuits.

coaxial speaker — 1. A single speaker comprising a high- and a low-frequency unit plus an electrical crossover network. 2. A tweeter mounted on the axis of and inside the cone of a woofer.

coaxial stub — A short length of coaxial cable joined as a branch to another coaxial cable. Frequently a coaxial stub is short-circuited at the outer end, and its length is so chosen that a high or low impedance is presented to the main coaxial cable at a certain frequency range.

coaxial transistor — A diffused base-alloy emitter, epitaxial mesa germanium semiconductor device with a bandwidth product up to 3 gigahertz, capable of being operated at medium power.

coaxial transmission line — *See* coaxial line.

cobalt doped — Utilizing a combination of "standard" gamma ferric oxide and cobalt as the magnetically active portion of the coating of a recording tape in order to improve maximum output level at low and high frequencies.

COBOL — Acronym for common business-oriented language. A higher-level computer language developed for programming business problems. Used to express problems of data manipulation and processing in English narrative form.

co-channel interference — Interference between two signals of the same type from transmitters operating on the same channel.

Cockcroft-Walton accelerator — A device for accelerating charged particles by the action of a high direct-current voltage on a stream of gas ions in a straight insulated tube. The voltage is generated by a voltage-multiplier system consisting essentially of a number of capacitor pairs connected through switching devices (vacuum tubes). The particles (which are nuclei of an ionized gas, such as protons from hydrogen) gain energies of up to several million electronvolts from the single acceleration so produced. Named for the British physicists J. D. Cockcroft and E. T. S. Walton, who developed this machine in the 1930s.

codan — Acronym for carrier-operated device, anti-noise. An electronic circuit that keeps a receiver inactive except when a signal is received.

codan lamp — A visual indication that a usable transmitted signal has been received by a particular radio receiver.

code — 1. A communications system in which arbitrary groups of symbols represent units of plain text of varying length. Codes may be used for brevity or security. 2. System of signaling by utilizing dot-dash-space, mark-space, or some other method in which each letter or figure is represented by prearranged combinations. 3. A system of characters and rules for representing information in a language capable of being understood and used by a computer. Code can be in the form of alphanumeric characters or binary data that can be executed directly by a computer. 4. The language that translates bits into characters in the word processor; it is basically the number of bits the code uses to make up a character and the pattern in which those bits are arranged. For example, ASCII, the industry standard for non-IBM machines, uses seven bits to make up a character, whereas EBCDIC, the IBM system, uses eight bits to make up a character. Systems with different codes will interpret each other's communications incorrectly. 5. A system of using symbols to represent other information. 6. A specific way of using symbols and rules to represent information.

codec — Abbreviation for coder/decoder. 1. A combination device for either digitally coding an analog signal (such as a voice signal) preliminary to transmission operations or decoding a digital signal preparatory to analog processing. 2. A coding-decoding device that converts voice signals into digital bit streams and back again. 3. A complex mixture of analog and digital functions, including (as a minimum) a band-limiting filter, S/H circuit, and nonlinear coder/decoder. 4. The related pair of digital coding and decoding functions in a pulse code modulation (PCM) telephone channel. Two codec pairs are required to complete a bidirectional channel. Each pair permits analog voice and tone signals to be multiplexed, switched, and transmitted in digitally encoded form. The subscriber loop is separated into individual transmit and receive paths by a two- to four-wire hybrid that is part of the subscriber loop interface circuit (SLIC). 5. A single integrated circuit that contains both an analog-to-digital converter (ADC) and a digital-to-analog converter (DAC). A codec performs both pulse code modulation and demodulation.

code character — One of the elements that makes up a code and that represents a specified symbol or value to be encoded. For example, dot-dot-dot-dash is the Morse code character for the letter V.

code conversion — The changing of the bit grouping for a character in one code into the corresponding bit grouping in another code.

code converter — 1. A device for translating one code to another. Examples: ASCII to EBCDIC, Gray to BCD, Hollerith to EBCDIC, etc. 2. A decision-making type of digital building block that converts information received at its inputs to another digital code that is transmitted at its outputs. (Also called encoder or decoder.) 3. A ROM or equivalent device that changes inputs into equivalent forms in a desired output format, such as ASCII, IBM Selectric, or Baudot.

coded-alarm system — An alarm system in which the source of each signal is identifiable. This is usually accomplished by means of a series of current pulses that operate audible or visible annunciators or recorders or both to yield a recognizable signal. This technique is usually used to allow the transmission of multiple signals on a common circuit.

coded decimal digit — A decimal digit expressed in terms of four or more 1s and 0s.

code-division multiple access — Abbreviated CDMA. A spread-spectrum scheme, transmitting on a bandwidth much larger than needed but at a much lower level. PN (pseudorandom noise) code is used for retrieving the information. *See also* CDMA.

coded passive reflector antenna — An object intended to reflect Hertzian waves and having variable reflecting properties according to a predetermined code for the purpose of producing an indication on a radar receiver.

coded program — A description of a procedure for solving a problem with a digital computer. It may vary in detail from a mere outline of the procedure to an explicit list of instructions coded in the machine's language.

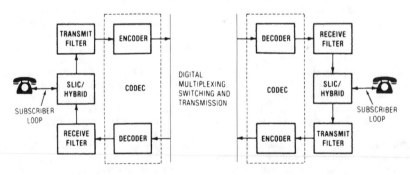

Codec.

code element — 1. One of the finite set of parts of which the characters in a code may be composed. 2. One of the discrete conditions or events in a code, such as the presence or the absence of a pulse.

code holes — The holes in perforated tape that represent information, as opposed to the feed holes or other perforations.

code-practice oscillator — An audio oscillator with a key and either headphones or a speaker, used to practice sending and receiving Morse code.

coder — 1. A device that sets up a series of signals in code form. 2. A beacon circuit that forms the trigger-pulse output of a discriminator into a series of pulses and then feeds them to a modulator circuit. 3. A person who prepares instruction sequences from detailed flowcharts and other algorithmic procedures prepared by others, as contrasted with a programmer, who prepares the procedures and flowcharts.

coder/decoder — *See* codec.

code ringing — In a telephone system, a method of ringing in which the number and/or duration of rings indicate which station on a party line is being called.

code set — 1. The entire set of unique codes that represent specific characters. Different code sets are employed in different equipment. 2. A specific set of symbols and rules used to represent information. 3. All possible values of a code. For an n-bit code, the code set consists of 2^n values. For example, EBCDIC, which is an eight-bit code, has 256 (2^8) possible values. The number of characters a code set can represent may exceed the total number of possible values.

code translation — In telephone operation, the changing of a directory code or number into a predetermined code that controls the selection of an outgoing trunk or line.

coding — 1. Converting program flowcharts into the language used by the computer. 2. The assignment of identification codes to transactions, such as a customer code number. 3. A method of representing characters within a computer. 4. Changing a communications signal into a form suitable for transmission or processing. 5. The preparation of a set of instructions or symbols that, when used by a programmable controller, have a special external meaning.

coding delay — Arbitrary time delay between pulse signals sent by master and slave transmitters.

coding disk — A disk with small projections that operate contacts to generate a predetermined code.

coding form — A form on which the instructions for programming a computer are written. Also called a coding sheet.

coding line — A single command or instruction that directs a computer to solve a problem, usually written on one line.

codiphase radar — A radar system including a phased-array radar antenna and signal-processing and beam-forming techniques.

codistor — A multijunction semiconductor device that provides noise-rejection and voltage-regulation functions.

coefficient — 1. The ratio of change under specified conditions of temperature, length, volume, etc. 2. A number (often a constant) that expresses some property of a physical system in a quantitative way.

coefficient of coupling — *See* coupling coefficient.

coefficient of expansion — The fractional change in dimension of a material for a unit change in temperature.

coefficient of reflection — The square root of the ratio of the power reflected from a surface to the power incident on the same surface.

coefficient of thermal expansion — The average expansion per degree over a specified temperature range, expressed as a fraction of the original dimension. The coefficient may be linear or volumetric.

coercive force — Symbolized by H. The magnetizing force that must be applied to a magnetic material in a direction opposite the residual induction in order to reduce the induction to zero.

coercivity — 1. The property of a magnetic material measured by the coercive force corresponding to the saturation induction for the material. 2. A measure of the amount of applied magnetic field (of opposite polarity) that is necessary to restore a magnetized tape to a state of zero magnetism. High-coercivity tapes exhibit less tendency toward self-erasure and thus have enhanced high-frequency response characteristics, but they require more current through the erase head for full erasure of a recorded signal.

cofire — To place circuits onto an unfired ceramic and fire both circuits and ceramic simultaneously.

cofired ceramic — A material used in making IC packages; its construction is of layers of alumina laminated and fired together to produce a high-strength monolithic structure.

cofiring — Processing the thick-film conductors and resistors through the firing cycle at the same time.

cogging — 1. Nonuniform angular velocity. The armature coil of a motor tends to speed up when it enters the magnetic field produced by the field coils, and to slow down when leaving it. This becomes apparent at low speeds; the fewer the coils, the more noticeable it is. 2. In a motor, the effect caused by improper ratio of stator to rotor slots for a particular speed. It is caused by a magnetic interaction between rotor and stator teeth. In dc motors it is caused by the rich ripple content of the power supply rectifier. The term *cogging torque* is usually applied to reluctance-type synchronous motors. This is the maximum torque of such a synchronous motor before starting to cog or slip, as if a gear were slipping one tooth at a time.

coherence — 1. The property of a set of waves by which their phases are completely predictable along an arbitrarily specified surface in space; also, the relation between a set of sources by which the phases of their respective radiations are similarly predictable. 2. A term used to denote various forms of temporal or statistical phase correlations of electromagnetic fields at different spatial positions; the more extensive the correlations, the greater the coherence. *See* laser; maser. 3. The property of laser light in which the beam emitted is largely of one color or frequency. In the case of the helium-neon laser, this color is red, at 632.9 nm.

coherent bundle — 1. A fiber-optics bundle in which each fiber maintains its relative location throughout the bundle. Thus, an image introduced at one end is transmitted to the other without being scrambled. 2. A bundle of optical fibers in which the spatial coordinates of each fiber are the same or bear the same spatial relationship to each other at the two ends of the bundle. Also called aligned bundle.

coherent carrier — A carrier, derived from a cw signal, the frequency and phase of which have a fixed relationship to the frequency and phase of the reference signal.

coherent carrier system — A transponder system in which the interrogating carrier is raised to a definite multiple frequency and retransmitted for comparison.

coherent detection — A method of deriving additional information from the phase of the carrier.

coherent detector — A detector that gives an output signal amplitude dependent on the phase rather than the strength of the echo signal. It is required for a radar display that shows only moving targets.

coherent display — In random-sampling oscilloscope technique, a plot of a group of samples in which the time sequence of signal events is maintained.

coherent echo — A radar echo that has relatively constant phase and amplitude at a given range.

coherent electroluminescence device — *See* diode laser.

coherent emitter — A source of power that provides a high degree of spectral purity, near-perfect beam collimation, and enormous power densities. Lasers are coherent emitters. *See* incoherent emitter.

coherent fiber bundle — A rigid or flexible fiber bundle that is capable of transmitting an image from one end of the bundle to the other.

coherent interrupted waves — Interrupted continuous waves occurring in wave trains in which the phase of the waves is maintained through successive wave trains.

coherent light — 1. A single frequency of light. Light having characteristics similar to a radiated radio wave that has a single frequency. 2. Light of but a single frequency that travels in intense, nearly perfect, parallel rays without appreciable divergence. 3. Light that has the property that, at any point in time or space, particularly over an area in a plane perpendicular to the direction of propagation, or over time at a particular point in space, all the parameters of the wave are predictable and are correlated. 4. Light in which all waves are of exactly the same frequency and exactly in phase. It can therefore act as a carrier and can be modulated for the transmission of information. 5. Light in which the phase relationship between successive waves is such that the beam consists of parallel rays that provide a high concentration of energy.

coherent light communications — Communications using amplitude or pulse-frequency modulation of a laser beam.

coherent light detection and ranging — *See* colidar.

coherent light source — A light source that is capable of producing radiation with waves vibrating in phase. The laser is an example of a coherent light source.

coherent oscillator — An oscillator within some radar sets that furnishes phase references for target returns during intervals between transmitter pulses and that has its output compared with the returns so that the echo becomes coherent video. Coherent video is applied to a cancellation circuit that eliminates nonmoving targets, and only moving targets are supplied to the indicator.

coherent-pulse operation — The method of pulse operation in which a fixed phase relationship is maintained from one pulse to the next.

coherent radar — A type of radar containing circuits that make possible comparison of the phase of successive received target signals.

coherent radiation — 1. A form of radiation in which definite phase relationships exist between radiation at different positions in a cross section of the radiant beam. 2. Radiation in which the phase between any two points in the radiation field has a constant difference, or is exactly the same throughout the duration of the radiation. 3. Single-frequency energy such that there is reinforcement when portions of a signal coincide in phase and cancellation when they are in phase opposition.

coherent reference — The reference signal, usually of stable frequency, to which other signals are phase-locked to establish coherency throughout a system.

coherent reflector — Simple or complex surface (such as a corner reflector) from which reflected wave components are coherent with respect to each other and

thus combine to yield larger effective power than would be observed from a diffuse scattering surface of the same area.

coherent source — A theoretically ideal light source (fiber optic) that emits a very narrow, unidirectional beam of light of one wavelength (monochromatic). All light emitted from a coherent source is in phase. A laser approximates a coherent source. Contrast *incoherent source*.

coherent system of units — Also called absolute system of units. A system of units in which the magnitude and dimensions of each unit are related to those of the other units by definite simple relationships in which the proportionality factors are usually chosen to be unity.

coherent transponder — A transponder, the output signal of which is coherent with the input signal. Fixed relations between frequency and phase of input and output signals are maintained.

coherent video — A video signal resulting from the combination of a radar echo signal with the output of a continuous-wave oscillator. After delay, the signal so formed is detected, amplified, and subtracted from the next pulse train to give a signal that represents only moving targets.

coherer — An early form of detector used in wireless telegraphy.

coil — Also called inductance and inductor. 1. A number of turns of wire wound around an iron core or onto a form made of insulating material, or one that is self-supporting. A coil offers considerable opposition to the passage of alternating current, but very little opposition to direct current. 2. A number of turns of wire used to introduce inductance into an electric circuit, to produce magnetic flux, or to react mechanically to a charging magnetic flux. In high-frequency circuits, a coil may be only a fraction of a turn. The electrical size of a coil is called inductance and is expressed in henrys. The opposition that a coil offers to alternating current is called impedance and is expressed in ohms. The impedance of a coil increases with frequency.

coil dissipation — The amount of electrical power consumed by a winding. For most practical purposes, this equals the I^2R loss.

coil effect — The inductive effect exhibited by a spiral-wrapped shield, especially above audio frequencies.

coil form — An insulating support of ceramic, plastic, or cardboard onto which coils are wound.

coil loading — As commonly understood, the insertion of coils into a line at uniformly spaced intervals. However, the coils are sometimes inserted in parallel.

coil neutralization — *See* inductive neutralization.

coil rating — *See* input-power rating.

coil resistance — The total terminal-to-terminal resistance of a coil at a specified temperature.

coil serving — A covering, such as thread or tape, that serves to protect a coil winding from mechanical damage.

coil temperature rise — The increase in winding temperature above the ambient temperature when energized under specified conditions for a given period of time, usually that required to reach a stable temperature.

coil tube — A tubular coil form. *See also* spool.

coincidence amplifier — An amplifier that produces no output unless two input pulses are applied simultaneously to the circuit.

coincidence circuit — 1. A circuit that produces a specified output pulse when and only when a specified number (two or more) or combination of input terminals receives pulses within a specified time interval. 2. A circuit that has an output signal only when all input signals are present. 3. An AND circuit.

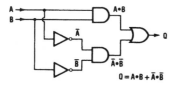

Coincidence circuit.

coincidence counting — The use of electronic devices to detect when two or more pulses from separate counters occur within a given time interval. This is done to determine whether the pulses were produced by the same particle, for example, in scintillation counting, or whether they correspond to the same event. Also called coincidence setting.

coincidence gate — A circuit with the ability to produce an output that is dependent on a specified type or the coincident nature of the input.

coincidence setting — The use of electronic devices to detect when two or more pulses occur within a given time interval. This is done to determine whether the pulses were produced by the same particle, for example, in scintillation counting, or whether they correspond to the same event. Also called coincidence counting.

coincident-current selection — The selection of a magnetic core, for reading or writing, by the simultaneous application of two or more currents.

coincident resonance — A condition in which two systems are joined together (or coupled) and both systems are vibrating near the resonant frequencies at the same time. It often happens that the amplified output from the first system turns out to be the input to the second system, which amplifies that input a second time.

cold — Idiomatic term generally used to describe electrical circuits that are disconnected from voltage sources and are at ground potential. Opposite of the term *hot*.

cold-blade stripper — Device for removing insulation utilizing a cold knife blade.

cold boot — Starting a computer by turning on the power.

cold cathode — 1. A cathode whose operation does not depend on the temperature being above the ambient temperature. 2. A cathode that emits electrons, not with the influence of heat radiation, but by means of high-voltage gradient at its surface.

cold-cathode tubes — Tubes in which no external source is used for heating the cathode. These include tubes such as photoelectric cells, gas glow tubes, and mercury rectifiers.

cold cleaning — An organic-solvent cleaning process in which liquid contact accomplishes the solution and removal of residues after soldering.

cold emission — *See* field emission.

cold light — Light that is produced without attendant heat, as from the ionization of a gas by high voltage (neon bulbs, fluorescent lamps), or by electroluminescence, bioluminescence, or cathodoluminescence.

cold-pressure welding — A method of making an electrical connection in which the members to be joined are compressed to the plastic range of the metals.

cold rolling — Rolling a magnetic core alloy into the form of a rod so that the metallic grains are oriented in the long direction of the rod.

cold solder joint — A solder connection exhibiting poor wetting and a grayish, porous appearance due to insufficient heat, inadequate cleaning prior to soldering, or to excessive impurities in the solder solution.

cold weld — A joint between two metals (without an intermediate material) produced by the application of pressure only.

colidar — Acronym for coherent light detection and ranging. An optical radar system that uses the direct output from a ruby laser source without further pulse modulation.

collate — To combine two or more similarly ordered sets of items to produce another ordered set. Both the number and the size of the individual items may differ from those of the original sets or their sums.

collating sequence — 1. In digital computers, the sequence in which the characters acceptable to a computer are ordered. The British term is *marshalling sequence*. 2. An ordering assigned to a set of items such that any two sets in that assigned order can be collated.

collator — 1. A device to collate sets of punched cards or other documents into a sequence. 2. A device for determining and indicating the coincidence or noncoincidence of two signals.

collection — The mechanism whereby the high-potential gradient and intense electric (drift) field present within the depletion layer of a reverse-biased pn junction can cause the depletion layer to collect any carriers of appropriate type that happen to diffuse it from the adjacent semiconductor regions.

collector — 1. In a transistor, the region into which majority carriers flow from the base under the influence of a reverse bias across the two regions. 2. The external terminal of a transistor that is connected to this region. 3. In certain electron tubes, an electrode to which electrons or ions flow after they have completed their function.

collector capacitance — Depletion-layer capacitance associated with the collector junction of a transistor.

collector-coupled logic — *See* current-sourcing logic.

collector current — The direct current flowing in the collector of a transistor.

collector cutoff — The operating condition of a transistor when the collector current is reduced to the leakage current of the collector-base junction.

collector cutoff current — 1. The minimum current that will flow in the collector circuit of a transistor with zero current in the emitter circuit. 2. The reverse current of the base-collector junction when the emitter is open circuited, the reverse voltage being specified.

collector-diffusion isolation — Abbreviated CDI. A technique for fabrication of bipolar ICs; the collector diffusion is used to isolate transistors on the same silicon chip electrically, thus reducing the number of photolithographic masking steps required.

collector efficiency — The ratio, usually expressed in percentage, of useful power output to final-stage power-supply power input of a transistor.

collector family — Set of transistor characteristic curves in which the collector current and voltage are variables.

collector junction — The semiconductor junction between the base and collector regions of a transistor. In normal transistor operation, it is reverse biased to collect carriers injected by the emitter (base-to-emitter) junction. In general, the collector junction is designed for a high breakdown voltage, and the emitter junction is designed for a high emitting efficiency.

collector resistance — Resistance in series with the collector lead in the common-T equivalent circuit of a transistor.

collector ring — The collector electrode in an iconoscope.

collector rings — Metal rings suitably mounted on an electric machine that serve, through stationary brushes bearing thereon, to conduct current into or out of the rotating member. *See also* slip ring.

collector transition capacitance — The capacitance across the collector-to-base transition region of a transistor.

collector voltage — The dc collector supply voltage applied between the base and collector of a transistor.

collimated light — 1. Parallel light rays, as opposed to converging or diverging rays. 2. A bundle of light rays in which the rays emanating from any single point in the object are parallel to one another, such as the light from an infinitely distant real source or apparent source, such as a collimator reticle.

collimated transmittance — Transmittance of an optical waveguide, such as an optical fiber or integrated optical circuit, in which the light wave at the output has coherency related to the coherency at the input.

collimation — 1. The process of adjusting an instrument so that its reference axis is aligned in a desired direction within a predetermined tolerance. 2. The process of making light rays parallel. 3. The property of laser light that keeps the beam from spreading out as it moves away from the laser.

collimation equipment — Equipment designed specifically for aligning optical equipment.

collimation tower — A tower supporting a visual and a radio target used to check the electrical axis of an antenna.

collimator — 1. An optical device that creates a beam made up of parallel rays of light, used in testing and adjusting certain optical instruments. It may be used to simulate a distant target, align the optical axes of instruments, or prepare rays for entry into the end of an optical fiber, fiber bundle, or optical thin film. 2. A lens or lens assembly that focuses light into a beam.

collinear array — An antenna array in which half-wave elements are arranged end to end on the same vertical or horizontal line.

collision — Overlapping transmissions when two or more nodes attempt to transmit messages simultaneously or almost simultaneously.

collision-avoidance system — A group of sensors and related instruments placed on board an aircraft to help the pilot detect points of collision and take the appropriate maneuvers to avoid them.

collision detection — The ability of a transmitting node to detect simultaneous transmission attempts on a shared medium.

colocation — The placing of several satellites near each other in orbit. Colocation allows a single fixed receiving antenna to receive signals from all the satellites without moving from one satellite to another (tracking).

color — 1. The everyday term for hue. One of the two constituents of the quality chroma (the other being white light). The mind-eye sensation produced by different wavelengths within the visible spectrum produces the colors, some of which are simply identifiable by such names as violet, blue, green, yellow, orange, and red. (Black, white, and gray are not properly described as colors.) 2. Embodies the characteristics of light, other than brightness or luminance, by which a human observer may distinguish between two structure-free patches of light of the same size and shape.

color balance — The adjustment of electron-gun emissions to compensate for the difference in the light-emitting efficiencies of the three phosphors on the screen of the color picture tube.

color bar-dot generator — An rf signal generator that develops a bar or dot pattern on the screen of a color TV picture tube used for test and alignment purposes.

color bars — A test pattern of specifically colored vertical bars used as a reference to test the performance of a color television and transmission paths.

color-bar signal — A test signal used in checking chrominance functions of color TV systems. Typically, it contains bars of six colors, yellow, cyan, green, magenta, red, and blue.

color breakup — Any fleeting or partial separation of a color picture into its display primary components because of a rapid change in the condition of viewing. For example, fast movement of the head, abrupt interruption of the line of sight, and blinking of the eyes are illustrations of rapid changes in the conditions of viewing.

color burst — Also called reference burst. In NTSC color systems, normally refers to a burst of approximately 9 Hz of the 3.579545-MHz unmodulated color subcarrier on the back porch (during the blanking pulse) of the composite video signal. This serves as a color synchronizing signal to establish a frequency and phase reference for the chrominance signal.

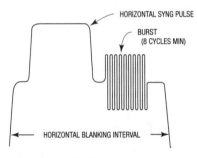

Color burst.

color carrier — *See* chrominance subcarrier.

color-carrier reference — A continuous signal of the same frequency as the color subcarrier and having a fixed phase relationship to the color burst. It is used for modulation at the transmitter and demodulation at the receiver.

colorcast — A color television broadcast.

color code — 1. A system of colors for specifying the electrical value of a component part or for identifying terminals and leads. Also used to distinguish between cable conductors. 2. A color system for component identification by use of solid colors, tracers, braids, surface printing, etc.

color coder — Also called color encoder. In a color TV transmitter, that circuit or section which combines the camera signals and the chrominance subcarrier to form the transmitted color picture signal.

color coding — A system of identification of terminals and related devices through the use of colored markings.

color contamination — An error in color rendition due to incomplete separation of the paths carrying different color components of the picture. Such errors can arise in the optical, electronic, or mechanical portions of a color television system.

color-coordinate transformation — Computation of the tristimulus values of colors in terms of one set of primaries from the tristimulus values of the same colors in another set of primaries. Such computation may be performed electrically in a color television system.

color decoder — A section or circuit of a color television receiver used for deriving the signals for the color display device from the color picture signal and the color burst.

color-difference signal — The signal produced when the amplitude of a color signal is reduced by an amount equal to the amplitude of the luminance signal. Color-difference signals are usually designated R-Y, B-Y, and G-Y. In a sense, I and Q signals are also color-difference signals because they are formed when specific proportions of R-Y and B-Y color-difference signals have been combined.

color edging — 1. Spurious color at the boundaries of differently colored areas in a picture. 2. Extraneous colors that appear along the edges of video pictures but have no color relationship to those areas.

color encoder — A device that produces an NTSC color signal from separate R, G, and B video inputs. *See* color coder.

color facsimile transmission — The transmission of a facsimile of a color photograph by separating the colors into varying intensities of red, blue, and green and then sending separate transmissions of the three-color reading to a receiving station, which recombines the three signals into the original print.

color fidelity — The degree to which a color television system is capable of faithfully reproducing the colors in an original scene.

color flicker — The flicker that results from fluctuation of both the chromaticity and the luminance.

color fringing — Spurious chromaticity at the boundaries of objects in a color TV picture. It can be caused by the change in relative position of the televised object from field to field, or by misregistration. Color fringing may cause small objects to appear separated into different colors.

color generator — A special rf signal generator for adjusting or troubleshooting a color TV receiver.

colorimeter — An optical instrument designed to compare the color of a sample with that of a standard sample or a synthesized stimulus. (In a three-color colorimeter, the synthesized stimulus consists of three colors of contrast chromaticity but variable luminance.)

colorimetric — Pertaining to the measurement of color characteristics, particularly wavelength and primary-color content.

colorimetric photometer — A photometer that uses a set of color filters to measure the intensity of light in various regions of the spectrum.

colorimetry — The technique of measuring color and interpreting the results.

color killer — A stage designed to prevent signals in a color receiver from passing through the chrominance channel during monochrome telecasts.

color match — The condition in which the two halves of a structureless photometric field look exactly the same.

color media — Transparent colored materials that can be placed in front of an instrument to color the emitted light. These materials are often referred to as gels (for gelatin), but glass or plastic also may be used.

color mixture — Color produced by the combination of lights of different colors. The combination may be accomplished by successive presentation of the components, provided the rate of alternation is sufficiently high; or the combination may be accomplished by simultaneous presentation, either in the same or in adjacent areas, provided the components are small enough and close enough together to eliminate pattern effects.

color oscillator — In a color television receiver, the oscillator operating at the burst frequency of 3.579545 MHz. Its frequency and phase are synchronized by the master oscillator at the transmitter.

color phase — 1. The difference in phase between a chrominance primary signal (I or Q) and the chrominance carrier reference. 2. The proper timing relationship with

a color signal. Color is considered to be in phase when the hue is reproduced correctly on the screen.

color-phase alternation — Periodic changing of the color phase of one or more components of the color television subcarrier between two sets of assigned values.

color-phase diagram — A vector diagram that denotes the phase difference between the color-burst signal and the chrominance signal for each of the three primary and complementary colors. This diagram also designates vectorially the peak amplitude of the chrominance signal for each of these colors, and the polarities and peak amplitudes of the in-phase and quadrature portions required to form these chrominance signals.

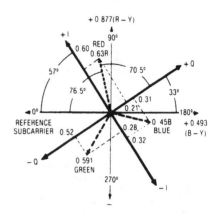

Color-phase diagram.

color picture signal — A signal that represents electrically the three color attributes (brightness, hue, and saturation) of a scene. 2. A combination of the luminance and chrominance signals, excluding all blanking and synchronizing signals.

color picture tube — An electron tube that provides an image in color by scanning a raster and by varying the

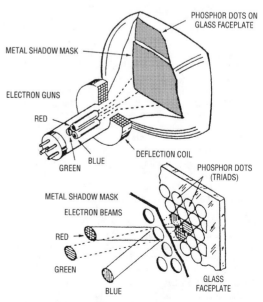

Color picture tube.

intensity at which it excites the phosphors on the screen to produce light of the chosen primary colors.

color primaries — In the color receiver, the saturated colors of definite hue and variable luminance produced by the receiver. These color primaries, when mixed in proper proportions, form other colors.

color purity — Freedom of a color from white light or any colored light not used to produce the desired color. In reference to the operation of a tricolor picture tube, the term refers to the production of pure red, green, or blue illumination of the phosphor-dot faceplate.

color-purity magnet — A magnet placed in the neck region of a color picture tube to alter the electron-beam path and thereby improve color purity.

color registration — The accurate superimposing of the red, green, and blue images that are used to form a complete color picture in a color television receiver.

color rendering index — Abbreviated CRI. A number that approximately represents the effect of a light source on the appearance of colored surfaces. A measure of the degree to which the perceived colors of objects illuminated by the source conform to those of the same objects illuminated by a reference source for specified conditions.

color sampling rate — In a color television system, the number of times per second that each primary color is sampled.

color saturation — 1. The degree to which white light is absent in a particular color. A fully saturated color contains no white light. If 50 percent of the light intensity is due to the presence of white light, the color is said to have a saturation of 50 percent. 2. A measure of the amount of white light in a hue. High saturation means that there is no white-light component and that the color is of good quality.

color sensitivity — The spectral sensitivity of a light-sensitive device such as a phototube or camera tube.

color signal — Any signal at any point in a color television system, used for wholly or partially controlling the chromaticity values of a color television picture. This is a general term encompassing many specific connotations, such as are conveyed by the words "color picture signal," "chrominance signal," "carrier color signal," "monochrome signal" (in color television), etc.

color subcarrier — 1. A subcarrier that is added to the main video signal to convey the color information. In NTSC systems, the color subcarrier is centered on a frequency of 3.579545 MHz, referenced to the main video carrier. 2. A monochrome signal to which modulation sidebands have been added to convey color information.

color sync burst — A burst of 8 to 11 cycles in the 4.43361875 MHz (PAL) or 3.579545 MHz (NTSC) color subcarrier frequency. This waveform is located on the back porch of each horizontal blanking pulse during color transmissions. It serves to synchronize the color subcarrier's oscillator with that of the transmitter in order to recreate the raw color signals. *See also* color burst.

color sync signal — The series of color bursts (pulses of subcarrier reference signal) applied to the back porch of the horizontal-sync pedestal in the composite video signal.

color television receiver — A standard monochrome receiver to which special circuits have been added. Phosphors capable of glowing in the three primary colors are used on the special screen. By using these primary colors and mixing them to produce complementary colors, and by varying their intensity, it is possible to reproduce an image in somewhat the original colors.

color television signal — The complete signal used to transmit a color picture. Included are horizontal-, vertical-, and color-sync components.

color temperature — 1. A way of describing the color of a radiating source in terms of the temperature (in degrees Kelvin) of a blackbody radiating with the same dominant frequency as the source. Certain high-end monitors offer the possibility of setting the color temperature to any desired value. By setting the color temperature, one can often achieve more realistic screen colors. 2. The temperature to which a perfectly black body must be heated to match the color of the source being measured. Color-temperature measurements begin at absolute zero and are expressed in kelvins.

color temperature of a light source — The absolute temperature at which a blackbody radiator must be operated to have a chromaticity equal to that of the light source.

color transmission — 1. The transmission of color-television signals that can be reproduced with different values of hue, saturation, and luminance. 2. The transmission of a signal that represents both the brightness values and the color values in a video picture.

color triad — One cell of a three-color phosphor-dot screen of a phosphor-dot color picture tube. Each triad contains one dot of each of the three color-producing phosphors.

color triangle — A triangle drawn on a chromaticity diagram to represent the entire range of chromaticities obtainable when the three prescribed primaries are added. These are represented by the corners of the triangle.

color video — Usually, the technique of combining the three primary colors of red, green, and blue to produce color pictures within the usual spectrum

Colpitts oscillator — A sinusoidal oscillator using a three-terminal active element, such as a tube, transistor, etc., and a feedback loop containing a parallel *LC* circuit. The capacitance of the *LC* circuit consists of two capacitors in series, forming a voltage divider that serves to match the input and output impedance of the active device.

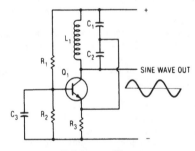

Colpitts oscillator.

column — Also called place. In positional notation, a position corresponding to a given power of the radix. A digit located in any particular column is a coefficient of a corresponding power of the radix.

column binary code — A punched-card code in which successive bits are represented by the presence or absence of punches in adjacent positions in successive columns rather than successive rows. It is used with 36-bit-word computers, in which each group of three columns is used to represent a single word.

column speaker — 1. A loudspeaker cabinet of long columnar shape. Usually the loudspeaker is at one end so that the rear of the drive unit is loaded by a column of air. Often columns are made from drain pipes. Also, the name given to a long speaker cabinet containing several loudspeakers for public address work. 2. An array of loudspeakers arranged in a vertical line, having the property of spreading its radiation through a wide angle in the horizontal plane while keeping it in a beam with respect to the vertical plane.

COM — *See* computer-output microfilm printer.

coma — 1. An aberration of spherical lenses, occurring in the case of oblique incidence, when the bundle of rays forming the image is unsymmetrical. The image of a point is comet shaped, hence the name. 2. A cathode-ray tube image defect that makes the spot on the screen appear comet shaped when away from the center of the screen.

coma lobe — A side lobe that occurs in the radiation pattern of a microwave antenna when the reflector alone is moved to sweep the beam. The lobe appears because the feed is not always at the center of the reflector. This scanning method is used to eliminate the need for a rotary joint in the feed waveguide.

comb amplifier — Several sharply tuned bandpass amplifiers whose inputs are connected in parallel and whose outputs are separate. The amplifiers separate various frequencies from a multifrequency input signal. The name is derived from the comblike appearance of the response pattern of various output peaks displayed along a frequency-base axis.

comb filter — 1. A filter for high resolution and picture sharpness that eliminates "dot crawl" and "hanging dots" caused by cross-color and cross-luminance distortion. Comb filters provide exceptional isolation of chrominance and luminance information and increased picture resolution for sharper, clearer, more detail-perfect color picture reproduction. Capable of delivering 700 lines of horizontal resolution, at the S-video input, it greatly exceeds the maximum clarity of the 330 lines of horizontal resolution delivered by broadcast television and easily supports the demands of videodisc, CD-I, and other advanced high-resolution video media. 2. A type of filter network that is, in effect, a multiple-bandpass design that passes only frequencies within a number of narrow bands, or provides outputs corresponding to each of its passbands. Comb filters are so named because their response characteristics have the appearance of a comb. 3. A filter that has a series of very narrow, deep notches where signals are attenuated.

comb generator — Usually a step-recovery diode circuit that converts a single frequency rf input into an rf output signal that contains a large number of spectral lines, each of which is harmonically related to the input frequency.

combination cable — A cable in which the conductors are grouped in combinations such as pairs and quads.

combination microphone — A microphone consisting of two or more similar or dissimilar microphones combined into one.

combination tones — Frequencies produced in a nonlinear device, such as in an audio amplifier, having appreciable harmonic distortion.

combinatorial logic — Digital circuitry in which the states of the outputs from a device depend only on the states of the inputs. *See also* sequential logic.

combined-gate IC — A single IC chip in which several gate circuits are interconnected to form a more complex circuit.

combiner — A circuit for mixing video, trigger, and scan data from the synchronizer for the modulation of a link.

combiner circuit — A circuit that combines the luminance and chroma channels with the sync signals in color-television cameras.

combustible — *See* flammable.

come-back — A point in the stopband of a filter where a spurious response occurs beyond points at which there is proper attenuation. Come-backs usually occur at frequencies much higher than the passband frequencies because of feed-through in parasitic elements.

command — 1. In a computer, one or more sets of signals that occur as the result of an instruction. *See also* instruction, 1. 2. An independent signal from which the dependent signals in a feedback control system are controlled according to the prescribed system relationships. 3. A signal that initiates or triggers an action in the device that receives the signal. *See* operation code. 4. An issue to the PC to execute a function, such as print, file, erase, or send a document.

commandable bug — A surreptitious listening device that can be turned on and off remotely.

command code — *See* operation code.

COMMAND.COM — A disk file that contains the command processor and must be present on the start-up disk for DOS to run.

command control — A system whereby functions are performed as the result of a transmitted signal.

command destruct signal — A radio signal for destroying a missile in flight.

command file — A computer file that will execute a program when the filename is entered.

command guidance system — A missile guidance system in which both the missile and the target are tracked by radar. The missile is guided by signals transmitted to it while it is in flight. *See also* command link.

command language — A computer source language that consists primarily of procedural operators, each of which is capable of invoking a function to be executed.

command link — The portion of a command guidance system used to transmit steering commands to the missile. *See also* command guidance system.

command module — *See* module.

command net — A communication network that connects an echelon of command with some or all of its subordinate echelons for the purpose of command and control.

command pointer — A multiple-bit register that indicates the memory location being accessed in the control store.

command reference — In a servo or control system, the voltage or current to which the feedback signal is compared. As an independent variable, the command reference exercises complete control over the system output.

comment — An expression that explains or identifies a particular step in a routine but which has no effect on the operation of the computer in performing the instructions for the routine.

comment field — In a computer, an area in a record assigned for entry of explanatory comments about a program.

commercial test (measuring) equipment — Devices used as working instruments selected from suppliers' catalogs as suitable for the user's measuring needs and procured as standard (off-the-shelf) items. This includes measuring devices that are installed, intact, and in consoles. Devices procured from suppliers but modified by user-imposed specifications in such a manner as to affect console performance do not fall into this category.

commercial time-sharing — A type of computer use with remote terminal, interactive, and time-sharing characteristics. The user of the machine pays an amount determined by the time used.

committed power and ground — Patterns that have been permanently established to input/output connectors for power and ground plane busing.

common — 1. Shared by two or more circuits. Used to designate the terminal of a three-terminal device that is shared by the input and output circuits. (Thus, a transistor may be operated in a common-base configuration, a common-collector configuration, or a common-emitter configuration.) Vacuum tube connections may be characterized in a similar way, but *grounded* is normally used instead of *common*. 2. A point that acts as the reference potential for several circuits — a ground. 3. *See* tailor-made.

common base — A circuit configuration in which the base terminal is common to the input circuit and to the output circuit and in which the input terminal is the emitter terminal and the output terminal is the collector terminal.

common-base amplifier — Also called a grounded-base amplifier. A transistor amplifier in which the base element is common to both the input and the output circuit. It is comparable to the grounded-grid configuration of a triode electron-tube amplifier.

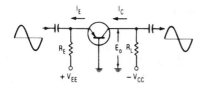

Common-base amplifier.

common-base circuit — A transistor circuit in which the base electrode is common to both input and output circuits.

common-base feedback oscillator — A common-base bipolar transistor amplifier with a feedback network between the collector (output) and the emitter (input) to produce oscillations at a desired frequency.

common-base transistor — Circuit configurations in which the base terminal is common to the input circuit and to the output circuit and in which the input terminal is the emitter terminal and the output terminal is the collector terminal.

common battery — A system of current supply in which all direct-current energy for a unit of a telephone system is supplied by one source in a central office or exchange.

common-battery office — A central office that supplies transmitter and signal current for its associated stations, and for signaling by the central office equipment, from a power source located in the central office.

common business-oriented language — Specific language by which business data-processing procedures may be precisely described in a standard form. Intended not only as a means for directly presenting any business program to any suitable computer for which a compiler exists, but also as a means of communicating such procedures among individuals. Also called COBOL.

common bus system — A set of standard data, address, and control lines available to all computer modules. The use of bus interface circuits makes it possible for a user to tie in and communicate with other users.

common carrier — 1. A government-regulated (by the FCC) private company that furnishes the general public with telecommunications service facilities, for example, a telephone or telegraph company. Specialized common carriers offer private line services. 2. A commercial company that sells communications services to any member of the public. A common carrier cannot provide programming for the channels itself.

common-carrier fixed station — A fixed station that is open to public correspondence.

common-channel interference — Radio interference resulting from two stations transmitting on the same channel. Characterized principally by beat-note generation (heterodyne whistle) and the suppression or capture of the weaker signal by the stronger one.

common-channel signaling — A method of using a single signaling channel to carry signaling information relating to a number of information channels. Common-channel signaling information is sent in packet form.

common collector — A circuit configuration in which the collector terminal is common to the input circuit and to the output circuit and in which the input terminal is the base terminal and the output terminal is the emitter terminal.

common-collector amplifier — Also known as an emitter-follower and a grounded-collector amplifier. A transistor amplifier in which the collector element is common to both the input and the output circuit. This configuration is comparable to an electron-tube cathode follower.

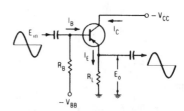

Common-collector amplifier.

common-collector circuit — A bipolar transistor circuit in which the collector is the common (grounded) electrode.

common-collector transistor — Circuit configuration in which the collector terminal is common to the input circuit and to the output circuit and in which the input terminal is the base terminal and the output terminal is the emitter terminal.

common communications carrier — A company recognized by an appropriate regulatory agency as having a vested interest in furnishing communications services.

common emitter — A circuit configuration in which the emitter terminal is common to the input circuit and to the output circuit and in which the input terminal is the base terminal and the output terminal is the collector terminal.

common-emitter amplifier — Also called grounded-emitter amplifier. A transistor amplifier in which the emitter element is common to both the input and the output circuit. This configuration is comparable to a conventional electron-tube amplifier.

common-emitter circuit — A bipolar transistor circuit in which the emitter is the common (grounded) electrode.

common-emitter transistor — Circuit configuration in which the emitter terminal is common to the input circuit and to the output circuit and in which the input terminal is the base terminal and the output terminal is the collector terminal.

common language — A form representing information that a machine can read and that is common to a group of computers and data-processing machines.

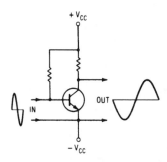

Common-emitter amplifier.

common mode — 1. Signals that are identical with respect to both amplitude and time. Also used to identify the respective parts of two signals that are identical with respect to amplitude and time. 2. A high-speed-modem interface name.

common-mode characteristics — The characteristics pertaining to performance of an operational amplifier in which the inverting and noninverting inputs have a common signal.

common-mode coupling — Coupling that results in similar signals with respect to ground on different circuit leads.

common-mode gain — The ratio of the output voltage of a differential amplifier to the common-mode input voltage. The common-mode gain of an ideal differential amplifier is zero. Typical values in op amps are around −30 dB.

common-mode impedance input — The internal impedance between either one of the input terminals of a differential operational amplifier and signal ground.

common-mode input — That signal applied in phase (i.e., common mode) equally to both inputs of a differential amplifier.

common-mode input capacitance — The equivalent capacitance of both inverting and noninverting inputs of an operational amplifier with respect to ground.

common-mode input voltage — The maximum voltage that can be applied simultaneously between the two inputs of a differential amplifier and ground without causing damage.

common-mode interference — Interference that appears between the terminals of the measuring circuit and ground.

common-mode output voltage — The output voltage of an operational amplifier resulting from the application of a specified voltage common to both inputs.

common-mode range — Maximum voltage that can be applied to differential inputs with respect to ground. The maximum difference between inputs is the full-scale input range

common-mode rejection — Abbreviated CMR. 1. Also called in-phase rejection. A measure of how well a differential amplifier ignores a signal that appears simultaneously and in phase at both input terminals (called a common-mode signal). Usually and preferably stated as a voltage ratio, but more often stated in the dB equivalent of said ratio at a specified frequency, e.g., "120 dB at 60 Hz with a source impedance of 1000 ohms." 2. A measure of the ratio of differential-mode gain to the common-mode gain present in all practical amplifiers. When both inputs to the amplifier are raised by the same voltage, the output should, ideally, be unaffected.

common-mode rejection in decibels — Twenty times the log of the common-mode rejection ratio.

common-mode rejection ratio — Abbreviated CMRR. 1. The ratio of the common-mode input voltage to the output voltage expressed in dB. The extent to which a differential amplifier does not provide an output voltage when the same signal is applied to both inputs. 2. The ratio of differential-mode gain to common-mode gain. 3. The ratio of the change of input offset voltage of an operational amplifier to the change in common-mode voltage producing it.

common-mode resistance — The resistance between the input- and output-signal lines and circuit ground. In an isolated amplifier, this is its insulation resistance. (Common-mode resistance has no connection with common-mode rejection.)

common-mode signal — 1. The instantaneous algebraic average of two signals applied to a balanced circuit (i.e., two ungrounded inputs of a balanced amplifier), with all signals referred to a common reference. 2. In an amplifier with a differential input, a signal (referred to ground) that appears at both inverting and noninverting inputs with the same phase, amplitude, and frequency. Power-line hum is the most frequently encountered common-mode signal.

common-mode voltage — Abbreviated CMV. 1. The amount of voltage common to both input lines of a balanced amplifier. Usually specified as the maximum voltage that can be applied without breaking down the insulation between the input circuit and ground. (Common-mode voltage has no connection with common-mode rejection.) 2. The voltage component of a two-wire input (control) signal that is common to both lines. The actual control signal is the difference between the two input voltages; for example, if the two input signals are (with respect to a common ground) 110 volts and 100 volts, then the common-mode voltage is 100 volts, and the control voltage is 10 volts. 3. An undesirable signal picked up in a transmission line by both wires making up the circuit, to an equal degree, with respect to an arbitrary "ground."

common-mode voltage gain — The ratio of ac voltage with respect to ground at the output terminal of an amplifier (or between the output terminals of an amplifier with differential outputs) to the common-mode input voltage.

common-mode voltage range — The range of voltage that may be applied to both inputs of an operational amplifier without saturating the input stage. This may limit the output capabilities in the voltage-follower connection.

common pool — A dedicated area of memory used as storage and shared by various processes.

common-user channels — Communication channels that are available to all authorized agencies for transmission of command, administrative, and logistic traffic.

common-user circuit — A circuit shared by two or more services, either concurrently or on a time-sharing basis. It may be a unilateral, bilateral, or joint circuit.

communal chained memory — A technique employed in dynamic storage allocation in a computer.

communicating word processor — *See* electronic mail.

communication — 1. The transmission of information from one point, person, or equipment to another. 2. The sensing of a measurement signal or phenomena for display, recording, amplification, transmission, computing, or processing into useful information. 3. Transmission of intelligence between points of origin and reception without alteration of the sequence or structure of the information content.

communication band — The band of frequencies due to the modulation (including keying) necessary for a given type of transmission.

communication channel — Part of a radio or wire circuit, or a combination of wire and radio that connects two or more terminals.

communication control character — A character whose purpose is to control or facilitate data transmission over communication networks.

communication engineer — An electrical engineer who specializes in the design, construction, operation, or maintenance of communication circuits, equipment, or systems whether by wire or radio (wireless).

communication facilities — Installations, equipment, and personnel used to provide telecommunication.

communication-line controller — A hardware unit that performs line-control functions with a modem.

communication link — The physical means of connecting one location to another for the purpose of transmitting and receiving information.

communications common carriers — Companies that furnish communications services to the public, regulated by the FCC or appropriate state agencies.

communications port — A connection on a terminal through which data is input and/or output.

communications program — Software that gives a computer the ability to communicate with other computers over telephone lines.

communications protocol — A set of rules that govern the communications between computers over telephone lines. Both computers must have the same settings and follow the same standards for communication to be successful.

communications receiver — A receiver designed for reception of voice or code signals from stations operating in the communications service.

communications satellite — An orbiting space vehicle that actively or passively relays signals between communications stations.

communications security — The protection resulting from all measures designed to deny unauthorized persons information of value that might be derived from the possession and study of telecommunications or to mislead unauthorized persons in their interpretations of the results of such possession and study.

communication switch — A device used to execute repetitive sequential switching.

communication zone indicator — A device that indicates whether or not long-distance high-frequency broadcasts are successfully reaching their destination.

community antenna television — Abbreviated CATV. A television system that receives and retransmits television broadcasts. Microwave transmitters and coaxial cables are used to bring the television signals to subscribers in a community.

community bulletin board — A dial-up computer used to exchange messages.

community dial office — A small dial-telephone office that serves an exchange area and that operates with no employees located in the building.

community television system — A receiving system by means of which television signals may be distributed over coaxial cables to homes in an entire community.

community TV cable — Coaxial cables used to transmit television signals from a master antenna to a group of receivers in a community.

commutating capacitor — Also called speedup capacitor. 1. In a flip-flop circuit, a capacitor connected in parallel with the cross-coupling resistor to accelerate the transition from one stable state to another. 2. The capacitor connected in parallel between SCR stages to momentarily reverse the current through the SCR so as to cut it off (commutate).

commutating filter — 1. A bandpass filter whose center frequency depends only on the frequency with which it is driven. 2. A clocked, switched-capacitance digital filter that uses periodic signal sampling techniques to synthesize discrete approximations of all the classic analog filters.

commutation — 1. A mechanical process of converting the alternating current in the armature of direct-current generators into the direct-current generator output. 2. Sampling of various quantities in a repetitive manner for transmission over a single channel. 3. The switching of currents back and forth between various paths as required for operation of some system or device. In particular, a switching of current to or from the appropriate armature coils of a motor or generator. The turning off of an active element at the correct time as in an inverter or power controller. 4. The transfer of load current from one thyristor to another; sometimes used as a means of turning off a thyristor by rapidly reducing its current to zero.

commutation capacitors — 1. Cross-connected capacitors in a thyratron inverter. They provide a path such that the start of conduction in one thyratron causes an extinguishing pulse to be applied to the alternate thyratron. Also used in inverter circuits employing semiconductor devices. 2. A specially designed capacitor used in the turn-off (commutation) circuit of an SCR, where it is subjected to exceedingly fast rise time pulses. Thus the capacitor must be capable of discharging large peak currents in very short periods of time.

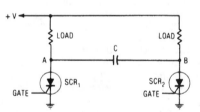

Commutation capacitor.

commutation switch — A device used to carry out repetitive sequential switching.

commutator — 1. The part of the armature to which the coils of a motor are connected. It consists of wedge-shaped copper segments arranged around a steel hub and insulated from it and from one another. The motor brushes ride on the outer edges of the commutator bars and thereby connect the armature coils to the power source. 2. Device used in a direct-current generator to reverse the direction of an electric current and maintain a current flowing in one direction. 3. A switch or equivalent device that permits the reversal or exchange of external connections of a transducer to provide a desired sequencing of signals.

compact cassette — A small ($4 \times 2\frac{1}{2} \times \frac{1}{2}$ inch or $10.2 \times 6.3 \times 1.3$ cm) tape cartridge developed by Philips, containing tape about $\frac{1}{8}$-inch (3.2 mm) wide, running at $1\frac{7}{8}$ ips (12.1 cm/s). Recordings are bidirectional, with both stereo tracks adjacent for compatibility with monophonic cassette recorders, whose heads scan both stereo tracks at once.

compact disc (CD) — 1. A read-only plastic disc that uses optical storage techniques to store large amounts of music (audio) or digitally encoded data. 2. A stereo digital audio system based upon a 12 cm single-sided polycarbonate disc with an internal reflective layer of aluminum or gold. Information is read by sensing the presence or absence of reflected light from a tightly focused laser beam pointing at the pits and bumps in the reflective surface within the disc. The digital audio signal is sampled to 16 bits per channel at a rate of 44.1 kHz. 3. Small plastic disc on which digital information is recorded. The data thus recorded can be played back using a laser. The packing density on the disc is very high, enabling hundreds of megabytes of data to be stored.

compactron — An electron tube based on a building-block concept that involves the standardizing of basic tube sections, diodes, triodes, and pentodes; clipping them together as required; and sealing them in a single envelope.

compander — 1. A combination consisting of a compressor at one point in a communication path to reduce the volume range of signals, followed by an expander at another point to improve the ratio of the signal to the interference entering the path between the compressor and expander. 2. A device consisting of a compressor at the transmitting end and an expander at the receiving end that operate as nonlinear amplifiers to obtain a more advantageous amplitude-quantizing relationship for the reduction of noise. The process of companding is particularly important with audio signals. The principles have also been effectively used in the processing of picture information. 3. A complementary noise-reduction system in which both pre- and postprocessing are used to provide no alteration of signal while reducing noise. It improves the signal-to-noise ratio in systems in which an analog signal is passed through a noisy transmission medium. 4. An electronic circuit that amplifies low-input signal amplitudes more than high amplitudes so that all signal amplitudes are compressed into a narrow range. This compressed signal can be expanded back into the original signal by amplifying large amplitudes more than smaller amplitudes. 5. A device used on some telephone channels to improve transmission performance. The equipment compresses the outgoing speech volume range and expands the incoming volume range on a long-distance telephone circuit.

companding — A process in which compression is followed by expansion. Companding is often used for noise reduction, in which case the compression is applied at the transmitter before the noise exposure and the expansion at the receiver.

compandor — 1. Abbreviation for compressor-expander. An audio-frequency circuit that reduces (compresses) loud sounds and increases (expands) quiet sounds so that the output level stays almost constant. A compressor is used in the transmitter. In the receiver, an expander would follow the demodulator and precede the audio power amp. 2. *See* compander.

companion keyboard — A remote keyboard connected by a multiwire cable to an ordinary keyboard and able to operate it.

comparator — 1. A circuit that compares two signals and supplies an indication of agreement or disagreement. 2. In a computer, a circuit that determines whether the absolute difference between a data sample and the previous sample passed is greater than or equal to a redundancy criterion (which may be a tolerance or a limit). 3. A device that compares two inputs for equality. One type compares voltages and gives one of two outputs: less than or greater than. Another type compares binary numbers and has three outputs: less than, equal to, or

greater than. A third type compares phase or frequency and gives a variable voltage depending on the relationship between the inputs. 4. A unit often found in audio showrooms, which, by switch selection, will connect up a combination of speakers, amplifier, tuner, pickup, tape player, etc., for comparing different types. 5. A circuit that compares two signals and provides a difference signal. 6. An active device that provides a logical 0 when its input is below a preset reference value and a logical 1 when its input is above that value. 7. A device that compares two different signals and provides an output when they differ in frequency, phase, voltage, or power level. 8. A circuit that evaluates an output parameter to determine if it falls within some predetermined limits. 9. A device for checking the accuracy of one set of data by comparing it with a second set of data and then noting any variation between the two.

compare — 1. A computer operation in which two quantities are matched for the purpose of discovering their relative magnitudes or algebraic values. 2. A computer instruction that effectively subtracts one word from another and indicates which of the two is larger.

comparison — The examination of how two similar items of data are related. The comparison is usually followed by a decision.

comparison bridge — A type of voltage-comparison circuit resembling a four-arm electrical bridge. The elements are so arranged that if a balance exists in the circuit, a zero error signal is derived.

comparison testing — Real-time comparison between the actual output responses of the device under test and those of a known good reference device when the same input stimulus patterns are applied to both devices in parallel.

compatibility — 1. That property of a color television system that permits unaltered monochrome receivers to receive substantially normal monochrome from the transmitted signal. 2. The property that makes possible use of a stereo system with a monophonic program source, or reproduction of a stereo program monophonically on a monophonic system. 3. The ability of one unit to be used with another without detrimental effect on the signal through mismatch. For example, a compatible pickup will play both mono and stereo records. 4. The ability to run the same software programs and connect the same peripherals and add-on equipment (e.g., boards, printers, modems) as another PC. When a machine is said to be compatible, it is more often than not compatible with the IBM PC. Compatible PCs are also referred to as clones.

compatible — A term applied to a computer system that implies that it is capable of handling both data and programs devised for some other type of computer system.

compatible color — A TV broadcast system that produces a color signal that can be received by either a black-and-white or a color set. The luminance values (the basis of black-and-white reception) and the chrominance values (the basis of color reception) are broadcast as different portions of the total signal so that the luminance values are not dependent on the chrominance values for reproduction.

compatible integrated circuit — A hybrid IC in which the active circuit element is within the silicon planar integrated structure. A passive network, which may be separately optimized, is deposited onto its insulating surface to complete the IC device.

compatible monolithic integrated circuit — A device in which passive components are deposited by thin-film techniques on top of a basic silicon-substrate circuit containing the active components and some passive parts. *See also* all-diffused monolithic integrated circuit.

compensated amplifier — 1. A broadband amplifier whose frequency range is extended by the proper choice of circuit constants. 2. A wideband amplifier made so by the addition of low- and high-frequency compensation.

compensated-impurity resistor — A diffused-layer resistor into which are introduced additional n- and p-type impurities.

compensated-loop direction finder — A direction finder employing a loop antenna and a second antenna system to compensate for polarization error.

compensated semiconductor — A semiconductor in which one type of impurity or imperfection (donor) partially cancels the electrical effects of the other (acceptor).

compensated volume control — *See* loudness control.

compensating filter — A filter used to alter the spectral emission of an emulsion to a specified response to different wavelengths.

compensation — 1. The controlling elements that compensate for, or offset, the undesirable characteristics of the process to be controlled in the system. 2. The shaping of an op-amp frequency response in order to achieve stable operation in a particular circuit. Some op amps are internally compensated, whereas others require some external compensation components in some circuits. 3. The phenomenon whereby extremely small quantities of donor and acceptor impurities present in a semiconductor crystal tend to cancel out each other, so that the material tends to behave according to the dominant impurity only. If both types of impurity are present to an equal extent, the material tends to behave as if it were an intrinsic material.

compensation signal — A signal recorded on a tape, along with the computer data and on the same track as the data; this signal is used during the playback of data to electrically correct for the effects of tape-speed errors.

compensation theorem — An impedance in a network may be replaced by a generator of zero internal impedance, the generated voltage of which at any instant is equal to the instantaneous potential difference produced across the replaced impedance by the current flowing through it.

compensator — 1. In a direction finder, the portion that automatically applies to the direction indication all or part of the correction for the deviation. 2. An electronic circuit for altering the frequency response of an amplifier system to achieve a specified result. This refers to record equalization or loudness correction.

compile — 1. To bring digital-computer programming subroutines together into a main routine or program. 2. To produce a binary coded program in a computer from a program written in source (symbolic) language by selecting appropriate subroutines from a subroutine library as directed by the instructions or other symbols of the source program. The linkage is supplied for combining the subroutines into a workable program, and the subroutines and linkage are translated into binary code.

compiler — 1. An automatic coding system in a computer that generates and assembles a program from instructions written by a programmer. 2. A unit that converts computer programs written in higher-level languages, such as BASIC and C++, into the machine language (object code) of the computer. It is necessary to write an entire program into a compiler's memory before the compiler executes or performs a translation on it. 3. Computer routine that translates symbolic instructions to machine instructions and replaces certain items of input with series of instructions, called subroutines.

4. A high-level, Englishlike programming language that converts instructions into executable machine code. Two examples of such programming languages are COBOL and FORTRAN. 5. A high-level language processor that converts or translates a sequence of source language statements into a corresponding sequence of machine language instructions that may be later loaded into memory and executed by the processor to perform the desired functions. No matter how many times a section of code is used in the assembled program, it will be translated only once and put in its proper place. 6. A program that converts a high-level language into machine language for a specific microprocessor/computer. 7. A program that translates high-level language programs into a series of machine-code instructions for a computer to execute; it may also check the programs' semantic consistency.

compiler language — A computer language system consisting of various subroutines that have been evaluated and compiled into one routine that can be handled by the computer. FORTRAN, COBOL, and ALGOL are compiler languages. Compiler language is the third level of computer language. *See* machine language, 3, for other levels. *See also* high-level language.

compiler program — Software, usually supplied by the manufacturer, to convert an application program from compiler language to machine language.

compiling routine — A routine by means of which a computer can itself construct the program used to solve a problem.

complement — 1. In an electronic computer, a number whose representation is derived from the finite positional notation of another by one of the following rules.

True complement: Subtract each digit from 1 less than the base, then add 1 to the least significant digit and execute all required carries.
Base-minus-1s complement: Subtract each digit from 1 less than the base. (For example, 9s complement in base 10, 1s complement in base 2, etc.)

2. To form the complement of a number. (In many machines, a negative number is represented as a complement of the corresponding positive number.) The binary opposite of a variable or function. The complement of 1 is 0 and the complement of 0 is 1; thus, for example, the complement of 011010 is 100101.

complementary — 1. A term describing integrated circuits that employ components of both polarity types connected in such a way that operation of either is complemented. A complementary bipolar circuit would employ both npn and pnp transistors, and a complementary MOS circuit (CMOS) would employ both n-channel and p-channel devices. In general, complementary devices operate with opposite polarity voltages and currents, which is advantageous in many circuit applications. 2. Two driving-point functions whose sum is a positive constant.

complementary binary or inverted binary — The negative true binary system. It is similar to the binary code except that all binary bits are inverted. Thus, zero scale is all 1s while full scale is all 0s.

complementary circuit — A circuit that provides push-pull operation (sink and source capability) with a single input.

complementary clocks — Two clock signals with opposite phase.

complementary colors — Two colors are complementary if, when added together in proper proportion such as by projection, they produce white light.

complementary constant current logic (C³L) — A high-density approach to bipolar LSI that has switching speeds of 3 nanoseconds.

complementary metal-oxide semiconductor — Device formed by the combination of a PMOS and an NMOS (p-type and n-type channel semiconductors), exhibiting very low power consumption and high noise immunity. Strictly speaking, CMOS refers to an IC manufacturing technology; the term is almost always used to describe an IC logic family. The CMOS logic family is characterized by very low power dissipation, low circuit density per chip, and moderate speed of operation when compared with other IC logic families. *See also* complementary MOS; ECL; I²L; Schottky TTL; TTL.

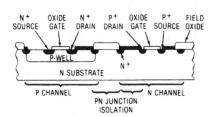

Conventional CMOS.

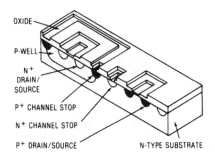

CMOS process.

Complementary metal-oxide semiconductor (CMOS).

complementary MOS — Abbreviated CMOS. 1. Pertaining to n- and p-channel enhancement-mode devices fabricated compatibly on a silicon chip and connected into push-pull complementary digital circuits. These circuits offer low quiescent power dissipation and potentially high speeds, but they are more complex than circuits in which only one channel type (generally p-channel) is used. 2. A digital inverter consisting of a p-channel and an n-channel enhancement-mode field-effect transistor. The transistors are connected in series across the power supply with gates linked together as the input.

complementary operator — The logic operator whose result is the NOT of a given logic operator.

complementary push-pull — A power amplifier in which the output transistors are of complementary polarities (i.e., pnp and npn). In some amplifiers of this kind the driven transistors also constitute a complementary pair.

complementary rectifier — Half-wave rectifying circuit elements that are not self-saturating rectifiers in the output of a magnetic amplifier.

complementary silicon-controlled rectifier — A pnpn semiconductor device that is the polarity complement of the silicon-controlled rectifier.

complementary-symmetry circuit — An arrangement of pnp and npn transistors that provides push-pull operation from one input signal.

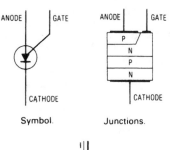

Symbol.   Junctions.

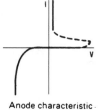

Anode characteristic.

Complementary silicon-controlled rectifier.

complementary tracking — A system of interconnection of two or more devices in which one (the master) operates to control the others (the slaves).

complementary transistor amplifier — An amplifier that utilizes the complementary symmetry of npn and pnp transistors.

complementary transistor logic — A digital logic circuit configuration making use of a complementary transistor emitter coupled AND-OR gate. Basically, a two-level diode gate using simultaneous npn and pnp action.

complementary transistors — Two transistors that have opposite conductivity (pnp and npn) and usually have matching electrical characteristics; npn and pnp pairs with similar electrical characteristics.

complementary unijunction transistor — Abbreviated CUJT. 1. An integrated semiconductor structure with characteristics similar to those of a unijunction transistor, but complementary to other unijunction transistors in the way that pnp transistors are complementary to npn transistors. 2. A silicon planar device similar to a unijunction transistor except that the operating currents and voltages are of the opposite polarity. The electrical characteristics are stable, consistent, and predictable over a wide temperature range. The CUJT will operate from a 5-volt supply and is therefore compatible with integrated circuits. Typically, the case is electrically connected to the substrate and must be isolated from the circuit.

complementary wave — A wave brought into existence at the ends of a coaxial cable or two-conductor transmission line, or any discontinuity along the line.

complementary wavelength — The wavelength of light of a single frequency. When combined with a sample color in suitable proportions, the wavelength matches the reference-standard light.

complementing — Changing each binary 1 into a 0 and each 0 into a 1.

complement number — A number that, when added to another number, gives a sum equal to the base of the numbering system. For example, in the decimal system, the complement of 2 is 8. Complement numbers are used in some computer systems to facilitate arithmetic operations.

complement number system — A system of number handling in which operations are performed on the complement of the actual number. The system is used in some computers to facilitate arithmetic operations.

complementor — A circuit or device that produces a Boolean complement. A NOT circuit.

complete carry — A system of executing the carry process in a computer. All carries, and any other carries to which they give rise, are allowed to propagate to completion in this system.

complex components — Indivisible and nonrepairable components having more than one function.

complex function — An integrated device in which three or more circuits are integral to a single silicon chip. In addition, the circuits are interconnected on the chip itself to form some electronic function at a higher level of organization than a single circuit. The interconnection pattern for the function is predetermined by the fixed mask; no wiring discretion is available for yield purposes. The input and outputs of all the circuits are not normally exposed to the package terminals. An example of a complex function would be a full adder or a multibit serial shift register.

complex instruction set computer — *See* CISC.

complex parallel permeability — The complex relative permeability measured under stated conditions on a core with the aid of a coil. The parameter characterizing the induction is the impedance of the coil when placed on the core, expressed as a parallel connection of reactance and resistance. The parameter characterizing the field strength is the reactance the coil would have if placed on a core of the same dimensions but with unity relative permeability, the distribution of the magnetic field being identical in both cases. The coil should have negligible copper losses.

complex permeability — Under stated conditions, the complex quotient of the moduli of the parallel vectors representing induction and field strength in a material. One of the moduli varies sinusoidally with time, and the component chosen from the other modulus varies sinusoidally at the same frequency.

complex series permeability — The complex relative permeability measured under stated conditions on a core with the aid of a coil. The parameter characterizing the induction in the core is the impedance of the coil when placed on the core, expressed as a series connection of reactance and resistance. The parameter characterizing the field strength is the reactance this coil would have if placed on a core of the same dimensions but with unity relative permeability. The coil should have negligible copper losses.

complex steady-state vibration — A periodic vibration of more than one sinusoid. It includes repeating square waves, sawtooth waves, etc., because these waveforms can be expressed in terms of a Fourier series of sinusoidal terms.

complex target — A radar target made up of a number of reflecting surfaces that, taken together, are smaller in all dimensions than the resolution capability of the radar.

complex tone — A sound wave produced by the combination of simple sinusoidal components of different frequencies. A sound sensation characterized by more than one pitch.

complex wave — A periodic wave made up of a combination of several frequencies or several sine waves superimposed on one another.

complex-wave generator — A device that generates a nonsinusoidal signal having a desired repetitive characteristic and waveform.

compliance — 1. The reciprocal of stiffness; that is, the ability to yield or flex. 2. The ease with which a phonograph stylus can be deflected by the groove wall.

Expressed in microcentimeters per dyne (10^{-8} m/dyne) as the distance through which the stylus will be deflected by a force of 1 dyne. Typical values are from 10 to 50 microcentimeters per dyne. In general, the higher the compliance, the better the low-frequency tracking at a given tracking force. Compliance can be measured in several ways — static and dynamic. 3. The mechanical and acoustical equivalent of capacitance. 4. The flexibility of a speaker cone's suspension. High compliance is important in a woofer for accurate reproduction of low-frequency signals of large amplitude.

compliance extension — A form of master/slave interconnection of two or more current-regulated supplies to increase their output voltage range through series connection.

compliance range — The range of voltage needed to sustain a given constant current throughout a range of load resistances.

compliance voltage — The output voltage of a dc power supply operating in constant-current mode.

compliance voltage range — The output voltage range of a dc power supply operating in a constant-current mode.

component — 1. An essential functional part of a subsystem or equipment. It may be any self-contained element with a specific function, or it may consist of a combination of parts, assemblies, accessories, and attachments. 2. In vector analysis, one of the parts of a wave, voltage, or current considered separately. 3. A packaged functional unit consisting of one or more circuits made up of devices, which in turn may be part of an operating system or subsystem. A part of, or division of, the whole assembly or equipment. Normally interchangeable with *unit*. 4. In high fidelity, a specialized item of equipment designed to do a particular part of the work in a sound system. 5. Any of the basic parts used in building electronic equipment, such as a resistor or capacitor.

component density — The number of components contained in a given volume or within a given package or chip. The quantity of components on a printed board per unit area.

component layout — The physical arrangement of the components in a chassis or printed circuit.

component level bus — The set of input and output pins, with defined functions and timing, through which a microprocessor sends and receives signals.

component operating hours — A unit of measurement for the period of successful operation of one or more components (of a specified type) that have endured a given set of environmental conditions.

component part — A term sometimes used to denote a passive device.

component placement equipment — Automatic systems for sorting and placing components onto hybrid circuit substrates or printed circuit boards, consisting of indexing conveyor, sorter, placement heads, missing component detector, programmable electropneumatic control, and options to handle special requirements.

component population — The variety and number of components (transistors, resistors, transformers, etc.) necessary to perform the desired electrical function.

component side — That side of a printed board on which most of the components will be mounted.

component stress — Those factors of usage or test, such as voltage, power, temperature, frequency, etc., that tend to affect the failure rate of component parts.

composite cable — A cable in which conductors of different gages or types are combined under one sheath.

composite circuit—A circuit that can be used simultaneously for telephony and direct-current telegraphy or signaling, the two being separated by frequency discrimination.

composite color signal—The color picture signal plus all blanking and synchronizing signals. Includes luminance and chrominance signals, vertical- and horizontal-sync pulses, vertical- and horizontal-blanking pulses, and the color-burst signal.

composite color sync—The signal comprising all the sync signals necessary for proper operation of a color receiver. Includes the deflection sync signals to which the color sync signal is added in the proper time relationship.

composite conductor—Two or more strands of different metals, such as aluminum and steel or copper and steel, assembled and operated in parallel.

composite controlling voltage—The voltage of the anode of an equivalent diode, combining the effects of all individual electrode voltages in establishing the space-charge limited current.

composited circuit—A circuit that can be used simultaneously for telephony and for direct-current telegraphy or signaling, separation between the two being accomplished by frequency discrimination.

composite dialing—In telephone operations, a method of dialing between distant offices over one leg of a composite set.

composite filter—A filter with two or more sections.

composite guidance system—A guidance system using a combination of more than one individual guidance system.

composite picture signal—The television signal produced by combining a blanked picture signal with the sync signal.

composite signal—The stereo FM broadcast modulation signal consisting of a 19-kHz pilot tone, L + R information and L − R information modulated on a suppressed 38-kHz carrier and (if any) a 67-kHz FM carrier with a ±6-kHz deviation SCA channel.

composite sync signal—The position of the composite video signal that synchronizes the scanning process.

composite TV signal—A combination of video picture, color, audio, and synchronization information.

composite video—A standard type of video in which red, green, and blue signals are mixed together.

composite video signal—1. The complete video signal, containing both picture and sync information. For monochrome, it consists of the picture signal and the blanking and synchronizing signals. For color, color-synchronizing signals and color-picture information are added. 2. The combined signals in a television transmission, including the picture signal, vertical and horizontal blanking, and synchronizing signals. 3. The complete video signal, consisting of the chrominance and luminance information as well as all sync and blanking pulses.

composite wave filter—A combination filter consisting of two or more low-pass, high-pass, bandpass, or band-elimination filters.

composite wire clad—A wire having a core of one metal to which is fused an outer shell of one or more different metals.

composition—The conversion of computer files containing text and typesetting commands into a format for input to an imaging system, such as a phototypesetter.

composition resistor—*See* carbon resistor.

compound-connected transistor—Two transistors that are combined to increase the current amplification factor at high emitter currents. This combination is generally employed in power-amplifier circuits.

compound-filled transformer—A transformer that is contained within a case and in which the structural insulating material is supplemented by submergence in a solid or semisolid insulating material introduced into the case in a fluid state.

compound horn—An electromagnetic horn of rectangular cross section. The four sides of the horn diverge in such a way that they coincide with or approach four planes, with the provision that the two opposite planes do not intersect the remaining planes.

compounding—A form of noise reduction using compression at the transmitting end and expansion at the receiver. A compressor is an amplifier that increases its gain for lower-power signals. The effect is to boost these components into a form having a smaller dynamic

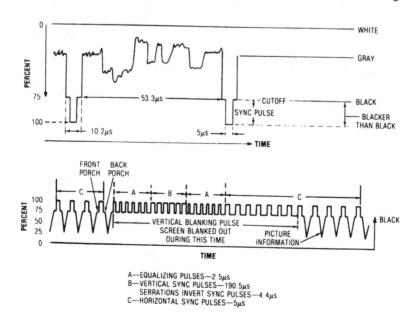

Composite picture signal.

range. A compressed signal has a higher average level and therefore less apparent loudness than an uncompressed signal, even though the peaks are no higher in level. An expander reverses the effect of the compressor to restore the original signal.

compound modulation—*See* multiple modulation.

compound-wound motor—A dc motor having two separate field windings. One, usually the predominant field, is connected in parallel with the armature circuit, and the other is connected in series.

compress—To reduce some parameter of a signal, such as bandwidth, amplitude variation, duration, etc., while preserving its information content.

compressed-air loudspeaker—A loudspeaker that has an electrically actuated valve to modulate a stream of compressed air.

compressed file—A file whose contents have been compressed by a special utility program so that it occupies less space on a disk or other storage device than in its uncompressed (normal) state.

compressed speech—A representation of speech in which some redundant features of the digitized speech have been removed.

compression—1. A process in which the effective amplification of a signal is varied as a function of the signal magnitude, the effective gain being greater for small than for large signals. In television, the reduction in gain at one level of a picture signal with respect to the gain at another level of the same signal. 2. Electronic reduction of the dynamic range so that quiet sounds are raised and loud sounds lowered. The most common application is an "automatic" recording in which it is important that all sounds recorded are made intelligible when played back. Also used when necessary to avoid overrecording and distortion, or to lift the signal level clear of background noise or hum. 3. A technique used to increase the number of bits per second sent over a data link by replacing often-repeated characters, strings, and command sequences with electronic code. When this compressed data reaches the remote end of the transmission link, data decompression is used to restore the data to its normal form for display. 4. The conversion of information to a format that requires fewer bits and can be reversed to its original state once transferred to a new location.

compressional wave—In an elastic medium, a wave that causes a change in volume of an element of the medium without rotation of the element.

compression driver unit—A speaker driver unit that does not radiate directly from the vibrating surface. Instead, it requires acoustic loading from a horn that connects through a small throat to an air space adjacent to the diaphragm.

compression ratio—The ratio between the magnitude of the gain (or amplification) at a reference signal level and its magnitude at a higher stated signal level.

compression seal—A seal made between an electronic package and its leads. The seal is formed as the heated metal, when cooled, shrinks around the glass insulator, thereby forming a tight joint.

compressor—1. A device that performs analog compression. 2. A transducer that, for a given amplitude range of input voltages, produces a smaller range of output voltages. In one important type of compressor, the envelope of speech signals is used to reduce their volume range.

compressor expander—*See* compander.

compromise network—In a telephone system, a network used in conjunction with a hybrid coil to balance a subscriber's loop. The network is adjusted for an average loop length, an average subscriber's set, or both,

and gives compromise (not precision) isolation between the two directional paths of the hybrid coil.

Compton diffusion—An elastic shock between a photon and an electron. The photon is diffused with a lesser energy and the electron acquires a kinetic energy equal to the energy decrease of the photons.

Compton effect—The elastic scattering of photons by electrons. Because the total energy and momentum are conserved in the collisions, the wavelength of the scattered radiation undergoes a change that depends on the scattering angle.

computational stability—The degree to which a computational process remains valid when subjected to such effects as errors or malfunctions.

compute bound—A program that is speed-limited by the computations being performed.

computer—1. Any device capable of accepting information, applying prescribed processes to the information, and supplying the results of these processes; sometimes, more specifically, a device for performing sequences of arithmetic and logical operations; sometimes, still more specifically, a stored-program digital computer capable of performing sequences of internally stored instructions, as opposed to calculators on which the sequence is impressed manually (desk calculator) or from tape or cards (card-programmed calculator). 2. A tool for managing data. It can work with numbers and alphanumeric data such as names, words, addresses, and stock numbers. It can be programmed to repeat the same function over and over. It can logically evaluate information given to it, and act on its own findings. It can store huge volumes of data for future use, reference, and updating, and even "converse" with its operator. 3. A machine in which stored instructions operate on other instructions to modify or alter them. Data words and instruction words of the same size, stored in the same medium, differ only in their function. The same word can be both a data word and an instruction word at different times during the execution of a program. 4. An electrical/electronic device that can accept information, process it mathematically (in accordance with previous instructions), and then provide the results of this processing. 5. An electronic device that uses programmed instructions to monitor and control various types of data in order to solve mathematical problems or control industrial applications. Its instructions are executed in various sequences, as required. 6. A calculating device that processes data represented by a combination of discrete data (in digital computers) or continuous data (in analog computers). 7. A device that manipulates data and makes comparisons according to a series of instructions stored in its memory. By changing the instructions the computer can be made to do a completely different

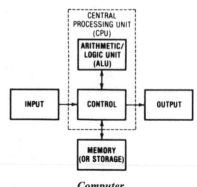

Computer.

task. Instructions and data are stored in the same memory and both can be manipulated by the computer with equal ease. 8. A device that is capable of solving problems or manipulating data by accepting data, performing prescribed operations (mathematical or logical) on the data, and then delivering or applying the results of these operations.

computer access device input — A device that automatically routes to the computer all teletypewriter observation reports that are received in a standard format.

computer-aided design — See CAD.

computer-aided engineering — Abbreviated CAE. An umbrella term that covers all uses of computers in engineering applications. Thus, computer-aided design and computer-aided manufacturing are branches of computer-aided engineering. The subject area is not usually considered to include software engineering.

computer-aided manufacturing — Abbreviated CAM. The use of computer technology to manage, control, and operate manufacturing either through direct or indirect computer interface with the physical and human resources of the company.

computer-aided software engineering — See CASE.

computer-aided tomography — See CAT.

computer architecture — That set of a computer's attributes (such as registers, addressing modes, and instruction set) that are visible to the programmer.

computer assisted tomography — See CAT.

computer code — Also called machine language. The code by which data is represented within a computer system. An example is binary-coded decimal.

computer control — The parts of a digital computer that have to do with the carrying out of instructions in the proper sequence, the interpretation of each instruction, and the application of signals to the arithmetic unit and other parts in accordance with this interpretation.

computer control counter — 1. A counter that stores the next required address. 2. Any counter that provides information to the control unit.

computer diagnosis — The use of data processing systems for evaluation of raw data.

computer entry punch — A combination card reader and keypunch used to enter data directly onto the memory drum of a computer.

computer-generated hologram — A synthetic hologram produced using a computer plotter. The binary structure is formed on a large scale and is then photographically reduced. The holograms are finally etched into a medium.

computer graphics — 1. Computer output in the form of pictorial representation (graphs, charts, drawings, etc.) that is displayed visually, usually by a cathode-ray tube. 2. A person-oriented system that uses a computer to create, transform, and display pictorial and symbolic data.

computer-integrated manufacturing — See CIM.

computer interface — 1. Peripheral equipment for attaching a computer to scientific or medical instruments. 2. A device designed for data communication between a central computer and another unit such as a PC processor.

computer interfacing — The synchronization of digital data transmission between a computer and one or more external I/O devices.

computerized axial tomograph — See CAT.

computerized robot — A servo model run by a computer. The computer controller does not have to be taught by leading the arm-gripper through a routine: new instructions can be transmitted electronically. The programming for such "smart" robots may include the ability to optimize, or improve, its work-routine instructions.

computer language — 1. A system of commands used to develop software for computers (e.g., DOS). 2. The method or technique used to instruct a computer to perform various operations. See high-level language; machine language.

computer-limited — Having to do with the condition in which the time required for computation is greater than the time required to read inputs and write outputs.

computer literacy — 1. Computer and information system comprehension. 2. The ability to use computer technology in a particular discipline.

computer network — Two or more connected computers that have the ability to exchange information.

computer numerical control — Abbreviated CNC. The use of a dedicated computer within a numerical-control unit to perform some or all of the basic numerical-control functions.

computer-output microfilm printer — Abbreviated COM printer. A microfilm printer that will take output directly from the computer, thus substituting for line printer or tape output.

computer polarization holography — A technique used to store wavefront information on thin polarization information-recordable materials (i.e., photochromic crystals) by controlling the polarization angle of a small illuminating spotlight in each sampling cell on a crystal.

computer port — The physical location at which the communication line interfaces to the computer.

computer program — A series of instructions or statements prepared in a form acceptable to the computer, the purpose of which is to achieve a certain result. See software.

computer programmer — A person who designs, writes, debugs, and documents computer programs.

computer programming language — A set of precisely defined structures and syntax (representation, conventions, and rules of use and interpretation) devised to simplify communication with a computer, such as BASIC, FORTRAN, C++, and Java. The greater the power of a higher-level language, the greater is the complexity of information that can be precisely conveyed in an efficient manner.

computer science — 1. The field of knowledge that involves the design and use of computer equipment, including software development. 2. The science of solving problems with computers.

computer system — The computer and its attached peripherals, such as disk drives, monitor, keyboard, and printer.

computer tape — A high-quality magnetic digital recording tape that must be rated at 1600 fci (flux changes per inch) or 530 flux changes per centimeter, or greater.

computer terminal — Peripheral computer equipment for entering and retrieving data. Sometimes incorporates cathode-ray tube for display.

computer user tape system — See CUTS.

computer utility — A network of central computers linked through data communications facilities to remote terminal systems.

computer word — A sequence of bits or characters that is treated as a unit and that can be stored in one computer location. Same as machine word.

computing — Performing basic and more involved mathematical processes of comparing, adding, subtracting, multiplying, dividing, integrating, etc.

computing device — Any electronic device or system that generates and uses timing signals or pulses of more than 10,000 pulses (cycles) per second and uses digital techniques; inclusive of telephone equipment that uses digital techniques or any device or system that generates and uses rf energy for data processing functions, such as electronic computations, operations, transformations, recording, filing, sorting, storage, retrieval, or transfer. (Radio transmitters and receivers; industrial, scientific, and medical equipment; and any other radio-frequency devices specifically subject to an examination requirement are excluded from this definition.)

computing machine — An automatic device that carries out well-defined mathematical operations.

COMSTAR I, II, and III — Three American satellites that can carry video and are operated by AT & T. They are located at 128°, 95°, and 87° west longitude.

concatenate — To unite in a series; to connect together; to chain. Used to describe the action of relating data used by a computer program in some organized manner.

concave — Curved inward.

concentrated-arc lamp — A type of low-voltage arc lamp having nonvaporizing electrodes sealed in an atmosphere of inert gas and producing a small, brilliant, incandescent cathode spot.

concentration gradient — A difference in carrier concentration (holes or free electrons) from point to point in a semiconductor.

concentrator — 1. A device that feeds the signals from several data terminals into a single transmission line for input to a computer, or vice versa. 2. An analog or digital buffer switch used to reduce the required number of trunks. 3. A device for combining many low-speed data lines into one high-speed data line. 4. A device that uses hardware and software to perform computer communication functions. (A term first applied to telephone-switching systems that permitted greater economy in use of facilities by combining many phone circuits into one.) 5. An electronic device that interfaces in a store-and-forward mode with multiple low-speed communication lines at a message level and then retransmits those messages to a processing site via one or more high-speed communication lines.

concentric cable — See coaxial line.

concentric groove — See locked groove.

concentric-lay cable — 1. A concentric-lay conductor. 2. A multiple-conductor cable composed of a central core surrounded by one or more layers of helically laid insulated conductors.

concentric-lay conductor — A conductor composed of a central core surrounded by one or more layers of helically laid wires. In the most common type of concentric-lay conductor, all wires are of the same size, and the central core is a signal wire.

concentric line — See coaxial line.

concentric strand — A strand that consists of a central wire or core surrounded by one or more layers of spirally laid wires. Each layer after the first has six more strands than the preceding layer and is applied in a direction opposite to that of the layer under it.

concentric stranding — A method of stranding wire in which the final wire is built up in layers such that the inner diameter of a succeeding layer always equals the outer diameter of the underlying layer.

concentric-wound coil — A coil with two or more insulated windings that are wound one over the other.

concurrent processing — The ability of a computer to work on more than one program at the same time.

condensation soldering — The immersion of a part to be reflow soldered into a reservoir of hot saturated vapor. As the vapor condenses on the part, the latent heat of vaporizaton is released to heat the part.

condensed mercury temperature — The temperature of a mercury-vapor tube, measured on the outside of the tube envelope, in the region where the mercury is condensing in a glass tube or at a designated point on a metal tube.

condenser — Obsolete term for capacitor in electronics. A device in an ignition circuit that is connected across the contact points to reduce arcing by providing a storage for electricity as points open. Note: The term condenser is still used in the automotive field.

condenser antenna — See capacitor antenna.

condenser microphone — See electrostatic microphone.

condenser speaker — See electrostatic speaker.

condenser tissue — Kraft paper of 0.002 inch (51 μm) or less nominal thickness used in the manufacture of capacitors with paper or paper/film dielectrics.

conditional — 1. In a computer, subject to the result of a comparison made during computation. 2. Subject to human intervention.

conditional breakpoint instruction — A conditional jump instruction that causes a computer to stop if a specified switch is set. The routine then may be allowed to proceed as coded, or a jump may be forced.

conditional jump — Also called conditional transfer of control. 1. An instruction causing a program transfer to an instruction location other than the next sequential instruction only if a specific condition tested by the instruction is satisfied. If the condition is not satisfied, the next sequential instruction in the program line is executed. 2. A computer program transfer that occurs only when the instruction that specifies it is executed and other specified conditions are satisfied.

conditional statement — Also called IF statement. A statement that causes the computer to check something and use that as a basis for choosing among alternative courses of action. Same as a branch.

conditional transfer — A program instruction that causes central control either to process the next instruction in sequence or to jump to some other indicated instruction, depending on the results of some previous operation.

conditional transfer of control — See conditional jump.

condition code — In a computer, a limited group of program conditions, such as carry, borrow, overflow, etc., that are pertinent to the execution of instructions. The codes are contained in a condition codes register.

conditioned line — A telephone circuit that has had its frequency response and/or delay characteristics optimized.

conditioning — 1. Equipment modifications or adjustments required to provide matching of transmission levels and impedances or to provide equalization between facilities. 2. The addition of equipment to a leased voice-grade channel to provide minimum line characteristics necessary for data transmission. 3. Time-limited exposure of a test specimen to a specified environment(s) prior to testing. 4. Applying electronic filtering elements to a communication line to improve its ability to support higher transmission data rates. See also equalization.

condition queue — A queue, declared in the monitor, that lines up blocked processes.

Condor — A cw navigational system, similar to Benito, that automatically measures bearing and distance from a single ground station and displays them in a cathode-ray indicator. The distance is determined by phase comparison, and the bearing by automatic direction finding.

conductance — Symbolized by G or g. 1. In an element, device, branch, network, or system, the physical property that is the factor by which the square of an instantaneous voltage must be multiplied to give the corresponding energy lost by dissipation as heat or other permanent radiation, or by loss of electromagnetic energy from the circuit. 2. The real part of admittance. 3. Reciprocal of resistance; measured in siemens. It is the ratio of current through a material to the potential difference at its ends. 4. The component of sinusoidal current in phase with the terminal voltage of a circuit, divided by that voltage

conducted heat — Thermal energy transferred by thermal conduction.

conducted interference — Any unwanted electrical signal conducted on the power lines supplying the equipment under test, or on lines supplying other equipment to which the one under test is connected.

conducted signals — Electromagnetic or acoustic signals propagated along wire lines or other conductors.

conducted spurious transmitter output — A spurious output of a radio transmitter that is conducted over a tangible transmission path such as a power line, control circuit, radio-frequency line, waveguide, etc.

conductimeter — *See* conductivity meter.

conduction — The transmission of heat or electricity through or by means of a conductor.

conduction band — 1. A partially filled energy band in which electrons can move freely, allowing the material to carry an electric current (with electrons as the charge carriers). 2. The band of energy levels occupied by a valence electron when it is liberated from an atom. Electrical conduction in a semiconductor crystal takes place through the transport of electrons in the conduction band.

conduction current — The power flow parallel to the direction of propagation, expressed in siemens per meter.

conduction-current modulation — 1. Periodic variations in the conduction current passing a point in a microwave tube. 2. The process of producing such variations.

conduction electrons — The electrons that are free to move under the influence of an electric field in the conduction band of a solid.

conduction error — The error in a temperature transducer due to heat conduction between the sensing element and the mounting to the transducer.

conduction field — Energy that surrounds a conductor when an electric current is passed through the conductor, and that, because of the difference in phase between the electrical field and magnetic field set up in the conductor, cannot be detached from the conductor.

conductive adhesive — An adhesive material that has metal powder added to increase electrical conductivity.

conductive epoxy — An epoxy material (polymer resin) that has been made conductive by the addition of a metal powder, usually gold or silver.

conductive gasket — A special highly resilient gasket used to reduce rf leakage in shielding that has one or more access openings.

conductive level detector — A device with single or multiple probes. A change in level completes an electrical circuit between the container and/or probes.

conductive material — A material in which a relatively large conduction current flows when a potential is applied between any two points on or in a body constructed from the material. Metals and strong electrolytes are examples of conductors.

conductive pattern — 1. The arrangement or design of the conductive lines on a printed circuit board. 2. The configuration or design of the conductive material on the base material. Includes conductors, lands, and through connections when these connections are an integral part of the manufacturing process.

conductive pattern-to-board outline — The location of the printed pattern relative to the overall outline dimensions of the printed circuit board.

conductive plastic potentiometer — A potentiometer in which the resistive element consists of a blend of resin (epoxy, polyester, etc.) and processed carbon powder applied to a plastic or ceramic substrate.

conductivity — 1. The conductance between opposite faces of a unit cube of material. The volume conductivity of a material is the reciprocal of the volume resistivity. 2. The ability of a material to conduct electric current. It is expressed in terms of the current per unit of applied voltage. It is the reciprocal of resistivity. 3. The ability to conduct or transmit heat or electricity. 4. The ability of a material to allow electrons to flow, measured by the current per unit of voltage applied. It is also the reciprocal of resistivity. 5. Synonym for conductance. 6. The parameter of a material that indicates the extent to which it permits a net electrical current; normally measured in terms of the conductance in reciprocal ohms (siemens) between opposite faces of a cube of the material measuring 1 centimeter on each side. The conductivity of a material is the reciprocal of its resistivity. 6. Property of a material to allow electrical current to flow with very little loss. For natural surfaces, conductivity in general is increased with increased moisture content.

conductivity meter — Also called conductimeter. An instrument that measures and/or records electrical conductivity.

conductivity modulation — 1. The change in conductivity of a semiconductor as the charge-carrier density is varied. 2. The process whereby the effective electrical conductivity of a semiconductor region is modified by the injection of excess carriers. Thus, excess majority carriers injected into a lightly doped region can cause the effective conductivity to be increased simply by providing further carriers for current conduction. Conversely, excess minority carriers injected into a heavily doped region can cause the effective conductivity to be reduced by increasing the incidence of recombination and, hence, reducing the number of carriers available for conduction.

conductivity-modulation transistor — A transistor in which the active properties are derived from minority-carrier modulation of the bulk of resistivity of the semiconductor.

conductor — 1. A bare or insulated wire or combination of wires not insulated from one another, suitable for carrying an electric current. 2. A body of conductive material so constructed that it will serve as a carrier of electric current. 3. A material (usually a metal) that conducts electricity through the transfer of orbital electrons. 4. A material, such as copper or aluminum, that offers low resistance or opposition to the flow of electric current. 5. A medium for transmitting electrical current. A conductor usually consists of copper, aluminum, steel, silver, or other material. 6. A solid, liquid, or gas that offers little opposition to the continuous flow of electric current. 7. Any material whose valence energy band is only partially filled with electrons, so that empty levels are immediately available for a net electron movement. Such materials conduct electricity readily, even at extremely low temperatures. 8. A substance or body that allows a current of electricity to pass continuously and easily through it. A member of a class of materials that conduct electricity easily, i.e., have a low resistivity (10^{-4} ohm/cm). 9. A single conductive path in a conductive pattern.

conductor side — The side of a single-sided printed board containing the conductive pattern.

conductor spacing — The distance between adjacent edges (not centerline to centerline) of isolated conductive patterns in a conductor layer of a printed circuit.

conductor-to-hole spacing — The distance between a conductor edge and the edge of a component hole.

conduit — 1. A tubular raceway designed for holding wires or cables designed and used expressly for this purpose. It may be a solid or flexible tube in which insulated electrical wires are run. 2. Metal sleeve through which electrical wires pass.

conduit wiring — Wiring carried in conduits and conduit fittings.

cone — The diaphragm that sets the air in motion to create a sound wave in a direct-radiator loudspeaker. Usually it is conical in shape.

cone breakup — The inability of a speaker cone to work as a piston at high frequencies, the effect being that the cone is not under the complete control of the voice coil, certain parts of it moving in opposition to other parts like a rippled rope. Responsible for uneven frequency response.

cone of nulls — A conical surface formed by directions of negligible radiation.

cone of silence — An inverted cone-shaped space directly over the aerial towers of some radio beacons. Within the cone, signals cannot be heard or will be greatly reduced in volume.

conference call — A telephone call that interconnects three or more telephones and permits all parties to converse at random.

confetti — Flecks or streaks of color caused by tube noise in the chrominance amplifier. Because of its colors, confetti is much more noticeable than snow in a black-and-white picture. The chrominance amplifier is therefore cut off during a monochrome program.

confidence — 1. The likelihood, expressed as a percentage, that a measurement or statement is true. 2. The degree of assurance that the stated failure rate has not been exceeded.

confidence factor — The percentage figure expressing confidence level.

confidence interval — A range of values believed to include, with a preassigned degree of confidence, the true characteristic of the lot.

confidence level — 1. The probability (expressed as a percentage) that a given assertion is true or that it lies within certain limits calculated from the data. 2. A degree of certainty.

confidence limits — Extremes of a confidence interval within which there is a designated chance that the true value is included.

configuration — 1. The relative arrangement of parts (or components) in a circuit. 2. A listing of the names and/or serial numbers of the assemblies that make up an equipment. 3. The hardware and/or software making up a system. 4. Combination of computer and peripheral devices at a single installation. 5. A general-purpose computer term that can refer to the way a computer is set up. It is also used to describe the total combination of hardware components that make up a computer system and the software settings that allow various hardware components of a computer system to communicate with one another.

configuration file — A file that contains information on the way a system is set up.

configure — The act of changing software or hardware actions by changing the settings in a computer.

confocal resonator — A wavemeter for millimeter wavelengths. It consists of two spherical mirrors that face each other; a change in the spacing between the mirrors affects the propagation of electromagnetic energy between them, making possible direct measurement of free-space wavelengths.

conformal coating — 1. A thin nonconductive coating, either plastic or inorganic, applied to a circuit for environmental and/or mechanical protection. 2. A protective coating applied to completed printed circuit boards that conforms to the shape of the components and provides complete electrical as well as environmental insulation.

conformance error — The deviation of a calibration curve from a specified curve line.

confusion jamming — An electronic countermeasure by means of which a radar may detect a target, but the radar operator is denied accurate data regarding range, azimuth, and velocity of the target. This result is accomplished through amplification and retransmission of an incident radar signal with distortion to create a false echo. Also called deception jamming.

confusion reflector — A device that reflects electromagnetic radiation to create echoes for purposes of causing confusion of radar, guided missiles, and proximity fuses.

congestion — A condition in which the number of calls arriving at the various inputs of a communications network are too many for the network to handle at once and are subject to delay or loss. (The concept applies in an analogous way to any system in which arriving traffic can exceed the number of servers.)

conical horn — A horn whose cross-sectional area increases as the square of the axial length.

conical scanning — A form of scanning in which the beam of a radar unit describes a cone, the axis of which coincides with that of the reflector.

conjugate — Either of a pair of complex numbers that are mutually related in that their real parts are identical and the imaginary part of one is the negative of the imaginary part of the other, that is, if $a = x + iy$, then $a = x - iy$ is its conjugate.

conjugate branches — Any two branches of a network in which a driving force impressed on one branch does not produce a response in the other.

conjugate bridge — A bridge in which the detector circuit and the supply circuits are interchanged, compared with a normal bridge.

conjugate impedance — An impedance whose value is the conjugate of a given impedance. For an impedance associated with an electric network, the conjugate is an impedance with the same resistance component as the original and a negative reactive component.

conjugate matching — A condition of source-and loading-impedance matching in which the source impedance and the load impedance have equal resistive parts and equal reactance values with opposite signs. This results in maximum power transfer.

connected — A network is connected if, between every pair of nodes of the network, there exists at least one path composed of branches of the network.

connecting block — A cable-termination block in which access to circuit connections is available.

connection — 1. The attachment of two or more component parts so that conduction can take place between them. 2. The point of such attachment.

connection diagram — 1. A diagram showing the electrical connections between the parts that make up an apparatus. 2. A pattern illustrating the connections needed to place an electronic system in operation when such a system includes one or more assemblies, power supplies, and devices being controlled.

connector — 1. A coupling device that provides an electrical and/or mechanical junction between two cables, or between a cable and a chassis or enclosure. 2. A device that provides rapid connection and disconnection of electrical cable and wire terminations. 3. A plug or receptacle that can be easily joined to or separated from its mate. Multiple-contact connectors join two or more conductors with others in one mechanical assembly. 4. A device consisting of a mating plug and receptacle. Various types of connectors include DIP, card-edge, two-piece, hermaphroditic, and wire-wrapping configurations. 5. Devices designed to provide separable through connections in cable-to-cable, cable-to-chassis, or rack and panel applications. 6. A device that holds two parts of a circuit together so that they make electrical contact.

connector assembly— The combination of a mated plug and receptacle.

connector discontinuity— An ohmic change in contact resistance.

connector flange— A projection that extends from or around the periphery of a connector and incorporates provisions for mounting the connector to a panel.

connector receptacle— 1. An electrical fitting with contacts constructed to be electrically connected to a cable, coaxial line, cord, or wire to join with another electrical connector mounted on a bulkhead, wall, chassis, or panel.

connect time— 1. The total time required for establishing a connection between two points. 2. In a computer-based data communications assembly, the switching time required to set up a connection between two terminal points.

conoscope— An instrument for determining the optical axis of a quartz crystal.

consequent poles— Additional magnetic poles present at other than the ends of a magnetic material.

consol— *See* sonne.

console— 1. A cabinet for a radio or television receiver that stands on the floor rather than on a table. 2. Main operating unit in which indicators and general controls of a radar or electronic group are installed. 3. A part of a computer that may be used for manual control of the machine. The computer operator's control panel or terminal. 4. An array of controls and indicators for the monitoring and control of a particular sequence of actions, as in the checkout of a rocket, a countdown, or a launch.

console operator— A person who monitors and controls an electronic computer by means of a central control unit or console.

consonance— Electrical or acoustical resonance between bodies or circuits not connected directly together.

constant— 1. An unvarying or fixed value or data item. 2. Any number not expected to change.

constant-amplitude recording— In disc recording, a relationship between the modulations in the groove and the electrical signals making them so that the width of the groove (the excursions of the cutting stylus) is proportional to the amplitude, or power, of the signal. In playback, a similar relation between the record and the motion of the stylus so that the cartridge produces equal voltage regardless of frequency. Crystal and ceramic pickups have a constant amplitude characteristic on playback.

constant current— 1. A current that does not undergo a change greater than the required precision of the measurement when the impedance of the generator is halved. 2. Having to do with a type of power-supply operation in which the output current remains at a preset value (within specified limits) while the load resistance varies, resulting in an output-voltage variation within the voltage range of the power supply.

constant-current characteristic— The relationship between the voltages of two electrodes, the current to one of them as well as all other voltages being maintained constant.

constant-current/constant-voltage supply— A power supply that behaves as a constant-voltage source for relatively large values of load resistance and as a constant-current source for relatively small values of load resistance. The crossover point between these two modes of operation occurs when the value of the critical load resistance equals the value of the supply voltage setting divided by the supply current setting.

constant-current modulation— Also called Heising modulation. A system of amplitude modulation in which the output circuits of the signal amplifier and carrier-wave generator or amplifier are directly and conductively coupled by a common inductor. The inductor has an ideally infinite impedance to the signal frequencies and therefore maintains the common plate-supply current of the two devices constant. The signal-frequency voltage thus appearing across the common inductor also modulates the plate supply to the carrier generator or amplifier, with corresponding modulation of the carrier output.

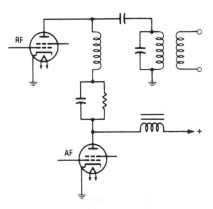

Constant-current modulation.

constant-current power supply— A regulated power supply that acts to keep its output current constant in spite of changes in load, line, or temperature. Thus, for a change in load resistance, the output current remains constant to a first approximation, while the output voltage changes by whatever amount necessary to accomplish this.

constant-current transformer— 1. A transformer that automatically maintains a constant current in its secondary circuit under varying conditions of load impedance when supplied from a constant-potential source. 2. A transformer that supplies a constant current to a varying load. Regulation is accomplished by either varying the separation between primary and secondary windings or by the use of a resonant network.

constant-delay discriminator— *See* pulse demoder.

constant K-filter— An image-parameter filter comprising a tandem connection of a number of identical prototype L-section filters. Each adjacent pair of L-sections together forms either a T- or π-network. The product of the series and shunt impedances is a constant that is independent of frequency.

constant K-network— A ladder network in which the product of its series and shunt impedances is independent of frequency within the range of interest.

constant-luminance transmission— A type of transmission in which the transmission primaries are a luminance primary and two chrominance primaries.

constant-power-dissipation line — A line super-imposed on the output static characteristic curves and representing the points of collector voltage and current, the product of which represents the maximum collector power rating of a particular transistor.

constant-ratio code — A code in which the combinations that represent all characters contain a fixed ratio of ones to zeros.

constant-resistance network — A network that will reflect a constant resistance to the output circuit of the driving amplifier when terminated in a resistive load. Loudspeakers do not reflect a constant impedance; therefore, an amplifier does not see a constant resistance. This disadvantage may be somewhat compensated for by the use of negative feedback by the amplifier.

constant-ringing drop — Abbreviated CRD. A relay that when activated even momentarily will remain in an alarm condition until reset. A key is often required to reset the relay and turn off the alarm.

constant-velocity recording — In disc recording, a relationship between the wiggles in the groove and the electrical signals making them, whereby the frequency of the signal determines the degree of excursion of the cutter. In playback, a similar relation between the recorded wiggles and the motion of the stylus so that the cartridge produces voltages that vary in strength, or amplitude, as the frequency in the groove varies. Magnetic cartridges have a constant velocity characteristic and must be equalized by special networks during playback.

constant voltage — 1. Voltage that does not undergo a change greater than the required precision of the measurement when the impedance of the generator is doubled. 2. Having to do with a type of power-supply operation in which the output voltage remains at a preset value (within specified limits) while the output current is varied within the range of the power supply.

constant-voltage charge — A charge method for a rechargeable battery (storage battery) in which voltage at the battery terminals is held at a constant value throughout the charge cycle.

constant-voltage charger — A battery charger that maintains a constant output voltage so that the charging current tapers off as the battery becomes charged. When fully charged, the battery and the charger supply only minor variations in the load current.

constant-voltage/constant-current cross over — Behavior of a power supply in which there is automatic conversion from voltage stabilization to current stabilization (and vice versa) when the output current reaches a preset value.

constant-voltage/constant-current (cv/cc) output characteristic — A regulated power supply that acts as a constant-voltage source for comparatively large load resistances and as a constant-current source for comparatively small load resistances.

constant-voltage power supply — 1. A regulated power supply that acts to keep its output voltage constant in spite of changes in load, line, or temperature. Thus, for a change in load resistance, the output current changes by whatever amount necessary to accomplish this. 2. A power supply capable of maintaining a fixed voltage across a variable load resistance and over a defined input voltage and frequency change. The output is automatically controlled to maintain constant the product of output current times load resistance.

constant-voltage transformer — A transformer delivering a fixed predetermined voltage over a limited range of input-voltage variations (e.g., 95–125 volts).

constructive synthesis — Synthetic voice generation system that builds words from a prescribed set of linguistic or phonetic sound segments, such as phonemes. Each language has its own set of such sound segments.

contact — 1. One of the current-carrying parts of a relay, switch, or connector that is engaged or disengaged to open or close the associated electrical circuits. 2. To join two conductors or conducting objects in order to provide a complete path for current flow. 3. The juncture point to provide the complete path. 4. The conducting part of a connector that acts with another such part to complete or break a circuit; contacts provide a separable through connection in a cable-to-cable, cable-to-box, or box-to-box situation. 5. The disc or bar of precious metal on a key, jack, or relay spring that touches another similar contact, thus making a temporary, low-resistance electrical connection through which current can flow.

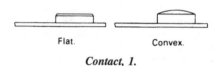

Flat.   Convex.

Contact, 1.

contact-actuation time — The time required for any specified contact on a relay to function. When not otherwise specified, it is the initial actuation time. For some purposes, it is preferable to use either the final or effective actuation time.

contact arc — The electrical (current) discharge that occurs between mating contacts when the circuit is being disestablished.

contact area — The common area between two conductors or a conductor and a connector through which the flow of electricity takes place.

contact arrangement — 1. The combination of contact forms that make up the entire relay-switching structure. 2. The number, spacing, and positioning of contacts in a connector.

contact bounce — 1. The uncontrolled making and breaking of contact when relay contacts are closed. 2. Internally caused intermittent and undesired opening of closed contacts of a relay, caused by one or more of the following: (*a*) Impingement of mating contacts. (*b*) Impact of the armature against the coil core on pickup or against the backstop on dropout. (*c*) Momentary hesitation, or reversal, of the armature motion during the pickup or dropout stroke.

contact bounce time — The time interval from initial actuation of a relay contact to the end of bounce brought about during pickup or dropout or from external causes.

contact chatter — 1. The undesired vibration of mating contacts during which there may or may not be actual physical contact opening. If there is no actual opening but only a change in resistance, it is referred to as dynamic resistance and appears as "grass" on the screen of an oscilloscope having adequate sensitivity and resolution. 2. A sustained rapid physical opening and closing of contact points caused by mechanical vibrations.

contact combination — 1. The total assembly of contacts on a relay. 2. Sometimes used for contact form.

contact current — The peak current through the contacts of a contact-triggered system at the instant the contacts open.

contact device — A device that when actuated opens or closes a set of electrical contacts; a switch or relay.

contact emf — A small voltage established whenever two conductors of different materials are brought into contact.

contact follow — The displacement of a stated point on the contact-actuating member following the initial closure of a contact.

contact force — 1. The amount of force exerted by one of a pair of closed contacts on the other. 2. The force exerted by the moving mercury on a stationary contact or electrode in a mercury switch.

contact gap — Also called contact separation. The distance between a pair of mating relay contacts when they are open.

contact length — The length of travel of one contact while touching another contact during the assembly or disassembly of a connector.

contactless vibrating bell — A vibrating bell whose continuous operation depends on application of an alternating current without circuit-interrupting contacts, such as those used in vibrating bells operated by direct current.

contact load — The electrical power demands encountered by a contact set in any particular application.

contact microphone — 1. A microphone designed to pick up mechanical vibrations directly from the sound source and convert them into corresponding electrical currents or voltages. 2. A microphone designed for attachment directly to a surface of a protected area or object; usually used to detect surface vibrations.

contact miss — Failure of a contact mating pair to establish the intended circuit electrically. This may be a contact resistance in excess of a specified maximum value.

contact modulator — Also called electromechanical chopper. A switch used to produce modified square waves having the same frequency as, and a definite phase relationship to, a driving sine wave.

contact noise — The random fluctuation of voltage across a junction through which current is flowing from one solid to another.

contactor — 1. A device for the purpose of repeatedly establishing or interrupting an electric power circuit. 2. A heavy-duty relay used to control electrical circuits.

contactor alarm — A signal calling attention to lowered pressure in a cable gas-pressure system.

contact potential — Also called Volta effect. 1. The difference of potential that exists when two dissimilar, uncharged metals are placed in contact. One becomes positively charged and the other negatively charged, the amount of potential depending on the nature of the two metals. 2. The potential difference between the contacting surfaces of two metals that have different work functions.

contact-potential difference — The difference between the work functions of two materials, divided by the electronic charge generated by them.

contact pressure — The amount of pressure holding a set of contacts together.

contact printing — A method of screen printing in which the screen is almost in contact with the substrate. Used for printing with metal mask.

contact rating — The electrical power-handling capability of relay or switch contacts under specified environmental conditions and for a prescribed number of operations.

contact rectifier — A rectifier consisting of two different solids in contact. Rectification is due to the greater conductivity across the contact in one direction than in the other.

contact resistance — 1. Total electrical resistance of a contact system, such as the resistance of a relay or a switch measured at the terminals. Usually this resistance is only a fraction of an ohm. 2. The ohmic resistance between the contacts of a switch connector or relay. It may an be extremely small value — typically in the milliohm range. Contact resistance is normally measured from terminal to terminal. 3. The resistance between the wiper terminal and the resistive element of a potentiometer. 4. Electrical resistance of pin and socket contacts when assembled in a connector under typical service use. Electrical resistance of each pair of mated pin and socket contacts in the connector assembly is determined by measuring from the pin to the extreme terminal end of the socket (excluding both crimps) when carrying a specified test current. (Overall contact resistance includes wire-to-wire measurement.) 5. In electronic elements, such as capacitors or resistors, the apparent resistance between the terminating electrode and the body of the device.

contact resistance variation — Abbreviated crv. The maximum instantaneous charge in contact resistance that results from moving the wiper of a potentiometer from one position to another. It is expressed as a percentage of the potentiometer's total resistance.

contact retainer — A device used to retain a contact in an inset or body; it may be either on the contact or in the insert.

contact retention — The minimum axial load a contact in a connector can withstand in either direction while remaining firmly fixed in its normal position in the insert.

contacts — In a relay, the current-carrying parts that engage or disengage to open or close electrical circuits.

contact separation — The maximum distance between the stationary and movable contacts when the circuit is broken.

contact spring — 1. A current-carrying spring to which the contacts are fastened. 2. A non-current-carrying spring that positions and tensions a contact-carrying member.

contact symbology diagram — Commonly referred to as a ladder diagram, it expresses the user-programmed logic of the controller in relay-equivalent symbology.

contact wetting — The coating of a contact surface with an adherent film of mercury.

contact wipe — The distance of travel (electrical engagement) of one contact during its engagement with or separation from another or during mating or unmating of the connector halves.

contaminated — Made radioactive by addition of a radioactive material.

content-addressable memory — Memory in which information is retrieved by specifying the data rather than the address at which the data is stored.

content-addressed storage — See associative storage.

content indicator — A display device that indicates the content in a computer, and the program or mode in use.

contention — 1. A condition that occurs on a multidrop communication channel when two or more locations attempt to transmit simultaneously. 2. Unregulated bidding for a line by multiple users. 3. Competition for use of the same communication facilities; a line-control method in which terminals request or bid to transmit.

contents — The information stored in any part of the computer memory.

contiguous allocation — An allocation method that assigns adjacent sectors to a file.

Continental code — See International Morse code.

continuity — 1. A continuous path for the flow of current in an electric circuit. 2. In radio broadcasting, the prepared copy from which the spoken material is presented.

continuity check — A test performed on a length of finished wire or cable to determine if electrical current flows continuously throughout the length. Conductors may also be checked against each other to ascertain that there are no shorts.

continuity test — An electrical test for determining whether a connection is broken.

continuity writer — In radio broadcasting, the person who writes the copy from which the spoken material is presented.

continuous carrier — A carrier over which transmission of information is accomplished by means which do not interrupt the carrier.

continuous commercial service — See CCS.

continuous-current rating — The designated rms alternating or direct current that a device can carry continuously under specified conditions.

continuous data — Any set of data whose information content can be ascertained continuously in time.

continuous duty — The ability of a device or a control to operate continuously with no off or rest periods.

continuous duty rating — The rating applied to equipment if operated for an indefinite length of time.

continuous load — A load in which the maximum current is expected to continue for three hours or more.

continuously loaded cable — A submarine cable in which the conductors are continuously loaded.

continuous output power — The maximum power (in watts) that an amplifier will deliver from each channel (with all channels operating) without exceeding its rated harmonic distortion. Measured with a 1-kHz signal. Power ratings should include harmonic distortion and the load impedance (4, 8, or 16 ohms). For example, continuous output power 40 W/40 W (at less than 1 percent harmonic distortion, into 8-ohm load).

continuous power — The power an amplifier is capable of delivering for at least 30 seconds with a sine-wave signal.

continuous power output — See rated power output.

continuous rating — The rating that defines the load that can be carried for an indefinite length of time.

continuous recorder — A recorder that makes its record on a continuous sheet or web rather than on individual sheets.

continuous scan thermograph — Equipment for presenting a continuous scan image of the thermal pattern (thermogram) of a patient or an object on a cathode-ray tube.

continuous spectrum — The spectrum that exhibits no structure and appears to represent a continuous variation of wavelength from one to the other.

continuous variable — A variable that may assume any value within a defined range.

continuous-wave radar — A system in which a transmitter sends out a continuous flow of radio energy to the target, which reradiates (scatters) the energy intercepted and returns a small fraction to a receiving antenna.

continuous waves — Abbreviated cw. Electromagnetic sine waves generated as a continuous train of identical oscillations. They can be interrupted according to a code, or modulated in amplitude, frequency, or phase in order to convey information.

continuous-wave tracking system — A tracking system that operates by keeping a continuous radio beam on a target and determining its behavior from changes in the antenna necessary to keep the beam on the target.

contour control system — In automatic control of machine tools, a system in which the cutting path of a tool is controlled along two or more axes.

contourograph — A device in which a cathode-ray oscilloscope is used to produce images that have a three-dimensional appearance.

contrahelical — In the wire and cable industry, the term is used to mean the direction of a layer with respect to the previous layer. Thus, it would mean a layer spiraling in an opposite direction from the preceding layer within a wire or cable.

CONTRAN — A computer-programming language in which instructions are written at a compiler level, thereby eliminating the need for translation by a compiling routine.

contrast — 1. The actual difference in density between the highlights and the shadows. Contrast is not concerned with the magnitude of density, but only with the difference in densities. 2. Amplitude ratio between picture white and picture black. 3. Ratio between the maximum and minimum brightness values in a picture. 4. In optical character recognition, the differences between the color or shading of the printed material on a document and the background on which it is printed. 5. A noticeable difference in color, brightness, or other characteristics in a side-by-side comparison. 6. The difference in tone between the lightest and darkest areas of a photographic print or television image. 7. The ratio between the brightest and darkest parts of a picture.

contrast control — 1. A method of adjusting the contrast in a television picture by changing the amplitude of the video signal. 2. With respect to television, a potentiometer that allows variation of the intensity of the different elements of an image and that can be used to accentuate the highlights and shadows in an image. In a color television system, saturation and hue may also be controlled.

contrast range — The ratio between the whitest and blackest portions of a television image.

contrast ratio — 1. Ratio of the maximum to the minimum luminance values in a television picture or a portion thereof. 2. The ratio of total display element of luminance to the background luminance.

control — Also called a control circuit. 1. In a digital computer, those parts that carry out the instructions in proper sequence, interpret each instruction, and apply the proper signals to the arithmetic unit and other parts in accordance with the interpretation. 2. Sometimes called a manual control. In any mechanism, one or more components responsible for interpreting and carrying out manually initiated directions. 3. In some business applications, a mathematical check. 4. In electronics, a potentiometer or variable resistor. 5. In an alarm system, any mechanism that sequences the interrogation of protected site units, resets latched alarms, and performs similar functions.

control ampere turns — The magnitude and polarity of the control magnetomotive force required for operation of a magnetic amplifier at a specified output.

control amplifier — See preamplifier.

control block — A storage area through which information of a particular type required for control of the operating system is communicated among the parts of the system.

control cable — A multiconductor cable made for operation in control or signal circuits, usually flexible, relatively small in size, and with relatively small current ratings.

control card — In computer programming, a card containing input data or parameters for a specific application of a general routine.

control center — See preamplifier.

control character — 1. A character whose occurrence in a particular context initiates, modifies, or halts operation. 2. An element of a character set that may produce some action in a device other than a printed or displayed character. A character may become a control character in some systems by being preceded by a special character or set of characters. 3. Within a code set, a character intended to initiate, modify, or stop a control function.

control characteristic — 1. A plot of the load current of a magnetic amplifier as a function of the control ampere turns for various loads and at the rated supply voltage and frequency. 2. The relationship between the critical grid voltage and the anode voltage of a tube.

control circuit — *See* control.

control circuits — In a digital computer, the circuits that carry out the instruction in proper sequence, interpret each instruction, and apply the proper commands to the arithmetic element and other circuits in accordance with the interpretation.

control-circuit transformer — A voltage transformer utilized to supply a voltage suitable for the operation of control devices.

control-circuit voltage — The voltage provided for the operation of shunt-coil magnetic devices.

control compartment — A space within the base, frame, or column of a machine used for mounting the control panel.

control counter — In a computer, a device that records the storage location of the instruction word to be operated on following the instruction word in current use.

control current — Current that occurs in the control circuit when control voltage is applied.

control data — In a computer, one or more items of data used to control the identification, selection, execution, or modification of another routine, record file, operation, data value, etc.

CONTROL DATA or Control Data — A trademark and service mark of Control Data Corporation in respect to data processing equipment and related services.

control electrode — An electrode on which a voltage is impressed to vary the current flowing between other electrodes.

control field — In a sequence of similar items of computer information, a constant location where control information is placed.

control-flow machine — A parallel-processing architecture with a single central sequence of instruction, carried out by many processors.

control grid — The electrode of a vacuum tube, other than a diode, upon which a signal voltage is impressed to regulate the plate current.

control-grid bias — The average direct-current voltage between the control grid and cathode of a vacuum tube.

control-grid plate transconductance — The ratio of the amplification factor of a vacuum tube to its plate resistance, combining the effect of both into one term.

controlled avalanche — A predictable, nondestructive avalanche characteristic designed into a semiconductor device as protection against reverse transients that exceed its ratings.

controlled-avalanche device — A semiconductor device that has very specific maximum and minimum avalanche-voltage characteristics and is also able to operate and absorb momentary power surges in this avalanche region indefinitely without damage.

controlled-avalanche silicon rectifier — A silicon diode manufactured with characteristics such that, when operating, it is not damaged by transient voltage peaks.

controlled-carrier modulation — Also called variable-carrier or floating-carrier modulation. A modulation system in which the carrier is amplitude modulated by the signal frequencies, and also in accordance with the

envelope signal, so that the modulation factor remains constant regardless of the amplitude of the signal.

controlled-impedance cable — Package of two or more insulated conductors in which impedance measurements between respective conductors are kept essentially constant throughout the entire length.

controlled rectifier — 1. A rectifier employing grid-controlled devices such as thyratrons or ignitrons to regulate its own output current. 2. Also called an SCR (silicon-controlled rectifier). A four-layer pnpn semiconductor that functions like a grid-controlled thyratron.

controller — 1. An instrument that holds a process or condition at a desired level or status as determined by comparison of the actual value with the desired value. 2. A device or group of devices that serves to govern, in some predetermined manner, the electric power delivered to the apparatus to which it is connected. 3. A hardware interface that accepts instructions from a computer and reformats them to program an instrument or peripheral.

controller function — Regulation, acceleration, deceleration, starting, stopping, reversing, or protection of devices connected to an electric controller.

control-line timing — Clock signals between a modem and a communication-line controller unit.

control link — Apparatus for effecting remote control between a control point and a remotely controlled station.

control locus — A curve that shows the critical value of grid bias for a thyratron.

control operator — An amateur radio operator designated by the licensee of an amateur radio station to also be responsible for the emissions from that station.

control panel — A panel having a systematic arrangement of terminals used with removable wires to direct the operation of a computer or punched-card equipment.

control point — 1. A point that may serve as a reference for all incremental commands. 2. The operating position of an amateur radio station where the control operator's function is performed.

control-power disconnecting device — A disconnective device, such as a knife switch, circuit breaker, or pullout fuse block, used for the purpose of connecting and disconnecting the source of control power to and from the control bus or equipment.

control program — A computer program that places another program and its environment in core memory in proper sequence and retains them there until it has finished operating.

control ratio — 1. The ratio of the change in anode voltage to the corresponding change in critical grid voltage of a gas tube, with all other operating conditions maintained constant. 2. Also called programming coefficient. The required range in control resistance of a regulated power supply to produce a 1-volt change in output voltage. Expressed in ohms per volt.

control read-only memory — Abbreviated CROM. A major component in the control block of some microprocessors. It is a ROM that has been microprogrammed to decode control logic.

control rectifier — A silicon rectifier capable of switching or regulating the flow of a relatively large amount of power through the use of a very small electrical signal. These solid-state devices can take the place of mechanical and vacuum tube switches, relays, rheostats, variable transformers, and other devices used for switching or regulating electric power.

control register — Also called instruction register. In a digital computer, the register that stores the current instruction governing the operation of the computer for a cycle.

control section — *See* control unit.

control sequence — In a computer, the normal order of execution of instructions.

control station — Station licensed to conduct remote control of another amateur radio station.

control store — A memory circuit designed to hold the sequence of commands that determines operation of the sequential-state machine. Sometimes referred to as the microprogram store.

control tape — In a computer, a paper or plastic tape used to control the carriage operation of some printing output devices. Also called carriage tape.

control unit — 1. That section of an automatic digital computer that directs the sequence of operations, interprets coded instructions, and sends the proper signals to the other computer circuits to carry out the instructions. Also called control section. 2. A preamplifier unit in an audio setup. Signals from audio sources, e.g., tuner or pickup microphone, are fed into it. Equalization (where necessary) is applied, then the signal is fed to the main amplifier. Volume and tone controls are usually incorporated, together with any necessary program-selection switch.

control variable — The plant inputs and outputs that a control system manipulates and measures to properly control them.

control voltage — Voltage applied to control input terminals of a relay.

control-voltage winding — The motor winding that is excited by a varying voltage at a time phase difference from the voltage applied to the fixed-voltage windings of a servomotor.

control winding — In a saturable reactor, the winding used for applying a controlling magnetomotive premagnetization force to the saturable-core material.

control wire and cable — Any wire that carries current to control a tube, device, relay, or to cause any event without actually carrying the energy controlled in the event.

convection — 1. The motion in a fluid as a result of differences in density and the action of gravity. 2. The transfer of heat from a high-temperature region in a gas or a liquid as a result of movement of masses of the fluid. 3. A conveying, or transference, of heat or electricity by moving particles of matter.

convection cooling — A method of heat transfer that depends on the natural upward movement of the air warmed by the heat dissipated from the device being cooled.

convection current — The amount of time required for a charge in an electron stream to be transported through a given surface.

convection-current modulation — 1. The time variation in the magnitude of the convection current passing through a surface. 2. The process of producing such a variation.

convenience receptacle — An assembly consisting of two or more stationary contacts mounted in a small insulating enclosure that has slots to permit blades on attachment plugs to enter and make contact with the circuit.

convention — A definite formatting method used in electronic diagrams to present the clearest picture of the circuit function. Some common conventions are as follows: (*a*) circuit signal flow from left to right, with inputs on the left and outputs on the right; (*b*) locating various circuit functional stages in the same sequence as the signal flow; (*c*) placing the highest voltage sources at the top of the sheet and the lowest at the bottom; and (*d*) showing auxiliary circuits that are included but are not a main part of the signal flow, such as oscillators and power supplies, on the lower half of the drawing.

convergence — 1. The condition in which the electron beams of a multibeam cathode-ray tube intersect at a specified point. 2. Orientation of the three electron beams in a color TV picture tube so they pass through the same hole in a shadow mask at the same time. 3. In optics, the bending of light rays toward each other, as by a convex or positive lens.

convergence coil — One of the two coils associated with an electromagnet, used to obtain dynamic beam convergence in a color television receiver.

convergence control — A variable resistor in the high-voltage section of a color television receiver. It controls the voltage applied to the three-gun picture tube.

convergence electrode — An electrode whose electric field causes two or more electron beams to converge.

convergence magnet — A magnet assembly whose magnetic field causes two or more electron beams to converge.

convergence phase control — A variable resistor or inductance for adjusting the phase of the dynamic convergence voltage in a color TV receiver employing a three-gun picture tube.

convergence surface — The surface generated by the point at which two or more electron beams intersect during the scanning process in a multibeam cathode-ray tube.

conversation — An interactive exchange of information between two systems or systems users.

conversational mode — A mode of computer operation in which a user is in direct contact with a computer, and interaction is possible between human and machine without the user being conscious of any language or communications barrier.

conversational operation — A type of operation similar to the interactive mode, except that the computer user must wait until a question is posed by the computer before interacting.

conversational system — *See* interactive system.

conversion — 1. The process of changing from one data-processing method or system to another. 2. The process of changing from one form of representation to another. 3. *See* encode, 2.

conversion efficiency — 1. The ratio of ac output power to the dc input power to the electrodes of an electron tube. 2. The ratio of the output voltage of a converter at one frequency to the input voltage at some other frequency. 3. In a rectifier, the ratio of dc output power to ac input power. 4. The ratio of maximum available luminous or radiant flux output to total input power. 5. Of a solar cell, the ratio of the electrical power obtained from the cell to the radiant power falling on the cell.

conversion gain — 1. The ratio of the intermediate-frequency output voltage to the input-signal voltage of the first detector of a superheterodyne receiver. 2. The ratio of the available intermediate-frequency power output of a converter or mixer to the available radio-frequency power input.

conversion loss — The ratio of available input power to available output power under specified test conditions.

conversion rate — The speed at which an analog-to-digital converter or digital-to-analog converter can make repetitive data conversions, or the number of conversions performed per second. It is affected by propagation delay in counting circuits, ladder switches, and comparators; ladder *RC* and amplifier settling times; and amplifier and comparator slew rates and integrating time of dual-slope converters. Conversion rate is specified as a number of conversions per second, or conversion time is specified as a number of microseconds to complete one conversion (including the effects of settling time). Sometimes conversion rate is specified for less than full resolution, thus showing a misleading (high) rate.

conversion speed — The measure of how long it takes an analog-to-digital converter to arrive at the proper output code. It is the time delay between the edge of the pulse that starts conversion and the edge of the signal that indicates completion of the conversion.

conversion time — 1. The length of time required by a computer to read out all the digits in a given coded word. 2. The time required for a complete conversion or measurement by an analog-to-digital converter, starting from a reset condition. In successive-approximation converters, it ranges typically from 0.8 microsecond to 400 microseconds. 3. Time required for an a/d converter to digitize an input signal. Throughput, the reciprocal of conversion time plus acquisition time, is expressed in channels per second.

conversion transconductance — The magnitude of the desired output-frequency component of current divided by the magnitude of the input-frequency component of voltage when the impedance of the output external termination is negligible for all frequencies that may affect the result.

conversion transducer — A transducer in which the signal undergoes frequency conversion. The gain or loss is specified in terms of the useful signal.

conversion voltage gain (of a conversion transducer) — With the transducer inserted between the input-frequency generator and the output termination, the ratio of the magnitude of the output-frequency voltage across the output termination to the magnitude of the input-frequency voltage across the input termination of the transducer.

convert — 1. To change information from one form to another without changing the meaning, e.g., from one number base to another. 2. In computer terminology, to translate data from one form of expression to a different form.

converted data — The output from a unit that changes the language of information from one form to another so as to make it available or acceptable to another machine, e.g., a unit that takes information punched on cards to information recorded on magnetic tape, possibly including editing facilities.

converter — 1. In a superheterodyne receiver, the section that converts the desired incoming rf signal into a lower carrier frequency known as the intermediate frequency. 2. A rotating machine consisting of an electric motor driving an electric generator, used for changing alternating current to direct current. 3. A facsimile device that changes the type of modulation delivered by the scanner. 4. Generally called a remodulator. A facsimile device that changes amplitude modulation to audio-frequency-shift modulation. 5. Generally called a discriminator. A device that changes audio-frequency-shift modulation to amplitude modulation. 6. A conversion transducer in which the output frequency is the sum or difference of the input frequency and an integral multiple of the local-oscillator frequency. 7. A device that accepts an input that is a function of maximum voltage and time and converts it to an output that is a function of maximum voltage only. 8. *See* shaft position encoder. 9. A device capable of converting impulses from one mode to another, such as analog to digital, or parallel to serial, or one code to another. 10. Device in a digital system that transforms information coded in one number system to its equivalent in another number system. Typically, conversion is either decimal-to-binary or binary-to-decimal.

converter tube — A multielement electron tube that combines the mixer and local-oscillator functions of a heterodyne conversion transducer.

converter unit — The unit of a radar system in which the mixer of a superheterodyne receiver and usually two stages of intermediate-frequency amplification are located. Performs a preamplifying operation.

converting — Changing data from one form to another to facilitate its transmission, storage, or the manipulation of information.

convex — Curved outward.

Cook system — An early stereo-disc recording technique in which the two channels were recorded simultaneously with two cutters on different portions (bands) of a record as concentric spirals. The playback equipment consisted of two pickups mounted side by side so that each played at the correct spot on its own band.

Coolidge tube — An X-ray tube in which the electrons are produced by a hot cathode.

coordinate digitizer — A device that transcribes graphic information in terms of a coordinate system for subsequent processing.

coordinated indexing — 1. In a computer, a system in which individual documents are indexed by descriptors of equal rank so that a library can be searched for a combination of one or more descriptors. 2. A computer indexing technique in which the coupling of individual words is used to show the interrelation of terms.

coordinated transpositions — Transpositions that are installed in either electric supply or communication circuits, or in both, for the purpose of reducing induction coupling, and which are located effectively with respect to the discontinuities in both the electric supply and communication circuit.

coordinate system — A way by which a pair of numbers is associated with each point in a plane (or a triplet of numbers is associated with each point in three-dimensional space) without ambiguity.

coordination — A term describing the ability of the lower rating of two breakers in series to trip before the higher-rating one trips.

coordinatograph — A precision drafting instrument used in the preparation of artwork for mask making.

copperclad — A thin coating of copper fused to an aluminum core. Used in some building wires (No. 12 and larger).

copper-covered steel wire — A wire having a steel core to which is fused an outer shell of copper.

copper loss — *See* I^2R loss.

copper-oxide photocell — An early type of nonvacuum photovoltaic cell consisting of a layer of copper oxide on a metallic substrate, with a thin transparent layer of a conductor over the oxide. Light falling on the cell produces a small voltage between the substrate and the conducting layer. This type of cell is extensively used in exposure meters for cameras because it requires no external source of electric power.

copper-oxide rectifier — A metallic rectifier in which the rectifying barrier is the junction between metallic copper and cuprous oxide. A disc of copper is coated with cuprous oxide on one side, and a soft lead washer is used to make contact with the oxide layer.

copper-sulfide rectifier — A semiconductor rectifier in which the rectifying barrier is a junction between magnesium and copper sulfide.

copperweld — A thin coating of copper fused to a steel core. Used in line wire and cable messengers, and stranded with copper for strength or extending flex life.

coprocessor — 1. A device that performs specialized processing in conjunction with the main microprocessor of a system. 2. In a computer, a device that performs specialized processing in conjunction with the main microprocessor. It works in tandem with another central processing unit to increase the computing power of a system. An extra microprocessor to handle certain tasks faster than the main processor. Same as math coprocessor, numeric coprocessor.

copy — 1. To hear a transmission. *See* subject copy. 2. To duplicate a file or program so that one can retain the original and work on the duplicate. Usually refers to duplicating one disk to another. *See also* backup.

copyguard — Also called stop-copy. Trademarked names for processing applied to a prerecorded video tape to prevent unauthorized copying of the recording. Typically the 60-Hz vertical-sync pulses are weakened, with the expectation that when they are further weakened in copying the tape, the image will roll vertically in playback.

copying telegraph — An absolute term for a facsimile system for the transmission of black-and-white copy only.

copy the mail — CB radio term for just listening to the radio without talking much.

Corbino effect — A special case of the Hall effect that occurs when a disc carrying a radial current is placed perpendicularly into a magnetic field.

cord — 1. One or a group of flexible insulated conductors covered by a flexible insulation and equipped with terminals. 2. A small, very flexible insulated cable constructed to withstand mechanical abuse. Generally, a cord is considered to be a size No. 10 and smaller.

cord circuit — A circuit, terminated in a plug at one or both ends, used at a telephone switchboard position in establishing connections.

cordless phone — A communication system that consists of two pieces: a transponder and a portable wireless handset. The transponder answers the telephone call, or processes an outgoing call, and is connected directly to the telephone line. Typically the transponder transmits to the handset on a frequency that is nominally 1.6 MHz (the high end of the broadcast band) by feeding rf into the ac power line. The handset receives the signal through a ferrite bar or loop antenna built into the handset. The handset transmits to the transponder on the radio control band at 49 MHz. The modulation is NBFM (narrow-band FM), which provides essentially noise-free reception both ways over an (approximately) 300-foot operating range.

cordless switchboard — A telephone switchboard in which manually operated keys are used to make connections.

cord sets — Portable cords fitted with any type of wiring device at one or both ends.

cordwood — 1. A sandwich-type construction wherein components lie in a vertical cordwood pattern between horizontal layers. 2. The technique of producing modules by bundling parts as closely as possible and interconnecting them into circuits by welding or soldering leads together.

cordwood module — 1. A high-density circuit module in which discrete components are mounted between and perpendicular to two small, parallel printed circuit boards to which their terminals are attached. 2. A module formed by bundling or stacking parts between a pair of end plates and interconnecting them into circuits by welding or soldering leads together.

core — 1. A magnetic material placed within a coil to intensify the magnetic field. 2. Magnetic material inside a relay or coil winding. 3. In fiber optics, the light-conducting portion of the fiber, defined by the high refractive index region. The core is normally in the center of the fiber, bounded by the cladding. 4. The central, light-transmitting portion of a fiber-optic cable. It must have a higher index of refraction than the cladding. 5. A small magnetic torus of ferrite used to store a bit of information. Core memories can be strung on wires so that organizations of 32K by 18 are possible in a size of $1/2$ inch high by 6 by 6 inches (1.27 by 15.24 by 15.24 cm). Advantages of core memory are that it is nonvolatile and the oldest main storage technology. (This technology is no longer used.)

core instruction set — A complete set of the operators of the instructions of a computer and the types of meanings associated with their operands.

coreless-type induction heater — A device in which an object is heated by induction without being linked by a magnetic core material.

core loss — Also called iron loss. Loss of energy in a magnetic core as the result of eddy currents, which circulate through the core and dissipate energy in the form of heat.

core memory — 1. A magnetic type of memory made up of miniature ferrite toroids, each of which can be magnetized in one direction to represent a 0 and in the other direction to represent a 1. It is a permanent memory, since if the power is removed the stored information remains. Core memory is characterized by low-cost storage and relatively slow memory operating speed. Core memories are nonvolatile, but have destructive readouts. 2. An array of doughnut-shaped ferrite cores whose diameter ranges from 9 to 18 mils (thousandths of an inch) or 37.5 to 75 μm. Core memories are arranged in a stack configuration in which fine wires are strung through the center of the cores, usually by manual methods. The wires supply current, which causes data to be written into and read out from the core. The storage property is a magnetic one in which the orientation of the core molecules is changed to read or write data. Core stacks can contain as few as 1024 bits or as many as 10^6 bits (one megabit). 3. A memory that is characterized by low-cost storage and relatively slow memory operating speed. Core memories are nonvolatile, but have destructive readouts. *See also* internal storage, 1. (This technology is no longer used.)

core plane — A horizontal network of magnetic cores that contains a core common to each storage position.

core rope storage — Direct-access storage in which a large number of doughnut-shaped ferrite cores are arranged on a common axis, and sense, inhibit, and set wires are threaded through individual cores in a predetermined manner to provide fixed storage of digital data. Each core stores one or more complete words instead of a single bit.

core storage — 1. In a computer, a form of high-speed storage that uses magnetic cores. 2. In a calculator, a storage register in which the contents will remain even after the machine has been switched off.

core store — A matrix of small magnetic rings or cores upon which electrical pulses may be stored. The presence of a pulse in a train is recorded by magnetizing a core, the absence of a pulse by leaving a core unmagnetized.

core transformer — A transformer in which the windings are placed on the outside of the core.

core tuning — Adjusting the inductance and thereby the frequency of resonance of a coil by moving a powdered iron or ferrite core in or out of the coil.

core-type induction heater — A device in which an object is heated by induction. Unlike the coreless type, a magnetic core links the induction winding to the object.

core wrap — Insulation placed over a core before the addition of windings.

corner — 1. An abrupt change in direction of the axis of a waveguide. 2. A neighborhood or point at which a curve makes a sharp or discontinuous change of slope.

corner cut — A corner removed, for orientation purposes, from a card to be used with a computer.

corner effect — The rounding off of the attenuation versus frequency characteristic of a filter at the extremes (or corners) of the passband.

corner frequency — 1. The frequency at which the open-loop gain-versus-frequency curve changes slope. For a servo motor, the product of the corner frequency in radians per second and the time constant of the motor is unity. 2. The frequency at which the two asymptotes of the gain-magnitude curves of an operational amplifier intersect. 3. The upper frequency at which 3-dB attenuation occurs in a high-gain amplifier. A cornering circuit usually is introduced to attenuate the high-frequency signals before the natural phase shift of the amplifier becomes greater than 90°. When properly designed, the cornering circuit prevents high-frequency oscillations in feedback amplifiers. The corner frequency is sometimes erroneously referred to as the cutoff frequency.

corner reflector — A reflecting object consisting of two (dihedral) or three (trihedral) mutually intersecting conducting surfaces. Trihedral reflectors are often used as radar targets.

corner-reflector antenna — An antenna consisting of a primary radiating element and a dihedral corner reflector formed by the elements of the reflector.

Corner-reflector antenna.

corona — 1. A luminous discharge of electricity, due to ionization of the air, appearing on the surface of a conductor when the potential gradient exceeds a certain value but is not sufficient to cause sparking. 2. Any electrically detectable, field-intensified ionization that occurs in an insulating system but does not result immediately in catastrophic breakdown. (Corona always precedes dielectric breakdown.) 3. The ionization of gases about a conductor that results in a bluish-purple glow due to the voltage differential between a high-voltage conductor and the surrounding atmosphere. 4. A device used in an electrostatic copier to impart an electrical charge (in the dark) to the photoconductive material (zinc-oxide-coated paper) to make it sensitive to the action of light. 5. The small, erratic current pulses resulting from discharges in voids in a dielectric during voltage stress.

corona discharge — A phenomenon that occurs when an electric field is sufficiently strong to ionize the gas between electrodes and cause conduction. The effect is usually associated with a sharply curved surface, which concentrates the electric field at the emitter electrode. The process operates between an inception voltage and a spark breakdown voltage. These potentials and the current-voltage characteristics within the operating range are affected by the polarity of the corona electrodes as well as the composition and density of the gas in which the discharge occurs.

corona effect — The glow discharge that occurs in the neighborhood of electric conductors where the insulation is subject to high electric stress. With an alternating current, the effect produced when two wires, or other conductors having a great difference of voltage, are placed near each other.

corona endurance — Resistance to corona cutting.

corona extinction voltage — Abbreviated cev. The voltage at which discharges preceded by corona cease as the voltage is reduced. The corona extinction voltage is always lower than the corona start voltage.

corona failure — Failure due to corona degradation at areas of high voltage stress.

corona loss — A loss or discharge that occurs when two electrodes having a great difference of pressure are placed near each other. The corona loss takes place at the critical voltage and increases very rapidly with increasing pressure.

corona resistance — Also called ionization resistance, brush-discharge resistance, slot-discharge resistance, or voltage endurance. 1. That length of time that an insulation material withstands the action of a specified level of field-intensified ionization that does not result in the immediate, complete breakdown of the insulation. 2. The ability of a material to withstand sustained high applied voltage.

corona shield — A shield placed around a high-potential point to redistribute electrostatic lines of force and prevent corona.

corona start voltage — Abbreviated csv. The voltage at which corona discharge begins in a given system.

corona voltage level — The voltage at which corona discharge does not exceed a specified level following the application of a specified higher voltage.

corona voltmeter — A voltmeter in which the peak voltage value is indicated by the beginning of corona at a known and calibrated electrode spacing.

correction — An increment that, when added algebraically to an indicated value of a measured quantity, results in a better approximation to the true value of the quantity.

corrective equalization — *See* frequency-response equalization.

corrective maintenance — The maintenance performed on a nonscheduled basis to restore equipment to satisfactory condition.

corrective network — Also called shaping network. An electric network designed to be inserted into a circuit to improve its transmission or impedance properties, or both.

correed — A glass-enclosed miniature reed switch. It is similar to the ferreed except that it is operated only when there is current through its surrounding winding, and releases when current stops.

correed relay — A device consisting of a hermetically sealed reed capsule surrounded by a coil. It is used as a switching device in telephone equipment.

correlated characteristic — A characteristic known to be reciprocally related to some other characteristic.

correlation — 1. The relationship, expressed as a number between −1 and 1, between two sets of data, etc. 2. A relationship between two variables; the strength of the linear relationship is indicated by the coefficient of correlation. 3. A measure of the similarity of two signals.

correlation detection — A method of detection in which a signal is compared, point to point, with an internally generated reference. The output of such a detector is a measure of the degree of similarity of the input and reference signal. The reference signal is constructed in such a way that it is at all times a prediction, or best guess, of what the input signal should be at that time.

correlation direction finder — A satellite station separated from radar to receive a jamming signal. By correlating the signals received from several such stations, the range and azimuth of many jammers may be obtained.

correlation distance — A term used in tropospheric propagation. The minimum spatial separation between antennas that will give rise to independent fading of the received signals.

correlation orientation tracking and range system — A system generally using a parabolic antenna for the analysis of a narrow band of radar energy for tracking and ranging purposes.

correlation tracking and ranging — A nonambiguous short-base-line, single-station, cw phase comparison system measuring two direction cosines and a slant range, from which space position can be computed.

correlation tracking and triangulation — A trajectory-measuring system composed of several antenna base lines separated by large distances and used to measure direction cosines to an object. From these measurements, the space position is computed by triangulation.

correlation tracking system — A system utilizing correlation techniques in which signals derived from the same source are correlated to derive the phase difference between the signals. This phase difference contains the system data.

corrosion — 1. A chemical action that causes gradual destruction of the surface of a metal by oxidation or chemical contamination. Also caused by reduction of the electrical efficiency between the metal and a contiguous substance or the disintegrating effect of strong electrical currents or ground-return currents in electrical systems. The latter is known as electrolytic corrosion. 2. In semiconductors, a defect in or on the aluminum metallization, usually a white crystalline growth. 3. A material's chemical alteration by electrochemical interaction with its environment. Corrosion reflects a metal's proclivity to return to the more stable compound state from which it was refined. Corrosive reaction represents an essential division of chemical kinetics.

corrosive fluxes — Also called acid fluxes. Fluxes consisting of inorganic acids and salts; they are generally required where the condition of the surface is well below the ideal for rapid wetting by molten solder.

cosecant-squared antenna — An antenna that emits a cosecant-squared beam. In the shaped-beam antenna used, the radiation intensity over part of its pattern in some specified plane (usually the vertical) is proportionate to the square of the cosecant of the angle measured from a specified direction in that plane (usually the horizontal).

cosecant-squared beam — A radar-beam pattern designed to give uniform signal intensity in echoes from distant and nearby objects. It is generated by a spun-barrel reflector. The beam intensity varies as the square of the cosecant of the elevation angle.

cosine law — The law which states that the brightness in any direction from a perfectly diffusing surface varies in proportion to the cosine of the angle between that direction and the normal to the surface.

cosmic noise — Radio static whose origin is due to sources outside the earth's atmosphere. The source may be similar to sunspots, or spots on other stars.

cosmic rays — Any rays of high penetrating power produced by transmutations of atoms in outer space. These particles continually enter the earth's upper atmosphere from interstellar space.

COS/MOS — *See* complementary metal-oxide semiconductor.

coulomb — 1. The quantity of electricity that passes any point in an electric circuit in 1 second when the current is maintained constant at 1 ampere. The coulomb is the unit of electric charge in the mksa system. 2. The measure of electric charge, defined as a charge equivalent to that carried by 6.281×10^{18} electrons.

Coulomb's law — Also called law of electric charges or law of electrostatic attraction. The force of attraction or repulsion between two charges of electricity concentrated at two points in an isotropic medium is proportionate to the product of their magnitudes and is inversely proportionate to the square of the distance between them. The force between unlike charges is an attraction, and the force between like charges is a repulsion.

coulometer — An electrolytic cell that measures a quantity of electricity by the amount of chemical action produced.

Coulter counter — An electronic cell-counting instrument operating on the ion-conductivity principle. Designed by J. R. Coulter. *See also* cell counter.

count — In radiation counters, a single response of the counting system.

countdown — A decreasing tally that indicates the number of operations remaining in a series.

counter — 1. A circuit that counts input pulses. One specific type produces one output pulse each time it receives some predetermined number of input pulses. The same term may also be applied to several such circuits connected in cascade to provide digital counting. Also called divider. 2. In mechanical analog computers, a means for measuring the angular displacement of a shaft. 3. Sometimes called accumulator. A device capable of changing from one to the next of a sequence of distinguishable states upon receipt of each discrete input signal. 4. An arrangement of flip-flops producing a binary word that increases in value by 1 each time an input pulse is received. It may also be called a divider, since successive counter stages divide the input frequency by 2. A counter has a maximum count, depending on its size, called a modulus or mod. For example, a mod-8 counter can count up to 7, and on the eighth input it resets itself back to a count of 0. When it resets, it also provides an output pulse, which could be counted by another counter. 5. A device capable of changing stages in a specified sequence upon receiving appropriate input signals; a circuit that provides an output pulse or other indication after receiving a specified number of input pulses. 6. A memory-type digital building block that counts pulses received at its input and transmits the cumulative total at its output. 7. In relay-panel hardware, an electromechanical device that can be wired and preset to control other devices according to the total cycles of one on and off function. 8. An instrument that detects and records the occurrence of events either for as long as the counter remains energized or over some predetermined period.

counterbalance — A weight, usually adjustable, fitted at the pivot end of a pickup arm. It counters the weight of the pickup head and cartridge unit and allows adjustment of the stylus pressure to the desired value.

counter circuit — A circuit that receives uniform pulses representing units to be counted and produces a voltage in proportion to their frequency.

counterclockwise polarized wave — *See* left-handed polarized wave.

counter-countermeasures — Use of anti-jamming techniques and circuits designed to decrease the effectiveness of electronic countermeasure activities on electronic equipment.

counterelectromotive cell — A cell of practically no ampere-hour capacity used to oppose the line voltage.

counterelectromotive force — Abbreviated counter EMF. A voltage developed in an inductive circuit by an alternating or pulsating current. The polarity of this voltage is at every instant opposite that of the applied voltage.

countermeasures — That part of military science dealing with the employment of devices and/or techniques intended to impair the operational effectiveness of enemy activity.

counterpoise — A system of wires or other conductors, elevated above and insulated from ground, forming a lower system of conductors of an antenna. Used to capacitively couple a radio transmitter to the ground when the ground resistance is high.

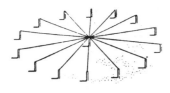

Counterpoise.

counters (Geiger and scintillation) — Instruments used to detect ionizing radiation having very short wavelength (about one-thousandth the wavelength of visible light). Natural sources of this radiation are radium, uranium isotopes, cosmic rays, and ores in which these elements are present; man-contrived sources are the atomic bomb, nuclear reactors used for generating electric power, high-voltage radar CRTs, and X-ray machines.

counter tube — Also called radiation counter tube. An electron tube that converts an incident particle or burst of incident radiation into a discrete electric pulse. This is generally done by utilizing the current flow through a gas that is ionized by the radiation.

counting efficiency — In a scintillation counter, the ratio, under specified conditions, of the average number of photons or particles of ionizing radiation that produce counts to the average number of photons or particles incident on the sensitive area.

counting-rate meter — A device for indicating the time rate of occurrence of input pulses averaged over a time interval.

counting-type frequency meter — An instrument for measuring frequency. Its operation depends on the use of pulse-counting techniques to indicate the number and/or rate of recurring electrical signals applied to its input circuits.

counts — Clicking noises made by a radiation-detecting instrument in the presence of radiation. *See* scintillation counter.

counts per turn — The total number of code positions per 360° of encoder shaft rotation.

couple — Two or more dissimilar metals or alloys in electrical contact with each other that act as the electrodes of an electrolytic cell when they are immersed in an electrolyte.

coupled circuit — Any network containing only resistors, inductors (self and mutual), and capacitors and having more than one independent mesh.

coupled modes — In a waveguide, such as an optical fiber, coaxial cable, or metal pipe, coexisting propagation modes whose fields are interrelated and whose energies are mutually interchanged.

coupler — 1. A passive device that divides an antenna signal to feed two or more receivers, or combines two or more antenna signals to feed a single down lead. A coupler provides some interset isolation and maintains an impedance match between the antenna and receiver. 2. A component that interconnects a number of optical waveguides (fibers) and provides an inherently bidirectional system by mixing and splitting all signals within the component.

coupling — The association or mutual relationship of two or more circuits or systems in such a way that power may be transferred from one to another.

coupling angle — In connection with synchronous motors, the mechanical-degree relationship between the rotor and the rotating field.

coupling aperture — Also called coupling hole or coupling slot. An aperture, in the wall of a waveguide or cavity resonator, designed to transfer energy to or from an external circuit.

coupling capacitor — 1. Any capacitor used to couple two circuits together. Coupling is accomplished by means of the capacitive reactance common to both circuits. 2. A capacitor that allows alternating currents to pass but blocks direct currents. 3. A capacitor intended for the coupling of two or more ac circuits with different dc levels.

coupling coefficient — Also called coefficient of coupling. The degree of coupling that exists between two circuits. It is equal to the ratio between the mutual impedance and the square root of the product of the total self-impedances of the coupled circuits, all impedances being of the same kind.

coupling hole — *See* coupling aperture.

coupling loop — A conducting loop projecting into a waveguide or cavity resonator and designed to transfer energy to or from an external circuit.

coupling loss — 1. *See* fiber optics. Those signal losses due to small differences in numerical aperture, core diameter, core concentricity, and tolerances in splicing connectors when two fibers are aligned. Also known as splicing loss and transfer loss. 2. The total optical power loss within a junction, expressed in decibels, attributed to the termination of the optical conductor. 3. Attenuation of the optical signal due to coupling inefficiencies between the flux source and the optical fiber, or between fibers, or between the fiber and the detector in a receiver. Expressed in decibels.

coupling probe — A probe projecting into a waveguide or cavity resonator and designed to transfer energy to or from an external circuit.

coupling slot — *See* coupling aperture.

coupling transformer — A transformer that couples two circuits together by means of its mutual inductance.

courtesy tone — An audible signal transmitted by a repeater that lets users know the repeater had reset at the end of one person's transmission and is available for use by the next person.

covalent bond — A type of linkage between atoms. Each atom contributes one electron to a shared pair that constitutes an ordinary chemical bond.

coverage — A percentage of the completeness with which a braid or shield covers the surface of an underlying insulated conductor or conductors.

covered relay — A relay contained in an unsealed housing.

Covington and Broten antenna — A compound interferometer in which a long line source is adjacent to a two-element interferometer of comparable aperture, in the same straight line.

CPC — Abbreviation for calling party control. A telephone signaling system that notifies the terminating office and any line-connected auxiliary equipment when the calling party has disconnected. This system permits more efficient use of telephone trunk lines by removing the called party from the line as soon as the calling party disconnects. The CPC signal is a pulse to ground potential, usually 100 ms long, equivalent to shorting the two wires of the phone line together.

cpm — Abbreviation for cycles per minute.

C power supply — A device connected in the circuit between the cathode and grid of a vacuum tube to apply grid bias.

cps — 1. Abbreviation for cycles per second, an obsolete term. Replaced by the term *hertz*, abbreviated Hz. 2. The number of times per second an electronic event is repeated. 3. Abbreviation for characters per second when speaking of data transmission. A data-rate unit, not to be confused with cycles per second.

cps/bps — The number of characters or bytes per second (bits per track per second) written to or read from a magnetic tape.

CPU — Abbreviation for central processing unit. A primary unit of the computer system that controls interpretation and execution of instructions.

CPU portion — *See* chip sets.

crash — 1. A computer condition that causes it to stop working for some reason and need to be restarted by the operator. 2. Abrupt computer failure.

crash-locator beacon — Airborne equipment consisting of various transmitters, collapsible antennas, etc., designed to be ejected from a downed aircraft and to transmit beacon signals to help searching forces to locate the crashed aircraft.

crater lamp — 1. A glow-discharge type of vacuum tube whose brightness is proportional to the current passing through the tube. The glow discharge takes place in a cup or crater rather than on a plate as in a neon lamp. 2. A gaseous lamp usually containing neon. Provides a point source of light that can be modulated with a signal.

crazing — Checking of an insulation material when it is stressed and in contact with certain solvents or their vapors.

CRC — Abbreviation for cyclic redundancy check. 1. A method of error detection consisting of a character generated at the transmitting terminal that is matched with a character at the receiving terminal. Matched characters signify correct character reception; unmatched characters indicate an error. 2. An error-checking control technique utilizing a binary prime divisor that produces a unique remainder.

credence — A measure of confidence in a radar target detection; generally it is proportional to the target-return amplitude.

credit balance indicator — In a calculator, warning light to indicate a negative answer.

creepage — The conduction of electricity across the surface of a dielectric.

creepage distance — The shortest distance between conductors of opposite polarities, or between a live part and ground, measured over the surface of the supporting material.

creepage path — The path across the surface of a dielectric between two conductors. Lengthening the creepage path reduces the possibility of arc damage or tracking.

creepage surface — An insulating surface that provides physical separation between two electrical conductors of different potential.

creep-controlled bonding — A method of diffusion bonding in which enough pressure is exerted to cause significant creep deformation at the joint interfaces. The method is characterized by use of intermediate and low unit loads for a period of hours.

creep distance — The shortest distance on the surface of an insulator between two electrically conductive surfaces separated by the insulator.

creep recovery — The change in no-load output occurring with time after removal of a load that had been applied for a specific period of time.

crest factor — 1. The ratio of the peak voltage to the rms voltage of a waveform (with the dc component removed). 2. An instrument's dynamic range and ability to respond faithfully to waveform peaks as the rms value approaches full scale. Can also refer to the quality of rms-conversion techniques in general.

crest value — Also called peak value. The maximum absolute value of a function.

crest voltmeter — A peak-reading voltmeter.

crest working off-stage voltage — The highest instantaneous value of the off-state voltage that occurs across a thyristor, excluding all repetitive and nonrepetitive transient voltages.

crest working reverse voltage — The highest instantaneous value of the reverse voltage that occurs across a semiconductor diode or reverse-blocking thyristor, excluding all repetitive and nonrepetitive transient voltages.

CRI — *See* color rendering index.

crimp — 1. To compress or deform a connector barrel around a cable so as to make an electrical connection. 2. Final configuration of a terminal barrel formed by the compression of terminal barrel and wire.

crimp contact — A contact whose back portion is a hollow cylinder to allow it to accept a wire. After a bared wire is inserted, a swedging tool is applied to crimp the contact metal firmly against the wire. An excellent mechanical and electrical contact results. A crimp contact often is referred to as a solderless contact.

crimping — A method of attaching a terminal, splice, or contact to a conductor through the application of pressure.

crimping tool — A device used to apply solderless teminals to a conductor.

crimp terminal — A point at which the bared portion of the hookup wire is crimped to either the contact or a tab or pin that mates with the contact terminal.

crimp termination — Connection in which a metal sleeve is secured to a conductor by mechanically crimping the sleeve with pliers, presses, or automated crimping machines. Splices, terminals, and multicontact connectors are typical terminating devices attached by crimping. Suitable for all wire types.

crimp-type termination — Open-barrel or closed-barrel termination in which a stripped wire is inserted into or through a tube that is crimped to the wire with an appropriate tool.

crippled leapfrog test — In a computer, a variation of the leapfrog test in which the test is repeated from a single set of storage locations rather than from a changing set of storage locations. *See also* leapfrog test.

critical area — *See* picture element, 2.

critical angle — 1. The maximum angle at which a radio wave may be emitted from an antenna and will be returned to the earth by refraction in the ionosphere. 2. The maximum angle of incidence for which light will be transmitted from one medium to another. Light approaching the interface at angles greater than the critical angle will be reflected back into the first medium. 3. The maximum angle at which light can be propagated within a fiber. The sine of the critical angle equals the ratio of the numerical aperture to the index of refraction of the fiber core. 4. Basically, the least angle of incidence at which total reflection takes place. The angle of incidence in a denser medium, at an interface between the denser and less dense medium, at which the light is refracted along the interface. When the critical angle is exceeded, the light is totally reflected back into the denser medium. The critical angle varies with the indexes of refraction of the two media with the relationship

$$\sin l_c = n'/n$$

where

l_c is the critical angle,
n' is the refractive index of the less dense medium, and
n is the refractive index of the denser medium.

critical characteristic — A characteristic not having the normal tolerance to variables.

critical coupling — Also called optimum coupling. Between two circuits independently resonant to the same frequency, the degree of coupling that transfers the maximum amount of energy at the resonant frequency.

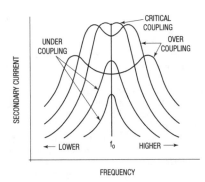

Critical coupling.

critical current — That current, at a specified temperature and in the absence of external magnetic fields, above which a material is normal and below which it is superconducting.

critical damping — The value of damping that provides the most rapid transient response without overshoot. Operation between underdamping and overdamping.

critical dimension — The dimension of a waveguide cross section that determines the cutoff frequency.

critical failure — A failure that causes a system to operate outside designated limits.

critical field — Also called cutoff field. Of a magnetron, the smallest theoretical value of a steady magnetic-flux density, at a steady anode voltage, that would prevent an electron emitted from the cathode at zero velocity from reaching the anode.

critical flicker frequency — The frequency at which a flickering light source is perceived as changing from pulsating to continuous.

critical frequency — Also called penetrating frequency. The limiting frequency below which a magneto-ionic wave component is reflected by an ionospheric layer and above which the component penetrates the layer at vertical incidence. *See also* waveguide cutoff frequency.

critical grid current — The instantaneous value of grid current in a gas tube when the anode current starts to flow.

critical grid voltage — The instantaneous value of grid voltage at which the anode current starts to flow in a gas tube.

critical high-power level — The radio-frequency power level at which ionization is produced in the absence of a control-electrode discharge.

critical inductance — In an inductor input power supply filter, the minimum value of the input inductor needed to ensure that the current drawn through the rectifier never goes to zero.

critical item — *See* critical part.

critical loads — Those loads that will not tolerate a loss of power without incurring major losses in the form of product or equipment damage, or without creation of safety hazards. Critical loads can be subclassified as those that are uninterruptible or those that are short-term interruptible for the time it takes transfer switches to operate.

critical magnetic field — That field intensity below which, at a specified temperature and in the absence of current, a material is superconducting and above which it is normal.

critical part — A part whose failure to meet specified requirements results in the failure of the product to serve its intended purpose. Also called critical item.

critical path — The path that determines a circuit's overall clock rate, typically the longest path in the circuit.

critical potential — *See* ionization potential.

critical race — *See* race

critical rate of rise of off-state voltage — The minimum value of the rate of rise of principal voltage that will cause a semiconductor switching device to switch from the off state to the on state under specified conditions.

critical rate of rise of on-state current — The maximum value of the rate of rise of on-state current that a thyristor can withstand without deleterious effect.

critical resistance value — For a given voltage rating and a given power rating, the one value of resistance that will dissipate full rated power at rated voltage. For values of resistance below the critical value, the maximum (element) voltage is never reached; for values of resistance above the critical value, the power dissipated becomes lower than rated.

critical temperature — That temperature below which, in the absence of current and external magnetic fields, a material is superconducting and above which it is normal.

critical voltage — Also called cutoff voltage. In a magnetron, the highest theoretical value of steady anode voltage, at a given steady magnetic-flux density, at which electrons emitted from the cathode at zero velocity will fail to reach the anode.

critical wavelength — The free-space wavelength corresponding to the critical frequency.

CRO — Abbreviation for cathode-ray oscilloscope.

CROM — Abbreviation for control read-only memory.

Crookes' dark space — *See* cathode dark space.

Crosby system — A compatible multiplex FM stereo broadcast technique in which the right and left signals are combined in phase (sum signal) and transmitted on the main carrier, and also combined out-of-phase (difference signal) and transmitted on the subcarrier. The two signals are combined (matrixed) in the receiving apparatus to restore the right and left channels.

cross alarm — 1. An alarm condition signaled by crossing or shorting an electrical circuit. 2. The signal produced due to a cross alarm condition.

cross assembler — 1. A symbolic language translator that runs on one type of computer to produce machine code for another type of computer. 2. A type of assembler program that generates binary code of a program for a computer other than the model with which it is being used; e.g., an 8080 cross assembler might operate on a PDP-8 minicomputer. 3. Frequently used in conjunction with a down-line load capability for remote control of an unattended microprocessor. *See* assembler; resident assembler.

crossbanding — The use of combinations of interrogation and reply frequencies such that either one interrogation frequency is used with several reply frequencies or one reply frequency is used with several interrogation frequencies.

crossbar — A type of telephone control switching system using a crossbar or coordinate switch. Crossbar switching systems suit data switching because they have low noise characteristics and can handle Touch Tone® dialing.

crossbar switch — 1. A switch having a number of vertical paths, a number of horizontal paths, and an electromagnetically operated mechanism for interconnecting any one vertical path with any one horizontal path. An automatic telephone switching system that uses a crossbar switch. 2. A switch with a plurality of vertical paths and electromechanically operated mechanical means for interconnecting any one of the vertical paths with any of the horizontal paths.

crossbar switching system — A method of switching that, when directed by a common control unit, will select and close a path through a matrix arrangement of switches.

crossbar system — An automatic telephone switching system in which, generally, the selecting devices are crossbar switches. Common circuits select and test the switching paths and control the selecting mechanisms. The method of operation is one in which the switching information is received and stored by controlling mechanisms that determine the operation necessary to establish a connection.

cross beat — A spurious frequency that arises as a result of cross modulation.

crosscheck — To check a computation by two different methods.

cross color — The interference in a color-television receiver chrominance channel caused by crosstalk from monochrome signals.

cross-compiler — A compiler that runs on one computer system but generates machine code for another computer system. Typically it runs on a large computer and generates code for a microcomputer, speeding up software development.

cross-correlation function — A measure of the similarity between two signals when one is delayed with respect to the other.

cross coupling — Unwanted coupling between two different communication channels or their components.

cross-current conduction — Transistor turn-off delay that stems mainly from storage in a converter's switching transistors. The phenomenon appears as a stretching out of the saturated conduction period after base drive is removed.

crossed-pointer indicator — A two-pointer indicator used with instrument landing systems to indicate the position of an aircraft with respect to the glide path.

cross-field device — An electronic device in which electrons from the cathode are influenced by a magnetic field that acts at right angles to the applied electric field. When electrons leave the cathode in a direction perpendicular to the magnetic field, this field causes a force to act at right angles to the electron motion. The electrons then spiral into orbit around the cathode instead

of moving colinearly with the electric field. Most of the electrons gradually move toward the anode, giving up potential energy to the rf field as they interact with the anode slow-wave structure. The tube structure may be either cylindrical or linear.

crossfield recording — A system in which the bias is not applied to the tape by a recording head but by a separate head on the tape's backing side, so that the bias signal will not partially erase high frequencies as they are being recorded.

crossfire — Interfering current in one telegraph or signaling channel resulting from telegraph or signaling currents in another channel.

crossfoot — 1. In a computer, to add or subtract numbers in different fields of the same punch card and punch the result into another field of the same card. 2. To compare totals of the same numbers obtained by different methods.

cross hairs — On a cursor, a horizontal line intersected by a vertical line to indicate a point on the display whose coordinates are desired.

crosshatching — 1. In a printed circuit board, the breaking up of large conductive areas where shielding is required. 2. Process of filling in an outline with a series of symbols to highlight part of a design.

cross magnetostriction — Under specified conditions, the relative change of dimension in a specified direction perpendicular to the magnetization of a body of ferromagnetic material when the magnetization of the body is increased from zero to a specified value (usually saturation).

cross modulation — 1. A spurious response or form of interference that occurs when the carrier of a desired signal intermodulates with the carrier of an undesired signal. This often happens in early stages of radio receivers, particularly when strong signals from local stations drive these stages into nonlinear operation. 2. A form of interference caused by the modulation of one carrier affecting that of another signal. It can be caused by overloading an amplifier as well as by signal imbalances at the headend.

cross modulation distortion — The amount of modulation impressed on an unmodulated carrier when a signal is simultaneously applied to the rf port of a mixer under specified operating conditions. The tendency of a mixer to produce cross modulation is decreased with an increase in conversion compression point and intercept point.

cross neutralization — A method of neutralization used in push-pull amplifiers. A portion of the plate-to-cathode ac voltage of each tube is applied to the grid-to-cathode circuit of the other tube through a neutralizing capacitor.

cross office switching time — The time required for connection of any input through the switching center to any selected output.

crossover — 1. The point where two conductors that are insulated from each other cross. 2. A connection formed between two elements of a circuit by depositing conductive material across the insulated upper surface of another interconnection or element. 3. A point in an integrated or MOS circuit at which an interconnect pattern passes over another conductive part of a circuit but is insulated from it by a thin dielectric layer. *See also* underpass.

crossover distortion — 1. Distortion that occurs in a push-pull amplifier at the points of operation where the input signals cross over (go through) the zero reference points. 2. The type of distortion resulting from class B push-pull power amplifiers owing to the lack of coincidence of the two transfer characteristics at the crossover point. The effect is reduced by applying a critical value of biasing to optimize the quiescent current and hence "linearize" the middle portion of the transfer characteristic. The situation is further improved by heavy negative feedback and by circuit design, such that the crossover distortion from hi-fi amplifiers is very small.

crossover frequency — 1. As applied to electrical dividing networks, the frequency at which equal power is delivered to each of the adjacent frequency channels when all channels are terminated in the specified load. *See also* transition frequency. 2. A frequency at which other frequencies above and below it are separated. In a two-way speaker system, for instance, the crossover frequency is the point at which woofer and tweeter response is divided.

crossover network — 1. An electrical filter that separates the output signal from an amplifier into two or more separate frequency bands for a multispeaker system. 2. A circuit (usually employing capacitors and coils) that feeds low notes to a low-frequency speaker (woofer) and high notes to a high-frequency speaker (tweeter). The crossover frequency is that at which frequency bands divide. Sometimes the audio spectrum is divided into more than two bands to drive more than two speakers. In a three-way system, midrange frequencies go to a midrange driver. Frequencies outside the range of each driver are attenuated at a rate determined by the network design. 3. A frequency at which each of two drivers is receiving half the amplifier's power; below or above that point, one speaker will receive more power than the other.

crossover spiral — *See* lead-over groove.

crosspoint — The operated contacts on a crossbar switch.

cross polarization — 1. The component electric field vector normal to the desired polarization component. 2. Describes signals of opposite polarization being transmitted and received. Cross-polarization discrimination refers to the ability of a feed to detect one polarity and reject the opposite-polarity signals.

cross-reference generator — A device that permits symbols (labels, variables, constants) to be correlated with their storage locations in a computer.

cross-sectional area of a conductor — The summation of all cross-sectional areas of the individual strands in the conductor, expressed in square inches or, more commonly, in circular mils.

cross software — Programs that permit a target system to be developed on a host computer with different CPU architecture.

crosstalk — 1. Interference caused by stray electromagnetic or electrostatic coupling of energy from one circuit to another. 2. Undesired signals from another circuit in the same system. 3. Breakthrough of the signal from one channel to another by conduction or radiation. 4. Transient noise induced on a switching signal by interaction with other switching transitions. 5. Audio interference from one track of a stereo tape to another. Poor head alignment often causes this. 6. Leakage of recorded signal from one channel of a stereo device into the adjacent channel or channels. Crosstalk between stereo channels impairs stereo separation; crosstalk between reverse-direction track signals can be heard, backward, during quieter parts of the desired program. 7. A measure of the amount of signal input to an off channel that appears at the output of a multiplexer, superimposed on the signal passed through the on channel. This is a direct function of the frequency of the signals, since semiconductor switches are capacitively coupled within the IC chip. The higher the frequency, the greater the crosstalk. This phenomenon is similar to the feedthrough problem of signal-handling devices. 8. In a transmission line, the

noise generated in a passive signal line due to information traveling down an active line in the cable. 9. Measurable leakage of optical energy from one optical conductor to another. 10. Optical coupling between a pair of optical fibers caused by light leakage.

crosstalk coupling — Also called crosstalk loss. Cross coupling between speech communication channels or their component parts. Note: Crosstalk coupling is measured between specified points of the disturbing and disturbed circuit and is preferably expressed in decibels.

crosstalk level — 1. The volume of crosstalk energy, measured in decibels, referred to a base. 2. The effective power of crosstalk, expressed in decibels below 1 mW.

crosstalk loss — *See* crosstalk coupling.

crossunder — 1. A connection of two elements of a circuit by a conductive path deposited or diffused into a substrate. 2. A point in an integrated or MOS circuit at which there is a crossing of two conductive paths, one of which is built into the active substrate for interconnection.

crowbar — 1. A term describing the action that effectively creates a high overload on the actuating member of the protective device. This crowbar action may be triggered by a slight increase in current or voltage. 2. A circuit that shorts the output terminals of a power supply if the output of the power supply rises above a certain preset limit. The crowbar thus brings the output voltage to zero. 3. To place a low-resistance short across the input to a circuit. 4. A protective circuit in a power distribution circuit that shorts the power supply output to ground when an overvoltage condition occurs.

crowbar circuit — 1. An electronic switching system used to protect high-voltage circuits from damage caused by arc currents. The system places a momentary short across the circuit to be protected. 2. A circuit that protects a circuit or system from dangerously high voltage surges by shutting down the power source.

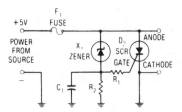

Crowbar overvoltage circuit (SCR).

crowbar protection circuit — A protection circuit that by rapidly placing a low resistance across the output terminals of a power supply initiates action that reduces the output voltage to a low value.

crowbar voltage protector — A separate circuit that monitors the output of a power supply and instantaneously throws a short circuit (or crowbar) across the output terminals of the power supply whenever a preset voltage limit is exceeded. An SCR is usually used as the crowbar device.

CRT — Abbreviation for cathode-ray tube.

CRT display — *See* cathode-ray-tube display.

CRT terminal — A visual-display terminal with an electrostatic deflection-type display. The term sometimes refers to all video "glass" terminals. These terminals' displays use either screen storage or memory-refreshed, raster-scanning techniques to generate visuals.

cryoelectrics — Technology having to do with the characteristics of electronic components at cryogenic temperatures. The branch of electronics that is concerned with applications of cryogenics. A contraction of cryogenic electronics

cryogenic — Of or having to do with temperatures approaching absolute zero.

cryogenic device — A device inteneded to function best at temperatures near absolute zero.

cryogenic motor — A motor that operates at a temperature below $-129°C$ and that uses a cryogenic fluid or gas to cool its windings and bearings.

cryogenics — 1. The subject of physical phenomena at temperatures below about $-50°C$. More generally, the term is used to refer to methods for producing very low temperatures. Also called cryogeny. 2. The study of the behavior of matter at supercold temperatures. 3. The science and technology applied to the creation of low temperatures (i.e., approaching absolute zero).

cryogenic temperature — A temperature close to absolute zero.

cryogeny — See cryogenics.

cryosar — A semiconductor device primarily intended for high-speed computer switching and memory applications. This device operates by the low-temperature avalanche breakdown produced by impact ionization of impurities.

cryosistor — A cryogenic semiconductor device in which the ionization between two ohmic contacts is controlled by means of a reverse-biased pn junction. After ionization, the device can act as a three-terminal switch, a pulse amplifier, an oscillator, or a unipolar transistor.

cryostat — A refrigerating unit such as that for producing or utilizing liquid helium in establishing extremely low temperatures (approaching absolute zero).

cryotron — 1. A superconductive four-terminal device in which a magnetic field, produced by passing a current through two input terminals, controls the superconducting-to-normal transition — and thus the resistance — between the two output terminals. 2. An electrical switching and binary storage instrument that depends on the oscillating of a superconducting component between low and high resistance levels due to alternation in the magnetic field.

cryotronics — A contraction of cryogenic electronics.

cryptanalyst — A person who solves cryptograms, that is, one who converts secret messages into their original plain-language form without having authorized knowledge of the cryptographic system.

crypto- — 1. A prefix used to form words that pertain to the transformation of data to conceal its actual meaning, usually by conversion to a secret code. 2. The science of encrypting data so that only the intended recipient can read it.

cryptochannel — Complete system of cryptocommunications between two or more holders.

cryptogram — See ciphertext.

cryptographic system — An algorithm for converting a message in ordinary language into a secret form.

cryptologic — Pertaining to communication intelligence and communication security.

cryptosecurity — Component of communication security that results from the provision of technically sound cryptosystems and their proper use.

crystal — 1. A solid in which the constituent atoms are arranged with some degree of geometric regularity. In communication practice, a piezoelectric crystal, piezoelectric crystal plate, or crystal rectifier. 2. A thin slab or plate of quartz ground to a thickness that causes it to vibrate at a specific frequency when energy is supplied. It is used as a frequency-control element in radio-frequency oscillators. 3. Quartz crystal whose

piezoelectric vibrational modes provide a highly accurate frequency for clock timing. 4. Solid material in which the atoms or molecules are arranged in a regular three-dimensional lattice array.

crystal anisotropy — A force that directs the magnetization of a single-domain particle along a direction of easy magnetization. To rotate the magnetization of the particle, an applied magnetic field must provide enough energy to rotate the magnetization through a difficult crystal direction.

crystal audio receiver — Similar to a crystal video receiver except for the path direction bandwidth, which is audio rather than video. See crystal video receiver.

crystal calibrator — A crystal-controlled oscillator used as a reference to check and set the frequency tuning of a receiver or transmitter.

crystal-can relay — A relay mounted in a can of a specific size and shape; called a crystal can because of its common usage as a mounting for quartz crystals used in frequency-control circuits.

crystal control — Control of the frequency of an oscillator by means of a specially designed and cut crystal.

crystal-controlled oscillator — See crystal oscillator.

crystal-controlled transmitter — A transmitter in which the carrier frequency is controlled directly by a crystal oscillator.

crystal counter — An instrument that is used to detect high-energy particles by the pulse of the current formed when a particle passes through a normally insulating crystal to which a potential difference is applied.

crystal cutter — A disc cutter in which the mechanical displacements of the recording stylus are derived from the deformities of a crystal having piezoelectric properties.

crystal detector — A mineral or crystalline material that allows electrical current to flow more easily in one direction than in the other. In this way, an alternating current can be converted to a pulsating current.

crystal diode — 1. A two-electrode semiconductor device that makes use of the rectifying properties of a pn junction (junction diode) or a sharp metallic point in contact with a semiconductor material (point-contact diode). Also called crystal rectifier, diode, and semiconductor diode. 2. A diode rectifier using a point contact on a silicon or germanium crystal. Because of its low capacitance, it will rectify at ultrahigh and superhigh frequencies.

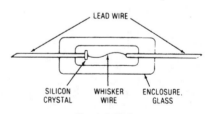

LEAD WIRE

SILICON CRYSTAL   WHISKER WIRE   ENCLOSURE, GLASS

Crystal diode.

crystal field — The electrostatic field acting locally within a crystal as a result of the microscopic arrangement of atoms and ions in the lattice.

crystal filter — 1. A highly selective circuit capable of discriminating against all signals except those at the center frequency of a crystal, which serves as the selective element. Resonant mechanical section consists of ceramic discs that vibrate at the band of frequencies to be removed. 2. Electrically coupled, two-terminal

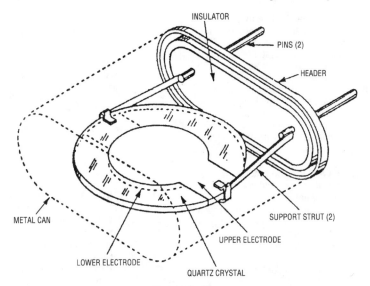

Crystal holder.

electroacoustic resonators (crystals) in ladder and lattice configurations. Monolithic filters with quartz substrates are also called crystal filters. 3. A bandpass filter with piezoelectric crystal components for the passage or impedance of electrical signals of various frequencies.

crystal headphones — Headphones using Rochelle-salt or other crystal elements to convert audio-frequency signals into sound waves.

crystal holder — A case of insulating material for mounting a crystal. External prongs allow the crystal to be plugged into a suitable socket.

crystal imperfection — Any deviation in lattice structure from that of a perfect single crystal.

crystal lattice — A periodic geometric arrangement of points that correspond to the locations of the atoms in a perfect crystal.

crystal loudspeaker — A loudspeaker in which piezoelectric action is used to produce mechanical displacements. Also called piezoelectric loudspeaker.

crystal microphone — Also called piezoelectric microphone. 1. A microphone that depends for its operation on the generation of an electric charge by the deformation of a body (usually crystalline) having piezoelectric properties. 2. A microphone whose generating element is a crystal or ceramic element, which generates a voltage when bent or stressed.

crystal mixer — 1. A mixer circuit with the frequency of the local oscillator being controlled by a crystal. Normally used in superheterodyne radio receivers. 2. A mixer that utilizes the nonlinear characteristic of a crystal diode to mix two frequencies. Frequently used in radar receivers to convert the received radar signal to a lower intermediate-frequency signal by mixing it with a local-oscillator signal.

crystal operation — Operation using crystal-controlled oscillators.

crystal orientation — For MOS devices, the terms ⟨100⟩ and ⟨111⟩ are commonly used. This refers to the angle with respect to crystal facets at which the silicon crystal is sliced. Each has a direct effect on MOS transistor characteristics.

crystal oscillator — Also called crystal-controlled oscillator. An oscillator in which the frequency of oscillation is controlled by a piezoelectric crystal.

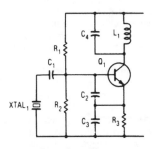

Crystal oscillator.

crystal oven — 1. A container, maintained at a constant temperature, in which a crystal and its holder are enclosed in order to keep their temperature constant and thereby reduce frequency drift. 2. A temperature-controlled container used to stabilize the temperature and resonant frequency of a crystal found in a crystal-controlled oscillator.

crystal pickup — Also called piezoelectric pickup. A phonograph pickup that depends for its operation on the generation of an electric charge by the deformation of a body (usually crystalline) having piezoelectric properties.

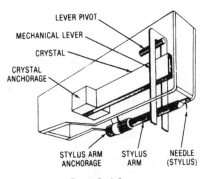

Crystal pickup.

crystal pulling — 1. A method of growing crystals in which the developing crystal is gradually withdrawn from a melt. 2. A technique (first developed by J. C. Czochralski) in which a monocrystalline seed is introduced into the top of a body of molten material and then withdrawn slowly to grow, or pull, a large single monocrystal. This technique is used in semiconductor manufacture to produce the uniformly doped monocrystal boules from which most devices are fabricated.

crystal rectifier — 1. An electrically conductive or semiconductive substance, natural or synthetic, that has the property of rectifying small radio-frequency voltages. *See* crystal diode.

crystal set — A simple type of radio receiver having no amplifier stages, and only a crystal-detector stage for demodulation of the received signal.

crystal shutter — A mechanical switch for shorting a waveguide or coaxial cable so that undesired rf energy is prevented from reaching and damaging a crystal detector.

crystal slab — A relatively thick piece of crystal from which crystal blanks are then cut.

crystal speaker — Also called piezoelectric speaker. A speaker in which the mechanical displacements are produced by piezoelectric action.

crystal-stabilized oscillator — Abbreviated CSO. A microwave rf source that uses a crystal oscillator operating at some low frequency (usually below 150 MHz) to drive a multiplier to obtain a microwave output frequency. A crystal oscillator can also be used for injection-locked stabilization of a free-running microwave oscillator.

crystal-stabilized transmitter — A transmitter employing automatic frequency control, in which the reference frequency is the same as the crystal-oscillator frequency.

crystal transducer — *See* piezoelectric transducer.

crystal video receiver — A radar receiver consisting only of a crystal detector and video amplifier.

CSA — Abbreviation for Canadian Standards Association, a testing and approval agency for products sold in Canada. This is the Canadian equivalent of Underwriters Laboratories in the United States.

C-scope — A rectangular radar display in which targets appear as bright spots with azimuth indicated by the horizontal coordinate and elevation angle by the vertical coordinate.

CSMA — Abbreviation for carrier sense multiple access. A contention scheme in which stations listen for the presence of a carrier on the channel before sending a packet.

csv — Abbreviation for corona start voltage.

CTCSS — Abbreviation for Continuous Tone-Coded Squelch System. Also called subaudible tones or PL tones (trademarked name by Motorola). This is a tone that is transmitted in addition to a voice signal. When it is equipped with a CTCSS decoder, a repeater will not function unless it receives both the CTCSS tone and the carrier signal.

CT-cut crystal — A natural quartz crystal cut to vibrate below 500 kHz.

CTL — Abbreviation for complementary transistor logic.

CTS — Abbreviation for clear to send. A control signal between a modem and a controller used to initiate data transmission over a communication line.

cubical antenna — An antenna array whose elements are positioned to form a cube.

cue — *See* address.

cue bus — In a mixing console, the bus or channel that is used to feed a program to a performer's headphones. Also called foldback.

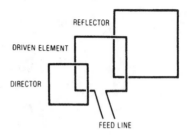

Cubical antenna.

cue circuit — A one-way communication circuit for conveying program control information.

cue control — A switch that temporarily disables a recorder's tape lifters during fast-forward and rewind, so the operator can judge what portion of the recording is passing the heads.

cuff electrode — An electrode in the shape of the letter "C" designed for application of potentials to small circular bodies, such as peripheral nerves.

cuing — 1. Locating a particular spot on a recorded tape, preparatory to playing through from that spot. 2. The marking or other identification of particular points or sections of tape to aid in the location of specific, desired selections or portions of the recording. This may be done with grease pencil or other markings directly on the tape, or by making use of a digital index counter, if one exists on the recorder.

cuing device — A lever or control that raises and lowers the tonearm without direct handling by the operator. Usually viscous damped for uniform rise and fall times, no matter how rapidly the control is moved.

CUJT — Abbreviation for complementary unijunction transistor.

cup — A single mechanical section of a potentiometer that may contain one or more electrical resistance elements.

cup core — A core that forms a magnetic shield around an inductor. Usually a cylinder with one end closed. A center core inside the inductor is normally used and may or may not be part of the cup core.

cupping — Curvature of a recording tape in a direction perpendicular to the length of the tape.

cup washer — A washer formed with a recess in one side to retain compression springs or, on binding-post terminals, to prevent escape of connecting wire strands.

curie — A unit of radioactivity; 1 curie (Ci) equals 3.7×10^{10} nuclear transformations per second.

Curie point — In ferroelectric dielectrics, the temperature or temperatures at which peak values of the dielectric constant occur. Also called Curie temperature. The (critical) temperature at which piezoelectric materials lose their polarization and therefore also their piezoelectric properties.

Curie temperature — Temperature in degrees Celsius at which a magnetized sample is completely demagnetized due to thermal agitation. *See* Curie point.

curl — The degree to which a wire tends to form a circle after removal from a spool. An indication of the ability of the wire to be wrapped around posts in long runs.

curls — Extruded material coming out from the edge of a bond.

Curpistor — A subminiature constant-current tube containing two electrodes and filled with radioactive nitrogen.

current — 1. The movement of electrons through a conductor. Current is measured in amperes, milliamperes, microamperes, nanoamperes, or picoamperes. 2. A movement of electrons, positive ions, negative ions, or holes. 3. The rate of transfer of electricity from one point to another.

current amplification — 1. The ratio of the current produced in the output circuit of an amplifier to the current supplied to the input circuit. 2. In photomultipliers, the ratio of the signal output current to the photoelectric signal current from the photocathode.

current amplifier — 1. A device designed to deliver a greater output current than its input current. 2. An amplifier that has a low output impedance and is capable of delivering a heavy current.

current antinode — Also called current loop. The point at which current is a maximum along a transmission line, antenna, or other circuit element having standing waves.

current attenuation — The ratio of the magnitude of the current in the input circuit of a transducer to the magnitude of the current in a specified load impedance connected to the transducer.

current-balance relay — A relay in which operation occurs when the magnitude of one current exceeds the magnitude of another current by a predetermined ratio.

current-carrying capacity — 1. The maximum current a conductor (or braid) can carry without heating beyond a safe limit. 2. The maximum current that can be continuously carried without causing permanent change in the electrical or mechanical properties of a device or conductor. (As applied to phone jacks, it refers to carrying current without interrupting the circuit.) 3. The maximum current an insulated conductor can safely carry without exceeding its insulation and jacket temperature limitations. 4. The maximum current that can be carried continuously without damage to a device, conductor, or machine. It is mainly determined by the temperature which the insulation can withstand and by the ambient temperature.

current-carrying rating — The current that can be carried continuously or for stated periodic intervals without impairment of the contact structure or interrupting capability.

current-controlled oscillator — Abbreviated CCO. A circuit that creates an ac output signal whose frequency is a function of the dc input current.

current density — 1. The amount of electric current passing through a given cross-sectional area of a conductor in amperes per square inch; i.e., the ratio of the current in amperes to the cross-sectional area of the conductor. 2. The ratio of current to surface area.

current echo — The signal that on a transmission line is reflected as the result of some discontinuity.

current feed — The excitation or feeding of an antenna by connecting the feeder at a point of maximum current, as at the center of a dipole or half-wave antenna.

current flicker — Current surges resulting from momentary shorts that can occur within a solid electrolyte capacitor. Under certain conditions, current flicker can avalanche to cause a short, which under low-impedance circuit conditions results in catastrophic destruction of the capacitor.

current foldback — In a dc power supply, a self-resetting protective method that reduces the output current to less than full-rated output under overload or short-circuit conditions.

current generator — A two-terminal circuit element with a terminal current independent of the voltage between its terminals.

current hogging — 1. A condition in which one of several parallel logic circuits takes the largest share of the available current because it has a lower resistance than the other circuits. 2. A condition that exists when several base-emitter junctions are driven from the same output and the input with the lowest base-emitter junction forward potential severely limits the drive current to the other transistor bases.

current-hogging injection logic — Abbreviated CHIL. A logic form that combines the input flexibility of current-hogging logic with the performance and packing density of injection logic.

current limiter — A device that detects current leakage and prevents potential shock hazard by minimizing or interrupting current flow.

current-limiter relay — A relay that opens its contacts when the current in a circuit exceeds a certain preset value.

current limiting (automatic) — 1. An overload-protection mechanism that limits the maximum output current of a power supply to a preset value and automatically restores the output when the overload is removed. 2. In a dc power supply, a self-resetting safeguard that protects the power supply against short-circuit and overload conditions by limiting output current to a predetermined maximum value. *See also* short-circuit protection.

current-limiting fuse — A protective device that anticipates a dangerous short-circuit current and opens the circuit, precluding the development of the peak available current.

current-limiting reactor — A form of reactor intended for limiting the current that can flow in a circuit under short-circuit conditions.

current-limiting resistor — A resistor inserted into an electric circuit to limit the flow of current to some predetermined value. Usually inserted in series with a fuse or circuit breaker to limit the current flow during a short circuit or other fault, to prevent excessive current from damaging other parts of the circuit.

current-limit sense voltage — The voltage across the current-limit terminals required to cause a regulator to current limit with a short-circuited output. This voltage is used to determine the value of the external current-limit resistor when external booster transistors are used.

current loop — 1. Means of communicating data via the presence or absence of current in a two-wire cable. 2. A two-wire transmit/receive interface in which the presence of a 20-mA current level indicates data (a binary 1 or mark) and its absence indicates no data (a 0 or space). Normally used with teletypes, and the only communication method that uses a current signal. *See* current antinode.

current margin — The difference between the steady-state currents flowing through a telegraph receiving instrument that correspond to the two positions of the telegraph transmitter.

current mirror — A circuit that relies on the collector current matching of two transistors (one strapped as a diode) when connected together base to base and emitter to emitter.

current-mode logic — Abbreviated CML. A non-saturating logic circuit that employs the characteristics of a differential amplifier circuit in its design. Because it is nonsaturating, it is a very fast switching logic design with low logic swings. The gate input element is the base of a transistor, with a separate transistor for each input.

current node — A point at which current is zero along a transmission line, antenna, or any other circuit element that has standing waves.

current noise — Also called excess noise. A low-frequency noise caused by current flowing in a resistor, particularly in film and carbon resistors. The amount of energy varies widely with the type and construction of the resistor. This low-frequency noise is generally measurable only in the region below 100 kHz; the noise power varies inversely with frequency and is a function of both the current in the resistor and the value of the resistor.

current penetration — The depth a current of a given frequency will penetrate into the surface of a conductor carrying the current.

current probe — A type of transformer, usually having a snap-around configuration, used for measuring the current in a conductor.

current pump — A circuit that drives, through an external load circuit, an adjustable, variable, or constant value of current regardless of the reaction of that load to the current, within rated limits of current, voltage, and load impedance.

current rating — 1. The maximum continuous current that can be carried by a conductor without degradation of the conductor or its insulation properties. 2. Maximum current that a device is designed to conduct for a specified time at a specified operating temperature. 3. The maximum continuous electrical current recommended for a given wire in a given situation. Expressed in amperes.

current regulator — 1. A device that functions to maintain the output current of a generator or other voltage source at a predetermined value, or varies the voltage according to a predetermined plan. 2. A regulating device that limits generator output and prevents excessive generator output by repeatedly inserting a resistance into the generator field circuit.

current relay — A relay that operates at a predetermined value of current. It can be an overcurrent relay, an undercurrent relay, or a combination of both.

current saturation — The condition in which the plate current of a vacuum tube cannot be further increased by increasing the plate voltage.

current-sensing resistor — A resistor of low value placed in series with a load to develop a voltage proportional to the output current. A regulated dc power supply regulates the current in the load by regulating the voltage across this sensing resistor.

current-sensitivity — The current required to give standard deflection on a galvanometer.

current sink — 1. A point toward which conventional current flows (electrons flow away from it). 2. An output type of sensor or analog device in which current flows from the load and into the output of the device at a low voltage when it is turned on.

current-sinking logic — Also called input-coupled logic. A logic form that requires that current flow out of the input of a circuit and back into the output of the preceding stage, which serves as a current sink instead of a source.

current source — 1. A point from which conventional current flows (electrons flow toward it). 2. An output type of switch or analog device in which current flows from it into the load at a high voltage when it is turned on.

current-sourcing logic — Also called collector-coupled logic. A logic form in which current flows from the output of a circuit and is forced into the input of a similar circuit to activate the circuit that drives.

current-stability factor — In a transistor, the ratio of a change in emitter current to a change in reverse-bias current between the collector and base.

current-transfer ratio — The ratio of an optocoupler's output current to input current at a specified bias. *See* beta.

current transformer — A transformer, intended for measuring or control purposes, designed to have its primary winding connected in series with a circuit carrying the current to be measured or controlled.

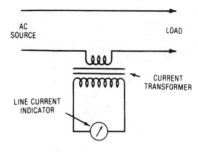

Current transformer.

current-type telemeter — A telemeter in which the magnitude of a single current is the translating means.

cursor — 1. A mechanically or electronically generated line that moves back and forth over another surface to delineate accurate readings. 2. A mechanical bearing line on a plan position indicator-type display for reading target bearing. 3. A visual movable pointer on a CRT used by the PC programmer to indicate where an instruction is to be added to the PC program. The cursor is also used during editing functions. 4. A small lighted line or square that appears on a video display screen at the point where the next character will appear. 5. A manually movable illuminated or flashing marker on a computer screen used to indicate the location of the next point at which an action will take place. 6. A means for indicating on a CRT screen the point at which data entry or editing will occur. The intensified element may be at constant high intensity or flashing (alternate high intensity and normal intensity). If flashing, additional data may be necessary to complete the instruction. 7. An indicator that marks the current working position on a display. 8. A blinking character that indicates where the next keyboard stroke will appear. 9. A position marker on a video display.

cursor and update — Circuitry that allows a user to add new material into a video terminal's memory and display. The cursor, which locates the current entry location, is generally a blinking underline, although some are overlines or boxes.

curve tracer — An instrument capable of producing a display of one current or voltage as a function of a second voltage or current, with a third voltage or current as a parameter.

customer set — *See* subscriber set.

cut and paste — To move graphics and/or text from one location to another.

Cutler antenna — A rear feed for a paraboloidal antenna reflector. It consists of a support waveguide with a terminating cavity containing two resonant slots, one on either side of the support waveguide, that face the reflector. Each slot is parallel to the broad faces of the feed waveguide.

Cutler feed — A resonant cavity, at the end of a waveguide, that feeds radio-frequency energy to the reflector of the antenna assembly of some airborne antennas.

cutoff — 1. Minimum value of bias that cuts off, or stops, the flow of plate current in a tube. 2. The frequency above or below which a selective circuit fails to respond. 3. The frequency of transmission at which the loss exceeds by 10 decibels that observed at 1000 hertz. 4. The condition when the emitter-base junction of a transistor has zero bias or is reverse biased and there is no collector current. 4. The frequency at which the modulus of measured parameter has decreased to $1/\sqrt{2}$ of its low-frequency value. (For a transistor, the cutoff frequency usually applies to the short-circuit small-signal forward current transfer ratio for either the common-base or common-emitter configuration.)

cutoff attenuator — A variable length of waveguide used below the cutoff frequency of the waveguide to introduce variable nondissipative attenuation.

cutoff current — Transistor collector current with no emitter current and normal collector-to-base bias.

cutoff field — *See* critical field.

cutoff frequency — 1. The frequency at which the gain of an amplifier falls below 0.707 times the maximum gain. 2. The frequency that marks the edge of the passband of a filter and the beginning of the transition to the stopband. 3. With respect to a line, the upper frequency limit, usually of a loaded transmission circuit, beyond which the attenuation rises very rapidly. 4. That frequency beyond which no appreciable energy is transmitted. It may refer to either an upper or lower limit of a frequency band.

cutoff limiting — Keeping the output of a vacuum tube below a certain point by driving the control grid beyond cutoff.

cutoff voltage — 1. The electrode voltage that reduces the dependent variable of an electron-tube characteristic to a specified value. *See also* critical voltage. 2. The voltage at which the discharge is considered complete. This need not be a very low voltage.

cutoff wavelength — The ratio of the velocity of electromagnetic waves in free space to the cutoff frequency of a waveguide.

cutout — 1. An electrical device that interrupts the flow of current through any particular apparatus or instrument, either automatically or manually. 2. Pairs brought out of a cable and terminated at some place other than at the end of the cable.

cutout relay — A device in the circuit between a generator and battery that closes to allow the generator to charge the battery and opens when the generator stops.

cut over — To transfer from one system to another.

CUTS — Abbreviation for computer user tape system. A standard method of recording data in serial form on an audiocassette recorder. Data is recorded at 300 baud by recording 8 pulses at 2400 Hz for a mark, or 4 pulses of 1200 Hz for a space.

cut-signal branch operation — In systems in which radio reception continues without cutting off the carrier, the cut-signal branch operation technique disables a signal branch in one direction when it is enabled in the other to preclude unwanted signal reflections.

cutter — Also called mechanical recording head. An electromechanical transducer that transforms an electrical input into a mechanical output (for example, the mechanical motion that a cutting stylus inscribes into a recording medium).

cut-through — The resistance of a solid material to penetration by an object under conditions of pressure, temperature, etc.

cut-through flow test — A test to measure the resistance to deformation of insulation subjected to heat and pressure.

cutting rate — The number of lines per inch the lead screw moves the cutting-head carriage across the face of a recording blank. Standard rates are 96, 104, 112, 128, 136, and greater, in multiples of 8 lines per inch. For microgroove recordings, 200 to 300 lines per inch are used.

cutting stylus — A special stylus used for cutting of phonograph records in the disc mastering process; it often has a built-in heater element and has special geometry.

CVD — Abbreviation for chemical vapor deposition.

CVD fiber — A process by which a heated gas produces an oxide deposit to fabricate a glass fiber preform. The deposited glass becomes the core.

cw — 1. Abbreviation for continuous wave. 2. Abbreviation for clockwise.

cw jamming — Transmission of a constant-amplitude, constant-frequency, unmodulated signal for the purpose of jamming a radar receiver by changing its gain characteristics.

cwp — Abbreviation for communication word processor. A word processor that can communicate, over special lines or regular telephone lines, with another cwp and with mainframe computers, Telexes and TWXs, photo-composers, optical character reading devices, intelligent copiers, and other terminals. The chief limitation is that the communicating systems must be compatible, meaning they must speak the same electronic language in order to understand each other.

cw reference signal — In color television, a sinusoidal signal used to control the conduction time of a synchronous demodulator.

cxr — Abbreviation for carrier. A communication signal used to indicate the intention to transmit data on a line.

cybernetics — 1. The study of systems of control and communications in humans and animals, and in electrically operated devices such as calculating machines. 2. The comparative study of the control and intracommunication of information-handling machines and nervous systems of animals and man in order to understand and improve communication. 3. The science that is concerned with the principles of communication and control, particularly as applied to the operation of machines and the functioning of organisms. 4. Study of multiple feedback loop, self-governing systems, usually of great complexity, as are found in living organisms and advanced human-made control systems. 5. A comparative study of the similarities as well as the differences between humans and machines, with respect to their ability to communicate and control.

cyberspace — 1. A general term referring to the virtual places that exist only on the Internet. 2. A term coined by William Gibson in his 1984 novel *Neuromancer* to describe a shared virtual environment whose inhabitants, objects, and spaces comprise data that is visualized, heard, and touched. The word *cyberspace* is currently used to describe the whole range of information resources available through computer networks and is often used as a synonym for the Internet. 3. The metaphoric space in which electronic communication takes place. Everything in cyberspace is virtual, that is, not physically real, but a shared experience nonetheless. 4. The "place" one goes to use a computer and modem to communicate with others. Being online puts one in cyberspace.

cyborg — Abbreviation for cybernetic organism. 1. An android capable of heuristic (learning by experience) updating of its own resident intelligence. 2. An organism superior in strength and/or perception to ordinary human beings. 3. A human altered or enhanced by mechanical means — somewhere between a man and a robot. The Borg in Star Trek are the best-known example.

cycle — 1. The change of an alternating wave from zero to a negative peak to zero to a positive peak and back to zero. The number of cycles per second (hertz) is called the frequency. *See also* alternation. 2. An off-on application of power. 3. A set of operations carried out in a predetermined manner. 4. In computer terminology, a regularly repeated sequence of operations, or the time required for one such sequence. 5. The complete sequence, including reversal, of an alternating electric current. 6. An interval of time in which an operation or set of events is completed.

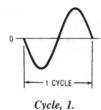

Cycle, 1.

cycle counter — A mechanism or device used to record the number of times a specified cycle is repeated.

cycle criterion — In computer terminology, the number of times a cycle is to be repeated.

cycle index — 1. In digital computer programming, the number of times a cycle has been executed. 2. The difference between the number of times a cycle has been executed and the number of times it is desired that the cycle be repeated.

cycle life — For rechargeable cells, the total number of discharge/charge cycles before the cell becomes inoperative.

cycle redundance check — Abbreviated CRC. 1. An error detection scheme, usually hardware implemented, in which a check character is generated by taking the remainder after dividing all the serialized bits in a block of data by a predetermined binary number. This remainder is then appended to the transmitted data and recalculated and compared at the receiving point to verify data accuracy. 2. Comparison of the checksum derived from data as it was originally written into storage with the checksum derived from the same data as it is being read out of storage. The first checksum is appended to the data as it enters storage. After reading this data, the controller computes a new checksum from it and compares the two. If the checksums match, the data is correct. A checksum error may indicate a damaged area on the memory, data that has changed since written, or erroneous reading of correct data for which a retry may work.

cycle reset — To return a cycle index to its initial value.

cycle shift — In a computer, the removal of the digits of a number of characters from a word from one end of the number or word and their insertion, in the same sequence, at the other end.

cycle stealing — 1. A memory cycle stolen from the normal CPU operation for a direct memory access (DMA) operation. 2. A computer memory access technique in which an I/O processor is synchronized to the general processor's memory cycles in such a way that it steals memory cycles between those of the general processor. This is possible on systems in which the memory cycle time is significantly shorter than the processor execution cycle. 3. A characteristic of DMA channels. An I/O device can delay CPU use of the I/O bus for one or more bus cycles while it accesses system memory.

cycle time — 1. The interval of time between the occurrence of corresponding parts of successive cycles. 2. The length of time required to obtain information from a memory and then write information back into the memory. Also called read-write cycle time, since it is normally equal to the sum of the write time and the read time. If system memory is core, the read cycle time includes a write-after-read (restore) subcycle. Cycle time is often used as a measure of computer performance, since this is a measure of the time required to fetch an instruction. 3. Time interval during which any set of operations is repeated regularly in the same sequence. 4. The time needed for a CPU to go through a complete operation cycle. 5. The total time required by a device

(usually a memory) to complete its cycle and become available again. Typically, the access time will be shorter than the cycle time, though they may sometimes be equal.

cycle timer — A controlling mechanism that opens or closes contacts according to a preset cycle.

cyclically magnetized condition — The condition of a magnetic material after being under the influence of a magnetizing force varying between two specific limits until, for each increasing (or decreasing) value of the magnetizing force, the magnetic-flux density has the same value in successive cycles.

cyclic code — Positional notation, not necessarily binary, in which quantities differing by one unit are represented by expressions that are identical except for one place or column, and the digits in that place or column differ by only one unit. Cyclic codes are often used in mechanical devices because no ambiguity exists at the changeover point between adjacent quantities.

cyclic decimal code — A four-bit binary code word only one digit of which changes state between one sequential code word and the next, and which translates to decimal numbers. It is categorized as one of a group of unit-distance codes.

cyclic memory — A memory that continuously stores information but provides access to any piece of stored information only at multiples of a fixed time called the cycle time.

cyclic shift — A shift in which the data moved out of one end of the storing register is reentered into the other end, as in a closed loop.

cycling — 1. A rhythmic variation, near the desired value, of the factor under control. 2. A periodic change from one value to another of the controlled variable in an automatic control system. 3. A periodic oscillation in an automatically controlled system between the high limit and the low limit over which the controls operate.

cycling vibration — Sinusoidal vibration applied to an instrument and varied in such a way that the instrument is subjected to a specified range of vibrational frequencies.

cycloconverter — A step-down static frequency converter that produces a constant or a precisely controllable output frequency from a variable-frequency ac power input. In general, the frequency ratio chosen is 3 to 1 or greater.

cyclogram — An oscilloscope display obtained by monitoring two voltages having a direct cyclic relationship to each other.

cyclograph — A device in which an electron beam moves in two directions, at right angles.

cyclometer register — A set of four or five wheels numbered from 0 to 9 inclusive on their edges, and enclosed and connected by gearing so that the register reading appears as a series of adjacent digits.

cyclotron — 1. A device consisting of an evacuated tank in which positively charged particles (for example, protons, deuterons) are guided in spiral paths by a static magnetic field while being highly accelerated by a fixed-frequency electric field. 2. Type of accelerator of nuclear particles (protons or deuterons) that uses an oscillating electric field and a fixed magnetic field to accelerate the particles. *See* accelerator.

cyclotron frequency — The frequency at which an electron traverses an orbit in a steady, uniform magnetic field and zero electric field. Given by the product of the electronic charge and the magnetic flux density, divided by 2π times the electron mass.

cyclotron-frequency magnetron oscillations — Those oscillations having substantially the same frequency as that of the cyclotron.

cyclotron radiation — The electromagnetic radiation emitted by charged particles orbiting in a magnetic field at speeds relatively slow compared with the speed of light. It arises from the centripetal acceleration of the particle moving in a circular orbit, as in a cyclotron.

cyclotron resonance — The effect characterized by the tendency of charged carriers to spiral around an axis in the same direction as an applied magnetic field, with an angular frequency determined by the value of the applied field and the ratio of the charge to the effective mass of the charge carrier.

cylindrical (or circular) connectors — Separable plugs and receptacles, both housed in cylindrically shaped metal or plastic shells. Shells may stand alone or may be mounted on panels, bulkheads, or walls. When mated, the shells are joined by a coupling ring of threaded, bayonet, or push-pull design. The receptacle (female) contacts are imbedded in an insert of plastic or vitreous insulating material (usually for hermetically sealed connectors) in the shell. The plug (male) contacts are cantilevered cylindrical pins, also partly imbedded in an insert of insulating material. Contact terminations are either soldered or crimp connected to lead-in wires.

cylindrical-film storage — A computer storage device, each storage element of which is a short length of glass tubing with a thin film of nickel-iron alloy on its outer surface. Wires running through the tubing act as bit and sense lines, and conducting straps at right angles to the tubing function as word lines.

cylindrical reflector — A reflector that is part of a cylinder, usually parabolic.

cylindrical wave — A wave whose equiphase surfaces form a family of coaxial cylinders.

Czochralski crystal — A crystal grown by slowly withdrawing a seed crystal from a melt while the melt is held slightly above the melting point of the material. High-quality, low-dislocation-density crystals of germanium and silicon are grown in this manner.

Czochralski technique — A method of growing large, single crystals by pulling them from a molten state. Usually used for growing single crystals of germanium and silicon.

D

D — Symbol for electrostatic flux density, deuterium, dissipation factor, or drain electrode.

DAA — Abbreviation for data access arrangement. 1. A direct interface attachment that connects a data communications device to a telephone line. It protects the public telephone network from a sudden surge of power or interference from the device that is coupled to the line. 2. A telephone-switching-system protective device used to attach uncertified non-telephone-company-manufactured equipment to the carrier network.

DABS — Abbreviation for discrete address beacon system. A sophisticated ground surveillance equipment and complementary airborne transponder. DABS includes a two-way automatic ground-to-air data link.

DAC — Abbreviation for digital-to-analog converter. Also abbreviated dac or d/a converter.

d/a converter — *See* digital-to-analog converter.

d/a decoder — A device that changes a digital word to an equivalent analog value.

Dag — Trademark of Acheson Industries, Inc. Abbreviation for Aquadag.

daisy chain — 1. In a computer, a bus line that is interconnected with units in such a way that the signal passes from one unit to the next in serial fashion. 2. A method for prioritizing interrupts. Each unit capable of requesting an interrupt can either pass on the processor acknowledge or block it. In this way the unit that is electrically closest to the processor has highest priority. 3. A method of propagating signals along a bus, often used in applications in which devices not requesting a daisy-chained signal respond by passing the signal on. The first device requesting the signal responds to it by performing an action and breaks the daisy-chained signal continuity. This scheme permits assignment of device priorities based on the electrical position of the device along the bus.

daisy wheel — 1. A letter-quality, impact-printing mechanism whose printing element consists of a flat, rimless, metal or plastic wheel with letters molded at the ends of the spokes. The wheel rotates until the required character is in position, and a hammer flies out and hits it, impacting the character onto paper. 2. The print element for a daisy-wheel printer. Print wheels are interchangeable, allowing the operator to select an appropriate font.

daisy-wheel printer — 1. An impact printer that prints fully formed characters one at a time by rotating a circular print element composed of a series of individual spokes, each containing two characters. 2. A printer that produces letter-quality type when characters, mounted in a circle, strike an inked ribbon. It is rarely used anymore.

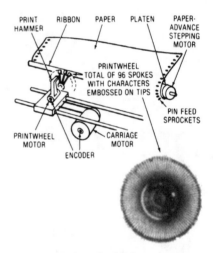

Daisy-wheel printer.

Damon effect — The change in susceptibility of a ferrite caused by a high rf power input.

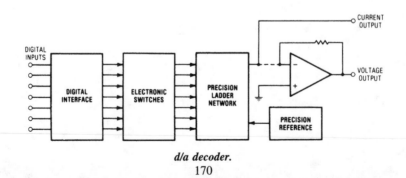

d/a decoder.

damped natural frequency — 1. The frequency at which a system with a single degree of freedom will oscillate, in the presence of damping, after momentary displacement from the rest position by a transient force. 2. The rate of free oscillation of a sensing element in the presence of damping.

damped oscillation — The oscillation that occurs when the amplitude of the oscillating quantity decreases with time. If the rate of decrease can be expressed mathematically, the name of the function describes the damping. Thus, if the rate of decrease is expressed as a negative exponential, the system is said to be an exponentially damped system.

damped waves — Waves in which successive cycles at the source progressively diminish in amplitude.

dampen — 1. To diminish progressively in amplitude; usually said of waves or oscillations. 2. To deaden vibrations.

damper diode — *See* freewheeling diode.

damper tube — The tube that conducts in the horizontal-output circuit of a television receiver when the current in the horizontal-deflecting yoke reaches its negative peak. This causes the sawtooth deflection current to decrease smoothly to zero instead of continuing to oscillator.

damper winding — Also called amortisseur winding. 1. In electric motors, a permanently short-circuited winding, usually uninsulated, arranged so that it opposes rotation or pulsation of the magnetic field with respect to the pole shoes. 2. A winding of copper bars or rods (squirrel cage) embedded in the pole face of synchronous motors and generators. Used as a starting winding on synchronous motors; acts as the squirrel cage of an induction motor. After the synchronous motor is up to speed, tends to prevent oscillations and hunting — a damping effect that gives it its name.

damping — 1. The reduction of energy in a mechanical or electrical oscillating system by absorption, conversion into heat, or by radiation. 2. Act of reducing the amplitude of the oscillations of an oscillatory system, hindering or preventing oscillation or vibration, or diminishing the sharpness of resonance of the natural frequency of a system. 3. The dissipation of kinetic energy in a system by a controlled energy-absorbing medium. A system can be described as being either critically damped, overdamped, or underdamped. 4. The manner in which the pointer settles to its steady indication after a change in the value of the measured quantity. Two general classes of damped motion are (*a*) periodic, in which the pointer oscillates about the final position before coming to rest; and (*b*) aperiodic, in which the pointer comes to rest without overshooting the rest position. Sometimes referred to as overdamping. 5. The energy-dissipating characteristic that, together with natural frequency, determines the upper limit of frequency response and the response time characteristics of a transducer. 6. The application of a mechanical resistance, such as a rubber or silicone material, to the cantilever pivot of a pickup to reduce the amplitude of the resonance between the tip mass and the compliance of the vinyl record material (which usually occurs between 15 and 30 kHz). 7. The suppression of oscillations, or ringing, in the rotary of a stepping motor caused by sudden changes in velocity. Damping can be accomplished by adding mechanical or viscous fluid dampers or by electronically shaping and controlling the energizing field power.

damping coefficient — The ratio of actual damping to critical damping.

damping constant — The Napierian logarithm of the ratio of the first to the second of two values of an exponentially decreasing quantity separated by a unit of time.

damping diode — A vacuum tube that damps the positive or negative half-cycle of an ac voltage.

damping factor — 1. For any underdamped motion during any complete oscillation, the quotient obtained by dividing the logarithmic decrement by the time required by the oscillation. 2. Numerical quantity indicating ability of an amplifier to operate a speaker properly. Values over 4 are usually considered satisfactory. 3. The ratio of rated load impedance to the internal impedance of an amplifier. 4. The ratio (larger to smaller) of the angular deviations of the pointer of an electrical indicating instrument on two consecutive swings from the equilibrium position. 5. *See* decrement, 1. 6. The ratio of load or speaker impedance to the amplifier's output impedance. Thus, the smaller the output impedance the greater the damping factor. The damping factor increases with increase of voltage negative feedback; with the large amounts of feedback applied to transistor hi-fi amplifiers, the source or output impedance can be as low as 0.1 ohm, giving a damping factor of 80, referred to as an 8-ohm load. 7. The standard damping constant of a second-order feedback system. In a phase-locked loop, it refers to the ability of the loop to respond quickly to an input signal frequency without excessive overshooting.

damping magnet — A permanent magnet and a movable conductor, such as a sector or disc, arranged in such a way that a torque (or force) is produced that tends to oppose any relative motion.

damping ratio — 1. The ratio of the degree of actual damping to the degree of damping required for critical damping. May be affected by changes in ambient temperature. 2. Of a galvanometer, the ratio, expressed as a positive number, of a given deflection to the next deflection in the opposite direction. The greater this ratio, the greater the degree of damping. The natural logarithm of this ratio is called the logarithmic decrement.

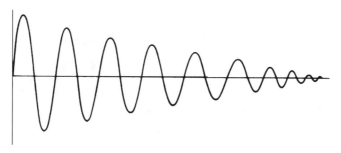

Damped oscillation.

dancer arm — A device that senses tape tension and signals the reel motor to take up or to supply tape.

Daniell cell — A cell having a copper electrode in a copper-sulfate solution and a zinc electrode in a diluted sulfuric acid or zinc-sulfate solution, with the two solutions separated by a porous partition. Generates an essentially constant electromotive force of about 1.1 volts.

daraf — The unit of elastance. It equals the reciprocal of capacitance and is actually farad spelled backward.

dark conduction — Residual electrical conduction in a photosensitive substance in total darkness.

dark current — *See* electrode dark current.

dark discharge — In a gas, an electric discharge that has no luminosity.

dark-field disc — A disc used in the optical electronic type of cell counter that controls light transmission.

dark noise — The current or pulses produced when the photocathode is shielded from all external optical radiation.

dark resistance — The resistance of a photoelectric device in total darkness.

dark satellite — A satellite that does not give information to friendly ground stations, either because it is controlled or because it carries inoperative radiating equipment.

dark space — A nonluminous region of a glow-discharge tube.

dark spot — A phenomenon sometimes observed in a reproduced television image. It is caused by the formation of electron clouds in front of the mosaic screen in the transmitter camera tube.

dark-spot signal — The signal existing in a television system while the television camera is scanning a dark spot.

dark-trace tube — A cathode-ray tube with a screen coated with a halide of sodium or potassium. The screen normally is nearly white, and whenever the electron beam strikes, it turns a magenta color that is of long persistence. The screen can be illuminated by a strong light source so that the reflected image may be made intense enough to be projected.

Darlington amplifier — Also called Darlington pair, double-emitter follower, or β multiplier. 1. A transistor circuit that, in its original form, consists of two transistors in which the collectors are tied together and the emitter of the first transistor is directly coupled to the base of the second transistor. Therefore, the emitter current of the first transistor equals the base current of the second transistor. This connection of two transistors can be regarded as a compound transistor with three terminals. 2. A two-transistor amplifier connected so that the amplification of the amplifier equals the product of the individual transistors' amplification. 3. A composite configuration of transistors that provides a high input impedance and a high degree of amplification.

Darlington-connected phototransistor — A phototransistor whose collector and emitter are connected to the collector and base, respectively, of a second transistor. The emitter current of the input transistor is amplified by the second transistor, and the device has a very high sensitivity to light.

Darlington connection — A form of compound connection in which the collectors of two or more transistors are connected together and the emitter of one is connected to the base of the next. Two transistors connected in this way constitute a Darlington pair.

Darlington pair — *See* Darlington amplifier.

Darlington transistor — A three terminal device that consists of two transistors connected so that the emitter of the first transistor is connected to the base of the second transistor. Such a direct coupled configuration has much higher current gain than can be achieved by a single transistor.

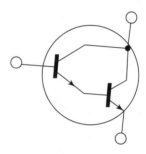

Darlington transistor.

D'Arsonval current — A high-frequency, low-voltage current of comparatively high amperage.

D'Arsonval galvanometer — A dc galvanometer consisting of a narrow rectangular coil suspended between the poles of a permanent magnet.

D'Arsonval instrument — *See* permanent-magnet moving-coil instrument.

D'Arsonval movement — A meter movement consisting essentially of a small, lightweight coil of wire supported on jeweled bearings between the poles of a permanent magnet. When the direct current to be measured is sent through the coil, its magnetic field interacts with that of the permanent magnet and causes the coil and attached pointer to rotate.

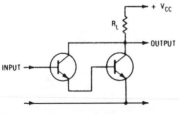

Darlington amplifier.

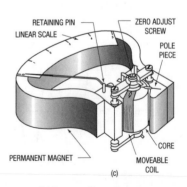

D'Arsonval movement.

DASD — Abbreviation for direct-access storage device. Any storage device utilizing addressing to let users enter or retrieve data without reference to their physical location. For example, a RAM.

dash — Term used in radiotelegraphy. It consists of three units of a sustained transmitted signal followed by one unit during which no signal is transmitted.

dashpot — 1. A device using a gas or liquid to absorb energy from or retard the movement of the moving parts of a circuit breaker or other electrical or mechanical device. 2. A cylinder and piston device using gas or a liquid to retard the movement of a relay or circuit breaker.

DAT — *See* digital audio tape.

data — 1. A general term used to denote any or all numbers, letters, symbols, or facts that refer to or describe an object, idea, condition, situation, or other factors. It connotes basic elements of information that can be processed or produced by a computer. Sometimes *data* is considered to be expressible only in numerical form, but *information* is not so limited. 2. A general term for any type of information. 3. Inputs in the form of a character string that may have significance beyond their numerical meaning. 4. Any representations, such as characters or analog quantities, to which meaning might be assigned.

data access arrangement — A protective connecting arrangement that serves as an interface between a customer-provided modem and the switched network. *See* DAA.

data acquisition — 1. The process by which events in the real world are translated to machine-readable signals. The term usually refers to automated systems in which sensors of one type or another are attached to machinery. 2. The simultaneous collection of data from external sensors, usually in analog form. 3. The function of obtaining data from sources external to a computer system, converting it to binary form, and processing it.

data acquisition and control systems — Assemblies of electronic and mechanical components used to monitor and control complex processes. These systems include the following:

- Process sensors that measure such parameters as temperature, pressure, voltage, and current
- Transmitters that convert measurement data to electrical or pneumatic signals and controls
- Digital computers that test set points, program sequential events, and perform calculations
- Software that provides the computer with instructions and routines
- Process actuators, such as solenoids, relays, valves, and motors, that modify the process in response to computer-generated commands
- Process interface devices, such as analog-to-digital converters, that link transmitters and actuators with digital computers
- Human interface devices, such as printers, keyboards, CRT terminals, switches, alphanumeric displays, chart recorders, and alarms, that facilitate human intervention

data acquisition and conversion system — A method of processing analog signals and converting them into digital form for subsequent processing or analysis by computer or for data transmission.

data acquisition system — 1. A system in which a computer at a central computing facility gathers data from multiple remote locations. 2. System for recording data, usually in digital form, from several sources; can include computing functions.

data bank — A comprehensive collection of libraries of data. For example, one line of an invoice may form an item, a complete invoice may form a record, a complete set of such records may form a file, the collection of inventory control files may form a library, and the libraries used by an organization are known as its data bank. Synonymous with database.

database — Also data base. 1. The entire body of data that has to do with one or more related subjects. Typically, it consists of a collection of data files (such as a company's complete personnel records concerning payroll, job history, accrued vacation time, etc.) stored in a computer system so that they are readily available. 2. A block of computer memory containing information about one given thing. 3. The collection of current variable data elements defined and maintained by the user. 4. A collection of data, consisting of at least one file, that is sufficient for a given purpose or for a given data-processing system. 5. A large and complete collection of information that covers a variety of subject areas. For instance, a medical diagnostic database might contain symptoms for all common diseases or injuries. 6. A collection of data fundamental to a system or to an enterprise. Made up of comprehensive files of information having predetermined structure and organization and able to be communicated, interpreted, or processed by humans or by automatic means. 7. A collection of related data that can be retrieved from memory at will, such as a mailing list or a list of accounts.

database management — 1. A systematic approach to the storage, updating, and retrieval of information stored as data items, usually in the form of records in a file, where many users, or even many remote installations, will use common data banks. 2. A program that enables a computer to store large amounts of information and then sort it in almost any manner. For example, a company's database could give a list of customers by ZIP code, by credit line, alphabetically by name, or by telephone number. The program takes care of managing the storage and retrieval of the data.

database management system — Abbreviated DBMS. A group of programs that allow users to store, alter, and retreive information from a database.

database relations — Linkages within a database that logically bind two or more elements in the database. For example, a nodal line (interconnect) is related to its terminal connection nodes (pins) because they all belong to the same electrical net.

data block — Typically, all the data for one item that is entered into a computer for processing, or the computer output that results from processing. An example of an input data block is an individual shipping list; an example of an output data block is a check to be sent.

data break — A facility that permits input/output transfers to take place on a cycle-stealing basis without disturbing execution of the program by a computer.

data bus — 1. A wire or group of wires used to carry data to or from a number of different locations. 2. The output pins of the MPU chip and associated circuitry used for the transmission of data from one point in the system to another. 3. In fiber optics, an optical waveguide used as a common trunk line to which a number of terminals can be interconnected through optical couplers. 4. A system incorporated into fiber-optic communications characterized by several spatially distributed terminals that are served with the same multiplexed signal.

data catalog — A software tool used to list all of the data elements in a database.

data channel (or communication) equipment — Abbreviated DCE. Equipment that interfaces a transmission facility to a transmitting/receiving device. A modem is a DCE.

data code — A structured set of characters used to stand for the data items of a data element, for example, the numerals 1, 2, ... 7 used to represent the data items Sunday, Monday, ... Saturday.

data collection — Also called data gathering. In a computer, the transferring of data from one or more points to a central point.

data communication equipment — *See* DCE.

data communications — 1. The technology covering the transfer of data over relatively long distances. 2. Transmission of data in both directions between a central location (host computer) and remote locations (terminals) through communication lines. To facilitate this, interfaces such as modems, multiplexers, concentrators, etc., are required at each end of the lines. 3. The movement of encoded information by means of electrical transmission systems.

data communications processor — A small computer used to control the flow of data between machines and terminals over communications channels. It may perform the functions of a concentrator, handshaking, and formatting, but does not include long-term memory or arithmetic functions.

data compression — 1. The process of reducing the number of recorded or transmitted digital data samples through the exclusion of redundant samples or the use of variable-length characters. 2. A technique that provides for the transmission of fewer data bits than originally required without information loss. The receiving location expands the received data bits into the original bit sequence. 3. A method of reducing the number of bits that are needed to represent information. Data compression allows higher communications speeds and also allows more information to be stored on a disk.

data control block — A control block through which the information required by access routines for storage and retrieval of data is communicated to them.

data conversion — 1. The changing of data from one form of representation to another. 2. Changing numerical information from one format to another via an IC processing chip or other means.

data display — Visual presentation of processed data by means of special electronic or electromechanical devices connected (either online or offline) with digital computers or component equipment. Although line printers and punch cards may display data, they usually are categorized as output equipment rather that as displays.

data distributor — An array of simple gates that can accept one or more input lines of data and route the data to specific outputs as determined by the levels of control inputs to the array.

data element — An element that converts data functions into a usable signal. *See also* element, 2.

data element dictionary — A software tool used to describe each data element, i.e., to tell what it is.

Data Encryption Standard — *See* DES.

data entry — 1. Data entered by an operator from a single data device, such as a card or badge reader, numeric keyboard, or rotary switch. 2. Converting data into a form suitable for entry into a computer system, such as by keying from a terminal onto magnetic disk or tape. 3. Entering data directly into a computer system.

data entry device — An electromechanical device to allow manual input of data to a display system. Examples of data entry devices are alphanumeric keys, data tablet, function keys, joystick, mouse, and trackball.

data file — A collection of related data records organized in a specific manner, such as a payroll or inventory file.

data-flow analysis — Graphical analysis of sequential data patterns, e.g., by tools that identify undefined variables.

data flowchart — A flowchart that represents the path of data used in a problem and that defines the major phase of the processing as well as the various data media used.

data-flow diagram — An illustration having a configuration such that it suggests a certain amount of circuit operation.

data-flow machine — A parallel-processing architecture in which each processor acts on instructions when the data needed become available.

data format — The structure and significance of data areas on a storage medium, without reference necessarily to the value of the data contained. Initial values, or limit values, are considered part of a data format definition. The data format itself may have been specified by parameter values at system generation time.

data gathering — *See* data collection.

data generators — Specialized word generators in which the programming is designed to test a particular class of device, in which the pulse parameters and timing are adjustable, and in which selected words may be repeated, reinserted later in the sequence, omitted, etc.

data-handling capacity — The number of bits of information that may be stored in a computer system at one time. The rate at which these bits may be fed to the input either by hand or with automatic equipment.

data-handling system — Semiautomatic or automatic equipment used in the collection, transmission, reception, and storage of numerical data.

data hierarchy — A data structure, made up of sets and subsets, in which every subset of a set is of lower rank than the data of the set.

data highway — A single-cable data link that provides communication among multiple stations that are separate program counters, computers, and data terminals. It eliminates the need for separate, independently wired data links. Whether communicating or not, all stations may function independently.

data identifier — The means of establishing the identity of data by a classification method so that data with close resemblance can be related on a one-to-one or a category basis. Typical identifiers are names, titles, limits or boundaries, classes, and types. Identifiers may also classify the manner of data generation, its completeness, or state of maturity.

data item — The simplest type of information dealt with by a computer system (e.g., a name or employee number). A collection of data items constitutes a record (e.g., payroll information on one employee), and a number of related records constitutes a file (e.g., payroll information on all employees of a company).

data link — 1. Electronic equipment that permits automatic transmission of information in digital form. 2. Equipment, especially transmission cables and interface modules, that permits the transmission of information. 3. A circuit designed to carry digital information, usually by time-division multiplex techniques. Often used to transfer information from one register to another. 4. Any information channel used for connecting data-processing equipment to any input, output, display device, or other data-processing equipment, usually at a remote location.

data logger — A system to measure a number of variables and make a written tabulation and/or record in a form suitable for computer input.

data management — Those control-program functions that provide access to data sets, enforcement of data-storage conventions, and regulation of the use of input/output devices.

data organization — Any of the data-management conventions for determining the arrangement of a data set.

dataphone — 1. A trademark of American Telephone and Telegraph Co. to identify the data sets made and supplied for use in the transmission of data over the regular telephone network. 2. A telephone equipped with a modem and appropriate switching for both voice and data transmission.

dataphone digital service — Abbreviated DDS. An all-digital transmission service offered by Bell Telephone that eliminates the need for special digital-to-analog modems.

data pointer — A register holding the memory address of the data (operand) to be used by an instruction. Thus, the register points to the memory location of the data.

data processing — 1. The handling of information in a sequence of reasonable operations. 2. The execution of a programmed sequence of operations upon data. A generic term for computing in business and other applications. 3. A term used in reference to operations performed by data-processing equipment.

data-processing center — An installation of computer equipment that provides computing services for users.

data-processing machine — A general name for a machine that can store and process numerical and alphabetical information.

data-processing system — A network of machine components that can accept information, process it in accordance with a plan, and produce the desired results.

data processor — 1. An electronic or mechanical machine for handling information in a sequence of reasonable operations. 2. Any device capable of performing operations on data; e.g., a desk calculator, a tape recorder, an analog computer, or a digital computer.

data rate — 1. The speed at which data is sent to a receiving computer or device, measured in bits per second. 2. The speed at which digital information is transmitted, expressed in hertz per second or bits per second.

data reduction — 1. The process of converting a large quantity of information into a more manageable form, usually including a reduction in volume and a simplification of format. 2. The conversion of raw data into a more useful form, such as graphic representations. 3. The process of transforming masses of raw test or experiment data, usually gathered by instrumentation, into a useful, ordered, or simplified form.

data-reduction system — Automatic equipment employed to simplify the use and interpretation of a large amount of data gathered by instrumentation.

data selector — A decision-making type of digital building block that routes data from any one of several inputs to its output.

data service unit — Abbreviated DSU. A device used on transmission facilities specifically designed to transmit digital data. It is equipped with local and remote testing capabilities (normally provided by a modem on an analog or voice transmission facility) and all other signals that are needed to provide a standard interface to terminals.

dataset — 1. A circuit-terminating device used to provide an interface between a circuit and terminal input/output equipment. 2. A modem. A device that converts the signal of a business machine to signals that are suitable for transmission over communications lines and vice versa. It may also perform other related functions. 3. The complete interface unit supplied by the carrier, including a network control and signaling unit, a

modem, and devices to protect the network from signals that might interfere with other users. 4. A device used to encode digital data onto voice phone lines.

data sheet — Also called spec sheet. A compilation of terminal information on a specific device defining the electrical and mechanical characteristics of that device.

data signaling rate — The data-transmission capacity, expressed in bits per second, of a set of parallel communication channels.

data sink — 1. A communications device that can accept data signals from a transmission device. Also, it may check the received signals and originate signals for error control. 2. A memory or recording device that can store information for future use. *See also* data source.

data source — A communications device that can originate data signals for a transmission device and may accept error control signals. *See also* data sink.

data stabilization — Stabilization of a radar display with respect to a selected reference, regardless of changes in the attitude of the vehicle that carries the radar, as in azimuth-stabilized PPI.

data stream — Data that is flowing from one point to another in a network. A data stream usually consists of a succession of messages (or data blocks).

data stream protocol — The part of a data stream that facilitates the time-critical delivery and synchronization of time-based media.

data switcher — A system used to connect network lines to a specific data-processing computer port.

data synchronizer — A device that controls and synchronizes the transmission of data between an input/output (I/O) device and the computer system.

data tablet — A data entry device consisting of a stylus and a graphic recorder with a coordinate grid similar to the number space of the screen. By pressing the stylus on the tablet, an interrupt is generated and the coordinates of the stylus are stored in special $X - Y$ input registers. The registers are then read by the host computer. The tablet generally replaces the light pen for cursor and tracking symbol movements, and is used extensively in storage-tube display systems, in which light-pen tracking and identification are impossible.

data terminal — 1. A common point at which data from various sources is collected and transferred; it may include or connect with several types of data-processing equipment. 2. Equipment at the end of a transmission system for the transmission or reception of data. 3. A class of devices characterized by keyboards and CRT displays.

data termination equipment — Abbreviated DTE. 1. Any device that generates information to be transmitted over a transmission facility. A terminal is a DTE. 2. Equipment that constitutes the data source and/or data sink and provides the communication control function protocol; it includes any piece of equipment at which a communication path begins or ends.

data tracks — Positions of information storage on drum storage devices. Information storage on the drum surface is in the form of magnetized and nonmagnetized areas.

data transfer — Moving data from one (or more) registers (or memory locations) to another, or interchanging data between registers (memory locations) in many different ways.

data transmission — The sending of information from one place to another or from one part of a system to another.

data-transmission equipment — The communications equipment used in direct support of data-processing equipment.

data transmission system — Means for transmitting data; for example, dataphone, radio, etc. *See also* telemetry.

data transmission utilization measure — In a data-transmission system, the ratio of the useful data output to the total data input.

data value — The information contained in data formats. Normally prepared for the generation of specific packages from generic packages.

data word — A computer word that contains or represents the data to be manipulated.

datum reference — A defined point, line, or plane used to locate the pattern or layer for manufacturing, inspection, or for both purposes.

daughterboard — 1. A board that mounts on, and connects to, the motherboard, in a computer. Sometimes used for memory upgrades. 2. A small circuit board that is mounted on and adds capability to another circuit board called a motherboard.

daughter card — Card or board interfaced with a motherboard or backplane.

dB — Abbreviation for decibel. Unit of measurement for power and voltage level.

dBa — Abbreviation for decibels adjusted. Used in conjunction with noise measurements. The reference level is −90 dBm, and the adjustment depends on the frequency-band weighting characteristics of the measuring device.

dBase — A database management program for PCs.

dBi — Decibels of gain relative to an isotropic antenna, or one that radiates equal power in all directions; used to measure antenna gain.

dBj — A unit used to express relative rf signal levels. The reference level is 0 dBj = 1000 microvolts. (Originated by Jerrold Electronics.)

dBk — Decibels referred to 1 kilowatt.

dBm — 1. Abbreviation for decibels above (or below) 1 milliwatt. A quantity of power expressed in terms of its ratio to 1 milliwatt. Power-level measurement unit in the telephone industry based on 600-ohm impedance and 1004-Hz frequency; 0 dBm is 1 mW at 1004 Hz, terminated by a 600-ohm impedance. 2. Decibels referenced to 1 milliwatt; used in communication work as a measure of absolute power values. Zero dBm equals 1 milliwatt.

dB meter — A meter having a scale calibrated to read directly in decibel values at a specified reference level (usually, 1 milliwatt equals 0 dB).

DBMS — *See* Database management system.

dBm0 — The power in dBm measured at, or referred to, a point of zero relative transmission level.

dBr — Decibels relative level. Used to define transmission levels at a point in a circuit, with respect to the level at the zero transmission level reference point.

dBRAP — Decibels above reference acoustical power, which is defined as 10^{-16} watt.

dBRN — Decibels above reference noise. This is a unit used to show the relationship between the interfering effect of a noise frequency, or band of noise frequencies, and a fixed amount of noise power commonly called reference noise. A tone of 1000 hertz having a power level of −90 dBm was selected as the reference noise power because it appeared to have negligible interfering effect and would permit the measurement of interfering effect in positive numbers.

DBS — Abbreviation for direct broadcast satellite.

dBV — The increase or decrease in voltage independent of impedance levels. A unit of electrical pressure. Decibels referred to a standard of 1 volt.

dBW — Decibels referenced to 1 watt.

dBx — Decibels above the reference coupling. Reference coupling is defined as the coupling between two circuits that would be required to give a reading of 0 dBa on a two-type noise-measuring set connected to the disturbed circuit when a test tone of 90 dBa (using the same weighting as that used on the disturbed circuit) is impressed on the disturbing circuit.

dc — Abbreviation for direct current.

D cable — Two-conductor cable, each conductor having the shape of the capital letter D, with insulation between the conductors and between the conductors and the sheath.

dc amplifier — *See* direct-current amplifier.

dc balance — An adjustment of circuitry to avoid a change in dc level when the gain is changed.

dc beta — The dc current gain of a transistor; the ratio of the collector current to the base current that caused it, measured at constant collector-to-emitter voltage.

dc block — 1. A (coaxial) component employed to prevent the flow of dc or video along a transmission line while allowing the uninterrupted flow of rf. The structure is a short section of coaxial line having a capacitance in series with the center and/or outer conductor. The rf flows with negligible reflection or attenuation, while the video frequencies or dc are blocked. 2. A device that stops the flow of dc power but permits passage of higher-frequency signals.

dc breakdown — Voltage at which ionization occurs at a slowly rising dc voltage.

dcc — Abbreviation for double cotton-covered.

dc capacitance — The capacitance of a capacitor calculated from the ratio of the capacitor charge and the direct voltage measured between the terminations.

dc capacitor — A capacitor designed essentially for application with a direct voltage.

dc circuit breaker — A device used to close and interrupt a dc power circuit under normal conditions or to interrupt this circuit under fault or emergency conditions.

dc component — The average value of a signal. In television it represents the average luminance of the picture being transmitted, and in radar it is the level from which the transmitted and the received pulses rise.

dc continuity — A circuit in which an impressed dc current — a reading on a conventional ohmmeter applied across the terminals of a circuit with dc continuity — will result in a deflection of the meter.

dc coupled — The connection by a device that passes the steady-state characteristics of a signal and that largely eliminates the transient or oscillating characteristics of the signal.

dc dump — The withdrawal of direct-current power from a computer. This may result in loss of the stored information.

DCE — Abbreviation for data communication equipment or data channel equipment. Equipment (such as a modem) installed at a user's premises that provides all the functions required to establish, maintain, and terminate a connection and provide signal conversion and coding between the data-terminal equipment and the common carrier's line.

dc generator — A rotating electric device for converting mechanical power into dc power.

dc inserter stage — A television transmitter stage that adds a dc component known as the pedestal level to the video signal.

dcl — Abbreviation for direct current leakage.

dc leakage current — Abbreviated dcl. 1. The relatively small direct current through a capacitor when dc voltage is impressed across it. 2. That direct current which passes through the capacitor dielectric when a capacitor is subjected to a polarized voltage potential across its terminals and has reached a state of charge equilibrium. Usually expressed in microamperes and associated mostly with electrolytic capacitors.

dcnm — Abbreviation for dc noise margin.

dc noise — The noise arising when reproducing a magnetic tape that has been nonuniformly magnetized by energizing the record head with direct current, either in the presence or absence of bias. This noise has pronounced long-wavelength components that can be as much as 20 dB higher than those obtained from a bulk-erased tape.

dc noise margin — Abbreviated dcnm. Noise margin is also called noise immunity. 1. The difference between the normal applied logic levels and the threshold voltage of a digital integrated circuit. 2. The difference between the output voltage level of a driving gate and the input threshold voltage of a driven gate for both the 1 and the 0 states.

dc operating point — The dc values of collector voltage and current of a transistor with no signal applied.

dc overcurrent relay — A device that functions when the current in a dc circuit exceeds a given value.

dc patch bay — Specific patch panels provided for termination of all direct-current circuits and equipment used in an installation.

dc picture transmission — Transmission of the dc component of the television picture signal. This component represents the background or average illumination of the overall scene and varies only with the overall illumination.

dc plate resistance — The value or characteristic used in vacuum-tube computations. It is equal to the direct-current plate voltage divided by the direct-current plate current and is given the symbol R_p.

dc reclosing relay — A device that controls the automatic closing and reclosing of a dc circuit interrupter, generally in response to load circuit conditions.

dc resistivity — The resistance of a body of ferromagnetic material having a constant cross-sectional area, measured under stated conditions by means of direct voltage, multiplied by the cross-sectional area, and divided by the length of the body.

dc restoration — The reestablishment by a sampling process of the dc and the low-frequency components of a video signal that have been suppressed by ac transmission.

dc restorer — Also called clamper or restorer. A clamping circuit that holds either amplitude extreme of a signal waveform to a given reference level of potential.

dc shift — An error in transient response, with a time constant approaching several seconds.

dc short — A coaxial component that provides a dc circuit between the center and outer conductors, while allowing the rf signal to flow uninterrupted. The unit has a high-impedance line shunted across the main coax line. This consequently makes the device frequency dependent.

dc signaling — A transmission method that utilizes direct current.

dc test — A general term for those tests that measure a static parameter such as leakage current.

DCTL — Abbreviation for direct-coupled transistor logic.

dc transducer — A transducer capable of proper operation when excited with direct current. Its output is given in terms of direct current unless otherwise modified by the function of the stimulus.

dc transmission — Transmission of a television signal in such a way that the dc component of the picture signal is still present. This is done to maintain the true level of background illumination.

dcwv — Abbreviation for direct-current working volts. The maximum continuous voltage that can be applied to a capacitor.

DDD — Abbreviation for direct distance dialing.

DDP — Abbreviation for distributed data processing.

DDS — *See* dataphone digital service.

deac — *See* deaccentuator.

deaccentuator — Abbreviated deac. A network or circuit employed in frequency-modulated receivers to deemphasize the higher frequencies in the received signal to restore their proper relative amplitude. *See also* deemphasis.

deactuate pressure — The pressure at which an electrical contact opens or closes as the pressure approaches the actuation level from the opposite direction.

dead — 1. Free from any electric connection to a source of potential difference and from electric charge; having the same potential as that of the earth. The term refers only to current-carrying parts that are sometimes alive, or charged. 2. *See* room acoustics.

dead band — Also called dead space, dead zone, or switching blank. 1. In a control system, the range of values through which the measurand can be varied without initiating an effective response. 2. A specified range of values in which the incoming signal can be altered without also changing the outgoing response. 3. The angle within which the rotor of a stepper motor can stop — the error band, or band of position uncertainty caused by bearing and load friction, and rotor and load inertia.

deadbeat — Coming to rest without vibration or oscillation; i.e., the pointer that a highly damped meter or galvanometer moves to a new position without overshooting and vibrating about its final position.

deadbeat instrument — A voltmeter, meter, or similar device in which the movement is highly damped to bring it to rest quickly.

deadbeat response — The response of a critically damped stepper motor that provides rotation from one step to another without overshoot or ringing.

dead break — An unreliable contact made near the trip point of a relay or switch at low contact pressure. As a result, the switch does not actuate, even though the circuit is interrupted.

dead-center position — The place on the commutator of a dc motor or generator at which a brush would be placed if the field flux were not distorted by armature reaction.

dead end — 1. In a sound studio, the end with the greater sound-absorbing characteristic. 2. In a tapped coil, the portion through which no current is flowing at a particular bandswitch position.

dead-end tower — An antenna or transmission-line tower designed to withstand unbalanced mechanical pull from all the conductors in one direction, together with the wind strain and vertical loads.

deadlock — 1. A situation that occurs when all tasks within a computer system are suspended, waiting for resources that have already been assigned to other tasks that are also waiting for additional resources. Thus, the system can perform no useful work unless tasks are destroyed and their resources reclaimed. 2. A condition in which two processes wait indefinitely for each other. The solution is to allocate resources on a priority basis or to have tie-breaking circuitry.

deadly embrace — A situation in which two processes each unknowingly wait for resources held by the other. *See also* deadlock.

dead range — *See* dead band.

dead room — 1. A room for testing the acoustic efficiency or range of electroacoustic devices such as speakers and microphones. The room is designed with an absolute minimum of sound reflection, and no two dimensions of the room are the same. A ratio of 3 to 4 to 5 is usually employed (e.g., 15 ft × 20 ft × 25 ft). The walls, floor, and ceiling are lined with a sound-absorbing material. 2. *See* anechoic.

dead short — A short circuit having minimum resistance.

dead space — 1. An area or zone, within the normal range of a radio transmitter, in which no signal is received. 2. *See* dead band, 2.

dead spot — 1. A geographic location in which signals from one or more radio stations are received poorly or not at all. 2. That portion of the tuning range of a receiver in which stations are heard poorly or not at all because of poor sensitivity.

dead time — 1. The minimum interval, following a pulse, during which a transponder or component circuit is incapable of repeating a specified performance. 2. Any definite delay intentionally placed between two related actions to avoid overlap that could result in confusion or permit another particular event, such as a control decision or switching event, to occur. 3. Time interval in which no clock phase is active.

dead volume — The total volume of the pressure port cavity of a pressure transducer at the rest position (i.e., with no stimulus applied).

dead zone — *See* dead band, 2.

debicon — A high-efficiency microwave generator in which use is made of crossed-field effects.

debouncing — The elimination of accidental bounce signals characteristic of mechanical switches that bounce repeatedly until the contact is finally closed or opened. Debouncing may be performed by hardware (latch) or software (delay).

de Broglie wavelength — The wavelength of radiation that corresponds to a photon whose energy is 1 electron volt: 1.24 micrometer.

debug — 1. To examine or test a procedure, routine, or equipment for the purpose of detecting and correcting errors. 2. To detect, locate, and remove mistakes from a program. Debugging programs are available that test for and isolate errors in another program. 3. A method of fault finding in programs, usually using data dumps and breakpoints. This would be handled in software by a debug routine. 4. Checking the logic of a software program to isolate and remove any mistakes. 5. To detect, locate, and remove all mistakes in a computer program and any malfunctions in the computing system itself.

debugger — 1. An essential program designed to facilitate software debugging. At a minimum, it provides breakpoints, dump facilities, and register and memory examine/modify. 2. Program that facilitates the testing of the object program on a microcomputer and its input-output devices. Debuggers usually accept commands from the user to perform such functions as (*a*) displaying or printing out the contents of the microcomputer memories, or the contents of the registers of the central processing unit, (*b*) modifying the RAM, (*c*) starting execution of the object program from a specific memory location, and (*d*) setting a breakpoint or stopping execution of the program when the instruction at a specific memory location is reached in the program or when a given condition is met. 3. A program that allows the user to observe the program flow and results of the program's operation in a step-by-step mode. It may be used to change data or instructions, alter registers, etc. 4. Program that helps track down and eliminate errors that occur in the normal course of program development.

debugging — 1. Isolating and removing all malfunctions (bugs) from a computer or other device to restore its operation. 2. A process of shakedown operation of each finished material that is performed prior to its being placed in use in order to exclude the early failure period. During debugging, weak elements are expected to fail and be replaced by elements of normal quality that are not subject to early failure. 3. Process of detecting, locating, and correcting mistakes in hardware (system wiring) or software (program).

debugging period — *See* early-failure period.

debugging routines — Programs that aid in the isolation and correction of malfunctions and/or errors in a unit of equipment or another program.

debug monitor — An interactive program that allows the design engineer to intercommunicate in a "friendly" manner (through a terminal device) with the microcomputer system under development and to control closely the execution of an untested microcomputer program in order to check its correct operation.

debug program — A special program used to find errors in a program that is being run on a computer. A debug program allows a programmer to correct programming errors in the program being run.

debunching — Space-charge effect that tends to destroy the electron bunching in a velocity-modulation vacuum tube by spreading the beam due to mutual repulsion of the electrons.

Debye effect — The selective absorption of electromagnetic waves by a liquid made up of molecules with permanent dipole moments.

Debye length — Also called Debye shielding distance or plasma length. A theoretical length that describes the maximum separation at which a given electron is influenced by the electric field of a given positive ion.

Debye shielding distance — *See* Debye length.

decade — 1. The interval between any two quantities having a ratio of 10:1. 2. A group or assembly of 10 units (e.g., a counter that counts to 10 in one column, or a resistor box that inserts resistance quantities in multiples of powers of 10). 3. Ten times a given quantity or range.

decade band — A band having frequency limits related by the equation $f_b - f_a = 10$.

decade box — A special assembly of precision resistors, coils, or capacitors. It contains two or more sections, each having 10 times the value of the preceding section. Each section is divided into 10 equal parts. By means of a 10-position selector switch or equivalent arrangement, the box can be set to any desired value in its range.

decade counter — A logic device that has 10 stable states and may be cycled through these states by the application of 10 clock or pulse inputs. A decade counter usually counts in a binary sequence from state 0 through state 9 and then cycles back to 0. Sometimes referred to as a divide-by-10 counter.

decade resistance box — A resistance box containing two or more sets of 10 precision resistors.

decade scaler — A decade counter, or scale-of-10 counter. A scaler with a factor of 10. It produces one output pulse for every 10 input pulses.

decametric waves — High-frequency band; 3 to 10 MHz.

decay — 1. Gradual reduction of a quantity. 2. Exponential reduction in amplitude. To be reduced in an exponential manner, as the current in a circuit decays when the source of potential is removed. 3. The decrease in the radiation intensity of any radioactive material with respect to time. 4. In a storage tube, a change in magnitude or configuration of stored information by any cause other then erasing or writing.

decay characteristic — *See* persistence characteristic (of a luminescent screen).

decay distance — The distance between an area of wave generation and a point of passage of the resulting waves outside the area.

Decca — A British long-range hyperbolic navigational system that operates in the 70- to 130-kilohertz frequency band. It is a continuous-wave system in which the receiver measures and integrates the relative phase difference between the signal received from two or more synchronized ground stations. One master station and three slave stations are usually arranged in star formation. Operational range is about 250 miles (402 km).

decelerated electrons — Electrons that, after traveling at a great rate of speed, strike a target, become quickly decelerated, and cause the target to emit X-rays.

decelerating electrode — In an electron-beam tube, an electrode to which a potential is applied to slow down the electrons in the beam.

deceleration — The act or process of moving, or of causing to move, with decreasing speed; the state of so moving.

deceleration time — 1. In a computer, the time interval between the completion of the reading or writing of a record on a magnetic tape and the time when the tape stops moving. 2. The time required to stop a motor, whether free running or with some braking means.

deception — Deliberate production of false or misleading echoes on enemy radar by the radiation of spurious signals synchronized to the radar or by the reradiation of the radar pulses from extraneous reflectors.

deception device — A device that works to make unfriendly signals either unusable or misleading.

deception jamming — See confusion jamming.

deci- — Prefix meaning one-tenth (10^{-1}).

decibel — 1. Abbreviated dB. The standard unit for expressing transmission gain or loss and relative power levels. One decibel is one-tenth of a bel. The term *dBm* is used when a power of 1 milliwatt is the reference level. Decibels indicate the ratio of power output (P_o) to power input (P_{in}):

$$dB = 10 \ \log_{10}(P_o/P_{in})$$

2. A unit of change in sound intensity. One decibel is approximately the smallest change that the ear can perceive. Larger decibel increments reflect the fact that sound intensity must be squared in order for the ear to perceive a doubling of intensity. An increase in intensity is expressed as a positive number of dBs, a decrease as a negative value. No change in intensity is 0 dB, and 0 is also used to indicate a starting point from which changes are measured. 3. A unit used to measure and compare signal levels on a logarithmic scale. 4. A measure of the ratio between two power levels. Doubling or halving the power corresponds to a 3-dB change, and 10 dB corresponds roughly to the audible effect of doubling or halving the loudness of a signal (although it represents a power ratio of 10:1). Decibels are frequently used to specify variation in signal level throughout a range of frequencies (i.e., 20–20,000 Hz ±1 dB) and to specify such other ranges as signal-to-noise ratio. 5. A relative measure of signal or sound intensity or volume. It expresses the ratio of one intensity to another. Can also express voltage and power ratios logarithmically.

decibel meter — Also called dB meter. 1. An instrument for measuring the electric power level, in decibels, above or below an arbitrary reference level. 2. Sound-level indicator.

decibels above or below 1 milliwatt — The unit used to describe the ratio of the power at any point in a transmission system to a reference level of 1 milliwatt. The ratio expresses decibels above or below this reference level of 1 milliwatt.

decibels above or below 1 watt — A measure of power expressed in decibels to a reference level of 1 watt.

decibels above reference noise — An expression used to describe the ratio of the circuit noise level in a transmission system, at any point, to some arbitrarily chosen reference noise. The expression signifies the reading of a noise meter. Where the circuit-noise meter has been adjusted to represent effect under specified conditions, the expression is in adjusted decibels.

decilog — A division of the logarithmic scale used for measuring the logarithm of the ratio of two values of any quantity. The number of decilogs is equal to 10 times the logarithm to the base 10 of the ratio. One decilog therefore corresponds to a ratio of $10^{0.1}$ (i.e., 1.25892+).

decimal — 1. Pertaining to a characteristic or property involving a selection, choice, or condition in which there are 10 possibilities. 2. Pertaining to the number representation system with a radix of 10. 3. Pertaining to a system of numerical representation in which there are ten symbols, 0, 1, 2, 3, . . . 9.

decimal attenuator — A system of attenuators arranged so that a voltage or current can be reduced decimally.

decimal-binary switch — A switch by means of which a single input lead is connected to appropriate combinations of four output leads (representing 1, 2, 4, and 8) for each of the decimal-numbered settings of the associated control knob. For example, with the knob in position 7, the input lead would be connected to output leads 1, 2, and 4.

decimal code — A code in which each allowable position has one of 10 possible states. The conventional number system with the base 10 is a decimal code.

decimal-coded digit — One of ten arbitrarily selected patterns of 1s and 0s that are used to represent decimal digits.

decimal digit — One of the numbers 0 through 9 used in the number system with the base 10.

decimal encoder — An encoder in which there are 10 output lines, one for each digit from 0 to 9, for each decade of decimal numbers.

decimal notation — The writing of quantities in the decimal numbering system.

decimal numbering system — The popular numbering system using the Arabic numerals 0 through 9 and thus having a base, or radix, of 10. For example, the decimal number 2345 can be derived in this way:

$$2000 + 300 + 40 + 5 = 2345$$

or

$$2(10^3) + 3(10^2) + 4(10^1) + 5(10^0) = 2345$$

In the decimal system, all numbers are obtained by raising the radix (total number of marks, or 10 in this system) to various powers.

decimal point — In a decimal number, the point that marks the place between integral and fractional powers of 10.

decimal-to-binary conversion — The mathematical process of converting a number written in the scale of 10 into the same number written in the scale of 2.

Decimal	Binary	Decimal	Binary
0	0	10	101
1	1	11	1011
2	10	12	1100
3	11	13	1101
4	100	14	1110
5	101	15	1111
6	110	16	10000
7	111	32	100000
8	1000	64	1000000
9	1001	128	10000000

decimetric waves — 1. Electromagnetic waves having wavelengths between 0.1 and 1 meter. 2. Ultrahigh frequency band; 300 MHz to 3 GHz.

decineper — One-tenth of a neper.

decinormal calomel electrode — A calomel electrode containing a decinormal potassium chloride solution.

decision — In a computer, the process of determining further action on the basis of the relationship of two similar items of data.

decision box — On a flowchart, a rectangle or other symbol used to mark a choice or branching in the sequence of programming of a digital computer.

decision element — In computers or data-handling systems, a circuit that performs a logical operation, such as AND, OR, NOT, or EXCEPT, on one or more binary digits in input information that represent "yes" or "no" and that expresses the result in its output. *See also* gate.

decision table — A table of all contingencies that are to be considered in the description of a problem, together with the actions to be taken. Decision tables are sometimes used in place of flowcharts for problem description and documentation.

deck — 1. In computer usage, a collection of cards, usually a complete set of cards punched for a definite purpose. 2. A term usually applied to a tape machine having no built-in power amplifiers or loudspeakers of its own, but intended rather for feeding a separate amplifier and speaker system, as in a component installation.

declination — The offset angle of an antenna from the axis of its polar mount as measured in the meridian plane between the equatorial plane and the antenna main beam.

declination offset angle — The adjustment angle of a polar mount between the polar axis and the plane of a satellite antenna used to aim at the geosynchronous arc.

decode — 1. In a computer, to obtain a specific output when specific character-coded input lines are activated. 2. To use a code to reverse a previous encoding. 3. To determine the meaning of characters or character groups in a message. 4. To determine the meaning of a set of pulses that describes an instruction, a command, or an operation to be carried out.

decoder — 1. A device for translating a combination of signals into one signal that represents the combination. It is often used to extract information from a complex signal. 2. In automatic telephone switching, a relay-type translator that determines from the office code of each call the information required for properly recording the call through the switching train. Each decoder has means, such as a cross-connecting field, for establishing the controls desired and readily changing them. 3. Sometimes called matrix. In an electronic computer, a network or system in which a combination of inputs is excited at one time to produce a single output. 4. A device that converts coded information into a more usable form, for example, a binary-to-decimal decoder. 5. A circuit that accepts coded input data and activates a specific output(s) in accordance with the code present at the input. 6. A circuit built into an FM tuner to enable it to translate stereo signal information into two matched audio outputs. 7. A means to extract and process recorded quadraphonic sound information from a complex signal into four matched outputs. 8. A device consisting of gates that is usually connected to the output of a counter. It provides an output when the counter is at a specific count or range of counts. 9. A logic device that breaks the code of an incoming binary signal; i.e., converts the coded information into a more usable form. 10. A logic device that converts data from one number system to another (e.g., an octal-to-decimal decoder). Decoders are also used to recognize unique addresses, such as a device address, and bit patterns. *See* code converter. 11. A circuit that restores a signal to its original form after it has been scrambled. 12. A device that reconstructs an encrypted signal so that it can be clearly received. 13. A television set-top device that enables the home subscriber to convert an electronically scrambled television picture into a viewable signal. (This should not be confused with a digital coder/decoder, known as a codec, which is used in conjunction with digital transmissions.)

decoding — 1. The process of obtaining intelligence from a code signal. 2. In multiples, a process of separating the subcarrier from the main carrier.

decoding matrix — A device for decoding many input lines into a single output line.

decoding network — A circuit made so that, when a particular combination of inputs is on, an output appears on one of a number of output lines.

decollimated light — In fiber optics, light rays made nonparallel by striae and boundary defects.

decommutation — The process of recovering a signal from the composite signal previously created by a commutation process.

decommutator — Equipment for separating, demodulating, or demultiplexing commutated signals.

decomposition — Breaking down a software specification, in depth and breadth, to determine all required functions and their relationships.

decoupler — A circuit for eliminating the effect of coupling in a common impedance.

decoupling — 1. The reduction of coupling. 2. To isolate two circuits on a common line. A decoupling network is a low-pass filter (RC or RLC) that does not isolate equally in both directions.

decoupling circuit — A circuit used to prevent interaction of one circuit with another.

decoupling network — A network of capacitors and chokes or resistors placed into leads that are common to two or more circuits, to prevent unwanted, harmful interstage coupling.

decoy — A reflecting object having reflective characteristics of a target, used in radar deception.

decrement — 1. Progressive diminution in the value of a variable quantity; also the amount by which a variable decreases. When applied to damped oscillations, it is usually called damping factor. 2. A specific part of an instruction word in some binary computers; thus, a set of digits. 3. To reduce the numerical contents of a counter. A decrement of one is usually assumed unless specified otherwise. 4. In an oscillating system with damping (each oscillation has less amplitude than the one preceding it), the ratio of the peak values (voltage, distance, etc.) of two successive half-cycles. It is expressed as a decimal fraction less than 1.

decremeter — An instrument for measurement of the logarithmic decrement (damping) of a wave train.

decryption — The process of "unscrambling" an encrypted or coded message.

dedicated — 1. To set apart for some special use. For example, a dedicated microprocessor is one that has been specifically programmed for a single application, such as weight measurement by scale, traffic light control, etc. (ROMs by their very nature [read-only] are dedicated memories.) 2. A piece of equipment that is assigned to one particular use only. Minicomputers are often dedicated. Microprocessors are intended to be dedicated.

dedicated computer — A computer whose use is reserved for a particular task.

dedicated line — 1. A communication line that isn't dialed, also called a leased or private line. 2. A communication line for voice and/or data rented from a communication carrier. 3. Full-term line allocated to one subscriber at a specific degree of conditioning. 4. Leased line, wired directly between communicating systems. It has faster transmission than regular telephone lines and does not require acoustic couplers (types of modems). 5. A permanent circuit for private use. The leased line physically connected between locations, or through a central office, without using the switching equipment. 6. A very common type of transmission facility for data communications. Each line is separate from the dial-up network (DDD) and its associated equipment and has a discrete, permanent path to its destination. Users pay a flat monthly fee for this service. 7. A telephone line that has a continuous connection, maintained by the telephone company.

dedicated machine — A PC or other microcomputer designed to handle a special, usually single, task. A dedicated server PC could only be used to service user requests, not as both server and workstation.

dedicated register — A register in a computer exclusively used to contain a specific item.

dee — A hollow, D-shaped accelerating electrode in a cyclotron.

dee line — A structural member that supports the dee of a cyclotron and together with the dee forms the resonant circuit.

deemphasis — Also called postemphasis or postequalization. 1. Introduction of a frequency-response characteristic that is complementary to that introduced in preemphasis. 2. Reduction of the level of the higher audio frequencies during FM reception or tape replay so that they compensate for the preemphasis applied to the transmission. This restores an overall uniform response. 3. A form of equalization used in FM tuners, complementary to a preemphasis used in transmission. The purpose is to improve the overall signal-to-noise ratio while maintaining a uniform frequency response. It is expressed in the form of a time constant or product of a resistance and capacitance. Standard FM broadcasts use a 75-μ time constant in the United States, and 50 μs in Europe, whereas Dolby B transmissions use a 25-μs time constant.

deemphasis network — A network inserted into a system to restore the preemphasized frequency spectrum to its original form.

deenergize — 1. To disconnect a device from its power source. 2. To stop the current in a circuit or to remove electrical potential from a circuit, as by opening a switch.

deep discharge — The withdrawal of all available electrical energy before recharging a cell or battery.

deep space net — A combination of radar and communications stations in the United States, Australia, and South Africa so located as to keep a spacecraft in deep space under observation at all times.

default — The value(s) or option(s) that are assumed during operation when not specified.

default value — The value that a database element assumes if the user does not specify another value.

defeat — The frustration, counteraction, or thwarting of an alarm device so that it fails to signal an alarm when a protected area is entered. Defeat includes both circumvention and spoofing.

defect — 1. A condition considered potentially hazardous or operationally unsatisfactory and therefore requiring attention. 2. Any nonconformance with the normally accepted characteristics for a unit. *See also* major defect; minor defect.

defect analysis — The process of examining technical or management (nontechnical) data, manufacturing techniques, or material to determine the cause of variations of electrical, mechanical, or physical characteristics outside the limitations established at any manufacturing checkpoint.

defect condition — Hole conduction in the valence band in a semiconductor.

Defense Electronic Supply Center — *See* DESC.

deferred addressing — An indirect addressing mode in which the directly addressed location contains the address of the operand, rather than the operand itself.

deferred entry — In a computer, an entry into a subroutine as a result of a deferred exit from the program that passes control to the subroutine.

deferred exit — In a computer, the transfer of control to a subroutine at a time controlled by the occurrence of an asynchronous event rather than at a predictable time

defibrillator — An electronic device that applies a brief high-voltage potential to the heart by means of electrodes placed on the chest wall. The defibrillator is used to restore regular rhythm to a heart in ventricular fibrillation.

definite-purpose relay — A relay with some electrical or mechanical feature that distinguishes it from a general-purpose relay.

definition — 1. The fidelity with which the detail of an image is reproduced. When the image is sharp (i.e., has definite lines and boundaries), the definition is said to be good. 2. The degree with which a communication system reproduces sound images or messages. 3. The fidelity with which the pattern edges in a printed circuit (conductors, inductors, etc.) are reproduced relative to the original master pattern. 4. The sharpness of a picture subjectively evaluated in terms of its resolution. 5. The sharpness of a screen-printed pattern; the exactness with which a pattern is printed.

deflecting coil — An inductor used to produce a magnetic field that will bend the electron beam a desired amount in the cathode-ray tube of an oscilloscope, television receiver, or television camera.

deflecting electrode — An electrode to which a potential is applied in order to deflect an electron beam.

deflecting torque — *See* torque of an instrument.

deflection — Movement of the electron beam in a cathode-ray tube as electromagnetic or electrostatic fields are varied to cause the light spot to traverse the face of the tube in a predetermined pattern.

deflection circuit — The circuit that regulates an electron beam's deflection in a CRT.

deflection coil — One of the coils in the deflection yoke.

deflection factor — *See* deflection sensitivity.

deflection focusing — The progressive defocusing of a cathode-ray-tube display image that occurs when the deflected electron beam impinges on the CRT screen at a slant.

deflection plane — A plane perpendicular to the cathode-ray-tube axis and containing the deflection center.

deflection plates — Two pairs of parallel electrodes, the pairs set one forward of the other and at right angles to each other, parallel to the axis of the electron stream

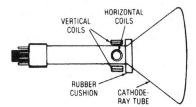

Deflection coils.

within an electrostatic cathode-ray tube. An applied potential produces an electric field between each pair. By varying the applied potential, this field may be varied to cause a desired angular displacement of the electron stream.

deflection polarity — The relationship between the direction of displacement of an oscilloscope trace and the polarity of the applied signal wave.

deflection sensitivity — Also called deflection factor. The displacement of the electron beam at the target or screen of a cathode-ray tube divided by the change in magnitude of the deflection field. Deflection sensitivity is usually expressed in millimeters (or inches) per volt applied between the deflecting electrodes, or in millimeters (or inches) per ampere in the deflection coil.

deflection voltage — The voltage applied to the electrostatic plates of a cathode-ray tube to control the movement of the electron beam.

deflection yoke — An assembly of one or more electromagnets for deflecting one or more electron beams.

defocus-dash mode — A method of storage of binary digits in a cathode-ray tube. Initially, the writing beam is defocused to excite a small circular area on the screen. For one kind of binary digit the beam remains defocused, and for the other kind of binary digit it is suddenly focused into a concentric dot, which traces out a dash on the screen during the interval of time before the beam is cut off and moved to the next position.

defocus-focus mode — A variation of the defocus-dash mode in which the focused dot is not caused to trace a dash.

defragmentation — A process in which all the files on a hard disk are rewritten so that all the parts of each file are written to contiguous sectors.

defruiting — Method of eliminating asynchronous returns in radar beacon systems.

degassing — The process of driving out and exhausting the gases of an electron tube occluded in its internal parts.

degauss — To neutralize the existing magnetic field. The term was coined during World War II to describe the neutralizing of a ship's magnetic field by a grid of cables generating an equal but opposite magnetic field.

degausser — Also called automatic degausser and bulk eraser. 1. A device to clarify the color picture by means of coils within the set. The coils deactivate the magnetization that builds up around a color TV set when it is moved around or when other electrical devices are brought too close to the receiver. 2. Any device for neutralizing magnetism, as in a recorder head or in a separate unit. Also called a tape eraser, for use with a complete tape recording on its reel. 3. Demagnetizer. *See* bulk eraser; head demagnetizer.

degaussing — Girdling a ship's hull with a web of current-carrying cable that sets up a magnetic field equal in value and opposite in polarity to that induced by the earth's magnetic field, thus rendering the ship incapable of actuating the detonator of a magnetic mine.

degeneracy — The condition in which two or more modes have the same resonant frequency in a resonant device.

degenerate modes — A set of modes having the same resonance frequency (or propagation constant). The members of a set of degenerate modes are not unique.

degenerate parametric amplifier — An inverting parametric device for which the two signal frequencies are identical and equal to one-half the frequency of the pump. (This exact but restrictive definition is often relaxed to include cases in which the signals occupy frequency bands that overlap.)

degeneration — *See* negative feedback.

degradation — 1. A gradual decline of quality or loss of ability to perform within required limits. The synonym *drift* is often used for electronic devices. 2. A condition in which the system continues to operate but at a reduced level of service. Unavailability of major equipment subsystems or components is the usual cause. 3. A gradual deterioration in performance as a function of time.

degradation failure — Failure of a device because a parameter or characteristic changes beyond some previously specified limit.

degree day — The measure of the deviation of the mean daily temperature from a given standard, with each variance from the standard during a single day recorded as one degree day.

degree of current rectification — The ratio between the average unidirectional current output and the root-mean-square value of the alternating-current input from which it was derived.

degree of membership — The confidence or certainty, expressed as a number from 0 to 1, that a particular value belongs in a fuzzy set.

degree of voltage rectification — The ratio between the average unidirectional voltage and the root-mean-square value of the alternating voltage from which it was derived.

degree rise — The amount of increase in temperature caused by the introduction of electricity into a unit.

degrees of freedom — In a vibrating system, the coordinates necessary to locate the position of the vibrating element at any time. For example, a single-degree-of-freedom system can move along only one axis, in both directions. A two-degrees-of-freedom system will require two coordinates to describe the position of the elements. A multiple-degree-of-freedom system generally has many elements that can move along many axes.

deinstall — To remove a program or hardware device from active service.

deion circuit breaker — A circuit breaker built so that the arc that forms when the circuit is broken is magnetically blown into a stack of insulated copper plates, giving the effect of a large number of short arcs in series. Each arc becomes almost instantly deionized when the current drops to zero in the alternating-current cycle, and the arc cannot reform.

deionization — The process by which an ionized gas returns to its neutral state after all sources of ionization have been removed.

deionization potential — The potential at which ionization of the gas within a gas-filled tube ceases and conduction stops.

deionization time — The time required for the grid of a gas tube to regain control after the anode current has been interrupted.

deionized water — Water that has been treated to remove ions. Deionized water is required in certain electronic applications to prevent contamination of parts coming in contact with the water. *See* demineralized water.

dekahexadecimal — *See* sexadecimal notation.

Dekatron — A cold-cathode counting tube.

delamination — 1. Separation of conductive patterns from the substrate, or separation of layers of the base material. 2. A separation between any of the layers of a base material or between the laminate and the conductive foil, or both.

delay — 1. The time required for a signal to pass through a device or conductor. 2. The time interval between the instants at which any designated point in a wave passes any two designated points of a transmission circuit.

delay circuit — 1. A circuit that delays the passage of a pulse or signal from one part of a circuit to another. 2. An electronic time-delay device that can introduce time delays from a few milliseconds to about 100 milliseconds without significantly degrading signal quality. Can be used to restore acoustics of a large auditorium to recorded programs heard in a normal-sized room.

delay coincidence circuit — A coincidence circuit actuated by two pulses, one of which is delayed a specific amount with respect to the other.

delay counter — In a computer, a device that can temporarily delay a program a sufficient length of time for the completion of an operation.

delay distortion — Also called envelope-delay distortion, phase-delay distortion, or phase distortion. 1. Phase-delay distortion; i.e., departure from flatness in the phase delay of a circuit or system over the frequency range required for transmission, or the effect of such departure on a transmitted signal. 2. Envelope distortion; i.e., departure from flatness in the envelope delay of a circuit or system over the frequency range required for transmission, or the effect of such departure on a transmitted signal. 3. The amount of variation in delay for various frequency components of the facsimile signal, usually expressed in microseconds from an average delay time. 4. The difference between the maximum and minimum phase delay within a specified band of frequencies. 5. Distortion caused by the fact that the higher-frequency components of a signal travel slower over a transmission facility than the lower-frequency components and therefore arrive later and out of phase. Numerically, it is the maximum difference in transmission time between any two frequencies in a specified frequency band, expressed in microseconds. Measured in microseconds of delay relative to the delay at 1700 Hz. Delay distortion doesn't affect voice communication but can seriously impair data transmissions.

delayed automatic volume control — Abbreviated delayed AVC. An automatic volume-control circuit that acts only on signals above a certain strength. It thus permits reception of weak signals even though they may be fading, whereas normal automatic volume control would make the weak signals even weaker.

delayed AVC — *See* delayed automatic volume control.

delayed contacts — Contacts that are actuated a predetermined time after the start of a (timing) cycle.

delayed PPI — A PPI (plan-position indicator) in which the initiation of the time base is delayed.

delayed repeater satellite — A satellite that stores information obtained from a ground terminal at one location and, upon interrogation by a terminal at a different location, transmits the stored message.

delayed sweep — 1. In a cathode-ray tube, a type of sweep that is not allowed to begin for a while after being triggered by the initiating pulse. 2. A sweep that has been delayed either by a predetermined period or by a period determined by an additional independent variable.

delay equalizer — 1. A device that adds delay at certain frequencies to a circuit in a way to reduce the delay distortion. 2. A corrective network that is designed to make the phase delay or envelope delay of a circuit or system substantially constant over a desired frequency range. 3. A network that introduces an amount of phase shift complementary to the phase shift in the circuit at all frequencies within the desired band. 4. Selective delaying of various frequency components of the received signal to match the delay of other frequency components caused by envelope delay distortion.

delay/frequency distortion — That form of distortion which occurs when the envelope delay of a circuit or system is not constant over the frequency range required for transmissions.

delay line — 1. A real or artificial transmission line or equivalent component that is used to delay a signal, either linear or digital, for a predetermined length of time. The delay time is defined as the duration of time between the leading edge of the input pulse and the 50-percent point on the leading edge of the output pulse. 2. A specially constructed cable used in the luminance channel of a color receiver to delay the luminance signal. 3. A sequential logic element that has one input channel and in which the state of an output channel at any instant is the same as the state of the input channel at the instant $t - n$, where n is a constant time interval for a given output channel (the input sequence undergoes a delay of n time units). 4. A device that can cause the transmission of one unit of information to be retarded until another unit can synchronize with it. 5. A device capable of causing an energy impulse to be retarded in time from point to point, thus providing a means of storage by circulating intelligence-bearing pulse configurations and patterns. Examples of delay lines are material media such as mercury, in which sonic patterns may be propagated in time; lumped constant electrical lines; coaxial cables; transmission lines; and recirculating magnetic drum loops. 6. A cable made to provide a very low velocity of propagation with long electrical delay for transmitted signals.

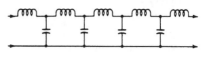

Lumped-constant line.

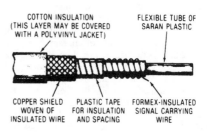

COTTON INSULATION
(THIS LAYER MAY BE COVERED
WITH A POLYVINYL JACKET)

FLEXIBLE TUBE OF
SARAN PLASTIC

COPPER SHIELD
WOVEN OF
INSULATED WIRE

PLASTIC TAPE
FOR INSULATION
AND SPACING

FORMEX-INSULATED
SIGNAL CARRYING
WIRE

Early-model delay line.

FIBRE INSULATION   FOIL STRIP   PLASTIC TUBE

Modern delay line.

Delay lines, 1.

delay-line memory — *See* delay-line storage.

delay-line register — An acoustic or electric delay line in an electronic computer, usually one or an integral number of words in length, together with input, output, and circulation circuits.

delay-line storage — Also called delay-line memory. In an electronic computer, a storage or memory device consisting of a delay line and a means for regenerating and reinserting information into it.

delay multivibrator — A monostable multivibrator that produces an output pulse a predetermined time after it is triggered by an input pulse.

delay on break — A term used to describe a mode of operation relative to timing devices. The delay begins when the initiate switch is opened (delay on break of initiate switch).

delay on energization — Also known as delay on make. A term used to describe a mode of operation relative to timing devices. The delay begins when the initiate switch is closed, or on application of power to the input.

delay on make — Same as delay on energization.

delay PPI — A radar indicator in which the start of the display sweep is delayed after the trigger so that distant targets are displayed on a short-range scale that gives an expanded presentation.

delay relay — Also called time-delay relay. A relay in which there is a delay between the time it is energized or deenergized and the time the contacts open or close.

delay time — 1. The amount of time one signal is behind (lags) another. 2. Of a switching transistor, the time interval between the application at the input terminals of a pulse that is switching the transistor from a nonconducting to a conducting state and the appearance at the output terminals of the pulse induced by the charge carriers. The time is usually measured between points corresponding to 10 percent of the amplitude of the applied pulse and of the output pulse. 3. Measurement of the interval between direction of signal to an LED and attainment of 10-percent output current in the photodetector.

delay timer — A term sometimes used to designate a timer that is primarily used for energizing (or deenergizing) a load at the end of a timed period. *See* time-delay relay.

delay unit — The unit of a radar system in which pulses may be delayed a controllable amount.

deleting — Removing something from a computer. It could be text from a word processing window, or a file from a hard disk drive.

deletion record — In a computer, a new record to replace or remove an existing record in a master file.

delimiter — Also called separator. 1. In a computer, a character that limits a string of characters and therefore cannot be a member of the string. 2. A flag that separates and organizes items of data. 3. Any means used to separate data items at input. Most frequently used are spaces or commas. The delimiter for the two integers 123 567 is the space between them. 4. A character that separates and organizes elements of a program in a computer. 5. A punctuation character, such as blackslash or comma, that separates one section of a computer command from another. 6. A text character that marks the beginning and/or end of a unit of data or separates different data components. For example, periods are used as delimiters in domain names, hyphens and parentheses are used in phone numbers and social security numbers, and blank spaces and commas are used in written text.

Dellinger effect — *See* radio fadeout.

delta — 1. The Greek letter delta (Δ) represents any quantity that is much smaller than any other quantity of the same units appearing in the same problem. 2. In a magnetic cell, the difference between the partial-select outputs of the same cell in a 1 state and in a 0 state. 3. Brainwave signals whose frequency is approximately 0.2 to 3.5 Hz. The associated mental state is usually a deep sleep or a trancelike state.

delta circuit — A three-phase circuit in which the windings of the system are connected in the form of a closed ring, and the instantaneous voltages around the ring equal zero. There is no common or neutral wire, so the system is used only for three-wire systems or generators.

delta connection — In a three-phase system, the terminal connections. So called because they are triangular, like the Greek letter delta.

delta-delta monitor — A monitor in which the red, green, and blue color guns are arranged in a triangle, and a shadow mask with round holes aligns each gun to the proper phosphor dots.

delta match — *See* Y match.

delta matched antenna — Also called Y antenna. A single-wire antenna (usually one half-wavelength long) to which the leads of an open-wire transmission line are connected in the shape of a Y. The flared part of the Y matches the transmission line to the antenna. The top of the Y is not cut, giving the matching section its triangular shape of the Greek letter delta, hence the name.

delta matching transformer — An impedance device used to match the impedance of an open-wire transmission line to an antenna. The two ends of the transmission line are fanned out so that the impedance of the line gradually increases. The ends of the transmission line are attached to the antenna at points of equal impedance, symmetrically located with respect to the center of the antenna.

delta modulation — 1. A means of encoding analog signals in control and communication systems. The output of the delta encoder is a single weighted digital pulse train that may be decoded at the receiving end to reconstruct an original analog signal. 2. A type of waveform encoding in which the differences between individual digitized speech samples are encoded. Data point values are determined by changes from preceding data point amplitudes.

delta modulator — A closed-loop sampled data control system that transmits binary output pulses whose polarity depends on the difference between the input signal being sampled and a quantized approximation of the preceding input signal. Delta modulation affords a simple, efficient method of digitizing voice for secure, reliable communications and for voice I/O in data processing.

delta network — A set of three branches connected in series to form a mesh.

delta pulse code modulation — A modulation system that converts audio signals into corresponding trains of digital pulses to provide greater freedom from interference during transmission over wire or radio channels.

delta tune — A control or switch similar in function to a clarifier, found on many AM transceivers. It compensates for signals off the center frequency of a CB channel. Although its effective range is about the same as that of a clarifier, adjustment is not as critical.

delta wave — A brain wave whose frequency is below 4 hertz.

dem — Abbreviation for demodulator.

demagnetization — Partial or complete reduction of residual magnetism.

demagnetization coefficient — *See* permeance coefficient.

demagnetization curve — In the second quadrant of a hysteresis loop, the portion that lies between the residual induction point, B_r, and the coercive force point, H_c.

demagnetization effect — 1. A decrease in internal magnetic field caused by uncompensated magnetic poles at the surface of a sample. 2. The portion of the normal hysteresis loop in the second quadrant showing the induction, B (gauss), in a magnetic material as related to the demagnetizing field, H (oersted). Points on this curve are usually designated by the coordinates B_d and H_d. This curve describes the characteristics of a permanent magnet (as contrasted with an electromagnet).

demagnetizer — A device that removes residual magnetism from recording or playback tape heads. This magnetism, if not removed, can introduce noise on recordings and cause high-frequency loss.

demagnetizing coefficient — See permeance coefficient.

demagnetizing field — A magnetizing force applied in such a direction as to reduce the remanent induction in a magnetized body.

demagnetizing force — A magnetizing force applied in such a direction that it reduces the residual induction in a magnetized body.

demand-driven machine — A parallel-processing architecture in which processors carry out instructions when the results of a processing step are demanded.

demand factor — The ratio of the maximum demand of an electrical system, or part of a system, to the total connected load of a system or that part of a system under consideration.

demand load — The load that is drawn from the source of supply at the receiving terminals, averaged over a suitable and specified interval of time, expressed in kilowatts, amperes, etc.

demarcation strip — A physical interface, usually a terminal board, between a business machine and a common carrier. See also interface, 1.

Dember effect — Also known as the photodiffusion effect. The production of a potential difference between two regions of a semiconductor specimen when one is illuminated. This phenomenon is related to the photo-electromagnetic effect, except there is no magnetic field. H. Dember discovered that when an illuminated metal plate producing electrons is bombarded by other electrons from an outside source, the photoelectric emission increases because, in addition to photoelectrons, secondary electrons are also knocked out by bombardment.

demineralized water — Water that has been treated to remove the minerals that are normally present in hard water. Demineralized water is required in some electronic applications in which extreme precautions must be taken to prevent contamination. See deionized water.

demodulation — Also called detection. 1. Operation on a previously modulated wave in such a way that it will have substantially the same characteristics as the original modulating wave. 2. The process by which a wave corresponding to the modulating wave is obtained from a modulated wave. 3. The process of retrieving digital (computer) data from a modulated analog (telephone) signal.

demodulator — Abbreviated dem. 1. A device that operates on a carrier wave to recover the wave with which the carrier was originally modulated. 2. A facsimile device that detects an amplitude-modulated signal and produces the modulating frequency as a direct current of varying amplitude. This type of unit is used to provide a keying signal for a frequency-shift exciter unit for radio facsimile transmission. 3. A device that receives tones from a transmission circuit and converts them to electrical pulses, or bits, that may be accepted by a business machine. 4. Circuitry that plays back a CD-4 disc's four signals after reprocessing the base and carrier bands inscribed in each side of the record groove. 5. A

functional section of a modem that converts received analog line signals to digital form. 6. A device that separates information from the carrier.

demodulator probe — A probe designed for use with an oscilloscope, for displaying modulated high-frequency signals.

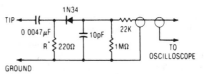

Demodulator probe.

demon — A computer program that waits until an event occurs before running; widely used to circumvent copy-protection procedures.

de Morgan's theorem — A theorem that states that the inversion of a series of AND implications is equal to the same series of inverted OR implications, or the inversion of a series of OR implications is equal to the same series of inverted AND implications. In symbols, $\overline{A \cdot B \cdot C} = \bar{A} + \bar{B} + \bar{C}$, or $\overline{A + B + C} = \bar{A} \cdot \bar{B} \cdot \bar{C}$.

demountable tube — A high-power electron tube having a metal envelope with porcelain insulation. Can be taken apart for inspection and for renewal of electrodes.

demultiplexer — 1. A device used to separate two or more signals that were previously combined by a compatible multiplexer and transmitted over a single channel. 2. A circuit that directs information from a single input to one of several outputs at a time in a sequence that depends on the information applied to the control inputs. 3. A device that reverses the action of a multiplexer, deriving a group of separate channels from the complex multiplex signal. 4. A logic circuit that can route a single line of digital information to other lines. The demultiplexer acts to switch information to many different points. 5. A circuit that applies the logic state of a single input to one of several outputs.

demultiplexing — Abbreviated demux. The process of separating a multiplexed signal into its separate intelligence signals.

demultiplexing circuit — A circuit that is used to separate the signals that have been combined for transmission by multiplex.

demux — See demultiplexing.

denary band — A band having frequency limits with the ratio of $f_b/f_a = 10$.

dendrite — A semiconductor crystal with a heavily branched, treelike structure that grows from the nucleus as the metal becomes solidified.

dendritic growth — 1. A technique of producing semiconductor crystals in long, uniform ribbons with optically flat surfaces. 2. The electrolytic transfer of metal from one conductor to another, similar to electroplating except that the dendritic growth usually, though not always, forms from cathode to anode. The dendrite resembles a tree in appearance, and when it touches the opposite conductor, there is an abrupt rise in current.

dense binary code — A binary code in which all the possible states of the pattern are used.

densitometer — An instrument for measuring the optical density (photographic transmission, photographic reflection, visual transmission, and so forth) of a material.

density — 1. A measure of the light-transmitting or reflecting properties of an area expressed by the common logarithm of the ratio of incident to transmitted or reflected light flux. 2. The mass per unit volume. The specific gravity of a body is the ratio of a density to the density of a standard substance. Water and air are commonly used as the standard substances. 3. Amount per unit cross-sectional area (e.g., current, magnetic flux, or electrons in a beam). 4. The logarithm of the ratio of incident to transmitted light. *See* opacity.

density modulation — Modulation of an electron beam by varying the density of the electrons in the beam with time.

density packing — The number of magnetic pulses (representing binary digits) stored on tape or drum per linear inch on a single head.

density step tablet — A facsimile test chart consisting of a series of areas that increase in steps from a low value of density to a maximum value of density.

dentonphonics — The technique of using electronics in broadcasting speech from the mouth. The principle is the same as that of a throat microphone, in which a transducer responds to sound energy transmitted through the tissues as a person speaks.

dependent linearity — Nonlinearity errors expressed as a deviation from a desired straight line of fixed slope and/or position.

dependent mode — In network analysis, a node having one or more incoming branches.

deperming — Another name for demagnetization.

depletion field-effect transistor — An active semiconductor device in which the main current is controlled by the depletion width of a pn junction.

depletion layer — Also called barrier layer. 1. In a semiconductor, the region in which the mobile-carrier charge density is insufficient to neutralize the net fixed charge density of donors and acceptors. 2. That region in the immediate vicinity of a semiconductor pn junction that becomes exhausted or *depleted* of current carriers in order to set up the internal potential barrier involved in either the balance between diffusion and drift currents present in the equilibrium case, or the imbalance between these currents present in a nonequilibrium situation. Being depleted of carriers, the depletion-layer region is virtually composed of "intrinsic" material, irrespective of the doping levels of the p-type and n-type materials from which it is formed. 3. A zone, several atoms thick, at the junction of n-type and p-type semiconductor materials, in which there are no current carriers, either free electrons or holes, unless biased by a direct voltage. Free electrons in the n-type material are repelled by negative charges in the p-type material, and the holes in the p-type material are repelled by the positive nucleus of atoms in the n-type material.

depletion-layer capacitance — Also called barrier capacitance. Capacitance of the depletion layer of a semiconductor. It is a function of the reverse voltage.

depletion-layer rectification — Also called barrier-layer rectification. The rectification that appears at the contact between dissimilar materials, such as a metal-to-semiconductor contact or a pn junction, as the energy levels on each side of the discontinuity are readjusted.

depletion-layer transistor — Any of several types of transistors that rely directly for their operation on the motion of carriers through depletion layers (for example, a spacistor).

depletion-mode device — A field-effect transistor or IC that passes maximum current at zero gate potential, and a decreasing current with applied gate potential.

depletion-mode field-effect transistor — 1. A field-effect transistor that exhibits substantial device current (I_{DSS}) with zero gate-to-source bias ($V_{GS} = 0$ V). 2. An MOS transistor normally on with zero gate voltage applied (channel formed during processing). A voltage of the correct polarity applied to the gate will force majority carriers from the channel, thus depleting it and turning the transistor off.

depletion-mode operation — The operation of a field-effect transistor such that changing the gate-to-source voltage from zero to a finite value decreases the magnitude of the drain current.

depletion region — Also referred to as space-charge, barrier, or intrinsic region. The region, extending on both sides of a reverse-biased semiconductor junction, in which all carriers are swept from the vicinity of the junction; that is, the region is depleted of carriers. This region takes on insulating characteristics and is capable of isolating semiconductor regions from each other. Depletion regions make planar bipolar integrated circuits possible.

depletion-type field-effect transistor — A field-effect transistor having appreciable channel conductivity for zero gate-source voltage; the channel conductivity may be increased or decreased according to the polarity of the applied gate-source voltage.

depolarization — The process of preserving the activity of a primary cell by the addition of a substance to the electrolyte. This substance combines chemically with the hydrogen gas as it forms, thus preventing excessive buildup of hydrogen bubbles.

depolarize — To make partially or completely unpolarized.

depolarizer — A chemical substance, usually manganese dioxide, added to a dry or primary cell to remove the polarizing chemical products resulting from discharge, and thus to keep the discharge rate constant; to prevent formation of hydrogen bubbles at the positive electrode.

deposited carbon — Resistive element made of a thin film of crystalline carbon or a carbon alloy sputtered onto a ceramic rod.

deposition — The application of a material to a substrate through the use of chemical, vapor, electrical, vacuum, or other processes.

depth of cut — The depth to which the recording stylus penetrates the lacquer of a recording disc.

depth of field — 1. The in-focus range of a lens or optical system. It is measured from the distance behind an object to the distance in front of the object when the viewing lens shows the object to be in focus. 2. The distance between the first object in focus and the last object in focus within a scene as viewed by a particular lens; it can vary with the quality and focal length of the lens or with its f-stop setting.

depth of heating — The depth at which effective dielectric heating can be confined below the surface of a material when the applicator electrodes are placed adjacent to only one surface.

depth of modulation — In a radio-guidance system obtaining directive information from the two spaced lobes of a directional antenna, the ratio of the difference in total field strength of the two lobes to the field strength of the greater lobe at a given point in space.

depth of penetration — The thickness of a layer extending inward from the surface of a conductor and having the same resistance to direct current as the whole conductor has to alternating current of a given frequency. *See also* skin depth.

de-Q — To reduce the Q of a tuned circuit, as generally applied to carrier-current transmission systems.

derate — 1. To use a device at a lower current or voltage than it is capable of handling in order to reduce the probability of failure or to permit its use under a condition of high ambient temperature. 2. To change the rating or value to a lower rating in consideration of other affecting factors.

derating — 1. The reduction in rating of a device or component, especially the maximum power-dissipation rating at higher temperatures. 2. Deliberately understressing components so as to provide increased reliability. (This requires the selection of components of higher stress than is required for normal operation.) For example, using a ½-watt resistor in circuit conditions demanding a ¼-watt dissipation. This process is recognized as an effective and well-established method of achieving reliable designs. The ratio of applied stress to rated stress is the stress ratio; thus, in the example above this would be 0.25/0.5, or 50 percent derated in terms of power. 3. The intentional reduction of stress-to-strength ratio in the application of an item, usually for the purpose of reducing the occurrence of stress-related failures.

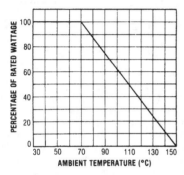

Derating curve.

derating factor — The factor by which the ratings of component parts are reduced to provide additional safety margins in critical applications or when the parts are subjected to extreme environmental conditions for which their normal ratings do not apply.

derivative action — *See* rate action.

derivative control — Automatic control in which the rate of correction is determined by the rate at which the error producing it changes.

derived center channel — A monophonic composite signal derived from the sum or difference of the left and right stereo channels, often fed to an extra speaker to fill in an aural hole between the left and right speakers. The signal from a voltage-derived center-channel output must be fed to an external power amplifier before it can drive a speaker. A power-derived center channel can drive a speaker directly.

DES — Abbreviation for Data Encryption Standard. A scheme approved by the National Bureau of Standards that encrypts data for security purposes. For use within the United State by the NSA (National Security Agency); DES is the data-communications encryption standard specified by Federal Information Processing Systems (FIPS) Publication 46.

DESC — Abbreviation for Defense Electronic Supply Center. 1. The agency that sets military specifications for all electronic components and verifies performance to these requirements. 2. Goverment agency that controls procurement policies and monitors quality for military electronics contracts.

desensitization — 1. The saturation of one component (an amplifier, for instance) by another so that the

first cannot perform its proper function. 2. The effect on a receiver section tuned to one channel that is caused by a strong signal on another channel. It is an AGC-type effect whereby the desired signal's strength appears to be decreased by the presence of a nearby signal. This effect influences a receiver's overall selectivity. 3. The tendency of a receiver to fail to recognize valid DTMF signals in the presence of such factors as dial tone, pilot signals, or data signals. 4. The reduction in sensitivity of a receiver caused by either noise or rf overload from a nearby transmitter.

desiccant — A substance used as a drying agent because of its affinity for water.

design-center rating — Limiting values of operating and environmental conditions that should not be exceeded under normal conditions in a bogey electronic device of a specified type as defined by its published data. These values are chosen by the device manufacturer to provide acceptable serviceability of the device in average applications, taking responsibility for normal changes in operating conditions due to rated supply-voltage variation, equipment component variation, equipment control adjustment, load variation, signal variation, environmental conditions, and variations in the characteristics of all electronic devices.

design compatibility — Electromagnetic compatibility achieved by incorporating in all electromagnetic radiating and receiving apparatus (including antennas) characteristics or features for elimination or rejection of undesired self-generated or external signals in order to enhance operating capabilities in the presence of natural or human-made electromagnetic noise.

design engineer — An engineer who has been assigned to design a specific product for a specific application.

design for maintainability — Those features and characteristics of design of an item that reduce requirements for tools, test equipment, facilities, spares, highly skilled personnel, etc., and improve the capability of the item to accept maintenance actions.

design-maximum ratings — Limiting values of operating and environmental conditions applicable to a bogey electronic device of a specified type as defined by its published data; they should not be exceeded under the worst probable conditions.

design-maximum rating system — *See* design-center rating. These values are chosen by the device manufacturer to provide acceptable serviceability of the device, taking responsibility for the effects of changes in operating conditions due to variations in the characteristics of the electronic device under consideration.

Desk-Fax — Trade name of Western Union Telegraph Co. for a small facsimile transceiver employed principally in short-line telegraph service.

desktop computer — 1. A self contained, totally integrated system that comes complete with central processor, read/write memory, external mass storage in the form of a cassette tape or diskette system, keyboard, computer language and operating system in firmware (usually in the form of read-only memory), connectors for external devices, and an output display such as a CRT, LED readout, or printer. When the computer is turned on, it is ready to solve problems. The operating system and language do not have to be loaded. Compiling is done automatically. Because of this integrated structure, a user can edit and execute complex commands by way of short, simple instructions. The internal operating system will recognize typing or syntax errors and give an easy-to-understand explanation on the display or printout. Also, the computer can prompt the user to supply subsequent inputs so that steps are not missed or implemented out of sequence.

Especially significant in design applications is the unit's ability to accept changes in problem parameters while a program is running. 2. A complete microprocessor-based system that is fairly easy to use by noncomputer experts. 3. A computer designed for scientific/engineering problems that are too complex for personal calculators or that require peripherals or interfacing capabilities for data acquisition or instrumentation control. Desktop computers have full, high-level programming languages, much larger memories than personal calculators, built-in mass storage devices, and I/O ports for interfacing to peripherals or other instruments. Some have impressive graphics capabilities for plotting, drawing, graphing, and lettering. 4. A complete, highly integrated system ready to use as it comes out of the box. It has an input device (a keyboard), a display device (a cathode-ray tube or single-line display), mass storage (magnetic cards, cassettes, floppy discs, CD-ROMs, etc.), a processor, memory, connectors for external I/O (input/output) devices, a power supply, and a language that resides in ROM (read-only memory) all housed in a single package that literally sits on a desk. Of course, the desktop computer may also have special peripherals, such as plotters, digitizers, and other instruments. One can simply connect the external I/O devices and peripherals, if any, turn the power on, and use the machine.

desktop publishing — Abbreviated DTP. The creation on personal computers of publication-quality printed documents that combine text and graphics.

desoldering — Process of disassembling soldered parts in order to repair, replace, inspect, or salvage them. Typical desoldering methods are wicking, pulse vacuum (solder sucker), heat and pull, and solder extraction.

destaticization — Treatment of a material to minimize the accumulation of static electricity and, as a result, the amount of dust that adheres to the material because of such static charges.

destination address — In computer systems having a source-destination architecture, the destination address is the address of the device register or memory location to which data is being transferred.

destination register — In a computer, a register into which data is being placed.

Destriau effect — Sustained emission of light by suitable phosphor powders embedded in an insulator and subjected only to the action of an alternating electric field.

destructive readout — 1. The destruction of data in a storage device by the act of reading the data. For example, reading data from a core memory clears the addressed location. When destructive readouts are used, data (modified or unmodified) is written back into the same location. 2. A characteristic of a memory. The memory is said to have a destructive readout if information retrieved from memory must be written back in immediately after it is used or else it is lost. A core memory has destructive readout. Computers with destructive readouts contain special circuits to write information back into memory after readout.

destructive-readout memory — See DRO memory.

destructive test — Any test resulting in the destruction or drastic deterioration of the test specimen.

detail — A measure of the sharpness of a recorded facsimile copy or reproduced image. Generally related to the number of lines scanned per inch. Defined as the square root of the ratio between the number of scanning lines per unit length and the definition in the direction of the scanning line.

detail contrast — The ratio of the amplitude of the high-frequency components of a video signal to the amplitude of the reference low-frequency component.

detail enhancement — Also called image enhancement. A system in which each element of a video picture is analyzed in relation to adjacent horizontal and vertical elements. When differences are detected, a detail signal is generated and added to the luminance signal to enhance it.

detection — See demodulation.

detection range — The greatest distace at which a sensor will consistently detect an intruder under a standard set of conditions.

detectophone — An instrument for secretly listening in on a conversation. A high-sensitivity, nondirectional microphone is concealed in the room and connected to an amplifier and headphones or recorder remotely located. Sometimes the microphone feeds into a wireless transmitter that broadcasts over power lines, to permit the listener to be farther away.

detector — 1. A device for effecting the process of detection or demodulation. 2. A mixer or converter in a superheterodyne receiver; often referred to as a first detector. 3. A device that produces an electrical output that is a measure of the radiation incident on the device. 4. A rectifier tube, crystal, or dry disc by which a modulation envelope on a carrier or the simple on-off state of a carrier may be made to drive a lower-frequency device. 5. A device that converts light signals from optical fibers to electrical signals that can be further amplified to allow reproduction of the original signal. 6. A device that converts optical power to other forms. See photodetector.

detector balance bias — A controlling circuit used in radar systems for anticlutter purposes.

detector circuit — That portion of a receiver which recovers the modulation signal from the rf carrier wave.

detector diode — A diode, often associated with microwave circuits, that converts rf energy into dc or video output.

detector power efficiency — The ratio of the change in dc power in the load resistance produced by the ac signal to the available power from a sinusoidal voltage generator when the diode is operated under specified conditions.

detector probe — A probe containing a high-frequency rectifying element such as a crystal diode or a tube. Used with an oscilloscope, vacuum-tube voltmeter, or signal tracer for recovering the modulation from a carrier.

detector quantum efficiency — The ratio of the number of carriers generated to the number of photons absorbed.

detector voltage efficiency — The ratio of the dc load voltage to the peak sinusoidal input voltage under specified circuit conditions.

detent — 1. A stop or other holding device, such as a pin, lever, etc., on a ratchet wheel. 2. Switch action typified by a gradual increase in force to a position at which there is an immediate and marked reduction in force. 3. A bump or raised section projecting from the surface of a spring or other part. 4. A device that holds a part, control, or assembly in a given position. For example, some connectors use locking detents on the plug half and indents on the cap half to hold the halves together in proper mated position.

detent torque — A measure of the maximum torque that can be applied to the shaft of a deenergized stepper motor before it begins to rotate.

deterministic signal — A signal whose future behavior can be predicted precisely.

detritus — Loose material dislodged during resistor trimming but remaining in the trimmed area.

detune — 1. To change the inductance and/or capacitance of a tuned circuit and thereby cause it to be resonant at other than the desired frequency. 2. To adjust a circuit so that it does not respond to (is not resonant at) a particular frequency.

detuning stub — A quarter-wave stub for matching a coaxial line to a sleeve-stub antenna. The stub tunes the antenna itself and detunes the outside of the coaxial feed line.

Deutsches Institut für Normung — *See* DIN.

deviation — 1. The difference between the actual and specified values of a quantity. 2. The difference, usually the absolute difference, between a number and the mean of a set of numbers, or between a forecast value and the actual datum. 3. In FM transmissions and reception, the increase or decrease of signal carrier frequency from the nominal; also applied to drifting. Standard maximum deviation rating is ±75 kHz for FM radio. 4. A departure from specification requirements for which approval is obtained from the consumer prior to occurrence of the departure.

deviation absorption — Absorption that occurs at frequencies near the critical frequency. Occurs in conjunction with the slowing up of radio waves near the critical frequency, upon reflection from the ionosphere.

deviation distortion — Distortion caused by inadequate bandwidth, amplitude-modulation rejection, or discriminator linearity in an FM receiver.

deviation ratio — In frequency modulation, the ratio of the maximum change in carrier frequency to the highest modulating frequency.

deviation sensitivity — The smallest frequency deviation that produces a specified output power in FM receivers.

device — Also called item. 1. A single discrete conventional electronic part such as a resistor or transistor, or a microelectronic circuit. 2. Any subdivision of a system. 3. A mechanical, electrical, and/or electronic contrivance intended to serve a specific purpose. 4. The physical realization of an individual electrical element in a physically independent body, which cannot be further reduced or divided without destroying its stated function. This term is commonly applied to active devices. Examples are transistors, pnpn structures, tunnel diodes, and magnetic cores, as well as resistors, capacitors, and inductors. It is not, for example, an amplifier, a logic gate, or a notch filter. 5. An electronic part consisting of one or more discrete active or passive elements. 6. A unit of processing equipment in a computer system external to the CPU; synonymous with the term *peripheral*.

device adaptor — *See* interface adaptor.

device channel — A dedicated channel associated with a device; connects a file variable to that device.

device complexity — The number of circuit elements within an integrated circuit.

device cutoff — The condition of an electronic device in which its conduction is either zero or relatively insignificant. With semiconductor devices such as FETs, bipolar transistors, and thyristors, cutoff is normally that condition in which the device passes only leakage currents.

device independence — In a computer, the ability to request input/output operations without regard to the nature of the input/output devices.

device register — An addressable register used to store status and control information or data for transfer into or out of a device.

device under test — Abbreviated DUT. A finished assembled device, or an untested die on a wafer.

Dewar flask — A container with double walls. The space between the walls is evacuated, and the surfaces bounding this space are silvered.

dewetted surface — 1. A surface that was initially wetted, i.e., a surface on which the solder flowed uniformly. 2. A condition that results when molten solder has coated a surface and then receded, leaving irregularly shaped mounds of solder separated by areas covered with a thin solder film; base metal is not exposed.

dew point — 1. The temperature at which condensation first occurs when a vapor is cooled. 2. The temperature at which moisture begins to condense out of a vapor. The relative humidity is then 100 percent.

df — Abbreviation for direction finder or dissipation factor.

df antenna — Any antenna combination included in a direction finder for obtaining the phase or amplitude reference of the received signal. May be a single or orthogonal loop, an adcock, or spaced differentially connected dipoles.

df antenna system — One or more df antennas and their combining circuits and feed systems, together with the shielding and all electrical and mechanical items up to the receiver input terminals.

D flip-flop — A flip-flop whose output is determined by the input that appeared one pulse earlier; for example, if a 1 appeared at the input, the output after the next clock pulse would be 1.

DHG — Abbreviation for digital harmonic generation.

DI — Abbreviation for dielectric isolation. A fabrication technique by which components in an integrated circuit are electrically isolated from each other by an insulator (dielectric material). DI surrounds the sides and bottom of each transistor with a layer of silicon dioxide (glass). DI has proven particularly advantageous for fabricating high-performance analog ICs. The conventional DI fabrication process for bipolar ICs begins with a wafer of n-type silicon. The side of the wafer that will eventually be the bottom is deeply etched (in V-shaped grooves) to form the sidewall pattern, then silicon dioxide and polycrystalline silicon are grown to fill the etched moats and to thicken the eventual DI substrate. The opposite side of the wafer is polished until the insulating sidewalls appear at the wafer surface. Conventional diffusion and metallization processes follow to complete the IC.

diac — 1. Two-lead alternating-current switching semiconductor. 2. *See* three-layer diode. 3. A bidirectional breakdown diode that conducts only when a specified breakdown voltage is exceeded.

diagnostic — 1. Having to do with the detection and isolation of a malfunction or error in a computer. 2. A message output by a compiler or assembler indicating that a computer program contains a mistake. 3. Software designed to locate either a fault in the equipment or an error in programming.

diagnostic code — An alphanumeric or word display that indicates a system condition such as a malfunction. The code is either self-explanatory or used to refer to further instructions that are explained in an operator guide.

diagnostic function test — A program for testing overall system reliability.

diagnostic program — 1. Special program for checking computer's hardware for proper operation. For example, there are CPU diagnostic checks, memory diagnostic checks, and so on. 2. A test program to help isolate hardware malfunctions in the programmable controller and application equipment. 3. A troubleshooting aid for locating hardware malfunctions in a system, or a program to aid in locating coding errors in newly developed programs.

diagnostic routine — 1. An electronic-computer routine designed to locate a malfunction in the computer, a mistake in coding, or both. 2. *See* debug.

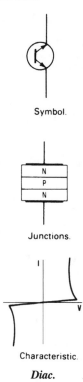

Symbol.

Junctions.

Characteristic.

Diac.

diagnostics — 1. Information on what tests a device failed and how they were failed, used to aid in troubleshooting. 2. An output from a tool, indicating software discrepancies and other attributes. 3. Methods used for detecting and isolating faults in a unit under test. 4. Computer system support tasks (usually supplied with an operating system) that test memory, interrupts, timers, and peripheral devices. 5. Programs or procedures used to test a piece of equipment, a communications link or network, or any similar system.

diagnostic test — A test performed for the purpose of isolating a malfunction in the unit under test.

diagnostic testing — Testing designed to locate and identify errors in a computer routine, hardware component, or communication network.

diagonal horn antenna — A horn antenna whose cross sections are all square and in which the electric vector is parallel to a diagonal. The radiation pattern in the far field has almost perfect circular symmetry.

diagonal pliers — Pliers with cutting jaws at an angle to the handles to permit cutting off wires close to terminals or printed circuit boards.

diagram — 1. Schematics, prints, charts, or any other graphical representation, the purpose of which is to explain rather than to represent. 2. A schematic representation of a sequence of subroutines designed to solve a problem. 3. A coarser and less symbolic representation than a flowchart, frequently including descriptions in English words. 4. A schematic drawing showing an electrical circuit, or a logical drawing showing logical arrangements within a circuit.

dial — 1. A means for indicating the value to which a control knob has been set. 2. A calling device that generates the required number of pulses in a telephone set and thereby establishes contact with the party being called.

dial cable — *See* dial cord.

dial central office — A telephone or teletypewriter office at which is located the automatic equipment necessary for connecting two or more user stations together by wires for communication purposes.

dial cord — Also called dial cable. A braided cord or flexible wire cable connected to a tuning knob so that turning the knob will move the pointer or dial that indicates the frequency to which an analog radio receiver is tuned. Also used for coupling two shafts together mechanically.

dialer — 1. A device that detects and reports emergencies by automatic dialing of telephone numbers. When an emergency is detected, the dialer usually begins playing a prerecorded tape containing the telephone number or numbers to be called (in the form of a series of pulses) and the emergency message. When the number has been dialed, the tape continues to play the prerecorded message. 2. *See* automatic dialer.

dialing area — The area within which a telephone company customer may make calls by dialing without using area codes.

dialing key — A dialing method in which a set of numerical keys instead of a dial is used to originate dial pulses. Generally, it is used in connection with voice-frequency dialing.

dial jacks — Strips of jacks associated with and bridged to a regular outgoing trunk jack circuit so that connections between the dial cords and the outgoing trunks can be made.

dial key — The key unit of the subscriber's cord circuit that is used to connect the dial to the line.

dial leg — The circuit conductor brought out for direct-current dial signaling.

dial light — A small pilot lamp that illuminates the tuning dial of a radio receiver.

diallyl phthalate — A thermosetting resin that has excellent electrical insulation properties.

dial office — Telephone central office operating on dial signals.

dialog — Interchange of information between program and user. The program gives prompts or messages, and the user responds by entering appropriate data.

dialog box — 1. An on-screen message box in a graphical user interface that allows users to input more specific information than standard commands. 2. An on-screen box that conveys or requests information from the user.

dial pulse — A momentary interruption in the direct current in the loop of a calling telephone, produced by the opening and closing of the dial pulse springs of a calling telephone in response to the dialing of a digit. The current in the calling-line loop is interrupted as many times as there are units in the digit dialed; i.e., dialing the digit 7 generates seven dial pulses (interruptions) in current flowing through the loop of the calling telephone.

dial pulsing — The transmission of telephone address information by the momentary opening and closing of a dc circuit a specified number of times, corresponding to the decimal digit that is dialed. This is usually accomplished, as with an ordinary telephone dial, by manual operation of a finger wheel.

dial register — *See* standard register of a motor meter.

dial telephone system — Telephone system in which telephone connections between customers are ordinarily established by electronic and mechanical apparatus controlled by manipulations of dials operated by the calling parties.

dial tone — 1. A hum or other tone employed in a dial telephone system to indicate that the line is not busy and that the equipment is ready for dialing. 2. A tone indicating that automatic telephone switching equipment is ready to receive dial signals, tones, or pulses.

dial-up — 1. The use of a dial or push-button telephone for initiating a station-to-station call. 2. Two simplex transmission paths between central offices and a full-duplex path in each subscriber loop. 3. Use of a phone to initiate a call.

dial-up connection — A connection from a computer using a modem to a host computer over standard telephone lines.

dial-up line — Also called switched line. A communication line accessible via dial-up facilities, typically the public telephone network.

diamagnetic — 1. Term applied to a substance with a negative magnetic susceptibility. 2. Bars of certain elements, such as zinc, copper, lead, and tin, when freely suspended in a magnetic field arrange themselves at right angles to the lines of force of the magnetic field, i.e., they are magnetized in the opposite direction to the magnetizing field. These elements are said to be diamagnetic.

diamagnetic material — 1. A material that is less magnetic than air, or in which the intensity of magnetization is negative. There is no known material in which this effect has more than a very feeble intensity. Bismuth is the leading example of materials of this class. 2. A material having a permeability less than that of vacuum: $\mu < 1$ gauss/oersted. For practical evaluation, a nonmagnetic material.

diamagnetism — 1. A phenomenon whereby the magnetization induced in certain substances opposes the magnetizing force. 2. The negative susceptibility exhibited by certain substances. The permeability of such substances is less than unity.

diamond antenna — Also called a rhombic antenna. A horizontal antenna having four conductors that form a diamond, or rhombus.

diamond lattice — The crystal structure of germanium and silicon (as well as a diamond).

diamond stylus — A phonograph pickup with a ground diamond as its point.

diapason — The unique fundamental tone color of organ music.

diaphragm — 1. A flexible membrane used in various electroacoustic transducers for producing audio-frequency vibrations when actuated by electric impulses, or electric impulses when actuated by audio-frequency vibrations. 2. In electrolytic cells, a porous or permeable membrane, usually flexible, separating the anode and cathode compartments. 3. In waveguide technique, a thin plate, or plates, placed transversely across the waveguide, not completely closing it, and usually introducing a reactance component. *See also* iris. 4. A sensing element consisting of a membrane placed between two volumes. The membrane is deformed by the pressure differential applied across it.

diathermal apparatus — Apparatus for generating heat in body tissue by high-frequency electromagnetic radiation.

diathermy — 1. The therapeutic use of an oscillating electric current of high frequency to produce localized heat in body tissues. 2. The use of radio-frequency fields to produce deep heating in body tissues. The output of a powerful rf oscillator is applied to a pair of electrodes, known as pads, between which the portion of the body to be treated is placed. The body tissues thus become the dielectric of a capacitor, and dielectric losses cause heating of the tissues.

diathermy interference — A form of television interference caused by diathermy equipment, resulting in a horizontal herringbone pattern across the picture.

diathermy machine — A medical apparatus consisting of an rf oscillator frequently followed by rf amplifier stages, used to generate high-frequency currents that produce heat within some predetermined part of the body for therapeutic purposes.

dibit — A group of two bits. In four-phase modulation, each possible dibit is encoded in the form of one of four unique phase shifts of the carrier. The four possible states for a dibit are 00, 01, 10, and 11.

DIC — Dielectrically isolated integrated circuit. Also abbreviated DIIC.

dice — The plural of die.

dichroic — Pertaining to the quality of dichroism.

dichroic filter — 1. An optical filter capable of transmitting all frequencies above a certain cutoff frequency and reflecting all lower frequencies, thus being either a high-pass or a low-pass filter. 2. A filter used to selectively transmit light according to its wavelength and not its plane of vibration.

dichroic mirror — 1. A special mirror through which all light frequencies pass execpt those for the color that the mirror is designed to reflect. 2. A semitransparent mirror used to selectively reflect light according to its wavelength and not its plane of vibration.

dichroism — Also called dichromatism or polychromatism. 1. A property of an optical material that causes light of some wavelengths to be absorbed when the incident light has its electric-field vector in a particular orientation and not absorbed when the electric-field vector has other orientations. 2. In anisotropic materials, such as some crystals, the selective absorption of light rays vibrating in one particular plane relative to the crystalline axes, but not those vibrating in a plane at right angles thereto. As applied to isotropic materials, this term refers to the selective reflection and transmission of light as a function of wavelength regardless of its plane of vibration. The color of such materials, as seen by transmitted light, varies with the thickness of material examined.

dichromatism — *See* dichroism.

dicing — The process of sawing a crystal wafer into blanks.

dictionary — In digital computer operations, a list of mnemonic code names together with the addresses and/or data to which they refer.

diddle — Automatic transmission of letter or figure characters by a terminal unit if no characters are ready for transmission (most often used with a FIFO memory).

die — 1. Sometimes called chip. A tiny piece of semiconductor material, broken from a semiconductor slice, on which one or more active electronic components are formed. (Plural: dice.) 2. A portion of a wafer bearing an individual circuit, or device cut or broken from a wafer containing an array of such circuits or devices. 3. An uncased discrete or integrated device obtained from a semiconductor wafer. *See* chip. 4. A single miniature active or passive component. So named because the circuits are batch fabricated by diffusion processes on a silicon wafer, which is then cut into individual components. Examples: transistors, diodes, integrated circuits, diffused resistors.

die attach — The operation of mounting chips to a substrate. Methods include gold-silicon eutectic bonding, various solders, and conductive (and nonconductive) epoxies.

die bond — 1. Attachment of a die or chip to the hybrid substrate. 2. A process in which chips are attached to a substrate (gold, epoxy, wax, etc.). The joint between a die and the substrate.

die bonding — 1. The method by which a semiconductor die, or chip, is attached to a mechanical support. 2. Attaching a semiconductor chip to the substrate with an epoxy, eutectic, or solder alloy. 3. The attachment of a die to a gold base, such as a substrate pad, or to a header. Heat, pressure, and a mechanical scrubbing action are used to create a gold-silicon eutectic bond between the die and base.

dielectric — 1. The insulating (nonconducting) medium between the two plates of a capacitor. Typical dielectrics are air, wax-impregnated paper, plastic, mica, and ceramic. A vacuum is the only perfect dielectric. 2. A medium capable of recovering, as electrical energy, all or part of the energy required to establish an electric field (voltage stress). The field, or voltage stress, is accompanied by displacement or charging currents. 3. The insulating material between the metallic elements of an electromechanical component or any of a wide range of thermoplastics or thermosetting plastics. 4. Any insulating medium that intervenes between two conductors and permits electrostatic attraction and repulsion to take place across it. 5. A material having the property that energy required to establish an electric field is recoverable, in whole or in part, as electric energy. 6. A material medium in which an electric field can exist in the stationary state. 7. Characteristic of materials that are electrical insulators or in which an electric field can be sustained with a minimum dispersion of power. Such materials exhibit nonlinear properties, such as anisotropy of conductivity or polarization, or saturation phenomena. 8. An insulator. Localized regions of dielectric materials are used in semiconductor devices, for example, to provide electrical isolation between dice, between metal interconnect layers, and between the gate electrode and the channel.

dieletric absorption — Also called dielectric hysteresis (short-term effect), or dielectric soak (long-term effect). 1. A characteristic of dielectrics that determines the length of time a capacitor takes to deliver the total amount of its stored energy. It manifests itself as the reappearance of potential on the electrodes after the capacitor has been discharged. Its magnitude depends on the charge and discharge time of the capacitor. 2. That property of an imperfect dielectric as a result of which all electric charges within the body of the material caused by the application of an electric field are not returned to the field. 3. Reluctance of a capacitor to give up all the electrons stored when the capacitor is discharged. Primarily caused by a polarization effect of dielectric dipoles and to a lesser extent by free electrons in the dielectric requiring a finite time to move to the electrode. The recovery voltage appearing after discharge divided by the charging voltage and expressed as a percentage is called the percent dielectric absorption. 4. The property of a capacitor with slow polarization of its dielectric that results in voltage appearance on the capacitor electrodes after its short-term discharge through a low resistance.

dielectric amplifier — An amplifier employing a device similar to an ordinary capacitor, but with a polycrystalline dielectric that exhibits a ferromagnetic effect.

dielectric analysis — Method of directly monitoring resin cooking, resin staging, and resin curing. Such analysis eliminates many of the variables influencing the selection of a given set of fabrication conditions. Dielectric analyzers consist of a press with heated platens, a clamshell autoclave, and a DTA/dielectric cell.

dielectric anisotropy — The difference between the dielectric along the director and the dielectric perpendicular to the director of a liquid crystal system.

dielectric antenna — An antenna in which a dielectric is the major component producing the required radiation pattern.

dielectric breakdown — 1. An abrupt increase in the flow of electric current through a dielectric material as the applied electric field strength exceeds a critical value. 2. A complete failure of a dielectric material characterized by a disruptive electrical discharge through the material due to a sudden and large increase in voltage.

dielectric breakdown voltage — Also called electric breakdown voltage, breakdown voltage, or hi-pot. 1. The voltage between two electrodes at which electric breakdown of the specimen occurs under prescribed test conditions. 2. The voltage at which a dielectric material punctures. 3. The voltage required to cause electrical failure or breakthrough of insulation. Usually expressed as a voltage gradient (volts per mil).

dielectric capacity — The inductivity or specific inductive capacity of a substance, being its ability to convey the influence of an electrified body.

dielectric constant — Also called permittivity, specific inductive capacity, or capacitivity. 1. The ratio of the capacitance of a capacitor with the given dielectric to the capacitance of a capacitor having air for its dielectric but otherwise identical. Symbol: K. 2. That property of a dielectric that determines the electrostatic energy stored per unit volume for a unit potential gradient. 3. The property of a material that determines the amount of electrostatic energy that can be stored when a given voltage is applied.

dielectric crystal — A crystal that is characterized by its relatively poor electrical conductance.

dielectric current — The current flowing at any instant through the surface of an isotropic dielectric that is in a changing electric field.

dielectric dissipation — See loss tangent.

dielectric dissipation factor — The cotangent of the dielectric phase angle of a material.

dielectric fatigue — The property of some dielectrics in which the insulating quality decreases after a voltage has been applied for a considerable length of time.

dielectric guide — A waveguide made of a solid dielectric material through which the waves travel.

dielectric heating — A method of raising the temperature of a nominally insulating material by sandwiching it between two plates to which an rf voltage is applied. The material acts as a dielectric, and its internal losses cause it to heat up.

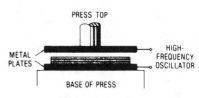

Dielectric heating.

dielectric hysteresis — 1. Short-term effect of dielectric absorption. 2. A lagging of an electric field in a dielectric behind the alternating voltage that produces it. It causes a loss similar to that of magnetic hysteresis. 3. See also dielectric absorption.

dielectric isolation — 1. The electrical isolation of monolithic integrated circuit elements from each other by dielectric material rather than by reverse-biased pn junctions. 2. The use of silicon dioxide barriers created during silicon IC processing to provide isolation between components on a chip.

dielectric lens — A lens used with microwave antennas; it is made of dielectric material so that it refracts radio waves similar to the way an optical lens refracts light waves.

dielectric loss— 1. The power dissipated by a dielectric as the friction of its molecules opposes the molecular motion produced by an alternating electric field. 2. The time rate at which electric energy is transformed into heat in a dielectric when it is subjected to a changing electric field.

dielectric loss angle— Also called dielectric phase difference. The complement of the dielectric phase angle (i.e., the dielectric phase angle minus 90°).

dielectric loss factor— Also called dielectric loss index. The product of the dielectric constant of a material times the tangent of the dielectric loss angle.

dielectric loss index— The product of a medium's relative permittivity and the tangent of its dielectric loss angle. See dielectric loss factor.

dielectric matching plate— In waveguide technique, a dielectric plate used as an impedance transformer for matching purposes.

dielectric mirror— A highly frequency-selective, multilayer dielectric reflector acting by partial reflection of light at the interface between materials of unequal refractive indices.

dielectric phase angle— The angular difference in phase between the sinusoidal alternating voltage applied to a dielectric and the component of the resultant alternating current having the same period.

dielectric phase difference— See dielectric loss angle.

dielectric polarization— See polarization, 3.

dielectric power factor— The cosine of the dielectric phase angle, or sine of the dielectric loss angle.

dielectric process— Also called electrographic process. A nonimpact printing technique in which specially treated paper consisting of a conductive base layer coated with a nonconductive thermoplastic material is used to hold an electric charge applied directly by a set of electrode styli. The electric charge corresponds to the latent image of the original. After charging, the image is produced by a toner system similar to that used in electrostatic copying devices. The dielectric process employed on general-purchase nonimpact printers, facsimile devices, and some photocopiers.

dielectric rating— Standard test voltages and frequencies above which failure occurs between specified points in a relay structure.

dielectric resonator— A high-Q, temperature-stable ceramic microwave resonator that is used in microwave oscillator circuits. It can exist in any regular geometrical form and resonates in various modes at frequencies determined by its dimensions and shielding conditions.

dielectric-rod antenna— An antenna in which propagation of a surface wave on a tapered dielectric rod produces an endfire radiation pattern.

dielectric soak— Long-term effect of dielectric absorption. See dielectric absorption.

dielectric strength— Also called electric strength, breakdown strength, electric field strength, and insulating strength. 1. The maximum voltage a dielectric can withstand without rupturing. 2. Maximum potential gradient that a dielectric can withstand before it ruptures or a conducting path forms through it. Normally expressed as the ratio of breakdown voltage to the dielectric thickness (volts per mil). 3. The property of a dielectric to withstand an electrical stress. 4. The ratio of dielectric breakdown to the thickness of the insulating materials. 5. The ultimate breakdown voltage of the dielectric or insulation of the resistor when the voltage is applied between the case and all terminals that are tied together. Dielectric strength is usually specified at sea level and simulated high-altitude air pressures.

dielectric susceptibility— The ratio of the polarization in a dielectric to the electric intensity responsible for it.

dielectric tests— 1. Tests that consist of the application of a voltage higher than the rated voltage for a specified time for the purpose of determining the adequacy against breakdown of insulating materials and spacings under normal conditions. 2. The testing of insulating materials by the application of a constantly increasing voltage until failure occurs.

dielectric waveguide— A waveguide constructed from a dielectric (nonconductive) substance.

dielectric wedge— A wedge-shaped piece of dielectric material used in one waveguide to match its impedance to that of another waveguide.

dielectric wire— A dielectric waveguide used for short-distance transmission of UHF radio waves between parts of a circuit.

dielectric withstanding voltage— Maximum potential gradient that a dielectric material can withstand without failure.

difference— The signal energy representing the differences in information between the signals in two or more stereo channels. A difference signal is produced when stereo signals differing in electrical polarity or in intensity are mixed together in opposing polarity.

difference amplifier— The basic input stage of most operational amplifiers. An amplifier with an inverting and a noninverting input. The output voltage is a function of the voltage difference between the two inputs. See differential amplifier.

difference channel— In a stereophonic sound system, an audio channel that handles the difference between the signals in the left and right channels.

difference detector— A detector circuit in which the output is a function of the difference between the peak or rms amplitudes of the input waveforms.

difference frequency— 1. A signal representing, in essence, the difference between the left and right sound channels of a stereophonic sound system. 2. One of the output frequencies of a converter. It is the difference between the two input frequencies.

difference in depth modulation— In directive systems employing overlapping lobes with modulated signals, a ratio obtained by subtracting from the percentage of modulation of the larger signal the percentage of modulation of the smaller signal and dividing by 100.

difference of potential— The voltage or electrical pressure existing between two points. It will result in a flow of electrons whenever a circuit is established between the two points.

difference signal— In a quadraphonic sound system, a signal arrived at by subtracting left or right back-channel signal from its respective front-channel signal. Left front and left back signals, if added to left front minus left back signals, yield a left front channel; if subtracted, they yield a left back channel.

differential— 1. A planetary gear system that adds or subtracts angular movements transmitted to two components and delivers the answer to a third. Widely used for adding and subtracting shaft movements in servo systems and for addition and subtraction in computing machines. 2. In electronics, the difference between two levels. 3. A method of signal transmission through two wires that always have opposite states. The signal data is the polarity difference between the wires: whenever either is high, the other is low. Neither wire is grounded. 4. Describing any device whose operation is dependent on the difference between two quantities.

differential amplifier—1. An amplifier with two similar input circuits so connected as to respond to the difference between two voltages (or currents) and effectively suppress voltages or currents that are alike in the two input circuits. 2. A circuit that amplifies the difference between two input signals.

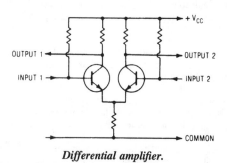

Differential amplifier.

differential analyzer—A mechanical or electrical device primarily designed and used to solve differential equations.

differential angle—The total angle from the operation to the releasing position in a mercury switch.

differential capacitance—The derivative with respect to voltage of a capacitor charge characteristic at a given point on the curve.

differential capacitance characteristic—The function that relates differential capacitance to voltage.

differential capacitor—A variable capacitor having two similar sets of stator plates and one set of rotor plates. When the rotor is turned, the capacitance of one section is increased and the capacitance of the other section is decreased. The sum of the two capacitance values remains substantially constant at all settings, however.

differential comparator—1. A circuit in which differential-amplifier design techniques are applied to the comparison of an input voltage with a reference voltage. When the input voltage is lower than the reference voltage, the circuit output is in one state; when the input voltage is higher than the reference voltage, the output is in the opposite state. Commonly used for pulse-amplitude detector circuits, a-d conversion, and differential receivers for data transmission in noisy environments over a twisted-pair line. 2. A differential circuit for indicating when two input signals are essentially equal, as in a differential pair.

differential cooling—A lowering of temperature that takes place at a different rate at various points on an object or surface.

differential delay—The difference between the maximum and minimum frequency delays occurring across a band.

differential discriminator—A discriminator that passes only pulses having amplitudes between two predetermined values, neither of which is zero.

differential duplex system—A duplex system in which the sent currents divide through two mutually inductive sections of the receiving apparatus. These sections are connected to the line and to a balancing artificial line in opposite directions. Hence, there is substantially no net effect on the receiving apparatus. The received currents pass mainly through one section, or through the two sections, in the same direction and operate the apparatus.

differential equation—A generalization of the algebraic equation in which the unknowns are not simply numbers, but functions of one or more independent variables. Not only the unknown function or functions appear explicitly in the differential equation, but also the first- and possibly higher-order derivatives with respect to the independent variable or variables.

differential flutter—Speed-change errors that occur at different magnitudes, frequencies, or phases across the width of a magnetic tape.

differential gain—1. The ratio of the differential output signal of a differential amplifier divided by the differential input signal causing that output. 2. Variation in the gain of a transmission system with changing modulation.

differential gain control—Also called gain sensitivity control. A device for altering the gain of a radio receiver in accordance with an expected change of signal level in order to reduce the amplitude differential between the signals at the receiver output.

differential galvanometer—A galvanometer having two similar but opposed coils, so that their currents tend to neutralize each other. A zero reading is obtained when the currents are equal.

differential gap—1. The difference between two target values, one of which applies to an upswing of conditions and the other to a downswing. 2. The span between on and off switching points. For example, a room thermostat set for 70° might switch the furnace on at 68° and off at 72°, resulting in a 4° differential. *See also* dead band.

differential gear—In an analog computer, a mechanism that relates the angles of rotation of three shafts. Usually it is designed so that the algebraic sum of the rotations of two shafts is equal to twice the rotation of the third. The device can be used for addition or subtraction.

differential generator—A synchro differential generator driven by a servo system.

differential GPS—A technique using the global positioning satellite (GPS) network, in which a fixed ground GPS receiver (for precisely known three-dimensional coordinates locally) determines corrections to be applied to local aircraft (or other mobile units) using satellite guidance.

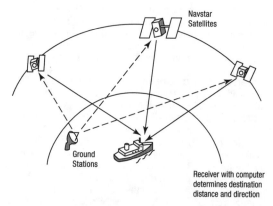

Differential global positioning system (GPS).

differential impedance—The internal impedance observed between the input terminals of an operational amplifier.

differential input — 1. An input circuit that rejects voltages that are the same at both input terminals and amplifies the voltage difference between the two input terminals. May be either balanced or floating and may also be guarded. 2. An input applied between two terminals of an operational amplifier, neither of which is at ground (earth) potential.

differential-input amplifier — An amplifier in which the output is ideally a function only of the difference between the signals applied to its two inputs, both signals being measured with respect to a common low or ground reference point.

differential-input capacitance — The capacitance between the inverting and noninverting input terminals.

differential-input impedance — 1. The impedance between the inverting and noninverting input terminals of a differential amplifier. 2. The impedance measured between the positive and the negative input terminals of an operational amplifier.

differential-input measurement — Also called floating input. A measurement in which the two inputs to a differential amplifier are connected to two points in a circuit under test and the amplifier displays the difference voltage between the points. In this type of measurement, each input of the amplifier acts as a reference for the other, and ground connections are used only for safety reasons.

differential-input rating — The maximum differential input that may be applied between the two terminals of an operational amplifier.

differential-input resistance — The resistance between the inverting and noninverting input terminals of a differential amplifier. *See* input resistance.

differential-input voltage — The maximum voltage that can be applied across the input terminals of a differential amplifier without damaging the amplifier.

differential-input voltage range — The range of voltages that may be applied between input terminals without forcing the circuit to operate outside its specifications.

differential-input voltage rating — The maximum allowable signal that may be applied between the inverting and noninverting inputs of a differential amplifier without damaging the amplifier.

differential instrument — A galvanometer or other measuring instrument having two circuits or coils, usually identical, through which currents flow in opposite directions. The difference, or differential effect, of these currents actuates the indicating pointer.

differential keying — A method of obtaining chirp-free break-in keying of a cw transmitter by turning the oscillator on quickly before the keyed amplifier stage can pass any signal, and turning it off quickly after the keyed amplifier stage has cut off.

differential linearity — The measure of linearity among digital states in a/d and d/a converters. If the differential linearity is specified as $\pm^1/_2$ lsb, the step size from one state to the next may range from $^1/_2$ to $^3/_2$ of an ideal 1-lsb step.

differential microphone — *See* double-button carbon microphone.

differential-mode gain — Abbreviated DMG. 1. The ratio of the output voltage of a differential amplifier to the differential-mode input voltage. 2. The voltage gain exhibited by an operational amplifier in response to differential-mode signals.

differential-mode input — The voltage difference between the two inputs of a differential amplifier.

differential-mode signal — 1. A signal that is applied between the two ungrounded terminals of a balanced three-terminal system. 2. In an amplifier with a differential input, a signal that appears at inverting and noninverting inputs with opposite phase but identical frequency and amplitude. It is not necessarily referred to ground.

differential modulation — A type of modulation in which the choice of the significant condition for any signal element is dependent on the choice for the previous signal element.

differential nonlinearity — The difference between actual analog voltage change and the ideal (1 1sb) voltage change at any code change of a digital-to-analog converter.

differential output voltage — The difference between the values of the two ac voltages that are present in phase opposition at the output terminals of an amplifier when a differential voltage is applied to the input terminals of the amplifier.

differential pair — A pair of transistors sharing a common emitter circuit but with two independent base inputs.

differential permeability — 1. The ratio of the positive increase of normal induction to the positive increase of magnetizing force when these increases are minute. 2. The slope of the normal induction curve.

differential phase — 1. The difference in phase shift through a television system for a small, high-frequency sine-wave signal at two stated levels of a low-frequency signal on which the first signal is superimposed. 2. In a color TV signal, the phase change of the color subcarrier introduced by the overall circuit, measured in degrees as the picture signal on which it rides is varied from blanking to white level.

differential phase-shift keying — Abbreviated DPSK. A modulation scheme in which the information is conveyed by changes in carrier phase during one interval relative to the preceding interval.

differential pressure transducer — A pressure transducer that accepts simultaneously two independent pressure sources, and the output of which is proportional to the pressure difference between the sources.

differential protective relay — A protective device that functions on a percentage of phase angle or other quantitative difference of two currents or of some other electrical quantities.

differential relay — A relay with multiple windings that functions when the voltage, current, or power difference between the windings reaches a predetermined value. The power difference may result from the algebraic addition of the multiple inputs.

differential resistance — The resistance measured between the terminals of a diode under small-signal and specified bias conditions.

differential resolver — A servo unit with a two-phase stator and a three-phase rotor that is used as a transolver, with the advantage that when connected as a control transformer, the signal does not travel through slip rings.

differential selsyn — A selsyn in which both the rotor and the stator have similar windings that are spread 120° apart. The position of the rotor corresponds to the algebraic sum of the fields produced by the stator and rotor.

differential stage — A symmetrical amplifier stage in which two inputs are balanced against each other so that when there is no input signal, or equal input signals, there is no output signal. An input-signal unbalance, including a signal to only one input, produces an output signal proportional to the difference between the input signals.

differential synchro — *See* synchro differential generator; synchro differential motor.

differential transducer — A device capable of simultaneously measuring two separate stimuli and providing an output proportionate to the difference between them.

differential transformer — Also called linear variable-differential transformer. 1. A transformer used to join two or more sources of signals to a common transmission line. 2. An electromechanical device that continuously translates displacement of position change into a linear ac voltage.

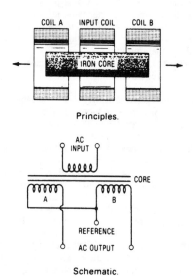

Principles.

Schematic.

Cutaway.

Differential transformer, 1.

differential voltage — For a glow lamp, the difference between the breakdown and maintaining voltage.

differential voltage gain — 1. The ratio of the change in output-signal voltage at either terminal of a differential device to the change in signal voltage applied to either input terminal, all voltages measured to ground. 2. The ratio of the differential output voltage of an amplifier to the differential input voltage of the amplifier. If the amplifier has one output terminal, the differential voltage gain is the ratio of the ac output voltage (with respect to ground) to the differential input voltage.

differential winding — A coil winding so arranged that its magnetic field opposes that of a nearby coil.

differential-wound field — A type of motor or generator field having both series and shunt coils connected so they oppose each other.

differentiate — 1. To distinguish. 2. To find the derivative of a function. 3. To deliver an output that is the derivative with respect to time of the input.

differentiating circuit — Also called differentiating network and differentiator. A circuit whose output voltage is proportional to the rate of change of the input voltage. The output waveform is then the time derivative of the input waveform, and the phase of the output waveform leads that of the input by 90°. An RC circuit gives this differentiating action.

differentiating network — See differentiating circuit.

differentiator — See differentiating circuit.

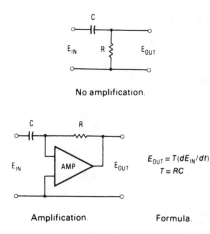

No amplification.

$$E_{OUT} = T(dE_{IN}/dt)$$
$$T = RC$$

Amplification. Formula.

Differentiator.

diffracted wave — A radio, sound, or light wave that has struck an object and been bent or deflected other than by reflection or refraction.

diffraction — 1. The bending of radio, sound, or light waves as they pass through an object or barrier, thereby producing a diffracted wave. 2. The phenomenon whereby waves traveling in straight paths bend around an obstacle. See X-ray crystallography.

diffuse — 1. To undergo or cause to undergo diffusion. 2. Light that has been either inadvertently or purposely scattered. Such diffused light propagates in many directions and is not intensely polarized when it illuminates surfaces. With diffused light, a high brightness level may be achieved with minimal glare.

diffused-alloy transistor — Also called drift transistor. A transistor in which the semiconductor wafer is subjected to gaseous diffusion to produce a nonuniform base region, after which alloy junctions are formed in the same manner as for an alloy-junction transistor. It may also have an intrinsic region to give a pnip unit.

diffused-base transistor — Also called graded-base transistor. A type of transistor made by combining diffusion and alloy techniques. A nonuniform base region and the collector-to-base junction are formed by gaseous dissemination into a semiconductor wafer that constitutes the collector region. Then the emitter-to-base junction is formed by a conventional alloy process on the base side of the diffused wafer.

diffused device — A semiconductor device in which a base, usually of silicon, has successive layers of p and n characteristics diffused upon and into the base by means of a series of masks and around which p and n materials, usually phosphorous and boron, adhere to the base by gaseous diffusion in a high-temperature furnace. It is possible to build areas of resistance, capacitance, and active diodes and transistors into the base, creating an entire circuit. Performance is poor in the presence of radiation.

diffused-emitter-and-base transistor — Also called double-diffused transistor. A semiconductor wafer that has been subjected to gaseous dissemination of both n- and p-type impurities to form two pn junctions in the original semiconductor material.

diffused-emitter-collector transistor — A transistor whose emitter and collector are both produced by diffusion.

diffused junction — 1. Type of pn junction, made by using masks to control the diffusion of impurities into monocrystalline semiconductor material. 2. A junction formed by the diffusion of an impurity within a semiconductor crystal.

diffused-junction rectifier — A semiconductor diode in which the pn junction is produced by diffusion.

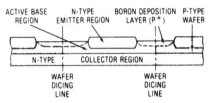

Diffused-junction silicon transistor wafer.

diffused-junction transistor — A transistor in which the emitter and collector electrodes have been formed by diffusion of an impurity into the semiconductor wafer without heating.

diffused-layer resistor — A resistor formed by including an appropriate pattern in the photomask to define diffusion areas.

diffused mesa transistor — A transistor in which the collector-base junction is formed by gaseous diffusion, and the emitter-base junction is formed either by gaseous diffusion or by an evaporated metal strip. The collector-base junction is then defined by etching away the undesired parts of the emitter and base regions, thus producing a mesa.

diffused metal-oxide semiconductor — *See* DMOS.

diffused planar transistor — A transistor made by two gaseous diffusions, but in which the collector-base junction is defined by oxide masking. Junctions are formed beneath this protective oxide layer, with the result that the device has lower reverse currents and good dc gain at low currents.

diffused sound — Sound that has uniform energy density, meaning that the energy flux is equal in all parts of a given region.

diffused transistor — A transistor in which the emitter and collector junctions are both formed by diffusion.

diffused transmission — The total net transmission, by a medium or device, of light that is neither perfectly Lambertian nor parallel. Often used interchangeably with the term *gross transmission*.

diffuser — A device used to scatter or disperse light emitted from a source, usually by the process of diffuse transmission.

diffuse scan — A reflective scan technique in which reflection from a nearby nonshiny surface illuminates a photosensor.

diffuse transmission — Transmission accompanied by diffusion or scatter to the extent that there is no regular or direct transmission.

diffusion — 1. The movement of carriers from a region of high concentration to regions of lower concentration. 2. The tendency of entities such as current carriers to diffuse themselves, or move in directions that increase the uniformity with which their number occupy the available space. Hence, carrier diffusion is a mechanism whereby carriers tend to move downhill along concentration gradients, away from regions of high concentration and toward regions of low concentration. 3. A thermally induced process in which one material permeates another. In silicon processing, doping impurities diffuse into the silicon at elevated temperatures to form the desired junctions. The same impurities penetrate silicon dioxide much more slowly, and therefore silicon dioxide on the surface of the silicon acts as a mask to determine the areas into which diffusion occurs. 4. A high-temperature process involving the movement of controlled densities of n-type or p-type impurity atoms into the solid silicon slice in order to change its electrical properties. 5. The process of adding impurities to a semiconductor material in order to affect its characteristics. 6. One method of modifying the impurity doping of a semiconductor crystal, which makes use of the fact that excited dopant atoms, like carriers, have a tendency to diffuse away from regions of high concentration and toward regions of low concentration. The technique involves prolonged exposure of the semiconductor crystal to a concentrated vapor of the dopant at elevated temperatures, whereupon dopant atoms diffuse into the crystal structure. The resulting doping gradient is roughly exponential, with highest density at the surface.

diffusion and oxidation systems — Equipment in which nonconductive materials are made semiconductive by diffusing controlled amounts of selected impurities into the surface and oxidizing the surface of silicon selectively to provide a protective or insulative layer. Diffusion and oxidation are accomplished by exposing the silicon wafer to specific atmospheres in a high temperature furnace.

diffusion bonding — Formation of a metallurgical joint between similar or dissimilar metals by the process of interdiffusion of atoms across the joint interface in either the solid or liquid state. The term generally is applied to, but is not limited to, solid-state diffusion. The joining surfaces must be brought within atomic distances through the application of pressure.

diffusion capacitance — The capacitance of a forward-biased pn junction.

diffusion constant — The quotient of diffusion-current density in a homogeneous semiconductor, divided by the charge-carrier concentration gradient. It is equal to the drift mobility times the average thermal energy per unit charge of carriers.

diffusion current — 1. The current produced when charges move by diffusion. 2. The flow of a particular type of carrier in a semiconductor due to a concentration difference in that type of carrier. Carriers will flow from an area of high concentration to an area of low concentration.

diffusion furnace — System designed for enclosed elevated temperature processing of solid-state devices and systems in gaseous atmospheres. Diffusion furnaces are operated at temperatures from 1000 to 1300°C to achieve doping of semiconductor substrates by one of a number of processes. Oxidation is a process that puts a protective layer of silicon oxide on the wafer that is used either as an insulator or to mask out certain areas when doping. Deposition systems, of which there are three (liquid, gaseous, solid), are used to deposit impurities on the silicon wafer. Other systems include a drive-in system, used to diffuse impurities into the wafer to a specified level, and an alloy system that is used in a final step of the metallization process.

diffusion length — In a homogeneous semiconductor, the average distance the minority carriers move between generation and recombination.

diffusion process — Doping of a semiconductor material by injection of an impurity into the crystal lattice at an elevated temperature. Usually, the semiconductor crystal is exposed to a controlled surface concentration of dopants.

diffusion transistor — A transistor in which current depends on the diffusion of carriers, donors, or acceptors, as in a junction transistor.

diffusion under the epitaxial film — *See* DUF.

diffusion window — In a semiconductor, the opening etched through the oxide to permit the diffusion of the emitter and base.

digiralt — A system of high-resolution radar altimetry in which pulse-modulated radar and high-performance time-to-digital conversion techniques are combined.

digit — 1. One of the symbols, 0, 1, 2, 3, 4, 5, 6, 7, 8, and 9, used in numbering in the scale of 10. One of these symbols, when used in a scale of numbering to the base n, expresses integral values ranging from 0 to $n - 1$ inclusive. 2. A character used to represent a nonnegative integer smaller than the radix, e.g., either 0 or 1 in binary notation or 0 to F in hexadecimal. 3. In a dial telephone system, one of the successive series of pulses incoming from a dial for operation of a switching train.

digital — 1. Using numbers expressed in digits and in a certain scale of notation to represent all the variables that occur in a problem or calculation. 2. Of or pertaining to the class of devices or circuits in which the output varies in discrete steps (i.e., pulses or on/off operation). 3. Of or pertaining to an element or circuit whose output is utilized as a discontinuous function of its input. 4. Circuitry in which data-carrying signals are restricted to either of two voltage levels, corresponding to logic 1 or 0.5. The representation of numerical quantities by means of discrete numbers. It is possible to express in binary digital form all information stored, transferred, or processed by dual-state conditions; e.g., on/off, open/closed, octal, and BCD values. 5. Referring to communications procedures, techniques, and equipment by which information is encoded as either a binary 1 or a binary 0; the representation of information in discrete binary form, discontinuous in time.

digital absorbing selector — A dial switch that sets up and then falls back on the first of two digits dialed; it then operates on the next digit dialed.

digital audio tape — Abbreviated DAT. 1. A digital recording audio format that combines the 16-bit audio quality of compact discs with the recording capability of analog cassettes. 2. A stereo tape format that records sound in digital code and is smaller in size than an audio cassette. A 4-mm DAT drive holds over 1 gigabyte of data and is used as a high-capacity backup medium for computers and as a master source medium for sending data to a CD manufacturer.

digital broadband technology — The technology used in the transmission of digital signals through fiber-optic cables.

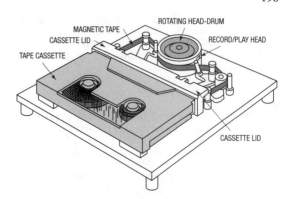

Digital audio tape (DAT) transport system.

digital circuit — 1. A circuit that operates like a switch (it is either on or off) and can make logical decisions. It is used in computers or similar decision-making equipment. The more common families of digital integrated circuits (called logic forms) are RTL, DTL, HTL, ECL, and TTL. 2. A circuit that has only two stable states, operating in the manner of a switch, that is, it is either on or off, or high or low (i.e., high voltage or low voltage).

digital clock — A series of synchronized pulses that determine the bit times (data rate) of a digital pattern.

digital communication — 1. The transmission of intelligence by the use of encoded numbers — usually uses the binary rather than decimal number system. 2. A system of telecommunications employing a nominally discontinuous signal that changes in frequency, amplitude, or polarity.

digital communications interface equipment — Line interface equipment, including modems.

digital computer — 1. An electronic calculator that operates with numbers expressed directly as digits, as opposed to the directly measurable quantities (voltage, resistance, etc.) in an analog computer. In other words, the digital computer counts (as does an adding machine); the analog computer measures a quantity (as does a voltmeter). 2. A computer that processes information in numerical form. Electronic digital computers generally use binary or decimal notation and solve problems by repeated high-speed use of the fundamental arithmetic

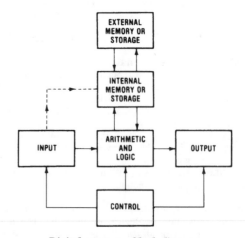

Digital computer block diagram.

processes of addition, subtraction, multiplication, and division. 3. A computer system in which circuit operation is based on specific signal levels. In a binary digital computer, there are two such signal levels, one at or near zero and the other at a defined voltage. 4. A device that performs sequences of arithmetic and logic operations on discrete data. 5. A type of data-processing equipment that counts, utilizing numbers to express the values and quantities. General-purpose digital computers include central storage units and peripheral control units and are designed to solve a wide class of problems. A common feature of general-purpose equipment is the ability to externally modify the program of instructions. Special-purpose digital computers are not intended for a typical commercial physical environment and include rugged computers for military and space applications. An analog computer measures cost or conditions. Hybrid computers utilize both modes. 6. A computer that solves problems by operating on discrete representing variables by performing arithmetic and logic processes on this data.

digital data — 1. Data represented in discrete, discontinuous form, as contrasted with analog data represented in continuous form. Digital data is usually represented by means of coded characters (e.g., numbers, signs, symbols, etc.). 2. Any data that is expressed in digits. The term usually implies the use of binary digits.

digital data-handling system — The electronic equipment that receives digital data, operates on it in a suitable manner, records it in a suitable manner on a suitable medium, and presents it directly to a computer or a display.

digital delay line — *See* active delay line.

digital delay module — *See* active delay line.

digital delay unit — *See* active delay line.

digital device — 1. Typically, an IC that switches between two exclusive states or levels, usually represented by logical 1 or 0. 2. An electronic device that processes electrical signals that have only two states, such as on or off, high or low voltages, or positive or negative voltages. In electronics, *digital* normally means binary or two-state.

digital differential analyzer — A special-purpose digital computer that performs integration and that can be programmed for the solution of differential equations in a manner similar to that of an analog computer.

digital disc recording — An analog disc recording that has been made from a master tape recording that was digitally recorded.

digital filter — 1. A linear computation or algorithm performed on a selected series in the form of an input signal that produces a new series as output. The computational device may be a specifically designed electronic system or a conventional computer. 2. Network that operates on discrete samples of a signal to achieve a desired transfer-function operation on that signal. Digital filters divide into two classes: nonrecursive filters produce an output that is a function of only the previous and present inputs; recursive filters produce an output that is a function of both the past and present inputs and outputs.

digital filtering — 1. A computational process or algorithm by which a sampled signal or sequence of numbers, acting as input, is transformed into a second sequence of numbers called the output. The computational process may correspond to high-pass, low-pass, bandpass, or bandstop filtering, integration, differentiation, or something else. The second sequence can be used for further processing, as in a fast-Fourier-transform analyzer, or it can be converted to an analog signal, producing a filtered version of the original analog signal. 2. The process of smoothing, spectrally shaping, or removing noise from a signal. Digital filters are basically mathematical functions that are performed on the digital data stream; their characteristics can be altered under software control, which adds to their overall flexibility. Finite impulse response (FIR) and infinite impulse response (IIR) are examples of digital filter functions.

digital frequency monitor — A special-purpose digital counter that permits a train of pulses to pass through a gate for a predetermined time interval, counts them, and indicates the number counted.

digital harmonic generation — Abbreviated DHG. The use of circuit elements whose outputs are discontinuous functions of their inputs to produce signals that are an integral multiple of the (fundamental) input signal.

digital image analysis — Technology to measure and standardize the output of a computer-interfaced vidicon system.

digital imaging — The process by which an image that is in electronic form (e.g., a bit-mapped graphic) is altered.

digital information display — The presentation of digital information in tabular form on the face of a digital information display tube.

digital integrated circuit — 1. A switching-type integrated circuit. 2. An integrated circuit that processes electrical signals that have only two states, such as on or off, high or low voltages, or positive or negative voltages. In electronics, *digital* normally means binary or two-state. 3. A monolithic group of logic elements. May be small-scale integration (e.g., SSI gates, flip-flops, latches), medium-scale integration (e.g., MSI decoders, adders, counters), or large-scale integration (e.g., LSI memories, microprocessors). 4. A class of integrated circuits that processes digital information (expressed in binary numbers). The processing operations are arithmetic (such as addition, subtraction, multiplication, and division) or logical (in which the circuit senses certain patterns of input binary information and indicates the presence or absence of those patterns by appropriate output binary signals).

digital integrator — Device for summing or totalizing areas under curves that gives numerical readout. *See also* integrator.

digital logic modules — Circuits that perform basic logic decisions (AND, OR, NOT); used widely for arithmetic and computing functions, flip-flops, half-adders, multivibrators, etc. *See also* logic system.

digitally programmable oscillator — A voltage-controlled oscillator designed to accept a digital tuning word instead of the usual analog signal. Internal digital-to-analog (d/a) converter circuits transform the digital input to an analog voltage. Tuning-curve linearization is usually accomplished through a digital memory. The frequency speed is primarily limited by the d/a circuits.

digital modulation — A method of transmitting human voice or other analog signals using a binary code (0s and 1s). Digital transmission offers a cleaner signal than analog technology.

digital multimeter — Abbreviated DMM. A test instrument used to measure voltage, current, and resistance. The readout of measured values is shown on a digital display which is typically a liquid crystal display (LCD).

digital optical processing — The scanning of photographs or transparencies of images, either by a vidicon camera or flying-spot scanner, for the conversion of the images to digital form for storage on magnetic tape.

digital output — An output signal that represents the size of a stimulus or input signal in the form of a series of discrete quantities that are coded to represent digits in a system of numerical notation. This type of output is to be distinguished from one that provides a continuous output signal.

digital panel meter — Abbreviated DPM. 1. A compact electronic measuring system capable of converting an electrical variable into an unambiguous, accurate numerical reading. Consists of a solid-state analog-to-digital converter and a logic-driven numerical display that can provide readings with accuracies of 0.1 to 0.005 percent of full scale, without the need for eye-straining interpolation. 2. A meter with digital-numerical readout, capable of indicating a range of values from zero to rated maximum, in increments of one least-significant digit.

digital phase shifter — A device that provides a signal phase shift by the application of a control pulse. A reversal of phase shift requires a control pulse of opposite polarity.

digital plotter — An output unit that graphs data via an automatically controlled pen. Data is normally plotted as a series of incremental steps.

digital position transducer — A device that converts motion or position into digital information.

digital programmable delay line — See active delay line.

digital radio — 1. Microwave radio in which one or more properties (amplitude, frequency, and phase) of the rf carrier are quantized by the modulating signal. 2. Radio whose instantaneous rf carrier can assume one of a discrete set of amplitude levels, frequency shifts, or phase shifts as a result of the modulating signal. 3. Radio in which the rf carrier, or even the baseband signal, is quantized by means of a modem. In such a radio the rf carrier is still quantized, but baseband or intermediate-frequency filtering can be used to provide some bandwidth control and thereby reduce rf filtering requirements. 4. Any radio that transmits a signal whose informational content is, in whole or in part, digital in format.

digital readout indicator — An indicator that reads directly in numerical form, as opposed to an analog indicator needle and scale.

digital recording — A technique for recording information as discrete points onto magnetic recording media.

digital rotary transducer — A rotating device utilizing an optical sensor that produces a serial binary output as a result of shaft rotation.

digital set top — Also referred to generically as a set-top unit (STU). Television set-top units that accept digital video as well as analog (traditional) video. Some digital set-top units enable interactivity and therefore support shopping, banking, games, etc.

digital signal — 1. An electrical signal with two states — on or off, high or low, positive or negative — such as could be obtained from a telegraph key or two-position toggle switch. *Digital* normally means binary or two-state. 2. Representation of information by a set of discrete values in accordance with a prescribed law. These values are represented by numbers.

digital signals — 1. Discrete or discontinuous signals whose various states are discrete intervals apart. 2. Signals made up of discontinuous pulses whose information is contained in their durations, periods, and/or amplitudes.

digital signature — 1. A numerical representation of a set of logic states, typically used to describe the logic-state history at one device under test output pin during the complete test program. 2. A personal authentication method based on encryption and secret authorization codes used for "signing" electronic documents.

digital simulation — See simulation.

digital speech communications — Transmission of voice signals in digitized or binary form.

digital status contact — A logical (on/off) input used mainly to sense the status of remote equipment in process control systems.

digital storage oscilloscope — A special oscilloscope that adapts analog monitoring and recording systems to the capture and analysis of all types of one-shot physical phenomena.

digital switch — 1. A means to interconnect two or more circuits whose information is represented in digital form, using a time-divided network consisting of nonlinear elements. 2. An automatic switching center capable of switching digital signals. It may be a circuit switch or a message switch.

digital switching — Switching of messages digitally by use of integrated electronic circuits for logic and memory, rather than by electromechanical switches.

digital synthesizer — A means of generating several different frequencies without using separate oscillators governed by crystals specially ground for each frequency. A digital synthesizer uses only one reference crystal, a phase-locked loop, and a digital counter to generate a large number of stable frequencies. These circuits are used to reduce dependence on individual crystals, which are relatively expensive.

digital telephone dialer — An automatic telephone dialer that uses a digital code.

digital television — Abbreviated DTV. 1. A television system in which reduction or elimination of picture redundancy is obtained by transmitting only the information needed to define motion in the picture, as represented by changes in areas of continuous white or black. 2. An umbrella term used to describe the digital television system adopted by the FCC in December 1996.

digital thermometer — Electronic temperature-measuring device that reads and/or prints out numerically.

digital-to-analog conversion — The generation of analog (usually variable-voltage) signals in response to a digital code.

digital-to-analog converter — Abbreviated DAC, dac, or d/a converter. 1. A computing device that changes digital quantities into physical motion or into a voltage (i.e., a number output into turns of a potentiometer). 2. A unit or device that converts a digital signal into a voltage or current whose magnitude is proportional to the numeric value of the digital signal. For example:

Digital Input	Analog Output
00101 (binary 5)	2 volts
01010 (binary 10)	4 volts
10100 (binary 20)	8 volts

3. A circuit that accepts the discrete, binary outputs of computers and changes them into continuous analog quantities. In general, DACs convert mathematical results into usable electrical quantities. Digital-to-analog converters are used to generate and modulate waveforms, stimulate devices under test, drive motors, or display information. They have applications in process and industrial control systems, cathode-ray-tube displays, and digitally programmed power supplies. Most DACs consist of three major blocks: a precision reference, a set of resistors forming a ladder network, and switches that connect or disconnect the resistor ladder and the reference. Some DACs also include an output amplifier that buffers the current from the ladder and interfaces it to the circuits at the DAC's output. 4. An interface that converts data in a digital form to data in analog form. Used to permit analog output from a digital computer.

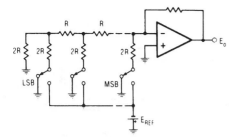

Digital-to-analog converter.

digital-to-disc recording — A recording technique in which a digital master tape is transferred to a conventional master lacquer disc for the manufacturing of phonograph records.

digital transmission — 1. A mode of transmission in which all information to be transmitted over the facility is first converted into digital form and then sent down the line as a stream of pulses. (Such transmission may imply a serial bit stream, but parallel forms are also possible.) When noise and distortion threaten to destroy the integrity of the pulse stream, the pulses are detected and regenerated. 2. The transmission of a signal in which information is represented by a code of discrete elements. Compare with *analog transmission*.

digital versatile disc — *See* DVD.

digital video disc — *See* DVD.

digital voltmeter — 1. An indicator that provides a digital readout of measured voltage rather than a pointer indication. 2. An electronic instrument that converts an analog voltage of unknown magnitude into a digital display of known value.

digit compression — In a computer, any of several techniques used to pack digits.

digitization — 1. The conversion of black-and-white artwork and continuous-tone photographs to a series of binary numbers that can be stored and processed in a computer system. 2. The process of converting analog video, images, or audio into digital format (i.e., 1s and 0s).

digitize — 1. To transform information from the analog (continuous wave-like signal) into a digital format, i.e., into a series of 1s and 0s. 2. To convert an image or signal into digital code for input into a computer; to scan an image, trace a picture on a graphics tablet, or convert camera images. 3. To convert an analog measurement of a physical variable into a number expressed in digits in a scale or notation. 4. To translate a quantitative measurement into a coded numerical equivalent. 5. To convert drawing or picture information into digital form.

digitized image — An image that has been converted into a series of discrete units that are represented in a computer by the binary digits 1 and 0; such an image can then be manipulated for various purposes.

digitized speech — A numerical representation of speech in which the amplitude of the speech waveform has been recorded at regular intervals. Speech is typically sampled from 8000 to 12,500 times per second.

digitizer — 1. A device that converts analog data into numbers expressed in digits in a system of notation. 2. A device that transforms graphical data into planar coordinate information that can be read and understood by a computer. These coordinates are usually presented as *x* and *y* coordinates based on the position of a cursor on the surface, or platen, of the digitizer. The cursor has a viewing area with a crosshair for alignment with the point of interest on the document, and is coupled either mechanically or electrically to a position-sensing device that provides the positional information to the computer. 3. A device that converts coordinate information into numeric form readable by a digital computer. 4. A computer peripheral device that converts an analog electrical signal into numeric form readable by a digital computer. 5. A device that translates input into digital form, to make it possible, for example, to enter sketches into a computer.

digitizing — 1. The process of converting an analog signal to a digital signal. 2. Any method of reducing feature locations on a flat plane to digital representation of *x* and *y* coordinates. 3. The process of converting graphic representations, such as pictures and drawings, into digital data that can be processed by a computer system.

digitron display — In a calculator, a type of display in which all digits appear in the same plane. Similar to mosaic lamp display.

digit selector — In a computer, a device for separating a card column into individual pulses that correspond to punched row positions.

digit-transfer bus — The main wire or wires used to transfer information (but not control signals) among the various registers in a digital computer.

diheptal base — Also called diheptal socket. A vacuum-tube base having 14 pins (such as the base of a cathode-ray tube).

diheptal socket — *See* diheptal base.

DIIC — Dielectrically isolated integrated circuits. Devices isolated from each other by a layer of dielectric insulation, usually glass, rather than by the more conventional reverse-biased pn junction. This "insulated substrate" structure is far more radiation resistant than junction-isolated units, making DIICs valuable in military and aerospace applications.

dimensional stability — 1. The ability of a body to maintain precise shape and size. 2. A measure of dimensional change caused by such factors as temperature, humidity, chemical treatment, age, or stress, usually expressed as units per unit.

dimension ratio (L/D) — The ratio of the length of a magnet in the direction of magnetization to its diameter. Or, the ratio of the length of the magnet to the diameter of a circle having an area equal to the cross-sectional area of the magnet. Used as figure of merit to find a magnet's composite permeance coefficient.

diminished-radix complement — *See* radix-minus-one complement.

dimmer — 1. A device for controlling the amount of light emitted by a luminaire. Common types employ resistance, autotransformer, magnetic amplifier, silicon-controlled rectifier or semiconductor, thyratron, or iris control elements. 2. An electric or electronic device that regulates the voltage going to a light source as a means of varying the intensity of the light emitted by the source.

dimmer curve — The performance characteristic of a light dimmer expressed as a graph of the light output of a dimmer-controlled lamp versus the setting of the control in terms of an arbitrary linear scale of 0 to 10.

DIN — 1. The abbreviation for the association in West Germany that determines the standards for electrical and other equipment in that country, Deutsches Institut für Normung (German Institute for Standards). Similar to the American National Standards Institute. 2. A set of standards and specifications promulgated by German manufacturers and covering such audio-related matters as connectors, frequency weighting, measurement techniques, and specifications.

D-indicator — A radar indicator that combines types B and C indicators. The signal appears as a bright spot, with azimuth angle as the horizontal coordinate and elevation angle as the vertical coordinate. Each horizontal trace is expanded vertically by a compressed time sweep to facilitate separation of the signal from noise and to give a rough range indication.

DIN jack — A system of multipin jacks and plugs allowing several connections to be made at once. Named after the German Institute for Standards (DIN).

diode — 1. An electron tube having two electrodes: a cathode and an anode. 2. *See* crystal diode. 3. A two-element electron tube or solid-state device. Solid-state diodes are usually made of either germanium or silicon and are primarily used for switching purposes, although they can also be used for rectification. Diodes are usually rated at less than one-half ampere. 4. A two-terminal electronic device that will conduct electricity much more easily in one direction than in the other. 5. A semiconductor device with two terminals and a single junction, exhibiting varying conduction properties depending on the polarity of the applied voltage. 6. A two-terminal semiconductor device exhibiting a nonlinear voltage-current characteristic; it has the asymmetrical voltage-current characteristic exemplified by a single pn junction.

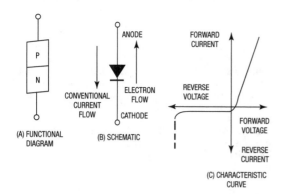

Diode.

diode amplifier — A parametric amplifier that uses a special diode in a cavity. Used to amplify signals at frequencies as high as 6000 MHz.

diode assembly — A single structure of more than one diode.

diode characteristic — The composite electrode characteristic of a multielectrode tube, taken with all electrodes except the cathode connected together.

diode demodulator — Also called diode detector. A demodulator in which one or more semiconductor or electron-tube diodes are used to provide a rectified output that has an average value proportional to the original modulation.

diode detector — *See* diode demodulator.

diode gate — An AND gate that uses diodes as switching elements.

diode isolation — A method in which a high electrical resistance between an IC element and the substrate is obtained by surrounding the element with a reverse-biased pn junction.

diode laser — Also called laser diode, injected laser, coherent electroluminescence device, semiconductor laser. A pn junction semiconductor electron device that converts direct forward-bias electrical input (pump power) directly into coherent optical output power via a process of stimulated emission in the region near the junction.

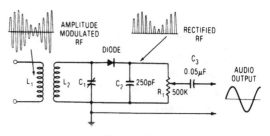

Diode detector.

diode limiter — A circuit employing a diode and used to prevent signal peaks from exceeding a predetermined value.

diode logic — An electronic circuit using current-steering diodes in an arrangement such that the input and output voltages have relationships that correspond to AND or OR logic functions.

diode matrix — 1. A two-dimensional array of diodes used for a variety of purposes, such as decoding and read-only memory. 2. A hardware pattern in which diode leads may be inserted to change solid-state control logic.

diode mixer — A diode that mixes incoming radio-frequency and local-oscillator signals to produce an intermediate frequency.

diode modulator — A modulator in which one or more diodes are employed to combine a modulating signal with a carrier signal. It is used chiefly in low-level signaling because it has inherently poor efficiency.

diode pack — A combination of two or more diodes integrated into a solid block.

diode peak detector — A diode used in a circuit to indicate when audio peaks exceed a predetermined value.

diode-pentode — A vacuum tube having a diode and a pentode combined in the same envelope.

diode rectification — The conversion of an alternating current into a unidirectional current by means of a two-element device such as a crystal, vacuum tube, etc.

diode switch — A diode in which positive and negative biasing voltages (with respect to the cathode) are applied in sucession to the anode in order to pass and block, respectively, other applied waveforms within certain voltage limits. In this way, the diode acts as a switch.

diode-transistor logic — Abbreviated DTL. 1. A logic circuit that uses diodes at the input to perform the electronic logic function that activates the circuit transistor output. In monolithic circuits, the DTL diodes are a positive-level logical AND function or a negative-level OR function. The output transistor acts as an inverter to result in the circuit becoming a positive NAND or a negative NOR function. 2. Any logic gate circuit that uses several diodes to perform the AND or OR function, followed by one or more transistors to add power to (and possibly invert) the output. Formerly very popular in digital systems, but now largely superseded by TTL circuits. 3. Logic employing diodes at the input with transistors used as amplifiers and resistor pull-up on the output.

diode-triode — A vacuum tube having a diode and triode combined in the same envelope.

diopter — 1. The unit of optical measurement that expresses the refractive power of a lens or prism. 2. A measure of lens power equal to the reciprocal of the lens focal length in meters.

dip — 1. A drop in the plate current of a class C amplifier as its tuned circuits are being adjusted to resonance. 2. The angle between the direction of the earth's magnetic field and the horizontal as measured in a vertical plane.

DIP — Abbreviation for dual in-line package.

dip coating — 1. A method of applying an insulating coating to a conductor by passing it through an applicator containing the insulating medium in liquid form. The insulation is then sized and passed through ovens to solidify. This medium can be used for magnet wire. 2. A process of applying a relatively thin (less than 50 mils) conformal coating to a part or assembly. The final coating thickness is determined by the viscosity of the coating material, rate of withdrawal, temperature, and the number of coats. 3. Method in which an object is coated by dipping into a plastisol or organosol.

dip encapsulation — A type of conformal coating. An embedding process in which the insulating material is applied by immersion and without the use of an outer container. The coating conforms generally with the contour of the embedding part or assembly.

diplexer — 1. A coupling unit that allows more than one transmitter to operate together on the same antenna. 2. A device that enables two (radio) transmitters operating at different frequencies to use the same antenna simultaneously.

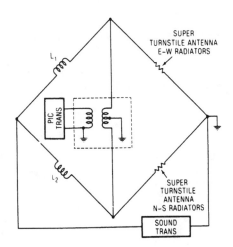

Diplexer circuit.

diplex operation — 1. The simultaneous transmission or reception of two messages from a single antenna or on a single carrier. 2. The operation of two radio transmitters on different frequencies into the same antenna simultaneously.

diplex radio transmission — Simultaneous transmission of two signals by using a common carrier wave.

diplex reception — The simultaneous reception of two signals having some feature in common, for example, a single receiving antenna or a single carrier frequency.

dipole — 1. A molecule that has an electric moment. For a molecule to be a dipole, the effective center of the positive charges must be at a different point than the center of the negative charges. 2. A form of speaker that radiates in approximately equal amounts to the rear and the front. 3. *See* dipole antenna.

dipole antenna — Also called dipole. A straight radiator usually fed in the center. Maximum radiation is produced in the plane normal to its axis. The length specified is the overall length.

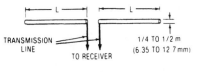

$$2L = 1/2 \text{ WAVELENGTH} \times 0.95$$
$$L(\text{ft}) = 234/\text{FREQ(MHz)}$$
$$L(\text{cm}) = 7224/\text{FREQ(MHz)}$$

Dipole antenna.

dipole disc feed — An antenna, consisting of a dipole near a disc, used to reflect energy to the disc.

dipping — The process of impregnating or coating insulating materials or windings by the simple method of immersion in the liquid insulating material. A step in the process of treating insulating materials or electrical components by immersion in a liquid insulation, followed by draining and curing to provide increased electrical and mechanical protection.

dip soldering — 1. The process of soldering component leads, terminals, and hardware to the conductive pattern on the bottom of a printed circuit board by dipping that side into molten solder or floating it on the surface. 2. A process of joining metals, previously cleaned and fluxed, by immersing them wholly or partially into molten solder. The filling of the joint is by capillary attraction. 3. A process whereby items to be soldered are brought in contact with the surface of a static pool of molten solder for the purpose of soldering the entire exposed conductive pattern in one operation. 4. Soldering by dipping fluxed and fixtured parts in a solder pot. 5. The simplest form of mass soldering. It involves the lowering of a prefluxed assembly onto a solder bath surface. The assembly is then submerged sufficiently for the solder to spread and form the required joints but not flow over the top surface of a printed wiring board.

dip solder terminal — The terminals on a connector that are inserted into holes in the printed circuit board and then soldered in place.

direct — A method of expressing an absolute address in an MPU instruction where the actual address would be specified in hexadecimal in the instruction.

direct access — The process of storing data in, or getting data from, a storage device in such a manner that surrounding data need not be scanned to locate the desired data. The time required to get desired data from the storage device is independent of the location of the data.

direct-access device — *See* random-access device.

direct-access file — A file in which each record may be accessed directly, regardless of its relative position in the file.

direct-access storage — Also called random-access storage. Pertaining to the process of obtaining data from or placing data into storage in which the time required for such access is independent of the location of the data most recently obtained or placed in storage.

direct-acting recording instrument — An instrument in which the marking device is mechanically connected to or directly operated by the primary detector.

direct address — An address that specifies the location in a computer of an instruction operand.

direct addressing — The standard addressing mode in a computer. It is characterized by an ability to reach any point in main storage directly. Direct addressing is sometimes restricted to the first 256 bits in main storage.

direct broadcast satellite — Abbreviated DBS. 1. A satellite that allows the use of inexpensive home reception dishes to receive its high-power signals, including software and data. 2. A term commonly used to describe Ku-band broadcasts via satellite directly to individual end users. The DBS band ranges from 11.7 to 12.75 GHz.

direct capacitance — 1. The capacitance between two conductors, excluding stray capacitance that may exist between the two conductors and other conducting elements. 2. The capacitance measured directly from conductor to conductor through a single insulating layer.

direct-connect modem — A modem that connects directly to a phone line via modular connectors, rather than going through a telephone headset and an acoustic coupler.

direct-coupled amplifier — 1. A direct-current amplifier in which the plate of one stage is coupled to the grid of the next stage by a direct connection or a low-value resistor. 2. An amplifier in which the output of one stage is connected to the input of the next stage without the use of intervening coupling components.

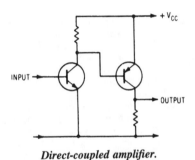

Direct-coupled amplifier.

direct-coupled transistor logic — Abbreviated DCTL. A NOR-gate type of bipolar logic in which the output of one gate is coupled directly to the input of the succeeding gate. This form of logic evolved into resistor-transistor logic because of the difficulty in mass producing transistors within the close tolerances necessary for direct coupling.

direct coupling — 1. The association of two or more circuits by means of an inductance, a resistance, a wire, or a combination of these so that both direct and alternating currents can be coupled. 2. Interstage coupling or speaker coupling with no intervening transformer or capacitor. To the speaker it ensures that the damping factor remains high at low frequencies (but increasing power supply impedance at low frequencies can influence this), while direct coupling generally minimizes low-frequency phase shift and encourages enhanced bass performance. 3. The connection of two circuits in such a way that both ac and dc currents can flow. Neither a transformer nor a capacitor can be used in series with the signal-carrying wires since these components do not pass dc.

direct current — Abbreviated dc. 1. An essentially constant-value current that flows in only one direction. 2. A flow of continuous electric current in one direction as long as the circuit is closed (as opposed to alternating current). 3. A current that flows in only one direction in an electric circuit. It may be continuous or discontinuous, and may be constant or varying.

direct-current amplifier — Also called dc amplifier. An amplifier capable of boosting dc voltages. Resistive coupling only is generally employed between stages, but sometimes will be combined with other forms.

direct-current erasing head — A head that uses direct current in magnetic recording to produce the magnetic field required for erasure. Direct-current erasing is achieved by subjecting the medium to a unidirectional field. Such a medium is therefore in a different magnetic state from one erased by alternating current.

direct-current generator — 1. A rotating machine that changes mechanical into electrical energy in the form of direct current. This is accomplished by commutating bars on the armature. The bars make contact with stationary brushes, from which the direct current is taken. 2. A rotary machine having a commutator that generates direct current when its armature is rotated in a magnetic field by an engine or motor. The commutator reverses the alternating current generated in the armature winding to produce direct current at the machine's output terminals.

direct-current resistance — Abbreviated DCR. The resistance offered by any circuit to direct current.

direct-current restorer — The means by which a direct-current or low-frequency component is reinserted after transmission. Used in a circuit incapable of transmitting slow variations, but capable of transmitting components of higher frequency.

direct-current transmission — Of television, that form of transmission in which a fixed setting of the controls makes any instantaneous value of signal correspond to the same value of brightness at all times.

direct data entry — The entry of data directly into a computer through machine-readable source documents or through the use of online terminals.

direct digital control — Time-sharing, or multiplexing, of a computer among many controlled loops.

direct distance dialing — Abbreviated DDD. 1. A telephone exchange service that enables the telephone user to call other subscribers outside his or her local area without operator assistance. 2. Direct distance dialing by subscribers over the nationwide intertoll telephone network. *See also* area code.

direct drive — A drive system used to rotate a turntable, in which the platter is driven directly by the motor shaft at the exact speed required. These designs usually include electronic motor control.

direct-drive torque motor — A servoactuator that can be directly attached to the load it is to drive. It converts electrical signals directly into sufficient torque to maintain the desired accuracy in a positioning or speed control system.

direct electromotive force — A unidirectional electromotive force in which the changes in values are either zero or so small that they may be neglected.

direct grid bias — The dc component of grid voltage; commonly called grid bias.

direct-insert subroutine — *See* open subroutine.

direct inward dialing — A service allowing outside parties to call directly to extensions on the customer's premises.

direction — The position of one point in space with respect to another.

directional — Having radiative characteristics that vary with direction.

directional antenna — An antenna that radiates radio waves more effectively in some directions than in others. (The term is usually applied to an antenna whose directivity is larger than that of a half-wave dipole.)

directional beam — An electromagnetic wave that is concentrated in a given direction.

directional coupler — 1. A junction consisting of two waveguides coupled together in such a manner that a traveling wave in either guide will induce a traveling wave in the same direction in the other guide. 2. A bilateral electrical network that may be used as a hybrid power splitter, power adder, or mixer.

directional filter — Also called directional separation filter. A filter used to separate the two frequency ranges in a carrier system in which one range of frequencies is used for transmission in one direction and another range of frequencies for transmission in the opposite direction.

directional gain — *See* directivity index.

directional homing — The procedure of following a path in such a way that the target is maintained at a constant relative bearing.

directional hydrophone — A hydrophone having a response that varies significantly with the direction of incidence of sound.

directional lobe — *See* lobe.

directional microphone — 1. A microphone whose response varies significantly with the direction of sound. *See also* bidirectional microphone; semidirectional microphone; unidirectional microphone. 2. A microphone that is more sensitive to sounds coming from certain directions than to sounds coming from other directions. Such microphones can be aimed so their most sensitive sides face the sound source, while their least sensitive sides face sources of noise or other undesired sound. *See* cardioid microphone; figure-8 microphone.

directional pattern — Also called radiation pattern. A graphical representation of the radiation or reception of an antenna as a function of direction. Cross sections are frequently given as vertical and horizontal planes, and principal electric and magnetic polarization planes.

directional phase shifter — A passive phase-shifting device in which the phase change for transmission in one direction is different from the phase change for transmission in the opposite direction.

directional power relay — A device that functions on a desired value of power flow in a given direction, or upon reverse power resulting from arcback in the anode or cathode circuits of a power rectifier.

directional relay — A relay that functions in conformance with the direction of power, voltage, current, pulse rotation, etc. *See also* polarized relay.

directional separation filter — *See* directional filter.

direction angle — The angle between the antenna base line and a line connecting the center of the base line with the target.

direction cosine — The cosine of the angle between the base line and the line from the center of the base line to the target.

direction finder — Abbreviated df. Also called radio compass. Apparatus for receiving radio signals and taking their bearings in order to determine their points of origin.

direction finding — The principle and practice of determining a bearing by radio means, using a discriminating antenna system and a radio receiver so that the direction of an arriving wave, and ostensibly the direction or bearing of a distant transmitter, can be determined.

direction of lay — 1. The lateral direction in which the strands of a conductor run over the top of the cable conductor as they recede from an observer looking along the axis of the conductor or cable. Also applies to twisted cable. 2. The direction, either right-hand (clockwise) or left-hand (counterclockwise), in which a conductor or group of conductors spiral around a cable core as they travel away from the observer.

direction of polarization — For a linearly polarized wave, the direction of the electrostatic field.

direction of propagation — At any point in a homogeneous, isotropic medium, the direction of the time-average energy flow. In a uniform waveguide, the

Direction finder.

direction of propagation is often taken along the axis. In a uniform lossless waveguide, the direction of propagation at every point is parallel to the axis and in the direction of time-average energy flow.

direction rectifier — A rectifier that supplies a direct-current voltage, the magnitude and polarity of which are determined by the magnitude and relative polarity of an alternating-current selsyn error voltage.

directive gain — In a given direction, 4π times the ratio of the radiation intensity to the total power radiated by the antenna.

directivity — 1. The property that causes an antenna to radiate or receive more energy in some directions than in others. 2. The value of the directive gain of an antenna in the maximum-gain direction. 3. The ability of an antenna to pick up signals from one general direction (usually from the front) and effectively reject those from other directions (usually from the back and sides). The front-to-back ratio is one measure of an antenna's directivity. 4. A tendency for some microphones to respond less strongly to sounds arriving from the sides and/or rear. Directional microphones are useful in discriminating on the basis of direction between wanted sounds (musical instruments) and unwanted sounds (audience noises). Directivity is typically graphed on a polar pattern, and is thus classed as nondirectional (omnidirectional), bidirectional (figure-8), or unidirectional (cardioid), supercardioid, or hyperdirectional.

directivity diagram of an antenna — The graphical representation of the gain of an antenna in the different directions of space.

directivity factor — 1. In acoustics, the directivity factor is equivalent to directivity, as applied to an antenna. 2. Of a transducer used for sound emission, the ratio of the intensity of the radiated sound at a remote point in a free field on the principal axis to the average intensity of the sound transmitted through a sphere passing through the remote point and concentric with the transducer. 3. Of a transducer used for sound reception, the ratio of the square of the electromotive force produced in response to sound waves arriving in a direction parallel to the principal axis of the transducer to the mean square of the electromotive force that would be produced if sound waves having the same frequency and mean square pressure were arriving at the transducer simultaneously from all directions with random phase. The frequency should be specified in both cases.

directivity index — Also called directional gain. A measure of the directional properties of a transducer. It is the ratio, in decibels, of the average intensity or response over the whole sphere surrounding the projector or hydrophone to the intensity or response on the acoustic axis.

directivity of a directional coupler — Ratio of the power measured at the forward-wave sampling terminals with only a forward wave present in the transmission line to the power measured at the same terminals when the forward wave reverses direction. This ratio is usually expressed in dB and would be infinite for a perfect coupler.

directivity of an antenna — The ratio of the maximum field intensity to the average field intensity at a given distance, implying a maximum value.

directivity pattern — A plot of the response of an electroacoustic transducer as a function of direction.

directivity signal — A spurious signal present in the output of any coupler because its directivity is not infinite.

direct light — Light from a luminous object such as the sun or an incandescent lamp, as opposed to reflected light.

direct lighting — A system of lighting that delivers a majority of light in useful directions without being deflected from the ceiling or walls. Any lamp equipped with a glass or metal reflector arranged to reflect the light toward the object to be illuminated is classified as direct lighting.

directly grounded — See solidly grounded.

directly heated cathode — A wire, or filament, designed to emit the electrons that flow from cathode to plate. This is done by passing a current through the filament; the current heats the filament to the point at which electrons are emitted. In an indirectly heated cathode, the hot filament raises the temperature of a sleeve around the filament; the sleeve then becomes the electron emitter.

direct material — A semiconductor material in which electrons move directly from the conduction band to the valence band to recombine with holes. The process of recombination conserves energy and momentum.

direct memory access — Abbreviated DMA. 1. A technique that permits a peripheral device to enter or extract blocks of data from a microcomputer memory without involving the central processing unit. In some cases, a CPU can perform other functions while the transfer occurs. 2. A method of I/O data transfer that does not alter minicomputer instruction-execution flow. The peripheral "steals" memory or CPU cycles to transfer data. 3. A mechanism that allows an input/output device to take control of the CPU for one or more memory cycles in order to write into memory or read from memory. The order of executing the program steps (instructions) remains unchanged. 4. Direct access to a block of memory by more than one system. 5. The technique generally used to transfer blocks of data between a peripheral and random-access memory. It is called direct because the host does not handle the data during the transfer operation. 6. A method of transferring blocks of data directly between an external device and system memory without the need for CPU intervention. This method significantly increases the data transfer rate and hence system efficiency. See also cycle stealing.

direct metal mask — A metal mask made by etching a pattern into a sheet of metal.

direct numerical control — A system connecting a set of numerically controlled machines to a common memory for part program or machine program storage with provision for on-demand distribution of data to the machines. Direct numerical control systems have additional provisions for collection, display, or editing of part programs, operator instructions, or data related to the numerical control process.

director — 1. A parasitic antenna element located in the general direction of the major lobe of radiation for the purpose of increasing the radiation in that direction. 2. Equipment in common-carrier telegraph message switching systems, used to make cross-office selection and connection from an input-line to an output-line equipment in accordance with addresses in the message. 3. A telephone switch that translates the digits dialed into the directing digits actually used to switch the call. 4. Electromechanical equipment that is used to track a moving target in azimuth and angular height and that, with the addition of other necessary information from an outside source such as a radar set or a range finder, continuously computes firing data and transmits them to the guns. 5. In a machine-tool or process control system, the part of the system that receives the command signals from a controller and converts and amplifies these signals to make them usable by the control devices in the machine or process. 6. In a liquid crystal system, a local symmetry axis around which the long-range order of the crystal is aligned. For the nematic phase, the molecular long axis is — on average — parallel to the director.

directory — Also called a catalog. 1. A table of contents designed to allow convenient access to specific files. 2. A file containing information concerning the other files on a mass-storage device. 3. In the logical format of a disk or disc, a branch of the information tree containing other directories (subdirectories) and/or files. 4. An index of the files on a disk. A directory can contain individual files in addition to other directories.

direct outward dialing — Abbreviated DOD. 1. Dialing of a call into the city system from a PAX/PABX extension without the help of an operator. Usually accomplished by first dialing the digit 9. 2. The dialing of a call from a local system into a toll network without the help of an operator.

direct pickup — Transmission of television images without resorting to an intermediate magnetic or photographic recording.

direct piezoelectricity — A name sometimes given to the piezoelectric effect in which an electric charge is developed on a crystal by the application of mechanical stress.

direct point repeater — A telegraph repeater in which the relay controlled by the signals received over a line sends corresponding signals directly into another line or lines without the use of any other repeating or transmitting apparatus.

direct radiative transition — A transition that involves photons alone. See transition, 1.

direct radiator — A speaker that is not horn loaded. The term usually refers to a cone-type speaker, as opposed to a compression driver/horn assembly.

direct-radiator speaker — A speaker in which the radiating element acts directly on the air instead of relying on any other element such as a horn.

direct recording — 1. The production of a visible record without subsequent processing, in response to received signals. 2. Analog recording in which continuous amplitude variations are recorded linearly through the use of ac bias.

direct-recording magnetic tape — A method of recording using a high frequency bias in which the electrical input signal is applied to the recording head without alteration.

direct/reflected speaker — A form of speaker in which a small part of the total output is radiated directly forward, with the major part reflected from the wall behind the speaker.

direct resistance-coupled amplifier — An amplifier in which the plate of one stage is connected either directly or through a resistor to the control grid of the next stage, with the plate-load resistor being common to both stages. Used to amplify small changes in direct current.

direct route— In wire communications, the trunks that connect two switching centers, regardless of the geographical path the actual trunk facilities may follow.

direct scanning— A scanning technique in which the object is illuminated the entire time, and in which picture elements of the object are viewed singly by the television camera.

direct sound wave— A wave emitted from a source in an enclosure prior to the time the wave has undergone its first reflection from a boundary of the enclosure. Frequently a sound wave is said to be direct if it contains reflections that have occurred from surfaces within about 0.05 second after the sound was first emitted.

direct synthesizer— 1. A frequency synthesizer producing an output frequency that is related to the reference frequency by the ratio of two integers. The primary advantage of this type of synthesizer lies in its ability to change output frequencies at a moderately fast rate (typically in the microsecond range) and in a random way. Principal applications include frequency-agile radars, secure communication links, and electronic countermeasures. 2. Derives an output from one or more fixed-frequency reference oscillators, using combinations of frequency division, multiplication, mixing, summing, and filtering.

direct-to-disc recording— A technique in which a live performance is recorded directly onto the master lacquer disc for manufacturing of phonograph records.

direct voltage— Also called dc voltage. A voltage that forces electrons to move through a circuit in the same direction and thereby produce a direct current.

direct wave— 1. A wave that is propagated directly through space, as opposed to one that is reflected from the sky or ground. 2. A radio wave that travels from the transmitting antenna to the point of reception without reflection or refraction.

direct Wiedemann effect— See Wiedemann effect.

direct-wire circuit— A supervised protective signaling circuit usually consisting of one metallic conductor and a ground return and having signal-receiving equipment responsive to either an increase or a decrease in current.

direct writing galvanometer recorder— Recorder using a pen attached directly to a galvanometer movement for direct writing of signals of frequencies up to about 300 Hz.

disable— To prevent the passage of binary signals by application of the proper signal to the disable terminal of a device.

disassemblers— Programs that do the opposite of compiler programs. Given a machine-code program listing, the disassembler turns it back into an assembly listing, with mnemonic representations, for troubleshooting purposes.

disassembly— 1. Retranslation of machine language into mnemonics during debugging. 2. Translation of binary machine code into assembly-language statements.

disc— 1. A phonograph record. 2. The blank used in a recorder. See also disk.

disc capacitor— A small disc-shaped capacitor with a ceramic dielectric, generally used for bypassing or for temperature compensation in tuned circuits.

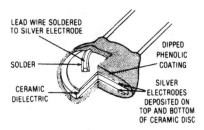

Disc (ceramic) capacitor.

disc files— A type of storage medium consisting of numbers of discs that rotate; each disc has a special coating for retaining stored information.

disc generator— A capacitive-charge type of voltage generator.

discharge— 1. In a storage battery, the conversion of chemical energy into electrical energy. 2. The release of energy stored in a capacitor when a circuit is connected between its terminals. 3. The conversion of dielectric stress of a capacitor into an electric current.

discharge breakdown— Breakdown of a material as a result of degradation due to gas discharges.

discharge key— A device for switching a capacitor suddenly from a charging circuit to a load through which it can discharge.

discharge lamp— A lamp containing a low-pressure gas or vapor that ionizes and emits light when an electric discharge is passed through it. Fluorescent materials are sometimes used on the inside of the glass envelope to

Direct waves.

increase the illumination, as in an ordinary fluorescent lamp.

discharge rate—The amount of current a battery will deliver over a given period of time. A slower discharge rate generally results in more efficient use of a battery.

discharge tube—A tube containing a low-pressure gas that passes a current whenever sufficient voltage is applied.

discharge voltage—Also called clamping voltage. The maximum peak voltage measured across suppressor device terminals when subjected to peak pulse current.

discone antenna—A special form of biconical antenna in which the vertex angle of one cone is 180°.

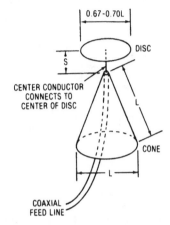

Discone antenna.

disconnect—Also called release. 1. To break an electric circuit. 2. To remove the power from an electrical device (colloquially, "to unplug the device"). 3. To disengage the apparatus used in a connection and to restore it to its ready condition when not in use. 4. A device or group of devices that removes electrical continuity from between the conductors of a circuit and the source of supply.

disconnecting means—A device whereby the current-carrying conductors of a circuit can be disconnected from their source of supply.

disconnector release—A device that disengages the apparatus used in a telephone connection to restore it to the condition in which it exists when not in use.

disconnect signal—1. A signal sent from one end of a trunk or subscriber line to indicate at the other end that the established connection should be released. 2. The on-hook signal in a telephone system by which the calling and called terminals notify the switching equipment that an established connection is no longer needed and should be released.

disconnect switch (motor circuit switch)—A switch intended for use in a motor branch circuit. It is rated in horse power and is capable of interrupting the maximum operating overload current of a motor of the same rating at the rated voltage.

discontinuity—1. A broken connection, or the loss of a specific connection characteristic. 2. The temporary interruption or variation in current or voltage. 3. A point of abrupt change in the impedance of a circuit, where wave reflections can occur.

discontinuous amplifier—An amplifier that reproduces an input waveform on some type of averaging basis.

disc pack—A set of magnetic discs that can be removed from a disc storage as one unit.

disc recorder—A recording device in which the sounds are mechanically impressed onto a disc; as opposed to a tape recorder, which impresses the sound magnetically on a tape.

discrete—1. An individual circuit component, complete in itself, such as a resistor, diode, capacitor, or transistor, and used as an individual and separable circuit element. 2. Pertaining to distinct elements, such as characters, or to representation by means of distinct elements. 3. A term applied to four channels when there are four electrically independent signals, as opposed to matrix. 4. A quad disc or record-playback method that keeps four signals separate, distinct, and independent from recording to playback. 5. Having an individual identity. Fabricated prior to installation and/or separately packaged, not part of an integrated circuit.

discrete circuit—1. A circuit built from separate components that are individually manufactured, tested, and assembled. 2. Electronic circuit built of separate components (transistors, resistors, etc.) connected by wiring or printed-circuit etched conductors.

discrete component—1. A component that has been fabricated prior to its installation (e.g., resistors, capacitors, diodes, and transistors). 2. A circuit component having an individual identity, such as a transistor, capacitor, or resistor.

discrete device—1. A class of electronic components, such as power MOSFETs, bipolar power transistors, MOVs, optoelectronic devices, rectifiers, power hybrid circuits, intelligent power discretes, and transistors. Typically, these devices contain one active element, such as a transistor or diode. However, hybrids, optoelectronic devices, and intelligent discretes may contain more than one active element. In contrast, integrated circuits typically contain hundreds, thousands, or even millions of active elements in a single die. 2. An individual electrical component, such as a resistor, capacitor, or transistor, as opposed to an integrated circuit, which is equivalent to several discrete components.

discrete element—An electronic element, such as a resistor or transistor, fabricated in such a way that it can be measured and transported individually.

discrete part—A separately packaged single circuit element supplying one fundamental property as a lumped characteristic in a given application. Examples: resistor, transistor, diode.

discrete sampling—The lengthening of individual samples so that the sampling process does not deteriorate the intelligence frequency response of the channel.

discrete thin-film component—An individually packaged electronic component having one or more thin films serving as resistive, conductive, and/or insulating elements. Resistors and potentiometers having thin-film metallic resistance elements are examples.

discretionary wiring—The use of a selective metallization pattern in the interconnection of large numbers of basic circuits on a slice of semiconductor material to form complex arrays. The metallization pattern connects only the "good" circuits on the wafers. Discretionary wiring requires a different interconnection pattern for each wafer.

discrimination—1. The difference between losses at specified frequencies, with the system or transducer terminated in specified impedances. 2. In a frequency-modulated system, the detection or demodulation of the imposed variations in the frequency of the carriers. 3. In a tuned circuit, the degree of rejection of unwanted signals.

discrimination ratio—The ratio of the width of the passband of a filter to the width of the stopband of the filter.

discriminator—1. A device in which amplitude variations are derived in response to frequency or phase variations. 2. A facsimile auxiliary device between the radio receiver and the recorder that converts an audio-frequency-shifted facsimile signal to an amplitude-modulated facsimile signal.

discriminator transformer—A transformer used in FM receivers to convert frequency changes directly to audio-frequency signals.

discriminator tuning unit—A device that tunes the discriminator to a particular subcarrier.

disc-seal tube—Also called lighthouse tube or megatron. An electron tube with disc-shaped electrodes arranged in closely spaced parallel layers to give a low interelectrode capacitance along with a high power output in the UHF region.

dish—1. A microwave antenna, usually shaped like a parabola, that reflects the radio energy leaving or entering the system. 2. A parabolic type of radio or radar antenna, roughly the shape of a soup bowl. 3. A colloquial expression for a parabolic antenna. 4. Common term for a parabolic microwave antenna.

dish illumination—The area of a dish as seen by the feedhorn.

disk—1. An electromagnetic storage medium for digital data. 2. High-capacity random-access magnetic storage medium. *See also* disc.

disk cartridge—The flat, round, removable disk pack, containing programs and data, that is placed into a disk drive.

disk drive—1. Identified floppy, removable, and nonremovable bulk storage for most minicomputer and mainframe systems, and microcomputer systems needing several megabytes of storage. 2. A disk player that rotates the disk, writes data onto it, and reads data from it as instructed by a program.

diskette—*See* floppy disk.

disk operating system—Abbreviated DOS. 1. The software that organizes how a computer reads, writes, and reacts with its disks and talks to its various peripherals (input/output devices), such as keyboards, screens, serial and parallel ports, printers, modems, etc. The most popular operating system for PCs is MS-DOS from Microsoft. 2. An operating system (set of programs) that instructs a disk-based computing system to manage resources and operate related equipment. 3. A set of programs that controls a computer. The DOS performs a variety of tasks, including managing communications between the computer and its peripherals. *See also* operating system.

disk pack—The vertical stacking of a series of magnetic disks in a removable self-contained unit.

disk storage—1. Random-access auxiliary memory device in which information is stored on constantly rotating magnetic disks. 2. The storage of data on the surface of magnetic disks. 3. A mass storage memory device employing a flat, rotating medium onto which data can be stored via magnetic recording techniques and retrieved by magnetic playback. 4. A method of high-speed bulk storage of programs and data. The medium is a rotating circular plate coated with a magnetic material, such as iron oxide. Data is written (stored) and read (retrieved) by fixed or movable read/write heads positioned over data tracks on the surface of the disk. Addressable portions can be selected for read or write operations.

dislocation—In a crystal, a region in which the atoms are not arranged in the perfect crystal-lattice structure.

dispatcher—In a digital computer, the section that transfers the words to their proper destinations.

dispenser—A device that automatically distributes radar chaff from an aircraft.

disperse—In data processing, to distribute grouped input items among a larger number of groups in the output.

dispersion—1. Separation of a wave into its component frequencies. 2. Scattering of a microwave beam as it strikes an obstruction. 3. The property of an optical material that causes some wavelengths of light to be transmitted through the material at different velocities, with the velocity a function of the wavelength. (This causes each wavelength of light to have a different refractive index.) 4. In a magnetostrictive delay line, the variation of delay as a function of frequency. 5. The frequency difference that can be analyzed in one sweep by a spectrum analyzer. Dispersion can be considered as that frequency width over which sampling can be performed, and is always equal to or less than the frequency range. 6. The extent to which a speaker distributes acoustical power widely and evenly into the listening area. 7. The undesirable effect of the broadening of optical pulses caused by lengthening of rise and fall times as the pulse travels along the fiber. Sometimes referred to as pulse spreading, it results from either modal or material effects in the fiber that reduce bandwidth. Expressed in nanoseconds per kilometer. 8. A fiber-optic phenomenon that causes pulse widths of transmitted data to lengthen. Dispersion is caused by the arrival of data at the far terminal at different times due to the varying lengths of optical paths in multimode fiber, and by inherent properties in the fiber. Dispersion increases with length of conductor and is caused by the difference in ray path lengths within the fiber core.

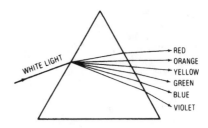

Dispersion, 3 (by a prism).

dispersive medium—A medium in which the phase velocity of a wave is related to the frequency.

displacement—1. The vector quantity representing change of position of a particle. 2. A number that a computer must add to a base address to form an effective address.

displacement current—A current that exists in addition to ordinary conduction current in ac circuits. It is proportional to the rate of change of the electric field. The current at right angles to the direction of propagation determined by the rate at which the field energy changes.

displacement of porches—The difference in level between the front and back porch of a television signal.

displacement transducer—A device that converts mechanical energy into electrical energy, usually by the movement of a rod or an armature. The amount of output voltage is determined by the amount the rod or armature is moved.

display—Also called readout. 1. Visual presentation of a received signal on a cathode-ray tube or video screen.

2. Row of digits across the top of a calculator, showing input or final answer. In printing-type calculators, referred to as printout. 3. The observable illustration of an image, scene, or data on a screen, such as a console or CRT screen, seen as a graph, report, or drawing. 4. The representation of data in visible form, e.g., on a cathode-ray tube, by lights or indicators on the console of a computer, or a printed report.

display console — A visual display used with a computer to give access to the many elements of data as an array of points. With the display console, an operator may check information in the computer and change it if required.

display-generation time — The time span between the output of data from the host computer and the moment at which the complete display can be viewed.

display generator — An electronic device that interfaces computer-graphics display information with a graphics-display device. Typically, the interface is made between a digital computer and a CRT. In general, a display generator for a rasterscan display contains four subsystems: display controller, display processor, refresh memory, and video driver.

display highlighting — The ability of the word processor to intensify or blink certain portions of the display screen — either the characters themselves or the screen area behind the characters — to emphasize a text segment designated for some special activity such as deleting or moving.

display information processor — A computer used in a combat operations center to generate situation displays.

display loss — See visibility factor.

display modes — Each display mode, such as vector, increment, character, point, vector continue, or short vector, specifies the manner in which points are to be displayed on the screen.

display panel — The substrate containing the media for creating an image, including electric connections but excluding the electronic interface.

display primaries — Also called receiver primaries. The red, green, and blue colors produced by a color television receiver and mixed in proper proportions to produce other colors.

display processor — A component of a display generator used to add intelligence. Typically, the device is a microcomputer with stored programs that perform high-level graphics functions.

display-storage tube — A special cathode-ray tube with a long and controllable image presistence and high luminescence.

display unit — A device used to provide a visual representation of data.

display window — The width of the portion of the frequency spectrum presented on panoramic presentation, expressed in frequency units, usually megahertz.

disruptive discharge — The sudden, large current through an insulating medium when electrostatic stress ruptures the medium and thus destroys its insulating ability.

dissector — In optical character recognition, a mechanical or electronic transducer that sequentially detects the level of light in different areas of a completely illuminated sample space.

dissector tube — A camera tube having a continuous photocathode on which a photoelectric emission pattern is formed. Scanning is done by moving the electron optical image of the pattern over an aperture. See also image dissector, 1.

dissipation — The undesired loss of electrical energy by conversion into heat.

dissipation constant — A constant of proportionality between the power dissipated and the resultant temperature rise in a thermistor at a specified temperature.

dissipation factor — 1. Symbolized by D. Ratio between the permittivity and conductivity of a dielectric. The reciprocal of the dissipation factor (df) is the storage factor, sometimes called the quality factor (Q). 2. A measure of the ac loss. Dissipation factor is proportional to the power loss (P_L) per cycle (f) per potential gradient squared (E^2) per unit volume (V) as follows:

$$\text{dissipation factor} = (P_L/kE^2 fV)$$

where k is a constant. Dissipation factor is approximately equal to power factor when the loss angle is small.

dissipation line — A length of stainless-steel or Nichrome wire used as a noninductive impedance for termination of a rhombic transmitting antenna when power of several kilowatts must be dissipated.

dissonance — The formation of maxima and minima by the superposition of two sets of interference fringes from light of two different wavelengths.

dissymmetrical network — See dissymmetrical transducer.

dissymmetrical transducer — Also called dissymmetrical network. A transducer with unequal input and output image impedances.

distance mark — Also called range mark. A mark that indicates, on a cathode-ray screen, the distance from the radar set to a target.

distance-measuring equipment — Abbreviated DME. A radio navigational aid for determining the distance from a transponder beacon by measuring the time of transmission to and from it.

distance protection — The effect of a device operative within a predetermined electrical distance on the protected circuit to cause and maintain an interruption of power in a faulty circuit.

distance relay — 1. A protective relay, the operation of which is a function of the distance between the relay and the point of fault. 2. A device that functions when the circuit admittance, impedance, or reactance increases or decreases beyond predetermined limits.

distance resolution — The ability of a radar to differentiate targets solely by distance measurement. Generally expressed as the minimum distance the targets can be separated and still be distinguishable.

distortion — 1. Undesired changes in the waveform of a signal so that a spurious element is added. All distortion is undesirable. Harmonic distortion disturbs the original relationship between a tone and other tones naturally related to it. Intermodulation distortion (IMD) introduces new tones caused by mixing of two or more original tones. Phase distortion, or nonlinear phase shift, disturbs the natural timing sequence between a tone and its related overtones. Transient distortion disturbs the precise attack and decay of a musical sound. Harmonic and IMD distortion are expressed in percentages; phase distortion in degrees; transient distortion is usually judged from oscilloscope patterns. 2. Unwanted changes in the purity of sound being reproduced or in rf signals. In audio, it generally implies intermodulation and/or harmonic distortion. These are derived from phase differences and/or amplitude distortion in which the amplitude of the output does not bear the same proportion to the input at all frequencies. 3. With a signal frequency (sine wave) signal, distortion appears as harmonics (multiples) of the input frequency. The rms (effective ac point) sum of all harmonic distortion components, plus hum and noise, is known as total harmonic distortion, or THD. When a two-tone test signal is used, distortion components appear at frequencies that are sums and differences of multiples of the input frequencies. Their magnitude is expressed as intermodulation distortion, which is more distressing to hear than THD. The lower the distortion in any form, the better. 4. Any difference in the waveshape after the signal has traversed the transmission circuit. 5. The unwanted changes in signal or signal shape that occur during transmission between two points.

distortion factor — *See* harmonic distortion.

distortion factor of a wave — The ratio of the effective value of the residue after the elimination of the fundamental to the effective value of the original wave.

distortionless line — A transmission line whose propagation constant is independent of frequency. (This is approached in a practical case by adjusting the line parameters, series inductance (l), shunt capacitance (c), series resistance (r), and shunt conductance (g) so that $r/g = 1/c$.)

distortion meter — 1. An instrument that measures the deviation of a complex wave from a pure sine wave. 2. An instrument that measures the harmonic content of a sine wave, usually calibrated to read in percent distortion.

distortion tolerance — Of a telegraph receiver, the maximum signal distortion that can be tolerated without error in reception.

distress frequency — A frequency reserved for distress calls, by international agreement. It is 500 kHz for ships at sea and aircraft over the sea.

distributed — Spread out over an electrically significant length, area, or time.

distributed amplifier — A multistage amplifier in which the high-frequency limitation, due to the input and output capacitances of the active element, is circumvented by making these capacitances the shunt elements of lumped-parameter device lines. In this way the overall gain is the sum of the gains of the individual stages rather than the product, thus allowing amplification even when the individual gains are less than unity.

distributed capacitance — Also called self-capacitance. Any capacitance not concentrated within a capacitor, such as the capacitance between the turns in a coil or choke, or between adjacent conductors of a circuit.

distributed computer network — A collection of computers and I/O devices that can communicate with each other. *See* distributed processing.

distributed constants — Constants such as resistance, inductance, or capacitance that exist along the entire length or area of a circuit, instead of being concentrated within circuit components.

distributed data processing — Abbreviated DDP. The functional distribution of certain data-processing activities along logical organizational lines.

distributed-emission photodiode — A broadband photodiode for use in detecting modulated laser beams at millimeter wavelengths.

distributed inductance — The inductance along the entire length of a conductor, as distinguished from the inductance concentrated within a coil.

distributed network — 1. An electrical-electronic device that for proper operation depends on physical size in comparison to a wavelength and physical configuration. 2. A network configuration in which all node pairs are connected either directly or by redundant paths through intermediate nodes.

distributed parameter network — A network in which the parameters of resistance, capacitance, and inductance cannot be taken as being concentrated at any one point in space. Rather, the network must be described in terms of its magnetic and electric fields and the quantities related to the distributed constants of the network.

distributed paramp — A paramagnetic amplifier consisting essentially of a transmission line shunted by uniformly spaced, identical varactors. The varactors are excited in sequence by the applied pumping wave to give the desired traveling-wave effect.

distributed pole — A motor has distributed poles when its stator or field windings are distributed in a series of slots located within the arc of the pole.

distributed processing — 1. A multiprocessing computer technique in which each processor has a specific task or set of tasks to perform. These processors transfer commands and data via a standard communication interface. 2. Performing a data-processing task by performing the needed calculations in a distributed computer network. The efficiency of the data-processing task is improved through the simultaneous performance of operations in several interconnected processors of a distributed computer network. 3. Data-processing tasks performed simultaneously in several interconnected processors of a computer network.

distributing amplifier — An amplifier, either radio frequency or audio frequency, having one input and two or more isolated outputs.

distributing cable — *See* distribution cable.

distributing frame — A structure for terminating permanent wires of a central office, private branch exchange, or private exchange and for permitting the easy change of connection between them by means of cross-connecting wires.

distributing terminal assembly — A frame situated between each pair of selector bays to provide terminal facilities for the selector bank wiring and facilities for cross connection to trunks running to succeeding switches.

distribution — Also called frequency distribution. The number of occurrences of the particular values of a variable as a function of those values.

distribution amplifier — 1. A power amplifier designed to energize a speech, music, or antenna distribution system. Its output impedance is sufficiently low that changes in the load do not appreciably affect the output voltage. 2. A device that provides several isolated outputs from one looping or bridging input, and has a sufficiently high input impedance and input-to-output isolation to prevent loading of the input source.

distribution cable — Also called distributing cable. 1. A cable extended from a feeder cable for the purpose of providing service to a specific area. 2. In a system, the transmission cable from the distribution amplifier to the drop cable.

distribution center — In an alternating-current power system, the point at which control and rotating equipment is installed.

distribution coefficients — Equal-powered tristimulus values of monochromatic radiations.

distribution switchboard — A power switchboard used for the distribution of electrical energy at the voltage common for each distribution within a building.

distributive sort — A sorting procedure that divides data elements into two or more distinct groups or subsets. The partition sort is an example.

distributor — 1. *See* memory register. 2. The electronic circuitry that acts as an intermediate link between the accumulator and drum storage.

distributorless — A semiconductor automotive ignition system that does not utilize breaker contacts to time or trigger the system, nor does it utilize a distributor distribution of the secondary voltage.

disturbance — 1. An irregular phenomenon that interferes with the interchange of intelligence during transmission of a signal. 2. An interruption of a quiet state. 3. Any form of interference with normal communications.

disturbed-one output — A "one" output of a magnetic core to which partial-read pulses have been applied since that core was last selected for writing.

disturbed-zero output — A "zero" output of a magnetic core to which partial-write pulses have been applied since that core was last selected for reading.

disturbing conductor — A conductor carrying energy that creates spurious signals in another conductor.

dither — 1. An oscillation introduced for the purpose of overcoming the effects of friction, hysteresis, or clogging. 2. A small electrical signal deliberately injected into an electromechanical device for the purpose of overcoming static friction in the device. In a recording instrument it makes the indicator ready to jump. 3. Constant vibration about a point. 4. The technique of adding controlled amounts of noise to a signal to improve overall system loop control, or to smear quantizing error in an analog-to-digital converter application.

dithering — 1. The application of intermittent or periodic acceleration forces sufficient to minimize the effect of static friction with a transducer, without introducing other errors. 2. The creation of additional colors or shades of gray to create special effects or to make hard edges softer.

divergence loss — The part of transmission loss that is caused by the spreading of sound energy.

diverging lens — A lens that is thinner in the center than at the edges. Such a lens causes light passing through to spread out, or diverge.

diversity — 1. A form of transmission and/or reception using several modes, usually in space or in time, to compensate for fading or outages in any one of the modes. In the space diversity system, the same signal is sent simultaneously over several different transmission paths, which are separated enough so that independent propagation conditions can be expected. With time diversity, the same path may be used, but the signal is transmitted more than once, at different times. There are other forms of diversity, using different frequencies or different polarizations to provide the separate transmission pole. *See* diversity reception. 2. The practice of constructing a portion of the system or backup system by using a different technology, component, or design, such that the two portions of the total system are not vulnerable to a common-cause failure.

diversity factor — The ratio of the sum of the individual maximum demands of the subdivisions of a system (or part of a system) to the maximum demand of the entire system (or part of the system).

diversity gain — The gain in reception as a result of the use of two or more receiving antennas.

diversity reception — 1. A method of minimizing the effects of fading during reception of a radio signal. This is done by combining and/or selecting two or more sources of received-signal energy that carry the same intelligence but differ in strength or signal-to-noise ratio in order to produce a usable signal. *See* frequency-diversity reception; space-diversity reception. 2. A technique for reducing the adverse effects of multipath fading by receiving the same signal on two or more diverse, or different, antenna-receiver combinations with a means of choosing the combination with the strongest signal. A two-channel system is known as dual diversity; a three-channel system, triple diversity. Diversity reception is widely and effectively used in commercial high-frequency installations.

diverter pole generator — A compound-wound direct-current generator with the series winding of the diverter pole opposing the flux generated by the shunt-wound main pole; provides a close voltage regulation.

divide-by-*N* counter — A group of counter stages that can be programmed to divide an input frequency by any number up to *N*.

divide-by-16 counter — A logic device in which four flip-flops count from 0 through 15 and then recycle to 0. All 16 states of the combination of four flip-flops are used. Sometimes referred to as a hexadecimal counter.

divide-by-10 counter — *See* decade counter.

divide-carrier modulation — The process by which two signals are added so that they can modulate two carriers of the same frequency but 90° out of phase. The resultant signal will have the same frequency as the carriers, but its amplitude and phase will vary in step with the variations in amplitude of the two modulating signals.

divide check — In a computer, an indicator that shows that an invalid division has occurred or has been attempted.

divider — *See* counter, 4.

dividing network — Also called speaker dividing network and crossover network. A frequency-selective network that divides the audio-frequency spectrum into two or more parts to be fed to separate devices such as amplifiers or speakers.

D layer — The lowest ionospheric layer, located between about 35 to 55 miles (56 to 88 km) above the earth. Its intensity is proportional to the height of the sun and is greatest at noon. Waves below approximately 3 MHz are absorbed by the D layer when it is present.

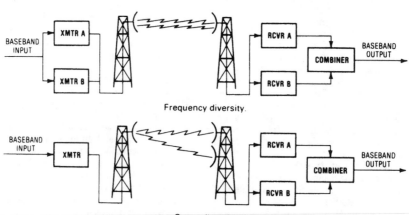

Frequency diversity.

Space diversity.

Diversity reception.

High-angle radiation may penetrate the D layer and be reflected by the E layer.

DMA—Abbreviation for direct memory access. A method in which a device in a computer other than the main processor can access main memory directly. It must first obtain control of the bus from the processor or other DMA devices. Then it can read and write to memory directly. This is a very fast method of transferring large amounts of data from a peripheral to main memory. Using the main processor to handle the data transfer requires more software overhead and reduces system throughput.

DMI—Abbreviation for dual-mode ignition. An adaptation of the Duraspark (Ford Motor) electronic ignition system that allows its use on smaller four-cylinder engines like the 2.3-liter size.

DMM—Abbreviation for digital multimeter.

DMOS—Abbreviation for double-diffused metal-oxide FET semiconductor. A process in which n and p atoms are diffused through the same mask opening to give precisely sized narrow channels. Used on discrete field-effect transistors (not MOS ICs) for ultrahigh gains and frequency performance. A very fast MOSFET fabricated with an extra diffusion step.

DNC—Abbreviation for direct numerical control.

DNL—*See* dynamic noise limiter.

DOC files—Document files. The default file extension for Microsoft Word. Most files with this extension are Microsoft Word files, but there are many that are plain text.

documentation—An orderly collection of recorded hardware and software data such as tables, listings, diagrams, etc.

document reader—A general term referring to OCR or OMR equipment that reads a limited amount of information (one to five lines). Generally operates from a predetermined format and is therefore more restricted in the location of information to be read. The forms involved are generally tab card size or slightly smaller or larger.

doghouse—A small enclosure located near the base of a transmitting-antenna tower and used to house antenna tuning equipment.

Doherty amplifier—A radio-frequency linear power amplifier divided into two sections, the inputs and outputs of which are connected by quarter-wave (90°) networks. As long as the input-signal voltage is less than half the maximum amplitude, section No. 2 is inoperative and section No. 1 delivers all the power to the load. The load presents twice the optimum impedance required for maximum output. At one-half the maximum input, section No. 1 is operating at peak efficiency but is beginning to saturate. Above this level, section No. 2 comes into operation and decreases the impedance presented to section No. 1. As a result, section No. 2 delivers more and more power to the load until, at maximum signal input, both sections are operating at peak efficiency and each section is delivering one-half the total output power.

Dolby—1. A technique that increases the signal-to-noise ratio of a recording medium by raising the volume of quiet passages prior to recording, and lowering them to their original levels during playback. The lowering process automatically reduces any noise that was introduced as a result of the recording or playback processes. 2. Noise-reduction circuit that boosts the recorded signal at the tape hiss frequencies for low levels and reduces the boost progressively as the signal becomes large enough to mask the noise. (The Dolby system has the important advantage that it is standardized and any Dolby tape can be replayed accurately on any other Dolby machine.) 3. Name of a noise-reduction system available as a special circuit on some stereo cassette tape decks. 4. A proprietary electronic device or circuit that reduces the amount

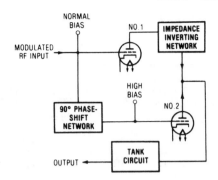

Doherty amplifier.

of noise (principally tape hiss) introduced during the recording process by boosting—in carefully controlled amounts—the strength of weak high-frequency signals before they are recorded. During playback the signals (and the noise) are cut back by an exactly equivalent amount. The original dynamics are restored, but the noise is reduced by 10 dB.

Dolby A—The original form of the Dolby noise-reduction device, intended for professional use. It has four independently controlled noise-reduction channels, to increase signal-to-noise ratio at low, middle, high and very high frequencies.

Dolby B—1. A simplified version of the original Dolby A, intended primarily for use by nonprofessional recordists. Dolby B functions identically to Dolby A, but has only one controlled frequency band, which is effective primarily on tape hiss. 2. A noise-reduction system widely used in cassette recorders, as well as in some open-reel and cartridge machines and in FM broadcasting. The high-frequency portions of signals being recorded are compressed, with the degree of compression being greater as signal level decreases. An opposite expansion process takes place in playback, restoring the original frequency response, but with a reduction in high-frequency hiss. 3. A complementary noise-reduction system designed to reduce tape (and FM) hiss. A Dolby B circuit boosts low-level high-frequency signals during recording and reduces them, along with the tape's added noise, in a complementary fashion during playback. Noise can be reduced up to 10 db above 5 kHz with the Dolby B system. It is now in virtually universal use in cassette decks.

Dolby B & C noise reduction (cassette deck)—Two systems of noise/hiss reduction invented by Ray Dolby. They work by boosting high frequencies during recording (also called encoding) and attenuating them during playback (also called decoding). Dolby B noise reduction boosts the level of the high-frequency range during recording and tapers the high frequencies during playback, reducing tape hiss by 8 to 10 dB. Dolby C noise reduction uses the same principle as Dolby B, with the addition of a second stage in which the frequencies affected are lower by about one octave. This results in a 15 to 18 dB reduction in tape hiss over an extended frequency range.

Dolby digital—The multi-channel audio encoding formed for DVD.

dolly—A wheeled platform or frame on which the tripod or frame supporting a television camera or other apparatus is mounted to give it wider mobility.

domain—1. In magnetic theory, that region of a magnetic material in which the spontaneous magnetization is all in one direction. In conventional magnetic-tape coatings, this corresponds to one oxide particle. 2. A region within a ferromagnetic substance where the atomic magnets of many atoms tend to orient themselves parallel to each other; the north poles pointing one way act spontaneously. The domains may be treated as small bar magnets of microscopic dimension. 3. In the Internet and other networks, an extension in a host name that identifies the type of host. The seven domains established by the Inter-NIC are .arpa (ARPANET), .com (company/commercial), .edu (educational institutions), .gov (government), .mil (military), .net (Internet access providers), and .org (organization). Outside the United States, the domain name is a two-letter country code (for example, .fr for France and .ca for Canada).

domain name — The unique name that identifies an Internet site. Domain names always have two or more parts, separated by dots. The part on the left is the most specific, and the part on the right is the most general. A given machine may have more than one domain name, but a given domain name points to only one machine. Usually, all of the machines on a given network will have the same thing as the right-hand portion of their domain names.

domains — *See* particles.

domestic induction heater — A home cooking utensil that is heated by induced currents within it. The unit contains a primary inductor, with the utensil itself acting as the secondary.

dominant mode — Also called fundamental mode or principal mode. In waveguide transmission, the mode with the lowest cutoff frequency. Designations for this mode are $TE_{0,1}$ and $TE_{1,1}$ for rectangular and circular waveguides, respectively.

dominant wave — The guided wave that has the lowest cutoff frequency. It is the only wave that will carry energy when the excitation frequency is between the lowest and the next higher cutoff.

dominant wavelength — 1. Of a color sample, the wavelength of light that matches it in chromaticity when mixed with white light. 2. The wavelength that is a quantitative measure of the apparent color of light as perceived by the human eye.

dominant wavelength of a color — The predominant wavelength of light in a color.

donor — Also called donor impurity. 1. An impurity atom that tends to give up an electron and thereby affects the electrical conductivity of a crystal. Used to produce n-type semiconductors. 2. A chemical that adds electrons to crystal lattices. 3. An impurity from column V of the periodic table, which adds a mobile electron to the conduction band of silicon, thereby making it more n-type. Commonly used donors are arsenic and phosphorous (compare with *acceptor*).

donor impurity — An element or compound whose atoms or molecules have more valence electrons than those of the intrinsic semiconductor material into which they are introduced in small quantities as an impurity or dopant. Because the donor impurity possesses more valence electrons, the material doped with a donor impurity is an n-type semiconductor. *See* donor.

donor-type semiconductor — An n-type semiconductor.

donut — *See* land, 2.

door cord — A short, insulated cable with an attaching block and terminals at each end used to conduct current to a device, such as foil, mounted on the movable portion of a door or window.

doorknob tube — A vacuum tube so called because of its shape designed for UHF transmitter circuits. It has a low electron-transit time and low interelectrode

capacitance because of the close spacing and small size, respectively, of its electrodes.

door trip switch — A mechanical switch mounted so that movement of a door will operate the switch.

dopant — 1. An impurity added to a semiconductor to improve its electrical conductivity; any material added to a substance to produce desired properties in the substance. 2. Selected impurity introduced into semiconductor substrates in controlled amounts, the atoms of which form negative (n-type) and positive (p-type) conductive regions. Phosphorus, arsenic, and antimony are n-type dopants for silicon; boron, aluminum, gallium, and indium are p-type dopants for silicon.

dope — To add impurities (called dopants) to a substance, usually a solid, in a controlled manner to cause the substance to have certain desired properties. For example, the number of electrical carriers in silicon can be increased by doping it with small amounts of other semimetallic elements. Ruby is aluminum oxide doped with chromium oxide.

doped junction — A semiconductor junction produced by the addition of an impurity to the melt during crystal growth.

doped region — A layer of an integrated circuit in which impurities have been introduced.

doped solder — Solder to which an element not normally found in solder has been intentionally added.

doping — The addition of controlled amounts of impurities to a semiconductor to achieve a desired characteristic, such as to produce an n-type or p-type material, accomplished through thermal diffusion or ion implantation. Common doping agents for germanium and silicon include aluminum, antimony, arsenic, gallium, and indium.

doping agent — An impurity element added to semiconductor materials used in crystal diodes and transistors. Common doping agents for germanium and silicon include aluminum, antimony, arsenic, gallium, and indium.

doping compensation — The addition of donor impurities to a p-type semiconductor or of acceptor impurities to an n-type semiconductor.

Doppler cabinet — A speaker cabinet in which either the speaker or a baffle board is rotated or moved to change the length of the sound path cyclically and thereby produce a vibrato effect mechanically.

Doppler effect — 1. The observed change of frequency of a wave caused by a time rate of change of the effective distance traveled by the wave between the source and the point of observation. As the distance between a source of constant vibration and an observer diminishes or increases, the received frequencies are greater or less. 2. The apparent change in the frequency of radio wave reaching an observer, due either to motion of the source toward or away from the observer, to motion of the observer, or both. 3. The apparent change in frequency of sound or radio waves when reflected from or originating from a moving object. Utilized in some types of motion sensors. 4. The radiation emitted from a source that moves away from an observer appears to be of lower frequency than the radiation emitted from a stationary source. The radiation emitted from a source moving toward the observer appears to be of a higher frequency than that from a stationary source.

Doppler principle — The theory established by Doppler in 1842 that states that the rate of change in distance between a perceiver and a radiation source determines the change in frequencies.

Doppler radar — A radar unit that measures the velocity of a moving object by the shift in carrier frequency of the returned signal. The shift is proportionate to the velocity of the object as it approaches or recedes.

Doppler ranging — Abbreviated as doran. A cw trajectory-measuring system that utilizes the Doppler shift to measure the distance between a transmitter, missile, transponder, and several receiving stations. From these measurements, trajectory data is computed. In contrast to a similar system, doran circumvents the necessity of continuously recording the Doppler signal by performing the distance measurements with four different frequencies simultaneously.

Doppler shift — 1. The change in frequency of a wave reaching an observer or a system, caused by a change in distance or range between the source and the observer or the system during the interval of reception. It is due to the Doppler effect. 2. The change in frequency with which energy reaches a receiver when the source of radiation or a reflector of the radiation and the receiver are in motion relative to each other. The Doppler shift is used in many tracking systems. 3. The magnitude, expressed in cycles per second, of the alteration in the wave frequency observed as a result of the Doppler effect.

Doppler signal — The signal, traveling from transmitter to receiver, that has an altered frequency due to the Doppler effect.

Doppler velocity and position — 1. Having to do with a beacon tracking system in which pulses are sent from a tracking station to a receiver in the object to be tracked, and returned to the station on a different frequency. 2. Having to do with a Doppler trajectory-measuring system for determining target position relative to transmitting and receiving stations on the ground.

doran — *See* Doppler ranging.

DOS — Abbreviation for disk operating system.

dosage meter — *See* dosimeter; intensitometer.

dose — A measure of energy actually absorbed in tissue as a result of ionizing radiation.

dosimeter — Also called intensitometer or dosage meter. 1. An instrument that measures the amount of exposure to nuclear or X-ray radiation utilizing the ability of such radiation to produce ionization of a gas. 2. Quartz fiber electrometer that is charged by a battery and discharges when exposed to radiation. Can be direct reading or indirect reading. Measures the total radiation dose received, in rem, and is carried by a person who works with radiation.

DOS prompt — The letter informing DOS system users what drive they're in, followed by the greater-than symbol (C>), which indicates that the system is ready to receive a command.

dot — 1. *See* button, 2. 2. Also called bubble. A symbol placed at the input of a logic symbol to indicate that the active signal input is negative. The absence of a dot indicates a positive active signal.

dot AND — *See* wired AND.

dot-bar generator — An instrument that generates a specified output pattern of dots and bars. Used for measuring scan linearity and geometric distortion of TV cameras, video monitors, and TV receivers. Also used for converging cathode-ray tubes as recommended by color monitor and receiver manufacturers.

dot cycle — One cycle of a periodic alternation between two signaling conditions, each condition having unit duration. Thus, two-condition signaling consists of a dot, or marking element, followed by a spacing element. In teletypewriter applications, one dot cycle consists of a mark and a space. The speed of telegraph transmission sometimes is stated in terms of dot cycles per second, or dot speed (half the speed of transmission expressed in bauds).

dot encapsulation — A packaging process in which cylindrical components are inserted into a perforated wafer to form a solid block with interconnecting conductors on both surfaces joining the components.

dot generator — An instrument used in servicing color television receivers. It produces a pattern of white dots so that convergence adjustments can be made on the picture tube.

dot matrix — 1. A pattern of dots in a fixed area used for formulation of characters. A method of display character generation in which each character is formed by a grid or matrix pattern of 5×7 to 7×9 dots, combinations of which form characters on a video or hard-copy medium. For very high quality, 11×13 patterns or greater are required. 2. The printing of characters by a matrix pattern of ink dots. 3. A technique for representing characters by composing them out of selected dots from within a rectangular matrix of dots.

dot-matrix character — 1. A printed character formed of dots so close together that it gives the impression of having been printed by uninterrupted strokes. The dots are formed by wire ends, jets of ink, electrical charge, or laser beams. 2. A character composed from a rectangular matrix of dots.

dot-matrix display — 1. A display format consisting of small light-emitting elements arranged as a matrix. Various elements are energized to depict a character. A typical matrix is 5×7. 2. A display composed of dots in close, rectangular array, capable of being individually illuminated to produce alphanumeric characters and graphic displays.

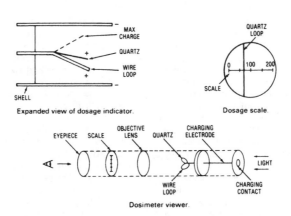

Expanded view of dosage indicator.

Dosage scale.

Dosimeter viewer.

Dosimeter.

Dot-matrix characters and numerals (5 × 7 matrix).

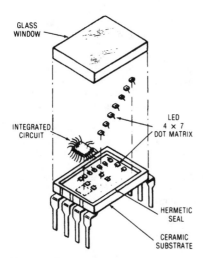

GLASS WINDOW

INTEGRATED CIRCUIT

LED 4 × 7 DOT MATRIX

HERMETIC SEAL

CERAMIC SUBSTRATE

Dot-matrix LED display with on-board IC.

dot-matrix printer — A printer that forms letters by striking the paper with small pins or other means of making a mark (such as a small jet of ink), forming each letter with a pattern of dots. More marks per letter result in a denser image. Some dot-matrix printers squeeze the dots so closely together that they look almost like letter-quality print, which is made by machines that form letters with a single impact, as do traditional typewriters. When dot-matrix printers produce letters that are close to the quality of traditional electric typewriters, the result is often called correspondence-quality or letter-quality print. 2. *See also* wireprinter.

dot pattern — Small dots of light produced on the screen of a color picture tube by the signal from a dot generator. If overall beam convergence has been obtained, the three color-dot patterns will merge into one white-dot pattern.

dot pitch — 1. A measure of picture quality or resolution in RGB color monitors. 2. The shortest distance between two phosphor dots of the same color on the screen. The smaller the dot pitch, the better the resolution of the monitor. High-resolution monitors usually have a dot pitch of 0.28 mm. 3. For printers, the number of dots per linear inch; for example, a desktop laser printer prints at least 300 dpi. The larger the number, the higher the resolution. 4. A measure of picture quality or resolution in RGB color monitors, which is more commonly known as pixel resolution. Dot pitch is the distance between screen dots (pixels) measured in millimeters. The shorter the distance, the better the resolution. It is specified in pixels/mm.

dot sequential — Pertaining to the association of the primary colors in sequence with successive picture elements of a color television system. Examples: dot-sequential pickup, dot-sequential display, dot-sequential system, dot-sequential transmission.

double-amplitude-modulation multiplier — A multiplier in which a carrier is amplitude modulated by one variable, and the modulated signal is again amplitude modulated by a second variable. The product of the two variables is obtained by applying the resulting double-modulated signal to a balanced demodulator.

double armature — An armature having two windings and commutators but only one core.

double-base diode — *See* unijunction transistor.

double-base junction transistor — Also called tetrode junction transistor. Essentially a junction triode transistor with two base connections on opposite sides of the central region of the transistor.

double-beam cathode-ray tube — A cathode-ray tube having two electron beams capable of producing on the screen two independent traces that may overlap. The beams may be produced by splitting the beam of one gun or by using two guns.

double-bounce calibration — A method of calibration used to determine the zero set error by using round-trip echoes. The correct range is the difference between the first and second echoes.

double-break contacts — A set of contacts in which one contact is normally closed and makes simultaneous connection with two other contacts.

double-break switch — A switch that opens the connected circuit at two points.

double bridge — *See* Kelvin bridge.

double-button carbon microphone — Also called differential microphone. A microphone with two carbon resistance elements or buttons, one on each side of a central diaphragm. They are connected in parallel to the current source in order to give twice the resistance change obtainable with a single button.

double-channel duplex — A method for simultaneous communication between two stations over two rf channels, one in each direction.

double-channel simplex — A method for nonsimultaneous communication between two stations over two rf channels, one in each direction.

double-checkerboard pattern — *See* worst-case noise pattern.

double-clocking — Incorrect setting of a flip-flop due to bounce in input signal.

double-conversion receiver — A receiver using a superheterodyne circuit in which the incoming signal frequency is converted twice, first to a high IF and then to a lower one.

double-current generator — A machine that supplies both direct and alternating current from the same armature winding.

double density — A type of diskette that allows twice as much data to be stored as single density.

double-diffused epitaxial mesa transistor — *See* epitaxial-growth mesa transistor.

double-diffused metal-oxide FET semiconductor — *See* DMOS.

double-diffused transistor — Also called double-emitter-and-base transistor. A transistor in which two pn junctions are formed in the semiconductor wafer by

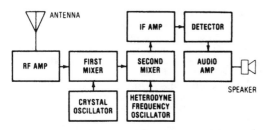

Double-conversion superheterodyne receiver block diagram.

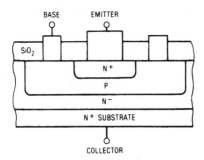

Double-diffused epitaxial planar bipolar npn transistor.

gaseous diffusion of both p-type and n-type impurities. An intrinsic region can also be formed.

double diode — Also called duodiode. A vacuum tube or semiconductor having two diodes in the same envelope.

double-diode limiter — A type of limiter used to remove all the positive signals from a combination of positive and negative pulses or to remove all the negative signals from such a combination of positive and negative pulses.

double-doped transistor — A transistor formed by growing a crystal and successively adding p- and n-type impurities to the melt while the crystal is being grown.

double-doublet antenna — An antenna composed of two half-wave doublet antennas criss-crossed at their centers; one is made shorter than the other to give broader frequency coverage.

double drop — An alarm-signaling method often used in central-station alarm systems, in which the line is first opened to produce a break alarm and then shorted to produce a cross alarm.

double-emitter follower — *See* Darlington amplifier.

double frequency-shift keying — A multiplex system in which two telegraph signals are combined and transmitted simultaneously by frequency shifting among four radio frequencies.

double-grip terminal — A solderless terminal with an extended flared barrel that permits a crimp to be made over the insulation of a wire as well as over the stripped portion.

double image — A television picture consisting of two overlapping images due to reception of the signal over two paths that differ in length so signals arrive at slightly different times.

double insulation — The insulation system resulting from a combination of functional and supplementary insulation.

double-junction photosensitive semiconductor — A semiconductor in which the current flow is controlled by light energy. It consists of three layers of a semiconductor material, with electrodes connected to the ends of each.

double-length number — Also called double-precision number. An electronic computer number having twice the normal number of digits.

double local oscillator — An oscillator mixing system that generates two rf signals accurately spaced a few hundred hertz apart and mixes these signals to give the difference frequency that is used as the reference. This equipment is used in an interferometer system to obtain a detectable signal containing the phase information of an antenna pair, and the reference signal to allow removal of the phase data for use.

double-make contacts — A set of contacts in which one contact is normally open and makes simultaneous connection with two other independent contacts when closed.

double moding — Changing from one frequency to another abruptly and at irregular intervals.

double modulation — 1. The process of modulation in which a carrier wave of one frequency is first modulated by a signal wave, and the resultant wave is then made to modulate a second carrier wave of another frequency. 2. A two-step modulation scheme in which an intelligence wave modulates a subcarrier, and then the modulated subcarrier is used to modulate a higher-frequency carrier.

double operand — An instruction type containing two address fields: source operand address field and destination operand address field.

double orthomode coupler — A dish-mounted device that allows reception of both vertically and horizontally polarized signals.

double photoresist — A technique for eliminating pinholes in the photoresist coating during fabrication of microelectronic integrated circuits. The method may consist of two separate applications and exposures of photoresist emulsions of the same or different types.

double-play tape — Tape having half the thickness, and hence double the running time (for a given reel size) of standard 1.5-mil tape.

double pole — A term applied to a contact arrangement to denote that it includes two separate contact forms (i.e., two single-pole contact assemblies).

double-pole, double-throw switch — Abbreviated dpdt. A switch that has six terminals and is used to connect one pair of terminals to either of the other two pairs.

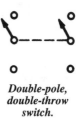

Double-pole, double-throw switch.

double-pole, single-throw switch — Abbreviated dpst. A switch that has four terminals and is used to connect or disconnect two pairs of terminals simultaneously.

double-pole switch — A switch that operates simultaneously in two separate electric circuits or in both lines of a single circuit.

*Double-pole,
single-throw
switch.*

double precision — 1. Having to do with the use of two computer words to represent one number. 2. The technique of allocating twice the data storage space for representing numeric information in order to achieve twice the accuracy.

double-precision arithmetic — The use of two computer words to represent a single number. This is done where it is necessary to obtain a greater accuracy than a single word of computer storage will provide. It effectively doubles the data word size.

double-precision number — *See* double-length number.

double-pulsing station — A loran station that receives two pairs of pulses and emits pulses at two pulse rates.

double pumping — A technique of pumping a laser for a relatively long time to store energy in subthreshold-level excited states, followed immediately by a very brief second pumping in which the threshold condition is exceeded in some region. This triggers laser oscillation throughout the entire active region and produces peak output powers several times larger than normally seen.

double rail — *See* dual rail.

double-rail logic — Pertaining to self-timing asynchronous circuits in which each logic variable is represented by two electrical lines that in combination can assume three meaningful states: zero, one, and undecided.

double screen — A three-layer screen consisting of a two-layer screen with the additional second long-persistence coating having a different color and different persistence from the first.

double-shield enclosure — A shielded enclosure or room whose inner wall is partially isolated electrically from the outer wall.

double sideband — Amplitude-modulated intelligence that is transmitted at frequencies both above and below the carrier frequency by the audio-frequency value of the intelligence.

double-sideband transmitter — A transmitter that transmits not only the carrier frequency, but also the two sidebands resulting from modulation of the carrier.

double-sided board — A printed board with a conductive pattern on both sides.

double-spot tuning — Superheterodyne reception of a given station at two different local-oscillator frequencies. The local oscillator is adjusted either above or below the incoming signal frequency by the intermediate-frequency value.

double-stream amplifier — A microwave traveling-wave amplifier in which amplification occurs through interaction of two electron beams having different average velocities.

double-stub tuner — An impedance-matching device consisting of two stubs, usually fixed three-eighths of a wavelength apart, in parallel with the main transmission lines.

double superheterodyne reception — Also called triple detection. The method of reception in which two frequency converters are employed before final detection.

double-surface transistor — A point-contact transistor, the emitter and collector whiskers of which are in contact with opposite sides of the base.

doublet — The output voltage waveform of a delay line under linear operating conditions when the input to the line is a current step function.

doublet antenna — An antenna consisting of two elevated conductors substantially in the same straight line and of substantially equal length, with the power delivered at the center.

double tape mark — A delimiter, consisting of two consecutive tape marks, that is used to indicate the end of a volume or of a file set.

double throw — A term applied to a contact arrangement to denote that each contact form included is a break-make.

double-throw circuit breaker — A circuit breaker by means of which a change in the circuit connections can be obtained by closing either of two sets of contacts.

double-throw switch — A switch that alternately completes a circuit at either of its two extreme positions. It is both normally open and normally closed.

double-track recorder — *See* dual-track recorder.

double trigger — A trigger signal consisting of two pulses spaced by a fixed amount for coding.

double triode — *See* duotriode.

double-tuned amplifier — An amplifier in which each stage utilizes coupled circuits having two frequencies of resonance for the purpose of obtaining wider bands than are possible with single tuning.

double-tuned circuit — A circuit in which two circuit elements are available for tuning.

double-tuned detector — A type of FM discriminator in which the limiter output transformer has two secondary windings, one tuned a certain amount above the center frequency and the other tuned an equal amount below the center frequency.

double-V antenna — Also called fan antenna. A modified single dipole that has a higher input impedance and broader bandwidth than an ordinary dipole.

Double-V antenna.

double-winding synchronous generator — A synchronous generator that has two similar windings in phase with one another, mounted on the same magnetic structure but not connected electrically, designed to supply power to two independent external circuits.

doubling — The generation of large amounts of second-harmonic distortion by nonlinear motion of a loudspeaker cone.

doubly balanced modulator — A modulator circuit in which two class A amplifiers are supplied with modulating and carrier signals of equal amplitudes and opposite polarities. Carrier suppression takes place because the two amplifiers share a common plate circuit and only the sidebands appear at the output.

downconversion — The process of converting microwave signals down into a frequency range in which signal processing components are less expensive. Typically, this is a VHF frequency of 70 MHz.

downconverter — 1. A type of converter whose input is a radio frequency and whose output is an intermediate frequency. 2. A microwave system (consisting of local oscillators, mixers, and bandpass filters) that accomplishes downconversion. 3. The front end of the satellite TV receiver. 3. The part of a satellite receiving system that converts the downlink signals to a 70-MHz intermediate frequency that is used by the receiver. Although it is sometimes part of the receiver, it is more often externally mounted directly at the LNA so that inexpensive coaxial cable can bring the signal to the receiver.

down lead — Also called a lead-in wire. The wire that connects an antenna to a transmitter, receiver, or downconverter.

downlink — 1. A satellite-to-earth microwave channel and related components, such as the earth station receiving equipment. The satellite contains a downlink transmitter; downlink components in the earth station are involved with the reception and processing of satellite-transmitted signals. 2. The communication path from a TV satellite to its ground (earth) stations.

download — 1. The process of sending communications instructions, operating software or data from a central computer to individual terminals, including personal computers. To electronically copy a file from one computer to another computer. 2. To receive a file from a remote computer. 3. To transfer information stored in a remote computer system to the user's system. The reverse process is an upload.

downloading — 1. The electronic transfer of information from one computer to another, generally from a larger computer to a smaller one, such as a microcomputer. 2. Transferring information from a host computer to another computer. 3. To transfer data stored in one computer to a storage device in another computer. 4. The process of sending configuration parameters, operating software, or related data from a central source to remote stations. 4. Direct transfer of code from a host system (MDS) into a target system or a PROM programmer.

downstream — Outlet side of an instrument.

downtime — 1. Any period of time during which a system or device cannot be used as a result of a failure or routine maintenance, but not because of a lack of work or the absence of an operator. 2. The period during which computer or network resources are unavailable to users because of a system or component failure.

downward compatibility — Software that can run on older and/or less powerful versions of a computer it was designed to run on.

downward modulation — Modulation in which the instantaneous amplitude of the modulated wave is never greater than that of the unmodulated carrier.

dpdt — Abbreviation for double-pole, double-throw.

dpi (dots per inch) — A measurement of resolution, usually used with printers and scanners.

DPM — See digital panel meter.

dpst — Abbreviation for double-pole, single-throw.

draft mode — A low-quality printing mode available on some printers.

drag — Selecting and moving an on-screen icon via a mouse.

drag angle — A stylus cutting angle of less than 90° to the surface of the record. So called because the stylus drags over the surface instead of digging in. It is the opposite of dig-in angle.

drag cup — A nonmagnetic metal rotated in a magnetic field to generate a torque or voltage proportional to its speed.

drag-cup motor — A small, high-speed, two-phase, alternating-current electric motor having a two-pole, two-phase stator. The rotating element consists only of an extremely light metal cup attached to a shaft rotating on ball bearings. Reversal is accomplished by reversing the connections to one phase. Used in applications requiring quick starting, stopping, and reversal characteristics.

dragging — An interactive technique for repositioning an image on a display screen.

drag magnet — See retarding magnet.

drag soldering — A form of mass soldering in which a printed circuit board is mounted on a conveyor and contacts the surface of a static pool of molten solder at a slight angle when entering and exiting the solder bath. The board remains in contact with solder for a defined length (the dragging length) of time, travelling horizontally across the solder surface.

drain — 1. The current taken from a voltage source. 2. The working-current terminal (at one end of the channel in a FET) that is the drain for holes or free electrons from the channel. Corresponds to a collector of a bipolar transistor. 3. Terminal that receives carriers from the MOS channel. 4. One of the three regions that form a field-effect transistor. Majority carries that originate at the source and traverse the channel are collected at the drain to complete the current path. The flow between source and drain is controlled by the voltage applied to the gate.

drainage equipment — Equipment used to protect connected circuits from transients produced by the operation of protection equipment.

drain conductor — A conductor in continuous contact with a shield for ground termination.

drain cutoff current — The current into the drain terminal of a depletion-type transistor with a specified reverse gate-to-source voltage applied to bias the device to the off state.

drain terminal — The terminal electrically connected to the region into which majority carriers flow from the channel.

drain wire — 1. An uninsulated solid or stranded tinned copper wire that is placed directly under a shield. It touches the shield throughout the cable, and therefore may be used in terminating the shield to ground. It is completely necessary on spiral shielded cables because it eliminates the possibility of induction in a spiral shield. 2. An uninsulated wire, usually placed directly beneath and in electrical contact with a shield. It is used for making shield connections through terminal strips and to ground.

DRAM — Abbreviation for dynamic random-access memory (pronounced "dee-ram"). 1. The least expensive and most popular type of semiconductor read/write memory chip, in which the presence or absence of a capacitive charge represents the state of a binary storage element (0 or 1). The charge must be periodically refreshed. 2. Memory that requires periodic refreshing because of charge leaking from capacitors in the cell circuit. 3. Main memory system of large computers, minicomputers, and even some large microcomputers.

drawing — In the manufacture of wire, pulling the metal through a die or series of dies for reduction of diameter to a specified size.

D-region—The region of the ionosphere up to about 90 kilometers above the earth's surface. It is below the E-region.

dress—1. The exact placement of leads and components in a circuit to minimize or eliminate undesirable feedback and other troubles. 2. To arrange wire connections, cable ends, or cables so that they present a neat and orderly appearance.

dressed contact—A contact that has a locking spring member permanently attached.

drift—1. Movement of carriers in a semiconductor as voltage is applied. 2. A change in either absolute level or slope of an input-output characteristic. 3. *See* flutter, 1. 4. *See* degradation, 1. 5. An undesired change in one of the output parameters of a power supply (voltage, current, frequency, etc.) over a period of time. The change is unrelated to all other variables, such as load, line, and environment. Drift is measured over a period of time by keeping all variables (such as line, load, and environment) constant. Specifications usually apply only after a warm-up period. 6. The angular displacement of an aircraft by the wind, generally expressed in degrees. 7. In a dc amplifier, the change in output with constant input, usually measured in terms of the dc input signal required to restore normal output; may be called out as microvolts or millivolts per hour. 8. A change in output attributable to any cause. 9. A change in the properties of an electrical circuit, as a result of aging or temperature changes.

drift current—1. The flow of carriers in a semiconductor due to an electric field. In the same electric field, holes and electrons will flow in opposite directions due to their opposite charge. 2. The relatively small directional bias that becomes superimposed on the random motion of carriers in an excited crystal lattice under the influence of an applied electric field (drift field).

drifting—An instability in a preset voltage, frequency, or other electronic circuit parameter.

drift mobility—The average drift velocity of carriers in a semiconductor per unit electric field. In general, the mobilities of electrons and holes are not the same.

drift space—1. In an electron tube, a region substantially free of alternating fields from external sources, in which relative repositioning of the electrons depends on their velocity distributions and the space-charge forces. 2. The distance between the buncher and catcher in a velocity-modulated vacuum tube.

drift speed—Average speed at which electrons or ions progress through a medium.

drift transistor—A type of transistor manufactured with a variable-conductivity base region. Such a base sets up an electric field that speeds up the carriers, thus reducing the transit time and improving high-frequency operation. *See also* diffused-alloy transistor.

drift velocity—Net velocity of charged particles in the direction of the applied field.

drip loop—A loop formed in a transmission line at a point where it enters a building. Condensation of moisture and water that may form on the line will drip off at the loop and thus will not enter the building.

drip-proof motor—A motor in which the ventilating openings are such that foreign matter falling on the motor at any angle not exceeding 15° from the vertical cannot enter the motor either directly or indirectly.

driptight enclosure—An enclosure that is intended to prevent accidental contact with the enclosed apparatus and, in addition, is so constructed as to exclude falling moisture or dirt.

drive—1. A unit used in mass storage applications to hold and operate the medium (e.g., disk or tape) being used to store data or programs. 2. Also called excitation.

The signal applied to the input of a power amplifier. *See* excitation, 2.

drive belt—A belt used to transmit power from a motor to a driven device.

drive circuit—A circuit, usually a printed circuit card or an encapsulated module, that converts an input pulse to the appropriate winding excitation sequence to produce one step of the motor shaft.

drive control—*See* horizontal drive control.

driven element—1. An antenna element connected directly to the transmission line. 2. An element of an antenna, such as of a Yagi antenna, that is energized directly from the antenna feed line.

driven sweep—A sweep signal triggered by an incoming signal only.

drive pattern—In a facsimile system, an undesired pattern of density variations that result from periodic errors in the position of the recording spot.

drive pin—In disc recording, a pin similar to the center pin but located at one side of it and used to prevent a disc record from slipping on the turntable.

drive pulse—A pulsed magnetomotive force applied to a magnetic cell from one or more sources.

driver—1. An electronic circuit that supplies input to another electric circuit. 2. A stage of amplification that precedes the power output stage. 3. In a radar transmitter, a circuit that produces a pulse to be delivered to the control grid of the modulator tube. 4. An element coupled to the output stage of a circuit to increase the power- or current-handling capability or fanout of the stage; for example, a clock driver is used to supply the current necessary for a clock line. 5. A device in a logic family controlled with normal logic levels whose output has the capability of sinking or sourcing high current. The output may control a lamp, relay, or a very large fanout of other logic devices. Also a device driving a higher-output device or transistor by supplying power, voltage, or current to it. 6. Any individual speaker within an audio system, such as the woofer, tweeter, etc. 7. A transistor output circuit that has an emitter-follower configuration. 8. A dc driver output module that contains driver output circuits. Each load must be connected between the output and ground as specified in the module data sheet. 9. Amplifier circuit used to reshape signals on a bus when more than one TTL load is present. 10. The low-power oscillator-modulator-amplifier unit that supplies the excitation to a power amplifier. *See also* exciter. 11. A program or routine that controls either external devices or other programs. 12. Typically, an electronic function used to provide amplification to drive high current loads. Term often used to denote bus drivers that rapidly charge and discharge capacitance. Also used to denote the ability to control power, such as when driving a solenoid or other high-current device.

driver element—An antenna array element that receives power directly from the transmitter.

driver gate—An analog switch, usually including two parts: the switch gate and a driver that controls the switch.

driver stage—The amplifier stage preceding the power-output stage.

driving-point admittance—The complex ratio of the alternating voltage for an electron tube, network, or other transducer.

driving-point impedance—1. At any pair of terminals in a network, the driving-point impedance is the ratio of an applied potential difference to the resultant current at these terminals, all terminals being terminated in any specified manner. 2. The input impedance of a transmission line or of an antenna.

driving power—The power supplied to the grid circuit of a tube where the grid swings positive and draws current for part of each cycle of the input signal.

driving-range potential—The voltage difference between the potential of the electrochemically more active anode and the less active protected metal or cathode. One example of driving potential is the electromotive force in a cathodic protection system that causes current between the protected structure (cathode) and the anode. The driving potential decreases as the electrodes become polarized.

driving signal—Television signals that time the scanning at the pickup point. Two kinds of driving signals are usually available from a central sync generator, one composed of pulses at the line frequency and the other of pulses at the field frequency.

driving spring—The spring driving the wipers of a stepping relay.

DRO memory—Destructive readout memory. A memory in which the contents of a storage location are destroyed in being read. Information must be rewritten after reading, if it is to be returned. An example of a DRO memory is the common computer core memory.

drone cone—An undriven speaker cone mounted in a bass-reflex enclosure.

droop—The decrease in mean pulse amplitude, expressed as a percentage of the 100-percent amplitude, at a specified time following the initial attainment of 100-percent amplitude.

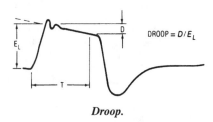

$$DROOP = D/E_L$$

Droop.

drop—1. To develop a specified difference of potential between a pair of terminals as the result of a flow of current. 2. *See* voltage drop.

drop bar—A protective device used to ground a high-voltage capacitor when opening a door.

drop bracket transposition—Reversal of the relative positions of two parallel wire conductors while depressing one, so that the crossover is in a vertical plane.

drop cable—In a cable TV system, the transmission cable from the distribution cable to a dwelling.

drop channel—A type of operation in which one or more channels of a multichannel system are terminated (dropped) at some intermediate point between the end terminals of the system.

drop-down menu—A type of computer menu that drops from the menu bar when requested and remains open without further action until the user closes it or chooses an item from the menu. Same as pull-down menu.

drop-in—The reading of a spurious signal of amplitude greater than a predetermined percentage of the nominal signal amplitude.

drop indicator—An indicator for signaling, consisting of a hinged flap normally held up by a catch. The catch is released by an electromagnet, allowing the flap to drop when a signal is received.

drop-in mike—A surreptitious listening device that transmits phone conversations only, by rf, to a radio receiver. So called because it is "dropped in," replacing the standard carbon microphone in the ordinary telephone handset.

dropout—1. A momentary loss of volume or treble response due to a brief separation of the tape from the surface of the record or play head. A very slight separation causes a treble dropout; more severe loss of head-to-tape contact causes the whole signal to drop out. Dropouts can be caused by buckled or crinkled tape, lumps or pits in the magnetic coating, or detached clumps of oxide passing across the head surface. 2. Short pause in tape replay due to bad tape coating. 3. Momentary loss of signal in a transmission channel. 4. Momentary signal losses due to imperfections in the surface of recording tape or phonograph record.

dropout compensator—Circuitry that senses a signal loss produced by dropout and substitutes missing information with signal from the preceding line; if one line drops out of a picture, it is filled in with the preceding line, resulting in no dropout on the screen.

dropout error—An error, such as loss of a recorded bit, that occurs in recorded magnetic tape because of foreign particles on or in the magnetic coating or because of defects in the backing.

dropouts—Also called keys. Special images inserted at certain points in the array on a photomask used in the production of monolithic circuits.

dropout value—The maximum value of current, voltage, or power that will deenergize a previously energized relay. *See also* hold current; pickup value (voltage, current, or power).

dropout voltage—The input-output voltage differential at which a regulator circuit ceases to regulate against further reductions in input voltage.

dropping resistor—1. A resistor used to decrease a given voltage by an amount equal to the potential drop across the resistor. 2. A resistor placed in series between a voltage source and a load to reduce the voltage supplied to the load.

drop rate—The rate of discharge or decay rate of the sample-and-hold capacitor of a sample-and-hold device. The rate is a function of switch leakage current and the current required by other circuit elements connected to the capacitor. It is expressed as millivolts per second.

drop relay—A relay activated by incoming ringing current to call the attention of an operator to the subscriber's line.

drop repeater—A microwave repeater station equipped for local termination of one or more circuits.

dropsonde—A parachute-carried radiosonde dropped from a high-flying aircraft to measure weather conditions and report them to the aircraft. It is used over water or other areas where ground stations cannot be maintained.

drop wire—A wire suitable for extending an open wire or cable pair from a pole or cable terminal to a building.

dross—Oxide and other contaminants that form on the surface of molten solder.

drum—1. A random-access auxiliary memory device in which information is stored on a revolving drum that is coated with a magnetic material. 2. Rotating magnetic memory that uses the surface of a cylinder.

drum controller—1. A device in which electrical contacts are made on the surface of a rotating cylinder or sector. 2. A device in which contacts are made by the operation of a rotating cam.

drum memory—A rotating cylinder or disk coated with magnetic material so that information can be stored in the form of magnetic spots.

drum parity—A parity error that occurs during the transfer of information to or from drums.

drum plotter—Plotter that draws an image on paper or film mounted on a drum.

drum printer—A type of printer that employs a rotating cylinder. A complete set of characters is embossed on the circumference of the drum for each print position. A set of hammers is used to strike the drum (through the paper and ribbon) and print the proper character each time the drum rotates. (No longer used.)

drum programmer—An electromechanical device that provides stored program logic for control of a sequential operation such as batch processing or machine cycling. It ranks between relay and solid-state systems in the cost/complexity scale.

drum recorder—A facsimile recorder in which the record sheet is mounted on a rotating cylinder. (No longer used.)

drum sequencer—Mechanical programming device that can be used to operate switches or valves.

drum speed—The number of revolutions per minute made by the transmitting or receiving drum of a facsimile transmitter or recorder.

drum storage—A storage device in which information is recorded magnetically on a rotating cylinder; a type of addressable storage associated with some computers.

drum switch—A switch in which the electrical contacts are made on pins, segments, or surfaces on the periphery of a rotating cylinder.

drum transmitter—A facsimile transmitter in which the copy is mounted on a rotating cylinder. (No longer used.)

drum-type controller—1. A multicircuit timing device, with or without a motor, using a cylindrical carriage into which pins are inserted to program events. 2. A multicircuit timing device intended to be driven from an external rotary power source.

drunkometer—A device measuring the degree of alcoholic intoxication by analyzing the subject's breath.

dry—1. A condition in which the electrolyte in a cell is immobilized. The electrolyte may be either in the form of a gel or paste or absorbed in the separator material. 2. Said of circuits or contacts that do not carry direct current.

dry battery—Two or more dry cells arranged in series, parallel, or series-parallel within a single housing to provide desired voltage and current values.

dry cell—1. A voltage-generating cell having an immobilized electrolyte. The commonest form has a positive electrode of carbon and a negative electrode of zinc in an electrolyte of sal ammoniac paste. 2. A source of energy produced by the reaction of an acid or alkaline paste on dissimilar metals or on a metal and a carbon electrode. Normal open-circuit voltage is 1.5 volts. The paste is sealed in a container in normal use, and the cell cannot be recharged.

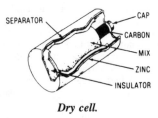

Dry cell.

dry-charged battery—A battery with the plates charged but lacking electrolyte. When ready to be placed in service, the electrolyte is added.

dry circuit—1. A circuit in which current and voltage are so low that there is no arcing to roughen the contacts. As a result, an insulating film can develop that prevents an electrical closing of the circuit when contacts are brought together, if the proper means are not employed to prevent the initial formation of the film. 2. A circuit in which the open-circuit voltage is 0.03 V or less and the current 200 mA or less. (The voltage is most important because at such a low level it is not able to break through most oxides, sulfides, or other films that can build up on contacting surfaces.) 3. A telephone circuit over which voice signals are transmitted and which carries no direct circuit.

dry-circuit contact—A contact that carries current but neither makes nor breaks while its load circuit is energized. Sometimes erroneously used if referring to low level.

dry contacts—Contacts through which there is no direct current.

dry-disc rectifier—A rectifier consisting of discs of metal and other materials in contact under pressure. Examples are the copper-oxide and the selenium rectifier.

dry dross—Nonmetallic components normally found on still solder pots and drag machines. A dry dross may be mixed in with lux, oil, or any other organic material. The term *dry* refers basically to the absence of any metallic solder.

dry-electrolytic capacitor—An electrolytic capacitor with a paste rather than liquid electrolyte. By eliminating the danger of leakage, the paste electrolyte permits the capacitor to be used in any position.

dry flashover voltage—The voltage at which the air surrounding a clean, dry insulator or shell completely breaks down between electrodes.

dry-reed contact—An encapsulated switch containing two metal wires that serve as the contact points for a relay.

dry-reed relay—A relay that consists of one or more capsules containing contact mechanisms that are generally surrounded by an electromagnetic coil for actuation. The capsule consists of a glass tube with a flattened ferromagnetic reed sealed in each end. These reeds, which are separated by an air gap, extend into the tube so as to overlap. When placed in a magnetic field, they are brought together and close a circuit.

dry shelf life—The length of time that a cell can stand without electrolyte before it deteriorates to a point at which a specified output cannot be obtained.

dry-type forced-air-cooled transformer (class AFA)—A transformer that is not immersed in oil and that derives its cooling by the forced circulation of air.

dry-type self-cooled/forced-air-cooled transformer (class AA/FA)—A transformer that is not immersed in oil and that has a self-cooled rating with cooling obtained by the forced circulation of air.

dry-type self-cooled transformer (class AA)—A transformer that is cooled by the natural circulation of air and that is not immersed in oil.

dry-type transformer—A transformer that is cooled by the circulation of air and that is not immersed in oil.

dsc—Abbreviation for double silk-covered.

D-scope—A radar display similar to a C-scope except that the blips extend vertically to give a rough estimate of the distance.

"D" service—FAA service pertaining to radio broadcast of meteorological information, advisory messages, and notices to airmen.

DSP — Abbreviation for digital signal processor. A specialized chip and/or system that is dedicated to processing real-time signals. Typically used for modems, audio, imaging, and video applications.

DSU — *See* data service unit.

DT-cut crystal — A crystal cut to vibrate below 500 kHz.

DTE — *See* data termination equipment.

DTL — Abbreviation for diode-transistor logic.

DTMF — Abbreviation for dual-tone multifrequency (signaling).

DTP — Abbreviation for desktop publishing.

DTV — Abbreviation for digital television.

D-type flip-flop — A flip-flop that, on the occurrence of the leading edge of a clock pulse, propagates to the 1 output whatever information is at its D (data) conditioning input prior to the clock pulse.

dual — Either of a pair of systems, circuits, etc., that are described by equations of the same form in which the same functional relationships hold provided that the dependent and independent dynamic variables are interchanged between these equations.

dual-beam oscilloscope — An oscilloscope in which the cathode-ray tube produces two separate electron beams that may be individually or jointly controlled.

dual capacitor — Two capacitors within a single housing.

dual capstan — *See* closed loop drive.

dual-channel amplifier — An amplifier that has two channels independent of each other, but similar in design, construction, and output.

dual coaxial cable — Two individually insulated conductors laid parallel or twisted and placed within an overall shield and sheath.

dual cone — Speaker unit containing a main cone for bass and middle frequencies and a smaller, stiffer inner cone for treble frequencies, sometimes called a full-range speaker unit.

dual-diversity receiver — 1. A radio receiver that receives signals from two different receiving antennas and uses whichever signal is the stronger at each instant to offset fading. In one arrangement, two identical radio-frequency systems, each with its own antenna, feed a common audio-frequency channel. In another arrangement, a single receiver is changed over from one antenna to the other by electronic switching at a rate fast enough to prevent loss of intelligibility. 2. The operation of combining two identical signals received over diverse paths to obtain an improvement of up to 3 dB in signal-to-noise ratio.

dual-emitter transistor — A passivated pnp silicon planar epitaxial transistor with two emitters; used as a low-level chopper.

dual feedhorn — A feedhorn that can simultaneously receive both horizontally and vertically polarized signals.

dual-frequency induction heater — A type of induction heater in which work coils operating at two different frequencies induce energy, either simultaneously or successively, to material within the heater.

dual-groove record — *See* Cook system.

dual in-line package — Abbreviated DIP. 1. A type of housing for integrated circuits. The standard form is a molded plastic container about $^3/_4$ inch long and $^1/_3$ inch wide (1.9×0.8 cm), with two rows of pins spaced 0.1 inch (2.5 cm) between centers. This package is more popular than the flat pack or TO-type can for industrial use because it is relatively inexpensive and is easily dip-soldered into printed circuit boards. 2. Carrier in which a semiconductor integrated circuit or other components, such as transistors, diodes, capacitors, inductors, resistors,

or film hybrid circuits, are assembled and sealed. Package consists of a plastic or ceramic body with two rows of seven or more vertical leads that are inserted into a circuit board and secured by soldering. 3. The most common type of integrated-circuit package, which can be either plastic (DIP-plastic) or ceramic (CERDIP). Circuit leads or pins extend symmetrically outward and downward from opposite sides of the rectangular package body. "DIP, side-brazed" is a dual in-line package with leads brazed externally, on the sides of the package.

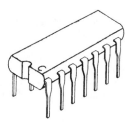

Dual in-line package.

dual meter — A meter constructed so that two aspects of a circuit may be read simultaneously.

dual-mode ignition — *See* DMI.

dual-mode phone — A phone that operates on both analog and digital networks.

dual modulation — The use of two different types of modulation, each conveying separate intelligence, to modulate a common carrier or subcarrier wave.

dual operation — A logic operation whose result is the negation of the result of an original operation when applied to the negation of its operands; for example, the OR operation is the dual of the AND operation. A dual operation is represented by writing 0 for 1 and 1 for 0 in the tabulated values of P, Q, and R for the original operation.

dual pickup — *See* turnover pickup.

dual rail — Also called double rail. Pertaining to a method of transferring data in which the data and the complement of the data are available on different input or output lines or wires.

dual slope converter — An integrating analog-to-digital converter in which the unknown signal is converted to a proportional time interval that is then measured digitally.

dual-tone multifrequency — Abbreviated DTMF. 1. A signaling method in which are employed set pairs of specific frequencies used by subscribers and PBX attendants, if their switchboard positions are so equipped, to indicate telephone address digits, precedence ranks, and end of signaling. 2. A portion of the touch-tone telephone dialing system developed by AT&T that combines two of a set of standard frequencies. The result of the combination is a third or beat frequency (signal) that is the desired or useable signal. DTMF signaling transports precisely defined and matched tone pair signals over sharply tuned amplitude-guarded channels to achieve highly error-immune operation.

dual-tone multifrequency signaling — Abbreviated DTMF signaling. Sending numerical address information from a telephone by sending, simultaneously, a combination of two tones out of a group of eight. The eight frequencies are 697, 770, 852, 941, 1209, 1336, 1477, and 1633 Hz.

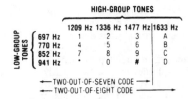

LOW-GROUP TONES	HIGH-GROUP TONES			
	1209 Hz	1336 Hz	1477 Hz	1633 Hz
697 Hz	1	2	3	A
770 Hz	4	5	6	B
852 Hz	7	8	9	C
941 Hz	*	0	#	D

←— TWO-OUT-OF-SEVEN CODE —→
←— TWO-OUT-OF-EIGHT CODE —→

Dual-tone multifrequencies.

dual trace — A mode of operation in which a single beam in a cathode-ray tube is shared by two signal channels. *See also* alternate mode; chopped mode.

dual-track recorder — Normally a monophonic recorder in which the recording head covers slightly less than half the width of a standard quarter-inch tape, making it possible to record one track on the tape in one direction and, after turning the reels over, a second track in the opposite direction. Known also as half-track or two-track recorder.

dual-use line — A communications link that normally is used for more than one mode of transmission (e.g., for voice and data).

dub — Also called rerecording. 1. A copy of a recording. 2. To make a copy of a recording by recording on one machine while another machine is playing.

dubbing — 1. In radio broadcasting, the addition of sound to a prerecorded tape or disc. 2. Copying of already recorded material. In tape recording, playing a tape or disc on one machine while recording it on another. The copy is called a dub.

duct — 1. An overhead or underground protective tube or pipe used for carrying electrical conductors. 2. In microwave transmission, atmospheric conditions may cause radio waves to follow a narrower path than usual. The narrower path is called a duct. The presence of ducting sometimes causes unusual transmission because the transmission waves do not follow the intended path.

ducting — The trapping of an electromagnetic wave, in a waveguide action, between two layers of the earth's atmosphere, or between a layer of the atmosphere and the earth's surface.

DUF — Abbreviation for diffusion under the epitaxial film. A method for providing a low-resistance path between the active region of an IC transistor and the contact electrode at the surface. A region of high conductance is formed by selective diffusion in the required location prior to deposition of the epitaxial layer.

dumb terminal — 1. Conversational slave to a host computer in a communication network. 2. An inexpensive means of interactive computer control that is good for on-line fixed-program applications. All software is in the host computer mainframe. A dumb terminal contains no user programming or memory for data manipulation. 3. The minimum equipment necessary to communicate with a computer. It consists of a monitor and a keyboard and will do little more than send and receive data. 4. A display terminal with no processing capabilities other than that associated with screen formatting. For processing, the terminal is entirely dependent on the main computer with which it communicates.

dummy — 1. A device that can be substituted for another, but which has no operating features. 2. A telegraphy network used to simulate a customer's loop for adjustment of a telegraph repeater. The dummy side of the repeater is the side toward the customer. 3. In a computer, an artificial address, instruction, or other unit of information inserted solely for the purpose of fulfilling such prescribed conditions as word length or block length

without affecting operations. 4. A simulating circuit that has no features.

dummy antenna — *See* artificial antenna.

dummying — The process of removing metallic impurities from plating solution with a large-area dummy cathode.

dummy instruction — An artificial instruction or address inserted in a list of instructions to a computer solely to fulfill prescribed conditions (such as word length or block length) without affecting the operation.

dummy load — *See* artificial load.

dummy variable — In a computer program, a symbol inserted at definition time, which will later be replaced by the actual variable.

dump — Also called power dump. 1. To withdraw all power from a computer, either accidentally or intentionally. 2. To transfer all or part of the contents of one section of a digital-computer memory into another section. 3. To transfer information from a register word to a memory position. 4. To transfer all of the information contained in a record into another storage medium. For example, a disc record could be dumped onto tape. 5. Copying contents of memory from one place to another. Same as memory dump.

dump check — Checking a computer by adding all digits as they are dumped (transferred) to verify the sum to make sure no errors exist as the digits are retransferred.

dump circuit — One form of transient suppression circuit, a self-biased snubber. The dump circuit is virtually inactive until the voltage across it exceeds the recent (slow) average value.

dumping resistor — A resistor whose function is to discharge a capacitor or network for safety purposes.

Dunmore cell — *See* lithium chloride sensor.

duodecal socket — A vacuum-tube socket having 12 pins. Used for cathode-ray tubes.

duodecimal — 1. Pertaining to a characteristic or property involving a selection, choice, or condition in which there are 12 possibilities. 2. Pertaining to the numbering system with a radix of 12.

duodiode — Also called dual diode. A vacuum tube or semiconductor having two diodes within the same envelope.

duodiode-pentode — An electron tube containing two diodes and a pentode in the same envelope.

duodiode-triode — An electron tube containing two diodes and a triode in the same envelope.

duolateral coil — *See* honeycomb coil.

duopole — An all-pass action with two poles and two zeros.

duotriode — Also called double triode. An electron tube containg two triodes in the same envelope.

duplex — 1. The method of operation of a communication circuit in which each end can simultaneously transmit and receive. (Ordinary telephones are duplex. When used on a radio circuit, duplex operation requires two frequencies.) 2. Two-in-one, as two conductors with a common overall insulation or two telegraph transmission channels over one wire. 3. Two conductors twisted together, usually with no outer covering. This term has a double meaning; it is possible to have parallel wires and jacketed parallel wires, and still refer to them as duplex. 4. Two-way data transmission. *Full duplex* describes two data paths that allow simultaneous data transmission in both directions. *Half duplex* describes one data path that allows data transmission in either of two directions, but only one direction at a time.

duplex artificial line — A balancing network simulating the impedance of the real line and distant terminal apparatus; it is employed in a duplex circuit for the purpose of making the receiving device unresponsive to outgoing signal currents.

duplex cable — A cable composed of two insulated stranded conductors twisted together. They may or may not have a common insulating covering.

duplex channel — A communication channel providing simultaneous transmission in both directions.

duplexer — 1. A radar device that, by using the transmitted pulse, automatically switches the antenna from receive to transmit at the proper time. 2. Highly selectable, tunable filters that allow a transmitter and receiver to use one common antenna.

duplexing assembly (radar) — *See* transmit-receive switch.

duplex operation — Simultaneous operation of transmitting and receiving apparatus at two locations.

duplex system — A system with two distinct and separate sets of facilities, each of which is capable of assuming the system function while the other assumes a standby status. Usually both sets are identical in nature.

duplex tube — A combination of two vacuum tubes on one envelope.

duplicate — To copy in such a way that the result has the same physical form as the source. For example, to make a new punched card that has the same pattern of holes as an original punched card.

duplication check — A computer check in which the same operation or program is checked twice to make sure the same result is obtained both times.

Duraspark — Conventional (Ford Motor) electronic ignition for use with high-voltage, high-energy spark timing control.

duration control — A control for adjusting the time duration of reduced gain in a sensitivity time control circuit.

duress alarm device — A device that produces either a silent alarm or local alarm under a condition of personnel stress such as holdup, fire, illness, or other panic or emergency. The device is normally manually operated and may be fixed or portable.

duress alarm system — An alarm system that employs a duress alarm device.

during cycle — The interval while a timer is operating for its preset time period.

dust core — A pulverized iron core consisting of extremely fine iron particles mixed with a binding material for use in radio-frequency coils.

dust cover — A device specifically designed to cover the mating end of a connector so as to provide mechanical and/or environmental protection.

DUT — Abbreviation for device under test.

duty cycle — 1. Ratio of working time to total time for intermittently operated devices. 2. The ratio of on-time to off-time in a periodic on-off cycle. 3. The ratio of operating time to total elapsed time of a device that operates intermittently, expressed as a percentage. 4. In percent, $100(t_o/T)$, where T is the period between pulses and t_o is the pulse width.

duty cyclometer — A test meter that gives a direct reading of duty cycle.

duty factor — 1. In a carrier composed of regularly recurring pulses, the product of their duration and repetition frequency. 2. Ratio of average to peak power. 3. Same as duty cycle except it is expressed as a decimal rather than a percentage. Usually calculated by multiplying pulses per second times pulse width.

duty ratio — In a pulsed system, such as radar, the ratio of average to peak power.

DVD — Abbreviation for the digital versatile disc and digital video disk. A high capacity optical storage medium with improved capacity and bandwidth over the CD for prerecorded multimedia program material with applications in video playback and PC areas.

dv/dt — The rate of change of voltage with respect to time. Proportional to current in a capacitor.

Dvorak keyboard — A keyboard arrangement designed by August Dvorak for increased speed and comfort. It reduces the rate of errors by placing the most frequently used letters in the center of the keyboard for use by the strongest fingers.

DX — 1. Abbreviation for distance. 2. Reception of distant stations. 3. Distant and/or difficult to hear radio stations.

DXer — One who listens to distant or hard-to-hear stations as a hobby.

DX hound — An amateur who specializes in making distant contacts.

DXing — The hobby of listening to distant or otherwise hard-to-hear stations.

dyadic Boolean operator — A Boolean operator that has two operands. The dyadic Boolean operators are AND, exclusive OR, NAND, NOR, and OR.

dyadic operation — An operation on two operands.

dye laser — A laser using a dye solution as its active medium. Its output is a short pulse of broad spectral content, and its achievable gain is high. Dye lasers function at room temperature. Synchronous pumping can be used to produce a continuous train of tunable picosecond pulses for sustained periods.

dynamic — 1. Of, concerning, or dependent on conditions or parameters that change, particularly as functions of time. 2. A speaker drive principle using the interaction between the magnetic field surrounding a voice coil carrying a signal current and a fixed magnetic field to move the coil and the cone to which it is attached. 3. A headphone driver using a voice coil in a magnetic field driving a paper or plastic diaphragm, as in a speaker.

dynamic acceleration — Acceleration in a constantly changing magnitude and direction, either simple or complex motion, usually called vibration. Also measured in gravity units.

dynamically balanced arm — A type of tonearm whose masses are balanced about its pivot, with tracking force applied by a spring. This type of arm does not require that the turntable be level for proper tracking.

dynamic analogies — The similarities in form between the differential equations that describe electrical, acoustical, and mechanical systems that allow acoustical and mechanical systems to be reduced to equivalent electrical networks, which are conceptually simpler than the original systems.

dynamic analysis — Execution of an instrumented program to collect information on its behavior and correctness.

dynamic behavior — The way a system or individual unit functions with respect to time.

dynamic braking — 1. A system of braking of an electric drive in which the motor is used as a generator, and the kinetic energy of the motor and driven machinery is employed as the actuating means of exerting a retarding force. 2. A type of motor braking caused by current being applied to the windings after the power is shut off. This is accomplished either by self-excitation (dc motors) or by separate excitation (ac motors).

dynamic burn-in — High-temperature test with devices subject to actual or simulated operating conditions.

dynamic cell — A memory cell that stores data as charge (or absence of charge) on a capacitor. A typical cell isolates the capacitor from the data line (bit line) with a transistor switch. Thus, when no read or write operation is desired, there is essentially no power required to maintain data. However, normal leakage requires that the charge be periodically restored by a process called refresh. Characteristics of a dynamic cell are very low data retention power, fewer transistors per bit (a one-transistor cell is common), and, usually, less area and lower cost per bit than for static cells.

dynamic characteristics — Relationship between the instantaneous plate voltage and plate current of a vacuum tube as the voltage applied to the grid is varied.

dynamic check — A check used to ascertain the correct performance of some or all components of equipment or a system under dynamic or operating conditions.

dynamic contact resistance — 1. In a relay, a change in contact electrical resistance due to a variation in contact pressure on mechanically closed contacts. For example, during wiping motion of sliding contacts during make or prior to break. Also when contact members no longer actually open, as in contact bounce, but members are still vibrating and varying the contact pressure and hence its resistance. 2. A varying contact resistance on contacts mechanically closed.

dynamic convergence — 1. The condition in which the three beams of a color picture tube come together at the aperture mask as they are deflected both vertically and horizontally. 2. A composite horizontal and vertical voltage used to ensure correct convergence of the three beams of a tricolor picture tube over the entire surface of the phosphor-dot faceplate. *See also* horizontal dynamic convergence; vertical dynamic convergence.

dynamic crosstalk — A condition (in an amplifier utilizing a single power supply) in which the demands made on one channel will effectively modulate the output of the other channel because the power supply feeding both is pumping up and down. To completely eliminate dynamic crosstalk may require separate and well-shielded power supplies for each channel.

dynamic decay — In a storage tube, decay caused by an action such as that of ion charging.

dynamic demonstrator — A three-dimensional schematic diagram in which the components of the radio, television receiver, etc., are mounted directly on the diagram.

dynamic deviation — The difference between the ideal output value and the actual output value of a device or circuit when the reference input is changing at a specified constant rate and all other transients have expired.

dynamic dump — A dump performed while a program is being executed.

dynamic electrode potential — The electrode potential measured when current is passing between the electrode and the electrolyte.

dynamic equilibrium of an electromagnetic system — 1. The tendency of any electromagnetic system to change its configuration so that the flux of magnetic induction will be maximum. 2. The tendency of any two current-carrying circuits to maintain the flux of magnetic induction linking the two at maximum.

dynamic error — An error in a time-varying signal resulting from inadequate dynamic response of a transducer.

dynamic focus — The application of an ac voltage to the focus electrode of a color picture tube to compensate for the defocusing caused by the flatness of the screen.

dynamic headroom — The ability of an amplifier to produce more than its rated power for very short periods of time. An amplifier rated at 100 watts per channel with 3 dB of dynamic headroom can briefly produce 200 watts per channel.

dynamic magnetic field — A field whose intensity is changing and whose lines of force are expanding or

contracting. Such change can be periodic or random. Unlike the static field, the dynamic field can transfer energy from one point to another without relative motion between the points.

dynamic memory — 1. A type of semiconductor memory in which the presence or absence of an electrical or capacitive charge represents the two states of a binary storage element. Without refresh, the data represented by the electrical charge would be lost. 2. An MOS RAM memory using dynamic circuits. Each bit is stored as a charge on a single MOS transistor. This results in very high-density storage (only one transistor per bit). The charge leaks; therefore, a typical dynamic memory must be refreshed every 2 ms by rewriting its entire contents. In practice, this does not slow down the system, but requires additional memory refresh logic. (Dynamic chips are inexpensive and generally preferred to static ones for sizes over 16 K.)

dynamic memory allocation — Allocation of a limited main memory to successive programs in function of an allocation strategy based on priority, availability, or size.

dynamic microphone — *See* moving-coil microphone.

dynamic MOS array — A circuit made up of MOS devices that requires a clock signal. The circuit must be tested at its rated (operating) speed. Known as clock-rate testing.

dynamic mutual-conductance tube tester — *See* transconductance tube tester.

dynamic noise limiter — Abbreviated DNL. A compatible circuit designed primarily for use with tape recorders. It improves the effective signal-to-noise ratio during replay by selective filtering at low signal levels.

dynamic noise suppressor — An audio filter whose bandpass is adjusted automatically to the signal level. At low signal levels, filtering is highest; at high signal levels, all filter action is removed.

dynamic output impedance — *See* output impedance, 2.

dynamic pickup — A phonograph pickup whose electrical output is the result of the motion of a conductor in a magnetic field.

dynamic plate impedance — The internal resistance to the flow of alternating current between the cathode and plate of a tube.

dynamic plate resistance — *See* ac plate resistance.

dynamic power — *See* music power.

dynamic printout — In a computer, a printout of data that occurs as one sequential operation during the machine run.

dynamic problem check — A dynamic check used to ascertain that the solution determined by an analog computer satisfies the given system of equations.

dynamic programming — 1. A procedure used in operations research for optimization of a multistage problem solution in which a number of decisions are available at each stage of the process. 2. A method of sequential decision making in which the result of the decision in each stage affords the best possible answer to exploit the expected range of likely (yet unpredictable) outcomes in the following decision-making stages.

dynamic RAM — *See* DRAM.

dynamic range — 1. The difference, in decibels, between the overload level and the minimum acceptable signal level in a system or transducer. 2. The span of volume between the loudest and softest sounds, either in an original signal (original dynamic range) or within the span of a recorder's capability (recorded dynamic range). Dynamic range is expressed in decibels. *See* signal-to-noise ratio. 3. The range of signal amplitudes, from the loudest to the quietest, that can be reproduced effectively by an equipment. Limited by the intrinsic noise of the amplifier and the ambient background noise level of the listening environment and by the power capacity of the amplifier and speaker system. 4. The ratio between the maximum recorded level (usually that which results in 3-percent playback distortion) and the playback noises from a tape recorded with no signal input. Expressed in decibels. 5. The difference between the maximum acceptable signal level and the minimum acceptable signal level. 6. The ratio of the largest to the smallest values of range, often expressed in decibels. 7. The ratio of the maximum output signal to the smallest output signal that can be processed in a system, usually expressed logarithmically in dB. (Dynamic range can be specified in terms of harmonic distortion, signal-to-noise ratio, spurious-free dynamic range, or other ac input-based performance criteria.)

dynamic register — A memory in which the storage takes the form of capacitively charged circuit elements and therefore must be continually refreshed, or recharged, at regular intervals.

dynamic regulator — A transmission regulator in which the adjusting mechanism is in self-equilibrium at only one or a few settings and requires control power to maintain it at any other setting.

dynamic relocation — The ability to move computer programs or data from auxiliary memory to any convenient location in the memory. Normally the addresses of programs and data are fixed when the program is compiled.

dynamic reproducer — *See* moving-coil pickup.

dynamic resistance — Incremental resistance measured over a relatively small portion of the operating characteristic of a device.

dynamic router — A router that automatically broadcasts routing information throughout the Internet at regular intervals. Other dynamic routers use this information to update their routing tables in case any changes have been made to the network.

dynamic run — Also called dynamic test. 1. The test performed on an instrument to obtain the overall behavior and to establish or corroborate specifications such as frequency response, natural frequency of the device, etc. 2. Test based on a time-interval measurement as, for example, the rise time or fall time of a pulse. 3. A test of one or more of the signal properties or characteristics of equipment that is energized or in a nonquiescent state.

dynamic sequential control — A method of operation in which a digital computer can alter instructions or their sequence as the computation proceeds.

dynamic shift register — A shift register in which information is stored by means of temporary charge storage techniques. The major disadvantage of this method is that loss of the information occurs if the clock repetition rate is reduced below a minimum value.

dynamic skew — Short-term misalignment of the read head of a tape player as referenced to a master skew tape. It results from variations in tape-path geometry and tape-path alignment and slitting and "snaking" tolerances of magnetic tape.

dynamic speaker — *See* moving-coil speaker.

dynamic storage — 1. A data-storage device in which the data is permitted to move or vary with time in such a way that the specified data is not always immediately available for recovery. Magnetic drums and disks are permanent dynamic storage; acoustic delay line is a volatile dynamic storage. 2. Information storage using temporary charge storage techniques. It requires a clock repetition rate high enough to prevent loss of information.

dynamic storage allocation — A storage-allocation technique in which program and data locations are determined by criteria applied at the moment of need.

dynamic subroutine — In digital-computer programming, a subroutine that involves parameters (such as decimal point position) from which a properly coded subroutine is derived. The computer itself adjusts or generates the subroutine according to the parametric values chosen.

dynamic test — *See* dynamic run.

dynamic transfer-characteristic curve — A curve showing the variation in output current as the input current changes.

dynamo — 1. Normally called a generator. A machine that converts mechanical energy into electrical energy by electromagnetic induction. 2. In precise terminology, a generator of direct current — as opposed to an alternator, which generates alternating current.

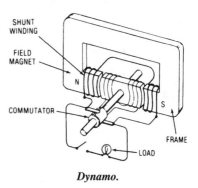

Dynamo.

dynamoelectric — Pertaining to the relationship between mechanical force and electrical energy or vice versa.

dynamometer — 1. An instrument in which the force between a fixed and a moving coil provides a measure of current, voltage, or power. 2. Equipment designed to measure the power output of a rotating machine by determining the friction absorbed by a hand brake opposing the rotation.

dynamotor — Also called a rotary converter or synchronous inverter. 1. A rotating device for changing a dc voltage to another value. It is a combination electric motor and dc generator with two or more armature windings and a common set of field poles. One armature winding receives the direct current and rotates (thus operating as a motor), while the others generate the required voltage (and thus operate as a dynamo or generator). 2 A rotary electrical machine used to convert direct current to alternating current. The machine has a single field structure and a single rotating armature having two windings, one equipped with a dc commutator and the other with ac slip rings.

dynaquad — A germanium pnpn semiconductor switching device that is base controlled and has three terminals. Its operation is similar to that of a flip-flop circuit or latching relay.

dynatron — Also called negatron. A type of vacuum tube in which secondary emission of electrons from the plate causes the plate current to decrease as the plate voltage increases, with the result that the device exhibits a negative-resistance characteristic. Used in oscillator circuits. *See* tetrode.

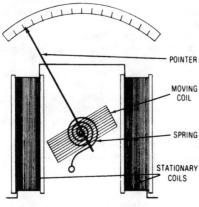

Dynamometer, 1.

POINTER

MOVING COIL

SPRING

STATIONARY COILS

dynatron oscillator—A negative-resistance oscillator with negative resistance derived between the plate and cathode of a screen-grid tube operating such that secondary electrons produced at the plate are attracted to the higher-potential screen grid.

dyne—The fundamental unit of force in the cgs system that, if applied to a mass of 1 gram, would give it an acceleration of 1 cm/s^2.

dyne per square centimeter—Also called microbar. The unit of sound pressure. One dyne per square centimeter was originally called a bar in acoustics, but the full expression is used in this field now because the bar is defined differently in other applications.

dynistor—A nonlinear semiconductor having the characteristics of a small current flow as voltage is applied. As the applied voltage is increased, a point is reached at which the current flow suddenly increases radically and will continue at this rate even though the applied voltage is reduced.

dynode—1. An electrode having the primary function of supplying secondary electron emission in an electron tube. 2. The auxiliary electrode that, when functioning within a photomultiplier tube and when bombarded by photoelectrons, gives rise to secondary emission and amplification.

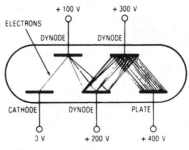

Dynode.

E

E — 1. Symbol for voltage or emitter. 2. Abbreviation for illumination.

E and M leads — In a signaling system, the output and input leads, respectively.

E- and M-lead signaling — Communications between a trunk circuit and a separate signaling unit by way of two leads: an M lead over which battery or ground signals are transmitted to the signaling equipment, and an E lead over which open or ground signals are received from the signaling unit.

early failure — A failure that occurs during the initial life phase and is generally caused by initial production, assembly, test, installation, or commissioning errors.

early-failure period — Also called debugging period, burn-in period, or infant-mortality period. The period of equipment life, starting immediately after final assembly, during which equipment failures initially occur at a higher than normal rate due to the presence of defective parts and abnormal operating procedures.

early-warning radar — A radar that usually scans the sky in all directions in order to detect approaching enemy planes and/or missiles at distances far enough away that interceptor planes can be in the air to meet their approach before they are near their target.

EAROM — Abbreviation for electrically alterable ROM. 1. A read-only memory (ROM) that can be erased and reprogrammed any number of times. 2. Device that resembles an EPROM except that electric current rather than ultraviolet light does the erasing. This is an expensive type of PROM because it requires complete read/write control logic. Use is restricted mainly to applications such as machine controllers in which operators must change programs regularly. 3. Similar to the EPROM, the EAROM can be erased by a sort of reversed programming with a high voltage. The EAROM is thus like a RAM that will not lose its data if power is removed.

earphone — Also called receiver. 1. An electroacoustic transducer intended to be placed in or over the ear. 2. An electroacoustic device that transforms electric waves into sound waves. It is intended to be closely coupled, acoustically, to the ear.

earth — Term used in Great Britain for ground.

earth conductivity — The conductance between opposite faces of a unit cube (usually cubic meter) of a given earth material (loam, clay, sand, rock, etc.). The volume conductivity of this earth sample is the reciprocal of its volume (not to be confused with surface) resistivity.

earth current — Also called ground current. 1. Current in the ground as a result of natural causes and affecting the magnetic field of the earth, sometimes causing magnetic storms. 2. Return, fault, leakage, or stray current passing from electrical equipment through the earth.

earthed — A British term meaning grounded.

earth ground — 1. A connection from an electrical circuit or equipment to the earth through a water pipe or a metal rod driven into the earth. This connection reduces shock hazards from faulty equipment. Water pipes may no longer be reliable grounds because of the use of transite pipe, neoprene gaskets, and other nonconducting links. Any ground rods driven under the interior of a large building may gradually become ineffective because the building may drive the local water table down so far that the rod is essentially surrounded by dry soil. 2. An actual connection into the surface of the earth by way of a metal or chemical rod or a wire connected to such a conductor.

earth inductor — *See* generating magnetometer.

earth-layer propagation — 1. Propagation of electromagnetic waves through layers in the atmosphere of the earth. 2. Propagation of electromagnetic waves through layers below the surface of the earth.

earth oblateness — The slight departure from a perfect spherical shape of the earth and the form of its gravity field.

earth permittivity — The ratio of a capacitor's capacitance using our earth sample as a dielectric to that with air as a dielectric. Permittivity has also variously been termed dielectric constant, specific inductive capacity, and capacitivity.

earth station — The term used to describe the combination of antenna, low-noise amplifier (LNA), downconverter, and receiver electronics; used to receive a signal transmitted by a satellite. Earth station antennas vary in size from the 2-foot to 12-foot (65 centimeters to 3.7 meters) diameter size used for TV reception to those as large as 100 feet (30 meters) in diameter sometimes used for international communications. The typical antenna used for INTELSAT communication today is 13 to 18 meters or 40 to 60 feet in diameter.

EAS — Abbreviation for extensive area service.

EAX — Abbreviation for electronic automatic exchange.

EBCDIC — Abbreviation for extended binary coded decimal interchange code. A coding scheme wherein letters, numbers, and special symbols are represented as unique eight-bit values, allowing for standardization between data communications devices; popularized by IBM. The code can accommodate 256 characters.

E-bend — *See* E-plane bend.

ebiconductivity — Conductivity induced as the result of electron bombardment.

ebmd — Abbreviation for electron-beam mode discharge.

EBS amplifier — Abbreviation for electron-bombarded semiconductor amplifier.

ec — Abbreviation for enamel covered.

eccentric circle — *See* eccentric groove.

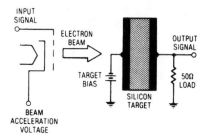

EBS amplifier.

eccentric groove — Also called eccentric circle. An off-center locked groove for actuating the trip mechanism of an automatic record changer at the end of a recording.

eccentricity — In disc recording, the displacement of the center of the recording-groove spiral with respect to the record center hole.

Eccles-Jordan circuit — A flip-flop consisting of a two-stage, resistance-coupled amplifier. Its output is coupled back to its input, two separate conditions of stability being achieved by alternately biasing the two stages beyond cutoff.

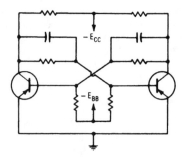

Eccles-Jordan multivibrator.

ECCM — Abbreviation for electronic counter-countermeasures.

ECG — *See* electrocardiogram.

echelon — One of a series of levels of accuracy of calibration, the highest of which is represented by an accepted national standard. There may be auxiliary levels between two successive echelons.

echo — 1. In radar, that portion of the energy reflected to the receiver from a target. 2. A wave that has been reflected or otherwise returned with sufficient magnitude and delay to be distinguishable from the directly transmitted wave. 3. In facsimile, a multiple reproduction on the record sheet caused by the arrival of the same original facsimile signal at different times over transmission paths of different lengths. 4. In a radio system, an electronic condition that causes a signal such as a voice signal to be reflected from some point or points in the circuit back to the point of origination of the signal. 5. A delayed repetition (sometimes several rapid repetitions) of the original sound or signal. 6. In tape recording, refers to a provision for picking up some of the sound from a play head while recording, and feeding it back to the record head to produce a rapidly periodic repetition of each sound. Correct echo-volume adjustment causes a decay of the repeated sounds to simulate acoustical reverberation. 7. A special recording effect, in which a portion of the recorded program is taken from the playback head, a short interval after being recorded, and mixed with the incoming program. Principally used at tape speeds greater than $3\frac{3}{4}$ ips (9.5 cm/s), where the delayed signal is not heard as a separate sound. 8. An instantaneous repetition of the sound heard in playing some tapes or other recordings. This is caused by print through. 9. A portion of the transmitted signal returned from a distant point to the transmitted source with sufficient time delay to be received as interference. 10. A signal that has been reflected at one or more points during transmission with sufficient magnitude and time difference as to be detected as a signal distinct from that of the primary signal. Echoes can either lead or lag the primary signal and appear as reflections or ghosts. 11. Displaying information sent or received on a terminal, to visually detect transmission errors. Remote echo comes from the host computer. Local echo comes from the sender's transmission. 12. The distortion created when a signal is reflected back to the originating station.

echo area — Equivalent echoing area of a radar target (i.e., the relative amount of radar energy the target will reflect).

echo attenuation — In a four-wire (or two-wire) circuit equipped with repeater or multiplex equipment in which the two directions of transmission can be separated from each other, the attenuation of the echo currents (which return to the input of the circuit under consideration) is determined by the ratio of the transmitted power to the echo power received.

echo box — Also called phantom target. A device for checking the overall performance of a radar system. It comprises a resonant cavity that receives a portion of the pulse energy from the transmitter and retransmits it to the receiver as a slowly decaying transient. The time required for this transient response to decay below the minimum detectable level on the radar indicator is known as the ring time and is indicative of the overall performance of the radar set.

echo cancellation — 1. A circuit that uses digital signal processing technology in a full-duplex communications node to remove echoes of the transmitted signal from the received signal. 2. The technique used in modems to filter out unwanted signals.

echo canceller — An electronic circuit that attenuates or eliminates the echo effect on satellite telephony links. Echo cancellers are largely replacing obsolete echo suppressors.

echocardiogram — Ultrasound image of the structure and motions of the heart.

echocardiography — A sonarlike noninvasive method of diagnosing cardiac malfunctions. A pulsed ultrasound beam is directed through the chest wall and echoes reflected from differing tissue interfaces (e.g., soft tissue and blood; tissue and bone, etc.) are detected by the barium titanate transducer that originated the pulse. Echoes are recorded against a linear time base to provide a hard-copy record of cardiac structure movement in a given time span, generally defined by the ECG.

echo chamber — A reverberant room or enclosure used for adding hollow effects or actual echoes to radio or television programs.

echo check — A method of checking the accuracy of transmission of data in which the received data is returned to the sending end for comparison with the original data.

echo checking — A method of checking in which transmitted information is reflected back to the transmitting point and compared with what was sent.

echo distortion — 1. A modulation-related impairment created when one or more delayed echo signals introduce a phase-sensitive delay ripple across the FM portion of a system. The amplitude of this distortion is a complex function involving the relative magnitude and delay time of the echo signal with respect to the main signal, the level of baseband loading present, and the relative position of the channel in the baseband. 2. A telephone-line impairment caused by electrical reflections at distant points where line impedances are dissimilar.

echoencephaloscope — An ultrasonic instrument for use in brain studies. A transducer that generates a series of ultrasonic pulses and detects the returning echoes is placed against the patient's head. Each pulse is displayed together with its echoes on a cathode-ray tube.

echo intensifier — A device, located at the target, that is used to increase the amplitude of the reflected energy to an abnormal level.

echo matching — Rotating an antenna to a position in which the pulse indications of an echo-splitting radar are equal.

echo splitting — In certain radar equipment, the echo return is split and appears as a double indication on the screen of the radar indicator. This splitting is accomplished by special electronic circuits associated with the antenna lobe switching mechanism. When the two echo indications are of equal height, the target bearing is read from a calibrated scale.

echo suppression — 1. A control used to disable a responder for a short interval of time so that reception of echoes of the interrogator pulse from nearby targets is prevented. 2. A circuit used to eliminate reflected waves.

echo suppressor — 1. A voice-operated device that is connected to a two-way telephone circuit to attenuate echo currents in one direction caused by telephone currents in the other direction. 2. In navigation, a circuit that desensitizes the equipment for a fixed period after the reception of one pulse, for the purpose of rejecting delayed pulses arriving from indirect reflection.

echo talker — A portion of the transmitted signal returned from a distant point to the transmitting source with sufficient time delay to be received as interference.

ECL — Abbreviation for emitter-coupled logic. 1. A type of unsaturated logic performed by emitter-coupled transistors. Higher speeds may be achieved with ECL than are obtainable with standard logic circuits. 2. An IC logic family characterized by its very high speed of operation, low circuit density per chip, and very high power dissipation when compared with other IC logic families. Used mainly in large, very high-speed digital computers and sold mainly on a custom-designed basis. 3. A type of current-mode logic in which the circuits are coupled with one another through emitter followers at the input or output of the logic circuit. 4. A bipolar digital IC family that uses a more complex design than TTL to speed up IC operations. Emitter-coupled logic is costly, power hungry, and difficult to use, but it is four times faster than TTL. 5. A family of nonsaturated logic devices that operate at very high speed. ECL logic dissipates relatively large amounts of power, requires a bias supply, and is characterized by low component density.

ECL bipolar memories — Very high-speed cache, writable control stores, and processing sections of large computers.

ECM — Abbreviation for electronic countermeasures.

ECO — Abbreviation for electron-coupled oscillator.

E-core — The laminated configuration resembling the capital letter E in some transformers and inductive transducers.

ECU — Abbreviation for electronic control unit. 1. Electronic ignition system (Chrysler Corp.) that replaced the breaker-point ignition system. Provides spark timing to the high-energy coil and spark plugs. 2. A high-energy ignition system using a Hall effect semiconductor pickup in the distributor instead of the conventional reluctance pickup.

EDC — Abbreviation for error detection code. 32 bits in each sector that are used to detect errors in the sector data of a CD.

EDCT — Abbreviation for electrochemical diffused-collector transistor. A pnp transistor in which all the mass of p material is etched off and replaced with metal, which acts as a heat sink. It is suitable for high-current, high-speed core driver and computer-memory applications.

eddy current clutch — A device that permits connection between a motor and a load by electrical (magnetic) means — no physical contact is involved. This method is also used for speed control by clutch slippage.

eddy-current heating — Synonym for induction heating.

eddy-current loss — The core loss that results when a varying induction produces electromotive forces that cause a current to circulate within a magnetic material.

eddy currents — Also called Foucault currents. 1. Those currents induced in the body of a conducting mass by a variation in magnetic flux. 2. Circulating currents induced in conducting materials by varying magnetic fields. Usually undesirable because they represent loss of energy and cause heating. 3. Induced currents through an iron core in a transformer. They cause a waste of power.

edge-board connector — Also called card-edge connector. A connector that mates with printed wiring leads running to the edge of a printed circuit board.

edge-board contact — A series of contacts printed on or near any edge of a printed board and intended for mating with an edge connector.

edge connector — 1. A one-piece receptacle, containing female contacts, designed to receive the edge of a printed circuit board and interconnect on which the male contacts are etched or printed. The connector may contain either a single or double row of female contacts. Both thermoplastic and thermosetting insulating materials are used. 2. A connector designed to mate with printed circuit boards. May be equipped with a polarizing pin or a key to ensure correct polarity. 3. A row of etched lines on the edge of a printed circuit board that is inserted into a motherboard or an expansion slot of a computer.

Edge connector.

edge effect — 1. *See* following blacks; following whites; leading blacks; leading whites. 2. Nonuniformity of electric fields between two parallel plates caused by an outward bulging of electric flux lines at the edges of the plates. 3. The bulging out of the lines of the electric field at the edges of two parallel electrodes, created by a potential difference between the electrodes.

edge-triggered flip-flop — A type of flip-flop in which some minimum clock-signal rate of change is one necessary condition for an output change to occur.

edging — Undesired coloring around the edges of different-colored objects in a color television picture.

EDI — Abbreviation for electronic data interchange. A format in which business data is represented using national or international standards.

Edison base — Standard screw-thread base used for ordinary electric lamps.

Edison distribution system — A three-wire direct-current distribution system, usually 120 to 240 volts, for combined light and power service from a single set of mains.

Edison effect — Also called Richardson effect. The phenomenon wherein electrons emitted from a heated element within a vacuum tube will flow to a second element that is connected to a positive potential.

Edison storage cell — A storage cell having negative plates of iron oxide and positive plates of nickel oxide immersed in an alkaline solution. An open-circuit voltage of 1.2 volts per cell is produced.

E-display — In a radar, a rectangular display in which targets appear as blips with distance indicated by the horizontal coordinate and elevation by the vertical coordinate.

edit — 1. To arrange or rearrange output information from a digital computer before it is printed out. Editing may involve deleting undesired information, selecting desired information, inserting invariant symbols, such as page numbers and typewriter characters, and applying standard processes such as zero suppression. 2. To deliberately modify the user program in a memory.

editing — 1. The rearrangement of recorded material to provide a change of content or form, or for replacement of imperfect material. Usually accomplished by cutting and splicing the tape. 2. Revising text with a word processor to create an updated document.

editor — 1. A program allowing such text editing functions as addition of a line or character, insertion, or deletion to permit altering of a program. Input data can be anything from programs or reports to raw instrument data. 2. Program that permits data or instructions to be manipulated and displayed. Most common use is in the preparation of new programs. 3. Program that takes the source program, written by the programmer in assembly or high-level language and entered through a keyboard or paper tape, and transfers it to a file in the computer's auxiliary memory, such as magnetic disk or tape. The editor also acts on special commands from the user to add, delete, or replace portions of the source program in the auxiliary memory. Editors can vary significantly in the ease with which they permit a user to make changes in the program. For example, some editors can operate only on entire lines in a program, whereas others can add, delete, or replace arbitrary character strings in the program. However, the less-sophisticated editors are usually easier to learn to use. 4. An interactive software subsystem that allows users to modify test programs directly on an automatic system. 5. A program that permits a user to create new files in symbolic form or to modify existing files. 6. A program that permits a series of mnemonic instructions comprising a user program to be displayed, analyzed, corrected, and otherwise modified quickly and easily on a CRT screen or other terminal device.

EDP — Abbreviation for electronic data processing.

EDP center — *See* electronic data-processing center.

EDPM — Abbreviation for electronic data-processing machine.

EEC — Abbreviation for electronic engine control. Precision control of engine spark timing and exhaust gas recirculation for emissions control and good fuel mileage (Ford Motor). Uses a microprocessor chip.

EEC IV — The fourth progressive version of the electronic engine control (Ford Motor) using the latest technology of microprocessor chips to control engine spark timing, EGR, and feedback carburetor fuel management.

EED — Abbreviation for electroexplosive device. An electroexplosive device for use with an air cushion that automatically inflates to protect the driver when an automobile equipped with such a cushion is involved in a collision.

EEG — *See* electroencephalograph and electroencephalogram.

EEG electrode — Electrode that attaches to scalp for detecting brain waves.

EEPROM or E²PROM — Abbreviation for electrically erasable programmable read-only memory. 1. A field-programmable read-only memory in which cells may be erased electrically and each cell may be reprogrammed electrically. The number of times the E^2PROM can be reprogrammed (write/erase cycles) ranges from 10 times to 10^6 times. 2. Similar to PROM, but with the capability of selective erasure of information through special electrical stimulus. Information stored in EEPROM chips is retained when the power is turned off (compare with PROM). *See also* electronically programmable read-only memory.

effective acoustic center — Also called apparent source. The point from which the spherically divergent sound waves from an acoustic generator appear to diverge.

effective actuation time — The sum of the initial actuation time and the contact chatter intervals of a relay following such actuation.

effective address — The address that is actually used in carrying out a computer instruction.

effective ampere — That alternating current which, when flowing through a standard resistance, produces heat at the same average rate as 1 ampere of direct current flowing in the same resistance.

effective antenna length — The length that, when multiplied by the maximum current, will give the same product as the length and uniform current of an elementary electric dipole at the same location, and the same ratio field intensity in the direction of maximum radiation.

effective aperture delay — In a sample-and-hold circuit, the time difference between the hold command and the time at which the input signal is at the held voltage.

effective area — The effective area of an antenna in any specified direction is equal to the square of the wavelength multiplied by the power gain (or directive gain) in that direction, divided by 4π.

effective bandwidth — For a bandpass filter, the width of an assumed rectangular bandpass filter having the same transfer ratio at a reference frequency and passing the same mean-square value of a hypothetical current and voltage having even distribution of energy over all frequencies.

effective capacitance — The total capacitance existing between any two given points of an electric circuit.

effective conductivity — The conductance between the opposite parallel faces of a portion of a material having unit length and unit cross section.

effective confusion area — Amount of chaff whose radar cross-sectional area equals the radar cross-sectional area of a particular aircraft at a particular frequency.

effective current — That value of alternating current that will give the same heating effect as the corresponding value of direct current. For sine-wave alternating currents, the effective value is 0.707 times the peak value.

effective cutoff — *See* effective cutoff frequency.

effective cutoff frequency — Also called effective cutoff. The frequency at which the insertion loss of an electric structure between specified terminating impedances exceeds the loss at some reference point in the transmission band.

effective facsimile band — A frequency band equal in width to the difference between zero frequency and the maximum keying frequency of a facsimile signal.

effective field intensity — Root-mean-square value of the inverse distance fields 1 mile from the transmitting antenna in all directions horizontally.

effective height — 1. The height of the antenna center of radiation above the effective ground level. 2. In loaded or nonloaded low-frequency vertical antennas, a height equal to the moment of the current distribution in the vertical section divided by the input current.

effective irradiance to trigger — The minimum effective irradiance required to switch a light-activated silicon-controlled rectifier from the off state to the on state.

effective isotropically radiated power — *See* EIRP.

effectively grounded — Grounded through a ground connection of sufficiently low impedance (inherent and/or intentionally added) so that fault grounds that may occur cannot build up voltages that are dangerous to connected equipment.

effectiveness — The capability of the system or device to perform its function.

effective parallel resistance — The resistance considered to be in parallel with a pure dielectric.

effective percentage modulation — For a single sinusoidal input component, the ratio between the peak value of the fundamental component of the envelope and the direct-current component in the modulated conditions, expressed as a percentage.

effective radiated power — 1. The product of the radio-frequency power, expressed in watts, delivered to an antenna and the relative gain of the antenna over that of a half-wave dipole antenna. 2. The product of the antenna power (transmitter power less transmission-line loss) times either the antenna power gain or the antenna field gain squared. Where circular or elliptical polarization is employed, the term *effective radiated power* is applied separately to the horizontal and vertical components of radiation. For allocation purposes, the effective radiated power authorized is the horizontally polarized component of radiation only. If specified for a particular direction, it is the antenna power gain in that direction only.

effective radius of the earth — A value used in place of the geometrical radius to correct the atmospheric refraction when the index of refraction in the atmosphere changes linearly with height. Under conditions of standard refraction, the effective radius is one and one-third the geometrical radius.

effective resistance — 1. The average rate of dissipation of electric energy during a cycle divided by the square of the effective current. 2. The equivalent pure dc resistance that, when substituted for the winding of a motor being checked, will draw the same power. It is also equivalent to the impedance of a circuit having a capacitor connected in parallel with the winding and the capacitor adjusted to unity power factor for the circuit.

effective series resistance — A resistance considered to be in series with an assumed pure capacitance inductance.

effective sound pressure — The root-mean-square value of the instantaneous sound pressure at one point over a complete cycle. The unit is the dyne per square centimeter.

effective speed — The speed (less than rated) that can be sustained over a significant period of time and that reflects slowing effects of control codes, timing codes, error detection, retransmission, tabbing, hand keying, etc.

effective speed of transmission — Also called average rate of transmission. The average rate over some specified time interval at which information is processed by a transmission facility. Usually expressed as average characters or average bits per unit time.

effective thermal resistance — Of a semiconductor device, the effective temperature rise per unit power dissipation of a designated junction above the temperature of a stated external reference point under conditions of thermal equilibrium.

effective value — Also called the rms (root-mean-square) value. The value of alternating current that will produce the same amount of heat in a resistance as the corresponding value of direct current. For a sine wave, the effective value is 0.707 times the peak value.

effective wavelength — The wavelength corresponding to the effective propagation velocity and the observed frequency.

efficiency — 1. Ratio of the useful output of a physical quantity that may be stored, transferred, or transformed by a device to the total input of the device. 2. Ratio of the output power to the input power of a power supply, usually expressed as a percentage of the input power measured at nominal line and load conditions. That part of the input not appearing in the output is converted into heat, which must be conducted away from the power supply circuitry. Efficiency is determined to a great extent by the method of regulation and is expressed as a percentage. In the absence of statements to the contrary, it is assumed to be taken at nominal input and output levels and full load conditions. 3. The percentage of the electrical input power to a speaker that is converted to acoustic energy. Varies from a small fraction of 1 percent to as much a 10 percent or more, depending on the design of the speaker. Higher efficiency means that less electrical amplifier power is required for a given listening volume, but it is not directly related to sound quality.

efficiency of a source of light — The ratio of the total luminous flux to the total power consumed. In the case of an electric lamp, it is expressed in lumens per watt.

efficiency of rectification — Ratio of direct-current power output to alternating-current power input of a rectifier.

***E*-field sensor** — A passive sensor that detects changes in the earth's ambient electric field caused by the movement of an intruder. *See also H*-field sensor.

EFTS — Abbreviation for electronic funds transfer system. A payments system in which the processing and communications necessary to effect economic exchange and the processing and communications necessary for the production and distribution of services incidental or related to economic exchanges are dependent wholly or in large part on the use of electronics.

EGA — Abbreviation for enhanced graphics adapter. A color graphics system for IBM PCs and compatibles that supports 16 colors. Provides higher resolution than CGA, lower than VGA.

EHF — Abbreviation for extremely high frequency.

E-H tee — A waveguide junction composed of a combination of E- and H-plane tee junctions that intersect and the main guide at a common point.

E-H tuner — An E-H tee having two arms terminated in adjustable plungers. It is used for impedance transformation.

EIA — *See* Electronic Industries Association.

EIA interface — 1. A set of signal properties (time duration, voltage, and current) specified by the Electronic Industries Association for business machine/data set connections. 2. A standardized set of signal characteristics for connection of terminals to modem units.

EIA RS-232-C standard — A set of specifications used throughout the data communications industry to define the interconnection of data terminal equipment (DTE) and data communication equipment (DCE) for the exchange of serial binary data. This standard defines electrical signal characteristics, mechanical interface characteristics, and circuits.

EIC — Abbreviation for electronic instrument cluster (Chrysler Corp.). Digital instrument display for speed, miles, fuel level, fuel consumption clock, etc.

eight-bit chip — A CPU chip that processes data eight bits at a time.

eight-bit color — The color range possible with an eight-bit-graphics system. In such a system, each pixel can display one of 256 colors at any given time.

eight-level code — A code in which eight impulses are utilized for describing a character. Start and stop elements may be added for asynchronous transmission. The term is often used to refer to the U.S. ASCII code.

eight-track — Most commonly, a cartridge tape system having eight narrow tracks on $1/4$-inch (6.35-mm) tape wound in a continuous loop around a single hub.

eight-track tape-recording format — Either of two professional tape recording formats (half-track or quarter-track) in which eight independent channels can be recorded in the same direction.

E-indicator — A rectangular radar display in which the horizontal coordinate of a target blip represents range and the vertical coordinate represents elevation.

einstein — A unit of energy equal to the amount of energy absorbed by one molecule of material undergoing a photochemical reaction, as determined by the Stark-Einstein law.

Einthoven string galvanometer — A moving-coil type of galvanometer in which the coil is a single wire suspended between the poles of a powerful electromagnet.

E-I pick-off — An assembly of transformer-like laminations, the output coils of which develop a voltage proportional to the displacement of a magnetic element from the neutral position for limited rotary as well as angular travel.

EIRP — Abbreviation for effective (or equivalent) isotropically radiated power. 1. A measure of the signal strength that a satellite transmits toward the earth below. The EIRP is highest at the center of the beam and decreases at angles away from the boresight. 2. The gain of an antenna in a given direction, multiplied by the net power accepted by the antenna from the transmitter.

EKG — Abbreviation for electrocardiograph.

elastance — 1. Symbolized by S. In a capacitor, the ratio of potential difference between its electrodes to the charge in the capacitor. It is the reciprocal of capacitance. The unit of measure is the daraf.

$$S\,(\text{daraf}) = V/Q$$

2. A measure of the difficulty of placing an electric charge in a capacitor.

elasticity — The resistance of an electrostatic field. It is the reciprocal of permittivity.

elastic wave — A pure acoustic wave; a moving lattice distortion without a magnetic component.

elastomer — 1. A material that has the ability to recover from extreme deformation, in the order of hundreds of percent. It may be thermosetting or thermoplastic.

2. A material that at room temperature stretches under low stress to at least twice its length and snaps back to the original length on release of stress.

E layer — 1. One of the regular ionospheric layers, with an average height of about 100 kilometers or 60 to 70 miles. This layer occurs during daylight hours, and its ionization is dependent on the sun's angle. The principal layer corresponds roughly to what was formerly called the Kennelly-Heaviside layer. 2. The lowest layer of the ionosphere that supports long-distance radio communication.

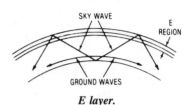

E layer.

elbow — In a waveguide, a bend with a relatively short radius and an angle normally of 90° but sometimes, for acute angles, down to 15°.

el casel — A tape system that uses a cassette similar in plan to the compact cassette, but holding $1/4$-inch (6.35-mm) tape running at $3\,3/4$ inches per second (9.5 cm/s); the tape is looped out of the cassette to reach the heads and capstan during recording and playback. In other cassette systems, the heads contact the tape through windows in the cassette shell.

electra — A specific radionavigational aid that provides a number (usually 24) of equasignal zones. Electra is similar to sonne except that in sonne the equasignal zones as a group are periodically rotated in bearing.

electret — 1. A permanently polarized piece of dielectric material produced by heating the material and placing it in a strong electric field during cooling. Some barium titanate ceramics can be polarized in this way, and so can carnauba waxes. The electric field of an electret corresponds somewhat to the magnetic field of a permanent magnet. 2. A special plastic piezoelectric element, polarized during manufacture to become the equivalent of a permanently charged capacitor. Generates an amplitude-responsive output voltage, like a ceramic element, but requires less energy from the stylus system. By loading with an appropriate resistance, its output can be converted to the equivalent of a magnetic cartridge's velocity-responding output characteristic.

electric — Containing, producing, arising from, actuated by, or carrying electricity, or designed to carry electricity and capable of so doing. Examples: electric eel, energy, motor, vehicle, wave.

electrical — Related to, pertaining to, or associated with electricity but not having its properties or characteristics. Examples: electrical engineer, handbook, insulator, rating, school, unit.

electrical angle — A quantity that specifies a particular instant in a cycle of alternating current. One cycle is considered to be 360°, so a half cycle is 180° and a quarter cycle is 90°. If one voltage reaches its peak value a quarter cycle after another, the phase difference, or electrical angle between the voltages, is 90°.

electrical bail — A switch action in which, upon actuation of one station, the switch changes the contact position, electrically locks the switch in that position, and releases any station previously actuated.

electrical bias— An electrically produced force tending to move the armature of a relay toward a given position.

electrical boresight— The tracking axis as determined by an electrical indication, such as the null direction of a conical scanning or monopulse antenna system or the beam maximum direction of a highly directive antenna.

electrical bridging— The formation of a conductive path between conductors.

electrical center— 1. The point approximately midway between the ends of an inductor or resistor. This point divides the inductor or resistor into two equal electrical values (e.g., voltage, resistance, inductance, or number of turns). 2. The center established by the electromagnetic field distribution within a test coil. A constant-intensity signal, irrespective of the circumferential position of a discontinuity, is indicative of electrical centering. The electrical center may be different from the physical center of the test coil.

electrical charge— The excess on (or in) a body of one kind (polarity) of electricity over the other kind. A plus sign indicates that positive electricity predominates, and a minus sign indicates that negative electricity predominates. Symbol: Q or q.

electrical conductivity— 1. The reciprocal of the resistance in ohms measured between opposite faces of a centimeter cube of an aqueous solution at a specified temperature. 2. The property of a fluid or solid that permits the passage of an electrical current as a result of an impressed emf. It is measured by the quantity of electricity transferred across unit area per unit potential gradient per unit time. In sampling and analysis, changes in this property are utilized to measure the presence of certain ions and compounds such as sulfur dioxide.

electrical coupling— Coupling discrete elements with either electrical conductors or reactances.

electrical degree— One 360th part of a cycle of alternating current.

electrical discharge machining— Machining in which metal is removed by a controlled electrical spark in a dielectric.

electrical distance— The distance between two points, expressed as the length of time an electromagnetic wave in free space takes to travel between them.

electrical element— The concept in uncombined form of any of the individual building blocks from which electronic circuits are synthesized. Examples of basic electrical elements are insulation, conductance, resistance, capacitance, and inductance.

electrical erosion— The loss of contact material due to action of an electrical discharge.

electrical forming— The application of electric energy to a semiconductor device in order to permanently modify the electrical characteristics.

electrical gearing— A term used to describe the action of a system in which the output shaft rotates at a different speed from the input shaft, the ratio being established by electrical means.

electrical generator— A machine that converts mechanical energy into electrical energy.

electrical glass insulation— Insulating materials made from glass fibers of varying diameters, lengths, compositions, etc., including yarns, rovings, slivers, cords, and sheets or mats, bounded or treated only as necessary to their manufacture.

electrical ground— The zero voltage reference for the power supply in an electronic device, usually connected to the equipment chassis. May also be connected to the power mains or earth ground.

electrical-impedance cephalography— A method of evaluating blood circulation in the brain by measuring changes in the impedance between two surface electrodes attached to the head. This impedance decreases when the blood volume in the brain increases. The technique is also known as rheoencephalography.

electrical inertia— Inductance that opposes any change in current through an inductor.

electrical initiation— Any source of electrical power used to start a function or sequence.

electrical interlocks— Switches mounted on contactors or other devices and operated by rods or levers. These interlocks open or close depending on the open or closed position of the contact or device with which they are associated, and are used to govern succeeding operations of the same or allied devices.

electrical length— Length expressed in wavelengths, radians, or degrees. Distance in wavelengths × 2π = radians; distance in wavelengths × 360 = degrees.

electrical load— A device (e.g., a speaker) comprising resistive and/or reactive components into which an amplifier, generator, etc., delivers power.

electrically alterable read-only memory— Abbreviated EAROM. An electrically erasable programmable read-only memory built using metal-nitride-oxide-semiconductor (MNOS) technology. The term electrically alterable read-only memory (EAROM) is being replaced by electrically erasable programmable read-only memory (EEPROM or E²PROM). However, the MNOS technology devices should always be referred to as EAROMs.

electrically connected— Joined through a conducting path or a capacitor, as distinguished from being joined merely through electromagnetic induction.

electrically erasable programmable read-only memory— Abbreviated EEPROM or E²PROM.

electrically operated rheostat— A rheostat used to vary the resistance of a circuit in response to some means of electrical control.

electrically operated valve— A solenoid- or motor-operated valve used in vacuum, air, gas, oil, water, or similar lines.

electrically variable inductor— An inductor in which the inductance can be controlled by a current or a voltage. It is usually made in the form of a saturable reactor with two windings. One is called the signal, or tuned winding, corresponding to the ac or load winding of a power-handling saturable reactor; the other is the control winding and corresponds to the dc winding of the saturable reactor.

electrical noise— Unwanted electrical energy other than crosstalk in a transmission system.

electrical overstress— The operation of an electronic device beyond its normal range of voltage, current, and power when abnormal electric signals are presented to it.

electrical radian— 57.296°, or $1/2\pi$ (or 1/6.28) of a cycle of alternating current or voltage.

electrical ratings— The combinations of voltage and current under which a device or component will operate satisfactorily in specified circuits under standard atmospheric conditions.

electrical reset— A term applied to a relay to indicate that it is capable of being electrically reset after an operation.

electrical resistivity— The resistance of a material to passage of an electric current through it. Expressed as ohms (units of resistance) per mil foot or as microhms (millionths of an ohm) per centimeter cubed (cm^3) at a specified temperature.

electrical resolver — Special type of synchro having a single winding on the stator and two windings whose axes are 90° apart on the rotor.

electrical scanning — Scanning accomplished through variation of the electrical phases or amplitudes at the primary radiating element of an antenna system.

electrical service entrance — A combination of intake wires and equipment including the service entrance wires, electric meter, main switch or circuit breaker, and main distribution or service panel through which the supply of power enters the home.

electrical sheet — Iron or steel sheets from which laminations for electric motors are punched.

electrical shielding — Copper screen, a wire braid, or any conducting material that surrounds a circuit or cable conductors to exclude electrostatic or radio-frequency noises.

electrical switch — A device that makes, breaks, or changes the connections in an electric circuit.

electrical system — The organized arrangement of all electrical and electromechanical components and devices in a way that will properly control the particular machine tool or industrial equipment.

electrical twinning — *See* twinning.

electrical zero — A standard synchro position at which electrical outputs have defined amplitudes and time phase.

electric arc — A discharge of energy through a gas.

electric bell — An audible signaling device consisting of one or more gongs and an electromagnetically actuated striking mechanism.

electric brazing — A brazing (alloying) process in which the heat is furnished by an electric current.

electric breakdown voltage — *See* dielectric breakdown voltage.

electric breeze or wind — The emission of electrons from a sharp point of a conductor that carries a high negative potential.

electric charge — 1. Electric energy stored on the surface of an object. 2. A property of electrons and protons. Similarly charged particles repel one another. Particles having opposite charges attract one another. 3. Electric energy that is stored as stress on the surface of a dielectric.

electric chronograph — A highly accurate apparatus for measuring and recording time intervals.

electric circuit — A continuous path consisting of wires and/or circuit elements over or through which an electric current can flow. If the path is broken at any point, current can no longer flow and there is no circuit.

electric coil — Successive turns of insulated wire that create a magnetic field when an electric current is passed through them. It may also consist of a number of separately insulated sections that lie side by side around the same magnetic circuit.

electric contact — A separable junction between two conductors that is designed to make, carry, or break (in any sequence or singly) an electric circuit.

electric controller — A device that governs the amount of electric power delivered to an apparatus.

electric current — Electricity in motion. In the atoms of metallic substances, there are a number of free electrons, or negatively charged particles that wander in the spaces between the atoms of the metal. The electron movement is normally without any definite direction and cannot be detected. The connection of an electric battery produces an electric field in the metal and causes the free electrons to move or drift in one direction; it is this electron drift that constitutes an electric current. Electrons, being of negative polarity, are attracted to the positive terminal of the battery, and so the actual direction of flow of electricity is from negative to positive, that is, opposite to the conventional direction usually adopted.

electric delay line — A delay line using properties of lumped or distributed capacitive and inductive elements. Can be used as a storage medium by recirculating the information-carrying signal.

electric dipole — Also called a doublet. A simple antenna comprising a pair of oppositely charged conductors capable of radiating an electromagnetic wave in response to the movement of an electric charge from one conductor to the other.

electric-discharge lamp — 1. A sealed glass enclosure containing a metallic vapor or an inert gas through which electricity is passed to produce a bright glow. 2. A lamp in which light (or radiant energy) is produced by the passage of an electric current through a vapor or a gas. Electric discharge lamps may be named after the filling gas or vapor that is responsible for the major portion of the radiation, e.g., mercury lamps, sodium lamps, neon lamps, and argon lamps.

electric displacement — *See* electric-flux density.

electric-displacement density — *See* electric-flux density.

electric eye — 1. The layman's term for a photoelectric cell. 2. The cathode-ray, tuning-indicator tube used in some radio receivers.

electric field — 1. The region about a charged body. Its intensity at any point is the force that would be exerted on a unit positive charge at that point. 2. A condition detectable in the vicinity of an electrically charged body such that forces act on other electric charges in proportion to their magnitudes. 3. Field of force that exists in the space around electrically charged particles. Lines of force are imagined to originate at the protons or positively charged particles and to terminate on electrons or negatively charged particles.

electric-field intensity — A measure of the force exerted at a point by a unit charge at that point.

electric-field strength — The magnitude of the electric field in an electromagnetic wave. Usually stated in volts per meter. *See also* dielectric strength.

electric-field vector — At a point in an electric field, the force on a stationary positive charge per unit charge. May be measured in either newtons per coulomb or volts per meter. This term is sometimes called the electric-field intensity, but such use of the word *intensity* is deprecated in favor of *field strength*, since *intensity* denotes power in optics and radiation.

electric-filament lamp — A glass bulb either evacuated or filled with an inert gas and having a resistance element electrically heated to, and maintained at, the temperature necessary to produce incandescence.

electric filter — 1. Device for rejecting or passing a specific band of signal frequencies. 2. *See* electric-wave filter.

electric-flux density — Also called electric-displacement density or electric displacement. At a point, the vector equal in magnitude to the maximum charge per unit area that would appear on one face of a thin metal plate introduced in the electric field at that point. The vector is normal to the plate from the negative to the positive face.

electric force — Electric field intensity measured in dynes.

electric furnace — A furnace in which electric arcs provide the source of heat.

electric generator — A machine that transforms mechanical power into electrical power.

electric governor-controlled series-wound motor — A series-wound motor having an electric speed governor connected in series with the motor circuit. The governor is usually built into the motor.

electric-heat soldering — Soldering by heating the joint with an electric current.

electric hygrometer — An instrument for indicating humidity by electric means. Its operation depends on the relationship between the electric conductance and moisture content of a film of hygroscopic material.

electric hysteresis — Internal friction in a dielectric field (e.g., the paper or mica dielectric of a capacitor in an ac circuit). The resultant heat generated can eventually break down the dielectric and cause the capacitor to fail.

electrician — A person engaged in designing, making, or repairing electric instruments or machinery. Also, one who sets up an electrical installation.

electric image — The electrical counterpart of an object; i.e., the fictitious distribution of the same amount of electricity that is actually distributed on a nearby object.

electricity — 1. The property of certain particles to possess a force field that is neither gravitational nor nuclear. The type of force field associated with electrons is defined as negative and that associated with protons and positrons as positive. The fundamental unit is the charge of an electron: 1.60203×10^{-19} coulomb. Electricity can be further classified as static electricity or dynamic electricity. Static electricity in its strictest sense refers to charges at rest, as opposed to dynamic electricity, or charge in motion. Static electricity is sometimes used as a synonym for triboelectricity or frictional electricity. 2. A basic property of all matter, which consists of negative and positive charges (electrons and protons) that attract each other. 3. The potential energy of electrons at rest. 4. The kinetic energy of electrons in motion. 5. A manifestation of free electrons that can be generated by induction, friction, or chemical action. It is recognized by its magnetic, chemical, and radiant effects.

electric lamp — Any lamp whose emission of radiant energy is dependent on the passage of an electrical current through the emissive medium.

electric light — Light produced by an electric lamp.

electric lines of force — In an electric field, curves whose tangents at any point give the direction of the fields at that point.

electric meter — A device that measures and registers the amount of electricity consumed over a certain period of time.

electric mirror — *See* dynode.

electric moment — For two charges of equal magnitude but opposite polarities, a vector equal in magnitude to the product of the magnitude of either charge by the distance between the centers of the two charges. The direction of the vector is from the negative to the positive charge.

electric motor — A device that converts electrical energy into rotating mechanical energy.

electric network — A combination of any number of electric elements, having either lumped or distributed impedances, or both.

electric oscillations — The back-and-forth flow of electric charges whenever a circuit containing inductance and capacitance is electrically disturbed.

electric potential — A measure of the work required to bring a unit positive charge from an infinite distance or from one point to another (the difference of potential between two points).

electric precipitation — The collecting of dust or other fine particles floating in the air. This is done by inducing a charge in the particles, which are then attracted to highly oppositely charged collector plates.

electric probe — A rod inserted into an electric field during a test to detect dc, audio, or rf energy.

electric reset — A qualifying term indicating that the contacts of a relay must be reset electrically to their original positions following an operation.

electric robot — Programmable machine that is powered by servomotors or stepping motors.

electric service panel — The main cabinet where electricity is brought into a building, then distributed to branch circuits. The panel usually contains the main circuit breaker for shutting down the entire system, and circuit breakers or fuses for shutting down each independent circuit.

electric shield — A housing, usually aluminum or copper, placed around a circuit to provide a low-resistance path to ground for high-frequency radiations and thereby prevent interaction between circuits.

electric strain gage — A device that detects the change in shape of a structural member under load and causes a corresponding change in the flow of current through the device.

electric strength — The maximum electric charge a dielectric material can withstand without rupturing. *See also* dielectric strength; insulating strength. The value obtained for the electric strength will depend on the thickness of the material and on the method and conditions of test.

electric stroboscope — An instrument for observing or for measuring the speed of rotating or vibrating objects by electrically producing periodic changes in the intensity of light used to illuminate the object.

electric tachometer — A tachometer (rpm indicator) that utilizes voltage or electrical impulses.

electric telemeter — A system consisting of a meter that measures a quantity, a transmitter that sends the information to a distant station, and a receiver that indicates or records the quantity measured.

electric transcription — In broadcasting, a disc recording of a message or a complete program.

electric transducer — A device actuated by electric waves from one system and supplying power, also in the form of electric waves, to a second system.

electric tuning — A system by which a radio receiver is tuned to a station by pushing a button (instead of, say, turning a knob).

electric vector — 1. A component of the electromagnetic field associated with electromagnetic radiation. The component is of the nature of an electric field. The electric vector is supposed to coexist with, but act at right angles to, the magnetic vector. 2. The electric field associated with an electromagnetic wave and thus with a light wave. The electric vector specifies the direction and amplitude of this electric field.

electric watch — A timepiece in which a battery replaces the mainspring as the prime energy source of the watch, and in which an electromagnet impels the balance wheel through a mechanical switching-contact arrangement.

electric wave — Another term for the electromagnetic wave produced by the back-and-forth movement of electric charges in a conductor.

electric-wave filter — Also called electric filter. 1. A device that separates electric waves of different frequencies. 2. A frequency-selective device, usually passive, made up of resistance, capacitance, and inductance elements having signal transmission characteristics that are a function of frequency. Basically, filters are used to pass desired signals and reject unwanted or interfering signals. Bandpass filters are used when the desired signals encompass a frequency band that does not contain any unwanted signals.

electrification — 1. The process of establishing an excess of positive or negative charges in a material. 2. The process of applying a voltage to a component or device.

electrification time — The time during which a steady direct potential is applied to electrical insulating materials or before the current is measured.

electroacoustic — Pertaining to a device (e.g., a speaker or a microphone) that involves both electric current and sound-frequency pressures.

electroacoustic device — One that employs phonon propagation or vibrations of a material's crystal lattice structure as the basic energy transport mechanism. Electrical energy is converted into acoustic energy by the material's piezoelectric properties.

electroacoustic transducer — A device that receives excitations from an electric system and delivers an output to an acoustic system, or vice versa. A speaker is an example of the first, and a microphone is an example of the second.

electroanalysis — The process of determining the quantity of an element or compound in an electrolyte solution by depositing the element or compound on an electrode by electrolysis.

electrobiology — The science concerned with electrical phenomena of living creatures.

electrobioscopy — The application of a voltage to produce muscular contractions.

electrocardiogram — Abbreviated EKG or ECG. 1. Essentially an electromyogram of the heart muscle. All muscular activity in the body is characterized by the discharge of polarized cells, the aggregate current from which causes a voltage drop that can be measured on the skin. A changing emf will appear between electrodes connected to the arms, legs, and chest, which rises and falls with heart action such that the period of the resulting waveform is the time between heartbeats. Various positive and negative peaks within one cycle of this waveform have been lettered P, Q, R, S, and T, a notation that aids in subsequent analysis and diagnosis. 2. A hardcopy record of heart action potentials obtained by measuring instantaneous potential differences at the surface of the body. In general, the recording describes the depolarization of myocardial muscle cell masses, providing a graphic, but indirect, view of the heart's competence. 3. Graphic tracing of the electric current that is produced by the rhythmic contraction of the heart muscle. Visually, a periodic wave pattern is produced. Changes in the wave pattern may appear in the course of various heart diseases; the tracing is obtained by applying electrodes on the skin of the chest and limbs.

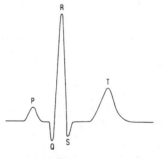

Electrocardiogram for one heartbeat.

electrocardiograph — A medical instrument for detecting irregularities in the action of a human heart. It measures the changes in voltage occurring in the human body with each heartbeat. Abbreviated EKG or ECG.

electrocardiography — Recording and interpretation of the electrical activity of the heart. The voltage generated by the heart is picked up by surface electrodes on the limbs and chest, amplified, and applied to a strip-chart recorder.

electrocardiophonograph — An instrument that records heart sounds.

electrochemical cell — An electrochemical system consisting of an anode and a cathode in metallic contact and immersed in an electrolyte. The anode and cathode may be different metals or dissimilar areas on the same metal surface.

electrochemical deterioration — A process in which autocatalytic electrochemical reactions produce an increase in conductivity and in turn ultimate thermal failure.

electrochemical device — A device that operates on both electrical and chemical principles, e.g., a lead-acid storage battery.

electrochemical diffused-collector transistor — *See* EDCT.

electrochemical equivalent — The weight of an element, compound, radical, or ion involved in a specified electrochemical reaction during passage of a specified quantity of electricity such as a coulomb.

electrochemical junction transistor — A junction transistor produced by etching an n-type germanium wafer on opposite sides with jets of a salt solution such as indium chloride.

electrochemical potential — Also called electrochemical tension. The partial derivative of the total electrochemical free energy of the system with respect to the number of moles of the constituent except that it includes the electrical as well as the chemical contributions to the free energy.

electrochemical recording — A recording made by passing a signal-controlled current through a sensitized sheet of paper. The paper reacts to the current and thereby produces a visual record.

electrochemical tension — *See* electrochemical potential.

electrochemical transducer — A device that uses a chemical change to measure the input parameter, and the output of which is a varying electrical signal proportional to the measurand.

electrochemical valve — Electric valve consisting of a metal in contact with a solution or compound, across the boundary of which current flows more readily in one direction than in the other direction and in which the valve action is accompanied by chemical changes.

electrochemistry — 1. That branch of science concerned with reciprocal transformations of chemical and electrical energy. This includes electrolysis, electroplating, the charge and discharge of batteries, etc. 2. The study of the reversible conversion of chemical energy into electrical energy. Electroplating is an electrochemical process.

electrochromic display — A passive solid-state display that is made from a material whose light-absorption properties are changed by an externally applied electric field. Ordinarily electrochromic materials do not absorb light in the visible range of the spectrum, so they are completely transparent. When a moderate electric field is applied, the material develops an absorption band in the visible spectrum and takes on a color that remains even after the electric field is removed and lasts from minutes to months. The color change can be reversed and the display returned to its original state when the polarity of the applied electric field is simply reversed.

electrocoagulation — The process of solidifying tissue by means of a high-frequency electrical current.

electrocution — Killing by means of an electric current.

electrode — 1. In an electronic tube, the conducting element that does one or more of the following: emits or collects electrons or ions, or controls their movement by means of an electric field on it. 2. In semiconductors, the element that does one or more of the following: emits or collects electrons or holes, or controls their movements by means of an electric field on it. 3. In electroplating, the metal being plated. 4. A conductor by means of which a current passes into or out of a fluid or an organic material, such as human skin; often one terminal of a lead. 5. A metallic conductor such as in an electrolytic cell, in which conduction by electrons is changed to conduction by ions or other charged particles. 6. A conductor, not necessarily metal, through which a current enters or leaves an electrolytic cell, arc, furnace, vacuum tube, gaseous discharge tube, or any conductor of the nonmetallic class. 7. That part of a semiconductor device providing the electrical contact between the specified region of the device and the lead to its terminal. 8. In a spark plug, the center rod passing through the insulator forms one electrode. The rod welded to the shell forms another. They are referred to as the center and side electrodes. 9. A conducting element at whose surface electricity passes into another conducting medium.

electrode admittance — The alternating component of the electrode current divided by that of the electrode voltage (all other electrode voltages maintained constant).

electrode capacitance — The capacitance between one electrode and all the other electrodes connected together.

electrode characteristic — The relationship, usually shown by a graph, between the electrode voltage and current, all other electrode voltages being maintained constant.

electrode conductance — The quotient of the in-phase component of the electrode alternating current divided by the electrode alternating voltage, all other electrode voltages being maintained constant. This is a variational and not a total conductance.

electrode current — Current passing into or out of an electrode.

electrode dark current — Also called dark current. 1. In phototubes, the component of electrode current that flows in the absence of ionizing radiation and optical photons. 2. The current that flows in a photodetector when there is no incident radiation on the detector.

electrode dissipation — The power that an electrode dissipates as heat when bombarded by electrons and/or ions and radiation from nearby electrodes.

electrode drop — The voltage drop produced in an electrode by its resistance.

electrode impedance — The reciprocal of electrode admittance.

electrode inverse current — Current through a tube electrode in the direction opposite to that for which the tube was designed.

electrodeless discharge — A luminous discharge produced by means of a high-frequency electric field in a gas-filled glass tube that has no internal electrodes.

electrodeless discharge tube — Abbreviated EDT. A device consisting of an airtight quartz tube that holds the material to be analyzed. When a high-frequency electrostatic field, generated by microwaves, is applied to the tube, it emits energy of a wavelength identical with that of the contained material.

electrodeposition — Also called electrolytic deposition. *See also* electroplating.

electrode potential — 1. The instantaneous voltage on an electrode. Its value is usually given with respect to the cathode of a vacuum tube. 2. The difference in potential between an electrode and the immediately adjacent electrolyte referred to some standard electrode potential as zero. 3. The potential in volts that an electrode has when immersed in an electrolyte, compared to the zero potential of a hydrogen electrode. The potential depends on the material of which the electrode is made.

electrode reactance — The imaginary component of electrode impedance.

electrode resistance — The reciprocal of electrode conductance. It is the effective parallel resistance, not the real component of electrode impedance.

electrodermography — The recording of the electrical resistance of the skin, which is a sensitive indicator of the activity of the autonomic nervous system.

electrode voltage — The voltage between an electrode and the cathode or a specified point of a filamentary cathode. The terms *grid voltage, anode voltage, plate voltage*, etc., designate the voltage between these electrodes and the cathode. Unless otherwise stated, electrode voltages are measured at the available terminals.

electrodialytic process — A process for producing fresh water by using a combination of electric current and two types of chemically treated membranes.

electrodynamic — Pertaining to electric current, electricity in motion, and the actions and effects of magnetism and induction.

electrodynamic braking — A method of stopping a tape-deck motor gently by the application of a predetermined voltage to the motors.

electrodynamic instrument — An instrument that depends for its operation on the reaction between the current in one or more moving coils and the current in one or more fixed coils.

electrodynamic machine — Electric generator or motor in which the output load current is produced by magnetomotive currents generated in a rotating armature.

electrodynamics — 1. The science dealing with the various phenomena of electricity in motion, including interactions of currents with each other, with their associated magnetic fields, and with other magnetic fields. 2. The study of the generation of electromagnetic power by radiation from high-energy beams.

electrodynamic speaker — A speaker consisting of an electromagnet called the field coil, through which a direct current flows.

electrodynamometer — 1. An instrument for detecting or measuring an electric current by determining the mechanical reactions between two parts of the same circuit. 2. A meter movement consisting of a rotatable (moving) wire coil suspended between two fixed (field) wire coils. The three coils can be connected in various configurations, so that rotation of the moving coil is proportional to applied ac or dc voltage or current, to power, power factor, etc.

electroencephalogram — 1. A waveform obtained by plotting brain voltages (available between two points on the scalp) against time. An electroencephalogram is not necessarily a periodic function, although it can beparticularly if the patient is unconscious. These voltages are of extremely low level and require recording apparatus that displays excellent noise rejection. 2. The tracing of brain waves made by an electroencephalograph.

electroencephalograph — Abbreviated EEG. An instrument for measuring and recording the rhythmically varying potentials produced by the brain by the use of electrodes applied to the scalp.

electroencephalography — 1. Recording and interpretation of the electrical activity of the brain. Voltage (typically 50 microvolts) picked up by electrodes on the scalp is amplified and applied to a strip-chart recorder. 2. Recording of electric currents developed in the brain by means of electrodes applied to the scalp, to the surface of the brain, or placed within the substance of the brain.

electroencephaloscope — An instrument for detecting brain potentials at many different sections of the brain and displaying them on a cathode-ray tube.

electroexplosive device — *See* EED.

electrofluid dynamics generator — Abbreviated EFD. A generator in which the only moving parts are wind-driven charged particles. Their movement from electrode to electrode is analogous to the spinning of an armature. Ideally suited for use at sea or at the seaside, bladeless windmills could be used anywhere there is a source of moisture. Principal advantages of the concept include efficiency and low cost.

electroforming — Also called electrodeposition and electroplating. 1. Making a metal object by using electrolysis to deposit a metal on an electrode. 2. Creating a pn junction by passing a current through point contacts on a semiconductor. 3. The production or reproduction of articles by electrodeposition on a mandrel or mold that is subsequently separated from the deposit. 4. The process of depositing a substance on an electrode by electrolysis, as in electroplating, electroforming, electrofining, or electrotinning.

electroforming process — An electrochemical process of metal fabrication using an electrolyte, an anode to supply the metal, and a control of the electrical current and of the deposition of metal on the matrix of a reflector.

electrogalvanizing — Electrodeposition of zinc coatings.

electrogastrogram — The graphic record that results from synchronous recording of the electrical and mechanical activity of the stomach.

electrograph — 1. A plot, graph, or tracing made by means of the action of an electric current on sensitized paper or other material, or by means of an electrically controlled stylus or pen. 2. Equipment for facsimile transmission.

electrographic process — *See* dielectric process.

electrographic recording — Also called electrostatography. The producing of a visible record by using a gaseous discharge between two or more electrodes to form electrostatically charged patterns on an insulator. *See also* electrostatic electrography.

electrokinetics — The branch of physics concerned with electricity in motion.

electroless deposition — The deposition of conductive material from an autocatalytic plating solution without application of electrical current.

electroless plating — 1. A method of metal deposition by means of a chemical reducing agent present in the processing solution. The process is further characterized by the catalytic nature of the surface, which enables the metal to be plated to any thickness. 2. A chemical process by which certain metals can be plated without electrical current. Tin may be plated onto copper in this manner. 3. The controlled autocatalytic reduction of a metal ion on certain catalytic surfaces.

electroluminescence — 1. Luminescence resulting from a high-frequency discharge through a gas or from application of an alternating current to a layer of phosphor. 2. Direct conversion of electrical energy into light energy in a liquid or solid; for example, photoemission as a result of electron-hole recombination in a pn junction. This is the mechanism employed by the injection laser. The standard abbreviation for the effect is written

EL. (This process is not to be confused with the ordinary tungsten filament bulb, where there is an intermediate stage of heat, making the process thermoluminescent.) 3. Light produced in a phosphor that is in an alternating electric field. Consists of a phosphor a few mils thick placed between two metal films, one of which is transparent. Alternating current is applied to the plates through a current-limiting resistor.

electroluminescent display — 1. A display, designated EL, whose segments or elements consist of transparent conductive electrodes separated by a dielectric containing a luminescent phosphor. Application of ac voltage to opposing electrodes causes the dielectric between them to glow with a characteristic blue-green light. 2. The utilization of the light produced when electrical energy is directly converted into light within devices used for visual readout displays or as complex logic-circuit elements. The display devices may be flat, giving a wide viewing angle without parallax, and may have low power needs. Can yield blue, green, yellow, and white colors.

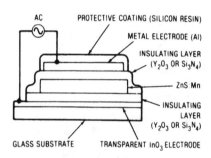

Electroluminescent display.

electroluminescent lamp — A lamp in the shape of a panel that is decorative as well as illuminative. It consists primarily of a capacitor having a ceramic dielectric with electroluminescent phosphor. The amount of illumination is determined by the voltage across the layer and by the frequency applied to it.

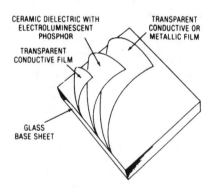

Electroluminescent lamp.

electroluminescent-photoconductive image intensifier — A panel, made up of electroluminescent and photoconductive (EL-PC) layers, used as either a positive or negative image intensifier, depending on amplitude and phase of its two power supply voltages. The photoconductive layer receives the image and converts it to electrical signals; subsequently, the electroluminescent layer converts the signals to visible, brighter light.

electrolysis — 1. The process of changing the chemical composition of a material (called the electrolyte) by sending an electric current through it. 2. The decay of an underground structure by chemical action due to stray electrical currents. 3. Production of chemical changes of the electrolyte by the passage of current through an electrochemical cell. 4. Conduction of an electric current through a chemical compound in its natural state, solution or as a molten, to decompose the compound.

electrolyte — 1. A substance in which the conduction of electricity is accompanied by chemical action. 2. The paste that forms the conducting medium between the electrodes of a dry cell, storage cell, or electrolytic capacitor. 3. A substance that, when dissolved in a suitable liquid (often water), dissociates into ions, thus rendering the liquid electrically conducting. 4. The current-conducting substance (liquid or solid) between two capacitor electrodes, at least one of which is covered by a dielectric film. 5. A conducting medium in which current is accompanied by movement of matter. Most often an aqueous solution of acids, bases, or salts, but includes many other media, such as fused salts, ionized gases, some solids, etc. 6. A substance that is capable of forming a conducting liquid medium when dissolved or melted.

electrolyte conductivity — Also called specific conductance. A measure of the ability of a solution to carry an electric current. Defined as the reciprocal of the resistance in ohms of a 1-cm cube of the liquid at a specified temperature. The units of specific conductance are the reciprocal ohm-cm (or siemens/cm) and one millionth of this. High-quality condensed steam and distilled or demineralized water have specific conductances at room temperatures as low as or lower than 1 microsiemens/cm.

electrolyte recording — A form of facsimile recording in which ionization causes a chemically moistened paper to undergo a change.

electrolytic — 1. Pertaining to or made by electrolysis; deposited by electrolysis; pertaining to or containing an electrolyte. 2. Said of an electrical device that contains an electrolyte.

electrolytic capacitor — 1. A capacitor consisting of two conducting electrodes, with the anode having a metal oxide film formed on it. The film acts as the dielectric or insulating medium. The capacitor is operable in the presence of an electrolyte, usually an acid or salt. Generally used for filtering, bypassing, coupling, or

decoupling. 2. A capacitor in which the dielectric is a film of oxide electrolytically deposited on a plate or slug of aluminum or tantalum. The thinness of the film permits a high capacitance-to-volume ratio. The oxide acts as a dielectric in one direction only. The device is, therefore, polarized. (A nonpolarized electrolytic capacitor is, in effect, two polarized types in series with their like terminals connected together.) 3. A fixed capacitor, having a relatively high capacitance-to-volume ratio due to a very thin electrically formed, nonconducting chemical dielectric (oxide) film. 4. A capacitor in which the electrolytically formed oxide layer on the surface of the anode serves as a dielectric, with a solid or nonsolid electrolyte forming the cathode, thus giving the capacitor polar properties.

electrolytic capacitor paper — Very pure, porous paper, 17 to 100 micrometers thick, used to separate the metallic electrodes in electrolytic capacitors.

electrolytic cell — 1. In a battery, the container, two electrodes, and the electrolyte. 2. A unit apparatus in which electrochemical reactions are produced by applying electrical energy, or which supplies electrical energy as a result of chemical reactions and which includes two or more electrodes and one or more electrolytes contained in a suitable vessel.

electrolytic cleaning — A process of removing soil, scale, or corrosion products from a metal surface by subjecting it as an electrode to an electric current in an electrolytic bath.

electrolytic conduction — The flow of current between electrodes immersed in an electrolyte. It is caused by the movement of ions from one electrode to the other when a voltage is applied between them.

electrolytic corrosion — Corrosion by means of electrochemical erosion. *See* corrosion.

electrolytic deposition — *See* electrodeposition.

electrolytic development — The method of developing a photographic image by means of an applied electric field. The systems used include electrolysis and photoconductive systems.

electrolytic dissociation — The breaking up of molecules into ions in a solution.

electrolytic interrupter — A device that is tilted to change the current through it.

electrolytic iron — Iron obtained by an electrolytic process. The iron possesses good magnetic qualities and is exceptionally free of impurities.

electrolytic plating — A method of metal deposition employing the work or cathode, the anode, the electrolyte, a solution containing dissolved salts of the metal to be plated, and a source of direct current. The anode metal is dissolved by chemical and electrical means; subsequently, cations are deposited onto the cathode. Electric current is the reducing agent. Copper, nickel, chromium, zinc, brass, cadmium, tin, gold, and silver are the metals most commonly electroplated.

electrolytic potential — The difference in potential between an electrode and the immediately adjacent electrolyte, expressed in terms of some standard electrode difference.

electrolytic rectifier — A rectifier consisting of metal electrodes in an electrolyte, in which rectification of alternating current is accompanied by electrolytic action. A polarization film formed on one of the electrodes permits current in one direction but not in the other.

electrolytic refining — The refining or purifying of metals by electrolysis.

electrolytic shutter — A high-speed shutter, similar to a Kerr cell, that uses the birefringence produced in a liquid during the passage of an electric current through it to change the liquid's optical transmission characteristics.

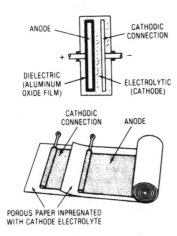

ANODE — CATHODIC CONNECTION

+ −

DIELECTRIC (ALUMINUM OXIDE FILM) — ELECTROLYTIC (CATHODE)

CATHODIC CONNECTION   ANODE

POROUS PAPER INPREGNATED WITH CATHODE ELECTROLYTE

Electrolytic capacitor.

electrolytic switch — A switch having two electrodes projecting into a chamber containing a precisely measured quantity of a conductive electrolyte, leaving an air bubble of predetermined width. When the switch is tilted from true horizontal, the bubble shifts position and changes the amount of electrolyte in contact with the electrodes, thereby changing the amount of current passed by the switch. Used as a leveling switch in gyro systems.

electrolyzer — An electrolytic cell that produces alkalies, metals, chlorine, or other allied products.

electromagnet — 1. A temporary magnet consisting of a solenoid with an iron core. A magnetic field exists only while current flows through the solenoid. 2. A magnet, consisting of a solenoid with an iron core, that has a magnetic field existing only during the time of current flow through the coil. 3. A coil of wire, usually wound on an iron core, that produces a strong magnetic field when current is sent through the coil. 4. A magnet created by inserting a suitable metal core within or near a magnetizing field that is usually formed by passing electric current through a coil of insulated wire. 5. A soft iron core that becomes a magnet temporarily when current flows through a coil of wire that surrounds it.

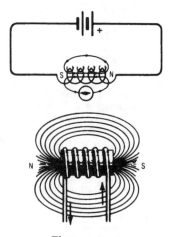

Electromagnet.

electromagnetic — 1. Having both magnetic and electric properties. 2. Pertaining to the mutually perpendicular electric and magnetic fields associated with the movement of electrons through conductors, as in an electromagnet. 3. Pertaining to the combined electric and magnetic fields associated with radiation or with movements of charged particles. 4. Pertaining to or caused by the combined electric and magnetic fields that are always associated with an electric current. 5. Pertaining to the relationship between currents and magnetic fields.

electromagnetic amplifying lens — A system made up of a large number of waveguides symmetrically arranged with respect to an excitation medium so that they are excited with equal amplitude and phase in order to provide an effective gain in energy.

electromagnetic bonding — Method for joining thermoplastics in which a metallic preform is placed in the joint area to convert electromagnetic energy into heat for fusion bonding.

electromagnetic cathode-ray tube — A cathode-ray tube that uses electromagnetic deflection to deflect the electron beam.

electromagnetic communications — The electromagnetic wave conductor is space itself. The electromagnetic frequencies available today for communications fall into two categories: frequencies that form "wireless" communications (such as visual light of fairly high frequency), and frequencies that humans use for wireless communications (such as radio, shortwave, and microwave transmitting, of relatively lower frequencies). In communicating by radio, shortwave, and microwave frequencies, translators similar in principle to those used in electrical communications are needed, although the equipment requirement increases.

electromagnetic compatibility — Abbreviated EMC. 1. The ability of electronic devices and communications equipment, subsystems, and systems to operate in their intended environments without suffering or causing unacceptable degradation of performance as a result of unintentional electromagnetic radiation or response. 2. A directive that specifies the acceptable limits for electromagnetic emissions from an electronic device, and how much electromagnetic interference the device should tolerate. 3. Abbreviation for electronic message center (Ford Motor). Digital dashboard electronics displaying digital readouts of speed, miles, fuel, clock, etc.

electromagnetic complex — The electromagnetic configuration of an installation, including all radiators of significant amounts of energy.

electromagnetic coupling — The mutual relationship between two separate but adjacent wires when the magnetic field of one induces a voltage in the other.

electromagnetic crack detector — An instrument for detecting hidden cracks in iron or steel objects by magnetic means.

electromagnetic deflection — The deflection of an electron stream by means of a magnetic field. In a television receiver, the magnetic field for deflecting the electron beam horizontally and vertically is produced by two pairs of coils, called the deflection yoke, around the neck of the picture tube.

electromagnetic deflection coil — A coil around the neck of a CRT, for deflecting the electron beam.

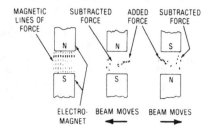

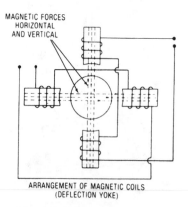

Electromagnetic deflection coil.

electromagnetic delay line— A delay line whose operation is based on the time of propagation of electromagnetic waves through distributed or lumped capacitance and inductance.

electromagnetic energy— Forms of radiant energy, such as radio waves, heat waves, light waves, X-rays, gamma rays, and cosmic rays.

electromagnetic environment— The rf field or fields existing in an area or desired in an area to be shielded.

electromagnetic field— 1. The field of influence produced around a conductor by the current flowing through it. 2. A rapidly moving electric field and its associated magnetic field. The latter is perpendicular to both the electric lines of force and their direction. 3. The field associated with radio or light waves, consisting of a magnetic and an electric field at right angles to each other and to the direction of wave propagation.

electromagnetic focusing— In a television picture tube, the focusing produced by a coil mounted on the neck. Direct current through the coil produces magnetic field lines parallel to the tube axis.

electromagnetic horn— A horn-shaped structure that provides highly directional radiation of radio waves in the 100-megahertz or higher frequency range.

electromagnetic induction— The voltage produced in a coil as the number of magnetic lines of force (flux linkages) passing through the coil changes.

electromagnetic inertia— 1. The characteristic delay of a current in an electric circuit in reaching its maximum or zero value after application or removal of the source voltage. 2. The property of self-induction.

electromagnetic interference— Abbreviated EMI. 1. Unintentional interfering signals generated within or external to electronic equipment. Typical sources could be power-line transients, noise from switching-type power supplies, and/or spurious radiation from oscillators. EMI is suppressed with power-line filtering, shielding, etc. EMI suppression requirements are frequently specified for military equipment. 2. Electromagnetic phenomena which, either directly or indirectly, can contribute to a degradation in performance of an electronic receiver or system. (The terms *radio interference, radio-frequency interference, noise, emi,* and *rfi* have been employed at various times in the same context.) 3. Disturbances caused by electromagnetic waves (radio, heat, light, etc.) that can impair the reception of the desired transmitted signal. 4. Unwanted electromagnetic emissions, generated by lightning or by electronic or electrical devices, that degrade the performance of another electronic device. Interference may be reduced by shielding. Maximum acceptable levels of EMI from electronic devices are detailed by the Federal Communications Commission.

electromagnetic lens— 1. An electron lens in which the electron beams are focused electromagnetically. 2. An electromagnet that produces a suitably shaped magnetic field for the focusing and deflection of charged particles in electron-optical systems. 3. An electron lens consisting of a homogeneous axial electric field and a magnetic field, used in high-quality image tubes for high MTF and small geometrical distortion requirements.

electromagnetic mirror— A surface or region capable of reflecting radio waves, such as one of the ionized layers in the upper atmosphere.

electromagnetic oscillograph— An oscillograph in which a mechanical motion is derived from electromagnetic forces to produce a record.

electromagnetic pollution— The effects of electromagnetic interference (EMI) produced by human-made apparatus. The seriousness of this interference ranges from annoying interference that affects a radio or television channel to interference that causes failure of an important communication channel or a cardiac pacemaker.

electromagnetic pulse— Abbreviated EMP. A reaction of large magnitude resulting from the detonation of nuclear weapons.

electromagnetic radiation— 1. That form of energy which is characterized by transversely oscillating electric and magnetic fields and which propagates at velocity c in free space. At a sufficient distance from the source, the electric-field vector and the magnetic field vector are at right angles to each other, forming a right-handed (coordinate) system. In an ionized medium, a longitudinal component may be present. 2. A form of power emitted from vibrating charged particles. A combination of oscillating electric and magnetic fields, electromagnetic radiation propagates through otherwise empty space with the velocity of light. This (constant) velocity equals the alternation frequency multiplied by the wavelength; hence the frequency and wavelength are inversely proportional to each other. The spectrum of electromagnetic radiation is continuous over all frequencies. 3. Abbreviated EMR. When discussing shielding, describes radiation generated by electrical means, ranging from a stationary magnetic or electrostatic field to high-frequency changing fields and transmitted plane waves of radio frequency.

electromagnetic reconnaissance— Activity conducted to locate and identify potential hostile sources of electromagnetic radiation, including radar, communication, missile-guidance, and air-navigation equipment.

electromagnetic relay— 1. Device that opens or closes contacts by setting "moving" contacts against "fixed" contacts when current passes through an electromagnet. Current sets up a magnetic attraction between the core of the electromagnet and a hinged arm to the tip of which is attached the moving contact. The movement of the arm toward the core of the electromagnet brings moving and fixed contacts together. When current is withdrawn, a spring returns the arm to its original position and the contacts separate. 2. A mechanical switch operated by electric power.

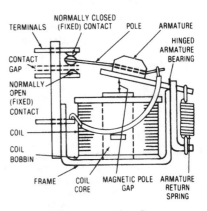

Electromagnetic relay.

electromagnetic repulsion— The repelling action between like poles of electromagnets.

electromagnetics — In physics, the branch concerned with the relationships between electric currents and their associated magnetic fields.

electromagnetic spectrum — 1. A chart or graph showing the relationships among all known types of electromagnetic radiation classified by wavelengths. 2. The continuous range of frequencies, from 0.1 to 10^{22} hertz, of which a radiated signal is composed. Spectral dimensions are more conveniently described in terms of wavelength (angstroms), where 1 angstrom is equivalent to 10^{-7} mm. The electromagnetic spectrum includes radio-frequency waves, light waves, microwaves, infrared, X-rays (Roentgen rays), and gamma rays. 3. The ordered array of known electromagnetic radiations, extending from the shortest wavelengths, cosmic rays, through gamma rays, X-rays, ultraviolet radiation, visible radiation, and infrared, and including microwave and all other wavelengths of radio energy. 4. The entire range of wavelengths, extending from the shortest to the longest or conversely, that can be generated physically. This range of electromagnetic wavelengths extends almost from zero to infinity and includes the visible portion of the spectrum known as light. *See also* visible spectrum. 5. The total range of wavelengths or frequencies of electromagnetic radiation, extending from the longest radio waves to the shortest known cosmic rays.

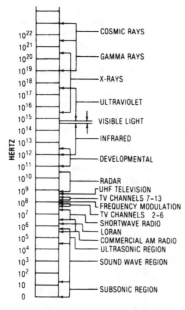

Electromagnetic radiation spectrum chart.

electromagnetic tester — A nondestructive test method for engineering materials, including magnetic materials, that uses electromagnetic energy having frequencies less than those of visible light to yield information regarding the quality of the tested materials.

electromagnetic theory — The theory of propagation of energy by combined electric and magnetic fields included in Maxwell's equations.

electromagnetic theory of light — The theory that states that electromagnetic and light waves have identical properties.

electromagnetic transduction — 1. Conversion of the measurand into the output induced in a conductor by a change in magnetic flux. 2. A wave produced by the oscillation of an electric charge. 3. A wave in which there are both electric and magnetic displacements. 4. A transverse wave associated with the transmission of electromagnetic energy.

electromagnetic-type microphones — Microphones in which the voltages are varied by an electromagnet (namely, ribbon or velocity, dynamic or moving-coil, and reluctance or moving-vane microphones).

electromagnetic unit — Abbreviated emu. A unit of electricity based primarily on the magnetic effect of an electric current. The fundamental centimeter-gram-second unit is the abampere. Now considered obsolete.

electromagnetic vibrator — A mechanical device for interrupting the flow of direct current and thereby making it a pulsating current. This is done where a circuit requires an alternating current to operate. A reed within the vibrator is alternately attracted to two electromagnets.

electromagnetic wave — 1. The radiant energy produced by oscillation of an electric charge. It includes radio, infrared, visible and ultraviolet light waves, and X-, gamma, and cosmic rays. 2. A wave in which both electric and magnetic displacement are present. 3. Waves of radiation identified by individual fluctuations of electric and magnetic fields. All such waves propagate at the speed of light in free space, which includes most realistic atmospheric conditions. Three material parameters are necessary and sufficient to describe electromagnetic waves in a given medium: dielectric constant (or permittivity), permeability, and conductivity.

electromagnetism — 1. The magnetic field around a wire or other conductor when, and only when, current passes through it. 2. Magnetism caused by an electric current in a conductor.

electromanometer — Instrument used for measuring pressure of gases or liquids by electronic methods.

electromechanical — Any device using electrical energy to produce mechanical movement.

electromechanical bell — A bell with a prewound spring-driven clapper that is tripped electrically to ring the bell.

electromechanical breakdown — A mechanical runaway that occurs when the mechanical restoring force fails to balance the electrical compressive force.

electromechanical chopper — *See* contact modulator.

electromechanical diffused-collector transistor — *See* EDCT.

electromechanical energy — Energy present in an induction coil or solenoid.

electromechanical frequency meter — A meter that uses the resonant properties of mechanical devices to indicate frequency.

electromechanical recorder — A device that transforms electrical signals into equivalent mechanical motion that is transferred to a medium by cutting, embossing, or writing.

electromechanical timer — Usually refers to a motor-driven timer, with or without an electrically operated clutch. Can also apply to pneumatic and thermal timers, or slow pull-in or drop-out relays.

electromechanical transducer — A device that transforms electrical energy into mechanical energy and vice versa. A speaker is an example of the first, and a microphone of the second.

electromechanics — That branch of electrical engineering concerned with machines producing or operated by electric currents.

electrometallurgy—That branch of science concerned with the application of electrochemistry to the extraction or treatment of metals.

electrometer—1. An electrostatic instrument that measures a potential difference or an electric charge by the mechanical force exerted between electrically charged surfaces. 2. A dc voltmeter with an extremely high input resistance, usually around 10^{10} megohms, as opposed to 10 megohms or less for a conventional type.

electrometer amplifier—An amplifier circuit having sufficiently low current drift and other noise components, sufficiently low amplifier input-current offsets, and adequate power and current sensitivities to be usable for measuring current variations of considerably less than 10^{-12} A.

electrometer tube—A vacuum tube having a very low control-electrode conductance, to facilitate the measurement of extremely small direct currents and voltages.

electromigration—Motion of ions of a metal conductor (such as aluminum) in response to the passage of high current through it. Such motion can lead to the formation of "voids" in the conductor, which can grow to a size such that the conductor is unable to pass current. Electromigration is aggravated at high temperature and high current density and therefore is a reliability "wearout" process. Electromigration is minimized by limiting current densities and by adding metal impurities such as copper or titanium to aluminum.

electromotive force—Abbreviated emf. 1. The force that causes electricity to flow when there is a difference of potential between two points. The unit of measurement is the volt. 2. Electrical pressure at the source. Not to be confused with potential difference, which is the voltage developed across a resistance or impedance due to current flowing through it. Both are measured in volts. 3. Electric pressure that causes a current to flow in a circuit; it is the energy put into the circuit by the source per unit electric charge that it supplies to the circuit. The unit of emf is the volt, being the electromotive force required to cause a current of 1 ampere to flow in a resistance of 1 ohm. 4. The difference of electrical potential found across the terminals of a source of electrical energy; more precisely, the limit of the potential difference across the terminals of a source as the current between the terminals approaches zero.

electromotive force series—A list of elements arranged according to their standard electrode potentials, with noble metals, such as gold, being positive and active metals, such as zinc, being negative.

electromotive series—A list of metals arranged in decreasing order of their tendency to pass into ionic form by losing electrons.

electromyogram—A waveform of the contraction of a muscle as a result of electrical stimulation. Usually the stimulation comes from the nervous system (normal muscular activity). The record of potential difference between two points on the surface of the skin resulting from the activity or action potential of a muscle.

electromyograph—An instrument for measuring and recording potentials generated by muscles.

electromyography—Abbreviated EMG. Recording and interpretation of the electrical activity of muscle tissue. A single electrical spike potential is generated when a muscle fiber contracts. The magnitude of the spike potentials is roughly proportional to the amount of muscular tension. Surface detecting electrodes (for many muscle fibers) or needle electrodes (for one or a few fibers) provide a signal that is amplified and displayed on a cathode-ray tube.

electron—An elementary atomic particle that carries the smallest negative electric charge (1.6×10^{-19} coulombs). Electrons are light in mass (1/1840 of the mass of the hydrogen atom, or 9.107×10^{-28} gram), highly mobile, and orbit the nucleus of an atom. Electrons are responsible for the bonds between atoms. Positive electrons, or positrons, also exist.

electronarcosis—1. The induction of unconsciousness by passage of a weak current through the brain. 2. Anesthesia induced by the passage of a precisely controlled electric current through the brain.

electron attachment—Process by which an electron is attached to a neutral molecule to form a negative ion. Often characterized by the attachment coefficient η, which is the number of attachments per centimeter of drift. Also characterized by the ratio $h = \sigma/\theta$, where σ is the attachment cross section and θ the total cross section.

electron avalanche—The chain reaction started when one free electron collides with one or more orbiting electrons and frees them. The free electrons then free others in the same manner, and so on.

electron band—A spectrum band composed of molecules that is usually found in the visible or the ultraviolet because of the electron transition taking place with the molecule.

electron beam—1. A narrow stream of electrons moving in the same direction under the influence of an electric or magnetic field. 2. A stream of electrons, emitted by a single source, that move in the same direction and at the same speed. 3. The electrons emitted by the cathode in a picture tube and focused into a beam that is deflected line by line across the phosphor screen to produce an image.

electron-beam bonding—Process using a stream of electrons to heat and bond two conductors within a vacuum.

electron-beam evaporation—An evaporation technique in which the evaporant is heated by electron bombardment.

electron-beam generator—A velocity-modulated generator, such as a klystron tube, used to generate extremely high frequencies.

electron-beam gun—A device generally used in a cathode-ray or camera tube to emit a stream of electrons moving at uniform velocity in a straight line. It consists of an emitting cathode and an anode, with an aperture for passage of some of the electrons.

electron-beam instrument—Also called a cathode-ray instrument. An instrument in which a beam of electrons is deflected by an electric or magnetic field (or both). Usually the beam is made to strike a fluorescent screen so the deflection can be observed.

electron-beam machining—A process in which controlled electron beams are used to weld or shape a piece of material.

electron-beam magnetometer—An instrument that measures the intensity and direction of magnetic forces by the immersion of an electron beam into the magnetic field.

electron-beam mode discharge—Abbreviated ebmd. A form of discharge produced by a perforated-wall hollow cathode operating under conditions of pressure, voltage, and geometry usually associated with the abnormal glow discharge.

electron-beam recording—The recording of the information contained in a modulated electron beam onto photographic or silicon-resin-coated materials.

electron-beam tube—An electron tube that depends for its operation on the formation and control of one or more electron beams.

electron-beam welding — 1. The process of using a focused beam of electrons to heat materials to the fusion point. 2. Process in which a welder generates a stream of electrons traveling at up to 60 percent of the speed of light. It focuses the beam to a small, precisely controlled spot in a vacuum and converts the kinetic energy into an extremely high temperature on impact with the work piece.

electron-bombarded semiconductor amplifier — Abbreviated EBS amplifier. An amplifier consisting of an electron-gun modulation system, semiconductor target, and output coupling network all within a glass or ceramic envelope. The semiconductor target is a pair of silicon diodes, each consisting of two metallic electrodes with a pn junction under the top contact. Amplifier operation is based on the fact that a modulated electron beam can control the current in a reverse-biased semiconductor junction.

electron-bombardment-induced conductivity — In a multimode display storage tube, a process by which the image on the surface of the cathode-ray tube is erased by the use of an electron gun.

electron charge — Also called elementary charge. The charge of a single electron. Its value is 1.602189×10^{-19} coulomb. The fundamental unit of electrical charge.

electron-coupled oscillator — Abbreviated ECO. A circuit using a multigrid tube in which the cathode and two grids operate as a conventional oscillator and the electron stream couples the plate-circuit load to the oscillator.

electron coupling — In vacuum (principally multigrid) tubes, the transfer of energy between electrodes as electrons leave one and go to the other.

electron device — Any device in which the passage of electrons through a vacuum, gas, or semiconductor is the device's principal means of conduction.

electron diffraction — 1. The phenomenon or the technique of producing diffraction patterns through the incidence of electrons on matter. 2. The bending of an electron stream that occurs when the stream travels through a medium such as very thin metal foil.

electron-diffraction camera — A special evacuated camera equipped with means for holding a specimen and bombarding it with a sharply focused beam of electrons. A cylindrical film placed around the specimen records the electrons that may be scattered or diffracted by it.

electron drift — The movement of electrons in a definite direction through a conductor, as opposed to the haphazard transfer of energy from one electron to another by collision.

electronegative — Having an electric polarity that is negative.

electronegative developer — A developer containing negatively charged toner particles.

electron emission — The freeing of electrons into space from the surface of a body under the influence of heat, light, impact, chemical disintegration, or a potential difference.

electron emitter — In a cathode tube, the electrode that serves as a source for electrons.

electron filter lens — An electrostatic device that uses an electric potential barrier to allow the transmittance of electrons at or above a set level of energy while stopping the passage of those below it.

electron flow — The movement of electrons from a negative to a positive point in a metal or other conductor, or from a negative to a positive electrode through a liquid, gas, or vacuum.

electron gun — 1. An electrode structure that produces and may control focus and may deflect and converge one or more electron beams. 2. A device for producing and accelerating a beam of electrons. 3. The portion of a TV picture tube or cathode-ray tube that produces the stream of electrons and may also focus and center the stream. 4. The source of the electron beam in a picture tube, comprising a cathode plus several focusing electrodes that collimate and focus the electron beam into a spot on the screen. In a color tube there may be three electron guns usually integrated into a single unit (unitized gun), or a single gun for the three colors.

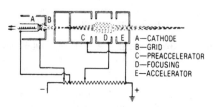

A—CATHODE
B—GRID
C—PREACCELERATOR
D—FOCUSING
E—ACCELERATOR

Electron gun.

electronic — 1. Pertaining to that branch of science which deals with the motion, emission, and behavior of currents of free electrons, especially in vacuum, gas, or phototubes and special conductors or semiconductors. This is contrasted with electric, which pertains to the flow of large currents in metal conductors. 2. Of or pertaining to devices, circuits, or systems using the principle of electron flow through a conductor. Examples: electronic control, electronic equipment, electronic instrument, electronic circuit.

electronic autopilot — An arrangement of gyroscopes, electronic amplifiers, and servomotors for detecting deviations in the flight of an aircraft and applying the required corrections directly to its control cables.

electronic balance — Weighing balance that uses forces produced by known currents to balance unknown currents, and thereby unknown weights, very accurately to within parts of a microgram.

electronic "bug" — A keying system that converts the Morse signals from a hand key into correctly proportioned and spaced dots and dashes.

electronic bulletin board — A shared file where users can enter information for other users to read or download. Many bulletin boards are set up according to general topics and are accessible throughout a network.

electronic calculator — Electronic device for arithmetic and logarithmic computations; may also include digital printer and computer.

electronic camouflage — Use of electronic means or exploitation of electronic characteristics to reduce, submerge, or eliminate the radar-echoing properties of a target.

electronic carburetor — A fuel-metering actuator in which the air/fuel ratio is controlled by continual variations of the metering rod position in response to an electronic control signal.

electronic charge — The quantity of charge represented or possessed by one electron. It is equal to 1.602189×10^{-19} coulomb.

electronic circuit — A circuit containing one or more electron tubes, transistors, integrated circuits, magnetic amplifiers, etc.

electronic commutator — A type of switch that provides a continuous switching or sampling of a number of circuits by means of a radial-beam electronic tube or electronic switching circuit.

electronic confusion area — Amount of space that a target appears to occupy in a radar resolution cell, as it appears to that radar beam.

electronic control — Also called electronic regulation. The control of a machine or condition by electronic devices.

electronic control unit — See ECU.

electronic counter — An instrument capable of counting up to several million electrical pulses per second.

electronic counter-countermeasures — Abbreviated ECCM. 1. Equipment and techniques that allow electronic systems such as radar and communication systems to operate effectively while attempts are made to disrupt or jam their operation. 2. That division of electronic warfare involving actions taken to ensure friendly and effective use of the electromagnetic spectrum despite the enemy's use of electronic warfare. 3. Retaliatory tactics used to reduce the effectiveness of electronic countermeasures.

electronic countermeasures — Abbreviated ECM. 1. All measures taken to reduce the effectiveness of enemy electronic systems such as radar and communications. There are two distinct areas: passive measures, or reconnaissance, and active measures, such as jamming. 2. That division of electronic warfare involving actions taken to prevent or reduce an enemy's effective use of the electromagnetic spectrum. Includes techniques such as chaff and barrage jamming as well as sophisticated methods to deceive the systems without indication to the opposing operators that their systems are being affected. 3. Methods of jamming or otherwise hindering the operation of enemy electronic equipment.

electronic countermeasures control — 1. Collection and sorting of large quantities of data for the purpose of measuring and defining radar signals. 2. Examination of the data received in order to determine selection and switching of countermeasure devices with little or no time delay.

electronic coupling — The method of coupling electrical energy from one circuit to another through the electron stream in a vacuum tube.

electronic crowbar — An electronic switching device generally used in a power supply to divert a fault current from more delicate components until a fuse, circuit breaker, or the like has time to respond.

electronic data exchange — See EDI.

electronic data processing — Abbreviated EDP. 1. Operations on data carried out mainly by electronic equipment. 2. Use of electronic memories to store, update, and read information automatically, and using that information in accounting, filing, etc. 3. Any computerized information system and the equipment used in that system.

electronic data-processing center — Abbreviated EDP center. A place in which is kept automatically operated equipment, including computers, designed to simplify the interpretation and use of data gathered by instrumentation installations or information-collection agencies.

electronic data-processing machine — Abbreviated EDPM. A machine or its device and attachments used primarily in or with an electronic data-processing system.

electronic data-processing system — Any machine or group of automatically intercommunicating machines capable of entering, receiving, sorting, classifying, computing, and/or recording alphabetical or numerical accounting or statistical data (or all three).

electronic deception — Deliberate radiation, reradiation, alteration, absorption, or reflection of electromagnetic radiations in a manner intended to cause the enemy to obtain misleading data or false indications from his electronic equipment. There are two categories of electronic deception: (a) Manipulative deception — the alteration or simulation of friendly electromagnetic radiation to accomplish deception. (b) Imitative deception — the introduction into enemy channels of radiations that imitate the enemy's own emissions.

electronic device — 1. A device in which conduction is principally by the movement of electrons through a vacuum, gas, or semiconductor. 2. An electronic tube or valve, transistor, or other semiconductor device. This definition excludes inductors, capacitors, resistors, and similar components.

electronic differential — An input or output type of circuit that only amplifies or responds to the difference of two signals, and does not respond to the signal with respect to ground or a supply voltage.

electronic differential analyzer — A form of analog computer using interconnected electronic integrators to solve differential equations.

electronic digital computer — A machine that uses electronic circuitry in the main computing element to perform arithmetic and logical operations on digital data (i.e., data represented by numbers or alphabetic symbols) automatically by means of an internally stored program of machine instructions. Such devices are distinguished from calculators, on which the sequence of instructions is externally stored and is impressed manually (desk calculators) or from tape or cards (card-programmed calculators).

electronic efficiency — The ratio of (a) the power at the desired frequency delivered by the electron stream to the oscillator or amplifier circuit to (b) the direct power supplied to the stream.

electronic engine control — See EEC.

electronic engineering — A branch of electrical engineering that applies the principles of electronics to the solution of practical problems. See also electronics.

electronic flash — Also called strobe. 1. The firing of special light-producing, high-voltage, gas-filled glass tubes with a high instantaneous surge of current furnished by a capacitor or bank of capacitors that have been charged from a high-voltage source (usually 450 volts or higher). 2. A device that upon command produces a pulse of luminous energy caused by a discharge of electrical energy through a gas. The term usually implies the use of a flash tube and associated power source and trigger circuit.

electronic flash tube — See flash tube.

electronic flash unit — A small xenon-filled tube with metal electrodes fused into the ends. The gas flashes brilliantly when a capacitor is discharged through the tube. The duration of the flash primarily depends on the capacitance of the capacitor. As a rule of thumb, flash time is approximately equal in microseconds to capacitance in microfarads. Sometimes the flash recurs at a specified frequency, which may reach many thousands per second; such a device is called a strobe unit because it

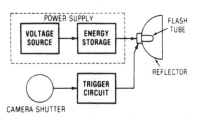

Electronic flash unit.

produces stroboscopic effects and makes rapidly moving parts appear to stand still.

electronic frequency synthesizer — A device that generates two or more selectable frequencies from one or more fixed-frequency sources.

electronic funds transfer system — *See* EFTS.

electronic gate — A device in which diodes and/or transistors provide input-output relations that correspond to a Boolean-algebra function (AND, OR, etc.).

electronic heating — Also called high-frequency heating. A method of heating a material by inducing a high-frequency current in it or having the material act as the dielectric between two plates charged with a high-frequency current.

electronic hookup wires — Wires used to make the internal connections between the various electrical parts of electronic assemblies.

electronic industries — Industrial organizations engaged in the manufacture, design, development, and/or substantial assembly of electronic equipment, systems, assemblies, or the components thereof.

Electronic Industries Association — Abbreviated EIA. A trade association of the electronics industry. Some of its functions are the formulation of technical standards, dissemination of marketing data, and the maintenance of contact with government agencies in matters relating to the electronics industry. The association was originally known as the Radio Manufacturers Association (RMA), and later as the Radio-Electronics-Television Manufacturers Association (RETMA).

electronic instrument — Any instrument that depends for its operation on the action of either one or more electron devices.

electronic instrument cluster — *See* EIC.

electronic intelligence — The technical and intelligence information derived from foreign noncommunications electromagnetic radiations emanating from other than nuclear detonations or radioactive sources.

electronic interference — Electrical or electromagnetic disturbances that result in undesired response in electronic equipment.

electronic jamming — Intentional radiation, reradiation, or reflection of electromagnetic energy for the purpose of reducing the effectiveness of enemy electromagnetic devices or impairing the use of any electronic devices, equipment, or systems being used by an enemy.

electronic keyboard — A keyboard that is used to generate characters through electronic means rather than through mechanical linkages.

electronic keying — A method of keying whereby the dots and dashes are produced solely by electronic means.

electronic line scanning — Facsimile scanning in which a spot on a cathode-ray tube moves across the copy electronically while the record sheet or subject copy is moved mechanically in a perpendicular direction.

electronic mail — Abbreviated e-mail or email. 1. Electronic messages that can be sent over a communications network from one computer to another. 2. Sending messages electronically between computers or terminals.

electronic microphone — A device that depends for its operation on the generation of a voltage by the motion of one of the electrodes in a special electron tube.

electronic mine detector — *See* mine detector.

electronic multimeter — A device employing the characteristics of an electron-tube circuit for the measurement of electrical quantities, at least one of which is voltage or current, on a single calibrated scale.

electronic music — The electronic generation and processing of audio signals, or the electronic processing of natural sound, and the manipulation and arrangement of these signals via tape recorders into a finished musical composition.

electronic music synthesizer — An audio signal processor that contains sound generators (oscillators) and additional circuitry such as filters to produce familiar sounds, such as those produced by conventional musical instruments, or to create unique sounds and effects.

electronic news-gathering — Abbreviated ENG. The use of video cameras, recording, and other ancillary electronic gear to collect news stories for TV airing.

electronic organ — The electronic counterpart of the pipe organ. All tones and tone variations, such as vibrato, tremolo, etc., are produced by electronic circuits instead of by pipes.

electronic pacemaker — Also called a pacemaker. An electrical device, usually with electrodes planted in the myocardium, that performs the pacing function in a diseased heart no longer capable of pacing itself. Electronic pacemakers can receive power from implanted batteries, radio-frequency signals, biological energy sources, etc.

electronic packaging — The coating or surrounding of an electronic assembly with a dielectric compound.

electronic part — A basic circuit element that cannot be disassembled and still perform its intended function. Examples of electronic parts are capacitors, connectors, filters, resistors, switches, relays, transformers, crystals, electron tubes, and semiconductor devices.

electronic photometer — Also called photoelectric photometer. A photometer with a photocell, phototransistor, or phototube for measuring the intensity of light.

electronic power supply — A circuit that transforms electrical input energy, that is, alternating or direct current. (Sources operating on rotating machine principles, or deriving electrical power from other energy forms such as batteries and solar cells, are excluded.) Supplies covered by this definition fall into one of four groups: 1) ac in, dc out — most common supplies. 2) ac in, ac out — line regulators, variable-frequency supplies. 3) dc in, dc out — converters. 4) dc in, ac out — inverters.

electronic products — Materials, parts, components, subassemblies, and equipment that employ the principles of electronics in performing their major functions. These products may be used as instruments and controls in communications, detection, amplification, computation, inspection, testing, measurement, operation, recording, analysis, and other functions employing electronic principles.

electronic profilometer — An electronic instrument for measuring surface roughness. The diamond-point stylus of a permanent-magnet dynamic pickup is moved over the surface being examined. The resultant variations in voltage are amplified and then measured with a meter calibrated to read directly in microinches of deviation from smoothness.

electronic raster scanning — Scanning by electronic means so that substantially uniform coverage of an area is provided by a predetermined pattern of scanning lines.

electronic reconnaissance — Search for electromagnetic radiations to determine their existence, source, and pertinent characteristics for electronic warfare purposes.

electronic rectifier — A rectifier using electron tubes or equivalent semiconductor elements as rectifying elements.

electronic regulation — *See* electronic control.

electronic relay — An electronic circuit that provides the functional equivalent of a relay but has no moving parts.

electronics — 1. The field of science and engineering concerned with the behavior of electrons in devices and the utilization of such devices. 2. Of or pertaining to the field of electronics, such as electronics engineer, course, laboratory, committee. 3. Name given to that branch of electrical engineering which deals with devices whose operation depends upon the movement of electrons in space as opposed to the movement of electrons in liquids or solid conductors, e.g., radio tubes, photoelectric cells. It includes the study of radio, radar, television, sound films, and control of industrial processes. 4. That branch of science involved in the study and utilization of the motion, emissions, and behaviors of currents of electrical energy through gases, vacuums, semiconductors, and conductors; not to be confused with electrics, which deals primarily with the conduction of large currents of electricity through metals. 5. That branch of science and technology which deals with the study, application, and control of the phenomena of conduction of electricity in a vacuum, in gases, in liquids, in semiconductors, and in conducting and superconducting materials.

electronic search reconnaissance — The determination of the presence, source, and significant characteristics of electromagnetic radiations.

electronic security — Protection resulting from measures designed to deny to unauthorized persons information of value that might be obtained by interception and analysis of noncommunications electromagnetic radiations.

electronic shock absorption — An integrated data bit storage buffer inside a CD portable, which receives information at twice the normal speed but supplies information to the digital-to-analog converter at normal speed, ensuring that any interruption of the data flow caused by shocks or bumps does not result in interruption of play.

electronic shutter — A mechanical shutter with an electronic timing circuit. This circuit allows a wider range of exposure times, can be more accurate, and, placed in a circuit with a photoconductive cell, allows automatic setting of shutter speeds.

electronic sky screen equipment — An electronic device for indicating the departure of a missile from a predetermined trajectory.

electronic speed control — 1. A system whereby a motor's speed is controlled by feedback from a frequency-sensing circuit attached to the device being powered; changes from the desired speed cause corrective signals to speed up or slow down the motor. 2. Changes in speed in a record player, whether gross (as from $33\frac{1}{3}$ to 45 rpm) or small (as an order of ±3 percent), can be made by alternating components in the external speed-regulation circuit, rather than by mechanically shifting belts or idler wheel.

electronic sphygmomanometer — Device that measures and/or records blood pressure electronically.

electronic stethoscope — An electronic amplifier of sounds within a body. Its selective controls permit tuning for low heart tones or high pulmonary tones. It has an auxiliary output for recording or viewing audio patterns.

electronic stimulator — A device for applying electronic pulses or signals to activate muscles, or to identify nerves, or for muscular therapy, etc.

electronic surge arrestor — A device used to switch high-energy surges to ground so as to reduce the transient energy to a level that is safe for secondary protectors (e.g., zener diodes, silicon rectifiers).

electronic switch — 1. A circuit element causing a start and stop action or a switching action electronically, usually at high speeds. 2. An electronic circuit used to perform the function of a high-speed switch. Applications include switching a cathode-ray oscilloscope back and forth between two inputs at such high speed that both input waveforms appear simultaneously on the screen.

electronic switching — Electronic circuits and solid-state devices used to perform most telephone central office switching functions.

electronic switching system — Abbreviated ESS. 1. A telephone switching system that uses a computer with a storage containing program switching logic. The output of the computer actuates reed or electronic switches that establish telephone connections automatically. 2. A system that uses solid-state switching devices and computerlike operations to accomplish switching of telephone calls. 3. A type of telephone switching system that uses a special-purpose digital computer to direct and control the switching operation. ESS permits custom-calling services such as speed dialing, call transfer, and three-way calling.

electronic thermal conductivity — The part of thermal conductivity due to the transfer of thermal energy by means of electrons and holes.

electronic timer — 1. A synchronizer, pulse generator, modulator, or keyer that originates a series of continuous control pulses at an unvarying repetition rate known as the pulse-recurrence frequency. 2. A timer using electronic circuits (either tube or transistor type) to control a time period, in place of a motor or other means.

electronic tube relay — A relay that employs electronic tubes as components.

electronic tuning — 1. Altering the frequency of a reflex klystron oscillator by changing the repeller voltage. 2. Frequency changing in a transmitter or receiver by changing a control voltage rather than circuit components.

electronic video recording — The recording of video images by means of photographic film, or magnetic tape or disk, so that the image's record can be played back in a video format at a later time.

electronic viewfinder — Also called viewfinder monitor. 1. A small TV screen attached to a video camera that allows the operator to view a given scene exactly as it is being viewed by the camera. 2. A small television camera that replaces the reflex viewfinder of a motion picture camera. This permits the image photographed to be viewed simultaneously by a number of people, since the TV image may be transmitted to several receivers.

electronic voltmeter — Also called vacuum-tube voltmeter. A voltmeter that utilizes the rectifying and amplifying properties of electron tubes or semiconductors and their circuits to secure such characteristics as high input impedance, wide frequency range, peak-to-peak indications, etc.

electronic volt-ohmmeter — A device employing the characteristics of an electron-tube or semiconductor circuit for the measurement of voltage and resistance on a single-calibrated scale.

electronic warfare — Abbreviated EW. 1. Military usage of electronics to reduce an enemy's effective use of radiated electromagnetic energy and to ensure our own effective use. 2. Military action involving the use of electromagnetic energy to determine, exploit, reduce, or prevent hostile use of the electromagnetic spectrum, and action that retains friendly use of the electromagnetic spectrum. There are three divisions within electronic warfare: electronic warfare support measures (ESM), electronic countermeasures (ECM), and electronic counter-countermeasures (ECCM).

electronic warfare support measures — Abbreviated ESM. That division of electronic warfare involving actions taken to search for, intercept, locate, and immediately identify radiated electromagnetic energy for the purpose of immediate threat recognition. Thus, ESM provides a source of information required for immediate action involving electronic countermeasures, electronic counter-countermeasures, avoidance, targeting, and other tactical employment of forces.

electronic watch — A timepiece in which a battery replaces the mainspring, and semiconductor elements replace the mechanical switching-contact arrangement.

electronic waveform synthesizer — An instrument using electron devices to generate an electrical signal of a desired waveform.

electron image — A representation of an object formed by a beam of electrons focused by an electron optical system.

electron image tube — 1. A cathode-ray tube having a photoemissive mosaic upon which an optical image is projected, and an electron gun to scan the mosaic and convert the optical image into corresponding electrical current. 2. A cathode-ray tube that increases the brightness or size of an image, or forms a visible image from invisible radiation. The focal plane for the optical image is a large, light-sensitive cold cathode. The emission from the cathode is first accelerated through a suitable lens system and then strikes a fluorescent screen, where an image is formed that is an enlarged and brightened reproduction of the original image.

electron lens — 1. The convergence of the electrons into a narrow beam in a cathode-ray tube by deflecting them electromagnetically or electrostatically. So called because its action is analogous to that of an optical lens. 2. A system of deflecting electrodes or coils designed to produce an electric field that influences a beam of electrons in the same manner in which a lens affects a light beam.

electron metallurgy — That branch of metallurgy that uses electron microscopic techniques in the examination of the nature of metals.

electron micrograph — A reproduction of an image formed by the action of an electron beam on a photographic emulsion.

electron micrography — The photographic recording of images produced by the electrons from an electron microscope. The electron beam carries the images through an array of lenses, and an enlarged electron image is used to stimulate a fluorescent screen that is photographed by a camera system.

electron microradiography — The photographic recording and later enlarging of very thin specimens, using an electron beam to form the image.

electron microscope — A device utilizing an electron beam for the observation and recording of submicroscopic samples with the aid of photographic emulsions or other short-wavelength sensors. Useful magnification is over 300,000.

electron microscopy — The study of materials by means of the electron microscope.

electron mirror — An electron instrument used to totally reflect an electron beam.

electron multiplier — A vacuum tube in which electrons liberated from a photosensitive cathode are attracted to a series of electrodes called dynodes. In doing so, each electron liberates others by secondary emission and thereby greatly increases the number of electrons flowing in the tube.

electron-multiplier section — A section of an electron tube in which an electron current is amplified by one or more successive dynode stages.

electron optical system — A combination of parts capable of producing and controlling a beam of electrons to produce an image of an object.

electron optics — 1. The branch of electronics concerned with the behavior of the electron beam under the influence of electrostatic and electromagnetic forces. 2. The control of free electron movement through the use of electric or magnetic fields, and use of the electron movement in research investigation of electronic diffraction phenomena, directly analogous to the control of light through use of lenses. 3. The area of science devoted to the directing and guiding of electron beams using electric fields in the same manner as lenses are used on light beams. 4. Pertaining to devices whose operation relies on modification of a material's refractive index by electric fields, for example, image-converter tubes and electron microscopes.

electron-pair bond — A valence bond formed by two electrons, one from each of two adjacent atoms.

electron paramagnetic resonance — A condition in which a paramagnetic solid subjected to two magnetic fields, one of which is fixed and the other normal to the first and varying at the resonance frequency, emits electromagnetic radiation associated with changes in the magnetic quantum number of the electrons.

electron probe — A narrow beam of electrons used to scan or illuminate an object or screen.

electron-ray tube — Also called a magic eye. 1. A tube that indicates visibly on a fluorescent target the effects of changes in control-grid voltage applied to the tube. Used as a tuning indicator in receivers. 2. A type of recording-level indicator using a luminous display in a special tube. The display is typically like an "eye" with a keyhole in the middle, and maximum recording level corresponds to the closing-up of a slot at the bottom of the keyhole (largely superseded by meters in current-model recorders).

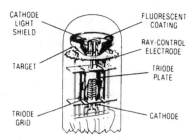

Electron-ray tube.

electron scanning — 1. The moving of an electron beam back and forth and/or up and down by deflecting the beam electromagnetically or electrostatically. 2. A deflection of a beam of electrons, at regular intervals, across a CRT screen, according to a definite pattern.

electron spectroscopy — The theory and interpretation of spectra produced by the electron emissions of substances after their irradiation by X-rays.

electron spin — The twirling motion of an electron, independent of any orbital motion.

electron-stream potential — The time average of the difference in potential between a point in an electron stream and the electron-emitting surface.

electron-stream transmission efficiency — With respect to an electrode through which an electron stream passes, the ratio of the average stream current through the electrode to the stream current approaching the electrode. (In connection with multitransit tubes, the electron stream is considered to include only those electrons approaching the electrode for the first time.)

electron telescope — 1. An apparatus for seeing through haze and fog. An infrared image is formed optically on the photoemissive mosaic of an electron-image tube and then made visible by the tube. 2. An instrument that serves to produce an enlarged electron image on a fluorescent screen by focusing an infrared image of a distant object on a photosensitive cathode.

electron trajectory — The path of an electron.

electron transit time — The time required for electrons to travel between two electrodes in a vacuum tube. This time is extremely important in tubes designed for ultrahigh frequencies.

electron tubes — Devices used to control the flow of electrons. They may be either gas filled or partially or fully evacuated (vacuum). Common tubes include vacuum tubes, cathode-ray tubes, phototubes, mercury vapor tubes, thyratrons, and microwave tubes.

electron-tube static characteristic — The relationship between two variables of an electron tube, such as the voltage and current of an electrode, with all other variables maintained constant.

electron unit — The unit of charge (negative or positive) equal to the charge on an electron.

electron velocity — The rate of motion of an electron.

electronvolt — Abbreviated ev or eV. The amount of kinetic energy gained by an electron when it is accelerated through an electric potential difference of 1 volt. It is equivalent to 1.602189×10^{-12} erg, or 1.602189×10^{-9} J. It is a unit of energy or work, not of voltage.

electron-wave tube — An electron tube in which streams of electrons having different velocities interact and cause a progressive change in signal modulation along their length.

electro-oculography — Recording and interpretation of the voltages that accompany eye movements. Eye-position voltages from electrodes placed on the skin near the eye are amplified and applied to a strip-chart recorder.

electro-optical detector — A device that detects radiation by utilizing the influence of light in forming an electrical signal. The detector may be a phototube; a photoconductive, photovoltaic, or photojunction cell; a phototransistor; or a thermal detector, such as a thermocouple or bolometer.

electro-optical transistor — A transistor capable of responding in nanoseconds to both light and electrical signals.

electro-optic coefficient — A measure of the extent to which the index of refraction changes with applied high electric fields, such as several parts per ten thousand for applied fields of the order of 20 volts per centimeter. Since the phase shift of a light wave is a function of the index of refraction of the medium in which it is propagating, the change in index can be used to phase modulate the light wave by shifting its phase at a particular point along the guide by changing the propagation time to the point.

electro-optic effect — The change in the index of refraction of a material when subjected to an electric field. The effect can be used to modulate a light beam in a material since many properties — such as light-conducting velocities, reflection and transmission coefficients at interfaces, acceptance angles, critical angles, and transmission modes — are dependent on the refractive indexes of the media in which the light travels.

electro-optic material — A material having refractive indexes that can be altered by an applied electric field.

electro-optic modulator — A device that uses an applied electric field to alter the polarization properties of light.

electro-optic phase modulation — Modulation of the phase of a light wave, such as by changing the index of refraction and thus the velocity of propagation and hence the phase at a point in the medium in which the wave is propagating, in accordance with an applied field serving as the modulating signal.

electro-optic radar — A radar system in which electro-optic instead of microwave techniques and equipment are used to perform the acquisition and tracking operation.

electro-optics — The study of the effects of electric fields on optical phenomena.

electro-optic shutter — A device used to control or block a light beam by means of the Kerr electro-optical effect.

electropad — The part of an electrocardiograph body electrode that makes contact with the skin.

electrophonic effect — The sensation of hearing produced when an alternating current of suitable frequency and magnitude from an external source is passed through an animal or human body.

electrophoresis — 1. The movement of particles or ions in solution caused by applying an electric field, as reported by O. Lodge in 1886. 2. The migration of colloidal particles under the influence of an applied electrical field. A colloidal particle, such as a protein molecule, has large numbers of positive and negative radicals that act as if they were on the surface. Thus, since protein molecules carry electric charges, they will migrate when subjected to an electric field. The fractional nature of the net charge makes possible a wide variety of electrophoretic patterns at a given pH. 3. The migration of molecules under the influence of an electric field. 4. The migration of dispersed solid, liquid, or gaseous material to one of two electrodes under the influence of an impressed direct-current voltage. 5. The motivation of particles or ions, suspended in a solution, toward the electrode having the opposite sign, due to the application of an electrical field.

electrophoresis apparatus — An apparatus for causing migration of charged particles (ions) in solution in an electric field. Types include paper, cascading electrodes, high voltage, gel, and thin layer.

electrophoresis scanner — An instrument for reading bands on paper strips or gel, for the purpose of measuring particle movement due to electrophoresis.

electrophoretic display — 1. A reflective display that offers a wide choice of colors and has a short-to-medium-term memory that consumes no power. The heart of the display is a suspension of charge pigment particles in a liquid of another color. The suspension, a layer typically 50 micrometers thick, is sandwiched between a pair of electrodes, one of which is transparent. When direct current of the right polarity is applied to the electrodes, the particles are pulled toward the transparent electrode, thus displacing the contrasting liquid and showing their own coloration. When the polarity is reversed, they move to the other electrode and are hidden by the liquid. 2. The movement of suspended particles through a fluid by an electromotive force.

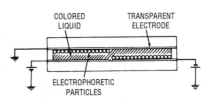

Electrophoretic display.

electrophorus — 1. An early type of static-electricity generator. 2. Simple piece of apparatus used in the laboratory to obtain a number of charges of static electricity from a single initial charge. Typically, it consists of a thick ebonite disc held in a brass sole, and a brass disc with insulated handle. The ebonite disc is charged by rubbing with fur, and the metal disc is brought near and allowed to pick up an induced charge that can be lifted and conveyed where required. 3. A device in which the electric field of an object that has been electrified by friction is used to induce charges in conductors.

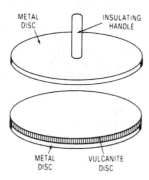

Electrophorus.

electrophotograph — The image formed in electrophotography.

electrophotographic process — The process in which images are formed by various electrical and photographic means. Examples are processes employing selenium-coated drums or zinc-oxide-coated paper.

electrophotography — 1. A term referring to a photographic process in which electrical energy is used to make materials sensitive to light. 2. The photographic recording of an image formed by the alteration in electrical properties of the sensitive materials, induced by the action of light.

electrophotometer — An instrument using a photoelectric sensor for colorimetric determinations.

electrophysiology — The science of physiology as related to electric reactions of the body.

electroplaques — Individual electricity-producing cells in eels and other electric fishes connected in series-parallel arrays, like miniature elements of a battery. They are usually thin waferlike cells, the two surfaces of which differ markedly.

electroplate — 1. To deposit a metal on the surface of certain materials by means of electrolysis. 2. To effect the transfer of one metal to another by means of electrolysis. 3. To apply a metallic coating on a conductive surface by means of electrolytic action.

electroplating — The electrodeposition of an adherent metal coating on a conductive object for protection, decoration, or other purposes, such as securing a surface with properties or dimensions different from those of the basis metal. The object to be plated is placed in an electrolyte and connected to one terminal of a dc voltage source. The metal to be deposited is similarly immersed and connected to the other terminal. Ions of the metal provide transfer to the metal as they make up the current between the electrodes. *See also* electroforming.

electropolishing — 1. The process of producing a smooth, lustrous surface on a metal by making it the anode in an electrolytic solution and preferentially dissolving

the minute protuberances. 2. The improvement in surface finish of a metal effected by making it anodic in an appropriate solution.

electropositive developer — A developer containing positively charged toner particles.

electrorefining — 1. The removal of impurities from a metal by electrolysis. 2. The process of anodically dissolving a metal from an impure anode and depositing it cathodically in a purer form.

electroretinograph — An instrument for measuring the electrical response of the human retina to light stimulation.

electroretinography — Recording and interpretation of the voltage generated by the retina of the eye. An electrode fitted to a plastic contact lens is used to pick up voltage from the surface of the eyeball.

electroscope — An electrostatic instrument for measuring a potential difference or an electric charge by means of the mechanical force exerted between electrically charged surfaces.

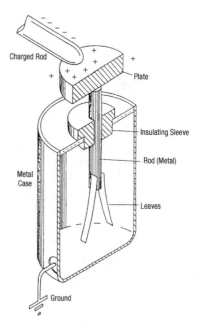

Electroscope.

electrosection — A surgical cutting technique that makes use of a radio-frequency arc.

electrosensitive paper — A paper that turns blue, brown, or black wherever a direct current passes through it. Used with facsimile and high-speed printers.

electrosensitive processor — A facsimile printing process whereby imaging is based on a two-layer paper composed of a white titanium oxide coating and dark underlayer. The paper (which may be sheet or roll fed) is imaged via contact with an electric stylus; as the charged wire touches the paper, the white coating is burned off, line by line, to correspond to the dark image areas of the original.

electrosensitive recording — 1. The passage of electric current into a sheet of sensitive paper to produce a permanent record. 2. A technique that uses the passage of an electric current through a recording medium to produce a permanent image on that medium.

electroshock — A state of shock produced by passing an electric current through the brain. It is useful in the treatment of certain mental disorders.

electroshock therapy — Treatment of certain mental disorders by passing an electric current through the brain.

electrospinograph — A device for detecting and recording electric signals of the spinal cord.

electrostatic — 1. Pertaining to static electricity — that is, electricity, or an electric charge, at rest. A constant-intensity electric charge. 2. Applied to speakers and microphones (capacitance type). An electrostatic force is used to activate the diaphragm. The charged diaphragm is suspended between two perforated plates. As an ac signal is applied to the outer plates, the diaphragm vibrates. 3. A form of electrical energy that has the capability of attracting and holding small particles having an opposite electrical charge. 4. A headphone drive system using a thin plastic membrane in a high-voltage electrostatic field, whose variation by the signal voltage moves the entire diaphragm to create a sound pressure wave. 5. The effects produced by electrical charges or fields, alone, without interaction with magnetic influence.

electrostatic actuator — An apparatus comprising an auxiliary external electrode that permits known electrostatic forces to be applied to the diaphragm of a microphone for the purpose of obtaining a primary calibration.

electrostatic capacitor — Two conducting electrodes separated by an insulating material such as air, ceramic, mica, gas, paper, plastic film, or glass. These are generally high-impedance devices.

electrostatic charge — 1. An electric charge stored in a capacitor or on the surface of an insulated object 2. The algebraic sum of all positive and negative electric charges present in a specific volume or surface element. 3. An electric charge that is in a state of equilibrium.

electrostatic charge mobility — The property of a barrier material that facilitates or impedes the movement of electrostatic charges internally or on the surface.

electrostatic coating — Process in which the coating material is electrically charged as it leaves the spray gun and is attracted to the part, which has an opposite charge.

electrostatic component — The portion of radiation due to electrostatic fields.

electrostatic-convergence principle — The principle of electron-beam convergence through use of an electrostatic field.

electrostatic copier — A type of copier that employs the principles of photoconductivity and electrostatic attraction.

electrostatic coupling — Method of coupling by which charges on one surface influence those on another through capacitive action.

electrostatic deflection — 1. In a cathode-ray tube, the deflection of the electron beam by means of pairs of charged electrodes on opposite sides of the beam. The electron beam is bent toward a positive electrode and bent away from a negative electrode. 2. The deflection of an electron beam by the action of an electrostatic field that has a component perpendicular to the direction of the beam.

elecrostatic discharge — See ESD.

electrostatic electrography — That branch of electrostatography which produces a visible record by employing an insulating medium to form latent electrostatic images with the aid of electromagnetic radiation.

electrostatic electrophotography — That branch of electrostatography which produces a visible record by employing a photoresponsive medium to form latent electrostatic images with the aid of electromagnetic radiation.

electrostatic energy — The energy contained in electricity at rest, such as in the charge of a capacitor.

electrostatic field — The vector force field set up in the vicinity of nonmoving electrical charges. The strength of this static field at a point is defined as the force per unit charge on a stationary positive test charge, provided the test charge is so small that it does not disturb the original charge distribution.

electrostatic flux — The electrostatic lines of force existing between bodies at different potentials.

electrostatic focusing — 1. The focusing of an electron beam by the action of an electric field. 2. A method of focusing the cathode-ray beam to a fine spot by application of electrostatic potentials to one or more elements of an electron lens system.

electrostatic galvanometer — Galvanometer operated by the effects of two electric charges on each other.

electrostatic generator — A device for the production of electric charges by electrostatic action.

electrostatic headphones — A device held against the ear that reproduces incoming electrical signals as sound. It relies on changes in electrical charge across a diaphragm stretched between two perforated, polarized plates. All parts of the diaphragm experience equal force, and the sound is inherently more linear.

electrostatic induction — 1. The process of inducing stationary electric charges on an object by bringing it near another object that has an excess of electric charges. A positive charge will induce a negative charge, and vice versa. 2. Capacitive induction of interfering signals over an air gap separating an instrument (e.g., from its wiring or housing).

electrostatic instrument — An instrument that depends for its operation on the attraction and repulsion between electrically charged bodies.

electrostatic latent image — In an electrostatic copier, the invisible image formed on the zinc-oxide-coated paper by the action of light.

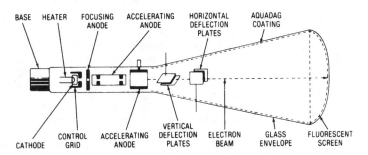

Electrostatic deflection.

electrostatic lens — A lens producing a potential field capable of deflecting electron rays to form an image of an object.

electrostatic loudspeaker — Loudspeaker in which the mechanical forces are produced by the action of electrostatic fields.

electrostatic memory — Also called electrostatic storage. A memory device in which information is retained by an electrostatic charge. A special type of cathode-ray tube is usually employed, together with associated circuitry.

electrostatic memory tube — Also called storage tube. An electron tube in which information is retained by electric charges.

electrostatic meter — A meter with a movement consisting of fixed and movable metal plates interleaved to form a capacitor. Rotation of the movable plates is proportional to the dc or ac voltage applied across the capacitor.

electrostatic microphone — Also called capacitor microphone or condenser microphone. 1. A microphone that contains a metal plate and a thin metal diaphragm set close together. The capacitance of the microphone is thus affected by movement of the diaphragm from air pressure waves. A polarizing voltage is applied to the plates. 2. A microphone whose transduction principle is based on the varying electrical charge across a sound-modulated capacitor. 3. A microphone whose capacitance varies with sound pressure; electronic circuits within the microphone convert this change in capacitance to a varying voltage signal. Electrostatic microphones, unlike other types, require a battery or other voltage source. 4. A type of microphone characterized by its wide frequency range and low distortion. Used for precision measurements and high-quality recording. Can be omnidirectional or cardioid.

electrostatic potential — The voltage that can be measured between any two objects that have different static charges.

electrostatic precipitation — The process of removing smoke, dust, and other particles from the air by charging them so that they can be attracted to and collected by a properly polarized electrode.

electrostatic printer — 1. A nonimpact printing technique that forms a copy by attracting toner particles to a static charge on the surface of a photoconductor, then transferring the toner image to the surface of a sheet of copy paper. The image is formed by a laser that develops an electrostatic image charge on the photoconductor according to information being supplied through the input data stream. Each bit of data can be related to a character shape in the memory of the printing system; in most cases, characters are formed by a dot-matrix method similar in concept to that of the matrix printer. Paper is sheet or roll fed. 2. A nonimpact printer that prints dot-matrix characters one at a time by means of wires or pins that supply an electrical charge in the desired patterns onto an aluminum-coated paper; particles of dry ink adhere to the magnetized areas and are then fixed by heat.

electrostatic process — 1. A reproduction method in which image formation depends on electrical rather than chemical changes induced by light. 2. A nonchemical, nonimpact imaging process in which a light source, corresponding to the image to be formed, discharges a charged dielectric photoconductive surface to form a latent image. This surface (a photoconductor) containing the latent image is then dusted with dielectric toner powder, which adheres to the charged areas, rendering the image visible.

electrostatic recording — Recording by means of a signal-controlled electrostatic field.

electrostatic relay — A relay in which two or more conductors that are separated by insulating material move because of the mutual attraction or repulsion produced by electric charges applied to the conductors.

electrostatics — The branch of physics concerned with electricity at rest.

electrostatic separator — An apparatus in which a finely pulverized mixture of the materials to be separated is passed through the powerful electrostatic field between two electrodes.

electrostatic series — *See* triboelectric series.

electrostatic shield — 1. A shield that prevents electrostatic coupling between circuits, but permits electromagnetic coupling. 2. A metallic enclosure or screen placed around a device so it will not be affected by external electric fields.

electrostatic speaker — Also called capacitor or condenser speaker. A speaker in which the mechanical forces are produced by the action of electrostatic fields.

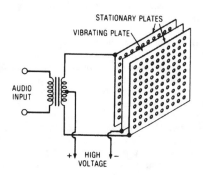

Electrostatic speaker.

electrostatic spraying — A technique of spraying wherein the material being sprayed is given a high electrical charge, while the test piece is grounded.

electrostatic storage — 1. The storage of changeable information in the form of charged or uncharged areas, usually on the screen of a cathode-ray tube. 2. Information storage on a dielectric medium that represents the data as those spots on the medium having electrostatic charges, forming an observable record of the data.

electrostatic transducer — A transducer that consists of a capacitor, at least one plate of which can be set into vibration. Its operation depends on the interaction between its electric field and a change in its electrostatic capacity.

electrostatic tweeter — A speaker with a movable flat metal diaphragm and a nonmovable metal electrode capable of reproducing high audio frequencies. The diaphragm is driven by the varying high voltages applied across it and the electrode.

electrostatic unit — An electric unit based primarily on the dynamic interaction of electric charges. Defined as a charge that, if concentrated on a small sphere, would repel with a force of 1 dyne a similar charge 1 centimeter away in a vacuum.

electrostatic voltmeter — A voltmeter depending for its action on electrostatic forces. Its scale is usually graduated in volts or kilovolts.

electrostatography — The process of recording and reproducing visible patterns by the formation and utilization of latent electrostatic charge patterns.

electrostriction — 1. A mechanical deformation caused by the application of an electric field to any dielectric material. The deformation is proportional to the square of the applied field. This phenomenon results from the induced dipole movement caused by the applied field, resulting in the mechanical distortion. 2. Elastic deformation of a dielectric caused by volume force when the dielectric is placed in an inhomogeneous electric field.

electrostrictive effect — The elastic deformation of a dielectric by an electrostatic field.

electrostrictive relay — A relay whose operation is produced by an electrostrictive dielectric actuator.

electrosurgery — The surgical use of electricity in such applications as dissection, coagulation, laser heating, laser welding, diathermy, desiccation of tumors, and hemostasis.

electrosurgical unit — An rf generator whose output is applied to a blade or wire loop used instead of a conventional scalpel for surgical incision or excision.

electrotape — An electronic distance-measuring device.

electrotherapeutics — *See* electrotherapy.

electrotherapy — Also known as electrotherapeutics. 1. The medical science or use of electricity to treat a disease or ailment. 2. Applying electric current to the body for massage or heat treatment.

electrotherapy apparatus — Equipment for applying electric current to the body for massage or heat treatment.

electrothermal — The heating effect of electric current, or the electric current produced by heat.

electrothermal expansion element — An actuating element consisting of a wire strip or other shape and having a high coefficient of thermal expansion.

electrothermal recorder — A recorder in which heat produces the image on the recording medium in response to the received signals.

electrothermal recording — *See* electrothermal recorder.

electrothermic instrument — An instrument that depends for its operation on the heating effect of a current. Examples are the thermocouple and bolometric, hot-wire, and hot-strip instruments.

electrothermics — The branch of science concerned with the direct transformation of electric energy into heat.

electrotinning — Electroplating tin on an object.

electrotyping — The production of printing plates by electroforming.

electrowinning — The process by which metals are recovered from a solution by electrolysis.

element — 1. One of the 104 known chemical substances that cannot be divided into simpler substances by chemical means. A substance whose atoms all have the same atomic number (e.g., hydrogen, lead, uranium). 2. In a computer, the portion or subassembly that constitutes the means of accomplishing one particular function, such as the arithmetic element. 3. Any electrical device (such as an inductor, resistor, capacitor generator, line, or electron tube) with terminals at which it may be connected directly to other electrical devices. 4. The dot or dash of an International Morse character. 5. A radiator, either active or parasitic, that is part of an antenna. 6. The smallest portion of a televised picture that still retains the characteristics of the picture. 7. A portion of a part that cannot be renewed without destruction of the part. 8. A part of an integrated circuit that contributes directly to its electrical characteristics. An active element exhibits gain, such as a transistor; a passive element does not have gain, such as a resistor or capacitor. 9. Lowest level design entity having an identifiable logical, electrical, or mechanical function.

elemental area — *See* picture element.

elemental semiconductor — A semiconductor containing only one element in the undoped state.

elementary charge — A natural unit or quantum into which both positive and negative charges appear to be subdivided. It is the charge on a single electron and has a value of about 4.8037×10^{-10} electrostatic units.

element error rate — The ratio of the number of elements incorrectly received to the total number of elements sent.

elevation — The angular position perpendicular to the earth's surface.

elevation-position indicator — A radar display that simultaneously shows angular elevation and slant range of detected objects.

elevator leveling control — A positioning control used to align the platform of an elevator with the floor level of the building. Metal vanes are mounted in the elevator shaft at each floor level, and an oscillator is mounted on the elevator car. When the elevator is properly leveled, the metal vane is between the plate and the grid coils of the oscillator. A relay connected in the oscillator circuit now energizes. The contacts of this relay are connected in the motor-control circuit of the elevator so that the elevator stops in alignment with the floor level.

ELF — Abbreviation for extremely low frequency.

eliminator — Also called a battery eliminator. A device operated from an ac or dc power line and used for supplying direct current and voltage to a battery-operated circuit.

E-lines — Contour lines of constant electrostatic field strength with respect to some reference base.

elliptically polarized wave — An electromagnetic wave whose electric intensity vector describes an ellipse at one point.

elliptical polarization — Polarization in which the wave vector rotates in an elliptical orbit about a point.

elliptical stylus — A stylus whose cross section, as seen from above, is an ellipse placed across the record groove. Elliptical styli can more readily trace the finer high-frequency modulations of the groove than can spherical styli. Such styli have two radii (e.g., 0.4×0.7 mil). *See* biradial stylus.

elliptic function — A mathematical function employed in obtaining the squarest possible amplitude response, or the sharpest passband magnitude rolloff, of a filter with a given number of circuit elements. The elliptic function has a Tchebychev response in both the passband and the stopband. The phase response and transient response of an elliptic-function filter are poorer than for any of the classical transfer functions.

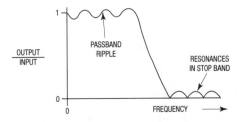

Elliptic-function filter passband.

elongation — Extension of the envelope of a signal as a result of the delayed arrival of certain of the multipath components.

e/m — The ratio of the electric charge to the mass for particles such as electrons and positive rays. For slow-moving electrons, the value of e/m is 1.7588×10^4 coulombs per gram. The value decreases with increasing velocity, however, because of an increase in effective mass.

e-mail — Abbreviation for electronic mail. 1. Messages, usually text, sent from one person to another via computer. E-mail can also be sent automatically to a large number of addresses (mailing list). 2. A system for transmitting messages from one computer terminal to another, where it can be displayed at the receiver's convenience.

embedded code — Machine instructions for checking copy protection that are interspersed with code for other purposes.

embedded software — Computer code that is not visible to the end user.

embedded system — A system into which one or more computing devices (which may be microprocessors or microcontrollers) are incorporated in such a way that the embedded device or devices are not directly accessible to the user of the system.

embedding — A general term for all methods of surrounding or enclosing components and assemblies with a substantial thickness of electrically insulating solid or foam material that substantially fills the voids or interstices between parts.

embedment — The complete encasement of a part or assembly to some uniform external shape. A relatively large volume of a complete package consists of the embedment material.

embossed-foil printed circuits — A printed circuit formed by indenting the desired pattern of metal foil into an insulating base and then mechanically removing the remaining unwanted raised portion.

embossed-groove recording — A method of recording sounds on discs or film strips by embossing sound grooves with a blunt stylus rather than by cutting into them with a sharp stylus. Embossing throws the material up in furrows on each side of the sound groove without actually removing any of the material in the disc or strip.

Embossed-groove recording.

embossing stylus — A recording stylus with a rounded tip that forms a groove in the recording medium by merely displacing the material instead of removing it completely.

EMC — Abbreviation for electromagnetic compatibility.

emergency communication — The transmission or reception of distress, alarm, urgent, or safety signals or messages relating to the safety of life or property, or the occasional operation of equipment to determine whether it is in working condition.

emergency radio channel — Any radio frequency reserved for emergency use, particularly for distress signals.

emergency receiver — Receiver immediately available in a station for emergency communication and capable of being energized solely by a self-contained or emergency power supply.

emergency service — The radiocommunication service carrier used for emergency purposes.

emf — Abbreviation for electromotive force.

EMG — Abbreviation for electromyography.

EMI — Abbreviation for electromagnetic interference.

emission — 1. The waves radiated into space by a transmitter. 2. The ejection of electrons from the surface of a material (under the influence of heat, for example).

emission characteristic — The relationship between the emission and the factor controlling it, such as temperature, voltage, or current of the filament or heater. This relationship is usually shown on a graph.

emission current — The current produced in the plate circuit of a tube when all the electrons emitted by the cathode pass to the plate.

emission efficiency — The rating of a hot cathode. Expressed in milliamperes per watt.

emission power — The time rate at which radiant energy is given off in all directions per unit surface area of a radiating body at a given temperature.

emission spectrum — 1. The spectrum showing the radiation emitted by a substance, such as the light emitted by a metal when placed in an electric arc, or the light emitted by an incandescent filament. 2. The spectrum formed by radiation from an emitting source, in contrast to absorption spectra.

emission types — The classification of modes of radio transmission adopted by international agreement. The AM designations are as follows:

Type A0: Unmodulated continuous-wave transmission
Type A1: Telegraphy or pure continuous waves
Type A2: Modulated telegraphy
Type A3: Telephony
Type A4: Facsimile
Type A5: Television

emission-type tube tester — Also called an English-reading tube tester. A tube tester for checking the electron emission from the filament or cathode. The indicating meter is generally calibrated to read "good" or "bad." The tester connects all elements, such as the plate and screen, suppressor, and control grids, together and uses them as an anode.

emission velocity — The initial velocity at which electrons emerge from the surface of a cathode, ranging from zero up to a few volts (attained by very few electrons). This effect accounts for the existence of virtual cathodes and also for the shape of the cutoff region of plate current.

emissive power — The emissivity of a body times the emissive power of a blackbody at the same temperature. For a blackbody, it is the total radiation per unit area of radiating surface.

emissivity — 1. The ratio of the radiant energy emitted by a radiation source to the radiant energy of a perfect (blackbody) radiator having the same area and at the same temperature and conditions. 2. The percentage of infrared energy emitted from a surface for a given temperature compared with the total energy it should emit for that temperature. Instead of percentage, emissivity (E) is expressed in terms of unity. A perfect blackbody has an $E = 1.0$. When the surface has less than 100-percent radiating efficiency, the difference between the E of the surface and 1.0 represents the approximate reflectivity of the surface. Reflectance varies inversely with emissivity. This allows surrounding ambient surfaces to add to or subtract from the temperature.

emitron — A cathode-ray tube developed in England by J. H. Hixenweaver in 1955.

emitron camera — A British television camera tube resembling an iconoscope.

emittance — The power per unit area radiated by a source of energy.

emitter — Also called source. 1. One of the three regions that form a bipolar transistor. Under forward bias of the emitter-base pn junction, the emitter injects minority carriers (electrons or holes) into the base region, where they either recombine or diffuse into the collector. The flow of minority carriers from the emitter to the collector is controlled by the base-emitter pn junction, thereby giving rise to signal amplification. 2. An electrode within a transistor from which carriers are usually minority carriers; when they are majority carriers, the emitter is referred to as a majority emitter. 3. In fiber optics, the source of optical power. *See* coherent emitter; incoherent emitter.

emitter-base and collector-base junction — In a semiconductor, the region where the base and collector and the emitter and base meet. These junctions are defined on the surface of the chip as an oxide step.

emitter bias — The bias voltage applied to the emitter of a transistor.

emitter-coupled logic — *See* ECL.

emitter current — The direct current flowing in the emitter circuit of a transistor.

emitter cutoff frequency — That frequency at which the β of a transistor is down 3 dB from the low-frequency value.

emitter depletion-layer capacitance — The part of the capacitance across an emitter-base junction of a semiconductor that is associated with its depletion layer. The emitter depletion-layer capacitance is a function of the total potential drop across the depletion layer.

emitter follower — A transistor amplifier circuit configuration analogous to a vacuum-tube follower. The circuit is characterized by relatively high input impedance, low output impedance, and a voltage gain of less than unity.

emitter junction — 1. A semiconductor junction normally biased in the low-resistance direction so that minority carriers are injected into the interelectrode region. 2. A junction between the base and emitter regions of a semiconductor normally biased in the forward direction, and through which the charge carriers flow from a region in which they are majority carriers to one in which they are minority carriers.

emitter region — That part of a transistor lying between the emitter junction and the emitter electrode from which carriers flow across the emitter junction.

emitter resistance — The resistance in series with the emitter lead in the common-T equivalent circuit of a transistor.

emitter semiconductor — A junction normally biased in the low-resistance direction to inject minority carriers into an interelectrode region.

emitter series resistance — The resistance between the emitter terminal of a semiconductor and the internal inaccessible emitter point in an equivalent circuit.

emitter terminal — The specified externally available point of connection to the emitter region.

emitter voltage — The voltage between the emitter terminal and a reference point.

EMP — Abbreviation for electromagnetic pulse.

emphasizer — A circuit or device that provides an intentional increase in signal strength at certain audio frequencies.

empire cloth — A cotton or linen cloth coated with varnish and used as insulation on coils and other parts of electrical equipment.

empirical — 1. Based on actual measurement, observation, or experience, as opposed to theoretical determination. 2. Based solely on experiment or observation, rather than on scientific theory. 3. Pertaining to a statement or formula based on experience or observation rather than on deduction or theory.

EMR — Abbreviation for electromagnetic radiation.

emu — Abbreviation for electromagnetic unit.

emulate — 1. To imitate one system with another, such that the imitating system accepts the same data, executes the same programs, and achieves the same results as the imitated system. 2. To imitate a computer system by a combination of hardware and software that allows programs written for one computer to be run on another. *See* simulate.

emulation — 1. The imitation of all or part of one device, terminal, or computer by another, so that the emulating device accepts the same data, performs the same functions, and appears to other network devices as if it were the emulated device. 2. The imitation of a computer system, performed by a combination of hardware and software, that allows programs to run between incompatible systems. 3. For PCs, the process of imitating the behavior of one operating system using a completely different operating system. 4. The generation of one system's code set by another so that the two may communicate. For example, a system with TTY emulation appears like a Teletype system when communicating with another Teletype. 5. The use of hardware or software to generate in real time the expected correct output responses for comparison to the device under test. 6. A hardware model of the target microprocessor used to check out the target system. This can be either the same microprocessor model as used in a target system, or bit-slice architecture that mimics the target microprocessor's function. Using the target microprocessor is called substitutional emulation or in-circuit emulation. *See also* in-circuit emulation.

emulator — 1. A device that is capable of operating in such a manner that it appears to have all of the characteristics of another device. For example, a hardware and software combination that enables one computer to execute programs written for another computer, or a device that produces the same set of outputs for a given set of inputs as does another device. 2. The combination of programming techniques and special machine features that permit a given computing system to execute programs written for another system. 3. A program or a hardware device that duplicates the instruction set of one computer on a different computer, allowing program development for the emulated computer without that computer being available.

emulsion — A suspension of finely divided photosensitive chemicals in a viscous medium, used in semiconductor processing for coating glass masks.

enable — To permit a circuit to be activated by the removal of a suppression signal.

enabling gate — A circuit that determines the start and length of a generated pulse.

enabling pulse — 1. A pulse that opens a normally closed electric gate, or otherwise permits occurrence of an operation for which it is a necessary but not sufficient condition. 2. A pulse that prepares a circuit for some subsequent action.

enameled wire — Wire coated with a layer of baked-enamel insulation.

encapsulant — A material, usually epoxy, used to encase and seal all components in an electronic circuit.

encapsulate — To embed electronic components or other entities in a protective coating, usually done when the plastic encapsulant is in fluid state so that it will set in solid form as an envelope around the work.

encapsulated relay — A relay embedded in a suitable potting compound.

encapsulating — 1. Coating by dipping, brushing, spreading, or spraying an electronic component or assembly. An encapsulated unit usually retains its original geometry. 2. Enclosing an article in an envelope of plastic by immersing the object in a casting resin and allowing the resin to polymerize or, if hot, cool.

encapsulating material — A composition primarily adapted for use on or around an electrical device to provide protection from the surrounding environment.

encapsulation — 1. A protective coating of cured plastic placed around delicate electronic components and assemblies. It is similar to potting, except the cured plastic is removed from the mold. The plastic therefore determines the color and surface hardness of the finished part. The molds may be made of any suitable material. 2. An embedding process using removable molds or other techniques in which the insulating material forms the outer surfaces of the finished unit. 3. The process of either (a) applying a conformal coating by dipping an object in a high viscosity or thixotropic material, or (b) using containment and a low viscosity material to provide a relatively thin protective encasement (50 to 100 mils or 1.27 to 2.54 mm) to a part or assembly.

encased control — A self-contained motor speed/torque control completely housed in an enclosure. Switching, indicating, and adjusting devices are provided on the outside of the enclosure. Unit portability, safety, and component protection are leading assets of this design.

encipher — To convert a message from ordinary language into a secret form. *See also* encode

enciphered facsimile communications — Communications in which security is provided by mixing pulses from a key generator with the output of a facsimile converter. Plain text is recovered at the receiving terminal by subtracting identical key pulses. Unauthorized persons are unable to reconstruct the plain text unless they have an identical key generator and they know the daily key setting.

enclosed relay — A relay in which both the coil and the contacts are protected from the environment.

enclosed switch — Switch having internal parts protected by a housing. The enclosed switch can be dust proof, moisture proof, oil or contamination proof, or hermetically sealed.

enclosure — 1. An acoustically designed housing or structure for a loudspeaker; also any cabinet for a component, electrical, or electronic device. 2. A surrounding case designed to provide a degree of protection for equipment against a specified environment and to protect personnel against accidental contact with the enclosed equipment.

encode — Also called encipher. 1. To use a code, frequently one composed of binary numbers, to represent individual characters or groups of characters in a message. 2. To change from one digital code to another. If the codes are greatly different, the process usually is called conversion. 3. To substitute letters, numbers, or characters, usually with the intention of hiding the meaning of the message except from persons who know the encoding scheme. 4. The process of converting an event such as a switch closure into a form suitable for transmission over a communication channel.

encoder — 1. A device used to electronically alter a signal so that it can only be viewed on a receiver equipped with a special decoder. 2. Any device that modifies information into the desired pattern or form for a specific method of transmission. 3. An electromechanical device that can be attached to a shaft to produce a series of pulses to indicate shaft position; when the output is differentiated, the device is an accurate tachometer. (It is fundamentally oriented to digital rather than analog techniques.) An encoder contains a disc with a printed pattern; as the disc rotates, it makes and breaks a circuit. The more make-and-break cycles per revolution, the better the resolution. 4. A digital-to analog converter. 5. Circuitry in a quadriphonic sound system that, by matrixing in the recording process, turns four signals into two for inscribing, stereo style, on each wall of the record groove. 6. Electromechanical device that transforms analog motion into digital electrical signals. The outputs are incrementally constant for uniform motion characterized by a staircase function, where the output remains constant for a small range of input values. 7. A digital device for converting an input digital signal into its equivalent binary code. *See also* code converter.

encoder accuracy — The maximum positional difference between the input to an encoder and the position indicated by its output; includes both deviation from theoretical code transition positions and quantizing uncertainty caused by converting from a scale having an infinite number of points to a digital representation containing a finite number of points.

encoding — 1. Translation of information from an analog or other easily recognized form to a coded form without a significant loss of information. 2. The process of converting an event such as a switch closure into a form suitable for transmission over a communication channel. 3. The scrambling of a signal to prevent viewing of a program by nonsubscribers.

encryption — 1. A change made to data, code, or a file such that it can no longer be read or accessed without processing (or unencrypting). 2. The technique of modifying a known bit stream on a transmission line to make it appear like a random sequence of bits to an unauthorized observer.

end-around carry — A computer operation in which the carried information from the left-most bit is added to the results of the right-most addition. It is used for ones complement and nines complement arithmetic.

end-around shift — In a computer, the movement of characters from one end of the register to the other end of the same register.

end bell — An accessory that is similar to a cable clamp and attaches to the back of a plug or receptacle. It serves as an adapter for the rear of connectors. Some angular end bells have built-in cable clamps. Angular end bells up to 90° are available. *See also* end shield.

end bracket — *See* end shield.

end-cell rectifier — A small trickle-charge rectifier for maintaining the voltage of storage-battery end cells.

end cells — Cells that can be switched in series with a storage battery to maintain the output voltage of the battery when it is not being charged.

end central office — The local central telephone office that interconnects customer lines and trunks. It is designated a Class 5 office in the DDD or intertoll network.

end distortion — A shifting of the ends of all marking pulses of start-stop teletypewriter signals from their proper positions relative to the beginning of the start pulse.

end effect — The capacitive effect at the ends of a half-wave antenna. To compensate for this effect, a dipole is cut slightly shorter than a half wave.

end effector — Terminal on a robot arm that carries a hand, welding gun, painting nozzle, or other tool.

end finish — Surface condition at the optical conductor face.

end-fire array — A linear or cylindrical antenna having its direction of maximum radiation parallel to the long axis of the array.

end instrument — A device connected to one terminal of a loop and capable of converting usable intelligence into electrical signals or vice versa. Includes all generating, signal-converting, and loop-terminating devices at the transmitting and/or receiving location.

endless loop recorder — A dictation system in which a nonremovable magnetic tape is sealed in a tank and the tape runs in a continuous loop.

end mark — In a computer, a code or signal used to indicate the termination of a unit of information.

endocardiac electrodes — *See* implantable pacemaker.

endodyne reception — A British term applying to reception of unmodulated code signals. A vacuum-tube circuit having a local oscillator whose frequency is slightly different from that of the carrier signal. Thus, a beat signal in the audio range is produced.

end-of-block signal — A symbol or indicator that defines the end of a block of data.

end of file — Abbreviated EOF. A code placed by a program after the last byte of a file to tell the computer's operating system that no additional data follows.

end-of-file mark — In a computer, a code instruction indicating that the last record of a file has been read.

end of message — The end of data to be transmitted. It can be indicated by a special control code, as in the ASCII code set; by an absence of data for a specified time interval; or by a particular sequence of block gaps and data, as is done on magnetic tape.

end of tape — The point on a computer tape at which the system or operator is given a warning that the physical end of the tape is approaching. It is approximately 25 feet (7.62 m) from the actual end of the tape on $1/2$-inch (12.7-mm) computer tape and approximately 50 feet (15.24 m) from the halt marker on $1/4$-inch (6.35-mm) tape.

end-of-tape marker — A marker placed on a magnetic tape to indicate the end of the permissible recording area. It may be a photoreflective strip, a transparent section of tape, or a particular bit pattern.

end-of-transmission card — Last card of each message; used to signal the end of a transmission. Contains the same information as the header card, plus additional data for traffic analysis.

endogeneous variable — A variable whose value is determined by relationships included within the model. *See also* exogenous variable.

end-on armature — Of a relay, an armature that moves in the direction of the core axis, with the pole face at the end of the core and perpendicular to this axis.

end-on directional antenna — A directional antenna that radiates chiefly toward the line on which the antenna elements are arranged.

endoradiograph — Equipment for X-ray examination of internal organs and cavities by means of radiopaque materials.

endoradiosonde — Also called radio pill. A device for detecting and transmitting physiological data from the gastrointestinal tract or other inaccessible body cavities.

endothermic — A term describing a chemical reaction in which heat is absorbed.

endothermic reaction — A reaction that is accompanied by the absorption of heat.

end point — The shaft positions immediately before the first and after the last measurable change(s) in output ratio after wiper continuity has been established, as the shaft of a precision potentiometer moves in a specified direction.

end-point control — Quality control by means of continuous automatic analysis. In highly automatic processes, the final product is analyzed, and if any undesirable variations are detected, the control system automatically brings about the necessary changes.

end-point sensitivity — The algebraic difference in electrical output between the maximum and minimum value of the measurand over which an instrument is calibrated.

end-point voltage — The terminal voltage of a cell below which equipment connected to it will not operate or should not be operated.

end resistance — The resistance of a precision potentiometer measured between the wiper terminal and an end terminal, with the shaft positioned at the corresponding end point.

end-resistance offset — In potentiometers, the residual resistance between a terminal and the moving contact, at a position corresponding to full rotation against that terminal.

end-scale value — The value of the actuating electrical quantity that corresponds to end-scale indication of an instrument. When zero is not at the end or at the electrical center of the scale, the higher value is taken. Certain instruments such as power-factor meters, ohmmeters, etc., are necessarily excepted from this definition.

end setting — In a potentiometer, the minimum resistance that is measured between one end of a potentiometer and the wiper, with the wiper mechanically positioned at that end.

end shield — 1. Frequently called end bracket or end bell. In a motor housing, the part that supports the bearing and also guards the electrical and rotating parts inside the motor. 2. In a magnetron, the shield that confines the space charge to the interaction space.

end spaces — In a multicavity magnetron, the two cavities at either end of the anode block that terminate all the anode-block cavity resonators.

end-to-end check — Tests conducted on a completed wire and/or cable run to ensure electrical continuity.

end use — The way the ultimate consumer uses a device.

energize — To apply the rated voltage to a circuit or device, such as to the coil of a relay, in order to activate it.

energized — Also called alive, hot, and live. Electrically connected to a voltage source. Turned on, alive.

energized part — A part at some potential with respect to another part, or the earth.

energy — 1. The capacity for performing work. A particle or piece of matter may have energy because it is moving or because of its position in relation to other particles or pieces of matter. A rolling ball is an example of the first; a ball at rest at the top of an incline is an example of the second. 2. The capacity for doing work and overcoming resistance.

energy conversion — The change of energy from one form to another, e.g., from chemical energy to electrical energy.

energy conversion devices — Devices including primary and secondary cells; fuel cells; photovoltaic systems; electrochemical energy converters; radiation conversion devices; thermionic converters; converters using solar, ionic, or nuclear energy sources; devices for creating a plasma in an interaction space between an emitter and a collector; electrostatic generators for creating an electrical output; organic and inorganic ion exchange and membrane devices; electron volt energy devices; devices for direct conversion of fuel to electricity; and electrical-energy storage-unit devices capable of delivering a power output.

energy density — 1. The energy output of a battery, expressed in watt-hours, per unit weight or volume of the battery. 2. The energy per unit volume of a medium.

energy dispersal waveform — A triangular-shaped signal at 30 Hz synchronized with the vertical blanking interval in the TV signal from the satellite, which ensures that the signal will average its power out over the whole channel, even when just the carrier is present. This waveform is removed by the receiver after FM demodulation.

energy efficiency ratio — Term used to gauge the relative electrical efficiency of appliances; found by dividing btu-per-hour output by the number of electrical watts used.

energy gap — The energy range between the bottom of the conduction band and the top of the valence band of a semiconductor.

energy level — A particular value of energy of a physical system, such as a nucleus, that the system can maintain for a reasonably long length of time. Systems on an atomic scale have only certain discrete energy levels and cannot occupy values between these levels.

energy-level diagram — A line drawing that shows the increase or decrease in electrical power as current intensities rise and fall along a channel of signal communications.

energy-measuring equipment — Equipment used to measure energy in electrical, electronic, acoustical, or mechanical systems.

energy of a charge — Represented by $E = 1/2QV$, given in ergs, when the charge Q and the potential V are in electrostatic units.

energy product — The product of the magnetic flux density B in gauss times the magnetic field strength H in oersteds. Used as an index of magnet quality. The larger the maximum energy product, the smaller the required magnet for a given job.

energy-product curve — A curve obtained by plotting the product of the value of magnetic induction B and demagnetizing force H for each point of the demagnetization curve of a permanent magnetic material. Usually shown together with the demagnetization curve.

energy redistribution — A method of finding the duration of an irregularly shaped pulse by considering it as a power curve. The area under the curve can be represented by an equivalent rectangle of the same area and peak amplitude. The original-pulse duration is equal to the rectangle width.

energy state — The position and speed of an electron relative to the position and speed of other electrons in the same atom or adjoining atoms.

energy storage capacitor — Specifically designed capacitor for use in applications wherein the capacitor can be charged over a relatively long period and discharged in a short period, thus increasing the instantaneous power in energy storage systems. Provisions are made to permit the very high currents that accompany the high rate of energy discharge. When the number of discharges is very large, the actual total time during which the capacitor may be subjected to the maximum operating voltage is relatively short.

energy-variant sequential detection — A technique for sequential detection in which a fixed number of transmitted pulses of varying energy are received with a single (upper) threshold device.

ENG — See electronic news-gathering.

engineered military circuit — 1. Leased long lines of which only the station equipment, local loops, and reserved positions of interexchange channels are paid for continuously. The unresevered portions of leased long lines or interexchange channels are on a steady status and are placed in an operational status and paid for only when required by the command concerned. 2. A standby or on-call circuit that is engineered specifically to meet military criteria.

engineering — A profession in which a knowledge of the natural sciences is applied with judgment to develop ways of utilizing the materials and forces of nature.

English-reading tube tester — See emission-type tube tester.

enhanced carrier demodulation — An amplitude-demodulation system in which a synchronized local carrier of the proper phase is added to the demodulator. This has the effect of materially reducing the distortion produced in the demodulation process.

enhancement — Modification of the subjective features of an image to increase its impact on the observer.

enhancement mode — 1. An MOS transistor that is normally off with zero gate voltage applied. A gate voltage of the correct polarity attracts majority carriers to the gate area, thus "enhancing" it and forming a current-conducting channel. 2. A device type that is normally off with zero gate voltage. A threshold voltage is then required to turn the device on.

enhancement-mode field-effect transistor — A field-effect transistor in which no device current flows (leakage only) when V_{GS} is zero volts. Conduction does not begin until V_{GS} reaches the threshold voltage.

enhancement mode operation — The operation of a field-effect transistor such that changing the gate-to-source voltage from zero to a finite value increases the magnitude of the drain current.

enhancement MOS transistor — A type of MOS transistor in which no current flows in the absence of an input control signal on the control terminal (called the gate) of the transistor; i.e., a control signal input is required to turn on the device. This reduces power dissipation (power dissipation occurs only when an input signal is present) and results in excellent logic state recognition (full off being one state and on the other).

ensemble — A collection of sample functions of a random process, all of which start from the same zero time.

enterprise number — Also called toll-free number. Unique telephone exchange number that permits the called party to be automatically billed for incoming calls.

entrance box — A metal box that houses overcurrent-protection devices and serves as the point of distribution for the various electrical circuits in a structure.

entrance cable — A cable by means of which electrical power is brought from an outside power line into a building.

entrance delay — The time between actuating a sensor on an entrance door or gate and the sounding of a local alarm or transmission of an alarm signal by the control unit. This delay is used if the authorized access switch is located within the protected area, and permits a person with the control key to enter without causing an alarm. The delay is provided by a timer or timing circuit within the control unit.

entrapped material — Gas or particles bound up in an electrical package so that they cannot escape the package.

entropy — 1. A measure of the unavailable energy in a thermodynamic system. 2. The unavailable information in a set of documents. 3. An inactive or static condition (total entropy). 4. A measure of the amount of information in a communication signal, equal to the average number of bits per symbol.

entry — Each statement in a computer programming system.

entry point — In a computer, the programmer-defined instruction at which a task is to begin execution.

enunciation — The act of pronouncing words clearly and distinctly. Articulation.

envelope — 1. Also referred to as a bulb. The glass or metal housing of a vacuum tube. The glass housing that encloses an incandescent source. 2. The curve passing through the peaks of a graph and showing the waveform of a modulated radio-frequency carrier signal.

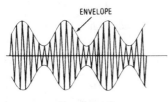

Envelope, 2.

envelope delay — 1. The time that elapses as a transmitted wave passes any two points of a transmission circuit. Such delay is determined primarily by the constants of the circuit and is measurable in milliseconds or microseconds. 2. Sometimes called time delay or group delay. The propagation time delay undergone by the envelope of an amplitude-modulated signal as it passes through a filter. Envelope delay is proportional to the slope of the curve of phase shift as a function of frequency. Envelope-delay distortion is introduced when the delay is not the same at all frequencies in the passband. 3. The time difference between the longest delay and the shortest delay for a given band of frequencies. 4. A type of distortion on an analog line in which the signal delay varies with signal frequency. 5. Characteristics of a circuit that result in some frequencies arriving ahead of others, even though they were transmitted together.

envelope-delay distortion — The distortion that occurs during transmission when the phase shift of a circuit or system is not constant over the frequency range.

envelope generator — A circuit in a synthesizer that produces a single, carefully defined waveform. Useful in creating attacks and decays to define notes or for special effects.

environment — The aggregate of all conditions that externally influence a device's performance.

environmental chambers — Test chambers designed to expose the subject being tested to external conditions, such as heat, shock, pressure, and moisture, for the study of their effects of the subject.

environmental conditions — External conditions of heat, shock, vibration, pressure, moisture, etc.

environmentally sealed — Provided with gaskets, seals, potting, or other means to keep out contamination that might reduce performance.

environmental testing — The testing of a system or component under controlled environmental conditions, each of which tends to affect its operation or life.

environmentproof switch — A switch that is completely sealed to ensure constant operating characteristics. Sealing normally includes an O-ring on the actuator shaft and fused glass-to-metal terminal seals or complete potting and an elastomer plunger-case seal.

EOG — Abbreviation for electro-oculography.

episcotister — A device consisting of alternate opaque and transparent discs that rotate at a speed that interrupts light beams at an audio-frequency rate. It modulates the light beam used to excite a photoelectric element.

epitaxial — Pertaining to a single crystal layer on a crystalline substrate, oriented the same as the substrate. In certain semiconductor processes, an expitaxial layer is grown on a silicon substrate during the fabrication of transistors and integrated circuits.

epitaxial deposition — 1. The growth of additional material, usually in a thin film, on a substrate. Often the added material has a crystal structure and orientation controlled by matching that of the substrate. 2. Epitaxy. The technique of growing a semiconductor layer on an existing crystal by depositing it directly from reactant vapors, so that the structure of the new layer is isomorphic with, or simply an extension of, that of the original crystal.

epitaxial device — A device constructed in such a manner that the crystalline structure of successive layers is oriented in the same direction as that of the original base material.

epitaxial film — 1. A film of single-crystal semiconductor material that has been deposited onto a single-crystal substrate. 2. Any deposited film, provided the orientation of its crystal is the same as that of the substrate material.

epitaxial growth — 1. A semiconductor fabrication process in which single-crystal p or n material is deposited and grows on the surface of a substrate. Usually, this material has a different conductivity than the substrate. 2. Crystal growth obtained by depositing a film of monocrystalline semiconductor material on a monocrystalline substrate. 3. The process of producing an additional crystal layer of semiconductor material on a semiconductor substrate. The crystalline structure of the substrate is continued into the epitaxial layer; however, the impurity concentration can be made to differ greatly. 4. The deposition of a single-crystal film on the surface of a single-crystal substrate so that the crystal orientations of the two layers are alike. 5. A process of growing layers of material on a selected substrate. Usually silicon is grown in a silicon substrate. Silicon and other semiconductor materials may be grown on a substrate with compatible crystallography, such as sapphire (silicon-on-sapphire).

epitaxial-growth mesa transistor — A transistor made by overlaying a thin mesa crystal over another mesa crystal.

epitaxial growth process — The process of growing a semiconductor material by depositing it in vaporized form on a semiconductor seed crystal. The deposited layer continues the single-crystal structure of the seed.

epitaxial layer — 1. A grown or deposited crystal layer with the same crystal orientation as the parent material and, in the case of semiconductor circuits, of the same basic material as the original substrate. 2. A single-crystal p-type or n-type material deposited on the surface of a substrate. 3. A thin, precisely doped monocrystalline silicon layer grown on a heavily doped thick wafer, into which are diffused semiconductor junctions. In conventional processing of an integrated circuit, the thick wafer is p doped and the epitaxial layer is n doped. 4. A single crystal layer that has been deposited or grown on a crystalline substrate having the same structural arrangement. 5. A layer of silicon grown atop single-crystal silicon that reproduces the same crystallographic orientation as the single-crystal material.

epitaxial material — A material whose atoms are arranged in single-crystal fashion upon a crystalline substrate so that its lattice structure duplicates that of the substrate.

epitaxial planar transistor — A transistor in which a thin collector region is epitaxially deposited on a low-resistivity substrate, and the base and emitter regions are produced by gaseous diffusion with the edges of the junction under a protective oxide mask.

epitaxial process — The process of growing from the vapor phase a single-crystal semiconductor material with controlled resistivity and thickness.

epitaxial transistor — A transistor with one or more epitaxial layers.

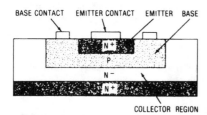

Epitaxial transistor (triple diffused).

epitaxy — 1. The controlled growth on a crystalline substrate of a crystalline layer, called an epilayer. In homoepitaxy (e.g., silicon layers on a silicon substrate), the epilayer exactly duplicates the properties and crystal structure of the substrate. In heteroepitaxy (e.g., silicon on sapphire), the deposited epilayer is a different material with a different crystalline structure than that of the substrate. 2. The growth of a crystal on the surface of a crystal of another substance in such a way that the orientation of the atoms in the original crystal controls the orientation of the atoms in the grown crystal.

E-plane — The plane of an antenna containing the electric field. The principal E-plane also contains the direction of maximum radiation.

E-plane bend — Also called E-bend. The smooth change in direction of the axis of a waveguide. The axis remains parallel to the direction of polarization throughout the change.

E-plane T-junction — Also called series T-junction. A waveguide T-junction in which the structure changes in the plane of the electric field.

epoxy — 1. Pertaining to a family of thermosetting materials that are widely used for casting and potting and as adhesives. 2. A family of thermosetting resins used in the packaging of semiconductor devices. Epoxies form a chemical bond to many metal surfaces.

EPROM — Abbreviation for erasable programmable read-only memory. A type of nonvolatile memory device whose contents can be erased by exposure to ultraviolet light. *See also* EEPROM; PROM.

epsilon — The Greek letter E, or ε, frequently used to represent 2.71828, which is the base of the natural system of logarithms.

equal-energy source — A source of electromagnetic or sound energy that emits the same amount of energy at each frequency in the spectrum.

equal-energy white — The light produced by a source that radiates equal energy at all visible wavelengths.

equalization — 1. The process of reducing the frequency and/or phase distortion of a circuit by the introduction of networks to compensate for the difference in attenuation and/or time delay at the various frequencies in the transmission band. 2. A process of compensating for increases in attenuation (signal loss) with frequency. Different signal frequencies are attenuated differently over a given distance. 3. An intentional departure from response flatness to compensate for complementary characteristics introduced elsewhere in the system (as with discs, tape, and FM broadcasting). Also used to correct for response deficiencies in speakers and other components. 4. Different equalization characteristics are used in the recording and playback amplifiers of a tape recorder, to compensate for the magnetic characteristics of the tape and the heads. Playback equalization is standardized to give flat frequency response with any properly recorded tape, while recording equalization is a property of a particular machine, depending on its head design and the tape for which it was meant. 5. Reshaping the playback characteristics of a recording during playback mode. The simplest way is to adjust the treble and bass controls, but true equalization requires continuous adjustment of the playback frequency response curve at several points. A graphic equalizer is often used for this. 6. The selective amplification or attenuation of certain frequencies. Also refers to recognized industry standards for recording and reproducing characteristics, such as the NAB Standard. 7. The intentional increase in level of certain portions of the audio-frequency spectrum. The term is sometimes misapplied when actually referring to the attenuation of portions of the audio-frequency spectrum. 8. A technique used to compensate for distortions present on a communication channel. Equalizers add loss or delay to signals in inverse proportion to the channel characteristics. The signal response curve is then relatively flat and can be amplified to regain its original form.

equalize — To apply to a circuit an electrical network whose transmission characteristics are complementary to those of the line, so that when the loss (or delay) in the line and that in the equalizer are combined, the overall loss (or delay) is almost the same at all frequencies.

equalizer — 1. A passive device designed to compensate for an undesired amplitude-frequency and/or phase-frequency characteristic of a system or component. 2. A series of connections made in paralleled, cumulatively compound direct-current generators to give the system stability. 3. A network, usually adjustable, that corrects the transmission-frequency characteristics of a circuit to permit it to transmit all the frequencies that it passes in a uniform manner. 4. An electronic circuit that introduces compensation for frequency-discriminative effects of elements within a television system. 5. An electronic device that amplifies (boosts) and/or attenuates certain portions of the audio-frequency spectrum. There are many different types of equalizers. 6. A device to allow the frequency response of an audio-signal path to be adjusted in some way.

equalizer circuit breaker — A breaker that serves to control or to make and break the equalizer or the current-balancing connections for a machine field, or for regulating equipment, in a multiple-unit installation.

equalize voltage — A voltage applied to a battery for charging at installation and as a periodic boost charge, common in lead-antimony lead-acid battery systems.

equalizing current — A current circulated between two parallel-connected compound generators to equalize their output.

equalizing network — A network connected to a line to correct or control its transmission frequency characteristics.

equalizing pulses — 1. A series of pulses (usually six) occurring at twice the line frequency before and after the serrated vertical TV synchronizing pulse. Their purpose is to cause vertical retrace to occur at the correct instant for proper interlace. 2. In the standard television signal, pulses that minimize the effect of line-frequency pulses on the interlace.

equal-loudness contours— *See* Fletcher-Munson curves.

equation function— As applied to microelectronic circuitry, a combination of electronic elements or circuits capable of solving the electronic-counter portion of a mathematical or Boolean equation. In obtaining the solution, it performs the necessary function within an electronic or electromechanical system.

equation solver— A computer, usually of the analog type, designed to solve systems of linear simultaneous (nondifferential) equations or to find the roots of polynomials.

equilibrium— In a semiconductor context, that state of a semiconductor crystal when there is no net current through the crystal. A crystal is normally in this state when no external voltages or current are impressed on it.

equilibrium brightness— The brightness of the viewing screen when a display storage tube is in a fully written condition.

equilibrium electrode potential— A static electrode potential when the electrode and the electrolyte are in equilibrium with respect to a specified electrochemical reaction.

equiphase surface— In a wave, any surface over which the field vectors at the same instant are either in phase or 180° out of phase.

equiphase zone— In radionavigation, the region in space within which the difference in phase between two radio signals is indistinguishable.

equipment— 1. An item having a complete function apart from being a substructure of a system. Sometimes called a set. 2. A general term referring to practically every part of an electrical system, including the parts consuming electrical energy. (Devices are also included in this category.)

equipment augmentation— 1. Procuring additional automatic data-processing equipment capability to accommodate increased workload within an established data system. 2. Obtaining additional automatic data-processing equipment capability to extend an established data system to additional sites or locations.

equipment bonding jumper— The connection between two or more portions of the equipment grounding conductor.

equipment chain— A group of units of equipment that are functionally in series. The failure of one or more individual units results in loss of the function.

equipment characteristic distortion— A repetitive display or disruption peculiar to specific portions of a teletypewriter signal. Normally, it is caused by improperly adjusted or dirty contacts in the sending or receiving equipment.

equipment ground— A connection from earth ground to a non-current-carrying metal part of a wiring installation of electric equipment. It reduces shock hazard and provides electrostatic shielding.

equipment life— The arithmetic mean of the cumulative operating times of identical pieces of equipment beginning with the time of acceptance by the ultimate consumer and ending when the equipment is no longer serviceable.

equipotential— A conductor having all parts at a single potential. The cathode of a heater-type tube is equipotential, whereas the filament is not because its voltage varies from one end to the other.

equipotential cathode— *See* indirectly heated cathode.

equipotential line— An imaginary line in space having the same potential at all points.

equipotential surface— A surface or plane passing through all points having the same potential in a field of flow.

equisignal localizer— Also called tone localizer. A type of localizer in which lateral guidance is obtained by comparing the amplitudes of two modulation frequencies.

equisignal radio-range beacon— A radio-range beacon used for aircraft guidance. It transmits two distinctive signals, which are received with equal intensity only in certain directions called equisignal sectors.

equisignal surface— The surface formed around an antenna by all points that have a constant field strength (usually measured in volts per meter) during transmission.

equisignal zone— In radionavigation, the region in space within which the difference in amplitude between two radio signals is indistinguishable.

equivalence— A logic operator having the property that if P is a statement, Q is a statement, R is a statement, etc., then the equivalence of P, Q, R... is true if and only if all statements are true or all statements are false.

equivalent absorption— The rate at which a surface will absorb sound energy, expressed in sabins. Defined as the area of a perfect absorption surface that will absorb the same sound energy as the given object under the same conditions.

equivalent absorption area— Area of perfectly absorbing surface that will absorb sound energy at the same rate as the given object under the same conditions. The acoustic unit of equivalent absorption is the sabin.

equivalent binary digits— 1. The number of binary digits equivalent to a given number of decimal digits or other characters. 2. The number of binary places required to count the elements of a given set.

equivalent circuit— 1. An arrangement of common circuit elements that has characteristics over a range of interest electrically equivalent to those of a different or more complicated circuit or device. 2. A simplified circuit that has the same response to changing voltage and frequency as a more complex circuit. Used to facilitate mathematical analysis.

equivalent circuit of a piezoelectric crystal unit— The electric circuit that has the same impedance as the unit in the frequency region of resonance. It is usually represented by an inductance, capacitance, and resistance in series, shunted by the direct capacitance between the terminals of the crystal unit.

equivalent component density— In circuits in which discrete components are not readily identifiable, the volume of the circuit divided by the number of discrete components necessary to perform the same function.

equivalent conductance— The normal conductance of an atr tube in its mount, measured at its resonance frequency.

equivalent dark-current input— The incident luminous flux required to give an output current equal to the dark current.

equivalent differential input capacitance— The equivalent capacitance looking into the inverting or non-inverting inputs of a differential amplifier with the opposite input grounded. *See also* equivalent differential input impedance.

equivalent differential input impedance— The equivalent impedance looking into the inverting or non-inverting input, with the opposite input grounded and the operational amplifier operated in the linear amplification region.

equivalent differential input resistance — The equivalent resistance looking into the inverting or non-inverting input of a differential amplifier with the opposite input grounded. *See also* equivalent differential input impedance.

equivalent diode — An imaginary diode consisting of the cathode of a triode or multigrid tube and a virtual anode to which is applied a composite controlling voltage of such a value that the cathode current would be the same as the current in the triode or multigrid tube.

equivalent faults — Two or more faults that cause the same output responses and that cannot be isolated from the board output pins and internal nodes being monitored by the tester.

equivalent four-wire system — A transmission system using frequency division to obtain full-duplex operation over only one pair of wires.

equivalent grid voltage — The grid voltage plus plate voltage divided by the mu of the tube.

equivalent height — The virtual height of an ionized layer of the ionosphere.

equivalent input noise current — The equivalent input noise current that would reproduce the noise seen at the output of an operational amplifier if all amplifier noise sources were set to zero and the source impedances were large compared with the optimum source impedance.

equivalent input noise voltage — The equivalent input noise voltage that would reproduce the noise seen at the output of an operational amplifier if all amplifier noise sources and the source resistances were set at zero.

equivalent input offset current — The difference between the two currents flowing into the inverting and noninverting inputs of a differential amplifier when the output voltage is zero.

equivalent input offset voltage — The amount of voltage required at the input to bring the output to zero. Usually this voltage is adjustable to zero by using either a built-in or an external variable resistor (balance control).

equivalent input wideband noise voltage — The output noise voltage of a differential amplifier with the input shorted, divided by the dc voltage gain of the amplifier. This voltage is measured with a true rms voltmeter and is limited to the combined bandwidth of the amplifier and meter.

equivalent loudness — The intensity level of a sound relative to some arbitrary reference intensity, such as a 1000-hertz pure tone, that is judged by the listeners to be equivalent in loudness.

equivalent network — One network that replaces another in a system without altering in any way the electrical operation of the system external to the network.

equivalent noise conductance — The spectral density of a noise-current generator expressed in conductance units at a specified frequency.

equivalent noise input — In a photosensitive device, the value of incident luminous flux that produces an rms output current equal to the rms noise curent within a specified bandwidth when the flux is modulated in a stated manner.

equivalent noise pressure — *See* transducer equivalent noise pressure.

equivalent noise resistance — A measure of the residual noise output of a potentiometer while the slider is being actuated. (The residual noise consists of active components in the form of self-generated voltages arising in the slider contact interface, and passive components in the form of ohmic contact resistance at the point of slider contact.)

equivalent noise temperature — The absolute temperature at which a perfect resistor with the same resistance as the component would generate the same noise as the component at room temperature.

equivalent open-circuit rms noise current — That noise which occurs at the input of the noiseless amplifier due only to noise currents. It is expressed in picoamperes per hertz at a specified frequency or in nanoamperes in a given frequency band.

equivalent periodic line — A periodic line that, when measured at its terminals or at corresponding section junctions, has the same electrical behavior at a given frequency as the uniform line with which it is compared.

equivalent permeability — The relative permeability that a component would have under specified conditions if it had the same reluctance as a component of the same shape and size but different materials.

equivalent plate voltage — The plate voltage plus mu times the grid voltage.

equivalent resistance — The concentrated or lumped resistance that would cause the same power loss as the actual small resistances distributed throughout a circuit.

equivalent series resistance — Abbreviated ESR or R. 1. In a circuit or component, the square root of the difference between the impedance squared and the reactance squared. All internal series resistance of the circuit or component treated as being concentrated in a single resistance at one point. 2. All internal ac series resistance of a capacitor treated as a single resistor. 3. An effective resistance that, if connected in series with an ideal capacitor of a capacitance value equal to that of the capacitor in question, would result in a power loss equal to the active power dissipated in that capacitor at a given frequency.

equivalent time — In random-sampling oscilloscope operation, the time scale associated with the display of signal events.

equivalent time sampling — A method of allowing the storage of repetitive events that occur faster than the maximum digital oscilloscope sampling frequency. The scope acquires the waveform during multiple sweeps by taking samples at various times until scope memory is filled.

equivocation — In a computer, the conditional information contained in an input symbol given an output symbol, averaged over all input-output pairs.

erasable programmable read-only memory — Abbreviated EPROM. A field-programmable read-only memory that can have the data content of each memory cell altered more than once. An EPROM is bulk-erased by exposure to high-intensity ultraviolet light. Sometimes referred to as a reprogrammable read-only memory.

erasable storage — Storage media in a computer that hold information that can be changed.

erase — 1. To replace all the binary digits in a storage device by binary zeros. In a binary computer, erasing is equivalent to clearing; in a coded decimal computer, in which the pulse code for decimal zero may contain binary ones, clearing leaves decimal zero, whereas erasing leaves all-zero pulse codes in all storage locations. 2. To remove all information from a register or a memory. Sometimes this consists of writing a zero into all memory positions.

erase head — 1. A head on a tape recorder that applies a strong high-frequency alternating magnetic field to the tape so that earlier recordings may be erased as the tape runs past the head. 2. A magnetic tape head that removes previously recorded signals from a tape, usually by applying inaudible high-level, high-frequency bias signals.

erasing speed — In charge-storage tubes, the rate of erasing successive storage elements.

erasure — 1. A process by which a signal recorded on a tape is removed and the tape made ready for rerecording. This may be accomplished by ac erasure, in which the tape is demagnetized by an alternating field that is reduced in amplitude from an initially high value, or by dc erasure, in which the tape is saturated by applying a primarily unidirectional field. 2. The neutralization of the magnetic pattern stored on tape.

E region — The region of the ionosphere about 50 to 100 miles (80 to 160 km) above the earth's surface.

E register — The extension of the computer A-register for use in double-precision arithmetic or logic-shift operations.

$E_r - E_y$ — The resultant color television signal when E_y is subtracted from the original full red signal.

erg — 1. The absolute centimeter-gram-second unit of energy and work. The work done when a force of 1 dyne is applied through a distance of 1 centimeter. 2. Measure of energy. 1 erg $= 10^7$ joules $= 6.25 \times 10^{10}$ eV.

ergonomics — The science of designing office systems to meet the needs of the human body.

ERP — Abbreviation for effective radiated power. The amount of power radiated by an antenna, which may be more or less than the power absorbed by it from the transmitter.

error — 1. In mathematics, the difference between the true value and a calculated or observed value. A quantity (equal in absolute magnitude to the error) added to a calculated or observed value to obtain the true value is called a correction. 2. Any discrepancy between a computed, observed, recorded, or measured quantity and the true, specified, or theoretically correct value or condition. 3. In a computer or data-processing system, any incorrect step, process, or result. In addition to the mathematical usage in the computer field, the term also commonly refers to machine malfunctions, or machine errors, and to human mistakes, or human errors. 4. In data communications, *error* means that a bit transmitted as a 1 was received as a 0 (or vice versa). 5. A deviation occurring during transmission such that a mark signal is received instead of a space signal, or vice versa. 6. A situation in which the data readout of a memory location is different from the data originally stored in that location. The term *error* is normally not applicable to circuits that do not contain memory elements. It can be applied, however, to complex logic arrays that contain memory elements, such as flip-flops or small arrays of RAM. Errors can be classified into three categories: hard error, medium error, and soft error.

error-correcting code — 1. A code in which each acceptable expression conforms to specific rules of construction that also define one or more equivalent nonacceptable expressions, so that if certain errors occur in an acceptable expression, the result will be one of its equivalents and thus the error can be corrected. 2. A computer code using extra bits that automatically detect and correct errors. 3. Code for information transmission that makes it possible to detect errors and to fix them. This involves sending extra information along with each word, with a corresponding reduction in the transmission rate.

error-correcting telegraph system — A system that employs an error-detecting code in such a way that any false signal initiates retransmission of the character incorrectly received.

error-correction routine — A series of computer instructions programmed to correct a detected error condition.

error detecting and feedback system — A system employing an error-detecting code and so arranged that a signal detected as being in error automatically initiates a request for retransmission of that signal.

error-detecting code — In a digital computer, a system of coding characters such that any single error produces a forbidden or impossible code combination.

error detection — 1. An arrangement that senses flaws in received data by examining parity bits, verifying block check characters, or using other techniques. 2. System that detects errors occasioned by transmission equipment or facilities.

error detector — That portion of an automatic control system that determines when the regulated quantity has deviated outside the dead zone.

error rate — 1. A measure of quality of a digital circuit or equipment item. 2. The number of erroneous bits or characters in a sample; it is frequently taken as the number of errors per 100,000 characters. 3. A measure of the ratio of the number of characters of a message incorrectly received to the number of characters of the message received. 4. The ratio of incorrectly received data (bits, elements, characters, or blocks) to the total amount of data transmitted.

error-rate damping — A damping method in which a signal proportional to the rate of change of error is added to the error signal for anticipatory purposes.

error signal — In an automatic control device, a signal whose magnitude and sign are used to correct the alignment between the controlling and the controlled elements.

error tape — A special tape developed and used for writing out errors in order to correct them by study and analysis after printing.

error voltage — A voltage that is present in a servo system when the input and output shafts are not in correspondence. The error voltage, which actuates the servo system, is proportional to the angular displacement between the two shafts.

ESA — Abbreviation for electronic shock absorption.

Esaki diode — *See* tunnel diode.

ESC — Abbreviation for electronic spark control. For General Motors turbo-charged engines, a system that controls timing and engine knock.

escape character — A code-extension character used, sometimes with one or more succeeding characters, to form an escape sequence, which indicates that the interpretation of the succeeding characters is to be different from the code currently in use.

E-scope — A radar display in which targets appear as blips with distance indicated by the horizontal coordinate and elevation by the vertical coordinate.

escutcheon — A backing plate around an opening. Commonly the ornamental metal, wood, plastic, or other framework around a radio tuning dial, control knob, or other panel-mounted part in a radio receiver or television receiver, audio-frequency amplifier, etc.

ESD — Abbreviation for electrostatic discharge. Discharge of a static charge on a surface or body through a conductive path to ground. An electronic component may suffer irreparable damage when it is included in the discharge path.

ESM — *See* electronic warfare support measures.

ESR — Abbreviation for equivalent series resistance.

ESS — 1. Abbreviation for electronic spark selection. An electronic system (General Motors) that controls timing on fuel-injected engines to increase fuel economy and improve emissions. Operates on inputs from engine temperature, rpm, and vacuum sensors. 2. Abbreviation for electronic switching system.

essential loads — Those loads that must be served to keep plant operations at an acceptable level during a prolonged commercial power outage. Such loads might be interruptible for periods of a few seconds to several minutes.

established reliability — A quantitative maximum failure rate demonstrated under controlled test conditions in accordance with a military specification; usually expressed as percent failures per thousand hours of test.

Estiatron — A special type of electrostatically focused traveling-wave tube.

etch — The process of removing material from a wafer (such as oxides or other thin films) by chemical, electrolytic, or plasma (ion bombardment) means. Examples: nitride etch, oxide etch.

etchant — 1. A chemical agent that can remove a solid material. For example, a highly selective etchant that acts on silicon dioxide, silicon, or both is employed in semiconductor processing to pattern the silicon surface for diffusion masking. 2. A solution used, by chemical reaction, to remove the unwanted portion of a conductive material bonded to a base. 3. A liquid solution of a chemical or combination of chemicals used to preferentially dissolve metal.

etchback — The controlled removal of all components of base material by a chemical process on the side wall of holes in order to expose additional internal conductor areas.

etched metal mask — A metal mask used for screening wherein the pattern is created in a sheet of metal by the etching process.

etched printed circuit — A type of printed circuit formed by chemically or electrolytically (or both) removing the unwanted portion of a layer of material bonded to a base.

etch factor — A ratio of etched depth to the lateral etch at the boundary of the interface of the photoresist and the substrate.

etching — 1. The selective removal of unwanted material from a surface, usually by chemical means. 2. A process using either acids or a gas plasma to remove unwanted material from the surface of a wafer.

etching resist — Material deposited on the surface of a copper-clad base material that prevents the removal by etching of the conductive areas the material covers.

etching to frequency — Finishing a crystal blank to its final frequency by etching it in hydrofluoric acid.

ether — A hypothetical medium that pervades all space (including vacuum) and all matter; assumed to be the vehicle for propagation of electromagnetic radiations.

Ethernet — 1. A high-speed network connection. 2. A network standard first developed by Xerox, and refined by DEC and Intel. Ethernet interconnects personal computers and transmits at 10 megabits per second. It uses a bus topology that can connect up to 1024 PCs and workstations within each main branch, and uses a protocol known as carrier sense multiple access with collision detection (CSMA/CD) to regulate communication line traffic. Nodes are linked using coaxial cable, fiber-optic cable, or twisted-pair wiring. Ethernet is codified as the IEEE 802.3 standard. 3. A very common method of networking computers in a LAN. Ethernet will handle about 10 megabits per second and can be used with almost any kind of computer.

E-transformer — A special form of differential transformer employing an E-shaped core. The secondaries of the transformer are wound on the outer legs of the E, and the primary is on the center leg.

ETSI — *See* European Telecommunications Standards Institute.

Ettingshausen effect — Analogous to the Hall effect. The different temperatures found on opposite edges of a metal strip that is perpendicular to a magnetic field and through which an electric current flows longitudinally.

eTV — Abbreviation for educational TV. A nonprofit television station operating to serve community needs in the areas of instruction and cultural development.

Eureka — A ground transponder in the British Rebecca-Eureka radar navigational system.

Euro connector — *See* SCART

European Telecommunications Standards Institute — Abbreviated ETSI. One of the European organizations responsible for establishing common industry-wide standards for telecommunications.

eutectic — 1. An isothermal reversible reaction in which a liquid solution is converted into two or more intimately mixed solids on cooling, the number of solids formed being the same as the number of components in the system. 2. An alloy having the composition indicated by the eutectic point on an equilibrium diagram. 3. An alloy structure of intermixed solid constituents formed by a eutectic reaction. 4. Referring to an alloy or solid solution that has the lowest possible melting point, usually below that of its components. 5. The most fusible series of alloys (e.g., 63/37 tin-lead solder). 6. The specific proportions of the constituents of an alloy having the lowest melting point. The system goes from totally molten to totally solid without going through a slushy range at the eutectic composition.

eutectic alloy — 1. A combination of two or more metals that has a sharply defined melting point and no plastic range. 2. An alloy with a low and sharp melting point that converts from a solid to a liquid state at a specific recurring point. Used in thermal overload devices.

eutectic bonding — Formation of a metallurgical joint in similar or dissimilar metals through the introduction of a thin film of another metal at the joint interface. Upon application of heat and moderate pressure, the intermediate film and the metals to be joined form a molten eutectic phase, which is then eliminated from the joint by thermal diffusion into the base metals.

eutectic solder — Solder that has the lowest possible melting point for its combination of elements. Eutectic tin-lead solder is composed of 63 percent tin and 37 percent lead. It melts at 361°F. Eutectic tin-silver solder has 96.5 percent tin with 3.5 percent silver, and melts at 430°F.

eV or ev — Abbreviation for electron-volt.

evaporation — The deposition in high vacuum of insulation that is thermally liberated from a parent source. Silica films of low optical absorption have been produced by electron bombardment of the parent oxide.

evaporation materials — Metals used for evaporation charges and sputtering targets, including chromium

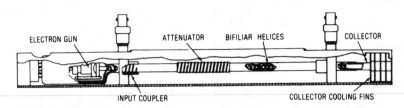

Estiatron.

and its alloys, for (*a*) a thin adhesive layer on IC substrates to allow better deposition of gold or other metal, (*b*) resistor material, and (*c*) vacuum deposition in mask production.

evaporation of electrons — The cooling that occurs on the surface of a cathode during emission. It is analogous to the cooling of a liquid or solid as it evaporates.

evaporation sources — Boats and filaments used as heat sources for vacuum evaporation to form thin layers on substrates. The process is frequently done by resistively heating the evaporant in a ceramic crucible or by self-heating of boats constructed of tungsten, molybdenum, or tantalum.

evaporative deposition — The process of condensing a thin film of evaporated material upon a substrate. Evaporation usually is produced by heating a material in a high vacuum.

E-vector — A vector representing the electric field of an electromagnetic wave. In free space it is perpendicular to the direction of propagation.

even harmonic — Any harmonic that is an even multiple (2, 4, 6, etc.) of the fundamental frequency. The even harmonics of 60 Hz are 120 Hz, 240 Hz, 360 Hz, etc.

even parity — 1. Parity bit that is added to a word so that the total number of 1s is even. 2. The condition that occurs when the sum of the number of 1s in a binary word is always even. 3. A dumb-terminal data verification method in which each character must have an even number of 1 bits.

event counter — Instrument that records and totalizes occurring events; can include time of occurrence of events.

event flag — In a computer, an easily implemented synchronization mechanism that can be used for passing messages and data buffers between two cooperating tasks.

EW — Abbreviation for electronic warfare.

E-wave — Designation for TM (transverse magnetic) wave, one of the two classes of electromagnetic waves that can be sent through waveguides.

EX — In a calculator, it is the abbreviation for exchange key. Interchanges the last entry with the preceding value in the calculator.

exalted-carrier receiver — A receiver that counteracts selective fading by maintaining the carrier at a high level at all times.

exalted-carrier reception — A method of receiving either amplitude- or phase-modulated signals in which the carrier is separated from the sidebands, filtered and amplified, and then recombined with the sidebands at a higher level prior to demodulation.

except gate — A gate in which the specified combination of pulses producing an output pulse is the presence of a pulse on one or more input lines and the absence of a pulse on one or more other input lines.

exception — In a computer, a condition that is out of the ordinary in normal task execution; e.g., arithmetic overflow.

excess carriers — Any carriers present in a semiconductor material or region in addition to those present in equilibrium.

excess conduction — Conduction by excess electrons in a semiconductor.

excess electron — An electron introduced into a semiconductor by a donor impurity and available to promote conduction. An excess electron is not required to complete the bond structure of the semiconductor.

excess fifty — In a computer, a representation in which a number N is denoted by the equivalent of $(N + 50)$.

excess meter — An electricity meter that measures and registers the integral, with respect to time, of those portions of the active power in excess of the predetermined value.

excess minority carriers — In a semiconductor, the number of minority carriers that exceed the normal equilibrium number.

excess modified index of refraction — *See* refractive modulus.

excess noise — Also called current noise, bulk noise, and $1/f$ noise. Noise resulting from the passage of current through a semiconductor material.

excess sound pressure — The total instantaneous pressure at a point in a medium containing sound waves, minus the static pressure when no sound waves are present. The unit is the dyne per square centimeter.

excess-three BCD — Abbreviation for excess-three binary-coded decimal. Pertaining to a code based on adding 3 to a decimal digit and then converting the result directly to binary form. Use of this code simplifies the execution of certain mathematical operations in a binary computer that must handle decimal numbers.

exchange — To remove the contents of one storage unit of a computer and place it in a second, at the same time placing the contents of the second storage unit into the first.

exchange cable — A lead-covered, nonquadded, paper-insulated cable used in providing cable pairs between local subscribers and a central office.

exchange code — The three digits following an area code in a telephone number.

exchange key — *See* EX.

exchange line — A line that joins a subscriber or switchboard to a commercial exchange.

exchange plant — Facilities used to serve the needs of subscribers, as distinguished from facilities used for long-distance communication.

exchange register — *See* memory register.

exciplex — From *exci*ted state com*plex*. A chemical reaction occurring in certain lasing materials known as organic dyes and used for adjusting a laser so that it emits light in a color range from near ultraviolet to yellow.

excitation — 1. Also called stimulus. An external force or other input applied to a system to cause it to respond in some specified way. 2. Also called drive. A signal voltage applied to the control electrode of an electron tube. 3. In electric or electromagnetic equipment, supplying with a potential, a charge, or a magnetic field. 4. The addition of energy to a system so as to transfer the system from its ground state to an excited state. 5. That energy which is present in a crystalline material as a result of its dynamic interaction with the external environment. This includes the energy acquired by the material in the form of sound, heat, light, and other forms of radiation. 6. Current supplied to energize the field coils of a generator. 7. The signal applied to a power amplifier. 8. The signal applied to a transmitting antenna.

excitation anode — An auxiliary anode of a pool-cathode tube, used to maintain a cathode spot when the output current is zero.

excitation current — The resultant current in the shunt field of a motor when voltage is applied across the field.

excitation energy — The external electrical energy required for proper operation of a transducer.

excitation purity — Also called purity. The ratio between the distance from the reference point to the point representing the sample and the distance along the same straight line from the reference point to the spectrum locus or to the purple boundary, both distances being measured (in the same direction from the reference point) on the CIE chromaticity diagram.

excitation voltage — The voltage required for excitation of a circuit.

excited-field speaker — A speaker in which the steady magnetic field is produced by an electromagnet.

exciter — 1. In a directional transmitting antenna system, the part connected directly to the source of power, such as to the transmitter. 2. A crystal or self-excited oscillator that generates the carrier frequency of a transmitter. 3. A small, auxiliary generator that provides field current for an ac generator.

exciter lamp — 1. A high-intensity incandescent lamp having a concentrated filament. It is used in making variable-area, sound-on-film recording and in reproducing all types of sound tracks on film, as well as in some mechanical television systems. 2. A light source used in a facsimile transmitter to illuminate the subject copy being scanned. 3. Small incandescent lamp whose intense beam is focused on the optical sound track of a motion picture film. The sound track modulates the beam, which in turn is detected by a photocell that produces an electrical output that is eventually converted into audible sounds.

exciter or dc-generator relay — A device that forces the dc machine-field excitation to build up during starting, or which functions when the machine voltage has built up to a given value.

exciter response — In rotating electrical machinery, the rate of change of the main exciter voltage when the resistance in the main exciter field circuit is suddenly changed. The exciter response may be expressed in volts per second or by the numerical value obtained by dividing the volts per second by some designated value of voltage, such as the nominal collector-ring voltage.

exciting current — Also called magnetizing current. 1. The current that flows in the primary of a transformer when the secondary is open-circuited. This current produces a flux that generates a back emf equal to the applied voltage. 2. The current that passes through the field windings of a generator.

exciton — A mobile, electrically neutral, excited state of holes and electrons in a crystal. An example is a weakly bound electron-hole pair; when such a pair recombines, the energy yielded is the bandgap reduced by the binding energy of the pair.

excitron — A type of rectifier tube used in applications with heavy power requirements and in power distribution systems. It has a single anode and a mercury-pool cathode and is provided with a means for maintaining a continuous cathode spot.

exclusion principle — The principle that states that if particles are considered to occupy quantum states, then only one particle of a given kind can occupy any one state. Particles differ in kind due to their direction of spin momentum, orbit, etc.

exclusive OR — A function that is valid (its value is 1) if one and only one of the input variables is valid. The exclusive OR applied to two variables is valid, or 1, if the binary inputs are different. The term *half-add* is often applied to the exclusive OR with two input variables.

exclusive OR function — A logic operation in which the result is logically true when only one input function is true and false when both inputs are true or false.

exclusive OR gate — 1. A type of gate that produces an output when the inputs are the same, but not when they are different. 2. An electronic circuit whose output is a logical 1 if its two inputs are different, and a logical 0 if they are the same. This follows the rules for binary addition if carries are disregarded. For this reason, the exclusive OR gate is sometimes known as a half-adder.

excursion — A single movement away from the mean position in an oscillating or alternating motion.

execute — 1. The process of interpreting an instruction and performing the indicated operation (s). 2. To perform a specified computer instruction. To run a program. 3. The third cycle of the three cycles for program instruction execution. During this time, the operation instruction is performed. *See also* decode; fetch.

execution — 1. The performance of an operation or instruction. 2. The performance of a specific operation such as would be accomplished through processing one instruction, a series of instructions, or a complete program.

execution time — 1. The time required for a computer to execute an instruction, usually several machine cycles. 2. The total time required for the execution of one specific operation, including fetch, decode, and execute steps.

executive instruction — Instruction to determine how a specially written computer program is to operate.

executive routine — In computer operation, a set of coded instructions that controls loading and relocation of other routines, and in some cases employs instructions not known to the general programmer. Effectively, an executive routine is a nonhardware part of the computer itself, except that it is superimposed on all lower-level programs and instructional sets.

exercisers — 1. Multioutput data generators; e.g., a memory exerciser produces, at different sets of output terminals, data-input words, addresses, and coincidence/complement signals for an error detector (discriminator), as well as appropriate write and read commands. These instruments are actually small test systems, and might properly be classed with them. 2. A test system or program for a device, such as memory, disk, or tape, designed to detect malfunctions prior to use.

exhaustion — The removal of gases from a space, such as the bulb of a vacuum tube, by means of vacuum pumps.

exit — In a computer, the means of halting a repeated cycle of operation in a program.

exitance — The flux per unit area leaving (diverging) from a source of finite area.

exit angle — The exit angle is the angle between the output radiation vector and the axis of the fiber or fiber bundle.

exit delay — The time between turning on a control unit and the sounding of a local alarm or transmission of an alarm signal on actuation of a sensor on an exit door. This delay is used if the authorized access switch is located within the protected area and permits a person with the control key to turn on the alarm system and to leave through a protected door or gate without causing an alarm. The delay is provided by a timer within the control unit.

exogeneous variable — A variable whose values are determined by considerations outside the model in question. *See also* endogenous variable.

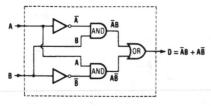

Exclusive OR.

exosphere—The outermost region of the earth's atmosphere, where the atoms and molecules move in dynamic orbits under the influence of the gravitational field.

exothermic—A chemical reaction in which heat is produced.

exothermic reaction—A reaction in which heat is given off.

expand—To spread out part or all of the trace of a cathode-ray display.

expandable breadboarding system—*See* microcomputer development system.

expandable gate—A logic gate whose number of inputs can be increased by the simple addition of an expander block.

expanded contact—In a semiconductor, any pattern that has metallization crossing a diffused junction.

expanded-position-indicator display—An expanded display of a sector from a plan-position-indicator presentation.

expanded-scale meter—1. A meter in which the ratio of deflection per unit of applied energy becomes greater as the energy approaches a specified value. 2. A meter that confines its measurements to a selected narrow range, which occupies the entire scale of the meter. Lesser values are ignored, and no deflection is exhibited until the minimum scale value is reached or exceeded.

expanded scope—A magnified portion of a cathode-ray-tube presentation.

expanded sweep—A preselected portion of a sweep, during which time the electron beam is speeded up in a cathode-ray tube.

expander—1. A transducer that produces a larger range of output voltages for a given amplitude range of input voltages. One important type expands the volume range of speech signals by employing their envelope. 2. A logic block that may be connected easily to an expandable gate to make a larger number of logic inputs available. 3. The part of a compander used at the receiving end of a circuit to restore the compressed signal to its original form. It reduces the amplitude of weak signals and amplifies strong signals. 4. A device used to restore natural dynamic range by counteracting the compression of dynamic range used in the making of recordings and in broadcasting.

expander inputs—Gates used for increasing the number of logic-performing inputs.

expandor—A device that reverses the effect of analog compression. *See also* compander.

expansion—1. A process in which the effective gain applied to a signal is increased for larger signals and decreased for smaller signals. 2. In facsimile transmission, an increase in the contrast between light and dark portions of the transmitted picture.

expansion board—1. A circuit board designed to fit the bus of a computer in order to add additional ports, memory, or functions. 2. A PC board added to a computer and giving it added capabilities, such as increasing the amount of main memory.

expansion slot—A socket on the main board of a personal computer where an accessory card, or expansion board, can be inserted.

experimental model—An equipment model that demonstrates the technical soundness of a basic idea but does not necessarily have the same form or parts as the final design.

experimental station—Any station (except amateur) utilizing electromagnetic waves between 10 kHz and 3000 GHz in experiments, with a view toward the development of a science or technique.

exploring coil—*See* magnetic test coil.

explosion-proof motor—A motor designed and constructed so as to withstand an internal explosion of a specified gas or vapor and to prevent the ignition of the specified gas or vapor surrounding the motor by sparks, flashes, or explosions of the specified gas or vapor inside the motor casing.

explosion-proof switch—A switch (UL listed) that can withstand an internal explosion of a specified gas without causing ignition of surrounding gases.

explosive atmosphere—The condition in which air is mixed with dust, metal particles, or inflammable gas in such proportion that it is capable of igniting or exploding.

exponential—Pertaining to exponents or to an expression having exponents. A quantity that varies in an exponential manner increases by the square or some other power of a factor, instead of linearly.

exponential curve—A curve representing the variation of an exponential function.

exponential damping—Damping that follows an exponential law.

exponential decay—The decay of signal strength, radiation, charge, or some other quantity at an exponential rate.

exponential horn—1. A horn whose cross-sectional area increases exponentially with axial distance. 2. Speaker horn (low or high frequency) in which the flare rate of the horn follows an exponential curve.

exponential quantity—A single quantity that increases or decreases at the same rate as the quantity itself (e.g., the discharge current of a capacitor through a noninductive resistor).

exponential sweep—An electron-beam sweep that starts rapidly and slows down exponentially.

exponential transmission line—A two-conductor transmission line whose characteristic impedances vary exponentially with the electrical length of the line.

exponential waveform—A waveform that is characterized by smooth curves but which possesses pulse properties because it contains numerous constituent frequencies. The exponential waveform undergoes a rate of amplitude change that is either inversely or directly proportional to the instantaneous amplitude.

exposure—1. A measure of the X-ray or gamma radiation at any point; it is used to describe the energy of the radiation field outside the body. 2. The act of subjecting photosensitive surfaces or matter to radiant energy, such as light, to produce an image.

exposure meter—In photography, an instrument that measures scene brightness and indicates proper lens opening and exposure time.

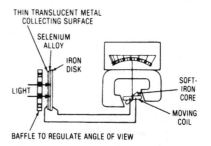

Exposure meter.

expression—A valid series of constants, variables, and functions that can be connected by operation symbols and punctuated, if necessary, to describe a desired computation.

expression control — In an organ, the control that regulates the overall volume. Usually operated with the right foot.

extended addressing — An addressing mode in a computer that can reach any place in memory. *See also* direct addressing.

extended-area service — Telephone exchange service, without toll charges, that extends over an area where there is a community of interest, in return for a somewhat higher exchange service rate.

extended-cutoff tube — *See* remote-cutoff tube.

extended foil capacitor — *See* noninductive capacitor.

extended-foil construction — A method of fabricating capacitors in which two foil electrodes, separated by a dielectric, are offset so that they may be wound with one foil extending from one end of the winding and the other foil extending from the opposite end.

extended-interaction tube — A microwave tube in which a moving electron stream and a traveling electric field interact in a long resonator. The bandwidth of such a tube is between that of a klystron and a traveling-wave tube.

extended octaves — In an organ, tones above or below the notes on the regular keyboard that can be sounded only when certain couplers are on.

extended play — Abbreviated EP. A 45-rpm record on a 7-inch disc. It provides 8 minutes of playing time, instead of the 5 minutes of a standard 45-rpm disc.

extended-play tape — Recording tape made with thinner base material so that more tape can be spooled onto a given size reel. Generally refers to 1-mil (25.4-μm) or 0.5-mil (12.75-μm) thick tapes, as opposed to standard 1.5-mil (38.1-μm) tape. Extended-play tapes are more susceptible to stretching, breaking, and print-through than standard tapes.

extended speed range — An extension of the basic speed range of a motor by means of field voltage control. When operating in this range, the load must be reduced to maintain a constant horsepower output to prevent overheating of the motor.

extension — 1. A small computer program that can be plugged into a larger one, usually to enhance the main program's capabilities. All operating systems use extensions to enable them to read compact discs. 2. Filenames often end with a period followed by additional characters known as the file extension. An extension is generally a standard abbreviation for a type of file. For example, .text is often used for ASCII files, and .ps for PostScript files. 3. An additional telephone bridged on the same line with the main phone.

extension number — The PBX or Centrex telephone number associated with a particular caller.

extent — The physical positions on input-output devices occupied by or reserved for a particular data set.

external armature — A ring-shaped armature that rotates around the field magnets of a generator or motor.

external circuit — All wires and other conductors that are outside the source.

external control devices — All control devices mounted external to the control panel.

external critical damping resistance — The value of resistance that must be placed in series or in parallel with a galvanometer in order to produce the critically damped condition.

external device — 1. An input or output device, such as a paper-tape reader, line printer, or magnetic tape recorder, under control of a computer. 2. A unit of processing equipment in a computer system external to the CPU.

external feedback — In a magnetic amplifier, the ampere-turns of auxiliary windings that assist in equalizing the load-winding ampere-turns, thereby reducing the control ampere-turns required for control. External feedback may be degenerative to reduce the gain and improve the stability of an amplifier.

externally adjustable timer — A time-delay unit that can be adjusted by varying the resistance of the timer resistor externally, usually by means of a screwdriver.

externally caused contact chatter — That chatter resulting from shock or vibration imposed on the relay by external action.

externally caused failure — A failure caused by an environment outside the design limitations, such as excessive loads, voltages, etc., resulting from operator error, accident, or failure of another part.

externally quenched counter tube — A radiation counter tube that requires an external circuit to prevent it from reigniting.

external memory — An auxiliary storage unit apart from the internal memory of a computer (e.g., magnetic tape).

external photoeffect — The emission of photon-excited electrons from the surface of a material after overcoming the energy barrier at the surface of a photoemissive material.

external photoeffect detector — A photodetector in which the energy of each photon incident on the detector surface is sufficient to liberate one or more electrons; i.e., Planck's constant times the frequency, which is the energy of the photon, is sufficient to overcome the work function of the material, and the liberated electrons move under the influence of an applied electric field. Photoemissive devices make use of the external photoeffect.

external photoelectric effect — The ejection of electrons from the surface of a solid by the absorption of a sufficient amount of photons.

external processor loop — *See* tape monitor.

external Q — In a microwave tube, the reciprocal of the difference between the reciprocal of the loaded Q and the reciprocal of the unloaded Q.

external storage — Storage facilities separate from the computer itself but holding information in a form acceptable to the computer, e.g., magnetic tapes, optical discs, etc.

external wiring — Category of electronic wiring that interconnects various subsystems within the weapons system. It is frequently subjected to severe environments.

extinction potential — The lowest value to which the plate voltage of a gaseous tube can be reduced without cutting off the flow of plate current.

extinguishing voltage — 1. The voltage across a glow lamp when the lamp ceases to glow. 2. The lowest anode voltage at which a gas tube can sustain a discharge.

extraband spurious transmitter output — A spurious transmitter output that lies outside the specified band of transmission.

extra-class license — The highest classification of United States amateur license. Requirements include a code sending and receiving ability of 20 words per minute, a knowledge of advanced theory, and the holding of a general- or conditional-class license for two years.

extract — 1. To remove from a set of items of information all those items that meet some arbitrary criterion. 2. In computer operations, to obtain specific digits from a stored word. 3. To form a new word from selected segments of given words.

extract instruction — In a digital computer, the instruction to form a new word by placing selected segments of given words side by side.

extractor — *See* filter, 3.

extra-high performance macros — Computers that have a throughput of 160 gigabits per second, and a memory size of 256 megabytes.

extramural absorption — The absorption of light, transmitted radically through the cladding of an optical fiber, by means of a dark or opaque coating placed over the cladding.

extramural cladding — A layer of dark or opaque absorbing coating placed over the cladding of an optical fiber to increase internal reflection, protect the smooth reflecting wall of the cladding, and absorb scattered or escaped stray light that might penetrate the cladding.

extraneous emission — Any emission of a transmitter or transponder other than the carrier and those sidebands intentionally added to convey intelligence.

extraneous response — Any undesired response of a receiver, recorder, or other susceptible device due to the desired signals, undesired signals, or any combination or interaction among them.

extraordinary wave — One of two components into which a sky wave is split in the ionosphere. When viewed below the ionosphere in the direction of propagation, it has clockwise or counterclockwise elliptical polarization, depending on whether the magnetic field of the earth has a positive or negative component in the same direction. The extraordinary wave is designated by the letter X and is called the X-wave. The other component is the ordinary wave, or O-wave.

extra play — Tape recording term. Originally, all recording tapes were 1.5-mils (38.1-μm) thick, and a 7-inch (17.8-cm) reel would accommodate about 1200 feet (365 m) of this tape, for a half-hour of continuous recording at $7\frac{1}{2}$ ips. The later extra-play tapes are 1-mil (25.4-μm) thick and allow for 45 minutes of recording, or 15 minutes extra.

extrapolate — 1. To estimate the value of a function for variables lying outside the range in which values of the function are known (e.g., to extend the graph of the function beyond the plotted points). 2. To estimate the values of a function that are less than or greater than those already known.

extrapolation — Estimating the future value of some data series based on past observations. Statistical forecasting represents a common example.

extraterrestrial noise — Radio disturbances originating from sources other than those related to the earth; cosmic or solar noise.

extremely high frequency — Abbreviated EHF. 1. The frequency band extending from 30 to 300 GHz. 2. Any of the radio frequencies in the band 30,000 to 300,000 MHz.

extremely low frequency — Abbreviated ELF. A frequency below 300 hertz.

extrinsic base resistance — The resistance between the base terminal of a transistor and the internal inaccessible base point in an equivalent circuit.

extrinsic base resistance-collector capacitance product — The product of the base resistance and collector capacitance of a transistor. It is expressed in units of time, since it is an *RC* time constant, and affects the high-frequency operation of a transistor.

extrinsic conductance — The conductance resulting from impurities or external causes.

extrinsic detector — A photodetector composed of semiconductor material whose responsive properties can be altered by the addition of impurities to the basic material. Copper and mercury-doped germanium are both examples of this semiconductor material.

extrinsic photoconductivity — Photoconductivity due to the addition of impurities or external causes.

extrinsic properties — The properties exhibited by a semiconductor as the result of its modification by imperfections and impurities in the crystal.

extrinsic semiconductor — 1. A semiconductor with charge carrier concentration dependent on impurities or other imperfections. 2. The resulting semiconductor produced when impurities are introduced into an otherwise nonsemiconductor crystal. The electrical properties depend on the impurities. 3. A semiconductor whose electrical properties are dependent on impurities.

extrinsic semiconductor material — A semiconductor material with charge carrier concentration dependent on impurities or other imperfections.

extrinsic transconductance — The quotient of a small change in collector current divided by the small change in emitter-to-base voltage producing it, under the condition that other voltages remain unchanged. Thus, if an emitter-to-base voltage change of 0.1 volt causes a collector-current change of 3 milliamperes (0.003 ampere), with other voltages constant, the transconductance is 0.003 divided by 0.1, or 0.03 siemens. For convenience, a millionth of a siemens (which was formerly termed mho), or a microsiemens, is used to express transconductance. Thus, in the example, 0.03 siemens is 30,000 microsiemens.

extruded cables — Cables with conductors that are insulated and formed in a uniform configuration by the application of a homogeneous insulation material in a continuous extrusion process.

extrusion — A method of forcing plastic, rubber, or elastomer material through an orifice in a more or less continuous fashion to apply insulation or jacketing to a conductor or cable.

eyelet — A tubular metal piece having one end (and possibly a second) headed or rolled over at a right angle.

E zone — In the making of frequency predictions, one of three zones into which the earth is divided to show the variations of the F_2 layer with respect to longitude. This zone includes Asia, Australia, the Philippines, and Japan.

F

f—1. Letter symbol for femto (10^{-15}). 2. Symbol for focal length, frequency.

F—1. Symbol for filament, fuse. 2. Abbreviation for Fahrenheit. 3. Letter symbol for farad.

F− (**F-minus** or **F-negative**)—See A− (A-minus or A-negative).

F+ (**F-plus** or **F-positive**)—See A+ (A-plus or A-positive).

fA—Letter symbol for femtoampere (10^{-15} ampere).

fabrication holes—See pilot holes.

fabrication tolerance—In the construction or assembly of an equipment or portion of an equipment, the maximum variation in the characteristics of a part that, considering the defined variations of the other parts in the equipment, will permit the equipment to operate within specified performance limits.

Fabry-Perot interferometer—A resonant cavity bounded by two end mirrors separated so as to produce interference for certain allowed optical frequencies. The enhancements and cancellations produced by the internal reflections are, in fact, optical standing waves. The structure can be used as a laser cavity, mode filter, or frequency selector.

face—1. A plane surface on a crystal that stands in a particular and invariable relation to the axes and planes of reference and to other faces. 2. Front, or viewing, surface of a cathode-ray tube. 3. The portion of a meter bearing the scale markings.

face bonding—1. A method of attaching active devices to thin-film passive networks. The semiconductor chips are provided with small mounting pads turned face down and bonded directly to the end of the thin-film conductors on the passive substrate. The term includes ultrasonic, solder-reflow, and solder-ball techniques. 2. Process of bonding a semiconductor chip so that its circuitry side faces the substrate. Flip-chip and beam-lead bonding are two common methods. (Opposite of back bonding.)

faced crystal—A single or twinned mass of quartz bonded in part or entirely by the original growth faces.

face-parallel cut—A Y-cut for a quartz crystal.

face-perpendicular cut—An X-cut for a quartz crystal.

facility—1. A transmission path between two or more locations without terminating or signaling equipment. 2. Anything used or available for use in providing communication service; a communications path.

facom—A long-distance measuring or radionavigational system that derives information of distances by comparing the phases of received and locally generated signals. It is a base-line system operating in the low-frequency band and will work under adverse propagation and noise conditions at ranges of up to 3000 miles (4827 km) from the signal source.

facsimile—1. A process or the result of a process by which fixed graphic material, including pictures or images, is scanned and the information converted into electrical signal waves, which are used either locally or remotely to produce in record form a likeness (facsimile) of the subject copy. 2. Electronic transmission of photographs or documents over a telephone channel. 3. Technology that allows a paper message to be scanned optically, translated into digitally encoded pixels, and sent across the public telephone network to a receiving facsimile (fax) machine, or computer which then reconstructs the original image.

facsimile broadcast station—A station licensed to transmit images of still objects for reception by the general public.

facsimile receiver—An instrument designed to receive facsimile transmissions and translate them into a reproduced image.

facsimile recorder—Apparatus that reproduces on paper the image transmitted by a facsimile system.

facsimile-signal level—An expression of the maximum signal power or voltage created by a scanning of the subject copy as measured at any point in a facsimile system. According to whether the system employs positive or negative modulation, this will correspond to picture white or black, respectively. It may be expressed in decibels with respect to some standard value, such as 1 milliwatt or 1 volt.

facsimile system—An integrated assembly of the elements used for facsimile transmission and reception.

facsimile transmitter—An apparatus employed to convert the subject copy into suitable facsimile signals.

factory calibration—The tuning or altering of a circuit by the manufacturer to bring the circuit into specification. Normally stated as a percent deviation.

fade—1. The gradual lowering in amplitude of a signal. 2. A gradual change of signal strength. 3. To change the strength of a signal gradually. 4. Reducing signal level until it is largely attenuated or completely inaudible (faded away). Short fades occur almost immediately; long fades take from 5 to 30 seconds. Fades are most often done at the end of a recording.

fade chart—A graph of the null areas of an air-search radar antenna; it is used as an aid in estimating target altitude.

fade in—To increase the signal strength gradually in a sound or television channel.

fade out—1. The gradual decrease in signal strength in a sound or television channel. 2. The cessation or near cessation of radio-wave propagation through parts of the ionosphere due to a sudden atmospheric disturbance.

fader—1. A multiple-unit control used in radio for gradual changeover from one microphone or audio channel to another; in television, from one camera to another; and in motion-picture projection, from one projector to another. 2. A control or group of controls for effecting fade in and fade out of video or audio signals.

fading — 1. A drift in the level of received radio signals beyond intelligibility. It is often caused by changes in the upper atmosphere. 2. Variations in intensity of some or all components of a received (radio) signal due to changes in the propagation path. 3. Deliberate slow reduction of signal level by means of the volume control.

fading margin — The number of decibels of attenuation that can be added to a specified radio-frequency propagation path before the signal-to-noise ratio of the channel falls below a specified minimum.

Fahnestock clip — A spring-type terminal to which a temporary connection can readily be made.

Fahrenheit temperature scale — A temperature scale in which the freezing point of water is defined as 32° and the boiling point as 212° under normal atmospheric pressure (760 mm of mercury).

fail hardover — Failure that results in a steady-state maximum system output with no input signal.

fail-safe — 1. Describing a circuit or device that fails in such a way as to maintain circuit continuity or prevent damage. 2. A feature of a system or device that initiates an alarm or trouble signal when the system or device either malfunctions or loses power. 3. A control system that either continues to function safely after a failure or lapses into a predefined condition known to be safe.

fail-safe circuit — A circuit that has an output state that indicates that either a circuit input or the circuit itself has failed. Finds circuit application in complex systems where self-healing subsystems exist. When a subsystem failure is detected, a backup subsystem is automatically inserted.

fail-safe control — A system of remote control for preventing improper operation of the controlled function in event of circuit failure.

fail-safe operation — An electrical system so designed that the failure of any component in the system will prevent unsafe operation of the controlled equipment.

fail soft — 1. An active failure that can be compensated for by the system, or one in which the operator can safely assume control. 2. A method of system implementation that prevents irrecoverable loss of computer usage due to failure of any system resource. It provides for graceful degradation of service.

fail-soft system — A system in a computer that continues to process data despite the failure of parts of the system. Usually accompanied by a deterioration in performance.

failure — 1. The inability of a system, subsystem, component, or part to function in the required or specified manner. 2. The termination of a capability of an item to perform its required function.

failure-activating cause — A stress or force (e.g., shock or vibration) that induces or activates a failure mechanism.

failure analysis — 1. Examination of electronic parts to determine the cause of performance variations outside previously established limits, for the purpose of identifying failure modes and failure-activating causes. 2. The logical, systematic examination of an item or its diagram(s) to identify and analyze the probability, causes, and consequences of potential and real failures.

failure indicator — The observed characteristic that shows that an item is defective.

failure mechanism — 1. A structural or chemical defect, such as corrosion, a poor bond, or surface inversion, that causes failure. 2. The process of degradation or chain of events that results in a particular failure mode. 3. The basic chemical or physical change that

results in a catastrophic, degradation, or intermittent failure. 4. Underlying physical or chemical phenomenon that causes a failure. *See* failure mode, 1.

failure mode — 1. Also called failure mechanism. The manner in which a failure occurs, including the operating condition of the equipment or part at the time of the failure. 2. An electrical parameter that is sensitive to degradation and therefore is useful in the identification of failure mechanisms. 3. A catastrophic, degradation, or intermittent failure, usually in the form of opens, shorts, or parameters out of specification.

failure rate — Also called hazard. 1. The average proportion of units failing per unit of time, normally expressed in percent per thousand hours. It is used in assessing the life expectancy of a device. To be meaningful, a statement of the failure rate must be accompanied by complete information regarding the testing conditions, failure criteria, parameters monitored, and confidence level. 2. The statistical probability of the incidence of catastrophic failure in a large number of identical components operated continuously for a given period under stated environmental conditions. Usually expressed as maximum percentage of units predicted to fail under continuous service at maximum rated power (or voltage, if that limit is reached first) at a stated ambient temperature. 3. Expected number of failures in a given time interval under specified conditions.

failure unit — One failure in 10^9 device operating hours.

fall-in — In a synchronous motor, the point at which synchronous speed is reached.

fallouts — *See* transistor seconds.

fall time — 1. The length of time during which a pulse is decreasing from 90 to 10 percent of its maximum amplitude. 2. A measure of time required for a circuit to change its output from a high level (1) to a low level (0). 3. The time required for the pointer of an electrical indicating instrument to move from a steady full-scale deflection to 0.1 (± specified tolerance) of full scale when the instrument is short circuited. 4. The time interval between the instants at which the magnitude of the pulse at the output terminals of a switching transistor reaches specified upper and lower limits, respectively, when the transistor is being switched from its conducting to its nonconducting state. The upper and lower limits are usually 90 and 10 percent, respectively, of the amplitude of the output pulse.

false add — To form a partial sum; that is, to perform an addition without carries.

false alarm — 1. A radar indication of a detected target even though one does not exist. It is caused by noise or interference levels that exceed the set threshold of detection. 2. An alarm signal transmitted in the absence of an alarm condition. These may be classified according to causes: environmental, e.g., rain, fog, wind, hail, lightning, temperature; animals, e.g., rats, dogs, cats, insects; human-made disturbances, e.g., sonic booms, electromagnetic interference, vehicles; equipment malfunction, e.g., transmission errors, component failure; operator error; and unknown.

false course — In navigation normally providing one or more course lines, a spurious additional course-line indication due to undesired reflections or maladjusted equipment.

false-echo device — A device for producing an echo different from that normally observed.

false statement — A statement having a value of 0 in Boolean algebra.

falsing — Extraneous signal or signals that cause a decoding device to operate without the normal input of proper encoding signals.

FAMOS — Abbreviation for floating-gate avalanche-injection MOSFET. A type of MOSFET capable of long-term memory storage; used in EPROMs.

fan — 1. The volume of space energized periodically by a radar beam(s) as it repeatedly traverses an established pattern. 2. An air-moving machine used to cool electronic circuitry in applications in which natural convection and radiation cooling are insufficient. Employed in electronics enclosures, fans are capable of moving air past the heat-radiation surfaces of the various active components or their associated heat sinks. The terms *fan* and *blower* are sometimes used interchangeably. In practice, however, the fan is a machine with an axially mounted impeller or fan blade, whereas the blower is one with a paddle-wheel-shaped impeller that draws air in at the side and exhausts it at the end. 3. An axial low-pressure device capable of providing high air flow parallel to the axis of the motor shaft. Available in several types, including the tube-axial with integral venturi porting and, in open construction, without venturi porting.

fan antenna — *See* double-V antenna.

fan beam — A field pattern having an elliptically shaped cross section in which the ratio of major to minor axes usually exceeds 3 to 1.

fan-in — 1. The number of inputs that can be connected to a logic circuit. 2. The number of operating controls in a single device that individually or in combination result in the same output from the device. 3. Number of units that can be input-connected to a single similar unit.

fan-in circuit — A circuit that has many inputs feeding to a common point.

fan marker — A radio signal having a vertically directed fan beam that tells the pilot the location of his aircraft while flying along a radio range.

fanned-beam antenna — A unidirectional antenna so designed that transverse cross sections of the major lobe are almost elliptical.

fanning beam — A narrow antenna beam that is scanned repeatedly over a limited arc.

fanning strip — An insulated board, often made of wood, that spreads out the wires of a cable for distribution to a terminal board.

fan-out — 1. The number of parallel loads within a given logic family that can be driven from one output mode of a logic circuit. 2. To spread a group of conductors, such as a cable end, apart so that each can be individually tested or identified. 3. The number of standard loads that can be driven by a circuit output in a logic family. A standard load is the current required to switch the basic gate of the family. 4. In a digital circuit, the number of other ICs that can be driven by the device. 5. The number of elements that can be operated in parallel from a single similar element. "Similar" in this case refers to operating parameters such as impedance and does not mean identical function devices.

fan-out circuit — A circuit that has a single output point that feeds many branches.

"fantasy" decoder — A type of descrambler used for satellite descrambling. It uses a 94-kHz sine-wave scrambling plus video inversion. Audio is encoded on a 15.7-kHz carrier that is one-sixth that of the scrambling sine wave. The 94-kHz sine wave is not an exact multiple of the scan rate, in general.

farad — The capacitance of a capacitor in which a charge of 1 coulomb produces a change of 1 volt in the potential difference between its terminals. The farad is the unit of capacitance in the mksa system.

faraday — A unit equal to the number of coulombs (96,500) required for an electrochemical reaction involving one electrochemical equivalent. In an electrolytic process, 1 gram equivalent weight of matter is chemically altered at each electrode for 1 faraday of electricity passed through the electrolyte.

Faraday cage — *See* Faraday shield.

Faraday dark space — The relatively nonluminous region between the negative glow and the positive column in a glow-discharge cold-cathode tube. *See also* glow discharge.

Faraday effect — 1. The rotation of the plane of polarization of radio waves as they pass though the ionosphere in the earth's magnetic field. This effect produces a decoupling loss between linearly polarized antennas. 2. The rotation of the plane of polarization that occurs when a plane-polarized beam of light passes through certain transparent substances in a direction parallel to the lines of force of a strong magnetic field. This effect also governs the action of a ferrite rotator in a waveguide.

Faraday rotation — 1. Rotation of a signal's polarization caused by the atmosphere's E- and F-layers. 2. The apparent rotation of the plane of polarization of a linearly polarized wave as it passes through a medium (e.g., a ferrite material) that has a different propagation constant for each of the two component waves of opposite rotational sense.

Faraday screen — *See* Faraday shield.

Faraday shield — Also called Faraday screen or Faraday cage. 1. A network of parallel wires connected to a common conductor at one end to provide electro-static shielding without affecting electromagnetic waves. The common conductor is usually grounded. 2. An electrostatic shield between input and output windings of a transformer. This is done to reduce capacitive coupling between the input and output of the power supply.

Faraday's laws — 1. The mass of a substance liberated in an electrolytic cell is proportionate to the quantity of electricity passing though the cell. 2. When the same quantity of electricity is passed through different electrolytic cells, the masses of the substances liberated are proportionate to their chemical equivalents. 3. Also called the law of electromagnetic induction. When a magnetic field cuts a conductor, or when a conductor cuts a magnetic field, an electric current will flow through the conductor if a closed path is provided over which the current can circulate.

faradic current — An intermittent and nonsymmetrical alternating current, such as that obtained from the secondary winding of the induction coil.

faradmeter — An instrument for measuring electric capacitance.

far-end crosstalk — Crosstalk that travels along the disturbed circuit in the same direction as the signals in that circuit. To determine the far-end crosstalk between two pairs, 1 and 2, signals are transmitted on pair 1 at station A, and the crosstalk level is measured on pair 2 at station B.

far field — The space beyond the near field of an antenna, in which radiation is essentially confined to a fixed pattern, and power density along the axis of the pattern falls off inversely with the square of the distance.

far-field region — That region of the field of an antenna where the angular field distribution is essentially independent of the distance from the antenna.

far-infrared radiation — That radiation composed of the wavelengths falling between light and microwaves, ranging from 10 to 2000 micrometers. The portion of the infrared spectrum that contains the longest wavelengths.

Farnsworth image-dissector tube — A special cathode-ray tube for use in television cameras.

far-ultraviolet radiation — That radiation characterized by wavelengths ranging from 0.2 to 0.3 micrometer in the electromagnetic spectrum.

far zone — The region distant from a radio transmitting antenna in which the radiation field is stronger than the induction field and in which the intensity of the radiation field varies inversely with the square of the distance from the antenna.

FAST — Acronym for Fairchild advanced Schottky TTL (Fairchild Camera & Instrument Corp.). A family of high-speed TTL logic that uses very small, low-junction-capacitance transistors.

fast-access storage — In a computer memory or storage, the section from which information may be obtained most rapidly.

fast answerback — *See* calculator mode.

fast automatic gain control — A radar AGC method in which the response time is long compared with a pulse width, but short compared with the time on target.

fastener — A device used to secure a conductor (or other object) to the structure that supports it.

fast forward — 1. The provision on a tape recorder permitting tape to be run rapidly through the recorder in the play direction, usually for search or selection purposes. 2. High-speed winding (shuttling) of tape from the supply reel onto the takeup reel.

fast-forward control — A tape-recorder control that permits running the tape through the machine rapidly in the forward direction.

fast Fourier transform — Abbreviated FFT. A computationally efficient mathematical technique that converts digital information from the time domain to the frequency domain for rapid spectral analysis. FFTs generally utilize a time-weighting function to compensate for data records with a noninteger number of samples.

fast groove — Also called fast spiral. In disc recording, an unmodulated spiral groove having a much greater pitch than the recorded grooves.

fast-operate, fast-release relay — A high-speed relay designed specifically for both short operate and short release times.

fast-operate relay — A high-speed relay designed specifically for short operate and long release times.

fast-operate, slow-release relay — A relay designed specifically for short operate and long release times.

fast-release relay — A high-speed relay designed specifically for short release but not short operate time.

fast spiral — *See* fast groove.

fast time constant — An antijamming device used in radar video-amplifier circuits. It differentiates incoming pulses so that only the leading edges of the pulses are used.

fatal error — An error that causes a program to abort.

fatigue — The weakening of a material under repeated stress.

fault — 1. A defect in a wire circuit due to unintentional grounding, a break in the line, or a crossing or shorting of the wires. 2. A disturbance (such as lightning) that impairs normal operation. 3. A failure of a hardware or software component in a system that may lead to a system failure or error or some other manifestation that can be detected by a user. 4. An anomaly that prevents the correct operation of the device.

fault current — 1. The current that may flow in any part of a circuit or amplifier under specified abnormal conditions. 2. The current in a circuit that results from loss of insulation between conductors or between a conductor and ground.

fault dictionary — A set of fault signatures, each of which indicates the probable faults that could cause the error message matching the signature.

fault electrode current — The peak current that flows through an electrode during a fault, such as an arcback or a load short circuit.

fault finder — A test set for locating troubles in a telephone system.

fault indicator — Equipment that provides an instantaneous alarm, both visual and audible, of failures detected in the various components of its assorted equipment.

fault isolation — 1. The process of identifying and locating failures in a unit under test. 2. Determining the cause of a test failure, typically by identifying a defective component or process failure on a board.

fault-isolation resolution — The average number of components to which a fault can be isolated.

fault model — A set of data that logically describes the operations of a device or circuit containing one or more faults.

fault signature — 1. Data representing the outputs of a known good unit and used to compare against outputs of a unit under test. 2. A particular output response or set of responses generated when a test program is executed on a device containing a fault. A typical fault signature consists of the incorrect output pin numbers and the test step number at which a test program first detects a fault.

fault simulation — The process of injecting faults during a simulation run to determine the test comprehensiveness of an input pattern.

fault-tolerant — Refers to a computer program or system in which some parts may fail and the program or system will still execute properly.

fault-tolerant circuits — Circuits that are designed so that they could continue to function properly even though part of the circuit failed.

Faure plate — A storage-battery plate consisting of a conductive lead grid filled with active paste material.

fax or FAX — 1. Abbreviation for facsimile. 2. A scanner/printer combination that transmits text and graphics over telephone lines. It uses CCITT Group 3 data compression techniques. Small paper documents can be transmitted over long distances very quickly, but the information is not represented as structured data elements, as in EDI.

fax modem — A device connected to a personal computer, giving it the ability to send or receive electronic messages or images over telephone lines.

fc — Letter symbol for footcandle.

FCC — Abbreviation for Federal Communications Commission.

F-connector — A small, metallic, male-type connecting device with internal threads that attach to the end of a coaxial cable to secure and electrically connect the coax to a female F-fitting. The internal threads of the male connector screw onto the external threads of the female connector.

F/D — *See* focal distance-to-diameter ratio.

FDDI — *See* fiber distributed data interface.

F-display — Also called F-scan or F-scope. In radar, a rectangular display in which a target appears as a centralized blip when the radar antenna is aimed at it. Horizontal and vertical aiming errors are indicated by the horizontal and vertical displacement of the blip.

FDM — *See* frequency-division multiplexer and frequency-division multiplexing.

FDMA — Abbreviation for frequency-division multiple access. A method of sharing the capacity of a satellite transponder by assigning each earth terminal a portion of the transponder's bandwidth, into which the terminal places a carrier modulated by its information. The carrier can be received by all other earth terminals.

FDX — *See* full duplex.

feasibility study — An investigation of the advantages and disadvantages of using an alternative approach over the presently used approach.

FEC — Abbreviation for forward error correction. 1. Used to describe equipment that corrects transmission errors at a receiver. The technique provides for transmission of additional information with the original bit stream so that if an error is detected, the receiver can recreate the correct information without a retransmission. 2. A system of data transmission in which redundant bits generated at the transmitter are used at the receiving terminal to detect, locate, and correct any transmission errors before delivery to the data sink. The advantage of this system is that it does not require a feedback channel; therefore, it can be used with a one-way transmission system.

Federal Communications Commission — Abbreviated FCC. A U.S. federal government agency made up of a board of seven commissioners appointed by the president under the Communications Act of 1934, having the power to regulate all interstate and foreign radio communication originating in the United States, including radio, television, facsimile, telegraph, telephone, and cable systems.

federal telecommunications system — System of commercial telephone lines, leased by the government, for use between major government installations for official telecommunications.

feedback — 1. In a transmission system or a section of it, the returning of a fraction of the output to the input. 2. In a magnetic amplifier, a circuit connection by which an additional magnetomotive force (which is a function of the output quantity) is used to influence the operating condition. 3. In a control system, the signal or signals returned from a controlled process to denote its response to the command signal. Feedback is derived from a comparison of actual response to desired response, and any variation is used as an error signal combined with the original control signal to help attain proper system operation. Systems employing feedback are termed closed-loop systems; feedback closes the loop. 4. Squeal or howl from a speaker caused by speaker sound entering microphone of same recorder or amplifier. 5. The return of a portion of the output of a circuit or device to its input. With positive feedback, the signal fed back is in phase with the input and increases amplification, but may cause oscillation. With negative feedback, the signal is 180° out of phase with the input and decreases amplification but stabilizes circuit performance and tends to lower an amplifier's output impedance, improve signal stability, and minimize noise and distortion. 6. The process of coupling some of the output of an amplifier back to its input. Negative feedback reduces the gain of an amplifier, but has compensating beneficial results. Positive feedback can be used to boost

gain (regeneration), but usually results in oscillation. 7. The flow of information back into the control system so that actual performance can be compared with planned performance. 8. The transfer of a portion of energy from one point in an electrical system to a preceding point. The transfer may be either electrical or acoustical.

feedback admittance — In an electron tube, the short-circuit transadmittance from the output electrode to the input electrode.

feedback amplifier — An amplifier that uses a passive network to return a portion of the output signal to modify the performance of the amplifier.

feedback attenuation — In the feedback loop of an operational amplifier, an attenuation factor by which the output voltage is attenuated to produce the input error voltage.

feedback circuit — A circuit that permits feedback in an electronic device.

feedback compensation — The placement of a device, or an additional circuit, into a feedback control system to improve its response in relation to a specific characteristic of a system.

feedback control — 1. A type of system control obtained when a portion of the output signal is operated upon and fed back to the input in order to obtain a desired effect. 2. An automatic means of sensing speed variations and correcting to maintain a constant speed or close speed regulation. 3. Guidance technique used by robots to bring the end effector to a programmed point.

feedback control loop — A closed transmission path that includes an active transducer and consists of a forward path, a feedback path, and one or more mixing points arranged to maintain a prescribed relationship between the loop input and output signals.

feedback control signal — That portion of the output signal that is returned to the input in order to achieve a desired effect, such as fast response.

feedback control system — 1. A control system comprising one or more feedback control loops; it combines the functions of the controlled signals and commands, tending to maintain a prescribed relationship between the two. 2. A system designed to control the output quantity of a device by returning a portion of its output signal back to its input. This results in the manipulation of the input quantity so that the desired relationship between the input and output signals can be maintained.

feedback cutter — An electromechanical transducer that performs like a disc cutter except it is equipped with an auxiliary feedback coil in the magnetic field. Signals exciting the cutter are induced into the feedback coil, the output of which is fed back in turn to the input of the cutter amplifier. The result is a substantially uniform frequency response.

feedback diode — *See* freewheeling diode.

feedback loop — The components and processes involved in using part of the output as an input for correction or control of the operation of a system.

feedback oscillator — An oscillating circuit, including an amplifier, in which the output is coupled in phase with the input. The oscillation is maintained at the frequency determined by the parameters of the amplifier and the feedback circuits, such as *LC*, *RC*, and other frequency-selective elements.

feedback path — In a feedback control loop, the transmission path from the loop output signal to the loop feedback signal.

feedback regulator — A feedback control system that tends to maintain a prescribed relationship between certain system signals and other predetermined quantities.

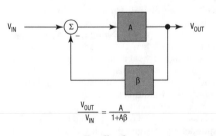

$$\frac{V_{OUT}}{V_{IN}} = \frac{A}{1+A\beta}$$

Feedback.

feedback sense voltage — The voltage, referred to ground, on the feedback terminal of a regulator while it is operating in regulation.

feedback transfer function — In a feedback control loop, the transfer function of the feedback path.

feedback winding — In a saturable reactor, the control winding to which a feedback connection is made.

feeder — 1. A conductor or group of conductors connecting two generating stations, two substations, a generating station and a substation or feeding point, a substation and a feeding point, or a transmitter and antenna. 2. A set of conductors between a main distribution center and secondary ones. A radial feeder supplies energy from one source to a substation or a feeding point that does not receive energy from any other source. The normal flow of energy in such a feeder is in one direction only. 3. The circuit conductors between the service equipment and the branch circuit overcurrent device. 4. A transmission line between an antenna and a radio transmitter or receiver.

feeder cable — Also called trunk cable. 1. A communication cable extending from the central office along a primary route (main feeder cable), or from a main feeder cable along a secondary route (branch feeder cable), and providing connections to one or more distribution cables. 2. In a CATV system, the transmission cable from the head end (signal pickup) to the trunk amplifier.

feeder link — A radio link between an earth station and a satellite, conveying information for a space radio-communication service other than a fixed-satellite service. In the broadcasting-satellite service, all feeder links are uplinks (from the earth to the satellite), but in the mobile-satellite service, feeder links can be both uplinks and downlinks.

feed-forward — A frequency-compensation technique in operational amplifiers. A small-value capacitor is used to bypass a gain stage that has poor performance at high frequency.

feed function — In automatic control of machine tools, the relative motion between the work and the cutting tool (excluding the motion provided for removal of material).

feed holes — A series of small holes in perforated paper tape that convey no information, but are solely for the purpose of engaging the feed pawls or sprocket that transports the tape over the sensing pins of various reading devices.

feedhorn — 1. A satellite TV receiving antenna component mounted at the focal point of a parabolic dish that collects the signal reflected from the main surface reflector and channels this signal into the low-noise amplifier (LNA). 2. A device that collects microwave signals reflected from the surface of an antenna. It is mounted at the focus of all prime focus parabolic antennas.

feed pitch — The distance between the centers of adjacent feed holes in a tape.

feed reel — Also called stock, supply, or storage reel. On a tape recorder, the reel from which the tape unwinds while playing or recording.

feedthrough — 1. The accidental or unintentional transfer of a signal from one track to another on a multitrack tape. 2. The use of special connectors to pass conductors through bulkheads or panels. Contacts can be male on one side and female on the other, or they can be male on both sides or female on both sides. Feedthrough connectors differ from rack-and-panel types in that connection can be made on both sides of the panel. 3. A conductor through the thickness of a substrate, thereby electrically connecting both surfaces. 4. The change in the output voltage of a sample/hold circuit in the hold mode caused by a voltage change in the input. Measured in decibels of attenuation.

feedthrough capacitor — 1. A feedthrough insulator that provides a desired value of capacitance between the feedthrough conductor and the metal chassis or panel through which the conductor is passing. Used chiefly for bypass purposes in UHF. 2. A coaxial capacitor with a central current-carrying conductor, or a conductor connected with a current-carrying rod surrounded by a capacitor element, that is symmetrically bonded to the center conductor and to the outer casing to form a coaxial construction.

feedthrough insulator — A type of insulator that permits wire or cable to be fed through walls, etc., with minimum current leakage.

feed-thru connection — *See* thru-hole connection.

female — Pertaining to the recessed portion of a device into which another part fits.

female contact — A contact located in an insert or body in such a manner that the mating contact is inserted in the unit. This is similar in function to a socket contact.

femto- — Prefix meaning 10^{-15}. Letter symbol: f.

femtoampere — A unit of current equal to 10^{-15} ampere. Letter symbol: fA.

femtovolt — A unit of voltage equal to 10^{-15} volt. Letter symbol: fV.

fence — 1. A line or system of early-warning radar stations. 2. A concentric steel fence placed around a ground radar transmitting antenna to act as an artificial horizon and suppress ground clutter that would otherwise mask weak signals returned from a target at a low angle.

fence alarm — Any of several types of sensors used to detect the presence of an intruder near a fence or any attempt by an intruder to climb over, go under, or cut through the fence.

Fermi-Dirac distribution — A mathematical description of the way in which the current carriers present in a crystalline material have energies distributed above and below the Fermi level; this distribution is a function of the excitation of the material.

Fermi level — The value of electron energy at which the Fermi distribution function is one-half. The average carrier level of a semiconductor region. Hence, a semiconductor crystal in equilibrium has a constant Fermi level throughout.

ferpic — *Fer*roelectric ceramic *pic*ture device. A sandwichlike structure made up of transparent electrodes, a photoconductive film, and a thin plate of fine-grained ferroelectric ceramic. The device stores images in the form of a variation of the birefringence of the ceramic plane (that is, as a variation in the way the plate transmits polarized light).

ferreed — An electromechanical switch that combines the rapid switching of bistable magnetic material with metallic contacts to produce output indications that persist as long as desired without further application of power. Describes any relay whose contact assembly of individual magnetic reeds is sealed in an evacuated glass tube, operated by an external winding. When operated, it is held by magnetism until released by a current pulse.

ferret — An aircraft, ship, or vehicle especially equipped for ferret reconnaissance.

ferret reconnaissance — A form of reconnaissance that detects, locates, and analyzes enemy radars. It is a passive technique that listens for signals transmitted by enemy radars and thus cannot be jammed. The maximum effective range is limited only by the radio horizon. Ferret systems flown at altitudes of a few miles can accurately locate and analyze radars 300 miles (483 km) away. Ferret is an all-weather reconnaissance technique, and no camouflage system works against a properly designated ferret.

ferri- — Prefix indicating a material having a net dipole moment.

ferric oxide (Fe_2O_3) — A red, iron oxide coating for magnetic recording tapes.

ferrimagnetic amplifier — A microwave amplifier utilizing ferrite material in the coupling inductors and transformers.

ferrimagnetic limiter — A power limiter used to replace tr tubes in microwave systems. Its operation is based on a ferrimagnetic material, such as a piece of ferrite or garnet, that exhibits nonlinear properties.

ferrimagnetic materials — Those materials in which spontaneous magnetic polarization occurs in nonequivalent sublattices; the polarization in one sublattice is aligned antiparallel to the other.

ferrimagnetism — A type of magnetism that, because the magnetic moment of neighboring ions tends to align antiparallel, appears microscopically similar to antiferromagnetism. The fact that these moments may be of different magnitudes allows a large resultant magnetization that macroscopically resembles ferromagnetism.

ferristor — A two-winding ferroresonant magnetic amplifier that operates on a high carrier frequency. A two-winding saturable reactor that may be connected as a coincidence gate, current discriminator, free-running multivibrator, oscillator, or ring counter.

ferrite — 1. Also called ferrospinel. A powdered, compressed, and sintered magnetic material having high resistivity, consisting chiefly of ferric oxide combined with one or more other metals. The high resistance makes eddy-current losses extremely low at high frequencies. Examples of ferrite compositions include nickel ferrite, nickel-cobalt ferrite, manganese-magnesium ferrite, yttrium-ion garnet, and single-crystal yttrium-ion garnet. 2. The generic term for a class of nonmetallic, ceramic, ferromagnetic materials (MFe_2O_4) having a spinel-crystal structure. The materials are noted for their high resistivity relative to ferromagnetic metals. Although in its simplest form the M in the formula is iron (Fe), iron is seldom used in actual ferrite manufacture. Manganese, nickel, zinc, magnesium, or copper are more common, and MnZn, NiZn, and MgZn are the most popular combinations.

ferrite bead — A magnetic device for storage of information. It is made of ferrite powder mixtures in the form of a bead fired on the current-carrying wires of a memory matrix.

ferrite circulator — A nonreciprocal microwave network that transmits power from one terminal to another in sequence. Can replace a conventional duplexer, provide isolation of transmitter from receiver, eliminate the requirement for an atr, and isolate the transmitter from antenna reflections.

ferrite core — A core made from iron and other oxides and usually shaped like a doughnut. It is used in circuits and magnetic memories and can be magnetized and demagnetized very rapidly.

ferrite-core memory — A magnetic memory in which read-in and read-out wires are threaded through a matrix of very small toroidal cores molded from a square-loop ferrite.

ferrite isolator — A device either in a waveguide or coax that allows power to pass through in one direction with very little loss, while the rf power in the reverse direction is absorbed. It is useful for maintaining signal source stability and eliminating long-line and frequency-pulling effects in all types of low-power microwave signal sources.

ferrite limiter — A passive low-power microwave limiter that provides an insertion loss of less than 1 dB, with minimum phase distortion, when operating in its

Ferrite cores.

linear range. It is used for protecting sensitive receivers from burnout and from blocking by a strong interfering signal.

ferrite phase-differential circulator — A combination microwave duplexer and load isolator that serves as a switching device between a radar antenna and the associated high-power radar magnetron and radar receiver.

ferrite-rod antenna — Also called ferrod or loopstick antenna. An antenna used in place of a loop antenna in a radio receiver. It consists of a coil wound around a ferrite rod.

ferrite rotator — A gyrator, composed of a ferrite cylinder surrounded by a ring-type permanent magnet, that is inserted in a waveguide to rotate the plane of polarization of electromagnetic waves that travel through the waveguide.

ferrites — 1. Chemical compounds of iron oxide and other metallic oxides combined with ceramic material. They have ferromagnetic properties but are poor conductors of electricity. Hence, they are useful where ordinary ferromagnetic materials (which are good electrical conductors) would cause too great a loss of electrical energy. 2. Ceramic structures made by mixing iron oxide (Fe_2O_3) with oxides, hydroxides, or carbonates of one or more of the divalent metals, such as zinc, nickel, manganese, copper, cobalt, magnesium, cadmium, or iron.

ferrite switch — A ferrite device that obstructs the flow of energy through a wavelength by causing a 90° rotation of the electric field vector.

ferroacoustic storage — A delay-line type of storage comprising a thin tube of magnetostrictive material, a central conductor that passes through the tube, and an ultrasonic driver transducer at one end of the tube.

ferrod — *See* ferrite-rod antenna.

ferrodynamic instrument — An electrodynamic instrument in which the measuring forces are materially increased by the presence of ferromagnetic material.

ferroelectric — 1. Pertaining to a phenomenon exhibited by certain materials in which the material is polarized in one direction or the other or reversed in direction by the application of a positive or negative electric field of magnitude greater than a certain amount. The material retains the electric polarization unless it is disturbed. The polarization can be sensed by the fact that a change in the field induces an electromotive force that can cause a current. 2. That property of certain materials that determines that they will be polarized in one direction or the other, or reversed in direction of polarization when a positive or negative electric field is applied. The material will remain ferroelectric until it is disturbed.

ferroelectric converter — A device that generates high voltage when heat is applied to it. Its operation is based on the change in the dielectric constant or the permittivity of certain materials, such as barium titanate, when heated. This change reaches maximum at the Curie point.

ferroelectric crystal — A crystal that can be polarized in the opposite direction by applying an electric field weaker than the breakdown strength of the material.

ferroelectric domain — The region of a ferroelectric crystal where spontaneous polarization is uniformly directed.

ferroelectric film — A film in which electric polarization is reversible when influenced by an electric field.

ferroelectricity — A property of certain crystalline materials whereby they exhibit a permanent, spontaneous electric polarization (dipole moment) that is reversible by means of an electric field; the electric analog of ferromagnetism. Materials that show this effect are piezoelectric as well.

ferroelectric materials — Those materials in which the electric polarization is produced by cooperative action between groups or domains of collectively oriented molecules.

ferroelectrics — Pyroelectric materials whose direction of polarization can be reversed by application of an electric field.

ferromagnetic — Pertaining to a phenomenon exhibited by certain materials in which the material is polarized in one direction or the other or reversed in direction by the application of a positive or negative magnetic field of magnitude greater than a certain amount. The material retains the magnetic polarization unless it is disturbed. The polarization can be sensed by the fact that a change in the field induces an electromotive force that can cause a current.

ferromagnetic amplifier — A parametric amplifier based on the nonlinear behavior of ferromagnetic resonance at high rf power levels. In one version, microwave pumping power is supplied to a garnet or other ferromagnetic crystal mounted in a cavity containing a strip line. A permanent magnet provides sufficient field strength to produce gyromagnetic resonance in the garnet at the pumping frequency. The input signal is applied to the crystal through the strip line, and the amplified output signal is extracted from the other end of the strip line. Sometimes incorrectly called a garnet maser, but the operating principle differs from that of the maser.

ferromagnetic material — 1. A material having a specific permeability greater than unity, the amount depending on the magnetizing force. A ferromagnetic material usually has relatively high values of specific permeability, and it exhibits hysteresis. The principal ferromagnetic materials are iron, nickel, cobalt, and certain of their alloys. 2. A paramagnetic material that exhibits a high degree of magnetizability. 3. A material that exhibits hysteresis phenomena and whose permeability, greater than 1, is dependent on the magnetizing field.

ferromagnetic oxide parts — Parts consisting primarily of oxides that display ferromagnetic properties.

ferromagnetic resonance — A condition under which the apparent permeability of a magnetic material reaches a sharp maximum at a microwave frequency.

ferromagnetics — The science that deals with the storage of information and the control of pulse sequences through use of the magnetic polarization properties of materials.

ferromagnetic tape — Tape made of magnetic material and used for winding closed cores for toroids and transformers.

ferromagnetism — 1. A high degree of magnetism in ferrites and similar compounds. The magnetic moments of neighboring ions tend not to align parallel with each other. The moments are of different magnitudes, and the resultant magnetization can be large. 2. Strong magnetic property of such substances as iron, cobalt, nickel, and certain alloys. Ferromagnetic substances are essential to the construction of such pieces of equipment as speakers, transformers, electric generators, etc. 3. A property of certain metals, alloys, and compounds whereby below a certain critical temperature (the Curie point) the magnetic moments of the atoms tend to align, giving rise to a spontaneous, permanent magnetism (dipole moment) that is reversible by means of a magnetic field. 4. The properties of certain materials that cause them to have relative permeabilities that exceed unity. This permeability permits the materials to exhibit hysteresis.

ferromanganese — An alloy of iron and manganese.

ferrometer — An instrument for making permeability and hysteresis tests of iron and steel.

ferroresonance — Resonance associated with circuits in which at least one of the circuit elements is nonlinear and contains iron.

ferroresonant circuit — A resonant circuit in which one of its elements is a saturable reactor.

ferroresonant transformer — 1. A voltage-regulating device that gives regulated ac voltages and incorporates some special advantages. Capable of acting as a step-up or step-down voltage transformer as well as an ac voltage regulator, this component delivers a more or less constant ac output voltage even if the magnitude of the input voltage changes. In addition, the ferroresonant transformer is efficient, inexpensive, rugged, and requires no heat sink. It generates no high levels of electrical noise and provides a degree of protection from transients riding on the ac power line. 2. The principal elements of a constant-voltage power supply. The transformer is specially wound to be tuned to line frequency. The resonant condition established is the key to its operation. Only a simple filter is needed to form a highly reliable and very efficient power supply. It does not require a dissipative regulator. A typical ferroresonant power supply offers line regulation of 2 percent and load regulation of 5 percent; efficiency may reach 80 percent. However, the ferroresonant transformer is considerably larger than the transformer for a linear supply and may be twice as heavy.

ferrospinel — A ceramic-like material containing iron and other elements combined with oxygen. A poor conductor of electricity, it is used in transformers, antenna, loops and television deflecting yokes. *See also* ferrite.

ferrous — Composed of and/or containing iron. A ferrous metal exhibits magnetic characteristics, as opposed to a nonferrous metal, such as aluminum, which does not.

ferrous oxide — The substance on magnetic recording disks and tapes. It can be magnetized, thereby permitting information to be recorded on it magnetically.

ferrule — 1. A short tube used to make solderless connections to shielded or coaxial cable. Also molded into the plastic inserts of multiple-contact connectors to provide strong, wear-resistant shoulders on which contact retaining springs can bear. 2. The metal cap around the end of a cartridge-fuse tube that serves as a contact for the fuse.

ferrule resistor — A resistor having ferrule terminals for mounting in standard fuse clips.

FET — Abbreviation for field-effect transistor. Also called unipolar transistor. 1. A transistor controlled by voltage rather than current. The flow of working current through a semiconductor channel is switched and regulated by the effect of an electric charge in a region close to the channel called the gate. A FET has either p-channel or n-channel construction. 2. A transistor whose internal operation is unipolar in nature. The metal-oxide semiconductor FET (MOSFET) is widely used in integrated circuits because the devices are very small and can be manufactured with few steps. 3. A solid-state device in which current is controlled between source terminal and drain terminal by voltage applied to a nonconducting gate terminal. 4. Semiconductor device in which resistance between source and drain terminals is modulated by a field applied to the third (gate) terminal.

fetal cardiotachometer — *See* fetal monitor.

fetal electrocardiograph — *See* fetal monitor.

fetal monitor — An instrument that displays or records the fetal electrocardiogram or other indication of heart action. In some instruments, the maternal electrocardiogram is recorded simultaneously. The instrument may be referred to more definitely as a fetal electrocardiograph, fetal cardiotachometer, or fetal phonocardiograph, depending on its primary purpose.

fetal phonocardiograph — An instrument that provides continuous instantaneous recording of beat-to-beat changes in the fetal heart rate. *See also* fetal monitor.

fetch — 1. To go after and return with things. In a microprocessor, the "objects" fetched are instructions that are entered in the instruction register. The next, or a later, step in the program will cause the machine to execute what it was programmed to do with the fetched instructions. Often referred to as an instruction fetch. 2. In a computer, the collective actions of acquiring a memory address and then an instruction or data byte from memory. 3. Reading out an instruction at a particular memory location into the CPU. 4. To obtain a quantity of data from a place of storage. 5. The action of obtaining an instruction from a stored program and decoding that instruction. Also refers to that portion of a computer's instruction cycle during which that action is performed.

fetch cycle — The first cycle in the fetch-decode-execute sequence of instruction execution. During the fetch cycle, the contents of the program counter are placed on the address bus, a read signal is generated, and the program counter is incremented. The data word coming from the memory (the instruction that has been fetched) will then be gated into the instruction register of the control unit.

FET resistor — A field-effect transistor in which, generally, the gate is tied to the drain, and the resultant structure is used in place of a resistor load for a transistor.

"f" factor — The slope of the straight line from which the nonlinearity of a displacement transducer is calculated in microvolts output per volt excitation per unit stimulus.

FFC — Abbreviation for flat flexible cable.

FFT — Abbreviation for fast Fourier transform.

fiber — 1. A tough insulating material, generally of paper and cellulose, compressed into rods, sheets, or tubes. 2. A clear glass or plastic optical cable, consisting of a core and cladding, designed to propagate optical energy. The diameter of a fiber can vary from about 10 to 1000 micrometers, depending on type. 3. Glass, silica, or plastic cable by which light is conducted or transmitted. Can be multimode (capable of propagating more than one mode of given wavelength) or single mode (one that supports propagation of only one mode of given wavelength). 4. A thread or threadlike structure such as composes cellulose, asbestos, or glass yarn. Also, a single discrete element used to transmit optical (lightwave) information. Analogous to a single wire used to transmit electrical information. Usually consists of a core

that transmits the information and a cladding around the core. 5. The material path along which light propagates; a single discrete optical transmission element.

fiber bundle — 1. A consolidated group of single fibers used to transmit a single optical signal. 2. A rigid or flexible concentrated assembly of glass or plastic fibers used to transmit optical images or light. *See* coherent fiber bundle. 3. An assemblage of transparent glass fibers all bundled together parallel to one another. The length of each fiber is much greater than its diameter. By a process of total reflection, this bundle of fibers can transmit a picture from one of its surfaces to the other around curves and into otherwise inaccessible places with an extremely low loss of definition and light.

fiber cable — A cable composed of a fiber bundle or a single fiber, strength members, and a cable jacket.

fiber dispersion — In fiber optics, pulse spreading in a fiber caused by differing transit times of various modes of electromagnetic waves.

fiber distributed data interface — Abbreviated FDDI. A standard for transmitting data on optical fiber cables at a rate of around 100,000,000 bits per second (10 times as fast as Ethernet and about twice as fast as T-3).

fiber metallurgy — The growing of superfine crystal whiskers whose characteristic is relatively great strength in their length-to-diameter ratio.

fiber needle — A playback point or phonograph needle made from fiber. Being softer than a metal or diamond needle, it is less scratchy; however, it has an extremely short life.

fiber-optic field flattener — A plate consisting of fused optical fibers with both surfaces ground and polished, and having the entrance surface curved to match the image curvature of the input system. The plate transmits to the flat exit surface.

fiber-optic rod multiplexer-filter — A graded-index, cylindrically shaped section of optical fiber or rod with a length corresponding to the pitch of the undulations of light waves caused by the graded refractive index, the light beam being injected via fibers at an off-axis end point on the radius, with the undulations of the resulting wave varying periodically from one point to another along the rod with interference layers at the $1/4$-pitch point of the undulations, providing for multiplexing or filtering.

fiber optics — Abbreviated FO. 1. A technology that uses light as a digital information carrier. The transmission medium is made up of small strands of glass, each of which provides a path for light rays that carry the data signal. Fiber-optic technology offers large bandwidth, very high security, and immunity to electrical interference. The glass-based transmission facilities also occupy far less space than other high-bandwidth media, which is a major advantage in crowded underground ducts. 2. Also called optical fibers or optical fiber bundles. An assemblage of transparent glass fibers all bundled together parallel to one another. The length of each fiber is much greater than its diameter. This bundle of fibers has the ability to transmit a picture from one of its surfaces to the other around curves and into otherwise inaccessible places with an extremely low loss of definition and light, by a process of total reflection. 3. The technique of conveying information in the form of light signals through a particular configuration of glass or plastic fibers. 4. A general term describing a light-wave or optical communications system. In such a system, electrical information is converted to light energy, transmitted to another location through optical fibers, and is there converted back into electrical information. 5. A method of communicating analog or digital signals through a noninductive/nonconductive dielectric medium such as glass- or plastic-core cables. The fiber-optic system consists of a light source and

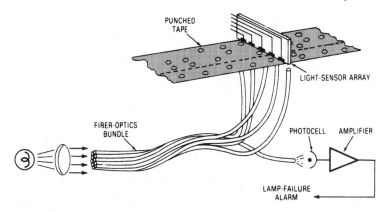

Fiber optics.

light detector, cable-to-semiconductor connectors, and the fiber-optic cable. The transmitter converts the input signal to an analog- or digital-modulated light signal that is communicated through the fiber-optic cable to a light detector. The light detector converts the optical energy back into usable electric form. 6. The technology of guidance of optical power, including rays and waveguide modes of electromagnetic waves, along conductors of electromagnetic waves in the visible and near-visible region of the frequency spectrum, specifically when the optical energy is guided to another location through thin transparent strands. Techniques include conveying light or images through a particular configuration of glass or plastic fibers. Incoherent optical fibers will transmit light, as a pipe will transmit water, but not an image. Coherent optical fibers can transmit an image through perfectly aligned, small (10–12 micrometers) clad optical fibers. Specialty fiber optics combine coherent and incoherent aspects.

fiber-optics bundles — 1. Assemblies of optical fibers. 2. Very fine transparent glass or plastic threads, each of which transmits light.

fiber-optic scanner — A scanner in which a fiber-optic assembly replaces a lens system.

fiber-optics computer interconnection — A means of connecting a computer with a terminal or another computer to transmit electrical signals via fiber-optics cable instead of wire.

fiber-optics multiport coupler — An optical unit, such as a scattering or diffusion solid "chamber" of optical material, that has at least one input and two outputs, or at least two inputs and one output, that can be used to couple various sources to various receivers. The ports are usually optical fibers. If there is only one input and one output port, it is simply a connector.

fiber-optics probe — A flexible probe made up of a bundle of fine glass fibers optically aligned to transmit an image, light, or both.

fiber-optics scrambler — 1. Similar to a fiberscope except that the middle section of loose fiber is deliberately disoriented as much as possible, then potted and sawed. Each half is then capable of coding a picture, which can be decoded by the other half. 2. A device used for coding messages that has a fiber bundle that is aligned at both ends and scrambled in the middle. The resulting halves of the bundles serve as encoders or decoders and the message appears as a random scattering of black dots.

fiber-optics splice — A nonseparable junction joining one optical conduction to another.

fiber-optics system — A light source and light detector; transmitter and receiver; and fiber-optic cable with connectors. The transmitter converts the encoded input signal to dc current that is converted to light energy by the light-emitting diode. This light energy is coupled from the LED into the fiber-optic cable with connectors and is then transmitted through the cable, which is connected to a light detector such as a photodiode (PIN diodes), phototransistor, or Darlington phototransistor. These devices detect the light energy and convert it to a current that may be amplified and decoded to faithfully reproduce the original input signal.

fiber-optics terminus — A device used to terminate an optical conductor that provides a means to locate and contain an optical conductor within a connector.

fiber-optics transmission system — Abbreviated FOTS. A transmission system utilizing small-diameter transparent fibers through which light is transmitted. Information is transferred by modulating the transmitted light. These modulated signals are detected by light-sensitive devices, such as photodetectors.

fiber-optics waveguide — A relatively long, thin strand of transparent substance, usually glass, capable of conducting an electromagnetic wave of optical wavelength (visible or near-visible region of the frequency spectrum) with some ability to confine longitudinally directed, or near-longitudinally directed, light waves to its interior by means of internal reflection.

fiber scattering — In an optical fiber, the coupling, or leaking, of light-wave power out of the core of the fiber by Rayleigh scattering or guide imperfections such as dielectric strain, compositional or physical discontinuities in the core or cladding, irregularities and extraneous inclusions in the core-cladding interface, curvature of the optical axis, or tapering. Scattering losses are measured in all directions as an integrated effect and expressed in decibels per kilometer.

fiberscope — Optical glass fibers, when systematically arranged in a bundle, transmit a full-color image that remains undisturbed when the bundle is bent. By mounting an objective lens on one end of the bundle, and an eyepiece at the other, the assembly becomes a flexible fiberscope that can be used to view objects that would be inaccessible for direct viewing.

fibrillation — A result of loss of synchronization of the heart muscle, causing the individual muscle fibers to contract in a random, uncoordinated sequence. As a result, the heart muscle merely quivers instead of pumping forcefully.

fidelity — 1. The accuracy with which a system or portion of a system reproduces at its output the essential characteristics of the signal impressed on its input. 2. A measure of the exactness with which sound is duplicated or reproduced.

field — 1. One of the two equal parts into which a frame of a television image is divided in interlaced scanning. With present U.S. standards, pictures are transmitted in two fields of 262.5 lines each, which are interlaced to form 30 complete frames, or images, per second. 2. That area or space in which a particular geophysical effect, such as gravity or magnetism, occurs and can be measured. 3. A group of characters in a computer that is treated as a single unit of information. 4. A region near an electric charge, a source of electromagnetic radiation, or a magnet in which components or materials may be affected. 5. That silicon area on a chip not used or occupied by active transistors. 6. A group of adjacent bits in a microinstruction. 7. A portion of a microprogram word that represents a group of bits dedicated to controlling a specific piece of hardware. 8. A subdivision of a record, usually consisting of a single item of information related to the rest of the record and serving a similar function in all records of that group. The smallest unit normally manipulated by a database management system.

field application relay — A device that automatically controls the application of the field excitation to an ac motor at some predetermined point in the slip cycle.

field circuit breaker — A device that functions to apply, or to remove, the field excitation of a machine.

field control of speed — The varying voltage applied to the field of a shunt-wound motor to control the motor's speed over the extended range.

field density — *See* magnetic induction.

field-discharge protection — A control function or device to limit the induced voltage in the field when the field current attempts to change suddenly.

field distortion — Distortion between the north and south poles of a generator due to the counterelectromotive force in the armature winding.

field-effect tetrode — A semiconductor device consisting basically of a thin n region adjacent to a similarly thin p region. Two contacts are made to the n side and two to the p side so that currents can be passed through each thin region parallel to the single junction. The two currents remain separate because reverse bias is maintained on the junction. A current in either side affects the resistance of the other side and hence the current in the other side.

field-effect transistor — Abbreviated FET. 1. A transistor in which current carriers (holes or electrons) are injected at one terminal (the source) and pass to another (the drain) through a channel of semiconductor material whose resistivity depends mainly on the extent to which it is penetrated by a depletion region. The depletion region is produced by surrounding the channel with semiconductor material of the opposite conductivity and reverse-biasing the resulting pn junction from a control terminal (the gate). The depth of the depletion region depends on the magnitude of the reverse bias. Because the reverse-biased junction draws negligible current, the characteristics of the device are similar to those of a vacuum tube. 2. A transistor in which the current through a conducting channel is controlled by an electric field arising from a voltage applied between the gate and source terminals.

field-effect tube — A triode with its grid replaced by a nonintercepting control gate. A high positive voltage is applied to this gate in order to draw sufficient current from the cathode. The result is a strong concentration of the electric field at the gap between the gate and the cathode, producing an electron beam passing through the gate to the anode.

field-effect varistor — A passive, nonlinear, two-terminal semiconductor device that maintains a constant current over a wide range of voltage.

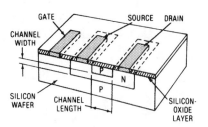
Field-effect transistor.

field emission — Also called cold emission. The liberation of electrons from a solid or liquid by application of a strong electric field at the surface.

field-enhanced photoelectric emission — Increased photoelectric emission resulting from the action of a strong electric field on the emitter.

field-enhanced secondary emission — Increased secondary emission resulting from the action of a strong electric field on the emitter.

field forcing — The effect of a control function or device that temporarily overexcites or underexcites the field of an electrical machine in order to increase the rate of change of flux.

field-free emission current — Also called zero-field emission. The electron current emitted by a cathode when the electric field at the surface of the cathode is zero.

field frequency — Also called field repetition rate. In television, the frame frequency multiplied by the number of fields contained in one frame. In the United States the field frequency is 60 per second, or twice the frame frequency.

field intensity — *See* field strength.

field inversion — Also called parasitic field turn-on. The creation of a channel between two nonassociated diffused beds in the field by voltages on conductors passing over.

field loss relay — *See* motor field-failure relay.

field magnet — An electromagnet or permanent magnet that produces a strong magnetic field in a speaker, microphone, phonograph pickup, generator, motor, or other electrical device.

field-neutralizing coil — A coil encircling the faceplate of a color picture tube. The current through it produces a magnetic field that offsets any effects of the earth's and other stray magnetic fields on the electron beams.

field-neutralizing magnet — Also called rim magnet. A permanent magnet mounted near the edge of the faceplate of a color picture tube to prevent stray magnetic fields from affecting the path of the electron beam.

field of view — 1. The solid angle from which objects can be acceptably viewed, photographed, or otherwise detected. 2. The maximum angle of view that can be seen through a lens or optical instrument.

field oxide — That portion of the oxide on an MOS device that is the thickest when measured perpendicular to the bulk silicon, usually 12 K to 20 K A.

field period — The time required to transmit one television field. In the United States, it is $\frac{1}{60}$th of a second.

field pickup — Also called a remote or nemo. A radio or television program originating outside the studio.

field pole — A structure, made of magnetic material, on which may be mounted a field coil.

field relay — A device that functions on a given or abnormally low value or failure of machine field current, or on an excessive value of the reactive component of armature current in an ac machine, indicating abnormally low field excitation.

field-repetition rate — *See* field frequency.

field resistor — A component in which the resistance element is a thin layer of conductive material on an insulated form. The conductive material does not contain either binding or insulating material.

field rheostat — A variable resistance connected to the field coils of a motor or generator and used for varying the field current.

field ring — The part that supports the field of a dc or series-wound motor housing. The motor end shields are attached to the ends of the field ring.

field scan — In a television system the downward excursion of an electron beam across the face of a cathode-ray tube, resulting in the scanning of alternate lines.

field selection — In a computer, the isolation of a particular data field within one computer word without isolating the word.

field sequential — Pertaining to the association of individual primary colors with successive fields in a color television system (e.g., field-sequential pickup, display, system, transmission).

field-sequential color television — A color television system in which the individual primary colors (red, blue, and green) are produced in successive fields.

field shield — A process whereby a conducting layer covers an entire MOS chip (except at transistor terminals) between the doped substrate and interconnecting conductors to control field inversion problems.

field-simultaneous system — A color television system in which a succession of full-color images is produced rather than a succession of primary-color fields.

field strength — Also called field intensity. 1. The value of the vector at a point in the region occupied by a vector field. In radio, it is the effective value of the electric-field intensity in microvolts or millivolts per meter produced at a point by radio waves from a particular station. Unless otherwise specified, the measurement is assumed to be in the direction of maximum field intensity. 2. The amount of magnetic flux produced at a particular point by an electromagnet or permanent magnet. 3. The strength of radio waves at a distance from a transmitting antenna, usually expressed in microvolts per meter. This is not the same as the strength of a radio signal at the antenna terminals of a receiver. 4. The intensity of a signal emitted by an antenna. It is proportional to the current in the antenna.

field-strength meter — A calibrated measuring instrument for determining the strength of radiated energy (field strength) being received from a transmitter.

field telephone — A durable, portable telephone designed for use in the field.

field weakening — The introduction of a resistance in series with the shunt field of a motor to reduce the voltage and current and increase the motor speed.

field wire — A flexible insulated wire used in field telephone and telegraph systems.

FIFO — Acronym for first in, first out. 1. A special integrated circuit memory device that will accept parallel data and retransmit it on a first in, first out basis. 2. First in, first out method of storing and retrieving items from a stack, table, or list. Compare with *LIFO*. 3. A method of coordinating the sequential flow of data through a buffer.

FIFO (first in, first out) buffer or shift register — A shift register with an additional control section that permits input data to fall through to the first vacant stage so that if there is any data contained, it is available at the output even though all the stages are not filled. In effect, it is a variable-length shift register whose length is always the same as the data stored therein.

FIFO/LIFO — Acronyms describing a method of data storage and retrieval. FIFO stands for first in, first out; LIFO for last in, first out. Both describe data input and output order.

figure-eight microphone — A microphone (usually a ribbon type) whose sensitivity is greatest to the front and rear, and weakest to both sides. *See* bidirectional microphone.

figure of merit — 1. The property or characteristic that makes a tube, coil, or other electronic device suitable for a particular application. It is a quality to look for in choosing a piece of equipment. 2. In a magnetic amplifier, the ratio of the power gain to the time constant. 3. For a thermoelectric material, the quotient of the square of the absolute Seebeck coefficient (α) divided by the product of the electrical resistivity (ρ) and the thermal conductivity. 4. In a sample-and-hold circuit, the ratio of the available charging current during the sample mode to the leakage circuit during the hold mode.

filament — Also known as a filamentary cathode. 1. The cathode of a thermionic tube, usually a wire or ribbon, which is heated by passing a current through it. 2. In tubes employing a separate cathode, the heating element. 3. A slender thread of material, such as carbon or tungsten, that emits light when raised to a high temperature by an electric current (as in an incandescent light bulb).

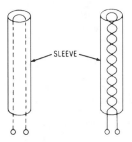

Filaments.

filamentary display — A numerical or alphanumerical display whose segments are composed of individual incandescent-type filament wires, which emit white light when energized.

filamentary transistor — A conductivity-modulation transistor that is much longer than it is wide.

filament battery — The source of energy for heating the filament of a vacuum tube.

filament circuit — The complete circuit through which filament current flows.

filament current — The current supplied to a filament to heat it.

filament emission — The freeing of electrons from a filament in an electron tube as the result of the filament being heated by an electric current.

filament power supply — The source of power for the filament or heater of a vacuum tube.

filament resistance — The resistance (in ohms) of the filament of a vacuum tube or incandescent lamp.

filament rheostat — A variable resistance placed in series with the filament of a vacuum tube to regulate the filament current.

filament sag — The bending of a filament when it heats up and expands.

filament saturation — Also called temperature saturation. The condition whereby a further increase in filament voltage will no longer increase the plate current at a given value of plate voltage.

filament transformer — A transformer used exclusively to supply filament voltage and current for vacuum tubes.

filament voltage — The voltage value that must be applied to the filament of a vacuum tube to obtain the rated filament current.

filament winding — A secondary winding provided on a power transformer to furnish alternating filament voltage for one or more vacuum tubes.

file — 1. A collection of related records. For example, in inventory control, one line of an invoice containing data on the material, the quantity, and the price forms an item; a complete invoice forms a record; and the complete set of such records forms a file. 2. To insert an item into such a set. 3. A user-defined collection of information of variable length. 4. A list. Usually, a file is a list of instructions plus data and comments. 5. A collection of information consisting of records pertaining to a single subject. A file may be recorded on all or on part of a volume or on more than one volume. 6. A logical block of computer information, designated by name, and considered as a unit by a user. A file may be physically divided into records, blocks, or other units required by the memory device. 7. A collection of related information stored on a disk.

filed coil — A coil of insulated wire wound around an iron core. Current flowing in the coil produces a magnetic field.

file gap — On a data medium, an area intended to be used to mark the end of a file and, possibly, the start of another. A file gap frequently is used for other purposes, in particular for indicating the end or beginning of some other group of data.

file layout — 1. The organization and structure of data in a file, including the sequence and size of the components. 2. By extension, the description thereof.

file maintenance — The processing of a computer file in order to bring it up to date.

file management — An operating system facility for the manipulating of data files to and from secondary storage devices (usually disk files or magnetic tapes); it is used for building files, retrieving information from them, or modifying the information.

file mark — Also termed tape mark or end-of-file mark. A specially recorded block containing no data but acting as a data-block separator.

file-protection device — 1. A device by which the existence and integrity of a file are maintained. 2. A ring that must be in place in the hub of a reel before data can be recorded on the tape contained by the reel. A reel of tape not provided with a file-protection device can be read but not written.

file section — That part of a file which is recorded on any one volume. The file sections may not have sections of other files interspersed.

file set — A collection of one or more related files, recorded consecutively on a volume set.

file transfer — A procedure that calls for a communication link (typically over telephone lines) to be established between two or more PCs using modems. This connection allows data files to be transferred from one computer's storage device (usually a floppy or hard-disk drive) to the other's.

File Transfer Protocol — Abbreviated FTP. 1. The protocol used for copying files to and from remote computer systems on a network using TCP/IP, such as the Internet. 2. A very common method of moving files between two Internet sites. FTP is a special way to log in to another Internet site for the purpose of retrieving and/or sending files. There are many Internet sites that have established publicly accessible repositories of material that can be obtained using FTP, by logging in using the account name "anonymous." These sites are called anonymous FTP servers. *See also* FTP.

fill — 1. The number of working lines in a particular cable or cable center. 2. The number of working lines as a percentage of the total pairs provided.

filler — 1. In mechanical recording, the inert material of a recording compound (as distinguished from the binder). 2. Nonconducting component cabled with insulation conductors to impart roundness, flexibility, tensile strength, or a combination of all three, to the cable.

film — Single or multiple layers or coatings of thin or thick material used to form various elements (resistors, capacitors, inductors) or interconnections and cross-overs (conductors, insulators). Thin films are deposited by vacuum evaporation or sputtering and/or plating. Thick films are deposited by screen printing.

film badge — A type of dosimeter consisting of a small piece of film sensitive to radiation, placed in a light-tight holder and carried by a person who works with radiation. When the film is developed, the amount of darkening can be measured to determine the total dose of ionizing radiation to which the badge has been subjected.

film capacitor — 1. A capacitor with a dielectric consisting of a plastic film. 2. A capacitor that is made by winding metal and dielectric (such as polyester, polycarbonate, polystyrene, polypropylene, or polysulfone) ribbons into a tubular shape. The metal electrodes can be separate metal foil, or can be vacuum-deposited onto the dielectric.

film chain — An arrangement of a film projector or projectors and a CCTV camera for transmitting moving pictures over a television system.

film conductor — 1. A conductor formed *in situ* on a substrate by depositing a conductive material by screening, plating, or evaporation techniques. 2. Electrically conductive material formed by deposition on a substrate.

film integrated circuit — Also called film microcircuit. 1. A circuit made up of elements that are films all formed in place upon an insulating substrate. To further define the nature of a film integrated circuit, additional modifiers may be prefixed. Examples: thin-film integrated circuit, thick-film integrated circuit. 2. Thin- or thick-film network forming an electrical interconnection of numerous devices.

film microcircuit — *See* film integrated circuit.

film pickup — A film projector combined with a television camera for telecasting scenes from a motion-picture film.

film reader — A computer input device that scans opaque and transparent patterns on photographic film and relays the corresponding information to the computer.

film recorder — An instrument designed to place nongraphic information, usually generated by a computer, onto photographic film. The information is generally encoded as a series of opaque and translucent spots, or light and dark spots.

film reproducer — An instrument that reproduces a recording on film.

film resistor — 1. A fixed resistor whose resistance element is a very thin layer of conductive material on an insulated form. Some sort of mechanical protection is placed over this layer. 2. A resistor whose characteristics depend on film rather than bulk properties. 3. A device whose resistive material is a film on an insulator substrate; final resistance value may be determined by trimming.

film scanning — 1. The process of converting movie film into corresponding electrical signals that can be transmitted by a television system. 2. The process by which the light from the images of photographic film is encoded into electrical signals for video transmission.

filter — Also called extractor or mask. 1. A selective network of resistors, inductors, or capacitors that offers comparatively little opposition to certain frequencies or to direct current, while blocking or attenuating other frequencies. *See also* wave filter. 2. A device or program that separates data, signals, or materials in accordance with specified criteria. 3. A machine word that specifies which parts of another machine word are to be operated on. 4. A circuit that attenuates signals above or below specific frequency without materially affecting signals in its passband. The action of a filter is usually defined by its slope (in decibels per octave; usually some multiple of 6 dB/octave) and by its turnover frequency. 5. Electrical device used to suppress undesirable electrical noise. 6. With respect to radiation, a device used to attenuate particular wavelengths or frequencies while passing others with relatively no change. 7. A circuit that passes signals above, below, or within a particular frequency band while rejecting all other signals. Active filters incorporate amplifier circuits, whereas passive filters are networks of capacitors, inductors, and resistors.

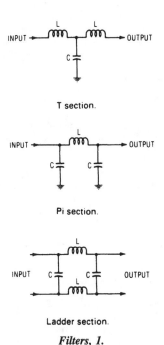

T section.

Pi section.

Ladder section.

Filters, 1.

filter attenuation — A loss of power through a filter as a result of absorption in resistive materials, of reflection, or of radiation. Usually expressed in decibels.

filter attenuation band — Also called a filter stopband. A frequency band in which the attenuation constant is not zero if dissipation is neglected. In other words, a frequency band of attenuation.

filter capacitor — A capacitor used in a filter circuit. The term is usually reserved for electrolytic capacitors in a power-supply filter circuit.

filter center — In an aircraft control and warning system, a location at which information from observation posts is filtered for further dissemination to air-direction centers.

filter choke — Normally, an iron-core coil that allows direct current to pass while opposing the passage of pulsating or alternating current.

filter crystal or plate — A quartz plate or crystal used in an electrical circuit designed to pass energy only at certain frequencies.

filter discrimination — The difference between the minimum insertion loss at any frequency in the attenuation band of a filter and the maximum insertion loss at any frequency in the transmission band of the filter. The loss is determined under the conditions of normal use of the filter.

filtered radar data — Radar data from which undesired returns have been removed by mapping.

filter-impedance compensator — An impedance compensator that is connected across the common terminals of electric-wave filters, when the latter are used in parallel, in order to compensate for the effects of the filters on each other.

filter passband — *See* filter transmission band.

filter section — Any of various simple networks that may be connected in cascade to form a filter. The simplest is the half section, consisting of a series impedance (Z) followed by a shunt admittance (Y). A full section is either a T-network in which the shunt arm is Y and the series arms are $Z/2$, or a pi network in which the series arm is Z and the shunt arms $Y/2$. Full sections, unlike half sections, have equal input and output impedances.

filter slot — A choke, in the form of a slot, designed to suppress unwanted modes in a waveguide.

filter stopband — *See* filter attenuation band.

filter transmission band — Also called filter passband. A frequency band in which the attenuation constant is zero if dissipation is neglected. In other words, a frequency band of free transmission.

fin — A metal disc or a thin, projecting metal strip attached to a semiconductor to dissipate heat.

final actuation time — The time of termination of the chatter of a relay following contact actuation.

final amplifier — The stage that feeds the antenna in a transmitter.

final control element — The part of a control system that actually changes the amount of energy or fuel to the process. For example, in an industrial oven the final control element could be a valve that controls the amount of fuel reaching the burner.

final seal — The hybrid microelectronic packaging step that encloses the circuit so that further internal processing cannot be performed without disassembly.

final wrap — The outer layer of insulation around a coil, covering the saddle and splice insulation.

finder — In a telephone switching system, a name applied to the switch or relay group that selects the path that the call is to follow through the system.

fine-chrominance primary — Also called the I signal. In the color television system presently standardized for broadcasting in the United States, the chrominance primary associated with the greater transmission bandwidth.

fine-tuning control — A receiver control that varies the frequency of the local oscillator over a small range to compensate for drift and permit fine adjustment to the carrier frequency of a station.

finger plethysmograph — An instrument for detecting and displaying changes in the volume of blood in the finger during the cardiac cycle. In some types, a light source and a photocell are placed on opposite sides of the finger; the volume of the blood in the finger determines the amount of light reaching the photocell. In another type, the finger is placed between two electrodes. The increased blood volume during each contraction of the heart reduces the impedance of the finger, and the resulting change in the current between the electrodes is recorded or displayed on a cathode-ray tube.

fingers — Nonpreferred term for edge-board contact.

finished blank — A crystal product after completion of all processes. It may also include the electrodes adherent to the crystal blank.

finishing — The process of repeated hand lapping and electrical testing by which a finished crystal blank is brought up to specifications.

finishing rate — Expressed in amperes, the rate of charge to which the charging current of a battery is reduced near the end of the charge to prevent excessive gassing and temperature rise.

finish lead — The lead connected to the finish, or outer end, of a coil.

finite — Having fixed and definite limits.

finite-element method — A procedure for solving electromagnetic field problems by dividing the domain of interest into small, basic elements and by solving differential equations in each of those elements.

finite-state machine — A computer in which a set of inputs determines both the set of outputs and also the internal state of the computer.

fins — Radial sheets or discs of metal attached to metal parts of a power tube, power transistor, or other component for the purpose of dissipating heat.

fin waveguide — A waveguide in which a thin longitudinal metal fin is placed to increase the range of wavelengths over which the waveguide can transmit signals efficiently. The method usually is used with circular waveguides.

FIPS — Abbreviation for Federal Information Processing Standard. A standard approved for use by U.S. government agencies.

fire — 1. To change from a blocked condition, in which negligible current flows, to a saturated condition, in which heavy current flows. 2. The term used to describe the act of heating a thick-film circuit so that the resistors, conductors, capacitors, etc., will be transformed into their final form.

fire-control equipment — Equipment that takes in target indications from optical or radar devices and, after calculating the motion of the target and firing vehicle, properties of air, etc., puts out directions of bearing, elevation, and timing for aiming and firing the guns.

fire-control radar — Radar employed for directing gunfire against the targets it observes.

fired tube (tr, atr, and pre-tr tubes) — The condition of a tube while a radio-frequency glow discharge exists at the resonant gap, resonant window, or both.

firewall — 1. A combination of hardware and software that separates a local area network (LAN) into two or more parts for security purposes. 2. A method of partially or totally blocking access (from machines not on a LAN) or of filtering/monitoring incoming packets.

firing — 1. In any gas- or vapor-filled tube, the ionization of the gas and the start of current flow. 2. The excitation of a device during a brief pulse. 3. In a magnetic amplifier, the transition from the unsaturated to the saturated state of the saturable reactor during the conducting or gating alternation. 4. An adjective modifying phase or time, to designate when firing occurs.

firing angle — 1. The electrical angle of the plate-supply voltage at which ionization of a gaseous tube occurs. 2. In a magnetic amplifier, the point on a sine-wave control voltage at which the control ampere-turns are sufficient to saturate the core. This is the point at which the secondary winding (load) impedance drops to zero, and almost all of the supply voltage appears across the load.

firing furnace — Furnace used for the curing of multilayer ceramics for integrated electronics and for the firing of thick-film materials on substrates.

firing point — The point at which the gas or vapor in the tube ionizes and current begins to flow.

firing potential — The controlled potential at which conduction through a gas-filled tube begins.

firing profile — A graph of time versus temperature or, in a continuous thick-film furnace, of position versus temperature.

firmware — 1. A computer program or software stored permanently in PROM or ROM or semipermanently in EPROM. 2. Programs or instructions that are stored in read-only memories; firmware is analogous to software in a hardware form. These instructions are for internal processor functions only, and are transparent to the user. 3. The internal interconnections that permanently determine what functions a device or system can perform. Also called microprogram. 4. Sets of instructions cast into user-modifiable hardware. 5. A series of instructions in ROM (read-only memory). 6. Data stored in a nondestructive form such as hard-wired or in a ROM. 7. An extension to a computer's basic command repertoire to create micro-programs for a user-oriented instruction set. This extension to the basic instruction set is done in ROM and not in software. The ROM converts the extended instructions to the basic instructions of the computer.

first article — A sample part or assembly manufactured prior to the start of production for the purpose of ensuring that the manufacturer is capable of manufacturing a product that will meet the requirements.

first audio stage — The first stage in an audio amplifier.

first detector — Now called the mixer. In a super-heterodyne receiver, the stage in which the local-oscillator signal is combined with the modulated incoming radio-frequency signal to produce the modulated intermediate-frequency signal.

first fit — In a computer, an algorithm for memory allocation that searches the free list only long enough to find an unused memory block that is large enough to satisfy the requesting task.

first Fresnel zone — In optics and radio communications, the circular portion of a wave front intersecting the line between an emitter and a more distant point where the resultant disturbance is being observed. The center intersects the front with the direct ray, and the radius is such that the shortest path from the emitter through the periphery to the receiving point is one half-wave longer than the ray.

first-generation computer — A computer in which vacuum-tube components are used.

first-in, first-out (FIFO) memory — 1. A type of memory with separate input and output ports. The first data to enter the input port are the first to exit the output port. One use of FIFO memory is as a buffer between a terminal and a LAN in a network interface controller. 2. A data access mechanism that implements a queue. Data elements are always extracted from the data structure in the same order that they are entered (the first element in is the first element out).

first selector — The selector that immediately follows a line-finder in a switch train. It responds to the dial pulses that represent the first digit of the called telephone number.

first-shot effect — A term used relative to solid-state (electronic) timers using a resistor-capacitor (RC) single time-constant circuit. Due to the forming effect leakage found in electrolytic capacitors when stored, the first and sometimes second and third operations have longer time delays. The repeat accuracy specification does not include this condition. Digital-type circuits do not exhibit this condition.

fishbone antenna — 1. An antenna consisting of a series of coplanar elements arranged in collinear pairs and loosely coupled to a balanced transmission line. 2. Directional antenna in the form of a plane array of doublets arranged transversely along both sides of a transmission line.

fishpaper — A tough fiber used in sheet form for insulating transformer windings from the core, field coils from field poles, or conductors from the armature.

fission — Also called atomic fission or nuclear fission. The splitting of an atomic nucleus into two parts. Fission reactions occur only with heavy elements such as uranium and plutonium and are accompanied by large amounts of radioactivity and heat.

fissionable — Capable of undergoing fission.

fission products — The elements that result from atomic fission. They may consist of more than forty different radioactive elements, such as arsenic, silver, cadmium, iodine, barium, tin, cerium, and others.

fissuring — The cracking of dielectrics or conductors. Often dielectrics, if incorrectly processed, will crack in the presence of conductors because of stresses occurring during firing.

FIT — Failure in 10^9 device-hours.

FITs — The measure of a semiconductor's reliability is often expressed in failure units (FITs)the number of failures per 10^9 device-hours. For example, if 100 units operate for 1000 hours with one failure, the failure rate in FITs is $(1/100) \times (10^9/1000)$ or 10^4 FITs. FITs are used as a measure of reliability because the units are particularly useful for expressing the low failure rates encountered in electronic devices. For example, it is much less cumbersome to work with a figure like 10 FITs rather than with 0.001%/1000 hours. FITs are also useful because they can be added when calculating the total failure rate for a system. FITs can also be used to calculate the mean time between failure, which is the reciprocal of the FIT rate times the total number of components in a system. For example, a system with 10,000 components with FIT rates of 100 would have a mean time between failure of $10^9/(100)$ (10,000), or 1000 hours.

fitting — An accessory, such as a locknut or bushing, to a wiring system. Its function is primarily mechanical rather than electrical.

five-layer device — A semiconductor, as a diac, triac, etc., in which there are four pn junctions.

five-level code — A telegraph code in which five impulses are utilized for describing a character. For asynchronous transmission, start and stop elements may be added. A common five-level code is the Baudot code.

five-level start-stop operation — Simplex mode of teletypewriter operation. Each code character consists of five electrical units. The distributor unit of the machine makes a positive start and stop for the transmission of each character.

fix — A position determined without reference to any former position.

fixed bias — A constant value of bias voltage.

fixed capacitor — A capacitor designed with a definite capacitance that cannot be adjusted.

fixed-composition (carbon-composition) resistor — Resistive element consisting of a carbon composition that is molded under extreme pressure, then enclosed in an insulating sleeve.

fixed contact — A contact that is permanently included in the insert material. It is permanently locked, cemented, or imbedded in the insert during molding.

fixed crystal — A crystal detector with a nondefinite contact position.

fixed-cycle operation — 1. A type of computer performance whereby a fixed amount of time is allocated to an operation. 2. Synchronous or clock-type arrangement in a computer, in which events occur as functions of measured time.

fixed decimal — Calculator that is limited to established decimal category; can be preset for specified number of places in answer, or preset so that numbers are entered as they would be written.

fixed decimal point — Location of the decimal point in the display of a calculator chosen by a selector switch. For example, if the switch is set to position six on an eight-digit machine, the numbers between 99 and 0.001 can be used. In some machines no selector is provided, and a calculation such as 123/456 yields the answer 0.27 instead of 0.2697368.

fixed echo — A stationary echo indication on a radar PPI display, indicating a fixed target.

fixed-frequency IFF — A class of IFF (identification, friend or foe) equipment that responds immediately to every interrogation, thus permitting the response to be displayed on plan-position indicators.

fixed-frequency transmitter — A transmitter designed for operation on a single carrier frequency.

fixed-instruction computer — A computer having an instruction set that is fixed by the manufacturer. Users must design applications programs using this instruction set. Contrasted with microprogrammable computer.

fixed-length record — Pertaining to a file in which all records are constrained to be of the same predetermined length. (Opposite of variable-length record.)

fixed logic — Circuit logic computers or peripheral devices that cannot be changed through operation of external controls. Connections must be physically changed to rearrange the logic.

fixed logic levels — Digital data with high and low levels that are programmable or adjustable.

fixed memory — 1. A nondestructive-readout computer memory that is alterable only by mechanical means. 2. A memory into which information normally can be written only once. The ROM is a fixed program memory. Programs are usually stored in fixed memories.

fixed operation — Radiocommunication conducted from the specific geographical land location shown on the station license.

fixed point — Pertaining to notation or a system of arithmetic in which all numeric quantities are expressed with a predetermined number of digits and the point is located implicitly at some predetermined position.

fixed-point arithmetic — 1. Calculations in which the computing device is not concerned with the location of the point. An example is a slide-rule calculation, since the human operator must locate the decimal point. 2. A type of arithmetic in which all figures must remain within certain fixed limits. 3. Arithmetic in which the binary point that separates the integer and fractional portions of numerical expressions is either explicitly stated for all expressions or is fixed with respect to the first or last digit of each expression.

fixed-point system — A system of notation in which a number is represented by a single set of digits and the position of the radix point is not numerically expressed. *See also* floating-point system.

fixed-program computer — *See* wired-program computer.

fixed resistor — 1. A resistor designed to introduce only a predetermined amount of resistance into an electrical circuit; it is not adjustable. 2. A basic, universal electronic component designed to impede the flow of electric current; classed according to the materials technology used to form the resistive element. Carbon, nickel chromium, tin oxide, and various other alloys and bonded mixtures are commonly used for resistive elements that are molded, imprinted, or deposited. Most discrete resistors are leaded and coated to form compact, uniform units ready for assembly.

fixed screen — Application of a potential to a screen grid that is unaffected by other operating conditions within the tube.

fixed service — 1. Any service communicating by radio between fixed points, except broadcasting and special services. 2. A point-to-point radiocommunication service between specific fixed stations on the earth.

fixed station — 1. A station in the fixed service. (A fixed station may, as a secondary service, transmit to mobile stations on its normal frequencies.) 2. A permanent station that communicates with other fixed stations.

fixed transmitter — A transmitter operated from a permanent location.

fixed-voltage winding — The motor winding that is excited by a fixed voltage.

fixed word length — The condition in which a machine word always contains a fixed number of bits, characters, bytes, or digits.

fixer network or system — A combination of radio or radar direction-finding installations that, when operated in combination, can determine the position relative to the ground of an aircraft in flight.

fixturing — An assortment of electronic switches, wiring, black boxes, etc., to connect two systems, e.g., a test system and a board under test.

fL — Letter symbol for footlambert.

flag — 1. A large sheet of metal or fabric for shielding television camera lenses from light. 2. In a computer, an indication that a particular operation has been completed and may be skipped by the program. 3. A single flip-flop that indicates that a certain condition has arisen as, for example, during the course of an arithmetic or logical operation in a computer program. 4. Usually a flip-flop storing one bit that indicates some aspect of the status of the central processing unit. For example, a carry flag is set to 1 when an arithmetic operation produces a carry, and to 0 when the result is zero. These flags aid in interpreting the results of certain calculations. Others are sometimes provided to permit access by interrupt request lines; for example, if a CPU is engaged in the highest priority of calculation, it may set all status flags to zero, which, loosely translated, means "don't bother me now." If only some of these flags are set, then only certain interrupt lines will be able to get through, according to their priority. 5. A "permanent" status signal, normally stored in a flip-flop or a register of a computer to indicate a special condition. Typically, every microprocessor provides at least the following status flags: carry, zero, sign, overflow, half-carry. 6. A delimiting bit field used to separate portions of data. 7. In a computer, an indicator, usually a single binary bit, whose state is used to inform a later section of a program that a condition, identified with the flag and designated by the state of the flag, has occurred. A flag can be both software and hardware implemented. 8. Any of various types of indicators used for identification of a condition or event; for example, a character that signals the termination of a transmission.

flag bit — A processor memory bit, controlled through firmware, to signify a certain condition (e.g., battery low). A flag bit may be monitored by user-programmed instructions.

flag lines — Inputs to a microprocessor controlled by I/O devices and tested by branch instructions.

flag terminal — A type of solderless terminal in which the tongue projects from the side rather than the end of the terminal barrel.

flame — To send angry or critical messages to someone via e-mail or a newsgroup.

flame-failure control — A system that automatically stops the fuel supply to a furnace if the pilot burner accidentally goes out.

flame microphone — A microphone in which the action of sound waves on a flame changes the resistance between two electrodes in the flame.

flameoff — The procedure in which a wire is severed by passing a flame across the wire, thereby melting it. The procedure is used in gold wire thermocompression bonding to form a ball for making a ball bond.

flameproof — Said of insulated wire or other material that has been chemically treated so it will not aid the spread of flames.

flameproof wire — Wire having insulation that is chemically treated so that it will not support combustion.

flame resistance — The characteristic of a material that prevents it from flaming when the source of heat is removed.

flame-resistant — See flame-retardant.

flame-retardant — Also called flame-resistant. 1. Retarding ignition and the spread of flames, either inherently or because of special treatment. 2. Constructed or treated so as not to be able to convey flame.

flammability — Measure of a material's ability to support combustion.

flammable — Also called combustible. Term applied to material that readily ignites and burns when exposed to flames or elevated temperatures.

flange — 1. Also called waveguide flange. A fitting used at the end of a waveguide for making attachment to a microwave component or to another waveguide. 2. The side of a tape reel, which prevents the tape on the hub from slipping sideways off the "pie." 3. A projection extending from, or around the periphery of, a connector and provided with holes to permit mounting the connector to a panel or to another mating connector half.

flange connector — A mechanical joint employing plane flanges bolted together in a waveguide.

flange coupling — A connection utilizing flanges not in mechanical contact between two parts of a waveguide, yet introducing no discontinuity in the flow of energy along the guide.

flange focus — The distance from the mounting flange or reference surface of a lens to the focal plane for a subject at infinity.

flanking effect — The effect on filter characteristics of connecting additional filters in parallel.

flap attenuator — A form of waveguide attenuator in which a variable amount of loss is introduced by insertion of a sheet of resistive material, usually through a nonradiating slot.

flare — An enlarged and distorted radar-screen target indication due to excessive brightness.

flare angle — The continuous change in cross section of a waveguide.

flare factor — A number expressing the degree of outward curvature of a speaker horn.

flash — Sometimes called hit. Momentary interference to a television picture, lasting approximately one field or less and of sufficient magnitude to totally distort the picture information. In general, this term is used only when the impairment is so short that the basic impairment cannot be recognized.

flashback voltage — The inverse peak voltage at which ionization takes place in a gas tube.

flasher — 1. A device that is designed to automatically turn electric lamps on and off in a rapidly repeating sequence. The device may use a motor-driven mechanism, a combination heater filament and bimetallic strip, to stop and start the current. 2. A control whereby the output to the load (normally a lamp) is turned on and off repeatedly at a given rate of operation or flashes per minute (fpm).

flasher relay — A self-interrupting relay.

flashing — The application of a high-frequency electromagnetic field to an electron tube through the envelope to flash its getter during evacuation.

flashing of a dynamo — The flashing or sparking that is likely to take place at the brushes of a commutator.

flash lamp — 1. A device in which a large amount of stored electrical energy is converted into light by means of a sudden electrical discharge. The flash is obtained by storing electrical energy in a capacitor and allowing the capacitor to discharge through the lamp. 2. A device or electronic circuitry that converts a large amount of stored electrical energy into light by means of a sudden electrical discharge. *See also* flash tube.

flash magnetization — Magnetization of a ferromagnetic object by an abrupt current impulse.

flashover — 1. A disruptive discharge through air, around or over the surface of insulation, or between parts of different potential or polarity produced by the application of voltage, wherein the breakdown path becomes sufficiently ionized to maintain an electric arc. 2. An electric discharge that occurs around an edge or across a surface. Can result in a carbon track and permanent degradation.

flashover voltage — 1. The highest value attained by any voltage impulse that caused a flashover. 2. The voltage at which insulation fails by discharge between electrodes across the insulation surface.

flash plating — The application of extremely thin deposits of a plating material for environmental protection or as a base for a subsequent layer of plating material.

flashpoint of impregnate — The temperature to which a liquid or solid impregnate must be heated before it gives off sufficient vapor to form a flammable mixture.

flash pulsing — Transmission of short bursts of radiation at irregular intervals by a mechanically controlled keyer.

flash radiography — A technique used in radiography to obtain an unblurred image of a moving object by the use of very short X-ray exposures, such as 1 microsecond, to record the image.

flash spectroscopy — The study and interpretation of the spectra of substances after they have absorbed the radiant energy emitted by a brief, intense light source.

flash test — Also known as high-potting, high-pot, or short check. 1. A method of testing insulation by momentarily applying a voltage much higher than the working voltage. 2. Method of testing capacitors for shorts (or potential shorts) between electrodes by momentarily applying a voltage much higher than the rated working voltage.

flash tube — Also called electronic flash tube and photoflash tube. A gas-discharge tube for producing high-intensity, short-duration flashes of light. It consists of a glass tube bent in a U, a helix, or a combination of the two and filled with a rare gas. The tube has an anode, a cold cathode, and a trigger electrode. It is flashed by applying a high-voltage pulse to the trigger electrode. *See also* flash lamp.

flash welding — Welding in which an arc is first struck between the pieces to be welded. After the ends are thus heated, the weld is completed by bringing them together under pressure and cutting off the current.

flat — Having a slope of zero at all points, as a graph, curve, etc.

flat back paper — A flat kraft paper tape used in splicing electrical cable.

Flash tube.

flatbed plotter — 1. Plotter that draws an image on paper, glass, or film mounted on a flat table. 2. Type of plotter in which the paper is held flat against a table electrostatically.

flatbed scanner — An optical scanner with a flat surface that can copy a page into an electronic file.

flat braid — A woven braid, composed of tinned copper strands, that is rolled flat at the time of manufacture to a specific width depending on construction. It is generally used as a high-current conductor at low voltages.

flat cable — Also called flexible flat cable or flat conductor cable. 1. Any cable with two or more parallel round or flat conductors in the same plane encapsulated by an insulating material. 2. Any cable with two smooth or corrugated but essentially flat surfaces.

flat-compounded generator — A compound-wound generator in which the series field winding is adjusted so that the output voltage is virtually constant for currents between no load and full load.

flat conductor — 1. A conductor with a width-to-thickness ratio of arbitrarily 5 to 1 or greater. 2. A cable with a plurality of flat conductors. Flexible flat cable with conductors that have rectangular, rather than round, cross sections.

flat fading — That type of fading in which all components of the received radio signal fluctuate in the same proportion simultaneously.

flat flexible cable — *See* tape cable.

flat frequency response — The response of a system to a constant-amplitude function that varies in frequency is flat if the response remains within specified limits of amplitude, usually specified in decibels from a reference quantity.

flat leakage power (tr and pre-tr tubes) — The peak radio-frequency power transmitted through the tube after establishment of a steady-state radio-frequency discharge.

flat line — A radio-frequency transmission line or part of a line having a low standing wave ratio.

flat pack — 1. A flat, rectangular integrated circuit or hybrid circuit package with coplanar leads. 2. Semiconductor network encapsulated in a thin rectangular package, with the necessary connecting leads projecting from the edges of the unit. 3. A slab-shaped, very low profile package for electronic components. Often used when printed circuit boards must be stacked close together. 4. An integrated circuit package that has leads extending from the package in the same plane as the package so that leads can be spot welded to terminals on a substrate or soldered to a printed circuit board. The small size and low profile of the flat pack contribute to high-density circuit packaging. 5. A small, flat IC package formed by sandwiching a hermetically sealed chip between two layers of metal or ceramic; the leads protrude from either two or all four of its edges.

flat response — Ability of a sound system to reproduce all tones (from the lowest to the highest) in their proper proportions. (For example, a specification of response within ±1 dB from 30 to 15,000 Hz would be considered flat.)

flat-square — Refers to cathode-ray tubes that are both full-square and have relatively flat screen surfaces also.

flat top — The horizontal portion of an antenna.

flat-top antenna — An antenna having two or more lengths of wire parallel to each other and to the ground.

flat-top response — Response characteristic in which a definite band of frequencies is transmitted uniformly.

flat-type armature — Of a relay, an armature that rotates about an axis perpendicular to that of the core, with the pole face on a side surface of the core.

flat-type relay — A relay having a flat-type armature.

flaw — In a material, any discontinuity that would be harmful to proper functioning of the material.

flaw detection — The process of using sonic or ultrasonic waves to locate imperfections in a solid material. This is done by transmitting the waves through the material and listening for reflections or variations in transmission when they strike an imperfection in the material.

F layer — 1. An ionized layer in the F region, existing in the night hemisphere and in the weakly illuminated portion of the day hemisphere. 2. The layer of the ionosphere which causes most long-distance communication. It sits about 130 to 160 miles (210 to 260 km) above the earth.

F₁ layer — One of the regular ionospheric layers, having an average height of about 225 kilometers, which occurs during the daylight hours.

F₂ layer — The most useful of the ionospheric layers for radio-wave propagation. It is the most highly ionized and highest of the layers, having an average night height of 225 kilometers and a midday height of about 400 kilometers.

Fleming's rule — Also called the right-hand or left-hand rule. 1. If the thumb and the first and second fingers are extended at right angles to one another, with the thumb representing the direction of the wire motion, the first finger representing the direction of magnetic lines of force (from the north pole to the south pole), and the second finger representing the direction of the current, then the right hand will give the correct relationships for a conductor in the armature of a generator, and the left hand will give the correct relationships for a conductor in the armature of a motor. This rule is applied to the so-called conventional current flow, which is the opposite of electron flow. 2. A rule stating that if the fingers of the right hand are placed around a current-carrying wire so that the thumb points in the direction of the conventional current, the fingers will point in the direction of the magnetic field.

Fleming valve — An early name for a diode, or two-electrode thermionic vacuum tube used as a detector.

Fletcher-Munson curves — Also called equal-loudness contours. A group of sensitivity curves showing the characteristics of the human ear for different intensity levels between the threshold of hearing and the threshold of feeling. The reference frequency is 1000 Hz.

Flewelling circuit — An early radio circuit in which one tube served as a detector, amplifier, and local oscillator.

flexible coupling — 1. A device for connecting two shafts end to end so that they can be rotated even though not exactly aligned. 2. Mechanical connection between two lengths of a waveguide normally lying in a straight line; designed to allow a limited angular movement between axes.

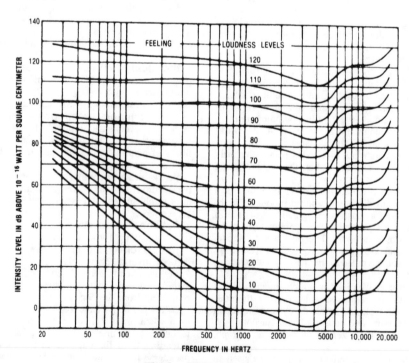

Fletcher-Munson curves.

flexible disk— A magnetic storage medium constructed of thin plastic. *See* floppy disk.

flexible printed circuit— A random arrangement of printed wiring and components utilizing flexible base materials with or without flexible cover layers.

flexible printed wiring— A random arrangement of printed wiring utilizing flexible base materials with or without flexible cover layers.

flexible resistor— A wirewound resistor that looks like a flexible lead. It is made by winding Nichrome or any other type of resistance wire around asbestos or other heat-resistant cord. The wire is then covered with braided insulation, which is color coded to indicate the resistor value.

flexible shaft— A flexible core made up of layers of wire that rotate inside a metal or rubber-covered flexible casing. The casing not only supports the core, but also acts as the bearing surface for the core.

flexible substrate— Thick- and thin-film circuits have generally been deposited on rigid substrates, but it is possible to deposit these circuits on some plastic substrates.

flex life— 1. A measure of the resistance of a conductor or other device to failure due to fatigue from repeated bending. 2. The time of heat aging that an insulating material can withstand before failure when bent around a specific radius (used to evaluate thermal endurance).

flexode— A flexible diode containing a junction that may be altered at will from a pn junction in one direction, to no junction at all, to a pn junction in the opposite direction. Thus, the direction of easy current may be reversed without reversing the leads to the diode, and the resistance of the diode may be continuously varied from the back-resistance value to the forward-resistance value. It may be set to behave as a simple resistor, with the same value for both directions of current.

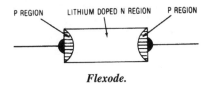

Flexode.

Flexowriter— A trade name for the typewriter that can provide input data to a computer or type outputs from the computer. (No longer used.)

flexure failure— A conductor failure due to repeated flexing that is indicated by an increase of resistance to a specified value for a specified time.

flicker— Also called jitter or wobble. 1. In television, the flickering produced in the picture when the field frequency is insufficient to completely synchronize the visual images. 2. In a regulated power supply, a phenomenon due primarily to sudden, minute changes of brief duration in the reference voltage or the input-stage junctions of the correction amplifier in the regulator. 3. Noise in an amplifier, of higher frequency than drift, but lower than power-line or chopper-drive frequency noise. 4. The sensation of image intermittence or of brightness or color variation. Flicker occurs when the frequency of the observed variation is less than the screen's flicker fusion frequency. 5. The fluctuation in apparent illumination that has a rate comparable to the reciprocal of the period of persistence in vision.

flicker effect— Small variations in the plate current of a thermionic vacuum tube, believed to be due to random emission of positive ions by the cathode.

flicker noise— Also called $1/f$ noise. One of the sources of noise associated with solid-state devices such as mixers or diode detectors, the amplitude of which varies inversely with frequency. It has a random amplitude similar to shot and thermal noise but with a $1/f$ spectral power density. This means that the noise increases at low frequencies and is associated with the level of direct current in the device.

flicker photometer— A device for measuring the intensity of a light source. Illumination from the light source being measured and a standard light source are observed alternately in rapid succession. When the standard source is equal to the other, the flickering disappears.

flight control— Real-time calculations for the control of a vehicle in flight; includes stabilization, fuel monitoring, cruise control, etc.

flight path— A planned course for an airborne vehicle.

flight-path computer— A computer that includes all the functions of a course-line computer and also controls the altitude of an aircraft in accordance with a desired plan of flight.

flight-path deviation— The difference between the flight path of an aircraft and the actual flight track, expressed in terms of either angular or linear measurement.

flight-path deviation indicator— An instrument that provides a visual indication of deviation from a flight path.

flight track— The three-dimensional path in space actually traced by a vehicle.

Flinders bar— A bar of soft iron placed near a compass to correct errors due to variation of the vertical component of the earth's magnetism in different parts of the world.

flip chip— 1. An unencapsulated semiconductor device in which bead-type pads terminate on one face to permit flip (face-down) mounting of the device by contact of the leads to the required circuit interconnectors. 2. A mounting approach in which the chip (die) is inverted and connected directly to the substrate rather than using the more common wire bonding technique. Examples of this kind of flip-chip mounting are beam lead and solder bump. 3. A generic term describing a semiconductor device having all terminations on one side in the form of bump contacts. After the surface of the chip has been passivated or otherwise treated, it is flipped over for attaching to a matching substrate. 4. A semiconductor die that is fabricated so it can be bonded to the next higher assembly without the use of flywires. The termination points are typically raised pads or solder balls that are attached to the substrate metallization by ultrasonic scrubbing or thermocompression bonding. 5. A leadless monolithic structure, containing circuit elements, that is designed to electrically and mechanically interconnect to a hybrid circuit by means of an appropriate number of bumps, located on its face, which are covered with a conductive bonding agent.

flip-chip bonding— Method of interconnecting ICs in a circuit by bonding bumps, located on the IC chip's back surface, to the circuit's conducting paths.

flip-chip mounting— A method of mounting flip chips on thick- or thin-film circuits without the need for subsequent wire bonding.

flip coil— A small coil used for measuring a magnetic field. When connected to a ballistic galvanometer or other instrument, it gives an indication whenever the magnetic field of the coil or its position in the field is suddenly reversed.

flip-flop — Also called bistable multivibrator, Eccles-Jordan circuit, or trigger circuit. 1. A two-stage multivibrator circuit having two stable states. In one state, the first stage is conducting and the second is cut off. In the other state, the second stage is conducting and the first stage is cut off. A trigger signal changes the circuit from one state to the other, and the next trigger signal changes it back to the first state. For counting and scaling purposes, a flip-flop can be used to deliver one output pulse for each two input pulses. 2. A similar bistable device with an input that allows it to act as a single-stage binary counter. 3. An electronic circuit having two static states and the ability to change from one state to the other on application of a signal in a special manner. 4. A type of digital circuit that can be in either of two states, depending both on the input received and on which state it was in when the input was received.

flip-flop calculator — Calculator that can display double its digital capacity in two steps by depressing a flip-flop key.

flip-flop circuit — An electronic circuit that has two conditions of permanent stability and a means for changing from one to the other in response to an external stimulus. *See also* Eccles-Jordan circuit.

flip-flop equipment — An electronic or electromechanical device that causes automatic alternation between two possible circuit paths.

flip-flop multivibrator — Also called start-stop multivibrator. A biased rectangular wave generator that operates for one cycle when a synchronizing trigger signal is applied.

flipover cartridge — A phonograph cartridge having separate needles for playing microgroove and standard records. It may be turned to bring the proper needle into playing position.

flippies — Floppy disks that flip over; two-sided diskettes.

flippy — A double-sided diskette.

float — 1. To be connected to no source of electrical potential. (Often used with respect to a particular point.) 2. To be maintained in a constant state of charge by being connected to a source of constant voltage, as a storage battery. 3. To operate a storage battery in parallel with a charger and a load at such a voltage that the charger supplies the load current and the battery supplies only transient peaks above the normal load.

float-charging — Charging a storage battery at about the same rate that it is being discharged by the load.

floated battery — A storage battery kept fully charged across the leads of a generator. The generator carries the normal load, and the battery assists during peaks.

floating — 1. Keeping a storage battery connected in parallel with an electric supply to serve as a standby in case of supply failure and to assist in handling peak loads. 2. The condition of a device or circuit that is not grounded and not tied to any established potential.

floating address — *See* symbolic address.

floating battery — A direct current supply from a constant-voltage source (generator or rectifier) paralleled with a storage battery. If the constant-voltage source is interrupted, the storage battery maintains power to the load. Minor variations in load current are supplied from the battery.

floating-carrier modulation — *See* controlled-carrier modulation.

floating carrier system — Method of radio transmission in which the percentage modulation is held constant by varying the amplitude of the carrier wave

to offset variations in the strength of the modulating wave.

floating charge — Continuous charging of a storage battery with a low current to keep the battery fully charged while idle or on light duty.

floating decimal — A calculator function that allows the user to calculate any decimal category. The decimal may or may not be present; if present, it automatically positions itself correctly in the answer.

floating-decimal arithmetic — *See* floating-point arithmetic.

floating decimal point — Calculator entry that may contain the decimal point in any position. The number and decimal point will be properly positioned automatically when displayed.

floating gate — A technique used for ultraviolet-erasable EPROMs, in which a silicon gate is isolated inside the silicon dioxide.

floating grid — A vacuum-type grid that is not connected to any circuit. It assumes a negative potential with respect to the cathode.

floating ground — A reference ground that is not earthed. A reference point or voltage in a circuit that is not tied to an actual external ground.

floating in — Decimal-point position need not be preset; numbers in a calculator are entered as they would be written.

floating input — 1. An isolated input circuit not connected to ground at any point (the maximum permissible voltage to ground is limited by electrical design parameters of the circuit involved). It is understood that in a floating-input circuit, both conductors are equally free from any reference potential, a qualification that limits the types of signal sources that can be operated floating. 2. *See* differential-input measurement.

floating junction — A semiconductor junction through which no net current flows.

floating neutral — A circuit in which the voltage to ground is free to vary with circuit conditions.

floating out — Decimal point in a calculator is automatically aligned in the answer.

floating point — 1. A method of representing a numeric value that contains a decimal point, i.e., not necessarily a whole number. 2. The representation of numbers in scientific notation, with the exponent and mantissa given separately, so as to be able to accommodate a very wide dynamic range. 3. Pertaining to a form of number representation in which quantities are expressed in terms of a bounded number (mantissa) and a scale factor (characteristic or exponent) consisting of a power of the number base. For example, $127.6 = 0.1276 \times 10^3$, where the bounds are 0 and 1.

floating-point arithmetic — 1. Computer handling of data in which the point is not always in the same position. Floating-point numbers are expressed in terms of digits and exponents. 2. In a digital computer, a form of arithmetic in which each number is represented by several significant digits, with an explicitly placed decimal point, multiplied by the base of the number system raised to a power, as for instance 6.3542×10^5. In computations of this kind the decimal point and exponent are adjusted automatically.

floating-point calculation — In a computer, a calculation taking into account the varying location of the decimal point (if base 10) or binary point (if base 2). The sign and coefficient of each number are specified separately.

floating-point mathematics — Calculations on data elements represented as a fixed-point or fractional component and an exponent; such calculations assure a specific degree of accuracy for values over a wide numerical range.

floating-point routine — Coded instructions in proper sequence to direct a computer to perform a calculation with floating-point operation.

floating-point system — A system of numbering in which an added set of digits is used to denote the location of the radix point. *See also* fixed-point system.

floating-point unit — In programming languages, a constant of type integer, real, double precision, or complex. Relates to a mathematical coprocessor.

floating potential — The dc voltage between an open-circuited terminal of a circuit and a reference point when a dc voltage is applied to the other circuit terminals as specified.

floating zero — In a machine-tool control system, the characteristic that allows the reference-point zero to be located readily anywhere along an axis of travel. Previously established reference points are eliminated from the control memory.

float switch — A switch actuated by a float on the surface of a liquid.

float-zone crystal — A crystal grown by passing a molten zone through a cylinder of material. No other material, with the possible exception of a gas, contacts the molten zone. When the crystal is grown in a vacuum, the term *vacuum float-zone crystal* is frequently used.

flock — Finely divided felt used on phonograph turntables, underneath microphone stands, or wherever a non-scratching surface is desired.

F/logic — A computer program developed by Bell Northern Research that can simulate large digital circuits with up to 32,000 gates. The program can simulate and detect faulty components, and trace and measure the circuit response. F/logic is a gate-level simulator, which means the simulator recognizes a circuit in terms of its constituent gate elements, such as NAND, NOR, and inverter gates. It is applicable in four areas of the design process: conceptual logic verification, completed design verification, fault simulation, and logic documentation.

flood projection — In facsimile transmission, an optical method in which all of the subject to be transmitted is illuminated and the scanning spot is defined by an aperture between the subject and the light-sensitive device.

floor trap — A trap, such as a trip wire switch or a mat switch, installed so as to detect the movement of a person across the floor space.

floppy disk — Also called a diskette. 1. A small, flexible disk carrying a magnetic medium in which digital data is stored for later retrieval and use. 2. A double-sided flexible vinyl disk that is coated with a magnetic oxide that serves as a memory medium for personal computers. Each side is organized in concentric circles called tracks, and each track is divided into sectors. The standard 3.5-in. diameter double-density disk is contained in a rigid square protective enclosure and has a formatted capacity of 1.44 Mbytes. The 5-1/4 inch and older 8-inch disks are now obsolete.

flow — 1. The passage of electrons (a current) through a conductor or through the space between electrodes. 2. A general term to indicate a sequence of events.

flow amplification — The rate of change of the flow in a specified load impedance, connected to a device, with respect to the change in the flow applied to the controls of the device.

flow amplifier — A device that causes a change in output power following a change in control power of sufficient magnitude.

flowchart — Also called flow diagram. 1. A graphic presentation of the major steps of work in process with

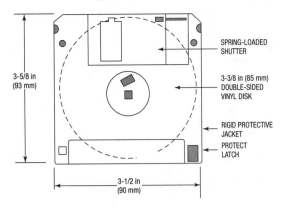

Floppy disk.

accent on how the work flows through the process rather than on how the steps are done. 2. A graphical representation of the definition, analysis, or method of solution of a problem, particularly a problem to be solved on a computer. Symbols are used to represent operations, data flow, a process or sequence of decisions, and events. 3. A graphical representation used primarily to help in the development of a computer program by illustrating how a computer program logic is laid out, and to provide documentation for the program. 4. Symbolic representation of a program sequence, where boxes represent orders or computations and diamonds represent tests and decisions (branches). A flowchart is the recommended step between algorithm specification and program writing since it greatly facilitates understanding and debugging by breaking down the program into logical sequential modules.

flowcharting — A means of illustrating the steps required to solve a computer problem. It helps to clearly visualize each step in the solution of a problem and also provides a schematic diagram of the steps used. The shape of the symbol indicates its use.

flowchart symbol — A symbol used on a flowchart to represent data flow, equipment, or an operation.

flow control — 1. A mechanism that allows a receiver to limit the amount of data a sender may transmit at any time. 2. The method used to regulate the rate of data exchange between the end users of a packet network in order to prevent system overloading. In general, the input is slowed down or stopped until the network handles the previous input.

flow diagram — Also called flowchart. A chart showing all the logical steps of a computer program. A program is coded by writing down the successive instructions that will cause the computer to perform the logical operations necessary for solving the problem, as represented on a flowchart.

flow direction — In flowcharting, the antecedent-to-successor relation between operations on a flowchart; it is indicated by arrows or other means.

flowed wax — A mechanical recording disc prepared by melting and flowing wax onto a metal base.

flowline — On a flowchart, a line that represents a connecting path between flowchart symbols, such as a line indicating transfer of data or control.

flowmeter — A device for measuring the rate of flow of liquids or gases.

flow soldering — Also called wave soldering. A method of soldering printed circuit boards by moving them over a flowing wave of molten solder in a solder bath.

fluctuating current — A direct current that changes in value, but not at a steady rate.

fluctuation noise — Noise resulting from undesired fluctuations in quantity and/or velocity of electron (or hole) flow. *See* random noise; shot noise.

fluctuation voltage — Small voltage variations in a thermionic tube due to thermal agitation, shot effect, flicker effect, etc.

fluid computer — A digital computer constructed totally from fluid logic elements. All logic functions are carried out by interaction between jets of air or liquid, and the device contains no moving parts or electronic circuits.

fluid damping — Damping obtained through displacement of a viscous fluid and the accompanying dissipation of heat.

fluidic — Of or pertaining to devices, systems, assemblies, etc., utilizing fluidic components.

fluidics — 1. The branch of engineering and technology concerned with the design and production of logic elements, amplifiers, and the like, that depend for their operation on interactions between jets of fluid rather than on electrical phenomena. (While slower than electronic logic systems, fluid logic systems can operate in environments that would damage electronic systems.) 2. The technology wherein sensing, control, information processing, and/or actuation functions are performed solely through utilizing fluid dynamic phenomena.

fluidized bed coating — A method of applying a resin coating to an article. The heated article is immersed in a dense-phase aerated bed of powdered resin, and then is heated in an oven to obtain a smooth, pinhole-free coating.

fluorescence — The emission of light (or other electromagnetic radiation of longer wavelengths) by a substance as a result of the absorption of some other radiation of shorter wavelengths only as long as the stimulus producing it is maintained. Luminescence persists for less than about 10^{-8} second after excitation is stopped.

fluorescence spectroscopy — The spectroscopic study of radiation emitted by the process of fluorescence.

fluorescent — Having the property of giving off light when activated by electronic bombardment or a source of radiant energy.

fluorescent display — A numerical or alphanumerical display whose segments are composed of fluorescent material that glows with a blue-green light when bombarded by electrons.

fluorescent lamp — 1. An electric discharge lamp in which a gas ionizes and produces radiation that activates the fluorescent material inside the glass tubing. The phosphors in the fluorescent material transform the radiant energy from the electric discharge into wavelengths giving more light (higher luminosity). 2. A low-pressure mercury electric-discharge lamp in which a fluorescing coating (phosphor) transforms some of the ultraviolet energy generated by the discharge into light.

fluorescent light source — A tube containing mercury vapor, lined with phosphor. When current is passed through the vapor, the strong ultraviolet emission excites the phosphor, which emits visible light. The ultraviolet itself cannot emerge becasue it is absorbed by the glass.

fluorescent material — A material that fluoresces readily when exposed to electron beams, X-rays, or other radiation.

fluorescent screen — A sheet of material coated with a fluorescent substance so as to emit visible light when struck by ionizing radiation such as X-rays or electron beams.

fluorometer — An instrument for measuring fluorescence.

fluoroscope — 1. An instrument with a fluorescent screen suitably mounted with respect to an X-ray tube, used for immediate indirect viewing of internal organs of the body, or internal structures in apparatus, or masses of metals, by means of X-rays. A fluorescent image (really a kind of X-ray shadow picture) is produced. 2. An X-ray device in which the image appears on a fluorescent screen rather than on a photographic film.

fluoroscopy — The use in diagnosis, testing, etc., of a fluorescent screen activated by X-rays.

flush receptacle — A receptacle recessed into a wall, with only the plate extending beyond the surface.

flush-type instrument — An instrument designed to be mounted with its face projecting only slightly from the front of the panel.

flutter — Also called wow and drift. 1. The frequency deviations produced by irregular motion of a turntable or tape transport during recording, duplication, or reproduction. The term *flutter* usually refers to relatively high cyclic deviations (for example, 10 Hz), and the term *wow* to relatively low ones (for example, a variation of once per turntable revolution). The term *drift* usually refers to a random rate close to 0 Hz. 2. In communications, (*a*) distortion due to variations in loss resulting from simultaneous transmission of a signal at another frequency, or (*b*) a similar effect due to phase distortion. 3. Rapidly repeated fluctuations in tape speed that introduce spurious burbling, quivering, or shimmering variations in the pitch of the reproduced sound. 4. A fast change in pitch (about 10 Hz) caused by a change in the speed of a turntable specified as a percentage of the test frequency (usually 3000 or 3150 Hz); figures below about 0.1 percent are good. 5. The audible effect of short-term recording speed fluctuations, occurring at a low audio or an infrasonic rate (0.5 to 200 Hz). This causes a frequency modulation of the program material, heard as a wavering or roughness of the sound. It is described as a percentage of rated speed; the smaller this percentage, the less audible the flutter. The percentage is generally combined with wow. It is often weighted (wrms) so that it corresponds to the average human hearing response. Flutter is particularly noticeable on piano or oboe, and may make the music sound watery or sour. 6. A rapid, extraneous variation in the pitch or frequency of a sound, usually caused by mechanical deviation in an element that should maintain constant speed. In tape recording, this may be caused by a faulty mechanism or by momentary sticking of the tape as it feeds through the transport and past the head.

flutter bridge — An instrument for measuring the irregularities in a constant-speed device such as a film, disc, or tape recorder.

flutter echo — 1. A rapid succession of reflected pulses resulting from a single initial pulse. 2. A multiple echo in which the reflections occur in rapid succession. If periodic and audible, it is referred to as a musical echo.

flutter rate — The number of times per second the flutter varies.

flux — 1. Number of particles crossing a unit area per unit time. The common unit of flux is particles/cm^2/s. Integrated flux, after an exposure of time T, is equal to the total number of particles that have traversed a unit area during time T. 2. The number of photons that pass through a surface per unit time. Expressed in lumens or watts. 3. The lines of force that make up an electrostatic field. 4. The rate of energy flow passing to, from, or through a surface or other geometric entity. Radiant flux is expressed in watts; luminous flux is expressed in lumens. Flux is sometimes erroneously referred to as optical power. 5. A substance used to promote or facilitate fusion, such as a material that removes oxides from surfaces to be joined by soldering, brazing, or welding. The flux also reduces surface tension of molten solder and metal to be soldered, and it covers the material being soldered to prevent reoxidation of the surface during the soldering operation. Rosin is widely used in electronics soldering. 6. The total amount of energy radiated in all directions per unit time from an electromagnetic source.

flux changes per inch — The number of polarity reversals possible in 1 inch (2.54 cm) of magnetic tape.

flux concentration — The intensity of radiation transmitted to a receiver.

flux concentrator — Any ferrous material attached to the sensor package to concentrate more of the available flux into the sensing area, thereby increasing the flux density at the chip.

flux-cored solder — Hollow-wire solder containing flux.

flux density — 1. A measure of the strength of a wave; flux per unit area normal to the direction of the flux; number of photons passing through a surface per unit time per unit area. Expressed in watts/cm^2 or lumens/ft^2. 2. The number of lines or maxwells per unit area in a section normal to the direction of the flux.

fluxgate — A magnetic azimuth-sensitive element of the fluxgate-compass system activated by the earth's magnetic field.

fluxgate compass — A gyrostabilized, remote-indicating compass and azimuth-control system used with automatic pilots.

fluxgraph — A machine that automatically plots on paper the magnetic field strength at various points in the vicinity of a coil.

flux guide — In induction heating, a magnetic material used for guiding the electromagnetic flux to the desired location or for confining it to definite regions.

flux intensity — Flux per unit solid angle.

flux linkage — 1. Magnetic lines of force that link a coil of wire. Whenever the flux linkage changes, an emf is generated in the coil. 2. The product of a number of turns in an electrical circuit by the average value of flux linked with the circuit.

fluxmeter — An instrument used with a test coil for measuring magnetic flux. It consists usually of a moving-coil galvanometer in which the torsional control is either negligible or compensated.

flyback — Also called retrace. 1. The shorter of the two time intervals comprising a sawtooth wave. 2. As applied to a cathode-ray tube, the return of the spot to

its starting point after having reached the end of its trace. This portion of the wave is usually not seen because of blanking circuits or the shortage of time.

flyback checker — An instrument used to check flyback or other transformers or inductors for open windings or shorted turns.

flyback diode — *See* freewheeling diode.

flyback power supply — The power supply that generates the high dc voltage required by the second anode of a picture tube. This voltage is produced during the flyback period, the current in the horizontal-deflecting coils reversing and inducing a sharp pulse in the primary of the transformer supplying the deflection circuit. This pulse is stepped up by an autotransformer and rectified. After suitable filtering, it becomes a very high dc voltage.

flyback tester — An instrument that tests flyback transformers and sometimes also deflection yokes.

flyback time — The period during which the electron beam is returning from the end of a scanning line to begin the next line.

flyback transformer — Also called horizontal-output transformer. A transformer used in the horizontal-deflection circuit of a television receiver to provide the horizontal scanning and accelerating anode voltages for the cathode-ray tube. It also supplies the filament voltage for the high-voltage rectifier.

flycutter — An accessory used with a drill press to cut out large round holes in metal or wood.

flying erase head — An erase head in a camcorder that allows the user to edit while shooting and achieve perfectly clean splices between the edited segments. The flying erase head follows a path in its erasing process that matches that of the recording heads, completely eliminating the unerased blanks and rainbow streaking effect left by fixed erase heads.

flying spot — 1. A small, rapidly moving spot of light, usually generated by a cathode-ray tube, used to scan an image field for television transmission. 2. The quick, mobile spot of light emitted by a source, generally a cathode-ray tube, to illuminate specific points of an area carrying light and dark regions according to a specific pattern.

flying-spot scanner — Also called light-spot scanner. 1. A television scanning device embodying a small beam that is moved over a scene or film and translates the highlights and shadows into electrical signals. 2. In optical character recognition, a device employing a moving spot of light to scan a sample space, the intensity of the transmitted or reflected light being sensed by a photoelectric transducer.

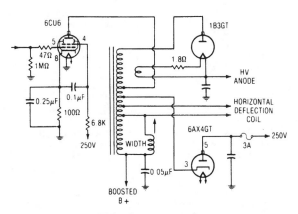

Flyback power supply.

fly's-eye lens — A multiple lens made up of hundreds of small, closely spaced lenses. It forms many images of the same subject and eliminates the need for step-and-repeat techniques in the fabrication of microelectronic circuits.

flywheel diode — *See* freewheeling diode.

flywheel effect — The maintaining of oscillations in a circuit in the intervals between pulses of excitation energy. The action is analogous to the rotation of a flywheel due to its stored mechanical energy.

flywheel synchronization — Automatic frequency control of a scanning system based on the average timing of the incoming sync signals rather than triggering of the scanning circuit by each pulse. It is used in high-sensitivity television receivers for fringe-area reception, in which noise pulses might otherwise trigger the sweep circuit prematurely.

flywheel tuning — A tuning-dial mechanism that uses a heavy flywheel on the control shaft for added momentum, to obtain a smoother tuning action.

flywire — A fine [0.001- to 0.003-inch (25- to 75-μm) diameter] gold or aluminum wire used for circuit interconnection.

FM — Abbreviation for frequency modulation.

FM/AM — A system in which information subcarriers are frequency modulated and are used to amplitude modulate the carrier.

FM/AM multiplier — A multiplier in which a carrier is modulated so that its frequency deviation from the center value is proportional to one variable, and its amplitude is proportional to another variable. The modulated carrier is then consecutively demodulated for FM and for AM. The final output is proportional to the product of the two variables.

FM broadcast band — The band of frequencies extending from 88 to 108 MHz, which includes those assigned to noncommercial educational broadcasting.

FM broadcast channel — A band of frequencies 200 kHz wide and designated by its center frequency. Channels for FM broadcast stations begin at 88.1 MHz and continue in steps of 200 kHz through 107.9 MHz. The portion of the band from 88.1 to 91.9 MHz is reserved for educational broadcasts.

FM broadcast station — A station employing frequency modulation in the FM broadcast band and licensed primarily for the transmission of radio emissions intended to be received by the general public.

FM discriminator — A device that converts frequency variations to proportional variations in the amplitude of an electrical signal. Discriminators may be of several basic types, such as pulse averaging, Foster-Seely, ratio detector, or phase-lock correlation detector.

FM discriminator (subcarrier) — The same as an FM discriminator except that it is used to convert subcarrier frequency variations into proportional voltage or current signals.

FM Doppler — Type of radar involving frequency modulation of both carrier and modulation on radial sweep.

FM-FM — Frequency modulation of a carrier by one or more subcarriers that are themselves frequency modulated by information.

FM laser — A conventional laser with a phase modulator inside its Fabry-Perot cavity. It is characterized by a lack of noise resulting from random phase fluctuation in the various modes.

FM multiplex — *See* FM stereo.

FM noise level — Residual frequency modulation of an aural transmitter as a result of disturbances in the frequency range between 50 and 15,000 Hz.

FM noise/phase noise — The short-term frequency variations in the output frequency that appear as energy at frequencies other than the carrier. It is usually expressed in terms of dBc or as an rms frequency deviation in a specified frequency removed from the carrier.

FM/PM — A system in which information subcarriers are frequency modulated and used to phase modulate the carrier.

FM radar — *See* frequency-modulated radar.

FM receiver deviation sensitivity — The smallest frequency deviation that results in a specified output power.

FM recording (magnetic) tape — A method of recording in which the input signal modulates a voltage-controlled oscillator, the output of which is delivered to the recording head.

FM stereo — Also called FM multiplex. 1. A means by which FM radio stations are able to transmit stereophonic program material to specially designed receivers and which is at the same time compatible with monophonic equipment. 2. FM broadcasting in which two channels of sound are transmitted, offering a signal similar to the stereo available from records and tapes. To hear FM stereo requires either a stereo FM tuner or a monophonic FM tuner fitted with an FM stereo adapter. The technical means for transmitting FM stereo is known as multiplexing.

FM stereophonic broadcast — The transmission of a stereophonic program by a single FM broadcast station utilizing the main channel and a subchannel to carry the signals required to produce the stereophonic effect.

f-number — In optical terminology, a number that describes a lens; ratio of focal length to lens diameter.

FO — Abbreviation for fiber optics.

foam fluxing — A commonly used wave-solder fluxing method in which flux foam is generated from a liquid flux by means of a porous diffuser, such as a hollow cylindrical stone. Low compressed air forced through the pores of the stone, immersed in the flux, generates fine bubbles of foam, which are guided to the surface by a chimney nozzle. *See also* brush fluxing; spray fluxing; wave fluxing.

focal length — 1. Symbolized by f. The distance from the principal focus (focus of parallel rays of light) to the surface of a mirror or the optical center of a lens. 2. Of a lens, the distance from the focal point to the principal point of the lens.

focal plane — A plane (through the focal point) at right angles to the principal axis of a lens. That surface on which the best image is formed.

focal point — The point at which a lens or mirror will focus parallel incident radiation.

focal distance-to-diameter ratio (F/D) — The ratio of feedhorn distance to the center of an antenna divided by the diameter of the antenna.

focometer — An instrument for measuring the focal length of a lens or an optical system.

focus — 1. The convergence of light rays or an electron beam at a selected point. 2. The sharp definition of a scanning beam in television receivers or optical systems. 3. The point at which light rays or an electron beam form a minimum-size spot. 4. The action of bringing light or electron beams to a fine spot.

focus control — 1. On a television receiver, a potentiometer control used for fine focusing of the electron beam. The control varies the first-anode voltage of an electrostatic tube or the focus-coil current of a magnetic tube. 2. A manual adjustment for electrostatically bringing the electron beam of a vidicon or picture tube to a minimum-size spot, producing the sharpest image.

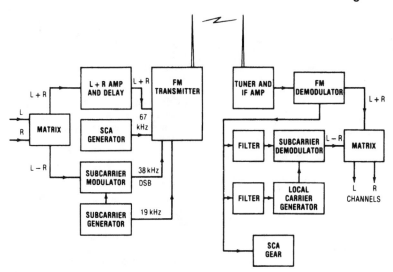

FM stereophonic broadcast system.

focusing — The process of controlling the convergence and divergence of an electron or light beam.

focusing anode — One of the electrodes used to focus the electron beam in a cathode-ray tube. As its potential changes, so does the electric field, thereby altering the path of the electrons.

focusing coil — The coil around the neck of a cathode-ray tube. It provides a magnetic field, parallel to the electron beam, for controlling the cross-sectional area of the beam on the screen.

focusing electrode — An electrode to which a potential is applied to control the cross-sectional area of the electron beam.

focusing magnet — A permanent magnet assembly that produces a magnetic field for focusing the electron beam in a cathode-ray tube.

focus projection and scanning — A method of magnetically focusing and electrostatically deflecting the electron beam in a hybrid vidicon. A transverse electrostatic field deflects the beam, and an axial magnetic field focuses the beam.

foil — 1. A thin continuous sheet of metal, usually copper or aluminum, used as the conductor for printed circuits. Foils used for printed circuits are commonly 1 or 2 ounces per square foot (30 or 60 g/cm^2); the thinner the foil, the lower the required etch time. Thinner foils also permit finer definition and spacing. 2. The thin metal shield in a shielded cable. It is equivalent to a long tubular capacitor that surrounds one or more signal wires. This electrostatic shield must be grounded at only one end of the cable by means of the drain wire. *See* shield. 3. A very thin sheet of metal, such as tin or aluminum. Used in the construction of fixed capacitors. 4. Thin metallic strips that are cemented to a protected surface (usually glass in a window or door), and connected to a closed electrical circuit. If the protected material is broken so as to break the foil, the circuit opens, initiating an alarm signal. Also called tape. A window, door, or other surface to which foil has been applied is said to be taped or foiled.

foil connector — An electrical terminal block used on the edge of a window to join interconnecting wire to window foil.

foil electret — A polymer plastic film about 1 mil (25 μm) thick with a very thin metal layer evaporated on one surface and having permanent electrostatic polarization created by electron bombardment or by heating while exposed to a powerful electric field. Used to make electret microphones.

foldback — British term synonymous with talkback. A technique for protecting voltage regulators from short circuits. After a certain output-current level is reached, any further load on the regulator results in less, rather than more, current. *See* cue bus.

foldback characteristic — *See* current limiting (automatic).

foldback current limiting — 1. An overload protection method whereby the output current of a power supply is decreased as the load approaches short circuit. Under output short circuit, the output current is, therefore, less than the rated output current. This technique minimizes internal power dissipation under overload conditions. 2. A protective circuit in a power supply that monitors the output current drain, and automatically reduces the output voltage to very low levels when the drain current exceeds a preset level. The output voltage is usually reduced to a level at which approximately 25 percent of the nominal current flows.

foldback operation — In a power supply, a technique similar to current limiting except that when the load demands too much current, the power supply reacts by reducing both its output current and output voltage.

folded cavity — An arrangement used for producing a cumulative effect in a klystron repeater. This is done by making the incoming wave act in several places on the electron stream from the cathode.

folded-dipole antenna — An antenna comprising two parallel, closely spaced dipole antennas. Both are connected together at their ends, and one is fed at its center.

folded heater — A strand of bent, coated wire inserted into a cathode sleeve.

folded horn — 1. An acoustic horn that is curled to permit more efficient use of the space it occupies. 2. A type of speaker enclosure employing a horn-shaped passageway for aiding the bass response.

folding frequency — The frequency that is one-half the sampling rate when samples are made continuously at equal intervals.

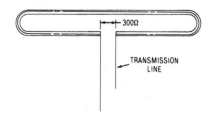

Folded-dipole antenna.

foldover — A distorted television picture that appears to overlap horizontally or vertically. It is due to nonlinear horizontal- or vertical-sweep circuits.

follow current — 1. That line current which tends to follow a lightning discharge through an arrester to ground. 2. The current through a lightning protector from a connected steady-state power source that flows during and following the discharge of a surge or transient current.

follower — A circuit in which the output of a high-gain amplifier is fed directly back to its negative input. The input signal is reproduced without polarity reversal.

follower drive — Also called slave drive. A drive in which the reference input and operation are direct functions of a master drive.

follower with gain — A follower in which only a part of the output voltage is fed back in series opposition to the input signal. Hence, closed-loop gain greater than unity is obtained over the rated range of operation.

following blacks — Also called edge effect, trailing reversal, or trailing blacks. A picture condition in which the edge following a white object is overshaded toward black (i.e., the object appears to have a trailing black border).

following whites — Also called edge effect, trailing reversal, or trailing whites. A picture condition in which the edge following a black or dark gray object is shaded toward white (i.e., the object appears to have a trailing white border).

follow-on current — The current from power sources that flows through a transient suppressor during and after discharge. Spark gaps, gas-discharge tubes, and SCR/thyristor crowbars offering a low impedance path to transient energy provide a low-impedance path for line current, as well. This follow-on current must be handled by the device until the follow-on voltage is removed.

font — 1. The characteristic style of a set of alphanumerics, e.g., gothic. 2. An alphabetic, numeric, or other graphic shape, i.e., 10-point Times Roman font, 1428E font, ocr (A) font, etc. 3. A family or assortment of characters of a given size and style. 4. A character set in a particular style and size of type, including all alphabet characters, numerics, punctuation marks, and special symbols. 5. A complete set of letter, numbers and symbols with a common design.

footcandle — Letter symbol: fc. The unit or measure of illumination in which the foot is taken as the unit of length. It is the illumination on a surface 1 square foot in area on which there is a uniformly distributed flux of 1 lumen, or the illumination produced on a surface all points of which are at a distance of 1 foot from a directionally uniform point source of 1 candela.

foot control — A foot-actuated start-stop switch, usually used for dictating and transcribing via tape.

footlambert — Letter symbol: fL. A unit of luminance (photometric brightness) equal to $1/\pi$ candela per square foot or to the uniform luminance of a perfectly diffusing surface emitting or reflecting light at the rate of 1 lumen per square foot or to the average luminance of any surface emitting or reflecting light at that rate. The average luminance of any reflecting surface in footlamberts is, therefore, the product of the illumination in footcandles by the luminous reflectance of the surface. Thus, a desk may be receiving 100 footcandles of illumination, but if it is a dark color it may be reflecting, say, 20 footlamberts. We "see" footlamberts, not footcandles. For a perfectly reflecting and perfectly diffusing surface, the number of lumens per square foot is equal to the number of footlamberts.

foot-pound — A unit of measurement equivalent to the work of raising one pound vertically a distance of 1 foot.

footprint — 1. The area covered by downlink signals transmitted from a satellite. 2. The space a device occupies on a desk or in a workplace. 3. The geographic area toward which a satellite downlink antenna directs its signal. The measure of strength of this footprint is the EIRP.

foot rail — A holdup alarm device, often used at cashiers' windows, in which a foot is placed under the rail, lifting it, to initiate an alarm signal.

forbidden band — The energy band lying between the conduction and valence bands. The energy difference across it determines whether a solid acts as a conductor, semiconductor, or insulator.

forbidden combination — A combination of bits or other representations that is invalid according to some criterion.

forbidden-combination check — A test, usually automatic, for the occurrence of a code expression that is not permissible. A self-checking code (or error-detecting code) uses code expressions such that errors result in a forbidden combination. A parity check uses a self-checking binary-digit code in which the total number of 1s (or 0s) in each permissible code expression is always even or always odd. A check may be made for either even parity or odd parity. A redundancy check makes use of a self-checking code that employs redundant digits called check digits.

forbidden energy gap — The energy range of a semiconductor between the bottom of the conduction band and the top of the valence band. Electrons cannot exist at energies within this range.

force — 1. Any physical action capable of moving a body or modifying its motion. 2. In computer programming, manual intervention that directs the computer to execute a jump instruction.

force-balance transducer — A transducer in which the output from the sensing member is amplified and fed back to an element that causes a force-summing member to return it to its rest position. The magnitude of the signal fed back determines the output of the divide, like the error signal in a servo signal.

forced coding — *See* minimum-access programming.

force differential — The difference between the operating force and the release force of a momentary contact switch.

forced oscillation — In a linear constant-parameter system, the response to an applied driving force, excluding the transient that results from energy at the time the driving force is applied.

force factor (of an electromechanical or electroacoustic transducer) — 1. The complex quotient of the force required to block the mechanical or acoustic system, divided by the corresponding current in the electrical system. 2. The complex quotient of the resultant open-circuit voltage in the electric system, divided by the velocity in the mechanical or acoustic system.

force feedback — Sensing technique using electrical or hydraulic signals to control a robot end effector.

force-summing device — In a transducer, the element directly displaced by the applied stimulus.

foreground processing — The automatic execution of computer programs that have been designed to preempt the use of the computing facilities.

fore pump — An auxiliary vacuum pump used as the first stage in evacuating vacuum systems.

foreshortened addressing — A feature of control computers that makes it possible to use simpler instructions when addressing the computer; hence less of the available computer storage is used for this purpose.

fork oscillator — An oscillator in which a tuning fork is the frequency-determining element.

fork tines — The projecting ends of a tuning fork. When vibrated, they produce a constant frequency.

form — To apply a voltage to an electrolytic capacitor, semiconductor, or other component as part of a manufacturing process, in order to cause a desired change in its characteristics.

formal logic — The study of the structure and form of a valid argument without regard to the meaning of the terms in the argument.

formant — 1. The particular frequency region in which the energy of a vowel sound is concentrated most strongly. 2. In an organ, an electrical circuit whose purpose is to alter the tone quality of sound amplified by it. A formant filter is applied to the entire output from a manual, rather than to individual tones.

formant filter — A waveshaping network used in an organ to modify the signal from the tone generator so it will assume the waveshape of the desired tone.

formants — Resonances in the frequency spectra of voiced speech. Formants appear as bands in a spectrographic display of voiced speech. Formants help to distinguish one sound from another.

formant synthesis — 1. A technique for modeling the natural resonances of the vocal tract. For recognizable speech, at least three formants should be used for each voice utterance. Voiced sounds are generated from an impulse source that is modulated in amplitude to control intensity. The resulting signal is passed through two levels of filtering. The first is a time-varying filter composed of cascaded resonators that correspond to the source-spectrum and mouth-radiation characteristics of the speech waveform. Unvoiced sounds, generated as white noise, are passed through a variable-pole-zero filter. The second filter used for voiced sounds can be reused for the unvoiced sounds. The coefficients for these filters are stored in ROM. An approximate number of memory bits required for a second of speech is 400. 2. A parameter-encoding technique that models speech information by tracking the formants of the frequency spectrum.

format — 1. In a computer, a specified grouping of data to facilitate storage and movement of the data in the system. A given format may include control codes, record marks, block marks, and tape marks in a prearranged sequence. The format tells the operator or the system how the transfer, processing, and printing of data are to be controlled. The term *format* also describes the layout of characters on printed copy, which is directly related to the data format. 2. A model for a microinstruction consisting of fields that contain constants, variables, and don't cares. 3. A message or data structure that allows identification of specific control codes or data by their position during processing. 4. The orderly structured arrangement of data elements (bits, characters, bytes, and/or files) to form a larger entity, such as a list, table, record, file, or dictionary. 5. A contraction meaning the *form* of *mat*erial, designating the predetermined arrangement of text/data

for output. 6. The layout, presentation, or arrangement of data on a screen, file, or paper. 7. A specified arrangement of data that permits identification of control and information content. 8. To prepare a disk for reading, writing, and accepting files.

formatted diskette — 1. A computer disk that has been initialized with DOS. 2. A diskette on which track and sector control information has been written.

formatted display — Standardized data arrangement to make data entry faster and more organized.

formatter — *See* buffer.

formatting — 1. Preparing a diskette for use so that the operating system can write information on it. The formatting process erases any previous information on the diskette. 2. The arranging in a predefined order of code characters within a record. 3. The division of tracks into sections to make it easier to retrieve and update data. In each sector, the block of data is preceded by an identifying header. Gaps are inserted between sectors and between the header and data block within each sector to allow time for control logic functions.

form factor — 1. Shape (diameter/length) of a coil. 2. Ratio of the effective value of a symmetrical alternating quantity to its half-period average value. 3. A figure of merit that indicates how much the current departs from pure dc or from a continuous, nonpulsating current. Unity represents pure dc. Values greater than 1 indicate an increasing departure from pure dc. A departure from unity form factor increases the heating effect in a motor and reduces brush life.

Formica — Trade name for a phenolic compound having good insulating qualities.

forming — The application of voltage to an electrolytic capacitor, electrolytic rectifier, or semiconductor device to produce a desired permanent change in electrical characteristics as a part of the manufacturing process.

formula translation — *See* FORTRAN.

form-wound coil — An armature coil that is formed or shaped over a fixture before being placed on the armature of a motor or generator. Any coil wound on a fixture or dummy form.

FORTRAN — 1. Acronym for *fo*rmula *tran*slation. A procedure-oriented computer language designed to be used with problems expressible in algebraic notation. There are several forms: FORTRAN II, FORTRAN IV, etc. 2. A computer-programming language designed mainly for scientific problems. 3. A higher-level programming language designed for programming scientific-type problems.

fortuitous conductor — Any conductor that may provide an unintended path for intelligible signals; for example, water pipe, wire or cable, metal structural members, and so forth.

fortuitous telegraph distortion — 1. Distortion other than bias or characteristic. It occurs when a signal pulse departs from the average combined effects of bias and characteristic distortion for one occurrence. Since fortuitous distortion varies from one signal to another, it must be measured by a process of elimination over a long period. It is expressed in a percentage of unit pulse. 2. Distortion of telegraph signals that does not follow any pattern and is not predictable.

forty-five/forty-five — Also called the Westrex system. A system of disc recording in which signals originating from two microphones are impressed on each side of a groove. The two sides are cut 45° from the surface of the record.

forty-five record — A 7-inch (17.8-cm) record with a 1$\frac{1}{2}$-inch (3.8-cm) center hole. It is recorded at 45 rpm and played at the same speed. (No longer in use.)

forty-four-type repeater — Type of telephone repeater used in a four-wire system. It employs two amplifiers and no hybrid arrangements.

forward — In or of the direction in which a nonlinear element, like a pn junction, conducts most easily.

forward-acting regulator — A transmission regulator that makes an adjustment without affecting the quantity that caused the adjustment.

forward-backward counter — A counter having both an add and subtract input and thus capable of counting in either an increasing or a decreasing direction.

forward bias — 1. A voltage applied across a rectifying junction with a polarity that provides a low-resistance conducting path. By contrast, reverse bias causes the junction to block normal current. 2. An external voltage applied in the conducting direction of a pn junction. The positive terminal is connected to the p-type region, and the negative terminal to the n-type region.

forward-biased second breakdown — A local thermal runaway phenomenon in a semiconductor characterized by high local temperature and uneven current density. It is strongly a function of breakdown voltage and is affected by the structure used.

forward coupler — A directional coupler used for sampling incident power.

forward current — 1. The current that flows across a semiconductor junction when a forward-bias voltage is applied. 2. Of a diode, the current through a diode in the forward direction.

forward dc resistance — Of a diode, the quotient of forward voltage across a diode and the corresponding forward current.

forward direction — 1. The direction of easy current flow through a semiconductor device when a given voltage within the ratings of the device is applied. In a conventional rectifier or diode, the forward direction is from anode to cathode when the anode is at a positive voltage with respect to the cathode. 2. Of a pn junction or a semiconductor diode, the direction of the unidirectional current in which the junction or diode has the lower resistance. 3. Of a tunnel diode, the direction of current within the diode for which the characteristic includes negative differential conductance.

forward error correction — Abbreviated FEC. A technique for improving the accuracy of data transmission. Excess bits are included in the outgoing data stream so that error correction algorithms can be applied upon reception.

forward gate current — The current into the gate terminal of a field-effect transistor with a forward gate-to-source voltage applied. The gate current corresponding to the forward gate voltage.

forward gate-to-source breakdown voltage — The breakdown voltage between the gate and source terminals of an insulated-gate field-effect transistor with a forward gate-to-source voltage applied and all other terminals short-circuited to the source terminal.

forward gate voltage — The negative gate-to-anode voltage for n-gate thyristors. The positive gate-to-cathode voltage for p-gate thyristors.

forward path — In a feedback control loop, the transmission path from the loop-actuating to the loop-output signal.

forward propagation by ionospheric scatter — A radiocommunication technique using the scattering phenomenon exhibited by electromagnetic waves in the 30- to 100-megahertz region when passing through the ionosphere at an elevation of about 85 kilometers.

forward propagation by tropospheric scatter — A method of communication by means of ultrahigh-frequency FM radio. It provides reliable multichannel telephone, teletype, and data transmission without line-of-sight restrictions or the necessity of using wire or cables.

forward recovery time — 1. In a semiconductor diode, the time required for the current or voltage to arrive at a specified condition after instantaneous switching from zero or a specified reverse voltage to a specified forward voltage. 2. The time required for the current or voltage to recover to a specified value after instantaneous switching of a semiconductor from zero or a specified reverse voltage to a specified forward bias condition.

forward resistance — The resistance measured at a specified forward voltage drop or forward current in a rectifier.

forward scatter — 1. Propagation of electromagnetic waves at frequencies above the maximum usable high frequency through use of the scattering of a small portion of the transmitted energy when the signal passes from a nonionized medium into a layer of the ionosphere. 2. A term referring collectively to very-high-frequency forward propagation by ionospheric scatter and ultrahigh-frequency forward propagation by tropospheric scatter communication techniques.

forward scattering — The reflected radiation of energy from a target away from the illuminating radar.

forward short-circuit current amplification factor — In a transistor, the ratio of incremental values of output to input current when the output circuit is ac short-circuited.

forward-transfer function — In a feedback control loop, the transfer function of the forward path.

forward voltage — 1. Voltage of the polarity that produces the larger current. 2. The voltage drop across a device after breaking over into conduction at some specified current. 3. Of a diode, the voltage across the terminals that results from the current in the forward direction. 4. Of a thyristor, a positive anode voltage.

forward voltage drop — The resultant voltage drop when current flows through a rectifier in the forward direction.

forward wave — In a traveling-wave tube, a wave with a group velocity in the same direction the electron stream moves.

Foster-Seeley discriminator — A type of frequency discriminator that converts a frequency-modulated signal into an audio signal. It requires a limiter to prevent random amplitude variations of the FM signal from appearing in its output.

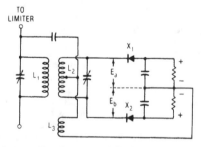

Foster-Seeley discriminator.

Foster's reactance theorem — The driving-point impedance of a finite two-terminal network composed of pure reactances is a reactance that is an odd rational function of frequency and that is completely determined, except for a constant factor, by assigning the resonant and antiresonant frequencies. In other words, the driving-point impedance consists of segments going from minus infinity to plus infinity (except that at zero, or infinite, frequency, a segment may start or stop at zero impedance). The frequencies at which the impedance is infinite are termed poles, and those at which it is zero are termed zeros.

FOTS — *See* fiber-optics transmission system.

Foucault currents — *See* eddy currents.

four-address code — An artificial language for describing or expressing the instructions carried out by a digital computer. In automatically sequenced computers, the instruction code is used for describing or expressing sequences of instructions. Each instruction word then contains a part specifying the operation to be performed, plus one or more addresses that identify a particular location in storage.

four-channel sound — *See* quadraphonic.

four-frequency diplex telegraphy — A method of frequency-shift telegraphy in which a separate frequency is used to represent each of the four possible signal combinations corresponding to two telegraph channels.

four-horn feed — A cluster of four rectangular horn antennas used as the radiating and receiving elements of parabolic or lens-type radar antennas. The four segments of the horn assembly define the four quadrants of information for direction to target sensing. Used on monopulse-type radar systems such as the AN/FPS-16.

Fourier analysis — 1. The process of analyzing a complex wave by separating it into a plurality of component waves, each of a particular frequency, amplitude, and phase displacement. 2. The representation of arbitrary functions as the superposition of sinusoidal functions, whereby the representations themselves are referred to as Fourier series or Fourier integers.

Fourier series — A mathematical analysis that permits any complex waveform to be resolved into a fundamental, plus a finite number of terms involving its harmonics.

Fourier transform — 1. A mathematical relationship that provides a connection between information in the frequency domain and the time domain. The Fourier transform of correlation functions yields the power spectra. 2. A mathematical operation that decomposes a time-varying signal into its complex frequency components (amplitude and the phase, or real and imaginary components). 3. Any of the various methods of decomposing a signal into a set of coefficients of orthogonal waveforms (trigonometric functions). 4. Mathematical operation used to derive the frequency domain description of a distribution. An efficient digital implementation is the fast Fourier transform, or FFT. The inverse Fourier transform returns a frequency domain description to the original distribution. The digital inverse form is known as the IFFT.

four-layer diode — 1. A semiconductor diode that has three junctions, with connections made only to the two outer layers that form the junctions. A Schottky diode is an example. 2. A pnpn two-terminal thyristor exhibiting a negative resistance characteristic in one direction. It has two stable states: an off state in which it displays a high series resistance, and an on state in which the series resistance is quite low. Switching time for the four-layer diode is in the nanosecond region. A very high ratio of hold current to switching makes it ideal for oscillator application.

four-layer transistor — A junction transistor that has four conductivity regions, but only three terminals. A thyristor is an example.

four-level laser — A type of laser that differs from a three-level type in that it has a terminal (lower level) for the laser transition that itself is an excited state of the system rather than ground level. Ordinarily, less energy is required to obtain the necessary population inversion in a four-level laser because the terminal level may be almost empty initially.

four-level system — A laser involving four electronic energy levels. The ground state (level 1) is pumped to level 4, from which the excited electrons make a downward transition to the upper laser level 3 (or metastable level 3). Then, stimulated transition to the lower laser level 2 occurs, followed by rapid decay to the ground state. The four-level system has the advantage that the pump level and ground state are isolated from the laser action.

four-pole network — *See* two-terminal-pair network.

four-quadrant multiplier — In analog computers, a multiplier in which operation is not restricted with regard to the signs of the input variables.

four-track — A quarter-track tape format in which the width of the tape is recorded in four parallel magnetic tracks, separated by narrow unrecorded guard bands.

four-track recorder — *See* track configuration.

four-track recording — Also called quarter-track. On quarter-inch-wide tape, the arrangement by which four different channels of sound may be recorded on adjacent tracks. These may be recorded as four separate and distinct tracks (monophonic), or two related (stereo) pairs of tracks. By convention, tracks 1 and 3 are recorded in the forward direction of a given reel, and tracks 2 and 4 are recorded in the reverse direction.

four-track tape — Also known as quarter-track. Tape on which four separate sound paths are recorded. The use of four tracks permits stereo in both directions of tape movement, or alternately, monophonic recording across four times the length of a given tape.

four-track tape recording format — Either of two professional tape recording formats (four-track, half-track, or four-track quarter track) in which four channels (tracks) can be recorded in the same direction.

four-wire circuit — 1. A two-way circuit with two paths. Each path transmits the electric waves in one direction only. The transmission paths may or may not employ four wires. 2. A full-duplex communications channel in which transmission occurs over one pair of wires and reception occurs over a separate pair.

four-wire line — A two-way transmission circuit using separate paths for the two directions of transmission. For voice-frequency transmission, this requires two pairs (four wires).

four-wire modem — A modem, using two pairs of wires, capable of simultaneous data transmission in both directions (i.e., full-duplex modem).

four-wire repeater — A telephone repeater used in a four-wire circuit. It has two amplifiers; one amplifies the telephone currents in one side of the four-wire circuit, and the other in the other side.

four-wire resistance — Resistance measurement method that compensates for the resistance in the measurement leads as well as those in the meter's input terminals (and anywhere else in the measurement circuit, such as junctions).

four-wire terminating set — A hybrid arrangement involving termination of four-wire circuits on a two-wire basis for interconnection with two-wire circuits.

fox message — A diagnostic test message that includes all the alphanumerics on a teletypewriter, as well as most of the function characteristics such as space, figure shift, letter shift, etc. It is: THE QUICK BROWN FOX JUMPED OVER A LAZY DOG'S BACK 1234567890- - -SENDING. The sending station's identification is inserted in the three space blanks that precede the word "sending."

fox test — *See* fox message.

FPGA — Abbreviation for field-programmable gate array. An array of gates on a chip whose interconnections can be arranged electronically by the user.

FPLA — Abbreviation for field-programmable logic array. A PLA that can be programmed by the user. FPLAs are used in particular to implement the control section of bit-slice processors.

fractal — 1. A form of computer-generated art-making process that creates complex, repetitive, mathematically based geometric shapes and patterns that resemble those found in nature. 2. An object (or set of points, curves, or patterns) that exhibits increasing detail with increasing magnification. In computer graphics applications, a technique for attaining a degree of complexity analogous to that in nature from a handful of data points.

fractional arithmetic units — Arithmetic units in a computer that is operated with the decimal point at the extreme left so that all numbers have a value less than 1.

fractional frequency offset — *See* frequency offset.

fractional-horsepower motor — Any motor having a continuous rating of less than 1 horsepower.

fragmentation — 1. Condition in which a mass memory has many separate holes (available spaces) and needs compacting. 2. In a computer, the division of a contiguous storage area such as a main memory or secondary storage in a way that causes areas to be wasted. 3. The existence of small increments of unused space throughout disk storage. The uneven distribution of data on a disk that occurs whenever files on a disk are deleted and new files are added.

Frahm frequency meter — A meter that measures the frequency of an alternating current. It consists of a row of steel reeds, each with a different natural frequency. All are excited by an electromagnet fed with the current to be measured. The reed that vibrates is the one with a frequency corresponding most nearly to that of the current.

frame — 1. In television, the total area occupied by the picture. In the United States, each frame contains 525 horizontal scanning lines, and 30 complete frames are shown per second. 2. One cycle of a recurring number of pulses. 3. In pulse-amplitude modulation and pulse-duration modulation, one complete commutator revolution or sweep. In pulse-code modulation, a recurring group of words that includes a single synchronizing signal. 4. The array of binary digits across the width of magnetic or paper tape. 5. The time period needed to transmit either bits or bytes of data along with the parity and other control information. 6. To center an image or place it in any part of the TV screen desired. Also applies to stills. 7. A single image of the connected multiple images on motion-picture film. 8. The size of the copy produced by a facsimile system. 9. In a computer, a logical block of consecutive addresses or lines within a structure such as a record or a file. 10. The total area occupied by a television image that is scanned, equivalent to one frame of moving-picture film.

frame buffer — Hardware to hold a bit-map picture, especially when hooked up to a computer so that a user may add effects one by one and see the result.

frame capture — A method of acquiring images displayed on a computer monitor into memory, so that they can be imported into an application program, manipulated, and printed.

frame frequency — 1. The number of times per second the picture area is completely scanned (30 per second in the United States television system). 2. In a computer, the number of frames per unit time. 3. In telemetry, the number of times per second that a frame of pulses is sent or received.

frame grabber — An electronic circuit in a computer used to capture a still video image for storage or processing.

frame grid — The grid of a vacuum tube consisting of a rigid welded frame on which tungsten wire is wound under tension, resulting in a firm precision structure that can be positioned accurately. It also allows the use of much finer grid wire, which reduces electron interception and power dissipation in the grid.

frame-grounding circuit — A conductor that is electrically bonded to the machine frame and/or to any conduction parts that are normally exposed to operating personnel. This circuit may further be connected to external grounds as may be required by applicable Underwriters code.

frame of reference — A set of points, lines, or planes used for defining space coordinates.

frame pulse synchronization — Synchronization of the local-channel rate oscillator by comparison and phase lock with the separate frame-synchronizing pulses.

framer — A device for adjusting facsimile equipment so that the recorded elemental area bears the same relationship to the record sheet as the corresponding transmitted elemental area bears to the subject copy as the line progresses.

frame rate — *See* frame frequency.

frame roll — A momentary roll, or flip-flop, of a television picture.

frame synchronization signal — In pulse-amplitude modulation, a coded pulse or interval to indicate the start of the commutation frame period. In pulse-code modulation, any signal used to identify a frame of data.

frame-synchronizing pulse — A recurrent signal that establishes each frame.

frame-synchronizing pulse separator — A circuit for separating frame-synchronizing pulses or intervals from commutated signals.

framing — 1. Adjusting the picture to a desired position in the direction of line progression. 2. The process of selecting the bit groupings representing one or more characters from a continuous stream of bits.

framing bits — Also called sync bits. Non-information-carrying bits used to make possible the separation of characters in a bit stream.

framing control — More often called centering control. A knob (or knobs) for centering and adjusting height and width of a television picture.

framing magnet — *See* centering magnet.

Franklin antenna — A base-fed vertical antenna that is several wavelengths high and that gives broadside radiation as a result of the elimination of phase reversals by means of loading coils or wire folds.

Franklin oscillator — A two-terminal feedback oscillator using two tubes or transistors and having sufficient loop gain to permit extremely loose coupling to the resonant circuit.

Fraunhofer region — The region in which the energy from an antenna proceeds essentially as though coming from a source located in the vicinity of the antenna.

fraying — The unraveling of a fibrous braid.

free-carrier absorption — The phenomenon whereby an electron within a band absorbs radiation by transferring from a low-energy level to an empty high-energy level.

free-carrier photoconductivity — Photoconductivity that may be extended as far as the microwave region, due to the absorption of photons by electrons.

free electrons — Electrons that are not bound to a particular atom, but circulate among the atoms of a substance.

free energy — The available energy in a thermodynamic system.

free field — 1. Theoretically, a field (wave or potential) that is free from boundaries in a homogeneous, isotropic medium. In practice, a field in which the effects of the boundaries are negligible over the region of interest. 2. A property of information-processing recording media that permits recording of information without regard to a preassigned or fixed field; e.g., information-retrieval-devices information may be dispersed in the record in a sequence or location.

free-field emission — Electron emission that occurs when the electric field at the surface of an emitter is zero.

free grid — A grid electrode that is left unconnected in a vacuum tube. Its potential exerts a control over the plate current.

free impedance — Also called normal impedance. The input impedance of a transducer when the load impedance is zero.

free list — A list of computer memory locations that are currently unused and may be allocated by the memory manager to requesting tasks. Free lists are usually organized as linked lists of memory blocks, in which each block contains the size of the block and a pointer to the next block in the list.

free magnetic pole — A magnetic pole so far from an opposite pole that it is free from the effect of the other pole.

free motional impedance — The complex remainder after the blocked impedance of a transducer has been subtracted from the free impedance.

free net — A net in which any station may communicate with any other station in the same net without first obtaining permission from the control station.

free oscillations — Commonly referred to as shock-excited oscillations. Oscillations that continue in a circuit or system after the applied force has been removed. The frequency of the oscillations is determined by the parameters of the system or circuit.

free-point tube tester — A tester instrument that permits transferring a tube from a circuit to a test panel at which either voltage or current measurement for any electrode of the tube is readily made by plugging a meter into appropriate jacks. Connections to the receiver are made by means of a cord and plug inserted into the socket from which the tube was removed.

free position — The initial position of the actuator of a momentary-contact switch when there is no external force (other than gravity) applied on the actuator, and the switch is in the specified position.

free progressive wave — Also called free wave. A wave free from boundary effects in a medium. In other words, there are no reflections from nearby surfaces. A free wave can only be approximated in practice.

free radicals — Atoms, ionized fragments of atoms, or molecules that combine and release enormous amounts of energy.

free reel — The reel that supplies the magnetic tape on a recorder.

free-rotor gyro — A gyro whose rotor is supported by a gas-lubricated spherical bearing.

free routing — A method of traffic handling in which a message is sent toward its destination over any available channel without dependence on a predetermined routing doctrine.

free-running frequency — The frequency at which a normally synchronized oscillator operates in the absence of a synchronizing signal. *See* center frequency.

free-running local synchronizer oscillator — A free-running oscillator circuit in the decommutator normally triggered by separated channel synchronizing pulses. It supplies substitute pulses for missing channel pulses.

free-running multivibrator — A multivibrator that oscillates without triggering pulses. *See* astable multivibrator (free-running).

free-running sweep — A sweep operating without synchronizing pulses.

free sound field — A field in a medium free from discontinuities or boundaries. In practice it is a field in which the boundaries cause negligible effects over the region of interest.

free space — 1. Empty space, or space with no free electrons or ions. It has approximately the electrical constants of air. 2. Having to do with a condition in which the radiation pattern of an antenna is not affected by surrounding objects such as the earth, buildings, vegetation, etc. 3. The amount of unused space available on a hard disk or other device. DOS chkdsk or another utility determines free space.

free-space field intensity — The radio-field intensity that would exist at a point in a uniform medium in the absence of waves reflected from the earth or other objects.

free-space loss — 1. The theoretical radiation loss that would occur in radio transmission if all variable factors were disregarded. 2. The theoretical transmission loss between two isotropic radio antennas dependent only on distance and frequency, with all variable factors eliminated.

free-space propagation — Electromagnetic radiation over a straight-line path in a vacuum or ideal atmosphere, sufficiently removed from all objects that affect the wave in any way.

free-space radar equation — The equation for determining the characteristic of a radar signal propagated between the radar set and a reflecting target in free space.

free-space radiation pattern — The radiation pattern of an antenna in free space, where there is nothing to reflect, refract, or absorb the radiated waves.

free-space transmission — Electromagnetic radiation over a straight line in a vacuum or ideal atmosphere sufficiently removed from all objects that affect the wave.

free-space wave — That portion of a radio wave which travels in a direct path between transmitting and receiving antennas, without reflections or refractions.

free speed — The angular speed of an energized motor under no-load conditions. *See also* angular velocity.

freeware — A computer file that is made available to the public free of charge from the author.

free wave — *See* free progressive wave.

freewheeling circuit — A motor arrangement in which the field is shunted by a half-wave rectifier that discharges the energy stored in the field during the negative half-cycles.

freewheeling diode — Also called damper diode, flyback diode, feedback diode, and flywheel diode. A fast recovery rectifier connected across an inductive load so it conducts current proportional to the energy stored in the inductance. This current flows when no power is supplied to the load and continues until all energy stored in the inductor has been removed or until energy is again supplied to the inductance from the power source.

freeze-out — A short-time denial of a telephone circuit to a subscriber by a speech-interpolation system.

F region — The region of the ionosphere above 100 miles (160 km).

freq — Abbreviation for frequency.

frequency — 1. Symbolized by f. The number of recurrences of a periodic phenomenon in a unit of time. Electrical frequency is specified as so many hertz. Radio frequencies are normally expressed in kilohertz at and below 30,000 kilohertz and in megahertz above this frequency. 2. The number of complete cycles in 1 second of alternating current, voltage, or electromagnetic or sound pressure waves. 3. Number of alternations or repetitions per second in any recurring action. In the case of alternating current and other forms of wave motion, it is expressed in hertz. 4. With reference to electromagnetic radiation, the number of crests of waves that pass a fixed point in a given unit of time, in light or other wave motion.

frequency agile — The ability of a satellite TV receiver to select or tune all channels (transponders) from a satellite. Receivers not frequency agile are dedicated to a single channel, and are most often used in the CATV industry. Frequency agility can be via continuously variable tuning or discrete-step (channel selection) tuning.

frequency agility — The rapid and continual shifting of a radar frequency to avoid jamming by the enemy, reduce mutual interference with friendly sources, enhance echoes from targets, or provide necessary patterns of ECM (electronic countermeasures) or ECCM (electronic counter-countermeasures) radiation.

frequency allocation — 1. The assignment of available frequencies in the radio spectrum to specific stations, for specific purposes. This is done to yield maximum utilization of frequencies with minimum interference between stations. Allocations in the United States are made by the Federal Communications Commission. 2. A band of radio frequencies identified by an upper and lower frequency limit ear-marked for use by one or more of the 38 terrestrial and space radiocommunication services defined by the International Telecommunication Union under specified conditions.

frequency allotment — The designation of portions of an allocated frequency band to individual countries or geographical areas for a particular radiocommunication service; for a satellite service, specific orbital positions may also be allotted to individual countries.

frequency assignment — Authorization given by a nation's government for a station or operator in that country to use a specific radio frequency channel under specified conditions.

frequency authorization — The document of power that legalizes the assignment of a frequency or a frequency band.

frequency-azimuth-intensity — Pertaining to a type of radar display in which frequency, azimuth, and strobe intensity are correlated.

frequency band — A continuous and specific range of frequencies. A range of frequencies between a lower and an upper limit.

frequency band of emission — The frequency band required for a specific type of transmission and speed of signaling.

frequency bias — A constant frequency purposely added to the frequency of a signal.

frequency changer — *See* frequency converter.

frequency-change signaling — A telegraph signaling method in which one or more particular frequencies correspond to each desired signaling condition of a telegraph code. The transition from one set of frequencies to the other may be either a continuous or a discontinuous change in frequency or in phase.

frequency-changing circuit — A circuit comprising an oscillator and a mixer and delivering an output at one or more frequencies other than the input frequency.

frequency channel — A continuous portion of the appropriate frequency spectrum for a specified class of emission.

frequency compensation — 1. The technique of modifying an electronic circuit or device for the purpose of improving or broadening the linearity of its response with respect to frequency. 2. The compensation required in feedback amplifiers to ensure stability and prevent unwanted oscillations.

frequency constant — The number relating the natural vibration frequency of a piezoid (finished crystal blank) to its linear dimension.

frequency conversion — 1. The process of converting a signal to some other frequency by combining it with another frequency. 2. Of a heterodyne receiving system, converting the carrier frequency of a received signal from its original value to the intermediate-frequency (IF) value in a superheterodyne receiver.

frequency converter — Also called frequency changer. A circuit, device, or machine that changes an alternating current from one frequency to another, with or without a change in voltage or number of phases. In a superheterodyne receiver, the oscillator and mixer first-detector stages together serve as a frequency converter.

frequency correction — Compensation, by means of an attenuation equalizer, for unequal transmission of various frequencies in a line.

frequency counter — An instrument in which frequency is measured by counting the number of cycles (pulses) occurring during a precisely established time interval.

frequency cutoff — The frequency at which the current gain of a transistor drops 3 dB below the low-frequency gain.

frequency demodulation — Removal of the intelligence from a modulated carrier.

frequency departure — The amount a carrier or center frequency deviates from its assigned value.

frequency deviation — 1. In frequency modulation, the peak difference between the instantaneous frequency of the modulated wave and its carrier frequency. 2. A measure of the output frequency excursion around the carrier caused by modulating the oscillator's tuning input, which produces a frequency-modulated output signal. 3. The measure of the percentage modulation of a frequency-modulated wave. It is the peak difference between the instantaneous frequency of a frequency-modulated wave and the carrier frequency.

frequency-deviation meter — An instrument that indicates the number of hertz a transmitter has drifted from its assigned carrier frequency.

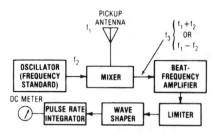

Frequency-deviation meter.

frequency discrimination — A term applied to the operation of selecting a desired frequency or frequencies from a spectrum of frequencies.

frequency discriminator — A circuit that converts a frequency-modulated signal into an audio signal.

frequency distortion — 1. The distortion that results when all frequencies in a complex wave are not amplified or attenuated equally. 2. The unequal amplification of all frequencies over the passband of an amplifier. 3. Distortion in which there is change in the relative magnitudes of the different frequency components of a complex wave, provided that the change is not caused by nonlinear distortion. *See also* frequency response.

frequency distribution — The number of occurrences of particular values plotted against those values.

frequency diversity — *See* frequency-diversity reception.

frequency-diversity reception — Also called frequency diversity. The form of diversity reception that utilizes transmission at different frequencies.

frequency divider — 1. A device delivering an output voltage that is at an integral submultiple or proper fraction of the input frequency. 2. A counter that has a gating structure added that provides an output pulse after a specified number of input pulses are received.

frequency-division data link — A data link in which frequency-division techniques are used for channel spacing.

frequency-division multiplexer — Abbreviated FDM. 1. A device or process for transmitting two or more signals over a different frequency band. 2. A multiplex system in which the total transmission bandwidth is divided into narrower bands, each used for a single, separate channel.

frequency-division multiplexing — Abbreviated FDM. 1. Taking the frequency spectrum of one leased line and subdividing it into a series of low-frequency bands, each of which will transmit the data of an associated low-speed device. 2. The multiplexing technique that assigns to each signal a specific set of frequencies (called a channel) within the larger block of frequencies available on the main transmission path in much the same way that many radio stations broadcast at the same time but can be separately received. 3. The transmission of two or more signals over a common path by using a different frequency band for each signal. 4. Multiplexing by splitting the bandwidth into some number of low-speed channels or subbands. 5. A technique in which a data line's bandwidth is divided into different frequency subchannels. It permits several terminals to share the same line. 6. A technique in which an analog communication channel's bandwidth is divided into frequency subchannels to permit several circuits to share the same channel.

frequency domain — A way of looking at waveforms in terms of the frequency components of the waveforms. An analysis in the frequency domain of a simple sine wave would be described as a pure sine wave of a single frequency. An analysis of the square wave would show that the square wave could be described as a sum of sine waves of different frequencies and magnitudes. In fact any waveform, no matter how complex, may be described as a sum of sine waves in the frequency domain.

frequency doubler — An electronic stage having a resonant output circuit tuned to the second harmonic of the input frequency. The output signal will then have twice the frequency of the input signal.

frequency-doubling transponder — A transponder that doubles the frequency of the interrogating signal before retransmission.

frequency drift — Any undesired change in the frequency of an oscillator, transmitter, or receiver.

frequency-exchange signaling — The method in which the change from one signaling condition to another is accompanied by a decay in amplitude of one or more frequencies and by a build-up in amplitude of one or more other frequencies.

frequency frogging — The interchanging of the frequency allocations of carrier channels to prevent singing, reduce crosstalk, and correct for line slope. It is accomplished by having the modulation in a repeater translate a low-frequency group to a high-frequency group and vice versa.

frequency hopping — Carrier-frequency shifting in discrete increments in a pattern dictated by a code sequence. The transmitter jumps from frequency to frequency within some predetermined set: the order of frequency hops is determined by a code sequence that, in turn, is detected and followed by the receiver.

frequency indicator — A device that shows when two alternating currents have the same phase or frequency.

frequency influence — In a measuring instrument other than a frequency meter, the change, expressed as a percentage of the full-scale value, in the indicated value as a result of a departure of the measured quantity from a specified reference frequency.

frequency interlace — 1. In television, the relationship of intermeshing between the frequency spectrum of an essentially periodic interfering signal and the spectrum of harmonics of the scanning frequencies. Such a relationship minimizes the visibility of the interfering pattern by altering its appearance on successive scans. 2. The method by which color and black-and-white sideband signals are interwoven within the same channel bandwidth.

frequency keying — A method of keying in which the carrier frequency is shifted between two predetermined frequencies.

frequency-measuring equipment — Equipment for indicating or measuring the frequency or pulse-repetition rate of an electrical signal.

frequency meter — 1. An instrument for measuring the frequency of an alternating current. 2. Instrument for measuring the repetition rate of a recurring phenomenon, as the cycles per second of a sinusoidal waveform. 3. An instrument that measures the number of periods per unit of time of a signal in hertz.

frequency-modulated broadcast band — The band of frequencies from 88 to 108 megahertz. It is divided into 100 channels, each 200 kilohertz in width, and set aside for frequency-modulated broadcasting. *See* FM broadcast channel.

frequency-modulated carrier-current telephony — A form of telephony in which a frequency-modulated carrier signal is transmitted over power lines or other wires.

Frequency-division multiplex.

frequency-modulated cyclotron — A cyclotron in which the frequency of the accelerating electric field is modulated in order to hold the positively changed particles in synchronism with the accelerating field despite their much greater mass at very high speeds.

frequency-modulated jamming — A jamming technique in which an rf signal of constant amplitude is varied in frequency about a center value to produce a signal that covers a band of frequencies.

frequency-modulated output — A transducer output that is obtained in the form of a deviation from a center frequency, where the deviation is proportional to the applied stimulus.

frequency-modulated radar — Also called FM radar. A form of radar in which the radiated wave is frequency modulated. The range is measured by beating the returning wave with the one being radiated.

frequency-modulated transmitter — A transmitter in which the frequency of the wave is modulated.

frequency-modulated wave — A carrier wave whose frequency is varied by an amount proportionate to the amplitude of the modulated signal.

frequency modulation — Abbreviated FM. 1. Modulation of a sine-wave carrier so that its instantaneous frequency differs from the carrier frequency by an amount proportionate to the instantaneous amplitude of the modulating wave. Combinations of phase and frequency modulation also are commonly referred to as frequency modulation. *See also* frequency-modulated wave. 2. One of three ways of modifying a sine-wave signal to make it carry information. The sine wave, or carrier, has its frequency modified in accordance with the information to be transmitted. The frequency function of the modulated wave may be continuous or discontinuous. In the latter case, two or more partial frequencies may correspond to one significant condition. 3. A method of transmitting digital information on an analog line by the carrier frequency.

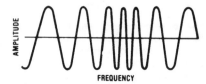

Frequency modulation.

frequency-modulation deviation — The peak difference between the instantaneous frequency of a modulated wave and the carrier or reference frequency.

frequency-modulation frequency modulation — A system in which frequency-modulated subcarriers are used to frequency modulate a second carrier.

frequency-modulation phase modulation — A system in which frequency-modulated subcarriers are used to phase modulate a second carrier.

frequency monitor — An instrument for indicating the amount a frequency deviates from its assigned value.

frequency multiplex — A technique for the transmission of two or more signals over a common path. Each signal is characterized by a distinctive reference frequency or band of frequencies.

frequency multiplier — A device for delivering an output wave whose frequency is a multiple of the input frequency (e.g., frequency doublers and triplers).

frequency offset — 1. The amount by which a frequency lies above or below a reference frequency. For example, if a frequency measures 1.000 001 MHz when compared against a reference frequency of 1.000 000 MHz, then its fractional frequency offset is 1 Hz/MHz, or 1 part in 10^6. 2. Analog-line frequency change, an impairment encountered on a communication line.

frequency-offset transponder — A transponder that changes the interrogating signal frequency by a fixed amount before retransmission.

frequency output (transducer) — An output in the form of frequency that is a function of the applied measurand (e.g., angular speed and flow rate).

frequency overlap — That part of the frequency band which is shared as a result of interleaving.

frequency-prediction chart — A graph that shows the maximum usable frequency, optimum working frequency, and lowest usable frequency between two specific points for various times throughout a 24-hour period.

frequency pulling — A change in oscillator frequency due to a change in the load impedance.

frequency pushing — A source-frequency change caused by a change in electron flow within the source oscillator.

frequency range — 1. In a transmission system, those frequencies at which the system is able to transmit power without attenuating it more than an arbitrary amount. 2. In a receiver, the frequency band over which the receiver is designed to operate, covering those frequencies the receiver will readily accept and amplify. 3. A designated portion of the frequency spectrum.

frequency record — A recording of various known frequencies at known amplitudes, usually for testing purposes.

frequency regulator — A regulator that maintains the frequency of the frequency-generating equipment at a predetermined value or varies it according to a predetermined plan.

frequency relay — A relay that functions at a predetermined value of frequency. It may be an overfrequency or underfrequency relay, or a combination of both.

frequency response — 1. A measure of how effectively a circuit or device transmits the different frequencies applied to it. 2. The portion of the frequency spectrum that can be sensed by a device within specified limits of amplitude error. 3. A graphical characteristic showing relative signal levels at different frequencies with respect to a given reference level. A flat frequency response is one that has a uniform level at all frequencies within a given bandwidth. 4. A measure of the ability of a device to take into account, follow, or act upon a rapidly varying condition, e.g., as applied to amplifiers. 5. The measure of any component's ability to pass signals of different frequency without affecting their relative strengths. This is shown as a graph, or curve, that assumes input signals equally strong at all frequencies and plots their output intensities against a decibel scale. The ideal curve is a straight line. Frequency response may also be stated as a frequency range but with specified decibel limits indicating the maximum deviations from flat response. For instance, 30 to 20,000 Hz ±2 dB means the component will not change the relative intensities of any frequencies within that range by more than 2 dB above or 2 dB below the ideal 0 dB (volume unchanged) point. 6. The range of frequencies over which an amplifier responds within defined limits of amplification (or signal output). 7. The range or band of frequencies to which a unit of electronic equipment will offer essentially the same characteristics.

frequency-response analysis — The use of alternating or pulsating signals to excite a control system so that the response of the system to different frequencies can be ascertained to permit analysis of its operating characteristics.

frequency-response analyzer—An instrument that analyzes the output amplitude of a signal waveform passing through a circuit over a specified band of frequencies.

frequency-response characteristic—The amount by which the gain or loss of a device varies with the frequency.

frequency-response curve—A graphical representation of the way a circuit responds to different frequencies within its operating range.

frequency-response equalization—Also called equalization or corrective equalization. The effect of all frequency-discrimination means employed in a transmission system to obtain the desired overall frequency response.

frequency reuse—A method that allows two different TV channels to be broadcast simultaneously on the same transponder by vertically polarizing one channel and horizontally polarizing the other. Another method of frequency reuse is to space satellites about 4° apart. A TVRO pointed at one satellite will not detect any signal from the other satellite, even if it is operating at the same frequency.

frequency run—A series of tests for determining the frequency-response characteristics of a transmission line, circuit, or device.

frequency-scan antenna—A radar antenna, similar to a phased-array antenna, in which scanning in one dimension is accomplished through frequency variation.

frequency scanning—A technique in which the output frequency is made to vary over a desired range at a specified rate.

frequency selectivity—The degree to which a transducer is capable of differentiating between the desired signal and signals or interference at other frequencies.

frequency-sensitive relay—A relay that operates only when energized with voltage, current, or power within specific frequency limits. A resonant reed relay is one example of this type.

frequency-separation multiplier—A multiplying device in which each variable is split into low-frequency and high-frequency parts that are multiplied separately to obtain results that are added to give the required product. The system makes possible high accuracy and broad bandwidth.

frequency separator—The circuit that separates the horizontal-scanning from the vertical-scanning synchronizing pulses in a television receiver.

frequency shift—1. Pertaining to radio-teletypewriter operation in which the mark and space signals are transmitted as different frequencies. 2. A change in the frequency of a radio transmitter or oscillator. 3. Pertaining to a modulation system in which one radio frequency represents picture black and another represents picture white; frequencies between the two limits represent shades of gray. 4. The frequency difference in a frequency-shift modulation system.

frequency-shift converter—A device that limits the amplitude of the received frequency-shift signal and then changes it to an amplitude-modulated signal.

frequency-shifted keyed filter—Abbreviated FSK filter. A highly selective bandpass filter that passes two closely spaced FSK frequencies corresponding to binary 0s and 1s, while rejecting voice; this allows transmission of digital data over voice-frequency channels.

frequency-shift indicator—In automatic code transmission, a device that designates marks and spaces

by shifting the carrier back and forth between two frequencies instead of keying it on and off.

frequency-shift keying—Abbreviated FSK. 1. A form of frequency modulation in which the modulating wave shifts the output frequency between predetermined values and the output wave has no phase discontinuity. 2. A method of modulating a carrier frequency. A binary 1 shifts the frequency above the center carrier frequency; a binary 0 shifts the frequency below the center carrier frequency. 3. A frequency-modulation method in which the frequency is made to vary at the significant instants as follows: (a) by smooth transitions—the modulated wave and the change in frequency are continuous at the significant instants; (b) by abrupt transitions—the modulated wave is continuous, but the frequency is discontinuous at the significant instants. 4. A form of frequency modulation in which the carrier frequency is made to vary or change in frequency precisely when a change in the state of a transmitted signal occurs. 5. A modulation scheme that shifts between two frequencies to represent a 1 or 0 state of data transmission.

frequency-shift telegraphy—Telegraphy by frequency modulation in which the telegraph signal shifts the frequency of the carrier between predetermined values. There is phase continuity during the shirt from one frequency to the other.

frequency-shift transmission—A method of transmitting the mark and space elements of a telegraph code by shifting the carrier frequency slightly, usually about 800 Hz.

frequency-slope modulation—A method of modulation in which the carrier is swept periodically over the entire band. Modulation of the carrier with a communications signal changes the system bandwidth without affecting the uniform distribution of energy over the band. Thus, the desired information can be recovered from any part of the system bandwidth, and portions of the band that have interference can be filtered out without loss of desired information.

frequency spectrum—The entire range of frequencies of electromagnetic radiations.

frequency splitting—One condition of magnetron operation in which rapid alternation occurs from one mode of operation to another. This results in a similar rapid change in oscillatory frequency and consequent loss of power at the desired frequency.

frequency stability—The ability of electronic equipment to maintain the desired operating frequency.

frequency stabilization—The controlling of the center or carrier frequency so that it does not differ more than a prescribed amount from the reference frequency.

frequency standard—A stable low-frequency oscillator used for frequency calibration. It can generate a fundamental frequency of 50 to 100 kilohertz with a high degree of accuracy. Harmonics of this fundamental are then used as reference points for checking throughout the radio spectrum at 50- or 100-kilohertz intervals.

frequency swing—The instantaneous departure of the emitted wave from the center frequency when its frequency is modulated.

frequency synthesizer—1. A frequency source of high accuracy generally characterized by the fact that the output frequency is composed of two components. The frequency steps, mostly decadic, are derived from the crystal-stabilized frequency standard and the variable frequency of a free-running oscillator, which fills in between these steps. The simplest method of frequency synthesis is to derive standard frequencies from a crystal-controlled frequency by harmonic generation and frequency division. 2. A circuit capable of producing a multitude of

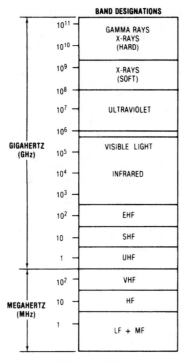

Frequency spectrum.

300,000,000 divided by the frequency in hertz, or 300 divided by the frequency in megahertz.

Fresnel — A little-used unit of frequency equal to 10^{12} hertz.

Fresnel lens — 1. A lens similar in action to a plano-convex lens but made thinner and lighter because of the presence of steps on the convex side. Often the flat side has a rough surface to diffuse the light slightly and thereby smooth the light beams. 2. A lens resembling a structure that is cut into narrow rings and flattened out. If the steps are narrow, the surface of each step is generally made conical and not spherical. Fresnel lenses can be large glass structures, as in lighthouses, floodlights, or traffic signals, or a thin molded plastic plate with fine steps.

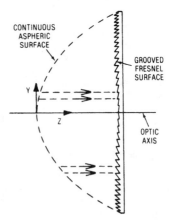

Fresnel lens.

output frequencies from a single input frequency. 3. An rf source that can provide, by external command, any discrete and precise frequency within its range and resolution. The output signal is stabilized to a fixed frequency reference, which may be internal or external to the synthesizer. Primary applications include automatic test equipment, electronic warfare, and communications systems. 4. A system utilizing the phase-locked loop (PLL) principle in conjunction with a programmable digital frequency divider to generate any of a number of discrete frequencies. (May replace a number of crystals or other timing elements with only one timing element.)

frequency-time-intensity — Pertaining to a type of radar display in which frequency, time, and strobe intensity are correlated.

frequency tolerance — 1. The maximum permissible deviation with respect to the reference frequency of the corresponding characteristic frequency of an emission. Expressed in percent or in hertz. 2. The extent to which the carrier frequency of a transmitter may be permitted to depart from the frequency assigned.

frequency translation — The transfer *en bloc* of signals occupying a definite frequency band, such as a channel or group of channels, from one position in the frequency spectrum to another in such a way that the arithmetic frequency difference of the signals within the band is unaltered.

frequency tripler — An amplifier whose output circuit is resonant to the third harmonic of the input signal. The output frequency is three times the input frequency.

frequency-type telemeter — A telemeter that employs the frequency of a periodically recurring electric signal as the translating means.

frequency-wavelength relation — For radio waves, the frequency in hertz is equal to approximately 300,000,000 divided by the wavelength in meters. The wavelength in meters is equal to approximately

Fresnel loss — *See* surface reflection.

Fresnel number — The square of the radius of a lens aperture divided by the product of the focal length and the wavelength. It provides a measure of the importance of diffraction in the image formed by the lens: a small Fresnel number indicates greater diffraction effects.

Fresnel reflection losses — Losses that are incurred at the optical conductor interface due to refractive index differences.

Fresnel reflections — Losses due to reflections at the input and output ends of an optical fiber caused by differences in the reflective indexes of the fiber core and the immersion medium.

Fresnel region — 1. The region adjacent to the region in which the field of an antenna is focused (that is, just outside the Fraunhofer region). 2. The region between an antenna and the beginning of the Fraunhofer region.

Fresnel zone — 1. An area selected in the aperture of a radiating system so that radiation from all parts of the system reaches some point at which it is desired at a common phase within 180°. 2. A circular zone about the direct path between a transmitter and a receiver at each radius such that the distance from a point on this circle to the receiving point has a path length that is some multiple of a half-wavelength longer than the direct path.

fretting — A condition whereby mated surfaces of a connector move slightly and continually expose fresh metal. The exposed metal oxidizes and builds up until electrical continuity of the system is broken.

frictional electricity — Electric charges produced by rubbing one material against another.

frictional error— Applied to pickups, the difference in values measured in percent of full scale before and after tapping, with the measurand constant.

frictional loss— The loss of energy due to friction between moving parts.

frictional machine— Also called a static machine. A device for producing frictional electricity.

friction effects— The difference in resistance or output between readings obtained prior to and immediately after tapping an instrument while applying a constant stimulus. Particularly applicable to potentiometric transducers.

friction error— A change in a reading originally taken in the absence of vibration that occurs after a transducer is tapped or dithered to remove internal friction.

friction-free calibration (transducer)— Calibration under conditions minimizing the effect of static friction often obtained by dithering.

friction tape— A fibrous or plastic tape impregnated with a sticky, moisture-resistant compound that provides a protective covering or insulation.

friendly environment— A software environment in which all software is adequately tested; therefore, one task will not interfere with or cause errors in the execution of another task.

friendly terminal— A terminal that is relatively easy and comfortable to use and whose features are designed with the needs of their operator in mind.

fringe— A unit of linear measurement equal to half the wavelength of thallium green light (approximately 0.01 mil or 0.25 μm). It is used in measuring the depth of diffusion in silicon.

fringe area— The area just beyond the limits of the reliable service area of a television transmitter. Signals are weak and erratic, requiring the use of high-gain directional receiving antennas and sensitive receivers for satisfactory reception.

fringe effect— The extension of the flux in a field beyond the edges of a gap, as electric flux at the edges of the plates of a capacitor or magnetic flux at the edges of an air gap in a magnetic circuit.

fringe howl— A squeal or howl heard when some circuit in a receiver is on the verge of oscillation.

frit— 1. Metallic powders fused in a glass binder. 2. To melt and fuse together, as a set of electrical contacts that are subject to repeated discharges. 3. Glass composition ground up into a fine powder form and used in thick-film compositions as the portion of the composition that melts upon firing to give adhesion to the substrate and hold the composition together. 4. In fiber optics, finely ground glass used to join glass to metal or other glasses. Also called solder glass, it may or may not devitrify during temperature cycles. 5. A term used interchangeably with "glass," as in frit- or glass-sealed packages such as CERDIP and CERPACK.

fritting— A type of contact erosion in which an electrical discharge makes a hole through the contact film and produces molten matter that is drawn through the hole by electrostatic forces and then solidifies and forms a conducting bridge.

frogging— At an intermediate carrier repeater, changing low-group frequencies on the input to high-group frequencies on the output, and vice versa, as a means of reducing crosstalk.

frogging repeater— A carrier repeater that has provisions for frequency frogging to make possible the use of a single multipair voice cable without excessive crosstalk.

Frolich high-temperature breakdown theory— A thermal mechanism for breakdown in which electrons rather than ions carry the current. The necessary number of conduction electrons is produced by thermal excitation of the electrons in impurity and imperfection levels.

Frolich low-temperature breakdown theory— Also called the high-energy criterion. Similar to the Von Hippel theory, except that an electron energy distribution is assumed. Only a few electrons on the high-energy tail of the distribution must gain the necessary critical energy.

front contact— A movable relay contact that closes a circuit when the associated device is operated.

front end— 1. The section of a tuner or receiver that is used to select the desired station from either the AM, FM, or TV band, and to convert the rf signal to intermediate frequency. To do its job properly, a front end requires a high-gain, low-noise rf stage, a mixer, and an oscillator. The degree to which a desired station can be received without interference and without adding noise is expressed by sensitivity and signal-to-noise ratio. 2. An auxiliary computer system that performs network-control operations, releasing the host computer system to process data. 3. In a process control system, the input end at which raw signals are converted to digital information for further processing.

front-end overload— Distortion or interference caused by an FM tuner's inability to handle strong signals from a nearby transmitter. Front-end overload can cause a station to appear at more than one place on the dial.

front-end processor— In a computer, a processor in charge of interfacing with a user or a process. May perform preprocessing (translations) and file handling. The main processor performs interpretation, execution, or number crunching.

front-end rejection— A dB expression of the relative ability of a receiver network to reject signals outside the tuned bandwidth.

front porch— In the composite television signal, that portion of the synchronizing signal (at the blanking or black level) preceding the horizontal-sync pulse at the end of each active horizontal line. The standard EIA signal is 1.27 microseconds in duration.

front projection— A system of picture enlargement using an opaque reflective screen. The projector and viewers are on the same side of the screen.

front-surface mirror— An optical mirror on which the reflecting surface is applied to the front of the mirror instead of the back.

front-to-back ratio— Also called front-to-rear ratio. 1. The ratio of power gain between the front and rear of a directional antenna. 2. Ratio of signal strength transmitted in a forward direction to that transmitted in a backward direction. For receiving antennas, the ratio of received-signal strength when the source is in the front of the antenna to the received-signal strength when the antenna is rotated 180°. 3. The ratio between a cardioid microphone's sensitivity to sounds arriving from the front and from the rear, a measure of its directionality. 4. On a printed circuit board, the location of the printed pattern on one side relative to the printed pattern on the opposite side. 5. A measure of the directivity of an antenna that is based on the difference between the strengths of signals received from the antenna front and those received from the back. The difference usually is expressed in decibels (dB). For example, a front-to-back ratio of 40 dB indicates that the power of signals received from the antenna front will be 10,000 times greater than those received from the back. Generally, the higher the rating in dB, the greater the directivity of the antenna.

front-to-back registration— On a printed circuit board, the location of the printed pattern on one side relative to the printed pattern on the opposite side.

front-to-rear ratio— *See* front-to-back ratio.

fruit — *See* fruit pulse.

fruit pulse — Also called fruit. Radar beacon system video display of a synchronous beacon return that results when several interrogator stations are located within the same general area. Each interrogator receives its own synchronous reply as well as many synchronous replies resulting from interrogation of the airborne transponders by other ground stations.

F-scan — *See* F-display.

F-scope — *See* F-display.

FSK — *See* frequency-shift keying.

FSK filter — *See* frequency-shift keyed filter.

FSS — Abbreviation for fixed-satellite service. A radiocommunication service between earth stations at given fixed positions via one or more satellites.

f-stop — Also called f-number and f-system. Refers to the speed or ability of a lens to pass light. It is calculated by dividing the focal length of the lens by its diameter.

FTP — Abbreviation for File Transfer Protocol. 1. A way to download remote files over the Internet. *See* File Transfer Protocol. 2. A protocol used for the transfer of data files consisting of one or more segments.

fuel cell — An electrochemical generator in which the chemical energy from the reaction of air (oxygen) and a conventional fuel is converted directly into electricity. A fuel cell differs from a battery in that it uses hydrocarbons (or some derivative such as hydrogen) for fuel, and it operates continuously as long as fuel and air are available.

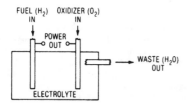

Fuel cell.

full adder — A circuit that provides an output equal to the sum of three binary-digit inputs (two digits to be added and a carry digit from a previous stage). Sum and carry outputs are provided.

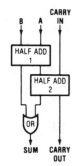

TRUTH TABLE

CARRY IN	B	A	CARRY OUT	SUM
0	0	0	0	0
0	0	1	0	1
0	1	0	0	1
0	1	1	1	0
1	0	0	0	1
1	0	1	1	0
1	1	0	1	0
1	1	1	1	1

Full adder.

full-differential input — Input configuration of an analog amplifier in which neither input is connected directly to ground or to other channels. This configuration has better noise immunity than a single-ended input.

full duplex — Abbreviated FDX. 1. A method of operation of a communication circuit in which each end can simultaneously transmit and receive. When used on a radio circuit, duplex operation requires two frequencies. 2. A data circuit that is capable of both sending and receiving data simultaneously. 3. A mode of data transmission that is the equivalent of two paths — one in each direction simultaneously. 4. Simultaneous two-way transmission.

full-duplex operation — 1. Simultaneous operation in opposite directions in a telegraph system. 2. A method of operation that provides simultaneous two-way communications between two points.

full excursion — The application of a measurand, in a controlled manner, over the entire range of a transducer.

full/full duplex — A protocol for a multidrop line that permits transmission from a master location to a slave site; the master location can also simultaneously receive a transmission from another slave site on that line.

fullhouse — A multichannel radio control system for model airplanes in which all controls work to allow the model to fly a complete flight pattern.

full load — The greatest load a piece of equipment is designed to carry under specified conditions.

full-period allocated circuit — A communication link (allocated circuit) assigned exclusively for the use of previously defined users at two or more terminal points.

full-pitch winding — A type of armature winding in which the number of slots between the sides of the coil equals the pole-pitch measure in the slots.

full-range speaker unit — *See* dual cone.

full scale — The total interval over which an instrument is intended to be operated. Also, the output from a transducer when the maximum rated stimulus is applied to the input.

full-scale cycle — The complete range of an instrument, from minimum reading to full scale and back to minimum reading.

full-scale error — The difference between the actual voltage or current that produces full-scale deflection of an electrical indicating instrument and the rated full-scale input of the instrument.

full-scale output — The algebraic difference in electrical output between the maximum and minimum values of measurand over which an instrument is calibrated. When the sensitivity slope is given by any other line than the end-point sensitivity, full scale expresses the algebraic difference, for the span of the instrument, which is calculated from the slope of the straight line from which nonlinearity is determined.

full-scale range — The continuum of input values (and resultant readings) over which a meter has been designed to function before exceeding the maximum scale or digital reading (or before entering the overrange region in digital meters providing overranging).

full-scale sensitivity — *See* full-scale output.

full-scale value — The largest value of applied electrical energy that can be indicated on a meter scale. When zero is between the ends of the scale, the full-scale value is the arithmetic sum of the values of the applied electrical quantity corresponding to the two ends of the scale. On a suppressed meter, the full-scale value is the largest value of applied electrical input less the smallest value of applied electrical energy input.

full-scale value of an instrument — The largest actuating electrical quantity that can be indicated on the scale; or, for an instrument having its zero between the ends of the scale, the sum of the values of the actuating electrical quantity corresponding to the two ends.

full subtractor—A device that can obtain the difference of two input bits, subtract a borrow from these two bits, and provide difference and borrow outputs.

full-track recording—1. Defines the track width as essentially equal to the tape width. Applies to inch-wide (or less) tape only. 2. A recorded signal occupying the full width of a $\frac{1}{4}$-inch (6.35-mm) tape. 3. Recording monophonically on one track whose width is essentially the same as the tape's. *See* track configuration.

full-track tape recording—A professional tape recording standard in which one track occupies the entire width of a $\frac{1}{4}$-inch (6.35-mm) wide recording tape.

full-voltage starting—*See* across-the-line starting.

full-wave rectification—The process of inverting the negative half-cycle of current of an alternating input so that it flows in the same direction as the positive half-cycle. A way to accomplish this is to use four diodes placed in a bridge configuration.

full-wave rectifier—A circuit, electron tube, or other device that uses both positive and negative alternations in an alternating current to produce direct current.

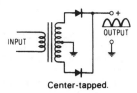

Center-tapped.

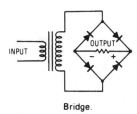

Bridge.

Full-wave rectifier.

full-wave rectifier tube—A tube containing two sets of rectifying elements to provide full-wave rectification.

full-wave vibrator—A vibrator having an armature that moves back and forth between two fixed contacts, so as to change the direction of the direct current flow through a transformer at regular intervals and thereby permit voltage step-up by the transformer. Used in battery-operated power supplies for mobile and marine radio equipment.

function—1. A quantity of value that depends on the value of one or more other quantities. 2. A specific purpose of an entity, or its characteristic action. 3. A means of referring to a type or sequence of calculations within an arithmetic statement. 4. An integral unit of computational work (add, subtract, sine, cosine, read, write). It cannot be subdivided further without destroying its nature—it is at its optimum granularity.

functional block—Also called molecular electronic circuit. A more or less homogeneous combination of several solid-state materials to perform a desired circuit function. The quartz crystal is often represented as a natural expression of the functional block combination of resistance, capacitance, and inductance.

functional board tester—A tester that verifies the correct logical operation of a logic board by applying test patterns at the board-edge connector. The output responses usually are monitored at the connector, although some test points may be used.

functional design—The specification of the working relations between the parts of a system in terms of their characteristic action.

functional device—*See* integrated circuit.

functional diagram—A diagram showing the functional relationships among the parts of a system.

functional electronic block—A fabricated device serving a complete electronic function, such as amplification, without other individual components or conducting wires except those required for input, power, and output.

functional insulation—The insulation necessary for the proper functioning of a device and for basic protection against electrical shock hazard.

functional interface—An interface between the operating characteristics of equipment such as electrical power and signal characteristics, signal timing, and environmental coupling.

functional language—A programming language that uses exclusively expressions to be evaluated.

functional parts—Discrete items defined by functional characteristics and dimensions that are not repairable with the use of spare parts (e.g., resistors, capacitors, diodes, potted transformers, permanently sealed batteries, etc.).

functional testing—Testing to determine whether the device under test reacts correctly to inputs, qualitatively.

functional tests—The application of functional input vectors and the corresponding responses that assure proper operation of a digital IC.

functional trimming—Trimming of a circuit element (usually resistors) on an operating circuit to set a voltage or current on the output.

function characteristic—The relationship between the output ratio and the shaft position of a precision potentiometer.

function codes—Codes that appear in tape or cards to operate machine functions, such as carriage returns, space, shift, skip, tabulate, etc.

function digit—A coded instruction used in a computer for setting a branch order to link subroutines into the main program.

function generator—1. A device capable of generating one or more desired waveforms. 2. An electrical network that can be adjusted to make its output voltage (or current) a desired function of time. Used in conjunction with analog computers. 3. A character, conic, or vector generator in a display controller. Some display controllers have additional function generators, such as sweep generators for displaying television-type pictures on the display screen. 4. A signal source that can produce a variety of output waveshapes that, if completely free from distortion, would represent a readily definable mathematical relationship between the time base and the output voltage. Thus, square, triangle, sawtooth, and equivalent waveshapes constitute the basic outputs of a function generator. If the instrument incorporates adequate signal-modifying circuitry, it can alter triangle or square waveforms to closely approximate sine-haversine and havertriangle or haversquare outputs. With circuitry for amplitude hold, internal gating, slope control, dc offset, signal inversion, and horizontal and limit control, it becomes simple to produce trapezoids, forward or backward skewed sines, or almost any other combination or variation of internally generated waveforms derived from basic mathematical functions.

functioning time — In a relay, the time that elapses between energization and operation or between deenergization and release.

functioning value — In a relay, the value of applied voltage, current, or power at which operation or release occurs.

function keyboard — An input device for an interactive display terminal that consists of various function keys.

function keys — 1. Data energy devices usually programmed to initiate or terminate a particular function or process in the graphic system. Function keys may or may not generate interrupts when depressed and/or released. Some systems will generate an interrupt upon depression and release, some just on depression, and others do not generate any interrupt but rather set a bit in a register. The latter type of function key must have the register periodically read by the host computer to determine when a key is depressed. The register bit is cleared when the key is released. Function keys may be mounted on the alphanumeric keyboard or on a separate box on the console. 2. Keys on a keyboard or a control panel that, when depressed, activate a particular machine function. The functional operation can usually be programmed or defined dynamically. 3. Keys on a keyboard that cause an operation, such as LETTERS, FIGURES, CARRIAGE RETURN, and LINE FEED, but not the printing of a character.

function switch — 1. A network of systems having a number of inputs and outputs. When signals representing information expressed in a certain code are applied to the inputs, the output signals will represent the input information in a different code. 2. In adapters or control units, the switch that determines whether the system plays as a monophonic or stereophonic unit; it may parallel the speakers or cut out one or the other, switch amplifiers from one speaker to the other, reverse channels, etc.

function table — 1. A table of values for a mathematical function. 2. A hardware device or a computer program that translates one representation of information into another. 3. A routine by means of which a dependent variable is derived from the values of independent variables. 4. A subroutine that can be used either to decode multiple inputs into a single output or to encode a single input into multiple outputs.

function test — A check for correct logical operation of the device. Basically, a function test is a device truth table verification.

function unit — A device that can store a functional relationship and release it continuously or sporadically.

fundamental — *See* fundamental frequency.

fundamental component — The fundamental frequency component in the harmonic analysis of a wave.

fundamental frequency — 1. The principal component of a wave; i.e., the component with the lowest frequency or greatest amplitude. It is usually taken as a reference. 2. The lowest frequency component of a complex sound or electrical signal.

fundamental group — In wire communications, a group of trunks by means of which each local or trunk switching center is connected to a trunk switching center of higher rank on which it "homes." The term also applies to groups that interconnect zone centers.

fundamental harmonic — The harmonic component with the lowest frequency.

fundamental mode — 1. Of vibration, the mode having the lowest natural frequency. 2. The waveguide mode that has the lowest critical frequency. *See* dominant mode.

fundamental piezoelectric crystal unit — A unit designed to use the lowest resonant frequency for a particular mode of vibration.

fundamental tone — 1. In a periodic wave, the component corresponding to the fundamental frequency. 2. In a complex tone, the component tone of lowest pitch.

fundamental units — Units arbitrarily selected as the basis of an absolute system of units.

fundamental wavelength — The wavelength corresponding to the fundamental frequency. In an antenna, the lowest resonant frequency of the antenna alone, without inductance or capacitance.

fungusproof — To chemically treat a material, component, or unit to prevent the growth of fungus spores.

furcation coupling — The mixing of signals from several separate optical fibers by passing them through a common single fiber rod, thus obtaining a signal containing all the components of the several signals. The mixing of several colors can take place in this manner.

furnace soldering — Soldering by placing clamped and fluxed assemblies, on which solder has been positioned, in a furnace or oven.

fuse — 1. A protective device, usually a short piece of wire but sometimes a chemical compound, that melts and breaks the circuit when the current exceeds the rated value. 2. To equip with a fuse. 3. A device used for protection against excessive currents. Consists of a short length of fusible metal wire or strip that melts when the current through it exceeds the rated amount for a definite time. Placed in series with the circuit it is to protect. 4. A replaceable protective device that will break the current when the current exceeds the capacity of the fuse.

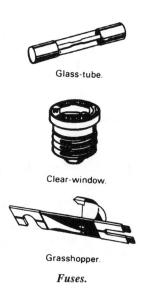

Glass-tube.

Clear-window.

Grasshopper.

Fuses.

fuse alarm — A circuit that produces a visual and/or audible signal to indicate a blown fuse.

fuse block — An insulating base on which fuse clips or other contacts for holding fuses are mounted.

fuse box — An enclosed box containing fuse blocks and fuses.

fuse clips — Contacts on the fuse support for connecting the fuse holder into the circuit.

fuse cutout — An assembly consisting of a fuse support and a fuse holder that may or may not include the fuse link.

fused arrays of fibers — Optical fibers fused together to form solid, vacuum-tight assemblies in the form of slabs or rods. Disks or rectangular shapes having ground and polished surfaces perpendicular to the fiber lengths will transmit image information from one surface to the other. Rods of this type, either straight or tapered, will transmit images from one end of the rod to the other.

fused coating — A metallic coating (usually tin or solder alloy) that has been melted and solidified, forming a metallurgical bond to the base material.

fused conductors — Individual strands of heavy tinned copper wire stranded together and then bonded together by induction heating.

fused disconnect — Generally an air-break switch with a fusing unit in the blade. Used for opening and closing high-voltage circuits.

fused junction — *See* alloy junction.

fused quartz — 1. A glasslike insulating material having exceptional resistance to the action of heat and acid. 2. Crystal quartz that is melted at a white heat and cooled to form an amorphous glass. It is not birefringent, and the refractive index is much lower than that of crystal quartz. Fused quartz of optical quality can be prepared by suitable techniques.

fused semiconductor — The junction formed by recrystallization on a base crystal from a liquid phase of one or more components and the semiconductor.

fuse filler — Material placed within the fuse tube to aid circuit interruption.

fuse holder — A device for supporting a fuse and providing connections for its terminals.

Fuse holder.

fuse link — In a fuse, the current-carrying portion, which melts when the current exceeds a predetermined value.

Fusestat — Trade name for a time-delay fuse similar to a Fusetron. It has a sized base requiring a permanent socket adapter, which prevents insertion of a fuse or Fusetron of an incorrect rating.

fuse terminal block — A block designed for cartridge fuses. Some are fitted with a neon or incandescent fuse failure indicator and may be mixed with feedthrough models on the same standard assembly vail.

Fusetron — Trade name for a screw-plug time-delay fuse that permits up to 50-percent overload for short periods without blowing.

fuse tube — The insulated tube enclosing a fuse link.

fuse unit — An assembly consisting of a fuse link mounted in a fuse holder, which contains parts and materials essential to the operation of the fuse link.

fuse wire — A wire made from an alloy that melts at a relatively low temperature.

fusible line — The current-carrying portion of a fuse, which is designed to melt when the rated current is exceeded for a specified length of time. Some fuses are made so that the fusible line can be replaced.

fusible link — A type of programmable read-only memory integrated circuit in which circuits form bit patterns by being "blasted" open (by a destructive current) or left closed (intact).

fusible-link diode-matrix integrated circuit memory — A programmable read-only memory (PROM) in which shorting fusible links are blown by applying overcurrent to configure a diode matrix with the proper program.

fusible-link readout memory — A large semiconductor array of prediffused cells interconnected by a fixed metallization pattern that can be tailored to a particular need simply by burning out selected interconnections.

fusible resistor — A resistor designed to protect a circuit against overload by opening when the current drain exceeds the design limits.

fusible wire — A wire used in fire-alarm circuits. It is made of an alloy with a low melting point.

fusing — 1. The melting of a metallic coating (usually electrodeposited) followed by solidification. 2. Melting and cooling two or more powder materials together so that they bond together in a homogeneous mass.

fusion — 1. Also called atomic fusion or nuclear fusion. The melting of atomic nuclei, under extreme heat (millions of degrees), to form a heavier nucleus. The fusion of two nuclei of light atoms is accompanied by a tremendous release of energy. 2. Melting, usually as the result of interaction of two or more materials.

fusion welding — Welding technique used to join metals by melting and fusing them at high temperatures using an electric arc or combustible gases.

fuzz — Also called fuzz tone. An intentional distortion of the natural tone of an electric guitar.

fuzz box — A special-effects device, usually used with electric guitars, which creates a great amount of harmonic distortion. The lack of clarity that results is called fuzz. A typical fuzz box operates by overdriving a preamplifier and then attenuating the preamplifier output so the output signal level is not necessarily higher than the output level.

fuzz tone — *See* fuzz.

fuzzy computer — A specially designed computer that uses fuzzy logic. Fuzzy logic computers are designed for artificial intelligence applications.

fuzzy logic — 1. A branch of logic that uses degrees of membership in sets rather than a strict true/false membership. 2. A kind of logic using graded or qualified statements rather than ones that are strictly true or false. The results of fuzzy reasoning are not as definite as those derived by strict logic, but they cover a larger field of discourse. 3. A method of handling imprecision and uncertainty that attaches various measures of credibility to propositions. A form of logic used in some expert systems and other artificial intelligence applications.

fuzzy sets — Sets that do not have a crisply defined membership, but rather allow objects to have grades of membership from 0 to 1.

fV — Letter symbol for femtovolt (10^{-15} volt).

G

g — Also called G force. Symbol for the acceleration of a free-falling body due to the earth's gravitational pull. Equal to 32.17 feet per second per second.

G — 1. Symbol for conductance, a grid of a vacuum tube, a generator, or ground. 2. Letter symbol for giga (10^9). 3. Abbreviation for the gate of a field-effect transistor.

g-a — Abbreviation for ground-to-air. Communication with airborne objects from the ground.

GaAs FET — *See* gallium arsenide field-effect transistor.

GA coil — A coil wound with air spaces between its turns and layers to reduce the capacitance.

gage — Also spelled gauge. 1. An instrument or means for measuring or testing. By extension, the term is often used synonymously with *transducer*. 2. A system for specifying wire size. The American wire gage (AWG), also known as Brown & Sharpe gage, is used for copper. An increase of three gage numbers doubles area and weight, and halves dc resistance.

gage pressure — 1. A differential pressure measurement using the ambient pressure as a reference. 2. A pressure in excess of a standard atmosphere at sea level (i.e., 14.7 pounds per square inch or 1.033×10^4 kg/m^2).

gage pressure transducer — A pressure transducer that uses ambient pressure as the reference pressure. The sensing element is normally vented to the ambient pressure.

gain — Also called transmission gain. 1. Any increase in power when a signal is transmitted from one point to another. Usually expressed in decibels. Widely used for denoting transducer gain. 2. The ratios of voltage, power, or current with respect to a standard or previous reading. 3. Any increase in the strength of an electrical signal, as takes place in an amplifier. Gain is measured in terms of decibels or number of times of amplification; for example, 6 dB (a gain of 2) increases an input voltage to an output twice as large. 4. The change in source-drain current per unit change in gate voltage. Thus, higher gain gives faster devices. 5. The degree to which a signal's amplitude is increased. The amount of amplification realized when a signal passes through an amplifier or repeater, normally measured in decibels. 6. The ratio of output power to the input power for a system or component. 7. Change in signal level due to processing functions that increase the magnitude of the signal. Examples include signal amplification in a radar receiver; processing gain in the processor; and antenna gain, a result of the directivity of the pattern.

gain-bandwidth product — 1. The product of the closed-loop gain of an operational amplifier and its corresponding closed-loop bandwidth. This product is often constant in operational amplifiers. 2. The product of the gain of an active device and a specified bandwidth. For an avalanche photodiode, the gain-bandwidth product

is the gain times the frequency of measurement when the device is biased for maximum obtainable gain.

gain block — A single stage of gain or a cascaded series of gain stages.

gain control — A device for varying the gain of a system or component.

gain function — A transfer function that relates either a pair of voltages or a pair of currents.

gain margin — The amount of gain change of an operational amplifier at 180° phase-shift angle frequency that would produce instability.

gain nonlinearity — The degree to which gain of an amplifier or instrument varies over the range of permissible input levels for a given gain setting, generally expressed as a percentage of deviation from the desired gain.

gain-sensitivity control — *See* differential gain control.

gain stability — The extent to which the sensitivity of an instrument remains constant with time. (The property reported in specifications should be instability, which is the maximum change in sensitivity from the initial value over a stated period of time under stated conditions.)

gain-time control — *See* sensitivity-time control.

galactic noise — All noise that originates in space as a result of radiation of celestial bodies other than the sun.

galena — A bluish-gray, crystalline form of lead sulfide often used as the crystal in a variable-crystal detector.

gallium arsenide field-effect transistor — Abbreviated GaAs FET. A field-effect transistor with a reverse-biased Schottky-barrier gate fabricated on a gallium arsenide substrate. Roughly equivalent to a silicon MOSFET, GaAs FETs are depletion-mode devices. Because charge carriers reach approximately twice the velocity as in silicon, for a given geometry a given gain can be reached at about twice the frequency. Also called GaAs MESFET, for metal epitaxial semiconductor FET.

gallium arsenide (GaAs) injection laser — A laser system consisting of a planar pn junction within a single crystal of gallium arsenide. The pair of parallel, semireflective end faces produces a Fabry-Perot resonant cavity, whereas the other faces are sawed to suppress all except the modes that propagate between the end faces. The lasing occurs in the modes that have the maximum optical gain and the least optical loss.

galloping ghost — *See* proportional control, 2.

galvanic — 1. An early term for current resulting from chemical action, as distinguished from electrostatic phenomena. 2. Describing any substance from which, or through which, direct current occurs as a result of chemical action.

galvanic anode — A source of emf for cathodic protection provided by a metal less noble than the one to be protected (i.e., magnesium, zinc, or aluminum as used for cathodic protection of steel).

galvanic cell — An electrolytic cell capable of producing electric energy by electrochemical action.

galvanic corrosion — Accelerated electrochemical corrosion produced when one metal is in electrical contact with another more noble metal, both being in the same corroding medium, or electrolyte, with a current between them. (Corrosion of this type usually results in a higher rate of solution of the less noble metal and protection of the more noble metal.)

galvanic current — An electrobiological term for unidirectional current such as ordinary direct current.

galvanic series — A list of metals and alloys arranged in the order of their relative potentials (ability to go into solution) in a given environment. The table of potentials is arranged with the anodic, or least noble, metals at one end and the cathodic, or more noble metals, at the other. (For marine use, the potentials listed are related to a seawater environment.)

Galvanic Series

(Anodic)
Magnesium
Zinc
Cadmium
Steel or iron
Cast iron
Chromium iron (active)
Lead-tin solders
Lead
Tin
Nickel
Brasses
Copper
Bronzes
Copper-nickel alloys
Monel
Silver
Graphite
Gold
Platinum
(Cathodic)

galvanizing — The coating of steel with zinc to retard corrosion.

galvanoluminescence — The emission of radiant energy produced by the passage of an electrical current through an appropriate electrolyte in which an electrode made of certain metals such as aluminum or tantalum has been immersed.

galvanometer — 1. An instrument for measuring an electric current. This is done by measuring the mechanical motion produced by the electromagnetic or electrodynamic forces set up by the current. 2. An instrument for detecting or measuring a small electric current by movements of a magnetic needle of a coil in a magnetic field. 3. An instrument used to measure the presence, amount of, and direction of an electric current.

galvanometer constant — The factor by which a certain function of a galvanometer reading must be multiplied to obtain the current in ordinary units.

galvanometer lamp — The lamp that illuminates a movable mirror of the galvanometer in some spectrophotometers. The angle at which the light is reflected to the galvanometer scale depends on the amount of current through the galvanometer coil.

galvanometer recorder (for photographic recording) — A combination of a mirror and coil suspended in a magnetic field. A signal voltage, applied to the coil, causes a light beam from the mirror to be reflected across a slit in front of a moving photographic film.

galvanometer shunt — A resistor connected in parallel with a galvanometer to increase the range of the instrument. The resistor limits the current to a known fraction and thus prevents excessive current from damaging the galvanometer.

galvanometric controller — A temperature indicator that has been converted to a temperature controller. The indicator operates directly off the sensor's input signal. The controller portion detects the mechanical position of the pointer and varies the output to keep the pointer at the desired temperature.

game controller — Joysticklike device that plugs into a computer and make it easier to play games.

game paddle — A game controller that moves an object in one of two directions.

game port — Interface on an adapter card used to connect a game peripheral and the computer.

game theory — A branch of mathematics that aims to analyze various problems of conflict by abstracting common strategic features for study in theoretical models termed games because they are patterned on actual games such as bridge and poker.

gamma — 1. A unit of magnetic intensity, equal to 10^{-5} oersted. 2. A number indicating the degree of contrast in a photograph, facsimile reproduction, or received television picture. This quantity is the exponent of that power law which is used to approximate the curve of output magnitude versus input magnitude over the region of interest.

gamma correction — Introduction of a nonlinear output-input characteristic for the purpose of changing the effective value of gamma.

gamma ferric oxide — The magnetic constituent of practically all present-day tapes, in the form of a dispersion of fine acicular particles within the coating.

gamma radiation — A highly penetrating electromagnetic disturbance (photons) that emanates from the nucleus of an atom. This type of radiation travels in waveform much like X-rays or light, but it has a shorter wavelength of approximately 1 angstrom or 10^{-7} millimeter.

gamma radiography — Radiography using the emission of gamma rays to form an image of the structure penetrated by the radiation.

gamma rays — 1. The emission from certain radioactive substances. They are electromagnetic radiations similar to X-rays, but with a shorter wavelength from about 10^{-12} to 10^{-9} centimeter. 2. The spontaneous emittance of electromagnetic radiation by the nucleus of certain radioactive elements during their quantum transition between two energy levels.

gang — To mechanically couple two or more variable capacitors, switches, potentiometers, or other components together so they can be operated from a single control knob.

gang capacitor — Also called gang tuning capacitor. Two or more variable tuning capacitors mounted on the same shaft and controlled by a single knob, but each capacitor tuning a different circuit. Thus, more than one circuit can be tuned simultaneously by a single control.

gang control — Simultaneous control of several similar pieces of apparatus with one adjustment knob or other control.

ganged — Describing a group of devices, such as variable capacitors or resistors, that are coupled mechanically so that all are adjusted simultaneously from one control.

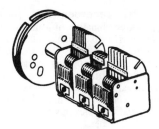

Gang capacitor.

ganged tuning — Simultaneous tuning of two or more circuits with a single mechanical control.

gang punch — To punch information that is identical or constant into all of a group of punch cards.

gang switch — A number of switches mechanically coupled for simultaneous operation, but electrically connected to different circuits. In one common form, two or more rotary switches are mounted on the same shaft and operated by a single control.

gang tuning capacitor — *See* gang capacitor.

gap — 1. In a magnetic circuit, the portion that does not contain ferromagnetic material (i.e., an air space). 2. The space between two electrodes in a spark gap. 3. The tiny space between the pole pieces of a record or playback head, across which the magnetic fields appear (when the tape is running) for transduction into alternating electrical signals. 4. A space between radar-antenna radiation lobes where the field strength is low, resulting in incomplete radar coverage. 5. A space in which the radiation fails to meet minimum coverage requirements, either because the space is not covered or because the minimum specified overlap is not obtained. 6. An interval of space or time used to indicate automatically the end of a word, record, or file of data on a tape. 7. Space between two records or two blocks of information on a magnetic memory that is usually set to predetermined value, such as all zeros. It allows rewriting blocks in slightly expanded or reduced format, due to speed variations of the drive.

gap arrester — A type of lightning arrester comprising a number of air gaps in series between metal cylinders or cones.

gap coding — 1. A means for inserting periods of no transmission in a system in which transmission is normally continuous. The spacing and duration of the periods of silence from the code are variable. 2. Subdividing the response of a transponder into long and short groups of pulses (analogous to Morse code) for purposes of recognition.

gap depth — The dimension of the gap in a recording head, measured perpendicular to the surface of the head.

gap factor — In a tube that employs electron-accelerating gaps (a traveling-wave tube), the ratio of the maximum energy gained (expressed in volts) to the maximum gap voltage.

gap filler — 1. A lightweight radar set used to fill in gaps in the coverage pattern of an early-warning radar net. 2. An auxiliary radar antenna used to fill in gaps in the pattern of the main radar antenna.

gap filling — Electrical or mechanical rearrangement of an antenna array, or the use of a supplementary array, to produce lobes where gaps previously occurred.

gap insulation — Insulation wound in a gap.

gap length — The dimension of the gap of a recording head, measured from one pole face to the other. In longitudinal recording, the gap length is the dimension of the gap in the direction of tape travel.

gap loss — 1. The loss in output attributable to the finite gap length of the reproduce head. The loss increases as the wavelength decreases, amounting to approximately 4 dB when the wavelength is equal to twice the gap length, and subsequently increases rapidly toward the complete extinction of output when the wavelength is approximately equal to 1.15 times the gap length. 2. In fiber optics, a power loss caused by deviation from optimum spacing between the elements to the fiber junction, fiber-to-fiber junction, or fiber-to-element junction. Gap loss is expressed in decibels.

gap motor — A spark-gap drive motor.

gap scatter — In a computer, the deviation from true vertical alignment of the gaps of the magnetic readout heads for the several parallel tracks. The mechanical misalignment of a head's read/write gaps in the direction of tape travel.

gap width — The dimension of the gap, measured in the direction parallel to the head surface and pole faces. The gap width of the record head governs the track width. The gap widths of reproduce heads are sometimes made appreciably less than those of the record heads to minimize tracking errors.

garbage — 1. In a computer, a slang term for unwanted and meaningless information carried along in storage. Sometimes called hash. 2. Undecipherable or meaningless sequences of characters produced in computer output or retained within storage. 3. An informal term for corrupted data.

garbage collection — Technique for collecting empty spaces in a mass memory and then compacting them.

garbage in, garbage out — *See* GIGO.

garble — Faulty transmission, reception, or encoding that renders the message incorrect or unreadable.

garnet maser — *See* ferromagnetic amplifier.

garter spring — In facsimile, the spring fastened around the drum to hold the record sheet or copy in place.

gas — One of the three states of matter. An aeriform fluid having neither independent shape nor volume, but tending to spread out and occupy the entire enclosure in which it is placed. Gases are formed by heating a liquid above its boiling point.

gas amplification — Ratio of the charge collected to the charge liberated by the initial ionizing of the gas in a radiation counter.

gas-amplification factor — Ratio of radiant or luminous sensitivities with and without ionization of the gas in a gas phototube.

gas cell — A cell whose action is dependent on the absorption of gases by the electrodes.

gas cleanup — The tendency of many gas-filled tubes to lose their gas pressure and hence become inoperable. This occurs when the ions of gas are driven at high velocity into the metal parts or the glass envelope of the tube, where they form stable compounds and are lost as far as the tube is concerned.

gas current — 1. The current in the grid circuit of a vacuum tube when the gas ions within the tube are attracted by the grid. 2. A flow of positive ions to an electrode, the ions having been produced as a result of gas ionization by an electron current between other electrodes. 3. The positive-ion current created in an electron tube as the result of the collisions between electrons and residual gas molecules.

gas detector — An instrument used to indicate the concentration of harmful gases in the air.

gas diode — A tube having a hot cathode and an anode in an envelope containing a small amount of an inert gas or vapor. When the anode is made sufficiently positive, the electrons flowing to it collide with gas atoms and ionize them. As a result, the anode current is much greater than that for a comparable vacuum diode.

gas-discharge device — A device utilizing the conduction of electricity in a gas due to movements of electrons and ions produced by collision.

gas-discharge display — A device containing an inert gas that gives off orange light when a high voltage is applied to break down (ionize) the gas. Gas-discharge displays have a seven-segment format.

gas-discharge laser — *See* gas laser.

gas-electric drive — A self-contained power-conversion system comprising an electric generator driven by a gasoline engine. The generator in turn supplies power to the driving motor or motors.

gaseous discharge — The state of a gas or mixture of gases in which a conduction current can be maintained by ionization. The ionization results from collisions between electrons and atoms or molecules of the gas, the energy being furnished by an external source such as an electric field.

gaseous electronics — The field of study involving the conduction of electricity through gases and a study of all atomic-scale collision phenomena.

gaseous tube — An electronic tube into which a small amount of gas or vapor is introduced after the tube has been evacuated. Ionization of the gas molecules during operation of the tube affects its operating characteristics.

gaseous-tube generator — A power source comprising a gas-filled electron-tube oscillator and a power supply, plus associated controls.

gas-filled cable — A coaxial or other type of cable containing gas under pressure, which serves as insulation and prevents moisture from entering.

gas-filled lamp — A tungsten-filament lamp containing nitrogen or an inert gas such as argon.

gas-filled radiation-counter tube — A gas tube used for the detection of radiation. It operates on the principle that radiation will ionize a gas.

gas-filled tube rectifier — A rectifier tube in which a unidirectional flow of electrons from a heated electrode ionizes the inert gas within the tube. In this way, rectification is accomplished.

gas focusing — Also called ionic focusing. The use of an inert gas to focus the electron beam in a cathode-ray tube. Beam electrons ionize the gas molecules, forming a core of positive ions along the beam path that tends to attract beam electrons, making the beam more compact.

gasket — A device used to retain fluids under pressure or seal out foreign matter. Also a static seal to reduce electromagnetic interference or susceptibility.

gasket-sealed relay — A relay in an enclosure sealed with a gasket.

gas laser — Also called gas-discharge laser. A laser in which the active medium is a discharge in gas, vapor, or mixture within a glass or quartz tube that has a Brewster window at each end. This gas can be excited by a high-frequency oscillator or by a direct current between electrodes inside the tube to pump the medium so as to obtain a population inversion.

gas magnification — The increase in current through a phototube due to ionization of the gas within the tube.

gas maser — A maser in which the microwave electromagnetic radiation interacts with the molecules of a gas such as ammonia. Use is limited chiefly to highly stable oscillator applications, as in atomic clocks.

gas noise — Electrical noise produced by erratic motion of gas molecules in gas or partially evacuated vacuum tubes.

gas-phase laser — A continuous-wave device for general experimental work with coherent light. It employs a resonator made up of a fused-silica plasma tube 60 cm long having internal, multilayer, dielectric-coated confocal reflectors of optical-grade fused silica.

gas photocell — A photoemissive cell having an inert gas added to its envelope. Subsequent ionization of the gas increases the responsivity of the photocell.

gas phototube — A phototube into which a quantity of gas has been introduced, usually to increase its sensitivity.

gas ratio — The ratio of the ion current in a tube to the electron current that produces it.

gassiness — The presence of unwanted gas in a vacuum tube, usually in relatively small amounts. It is caused by leakage from outside the tube or by evolution from its inside walls or elements.

gassing — 1. Evolution of a gas from one or more electrodes during electrolysis. 2. The production of gas in a storage battery when the charging current is continued after the battery has been completely charged.

gassy — Having operating characteristics that are impaired as a result of an excessive amount of gas inside its envelope, as a vacuum tube.

gassy tube — *See* soft tube.

gas-tight connection — A contact or joint that prevents contaminant gases from reaching the contact area.

gaston — A modulator that produces a random-noise modulation signal from a gas tube. It may be attached to any standard aircraft communications transmitter to provide a counterjamming modulation.

gas tube — 1. A partially evacuated electron tube containing a small amount of gas. Ionization of the gas molecules is responsible for the current. 2. An electron tube whose current is affected by the pressure on the gas or vapor contained in the tube.

gas-tube relaxation oscillator — A relaxation oscillator in which the abrupt discharge is provided by the breakdown of the gas in the tube.

gas X-ray tube — An X-ray tube in which electron emission from the cathode is produced by bombarding it with positive ions.

gate — 1. A circuit having two or more inputs and one output, the output depending upon the combination of logic signals at the inputs. There are four gates, called

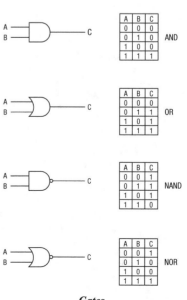

Gates.

AND, OR, NAND, and NOR. The definitions assume positive logic is used. In computer work, a gate is often called an AND circuit. 2. A signal used to trigger the passage of other signals through a circuit. 3. One of the electrodes of a field-effect transistor; it is analogous to the base of a transistor or the grid of a vacuum tube. Symbol: G. 4. An output element of a cryotron. 5. A circuit in which one signal (usually a square wave) switches another signal on or off. 6. To control the passage of a pulse or signal. 7. Voltage-actuated control terminal of an MOS transistor. 8. A circuit that admits and amplifies or passes a signal only when a gating (triggering) pulse is present. 9. A digital logic element usually with one output and several inputs, where the output is a function of a combination of the inputs. 10. A single logic function, such as NAND, NOR, AND, OR, XOR, or NOT.

gate array — 1. A geometric pattern of basic gates contained in one chip. It is possible to interconnect the gates during manufacture to form a complex function that may be used as a standard production. 2. A cellular arrangement of logic elements with a custom interconnection. Arrays offer complexities of several hundred to more than 2000 gates, performance that equals or betters that of standard TTL or ECL, and as little as 12-week turnaround from design to prototype. 3. Also called a semicustom large-scale integrated or very large-scale integrated circuit (VLSI). A matrix of unconnected transistors and resistors that become a dedicated array only after they are interconnected by one or more masking and interconnection steps. The term *gate array* is a holdover from the days when functional logical cells, equivalent to small- and medium-scale integrated circuits, were made by first forming equivalent gates. 4. Cells of transistors stacked in columns and separated by conductive channnels. The cells are selectively connected to the channels in the final stages of manufacturing to provide logic functions. Like PLAs, gate arrays are less space efficient and more expensive than dedicated circuits. 5. A semicustom IC consisting of a regular arrangement of gates that are interconnected through one or more layers of metal to provide custom functions. Generally, gate arrays are preprocessed up to the first interconnect level so they can be quickly processed with final metal to meet a customer's specified logic function.

gate circuit — A circuit that passes a signal only when a gating pulse is present.

gate-controlled switch — Also called gate turn-off switch. A three-junction, three-terminal, solid-state device constructed very much like a silicon-controlled rectifier except that it has a turn-off ability, which is controlled by a negative current pulse applied to the gate.

gate-controlled turn-off time — 1. The time interval during which a thyristor is switching from the on state to the off state as a result of the application of a gate trigger pulse. The interval is usually measured from the 10-percent point on the gate pulse to the instant at which the principal voltage has increased to 90 percent of the off-state voltage. 2. The time interval during which a thyristor is switching from the off state to the on state as a result of the application of a gate trigger pulse. The interval is usually measured from the 10-percent point on the gate pulse to the instant at which the principal voltage has dropped to 10 percent of its initial value. The turn-on time is the sum of the delay time and the rise time.

gate-controlled turn-on time — The time interval between the 10-percent rise of the gate pulse and the 90-percent rise of the principal current pulse during switching of a thyristor from the off state to the on state.

gate current — Instantaneous current flowing between the gate and cathode of a silicon-controlled rectifier.

gate current for firing — Gate current required to fire a silicon-controlled rectifier when the anode is at a fixed dc voltage with respect to the cathode and with the device at stated temperature conditions.

gated-beam detector — A single-stage FM detector using a gated-beam tube.

gated-beam tube — A five-element tube in which the electrons flow in a beam between the cathode and plate. A small increase in voltage on the limiter grid will cut off the plate current, and further increases will have a negligible effect on it.

gated buffer — A low-impedance inverting driver circuit that may be used as a line driver for pulse differentiation or in multivibrators.

gated flip-flop — A flip-flop that has a steering circuit that prevents both flip-flop outputs from becoming 0 at the same time. Alternating-current input pulses must be used to prevent the flip-flop from oscillating. This situation can arise if both input lines are made high simultaneously.

gated sweep — Sweep in which the duration as well as the starting time is controlled to exclude undesired echoes from the indicator screen.

gated transistor — A transistor in which a gate electrode covers the emitter and collector junctions. This allows the application of an electric field at the surface of the base region.

gate electrode — A control electrode to which trigger pulses are applied.

gate equivalent circuit — A basic unit for describing relative digital circuit complexity. The number of gate equivalent circuits is that number of individual logic gates that would have to be interconnected to perform the same function.

gate generator — A circuit or device used to produce one or more gate pulses.

gate impedance — The impedance of a gate winding in a magnetic amplifier.

gate multivibrator — A rectangular-wave generator designed to produce a single positive or negative gate voltage upon being triggered and then to become inactive until the arrival of the following trigger pulse.

gate nontrigger current — The highest gate current that will not cause a thyristor to switch from the off state to the on state.

gate nontrigger voltage — The maximum gate voltage that will not cause a thyristor to switch from the off state to the on state.

gate power dissipation — The power dissipated between the gate and cathode terminals of a silicon-controlled rectifier.

gate-producing multivibrator — A rectangular-wave generator that produces a single positive or negative gate voltage only when triggered by a pulse.

gate pulse — A pulse that enables a gate circuit to pass a signal. The gate pulse generally has a longer duration than the signal to ensure time coincidence.

gate region — Of a field-effect transistor, a region associated with the gate electrode in which the electric field due to the control voltage is effective.

gate resistance — The dc resistance between gate and source terminals of a field-effect transistor at specified gate voltage with drain short-circuited to the source.

gate signal — That signal generated by some form of delay circuit required in connection with beam switching, automatic following, the application of AGC to a selected echo, and many other purposes.

gate terminal — 1. The terminal in a field-effect transistor electrically connected to the electrode associated with the region in which the electric field, due to the control voltage, is effective. 2. A terminal to or from which only control current, usually called the gate current, flows.

gate-to-source leakage current — A current through gate and source terminals of a field-effect transistor at specified drain circuit conditions.

gate trigger current — In a controlled rectifier, the minimum gate current, for a given anode-to-cathode voltage, required to switch the rectifier from the off state to the on state.

gate trigger voltage — In a controlled rectifier, the gate voltage that produces the gate trigger current.

gate tube — A tube that does not operate unless two signal voltages, derived from two independent circuits, are applied simultaneously to two separate electrodes.

gate turn-off current — In a controlled rectifier, the minimum gate current, for a given collector current in the on state, required to cause the rectifier to switch off. Not all thyristors can be turned off by the gate.

gate turn-off switch — *See* gate-controlled switch.

gate turn-off voltage — The gate voltage required to produce the gate turn-off current in a thyristor. Not all thyristors can be turned off by the gate.

gate voltage — 1. The voltage across the gate-winding terminals of a magnetic amplifier. 2. The instantaneous voltage between gate and cathode of a silicon-controlled rectifier with anode opening. 3. The voltage between a gate terminal and a specified main terminal.

gateway — 1. A link, or bridge, from one type of communication network to another. 2. A machine or set of machines used to relay packets from one network to another network. *See also* router. 3. Equipment used to connect network architectures that use different protocols by providing protocol translation. 4. A link from one computer system to a different computer system. Some gateways are the Internet and Stocklink.

gateway station — An earth station that provides mobile units with access to the public telephone network.

gate winding — The reactor winding that produces the gating action in a magnetic amplifier.

gating — 1. Selecting those portions of a wave that exist during certain intervals or that have certain magnitudes. 2. Applying a rectangular voltage to the grid or cathode of a cathode-ray tube, to sensitize it during the sweep time only. 3. Application of a specific waveform to perform electronic switching. *See also* blanking.

gating circuit — A circuit that operates as a selective switch and allows conduction only during selected time intervals or when the signal magnitude is within specified limits.

gating pulse — A pulse that modifies the operation of a gate circuit.

gauge — *See* gage.

gauge pressure — *See* gage pressure.

gauss — 1. The centimeter-gram-second electromagnetic unit of magnetic induction. One gauss represents one line of flux (one maxwell) per square centimeter. Letter symbol: G.

Gaussian distribution — Also called normal distribution. A density function of a population that is bell-shaped and symmetrical and that is completely defined by two independent parameters: the mean and the standard deviation.

Gaussian elimination — A method for solving systems of linear equations based on manipulation of the matrix representing those equations.

Gaussian function — A mathematical function used in designing a filter to pass a step function with zero overshoot and minimum rise time (similar to a Bessel-function filter).

Gaussian noise — 1. Unwanted electrical disturbances or perturbations described by a probability density function that follows a normal law of statistics. This normal distribution is the well-known symmetrical bell-shaped density function of a population that is completely defined by two independent parameters, viz, the mean and the standard deviation. The Gaussian (normal) amplitude distribution is of fundamental significance in statistical theory, describing numerous natural phenomena. The central-limit theorem of statistics states, in essence, that the sum of a number of independent random variables approaches the Gaussian distribution as the number of said variables increases, regardless of the distribution of the individual variables. A significant facet of Gaussian random noise is that it exceeds its positive root-mean-square amplitude and attains twice that value for 16 and 20 percent of the time, respectively. 2. A random-noise signal whose frequency components have a Gaussian distribution centered on a specified frequency. 3. Noise whose amplitude is characterized by the Gaussian distribution, e.g., white noise, ambient noise, and hiss.

Gaussian random vibration — *See* random vibration.

Gaussian waveform — In pulse-compression systems, a waveform that produces very low transmitted side lobes.

gaussmeter — An instrument that provides direct readings of magnetic field density (flux density) by virtue of the interaction with an internal magnetic field.

Gauss's theorem — The summation of the normal component of the electric displacement over any closed surface is equal to the electric charge within the surface.

GB — Abbreviation for gigabyte, one billion bytes.

GCA — Abbreviation for ground-controlled approach.

GCI — Abbreviation for ground-controlled interception.

GCS — Abbreviation for gate-controlled switch.

GCT or Gct — Abbreviation for Greenwich civil time.

G-display — Also called G-scan or G-scope. In radar, a rectangular display in which a target appears as a laterally centralized blip on which wings appear to grow as the target approaches. Horizontal and vertical aiming errors are indicated by horizontal and vertical displacement of the blip.

gear — An element shaped like a toothed wheel that engages one or more similar wheels. The energy transmitted can be stepped up or down by making the driven gears of different sizes.

geared synchro system — A system in which the transmitting and receiving synchros turn at a higher speed than the input and output shafts. Geared systems are generally used when a high degree of accuracy is required.

gearmotor — A train of gears and a motor used for reducing or increasing the speed of the driven object.

Geiger counter — Also called Geiger-Mueller or G-M counter. 1. A radiation detector that uses a Geiger-Mueller counter tube, an amplifier, and an indicating device. The tube consists of a thin-walled gas-filled metal cylinder with a projecting electrode. Nuclear particles enter a window in the metal cylinder and temporarily ionize the gas, causing a brief pulse discharge. These pulses, which appear at the projecting electrode, are amplified and indicated visibly or audibly. 2. Gas-chamber-type radiation counter in which the chamber operates in avalanche region for high amplification and sensitivity.

Geiger-Mueller counter — *See* Geiger counter.

Geiger-Mueller counter tube — A radiation-counter tube designed to operate in the Geiger-Mueller region.

Geiger-Mueller region — Also called Geiger region. The voltage interval in which the pulse size is independent of the number of primary ions produced in the initial ionizing event.

Geiger-Mueller threshold — Also called Geiger threshold. The lowest voltage at which all pulses produced in the tube by any ionizing event are of the same size regardless of the size of the primary ionizing event. This threshold is that start of the Geiger region in which the counting rate does not substantially change with applied voltage.

Geiger region — *See* Geiger-Mueller region.

Geiger threshold — *See* Geiger-Mueller threshold.

Geissler tube — A gas-filled dual-electrode discharge tube that glows when electric current passes through the gas.

gel — A material composed of a solid held in a liquid.

genemotor — A type of dynamotor having two armature windings. One winding serves as the driving motor and operates from the vehicle battery. The other winding functions as a high-voltage dc generator for operation of mobile equipment.

general address — Group of characters included in the heading of a message to cause routing of the message to all addresses included in the general address category.

general background lighting — Overall lighting level in a room or area, exclusive of local lighting.

general class license — A license issued by the FCC to amateur radio operators who are able to send and receive code at the rate of 13 words per minute and who are familiar with general radio theory and practice. Holder enjoys all authorized amateur privileges except those reserved for higher license classes.

generalized network — A set of elements, commonly called nodes, that are interconnected in some way. In some cases it is sufficient to assign some meaning to the nodes and interconnections for the network to have some useful purpose. For example, a PERT network is simply a planning model of a complex set of tasks to be performed. Each node represents the completion of a task. The interconnecting lines represent precursor and successor relationships among tasks. Such models are commonly used to determine the critical path — the most complex sequence of events that will complete the task.

general light — The total light or light level in a room or area, including all sources.

general-purpose computer — A computer designed to solve a wide variety of problems, the exact nature of which may have been unknown before the computer was designed.

general-purpose digital computer — A digital computer designed to solve a large variety of problems; that is, a computer that can be adapted to a large class of applications (as opposed to a computer designed specifically to control a manufacturing process). A typical general-purpose digital computer consists of four subsystems: (1) input/output, which permits communication with the outside world; (2) memory, which stores data

and instructions; (3) central processing unit (CPU), or arithmetic unit, which performs the arithmetic and data processing operations; and (4) control, which ties all of the subsystems together so that they operate in a fully automated way.

general-purpose interface bus — *See* GPIB.

general-purpose motor — A motor of 200 hp or less and 450 rpm or more, rated for continuous operation, having standard ratings, and suitable for use without restriction to a particular application.

general-purpose relay — A term that covers a wide variety of electromechanical relays that typically are of the common coil-and-armature type, the most common styles having the ability to switch either ac or dc with dc ratings from 2 to 10 amperes. These relays are usually employed in switching power and can be found in business machines, process controls, fractional-horsepower motor controls, and many other applications. They may be plug-in, bracket, stud, or screw mounted. Plug-in choices are square, octal, miniature, and multiple-pin. A relay that is adaptable to a variety of applications.

general rate — The amount of time taken by the creation of electron-hole pairs in a semiconductor.

general register — One of a specified number of internal addressable registers in a CPU that can be used for temporary storage, as an accumulator, an index register, a stack pointer, or for any other general-purpose function.

general routine — A computer routine designed to solve a general class of problems, but when appropriate parametric values are supplied, it specializes in a specific problem.

generated noise — In potentiometric transducers, the noise that is attributable to causes such as the generation of emf when dissimilar metals are rubbed against each other, or the emf resulting from the thermocouple effects at points where dissimilar metals are joined.

generating electric field meter — Also called a gradient meter. A device for measuring the potential gradient at the surface of a conductor. A flat conductor is alternately exposed and then shielded from the electric field to be measured. The resultant current in the conductor is then rectified and used as the measure.

generating magnetometer — Also called earth inductor. A magnetometer that measures a magnetic field by the amount of emf generated in a coil rotated in the field.

generating station — An installation that produces electric energy from chemical, mechanical, hydraulic, or some other form of energy.

generating voltmeter — Also called a rotary voltmeter. A device that measures voltage. A capacitor is connected across the voltage, and its capacitance is varied cyclically. The resultant current in the capacitor is then rectified and used as a measure.

generation — The number of dubbing steps between a master recording and a given copy of the master. Thus, a second-generation dub is a copy of a copy of the original master.

generation data group — A collection of successive data sets that are historically related.

generation rate — The time rate of creation of electron-hole pairs in a semiconductor.

generator — Symbolized G. Also called dynamo. 1. A machine that converts the mechanical energy of a spinning rotor into electrical energy. 2. An electronic device that converts dc voltage to alternating current of the desired frequency and waveshape. 3. Any device that generates electricity. 4. In computer operation, a routine for producing specific routines from specific input parameters and skeletal coding. 5. A machine that

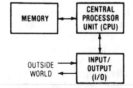

General-purpose digital computer.

converts mechanical energy into electric energy. In its commonest form, a large number of conductors are mounted on an armature that is rotated in a magnetic field produced by field coils. 6. A vacuum-tube oscillator or any other nonrotating device that generates an alternating voltage at a desired frequency when energized with dc power or low-frequency ac power. Such generators are used to produce large amounts of rf power, such as for high-frequency heating and ultrasonic cleaning. 7. A circuit that generates a desired repetitive or nonrepetitive waveform, such as a pulse generator.

generator efficiency — 1. In a generator, the ratio between the power required to drive the generator and the output power obtained from it. 2. In a thermoelectric couple, the ratio of the electrical power output to the thermal power input. It is an idealized efficiency assuming perfect thermal insulation of the thermoelectric arms.

generator field control — Regulation of the output voltage of a generator by control of the voltage that excites the field winding of the generator.

generator lock — See genlock.

generator voltage regulator — A regulator that maintains or varies the voltage of a synchronous generator, capacitor motor, or direct-current generator at or within a predetermined value.

generic package — A collection of software items from which more than one installation-dependent specific package may be generated by a software manufacturing process. Generic packages may be in source-code form (generic source package) or in object-code form (generic object package).

genlock — Abbreviation for generator lock. Also abbreviated GENLOCK. 1. When two composite video signals are mixed, or one is fed as input to another, their vertical, horizontal and subcarrier signals must be synchronized exactly or with a fixed offset. Genlock is achieved by means of locking the originating signal with the second with a digital or analog phase-locked loop circuit. 2. A process of sync generator locking, that makes it possible for two sync generators to run at the same frequency and phase. Used in video processing.

geodesic — The shortest line between two points on a given surface.

geometric distortion — In television, any geometric dissimilarity between the original scene and the reproduced image.

geometric mean — The square root of the product of two quantities.

geometric symmetry — Filter response in which there is mirror-image symmetry about the center frequency when frequency is plotted on a logarithmic scale. (This is the natural response of many electrical circuits.) See also arithmetic symmetry.

geometry — The shape or configuration of a die or IC.

geophysical cable — Cable used in exploring for underground oil deposits.

george box — An amplitude-sensitive device employed in an intermediate-frequency amplifier. It rejects jamming signals of insufficient amplitude to operate its circuits; however, jamming signals having sufficient amplitude are not affected.

geostationary — Refers to a geosynchronous satellite angle with zero inclination so that the satellite appears to hover over one spot on the earth's equator. A satellite in the Clarke belt is geostationary.

geostationary orbit — Also called Clarke orbit, in honor of Arthur C. Clarke, who first described it. This circular orbit above the equator is precisely the altitude (35,863 km or 22,238 miles) at which any size satellite will revolve around the earth once every 24 hours.

From the ground below it thus appears parked in space overhead, and, from above, one-third of the earth's surface can always be seen. Television satellites are separated by 4° intervals on this orbit to avoid mutual interference.

geosynchronous — Within the Clarke circular orbit above the equator. For a planet the size and mass of the earth, this point is 22,238 miles above the surface.

geosynchronous satellite — A satellite positioned at the proper distance from the earth and moving at the proper relational speed to appear stationary to an observer on earth.

germanium — Symbol Ge. A brittle, grayish-white metallic element having semiconductor properties. Widely used in transistors and crystal diodes. Its atomic number is 32.

germanium detector — A type of photoconductive detector in which germanium, usually doped with boron, gallium, and indium, serves as a semiconductor and can detect frequencies up to and beyond 100 micrometers.

germanium diode — A semiconductor diode in which a germanium crystal pellet is used as the rectifying element.

germanium transistor — A transistor in which germanium is the semiconducting material.

German silver — Usually called nickel silver. A silverish alloy of copper, zinc, and nickel.

getter — An alkali metal introduced into a vacuum tube during manufacture. It is fired after the tube has been evacuated, to react chemically with and eliminate any remaining gases. The getter then remains inactive inside the tube. The silvery deposit sometimes seen on the inside of the glass envelope is due to getter firing.

gettering — A semiconductor manufacturing process that aims to remove defects from the neighborhood of the devices, either removing them completely from the silicon wafer or transporting them into regions where they will have no effect on device performance. Defects can be introduced at almost any time during processing, and the defects can take on many forms. Consequently, a large number of different types of gettering processes are presently used. Several different gettering processes may be used during the fabrication of one particular device.

G force — See g.

g-g — Abbreviation for ground-to-ground. Communication between two points on the ground.

ghost — 1. A spurious video image resulting from an echo. 2. An undesired duplicate image in a television picture, fainter and to one side of the normal picture, due to multipath transmission. See ghost image.

ghost image — Also called ghost. An undesired duplicate image offset somewhat from the desired image as viewed on a television screen. It is due to a reflected signal traveling over a longer path and, hence, arriving later than the desired signal. It may be eliminated by the use of a directional antenna array that receives signals over only one path.

ghosting — The appearance of multiple TV images, which is usually caused by reception of a signal via two different paths.

ghost mode — A waveguide mode in which there is a trapped field associated with an imperfection in the waveguide wall. A ghost mode can cause difficulty in a waveguide operated near the cutoff frequency of a propagation mode.

ghost pulse — See ghost signal.

ghost signal — Also called ghost pulse. An unwanted signal on the screen of a radar indicator. Echoes that experience multiple reflections before reaching the receiver are an example. The term also is applied to a reflected television signal. See ghost image.

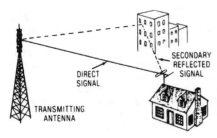

Ghost signal.

giant grid — An extensive regional or national system of backbones and networks.

giant ties — *See* interconnection, 1.

Gibson girl — A portable, hand-operated transmitter used by pilots forced down at sea.

GIF — Abbreviation for Graphics Interchange Format. A standard color image format commonly encountered on the Internet. Other common formats are TIFF, PICT, and JPEG.

giga- — A prefix meaning one billion, or 10^9.

gigacycle — One kilomegacycle, or one billion cycles. An obsolete term. The currently preferred term is gigahertz.

gigahertz (GHz) — One billion (10^9) cycles per second. Used to replace the more cumbersome and obsolete term *kilomegacycle*.

gigawatt — One thousand megawatts (10^9 watts). Letter symbol: GW.

GIGO or gigo — 1. An acronym formed from the phrase "garbage in, garbage out." It is used to describe a computer whose operation is suspect. 2. A term used to describe data moving into and out of a computer system. If the input is bad (garbage in), then the output will also be bad (garbage out).

gigohm — One thousand megohms (10^9 ohms).

gilbert — A cgs unit of the magnetomotive force required to produce one maxwell of magnetic flux in a magnetic circuit of unit reluctance; 1 gilbert = $10/4\pi$ ampere-turn. Letter symbol: Gb.

gilbert per centimeter — The practical cgs unit of magnetic intensity. Gilberts per centimeter are the same as oersteds.

Gill-Morrell oscillator — A retarding-field oscillator in which the oscillation frequency depends not only on the electron transit time within the tube, but also on the associated circuit parameters.

gill selector — A slow-acting telegraph sender and calling key for selective signaling.

gimbal — A mechanical frame having two perpendicularly intersecting axes of rotation.

gimmick — 1. A capacitor with a value of a few picofarads, improvised by twisting together two insulated wires. 2. Length of twisted two-conductor cable used as a variable-capacity load, in which the capacity is varied by untwisting and separating the individual conductors.

gimp — A slang name given to the extremely flexible wire that was used in telephone cords and similar equipment. This wire cannot be directly soldered to, as it is a metallic cloth-type material.

Giorgi system — *See* mksa electromagnetic system of units.

glass — In fiber optics, an amorphous transparent or translucent brittle material usually made by fusion of silica, soda ash, lime, and salt cake or similar materials. Used as a fiber-optic cable, glass offers resistance to high temperatures. It is not subject to corrosion, and eliminates fire hazards and problems caused by short circuits.

glass-ambient technology — The technique by which glass is applied directly to the surface of a semiconductor material. Typically, glass is placed on the surface of a microelectronic device by means of pyrolytic deposition, vapor deposition, or the firing of a glass powder to the surface.

glass binder — The glass powder added to a resistor or conductor ink to bind the metallic particles together after firing.

glassbreak vibration detector — A vibration-detection system that employs a contact microphone attached to a glass window to detect cutting or breakage of the glass.

glass electrode — In electronic pH measurement, an electrode used for determining the potential of a solution with respect to a reference electrode. The calomel type is the most common.

glassivation — Also called passivation. 1. A method of transistor passivation by a pyrolytic glass-deposition technique, whereby silicon semiconductor devices, complete with metal contact systems, are fully encapsulated in glass. 2. The deposition of glass on a chip to give protection to underlying device junctions. 3. A process in which a dielectric material is diffused over the entire wafer to provide mechanical and environmental protection for the circuits. 4. A method of semiconductor passivation by coating the element with a pyrolytic glass deposition.

glass-plate capacitor — A high-voltage capacitor in which the metal plates are separated by sheets of glass for dielectric. The complete assembly is generally immersed in oil.

glass-to-metal seal — An airtight seal between glass and metal parts of an electron tube, made by fusing together a special glass and special metal alloy having nearly the same temperature coefficients of expansion.

glass tube — A vacuum or gaseous tube that has a glass envelope.

glazed substrate — Ceramic substrate with a glass coating to effect a smooth and nonporous surface.

glide path — The approach path used by an aircraft making an instrument landing.

glide-path localizer — In an aircraft instrument-landing system, the part that indicates the altitude of the plane and creates a glide path for a blind landing.

glide-path transmitter — A transmitter that produces signals for vertical guidance of aircraft along an inclined surface that extends upward from the desired point of ground contact.

glide slope — A radio beam used by pilots to determine the altitude of the aircraft during a landing.

glide-slope facility — A radio transmitting facility that provides the glide-slope signals.

glidetone — A device used in electronic music that produces a continuous shift in the frequency of an audio signal.

G-line — 1. A round wire coated with a dielectric and used to transmit microwave energy. 2. A signal insulated wire that can be strung in the open and used as a surface-wave radio-frequency transmission line. A G-line acts like a coaxial cable without the outer tubular conductor.

glint — Also called glitter. 1. A distorted radar-signal echo that varies in amplitude from pulse to pulse because the beam is being reflected from a rapidly moving object such as an airplane propeller. 2. An electronic-countermeasures technique in which the scintillating effect of shuttered or rotating reflectors is used to degrade the tracking or seeking functions of an enemy weapons system.

glissando — A tone that changes smoothly from one pitch to another.

glitch — 1. A form of low-frequency interference appearing as a narrow horizontal bar moving vertically through the television picture. This is also observed on an oscilloscope at the field or frame rate as an extraneous voltage pip moving along the signal at approximately the reference-black level. 2. An unwanted transient condition. 3. An undesired pulse or burst of noise that causes crashes and failures in computers. A small pulse of noise is called a snivitz. 4. A temporary or random error, or a problem or malfunction in hardware, such as a malfunction that can be caused by a power surge. 5. A spike caused by the skew (difference in turn-on/turn-off time) of switches or logic. Glitches are a troublesome source of error in high-speed D/A converters and are most prevalent at the midscale switching location, when all digital input bits are switching. Glitch energy is specified in picovolt-seconds, which describes the area under the voltage-time curve at its worst-case occurrence.

glitter — See glint.

global beam — 1. An INTELSAT antenna downlink pattern covering a third of the earth's surface. These patterns are bore-sighted at the middle of an ocean to provide services to nations all the way around the ocean basin. 2. A broad pattern of signal radiation from a satellite that covers one-third of the earth's surface. (Type of beam used by INTELSAT satellites.)

Global Systems for Mobile Communications — Originally called the Groupe Speciale Mobile. Abbreviated GSM. The digital cellular standard for Europe.

globule test — A solderability test specifically for component leads. The time required for a globule of solder to completely wet around a component lead is measured and recorded and then compared against a known standard. This particular test requires a certain amount of human evaluation.

glossmeter — A photoelectric instrument for determining the gloss factor of a surface (i.e., the ratio of light reflected in one direction to the light reflected in all directions).

glow discharge — A discharge of electricity through a gas in an electron tube. It is characterized by a cathode glow resulting from a space potential much higher than the ionization potential of the gas in the vicinity of the cathode.

glow-discharge microphone — A microphone in which the sound waves cause corresponding variations in the current forming a glow discharge between two electrodes.

glow-discharge tube — A gas tube that depends for its operation on the properties of a glow discharge.

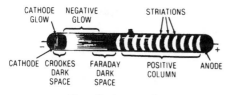

Glow-discharge tube.

glow-discharge voltage regulator — A gas tube used for voltage regulation. The resistance of the gas within the tube varies in step with the voltage applied across the tube.

glow lamp — 1. A lamp containing a small amount of gas or vapor. Current between the two electrodes ionizes the gas and causes the lamp to glow but does not provide rectification. Neon gives a red-orange glow, mercury vapor a blue glow, and argon a purple glow. 2. Gas-discharge tube serving as a concentrated source of light whose brightness varies in proportion to current flow. When an audio-frequency signal is combined with the lamp current, the brightness of the glow discharge varies according to the audio-frequency signal variations. 3. Glow-discharge type of tube whose light brightness is proportional to the current passing through the tube; used for photographic recording of facsimile signals. 4. A lamp in which the ionization of the inert gas contained in it produces a glow in the space close to the negative electrode.

glow potential — The voltage at which a glow discharge begins in a gas-filled electronic tube as the voltage is gradually increased.

glow-tube rectifier — Also called a point-plane rectifier. A cold-cathode gas-discharge tube that provides a unidirectional current flow.

glow switch — An electron tube used in some fluorescent-lamp circuits. It contains two bimetal strips that are closed when heated by the glow discharge.

glue-line heating — An arrangement of electrodes designed to heat a thin film of material having a high loss factor between alternate layers of materials having a low loss factor.

g_m — Symbol for the mutual conductance or transconductance of a vacuum tube.

G-M counter — Abbreviation for Geiger-Mueller counter.

GMT or Gmt — Abbreviation for Greenwich mean time.

gobo — A dark mat used to shield the lens of a television camera from stray lights.

gold — A very soft, ductile material that is noted for its resistance to corrosive media. It is used primarily as a coating or plating.

gold-bonded diode — A semiconductor diode in which a preformed whisker of gold contacts an n-termination substrate as the junction is formed by millisecond electrical pulses.

gold doping — 1. A technique used to control the lifetime of minority carriers in a diffused-mesa transistor. Gold is diffused into the base and collector regions to reduce the storage time. 2. A process sometimes used in the manufacture of integrated circuits, in which gold is diffused into the semiconductor material, resulting in higher operating speeds.

gold-leaf electroscope — An apparatus comprising two pieces of gold leaf joined at their upper ends and suspended inside a glass jar. When a charge is applied to the terminal connected to the leaves, they spread apart due to repulsion of the like charges on them.

Goldschmidt alternator — An early radio transmitter. It is a rotating machine employing oscillating circuits in connection with the field and the armature to introduce harmonics in the generated fundamental frequency. Interaction between the stator and rotor harmonics gives a cumulative effect and thereby provides very high radio frequencies.

goniometer — 1. In a radio-range system, a device for electrically shifting the directional characteristics of an antenna. 2. An electrical device for determining the azimuth of a received signal by combining the outputs of individual elements of an antenna array in certain phase relationships.

go/no-go test — A test that determines whether a unit under test is functioning in accordance with specifications, but does not perform any diagnostic tests to determine the cause of an incorrect output.

Gopher — 1. A menu-based system for organizing and distributing information on the Internet, which allows users to browse or download files and directories. Simpler to use but similar in functionality to FTP. A key feature is the ability to include menu items that connect the user to other Gopher servers. Developed at the University of Minnesota, and partially named after their mascot. 2. Internet databases that can be accessed by the World Wide Web (WWW) as well as other Gopher clients.

Gopherspace — All Gopher sites are at some point interconnected, and this network is known as Gopherspace. Gopherspace results from the ability to link different Gopher sites together.

goto circuit — A circuit capable of sensing the direction of current. It can be used in majority logic circuits, in which the output is either positive or negative depending on whether the majority of the inputs is positive or negative.

goto pair — Two tunnel diodes connected in series in a way such that one is in the reverse tunneling region when the other is in the forwarded conduction region. This arrangement is used in high-speed gate circuits.

governed series motor — A motor used with teletypewriter equipment. It has a governor for regulating the speed.

governor — 1. A motor attachment that automatically controls the speed at which the motor rotates. 2. The equipment that controls the gate or valve opening of a prime mover.

GPI — Abbreviation for ground-position indicator.

GPIB — Abbreviation for general-purpose interface bus. Also called HPIB (Hewlett Packard interface bus) and, more formally, IEEE Standard 488.2 interface. 1. A byte-serial bus created to interconnect instruments and computers. 2. The standard interface hardware used in computer systems. It has been adopted by the IEEE as the IEEE-488 instrument bus standard, which is intended to define standard interface techniques for limited transmission distances (up to 20 meters). It is an eight-bit wide digital interface applicable for both programmable and nonprogrammable components. The bus provides compatibility between interfacing components, offers noise immunity, and covers the use of interfacing with special connectors.

grabber hand — An on-screen image of a hand that is controlled with a mouse.

graceful degradation — A computer programming technique whose purpose is to prevent catastrophic system failure by permitting the machine to operate, although in a degraded mode, in spite of failures or malfunctions in several integral units or subsystems.

graded-base transistor — *See* diffused-base transistor.

graded-core glass optical fiber — For most applications of high information rate (bandwidth), a fiber that has a core in which the highest optical density is at the center. Optical density decreases with distance from the center until it is the same as the cladding. That is, the optical density of the core is graded downward from the center of the core to the edge of the core, at which point it is equal to the cladding optical density.

graded filter — A power-supply filter in which the output stage of a receiver or audio amplifier is connected at or near the filter input so that the maximum available

dc voltage will be obtained. The output stage has low gain; therefore, ripple is not too important.

graded-index fiber — 1. An optical fiber made with a refractive index that gets progressively lower as the diameter increases. 2. An optical fiber in which the index of refraction varies in the fiber, usually decreasing approximately parabolically from the center to the surface. 3. A fiber whose index of refraction decreases with increasing radial distance from the center of the core.

graded insulation — A combination of insulation proportioned so as to improve the distribution of the electric field to which the combination is subjected.

graded-junction transistor — *See* rate-grown transistor.

graded thermoelectric arm — A thermoelectric arm having a composition that changes continuously in the direction of the current.

gradient — The rate at which a variable quantity increases or decreases. For example, potential gradient is the difference of potential along a conductor or through a dielectric.

gradient-index fibers — Optical fibers that keep a pulse of photons together for a longer time by gradually bending the photon paths back to the core of the fiber before they get all the way to the surface. While the weaving photon travels a longer distance than those traveling in a straight line down the center of the fiber, the glass in the core is treated to slow down the light. This keeps the packets of photons in proximity for a longer time. This fiber is being used in almost all phone company installations today.

gradient meter — *See* generating electric field meter.

gradient microphone — A microphone in which the output rises and falls with the sound pressure. *See also* pressure microphone.

grain growth — The increase in the size of the crystal grains in a glass coating or other material over a period.

gram — A unit of mass and weight in the metric system. Letter symbol: g.

gramme ring — A ring-shaped iron armature around which the coils are wound. Each turn is tapped from the inside diameter of the ring to a commutator segment.

grandfather cycle — The period during which magnetic-tape records are retained before reusing so that records can be reconstructed in the event of loss of information stored on a magnetic tape.

grand-scale integration — *See* GSI.

granular carbon — Small particles of carbon used in carbon microphones.

granularity — A characteristic of the output data of a measuring instrument. The measure of granularity is the smallest increment of the output data when it is in a digital form. The smallest increment is also called least count.

graph — A pictorial presentation of the relationship between two or more variables.

graphechon — A specially designed electron memory tube, based on iconoscope principles, in which electrical signal information is stored and recovered at different

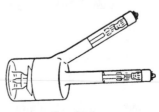

Graphechon.

scanning rates. It is used in radar and computer applications.

graphic — A symbol produced by a process such as handwriting, drawing, or printing.

graphical analysis — The use of diagrams and other graphic methods to obtain operating data and answers to scientific or mathematical problems.

graphical data operations — Manipulations that a system can perform on points, lines, symbols, angles, and other graphical representations. These operations include delete, insert, replace, move, rotate, expand, contract, and extrapolate.

graphical language — A programming language that expresses programs in a graphical form resembling flowcharts.

graphic equalizer — 1. An equalizer that functions simultaneously at a number of preset frequencies, any of which may be boosted or cut independently of all others. Often done at standard 1, $\frac{1}{2}$, $\frac{1}{3}$, or $\frac{1}{6}$ octave center frequencies. Graphic equalizers are generally peaking-type equalizers. 2. A multiband equalizer whose controls are sliders, so that their settings can be seen as a rough graph of their frequency response characteristics. 3. Tone control device that uses separate adjustments to cut or boost different frequencies within the audio band. A graphic equalizer provides greater control over tone than single-knob tone controls.

graphic instrument — See recording instrument.

graphics — 1. In communications systems, an information mode in which intelligence is reproduced by use of a graphic system (a variation of facsimile). 2. Nonvoice analog information modes and devices such as facsimile and television. 3. Using computer technology to create a drawing that is usually displayed on a terminal or plotter.

graphics board — A hardware add-on that boosts a computer's graphics capabilities. Common graphics adapters include CGA, EGA, VGA, and Super VGA.

graphic symbol — An electronic circuit diagram symbol formed using one or more basic elements such as lines, circles, arcs, and squares. The meaning of a symbol is not changed by its orientation, although some symbols are generally shown in one direction by convention. Line width and symbol size do not affect a graphic symbol's meaning.

graphic tablet — A surface through which coordinate points can be transmitted by identification with a cursor or stylus.

graphic terminal — 1. A cathode-ray-tube display. 2. An XY plotter.

graphite — A finely divided carbon used as a lubricant and in the construction of some carbon resistance elements. The most common use is in so-called lead pencils.

grass — The pattern produced by random noise on an A-scope; it appears as closely spaced, sharp, constantly moving pulses on the base line. See also random noise.

grasshopper fuse — A small fuse with a spring-loaded fusible wire. When the wire fuses to open the circuit, the spring shows a visible signal and closes an auxiliary circuit to actuate an alarm.

graticule — A calibrated screen placed in front of a cathode-ray tube for measurement purposes.

grating — A device for spreading out light or other radiation. It consists of narrow parallel slits in a plate or narrow parallel reflecting surfaces made by ruling grooves on polished metal. The slits or grooves break up the waves as they emerge. See also ultrasonic cross grating; ultrasonic space grating.

grating reflector — An antenna reflector consisting of an openwork metal structure that resembles a grating.

Gratz rectifier — An arrangement of two rectifiers per phase connected into a three-phase bridge circuit to provide full-wave rectification.

gravity — The force that tends to pull bodies toward the center of the earth, thereby giving them weight. See also g.

gravity cell — A primary cell in which two electrolytes are kept separated by differences in specific gravity. It is a modification of the Daniell cell and is now obsolete.

gray body — A radiating body whose spectral emissivity remains the same at all wavelengths. It is in constant ratio of less than unity to the radiation of a blackbody radiator at the same temperature.

Gray code — 1. A positional binary number notation in which any two numbers whose difference is 1 are represented by expressions that are the same except in one place or column and differ by only one unit in that place or column. 2. A numeric code composed of a number of bits, assigned in such a way that only one bit changes at each increment (or decrement). 3. A modified binary code. Sequential numbers are represented by binary expressions in which only one bit changes at a time; thus, errors are easily detected.

Gray-code test patterns — A sequence of input patterns in which only one input pin changes state at each test step.

gray image — Any image composed of the full spectrum of gray shades ranging from black to white.

gray scale — 1. A series of regularly spaced tones ranging from white to black through intermediate shades if gray used as a reference scale for control purposes in photography or TV. 2. The discrete levels of the video signal between reference-black and reference-white levels.

gray-scale capability — The ability to accurately reproduce different light levels. At present there are several standards for judging gray-scale display. Some define each light "level" that makes up the scale as the brightness change discernible by the eye (typically a 3- to 5-percent change). Another standard, less subjective, defines gray scale as the difference in brightness required to produce a specified density change on processed film.

gray scale image — An image consisting of an array of pixels that can have more than two values.

great manual — Also called the accompaniment manual or lower manual. In an organ, the keyboard normally used for playing the accompaniment to the melody.

green-gain control — A variable resistor used in the matrix of a three-gun color television receiver to adjust the intensity of the green primary signal.

green gun — The electron gun whose beam, when properly adjusted, strikes only the green phosphor dots in the color picture tube.

green restorer — A dc restorer used in the green channel of a three-gun color-television picture-tube circuit.

green video voltage — The signal voltage that controls the grid of the green gun in a three-gun picture tube.

Greenwich civil time — See universal time.

Greenwich mean time — Abbreviated GMT or Gmt. The mean solar time at the meridian of Greenwich (zero longitude). It is used as a world-wide reference time. Also called zulu time, because of the Z time zone. This widely used standard time reference is equivalent to EDT plus 4 hours, EST or CDT plus 5 hours, CST or MDT plus 6 hours, MST or PDT plus 7 hours, and PST plus 8 hours.

Gregorian antenna—A satellite antenna with a sub-reflector mounted near the focal point, for improved focusing of satellite signals.

grid—1. An electrode having one or more openings for the passage of electrons or ions. *See also* control grid; screen grid; shield grid; space-charge grid; suppressor grid. 2. An interconnected system in which high-voltage, high-capacity backbone lines overlay and are connected with networks of lower voltages. 3. A two-dimensional network consisting of a set of equally spaced parallel lines superimposed upon another set of equally spaced parallel lines so that the lines of one set are perpendicular to the lines of the other, thereby forming square areas. The intersections of the lines provide the basis for an incremental location system. 4. An arrangement of electrically conducting wire, screen, or tubing placed in front of doors or windows or both, which is used as a part of a capacitance sensor. 5. A lattice of wooden dowels or slats concealing fine wires in a closed circuit that initiates an alarm signal when forcing or cutting the lattice breaks the wires. Used over accessible openings. Sometimes called a protective screen. 6. A screen or metal plate, connected to earth ground, sometimes used to provide a stable ground reference for objects protected by a capacitance sensor. If placed against the walls near the protected object, it prevents the sensor sensitivity from extending through the walls into areas of activity. 7. Network of uniformly spaced points on an input device used for locating position.

grid battery—Sometimes called a C battery. A source of energy for supplying a bias voltage to the grid of a vacuum tube.

grid bearing—A bearing made with the reference line to grid north.

grid bias—Also called C-bias. A constant potential applied between the grid and cathode of a vacuum tube to establish an operating point.

grid-bias cell—A small cell used in a vacuum-tube circuit to make the grid more negative than the cathode. It provides a voltage, but cannot supply an appreciable amount of current.

grid blocking—1. Blocking of capacitance-coupled stages in an amplifier because of an accumulated charge on the coupling capacitor as the result of current flow during the reception of large signals. 2. A method of keying a circuit by application of a negative grid voltage several times during key-up conditions; when the key is down, the blocking bias is removed, and normal current through the keyed circuit is restored.

grid cap—At the top of some vacuum tubes, the terminal that connects to the control grid.

grid capacitor—A capacitor in parallel with the grid resistor or in series with the grid lead of a tube.

grid-cathode capacitance—Capacitance between the grid and the cathode in a vacuum tube.

grid characteristic—The curve obtained by plotting grid-voltage values of a vacuum tube as abscissas against grid-current values as ordinates on a graph.

grid circuit—The circuit connected between the grid and cathode and forming the input circuit of a vacuum tube.

grid-circuit tester—A tester designed to measure the grid resistance of vacuum tubes without discriminating between the type or polarity of impedance.

grid clip—A spring clip used for making a connection to the top-cap terminal of some vacuum tubes.

grid-conductance—The in-phase component of the alternating grid current divided by the alternating grid voltage, all other electrode voltages being maintained constant.

grid control—Control of the anode current of an electron tube by means of changes in the voltage between the control grid and cathode of the tube.

grid-controlled mercury-arc rectifier—A mercury-arc rectifier employing one or more electrodes exclusively for controlling start of the discharge.

grid-controlled rectifier—A triode mercury-vapor rectifier tube in which the grid determines the instant at which plate current starts to flow during each cycle, but does not determine how much current will flow.

grid-control tube—A mercury-vapor-filled thermionic vacuum tube with an external grid control.

grid current—The current that flows in the grid-to-cathode circuit of a vacuum tube. It is usually a complex current made up of several currents having a variety of polarities and impedances.

grid detection—Detection by rectification in the grid circuit of a vacuum tube.

grid-dip meter—A multiple-range oscillator incorporating a meter in the grid circuit to indicate grid current. The meter is so named because its reading dips (reads a lower grid current) whenever an external resonant circuit is tuned to the oscillator frequency.

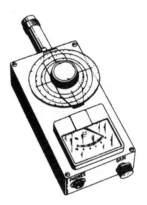

Grid-dip meter.

grid dissipation—The power lost as heat at the grid of a tube.

grid-drive characteristic—The relationship between the electrical or light output of an electron tube and the control-electrode voltage measured from cutoff.

grid driving power—The average product of the instantaneous value of grid current and the alternating component of grid voltage over a complete cycle.

grid emission—Electron or ion emission from the grid of an electron tube.

grid-glow tube—A glow-discharge cold cathode tube in which one or more control electrodes initiate the anode current but do not limit it except under certain conditions.

gridistor—A field-effect transistor that uses the principle of centripetal striction and has a multichannel structure, combining the advantages of both field-effect transistors and minority-carrier injection transistors.

grid leak—A high resistance connected across the grid capacitor or between the grid and cathode. It provides a direct-current path, to limit the accumulation of a charge on the grid.

grid-leak capacitor—A small capacitor connected in a vacuum-tube grid circuit, together with a resistor, to produce grid bias.

grid-leak detector — A triode or multielectrode tube in which rectification occurs because of electron current through a high resistance in the grid circuit. The voltage associated with this flow appears in amplified form in the plate circuit.

grid limiting — The use of grid-current bias derived from the signal, through a large series grid resistor, in order to cut off the plate current and consequently level the output wave for all input signals above a critical value.

grid locking — Faulty tube operation in which excessive grid emission causes the grid potential to become continuously positive.

grid modulation — Modulation produced by application of the modulating voltage to the control grid of any tube in which the carrier is present. Modulation in which the grid voltage contains externally generated pulses is called grid-pulse modulation.

grid neutralization — A method of neutralizing an amplifier. A portion of the grid-to-cathode alternating-current voltage is shifted 180° and applied to the plate-to-cathode circuit through a neutralizing capacitor.

grid north — An arbitrary reference direction used with the grid system of navigation.

grid-plate capacitance — The direct capacitance between the grid and plate of a vacuum tube.

grid-plate transconductance — Mutual conductance, which is the ratio of a plate-current change to the grid-voltage change that produces it.

grid-pool tank — A grid-pool tube having a heavy metal envelope somewhat resembling a tank in appearance.

grid-pool tube — A gas-discharge tube having a mercury-pool cathode, one or more anodes, and a control electrode or grid to control the start of current. *See also* excitron; ignitron.

grid-pulse modulation — Modulation produced in an amplifier or oscillator by application of one or more pulses to a grid circuit.

grid pulsing — Method of controlling the operation of a radio-frequency oscillator. The oscillator-tube grid is biased so negatively that oscillation occurs, even at full plate voltage, except when this negative bias is removed by application of a positive voltage pulse to the grid.

grid resistor — A general term that denotes any resistor in the grid circuit.

grid return — An external conducting path for the return of grid current to the cathode.

grid suppressor — A resistor, sometimes connected between the control grid and the external circuit of an amplifier, to prevent parasitic oscillations caused by stray-capacitance feedback.

grid swing — The total variation in grid-to-cathode voltage from the positive peak to the negative peak of the applied-signal voltage.

grid-to-cathode capacitance — The direct capacitance between the grid and cathode of a vacuum tube.

grid-to-plate capacitance — Designated C_{gp}. The direct capacitance between the grid and plate in a vacuum tube.

grid-to-plate transconductance — The mutual conductance, or ratio of plate-current to grid-voltage changes, in a vacuum tube.

grid voltage — The voltage between the grid and cathode of a tube.

grid-voltage supply — The means for supplying, to the grid of an electron tube, a potential that is usually negative with respect to the cathode.

grille cloth — A loosely woven fabric that is virtually transparent to sound, often stretched across the opening in a speaker enclosure that seats the radiating side of the speaker.

grommet — An eyelet of rubber or neoprene placed in a hole in sheet metal, such as a terminal entrance, to insulate and protect wires that pass through.

groove — In mechanical recording, the track inscribed in the record by a cutting or embossing stylus, including undulations or modulations caused by vibration of the stylus. In stereo discs its cross section is a right-angled triangle, with each side at a 45° angle to the surface of the record; information is cut on both sides of the groove. In a long-playing record, groove dimensions could be: width 2.5 mils (63.5 μm); depth 1 mil (25.4 μm), and pitch 250–350 groove revolutions per inch (98–138 groove revolutions per centimeter).

groove angle — In disc recording, the angle between the two walls of an unmodulated groove in a radial plane perpendicular to the surface of the recording medium.

groove shape — In disc recording, the contour of the groove in a radial plane perpendicular to the surface of the recording medium.

groove speed — In disc recording, the linear speed of the groove with respect to the stylus.

groove velocity — The speed with which the record groove moves under the cartridge. An lp record rotates at a constant $33\frac{1}{3}$ rpm with grooves cut at diameters that decrease gradually from $11\frac{1}{2}$ to $4\frac{3}{4}$ inches (29.2 to 12.1 cm). Groove velocity therefore ranges from 20 in/s (50.8 cm/s) at the outside of the record to 8.3 in/s (21.1 cm/s) in the innermost groove.

gross information content — A measure of the total information, including redundant portions, contained in a message. It is expressed as the number of bits or hartleys necessary to transmit the message with specified accuracy by way of a noiseless medium without coding.

ground — 1. A connection to the earth for conducting electrical current to and from the earth. 2. The voltage reference point in a circuit. There may or may not be an actual connection to earth, but it is understood that a point in the circuit said to be at ground potential could be connected to earth without disturbing the operation of the circuit in any way. 3. A point in an electrical system that has zero voltage. Usually, the chassis of an electrical component is at ground potential and thus serves as the return path for signals as well as for power circuits. The shield in coaxial signal cable is, or should be, at ground potential to avoid hum pickup. Ground also designates the earth, literally, which is used as a return path for radio waves from an antenna. In British terminology *earth* is used to designate all ground connections. 4. A conducting connection through which a circuit or electrical equipment is connected to the earth or to a conducting body that is at earth potential. A ground may be accidental or intentional. 5. To connect to a ground. 6. Connection (intentional or accidental) between an electrical circuit and the earth or its electrical equivalent.

ground absorption — The loss of energy during transmission because of the radio waves dissipated to ground.

ground bus — A conductor, usually large-diameter wire, that connects a number of points to one or more ground electrodes.

ground check — Also known as base-line check. 1. A procedure followed prior to the release of a radiosonde in order to obtain the temperature and humidity correction for the radiosonde system. 2. Any instrumental check prior to the ground launch of an airborne experiment.

ground clamp—A clamp used for connecting a grounding conductor (ground wire) to a grounded object such as a water pipe.

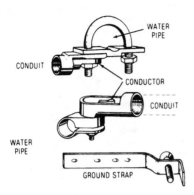

Ground clamps.

ground clutter—The pattern produced on the screen of a radar indicator by undesired ground return.

ground conductor—A conductor in a transmission cable or line that is grounded.

ground conduit—A conduit used solely to contain one or more grounding conductors.

ground control—To guide or direct an aircraft or missile by control exercises from the ground.

ground-controlled approach—Abbreviated GCA. 1. The radar system developed to give direction, distance, and elevation along a fixed approach path to an airport. The ground controller at the radarscope communicates instructions to the pilot to direct the aircraft along the approach line. 2. Technique or procedure for "talking down," through the use of both surveillance and precision approach radar, an aircraft during its approach so as to place it in a position for landing.

ground-controlled interception—Abbreviated GCI. A radar system used for directing an aircraft to intercept enemy aircraft.

ground controller—Aircraft controller stationed on the ground. Generic term applied to the controller in ground-controlled approach, ground-controlled interception, etc.

ground current—Current in the earth or grounding connection.

ground detector—1. An instrument or equipment that indicates the presence of a ground on a normally ungrounded system. 2. Device that indicates ground faults in electrical circuits.

ground distance—The great-circle component of distance from one point to another at mean sea level.

grounded—1. Connected to the earth, to a rod or pipe that makes a good electrical connection with the earth, or to some conduction body in place of the earth. 2. A system, circuit, or apparatus that is provided with a ground.

grounded-base amplifier—See common-base amplifier.

grounded cable bond—A cable bond used for grounding the armor and/or sheaths of cables.

grounded capacitance—In a system having several conductors, the capacitance between a given conductor and the other conductors when they are connected together and to ground.

grounded-cathode amplifier—The conventional amplifier circuit. It consists of a tube amplifier in which the cathode is at ground potential at the operating frequency. The input is applied between the control grid and ground, and the output load is between the plate and ground.

grounded circuit—1. A circuit in which one conductor or point (usually the neutral conductor or neutral point of transformer or generator windings) is intentionally grounded (earthed) either solidly or through a grounding device. 2. A circuit that is connected to earth at one or more points.

grounded-collector amplifier—See common-collector amplifier.

grounded conductor—A conductor that is intentionally grounded, either directly or through a current-limiting device.

grounded dielectric constant—The dielectric constant of the earth at a given location.

grounded-emitter amplifier—See common-emitter amplifier.

grounded-gate amplifier—A FET amplifier circuit in which the gate electrode is connected to ground. The input signal is applied to the source electrode, and the output is taken from the drain electrode.

grounded-grid amplifier—An electron-tube amplifier circuit in which the control grid is at ground potential at the operating frequency. The input is applied between the cathode and ground, and the output load is between the plate and ground. The grid-to-plate impedance of the tube is in parallel with the load, instead of acting as a feedback path.

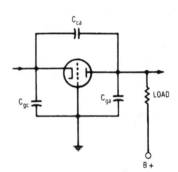

Grounded-grid amplifier.

grounded-grid triode—A type of triode designed for use in a grounded-grid circuit.

grounded-grid triode circuit—A circuit in which the input signal is applied to the cathode and the output is taken from the plate. The grid is at rf ground and serves as a screen between the input and output circuits.

grounded-grid triode mixer—A triode in which the grid forms part of a grounded electrostatic screen between the anode and cathode. It is used as a mixer for centimeter wavelengths.

grounded neutral—The neutral wire in an electrical power system metallically connected to ground.

grounded outlet—An outlet equipped with a receptacle of the polarity type having, in addition to the current-carrying contacts, one ground contact, which can be used for the connection of an equipment-grounding conductor.

grounded parts—Parts of a completed installation that are so connected that they are substantially at the same potential as the earth.

grounded-plate amplifier—Also called cathode follower. An electron-tube amplifier circuit in which the plate is at ground potential at the operating frequency. The input is applied between the control grid and ground, and the output load is between the cathode and ground.

grounded system — A system of electrical conductors in which at least one conductor or point (usually the middle wire or neutral point of transformer or generator windings) is intentionally grounded, either solidly or through a current-limiting device.

ground environment — 1. The environment surrounding and affecting a system or item of equipment that operates on the ground. 2. A system or part of a system that functions on the ground. 3. The aggregate of all ground-installed equipment that makes up a communications-electronics system, facility, station, set, etc. 4. The portion of an air-defense system that provides for the detection, surveillance, and control of airborne objects. It includes ground-based facilities and overwater facilities, such as picket vessels and airborne early warning and control aircraft.

ground-equalizer inductors — Relatively low-inductance coils inserted in the circuit to one or more of the grounding points of an antenna to obtain a desired distribution of the current to the various points.

ground fault — 1. An unintentional electrical path between a part operating normally at some potential to ground and ground. 2. A current leak from the hot side of the line through a path that bypasses the load to ground. This current leak causes an imbalance between the hot and neutral wires to the load.

ground fault current — A fault current that flows to ground rather than between conductors.

ground-fault interrupter — A device that senses the imbalance caused by a ground fault with a differential-current transformer. Two primary windings, each in series with one side of the line, are wound on the transformer core. Equal currents in both lines cancel each other to produce no transformer signal in a third sensor winding. However, current imbalance generates a sensor output, which is amplified to trip a circuit-opening device. *See also* current limiter.

ground-gate amplifier — A FET amplifier circuit in which the gate electrode is connected to the ground. The input signal is applied to the source electrode, and the output is taken from the drain electrode.

ground gating — The conversion of pulse-amplitude modulated signals at a telemetry ground station to 50-percent duty-cycle signals.

ground grid — A system of grounding electrodes interconnected by bare cables buried in the earth to provide lower resistance than a single grounding electrode.

ground indication — An indication of the presence of a ground on one or more of the normally ungrounded conductors of a system.

grounding — 1. Connecting to ground, or to a conductor that is grounded 2. A means of referencing electrical circuits to the well-bonded equipotential surface. 3. A connection between the electrical system and the earth to prevent shock. Ground wires are usually bare; grounding connections are normally green in color. A safety measure to avoid electric shock.

grounding conductor — A conductor that, under normal conditions, carries no current but serves to connect exposed metal surfaces to earth ground, to prevent hazards in case of breakdown between current-carrying parts and the exposed surfaces. The connector, if insulated, is colored green with or without a yellow stripe.

grounding connection — A connection used to establish a ground, consisting of a grounding conductor, a grounding electrode, and the earth surrounding the electrode.

grounding electrode — A conductor or network of conductors, usually embedded in the earth, used for maintaining ground potential on conductors connected to it or for dissipating into the earth any current conducted to it.

grounding outlet — An alternating-current receptacle that has a third contact connected to ground. Used with three-wire plugs and cords to ground portable electric tools and appliances safely.

grounding plate — An electrically grounded metal plate on which a person stands in order to discharge any static electricity that may be picked up by his or her body.

grounding switch — A form of air switch for connecting a circuit or apparatus to ground.

grounding transformer — A transformer intended primarily for the purpose of providing a neutral point for grounding purposes.

ground insulation — The major insulation used between a winding and structural parts at ground potential.

ground junction — In a semiconductor, a junction formed during the growth of a crystal from a melt.

ground level — *See* ground state.

ground loop — 1. An unwanted feedback condition in which power current in a single ground wire causes instability or errors. 2. A potentially detrimental condition produced when two or more points in an electrical system that are nominally at ground potential are connected by a conducting path. The term usually is applied when, because of improper design or by accident, unwanted noise signals are generated in the common return of relatively low-level signal circuits by the return currents or by magnetic fields produced by relatively high-power circuits or components. 3. The electrical path between two separate grounds. 4. A condition occurring when two or more paths to ground exist and a voltage is induced unequally in these paths, causing interference like hum, buzz, or noise. 5. A term used to describe situations occurring in ground systems in which a difference in potential exists between two ground points due to the resistance of ground conductors.

ground-loop disturbances — Detrimental interference formed when two or more points in an electrical system that are nominally at ground are connected by a conducting path such that either or both are not at the same ground potential.

ground lug — A lug for connecting a grounding conductor to a grounding electrode.

ground mat — A system of bare conductors, on or below the surface of the earth, connected to a ground or ground grid to provide protection from dangerous touch voltage.

ground noise — In recording and reproducing, the residual noise in the absence of a signal. It is usually caused by dissimilarities between the recording and reproducing media, but may also include amplifier noise such as from a tube or noise generated in resistive elements at the input of the reproducer amplifier system.

ground-noise margin — The voltage that may be applied at the ground connection of a logic circuit without causing the circuit to malfunction. It is usually measured by increasing the static ground voltage on a single gate until the logic fails to operate properly.

ground outlet — An electrical outlet equipped with a polarized receptacle that has, in addition to the current-carrying contacts, a grounded contact to which can be connected an equipment-grounding conductor.

ground plane — 1. Copper or brass sheet used in interference testing to simulate missile, aircraft, or vehicle frame or skin so that actual installation and grounding conditions may be approximated. 2. A conductor layer, or portion of a conductor layer, used as a common reference point for circuit returns, shielding, or heat sinking. 3. Expanded copper mesh that is laminated into some flat cable constructions as a shield. May be supplied with one, two, or no drain wires. A common ground electrical path for power and/or signals.

ground-plane antenna — A vertical antenna combined with a turnstile element to lower the angle of radiation. It has a concentric base support and a center conductor that place the antenna at ground potential, even though located several wavelengths above ground.

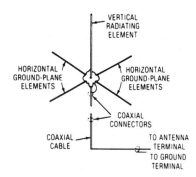

Ground-plane antenna.

ground plate — A plate of conductive material buried in the earth to serve as a grounding electrode.

ground-position indicator — Abbreviated GPI. A dead-reckoning computer, similar to an air-position indicator, with provision for taking drift into account.

ground potential — Zero potential with respect to ground or the earth.

ground power cable — A cable assembly fitted with appropriate terminations to supply power to an aircraft from a ground power unit.

ground protection — Protection of a circuit by means of a device that opens the circuit when a fault to ground occurs.

ground protective relay — A device that functions on failure of the insulation of a machine, transformer, or other apparatus to ground, or on flashover of a dc machine to ground.

ground range — In range measurements related to airborne radar, the distance on the surface of the earth between the object under consideration and a point directly below the aircraft that carries the radar.

ground-reflected wave — 1. In a ground wave, the component reflected from the earth. 2. A radio wave reflected one or more times from the earth's surface before reaching the point of reception.

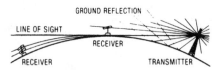

Ground-reflected waves.

ground resistance — 1. The opposition of the earth to the flow of current through it. Its value depends on the nature and moisture content of the soil; the material, composition, and physical dimensions of the connection to earth; and the electrolytic action present. 2. The ohmic resistance between a grounding electrode and a remote or reference grounding electrode so spaced that their mutual resistance is essentially zero.

ground return — 1. In radar, the echoes reflected from the earth's surface and fixed objects on it. 2. A lead from an electronic circuit, antenna, or power line to ground.

ground-return circuit — A circuit that has a conductor (or two or more in parallel) between two points and which is completed through ground or the earth. A circuit using the earth as one side of the complete circuit.

ground rod — 1. A steel or copper rod driven into the earth to make an electrical contact with it. 2. A long metal rod that is driven into the ground near an antenna installation and to which are attached the grounding wires from the mast and antenna discharge unit to discharge static electricity to ground before it can enter and damage the TV receiver.

ground-scatter propagation — Multihop ionospheric propagation of radio waves along other than the great-circle path between the transmitting and receiving points. Radiation from the transmitter is returned from the ionosphere to the surface of the earth, from which it is then scattered in many directions.

ground shift — The variation in signal amplitude at different grounding points because of a voltage drop along a ground line.

ground speed — In navigation, the speed of a vehicle with reference to ground.

ground state — Also called ground level. The lowest energy level or state of an atom or atomic system; all other states of the system are called excited states.

ground support cable — Cable construction, usually rugged and heavy, for use in ground support control or power systems.

ground-support equipment — All ground equipment that is part of a complete weapons system and that must be furnished to ensure complete support of the weapons system.

ground system of an antenna — The portion of an antenna system that includes an extensive conducting surface, which may be the earth itself, and those parts of the antenna closely associated with that surface.

ground-to-air communication — One-way communication from ground stations to aircraft.

ground wave — 1. A radio wave that travels along the earth's surface rather than through the upper atmosphere. 2. That portion of a radio wave traveling between transmitting and receiving antennas that is associated with currents induced in the ground or water surface of the earth. Important only below about 10 MHz.

ground wire — 1. A conductor leading to an electric connection with the earth. 2. A heavy copper conductor, usually insulated, that is used to connect protectors or other equipment to a ground rod or cold water pipe.

group — 1. In carrier telephony, a number of voice channels multiplexed together and treated as a unit. Commonly, a group contains 12 channels, each with a bandwidth of 4 kHz, frequency multiplexed and occupying the band from 60 to 180 kHz. 2. The second-highest stratum of an organizational hierarchy, usually identifying departments, divisions, or regions.

group busy tone — A high tone fed to the jack sleeves of an outgoing trunk group to serve as an indication that all trunks in the group are busy.

group channel — A unit or method of organization on telephone carrier (multiplex) systems. A full group is a channel equivalent to 12 voice-grade channels (48 kHz).

A half group has the equivalent bandwidth of 6 voice-grade channels (24 kHz). (When not subdivided into voice facilities, group channels can furnish high-speed data communication.)

group delay — Also called envelope delay. 1. The delay in transmission of information modulated on a carrier. 2. Distortion resulting from nonuniform speed of transmission of the various frequency components of a signal through a transmission medium; specifically, the propagation delay of a lower frequency is different from that of a higher frequency. This creates a time-related delay-distortion error.

grouped-frequency operation — A method in which different frequency bands are used for channels in opposite directions in a two-wire carrier system.

group frequency — The number of sets or groups of waves passing a given point in one second.

grouping — 1. Nonuniform spacing between the grooves of a disc recording. 2. Periodic error in the spacing between recorded lines in a facsimile system.

grouping circuits — Circuits used to interconnect two or more positions of a switchboard so that one operator may handle the several positions from one operator's set.

group loop — A source of interference when a system is grounded improperly at several points.

group mark — A mark used to identify the beginning or the end of a set of data, which could include words, blocks, or other items.

group modulation — The process by which a number of channels, already separately modulated to a specific frequency range, are again modulated to shift the group to another range.

Group I fax — An analog facsimile device that transmits or receives a standard page in four to six minutes. Group I machines are no longer being manufactured and are rarely in use today.

group technology — Facilitation of processing through combination of similar parts into production families.

Group III fax — The standard for current facsimile devices. Most current facsimile systems are digital devices offering operating speeds of one minute or less. When equipped with automatic speed recognition, these systems can be compatible with Group I and II units, although a number of the lower-cost models are strictly Group III compatible. Machines that can recognize speed automatically can select the fastest speed available when sending to or receiving from Group I or Group II devices.

Group II fax — An analog device that transmits or receives a page in two or three minutes. These systems offer some data-compression techniques for faster transmission and can be compatible with Group I devices. Like the Group I units, Group II models are not actively marketed today.

group velocity — 1. Of a traveling plane wave, the velocity of propagation of the envelope delay is approximately constant. It is equal to the reciprocal of the envelope delay per unit length. (Group velocity differs from phase velocity in a medium in which the phase velocity varies with the frequency.) 2. The velocity of the envelope of an electromagnetic wave as it travels in a medium, usually identified with the velocity of energy propagation.

groupware — Software that is designed to be used by a group of people working on the same information, whether in the same room, building, or across town or the globe.

Grove cell — A primary cell with a platinum electrode submerged in an electrolyte of nitric acid within a porous cup, surrounded by a zinc electrode in an electrolyte of sulfuric acid. This cell normally operates on a closed circuit.

growler — 1. An electromagnetic device consisting essentially of two field poles arranged as in a motor, energized with ac and used for locating short-circuited coils in a generator or motor armature. A growling noise indicates a short-circuited coil. 2. An electromagnetic device for magnetizing or demagnetizing objects.

grown-diffused transistor — A transistor made by combining the diffusion and double-doped techniques. Suitable n- and p-type impurities are added simultaneously to the melt while the crystal is being grown. Subsequently, the base region is formed by diffusion as the crystal grows.

grown injunction — The boundary between p- and n-type semiconducting materials. It is produced by varying the impurities during the growth of a crystal from the melt. Such junctions have strong rectifying properties, the forward current being obtained when p is positive to n.

grown-junction photocell — A photodiode made of a small bar of semiconductor material that has a pn junction at right angles to its length and an ohmic contact at each end of the bar.

grown-junction transistor — A transistor in which junctions are formed by adding impurities to the melt while the crystal is being grown.

grown-junction wafer — A semiconductor wafer on which pn junctions are formed during manufacture.

grown semiconductor junction — A junction formed during the growth of a crystal from a melt.

G-scan — *See* G-display.

G-scope — *See* G-display.

GSI — Abbreviation for grand-scale integration. Monolithic integrated circuits with a typical complexity in excess of 1000 or more gates or gate equivalent circuits.

GSO — Abbreviation for geostationary satellite orbit. A circular orbit 35,863 km above the earth, in the plane of the earth's equator, in which a satellite revolves around the earth in the same time that the earth rotates on its axis; thus, the satellite appears approximately stationary over one point on the earth.

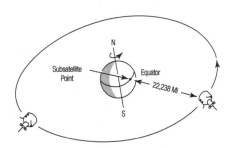

Geostationary satellites.

GTAW — Abbreviation for gas tungsten arc welding. *See also* TIG.

GTO SCR — Abbreviation for gate turn-off SCR. A silicon-controlled rectifier whose output-current switching can be turned off by a voltage at the gate.

guard — A mechanism to terminate program execution (real or simulated) upon access to data at a specified memory location. Used in debugging.

guard arm — 1. A crossarm placed across and in line with cable to protect it from damage. 2. A crossarm placed over wires to prevent other wires from falling into them.

guard band — 1. Also called interference guard band. A frequency band left vacant between two channels to safeguard against mutual interference. 2. The unused chip surface area that, by virtue of physical spacing, serves to isolate functional elements in a printed circuit or an integrated circuit. Also refers to the consideration given instrumentation precision in electrical testing. 3. The unused bandwidth separating channels to prevent crosstalk.

guard channel — 1. Unused portions of the frequency spectrum that are located between program channels to prevent adjacent channel interference. *See also* guard band. 2. One or more auxiliary parallel processing channels to control the main processing channel in order to reject interference that is partly in, but not centered on, the main channel. Guard channels may be displaced in time (range), Doppler frequency, carrier frequency, or angle.

guard circle — An inner concentric groove on disc records. It prevents the pickup from being thrown to the center of the record and possibly damaged.

guarded input — An input that has a third terminal that is maintained at a potential near the input-terminal potential for a single-ended input, or near the mean input potential for the differential input. It is used to shield the entire input circuit.

guarded motor — An open motor in which all openings given direct access to live or rotating parts (except smooth shafts) are limited in size by the structural parts or by screens, grilles, expanded metal, etc., to prevent accidental contact with such parts. Such openings shall not permit the passage of a cylindrical rod $1/2$ inch (12.7 mm) in diameter.

guard frequency — The frequencies between subchannels in frequency-division multiplexed systems used to guard against subchannel interference.

guarding — 1. The introduction of conducting surfaces at critical points in a circuit to intercept and divert leakage currents that otherwise would cause undesired effects or measurement errors. 2. A method of protecting the inputs to a high-gain op amp by surrounding the input terminals with a conducting ring of printed circuit board conductors. This isolates the inputs from potential leakage currents from other parts of the circuit. 3. The use of special circuitry, insulated from ground, to provide freedom from adverse effects of leakage currents. The stray current is bypassed through a noncritical path so that it does not affect the accuracy of measurement. Guarding is especially important when measuring low voltages and when measuring high-value resistances, particularly when high humidity causes a reduction in normally high insulation resistance.

guard relay — A relay used in the linefinder circuit to prevent more than one linefinder from being connected to any line circuit when two or more line relays are operated simultaneously.

guard ring — 1. A metal ring placed around a charged terminal or object to distribute the charge uniformly over the surface of the object. 2. A ring-shaped electrode intended to limit the extent of an electric field, as, for instance, in elimination of the fringe effect at the edges of the plates of a capacitor.

guard-ring capacitor — A capacitor with parallel electrodes, one of which is surrounded by a ring held at the potential of that electrode in order to reduce the edge effect.

guard shield — An internal floating shield surrounding the input section of an amplifier. Effective shielding results only when the absolute potential of the guard is stabilized relative to the incoming signal.

guard-well capacitor — A primary standard capacitor, fixed or variable, for values of capacitance below 1 picofarad. The guard ring forms a well in which a Pyrex disk is mounted for the accurate location of the electrode assembly.

guard wire — A grounded wire used frequently where high-tension lines cross a thoroughfare. Should a line break, it will contact the guard wire and be grounded.

Gudden-Pohl effect — The momentary illumination produced when an electric field is applied to a phosphor previously excited by ultraviolet radiation.

guidance — Control of a missile or vehicle from within by a person, a preset or self-reacting automatic device, or a device that reacts to outside signals.

guidance system — A system that measures and evaluates flight information, correlates it with target data, converts the resultant into the parameters necessary to achieve the desired flight path, and communicates the appropriate commands to the flight-control system.

guidance tapes — Magnetic or paper tapes that are placed in a missile or computer and that contain previously entered information necessary for directing the missile to the selected target.

guide — In a tape recorder, a grooved or flanged pin or roller that guides the tape in a straight line between the reels and the heads, to keep it perfectly in line with the pole pieces.

guided ballistic missile — A ballistic missile that is guided during the powered portion of the trajectory and follows a free ballistic path during the remainder.

guided clip — *See* guided probe.

guided missile — An unmanned vehicle moving above the surface of the earth, the trajectory or flight path of which is capable of being altered by an external or internal mechanism.

guided probe — 1. A fault-isolation technique in which the test system automatically displays the next mode or IC that the operator should probe or clip. The system leads the operator along a path back from a faulty output pin to the location of the fault. A software algorithm uses stored interconnection information and expected responses at each node to determine the next node to be probed. 2. A hand-held probing device (single-point or multipin clip) guided by an operator with instructions from a computer-controlled algorithm.

guided propagation — A type of radiowave propagation in which radiated rays are bent excessively by refraction in the lower layers of the atmosphere. This bending creates an effect much as if a duct or a waveguide has been formed in the atmosphere to guide part of the radiated energy over distances far beyond normal range.

guided spark — An electrical discharge between two electrodes that has its path guided or constrained by the presence of a dielectric material or a gas jet.

guided wave — A wave in which the energy is concentrated near a boundary (or between substantially parallel boundaries) separating materials of different properties. The direction of propagation is parallel to the boundary.

guide pin — A pin or rod that extends beyond the mating faces of a connector in such a way that it guides the closing or mating of the connector and ensures proper engagement of the contacts.

guide wavelength — *See* waveguide wavelength.

Guillemin line — A special type of artificial transmission line or pulse-forming network used in radar sets to control the duration of the pulses. It generates a nearly square pulse for use in high-level pulse modulation.

guillotine capacitor—A translatory motion tuning capacitor consisting of a pair of stators and a sliding plunger in place of a rotor.

gullwing—A common lead form used to interconnect surface-mounted packages to a printed circuit board.

gun-directing radar—Radar used for directing antiaircraft or similar artillery fire.

Gunn diode—1. A tiny wafer of n-type gallium arsenide consisting of a thin active layer of n-type gallium arsenide grown on a low-resistivity substrate of the same material. The substrate is bonded to the anode terminal of the encapsulation, and the other face of the wafer has an evaporated cathode contact connected by a bonded gold wire. The diode has no pn junction and cannot be used for rectification. When a few volts dc are applied to make the anode positive with respect to the cathode, the current that flows is dc with superimposed pulses. 2. A microwave diode that exhibits negative resistance arising from the bulk negative differential conductivity that occurs in several compound semiconductors, such as gallium arsenide, and that operates at a frequency determined by the transit time of charge bunches that are formed due to this negative differential conductivity.

Gunn effect—Current oscillations that occur at an rf rate when an electric field of about 3000 V/cm is applied to a short (127 μm or 0.005 in or less) specimen of n-type gallium arsenide. This effect takes place because electrons under the influence of sufficiently high fields are transferred from high- to low-mobility valleys in the conduction band of GaAs.

Gunn oscillator—An oscillator in which the active element is a Gunn diode operating in the negative resistance mode. This type of oscillator is one of the simplest means of generating microwave signals because only a microwave-tuned circuit, Gunn diode, bias network, and low-voltage power supply are required. Presently available units are restricted to operation above 4 GHz, and have dc to rf conversion efficiencies of less than 10 percent.

gutta-percha—A natural vegetable gum, similar to rubber, used principally as insulation for wire and cables.

guy ring—A circular metal collar with attachment holes (eyes) that is slipped on and clamped to an antenna mast. Guy wires are then attached to the mast through the holes in the guy ring.

guy wire—A wire used to brace the mast or tower of a transmitting or receiving antenna system.

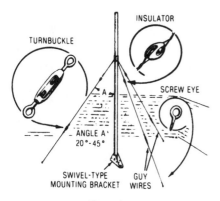

Guy wire.

gyrator—1. A two-port circuit element that exhibits a 180° differential, or nonreciprocal, phase shift. The gyrator circuit symbol indicates that an rf signal transmitted from port 1 to port 2 will undergo a 180° phase shift relative to an rf signal transmitted in the reverse direction. 2. A negative-impedance device that can change the sign of a reactance, thus allowing a capacitor to act as an inductor. Usually consists of a ferrite section in a waveguide. Through the use of such a device, some of the shortcomings of inductors can be eliminated. These shortcomings are large physical size, low Q, nonlinearity, and interwinding capacitance. A properly designed gyrator will provide a synthetic inductor with Q, wide bandwidth, inductance value independent of frequency, and good stability. Filters for frequencies up to 50 kHz can be designed using gyrators. 3. A device providing 180° phase shift in one direction relative to the other direction of signal passage.

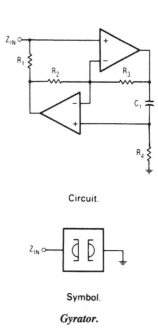

Circuit.

Symbol.

Gyrator.

gyro—Abbreviation for gyroscope.

gyrocompass—*See* gyroscope.

gyrofrequency—The natural frequency at which charged particles rotate around the lines of force of the earth's magnetic field. For electrons, it is 700 to 1600 kilohertz; for ions, it is in the audio-frequency range.

gyromagnetic—The magnetic properties of rotating electric charges, such as electrons spinning within atoms.

gyromagnetic effect—The change in the angular momentum of a body as a result of being magnetized, arising as a result of the fact that the magnetic moments of its electrons are associated with their spins or orbital angular momentum.

gyropilot—*See* autopilot.

gyroscope—Abbreviated gyro. A rotating device whose axle will maintain a constant direction, even though the earth is turning under it. It consists of a wheel mounted so that its spinning axis is free to rotate around either of two other axes perpendicular to itself and each other. When its axle is pointed north, it can be used as a gyrocompass.

gyroscopic action—An action that causes a mass to turn on an axis perpendicular to the applied torque and to the axis of spin.

gyrostabilized platform—*See* stable platform.

gyrotron—A microwave vacuum tube based on the interaction between an electron beam and microwave fields in which coupling is achieved by the cyclotron resonance condition. This type of coupling allows the beam and microwave circuit dimensions to be large compared with a wavelength. Thus, the power density problems encountered in conventional traveling-wave tubes and klystrons at millimeter wavelengths are avoided in the gyrotron.

G–Y signal—In color television, the green-minus-luminance signal, representing primary green minus the luminance, or Y, signal. It is combined with a luminance, or Y, signal outside or inside the picture tube to yield a primary green signal.

H

H — 1. A radar air-navigation system using an airborne interrogator to measure the distance from two ground responder beacons. *See also* shoran. 2. Symbol for heater, magnetic field strength, or henry.

hacker — 1. A person with computer expertise intrinsically interested in the exploration of computer systems and their capabilities. 2. A person who accesses computer systems without authorization. 3. One who deliberately tries to penetrate the security of other computers. 4. A skilled computer enthusiast who is obsessed with learning about programming and exploring the capabilities of computer systems. 5. A person who gains unauthorized access to a computer system.

hacking — Using a microcomputer system or terminal to bypass the security of a large computer system.

hairpin pickup coil — A hairpin-shaped, single-turn coil for transferring UHF energy.

hairpin tuning bar — A sliding hairpin-shaped metal bar inserted between the two halves of a doublet antenna to vary its electrical length.

halation — 1. Distortion seen as blurred images and caused by reflection of the image rays off the back of a fluorescent screen that is too thick. 2. The spreading of light in a photographic emulsion outside the intended area of exposure by reflection from the rear surface of the material supporting the emulsion; this is distinguished from the diffusion that takes place within the emulsion layer. 3. A glow or diffusion that surrounds a bright spot on a television picture tube screen. A defect in picture tube quality is indicated. 4. In a cathode-ray tube, the glow surrounding a bright spot that appears on the fluorescent screen as the result of the screen's light being reflected back by the front and rear surfaces of the tube's face.

half-add — In a computer, an operation that is performed first in carrying out a two-step binary addition. It consists of addition of corresponding bits in two binary numbers, with any carry information being ignored. *See also* exclusive OR.

half adder — 1. A circuit that will accept two binary input signals and produce corresponding sum and carry outputs. So called because, above the first order, two half adders per order are required when adding two quantities. 2. Building-block circuit used in digital computers. A combination of logic gates adds two bits and delivers an answer — two bits called sum and carry. Half adders can be combined to add numbers of any length. Two half adders make up a full adder. 3. A logic element that adds two input bits, but does not have provision for adding the carry from a previous addition. *See also* full adder.

half cell — An electrode, submerged in an electrolyte, for measuring single electrode potentials.

half cycle — The time interval required for the operating frequency to complete one-half, or 180°, of its cycle.

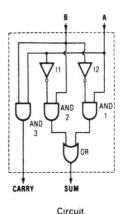

Circuit.

B	A	CARRY	SUM
0	0	0	0
0	1	0	1
1	0	0	1
1	1	1	0

Truth table.

Symbol.

Half adder.

half-digit — In digital meters provided with 100 percent overranging, an additional most-significant digit provided at the left of the readout, capable of displaying the numeral 1 when the measurement exceeds the full-scale range.

half duplex — 1. A communication system in which information can be transmitted in either direction, but only in one direction at a time. 2. In communications, pertaining to an alternate independent transmission made in one direction at a time. 3. A mode of data transmission capable of communicating in one of two directions, but in only one direction at a time. 4. Refers to a communication channel that can receive and transmit, but not simultaneously.

half-duplex circuit — Abbreviated HDX. A circuit that permits one-direction electrical communications between stations. Technical arrangements may permit operation in either direction, but not simultaneously. Therefore, this term is qualified by one of the following suffixes: s/o for send only, r/o for receive only, or s/r for send or receive.

half-duplex modem — A modem with a single wire pair that can transmit in both directions but not simultaneously.

half-duplex operation — A duplex telegraph system capable of operating in either direction, but not in both simultaneously.

half-duplex repeater — A duplex telegraph repeater provided with interlocking arrangements that restrict the transmission of signals to one direction at a time.

half-hertz transmission — A data transmission and control system in which synchronized sources of 60-Hz power are used at transmitting and receiving ends. Either of two relays at the receiver can be actuated by proper choice of the half-hertz polarity of the transmitter power supply.

half-life — 1. The time in which half the atoms in a radioactive substance decay. In the first half-life interval, the amount of radioactive material left unchanged is one-half the original amount; in the next half-life interval, half of the remaining amount, or one-fourth of the original amount, remains. Thus, by determining the remaining radioactivity of a fossil and comparing it with the half-life of the material, scientists can estimate the age of the fossil. The half-life of various materials varies greatly — from millionths of a second to billions of years. 2. Time required for a source to decay to one-half its initial millicurie value. Typical values are Cs-137, 30 years; Co-60, 5.3 years; Sr-90, 28 years; Kr-85, 10.7 years; Am-241, 457 years; and Ra, 1622 years.

half-nut — A feed nut that engages half the circumference or less of a lead screw, so that it can be withdrawn from the lead screw to stop the lateral scanning movement.

half-power frequency — Either a high frequency or a low frequency at which the output of an amplifier, network, transducer, etc., falls to one-half (±3 dB) of its maximum or nominal response.

half-power point — On an amplitude response characteristic or other curve of the magnitude of a network quantity versus frequency, distance, angle, or other variable, the point that corresponds to half the power of a neighboring point having maximum power.

half-power width of a radiation lobe — In a plane containing the direction of the maximum of the lobe, the full angle between the two directions in that plane in which the radiation intensity is one-half the maximum value of the lobe.

half-shift register — 1. A logic circuit that consists of a gated input storage element with or without an inverter. 2. A logic device equivalent of half of a full master-slave flip-flop.

half step — See semitone.

half tap — A bridge that can be placed across conductors without disturbing their continuity.

half-time emitter — A device that produces synchronous pulses midway between the row pulses of a punched card.

half-tone characteristic — In facsimile, the fidelity of the recorded density shadings in comparison with the original transmitted subject copy. Also used to express the relationship between the facsimile signal and the subject or recorded copy.

half track — See two track.

half-track recorder — See dual-track recorder.

half-track tape — Also called two-track tape. Quarter-inch magnetic tape on which half the width of the tape is used for one sound path. Such a tape provides stereo in one direction of tape travel, or mono sound in both directions.

half-track tape-recording format — A professional tape recording standard format in which independent tracks (channels) are recorded in the same direction. Two tracks on $1/4$-inch (6.35-mm) wide tape or four tracks on $1/2$-inch (12.7-mm) wide tape or eight tracks on 1-inch (25.4-mm) wide tape or 16 tracks on 2-inch (50.8-mm) wide tape are used.

half wave — A wave with an electrical length of half a wavelength.

half-wave antenna — An antenna having an electrical length equal to half the wavelength of the signal being transmitted or received.

half-wave dipole — A straight, ungrounded antenna measuring substantially one-half wavelength.

half-wave rectification — 1. The production of a pulsating direct current by passing only half the input cycle of an alternating current. The other half is blocked by the rectifier. 2. The process of blocking the negative half cycle of current of an alternating input. This is accomplished by a single diode.

half-wave rectifier — A rectifier utilizing only one-half of each cycle to change alternating current into pulsating direct current.

half-wave transmission line — A piece of transmission line having an electrical length equal to half the wavelength of the signal being transmitted or received.

half-wave vibrator — A vibrator used mainly in battery-operated mobile power supplies. It has only one pair of contacts, and supplies an intermittent unidirectional current at its output (usually connected to a half-wave rectifier).

half word — A nonbroken sequence of bits or characters that makes up half a computer word and that can be addressed as a unit.

Hall constant — The constant of a proportionality in the equation for a current-carrying conductor in a magnetic field. The constant is equal to the transverse electric field (Hall field) divided by the product of the current density and the magnetic field strength. The sign of the majority carrier can be inferred from the sign of the Hall constant.

Hall effect — 1. In a current-carrying semiconductor bar located in a magnetic field that is perpendicular to the direction of the current, the production of a voltage perpendicular to both the current and the magnetic field. 2. The description given to the following phenomenon:

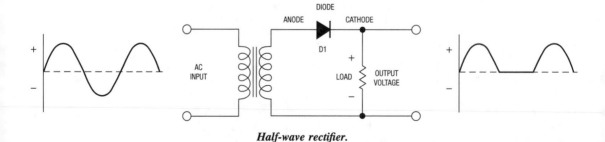

Half-wave rectifier.

when a conductor, through which a current is flowing, is placed in a magnetic field, a potential difference is generated between the two opposed edges of the conductor in the direction mutually perpendicular to both the field and the conductor.

Hall-effect generator — Also called Hall sensor. 1. A magnetic sensor using the Hall effect to give an output voltage proportional to magnetic field strength. 2. A device made of compounds such as indium arsenide, indium antimonide, or silicon that produces a useful Hall-effect potential. Its output is proportional to magnetic induction through the semiconductor.

Hall-effect modulation — Use of a Hall-effect multiplier as a modulator to produce an output voltage proportional to the product of two input voltages or currents.

Hall-effect switch — A keyboard switch that incorporates an IC chip containing a Hall generator, trigger circuit, and amplifier. Depressing the key moves the magnet shunt member across the chip. This increases the magnetic flux through the chip and causes the analog voltage generated by the Hall element to switch the trigger circuit to its on state.

Hall generator — A thin wafer of semiconductor material used for measuring ac power and magnetic field strength. Its output voltage is proportional to the current passing through it times the magnetic field perpendicular to it.

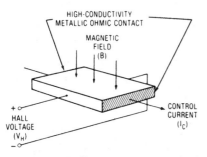

Hall generator.

Hall mobility — The product of conductivity and the Hall constant for a conductor or semiconductor. It is a measure of the mobility of the electrons or holes in a semiconductor.

Hall sensor — *See* Hall-effect generator.

halo — 1. The undesirable ring of light around a spot on the fluorescent screen of a cathode-ray tube. 2. The appearance of a blank border around unusually bright objects in a televised scene.

halogen — A general name applied to four chemical elements, fluorine, chlorine, bromine, and iodine, that have similar chemical properties.

halogen quenching — A method of quenching the discharge in a counter tube by the introduction of a small quantity of one of the halogens.

halt — The state in which a computer stops and does nothing.

ham — Also called amateur. Slang for a licensed radio operator who operates a station as a hobby rather than a business.

Hamming code — 1. One of the error correction code systems used in data transmission. 2. A forward error correction technique named for its inventor. It corrects single-bit errors. 3. A seven-bit error-correcting computer code.

hand capacitance — The capacitance introduced when one's hand is brought near a tuning capacitor or other insufficiently shielded part of a tuned circuit.

handheld — An amateur radio transceiver small enough to be carried in one hand (often abbreviated HT). Typically, amateur handhelds are for VHF/UHF use.

Handie-Talkie — Trade name of the Motorola Communications Division for a two-way radio small enough to be carried in one's hand.

handler — Also called device handler. 1. A section of a computer program used to control or communicate with an external device. 2. A software routine that controls the operation of a peripheral.

handoff — The transfer of a cellular phone call to a new cell, designed to be transparent to the cellular phone user. During a cellular conversation, when the user reaches the edge of the service area of a cell, computers in the network assign another tower in the next cell to provide the phone with continuing service.

hand receiver — An earphone held to the ear by hand.

hand reset — A relay in which the contacts must be reset manually to their original positions after normal conditions are resumed.

handset — 1. A telephone-type receiver and transmitter mounted on a single frame. 2. That portion of the telephone containing the transmitter and receiver that is hand-held when the telephone is in use. Consists of a receiver and transmitter about 6 inches (15.2 cm) apart at the ends of a common handle, connected by an electrical cord to the handset mounting. Sometimes includes a dial or Touchtone pad.

handset telephone — *See* hand telephone set.

hands-free telephone — A desk telephone containing a microphone and voice-switched amplifiers with a separate speaker unit, permitting telephone conversation without use of a hand-held handset.

handshake — 1. An interface procedure that is based on status/data signals that assure orderly data transfer as opposed to asynchronous exchange. 2. In communications, a preliminary exchange of predetermined signals performed by modems and/or terminals to verify that communication has been established and can proceed.

handshake cycle — The process whereby digital signals effect the transfer of each data byte across the interface by means of an interlocked sequence of status and control signals. "Interlocked" denotes a fixed sequence of events in which one event in the sequence must occur before the next event may occur.

handshaking — 1. A process in which predetermined arrangements of characters are exchanged by the receiving and transmitting equipment to establish sychronization. 2. The exchange of predetermined signals between machines connected by a communications channel to assure each that it is connected to the other. May also include the use of passwords and codes by an operator. 3. A colloquial term that describes the method used by a modem to establish contact with another modem at the other end of a telephone line. Often used interchangeably with buffering and interfacing, but with a fine line of difference in which handshaking implies a direct package-to-package connection regardless of functional circuitry. 4. A CPU-terminal interface process that prevents overrun and ignores signaling. The transmitter sends a signal in response to a request from the receiver, which then sends an acknowledgment signal to the transmitter. 5. The exchange of predetermined signals for control purposes during establishment of a connection between two data sets or modems. 6. Refers to the sequence of signals exchanged when a connection is established between two modems or between a DCE and a DTE. 7. Line-termination interplay to establish a data-communication path. 8. Exchange of predetermined signals establishing contact between two data sets. *See* answerback.

hand telephone set — Also called a handset telephone. A telephone set having a handset and a mounting that supports the handset when not in use.

hangover — 1. Also called tailing. The smeared or blurred bass notes reproduced by a poorly damped speaker or one mounted in an improperly vented enclosure. 2. In television, overlapping and blurring, in the direction opposite to subject motion, of successive frames as a result of improper transient response. 3. In facsimile, distortion that occurs when the signal changes from maximum to minimum at a slower rate than required, with the result that there is tailing on the lines in the copy.

hangup — 1. A condition in which the central processor of a computer is trying to perform an illegal or forbidden operation or in which it is continually repeating the same routine. 2. Commutation failure in which the load controlled by a solid-state relay cannot be turned off because the thyristor current does not reach zero or stay near zero for long enough for the gate circuit to regain control.

hard — 1. Indicating an electron tube that has been evacuated to a high degree. 2. Indicating X-rays of relatively high penetrating power.

hard automation — Production technique in which equipment is engineered specifically for a unique manufacturing sequence. Hard automation implies programming with hardware in contrast to soft automation, which uses software or computer programming.

hard contacts — Any type of physical switch contacts. Contrasted with electronic switching devices, such as triacs and transistors.

hard contact switch — A keyboard switch in which switching action is accomplished by the movement of one gold-plated bar against another at right angles — the classic cross-bar switching. The knife-edge contact area is extremely small (typically about 9×10^{-6} square inch or 5.8×10^{-5} cm^2), resulting in an extremely high contact pressure (typically about 5000 psi or 3.45×10^7 Pa).

hard copy — Also hardcopy. 1. Typewritten or printed characters on paper produced by a computer at the same time information is copied or converted into machine language that is not easily read by a human. 2. A printed copy of a machine output. 3. Data in a permanent and tangible form, such as printed, punched, or even handwritten. 4. The printed original copy of a message. 5. Output in printed form, such as the output of a teleprinter. Used as an adjective, signifies that the device involved produces such output. 6. Computer or machine output in a permanent, visually readable form. For example, printed reports, listings, translation lists, documents, and summaries. 7. A tangible, printed copy of a message, such as that obtained from a teletypewriter or computer, as opposed to a volatile display on a video terminal. *See also* printout.

hard-copy printer — An automatic device that produces intelligible symbols in a permanent form.

hard disk — A rigid disk of magnetic or magnetically coated material, rotating in a sealed housing and used as a recording and playback system for computer programs and data. Hard disks can store far more information than floppy disks and can write and read information more quickly.

hard-drawn copper wire — Copper wire that is not annealed after work hardening during drawing, thus providing increased tensile strength.

hardened links — 1. Transmission links for which special construction or installation is necessary to assure a high probability of survival under nuclear attack. 2. Passive protection to aid survival.

hardener — Also called curing agent. A chemical added to a thermosetting resin to stimulate curing.

hard error — 1. An error in magnetic media, electromechanical devices, or electronics that is repeatable. 2. Semiconductor memory condition in which the cell or cells will not properly store data under any condition, test, or operation.

hard firing — A condition in which the gate signal of an SCR is several times the dc triggering current and in which the rise time of the gate current is short relative to the turn-on time.

hard limiting — The condition in which limiting takes place for at least 20 dB into the signal noise.

hardline — 1. The intelligence link between two objects that consists of a wire or wires, as opposed to a radio or radar link. 2. A low-loss coaxial cable that has a continuous hard metal shield instead of a conductive braid around the outer perimeter.

hard magnetic materials — Magnetic materials that are not easily demagnetized.

hardness — 1. Referring to X-rays, the quality that determines their penetrating ability. The shorter the wavelength, the harder and hence more penetrating they are. 2. Property of an installation, facility, transmission link, or equipment that will prevent an unacceptable level of damage.

hardness tester — Equipment for determining the force required to penetrate the surface of a solid.

hard rubber — A material formerly widely used for insulation. It is formed by vulcanizing rubber at high temperature and pressure to give it the desired hardness.

hard scrambling — An encryption method that uses proprietary, highly secure technology (i.e., digital), such as that used by VideoCipher II.

hard-sectored — A disk whose sectors are marked by holes in the disk itself.

hard solder — Solder composed principally of copper and zinc. It must be red hot before it will melt. Hard soldering is practically equivalent to brazing.

hard soldering — Process of joining two metals by utilizing an alloy with a melting temperature higher than 800°F (427°C). *See* soft soldering.

hard tube — A high-vacuum electronic tube.

hardware — 1. Mechanical, magnetic, electrical, or electronic devices; physical equipment from which a system is fabricated. (Contrasted with software.) 2. Particular circuits or functions built into a system. 3. The physical components of a computer or a system. (Software is the term used to describe the programs and instructions for a computer.) 4. The electronic components, such as gates, inverters, and storage devices, that make up a system (as opposed to software and firmware). 5. Items of equipment used in a communications or data processing system. 6. The physical equipment components of a computer system, e.g., mechanical, magnetic, electrical, or electronic devices.

hardware buffer — A register or set of registers used to store information temporarily, usually to act as a transition medium between a fast and a slow device.

hardware independent — Computer software that is not dependent on a certain make of computer.

hardwire — A colloquialism meaning a circuit evidencing dc continuity.

hardwired — 1. Electrical devices interconnected through physical wiring. 2. The implementation of a function with logic gates; i.e., hardware as opposed to software. 3. Electronic-programming technique using soldered connections; hence, not readily reprogrammable. 4. Physically interconnected, usually for a specific purpose. Hardwired logic is essentially unalterable. 5. Pertaining to the physical connection of two pieces of electronic equipment by means of a cable or wires.

hard-wired logic — 1. A group of solid-state logic modules mounted on one or more circuit boards and interconnected by electrical wiring. The logic control functions are determined by the way in which the modules are interconnected. (As contrasted with a programmable controller or microprocessor, in which the logic is in program form.) 2. A group of solid-state logic modules mounted on one or more circuit boards and interconnected by electrical wiring. The logic control functions are determined by the way in which the modules are interconnected. 3. A group of logic circuits permanently interconnected to perform a special function. Permanently assigned device addresses, memory block assignments, and interrupt vector addresses.

hard X-rays — Highly penetrating X-rays, as distinguished from less penetrating, or soft, X-rays.

harmful interference — Any radiation or any induction that disrupts the proper functioning of an electromagnetic system.

harmonic — A sinusoidal wave having a frequency that is an integral multiple of the fundamental frequency. For example, a wave with twice the frequency of the fundamental is called the second harmonic.

harmonica bug — *See* infinity device.

harmonic analysis — 1. A method of identifying and evaluating the harmonics that make up a complex waveform of voltage, current, or some other varying quantity. 2. The expression of a given function as a series of sine and cosine terms that are approximately equal to the given function, such as a Fourier series. 3. Defining a complex wave as the sum of several harmonics of the fundamental wave, each harmonic having a specified magnitude and phase.

harmonic analyzer — Also called harmonic-wave analyzer. A mechanical or electronic device for measuring the amplitude and phase of the various harmonic components of a wave from its graph.

harmonic antenna — An antenna whose electrical wavelength is an integral multiple of a half wavelength.

harmonic attenuation — Elimination of a harmonic frequency by using a pi network and tuning its shunt resistances to zero for the frequency to be eliminated.

harmonic component — Of a periodic quantity, any one of the simple sinusoidal quantities of the Fourier series into which the periodic quantity may be resolved.

harmonic content — 1. The degree of distortion in the output signal of an amplifier. 2. The components remaining after the fundamental frequency has been removed from a complex wave.

harmonic conversion transducer — A conversion transducer in which the useful output frequency is a multiple or submultiple of the input frequency.

harmonic detector — A voltmeter circuit that measures only a particular harmonic of the fundamental frequency.

harmonic distortion — 1. The production of harmonic frequencies at the output by the nonlinearity of a transducer when a sinusoidal voltage is applied to the input. The amplitude of the distortion is usually a function of the amplitude of the input signal. 2. The voltages of harmonics resulting from amplitude distortion expressed as a percentage of the voltage of the fundamental. A common measurement is total harmonic distortion (THD), in which the fundamental of a very low distortion sine-wave test signal is removed by a steep notch filter. The summed harmonics that remain are then measured as a voltage and expressed as a percentage (or dB value) of the voltage of

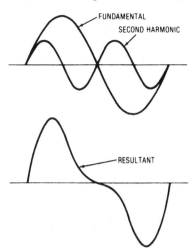

Harmonic content, 1.

the fundamental at the required test power of the amplifier. (When measured in this manner, the term THD is not really correct since the distortion also has the amplifier noise added to it within the test passband. The correct term is *distortion factor*.) 3. The sum of all signals in an output that are multiples of the input signal frequencies (harmonics). Their intensities are expressed as a percentage of the total output intensity. 4. The production of spurious frequencies, not present in the original sound, that are multiples of the original sound frequency. For example, a 100-Hz tone in the original may produce spurious tones at 200 Hz, 300 Hz, and so on. The result is an audible blurring or loss in the reproduced sound. Total harmonic distortion of no more than 1 percent is considered to be inaudible. 5. A data-communication-line impairment caused by erroneous frequency generation along the line.

harmonic filter — A combination of inductance and capacitance tuned to an undesired harmonic to suppress it.

harmonic generator — A vacuum tube transistor, or other generator operated so that it generates strong harmonics in the output.

harmonic interference — Interference between radio stations because harmonics of the carrier frequency are present in the output of one or more stations.

harmonic-leakage power (tr and pre-tr tubes) — The total radio-frequency power transmitted, through the fired tube in its mount, at other than the fundamental frequencies generated by the transmitter.

harmonic motion — Back and forth motion, such as that of a pendulum, in which the distance on one side of equilibrium always equals the distance of the other side; the acceleration is toward the point of equilibrium and directly proportional to the distance from it. Graphically, harmonic motion is represented by a sine wave.

harmonic oscillator — 1. A circuit in which the oscillating frequency of the active device and the output frequency are not the same. For example, in a push-push configuration, each transistor oscillates at f_o but the output is combined to provide $2f_o$. 2. An oscillator whose output is very nearly a sine wave and whose output amplitude and frequency are very nearly constant.

harmonic producer — A tuning-fork-controlled oscillator used to provide carrier frequencies for broadband carrier systems. It is capable of producing odd and even harmonics of the fundamental tuning-fork frequency.

harmonic ringing—A system of selectively signaling several parties on a subscriber's line. The different rings are produced by currents that are harmonics of several fundamental frequencies.

harmonics—Undesired signals that appear at multiples (2, 3, etc.) of the desired fundamental frequency. They are produced by nonlinear amplifiers and can cause interference and other phenomena.

harmonic selective ringing—Selective ringing that employs currents of several frequencies and ringers, each tuned mechanically or electrically to the frequency of one of the ringing currents, so that only the desired ringer responds.

harmonic series of sounds—A series in which each basic frequency is an integral multiple of a fundamental frequency.

harmonic telephone ringer—A ringer that responds only to alternating current within a very narrow frequency band. A number of such ringers, each responding to a different frequency, are used in one type of selective ringing.

harmonic-wave analyzer—See harmonic analyzer.

harness—1. Wires and cables arranged and tied together so they can be connected or disconnected as a unit. 2. A group of conductors laid parallel or twisted by hand, usually with many breakouts, laced or bundled together, or pulled into a rubber or plastic sheath. Used to interconnect electrical circuits.

hartley—In computers, a unit of information content equal to one decimal decision, or the designation of one of ten possible and equally likely values or states of anything used to store or convey information. One hartley equals $\log_2 10 = 3.23$ bits.

Hartley oscillator—An oscillator in which a parallel-tuned tank circuit is connected between the grid and plate of an electron tube or between the base and collector of a junction transistor, the inductive element of the tank having an intermediate tap at the cathode or emitter potential.

hash—1. Electrical noise generated within a receiver by a vibrator or a mercury-vapor rectifier. See also grass. 2. A completely random interfering signal usually caused by arcing and occasionally by natural environmental disturbances.

hash-mark stripe—A noncontinuous helical stripe applied to a conductor for identification.

hash total—In a computer, a total for checking purposes. It is determined by adding all the digits or all the numbers in a particular field in a batch of unit records to be processed or manipulated, with no attention paid to the meaning or significance of the total. After processing, the hash total is recalculated and compared with the original total. If the two do not match, the original data has been changed in some way.

hat—To arrange a fixed number of symbols or groups of symbols in a random sequence, as if they had been drawn from a hat.

hatted code—A randomized code consisting of an encoding section. The plain-text groups are arranged in a significant order, accompanied by their code groups arranged in a random order.

haul—An arbitrary classification of telephone toll calls, as follows: short haul, less than 30 miles or 48 km; medium haul, 30 to 1000 miles or 48 to 1609 km; long haul, over 1000 miles or 1609 km.

Hay bridge—A four-arm, alternating-current bridge used for measuring inductance in terms of capacitance, resistance, and frequency. The arms adjacent to the unknown impedance are nonreactive resistors, and the opposite arm is composed of a capacitor in series with a resistor (unlike the Maxwell bridge, where it is in parallel). Usually the bridge is balanced by adjustment of the resistor, which is also in series with the capacitor and one of the nonreactive arms. The balance depends upon the frequency.

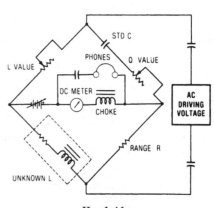

Hay bridge.

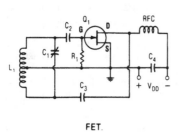

FET.

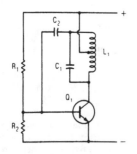

Bipolar transistor.

Hartley oscillator.

Hayes-compatible—A modem that recognizes the commands of a Hayes-manufactured modem.

hazard—See failure rate.

hazard rate—At a particular time, the rate of change of the number of items that have failed divided by the number of items surviving.

Hazeltine neutralizing circuit—An early form of neutralized radio-frequency amplifier circuit.

H-beacon—A nondirectional homing beacon with an output power of 50 to 2000 watts.

H-bend—Also called H-plane bend. In waveguide technique, a smooth change in the direction of the axis of the waveguide. Throughout the change, the axis remains perpendicular to the direction of polarization.

HBO — Abbreviation for Home Box Office. The most popular pay-TV network, which is distributed on SAT-COM F1.

HCD — Abbreviation for hot-carrier diode.

H-display — Also called H-scan. In radar, a B-display modified to indicate the angle of elevation. The target appears as two closely spaced blips that approximate a short, bright line that slopes in proportion to the sine of the angle of target elevation.

HDLC — Abbreviation for high-level data link control. 1. A packet transmission protocol developed by the International Standards Organization (ISO) derived from IBM's Synchronous Data Link Control (SDLC). Messages are transmitted in units called frames, which can contain differing amounts of data but must be organized in a particular way. 2. A CCITT standard data-communication line protocol.

HDTV — *See* high-definition television.

HDX — *See* half-duplex circuit.

head — 1. A device that reads, records, or erases data on a storage medium. For example, a small electromagnet used to read, write, or erase data on a magnetic drum or tape, or the set of perforating reading or marketing devices used for punching, reading, or printing on paper tape. 2. In a tape recorder, any device intended to change the magnetic state of the tape. Specialized types of heads are used for erasing the tape, recording a signal on it, and playing back the signal from it. In many recorders, the recording and playback functions are both performed (at different times) by the same head. 3. An electromagnetic device, usually consisting of a ring-shaped metal core wound with coils of wire, in which the continuity of the core is broken at one place, called the gap. Tape touches the head at the gap as it moves past it. A reproducing or playback head senses signals already magnetized (recorded) on tape and transforms them into electrical impulses that are then amplified and fed to a loudspeaker. A recording head accepts electrical signals and transforms them into magnetic impulses that are deposited on tape as it passes the head. Most cassette recorders use a single, combination record/playback head.

head alignment — Also called azimuth alignment. 1. Positioning the record/playback head on a tape recorder so that its gap is perpendicular to the tape. 2. On a VTR or VCR, the positioning of the audio or video heads so that they describe the current path at the correct angle across the videotape; heads that are out of alignment won't record or play back properly. 3. Adjustment of the recording or reproducing head so that it's at right angles to the longitudinal axis of tape. 4. Mechanical adjustment of the spatial relationships between the head gaps and the tape.

head amplifier — An audio-frequency amplifier mounted on or near the sound head of a motion-picture projector to amplify the extremely weak output of the phototube.

head demagnetizer — Also called degausser. 1. A special demagnetizer with elongated pole pieces enabling them to be brought into proximity to head surfaces for elimination of the built-up magnetic charge that develops over a period of time as a result of asymmetrical electrical input signals. 2. A device used to neutralize possible residual or induced magnetism in heads or tape guides.

headend — 1. In a broadband transmission network, a group of active and/or passive components that translate one range of frequencies (Transmit) to a different frequency band (Receive); allows devices on a single cable network to send and receive signals without interference. 2. The master distribution center of a satellite TV

(SATV) system in which the incoming television signals from space and distant broadcast stations are received, amplified, and remodulated onto television channels for transmission down the SATV coaxial cable. 3. In a broadband local area network, the central location that has access to signals traveling in both inbound and outbound directions.

header — 1. The part of a sealed component or assembly that provides support and insulation for the leads passing through the walls. 2. A feedthrough device that forms a conductive path through an insulating plate or surface. 3. The part of a semiconductor device package to which the actual chip or die is mounted. May consist of metal, ceramic, or one of a number of plastics, such as epoxy resin. 4. The portion of a device package from which the external leads extend. Examples include the TO-5 header and flat-pack case. 5. An information structure that precedes and identifies the information that follows, such as a block of bytes in communications, a file on a disk, a set of records in a database, or an executable program.

header card — A card containing information about the data in other cards that follow.

header record — A computer input record that contains common, constant, or identifying information for other records that follow.

head gap — A space inserted intentionally into the magnetic circuit of a magnetic recorder head to force or direct the flux into the recording medium.

head guy — The messenger cable and attachments placed so they pull toward the pole line.

heading — The direction of a ship, aircraft, or other object with reference to true, magnetic, compass, or grid north.

headlight — An aircraft radar antenna small enough to be housed in the wing, like an automobile headlight. The beam operates like a searchlight.

headphone — Also called a head receiver or phone. 1. A device held against the ear and having a diaphragm that vibrates according to current variations. It reproduces the incoming electrical signals as sound. Thus it permits private listening to a receiver amplifier or other device. 2. Small sound reproducers, superficially resembling miniature loudspeakers, set in a suitable frame for wearing about the head and listening to by close coupling to the ears. Recent headphones, improved greatly in fidelity, have become increasingly popular among audiophiles for private listening without disturbing others, as well as to prevent outside noises from interfering with the listening. Headphones are available in mono or stereo. 3. A telephone receiver held against the ear by a headband. *See also* headset.

head receiver — *See* headphone.

head room — 1. The safety margin that is normally provided between the maximum recording level as indicated on a recorder level indicator and the actual point of severe tape overload. Most good recorders provide 6 to 8 dB of head room above the indicated 0-vu or normal maximum indicated recording level, to allow for the inability of the needle of the VU meter needle to respond fully to sudden, intense bursts of signal energy. 2. The difference between the nominal operating level and the maximum level at any point in a system or device. Usually expressed in decibels. 3. The margins between an actual signal operating level and the level that would cause substantial distortion. For a tape recorder this would be the level above 0 vu that gives a (specified) distortion.

headset — 1. A headphone (or a pair of headphones) and its associated headband and connecting cord. 2. Small portable telephone receivers, usually in pairs, with a connecting clamp to support the phones against the ears, for operators of receiving equipment. 3. An operator's head telephone set.

headset cord — A very flexible cord used for communication equipment, usually 24 AWG to 22 AWG multiconductor. Usually made with Buna insulation, rubber, or neoprene jacket; sometimes the outer jacket is a cotton braid. The conductor may be bare copper or cadmium bronze.

headshell — The end of a pickup arm where the cartridge fits. Sometimes bonded to the arm, though often detachable.

head stack — A group of two or more heads mounted in a single unit, used to provide multiple-track recording or reproduction.

head-to-tape contact — The degree to which the surface of the magnetic coating approaches the surface of the record or replay heads during normal operation of a recorder. Good head-to-tape contact minimizes separation loss and is essential in obtaining high resolution.

heap — A storage area used for dynamically allocated variables created by a running process without correlation to the static structure of the program.

hearing aid — A small audio-reproducing system for the hard of hearing. It consists of a microphone, amplifier, battery, and earphone and is used to increase the sound level normally received by the ear.

hearing loss — Also called deafness. 1. The hearing loss of an ear at a specified frequency — i.e., the ratio, expressed in decibels, of its threshold of audibility to the normal threshold. 2. The difference in level, expressed in decibels, between the weakest sound a particular human ear can hear and the weakest sound heard by an average, normal ear.

hearing loss for speech — The difference in decibels between the speech levels at which the average normal ear and the defective ear, respectively, reach the same intelligibility. It is often arbitrarily set at 50 percent.

heart pacer — *See* pacemaker.

heat aging — A test used to indicate the relative resistance of various insulating materials to heat degradation.

heat coil — A protective device that grounds or opens a circuit, or both, when the current rises above a predetermined value. A mechanical element moves when the fusible substance that holds it in place is heated above a certain point by current through the circuit.

heater — Also called filament. 1. An element that supplies the heat to an indirectly heated cathode. 2. A resistor that converts electrical energy into heat.

heater biasing — Application of a dc potential to the heater of a vacuum tube to eliminate diode conduction between it and some other element within the tube.

heater cord — Flexible stranded copper conductor, cotton wrapped, with rubber insulation and asbestos roving. For indoor use on household appliances.

heater current — The current flowing through a heater in a vacuum tube.

heater voltage — The voltage between the terminals of a heater.

heater-voltage coefficient — In a klystron, the frequency change per volt of heater voltage change when the reflector voltage is adjusted for the peak of a reflector voltage mode.

heat-eye tube — A cathode-ray tube powered by a midget generator. It is used as an infrared instrument that can "see" in the dark.

heat gradient — The difference in temperature between two parts of the same object.

heating effect of a current — Assuming a constant resistance, the amount of heat produced by the current through it. It is proportionate to the square of the current.

heating element — The wirewound resistor, terminals, and insulating supports used in electric cooking and heating devices.

heating pattern — In induction or dielectric heating, the distribution of temperature in a load or charge.

heating station — In induction or dielectric heating, the work coil or applicator and its associated production equipment.

heat loss — The loss due to conversion of part of the electric energy into heat.

heat of emission — Additional heat energy that must be supplied to an electron-emitting surface to keep its temperature constant.

heat of radioactivity — Heat generated by radioactive disintegration.

heat sealing — A method of joining plastic films by simultaneous application of heat and pressure to areas in contact. Heat may be supplied conductively or dielectrically.

heatseeker — A guided missile that uses an infrared sensor to detect and home in on an enemy target. The missile is guided by the high infrared emissions produced by a target, such as the heat from an aircraft or tank engine.

heat sensor — 1. A sensor that responds to either a local temperature above a selected value or a local temperature increase that is at a rate of increase greater than a preselected rate. 2. A sensor that responds to infrared radiation from a remote source, such as a person.

heat shock — Test to determine stability of material by sudden exposure to a high temperature for a short period.

heat sink — 1. A mounting base, usually metallic, that dissipates, carries away, or radiates into the surrounding atmosphere the heat generated within a semiconductor device. The package of the device often serves as a heat sink, but, for devices of higher power, a separate heat sink on which one or more packages are mounted is required to prevent overheating and consequent destruction of the semiconductor junction. 2. A mass of metal that is added to a device for the absorption or transfer of heat away from critical parts. Generally made from aluminum to achieve high heat conductivity and light weight, most heat sinks are of one-piece construction. They may also be designed for mounting on printed circuit boards. 3. A method used to transfer a rise in temperature. A metal plate or fin-shaped object with good heat-transfer efficiency that helps dissipate heat into the surrounding air, into a liquid, or into a larger mass. 4. A material capable of absorbing heat; a device utilizing such material for the thermal protection of components or systems. 5. A metal part with maximized surface areas to remove heat from electronic components, such as transistors, integrated circuit amplifiers, etc. 6. A mass of metal, often with fins, mounted on or under a circuit component that produces heat, such as a power transistor, silicon rectifier, etc. The heat sink absorbs and then radiates the heat to maintain a safe working temperature for the component.

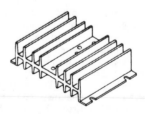

Heat sink.

heat sink compound—A silicon compound filled with alumina or other heat-conductive oxide. Used to fill voids and irregularities in surfaces between two mating objects to permit optimum heat transfer.

heat soak—Heating a circuit over a period to allow all parts of the package and circuit to stabilize at the same temperature.

heat waves—Infrared radiation similar to radio waves but of a higher frequency.

heat-writing recorder—A type of stripchart recorder in which a heated stylus writes on a strip of chemically treated paper. The paper is discolored by the heat, and the path followed by the pen over the surface of the paper is thus made visible.

Heaviside-Campbell mutual-inductance bridge—A Heaviside mutual-inductance bridge in which one inductive arm contains a separate inductor that is included in the bridge arm during the first of a pair of measurements and is short-circuited during the second. The balance is independent of frequency. *See also* Heaviside mutual-inductance bridge.

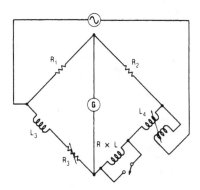

Heaviside-Campbell mutual-inductance bridge.

Heaviside layer—Also called the Kennelly-Heaviside layer. The region of the ionosphere that reflects radio waves back to earth.

Heaviside mutual-inductance bridge—An alternating-current bridge normally used for the comparison of self- and mutual inductances. Each of the two adjacent arms contains self-inductance, and one or both of them have mutual inductance to the supply circuit. The other two arms normally are nonreactive resistors. The balance is independent of frequency.

heavy hydrogen—Another term for deuterium ($_1H^2$) or tritium ($_1H^3$).

hecto-—A prefix meaning 100.

hectometric wave—An electromagnetic wave between the wavelength limits of 100 and 1000 meters, corresponding to the frequency range 300 kHz to 3 MHz.

heelpiece—1. Part of a relay magnetic structure at the end of a coil, opposite the armature. It generally supports the armature and completes the magnetic path between it and the core of the coil. 2. The base of a relay, on which one or more contact spring assemblies are mounted, and to which the core of a relay is fastened.

Hefner lamp—A standard source that gives a luminous intensity of 0.9 candlepower.

HEI—Abbreviation for high-energy ignition. Conventional (General Motors) automotive electronic ignition

that replaces the distributor point system. Provides high voltage for high-energy spark timing control.

height control—In a television receiver, the adjustment that determines the amplitude of the vertical-scanning pulses and hence the height of the picture.

height finder—A radar that measures the altitude of an airborne object.

height input—Information regarding target height received by a computer from a height finder and relayed by way of a ground-to-ground data link or telephone.

height overlap coverage—A region of height-finder coverage within which there is duplicated coverage from adjacent height finders of other radar stations.

height-position indicator—A radar display that simultaneously shows the angular-elevation slant range and height of objects.

height-range indicator—A cathode-ray tube from which altitude and range measurements of airborne objects may be viewed.

Heising modulation—*See* constant-current modulation.

helical—Spiral-shaped.

helical antenna—Also called a helical-beam antenna. 1. A spiral conductor wound around a circular or polygonal cross section. The axis of the spiral normally is mounted parallel to the ground and fed at the adjacent end. The radiation produced has approximately a circular polarization and is confined mainly to a single lobe located along the axis of the spiral. 2. An antenna made of wire wound as a coil, usually on a Fiberglas rod and with the wire usually within Fiberglas.

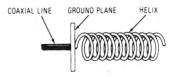

Helical antenna.

helical-beam antenna—*See* helical antenna.

helical potentiometer—A precision potentiometer that requires several turns of the control knob to move the contact arm from one end of the spiral-wound resistance element to the other end.

helical scanning—1. Radar scanning in which the rf beam describes a distorted spiral motion. The antenna rotates about the vertical axis while the elevation angle rises slowly from 0° to 90°. 2. Method of facsimile scanning in which the elemental area sweeps across the copy in a spiral motion.

helical stripe—A continuous, colored, spiral stripe applied to a conductor for circuit identification.

helicone—A circularly polarized antenna that produces a low side-lobe level. It consists of a helix excited in the axial mode and placed inside a conical horn. The axial length of the helix is approximately equal to the altitude of the truncated cone.

helionics—The conversion of solar heat to electric energy.

helitron oscillator—An electrostatically focused, low-noise, backward-wave microwave oscillator. The frequency of the output signal can be swept rapidly through a wide range by variation of the voltage applied between the cathode and the associated rf circuit.

helium tight—*See* hermetic.

helix—A spiral winding, such as a coil.

helix recorder — A recorder in which helical scanning is used.

Helmholtz coil — A phase-shifting network used for determining the range in certain types of radar equipment. It consists of fixed and movable coils. The phase is kept constant at the input, but may be continually shifted from 0° to 360° at the output.

Helmholtz resonator — An acoustic enclosure with a small opening that causes the enclosure to resonate. The frequency at which it does depends on the geometry of the resonator.

hemimorphic — Terminated at the two ends by dissimilar sets of faces.

HEM wave — *See* hybrid electromagnetic wave.

henry — The cgs electromagnetic unit of inductance or mutual inductance. The inductance of a closed circuit in which an electromotive force on 1 volt is produced when the electric current in the circuit is varied uniformly at a rate of 1 ampere per second. Letter symbol: H.

heptode — A vacuum tube that contains seven electrodes: an anode, a cathode, a control electrode, and four additional electrodes, usually grids.

hermaphrodite connector — A cable connector in which the jack and the plug are identically shaped.

hermaphroditic connector — 1. A connector in which both mating members are exactly alike at their mating face. There are no male or female members, but provisions have been made to maintain correct polarity, hot lead protection, sealing, and coupling. 2. Either of a pair of coaxial connectors whose mating faces are alike. 3. A connector design in which pin and socket contacts are arranged in a balanced manner such that both mating connectors are identical. The contacts may also be hermaphroditic and arranged as male and female contacts, as for pins and sockets. Hermaphroditic contacts may also be used in a manner such that one half of each contact mating surface protrudes beyond the connector interface, and both mating connectors are identical.

hermaphroditic contact — 1. A contact designed so that it is neither pin nor socket and can be mated with any other contact of the same design. 2. Slotted contacts with beveled contacting surfaces in which both mating portions are exactly alike in shape, but mate at a 90° angle from each other.

hermetic — Also called helium tight, leak tight, and vacuum tight. 1. Pertaining to permanent sealing, by fusion, soldering, or other means, to prevent the transmission of gases. 2. A characteristic of packages providing an absolute seal against moisture to prevent degradation. These packages are generally metal cans, dual-in-line packages (DIPs) with solder seals, or ceramic packages. Plastic packages, although less expensive, are not hermetic.

hermetically sealed — Contained within an enclosure that is sealed by fusion or other comparable means to ensure a low rate of gas leakage over a long period. This generally refers to metal-to-metal or metal-to-glass seal.

hermetic seal — 1. An airtight seal between two parts of a container, such as between the can and header of a metal component package. 2. A mechanical or physical closure that is impervious to moisture or gas, including air. Usually pertains to an envelope or enclosure containing electronic components or parts, or to a header. 3. An IC package enclosure technique by which the IC chip is completely protected from environmental contaminants.

herringbone pattern — Television interference seen as one or more horizontal bands of closely spaced V- or S-shaped lines.

hertz — The standard unit for frequency, equivalent to one cycle per second. Letter symbol: Hz. Named after H. R. Hertz, a 19th-century German physicist.

Hertz antenna — 1. An antenna system that does not depend for its operation on the presence of ground. Its resonant frequency is determined by its distributed capacitance, which varies according to its physical length. 2. An elementary linear dipole radiator; it may or may not have spherical or flat-plate ends.

Hertz effect — The ionization and spark emission due to exposure to ultraviolet radiation.

Hertzian oscillator — A type of oscillator for producing ultrahigh-frequency oscillations. It consists of two metal plates or other conductors separated by an air gap. The capacitor formed has such a small capacitance that ultrahigh-frequency oscillations can occur.

Hertzian waves — Electromagnetic waves of frequencies between 10 kHz and 30,000 GHz. Radio waves.

hertz-matching loran — *See* low-frequency loran.

Hertz vector — A vector that specifies the electromagnetic field of a radio wave. Both the electric and the magnetic intensities can be specified in terms of it.

heterodyne — Also called beat. To mix two frequencies together in a nonlinear component in order to produce two other frequencies equal to the sum and difference of the first two. For example, heterodyning a 100-kHz and a 10-kHz signal will produce a 110-kHz (sum frequency) and a 90-kHz (difference frequency) signal.

heterodyne conversion transducer (converter) — A conversion transducer in which the output frequency is the sum or difference of the input frequency and an integral multiple of the frequency of another wave.

heterodyne detection — Detection (or conversion) by mixing two signals together to generate the intermediate frequency in a superheterodyne receiver or to make cw signals audible.

heterodyne detector — A detector that converts an incoming rf signal to an audible tone by heterodyning. It incorporates a local oscillator (called a beat-frequency oscillator).

heterodyne frequency — 1. The sum or difference frequency produced by combining two other frequencies. 2. Either of the two frequencies, the sum or the difference, that result from an amplitude-modulation process.

heterodyne frequency meter — *See* heterodyne wavemeter.

heterodyne oscillator — An oscillator that produces a desired frequency by combining two other frequencies (e.g., two radio frequencies to produce an audio frequency, or the incoming and local-oscillator frequencies to produce the intermediate frequency of a superheterodyne receiver).

heterodyne principle — *See* heterodyne.

heterodyne reception — Also called beat reception. 1. Reception by combining a received high-frequency wave with a locally generated wave in a nonlinear device to produce sum and difference frequencies at the output. 2. A form of reception in which the receiver combines the incoming signal with a locally generated signal of different frequency. The combination creates a new signal, the intermediate frequency, at the difference, or beat, frequency between the two components.

heterodyne repeater — A radio repeater in which the incoming radio signals are converted to an intermediate frequency, amplified, and reconverted to another frequency band before being transmitted over the next repeater section.

heterodyne-type frequency meter — An instrument for measuring frequency by producing a zero difference frequency (zero beat) between the signal under test and an internally generated signal.

heterodyne wavemeter — A wavemeter employing the heterodyne principle to compare the frequency being measured with a frequency being generated in a calibrated oscillator circuit.

heterodyne whistle — A steady squeal heard in a radio receiver when the signals from stations having nearly equal frequencies beat together.

heterodyning — *See* heterodyne.

heterogeneity — A state or condition of being unlike in nature, kind, or degree.

heterogeneous — Composed of different materials (opposite of homogeneous).

heterojunction — An interface between two semiconductors of different chemical compositions (for example, between indium gallium arsenide phosphide and indium phosphide) but not necessarily different majority carrier types (n-type or p-type).

heterosphere — The portion of the upper atmosphere in which the relative proportions of oxygen, nitrogen, and other gases are unfixed and radiation particles and micrometeoroids are mixed with the air particles.

heuristic — 1. Pertaining to exploratory problem-solving methods in which solutions are discovered through evaluation of the progress made toward the final result (as opposed to algorithmic methods). 2. Empirical. Referring to knowledge or procedures determined by experience, but difficult to prove. 3. A problem-solving technique in which general principles and rules of thumb are used to approach a solution. Unlike an algorithm, the heuristic approach does not guarantee a solution. However, heuristic methods can result in a faster and simpler solution, and where algorithms are not available they are often the only resource. 4. A rule of thumb or an educated guess that simplifies and limits a search for solutions and applications that are difficult or poorly understood.

heuristic program — A set of computer instructions that simulate the behavior of human operators in approaching similar problems.

Hewlett-Packard Interface Bus — Abbreviated HP-IB. The Hewlett-Packard implementation of the IEEE-488 bus used to interface multiple devices by a well-defined hardware protocol. *See* GPIB.

hex — The hexadecimal numbering system. Since 16 is a power of 2, binary numbers are easily converted into hex, so machine-language computer programs are often written in hex to save space. For example, binary 11010110 (decimal 214) could be written in hex as D6.

hexadecimal — 1. Number system using 0, 1, . . . , A, B, C, D, E, F to represent all the possible values of a four-bit digit. The decimal equivalent is 0 to 15. Two hexadecimal digits can be used to specify a byte. 2. A counting system similar to BCD. 3. A number system using the equivalent of the decimal number 16 as base. *Compare* binary number system.

hexadecimal counter — *See* divide-by-16 counter.

hexadecimal display — A solid-state display capable of exhibiting numbers 0 through 9 and alphabet characters A through F.

hexadecimal number system — A number system having as its base the equivalent of the decimal number 16.

hex inverter — A group of six logic inverters contained in a single package.

hexode — A vacuum tube containing six electrodes: an anode, a cathode, a control electrode, and three additional electrodes, usually grids.

HF — Abbreviation for high frequency.

Hexadecimal system

Decimal	Binary	Octal	Hexadecimal
0	0	0	0
1	1	1	1
2	10	2	2
3	11	3	3
4	100	4	4
5	101	5	5
6	110	6	6
7	111	7	7
8	1000	10	8
9	1001	11	9
10	1010	12	A
11	1011	13	B
12	1100	14	C
13	1101	15	D
14	1110	16	E
15	1111	17	F

H-**field sensor** — A passive sensor that detects changes in the earth's ambient magnetic field caused by the movement of an intruder. *See also E*-field sensor.

HH beacon — A nondirectional radio homing beacon with a power output of 2000 watts or more.

HIC — Abbreviation for hybrid integrated circuit.

hidden codes — Formatting codes embedded into a document that are not visible on the screen.

hidden file — Also called invisible file. A file in a computer that occupies disk space but does not appear in directory listings. Files are hidden to prevent their display or change. Users cannot display, erase, or copy hidden files.

hierarchical network — Also called a tree network. A network topology organized in the form of a pyramid with one terminal at the top and increasing numbers of terminals at each lower level.

hierarchy — 1. A series of items classified according to rank or order. 2. The order in which arithmetic operations within a formula or a statement will be executed. 3. An arrangement into a graded series.

hi-fi — *See* high fidelity.

high-altitude electromagnetic pulse — An electromagnetic radiation of very short rise time, large amplitude, and brief duration that follows a nuclear explosion above the atmosphere.

high band — Television channels 7 to 13, covering a frequency range of 174 to 216 MHz.

high boost — *See* high-frequency compensation.

high-contrast image — A picture in which strong contrast between light and dark areas is visible. Intermediate values, however, may be missing.

high definition — The condition of a reproduced television or facsimile image in which it contains sufficient accurately reproduced elements for the picture details to approximate those of the original scene.

high-definition television — Abbreviated HDTV. A television picture of higher quality than standard NTSC video. This usually involves 720 lines, progressive scan more horizontal lines (e.g., 1080 or higher rather than 525), a different aspect ratio (16 : 9 rather than 4 : 3), and Dolby digital audio.

high-density disk — A floppy disk that holds more information than a double-density disk. A 3.5-inch high-density disk holds 1.44 megabytes.

high-energy materials — Also called hard magnetic materials. Magnetic materials having a comparatively high-energy product; e.g., materials used for permanent magnets.

higher-level language — A programming language that closely resembles natural language. A statement in a higher-level language will produce many machine-language instructions. The higher-level languages are usually independent of the computer.

highest probable frequency — Abbreviated HPF. An arbitrarily chosen frequency value 15 percent above the F_2 layer MUF (maximum usable frequency) for the radio circuit. For the E-layer, the HPF is equal to the MUF.

high fidelity — Popularly called hi-fi. 1. The characteristic that enables a system to reproduce sound as nearly like the original as possible. 2. Reproduction of audio so perfect that listeners hear exactly what they would have heard if present at the original performance. 3. The reproduction of audio sounds so perfectly that a listener is not aware of any loss of naturalness.

high-fidelity receiver — A radio receiver capable of receiving and reproducing, without noticeable distortion, the original modulation impressed on the carrier waves.

high filter — An audio circuit designed to remove undesired high-frequency noise from the program material. Such noise includes record scratch, tape hiss, AM whistles, etc.

high frequency — Abbreviated HF. The frequency bands from 3 to 30 MHz (100 meters to 10 meters).

high-frequency alternator — An alternator capable of generating radio-frequency carrier waves.

high-frequency band — The band of frequencies extending from 3 to 30 MHz.

high-frequency bias — In a tape recorder, a sinusoidal voltage that is mixed with the signal being recorded to improve the linearity and dynamic range of the recorded signal. In practice, the bias frequency is three to four times the highest information frequency to be recorded.

high-frequency carrier telegraphy — Carrier telegraphy with the carrier currents above the frequencies transmitted over a voice telephone channel.

high-frequency compensation — Also called high boost. An increase in the amplification of the high frequencies with respect to the low and middle frequencies within a given band of frequencies.

high-frequency heating — See electronic heating.

high-frequency induction heater or furnace — An induction heater or furnace using frequencies much higher than the standard 60 hertz.

high-frequency resistance — Also called rf or ac resistance. The total resistance offered by a device in a high-frequency ac circuit. This includes the dc and all other resistances due to the effects of the alternating current.

high-frequency treatment — Therapeutic use of intermittent and isolated trains of heavily damped oscillations having a high frequency and voltage and a relatively low current.

high-frequency trimmer — A trimmer capacitor that is used to calibrate the high-frequency end of the tuning range in a superheterodyne receiver.

high-frequency unit — See tweeter.

high-frequency welding — See radio-frequency welding.

high-intensity discharge lamps — Abbreviated HID. A general group of lamps, consisting of mercury, metal halide, and high-pressure sodium lamps.

high-K ceramic — A ceramic dielectric composition (usually $BaTiO_3$) that exhibits large dielectric constants

and nonlinear voltage and temperature response characteristics.

high level — 1. In digital logic, the more positive of the two binary-system logic levels. 2. Commands for computer systems in which each instruction is actually equated to many machine-code instructions strung together. *See also* low level; negative logic; positive logic. 3. Term describing a computer language in which the statements are closer to human communication than to machine code. Usually high-level languages can be easily read and understood by people since they consist of pseudo-English statements. To be used by a processor, high-level statements must be translated into machine code.

high-level crossover network — A crossover network designed to operate at high levels and which is placed between the power amplifier and the speakers (this type of crossover is normally built into a speaker system).

high-level data link control — See HDLC.

high-level detector — A linear power detector with a voltage-current characteristic that may be treated as a straight line or two intersecting lines.

high-level firing time — The time required to establish a radio-frequency discharge in a switching tube after radio-frequency power is applied.

high-level language — 1. An application or problem-oriented programming language, as distinguished from a machine-oriented programming language. The instruction approach is closer to the needs of the problems to be solved than it is to the language of the machine on which it is to be run. Examples are Ada, C, COBOL, FORTRAN, Lisp, and Pascal. 2. Any programming language that allows a person to give instructions to a computer in English-like text rather than in the numerical (binary) code of 1s and 0s that the computer understands. FORTRAN and BASIC are high-level languages. 3. A step above assembly language. It can be translated to a lower level but not into English except by a programmer. 4. A sophisticated, though easy to use, computer language (written words and special punctuation) that allows a programmer to write software without being concerned with housekeeping functions (e.g., register allocation) or optimization.

high-level modulation — A system in which the modulation is introduced at a point where the power level approximates the output power.

high-level radio-frequency signal (tr, atr, and pre-tr tubes) — A radio-frequency signal with sufficient power to fire the tube.

high-level VSWR (switching tubes) — The voltage standing-wave ratio caused by a fired tube located between a generator and the matched termination in a waveguide.

highlight — The brightest portion of a reproduced image.

high-mu tube — A vacuum tube with a high amplification factor.

high-noise-immunity logic — Abbreviated HNIL. A special type of logic designed specifically to provide very high resistance to electrical noise. Sometimes called HTL (high-threshold logic).

high-order language — A programming language that is independent of the computer. Usually, it resembles natural languages, and a compiler is required for translation into machine language. Examples are FORTRAN and ALGOL.

high-pass filter — 1. A wave filter having a single transmission band extending from some critical, or cutoff, frequency other than zero, up to infinite frequency. 2. A filter that, above a critical frequency, allows the unrestricted passage of high-frequency signals.

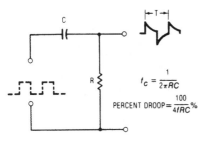

$$f_c = \frac{1}{2\pi RC}$$

$$\text{PERCENT DROOP} = \frac{100}{4fRC}\%$$

High-pass RC filter.

Reciprocally, a bass-cut filter. 3. A filter that passes frequency components above some limited frequency and rejects components below that limit.

high-performance equipment — Equipment having sufficiently exacting characteristics to permit their use in trunk or link circuits.

high pot — *See* flash test.

high-potential test — A test for determining the breakdown point of insulating materials and spacings. It consists of applying a voltage higher than the rated voltage between two points or between two or more windings. However, it it not a test of conductor insulation.

high potting — *See* flash test.

high-power silicon rectifiers — A group of rectifiers with continuous ratings exceeding 50 average amperes per section in a single-phase, half-wave circuit.

high Q — Having a high ratio of reactance to effective resistance. The factor determining the efficiency of a reactive component.

high-rate discharge — The storage-battery discharge equivalent to the heaviest possible duty in service.

high-recombination rate contact — A semiconductor contact at which thermal equilibrium charge-carrier concentrations are maintained substantially independent of current density.

high-resistance joint — A faulty union of conductors or conductor and terminal. The result is less current flow and a drop in voltage at the union.

high-resistance voltmeter — A voltmeter having a resistance considerably higher than 1000 ohms per volt. As a result, it draws very little current from the circuit being measured.

high resolution — Descriptive of a camera or monitor capable of displaying a great number of scanning lines (1000–2000), which produces a picture that is very detailed, defined, and sharp.

high-speed bus — *See* memory register.

high-speed carry — Also called standing-on-nines carry. 1. A carry into a column results in a carry out of that column, because the sum without carry in that column is 9. 2. Instead of a normal adding process, a special process is used which takes the carry at high speed to the actual column where it is added.

high-speed data rate — Data transmission at a rate between 2401 bauds and 500 kilobauds.

high-speed dc circuit breaker — A device that starts to reduce the current in the main circuit in 0.01 second or less after the occurrence of a dc overcurrent or an excessive rate of current rise.

high-speed excitation system — An excitation system that can change its voltage rapidly in response to a change in the field circuit of the excited generator.

high-speed pattern board — Board completely covered with copper foil on both sides with the exception of small annular rings etched away to provide isolated interconnections. High V_{CC} and ground-plane coverage

enhance high-speed logic operation due to afforded control of signal-line impedance and reduction of noise and crosstalk.

high-speed printer — 1. A printer that has a speed of operation compatible with the speed of computation and data processing so that it may operate online. 2. A signal-responsive alphanumeric printer capable of printing computer output signals at rates on the order of 300 characters per second or greater.

high-speed reader — A reading device that can be connected to a computer so as to operate online without seriously slowing the operation of the computer.

high-speed relay — A relay designed specifically for short operate or short release time, or both.

high-speed storage — *See* rapid storage.

high-speed telegraph transmission — Transmission of code at higher speeds than are possible with hand-operated keys.

high state — The relatively more positive signal level used to assert a specific message content associated with one of two binary logic states.

high-tech — A general term that refers to sophisticated technical innovation; cutting-edge technology, often involving computers and other electronic devices.

high-temperature reverse bias — Burning-type test of diodes and transistors, conducted with the junctions reverse biased to effect any failure due to ion migration in bonds of dissimilar metals.

high-tension — Lethal voltages, on the order of thousands of volts.

high-tension magneto — A self-contained generator in which the required high potential is generated directly; no induction coil is needed.

high-threshold logic — Abbreviated HTL. 1. Logic with a high noise margin, used primarily in industrial applications. It closely resembles diode-transistor logic (DTL), except that in HTL a reverse-biased emitter junction is used as a threshold element operating as a zener diode. A typical noise margin is 6 volts with a 15-volt supply. 2. Logic that allows for higher degree of inherent electrical noise immunity. A considerably larger input threshold characteristic is exhibited by HTL devices by using a reversed-biased base-emitter junction that operates in the breakdown avalanche mode. A higher input signal is required to turn on the HTL output-inverting transistor than the DTL.

high-vacuum phototube — A phototube that is highly evacuated so that its electrical characteristics are essentially unaffected by gaseous ionization. In a gas phototube, some gas is intentionally introduced.

high-vacuum rectifier — A vacuum-tube rectifier in which conduction is entirely by electrons emitted from the cathode.

high-vacuum tube — Also known as a hard tube. An electron tube whose electrical characteristics will not be affected by gaseous ionization because of its high degree of evacuation.

high-velocity scanning — The scanning of a target with electrons of such velocity that the secondary-emission ratio is greater than unity.

high voltage — The accelerating potential that speeds up the electrons in a beam of a cathode-ray tube.

high-voltage probe — A probe with a high internal resistance, for measuring extremely high voltages. It is used with a voltmeter having an internal resistance of 20,000 ohms per volt or more. *See* Figure on p. 352.

hill-and-dale recording — *See* vertical recording.

hinge — A joint in a relay that permits movement of the armature relative to the stationary parts of the relay structure.

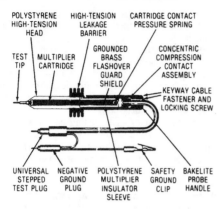

High-voltage probe.

H-lines — Imaginary lines that represent the direction and strength of magnetic flux on a diagram.

HNDT — *See* holographic nondestructive testing.

H-network — A network composed of five impedance branches. Two are connected in series between an input terminal and an output terminal, and two are connected between another input and another output terminal. The fifth is connected from the junction points of the two branches.

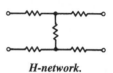

H-network.

hinged-iron ammeter — A moving-iron ammeter in which the fixed portion of the magnetic circuit is placed around the conductor to measure the current through it.

HIPERNAS — Acronym for *high-per*formance *navi*gation *system.* A self-compensated, pure-inertial guidance system.

hipot — 1. *See* dielectric breakdown voltage. 2. Contraction of high potential. Commonly refers to a device used, with high voltages, for testing insulation breakdown or leakage. High potting is the verb. 3. High-potential voltage applied across a conductor to test the insulation. 4. A test designed to determine the highest potential that can be applied to a conductor without breaking through the insulation.

hipot tester — A *high-pot*ential test instrument that applies a high-voltage source to the insulating material of a device or cable to determine the ability of the unit to withstand the voltage without breakdown.

hi-rel — A contraction of high reliability; refers to products that are assembled and inspected under rigid standards, given extra testing and conditioning, and typically used in military, space, or medical applications.

hiss — 1. Random noise characterized by prolonged sibilant sounds in the audio-frequency range. 2. The primary background noise in tape recording, stemming from circuit noise in the playback amplifiers or from residual magnetism of the tape.

histogram — 1. A graphical representation of a frequency distribution by a series of rectangles that have for one dimension a distance proportional to a definite range of frequencies, and for the other dimension a distance proportional to the number of frequencies appearing within range. 2. A description of one (or all) parameters, showing distribution, standard deviation, mean-value failure limits, and sample lot size for all samples within the lot. 3. A graph of contiguous vertical bars representing a frequency distribution in which the groups or classes of items are marked on the *x*-axis, and the number of items in each class is indicated by a horizontal line segment drawn above the *x*-axis at a height equal to the number of items in the class.

hit — 1. *See* flash. 2. Momentary surge of voltage on a transmission channel. 3. An impulse noise having a duration of about 1 millisecond.

hit-on-the-fly printer — 1. A mechanical printer in which the printing head is in continual motion. 2. A printer in which the paper and/or the printing mechanism are in constant motion so that starts and stops are not needed.

hits — Momentary line disturbances that could result in mutilation of characters being transmitted.

HNIL — Abbreviation for high-noise-immunity logic.

hobby computer — A computer that is not used for profit.

hog horn — A microwave feed horn shaped so that the input energy from the waveguide approaches from the same direction as the horn opening.

hold — Opposite of clear. 1. To maintain storage elements in charge storage tubes at equilibrium potentials by electron bombardment. 2. To retain the information contained in one storage location for a computer after copying the information into another storage location, as opposed to clearing or erasing the information. 3. To maintain an established telephone connection, possibly while disconnecting to answer another call.

hold control — In a television receiver, the adjustment that controls the frequency of the vertical or horizontal scanning pulses and hence the stability of the picture.

hold current — Also called the electrical hold value. The minimum current that will keep the contact springs energized in a relay.

hold electrode — In a mercury switch, the electrode that remains in contact with the mercury pool while the circuit is being closed or opened.

holding anode — In a mercury-arc rectifier, a small auxiliary anode that maintains the ionization while the main anode current is zero.

holding beam — A diffused beam of electrons for regenerating the charges retained on the dielectric surface of an electrostatic memory or storage tube.

holding circuit — Also called a locking circuit. An alternate operating circuit that, when completed, maintains sufficient current in a relay winding to keep the relay energized after the initial current has ceased.

holding coil — A separate relay coil that keeps the relay energized after the original current has been removed.

holding current — 1. That value of average forward current (with the gate open) below which a silicon-controlled rectifier returns to the forward blocking state after having been in forward conduction. 2. The minimum current that must pass through a device such as a silicon-controlled rectifier, thyratron, neon glow tube, etc., to maintain it in a conducting condition. 3. The minimum current that will hold a relay in its operated position. The holding current is less than the operating current. 4. The minimum principal current required to maintain a thyristor in the on state.

holding gun — In a storage tube, the source of electrons constituting the holding beam.

holding time — The total time a trunk or circuit is in use on a call, including both operator's and user's time.

holding torque — Also called restoring or stall torque. 1. Force moment required to deflect the rotor of a stepper motor a full step with the motor energized but at a standstill. 2. A measure of the maximum torque that can be applied to the shaft of a stepper motor with one or more of its phases energized before the shaft begins to rotate.

hold lamp — An indicating lamp that stays lighted while a telephone connection is being held.

hold mode — In integrators or other charge-storage circuits, a condition or time interval in which input(s) are removed and the circuit is commanded (or expected) to maintain a constant output.

hold-mode drop — In a sample-and-hold circuit, the output voltage change per unit of time while in hold. Commonly specified in volts per second, microvolts per microsecond, and other convenient units.

hold-mode feedthrough — In a sample-and-hold circuit, the percentage of an input sinusoidal signal that is measured at the output of a sample hold when it is in hold mode.

hold-mode settling time — In a sample-and-hold circuit, the time from the hold-command transition until the output of the sample hold has settled within the specified error band. It includes aperture delay time.

hold-off voltage — The maximum voltage an electronic flash tube will stand without self-flashing. Normal hold-off voltage is reduced at the end of lamp life and in the presence of high temperatures or rf fields.

holdover — The condition that occurs when a lightning-protector gap continues to conduct follow current.

holdover time — Also called holdup time. The length of time a power supply will maintain its rated output after the input has been lost. In nonuninterruptible power supplies, the holdover time is usually measured in milliseconds.

holdover voltage — The steady-state voltage at which a gap just fails to clear a given value of follow current.

hold time — Also called release time. 1. In resistance welding, the time that is allowed for the weld to harden. 2. The length of time after the clocking of a flip-flop that data must remain unchanged.

holdup — The ability of a power supply to provide energy for approximately 30 to 50 milliseconds after an ac power loss. This provides an adequate amount of time to make a transfer to a standby power system. Holdup time is especially useful in computer systems, in which data in volatile memories may be lost in nuisance shutdowns.

holdup button — A manually actuated mechanical switch used to initiate a duress alarm signal; usually constructed to minimize accidental activation.

holdup time — See holdover time.

hole — 1. In the electronic valence structure of a semiconductor, a mobile vacancy that acts like a positive electronic charge with a positive mass. 2. In a semiconductor, the term used to describe the absence of an electron; this absence has the same electrical properties as an electron except that it carries a positive charge. 3. A mobile vacancy or electron deficiency in the valence structure of a semiconductor. It is equivalent to a positive charge. 4. A defect in the valence electron system of a semiconductor crystal lattice, equivalent to the absence of a single valence electron. Like a conduction electron, a hole is capable of moving through the crystal and thus forms an effective current carrier having a positive charge. However, unlike a conduction electron, the hole must remain in the valence-bonding system of the crystal and thus it has a lower mobility.

hole conduction — Conduction occurring in a semiconductor when electrons move into holes under the influence of an applied voltage and thereby create new holes. The apparent movement of such holes is toward the more negative terminal, and is hence equivalent to a flow of positive charges in that direction.

hole current — Conduction in a semiconductor when electrons move into holes, creating new holes. The holes appear to move toward the negative terminal, giving the equivalent of positive charges flowing to the terminal.

hole density — In a semiconductor, the density of holes in an otherwise full band.

hole-electron pair — A positive charge carrier (hole) and a negative charge carrier (electron) considered together as one entity.

hole injection — The production of mobile vacancies in an n-type semiconductor when a voltage is applied to a sharp metal point in contact with the surface of the material.

hole injector — A pointed metallic device for injecting holes into an n-type semiconductor.

hole-in-the-center effect — Also called hole-in-the-middle effect. The lower volume or absence of sound between the left and right speakers of a stereo system.

hole mobility — The ability of a hole to travel easily through a semiconductor.

hole site — The area on a computer punch card or paper tape where a hole may or may not be punched. It can be a form of binary storage, in which a hole represents a 1 and the absence of a hole represents a 0.

hole storage factor — In a transistor, the excess stored charge (when the transistor is in saturation) per unit excess base current. Excess base current is defined as the amount of current supplied to the base in excess of the current required to just keep the transistor in saturation.

hole trap — A semiconductor impurity that can trap holes by releasing electrons into the conduction or valence bands.

Hollerith — Pertaining to a particular type of code or punched card utilizing 12 rows per column and usually 80 columns per card.

Hollerith code — 1. A code based on the punching of holes in cards at specified locations. From one to three punches may be made in each column of the card and up to 80 columns may be punched in each card. Each column corresponds to one character; the specific character is determined by the number and location of the punches in that column. 2. A 12-level (12 bits per character) code that defines the relation between an alphanumeric character and the punched holes in an 80-column data card.

hollow-cathode tube — A gas discharge tube with a hollow cathode closed at one end. Almost all the radiation is from the cathode glow within the hollow cathode.

hollow core — A plain ferrite core having a center hole for mounting purposes.

hologram — 1. A photograph, made with laser light, that appears to have three dimensions. 2. A recording of the two-dimensional intensity distribution of the interference pattern produced by the interaction of two or more monochromatic waves that have phases derived from the same source. One of the waves is reconstructed when a replica of the other wave is diffracted from the hologram. 3. An interference pattern recorded on photographic film or similar media. This pattern is created by directing two beams of coherent light into the film. One, called the reference beam, strikes the film directly. The second, called the object beam, bounces off, or passes through the test specimen, then strikes the film. The interaction of these two beams makes up the interference pattern called a hologram. To "decode" the swirls and dots of the pattern and create a visible three-dimensional image, a coherent light beam is directed onto the hologram.

holographic cinematography — A technique used to create a succession of interrelated holographic images that give an appearance of motion when projected in sequence. A flashing laser is used in both the recording and projecting of the images.

holographic display — A three-dimensional display created by using lasers.

holographic lenses — Photographic recordings of interference patterns between a plane wave and a spherical wave on a high-resolution photographic emulsion.

holographic memory — The storage of data as bits in memory by holographic processes. A laser beam is divided into reference and object beams, and bit information is stored as a hologram.

holographic nondestructive testing — Abbreviated HNDT. The application of coherent wavefront techniques to the determination of the physical state of a system without appreciably altering that state.

holography — 1. The optical recording of an object wave formed by the resulting interference pattern of two mutually coherent component light beams. A coherent beam is first split into two component beams, one of which irradiates the object, the second of which irradiates a recording medium. The diffraction or scattering of the first wave by the object forms the object wave that proceeds to and interferes with the second coherent beam, or reference wave, at the medium. The resulting pattern is a three-dimensional record (hologram) of the object wave. 2. The recording of an object wave (usually optical) in such a way that an identical wave can subsequently be reconstructed. Whereas a conventional photograph records only the intensity of the light incident on it, a hologram records both the amplitude and phase. The additional phase information is contained in an interference pattern that is formed from the object wave and a reference wave. 3. The science and technique of producing holograms.

holtite contact system — A pluggable solderless means of connecting DIP, SIP, and discrete-component packages to printed circuit boards with none of the socket material showing above the board surface.

Home Box Office — *See* HBO.

home loop — An operation involving only those input and output units associated with the hole terminal.

home-on-jam — A radar feature that permits angular tracking of a jamming source.

home page — 1. An HTML document associated with an individual or organization. It contains text, pictures, sounds, and links to other sites that appear (generally in blue) as underlined words or phrases. Clicking on these underlined words opens a network connection to other HTML documents, which can be anywhere on the Internet, or spawns an application on the host computer. 2. An HTML page that is the primary or index document representing an entity such as a company or individual. It is usually the first page received from a web server and, as such, serves as an introduction to the entity or content being served. 3. The first screen of a collection of web sites particular to one person or business.

hometaxial-base transistor — A transistor manufactured by a single-diffusion process so that both the emitter and collector junctions are formed in a uniformly doped silicon slice. The homogeneously doped base region that results is free from accelerating fields in the axial (collector-to-emitter) direction; such fields could cause undesirable high flow and destroy the transistor.

homing — 1. Approaching a desired point by maintaining some indicated navigational parameter constant (other than altitude). 2. In missile guidance, the use of radiation from a target to establish a collision course.

homing adapter — A device used with an aircraft radio receiver to produce aural and/or visual signals that indicate the direction of a transmitting radio station.

homing antenna — A type of directional-antenna array used for pinpointing a target.

homing beacon — A radio transmitter that emits a distinctive signal for determining bearing, course, or location.

homing device — 1. An automatic device that moves or rotates in the correct direction without first having to go to the end of its travel in the opposite direction. 2. A radio device that guides an aircraft to an airport or transmitter site.

homing guidance — A missile-guidance system in which the missile steers itself toward a target by means of a self-contained mechanism (infrared detectors, radar, etc.). It is activated by some distinguishing characteristic of the target. Homing guidance may be active, semiactive, or passive.

homing guidance system — A system of sensors and related instrumentation that allows a navigable object (usually a missile) to locate its destination by some distinguishing characteristics of that target, and then calculate and alter its course so that the destination is reached.

homing relay — A stepping relay that returns to a specified starting position prior to each operating cycle.

homing station — A radionavigational aid incorporating direction-finding facilities.

homodyne reception — Also called zero-beat reception. 1. A form of reception in which the receiver generates a signal at the original carrier frequency and combines it with the incoming signal. 2. A system of reception using a locally generated voltage at the carrier frequency.

homogeneity — The state or condition of being similar in nature, kind, or degree.

homogeneous — Of the same nature (the opposite of heterogeneous). That property of a substance that determines that all components of volume are the same relative to composition and other properties.

homogeneous crystal — Crystalline material having a uniform composition. In the context of impurity semiconductor materials, a homogeneous crystal is one having a uniform doping concentration.

homogeneous multiprocessor — A multiprocessor in which all processors of significance are functionally identical.

homologous field — A field in which the lines of force in a given plane all pass through one point (e.g., the electric field between two coaxial charged cylinders).

homopolar — Electrically symmetrical, i.e., having equally distributed charges.

homopolar generator — A dc generator in which all the poles presented to the armature are of the same polarity, so that the armature conductor always cuts the magnetic lines of force in the same direction. A pure direct current can thus be produced without commutation.

homopolar magnet — A magnet with concentric pole pieces.

homosphere — That part of the atmosphere which is made up mostly of atoms and molecules found near the earth's surface and which retains the same relative proportions of oxygen, nitrogen, and other gases throughout.

homotaxial — A term coined by RCA from *homogeneous* and *axial* to describe a single-diffused transistor with a base region of homogeneous resistivity silicon on the axial (emitter-to-collector) direction.

honeycomb coil — An air-core radio-frequency inductance wound in a crisscross lattice to reduce its distributed capacitance.

honeycomb winding — A method of winding a coil with crisscross turns to minimize distributed capacitance.

hood — 1. A shield placed over a cathode-ray tube to eliminate extraneous light and thus make the image on the screen appear more clearly. 2. An enclosure, attached to the back of a connector, to contain and protect wires and cable attached to the terminals of a connector. A cable clamp is usually an integral part of the hood.

hood contact — A switch that is used for the supervision of a closed safe or vault door. Usually installed on the outside surface of the protected door.

hook — 1. Hidden capacitance between conductors on a printed wiring board. Molecular structure causes capacitance change with frequency. Hook is responsible for signal-waveshape distortion and timing problems. 2. The effect on a signal's voltage caused by a change in printed circuit board capacitance with frequency. Board capacitance is created between printed circuit board conductors separated by dielectric material. It can change the response time of a square wave and can bring about erroneous responses at certain frequencies of sine waves.

hookswitch — The device on which a telephone receiver hangs or on which a telephone handset hangs or rests when not in use. The weight of the receiver or handset operates a switch that opens the telephone circuit, leaving only the bell connected to the line. When the handset is lifted, the switch closes the telephone circuit or loop.

hook terminal — Terminal with a hook-shaped tongue.

hook tongue — Type of terminal with a tongue that opens from the side rather than from the end.

hook transistor — A transistor having four alternating p-type and n-type layers, with one layer floating between the base layer and the collector layer. This arrangement gives high emitter-input current gains. The pnpn transistor has a p-type floating layer, while the npnp transistor has an n-type floating layer.

hookup — 1. Method of connection between the various units in a circuit. 2. The diagram of connections used. 3. An interconnection of circuit components for a particular purpose.

hookup wire — 1. The wire used in coupling circuits together. It may be solid or stranded, and is usually tinned and insulated No. 18 or 20 soft-drawn copper. 2. Wire used for point-to-point connection within electronic equipment, usually carrying low voltages (under 1000 V) and currents. 3. Wire used to make the internal connections between the various electrical parts of electronic assemblies.

hop — An excursion of a radio wave from the earth to the ionosphere and back. It is usually expressed as single-, double-, and multihop. The number of hops is called the order of reflection.

hopoff — In a potentiometer, the sudden jump in resistance as the contact is rotated over the junction of two resistance slopes. The magnitude of the hopoff is dependent on the ratio of the slopes and on the junction blending characteristic.

hopper — An area in a temporary memory unit, such as a call store, used to record a list of items for subsequent communications with processing programs or input-output programs sent to central control.

horizon — An apparent or visible junction of earth and sky as seen on or above the earth. It bounds the part of the earth's surface that can be reached by the direct wave

of a radio station. The distance to the horizon is affected by atmospheric refraction.

horizon distance — The space between the farthest visible point and the transmitter antenna. It is the distance over which ultrahigh-frequency transmission can be received under ordinary conditions with an unelevated receiving antenna.

horizontal — 1. Perpendicular to the direction of gravity. 2. In the direction of or parallel to the horizon. 3. On a level.

horizontal angle of deviation — The horizontal angle between the great-circle path from the transmitter to the receiver and the direction of departure or arrival of the wave along the line of propagation.

horizontal axes — The three horizontal axes of crystallographic reference.

horizontal blanking — Cutting off the CRT electron beam between successive active horizontal lines during retrace. Blanking of the picture during the period of horizontal retrace.

horizontal blanking interval — The brief time between scan lines required for the scanning electron gun to retrace from the right edge of the image back to the left to begin the next scan line.

horizontal blanking pulse — A rectangular pedestal in the composite television signal. It occurs between active horizontal lines and cuts off the beam current of the picture tube during retrace.

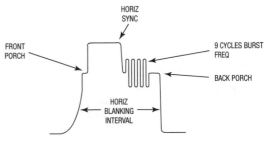

Horizontal blanking pulse.

horizontal centering control — In a television receiver or cathode-ray oscilloscope, the adjustment for moving the entire display back and forth.

horizontal-convergence control — In a color television receiver, the control that adjusts the amplitude of the horizontal dynamic convergence voltage.

horizontal definition — See horizontal resolution.

horizontal-deflecting electrodes — A pair of electrodes that move the electron beam from side to side on the screen of a cathode-ray tube employing electrostatic deflection.

horizontal-discharge tube — A vacuum tube used in the horizontal-deflection circuit to discharge a capacitor and thereby form the sawtooth scanning wave. See also discharge tube.

horizontal-drive control — In an electromagnetically deflected television receiver, the control that adjusts the ratio of the pulse amplitude to the linear portion of the scanning-current wave.

horizontal dynamic convergence — Convergence of the three electron beams in a color picture tube at the aperture mask during scanning of a horizontal line.

horizontal field-strength diagram — A representation of the field strength in a horizontal plane and at a constant distance from an antenna. Unless otherwise specified, the plane passes through the antenna.

horizontal frequency — *See* line frequency, 1.

horizontal hold control — A synchronization control that varies the free-running frequency of the horizontal deflection oscillator so it will be in step with the scanning frequency at the transmitter.

horizontal hum bars — Broad, horizontal, moving or stationary bars, alternately dark and light, that extend over an entire television picture. They are caused by interference at approximately 60 Hz or a harmonic of 60 Hz.

horizontal-linearity control — In a television receiver, the control for adjusting the width at the left side of the screen.

horizontal line frequency — *See* line frequency, 1.

horizontal lock — The circuit that maintains horizontal synchronization in a television receiver.

horizontally polarized wave — A linearly polarized wave with a horizontal electric-field vector.

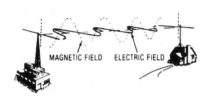

Horizontally polarized wave.

horizontal-output transformer — *See* flyback transformer.

horizontal parabola control — *See* phase control, 1.

horizontal polarization — 1. Transmission in which the electrostatic field leaves the antenna in a horizontal plane. Elements of the transmitting and receiving antennas likewise are horizontal. Horizontal polarization is standard for television in the United States. 2. Transmission of radio waves whose undulations vary horizontally with respect to the earth. (Horizontally polarized antennas are used mainly for base-to-base transmission.)

horizontal redundancy checking — *See* HRC.

horizontal repetition rate — Also called horizontal scanning frequency. The number of horizontal lines per second (15,750 hertz in the United States).

horizontal resolution — Also called horizontal definition. 1. The number of individual picture elements that can be distinguished in a horizontal scanning line. 2. The capability of a TV system to resolve detail in a horizontal direction across the screen. The higher the resolution number, the sharper the picture will be. 3. The number of vertical lines that can be displayed in a picture width equal to the picture height. This measurement is usually done via direct (RGB or Y/C) input in order to bypass limiting factors such as transmission standards (bandwidth, color carrier, etc.). For standard NTSC (60 Hz, 525 lines, 4 : 3 aspect ratio) and PAL (50 Hz, 625 lines, 4 : 3 aspect ratio), the picture is in practice the same: 80 lines per MHz bandwidth.

horizontal retrace — 1. The return of the electron beam from the right to the left side of a CRT raster after the scanning of one line. 2. The line that would be seen on a CRT screen while the spot is returning from right to left, if retrace blanking were not used.

horizontal ring-induction furnace — A furnace for melting metal. It comprises an open trough or melting channel, a primary inductor winding, and a magnetic core that links the melting channel to the primary winding.

horizontal-scanning frequency — *See* horizontal repetition rate.

horizontal sweep — Movement of the electron beam from left to right across the screen or the scene being televised.

horizontal-sync discriminator — A circuit employed in the flywheel method of synchronization to compare the phase of the horizontal-sync pulses with that of the horizontal-scanning oscillator.

horizontal-sync pulse — The rectangular pulse that occurs above the pedestal level between each active horizontal line. The pulses keep the horizontal scanning at the receiver in step with that at the transmitter.

horn — Also called an acoustic horn. 1. A tubular or rectangular enclosure for radiating or receiving acoustic waves. 2. A primary element consisting of a section of metal waveguide in which one or both of the cross-sectional dimensions increase toward the open end. 3. Any flared or funnel-shaped passage used to couple a speaker efficiently to the air in the environment. It also provides control over sound dispersion pattern over a specified frequency range. The driver is connected to the throat of the horn, and the sound emerges from its mouth. *See also* horn antenna.

horn antenna — Also called a horn. 1. A tubular or rectangular microwave antenna that is wider at the open end and through which radio waves are radiated into space. 2. A microwave antenna formed by flaring the end of a waveguide into the shape of a horn.

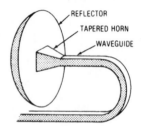

Horn antenna.

horn arrester — A lightning arrester that has a spark gap with upward-projecting diversion horns of thick wire. When the arc is formed, it travels up the gap and is extinguished upon reaching the widest part of the gap.

horn cutoff frequency — A frequency below which an exponential horn will not function correctly because it fails to provide for proper expansion of the sound waves.

horn gap — A type of spark gap with divergent electrodes.

horn-gap switch — A form of air switch with arcing horns.

horn loading — A method of coupling a speaker diaphragm to the listening space by an expanding air column having a small throat and large mouth.

horn loudspeaker — A very directional speaker in which the driver is fed into a metal horn, whose flare is usually an exponential curve.

horn mouth — An open-ended metallic device for concentrating energy from a waveguide and directing this energy into space.

horn speaker — A speaker in which a horn couples the radiating element to the medium.

horn throat — The narrow end of a horn.

horsepower — Abbreviated hp. A unit of power, or the capacity of a mechanism to do work. It is the equivalent of raising 33,000 pounds 1 foot in 1 minute, or 550 pounds 1 foot in 1 second. One horsepower equals 746 watts.

horseshoe magnet — A permanent magnet or electromagnet shaped like a horseshoe or U to bring the two poles close together.

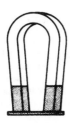

Horseshoe magnet.

host — 1. A node computer on a network. The etymology of the term *host* may be from the days when a computer on a network hosted multiple users and services. 2. Any computer on a network that is a repository for services available to other computers on the network. It is quite common to have one host machine provide several services, such as WWW and Usenet. 3. The primary or controlling computer in a multiple-part system.

host computer — 1. In the context of networks, a computer that directly provides service to a user. In contrast to a network server, which provides services to a user through an intermediary host computer. 2. The master or controlling computer in a multicomputer network. 3. A computer that prepares programs to be run on another computer system. 4. A computer that monitors and controls other computers and programmable controllers. 5. The central computer (or one of a collection of computers) that provides functions such as computation, database access, or special programs or programming languages; often shortened to *host*. 6. The central computer (or one of a collection of computers) in a data-communications system, which provides the primary data-processing functions such as computation, database access, or special programs or programming languages.

host system — The computer system in which an emulation program or accessory card is used to imitate another system.

hot — 1. Connected, alive, energized; pertains to a terminal or any ungrounded conductor. 2. Not grounded. 3. Strongly radioactive. 4. Excited to a relatively high energy level. 5. Idiomatic term generally used to describe conductors carrying an electrical charge.

hot-carrier diode — Abbreviated HCD. A diode in which a closely controlled metal-semiconductor junction provides virtual elimination of charge storage. The device has extremely fast turn-on and turn-off times, excellent diode forward and reverse characteristics, lower noise characteristics, and wider dynamic range. *See* Schottky barrier diode.

hot carriers — In barrier diodes, carriers that have energies greater than those that are in terminal equilibrium with the metal. Thus, electrons that cross the junction from semiconductor to metal must be energetic enough to surmount the barrier. Therefore, electrons that are energetic enough to cross the junction are called hot electrons.

hot cathode — Also called thermionic cathode. 1. A cathode that supplies electrons by thermionic emission. (As opposed to a cold cathode, which has no heater.) 2. An electron-tube cathode in which electron emission is produced by heat.

hot-cathode tube — Also called thermionic tube. Any electron tube containing a hot cathode.

hot-cathode X-ray tube — A high-vacuum X-ray tube in which a hot rather than cold cathode is used.

hot electrons — *See* hot carriers.

hot-electron triode — A solid-state, evaporated-thin-film structure that is directly equivalent to a triode vacuum tube.

hotkey — A single key or a simple key combination that can be used to activate a computer program, or a function in a program.

hot line — Communications channel between two points available for immediate use without patching or switching.

hot plate — Electrically heated flat surface, sometimes combined with auxiliary equipment such as a magnetic stirrer.

hot spot — The point of maximum temperature on the outside of a device or component.

hot-spot temperature — The maximum temperature measured on a resistor due to both internal heating and the ambient operating temperature. Maximum hot-spot temperature is predicted on thermal limits of the materials and the design. The hot-spot temperature is also usually established as the top temperature on the derating curve at which the resistor is derated to zero power.

hot stamping — Method of imprinting letters, numbers, and symbols with a heated die.

hot tin dip — A process of passing a bare wire through a bath of molten tin to provide a coating.

hot-wire ammeter — Also called thermal ammeter. An ammeter in which the expansion of a wire moves a pointer to indicate the amount of current being measured. The current flows through the wire and changes its length in proportion to I^2. Instability because of wire stretching, and the lack of ambient temperature compensation, make the hot-wire ammeter commercially unsatisfactory.

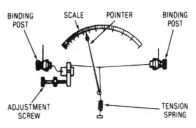

Hot-wire ammeter.

hot-wire anemometer — An instrument that measures the velocity of wind or a gas by its cooling on an electrically heated wire.

hot-wire instrument — 1. An electrothermic instrument operated by expansion of a wire heated by the current it is carrying. 2. A measuring device or transducer whose operation depends either on the expansion of a wire due to its being heated by an electric current or on the change in electrical resistance on the part of a wire that is heated or cooled.

hot-wire microphone — A microphone in which the cooling or heating effect of a sound wave changes the resistance of a hot wire and thus the current through it.

hot-wire relay — A form of linear-expansion time-delay relay in which the longitudinal expansion of a wire, when heated, provides the mechanical motion to open or close contacts. The time required to heat the wire constitutes the delay.

hot wires — In an electrical system, those wires that carry the live current through the electrical system. Hot wires are usually black or red.

hot-wire transducer — A unilateral transducer in which the cooling or heating effect of a sound wave changes the resistance of a hot wire and thus the current.

house cables — Conductors inside a building used to connect communication equipment to outside lines.

housekeeping — In a computer routine, those operations, such as setting up constants and variables for use in the program, that contribute directly to the proper operation of the computer but not to the solution of the problem.

howl — An undesirable prolonged wail produced in a speaker by electric or acoustic feedback.

howler — 1. An electromechanical device that produces an audio-frequency tone. 2. A unit by which the operator at a telephone test desk may apply a high tone of varying loudness to a line to call a subscriber's attention to the fact that his or her receiver is off the hook.

howling — System instability caused by acoustic feedback from loudspeaker to microphone.

howl repeater — In the operation of telephone repeaters, a condition in which more energy is returned than is sent, with the result that an oscillation is set up on the circuit.

hp — Abbreviation for horsepower.

H-pad — An attenuation network in which the elements are arranged in the form of the letter H.

h-parameters — *See* hybrid parameters.

h-particle — The positive hydrogen ion or proton resulting from bombardment of the hydrogen atom by alpha rays or fast-moving positive ions.

HPF — Abbreviation for highest probable frequency.

H-plane — The plane in which the magnetic field of an antenna lies. It is perpendicular to the E-plane. The principal H-plane of an antenna is the H-plane that also contains the direction of maximum radiation.

H-plane bend — *See* H-bend.

H-plane T-junction — Also called shunt T-junction. A wavelength T-junction in which the structure changes in the plane of the magnetic field.

HRC — Abbreviation for horizontal redundancy checking. A validity-checking technique used on data-transmission blocks, in which redundant information is included with the information to be checked.

H-scan — *See* H-display.

HTL — Abbreviation for high-threshold logic.

HTML — Abbreviation for Hypertext Markup Language. The coding language used to create hypertext documents for use on the World Wide Web. HTML allows a document to contain links to another document, giving WWW its hypertext capabilities.

HTML page — A single file document written in HTML. Although this page may require multiple pages of paper if printed, it is typically viewed as a single scrolled page in a web browser.

HTTP — Abbreviation for Hypertext Transfer Protocol. A client/server networking protocol for retrieving HTML documents on the World Wide Web. The client software application used in HTTP transactions is a web browser.

hub — 1. On a control panel or plugboard, a socket or receptacle into which an electrical lead or plug wire may be connected for the purpose of carrying signals. 2. The narrow spindle around which the tape is wound on a reel or in a cassette. 3. Local distribution center where signals from a master feed are relayed to other inputs.

hue — 1. Often used synonymously with the term *color*, but does not include gray. It is the dominant wavelength — i.e., the one that distinguishes a color as red, yellow, etc. Varying saturations may have the same hue. 2. A more inclusive and more precise identification of the optical wavelengths than the commonly used word *color*. Rather than state that the maximum daylight sensitivity of the eye is to green-yellow or yellow-green, it is technically more rigorous to state that it occurs at a hue of 555 nanometers. 3. The perceptual term for that aspect of color described by words such as red, yellow, or blue. Achromatic colors, such as white, gray, and black, do not exhibit hue. 4. In color TV, one of the three characteristics of color. Defines color on the basis of its position in the frequency spectrum, i.e., whether red, blue, green, or yellow, etc. *See also* luminance; saturation.

hue control — On a color television receiver, the operating control that changes the hue (color) of the picture.

hull potential — The voltage difference between a reference electrode and an immersed metallic hull, or the bonded underwater metallic appendages of a nonmetallic hull.

hum — 1. In audio-frequency systems, a low-pitched droning noise consisting of several harmonically related frequencies. It results from an alternating-current power supply, ripple from a direct-current power supply, or induction from exposure to a power system. By extension, the term is applied in visual systems to interference from similar sources. 2. A pattern produced on a facsimile record sheet when a signal at the power-line frequency or a harmonic of the power-line frequency is mixed with or modulates the facsimile signal. 3. A continuous low-frequency interference caused by inadvertent pickup of 60-Hz or 120-Hz energy from nearby ac power sources. Most likely to originate in devices (like microphones) requiring substantial amplification. 4. A background tone caused by improper shielding of audio components or inadequate filtering of line voltage entering the equipment. 5. Electrical disturbance at the power supply frequency or harmonics thereof.

human engineering — 1. The science and art of developing machines for human use, giving consideration to the abilities, limitations, habits, and preferences of the human operator. 2. The determination of man's capabilities and limitations as they relate to the equipment or systems he will use, and the application of this knowledge to the planning, design, and testing of man-machine combinations to obtain optimum performance, operability, reliability, efficiency, safety, and maintainability. 3. The study of the behavioral properties of humans in interaction with machines, and of total human-machine systems; the structuring of human-machine systems to enhance system performance.

human factors — A body of scientific facts about human characteristics. The term covers biomedical and psychosocial considerations in the areas of human engineering, personnel selection, training, life support, job performance aid, and human-performance evaluation.

humanoid — Robot in the form of a person.

hum-balancing pot — A potentiometer usually placed across the heater circuit. Its arm is grounded so that the heater voltage is balanced with respect to ground.

hum bar — A dark band extending across the picture. It is caused by excessive 60-Hz hum (or harmonics) in the signal applied to the picture-tube input.

hum bars — Relatively broad horizontal bars, alternately black and white, that extend over an entire TV picture. They may be stationary or may move up and down. Sometimes referred to as a venetian-blind effect, they are caused by an approximate 60-Hz interfering frequency or one of its harmonic frequencies.

hum-bucking — The introduction of a small amount of voltage, at the power-line frequency, into a circuit to cancel unwanted power.

hum-bucking coil — A coil wound around the field coil of a dynamic speaker and connected in series opposition with the voice coil. In this way, any hum voltage induced in the field coil will be induced in the voice coil in the opposite direction and will buck, or cancel, the effects of the hum.

humidity — 1. An indication of the water-vapor content of a gas mixture. 2. The amount of moisture in the air. Measured in percent relative humidity.

humidity transducer — A layer of hygroscopic (moisture-absorbing) substance deposited between two metal electrodes. These electrodes establish electrical contact with the hygroscopic chemical, which serves as a resistance element. Since the chemical coating tends to absorb moisture from the surrounding air, its resistance decreases as the humidity increases. In this manner, humidity variations are converted to resistance variations.

hum loop — A condition arising from the connection of two or more grounds to an amplifier system whereby circulating currents of low value at power-line frequency and harmonics are added to the program signals, causing hum to appear in the background.

humming — A sound produced by transformers having loose laminations or by magnetostriction effects in iron cores. The frequency of the sound is twice the power-line frequency.

hum modulation — Modulation of an rf signal or detected audio-frequency signal by hum. This type of hum is heard only when the receiver is tuned to a station.

hunting — 1. Continuous, cyclical searching by a control system for a desired or ideal value. Rapid hunting usually is termed oscillation; slower cycling is called bird-dogging. 2. Movements of a selector from terminal to terminal until an idle one is found.

HV — Abbreviation for high voltage.

H-vector — A vector that represents the magnetic field of an electromagnetic wave. In free space, it is perpendicular to the **E**-vector and the direction of propagation.

H-wave — A mode in which electromagnetic energy can be transmitted in a waveguide. An H-wave has an electric field perpendicular to the length of the waveguide, and a magnetic field parallel as well as perpendicular to the length.

hybrid — 1. An electronic circuit that contains both vacuum tubes and transistors. 2. A mixture of thin-film and discrete integrated circuits. 3. A computer that has both analog and digital capabilities. 4. *See* hybrid junction. 5. A transformer or combination of transformers or resistors that affords paths to three branches, A, B, and C, so arranged that A can send to C, and B can receive from C, but A and B are effectively isolated. 6. A mixture or combination of two different technologies. 7. Made up of several different components. 8. A telephone circuit that joins a two-wire line to a four-wire line. Originally, hybrids were transformers, but today they are electronic circuits. 9. In telecommunications, a circuit that divides a signal transmission channel into two channels (i.e., one for each direction) or, conversely, combines two channels into one.

hybrid arrangement — On a telephone system, the use of hybrid-type transformers with carrier circuits or two-wire repeaters to amplify conversations in both directions without causing feedback or singing effects.

hybrid balance — A measure of the degree of balance between two impedances connected to two conjugate sides of a hybrid set. Given by the formula for return loss. The better the balance, the greater is the transhybrid loss.

hybrid circuit — 1. A circuit that combines the thin-film and semiconductor technologies. Generally, the passive components are made by thin-film techniques, and the active components by semiconductor techniques. The active devices are attached to the thin-film passive components by a suitable bonding process. 2. Also called two-wire–four-wire terminating set. In telephone transmission circuits, a circuit for interconnecting two-wire and four-wire circuits through a differential balance or bridge circuit in which the two sides of the four-wire circuit form conjugate arms. 3. Any circuit made by using a combination of the following component manufacturing technologies: monolithic IC, thin film, thick film, and discrete component. 4. An integrated microelectronic circuit in which each component is fabricated on a separate chip or substrate, interconnected by means of lead wires so that each component can be independently optimized for performance. Example: a circuit package composed of transistor and diode dice, capacitor chips, and thick-film resistors and conductors.

hybrid coil — Also called bridge transformer. 1. A single transformer that has, effectively, three windings and that is designed to be connected to four branches of a circuit so as to render these branches conjugate in pairs. 2. A four-winding transformer used at the junction between a two-wire and a four-wire circuit. It effectively separates the transmit and receive paths. 3. A hybrid consisting of a three-winding tapped transformer used with a balancing network to convert a four-wire telephone line to a two-wire line.

hybrid computer — 1. A computer that results from the interconnection of an analog computer and a digital computer, plus conversion equipment, each contributing its special advantages to an assigned part of the solution of a class of complex problems. 2. A computer that combines both analog and digital equipment for purposes of solving problems that cannot be adequately or economically handled by either type of computer operating independently. The term does not denote the use of some analog equipment to preprocess data that is then converted to digital form and subsequently entered into a conventional digital computer. Rather, there is usually a continual flow of data in both directions between analog and digital equipment. 3. A computer for data processing in which both analog and discrete representations of data are used. 4. A computer designed with both digital and analog characteristics, combining the advantages of analog and digital computers when working as a system.

hybrid electromagnetic wave — Abbreviated HEM wave. A wave in the electromagnetic spectrum that has both electric and magnetic field vectors in the direction of propagation.

hybrid electromechanical relay — A relay with isolated input and output in which electromechanical and electronic devices such as a solid-state amplifiers are combined to perform a switching function with an electromechanical output.

hybrid integrated circuit — Abbreviated HIC. 1. An integrated circuit combining parts made by a number of techniques, such as diffused monolithic portions, thin-film elements, and discrete devices. 2. An arrangement consisting of one or more integrated circuits in combination with one or more discrete passive devices. Alternatively, the combination of more than one type of integrated circuit into a single integrated component. Hybrid ICs offer more circuit complexity than can be achieved with present-generation monolithic ICs. 3. A composite of either monolithic integrated circuits or discrete semiconductor device circuits, in a unit-packaging configuration. 4. The physical realization of electronic circuits or subsystems from a number of extremely small circuit elements electrically and mechanically interconnected on a substrate. 4. The combination of thin-film or thick-film circuitry deposited on a substrate with chip transistors, capacitors, and other components. Thin-film construction is used for microwave integrated circuits (MICs).

hybrid junction — 1. A transformer or waveguide circuit having four terminals (or four ports) so arranged that a signal entering at one terminal will divide and emerge from the two adjacent terminals but will be unable to reach the opposite terminal. Hybrid junctions (quadrature hybrids) are widely used in microwave circuits as power dividers and combiners (e. g., in balanced amplifiers, double-balanced mixers). 2. Also called hybrid tee or magic tee. A waveguide arrangement with four branches. When they are properly terminated, energy is transferred from any one branch into two of the remaining three branches. In common usage, this energy is divided equally between the two. 3. Any network or device that provides a low impedance path and impedance matching between adjacent circuits, but maintains a high degree of isolation between opposite circuits. Several types are common: (*a*) a three-winding hybrid transformer, (*b*) a resistance bridge circuit, and (*c*) a waveguide device known as a hybrid tee.

hybrid loss — The transmission loss incurred when a signal goes through a hybrid coil. This loss is about 3.6 dB (3 dB because the current divides into two equal halves, and 0.6 dB for coil loss).

hybrid microcircuit — A circuit produced by the combination of several different components in a single package. Hybrids form a middle ground between boards or modules using packaged components and monolithic ICs, which may not offer sufficient performance. Hybrids are commonly formed from a combination of chip or packaged active devices and thin- or thick-film passive devices.

hybrid microelectronics — The entire body of electronic art that is connected with or applied to the realization of electronic systems using hybrid circuit technology.

hybrid network — A nonhomogeneous communication network required to operate with signals of dissimilar characteristics (such as analog and digital modes).

hybrid parameters — Also called h-parameters. The resultant parameters of an equivalent transistor circuit when the input current and output voltage are selected as independent variables.

hybrid ring — Also called a rat race. A hybrid junction commonly used as an equal power divider. It consists of a reentrant line (waveguide) to which four side arms are connected. The line is of the proper electrical length to sustain standing waves.

hybrids — A particular type of circuit or module consisting of a combination of two or more integrated circuits, or one integrated circuit and discrete elements. *See* hybrid integrated circuit.

hybrid set — Two or more transformers interconnected to form a network having four pairs of accessible terminals. Four impedances may be connected to the four terminals, so that the branches containing them may be made conjugate in pairs.

hybrid solid-state relay — A relay with isolated input and output in which electromechanical and electronic devices are combined to perform switching functions with a solid-state output. Typically, a reed switch is used to trigger a solid-state output device.

hybrid tee — *See* hybrid junction, 2.

hybrid thin-film circuit — A microcircuit formed by attaching discrete components and semiconductor devices to networks of passive components and conductors that have been vacuum deposited on glazed ceramic, sapphire, or glass substrates.

hybrid transformer — *See* hybrid coil.

hybrid-type circuit — *See* hybrid integrated circuit; multichip circuit.

hydraulic robot — Programmable machine that uses hydraulic motors and cylinders much as pneumatic robots use pneumonic motors and cylinders. In most cases these motors and cylinders are controlled by servo valves, a system that permits smooth motion as well as high lifting capacity. However, some hydraulic robots control their motors and cylinders with mechanical stops.

hydroelectric — The production of electricity by water power.

hydrogen electrode — A platinum electrode covered with platinum black, around which a stream of hydrogen is bubbled. The hydrogen electrode furnishes a standard against which other electrode potentials can be compared.

hydrogen lamp — A special light source, used in some spectrophotometers, that produces invisible light energy. It is used in finding the light-energy frequency of test solutions.

hydrogen thyratron — A thyratron containing hydrogen.

hydrolysis — The chemical decomposition of a substance in the presence of water. Usually, it is considered in the sense of chemical degradation of insulating materials under the influence of heat or pressure and in contact with moisture (for example, hydrolysis of polyester films and coatings).

hydromagnetics — *See* magnetohydrodynamic power generator.

hydromagnetic waves — Waves in which the energy oscillates between the magnetic field energy and kinetic energy of the hydrodynamic motion, the reservoirs being the self-inductance of the conductive matter and the mass inertia of the moving fluid.

hydrometer — An instrument for determining the specific gravity of liquids, especially of a storage-battery electrolyte. It consists of a weighted glass float having a graduated stem that sinks into the liquid to a point determined by the specific gravity of the liquid. The float is usually contained in a glass, and a rubber syringe is used to withdraw a sample of the liquid.

hydrophone — An electoacoustic transducer that responds to waterborne sound waves and delivers essentially equivalent electric waves.

hydrostatic pressure — *See* static pressure.

hygrometer — An instrument that measures the relative humidity of the atmosphere.

hygroscopic — Readily capable of absorbing and retaining moisture from the atmosphere. The opposite term is nonhygroscopic.

hygrostat — A device that closes a pair of contacts when the humidity reaches a prescribed level.

hyperbola — 1. A curve that is the locus of points having a constant difference of distance from two fixed points. 2. In hyperbolic guidance systems, a path along which the difference between the arrival times of pulses from two transmitters is constant. *See also* hyperbolic guidance system.

hyperbolic error — The error in an interferometer system arising from the assumption that the directions of the wavefronts incident at two antennas of a base line are parallel, whereby the equiphase path is a cone. Mathematically the equiphase path is a hyperbola.

hyperbolic grind — A shape of tape playback and record heads. It permits good head contact and better response at high frequencies.

hyperbolic guidance system — A method of guidance in which sets of ground stations transmit pulses from which a hyperbolic path can be derived to give range and course information for steering. *See also* hyperbola.

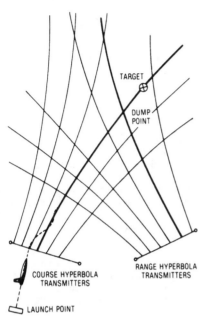

Hyperbolic guidance.

hyperbolic head — A recording head whose pole-piece surfaces (when viewed from the edge of the tape) are shaped like the graph of a mathematical hyperbolic function. This shape offers a good compromise between intimate tape contact at the gap and proximity to the tape of the rest of the pole-piece face (the latter is necessary for good low-frequency response).

hyperbolic horn — A horn in which the equivalent cross-sectional radius increases according to a hyperbolic law.

hyperbolic navigation system — A method of radionavigation (e.g., loran) in which pulses transmitted by two ground stations are received by an aircraft or ship. The difference in arrival time from each station is a measure of the difference in distance between the aircraft or ship and each station. This distance is plotted on one of many hyperbolic curves on a map. A second reading from another pair of stations (or from the same master and a different slave) establishes another point on a different hyperbolic curve. The intersection of the two curves gives the position of the aircraft or ship.

hyperboloidal subreflector — The secondary reflector used in a Cassegrain antenna system. The surface has a hyperbolic shape.

HyperCard — A software tool introduced by Apple computer in 1987 to provide new ways to organize, display, and navigate through data that broadened the capability of the Macintosh computer so noncomputer programmers could design and write their own computer applications.

hyperfocal distance — That object distance at which a camera must be focused so that the far depth of field just extends to infinity. The near limit of the depth of field is then half the hyperfocal distance. For normal photographic work this distance equals 1000 times the lens aperture diameter.

hyperfrequency waves — Microwaves having wavelengths in the range from 1 centimeter to 1 meter.

hyperlink — On a computer screen, a colored section of text (usually blue) that, when clicked, takes one to another web page.

hypersensor — A single-component, reset-table circuit breaker that operates as a majority-carrier tunneling device. It is used to provide overcurrent or overvoltage protection of integrated circuits.

hypersonic — Having five or more times the speed of sound.

hypertext — 1. A concept for organizing information made possible by computers, in which keywords or phrases not only reference additional resources but also serve as software links to these resources. When viewing an HTML document with a browser, hypertext links or anchors are displayed (usually in blue) as underlined text. Clicking on this text immediately establishes a network connection to another file (or another place in the same file) containing more information on the underlined subject, and causes the browser to display this information. The other file can be anywhere on the Internet, and may contain almost anything including text, images, movies, or sounds. 2. A system that allows documents to be cross-linked in such a way that the reader can explore related documents by clicking on a highlighted word or symbol. 3. Generally, any text that contains links to other documents — words or phrases in the document that, when selected by a reader, cause another document to be retrieved and displayed.

hyspersyn motor — A synchronous motor that combines the desirable features of the induction, hysteresis, and dc-excited synchronous motor, resulting in high efficiency and power factor. It possesses the vigorous starting torque of an induction motor, the synchronization torque of a hysteresis motor, and the stiffness of an externally dc-excited synchronous motor.

hysteresigraph — A device for experimentally presenting or recording the hysteresis loop of a magnetic specimen.

hysteresis — 1. A property of all magnetic materials that causes the value of magnetic flux density to lag behind the change in value of the magnetizing force that produces the flux. It is caused by the reluctance of the molecules to change their orientation. Work done to move the molecular magnets is a loss and appears in the form of heat. 2. A type of oscillator behavior in which multiple values of the output power and/or frequency correspond to given values of an operating parameter. 3. The temporary change in the counting rate-versus-voltage characteristic of a radiation-counter tube (caused by previous operation). 4. The difference between the response of a unit or system to an increasing and a decreasing signal. 5. A form of nonlinearity in which the response of a circuit to a particular set of input conditions depends not only on the instantaneous values of those conditions, but also on the immediate past (recent history) of the input and output signal. Hysteretical behavior is characterized by inability to retrace exactly on the reverse swing a particular locus of input/output conditions. 6. The lag in the response of an instrument or process when a force acting on it changes abruptly. 7. The property of a magnetic material by virtue of which the magnetic induction for a given magnetizing force depends on the previous conditions of magnetization. 8. This term literally means to lag behind.

It is quite often used to describe the residual effect that remains after the primary effect has been removed, or the lag that exists between the responding parameter and the changing parameter. It can be seen in stress-strain and magnetizing-force magnetic-field relationships. 9. A tendency for a display element to stay in either the on or off condition once it has been switched. With hysteresis, for example, a sustaining voltage can be applied to a display to keep all lighted pixels glowing without lighting any that are supposed to be off.

hysteresis brake — *See* hysteresis clutch.

hysteresis clutch — Also called hysteresis brake. A proportional torque-control device that employs the hysteresis effect in a permanent-magnet rotor to develop its output torque. It is capable of synchronous driving or continuous slip, provided heat can be removed, with almost no torque variation at any slip differential. Its control-power requirement is small enough for vacuum-tube or transistorized drive.

hysteresiscope — An instrument used to obtain hysteresis loops on a cathode-ray oscilloscope screen without the need for specially prepared ring samples. It is used in the inspection of magnetic material.

hysteresis curve — 1. A curve showing the relationship between a magnetizing force and the resultant magnetic flux. 2. A graph showing the amount of magnetism imparted to a magnetizable material as the result of a varying magnetic field. This coincides with the variations of the applied field only through a relatively narrow range between zero magnetism and saturation, but the addition of a bias allows an audio signal to be recorded on a magnetic tape within this linear range, for minimum distortion.

hysteresis distortion — Distortion of waveforms in circuits containing magnetic components. It is due to the hysteresis of the magnetic cores.

hysteresis error — The difference in the reading obtained on a measuring instrument containing iron when the current is increased to a definite value and when the current is reduced from a higher value to the same definite value.

hysteresis heater — An induction device in which a charge (or a muffle around the charge) is heated by hysteresis losses due to the magnetic flux produced in the charge.

hysteresis loop — 1. A curve (usually with rectangular coordinates) that shows, for a magnetic material in a cyclically magnetized condition, two values of the magnetic induction for each value of the magnetizing force: one when the magnetizing force is increasing, the other when it is decreasing. 2. The graphical representation of

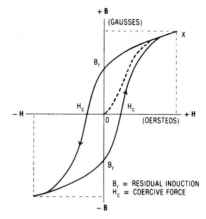

Hysteresis loop.

relationship between the magnetizing force and the resultant induced magnetization of a ferromagnetic material when the magnetizing field is carried through a complete cycle of equal and opposite values under cyclic conditions.

hysteresis loss — 1. The power expended in a magnetic material as a result of magnetic hysteresis. 2. The power dissipated in a ferromagnetic core as a result of its hysteresis; proportional to the product of the area of the loop, times the frequency, times the volume of the core. 3. Loss that occurs when a rapidly changing current, such as ac, is forced to continually supply energy to an iron core that tends to "memorize" previous magnetic states. 4. The power loss in a magnetic core, such as in a transformer energized by an alternating current, that is due to hysteresis.

hysteresis meter — An instrument for determining the hysteresis loss in a ferromagnetic material. It measures the torque produced when the test specimen is placed in a rotating magnetic field or is rotated in a stationary magnetic field.

hysteresis motor — A synchronous motor without salient poles or direct-current excitation. It is started by the hysteresis losses induced in its secondary by the revolving field.

hystoroscope — An instrument used to observe, measure, and record the magnetic characteristics of both easy and hard axes of magnetic materials.

Hz — Letter symbol for hertz, meaning cycles per second (of any periodic phenomenon).

I

I — 1. Symbol for current. 2. Abbreviation for luminous intensity.

IACS — Abbreviation for International Annealed Copper Standard. A standard of copper conductivity by specifying resistivity and temperature.

IBM card — A type of paper card that may have information recorded on it by means of punched holes and that may be read by a computer.

IC — 1. Abbreviation for integrated circuit. 2. Abbreviation for internal connection.

I_{CBO} — The reverse current that occurs when a specific dc voltage is applied in the nonconducting direction to the collector junction of a transistor while the emitter is open-circuited.

ICE — Abbreviation for in-circuit emulator and in-circuit emulation.

ice loading — The weight of ice an antenna can accumulate without being damaged.

icon — A small picture displayed on a computer screen that represents a command or an object that can be manipulated by the user. Usually, the picture shows what the icon does. For example, the PRINT icon generally looks like a printer.

iconoscope — A camera tube in which a beam of high-velocity electrons scans a photoemissive mosaic capable of storing an electrical charge pattern.

IC socket — Female contact that provides pluggable electrical engagement on its inner surface for integrated circuit components to achieve interfacing to a printed circuit board.

ICW — Abbreviation for interrupted continuous wave.

ID — Abbreviation for inside diameter.

ideal bunching — A theoretical condition in which bunching of the electrons in a velocity-modulated tube would give an infinitely large current peak during each cycle.

ideal capacitor — A capacitor having a single-valued transferred-charge characteristic.

ideal crystal — A crystal having no mosaic structure and capable of X-ray reflection in accordance with the Darwin-Ewald-Prins law.

ideal dielectric — A dielectric in which all the energy required to establish an electric field in the dielectric is returned to the source when the field is removed. (A perfect dielectric must have zero conductivity. Also, all absorption phenomena must be lacking. A vacuum is the only known perfect dielectric.)

ideal filter — Any filter in which the range of frequencies within a chosen radius suffers no attenuation and the range of frequencies outside the radius is entirely attenuated.

ideal-noise diode — A diode that has an infinite internal impedance and in which the current exhibits full shot-noise fluctuations.

ideal transducer — Theoretically, any linear passive transducer that — if it dissipated no energy and, when connected to a source and load, presented its combined impedance to each — would transfer maximum power from source to load.

ideal transformer — A hypothetical transformer that would neither store nor dissipate energy. Its self-inductances would have a finite ratio and unity coefficient of coupling, and its self- and mutual impedances would be pure inductances of infinitely great value.

I demodulator — A demodulator circuit whose inputs are the chrominance signal and the signal from the local 3.58-MHz oscillator. The output of this demodulator is a video signal representing color in the televised scene. The Q demodulator is similar except that its input from the local oscillator is shifted 90°.

identification — 1. In radar, determining the identity of a displayed target (i.e., which one of the blips in the display represents the target). 2. In a computer, a code number or code name that uniquely identifies a record, block, file, or other unit of information.

identification beacon — A code beacon used for positively identifying a particular point on the earth's surface.

identification, friend or foe — Abbreviated IFF. A system using radar transmissions to which equipment carried by friendly forces automatically responds, for example, by emitting pulses, thereby distinguishing themselves from enemy forces. It is the primary method of determining the friendly or unfriendly character of aircraft and ships by other aircraft or ships and by ground forces employing radar-detection equipment and associated IFF units.

identifier — 1. A symbol whose purpose is to identify, indicate, or name a body of data. 2. A mnemonic code used to identify or name an item of data or data format in a computer.

identify — In a computer, to attach a unique code or code name to a specific unit of information.

idiochromatic — Having photoelectric properties characteristic of the pure crystal itself and not due to foreign matter.

I-display — In radar, a display in which a target appears as a circle when the radar antenna is pointed directly at it. The radius of the circle is proportionate to the target distance. When the antenna is not pointing at the target, only a segment of the circle appears. Its length is inversely proportional to the magnitude of the pointing error, and the segment points away from the direction of error.

idle characters — Control characters interchanged by a synchronized transmitter and receiver to maintain synchronization during a nondata period.

idle noise — 1. Noise that exists in a communication system when no signals are present. 2. Unwanted, random electrical energy present in a transmission system under unmodulated conditions.

idler — 1. A rubber-tired wheel that transfers power from a phonograph motor to the turntable rim. 2. An intermediate drive wheel, usually with a rubber or neoprene "tire," that transfers rotational energy from a driven wheel to a third wheel. Often used for speed reduction between a drive motor and capstan shaft. *See also* pinch roller.

idler drive — 1. A drive system used to rotate a turntable, which consists of a drive-shaft that is turned by the motor pulley and that drives the inside rim of the turntable platter. 2. A system for transferring power from a motor to a turntable through a rubber wheel that contacts the motor shaft and the inside rim of the platter.

idler frequency — A sum or difference frequency, other than the input, output, or pump frequencies, generated within a parametric device and requiring specific circuit consideration to achieve the desired performance of the device.

idler pulley — A pulley used only for tightening a belt or changing its direction. The shaft does not drive any other part.

idle time — That portion of available time during which the hardware is not in use.

idle-trunk lamp — A signal lamp that indicates that the outgoing trunk with which it is associated is not busy.

idling current — Also called quiescent current. The zero-signal power supply current drawn by a circuit or by a complete amplifier.

IDT — Abbreviation for interdigital transducer.

IEC — 1. Abbreviation for integrated electronic component. 2. Abbreviation for International Electrotechnical Commission. An organization that cooperates with the ISO for technology standards.

IEEE — Abbreviation for Institute of Electrical and Electronic Engineers. A professional organization of scientists and engineers whose purpose is the advancement of electrical engineering, electronics, and allied branches of engineering and science. (The IEEE resulted from the merger of the IRE and the AIEE.)

IEEE-488 bus — Also known as the general-purpose interface bus (GPIB). 1. An interface standard that defines digital data exchange between up to 15 instruments; a bus widely used to connect test instrumentation. 2. An industry-standard bus that defines a digital interface for programmable instrumentation; it uses a byte-serial, bit-parallel technique to handle 8-bit-wide data words. Published by the IEEE in 1975, revised 1978.

IEM — Abbreviation for illuminated entry module. An electronic convenience control (offered by Ford Motor and General Motors) that lights up a car's interior and door keyhole slots for a timed interval of about 20 seconds while the car owner unlocks the door.

IF — Abbreviation for intermediate frequency.

IF amplifier — *See* intermediate-frequency amplifier.

IF bandwidth — The range of frequencies centered about the intermediate frequency limited by the −3-dB amplitude points.

IF canceler — In radar, a moving-target-indicator canceler operating at intermediate frequencies.

IFF — *See* identification, friend or foe.

IF rejection — The ability of a superheterodyne AM or FM tuner's IF circuits to reject external interference at the intermediate frequency. Measured in decibels (the higher the better), it is of more significance in AM than in FM reception because the lowest broadcast AM frequency, now 530 kHz, is so close to the standard AM intermediate frequency of 455 kHz.

IFRU — *See* interference-rejection unit.

IF selectivity — The ability of the IF stages of a superheterodyne receiver to accept the signal from one station while rejecting the signal of the adjacent stations; it is the ratio of desired to undesired signal required for 30-dB suppression of the undesired signal (IHF standard).

if statement — *See* conditional statement.

IF strip — *See* intermediate-frequency strip.

IF transformer — *See* intermediate-frequency transformer.

IGFET — Abbreviation for insulated-gate field-effect transistor. Though a less popular term than MOS (metal-oxide semiconductor), it more precisely defines devices made by various MOS processes. *See also* metal-oxide semiconductor field-effect transistor.

ignition cable — Cable designed primarily for automotive ignition systems.

ignition coil — 1. An iron-core transformer that converts a low direct voltage to the 20,000 volts or so required to produce an ignition spark in gasoline engines. It has an open core, a heavy primary winding connected to the battery or other source through a vibrating armature contact, and a secondary winding with many turns of fine wire. 2. That part of the ignition system which acts as a transformer to step up the battery voltage to many thousands of volts. The high-voltage surge then produces a spark at the spark-plug gap.

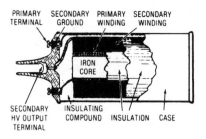

Ignition coil.

ignition control — Control of the instant that static current begins to flow in the anode circuit of a gas tube.

ignition interference — Noise produced by sparks or other ignition discharges in a car, motor, or furnace ignition, or by equipment with loose contacts or connections.

ignition reserve — In a gasoline engine, the difference between the available voltage and the required ignition voltage.

ignition system — In an automobile, the system that furnishes high-voltage sparks to the engine cylinders to fire the compressed air-fuel charges. It consists of battery, ignition coil, ignition distributor, ignition switch, wiring, and spark plugs.

ignition terminal — Solderless terminal designed for use in automotive ignition work.

ignition voltage — In a gasoline engine, the peak voltage required to produce a spark across the plug electrodes.

ignitor discharge — In switching tubes, a dc glow discharge between the ignitor electrode and a suitably located electrode. It is used to facilitate radio-frequency ionization.

ignitor electrode — An electrode (which is partly immersed in the mercury-pool cathode of an ignitron) used to initiate conduction at the desired points in each cycle.

ignitor firing time — In switching tubes, the interval between application of a dc voltage to the ignitor electrode and start of current flow.

ignitor interaction — In a tr, pre-tr, or attenuator tube, the difference between the insertion loss measured at a specified level of ignitor current and that measured at zero ignitor current.

ignitor leakage resistance — In a switching tube, the insulation resistance measured between the ignitor electrode terminal and the adjacent rf electrode in the absence of an ignitor discharge.

ignitor oscillation — A relaxation type of oscillation in the ignitor circuit of a tr, pre-tr, or attenuator tube.

ignitor voltage drop — In switching tubes, the dc voltage between the cathode and anode at a specified ignitor current.

ignitron — A type of mercury-pool rectifier that has only one anode. The arc is started for each cycle of operation by an ignitor that dips into the mercury pool. The mercury pool serves as the cathode of the rectifier. The ignitron is characterized by the ability to withstand tube currents several times as high as rated values for a few cycles.

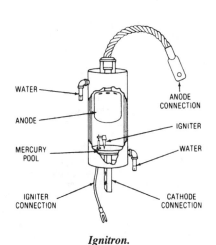

Ignitron.

ignore — In a computer, a character code indicating that no action is to be taken.

IGY — Abbreviation for international geophysical year.

IHFM — Abbreviation for Institute of High-Fidelity Manufacturers, an association of manufacturers that publishes ratings and standards for high-fidelity equipment.

I²L — Abbreviation for integrated injection logic (pronounced "I-squared L"). 1. A bipolar logic circuit that has a much higher packing density than conventional transistor-transistor logic or complementary metal-oxide semiconductor (CMOS), dissipates low power, and has high-speed characteristics. To an extent it combines bipolar speed, MOS circuit density, and the low-power dissipation of CMOS.

I³L® — Abbreviation for isoplanar integrated injection logic (Fairchild Camera & Instrument Corp.).

illegal character — A character or combination of bits that does not have validity according to some criterion; for example, a character that is not a member of a specified alphabet.

illegal operation — An impossible-to-execute computer instruction.

illuminance — 1. The density of luminous flux at a given distance from the center of a source. It is equal to the total flux divided by the surface area over which it is uniformly spread. The units of illuminance are lumens/cm², lumens/ft², etc. One lumen/ft² is the same as a footcandle; one lumen/cm² is the same as one phot; and one lumen/m² is equal to one lux, or one meter-candle. 2. Luminous flux incident per unit area of a surface; luminous incidence. (The use of the term *illumination* for this quantity conflicts with its more general meaning.)

illuminant-C — The reference white of color television—i.e., light that most nearly matches average daylight.

illuminate — 1. To expose to light. 2. In radar, to strike with a radar signal so that reflection returns the signal to the source for interpretation.

illuminated — Characteristic of a surface or object that has luminous flux incident on it.

illuminated entry module — *See* IEM.

illumination — 1. The light flux incident on a unit projected area; it is the photometric counterpart of irradiance and is expressed in footcandles. 2. The density of the luminous flux incident on a surface; it is the quotient of the luminous flux by the area of the surface, when the latter is uniformly illuminated (SI unit = lux or lx). 3. The general term meaning the application of light to a subject. Should not be used in place of the specific quantity *illuminance*.

illumination control — A photorelay circuit that turns on artificial lighting when natural illumination decreases below a predetermined level.

illumination sensitivity — The output current of a photosensitive device divided by the incident illumination at constant electrode voltages.

illuminometer — A photometric instrument used to measure the illumination falling on a surface. May be photoelectric or visual.

ILO — *See* injection-locked oscillator.

ILS — Abbreviation for instrument landing system. A radio beacon system forming a straight pencil beam from the runway that planes can follow from 5 to 7 miles (8 to 12 kilometers) away. An alternative is a microwave landing system.

image — 1. The instantaneous illusion of a picture on a flat surface. 2. The unused one of the two groups of sidebands generated in amplitude modulation. 3. A spatial distribution of some physical property (e.g., radiation, electric charge, conductivity, or reflectivity) made to correspond with another distribution of the same or another physical property. 4. A two-dimensional representation of an object or a scene formed by creating a pattern from the light received from the scene.

image admittance — Reciprocal of image impedance.

image antenna — The imaginary counterpart of an actual antenna. For mathematical purposes it is assumed to be located below the ground and symmetrical with the actual antenna.

image-attenuation constant — The real part of the transfer constant.

image compression — The translation of data in any format (such as video or graphics) to a more compact form for storage or transmission such that it takes up less space in a computer's memory.

image converter — 1. A solid-state optoelectric device capable of changing the spectral characteristics of a radiant image. Examples of such changes are infrared-to-visible and X-ray-to-visible. 2. An electron tube that employs electromagnetic radiation to produce a visual replica of an image produced on its cathode. Electrons ejected from the photosensitive cathode by the incident radiation are accelerated to and focused upon a fluorescent phosphor screen, thus forming the visual replica. Image converters can be used in the infrared, ultraviolet, and

X-ray regions as well as in the visible. An example of an infrared-sensitive image converter is the snooperscope.

image-converter tube — *See* image tube.

image dissection — An optical, mechanical, or electronic process, or a combination of such processes, in which an optical image is divided into discrete segments prior to being photographed, recorded, transmitted, or processed in some other way.

image dissector — 1. A television camera tube in which the image is swept past an aperture in a series of 525 interlaced lines 30 times per second. Instead of a beam scanning the image, the entire image is scanned past the aperture, which "dissects" the image — hence the name. *See also* dissector tube. 2. In OCR, a mechanical or electronic transducer that detects in sequence the light levels in different areas of a completely illuminated sample space.

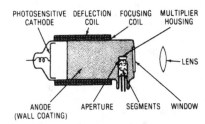

Image dissector, 1.

image dissector tube — An electron tube that is used as a camera tube for a television system. When the picture to be transmitted is focused on a photosensitive surface, electrons are emitted from each section of the surface in proportion to the amount of light in that certain part of the picture, and are then drawn down the tube in a positive anode. However, the electrons are still focused in an electric field. The focusing fields are changed regularly to sweep the electron picture horizontally and vertically as it travels down the tube. The picture then passes through an aperture into an electron multiplier, where the output of the multiplier varies with the different parts of the picture. This output represents the dissecting of the picture into ordered parts.

image distortion — Failure of the reproduced image in a television receiver to resemble the original scene scanned by the camera.

image effect — An effect produced on the field of an antenna as the electromagnetic waves are reflected from the earth's surface.

image enhancement — Also called detail enhancement. The process by which the image is manipulated to increase the information extracted by the human visual system.

image-enhancing equipment — An elaborate device, often involving a computer, in which a photograph is scanned by a point of light, the amplitude of the electrical signal being modified electronically before being rerecorded on another film.

image force — The force on a charge due to that charge or polarization which it induces on neighboring conductors or dielectrics.

image frequency — 1. In heterodyne frequency converters, an undesired input frequency capable of producing the selected frequency by selecting one of the two sidebands produced by beating. The word *image* implies the mirrorlike symmetry of signal and image frequencies about the beating oscillator frequency or intermediate

frequency, whichever is higher. 2. An undesired signal obtained in frequency conversion using a mixing or heterodyning process.

image-frequency rejection ratio — Of a super heterodyne receiver, the ratio of the response at the desired frequency to the response at the image frequency

image iconoscope — 1. An iconoscope in which greater sensitivity is obtained by separating the function of charge storage from that of photoelectric emission. An optical image is projected on a continuous photosensitive screen, and the electron emission from the back of this screen is focused electromagnetically onto a mosaic screen that is scanned by an electron beam as the original emitron cathode-ray tube. The British term is super-emitron. 2. A camera tube similar in design to the iconoscope. However, the image formed in the image iconoscope is projected on a photocathode that emits photoelectrons to be focused on a material, forming the charge image.

image impedance — The impedances that will simultaneously terminate all inputs and outputs of a transducer so that at each of its input and outputs the impedances in both directions are equal.

image intensifier — 1. A system for increasing the sensor response to a radiation pattern or image by interposing active elements between the sensor and the image, and supplying power to the active element. This is normally done by focusing the scene to be imaged on the photocathode of the tube, giving rise to a photoelectron pattern corresponding to the optical image. This pattern is accelerated and focused onto a phosphor, which emit light to reproduce a visual image of the scene. 2. Device used in X-ray techniques for brightening the fluoroscopic image several hundred times and reducing radiation exposure. 3. An electronic tube equipped with a light sensitive electron emitter at one end and a phospho screen at the other end; an electron lens inside the tube relays the image. This device is used in astronomy and in military and surveillance systems to provide night vision

image interference — In a receiver, a response due to signals of a frequency removed from the desired signal by twice the intermediate frequency.

image-interference ratio — In a superheterodyne receiver, the effectiveness of the preselector in rejecting signals at the image frequency.

image inverter — A fiber-optic device that rotates an image through a predetermined angle.

image orthicon — 1. A camera tube in which a photoemitting surface produces an electronic image and focuses it on one side of a separate storage target. The opposite side of the target is then scanned by low-velocity electrons to produce the output. 2. A camera tube widely used in TV broadcasting. It consists of three sections all included in a single vacuum envelope. The three parts are as follows: (*a*) A photosensitive film sometimes called the photocathode. The scene to be televised is focused on this film by an outside camera lens. (*b*) The scanning beam provided by an electron gun, which scans lines of the target film. The gun's beam is deflected by electromagnet within the tube (the whole picture is scanned by the beam in $1/30$ of a second). (*c*) A return beam of electrons falls on a multiplier section, where an electron current of sufficient magnitude is developed to be sent out from the broadcast transmitter as a video signal. *See* Figure on p. 367.

image phase constant — The imaginary part of the transfer constant.

image plane holography — A type of holographic process in which a lens is used to image the subject in the film plane.

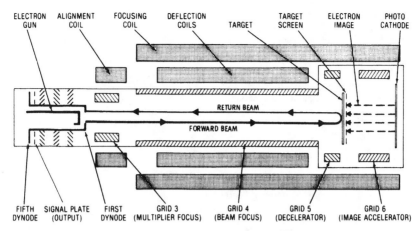

Image orthicon.

image ratio — In a heterodyne receiver, the ratio of the image-frequency signal input to the desired signal input for identical amplitude outputs.

image redundancy — The multiple storage of a single image.

image rejection — 1. The suppression of image-frequency signals in a superheterodyne receiver. 2. The rejection by the tuner of a signal at the image (second-channel) frequency, corresponding to the tuned (real) frequency plus twice the intermediate frequency when the local oscillator is working above the signal frequency, or minus twice the intermediate frequency when the local oscillator is working below the signal frequency. 3. The ability of a superheterodyne receiver to ignore signals removed from the desired frequency by twice the intermediate frequency (10.7 MHz in home FM receivers, 455 kHz in home AM receivers). Image response can be reduced by using selective tuned stages in the input circuits of the tuner. As with most tuner rejection or suppression specifications, it is measured in decibels, with higher numbers indicating more suppression.

image-rejection ratio — The ratio (in decibels) of the signal required for a 30-dB signal-to-noise ratio to that required for the same ratio but at the image frequency. An increase in front-end selectivity increases the ratio.

image-reject mixer — A combination of two balanced mixers and associated hybrid circuits designed for separation of the image channel from the signal channels normally present in a conventional mixer. The arrangement makes possible image rejection of up to 30 dB without the use of filters.

image response — The response of a heterodyne receiver to a signal that is separated by twice the intermediate frequency from the frequency to which the receiver is tuned. Unless there is some preselection, images will cause spurious unwanted responses when the spectrum occupied by a signal is greater than twice the frequency of the first IF stage of the receiver.

image retaining panel — A type of electroluminescent display that records and maintains an irradiated image on its phosphor screen, provided a dc potential is applied to the screen. Used to record X-ray images, the display can retain an image for up to 30 minutes if the dc potential is maintained.

image retention — The vidicon pickup tube's tendency to retain an image on its target area after it has stopped scanning that image. Extreme image retention results in the image being burned into the target area.

image storage panel — A modified form of an image retaining panel that can be used in subdued daylight. This is achieved by the addition of a layer of zinc oxide between the panel's phosphor layer and its rear electrode. The zinc oxide cannot be made photoconductive by low-intensity daylight, but does become so when exposed to X-rays. An electroluminescent image is obtained by the application of ac voltage to the panel following its exposure to the X-rays.

image transducer — Any arrangement of a bundle of optical fibers that alters the shape of the image. For example, by a systematic regulation of the spacing of the fibers from the entrance end to the exit end, a distortion of the image can occur that may be used to neutralize or compensate for the distortion introduced by the lens, prism, or mirror components in an optical system.

image-transfer constant — *See* transfer constant.

image tube — Also called an image-converter tube. An electron tube that reproduces on its fluorescent screen an image of an irradiation pattern incident on its photosensitive surface.

image-tube camera — A camera system in which the image formed on the fluorescent screen of an image converter tube in the system is recorded by photography, or direct-contact printing from the face of the tube.

imaging — The process of creating and manipulating data for visual presentation and storage.

imbedded layer — A conductor layer deposited between insulating layers.

IMD — *See* intermodulation distortion.

imitative deception — The transmission of messages in the enemy's communication channels for the purpose of deceiving him or her.

immediate access — The ability of a computer to put data in storage or remove it from storage without delay.

immediate-access store — A store whose access time is negligible compared with other operating times.

immediate addressing — 1. In a computer, a mode of addressing in which the operand contains the value to be operated on, and no address reference is required. 2. An addressing mode in which the data for an instruction is the next sequential byte in the instruction stream.

immediate data — Data that immediately follows an instruction in a memory and is used as an operand by that instruction.

immersion plating — 1. A method of metal deposition that depends on a galvanic displacement of the metal being plated by the substrate. Thickness of the plating is limited to 10 to 50 microinches (0.254 to 1.27 µm). 2. The chemical deposition of a thin metallic coating over certain base metals by a partial displacement of the base metal.

immersion pyrometer — An instrument for determining molten-steel temperature and normally consisting of a platinum-platinum rhodium bimetal thermocouple junction and a recording device for transposing the millivoltage into degrees of temperature.

immittance — A term that denotes both impedance and admittance. It is commonly applied to transmission lines, networks, and certain kinds of measuring instruments.

IMOS — *See* ion-implanted MOS.

impact excitation — The starting of damped oscillations by a sudden surge, such as by a spark discharge.

impact modulator amplifier — A fluidic device in which the impact plane position of two opposed streams is controlled to alter the output.

impact predictor — A device that can determine, in real time, the point on the earth's surface where a ballistic missile will impact if thrust is instantaneously terminated.

impact printer — 1. Any type of printer that generates characters by using some form of stamping or inking through a ribbon by some sort of character slug, element, or hammer-needle. (Daisy-wheel printers are impact printers.) 2. Any mechanical imprinting device that forms characters by striking characters against a ribbon onto paper. 3. A mechanical printer operating at relatively low speeds — from 150 lines per minute to 1800 lines per minute. Has multiple copy capability and normally is capable of producing an original plus three copies. Impact printers are identified as letter-quality, dot-matrix, and line-printing types. No longer in common use.

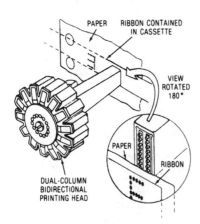

Impact printer (dot-matrix).

impact printing — Printing that is produced by the impact of a key on an inked ribbon, transferring the impression of a character onto the paper located behind the ribbon.

impact resistance — Resistance to fracture under shock force.

IMPATT — Acronym for impact avalanche and transit time.

IMPATT diode — Impact avalanche and transmit time diode. 1. A pn-junction diode operated with heavy back bias so that avalanche breakdown occurs in the active region. To prevent burnout, the device is so constructed that the active region is very close to a good heat sink. For the same reason, the bias supply must be a constant-current type. 2. A device whose negative resistance characteristic is produced by a combination of impact avalanche breakdown and charge-carrier transit-time effects. Avalanche breakdown occurs when the electric field across the diode is high enough for the charge carriers (holes or electrons) to create electron-hole pairs. With the diode mounted in an appropriate cavity, the field patterns and drift distance permit microwave oscillations or amplification. 3. A semiconductor microwave diode that, when its junction is biased onto avalanche, exhibits a negative resistance over a frequency range determined by the transit time of charge carriers through the depletion region.

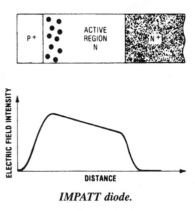

IMPATT diode.

IMPATT oscillator — An oscillator in which the active element is an IMPATT diode operating in a negative resistance mode. Dc to rf conversion efficiencies are normally less than 20 percent. Present devices operate above 5 GHz.

impedance — 1. The total opposition (i.e., resistance and reactance) a circuit offers to the flow of alternating current at a given frequency; the ratio of the potential difference across a circuit or element of a circuit to the current through the circuit or element. It is measured in ohms, and its reciprocal is called admittance. Symbol: Z. 2. The combination of resistance and reactance 3. Combined opposition to current resulting from resistance, capacitance, and inductance. 4. The sinusoidal terminal voltage of a circuit divided by the current through it. 5. A speaker's resistance to an alternating current, which varies with frequency. A speaker's rated impedance is usually the value measured at 400 Hz. 6. The opposition to alternating current in a circuit, generally categorized as either high or low, and measured in ohms. 7. The total opposition that a circuit offers to the flow of alternating current or any other varying current at a particular frequency. It is a combination of resistance, R, and reactance, X, measured in ohms.

impedance angle — Angle of the impedance vector with respect to the resistance vector. Represents the phase angle between voltage and current.

impedance at the intermediate frequency — In a mixer semiconductor diode, the impedance measured at the output terminals of a mixer circuit when the device is driven by a local oscillator under specified conditions.

impedance bridge — A device for measuring the combined resistance and reactance of a component part of a circuit.

impedance characteristic — A graph of impedance versus frequency of a circuit or component.

impedance coil — A coil whose inductive reactance is used to hinder the flow of alternating current in or between circuits.

impedance compensator — 1. An electric network used with a line or another network to give the impedance of the combination a certain characteristic over a desired frequency range. 2. A circuit that ensures that a transmission line is a proper electrical load for its communicating devices. It is connected in parallel with the devices.

impedance coupling — A method of coupling using an impedance as the coupling device common to both the primary and secondary circuits. This type of coupling is usually limited to audio systems, where high gain and limited bandpass are required.

impedance drop — The vector sum of the resistance drop and the reactance drop. (For transformers, the resistance drop, the reactance drop, and the impedance drop are, respectively, the sum of the primary and secondary drops reduced to the same terms. They are usually expressed in percent of the secondary-terminal voltage.)

impedance ground — An earth connection made through an impedance of predetermined value usually chosen to limit the current of a short-circuit to ground.

impedance irregularities — Breaks or abrupt changes that occur in an impedance-frequency curve when unlike sections of a transmission line are joined together or when there are irregularities on the line.

impedance match — The condition in which the impedance of a component or circuit is equal to the internal impedance of the source or the surge impedance of a transmission line, thereby giving maximum transfer of energy from sources to load, minimum reflection, and minimum distortion.

impedance matching — 1. The connection across a source impedance of another impedance having the same magnitude and phase angle. If the source is a transmission line, reflection is thereby avoided. 2. The process of adjusting the impedances of a load and of its power source so that they are equal. This permits the greatest possible transfer of power. 3. Making the impedance of a terminating device equal to the impedance of the circuit to which it is connected in order to achieve optimum signal transfer.

impedance-matching transformer — A transformer used to match the impedance of a source and load.

impedance plethysmograph — An instrument used to detect the increased blood volume in the tissues of the body during a contraction of the heart. *See also* electrical-impedance cephalography; finger plethysmograph.

impedance transformer — A transformer that transfers maximum energy from one circuit to another.

impedance triangle — A diagram consisting of a right triangle. The sides are proportional to the resistance and reactance in an ac circuit, with the hypotenuse representing the impedance.

imperfect dielectric — A dielectric in which part of the energy required to establish its electric field is converted into heat instead of being returned to the electric system when the field is removed.

imperfection — In a crystalline solid, any deviation in structure from an ideal crystal (one that is perfectly periodic in structure and contains no foreign atoms).

implantable pacemaker — A miniature pulse generator surgically implanted beneath the skin and provided with output leads that connect directly to the heart muscle. The electrodes may contact either the outer wall of the heart muscle (myocardial electrodes) or the inner surface of the heart chamber (endocardiac electrodes).

implied AND — Also called dot AND or wired AND. A logic element in which the combined outputs are true if and only if all outputs are true. (Sometimes improperly called dot OR or wired OR.)

implied OR — Also called wired OR. A logic element in which the combined outputs are true if one or more of the outputs are true.

implode — The inward bursting of a picture tube due to its high vacuum.

import — To copy data created by one computer program or file into another.

impregnant — 1. A substance, usually a liquid, used to saturate the paper dielectric of a capacitor and replace the air between its fibers, thereby increasing the dielectric strength and the dielectric constant of the capacitor. 2. A substance intended to replace the air as dielectric between the electrodes of a capacitor.

impregnate — 1. To fill voids and air spaces (of a capacitor or transformer) with a material having good insulating properties commonly called an impregnant. 2. To fill the voids and interstices of a material with a compound. This does not imply complete fill or complete coating of the surfaces by a hole-free film.

impregnated coils — Coils that have been permeated with an electric grade varnish or other protective material to protect them from mechanized vibration, handling, fungus, and moisture.

impregnating — Complete filling of even the smallest voids in a component or closely packed assembly of parts. Low-viscosity compounds, usually liquids, are used. The process is frequently accomplished by a vacuum process in which all air is removed before introducing the impregnating material. Typical examples of impregnating are the filling of capacitors or transformer windings.

impregnation — 1. The process of coating the insides of coils and closely packed electronic assemblies by dipping them into a liquid and letting it solidify. 2. The process of completely filling all interstices or a part or assembly with a thin, liquid, electrically insulating material. The process is best accomplished by first removing all air (creating a vacuum), then introducing the impregnant, and finally applying atmospheric or elevated pressures to completely force-fill the system.

impressed voltage — The voltage applied to a circuit or device.

improvement threshold — A characteristic of FM radio receivers that determines the minimum rf signal power required to overcome the inherent thermal noise. For increasing values of rf power above this point, an improvement of signal-to-noise ratio is obtained.

impulse — 1. A pulse that begins and ends within so short a time that it may be regarded mathematically as infinitesimal. The change produced in the medium, however, is generally of a finite amount. 2. A current surge of unidirectional polarity. *See also* pulse.

impulse bandwidth — The area divided by the height of the voltage-response selectivity as a function of frequency. It is used in the calculation of broadband interference.

impulse-driven clock — An electric clock in which the hands are moved forward at regular intervals by current impulses from a master clock.

impulse excitation — Also called shock excitation. 1. A method of producing oscillatory current in which the duration of the impressed voltage is relatively short compared with that of the current produced. 2. The sudden application of a momentary steep-wavefront voltage to a resonant circuit, resulting in a damped oscillation.

impulse frequency — The number of pulse periods per second generated by the dial-pulse springs in a telephone as they rapidly open and close in response to the dialing of a digit.

impulse generator — Also called surge generator. 1. An electric apparatus that produces high-voltage surges for testing insulators and for other purposes. 2. A device that generates a broad energy spectrum by means of a very narrow impulse. Usually generated by discharging a short coaxial or waveguide transmission line. The pulses are discrete and regularly spaced, and are generally variable at a repetition rate from a few pulses per second to a few thousand pulses per second. The output of an impulse generator is specified as the rms equivalent of the peak voltage in dB above 1 microvolt per megahertz. 3. An oscillator circuit that generates electric impulses for synchronizing purposes in a television system. 4. A circuit, typically using a step-recovery diode, used to convert a sinusoidal input to a voltage impulse output. The basic circuit block in both step recovery diode multiples and comb generators.

impulse noise — 1. Noise due to disturbances having abrupt changes and of short duration. These noise impulses may or may not have systematic phase relationships. The noise is characterized by nonoverlapping transient disturbance. The same source may produce impulse noise in one system and random noise in a different system. 2. A type of communication-line interference characterized by high amplitude and short duration. 3. An unwanted signal characterized by a steep wavefront.

impulse-noise generator — Equipment for generating repetitive pulses that provide random noise signals uniformly spread over a wide band of frequencies.

impulse pay-per-view — Abbreviated IPPV. A feature of a decoder that allows an authorized subscriber to purchase a one-time scrambled program at will. IPPV shows are selected by a button on the decoder or its remote control.

impulse period — See pulse period.

impulse ratio — The ratio of the flashover, sparkover, or breakdown voltage of an impulse to the crest value of the power-frequency flashover, sparkover, or breakdown voltage.

impulse relay — 1. A relay that stores enough energy from a brief impulse to complete its operation after the impulse ends. 2. A relay that can distinguish between different types of impulses, operating on long or strong impulses and not operating on short or weak ones. 3. An integrating relay.

impulse response of a room — The time sequence of signals received at some point in a room due to a sound pulse generated at some other point in the room. It defines the arrival of a sound that has transversed the direct path between source and microphone and the arrivals of the various reflections.

impulse sealing — A heat-sealing technique in which a pulse of intense thermal energy is applied to the sealing area for a very short time, followed immediately by cooling. It is usually accomplished by using an rf heated metal bar that is cored for water cooling, or is of such a mass that it will cool rapidly at ambient temperatures.

impulse separator — Normally called sync separator. In a television receiver, the circuit that separates the synchronizing impulses from the video information in the received signal.

impulse sparkover voltage — The highest value of spark-gap or gas-discharge tube terminal voltage prior to ionization and the flow of discharge current.

impulse speed — The rate at which a telephone dial mechanism makes and breaks the circuit to transmit pulses.

impulse strength — A measure of the ability of insulation to withstand voltage surges on the order of microseconds in duration.

impulse timer — A timing device electrically powered by a synchronous motor, featuring a mechanical stepping device that enables it to advance a predetermined number of degrees within a predetermined time interval, controlling a multiple number of circuits. Said circuits are controlled by individual cams, which program their activity.

impulse train — See pulse train.

impulse transmission — The form of signaling used principally to reduce the effects of low-frequency interference. Impulses of either or both polarities are employed for transmission, to indicate the occurrence of transitions in the signals.

impulse-transmitting relay — A relay in which a set of contacts closes briefly when the relay changes from the energized to the deenergized position, or vice versa.

impulse-type telemeter — A telemeter that employs the characteristics of intermittent electric signals, other than their frequency, as the translating means.

impurity — Also called dopant. 1. A material such as boron, phosphorus, or arsenic added to a semiconductor such as germanium or silicon to produce either p-type or n-type material. Impurities that provide free electrons are called donors and cause the semiconductor material to be n-type. Impurities that accept electrons are called acceptors and cause the material to be p-type. 2. A foreign material present in a semiconductor material, usually in small quantities. Some impurities are unwanted, and great pains are taken to extract them from the material. Others are intentionally added to semiconductor materials as dopants in order to modify their electrical behavior. 3. In semiconductor technology, a material such as boron, phosphorus, or arsenic added in small quantities to a crystal to produce an excess of electrons (donor impurity) or holes (acceptor impurity).

impurity density — The amount of impurity material diffused into a certain volume of semiconductor material used in manufacturing semiconductor devices.

impurity ions — An alien, electrically charged atomic system in a solid; an ion substituted for a constituent atom or ion in a crystal lattice, or located in an interstitial site in the crystal.

impurity level — The energy level existing in a substance because of impurity atoms.

inaccuracy — 1. The difference between the input quantity applied to a measuring instrument and the output quantity indicated by that instrument. The inaccuracy of an instrument is equal to the sum of its instrument error and its uncertainty. 2. The term sometimes used to indicate the deviation from an indicated or recorded value or the measure of conformity to an accepted standard.

inactive leg — Within a transducer, an electrical element that does not change its electrical characteristics with the applied stimulus. Applied specifically to elements that complete a Wheatstone bridge in certain transducers.

in-band signaling — 1. The transmission of signaling tones at some frequency or frequencies within the channel normally used for voice transmission. 2. A signaling scheme that uses the same path for both data and signaling information.

in-band spurious transmitter output — A spurious transmitter output that lies within the specified band of transmission.

incandescence — 1. The state of a body with such a high temperature that it gives off light. 2. The generation of light caused by passing an electric current through a wire filament. The resistance of the filament to the current causes the filament to heat up and emit radiant energy, some of which is in the visible range.

incandescent lamp — 1. An electric lamp in which electric current flowing through a filament of resistance material heats the filament until it glows. 2. A lamp that emits light when an electric current passes through a resistant metallic wire situated in a vacuum tube. 3. An electric lighting and signaling device operating on the principle of heating a fine metal wire filament to a white heat by passing an electric current through it. The filament wire has a positive temperature coefficient, which results in high inrush currents, up to 10 times the steady-state current.

INCH — Acronym for integrated chopper. It is a device designed to operate as a chopper, commutator, modulator, demodulator, or mixer, depending on circuit requirements.

inching — *See* jogging.

incidence angle — The angle between an approaching light ray or emission and the perpendicular (normal) to the surface in the path of the ray.

incidental FM — Also called residual FM. 1. The short-term jitter or undesired FM deviation of a local oscillator. It limits resolution when it approaches the IF bandwidth in magnitude. 2. Peak-to-peak variations of a carrier frequency caused by external variations not a part of normal action of the carrier-tuned circuits. 3. In a klystron, frequency modulation of the fundamental frequency due to shot and ion noises, ac heater voltage, etc.

incident field intensity — The field strength of a down-coming sky wave, not including the effects of earth reflections at the receiving location.

incident light — The light that falls directly on an object.

incident-light meter — An exposure meter designed to measure the light striking an object and used at a suitable location in the scene.

incident power — The product of the outgoing current and voltage traveling from a transmitter down a transmission line to an antenna.

incident ray — A ray of light that falls on or strikes a surface of an object, such as a lens. It is said to be incident to the surface.

incident wave — In a medium of certain propagation characteristics, a wave that strikes a gap in the medium or strikes a medium having different propagation characteristics.

incipient failure — A degradation failure in its beginning stages.

in-circuit emulation — Abbreviated ICE. A capability provided on some microcomputer development systems that enables a system designer to use the facilities of the development system to debug prototype hardware and software. This is accomplished via an umbilical cable from the development system that plugs into the microprocessor socket in the prototype system. *See also* emulation.

in-circuit emulator — Abbreviated ICE. A microcomputer development system that can be plugged into a microcomputer system to control, alter, interrogate, and debug that system using a known good environment.

in-circuit tester — Also called a bed-of-nails tester or an in-situ tester. A tester that checks the individual components on a board using a fixture that provides access to each node of each component. Used to test for short and open circuits on bare boards, correct values of analog components (using a guarding technique), and correct functions of individual ICs (using a pulsing technique).

inclination — 1. The angle between the orbital plane of a satellite and the equatorial plane of the earth. 2. The angle that a line, surface, or vector makes with the horizontal.

inclinometer — An instrument for measuring the magnetic inclination of the earth's magnetic field. It uses a magnetic needle that pivots vertically to indicate the inclination.

inclusive AND — A logic element whose output is true if all inputs are true, all inputs are false, or all inputs but one are false.

inclusive NAND — A logic element whose output is true if one and only one of the inputs is false.

incoherent — Denotes the lack of a fixed phase relationship between two waves. If two incoherent waves are superposed, interference effects cannot last longer than the individual coherent times of the waves.

incoherent detection — Detection wherein the information contained in the phase of the carrier is discarded.

incoherent emitter — A fiber-optic source of radiation that has been used for short-length optical transmission lines. Light-emitting diodes are incoherent emitters. *See* coherent emitter.

incoherent scattering — The disordered change in their direction of propagation that occurs when radio waves encounter matter.

incoherent source — A fiber-optic light source that emits wide, diffuse beams of light of many wavelengths. The light waves emitted from an incoherent source are out of phase. *Contrast* coherent source.

incoming — Describing a telecommunication trunk that is used only for calls coming in from another office.

incoming selector — In a telephone central office, a selector associated with trunk circuits from another central office.

incomplete sequencer relay — A device that returns the equipment to the normal, or off, position and locks it out if the normal starting, operating, or stopping sequence is not properly completed within a predetermined time.

increductor — A controllable inductor similar to a saturable reactor, except that it is capable of operating at high frequency (e.g., up to 400 MHz).

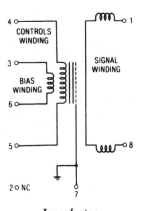

Increductor.

increment — 1. A small change, either positive or negative, in the value of a variable quantity. 2. *See* step angle.

incremental compiler — In a computer, a compiler capable of compiling additional statements without a complete recompilation.

incremental computer — 1. A computer in which the use of incremental representation of data is predominant. 2. A special-purpose computer designed specifically to process changes in the variables as well as absolute values of the variables.

incremental digital-position transducer equalizer — Digital-position transducer that, without absolute position reference, produces a digital signal by motion.

incremental digital recorder — A magnetic tape recorder that advances the tape across the recording head step by step, as in a punched-tape recorder. It is used for economical and reliable recording of an irregular flow of data.

incremental frequency shift — A method by which incremental intelligence may be superimposed on other intelligence by shifting the center frequency of an oscillator a predetermined amount.

incremental hysteresis loss — Losses in a magnetic material that has been subjected to a pulsating magnetizing force.

incremental induction — One-half the algebraic difference between the maximum and minimum magnetic induction at a point in a material that has been subjected simultaneously to a polarizing and a varying magnetizing force.

incremental integrator — A digital integrator modified so that the output signal is maximum negative, zero, or maximum positive when the value of the output is negative, zero, or positive, respectively.

incremental permeability — 1. The ratio of the cyclic change in magnetic induction to the corresponding cyclic change in magnetizing force when the mean induction differs from zero. For small changes in magnetizing force, the incremental permeability is approximately equal to the slope of the minor hysteresis loop generated. 2. The ratio of incremental change in flux density to the incremental change in magnetizing force at any point on the hysteresis loop.

incremental sensitivity — The smallest change that can be detected by a particular instrument in a quantity under observation.

incremental tape — Magnetic tape written one character at a time instead of the usual method of continuous recording.

incremental tuner — A television tuner in which antenna, rf amplifier, and rf oscillator inductors are continuous or in small sections connected in series. Rotary switches, connected to taps on the inductors, provide the portion of total inductance required for a channel, or short-circuit all remaining inductance except that required for the channel.

independent failure — A failure that has no significant relationship to other failures in a given device and can occur without interaction with other component parts in the equipment.

independent load contacts — Contacts that can control electrical loads that must be isolated from the timer clutch solenoids and motor circuit.

independent variable — One of several voltages and currents chosen arbitrarily and considered to vary independently.

indeterminacy — The time coincidence uncertainty between an external input trigger pulse and an independently clock-derived output pulse from a timing unit such as a digital delay generator. It stems from a clock pulse versus trigger pulse phase uncertainty and is usually expressed in terms of nanoseconds or microseconds. For example, the $\pm\frac{1}{2}$ clock pulse uncertainty of a 10-MHz source would be ±50 ns.

indeterminate state — The unknown logic state (X) of a memory element caused by critical races or oscillations, or existing after power is applied and before initialization. Some simulators can model indeterminate states and typically assign an X to indicate an indeterminate state.

index counter — An odometer-type cumulative-digit indicator for keeping track of the amount of tape that has passed through a tape machine. The counter is generally driven by the takeup-reel turntable and thus registers rotation rather than tape footage, although the accuracy is generally good enough to allow for locating specific recorded segments according to previously noted index counter numbers.

indexed address — An address that is altered by the content of an index register before or during the execution of a computer instruction.

indexed addressing — 1. An addressing mode in which a computer finds the address of the desired memory location by referring to an index register. By successively adding or subtracting 1 to this index register, the computer can be made to step through a list or table. 2. An addressing system in which the address of the data is expressed as relative to the address stored in an index or pointer register. To obtain the absolute address, the offset address is added to the pointer address. This system is useful in processing tables or matrices of data.

index hole — A hole punched in a floppy disk to indicate the beginning of the first sector.

indexing — In a computer, a technique of address modification that is often implemented by means of index registers.

indexing mechanism — A mechanical device on a rotary switch to locate and to maintain each position of the rotor.

indexing slots — *See* polarizing slots.

index matching fluid — In fiber optics, a fluid with a refractive index the same as a fiber core; used to fill the air gap between the fiber ends at connectors.

index matching materials — Materials used in intimate contact between the ends of optical conductors to reduce coupling losses by reducing Fresnel loss.

index of cooperation — In rectilinear scanning or recording, the product of the total length of a line and the number of lines per unit length.

index of modulation — The modulation factor.

index of refraction — 1. Ratio of the speeds of light or other radiation in two different materials. This determines the amount the ray will be refracted or bent when passing from one material to the other, such as from air to water. The index of refraction of air is generally taken as unity. 2. The physical property of a material that describes the behavior of optical energy passing through it. It is defined as the ratio of the velocity of light in a vacuum to the velocity of light in the material. It varies with wavelength.

index register — 1. In a computer, a register that holds a quantity that may be used for modifying addresses or for other purposes, as directed by the program. 2. In a computer, a register whose contents can be added automatically to an address field contained in an instruction, when the indexing mode is specified. 3. A user-accessible register implicitly used for address computation by many instructions that reference main memory.

indicating demand meter — A meter equipped with a scale over which a pointer is advanced to indicate maximum demand.

indicating fuse — A protective device placed in a telephone circuit to provide visual and audible indication of a fault in the line. It consists of a fuse, pilot lamp, relay, and buzzer. When a line fault blows the fuse, the lamp lights and the buzzer sounds.

indicating instrument — An instrument that visually indicates only the present value of the quantity being measured.

indicating lamp — A lamp that indicates the position of a device or the condition of a circuit.

indicating meter — A meter that gives a visual indication of only the present or short-time average value of the measured quantity.

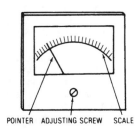

POINTER   ADJUSTING SCREW   SCALE

Indicating meter.

indication — The display to the human senses of information concerning a quantity being measured.

indicator — An instrument that makes information available but does not store it.

indicator gate — A rectangular voltage waveform applied to the grid or cathode circuit of an indicator cathode-ray tube to sensitize or desensitize the tube during the desired portion of the operating cycle.

indicator tube — An electron-beam tube that conveys useful information by the variation in cross section of the beam at a luminescent target.

indirect-acting recording instrument — An instrument in which the marking device is actuated by raising the level of measurement energy of the primary detector. This is done mechanically, electrically, electronically, photoelectrically, or by some other intermediate means.

indirect address — An address in a computer instruction that indicates a location where the address of the referenced operand is to be found.

indirect addressing — 1. A method of computer cross reference in which one memory location indicates where the correct address of the main fact can be found. 2. A method of storage addressing in which an addressed location contains an address rather than data. Quite often several levels of indirect addressing may occur before the sought-after data item is obtained.

indirect light — Light from an object that has no self-luminous properties. Instead, it reflects light from another source.

indirect lighting — A system of lighting in which all the light is directed to the ceiling or walls, which in turn reflect it to the objects to be illuminated. Lighting by luminaries distributing 90 to 100 percent of the emitted light upward.

indirectly controlled variable — A variable that is related to and influenced by the directly controlled variable but is not directly measured for control.

indirectly heated cathode — Also called equipotential or unipotential cathode. A cathode that is heated by an independent heater.

indirectly heated thermistor — A thermistor that incorporates, as part of its composite structure, an electrical heater. A thermistor whose body temperature in use is significantly higher than the temperature of its surrounding medium as a result of current passing through its heater.

indirect material — A type of semiconductor material in which electrons do not drop directly from the conduction band to the valence band, but drop in steps as a result of the trapping levels in the forbidden gap.

indirect piezoelectricity — The production of a mechanical strain in a crystal by applying a voltage to it (as opposed to the more common piezoelectric effect of applying a strain to the crystal in order to produce a voltage).

indirect radiative transition — *See* transition, 1.

indirect scanning — A television technique used in early mechanical systems, and today in the flying-spot scanning of films. A small beam of light is moved across the subject and then reflected to a battery of phototubes.

indirect synthesizer — A synthesizer employing phase-locked loops, digital dividers, and high-Q varactor-tuned oscillators. The discrete output frequencies are not limited to integer ratios of the reference frequency. Frequency step size or increments are primarily determined by the digital dividers. Switching speed between discrete output frequencies is usually limited by the phase-lock circuits. Applications include automatic test and satellite communications systems.

indirect wave — A wave reaching a given reception point by a path other than the direct-line path between the two (e.g., a sky wave received after deflection from the ionosphere layers).

individual gap azimuth — In a magnetic-tape record or reproduce head stack, the angle of an individual gap relative to a line perpendicular to the precision-milled mounting pads in a plane parallel to the surface of the tape.

individual line — A subscriber line that serves one main station and optional additional stations connected to the line as extensions; the line is not arranged for discriminatory ringing with respect to the stations.

indoor antenna — Any receiving antenna located inside a building but outside the receiver.

indoor transformer — A transformer that must be protected from the weather.

Indox — The trade name of Indiana Steel Products Company's barium ferrite permanent magnet alloy.

induced — Produced by the influence of an electric or magnetic field.

induced charge — 1. An electrostatic charge produced in one object by the electric field surrounding a nearby object. 2. An electrostatic charge produced on a conducting body when it is brought near to or connected to another body that bears an electric charge.

induced current — 1. The current that flows in a conductor that is moved perpendicularly to a magnetic field, or which is subjected to a magnetic field of varying intensity. The former takes place in an induction-motor rotor; the latter, in the secondary winding of a transformer. 2. In induction heating, the current that flows in a conductor when a varying electromagnetic field is applied. 3. The current that results when a conductor is cut by magnetic flux lines. 4. Current that flows in a conductor because of an induced voltage.

induced electromotive force — Represented by E; proportional to the rate of change of magnetic flux through the circuit ($d\phi/dt$).

induced environment — The temperatures, vibrations, shocks, accelerations, pressures, and other conditions imposed on a system due to the operation or handling of the system.

induced failure — A failure that is basically caused by a condition or phenomenon external to the item that fails.

induced voltage — The voltage produced in a conductor when the conductor is moved up and down through the magnetic field of a second conductor, or when the field varies in intensity and cuts across the first conductor. Even though there is no mechanical coupling between the two conductors, the one producing the field will produce a voltage in the other.

inductance — 1. Property of a circuit that tends to oppose any change of current because of a magnetic field associated with the current itself. Whenever an electric current changes in value — rises or falls — in a circuit, its associated magnetic field changes, and when this links with the conductor itself, an electromotive force is induced that tends to oppose the original current change. Self-inductance is the full name for this, but the term *inductance* is usually used. The unit of inductance is the henry. When a current changing at the rate of 1 ampere per second induces a voltage of 1 volt, the inductance of the circuit is 1 henry. 2. *See also* coil. 3. The property of a circuit or circuit element that opposes a change in current flow, thus causing current changes to lag behind voltage changes. It is measured in henrys.

inductance bridge — An instrument, similar to a Wheatstone bridge, for measuring an unknown inductance by comparing it with a known inductance.

inductance coil — *See* inductor.

inductance-tube modulation — A method of modulation employed in frequency-modulated transmitters. An oscillator control tube acts as a variable inductance in parallel with the tank circuit of the radio-frequency oscillator tube. As a result, the oscillator frequency varies in step with the audio-frequency voltage applied to the grid of the oscillator control tube.

induction — 1. The establishment of an electric charge or a magnetic field in a substance by the proximity of an electrified source, a magnet, or a magnetic field. 2. The setup of an electromotive force and current in a conductor by variation of the magnetic field affecting the conductor.

induction brazing — The electric brazing process in which heat is produced by an induced current.

induction coil — A device for changing direct current into high-voltage alternating current. Its primary coil contains relatively few turns of heavy wire; its secondary coil, wound over the primary, contains many turns of fine wire. Interruption of the direct current in the primary by a vibrating-contact arrangement induces a high voltage in the secondary.

induction compass — A compass in which the indications are produced by the current generated in a coil revolving in the magnetic field of the earth.

induction-conduction heater — A heating device through which electric current is conducted but is restricted by induction to a preferred path.

induction density — *See* flux density.

induction factor — In an alternating-current circuit, the ratio between that element of the current that does no work and the total strength of the current.

induction field — 1. That portion of the electromagnetic field of a transmitting antenna that acts as if it were permanently associated with the antenna, and into which energy is alternately stored and removed. 2. The electromagnetic field of a coil carrying alternating current, responsible for the voltage induced by that coil in itself or in a nearby coil. 3. The magnetic field that is predominant in the near zone of a radio transmitting antenna and that is directly proportional to the current in the antenna.

induction frequency converter — A slip-ring induction machine driven by an external source of mechanical power. Its primary circuits are connected to a source of electric energy having a fixed frequency. The energy delivered by its secondary circuits is proportionate in frequency to the relative speed of the primary magnetic field and the secondary member.

induction furnace — A furnace heated by electromagnetic induction.

induction hardening — The process of hardening the surface of a casting by heating it above the transformation range by electrical induction, followed by rapid cooling.

induction heating — The method of producing heat by subjecting a material to a variable electromagnetic field. Internal losses in the material then cause it to heat up.

induction instrument — An instrument operated by the reaction between the magnetic flux set up by one or more currents in fixed windings and the currents set up by electromagnetic induction in movable conductive parts.

induction loudspeaker — A speaker in which the current that reacts with the steady magnetic field is induced in the moving member.

induction motor — 1. An alternating-current motor in which the primary winding (usually the stator) is connected to the power source and induces a current into a polyphase secondary or squirrel-cage secondary winding (usually the rotor). Currents, and therefore magnetic poles, are induced into the secondary by the rotating primary magnetic field; thus, the armature rotates at slightly slower than synchronous speed, a condition that is called slip. The amount of slip increases with an increase in mechanical loading. 2. A motor that runs asynchronously; that is, not in step with the alternations of the alternating current.

induction-motor meter — A meter containing a rotor that moves in reaction to a magnetic field and the currents induced into it.

induction noise — The noise — other than thump, flutter, cross fire, or crosstalk — produced when two circuits are inductively coupled together.

induction-resistance welding — Welding in which electromagnetic induction alone causes the heating current to flow in the parts being welded.

induction-ring heater — A core-type induction heater adapted principally for heating round objects. The core is open or can be taken off to facilitate linking the charge.

induction soldering — A method of soldering in which the solder is reflowed or supplied by preforms. If the work is moved slowly through the energy field, the induction process may be made continuous.

induction speaker — A speaker in which the current that reacts with the steady magnetic field is induced into the moving member.

induction-voltage regulator — A device having a primary winding in shunt and a secondary winding in series with a circuit for gradually adjusting the voltage or the phase relation of the circuit by changing the relative positions of the exciting and series windings of the regulator.

inductive — Pertaining to inductance or to the inducing of a voltage through mutual or electrostatic induction.

inductive circuit — A circuit with more inductive than capacitive reactance. A circuit having a net inductive reactance; that is, a higher value of inductance reactance than of capacitive reactance.

inductive coordination — Location, design, construction, operation, and maintenance of electric supply and communication systems in a manner that prevents inductive interference.

inductive coupled circuit — A network with two meshes having only mutual inductance in common.

inductive coupling — 1. The association of one circuit with another through inductance common to both. When used without modifying words, the term commonly refers to coupling by means of mutual inductance, whereas coupling by means of self-inductance common to both circuits is called direct inductive coupling. 2. In inductive-coordination practice, the interrelation of neighboring electric supply and communication circuits resulting from electric and/or magnetic induction. 3. Coupling that exists between two circuits through a mutual inductance, such as that in a transformer. 4. Coupling between two circuits through an inductance that is common to the two circuits; direct inductive coupling.

inductive feedback — The transfer of energy from the output circuit to the input circuit of an amplifying device through an inductor or inductive coupling.

inductive interference — 1. Interference produced in communication systems by induced voltages within the system. 2. Effect arising from the characteristics and inductive relations of electric supply and communication systems of such character and magnitude as would prevent the communication circuit from rendering service satisfactorily and economically if methods of inductive coordination were not applied.

inductive kick — 1. The voltage, many times higher than the impressed voltage, produced by the collapsing field in a coil when the current through it is abruptly cut off. 2. A voltage surge that is induced in an inductance when the current through it is interrupted and the magnetic flux collapses suddenly.

inductive level detector — A level-measuring system incorporating an oscillator and electromagnetic field.

inductive load — Also called lagging load. A load that is predominantly inductive, so that the alternating load current lags behind the alternating voltage of the load. An electrical load that has a significant inductive reactance.

inductive microphone — See inductor microphone.

inductive neutralization — Also called shunt or coil neutralization. A method of neutralizing an amplifier, whereby the equal and opposite susceptance of an inductor cancels the feedback susceptance caused by interelement capacitance.

inductive pickup — Signals generated in a circuit or conductor due to mutual inductance between it and a disturbing source.

inductive post — A metal post or screw extended across a waveguide parallel to the E field to act as inductive susceptance in parallel with the waveguide for purposes of tuning or matching.

inductive reactance — The opposition to the flow of alternating or pulsating current by the inductance of a circuit. It is measured in ohms, and its symbol is X_L. It is equal to 2π times the frequency in hertz times the inductance in henrys.

inductive system — An ignition system that stores its primary energy in an inductor or coil.

inductive transducer — A transducer in which changes in inductance convey the stimulus information.

inductive transduction — The conversion of the measurand into a change in the self-inductance of a single coil.

inductive tuning — A method of tuning a radio by moving a core into and out of a coil to vary the inductance.

inductive winding — A coil through which a varying current is sent to give it an inductance.

inductive window — A conducting diaphragm extended into a waveguide from one or both sidewalls to act as an inductive susceptance in parallel with the waveguide.

inductometer — An inductor whose inductance can be varied, sometimes by a calibrated amount.

inductor — Also called inductance or retardation coil. 1. A conductor used for introducing inductance into an electric circuit. The conductor is wound into a spiral, or coil, to increase its inductive intensity. 2. A passive fluidic element that, because of fluid inertness, has a pressure drop that leads flow by essentially 90°. 3. See coil.

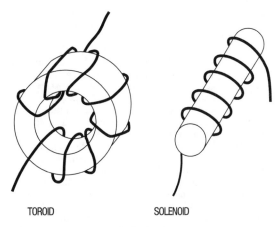

TOROID   SOLENOID

Inductors.

inductor microphone — Also called inductive microphone. A microphone in which the sound waves move a conductor back and forth, cutting magnetic lines of force and producing an electrical output of the same frequency and proportional to the amplitude of the sound waves.

inductor-type synchronous motor — A type of synchronous motor having field magnets that are fixed in magnetic position relative to the armature conductors, the torques being produced by forces between the stationary poles and salient rotor teeth. Such motors usually have permanent-magnet field excitation, are built in fractional-horsepower frames, and operate at low speeds (300 revolutions per minute or less).

Inductosyn — An extremely precise transducer based on the magnetic circuit of a conductor deposited on glass for stability, and operated at a relatively high frequency. Extremely accurate, but requires much auxiliary equipment (Farrand Controls Inc.).

industrial-grade IC — Typically, an integrated circuit whose performance is guaranteed over the temperature range 0 to 70°C.

industrial radio services — Radiocommunication services essential to, operated by, and for the sole use of those enterprises that require radiocommunications in order to function efficiently.

industrial television — The cameras and related instrumentation of a closed-circuit television system. Such equipment is designed to function in the different environments found in industrial processes. Used to monitor areas that are hazardous to personnel, such as high-radiation areas, areas that do not require steady supervision, industrial processes for surveillance or quality control, or areas requiring security measures.

industrial timer — A timing device, impulse or constant-speed type, used in industrial applications other than the appliance industry.

industrial tube — A vacuum tube designed for industrial electronic equipment.

inelastic collision — Collision resulting in excitation of a molecule.

inertance — Acoustical equivalent of inductance.

inert gas — *See* noble gas.

inertia — 1. The tendency of an object at rest to remain at rest, or of a moving object to continue moving in the same direction and at the same speed, unless disturbed by an outside force. Resulting from mass and inhibiting change in velocity. Important in pickup mechanics. 2. The resistance to change in speed or velocity. In stepper motors, inertia does not affect the maximum stepping rate, only the time required to attain it.

inertial guidance/navigation — A self-contained system for navigation in which position can be computed by knowing a craft's starting point and where it has been. Changes in acceleration are detected by gyroscopes for direction and attitude and by accelerometers for velocity. These signals are integrated to determine resulting velocity and distance. The system needs no outside reference and cannot be jammed.

inertial navigation — 1. A guidance technique in which airframe acceleration is first measured and then integrated twice with respect to time in order to determine the distance traveled. External aids such as radio and radar are not necessary. The acceleration or deceleration of the airframe is measured continuously with accelerometers oriented in some convenient frame of reference, usually corresponding to the earth's north-south, east-west coordinates. 2. A form of navigation that uses dynamic measurements of acceleration forces acting on a gyroscopically stabilized device as a basis for computing position and velocity information. This device, which is mechanized so as to be completely self-contained, has the inherent capability of providing continuously available navigation information in terms of conventional, directly usable latitude and longitude coordinates.

inertia relay — A relay having added weights or other modifications that increase its moment of inertia and either slow it or cause it to continue in motion after the energizing force is removed.

inertia switch — A switch capable of sensing acceleration, shock, or vibration. It is designed to actuate upon an abrupt change in velocity.

inertia welding — A forge-welding process in which stored kinetic energy is released as frictional heat when two parts are rubbed together under the proper conditions.

infant mortality — The occurrence of premature catastrophic-type failures of a component or equipment

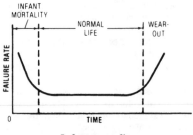

Infant mortality.

at a rate substantially greater than that observed during life prior to wearout.

infant-mortality period — *See* early-failure period.

inferential — The kind of instrumentation, especially its signal source, in which there is sampling of an entirely different quantity from the one of interest, upon the assumption that they vary in perfect proportion. Linearity or a perfectly repeatable relationship between the two is inferred for the sake of a more convenient signal-source arrangement.

infiltration — The process of filling the pores of a sintered compact with a metal or alloy of lower melting point.

infinite — Boundless; having no limits whatsoever.

infinite baffle — 1. An airtight speaker enclosure that completely absorbs or dissipates a speaker's rear sound waves. 2. A speaker mounting in which ideally there is no path of air between the back and front of the speaker diaphragm. An infinite baffle improves the forward radiation of sound at low frequencies and preferably should be a very large plane surface, like the wall of a room or a screen of very rigid material (e.g., $^3/_4$-inch or 2-cm wood), on which a speaker is mounted. (In practice, truly infinite baffles are rarely accomplished except in sealed boxes, but these give rise to problems of resonance.) 3. An airtight speaker enclosure containing a bass speaker with very low open-air resonance, plus a sealed midrange speaker and tweeter.

infinite-baffle speaker system — A speaker in which the bass driver is located in an almost airtight enclosure.

infinite-impedance detector — A detector circuit in which the load is a resistor connected in parallel with an rf bypass capacitor between the cathode and ground. Since the grid is always negative with respect to the cathode, the tube presents an infinite impedance to the input.

infinite line — A transmission line with the same characteristics as an ordinary line that is infinitely long.

infinite resolution — The capability of a device to provide continuous output over its entire range.

infinitesimal — Immeasurably small; approaching zero.

infinity — 1. A hypothetical amount larger than any assignable amount. 2. A number larger than any number a computer can store in any register. 3. Any distance of a subject from a lens for which the image no longer moves when the subject moves along the optical axis.

infinity device — Also called harmonica bug. A surreptitious listening device that uses the telephone as a sensor. Allows the bugger to listen to sounds near the phone while the phone is not in use. Does not allow him or her to hear telephone conversations.

inflection point — The point at which a curve changes direction.

infobond — An automated system of point-to-point wiring on the back of a two-sided printed wiring board (the components are on the front, or other side). The No. 38 AWG copper wire used is solder-bonded to terminations by an automatic soldering gun.

information — 1. In computing, the basic data and/or program entered into the system. 2. That property of a signal or message whereby it conveys something meaningful and unpredictable to the recipient, usually measured in bits. 3. Data that has been organized into a meaningful sequence.

information bits — In telecommunications, those bits originated by the data source and not used for error control by the data-transmission system.

information center — A facility specifically designed for storing, processing, and retrieving information to be disseminated at regular intervals, on demand, or selectively, according to the needs of users.

information channel — The transmission and intervening equipment involved in the transfer of information in a given direction between two terminals. An information channel includes the modulator and demodulator and any error-control equipment irrespective of its location, as well as the backward channel, when provided.

information extraction — An analysis of an image to recognize and isolate a specific feature or relationship among features.

information feedback system — In telecommunications, an information-transmission system in which an echo check is employed to verify the accuracy of the transmission.

information gate — A circuit that permits information or data pulses to pass when the circuit is triggered by an external source.

information handling — The storing and processing of information and its transmission from the source to the user. Information handling excludes the creation and use of information.

information processing — A term that encompasses both word processing and data processing. It describes the entire scope of operations performed by a computer.

information rate — In computers, the minimum number of binary digits per second required to specify the source messages.

information rate changer — A device that speeds up the playback of tape-recorded speech without pitch change or deterioration of characteristic resonances. This is accomplished by rotating the playback head in the direction of tape travel.

information retrieval — 1. A method for cataloging vast amounts of data related to one field of interest so that any part or all of this data can be called out at any time with accuracy and speed. 2. The recovery of data that has been stored at a particular address in a memory. 3. A technique of classifying and indexing useful data in mass storage devices in a format amenable to interaction with the user(s). 4. The art of storing information so that it may be recovered easily. Branches include abstracting, locating facts of interest, and language translation.

information-retrieval system — A system for locating and selecting on demand certain documents or other graphic records relevant to a given information requirement from a file of such material.

information separator — A control character used to identify a logical boundary of information. The name of the separator is not necessarily indicative of what it separates.

information superhighway — The concept of a high-bandwidth network that links everyone with everyone and can transport all media types. The information superhighway links the concepts of online services, Internet, and interactive TV together.

information system — A group of computer-based systems and data required to support the information needs of one or more business processes.

information theory — 1. The branch of learning that deals with the likelihood of accurate transmission of messages subject to transmission failure, distortion, and noise. 2. The mathematical theory that deals with the transmission of information and the effects of bandwidth, distortion, and noise.

infra- — Prefix meaning below; beneath; less than.

infradyne receiver — A superheterodyne receiver whose intermediate frequency is made higher than the signal frequency in order to obtain high selectivity.

infrared — Abbreviated IR. 1. Pertaining to or designating those radiations with wavelengths just beyond the red end of the visible spectrum, such as those emitted by a hot body. These wavelengths are longer than those of visible light and shorter than those of radio waves. 2. That section of the electromagnetic spectrum, invisible to the eye, lying between wavelengths of 750 nm and about 1 mm. Thermography utilizes waves in this region for recording changes in temperature. 3. Part of the electromagnetic spectrum between the visible light range and the radar range. 4. The electromagnetic wavelength region between approximately 0.75 and 1000 micrometers. For fiber-optic transmission, the near-infrared region between 0.75 and 1.3 micrometers is the most relevant region because glass, light sources, and detector techniques are most nearly matched in this wavelength region.

infrared alarm systems — A system that uses infrared detectors and related instrumentation to determine when abnormal amounts of infrared radiation, usually in the form of heat, are present in an area; used for the detection of fires or the presence of intruders in a restricted area.

infrared binoculars — An instrument, similar in design to regular binoculars, that can transmit and enlarge infrared images using electronic circuits.

infrared communications set — The collection of components necessary to operate a two-way electronic system in which infrared radiation is used to carry intelligence.

infrared counter-countermeasures — Action taken to employ infrared radiation equipment and systems in spite of enemy measures to counter their use.

infrared countermeasures — Action taken to reduce the effectiveness of enemy equipment employing infrared radiation.

infrared detector — A transducer that is sensitive to invisible infrared radiation (wavelengths between 0.75 and 1000 micrometers), usually using a semiconductor (photon), thermocouple bolometer, or pneumatic (pressure) device to detect the radiation. Also a device used to detect radiation from the infrared region.

infrared emitter — *See* infrared light-emitting diode.

infrared-emitting diode — 1. A semiconductor device with a semiconductor junction in which infrared radiant flux is nonthermally produced when a current flows as a result of applied voltage. 2. A pn diode in which a fraction of the injected minority carriers recombine by means of radiative transistors. When the junction is forward biased, electrons from the n region are injected into the p region, where they recombine with excess holes. In the radiative process, energy given up in recombination is in the form of photon emission. The generated photons travel through the lattice until they are either reabsorbed by the crystal or escape from the surface as radiant flux.

infrared guidance — A system using infrared heat resources for reconnaissance of targets or for navigation.

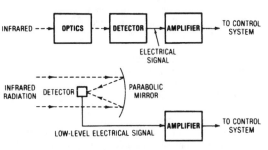

Infrared guidance.

infrared homing—A type of missile homing in which the guidance system tracks the target from the infrared radiation it emits. *See* heatseeker.

infrared instruments—Any of the photoelectric and thermal detectors, spectrographs and monochromators, thermographs, scanners, amplifier tubes, snooperscopes, and related equipment especially developed for use with infrared radiation.

infrared jamming—A countermeasure used against heatseeking missiles to reduce their effectiveness. Normally it involves the emittance of infrared radiation at a level that will overload the missiles' detectors.

infrared light—Light rays just below the red end of the visible spectrum.

infrared light-emitting diode—Also called infrared emitter. An optoelectronic device containing a semiconductor pn junction that emits radiant energy in the 0.75- to 100-micrometer wavelength region when forward biased.

infrared motion detector—A sensor that detects changes in the infrared light radiation from parts of the protected area. Presence of an intruder in the area changes the infrared light intensity from his or her direction.

infrared optics—Lenses, prisms, and other optical elements for use with infrared radiation (radiation with a wavelength between 0.75 and 1000 micrometers).

infrared ovens—Units that dry, cure, and preheat parts directly (i.e., without heating the oven air) via infrared energy.

infrared radiation—1. Invisible radiation with wavelengths in the range between 7500 angstroms (red) and about 1,000,000 angstroms (microwaves). 2. The electromagnetic wavelength region between approximately 0.75 and 1000 micrometers, longer than the wavelength of visible light.

infrared radiation sources—Almost any warm thing, from an electric blanket to a living human being, that acts as a source for the longer wavelength end of the IR range, which generally is regarded as extending from 0.75 to 1000 micrometers. Calibrated secondary sources are usually a heated cavity (blackbody) or a carbon filament lamp rated in wattage output or in ergs per second of radiation incident on a surface at a specified distance.

infrared sources—Emitters of radiation with a wavelength between 0.75 and 1000 micrometers.

infrared spectrum—That portion of the electromagnetic spectrum between the wavelengths of 0.75 and 1000 micrometers.

infrared thermometer—A temperature-measuring device that detects infrared radiation from an object and converts that measurement into a reading representing the temperature of the object.

infrared waves—Also called black light. Invisible waves longer than the longest visible red light waves but shorter than radio-frequency waves.

infrared window—A region of relatively high transmission in the infrared-frequency range.

infrasonic—Pertaining to frequencies below the range of human hearing, hence below about 15 hertz. Formerly called subsonic.

infrasonic frequency—A frequency below the audio range. Infrasonic vibrations can be felt but not heard. Replaces the obsolete term *subsonic frequency*.

inharmonic frequency—A frequency that is not a rational multiple of another frequency.

inherent delay—Delay between the insertion of information into a unit and presentation of the information at the output. For example, a delay inserted into the CRT vertical amplifier of pulse analyzers to allow the leading

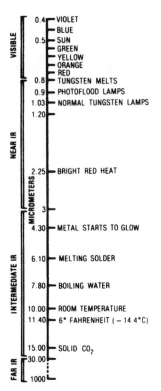

Infrared spectrum.

edge of the signal triggering the sweep to be seen. *See also* insertion delay.

inherent interference—A type of electromagnetic interference generated within a receiver by thermal agitation, shot effect, and nonlinear impedance.

inherent reliability—The potential reliability of an item present in its design.

inherited error—In a computer, the error in the initial values, especially that error accumulated from prior steps in a step-by-step integration.

inhibit—To prevent an action from taking place or data from being accepted by applying an appropriate signal (generally a logic 0 in positive logic) to the proper input.

inhibit gate—A circuit that provides an output only when certain signals are present and other signals are not present at the inputs.

inhibiting input—A computer gate input which, if in its prescribed state, prevents any output which might otherwise occur.

inhibiting signal—A signal whose presence prevents an operation from taking place.

inhibition gate—A gate circuit used as a switch and placed in parallel with the circuit it is controlling.

inhibitor—Also called inhibition gate. In a digital computer, a logic circuit that clamps a specified output to the zero level when energized.

inhibit pulse—A computer drive pulse that tends to prevent certain drive pulses from reversing the flux of a magnetic cell.

initial actuation time—The time of the first closing of a previously open contact of a relay or the first opening of a previously closed contact.

initial contact chatter—That chatter caused by vibration produced by opening or closing the contacts in a relay themselves, as by contact impact in closure.

initial differential capacitance — The differential capacitance of a nonlinear capacitor when the capacitor voltage is zero.

initial drain — The current supplied at nominal voltage by a cell or battery.

initial element — *See* primary detector.

initial erection — The mode of operation of a vertical gyro in which the gyro is being erected or slaved initially. The initial erection rate is usually relatively fast.

initial failure — The first failure that occurs in use.

initial inverse voltage — Of a rectifier tube, the peak inverse anode voltage immediately following the conducting period.

initial ionizing event — Also called primary ionizing event. An ionizing event that initiates a tube count.

initialization — 1. The process in which information (memory locations for data and results, tolerances, limits, etc.) is supplied to a computer prior to the running of a program. 2. Applying input patterns to a logic circuit so that all internal memory elements achieve a known logic state. 3. A process that takes place whenever the state of a device or program must be known at startup.

initialize — 1. To set counters, switches, and addresses to their starting values at the beginning of a computer routine or at prescribed points in the routine. 2. To establish an initial condition or starting state; for example, to set logic elements in a digital circuit or the contents of a storage location to a known state so that subsequent application of digital test patterns will drive the logic elements to another known state. 3. To reset a computer and its peripherals to a starting state before beginning a task. Done automatically by the disk operating system.

initializing — 1. The preliminary steps in arranging those instructions and data in a computer memory that are not to be repeated. 2. Setting flip-flops to known states prior to testing.

initial permeability — The slope of the normal induction curve at zero magnetizing force. Permeability at a field density approaching zero.

initial program loader — The procedure that results in loading of the initial part of an operating system or other program so that the program can then proceed under its own control.

initial reversible capacitance — In a nonlinear capacitor, the reversible capacitance at a constant bias voltage of zero.

initial-velocity current — A current that flows between an electrode, such as the grid of a vacuum tube, and its cathode as a result of electrons thrown off from the cathode because of heat alone. Their velocity is sufficient to allow the electrons to reach the grid unaided by an accelerating field.

injected laser — *See* diode laser.

injection grid — A vacuum-tube grid that controls the electron stream without causing interaction between the screen and control grids. In some superheterodyne receivers, the injection grid introduces the oscillator signal into the mixer stage.

injection laser — Also known as a pn junction laser. 1. An optical oscillator or amplifier that has as its active medium a forward-biased semiconductor diode in which a population inversion has been established between the conduction and valence bands. Radiation is emitted in the process of recombination across the bandgap. High frequency modulation of the output beam can be achieved by modulating the input current. Usually, the optical resonator is formed by cleaving or polishing opposite faces of the diode crystal. Typical dimensions of the device are 0.1 mm × 0.1 mm × 0.5 mm. 2. A semiconductor diode carrying a high current in the forward direction. Radiation is produced as electrons recombine with holes in the

junction region. For coherent emission, the current density must exceed a threshold commonly about 10,000 A/cm^2 for gallium arsenide diodes. 3. A solid-state semiconductor device with at least one pn junction capable of emitting coherent or stimulated radiation under specified conditions. Incorporates a resonant optical cavity. 4. A solid-state laser having at least one pn junction. Its energy level transitions are between energy bands of semiconductors and it can be tuned in frequency by temperature or pressure alterations and by the effect of a magnetic field.

injection laser diode — 1. A coherent radiant source LED consisting of an extremely flat junction area, end mirrors, and direct bandgap semiconductors, having a Fabry-Perot optical cavity. 2. In fiber optics, a semiconductor device in which lasing takes place within the pn junction. Light is emitted from the diode edge.

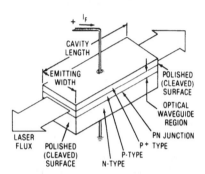

Injection laser diode.

injection-locked oscillator — Abbreviated ILO. A free-running microwave oscillator that is stabilized by injecting a reference signal into the oscillator's resonant circuitry. The required injected signal level is determined by the output signal characteristic requirements (i.e., noise, stability, etc.) and is typically in the range of 70 to 30 dB below the output level of the ILO.

injection luminescent diode — 1. A gallium arsenide diode, operating in either the laser or noncoherent mode, that can be used as a source of visible or near infrared light for use in triggering such devices as light-activated switches. 2. A semiconductor (gallium arsenide) diode operating in either a coherent or incoherent mode that is used as a near-infrared or visible source in triggering light-activated devices.

injector — An electrode on a spacistor.

ink — 1. One of several conductive materials used for chip bonding, electrostatic shielding, corona shielding, making connections, repairing on printed circuits, attaching leads, adhesive work, ignition cable sheath coating, and making electrodes, contacts, terminations, and surfaces receptive to plating, etc. 2. Synonymous with *composition* and *paste* when relating to screenable thick-film materials, usually consisting of glass frit, metals, metal oxide, and solvents. 3. In hybrid technology, the conductive paste used on thick-film materials to form the printed conductor pattern. Usually contains metals, metal oxide, glass frit, and solvent. 4. In thick film, composition of micrometer-size polycrystalline solids suspended in a thixotropic vehicle. The solids are chosen for their electrical characteristics (i.e., metals for conductives, metals and oxides for resistives, and glasses for glazes and dielectrics).

ink blending — *See* blending.

ink-jet printer — A nonimpact printer that forms letters and numbers by electrostatically aiming a jet of ink onto the paper.

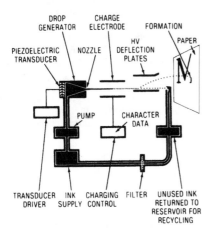

Ink-jet printer.

ink-jet printing — A nonimpact printing technique that utilizes droplets of ink to form copy images. As the print head moves across the surface of the copy paper, it shoots a stream of tiny, electrostatically charged ink drops at the page, placing them precisely to form individual print characters.

ink-mist recording — Also called ink-vapor recording. In facsimile, electromechanical recording in which particles of an ink mist are deposited directly onto the record sheet.

ink recorder — The ink-filled pen or capillary tube that produces a graphic record.

ink recording — A type of mechanical facsimile recording in which an inked helix marks the record sheet.

ink-vapor recording — *See* ink-mist recording.

inleads — Those portions of the electrodes of a device that pass through an envelope or housing.

in-line heads — *See* stacked heads.

in-line procedures — 1. In COBOL, the procedural instructions that are part of the main sequential and controlling flow of the program. 2. Short functions whose code is inserted by the compiler at the point of call, thereby avoiding the overhead of a normal function call.

in-line processing — The processing of data in random sequence not subject to preliminary sorting or editing.

in-line subroutine — A subroutine that is inserted directly into the linear operational sequence. Such a subroutine must be recopied at each point in a routine where it is needed.

in-line tuning — The method of tuning the intermediate-frequency strip of a superheterodyne receiver in which all the intermediate-frequency amplifier stages are made resonant to the same frequency.

inorganic electrolyte — A solution that conducts electricity due to the presence of ions of substances not of organic origin.

in phase — Two waves of the same frequency that pass through their maximum and minimum values of like polarity at the same instant are said to be in phase.

in-phase portion of the chrominance signal — That portion of the chrominance signal having the same phase as, or exactly the opposite phase from, that of the subcarrier modulated by the I signal. This portion of the chrominance signal may lead or lag the quadrature portion by 90 electrical degrees.

input — 1. The current, voltage, power or other driving force applied to a circuit or device. 2. The terminals or other places where current, voltage, power, or driving force may be applied to a circuit or device. 3. Data to be processed. 4. The process of transferring data from an external computer storage to an internal storage. 5. The terminals, jack, or receptacle provided for the introduction of an electrical signal or electric power into a device or system.

input admittance — 1. The reciprocal of the input impedance. 2. The admittance between the input terminals with the outputs shorted together.

input area — In a computer, the area of internal storage into which data from external storage is transferred.

input bias current — 1. The current that must be supplied to each input of an IC operational amplifier to assure proper biasing of the differential-input-stage transistors. In specification sheets, this term refers to the average of the two input bias currents. 2. One-half the sum of the separate currents entering the two input terminals of a balanced amplifier. 3. The average of the two input currents of an operational amplifier.

input block — In a computer, a section of the internal storage reserved for receiving and processing input data.

input capacitance — 1. The capacitance at the input terminals of a device. 2. The capacitance between gate and source terminals of a field-effect transistor at specified bias and frequency conditions, with the drain ac short-circuited to the source.

input channel — A channel through which a state is impressed on a device or logic element.

input common-mode range — The maximum input that can be applied to either input of an operational amplifier without causing damage or abnormal operation.

input common-mode rejection ratio — 1. The ratio of the change in input voltage to the corresponding change in output voltage, divided by the open-loop voltage gain. 2. The ratio of the full differential voltage gain to the common-mode voltage gain.

input common-mode voltage range — The range of voltages on the input terminals of an operational amplifier for which the amplifier is operational. Note that the specifications are not guaranteed over the full common-mode voltage range unless specifically stated.

input device — 1. The device or set of devices through which data is brought into another device. 2. A device such as a card reader or terminal keyboard that converts data from the form in which it has been received into electronic signals that can be interpreted by the computer.

input equipment — The equipment that introduces information into a computer.

input error voltage — The error voltage appearing across the input terminals of an operational amplifier when a feedback loop is applied around the amplifier.

input extender — A high-speed diode array used in a logic circuit when increased fan-in capability is required.

input formatting — The technique a system uses to put all entered data into a standard (or intelligible) format.

input gap — Also called buncher gap. In a microwave tube, the gap where the initial velocity modulation of the electron stream occurs.

input impedance — 1. The impedance a transducer presents to a source. 2. The effective impedance seen looking into the input terminals of an amplifier; circuit details, signal level, and frequency must be specified. 3. The impedance that exists between the input terminals of an amplifier or transmission line when the source is disconnected.

input impedance of a transmission line — The impedance between the input terminals with the generator disconnected.

input offset current — The difference between the input bias currents flowing into each input of an IC operational amplifier, when the output of the operational amplifier is at zero volts.

input offset voltage — That voltage which must be applied between the input terminals of an operational amplifier, through two equal resistances, to obtain zero output voltage.

input/output — Abbreviated I/O. 1. Pertaining to devices that accept data for transmission to a computer system (input) or that accept data from a computer system for transmission to a user or process. Devices that perform both functions are known as I/O devices (e.g., terminals). 2. A general term for the peripheral devices used to communicate with a digital computer, and the data involved in the communication. 3. The process of transmitting information from an external source to an equipment unit, or from an equipment unit to an external source. 4. Transferring information from an input to an accumulator, or from an accumulator to an output. 5. Interface circuits or devices offering access between external circuits and the central processing unit or memory.

input-output bound — *See* input-output limited.

input/output connector — A mating pair of connectors used to carry signals into and out of a panel-mounted subsystem. An example is a connector pair that interconnects the individual back panels in a large array of panels.

input-output limited — Also called input-output bound. Pertaining to a system or condition in which the time taken by input and output operations exceeds the time for other operations.

input pins — The terminals of the device to which input logic signals may be applied.

input-power rating — Also called coil rating. A statement of the allowable voltage, current, or power to the actuating element of a relay beyond which unsatisfactory performances will occur.

input process — 1. The process in which a device receives data. 2. The transmission of data from peripheral equipment or external storage to internal storage.

input recorder — Any device that makes a record of an input electrical signal.

input reflected current — In dc-dc converters, the peak-to-peak ac current generated by the switching transients. The value of this current should not exceed 1 percent of the nominal input current.

input reflected ripple — A dc-dc converter term that describes the voltage spike resulting from switching generated transient currents as measured at the dc input source.

input register — In a computer, the register of internal storage able to accept information from outside the computer at one speed and supply the information to the computer calculating unit at another, usually much greater, speed.

input resistance — Also called differential input resistance. The small signal resistance measured between the inverting and noninverting inputs of an operational amplifier. Input capacitance is the capacitance seen between the same two inputs.

input resonator — The buncher resonator in a velocity-modulated tube. It modifies the velocity of the electrons in the beam.

input sensitivity — The input signal level that will result in rated output of a piece of amplifying equipment. In preamplifiers, it is the signal that gives the rated voltage output of the preamplifier; in power amplifiers, the signal that gives the rated power output. (In preamplifiers, the phono sensitivity is commonly 1 millivolt; high-level inputs, such as tape and tuner, are commonly 250 millivolts. In power amplifiers, common values are between 0.5 and 1.0 volt.)

input transformer — A transformer that transfers energy from an alternating-voltage source to the input of a circuit or device. It usually provides the correct impedance match as well.

input uncertainty — In an operational amplifier, the algebraic sum of all the factors, including environmental and time effects, that contribute to the nonideal behavior of the input circuit.

input unit — In a computer, the unit that takes information from outside the computer into the computer.

input voltage drift — The change in output voltage of an operational amplifier divided by the open-loop gain, the quotient expressed as a function of temperature or time.

input voltage offset — The dc potential difference between the two inputs of a differential amplifier when the potential difference between the output terminals is zero.

input winding — *See* signal winding.

inquiry — 1. The withdrawal of stored information from an electronic data processing system by interrogating the contents of the storage of a computer. 2. A technique for initiating the interrogation of the contents of the storage of a computer.

inquiry station — A remote terminal from which an inquiry may be sent over a wire line to a computer.

inquiry unit — A device used to extract a quick reply to a random question regarding information in a computer storage.

inrush — The initial surge of current through a load when power is first applied. Lamp loads, inductive motors, solenoids, contactors, valves, and capacitive load types all have inrush or surge currents higher than the normal running or steady-state currents. Resistive loads, such as heater elements, have no inrush.

inrush current — 1. The current in a load circuit immediately following turn-on. In capacitive and tungsten-lamp loads, this exceeds steady-state current for some period following turn-on. 2. In a solenoid or coil, the steady-state current drawn from the line with the armature in its maximum open position.

inrush current limiting — Protective circuit in a power supply that prevents excessively large currents through a rectifier to charge the filter capacitors. To prevent unnecessary power loss, the circuit is usually inhibited after the capacitors attain full charge.

insert core — An iron core used generally for adjusting an inductor to a fixed frequency. It consists of a threaded metal insert molded or cemented into one or both ends of the core.

insert earphones — Small earphones that fit partially inside the ear.

insertion delay — Also called inherent delay, intrinsic delay, and propagation delay. The interval for a circuit or instrument to respond with an output after being triggered either internally or externally. It is usually implied that the numerically expressed period is an irreducible minimum.

insertion force — 1. Of a connector, the force needed to fully engage a connector plug and receptacle. Depending on the number, arrangement, and size of the contact pins, in addition to the strength of the springs, their surface finish, and the rigidity of their mountings, the required mechanical insertion force will be dictated to a great extent by the life expectancy of the connectors, the intended use, the speed of insertion, and their immunity to shock and vibration. 2. The effort, usually measured in ounces, required to engage mating components.

insertion gain — The gain resulting from the insertion of a transducer in a transmission system is the ratio of the power delivered to that part of the system following the transducer to the power delivered to that same part before insertion. (If more than one component is involved in the input or output, the particular component used must be specified. This ratio is usually expressed in decibels.)

insertion loss — 1. The difference between the power received at the load before and after the insertion of apparatus at some point in the line. 2. The loss in load power resulting from the insertion of a component, connector, or device. Insertion loss is expressed in decibels as the ratio of power received at the load before insertion to the power received at the load after insertion. 3. Signal-power loss resulting from connecting communication equipment with dissimilar impedance values. 4. A power loss that results from inserting a component into a previously continuous path or creating a splice in it.

insertion phase shift — The change in phase of an electric structure when inserted into a transmission system.

insertion switch — A process by which information is inserted into a computer by the manual operation of switches.

insertion tool — A small, hand-held tool used to insert contacts into a connector.

inside lead — See start lead.

inside spider — A flexible device placed inside a voice coil to center it with the pole pieces of a speaker.

in-situ tester — See in-circuit tester.

inspection chamber — In a spectrophotometer, the part in which the solution to be tested is placed for analysis.

inspectoscope — An instrument for viewing quartz crystals, while they are immersed in oil, to determine mechanical faults, the approximate direction of the optical axis, and regions of optical twinning.

instability — 1. The measure of the fluctuations or irregularities in the performance of a device, system, or parameter. 2. An undesired change that occurs over a period of time and that is not related to input, operating conditions, or load.

instantaneous automatic gain control — Abbreviated instantaneous AGC. A portion of a radar system that automatically adjusts the gain of an amplifier for each pulse so that there is a substantially constant output-pulse peak amplitude with different input-pulse peak amplitudes. The circuit is capable of acting during the time in which a pulse is passed through the amplifier.

instantaneous companding — Companding that varies the effective gain in response to instantaneous values of the signal wave.

instantaneous contacts — Contacts that are actuated immediately when a starting signal is applied to a timer.

instantaneous disc — A blank recording disc that can be played back on a phonograph immediately after being cut on a recorder.

instantaneous frequency — The rate at which the angle of a wave changes when the wave is a function of time. If the angle is measured in radians, the frequency in hertz is the rate of change of the angle divided by 2π.

instantaneous overcurrent relay — Also called rate-of-rise relay. A device that functions instantaneously on an excessive value of current or on an excessive rate

of current rise, thus indicating a fault in the apparatus of the circuit being protected.

instantaneous power — The power at the points where an electric circuit enters a region. It is equal to the rate at which the circuit is transmitting electrical energy into the region.

instantaneous power output — The rate at which energy is delivered to a load at a particular instant.

instantaneous readout — Readout by a radio transmitter at the instant the information to be transmitted is computed.

instantaneous recording — A recording intended for direct reproduction without further processing.

instantaneous sampling — The process of obtaining a sequence of instantaneous values of a wave. These values are called instantaneous samples.

instantaneous sound pressure — The total instantaneous pressure at a certain point, minus the static pressure at that point. The most common unit is the microbar.

instantaneous speech power — The rate at which the speaker is radiating sound energy at any given instant.

instantaneous speed variations — See ISV.

instantaneous start-stop rate — The maximum stepping rate that can be attained by an unloaded stepper motor from a standstill without losing synchronism with the field and without overshooting to the next step when coming to a stop.

instantaneous value — 1. The magnitude, at any particular instant, of a varying value. 2 The value of voltage or current at a particular instant. If the selected instant is the time when the polarity of the waveform changes, this value will be zero.

instruction — 1. Information that, when properly coded and introduced as a unit into a digital computer, causes the computer to perform one or more of its operations. All instructions commonly include one or more addresses. 2. A binary code applied to a logic circuit to affect its mode of operation. 3. A statement that specifies an operation and the values or locations of its operands. In this context, the term *instruction* is preferable to the terms *command* or *order*, which are sometimes used synonymously. 4. A set of bits that defines a computer operation and is a basic command understood by the CPU. It may move data, do arithmetic and logic functions, control I/O devices, or make decisions as to which instructions to execute next. 5. In a computer, a single order within a program. This order will be fetched from memory, decoded, and executed by the CPU. Instructions may be arithmetic or logical, and operate on registers, memory, I/O devices, or specify control operations. A sequence of instructions is a program. 6. A machine-language command executed by the microprocessor in a computer system.

instructional constant — Also called pseudoinstruction. In a computer, data stored in the program or instructional area that will be used only as a test constant.

instruction code — The list of symbols, names, and definitions of the instructions that are intelligible to a given computer or computing system.

instruction counter — A multiple-bit register that keeps track of the address of the current instruction. See control counter.

instruction cycle — The process of fetching an instruction from memory and executing it.

instruction deck — A set of punched cards containing a symbolic coded program to be read into a computer.

instruction fetch — See fetch.

instruction length — The number of words needed to store an instruction. It is one word in most computers, but some will use multiple words to form one instruction. Multiple-word instructions have different instruction execution times depending on the length of the instruction.

instruction modification — A change in the operation-code portion of a computer instruction or command such that, if the routine containing the instruction or command is repeated, the computer will perform a different operation.

instruction register — 1. In a computer, the register that temporarily stores the instruction currently being performed by the control unit of the computer. 2. A computer storage for the binary code for the operation to be performed. Usually this instruction represents the contents of the address just designated by the program counter. However, the contents of the instruction register or the program counter may be changed by the computations. This, of course, represents one of the key ideas of a stored-program computer — instructions, as well as data, can be operated on and subsequent operations will be determined by the results.

instruction repertoire — The instruction set for a computer.

instruction set — 1. A means of describing computer capability. It consists of a listing of all the instructions the computer can execute. The basic operations that can be performed by a CPU. Necessary instructions are arithmetic, logical, test and branch, and moves. 2. The set of general-purpose instructions available with a given computer. In general, different machines have different instruction sets. The number of instructions only partially indicates the quality of an instruction set. Some instructions may only be slightly different from one another; others rarely may be used. Instruction sets should be compared using benchmark programs typical of the application to determine execution times and memory requirements.

instruction storage — The storage medium that contains basic machining instructions in coded form.

instruction time — The time required to fetch an instruction from memory and then execute it.

instruction word — A computer word that causes the computer to execute a particular operation. *See* word.

instrument — A device capable of measuring, recording, and/or controlling.

instrument approach — A blind landing — i.e., solely by navigational instruments, without visual reference to the terrain.

instrument-approach system — In navigation, a system furnishing vertical and horizontal guidance to aircraft during descent. Touchdown requires some other guidance.

instrumentation — 1. The use of devices to measure the values of varying quantities, usually as part of a system for keeping the quantities within prescribed limits. 2. Adding code to a program for injecting data and collecting information, usually for dynamic analysis.

instrumentation amplifier — High accuracy analog amplifier with full differential inputs and gains ranging from 1 to 1000.

instrumentation bus — A dedicated bus tailored for the connection of measurement and control devices.

instrument chopper — A vibrating switch used for modulating, demodulating, and switching dc or low-frequency ac information in instrumentation. It is driven synchronously from an ac or pulsating dc source. The driven switching circuit is designed for low-level (0- to 10-volt) signal information.

instrument driver — Software module that converts the parameters in the object code to the specific instruction sequence needed to stimulate an instrument.

INSTRUMENTATION AMPLIFIER'S INVERTING INPUT

INSTRUMENTATION AMPLIFIER'S NONINVERTING INPUT

$$\frac{e_0}{e_1} = \left(1 + \frac{R_1}{R_G/2}\right)$$

Instrumentation amplifier.

instrument error — The inaccuracy of an instrument.

instrument flight — A blind flight — i.e., one in which the pilot controls the path and altitude of the aircraft solely by instrument.

instrument lamp — A lamp that illuminates or irradiates an instrument.

instrument landing station — A special radio station for aiding in landing aircraft.

instrument landing system — Abbreviated ILS. A radionavigation system intended to aid aircraft in landing. It provides lateral and vertical guidance, including distance from the landing point. Consists of four ground radio transmitting stations at and in the vicinity of an airport, which radiate direction and position signals to approaching aircraft that are received on an instrument in the aircraft and alert the pilot to any deviation from the safe approach path to the correct touchdown point.

instrument landing system localizer — System of horizontal guidance embodied in the instrument landing system that indicates the horizontal deviation of the aircraft from its optimum path of descent along the axis of the runway.

instrument multiplier — *See* voltage-range multiplier.

instrument relay — A relay that operates on the principles employed in such electrical measuring instruments as the electrodynamometer, iron-vane, and D'Arsonval meters.

instrument shunt — An internal or external resistor connected in parallel with the circuit of an instrument to extend its current range.

instrument switch — A switch disconnecting an instrument or transferring it from one circuit or phase to another.

instrument transformer — A transformer that reproduces, in its secondary circuit, the primary current (or voltage) with its phase relationship substantially preserved, suitable for utilization in measurement, control, or protective devices.

instrument-transformer correction factor — The factor by which a wattmeter reading must be multiplied to correct for the effect of the instrument-transformer ratio correction factor and phase angle.

instrument zero — The lower end of the measuring instrument's scale. Instrument zero may not coincide with zero value of the measured variable. Zero error is the error at instrument zero.

insulated — Separated from other conducting surfaces by a nonconductive material offering a high, permanent resistance to the passage of current and disruptive discharge.

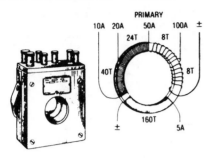

Instrument transformers.

insulated carbon resistor — A carbon resistor encased in fiber, plastic, or other insulation.

insulated clip — A clip terminating in an insulated eye through which flexible cords or wires may be run and supported.

insulated enclosure — A special shielded enclosure design providing insulation against weather or providing maximum temperature stability. Usually prefabricated as an exterior building panel in modular construction.

insulated-gate field-effect transistor — Abbreviated IGFET. In general, any field-effect transistor that has an insulated gate regardless of the fabrication process.

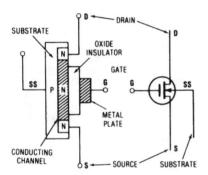

Insulated-gate field-effect transistor.

insulated-substrate monolithic circuit — An integrated circuit that may be either an all-diffused device or a compatible structure so constructed that the components within the silicon substrate are insulated from one another by a layer of silicon dioxide, instead of the reverse-biased pn junctions used for isolation in other techniques.

insulated terminals — Solderless terminals provided with an insulated sleeve over the barrel to prevent a short circuit.

insulated wire — A conductor covered with a nonconductive material.

insulating material — 1. A material on or through which essentially no current will flow. It is used to confine the flow of current within a conductor or to eliminate the shock hazard of a bare conductor. 2. Any composition primarily adapted for preventing the transfer of electricity therethrough, the useful properties of which depend on its chemical composition or atomic arrangement.

insulating sleeve — Tube or tape of insulating material placed around metal-enclosed capacitors to insulate the case electrically from other components and wiring.

insulating strength — The measure of the ability of an insulating material to withstand electrical stress without breaking down. It is defined in terms of the voltage per unit thickness necessary to initiate a disruptive discharge and usually is measured in volts per centimeter. *See also* dielectric strength; electric strength.

insulating tape — Tape that is wrapped around joints in insulated wires or cables. It is impregnated with an insulating material and covered with adhesive on one side.

insulating varnish — A varnish applied to coils and windings to improve their insulation (and, at times, their mechanical rigidity).

insulation — 1. A nonconductive material that prevents the leakage of electricity from a conductor, provides mechanical spacing or support, or protects against accidental contact. 2. The use of a material that passes negligible current to surround or separate a conductor to prevent loss of current. 3. A material that offers high electric resistance, making it suitable for covering components, terminals, and wires to prevent the possible future contact of adjacent conductors resulting in a short circuit. 4. Material used to cover electrical wires to prevent electrical leakage and short circuiting and to reduce the danger of shock.

insulation displacement termination — A connector that has insulated wire is forced into a channel constructed so that ridges or teeth in the channel cut through or displace the insulation and make an air-tight contact with the wire.

insulation piercing — A crimping method in which lances pierce wire insulation, enter into the strands, and make electrical contact without stripping the wire.

insulation rating — The dielectric-strength and insulation-resistance values required to ensure satisfactory performance.

insulation resistance — 1. The resistance offered by an insulating material to the flow of current resulting from an impressed dc voltage. 2. The ratio of the voltage applied between two electrodes in contact with a specific insulator to the total current between the electrodes. 3. Industrial specifications usually call for a certain minimum value (several thousand megohms) determined with a specific voltage applied. 4. The direct current resistance between the two terminals of a capacitor, or between either or both of the terminals and the capacitor case. 5. The ratio of dc voltage impressed across a capacitor to the resultant leakage current. For a particular capacitor design, the product of insulation resistance and capacitance (megohm-microfarad) is quite constant. 6. The electrical resistance of the insulating material (determined under specified conditions) between any pair of contacts, conductors, or grounding devices in various combinations.

insulation resistivity — The insulation resistance per unit volume of insulation.

insulation stress — The molecule separation pressure caused by a potential difference across an insulator. The practical stress on insulation is expressed in volts per mil.

insulation system — All of the insulation materials used to insulate a particular electrical or electronic product.

insulator — 1. A material in which the outer electrons are tightly bound to the atom and are not free to move. Thus, there is negligible current through the material when a voltage is applied. The resistivity is greater than 10^8 ohm-cm and generally decreases when the temperature rises. 2. A nonconducting substance such as porcelain, plastic, glass, rubber, etc. 3. A material of such low electrical conductivity that current through it can usually be neglected. 4. A material of low electrical conductivity designed for supporting a conductor, physically and electrically separating it from another conductor or object.

insulator arcing ring — A circular or oval metal part placed at one or both ends of an insulator to prevent current from arcing over and damaging it and/or the conductor.

insulator arcover — The flow of power current over an insulator in the form of an arc following a surface discharge.

insulazing — *See* surface insulation.

insulectrics — The science encompassing insulating materials in electrical insulation.

integer — A whole number, which may be positive, negative, or zero. It does not have a fractional part. Examples of integers: 1, 2, 48, −136, etc., but not 3.7 or $^3/_4$.

integral-cavity, reflex-klystron oscillator — A reflex-klystron oscillator in which tuning is accomplished by changing the physical dimensions of the resonant cavity. It is usually referred to as a diaphragm- or grid-gap-tuned klystron, since a flexible diaphragm is used to change the cavity dimension, i.e., the gap between the cavity grids.

integral circuit packages — Microcircuits assembled from discrete components and all circuits created essentially in an active or passive substrate.

integral contact — Current-carrying member of jack, switch, or relay. Usually a flat, flexible spring or other conducting member having no separate contacts attached at point of mating.

integral-external-cavity reflex oscillator — A reflex-klystron oscillator in which a fixed internal cavity is tightly coupled to a permanently attached external cavity. Tuning is achieved by varying a reactance probe in the external cavity.

integral-horsepower motor — A motor that is built into a frame and has a continuous rating of 1 horsepower.

integral resistor — An internal or external resistor preconnected to the electrical element and forming an integral part of the cup assembly to provide a desired electrical characteristic of a precision potentiometer.

integrated — A type of design in which two or more basic components or functions are physically, as well as electrically, combined — usually on one chassis, such as an integrated amplifier.

integrated amplifier — Also called, occasionally, a control amplifier, since it is an amplifier with controls. 1. An amplifier that embodies in a common housing the preamplifier and control section and the power amplifier. Some early amplifiers of large power were in two separate units, one the control unit with preamplifiers and the other the power amplifier. 2. A single component combining the functions and circuitry of a power amplifier and preamplifier. *See also* amplifier.

integrated circuit — Abbreviated IC. Also called functional device. 1. An electrical network — active or passive — composed of two or more circuit elements inextricably bound on a single semiconductor substrate. To further define the nature of an integrated circuit, additional modifiers may be prefixed. Examples: dielectric-isolated monolithic integrated circuit, beam-lead monolithic integrated circuit, or silicon-chip tantalum thin-film hybrid integrated circuit. *See also* film integrated circuit; hybrid integrated circuit; monolithic integrated circuit; multichip integrated circuit. 2. Any electronic device in which both active and passive elements are contained in a single package. The term frequently is used for circuits other than those containing semiconductors; for example, microwave

designers consider many types of waveguide assemblies to be integrated circuits. 3. A small chip of solid material (generally a semiconductor) upon which, by various techniques, an array of active and/or passive components has been fabricated and interconnected to form a functioning circuit. Integrated circuits, which are generally encapsulated with only input, output, power supply, and control terminals accessible, offer great advantages in terms of small size, economy, and reliability. 4. The physical realization of a number of electrical circuit elements inseparably associated on or within a continuous body of semiconductor material to perform the function of a circuit. 5. An electronic device containing several elements, active or passive, that perform all or part of a circuit function. 6. An interconnected array of conventional components — transistors, diodes, capacitors, and resistors — fabricated *in situ* within and on a single crystal of semiconductor material with the capability of performing a complete electronic circuit function. 7. An electronic circuit containing transistors, diodes, resistors, and perhaps capacitors and photocells, along with interconnecting electrical conductors processed and contained entirely within a single chip of silicon. 8. Multiple, interconnected circuit elements, contained on or in a common substrate, that function as a unit and not separately.

integrated-circuit array — Multiple integrated circuits formed on a common substrate and electrically interconnected during fabrication.

integrated-circuit package — The combined mounting and housing for an integrated circuit; the package protects the integrated circuit and permits external connections to be made to it.

integrated communication system — Communication system on either a unilateral or joint basis in which a message can be filed at any communications center in that system and be delivered to the addressee(s) by any other appropriate communication center in that system without reprocessing en route.

integrated component — A number of electrical elements comprising a single structure that cannot be divided without destroying its stated electronic function.

integrated console — Computer control console that is capable of controlling the operation of the switching center equipment of an integrated communications system.

integrated data processing — A method of transforming disjointed and repetitive paperwork tasks into a correlated and mechanized production of information for any purpose.

integrated electronic component — Abbreviated IEC. An assembly that consists of several integrated circuits interconnected on a single chip of silicon to provide a complete electronic function with a circuit content greater than 10 equivalent gates.

integrated electronics — That portion of electronic art and technology in which the interdependence of material, device, circuit, and system-design considerations is especially significant; more specifically, that portion of the art dealing with integrated circuits.

integrated electronic system — *See* integrated circuit.

integrated equipment components — Abbreviated IECs. Integrated-circuit chips that contain a complete logic function.

integrated injection logic — *See* I²L.

integrated microcircuit — *See* integrated circuit.

integrated morphology — The structural characterization of an electronic component in which the identity of the current- or signal-modifying areas, patterns, or volumes has become lost in the integration of electronic materials, in contrast to an assembly of devices performing the same function.

integrated optical circuit — Abbreviated IOC. Also called optical integrated circuit. A circuit, or group of interconnected circuits, consisting of miniature solid-state optical components, such as light-emitting diodes, optical filters, photodetectors (active and passive), and thin-film optical waveguides on semiconductor or dielectric substrates.

integrated optics — The interconnection of miniature optical components via optical waveguides on transparent dielectric substrates, using optical sources, modulators, detectors, filters, couplers, and other elements incorporated into circuits analogous to integrated electronic circuits, for the execution of various communication, switching, and logic functions.

integrated services digital network — Abbreviated ISDN. A switched network providing end-to-end digital connectivity for simultaneous transmission of voice and data over multiplexed communications channels. *See also* ISDN.

integrated software — An applications software package containing programs to perform more than one function.

integrated transducers — Semiconductor components that change the form of energy (e.g., piezoelectric devices, photogenerators, thermistors, etc.) and that are integrated into multifunction chips.

integrated voltage regulator — Abbreviated IVR. An integrated structure that serves as the reference, error amplifier, and shunt elements for shunt voltage regulation. A resistive voltage divider connected to its input provides adjustment of the output voltage.

integrating circuit — *See* integrator, 1.

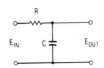

No amplification.

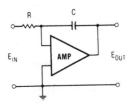

Amplification.

$$E_{OUT} = \frac{1}{T} \int_T E_{IN} dt$$

$$(T = RC)$$

Formula.

Integrating circuits.

integrating meter — A meter that adds up (integrates) the electrical energy used over a period of time. An ordinary electric watthour meter is an example.

integrating motor — A motor that maintains a constant ratio of output-shaft rotational speed to input signal.

Thus, the angle of rotation of the shaft is proportional to the time integral of the input signal.

integrating photometer — A photometer that, with a single reading, indicates the average candlepower from a source in all directions or at all angles in a single plane.

integrating relay — A relay that sums up the inputs of voltage or current supplied to it and opens or closes its contacts in response to the input so integrated.

integrating-sphere densitometer — A photoelectric instrument that measures the density of motion-picture film or its sound track.

integration time — That time during which all electrons formed by impinging photons are gathered in a potential well under an energized electrode.

integrator — 1. A device with an output proportionate to the integral of the input signal. 2. In certain digital machines, a device that numerically approximates the mathematical process of integration. 3. A device that determines on a continuous basis the total value of a quantity being measured, usually as a function of time. 4. Device for summing or totalizing counts, areas under curves, etc.

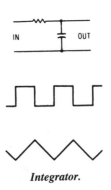

Integrator.

intelligence bandwidth — The total audio (or video) frequency bandwidths of one or more channels.

intelligence sample — Part of a signal taken as evidence of the quality of the whole.

intelligence signal — Any signal that conveys information (e.g., voice, music, code, or television).

intelligent controller — Device controller equipped with local interpreting functions, such as editing, input validity checks, and complex command decoding.

intelligent instruments — Devices that possess capabilities that raise them above the level of instruments that merely sense and display analog information. The following list presents one instrument intelligence rating system, from the lowest order of intelligence to the highest:

1. The ability to sense and display information.
2. Conversion of analog information into digital.
3. Mathematical manipulation of digital data.
4. Interpretation of results of mathematical manipulation.
5. Making of decisions on the basis of interpretation.

intelligent robot — A robot that can make decisions by itself through its sensing and recognizing capabilities.

intelligent terminal — 1. An input/output device in which a number of computer processing characteristics are physically built into, or attached to, the terminal unit. 2. Programmable control terminal that drives other terminals, peripherals, floppy disks, machines, program counters, etc. 3. A programmable data service, usually remote from the main computer, that unburdens the host computer by performing preliminary data processing such as formatting, verification, or validation. 4. A terminal that has editing and block-transmission capabilities, which allow manipulation of data in the terminal before transmission to the host computer. Some intelligent terminals can also perform checks on entered data. An intelligent terminal's programmability is restricted to formatting data on its screen. These formats can be called up through the keyboard and can have protected fields that allow entering of data (numeric, alphabetic, or alphanumeric) in the prescribed format. User programmability distinguishes the intelligent terminal from its less intelligent relatives. In addition to this programmability, an intelligent terminal has editing functions similar to those of a smart terminal. 5. An input/output device with built-in intelligence in the form of a microprocessor and able to perform functions that would otherwise require the central computer's processing power; sometimes called a stand-alone terminal. 6. A terminal with local processing power whose characteristics can be changed under program control.

intelligent time-division multiplexer — See ITDM.

intelligent voice terminal — Intelligent terminal operated by the human voice; software resident in the terminal is user-programmable. Best for applications suiting an intelligent terminal but where hands-free data entry is cost advantageous.

intelligibility — See articulation.

INTELSAT — Abbreviation for *In*ternational *Tele*communications *Sat*ellite Consortium. 1. A series of commercial communications satellites designed to relay telephony and television signals among the member INTELSAT nations. The INTELSAT global network employs an equatorial orbit above the Atlantic, Pacific, and Indian Oceans. Satellites in these three locations are capable of linking virtually all the inhabited areas of the world. 2. International (primarily noncommunist) satellite agency whose member nations lease transponder capacity on its satellite system to provide at least some TV in all parts of the world.

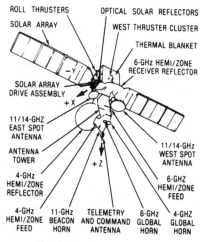

INTELSAT satellite.

intensifier electrode — Also called postaccelerating electrode. In some types of electrostatic cathode-ray tubes, an electrode that permits additional acceleration of the electron beam after it has been deflected. This electrode emits greater intensity of the trace without materially reducing the deflection sensitivity of the tube.

intensifying screen — A thin fluorescent screen placed next to a photographic plate to increase the effect of radiation on the plate.

intensitometer — Also called a dosage meter or dosimeter. An instrument that estimates the amount of X-ray radiation, for determining the duration of exposure during X-ray pictures or therapy.

intensity — 1. The strength of a quantity. 2. The relative strength, or amplitude, of electric, magnetic, or vibrational energy. 3. The brilliance of an image on the screen of a cathode-ray tube. 4. The strength of light or other electromagnetic energy being radiated or reflected per steradian. 5. The flux per unit solid angle radiating (diverging) from a source of finite area.

intensity control — Used with cathode-ray tubes to control the intensity of the electron beam and hence the amount of light generated by the fluorescent screen. Generally, the grid bias of the tube is regulated.

intensity level — Ratio of the intensity of the sound to a reference intensity of a free plane wave of 1 microwatt per square centimeter under normal conditions. Commonly expressed in decibels.

intensity modulation — 1. The process and/or effect of varying the electron-beam current in a cathode-ray tube, resulting in varying brightness or luminance of the trace. 2. The process in which the electron beam of a cathode-ray tube is varied in intensity in accordance with the magnitude of the signals it receives. See also Z-axis modulation.

intensity of radiation — The radiant energy emitted in a specified direction per unit time, per unit area of surface, per unit solid angle.

interaction — The effects two or more parts, components, etc., have on each other while each is performing a function.

interaction crosstalk — Crosstalk resulting from mutual coupling between two paths by means of a third path. For example, if a signal on pair 1 is coupled to pair 2, and then coupled from pair 2 to pair 3, where it is measured, it is known as interaction crosstalk.

interaction loss (of a transducer) — Expressed in decibels, it equals 20 times the logarithm (to the base 10) of the scalar value of the reciprocal of the interaction factor.

interaction space — In an electronic tube, the region where the electrons interact with an alternating electromagnetic field.

interactive — 1. Pertaining to an application in which each entry elicits a response. An interactive system may also be conversational, implying continuous dialog between the user and the system. 2. A program or system that can ask questions of the user and then take action based on his response. *Conversational* is often used to mean the same thing. 3. Refers to those applications in which a user communicates with a computer program via a terminal, entering data and receiving responses from the computer. 4. Processing of data on a two-way basis, with human intervention redirecting the processing in a predetermined manner.

interactive debugger — A computer system software utility that permits a user to examine his or her task while it executes by stopping it at given points (usually called breakpoints) and displaying and changing memory/register contents.

interactive environment — A situation in which a computer continually responds to the user on receipt of directives from his or her terminal.

interactive graphics — 1. The use of a large-screen, high-precision CRT and its associated circuitry — usually linked to a large-scale computer system through a small control computer — on which both alphanumeric and vector data are displayed and manipulated. By using the data entry device, two- and three-dimensional geometric designs can be created, deleted, and modified in real time to achieve desired results. 2. The use of a display terminal in a conversational or interactive mode. 3. Capability to perform graphics operations directly on the computer with immediate feedback.

interactive operation — Also called conversational mode. Online operation in which there is a give-and-take between person and machine.

interactive system — 1. A system in which it is possible for the human user or the device serviced by the computer to communicate directly with the operating program. For human users, this arrangement is termed a conversational system. 2. A system in which a computer or operating program communicates bilaterally with a user.

interactivity — Computer programs, online services, and interactive TV in which the user can make inputs that direct subsequent delivery of services. Channel surfing is a crude form of interactivity. Internet and online services provide simple forms of interactivity, i.e., point-and-click.

interaxis error — The deviation from 90° perpendicularity of one set of resolver windings when excitation is applied to one of the other windings. For rotor interaxis error, one stator winding is excited; for stator interaxis error, one rotor winding is excited.

interbase current — In a junction tetrode transistor, the current that flows from one base connection to the other through the base region.

interbase resistance — Resistance between base 2 and base 1 of a unijunction transistor measured at a specified interbase voltage with $I_E = 0$.

interblock space — See IRG.

intercarrier noise suppression — The means of suppressing the noise resulting from increased gain when a high-gain receiver with automatic volume control is tuned between stations. The suppression circuit automatically blocks the audio-frequency output of the receiver when there is no signal at the second detector.

intercarrier sound system — A television receiving system in which use of the picture carrier and the associated sound-channel carrier produces an intermediate frequency equal to the difference between the two carrier frequencies. This intermediate frequency is frequency modulated in accordance with the sound signal.

intercellular massage — The ultrasonic stimulation of body cells. Sometimes called micromassage.

intercepting — Routing of a call or message placed for a disconnected or nonexistent destination to an operator position or a specially designated terminal or machine answering device.

intercepting trunk — A trunk to which a call made to a vacant number, a changed number, or a line out of order is connected so that action may be taken by an operator.

intercept operator — The telephone operator who requests the number called, determines the reason for the intercept, and relays the information to the calling party.

intercept receiver — Also called search receiver. A specially calibrated receiver that can be tuned over a wide frequency range in order to detect and measure enemy rf signals.

intercept service — In a telephone system, a service provided to subscribers whereby calls to disconnected stations or dead lines are either routed to an intercept operator for explanation or the calling party receives a distinctive tone signal or recorded announcment to indicate that he or she has made such a call.

intercept tape — A tape used for temporary storage of messages intended for trunk channels and tributary stations in which there is equipment or circuit trouble.

intercept trunk — See intercepting trunk.

intercharacter space — In telegraphy, the space between characters of a word. It is equal to three unit lengths.

intercom — See intercommunication system.

intercommunication apparatus — Equipment and systems for paging and intercommunication within a building, including audio, bell systems, pillow systems, and pocket page systems.

intercommunication system — Also called intercom. 1. A two-way communication system without a central switchboard, usually limited to a single vehicle, building, or plant area. Stations may or may not be equipped to originate a call, but can answer any call. 2. A system that permits selective speaker voice communication via wires between any pair of several stations, usually in the same building. The stations may be either master stations, which may initiate calls to any of a group of stations, or slave stations, which may initiate calls only to their master station. 3. A communication system that bridges the gap between a regular telephone system and the public address (PA) system it locates people and permits communication with them. Most often, it is installed as an adjunct to telephone and PA systems.

intercom wire — Wire used to connect communications instruments, telephones, telegraph, etc.

interconnecting wire — 1. Wires used for connections between subassemblies, panels, chassis, and remotely mounted devices. Does not necessarily apply to the internal connection of these units. 2. The physical wiring between components (outside a module), between modules, between units, or between larger portions of a system or systems.

interconnection — 1. Also called tie line. A transmission line connecting two electric systems or networks and permitting energy to be transferred in either direction. Larger interconnections are often called interties, giant ties, or regional interconnections. 2. The conductive path required to achieve connection from a circuit element to the rest of the circuit. 3. Also called intraconnection. The physical wiring between components (outside a module), between modules, between units, or between larger portions of a system or systems.

interconnection diagram — Diagram showing the identity of all units in a piece of electronic equipment and the connections between them.

interconnections (microelectronic) — Those conductors and connections that are not in continuous integral contact with the substrate or circuit elements of an integrated circuit.

interconnection system — The electrical and mechanical interconnection of any one or all of the six levels of interconnections generally common to electronic equipment. The six levels of interconnection are intramodule, module to motherboard, intramotherboard, motherboard to back panel, backpanel wiring, and input/output.

interdiction — Techniques to prevent nonsubscribers access to information/programs on cable or drop cables. Interdiction electronics are outside the consumer's house and can be addressable.

interdigital magnetron — A magnetron with anode segments around the cathode. Alternate segments are connected together at one end, and remaining segments at the opposite end.

interdigital transducer — Abbreviated IDT. A number of interleaved metal electrodes whose width and spacing is equal and uniform throughout the transducer pattern. When a harmonic voltage is applied to the transducer terminals, the IDT pattern excites a periodic electric field that penetrates into the piezoelectric substrate. The substrate responds by periodically expanding and contracting in unison with these fields. With the proper choice of substrate orientation, this piezoelectric excitation gives rise to surface acoustic waves that propagate in the two directions normal to the IDT electrodes. The electric field can be produced by applying signals of opposite potential to two parallel metal electrodes formed in films deposited on the surface of the crystal. As the field is applied, the alternating signals send an acoustic wave across its surface; this wave is reconverted to an electric signal at a second pair of similar electrodes at the other end of the crystal. In practice, the transducer consists of several pairs of interleaved parallel electrodes, which give the transducer its name. The fingers, and the space between them, must be related to the size of a wave of the frequency desired. In certain applications the transducer fingers are less than a thousandth of a millimeter wide.

interelectrode capacitance — The capacitance between one electron-tube electrode and the next electrode toward the anode. The capacitance between the electrodes in an electron tube.

interelectrode coupling — Capacitive feedback from the plate of a tube to the grid. In triodes, this limits the maximum amplification possible without starting oscillation.

interelectrode leakage — The undesired current that flows between elements not normally connected in any way.

interelectrode transit time — The time required for an electron to travel between two electrodes.

interelement capacitance — The capacitance caused by the pn junctions between the regions of a transistor and measured between the external leads of the transistor.

interexchange channel — A channel connecting two different exchange areas.

interface — 1. A point or device at which a transition between media, power levels, modes of operation, etc., is made. 2. The two surfaces on the contact sides of mating connectors that face each other when mated. 3. A common boundary between two or more items. May be mechanical, electrical, functional, or contractual. 4. A common aspect at the boundary between two systems involving intersystem communication — e.g., the interaction between research and development, basic and applied science, or engineering and systems development. 5. The physical and space boundary surrounding the system, subsystem, equipment, or component, through which all environmental and operational stimuli essential to the device or affecting its proper operation must propagate or interact with other related devices or structures. 6. The hardware for linking two units of electronic equipment; for example, a hardware component to link a computer with its input (or output) device. 7. The means of connection between two logic elements, often elements that belong to two different "families." 8. The hardware or software required to be able to communicate with, sense, or control external equipment. 9. A circuit that controls the flow and format of data between a computer and a terminal or other peripheral. 10. An electrical connection that permits a peripheral device or communications channel to be attached to a system. 11. An electronic assembly that ties an external device to a computer. 12. A common boundary between automatic data-processing systems or parts of a single system. In communications and data systems, it may involve code, format, speed, or other changes as required. 13. The junction or point of interconnection between two systems or equipments with different characteristics. They may differ with respect to voltage, frequency, operating speed, type of signal, and/or type of information coding. 14. To interconnect different systems to resolve their incompatibilities. 15. A common or shared boundary between two or more instruments, devices, or systems, which enables exchange of information among interconnecting units or systems that may not be directly compatible.

interface adaptor — Also called device adaptor. A unit that provides a mechanical and electrical interconnection between the tester and the device under test. It may include special stimulus, measurement, load, and switching circuitry unique to a device or family of devices but which is not provided in the tester.

interface analysis — Checking the interfaces between program elements (modules) for consistency and proper data transfer.

interface card — A device that converts a computer I/O bus into some standard I/O configuration (8- or 16-bit parallel BCD, RS-232, IEEE 488, etc.).

interface circuit — 1. A circuit that links one type of logic family with another or with analog circuitry. 2. A circuit that allows two or more systems to be readily joined or associated. Examples include circuits that link linear and digital systems and those that enable communication between two circuits through a transmission line. 3. A circuit that links one type of device with another. Its function is to produce the required current and voltage levels for the next stage of circuitry from the previous stage. 4. An input/output circuit that permits different chips to communicate over a bus. 5. Linear or analog circuit with the prime function of supporting digital circuitry. Employed where the digital circuitry for the world of data processing and computation must meet the physical world of analog variables. Also used to mix functions from a number of digital logic families to obtain performance or economic benefits. There are at least nine different categories of interface circuits: bus interface, memory interface and control, line drivers and receivers, peripheral interface, numeric display interface, data converters, voltage reference, voltage comparators, and communications interface.

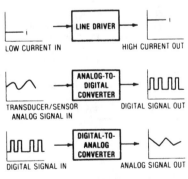

Interface circuits.

interface connection — Also called feedthrough. A conductor that connects patterns on opposite sides of a printed circuit board.

interface equipment — Equipment used between two other equipments that would otherwise be incompatible.

interface resistance — *See* cathode interface.

interface system — The device-independent mechanical, electrical, and functional elements of an interface necessary to effect communication among a set of devices. Cables, connector, driver and receiver circuits, signal lines, descriptions, timing and control conventions, and functional logic circuits are typical interface system elements.

interface unit — A device that translates incoming signals that are incompatible with the electrical characteristics of the computer without changing the information content. Also translates outgoing signals for the benefit of associated equipment that is designed to different electrical standards.

interfacial bond — An electrical connection between the two faces of a substrate.

interfacial connection — In a printed circuit board, a conductor that connects conductive patterns on opposite faces of the base.

interfacial junction — The junction that is formed by the faces of the two mating halves of a connector. This junction can be tightly compressed or loose, depending on the requirements of the application of the connector.

interfacial seal — Sealing of a mated connector pair over the whole area of the interface to provide sealing around each contact. This is usually done by providing a soft insert material on one or both halves of the connector, which are in compression when mated.

interfacing — 1. The joining of members of a group (people, instruments, and so on) in such a way that they are able to function in a compatible (synchronized) and coordinated fashion. 2. Interconnecting a program counter with its application devices and data terminals through various modules and cables. Interface modules convert program counter logic levels into external signal levels, and vice versa.

interference — 1. Any electrical or electromagnetic disturbance, phenomenon, signal, or emission, man-made or natural, which causes or can cause undesired response, malfunctioning, or degradation of the electrical performance of electrical and electronic equipment. 2. Any signal that degrades the accuracy of a system. Interfering signals may be separated into two classes: damaging and degrading. Damaging signals cause degradation of accuracy after the signal is removed (even permanently). 3. Any undesired electrical signal induced into a conductor by electrostatic or electromagnetic means. 4. The additive process whereby the amplitudes of two or more overlapping waves are systematically attenuated and reinforced. 5. The process whereby a given wave is split into two or more waves by, for example, reflection and refraction of beam splitters, and then possibly brought back together to form a single wave. 6. Extraneous energy that tends to interfere with the desired signal. 7. Unwanted occurrences on communication channels that result from natural or human-made noises and signals.

interference blanker — A device used with two or more pieces of radio or radar equipment to permit simultaneous operation without confusion of intelligence, or used with a single receiver to suppress undesired signals.

interference eliminator — A device designed for the purpose of reducing or eliminating interference.

interference fading — Fading produced by different wave components traveling slightly different paths in arriving at the receiver.

interference filter — A device added between a source of human-made interference and a radio receiver to attenuate or eliminate noise signals. It generally contains a combination of capacitance and inductance.

interference guard band — *See* guard band.

interference pattern — 1. The resultant space distribution of pressure, particle velocity, or energy flux when progressive waves of the same frequency and kind are superimposed. 2. The pattern produced on a radar scope by interference signals.

interference prediction — Estimation of the interference level of a particular item of equipment with respect to its future electromagnetic environment.

interference-rejection unit — Abbreviated IFRU. A tunable filter or wave trap capable of being adjusted to reject any frequency within the IF passband of a receiver while allowing the remainder of the passband curve to remain intact. It is adjusted to reject an interference signal and thus constitutes a form of antijamming.

interference source suppression — Techniques applied at or near a source of radiation to reduce its emission of undesired signals.

interference spectrum — The frequency distribution of the jamming interference in the propagation medium external to the receiver.

interferometer — An apparatus that shows interference between two or more wave trains coming from the same luminous area, and also compares wavelengths with observable displacements of reflectors or other parts.

interferometer homing — A homing guidance system in which the direction of the target is determined by comparing the phase of the echo signal as received at more than one antenna.

interferometer system — A method of determining the azimuth of a target through use of an interferometer to compare the signal phases at the output terminals of two antennas receiving a common signal from a distant source.

intergranular corrosion — A localized attack at metallic grain boundaries due to the presence of impurities and/or mechanical stress. On exposure to corrosive catalysis, the grain boundaries become anodic and the grains cathodic.

interior label — In a computer, a magnetically recorded sequence added to a tape to identify the contents.

interior-wiring-system ground — The ground connection to one of the current-carrying conductors of an interior wiring system.

interlace — 1. In a computer, to assign successive storage location numbers to physically separated storage locations on a magnetic drum. This serves to reduce access time. 2. To transmit different interrogation modes on successive sweeps.

interlaced scanning — Also called line interlace. 1. A system of scanning whereby the odd- and even-numbered lines of a picture are transmitted consecutively as two separate fields. These are superimposed to create one frame, or complete picture, at the receiver. The effect is to double the apparent number of pictures and thus reduce flicker. 2. A scanning process in which the distance from center to center of successively scanned

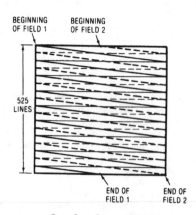

Interlaced scanning.

lines is two or more times the nominal line width, and in which the adjacent lines belong to different fields.

interlace factor — A measure of the degree of interlace of normally interlaced fields.

interlace operation — A type of computer operation in which data can be read out of or copied into the memory without causing interference to the other activities of the computer. *See also* interrupt; time sharing, 1.

interlacing — A method of scanning used in television, in which each picture is divided into two or more complete sets of interlacing lines to reduce flicker.

interlayer connection — An electrical connection between conductive patterns in different layers of multilayer printed board. *See* through connection.

interleave — 1. In a computer, to insert segments of one program into another program so that the two programs can be executed essentially simultaneously. 2. *See* interlace.

interleaving — 1. Placing between. For example, in the transmission of a composite color signal, the bands of energy of the chrominance signal are interleaved with, or placed between, those of the luminance signal. 2. Assigning successive memory locations to different physical memory modules.

interlock — A device actuated by the operation of some other device with which it is directly associated, to govern succeeding operations of the same or allied devices. Interlocks may be either electrical or mechanical.

interlock circuit — A circuit in which a given action cannot occur until after one or more other actions have taken place. The interlocking action is generally obtained through the use of relays.

interlocking — The forcing of a voltage of one frequency to be in step with a voltage of another frequency.

interlock relay — 1. A relay in which one armature cannot move or its coil be energized unless the other armature is in a certain position. 2. A relay with two sets of coils and respective armatures and contacts, so arranged that movement of one armature or energizing of its coil is dependent on the position of the other armature.

interlock switch — A safety switch that deenergizes a high-voltage supply when a door or other access cover is opened.

intermediate code — Machine input in a form between source and machine code; for example, pseudocode.

intermediate current — The range of current (milliamperes) at which formulation of carbonaceous material may significantly affect contact resistance.

intermediate fluxes — Fluxes consisting of mild organic acids and certain of their derivatives, such as the hydrohalides. As a class, they are weaker than the inorganic salt types.

intermediate frequency — Abbreviated IF. 1. A frequency to which a signal wave is shifted locally as an intermediate step in transmission or reception. 2. The fixed frequency resulting from heterodyning (i.e., beating or modulating to develop the sum or difference frequency signal) the incoming signal with a signal from the local oscillator. The IF used in FM tuners is commonly 10.7 MHz, and in AM tuners is 455 kHz. The IF signal is amplified in the IF channel, and it is here where most of the selectivity is introduced by tuned bandpass transformers and/or crystal or ceramic filters. 3. In superheterodyne receiving systems, the frequency to which all selected signals are converted for additional amplification, filtering, and eventual detection.

intermediate-frequency amplifier — 1. An amplifier tuned to a fixed frequency, or capable of single-control tuning over a range of frequencies, for the purpose of selecting one of the frequency components generated in a mixer circuit. 2. The central stages of a superheterodyne radio receiver that amplify the signals after they have been converted to a fixed intermediate frequency by a mixer (frequency converter).

intermediate-frequency harmonic interference — Interference caused in superheterodyne receivers by the radio-frequency circuit accepting harmonics of the intermediate-frequency signal.

intermediate-frequency interference ratio — *See* intermediate-frequency response ratio.

intermediate-frequency jamming — A form of jamming in which two cw signals are transmitted at frequencies separated by an amount equal to the center frequency of the IF amplifier in the radar receiver.

intermediate-frequency response ratio — Also called intermediate-frequency interference ratio. In a heterodyne receiver, the ratio of intermediate-frequency signal input at the antenna to the desired signal input for identical outputs.

intermediate-frequency strip — Also called IF strip. A subassembly containing the intermediate-frequency stages in a receiver.

intermediate-frequency transformer — Also called IF transformer. A transformer designed for use in the intermediate-frequency amplifier of a superheterodyne receiver.

intermediate-frequency transformer-lead color code — Transformer leads in many radio receivers are identified by the following standard EIA colors:

Blue: Plate   Green: Grid or diode
Red: B +   Black: Grid return
Green-black: Second diode (full-wave transformers only)

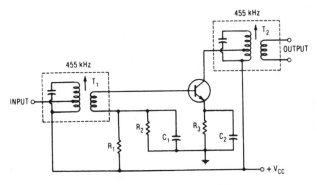

Intermediate-frequency amplifier.

intermediate horizon — A screening object (such as a hill, mountain, ridge, building, etc.) similar to the radar horizon but nearer to the radar site. For example, a distant mountain range might be the radar horizon on a given azimuth, while a nearer, lower ridge might screen a valley between it and the mountain range; the ridge would be an intermediate horizon.

intermediate means — All system elements needed to perform distinct operations in the measurement sequence between the primary detector and the end device.

intermediate repeater — A repeater used other than at the end of a trunk or line.

intermediate state — The partial superconductivity that occurs when a magnetic field of appropriate strength is applied to a sphere of material below its critical temperature (i.e., the temperature below which the material would superconduct if no magnetic field were present).

intermediate storage — The portion of the computer storage facilities in which information in the processing stage usually is stored.

intermediate subcarrier — A carrier used for modulating a carrier or another intermediate subcarrier. It also may have been modulated by one or more subcarriers.

intermediate switching region — In a relay, an area between low level (including dry circuit) and power switching (including full rated load) where the contact arc does not destroy deposits that are byproducts of the switching function.

intermediate trunk distributing frame — A frame in which are mounted terminal blocks for connecting linefinders and first selectors.

intermetallic bond — The ohmic contact made when two metal conductors are welded or fused together.

intermetallic compound — A compound of two or more metals that has a characteristic crystal structure that may have a definite composition corresponding to a solid solution, often refractory.

intermittent — 1. Occurring at intervals. 2. Electrical connections when conducting paths alternately open and close at some essentially uncontrolled rate. (Intermittents are undesirable since continuous connections are normally required.) 3. Not continuously present; disappearing and reappearing.

intermittent current — A unidirectional current that is interrupted at intervals.

intermittent defect — A defect that depends on variable conditions in a circuit. Hence, it is not present at all times.

intermittent duty — Operation for specified alternate intervals of load and no-load; load and rest; or load, no-load, and rest.

intermittent-duty rating — The output rating of a device operated for specified intervals rather than continuously.

intermittent-duty relay — A relay that must be deenergized at intervals to avoid excessive temperature, or a relay that is energized at regular or irregular intervals, as in pulsing.

intermittent pulsing — The transmission of short bursts of radiation at irregular intervals.

intermittent rating — The permissible output of a piece of apparatus when it is operated for alternate periods of load and rest that have a definite ratio to each other, or when it is run for a stated period of time that is not long enough to produce the final temperature.

intermittent reception — A defect in which the receiver operates normally for a while, at regular or irregular intervals.

intermittent scanning — One or two 360° scans of an antenna beam at irregular intervals to make detection by intercept receivers more difficult.

intermittent-service area — An area still receiving the ground wave of a broadcast station, but subject to interference and fading.

intermod — Abbreviation for intermodulation distortion (IMD). Interference that results when strong signals from a nearby transmitter mix with the desired signal in a radio receiver.

intermodulation — Sometimes called intermod. 1. In a nonlinear transducer element, the production of frequencies corresponding to the sums and differences of the fundamentals and harmonics of two or more frequencies transmitted through the transducer. 2. An error form that occurs in chopper-stabilized amplifiers when a beat component forms between the chopper drive frequency and normal signals that have frequencies near that of the chopper. 3. Mixing of two radio signals to produce a third signal that interferes with the reception of a desired signal.

intermodulation distortion — Abbreviated IMD. 1. Nonlinearity characterized by the appearance of frequencies in the output equal to the sums and differences of integral multiples of the component frequencies in the input signal. (Harmonics are usually not included.) 2. The introduction of unwanted signal energy as the result of interaction between two or more simultaneously reproduced tones, causing a smearing or veiling of the sound. All recording and amplifying equipment produces a certain amount of intermodulation distortion, but it can be held to sufficiently low levels to be below the threshold of audibility. 3. An analog-line impairment in which two frequencies interact to create an erroneous frequency, which in turn distorts the data-signal representation. 4. The production of spurious frequencies, not present in the original sound, that result from the interference or interaction of two (or more) sound signals that simultaneously occur in the original. These generally are sum and difference frequencies. For example, a 200-Hz and 75-Hz signal may occur at the same instant. If the equipment is prone to intermodulation distortion, these two may interact to produce a spurious 125-Hz tone. 5. The distortion caused by the addition of sum and difference modulation products when a complex wave (composed of two or more sine waves) passes through a nonlinear circuit. (When the modulation products are few, distortion is created; when they are many, noise is created.) 6. Distortion that results when two or more pure tones produce new tones with frequencies representing the sum and difference of the original tones and their harmonics.

intermodulation frequencies — The sum and difference frequencies generated in a nonlinear element.

intermodulation interference — The combination-frequency tones produced at the output by a nonlinear amplifier or network when two or more sinusoidal voltages are applied at the input. Generally expressed as the ratio of the root-mean-square voltage of one or more combination frequencies to that of one of the parent frequencies measured at the output.

intermodulation noise — Noise introduced in the channel of interest by signals being transmitted in other channels.

internal arithmetic — Any computations performed by the arithmetic unit of a computer, as distinguished from those performed by peripheral equipment.

internal calibration — Calibration by an internal voltage source (provided with the instrument) rather than an external standard.

internal connection — Abbreviated IC. In a vacuum tube, a base-pin connection designed not to be used for any circuit connections.

internal correction voltage — The voltage added to the composite controlling voltage of an electron tube.

It is the voltage equivalent of those effects produced by initial electron velocity, contact potential, etc.

internal graticule — A graticule whose rulings are a permanent part of the inner surface of the cathode-ray tube faceplate.

internal input impedance — The actual impedance at the input terminals of a device.

internally caused contact chatter — That chatter resulting from the operation or release of the relay. It may be classified as initial, armature-impact, armature-rebound, or armature-hesitation chatter.

internally stored program — A sequence of instructions (program) stored inside a computer in the same storage facilities as the computer data, as opposed to being stored externally on punched paper tape, pin boards, etc.

internal magnetic recording — Storage of information within the material itself, such as used in magnetic cores.

internal memory — Also called internal storage. The total memory or storage that is automatically accessible to a computer. It is an integral physical part of the computer and is directly controlled by it.

internal node — A junction between internal logic elements within an integrated circuit.

internal output impedance — The actual impedance at the output terminals of a device.

internal photoelectric effect — The creation of free electrons within a solid by the absorption of a sufficient amount of photons. The effect produces an increase in the conductivity of the solid.

internal resistance — 1. The effective series resistance in a source of voltage. 2. The resistance of a voltage source, such as a generator, battery or power supply, which acts to reduce the terminal voltage of the source as current is drawn.

internal storage — Also called main memory and core memory. 1. Storage facilities in a computer forming an integral physical part of and directly controlled by the computer. 2. The total storage automatically available to the computer. *See also* internal memory.

internal timer — In a computer, the internal clock equipped with multiple registers that can monitor the duration of external events or generate a pulse after a fixed time.

international broadcast station — A station licensed for transmission of broadcast programs for international public reception. By international agreement, frequencies are allocated between 6000 and 26,600 kHz.

international call sign — The identifying letters and numbers assigned to a radio station in accordance with the International Telecommunications Union. The first character, or the first two, identify the nationality of the station.

international code signal — A code, adopted by many nations for international communication, in which combinations of letters are used in lieu of words, phrases, and sentences.

international communication service — A telecommunication service between offices or stations (including mobile) belonging to different countries.

international control station — A fixed station in the Fixed Public Control Service, directly associated with the International Fixed Public Radio Communication Service.

international coulomb — The quantity of electricity passing any section of an electric circuit in 1 second when the current is 1 international ampere. One international coulomb equals 0.99985 absolute coulomb.

international farad — The capacitance of a capacitor when a charge of 1 international coulomb produces a potential difference of 1 international volt between the terminals. One international farad equals 0.99952 absolute farad.

International Fixed Public Radiocommunication Service — A fixed service, the stations of which are open to public correspondence, intended to provide radiocommunication between the United States or its territories and foreign points.

international henry — The inductance that produces an electromotive force of 1 international volt when the current is changing at a rate of 1 international ampere per second. One international henry equals 1.00018 absolute henrys.

international joule — The energy required to transfer 1 international coulomb between two points having a potential difference of 1 international volt. One international joule equals 1.00018 absolute joules.

International Morse code — Also called continental code. 1. A system of dot-and-dash signals used chiefly for international radio and wire telegraphy. It differs from American Morse code in certain code combinations only. 2. The Morse code that matches the English-language alphabet.

international ohm — The resistance at 0°C of a column of mercury of uniform cross section 106.300 centimeters in length and with a mass of 14.4521 grams. One international ohm equals 1.00048 absolute ohms.

International Radio Consultative Committee — Abbreviated CCIR. An international committee that studies technical operating and tariff questions pertaining to radio, broadcast television, and multichannel video transmissions and that issues recommendations. It reports to the International Telecommunications Union.

international radio silence — Three-minute periods of radio silence, commencing 15 and 45 minutes after each hour, on a frequency of 500 kHz only. During this time all radio stations are supposed to listen on that frequency for distress signals of ships and aircraft.

international system (of electrical and magnetic units) — A system for measuring electrical and magnetic quantities by using four fundamental quantities. Resistance and current are arbitrary values that correspond approximately to the absolute ohm and the absolute ampere. Length and time are arbitrarily called centimeter and second. The international system of electrical units was used between 1893 and 1947. By international agreement, it was discarded on January 1, 1948, in favor of the mksa (Giorgi) system.

international telecommunication service — A telecommunication service between offices or stations in different states, or between mobile stations that are not in the same state or that are subject to regulation by different states.

International Telecommunications Satellite Consortium — *See* INTELSAT.

International Telecommunication Union — The United Nations specialized agency that deals with telecommunications. Its purpose is to provide standardized communications procedures and practices, including frequency allocation and radio regulations on a worldwide basis.

International Telegraph Consultative Committee — Abbreviated CCIT. An international committee responsible for studying technical operating and tariff questions pertaining to telegraph and facsimile and issuing recommendations. It reports to the International Telecommunications Union.

international telephone address — A code not exceeding 12 digits that specifies a unique address for any telephone in the world. It consists of (*a*) a country or regional identity code of one, two, or three digits; (*b*) a three-digit numbering plan area code; (*c*) a two- or three-digit central office code; plus (*d*) a four-digit station number.

International Telephone Consultative Committee — Abbreviated CCIF. An international committee responsible for studying and issuing recommendations regarding technical operations and tariff questions pertaining to ordinary telephones; carrier telephones; and music, picture, television, and multichannel telegraph transmission over wire line. It reports to the International Telecommunication Union.

international temperature scale — A temperature scale adopted in 1948 by international agreement. Between the boiling point of oxygen (−182.97°C) and 630.5°C it is based upon the platinum resistance thermometer. From 630.5°C to 1063.0°C it is based on the platinum rhodium thermocouple, and above 1063.0°C on the optical pyrometer.

international volt — The voltage that will produce a current of 1 international ampere through a resistance of 1 international ohm. One international volt equals 1.00033 absolute volts.

international watt — The power expended when 1 international ampere flows between two points having a potential difference of 1 international volt. One international watt equals 1.00018 absolute watts.

Internet — A vast collection of many computer networks that span the entire world and communicate across dedicated high-speed phone lines using a single protocol family called TCP/IP. The network used for electronic mail, file transfers, chat services, and general data communications. It consists of a backbone connected via gateways to many smaller networks such as LANs and WANs, and thus each network can access every other network.

Internet gateway — An online service that allows access to the Internet.

Internet protocol — *See* IP.

Internet service provider — *See* ISP.

internetworking — The technique of connecting individual LANs to form a larger network.

internode — Communication paths that originate in one node and terminate in another.

interoffice trunk — The telephone channel between two central offices.

interoperability — The ability of computer equipment from different manufacturers to work together compatibly.

interphase transformer — An autotransformer or a set of mutually coupled reactors used with three-phase rectifier transformers to modify current relationships in the rectifier system and thereby cause a greater number of rectifier tubes to carry current at any instant.

interphone — A telephone communication system wholly contained within an aircraft, ship, or activity.

interphone system — An intercommunication system like that in an aircraft or other mobile unit.

interpolate — To estimate the values of a function that are intermediate between those already known.

interpolation — 1. The process of finding a value of a function between two known values. Interpolation may be performed numerically or graphically. 2. The process of determining a reading falling between two adjacent gradations on an analog meter scale.

interpole — A small auxiliary pole placed between the main poles of a direct-current generator or motor to reduce sparking at the commutator.

interposition trunk — A trunk connecting two positions of a large switchboard so that a line on one position can be connected to a line on the other position.

interpreter — 1. A system program that converts and executes each instruction of a high-level language program into machine code as it runs, before going on to the next instruction. An interpreted program is slow — as much as 20 times slower than an assembled program — but speeds up program development because the effect of source changes can be seen immediately. 2. A computer executive routine by which a stored program expressed in pseudocode is translated into machine code as a computation progresses, and the indicated operation is performed by means of subroutines as they are translated. 3. Programs that translate an assembled program into a complete machine-code listing on a line-by-line basis. If a statement is used in a program 10 times, it will be translated 10 times. 4. In a computer, a language translator that accepts high-level language (e.g., BASIC or Pascal) input text and translates this text into a special intermediate code that is simulated (interpreted) by a system program. Usually this intermediate code cannot be directly executed on a general-purpose processor. 5. A program that calls on subroutines to execute a program. Example: add two numbers, then divide by another number. BASIC is usually implemented as an interpreter. 6. A punch-card machine that will read the information conveyed by holes punched in a card and print its translation in characters arranged in specified rows and columns on the card.

interpreter code — A computer code that an interpretive routine can use.

interpretive programming — The writing of computer programs in a pseudo machine language, which the computer precisely converts into actual machine-language instructions before performing them.

interpretive routine — Computer routine designed to transfer each pseudocode and, using function digits, to set a branch order that links the appropriate subroutine into the main program.

interrecord gap — Also called interblock space. *See* IRG.

interrogate — 1. To determine the state of a device or circuit. 2. To retrieve information from computer files by use of predefined inquiries or unstructured queries handled by a high-level retrieval language.

interrogation — The triggering of one or more transponders by transmitting a radio signal or combination of signals.

interrogation signal — A pulsed or cw signal emitted to initiate a reply signal from a transponder or responder.

interrogation suppressed time delay — The overall fixed time that elapses between transmission of an interrogation and reception of the reply to this interrogation at zero distance.

interrogator — Also called challenger. A radio transmitter used to trigger a transponder.

interrogator-responser — A combined radio transmitter and receiver for interrogating a transponder and displaying the replies.

interrupt — 1. A processor feature that allows the currently executing program to be deferred in favor of servicing another. 2. A special instruction from a computer that temporarily disrupts normal operation and changes the normal flow of instruction execution. 3. In a computer, a break in the normal flow of a system or routine such that the flow can be resumed from that point at a later time. The source of the interrupt may be internal or external. 4. A method of stopping a process and identifying that a certain condition exists. In graphic systems, interrupts can originate from data entry devices, the display list, the host computer, the refresh clock, and display error conditions. When an interrupt occurs, the host computer and display refresh cease until the interrupt is answered and processed. At that time, the host computer will restart the refresh — usually from where it was halted.

If a new display list is to be presented, the display starts at the beginning of the list. 5. To stop a process in such a way that it can be resumed.

interrupted continuous waves — Abbreviated ICW. Continuous waves that are interrupted at an audio-frequency rate.

interrupt enable — *See* interrupt mark.

interrupter — 1. A magnetically operated device used for rapidly and periodically opening and closing an electric circuit in doorbells and buzzers and in the primary circuit of a transformer supplied from a dc source. 2. A device used to produce interrupted ringing cycles. It also may be employed with the release alarm to start signal alarm circuits of the switching equipment and thereby provide timed delay in the sounding of a failure alarm. 3. An electrical, electronic, or mechanical device that periodically interrupts a continuous current to produce pulses. 4. A device that produces an electrical output signal when an object breaks the light (visible or invisible) between a light source and a photodetector.

interrupter contacts — On a stepping relay, an additional set of contacts operated directly by the armature.

interrupting capacity — The maximum power in the arc that can be interrupted by a circuit breaker or fuse without the occurrence of restrike or violent failure. Rated in volt-amperes for ac circuits and in watts for dc circuits.

interrupting rating — 1. Conditions under which the contact of a relay must interrupt with a prescribed duty cycle and contact life. 2. The amount of current that a fuse can interrupt without the fuse cartridge rupturing or showing any external sign of damage.

interrupting time — In a circuit breaker, the interval between the energizing of the trip coil and the interruption of the circuit, at the rated voltage.

interrupt latency — The delay between an interrupt request and acknowledgment of the request.

interrupt mark — A mechanism that allows a program to specify whether or not an interrupt request will be accepted by the computer (sometimes called interrupt enable).

interrupt request — A signal to a computer that temporarily suspends the normal sequence of a routine and transfers control to a special routine. Operation can be resumed from this point later. Ability to handle interrupts is very useful in communication applications, where it allows the microprocessor to service many channels.

interrupt vector — 1. Typically, two memory locations assigned to an interrupting device and containing the starting address and processor status word for its service routine. 2. A function that facilitates fast handling of external interrupts by having the hardware supply a value corresponding to the device causing the interrupt. This value then becomes an index into the interrupt vector that contains a pointer to the appropriate interrupt service routine.

interstage — Between stages.

interstage coupling — Coupling between stages.

interstage transformer — A transformer that couples two stages together.

interstation noise suppression — Canceling of the noise that occurs when a high-gain radio receiver with automatic volume control is tuned between stations.

interstitial site — A position that is inside a crystal lattice but is not one of the proper sites ordinarily occupied by the atoms of the crystalline material. Impurity ions of the proper size can occupy such positions in a lattice that is otherwise regular.

intersymbol interference — 1. In a transmission system, extraneous signal energy during one or more keying intervals that tends to hinder the reception of the signal in another keying interval. 2. The disturbance that results from this condition.

interties — *See* interconnection, 1.

intertoll trunk — A trunk linking toll offices in different telephone exchanges.

interval — A period from one event to another. An interval timer controls the time a load is energized or deenergized.

interval calibration — *See* step calibration.

interval circuit — A circuit that is energized during timing only. This can be accomplished by using a timer with interval contacts, or by using a timer with delayed contacts in series with the start switch, or one with instantaneous contacts in series with delayed contacts.

interval contacts — In a timer, contacts that are actuated only for the duration of the preset time interval.

interval timer — 1. A device for measuring the time interval between two actions. 2. A timer that switches electrical circuits on or off for the duration of the preset time interval. 3. A hardware or software clock that generates an interrupt after a specified period has elapsed.

interword space — In telegraphy, the space between words or coded groups. It is equal to seven unit lengths.

intonation — The slight modification of pitch, or frequency, that makes a note sound flat or sharp compared with the natural frequency of the note played.

intracardiac — Pertaining to instruments whose pick-up element is inserted through a vein directly into the heart chambers.

intraconnections (microelectronics) — 1. Those conductors and connections that are in continuous integral contact with the substrate or circuit elements of an integrated circuit. 2. *See* interconnection.

intraoffice trunk — Telephone channel used to interconnect two customer lines within the same central office.

intrinsically safe — Incapable of releasing sufficient electrical energy to ignite a specific atmospheric mixture under normal conditions or such abnormal conditions as accidental damage to any part of the equipment or wiring insulation, failure of electrical components, application of overvoltage, or improper adjustment or maintenance operations.

intrinsic-barrier diode — A pin diode in which a thin region of intrinsic material separates the n-type and p-type regions.

intrinsic-barrier transistor — A pnip or npin transistor in which a thin region of intrinsic material separates the collector from the base.

intrinsic brightness — The luminous intensity measured in a given direction per unit of apparent (projected) area when viewed from that direction.

intrinsic characteristics — Characteristics of a material that depend on the material itself and do not result from impurities.

intrinsic coercive force — The magnetizing force that, when applied to a magnetic material in a direction opposite to that of the residual induction, reduces the intrinsic induction to zero.

intrinsic coercivity — The measurement (in oersteds) of the force required to reduce the intrinsic induction of the magnetized material to zero.

intrinsic concentration — In a semiconductor, the number of minority carriers that exceeds the normal equilibrium number.

intrinsic conduction — In an intrinsic semiconductor, the conduction associated with the directed movement of electron-hole pairs under the influence of an electric field.

intrinsic contact potential difference — The true potential difference between two spotlessly clean metals in contact.

intrinsic delay — *See* insertion delay.

intrinsic detector — A photodetector composed of a photoconductive material that, when exposed to radiation, conducts without the aid of added impurities and does not have to be cooled to the level of extrinsic material. Examples of intrinsic materials are silicon and cadmium sulfide.

intrinsic electric strength — The characteristic electric strength of a material.

intrinsic flux — The product of the intrinsic flux density and the cross-sectional area of a uniformly magnetized sample of material.

intrinsic flux density — In a sample of magnetic material, the excess of the normal flux density over the flux density in a vacuum for a given magnetizing field strength. In the cgs system, the intrinsic flux density is numerically equal to the difference between the ordinary flux density and the magnetizing field strength.

intrinsic hysteresis loop — A curve that shows intrinsic flux density as a function of magnetizing field strength, where the magnetizing field is cycled between equal values of opposite polarity. Hysteresis is indicated by the fact that the ascending and descending portions of the curve do not coincide.

intrinsic induction — The excess magnetic induction produced in a magnetic material by a given magnetizing force, over the induction that would be produced by the same magnetizing force in a vacuum.

intrinsic insulation — Any method of isolating fields or current within a semiconductor or other substrate. Among these are thermal oxidation, fabrication of an isolating layer of intrinsic semiconductor material, and creation of one or more back-based functions.

intrinsic-junction transistor — *See* intrinsic-region transistor.

intrinsic layering — The method of separating two regions of conductive semiconductor by a region of near-intrinsic semiconductor material. This material differs enough in resistivity from the adjacent regions to serve as an insulator.

intrinsic material — A semiconductor material in which there are equal numbers of holes and electrons, i.e., no impurities.

intrinsic mobility — The mobility of electrons in an intrinsic semiconductor or in a semiconductor having a very low concentration of impurities.

intrinsic noise — Noise that is due to the device or transmission path and is independent of modulation.

intrinsic permeability — Ratio of intrinsic normal induction to the corresponding magnetizing forces.

intrinsic photoconductivity — The absorption of a photon raising an electron across the forbidden gap from valence to conduction band of the semiconductor, whereby conductivity is increased and incident radiation may be measured.

intrinsic photoemission — The photoemission that would occur if a crystal were pure and its structure perfect.

intrinsic properties — The semiconductor properties that are characteristic of the pure, ideal crystal.

intrinsic Q — *See* unloaded Q (switching tubes).

intrinsic region — *See* i region; depletion region.

intrinsic-region transistor — Also called intrinsic-junction transistor. A four-layer transistor with an intrinsic region between the base and collector. Examples are npin, pnip, npip, and pnin transistors.

intrinsic reliability — The probability that a device will perform its specified function, determined by statistical analysis of the failure rates and other characteristics of the parts and components the device comprises.

intrinsic semiconductor — 1. A semiconductor in which some hole and electron pairs are created by thermal energy at room temperature, even though there are no impurities in it. 2. A semiconductor with substantially the same electrical properties as those of the ideal crystal. 3. An element or compound that has the same electron energy band configuration as an insulator, but has a forbidden energy gap that is sufficiently narrow to permit transfer of electrons from the valence band to the conduction bands at normal temperatures. Conduction in an intrinsic semiconductor takes place via equal number of conduction band electrons and valence band holes.

intrinsic standoff ratio — In a unijunction transistor, the difference between the emitter voltage at the peak point with a specific interbase voltage and the forward voltage drop of the emitter junction, divided by the voltage on base 2 with respect to base 1.

intrinsic temperature range — The temperature range at which impurities or imperfections within the crystal do not modify the electrical properties of a semiconductor.

intrusion alarm system — An alarm system for signaling the entry or attempted entry of a person or an object into the area or volume protected by the system.

Invar — An alloy containing 63.8 percent iron, 36 percent nickel, and 0.2 percent carbon. Has a very low thermal coefficient of expansion. Used primarily as resistance wire in wirewound resistors.

inverse beta — The transistor gain that results when the emitter and collector loads are physically reversed in the operation of a circuit.

inverse common base — Transistor circuit configuration in which the base terminal is common to the input circuit and to the output circuit and in which the input terminal is the collector terminal and the output terminal is the emitter terminal.

inverse common collector — Transistor circuit configuration in which the collector terminal is common to the input circuit and to the output circuit and in which the input terminal is the emitter terminal and the output terminal is the base terminal.

inverse common emitter — Transistor circuit configuration in which the emitter terminal is common to the input circuit and to the output circuit and in which the input terminal is the collector terminal and the output terminal is the base terminal.

inverse direction of operation — A mode of operating a transistor in which the nominal collector region acts as an emitter and in which the net flow of minority carriers is from the nominal collector region to the base region.

inverse electrical characteristics — In a transistor, those characteristics obtained when the collector and emitter terminals are interchanged and the transistor is then tested in the normal manner.

inverse electrode current — The current flowing through an electrode in the opposite direction from that for which the tube was designed.

inverse feedback — *See* negative feedback.

inverse-feedback filter — A tuned circuit at the output of a highly selective amplifier having negative feedback. The feedback output is zero for the resonant frequency, but increases rapidly as the frequency deviates.

inverse Fourier transform — A mathematical operation that synthesizes a time-domain signal from its complex spectrum components. If a time-domain signal is Fourier transformed and then inverse Fourier transformed, the original time function is reconstructed.

inverse limiter — A transducer with a constant output for inputs of instantaneous values within a specified range. Above and below that range, the output is linear or some other prescribed function of the input.

inverse networks — Any two two-terminal networks in which the product of their impedances is independent of frequency within the range of interest.

inverse neutral telegraph transmission — A form of transmission in which zero-current intervals are used as marking signals, and current pulses of either polarity are used as spacing signals.

inverse-parallel connection — *See* back-to-back circuit.

inverse peak voltage — The peak instantaneous voltage across a rectifier tube during the nonconducting half-cycle.

inverse photoelectric effect — The transformation of the kinetic energy of a moving electron into radiant energy, as in the production of X-rays.

inverse piezoelectric effect — Contraction or expansion of a piezoelectric crystal under the influence of an electric field.

inverse ratio — The seesaw effect whereby one value increases as the other decreases or vice versa.

inverse-square law — 1. The strength of a field, or the intensity of radiation, decreases in proportion to the square of the distance from its source. 2. The law stating that the illuminance (or irradiance) from a point source varies as the inverse square of the distance between the source and the receiver.

inverse time — A qualifying term applied to a relay, indicating that its time of operation decreases as the magnitude of the operating quantity increases.

inverse voltage — The effective voltage across a rectifier tube during the half-cycle when current does not flow.

inverse Wiedemann effect — *See* Wiedemann effect.

inversion — 1. The bending of a radio wave because the upper part of the beam is slowed down as it travels through denser air. This may occur when a body of cold air moves in under a moisture-laden body of air. 2. The producing of inverted or scrambled speech by beating an audio-frequency signal with a fixed band of the resultant beat frequencies. The original low audio frequencies then become high frequencies and vice versa.

inversion layer — A layer of doped semiconductor material that has changed to the opposite type, such as a p layer at the surface of an n-doped region. Surface inversion layers may be the result of surface ions or dopant gettering by surface passivation material or the action of induced electric fields. *See also* channel, 6.

inverted amplifier — An amplifier stage containing two vacuum tubes. The control grids are grounded, and the driving excitation is applied between the cathodes. The grids then serve as a shield between the input and output circuits. Thus, the output-circuit capacitance is greatly reduced.

inverted-L antenna — An antenna consisting of one or more horizontal wires with a vertical wire connected at one end.

inverted speech — *See* scrambled speech.

inverter — 1. A circuit that takes in a positive signal and puts out a negative one, or vice versa. 2. A device that changes alternating current to direct current or vice versa. It frequently is used to change 6-volt or 12-volt direct current to 110-volt alternating current. 3. A device that accepts an input that is a function of the maximum voltage and changes it into an output that is a function of both the maximum voltage and time. 4. A circuit with one input and one output, and whose function is to invert the

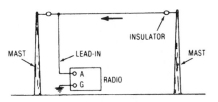

Inverted-L antenna.

input. When the input is high, the output is low, and vice versa. The inverter is sometimes called a NOT circuit, since it produces the reverse of the input. 5. A device or circuit that complements a Boolean function.

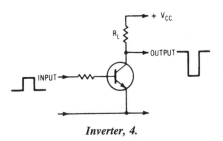

Inverter, 4.

inverter circuit — *See* NOT circuit.

inverting amplifier — An amplifier in which the output polarity is the opposite of the input polarity. Such an amplifier obtains negative feedback through a connection from the output to the input, and with high gain it is widely used as an operational amplifier.

inverting connection — The closed-loop connection of an operational amplifier when the forward gain is negative for dc signals. A 180° phase shift.

inverting input — An input terminal of a differential amplifier that produces an output signal of opposite phase (shifted 180°) than that of the input signal.

inverting parametric device — A parametric device whose operation depends essentially upon three frequencies: a harmonic of the pump frequency, and two signal frequencies, of which the higher signal frequency is the difference between the pump harmonic and the lower signal frequency.

invisible file — *See* hidden file.

invister — A high-frequency, high-transconductance unipolar structure made by means of lateral diffusion.

inward-outward dialing system — A dialing system by means of which calls within the local exchange area may be dialed directly to or from base private branch exchange telephone stations without the assistance of an operator at the base private branch exchange.

inward WATS — Inward wide area telephone service that permits a customer, for a monthly charge, to receive incoming station-to-station calls from telephones within prescribed service areas or in six interstate bands, without charge to the calling party. A telephone service similar to WATS but applicable to incoming calls only.

I/O or **i/o** — Abbreviation for input/output. 1. The components of a computer system's architecture that control the flow of data and instructions to and from the CPU. 2. Any equipment that introduces data into or extracts data from a data communications system.

IOC — *See* integrated optical circuit.

I/O device — Abbreviation for input/output device. 1. A disk drive, magnetic tape unit, printer, or similar device that transmits data to or receives data from a computer or secondary storage device. 2. A general term for equipment used to communicate with a computer. 3. A device that supplies data and instructions to the computer and receives the computer's output, or a circuit that interfaces with such a device. 4. Input/output equipment used to communicate with a system.

I/O electrical isolation — Separation of the field wiring circuits from the logic level circuits of the PC, typically done with optical isolation.

I/O modules — General-purpose circuit modules that interface a microprocessor to relays, switches, and transducers. Each module contains almost all necessary interface circuitry in one package either to sense a load, control a switch, or condition a transducer signal so that practically no other interface design is required.

ion — 1. An atom that has become charged by gaining or losing one or more orbiting electrons. A completely ionized atom is stripped of all electrons. 2. An atom or molecule with an electrostatic charge. 3. The charged particle formed when one or more electrons are taken from or added to a previously neutral atom or molecule. An atom that has gained additional electrons is thus a negative ion, whereas an atom that has lost an electron or electrons is a positive ion.

ion charging — In a storage tube, spurious charging or discharging caused by ions striking the storage surface.

ion counter — A tubular chamber for measuring the ionization of air.

ion-exchange electrolyte cell — A fuel cell that uses the reaction of hydrogen with oxygen from the air. It is similar to the standard hydrogen-oxygen fuel cell, except that an ion-exchange membrane replaces the liquid electrolyte. Operation is at atmospheric pressure and room temperature.

ionic conductivity — The transit phenomenon that occurs as a result of positive and/or negative charge movement in an electrolyte when placed in an electric field.

ionic focusing — Focusing the electron beam in a cathode-ray tube by varying the filament voltage and temperature to change the electrostatic focusing field automatically produced by the accumulation of positive ions in the tube.

ionic-heated cathode — A cathode that is heated primarily by bombardment with ions.

ionic-heated cathode tube — An electron tube containing an ionic-heated cathode.

ionic tweeter — A type of speaker in which a varying electrostatic field activates a mass of air ionized by a high-voltage radio-frequency field. Ionic speakers are capable of extremely extended high-frequency response (up to 100 kHz or so) because of the extreme lightness of the ionic "diaphragm."

ion implantation — 1. A precise and reproducible method of semiconductor doping in which selected dopants are ionized and accelerated at high velocity to penetrate the semiconductor substrate and become deposited below the surface. Charged atoms (ions) of elements such as boron, phosphorus, or arsenic are accelerated by an electric field into the semiconductor material. This technique ensures uniform, accurately controlled depth of implantation and ionic diffusion in the water. 2. A processing step by which standard p-channel diffused MOS devices are made directly compatible with TTL/DTL logic. It is a highly controllable process that allows the adjusting of gate threshold voltages and also allows the fabrication of both enhancement-mode and depletion-mode transistors on the same chip. 3. A process that uses accelerated atoms to implant source and drain regions in metal-oxide semiconductors. It offers higher

speed and lower threshold voltages and can also be used with PMOS, NMOS, and CMOS.

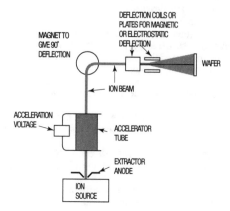

Ion implantation.

ion-implanted MOS — Abbreviated IMOS. A method for doping substrates with a stream of ionized dopant atoms. Ions are electrically shot into the substrate instead of diffusing atoms at high temperatures.

ionization — 1. The dissociation of inert-gas atoms into positive and negative ions in an electric field, resulting in the emission of light. 2. The state of an insulator in which it facilitates the passage of current because of the presence of charged particles (usually induced artificially). 3. The electrically charged particles produced by high-energy radiation (such as light or ultraviolet rays) or by the collision of particles during thermal agitation. 4. The formation of ions. The process of giving a net charge to a neutral atom or molecule by adding or subtracting an electron. This can be accomplished by radiation or by a strong electric field.

ionization arcover — 1. Formation of an electrical arc between terminals or contacts as a result of ionization of the adjacent air or gas. 2. Formation of an arc between the terminals of a satellite antenna as the satellite passes through the ionized regions of the ionosphere.

ionization chamber — 1. An enclosure containing two or more electrodes between which an electric current may pass when the gas within is ionized. The current is a measure of the total number of ions produced in the gas by externally induced radiations. 2. A chamber containing a gas through which ionizing particles pass. A voltage is applied across the chamber so as to collect the ions produced and permit the ion current to be measured.

ionization current — The current resulting from the movement of electric charges, under the influence of an applied electric field, in an ionized medium.

ionization energy — Sometimes called ionization potential. The minimum amount of energy (usually expressed in electronvolts) required to eject an electron from a molecule.

ionization factor — The difference between percent power factors of a dielectric at two specified values of electrical stress. The lower of the two stresses is usually selected so that the effect of ionization on power factor at this stress is negligible.

ionization gage — 1. A gage that measures the degree of a vacuum in an electron tube by the amount of ionization current in the tube. 2. A type of radiation detector that depends on the ionization produced in a gas by the passage of a charged particle through it. One of the best known is the Geiger-Mueller counter. Cloud chambers and spark chambers can also be included in this category.

ionization-gage tube — An electron tube that measures low gas pressure by the amount of ionization current produced.

ionization potential — 1. The energy, expressed in electronvolts, needed to remove one electron from a neutral atom or molecule in its ground state. 2. The amount of energy, for a particular kind of atom, required to remove an electron from the atom to infinite distance. The ionization potential is usually expressed in volts.

ionization pressure — An increase in the pressure in a gaseous discharge tube due to ionization of the gas.

ionization resistance — *See* corona resistance.

ionization smoke detector — A smoke detector in which a small amount of radioactive material ionizes the air in the sensing chamber, thus rendering it conductive and permitting a current through the air between two charged electrodes. This effectively gives the sensing chamber an electrical conductance. When smoke particles enter the ionization area, they decrease the conductance of the air by attaching themselves to the ions, causing a reduction in mobility. When the conductance is less then a predetermined level, the detector circuit responds.

ionization time — 1. The time interval between the initiation and the establishment of conduction in a gas tube at some stated voltage drop for the tube. 2. The elapsed time to achieve normal glow after a voltage greater than the breakdown voltage is applied to a glow lamp.

ionization transducer — A transducer in which displacement of the force-summing member is sensed by the induced changes in differential ion conductivity.

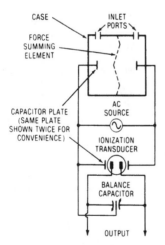

Ionization transducer.

ionization vacuum gage — A gage in which the operation depends on the positive ions produced in a gas by electrons as they accelerate between a hot cathode and another electrode in a vacuum. It ordinarily covers a pressure range of 10^{-4} to 10^{-10} mm of mercury.

ionization voltage (corona level) — The minimum value of falling rms voltage that sustains electrical discharge within the vacuous or gas-filled spaces in the cable construction or insulation.

ionize — To free an electron from an atom or molecule (e.g., by X-ray bombardment) and thus transform the atom

or molecule into a positive ion. The freed electron attaches itself to another atom or molecule, which then becomes a negative ion.

ionized layers — Layers of increased ionization within the ionosphere. They are responsible for absorption and reflection of radio waves and are important for communication and for tracking satellites and other space vehicles.

ionizing event — Any interaction by which one or more ions are produced.

ionizing radiation — 1. Radiant electromagnetic energy and high-energy particles that cause the division of a substance into parts carrying positive and negative charges. High-energy particles can directly ionize substances, whereas electromagnetic radiation sets in motion charged particles that then produce ions. 2. Generally, any radiation having sufficient energy to dislodge electrons from atoms or molecules while traveling through a substance, thereby producing ions.

ion migration — 1. Movement of the ions produced in an electrolyte by application of an electric potential between electrodes. 2. The movement of free ions within a material or across the boundary between two materials under the influence of an applied electric field.

ionophone — A high-frequency speaker in which the audio-frequency signal modulates an rf supply to maintain an arc in the mouth of a quartz tube. The resultant modulated wave acts directly on the ionized air under pressure and thus creates sound waves.

ionosphere — The part of the earth's outer atmosphere where sufficient ions and electrons are present to affect the propagation of radio waves.

ionospheric absorption — The loss of energy of a radio wave as it travels through the atmosphere, setting particles in motion. The energy needed to create this motion is equal to the loss.

ionospheric disturbance — 1. The variation in the state of ionization of the ionosphere beyond the normal observed random day-to-day variation from the average value for the location, date, and time of day under consideration. 2. Change in the part of the earth's outer atmosphere (ionosphere) that affects transmission and reception of radio signals.

ionospheric D scatter meteor burst — A phenomenon in which the penetration of meteors through the D region of the ionosphere affects ionospheric scatter communications.

ionospheric error — Also called sky error. In navigation, the total systematic and random error resulting from reception of the navigational signal after it has been reflected from the ionosphere. It may be due to variations in the transmission path, uneven height of the ionosphere, or uneven propagation within the ionosphere.

ionospheric prediction — The forecasting of ionospheric conditions and the preparation of radio propagation data derived from it.

ionospheric scatter — *See* forward scatter.

ionospheric storm — An ionospheric disturbance associated with abnormal solar activity and characterized by wide variations from normal, including turbulence in the F region and increases in absorption. Often the ionization density is decreased and the virtual height is increased. The effects are most marked in high magnetic latitudes.

ionospheric wave — Also called a sky wave. A radio wave that is propagated by way of the ionosphere.

ion sheath — A positive-ion film that forms on or near the grid of a gas tube and limits its control action.

ion spot — 1. In camera or image tubes, the spurious signal resulting from bombardment of the target or photocathode by ions. 2. On a cathode-ray-tube screen, an area where the luminescence has been deteriorated by prolonged bombardment with negative ions. 3. A spot on the fluorescent surface of a cathode-ray tube that is somewhat darker than the surrounding area because of bombardment by negative ions, which reduce the sensitivity.

ion trap — Also called a beam bender. 1. An electron-gun structure and magnetic field that diverts negative ions to prevent their burning a spot in the screen, but permits electrons to flow toward the screen. 2. An arrangement of magnetic fields and apertures that will allow an electron beam to pass through but will obstruct the passage of ions.

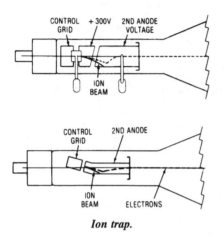

Ion trap.

I/O rack — A chassis that contains I/O modules.

I/O scan — The time required for the PC processor to monitor all inputs and control all outputs. The I/O scan repeats continuously.

IP — Abbreviation for Internet Protocol. The Internet standard protocol that provides a common layer over dissimilar networks, used to move packets among host computers and through gateways if necessary.

I_p — Symbol for the plate current of a vacuum tube.

I-phase carrier — Also called in-phase carrier. A carrier separated in phase by 57° from the color subcarrier.

ips — Abbreviation for inches per second. Used for specifying the speed of a tape traveling past the heads of a tape recorder. The most common speeds are $1\frac{7}{8}$ ips (4.75 cm/s); $3\frac{3}{4}$ ips (9.5 cm/s) and $7\frac{1}{2}$ ips (19 cm/s).

IR — 1. Abbreviation for interrogator response. 2. Abbreviation for infrared. 3. Abbreviation for insulation resistance. That resistance offered by an insulation to an impressed dc voltage, tending to produce a leakage current through the insulation.

I^2R — Power in watts expressed in terms of the current (I) and resistance (R).

IRAC — Acronym for Interdepartmental Radio Advisory Committee. It is composed of representatives of eleven government agencies: The FCC; Army; Navy; Air Force; Maritime Commission; and the Treasury, State, Commerce, Agriculture, Interior, and Justice Departments.

IR compensation — A control device that compensates for voltage drop due to current flow.

IR drop — 1. The voltage produced across a resistance (R) when there is a current (I) through the resistor. 2. The voltage drop that exists across a resistance when a current is flowing through it.

IRE — Abbreviation for Institute of Radio Engineers, an organization now merged with AIEE. *See* IEEE.

IRED — An infrared-emitting diode that emits photons in the infrared spectrum when forward biased. *See also* light-emitting diode.

i region — Abbreviation for intrinsic region. In silicon, a pure region of the group IV element (i.e., having neither excess holes or excess electrons, and therefore having very high resistivity).

IRG — Abbreviation for interrecord gap. 1. Erased area between records that allows stop/start and speed standardization when writing or reading data blocks on magnetic tape. 2. The space between records on magnetic tape caused by delays involved in starting and stopping the tape motion. This gap is used to signal that the end of a record has been reached.

iridescence — The rainbow exhibition of colors in a body, usually caused by interference of light of different wavelengths reflected from superficial layers in the surface of a material.

iris — Also called diaphragm. 1. In a waveguide, a conducting plate (or plates) that is very thin (compared with the wavelength) and occupies part of the cross section of a waveguide. When only a single mode can be supported in the waveguide, an iris appears substantially as a shunt admittance. 2. An adjustable aperture built into a camera lens to permit control of the amount of light passing through the lens.

iris diaphragm — A simple mechanism used to vary the diameter of an aperture. Consists of a number of thin, arc-shaped, metal blades that surround the aperture, each blade having a lower stud at one end and an upper stud at the other end. The lower studs fall into holes in a fixed ring surrounding the aperture, while the upper studs are held by radial slots in a rotatable control ring. Iris openings are measured in f-stops.

I^2R loss — The power lost in transformers, generators, connecting wires, and other parts of a circuit because of the current I through the resistance R of the conductors.

iron constantan — A combination of metals used in thermocouples, thermocouple wires, and thermocouple lead wires. Constantan is an alloy of copper, nickel, manganese, and iron. The iron wire is the positive, the constantan the negative wire.

iron-core coil — A coil in which iron forms part or all of the magnetic circuit, linking its winding. In a choke coil, the core is usually built up of laminations of sheet iron.

iron-core transformer — A transformer in which iron forms part or all of the magnetic circuit, linking transformer windings.

ironless rotor motor — A motor that has a stationary permanent magnet located in the center of the rotor. The housing is made of soft steel and provides a return path for magnetic flux. Rotor coils are wound on a non-magnetic cage. The rotating member consists only of the hollow cage, the winding shaft, and commutator. Advantages include the elimination of cogging, lower starting voltages, and smoother starting and running. The motors are lightweight, hence inertia is low, providing a high torque-to-inertia ratio and high acceleration. The main disadvantage is that the complex construction is more expensive.

iron loss — *See* core loss.

iron-vane instrument — An indicating instrument whose operating portion consists of two iron bars, one fixed, one pivoted, placed parallel to each other inside a signal coil. Current through the coil magnetizes the bars in the same direction, and they repel each other, causing the pointer to pivot against the force of a hairspring. A damping vane may be used to slow the movement of the pointer. Deflection is the same for ac or dc; the meter does not have polarity. The instrument has a nonlinear scale, and readings below ⅕ scale are extremely difficult to make. Because of inductance effects, use of this type of meter is limited to power-line frequencies.

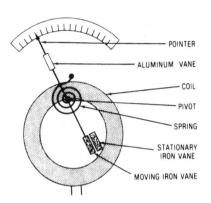

Iron-vane instrument.

irradiance — The incident radiated power per unit area of a surface; the radiometric counterpart of illumination, usually expressed in watts per square centimeter.

irradiation — 1. The application of X-rays, radium rays, or other radiation. 2. The amount of radiant energy per unit area received during a given time interval. This term is used in radiological therapy as well as in describing accidental exposure to radiation. It also can be used to denote radiant energy that ultimately passes through the skin to supply power to implanted electronic devices. 3. The exposure of a material to high-energy emissions. In insulations, done for the purpose of favorably altering the molecular structure.

irregularity — A change from normal.

ISA — Abbreviation for Industry Standard Architecture. The original bus architecture developed by IBM for its AT computer and opened up for use by other manufacturers.

I-scan — A radar display in which a target appears as a complete circle when the radar antenna is correctly pointed at it, the radius of the circle being proportional to the target distance.

I-scope — *See* I-scan.

ISDN — Abbreviation for integrated services digital network. 1. A digital telephone service network that allows transmission of data, voice, and images over one phone line simultaneously. It can provide speeds of roughly 128,000 bits per second over regular phone lines. In practice, most users are limited to 56,000 or 64,000 bits per second. 2. Telephone technology that provides digital access to voice and data network services for sending speech, data, and video across one line simultaneously.

ISHM — The International Society of Hybrid Microelectronics.

I signal — Also called the fine-chrominance primary. 1. A signal formed by the combination of +.74 of an R − Y signal and −0.27 of a B − Y signal. One of the two signals used to modulate the chrominance subcarrier, the other being the Q signal. 2. Also known as the in-phase signal. The color sidebands produced by modulating the color subcarrier at a phase 57° removed from the burst reference phase. This signal is capable of reproducing the range of colors from orange to cyan (bluish-green).

ISM equipment — A Federal Communications Commission designation for industrial, scientific, and medical equipment.

isobar — 1. On meteorological maps, a line denoting places having the same atmospheric pressure at a given time. 2. One of a group of atoms or elements having the same atomic weights but different atomic numbers.

isochromatic — Having the same color, as with the lines of the same tint in the interference figure of anisotropic crystals.

isochrone — On a map or chart, a line joining points associated with a constant time difference in the reception of radio signals.

isochrone determination — A radio location in which a position line is determined by the difference in transit times of signals along two paths.

isochronous — 1. Equal in length of time. 2. Occurring at equal intervals of time. 3. Describes modems, terminals, and transmissions in which all bits are of equal duration. There are no start or stop bits as in asynchronous transmissions, and no clocking signals as in synchronous transmissions.

isochronous circuits — Circuits having the same resonant frequency.

isochronous distortion — A measure of a modem's dynamic operation.

isochronous multiplexer — A multiplexer that can interleave two time-independent data streams into one higher-speed stream independent of the master timing control required by a synchronous multiplexer.

isoclinic line — *See* aclinic line.

isodynamic lines — On a magnetic map, lines passing through points of equal strength of the earth's magnetic field.

isoelectric — Uniformly electric throughout or having the same electric potential, and therefore producing no current.

isoelectronic — Pertaining to atoms having the same number of electrons outside the nucleus of the atom.

isolated — Utterly cut off; refers to that condition in which a conductor, circuit, or device is not only insulated from another (or others), but the two are mutually unable to engender current, emf, or magnetic flux in each other. As commonly used, insulation is associated predominantly with direct current, whereas isolation implies additionally a bulwark against ac fields.

isolated amplifier — A differential amplifier in which the input-signal lines are conductively isolated from the output-signal lines and chassis ground.

isolated I/O module — A module that has each input or output electrically isolated from every other input or output on that module, i.e., each input or output has a separate return wire.

isolating diode — A diode that passes signals in one direction through a circuit but blocks signals and voltages in the opposite direction.

isolating switch — A switch intended for isolating an electric circuit from the source of power. It has no interrupting rating and is intended to be operated only after the circuit has been opened by some other means.

isolation — 1. Electrical or acoustical separation between two locations. 2. The technique for producing a high electrical resistance between an integrated-circuit component and the substrate in which it is formed. 3. A reduction in the ability of a system to respond to an excitation or to generate an excitation. 4. A method for implementing electrical independence of devices integrated on the same integrated circuit.

isolation amplifier — Also called buffer amplifier. 1. An amplifier that is used to minimize the interaction between the circuitry which precedes and that which follows. 2. An amplifier designed to have a galvanic discontinuity between its input and output pins. This discontinuity (called an isolation barrier) must have high breakdown voltage, low dc leakage (high barrier resistance), and low ac leakage (low barrier capacitance). Three-port isolation amplifiers have an additional isolation barrier between the power-supply connection and the signal connections that allows the user to connect power in common with either the amplifier's input or its output. Isolation amplifiers generally serve the following functions not achievable with operational or instrumentation amps: sensing small signals in the presence of very high (10 volts) or unknown common-mode voltages, protecting patients undergoing medical monitoring or diagnostic measurements, and completely breaking ground loops. The isolation amplifier, as well as offering isolation, increases accuracy because of its floating input. In contrast to the instrumentation amplifier, it not only eliminates ground loop errors but further reduces the total system error, because its isolation-mode rejection ratio is generally one or two orders of magnitude higher than the common-mode rejection of an instrumentation amplifier. 3. An amplifier with input circuitry and output circuitry designed to eliminate the effects of changes made at either on the other. 4. An amplifier that has an input circuit that is galvanically (no ohmic connection) isolated from the output stage and the power-supply terminals. This isolation is provided by magnetic, optical, or mechanical coupling techniques. Isolation amplifiers are used in applications that require accurate and safe measurement of dc and low-frequency signals. 5. A circuit that amplifies a signal without needing a galvanic path between its input and output terminal. The circuit can be used to protect individuals and equipment from high voltages and to break ground loops in measurement systems. The high resistance and low capacitance of the isolation region permit little current leakage across its barrier and it withstands a specified high voltage without breakdowns or arcs.

isolation barrier — *See* isolation amplifier.

isolation diffusion — 1. A technique for separation of the individual components within a monolithic silicon n structure; p diffused isolation zones are used to form pn junctions that act as reverse-biased diodes. The transistors are double diffused; that is, they are processed by two diffusion steps after the isolation diffusion. 2. In monolithic integrated circuit technology, the diffusion step that generates back-to-back junctions to isolate active devices from one another.

isolation network — A network inserted into a circuit or transmission line to prevent interaction between circuits on each side of the insertion point.

isolation transformer — A transformer designed to provide magnetic coupling (flux coupling) between one or more pairs of isolated circuits, without introducing significant coupling of any other kind between them — i.e., without introducing either significant conductive (ohmic) or significant electrostatic (capacitive) coupling.

isolator — 1. A device that permits microwave energy to pass in one direction while providing high isolation to reflected energy in the reverse direction. Used primarily at the input of communications-band microwave amplifiers to provide good reverse isolation and minimize VSWR. Consists of microwave circulator with one port (port 3) terminated in the characteristic impedance. 2. A device that acts as a one-way valve for microwave signals to prevent stray receiver signals from leaking out past the LNA onto the antenna. 3. A device that allows the transmission of signals in one direction while blocking or attenuating them in the other.

isolator ferrite — A microwave device that allows rf energy to pass through in one direction with very little loss but that absorbs rf power in the other direction.

isolith — An integrated circuit of components formed on a single silicon slice, but with the various components interconnected by beam leads and with circuit parts isolated by removal of the silicon between them.

isomer — One of two or more substances composed of molecules having the same kinds of atoms in the same proportions but arranged differently. Hence, the physical and chemical properties are different.

isoplanar — A bipolar fabrication process that replaces conventional planar p+ isolation diffusion with an insulating oxide to provide isolation between active elements of a silicon integrated circuit. Circuit elements can be fabricated in less space than conventional isolation techniques with improved speed and power performance.

isopulse system — In adaptive communications, a pulse coding system in which special inserted pulses indicate the number of information pulses that are transmitted.

isostatic — Being subjected to equal pressure from every side.

isothermal region — The stratosphere considered as a region of uniform temperature.

isotones — A group of atoms whose nuclei have the same number of neutrons.

isotope — A species of matter whose atoms contain the same number of protons as some other species, but a different number of neutrons. The atomic numbers of isotopes are identical, but the mass numbers (atomic weights) differ. *See* radioisotope.

isotropic — 1. Having properties with the same values along axes in all directions. 2. Term applied to substances having certain properties that are manifest in every direction, e.g., electrical conductivity in metals. 3. The ability to react the same regardless of direction of measurement. Isotropic materials will react consistently even if stress is applied in different directions. Stress/strength ratio is uniform throughout material.

isotropic antenna — Also called unipole. A hypothetical antenna radiating or receiving equally in all directions. A pulsating sphere is a unipole for sound waves. In the case of electromagnetic waves, unipoles do not exist physically, but represent convenient reference antennas for expressing directive properties of actual antennas.

isotropic gain of an antenna — *See* absolute gain of an antenna.

isotropic magnet — A magnetic material having no preferred axis of magnetic characteristics.

isotropic material — 1. A material having the same magnetic characteristics along any axis. 2. A substance whose properties are similar when tested in any direction.

isotropic radiator — A radiator that sends out equal amounts of energy in all directions.

ISP — Abbreviation for Internet service provider. A general term for any company that provides a connection to the Internet for individuals or businesses.

ISV — Abbreviation for instantaneous speed variations. Short-term speed changes resulting from nonuniform capstan speed, tension changes caused by start/stop accelerations, and longitudinal vibrations that are caused by a magnetic tape's sliding over the heads and tape guides.

ITDM — Abbreviation for intelligent time-division multiplexer. Also called a statistical multiplexer. A multiplexer that assigns time slots on demand rather than on a fixed subchannel-scanning basis.

item — 1. A general term denoting one of a number of similar units, assemblies, objects, etc. 2. Any unique manufactured or purchased part or assembly, that is, end product, assembly, subassembly, component, or raw material.

iterations per second — The number of approximations per second in iterative division in a computer; the number of times a cycle of operation can be repeated in one second.

iterative — 1. Recurring an infinite number of times. 2. Characteristic of a network with an infinite number of identical sections, or of the impedance looking into such a network.

iterative array — In a computer, a large number of identical, interconnected processing modules used, with appropriate driver and control circuits, to perform simultaneous parallel operations.

iterative division — In computers, a method of performing division by use of addition, subtraction, and multiplication operations. A quotient of specified precision is obtained by a series of progressively closer approximations.

iterative filter — A four-terminal filter that provides iterative impedance.

iterative impedance — 1. An impedance that, when connected to one pair of terminals of a four-terminal transducer, causes the same value of impedance to appear between the other two terminals. The iterative impedance of a uniform transmission line is equal to the characteristic impedance of the line. In a symmetrical four-terminal transducer, the iterative impedances for the two pairs of terminals are equal and the same as the image impedances and the characteristic impedance. 2. The impedance that will terminate the output of a line or network such that the impedance then measured at the input of the line or network will be equal to the (iterative) terminating impedance.

iterative process — The calculating of a desired result by means of a repeating cycle of operations that comes closer and closer to the desired result.

iterative routine — A computer routine composed of repetitive computations, so that the output of every step becomes the input of the succeeding step.

ITU — Abbreviation for International Telecommunication Union.

iTV — Abbreviation for industrial television.

i-type — Intrinsic semiconductor.

i-type semiconductor material — A semiconductor material in which the electron and hole densities are effectively equal under conditions of thermal equilibrium.

IVR — Abbreviation for integrated voltage regulator.

IWW — Abbreviation for intermittent windshield wiper (General Motors, Ford Motor, and Chrysler Corp.). In an automobile, electronic control of the windshield wiper motor to vary the time interval between wipes.

J

J — Letter symbol for joule.

jack — 1. A socket to which the wires of a circuit are connected at one end, and into which a plug is inserted at the other end. 2. A type of two-way, or more, concentric contact socket for carrying audio signals. 3. A receptacle into which a mating connector may be plugged. 4. The receptacle that accepts a plug, specifically a phone plug. 5. A plug-in type terminal widely used in electronic apparatus for temporary connections or those requiring frequent connections and disconnections. A connection is made to a jack simply by plugging a probe or plug attached to a flexible insulated wire or cable into the jack. 6. Receptacle for a plug connector leading to the input or output circuit of a tape recorder or other piece of equipment. A jack matches a specific plug.

jack bay — Also called patch bay. A panel containing a number of signal jacks (usually standard phone jacks or mini phone jacks), commonly used in studio recording consoles and in equipment racks to provide flexibility in rerouting signals (beyond that provided by normal switches or controls). The jack bay also offers convenience when temporarily connecting certain equipment to a system, or troubleshooting and aligning equipment.

jacket — 1. Pertaining to wire and cable, the outer sheath that protects against environment and may also provide additional insulation. 2. An outer, nonmetallic, protective covering applied over an insulated wire or cable.

jack panel — An assembly composed of a number of jacks mounted on a board or panel.

jackscrew — A screw attached to one half of a two-piece, multiple-contact connector and used to draw both halves together and to separate them.

jaff — Slang for the combination of electronic and chaff jamming.

jag — In facsimile, distortion caused in the received copy by a momentary lapse in synchronism between the scanner and recorder.

jaggies — Irregular edges on something that should look smooth, a byproduct of the method of searching a scene and of too coarse a bit map.

jam — 1. In punch-card machines, a condition in the card feed that interferes with the normal travel of the punch cards through the machine. 2. To interfere electronically with the reception of radio signals.

jam input — The presetting or loading of a counter, using inputs provided for the purpose. Also, the establishment of a desired logic state or logic line by the direct application of the appropriate voltage level to the line, regardless of the outputs of other devices connected to it.

jammer — An electronic device for intentionally introducing unwanted signals into radar sets to render them ineffective.

jammer band — The radio-frequency band where the jammer output is concentrated. It is usually the band between the points where the intensity is 3 dB down from maximum.

jammer finder — Also called burnthrough. Radar that attempts to obtain the range of the target by training a highly directional pencil beam on a jamming source.

jammers tracked by azimuth crossings — Semi-automatic strobe processing and tracking that permits automatic detection and tracking on the basis of azimuth information obtained from the jamming signals emanating from an airborne vehicle.

jamming — Also called active electronic countermeasures. 1. The intentional transmission of radio signals in order to interfere with the reception of signals from another station. 2. Interference with hostile radio or radar signals for the purpose of deceiving or confusing the operator. It may be accomplished by saturating a receiver with sufficient noise to prevent detection and location of a target, or by deceiving the operator with intentionally misleading signals or false echoes without his knowing that such signals are present. 3. The deliberate radiation, reradiation, or reflection of electromagnetic energy to impair the use of electronic devices, equipment, or systems by the enemy. Equipment may consist of rudimentary cw or noise transmitters, broadband transmitters, or complex systems that generate deceptive signals.

jamming effectiveness — The jamming-to-signal ratio, that is, the percentage of information incorrectly received in a test message.

JAN specification — Joint Army-Navy specification. The forerunner of present military specifications; generally superseded by the designation MIL.

J-antenna — A half-wave antenna fed at one end by a parallel-wire, quarter-wave section having the configuration of a J.

Java — A platform-independent programming language invented by Sun Microsystems that is specifically designed for writing programs that can be safely downloaded to a computer through the Internet and immediately run without fear of viruses or other harm to the computer or to files. Using small Java programs (called applets), web pages can include functions such as animations, calculators, and other fancy tricks.

J-carrier system — A broadband carrier system that provides 12 telephone channels and utilizes frequencies up to about 140 kilohertz by means of four-wire transmission on a single open-wire pair.

J-display — Also called J-scan. In radar, a modified A-display in which the time base is a circle and the target signal appears as a radial deflection from it.

JEDEC — Acronym for Joint Electron Device Engineering Council. An industry-sponsored organization whose function is to provide a means of standardization for the industry. This encompasses numbering systems, testing methods and techniques, specifications uniformity,

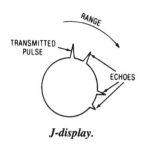

J-display.

and similar attempts on the part of the component manufacturers to assist the users of electronic components. In addition, JEDEC attempts to establish a code of ethics for the industry and to offer a set of standards for nonmilitary parts.

JETEC — Acronym for Joint Electron Tube Engineering Council.

jewel bearing — A natural or synthetic jewel, usually sapphire, used as a bearing for a pivot or other moving parts of a delicate instrument.

jezebel — A system for the detection and classification of submarines.

JFET — Abbreviation for junction field-effect transistor. A semiconductor device that operates by altering the conductivity of a region of the semiconductor (the channel) between two contacts (source and drain) by application of a voltage to a third terminal (gate). The current flow between source and drain is controlled by the gate voltage. In a JFET device, the gate voltage is applied to the channel across a pn junction, in contrast to its application across an insulator in a conventional MOSFET. JFETs are of two types: p channel and n channel, depending on whether the channel is n type or p type.

JHG — Abbreviation for Joule heat gradient.

jitter — 1. Instability of a signal in either its amplitude or its phase, or both, due to mechanical disturbances or to changes in supply voltage, component characteristics, etc. 2. In relation to cathode-ray tube displays, error in the signal amplitude, phase, or both that results in small, rapid abberations in size or position of the image. 3. Error of synchronization between a facsimile's transmitter and receiver characterized by a raggedness in the copy. 4. An aberration of a repetitive display, indicating instability of the signal or of the oscilloscope. May be random or periodic, and is usually associated with the time axis. 5. A loss of synchronization caused by electrical or mechanical malfunctions. 6. Type of analog-communication-line distortion caused by a signal's variation from its reference timing position, which can cause data-transmission errors, particularly at high speeds. This variation can be in amplitude, time, frequency or phase. *See* flicker, 3; fortuitous telegraph distortion.

jittered pulse recurrence frequency — The random variation of the pulse-repetition period. Provides a discrimination capability against repeater-type jammers.

J-K flip-flop — A flip-flop with two conditioning inputs (J and K) and one clock input. If both conditioning inputs are disabled prior to a clock pulse, the flip-flop does not change condition when a clock pulse occurs. If the J input is enabled and the K input is disabled, the flip-flop will assume the 1 condition upon arrival of a clock pulse. If the K input is enabled and the J input is disabled, the flip-flop will assume the 0 condition when a clock pulse arrives. If both the J and K inputs are enabled prior to the arrival of a clock pulse, the flip-flop will complement, or assume the opposite state, when the clock pulse occurs. The J-K flip-flop is a refinement of the R-S flip-flop with the advantage that it has a determinate state when signals

appear on both input terminals; it changes state when all inputs are activated. J and K have no particular meaning, but were selected to avoid conflict with other commonly used symbols.

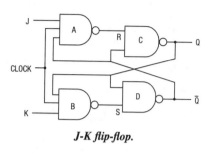

J-K flip-flop.

job — 1. A group of tasks specified as a unit of work for a computer. Usually by extension, a job includes all necessary programs, linkages, files, and instructions for the operating system. 2. In a computer, a collection of tasks, grouped and run together in order to perform a specific function.

job control language — A special computer command language designed for use in batch systems to inform the systems software and computer operator of unique requirements for the running of a computer program.

job library — A related series of user-identified, partitioned data sets that serve as the primary source of load modules for a given job.

job statement — A control statement that identifies the start of a series of job control statements for a single job.

job step — The carrying out of a computer program explicitly identified by a job control statement. The execution of several job steps may be specified by a job.

jogging — Also called inching. Quick and repeated opening and closing of a motor starting circuit to produce slight movements of the motor.

Johnson counter — Also called Mobius counter or twisted-ring counter. A counter composed of an N-stage shift register with the complement of the last stage returned to the input. It normally has 2N states through which it cycles. It has the distinguishing characteristic that only one stage changes state at each count. *See* ring counter, 2.

Johnson noise — Also called thermal noise. 1. The noise generated by any resistor at a temperature above absolute zero. It is proportionate to the absolute temperature and the bandwidth according to the following formula:

$$N = KTB$$

where

N is the noise power in watts
K is Boltzmann's constant, or 1.38047×10^{-23}
T is he absolute temperature in kelvins
B is he bandwidth in hertz

2. A frying or sizzling sound produced by thermal agitation voltages generated in amplifier circuits. It usually occurs in the input circuit (or front end) of an amplifier. Discovered by J. B. Johnson in the late 1920s, this thermal noise may be generated by a resistor at a temperature above absolute zero. It is random noise engendered by thermal agitation.

joined actuator — A multiple breaker such that when one pole trips, all trip, but whereas the faulted pole is tripfree, the other poles may be kept maintained by a restraining actuator.

joint — A connection between two or more conductors, or the connecting point of two conductors.

joint circuit — A communication link in which there is participation by the elements of more than one service, through control, operations, management, etc.

joint communications — The common use of communication facilities by two or more services of the same country.

Joint electron device Engineering Council — See JEDEC.

Joint electron tube Engineering Council — See JETEC.

joint pole — Pole used in common by two or more utility companies.

joint use — The simultaneous use of pole, line, or plant facilities by two or more kinds of utilities.

Jones plug — A type of polarized connector designed in the form of a receptacle and having several contacts.

Josephson effect — 1. The phenomenon described by Brian Josephson to explain the action of currents through and voltages across hairlike gaps in superconductors. On the basis of theoretical considerations, it is predicted that if two superconductors would be brought close enough together, a current could be made to flow across the gap between them. Under certain conditions, a voltage appears across the gap, and high-frequency radiation emanates from it. This predicted radiation would have a frequency precisely equal to $2eV/h$, where V is the measured voltage across the gap. 2. Characteristic of radiation detectors that produce energy that is similar to the energy of superconductive gaps when interacting with photons.

joule — 1. The work done by a force of 1 newton acting through a distance of 1 meter. The joule is the unit of work and energy in the mksa system. 2. The energy required to transport 1 coulomb between two points having a potential difference of 1 volt. The joule is 10^7 ergs. The kilowatt-hour is 3.6×10^6 joules. 3. A unit of energy of work $CV^2/2$ equal to 1 wattsecond. Energy stored in a capacitor is equal to $CV^2/2$ joules or wattseconds, where C is capacitance in farads and V is voltage in volts at the capacitor's terminals. The letter symbol for joule is J.

Joule effect — In a circuit, electrical energy is converted into heat by an amount equal to I^2R. Half of this heat flows to the hot junction and the other half to the cold junction.

Joule heat — The thermal energy produced as a result of the Joule effect.

Joule heat gradient — Abbreviated JHG. The rate at which the thermal heat produced by the Joule effect increases or decreases.

Joule's law of electric heating — The amount of heat produced in a conductor is proportional to the resistance of the conductor, the square of the current, and the time.

joystick — 1. A control device consisting of a handle with freedom of motion in all directions of a plane, connected to potentiometers or other control devices through suitable linkage permitting natural human input of positioning or other information. The term is derived from the joystick of aircraft. 2. A data-entry device used to manually enter coordinate values in special X-, Y-, and Z-input registers. The device consists of a vertically mounted stick or column that can be moved and twisted. When it is moved backward or forward or sideways, coordinate values are stored in the X- and Y-input registers. The Z-input register is varied whenever the

joystick is twisted clockwise and counterclockwise. These registers must be scanned by the host computer since joysticks normally do not generate interrupts when they are activated. Usually the joystick is used to move a cursor and/or tracking symbol on the face of the CRT screen. Used mainly for computer games. 3. A type of four-channel pan potentiometer that has a shaft with a handle that can be moved forward and back, left and right, or anywhere in between to direct a single input signal to any of four output channels.

J-scan — See J-display.

J-scope — Also called class J oscilloscope. A cathode-ray oscilloscope that presents a J-display.

J/S ratio — A ratio, normally expressed in dB, of the total interference power to the signal-carrier power in the transmission medium at the receiver.

juice — Slang for electric current.

jump — 1. To cause the next instruction to be selected from a specific storage location in a computer. 2. A departure from the normal one-step incrementing of the program counter. By forcing a new value (address) into the program counter, the next instruction can be fetched from an arbitrary location (either further ahead or back). For example, a program jump can be used to go from the main program to a subroutine, from a subroutine back to the main program, or from the end of a short routine back to the beginning of the same routine to form a loop. 3. Transfer of program logic flow by bypassing a number of instructions. The jump can be forward over a positive number of bytes or backward by expressing a negative number of bytes. The jump can be conditional on the status of the accumulator or other registers.

jumper — 1. A short length of wire used to complete a circuit temporarily or to bypass a circuit. 2. A direct electrical connection, which is not a portion of the conductive pattern, between two points in a printed circuit. 3. An electrical connection between two points on a printed board added after the intended conductive pattern is formed. 4. A short length of cable used to make electric connections within, between, among, and around circuits and their associated equipment. 5. A length of conductor used to establish connections (often temporary) between two points or to provide a path around a break in a circuit. A jumper can interconnect board to connector, board to board, or power supply to black box unit. 6. A direct electrical connection between two points on a film circuit. Jumpers are usually portions of bare or insulated wire mounted on the component side of the substrate. 7. A patch cable or wire used to establish a circuit, often temporarily, for testing or diagnostics.

jumper cable — A short flat cable interconnecting two printed wiring boards or devices.

junction — 1. A connection between two or more conductors or two or more sections of a transmission line. 2. A contact between two dissimilar metals or materials (e.g., in a rectifier or thermocouple). 3. A region of transition between p- and n-type semiconductor material. The controllable resultant asymmetrical properties are exploited in semiconductor devices. There are diffused, alloy, grown, and electrochemical junctions. 4. A joining of two different semiconductors or of semiconductor and metal. 5. Optical interface.

junction barrier — The opposition to the diffusion of majority carriers across a pn junction due to the charge of the fixed donor and acceptor ions.

junction battery — A nuclear type of battery in which radioactive strontium 90 irradiates a silicon pn junction.

junction box — 1. A box for joining different runs of raceway or cable, plus space for connecting and branching the enclosed conductors. 2. An enclosed distribution panel for connecting or branching one or more corresponding electric circuits without the use of permanent splices.

junction capacitance — The total small-signal capacitance between the contacts of an uninstalled semiconductor die.

junction capacitor — A capacitor in which the capacitance is that of a reverse-biased pn junction.

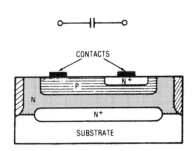

Junction capacitor.

junction diode — 1. A two-terminal device containing a single crystal of semiconducting material that ranges from p type at one terminal to n type at the other. It conducts current more easily in one direction than in the other and is a basic element of the junction transistor. Such a diode is the basic part of an injection laser; the region near the junction acts as a source of emitted light. When fabricated in a suitable geometrical form, the junction diode can be used as a solar cell. 2. A pn junction characterized by slower switching speed and higher operating voltage and temperature than the Schottky diode. The fast recovery version of the junction diode turns off faster than the conventional pn junction diode, usually in hundredths of a nanosecond. 3. A semiconductor diode having the property of conducting current more easily in one direction than the other. This device may be made by diffusion of an impurity into a semiconductor crystal to form a junction. Such diodes are the basic elements of an injection laser. Light is emitted from the area near the junction.

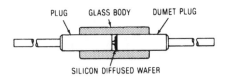

Junction diode assembly.

junction FET — A field-effect transistor having one or more gates that form pn junctions with the channel. *See also* junction field-effect transistor.

junction field-effect transistor — Abbreviated JFET. 1. A transistor made up of a gate region diffused into a channel region. The gate element is a region of semiconductor material (ordinarily, the substrate) insulated by a pn junction from the channel, which is material of opposite polarity. When a control voltage is applied to the gate, the channel is depleted or enhanced, and the current between source and drain is thereby controlled. There is no current when the channel is pinched off. 2. A field-controlled majority carrier

device where the conductance in the channel between the source and the drain is modulated by a transverse electric field. The field is controlled by a combination of gate-source bias voltage, V_{GS}, and the net drain-source voltage, V_{DS}.

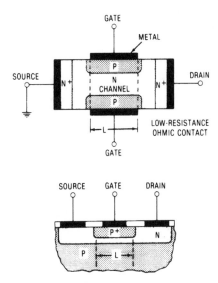

Junction field-effect transistor.

junction filter — A combination of a high-pass and a low-pass filter that is used to separate frequency bands for transmission over separate paths. For example, junction filters are used to separate voice and carrier frequencies at the junction between open-wire and cable so that the carrier frequencies and voice frequencies can be sent over nonloaded and VF-loaded cable pairs, respectively.

junction-gate field-effect transistor — A field-effect transistor having one or more gate regions that form pn junction(s) with the channel.

junction loss — 1. In telephone circuits, that part of the repetition equivalent that can be assigned to interactive effects originating at trunk terminals. 2. The transmission loss due to the mismatch of impedance between two types of transmission facilities. *See* repetition equivalent.

junction point — *See* node, 1.

junction pole — Pole at the end of a transposition section of an open-wire line or the pole common to two adjacent transposition sections.

junction station — A microwave relay station that joins a microwave radio leg or legs to the main, or through, route.

junction transistor — A transistor having three alternate sections of p-type or n-type semiconductor material. *See also* npn transistor; pnp transistor.

junction transposition — Transposition located at the junction pole between transposition sections of an open-wire line.

junctor — In crossbar systems, a circuit extending between frames of a switching unit and terminated in a switching device on each frame.

justification — 1. The act of adjusting, arranging, or shifting digits to the left or right so that they conform to a prescribed pattern. 2. The automatic inserting of blank spaces within text to make the right margin even on a page.

justify — To align data about a particular reference or to produce a text with flush left and right margins (a more printlike appearance).

just-operate value — Also called dropout value. The measured functioning value at which a particular relay operates. *Contrast* just-release value.

just-release value — The measured functioning value at which a particular relay releases.

just scale — A musical scale formed by three consecutive triads (those in which the highest note of one is the lowest note of the other), each having the ratio $4:5:6$ or $10:12:15$.

jute — Cordage fiber (such as hemp) saturated with tar and used as a protective layer over cable.

jute-protected cable — A cable having its sheath covered by a wrapping of tarred jute or other fiber.

K

K—1. Symbol for cathode or dielectric constant. 2. Letter symbol for kelvin or kilo. 3. Abbreviation for luminosity factor. 4. In a calculator, a fixed number (a constant) that can be used repetitively. 4. Symbol for 10^3. When referring to bits or words, $K = 1024$. A 4 K chip is a 4 K-bit chip. A 4 K memory is a 4 K-word memory (typically 4 K bytes). 5. When referring to memory storage capacity, 2^{10}; in decimal notation, 1024.

k—Letter symbol for kilo-.

kA—Letter symbol for kiloampere.

Ka band—The frequency band from 18 to 31 GHz.

Karnaugh map—A display of a truth table in a way such that reduction (simplification) of a Boolean expression is facilitated. It consists of a rectangular or square array (depending on the number of variables) of "locations" whose coordinates correspond to truth-table inputs.

KB or **Kb**—Abbreviation for kilobyte. 1024 bytes (8192 bits).

K-band—A radio frequency band extending from 11 to 36 GHz and having wavelengths of 2.73 to 0.83 cm.

Kbps—Abbreviation for kilobits per second. A standard measurement of data rate and transmission capacity. One Kbps equals 1024 bits per second.

kc—Abbreviation for kilocycle. Now obsolete. Replaced by kHz.

K-carrier system—A broadband carrier system that provides 12 telephone channels and utilizes frequencies up to about 60 kHz by means of four-wire transmission on cable facilities.

K-display—Also called K-scan. Modification of a type-A scan, used for aiming a double-lobe system in bearing or elevation. The entire range scale is displaced toward the antenna lobe in use. One signal appears as a double deflection from the range and relative scales. The relative amplitudes of these two pips indicate the amount of error in aiming the antenna.

keep-alive anode—An auxiliary electrode that maintains a dc discharge in a mercury-pool tube. It has the disadvantage of reducing the peak inverse voltage rating.

keep-alive circuit—In a tr or anti-tr switch, a circuit for producing residual ionization in order to reduce the time for full ionization when the transmitter fires.

keep-alive voltage—A dc voltage that maintains a small glow discharge within one of the gap electrodes of a tr tube. This allows the tube to ionize more rapidly when the transmitter fires, thus preventing damage to the receiver.

keeper—A magnetic conductor, placed over the ends of a permanent magnet, used to complete the magnet circuit of a permanent magnet to protect it against demagnetizing influences.

Kel-f—Polymonochlorotrifluoroethylene—a plastic used as a high-temperature insulation ($-55°C$ to $+135°C$).

kelvin—Formerly degree Kelvin. A unit of absolute temperature equal to 1/273.16 of the Kelvin scale temperature of the triple point of water. Letter symbol: K.

Kelvin balance—An instrument for measuring current. This is done by sending a current through a fixed and a movable coil attached to one arm of a balance. The resultant force between the coils is then compared with the force of gravity acting on a known weight at the other end of the balance arm.

Kelvin bridge—Also called a double or Thompson bridge. A seven-arm bridge for comparing the resistances of two 4-terminal resistors or networks. Their adjacent potential terminals are spanned by a pair of auxiliary resistance arms of known ratio, and they are connected in series by a conductor joining their adjacent current terminals.

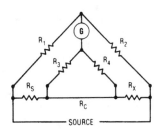

Kelvin bridge.

Kelvin scale—Also called absolute scale. A temperature scale using the same divisions as the Celsius scale, but with the zero point established at absolute zero ($-273°C$ or $-459°F$), theoretically the lowest possible temperature (the temperature at which all molecular motion stops).

Kendall effect—A spurious pattern or other distortion in a facsimile record. It is caused by unwanted modulation produced by transmission of a carrier signal. Such modulation appears as a rectified baseband that interferes with the lower sideband of the carrier.

Kennelly-Heaviside layer—*See* Heaviside layer.

kenopliotron—A diode-triode vacuum tube within one envelope. The anode of the diode also serves as the cathode of the triode.

kenotron—Also called a value tube. 1. A term used primarily in industrial and X-ray fields for a hot-cathode vacuum tube. 2. A high-vacuum, high-voltage thermionic diode, used primarily as a high-voltage rectifier.

Keplerian elements (Keps) — A collection of data relating to the position of a satellite in its orbit at any given time. This information is interpreted by satellite tracking programs to predict time and duration of satellite passes and directions in which to point antennas. Named for the 19th-century scientist Johannes Kepler.

keraunophone — A radio circuit device for audibly demonstrating the occurrence of distant lightning flashes.

kerf — The slit or channel or "L" cut in a resistor during trimming by a laser beam or abrasive jet.

kernel — 1. A line within a current-carrying conductor along which the magnetic intensity due to the current is zero. 2. The most basic portion of an operating system, usually supporting only task synchronization, scheduling, communication, and the most rudimentary of memory allocation capabilities. It implements software processes and furnishes the means of interprocess communication; usually written in assembly language. 3. The central portion of a program, containing the bulk of its calculations, that consumes the most execution time.

Kerr cell — A container filled with a transparent material that, when subjected to a strong electric field, exhibits double refraction. Because the two polarized components of an incident light beam have different phase velocities in the medium, this device can rotate the plane of polarization. When placed between crossed polarizers, the Kerr cell, usually containing nitrobenzene, can act as an extremely high-speed shutter for light beams.

Kerr effect — 1. An electro-optical effect in which certain transparent substances become double refracting when subjected to an electric field perpendicular to a beam of light. 2. The conversion of plane into elliptically polarized light when reflected from the polished end of a magnet.

Kettering ignition system — Inductive system commonly used for internal combustion engines. Employs induction coil, breaker contacts, capacitor, and a suitable power supply, such as a battery.

keV — Letter symbol for kiloelectronvolt.

key — 1. A hand-operated switching device for switching one or more parts of a circuit. It ordinarily consists of concealed spring contacts and an exposed handle or push button. 2. A projection that slides into a mating slot or groove so as to guide two parts being assembled and assure proper polarization.

keyboard — 1. In a calculator, keys for digits 0 through 9, plus additional keys for various functions, such as add, multiply, divide, subtract, clear, memory, etc. 2. That portion of a terminal used to generate the character stream to the computer or other communications device. 3. An array of push-button switches with related functions. Keyboards use standard push-button contact switches, Hall-effect switches, capacitive switches, and many others. Any type of momentary-contact switch suitable in terms of size, current switching capacity, etc., may be used. The electrical signal generated by switch closure is routed to the appropriate circuitry where the action indicated by closure of the keyboard switch is initiated. 4. A series of switches, usually in the form of a typewriter keyboard, that a computer operator uses to communicate with the computer.

keyboard computer — A computer whose input employs a keyboard, e.g., an electric typewriter.

keyboarding — Entering information (to a word processor) via a keyboard.

keyboard-operated transmission — See KOX.

keyboard perforator — A mechanism that punches a paper tape from which messages are automatically transmitted by a transmitter distributor. The keyboard is similar to that of a typewriter and can be operated by any trained typist after a few hours' instruction. As each key is depressed, the tape is punched with corresponding code symbols.

keyboard send/receive — A combination teletypewriter transmitter and receiver with transmission capability from a keyboard only. See KSR.

key cabinet — A case installed on a customer's premises and providing facilities so that different lines to the control office can be connected to various telephone stations. It has signals that indicate originating calls and busy lines.

key click — A transient signal sometimes produced when the key of a radiotelegraph transmitter is opened or closed. The transient is heard in a speaker or headphone as a click.

key-click filter — Also called a click filter. A filter that attenuates the surges produced each time the keying circuit of a transmitter is opened or closed.

keyed adapter — A device that detects a modulated signal and produces a dc output signal whose amplitude varies in accordance with the modulation. In radio facsimile transmission, it is used to provide the keying signal for a frequency-shift exciter unit.

keyed AGC — Abbreviation for keyed automatic gain control.

keyed automatic gain control — Abbreviated keyed AGC. A television automatic gain control in which the AGC tube is kept cut off except when the peaks of the positive horizontal-sync pulse act on its grid. The AGC voltage is therefore not affected by noise pulses occurring between the sync pulses.

keyed clamp — A clamping circuit in which a control signal determines the time of clamping.

keyed interval — In a periodically keyed transmission system, an interval that starts from a change in state and has a length equal to the shortest time between changes in state.

keyed rainbow generator — A color television test instrument that displays the individual colors of the spectrum, separated by black bars, on the picture tube.

keyed rainbow signal — A 3.563795-MHz (3.56-MHz) continuous sine-wave signal from a color-bar generator that is pulsed on and off. This signal creates a series of different color bars on the screen of the color picture tube. A typical pulse rate (for 10 color bars) is 12 times per 1 horizontal line.

keyer — 1. In telegraphy, a device that breaks up the output of a transmitter or other device into the dots and dashes that are used in the code. 2. A radar modulator.

keying — 1. The forming of signals, such as those employed in a telegraph transmission, by an abrupt modulation of the output of a director by an alternating-current source (e.g., by interrupting it or by suddenly changing its amplitude, frequency, or some other characteristic). 2. Mechanical arrangement of guide pins and sockets, keying plugs, contacts, bosses, slots, keyways, inserts, or grooves in a connector housing, shell, or insert that allows connectors of the same size and type to be lined up without the danger of making a wrong connection.

keying chirps — Sounds accompanying code signals when the transmitter is unstable and shifts slightly in frequency each time the sending key is closed.

keying frequency — In facsimile, the maximum number of times a second a black-line signal occurs while scanning the subject copy.

keying plug contact — A component that is inserted into a cavity of a connector housing or insert to ensure engagement of identically matched components.

keying wave — Also called marking wave. The emission that takes place in telegraphic communication while the information portion of the code characters is being transmitted.

keyless ringing — A type of machine ringing on a manual switchboard. Ringing is started automatically when the calling plug is inserted into the jack of the called line.

keypad — Device with a matrix of keys enabling a user to input (usually numeric) information into another device, as in dialing a touch-tone telephone.

key pulse — A telephone signaling system in which numbered keys are depressed instead of a dial being turned.

key pulsing — A switchboard arrangement using a nonlocking keyset for the transmission of pulse signals corresponding to the key depressions.

key-pulsing signal — The signal that indicates a circuit is ready for pulsing, in multifrequency and direct-current key pulsing.

keypunch — A keyboard-operated device that punches holes in punch cards to represent data to be input to a computer by a card reader.

keyshelf — A shelf on which are mounted the keys by means of which the operator of a manual telephone switchboard performs switching of one or more of the switchboard circuits.

key signal — A pseudorandom sequence of two-level pulses used to accomplish enciphering or deciphering processes.

key station — The master station from which a network radio or television program originates.

keystone distortion — The distortion produced when a plane target area not normal to the average direction of the beam is scanned rectilinearly with constant-amplitude sawtooth waves.

Keystone distortion.

keystone shaped — Wider at the top than at the bottom, or vice versa.

keystoning — The keystone-shaped scanning pattern produced when the electron beam in the television camera tube is at an angle with the principal axis of the tube. *See also* keystone shaped.

keyswitch — In an organ, the switch that is closed to allow a tone from the tone generator to sound when a key is depressed.

key telephone — Also called key telephone set. Telephone set having six button keys and used with relay equipment to provide call holding, multiline pickup, signaling, intercommunication, and conference services.

key telephone set — A desk telephone with six (illuminated) push-button keys across the set below the dial. The keys may be connected to hold, pick up a central office line, or pick up an intercom line.

key telephone system — A versatile station switching system located on a customer's premises, consisting of one or more multibutton telephone sets and associated equipment. Permits mutual access and control of a number of central office lines, as well as pickup, hold, signal, and intercommunication capability.

key up — To turn on a repeater by transmitting on its input frequency.

keyway — The mating slot or groove in which a key slides.

keyword — 1. Characteristic word in a computer file used to retrieve its contents. 2. A word (with a special meaning) that is recognized by the computer or program.

kHz — Abbreviation for kilohertz. A unit of frequency measurement equivalent to 1000 cycles per second; formerly expressed as kilocycles per second, or kc/s.

kickback — The voltage developed across an inductance by the sudden collapse of the magnetic field when the current through the inductance is cut off.

kickback power supply — *See* flyback power supply.

kick-sorter — British term for pulse-height analyzer.

kidney joint — A flexible joint, or air-gap coupling, located in the waveguide and near the transmitting-receiving position of certain radars.

Kikuchi lines — A series of spectral lines obtained by the scattering of electrons when an electron beam is directed against a crystalline solid. The pattern may be interpreted to yield information on the structure of the crystal and its mechanical perfection.

killer circuit — 1. The vacuum tube or tubes and associated circuits in which are generated the blanking pulses used to temporarily disable a radar set. 2. In a transponder, a logic circuit that kills replies to side-lobe interrogations.

killer pulse — *See* suppression pulse.

kilo- — A prefix representing 10^3, or 1000. Letter symbol: k.

kiloampere — 1000 amperes. Letter symbol: kA.

kilobaud — Abbreviated kBd. One thousand baud.

kilobit — 1024 bits.

kilobyte — 1. A standard quantity measurement for disk and diskette storage and semiconductor circuit capacity: 1 kilobyte of memory equals 1024 bytes (8-bit characters) of computer memory. 2. 2^{10}, or 1024, bytes. Commonly thought of as 1000 bytes. Abbreviated K and used as a suffix to describe memory size.

kilocycle (kc) — 1000 cycles. Generally interpreted as meaning 1000 cycles per second. Obsolete term; replaced by kilohertz (kHz).

kiloelectronvolt — 1000 electron volts. The energy acquired by an electron that has been accelerated through a voltage difference of 1000 volts. Letter symbol: keV.

kilogauss — 1000 gauss.

kilogram — Unit of mass. The mass of a particular cylinder of platinum-iridium alloy, called the international prototype kilogram, that is preserved in a vault at Sevres, France, by the International Bureau of Weights and Measures. Letter symbol: kg.

kilohertz — 1000 hertz. Letter symbol: kHz.

kilohm — 1000 ohms. Abbreviated k or kohm.

kilohmmeter — A meter designed for measuring resistance in kilohms.

kilomega (kM) — Obsolete prefix for giga- (G), representing 10^9 or 1,000,000,000.

kilomegacycle — Now called gigahertz. One billion cycles per second.

kilometer — One thousand meters, or approximately 3280 feet.

kilometric waves — British term for electromagnetic waves between 1000 and 10,000 meters in length.

kilosecond — 1000 seconds.

kilovar — One reactive kilovoltampere, or 1000 reactive voltamperes. Letter symbol: kvar.

kilovar-hour — 1000 reactive volt-amperehours.

kilovolt — 1000 volts. Letter symbol: kV.

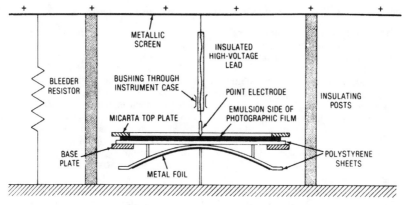

Klydonograph gradient recorder.

kilovoltampere — 1000 voltamperes. Letter symbol: kVA.

kilovoltmeter — A voltmeter that reads thousands of volts.

kilowatt — 1000 watts. Letter symbol: kW.

kilowatthour — The equivalent energy supplied by a power of 1000 watts for 1 hour. Letter symbol: kWh.

kine — Slang term for kinescope recording.

kine-klydonograph — An instrument that records the current-time characteristics of a lightning stroke. The instrument records a series of Lichtenberg figures in a manner similar to that of the field gradient recorder.

kinescope — 1. A cathode-ray tube that serves as a picture tube in a television receiver. The signal representing the picture intensity is transmitted to the electron gun grid so that the beam intensity varies with the intensity of the original scene. Deflecting voltages sweep the beam horizontally and vertically until it is synchronized with the scanning of the camera tube. The electron beam then strikes the fluorescent end screen of the tube and the scene is reproduced. 2. A film recording made from a television program on a picture tube and used as a permanent record or for subsequent rebroadcasting.

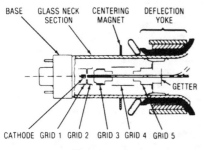

Kinescope, 1.

kinescope gun — The electron gun in the kinescope.

kinescope recorder — A camera that photographs television images directly from the picture tube onto motion-picture film.

kinetic energy — Energy that a system possesses by virtue of its motion.

Kirchhoff's laws — 1. The current flowing to a given point in a circuit is equal to the current flowing away from that point. 2. The algebraic sum of the voltage drops in any closed path in a circuit is equal to the algebraic sum of the electromotive forces in that path. (Laws 1 and 2 are also called laws of electric networks.) 3. At a given temperature, the emissive power of a body is the same as its radiation-absorbing power for all surfaces.

kit — A prepared package of parts with instructions for assembly and/or wiring a component or chassis (also a small accessory item).

klydonograph — A field gradient recording instrument that registers voltages on photographic film in the form of Lichtenberg figures.

klystron — An electron tube used as an oscillator or amplifier at ultrahigh frequencies. The electron beam is velocity modulated (periodically bunched) to accomplish the desired results.

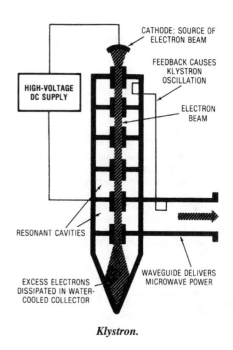

Klystron.

klystron control grid — An electrode that controls the emission, or beam current, of a klystron or other velocity-modulated tube.

klystron frequency multiplier — A two-cavity klystron that has the output cavity tuned to a multiple of the fundamental frequency.

klystron generator — A klystron tube used as a generator. Its cavity feeds energy directly into a waveguide.

klystron oscillator — An oscillator employing a klystron tube to generate radio-frequency power.

klystron repeater — A klystron tube operated as an amplifier and inserted directly into a waveguide in such a way that incoming waves velocity-modulate the electron stream emitted from a heated cathode. A second cavity converts the energy of the electron clusters into waves of a much higher amplitude and feeds them into the outgoing guide.

kM — Abbreviation for kilomega (an obsolete term). Replaced by giga.

kMc — Abbreviation for kilomegacycle. Now replaced by gigahertz.

knee — 1. An abrupt change in direction between two relatively straight segments of a curve, such as the region of a magnetization curve near saturation or the top bend of a vacuum-tube characteristic curve. 2. A section between two comparatively straight segments of a curve in which the magnitude of curvature, although of the same sign, is relatively high. 3. The region of maximum curvature.

knife-edge diffraction — In radio-wave propagation, an effect by which the atmospheric attenuation of a signal is reduced when the signal is diffracted as it passes over a sharp obstacle such as a mountain ridge.

knife-edge pointer (of a meter) — A pointer whose end is flattened and turned edgewise so the smallest dimension or edge is seen. Usually used with mirror-backed scales to eliminate parallax and increase the accuracy of reading.

knife switch — 1. A form of air switch in which a moving element is sandwiched between two contact clips. The moving element is usually a hinged blade; when it is not, it is removable. 2. A form of switch in which the moving element, usually a hinged blade, enters or embraces stationary contact clips.

knob — A round, polygonal, or pointer-shaped part that is fastened to one end of a control shaft so that the shaft can be turned more easily. The knob sometimes indicates the degree of rotation also.

knocker — A term used with some fire-control radars to indicate a subassembly comprising synchronizing and triggering circuits. It drives the rf pulse-generating equipment in the transmitter and also synchronizes the cycle of operation with the transmitted pulse in range units and indicators.

knockout — A removable portion in the side of a box or cabinet. During installation it can be readily taken out with a hammer, screwdriver, or pliers so the raceway, cables, or fittings can be attached.

knot — One nautical mile (6,080.20 feet, 1852 meters, or 1.15 statute miles) per hour.

Kooman antenna — A vertical array of horizontal full-wave dipoles that are driven by transposed two-conductor line and backed by a parasitic reflecting curtain or horizontal dipoles.

Kovar — An iron-nickel-cobalt alloy with a coefficient of expansion similar to that of glass and silicon and thermal characteristics similar to those of alumina. It is used as a material for headers and in glass-to-metal seals. Kovar/glass packages are used extensively in hybrid construction to preserve hermetic sealing.

KOX — Acronym for keyboard-operated transmission. A station equipped with a KOX system can turn on the transmitter and turn off the receiver simply by typing on the station's keyboard. When the operator ceases typing, the transmitter automatically turns off, and the receiver turns on after some preset delay.

kraft paper — Relatively heavy, high-strength sulfate paper used for electrical insulating material. (Capacitor tissue is kraft paper of normal thickness equal to 0.002 inch, or 50 μm, or less.)

K-scan — See K-display.

K-series — A series of frequencies in the X-ray spectrum of an element.

KSR — Abbreviation for keyboard send-receive unit. A combination transmitter and receiver with transmission capability from the keyboard only (teletypewriter term). Refers to a terminal device (teletype or similar) having only a keyboard for sending and a printer for receiving, i.e., no paper or magnetic tape equipment, but which is useful for conversational time-sharing and inquiry-response applications.

Ku band — 1. A band of microwave frequencies between 11 and 13 GHz. 2. Also called K band. A band of frequencies that extends from 11.7 to 12.7 GHz. The band is separated into two portions: 11.7 to 12.2 GHz (fixed satellite services — intended for point-to-point services) and 12.2 to 12.7 GHz (for broadcasting satellite service or DBS). 3. The frequency band from 10.9 to 17 GHz.

kurtosis — The degree of curvature of the peak of a probability curve.

kV — Letter symbol for kilovolt.

kVA — Letter symbol for kilovoltampere.

kvar — Letter symbol for kilovar.

kW — Letter symbol for kilowatt.

kWh — Letter symbol for kilowatthour.

kymograph — An instrument for recording wavelike oscillations of varying quantities for medical studies.

L

L — Symbol for coil, lambert, or inductance.

label — 1. A code name used to identify or classify a name, term, phrase, or document. 2. One or more characters that serve to identify an item of data. 3. A numerical value or a memory location in the programmable system of a computer. The specific absolute address is not necessary since the intent of the label is a general destination. Labels are a requisite for jump and branch instructions. 4. A number or letter or a name given to a program statement so that the computer can find it later. 5. A name used in source code to identify an instruction or executable statement in computer programs. 6. A block at the beginning or at the end of a volume or file that identifies, characterizes, and/or delimits that volume or file. A label is not considered to be part of a file.

label group — A collection of continuous label sets of the same label type.

labile oscillator — A local oscillator whose frequency is remote controlled by a signal received from a radio or over a wire.

laboratory power supply — A regulated dc source having (a) less than 10-kV output at up to 500 W, (b) output adjustable over a wide range, usually down to zero, (c) regulation on the order of ±0.01 percent static line and load.

labyrinth — Speaker enclosure with absorbing air chambers at the rear to eliminate acoustic standing waves. A mazelike construction extends the air column. Resonances are tamed by heavy damping material.

labyrinth loudspeaker — Loudspeaker mounted in an acoustic baffle having air chambers designed to prevent acoustic standing waves.

lacing — A network of fine wire surrounding or covering an area to be protected, such as a safe, vault, or glass panel, and connected into a closed-circuit system. The network of wire is concealed by a shield, such as concrete or paneling, in such a manner that an attempt to break through the shield breaks the wire and initiates an alarm.

lacing and harnessing — Also called bundling. A method of grouping wires by securing them in bundles or designated patterns.

lacing cord or twine — Cord used for lacing and tying cable forms, hookup wires, cable ends, cable bundles, and wire harness assemblies. Available in various materials and impregnants.

lacing tape — Flexible, flat, fabric tape for tying harnesses and wire bundles, securing of sleeves and other items, and general lacing and typing applications.

lacquer disc — Also called cellulose-nitrate disc. A mechanical recording disc, usually made of metal, glass, or paper and coated with a lacquer compound often containing cellulose nitrate.

lacquer master — See lacquer original.

lacquer original — Also called lacquer master. An original recording made on a lacquer surface to be used as a master.

lacquer recording — Any recording made on a lacquer medium.

ladder attenuator — A series of symmetrical sections used in signal generators and other devices in which voltages and currents must be reduced in known ratios. They are designed so that the required ratio of voltage loss per section is obtained with image-impedance operation. The impedance between any junction point and common ground in a ladder attenuator is half the image impedance.

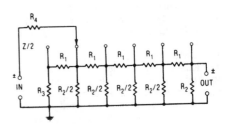

Ladder attenuator.

ladder diagram — 1. A diagram that shows actual component symbols and the basic wiring configuration of a relay logic circuit (as opposed to a logic diagram). 2. An industry standard for representing control logic relay systems. 3. Electrical engineering technique to schematically illustrate functions in an electrical circuit (relays, switches, timers, etc.) by diagramming them in a vertical sequence resembling a ladder. 4. See contact symbology diagram.

ladder network — Also called series-shunt network. 1. A network composed of H-, L-, T-, or pi networks connected in series. 2. A series of film resistors with values from the highest to the lowest resistor reduced in known ratios.

LAFOT — Coded weather broadcasts issued by the U.S. Weather Bureau for the Great Lakes region. They are broadcast every 6 hours by marine radiotelephone broadcasting stations on their assigned frequencies.

lag — 1. The displacement in time, expressed in electrical degrees, between two waves of the same frequency. 2. The time between transmission and reception of a signal. 3. In a television camera tube, the persistence of the electrical-charge image for a time interval equal to a few frames. 4. A time difference between the occurrence of two events.

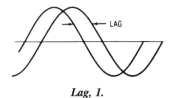

Lag, 1.

lagged-demand meter — A meter in which there is a characteristic time lag, by either mechanical or thermal means, before maximum demand is indicated.

lagging current — The current flowing in a circuit that is mostly inductive. If the circuit contains only inductance, the current lags the applied voltage by 90°. Because of the characteristics of an inductance, the current does not change direction until after the corresponding voltage does.

lagging load — A predominantly inductive load — i.e., one in which the current lags the voltage.

lag-lead — Also called lead-lag. A circuit whose response includes lag components and their derivatives.

lambda — Greek letter λ, used to designate wavelength measured in meters.

lambda diode — A simple two-terminal device consisting of a pair of complementary depletion-mode junction field-effect transistors that can be fabricated more easily than conventional negative-resistance devices. It can be integrated on a single chip or with bipolar and MOS devices on the same chip. Unlike tunnel diodes, which are limited to a narrow negative-resistance range, lambda diodes can be produced with a wide range of characteristics.

lambda wave — An electromagnetic wave propagated over the surface of a solid that has a thickness comparable to the wavelength of the wave.

lambert — A unit of luminance (photometric brightness) equal to $1/\pi$ candela per square centimeter and, therefore, equal to the uniform luminance of a perfectly diffusing surface emitting or reflecting light at the rate of 1 lumen per square centimeter. The lambert also is the average luminance of any surface emitting or reflecting light at the rate of 1 lumen per square centimeter. For the general case, the average must take account of variation of luminance with angle observation, also of its variation from point to point on the surface considered. Letter symbol: L.

lambertian — A radiance distribution that is uniform in all directions of observation.

Lambert's law of illumination — The illumination of a surface on which the light falls normally from a point source is inversely proportional to the square of the distance of the surface from the source. If the normal to the surface makes an angle with the direction of the rays, the illumination is proportional to the cosine of that angle.

laminar flow — A directed stream of filtered air moved constantly across a clean work station, usually parallel to the workbench surface.

laminate — A product made by bonding together two or more layers of material.

laminated — Made of layers.

laminated contact — A switch contact made up of a number of laminations, each making contact with an opposite conducting surface.

laminated core — An iron core for a coil, transformer, armature, etc. It is built up from laminations to minimize the effect of eddy currents. The sheet iron or steel laminations are insulated from each other by surface oxides or by oxides and varnish.

laminated record — A mechanical recording medium composed of several layers of material (normally a thin face of material on each side of a core).

lamination — A single stamping of sheet material used in building up a laminated object such as the core of a power transformer.

Lamont's law — The permeability of steel at any flux density is proportional to the difference between the saturation value of the flux density and its value at the point in question. This law is only approximately accurate and is not true for the initial part of the magnetization curve.

lamp — 1. A generic term for a human-made source of light. By extension, the term is also used to denote sources that radiate in regions of the spectrum adjacent to the visible range. 2. A device for producing light.

lamp bank — An arrangement of incandescent lamps commonly used as a resistance load during electrical tests.

lamp cord — 1. A twin conductor, either twisted or parallel, used for connecting floor lamps and other electric appliances to wall outlets. 2. Flexible stranded conductor cord, rubber or plastic insulated, used in wiring of lamps, household fans, and similar appliances. Not subject to hard usage.

lamp holder — A lamp socket.

lamp housing — A device designed to concentrate and direct a light source by enclosing the source in it and using a concave reflector to direct the light through its only opening.

lamp jack — Special electronic electromechanical component having a frame that holds a lamp and has the contact springs and terminals for applying power to the lamp. Used extensively in jack panels and other types of telephone equipment as a visual-indicating signal device.

Lampkin oscillator — A variation of the Hartley oscillator. Its distinguishing feature is that an approximate impedance match is effected between the tank and grid-cathode circuits.

lamp receptacle — A device that supports an electric lamp and connects it to a power line.

LAN — Abbreviation for local area network. 1. A data communications network spanning a limited geographical area, such as an office, an entire building, or industrial park. It provides communication between computers and peripherals. 3. A baseband or broadband interactive bidirectional communication system for voice, video, or data use on a common cable medium. The network uses some type of switching technology and does not use common-carrier circuits, although it may have gateways or bridges to other public or private networks.

land — 1. The surface between two adjacent grooves of a recording disc. 2. Also called boss, pad, terminal point, blivet, tab, spot, donut. In a printed circuit board, the conductive area to which components or separate circuits are attached. It usually surrounds a hole through the conductive material and the base material. 3. A portion of a conductive pattern usually, but not exclusively, used for the connection and/or attachment of components. 4. Widened conductor area on the major substrate used as an attachment point for wire bonds or the bonding of chip devices.

Landau damping — The damping of a space-charge wave by electrons moving at the phase velocity of the wave.

landing beacon — The radio transmitter that produces a landing beam for aircraft. *See also* landing beam.

landing beam — A highly directive radio signal projected upward from an airport to guide aircraft in making a landing during poor visibility.

landless hole — A plated-through hole without a land(s).

landline — A telegraph or telephone line passing over land, as opposed to submarine cables.

landline facilities — Domestic communications common-carrier's facilities that are within the continental United States.

landmark beacon — Any beacon other than an airport or airway beacon.

land mobile service — A radio service in which communication is between a base station and land mobile stations or between land mobile stations.

land mobile station — A two-way mobile station that operates solely on land.

land radio positioning station — A station in the radio positioning service, not intended to be operated while in motion.

land return — Radiation reflected from nearby land masses and returned to a radar set as an echo.

lands — Bonding points used in the manufacture of microelectronic circuits.

landscape — In word processing, printing a page horizontally across the width of the paper.

land station — A permanent, or fixed, station.

land transportation radio services — Radio-communication services whose transmitting facilities include fixed, land, or mobile stations, operated by and for the sole use of certain land transportation carriers.

Langevin ion — An electrified particle produced in a gas by an accumulation of ions on dust particles or other nuclei.

Langmuir dark space — The nonluminous region surrounding a negatively charged probe inserted into the positive column of a glow or arc discharge.

language — 1. A set of computer symbols, with rules for their combination. They form a code to express information with fewer symbols and rules than there are distinct expressible meanings. 2. A format for computer programs. Ultimately, computers receive their instructions in machine language, binary codes whose meanings are specific to each computer. Machine codes are usually written in hex or octal for easier use by humans. High-level languages, such as BASIC, allow programs to be written in fairly human terms (such as PRINT "NOW IS THE TIME") that are then translated into a sequence of machine codes. 3. A system for representing information and communicating it between people, or between people and machines. 4. A definition of the elements and syntax within which a computer program must be encoded. 5. The means by which people communicate with a computer.

language converter — A data-processing device designed to change one form of data, i.e., microfilm, strip chart, etc., into another (punch card, paper tape, etc.).

language translation — The process performed by an assembler, compiler, or other routine that accepts statements in one language and converts them to equivalent statements in another language.

language translator — A computer system program that translates text written in one language to another language. Assemblers, interpreters, and compilers are examples of language translators.

L-antenna — An antenna consisting of an elevated horizontal wire to which a vertical lead is connected at one end.

lanyard — A device that is attached to certain quick-disconnect connectors and that permits uncoupling and separation of connector halves by a pull on a wire or cable.

lap — 1. A rotation plate covered with liquid abrasive, used for grinding quartz crystals. 2. A fire-resistant, untwisted, ribbonlike form of asbestos felt made from slivers of asbestos fiber blended with cotton or other organic fibers. Used as a wrapping on wire and cable.

LAP — Abbreviation for Link Access Protocol. The data link layer protocol that is used in X.25-based networks in setting up channels between data termination equipment and data communication equipment.

lap computer — A battery-operated computer, small and light enough to be operated on the user's lap.

lap dissolve — In motion pictures or television, simultaneous transition in which one scene is faded down and out while the next scene is faded up and in.

lapel microphone — A microphone worn on the user's clothing.

lap joint — The connecting of two conductors by placing them side by side so that they overlap.

Laplace's law — The strength of the magnetic field at any given point due to any element of a current-carrying conductor is directly proportional to the strength of the current and the projected length of the element, and is inversely proportional to the square of the distance of the element from the point in question.

Laplace transform — A mathematical substitution whose use permits the solution of a certain type of differential equation by algebraic means.

lapping — 1. Bringing quartz crystal plates up to their final frequency by moving them over a flat plate over which a liquid abrasive has been poured. 2. Grinding and polishing such products as semiconductor blanks in order to obtain precise thicknesses or extremely smooth, flat, polished surfaces.

laptop — A small portable computer. Sometimes distinguished as larger than a notebook computer, but sometimes also used as a synonym for notebook computer.

lap winding — An armature winding in which opposite ends of each coil are connected to adjoining segments of the commutator so that the windings overlap.

lap wrap — Tape wrapped around an object in an overlapping condition.

large-scale integrated circuit — An integrated circuit that contains 100 gates or more in a single chip, resulting in an increase in the scope of the function performed by a single device.

large-scale integration — Abbreviated LSI. 1. The simultaneous achievement of large-area circuit chips and optimum density of component packaging for the express purpose of cost reduction by maximization of the number of system interconnections made at the chip level. 2. Monolithic integrated circuits of very high density. Such circuits typically have on a single chip the equivalent of about 200 to thousands of simple logic circuits. The term sometimes describes hybrid ICs built with a number of MSI or LSI chips. 3. A classification of ICs by size, applicable to chips containing more than 100 gates or circuits of equivalent complexity. 4. The technology that produces microcircuits with at least 100 active devices on a single chip. Functional blocks that include several op amps and other devices are examples of LSI devices.

large-signal characteristics — The characteristics of an amplifier when rated (full) output signals are produced.

large-signal dc current gain — The dc output current of a transistor with the dc output circuit shorted, divided by the dc input current producing the dc output current.

large-signal power gain — The ratio of the ac output power to the ac input power under specified large-signal conditions. Usually expressed in decibels (dB).

large-signal short-circuit forward-current transfer ratio — In a transistor, the ratio under specified test conditions of a change in output current to the corresponding change in input current.

large-signal voltage gain — The ratio of the output voltage swing of an operational amplifier to the change in input voltage required to drive the output from zero to this voltage.

Larmor orbit — The path of circular motion of a charged particle in a uniform magnetic field. The motion of the particle is unimpeded in the direction of the magnetic field, but motion perpendicular to the direction of the field is always accompanied by a force perpendicular to the direction of motion and the field.

laryngaphone — Also called a throat microphone. A microphone applied to the throat of a speaker to pick up voice vibrations directly. It is very useful in noisy locations because it picks up only the speaker's voice — no outside noises.

LASCR — Abbreviation for light-activated silicon-controlled rectifier. 1. A pnpn device in which incident light performs the function of gate current; three of the four semiconductor regions are available for circuit connections. A photoswitch. 2. A semiconductor device that is triggered into conduction when the light falling on the base-collector photodiode junction exceeds a given threshold level. Operation of the LASCR is similar to the silicon-controlled rectifier, with the major difference being that an external resistance between gate and cathode (in addition to bias voltage and current) determines light sensitivity. A positive electrical signal applied to the gate can be used to trigger the LASCR, as well as to modify the light sensitivity.

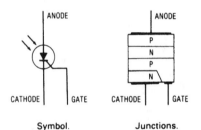

Symbol.   Junctions.

Anode characteristic.

LASCR.

LASCS — Abbreviation for light-activated silicon-controlled switch. 1. A device similar to an LASCR, except that all four semiconductor regions are accessible. 2. A semiconductor device that combines the LASCR and the planar silicon photoswitch (PSPS). Having four terminals, the LASCS can be triggered by light positive signals (at the gate terminal) and negative signals (at the anode gate terminal).

laser — 1. A device for transforming incoherent light of various frequencies of vibration into a very narrow, intense beam of coherent light. The name is derived from the initial letters of "light amplification by stimulated emission of radiation." In the emission of ordinary light, the molecules or atoms of the source emit their radiation

independently of each other, and consequently there is no definite phase relationship among the vibrations in the resultant beam. The light is incoherent. The laser, by means of an optical resonator, forces the atoms of the material of the resonator to radiate in phase. The emitted radiation is stimulated by the excitation of atoms to a higher energy level by means of energy supplied to the device. In the microwave region, the corresponding device is called a maser, and hence the laser is also known as a light maser. 2. A device for producing light by emission of energy stored in a molecular or atomic system when stimulated by an input signal. 3. A mechanically designed semiconductor junction that will optically pump (amplify light) short pulses of high-energy coherent radiation. 4. A device that produces a coherent monochromatic (single wavelength) collimated beam of concentrated light in which the subatomic particles that constitute the beam, known as photons, travel on a parallel axis. Scientists believe that a beam produced by a 5-megawatt laser would be able to melt objects in space at distances exceeding 5000 miles (8045 km).

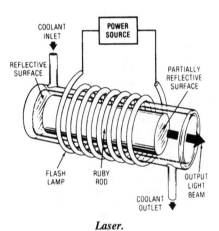

Laser.

laser basic mode — The primary or lowest-order fundamental transverse propagation mode for the emitted light wave of a laser, the emitted energy normally having Gaussian (bell-shaped) distribution in space and being in a single beam, with no side lobes.

laser bonding — 1. A process that forms a metal-to-metal fastened union, using a laser heat source to join conductors. 2. Effecting a metal-to-metal bond of two conductors by welding the two materials together using a laser beam for a heat source.

laser cavity — An optically resonant and hence mode-selecting low-loss structure in which laser action occurs through the buildup of electromagnetic field intensity upon multiple reflection.

laser diode — Abbreviated LD. A junction diode, consisting of positive and negative carrier regions with a pn transition region (junction), that emits electromagnetic radiation (quanta of energy at optical frequencies) when injected electrons under forward bias recombine with holes in the vicinity of the junction. In certain materials, such as gallium arsenide, there is a high probability of radiative recombination producing emitted light, rather than heat, at a frequency suitable for optical waveguides. Some light is reflected by the polished ends and is trapped to stimulate more emission. *See* diode laser; injection laser diode.

laser diode coupler — A coupling device that enables the coupling of light energy from a laser diode

(LD) source to an optical fiber or cable at the transmitting end of an optical fiber data link. The coupler may be an optical fiber pigtail expoxied to the LD. Synonym: LD coupler.

Laserdisc — A laser-read optical video disc system that can hold thousands of video images and hours of sound. A 12-inch plastic disc used to store video, audio, and other data for playback on a video disc player and video monitor.

laser Doppler velocimeter — A laser device utilizing either optical heterodyning or scanning interferometry to measure flow velocities by means of the Doppler shift.

laser drill — A system that uses a laser as the source in the evaporation of localized areas as small as 0.00025 cm in diameter of hard materials such as gemstones and tungsten. A pulsed ruby laser is most commonly used in this type of system.

laser dyes — Class of organic dyes that emit coherent radiation over a wide spectral range.

laser fiber-optic transmission system — A system consisting of one or more laser transmitters and associated fiber-optic cables. During normal operation the laser radiation is limited to the cable.

laser head — A module containing the active laser medium, resonant cavity, and other components within one enclosure, not necessarily including a power supply.

laser holographic camera — A camera system that has a laser, usually of the pulsed ruby type, as a light source; holographic optics; a plate; and a plate holder. It is used often in the laboratory or in production areas for analysis and detection. The laser's nanosecond exposure time is most efficient in freezing moving particles in a permanent hologram for convenient analysis.

laser linewidth — In the operation of a laser, the frequency range over which most of the laser beam's energy is distributed.

laser printer — A printer whose images are formed by laser light impinging on a light-sensitive drum like that of a photocopy machine; where the light strikes the drum, it will hold xerographic ink for transfer to a sheet of paper.

laser protective housing — A protective housing for a laser to prevent human exposure to laser radiation in excess of an allowable established or statutory emission limit for the appropriate class. Parts of the housing that can be removed or displayed and not interlocked may be secured in such a way that removal or displacement of the parts requires the use of special tools.

laser pulse length — Also called laser pulse width. The time duration of the burst of electromagnetic energy emitted by a pulsed laser. It is usually measured at the half-power points, i.e., on a plot of pulse power developed versus time, the time interval between the points that are at 0.5 of the peak of the power curve.

laser pulse width — See laser pulse length.

laser rangefinder — A portable instrument that measures the distance between itself and its target by determining the amount of time it takes for a pulsed laser beam to travel to the target and be reflected back to the instrument. A telescope is used to aim the beam, and a photomultiplier detects the reflected pulse.

laser ranger — A device similar to conventional radar but using high-intensity light rather than microwaves.

laser soldering — A selective soldering technique employing a programmable laser system. The system automatically makes and places solder preforms on one panel while simultaneously laser soldering pins on a second panel. The laser soldering system is effective for high-volume selective soldering of wire-wrapping pins to backplanes, power planes, and PCBs.

laser trim — The adjustment (upward in resistance) of a film resistor value by applying heat from a focused laser source to remove material, i.e., to cut a kerf.

laser trimming — An IC trimming technique that involves focusing a laser on the die to disconnect resistors.

laser welder — A system, similar to a laser drill, that uses the heat from a pulsed laser to weld metals. Because of the rapidity and localization with which the welding takes place, metals of vastly dissimilar melting points can be welded with this system.

laser welding — Process in which thermal energy released by a laser impinging on the surface of a metal is conducted into the bulk of the metal workpiece by thermal conduction, bonding component leads to highly conductive materials, such as copper printed circuitry.

lasing — 1. The phenomenon that occurs when resonant frequency controlled energy is coupled to a specially prepared material, such as a uniformly doped semiconductor crystal that has free-moving or highly mobile, loosely coupled electrons. As a result of resonance and the imparting of energy by collision or close approach, electrons are raised to highly excited energy states; when they move to lower states, they cause quanta of high-energy electromagnetic radiation to be released as coherent light waves. This action takes place in a laser. 2. A unique mode of light production by stimulated emission from excited atoms. The uniqueness of the lasing process is that the light thereby generated tends to be a single frequency, coherent in time and space.

lasing condition or state — The condition of an injection laser corresponding to the emission of predominantly coherent or stimulated radiation.

latch — 1. A feedback loop used in a symmetrical digital circuit (such as a flip-flop) to retain a state. 2. A simple logic storage element. The most basic form consists of two cross-coupled logic gates that store a pulse applied to one logic input until a pulse is applied to the other input; thus, the complementary information is stored in the latch. 3. A name commonly used to refer to a flip-flop (usually a D type) when used for data storage, as opposed to counting and logic functions. 4. To lock into a certain location or state.

latching — A technique for storing an event such as the momentary breaking of a perimeter circuit. The fact that the event has occurred will be available until the latched circuit has been reset. See alarm hold.

latching current — The minimum value of principal current required to maintain a thyristor in the on state after switching from the off state to the on state has occurred and the trigger signal has been removed.

latching relay — Also called bistable relay. 1. A relay with contacts that lock in either the energized or deenergized position, or both, until reset either manually or electrically. 2. A relay with two separate coils, one of which must be energized to change the state of the relay, which will remain in either state without the need for external power. 3. A relay that includes a means of holding the state of the relay in the last or latched position. In effect, the relay has a memory, because the contacts remain open or closed when the coil is not actuated. To change state, the latching relay coil must be reenergized.

latching sensor — A solid-state Hall-effect sensor that has a plus (south pole) maximum and minimum operate point, and a minus (north pole) maximum and minimum release point. Thus, when the sensor is operated with a south pole, it will stay in the operated condition even with removal of the south pole magnet and will release only in the presence of a north pole.

latch-in relay — Also called locking relay. A relay with contacts that remain energized or deenergized until reset manually or electrically.

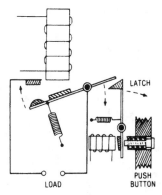

Latch-in relay.

latch mode — A mode of operation for a storage circuit in which all encoder contact closures, even momentary ones, are latched "on."

latch-up — 1. A condition in which the collector voltage in a given circuit does not return to the supply voltage when a transistor is switched from saturation to cutoff. Instead, the collector finds a stable operating point in the avalanche region of the collector characteristics. 2. An unintended stable circuit mode that will not revert to a previous intended circuit mode after removal of a stimulus such as a spurious signal or radiation. The effect is usually caused by parasitic circuit elements. 3. The characteristic of some op amps to remain in positive or negative saturation after their maximum differential input voltage is exceeded. 4. The switching of an electronic circuit to an unintended mode by improper voltage application. 5. An undesirable phenomenon in which either a pnpn or an npnp thyristor-type parasitic structure suddenly turns to an on state, thereby bypassing or shorting out portions of an IC.

latch voltage — The effective input voltage at which a flip-flop changes states.

late contacts — In a relay, contacts that open or close after other contacts when the relay operates.

latency — 1. In a serial storage computer system, the time necessary for the desired storage location to appear under the drum heads. 2. In computers, the time required to establish communication with a specific storage location, not including transfer time, i.e., access time less word time. 3. A state of seeming inactivity, such as that occurring between the instant of stimulation and the beginning of response. 4. A delay encountered in a computer when waiting for a specific response. Latency is caused by propagation delays and the queuing of disks or tapes when randomly addressed.

latency time — The time required to shift to any given bit (word) in a serial memory, such as in CCDs and bubble memories.

latent image — A stored image (e.g., the one contained in the charged mosaic capacitance in an iconoscope).

lateral chromatic aberration — Aberration that affects the sharpness of images off the axis. This occurs because different colors produce different magnifications.

lateral compliance — The force required to move the reproducing stylus from side to side as it follows the modulation on a laterally recorded record.

lateral-correction magnet — In a three-gun picture tube, an auxiliary component used for positioning the blue beam horizontally so that beam convergence will

be obtained. It operates on the principle of magnetic convergence and is used in conjunction with a set of pole pieces mounted on the focus element of the blue gun.

lateral forced-air cooling — A method of heat transfer that employs a blower to produce side to side circulation of air through or across the heat dissipators.

lateral loss — A power loss, expressed in decibels, due to the deviation from optimum coaxial alignment of the ends of separable optical conductors.

lateral recording — A mechanical recording in which the groove modulation is perpendicular to the direction of motion of the recording medium and parallel to its surface.

latex — Rubber material used for insulation of wire.

lattice — 1. In navigation, a pattern of identifiable intersection lines placed in fixed positions with respect to the transmitters that establish them. 2. The geometrical arrangement of atoms in a crystalline material.

lattice network — A network composed of four branches connected in series to form a mesh. Two nonadjacent junction points serve as input terminals, and the remaining two as output terminals.

lattice structure — In a crystal, a stable arrangement of atoms and their electron-pair bonds.

lattice-wound coil — *See* honeycomb coil.

launch angle — In an optical fiber or fiber bundle, the angle between the input radiation vector, i.e., the input light chief ray, and the axis of the fiber radiation vector, i.e., the axis of the fiber or fiber bundle. If the ends of the fibers are perpendicular to the axis of the fibers, the launch angle is equal to the angle of incidence when the ray is external and the angle of refraction when initially inside the fiber.

launch complex — The entire launch, control, and support system required for launching rockets.

launching — The transferring of energy from a coaxial cable or shielded paired cable in a waveguide.

lavalier microphone — A microphone with acoustical and vibration-isolation properties suiting it to use for speech pickup from a position on the speaker's chest. Lavalier mikes are fitted with a band or strap for hanging around the neck, and are frequently used when it is important that the mike not be conspicuously visible (as to a TV audience). The use of this mike frees the speaker's hands and allows a certain amount of freedom to move about.

lawn mower — 1. In facsimile, a term often used when referring to a helix-type recorder mechanism. 2. A type of rf preamplifier used with a radar receiver.

law of electric charges — Like charges repel; unlike charges attract. *See also* Coulomb's law.

law of electromagnetic induction — *See* Faraday's laws, 3.

law of electromagnetic systems — Every electromagnetic system tends to change its configuration so that the flux of magnetic induction will be a maximum.

law of electrostatic attraction — *See* Coulomb's law.

law of magnetism — Like poles repel; unlike poles attract.

law of normal distribution — The Gaussian law of the frequency distribution of any normal, repetitive function. It describes the probability of the occurrence of deviants from the average.

law of reflection — The angle of reflection is equal to the angle of incidence — i.e., the incident, reflected, and normal rays all lie in the same plane.

laws of electric networks — *See* Kirchhoff's laws, 1 and 2.

lay — 1. Pertaining to wire and cable, the axial distance required for one cable conductor or conductor strand to complete one revolution about the axis around which it is cabled. 2. The distance between successive points where the same strand (or insulated conductor) when twisted with one or more strands (or insulated conductors) presents itself in the same position. 3. The length measured along the axis of a wire or cable required for a single strand (in stranded wire) or conductor (in cable) to make one complete turn about the axis of the conductor or cable.

layer—1. The consecutive turns of a coil lying in a single plane. 2. One of several films in a multiple-film structure on a substrate.

layer-to-layer adhesion—The tendency for adjacent layers of recording tape in a roll to adhere, particularly after prolonged storage under conditions of high temperature and/or humidity.

layer-to-layer signal transfer—The magnetization of a layer of tape in a roll by the field from a nearby recorded layer. The magnitude of the induced signal tends to increase with storage time and temperature, and to decrease after the tape is unwound. These changes are a function of the magnetic instability of the oxide.

layer-to-layer spacing—The thickness of dielectric material between adjacent layers of conductive circuitry in a multilayer printed board.

layer winding—A coil-winding method in which adjacent turns are placed side by side and touch each other. Additional layers may be wound over the first and are usually separated by sheets of insulation.

layout—1. Diagram indicating the positions of parts on a chassis or panel. 2. The actual positions of the parts themselves. 3. The topological arrangement of conductors

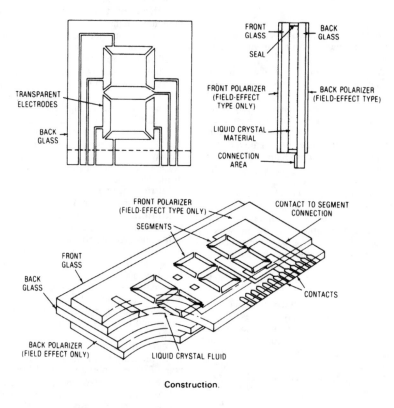

Construction.

Drive electronics.

LCD.

and components in the design of integrated circuits. It precedes the artwork. 4. A visual representation of a complete physical entity, usually to scale.

layout wiring drawing — A type of circuit diagram, made to show explicitly each wire, its gage, color coding, and terminations.

lay-up — The technique of registering and stacking layers of a multilayer board before the laminating cycle.

L-band — 1. Microwave band in which the wavelengths are at or near 23.5 cm. 2. A radio-frequency band of 390 to 1550 MHz and corresponding wavelengths of 77 to 19 cm. Used for mobile communications.

L-carrier system — A telephone carrier system employed on coaxial-cable systems and microwave line-of-sight and tropospheric-scatter radio systems. It occupies a frequency band extending from 68 kHz to over 8 MHz.

LCC — Abbreviation for leadless chip carrier. A surface-mounted package having metallized contacts (terminals) at its periphery. Usually made of ceramic material.

LCD — Abbreviation for liquid crystal display. 1. A seven-segment (typically) display device consisting basically of a liquid crystal hermetically sealed between two glass plates. One type of LCD (dynamic scattering) depends on ambient light for its operation, while a second type depends on a backlighting source. The readout is either dark characters on a dull white background or white on a dull black background. LCDs have very low power requirements. 2. A display whose segments or elements consist of transparent electrodes etched on glass separated by a liquid that has some crystalline properties, including orderly molecular alignment. Voltage applied to opposing electrodes alters the alignment in the liquid between them, affecting the passage of light through the region (reflected or transmitted), rendering the segments visible by contrast. 3. An optically passive device, in that it does not generate light to produce contrast, whose operation depends on the ability of the liquid crystal to rotate plane-polarized light relative to a pair of crossed polarizers attached to the outside of the display. Rotation of the plane of polarization is a function of the applied field and decreases with increasing field or voltage.

LC product — Inductance (L) in henrys multiplied by capacitance (C) in farads. See figure on p. 416.

L/C ratio — Inductance in henrys divided by capacitance in farads.

LCS — Abbreviation for loudness-contour selector.

L cut — A trim notch in a film resistor that is created by the cut starting perpendicular to the resistor length and turning 90° to complete the trim parallel to the resistor axis, thereby creating an L-shaped cut.

LD — See laser diode.

LD coupler — See laser diode coupler.

LDCS — Abbreviation for long-distance control system. A computer-based communication management system (Datapoint Corporation) that routes long-distance calls over least-cost lines and maintains accounting data on calls.

L-display — Also called L-scan. A radar display in which the target indication appears as two horizontal blips, one extending to the right and one to the left from a vertical time base. Azimuth pointing error is indicated by relative blip amplitude, and range is indicated by the position of the signal along the base line.

leaching — The process of dissolving and removing impurities and soluble components from plated items, tank materials, and the like.

lead — 1. A wire to or from a circuit element. 2. To precede (the opposite of lag). 3. A wire, with or without terminals, that connects two points in a circuit. 4. A conductive path that is usually self-supporting.

lead-acid cell — Also called lead cell. A cell in an ordinary storage battery. It consists of electrodes (plates) immersed in an electrolyte of dilute sulfuric acid. The electrodes contain certain lead oxides that change their composition as the cell is charged or discharged.

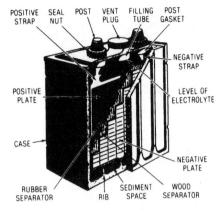

Lead-acid cell.

lead-acid storage cell — A storage cell in which both plates are lead-antimony or lead-calcium grids filled with spongy lead for the negative plate and lead peroxide for the positive plate. During discharge, the material in both plates is converted to lead sulfide. The electrolyte is a solution of sulfuric acid with a specific gravity of 1.200. The cell voltage is nominally 2 volts but rises to 2.15 volts on float and drops to 1.85 volts on discharge.

lead cell — See lead-acid cell.

lead-covered cable — A cable with a lead sheath. The sheath offers protection from the weather and mechanical damage to the wires contained therein.

lead dress — The placement or routing of wire and component leads in an electrical circuit.

leader — 1. Special nonmagnetic tape that can be spliced to either end of a magnetic tape to prevent damage and possible loss of recorded material and to indicate visually where the recorded portion of the tape begins and ends. 2. Tough, nonmagnetic tape spliced ahead of the recorded material on a tape that is expected to receive rough or frequent handling. Usually has one matte-finished surface for writing on, and often available in a variety of colors for coding purposes.

leader cable — A navigational aid in which the path to be followed is defined by a magnetic field around a cable.

leader tape — Also called timing tape. Plain nonmagnetic tape for splicing to either end of magnetic tape to facilitate threading and preserve recorded material, or for splicing between recorded tapes to separate selections or provide pauses.

lead frame — 1. A metal frame that holds the leads of a plastic encapsulated package (DIP) in place before encapsulation and is cut away after encapsulation. 2. The metal part of a solid-state device package that achieves electrical connection between the die and other parts of the system of which the IC is a component. Large-scale integrated circuits are welded onto lead frames in such a way that leads are available to facilitate making connections to and from the various solid-state devices to the packages. 3. The metallic portion of the device package that completes the electrical connection path from the die or dice and from ancillary hybrid circuit elements to external circuits.

lead-in — The conductor that provides the path for rf energy between the antenna and the radio/television receiver or transmitter.

leading blacks — Also called edge effect. In a television picture, the condition in which the edge preceding a white object is overshaded toward black (i.e., the object appears to have a preceding, or leading, black border).

leading current — 1. Current that reaches maximum before the voltage that produces it does. A leading current flows in any predominantly capacitive circuit. 2. In an alternating-current circuit in which the net reactance is capacitive, a current wave that precedes in phase the voltage wave that produces it.

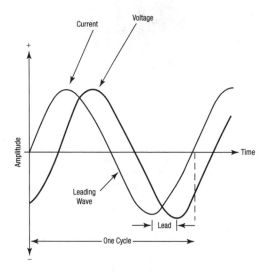

Leading current.

leading edge — That transition of a pulse which occurs first.

leading-edge pulse time — The time required by a pulse to rise from its instantaneous amplitude to a stated fraction of its peak amplitude.

leading ghost — A twin image appearing to the left of the original in a televised picture.

leading load — A predominantly capacitive load — i.e., one in which the current leads the voltage.

lead-in groove — Also called a lead-in spiral. A blank spiral groove around the outside of the record. Its pitch is usually much greater than the other grooves and is used to quickly lead the needle into the beginning of the recorded groove.

leading whites — Also called edge effect. In a television picture, the condition in which the edge preceding a black object is shaded toward white (i.e., the object appears to have a preceding, or leading, white border).

leading zeros — Zeros placed in front of a number to use up all blanks spaces in a data field.

lead-in insulator — A tubular insulator through which cables or wires are brought inside a building.

lead-in spiral — *See* lead-in groove.

lead-in wires — Wires that carry current into a building (e.g., from an antenna). *See* down lead.

lead-length compensation — In dc ammeters for use with external shunts, the leads that connect to the shunt become an integral part of the total instrument. An adjustable resistor is often included to compensate for the resistance of the leads and to improve overall accuracy.

leadless chip carrier — *See* LCC.

leadless inverted device — Abbreviated LID. A shaped metallized ceramic form used as an intermediate carrier for semiconductor chip devices, especially adapted for attachment to conductor lands of a thick- or thin-film network by reflow solder bonding.

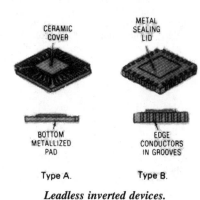

Leadless inverted devices.

lead network — A network, either ac or dc, designed to provide error-rate damping in the controlling device of a servo system.

lead-out groove — Also called a throw-out spiral. A blank spiral groove on the inside of a recording disc, next to the label. It is generally much deeper than the recording groove and is connected to either the locked or eccentric groove.

lead-over groove — Also called a crossover spiral. On disc records containing several selections, the groove in which the needle travels as it crosses from one selection to the next.

lead polarity of transformer — Also called polarity. A designation of the relative instantaneous directions of currents in the leads of the transformer. Primary and secondary leads are said to have the same polarity when, at a given instant, the current enters the primary lead in question and leaves the secondary lead in question in the same direction as though the two leads formed a continuous circuit.

lead screw — 1. In a recording, a threaded road that guides the cutter or reproducer across the surface of a disc. 2. In facsimile, a threaded shaft that moves the scanning mechanism or drum lengthwise.

lead selenide (PbSe) cell — A thin-film photoconductive cell that is sensitive to the infrared region. The photosensitive material of the cell is composed of lead selenide, and the cell is used in the detection of infrared radiation.

lead sulfide (PbS) cell — A photoconductive cell having its greatest sensitivity in the infrared region. The photosensitive material of the cell is lead sulfide, which is deposited on a glass plate.

lead time — In the display of a random sampling oscilloscope, the interval represented that occurs immediately before trigger recognition.

lead wires — Wire conductors for intraconnections or input/output leads.

leaf insulator — Leaf-spring-shaped insulator located in a switch stack adjacent to a contact spring or actuator spring to keep that spring from making electrical contact with an adjacent spring or other metallic surface.

leak — A condition that causes current to be shunted away from its destination through a low resistance.

leakage — 1. Undesired flow of electricity over or through an insulator. 2. The portion not utilized most effectively in a magnetic field (e.g., at the end pieces of an electromagnet).

leakage coefficient — Ratio of total to useful flux produced in the neutral section of a magnet.

leakage current — 1. An undesirable small-value stray current that flows through (or across the surface of) an insulator or the dielectric of a capacitor. 2. A current that flows between two or more electrodes in a tube other than across the interelectrode space. 3. Current prior to switching at a specified voltage. 4. Undesirable flow of current through or over a surface of an insulating material or insulator. 5. All currents, including capacitively coupled currents, that may be conveyed between energized parts of a circuit and ground or other parts. 6. The conduction current through a capacitor when a direct voltage is applied.

leakage flux — The flux that does not pass through the air gap, or useful part, of the magnetic circuit.

leakage inductance — A self-inductance due to the leakage flux generated in the winding of a transformer.

leakage power — In tr and pre-tr tubes, the radio-frequency power transmitted through a fired tube.

leakage radiation — Spurious radiation in a transmitting system — i.e., radiation from other than the system itself.

leakage rate — A laboratory procedure used to determine the amount and duration of resistance of an article to a specific set of destructive forces or conditions.

leakage reactance — 1. The reactance represented by the difference in value between two mutually coupled inductances when their fields are aiding and then opposing. 2. That portion of the reactance of a transformer primary which is due only to leakage flux.

leakage resistance — The normally high resistance of the path over which leakage current flows.

leakance — The reciprocal of insulation resistance.

leaktight — *See* hermetic.

leaky — Usually applied to a capacitor in which the resistance has dropped so far below normal that objectionably high leakage current flows.

leaky waveguide — A waveguide with a narrow longitudinal slot, permitting a continuous energy leak.

leaky waveguide antenna — An antenna constructed from a long waveguide with radiating elements along its length. It has a very sharp pattern.

leapfrogging — The process of phasing, or delaying, the ranging pulse of a tracking radar in order to move or shift (on the scope presentation) the tracking gate (at target blip) past the target blip from another radar.

leapfrog test — A computer check routine using a program that calls for performing a series of arithmetical or logical operations on one section of memory locations, transferring to another section, checking correctness of transfer, and repeating the series of operations. Eventually, all storage positions are checked by this process.

learning curve — The improvement that occurs in manufacturing processes with experience.

leased line — Also called dedicated or private line. A semipermanent leased telephone circuit that connects two or more points and is continuously available to the subscriber. *See also* dedicated line.

least maximum deviation — A manner of expressing nonlinearity as a deviation from a straight line for which the deviations for proportional or normal linearity are minimized.

least mechanical equivalent of light — The radiant power that is contained in 1 lumen at the wavelength of maximum visibility. It is equal to 1.46 milliwatts at a wavelength of 555 nanometers.

least significant bit — Abbreviated LSB. 1. In a system in which a numerical magnitude is represented by a series of binary (i.e., two-valued) digits, that digit (or bit) that carries the smallest value or weight; usually the rightmost bit. 2. The lowest-order bit or the bit with the least weight. Binary digit having a weight of 2^0, or 1. 3. 3. Smallest value that can be digitized; lowest-order digital output.

least significant digit — Abbreviated LSD. 1. The digit that has the lowest place value in a number; usually the rightmost digit. 2. Number at the extreme right of a group of numbers. Example: 6937. Digit 7 is the LSD.

least voltage coincidence detection — Abbreviated LVCD. A system that provides protection against interfering signals by blocking all signals except those having a pulse-repetition frequency the same as or some exact multiple of the radar-set pulse-repetition frequency.

Lecher line — *See* Lecher wire.

Lecher oscillator — A device for producing standing waves on two parallel wires called Lecher wires.

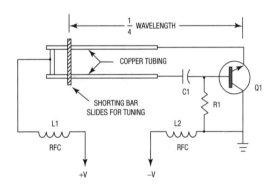

Lecher oscillator.

Lecher wire — 1. A type of transmission line used to measure wavelength, consisting of a pair of wires whose electrical length is adjustable. If a source of radio frequency is coupled to one end of the line and the line is adjusted until a set of standing waves is formed, the wavelength may be determined by measurement of the distance between adjacent nodes. 2. Two parallel wires with a movable shunt that are connected to the output of a radio-frequency source and are used mainly to measure wavelengths shorter than about 10 meters.

Leclanche cell — 1. Type of dry cell comprising a positive carbon pole contained in a porous vessel filled with manganese dioxide, the whole assembly standing in a container of an ammonium chloride solution that also contains the negative zinc pole. The electromotive force generated by a cell of this type is approximately 1.5 volts. 2. An ordinary dry cell. It is a primary cell with a positive electrode of carbon and a negative electrode of zinc in an electrolyte of sal ammoniac and a depolarizer of manganese dioxide.

LED — Abbreviation for light-emitting diode. 1. A pn junction semiconductor device specifically designed to emit light when forward biased. This light can be one of several visible colors — red, amber yellow, or green — or it may be infrared and thus invisible. Electrically, a LED is similar to a conventional diode in that it has a relatively low forward voltage threshold. Once this threshold is exceeded, the junction has a low impedance and conducts current readily. This current must be limited by an external circuit, usually a resistor. The amount of light emitted by a LED is proportional to the forward current over a broad range, thus it is easily controlled, either linearly or by pulsing. The LED is extremely fast in its light output response after the application of forward current. Typically, the rise and fall times are measured in nanoseconds. LEDs are constructed in either a multisegment (typically seven segments) display format or a dot-matrix display format. Red is used most often because of its lower cost. 2. A diode that operates similar to a laser diode, with the same output power level, the same output limiting modulation rate, and the same operational current densities, i.e., thousands of amperes per square centimeter, which causes catastrophic and graceful degradation, but with greater simplicity, tolerance, and ruggedness and about ten times the spectral width of its radiation.

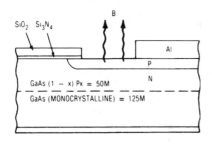

Layering.

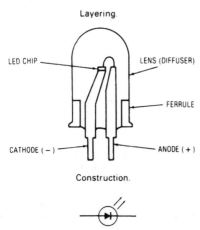

Construction.

Symbol.

LED.

ledger balance — A facility used with message switching equipment to ensure that no messages are lost within the center. It involves comparing the number of addresses received with the number of addresses transmitted.

left-handed polarized wave — Also called counterclockwise-polarized wave. An elliptically polarized transverse electromagnetic wave in which the electric intensity vector rotates counterclockwise (looking in the direction of propagation).

left-hand rule — *See* Fleming's rule, 1.

left-hand taper — The greater resistance in the counterclockwise half of the operating range of a rheostat or potentiometer than in the clockwise half (looking from the shaft end).

left (or right) signal — The electrical output of a microphone or combination of microphones placed so as to convey the intensity, time, and location of sound originating predominantly to the listener's left (or right) of the center of the performing area.

left (or right) stereophonic channel — The left (or right) signal as electrically reproduced in the reception of FM stereophonic broadcasts.

leg — A section or branch of a component or system (e.g., one of the windings of a transformer).

legend — A table of symbols or other data placed on a map, chart, or diagram to assist the reader in interpreting it.

Lenard rays — Cathode rays that emerge from a special vacuum tube through a thin glass window or metallic foil.

Lenard tube — An electron tube in which the beam can be taken through a section of the wall of the evacuated enclosure.

length of a scanning line — 1. The length of the path traced by the scanning or recording spot as it moves from line to line. 2. On drum-type equipment, the circumference of the drum. 3. The spot speed divided by the scanning-line frequency.

lens — 1. An optical device that focuses light by refraction. 2. An electrical device that focuses microwaves by refraction or diffraction. 3. An acoustic device that concentrates sound waves by refraction. 4. An electronic optical device that focuses electrons. 5. A transparent optical component consisting of one or more pieces of optical glass with surfaces so curved (usually spherical) that they serve to converge or diverge the transmitted rays of an object, thus forming a real or virtual image of that object.

lens antenna — A microwave antenna with a dielectric lens placed in front of the dipole or horn radiator so that the radiated energy is concentrated into a narrow beam.

lens disc — A television scanning disc having a number of openings arranged in a spiral, with a lens set into each opening.

lens speed — Refers to the ability of a lens to transmit light, represented as the ratio of the focal length to the diameter of the lens. A fast lens would be rated f/1.2; a much slower lens might be designated as f/8. The larger the f-number, the slower the lens.

lens turret — On a camera, an arrangement that accommodates several lenses and can be rotated to facilitate their rapid interchange.

Lenz's law — The current induced in a circuit due to a change in the magnetic flux through it or to its motion in a magnetic field is so directed as to oppose the change in flux or to exert a mechanical force opposing the motion. If a constant current flows in a primary circuit A and if, by motion of A or the secondary circuit B, a current is induced in B, the direction of the induced current will be such that, by its electromagnetic action on A, it tends to oppose the relative motion of the circuits.

LEO — Abbreviation for low earth orbit. Any orbit around the earth substantially below the geostationary satellite orbit, generally within several hundred kilometers above the earth's surface and usually inclined to the equatorial plane.

Lepel discharger — A quenched spark gap used in early radiotelegraph transmitters employing shock excitation.

letter quality — Resembling the output of a typewriter.

letter-quality printer — A machine that prints like a typewriter, by pressing complete characters through a ribbon onto paper.

letters shift — In the Baudot code, a control character following which all characters are interpreted as being in the group containing letters (lower case).

let-through current — The current that actually passes through a circuit breaker under short-circuit conditions.

level — 1. The magnitude of a quantity in relation to an arbitrary reference value. Level normally is stated in the same units as the quantity being measured (e.g., volts, ohms, etc.). However, it may be stated in units that express the ratio to a reference value (e.g., decibels). 2. A voltage that remains constant over a long period of time. 3. In describing codes or characters, a bit or element. 4. The intensity of an electrical signal.

level above threshold — Also called sensation level. The pressure level of a sound in decibels above its threshold of audibility for the individual listener.

level compensator — 1. An automatic gain control that minimizes the effect of amplitude variations in the received signal. 2. A device that automatically controls the gain in telegraph-receiving equipment.

level indicator — A device for showing visually the level of the audio signal, as a means of establishing the optimum amount of signal being fed to the tape.

level shifting — The process of changing a differential signal input to a single-ended output within an operational amplifier.

level translator — A circuit that accepts digital input signals at one pair of voltage levels and delivers output signals at a different pair of voltage levels. For example, a circuit to translate the -0.8-V "zero" and 1.6-V "one" of ECL to the -0.8-V "zero" and -4.2-V "one" suitable for COS/MOS.

level-triggered flip-flop — A flip-flop that responds to the voltage level rather than the rate of change of an input signal.

lever switch — Commonly referred to as a key lever or lever key. 1. A hand-operated switch for rapidly opening and closing a circuit. 2. A switch having a lever (toggle), whose movement results either directly or indirectly in the connection or disconnection of the switch terminations in a specified manner.

Lewis antenna — A microwave scanning antenna consisting of a lensed flat horn that tapers to a narrow rectangular opening across which a waveguide feed can be moved to scan the beam. The horn is folded by the incorporation of a 45° reflecting strip, and the thin rectangular end is formed into a circular annulus, around which the feed can be rotated. The deformed parallel-plate region that results has a conical shape with the feed circle as base.

Leyden jar — The original capacitor. It consists of metal foil sheets on the inside and outside of a glass jar. The foil serves as the plates and the glass as the dielectric.

LF — Abbreviation for low frequency (i.e., between 30 and 300 kHz).

LFM — A VHF fan-type marker. It is low powered (5 watts) and has a range of only 10 miles or less.

librarian — A system program that is responsible for creating, editing, and deleting software libraries.

library — 1. A collection, usually stored on magnetic tape, of computer programs or subroutines for special purposes. 2. A group of standard, proven computer routines

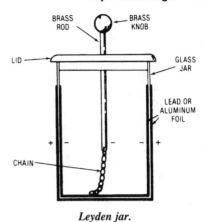

Leyden jar.

that can be incorporated into larger routines. 3. A collection of system and/or user tasks that may be executed by other tasks in the system. The major reason for libraries is to prevent software redesign each time a function is needed by a task. 4. A set of frequently used programs or program pieces.

LIC — Abbreviation for linear integrated circuit.

Lichtenberg figure camera — Also called klydonograph or surge-voltage recorder. A device for indicating the polarity and approximate crest value of a voltage surge by the appearance and dimensions of the Lichtenberg figure produced on a photographic plate or film. The emulsion coating of the plate or film contacts a small electrode coupled to the circuit in which the surge occurs. The film is backed by an extended plane electrode.

LID — *See* leadless inverted device.

lie detector — Also called a polygraph. An electronic instrument that measures the blood pressure, temperature, heart action, breathing, and skin moisture of the human body. Abrupt or violent changes in these variables are said to indicate that the subject is not telling the truth.

LIF connector — Abbreviation for low insertion force connector. Type of printed circuit board connector unit in which mating and unmating forces are reduced 70 to 90 percent. Typical engaging and separating forces in these devices are 0.5 oz/contact, contrasted with the 8 to 16 oz required for conventional printed circuit board connectors.

life — 1. The expected number of full excursions over which a transducer would operate within the limits of the applicable specification. 2. The number of performance hours, days, years, or actual operations an item is designated to meet.

life aging — 1. Burn-in test that moderates the elevation of temperature and extends the time period in order to test the overall device quality as opposed to infant mortality. 2. Long-period operation of items, components, or devices to test the consistency of device finality.

life cycle — A test that indicates the time span before failure; the test occurs in a controlled, usually accelerated environment.

life test — The test of a component or unit under the conditions that approximate, or simulate by acceleration, a normal lifetime of use. The test is performed to determine life expectancy or reliability throughout a predetermined life expectancy.

lifetime — 1. The average time interval between the introduction and recombination of minority carriers in a semiconductor. 2. The time it takes a thermally generated or photogenerated electron-hole pair to recombine.

LIFO—Last in, first out; method of storing and retrieving data in a stack, table, or list. *Compare* FIFO.

lifter—In a tape recorder, a movable rod or guide that draws the tape away from the heads during fast-forward or rewind modes to eliminate needless head wear. Lifters work automatically on most machines in either high-speed mode.

lifting magnet—A powerful electromagnet used on the end of a crane to lift iron and steel objects, which can be dropped instantly by merely cutting off the current.

light—1. Radiant energy within the wavelength limits perceptible by the average human eye (roughly, between 400 and 700 nanometers). Although ultraviolet and infrared emissions will excite some types of photocells, they are usually not considered light. 2. In combination with other terms, a device used as a source of luminous energy (e.g., a pilot light). 3. Radiant energy transmitted by wave motion with wavelengths from about 0.3 μm to 30 μm; this includes visible wavelengths (0.38 μm to 0.78 μm and those wavelengths such as ultraviolet and infrared, which can be handled by optical techniques used for the visible region. In more restricted usage, radiant energy within the limits of the visual spectrum.

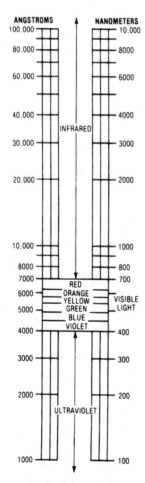

Light, 1 (spectrum).

light-activated silicon-controlled rectifier— *See* LASCR.

light-activated silicon-controlled switch—*See* LASCS.

light-activated switch—A semiconductor diode that is triggered into conduction by light irradiation of a light-sensitive part of the semiconductor pellet.

light amplifier—A device that serves to emit light of the same wavelength as the input light, only with an increase in intensity. It may be a solid-state device comprised of photoconductive and luminescent layers contained between two electrodes.

light-beam cathode-ray-tube recorder— Recorder using an electron beam to make multiple traces on a CRT screen. Traces are reflected from a fixed plane mirror onto moving photosensitive paper via an optical system.

light-beam galvanometer—A modified form of the D'Arsonval meter movement in which a small mirror is cemented to a moving coil mounted in the field of a permanent magnet. Current through the coil causes the coil to be deflected angularly, and the mirror reflects a beam of light onto a moving strip of photographic paper. The developed chart shows the waveform of the current through the coil.

light-beam instrument—An instrument in which a beam of light is the indicator.

light-beam oscillograph—Recorder using a mirror on a galvanometer to achieve recording response to 5 kHz.

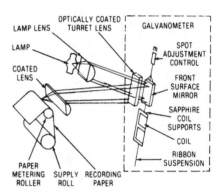

Light-beam oscillograph.

light-beam pickup—A phonograph pickup utilizing a beam of light as a coupling element of the transducer.

light chopper—A device for interrupting a light beam. It is frequently used to facilitate amplification of the output of a phototube on which the beam strikes.

light current—The current that flows through a photosensitive device, such as a phototransistor or a photodiode, when it is exposed to illumination or irradiance.

light-dimming control—A circuit, often employing a saturable reactor, used to control the brightness of the lights in theaters, auditoriums, etc.

light-emitting diode—Abbreviated LED. A pn junction that emits light when biased in the forward direction. *See* LED.

light-emitting diode coupler—Abbreviated LED coupler. A coupling device that enables the coupling of light energy from a light-emitting diode (LED) source to an optical fiber or cable at the transmitting end of an optical fiber data link. The coupler may be an optical fiber pigtail epoxied to the LED.

light flux — *See* luminous flux, 1.

light guide — 1. An assembly of optical fibers and other optical elements mounted and finished in a component that is used to transmit light. 2. A conduit made up of fibers randomly collected or bunched in a group; it conducts light and images.

light gun — A photoelectric cell used by computer operators to take specific actions in assisting and directing computer operation. So called because of its gun-like case.

lighthouse tube — An ultrahigh-frequency electron tube shaped like a lighthouse and having disc-sealed planar elements. *See also* disc-seal tube.

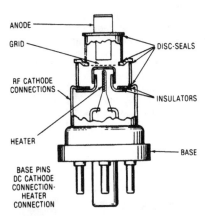

Lighthouse tube (cutaway view).

lighting outlet — An outlet for direct connection of a lamp holder, lighting fixture, or pendant cord termination in a lamp holder.

light intensity cutoff — In a photoelectric alarm system, the percent reduction of light that initiates an alarm signal at the photoelectric receiver unit.

light level — The amount of (or intensity of) light falling upon a subject.

light load — A fraction of the total load the device is designed to handle.

light meter — An electron device that contains a photosensitive cell and calibrated meter for the measurement of light levels.

light microsecond — The unit for expressing electrical distance. It is the distance over which light travels in free space in 1 microsecond (i.e., about 983 feet, or 300 meters).

light modualtion — Variation in the intensity of light, usually at audio frequencies, for communications or motion-picture sound purposes.

light modulator — The device for producing the sound track on a motion-picture film. It consists of a source of light, an appropriate optical system, and a means for varying the resulting light beam (such as a galvanometer or light valve).

light negative — Having a negative photoconductivity when subjected to light.

lightning arrester — A device to prevent damage to electrical equipment by transient overvoltages whether from lightning or switching. Spark gaps that can only be bridged by voltages above those used in the equipment allow the higher voltages to be discharged to ground.

lightning generator — A generator of high-voltage surge (e.g., for testing insulators).

lightning rod — A pointed metal rod carried above the highest point of a pole or building, and connected to earth by a heavy copper conductor, for the purpose of carrying a direct lightning discharge directly to earth without damage to the protected structure.

lightning surge — A transient disturbance in an electric circuit caused by lightning.

lightning switch — A switch for connecting a radio antenna to ground during electrical storms.

light pen — 1. A light-sensitive device used with a computer-operated CRT display for selecting a portion of the display for action by the computer. 2. A photosensor placed in the end of a penlike probe. It is used in conjunction with a CRT display for drawing, erasing, or location characters. Operation is by comparsion of the time it senses a light pulse to the scanning time of the display. 3. A hand-held data-entry device used only with refresh displays. It consists of an optical lens and photocell, with associated circuitry, mounted in a wand. Most light pens have a switch on the barrel that makes the pen sensitive to light from the screen. An activated light pen, when pointed at a vector or character on the screen, will generate an interrupt. It is then possible to identify the vector or character since the display stopped refreshing when the item was drawn that caused the interrupt. The most common uses of light pens are light-button selection and tracking.

light pencil — A narrow cone of light rays that diverges from a point source or converges to an image point.

light pipe — 1. A bundle of transparent fibers that can transmit light around corners with small losses. Each fiber transmits a portion of the images through its length, reflection being caused by the lower refractive index of the surrounding material, usually air. 2. Transparent matter that usually is drawn into a cylindrical or conical shape through which light is channeled from one end to the other by total internal reflections. Optical fibers are examples of light pipes.

light positive — Having positive photoconductivity — i.e., increasing in conductivity when subjected to light.

light-powered telephone — Technology that relies on a highly efficient photodetector that can detect incoming light signals at one frequency and transmit outgoing signals at another, thus permitting the sending and receiving of light signals over one fiber with a single device.

light ray — 1. A very thin beam of light. 2. A line, perpendicular to the wavefront of light waves, indicating their direction of travel and representing the light wave itself.

light relay — A photoelectric device that opens or closes a relay when the intensity of a light beam changes.

light-sail — A method of spacecraft propulsion using a giant sail to catch the solar wind, a nonfictional stream of ionized gas particles constantly emitted from the sun at speeds of up to 2 million miles an hour (3.2 million kilometers per hour).

light sensitive — Exhibiting a photoelectric effect when irradiated (e.g., photoelectric emission, photoconductivity, and photovoltaic action).

light-sensitive Darlington amplifier — Two stages of transistor amplification in one light-detector device. Darlingtons give much higher gain than single transistors.

light-sensitive tube — A vacuum tube that changes its electrical characteristics with the amount of illumination.

light source — Any object capable of emitting light. (In fiber optics, the light source is normally either an LED or a laser.)

light source power — The electrical power used to stimulate any light source. Power supplies may be step-up or stepdown transformers; rectifiers to convert ac to dc; ammeters and voltmeters to observe the input to the source; and regulators and variable resistors, for maintaining constancy of input.

light-spot scanner — Also called a flying-spot scanner. A television camera in which the source of illumination is a spot of light that scans the scene to be televised. The picture signal is generated in a phototube, which picks up light either transmitted through the scene or reflected from it.

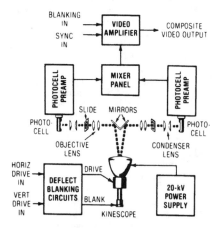

Light-spot scanner.

light valve — A device whose light transmission can be varied in accordance with an externally applied electrical quantity such as voltage, current, an electric or magnetic field, or an electron beam.

lightwave communications — Also called optical communications. 1. In fiber optics communications, using light, instead of an electric current, to carry the information. 2. That aspect of communications and telecommunications devoted to the development and use of equipment that uses electromagnetic waves in or near the visible region of the spectrum for communication purposes. Lightwave communication equipment includes sources, modulators, transmission media, detectors, converters, integrated optic circuits, and related devices, used for generating and processing light waves. The term *optical communications* is oriented toward the notion of optical equipment, whereas the term *lightwave communications* is oriented toward the signal being processed.

light-year — 1. The distance traveled by light in one year, or about 5,880,000,000,000 miles or 9,500,000,000,000 km, roughly 6 trillion miles (9.5 trillion kilometers). A parsec (parallex second) is equal to 3.26 light-years. Both are units of distance, not time.

limit bridge — A form of Wheatstone bridge used for rapid routine production testing. Conformity with tolerance limits, rather than exact value, is determined.

limit cycle — A mode of control system operation in which the controlled variable cycles between extreme limits, with the average near the desired value.

limited continuous word/speech — Voice recognition capability for certain sets of words uttered without pause (typically digits such as part numbers or postal zip codes), which can be trained into user-programmable

voice equipment designed with proper recognition processing algorithms.

limited signal — In radar, a signal that is intentionally limited in amplitude by the dynamic range of the system.

limited space-charge accumulation — A mode of oscillation for gallium arsenide diodes.

limited space-charge accumulation diode — *See* LSA diode.

limited stability — The property of a system that remains stable only as long as the input signal falls within a particular range.

limiter — 1. A device in which some characteristic of the output is automatically prevented from exceeding a predetermined value — e.g., a transducer in which the output amplitude is substantially linear (with regard to the input) up to a predetermined value and substantially constant thereafter. 2. A radio-receiver stage or circuit that limits the amplitude of the signals and hence keeps interfering noise low by removing excessive amplitude variations from the signals. 3. A device that reduces the intensity of very short duration peaks (transient peaks) in the audio signal without audibly affecting dynamic range. 4. A feedback element that acts to restrain a variable by modifying or replacing the function of the primary element when predetermined conditions have been reached.

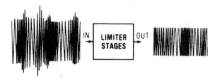

Limiter action.

limiting — The restricting of the amplitude of a signal so that interfering noise can be kept to a minimum.

limiting resolution — In television, the number of TV lines per picture height that can be just be resolved by visual inspection of a televised pattern. Wedge patterns are often used for this subjective test. These charts include five areas (center and four corners) for resolution measurement. Each area includes two identical sets of mutually orthogonal wedge patterns calibrated (up to 16) in hundreds of TV lines per picture height. 2. The details that can be distinguished on the television screen. Vertical resolution refers to the number of horizontal black and white lines that can be resolved in the picture height. Horizontal resolution refers to the black and white lines resolved in a dimension equal to the vertical height and may be limited by the video amplifier bandwidth.

limit of error — An accuracy index that indicates the expected maximum deviation of the measured value from the true value if all of the factors causing deviations act simultaneously and in the same direction.

limit ratio — The ratio of peak value to limited value.

limits — The minimum and maximum values specified for a quantity.

limit switch — 1. A mechanically operated contact-making or contact-breaking device mounted in the path of a moving object and actuated by its passage. 2. An electromechanical device that uses changes in mechanical motion to control electrical circuits. It functions as the interlocking link between a mechanical motion and an electrical circuit. 3. A switch that is actuated by some part or motion of a machine or equipment to alter the electrical circuit associated with it.

line — 1. In television, a single trace of the electron beam from left to right across the screen. The present United States standard is based on 525 lines to a complete picture. 2. A conductor of electrical energy. 3. The path of the moving spot in a cathode-ray tube. 4. A term used interchangeably for maxwell. 5. A row of actual or potential holes at right angles to the direction in which a punched tape advances. Line width is measured in terms of the maximum number of holes permissible, excluding the sprocket hole. 6. The interconnection between two electrical devices. Usually used with reference to a long run of interconnecting cable, as from a microphone to its tape-recorder input. 7. In communications, describes cables, telephone lines, etc., over which data is transmitted to and received from the terminal.

line advance — Also called line feed. The distance between the centers of the scanning lines.

line amplifier — 1. An amplifier that supplies a program transmission line or system with a signal at a specified level. 2. Also called line stretcher. An amplifier, usually remotely powered, used in a trunk line in a distribution system to increase the strength of the signal in order to drive an additional length of cable. 3. Also called program amplifier. An amplifier for audio or video signals that feeds a transmission line. 4. An audio amplifier that is used to provide preamplification of an audio alarm signal before transmission of the signal over an alarm line. Use of an amplifier extends the range of signal transmission. 5. An amplifier that supplies an audio system or an audio long cable with a signal at a specified level, usually between −10 and +4 dBv (245 millivolts to 1.23 volts rms). 6. An amplifier inserted in any part of the transmission line following the downconverter to compensate signal losses caused by long lengths of coaxial cable or the insertion of passive devices such as splitters. Line amplifiers are also used when the signal must drive a number of television receivers.

line and trunk group — A group consisting of four-wire line circuits, incoming trunks from private automatic branch exchanges, and intertoll trunk groups.

linear — 1. Having an output that varies in direct proportion to the input. 2. A ratio in which change in one of two related quantities is accompanied by a directly proportional change in the other.

linear acceleration — The rate of change in linear velocity.

linear accelerator — A device for speeding up charged particles such as protons. It differs from other accelerators in that the particles move in a straight line instead of in circles or spirals.

linear accelerometer — A transducer used to detect, measure, and record the rate of change in linear velocity of accelerative forces.

linear actuator — An actuator that produces mechanical motion from electrical energy.

linear amplification — Amplification in which the output is directly proportional to the input.

linear amplifier — 1. An amplifier that operates on the linear portion of its forward transfer characteristic so that its output signal is always an amplified replica of the input signal. 2. Amplifier whose gain is constant for a wide variation in amplitude of input signal — i.e., output signal is proportional to input signal. 3. Amplifier that has linear control characteristics and negligible response time in the active bandwidth, provides a wide speed range, and usually requires minimal external circuitry to prevent instability caused by phase-shifted feedback from reactive loads. Linear amplifiers also generate little electrical noise.

linear array — 1. An antenna array in which the elements are equally spaced and in a straight line. 2. A multielement antenna in which individual dipole elements are arranged end to end.

linear circuit — 1. A circuit in which the output voltage is approximately directly proportional to the input voltage; this relationship generally exists only over a limited range of signal voltages and often over a limited range of frequencies. 2. A circuit whose output is a continuous amplified version of its input. That is, the output is a predetermined variation of its input. 3. A circuit in which a proportional, or linear, relationship exists between the input and output. In manufacturers' circuit classifications the term often includes all analog circuits, both linear and nonlinear.

linear control — A rheostat or potentiometer having uniform distribution of graduated resistance along the entire length of its resistance element.

linear detection — Detection in which the output voltage is substantially proportionate to the input voltage over the useful range of the detector.

linear detector — A detector that produces an output signal directly proportionate in amplitude to the variations in amplitude (for AM transmission) or frequency (for FM transmission) of the rf input.

linear device — An amplifying-type analog device with a linear input/output relation, as opposed to a nonlinear digital device, which is either completely on or completely off over large ranges of input signals.

linear differential transformer — A type of electromechanical transducer that converts physical motion into an output voltage, the phase and amplitude of which are proportional to position. *See also* linear motion transducer.

linear distortion — Amplitude distortion in which the output and input signal envelopes are not proportionate, but no alien frequencies are involved.

linear electrical parameters of a uniform line — Frequently called the linear electrical constants. The series resistance and inductance, and the shunt conductance and capacitance, per length of a line.

linear electron accelerator — An evacuated metal tube in which electrons are accelerated through a series of small gaps (usually cavity resonators in the high-frequency range). The gaps are so spaced that, at a specific excitation frequency, the electrons gain additional energy from the electric field as they pass through successive gaps.

linear feedback-control system — A feedback-control system in which the relationship between the pertinent measure of the system signals is linear.

linear integrated circuit — Abbreviated LIC. 1. A circuit whose output is an amplified, linear version of its input or whose output is a predetermined variation of its input. A class of integrated circuits that process analog information expressed as voltages or currents. 2. An integrated circuit whose output remains proportional to the input level. Generally the term is taken to mean an analog IC, such as a voltage regulator, comparator, sense amplifier, driver, etc., as well as a linear amplifier. The operation of the circuit can be made nonlinear by connecting the basic linear amplifier to external circuit elements that have thresholds or other nonlinear characteristics.

linearity — 1. The relationship existing between two quantities when a change in a second quantity is directly proportionate to a change in the first quantity. 2. Deviation from a straight-line response to an input signal. 3. The ability of a meter to provide equal angular deflections proportional to the applied current. Usually expressed as a percent of the full-scale deflection. 4. The relationship between the actual electrical energy input and the deflection of a meter pointer, as referenced to a theoretical straight line. Linearity is often confused with tracking. 5. In a modulator, the ability to generate a modulation envelope that reproduces the modulating signal without distortion. 6. The state of an output that incrementally changes directly or proportionally as the input changes. 7. The closeness of a calibration curve to a specified straight line; the degree to which the output of a linear device is proportional to the input.

linearity control — A control that adjusts the variation of scanning speed through the trace interval.

linearity error — The deviation of a calibration curve from a specified straight line.

linear logarithmic intermediate-frequency amplifier — An amplifier used to avoid overload or saturation as a protection against jamming in a radar receiver.

linear magnetostriction — Under stated conditions, the relative change of length of a ferromagnetic object in the direction of magnetization when the magnetization of the object is increased from zero to a specified value (usually saturation).

linear mobility — The synchronized incremental mobility of functionally transitional electrons in a semiconductor.

linear modulation — Modulation in which the amplitude of the modulation envelope (or the deviation from the testing frequency) is directly proportional to the amplitude of the modulating wave at all audio frequencies.

linear modulator — A modulator in which the modulated characteristic of the output wave is substantially linear with respect to the modulating wave for a given magnitude.

linear motion transducer — An instrumentation component that translates straightline (linear) mechanical motion into an ac analog that is usable as a feedback signal for control or display. A transformer-type device in which a movable magnetic core is displaced axially by the moving component being monitored. When the core is moved in one direction from the center of its stroke, the output voltage is in phase with the excitation voltage, and when the core is moved in the opposite direction from the center, the output voltage is 180° out of phase. At the center, the output voltage is (virtually) zero. In either direction from center, the voltage increases as a precise linear function of probe displacement. Thus, the output signal has two basic analog components: phase relationship with the excitation voltage, indicating the direction of travel; and voltage amplitude, indicating the length of travel.

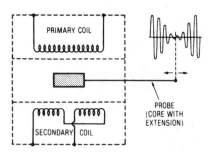

Linear motion transducer.

linear polarization — The polarization of a wave radiated by an electric vector that does not rotate but that alternates so as to describe a line. Normally the vector is oriented either horizontally or vertically.

linear polarized wave — At a point in a homogeneous isotropic medium, a transverse electromagnetic wave whose electric field vector lies along a fixed line.

linear power amplifier — A power amplifier in which the output voltage is directly proportionate to the input voltage.

linear predictive coding — 1. A method of analyzing and storing human speech by determining from speech patterns a description of a time-varying digital filter modeling the vocal tract. This filter is then excited by the proper type of input, depending on the sound to be synthesized. The output of the filter is passed through a digital-to-analog converter whose output is the desired synthetic speech. 2. Speech synthesis technique based in the frequency domain. The quality of the synthesis improves as the number of coefficients is increased. With ten coefficients, an approximate number of bits per second required for speech is 1200. 3. A parameter-encoding technique that models the human vocal tract with a digital filter whose controlling parameters change with time. Changes are based on previous speech samples.

linear programming — In computers, a mathematical method of sharing a group of limited resources among a number of competing demands. All decisions are interlocking because they must be made under a common set of fixed limitations.

linear pulse amplifier — A pulse amplifier that maintains the peak amplitudes of the input and output pulses in proportion.

linear rectification — The production, in the rectified current or voltage, of variations that are proportionate to variations in the input wave amplitude.

linear rectifier — A rectifier with the same output current or voltage waveshape as that of the impressed signal.

linear regression — A statistical function used when handling experimental data. It is especially used when performing an experiment to find a mathematical relationship between two variables. Linear regression is the name of the procedure that is used to find the line that best fits the set of data points that have been found experimentally. The procedure usually finds the equation of the straight line and also a parameter called the correlation coefficient, which indicates how well the data fits the line.

linear scan — A radar beam that traverses only one arc or circle.

linear scanning — Scanning in which a radar beam generates only one arc or circle.

linear sweep — In a television receiver, the movement of the spot across the screen at a uniform velocity during active scanning intervals.

Linearsyn — A linear displacement pickoff of the differential-transformer type consisting of a coil assembly and a movable magnetic core. Linear velocity units of high-coercive-force permanent magnetic cores that induce sizeable dc voltages while moving concentrically within shielded coils; the voltage varies linearly with the core velocity (Sanborn Co.).

line art — A computer-drawn graphic (without halftones) that can be clearly printed.

linear taper — A potentiometer that changes the resistance linearly as it is rotated through its range.

linear time base — In a cathode-ray tube, the time base in which the spot moves at a constant speed along the time scale. This type of time base is produced by application of a sawtooth waveform to the horizontal-deflection plates of a cathode-ray tube.

linear transducer — 1. A transducer for which the pertinent measures of all the waves concerned are related by a linear function (e.g., a linear algebraic differential or integral equation). 2. A transducer having its output at any given frequency proportional to the received input.

linear variable-differential transformer — *See* differential transformer.

linear varying parameter network — A linear network in which one or more parameters vary with time.

linear velocity transducer — A transducer that produces an output signal proportionate to the velocity of single-axis translational motion between two objects.

line-a-time printing — A type of computer output in which an entire horizontal row of characters is printed at the same time. *See also* line printer.

line balance — 1. The degree to which the conductors of a transmission line are alike in their electrical characteristics with respect to each other, other conductors, and ground. 2. Impedance equal to that of the line at all frequencies (e.g., in terminating a two-wire line).

line-balance converter — A device used at the end of a coaxial line to isolate the outer conductor from ground.

line characteristic distortion — Distortion experienced in teletypewriter transmission when the presence of changing current transitions in the wire circuit affects the lengths of the received signal impulses.

line circuit — In a telephone system, the relay equipment associated with each station connected to a dial or manual switchboard. The term is also applied to a circuit for interconnecting an individual telephone and a channel terminal.

line coordinate — In a matrix, a symbol (normally at the side) identifying a specific row of cells and, in conjunction with a column coordinate, a specific cell.

line cord — Also called a power cord. A two-wire cord terminating in a two-prong plug at the end that goes to the supply, and connected permanently into a radio receiver or other appliance at the other end.

line-cord resistor — An asbestos-enclosed, wire-wound resistance element incorporated into a line cord along with the two regular wires. It lowers the line voltage to the correct value for the series-connected tube filaments and pilot lamps of a universal ac/dc receiver.

line diffuser — A circuit used to produce small vertical oscillations of the spot on the screen of a television monitor or receiver to make the line structure of the image less noticeable to an observer close to the screen.

line driver — 1. An integrated circuit designed for transmitting logic information through long lines (normally at least several feet in length). 2. A buffer circuit with special characteristics (i.e., high current and/or low impedance) suitable for driving logic lines longer than normal interconnection length (greater than a few feet). It may have complementary (push-pull) outputs to work with the differential inputs of a line receiver. *See* line receiver. 3. A signal converter that conditions a digital signal to ensure reliable transmission over an extended distance.

line drop — A voltage loss occurring between any two points in a power or transmission line. Such a loss, or drop, is due to the resistance, reactance, or leakage of the line. An example is the voltage drop between a power source and load when the line supplying the power has excessive resistance for the amount of current.

line-drop signal — A signal associated with a subscriber line on a manual switchboard.

line-drop voltmeter compensator — A device using a voltmeter to enable it to indicate the voltage at some distant point in the circuit.

line equalizer — An inductance and/or capacitance inserted into a transmission line to correct its frequency-response characteristics.

line-equipment balancing network — A hybrid network designed to balance filters, composite sets, and other line equipment.

line-fault protection — A means of eliminating or reducing the effects of faults that occur on a transmission line such as a telephone circuit. Such faults include momentary losses of transmission due to signal outages and high noise levels.

line feed — *See* line advance.

line fill — The ratio of the number of main telephone stations connected to a line to the nominal main-station capacity of the same line.

line filter — 1. A device containing one or more inductors and capacitors. It is inserted between a transmitter, receiver, or appliance and the power line to block noise signals. In a radio receiver, it prevents power-line noise signals from entering the receiver. In other appliances it prevents their own electrical noises from entering the power line. 2. A filter associated with a transmission line. In some applications, line filter may imply a filter used to separate the speech frequencies. In other applications, it may imply directional separation, etc.

line-filter balance — A network designed to maintain phantom-group balance when one side of the group is equipped with a carrier system. Since the network must balance the phantom group for voice frequencies only, its configuration is much simpler than the filter it balances.

linefinder — 1. A switching mechanism that locates a calling telephone line among a group and connects it to a trunk, selector, or connector. 2. An electromechanical device that automatically line-feeds the platen of a printer to a predetermined line on a printed form.

linefinder shelf — A group (usually 20) of linefinders with the equipment required for routing the dial pulses from any of its associated calling telephones to a selector or connector.

linefinder switch — In a telephone system, an automatic switch for seizing the selector apparatus that provides the dial tone transmitted to the calling party.

line-focus tube — An X-ray tube in which the focal spot is roughly a line.

line fonts — Repetitive pattern used to give meaning to a line, e.g., solid, dashed, dotted, etc.

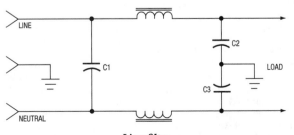

Line filter.

line frequency — Also called horizontal line frequency or horizontal frequency. 1. In television, the number of times per second the scanning spot crosses a fixed vertical line in the picture in one direction, including vertical-return intervals. 2. The nominal operating frequency of the power-line voltage used to supply operating power to instruments or equipment.

line-frequency regulation — The change in output (current voltage or power) of a regulated power supply for a specified change in line frequency.

line group — The frequency spectrum occupied by a group of carrier channels as they are applied to a transmission facility.

line hit — An electrical interference causing the introduction of spurious signals on a circuit.

line hydrophone — A directional hydrophone consisting of a single straight-line element, an array of adjacent electroacoustic transducing elements in a straight line, or the acoustic equivalent of such an array.

line impedance — The impedance measured across the terminals of a transmission line.

line input — Input channel of an amplifier designed to accept signal at a given level from a line at a specific impedance, usually 600 ohms.

line interlace — *See* interlaced scanning.

line leakage — Resistance existing through the insulation between the two wires of a telephone-line loop.

line lengthener — A device for altering the electrical length of a waveguide or transmission line, but not its physical length or other electrical characteristics.

line level — 1. The level of a signal at a certain point on a transmission line. Usually expressed in decibels. 2. Based roughly on the "standardized" signal intensity sent over a telephone line, this term refers to any audio signal having a maximum intensity of between $1/2$ and $1 1/2$ volts. Typically, this is the signal level put out by audio components that do not require preamplification (tuners, for instance).

line load — Usually a percentage of maximum circuit capability to reflect actual use during a period of time (e.g., peak-hour line load).

line loop — An operation performed over a communication line from an input unit at one terminal to output units at a remote terminal.

line loop resistance — Also called loop resistance. The metallic resistance of the pair of line wires that extend from a subscriber's telephone to the central office (does not include the resistance of the telephone).

line loss — The total of the various energy losses in a transmission line.

line microphone — A directional microphone consisting of a single straight-line element, an array of adjacent electroacoustic transducing elements in a straight line, or the acoustical equivalent of such an array.

line noise — Noise originating in a transmission line.

line of force — 1. Used in the description of an electric or magnetic field to represent the force starting from a positive charge and ending on a negative charge. 2. In an electric or magnetic field, an imaginary line in the same direction as the field intensity at each point. Sometimes called a maxwell when used as a unit of magnetic flux.

line of propagation — The path over which a radio wave travels through space.

line of sight — 1. The distance to the horizon from an elevated point, including the effects of atmospheric refraction. The line-of-sight distance for an antenna at zero height is zero. 2. A straight line between an observer or radar antenna and a target. 3. An unobstructed, or optical, path between two points. 4. The radio-propagation characteristic of a microwave. 5. The optical axis of a telescope or other observation system. The straight line connecting the object and the objective lens of the viewing device.

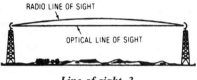

Line of sight, 3.

line-of-sight coverage — The maximum distance for transmission above the highest usable frequency. Radio waves at those frequencies do not follow the curvature of the earth and are not reflected from the ionosphere, but go off into space and are lost.

line-of-sight stabilization — In shipboard or airborne radar, compensating for the roll and pitch by automatically changing the elevation of the antenna in order to keep the beam pointed at the horizon.

line of travel — The path followed by an electromagnetic wave from one point to another

line oscillator — An oscillator in which the resonant circuit is a section of transmission line an integral number of quarter wavelengths in electrical length.

line output — Output channel of an amplifier designed to deliver signal at a given level to a line at a specific impedance, usually 600 ohms.

line pad — In radio broadcasting, a pad inserted between the final program amplifier and the transmitter to ensure a constant load on the amplifier.

line printer — 1. In computers, a high-speed printer that produces an entire line at one time. All characters of the alphabet are contained around the rim of a continuously rotated disc, and there are as many discs as there are characters in the line. The computer momentarily stops the discs at the right characters for each line, and stamps out an impression in a fraction of a second. 2. A high-speed printer capable of printing simultaneously a complete line (80 to 120 characters) at once. It is capable of printing as many as 3000 lines a minute.

line protocol — 1. A set of rules used to organize and control the flow of information between two or more stations connected by a common transmission facility. 2. A control program used to perform data-communication functions over network lines. Consists of both handshaking and line-control functions that move the data between transmit and receive locations.

line pulsing — A method of pulsing a transmitter by charging an artificial line over a relatively long period, and then discharging it through the transmitter tubes at a shorter interval determined by the line characteristics.

line radiator — A speaker enclosure that has several speakers arranged in a straight line to achieve a specific directional pattern.

line receiver — 1. A circuit to receive signals from a line, usually driven by a line driver and having features such as differential input, Schmitt trigger, and the like. 2. Used in conjunction with a line driver to detect signals at the receiving end of a long line. *See also* line driver.

line regulation — 1. The change in output (current, voltage, or power) of a regulated power supply for a specified change in line voltage. It may be stated as a percentage of the specified output and/or as an absolute value. 2. Percent change in output voltage at constant junction temperature for a specified change in input voltage. This determines output accuracy of a regulator for changes in input voltage. 3. The maximum deviation of the output voltage of power supply in percent as the input voltage is varied from nominal to high line and nominal to low line. Output load and ambient temperature are held constant.

line relay — A relay activated by the signals on a line.

line sensor — A sensor with a detection zone that approximates a line or series of lines, such as a photo-electric sensor that senses a direct or reflected light beam.

line-sequential color-television system — A color-television system in which the individual lines of green, red, and blue are scanned in sequence rather than simultaneously.

line side — 1. The contacts or terminals of an electrical device, designed to be connected to a conductor carrying line voltage from the source; the input side. 2. The side of equipment that "looks" toward the transmission path.

lines of force — In electric and magnetic fields, the electric and magnetic forces of repulsion or attraction, which are taken to follow certain imaginary lines radiating from the electric charge or the magnetic pole. (It is assumed that any unit electric charge or unit magnetic pole placed in the appropriate field will be acted upon so as to move in the direction of these imaginary lines.)

line spectra — Spectra that originate from atoms; they are composed of lines having irregular spacing and intensity.

line spectrum — The spectrum of a periodic, discrete signal consisting of one or more frequencies. For example, a square wave is characterized by a fundamental and odd-order harmonics.

line speed — 1. The maximum rate at which signals may be sent over a given channel, usually expressed in bauds or in bits per second. 2. The rate at which text is transmitted over a line, expressed in bits per second.

line splitter — An active or passive device that divides a signal into two or more signals that contain all of the original information. A passive splitter feeds an attenuated version of the input signal to the output ports. An active splitter amplifies the input signal to overcome the splitter loss.

line-stabilized oscillator — An oscillator in which the frequency is controlled by using one section of a line as a sharply selective circuit.

line stretcher — 1. A section of rigid coaxial line with telescoping inner and outer conductors that permit the section to be conveniently lengthened or shortened. 2. *See* line amplifier, 2.

line supervision — A means of determining that a transmission line is functional. Electronic protection of an alarm line accomplished by sending a continuous or coded signal through the circuit. A change in the circuit characteristics, such as a change in impedance due to the circuit's having been tampered with, will be detected by a monitor. The monitor initiates an alarm if the change exceeds a predetermined amount.

line switcher — Direct-current power supply that directly converts the ac line voltage to a lower dc value using switching techniques. Line switchers are highly efficient, but require care to contain electrical noise.

line switching — Also called circuit switching. A communications switching system that completes a circuit from sender to receiver at the time of transmission, as opposed to message switching.

line transformer — A transformer inserted into a system for such purposes as isolation, impedance matching, or additional circuit derivation.

line triggering — Triggering from the power-line frequency.

line unit — An electric device used in sending, receiving, and controlling the impulses of a teletypewriter.

line voltage — 1. The voltage existing in a cable or circuit, such as at a wall outlet or other terminals of a power line system. The line voltage is usually between 115 and 120 volts, with 117 as an average, but may vary at times as much as 5 volts above or below the 115- and 120-volt limits. It is also the voltage existing in a cable or circuit. 2. The voltage supplied by an electrical power line or source, measured at the point of supply or sometimes at the point of utilization.

line-voltage regulator — A device that counteracts variations in the power-line voltage and delivers a constant voltage to the connected load.

line voltage transient protection — A device, or circuit, generally located on the line side of the control, that will divert or dissipate the energy contained in an abnormal line voltage spike. The purpose of this is to prevent possible control damage, especially to solid-state components, as a result of excessive voltage surges.

linguistic — Pertaining to language or its study, including its origin, structure, phonetics, etc.

linguistic variable — Common-language terms used to describe a fuzzy set, such as "hot," "slow," "very cold," or "medium."

link — 1. A transmitter-receiver system connecting two locations. 2. In a digital computer, the part of a sub-program that connects it with the main program. 3. An interconnection. 4. In automatic switching, a path between two units of switching apparatus within a central office. *See* channel. 5. A communication channel between two adjacent signaling points that provides a path for messages to travel. 6. To connect two elements in a data structure by using index variables or pointer variables.

linkage — 1. A measure of the voltage that will be induced in a circuit by magnetic flux. It is equal to the flux times the number of turns linked by the flux. 2. A mechanical arrangement for transferring motion in a desired manner. It consists of solid pieces with movable joints. 3. In a computer, a technique used to provide interconnections for entry and exit of a closed subroutine to or from the main routine. 4. The instructions that connect one program to another, providing continuity of execution between the programs.

link blowing — An IC trimming technique that involves passing current pulses through points on the chip to open interconnection links between parallel resistors.

link circuit — A closed loop used for coupling purposes. It generally consists of two coils, each having a few turns of wire, connected by a twisted pair of wires or by other means, with each coil placed over, near, or in one of the two coils to be coupled.

link coupling — Inductive coupling between circuits. A coil in one circuit acts as the primary, and a coil in the second circuit as the secondary.

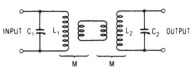

Link coupling.

linked list — A list formed by tying together (with pointers) several items on a disk.

linker — 1. A program that connects small sections of a program together so they can run as a whole. 2. A utility program that ties different software modules into one entity.

link fuse — An unprotected fuse consisting of a short, bare wire between two fastenings.

linking loader — 1. A computer program that takes program segments and places them one after the other in memory, adjusting jump and call instruction locations to match the new placement in the memory. 2. A routine that combines program segments and reassigns addresses to reflect the new memory locations. 3. A relocatable loader that links various object modules into a single load module, resolving external references in the process. This lets users load their programs into any memory area.

link neutralization — Neutralization of a tuned radio-frequency amplifier by means of an inductive coupling loop from the output to the input.

link transmitter — In broadcasting, a booster for a remote pickup or from a studio to main transmitter.

lin-log receiver — A radar receiver in which the amplitude response is linear for small-amplitude signals and logarithmic for large ones.

LiON — Abbreviation for lithium-ion.

lip microphone — A sensitive microphone that is placed in contact with the lip.

liquid-borne noise — Undesired sound characterized by fluctuations of pressure of a liquid about the static pressure as a mean.

liquid cooling — The use of a circulating liquid to cool components or equipment that heat up during operation.

liquid-core fiber — An optical fiber consisting of optical glass, quartz, or silica tubing filled with a higher-refractive-index liquid, such as tetrachloroethylene.

liquid-core optical fibers — Multimode straight fibers capable of transporting linearly polarized light with any incident polarization angle, and in which no form of internal stress can develop that could lead to birefringence.

liquid-crystal display — See LCD.

liquid crystals — Liquids that are doubly refracting and that display interference patterns in polarized light.

liquid-filled capacitor — A capacitor in which a liquid impregnant occupies substantially all of the case volume not required by the capacitor element and its connections. Space may be allowed for the expansion of the liquid under temperature variations.

liquid fuse unit — A fuse unit in which the fuse link is immersed in a liquid or the arc is drawn into the liquid when the fuse link melts.

liquid-impregnated capacitor — A capacitor in which a liquid impregnant is dominantly contained with the foil and paper winding, but does not occupy substantially all of the case volume not required by the capacitor element and its connections.

liquid laser — 1. A laser in which the active material is in the liquid state. Present types employ a chelated rare-earth ion dissolved in an organic liquid. 2. A laser whose active medium is in liquid form, such as organic dye and inorganic solutions. Dye lasers are commercially available; they are often called organic dye or tunable dye lasers.

liquid rheostat — A rheostat consisting of metal plates immersed in a conductive liquid. The resistance is changed by raising or lowering the plates or the liquid level to vary the area of the plates contacting the liquid.

liquidus — The lowest temperature at which a metal or alloy is completely liquid.

liser — A microwave oscillator of very high spectrum purity. Its emission consists of right circularly polarized waves of two different cavity resonant frequencies.

LISP — Acronym for list processing. A high-level programming language used primarily for list processing, symbol manipulation, and recursive operations; it can handle many different data types, treat programs as data, and provide for the self-modification of the program as it is executing; generally considered a difficult language to learn.

Lissajous figures — Patterns produced on the screen of a cathode-ray tube when sine-wave signal voltages of various amplitude and phase relationships are applied to the horizontal- and vertical-deflection circuits simultaneously.

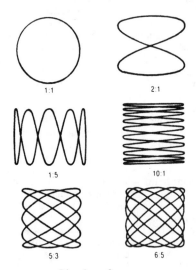

Lissajous figures.

list — Also called a chained or threaded list. A data structure in which each item of information has attached to it one or more links or pointers that refer to other items.

listening angle — The enclosed angle between the listener and the two speakers of a stereo reproducing system.

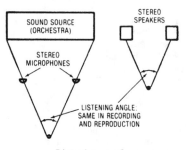

Listening angle.

listening post — The receiving terminus in a bugging or tapping operation. It may consist of recording equipment or a human operator or a combination of both.

listening sonar — See sonar.

list-processing language — A programming language that is widely used in artifical-intelligence research to manipulate categories or lists of items.

lithium — An alkali metal used in the construction of photocells and batteries.

lithium chloride sensor — Also called Dunmore cell. A hygroscopic element that has fast response, high accuracy, and good long-term stability and whose resistance is a function of relative humidity.

lithium-ion — Abbreviated LiON. A rechargeable battery technology that is able to produce considerably more charge than comparable size nickel-cadmium or nickel-metal hydride batteries.

lithography — A method of defining patterns for semiconductor device processing. Patterns are most frequently produced in thin films of materials called resists, which then resist a subsequent processing step being applied to an underlying material in accordance with that pattern. In typical semiconductor integrated-circuit fabrication, many different patterns are used to delineate features in a sequence of processing steps.

litz wire — Also called Litzendraht wire. A conductor composed of a number of fine, separately insulated strands that are woven together so that each strand successively takes up all possible positions in the cross section of the entire conductor. Litz wire gives reduced skin effect, hence, lower resistance to high-frequency currents.

live — 1. A term applied to a circuit through which current is flowing. 2. Connected to a source of an electrical voltage. 3. Charged to an electrical potential different from that of the earth. 4. Reverberant, as a room in which there are reflections of sound. 5. A program that is transmitted as it happens, with no delay.

live cable test cap — A protective cap placed over the end of a cable to insulate the cable and seal its sheath.

live end — The end of a radio studio where the reflection of sound is greatest.

live parts — Metallic portions of equipment that are at a potential different from that of the earth.

live room — A room with a minimum of sound-absorptive material, such as drapes, upholstered furniture, rugs, etc. Because of the many reflecting surfaces, any sound produced in the room will have a long reverberation time.

LLTV — Also LLLTV and L³TV. Abbreviation for low-light television and low-light-level television. A CCTV system capable of operating with scene illumination less than 0.5 lumen/ft².

LNA — *See* low-noise amplifier.

LNB — Abbreviation for low-noise block downconverter. A microwave amplifier that converts a block of frequencies to a lower frequency. LNBs for satellite TV typically convert C- and Ku-band signals to a frequency band of 950 to 1450 MHz for input to the receiver.

LNC — *See* low-noise converter.

L-network — A network composed of two impedance branches in series. The free ends are connected to one pair of terminals, and the junction point and one free end are connected to another pair.

LO — *See* local oscillator.

load — 1. The power consumed by a machine or circuit in performing its function. 2. A resistor or other impedance that can replace some circuit element. 3. The power delivered by a machine. 4. A device that absorbs power and converts it into the desired form. 5. The impedance to which energy is being supplied. 6. Also called work. The material heated by a dielectric or induction heater. 7. In a computer, to fill the internal storage with information obtained from auxiliary or external storage. 8. The resistance or impedance that the input of one device offers to the output of another device to which it is connected. *See* input impedance; termination, 1. 9. The circuit or transducer (e.g., speaker) connected to the output of an amplifier. The source (e.g., pickup) is loaded by the amplifier's input impedance. 10. The electrical demand placed on a circuit or a system by the utilization equipment connected to it. Also, any piece of electrical utilization equipment of any given rating so connected. 11. To feed a program into a computer system. A common means of loading the program is via a form of magnetic media. The media is inserted into the media drive and the program read into the system's memory.

load and go — In a computer, an operation and compiling technique in which the pseudo language is converted directly to machine language and the program is then run without the creation of an output machine-language program.

load balance — *See* load division.

loadbreak connector — A connector designed to close and interrupt current on energized circuits.

load cell — 1. Transducer that measures an applied load by a change in its properties, such as a change in resistance (strain-gage load cell), pressure (hydraulic load cell), etc. 2. A device that produces an output signal proportional to the applied weight or force.

load circuit — The complete circuit required to transfer power from a source to a load (e.g., an electron tube).

load-circuit efficiency — In a load circuit, the ratio between its input power and the power it delivers to the load.

load-circuit power input — The power delivered to the load circuit. It is the product of the alternating component of the voltage across the load circuit and the current passing through it (both root-mean-square values), times their power factor.

load coil — Also called a work coil. In induction heaters, a coil that, when energized with an alternating current, induces energy into the item being heated.

load curve — A curve of power versus time — i.e., the value of a specified load for each unit of the period covered.

load divider — A device for distributing power.

load division — Also called load balance. A control function that divides the load in a prescribed manner between two or more power sources supplying the same load.

loaded antenna — 1. An antenna to which extra inductance or capacitance has been added to change its electrical (but not its physical) length. 2. An antenna employing a loading coil at its base or above its base to achieve the required electrical length using physically shorter elements.

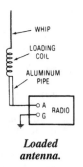

Loaded antenna.

loaded applicator impedance — In dielectric heating, the complex impedance measured at the point of application with the load material properly positioned for heating and at the specified frequency.

loaded impedance — In a transducer, the impedance at the input when the output is connected to its normal load.

loaded line — 1. A telephone line equipped with loading coils to add inductance in order to minimize amplitude distortion. 2. A transmission line that has lumped elements (inductance or capacitance) added at uniformly spaced intervals. Loading is used to provide a given set of characteristics to a transmission line.

loaded motional impedance — See motional impedance.

loaded Q — Also called the working Q. 1. The Q of an electric impedance when coupled or connected under working conditions. 2. In bandpass filters, a quantity that defines the percentages of 3-dB bandwidths. Numerically, it is equal to the center frequency divided by the 3-dB bandwidth.

loader — 1. A program in a minicomputer that takes a program from some input or storage device and places it in memory at some address. Programs loaded by a loader are then ready to run. 2. A program that places a binary (machine language) program into successive core memory locations for execution. Some can replace relocatable addresses with absolute addresses. 3. A program that accepts input from an external device and places it in memory. The simplest loaders do no more than this. Other loaders include relocating loaders, linking loaders, and bootstrap loaders. 5. A program that controls peripheral device operation when reading programs for execution into system memory. 6. A program that transfers the object program from an external medium, such as paper tape, to the microcomputer random-access memory. Some loaders also convert a relocatable version of the object program to a loadable version. A program might originally be assembled to reside in the microcomputer memory starting at address zero. If the compiler or assembler has allowed the object program to be relocatable, the programmer can specify to the loader the program's new base address and the loader will modify all addresses accordingly in the object program. Another feature that is sometimes available is linkage editing, which establishes the linkages between different object programs that make reference to one another. Linkage editing requires both a compiler or assembler and a loader program that can communicate the appropriate information.

load factor — The ratio of average power demand, over a stipulated period of time, to the peak or maximum demand for that same interval.

load impedance — The impedance that the load presents to a transducer.

loading — 1. In communication practice, the insertion of reactance into a circuit to improve its transmission characteristics. 2. Placing some material at the front or rear of a speaker so as to change its acoustic impedance and thus alter its radiation pattern.

loading coil — An inductor inserted into a circuit to increase its inductance and thereby improve its transmission characteristics.

loading-coil spacing — The line distance between the successive loading coils of a line.

loading disc — A metal disc placed on top of a vertical antenna to increase its natural wavelength.

loading error — The error introduced when more than negligible current is drawn from the output of a device.

loading factor — A number that represents the relative capacity of a gate for driving other gates or the relative load presented by a gate to the gate that drives it.

loading noise — Any unwanted signal caused by fluctuating contact resistance between the slider and the wire or film in a potentiometric transducer when current is drawn from the instrument.

loading routine — In a computer, a routine that, when in the memory, is able to read other information into the memory from external devices, tape, etc.

load isolator — A waveguide or coaxial device (usually ferrite) that provides a good energy path from a signal source to a load, but provides a poor energy path for reflections from a mismatched load back to the signal source.

load life — The ability of a device to withstand its full power rating over an extended period of time, usually expressed in hours.

load line — 1. A straight line drawn across a series of plate-current plate-voltage characteristic curves on a graph to show how plate current changes with grid voltage when a specified plate-load resistance is used. 2. A line drawn on the collector characteristic curves of a transistor on which the operating point of the transistor moves as the collector current changes. Called a load line because the slope of the line depends on the value of the collector load resistance. 3. The line used to locate the operating point of the permanent magnet on the demagnetization curve. The slope of the shearing line is equal to the permeance coefficient.

load matching — In induction and dielectric heaters, adjustment of the load-circuit impedance so that the desired energy will be transferred from the power source to the load.

load-matching network — An electrical impedance network inserted between the source and the load to provide for maximum transfer of energy.

load-matching switch — In induction and dielectric heaters, a switch used in the load-matching network to alter its characteristics and thereby compensate for a sudden change in the load characteristics (such as in passing through the Curie point).

load mode — In some variable-word-length computers, data transmission in which certain delimiters are moved with the data (in contrast with move mode).

load point — Sometimes called BOT (beginning of tape). The point on a magnetic tape at which writing and reading begin. It is indicated by a reflective marker placed on the tape.

load regulation — 1. For a constant-current supply, the change in the steady-state value of the output dc current due to a change in load resistance from a short-circuit current (zero resistance) to a value that results in the maximum rated output voltage. 2. For a constant-voltage supply, the change in the steady-state value of the output dc voltage due to a change in load resistance from an open-circuit condition (infinite resistance) to a value that results in the maximum rated output current. 3. The maximum deviation of the output voltage of a power supply in percent as the load is changed from minimum to maximum rated load. Input voltage is nominal value and ambient temperature is constant.

load resistor — A resistor connected in parallel with a high-impedance load so that the output circuit driving the load can provide at least the minimum current required for proper operation.

load sharing — A scheduling technique in multiprocessing systems whereby a task is executed by the next available processor. In order to make this technique operate successfully, all processors must be identical and have identical memory addressing capabilities.

load side — The contacts or terminals of an electrical device designed to be connected to conductors carrying voltage away from the device to the utilization equipment, or load, or another device; the output side.

loadstone — See lodestone.

load-transfer switch — A switch for connecting either a generator or a power source to one load circuit or another.

lobe — Also called directional, radiation, or antenna lobe. One of the areas of greater transmission in the pattern of a directional antenna. Its size and shape are determined by plotting the signal strength in various directions. The area with the greatest signal strength is known as the major lobe, and all others are called minor lobes.

lobe frequency — The number of times a lobing pattern is repeated per second.

lobe front — The major lobe of a directional antenna. The lobe in the direction of preferred reception or transmission.

lobe half-power width — In a plane containing the direction of the maximum energy of the lobe, the angle between the two directions in that plane about the maximum in which the radiation intensity is one-half the maximum value of the lobe.

lobe penetration — The penetration of the radar coverage of a station that is not limited by the pulse-repetition frequency, scope limitations, or the screening angle at the azimuth of penetration.

lobe switching — A form of scanning in which the maximum radiation or reception is periodically switched to each of two or more directions in turn.

lobing — The formation of maxima and minima at various angles of the vertical-plane antenna pattern by the reflection of energy from the surface surrounding the radar antenna. These reflections reinforce the main beam at some angles and tend to cancel it at other angles, producing fingers of energy.

local action — In a battery, the loss of otherwise usable chemical energy by currents that flow within regardless of its connections to an external circuit.

local alarm — An alarm that when activated makes a loud noise at or near the protected area or floods the site with light, or both.

local alarm system — An alarm system that when activated produces an audible or visible signal in the immediate vicinity of the protected premises or object. This term usually applies to systems designed to provide only a local warning of intrusion and not to transmit to a remote monitoring station. However, local alarm systems are sometimes used in conjunction with a remote alarm.

local area network — Abbreviated LAN. A data-communications system, usually owned by a single organization, that allows similar or dissimilar digital devices to talk to each other over a common transmission medium. Communications can also take place among diverse equipment types: mainframes, minicomputers, microcomputers, work processors, personal computers, intelligent terminals, workstations, printers, and disk drives. A local network provides such communications over a limited geographical area: a floor, a section of a building, an entire building or a cluster of buildings, or in a multistory building or factory complex. Distances can vary from a few hundred feet to several miles.

local battery — 1. A battery made of single dry cells located at the subscriber's station. 2. A battery that actuates the recording instruments at a telegraph station (as distinguished from the battery that furnishes current to the line). 3. A telephone circuit power source, usually in the form of dry cells, located at the customer's end of the line.

local-battery telephone set — A telephone set that obtains transmitter current from a battery or other current supply circuit individual to the telephone set. The signaling current may be obtained from a local hand-operated generator or from a central power source.

local cable — A handmade cable form for circuit terminations at an attendant's switchboard, at unit equipment, and at other locations where wiring is routed inside the section or unit.

local central office — A central office arranged for termination of subscriber lines and provided with trunks for making connections to and from other central offices.

local channel — 1. In private line services, that portion of a through channel within an exchange that is provided to connect the main station with an interexchange channel. 2. A standard broadcast channel in which several stations, with powers not in excess of 1000 watts daytime or 250 watts nighttime, may operate. 3. A channel connecting a communications subscriber to a central office.

local control — Also referred to as manual control. 1. Control of a radio transmitter directly at the transmitter, as opposed to remote control. 2. Manual control of a transmitter, with the control operator monitoring the operation on duty at the control point located at a station transmitter with the associated operating adjustments directly accessible. (Direct mechanical control, or direct wire control of a transmitter from a control point located on board an aircraft, vessel, or on the same premises on which the transmitter is located, is also considered local control.) 3. A method whereby a device is programmable by means of its local (front or rear panel) controls in order to enable the device to perform different tasks.

localizer — A radio facility that provides signals for guiding aircraft onto the center line of a runway.

localizer on-course line — A vertical line passing through a localizer. Indications of opposite sense are received on either side of the line.

localizer station — A ground radionavigation station that provides signals for the lateral guidance of aircraft with respect to the center line of a runway.

local loop — 1. The access line from either a user terminal or a computer port to the first telephone office along the line path. 2. The teletype circuit containing a power source, the selector magnets, and a keyboard. This connection allows local copy on the teleprinter. 3. A telephone circuit that connects a subscriber's station equipment to the switching equipment in the telephone company local office.

local memory — Also called buffer RAM or sequence processor. A high-speed random-access memory used to store sequential data patterns that cannot be generated by a hardware pattern generator. Local memory often includes the capability to process data stores in RAM as if they were instructions, thereby modifying data in the buffer.

local oscillator — Abbreviated LO. Also called beat oscillator. 1. An oscillator used in a superheterodyne circuit to reproduce a sum or difference frequency equal to the intermediate frequency of the receiver. This is done by mixing its output with the received signal. 2. The oscillator whose output is mixed with the incoming signal in superheterodyne receivers to produce an intermediate frequency for signal processing (i.e., filtering, amplifying, detecting, etc.).

local-oscillator injection — An adjustment used to vary the magnitude of the local-oscillator signal that is coupled into the mixer.

local-oscillator radiation — Radiation of the fundamental or harmonics of the local oscillator of a superheterodyne receiver.

local-oscillator tube — The vacuum tube that provides the local-oscillator signal in a superheterodyne receiver.

local program — A program originating at and released through only one broadcast station.

local side — The connections from a data terminal to input-output devices.

local trunk — A trunk between local and long-distance switchboards, or a trunk between local and private branch exchange switchboards.

location — 1. A unit-storage position in the main or secondary storage of a computer. 2. A storage position in memory.

location counter — In the control section of a computer, a register that contains the address of the instruction currently being executed.

lock — 1. To terminate the processing of a magnetic tape in a manner such that the contents of the tape are no longer accessible. 2. To synchronize or become synchronized with; follow or control precisely, as in frequency, phase, motion, etc., used with on, onto, in, etc.

lock byte — An entity used to represent a resource in synchronization schemes; also termed a semaphore.

locked groove — Also called concentric groove. The blank and continuous groove in the center, near the label, of a disc record. It prevents the needle from traveling farther inward, onto the label.

locked-rotor current — In a motor, the steady-state current taken from the line while the rotor is locked and the rated voltage (and frequency in alternating-current motors) is applied to the motor.

locked-rotor torque — Also called static torque. 1. The minimum torque a motor will develop at rest, for all angular positions of the rotor, when the rated voltage is applied at rated frequency. 2. Motor torque at zero speed.

lock-in — 1. The term used when a sweep oscillator is in synchronism with the applied sync pulse. 2. The shifting and automatic holding of one or both of the frequencies of two oscillating systems that are coupled together, so that the two frequencies have the ratio of two integral numbers.

lock-in amplifier — A form of synchronous detector having a balanced amplifier in which the signal is applied to the grids of two tubes as the control signal is applied to their plates or another grid. The difference of their output currents is then measured.

locking — 1. Controlling the frequency of an oscillator by means of an applied signal of constant frequency. 2. Automatic following of a target by a radar antenna. 3. Pertaining to code extension characters that indicate a change in the interpretation of an unspecified number of subsequent characters.

locking circuit — See holding circuit.

locking-in — In two oscillators that are coupled together, the shifting and automatic holding of one or both of their frequencies so that the two frequencies are in synchronism (i.e., have the ratio of two integral numbers).

locking-out relay — An electrically operated hand-reset or electrically reset device that functions to shut down and hold an equipment out of service on the occurrence of abnormal conditions.

locking relay — See latch-in relay.

lock-in range — See tracking range.

lock-on — 1. The instant at which radar begins to track a target automatically. 2. Signifies that a tracking or target-seeking system is continuously and automatically tracking a target in one or more coordinates.

lock-on range — The range from a radar to its target at the instant when lock-on occurs.

lockout — 1. In a telephone circuit controlled by two voice-operated devices, a condition in which excessive local circuit noise or continuous speech from either or both subscribers results in the inability of one or both subscribers to get through. 2. See receiver lockout system. 3. Mechanical function whereby not

more than one switch station can be fully depressed simultaneously — for switches with interlock, nonlock, or pushlock/push-release mechanical functions.

lock range — The range of frequencies over which a phase-locked loop will remain locked on. The related tracking, or hold-in, range refers to how much the loop frequency can deviate from the center frequency, and is one-half the lock range.

lock shaft — A mechanical locking arrangement to prevent rotation of an adjustment after it has once been set.

lockup — Electromechanical function whereby switch stations are immobilized by operation (either actuating or releasing) of a solenoid (sometimes manually). Actuated switch stations cannot be released, and unactuated switch stations cannot be actuated until lockup is released. Lockup release can be accomplished locally or remotely.

lock-up relay — A relay that is locked in the energized position magnetically or electrically rather then mechanically.

loctal base — See loktal base.

lodar — Also called lorad. A direction finder that compensates for night effect by observing the distinguishable ground- and sky-wave loran signals on a cathode-ray oscilloscope and positioning a loop antenna to obtain a null indication of the more suitable components.

lodestone — Also spelled loadstone. A natural magnet consisting chiefly of a magnetic oxide of iron called magnetite.

lodex — A Hitachi trade name for permanent magnets that consist of elongated, single-domain iron cobalt particles dispersed in a lead matrix. The fine-particle magnets and binder are mixed in powder form and then pressed to final shape.

Lodge antenna — A counterpoise antenna that consists of a vertical dipole provided with horizontal top and bottom plates or screens, the lower plate or screen being spaced from the ground. Other versions include the bow-tie antenna, in which two narrow triangular plates are connected at their smaller ends by a coil, and the umbrella antenna, in which the end plates are conical and made up of wires.

Loftin-White circuit — A type of direct-coupled amplifier.

log — 1. A listing of radio stations and their frequency, power, location, and other pertinent data. 2. A record of the station with which a radio station has been in communication. Amateur radio operators, as well as all commercial operators, are required by law to keep a log. 3. At a broadcast station, a detailed record of all programs broadcast by the station. 4. At a broadcast transmitter, a record of the meter readings and other measurements required by law to be taken at regular intervals. 5. Abbreviation for logarithm.

logarithmic amplifier — An amplifier whose output is a logarithmic (as opposed to linear) function of its input.

logarithmic converter — A device designed to convert linear change in the light state at input to logarithmic data at output.

logarithmic curve — A curve on which one coordinate of any point varies in accordance with the logarithm of the other coordinate of the point.

logarithmic decrement — 1. For an exponentially damped alternating current, the natural logarithm of the ratio of the first to the second of two successive amplitudes having the same polarity. 2. See damping ratio, 2.

logarithmic fast time constant — A constant false alarm rate system in which a logarithmic intermediate-frequency amplifier is followed by a fast time constant circuit.

logarithmic horn — A horn whose diameter varies logarithmically with the length.

logarithmic scale — A scale on which the various points are plotted according to the logarithm of the number with which the point is labeled.

logarithmic scale meter — A meter having deflection proportional to the logarithms of the applied energies.

logger — An instrument that automatically scans certain quantities in a controlled process and records readings of the values of these quantities for future record.

logging — Recording data, such as error events or transactions, for future reference.

logic — Also called symbolic logic. 1. The science dealing with the basic principles and applications of truth tables, switching, gating, etc. See logical design. 2. A mathematical approach to the solution of complex situations by the use of symbols to define basic concepts. The three basic logic symbols are AND, OR, and NOT. When used in Boolean algebra, these symbols are somewhat analogous to addition and multiplication. 3. In computers and information-processing networks, the systematic method that governs the operations performed on the information, usually with each step influencing the one that follows. 4. The systematic plan that defines the interactions of signals in the design of a system for automatic data processing. 5. Circuitry designed to enhance separation between a matrix disc's recovered channels in a quadraphonic sound system. Such circuits monitor the four decoded channels and adjust the decoder dynamically to favor the main, or dominant, channel. 6. A mathematical arrangement using symbols to represent relationships and quantities, handled in a microelectronic network of switching circuits, or gates, which perform certain functions; also, the type of gate structure used in part of a data processing system.

logical choice — In a computer, the correct decision where alternatives or different possibilities are open.

logical comparison — In computers, the consideration of two items to obtain a yes (1) if they are the same with regard to some characteristic, or a no (0) if they are not.

logical decision — 1. In a computer, the operation by which alternative paths of flow are selected depending on intermediate program data. 2. The ability of a computer to make a choice between alternatives; basically, the ability to answer yes or no to certain fundamental questions concerning equality and relative magnitude.

logical design — 1. The preplanning of a computer or data-processing system prior to its detailed engineering design. 2. The synthesizing of a network of logical elements to perform a specified function. 3. The result of 1 and 2 above, frequently called the logic of the system, machine, or network.

logical diagram — In logical design, a diagram that represents logical elements and the interconnection between them, without necessarily including construction or engineering details.

logical element — In a computer or data-processing system, the smallest building blocks that operators can represent in an appropriate system of symbolic logic. Typical logical elements are the AND gate and the flip-flop.

logical flowchart — A detailed, graphical presentation of workflow in its logic sequence; often, the built-in operations and characteristics of a given machine, with types of operations indicated by symbols.

logically equivalent circuits — Logic circuits that perform the same function even though details of the circuits differ.

logical manipulation — Performing functions of AND, OR, exclusive OR, complementing, or rotating data in registers or in memory.

logical 1 — The opposite of logical 0.

logical operations — Those operations considered to be nonarithmetical, such as selecting, searching, sorting, matching, and comparing.

logical state — Signal levels in logic devices are characterized by two stable states — the logical 1 state and the logical 0 state. The designation of the two states is chosen arbitrarily. Commonly the logical 1 state represents an on signal and the 0 state represents an off signal.

logical threshold voltage — At the output of a logic device, the voltage level at which the following logic device switches states.

logical 0 — One of the two values of a binary digit signal. If the signal is a voltage, logical 0 usually is the lower of the two voltage levels.

logic analyzer — 1. An instrument to monitor a number of test points to aid in troubleshooting. 2. An instrument that displays data or timing diagrams for designing, debugging, and troubleshooting digital equipment. 3. Equipment that displays program timing and response signals. The trigger is normally a match with a specified bit pattern, or a signal that fills the logic analyzer's buffer. Individual probes that can be attached to any desired signal line greatly enhance the analyzer's power. 4. An instrument that checks levels and/or timing of digital data.

logic array — An integrated device in which 50 or more circuits are integral to a single silicon chip. In addition, the circuits are interconnected on the chip to form some electronic function at a higher level of organization than a single circuit. Logic arrays are constructed by the unit cell method, in which a simple circuit (or function) is repeated many times on a slice. The interconnection pattern for converting groups of cells into large functions is determined after cell probe tests are completed. Each interconnection pattern may be unique to a single slice. In general, logic arrays are characterized by multiple levels of metallization to effect the large-scale function.

logic circuit — 1. A circuit (usually electronic) that provides an input-output relationship corresponding to a Boolean-algebra logic function. 2. An electronic device or devices used to govern a particular sequence of operations in a given system.

logic devices — Digital components that perform logic functions. They can gate or inhibit signal transmission in accordance with the application, removal, or combination of input signals.

logic diagram — 1. In logical design, a diagram representing the logical elements and their interconnections, but not necessarily their construction or engineering details. 2. A pictorial representation of interconnected logic elements using standard symbols to represent the detailed functioning of electronic logic circuits. The logic symbols in no way represent the types of electronic components used, but represent only their functions. 3. A drawing that depicts the multistate device implementation of logic functions with logic symbols and supplementary notation, showing details of signal flow and control, but not necessarily the point-to-point wiring. 4. A drawing that represents the logic functions AND, OR, NOT, etc.

logic element — 1. A device that performs a logic function; a gate or flip-flop, or in some cases a combination of these devices treated as a single entity. 2. Symbol that has logical meaning. May also be called a logic symbol, e.g., gate, flip-flop, etc.

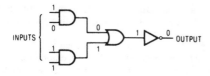

Logic diagram.

logic family — Group of digital integrated circuits sharing a basic circuit design with standardized input-output characteristics.

logic function — 1. A means of expression of a definite state or condition in a magnetic amplifier, relay, or computer circuit. 2. One of the Boolean algebra functions AND, OR, and NOT, or a combination of these.

logic ground — A level that is used as a reference for digital signals in a system. Not necessarily at the same potential as earth or safety ground.

logic instruction — An instruction for execution of an operation defined in symbolic logic, such as AND, OR, NOR, or NAND.

logic level — 1. One of two possible states, zero or one. 2. The voltage magnitude associated with signal pulses representing logical 1s and 0s in binary computation.

logic module — Circuit element comprising logic gates (AND, OR, NOT) and variations (NAND, NOR). Used in the design of binary arithmetic circuits, Boolean logic arrangements, etc.

logic-state analyzer — 1. A test instrument that synchronously captures and displays data valid at a system's clock edge. It traces the flow of digital data and, hence, is said to work in the data domain. 2. An instrument used to track state flow and record state sequences occurring during program execution. It can reveal errors in software execution.

logic swing — The difference in voltage between the voltage levels for 0 and 1 in a binary logic circuit.

logic switch — A diode matrix or other switching arrangement that can direct an input signal to a selected one of several outputs.

logic symbol — A symbol used to represent a logical element on a graph.

logic system — 1. A group of interconnected logic elements that act in combination to perform a relatively complex logic function. 2. Programming-recording system constructed of solid-state modules based on series of binary logic (go/no go) components.

logic timing analyzer — A test instrument that captures information that is asynchronous to a system's clock and displays the relative timing relationships of the traditional time domain. (Some logic analyzers display both data and timing diagrams.)

login — 1. The account name used to gain access to a computer system. The account name (in contrast to the password) is not secret. 2. The act of entering into a computer system.

log/linear preamplifier — A preamplifier whose electrical output is proportional to the common logarithm of the input, but which can be switched to a different mode wherein the output is proportional to the input. Any amplifier whose transfer characteristic or calibration curve can be arbitrarily switched from the $y = A \log x$ form to the form $y = Ax$ (where y is the output and x is the input).

log on — A mechanism by which a computer system user identifies himself or herself and gains access to system facilities.

log-periodic antenna — A type of directional antenna that achieves its wideband properties by geometric iteration. The radiating elements and the spacing between elements have dimensions that increase logarithmically from one end of the array to the other so that the ratio of element length to element spacing remains constant.

log receiver — A receiver in which the output amplitude is proportional to the logarithm of the input amplitude.

log scan — A spectrum display in which the frequency axis is calibrated logarithmically.

loktal base — Also spelled loctal. An eight-pin base for small vacuum tubes. It is designed so that it locks the tube firmly in a corresponding eight-hole socket. Unlike in other tubes, the pins are sealed directly into the glass envelope.

loktal tube — Also spelled loctal. A vacuum tube with a loktal base.

long base-line system — A system in which the distance separating ground stations approximates the distance to the target being tracked.

long-distance loop — A line that connects a subscriber's station directly to a long-distance switchboard.

long-distance navigational aid — A navigational aid usable beyond radio line of sight.

long-distance xerography — A facsimile system at the receiving terminal of which a lens projects a cathode-ray image onto the selenium-coated drum of a xerographic copying machine; a cathode-ray scanner is employed at the microwave transmitting terminal.

longevity — The normal operating lifetime of a piece of equipment, usually considered to be the period in which its failure rate is acceptably low and essentially constant.

longitudinal chromatic aberration — A lack of sharpness in the image because different colors come to a focus at different distances from the lens.

longitudinal circuit — A circuit in which one conductor is a telephone wire (or two or more wires in parallel) and the return is through the earth or any other conductors except those taken with the original wire or wires to form a metallic telephone circuit.

longitudinal coils — The coils that are used in transmission lines to suppress longitudinal currents.

longitudinal current — A current that flows in the same direction in both wires of a parallel pair. The earth is its return path.

longitudinal magnetization — In magnetic recording, magnetization of the recording medium in a direction essentially parallel to the line of travel.

longitudinal parity — Parity associated with bits recorded on one track in a data block to indicate whether there is an even or odd number of bits in the block.

longitudinal redundance — In a computer, a condition in which the bits in each track or row of a record do not total an even (or odd) number. The term is generally used to refer to records on magnetic tape, and a system can have either odd or even longitudinal parity.

longitudinal wave — A wave in which the direction of displacement at each point in the medium is perpendicular to the wavefront, as, for example, in a sound wave propagated in air.

long line — A communication line spanning a long distance relative to the local loop.

long-line effect — An effect that occurs when an oscillator is coupled to a transmission line with a serious mismatch. Oscillation may be possible at two or more frequencies, and the oscillator jumps from one side of these frequencies to another as its load changes.

long-lines engineering — Engineering with the objective of developing, modernizing, or expanding long-haul, point-to-point communications facilities that use radio, microwave, fiber optics, or wire circuits.

long-nose pliers — Pliers with long, narrow holding jaws suitable for wrapping wire around closely spaced terminals.

long-persistence screen — A fluorescent and phosphorescent screen in which the light intensity of its spots does not immediately die out after the beam has moved on. The generation of light at the instant the electron beam strikes the screen is due to fluorescence; the light that persists after the beam has moved on is due to phosphorescence. The time of persistence varies with the type of tube employed and the coating of the screen.

long-play record — Abbreviated LP record. Also called a microgroove record. A 10- or 12-inch (25.4- or 30.5-cm) record or transcription with finely cut grooves that give it a long playing time.

long-pull magnet — An electromagnet designed to exert a practically uniform pull for an extended range of armature movement. It consists of a conical plunger moving up and down inside a hollow core.

long-range navigation — *See* loran.

long-range radar — A radar installation capable of detecting targets 200 or more miles (320 km) away.

long-reach mike — *See* shotgun mike.

long shunt — A shunt field connected across the series field and the armature, instead of directly across the armature alone, of a motor or generator.

long-tailed pair — A two-tube circuit in which decreased plate current through one tube results in increased plate current through the other tube, and vice versa.

long-term stability (or long-term instability) — The slow changes in average frequency arising from changes in an oscillator. Statements of long-term stability for quartz oscillators often term this characteristic *aging rate* and specify it as parts per day (fractional frequency change over 24 hours). For cesium standards, this term commonly refers to the total fractional frequency drift for the life of the cesium beam tube.

long throw — A method of speaker design in which the woofer moves freely through long excursions, providing excellent low-frequency response with low distortion.

long wave — Wavelengths longer than about 1000 meters. They correspond to frequencies above 300 kHz.

long-wire antenna — 1. An antenna that has a length greater than one-half wavelength at the operating frequency. 2. A directional antenna consisting of a single straight wire whose length is several times greater than its operating wavelength.

look ahead — 1. A feature of the CPU of a computer that allows the machine to mask an interrupt request until the following instruction has been completed. 2. A feature of adder circuits and ALUs that allows these devices to look ahead to see that all carrys generated are available for addition.

lookthrough — 1. In jamming, sporadic interruption of the emission for extremely short periods in order to monitor the victim signal. 2. When a set is being jammed, the monitoring of the desired signal during lulls in the jamming signals.

loom — A flexible nonmetallic tubing placed around insulated wire for protection.

loop — 1. A complete electrical circuit. 2. In a computer, a series of instructions that are carried out repeatedly until a terminal condition prevails, e.g., until a predetermined count or other test is satisfied. 3. In automatic control, the path followed by command signals, which direct the actions to be performed, and feedback signals, which are returned to the command point to indicate what is actually happening. *See also* closed loop. 4. *See* mesh. 5. A length of tape having its ends spliced together to form an endless loop. Frequently used by film and radio/TV sound departments for prolonged backgrounds of continual or repetitive sound effects. The loop is the basis of the 8-track cartridge format. 6. A combination of one or more interconnected instruments arranged to measure or control a process variable, or both. 7. The two-wire circuit formed by a customer's telephone set, cable pair, and other conductors that connect it to the central office equipment. 8. An electric circuit consisting of several elements, usually switches, connected in series. 9. The curve or arc made by the wire between the attachment points at each end of a wire bond. 10. *See* antinodes.

loop actuating signal — The signal derived from mixing the loop input and loop feedback signals.

loop antenna — 1. An antenna used in radio direction-finding apparatus and in some radio receivers. It consists of one or more loops of wire. 2. An antenna consisting of several turns of wire in the same plane so arranged that it encloses an area in the electromagnetic field.

Loop antenna.

loopback — A diagnostic procedure used for transmission devices. A test message is sent to a device being tested. The message is then sent back to the originator and compared with the original transmission. Loopback testing may be performed with a locally attached device or conducted remotely over a communications circuit.

loopback test — A test in which signals are looped from a test center through a data set or loopback switch and back to the test center for measurement. *See also* busback.

loop circuit — A communication circuit that more than two parties share. In teletypewriter application, all machines print all data entered on the loop.

loop control — The maintaining of a specified loop of material between two sections of a machine by automatically adjusting the speed of at least one of the driven sections.

loop counter — In a computer, a register used to implement high-speed loop branching, including simple instruction loops (test and decrement must also be performed in the same instruction).

loop current — Flow of dc in the local loop of a telephone circuit. Indicates that a telephone is in use.

loop dialing — A return-path dialing method in which pulses are sent out over one side of the interconnecting line or trunk and returned over the other side. This arrangement is limited to short-haul traffic.

loop difference signal — The output signal from a summing point of a feedback control loop. It is a specific type of loop actuating signal produced by a particular loop input signal applied to that summing point.

loop disconnect dialer — A circuit used to convert ten-digit push-button telephone inputs to an output series of dialing pulses. This enables a push-button telephone to be connected into a telephone system that was designed to accept inputs from rotary-dial telephones only.

loop error — The desired value minus the actual value of the loop output signal.

loop error signal — The loop actuating signal, when it is the loop error.

loop feedback signal — The signal derived from the loop output signal and fed back to the mixing point for control purposes.

loop feeder — A feeder that follows along a circuit and distributes the voltage more evenly at different points.

loop gain — 1. The total usable power gain of a carrier terminal or two-wire repeater. Because any closed system tends to sing (oscillate), its usable gain may be less than the sum of the enclosed amplifier gains. The maximum usable gain is determined by — and cannot exceed — the losses in the closed path. 2. The increase in gain observed when the feedback path of an amplifier is opened, but with all circuit loads intact. 3. In an operational amplifier circuit, the product of the transfer characteristics of all of the elements (active or passive) encountered in a complete trip around the loop, starting at any point and returning to that point.

loop height — A measure of the deviation of the wire loop from the straight line between the attachment points of a wire bond. Usually it is the maximum perpendicular distance from this line to the wire loop.

looping — Repetition of instructions at delayed speeds until a final value is determined (as in a weight scale indication). The looped repetitions are usually frozen into a ROM memory location and then jumped when needed. Looping also occurs when the CPU of a computer is in a wait condition. *See also* loop through.

looping-in — Wiring method that avoids tee joints by carrying the conductor or cable to and from the point to be supplied.

loop input signal — An external signal applied to a feedback control loop.

loop-mile — The length of wire in a mile of two-wire line.

loop noise bandwidth — A phase-locked loop property related to damping and natural frequency that describes the effective bandwidth of the received signal. Noise and signal components outside this band are highly attenuated.

loop output signal — The controlled signal extracted from a feedback control loop.

loop pulsing — Also called dial pulsing. The regular, momentary interruption of the direct-current path at the sending end of a transmission line.

loop resistance — The total resistance of two conductors measured at one end (conductor and shield, twisted pair, conductor and armor, etc.). *See also* line loop resistance.

loop return signal — The signal returned, via a feedback control loop, to a summing point in response to a loop input signal applied to that summing point. The loop return signal is a specified type of loop input signal and is subtracted from it.

loopstick antenna — *See* ferrite-rod antenna.

loop test — A method of locating a fault in the insulation of a conductor when the conductor can be arranged to form part of a closed circuit.

loop through — Also called looping. The method of feeding a series of high-impedance circuits (such as multiple monitor displays in parallel) from a pulse or video source with a coaxial transmission line in such a manner that the line is bridged (with minimum length stubs) and that the last unit properly terminates the line in its characteristic impedance. This minimizes discontinuities or reflections on the transmission line.

loop transfer function — The transfer function of the transmission path. It is formed by opening and properly terminating a feedback loop. *See also* transfer function.

loose coupler — An obsolete tuning system consisting of two coils, one inside the other. Coupling was varied over a wide range by sliding one coil over the other.

loose coupling — Also called weak coupling. 1. In resonant systems, a degree of coupling that is considerably less than critical coupling. Hence, there is very little transfer of energy. 2. Coupling between two circuits so that (*a*) little power is transferred from one to the other and (*b*) an impedance change in one circuit has little effect on the other circuit.

LORAC — A trademark of Seismograph Service Corp. for a specific navigation system that determines a position fix by the intersection of lines of position. Each line is defined by the phase angle between two heterodyne beat-frequency waves. One wave is the beat frequency between two cw signals from two widely spaced transmitters. The other is a reference wave of the same frequency and is obtained by deriving the heterodyne beat of the same two cw signals at a fixed location and transmitting it, via a second radio-frequency channel, to the receiver being located.

lorad — *See* lodar.

loran — A contraction of long-range navigation. 1. A navigation aid sending out pulses at radio frequencies between 1800 and 2000 kHz. It defines lines of position that are based on the differences in travel time between radio waves from a master and a slave station. Airborne equipment utilizes a picture tube and measuring circuits to determine the time differences and relate them to time-difference lines drawn on a map. A navigation fix is established by the intersection of two of more lines of position. 2. A long-range electronic navigation system that uses the time divergence of pulse-type transmission from two or more fixed stations.

loran C — A hyperbolic navigation system that relies on the stability of electromagnetic propagational characteristics at a frequency of 100 kHz. Each loran-C station operates in a pulse mode in such a way that no two stations can be received simultaneously. During the transmission period of the master station, both slave stations remain silent. After a fixed delay time, slave X will transmit a similar pulse. After a second delay, slave Y will transmit a pulse.

loran D — A tactical loran system that employs the coordinate converter of loran C and can be operated in conjunction with inertial systems on board aircraft, without dependence on ground facilities and without radiation of rf energy that could reveal the location of the aircraft.

loran line — A line of position on a loran chart where each line is the locus of points, the distances from two fixed stations of which differ by a constant amount.

loran station — A radionavigational land station transmitting synchronized pulses.

loran tables — Tables giving terrestrial coordinates (latitude and longitude) of loran lines of position and values of the skywave correction.

Lorentz force — The force exerted by an electric field and a magnetic field on a moving electric charge.

Lorentz force equation — An equation relating the force on a charged particle to its motion in an electromagnetic field.

loss — 1. A decrease in power suffered by a signal as it is transmitted from one point to another. It is usually expressed in decibels. 2. Energy dissipated without accomplishing useful work. 3. A reduction in signal level or strength, usually expressed in decibels.

loss angle — The complement of the phase angle of an insulating material.

losser — A circuit having less power in the output as compared with the power applied to the input. This term is particularly applicable to mixers; a crystal mixer is a losser.

losser circuit—A resonant circuit having sufficient high-frequency resistance to prevent sustained oscillation at the resonant frequency.

Lossev effect—The resultant radiation when charge carriers recombine after being injected into a forward-biased pn or pin junction.

loss factor—*See* loss index.

loss index—Also called loss factor. 1. The product of the dielectric constant and the loss tangent. It is a measure of the power loss per cycle and receives extensive use in microwave and dielectric heating applications. 2. The product of the power factor and the dielectric constant. 3. The characteristic that determines the rate at which heat is generated in an insulating material. It is equal to the dielectric constant times the power factor.

lossless line—A theoretically perfect line—i.e., one that has no loss and hence transmits all the energy fed to it.

loss modulation—*See* absorption modulation.

loss tangent—Also called dielectric dissipation factor. The decimal ratio of the irrecoverable to the recoverable part of the electrical energy introduced into an insulating material by the establishment of an electric field in the material.

lossy—1. Insulating material that dissipates more than the usual energy. 2. Of, like, or made of an insulating material capable of damping out an unwanted mode of oscillation while having little effect on a desired mode.

lossy attenuator—A length of waveguide made from some dissipative material and used to deliberately introduce transmission loss.

lossy cable—A single-circuit cable (frequently of coaxial construction) deliberately constructed to have high transmission loss so that it can be used as an attenuator, an artificial load, or a termination.

lossy line—1. A transmission line designed with high attenuation. 2. A transmission line, usually a coaxial cable, designed to have a very high transmission loss per unit length.

lot—A group of similar components that have been either all manufactured in a continuous production run from homogeneous raw materials under constant process conditions or assembled from more than one production run and submitted for random sampling and acceptance testing. Specifically, a quantity of material all of which was manufactured under identical conditions and assigned an identifying lot number.

lot tolerance percent defective—Abbreviated LTPD. A statistically defined quality level, in percent defective that is rejected on an average 90 percent of the time. Thus, a lot that is truly 5 percent defective is rejected 90 percent of the time when a 5-percent LTPD is used.

Lotus 1-2-3—A flexible, easy to use spreadsheet program for IBM PCs and compatibles. It combines graphics, spreadsheet functions, and data management.

loudness—1. A measure of the sensitivity of human hearing to the strength of sound. Scaled in sones, it is an overall single evaluation resulting from calculations based on several individual band-index values. 2. Generally synonymous with volume, which is the intensity of perceived sound.

loudness compensation—1. A variety of equalization applied to a signal according to its volume in order to compensate for the tendency of the ear to change frequency responses at different listening levels. 2. A form of equalization, coupled with the volume control, that progressively emphasizes low frequencies (and sometimes also high frequencies) relative to the middle frequencies

as the volume is reduced. Intended to correct for the human ear's natural loss of hearing sensitivity at the frequency extremes when sound level is reduced, and thus to preserve proper frequency balance at different listening volume levels.

loudness contour—A curve showing the sound pressure required at each frequency to produce a given loudness sensation to a typical listener.

loudness-contour selector—Abbreviated LCS. A circuit that alters the frequency response of an amplifier so that the characteristics of the amplifier will more closely match the requirements of the human ear.

loudness control—Also called compensated volume control. A combined volume and tone control that boosts the bass frequencies at low volume to compensate for the inability of the ear to respond to them. Some loudness controls provide similar compensation at the treble frequencies.

loudness level—The sound-pressure level of a 1000-Hz tone judged by a listener to be as loud as the sound under consideration. It is measured in decibels relative to 0.0002 microbar.

loudspeaker—*See* speaker.

loudspeaker dividing network—*See* dividing network.

loudspeaker impedance—*See* speaker impedance.

loudspeaker system—*See* speaker system.

loudspeaker voice coil—*See* speaker voice coil.

louver—The grille of a speaker.

love wave—A dispersive or frequency-sensitive surface effect that also decays exponentially with depth. Love waves consist of particle motion in the plane of the surface and normal to the direction of wave propagation. These are generated when a substrate is covered with a film of solid material whose thickness is considerably smaller than the wavelength.

low-angle radiation—Radiation that proceeds at low angles above the ground.

low band—Television channels 2 through 6, covering frequencies between 54 and 88 MHz.

low-capacitance contacts—A type of contact construction providing small capacitance between contacts.

low-capacitance probe—A test probe with very low capacitance. It is connected between the input of an oscilloscope and the circuit under observation.

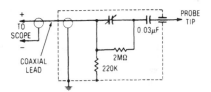

Low-capacitance probe circuit.

low-corner frequency—The frequency at which the output of a resolver is 3 dB below the midfrequency value and the phase shift is 45°.

low-definition television—A television system employing less than 200 scanning lines per frame.

low earth orbit—*See* LEO.

low-energy circuit—Also called a dry circuit. A circuit application that functions at low voltage (i.e., approximately 10 volts or less) and low current (i.e., approximately 1 mA or less).

low-energy criterion — *See* Von Hippel breakdown theory.

low-energy material — *See* soft magnetic material.

lower pitch limit — The minimum frequency at which a sinusoidal sound wave will produce a pitch sensation.

lower sideband — 1. The lower of two frequencies or groups of frequencies produced by the amplitude-modulation process. 2. In carrier transmission, the band of frequencies that is lower than the carrier frequency. It is the difference between the instantaneous values of the carrier frequency and the modulating frequency.

lowest effective power — The minimum product of the antenna input power in kilowatts and the antenna gain required for satisfactory communication over a particular radio route.

lowest useful frequency — Abbreviated LUF. The lowest frequency in the high-frequency radio band that can be used at a particular time of day for ionospheric propagation between two specified points.

lowest useful high frequency — In radio transmission, the lowest frequency effective at a specified time for ionospheric propagation of radio waves between any two points, excluding frequencies below several megahertz. It is determined by such factors as absorption, transmitter power, antenna gain, receiver characteristics, type of service, and noise conditions.

low filter — An audio circuit designed to remove low-frequency noises from the program material. Such noises include turntable rumble, tonearm resonance, etc.

low frequency — Abbreviated LF. The band of frequencies extending from 30 to 300 kHz (10,000 to 1000 meters).

low-frequency compensation — A technique for extending the low-frequency response of a broadband amplifier.

low-frequency distortion — Distortion effects that occur at low frequency. In television, generally considered as any frequency below the 15.75-kHz line frequency.

low-frequency impedance corrector — An electric network designed to be connected to a basic network, or to a basic network and a building network, so that the combination will simulate, at low frequencies, the sending-end impedance, including the dissipation, of a line.

low-frequency induction heater or furnace — A heater or furnace in which the charge is heated by inducing a current at the power-line frequency through it.

low-frequency loran — Also called hertz-matching loran. A modification of standard loran, with operation in the frequency range of approximately 100 to 200 kHz, for increased range over land and during daytime. Whereas the envelopes of the pulses are matched to obtain a line of position with ordinary loran, the hertz within the pulses are matched to provide a much more accurate fix with low-frequency loran.

low-frequency padder — In a superheterodyne receiver, a small adjustable capacitor connected in series with the oscillator tuning coil. During alignment it is adjusted to obtain correct calibration of the circuit at the low-frequency end of the tuning range.

low insertion force connector — *See* LIF connector.

low level — In digital logic, the more negative of the two binary-system logic levels. *See also* high level; negative logic; positive logic.

low-level contacts — Contacts that control only the flow of relatively small currents in relatively low-voltage circuits — e.g., alternating currents and voltages encountered in voice or tone circuits, direct current on the order of microamperes, and voltages below the softening

voltages on record for various contact materials (that is, 0.080 volt for gold, 0.25 volt for platinum, and the like). Also defined as the range of contact electrical loading where there can be no electrical (arc transfer) or thermal effects and where only mechanical forces can change the conditions of the contact interface. In general, loads below 5 volts dc and 10 milliamperes are considered low level.

low-level crossover network — A crossover network designed to operate at line levels and which is placed before the power amplifier(s); used in biamplified or tri-amplified speaker systems.

low-level modulation — 1. The modulation produced in a system when the power level at a certain point is lower than it is at the output. 2. Modulation introduced at a low-level point in a system, before amplification.

low-level radio-frequency signal (tr, atr, and pre-tr tubes) — A radio-frequency signal with insufficient power to fire the tube.

low-level signal — Very small amplitudes serving to convey information or other intelligence. Variations in signal amplitude are frequently expressed in microvolts.

low-loss dielectric — An insulating material, such as polyethylene, that has a relatively low dielectric loss, making it suitable for transmission of radio frequency.

low-loss insulator — An insulator with high radio-frequency resistance and hence slight absorption of energy.

low-loss line — A transmission line with relatively low losses.

low-noise amplifier — Abbreviated LNA. A device that receives and amplifies the weak satellite signal reflected by a dish via a feed. C-band LNAs typically have noise characteristics quoted as noise temperatures rated in degrees Kelvin. Ku-band LNA noise characteristics are usually expressed as a noise figure in decibels.

low-noise block downconverter — Abbreviated LNB. A low-noise microwave amplifier and converter that downconverts a block or range of frequencies at once to an intermediate frequency range, such as 950 to 1450 MHz, 950 to 1750 MHz, or 950 to 2050 MHz.

low-noise cable — Cable configuration specially constructed to eliminate spurious electrical disturbances caused by capacitance changes of self-generated noise induced by either physical abuse or adjacent circuitry.

low-noise converter — Abbreviated LNC. A combination low-noise amplifier and remote downconverter housed in one weatherproof box.

low-noise tape — Magnetic tape with a signal-to-noise ratio 3 to 5 dB better than conventional tapes, making it possible to record sound — especially music with a wide frequency range — at reduced tape speeds without incurring objectionable background noise (hiss) and with little compromise of fidelity.

low order — The least significant half of a word. Usually bits 0–7 of a 16-bit word.

low-pass filter — 1. A filter network that passes all frequencies below a specified frequency with little or no loss but discriminates strongly against higher frequencies. 2. Wave filter having a single transmission band extending from zero frequency up to some critical or cutoff frequency not infinite. 3. Filter that passes all frequencies below a certain cutoff point and attenuates all frequencies above that point.

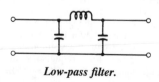

Low-pass filter.

low-power Schottky TTL (LS-TTL) circuit — A relatively high speed and low-power transistor-transistor logic (TTL) circuit using metal/semiconductor diodes (Schottky diodes) to prevent transistors from saturating.

low print-through tape — Special magnetic recording tape significantly less susceptible to print-through (the transfer of signal from one layer to another), which results when tape is stored for long periods of time. These tapes are especially useful for speech master recording (making an original recording from which copies will be made) on professional-quality equipment.

low-rate discharge — The withdrawal of a small current from a cell or battery for long periods of time.

low-speed data rate — Transmission of data at any speed up to 150 bauds.

low state — The relatively less positive signal level used to assert a specific message content associated with one of two binary logic states.

low tension — British term for low voltage. Generally refers to the heater or filament voltage.

low-velocity scanning — The scanning of a target with electrons having a velocity below the minimum required to give a secondary-emission ratio of unity.

low-voltage protection — *See* undervoltage protection.

L-pad — A volume control that has practically the same impedance at all settings. It consists essentially of an L-network in which both elements are adjusted simultaneously.

L + R, L − R — The sum and difference signals of the two stereo channels; the L + R signal combines the signals of both channels in phase; the L − R signal combines them out of phase. By combining in suitable circuitry, L + R and L − R can be added to obtain 2L, the signal from one channel; L − R can be subtracted from L + R to obtain 2R, the signal from the other channel.

LP record — Abbreviation for long-playing record.

LSA diode — Limited space-charge accumulation diode. A microwave diode similar to the Gunn diode except that by avoiding the formation of charge bunches or domains, it achieves higher output power at frequencies determined by a microwave cavity that are several times greater than the transit-time frequency.

LSB — Abbreviation for least significant bit.

L-scan — *See* L-display.

LSD — Abbreviation for least significant digit.

LSI — Abbreviation for large-scale integration. Integrated circuits containing between 100 and 5000 gate equivalents, or 1000 to 16,000 bits of memory. Over the years, integration levels have progressed from SSI (small-scale integration), MSI (medium-scale integration), and LSI to today's ULSI (ultra large scale integration).

Integration levels

Integration type	Chip area (sq-mils)	No. of gates	Memory (bits)
SSI	<10 K	<10	—
MSI	10–25 K	10–100	<1000
LSI	25–100 K	100–5000	1–16 K
VLSI	>100 K	>5000	>16 K
ULSI	>1000 K	>50,000	>256 K

LTPD — Abbreviation for lot tolerance percent defective.

LUF — Abbreviation for lowest useful frequency.

lug — 1. A device soldered or crimped to the end of a wire lead to provide an eye or fork that can be placed under the head of a binding screw. 2. A wire terminal. 3. An earlike projection of a terminal, to which an electrical connection can be made by soldering, crimping, or wrapping.

lumen — Letter symbol: lm. 1. The unit of luminous flux. It is equal to the flux through a unit solid angle (steradian) from a uniform point source of 1 candela (candle) or to the flux on a unit surface all points of which are at unit distance from a uniform point source of 1 candela. 2. A unit of light emitted from a point light source of 1 candle through a unit solid angle. All lamps are rated in their output in lumens of light.

lumen/ft^2 — A unit of incident light. It is the illuminance on a surface 1 square foot in area on which a flux of 1 lumen is uniformly distributed, or the illuminance at a surface all points of which are at a distance of 1 foot from a uniform source of 1 candela.

lumen-hour — A unit of the quantity of light delivered in 1 hour by a flux of 1 lumen.

lumen-second — The unit quantity of light equal to 1 lumen of luminous flux emitted for 1 second.

luminaire — 1. A complete lighting unit comprising a lamp or lamps and the parts required to distribute the light, position and protect the lamps, and connect the lamps to the power supply. 2. A complete unit containing a light source, globe, reflector, housing, socket, and other necessary components for lighting.

luminaire efficiency — The ratio of luminous flux (lumens) emitted by a luminaire to that emitted by the lamp or lamps used therein.

luminance — 1. The luminous flux emitted, reflected, or transmitted from the source. Usual units are the lumen per steradian per square meter, candle per square foot, meterlambert, millilambert, and footlambert. Formerly known as brightness. 2. In color television, the photometric quantity of light radiation. 3. The amount of light emitted or scattered by a surface, usually measured in footlamberts. One footcandle falling upon a perfectly diffusing white surface with no loss would produce 1 footlambert. 4. The luminous flux per unit solid angle emitted per unit projected area of a source. 5. The photometric equivalent of brightness. It is based on measurements made with a sensor having a spectral sensitivity curve corrected to that of the average human eye. The SI (international metric standard) units for luminance are candelas per meter squared, but the English footlamberts are still used extensively in the United States. One footlambert = 3.4263 candelas/m^2. 6. Luminous flux per unit solid angle and per unit area measured normal to the direction of propagation of the flux.

luminance channel — In a color television system, any path intended to carry the luminance signal. The luminance channel may also carry other signals such as the color carrier, which may or may not be used.

luminance channel bandwidth — The bandwidth of the path intended to carry the luminance signal.

luminance contrast — The observed brightness of a light-emitting element compared to the brightness of the surround, an integral part of the device and inseparable from it.

luminance factor — Ratio of the luminance of the specimen to that of a perfect reflecting or transmitting diffuser identically illuminated.

luminance flicker — The flicker resulting from fluctuation of the luminance only.

luminance meter — A type of photometer calibrated in luminance units (candelas per square unit, or lamberts). In photography, an exposure meter contains a luminance meter to record the average luminance of a scene.

luminance primary — The one of three transmission primaries whose amount determines the luminance of a color.

luminance range — An objective measure of an object's brightness that is derived from the ratio of the luminances of its lightest and darkest sections.

luminance ratio — The ratio between the luminances (photometric brightness) of any two areas in the visual field.

luminance signal — Also called Y signal. 1. The signal in which the amplitude varies with the luminance values of a televised scene. It is part of the composite color signal and is made up of 0.30 red, 0.59 green, and 0.11 blue. The luminance signal is capable of producing a complete monochromatic picture. 2. That portion of the NTSC color television signal which contains the luminance or brightness information.

luminescence — 1. The absorption of energy by matter, and its subsequent emission as light. If the light is emitted within 10^{-8} second after the energy is absorbed, the process is known as fluorescence. If the emission takes longer, the process is called phosphorescence. 2. The emission of light of certain wavelengths or limited regions of the spectrum in excess of that due to incandescence and the emissivity of the surface. This property is not exhibited by all materials 3. Emission of light due to any other cause than high temperature (incandescence). 4. A general term that is applied to the production of light, either visible or infrared, by the direct conversion of some other form of energy. The general term is then subdivided to denote the particular energy conversion involved (i.e., thermoluminescence, cathodoluminescence, photoluminescence, and electroluminescence). Luminescence includes both fluorescence and phosphorescence. Not to be confused with luminance (brightness).

luminescence threshold — Also called threshold of luminescence. The lowest radiation frequency that will excite a luminescent material.

luminescent — Any material that will give off light but not heat when energized by an external source such as a stream of electrons or radiant energy.

luminescent screen — A screen that becomes luminous in those spots excited by an electron beam. The screen of a cathode-ray tube is the best-known example.

luminosity — Ratio of luminous flux to the corresponding radiant flux at a particular wavelength. It is expressed in lumens per watt.

luminosity coefficients — The constant multipliers for the respective tristimulus values of any color. The sum of three products is the luminance of the color.

luminosity curve — A distribution curve showing luminous flux per element of wavelength as a function of wavelength.

luminosity factor — Abbreviated K. For radiation of a particular wavelength, the ratio of the luminous flux at that wavelength to the corresponding radiant flux. It is expressed in lumens per watt.

luminous — 1. An adjective used to indicate the production of light, e.g., "luminous source," to distinguish from electrical sources, etc. It is sometimes used before "intensity" or "flux." 2. Pertaining to electromagnetic radiation as perceived by the eye.

luminous efficiency — The ratio of luminous flux to radiant flux. It is usually expressed in lumens per watt of radiant flux and should not be confused with the term *efficiency* when applied to a practical source of light. The latter is based on the power supplied to the source — not the radiant flux from the source. For energy radiated at a single wavelength, luminous efficiency is synonymous with luminosity.

luminous emittance — The luminous flux emitted per unit area of the source; normal units of measurement are the lumen/cm^2 and the lumen/ft^2.

luminous energy — A measure of the time-integrated amount of flux. It has units of lumenseconds and could be used to describe such things as the radiant energy that the eye would receive from a photographic flash.

luminous flux — Also called light flux. 1. Time rate of flow of light (the total visible energy produced by a source per unit time). Usually measured in lumens. 2. Radiant power weighted at each wavelength in accordance with the ability of the eye to perceive it. 3. Radiant flux of wavelength within the band of visible light. 4. The radiant power of visible light modified by the eye response. It is the measure of the flow of visible light energy past any given point in space in a given period, and is defined as the amount of flux radiated by a source of 1 candela into a solid angle of 1 steradian.

luminous flux density (illuminance) — Radiation flux density of wavelength within the band of visible light. Measured in lumens per square foot or footcandles. At the wavelength of peak response of the human eye, 0.555 μm, 1 watt of radiative power is equivalent to 680 lumens.

luminous flux intensity — Luminous flux per unit solid angle (steradian).

luminous intensity — The ratio of luminous flux emitted by a source to the solid angle in which the flux is emitted. It is normally measured in terms of the lumen per steradian, which is the same as 1 candela.

luminous sensitivity — 1. In a phototube, the output current divided by the incident luminous flux at a constant electrode voltage. (The term "output current" here does not include the dark current.) 2. The sensitivity of an object to light from a tungsten-filament lamp operating at a color temperature of 2870 K. This definition is permissible, since luminous sensitivity is not an absolute characteristic but depends on the spectral distribution of the incident flux.

luminous transmittance — The ratio of the luminous flux transmitted by an object to the incident luminous flux.

lumped — Concentrated in single, discrete elements rather than being distributed throughout a system.

lumped constant — 1. The single quantity of a circuit property that is electrically equivalent to the total of that property distributed in a coil or circuit. 2. A resistance, inductance, or capacitance that is connected at a point, rather than being distributed uniformly throughout the circuit.

lumped-constant elements — Distinct electrical units smaller than a wavelength. They are calibrated and used, in conjunction with other electrical/electronic equipment, for controlling voltage and current.

lumped-constant oscillator — A microwave power-generation circuit that is realized by using discrete circuit elements such as inductors and capacitors.

lumped-constant tuned heterodyne-frequency meter — A device for measuring frequency. Its operation depends on the use of a tuned electrical circuit consisting of lumped inductance and capacitance.

lumped impedance — Impedance concentrated in a component, as distinguished from impedance due to stray or distributed effects.

lumped inductance — Inductance concentrated in a component, as opposed to stray or distributed inductance.

lumped loading — Inserting uniformly spaced inductance coils along the line, since continuous loading is impractical.

lumped parameter — Any circuit parameter which — for purposes of analysis — can be considered to represent a single inductance, capacitance, resistance, etc., throughout the frequency range of interest.

Luneberg lens — A type of lens used to focus radiated UHF electromagnetic energy to increase the gain of an antenna.

lux — Letter symbol: lx. 1. A practical unit of illumination in the metric system. It is equivalent to the metercandle and is the illumination on a 1 square-meter area on which there is a uniformly distributed flux of 1 lumen. 2. International system (SI) unit of illumination in which the meter is the unit of length. One lux equals 1 lumen per square meter.

Luxemburg effect — A nonlinear effect in the ionosphere. As a result, the modulation on a strong carrier wave will be transferred to another carrier passing through the same region.

luxmeter — 1. A type of illumination photometer employing a variable aperture and the contrast principle. 2. An illuminometer designed to measure illumination in terms of lux.

LVCD — Abbreviation for least voltage coincidence detection.

lx — Letter symbol for lux.

M

m — Letter symbol for the prefix milli-.

M — 1. Symbol for mutual inductance. 2. Letter symbol for the prefix mega-, or meg-, meaning one million.

mA — Letter symbol for milliampere.

MA — Letter symbol for megampere.

MAC-B — An acronym for mixed analog component signals, system B. It refers to placing TV sound into the horizontal line-blanking period and then separating the color and luminance for periods of 20 to 40 microseconds each during horizontal scan. In the process, luminance and chroma are compressed during transmission and expanded during reception, enlarging their bandwidths considerably. Transmitted via satellite as FM, this system provides considerably better TV definition and resolution. Its present parameters are within the existing NTSC format.

mach — Unit of speed measurement equal to speed of sound. Sound travels 759 mph (1221 km/hr) at sea level; it decreases about 2 percent for every thousand feet of altitude.

machine — A term that is often used as a synonym for computer, workstation, or host.

machine-available time — In a computer, power-on time less maintenance time.

machine code — 1. Code that is acceptable to the computing device on which the software is to be executed without further modification. 2. Commands for a computer or microprocessor system, often written in binary or hexadecimal format. Machine code is often referred to as machine language or object code. *See also* machine language. 3. In process planning, an alphanumeric code that identifies important machine-tool characteristics such as capacity, size, function, or capability. This number code can also be used in sorting routines to find similar machine types.

machine cycle — 1. The shortest complete process or action that is repeated in order. The minimum length of time in which the foregoing can be performed. 2. The basic CPU cycle. In one machine cycle an address may be sent to memory and one word (data or instruction) read or written, or in one machine cycle a fetched instruction can be executed.

machine error — A deviation from the correct result in computer-processed data as a result of equipment failure.

machine hardware — Circuits contained in the five parts of the computer: input, output, control, storage, and arithmetic sections; any other circuit is offline.

machine independent — 1. Pertaining to procedures or programs created without regard to the actual devices with which they will be used. 2. A term used to indicate that a program is developed in terms of the problem rather than in terms of the characteristics of the computer system.

machine instruction — An instruction that can be recognized and executed by a machine.

machine language — Also called object code, object language, or computer code. 1. A set of symbols, characters, or signs and the rules for combining them to convey instructions or data to a computer. 2. Programming language form directly executable by a computer. This language is nearly always expressed in numeric form, usually binary, octal, or hexadecimal. 3. A series of bits written as such for instruction of computers. The first level of computer language. *See* assembly language (second level); compiler language (third level). 4. Sets of numeric instructions that control the functioning of a computer. These numeric instructions execute the logic functions of the computer's logic circuits. For example, the instructions 010110 might tell the computer to clear the A register. To overcome the problems inherent in working with long strings of numbers, computer programs are more often written in assembly language.

machine learning — The ability of equipment to modify performance based on past experience stored in memory. Related to artificial intelligence.

machine ringing — Telephone ringing that is started either mechanically or by an operator and that continues automatically until the call is either answered or abandoned.

machine run — In a computer, the performance of one or more machine routines that are linked to form one operating unit.

machine sensible — Information in a form such that it can be read by a specific machine.

machine thermal relay — A device that functions when the temperature of an ac machine armature, or of the armature or other load-carrying winding or element of a dc machine, or converter, power rectifier, or power transformer (including a power-rectifier transformer), exceeds a predetermined value.

machine vision — The ability of an automated system to perform certain visual tasks normally associated with human vision, including sensing image formation, image analysis, and image interpretation or decision making.

machine word — *See* computer word.

macro — 1. A single pseudoinstruction that, when encountered by software or hardware, invokes the execution of a number of real instructions. 2. A stored set of keystrokes. They allow for quick execution of often repeated tasks, such as setting margins or tabs that are uniform to a particular business. 3. A symbol that the assembler expands into one or more machine language instructions, relieving the programmer of having to write out frequently occurring instruction sequences. 4. An instruction in an assembly language that is implemented in machine language by more than one machine language instruction.

macroassembler — 1. An assembler that facilitates definition of macros for frequently used code segments. Macroassemblers simplify program coding; however, unlike subroutine calls, they generate in-line code for each reference. 2. A computer assembly program that translates alphanumeric language into machine language.

macrocode — 1. A coding system that assembles groups of computer instructions into single code words and which therefore requires interpretation or translation so that an automatic computer can follow it. 2. An instruction in a source language that is equivalent to a specified sequence of machine instructions.

macrocommand — A computer program entity formed by a string of standard, but related, commands that are put into effect by means of a single macrocommand. Any group of frequently used commands can be combined into a macrocommand. The many become one.

macroelement — An ordered set of two or more elements used as one data element with one data use identifier. For example, the macroelement date could be the ordered set of the data elements year, month, or day.

macroinstruction — 1. In a computer, a source-language instruction that is equivalent to a specified sequence of machine instructions. 2. Either a conventional computer instruction (e.g., ADD MEMORY TO REGISTER, INCREMENT, SKIP) or device controller command (e.g., seek, read).

macroprogram — A computer program in the form of a sequence of instructions written in a source language.

macroprogramming — In a computer, the writing of machine-procedure statements in terms of macrocode instructions.

macros — Computers that have a throughput of 2 gigabits (2×10^9 bits) per second and a memory size of 2 megabytes.

macroscopic — Large enough to be observed by the unaided eye.

macrosonics — The utilization of high-amplitude sound waves for performing functions such as cleaning, drilling, emulsification, etc.

madistor — A semiconductor device that makes use of the effects of a magnetic field on a plasma current. The strength of a magnetic field determines the conductivity of the madistor material. A small change in the magnetic field produces a larger change in the madistor current.

MADT — Abbreviation for microalloy diffused-base transistor.

magamp — Abbreviation for magnetic amplifier.

magic eye — See electron-ray tube.

magic tee — See hybrid junction, 1.

magnaflux — A magnetic method of determining surface and subsurface defects in metals.

magnal base — An 11-pin base used on cathode-ray tubes.

magnal socket — An 11-pin socket used with cathode-ray tubes.

magnesium — An alkaline metal whose compounds are sometimes used for cathodes.

magnesium anode — A bar of magnesium, buried in the earth, connected to an underground cable to prevent cable corrosion due to electrolysis.

magnesium cell — A primary cell whose negative electrode is made of magnesium or one of its alloys.

magnesium-copper sulfide rectifier — A dry disc rectifier consisting of magnesium in contact with copper sulfide.

magnesium-silver chloride cell — A reserve primary cell that becomes activated when water is added.

Magnesyn — Trade name for a device made by Bendix Aviation Corp. It is a portion of a repeater unit consisting of a two-pole permanent-magnet rotor within a three-phase, two-pole, delta-connected stator. The rotor carries the indicating pointer and is free to rotate in any direction.

magnet — A body that has the property of attracting or repelling magnetic materials. In its natural form it is called a lodestone. It may also be produced by permanently magnetizing a piece of iron or steel. A temporary magnet — called an electromagnet — is produced by passing a current through a coil surrounding a piece of iron or steel; the magnetism persists only while current is flowing. When suspended freely, a magnet will turn and align its poles with the north and south magnetic poles of the earth. See also electromagnet; permanent magnet.

magnet brake — A friction brake controlled electromagnetically.

magnetic — Pertaining to magnetism.

magnetic aging — The normal change in the metallurgical change in the material. The term also applies when the metallurgical changes are accelerated by an increase or decrease in temperature. When used in reference to core loss, this term, unless otherwise modified, implies an increase in loss. When used in reference to permeability or remanence, the term, in a positive sense, indicates a decrease in these quantities.

magnetic air gap — The air space, or nonmagnetic portion, of a magnetic circuit.

magnetic alarm system — An alarm system that will initiate an alarm when it detects changes in the local magnetic field. The changes could be caused by motion of ferrous objects, such as guns or tools, near the magnetic sensor.

magnetically damped — A form of damping achieved by moving a metal vane through a magnetic field. This motion induces currents in the vane that set up magnetic fields opposing those of the stationary magnets, thus tending to bring the pointer to rest. This type of damping is found in many quality moving-iron and dynamometer-type instruments.

magnetically hard — A rather loose term not always synonymous with "hard" metallurgically. Used to designate a permanent-type magnetic material, i.e., an alloy having a high coercive force and residual induction.

magnetically isotropic — A material having the same magnetic characteristics along any axis or orientation. Could be considered the antonym of anisotropic.

magnetically polarized — The magnetic orientation of molecules in a piece of iron or other magnetizable material placed in a magnetic field, whereby the tiny internal magnets tend to line up with magnetic lines of force.

magnetically soft — A rather loose term not always synonymous with "soft" metallurgically. Used to designate a ferromagnetic material having a low coercive force.

magnetic amplifier — Often called magamp. 1. A device in which one or more saturable reactors are used, either alone or with other circuit elements, to obtain power gain. 2. A device in which a control signal applied to a system of saturable reactors modulates the flow of an alternating current in a power circuit.

magnetic analysis — The separation of a stream of electrified particles by a magnetic field in accordance with their mass, charge, or speed. This is the principle of the mass spectrograph.

magnetic anisotropy — The dependence of the magnetic properties of some materials on direction.

magnetic-armature speaker — Also called a magnetic-armature loudspeaker or magnetic speaker. A speaker comprising a ferromagnetic armature actuated by magnetic attraction.

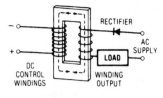

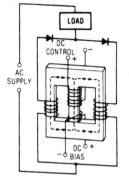

Magnetic amplifiers.

magnetic azimuth — Azimuth measured from magnetic north.

magnetic bearing — The position in which an object is pointing with respect to the earth's magnetic north pole. It is expressed in degrees clockwise from that pole.

magnetic bias — A steady magnetic field applied to the magnetic circuit of a relay.

magnetic biasing — The superimposing of another magnetic field on the signal magnetic field of a tape while a magnetic recording is being made.

magnetic bipolar sensor — Solid-state Hall-effect sensor that has a plus (south pole) maximum operate point and a minus (north pole) minimum release point. Because the operate and release points can be both positive or both negative, latching cannot be guaranteed. The sensor usually used in ring magnet applications. *Compare* latching sensor.

magnetic blowout — A magnet for establishing a field where an electrical circuit is broken. The field lengthens the arc and thus helps to extinguish it.

magnetic blowout switch — A switch that contains a small permanent magnet that provides a means of switching high dc loads. The magnet deflects the arc to quench it.

magnetic braking — Application of the brakes by magnetic force. The current for exciting the electromagnets is derived either from the traction motors acting as generators or from an independent source.

magnetic bremsstrahlung — *See* synchrotron radiation.

magnetic brush — In the electrostatic process, a device used to transport toner particles to the surface of the photoconductor. A brush is mounted on a roller mechanism. As it turns, its magnetic pull attracts toner particles and carries them to the drum surface, where they are deposited. *See* electrostatic process.

magnetic brush development — A type of development in which the material that forms the image is carried to the field of the electrostatic image by ferromagnetic particles that act as carriers under the influence of a magnetic field.

magnetic bubble — 1. A cylindrical magnetic domain with a polarization opposite to that of the thin magnetic film in which it is embedded. Domains form in a crystalline magnetic material in such a way as to minimize the total magnetic energy of the crystal. 2. Small, cylindrical domain formed in single-crystal thin films of synthetic ferrites or garnets, or in thin, amorphous magnetic-metal films when a stationary external magnetic bias field is applied perpendicularly to the plane of the films. These domains are mobile in the presence of a magnetic field gradient, and their direction of movement can be controlled by special structures deposited on top of the film and by a moving magnetic field. As the bias field increases in strength toward its optimum value, randomly distributed serpentine domains in the film layer shrink until they form the cylindrical domains, or bubbles. Once the bubbles are formed, they can be moved along a path defined by a deposited layer of metal on the surface of the magnetic film. The presence of a bubble corresponds to a logical 1 and absence to a logical 0.

magnetic bubble film — An amorphous film in which cylindrical bubbles of reverse magnetization can be formed to follow circuit paths usually made by depositing magnetic metal strips on the film surface. They are used in magneto-optic storage and processing of computer data.

magnetic bubble memory — A type of memory made of cylindrically shaped magnetic domains, called magnetic bubbles, formed in a thin-film layer of single-crystal synthetic ferrite or garnet when a magnetic field is applied perpendicular to the film's surface. A separate, rotating, magnetic field moves the bubbles through the film in shift register fashion. The presence of a bubble represents a digital 1 and the absence of a bubble, a digital 0. The basic bubble memory consists of a bubble memory chip, magnetic field coils that produce the rotating magnetic field, and two permanent magnets that maintain the magnetic bubble domains and nonvolatility.

magnetic bubble memory controller — A high-level interface between a microprocessor and a bubble memory. The controller performs parallel-to-serial conversion from the microprocessor to the bubble memory, and serial-to-parallel conversion from the bubble memory to the microprocessor. The controller's primary functions are to stop and start bubble movement, maintain page position, and raise or lower flags for such bubble memory functions as generate, swap, block replicate, and redundancy replicate. Control signals from the bubble memory controller are sent to the function timing generator — a monolithic IC that provides the precise timing signals necessary to operate the function driver, coil driver, and sense amplifier during each field cycle. (Obsolete technology).

magnetic card — A card on which data can be stored by selective magnetization of portions of a flat magnetic strip such as that found on credit cards.

magnetic card reader — A device that can retrieve information stored in a magnetic strip.

magnetic cartridge — *See* variable-reluctance pickup.

magnetic character — A character imprinted with ink having magnetic properties. These characters are unique in that they can be read directly by both humans and machines.

magnetic circuit — The path of the flux as it travels from the north pole, through the circuit components, and back to the north pole. In a generator, the magnetic-circuit components include the field yoke, field pole pieces, air gap, and armature core. The magnetic circuit may be compared to an electrical circuit, with magnetomotive force corresponding to voltage, flux lines to current, and reluctance to resistance.

magnetic circuit breaker—1. A circuit breaker that depends on the response of an electromagnetic coil for operation. 2. In their simplest form, a latching relay with normally closed contacts. When enough current flows through the coil, the armature is pulled in, transferring the contacts, which latch in their open position. In a magnetic breaker, the coil is wound so that the field generated by the rated current is not quite strong enough to pull in the armature. Any increase in current above the rated level increases the field strength enough to transfer the contacts and open the circuit.

magnetic coated disc—A magnetic disc-recording medium consisting of a coat of magnetizable material over a nonmagnetic base.

magnetic coating thickness—*See* coating thickness.

magnetic coil—The winding of an electromagnet.

magnetic compass—A device for indicating direction. It consists of a magnetic needle that pivots freely and points toward the earth's north magnetic pole.

magnetic conduction current—The rate of flow of magnetism through a magnetized body.

magnetic contactor—A contactor actuated electromagnetically.

magnetic controller—An electric controller that contains devices operated by means of electromagnets.

magnetic-convergence principle—The obtaining of beam convergence through the use of a magnetic field.

magnetic core—1. A configuration of magnetic material that is, or is intended to be, placed in a rigid special relationship to current-carrying conductors, and the magnetic properties of which are essential to its use. For example, it may be used to concentrate an induced magnetic field, as in a transformer, induction coil, or armature; to retain a magnetic polarization for the purpose of storing data; or for its nonlinear properties, as in a logic element. It may be made of such material as iron, iron oxide, or ferrite in such shapes as wires, tapes, toroids, or thin film. 2. The ferrous material in the center of an electromagnet.

magnetic core storage—A type of computer storage that employs a core of magnetic material surrounded by a coil of wire. The core can be magnetized to represent a binary 1 or 0. (Obsolete technology).

magnetic course—A course in which the reference line is magnetic north.

magnetic cutter—A cutter in which the mechanical displacement of the recording stylus is produced by the action of magnetic fields.

magnetic cycle—The sequence of changes in the magnetization of an object corresponding to one cycle of the alternating current producing the magnetization.

magnetic damping—1. The damping of a mechanical motion by means of the reaction between a magnetic field and the current generated in a conductor moving through that field. The resistance of this conductor converts excess kinetic energy to heat. 2. The slowing down of mechanical motion by the reaction between two magnetic fields. A common example is the metal disc that rotates between the poles of a permanent magnet in a watt-hour meter.

magnetic deflection—The moving of the electron beam by means of a magnetic field produced by a coil around the neck of the cathode-ray tube. Linear motion is produced by a sawtooth current through the coil.

magnetic delay line—A computer delay line in which magnetic energy is propagated, consisting in essence of a metallic medium along which the velocity of propagation of magnetic energy is small when compared with the speed of light. The storage of information,

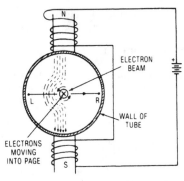

Magnetic deflection.

which is usually in a binary form, is accomplished by recirculation of wave patterns.

magnetic density—The number of lines of magnetic force passing through a magnet or magnetic field per unit area of cross section.

magnetic detecting device—A device for detecting cracks in iron or steel. This is done by introducing magnetic particles, which are attracted to the opposing magnetic poles created at the break.

magnetic device—Any device actuated electromagnetically.

magnetic digital versatile disc—Abbreviated M-DVD.

magnetic diode—A magnetism-sensitive semiconductor device that (like indium-antimony devices) has an internal resistance that varies as a function of an external magnetic field. Electrical signals may be obtained through alternation of the magnetic field, and, hence, nonelectrical quantities can be converted into electrical quantities.

magnetic dip—Also called magnetic inclination. The angle of the magnetic field of the earth with the horizontal at a particular location.

magnetic dipole—A pair of equal-strength north and south magnetic poles spaced close together, and so small that its directive properties are independent of its size and shape. It is the magnetic equivalent of an electrical dipole.

magnetic dipole antenna—A loop antenna that radiates an electromagnetic wave when electric current circulates in the loop.

magnetic direction indicator—Abbreviated MDI. An instrument that provides a compass indication, which it obtains electrically from a remote gyrostabilized magnetic compass (or its equivalent).

magnetic disc—Also disk. 1. A flat circular plate with a magnetic surface on which data can be stored by selective magnetization of portions of the flat surface. 2. A metal or plastic (floppy) disk looking something like a phonograph record whose surface can store data in the form of magnetized spots.

magnetic disk storage—1. A device or system in which information is stored in the form of magnetic spots arranged on the surfaces of magnetically coated disks to represent binary data. 2. A storage device or system made up of magnetically coated metal disks.

magnetic disk unit—A device used to read and write data on thin magnetic disks, the surfaces of which have been coated with a magnetizable material.

magnetic displacement—Magnetic flux density or magnetic induction.

magnetic domain — A volume within a magnetic material where the direction of magnetization is everywhere the same.

magnetic drum — 1. A storage device consisting of a rapidly rotating cylinder whose surface can be easily magnetized and that will retain the magnetization. Information is stored in the form of magnetized spots (or no spots) on the drum surface. 2. A right circular cylinder with a magnetic surface on which information can be stored by selective magnetization of parts of the curved surface.

magnetic-drum receiving equipment — Radar developed for detection of targets beyond line of sight using ionospheric reflection and very low power. To distinguish radar signals from ground noise, a noise-free standard system is employed whereby the amplitude and characteristics are reproduced and preserved for comparison with returned signals. Fluctuations in the amplitude of the returned signals indicate the position and velocity of the target accurately to within 10 miles (16 km).

magnetic drum storage — A device or system that uses a magnetic drum to store information.

magnetic drum unit — A device used to read and write data on a cylinder, the surface of which is covered with a magnetizable material.

magnetic electron multiplier — One type of electron multiplier in which the paths of the emitted secondary electrons are controlled by an externally applied magnetic field.

magnetic-energy product curve — The graphical representation of the energy per unit volume produced by a magnet. It is the product of the flux density and demagnetizing field as shown on the normal demagnetization curve. The maximum of this product as shown on such a curve is known as $B_d H_d$ max.

magnetic field — An area where magnetic forces can be detected around a permanent magnet, natural magnet, or electromagnet.

Magnetic field.

magnetic field strength — *See* magnetizing force.

magnetic figures — A pattern showing the distribution of a magnetic field. It is made by sprinkling iron filings on a nonmagnetic surface in the field.

magnetic flip-flop — A bistable amplifier using one or more magnetic amplifiers. The two stable output levels are determined by appropriate changes in the control voltage or current.

magnetic fluid — A fluid having three components: a carrier fluid, magnetite particles suspended by Brownian motion, and a stabilizer to prevent agglomeration of these fine particles. It is characterized by its viscosity and saturation magnetization value.

magnetic flux — 1. The magnetic induction in a material. An electromotive force will be induced in a conductor placed in a magnetic field whenever the magnitude of the flux changes. 2. A condition in a medium produced by a magnetomotive force such that, when altered in magnitude, a voltage is induced in an electric circuit linked with the flux. The cgs unit of magnetic flux is called the maxwell.

magnetic flux density — *See* magnetic induction.

magnetic focusing — The focusing of an electron beam by the action of a magnetic field.

magnetic freezing — In a relay, sticking of the armature to the core due to residual magnetism.

magnetic fusion — A fusion process that occurs when a burning plasma is confined by externally imposed magnetic fields.

magnetic gap — The nonmagnetic part of a magnetic circuit.

magnetic gate — A gate circuit used in a magnetic amplifier.

magnetic head — In magnetic recording, a transducer that converts electric variations into magnetic variations for storage on magnetic media or for reconverting such stored energy into electric energy. Stored energy can also be erased by this method.

magnetic heading — The heading of an aircraft with reference to magnetic north.

magnetic hysteresis — In a magnetic material, the property by virtue of which the magnetic induction for a given magnetizing force depends on the previous conditions of magnetization.

magnetic hysteresis loop — A closed curve that shows the induction of magnetization in a magnetic substance as a function of the magnetization force for a complete cycle of the magnetization force.

magnetic hysteresis loss — The power expended in a magnetic material as a result of magnetic hysteresis when the magnetic induction is cyclic.

magnetic inclination — *See* magnetic dip.

magnetic induction — Also called magnetic flux density (B). The flux per unit area (A) perpendicular to the direction of the flux (ϕ). The cgs unit of induction is called the gauss (plural, gausses) and is defined by the equation $B = d\phi/dA$.

magnetic ink — Visible ink containing magnetic particles. When printed on a document (e.g., a bank check), the ink can be read by a magnetic character sensor and also by humans.

magnetic-ink character recognition — Abbreviated MICR. 1. Machine recognition of characters printed with magnetic ink. (Contrast with OCR.) 2. A system of coding in which numeric and special characters are printed with magnetizable ink. 3. A mechanized method of data collection involving the electronic reading of data that has been printed in magnetic ink. Most checks have the bank transit number imprinted in magnetic ink at the lower-left part of the check.

magnetic instability — The property of a magnetic material that leads to variations in the residual flux density of a recording tape with time, mechanical flexing, or changes in temperature.

magnetic integrated circuit — An integrated component that utilizes one or more magnetic elements inseparably associated to perform all or at least a major portion of its intended function.

magnetic-latch relay — *See* polarized double-biased relay.

magnetic leakage — Passage of magnetic flux outside the path along which it can do useful work.

magnetic lens — 1. An apparatus that uses a nonuniform magnetic field to focus beams of rapidly moving electrons or ions in a cathode-ray or other tube. 2. An arranged series of coils, magnets, or electromagnets disposed in such a way that the resulting magnetic fields are used to focus beams of rapidly moving electrons or ions.

magnetic line of force—In a magnetic field, an imaginary line that has the direction of the magnetic flux at every point.

magnetic materials—Materials that show magnetic properties. Ferromagnetic materials are more strongly magnetic than paramagnetic materials.

magnetic medium—Any data-storage medium and related technology, including disks, diskettes, and tapes, in which different patterns of magnetization are used to represent bit values.

magnetic memory—A computer memory (or any portion) in which information is stored in the form of magnetism.

magnetic memory disk—A mass-memory storage element that is a large and rigid rotating circular plate coated with magnetic material, such as iron oxide. Data is written (stored) and read (retrieved) by fixed or movable read/write heads positioned over data tracks on the surface of the disk. Addressable portions are selectable for read and write operations.

magnetic memory plate—A magnetic memory that consists of a ferrite plate containing a grid of small holes through which the read-in and read-out wires are threaded. Printed wiring applied directly to the plate may replace conventionally threaded wires, making possible mass production of plates with a high storage capacity.

magnetic microphone—*See* variable-reluctance microphone.

magnetic mine—An underwater mine that detonates when near the steel hull of a ship. This is accomplished by relays that redistribute the magnetic field in the mine when it is near the ship.

magnetic modulator—Also called a magnettor. A modulator employing a magnetic amplifier as the modulating element.

magnetic moment—Ratio of the maximum torque exerted on a magnet to the magnetizing force of the field in which the magnet is situated.

magnetic needle—The magnetized needle used in a compass. When freely suspended, it will point to the earth's magnetic north and south poles.

magnetic north—The direction indicated by the north-seeking end of the needle in a magnetic compass.

magnetic phase modulator—A ferrite-core delay line in which the delay is varied by an external magnetic field.

magnetic pickup—*See* variable-reluctance pickup.

magnetic-plated wire—A wire with a nonmagnetic core and a plated surface of ferromagnetic material.

magnetic poles—Those portions of a magnet toward which the lines of flux converge. All magnets have two poles, called north and south, or north-seeking and south-seeking, poles.

magnetic pole strength—The magnetic moment of a magnetized body divided by the distance between the poles.

magnetic potential difference—The line integral of magnetizing force between two points in a magnetic field.

magnetic-powder-coated tape—Also called coated tape. A tape consisting of a coating of uniformly dispersed powdered ferromagnetic material on a nonmagnetic base.

magnetic printing—Also called magnetic transfer or crosstalk. The permanent transfer of a recorded signal from a section of one magnetic recording medium to another section of the same or a different medium when they are brought near each other.

magnetic probe—A loop-type conductor for detecting the presence of static, audio, or rf magnetic fields.

magnetic radio bearing—*See* radio bearing.

magnetic recorder—Equipment incorporating an electromagnetic transducer and a means for moving a magnetic recording medium past the transducer. Electric signals are recorded in the medium as magnetic variations. *See also* magnetic recording.

magnetic recording—Recording audio frequencies by magnetizing areas of a tape or wire. The magnetized tape or wire is played back by passing it through a reproducing head. Here the magnetized areas are reconverted into electrical energy, which headphones or speakers then change back into sound.

magnetic recording head—In magnetic recording, a magnetic head that transforms electric variations into magnetic variations for storage on a magnetic medium such as tape or disks.

magnetic recording medium—A wire, tape, cylinder, disk, or other magnetizable material that retains the magnetic variations imparted to it during magnetic recording.

magnetic reproducer—Equipment that picks up the magnetic variations on magnetic recording media and converts them into electrical variations.

magnetic reproducing head—In magnetic recording, the head that converts the magnetic variations into electric variations.

magnetic rod—A square-loop switching and storage element for digital systems. It consists of a silver-coated glass rod upon which a thin layer of iron and nickel is deposited. Conductors are wound around the rod for the drive, sense, enable, and inhibit currents.

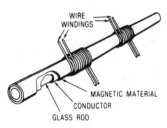

Magnetic rod.

magnetics—The branch of science concerned with the laws of magnetic phenomena.

magnetic saturation—1. In an iron core, the point at which application of a further increase in magnetizing force will produce little or no increase in the magnetic lines of force. 2. The condition under which all elementary moments have become oriented in one direction.

Magnetic poles.

magnetic separator — An apparatus for separating powdered magnetic ores from nonmagnetic ores, or iron filings and other small iron objects from nonmagnetic materials. An electromagnet is employed, which deflects the magnetic materials from the path taken by the nonmagnetic materials.

magnetic shield — A sheet or core of iron enclosing instruments of radio parts to protect them from stray magnetic fields. The shield provides a convenient path for the magnetic lines of force and thus diverts them from the component being protected.

magnetic shielding — Iron, steel, nickel, mu-metal, or other magnetic material used to exclude unwanted electromagnetic fields from circuitry, transformers, or conductors.

magnetic shift register — A register in which magnetic cores are used as binary storage elements. By means of pulses, the pattern of binary digital information can be shifted one position to the left or right in the register.

magnetic shunt — A piece of iron used during instrument calibration to divert a portion of the magnetic lines of force passing through an air gap in the instrument.

magnetic sound — Sound recording in which magnetic impulses are stored in a ferric emulsion bonded to plastic or film.

magnetic speaker — A speaker in which acoustic waves are produced by mechanical forces resulting from magnetic reaction.

magnetic starter — A starter actuated electromagnetically.

magnetic storage — Any storage system such as tape, diskette, or other form of magnetic media that makes use of the magnetic properties of materials to store information.

magnetic storm — A disturbance in the earth's magnetic field. It is associated with abnormal solar activity and is capable of disrupting both radio and wire transmission.

magnetic strain gage — Also spelled gauge. An instrument for measuring strain in rails or other structural members that bend only microscopically under a normal load. It does this by determining the change in reluctance of a magnetic circuit having a movable armature.

magnetic strip — Sheet or foil aluminum (either bare or insulated) used as a conductor in electric windings. Copper is also used sometimes.

magnetic susceptibility — Ratio of the intensity of magnetization to the corresponding value of the magnetizing force.

magnetic switch — A switch that consists of two separate units: a magnetically actuated switch and a magnet. The switch is usually mounted in a fixed position (door jamb or window frame) opposing the magnet, which is fastened to a hinged or sliding door, window, etc. When the movable section is opened, the magnet moves with it, actuating the switch.

magnetic tape — 1. The recording medium used in tape recorders. A paper or plastic tape on which a magnetic emulsion (usually ferric oxide) has been deposited. The most common width for home recorders is $1/4$ inch (6.35 mm). Some tapes are made of a magnetic material and hence need no magnetic-emulsion coating. 2. Tape made of plastic and coated with magnetic material; used to store information.

magnetic tape core — A toroidal core made by winding a strip of thin magnetic-core material around a form. A toroidal winder is then used to wind coils around the core.

magnetic tape reader — A computer device capable of converting the information recorded on magnetic tape into corresponding electric pulses.

magnetic tape storage — A storage system based on the use of magnetic spots (bits) on metal or coated-plastic tape; the spots are arranged so that the desired code is read out as the tape travels past the read-write head.

magnetic tape unit — A device used to read and write data in the form of magnetic spots on reels of tape coated with a magnetizable material.

magnetic test coil — Also called search or exploring coil. 1. A coil that is connected to a suitable device to measure a change in the magnetic flux linked with it. The flux linkage may be changed by either moving the coil or varying the magnitude of the flux. 2. The coil used with a ballistic galvanometer or flux meter to measure the magnetic flux.

magnetic thin film — A layer of magnetic material, usually less than 10,000 angstroms thick. In electronic computers, magnetic thin films may be used for logic or storage elements.

magnetic transducer — See variable-reluctance transducer.

magnetic transfer — See magnetic printing.

magnetic transition temperature — Also called the Curie point. In a ferromagnetic material, the point at which its transition to paramagnetic seems to be complete as its temperature is raised.

magnetic tubes of flux — Region in space whose sides are everywhere tangent to the magnetic induction and whose ends may meet to form closed rings.

magnetic units — Ampere-turn, gauss, gilbert, line of force, maxwell, oersted, and unit magnetic pole are examples of magnetic units — i.e., those used in measuring magnetic quantities.

magnetic-vane meter — Also called a moving-vane meter. A meter for measuring alternating current. It contains a metal vane that pivots inside a coil. Alternating current flows through the coil and sets up magnetic forces that rotate the vane and attached pointer in proportion to the value of the current.

magnetic variometer — An instrument for measuring the difference in a magnetic field with respect to space or time.

magnetic vector — A term denoting the amplitude and direction of the magnetic field associated with an electromagnetic wave.

magnetism — A property possessed by certain materials by which these materials can exert mechanical force on neighboring masses of magnetic materials and can cause voltages to be inducted in conducting bodies moving relative to the magnetized bodies.

magnetite — A mineral that exists in a magnetized condition in its natural state. It consists chiefly of a magnetic oxide of iron.

magnetization curve — A curve plotted on a graph to show successive states during magnetization of a ferromagnetic material. A normal magnetization curve is a portion of a symmetrical hysteresis loop. A virgin magnetization curve shows what happens to the material the first time it is magnetized.

magnetization intensity — At any point in a magnetized body, the ratio between the magnetic moment of the element of volume surrounding the point and an infinitesimal volume.

magnetize — To make magnetic.

magnetizing current — Also called exciting current. The current through the field windings of a generator.

magnetizing field — The magnetomotive force per unit length at any given point in the magnetic circuit. In the cgs system the unit is the oersted. It is defined by the quotient

$$\frac{\text{magnetomotive force in gilberts}}{\text{length in centimeters}}$$

magnet keeper—A bar of iron or steel placed across the poles of a horseshoe magnet when the magnet is not in use. The keeper prevents the magnet from becoming demagnetized, by completing the magnetic circuit to keep the flux from leaking off.

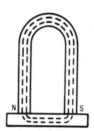

Magnet keeper.

magnet meter—Also called a magnet tester. An instrument for measuring the magnetic flux produced by a permanent magnet. It usually comprises a torque-coil or moving-magnet magnetometer with a particular arrangement of pole pieces.

magneto—1. An ac generator for producing ringing signals. 2. An ac generator for producing the ignition voltage in some gasoline engines. 3. An engine-driven unit that generates high voltage to fire the spark plugs. It needs no outside source of power such as a battery.

magnetodiode—A high-sensitivity magnetosensitive semiconductor that operates on the principle of controlled lifetime of injected carriers by an external magnetic field.

magnetoelastic coupling—Energy transfer between elastic and spin-wave modes of propagation. The strength of the effect varies with the material used.

magnetoelastic energy—In a crystal, energy of interaction between elastic (or lattice) strains and applied magnetization. If a ferromagnetic crystal is placed in a magnetic field, or if a magnetic field is varied about the crystal, the lattice structure of the crystal is distorted (magnetostriction occurs) and changes the amount of potential energy in the lattice. The energy associated with this change is magnetoelastic.

magnetoelastic wave—A hybrid of spin and elastic waves.

magnetoelectric generator—An electric generator with permanent-magnet field poles.

magnetoelectric surface waves—Extensions of bulk magnetoelastic waves phenomena, which combine acoustic and spin wave properties. The spin waves, created by oscillations of the angle between adjacent atomic moments in a ferromagnetic solid, have high dispersive characteristics that depend on a dc magnetic field. A strain of a crystal lattice can affect the magnetic movements, resulting in a coupling between acoustic and spin waves. This coupling is most effective when wavelengths and frequencies of the two waves are comparable.

magnetoelectric transducer—A transducer that measures the emf generated by the movement of a conductor relative to a magnetic field.

magnetofluid dynamics—*See* magnetohydrodynamics.

magnetofluidmechanics—*See* magnetohydrodynamics.

magnetogas dynamics—*See* magnetohydrodynamics.

magnetograph—A magnetometer that provides a continuous record of the changes occurring in the earth's magnetic field.

magnetohydrodynamic power generation—Also called hydromagnetic power generation. The generation of electric current by the motion of an ionized gas.

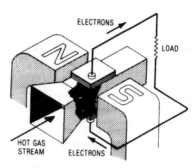

Magnetohydrodynamic generator.

magnetohydrodynamics—The study and application of the effects of electric and magnetic fields on conducting fluids and superheated ionized gases.

magnetoionic—Pertaining to the combined effect of atmospheric ionization and the magnetic field of the earth on electromagnetic wave propagation.

magnetoionic duct—A duct along the geomagnetic lines of force that exhibits waveguide characteristics for radio-wave propagation between conjugate points on the earth's surface.

magnetoionic wave component—Either of the two elliptically polarized wave components into which a linearly polarized wave in the ionosphere is separated by the earth's magnetic field.

magnetometer—An instrument for measuring the intensity or direction (or both) of a magnetic field or of a component of a magnetic field in a particular direction.

magnetomotive force—The force by which a magnetic field is produced, either by a current flowing through a coil of wire or by the proximity of a magnetized body. The amount of magnetism produced in the first method is proportional to the current through the coil and the number of turns in it. The cgs unit of magnetomotive force is called the gilbert.

magneto-optical switch—A device used for digital modulation that consists of a magnetic film deposited on a substrate. A small digital voltage impressed on the film creates a magnetic field in the film that polarizes light passing through the film. The polarized light beam can then be sent through a polarizing filter that, depending on the direction of the light polarization, will either pass or block the light beam.

magnetoplasmadynamic generator—A device that generates an electric current by shooting an ionized gas (plasma) through a magnetic field.

magnetoresistance—The change in electrical conductivity of a material when a magnetic field is applied. This change is quite pronounced in materials with a high carrier mobility.

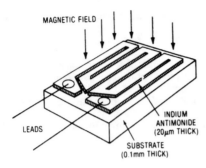

Magnetoresistor.

magnetoresistor—A semiconductor device in which the electrical resistance is a function of the applied magnetic field.

magnetosphere—A 900-mile thick belt in the upper atmosphere, composed primarily of helium gas.

magnetostatic field—A magnetic field that is neither moving nor changing direction. Such a field could be produced by a stationary magnetic pole or by a constant current flowing in a stationary conductor.

magnetostriction—That property of certain ferromagnetic metals, such as nickel, iron, cobalt, and manganese alloys, which causes them to shrink or expand when placed in a magnetic field. Conversely, if subjected to compression or tension, the magnetic reluctance changes, thus making it possible for a magnetostrictive wire or rod to vary a magnetic field in which it may be placed. This is true for lateral as well as longitudinal strains.

magnetostriction microphone—A microphone in which the deformation of a magnetostrictive material generates the required voltages.

magnetostriction oscillator—An oscillator in which the frequency is determined by the characteristics of the magnetostrictive element that inductively couples the plate circuit to the grid circuit.

magnetostriction speaker—A speaker in which the mechanical displacement is derived from the deformation of a magnetostrictive material.

magnetostriction transducer—A transducer comprising an element of magnetostrictive material inside a coil, and a force-summing member attached to one end of the element. Current flows through the coil, and the magnetic field around it expands and contracts the element to move the member back and forth.

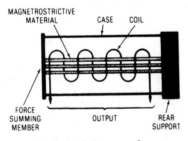

Magnetostriction transducer.

magnetostrictive delay line—A delay line made of nickel or certain other materials that become shorter when placed in a magnetic field.

magnetostrictive filter—A filter network that uses the magnetostrictive phenomenon to form high-pass, low-pass, bandpass, or band-elimination filters. The impedance characteristic is the inverse of that of a crystal.

magnetostrictive oscillator—An oscillator in which a magnetostrictive element controls the frequency.

magnetostrictive relay—A relay that functions because of dimensional changes occurring in a magnetic material under the influence of a magnetic field.

magnetostrictive resonator—A ferromagnetic rod that can be excited magnetically so that it will resonate (vibrate) at one or more desired frequencies.

magnetostrictive transducer—Sensor using contraction or expansion of an iron or nickel rod due to a magnetic field. Magnetostrictive elements are used in magnetostriction oscillators.

magneto switchboard exchange—A manual telephone exchange arranged so that calling and clearing by the subscribers and operators are done by means of magnetoelectric generators.

magneto telephone—A telephone equipped with a magneto (a hand-driven, two-pole, ringing-signal generator). Although obsolete for home and business telephones, it is still used in many field applications.

magnetron—An electric tube used to generate high power output in the UHF and SHF bands. The basis of its operation is the interaction of electrons with the electric field of a circuit element in crossed steady electric and magnetic fields to generate alternating-current power output.

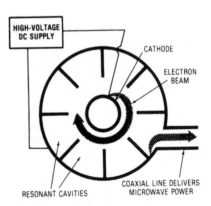

Magnetron.

magnetron effect—In a thermionic vacuum tube, the reduced electron emission due to the magnetic field of the filament current.

magnetron oscillator—An electron tube in which electrons are accelerated by a radial electric field between the cathode and one or more anodes and by an axial magnetic field that provides a high-energy electron stream to excite the tank circuits.

magnetron pulling—A shift in the frequency of a magnetron due to a change in the standing waves on the rf lines.

magnetron pushing—The shift in the frequency of a magnetron caused by faulty modulator operation.

magnetron rectifier—A gas-tube rectifier in which the electron stream is controlled by an optical element or instrument instead of by heated electrodes.

magnet steel—A special steel used in permanent magnets because of its high retentivity. In addition to steel, it also contains tungsten, cobalt, chromium, and manganese.

magnet tester — *See* magnet meter.

magnettor — *See* magnetic modulator.

magnet wire — 1. An insulated copper wire used for winding the coils of electromagnets. 2. Soft, insulated copper wire in the range of sizes used for winding relay and transformer coils. Usually has enamel insulation.

magnification ratio — The ratio of the size of an image produced by a lens to the size of the source.

magnified sweep — In an oscilloscope, a sweep whose time per division has been decreased by amplification of the sweep waveform rather than by changing the time constants used to generate it.

Magnistor — A saturable reactor for controlling electrical pulses or sine waves having frequencies of 100 kHz to 30 MHz and power levels ranging from microwatts to tens of watts.

magnitude — 1. Size; the quantity assigned to one unit so that it may be compared with other units of the same class (i.e., the ratio of one quantity to another). 2. The size of a quantity irrespective of its sign. For example, $+10$ and -10 have the same magnitude.

magnitude-controlled rectifier — A type of rectifier circuit in which a thyratron is used as the rectifying element. The load current is controlled by varying the basis on the grid of the thyratron.

mag-slip — A British term for synchro (i.e., a synchronous device such as the selsyn, autosyn, motortorque, and generator).

mailbomb — Sending a large amount of e-mail to a person's mailbox with the intent of harassing the person.

mailbox — 1. A computer location in a RAM storage area reserved for data addressed to specific peripheral devices as well as other microprocessors in the immediate environment. (Such an arrangement enables the coordinator CPU and the supplementary microprocessors to transfer data among themselves in an orderly fashion with minimal hardware.) 2. A system data structure that handles task communication. Tasks send messages to and receive messages from mailboxes. 3. A disk storage area assigned to a network user for receipt of e-mail messages.

mailer — A cassette or very small reel of tape sold in a cardboard box printed with name and address blanks, for sending through the mail.

mailer-daemon — An automated program that returns Internet mail if it is undeliverable for one reason or another. Usually, returned mail is the result of addressing the mail incorrectly.

mail merge — Taking information from a database and inserting it in a word processing document. Frequently used to create "personalized" form letters from mailing lists.

main anode — The anode that conducts the load current.

main bang — The transmitted pulse of a radar system.

main bonding jumper — The connection between the grounded circuit conductor and the equipment grounding conductor at the service.

main control unit — Transmitter or receiver controls for energizing, adjustment, etc., of the transmitter or receiver, but not for operating it while on the air.

main disconnect — The means provided in an electrical system to disconnect all conductors in a building from the service-entrance conductors. The means may consist of one or more, but not to exceed six, closely grouped switches or circuit breakers.

main distributing frame — In a telephone central office, a distributing frame, on one section of which terminate permanent outside lines, and on another section

of which terminate the subscriber-line multiple cabling, trunk multiple cabling, etc., used for associating an outside line with any desired terminal in such a multiple or with any other outside line. The main distributing frame usually carries the central-office protective devices and serves as a test point between line and office.

main entrance panel — A single enclosure housing the main disconnect, main overcurrent protection devices, branch circuit overcurrent devices, and branch circuit and feeder distribution means.

main-exciter — An exciter that supplies energy for the field excitation of another exciter.

mainframe — Also called central processing unit. 1. The central processor of the computer system. It contains the main storage, arithmetic unit, and special register groups. 2. The CPU of a computer plus the input/output unit and memory. As distinguished from peripheral equipment. 3. The main part of a computer system. 4. A large computer that stores information and distributes it to a network of micro- or minicomputers. Mainframes have largely given way to workstations, which serve fewer users and usually have the CPU and the terminal at the same location, but can be accessed remotely.

main gap — The conduction path between the principal cathode and anode.

main memory — *See* internal storage, 1.

main power — Power supplied to a complete system from a line.

mains — Interior wires extending from the service switch, generator bus, or converter bus to the main distribution center.

main service panel — The main electrical switch or circuit breaker and the circuit panel box that houses the circuit breakers or fuses for a branch circuit.

mains line — British word for ac power line.

main station — 1. A telephone station that has a distinct call number and a direct connection to a central office. 2. With respect to leased lines for customer equipment, the main point of interfacing of such equipment with the logic loop.

main storage — Usually the fastest storage device of a computer and the one from which instructions are executed.

main sweep — The longest-range scale available on some fire-control radars.

maintainability — A characteristic of design and installation expressed as the probability that an item will be retained in or restored to a specified condition within a given period when the maintenance is performed in accordance with prescribed procedures and resources.

maintained contact — Switch actuator that remains in a given position until it is actuated to another position, which is also maintained until further action.

maintained contact switch — A switch that remains in a given condition until actuated to another condition, which is also maintained until further actuation.

maintained switch — A switch that remains at the operated circuit condition when the actuating force is removed. It returns to the normal circuit condition, or moves to another position, when actuated a second time.

maintaining voltage — The voltage across a glow lamp after breakdown occurs.

maintenance — 1. All procedures necessary to keep an item in, or restoring it to, a serviceable condition, including servicing, repair, modification, modernization, overhaul, inspection, etc. 2. *See* file maintenance.

maintenance time — Time used for preventive and corrective maintenance of hardware.

major-apex face — In a natural quartz crystal, any one of the three larger sloping faces extending to the apex (pointed) end. The other three are called the minor-apex faces.

major cycle — 1. In a memory device that provides serial access to storage positions, the time interval between successive appearances of a given storage position. 2. A number of minor cycles.

major defect — A defect that could result in failure or significantly reduce the usability of a unit for its intended purpose.

major face — Any one of the three larger sides of a hexagonal quartz crystal.

major failure — 1. A noncritical failure that can degrade the system performance due to cumulative tolerance buildup. 2. A malfunction in a system that, while not causing complete breakdown, causes it to function uselessly.

major items — Items and components of communications electronics equipment that are designed to perform a specific function (e.g., radio, radar sets, transmitters, receivers, modulators, amplifiers, and assemblies) and for which the procurement and supply lead times are such that the items must be programmed and procured to be available at a given time. A programming term.

majority — A logical operator with the property that if P, Q, R, etc, are statements, the majority of P, Q, R, ... is true if more than half the statements are true, and false if half or fewer are true.

majority carrier — The predominant carrier in a semiconductor. Electrons are the majority carrier in n-type semiconductors, since there are more electrons than holes. Likewise, holes are the majority carrier in p-type semiconductors, since they outnumber the electrons.

majority-carrier contact — An electrical contact across which the ratio of majority-carrier current to applied voltage is substantially independent of the voltage polarity but the ratio of minority-carrier current to applied voltage is not.

majority emitter — The transistor electrode from which the majority carriers flow into the interelectrode region.

majority gate — A logic element whose output is true if more than half of the inputs are true, but whose output is false for other input conditions.

major lobe — 1. The antenna lobe in the direction of maximum energy radiation or reception. 2. The largest of the several lobes, or three-dimensional petals, that represent the radiation pattern of a directional antenna. The lobe in the direction of preferred reception or transmission.

major loop — A continuous network composed of all the forward elements and the primary feedback elements of a feedback control system.

major relay station — A tape relay station to which two or more trunk circuits are connected in order to provide an alternate route or to meet command requirements.

make — 1. The closing of a relay, key, or other contact. 2. A closed circuit or off-hook condition as determined by the dial of a telephone set.

make-before-break — The action of closing a switching circuit before opening another associated circuit.

make-before-break contacts — 1. Double-throw contacts so arranged that the moving contact establishes a new circuit before disrupting the old one. 2. Movable contact that makes the next circuit before breaking the first circuit. 3. A set of three contacts on a key or relay, so arranged that contact A makes with contact C before breaking from contact B.

make-before-break switch — *See* shorting switch.

make-break contacts — Also called continuity-transfer contact. A contact form of a relay, in which one contact closes connection to another contact and then opens its prior connection to a third contact.

make-break electrode — In a mercury switch, that electrode which serves the function of making and breaking contact with the mercury pool to close or open the electrical circuit.

make contact — 1. A contact that closes a circuit upon operation of the device of which the contact is a part. 2. Contact that closes when a key or relay is operated. *See also* normally open contact.

make percent — In pulse testing, the length of time, in comparison with the duration of the test signal, that a circuit stands closed.

make-up time — That portion of available time used for reruns made necessary by malfunctions or mistakes that occurred during a previous operating time.

male — Adapted so as to fit into a matching hollow part.

male contact — A contact located in the insert or body of a connector in such a way that the mating portion of the contact extends into the female contact. It is similar in function to a pin contact.

malfunction — Any incorrect functioning within electronic, electrical, or mechanical hardware. *See also* error, 3; fault.

malware — Malicious software, such as worms, viruses, and trojan horses.

Manchester encoding — A coding scheme used with several LANS. Manchester encoding has a logic transition in the center of each bit. A positive transition indicates a logical 1, and a negative transition indicates a logical 0.

mandrel — Also called an arbor. Shaft on which wound capacitors are wound.

mandrel test — A test used to determine the flexibility of insulation. In it a wire, with or without previous stretch, is wrapped around a mandrel.

manganin — An alloy wire used in precision wire-wound resistors because of its low temperature coefficient of resistance.

manipulated variable — The one variable (condition, quantity, etc.) of a process that is being controlled. The process can be controlled through manipulations of this variable.

man-made interference — A type of electromagnetic interference generated by electric motors, communication and broadcast transmitters, fluorescent lighting, and other electrical and electronic systems that radiate spurious signals.

man-made noise — 1. High-frequency noise signals caused by sparking in an electric circuit. 2. Noise created by voltages that result from the use of machinery, such as electric motors and internal-combustion engines.

man-made static — High-frequency noise signals created by sparking in an electric circuit. When picked up by radio receivers, it causes buzzing and crashing sounds from the speaker.

manpack — Also called packset. A portable radio-transmitting/receiving device that can be carried easily in a harness by a man.

man-rem — The absorbed dose of 1 rem by one human; also, the measure of population dosage, calculated by adding the doses received by each member of the population.

manual — 1. Hand operated. 2. In an organ, a group of keys played with the hand. In two-manual organs the upper manual, also referred to as the solo or swell manual, is normally used to play the melody. The lower manual, also referred to as the accompaniment manual or great manual, is normally used to play the accompaniment. 3. Operated by mechanical force applied directly by human intervention.

manual analysis — Also called manual test programming. The generation of input and output test patterns by a test engineer or technicians who study the function or structure of a logic circuit.

manual central office — A central office of a manual telephone system.

manual control — 1. The opening or closing of switches by hand. 2. The direction of a computer by means of manually operated switches.

manual controller — An electric controller in which all but its basic functions are performed by hand.

manual dimmer — A dimmer in which the only linkage between the control lever and the moving electrical contact that conducts the electrical power is mechanical.

manual direction finder — A radio compass that uses a rotatable loop that is operated manually.

manual exchange — A telephone exchange in which the lines are connected to a switchboard and interconnections are controlled by an operator.

manual input — 1. The entry of data into a system by hand at the time of processing. 2. Direct computer entry by means of manual intervention, or drum entry of manual data through card machines.

manual operation — Data processing in a system by means of direct manual techniques.

manual preset timer — A manual start timer, the cycle of which is initiated by turning a pointer to the desired setting.

manual rate-aided tracking — Radar tracking of individual targets by means of circuits that compute the velocity from manually inserted position fixes.

manual reset — A qualifying term used to indicate that a relay may be reset manually after an operation.

manual ringing — A method of ringing a telephone. The key must be held down for the ringing to continue. Nor does it stop when the receiver is lifted off the hook, unless the caller releases the key.

manual single play — A turntable operation in which records are played one at a time, and the arm is placed on the record and removed again manually.

manual start timer — An interval timer on which each cycle must be manually started at the timer.

manual switch — A switch that is actuated by an operator.

manual switchboard — A telephone switchboard in which the operator makes connections manually with plugs and jacks or with keys.

manual switching — A characteristic of a circuit breaker permitting manual opening and closing of the circuit by operation of the actuator.

manual telegraphy — Telegraphy in which an operator forms the individual characters of the alphabet in code.

manual telephone — A telephone without push button or rotary dial that rings the attendant automatically when the receiver is lifted.

manual telephone set — A set not equipped with a dial for securing the number to be called. Instead, lifting the receiver alerts the switchboard operator, who then connects the caller to the person being called.

manual telephone system — A system in which telephone connections between customers are ordinarily established manually by telephone operators, in accordance with orders given verbally by the calling parties.

manual test programming — *See* manual analysis.

manual tuning — Rotation by hand of a knob on a radio receiver to tune in a desired station.

manufacturing holes — *See* pilot holes.

manuscript — A form of storage medium, such as programming charts, in which is contained raw information in a sequential form suitable for translation.

MAR — Abbreviation for memory address register.

Marconi antenna — An antenna system in which one end of the signal source is connected to a radiating element and the other end is connected to ground.

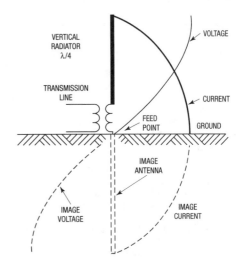

Marconi antenna.

margin — 1. The difference between an actual operating point and the point or condition where a failure to operate properly will occur. 2. Also called range or printing range. In telegraphy, the interval between limits on a scale, usually arbitrary, in which printing is error-free. 3. The distance that the electrode foil is indented from the edge of the dielectric when a capacitor is wound.

marginal — Operating at the border of permissible limits of voltage, current, distortion, etc. When operation is marginal, a very small impairment can cause the device or system to fail.

marginal checking — Also called marginal testing. Preventive maintenance in which certain operating conditions (e.g., supply voltage or frequency) are varied from normal in order to locate defects before they become serious.

marginal relay — A relay with a small margin between its nonoperative current value (maximum current applicable without operation) and its operative value (minimum current that operates the relay).

marginal testing — *See* marginal checking.

margin-punched card — A card in which holes representing data are punched only on the border, with the center left free for written or printed information.

marine broadcast station — A coastal station that regularly broadcasts the time and meteorological and hydrographic information.

marine radiobeacon station — A radionavigation land station whose emissions are used to determine the bearing or direction of a ship in relation to the marine radiobeacon station.

maritime mobile service — The radio service in which ships communicate with each other or with coastal and other land stations on specified frequencies.

maritime radionavigation service — A radio service intended to be used for the navigation of ships.

mark — 1. In telegraphy, the closed-circuit condition, i.e., the signal that closes the circuit at the receiver to produce a click of the sounder or to print a character on a teletypewriter. 2. The presence of signal. A mark impulse is equivalent to a binary 1. 3. A term that originated with telegraphy to indicate a closed key condition. Present usage implies the presence of current or carrier on a circuit or the idle condition of a teletypewriter. It also indicates the binary digit 1 in computer language. *See* flag. 4. In RTTY applications, the mark is one of two states. The mark is the condition characterized by a closed circuit. The space, the other state, is characterized by an open-circuit condition.

mark and space impulses — In neutral operation of a teletypewriter system, the mark impulse is the closed-circuit signal, and the space impulse is the open-circuit signal. In other than neutral operation, the mark impulse is the circuit condition that produces the same result in the terminal equipment that a mark impulse produces in neutral operation. Similarly, the space impulse is the circuit condition that produces the same result in the terminal equipment that a space produces in neutral operation.

marker — Also called marker beacon. A radio navigational aid consisting of a transmitter that sends a signal to designate the small area around and above it.

marker antenna — The transmitting antenna for a marker beacon.

marker beacon — *See* marker.

marker generator — An rf generator that injects one or more pips of specific frequency onto the response curve of a tuned circuit being displayed on the screen of a cathode-ray oscilloscope.

marker pip — The inverted V (Λ) or spot of light used as a frequency index mark in cathode-ray oscilloscopes for alignment of TV sets. It is produced by coupling a fixed-frequency oscillator to the output of a sweep-driven signal generator.

marker thread — A colored thread laid parallel and adjacent to the strands of an insulated conductor that identifies the wire manufacturer and often the specification under which the wire was constructed.

marking-and-spacing intervals — In telegraphy, the intervals corresponding to the closed and open positions, respectively, of the originating transmitting contacts.

marking bias — Bias that affects the results in the same direction they are affected by marking current.

marking current — The magnitude and polarity of line current when the receiving mechanism is in the operated condition.

marking pulse — The signal interval during which the selector unit of a teletypewriter is operated.

marking wave — Also called keying wave. In telegraphy, the emission while the active portions of the code characters are being transmitted.

mark sense — 1. To mark a position on a punch card, using a special pencil that leaves an electrically conductive deposit for later conversion to machine punching. 2. A mechanized technique of punching data into computer cards. A graphite line positioned on the card is read electronically and converted into holes by special equipment.

mark-sense card — A card designed to permit data to be entered on it with an electrographic pencil.

mark sensing — A technique for detecting special pencil marks entered in special places on a card and automatically translating the marks into punched holes.

mark-to-space transition — The change from a marking impulse to a spacing impulse.

marshalling sequence — *See* collating sequence, 1.

maser — Acronym for microwave amplification by stimulated emission of radiation. 1. A low-noise microwave amplifier in which a signal is boosted by changing the energy level of a gas or crystal (commonly, ammonia or ruby, respectively). 2. A means of focusing a stream of particles, which concentrates only on the high-energy particles. These are passed into a resonator that is resonating at the radiation frequency of the particles. The particles are raised to a strong oscillation in this state and can be used for control purposes. By reducing the flow of particles to the resonator to maintain oscillations, it can be used as an amplifier. (There are many other applications.) 3. Device for amplifying a microwave frequency signal by stimulated emission of radiation — i.e., the weak microwave signal causes electrons in an atom to change orbit in such a manner as to emit an amplified signal of the same frequency as the weak signal. 4. Amplification by a low-noise radio-frequency amplifier in which an input signal stimulates emission of energy stored in a molecular or atomic system by a microwave power supply.

mask — 1. A frame mounted in front of a television picture tube to limit the viewing area of the screen. 2. A device (usually a thin sheet of metal that contains an open pattern) used to shield selected portions of a base during a deposition process. 3. A device used to shield selected portions of a photosensitive material during photographic processing. 4. A logical technique in which certain bits of a word are blanked out or inhibited. 5. Template used to etch circuit patterns on semiconductor wafers. Images of the circuit patterns are produced on glass or metal photographically. The mask is then used to control the diffusion process, plus metallization. 6. A transparent (glass or quartz) plate covered with an array of patterns used in making integrated circuits. Each pattern consists of opaque and transparent areas that define the size and shape of all circuit and device elements. The mask is used to expose selected areas of photoresist, which define areas to be etched. Masks may use emulsion, chrome, iron oxide, silicon, or other material to produce the opaque areas. 7. Thin metals or other materials with an open pattern designed to mask off or shield selected portions of semiconductors or other surfaces during deposition processes. There also are photomasks or optical masks for contact or projection printing of wafers; these may use an extremely flat glass substrate with iron oxide, chrome, or emulsion coating. There also are thick-film screen masks. 8. The photographic negative that serves as the master for making thick-film screens and thin-film patterns. 9. The pattern, usually "printed" on glass, used to define areas of the chip or wafer. Masks are used for the diffusion, oxidation, and metallization steps used in manufacturing of semiconductors. 10. To hide, to obscure, to make less noticeable. For example, as noise masks crosstalk. 11. A material applied to enable selective etching, etching, plating, or the application of solder to a printed board. Also, the surface on which the master artwork of the circuit pattern is projected. 12. A thin steel arrangement with fine holes (shadow mask) or stripes (slot mask) that concentrates the electron beam at points on the CRT.

masked diffusion — The use of a mask pattern to obtain selective impregnation of portions of a semiconductor material with impurity atoms.

masked ROM — A regular read-only memory (ROM) produced by the usual masking process. (Contrasted with a PROM.)

masking — The process by which a sound is made audible by the addition of a second sound called the masking sound. The unit of measurement is usually the decibel.

masking audiogram — A graphical representation of the amount of masking by a noise. It is plotted in decibels as a function of the frequency of the masked tone.

mask microphone — A microphone designed for use inside an oxygen or other respiratory mask.

mask-programmed — A semiconductor device, most often a read-only memory, that is permanently programmed as a step in its manufacture. In contrast, *field-programmed* applies to memory devices that are programmed after manufacture.

mask set — A set of plates, usually glass, that are used to transfer a device topology in sequence to a wafer during fabrication.

Masonite — Trade name of the Masonite Corp. Fiberboard made from steam-exploded wood fiber. Its highly compressed forms are used for panels in electrical equipment.

mass — 1. The quantity of matter in an object. It is equal to the weight of a body divided by the acceleration due to gravity. 2. The bulk of matter, though not necessarily equal to its weight. 3. A mechanical unit whose electrical analog is inductance.

mass data — A larger amount of data than can be stored in the central processing unit of a computer at any one time.

mass-memory unit — A drum or disk memory that provides rapid-access bulk storage for messages being held until outgoing channels are available.

mass properties — Calculation of physical engineering information about a part, e.g., perimeter, area, volume, weight, and moments of inertia.

mass radiator — A spark radiator that generates a low-level broadband signal extending into and above the EHF band. Arcing occurs between fine metal particles suspended in a liquid dielectric.

mass spectrometer — An instrument that permits rapid analysis of chemical compounds. It consists of a vacuum tube into which a small amount of the gas to be studied is admitted. The gas is ionized by the electrons emitted from the cathode and speeded up by an accelerating grid. An electric field draws the ions out of the ionizing chamber. They are then sent through electric and magnetic fields that sort them according to their ratios of mass to charge.

mass spectrum — The spectrum obtained by deflecting a beam of electrons with an electric or magnetic field as they emerge from a tube containing a small quantity of the gas being investigated. The amount a particle is deflected depends on the ratio of its mass to its atomic charge. Hence, every element has a characteristic mass-spectrum line.

mass storage — 1. Refers to hardware devices providing massive amounts of online secondary storage, generally using strips of inexpensive magnetic media that can be accessed randomly, but with slower access times than those of conventional tape or disk devices. 2. Refers to peripheral devices into which programs and data are stored for immediate action. 3. In a computer, secondary, slower memory for bulky files. Mostly floppy disk, cassette, or tape.

mass storage device — Any device used to provide relatively inexpensive storage for large amounts of data.

mass termination — The simultaneous termination of several or all conductors of a cable. This process generally uses terminals that pierce the insulation without stripping to cold-flow mate with the conductors and form a gas-tight metal-to-metal joint. *See* insulation displacement connector.

mast — The pole on which an antenna is mounted.

master — 1. The mold from which other disc recordings are cast. It is made by electroforming from a disc recording, and is a negative of the disc (i.e., has ridges instead of grooves). 2. An original, or first special copy, of a recorded performance from which other copies may be made. 3. An original recording, made directly from recording microphones. A disc master is the lacquer original, usually cut from a tape from which stampers are made for vinyl pressings. 4. An element of a system that controls or initiates the action or responses of the other elements of the system.

master antenna television — Abbreviated MATV. 1. An antenna system that serves a concentration of television sets, such as in apartment buildings, hotels, or motels. 2. Broadcast receiving stations that use one or more high-quality centrally located UHF and/or VHF antennas that relay their signals to many televisions in a local apartment/condo or group-housing complex.

master brightness control — In a color television receiver, a variable resistor that adjusts the bias level on all three guns in the picture tube at the same time.

master clock — 1. In a computer, the primary source of timing signals. 2. A very accurate timer with an absolute time reference, providing controlled power to drive slave or auxiliary timers and display units. May also provide correcting pulses for slave devices.

master contactor — A device that is generally controlled by a master element or equivalent, and the necessary permissive and protective devices, and which serves to make and break the necessary control circuits to place an equipment into operation under the desired conditions and to take it out of operation under other or abnormal conditions.

master control — 1. In a studio, a central point from which sound or television programs are switched to one or more destinations. 2. An application-oriented routine usually applied to the highest level of a subroutine hierarchy.

master die — A substrate that contains unconnected active and passive elements in a predetermined pattern. Connection pads for each element and subelement are provided, and a variety of circuits may be obtained by appropriate choices of intraconnection patterns.

master drawing — A drawing showing the dimensional limits or grid locations applicable to any or all parts of a printed circuit, including the base.

master drive — A drive that determines the reference input for one or more follower drives.

master element — The initiating device, such as a control switch, voltage relay, float switch, etc., that serves either directly or through such permissive devices as protective and time-delay relays to place an equipment in or out of operation.

master file — In a computer, a file of relatively more permanent information that is usually updated periodically.

master gain control — 1. An amplifier control that permits adjusting the gain of two or more channels simultaneously. 2. Control of overall gain of an amplifying system as opposed to varying the gain of several individual inputs.

master instruction tape — A computer magnetic tape on which are recorded all programs for a system of runs.

master layout — The original layout of a circuit.

master mask — A chrome mask of a complete wafer's multiple images. It is used either in projection printing on a wafer or to contact-print additional masks.

master oscillator — 1. In a transmitter, the oscillator that establishes the carrier frequency of the output.

2. An oscillator that controls or provides modulator drive frequencies for a number of channels or channel groups.

master-oscillator power amplifier — Abbreviated MOPA. An oscillator followed by a radio-frequency buffer-amplifier stage.

master pattern — An accurately scaled pattern that is used to produce the printed circuit within the accuracy specified in the master drawing.

master plates — *See* photomask, 4.

master processor — The main processor in a master-slave configuration.

master reticle — A reticle plate properly aligned in a reticle frame and sealed in place. The master reticle is inserted into a photorepeater to make the final-sized photomask. Used in the production of monolithic circuits.

master routine — *See* subroutine, 2.

master scheduler — The control-program function that responds to operator commands, initiates requested actions, and returns information that is requested or required; the overriding medium for control of the use of the computing system.

master-slave — 1. A binary element consisting of two independent storage stages arranged with a definite separation of the clock function to enter information to the master stage and to transfer it to the slave stage. 2. A configuration in which one system, the master, always has control over another system, the slave.

master-slave flip-flop — A circuit that contains two flip-flops, a master and a slave. The master flip-flop receives information on the leading edge of a clock pulse, and the slave (output) flip-flop receives information on the trailing edge of the clock pulse.

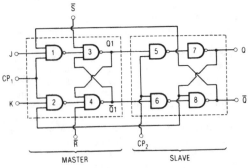

Master-slave J-K flip-flop.

master slice — A silicon wafer containing 30 or more groups of components. The elements can be interconnected with paths of aluminum to form desired circuits. The wafer is then diced into single devices.

master stamper — A master from which phonograph records are pressed.

master station — 1. The radio station to which the emission of other stations of a synchronized group are referred 2. In a hyperbolic navigation system, the one of a pair of transmitting stations that controls the transmission of the other (slave) station in the pair and maintains the time relationship between the pulses emitted from the two stations. 3. The device that controls all other devices for the purpose of organizing communications over a transmission line.

master switch — A switch located electrically ahead of a number of other switches.

master synchronization pulse — A pulse distinguished from other telemetering pulses by its different amplitude and/or duration and used to indicate the end of a sequence of pulses.

master TV system — A combination of components for providing multiple TV set operation from one antenna.

master wafer — A processed semiconductor wafer with unconnected active and passive circuit elements located in a standardized pattern. Different integrated circuits may be synthesized by using various interconnection paths.

MAT — Acronym for microalloy transistor.

match — 1. The similarity or equality of one thing to another. 2. To compare two or more items of data for identity.

matched filter — An optimum filter for separating a waveform of known shape from random perturbing noise.

matched impedance — The condition that exists when two coupled circuits are adjusted so that the impedance of one circuit equals the impedance of the other.

matched load — A device used to terminate a transmission line or waveguide so that all the energy from the signal source will be absorbed.

matched power gain — That power gain which is obtained when the impedance of the load is matched to the effective output impedance of the amplifier to which the load is connected.

matched pulse intercepting — Technique used in intercepting calls on party lines in a terminal-per-line office. A ground pulse is matched in time with the particular ringing frequency of the intercepted station.

matched symmetrical transistor — A special case of the bidirectional transistor in which not only are the requirements for a symmetrical transistor met, but actual matching specifications are also given. *See also* symmetrical transistor.

matched termination — A termination that causes no reflected wave at any transverse section of the transmission line. Its impedance is equal to the characteristic impedance of the line.

matched transmission line — A transmission line along which there is no wave reflection.

matched waveguide — A waveguide along which there is no reflected wave.

matching — 1. Connecting two circuits or parts together with a coupling device in such a way that the maximum transfer of energy occurs between the two circuits, and the impedance of either circuit will be terminated in its image. 2. The connection of a component's output to an input that provides the recommended value of load, or termination impedance. 3. Obtaining like impedances to provide a reflection-free transfer of signal.

matching diaphragm — A window consisting of an aperture (slit) in a thin piece of metal, placed transversely across a waveguide; used as a matching device. The orientation of the slit (whether parallel to the long or short dimensions of the waveguide) determines whether it is respectively capacitive or inductive.

matching impedance — The impedance value that must be connected to the terminals of a signal-voltage source for proper matching.

matching plate — In waveguides, a diaphragm used for matching.

matching stub — A device placed on a radio-frequency transmission line to vary its electrical length and hence its impedance.

matching transformer — A transformer used for matching impedances.

mate — Joining of two connector halves, or of a cable to a connector.

material dispersion — Broadening of light impulses arising from wavelength-dependent differential delay of light in a waveguide material.

material system — The designation of the number of basic metals (e.g., silver-antimony-telluride) making up thermoelectric materials.

math coprocessor — A special chip that is added to a computer to handle advanced mathematic functions, thereby freeing up the processing power of the main CPU.

mathematical check — A check making use of mathematical identities or other properties.

mathematical logic — Also called symbolic logic. Exact reasoning concerning nonnumerical relations by using symbols that are efficient in calculation.

mathematical model — 1. The general characterization of a process, object, or concept in terms of mathematics that enables the relatively simple manipulation of variables to be accomplished in order to determine how the process or concept would behave in different situations. 2. A mathematical representation that simulates the behavior of a process, device, or concept. 3. A mathematical equation that can be used to numerically compute a system's response to a particular input.

mating face — *See* interface.

matrices — Plural of matrix.

matrix — 1. A coding network or system in a computer. When signals representing a certain code are applied to the inputs, the output signals are in a different code. 2. In electronic computers, any logical network whose configuration is a rectangular array of intersections of its input-output leads, with logic elements connected at some of these intersections. The network usually functions as an encoder or decoder. 3. A computer network or system in which only one input is excited at a time and produces a combination of outputs. 4. In a color TV circuit, the section that combines the I, Q, and Y signals and transforms them into individual red, green, and blue signals that are applied to the picture-tube grids. 5. A rectangular array of scalar quantities, usually numbers or letters used to represent numbers. Rectangular array means that the elements are arranged into definite rows and columns. 6. An orderly two-dimensional array. An arrangement of circuit elements, such as wires, relays, diodes, etc., that can transform a digital code from one type to another. 7. The terminology applied to the several methods for encoding four channels onto two channels for later recovery back to four channels. Also referred to as 4-2-4. The actual electronics used to encode into two channels or decode back to four are known as matrixing electronics. 8. A rectangular array of elements, in cross-match fashion. Used to describe memory organization, character formation, diode layouts, and so forth. 9. A general process whereby several signals can be added together for recording or transmission on fewer channels, and later retrieved through a complementary process. 10. A mathematical array having height, width, and sometimes depth, into which collections of data may be stored and processed.

matrixer — Also called matrix unit. A device that transforms the color coordinates, usually by electrical or optical means.

matrixing electronics — *See* matrix, 7.

matrix life test — A test in which each test condition has two components. For example, in transistor matrix life testing each life-test condition is represented by an ambient temperature and a power dissipation. At each stress level of temperature, there are several dissipations; and at each stress level of dissipation, there are several temperatures. The test conditions are placed into blocks or groups.

matrix line printer — A printer in which each character is composed of a matrix of dots.

matrix printer — A printer that uses an impact process in which the desired character is created from small dots.

matrix storage — Storage in which the elements are arranged in such a way that access to any location requires the use of two or more coordinates, as, for example, in cathode-ray-tube storage and core storage.

matrix switcher — A combination or array of electromechanical or electronic switches that routes a number of signal sources to one or more designations.

matrix unit — *See* matrixer.

mat switch — A flat area switch used on open floors or under carpeting. It may be sensitive over an area of a few feet or several square yards.

matter — Any physical entity — i.e., having mass.

Matteucci effect — The ability of a twisted ferromagnetic wire to generate a voltage as its magnetization changes.

MATV — Abbreviation for master antenna television system.

max — Abbreviation for maximum.

maxima/minima — In radar, regions of maximum and minimum return from the transmitted pulse caused by additive and subtractive combinations of the direct and reflected wave. A plot of this data is usually known as a null pattern or fade chart.

maximize — In a Windows program, enlarging a window to its largest size, usually filling the entire screen.

maximum — Abbreviated max. The highest value occurring during a stated period.

maximum average power output — In television, the maximum radio-frequency output power, averaged over the longest repetitive modulation cycle.

maximum deviation sensitivity — Under maximum system deviation, the smallest signal input for which the output distortion does not exceed a specified limit.

maximum dissipation — The maximum average power a device can dissipate during operation while still remaining within published life specifications.

maximum frequency of oscillation — The maximum frequency at which a vacuum tube or transistor can be made to oscillate under specified conditions. This approximates to the frequency at which the maximum available power gain has decreased to unity.

maximum keying frequency — In a facsimile system, the frequency (in hertz) equal to half the number of critical areas of the subject copy scanned per second.

maximum luminous efficiency — The greatest luminosity possible for a specified chromaticity.

maximum luminous reflectance — The greatest luminous reflectance possible for a specified chromaticity.

maximum luminous transmittance — The greatest luminous transmittance possible for a specified chromaticity.

maximum modulating frequency — In a facsimile system, the maximum scanning frequency process that can be transmitted without degrading the recorded copy.

maximum output — The highest average output power into a rated load, regardless of distortion.

maximum overshoot — The maximum amplitude deviation from the average of the steady-state values that exist immediately before and after the transient.

maximum peak plate current — The highest instantaneous plate current a tube can safely carry.

maximum percentage modulation — The highest percentage of modulation permitted in a transmitter without producing excessive harmonics in the modulating frequency.

maximum permeability — The highest permeability reached as induction or magnetization is increased.

maximum power transfer theorem — The maximum power will be absorbed by one network from another joined to it at two terminals, when the impedance of the receiving network is varied, if the impedances (looking into the two networks at the junction) are conjugates of each other.

maximum record level — In direct recording, the amount of record-head current required to produce 3-percent third-harmonic distortion of the reproduced signal at the record-level set frequency. Such distortion must result from magnetic-tape saturation, not from electronic circuitry.

maximum response speed — The maximum pulse frequency that can be applied to a stepper motor at random and result in synchronized steps. The motor must not miss in step while operating within this range.

maximum retention time — The maximum time interval between writing into a storage element of a charge storage tube and reading an acceptable output.

maximum saturation — The highest value of saturation possible for a specified hue.

maximum sensitivity — The smallest signal input that produces a specified output.

maximum signal level — In an amplitude-modulated system, the level corresponding to copy black or copy white — whichever has the higher amplitude.

maximum sound pressure — For any given cycle of a periodic wave, the maximum absolute value of the instantaneous sound pressure. The most common unit is the microbar.

maximum storage time — In a storage tube, the length of time after writing during which an acceptable output can be read. *See also* maximum usable viewing time.

maximum system deviation — In a frequency-modulation system, the greatest permissible deviation in frequency.

maximum torque — *See* pull-out torque.

maximum undistorted output — Also called maximum useful output. The maximum power an amplifier can deliver without producing excessive harmonics.

maximum usable frequency — Abbreviated MUF. 1. In radio transmission by ionospheric reflection, the highest frequency that can be transmitted by reflection from regular ionized layers. 2. The highest frequency at which the ionosphere reflects signals to earth. Any signals of higher frequency than the MUF penetrate the ionosphere and enter space.

maximum usable viewing time — Also called maximum storage time. The length of time during which the visible output of a storage tube can be viewed, without rewriting, before a specified decay occurs.

maximum useful output — *See* maximum undistorted output.

maximum working voltage — The maximum voltage stress (dc or ac) that may be applied to a component under specified conditions of use.

maximum writing rate — The maximum spot speed that produces a line of specified density on a photographic negative or on the screen of a cathode-ray tube.

maxterm form — A function expanded into a product of sums, such as $(A + B) (C - D) (E + F)$.

maxwell — The cgs electromagnetic unit of magnetic flux, equal to 1 gauss per square centimeter, or one magnetic line of force.

Maxwell bridge — A four-arm ac bridge normally used for measuring inductance in terms of resistance and capacitance (or capacitance in terms of resistance and inductance). One arm has an inductor in series with a resistor, and the opposite arm has a capacitor in parallel with a resistor. The other two arms normally are nonreactive resistors. The balance is independent of frequency.

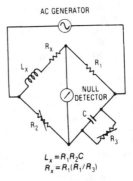

$$L_x = R_1 R_2 C$$
$$R_x = R_1 (R_1 / R_3)$$

Maxwell bridge.

Maxwell inductance bridge — A four-arm ac bridge for comparing inductances. Two adjacent arms have inductors, and the other two arms usually have nonreactive resistors. The balance is independent of frequency.

Maxwell mutual-inductance bridge — A four-arm ac bridge for measuring mutual inductance in terms of self-inductance. Mutual inductance is present between the supply circuit and the arm that includes one coil of the mutual inductor. The other three arms normally are nonreactive resistors. The balance is independent of frequency.

Maxwell's equations — Fundamental equations, developed by J. C. Maxwell, for expressing radiation mathematically and describing the condition at any point under the influence of varying electric and magnetic fields.

Maxwell's law — A movable portion of a circuit will always travel in the direction that gives maximum flux linkages through the circuit.

Maxwell triangle — A graph that defines the chromaticity values of a color in terms of three coordinates.

maxwell-turn — A unit of magnetic linkage equal to one magnetic line of force passing through one turn of a circuit.

mayday — International distress call for radiotelephone communication. It is derived from the French *m'aidez*, meaning "help me."

Mbps — Millions of bits per second (bps).

mc — Abbreviation for megacycle — an obsolete term superseded by MHz (megahertz).

MCM — 1. Abbreviation for multichip module. An electronic package containing several chips on a single substrate. 2. *See* Monte Carlo method.

McNally tube — A velocity-modulated vacuum tube that produces low-power UHF oscillations. It is used as a local oscillator in some radar receivers.

MCW — Abbreviation for modulated continuous wave.

MD — Abbreviation for mini-disc.

m-derived filter — A type of constant-k filter in which the constant-k elements are multiplied by the factor *m* or a function of *m*. Generally, the m-derived filter has more elements and provides sharper cutoff and more uniform attenuation in the pass region. It may provide high-pass, low-pass, bandpass, or band-stop filtering action.

m-derived L-section filter — A reactance network derived from the prototype L-section filter, so that the image-transfer coefficient and one image impedance are changed, but the other image impedance is left unchanged.

MDI — Abbreviation for magnetic direction indicator.

M-display — *See* M-scan.

MDS — Abbreviation for microcomputer development system.

M-DVD — Abbreviation for magnetic digital versatile disc. A 3.5-inch removable storage medium with 2.1 gigabytes of capacity which is achieved by using magneto-resistive technology.

Meacham-bridge oscillator — A crystal oscillator in which the crystal forms one arm in a bridge so as to obtain effective multiplication of the actual Q of the crystal.

meaconing — 1. The interception and rebroadcast of beacon signals. They are rebroadcast on the received frequency to confuse enemy navigation. As a result, aircraft or ground stations are given inaccurate bearings. 2. The clandestine generation or retransmission of a radionavigation signal in order to confuse navigation.

mean — 1. The arithmetic middle point of a range of values, obtained by adding the smallest value to the largest value and dividing that sum by two. 2. The arithmetic average of a group of values.

mean absolute deviation — Abbreviated MAD. The average of the absolute values of the deviations of some observed value from some expected value. MAD can be calculated based on observations and the arithmetic mean of those observations.

mean carrier frequency — The average carrier frequency of a transmitter (corresponding to the resting frequency in a frequency-modulated system).

mean charge — The arithmetic mean of the transferred charges corresponding to a given capacitor voltage, as determined from a specified alternating-charge characteristic curve.

mean charge characteristic — The function giving the relation of mean charge to capacitor voltage.

meander line — A transmission-line-matching section whose electrical length is dependent on frequency. The characteristics of a meander line are determined primarily by the width of the structure, the spacing between adjacent turns, and the angle of the line with respect to the ground plane. If the turns have sufficient separation that there is no space coupling between adjacent turns, the meander line becomes a simple length of transmission line.

mean free path — 1. The average distance that sound waves travel between successive reflections in an enclosure. 2. The average distance between collisions of atomic particles, which may be further specified according to type of collision (e.g., elastic, inelastic).

mean life — Also called average life. 1. In a semiconductor, the time taken by injected excess carriers to recombine with others of the opposite sign. 2. A measure of the probability that a part or equipment will function satisfactorily during its constant-failure-rate period. It is unrelated to longevity.

mean power of a radio transmitter — Power supplied to the antenna transmission line by a transmitter during normal operation, averaged over a time sufficiently long compared with the period of the lowest frequency encountered in the modulation. A time of 0.1 second during which the mean power is greatest will be selected normally.

mean pulse time — The arithmetic mean of the leading-edge and trailing-edge pulse times.

means of communications — The medium (i.e., electromagnetic or sound waves, visual messenger) by which a message is conveyed from one person or place to another.

mean time between failures — Abbreviated MTBF. 1. The average length of time between successive system failures. It is the reciprocal of the sum of the failure rates of all components and connections in the system. 2. The average time between failures of a continuously operating device, circuit, or system. 3. For a particular interval, the total functioning life of a population of an item divided by the total number of failures within the population during the measurement involved.

mean time between maintenance — Abbreviated MTBM. The mean of the distribution of the time intervals between maintenance actions (preventive, corrective, or both).

mean time to failure — Abbreviated MTTF. In a piece of equipment, its measured operating time divided by its total number of failures during that time. Normally this measurement is made between the early-life and wearout failures.

mean time to first failure — Abbreviated MTTFF. A special case of MTBF, where T is the accumulated operating time to first failure of a number of devices (failures).

mean time to repair — Abbreviated MTTR. 1. The total effective maintenance time during a given time interval divided by the total number of failures during the interval. 2. The arithmetic average of time required to complete a repair activity. 3. The total corrective maintenance time divided by the total number of corrective maintenance actions during a given period.

measurand — Also called stimulus. The physical quantity, force, property, or condition measured by an instrument.

measured pickup — The value of current or voltage at or below which the contacts of a relay must assume their fully operated position.

measured service — A type of telephone service for which a charge is made in accordance with the number of calls or message units during the billing period.

measurement — 1. The determination of the existence or magnitude of a variable. Measuring instruments include all devices used directly or indirectly for this purpose. 2. The act of determining the magnitude of something, in terms of a recognized standard.

measurement component — Those parts or subassemblies used primarily for the construction of measurement apparatus, excluding screws, nuts, insulated wire, or other stable materials.

measurement device — An assembly of one or more basic elements with other components needed to form a self-contained unit for performing one or more measurement operations. Included are the protecting, supporting, and connecting as well as functioning parts.

measurement energy — The energy required to operate a measurement device or system. Normally it is obtained from the measurand or the primary detector.

measurement equipment — Any assemblage of measurement components, devices, apparatus, or systems.

measurement inverter — *See* measuring modulator.

measurement range of an instrument — That part of the total range of measurement through which the accuracy requirements are to be met.

measurement voltage divider — Also called voltage ratio or volt box. A combination of two or more resistors, capacitors, or other elements arranged in series so that the voltage across any one is a definite, known fraction of the voltage applied to the combination (provided the current drain at the tap point is negligible or taken into account). The term *volt box* is usually limited to resistance voltage dividers intended for extending the range of direct-current potentiometers.

measuring modulator — Also called measurement inverter or chopper. An intermediate means of modulating a direct-current or low-frequency alternating-current input in a measurement system to give a proportionate alternating-current output, usually as a preliminary to amplification.

mechanical bail — Switch action in which, upon actuation of one station, the switch changes the contact position, mechanically locks the switch in that position, and releases any station previously actuated.

mechanical bandspread — A vernier tuning dial or other mechanical means of lengthening the rotation of a control knob. This permits more precise tuning in crowded shortwave bands.

mechanical bias — A mechanical force tending to move the armature of a relay toward a given position.

mechanical compliance — The displacement of a mechanical element per unit of force, expressed in centimeters per dyne. It is the reciprocal of stiffness and is analogous to capacitance.

mechanical coupling — See acoustic coupling.

mechanical differential analyzer — A form of analog computer in which interconnected mechanical surfaces are used for solving differential equations.

mechanical drum programmer — A sequencer that operates switches by means of pins placed on a rotating drum. The switch sequence may be altered by changing the pin pattern.

mechanical filter — 1. See mechanical wave filter. 2. Mechanical resonators coupled by mechanical means. Piezoelectric or magnetostrictive transducers are used to convert electrical and mechanical energy at input and output. Resonators are bars, discs, or electrode pairs; coupling elements are rods, wires, or nonelectroded regions. See monolithic filter.

mechanical impedance — The complex ratio of the effective force acting on a specified area of an acoustic medium or mechanical device to the resulting effective linear velocity through or of that area, respectively. The unit is newton-s/m or the mks mechanical ohm. (In the cgs system, the unit is the dyne-s/cm or the mechanical ohm.)

mechanical joint — A joint made by clamping cables or other conductors together mechanically rather than by soldering them.

mechanical life — The maximum number of complete cycles through which a device may be actuated without electrical or mechanical failure.

mechanically timed relay — A relay that is mechanically timed by such means as a clockwork, escapement, bellows, or dash-pot.

mechanically tuned oscillator — Any oscillator that is specifically designed for frequency tuning by mechanical means. Typically a cavity or discrete element.

mechanical ohm — The magnitude of a mechanical resistance, reactance, or impedance for which a force of 1 dyne produces a linear velocity of 1 centimeter per second (dyn-s/cm). When expressed in newton-seconds per meter it is called the mks mechanical ohm.

mechanical overtravel — The shaft travel of a precision potentiometer between each end point (or limit of theoretical electrical travel) and its adjacent corresponding limit of total mechanical travel.

mechanical phonograph — A phonograph whose playback stylus drives the diaphragm of an acoustic pickup that radiates acoustic energy without any further amplification.

mechanical phonograph recorder — Also called mechanical recorder. Equipment that converts electric or acoustic signals into mechanical motion and cuts or embosses it into a medium.

mechanical reactance — The magnitude (size) of the imaginary component of mechanical impedance.

mechanical reader — A reader that senses characters on a perforated tape by means of a contact closure caused by each hole.

mechanical recorder — See mechanical phonograph recorder.

mechanical recording head — See cutter.

mechanical rectifier — A rectifier in which its action is done mechanically (e.g., by making and breaking the electrical circuit at the correct times with a rotating wheel or vibrating reed).

mechanical register — An electromechanical device that records or indicates a count.

mechanical reproducer — See phonograph pickup.

mechanical resistance — The real part of the mechanical impedance. The cgs unit is the mechanical ohm. The mks unit is the mks mechanical ohm.

mechanical scanning — An obsolete type of scanning in which a rotating device, such as a disc or mirror, breaks up a scene into a rapid succession of narrow lines for conversion into electrical impulses.

mechanical shock — Shock that occurs when the position of a system is significantly changed in a relatively short time in a nonperiodic manner. It is characterized by suddenness and large displacements which develop significant internal displacements within the system.

mechanical switch — 1. Bringing together or separating the surfaces of two or more metallic contacts. Mechanical switches are operated manually or electromagnetically, as in relays. 2. A switch in which the contacts are opened and closed by means of a depressible plunger or button.

mechanical television system — A television system that uses mechanical scanning.

mechanical tilt — 1. Tilt of the mechanical axis of an antenna. 2. The angle of this tilt is shown by the tilt indicator dial.

mechanical timer — A timer that does not require electrical power to run or reset. Usually spring wound.

mechanical transducer — A device that transforms mechanical energy directly into acoustical energy.

mechanical transmission system — An assembly of gears, etc., for transmitting mechanical power.

mechanical tuning range — The frequency range of oscillation of a klystron that is obtainable by tuning mechanically while keeping the reflector voltage optimized for the peak of the reflector-voltage mode.

mechanical tuning rate — In a klystron, the frequency change per degree of rotation of the tuning apparatus while oscillation is maintained on the peak of the reflector-voltage mode.

mechanical wave filter — Also called mechanical filter. A filter that separates mechanical waves of different frequencies.

mechanical waveform synthesizer — A device that mechanically generates an electrical signal with the desired waveform.

mechanism of failure — See failure mode, 1.

mechanized assembly — The joining together of elements by operators using semiautomatic equipment as contrasted to fully automatic assembly.

media — Communications channels over which LANs operate. These include coaxial cable, twisted-pair telephone wiring, and fiber optic cabling.

median — 1. The middle, or average, value in a series (e.g., in the series 1, 2, 3, 4, and 5, the median is 3). 2. The middle value in a set of measured values when the items are arranged in order of magnitude. If there is no middle value, the median is the average of the two middle values. *Compare* mode and mean.

medical amplifier — Amplifier designed for receiving medical and biological signals (EEG, ECG, etc.) and increasing their magnitude.

medical diathermy — The production of heat in body tissues for therapeutic purposes by high-frequency currents that are insufficiently intense to damage tissues or to impair their vitality. Diathermy has been used in treating chronic arthritis, bursitis, fractures, gynecologic diseases, sinusitis, and other conditions.

medical electronics — 1. The branch of electronics concerned with its therapeutic or diagnostic applications in the field of medicine. 2. The application of the tools, techniques, and methods of electronic technology to the problems of medicine.

medical sonic applicator — An electromechanical transducer designed for the local application of sound for therapeutic purposes, for example, in the treatment of muscular ailments.

medium — Anything used for the propagation or transmission of signals, usually in the form of electrons or modulated radio, light, or acoustic waves; examples include optical fiber, cable, wire, dielectric slab, water, air, or free space.

medium error — Semiconductor memory condition in which the cell or cells will properly store data under some conditions or operations (voltage, temperature, and pattern) but will make errors under other conditions of operation.

medium frequency — The band of frequencies between 300 kHz and 3 MHz (100 to 1000 meters).

medium-power silicon rectifiers — Rectifiers with maximum continuous rating of 1 to 50 average amperes per section in a single-phase, half-wave circuit.

medium-scale integration — Abbreviated MSI. 1. Integrated circuits that function as simple, self-contained logic systems, such as decade counters or 5-bit shift registers. Such chips may contain up to 100 gates. 2. The accumulation of several circuits (usually less than 100) on a single chip of semiconductor. 3. The physical realization of a microelectronic circuit fabricated from a single semiconductor integrated circuit having circuitry equivalent to more than 10 individual gates or active circuit functions. 4. The technology that produces microcircuits with 15 or more active devices on a single chip. Most op amps are MSI devices. 5. A classification of ICs by size, applicable to chips containing between 12 and 100 gates or circuits of equivalent complexity. 6. A class of integrated circuits having a density between those of large-scale integration and small-scale integration.

medium speed data rate — Data transmission at a rate between 151 and 2400 bauds.

medium wave — The band of frequencies found on a regular AM radio dial, 540 to 1600 kHz.

meg — Abbreviation for megohm.

mega- — Prefix denoting 10^6 (one million). Letter symbol: M.

megabar — The absolute unit of pressure; equal to one million bars.

megabit — Abbreviated Mb or Mbit. 1,000,000 bits or, exactly, 1,048,576 bits; used in describing data transfer rates as a function of time (Mbps). *See* bit.

megabyte — Abbreviated Mbyte, MB, Meg, or M. 1,048,576 bytes, equal to 1024 kilobytes or 8,388,608 bits; basic unit of measurement of mass storage. *See* byte.

megacycle — One million cycles — obsolete term replaced by megahertz (MHz).

megaflop — Abbreviated Mflops. One million floating-point operations per second.

megaflops per second — Millions of floating-point operations per second; a measure of a computer's processing capability.

megahertz — One million hertz. Letter symbol: MHz.

megampere — One million amperes. Letter symbol: MA.

megarad — A unit for measuring radiation dosage. Equal to one million or 10^6 rads.

megatron — A tube having a high power output at high frequencies, but very low interelectrode capacitances because its electrodes are arranged in parallel layers. *See also* disc-seal tube.

megavolt — One million volts. Letter symbol: MV.

megavoltampere — One million voltamperes. Letter symbol: MVA.

megawatt — Abbreviated MW. One million watts.

megawatthour — One million watt-hours. Letter symbol: MWh.

Megger — A high-range ohmmeter having a built-in hand-driven generator as a direct voltage source, used for measuring insulation resistance values and other high resistances. Also used for continuity, ground, and short-circuit testing in general electrical power work.

megohm — Abbreviated meg. One million ohms.

megohm-farads — *See* megohm-microfarad.

megohm-microfarad — A term used to indicate the insulation resistance of capacitors. It is equal to the product of the insulation resistance in megohms and the capacitance in microfarads. For larger high-voltage capacitors, megohm-farads are used.

megohm sensitivity — The resistance in megohms that must be placed in series with a galvanometer in order that an applied emf of 1 volt shall produce the standard deflection. If the resistance of the galvanometer coil itself is neglected, the number representing the megohm sensitivity is equal to the reciprocal of the number representing the current sensitivity.

Meissner effect — The sudden loss of magnetism in superconductors as they are cooled below the temperature required for superconductivity. As a result, they become diamagnetic — i.e., the self-induced magnetization opposes the applied magnetic field to such an extent that there is no longer a magnetic field.

Meissner oscillator — An oscillator in which the grid and plate circuits are inductively coupled through an independent tank circuit, which determines the frequency.

mel — A unit of pitch. A simple 1000-hertz tone, 40 dB above a listener's threshold, produces a pitch of 1000 mels. The pitch of any sound that is judged by the listener to be n times that of a 1-mel tone is n mels.

melodeon — A broadband, panoramic countermeasures receiver. It displays all types of received electromagnetic radiation as vertical pips on a frequency-calibrated CRT indicator.

melt — Molten semiconductor material from which are drawn the basic single-crystal ingots.

meltback process — The making of junctions by melting a correctly doped semiconductor and allowing it to refreeze.

meltback transistor — A grown transistor produced by melting the tip of a double-doped pellet. Junctions are formed when the tip recrystallizes.

melting channel — The restricted portion of the charge in a submerged horizontal ring-induction furnace. The induced currents are concentrated here to effect high energy absorption and thereby melt the charge.

melting current — The minimum current that causes a fuse to melt before the end of a given period.

melt-quench transistor — A junction transistor made by quickly cooling a meltback region.

membrane potential — The electric potential that exists across the two sides of a membrane.

membrane switch — 1. A thin, flat, lightweight panel containing one or more individual touch-activated switches. An upper membrane with a flexible, conductive material on the lower side is separated by a shimlike spacer from a lower substrate that contains one or more sets of switch contact points. Depressing the membrane touches the conductive surfaces on the lower side of the membrane to the contact points to close the circuit. When the switch is released, the flexible membrane returns to its original position and breaks contact. 2. Contacts and interconnections deposited on two outside layers separated by a third layer that acts as a spacer and maintains the contact gap. A decorative graphic overlay is generally secured to the top layer of the switch with adhesive. The flexible top layer acts as a key cap. The bottom layer can be made of the same flexible material, or it can be a rigid printed circuit board that provides mechanical support. 3. A sandwich of three polyester films. The front layer is the membrane, with a moving contact on its inner surface. Stationary contacts are on the front surface of the rear layer or base. The middle layer contains openings that expose the moving contacts to the stationary ones. Pushing the front sheet over a cut-out area flexes the membrane toward the rear layer, closing the contacts. Only a few thousandths of an inch movement and a few ounces of pressure are required to make contact.

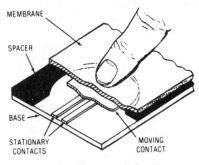

Membrane switch.

memistor — A nonmagnetic memory device composed of a resistive substrate in an electrolyte. When the device is used in an adaptive system, a dc signal deposits copper from an anode on the substrate, thus reducing the resistance of the substrate. Reversing the current reverses the process, increasing the resistance of the substrate.

memory — 1. The equipment and media used to hold machine-language information in electrical or magnetic form. Usually, the word *memory* means storage within a control system, whereas *storage* is used to refer to magnetic drums, MOS devices, disks, cores, tapes, punched cards, etc., external to the control system. Either term means collecting and holding pertinent information until it is needed by the computer. 2. The tendency of a material to return to its original shape after having been deformed. 3. Any device or circuit capable of storing a digital word or words. 4. The component of a computer, control system, guidance system, instrumented satellite, or the like designed to provide ready access to data or instructions previously recorded so as to make them bear upon an immediate problem. 5. That part of a computer that holds data and instructions. Each instruction or datum

is assigned a unique address that is used by the CPU when fetching or storing the information. 6. The storage capability or location in a computer system that receives and holds information for later use. Also, the storage arrangement, such as RAM or other type.

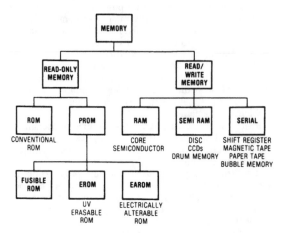

Memory types.

memory addressing modes — The method of specifying the memory location and an operand. Common addressing modes are direct, immediate, relative, indexed, and indirect. (These modes are important factors in program efficiency.)

memory address register — 1. The CPU register in a computer, which holds the address of the memory location being accessed. 2. A multiple-bit register that keeps track of where instructions are stored in the main memory.

memory allocation — In a computer, a technique of allocating memory to processes or devices.

memory array — In a computer, the memory cells arranged in a rectangular geometric pattern on a chip and organized in rows and columns.

memory buffer register — In a computer, a register in which a word is stored as it comes from memory (reading) or just prior to its entering memory (writing).

memory bus — A dedicated bus through which a central processing unit accesses the memory of a computer system.

memory capacity — *See* storage capacity.

memory cell — A single storage element of a memory, together with the associated circuits for inserting and removing one bit of information.

memory circuit — A circuit that, having been placed in a particular state by an input signal, will remain in that state after the removal of the input signal.

memory counter — Also called rewind. A system that allows recording tape to be rewound automatically to any predetermined point on the tape.

memory cycle — 1. In a computer, an operation consisting of reading from and writing into memory. 2. The operations required for addressing, reading, writing, and/or reading and writing data in memory.

memory dump — 1. In a computer, a process of writing the contents of memory consecutively in such a form that it can be examined for computer or program errors. 2. A computer printout showing the contents of memory.

memory expansion — Expansion of the horizontal and/or vertical memory contents in a digital scope for easy viewing of critical parts of a stored waveform. Typically, front-panel scope controls provide expansion over a 1 × to 10 × range.

memory fill — In a computer, the placing of a pattern of characters in the memory registers not in use in a particular problem to stop the computer if the program, through error, seeks instructions taken from forbidden registers.

memory hierarchy — A set of computer memories with differing sizes and speeds and usually having different cost-performance ratios. A hierarchy might consist of a very high-speed, small semiconductor memory, a medium-speed core memory, and a large, slow-speed core.

memory integrated circuit — In a computer, an integrated circuit consisting of memory cells and usually including associated circuits such as those for address selection and amplification. A class of ICs that store digital information being expressed in binary numbers. Examples of memory ICs are ROMs, dynamic and static RAMs, EPROMs, and EEPROMs.

memory light — In a calculator, indicates there is a number in the memory.

memory map — A listing of addresses or symbolic representations of addresses that defines the boundaries of the memory address space occupied by a program or a series of programs. Memory maps can be produced by a high-level language such as FORTRAN.

memory-mapped I/O — Operations performed by reading and writing to memory locations dedicated to a device.

memory module — A processor module consisting of memory storage and capable of storing a finite number of words (e.g., 4096 words in a 4 K memory module). Storage capacity is usually rounded off and abbreviated, with K representing each 1024 words.

memory + and − keys — In a computer, direct access to the memory for storing numbers. On machines without these, the memory has to be addressed first and the working register + and − keys used to store a number in the memory.

memory protection — 1. The mechanism used to prevent accidental writing into a computer memory location or area. 2. A method of ensuring that the contents of main memory within certain variable limits are not altered or inadvertently destroyed. *See* storage protection.

memory register — Also called high-speed bus, distributor, or exchange register. 1. In some computers, a register used in all data and instruction transfers between the memory, the arithmetic unit, and the control register. 2. A register in a calculator in which the contents can be added to or subtracted from without being recalled, or which can be recalled for other operations. The contents are available for repeated recall until the register is cleared.

memory relay — A relay in which each of two or more coils may operate independent sets of contacts, and another set of contacts remains in a position determined by the coil last energized. The term is sometimes erroneously used for polarized relay.

memory timer — A timing device wherein the cycle duration is infinitely variable within the specified overall cycle time and having the ability, once the cycle is selected, to repeat this cycle any number of times by a simple mechanical actuation of the time shaft.

memory typewriter — A typewriter that is capable of storing keyboarded material and playing it back automatically. Memory typewriters generally also have some text input features.

memory unit — That part of a digital computer in which information is stored in machine language, using electrical or magnetic techniques.

meniscus lens — A thin lens with one convex and one concave surface. The surface with the greatest radius of curvature is the convex surface for a positive lens and the concave one for a negative lens.

menu — 1. The list of available software functions available to the user of a computer software program, displayed on the computer screen and chosen by moving the cursor or entering an appropriate command. 2. A list of options or functions that is displayed on a CRT. The items in the menu are composed of light buttons for light-pen selection and/or of character strings that are alphanumeric key and/or function key selectable. 3. A display on a terminal device that lists options a user may choose. 4. Input device consisting of command squares on a digitizing surface. It eliminates the need for keyboard input for common commands. 5. A list of all the computer programs that can be used (within a package) or a list of all the tasks that can be performed by a program. The menu is displayed on the screen, and a letter or number represents each option. The option is selected by pressing the appropriate key.

mercuric-oxide-cadmium cell — A primary-cell electrochemical system. Its primary advantage is its long shelf life in the fully charged condition and its operation at low temperatures, far below 0°C. Nominal cell voltage is about 0.9 volt.

mercury — A silver-white metal that becomes a liquid above −38.87°C. In addition to thermometers, it is used in switches and many electronic tubes. When it is vaporized, mercury ionizes readily and conducts electricity.

mercury arc — A cold-cathode arc through ionized mercury vapor. A very bright bluish-green glow is given off.

mercury-arc converter — A frequency converter using a mercury-vapor tube.

mercury-arc rectifier — Also called mercury-vapor or simply mercury rectifier. A diode rectifier tube containing mercury vapor. The mercury vapor is ionized by the voltage across the tube, and a much greater current can flow. There is only a small voltage drop in the tube during conduction.

mercury barometer — An instrument for measuring atmospheric pressure.

mercury battery — A type of battery characterized by extremely uniform output throughout its life. Employs a zinc-powder anode; the cathode is mercuric oxide powder and graphite powder.

mercury cell — 1. A primary cell with a zinc anode and a mercuric oxide cathode in a potassium hydroxide

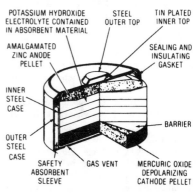

Mercury cell.

electrolyte. This cell offers a higher capacity than the alkaline or zinc-carbon cells and a much flatter voltage-discharge characteristic. 2. A small dry cell that provides a constant voltage of 1.35 volts. Used in hearing aids, watches, and calculators.

mercury-contact relay — A relay in which the contacts are mercury.

mercury delay line — *See* mercury memory.

mercury displacement relay — A relay in which the displacement of mercury, such as caused by a solenoid-actuated plunger, results in a mercury-to-mercury electrical contact.

mercury fence alarm — A type of mercury switch that is sensitive to the vibration caused by an intruder climbing on a fence.

mercury-hydrogen spark-gap converter — A spark-gap generator in which the source of radio-frequency power is the oscillatory discharge of a capacitor through an inductor and a spark gap. The latter comprises a solid electrode and a pool of mercury in hydrogen.

mercury-jet scanning switch — A commutating switch in which a stream of mercury performs the switching between the common circuit and those to be sampled. (Not in common use).

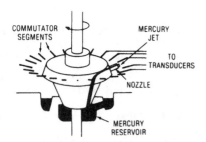

Mercury-jet scanning switch.

mercury lamp — An electric discharge lamp in which the major portion of the radiation is produced by the excitation of mercury atoms.

mercury memory — Also called mercury storage and mercury delay line. Delay lines using mercury as the medium for storage of a circulating train of waves or pulses.

mercury motor meter — A motor meter in which a portion of the rotor is immersed in mercury, which directs the current through conducting portions of the rotor.

mercury-pool cathode — The cathode of a gas tube consisting of a pool of mercury. An arc spot on the pool emits electrons.

mercury rectifier — *See* mercury-arc rectifier.

mercury relay — A relay in which the energized coil pulls a magnetic plunger into a tube containing mercury. The plunger moves the mercury in order to make connection between the contacts.

mercury storage — *See* mercury memory.

mercury switch — 1. An electric switch comprising a large globule of mercury in a metal or glass tube. Tilting the tube causes the mercury to move toward or away from the electrodes to make or break the circuit. 2. A switch operated by tilting or vibrating that causes an enclosed pool of mercury to move, making or breaking physical and electrical contact with conductors. It is used on tilting doors and windows and on fences.

mercury tank — In a computer, a container of mercury holding one or more delay lines for storing information.

mercury-vapor lamp — A glow lamp in which mercury vapor is ionized by the electric current, producing a bluish-green luminous discharge.

mercury-vapor rectifier — *See* mercury-arc rectifier.

mercury-vapor tube — A tube containing mercury vapor, which when ionized allows conduction and produces a luminous glow.

mercury-wetted contact relay — A special form of reed relay. It consists of a glass-encapsulated reed with its base mounted in a pool of mercury and the other end arranged so as to move between two sets of stationary contacts. By capillary action, the mercury flows up the reed to coat the movable and stationary contact surfaces, thus assuring mercury-to-mercury contact during make.

merge — 1. To produce a single sequence of items ordered according to some rule (i.e., arranged in some orderly sequence) from two or more sequences previously ordered according to the same rule, without changing the size, structure, or total number of the items. 2. To combine two or more files of data into one file in a predetermined sequence. 3. To combine two files into one while keeping the original order. 4. Also called a mail merge. Incorporating two individual documents to create a third.

merged technology switch — A means to interconnect two or more circuits using a series of analog and digital switching stages.

merging — Action of combining two computer files into a simple, ordered one.

meridial rays — The rays of light that propagate by passing through the axis of the fiber and that travel in one plane.

mesa — In certain transistors, the raised area (somewhat resembling a geological mesa) left when semiconductor material is etched away to allow access to the base and collector regions.

mesa diffusion — Technique used to manufacture semiconductors having diffused pn junctions. A single base region is diffused over the entire wafer, and an acid is used to etch a mesa configuration for the transistor elements.

mesa isolation — A process for isolating ICs in which transistors, resistors, and other components are fabricated before isolation.

mesa structure — A semiconductor whose structure is moundlike. During processing, material is etched away from the original chip in order to produce the final shape.

mesa transistor — A transistor that is produced by chemically etching away a transistor chip formed by either a double-diffused or diffused-alloy process. When the etching process is complete, the base and emitter regions appear as plateaus above the collector region. One result of the mesa construction is a reduction in the collector-base capacitance as a result of lowering the junction area.

MESFET — Abbreviation for metal epitaxial semiconductor field-effect transistor. A type of FET in which the channel is formed directly beneath a metal gate, which itself is in intimate contact with the semiconductor. Compare with MOSFET, in which the gate is separated from the semiconductor by a thin insulating oxide layer. *See also* gallium arsenide field-effect transistor.

mesh — Sometimes called a loop. A set of branches forming a closed path in a network — provided that if any branch is omitted, the remaining branches do not form a closed path.

mesh beat — *See* moiré, 1.

mesh current—The current assumed to exist over all cross sections of a closed path in a network. It may be the total current in a branch included in the path, or a partial current that, when combined with the others, forms the total current.

message—1. An ordered selection of an agreed set of symbols for the purpose of communicating information. 2. The original modulating wave in a communication system. 3. An arbitrary amount of information whose beginning and end are defined or implied. 4. One or more blocks of data that contain the total information to be transmitted. 5. A group of characters that have a meaning when taken together and that always are handled as a group.

message center—Communication agency charged with the responsibility for acceptance, preparation for transmission, receipt, and delivery of messages.

message circuit—A long-distance telephone circuit used in providing regular long-distance or toll service to the general public, as opposed to a circuit used for private-line service.

message exchange—A service used between a communications line and a computer to perform certain communications functions and free the computer for other tasks.

message interpolation—Insertion of data between syllables or during speech pauses on a busy voice channel without noticeably affecting the voice transmission.

message precedence—Designations employed to indicate the relative order in which a message of one precedence designation is handled with respect to all other precedence designations.

message switching—1. The technique of data transmission in which data may be received, stored until the proper line is available, then retransmitted. No direct connection is set up between the originator of the data and its destination. 2. Routing messages between three or more locations by store-and-forward techniques in a computer.

message unit—1. A unit of measurement used in charging for local telephone messages, based on time and distance between the parties. 2. Call measurement for a call within a local service area for which charges are accrued.

message-waiting lamp—A small lamp on a telephone set that can be lighted (or flashed) from the switchboard (or call waiting panel) to notify a hotel or motel guest that a message is being held for him or her.

metadyne—British term for amplidyne. A direct-current machine used for voltage regulation or transformation. It has more than two brushes for each pair of holes.

metal—A material that has high electrical and thermal conductivity at normal temperatures.

metal-base transistor—A transistor with a base of a thin metal film sandwiched between two n-type semiconductors, with the emitter doped more heavily than the base to give it a high electron-current-to-hole-current ratio.

metal detector—Also called metal locator. An electronic device for detecting concealed metal objects.

metal-etched mask—A mask formed by chemically etching openings in a metal film or plate where it is not protected by photoresist or other chemically resistant material.

metal film resistor—An electronic component in which the resistive element is an extremely thin layer of metal alloy vacuum-deposited on a substrate.

metal foil capacitor—A capacitor in which the electrodes consist of metal foils separated by a dielectric consisting of plastic film or paper.

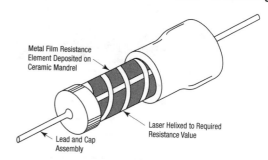

Metal film resistor.

metal gate—Refers to the use of aluminum as the gate conductor instead of silicon or refractory metals.

metal halide lamp—A discharge lamp in which the light is produced by the radiation from a mixture of metallic vapor (for example, mercury) and the products of the disassociation of halides (for example, halides of thallium, indium, or sodium).

metal-insulator silicon—See MIS.

metallic circuit—A circuit in which the earth itself is not used as ground.

metallic insulator—A shorted quarter-wave section of transmission line, which acts as an electrical insulator at the transmitted frequency.

metallic noise—Weighted noise current in a metallic circuit at a given point when the circuit is terminated at that point in the nominal characteristic impedance of the circuit.

metallic rectifier—A rectifier in which the asymmetrical junction between dissimilar solid conductors presents a high resistance to current flow in one direction and a low resistance in the opposite direction.

metallic rectifier cell—An elementary rectifying device having only one positive electrode, negative electrode, and rectifying junction.

metallic-rectifier stack—A single structure made up of one or more metallic rectifier cells.

metallization—1. The deposition of a thin-film pattern of conductive material onto a substrate to provide interconnection of electronic components or to provide conductive contacts (pads) for interconnections. 2. A film pattern (single or multilayer) of conductive material deposited on a substrate to interconnect electronic components, or the metal film on the bonding area of a substrate that becomes a part of the bond and performs both electrical and mechanical functions. 3. The selective deposition of metal film on a substrate to form conductive interconnection between IC elements and points for connections with the outside world.

metallized capacitor—A capacitor that is made with dielectric film that has had metal vacuum-deposited on it. This thin metallization restricts the maximum current capacity, but at the same time provides a very high volumetric efficiency and a unique self-healing property. Any internal arcover (which could be triggered by a transient voltage spike) will usually clear itself by vaporizing the deposited metal film in the immediate area, thus extending the arc path beyond the sustaining gap length limit. Foil capacitors cannot clear in this manner and may therefore sustain the arcovers and short out.

metallized resistor—A fixed resistor in which the resistance element is a thin film of metal deposited on the surface of a glass or ceramic substrate.

metallizing—Applying a thin coating of metal to a nonmetallic surface. This may be done by chemical

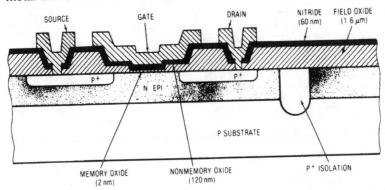

Metal-nitride-oxide semiconductor (MNOS).

deposition or by exposing the surface to vaporized metal in a vacuum chamber.

metal locator — *See* metal detector.

metal mask (screens) — A screen made not from wire or nylon thread but from a solid sheet of metal in which holes have been etched in the desired circuit pattern. Useful for precision and/or fine printing and for solder cream printing.

metal master — *See* original master.

metal negative — *See* original master.

metal-nitride-oxide semiconductor — Abbreviated MNOS. An MOS structure with a layer of silicon nitride added between the metal gate contact and the oxide protective layer to offset the permeability of silicon dioxide by contaminating sodium ions. In suitably fabricated devices, it is possible to generate a relatively nonvolatile memory element by charge storage at the nitride-oxide interface.

metal-on-glass mask — An optical mask comprising a glass substrate selectively covered by a thin opaque metal layer; a type of photomask.

metal-oxide resistor — A type of film resistor in which the material deposited on the substrate is tin oxide, which provides good stability.

metal-oxide semiconductor — Abbreviated MOS. 1. A field-effect transistor in which the gate electrode is isolated from the channel by an oxide film. 2. A capacitor in which semiconductor material forms one plate, aluminum forms the other plate, and an oxide forms the dielectric. 3. A circuit in which the active region is a metal-oxide semiconductor sandwich. The oxide acts as the dielectric insulator between the metal and the semiconductor. 4. A process that results in a structure of metal over silicon oxide over silicon. By appropriate topology this generates field-effect transistors, capacitors, or resistors. 5. Referring to a field-effect transistor (MOSFET) that has a metal gate insulated by an oxide layer from the semiconductor channel. A MOSFET is either enhancement-type (normally turned off) or depletion-type (normally turned on). MOS also refers to integrated circuits that use MOSFETs (virtually all enhancement-type). 6. Technology that employs field-effect transistors having a metal or conductive electrode that is insulated from the semiconductor material by an oxide layer of the substrate material. Whereas, bipolar devices permit current in only one direction, MOS devices permit bidirectional current. 7. A metal over silicon oxide over silicon arrangement that produces circuit components such as transistors. Electrical characteristics are similar to vacuum tubes. 8. A transistor whose characteristics are high packing density, low power dissipation, slow speed, and high input resistance.

metal-oxide semiconductor field-effect transistor — Abbreviated MOSFET. 1. A device consisting of diffused source and drain regions on either side of a p or n channel region, and a gate electrode insulated from the channel by silicon oxide. When a control voltage is applied to the gate, the channel is converted to the same type of semiconductor as the source and drain. This eliminates part of the pn junction and permits current to be established between the source and drain. Functionally, the main difference between a MOSFET and a bipolar transistor is that the source and drain of the MOSFET are interchangeable, unlike the emitter and collector of the bipolar transistor. 2. An insulated-gate field-effect transistor in which the insulating layer between each gate electrode and the channel is oxide material. 3. A class of voltage-driven devices that do not require the large input drive currents of bipolar devices. MOSFETs are a type of field-effect transistor that operates and functions similar to a junction field-effect transistor. The difference is that in the MOS device the controlling gate voltage is applied to the channel region across an oxide insulating material, rather than across a pn junction. The term can be applied either to transistors in an IC or to discrete power devices. The major advantage of a MOSFET is low power requirement due to its insulation from source and drain. Other advantages are its process simplicity, savings in chip real estate, and the ease of interconnection on chip. MOSFETs are of both p-channel and n-channel types.

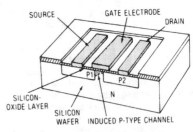

Metal-oxide semiconductor field-effect transistor.

metal-tank mercury-arc rectifier — A mercury-arc rectifier in which its anodes and mercury cathode are enclosed in a metal chamber.

metal-thick-nitride semiconductor — A device similar to an MTOS device except that a thick layer of silicon nitride or silicon nitride-oxide is used instead of a layer of oxide.

metal-thick-oxide semiconductor—Abbreviated MTOS. A device in which the oxide outside the desired active gate area is made much thicker so that problems with unwanted parasitic devices are reduced.

metal tube—A vacuum tube with a metal envelope. The electrode leads pass through glass beads fused into the metal housing.

metamers—Lights of the same color but of different spectral energy distribution.

meteorological aids service—A radio service in which emissions consist of special signals intended solely for meteorological uses.

meteorological radar station—A station, operating in the meteorological aids service, that employs radar and is not intended for operation while in motion.

meter—1. Any measuring device; specifically, any electrical or electronic measuring instrument. 2. The basic unit of linear measurement in the metric system. Originally based on the distance from the equator to the geographic north pole (10^5 meters). One meter is equal to 39.37 inches, 3.281 feet, or 1.094 yards. Letter symbol: m.

meterampere—A measure of the strength of a radio transmitter. It is equal to the antenna height in meters times the maximum antenna current in amperes.

metercandle—See lux.

meter-correction factor—The factor by which a meter reading must be multiplied to compensate for meter errors in order to obtain the true reading.

meter display—A display in which one or more pointer instruments give the indications.

meter-kilogram-second-ampere system of units—See mksa electromagnetic system of units.

meter rating—A manufacturer's designation used to indicate the operating limitations of the meter. The full-scale marking on a meter scale does not necessarily correspond to the rating of the meter.

meter resistance—The resistance of a meter as measured at the terminals at a given reference temperature. When applied to rectifier-type meters, the frequency and waveshape of the applied energy, as well as the indicated value at which the measurement is to be made, must be specified.

meter-type relay—A meter movement in which the armature function is performed by a contact-bearing pointer.

metrechon—A storage tube used in scanning converters (e.g., in radar and industrial TV).

metric—A measure for hardware or software performance, or of the extent to which a program exhibits certain characteristics.

metric system—The decimal system of weights and measures, used extensively by scientists.

metric waves—Waves with lengths between 1 and 10 meters (corresponding to frequencies between 300 and 30 MHz, respectively).

MeV—Usually pronounced M-E-V. Letter symbol for million electronvolts. 1 MeV equals 4.45×10^{-20} kilowatthours, or 1.6020×10^{-6} erg.

MEW—Abbreviation for microwave early warning.

MF—1. Abbreviation for medium frequency. 2. Abbreviation for multifrequency.

mf—Alternate abbreviation for microfarad.

mfd—Alternate abbreviation for microfarad.

mH—Letter symbol for millihenry.

MHD—Abbreviation for magnetohydrodynamics.

mho—The unit of conductance of a conductor when a potential difference of 1 volt between its ends maintains an unvarying current of 1 ampere. The mho was the unit of conductance in the mksa system. This term has now been replaced by siemens.

MHz—Letter symbol for megahertz.

MIC—1. Abbreviation for monolithic integrated circuit. 2. See microwave integrated circuit.

mica—1. A transparent mineral that can be split into thin sheets. Because of its excellent insulating and heat-resisting qualities, it is used to separate the plates of capacitors and to insulate electrode elements in vacuum tubes. 2. A silicate that separates into layers and has high insulation resistance, dielectric strength, and heat resistance. It is used as an insulation wrap in wires and cables to a limited degree where radiation resistance requirements are severe and for high-temperature work demanding good heat resistance.

mica capacitor—1. A fixed capacitor employing mica sheets as the dielectric material between adjacent plates. The complete units are usually encased in molded Bakelite. 2. A capacitor with a dielectric consisting of mica.

MICamp—Abbreviation for microwave integrated circuit amplifier

MICR—Abbreviation for magnetic-ink character recognition.

micro-—1. In the metric system, a prefix meaning one millionth (1/1,000,000). 2. A prefix meaning something very small.

microalloy diffused transistor—Abbreviated MADT. A transistor fabricated by etching emitter and collector pits into heavily doped germanium, depositing epitaxial layers of intrinsic germanium on both sides of the wafer, and adding electrodes that are plated and electrolayered in the pits.

microalloy transistor—Abbreviated MAT. A high-frequency transistor in which the emitter and collector are alloyed to a slight depth on opposite sides of the germanium base material by an electrochemical process.

microammeter—A meter having a scale that reads in microamperes, for measuring extremely small currents.

microampere—One millionth of an ampere.

microbar—A unit of pressure commonly used in acoustics. One microbar is equal to one dyne per square centimeter. (Originally the term *bar* denoted dyne per square centimeter. Therefore, to avoid confusion it is preferable to use *microbar* or *dynes per square centimeter* when speaking of sound pressures.)

micro B-display—A B-scope in which range and azimuth data are so expanded that only a small portion of the area under surveillance is presented, and, because of the degree of expansion, distortion of the presentation is negligible.

microbond—The realization of a very small fastened joint between conductors or between a conductor and a microelectronic chip device.

microcassette—A miniature cassette system originated by Olympus, allowing 30 minutes of recording per side on a capstan-driven tape, $1/7$ inch (3.63 mm) wide, running at $15/16$ ips (2.38 cm/s).

microcircuit—1. A small circuit with a high equivalent-circuit-element density, and considered as a single part, composed of interconnected elements on or within a single substrate, that performs an electronic-circuit function. (According to this definition, such structures as printed wiring boards, circuit card assemblies, and modules composed exclusively of discrete electronic parts are not considered to be microcircuits.) 2. An integrated circuit. 3. The physical realization of a (hybrid or monolithic) interconnected array of very small active and passive electronic elements. 4. A term used to describe the miniaturization of solid-state discrete components onto a continuous substrate.

microcircuit module — An assembly of microcircuits or of microcircuits and discrete parts, designed to perform an electronic-circuit function or functions and constructed in such a manner that it is considered to be a single entity for the purposes of specification, testing, commerce, and maintenance.

microcircuit wafer — A microwafer carrying one or more circuit functions such as a flip-flop or gate. Integrated-circuit chips may be bonded to deposited conductors.

microcode — 1. A computer-coding system that includes suboperations, such as multiplication and division, that ordinarily are inaccessible in programming. A list of very small program steps. 2. A set of control functions performed by the instruction decoding and execution logic of a computer that defines the instruction repertoire of that computer. Microcode is not generally accessible by the programmer. 3. Sequences of low-level steps, making up machine instructions, that are built into a microprocessor that directly control the interaction of the processor's computing elements — that is, machine instructions wired into the hardware that is being controlled. 4. Permanent basic subcommands, built into a computer, that are executed directly by the computer. Generally, these commands define the instruction set of a microprogrammable computer.

microcoding — In a computer, a system of coding that uses suboperations not ordinarily accessible in programming.

microcomponents — 1. Those components smaller than existing components by several orders of magnitude. 2. An assembly of very small, interconnected discrete components — active or passive — that forms an electronic circuit. Interconnection of the various leads is by soldering or welding. Microcomponents use no substrates.

microcomputer — 1. A general-purpose computer composed of standard LSI components built around a central processing unit (CPU). The CPU (or microprocessor) is program-controlled, featuring arithmetic and logical instructions and a general-purpose parallel I/O bus. The CPU is generally contained on a single chip. Generally intended for dedicated applications, the microcomputer also includes any number of ROMs and RAMs (for instruction and data storage) and, in some cases, one or more I/O devices. The simplest microcomputer consists of one CPU chip and one ROM. 2. A computer whose major sections — CPU, control, timing, and memory — are each contained on a single integrated-circuit chip, or, at most, a few chips. An LSI computer. 3. A computer that has a single or multichip LSI CPU, as distinguished from a minicomputer. 4. A class of computer having all major central processor functions contained on a single printed circuit board constituting a stand-alone module. 5. A computer whose central processing unit is a microprocessor. A microcomputer includes microprocessor, memory, and input/output controllers. 6. A device consisting of a central processor (usually a microprocessor, but sometimes a custom LSI chip set or a bit-slice processor) combined with a memory and input/output.

microcomputer development system — Abbreviated MDS. Also called expandable breadboarding system. 1. A key tool used in the development phase of microprocessor-based computers. It provides an efficient means of program coding, subprogram testing, and final testing for the completed program in a simulated operating condition without the use of the hardware that is contained in the final end product. The elements of the MDS combine to make up a complete microprocessor-based computer, which is composed of the central processing unit or microprocessor unit; a mass-storage memory section; either a floppy disk or tape cassette, complete with associated drive and control circuits; a CRT terminal (video readout), a keyboard; and, as there is in most systems, a printer. In addition, an emulator module is part of the system and is used to test the operation of the developed system under defined conditions without the use of final hardware. 2. A system designed exclusively to aid in the development of microprocessor systems. Microcomputer development systems enable a designer to develop software and hardware as if many standard operating system utilities were present in his or her final design. These utilities, however, actually reside in the development system and therefore do not require costly additions to every shipped system. 3. Microcomputer equipped with the hardware and software facilities required for efficient program development and hardware debugging.

microcontroller — A complete microprocessor system on a chip. Includes on-chip the CPU, local RAM, local ROM or EPROM, clock and control circuits, and serial and parallel I/O ports that can be programmed for various control functions. *See also* chip sets.

microcrack — A thin crack in a substrate or chip device, or in thick-film trim-kerf walls, that can only be seen under magnification and which can contribute to latent failure phenomena.

microdensitometer — An instrument used in spectroscopy to measure lines in a spectrum by light transmission measurement.

microelectrode — An extremely small electrode. Some microelectrodes are small enough to contact a single biological cell.

microelectronic circuit — Discrete electrical components assembled and connected in extremely small and compact form.

microelectronic device — An alternate term for integrated circuit.

microelectronics — Also called microsystems electronics. 1. The entire body of electronic art that is connected with or applied to the realization of electronic systems from extremely small electronic parts. 2. *See* integrated circuit. 3. All techniques for the manufacture of extremely small electronic circuits, generally including all types of silicon integrated circuits, thin-film circuits, and thick-film circuits. 4. The physical realization of electronic circuits or systems from a number of extremely small circuit elements inseparably associated on or within a continuous body. Microelectronics has developed along two basic technologies: monolithic integrated circuits and hybrid integrated circuits. 5. The field that deals with techniques for producing miniature circuits, e.g., integrated circuits, thin-film techniques, and solid-state logic modules. 6. The art of electronic equipment design and construction that uses microminiaturization schemes.

microelement — A resistor, capacitor, or transistor, diode, inductor, transformer, or other electronic element or combination of elements mounted on a small ceramic wafer 0.01 inch (or 0.25 mm) thick and about 0.3 inch (0.75 mm) square. Individual microelements are stacked, interconnected, and potted to form micromodules.

microelement wafer — A microwafer carrying one or more components or a simple network. The network can consist, for example, of several thin-film resistors deposited directly on the wafer.

microfarad — One millionth of a farad. Letter symbol: μF.

microfaradmeter — *See* capacitance meter.

microflash — An extremely short electronic flash, having a duration of about 1×10^{-6} second. Used in photographing rapidly moving subjects.

microflash lamp — A lamp that emits radiation pulses having a duration of approximately 1 microsecond.

micrographics — The use of microfilm and microfiche for filing basic documents that must be retained and made available for infrequent use. Special computer software can be used to maintain an index of such documents and to aid in their retrieval.

microgroove — In disc recording, the groove width of most long-play and 45-rpm records. Normally it is 0.001 inch (25.4 μm), or about half as wide as the groove on a 78-rpm record. (No longer manufactured.)

microgroove record — *See* long-play record.

microhenry — One millionth of a henry.

microhm — One millionth of an ohm.

microinstruction — Also called elementary operation, cycle, or function. 1. A very simple instruction (typically a register-to-register copy). 2. A bit pattern that is stored in a microprogram memory word and specifies the operation of the individual LSI computing elements and related subunits, such as main memory and input/output interfaces.

microinstruction sequence — The series of microinstructions that the microprogram control unit (MCU) selects from the microprogram to execute a single macroinstruction or control command. Microinstruction sequences can be shared by several macroinstructions.

microlock — A phase-lock loop system for transmitting and receiving information. Because the system reduces bandwidth drastically, it is used as a radar beacon for tracking, or to provide telemetering data.

micrologic elements — 1. A group of high-speed, low-powered integrated logic building blocks primarily intended to be used in building the logic section of a digital computer. 2. Semiconductor networks used in computer and other critical circuits.

micromanipulators — Devices that provide means for accurately moving miniscule tools over and onto the surface of a microscopic object.

micromassage — *See* intercellular massage.

micrometer — One millionth of a meter.

micromho — One millionth of a mho or of a siemens. Replaced by microsiemens.

micromicro- — An obsolete prefix meaning one millionth of a millionth, or 10^{-12}. Now called pico-.

micromicrofarad — Obsolete term for 10^{-12} farad. Now called picofarad.

micromicrowatt — Obsolete term for 10^{-12} watt. Now called picowatt.

microminiature lamp — Any incandescent lamp, usually rated in the milliwatt range, that operates on 3 volts or less. Diameters range from 0.01 to 0.06 inch, or 0.25 to 1.5 mm.

microminiaturization — 1. The producing of microminiature electronic circuits from individual miniature solid-state and other nonthermionic components. 2. A relative degree of miniaturization resulting in an equipment or assembly volume an order of magnitude smaller than that existing in subminiature equipment. 3. The technique of packaging a microminiature part of an assembly composed of elements radically different in shape and form. Electronic parts are replaced by active and passive elements through use of fabrication processes such as screening, vapor-deposition diffusion, and photoetching. 4. The process of packaging an assembly of microminiature active and passive electronic elements, replacing an assembly of much and different parts.

micromodule — 1. A tiny ceramic wafer made from semiconductive and insulative materials. It is capable of functioning as either a transistor, resistor, capacitor, or other basic component. 2. A microcircuit constructed of a number of components (e.g., microwafers) and encapsulated to form a block that is still only a fraction of an inch in any dimension.

micron — 1. An absolute unit of length equal to 10^{-6} meter. The term micrometer is now preferred. 2. A unit used in the measurement of very low pressures. It is equivalent to 0.001 mm (10^{-6} meter) of mercury at 32°F or 0°C.

microphone — 1. An electroacoustic transducer that responds to sound waves and delivers essentially equivalent electric waves. 2. A device for converting sound waves or sound-producing vibrations (as from the strings of a guitar) into corresponding electrical impulses. Microphones may use as transducing elements crystal or ceramic chips, ribbons, moving coils, or capacitors, and different recording applications may call for different transducers as well as for different directional patterns and impedances.

microphone amplifier — Also called a microphone preamplifier. An audio-frequency amplifier that boosts the output of a microphone before the signal reaches the main audio-frequency amplifier.

microphone boom — A movable crane from which a microphone is suspended.

microphone button — The resistance element of a carbon microphone. It is button-shaped and filled with carbon particles.

microphone cable — A shielded cable for connecting a microphone to an amplifier.

microphone mixer — An audio mixer that feeds the output from two or more microphones into a single input to an audio amplifier. The output from each microphone is adjustable by individual controls on the mixer.

microphone preamplifier — *See* microphone amplifier.

microphone sensitivity — The voltage that is produced by a microphone that is exposed to a specified sound pressure level. Usually specified in dBV in a 94 dB sound pressure level (spl) or 74 dB spl sound field, measured with no load on the microphone.

microphone stand — A stand that holds a microphone the desired distance above the floor or a table.

microphone transformer — An iron-core transformer used for coupling certain microphones to an amplifier or transmission line.

microphonics — 1. The generation of an electrical noise signal by mechanical motion of internal parts within a device. 2. Electrical disturbance (noise) due to mechanical disturbances of circuit elements. 3. A form of noise interference arising from the tendency for vibrations of certain objects to be converted into corresponding electrical signals. A microphonic device will cause a "bong" or "bing" in the signal when subjected to jarring. 4. Audio-frequency noise caused by the mechanical vibration of elements within a system or component. 5. Microphone noise that occurs in lasers when vibrations are transferred to the resonator structure.

microphonism — 1. The production of noise as a result of mechanical shock or vibration. 2. The quasiperiodic voltage output of a tube produced by mechanical resonance of its elements as a result of mechanical impulse excitation. 3. The periodic voltage output of a tube produced by mechanical resonances of its elements as a result of sustained mechanical excitation. 4. The output voltage of a tube acting as an electrical transducer of mechanical energy.

microphonograph — A device that amplifies and records weak sounds; used in training the deaf to speak.

microphonoscope — A binaural stethoscope using a membrane in the chest piece to accentuate the sound.

microphotograph — A small picture of a large subject. The microfilming of a check or other document produces a microphotograph.

microprobe — An extremely sharp and small exploring tool head attached to a positioning handle. Used for testing microelectronic circuits by establishing ohmic contact.

microprocessor — Also called MPU (microprocessor unit). 1. A central processing unit (CPU) fabricated on one or more chips, containing the basic arithmetic, logic, and control elements of a computer that are required for processing data. 2. An integrated circuit that accepts coded instructions, executes the instructions received, and delivers signals that describe its internal status. The instructions may be entered or stored internally. Widely used as control devices for household appliances, business machines, toys, etc., as well as for microcomputers. 3. An electronic integrated circuit, typically a single-chip package, capable of receiving and executing coded instructions.

microprocessor development system — A combination of hardware and software that acts as a tool for micro system design and debugging from concept to final production release. It contains assembler, compiler, and editor programs to assist with the original program writing. It also can simulate the system, both at the concept stage and during the final integration. The larger memory available makes it practical to document programs, that is, to add remarks within the program that will be ignored by the microprocessor and that will indicate to a subsequent user the purposes of specific instructions. These remarks do not, of course, go into the final ROM. The development system is also able to transfer the debugged program that it is already using into PROM or EROM.

microprocessor emulator — A software routine or device that imitates the functions of a specific microprocessor.

microprogram — 1. A computer program written in the most basic instructions or subcommands that can be executed by the computer. Frequently, it is stored in a read-only memory. *See also* firmware. 2. A special-purpose program, stored in a fixed memory, that is initiated by a single instruction in a system's main program. For example, one instruction in the main program may initiate a stored microprogram of six or seven instructions needed to execute the single main program instruction. 3. In a computer, a subelement of a conventional program built up of a sequence of even smaller operations called microinstructions. Each microinstruction is further subdivided into a collection of microoperations carried out in one basic machine cycle. (For example, the computer program consists of a sequence of instructions that are carried out in a specific order. Each instruction consists of a routine of one or more steps. This sequence of computer machine cycles necessary to execute a single instruction is called a microprogram.) 4. A sequence of instructions held in the control store that determines what operations the processor performs for each command given to it by the main memory. 5. A type of program that directly controls the operation of each function element in a microprocessor. 6. Stored routines in CPU control memory that define machine instructions as a series of elemental steps to be executed by the processor's control section. 7. A special computer program that sequences the control unit of a processor. It implements sequential instruction fetch, plus decoding and execution, by providing the appropriate signals to the required gates. Most MPUs are internally microprogrammed and can be microprogrammed by the user. Bit-slices are user microprogrammable. 8. A machine-executable description of the elementary steps that are executed in the course of what the software sees as an instruction.

microprogrammable computer — A computer in which the internal CPU control sequence for performing instructions is generated from a read-only memory (ROM). By changing the ROM contents, the instruction set can be changed (as contrasted with a fixed-instruction computer).

microprogramming — 1. The setting up of basic suboperations for a computer to handle, after which the programmer combines them, and they are presented to the computer again in a higher-level program. For example, if a computer has only basic instructions for addition, subtraction, and multiplication, the instruction for division would be defined by microprogramming. 2. A method of operating the control part of a computer in which each instruction is broken into several small steps (microsteps) that form part of a microprogram. 3. A method of organizing a general-purpose computer to perform desired functions, using instructions stored in a control array.

microprogram store — *See* control store.

microradiometer — Also called a radio micrometer. A thermosensitive detector of radiant power. It consists of a thermopile supported on and connected directly to the moving coil of a galvanometer.

microsecond — One millionth of a second: 1×10^{-6} or 0.000001 second. Letter symbol: μs.

Microsoft Windows — An operating environment for IBM PCs and compatibles that features icons, pull-down menus, desk accessories, and the ability to easily move text and graphics from one program to another via a clipboard. It can also operate more than one program at a time.

microstrip — Also called stripline. 1. A microwave transmission component in which a single conductor is supported above a ground plane. 2. Printed-wiring LC resonant circuits or transmission lines. Conductive patterns form inductors. Capacitance is developed between the inductive patterns and the ground plane. 3. (Microstripline) A transmission line consisting of a metallized strip and a solid ground plane metallization separated by a thin, solid dielectric. This transmission line configuration permits accurate fabrication of 50-ohm transmission line elements on a ceramic or printed circuit board substrate.

microsyn — A precise and sensitive pickoff device for converting angular displacement within a small range to an electrical signal.

microsystems electronics — *See* microelectronics.

microvolt — One millionth of a volt. Letter symbol: μV.

microvoltmeter — A highly sensitive voltmeter that measures millionths of a volt.

microvolts per meter — The potential difference in microvolts developed between an antenna system and ground, divided by the distance in meters between the two points.

microvolts/meter/mile — One method of stating the field strength of a radiated field. Radiation from industrial heating equipment, for example, must be suppressed so that the radiated field strength does not exceed 10 microvolts per meter at a distance of 1 mile (1.609 km) from the source.

microwafer — A basic microcircuit building block generally made of beryllia, alumina, or glass. Terminations on the edges are usually of gold on top of chromium, with a heavy nickel overlay for welding.

microwatt — One millionth of a watt. Letter symbol: μW.

microwave— 1. A term applied to waves in the frequency range of 1000 megahertz and upward. Generally defines operations in the region where distributed constant circuits enclosed by conducting boundaries are used instead of the conventional lumped-constant circuit components. 2. An electromagnetic wavelength in (or near) the span of 1–100 cm.

microwave alarm system— An alarm system that employs radio-frequency motion detectors operating in the microwave frequency region of the electromagnetic spectrum.

microwave amplification by stimulated emission of radiation— *See* maser.

microwave diode— A two-terminal device that is responsive in the microwave region of the electromagnetic spectrum, commonly regarded as extending from 1 to 300 GHz.

microwave discriminator— A tuned cavity that converts a frequency-modulated microwave signal into an audio or video signal.

microwave early warning— Abbreviated MEW. A high-power, long-range, early-warning radar. It has numerous indicators that give high resolutions and large traffic-handling capacity.

microwave filter— A filter built into a microwave transmission line to pass desired frequencies but reject or absorb all other frequencies.

microwave frequency— Any of the ultrahigh frequencies (approximately 1000 MHz and above) suitable for line-of-sight radiocommunication.

microwave holography— The holographic recording of the pattern formed by two sets of coherent microwaves that interfere at a scanning plane. A scanning device converts the microwave interference pattern into a light pattern that is recorded by a suitable camera.

microwave integrated circuit— Abbreviated MIC. 1. An electronic circuit fabricated by microelectronic techniques and capable of operating at frequencies above 1 gigahertz. Either hybrid or monolithic integrated circuit technology may be employed. 2. A hybrid type of construction in which thick- or thin-film technology is used to lay out a pattern of conducting lines on a ceramic substrate, and uncased active devices are then bonded to the conductor pattern. 3. A miniature microwave circuit usually using hybrid circuit technology to form the conductors and attach the chip devices. 4. In the microwave industry, a hybrid using thin- or thick-film conductors and passive components on a ceramic substrate combined with chip-form active and passive components.

microwave intruder detector— A radio-frequency device employing a Doppler frequency shift principle of detection. Microwave energy is generated in a resonant cavity and transmitted into the protection area. Objects within the protected area partially reflect the microwave energy back to the detector. If all objects

within the range are stationary, the reflected waves return at the same frequency; if they strike a moving object, they return at a different frequency. The difference in the transmitted and received frequency appears as a low-frequency signal, which is amplified and used to trip an alarm relay. The shape of the protection area is determined by the shape of the antenna. Range is controlled by adjusting the gain of the amplifier.

microwave level detector— A noncontacting level detector incorporating a microwave transmitter, oscillator, directional antenna, and receiver.

microwave mapping— The pattern of microwave field intensity that can be obtained by detecting the minute expansion of a microwave absorber slab when heated by the microwave field, accompanied by moiré interference of a set of optical fringe patterns on the slab surface pre- and post-deformation.

microwave phototube— A device designed to detect microwave modulation and to mix modulated and unmodulated laser beams. It consists of a photosensitive cathode and a microwave-electron-tube structure that amplifies, detects, and removes the modulation placed on the electron beam by the incident light beam signal.

microwave power transmission— A method of transmitting power through space from a microwave transmitting antenna to a remotely located receiving antenna.

microwave radio relay— The relaying of long-distance telephone calls and television broadcast programs by means of highly directional high-frequency radio waves that are received and sent on from one booster station to another.

microwave refractometer— A device for measuring the refractive index of the atmosphere at microwave frequencies, usually in the 3-cm region.

microwave region— The portion of the electromagnetic spectrum between the far infrared and conventional radio-frequency portion. Commonly regarded as extending from 1000 (30 cm) to 300,000 (1 mm) megahertz.

microwave relay system— A series of ultrahigh-frequency radio transmitters and receivers comprising a system for handling communications (usually multichannel).

microwave repeater— A repeater station in a microwave radio relay system. Repeaters are spaced 20 to 35 miles (32 to 56 km) apart in a typical microwave system.

microwave routes— Communications pathways in which microwave towers beam signals to and from each other as messages travel to telephones, data terminals, and the like. About 70 percent of interstate telephone messages travel this way.

microwaves— 1. Radio frequencies with such short wavelengths that they exhibit some of the properties of light. Their frequency range is from 1000 MHz up.

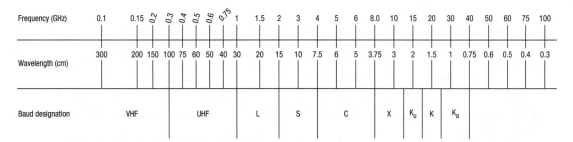

Frequency (GHz)	0.1		0.15	0.2	0.3	0.4	0.5	0.6	0.75	1		1.5	2	3	4	5	6	8.0	10	15	20	30	40	50	60	75	100
Wavelength (cm)	300		200	150	100	75	60	50	40	30		20	15	10	7.5	6	5	3.75	3	2	1.5	1	0.75	0.6	0.5	0.4	0.3
Baud designation		VHF				UHF					L		S		C			X	K_u	K	K_a						

Microwave frequency chart.

(Microwaves are preferred in point-to-point communications because they are easily concentrated into a beam.) 2. Short electromagnetic waves located between the television transmission and infrared frequency regions. For communication purposes, microwaves offer considerable appeal as they can be focused and directed like light and can be manipulated like electricity, providing a practical means of transmitting information great distances without the use of wires.

microwave spectrum — The spectrum of frequencies suitable for microwave radio communication (300 MHz to 300 GHz).

microwave tube — The source of power for generating microwave frequencies. Primary microwave generating tubes are klystrons, magnetrons, and traveling wave tubes. Other microwave devices include masers, parametric amplifiers, and backward-wave oscillators.

micro-Winchester drive — Also called microdisk drive. A memory storage device for personal computers that uses a $5\frac{1}{4}$-inch (13.3-cm) metal memory disk.

mid-band — The part of the frequency band that lies between television channels 6 and 7, reserved by the FCC for air, maritime, and land mobile units; FM radio; and aeronautical and maritime navigation. Mid-band frequencies, 108 to 174 MHz, can also be used to provide additional channels on cable television systems.

middle marker — In an instrument landing system, a marker located on a localizer course line, about 3500 feet (1067 m) from the approach end of the runway.

middle-side system — *See* mitte-seite stereo system.

MIDI — Abbreviation for musical instrument digital interface. A standard protocol that allows electronic instruments to communicate with each other and with computers.

midrange — The frequency range between bass and treble, from approximately 500 to 4000 Hz. This most important audio-frequency region includes all the frequencies necessary for intelligibility. Speaker systems often use a midrange unit that typically operates between about 400 and 3000 Hz.

migration — The movement of some metals, notably silver, from one location to another as a result of a plating action that takes place in the presence of moisture and an electrical potential.

mike — Slang for microphone.

MIL — Abbreviation for military. Pertains to a nation's armed forces, including its army, navy, and air force. Specifically, the Armed Forces of the United States.

mil — One thousandth of an inch. Used in the United States for measuring wire diameter and tape thickness.

mileage — Monthly user charges by a telephone company for line mileage based on the class of service the customer has and the air line distance from the base rate area.

military-grade IC — Typically, an integrated circuit whose performance is guaranteed over the temperature range from −55°C to +125°C.

military radio station — An amateur radio station licensed to the person in charge of a station at a land location provided for the recreational use of amateur radio operators under military auspices of the Armed Forces of the United States.

military specifications — Documents issued by the Department of Defense that define materials, products, or services used only or predominantly by military activities.

military standardization handbooks — Detailed handbooks describing a specific subject that is critical to military design. The title is comprised of the prefix letters MIL-HDBK followed by an assigned serial number. For example, MIL-HDBK-241 is the design guide for EMI reduction in power supplies.

military standards — Procedures for design, drawing, writing, and testing of components or equipment rather than giving a particular specification.

military temperature range — From −55°C to +125°C.

Miller bridge — A type of bridge circuit for measuring the amplification factor of vacuum tubes.

Miller capacitance — Feedback capacitance caused by gate metal overlapping source and drain regions.

Miller effect — The increase in the effective grid-to-cathode capacitance of a vacuum tube because the plate induces a charge electrostatically on the grid through grid-to-plate capacitance.

Miller oscillator — A crystal-controlled oscillator in which the crystal oscillates at its parallel resonant frequency due to the connection of negative resistance across its plates.

milli- — Prefix meaning one thousandth (1/1000, or 10^{-3}). 2. One-thousandth part of a whole. Letter symbol: m.

milliammeter — An electric current meter calibrated in milliamperes.

milliampere — One one-thousandth (.001) of an ampere. Letter symbol: mA.

millihenry — One one-thousandth (.001) of a henry. Letter symbol: mH.

millilambert — A unit of brightness equal to one one-thousandth (.001) of a lambert.

millimaxwell — One one-thousandth of a maxwell.

millimeter waves — Electromagnetic radiation in the frequency range of 30 to 500 gigahertz, with corresponding wavelengths of 10 millimeters to 0.6 millimeter.

millimicro- — Obsolete prefix for nano-, representing 10^{-9}.

millimicrometer — A unit of length equal to one ten-millionth of a centimeter (10^{-7} cm), or one one-thousandth of a micrometer.

milliohm — One-thousandth of an ohm. Letter symbol: mΩ.

millisecond — One-thousandth of a second: 10^{-3} or 0.001 second. Letter symbol: ms.

millitorr — One-thousandth of a torr.

millivolt — Abbreviated mV. One-thousandth of a volt.

millivoltmeter — A sensitive voltmeter calibrated in millivolts.

millivolts per meter — The potential difference in millivolts developed between an antenna system and ground, divided by the distance in meters between the two points.

milliwatt — Abbreviated mW. One one-thousandth of a watt. The reference level used for dB measurements.

Mills antenna — A combination of two independent fan-beam antennas placed at right angles in a cross formation, with a common center and common pencil-beam volume of low gain. Antennas are combined by switching from output phase addition to phase opposition at a constant rate to secure the angular resolution of the pencil-beam components.

MIL-STD-883B — The basic military standard for reliable semiconductors. Three classes are defined: A (aerospace), B (avionics), and C (ground).

mine detector — Also called electronic mine detector. An electronic device that indicates the presence of metallic or nonmetallic explosive mines under the ground or under water.

miniature lamp — 1. A small, filament-type lamp with an operating voltage less than 60 volts. 2. Very small tungsten lamps that are used where space is limited, as in surgical instruments such as cystoscopes. They are sometimes called grain-of-wheat lamps.

miniature sealed relay — A miniature hermetically sealed relay used in applications where space is at a premium and where high reliability is a requirement.

miniature tube — A small electron tube usually having a 7- or 9-pin base.

miniature wire — Insulated conductors of approximately 20 to 34 AWG with small overall diameters.

miniaturization — The process of reducing the minimum volume required by equipments or parts in order to perform their required functions.

minicassette — A miniature cassette system originated by Philips, allowing 15 minutes of recording per side on a narrow tape.

minicomputer — 1. A loosely used term for describing any small general-purpose digital computer in the low-to-moderate price range. The approximate ceiling price used to define a minicomputer is subject to wide interpretation. 2. A machine developed primarily for the processing of a single application or the processing of a number of small applications. For example, in a distributed processing network, a minicomputer could perform a specific operation and send the volume processing through communication lines to a large mainframe. 3. A class of small, digital process-control computers sized generally around a 32-bit word, with stored programs and various memory options for data acquisition and monitoring, supervisory, or direct digital control in systems having no more than 20 or 30 control loops. 4. A small- or medium-scale computer (also called a mini), usually operated with interactive dumb terminals. A mini can operate as a single powerful workstation or as a multiuser system.

mini-disc — Abbreviated MD. A 2.5-inch optical recording and playback system that approaches the sound quality of a compact disc.

minidisk — *See* floppy disk.

minifloppy handshake — A system of transferring data from one device to another. Device A will signal that it has data ready; device B will accept that data and signal that it has it to device A, which is now released to collect more data. In the meantime device B will set an indicator that will show that it is busy with the last data until this data has been processed. The action of setting and checking these various indicators is referred to as handshaking.

minigroove — A recording having more lines per inch than the average 78-rpm phonograph record but not as many as the extended-play, long-playing, or microgroove records.

minimum-access programming — Also called minimum-latency programming, or forced coding. Programming a digital computer so information is obtained from the memory in the minimum waiting time.

minimum-access routine — A computer routine coded in a way such that the actual waiting time to obtain information from a serial memory is much less than the expected random-access waiting time.

minimum detectable signal — The signal level that just exceeds the threshold.

minimum discernible signal — Abbreviated MDS. In a receiver, the smallest input signal that will produce a discernible signal at the output. The smaller the signal required, the more sensitive the receiver.

minimum firing power — The lowest radio-frequency power that will initiate a radio-frequency discharge in a switching tube at a specified ignitor current.

minimum flashover voltage — The crest value of the lowest voltage impulse of a given waveshape and polarity that causes flashover.

minimum-latency programming — *See* minimum-access programming.

minimum-phase network — A network for which the phase shift at each frequency equals the minimum value determined solely by the attenuation-frequency characteristic.

minimum reject number — Abbreviated MRN. A number that defines the maximum number of rejects allowed for each sample given for each LTPD (lot tolerance percent defective).

minimum reliable current — As applied to relay contacts, the range at which there is insufficient energy under arcing conditions at a mating contact surface to ensure good contacting for the kind of contact material, shape, and forces employed.

minimum resistance — The resistance of a precision potentiometer measured between the wiper terminal and any terminal with the shaft positioned to give a minimum value.

minimum shift frequency — The minimum frequency at which a flip-flop can be operated in a shift-register application.

minimum-signal level — In facsimile, the level corresponding to the copy white or copy black signal, whichever is the lower.

minimum starting voltage — The minimum voltage of rated frequency applied to the control-voltage winding of a servomotor necessary to start the rotor turning at no-load conditions with rated voltage and frequency on the fixed-voltage winding.

minimum toggle frequency — The minimum frequency at which a flip-flop can be toggled. In a typical device, the maximum toggle frequency usually is about 20 percent higher.

minis — Computers that have a throughput of 32 megabits per second, and a memory size of 256 bytes of real addressable memory.

minitrack — A satellite tracking system that uses a miniature pulse-type telemeter and a precise directional antenna system, with phase-comparison tracking techniques.

minor apex face — In a quartz crystal, any one of the three smaller sloping faces near but not touching the apex (pointed end). The other three are called the major apex faces.

minor bend — A rectangular waveguide bent so that one of its longitudinal axes is parallel to its narrow side throughout the length of the bend.

minor cycle — Also called word time. In a digital computer using serial transmission, the time required to transmit one word or space between words.

minor defect — A defect that is not likely to reduce the usability of a unit for its intended purpose. It may be a departure from established standards having no significant bearing on the effective use or operation of the unit.

minor face — One of the three smaller sides of a hexagonal quartz crystal.

minor failure — A failure that has no significant effect on the satisfactory performance of a system.

minority carrier — The less predominant carrier in a semiconductor. Conduction-band electrons are the minority carriers in p-type semiconductors, since there are fewer electrons than holes. Likewise, valence-band holes are the minority carriers in n types, since they are outnumbered by the electrons.

minority emitter — An electrode from which minority carriers flow into the interelectrode region.

minor lobe — Any of the lobes, except the major lobe, that represent the radiation pattern of a directional antenna.

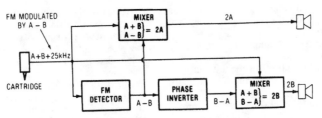

Minter stereo-disk playback.

minor loop—A continuous network composed of both forward and feedback elements, which is only part of the overall feedback system.

minor relay station—A tape relay station that has tape relay responsibility but does not provide an alternate route.

Minter stereo system—A stereo recording technique for producing the right and left channels. The two program channels are combined additively and recorded with a monophonic cutter. A 25-kHz note is also recorded and is frequency-modulated by the two channels combined subtractively. The sum and difference signals are then matrixed (combined) to produce the right and left channels.

MIPS—Abbreviation for millions of instructions per second. A measurement of data transmissions capacity. The more MIPS a computer has, the more powerful it is.

mirror galvanometer—A suspended-coil instrument that, instead of using a pointer to indicate the reading, employs a light beam reflected from a mirror attached to the moving coil.

mirror galvanometer oscillograph—An instrument that photographs the deflection of a light spot from a mirror attached to a moving coil. Used for recording small current variations.

mirror-reflection echoes—Multiple-reflection echoes produced when a radar beam is reflected from a large, flat surface (such as the side of an aircraft carrier) and strikes nearby targets.

mirror scale—Meter scale with a mirror arc used to align the eyeball perpendicular to the scale when taking a reading. By eliminating this human error in reading, accuracy can be improved by half.

MIS—Abbreviation for metal-insulator-silicon. Technology wherein a silicon dioxide layer is formed on a single crystal silicon substrate, and a polysilicon conductive is formed on the oxide. This layer is etched to form the electrode pattern and then doped with phosphorus to create the desired conductivity.

misfire—Failure of a mercury-pool cathode to establish an arc between the main anode and cathode during a scheduled conducting period.

mismatch—1. A condition whereby the coupled impedances inhibit optimum power transference or the source or output signal fails to match the amplifier sensitivity or the load's (i.e., speaker's) power capacity. 2. The condition in which the impedance of a load does not match the impedance of the source to which it is connected.

mismatch factor—See reflection factor, 1.

mismatch loss—The ratio between the power a device would absorb if it were perfectly matched to the source and the power it actually does absorb.

missile site radar—Phased-array radar located at a missile launch area to provide a guidance link with interceptor missiles enroute to their targets.

mission time—That element of uptime during which the item is performing its designated mission.

mistake—See error.

mistor—A magnetic-field sensing device whose resistance increases with an increase in magnetic-field intensity.

mitte-seite stereo system—German for middle-side stereo system. A technique of stereo pickup in which two directional microphones, placed close together and at right angles to each other, are oriented so that one picks up sound from directly ahead, and the other from the two sides, with maximum intensity.

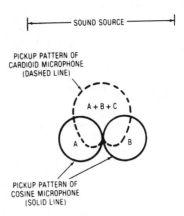

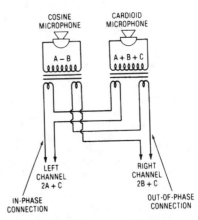

Mitte-seite stereo system.

mix—To combine two or more input signals in a transducer so as to produce a single output. If the transducer is linear, the output consists of a superposition of the input signals. If the transducer is nonlinear, the output consists of the heterodyne products of the input signals.

mixed analog component (MAC) transmissions — An innovative television transmission method that separates the data, chrominance, and luminance components and compresses them for sequential relay over one television scanline. There are presently five systems being designed: A-MAC; B-MAC; C-MAC; D-MAC; and E-MAC.

mixed-base notation — A number system in which a single base, such as 10 in the decimal system, is replaced by two number bases, such as 2 and 5, used alternately.

mixed calculation — In a calculator, calculation involving more than one arithmetic mode. Saves time by permitting calculations to be handled as a single problem.

mixed dielectric capacitor — Also called bifilm capacitor. A capacitor with a dielectric consisting of a combination of plastic film and paper.

mixed highs — In color television, the method of reproducing very fine picture detail by transmitting high-frequency components as part of the luminance signal for achromatic reproduction.

mixer — 1. In a sound transmission, recording, or reproducing system, a device having two or more inputs (usually adjustable) and a common output. The device combines the separate input signals linearly in the desired proportion to produce an output signal. 2. A circuit that generates output frequencies equal to the sum and difference of two input frequencies. 3. A device for blending two or more signals for special effects. 4. An audio control unit whose basic function is to combine two or more audio signals into a single, composite signal.

mixer diode — A diode, often associated with microwave circuits, that combines rf signals at two frequencies to generate an rf signal at a third frequency.

mixer ports — The input/output terminals of a mixer, identified as RF, LO, and IF. In most double balanced mixers, the LO and RF are either transformer or transmission line-coupled to the mixer diodes, and therefore have a limited low-frequency response, whereas the IF port is usually direct-coupled with an essentially unlimited low frequency response. In upconverting applications, the low frequency input signal is often applied to the IF port, with the higher-frequency output signal being taken from the RF port.

mixer tube — An electron tube that, when supplied with voltage or power from an external oscillator, performs the frequency-conversion function of a heterodyne conversion transducer.

mixing — 1. Combining two or more signals — e.g., the outputs of several microphones, or the received and local-oscillator signals in a superheterodyne receiver. 2. Blending of two or more signals for special effects, while exercising individual control over the volume of each. 3. Generating output frequencies that are the sum and difference of two input frequencies.

mixing amplifier — An amplifier that combines several signals, each with a different amplitude and waveshape, into a composite signal.

mixing point — In a block diagram of a feedback control loop, a symbol indicating that the output is a function of the inputs at any instant.

mksa — Meter-kilogram-second-ampere.

mksa electromagnetic system of units — Also called the Giorgi system. A system in which the fundamental units are the meter, kilogram, second, and ampere.

MLM — *See* multilayer metallization.

MLS — Abbreviation for microwave landing system. A microwave system that creates a fan-shaped beam over a wide volume of airspace so planes can use various approaches to a runway during bad weather.

mm — Letter symbol for millimeter.

mmf or **mmfd** — Abbreviation for micromicrofarad; this term is obsolete and has been replaced by picofarad.

mm Hg — Millimeters of mercury. A measure of absolute pressure, being the height of a column of mercury that the air or other gas will support. Standard atmospheric pressure will support a mercury column 760 millimeters high (760 mm Hg). Any value less than this represents some degree of vacuum.

MMIC — Abbreviation for monolithic microwave integrated circuit. A combination of active elements (diodes and transistors) and passive elements (resistors, capacitors, inductors, and transmission lines) on a single GaAs (gallium arsenide) substrate, MMICs replace conventional "chip and wire" microwave circuits. As amplifiers, attenuators, or switches at microwave frequencies, MMICs offer benefits of reduced size, lower unit cost, and higher reliability. An integrated circuit design.

mnemonic — 1. A term describing something used to assist the human memory. 2. A method of expressing complicated words, names, or phrases so as to render them easy to remember, usually by using the first letter or letters of each major syllable of the original. 3. The art of improving the efficiency of the memory (in computer storage). Abbreviation of instructions used in computer language programming. 4. An abbreviation or word that stands for another word or phrase.

mnemonic code — 1. Computer instructions written in a form the programmer can remember easily, but which must be converted into machine language later. 2. A memory jogger. 3. Computer instructions written in brief, easy-to-learn, symbolic or abbreviated form. Mnemonic code is also recognizable by the assembly program. For example, ADD, SUB, CLR, and MOV are mnemonic codes for instructions that will be executed as machine code.

mnemonic language — A programming language that is based on easily remembered symbols and that can be assembled into machine language by the computer.

mnemonic operation codes — Computer instructions that are written in a meaningful notation, for example: ADD, MPY, STO.

mnemonic symbol — A symbol chosen so that it assists the human memory; for example, the abbreviation MPY used for "multiply."

MNOS — Abbreviation for metal-nitride-oxide semiconductor.

mobile operation — Radiocommunication conducted while in motion or during halts at unspecified locations.

mobile radio service — Radio service between a fixed location and one or more mobile radio stations, or between mobile stations.

mobile radiotelephone — System providing telephone service to a station located in a mobile vehicle, using radio circuits to a base station that is connected to a central telephone office.

mobile receiver — The radio receiver in an automobile, truck, or other vehicle.

mobile-relay station — A type of base station in which the base-station receiver automatically turns on the base-station transmitter that then retransmits all signals received by the base-station receiver. Such a station is used to extend the range of mobile units and requires two frequencies for operation.

mobile service — A radiocommunication service between mobile and fixed stations, or between mobile stations. Depending on whether one or more of the earth stations are on land, sea, or air, the service is called land mobile, maritime mobile, or aeronautical mobile.

mobile station—A radio station intended for use while in motion or during halts at unspecified points. Included are hand- and pack-carried units.

mobile telemetering—Electric telemetering between moving objects, where interconnecting wires cannot be used.

mobile telephone service—1. Telephone service between a fixed base station and several mobile stations in vehicles. 2. Telephone service from mobile stations into the commercial telephone system. 3. Radiotelephone service provided to motor vehicles, railroad trains, and airplanes.

mobile transmitter—A radio transmitter installed and operated in a vessel, land vehicle, or aircraft.

mobility—Symbolized by the Greek letter mu (μ). 1. The ease with which carriers move through a semiconductor when they are subjected to electric forces. In general, electrons and holes do not have the same mobility in a given semiconductor. Also, their mobility is higher in germanium than in silicon. Normally expressed in terms of the average drift velocity attained by the type of carrier concerned per unit electric field intensity. 2. The velocity of a charged particle attained under the action of an applied electric field. Units are cm^2/V-Sec.

mobius counter—*See* Johnson counter.

modal dispersion—A fiber bandwidth-limiting factor caused by differences in the propagation characteristics of the various modes in a multimode fiber.

mod/demod—Abbreviated form of modulating and demodulating.

mode—1. One of several types of electromagnetic wave oscillation that may be sustained in a given resonant system. Each type of vibration is designated as a particular mode and has its own particular electric- and magnetic-field configurations. 2. One of several methods of exciting a resonant system. The term has also been used to describe the existence of a number of different input voltages that allow operation of a klystron at the same frequency. 3. A computer system of data representations (e.g., the binary mode). 4. The most frequent value in a series of measurements or operations. 5. The characteristic of a quantity being suitable for integer or floating-point computation. 6. The most common or frequent value in a group of values. 7. The pattern of an electromagnetic field within an optical fiber. A fiber that can propagate only one field pattern is referred to as a single-mode fiber; a fiber propagating more than one field pattern in a multimode fiber.

mode coupling—The interaction or exchange of energy between similar modes.

mode filter—A waveguide filter designed to separate waves of the same frequency but of different transmission modes.

mode hopping—The random shifting of laser output energy from one mode to another for short durations, usually in the microsecond range. The modes are generally closely spaced.

mode jump—A change in the mode of magnetron operation from one pulse to the next. Each mode represents a different frequency and power level.

model—1. An approximate representation of a process or system that tries to relate (usually mathematically) a part or all of the variables in the system so that a better understanding of the system is attained. 2. A geometrically accurate and complete representation of a real object stored in a CAD/CAM data base.

mode-locked laser—A laser that functions by modulation of the energy content of each mode internally to give rise selectively to energy bursts of high-peak power and short duration.

modem—Acronym for *mo*dulator/*dem*odulator, also known as a data set. 1. A device that transforms a typical two-level computer signal into a form suitable for transmission over the telephone network (for example, conversion of a two-level signal into a two-frequency sequence of signals). 2. A device that modulates and demodulates signals transmitted over communication facilities. 3. A device that performs modulation in the form of signal conversion, interfacing computers or computer peripheral equipment to the telephone line. Instead of trying to send a logic 0 or 1 dc voltage level over phone lines where voltage transients or noise pulses could be interpreted as false signals at the other end of the line and where transformer coupling is used, the modem changes the logic 0 or 1 into pulse audio tones. The tones travel over the phone line and enter a companion modem at the other end of the line and are converted back into 1s and 0s to properly interface and communicate with a computer or computer peripheral equipment. 4. A modulator-demodulator, whose primary function is to convert input digital data to a form compatible with basically analog transmission lines. Modulation is the digital-to-analog conversion process; demodulation is the reverse, whereby transmitted analog signals are reconverted to digital data compatible with the receiving data-handling equipment. In the process of carrying out this primary function, the modem must be compatible with the data communications equipment at its digital interface as well as with the telephone line at its analog interface. 5. A device that converts electrical signals to telephone tones and back again. The conversion occurs through acoustic coupling (placing speakers near the phone) or direct coupling to the line, which provides superior frequency response. Most modems use the RS-232C interface standard. 6. A device that converts signals from digital to analog and vice versa. The word modem is a contraction of the two main functions of the unit: modulator (converts digital to analog) and demodulator (converts analog to digital). 7. A device that modulates

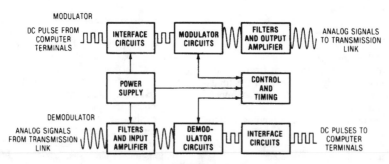

Modem.

and demodulates signals transmitted over communications facilities; that is, a device used to convert digital signals into analog (voice-like) signals for transmission over a telephone line. At the other end of the line another modem converts the analog signals back into digital form.

mode number — 1. In a reflex klystron, the number of whole cycles that a mean-speed electron remains in the drift space. 2. The number of radians of phase shift resulting from going once around the anode of a magnetron divided by 2π.

mode of failure — *See* failure mode.

mode of resonance — A form of natural electromagnetic oscillation in a resonator. It is characterized by an unvarying field pattern.

mode of transmission propagation — 1. A form of guided-wave propagation characterized by a particular field pattern that intersects the direction of propagation. The field pattern is independent of its position along the waveguide. For a uniconductor waveguide, it also is independent of frequency. 2. In a vibrating system, the state that corresponds to one of the resonant frequencies.

mode of vibration — The pattern formed by the movement of the individual particles in a vibrating body (e.g., piezoelectric crystal). This pattern is determined by the stresses applied to the body, the properties of the body, and the boundary conditions. The three common modes of vibration are flexural, extensional, and shear.

mode purity — The freedom of an atr tube from undesirable mode conversion while the tube is in its mount.

mode separation — In an oscillator, the difference in frequency between resonator modes of oscillation.

mode shift — In a magnetron, the change in mode during a pulse.

mode skip — Failure of a magnetron to fire during each successive pulse.

mode transducer — Also called mode transformer. A device that transforms an electromagnetic wave from one mode of propagation to another.

mode transformer — *See* mode transducer.

modification — The physical alteration of a system, subsystem, etc., for the purpose of changing its designed capabilities or characteristics.

modified constant-voltage charge — The charging of a storage battery in which the voltage of the charging circuit is held substantially constant, but a fixed resistance is inserted in the battery circuit, producing a rising voltage characteristic at the battery terminals as the charge progresses.

modified index of refraction — Also called modified refractive index. In the troposphere, the index of refraction at any height increased by h/a, where h is the height above sea level and a is the mean geometrical radius of the earth. When the index of refraction in the troposphere is horizontally stratified, propagation over a hypothetical flat earth through an atmosphere with the modified index of refraction is substantially equivalent to propagation over a curved earth through the real atmosphere.

modified refractive index — *See* modified index of refraction.

modifier — A device that alters an instruction but does not change the form of energy (for example, the changing of electrical input signals into electrical output signals).

moding — A defect of magnetron oscillation in which the magnetron oscillates in one or more undesired modes.

modular — 1. Made up of modules. 2. Dimensioned according to a prescribed set of size increments. 3. A partitioning scheme for a family of single-board microcomputer family circuit cards wherein system and I/O functions of various types are grouped and organized into modules (a memory module, for example). Each of the modular circuit board elements also includes a bus interface to allow intercommunication. The benefit to the system designer is that various functions can be included only to the degree needed for a specific application. 4. Having dimensions which are integral multiples of a unit of length called a module. 5. Composed of subunits. Computer systems and programs are made this way to make them easier to change or expand. 6. A design technique that permits a device or system to be assembled from interchangeable components; such a system or device can be expanded or modified simply by adding another module.

modular connector — A connector in which similar or identical sections can be assembled together to provide the best connector type or size for the application.

modulate — To vary the amplitude, frequency, or phase of a wave by impressing one wave on another wave of constant properties.

modulated amplifier — In a transmitter, the amplifier stage where the modulating signal is introduced and modulates the carrier.

modulated-beam photoelectric system — An intrusion-detector system that provides reliable beam ranges of several thousand feet. This is done by interrupting the light beam at the source with a rotating punched or slotted disc. In this way, the phototube output is converted to an ac signal, which can then be easily amplified.

modulated carrier — A radio-frequency carrier in which the amplitude or frequency has been varied by, and in accordance with, the intelligence to be conveyed.

modulated continuous wave — Abbreviated MCW. A wave in which the carrier is modulated by a constant audio-frequency tone. In telegraphy service, the carrier is keyed to produce the modulation.

modulated light — Light whose intensity has been made to vary in accordance with variations in an audio-frequency or code signal.

modulated oscillator — An oscillator whose output frequency is varied by an input signal.

modulated photoelectric alarm system — A photoelectric alarm system in which the transmitted light beam is modulated in a predetermined manner and in which the receiving equipment will signal an alarm unless it receives the properly modulated light.

modulated signal generator — A device that produces an output signal that may be changed in amplitude and/or frequency according to a desired pattern. It is calibrated in units of both power (or voltage) and frequency.

modulated stage — The radio-frequency stage to which the modulator is coupled and in which the continuous wave (carrier wave) is modulated in accordance with the system of modulation and the characteristics of the modulating wave.

modulated wave — A carrier wave in which the amplitude, frequency, or phase varies in accordance with intelligence signal being transmitted.

modulating amplifier using variable reactance — A very-high-frequency electron-beam parametric amplifier with bandpass characteristics that are independent of gain, and which is unconditionally stable.

modulating electrode — In a cathode-ray tube, an electrode to which a potential is applied to control the magnitude of the beam current.

modulating signal — *See* modulating wave.

modulating wave — Also called modulating signal, or simply signal. A wave that varies some characteristic (i.e., frequency, amplitude, phase) of the carrier.

modulation — 1. The process of modifying some characteristic of a wave (called a carrier) so that it varies in step with the instantaneous value of another wave (called a modulating wave or signal). The carrier can be a direct current, an alternating current (provided its frequency is above the highest frequency component in the modulating wave), or a series of regularly repeating, uniform pulses called a pulse chain (provided their repetition rate is at least twice that of the highest frequency to be transmitted). 2. The controlled variation of frequency, phase, and/or amplitude of a carrier wave of any frequency in order to transmit a message. 3. In fiber optics, the manner in which information is coded into light for transmission through a fiber. The modulation method may be either pulse modulation (digital) or intensity modulation (analog). 4. The process of varying one signal with another. 5. The imposing of a signal on some type of transmission or storage medium, such as a radio carrier or magnetic tape. 6. The process, or results of the process, whereby some characteristic of one signal is varied in accordance with another signal. The modulated signal is called the carrier and may be modulated in three fundamental ways: by varying the amplitude (amplitude modulation) by varying the frequency (frequency modulation) or by varying the phase (phase modulation). 7. The process by which some characteristic of a higher frequency wave is varied in accordance with the amplitude of a lower frequency wave.

modulation capability — The maximum percentage of modulation possible without objectionable distortion.

modulation code — A code used to cause variations in a signal in accordance with a predetermined scheme; normally used to alter or modulate a carrier wave to transmit data.

modulation distortion — Distortion occurring in the radio-frequency amplifier tube of a receiver when the operating point is at the bend of the grid-voltage/plate-current characteristic curve. As a result, the plate-current changes are greater on positive than on negative half-cycles. The effect is equivalent to an increase in the percentage of modulation.

modulation envelope — A curve, drawn through the peaks of a graph, showing how the waveform of a modulated carrier represents the waveform of the intelligence carried by the signal. The modulation envelope is the intelligence waveform.

Modulation envelope.

modulation factor — In an amplitude-modulated wave, the ratio of half the difference between the maximum and minimum amplitudes to the average amplitude. This ratio is multiplied by 100 to obtain the percentage of modulation.

modulation frequency — That signal which causes the output frequency of an oscillator to be modulated.

modulation index — 1. In frequency modulation with a sinusoidal modulating wave, the ratio of the frequency deviation to the frequency of the modulating wave. 2. A measure of the degree of modulation.

modulation indicator — A relative indicator that glows brighter and brighter as the modulation level approaches 100 percent. It gives the operator some idea of how fully he or she is modulating the carrier.

modulation meter — An instrument that measures the modulation percentage of an amplitude modulated signal and displays the results on a scale or dial.

modulation monitor — *See* modulation meter.

modulation noise — Also called noise behind the signal. 1. The noise caused by the signal, but not including the signal. 2. The noise produced on playback of a tape that is a function of the instantaneous amplitude of the signal. It is caused by poor particle dispersion and surface irregularities.

modulation percentage — The modulation factor multiplied by 100.

modulation plan — The arrangement by which the individual channels or groups of channels are modulated to a final frequency allocation.

modulation ratio — For an electrically modulated source, the number obtained by dividing the percentage of radiation modulation by the percentage of current modulation.

modulation rise — An increase in the modulation percentage. It is caused by a nonlinear tuned amplifier, usually the last intermediate-frequency stage of a receiver.

modulation transfer function — Abbreviated MTF. Also called sine-wave response, and contrast transfer function. 1. The ratio of input modulation to output modulation. 2. A plot of the video output amplitude as a function of TV line number for a sine-wave- modulated input, i.e., a bar pattern with sine-wave density variation serving as the object for the TV camera. The output video amplitude, measured by an oscilloscope, is calibrated at 100 percent for small line numbers (wide lines). 3. The function, usually a graph, describing the modulation of the image of a sinusoidal object as the frequency increases. The ratio of the modulation in the image to that in the object.

modulator — 1. A device that effects the process of modulation. 2. In radar, a device for generating a succession of short energy pulses that cause a transmitter tube to oscillate during each pulse. 3. An electrode in a spacistor.

modulator crystal — A crystal that is used to modulate a polarized light beam by the use of Pockel's effect. Useful as a modulator in laser systems.

modulator driver — A transmitter circuit that produces a pulse to be delivered to the control grid of the modulator stage.

modulator glow tube — A cold-cathode recorder tube used for facsimile and sound-on-film recording. It provides a modulated, high-intensity, point-of-light source.

modulator stage — The last stage through which the signal that modulates the radio-frequency wave is passed.

module — 1. A unit in a packaging scheme displaying regularity and separable repetition. It may or may not be separate from other modules after initial assembly. Usually all major dimensions are in accordance with a prescribed series of dimensions. 2. A packaging concept in which identical forms are used. 3. A complete sub-assembly of a larger system combined in a single package. 4. An interchangeable plug-in item containing electronic components, which may be combined with other interchangeable items to form a complete unit. 5. A collection of modules or a collection of software items in any combination in computer software. 6. A section of software that has well-defined inputs and outputs and may be tested independently of other software. 7. An encapsulated all-solid-state control circuit. 8. An equipment unit capable of being combined with other similar units to form a larger unit. 9. Any assembly of interconnected components that constitutes an identifiable item, device, instrument, or piece of equipment.

modulo — A term used to express the maximum number of states for a counter; this term is used to describe several packet-switched network parameters, such as packet number (usually set to modulo 8, counted from 0 to 7). When the maximum count is exceeded, the counter is reset to 0.

modulo-*n* counter — A counter with *n* unique states. *See also* programmable counter.

modulus — Abbreviated mod. 1. An integer designating the number of states through which a counter sequences during each cycle. 2. The number of distinct states that a counter has. For example, the modulo-10 counter has a modulus of 10 and therefore has 10 distinct states.

moiré — 1. A wavy or satiny effect produced by the convergence of lines. Usually appears as a curving of the lines in the horizontal wedges of the test pattern and is most pronounced near the center, where the lines forming the wedges converge. A moiré pattern is a natural optical effect when converging lines in the picture are nearly parallel to the scanning lines. To a degree this effect is sometimes due to the characteristics of color picture tubes and of image-orthicon pickup tubes (in the latter, it is termed mesh beat). 2. A coarse pattern of shading that occurs in a facsimile system when half-tone material is scanned. 3. In television, the spurious pattern in the picture resulting from interference beats between two sets of periodic structures in the image.

moiré effect — A fringe pattern arising from the interference between two superimposed line patterns. In a monitor it comes from the interference between the shadow mask pattern and the video information (video moiré) or between the shadow mask pattern and the horizontal line pattern (scan moiré). Autoscan (MultiSync) monitors, which operate over a range of scanning frequencies, may sometimes exhibit moiré in certain video modes.

moisture absorption — The amount of moisture (in percentage) that an insulation will absorb. The figure should be as low as possible when the insulation is to be used in a moist environment.

moisture repellent — Having properties such that moisture will not penetrate.

moisture resistance — The ability of a material not to absorb moisture, either from the air or from being immersed in water.

moisture resistant — Having characteristics such that exposure to a moist atmosphere will not readily lead to a malfunction.

mol — Abbreviation for molecular weight.

mold — In disc recording, a metal part derived from a master by electroforming. It is a positive of the recording (i.e., it has grooves similar to those of a recording and thus can be played).

molded capacitor — A capacitor that has been encased in a molded plastic insulation.

molded carbon potentiometer — A potentiometer in which a carbon resistance element and all other parts of the potentiometer are molded together.

molding — A process using pressure to apply accurately dimensional, specifically shaped jackets to parts or assemblies. This process usually uses costly heavy-machined metal molds.

molectronics — Abbreviation for molecular electronics.

molecular circuit — 1. *See* monolithic integrated circuit. 2. An electronic circuit demonstrating a measurable input and output, where the portions performing various functions such as resistance or capacitance are indistinguishable as discrete areas.

molecular circuitry — *See* morphological circuitry.

molecular clock — A device for time measurement based on an electromagnetic oscillation of extremely stable frequency from a beam of the rotation spectrum, the vibration or inversion of a particular molecule, and a counting device.

molecular electronics — Abbreviated molectronics. 1. The science of making a single block of matter perform the function of a complete circuit. This is done by merging the function with a material, using solid-state functional blocks. 2. Electronics on a molecular scale, dealing with the production of complex circuitry in semiconductor devices with integral elements processed by growing multizoned crystals in a furnace for the ultimate performance of electrical functions.

molecular integrated circuit — An integrated circuit such that the identity and location of specific electric elements cannot be determined even by microscopic disassembly of the material of which the circuit is formed. In contrast to a conventional microelectronic circuit, the molecular integrated circuit can be defined only by function, which in turn can be described only by mathematical models and incremental circuit representation.

molecular technique — A practice method of causing a single piece of material or a crystal to provide a complete circuit function.

molecular weight — Abbreviated mol. Also known as relative molecular mass. The sum of the weights of all the atoms in a molecule, expressed relative to $1/12$ of the mass of a carbon-12 atom.

molecule — In any substance, the smallest particle that still retains the physical and chemical characteristics of that substance. A molecule consists of one or more atoms of one or more elements. Sometimes two entirely different substances may have similar chemical elements, but their atoms will be arranged in a different order.

molybdenum — A metallic element (chemical symbol, Mo; atomic number, 42; atomic weight, 95.95) sometimes used for the grid and plate electrodes of vacuum tubes.

momentary contact — Closure of a normally open switch for a brief period.

momentary contact switch — A mercury switch designed to make contact for only a brief transitory interval while the mercury moves from one extreme position to another.

momentary loss of power — A short interruption of power to the total equipment.

momentary start — An electrical signal for starting a timer that is of any duration shorter than the time setting.

momentary switch — 1. A switch that returns to its normal circuit condition when the actuating force is removed. 2. A spring-loaded contact that, when pressed, closes two contacts. When pressure is removed, contacts open.

monaural — Describing sound reproduction using only one source of sound and giving a monophonic effect. *See* monophonic.

monaural channel unbalance — The ratio of the outputs from the right and left channels with a monaural signal applied to the input.

monaural recorder — Literally, a tape recorder intended for listening with one ear only; however, in popular usage refers to single-channel recorders, as distinguished from multichannel (stereophonic, binaural, etc.) types. More correctly but less universally called monophonic recorder.

monitor — 1. To listen to a communication service, without disturbing it, to determine its freedom from trouble or interference. 2. A device (e.g., a receiver, oscilloscope, teleprinter, etc.) used for checking signals. 3. A software package or a hardware device that can be used to measure the performance of a system or the utilization of specific devices. 4. Any device used to observe or measure a parameter. 5. Any device for listening incidentally to an audio signal that is primarily directed to some other purpose at that moment. A monitor loudspeaker is used for auditioning a recording or radio program incidentally to its committal to tape or its broadcast. 6. The operator of a television monitoring system who selects one out of several camera images for broadcasting. 7. A TV set without a tuner used to directly display the composite video signal from a camera, videotape recorder, or special-effects generator. 8. A controller of the operation of the various programs available, the monitor can access the editor, assembler, or other programs. 9. A program that controls a computer's basic operation, telling it how and where to acquire the programs and data, where to store them, and how to run them. 10. A device for video viewing connected directly to the camera output. A true monitor does not incorporate channel selector components or audio components.

monitored fast forward — A feature in a cartridge deck whereby the playback amplifier is left on at low volume during fast forward so the user can hear the program running through at the faster speed, to spot or cue up to the desired program.

monitor head — A playback head that is separated from the record head, enabling the recordist to listen to what is coming off the tape a fraction of a second after it has been recorded and while the recording is still in progress. Without a monitor head, a tape must be recorded to its end and then rewound and replayed before the recordist can evaluate the tape. On some cassette decks with monitor capability, the monitor head is not completely separate, but is built into the same shell as the record head.

monitoring — 1. Observing the characteristics of transmitted signals as they are being transmitted. 2. Listening to a communication service without disturbing it to determine its quality or freedom from trouble or interference.

monitoring amplifier — A power amplifier used primarily for evaluation and supervision of a program.

monitoring key — A key that, when operated, permits an attendant or operator to listen on a telephone circuit without causing appreciable impairment of transmission on the circuit.

monitoring radio receiver — A radio receiver for checking the operation of a transmitting station.

monitors — Programs that control the operation of an entire computer system. They often contain routines that tell the computer how to communicate with the outside world and how to allocate resources.

monitor systems — Programs that supervise other programs and keep computers functioning efficiently with a minimum of assistance from human operators.

monkey chatter — Garbled speech or music heard along with a desired program. This interference occurs when the side frequencies of an adjacent-channel station beat with the signal from the desired station.

monoboard microcomputer — See single-board microcomputer.

monobrid — A method of manufacturing an integrated circuit by using more than one monolithic chip within the same package.

monobrid circuit — An integrated circuit using a combination of monolithic and multichip techniques by means of which a number of monolithic circuits or a monolithic device in combination with separate diffused or thin-film components are interconnected in a single package.

monochromatic — 1. Pertaining to or consisting of a single color. 2. Radiation of a single wavelength.

monochromatic emissivity — See total emissivity.

monochromaticity — The degree of response to one color.

monochromatic light — Light consisting of just one wavelength. No light is completely monochromatic. The closest approach is particular lines in the mercury 198 spectrum excited in a discharge tube with no electrodes.

monochromatic sensitivity — The response of a device to light of a given color only.

monochromator — An instrument used to isolate narrow portions of the spectrum by making use of the dispersion of light into its component colors.

monochrome — Also called black-and-white in referring to television. 1. Having only one chromaticity — usually achromatic, or black and white and all shades of gray. 2. Black and white with all shades of gray.

monochrome channel — In a color television system, any path intended to carry the monochrome signal (although it may carry other signals also).

monochrome channel bandwidth — The bandwidth of the path that carries the monochrome signal.

monochrome signal — 1. In a monochrome TV transmission, the signal wave that controls the luminance values in the picture. 2. In a color-television transmission signal wave, the portion with major control of luminance, whether displayed in color or monochrome.

monochrome television — Also called black-and-white television. Television in which the final reproduced picture is monochrome. That is, it has only shades of gray between black and white.

monochrome transmission — Also called black-and-white transmission. 1. In television, the transmission of a signal wave that represents the brightness (luminance) values in the picture, but not the color (chrominance) values.

monoclinic — A crystal structure in which two of the three axes are perpendicular to the third, but not to each other.

monocord switchboard — A local-battery telephone switchboard in which each line terminates in a single jack and plug.

monocrystal — A crystal of material that has a continuous lattice structure and orientation throughout its volume, in contrast with the multigrain structure of a polycrystal. Almost all semiconductor devices are fabricated from monocrystalline material.

monocrystalline — Material made up of a single continuous crystal.

monoergic — A type of emission in which the particles or radiations are produced with a small energy spread (i.e., a "line spectrum").

monofier — A complete master oscillator and power amplifier system contained in a single evacuated envelope. It is equivalent electrically to a stable low-noise oscillator, an isolator, and a two- or three-cavity klystron amplifier.

monofilament — A single-strand filament, as opposed to a braided or twisted filament.

monogroove stereo — Also called single-groove stereo. A stereo recording in which both channels are contained in one groove.

monolithic — 1. Existing as one large, undifferentiated whole. 2. Refers to the single slice of silicon substrate on which an integrated circuit is built; hence, monolithic integrated circuit. 3. Elements or circuits formed within a single semiconductor substrate.

monolithic ceramic capacitor — An electrostatic capacitor that is constructed by cofiring alternate layers of metal (electrodes) with carefully selected ceramic (dielectric) materials.

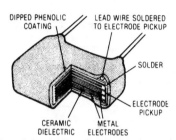

DIPPED PHENOLIC COATING — LEAD WIRE SOLDERED TO ELECTRODE PICKUP

SOLDER

ELECTRODE PICKUP

CERAMIC DIELECTRIC   METAL ELECTRODES

(Alternately deposited layers of ceramic dielectric material and metal electrodes fired into a single homogenous block)

Monolithic ceramic capacitor.

monolithic circuit — A monolithic semiconductor integrated circuit has all circuit components manufactured in or on top of a single crystal semiconductor material. Interconnections between the component parts within a given circuit are made by means of metallization patterns, and the individual parts are not separable from the complete circuit. Hence, the monolithic circuit is often referred to as a fully integrated circuit.

monolithic filter — A filter whose operations are based on the use of deposited electrode pairs acting as shear or thickness-mode resonators separated by nonelectrode regions that act as acoustic coupling elements. The entire filter is on a single quartz or ceramic substrate.

monolithic hybrid — Hybrid circuit built on a multilayer semiconductor substrate. Semiconductor elements are actually formed within the substrate in a process much like that used to build a monolithic IC. Additional components can be soldered or chemically bonded to the substrate, but components usually are fabricated within the substrate.

monolithic IC — *See* monolithic integrated circuit.

monolithic integrated circuit — Abbreviated MIC. 1. An integrated circuit whose elements are formed *in situ* upon or within a semiconductor substrate, with at least one of the elements formed within the substrate. *See also* integrated circuit, 1. To further define the nature of a monolithic integrated circuit, additional modifiers may be prefixed. Examples: pn junction-isolated monolithic integrated circuit, dielectric-isolated monolithic integrated circuit, or beamlead monolithic integrated circuit. 2. The physical realization of electronic circuits or subsystems from a number of extremely small circuit elements inseparably associated on or within a continuous body or a thin film of semiconductor material. 3. An integrated circuit that is fabricated completely on a single chip and which contains no thin film or discrete components. (As opposed to hybrid IC.) 4. A complete electronic circuit fabricated as an inseparable assembly of circuit elements in a single small structure. It cannot be divided without permanently destroying its intended electronic function.

5. An integrated circuit in which all components are created by pn junctions grown on a single chip of silicon crystal. 6. An electronic circuit formed within a single small chip of crystalline semiconductor material, usually silicon. Typically the chip is contained in a plastic or ceramic package. Electrical connection to package leads is made by fine wire, which is welded to metal pads on the chip and to the package leads. The advantages of integrated circuits include small size and extremely high reliability.

monolithic microcircuit — *See* monolithic integrated circuit.

monolithic microwave integrated circuit — *See* MMIC.

monolythic™ capacitor — A trademark of the Sprague Electric Company for a multilayer monolithic ceramic capacitor.

monomer — A single molecule that can join with another monomer or molecule to form a polymer or molecular chain.

monomode fiber — An optical fiber used for extremely long distance needs, with a core so tiny that all of the photons essentially have to travel in a straight line. These fibers may be used in transatlantic cable, with repeaters located at 12-mile (19-km) intervals. Since the photons have little chance to disperse, the light pulses fed into the fiber can in theory be so very rapid — 100 billion bits a second — that a single fiber could carry the digital equivalent of tens of thousands of voices simultaneously. (However, no laser now known can flash fast enough to exploit fully the capacity of a monomode fiber.)

monophonic — Also called by the older term monaural. 1. Pertaining to audio information on one channel (i.e., as opposed to binaural or stereophonic). Monophonic and monaural are usually, although not necessarily, associated with a one-speaker system. 2. Single-channel sound.

monophonic recorder — *See* monaural recorder.

monopinch — Antijam application of the monopulse technique in which the error signal is used to provide discrimination against jamming signals.

monopole — A stub antenna fed against a ground plane.

monopole antenna — A vertical antenna with a voltage node at the lower end and a current node at the top that is 0.25 wavelength at the operating frequency.

monopulse — A method of determining azimuth and elevation angles simultaneously.

monopulse radar — Radar using a receiving antenna system having two or more partially overlapping lobes in the radiation pattern. Sum and difference channels in the receiver compare the amplitudes or phases of the antenna outputs.

monopulse tracking — A form of telemetry tracking that compares the phase of the carrier received simultaneously at four points on a plane (two elevation and two azimuth references) and positions the antenna so that all four signals are in phase, or equidistant from the source of radiation. A variation of the monopulse tracking technique is the scan-coded or single-channel monopulse tracking system.

monorange speaker — A speaker that provides the full spectrum of audio frequencies.

monoscope — Also called phasmajector or monotron. An electron-beam tube in which the picture signal is generated by scanning an electrode, parts of which have different secondary-emission characteristics.

monoscope cathode-ray tube — A character-generator CRT that functions on the principle of secondary emission. The target holds a set of aluminum characters, select characters being scanned by the electron beam. The secondary emission from the target is gathered by the collector and becomes a video signal to be transmitted to and converted in other cathode-ray tubes.

monospacing — Method of printing in which each character printed takes up the same amount of space horizontally, irrespective of the size of the character.

monostable — Also called single-shot or one-shot. 1. A term used to describe a circuit that has one permanently stable state and one quasi-stable state. An external trigger causes the circuit to undergo a rapid transition from the stable state to the quasi-stable state, where it remains for a time and then spontaneously returns to the stable state. 2. A type of multivibrator that has one stable state. The integrated-circuit version usually includes input gating and sometimes a Schmitt trigger. 3. A system that has an at-rest bias causing one output condition consistently until appropriate input signaling occurs.

monostable circuit — A circuit capable of two states of operation in which only one such state is stable; i.e., if triggered into the unstable state, the circuit will (after a controllable time interval) return to the stable state without external intervention.

monostable multivibrator — Also called one-shot multivibrator, single-shot multivibrator, or start-stop multivibrator. 1. A circuit having only one stable state, from which it can be triggered to change the state, but only for a predetermined interval, after which it returns to the original state. 2. A multivibrator having one stable and one semistable condition. A trigger is used to drive the unit into the semistable state, where it remains for a predetermined time before returning to the stable condition. 3. A half-analog/half-digital circuit that produces a pulse on its output in response to a trigger signal at its input. The output pulse width is determined by a resistor-capacitor network. 4. A multivibrator that delivers one output pulse, of adjustable width, for each input pulse received.

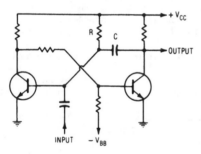

Circuit.

$$t_p = 0.69RC$$

Output.

Monostable multivibrator.

monostatic reflectivity — The characteristic of a reflector that reflects energy only along the line of the incident ray (e.g., a corner reflector).

monotone — A single musical tone unvaried in pitch.

monotonic DAC — A digital-to-analog converter that has an analog output that is a continuously increasing function of the input.

monotonicity — 1. A measure of the ability of a converter (either d/a or a/d) to produce an output in response to a continuously increasing input signal (a/d) or count (d/a) without decreasing in value or skipping codes. 2. Parameter for a d/a converter. It means that there is always an increase in the output level for each increase in digital input. If the output decreases, the converter is nonmonotonic and can seldom be used to its full resolution. 3. In a d/a converter, if the output analog signal either increases or stays the same for an increase in input digital code, it is termed monotonic. In an a/d converter, if the output digital code increases or stays the same for a 1-lsb increase in input voltage, it is termed monotonic. If a data converter's differential linearity lies within ±1 lsb, the device will be monotonic. Monotonicity is especially important in control loops where convergence is necessary. 4. In analog-to-digital conversion, indicates that the digital output never exhibits a decrease between one conversion and the next, as the analog input increases steadily from zero to rated maximum.

monotron — *See* monoscope.

Monte Carlo analysis — A mathematical technique that, by randomly changing component values, predicts the effect of statistical variations on circuit performance.

Monte Carlo method — Abbreviated MCM. 1. A computer technique in which a number of possible models under study are mathematically constructed from constituents selected at random from representative populations. 2. Any procedure that involves statistical sampling techniques in order to obtain an approximate solution of a mathematical or physical problem.

Monte Carlo simulation — A subset of digital simulation models based on random or stochastic processes.

MOPA — Acronym for master oscillator-power amplifier.

morphological circuitry — Also called molecular circuitry. A circuit made from a material in which the molecular structure has been arranged to perform a certain electrical function.

Morse code — A system of dot-and-dash signals developed by Samuel F. B. Morse and now used chiefly in wire telegraphy.

Morse code.

Morse sounder — 1. A telegraph receiving instrument that produces a sound at the beginning and end of each dot and dash. From these sounds, a trained operator can interpret the message. 2. A device consisting of an electromagnet with a spring-loaded armature that is used for the reception of Morse coded signals. The start of a mark signal produces a sharp "click" sound, and the end of the signal a "clack" sound.

Morse telegraphy — Telegraphy in which the Morse code or its derivative is used — specifically, the International (also called continental) or American Morse code.

MOS — Abbreviation for metal-oxide semiconductor.

mosaic — The light-sensitive surface of an iconoscope or other television camera tube. In one form, it consists of millions of tiny silver globules on a sheet of ruby mica. Each globule is treated with cesium vapor to make it sensitive to light. The globules retain a positive charge when bombarded with light and absorb electrons from a scanning beam in proportion to the amount of light they have received.

mosaic detector — A device in which a number of active elements are arranged in an array. It is generally used as an imaging device.

mosaic-lamp display — A form of display made up of seven straight line segments that can be used to form all digits. An individual lamp behind each segment lights to form each digit.

MOS capacitor — A capacitor in which a silicon-oxide dielectric layer and a metal top electrode are deposited on a conducting semiconductor region that acts as the bottom electrode.

MOS device — Semiconductor component (typically, in an IC) formed by metal-oxide semiconductor process. In an MOS device, current can flow in either of two directions, whereas in a bipolar semiconductor current flows in only one direction.

MOSFET — Abbreviation for metal-oxide semiconductor field-effect transistor. Sometimes called insulated-gate field-effect transistor (IGFET).

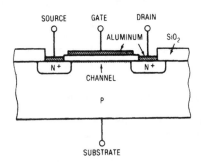

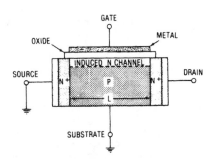

MOSFET (n-channel enhancement type).

MOS insulated-gate field-effect transistor — A semiconductor device consisting of two electrodes (source and drain) diffused into a silicon substrate and separated by a finite space, thus forming a majority-carrier conducting channel. A metal electrode (gate) is placed above and insulated from the channel.

MOS monolithic IC — A single-chip integrated circuit, consisting largely of interconnected unipolar active-device elements or MOS field-effect devices. This class of circuits results in higher equivalent parts density.

most significant bit — Abbreviated MSB. 1. The highest-order bit or the bit with the greatest weight. 2. The bit in the leftmost position.

most significant digit — Abbreviated MSD. 1. Number at the extreme left of a group of numbers. Example, 6937. Digit 6 is the MSD. Does not apply to number 0 when in the MSD position. 2. In a binary number, it is the digit with the highest weighting.

mother — In disc recording, a mold electroformed from the master.

motherboard — 1. The main board in a computer, into which the circuits are plugged. 2. The main board in a computer that holds the CPU chip, the ROM, the RAM, and sometimes coprocessors. 3. A circuit board that accommodates plug-in cards or daughterboards and makes appropriate interconnecting terminations between them. 4. A printed circuit board to which one or more other circuit boards may be assembled and connected. 5. The main printed circuit board equipped with female connectors in which all functional boards are inserted. It carries the system buses. *See also* backplane.

mother crystal — The quartz crystal found in nature. It has the characteristic geometric design of a crystal (i.e., flat faces at definite angles to each other), but all or some of the faces may be worn because of abrasion with stones or other objects.

motional feedback — Correction of a speaker's response by feeding information about its motion back to the amplifier. The amplifier then compares the speaker's motions with its own output and changes this output so as to counteract any changes (distortions) created by the speaker. Such speakers usually have special servo amplifiers built in, though a few can be used with other amplifiers if these are modified.

motional impedance — Also called loaded motional impedance. The complex remainder after the blocked impedance of a transducer has been subtracted from the loaded impedance.

motion detector — A device used in security systems that reacts to any movement on a CCTV monitor by automatically setting off an alarm when the monitor is not manned.

motion frequency — 1. The natural frequency of a servo system. 2. The frequency at which a given servo tends to oscillate.

motion-picture pickup — A television camera or technique for televising scenes directly from motion-picture film.

motion-sensing — A type of tape transport in which certain actions that could break or spill the tape are prevented or delayed until the instant the tape has come to a stop or reached a speed that allows the action to take place safely.

motion sensor — A sensor that responds to the motion of an intruder. *See also* infrared motion detector; radio-frequency motion detector; sonic motion detector; ultrasonic motion detector.

motor — 1. A device that moves an object. Specifically, a machine that converts electric energy into mechanical energy. 2. A device that converts continuous electric current into a regular rotary motion.

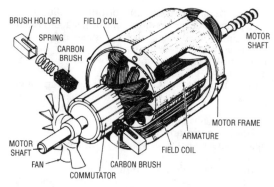

Motor.

motor board — Also called tape-transport mechanism. The platform or assembly of a tape recorder on which the motor (or motors), reels, heads, and controls are mounted. It includes parts of the recorder other than the amplifier, preamplifier, speaker, and case.

motorboating — 1. Interference heard as the characteristic "putt-putt" made by a motorboat. It is due to self-oscillation, usually pulsating, in an amplifier below or at a low audio frequency. 2. A generally periodic, relatively low-frequency pulse disturbance of the output voltage of a regulated power supply, frequently line- or load-dependent, unstable, and significantly large, as in the case of oscillations.

motor circuit switch — See disconnect switch.

motor controller — A device or group of devices that serves to govern, in a predetermined manner, the electrical power delivered to a motor.

motor converter — A device for converting an alternating current to a pulsating direct current. It consists of an induction motor to which an ac supply is connected. The armature of the induction motor is linked mechanically to the armature of a synchronous converter, which is connected to a dc circuit.

motor-driven relay — A relay in which the contacts are actuated by the rotation of a motor shaft.

motor effect — The repulsion force exerted between adjacent conductors carrying currents in the opposite direction.

motor element — That portion of an electroacoustic receiver that converts energy from the electrical system into mechanical energy.

motor-field control — The method of controlling the speed of a motor by changing the magnitude of its field current.

motor field-failure relay — Also called field loss relay. A relay that functions to disconnect the motor armature from the line in the event of loss of field excitation.

motor field induction heater — An induction heater in which the inducing winding typifies that of a rotary or linear induction motor.

motor-generator set — A motor-generator combination for converting one kind of electric power to another (e.g., alternating current to direct current). The two are mounted on a common base and their shafts are coupled together

motorized lens — A camera lens fitted with a small electric motor that can focus the lens, open the diaphragm, or, in the case of a zoom lens, change the focal length, all by remote control.

motor junction (conduit) box — An enclosure on a motor for the purpose of terminating a conduit run and joining the motor to power conductors.

motor lead wire — Wire that connects to the usually fragile and easily damaged magnet wire found in coils, transformers, and stator or field windings. General requirements are abrasion resistance, toughness, flexibility, dielectric strength, thermal resistance, and low percentage of extractables (where applicable, such as in hermetic wires).

motor meter — A meter comprising a rotor, one or more stators, and a retarding element that makes the speed of the rotor proportionate to the quantity being measured (e.g., power or current). A register, connected to the rotor by suitable gearing, counts the revolutions of the rotor in terms of the total.

motor-operated sequence switch — A multi-contact switch that fixes the operating sequence (or the major devices) during starting and stopping, or during other sequential switching operations.

motor-run capacitor — A capacitor that is left in the auxiliary motor winding, and which is in parallel with the main winding to obtain a higher power factor and efficiency.

motor-start capacitor — A capacitor that is in the circuit only during the starting period of a motor. The capacitor and its auxiliary winding are disconnected automatically by a centrifugal switch or other device when the motor reaches a predetermined speed, after which the motor runs as an induction motor.

motor starter — A device arranged to start an electric motor and accelerate it to normal speed; a motor starter has no running position other than fully on. It is a combination of all the switching means required to start and stop the motor, and it incorporates suitable overload protection.

mount — The flange or other means by which a switching tube, or a tube and cavity, are connected to a waveguide.

mount structure — The essential elements of a vacuum tube except the envelope.

mouse — A cigarette-pack-size hemispherical or rectangular input device moved with one hand on a horizontal surface that causes corresponding movements of a cursor on a computer's monitor. A mouse will usually have one or more finger-operated switches so that it can generate both on-off and positional input. The mouse may be connected to the computer by a cord or it may be wireless.

mousepad — A pad used under a mouse to provide more traction for the mouse ball.

mouth of a horn — The end having the larger cross section.

M-out-of-N code — A type of fixed-binary code in which M of the N digits are always in the same state.

MOV — Abbreviation for metal-oxide varistor. A varistor having a sintered zinc-oxide element and a symmetrical voltage-current characteristic. Such devices provide bidirectional transient suppression capability, enabling them to protect circuits against transient overvoltage occurring from opposite directions. These devices absorb very large amounts of energy — up to 10 K joules.

movable contact — The one of a pair of contacts that is moved directly by the actuating system.

movement differential — The distance or angle from the operating position to the releasing position of a momentary contact switch.

move mode — In some variable-word-length computers, data transmission in which certain delimiters are not moved with the data (as opposed to load mode).

moving average — An arithmetic average of the n most recent observations. As each new observation is added, the oldest one is dropped. The value of n, the number of periods to use for the average, reflects responsiveness versus stability in the same way that the choice of smoothing constant does in exponential smoothing.

moving-coil galvanometer — A galvanometer in which the moving element is a suspended or pivoted coil.

moving-coil meter — A meter in which a coil pivots between permanent magnets.

moving-coil microphone — Also called a dynamic microphone. 1. A moving-conductor microphone in which the diaphragm is attached to a coil positioned in a fixed magnetic field. The sound waves strike the diaphragm, moving it, and hence the coil, back and forth. An audio-frequency current is induced in the moving coil in the magnetic field and coupled to the amplifier. 2. A microphone using a permanent magnet and a vibrating coil or ribbon as its transducing system.

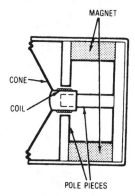

Moving-coil microphone.

moving-coil pickup — Also called dynamic pickup or reproducer. 1. A phonograph pickup in which a conductor or coil produces an electric output as it moves back and forth in a magnetic field. 2. A type of magnetic cartridge in which the coils, connected to the stylus, move within a stationary magnetic field. Output from such cartridges is low, and the stylus cannot usually be replaced by the user, but some users feel the sound quality outweighs these inconveniences.

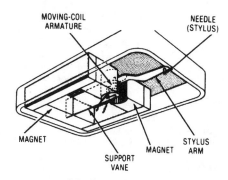

Moving-coil pickup.

moving-coil speaker — Also called a dynamic speaker. A speaker in which the moving diaphragm is attached to a coil, which is conductively connected to the source of electric energy and placed in a constant magnetic field. The current through the coil interacts with the magnetic field, causing the coil and diaphragm to move back and forth in step with the current variations through the coil.

moving-conductor microphone — A microphone that produces its electric output from the motion of a conductor in a magnetic field.

moving-conductor speaker — A speaker in which a conductor is moved back and forth in a steady magnetic field. The cone is moved by the reaction between the magnetic field and the current in the conductor.

moving contact — The portion of a relay or switch that moves toward or away from a fixed contact.

moving element — The portion of an instrument that moves as a direct result of a variation in the electrical quantity being measured by the instrument. One-half the weight of the springs (when used) is included in the weight of the moving element.

moving-head disk — An economical disk drive using a single head to access all the tracks.

moving-iron instrument — An instrument in which the current in one or more fixed coils acts on one or more pieces of soft iron or magnetically similar material, at least one of which is movable. The various forms of this instrument (plunger, vane, repulsion, attraction, repulsion-attraction) are distinguished chiefly by their mechanical construction. Otherwise, the action is the same.

moving-magnet instrument — An instrument in which a movable permanent magnet aligns itself in the field produced by another permanent magnet and an adjacent coil or coils carrying current, or by two or more current-carrying coils.

moving-magnet magnetometer — A magnetometer in which the torques act on one or more permanent magnets, which can turn in the field to be measured. Some types use auxiliary magnets (gaussian magnetometer); others use electric coils (sine or tangent galvanometer).

moving-target indicator — Abbreviated MTI. A device that limits the display of radar information primarily to moving targets.

moving-vane meter — *See* magnetic-vane meter.

moving-vane movement — Also known as moving-iron movement. A meter movement consisting of fixed and movable iron vanes, surrounded by a field coil. A magnetic field produced by current in the field coil causes repulsion between the two vanes. Deflection of the movable vane is proportional to the current.

MPC — Abbreviation for multimedia personal computer. Any workable combination of multimedia technologies.

MPU — Abbreviation for microprocessor unit.

MPX filter — Abbreviation for multiplex filter. 1. A circuit to remove 19-kHz tones from a signal to be recorded, in order to prevent audible interference between a tape recorder's bias signal and the 19-kHz pilot tone in the output signal from a stereo FM tuner or receiver. Some receivers and tuners have such filters built in. 2. A circuit that removes 19-kHz fm pilot tones from a signal.

MRN — Abbreviation for minimum reject number.

mR/min — Abbreviation for milliroentgens per minute.

ms — Letter symbol for millisecond.

MSB — Abbreviation for most significant bit.

M-scan — Also called M-display. An A-scan radar display in which the target distance is determined by moving a pedestal signal along the base line until it coincides with the horizontal position of the target-signal deflection. The control that moves the pedestal is calibrated in distance.

MSD — Abbreviation for most significant digit.

MS-DOS — Abbreviation for Microsoft Disk Operating System. The operating system for the IBM PC and compatible personal computers.

msec — Abbreviation for millisecond.

MSI — Abbreviation for medium-scale integration. A term generally applied to integrated circuit chips containing 10 or more gate equivalents, but less than 100. Also applies to memory devices with fewer than 1 K (1024) bits of memory.

MSI- or LSI-component processor — A processing unit built around medium-scale and/or large-scale integrated circuits, as opposed to one built around a monolithic processor, such as a microprocessor.

MSS — Abbreviation for mobile-satellite service. A service that links mobile earth stations with base stations and with one another via one or more satellites.

M-S stereo system — See mitte-seite stereo system.

MT — Also called MTST. The IBM magnetic tape Selectric typewriter.

MTBF — Abbreviation for mean time between failures.

MTE — Abbreviation for miles to empty. An electronic digital dashboard readout that indicates the number of miles to drive before the gas tank is empty (Ford Motor).

MTI — Abbreviation for moving-target indicator.

MTNS — Abbreviation for metal-thick-nitride semiconductor.

MTOS — Abbreviation for metal-thick-oxide semiconductor.

MTS — Abbreviation for message telecommunications service. 1. Services available to the public over the nationwide switched network. An interstate call is one example. 2. The official name for long-distance or toll service.

MTTF — Abbreviation for mean time to failure.

MTTFF — Abbreviation for mean time to first failure.

MTTR — Abbreviation for mean time to repair. Average time to repair a failure under the operating conditions encountered.

M-type backward-wave oscillator — A crossfield injected-beam oscillator. The electrons in this device interact with an rf wave traveling backward or opposite to the electron beam. It is efficient, broadband, and can be voltage-tuned. It is also insensitive to load variations.

mu — English spelling for the Greek letter μ.

μ — Greek letter mu. 1. Symbol for amplification factor. 2. Symbol for permeability. 3. Letter symbol for the prefix micro- (10^{-6})

μ**A** — Letter symbol for microampere.

mu-circuit — In a feedback amplifier, the circuit that amplifies the vector sum of the input signal and the feedback portion of the output signal in order to generate the output signal.

MUF — Abbreviation for maximum usable frequency.

mu-factor — Ratio of the changes between two electrode voltages, assuming the current and all other electrode voltages are maintained constant—i.e., it is a measure of the relative effect that the voltages on two electrodes have on the current in the circuit of a specified electrode.

μ**H** — Letter symbol for microhenry.

muldem — Acronym for *mu*ltiplexer/*dem*ultiplexer.

Muller tube — A thermionic vacuum tube having an auxiliary cathode or grid connected internally to the main cathode through a high-value resistor.

multiaddress — Pertaining to computer instructions that specify two or more addresses.

multianode microchannel array detector — A photon-counting array for use in both space-borne and ground-based photometric and spectroscopic instrumentation. The resulting tubes use opaque photocathodes, feedback-free microchannel plates, proximity-focused multianode readout arrays, and multilayer ceramic headers.

multianode tank — See multianode tube.

multianode tube — Also called multianode tank. An electron tube having two or more main anodes and a single cathode.

multiaperture reluctance switch — A two-aperture ferrite storage core that may be used to provide a nondestructive-readout memory for a computer.

multiband antenna — An antenna usable at more than one frequency band.

multicasting — Broadcasting a stereo program by using two FM stations. Two FM receivers are required.

multicavity magnetron — A magnetron in which the circuit has more than one cavity.

multicellular horn — A cluster of horns with juxtaposed mouths lying in a common surface. The cluster controls the directional pattern of the radiated energy.

multichannel radio transmitter — A radio transmitter having two or more complete radio-frequency portions capable of operating on different frequencies, either individually or simultaneously.

multichannel R/C — A radio-control installation that employs tuned reeds to supply several control functions. The basic carrier frequency remains the same, but different tones make possible a number of control channels.

multichannel sound — A system of stereo sound transmission for TV applications. Approved in early 1984 by the FCC, it is AM double-sideband for stereo L-R and operates on a 15,734-Hz pilot carrier, which is doubled to 31,468 Hz for stereo. Multichannel sound also contains higher-frequency carriers for SAP (second audio program) and professional channel(s).

multichip circuit — A microcircuit in which discrete, miniature active electronic elements (transistors and/or diode chips) and thin-film or diffused passive components or component clusters are interconnected by thermocompression bonds, alloying, soldering, welding, chemical deposition, or metallization.

multichip integrated circuit — 1. An integrated circuit whose elements are formed on or within two or more semiconductor chips that are separately attached to a substrate. See also integrated circuit. 2. Hybrid integrated circuit that includes two or more SIC, MSI, or LSI chips. 3. An electronic circuit in which two or more semiconductor wafers that contain single elements or simple circuits are interconnected and encapsulated in a single package to give a more complex circuit.

multichip microcircuit — A microcircuit whose elements are formed on or within two or more semiconductor chips that are attached separately to a substrate.

multiconductor — More than one conductor within a single cable complex.

multicoupler — A device for connecting several receivers to one antenna and properly matching their impedances.

multidrop — A telephone line configuration in which a single transmission facility is shared by several end stations.

multidrop line — Also called a multipoint line. 1. A communication system configuration using a single channel or line to serve multiple terminals. Use of this type of line normally requires some kind of polling mechanism, addressing each terminal with a unique identification. 2. A communication line with several subsidiary controllers sharing time on the line under a central site's control.

multielectrode tube — An electron tube containing more than three electrodes associated with a single electron stream.

multielement parasitic array — An antenna consisting of driven dipoles and parasitic elements arranged to produce a highly directive beam.

multiemitter transistor — A transistor that has more than one emitter. Used mainly in logic circuits.

multifiber (fiber optic) — A coherent bundle of fused single fibers that behaves mechanically as a single glass fiber.

multifrequency tones — Telecommunication signaling code utilizing pairs of frequencies in the 700- to 1700-Hz range.

multifrequency transmitter — A radio transmitter capable of operating on two or more selectable frequencies, one at a time, using preset adjustments of a single radio-frequency portion.

multifunction — Pertaining to an integrated device containing two or more circuits integral to a single silicon chip. In addition, each circuit has all inputs and outputs available at terminals for testing and interconnection with other packages. Typical multifunction devices are quadruple gates and dual flip-flops.

multifunction array radar — An electronic scanning radar that will perform target detection and identification, tracking, discrimination, and some interceptor missile tracking on a large number of targets simultaneously and as a single unit.

multigun tube — A cathode-ray tube having more than one electron gun. Used in color television receivers and multiple-presentation oscilloscopes.

multihop propagation — The bouncing of radio waves from the ionosphere to increase their range.

multilayer — A type of printed-circuit board that has several layers of circuit etch or pattern, one over the other and interconnected by electroplated holes. Since these holes can also receive component leads, a given component lead can connect to several circuit points, reducing the required dimensions of a printed circuit board.

multilayer board — A high-density printed-wiring board that consists of alternating conductive pattern layers and insulating layers bonded together. Interlayer connections are included as required by means of plate-through holes.

multilayer ceramic capacitor — A miniature ceramic capacitor manufactured by paralleling several thin layers or ceramic. The assembly is fired after the individual layers have been electroded and assembled.

multilayer circuit — 1. Three or more conductive patterns interspersed with layers of insulating base and laminated into sandwiches, with interconnection between layers provided by plate-through holes. 2. A composite circuit consisting of alternate layers of conductive circuitry and insulating materials (ceramic or dielectric compositions) bonded together with the conductive layers interconnected as required.

multilayer dielectric — A compound including glass and ceramic that is applied as an insulating barrier between conductors for multilayer and crossover work.

multilayer interconnection pattern — A technique used for the interconnection of arrays performing large, complicated electronic functions. This technique involves the use of alternating films of insulating and conducting materials as a precondition for the realization of multiple interconnection planes.

multilayer metallization — Abbreviated MLM. 1. A method used to increase the component density of a particular monolithic integrated circuit. Two or more layers of interconnecting metallization are stacked on top of the chip; the layers are separated by a thin dielectric (insulating) film except at the desired contact points. 2. An integrated circuit fabricating technique that makes economically feasible more complex monolithic logic subsystems or arrays: LSI MLM permits signal paths to be routed most efficiently and to cross without interaction. As a result, integrated circuit chips can be smaller or can contain more circuitry.

multilayer printed circuit — A type of printed circuitry wherein 2 to 14 or more printed circuit layers are fabricated as a complete assembly.

multilayer substrates — Substrates that have buried conductors so that complex circuitry can be handled. Assembled using processes similar to those used in multilayer ceramic capacitors.

multileaving — A technique for allowing simultaneous use of a communications line by two or more terminals.

multimedia or **multi-media** — 1. Term used to describe the use of more than one medium in a program or system, such as the use of audio, video, graphics, animation, and computer data together for a program. Video had been considered separate from audio and computers separate from video and audio. Multimedia means the joining of any two or more of these. Multimedia computers play back high-quality sound and video, as well as text and graphics. Multimedia combines multiple forms of media in the communication of information between users and machines. 2. The presentation of information on a computer using video, audio, animation, and graphics technologies combined in one interactive package.

multimedia communications — Communication that is made up of a combination of text, graphics, video, and audio.

multimeter — 1. Electronic device for measuring resistance and ac or dc current and voltage. *See also* circuit analyzer. 2. A portable test instrument that can be used to measure voltage, current, and resistance. 3. A meter having multiple scales. Usually means a volt-ohm-milliammeter.

multimode fiber — 1. An optical fiber that supports propagation of more than one mode of a given wavelength. 2. An optical fiber that propagates optical energy in more than one mode. 3. An optical fiber that transmits many modes.

multimode operation — The operation of a laser sufficiently above threshold so as to stimulate more than one mode. If the modes are not degenerate, the output pattern will contain measurable power at more than one frequency.

multimoding — The simultaneous generation of many frequencies instead of one discrete frequency.

multioffice exchange — A telephone exchange area in which there are several central office units.

multipactor — A high-power, high-speed microwave switching device in which an rf electric field drives a thin cloud of electrons back and forth between two parallel plane surfaces located in a vacuum. The device can be used as a switch in a waveguide or for high-speed switching of pulses in other microwave systems.

multipass sort — A computer program for sorting more data than can be contained by the internal computer storage. Intermediate storage, such as disk, tape, or drum, is required.

multipath — 1. The constructive and destructive combination of two or more out-of-phase versions of an FM signal at the receiver. It occurs when a building or similar structure reflects a portion of the signal. When the reflected signal arrives at the receiving antenna, its phase slightly lags the phase of the signal traveling directly from the transmitting site to the receiving antenna. 2. A condition in which a signal reaches the receiving antenna over two or more paths of different lengths The resulting interference causes distortion in the receiver, as well as loss of stereo channel separation. Multipath distortion can be minimized by using a directional receiving antenna, and by tuners having a low capture ratio and high AM suppression. Some tuners also have visual or audible multipath indicators that can be used as aids in adjusting the antenna for minimum multipath interference.

multipath cancellation — In effect, complete cancellation of signals because of the relative amplitude and phase differences of the components arriving over separate paths.

multipath delay — A form of phase distortion occurring most often in high-frequency layer-refracted or reflected signals, and also in VHF scattered signals. The existence of more than one signal path between transmitter and receiver causes the signal components to reach the receiver at slightly different times, causing echoes or ghosting.

multipath distortion/reception — Owing to reflection of the VHF FM signal by large buildings, hills, etc., a receiver sometimes receives not only the direct signal but also a reflected signal slightly later due to the greater path distance traveled. This results in high-order harmonic distortion and impairment to the stereo separation and quality.

multipath effect — The arrival of radio waves at slightly different times because all components do not travel the same distance.

multipath propagation — A transmission path anomaly that acts as a time-varying source of signal nonlinearity. Multipath propagation can distort or reduce a received signal to the point of unreliable reception. In television, multipath transmission is manifested as image ghosting.

multipath reception — Reception in which the radio signal from the transmitter travels to a receiver antenna by more than one route, usually because it is reflected from obstacles. The result is seen as ghosts in a TV picture.

multipath transmission — The phenomenon whereby signals reach the receiving antenna from two or more paths and usually have both amplitude and phase differences. It may cause jitter in facsimile (in Europe it is called echo).

multipattern microphone — A microphone that can be switch selected to provide two or more pickup patterns.

multiphonic organ — An electronic musical instrument in which there are only as many generators as notes you wish to play at the same time, as distinct from one generator for every note on the keyboard, which is the case with a polyphonic organ.

multiple — 1. A group of terminals arranged in parallel to make a circuit or group of circuits accessible at any of several points to which a connection can be made. 2. To render a circuit accessible as in (1) above by connecting it in parallel with several terminals.

multiple access — 1. The use of a communication satellite by more than one pair of ground stations at a time. Ideally it lets many independent ground stations use an active repeater without interfering with each other. 2. Pertaining to a computer system in which a number of online communication channels provide concurrent access to the common system.

multiple accumulating registers — In a computer, additional internal storage capacity that can contain loading, storing, adding, subtracting, and comparing factors up to four computer words in length.

multiple-address code — A computer instruction code that includes more than one address.

multiple-address instruction — In a computer, an instruction that contains more than one address.

multiple-aperture core — A magnetic core that has two or more holes through which may be passed wires and around which there may be magnetic flux. Such cores may be used for nondestructive reading.

multiple array — Two or more antennas mounted on the same mast with outputs coupled together. Multiple arrays are used to increase gain and directivity.

multiple-beam laser — A laser having a Q-switching method that allows separate parallel volumes of the lasing material to act independently of each other. This way, it is capable of producing several separate beams and is useful in high-speed holographic technology.

multiple break — In an electrical circuit, an interruption at more than one point.

multiple-break contacts — A contact arrangement such that a circuit is opened in two or more places.

multiple-chip circuit — *See* hybrid integrated circuit.

multiple-circuit layout — Layout of an array of identical circuits on a substrate.

multiple-conductor cable — A combination of two or more conductors cabled together and insulated from one another and from sheath or armor where used. Special cables are referred to as three-conductor cable, seven-conductor cable, fifty-conductor cable, etc.

multiple-conductor concentric cable — A cable composed of an insulated central conductor with one or more tubular stranded conductors laid over it concentrically and insulated from one another.

multiple-contact switch — A switch in which the movable contact can be set to any one of several fixed contacts.

multiple course — One of a number of lines of position defined by a navigational system. Any one of these lines may be selected as a course line.

multiple jacks — A series of jacks that appear on different panels and that have their tips, rings, and sleeves, respectively, connected in parallel.

multiple-length number — In a computer, a quantity or expression that occupies two or more registers.

multiple modulation — Also called compound modulation. A succession of modulation processes in which the modulated wave from one process becomes the modulating wave for the next.

multiple pileup — Also called multiple stack. An arrangement of contact springs composed of two or more pileups.

multiple-precision notation — A technique whereby two or more computer words represent a single numeric quantity.

multiple processing — Configuring two or more processors in a single system, operating out of a common memory. This arrangement permits execution of as many programs as there are processors.

multiple programming — In computer programming, simultaneous execution of two or more arithmetical or logical operations.

multiple-purpose tester — A single test instrument having several ranges, for measuring voltage, current, and resistance.

multiple-reflection echoes — Returned echoes that have been reflected from an object in the radar beam. Such echoes give a false bearing and range.

multiple sound track — Two or more sound tracks printed side by side on the same medium, containing the same or different material, but meant to be played at the same time (e.g., those used for stereophonic recording).

multiple stack — *See* multiple pileup.

multiple switchboard — A manual telephone switchboard in which each subscriber line is connected to two or more jacks so that they are within reach of more than one operator.

multiple tube counts — Spurious counts induced in a radiation counter by previous tube counts.

multiple-tuned antenna — A low-frequency antenna with one horizontal section and several tuned vertical sections.

multiple-twin quad — Quad cable in which the four conductors are arranged in two twisted pairs, and the two pairs are twisted together.

multiple-unit semiconductor device — A semiconductor device having two or more sets of electrodes associated with independent carrier systems. It is implied that the device has two or more output functions that are independently derived from separate inputs (e.g., a duotriode transistor).

multiple-unit steerable antenna — Abbreviated MUSA. An antenna unit composed of a number of stationary antennas, the major composite lobe of which is electrically steerable..

multiple-unit tube — Also called a duodiode, duotriode, diode-pentode, duodiode-triode, duodiode-pentode and triode-pentode. An electron tube containing two or more groups of electrodes associated with independent electron streams within one envelope.

multiplex — Abbreviated mux. 1. To carry out several functions simultaneously in an independent but related manner. 2. To interleave or transmit two or more messages simultaneously on a single channel. 3. A technique for transmitting two or more signals at the same time on the same carrier frequency. *See* FM stereo. 4. Transmission of two or more channels on a single carrier so that they may be recovered independently by a tuner. In FM stereo, transmission of left-plus-right (sum) signal and left-minus-right (difference) signal on main carrier and subcarrier, respectively. The multiplex decoder in the tuner recovers independent left and right stereo channels from the multiplex signal. 5. To combine two or more electrical signals into a single, composite signal. This may be done on a frequency basis (frequency-division multiplexing) or on a time basis (time-division multiplexing). 6. The system used to transmit two stereo program channels on a single FM carrier in such a form that the complete program (left plus right channels) can be heard on a mono FM tuner. A multiplex demodulator in the tuner converts the composite received program to its two-channel form.

multiplex adapter — A circuit incorporated in an FM tuner or receiver to permit two-channel, or stereophonic, reception from a station transmitting multiplex broadcasts. The other audio channel is produced by demodulating the transmitted subcarrier.

multiplex code transmission — The simultaneous transmission of two or more code messages in either or both directions over the same transmission path.

multiplexed line — A data-communication line equipped with multiplexers at each end.

multiplexer — Abbreviated mux. 1. A device for accomplishing simultaneous transmission of two or more signals over a common transmission medium. 2. An analog or linear device for selecting one of a number of inputs and switching its information to the output; the output voltage follows the input voltage with a small error. 3. A digital device that can select one of a number of inputs and pass the logic level of that input-channel. Selection usually is presented to the device in binary weighted form

and decoded internally. The device acts as a single-pole multiposition switch that passes digital information in one direction only. 4. A device that will interleave (time division) or simultaneously transmit (frequency division) two or more messages on the same communications channel. 5. A device that uses several communication channels at the same time and transmits and receives messages and controls the communication lines. This device itself may or may not be a stored-program computer. 6. A device that can combine several low-speed inputs into one high-speed output. Multiplexers can also function in reverse, a process called demultiplexing. 7. A decision-making type of digital building block that routes data from its one input to any one of several outputs. 8. The device or technique used to share a resource (usually a memory or a bus). 9. In television, a specialized optical device that makes it possible to use a single television camera in conjunction with one or more motion-picture projectors and/or slide projectors in a film chain. The camera and projectors are in a fixed relationship, and prisms or special (dichroic) mirrors are used to provide smooth and instantaneous nonmechanical transition from one program source to the other. 10. A device for selecting a single signal from one of many sources (similar to a multiposition switch). 11. A hardware device that allows the transmission of a number of different signals simultaneously over a single channel. 12. A combination of hardware and software that allows simultaneous transmission and reception of two or more data streams on a single channel.

multiplex filter — *See* MPX filter.

multiplexing — Abbreviated mux. 1. A method of signaling characterized by the simultaneous and/or sequential transmission and reception of multiple signals over a communication channel with means for positively identifying each signal. The signaling may be accomplished over a wire path or radio carrier or combination of both. 2. The division of a transmission facility into two or more channels either by splitting the frequency band transmitted by the channel into narrower bands, each of which is used to constitute a distinct channel (frequency-division multiplexing) or by allotting this common channel to several different information channels, one at a time (time-division multiplexing). 3. Combining two or more channels onto one transmission facility. 4. The time-shared scanning of a number of data lines into a single channel. Only one data line is enabled at any instant. 5. The combination of two or more measurements for transmission along a single wire, path, or carrier. The combination of information signals for transmission on one channel. 6. The act of combining a number of individual message circuits for transmission over a common transmission path. 7. A technique for the concurrent transmission of two or more signals in either or both directions, over the same wire, carrier, or other communication channel. The two basic multiplexing techniques are time-division multiplexing and frequency-division multiplexing.

multiplex operation — Simultaneous transmission of two or more messages in either or both directions over the same transmission path.

multiplexor — A device used for division of a transmission facility into two or more subchannels, either by splitting the frequency band into narrower bands (frequency division) or by allotting a common channel to several different transmitting devices one at a time (time division).

multiplex printing telegraphy — The form of printing telegraphy that uses a line circuit to transmit one character (or one or more pulses of a character) for each of two or more independent channels.

multiplex radio transmission — The simultaneous transmission of two or more signals over a common carrier wave.

multiplex stereo — A system of broadcasting both channels to a stereo program on a single carrier. This is commonly done by modulating an ultrasonic subcarrier — either with the signals of one of the stereo channels, or by a different signal composed of the two channels combined out of phase, or subtractively.

multiplex telegraphy — Telegraphy employing multiplex code transmission.

multiplex transmission — The simultaneous transmission of two or more signals within a single channel. Multiplex transmission as applied to FM broadcast stations means the transmission of facsimile or other signals in addition to the regular broadcast signals.

multiple x-y recorder — A recorder that plots a number of independent charts simultaneously, each showing the relation of two variables, neither of which is time.

multiplication point — A mixing point whose output is obtained by multiplication of its inputs.

multiplier — 1. A device in which the output represents the product of the magnitudes represented by the two or more input signals. 2. A series register placed in a voltmeter to increase the voltage range. 3. A fixed resistance, connected in series with the moving coil of a voltmeter, to enable measurements over a larger voltage range.

multiplier phototube — Also called photomultiplier. 1. Phototube with one or more dynodes between its photocathode and the output electrode. The electron stream from the photocathode is reflected off each dynode in turn, with secondary emission adding electrons to the stream at each reflection. 2. A tube consisting of an evacuated envelope with a photocathode that emits electrons when exposed to light. These electrons are accelerated by a positive electrostatic field and fall on a metal surface, where they emit secondary electrons that are again accelerated to generate more electrons at the next metal surface, and so on. The whole arrangement thus acts as a combination of a simple photocell with a high-gain amplifier in a self-contained unit.

multiplier-quotient register — In a computer, a register used for operations that involve multiplication and division.

multiplier resistor — *See* multiplier, 2.

multiplier traveling-wave photodiode — A device in which increased sensitivity is obtained by combining the construction of a traveling-wave tube with that of a multiplier phototube.

multiplier tube — A vacuum tube in which sequential secondary emission from a number of electrodes is used to obtain increased output current. The electron stream is reflected from the first electrode of the multiplier to the second, and so on.

multiplying factor — The number by which the reading of a meter must be multiplied to obtain the true value.

multipoint — Describes a network configuration in which several transmission facilities connect several end stations to a master station.

multipoint circuit — A circuit interconnecting several locations, wherein information transmitted is available at all locations simultaneously.

multipoint line — *See* multidrop line.

multipoint recorder — A data recorder that uses a single pen to record the values of several parameters by multiplexing the inputs, that is, by sequentially switching the inputs so that one parameter is recorded for a brief period, than the next, and so forth. Since the curves may cross each other and make identification

of parameters a problem, various methods of identifying each trace have been developed.

multipolar — Having more than one pair of magnetic poles.

multiport — Refers to the capability of communications equipment to accept more than one input/output data line.

multiposition relay — A relay having more than one operate or nonoperate position, e.g., a stepping relay.

multiprocessing — 1. The simultaneous or interleaved execution of two or more programs or instruction sequences by a computer or computer network. It may be accomplished by multiprogramming, parallel processing, or both. 2. The operation of more than one processing unit within a single system. Separate processors may take over communications or peripheral control, e.g., while the main processor continues program execution. 3. A complex technique that permits more than one computer program to be executed simultaneously on more than one processor in a shorter period of time than is required for the overall sequentially organized task, for which time may not be available. 4. The operation of several CPUs or similar programmable processors operating relatively independently of one another but sharing common resources, such as memory and I/O, over a common multiple-processor bus. This arrangement is well suited for larger overall processing tasks for which one processor is not a cost-effective solution. In general, multiprocessing increases system reliability, improves the use of resources, and permits greater system performance. Through parallel action, it often multiplies system performance by the number of processors used. Multiprocessing permits system performance to greatly surpass that of a single CPU. 5. Two or more CPUs operating together either in a tightly coupled system, in which all I/O and memory share a parallel bus, and access is interleaved cycle by cycle, or in a loosely coupled system, in which CPUs have independent memory and I/Os and share parallel or serial bus links. 6. Independent and simultaneous processing accomplished by a computer configuration consisting of more than one arithmetic and logic unit, each being capable of accessing a common memory. 7. Computer operation in which two or more similar computers are hooked together. The individual computers can work on different parts of the same problem simultaneously, or on entirely different problems. Or each computer can specialize in a particular kind of task, or two computers can work on the same program and check each other. 8. The ability of an operating system to support multiple processors. Generally, in order to make the operating system task manageable, all processors are equivalent and interchangeable, although this is not a requirement. 9. A processing method in which program tasks are logically and/or functionally divided among a number of independent CPUs, with the programming tasks being simultaneously executed.

multiprocessor — 1. A computer capable of executing one or more programs simultaneously using two or more processing units under integrated control of programs or devices. 2. A system architecture that permits complex operations, like computation, to be carried on simultaneously at different nodes.

multiprogramming — 1. A method in which many programs are interleaved or overlapped so that they can be operated on within the same interval of time. This technique is the basis for time-shared operation. 2. A programming technique in which two or more programs are operated on a time-sharing basis, usually under control of a monitor that determines when execution of one program stops and another begins. 3. The technique of utilizing several interleaved programs concurrently in a single computer system. 4. Pertaining to the concurrent execution of two or more programs by a computer. The programs operate in an interleaved manner within one computer system. 5. A mode of operation in which a computer divides its time among several programs, working on one for a while, then switching to another. 6. A word processing system that can simultaneously process different applications such as file sorting and text editing or even data-processing tasks. 7. Another term for multitasking.

multirate meter — A meter that registers at different rates or on different dials at different hours of the day.

multi-rf-channel transmitter — A radio transmitter having two or more complete radio-frequency portions that may be operated on different frequencies either individually or simultaneously.

multisegment magnetron — A magnetron with an anode divided into more than two segments, usually by parallel slots.

multispeed motor — A motor that can be operated at any one of two or more definite speeds, each practically independent of the load.

multistage LNA — Three or more transistor amplifier stages placed end to end (cascaded) so that the gain contribution of each will add up to total gain of approximately 50 dB. In most low-noise amplifiers (LNAs), the first stage (closest to the antenna feed probe) has the best noise characteristics needed to minimize the noise propagated along through the remaining stages.

multistage tube — An X-ray tube in which the cathode rays are accelerated by multiple ring-shaped anodes, each at a progressively higher potential.

multistate noise — Noise consisting of erratic switching that is generated within a device at various sharply defined levels of applied current.

multitasking — *See also* multiprogramming. 1. The capability of a personal computer to run more than one software program, or task, by interleaving segments of each task. 2. The concurrent execution of two or more tasks or applications by a computer. 3. A method of achieving concurrency by separating a program or programs into two or more interrelated tasks that share code, buffers, and files while running.

multithreading — The running of several processes in rapid sequence (multitasking) within a single computer program.

multitrack magnetic system — A magnetic recording system in which its medium has two or more tracks.

multitrack recording system — A recording system in which the medium has two or more recording paths, which may carry the same or different material but are played at the same time.

multiturn potentiometer — A potentiometer that must be rotated more than one turn for the slider to travel the complete length of the resistive element.

multiunit tube — An electron tube containing within one glass or metal envelope two or more groups of electrodes, each associated with separate electron streams.

multivibrator — A relaxation oscillator in which the in-phase feedback voltage is obtained from two electron tubes or transistors. Typically, their outputs are coupled through resistive-capacitive elements. The time constants of the coupling elements determine the fundamental frequency, which may be further controlled by an external voltage. When such a circuit is normally in a nonoscillating state and a trigger signal is required to start a cycle of operation, it is called a one-shot, flip-flop, or start-stop multivibrator.

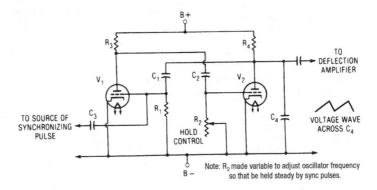

Tube circuit.

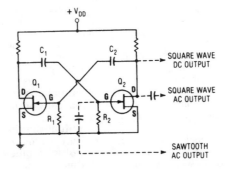

FET circuit.

Multivibrator.

multivoltage control — A method of controlling the voltage of an armature by successive impression of a number of substantially fixed voltages on the armature. The voltages are usually obtained from multicommutator generators common to a group of motors.

multiwire — A machine-controlled process for placing a customized interconnection pattern of insulated wires on a prepared epoxy glass laminate substrate.

MUM — Abbreviation for methodology for unmanned manufacturing. A Japanese program established to develop an unmanned factory in the early 1980s. A facility that depends heavily on robots.

Mumetal — A metallic alloy with high permeability and a low hysteresis loss. It is excellent for magnetic shielding.

Mumetal shield — A cone-shaped covering made of Mumetal. Placed over the flared portion of a picture tube, it acts as a shield to prevent outside magnetic fields from affecting the alignment of the electron beams in the tube.

μmho — Letter symbol for micromho.

Munsell Book of Color — A collection of color samples arranged in charts according to equal visually spaced steps in Munsell hue, value, and chroma.

Munsell chroma — Numerical scale of chroma devised by A. H. Munsell and exhibited in the Munsell Book of Color.

Munsell color system — A system of surface-color specification based on perceptually uniform color scales for the three variables: Munsell hue, Munsell value, and Munsell chroma. For an observer of normal color vision, adapted to daylight, and viewing the specimen when illuminated by daylight and surrounded with a middle gray to white background, the Munsell hue, value, and chroma of the color correlate well with the hue, lightness, and saturation of the perceived color.

Munsell notation — Numerical-alphabetical description of color according to Munsell hue, value, and chroma.

Munsell system — A color-specification system used principally in photography and color printing. It is based on sample cards containing the hue scale in five principal and five intermediate hues, and the brilliance scale in ten steps ranging from black to white. These represent visual, not physical, intervals.

Munsell value — 1. In the Munsell system of object color specification, the dimension that indicates the apparent luminous transmittance or reflectance of the object on a scale having approximately equal perceptual steps under the usual conditions of observation. 2. Numerical scale of lightness devised by A. H. Munsell and exhibited in the Munsell Book of Color.

$\mu\Omega$ — Letter symbol for micro-ohm.

Murray loop test — A method of localizing a fault in a cable. This is done by replacing two arms of a Wheatstone bridge with a loop formed by the cable under test and a good cable connected to the far end of the defective cable.

μs — Letter symbol for microsecond.

MUSA antenna — Acronym for multiple-unit steerable antenna.

muscle chip — An interface circuit that has high-voltage and/or high-current capability to drive relays displays, motors, printers, lamps, etc., from inputs provided from logic or other low-power signal sources (Sprague Electric Co.).

mush winding — A type of winding in an ac machine. The conductors are placed one by one in prepared slots and the end connections are separately insulated.

musical cushion — A musical selection added at the end of a program that is running short or over. By playing it slowly or cutting out portions, the program director can make the program come out on time.

musical echo — *See* flutter echo, 2.

musical instrument digital interface — *See* MIDI.

musical quality — *See* timbre.

music power — Also called dynamic power. 1. The short-term power available from an amplifier for the reproduction of program material. The music power output exceeds the rms power rating to a greater or lesser extent. Its measurement is standardized by the Institute of High Fidelity (IHF) and represents a practical means of stating the actual capabilities of an amplifier for the reproduction of program material. 2. The power that an amplifier (with both channels working, when stereo) is capable of delivering on a music type signal. Output is rated in watts into a specified load value, and the test signal is sometimes an interrupted sine-wave signal (i.e., tone bursts). This method of power expression fails to take account of the power supply impedance and regulation and the efficiency of the power transistor heat sinks. The power obtained is always greater than that measured when the test signal is continuous wave (i.e., steady-state sine-wave signal) and with both channels driven together. 3. A measurement of the peak output power capability of an amplifier performed either with a signal duration sufficiently short that the amplifier power supply does not sag during the measurement or with high-quality external power supplies. This measurement (an IHF standard) assumes that with normal music program material the amplifier power supplies will sag insignificantly.

must-operate value — A specified functioning value at which all relays meeting the specification must operate.

must-operate voltage — The minimum control voltage at which a relay will always turn on.

must-release value — The specified operating value at which all relays that meet a certain specification must release.

must-release voltage — The maximum control voltage at which a relay switches from on to off.

MUT — Abbreviation for module under test.

mute — To suppress the audio output of an amplifier in response to a command signal even though an input may be present. Used in stereo receivers to prevent the off-channel spurious response produced from reaching the speaker while tuning.

muting — Muffling or deadening a sound. *See also* quieting.

muting circuit — 1. A circuit that cuts off the receiver output when the rf carrier reaching the first detector is at or below a predetermined intensity. 2. A circuit for making a receiver insensitive while its associated transmitter is on.

muting switch — A switch used with automatic tuning systems to silence the receiver while it is being tuned.

mutual capacitance — Capacitance between two conductors with all other conductors connected together and to a grounded shield. (Does not apply to a shielded single conductor.)

mutual conductance — *See* transconductance.

mutual-conductance meter — *See* transconductance meter.

mutual-conductance tube tester — *See* transconductance tube tester.

mutual exclusion — A process synchronization rule that prohibits more than one task from using the same resource at the same time.

mutual impedance — 1. Between any two pairs of network terminals (all other terminals being open), the ratio of the open-circuit potential at either one of the two pairs to the current applied to the other pair of network terminals. 2. The impedance existing between the primary and secondary of a transformer. Numerically equal to the secondary voltage divided by the primary current.

mutual inductance — The common inductance of two coupled electrical circuits that determines, for a given rate of change of current in one of the circuits, the electromotive force that will be induced in the other.

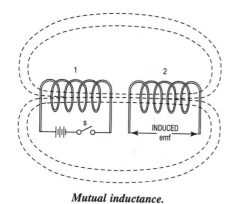

Mutual inductance.

mutual induction — The production of a voltage in one circuit by a changing current in a neighboring circuit, even though no apparent connection exists between the two circuits.

mutual inductor — An inductor for changing the mutual inductance between two circuits.

mutual information — *See* transinformation.

mutual interference — Human-made interference from two or more electrical or electronic systems that affects these systems on a reciprocal basis.

μV — Letter symbol for microvolt.

μW — Letter symbol for microwatt (10^{-6} watt).

mux — 1. Abbreviation for multiplex. 2. Abbreviation for multiplexer. 3. Abbreviation for multiplexing.

MV — Letter symbol for megavolt.

mV — Letter symbol for millivolt.

MVA — Letter symbol for megavolt-ampere.

MW — Letter symbol for megawatt.

mW — Letter symbol for milliwatt.

MWh — Abbreviation for megawatthour.

Mycalex — Trade name of the Mycalex Corp. for mica bonded with glass. Mycalex has a low power factor at high temperatures, and it is a good insulator at all frequencies.

Mylar — Trade name of E. I. duPont de Nemours and Co., Inc., for a highly durable, transparent plastic film of outstanding strength. It is used as a base for magnetic tape and as a dielectric in capacitors.

Mylar capacitor — A capacitor in which Mylar film, either alone or in combination with another film or paper, is the dielectric.

myocardial electrodes — *See* implantable pacemaker.

myoelectric signals — Complex pulse potentials of 10 to 1000 microvolts with durations between 1 and 10 milliseconds as generated by a muscular effort, recorded from the body surface.

myograph — A recorder of forces of muscular contraction. *See also* electromyograph.

myokinesimeter — An apparatus for measuring the response of a muscle to stimulation by electric current.

myophone — An instrument for making audible the sound of a muscle when it is contracting.

myriametric waves — Very-low frequency band; 3 kHz to 30 kHz (100 km to 10 km).

N

n—Letter symbol for the prefix nano- (10^{-9}).

N—Symbol for number of turns, or the north-seeking pole of a magnet.

NA—Abbreviation for numerical aperture.

nA—Letter symbol for nanoampere.

NAB—Abbreviation for National Association of Broadcasters.

NAB curve—Also called NAB equalization. 1. The standard playback equalization curve set by the National Association of Broadcasters. 2. In tape recording, this is used to describe anything for which standards have been established by the National Association of Broadcasters to ensure complete interchangeability from one recorder to another of the same speed.

NAB reel, NAB hub—Reels and hubs used in professional recording, having a large center hole and usually an outer diameter of $10\frac{1}{2}$ inches (26.7 cm).

NAEB—Abbreviation for National Association of Education Broadcasters.

nail-head bond—An alternate term for ball bond.

nail heading—The flared condition of copper on the inner conductor layers of a multilayer board caused by hole drilling.

NAND—A function of A and B that is true if either A or B is false.

NAND circuit—1. A circuit in which the normal output of an AND circuit is inverted; a NOT-AND circuit. 2. A circuit that has its output in the logical 0 state if and only if all of the control (input) signals assume the logical 1 state.

NAND gate—1. A combination of a NOT function and an AND function in a binary circuit that has two or more inputs and one output. The output is logical 0 only if all inputs are logical 1; it is logical 1 if any input is logical 0. With logic of the opposite polarity, this type of gate becomes a NOR gate. 2. A multiple-control device that evidences the simultaneous appearance of all positive-pressure control signals with a single zero-output signal. With the loss of one or more control signals, the zero NAND output signal ceases. The NAND gate can also be defined as a complemented AND or NOT AND device. 3. An AND circuit that delivers an inverted output signal.

nano-—Prefix meaning one-billionth (10^{-9}). Letter symbol: n.

nanoampere—One-thousandth of a microampere (10^{-9} ampere). Letter symbol: nA

nanocircuit—An integrated microelectronic circuit in which each component is fabricated on a separate chip or substrate so that independent optimization for performance can be achieved.

nanofarad—One-billionth of a farad, equal to 10^{-9} farad, 0.001 microfarad, or 1000 picofarads. Letter symbol: nF.

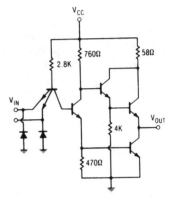

High-speed TTL.

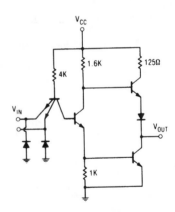

Standard TTL.

NAND gates.

nanohenry—One-thousandth of a microhenry, equal to 10^{-9} henry or 1000 picohenrys. Letter symbol: nH.

nanometer—A unit of length in the metric system equal to 10^{-9} meter. Formerly called a millimicron. Letter symbol: nm.

nanosecond—One-billionth of a second (10^{-9} second). Light travels approximately 1 foot (30.5 cm) in 1 nanosecond. Letter symbol: ns.

nanovolt—One-thousandth of a microvolt (10^{-9} volt). Letter symbol: nV.

nanovoltmeter—A voltmeter sufficiently sensitive to give readings in thousandths of microvolts.

Naperian logarithm — Also called hyperbolic or natural logarithm. A logarithm to the base 2.7128.

napier — *See* neper.

narrow band — 1. A band whose width is greater than 1 percent of the center frequency and less than one-third octave. 2. Pertaining to a communication channel of less than voice grade.

narrow-band amplifier — An amplifier designed for optimum operation over a narrow band of frequencies.

narrow-band axis — In phasor representation of the chrominance signal, the direction of the coarse-chrominance primary of a color TV system.

narrow-band FM — *See* narrow-band frequency modulation.

narrow-band FM adapter — An attachment that converts an AM communications receiver to FM.

narrow-band frequency modulation — Abbreviated NBFM. Frequency modulation that occupies only a small portion of the conventional FM bandwidth. Used mainly for two-way voice communication by police, fire, taxicabs, and amateurs.

narrow-band interference — Sharply tunable interference, having a spectrum that is small compared with the bandwidth of the measuring instrument. (Sinewave carriers both modulated and unmodulated are good examples.)

narrow-sector recorder — A radio direction finder with which atmospherics are received from a limited sector related to the position of the antenna. The antenna is usually rotated continuously, and the bearing of the atmospherics recorded automatically.

NARTB — Abbreviation for National Association of Radio and Television Broadcasters, the former name of the National Association of Broadcasters.

***n*-ary code** — A code in which each element can be any one of *n* distinct kinds or values.

***n*-ary pulse-code modulation** — A type of pulse-code modulation in which the code for each element of information can consist of any one of *n* distinct kinds or values.

NASA — Acronym for National Aeronautics and Space Administration. The federal agency charged with all scientific space missions.

National Association of Broadcasters — Abbreviated NAB. The official association of the radio and television broadcasting industry. Formerly called the NARTB.

National Association of Radio and Television Broadcasters — Abbreviated NARTB. A name used for a number of years by an association of broadcasters. In 1958 the name was changed back to National Association of Broadcasters, an earlier title.

National Electrical Code — Abbreviated NEC. 1. A recognized authority on safe electrical wiring. The code is used as a standard by federal, state, and local governments in establishing their own laws, ordinances, and codes on wiring specifications. 2. A set of regulations governing construction and installation of electrical wiring and apparatus, established by the National Fire Protection Association and suitable for mandatory application by governmental bodies exercising legal jurisdiction. It is widely used by state and local authorities within the United States and is incorporated in OSHA regulations. It has the force of law only when enforced by municipalities or states. 3. A compendium of articles pertaining to the general electrical wiring field, containing provisions considered essential for the practical safeguarding of persons and property from hazards arising from the use of electricity.

National Electrical Manufacturers Association — Abbreviated NEMA. 1. An organization of manufacturers of electrical products. 2. An industry association that standardizes specifications for cables, wires, and electrical components.

National Electrical Safety Code — A set of safety rules for the installation and maintenance of electric supply and communication lines, published as *National Bureau of Standards Handbook* and approved by the American Standards Association. It has the force of law only if enforced by municipalities or states.

national paging — Paging service provided on a national or regional basis, in which subscribers use a single pager that can operate in many different areas. This is usually achieved with a single frequency available nationwide (called nationwide paging) or through a form of networking that uses a pager that receives different frequencies in different areas.

National Television System Committee — Abbreviated NTSC. A committee organized in 1940 and comprising all United States companies and organizations interested in television. Between 1940 and 1941, it formulated the black and white television standards; between 1950 and 1953, it formulated the color television standards that were approved by the Federal Communications Commission.

nationwide paging — Method of national or regional paging in where a single frequency is used throughout the nation (or region) for sending messages to a paging system subscriber.

natural antenna frequency — The lowest resonant frequency of an antenna operated without external inductance or capacitance.

natural binary — 1. A number system to the base 2, in which the 1s and 0s have weighted value in accordance with their relative position in the binary word. Carries may affect many digits. (Contrasted with Gray code, which permits only one digit to change state.) 2. The usual $2n$ code with $2, 4, 8, 16, \ldots, 2n$ progression. An input or output high, or 1, is considered a signal, whereas a 0 is considered an absence of signal. This is a positive-true binary signal. Zero scale is then all 0s, while full scale is all 1s.

natural frequency — 1. The frequency at which a system with a single degree of freedom will oscillate from the rest position when displaced by a transient force. Sometimes used synonymously with damped natural frequency. 2. The lowest resonant frequency of a circuit or component without adding inductance or capacitance.

natural frequency of an antenna — The lowest resonant frequency of an antenna with no added inductance or capacitance.

natural interference — Electromagnetic interference caused by natural terrestrial phenomena (atmospheric interference) or by natural disturbances outside of the atmosphere of the earth (galactic and solar noise).

natural logarithm — *See* Naperian logarithm.

natural magnet — Magnetic ore (e.g., a lodestone) that exhibits the property of magnetism in its natural state.

natural period — The period of the free oscillation of a body or system. When the period varies with amplitude, the natural period is the period when the amplitude approaches zero.

natural radiation — *See* background radiation.

natural resonance — *See* periodic resonance.

natural wavelength — The wavelength corresponding to the natural frequency of an antenna or circuit.

Navaglobe — A long-distance navigational system of the continuous-wave, low-frequency type. Bearing information is provided by amplitude comparison.

navaids—Navigational aids. The electronic facilities provided on or in the immediate vicinity of an airport so that a pilot using compatible airborne equipment can execute the instrument approach or approaches authorized for the airport.

navar—A coordinated series of radar navigation and traffic-control aids utilizing transmissions at wavelengths of 10 and 60 centimeters. In an aircraft, it provides distance and bearing from a given point, display of other aircraft in the vicinity, and commands from the ground. On the ground, it provides a display of all aircraft in the vicinity, as well as their altitudes and identities, plus means for transmitting certain commands.

navarho—A continuous-wave, low-frequency navigation system that provides simultaneous bearing and range information over long distances.

Navascreen—A system for the display and computation of air-traffic-control data based on information from radar and other sources.

navigation—The process of directing an airplane or ship toward its destination by determining its position, direction, etc.

navigational parameter—In a navigational aid, a visual or aural output having a specific relationship to navigational coordinates.

navigation beacon—A light, radio, or radar beacon that provides navigational aid to ships and aircraft.

NBFM—Abbreviation for narrow-band frequency modulation.

nc—1. Abbreviation for normally closed. 2. Abbreviation for no connection.

NC—Abbreviation for numerical control.

nc contacts—Contacts that are normally closed when the control circuit is not energized.

n-channel—A device constructed on a p-type silicon substrate whose drain and source components are of n-type silicon.

n-channel FET—A field-effect transistor that has an n-type conduction channel.

n-channel MOS—1. A metal-oxide semiconductor (unipolar) transistor in which the working current consists of negative (n) electrical charges. 2. A type of metal-oxide silicon field-effect transistor using electrons to conduct current in the semiconductor channel. The channel has a predominantly negative charge.

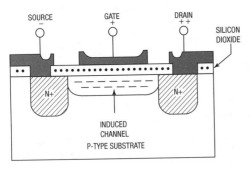

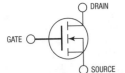

N-channel MOSFET.

n-conductor concentric cable—A cable composed of an insulated central conducting core with $n - 1$ tubular stranded conductors laid over it concentrically and separated by layers of insulation.

n-cube—Also called n-dimensional cube or n-variable cube. In switching theory, a term used to indicate two $n - 1$ cubes having corresponding points connected.

NC value—*See* noise-criteria value.

n-dimensional cube—*See* n-cube.

N-display—Also called N-scan or N-scope. A radar display similar to the K-display. The target appears as a pair of vertical deflections (blips) from the horizontal time base. Direction is indicated by the related amplitude of the vertical deflections. Target distance is determined by moving a pedestal signal (the control of which is calibrated in distance) along the base line until it coincides with the horizontal position of the vertical deflections.

near-end crosstalk—In a disturbed telephone channel, crosstalk propagated in the opposite direction from the current in the disturbing channel. The terminal in which the near-end crosstalk is present is ordinarily near, or coincides with, the energized terminal of the disturbing channel.

near field—1. The acoustic radiation field close to the speaker or some other acoustic source. 2. The electromagnetic field within a distance of 1 wavelength from a transmitting antenna.

near infrared—1. Name applied to the spectral region primarily comprising wavelengths between 3 and 30 meters. 2. That nonvisible radiant energy with wavelengths nearest the end of the visible spectrum.

near space—The space above the earth in which communication satellites travel, beginning at an altitude of about 120 miles (193 km).

near-ultraviolet—The longest wavelengths of the ultraviolet region, nominally 300 to 400 nanometers.

near zone—The region very close to a radio transmitting antenna in which the induction field predominates. The intensity of the induction field does not reduce in proportion to the square of the distance from the antenna.

NEC—Abbreviation for National Electrical Code. A consensus standard published by the National Fire Protection Association (NFPA) and incorporated in OSHA regulations.

necessary bandwidth—For a given class of emission, the minimum value of the occupied bandwidth sufficient to ensure the transmission of information at the rate and quality required for the system employed, under specified conditions. Emissions useful for the good functioning of the receiving equipment for example, the emission corresponding to the carrier of reduced carrier systems are included in the necessary bandwidth.

neck—The narrow tubular part of the envelope of a cathode-ray tube; it extends from the funnel to the base and houses the electron gun.

needle—1. A probe used on stacks of punched cards. 2. *See* stylus, 1.

needle chatter—*See* needle talk.

needle drag—*See* stylus drag.

needle electrode—Small instrument used for subcutaneous electrical recording and stimulating.

needle gap—A spark gap having needlepoint electrodes.

needle pressure—*See* stylus force.

needle scratch—*See* surface noise.

needle talk—Audible sounds from a record player pickup in the vicinity of the stylus.

needle test points—Sharp steel probes connected to test cords for making contacts with conductors.

neg—Abbreviation for negative.

negate — To invert the value of a function or variable; that is, to change it to 1 if it is 0 or to 0 if it is 1. The symbol for negation is a superscript bar.

negated-input OR gate — A gate that has a high output when one or both inputs are low. A negated-input OR gate is identical with a NAND gate in function.

negation — The NOT operation in Boolean algebra.

negative — Abbreviated neg. 1. Less than zero. 2. The opposite of positive (e.g., negative resistance, transmission, feedback, etc.). 3. A terminal or electrode having an excess of electrons. Electrons flow out of the negative terminal of a voltage source, toward a positive source. 4. An artwork, artwork master, or production master in which the intended conductive pattern is transparent to light, and the areas to be free from conductive material are opaque.

negative acceleration — A relative term often used to indicate that, referenced to zero acceleration, a negative electrical output will be obtained from a linear accelerometer when its sensitive axis is oriented normal to the surface of the earth and a given minus reference point along the center of the seismic mass.

negative acknowledgement character — Abbreviated NAK. A communication control character transmitted from a receiving point as a negative response to a sender. It may also be used as an accuracy control character.

negative bias — In a vacuum tube, the voltage that makes the control grid more negative than the cathode.

negative booster — A booster used with a ground-return system to reduce the difference of potential between two points to the grounded return. It is connected in series with a supplementary insulated feeder extending from the negative bus of the generating station or substation to a distant point on the grounded return.

negative charge — A condition in a circuit when the element in question retains more than its normal quantity of electrons.

negative conductor — A conductor connected to the negative terminal of a source of supply. Such a conductor is frequently used as an auxiliary return circuit in a system of electric traction.

negative differential conductance region — That part of the characteristic of a tunnel diode between the peak and valley points.

negative-effective-mass amplifiers and generators — A class of solid-state devices for broadband amplification and generation of microwave energy. Operation of these devices is based on the phenomenon by which the effective masses of charge carriers in semiconductors become negative when their kinetic energies are sufficiently high.

negative electricity — The type of electricity possessed by a body that has an excess of electrons.

negative electrode — The electrode from which the forward current flows.

negative electron — Also called negatron. An electron, distinguished from a positive electron or positron.

negative feedback — Also called degeneration, inverse feedback, or stabilized feedback. A process by which a part of the output signal of an amplifying circuit is fed back to the input circuit. The signal fed back is 180° out of phase with the input signal; therefore, the amplification is decreased, and distortion reduced.

negative-feedback amplifier — An amplifier in which negative feedback is employed to improve the stability or frequency response, or both.

negative ghost — A ghost that has the opposite shading from that of the original image (i.e., is black when the image is white, and vice versa).

negative glow — The luminous glow between the cathode and Faraday dark spaces in a glow-discharge cold-cathode tube. *See also* glow discharge.

negative-grid generator — A conventional oscillator circuit in which oscillation is produced by feedback from the plate circuit to a grid that is normally negative with respect to the cathode and that is designed to operate without drawing grid current at any time.

negative ground — The negative battery terminal of a vehicle is connected to the body and frame.

negative image — 1. A televised picture in which the whites appear black and vice versa. It is due to the picture signal having a polarity opposite to that of a normal signal. 2. A picture signal having a polarity that is opposite to normal polarity and which results in a picture in which the white and black area are reversed.

negative impedance — Also called negative resistance when there is no inductance or capacitance in the circuit. A characteristic of certain electrical devices or circuits — instead of increasing, the voltage decreases when the current is increased, and vice versa.

negative ion — An atom with more electrons than the normal number. Thus, it has a negative charge.

negative-ion generator — A device that bombards air molecules with electrons so that molecules are ionized negatively.

negative light modulation — In television, the process whereby a decrease in initial light intensity causes an increase in the transmitted power.

negative logic — A form of logic in which the more positive voltage level represents logical 0 and the more negative level represents logical 1.

negative modulation — In an AM television system, that modulation in which an increase in brightness corresponds to a decrease in transmitted power.

negative modulation factor — The maximum negative departure of the envelope of an AM wave from its average value, expressed as a ratio. This rating is used whenever the modulation signal wave has unequal positive and negative peaks.

negative phase-sequence relay — A relay that functions in conformance with the negative phase-sequence component of the current, voltage, or power of the circuit.

negative picture phase — 1. For a television signal, the condition in which an increase in brilliance makes the picture-signal voltage swing in a negative direction from the zero level. 2. Positioning the composite video signal so that the maximum level of the sync pulses is at 100% amplitude. The brightest picture signals are in the opposite, negative direction.

negative picture transmission — Television transmission system used in North America and other countries in which a decrease in illumination of the original scene causes an increase in percentage of modulation of the picture carrier. When modulated, signals with a higher modulation percentage have more positive voltages.

negative plate — The grid and active material connected to the negative terminal of a storage battery. When the battery is discharging, electrons flow from this terminal, through the external circuit, to the positive terminal.

negative resistance — 1. A resistance that exhibits the opposite characteristic from normal — i.e., when the voltage is increased across it, the current will decrease instead of increase. 2. A property of some circuits containing no reactance whereby the current increases as the voltage is decreased.

negative-resistance magnetron — A magnetron operated so that it acts as a negative resistance.

negative-resistance oscillator — An oscillator produced by connecting a parallel-tuned resonant circuit to a two-terminal negative-resistance device (one in which an increase in voltage results in a decrease in current). Dynatron and transitron oscillators are examples.

negative-resistance region — That operating region in which an increase in applied voltage results in a current decrease.

negative-resistance repeater — A repeater in which gain is provided by a series or shunt negative resistance, or both.

negative temperature coefficient — The amount of reduction in the value of a quantity, such as capacitance or resistance, for each degree of increase in temperature. *See also* temperature coefficient.

negative terminal — In a battery of other voltage source, the terminal having an excess of electrons. Electrons flow from it, through the external circuit, to the positive terminal.

negative torque — A torque developed in opposition to the normal torque of a motor. This may occur at starting — common to pole motors — or at some speed below the nameplate rpm. This causes "cusps" or "saddles" in the graphed torque curves.

negative-transconductance oscillator — An electron-tube oscillator whose output is coupled back to the input without phase shift, the phase condition for oscillation being satisfied by the negative transconductance of the tube. A transitron oscillator is an example.

negative transmission — The modulation of the picture carrier by a picture signal with a polarity such that the sync pulses occur in the blacker-than-black region and a decrease in initial light intensity increases the transmitted power.

negative-true logic — A logic system in which the voltage representing a logical 1 has a lower or more negative value than that representing a logical 0. Most parallel I/O buses use negative-true logic due to the nature of commonly available logic circuits.

negatron — 1. An electron. *See also* negative electron. 2. A four-electrode vacuum tube having a negative-resistance characteristic. 3. *See* dynatron.

NEMA — Abbreviation for National Electrical Manufacturers Association.

NEMA standards — Consensus standards for electrical equipment approved by the majority of the members of the National Electrical Manufacturers Association (NEMA).

nematic liquid — An organic chemical used in liquid crystal displays (LCDs) that simultaneously exhibits liquid and crystal properties. It is transparent until the threadlike molecules are disturbed by an electric field; then it takes on a milky-white appearance.

nematic liquid crystal — 1. A liquid crystal in such a state that the principal axis of all the molecules is aligned in one direction. This state is the result of the arrangement of the forces between the molecules and the shape of the molecules, keeping the molecules in a low-energy static state, aligned on one axis. 2. Ordered fluid whose molecules lie parallel to each other with their centers of mass randomly distributed. They are organic compounds whose molecules are rodlike in overall shape. The word *nematic* comes from the Greek *nema*, meaning "threadlike." A threadlike pattern is observed when nematics are viewed through a microscope under polarized light.

nematic phase — 1. An arrangement of parallel-oriented liquid crystal molecules. The nematic phase is turbid but much less viscous than the smectic phase, and the molecules are not arranged in layers. 2. A form of the liquid crystals with an appearance of moving, threadlike structures, particularly visible when observed, in thick specimens, with polarized light. During this phase the

molecules of the crystal are parallel and able to travel past each other following the direction of their longitudinal axes. This form has one optical axis that occupies the direction of an applied magnetic field.

nemo — *See* field pickup.

neon — An inert gas used in neon signs and in some electron tubes. It produces a bright red glow when ionized. Symbol, Ne; atomic number 10; atomic weight, 20.183.

neon bulb — A glass envelope filled with neon gas and containing two or more insulated electrodes. The tube will not conduct until the potential difference between two electrodes reaches the firing, or ionization, potential, and will remain conductive until the voltage is reduced to the extinction level.

Neon bulb.

neon-bulb oscillator — A simple relaxation oscillator in which a capacitor charges to the ionization potential of the neon gas within a bulb. Ionization rapidly depletes the charge on the capacitor, and a new charge cycle begins.

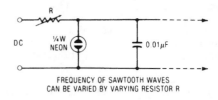

Circuit.

Output.

Neon-bulb oscillator.

neon indicator tube — A cold-cathode tube containing neon and designed to visually indicate a potential difference or field.

neon oscillator — *See* neon-bulb oscillator.

neon tube — An electron tube containing neon gas that uses the transmission of an electric current through the gas to ionize the neon and produce a luminous red discharge.

NEP — Abbreviation for noise equivalent power.

neper — Also called napier. The fundamental division of a logarithmic scale for expressing the ratio between two currents or voltages. The number of nepers denoting such a ratio is the natural (Naperian) logarithm of this ratio. One neper equals 0.8686 bels, or 8.686 decibels. Expressed as a formula:

$$N = \log_e(E_1/E_2)$$

where N is the number of nepers and E_1 and E_2 are the two voltages.

Nernst effect — The effect whereby a potential difference is developed across a heated metal strip placed perpendicular to a magnetic field.

Nernst-Ettinghausen effect — A thermomagnetic effect that occurs in certain pure crystals, whereby a temperature difference is produced in a direction perpendicular to an applied magnetic field and a longitudinal electric current.

Nernst lamp — An electric lamp consisting of a short, slender rod of zirconium oxide that is heated to incandescence by a current.

nest — To enclose a subroutine inside a larger routine, but not necessarily make it part of the outer routine. A series of looping instructions can be nested within each other.

nested subroutine — In a computer program, the subroutine called within another subroutine.

nesting — 1. A term used to indicate that a subroutine in a computer is enclosed inside a larger routine, but is not necessarily part of the outer routine. A series of looping instructions may be nested within each other. 2. A programming technique in which a segment of a larger program is executed iteratively (looping) until a specific data condition is detected or until a predetermined number of interactions has been performed. The nesting technique allows a program segment to be nested within a larger segment and that segment to be nested within an even larger segment. 3. A programming technique involving the embedding of routines within other routines.

nesting level — The number of times nesting can be repeated.

net — 1. Organization of stations capable of direct communications on a common channel, often on a definite schedule. 2. Also known as a signal. Collection of part and connector pins that must be connected. 3. One complete circuit connecting at least one output to at least one input. Must be some form of conductor, such as a wire. 4. The common abbreviation for the Internet.

net authentification — Identification used on a communications network to establish the authenticity of several stations.

net control station — Communications station having the responsibility of clearing traffic and exercising circuit discipline within a net.

net information content — A measure of the essential information contained in a message. It is expressed as the minimum number of bits or hartleys required to transmit the message with specified accuracy over a noiseless medium.

netlist — List of names, symbols, and their connection points that are logically connected in a net.

net loss — The algebraic sum of the gains and losses between two terminals of a circuit. It is equal to the difference in the levels at these points.

net reactance — The difference between the capacitive and inductive reactance in an ac circuit.

Netscape — A graphical web browser.

Netscape Communications — A company whose primary products are a graphical web browser called Netscape Navigator (sometimes known simply as Netscape) and HTTP server software named Netsite.

Netscape Navigator — The graphical web browser developed by Netscape Communications.

Netsite — The HTTP server software developed by Netscape Communications.

network — 1. A combination of electrical elements. 2. An interconnected system of transmission lines that provides multiple connections between loads and sources of generation. 3. An organization of stations with a capability for intercommunication, although not necessarily on the same channel. 4. Two or more interrelated circuits. 5. Decentralized computer architecture that retains the access-to-information characteristic of a centralized computer. The key feature of a network is that communication between the computers in the network is really communication between the programs running on those computers. 6. A complex of two or more interconnected computers. The hardware that supports it generally includes multiplexers, line adapters, modems, and computers with associated peripherals. Software products used consist of modules in the host computer's operating systems, front-end processors, and remote processors that handle services provided to users. 7. A structured connection of computer systems and/or peripheral devices, each remote from the others, exchanging data as necessary to perform the specific function of the connection. 8. The interconnection of a number of points by data communications facilities.

network analog — The expression and solution of a mathematical relationship between variables through the use of a circuit or circuits to represent those variables.

network analysis — 1. The obtaining of the electrical properties of a network (e.g., its input and transfer impedances, responses, etc.) from its configuration, parameters, and driving forces. 2. The process of creating a data model of transfer and/or impedance characteristics of a linear network through sine-wave testing over the frequency range of interest.

network analyzer — 1. A group of electric-current elements that can readily be connected to form models of electric networks. From corresponding measurements on the model, it is then possible to infer the electrical quantities at various points on the prototype system. 2. Also called network calculator. An analog device designed primarily for simulating electrical networks. 3. An instrument that evaluates the impedance characteristics of linear networks over a range of frequencies.

network architecture — 1. A descriptive phrase for the combination of hardware and software that comprises a computer network. 2. A set of rules, standards, or recommendations through which various computer hardware, operating systems, and applications software function together.

network calculator — An analog device designed primarily for simulating electric networks.

network constant — Any one of the resistance, inductance, mutual-inductance, or capacitance values in a circuit or network. When these values are constant, the network is said to be linear.

network filter — 1. A transducer for separating waves on the basis of their frequency. 2. A combination of electrical elements, e.g., interconnected resistors, coils, and capacitors, that presents relatively small attenuation to signals of certain frequency and great attenuation to all other frequencies.

network master relay — A relay that closes and trips an alternating-current, low-voltage network protector.

network phasing relay — A relay that functions in conjunction with a master relay to limit closure of the network protector to a predetermined relationship between the voltage and the network voltage.

network relay — A form of relay (e.g., voltage, power) used in the protection and control of alternating-current, low-voltage networks.

network synthesis — 1. The obtaining of a network from prescribed electrical properties such as input and transfer impedances, specified responses for a given driving force, etc. 2. The process of deriving the configuration and component values of a network with given performance specifications.

network topology — The physical and logical relationship of nodes in a computer network; the schematic arrangement of the links and nodes of a network; networks typically have a star, ring, tree, or bus topology, or some hybrid combination.

network transfer function — A frequency-dependent function, the value of which is the ratio of the output to the input voltage.

network transformer — A transformer suitable for use in a vault to feed a variable-capacity system of interconnected secondaries.

Neuman's law — Mutual inductance is a constant for a given relative physical position of coils, and is independent of the fact that the current flows in one or the other coil, and of frequency, current, and phase.

neural network — A set of linked microprocessors that can form associations and learn like the neurons in a human brain.

neuristor — 1. A two-terminal active device with some of the properties of neurons (e.g., propagation that suffers no attenuation and has a uniform velocity and a refractory period). 2. A device that is essentially an active transmission line designed to propagate signals without attenuation. It is an electrical analog of a nerve fiber.

neuroelectricity — The minute electric voltage generated by the nervous system.

neuron — The basic building block of the human nervous system. A specialized body cell that can conduct and code an electrical pulse.

neutral — In a normal condition; hence, neither positive nor negative. A neutral object has its normal number of electrons — i.e., the same number of electrons as protons.

neutral circuit — A teletypewriter circuit in which current flows in only one direction. The circuit is closed during the marking condition and open during the spacing condition.

neutral conductor — That conductor of a polyphase circuit or of a single-phase three-wire circuit that is intended to have a potential such that the potential differences between it and each of the other conductors are approximately equal in magnitude and are also equally spaced in phase.

neutral density filter — A light filter that reduces the intensity of light without changing the spectral distribution.

neutral ground — A ground connection to the neutral point or points of a circuit, transformer, rotating machine, or system.

neutralization — 1. The nullifying of voltage feedback from the output to the input of an amplifier through the interelectrode impedance of the tube. Its principal use is in preventing oscillation in an amplifier. This is done by introducing into the input a voltage equal in magnitude but opposite in phase to the feedback through the interelectrode capacitance. 2. The process that prevents unwanted self-oscillation in an amplifier by feeding back part of the amplified signal, out of phase, to the input.

neutralize — In an amplifier stage, to balance out the feedback voltage due to grid-plate capacitance, thus preventing regeneration.

neutralized radio-frequency stage — A stage in which a circuit is added for the purpose of feeding back, in the opposite phase, an amount of energy sufficient to cancel the energy that would otherwise cause oscillation.

neutralizing capacitor — A capacitor, usually variable, employed in a radio receiving or transmitting circuit to feed a portion of the signal voltage from the plate circuit back to the grid circuit.

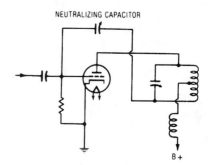

Neutralizing capacitor.

neutralizing circuit — In an amplifier circuit, the portion that provides an intentional feedback path from plate to grid. This is done to prevent regeneration.

neutralizing indicator — An auxiliary device (e.g., a lamp or detector coupled to the plate tank circuit of an amplifier) for indicating the degree of neutralization in an amplifier.

neutralizing tool — Also called a tuning wand. 1. A small screwdriver or socket wrench, partly or entirely nonmetallic, for making neutralizing or aligning adjustments in electronic equipment. *See also* alignment tool. 2. A plastic rod having a ferrite plug in one end and a brass plug in the other end. Used to check the tuning of radio-frequency circuits. When inserted in a coil, the ferrite increases the inductance while the brass decreases the inductance.

neutralizing voltage — The ac voltage fed from the grid circuit to the plate circuit (or vice versa). It is deliberately made 180° out of phase with and equal in amplitude to the ac voltage similarly transferred through undesired paths (usually the grid-to-plate interelectrode capacitance).

neutral operation — The system whereby marking signals are formed by current impulses of one polarity, either positive or negative, and spacing signals are formed by reducing the current to zero or nearly zero.

neutral point — The point that has the same potential as the point of junction of a group of equal nonreactive resistances connected at their free ends to the appropriate main terminals or lines of the system.

neutral relay — Also called a nonpolarized relay. 1. A relay in which the armature movement does not depend on the direction of the current in the controlling circuit. 2. A relay whose operation depends on the magnitude of the current through its operating winding, and not on the current's direction.

neutral section — That section through a magnet where the center line would pass.

neutral signals—Signals sent in the form of direct-current pulses for marks and an absence of current for spaces.

neutral transmission—Also called unipolar transmission. Transmission of teletypewriter signals in such a way that a mark is represented by current on the line, and a space is represented by the absence of current. By extension to tone signaling, a method of signaling involving two states, one of which represents both a space and the absence of signaling.

neutral zone—A range in the total control zone in which a controller does not respond to changes in the controlled process; a dead zone, usually in the middle of the control range.

neutrodyne—An amplifier circuit used in early tuned radio-frequency receivers. It is neutralized by the voltage fed back by a capacitor.

neutron—One of the three elementary particles (the electron and proton are the other two) of an atom. It has approximately the same mass as the hydrogen atom, but no electric charge. It is one of the constituents of the nucleus (the proton is the other one).

newsgroups—Discussion groups on Usenet. *See also* Usenet

newton—In the mksa system, the unit of force that will impart an acceleration of 1 meter per second to a mass of 1 kilogram. Letter symbol: N.

nF—Letter symbol for nanofarad.

n-gate thyristor—A thyristor in which the gate terminal is connected to the n region nearest the anode and which is normally switched to the on state by applying a negative signal between the gate and anode terminals.

nibble—Also nybble. 1. A sequence of 4 bits operated upon as a unit. *See* byte. 2. One half of a byte. 3. The first or last half of an 8-bit byte (4 bits). 4. Half of a byte (4 bits). Binary coded decimal (BCD) is packed into nybbles.

nibble mode—A form of DRAM addressing in which 4 sequential bits (a nibble) are accessed.

Nichrome—Trade name of Driver-Harris Co. for an alloy of nickel and chromium used extensively in wirewound resistors and heating elements.

nick—A small cut or notch in conductor strands or insulation.

nickel—A metal that offers combination of corrosion resistance, formability, and tough physical properties. For these reasons, nickel is used for alloying purposes and in nickel-clad copper wire.

nickel-cadmium cell—The most widely used rechargeable sealed cell, nickel-cadmium gives a flat voltage discharge characteristic, nominal cell voltage of 1.25 volts, and good low-temperature operation. A cell with a nickel and oxide positive electrode and a cadmium negative electrode. The plates are wrapped with a separator between them, and are immersed in a potassium-hydroxide electrolyte.

nickel-clad copper wire—A wire with a layer of nickel on a copper core, where the area of the nickel is approximately 30 percent of the conductor area. The nickel has been rolled and fused to the copper before drawing.

nickel-metal hydride—Abbreviated NiMH. A rechargeable battery technology that largely overcomes the "memory loss" problems of nickel-cadmium batteries. It can also store a greater charge than a comparable size Nicad battery.

nickel-oxide film diode—A solid-state diode made of nickel-oxide film. It may be switched from off to on by applying a 300-volt low-current pulse for 10 μs, and may be switched from on to off by applying a 30-volt

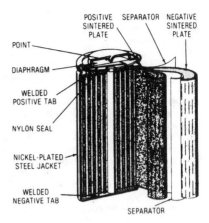

Nickel-cadmium cell.

high-current pulse for 10 μs. In effect it works like a dipole flip-flop.

nickel silver—*See* German silver.

NIF—Abbreviation for noise-improvement factor.

night-vision devices—Devices that use low-level visible radiation or infrared radiation to produce a visual image of a night scene as it would appear in daylight. These devices may rely on the amplification of existing visible light by photomultiplier tubes or infrared recording.

ni junction—A semiconductor junction between n-type and intrinsic materials.

NiMH—Abbreviation for nickel-metal hydride.

nines complement—1. A decimal digit that yields 9 when added to another decimal digit. 2. Pertaining to an arithmetic method of negating a decimal number so that subtraction can be performed by using addition techniques. The nines-complement negation of a number results from subtracting each decimal digit individually from 9. End-around carry must be used in performing a nines-complement addition.

Nipkow disc—A round plate used for scanning small elementary areas of an image in correct sequence for a mechanical television system. It has one or more spirals of holes around its outer edge, and successive openings are positioned so that rotation of the disc provides the scanning.

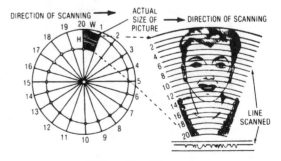

Nipkow disc TV system.

NIPO—Acronym for negative input, positive output.

N-ISDN—Abbreviation for narrowband integrated services digital network. Opposite of B-ISDN (broadband

ISDN). N-ISDN operates at channel speeds less than B-ISDN and has less bandwidth for asynchronous transfer mode-like transmission.

nit — 1. In a computer, a choice among events that are equally probable. One nit equals 1.44 bits. 2. The unit of luminance (photometric brightness) equal to 1 candela per square meter. Letter symbol: nt.

Nixie tube — A glow tube that converts a combination of electrical impulses into a visual display. *See also* numerical-readout tube.

Nixie-tube display — A form of display composed of a number of neon-filled tubes, each containing wires in the shape of the digits 0 to 9, with separate wires for each digit. On signal, the correct wires are energized, causing the neon gas around the wires to glow.

n-level logic — Pertaining to a collection of gates connected in such a way that not more than n gates appear in series.

NMOS — Abbreviation for n-channel metal-oxide semiconductor. 1. Pertaining to MOS devices made on p-type silicon substrates in which the active carriers are electrons that flow between n-type source and drain contacts. The opposite of PMOS, where n-type source and drain regions are diffused into p-type substrate to create an n-channel for conduction. NMOS is from two to three times faster than PMOS. 2. N-channel circuits, using currents made up of negative charges and producing devices at least twice as fast as PMOS and having the potential for high-density configurations. 3. A form of metal-oxide silicon transistor in which the charge flow from source to drain is electrons.

NMOS silicon-gate technology — A process used by National Semiconductor to design and fabricate high-density, high-performance microcomputer components. The process features scaled-geometry transistors (2-μm active length), of a variety of threshold voltages, to optimize the speed/power performance of the internal circuitry.

n-n junction — In an n-type semiconducting material, a region of transition between two regions having different properties.

no — Abbreviation for normally open.

no-address instruction — An instruction specifying an operation that the computer can perform without having to refer to its storage or memory unit.

noble — Chemically inert. A term often used to describe metals such as gold, platinum, etc.

noble gas — Also called inert gas or rare gas. One of the chemically inert gases, including helium, neon, argon, krypton, and xenon.

noble-metal paste — A soft, moist, smooth compound made up partially of precious metals such as gold, platinum, ruthenium, or others classed as noble metals, providing conductors in film circuitry.

noble system — Thick-film system using conductors of gold, platinum, and possibly palladium silver, or certain alloys of these precious metals.

no contacts — Contacts that are normally open when the control circuit is not energized.

noctovision — 1. A television system employing invisible rays (usually infrared) for scanning purposes at the transmitter. Hence, no visible light is necessary. 2. A television system used for seeing in the dark, particularly with the use of infrared rays.

nodal-point keying — Keying an arc transmitter at a point that is essentially at ground potential in the antenna circuit.

nodal points — 1. Two points on the axis of a lens such that a ray entering the lens in the direction of one leaves as if from the other and parallel to the original direction. 2. *See* node, 2.

node — Also called junction point, nodal point, branch point, or vertex. 1. A terminal of any branch of a network, or a terminal common to two or more branches. 2. The point, line, or surface in a standing-wave system where some characteristic of the wave field has essentially zero amplitude. The appropriate modifier should be used with the word *node* to signify the type that is intended (e.g., pressure node). 3. Provides data entry/exit point in computers and terminals, and switches or processes data. Smart terminals and smart programmable controllers — the ones based on microprocessors — also have true processing capability and qualify as nodes. Each node is potentially capable of performing application-oriented tasks. 4. A junction between interconnected integrated circuits on a logic board. 5. Any point in a communications network that is individually addressable, be it a terminal, computer, or communications controller. 6. A point of interconnection to a network common to two or more branches of a network. Normally, a point at which a number of terminals are located.

nodules — Clusters of oxide particles that protrude above the surface of magnetic tape.

noise — 1. Any unwanted disturbance within a dynamic electrical or mechanical system (e.g., undesired electromagnetic radiation in a transmission channel or device). 2. Any unwanted electrical disturbance or spurious signal that modifies the transmitting, indicating, or recording of desired data. 3. In a computer, extra bits or words that have no meaning and must be ignored or removed from the data at the time it is used. 4. Random electrical variations generated internally in electronic components. 5. In acoustics, a potentially distracting or distorting sensory pattern. 6. Any sounds that are not harmonically related to the signal and are added subsequently by recording or playback equipment (includes hiss, crackles, hum, rumblings, ticks, pops, etc.). Each recording system has its own set of inherent noise-producing components. 7. Any unwanted addition of frequencies unrelated to the signal that tends to obscure the signal information. In audio, noise is usually heard as hiss (random noise) or as hum (the power-line frequency and its harmonics). Since signals must be, in general, stronger than the noise level in order to convey information, noise defines the lower limits of a component's dynamic range. (The upper limit is imposed by the maximum tolerable distortion.) Noise may be defined in terms of absolute levels, but in audio equipment it is usually defined in decibels below the maximum tolerably distorted signal. 8. Random, unwanted electrical signals that can cause unwanted and false output signals in a circuit. For example, a random noise pulse might trigger a digital circuit, thereby producing a binary 1 at the circuit output where a binary 0 should be present. 9. An unwanted voltage fluctuation generated within a resistor. Total noise of a resistor always includes Johnson noise, which is dependent only on resistance value and temperature of the resistance element. Depending on type of element and construction, total noise may also include noise caused by current and noise caused by cracked bodies and loose end caps or leads. For variable resistors, noise may also be caused by jumping of contact over turns of wire and by an imperfect electrical path between contact and resistance elements. 10. The unwanted and unpredictable fluctuations that distort a received signal and, hence, tend to obscure the desired message. Noise disturbances, which may be generated in the devices of a communications system or which may enter the system from the outside, limit the range of the system and place requirements on the signal power necessary to ensure good reception. 11. A communication-line impairment inherent in the line design or induced by transient energy bursts. 12. Unwanted components, other than distortion, added to a signal or sound. 13. Random bursts of electrical energy or interference. In some cases, it will produce a salt-and-pepper pattern over a televised picture. Heavy noise is sometimes referred to as snow. 14. Random signals or other interference on a line. 15. The unpredictable or random difference between the observed data and the true process.

noise analysis — Determination of the frequency components that make up the noise being studied.

noise analyzer — An instrument used for determining the amplitude versus frequency characteristics of noise.

noise bandwidth — The width of the equivalent rectangular power gain versus frequency response of a network. It is the width of the rectangle whose area equals that of the actual power gain versus frequency response and whose height equals the maximum available power gain at a reference frequency f_0. It does not include any spurious responses. Any responses other than those in the useful channel are accounted for by a degraded noise figure.

noise behind the signal — See modulation noise.

noise blanker — 1. A circuit used in some receivers just before the detector to minimize ignition noise. 2. Circuit for reducing noise interference. It is usually placed in the rf section or at the beginning of the IF section of a receiver, before the high-selectivity circuits. It squelches the receiver for a very short period (the width of the noise spike). Noise blankers are more effective than noise limiters, but their circuitry is more complex.

noise-canceling microphone — A microphone that is built in such a way as to minimize transmission of any background noise at the transmitting site. See also close-talking microphone.

noise clipper — A circuit that automatically or manually clips the noise from the output of a receiver.

noise-criteria value — Also called NC value. A measure of background noise; it is a single overall value determined from the greatest level of sound pressure in several individual frequency bands. Sometimes values are stated for each band.

noise current — The equivalent open-circuit rms noise current that occurs at the input of a noiseless amplifier due to current at that input. It is measured by shunting an impedance across the input terminals and comparing this output noise with the output noise obtained when the input is shorted.

noise-current generator — A current generator in which the output is a random function of time.

noise diode — A standard electrical-noise source consisting of a diode operated at saturation. The noise is due to random emission of electrons.

noise equivalent bandwidth — The useful bandwidth of a thermistor bolometer for various frequencies in the input radiation, equal to $1/4t$ (where t is the time constant of the bolometer).

noise equivalent power — Abbreviated NEP. 1. The value of radiation that produces, in a detector, an rms signal-to-noise ratio of unity. It is usually measured with a blackbody radiation source at 500 K and a bandwidth of 1 or 5 hertz. The modulation rate varies with the type of detector and is usually between 10 and 1000 hertz. 2. The minimum detectable signal equal to the ratio of total detector and system noise to responsivity.

noise factor — Also called noise figure. 1. For a given bandwidth, the ratio of total noise at the output to the noise at the input. 2. A number expressing the amount by which a receiver falls short of equaling the theoretical optimum performance. 3. The ratio of the available signal-to-noise power ratio at the input to the available signal-to-noise power ratio at the output of a semiconductor device.

noise figure — Abbreviated NF. 1. The ratio of the total noise power at the output of an amplifier to that portion of the total output noise power attributable to the thermal agitation in the resistance of the signal source. See also noise factor. 2. In a network with a generator connected to its input terminal, the ratio of the available signal-to-noise-power ratio at the signal-generator terminals (weighted by the network bandwidth) to the available signal-to-noise-power ratio at the output. 3. The common logarithm of the ratio of the input signal-to-noise ratio to the output signal-to-noise ratio.

noise filter — 1. A combination of electrical components that prevents extraneous signals from passing into or through or out of an electronic circuit. 2. Combination of one or more coils and capacitors inserted between the power cord of a receiver and a wall outlet to block noise interference that might otherwise reach the receiver through the power line.

noise floor — The lowest input signal power level that will produce a detectable output signal from a microwave component, determined by the thermal noise generated within the microwave component itself. The noise floor limits the ultimate sensitivity to the weak signals of the microwave system, since any signal below the noise floor will result in an output signal with a signal-to-noise ratio of less than 1 and will be more difficult to recover.

noise-free environment — A space theoretically containing neither electrostatic nor electromagnetic radiation.

noise generator — An electronic device that generates wide-band white noise for use in testing. The generator output is obtained by amplifying low-level white noise obtained from a source such as a photomultiplier.

noise grade — A number that defines the noise at a particular location relative to the noise at other locations throughout the world.

noise immunity — 1. A measure of the insensitivity of a logic circuit to spurious or undesired electrical signals or noise. 2. See dc noise margin. 3. A measure of a receiver's ability to prevent valid signals from being rejected as noise. It is sometimes specified as the ability of a receiver to operate under conditions of Gaussian noise. 4. A device's ability to discern valid data in the presence of noise.

noise-improvement factor — Abbreviated NIF. In a receiver, the ratio of output signal-to-noise ratio to the input signal-to-noise ratio. (The term receiver is used in the broad sense and is taken to include pulse demodulators.)

noise index — The ratio of the rms current-noise voltage (over a 1-decade bandwidth) to the average (dc) voltage caused by a specified constant current passed through a resistor at a specified hot-spot temperature (usually expressed either in microvolts per volt or in decibels of voltage ratio).

noise jamming — See jamming.

noise killer — An electric network inserted into a telegraph circuit (usually at the sending end) to reduce interference with other communication circuits.

noise level — 1. The strength of extraneous audible sounds at a given location. 2. The strength of extraneous signals in a circuit. Noise level is referred to a specified base and usually measured in decibels. 3. A volume of noise energy, specified as so many decibels above a reference level.

noise limiter — Also called a series-gate noise limiter. 1. A circuit that cuts off all noise peaks stronger than the highest peak in the received signal. In this way, the effects of strong atmospheric or man-made interference are reduced. 2. A circuit that shaves off the noise spikes riding on the desired signal. Usually, diodes are used to obtain the clipping action. When the circuit operates at a predetermined threshold without any user activation, it is called an automatic noise limiter, or ANL. These diode noise limiters are usually in the intermediate-frequency or audio-frequency sections of the receiver.

noise margin — 1. The extraneous-signal voltage amplitude that can be added algebraically to the noise-free worst-case input level of a logic circuit before deviation of the output voltage from the allowable logic levels occurs. In this application, the term input generally refers to logic input terminals, ground reference terminals, or power supply terminals. 2. The difference between the operating voltage and the threshold voltage of a binary logic circuit.

noise-measuring set — *See* circuit-noise meter.

noise-modulated jamming — Random electronic noise that appears at a radar receiver as background noise and tends to mask the desired radar echo or radio signal.

noise pulse — Also called noise spike. A spurious signal of short duration and of a magnitude considerably in excess of the average peak value of the ordinary system noise.

noise quieting — The ability, usually expressed in decibels, of a receiver to reduce background noise in the presence of a desired signal.

noise ratio — The ratio of the available noise power at the output of a circuit to the noise power at the input.

noise-reducing antenna system — A receiving-antenna system so designed that only the antenna proper can pick up signals. It is placed high enough to be out of the noise-interference zone, and is connected to the receiver with a shielded cable or twisted transmission line that is incapable of picking up signals.

noise source — A device employed to generate random noise (e.g., photomultiplier and gaseous-discharge tubes).

noise spike — *See* noise pulse.

noise suppression — 1. The ability of a radio receiver to materially reduce the noise output when no carrier is being received. 2. A means of reducing surface noise during reproduction of a phonograph record.

noise suppressor — 1. A circuit that reduces high-frequency hiss or noise. It is utilized primarily with phonograph records. 2. In a receiver circuit, the portion that reduces noise automatically when no carrier is being received. 3. Any device intended to reduce the audibility of background noise.

noise temperature — 1. At a pair of terminals and a specific frequency, the temperature of a passive system having an available noise power per unit bandwidth equal to that of the actual terminals. 2. The temperature at which the thermal noise available from a device equals the total fluctuation noise actually available from the device. 3. A measure of the amount of thermal noise present in a system or a device. The lower the noise temperature, the better the performance.

noise voltage — The equivalent short-circuit input rms noise voltage that occurs at the input of a noiseless amplifier if the input terminals are shorted. It is measured at the output and divided by the amplifier gain and the square root of the bandwidth over which the measurement is made to yield units of nanovolts per square root of hertz.

noise-voltage generator — A voltage generator whose output is a random function of time.

noisy mode — In a computer, a floating-point arithmetic procedure associated with normalization. In this procedure, digits other than 0 are introduced in the low-order positions during the left shift.

no-load current — *See* exciting current, 1.

no-load losses — The losses in a transformer when it is excited at the rated voltage and frequency, but is not supplying a load.

nominal — The identifying value of a measurable property in a conductor or component. This standard value is halfway between maximum and minimum limits of the tolerance range.

nominal band — In a facsimile-signal wave, the frequency band equal to the width between the zero and the maximum modulating frequency.

nominal bandwidth — 1. The difference between the nominal upper and lower cutoff frequencies of a filter. It may be expressed in octaves, in hertz, or as a percentage of the passband center frequency. 2. The maximum band of frequencies, inclusive of guard bands, assigned to a channel.

nominal horsepower — The rated power of a motor, engine, etc.

nominal impedance — The impedance of a circuit under normal conditions. Usually it is specified at the center of the operating-frequency range.

nominal line pitch — The average separation between the centers of adjacent lines in a raster.

nominal line width — 1. In television, the reciprocal of the number of lines per unit length in the direction of line progression. 2. In facsimile transmission, the average separation between centers of adjacent scanning or recording lines.

nominal power rating of a resistor — The power that a resistor can dissipate continuously at a specified ambient temperature and for a stipulated length of time without excessive resistance drift.

nominal value — The stated or specified value, as opposed to the actual value.

nominal voltage — 1. The voltage of a fully charged storage cell when delivering rated current. 2. Average source voltage.

nomogram — Also called an alignment chart or nomograph. A computational aid consisting of two or more scales drawn and arranged so that the results of calculations may be found by the linear connection of points on them.

nomograph — *See* nomogram.

nomography — A method of representing the relation between any number of variables graphically on a plane surface such as a piece of paper.

nonblinking — The ability of a digital meter display to appear steady and free of blinking. Usually achieved by storing the reading during the same period.

nonbridging — A contact transfer in which the movable contact leaves one contact before touching the next.

nonbridging contacts — A contact arrangement in which the opening of one set of contacts occurs before the closing of another set.

noncoherent bundle — An assembly of optical fibers that will not transmit coherent images or information because the relationship of the fibers on either end of the assembly is different; can be applied to information scrambling.

noncoherent radiation — Radiation in which the waves are out of phase with respect to space and/or time.

noncombustible — *See* nonflammable.

noncomposite video — A video signal containing all information except sync.

nonconductor — 1. An insulating material — i.e., one through which no electric current can flow. 2. A substance that does not transmit certain forms of energy, such as sound, heat, and especially electricity.

noncontiguous allocation— An allocation method that assigns physically nonadjacent sectors to a file.

noncorrosive flux— A flux that does not contain acid and other substances that might corrode the surfaces being soldered.

nondegenerate amplifier— A parametric amplifier in which the pumping frequency is considerably higher than twice the signal frequency. The output is at the input-signal frequency. This type of amplifier has negative impedance characteristics, and therefore is capable of oscillation.

nondestructive readout— A type of memory operation or device in which reading out does not cause stored data to be lost from the memory. Such memories, therefore, do not require a write operation immediately after each read operation, as do destructive types.

nondeviated absorption— 1. Absorption that occurs without any appreciable slowing up of waves. 2. Normal sky-wave absorption.

nondirectional— *See* omnidirectional.

nondirectional antenna— *See* omnidirectional antenna.

nondirectional microphone— *See* omnidirectional microphone.

nondissipative stub— A lossless length of waveguide or transmission line coupled into the sides of a waveguide.

nonequivalence element— A logic element having an action that represents the exclusive-OR Boolean connective.

nonerasable storage— In a computer, a storage medium that cannot be erased and reused, e.g., punched cards, perforated paper tape, or a ROM.

nonessential loads— Electrical loads that are not necessary for plant operation, but may be desirable to maintain during longer than commercial power outages. Such loads include some lighting loads, some air-conditioning loads, and equipment used on a part-time basis.

nonfatal error— An error in a computer program that does not substantially affect the progress of the program.

nonferrous— Not made of or containing iron.

nonflammable— Also called noncombustible. Term applied to material that will not burn when exposed to flame or elevated temperatures, e.g., asbestos, ceramics, and structural metals.

nonhoming tuning system— A motor-driven automatic tuning system in which the motor starts in the direction of the previous rotation. If this direction is incorrect for the new station, the motor reverses, after turning to the end of the dial, and then proceeds to the desired station.

nonimpact printer— 1. A class of printers that forms images on paper without using a form of stamping or inking through a ribbon by an element, character slug, or hammer-needle. The shapes of the characters are stored in the system memory of the printer and are used to drive the imaging mechanism. 2. A high-speed printer using either thermal, optical, chemical, or electrical techniques. The speed of such a printer is from 30 to 21,000 lines per minute. Generally requires special paper to convert to readable copy. 3. A device that does not rely on mechanical force to print data. Methods such as heat or electricity are used instead.

nonimpact printing— Printing processes in which characters are transferred to paper by means other than physically striking the paper with a key-driven hammer; in nonimpact printing, characters are created by ink jets, thermal devices, or lasers.

noninductive— Having practically no inductance.

noninductive capacitor— Also called extended foil capacitor. A capacitor in which the inductive effects at high frequencies are reduced to a minimum. Foil layers are offset during winding so that an entire layer of foil (projecting at either end) is connected together for contact-making purposes. A current then flows laterally rather than spirally around the capacitor, and the inductive effect is minimized.

noninductive circuit— A circuit in which the inductance is reduced to a minimum or negligible value.

noninductive load— A load that has no inductance. It may consist entirely of resistance or capacitance.

noninductive resistor— 1. A wirewound resistor with little or no self-inductance. 2. A resistor constructed in such a manner that its inductance is negligible.

noninductive winding— A winding in which the magnetic fields produced by its two parts cancel each other and thereby provide a noninductive resistance.

non-interlaced— A method of sending a video image to a monitor that scans each successive video line of a picture so a full picture is painted onto the screen in one vertical sweep of the electron beam. At any given resolution, non-interlaced modes are preferable to interlaced modes but are more expensive to generate. Standard VGA (640×480) and SVGA (800×600) are always non-interlaced. *See* interlaced scanning.

noninverting connection— The closed-loop connection of an operational amplifier when the forward gain is positive for a dc signal ($0°$ phase shift).

noninverting input— A differential-amplifier input terminal for which the output signal has the same phase as the input signal.

noninverting parametric device— A parametric device whose operation depends essentially upon three frequencies — a harmonic of the pump frequency and two signal frequencies — of which one is the sum of the other plus the pump harmonics. Such a device can never provide gain at either of the signal frequencies. It is said to be noninverting because if either of the two signals is moved upward in frequency, the other will move upward in frequency.

nonionizing radiation— 1. Radiation that does not produce ionization (e.g., infrared, ultraviolet, and visible light). 2. Radiation that does not produce free electrons and ions, or electrically charged particles.

nonlinear— 1. Having an output that does not rise or fall in direct proportion to the input. 2. Characteristic of any device in which the output is not related to the input by a simple constant. Not proportional.

nonlinear capacitor— A capacitor that has a nonlinear mean-charge characteristic or peak-charge characteristic or a reversible capacitance that varies with bias voltage.

nonlinear coil— A coil with an easily saturable core. Its impedance is high at low or zero current, and is low when enough current flows to saturate the core.

nonlinear distortion— 1. Distortion that occurs when the output does not rise and fall directly in proportion to the input. The input and output values need not be of the same quantity — e.g., in a linear detector, these values are the signal voltage at the output and the modulation envelope at the input. 2. Distortion that occurs because the transmission properties of a system are dependent on the instantaneous magnitude of the transmitted signal. Note: Nonlinearity distortion gives rise to amplitude and harmonic distortion intermodulation and to flutter. 3. The generation of unwanted signals as a result of nonlinear operations on a desired signal or group of signals. (Harmonic distortion is one well-known example; if the main signal is a sine wave of frequency f, the unwanted signals appear as sine waves at $2f$, $3f$, and so on.) 4. Distortion in an electrical system in which the ratio of voltage to current in the system varies as a function of either the voltage or current.

nonlinear feedback control system — Feedback control system in which the relationships between the pertinent measures of the system input and output signals cannot be adequately described by linear means.

nonlinearity — The amount by which the measured output of a transducer, subjected to any load within its capacity, differs from an ideally linear output as defined by a straight line. The transducer output is measured on increasing load only.

nonlinear network — A network (circuit) not specifiable by linear differential equations with time as the independent variable.

nonlinear optical effects — Optical phenomena that can be observed only with the use of directional, nearly monochromatic light beams, such as those produced by a laser. Examples are optical mixing, harmonic generation, and the stimulated raman effect.

nonlinear programming — Similar to linear programming but incorporating a nonlinear objective function and linear constraints, or a linear objective function and nonlinear constraints, or both a nonlinear objective and nonlinear constraints.

nonloaded Q — Also called basic Q. The value of the Q of an electric impedance without external coupling or connection.

nonmagnetic — Material that is not attracted by a magnet and cannot be magnetized (e.g., paper, plastic, tin, glass). In a strict sense, having a permeability equal to that of air or 1.

nonmagnetic steel — A steel alloy that contains about 12 percent manganese, and sometimes a small quantity of nickel. It is practically nonmagnetic at ordinary temperatures.

nonmetallic sheathed cable — Two or more rubber- or plastic-covered conductors assembled in an outer sheath of nonconducting fibrous material that has been treated to make it flame and moisture resistant.

nonmultiple switchboard — A manual telephone switchboard in which each subscriber line is connected to only one jack.

nonnoble system — Thick-film system using conductors of copper, tungsten, nickel, molybdenum, and other nonnoble metals.

nonphysical primary — A color primary represented by a point outside the area of the chromaticity diagram enclosed by the spectrum locus and the purple boundary.

nonplanar network — A network that cannot be drawn on a plane without crossing branches.

nonpolar capacitor — See bipolar electrolytic capacitors.

nonpolar crystals — Crystals having the property that each lattice point is identical.

nonpolar electrolytic capacitor — An electrolytic capacitor that can be connected without regard to polarity. This is possible because the dielectric film is formed on both electrodes.

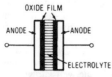

Nonpolar electrolytic capacitor.

nonpolarized relay — See neutral relay.

nonrecursive filter — See digital filter.

nonrenewable fuse unit — A fuse unit that cannot be readily restored for service after operation.

nonreset timer — A timer that cannot be reset by electrical means.

nonresonant line — A transmission line with a neutral resonant frequency different from that of the transmitted signal.

nonreturn-to-zero — A method of writing information on a magnetic surface in which the current through the write-head winding does not return to zero after the write pulse.

nonsaturated color — A color that is not pure — i.e., one that has been mixed with white or its complementary color.

nonsaturated logic — A type of logic circuit in which short delay times are achieved by preventing transistors from saturating.

nonshorting switch — Switch in which the width of the movable contact is less than the distance between contact clips so that one circuit is broken before another is completed.

nonsimultaneous transmission — Usually, transmission in which data can be moved in only one direction at a time by a device or facility.

nonsinusoidal wave — Any wave that is not a sine wave. It therefore contains harmonics.

nonstorage camera tube — A television camera tube in which the picture signal is always in proportion to the intensity of the illumination on the corresponding area of the scene being televised.

nonswitched line — A communications link that is permanently installed between two points.

nonsynchronous — Not related in frequency, speed, or phase to other quantities in a device or circuit.

nonsynchronous starting torque — The maximum load torque with which motors can start. However, the motor will not come to a synchronous speed.

nonsynchronous vibrator — A vibrator that interrupts a direct-current circuit at a frequency unrelated to the other circuit constants. It does not rectify the resulting stepped-up alternating voltage.

nonthermal radiation — The radiation given off by energetic particles not in thermal equilibrium.

nontrip-free circuit breaker — A breaker that can be maintained closed by manual override action while a tripping condition persists.

nonuniform field — A field in which the scalar (or vector) at that instant does not have the same value at every point in a given region.

nonvolatile memory — 1. A memory whose stored data is undisturbed by removal of operating power. 2. A type of computer system memory offering preservation of data storage during power loss or system shut-down. Magnetic-core read/write memory systems are typically nonvolatile and therefore do not require reloading to restore programs and data when system power is applied. See also volatile memory.

nonvolatile storage — 1. Storage media that retain information in the absence of power and that will make the information available when power is restored. 2. A storage medium that retains its data in the absence of power.

nonwetting — A condition whereby a surface has contacted molten solder, but has had none of the solder adhere to it.

nonwirewound trimming potentiometer — A trimming potentiometer characterized by the continuous nature of the surface area of the resistance element to be contacted. Contact is maintained over a continuous unbroken path. The resistance is achieved by using material compositions other than wire, such as carbon, conductive plastic metal film, and cement.

no-operation instruction — Also called blank instruction, waste instruction, or no-op. An instruction for a computer to do nothing but process the next instruction in sequence. *See also* skip.

NOR — 1. A function of A and B that is true if both A and B are false. 2. Contraction of NOT-OR, the logical negation of the OR function. A NOR B is false if either A or B is true; true otherwise.

NOR device — 1. A device that has its output in the logical 1 state if and only if all the control signals assume the logical 0 state. 2. An OR circuit that delivers an inverted output signal.

NOR element — A gate circuit having multiple inputs and one output that is energized only if all inputs are zero.

NOR gate — An OR gate followed by an inverter to form a binary circuit in which the output is logical 0 if any of the inputs is logical 1 and is logical 1 only if all the inputs are logical 0. With the opposite logic polarity, this type of gate is a NAND gate.

norm — 1. The mean or average. 2. A customary condition or degree.

normal condition — The deenergized condition of a relay.

normal contact — A contact that in its normal position closes a circuit and permits current to flow.

normal distribution — 1. The most common frequency distribution in statistics. The probability curve is bell-shaped, and the greatest probability occurs at the arithmetical average (i.e., at the top of the curve). The probability of occurrence of a particular value is shown by the areas between two abscissa values on the curve. *See also* Gaussian distribution. 2. A particular statistical distribution. For a distribution to be classified as a normal distribution, it must be unimodal — that is, most of the observations must fall fairly close to one mean — and symmetrical; that is to say, a deviation from the mean is as likely to be plus as it is likely to be minus. When graphed, the normal distribution takes the form of a bell-shaped curve.

normal electrode — A standard electrode used for measuring electrode potentials.

normal failure period — That period of time during which an essentially constant failure rate exists.

normal impedance — *See* free impedance.

normal induction — The limiting induction, either positive or negative, in a magnetic material that is under the influence of a magnetizing force that varies between two extremes.

normal induction curve — 1. The curve obtained by plotting B (induction) against H (magnetizing force), starting from a totally demagnetized state. 2. The locus of the tips of a series of hysteresis loops, obtained by cycling the magnetizing force, H, in successively increasing steps. Nearly coincident with the initial magnetization curve.

normalization — The transforming of signals to a common basis — e.g., adjusting two signals, representing the same spoken word but differing in loudness, to the same loudness.

normalize — 1. To adjust the representation of a quantity so that it lies within a prescribed range. 2. In a computer, to shift all digits of a word to the right or left to accommodate the maximum number of digits. 3. To divide a quantity (e.g., voltage, impedance, frequency) by a reference of the same dimension, thereby making the result dimensionless.

normalized admittance — The reciprocal of normalized impedance.

normalized impedance — An impedance divided by the characteristic impedance of a waveguide.

normalized plateau slope — The slope of the substantially straight portion of the counting-rate-versus-voltage characteristic of a radiation counter tube, divided by the quotient of the counting rate and the voltage at the Geiger-Mueller threshold.

normal linearity — A manner of expressing linearity as the deviation from a straight line in terms of a given percentage of the output at a certain stimulus value, usually the full-scale value.

normally closed — Symbolized by nc. Designation applied to the contacts of a switch or relay when they are connected so that the circuit will be completed when the switch is not activated or the relay coil is not energized.

normally closed contact — Also called break contact. A contact pair that is closed when the coil of a relay is not energized.

normally closed switch — 1. A switch that passes current until actuated. 2. A switch in which the contacts are closed when no external forces act on the switch.

normally high — A device in which the output is high in voltage in the rest condition.

normally low — A device in which the output is low in voltage in the rest condition.

normally open — Symbolized by no. Designation applied to the contacts of a switch or relay when they are connected so that the circuit will be broken when the switch is not activated or the relay coil is not energized.

normally open contact — A contact pair that is open when the coil of a relay is not energized.

normally open switch — 1. A switch that must be actuated to pass current. 2. A switch in which the contacts are open (separated) when no external forces act on the switch.

normal mode — The expected or usual operating conditions, such as the voltage that occurs between the two input terminals of an amplifier.

normal-mode interference — Interference that appears between the terminals of a measuring circuit.

normal mode of vibration — A characteristic distribution of vibration amplitudes among the parts of the system, each part of which is vibrating freely at the same frequency. Complex free vibrations are combinations of these simple vibration forms.

normal-mode voltage — 1. The actual signal voltage developed by a transducer. 2. The difference voltage between two input-signal lines.

normal operating period — The time interval between debugging and wear-out.

normal permeability — Ratio of the normal induction to the corresponding magnetizing force. In the cgs system, the flux density in a vacuum is numerically equal to the magnetizing force.

normal position — The position of the relay contacts when the coil is not energized.

normal propagation — The phenomenon of passing radio waves through space when atmospheric and/or ionospheric conditions are such as to permit the passage with little or no difficulty.

normal record level — In direct recording, the amount of record-head current required to produce 1 percent third-harmonic distortion of the reproduced signal at the record-level set frequency. Such distortion must result from magnetic-tape saturation, not from electronic circuitry.

normal-stage punching—In a computer, a card-punching system in which only the even-numbered rows are punched on the British standard card.

north pole—1. The pole of a magnet that points toward the north magnetic pole of the earth. (The lines of force internal to a magnet are assumed to leave from its north pole.) 2. A pole that attracts the south-seeking pointer of a field compass.

Norton's theorem—The current in any impedance Z_R, connected to two terminals of a network, is the same as though Z_R were connected to a constant-current generator whose generated current is equal to the current that flows through the two terminals when these terminals are short-circuited, the constant-current generator being in shunt with an impedance equal to the impedance of the network, looking back from the terminals in question.

NOT-AND circuit—An AND gating circuit that inverts the pulse phase.

notation—1. A manner of representing numbers. Some of the more important notation scales are as follows:

Base	Name
2	Binary
3	Ternary
4	Quaternary, tetral
5	Quinary
8	Octonary, octal
10	Decimal
12	Duodecimal
16	Hexadecimal, sexadecimal
32	Duotricenary
2.5	Biquinary

2. The act, process, or method of representing facts or quantities by a system or set of marks, signs, figures, or characters

notch—1. A rectangularly shaped depression that extends below the sweep line of the radar indicator in some equipment. 2. On a graph (a graph of frequency response, in particular), a point where a curve dips sharply and returns equally sharply to its original value.

notch antenna—An antenna that forms a pattern by means of a notch or slot in a radiating surface. Its characteristics are similar to those of a properly proportioned metal antenna and may be evaluated with similar techniques.

notch filter—1. An arrangement of electronic components designed to attenuate or reject a specific frequency band with sharp cutoff at either end. *See also* band-reject filter. 2. A band-elimination filter used to prevent the passage of specific frequencies. 3. A filter that attenuates a very narrow portion of the frequency spectrum, but will pass signals on either side. 4. A special filter, designed to reject a very narrow band of frequencies.

notch gate—The early and late gate display on the range CRT. It appears as a negative deflection of the range base line equal in width to the early and late gates.

notching—A term indicating that a predetermined number of separate impulses is required to complete operation of a relay.

notching circuit—A control circuit used in a cathode-ray oscilloscope to expand portions of the displayed image.

NOT circuit—Also called inverter circuit. A binary circuit with a single output that is always the opposite of the single input.

NOT device—A device that has its output in the logical 1 state if and only if the control signal assumes the logical 0 state. The NOT device is a single-input NOR device.

note—1. The pitch, duration, or both of a tone sensation. 2. The sensation itself. 3. The vibration causing the sensation. 4. The general term when no distinction is desired between the symbol, the sensation, and the physical stimulus. 5. A single musical tone. The notes of the musical scale are referred to by letters running alphabetically from A to G for the white keys. A black key may be called a sharp of the note below it or a flat of the note above it. Each note has a frequency exactly one-half that of the corresponding note in the next higher octave.

notebook computer—A personal computer that is easily portable and powered by a battery rather than an electrical current. Typical dimensions are "$8\frac{1}{2} \times 11 \times 2$ to 4" thick.

NOT gate—1. An inhibitory circuit equivalent to the logical operation of negation (mathematical complement). The output of the circuit is energized only when its single input is not energized, and there will be no output if the input is energized. 2. A single-control device that evidences the appearance of a positive-pressure control signal with a single zero output signal. With no positive control signal, the zero NOT output signal ceases. A NOR gate with a single control connection performs as a NOT gate.

NOT majority—A logic function in which the output is false if more than half the number of inputs are true; otherwise the output is true.

noval base—A nine-pin glass base for a miniature electron tube. For orientation of the tube into its socket, the spacing between pins 1 and 9 is greater than between the other pins.

novar—A beam-power tube that has a nine-pin base.

novice license—A class of radiocommunication amateur license. Its requirements are the easiest of all, but novice operation is limited. Transmitters must be crystal controlled, and the maximum permissible plate input power is 75 watts. Novice licensees are restricted to certain frequencies (3.70–3.75, 7.15–7.20, 21.10–21.25, and 145–147 MHz, 222–225 MHz, 902–928 MHz, and 1.27–1.295 GHz).

noys—A measure of perceived noisiness; the noys scale is linear.

npin transistor—An npn transistor with a layer of high-purity germanium added between the base and collector to extend the frequency range.

npip transistor—An intrinsic-region transistor in which the intrinsic region is located between two p-regions.

n-plus-one address instruction—In a computer, a multiple-address instruction in which one address serves to specify the location of the next instruction of the normal sequence to be executed.

n+-type material—Heavily doped n-type material, formed by introducing donor impurities into a silicon substrate. Conduction takes place by the movement of free electrons.

npnp transistor—A hook transistor with a p-type base, n-type emitter, and a hook collector.

npn semiconductor—Double junction formed by sandwiching a thin layer of p-type material between two layers of n-type material of a semiconductor.

npn transistor—A transistor with a p-type base and an n-type collector and emitter. In construction, it consists of a thin layer of p-type semiconductor material sandwiched between two layers of n-type semiconductor material.

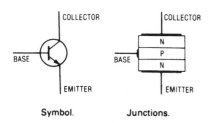

Symbol.   Junctions.

npn transistor.

NPO — A commonly used code that is synonymous with EIA code COG. Both indicate a temperature characteristic of ±30 ppm/°C or less for a capacitor.

npo body (negative positive zero) — A designation referring to a class of capacitors in the temperature-compensating series that have an essentially nonvariant dielectric constant over a specified temperature range (commonly +25°C to +85°C).

np semiconductor — Material in a region of transition between n- and p-type material.

N-quadrant — One of the two quadrants where the N-signal is heard in an A-N radio range.

n-region — Also called n-zone. The region where the conduction-electron density in a semiconductor exceeds the hole density.

NRZ — Abbreviation for nonreturn to zero. Data transmission and recording in which the transition represents the data, and the state remains unchanged until the next transition transmits more data on that line.

NRZI — Abbreviation for nonreturn-to-zero interchange. A recording technique using only negative and positive magnetic saturation states. Because NRZI does not use a neutral or unmagnetized state, usually called zero (0), there is never a return to zero in the recording. In the most common NRZI mode, a transition from one saturation to another coincides with a binary 1. No transition coincides with a binary 0.

ns — Letter symbol for nanosecond (10^{-9} s).

N-scan — *See* N-display.

N-scope — *See* N-display.

N-signal — A dash-dot signal heard in either a bisignal zone or an N-quadrant of a radio range.

NSOS — Same as PSOS (p-channel silican gate devices), except it denotes n-channel devices.

n-**terminal network** — A network with *n* accessible terminals.

*n*th **harmonic** — A harmonic whose frequency is *n* times the frequency of the fundamental component.

NTSC — Abbreviation for National Television Standard Committee. The color system established in the United States and North America and adopted by numerous other countries. The field rate for NTSC is 60 Hz with 525 lines per screen, and the subcarrier transmission method is a straight phase- and amplitude-modulation system for chroma using a subcarrier frequency of 3.579545 MHz.

NTSC color bar pattern — The standard television test pattern of six adjacent color bars including the three primary colors plus their three complementary shades.

NTSC signal — A 3.579545-MHz (3.58-MHz) signal, the phase of which is varied with the instantaneous hue of the televised color, and the amplitude of which is varied with the instantaneous saturation of the color, as specified by the National Television System Committee.

NTSC triangle — 1. On a chromaticity diagram, a triangle that defines the gamut of color obtainable through the use of phosphors. 2. The triangle in a chromaticity diagram joining the chromaticities of the NTSC phosphors, and containing all chromaticities that can be produced by additive mixture of their light.

n-type — 1. Refers to an excess of negative electrical charges in a semiconductor material. Natural silicon is made to be p-type by the addition of an acceptor impurity. 2. Semiconductor material whose impurities produce free electrons in the compound, leading to conduction.

n-type conductivity — The conductivity associated with conduction electrons in a semiconductor.

n-type crystal rectifier — A crystal rectifier in which forward current flows when the semiconductor is more negative that the metal.

n-type material — 1. A crystal of pure semiconductor material to which has been added an impurity (an electron donor such as arsenic or phosphorous) so that its characteristics are altered and electrons serve as the majority charge carriers. 2. A quadrivalent semiconductor material, with electrons as the majority charge carriers, that is formed by doping with donor atoms.

n-type region — Portion of semiconductor material containing a small number of dopant atoms that have an extra (free) electron in their outer orbit. The n-region is a source of mobile negative charges.

n-type semiconductor — 1. An intrinsic semiconductor in which the conduction-electron density exceeds the hole density. By implication, the net ionized impurity concentration is a donor type. 2. A semiconductor in which conduction is aided by the presence of more free electrons than holes. 3. Impurity semiconductor material containing a predominance of donor dopants, and in which conduction-band electrons normally form the principal current carriers. 4. A semiconductor material, such as germanium or silicon, with a small amount of impurity, such as antimony, arsenic, or phosphorous, added to increase the supply of free electrons. Such a material conducts electricity through the movement of electrons. 5. A semiconductor type in which the density of holes in the valence band is exceeded by the density of electrons in the conduction band. N-type behavior is induced by the addition of donor impurities, such as arsenic or phosphorus, to the crystal structure of silicon.

n-type semiconductor material — An extrinsic semiconductor material in which the conduction-electron density greatly exceeds the mobile-hole density.

nuclear battery — Also called an atomic battery. 1. A battery that converts nuclear energy into electrical energy. 2. Direct-conversion secondary cell or battery using radioisotopes as the energy source.

nuclear clock — A clock in which the time base is provided by the mean time interval between two successive disintegrations in a radioactive source.

nuclear magnetic resonance — The flipping over of a particle, such as a proton, as the result of the application of an alternating magnetic field at right angles to a steady magnetic field in which the particle is placed.

nucleation — The occurrence in an existing phase or state of a new phase or state.

nucleonics — The application of nuclear science in physics, chemistry, astronomy, biology, industry, and other fields.

nucleus — The core of an atom. It contains most of the mass and has a positive charge equal to the number of protons it contains. Its diameter is about one ten-thousandth that of the atom. Except for the ordinary hydrogen atom, the nuclei of all other atoms consist of protons and neutrons tightly locked together.

nude contact — A contact with a locking member that remains in the insert at all times.

null — 1. A balanced condition that results in zero output from a device or system. 2. To oppose an output that differs from zero so that it is returned to zero. 3. The minimum output amplitude (ideally, zero) in direction-finding systems in which the amplitude is determined by the direction from which the signal arrives or by the rotation in bearing of the system's response pattern. 4. In a computer, a lack of information, as opposed to a zero or blank for the presence of no information.

null balance — A condition in which two or more signals are summed and produce a result that is essentially zero.

null detection — A method of making DF measurements. The antenna is turned to the point where the received signal is weakest. The true bearing of the signal is then found by noting the antenna direction and using a correction factor.

null detector — 1. An apparatus that senses the complete balance, or zero-output condition, of a system or device. 2. A circuit to detect when there is no current or when no voltage is present.

null-detector circuit — A circuit that determines when two circuits are in exact electrical balance, i.e., having the same voltages or currents.

null-frequency indicator — A device that indicates frequency by heterodyning two electrical signals together to give a zero beat indication.

null indicator — A device that indicates when current, voltage, or power is zero.

null method — Also called zero method. A method of measurement in which the circuit is balanced, to bring the pointer of the indicating instrument to zero, before a reading is taken (e.g., in a Wheatstone bridge, or in a laboratory balance for weighing purposes).

null modem — A device connecting two asynchronous adapters, thus enabling communications.

null-spacing error — In a resolver, the difference between 180° and the angle between null positions of the output winding with respect to one input winding.

number — An abstract mathematical symbol for expressing a quantity. In this sense, the manner of representing the number is immaterial. Take 26, for example. This is its decimal form but it could be expressed as binary 011010 and still mean the same. Some common numbering systems are binary (base 2), quinary (base 5), octonary or octal (base 8), and decimal (base 10).

number crunching — 1. The act of performing complex numerical operations. 2. Refers to long or complex calculations. 3. Traditionally, using a computer for large or complex computational tasks; now applied to any task involving numbers.

number of overshoots — The total number of times that the transient response goes to a minimum or maximum value that is outside a specified range.

number of scanning lines — The ratio of the scanning frequency to the frame frequency.

number system — Any system for the representation of numbers. *See also* positional notation.

numeric — In computers, consisting entirely or partially of digits, as distinguished from alphabetic composition.

numerical analysis — 1. The providing of convenient methods for finding useful solutions to mathematical problems and for obtaining useful information from available solutions that are not expressed in easily handled forms. 2. Methods for finding useful solutions to problems once they have been stated mathematically. Numerical analysis also studies errors that arise during calculations and how they may be kept within limits.

numerical aperture — Abbreviated NA. 1. A measure of the light-gathering capability of a fiber. Mathematically, it is expressed as the sine of one-half of the acceptance angle. 2. The characteristic of an optic conductor in terms of its acceptance of impinging light. Degree of openness, light-gathering ability, and angular acceptance are terms describing this characteristic. 3. A number that indicates a fiber's ability to accept light and shows how much light can be off-axis and still be accepted by the fiber. 4. A measure of the maximum acceptance angle for light propagation in an optical fiber; at angles larger than this there is no longer internal reflection.

$$NA = \sin \theta = \sqrt{(n_1^2 - n_2^2)}$$

where

θ = one-half of the input cone angle
n_1 = index of refraction of core
n_2 = index of refraction of cladding

numerical control — Abbreviated NC. 1. Descriptive of systems in which digital computers are used for the control of operations, particularly of automatic machines (e.g., drilling or boring machines, wherein the operation control is applied at a discrete point in the operation or process). Contrasted with process control, in which control is applied continuously. 2. The technique of controlling a machine or process through the use of command instructions in coded numerical form. 3. Controlling a machine or process by means of digitally encoded numeric data. The control information can be prerecorded, or the equipment can be under the direct control of a computer. 4. A technique whereby prerecorded information in symbolic form represents the complete instructions for the operation of a machine. 5. The use of a dedicated controller to direct the operations of a machine tool.

numerical control system — A system controlled by direct insertion of numerical data at some point. The system must automatically interpret at least some portion of the data.

numerical data — Data expressed in terms of a set of numbers or symbols that can assume only discrete values or configurations.

numerical-readout tube — A gas-filled cold-cathode digital-indicator tube having a common anode and containing stacked metallic elements in the form of numerals. When a negative voltage is applied to one of them, the element then becomes the cathode of a simple gas-discharge diode.

numeric coding — 1. A system of abbreviation used in the preparation of information for machine acceptance in which all information is reduced to numerical quantities, in contrast to alphabetic coding. 2. System of coding, with digits as code characters, used in preparing information for input to a computer.

nutating feed — In a tracking radar, an oscillating antenna feed that produces an oscillating deflection of the beam without changing the plane of polarization.

nutation field — The time-variant, three-dimensional field pattern of a directional or beam-producing antenna having a nutating feed.

nuvistor — An electron tube in which all electrodes are cylindrical and are closely spaced, one inside the other, in a ceramic envelope.

nV — Letter symbol for nanovolt.

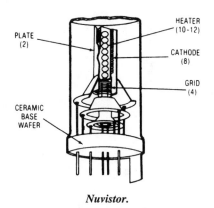

PLATE
(2)

HEATER
(10-12)

CATHODE
(8)

GRID
(4)

CERAMIC
BASE
WAFER

Nuvistor.

n-**variable cube** — *See n*-cube.

nW — Letter symbol for nanowatt.

nybble — *See* nibble.

nylons — Family of polymers with outstanding toughness and wear resistance, low coefficient of friction, excellent electrical properties, and chemical resistance.

Nyquist diagram — A plot, in rectangular coordinates, of the real and imaginary parts of factor $\mu\beta$ for frequencies from zero to infinity in a feedback amplifier — where μ is the amplification in the absence of feedback, and β is the fraction of the output voltage superimposed on the amplifier input.

Nyquist interval — The maximum separation in time that can be given to regularly spaced instantaneous samples of a wave of a given bandwidth for complete determination of the waveform of the signal. Numerically it is equal in seconds to one-half the bandwidth.

Nyquist rate — Of a channel, the maximum rate at which independent signal values can be transmitted over the specified channel without exceeding a specified amount of mutual interference.

Nyquist theorem — A theorem that states that if a continuous bandwidth-limited signal contains no frequency components higher than fC, then the original signal can be recovered without distortion if it is sampled at a rate of at least $2fC$. This theorem applies to A/D converter applications as well as data transmission density over limited-bandwidth channels. *See also* sampling theorem.

n-zone — *See* n-region.

O

OAO — Abbreviation for Orbiting Astronomical Observatory.

object — A data element that provides a well-defined external visibility but no insight into its internal structure.

object code — 1. The output of an assembler or compiler that will execute on the target processor. Linking and loading may be required before this code can execute directly on the processor. 2. Machine-language instructions that are actually executed by a computer system. They are produced by a compiler from an application program written in a higher-level language. 3. A series of binary 1s and 0s that can be used by a computer directly.

object computer — The computer of the receiving end of a communication attempt such as a fax or modem.

object language — The computer language in which the output from a compiler or assembler is expressed.

object module — 1. A module that serves as the output of an assembler or compiler and the input to a linkage editor. 2. An object program plus the control information a computer needs to load it and combine it with other object programs.

object-oriented programming — A programming environment in which symbols called objects communicate by passing messages. This approach permits the simulation of multiple unrelated, but simultaneously occurring, processes.

object program — 1. A computer program that is the output of an automatic coding system. The object program may be a machine-language program ready for execution, or it may be in an intermediate language. 2. The binary form of a source program produced by an assembler or a compiler. The object program is composed of machine-coded instructions that the computer can execute. 3. The machine-language program that a compiler produces from the programmer's source program in high-level language. It is the one program that the computer actually executes. 4. A set of problem-solving machine-language instructions obtained through the compilation or assembly of the related source program.

oblique-incidence transmission — Transmission by means of a radio wave that travels obliquely up to the ionosphere and down again.

Oboe — A radar navigation system consisting of two ground stations that measure the distance to an airborne transponder beacon and relay this information to the aircraft.

obsolescence-free — Not likely to become outdated within a reasonable time.

occlude — To absorb; e.g., some metals will occlude gases, which must be driven out before the metals can be used in the electrodes or supports of a vacuum tube.

occluded contaminants — Contaminants that have been absorbed by a material.

occluded gas — Gas that has been absorbed by a material (e.g., by the electrodes, supports, leads, and insulation of a vacuum tube).

occupied bandwidth — 1. The frequency bandwidth such that, below its lower and above its upper frequency limits, the mean powers radiated are each equal to 0.5 percent of the total mean power radiated by a given emission. 2. The band of frequencies that a given emission type uses.

OCR or ocr — Abbreviation for optical character recognition 1. A technology that enables a machine to automatically read and convert typewritten, machine-printed, mark-sensed, or hand-printed characters into electrical impulses for processing by a computer. OCR devices are designed to read man-readable data and convert it directly to a computer language. 2. The machine recognition of printed or written characters based on inputs from photoelectric transducers. 3. A device or scanner that can read printed or typed characters and convert them into a digital signal for input into a data or word processor. OCR units in word processing applications usually read special machine-readable type fonts. 4. A mechanized method of collecting data involving the reading of hand-printed or special-character fonts. If handwritten, the information must adhere to predefined rules of size, format, and location on the form.

octal — 1. A numbering system based on 8 and using the digits 0 through 7. Since 8 is a power of 2, binary programs are easily converted into octal. Binary 11010110 (decimal 214) is written in octal as 326. 2. *See* notation. 3. Pertaining to a characteristic or property involving a selection, choice, or condition in which there are eight possibilities.

octal base — An eight-pin tube base (although unneeded pins are often omitted without changing the position of the remaining pins). An aligning key in the center of the base assures correct insertion of the tube into the socket.

octal, binary coded — Pertaining to a binary-coded number system with the radix 8, in which the natural binary values of 0 through 7 are used to represent octal digits with values from 0 to 7.

octal debugging technique — Also called online debugging technique (ODT). A system program designed to help the user debug object programs interactively. All addresses, registers, and memory location contents are expressed in octal notation. Letters and symbols make up the command set for ODT.

octal digit — One of the symbols 0, 1, 2, 3, 4, 5, 6, and 7 when used in numbering in the scale of 8.

octal fraction — A shorthand expression of the binary contents of a half word.

octal fractional — A quantity less than a whole number referenced to a numbering system that has the radix eight.

octal loading program — A computer utility program with provision for making changes in programs and tables that are in core memory and drum storage, reading in words that are coded in octal notation on punched cards or tapes.

octal numbering system — A numbering system based on powers of 8. This system is used extensively in computer work because it is derived simply from binary numbering.

octal plug — An eight-pin male connector with a locating key for proper orientation.

octave — 1. The interval between two sounds having a basic frequency ratio of 2; by extension, the interval between any two frequencies having a ratio of 2 : 1. 2. A reference point for frequency or pitch; e.g., an 880-Hz tone is one octave higher than a 440-Hz tone and one octave lower than a 1760-Hz tone. An increase of one octave doubles the frequency. Dropping one octave reduces the frequency by one-half.

octave band — A band of frequencies whose limits have the ratio 2 : 1.

octave-band pressure level — Also called octave pressure level. The pressure level of a sound for the frequency band corresponding to a specified octave.

octave pressure level — *See* octave-band pressure level.

octode — An eight-electrode electron tube containing an anode, a cathode, a control electrode, and five additional grids.

octonary — *See* notation.

octonary signaling — A mode of communication in which information is represented by the presence and absence, or plus and minus variations, of eight discrete levels of a parameter of the signaling medium.

odd-even check — An automatic computer check in which an extra digit is carried along with each word to determine whether the total number of 1s in the word is odd or even, thus providing a check for proper operation. *See* parity check.

odd harmonic — Any harmonic whose frequency is the fundamental frequency multiplied by an odd number. The odd harmonics of 60 Hz are 180 Hz, 300 Hz, 420 Hz, etc.

odd-line interlace — The double-interlace system in which, since there are an odd number of lines per frame, each field contains a half line. In the 525-line television frame used in the United States, each field contains 262.5 lines.

odograph — Automatic electronic map tracer used in military vehicles for map making and land navigation. It automatically plots, on an existing map or on cross-sectional paper, the exact course taken by the vehicle. This is done by phototubes and thyratrons, which transfer the indication of a precision magnetic compass onto a plotting unit actuated by the speedometer drive cable. A pen then traces the course.

ODT — *See* octal debugging technique.

Oe — Letter symbol for oersted.

oersted — In the cgs electromagnetic system, the unit of magnetizing force equal to $1000/4\pi$ ampere-turns per meter. At any point in a vacuum, the magnetic intensity in oersteds is equal to the force in dynes exerted on a unit magnetic pole at that point. Letter symbol: Oe.

off bond — Bond that has some portion of the bond area extending off the bonding point.

off-center display — A PPI display whose the center does not correspond to the position of the radar antenna.

off-delay — A circuit that retains an output signal some definite time after the input signal is removed.

off-delay timers — A reset timer that is started by the opening of a circuit. Does not reset on power interruption.

off-ground — The voltage above or below ground at which a device is operated.

offhook or **off-hook** — 1. A condition that occurs when the telephone handset is lifted from its mounting, thus causing the hookswitch to operate (close) and closing the loop to the central office. The offhook condition indicates a busy condition to incoming calls. 2. The signal that is sent to the central office, i.e., loop closed, current flowing as a result of the offhook condition. 3. The condition that indicates a closed loop or the active state of a telephone customer's line. 4. A modem automatically answering a call on the dial network is said to go off-hook. The opposite condition is on-hook.

offhook service — Priority telephone service in which a connection from caller to receiver is afforded by removing the phone from its cradle or hook.

office code — The first three digits of a seven-digit telephone number assigned to a subscriber.

off-limit contacts — Contacts on a stepping relay used to indicate that the wiper has reached the limiting position on its arc and must be returned to normal before the circuit can function again.

offline — 1. Not physically connected to or using a communications medium. 2. The condition in which a user, terminal, or other device is not connected to a computer or is not actively transmitting via a network. Opposite of online. 3. Pertaining to equipment or devices not under direct control of the central processing unit. May also be used to describe the terminal equipment that is not connected to a transmission line. 4. Describing a communication system that has no direct connection between the communication line and the device which originates the message. 5. Disconnected from the computer system. 6. Equipment or devices in a system that are not under the direct control of the system.

offline cipher — Method of encryption that is not associated with a particular transmission system and in which the resulting cryptogram can be transmitted by any means.

offline equipment — Peripheral equipment that is not in direct communication with the computer central processing unit.

offline operation — 1. A control system in which the computer does not respond immediately or directly to the events of the controlled process; some time may elapse between the occurrence of an event and the reaction of the computer to it. 2. In a computer system, operation of peripheral equipment independent from the central processor, e.g., the transcribing of magnetic tape information to printed form. 3. Operation performed while the computer is not engaged in monitoring or controlling a process or operation. 4. Data-processing operation that is handled outside of the regular computer program. For example, the computer could generate a magnetic tape, which would then be used to generate a report offline while the computer was doing another job.

offline system — 1. A kind of teleprocessing system in which there must be human operations between the original recording functions and the ultimate data-processing function. 2. A system of processing information in which the data is submitted to some intermediate step (such as punched cards or magnetic tape) before it is processed by the computer.

offline unit — In a computer, the input/output device or auxiliary equipment not under direct control of the central processing unit.

off-normal contacts — Relay contacts that assume one condition when the relay is in its normal position and the reverse condition for any other position of the relay.

off-period—That portion of an operating cycle during which an electron tube or semiconductor is nonconducting.

offset—1. The measure of unbalance between halves of a symmetrical circuit. Generally caused by differences in transistor betas or in values of biasing resistors. 2. The change in input voltage necessary to cause the output voltage in a linear amplifier circuit to be zero. 3. In digital circuits, the dc voltage on which a signal is impressed. 4. The difference between the desired value or condition and the value or condition actually attained. 5. The voltage that must be applied between the inverting and noninverting inputs of an operational amplifier to cause the output voltage to go to zero. 6. A number that a computer adds to a base address to get a new effective address. 7. In a Hall-effect element, the output voltage at 0 gausses. 8. The difference between an amateur radio repeater's input and output frequencies. The offset on 2 meters is generally 600 kHz.

offset angle—In lateral-disc reproduction, the smaller of the two angles between the projections, into the plane of the disc, of the vibration axis of the pickup stylus and the line connecting the vertical pivot (assuming a horizontal disc) of the pickup arm with the stylus point.

offset binary—A natural binary code except that it is offset (usually ½ scale) in order to represent negative and positive values. Maximum negative scale is represented to be all 0s while maximum positive scale is represented as all 1s. Zero scale (actually center scale) is then represented as a leading 1 and all remaining 0s.

offset current—1. The difference in current into the two inputs of an operational amplifier required to bring the output voltage to zero. 2. The difference between the two bias currents drawn by the inputs of a differential input stage of an electronic meter.

offset error—1. The error by which the transfer function fails to pass through the origin, referred to the analog axis. This is adjustable to zero in available analog-to-digital converters. 2. The analog value by which the transfer function of a D/A or A/D converter fails to pass through zero; it is generally specified in millivolts or in percent of full scale.

offset nulling—The adjustment of an external resistor in an operational amplifier circuit so that the output voltage is made zero when both input potentials are zero.

offset stacker—In a computer, a card stacker having the ability to stack cards selectively under machine control so that they protrude from the balance of the stack, thus giving physical identification.

offset voltage—1. The difference in voltage at the two inputs of an operational amplifier required to bring the output voltage to zero. 2. The deviation from zero output voltage of an amplifier having zero input voltage.

off-state—The condition of a thyristor corresponding to the portion of the principal characteristic between the origin and the breakover point or points.

off-state current—The principal current when a thyristor is in the off state.

off-state voltage—The principal voltage when a thyristor is in the off state.

off-target jamming—Employment of a jammer away from the main units of the force. This is done to prevent the enemy from monitoring the jamming signals and using them to pinpoint the location of the force.

off-the-line supply—A supply in which raw dc is obtained by rectifying the line voltage directly without using an isolation transformer. Most switchers are off-the-line units.

OGO—Abbreviation for Orbiting Geophysical Observatory.

ohm—Symbolized by the Greek letter *omega* (Ω). 1. The unit of resistance. It is defined as the resistance, at 0°C, of a uniform column of mercury 106.300 cm long and weighing 14.451 grams. 2. The electrical resistance between two points of a conductor when a constant difference of potential of 1 volt, applied between these points, produces in this conductor a current of 1 ampere, the conductor not being the source of any electromotive force.

ohmic contact—1. Between two materials, a contact across which the potential difference is proportionate to the current passing through it. 2. A contact between two materials across which the voltage drop is the same regardless of the direction of current. 3. An electrical connection that passes current linearly in both directions. 4. In the context of semiconductor device design, a contact to a semiconductor crystal expressly designed so that it does not possess any of the unilateral properties of a normal metal semiconductor or p-type/n-type semiconductor junction. Usually all exterior electrode connections to a device chip are made by means of ohmic contacts.

ohmic heating—The energy imparted to charged particles as they respond to an electric field and make collisions with other particles. The name was chosen due to the similarity of this effect to the heat generated in an ohmic resistance due to the collisions of the charge carriers in their medium.

ohmic resistance—Resistance to direct current.

ohmic value—The resistance in ohms.

ohmmeter—A direct-reading instrument for measuring electric resistance. Its scale is usually graduated in ohms, megohms, or both. If the scale is graduated in megohms, the instrument is called a megohmmeter; if the scale is calibrated in kilohms, the instrument is a kilohmmeter.)

ohmmeter zero adjustment—In an ohmmeter, a potentiometer or other means of compensating for the drop in battery voltage with age. Usually a knob is rotated until the meter pointer is at zero on the particular scale being used.

Ohm's law—The voltage across an element of a dc circuit is equal to the current in amperes through the element, multiplied by the resistance of the element in ohms. Expressed mathematically as $E = IR$. The other two equations obtained by transposition are $I = E/R$ and $R = E/I$.

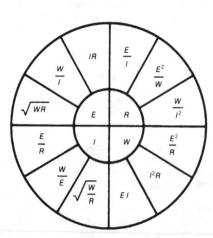

Ohm's law.

ohms per square — Also shown as ohms/square. 1. A unit of measurement of resistance by topological shape. A resistor topology can be considered to consist of continuous squares. The value of a resistor is equal to the number of squares times the ohms per square. 2. A unit of measure for thin- and thick-film resistors. The value of a film resistor depends on the sheet resistivity (ohms per square) and the aspect ratio (length to width). Resistors of different sizes but equal aspect ratios will have the same value. 3. The unit of sheet resistance or, more properly, of sheet resistivity.

ohms per square per mil — The unit of sheet resistivity. *See* sheet resistivity.

ohms per volt — A sensitivity rating for voltage-measuring instruments (the higher the rating, the more sensitive the meter). On any particular range, it is obtained by dividing the input resistance of the instrument (in ohms) by the full-scale voltage value of that range.

oil circuit breaker — A circuit breaker in which the interruption occurs in oil to suppress the arc and prevent damage to the contacts.

oiled paper — Paper that has been treated with oil or varnish to improve its insulating qualities.

oil-filled cable — Paper-insulated, lead sheathed cable into which high-grade mineral oil is forced under pressure, saturating the insulation. Main object is to prevent moisture and gases from entering. Also, it is easier to detect flaws due to leakage, as the oil is kept under constant pressure at all times.

oil fuse cutout — An enclosed fuse cutout in which all or part of the fuse support is mounted in oil.

oil-immersed forced-oil-cooled transformer (class FOA or FOW) — A transformer having its core and coils immersed in oil and cooled by the forced circulation of its oil through an external heat exchanger. If the heat exchanger utilizes forced circulation of air, the transformer is class FOA; if it utilizes forced circulation of water, the transformer is class FOW.

oil-immersed self-cooled/forced-air-cooled transformer (class OA/FA) — A transformer having its core and coils immersed in oil and having a self-cooled rating with cooling obtained by the natural circulation of air over the cooling surface (class OA) and a forced-air-cooled rating with cooling obtained by the forced circulation of air over this same cooling surface (class FA).

oil-immersed self-cooled transformer (class OA) — A transformer having its core and coils immersed in oil, the cooling being effected by the natural circulation of air over the cooling surface. (Smooth tank surfaces are generally sufficient to cool small transformers, but for about 25 kVA and larger, a supplementary heat-dissipating surface is provided in the form of external fins, tubes, or tubular radiators.)

oil-immersed transformer — A transformer whose core and coils are immersed in oil.

oil-immersed water-cooled transformer (class OW) — A transformer having its core and coils immersed in oil, the cooling being effected by the natural circulation of oil over the water-cooled surface. The water-cooled surface is provided by water flowing through copper cooling coils immersed in the oil and located inside the main transformer tank, through which the oil flows by thermosiphon action.

oil switch — A switch in which the interruption of the circuit occurs in oil to suppress the arc and prevent damage to the contacts.

Omega — A long-range hyperbolic navigation system that transmits interrupted cw signals from which phase differences are extracted. Transmissions occur from two or more locations on a single carrier frequency utilizing time-sharing techniques.

omnibearing — The bearing indicated by a navigational receiver on transmissions from an omnirange.

omnibearing converter — An electromechanical device that combines an omnirange signal with aircraft heading information to furnish electrical signals for operating the pointer of a radio magnetic indicator. *See also* omnibearing indicator.

omnibearing indicator — An omnibearing converter to which a dial and pointer have been added.

omnibearing line — One of an infinite number of straight, imaginary lines radiating from the geographical location of a VHF omnirange.

omnibearing selector — Instrument capable of being set manually to any desired omnibearing or its reciprocal in order to control a course-line deviation indicator.

omniconstant — A calculator that can add consecutively in any steps of any predetermined size, raising the power of any number in consecutive steps.

omnidirectional — Also called nondirectional. 1. All-directional; not favoring any one direction. Having no particular direction of maximum emission or sensitivity. An omnidirectional speaker is one that theoretically radiates equally in all directions. 2. Responding equally to sounds arriving from any direction. 3. Emitting sound equally in all directions. Frequently applied to speakers that are only "omni" in the forward or upward hemisphere.

omnidirectional antenna — Also called nondirectional antenna. 1. An antenna producing essentially the same field strength in all horizontal directions and a directive vertical radiation pattern. 2. An antenna that radiates or receives equally well in all directions in a horizontal plane.

omnidirectional hydrophone — A hydrophone having a response that is essentially independent of the angle of arrival of the incident sound wave.

omnidirectional microphone — Also called non-directional microphone. 1. A microphone that responds to a sound wave from almost any angle of arrival. 2. A microphone that picks up sound relatively evenly from all directions.

omnidirectional radio range — Radio aid to air navigation using a transmitter that radiates throughout 360° azimuth, providing aircraft with a direct indication of the bearing of the transmitter.

omnidirectional range — Also called omnirange. A radio facility providing bearing information to or from such facilities at all azimuths within its service area.

omnidirectional range station — In the aeronautical radionavigational service, a land station that provides a direct indication of the bearing (omnibearing) of that station from an aircraft.

omnigraph — An instrument that produces Morse-code messages for instruction purposes. It contains a buzzer circuit that is usually actuated by a perforated tape.

omnirange — *See* omnidirectional range.

OMR — Abbreviation for optical mark recognition.

on — Said of an electronic element that is conducting current.

on-board — Located on a circuit board. Most frequently refers to the motherboard of a computer.

on-call channels — Similar to allocated channels except that full-time exclusive use of the channel is not warranted.

on-course curvature — In navigation, the rate at which the indicated course changes with respect to the distance along the course path.

on-course signal — The monotone radio signal that indicates to the pilot that he is neither too far to the right nor to the left of the radio beam being followed.

on-delay — A circuit that produces an output signal some definite time after an input signal is applied.

on-demand system — A system from which the desired information or service is available at the time of request.

ondograph — An instrument for drawing alternating-voltage waveform curves. It employs a capacitor that is momentarily charged to the amplitude at a particular point on the curve and then discharged through a recording galvanometer. This is repeated at intervals of about once every hundred cycles, with the sample taken a little farther along on the waveform each time.

ondoscope — A glow-discharge tube used on an insulating rod to indicate the presence of high-frequency radiation near a transmitter. The radiation ionizes the gas in the tube, and the visible glow indicates the presence of radiation.

one-input terminal — Called the set terminal. The terminal which, when triggered, will put a flip-flop in the 1 (opposite of starting) condition.

one-line unit — In a computer, the input/output device or auxiliary equipment under direct control of the computer.

one-many function switch — A function switch in which only one input is excited at a time and each input produces a combination of outputs.

one output — *See* one state.

one-output terminal — The terminal that produces an output of the correct polarity to trigger a following circuit when a flip-flop circuit is in the 1 condition.

ones complement — 1. Having to do with arithmetic that provides a method of negating a binary number so that binary subtraction can be performed with the techniques of addition. The ones complement of a binary number is obtained by complementing all bits in that number. In ones-complement addition, end-around carry must be used. 2. A binary digit that when added to another binary digit yields 1; the inverse binary state of any given bit.

ones-complement arithmetic — A binary arithmetic system in which negative numbers are created by inverting individual bits in the binary representation of the positive number.

one-shot — 1. A circuit that produces an output signal of fixed duration when an input signal of any duration is applied. 2. *See* monostable. 3. A type of multivibrator that provides an output pulse of a specific length whenever it receives an input pulse. Often used to provide delays or stretch or expand pulses that are the wrong length or shape.

one-shot multivibrator — *See* monostable multivibrator.

one state — Also called one output. In a magnetic cell, the positive value of the magnetic flux through a specified cross-sectional area, determined from an arbitrarily specified direction. *See also* zero state.

one-third-octave band — A frequency band in which the ratio of the extreme frequencies is 1.2599.

one-to-partial-select ratio — In a computer, the ratio of a 1 output to a partial-select output.

one-to-zero ratio — In a computer, the ratio of a 1 output to the 0 output.

O-network — A network composed of four impedance branches connected in series to form a closed

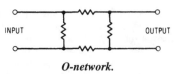

O-network.

circuit. Two adjacent junction points serve as input terminals, and the remaining two as output terminals.

one-way communication — Applied to certain radiocommunication or intercommunications systems in which a message is transmitted from one station to one or more receiving stations that have no transmitting apparatus.

one-way repeater — *See* repeater.

on-hook — 1. A condition that exists when the telephone handset is on its mounting, thus keeping the hook-switch open. The on-hook condition opens the dc loop, indicating that calls can be accepted. 2. The signal that is sent to the central office (i.e., loop open, no current) as a result of the on-hook condition. 3. The condition that indicates the idle state or open loop of a telephone customer's line. A modem not in use is said to be on-hook.

online — 1. Pertaining to a computer that is actively monitoring or controlling a process or operation, or to an operation performed by such a computer. 2. Pertaining to equipment or devices under direct control of the central processing unit. 3. Pertaining to a user's ability to interact with a computer. 4. Describes equipment or devices that are connected to the communications line. 5. A word or data processing operation that is performed on a local system connected to and sharing the facilities of a remote central processor. 6. The condition in which a user, terminal, or other device is actively connected with the facilities of a communications network or computer; opposite of offline. 7. Any equipment or process that sends information directly to a computer for immediate processing and immediate results is said to be online. This is in contrast with storing the information and having the computer process it later. 8. Pertaining to peripheral equipment or devices in direct communication with the central processing unit of a computer. May also be used to describe terminal equipment that is connected to a transmission line. 9. A direct connection between a remote terminal and a central processing site. 10. Refers to equipment or devices in a system that are directly connected to and under the control of the computer. 11. Physically connected to and using a communications medium or remote computer.

online computers — Incorporation of calculators or small computers into spectroradiometric systems to control wavelength intervals, wavelength ranges, and detector signals, and to produce computed absolute spectroradiometric values.

online data reduction — The processing of information as rapidly as the information is received by the computing system or as rapidly as it is generated by the source.

online debugging technique — *See* octal debugging technique.

online operation — 1. An operation carried on within the main computer system (e.g., computing and writing results onto a magnetic tape, printed report, or paper tape). 2. Operation in which information concerning a controlled process is fed directly into a computer, and the computer exercises direct control on the basis of this information. Since the computer reacts without appreciable time lag, results are said to be in real time. 3. A type of system application in which the input data to the system is fed directly from the measuring devices, and the computer results are obtained during the process of the event; e.g., a computer receives data from wind tunnel measurements during a run, and the computations of dependent variables are performed during the run, enabling a change in the conditions so as to produce particularly desirable results. 4. Operations in which the programmable controller is directly controlling the machine or process.

online processing — A data-processing approach in which transactions are entered into the computer directly as they occur.

online system — 1. A teleprocessing system in which the input enters the computer directly from the originating point and/or in which the output is transmitted directly to the point at which it is used. 2. A telegraph system that involves transmitting directly into the system. 3. A system of processing information in which the input data enters the computer directly from, and the output data is transmitted directly to, its point of origin.

on-off control — A simple control system in a switch, in which the device being controlled is either fully on or fully off and no operating positions are available.

on-off keying — Keying in which the output of a source is alternately transmitted and suppressed to form signals.

on-off ratio — Ratio of the duration (on) of a pulse to the space (off) between successive pulses.

on-off switch — See power switch.

on-off tests — The switching of various suspected interference sources on and off while the victim receiver is monitored so that the actual source can be identified.

on period — That portion of an operating cycle during which an electron tube is conducting.

on state — The condition of a thyristor corresponding to the low-resistance low-voltage portion of the principal characteristic.

on-state current — The principal current when a thyristor is in the on state.

on-state voltage — The principal voltage when a thyristor is in the on state.

on the air — Transmitting.

on-the-fly — Recording techniques in which reading, writing, and block spacing require no tape stoppage. These techniques require total control over data transmission. For long-length recording, coordination between data supply and tape demand is essential.

on-the-fly printer — A high-speed line printer in which continuously rotating print wheels and fast-acting hammers are used so that the successive letters in a line of text are printed so rapidly that they appear to be printed simultaneously.

on voltage — The voltage with respect to ground or the minus supply at a switch output when it is in the conducting, or on, state.

opacimeter — Also called turbidimeter. A photoelectric instrument for measuring the turbidity (amount of sediment) of a liquid. It does this by determining the amount of light that passes through the liquid.

opacity — 1. The degree of nontransparency of a substance — i.e., its ability to obstruct, by absorption, the transmission of radiant energy such as light. Opacity is the reciprocal of transmission. 2. The ratio of incident flux to transmitted flux. The inverse of transmission factor. 3. The characteristic of an object that prevents light from passing through.

opal lamp — A tungsten-filament lamp that uses an opal glass bulb to diffuse light.

op amp — Abbreviation for operational amplifier.

opaque — 1. The optical quality of a substance whereby light cannot pass through it. The meaning is directly opposite to that of transparent. Thus, steel is opaque to visible light. 2. A term describing a substance that is impervious to light; the characteristic of a substance that has no luminous transmittance.

op code — Also opcode. Abbreviation for operation code.

open — 1. A circuit interruption that results in an incomplete path for current (e.g., a broken wire, which opens the path of the current). 2. A break in the continuity of an electrical circuit that prevents current from flowing. 3. The computer command used to view a document or other file.

open architecture — 1. A system (usually a computer) whose characteristics comply with industry standards and can be connected to other systems that also comply with these standards. 2. Computer design practices that encourage other manufacturers to develop supportive peripherals, software, and internal components. *Compare* closed architecture.

open-center display — A PPI display on which zero range corresponds to a ring around the center of the display.

open circuit — 1. A circuit that does not provide a complete path for the flow of current. *See also* open. 2. A discontinuous electrical path or circuit through which no current can flow.

open-circuit impedance — The driving-point impedance of a line or four-terminal network when the far end is open.

open-circuit jack — A jack that has its through circuit(s) normally open. Circuits are closed by inserting mating plugs.

Open-circuit jack.

open-circuit parameters — The parameters of an equivalent circuit of a transistor that are the result of selecting the input current and output current as independent variables.

open-circuit signaling — Signaling in which no current flows under normal (i.e., inoperative) conditions.

open-circuit system — A system in which the sensors are connected in parallel. When a sensor is activated, the circuit is closed, permitting a current that activates an alarm signal.

open-circuit voltage — 1. The voltage at the terminals of a battery or other voltage source when no appreciable current is flowing. 2. The voltage at the terminals of a voltage source when no current is being drawn from the source. 3. The voltage of a cell or battery under a no-load condition and measured with a high-impedance voltmeter.

open collector — A type of output structure found in certain bipolar logic families. The output is characterized by an active transistor pulldown for taking the output to a low voltage level and no pull-up device. Resistive pull-ups are generally added to provide the high-level output voltage. Open-collector devices are useful when several devices are to be used together on one I/O bus, such as the IEEE-488 bus.

open collector output — A feature of some semiconductor memories and other circuits that permits the formation of larger arrays by wiring several units together.

open core — A core fitting inside a coil but having an external return path. The magnetic circuit thus has a long path through air.

open-delta connection — Two single-phase transformers connected so that they form only two sides of a delta, instead of the three sides with three transformers in a regular delta connection.

open-ended — Pertaining to a system or process that is receptive to augmentation.

open-entry contact — 1. A female contact designed for a certain size male pin, but having a spring construction that allows use of larger pins. 2. A female-opening contact unprotected from possible damage or distortion from a test probe or an oversized or misaligned male contact.

open-fuse cutout — An enclosed fuse cutout in which the fuse support and fuse holder are exposed.

open loop — 1. Pertaining to a control system that does not provide self-correcting action for errors in the desired operational condition. 2. Having to do with an operational amplifier that does not have feedback. 3. An arrangement in which a computer monitors a process or device and presents the results in real time so that an operator can make adjustments to the process or operation, if required.

open-loop bandwidth — Without feedback, and frequency limits at which the voltage gain of the device drops off 3 dB below the gain at some reference frequency.

open-loop control system — A control system in which there is no self-correcting action, as there is in a closed-loop system.

open-loop differential voltage gain — *See* differential voltage gain.

open-loop gain — The ratio of the (loaded) output of an amplifier without any feedback to its net input at any frequency. Usually implies voltage gain.

open-loop output impedance — The complex impedance seen looking into the output terminals of an operational amplifier with no external feedback and in the linear-amplification region. In closed-loop operation, the output impedance is equal to the open-loop impedance divided by the loop gain. If the open-loop impedance is not more than a few hundred ohms and the loop gain is high enough for good gain accuracy and stability, the closed-loop impedance will be on the order of an ohm or less, which can be neglected in most applications.

open-loop output resistance — The resistance looking into the output terminal of an operational amplifier operating without feedback and in the linear-amplification region.

open-loop system — A control system that does not have a means of comparing input and output for control purposes.

open-loop voltage gain — The ratio of the output signal voltage of an operational amplifier to the different input signal voltage producing it with no feedback applied.

open magnetic circuit — A magnet that has no closed external ferromagnetic circuit and does not form a complete conducting circuit itself (e.g., a permanent magnet ring interrupted by an air gap).

open motor — A motor having ventilating openings that permit passage of external cooling air over and around the windings of the machine. When such motors have an internal fan to aid the movement of ventilating air, they may be referred to as self-ventilated.

open-phase protection — The effect of a device operating on the loss of current in one phase of a polyphase circuit to cause and maintain the interruption of power in the circuit.

open-phase relay — A relay that functions when one or more phases of a polyphase circuit open and sufficient current is flowing in the remaining phase or phases.

open plug — A plug designed to hold jack springs in their open position.

open-reel — Also called reel-to-reel. Used to designate any tape format in which the running tape is wound onto a separate takeup reel. As distinguished from cartridges and cassettes, where the takeup reel is either nonexistent (cartridge) or is included in the tape package (cassette).

open relay — A relay not having an enclosure.

open repeater — A repeater whose access is not limited.

open routine — In a computer, a routine that it is possible to insert directly into a larger routine without a linkage or calling sequence.

open subroutine — Also called direct-insert subroutine. 1. A subroutine inserted directly into a larger sequence of instructions. Such a subroutine is not entered by a jump instruction; hence, it must be recopied at each point where it is needed. 2. In a computer, a separately coded sequence of instructions that is inserted in another instruction sequence directly in the line of flow.

open system — A computer processor or set of connected processors for which standards are published, thus allowing an application running in the system to communicate with other applications in the same or other systems.

open temperature pickup — A temperature transducer in which its sensing element is directly in contact with the medium whose temperature is being measured.

open wire — 1. A conductor separately supported above the earth's surface. 2. Communication lines that aren't insulated and formed into cables, but are instead mounted on aerial crossarms on utility poles.

open-wire circuit — A circuit made up of conductors separately supported on insulators.

open-wire loop — The branch line on a main open-wire line.

open-wire transmission line — A transmission line formed by two parallel wires. The distance between them, and their diameters, determines the surge impedance of the transmission line.

operand — 1. The quantity that is affected, manipulated, or operated on. It may be a number, a string, an address, etc. 2. Any of the quantities arising out of or resulting from the execution of a computer instruction. An operand can be an argument, a result of computation, a constant, a parameter, the address of any of these quantities, or the next instruction to be executed.

operate current — The minimum current required to trip all the contact springs of a relay.

operate time — 1. The time that elapses, after power is applied to a relay coil, until the contacts being checked have operated (i.e., first opened in a normally closed contact, or first closed in a normally open contact). 2. The phase of computer operation during which an instruction is being executed.

operate-time characteristic — The relation between the operate time of an electromagnetic relay and the operate power.

operating angle — The electrical angle (portion of a cycle) during which plate current flows in a amplifier or an electronic tube. Class A amplifiers have an operating angle of $360°$; class B, $180°$ to $360°$; and class C, less than $180°$.

operating code — Abbreviated opcode or op code. Source statement that generates machine codes after assembly.

operating conditions— Those conditions, excluding the variable measured by the device, to which a device is subjected.

operating cycle— The complete sequence of operations required in the normal functioning of an item of equipment.

operating frequency— The rated ac frequency of the supply voltage at which a device is designed to operate.

operating life— The minimum length of time over which the specified continuous and intermittent rating of a device, system, or transducer applies without change in performance beyond the specified tolerance.

operating-mode factor— A failure-rate modifier determined by the type of equipment environment, e.g., laboratory computer, nose-cone compartment, etc.

operating overload— The overcurrent to which electric apparatus is subjected in the course of the normal operating conditions that it may encounter.

operating point— Also called quiescent point. On a family of characteristic curves of a vacuum tube or transistor, a point whose coordinates describe the instantaneous electrode voltages and currents for the operating conditions under consideration.

operating position— The operator-attended terminal of a communications channel. It is usually used in a singular sense (e.g., a radio-operator's position, a telephone-operator's position), even when there is no more than one operating position.

operating power— The power actually reaching a transmitting antenna.

operating range— *See* receiving margin.

operating ratio— *See* availability, 1.

operating system— Also called a disk operating system. 1. Software that controls the carrying out of computer programs and that may provide scheduling, debugging, input/output control, accounting, compilation, storage assignment, data management, and related services. 2. The totality of software that describes the methods by which data is processed to obtain a desired result. 3. A collection of software that schedules jobs; assigns resources, such as peripherals; manages all data transfers between the computer and its peripherals; assigns places in memory to programs and data; processes interrupts; and performs various housekeeping functions. 4. A structured set of software routines whose function is to control the execution sequence of programs running on a computer, supervise the input/output activities of these programs, and support the development of new programs through such functions as assembly, compilation, editing, and debugging. 5. A collection of system software that permits user-written tasks to interface to the machine hardware and interact with other tasks in a straightforward, efficient, and safe manner. 6. An integrated collection of supervisory routines (usually user-transparent) responsible for allocation of system resources among user tasks. These routines may include memory management, I/O handling, logging, storage assignment, operator interaction, and job scheduling. Every operating system is different in what services it provides and the way in which the services are implemented. Programs written for one operating system won't normally run under other operating systems.

operating temperature— The temperature or range of temperatures at or over which a device is expected to operate within specified limits of error.

operating temperature range— The interval of temperatures in which a component device or system is intended to be used, specified by the limits of this interval.

operating time— The time period between turn-on and turn-off of a system, subsystem, component, or part during which the system, etc., functions as specified. Total operating time is the summation of all operating-time periods.

operating voltage— 1. The direct voltage applied to the electrodes of a vacuum tube under operating conditions. 2. The voltage at which a circuit element operates. 3. The voltage of the system on which a device is operated.

operation— 1. In mathematics, the determination of a quantity from two or more other quantities according to some rule. Examples are addition and multiplication in conventional arithmetic, and the AND and OR operations in Boolean algebra. 2. A specific action that a computer will perform whenever an instruction calls for it (e.g., addition, division).

operational amplifier— Also called op amp. 1. An amplifier that performs various mathematical operations. Application of negative feedback around a high-gain dc amplifier produces a circuit with a precise gain characteristic that depends on the feedback used. By the proper selection of feedback components, operational amplifier circuits can be used to add, subtract, average, integrate, and differentiate. An operational amplifier can have a single-input-single-output, differential-input-single-output, or differential-input-differential-output configuration. 2. Generally, any high-gain amplifier whose gain and response characteristics are determined by external components. 3. An exceptionally versatile linear amplifier (usually a linear IC) used extensively in control, computation, and measurement applications. 4. A stable, high-gain, direct-coupled amplifier that depends on external feedback from the output to the input to determine its functional characteristics. The term operational is derived from the fact that the first op amps were employed to perform addition, multiplication, division, integration, and differentiation, among other mathematical operations, in analog computers. 5. An active device characterized by high input impedance and very high voltage gain. Operational amplifiers generally have a differential input and a single-ended output. 6. A device having power gain down to zero frequency, for use in a feedback loop. It manipulates an output variable or variables in response to a local input variable or variables derived from an associated characterizing structure. This is done so as to enforce a prescribed functional relationship between one such output variable and a set of variables supplied to that structure, through reducing the local input variable or variables substantially to null or to equality. 7. An active, direct-coupled electrical subsystem having power gain, for use in a feedback circuit. In response to local input voltages or currents derived from an associated characterizing network, it manipulates an output voltage or current so as to enforce a prescribed functional relationship between a related output voltage or current and a set of currents or voltages supplied to that network, through bringing about a substantial balance. 8. A dc amplifier that has high gain, wide bandwidth, low noise, and low output impedance that can be combined with a passive network to provide an active analog device.

operational differential amplifier— An operational amplifier that utilizes a differential amplifier in the first stage.

operational readiness— The probability that, at any point in time and under stated conditions, a system will be either operating satisfactorily or ready to be placed in operation.

operational reliability— *See* achieved reliability.

operational transconductance amplifier— Abbreviated OTA. A device similar to a conventional operational amplifier, but with a current output and a means for controlling transconductance with a current input.

operation code — Also called op code, the operation part of an instruction, or a command. 1. The part of a machine language instruction or assembly language instruction specifying the operation to be performed. The other segments specify the registers, address, or input/output ports. (In the microprocessor world, the op code includes the first 8 bits.) 2. The list of operation parts occurring in an instruction code, together with the names of the corresponding operations (e.g., add, unconditional transfer, add and clear, etc.). 3. That part of a computer instruction word that designates the function performed by a given instruction (e.g., the op codes for arithmetic instructions include ADD, SUB, DIV, and MUL). 4. The symbols within an instruction that represent the particular operation to be performed (e.g., add). 5. The portion of an instruction which designates the operation to be performed by a computer (e.g., add, subtract, or move). 6. A code signifying a particular task to be done by the computer.

operation decoder — In a computer, circuitry that interprets the operation-code portion of the machine instruction to be executed and sets other circuitry for its execution.

operation number — In a computer, a number that designates the position of an operation or its equivalent subroutine in the sequence of operations that makes up a routine. A number that identifies each step in a program stated in symbolic code.

operation part — In a computer instruction, the part that specifies the kind of operation, but not the location of the operands. *See also* instruction code.

operation register — In a computer, the register that stores the operation-code portion of an instruction.

operations research — 1. The solving of large-scale or complicated problems that arise from an operation in which a major decision must be based on solutions of relationships involving a large number of variables. 2. The development and application of quantitative techniques to the solution of problems faced by managers of public and private organizations. More specifically, theory and methodology in mathematics, statistics, and computing are adapted and applied to the identification, formulation, solution, validation, implementation, and control of decision-making problems.

operation time — The amount of time required by a current to reach a stated fraction of its final value after voltages have been applied simultaneously to all electrodes. The final value is conventionally taken as that reached after a specified length of time.

operator — 1. Any person who operates equipment — specifically, a computer, radio transmitter, or other communications equipment. 2. A symbol that designates a mathematical operation, such as +, ×, etc.

operator license — The instrument of operator authorization, including the class of operator privileges.

opposition — The phase relationship between two periodic functions of the same period when the phase difference between them is half of a period.

optical ammeter — An electrothermic instrument for measuring the current in the filament of an incandescent lamp. It normally uses a photoelectric cell and indicating instrument to compare the illumination with that produced by a current of known magnitude in the same filament.

optical axis — 1. The Z-axis of a crystal. 2. The straight line passing through the centers of the curved surfaces of a lens or lens system.

optical bar-code reader — A device used with the data station to read coded information from particular documents at the rate of hundreds of characters per second.

optical character — A printed character, frequently used in utilities billing, that can be read by a machine without the aid of magnetic ink.

optical character reader — Abbreviated OCR. 1. A photoelectric device that scans printed or typed copy and produces electrical signals that allow recognition and/or reproduction of the copy being scanned. 2. A photosensitive device used to read numerical data at a rate of up to 480 characters per second.

optical character recognition — Identification of printed characters by a machine using light-sensitive devices. *See also* OCR.

optical communication cable — Fiber with a protective jacket around it. A cable may have one or more fibers within it. *See* optical communication fiber.

optical communication fiber — A term analogous to a single strand of electrical wire in that it carries information from point to point. *See* optical communication cable.

optical communications — 1. Communications through the use of beams of visible, infrared, or ultraviolet radiation, or, over much longer distances, through the use of laser beams. *See also* coherent light communications. 2. The transmission and reception of information by optical devices and sensors. *See also* lightwave communications.

optical conductors — In fiber optics, materials that offer a low optical attenuation to transmission of light energy. Types of optical conductors include the following: (*a*) single fiber — a discrete optical conductor; (*b*) bundle — a number of optical conductors in a random arrangement, grouped together and used as a single transmission medium (channel); (*c*) single-channel, single-bundle cable — a bundle of optical conductors with a protective covering; (*d*) multichannel, single-fiber cable — more than one single-fiber cable jacketed; (*e*) single-channel, single-fiber cable — a discrete optical conductor with a protective covering; (*f*) multichannel bundle cable — more than one single-bundle cable jacketed; and (*g*) multichannel cables — a combination of cables.

optical coupler — Also called optoisolator, optical isolator, photocoupler, or optocoupler. A device designed to transfer electrical signals by utilizing light waves to provide coupling with electrical isolation between input and output. A typical optical coupler uses a light-emitting diode (LED) to convert the electrical signal of the primary circuit into light and a phototransistor in the secondary circuit to reconvert the light signal back into an electrical signal.

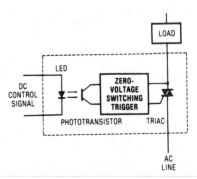

Optical coupler for solid-state relay.

optical coupling — Sometimes called crosstalk. In fiber optics, light leakage from one fiber to another by evanescent boundary wave interaction.

optical cropper — A mechanical or electrical electromagnetic device for passing and then interrupting a beam of light for a known brief interval. Examples include tuning forks, rotation shutters, and the more sophisticated Kerr cells.

optical damping — A damping ratio slightly less than unity in which the overshoot is less than the specified uncertainty of the instrument.

optical data link — A system consisting of a transmitter (i.e., a light source), a fiber-optic cable, and a receiver (i.e., a photodetector), connected together in such a manner that light waves from the source can be received at the receiver. Light from the transmitter is usually modulated by an intelligence-bearing signal.

optical density — The negative logarithm of the percent transmittance (or reflectance) of a transparent (or opaque) material.

optical detector — A device used in optical fiber communication systems, such as CATV and data links, for optical fiber measurements, to combine or split optical signals at desired ratios by insertion into a transmission line; for example, a three-port or four-port unit with precise connectors at each port to enable inputs to be coupled together and transmitted via multiple outputs.

optical disk — A very-high-density information storage medium that uses light to read and write digital information.

optical distortion — An aberration of spherical surface optical systems due to the variation in magnification with distance from the optical axis.

optical encoder — A device designed to accurately measure linear or rotary motion by detection of the movement of markings on a transparent medium past a fixed point of light. The encoder has a moving code plate, a glass disk with positional or angular information on it — generally bit information — and simple geometric optics for readout.

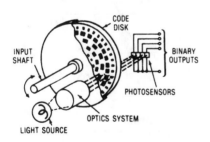

Optical encoder.

optical end-finish — The surface condition at the face of an optical conducting medium.

optical fiber — 1. An optical waveguide. 2. A core of transparent material surrounded by a transparent cladding that has a lower refractive index. Both core and cladding are usually made of glass. Because of this arrangement, light rays launched into the fiber are restricted by total internal reflection and are thus confined within the core. 3. A single discrete optical transmission element, usually consisting of a fiber core and a fiber cladding. As a light-guidance system (dielectric waveguide) that is usually cylindrical in shape, it consists either of a cylinder of transparent dielectrical material of given refractive index whose walls are in contact with a second dielectric material of a lower refractive index, or of a cylinder whose core has a refractive index that gets progressively lower away from the center. The length of a fiber is usually much greater than its diameter. The fiber relies on internal reflection to transmit light along its axial length. Light enters one end of the fiber and emerges from the opposite end with losses dependent on length, absorption, scattering, and other factors. A bundle of fibers has the ability to transmit a picture from one of its surfaces to another, around curves, and into otherwise inaccessible places with an extremely low loss of definition and light, by the process of total internal reflection. Each fiber transmits only one element of the composite emergent image. As the definition in the output image depends on the smallness of each element composing it, it is desirable to keep the cross section of the individual fibers as small as possible. If the spacing of the fibers increases toward the output end of the bundle, the image is magnified; if spacing is reduced, the image is reduced in size. By crossing the fibers systematically or randomly, the image is scrambled and can be recovered by retransmitting the scrambled image backward through the same or equivalent fiber bundle. 4. A cylinder or core of transparent dielectric material surrounded by a second dielectric. In order to propagate light, the refractive index of the core material is greater than that of the surrounding cladding material. To obtain satisfactory optical isolation of the core, the cladding material has a minimum thickness of one or two wavelengths of the light transmitted.

optical fiber bundle — Many optical fibers in a single protective sheath or jacket. The jacket is usually polyvinyl chloride (PVC). The number of fibers could range from a few to several hundred, depending on the application and the characteristics of the fibers.

optical horizon — The locus of points at which straight lines extended from a given point are tangent to the surface of the earth.

optical isolator — A photon-coupled device in which an electrical signal is converted into light that is projected through an insulating interface and reconverted to an electrical signal. *See also* optical coupler; optoisolator.

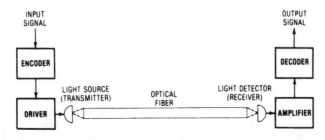

Optical fiber link.

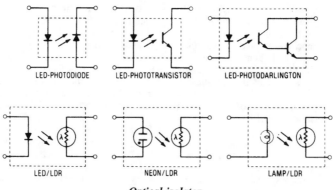

Optical isolator.

optical mark recognition—Abbreviated OMR. An information-processing technology that converts data into another medium for computer input. This is accomplished by the presence of a mark in a given position, each position having a value known to the computer and which may or may not be understandable to humans.

optical maser—*See* laser.

optical memory—The direct storage of data as bits in memory using optical systems and properties. The memory makes use of a laser beam that is divided by a beam splitter and controlled by a modulator and a deflector to transpose bits into a given area of storage in memory. On the other side of the memory plane, a laser and a deflector read the memory, bit by bit, the bits being read by a scanning photodetector. Erasure is accomplished by writing with the beam at a different wavelength.

optical microphone—Laser-powered telephone device for analog communications that employs a vibrating plastic membrane as a transmitter to modulate laser light piped via optical fibers from a central exchange, omitting the need for transducers or modulators.

optical mode—In a crystal lattice, a mode of vibration that produces an oscillating dipole.

optical mosaic—1. Grouping of fibers to form a cross-section area that will make a light pattern for acceptance by a detector or other receiving element. 2. A construction in which fibers are grouped and regrouped to build up an area, usually with some degree or type of imperfection developing at the boundaries of the subgroup. When this boundary condition becomes very noticeable, it is called whichen wire.

optical pattern—Also called Christmas-tree pattern. In mechanical recording, the pattern observed when the surface of the record is illuminated by a light beam of essentially parallel rays.

optical port—An opening through which optical energy can pass.

optical pumping—The process of changing the number of atoms or atomic system in a set of energy levels as a result of the absorption of light incident on the material. This process causes the atoms to be raised from certain lower to certain higher energy levels, and it may cause a population inversion between certain intermediate levels.

optical pyrometer—A temperature-measuring device comprising a comparison source of illumination, together with some convenient arrangement for matching this source—either in brightness or in color—against the source whose temperature is to be measured. The comparison is usually made by the eye.

optical receiver—An electro-optical module that converts an optical input signal to an electrical output signal.

optical repeater—A device in a fiber optics system that amplifies the optical signal to overcome previous attenuation.

optical scanner—A computer-system input device that reads a line of printed characters and produces a corresponding electronic signal for each character.

optical scanning—A technique for machine recognition of characters by their images. *Compare* optical character recognition.

optical sound—A system of recording and reproducing sound by using modulated light areas at the side of motion-picture film.

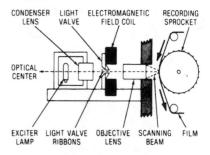

Optical sound recorder.

optical spectrometer—An instrument with an entrance slit, a dispersing device, and one or more exit slits, with which measurements are made at selected wavelengths within the spectral range, or by scanning over the range. The quantity detected is a function of radiant power.

optical storage—In data processing, the storage of information on photographic film in a manner that allows it to be retrieved in a nondestructive fashion, usually with a photosensing device.

optical transmitter—An electro-optical module that converts an electrical input signal to an optical output signal.

optical twinning—A defect occurring in natural quartz crystals. The right quartz and left quartz both occur in the same crystal. This generally results in small regions of unusable material, which are discarded when the crystal is cut up.

optical video disc — Abbreviated OVD. A disc on the surface of which digital data is recorded at high packing densities in concentric circles or in a spiral using a laser beam to record spots, which are read by means of a reflected laser beam of intensity lower than the recording intensity. Up to 10^{11} bits can be recorded on a single disc, which is thus suitable for an hour of TV programming playback.

optical waveguide — 1. A fiber used for optical communications. Analogous to a waveguide used for microwave communications. 2. An optical fiber with a high refractive index clad with a material having a lower index of refraction.

optic-electronic device — *See* optoelectronic device.

optics — That branch of physical science concerned with the nature and properties of the electromagnetic radiation known as light, including the infrared, visible, and ultraviolet regions.

optic scrambler — *See* fiber-optics scrambler.

optimization — 1. The continual adjustment of a process for the best obtainable set of conditions. 2. Achieving the best possible solution to a problem in terms of a specified objective function.

optimize — To arrange instructions or data in computer storage in such a way that a minimum of machine time is required for access when instructions or data are called out.

optimum bunching — The bunching condition required for maximum output in a velocity-modulation tube.

optimum coupling — *See* critical coupling.

optimum damping — The value of damping that permits fast response with some overshoot; this value is about 65 percent of critical damping.

optimum frequency — The most effective frequency at a specified time for ionospheric propagation of radio waves between two specified points.

optimum load — The value of load impedance that will transfer maximum power from the source to the load.

optimum plate load — The ideal plate-load impedance for a given tube and set of operating conditions.

optimum reliability — The value of reliability that yields a minimum total successful mission cost.

optimum working frequency — The frequency at which transmission by ionospheric reflection can be expected to be most effectively maintained between two specified points and at a certain time of day. (For propagation by way of the F_2 layer, the optimum working frequency is often taken as being 15 percent below the monthly median value of the F_2 maximum usable frequency for a specified path and time of day.)

optocoupler — Sometimes referred to as optoisolator. 1. A light source (input) and a light detector (output) both housed in a single package, sealed against outside light. An electrical signal applied to the light source changes the amount of light emitted. The emitted light falls upon, and is collected by, the detector. These input electrical signals are thus coupled to the output. From the output, the signals perform normal electronic functions, such as driving amplifiers, triggering a thyristor power supply, or switching logic levels. 2. A device that transmits electrical signals, without electrical connection, between a light source (input) and a light detector (output). The input is generally an LED; the output could be a photodiode, phototransistor, a photo-Darlington, etc.

optoelectronic device — Also called opticelectronic device. 1. A device responsive to electromagnetic radiation in the visible, infrared, or ultraviolet spectral regions of the frequency spectrum. It emits or modifies noncoherent or coherent electromagnetic radiation in these same regions, or utilizes such electromagnetic radiation for its internal operation. The wavelengths handled by these devices range from approximately 0.3 to 30 μm. Examples are LEDs, optical couplers, laser diodes, and photodetectors. 2. Electronic devices associated with light, serving as sources, conductors, or detectors. *See also* light.

optoelectronic integrated circuit — An integrated component that uses a combination of electroluminescence and photoconductivity in the performance of all or at least a major portion of its intended function.

optoelectronics — 1. Technology dealing with the coupling of functional electronic blocks by light beams. 2. Circuitry in which solid-state emitters and detectors of light are involved. 3. A term that applies to devices or circuits in which light is emitted or detected by a solid-state technique. Through usage, however, the term implies a close tie with logic-level signals for driving emitters and coupling isolated circuits. The three leading technologies in the field are light-emitting diode (LED), liquid-crystal display (LCD), and neon gas discharge. LED technology spans a range of devices: both visible and infrared light-emitting lamps, optocouplers or optoisolators, and various visible light-emitting displays.

optoelectronic transistor — A transistor that has an electroluminescent emitter, a transparent base, and a photoelectric collector.

optographics — The use of techniques of drafted artwork, diamond ruling, high-resolution photoreproduction and vacuum deposition to generate and replicate precise patterns. In turn, these patterns make up such instruments as resolution targets, reticles, code discs, scales, and step wedges, which are generally applied to glass, and sometimes to film, plastic, quartz, or other substrates.

optoisolator — 1. A coupling device consisting of a light sensor. Used for voltage and noise isolation between input and output while transferring the desired signal. 2. Any device that uses a light emitter and a photodetector to couple signals without any electrical connection. The use of the term is restricted by convention to devices employing an LED and semiconductor photodetector. 3. A small component that converts electrical signals to light signals, transmits the light signals across a short gap, and converts them back to electrical signals on the other side. Since no electrical connection exists across the gap, an optoisolator provides immunity to

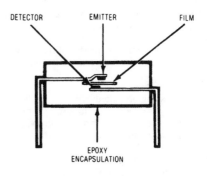

Cutaway view.

Optoisolator.

electromagnetic interference and ground-loop problems. *See also* optocoupler.

optophone — A photoelectric device that converts light energy into sound energy. Thus, a blind person can "read" by ear by using a selenium cell and a circuit for converting the resulting signals into sounds of corresponding pitch.

orange peel — A term applied to the surface of a recording blank that resembles an orange peel. Such a surface has a high background noise.

orbit — 1. The path of a body or particle under the influence of a gravitational or other force. For instance, the orbit of an celestial body is its path relative to another body around which it revolves. 2. To go around the earth or other body in an orbit.

orbital electron — An electron that is visualized as moving in an orbit around the nucleus of an atom or molecule, as opposed to a free electron.

orbital period — The time that it takes a satellite to complete one circumnavigation of its specific orbit.

OR circuit — *See* OR gate.

order — 1. In computer terminology, the synonym for instruction, command, and — loosely — operation part. These three usages, however, are losing favor because of the ambiguity between them and the more common meanings in mathematics and business. 2. To place in sequence according to some rules or standards.

ordering — In a computer, the process of sorting and sequencing.

orders of logic — A measure of the speed with which a signal can propagate through a logic network (commonly referred to as orders-of-logic capability).

order tone — A tone sent over trunks to indicate that the trunk is ready to receive an order, and to the receiving operator that it is about to arrive.

OR device — A device whose output is logical 0 if, and only if, all the control signals are logical 0.

ordinary differential equations — Differential equations in which there is only one independent variable. The unknowns are functions of this variable, and all derivatives in the equations are with respect to this variable.

ordinary wave — Sometimes called the O-wave. One of the two components into which the magnetic field of the earth divides a radio wave in the ionosphere. The other component is called the extraordinary wave, or X-wave. When viewed below the ionosphere in the direction of propagation, the ordinary wave has counterclockwise or clockwise elliptical polarization, depending on whether the magnetic field of the earth has a positive or negative component, respectively, in the same direction.

ordinate — 1. The vertical line, or one of the coordinates drawn parallel to it, on a graph. 2. The vertical scale on a graph. 3. A distance along the vertical or *y*-axis of a graph.

organ — In a computer subassembly, the portion that accomplishes some operation or function (e.g., arithmetic organ).

organic dye laser — A laser having a lasing material that is a fluorescing organic dye. It can produce, depending on the dye used, emission in any part of the visible spectrum into the near infrared and ultraviolet, and its most important property is continuous tunability over a broad spectrum.

OR gate — Also called an OR circuit. 1. A gate that performs the function of logical "inclusive OR." It produces an output whenever any one (or more) of its inputs is energized. 2. A logic circuit that requires that at least one input be in the on state to drive the output into the on state. 3. A digital logic circuit that produces

a logical 1 output when any one or all of its outputs are logical 1. It produces a logical 0 output only when all of its inputs are logical 0.

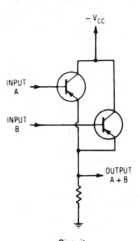

Circuit.

TRUTH TABLE

INPUTS		OUTPUT
A	B	A + B
0	0	0
0	1	1
1	0	1
1	1	1

Truth table.

OR gate (using transistors).

orient — To position or otherwise adjust with respect to some reference point (e.g., to orient an antenna for best reception).

orientation — An adjustment of the time, relative to the start transition, that teletypewriter receiving apparatus starts selection.

oriented — In crystallography, a crystal in which the axes of its individual grains are aligned so that they have directional magnetic properties.

orifice — An opening or window — specifically, in a side or end wall of a waveguide or cavity resonator, an opening through which energy is transmitted.

origin — The point in a coordinate system where the axes meet.

original lacquer — An original disc recording made on a lacquer surface for the purpose of producing a master.

original master — Also called metal master or metal negative. In disc recording, the master produced by electroforming from the face of a wax or lacquer recording.

O-ring — A circular piece of material with a round cross section; it effects a seal under pressure.

ORION — An acronym for Oak Restricted Information and Operational Network.

orthicon — Also called an orthiconoscope. A camera tube in which a beam of low-velocity electrons scans a photoemissive mosaic capable of storing an electrical-charge pattern. It is more sensitive than the iconoscope.

orthiconoscope — *See* orthicon.

orthocode — An arrangement of black and white bars that resembles a piano keyboard and that can be read by an electric eye device.

orthocore— A completely closed flux memory device designed to almost duplicate the geometry of the ferrite core memory, eliminate the wiring of memory cores, and provide a plurality of wires through the memory element. The concept involves the formation of a group of plastic rods around a suitable wiring array.

orthogonal— A term that signifies that two signals (or signal attributes) are mutually transparent and noninterfering with each other. Frequency and amplitude modulation are orthogonal signal attributes.

orthogonal antennas— A pair of radar transmitting and receiving antennas, or a single antenna for transmitting and receiving, designed to permit detection of a difference in polarization between the transmitted and returned energy.

orthogonal axes— Axes that are perpendicular to each other. In an instrument, these axes usually coincide with its axes of symmetry.

OS/2— One of the first personal computer operating systems to support true multitasking with multiple address spaces and processes. Introduced by IBM in 1990.

OSCAR— Acronym for orbiting satellite carrying amateur radio. Acronym describing amateur satellites generally; with a number attached (e.g., AMSAT-OSCAR-16, or AO-16), the name of a specific ham radio satellite.

osciducer— A transducer in which information pertaining to the stimulus is provided in the form of deviation from the center frequency of an oscillator.

oscillate— To repeat a cycle of motions or to pass through a cycle of state with strict periodicity.

oscillating current— An alternating current; specifically, one that changes according to some law.

oscillating quantity— A quantity that alternately increases and decreases in value, but always remains within finite limits — e.g., the discharge of current from a capacitor through an inductive resistance (provided the inductance is greater than the capacitance times the resistance squared).

oscillating transducer— A transducer in which information pertaining to the stimulus is provided in the form of deviation from the center frequency of an oscillator.

oscillation— 1. The state of a physical quantity when, in the time interval under consideration, the value of the quantity is continually changing in such a manner that it passes through maxima and minima (e.g., oscillating pendulum, oscillating electric current, and oscillating electromotive force). 2. Fluctuations in a system or circuit, especially those consisting of the flow of electric currents alternately in opposite directions; also, the corresponding changes in voltages. 3. *See* hunting, 1. 4. A periodic change in a variable, as in the amplitude of an alternating current.

oscillator— 1. An electronic device that generates alternating-current power at a frequency determined by the values of certain constants in its circuits. An oscillator may be considered an amplifier with positive feedback, with circuit parameters that restrict the oscillations of the device to a single frequency. 2. Something that oscillates. In particular, a self-excited electronic circuit whose output voltage or current is a periodic function of time. 3. A generator of an alternating signal, continuous, sinusoidal, or pulsed.

oscillator circuit— *See* oscillator.

oscillator coil— A radio-frequency transformer that provides the feedback required for oscillation in the oscillator circuit of a superheterodyne receiver or in other oscillator circuits.

oscillator harmonic interference— Interference caused in a superheterodyne receiver by the interaction of incoming signals with harmonics (usually the second harmonic) of the local oscillator.

oscillator-mixer-first-detector— A single stage that, in a superheterodyne receiver, combines the functions of the local oscillator and the mixer-first-detector.

oscillator padder— An adjustable capacitor placed in series with the oscillator tank circuit of a superheterodyne receiver. It is used to adjust the tracking between the oscillator and preselector at the low-frequency end of the tuning range.

oscillator radiation— The amount of voltage available across the antenna terminals of a receiver (or at a distance) traceable to any oscillators incorporated in the receiver.

oscillatory circuit— A circuit containing inductance and/or capacitance and resistance, so arranged or connected that a voltage impulse will produce a current that periodically reverses.

oscillatory current— A current that periodically reverses its direction.

oscillatory discharge— Alternating current of gradually decreasing amplitude which, under certain conditions, flows through a circuit containing inductance, capacitance, and resistance when a voltage is applied. *See also* damped waves.

oscillatory surge— A surge that includes both positive and negative polarity values.

oscillistor— A semiconductor bar that is subjected to a magnetic field and a direct current, and which generates oscillations believed to be due to diffusion of ions toward the surface of the semiconductor as a result of the magnetic field.

oscillogram— 1. The recorded trace produced by an oscillograph. 2. A photograph of the luminous trace or image produced by an oscilloscope. 3. A record formed when the luminous trace or image produced by an oscilloscope is photographed.

oscillograph— 1. An instrument primarily for producing a record of the instantaneous values of one or more rapidly varying electrical quantities as a function of time, or of another electrical or mechanical quantity. 2. An instrument used to record rapidly varying currents or voltages. May consist of a CRT oscilloscope with a camera attachment, or a mirror galvanometer with a lamp and optical system to trace rapid variations of electric current on a moving ribbon of photographic paper. The later type of oscillograph is often a multiple unit capable of recording 20 or more different current variations side by side on the same strip of paper.

oscillograph recorder— A form of mechanical oscillograph in which the waveform is traced on a moving strip of paper by a pen. It is used to record parameters that vary rapidly with time and can record variations in excess of 100 Hz. However, at these higher frequencies, the width of the trace is reduced. Typically such a device with a 40-mm trace width will record a signal up to 40 Hz at full trace width. If the maximum excursion of the recording pen is limited to 8 mm, it will record signals up to 100 Hz faithfully. Above 100 Hz, the signal width will be reduced. These recorders are available in multipen models to record several parameters on a single recorder.

oscillograph tube— *See* oscilloscope tube.

oscillography— The art and practice of utilizing the oscillograph.

oscillometer— An instrument for measuring oscillations (periodic variations) of any kind.

oscilloscope— 1. An instrument in which the horizontal and vertical deflection of the electron beam of a cathode-ray tube are, respectively, proportional to a pair of applied voltages. In the most usual application

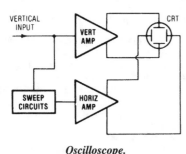

Oscilloscope.

of the instrument, the vertical deflection is a signal voltage and the horizontal deflection is a linear time base. 2. A cathode-ray tube with attendant amplifiers and control circuits for measuring and studying the waveforms of small currents and voltages. A CRT oscilloscope is particularly convenient for studying repetitive phenomena, but a tube with a long-delay phosphor can be used to analyze a single electrical pulse. An oscilloscope equipped with a camera (often of the instant type) becomes an oscillograph. 3. An electronic window that displays variations of voltage at any point in a circuit by displaying in graphic form on its screen the actual waveform of voltage plotted against time. In addition, an oscilloscope serves as an accurate ac/dc voltmeter and time-period counter. The typical scope is made up of five major interrelated parts: vertical amplifier section, horizontal amplifier section, sweep and synchronization circuits, picture tube (cathode-ray tube or CRT), and power supply. The electron beam strikes the fluorescent screen of the cathode-ray tube and temporarily produces a visible pattern or waveform of some fluctuating electrical quantity such as voltage. The pattern is employed to reveal the detailed variations in rapidly changing electric currents, potentials, or pulses.

oscilloscope differential amplifier — A device that amplifies and displays the voltage difference that exists at every instant between signals applied to its two inputs.

oscilloscope tube — Also called oscillograph tube. A cathode-ray tube that produces a visible pattern that is the graphical representation of electric signals. The pattern is seen as a spot or spots, which change position in accordance with the signals.

OSHA — Acronym for Occupational Safety and Health Act. A federal law that specifies the requirements an employer must follow in order to guard against employee illness and injury.

OSI — Abbreviation for Open Systems Interconnection. A layered architecture designed to permit interconnection between heterogeneous computer systems. Also, the international protocol for communications in a multiple-vendor environment.

OSO — Abbreviation for Orbiting Solar Observatory.

O-type backward-wave oscillator — A wideband, voltage-tunable microwave oscillator that uses a fundamental or space harmonic with phase and group velocity of different signs.

outage — 1. Loss of signal in a channel, usually the result of a dropout or a hit. 2. Status of equipment when it is out of service. Outages are termed *forced* when due to undesired occurrences, and *planned* when prescheduled, as for routine maintenance.

outconnector — In a flowchart, a connector indicating a point at which a flowline is broken to be continued at another point.

outdoor antenna — A receiving antenna located on an elevated site outside a building.

outdoor transformer — A transformer of weatherproof construction.

outer marker — In an instrument landing system, a marker located on a localizer course line at a recommended distance (normally about $4\frac{1}{2}$ miles or 7.2 km) from the approach end of the runway.

outgas — The release of gas from a material over a period.

outgassing — 1. A phenomenon in which a substance in a vacuum spontaneously releases absorbed and occluded constituents as vapors or gases. 2. De-aeration or other gaseous emission from a printed board assembly (printed board, component, or connector) when exposed to a reduced pressure or heat, or both.

outlet — 1. The point where current is taken from a wiring system. 2. Convenience receptacle used for supplying power in the home, shop, or laboratory from power-company mains. 3. A point on the wiring system that can be tapped to provide electrical current for appliances or lights.

outlet box — Metal box that houses a switch or receptacle.

outline drawing — A drawing showing approximately overall shape, but no detail.

out of phase — 1. Two or more waveforms that have the same shape, but do not pass through corresponding values at the same instant. 2. Relationship between periodic waves of the same frequency, but which do not pass through their maximum and minimum (or other corresponding) values at the same instant.

out-of-service jack — A jack, associated with a test jack, into which a shorted plug may be inserted to remove a circuit from service.

outphaser — In electronic organs, a circuit that changes a sawtooth wave to something approaching a square wave by adding to the sawtooth a second sawtooth of twice the frequency and half the amplitude in reverse phase, thus canceling the even harmonics.

outphasing — In electronic organs, a term applied to a method sometimes used for producing certain voices. Special circuitry, placed between the keying-system output and the formant filters, either adds or subtracts harmonics or subharmonics of the tone-generator signal.

output — 1. The current, voltage, power, or driving force delivered by a circuit or device. 2. The terminals or other places where the circuit or device may deliver the current, voltage, power, or driving force. 3. Information transferred from the internal to the secondary or external storage of a computer. 4. The electrical quantity produced by a transducer, which is a function of the measurand. 5. The useful energy delivered by a circuit or device. 6. In logic circuits, frequently used to mean a change in condition between conducting and nonconducting. (It is like calling the coil of a relay the input and the contacts the output.) 7. The signal level at the output of an amplifier or other device. 8. A port or set of terminals at which a system or component delivers useful energy or a useful signal. Also the energy or signal delivered. The useful signal delivered by a recorder using a particular type of tape, usually at an arbitrarily fixed level of harmonic distortion (1 or 3 percent) and relative to the performance of a tape with standard characteristics. 9. The transfer of information from an information process. 10. The act of providing information from a device to the outside world. Generally accompanied by a device that inputs the information being output by the first device.

output amplifier — A circuit that energizes high-power-level devices upon application of a low-power-level input signal.

output axis — The axis around which the spinning wheel of a gyroscope precesses after the wheel has received an input.

output block — 1. In a computer, a portion of the internal storage reserved for holding data that is to be transferred out. 2. A block of computer words treated as a unit and intended to be transferred from internal storage to an external location. 3. A block used as an output buffer.

output capability — The intensity of the strongest signal that a device can put out without exceeding certain limits of overload distortion.

output capacitance — 1. Of an *n*-terminal electron tube, the short-circuit transfer capacitance between the output terminal and all other terminals, except the input terminal, connected together. 2. The shunt capacitance at the output terminal of a device.

output capacitive loading — The maximum capacitance that can be placed on the output of an operational amplifier at unity gain without increasing the phase shift to the point of inducing oscillation. The limiting value increases in direct proportion to the closed-loop gain.

output capacity — The number of loads that can be driven by the output of a circuit.

output device — 1. The part of a machine that translates the electrical impulses representing data processed by the machine into permanent results such as printed forms, punched cards, and magnetic writing on tape. 2. Any device, such as a solenoid, motor starter, etc., that receives data from a programmable controller. 3. The unit of a computer, such as a card punch, that converts electrical signals into the form used by the output device, such as holes punched into cards.

output equipment — Equipment that provides information in visible, audible, or printed form from a computer.

output gap — An interaction gap with which usable power can be extracted from an electron stream.

output impedance — Also called dynamic output impedance. 1. The impedance measured at the output terminals of a transducer with the load disconnected and all impressed driving forces (including those connected to the input) taken as zero. 2. The impedance presented by a power supply to the load. It is calculated from the ratio of the change in output voltage (at the prescribed terminals) to the change in load current causing the change. The impedance is specified from dc to a stated maximum ac. 3. The impedance a device presents to its load.

output indicator — A meter or other device that indicates variations in the signal strength at the output circuits.

output limit — The maximum output signal available when an operational amplifier is operated in the saturation region.

output load current — The maximum current that the amplifier will deliver to, or accept from, a load. This rating includes the amount, however small, that is caused to flow in the feedback loop.

output meter — An alternating-current voltmeter that measures the signal strength at the output of a receiver or amplifier.

output-meter adapter — A device that can be slipped over the plate prong of the output tube of a radio receiver to provide a conventional terminal to which an output meter can be connected during alignment.

output offset voltage — 1. The difference between the dc voltages at the two output terminals (or at the output terminal and ground in an amplifier that has one output) when both input terminals are grounded. 2. The output voltage of a negative-feedback op-amp circuit when the input voltage to the circuit is zero. An ideal op amp has zero output offset voltage.

output port — In a fluidic device, the port at which the output signal appears.

output power — 1. The power that a system or component delivers to its load. 2. The maximum amount of power, limited by clipping or a specified value of distortion, that an amplifier is capable of delivering to a load of given value.

output resistance — The small-signal ac resistance of an operational amplifier seen looking into the output with no feedback applied and the output dc voltage near zero.

output saturation voltage — The lowest voltage level to which the collector of the output transistor can be reduced without degrading circuit performance.

output stage — The final stage in any electronic equipment. In a radio receiver, it feeds the speaker directly or through an output transformer. In an audio-frequency amplifier, it feeds one or more speakers, the cutting or recording head of a sound recorder, a transmission line, or any other load. In a transmitter, it feeds the antenna.

output transformer — A transformer used to couple the output stage of an amplifier to a load.

output tube — A power-amplifier tube designed for use in an output stage.

output unit — 1. A computer unit that transfers data from the computer to an external device or from internal storage to external storage. 2. A device capable of recording data coming from the internal storage unit of a computer, e.g., card punch, line printer, CRT display, magnetic disk, or teletypewriter.

output voltage — The maximum output voltage that an amplifier will develop in the linear operating region (i.e., before the onset of saturation).

output voltage swing — The peak output voltage swing of an operational amplifier referred to zero that can be obtained without clipping.

output winding — The winding of a saturatable reactor, other than a feedback winding, through which power is delivered to the load.

outside lead — *See* finish lead.

outward WATS — Outward wide-area telephone service that permits a customer, for a monthly charge, to place outgoing station-to-station paid calls to telephones within prescribed service areas.

oven — An enclosure and associated sensors and heaters for maintaining components at a controlled temperature.

overall bandwidth — The width of a receiver resonance curve which represents the ratio of signal strength required at various frequencies of resonance to the signal strength at resonance in order to give constant output.

overall electrical efficiency (induction- and dielectric-heating usage) — Ratio of the power absorbed by the load material to the total power drawn from the supply lines.

overall loudness level — A measure of the response of human hearing to the strength of a sound. It is scaled in phons and is an overall single evaluation calculated for the levels of sound pressure of several individual bands.

overall thermoelectric generator efficiency — The ratio of electrical power output to thermal power input to the thermoelectric generator.

overall ultrasonic system efficiency — The acoustical power output at the point of application, divided by the electrical power input into the generator.

overbiasing — A setup procedure whereby the bias current of a tape recorder is adjusted to slightly beyond the point that produces maximum or peak output from the tape. Overbiasing reduces differences due to imperfect uniformity from one tape to another, at the cost of some high-frequency response.

overbunching — The condition in which the buncher voltage of a velocity-modulation tube is higher than required for optimum bunching of the electrons.

overcharging — Continued charging of a battery after it has reached a charged condition. This action damages the battery and shortens its life.

overcoat — A thin film of insulating material, either plastic or inorganic (e.g., glass or silicon nitride), applied over integral circuit elements for the purposes of mechanical protection and prevention of contamination.

overcompounding — In a compound-wound generator, use of sufficient series turns to raise the voltage as the load increases, in order to compensate for the increased line drop. In a motor, overcompounding makes it run faster as the load increases.

overcoupled circuit — A tuned circuit in which the coupling is greater than the critical coupling. The result is a broadband response characteristic.

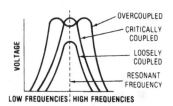

Overcoupled circuit.

overcurrent — In a circuit, the current that will cause an excessive or even dangerous rise in temperature in the conductor or its insulation.

overcurrent protection — *See* overload protection.

overcurrent protective device — 1. A device operative on excessive current that causes and maintains the interruption of power in the circuit. 2. A device designed to open an electric circuit automatically at a predetermined overcurrent level.

overcurrent relay — A protective device that separates one portion of an electric network from another when current exceeds a predetermined value.

overcutting — In disc recording, the cutting through of one groove into an adjacent one when the level becomes excessive.

overdamping — Any periodic damping greater than the amount required for critical damping. *See also* aperiodic damping.

overdriven amplifier — An amplifier stage designed to distort the input-signal waveform by permitting the driving signal to drive it beyond cutoff or even into saturation.

overflow — 1. The condition occurring whenever the result of an arithmetic operation exceeds the capacity of the number representation in a digital computer. 2. The carry digit arising from (1) above. 3. The generation of a quantity that exceeds the capacity of the computer storage facility. 4. That portion of the result of an operation that exceeds the capacity of the intended unit of storage. 5. In a calculator, when a number that is entered, or the answer to a calculation, exceeds the capacity of the working register, the excess digits overflow the register. In most machines, an overflow signal (either audible or visible) indicates when overflow occurs.

overflow indicator — 1. A bistable trigger that changes state on the occurrence of overflow in the computer register with which it is associated. The overflow indicator may be interrogated and/or restored to the original state. 2. In a calculator, a warning light or illuminated letter E to indicate existence of overflow condition.

overflow position — In a computer, an extra register position in which the overflow digit is developed.

overflow storage — Additional storage provided in a store and forward switching center of a computer to prevent the loss of messages (or parts of messages) offered to a completely filled line store.

overglaze — 1. A glass coating that is over another component or element, normally for physical or electrical protection purposes. 2. A glass compound in low-melting, vitreous form, used as a coating for passivate thick-film resistors and offering mechanical protection.

overglazed — Coated with a layer of printed and fired glass paste. Overglazing may be a solder barrier, a protective coating for resistors, or an insulator to prevent possible short circuits, as in the case of a wire bond crossing over a printed conductor.

overhang — 1. In terms of printed wiring, the inverted shelf of plating formed when conductor material is selectively removed from under the plating. The measurement between the base copper and the plating (generally refers to one side of the conductor only; thus, for a conductor width reduction, one must take two times the overhang). 2. The critical dimension by which the stylus overreaches the center spindle of a turntable when an offset tonearm is mounted for minimum tracking error.

overhead line — A conductor carried on elevated poles (e.g., telephone or telegraph wires).

overinsulation — In a coil, the insulating material placed over a wire that is brought from the center over the top or bottom wall.

overlap — 1. The amount by which the effective height of a scanning facsimile spot exceeds the nominal width of the scanning line. When the spot is rectangular, overlap may be expressed as a percentage of the nominal width of the scanning line. 2. To perform one operation at the same time that another operation is being performed; for example, to carry on input/output operations at the same time that instructions are being executed by the central processing unit. 3. The contact area between a film resistor and film conductor. 4. A mode of computer operation in which several processes take place seemingly simultaneously. In a multiprocessor system, simultaneous operation is truly possible. In a single-processor system, processes time-share the processor and appear to happen simultaneously but actually they occur in a time-sequential mode. Real time savings can be realized in either case, especially when extensive I/O to many devices of differing speeds is taking place.

overlapping contacts — Combinations of two sets of contacts actuated by a common means, each set closing in one of two positions and so arranged that the contacts of one set open after the contacts of the other set have been closed.

overlap radar — Long-range radar that is located in one sector but also covers a portion of another sector.

overlay — 1. In a computer, the technique of using the same blocks of internal storage for different routines during different stages of a problem, e.g., when one routine is no longer needed in internal storage, another routine can be placed in that storage location. 2. Memory management technique in which various routines occupy overlapping memory areas in succession. 3. A technique used to execute programs that are larger than the available memory size in systems without paging or segmentation capabilities. To utilize this method a program must be manually divided into a number of mutually exclusive groups of software modules.

overlaying—The technique of repeatedly using the same blocks of internal storage during different stages of a program, e.g., when one routine is no longer needed, another routine can replace all or part of it.

overlay load module—A computer load module that has been divided into overlay segments and has been provided by the linkage editor with information that permits the desired loading of segments to be implemented by the overlay supervisor when requested.

overlay supervisor—A routine for control of the sequencing and positioning of computer program segments in limited storage during their execution.

overlay transistor—A transistor containing a large number of emitters connected in parallel to provide maximum power amplification at extremely high frequencies.

overlay zone—An area of memory used to contain different programs or data at different times during the operation of a system.

overline—In teletypewriter operation, the printing of one group of characters over another.

overload—1. A load greater than that which an amplifier, other component, or a whole system is designed to handle. It is characterized by waveform distortion or overheating. 2. An electrically operated counter that indicates the number of times all trunks are busy between the various telephone office units. 3. In an analog computer, a condition within a computing element or at its output that causes a significant computing error as a result of the saturation of one or more parts of the computing element. 4. The amount beyond the specified maximum magnitude of the measurand which, when applied to a transducer, does not cause a change in performance beyond specified tolerance. 5. Any input that causes a meter to read beyond its rated range (full-scale range for analog meters or "full" range for digital meters).

overload capacity—The level of current, voltage, or power beyond which a device will be ruined. It is usually higher than the rated load capacity.

overload level—The level at which a system, component, etc., ceases to operate satisfactorily and produces signal distortion, overheating, damage, etc.

overload margin—In an amplifier, the safety margin prior to the onset of overload to avoid clipping on transients. This also enhances the reproduction, giving it a smoother quality in many cases.

overload operating time—The length of time a system, component, etc., may be safely subjected to a specified overload current.

overload protection—Also called overcurrent protection. 1. A device that automatically disconnects the circuit whenever the current or voltage becomes excessive. 2. Effect of a device operative on excessive current, but not necessarily on short circuit, to cause and maintain the interruption of current flow to the device governed. 3. A device or circuit that protects a power supply from damage due to excessive current demand by the load. Some schemes are also designed to protect the load. *See* foldback current limiting.

overload recovery time—The time required for an amplifier to regain its ability to amplify within stated specification limits after distortion of the output voltage amplitude by the application of a specified input voltage exceeding the rated amplitude.

overload relay—A relay designed to operate when its coil current rises above a predetermined value.

overmodulation—Modulation greater than 100 percent. Distortion occurs because the carrier is reduced to zero during certain portions of the modulating signal.

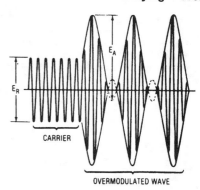

Overmodulation.

overpotential—Also called overvoltage. A voltage greater than the normal operating voltage of a device or circuit.

overpressure—Pressure greater than the full-scale rating of a pressure transducer.

overpunch—Also called zone punch. A hole punched in one of the three top rows of a punch card and which, in combination with a second hole in one of the nine lower rows, identifies an alphabetic or special character.

overrange—The situation in which the signal being measured exceeds the full-scale value of a digital panel meter. Digital panel meters are designed to indicate this condition with a blinking display or the actuation of an appropriate symbol.

overranging—A feature of a digital panel meter or digital voltmeter wherein some indication is given (usually by blinking or flashing display) that the quantity being measured is too high in value for the range selected.

override—To manually or otherwise deliberately overrule an automatic control system or circuit and thereby render it ineffective.

oversampling—A technology used in CD players for noise reduction after digital-to-analog conversion. The higher the rate, the lower the noise. An oversampling rate of 256x virtually eliminates any noise, even at low recording levels.

overscan—A video-display effect in which the image is enlarged, resulting in its edges being off the screen.

overscanning—In a cathode-ray tube, the deflection of the electron beam beyond the normal limits of the screen. This is the deflection of the beam of the tube over an angle that surpasses the angle that subtends the suitable area of the screen.

overscan recovery—A characteristic of differential comparators that states the time required for the amplifier to recover to within some amount of voltage after a return to the screen of a CRT.

overshoot—1. The initial transient response, which exceeds the steady-state response, to a unidirectional change in input. 2. Amplitude of the first maximum excursion of a pulse beyond the 100-percent amplitude level expressed as a percentage of this 100-percent amplitude. 3. A transient rise beyond regulated output limits, occurring when the ac power input is turned on or off, and for line or load step changes. 4. The amount that the indicator travels beyond its final steady deflection when a new constant value of the measured quantity is suddenly applied to the instrument. The overtravel and deflection are determined in angular measure, and the overshoot is expressed as a percentage of the change in steady deflection. 5. Reception of microwave signals at an unintended location because of an unusual atmospheric condition that sets up variations in the index of refraction. 6. Distance a motor shaft of a stepper motor can rotate beyond the step angle before it comes to rest at step-angle position.

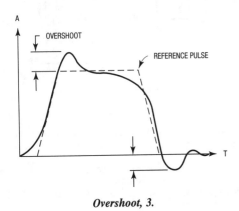

Overshoot, 3.

overshoot distortion — *See* overthrow distortion.

overtemperature protection — A thermal relay or other protective device that turns off the power automatically in the event of the occurrence of an overtemperature condition.

over-the-horizon radar — A type of high-powered radar used to "see" over the horizon by means of scatter propagation.

over-the-horizon transmission — *See* scatter propagation.

overthrow distortion — Also called overshoot distortion. The distortion that occurs in a signal wave when the maximum amplitude of the signal wavefront exceeds the steady-state amplitude.

overtone — 1. A component of a complex tone having a pitch higher than that of the fundamental component. The term *overtone* has frequently been used in place of *harmonic*, the nth harmonic being called the $(n-1)$st overtone. There is, however, ambiguity sometimes in the numbering of components of a complex sound when the word *overtone* is employed. Moreover, the word *tone* has many different meanings, so that it is preferable to employ terms that do not involve *tone* whenever possible. 2. A tone that is a harmonic of a fundamental tone.

overtone crystal — A quartz crystal cut so that it will operate at a harmonic of its fundamental frequency or at two frequencies simultaneously, as in a synthesizer.

overtravel — Distance in inches or degrees between the electrical operating position and the extreme position to which a switch actuator may be moved without switch damage.

overvoltage — 1. The amount by which the applied voltage in a radiation-counter tube exceeds the Geiger-Mueller threshold. 2. *See* overpotential.

overvoltage protection — Also called transient suppression. The built-in capability of an electrical circuit to dissipate or shunt electrical impulse energy at a voltage low enough to ensure the survival of circuit components.

overvoltage protector — A device or circuit that protects the load by automatically shutting down a supply

when its output voltage exceeds a preset level. A crowbar is one form of overvoltage protection.

overvoltage relay — A relay designated to operate when its coil voltage rises above a predetermined value.

Ovshinsky effect — The characteristic of a special thin-film solid-state switch that has identical response to both positive and negative polarities so that current can be made to have the same magnitude in both directions.

O-wave — *See* ordinary wave.

Owen bridge — A four-arm alternating-current bridge for measuring self-inductance in terms of capacitance and resistance. One arm, adjacent to the unknown inductor, comprises a capacitor and resistor in series. The arm opposite the unknown consists of a second capacitor, and the fourth arm is a resistor. Usually the bridge is balanced by adjusting the resistor in series with the first capacitor, and also the resistor in series with the inductor. The balance is independent of frequency.

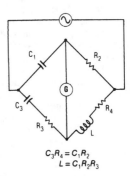

$$C_3 R_4 = C_1 R_2$$
$$L = C_1 R_2 R_3$$

Owen bridge.

oxalizing — *See* surface insulation.

oxidation — 1. Commonly known as rust when ferrous material is involved. The increase in oxygen or in an acid-forming element or radical in a compound. 2. The process of combining with oxygen. More generally, the process by which atoms lose valence electrons or begin to share them with more electronegative atoms. 3. The reaction of oxygen on a compound. Usually detected by a change in the appearance or feel of the surface or by a change in physical properties or both.

oxide — In magnetic recording, microscopic particles of ferric oxide dispersed in a liquid binder and coated on a recording-tape backing. These oxides are magnetically hard — i.e., once magnetized, they remain so permanently unless exposed to a strong magnetic field.

oxide breakdown voltage — That voltage which exceeds gate oxide dielectric breakdown, causing a gate-to-substrate short.

oxide-coated cathode — Also called Wehnelt cathode. A cathode that has been coated with oxides of alkaline earth metals to improve its electron emission at moderate temperatures.

oxide isolation — Electrical isolation of a circuit element by a layer of silicon oxide formed between the element and the substrate.

oximeter — An instrument for measuring the oxygenation of the blood, usually by measuring light transmission through the ear lobe. It uses a photoelectric cell and a source of illumination.

ozone — An extremely reactive form of oxygen, normally occurring around electrical discharges and present in the atmosphere in small but active quantities. It is faintly blue and has the odor of weak chlorine. In sufficient concentrations, it can break down certain rubber insulations under tension (such as a bent cable).

ozone-producing radiation—Ultraviolet energy shorter than about 220 nanometers, which decomposes oxygen (O), thereby producing ozone (O_3). Some ultraviolet sources generate energy at 184.9 nanometers, which is particularly effective in producing ozone.

P

p — 1. Abbreviation for power (combining form, as in pf for power factor), or plate of an electron tube. 2. Letter symbol for the prefix pico- (10^{-12}).

P — Symbol for permeance.

pA — Letter symbol for picoampere.

PABX or **pabx** — Abbreviation for private automatic branch exchange. 1. A small, local, user-owned automatic telephone exchange serving extensions in a business complex that accommodates the transmission of calls to and from the public telephone network. 2. A private branch exchange in which connections are made by remote-controlled switches.

pacemaker — Also called pacer or electronic pacemaker. An electronic instrument for starting and/or maintaining the heartbeat. The instrument is essentially a pulse generator with its output applied either externally to the chest or internally to the heart muscle. In cases requiring long-term application, the device is surgically implanted in the body, and its electrodes contact the heart directly.

pacer — *See* pacemaker.

pack — 1. In computer programming, to combine several fields of information into one machine word. 2. To compress data in a storage medium by taking advantage of known characteristics of the data in such a way that the original data can be recovered; e.g., to compress data in a storage medium by making use of bit or byte locations that would otherwise go unused.

package count — The number of packaged circuits in a system or subsystem.

packaged magnetron — An integral structure comprising a magnetron, its magnetic circuit, and its output matching device.

package lid — A flat cover plate that is used to seal a package cavity.

packaging — The physical process of locating, connecting, and protecting devices, components, etc.

packaging density — 1. The number of devices or equivalent devices per unit volume in a working system or subsystem. 2. In a computer, the number of units of information per dimensional unit. 3. Quantity of functions (components, interconnection devices, mechanical devices) per unit volume, usually expressed in qualitative terms, such as high, medium, or low.

packed data — Information that has been compressed to make optimal use of memory. Four BCD digits can be packed into a 16-bit memory location.

packed decimal — Numerical representation in which two or more BCD digits are present in every word.

packet — 1. A group of binary digits, including data and call control signals, that is switched as a composite whole. The data call-control signals, possible error-control signals, and possible error-control information are arranged in a specific format. 2. A group of ASCII characters (information) surrounded by control signals and error-detection features. The control signals help recognize the presence of a packet and tell any intervening switching equipment where the packet should be sent. 3. A digital communications technique involving the transmission of a short burst of data in a protocol format that contains addressing, control, and error-checking information, along with the field information, in each transmission burst. Packet can also refer to the fixed-length data unit sent over a communications network. A packet contains data plus the addresses of the sending and receiving terminals, control information, and error-checking information. 4. A unit of data to be routed from a source node to a destination node. 5. Common short form of "packet radio."

packet assembler/disassembler — *See* PAD.

packet-mode terminal — Data-terminal equipment that can control and format packets and transmit and receive them.

packet radio — 1. The time-division multiplexing of a radio channel so large numbers of users can share one channel without interfering with one another. Users don't know that they are sharing the channel with anyone else. The name *packet* is derived from the fact that each message is sent in a package. It has three parts: the address and return address, called the header; the data or message part; and the trailer, which is an error-detection scheme. 2. A method of communication that encodes information digitally and in such a manner as to virtually ensure error-free copy at the receiving station. A packet consists of binary data (which might be ASCII, Baudot, or some other code), and the modulation techniques may be essentially the same as for conventional ASCII or rtty, although the exact interpretation of the tones may be different. In a packet the individual characters or bytes are run together with no space at all between. This eliminates the need for both the start and stop bits as well as the dead time between characters. The analogs of start and stop bits are sent only for the beginning and end of the packet, and the transmitter is keyed only while information is actually being sent. 3. The most popular form of amateur radio digital communications, in which computers hooked to radios exchange data in packets.

packets — Data that is transmitted though networks and broken up into small packets (localized in time) rather than being sent as a continuous byte stream. This allows multiple transmissions to share the same line and also facilitates error detection.

packet-switched network — A data-communications network that transmits packets. Packets from different sources are interleaved and sent to their destination over virtual circuits. The term includes PDNs and cable-based LANs.

packet switching — 1. An efficient way to prevent transmission capacity consumption during the silent periods in voice conversation by compressing the conversations of a number of speakers onto a smaller number

534

of channels. In one such system, channel capacity is allocated only when appropriate hardware detects that a subscriber is actively speaking. 2. Transfer of data by means of addressed packets, whereby interim point-to-point channels are available only during the transmission of one packet. The channel then becomes available for the transfer of packets from the same or other messages. (Contrast with circuit switching, where the data network determines the end-to-end routing before the entire message transfer.) 3. The transmission of data by means of addressing packets, whereby a transmission channel is occupied for the duration of transmission of the packet only. The channel is then available for use by packets being transferred between different data-terminal equipment. 4. A method of digital communication in which messages are divided into packets of bit size determined by the needs of the transmission network, and are transferred to their destination in a store-and-forward manner over multiple virtual circuits, which are dedicated to the connection only for the duration of the packet's transmission. 5. A method of transmitting units of data (called packets) through a mesh network. There is no physical circuit established between end points; instead, each packet is individually relayed from one switching node to the next. Individual packets may take different routes through the switching network. 6. The use of software to route messages dynamically from source to destination within a communications network. 7. Data transmission method that divides messages into standard-sized packets for greater efficiency of routing and transport through a network.

packing — Excessive crowding of carbon particles in a carbon microphone. The abnormal pressure of the particles lowers their resistance. As a result, the current increases excessively and fuses some of the particles together, further lowering the resistance and raising the current. Packing causes the sensitivity of the microphone to decrease.

packing density — 1. In a digital computer, the number of units of desired information contained within a storage or recording medium. 2. The amount of digital information recorded along the length of a tape measured in bits per inch (bpi).

packing factor — The number of pulses or bits of information that can be written on a given length of magnetic surface.

packing fraction — 1. The fraction of total cross-sectional area composed of the fiber cores in a fiber-bundle assembly. 2. The ratio of the active core area of a fiber bundle to the total area at its light-emitting or receiving end.

pack unit — A term applied to a compact combination radio transmitter/receiver that can be carried or strapped on the back. Some pack units are popularly known as walkie-talkies.

PACSAT — Abbreviation for *packet satellite*. Satellite used to store and forward amateur digital (packet radio) messages.

pad — 1. A transducer capable of reducing the amplitude of a wave without introducing appreciable distortion. 2. A device inserted into a circuit to introduce transmission loss or to match impedances. 3. A metal electrode that is connected to the output of a diathermy machine and placed on the body over the region being treated. 4. In integrated circuit technology, the bonding area. 5. Also called land. The portion of the conductive pattern on printed circuits designated for mounting or attachment of components. 6. A metallized area on the surface of an active substrate as an integral portion of the conductive interconnection pattern to which bonds or test probes may be applied. 7. A passive resistor network that reduces the

power level of a signal. A pad may be utilized to match unequal input and output impedances for proper interface.

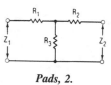

Pads, 2.

PAD — Abbreviation for packet assembler/disassembler. Equipment providing packet assembly and disassembly facilities.

pad character — A character introduced to consume time while a function (usually mechanical, such as carriage return, form eject, etc.) is being accomplished.

padder — Also called padder capacitor or padding capacitor. A relatively high-capacitance trimmer capacitor in the oscillator circuit of a superheterodyne receiver that permits calibration of the low-frequency end of the tuning range so that it tracks with the tuning dial markings.

padder capacitor — *See* padder.

padding — A technique used in digital information systems whereby an information block that is only partially filled is completed by the insertion of dummy data.

padding capacitor — *See* padder.

page — 1. A natural grouping of memory locations by higher-order address bits. In an 8-bit microprocessor, $2^8 = 256$ consecutive bytes often may constitute a page. The words on the same page only differ in the lower-order 8 address bits. 2. A full screen of information.

page printer — A high-speed unit that prints characters one at a time to full page format.

pager — Pocket-sized radio receiver that generates audible or physical signals when the user is paged.

page scrolling — The ability of the system to "flip" through the pages of a document, usually in both forward and backward directions, allowing access to all text of a multipage document.

pagination — The electronic makeup of complete page images, including all graphics, for automatic imaging onto photographic paper, negatives, or offset printing plates.

paging — 1. Methods for locating and exchanging segments to and from the main computer memory. 2. To summon a particular person over a public address system, or by selectively calling him or her on a pocket radio receiver that emits an alerting signal. 3. In the case of a CRT, switching from one page of information to the next. 4. In the case of a memory, a logical block of storage used for memory management (for example, 1 K words). An address is then specified by a page address (number) and a displacement (address within the page).

paging receiver — Small, lightweight FM radio receiver carried by persons who need to be paged when they are away from their phones.

paging system — Communications system for summoning individuals (doctors, nurses, hospital personnel, etc.) or making public announcements.

paint — 1. Vernacular for a target image on a radarscope. 2. To draw vectors when the beam is unblanked 3. To shade the interior of a closed graphical image with diagonal lines, cross hatch, points, etc.

pair — In electric transmission, two like conductors employed to form an electric circuit.

paired cable — 1. A cable in which all of the conductors are arranged in the form of twisted pairs, none of which is arranged with others to form quads. 2. Cable in which the conductors are combined in pairs. Two wires are twisted about each other, and each wire of the pair has a distinctive color of insulation.

pairing — 1. In television, the imperfect interlace of lines comprising the two fields of one frame of the picture. Instead of being equally spaced, the lines appear in groups of two — hence the name. 2. In television, an effect in which the lines of one field fail to fall exactly within the lines of the following field, both fields comprising one frame of the picture. The lines of the two fields fall directly over each other when the effect is more exaggerated, and the vertical definition is suitably decreased by half. 3. A faulty interlace scan in which the alternate scanning lines tend to overlap each other. The effect is a severe reduction in vertical resolution capability.

PAL — Abbreviation for phase alternating line. 1. Pertaining to a color television system in which the subcarrier derived from the color burst is inverted in phase from one line to the next in order to minimize hue errors that may occur in color transmission. 2. The German-developed TV color system used in Europe (except for France). It features 625 lines per frame and 50 fields per second. (Generally, this method gives higher-resolution TV pictures than the American NTSC 525-line color system.) PAL uses a similar transmission method as NTSC, but with the color information switched 180° on alternate scan lines. The subcarrier frequency is 4.43 MHz.

Palmer scan — A combination of circular and conical scans. The beam is swung around the horizon at the same time the conical scan is performed.

PALplus — An enhanced PAL transmission system used in Europe. It offers sharper images in wide screen format.

PAM — Abbreviation for pulse-amplitude modulation.

PAM/FM — Frequency modulation of a carrier by pulses that, in turn, are modulated by data.

PAM/FM/FM — Frequency modulation of a carrier by subcarriers modulated by pulses that, in turn, are modulated by data.

pan — 1. To move a television or movie camera slowly up and down or across a scene to secure a panoramic effect. 2. To move the camera up and down, or back and forth, in order to keep it trained on a moving object.

pan and tilt — An accessory upon which a camera is mounted to facilitate movement (panning and tilting) by the operator or by a remote control unit.

pancake coil — A coil shaped like a pancake, usually with the turns arranged in a flat spiral.

panel — 1. An electrical switchboard or instrument board. 2. A mounted plate of metal or insulation for the controls and/or other parts of equipment.

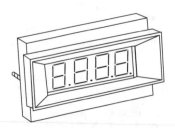

Panel meter.

panel code — Prearranged code designed for visual communications between ground units and friendly aircraft.

panel layout — The general physical layout of an electrical control panel with all relays, disconnect switch, control transformer, terminal strip, etc.

panning — *See* pan.

panoramic adapter — An attachment used with a search receiver to provide, on an oscilloscope screen, a visual presentation of the frequencies above and below the center frequency to which the search receiver is tuned.

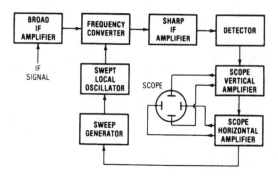

Panoramic adapter.

panoramic display — A display that shows at the same time all the signals received at different frequencies.

panoramic presentation — A presentation of signals as intensity pips (vertical deflections) along a line. The horizontal distance along the line represents frequency.

panoramic radar — A nonscanning radar that transmits signals omnidirectionally over a wide beam.

panoramic receiver — A radio receiver that displays, on the screen of a cathode-ray tube, the presence and relative strength of all signals within a wide frequency range. Used in communications for monitoring a wide band, locating open channels quickly, indicating intermittent signals or interference, and monitoring a frequency-modulated transmitter.

panoramic sonic analyzer — A heterodyne-type instrument that separates the frequency components of a complex waveform and displays them on an oscillographic screen, indicating both frequency and magnitude.

pan-pot — A potentiometer used to adjust the stereo balance of a monophonic signal, allowing it to be positioned anywhere across the stereo range.

pan-range — Intensity-modulated A-type radar indication with slow vertical sweep applied to video. Stationary targets give solid vertical deflection, and moving targets give broken vertical deflection.

pantography — A system for transmitting and automatically recording radar data from an indicator to a remote point.

pantophonic system — *See* ambisonic reproduction.

paper capacitor — 1. A fixed capacitor consisting of two strips of metal foil separated by oiled or waxed paper or other insulating material, the whole rolled together into a compact roll. The foil strips can be staggered so that one strip projects from each end, or tabs can be added. The connecting wires are attached to the strips or tabs. 2. A capacitor with a dielectric consisting of paper, usually impregnated.

537

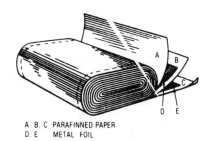

A, B, C PARAFINNED PAPER
D, E   METAL FOIL

Paper capacitor.

paper electrophoresis — Analytical instrument for a technique in which ions migrate along a strip of porous filter paper saturated with an electrolyte when a potential gradient is applied across the length of the strip. It is used to identify ion types in analysis of serums, proteins, biochemicals, inorganic ions, rare earths, etc.

paper tape — 1. A continuous strip of paper into which data is recorded as a series of holes along its length. Data is read by a paper-tape reader sensing the pattern of holes, which represent coded data. 2. One of the slowest but oldest and cheapest methods of storing archival information in a computer system. Data is stored in punched-hole sequences on a strip of tape.

paper-tape punch — A device that places binary characters on a paper tape in the form of holes punched in appropriate channels on the tape. A binary 1 is indicated by the presence of a hole, and a 0 is indicated by the absence of a hole.

paper-tape reader — 1. A device that senses and translates holes punches in a tape into electrical signals. 2. A device that senses the presence or absence of holes in punched paper tape, one character at a time, and produces electrical signals suitable for computer input.

PAR — *See* precision approach radar.

parabola — Locus of points equidistant from a fixed point and a straight line.

parabola controls — Sometimes called vertical-amplitude controls. Three controls in a color television receiver employing the magnetic-convergence principle. They are used for adjusting the amplitude of the parabolic voltages applied, at the vertical-scanning frequency, to the coils of the magnetic-convergence assembly.

parabolic antenna — 1. An antenna with a radiating element and a parabolic reflector that concentrates the radiated power into a beam. 2. A highly directional microwave antenna that uses a parabolic reflector. 3. The most frequently found satellite TV antenna, which takes

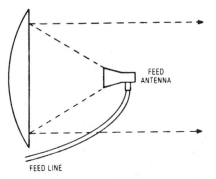

Parabolic antenna.

its name from the shape of the dish described mathematically as a parabola. The function of the parabolic shape is to focus the weak microwave signal hitting the surface of the dish (antenna) into a single focal point in front of the dish, where the feedhorn is usually located.

parabolic microphone — A microphone positioned at the focus of a parabolic sound reflector to give highly directional characteristics.

parabolic reflector — A metallic sheet formed so that its cross section is in the shape of a cylindrical parabola. The antenna elements are placed along the line that runs through the focal point of the parabola, parallel to the leading edge of the reflecting sheet.

parabolic-reflector microphone — A microphone employing a parabolic reflector for improved directivity and sensitivity.

parabolic/shotgun mike — A microphone with a reflector/director that concentrates sound energy so that specific sounds can be heard at a distance with minimum interference from other sound sources.

paraboloid — A reflecting surface of paraboloidal shape (the shape of a surface formed by rotating a parabola about its axis of symmetry).

paraboloidal reflector — A hollow concave reflector that is a portion of a paraboloid of revolution.

paradigm — In programming, an established coding model or structure.

paraffin — A vegetable wax having insulating properties.

paragraphic equalizer — A contraction of *para-metric* and *graphic*. A graphic equalizer in which the center frequency of each band is adjustable. May also have adjustable Q.

parallax — 1. An optical illusion that makes an object appear displaced when viewed from a different angle. Thus, a meter pointer will seem to be at different positions on the scale, depending from which angle it is read. To eliminate such errors, the eye should be directly above the meter pointer. 2. The apparent displacement of a meter pointer from its true position on the scale, when the observer's eye is not centered over the pointer.

parallel — Also called shunt. 1. Connected to the same pair of terminals, so that the current can branch out over two or more paths. 2. A method of connecting an electric circuit whereby each element is connected across the other. The addition of all the currents through each element equals the total current of the circuit. 3. In electronic computers, the simultaneous transmission of, storage of, or logical operations on a character or other subdivision of a word, using separate facilities for the various parts. 4. Indicating a type of computer in which several operations are performed on the same or different data at once. 5. A type of interface in which all bits of data in a given byte are transferred simultaneously, using a separate data line for each bit.

parallel access — Also called simultaneous access. The process of taking information from or placing information into computer storage whereby the time required for such access depends on simultaneously transferring all elements of a word from a given storage location.

parallel adder — A conventional technique for adding, in which two multibit numbers are presented and added simultaneously in parallel.

parallel addition — A form of addition in which the computer operates simultaneously on each set of corresponding digits of two numbers.

parallel arithmetic unit — In a computer, a unit in which separate equipment operates (usually simultaneously) on the digits in each column.

parallel buffer — An electronic device (for example, magnetic cores or flip-flops) used for temporary parallel storage of digital data.

parallel circuit — 1. A circuit in which all positive terminals are connected to a common point, and all negative terminals are connected to a second common point. The voltage is the same across each element in the circuit. 2. A circuit in which the current has two or more paths to follow. Two electrical elements are in parallel if (and only if) both terminals of both elements are electrically connected.

parallel computer — A computer in which the digits or data lines are handled at the same time by separate units.

parallel connection — Also called shunt connection. Connection of two or more parts of a circuit to the same pair of terminals, so that current divides between the parts; as contrasted with a series connection, in which the parts are connected end to end so that the same current flows through all.

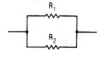

*Parallel-connected
resistors.*

parallel cut — A Y-cut in a crystal.

parallel digital computer — A computer in which the digits are handled in parallel. Mixed serial and parallel machines are frequently called serial or parallel according to the way arithmetic processes are performed. For example, a parallel digital computer handles decimal digits in parallel, although the bits that comprise a digit might be handled either serially or in parallel.

parallel feed — Also called shunt feed. Application of a dc voltage to the plate or grid of a tube in parallel with an ac circuit, so that the dc and ac components flow in separate paths.

parallel gap solder — Passing a high current through a high-resistance gap between two electrodes to remelt solder, thereby forming an electrical connection.

parallel gap welding — 1. A method of resistance welding in which both electrode tips are in close proximity to each other, being separated by a small gap or insulating material, approach the work from the same direction, and contact only one of the two materials being welded. 2. Type of resistance welding wherein electrodes contact the work from one side only. Mechanism by which bonding occurs is virtually always fusion. The

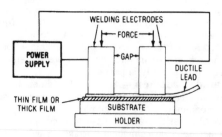

Parallel gap welding.

process is well suited to welding component leads to planar surfaces, such as IC leads to printed circuit conductors. 3. Passing a high current through a high-resistance gap between two electrodes that are applying force to two conductors, thereby heating the two work pieces to the welding temperature and effecting a welded connection.

paralleling reactor — A reactor for correcting the division of load between parallel-connected transformers with unequal impedance voltages.

parallel interface — 1. A multiline channel that transfers 8 parallel bits. 2. A port that sends or receives the 8 bits in each byte all at one time. Many printers likely to be used in homes use a parallel interface to connect to the computer. 3. A link between two devices in which all the information transferred between them is transmitted simultaneously over separate conductors.

parallel light — *See* collimated light.

parallel load — *See* shift, 2.

parallelogram distortion — In camera or image tubes, a form of distortion that amounts to a skewing of the reproduced image laterally across the CRT face.

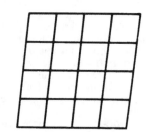

Parallelogram distortion.

parallel operation — Also called master/slave operation. 1. The connecting of two or more power supplies so that their outputs are tied together, permitting the accumulated current from all units to a common load. In regulated power supplies, interconnections other than the output terminals themselves may be required. For example, the amplifiers of all units but one may be made inoperable, and this single amplifier would control all regulating elements. 2. Pertaining to the manipulation of information within computer circuitry, in which the digits of a word are transmitted simultaneously on separate lines. Faster than serial operation, but requires more equipment. 3. Type of information transfer whereby all digits of a word are handled simultaneously.

parallel output — 1. An output arrangement in which two or more bits, channels, or digits are available simultaneously. 2. Simultaneous availability of two or more bits, channels, or digits.

parallel-plate oscillator — A push-pull, ultrahigh-frequency oscillator circuit that uses two parallel plates as the main frequency-determining elements.

parallel-plate waveguide — A pair of parallel conducting planes for propagating uniform cylindrical waves that have their axes normal to the plane.

parallel port — 1. Interface located on a host adapter card used to connect a disk drive or printer to the PC. 2. A port on a computer used to send several bits in parallel (usually 8 bits). Many printers connect to a parallel port. 3. A data communication channel that uses one wire for each bit in a single byte.

parallel processing — 1. In a computer, the processing of more than one program at a time through more than one active processor. 2. The simultaneous execution of two or more processes in multiple devices, such as channels or processing units.

parallel programming—A method of parallel operation of two or more power supplies in which the feedback terminals (voltage control terminals) of the units are also connected in parallel. Often, these terminals are connected to a separate programming source.

parallel recording—A technique in which the record heads in a head stack are energized simultaneously to record a specific set of bits.

parallel resonance—In a circuit comprising inductance and capacitance connected in parallel, the steady-state condition that exists when the current entering the circuit from the supply line is in phase with the voltage across the circuit.

parallel-resonant circuit—An inductor and capacitor connected in parallel to furnish a high impedance at the frequency to which the circuit is resonant.

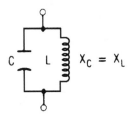

$$X_C = X_L$$

Parallel-resonant circuit.

parallel-rod oscillator—An ultrahigh-frequency oscillator circuit in which the tank circuits are formed by parallel rods or wires.

parallel-rod tank circuit—A tank circuit consisting of two parallel rods connected at their far ends. This is done to provide the small values of inductance and capacitance in parallel required for ultrahigh-frequency circuits.

parallel-rod tuning—A tuning method sometimes used at ultrahigh frequencies. The transmitter, receiver, or oscillator is tuned by sliding a shorting bar back and forth on two parallel rods.

parallel search storage—A type of computer storage in which one or more parts of all storage locations are queried at the same time. *See also* associative storage.

parallel-series circuit—Also called shunt-series circuit. Two or more parallel circuits connected together in series.

parallel shift—*See* shift, 2.

parallel splice—A device in which two or more conductors are joined and lie parallel and adjacent to each other.

parallel storage—Computer storage in which characters, words, or digits are accessed simultaneously.

parallel-T network—Also called twin-T network. A network composed of separate T-networks (usually two), the terminals of which are connected in parallel.

parallel-T oscillator—An *RC* sine-wave oscillator that provides phase inversion at one discrete frequency and is so connected that positive feedback results only when phase inversion occurs.

parallel-tracking arm—Pickup system that allows phonograph cartridge to track on the true radius of the record, as the recording was made, thereby minimizing lateral tracking error.

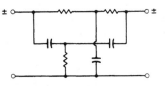

Parallel-T network.

parallel transfer—Data transfer in which all characters of a word are transferred simultaneously over a set of lines.

parallel transmission—1. In a computer, the system of information transmission in which the characters of a word are transmitted (usually simultaneously) over separate lines, as contrasted with serial transmission. 2. The simultaneous transmission of the bits making up a character, via either separate channels or different carrier frequencies on one channel. 3. The simultaneous transmission of a number of signal elements, either as tones or as dc pulses. 4. Transmission mode in which a number of bits of information are sent simultaneously (usually unidirectional) over separate lines (for example, eight bits over eight lines).

parallel-wire line—A transmission line consisting of two wires a fixed distance apart.

parallel-wire resonator—A resonator circuit consisting of two parallel wires connected at one end to the oscillator tube or transistor. The other end is short circuited and can be adjusted for the desired frequency.

Paraloc—A phase-shift amplifier whose positive feedback results in an oscillation that is frequency modulated by the change in a sensed variable. The sensor is one arm of a resistance bridge.

paramagnetic—1. Having a magnetic permeability greater than that of a vacuum but less than that of ferromagnetic materials. Unlike the latter, the permeability of paramagnetic material is independent of the magnetizing force. 2. Characteristic of a substance that is attracted by a magnetic field, but which has a permeability only slightly greater than 1.

paramagnetic amplifier—A parametric amplifying device in which a nonlinear element (varactor diode) is pumped at twice the frequency of the signal to be amplified. Amplification is obtained because the pumped state of the element can be returned to the normal state by an input signal of relatively low power, thus releasing the excess energy to an external circuit at the signal frequency.

paramagnetic material—A material having a permeability that is slightly greater than that of a vacuum and which is approximately independent of the magnetizing force, i.e., the material is broadly considered nonmagnetic.

paramagnetism—Magnetism that involves a permeability somewhat greater than unity.

parameter—1. A constant or element, the value of which characterizes the behavior of one or more variables associated with a given system. 2. A measured value that expresses performance. 3. A test variable used as an arbitrary constant. 4. A variable that is given a constant value for a specific purpose or process. 5. A measurable factor that indicates the degree of quality, performance, or capability of a device. 6. An element that determines the characteristics or behavior of the data communication equipment.

parameter extraction—A technique that reduces the bandwidth required to transmit a given data sample by means of an information-describing irreversible transformation. These transformations are considered irreversible in that, while they provide useful descriptions of the input signal, they so distort the signal that it is impossible to reconstruct the original waveform.

parameter format — The structure and significance of information that must be assigned values before a generic package may be generated.

parameters — Those indicators of device performance that relate one aspect of its behavior with another. Hence, input resistance is a parameter relating input voltage with input current, and current gain is a parameter relating output current with input current.

parameter spread — The inevitable variation in value of the parameters of a given device type due to manufacturing tolerances. Often expressed in terms of the various statistical distributions.

parameter tags — Constants used by several programs.

parametric amplification — A means of amplifying optical waves whereby an intense coherent pump wave is made to interact with a nonlinear optical crystal to produce amplification at two other optical wavelengths.

parametric amplifier — 1. A low-noise device for amplifying signals in the UHF and microwave regions of the electromagnetic spectrum. The essential element of the amplifier is a semiconductor crystal (varactor). By its very nature the diode contributes very little noise, unlike an amplifier utilizing electron tubes, where the random nature of the electron emission from the hot cathode produces large fluctuations in the current, called shot noise. Thus, the signals to be amplified can be of extremely low levels. 2. A microwave amplifier having a base element whose reactance can be varied periodically by an ac voltage at a pumping frequency. Operation is at room temperature. 3. An amplifier that uses a varying parameter, such as reactance, to take power from a local source of energy to amplify an input signal. (The local source of energy is usually a pump oscillator, operating at twice the frequency of the signal to be amplified.) 4. A low-noise amplifier that depends on a high-frequency source (Gunn-effect pump) rather than a dc source to amplify weak input signal. It works like a maser. The input signal modulates an electron beam, and the pump amplifies it while it travels to the output coupler.

parametric converter — An inverting or noninverting parametric device used to convert an input signal at one frequency into an output signal at a different frequency.

parametric device — A device whose operation depends essentially upon the time variation of a characteristic parameter usually understood to be a reactance.

parametric down-converter — A parametric converter in which the output signal is at a lower frequency than the input signal. An equalizer whose center frequency is continuously variable over a given frequency range, and whose Q (slope rate) is adjustable. If Q is nonadjustable, the equalizer is "quasi-parametric," "tunable," or "sweepable."

parametric excitation — A term referring to the method of exciting and maintaining oscillation in either an electrical or mechanical dynamic system, in which excitation results from a periodic variation in an energy storage element in a system such as a capacitor, inductor, or spring constant.

parametric frequency converter — A frequency converter that utilizes the variation of the reactance parameter of an energy-storage element for frequency conversion.

parametric modulator — A modulator that utilizes the variation in the reactance parameter of an energy-storage element to produce modulation.

parametric oscillator — A device using a parametric amplifier inside a resonant optical cavity to generate a frequency-tunable coherent beam of light from an intense laser beam of fixed frequency. It is tuned by varying the phase-matching properties of the nonlinear material.

parametric testing — Testing based on reasonably precise measurements of voltages or currents.

parametric tests — Tests that measure dc conditions of a chip, such as maximum current, leakage, and output drive.

parametric up-converter — A parametric converter in which the output signal is at a higher frequency than the input signal.

parametric value — The actual information that is assigned to a parameter format.

parametron — A digital circuit element utilizing the principle of parametric excitation. It is essentially a resonant circuit with a nonlinear reactive element that oscillates at half the driving frequency. The oscillation can be made to represent a binary digit by the choice between two stationary phases π radians apart.

paramistor — A digital logic circuit module containing several parametron elements.

paraphase amplifier — An amplifier that converts a single input signal into two out-of-phase signals for driving a push-pull stage.

parasite — Current in a circuit due to some unintentional cause such as inequalities of temperature or of composition; particularly troublesome in electrical measurement.

parasitic — An undesired low- or high-frequency signal in an electronic circuit.

parasitically excited — Said of an antenna element (such as a director or reflector) that is not directly connected to a source of rf energy, but instead is energized through radiation from a nearby element.

parasitic antenna — An antenna that is excited by radiation from other antennas rather than by electrical connection with them.

parasitic array — An antenna array containing one or more elements not connected to the transmission line.

parasitic components — In a monolithic integrated circuit, the capacitors and diodes that are formed between the planned circuit elements and the substrate during processing. The circuit design must allow for the functional effects of these parasitic components.

parasitic element — Also called passive element. 1. An antenna element (i.e., reflector, director, etc.) not connected to the transmission line or to any driven element. A parasitic element affects the gain and directivity pattern of an antenna, and also acts on a driven element by absorbing and returning energy from it. In a dipole reflector combination, the reflector is the parasitic element. 2. An element of a directional antenna that has no electrical connection to the active element(s) of the antenna, but which reflects or directs the radio waves so that they are additive along the directional axis. 3. An undesirable but inherent element in a circuit, such as wire resistance, core losses, winding capacitance, or leakage inductance.

parasitic field turn-on — *See* field inversion.

parasitic oscillation — 1. An undesired self-sustaining oscillation at a frequency other than the operating frequency. Parasitic oscillations occur chiefly in vacuumtube circuits. 2. Any undesired oscillation in an oscillator or amplifier stage.

parasitics — Parasitic oscillations.

parasitic suppressor — A parallel resistance, or a parallel combination of inductance and resistance, inserted into a grid or plate circuit to suppress parasitic oscillations.

paraxial ray — A ray that is parallel, or nearly parallel, to an optical axis.

PARD or **pard**—Acronym for *p*eriodic *a*nd *r*andom *d*eviation, which replaces the term "ripple and noise." 1. In an electronic power conversion unit, the periodic and random deviation of the output dc voltage, current, or power from its average value, with all external operational and environmental parameters maintained constant. The load impedance in particular must be held constant. Perturbations in the output that are induced by load impedance changes are dynamic load regulation or transient response. PARD may be defined as rms PARD or peak-to-peak PARD. 2. A broadband measurement of net ac variations in the dc output, usually measured within a 10- to 20-MHz bandwidth and expressed in millivolts rms. In a few special cases it is expressed as peak-to-peak variation.

parent population—Prototype or initial group of the articles under consideration.

parity—1. A method of checking the accuracy of binary numbers used in recorded, transmitted, or received data. An extra bit, called a parity bit, is added to a number. If even parity is used, the sum of all 1s in the number and its corresponding parity bit is always even. If odd parity is used, the sum of the 1s and the parity bit is always odd. 2. The number of 1s in a word (may be even or odd). When parity is used, an extra bit is used that indicates whether the data word has an odd or even number of 1s. ASCII uses seven bits for data and one bit for parity. Parity is one of the simplest error-detection techniques and will detect a single-bit failure. 3. A technique for testing transmitted data. Typically, a binary digit is added to the data to make the sum of all the digits of the binary data either always even (even parity) or always odd (odd parity). If any one bit is corrupted, the count will no longer agree and the error is detected.

parity bit—1. An additional bit used with a computer character or electronic channel data processor to provide a check for accuracy. 2. A binary digit appended to an array of bits to make the sum of all the bits always odd or always even. 3. An additional bit added to a memory word to make the sum of the number of 1s in a word always even (even parity) or always odd (odd parity). 4. A noninformation bit that is used to ensure that data has been transmitted accurately; a receiving device counts the "on" bits of every arriving byte; if odd parity is specified, an error condition will be flagged any time an even number of "on" bits is detected.

parity check—Synonymous with odd-even check. 1. An error-checking method that tests whether the number of 1s in a group of binary digits is odd or even. If a channel uses an odd parity scheme, an odd number of 1s implies that the data was received correctly, etc. 2. A technique used by the computer to check on the validity of data as it moves from one location to another. Parity for a given computer will be either even or odd; any data that contains an odd number of bits will be given one extra check bit in an even-parity computer. The computer can therefore recognize quickly whether any bit of information has been dropped or picked up as data has been moved. 3. A method of checking the correctness of binary data after that data has been transferred from or to storage. An additional bit, called the parity bit, is appended to the binary word or character to be transferred. The parity bit is the single-digit sum of all the binary digits in the work or character and its logical state can be assigned to represent either an even or an odd number of 1s making up the binary word. Parity is checked in the same manner in which it is generated. 4. The technique of detecting errors in a group of bits by adding a noninformation bit that will make the total number of 1s in the grouping either always even (even parity) or always odd (odd parity). 5. A check on the validity of a binary word by determining whether the number of 1s in the word is odd or even.

parity checking—An error-detection technique in which character bit patterns are forced into parity, so that the total number of 1 bits is always odd or always even. This is accomplished by the addition of a 1 or 0 bit to each byte, as the byte is transmitted; at the other end of the transmission, the receiving device verifies the parity (odd or even) and the accuracy of the transmission.

parity checks and checksums—Methods of verifying that short sequences of bits are identical with the sequences orignally transmitted.

parity error—An error occurring when the results of the parity calculations at the transmit and receive ends of a system don't agree.

parity tree—A group of exclusive OR gates that can be used to check a number of input bits for either odd or even parity. Parity trees are used both to check and generate parity wherever a redundant bit is added to a word in order to check for error.

part—1. The smallest subdivision of a system. 2. An item that cannot ordinarily be disassembled without destruction.

part failure—A breakdown that cannot be repaired and which ends the life of a part.

part-failure rate—The number of occasions, during a specified time period, on which a given quantity of identical parts will not function properly.

partial—1. A physical component of a complex tone. 2. A component of a sound sensation that can be distinguished as a simple tone that cannot be further analyzed by the ear and which contributes to the character of the complex sound.

partial carry—In parallel addition, a technique involving temporary storage of some or all of the carries instead of allowing them to propagate immediately.

partial dial tone—A high dial tone that notifies a calling party that he or she has not completed dialing within a specified period of time, or that not enough digits have been dialed.

partial motor—Also called a shell-type motor. A motor sold with rotor and stator only — no end bells and no containing frame.

partial node—The place in a standing-wave system at which some characteristic of the wave field has a minimum amplitude other than zero.

partial-read pulse—In a computer, any one of the applied currents that cause selection of a core for reading.

partial-select pulse—In a computer, the voltage response of an unselected magnetic cell produced by the application of partial-read pulses or partial-write pulses.

partial-write pulse—In a computer, any one of the applied circuits that cause a core to be selected for writing.

particle—An infinitesimal subdivision of matter — e.g., a molecule, atom, or electron.

particle accelerator—Any device for accelerating charged particles to high energies (e.g., cyclotron, betatron, Van de Graaff generator, linear accelerator).

particle orientation—The process by which acicular particles are positioned so that their longest dimensions tend to be parallel. Orientation is accomplished in magnetic tape by the combined effects of the sheer force applied during the coating process and a magnetic field applied to the coating while it is still fluid.

particles—Also known as domains. Small bits of oxide that are the recording media on magnetic tape. The smaller and more uniform they are, the better the tape's frequency response, provided they are evenly dispersed.

particle velocity—The velocity of a given infinitesimal part of a sound wave. The most common unit is centimeter per second.

partition — A portion of a hard disk with a size expressed in the disk sectors.

partitioning — Also called segmenting. 1. In a computer, subdividing a large block into smaller, more conveniently handled subunits. 2. Logical grouping of electrical functions within a given set of hardware components.

partition noise — A type of noise that appears in multielement vacuum tubes due to the random division of cathode current among the different electrodes. It has been called pseudoshot noise. It is more pronounced in pentodes and tetrodes than in triodes.

Part 95 rules — FCC rules and regulations governing the Citizens Radio Service.

part programmer — One who translates the physical operations for machining a part into a series of mathematical steps and then prepares the coded computer instructions for those steps.

parts density — The number of parts in a unit volume.

party line — A telephone line serving more than one subscriber, with discriminatory ringing for each. Usually either a two-party or a four-party line. Lines serving more than four parties are called rural lines.

Pascal — A block-structured high-level programming language in the style of ALGOL. It incorporates the control structures of structured programming (e.g., sequence, selection, repetition) and data structures (e.g., arrays, records, files, sets, and user-defined types). Named for the French physicist Blaise Pascal, it is in many ways simpler to use than BASIC. Pascal can be made machine-dependent or transportable.

pascal — The SI unit of pressure or stress. One pascal equals 1 newton per square meter. Letter symbol: Pa.

Paschen's law — The sparking potential between two terminals in a gas is proportional to the pressure times the spark length. For a given voltage, this means the spark length is inversely proportional to the pressure.

pass — One cycle of processing of a body of data.

passband — 1. The band of frequencies that will pass through a filter with essentially no attenuation. 2. The frequency range in which a filter is intended to pass signals.

passband filters — Filters used in modem design to allow only the frequencies within the communication channel to pass, while rejecting all frequencies outside the channel.

passband ripple — In a filter, the difference, in decibels, between the minimum loss point and the maximum loss point in a specified bandwidth.

pass element — An automatic variable-resistance device, either a vacuum tube or power transistor, in series with the source of dc power. The pass element is driven by the amplifier error signal to increase its resistance when the output needs to be lowered or to decrease its resistance when the output must be raised. *See also* series regulator.

passivate — To treat the surface of a semiconductor with a relatively inert material in order to protect it from contamination.

passivated region — Any region covered by glass, SiO_2, nitride, or other protective material.

passivation — 1. The growth of an oxide layer on the surface of a semiconductor to provide electrical stability by protecting the surface against moisture, contamination, particles, and mechanical damage. This reduces reverse-current leakage, increases breakdown voltages, and raises the power dissipation rating. 2. *See* glassivation. 3. A coating of electrically inert material, such as glass or silicon dioxide, used to protect semiconductors or resistors from environmental contamination. Unpassivated nickel-chromium resistors, for example, will open in the presence of water and a large applied dc potential. 4. The growth of an oxide layer on the surface of a semiconductor to provide mechanical protection by isolating the transistor surface from electrical, mechanical (scratching of metal), and chemical conditions in the environment. 5. The technique of providing a semiconductor device chip with an isolating layer or "skin" that protects it from contamination by unwanted impurity atoms or molecules. With silicon devices, the isolating layer is usually composed of silicon dioxide (quartz) or silicon nitride, grown on the chip at a high temperature. 6. Electrolytical treatment of a metal or semiconductor to create a chemically bonded oxide layer on the surface to protect it from corrosion. 7. The formation of an insulating layer directly over a circuit or circuit element to protect the surface from contaminants, moisture, or particles. 8. Surface oxidation that acts as a barrier to further oxidation and corrosion.

passive — 1. An inert component that may control, but does not create or amplify, energy. 2. Pertaining to a general class of device that operates on signal power alone. 3. Incapable of generating power or amplification. A nonpowered device that generally presents some loss to a system. 4. Describing a device that does not contribute energy to the signal it passes.

passive acoustic monitoring — The use of microphones and ancillary equipment to provide surveillance by monitoring the sounds in a protected premise.

passive communication satellite — A communication satellite that simply reflects a signal without amplification. In essence, it is a radio mirror. It requires a large reflecting surface and large, high-powered, complex ground stations.

passive component — 1. A nonpowered component generally presenting some loss (expressed in decibels) to a system. 2. A component that has no gain characteristics, such as a capacitor or a resistor. 3. An electrical component without gain or current-switching capability. Commonly used when referring to resistors, capacitors, and inductors.

passive decoder — A device that is set so that only one specific reply code will pass a decoder and give an output from one decoder for display.

passive detection — Detection of a target by reception of signals emitted by the target rather than by means involving a signal source independent of the target.

passive device — 1. A device that exhibits no transistance. It has no gain or control and does not require any input other than a signal to perform its function. Examples of passive devices are conductors, resistors, and capacitors. 2. A component that does not provide rectification, amplification, or switching, but reacts to voltage and current; e.g., resistor, capacitor. 3. An electronic component that does not require a bias voltage, e.g., a resistor, capacitor, or inductor.

passive electric network — An electric network with no source of energy.

passive element — 1. A parasitic element. 2. A circuit element with no source of energy (e.g., a resistor, capacitor, inductor). 3. An electronic circuit element that displays no gain or control, such as a resistor or capacitor.

passive film circuit — A thin- or thick-film circuit network consisting entirely of passive circuit elements and interconnections.

passive homing system — A guidance system based on the sensing of energy radiated by the target. *See also* active homing; homing guidance.

passive intrusion sensor — A passive sensor in an intrusion alarm system that detects an intruder within the range of the sensor. Examples are a sound-sensing detection system, a vibration detection system, an infrared-motion detector, and an *E*-field sensor.

passive network — A network with no source of energy.

passive probe — A test probe that is constructed entirely of passive components — resistors, capacitors, and, in some cases, inductors.

passive pull-up — A gate output circuit in which the charging current for a load capacitance is obtained through a resistor.

passive reflector — A reflector often used on microwave relay towers to change the direction of a microwave. This permits convenient location of transmitter, repeater, and receiver equipment on the ground rather than at the tops of towers.

passive repeater — 1. A radio-frequency device used to change the direction of a radio beam without amplification. 2. A large, flat metal or metal-screen surface that acts as a simple radio-frequency mirror.

passive satellite — A satellite that reflects, without amplification, communications signals from one ground station to another.

passive sensor — A sensor that detects natural radiation or radiation disturbances, but does not itself emit the radiation on which its operation depends.

passive solar energy — Energy gathered by natural convection and heating, rather than by mechanical means.

passive sonar — *See* sonar.

passive substrate — 1. A substrate that may serve as a physical support and thermal sink for a thick- or thin-film integrated circuit but does not exhibit transistance. Examples of passive substrates are glass, ceramic, alumina, etc. 2. A physical support and a thermal sink for circuits. The substrate itself performs no electrical function.

passive system — A system that emits no energy and therefore does not reveal its position or existence.

passive tracking system — Usually a system that tracks by reflected radiation from some external source, or by the jet emission of the vehicle (e.g., optical systems, use of commercial radio or television, reflection and infrared systems).

passive transducer — A transducer that does not require any local source of energy other than the received energy.

passive ultrasonic alarm system — An alarm system that detects the sounds in the ultrasonic frequency range caused by an attempted forcible entry into a protected structure. The system consists of microphones, a control unit containing an amplifier, filters, an accumulator, and a power supply. The unit's sensitivity is adjustable so that ambient noises or normal sounds will not initiate an alarm signal; however, noise above the preset level or a sufficient accumulation of impulses will initiate an alarm.

password — 1. A code used to gain access to a locked system. Good passwords contain letters and nonletters and are not simple combinations. A good password might be "Bit@1*6." 2. A string of characters known (supposedly) only to the user of a particular computer account, employed to identify that person when he logs into the system. 3. A unique string of characters required to gain access to a device, data channel, computer, or data files. In multiple-user systems, users identify themselves with a password, unique to each user, before the computer will let them use the system.

paste — 1. In batteries, the medium, in the form of a paste or jelly, containing an electrolyte. It is positioned adjacent to the negative electrode of a dry cell. In an electrolyte cell, the paste serves as one of the conducting plates. 2. Synonymous with *composition* and *ink* when relating to screenable thick-film materials.

paste blending — Mixing resistor pastes of different ohms per area value to create a third value in between those of the two original materials.

paste solder — 1. Finely divided particles of solder suspended in a flux paste. Used for screening application onto a film circuit and reflowed to form connections to chip components. 2. A mixture of flux and finely divided solder.

PA system — Abbreviation for public-address system.

patch — 1. To connect circuits together temporarily with a special cord known as a patch cord. 2. In a computer, to make a change or correction in the coding at a particular location by inserting transfer instructions at that location and by adding elsewhere the new instructions and the replaced instructions. This procedure is usually used during checkout. 3. The section of coding so inserted. 4. A group of instructions that have been inserted into a program to correct an error or deficiency. Patching a program rather than rewriting it is poor practice, for a program with numerous patches becomes very hard to understand.

patch bay — *See* jack bay.

patch board — A board or panel in which circuits are terminated in jacks for patch cords.

patch cable — A cable with plugs or terminals on each end of the conductor or conductors, used to temporarily connect circuits of equipment together.

patch cord — Sometimes called an attachment cord. 1. A short cord with a plug or a pair of clips on one end, for conveniently connecting two pieces of sound equipment such as a phonograph and tape recorder, an amplifier and speaker, etc. 2. A short audio cable with a male plug on each end. Commonly used for audio signal routing (patching) between nearby electronic devices, or between various jacks on a mixer or patch panel (patch bay). Patch cords often also route dc control signals as well as audio signals in electronic music synthesizers. 3. Cord, usually braid, with plugs or terminals on each end. Used to connect jacks or blocks in switchboards or programming systems. It is called a patch cord because it is used to temporarily "patch" a circuit.

patching — Connecting two lines or circuits together temporarily by means of a patch cord.

patching jack — A jack for interconnection of circuit elements.

patch panel — 1. In a computer, a panel that contains means for changing circuit configurations; usually, it consists of receptacles into which jumpers can be inserted. 2. A panel where circuits are terminated and facilities provided for interconnecting between circuits by means of jacks and plugs.

path — 1. In navigation, an imaginary line connecting a series of points in space and constituting a proposed or traveled route. 2. *See* channel, 2, 4, 5, 6, and 7.

path attenuation — The power loss between transmitter and receiver resulting from all causes.

pathfinding — The process of finding an idle path through a switching network, from entrance port to exit port.

path loss — Also known as space loss. The attenuation of a signal as it travels through space.

patient monitor — A system of instruments that permits remote monitoring at a central location in a hospital of such quantities as heart rate, blood pressure, temperature, etc.

pattern — 1. The means of specifying the character of a wave in a guide. This is done by showing the loops of force existing in the guide for that wave. The pattern identifies the order and mode of the wave and the cross-sectional shape of the guide. 2. A geometrical figure representing the directional qualities of an antenna array.

3. The configuration of conductive and nonconductive materials on a panel or printed board. Also the circuit configuration on related tools, drawings, and masters. 4. The outline of a collection of circuit conductors and resistors that defines the area to be covered by the material on a film circuit substrate.

pattern definition — The accuracy, relative to the original artwork, with which pattern edges are reproduced in integrated-circuit elements.

pattern recognition — In a computer, the examination of records for certain code-element combinations.

pattern-sensitive fault — A fault that appears in response to some particular data pattern.

pause control — A feature of some tape recorders making it possible to temporarily stop the movement of the tape without switching the machine from the play or record position. (Essential for a tape recorder used for dictation and generally helpful for editing purposes.)

PAX or **pax** — Acronym for private automatic exchange. 1. An automatic system used exclusively for interoffice dial communications and having no trunks to the central office. 2. A dial telephone exchange that provides private telephone service to an organization.

pay-per-view — A method of purchasing television programming on a per-program basis.

paystation — A coin-operated telephone.

pay television — Also called subscription television. 1. A system whereby viewers must insert coins or record cards into a decoding device in order to view a television program that has been deliberately scrambled to prevent unpaid viewing. 2. A service that provides the TV viewer, for an extra monthly fee, with special extra programming such as first-run movies, sporting events, news, and features. 3. Generally refers to additional channel service — over and above regular TV-network cable service — in exchange for an extra monthly charge.

P-band — A radio-frequency band extending from 225 to 390 MHz and having wavelengths from 133.3 to 76.9 cm.

PBX or **pbx** — Abbreviation for private branch exchange. 1. A private telecommunications exchange that includes access to a public telecommunications exchange. 2. A telephone exchange, having a switchboard and associated equipment, that serves a single organization and usually is located on the customer's premises. It provides for switching calls between any two extensions served by the exchange or between any extension and the national telephone system via a trunk to a central office. 3. An intelligent, programmable switch that can route telephone calls or digital data from computers and terminals throughout a building or to an outside line for local or long-distance transmission.

pC — 1. Letter symbol for picocoulomb.

pc — 1. Abbreviation for program counter. 2. Abbreviation for phase corrector. A part of synchronous modems that adjusts the local data-clocking signal to match the incoming receive data sent by the remote clocking signal. 3. Abbreviation for printed circuit.

PC — 1. Abbreviation for personal computer. Usually refers to an IBM-compatible microcomputer, but sometimes also is used to include Macintosh computers. 2. Abbreviation for programmable controller.

pcb — Abbreviation for printed circuit board.

PC-DOS — IBM's name for the disk operating system used in the IBM Personal Computer. (Similar to MS-DOS.)

p-channel — *See* PMOS.

p-channel device — A device constructed on an n-type silicon substrate, whose drain and source components are of p-type silicon. *See also* PMOS.

p-channel field-effect transistor or **p-channel FET** — A field-effect transistor that has a p-type conduction channel.

p-channel MOS — *See* PMOS.

PCM — 1. Abbreviation for pulse-code modulation. 2. Abbreviation for punched card machine.

PCM/FM — Frequency modulation of a carrier by pulse-code-modulated information.

PCM/FM/FM — Frequency modulation of a carrier by subcarrier(s) that is (are) frequency modulated by pulse-code-modulated information.

PCM/PM — Phase modulation of a carrier by pulse-code-modulated information.

PCS — Abbreviation for Personal Communication Services. A term for digital low-power mobile telephone service.

PD — Abbreviation for photodetector.

P-display — *See* plan-position indicator.

PDM — Abbreviation for pulse-duration modulation.

PDM/FM — Frequency modulation of a carrier by pulses that are modulated in duration by information.

PDM/FM/FM — Frequency modulation of a carrier by subcarrier(s) that is (are) frequency modulated by pulses that are time duration modulated by information.

PDM/pm — Phase modulation of a carrier by pulses that are duration modulated by information.

PE — Abbreviation for phase encoded.

peak — Also called crest. 1. A momentary high amplitude level occurring in electronic equipment. 2. A momentarily high volume level during a radio program. It causes the volume indicator at the studio or transmitter to swing upward. 3. The maximum instantaneous value of a quantity. 4. To increase or sharpen the peaks of a waveform. 5. To broaden the frequency response of an amplifier by including inductors in its coupling networks so as to cancel the input and output capacitances of its active elements.

peak alternating gap voltage — In a microwave tube, the negative of the line integral of the peak alternating electric field, taken along a specified path across the gap.

peak amplitude — The maximum deviation (e.g., of a wave) from an average or mean position.

peak anode current — The maximum instantaneous value of an anode current in an electron tube.

peak-charge characteristic — The function giving the relation of one-half the peak-to-peak value of transferred charge in the steady state to one-half the peak-to-peak value of a specified symmetrical alternating voltage applied to a nonlinear capacitor.

peak coil current — The peak current through the ignition coil primary winding of an inductive system at the instant the contacts open.

peak current — 1. The maximum current during a complete cycle. 2. Maximum amplitude of current an ionized device can pass without permanent change in breakdown ratings or published life specifications.

peak discharge energy — The maximum amount of energy a device can withstand during operation without permanent change in breakdown ratings or published life specifications.

peak distortion — The largest total distortion of signals noted during a period of observation.

peak electrode current — The maximum instantaneous current that flows through an electrode.

peak envelope power — Abbreviated PEP. 1. The power contained in the signal at the peak of the modulation envelope. 2. Of a radio transmitter, the average power supplied to the antenna transmission line by a transmitter during one radio-frequency cycle at the highest crest of the modulation envelope, taken under conditions of normal operation.

peak firing temperature — The maximum temperature seen by the resistor or conductor paste in the firing cycle as defined by the firing profile.

peak flux density — The maximum flux density in a magnetic material.

peak forward anode voltage — The maximum instantaneous anode voltage in the direction a tube is designed to pass current.

peak forward-blocking voltage — The maximum instantaneous value of repetitive positive voltage that may be applied to the anode of an SCR with its gate circuit open.

peak forward drop — The maximum instantaneous voltage drop measured when a tube or rectifier cell is conducting forward current, either continuously or during transient operation.

peak indicator — A visual (light) indicator showing when transient signal levels exceed a recorder's ability to handle them without distortion. Such indicators are often used to supplement recording-level meters, which usually indicate average signal levels.

peaking — 1. Adjusting a component so as to increase the response of a circuit at a desired frequency or band of frequencies. 2. To tune a circuit for a very sharp response at a particular frequency.

peaking circuit — A circuit capable of converting an input wave into a peaked waveform.

peaking control — In a television receiver, a fixed or variable resistor-capacitor circuit that controls the negative shape of the pulses originating at the horizontal oscillator. This is done to ensure a linear sweep.

peaking network — A type of interstage coupling network used to increase the amplification at the upper end of the frequency range. It consists of an inductance effectively in series (series peaking network) or shunt (shunt peaking network) with a parasitic capacitance.

peaking resistor — A resistor placed in series with the charging capacitor of the vertical sawtooth generator. By adding a negative peaking pulse to the sawtooth voltage, it creates the waveform required to produce a linear sawtooth current in the yoke.

peaking transformer — A transformer operated in such a way that its core is saturated in one direction or the other for most of a single ac cycle, with the result that the secondary voltage waveform is sharply peaked at each flux reversal. Sharpness of the peaking is enhanced by an approximately rectangular hysteresis loop in the core.

peak inverse anode voltage — The maximum instantaneous anode voltage in the direction opposite from that in which the tube is designed to pass current.

peak inverse voltage — The peak ac voltage that a rectifying cell or pn junction will withstand in the reverse direction.

peak level — The maximum instantaneous level that occurs during a specific time interval (i.e., in acoustics, the peak sound pressure level).

peak limiter — A device that automatically limits the magnitude of its output signal to approximate a preset maximum value by reducing its amplification when the instantaneous signal magnitude exceeds a preset value.

peak load — The maximum electrical power load consumed or produced in a stated period of time. It may be the maximum instantaneous load or the maximum average load over a designated interval of time.

peak magnetizing force — The upper or lower limiting value of a magnetizing force.

peak-or-valley readout memory — A circuit in which the output remains at the condition corresponding to the most positive (least negative) or vice versa input signal since the circuit was set to initial conditions, until reset to those conditions.

peak plate current — The maximum instantaneous current passing through the plate circuit of a vacuum tube.

peak point — The point on the characteristic curve of a tunnel diode corresponding to the lowest voltage in the forward direction for which the differential conductance is zero.

peak-point emitter current — The maximum emitter current that can flow without allowing a UJT to go into the negative-resistance region.

peak power — 1. The mean power supplied to the antenna of a radio transmitter during one radio-frequency cycle at the highest crest of the modulation envelope. 2. The maximum power of the pulse from a radar transmitter. Since the resting time of a radar transmitter is longer than its operating time, the average power output is much lower than the peak power. 3. Maximum instantaneous audio power available from a power amplifier.

peak power output — 1. The output power averaged over the radio-frequency cycle having the maximum peak value that can occur under any combination of signals transmitted. 2. Maximum instantaneous power output from any power amplifier. Usually related to the saturation power of the amplifier.

peak pulse amplitude — The maximum absolute peak value of the pulse, excluding unwanted portions such as spikes.

peak pulse power — The maximum power of a pulse, excluding spikes.

peak-reading meter — 1. A type of recording-level meter that responds to short transient signals. 2. A type of recording-level meter whose needle rises quickly and falls back at moderate speed, permitting the operator to judge the levels of transient peak waveforms.

peak response — The maximum response of a system to an input.

peak signal level — An expression of the maximum instantaneous signal power or voltage as measured at any point in a facsimile transmission system. This includes auxiliary signals.

peak sound pressure — The maximum absolute value of instantaneous sound pressure for any specified time interval. The most common unit is the microbar.

peak spectral emission — The wavelength at which the radiation from a lamp has the highest intensity.

peak speech power — The maximum instantaneous speech power over the time interval considered.

peak to peak — Abbreviated p-p. 1. The algebraic difference between the positive and negative maximum values of a waveform. 2. The amplitude (voltage) difference between the most positive and the most negative excursions (peaks) of an electrical signal.

peak-to-peak amplitude — The amplitude of an alternative quantity, measured from positive peak to negative peak. The sum of the absolute value of the peak positive and the peak negative excursions.

peak-to-peak output ripple — The periodic variations present in the output of a power supply. They have little effect on the average dc output but can clutter operation of circuits being powered.

peak-to-peak voltmeter — A voltmeter that indicates the overall difference between the positive and negative voltage peaks.

peak value — Also called crest value. 1. The maximum instantaneous value of a varying current, voltage, or power. For a sine wave, it is equal to 1.414 times the effective value of the sine wave. 2. The largest instantaneous amplitude of a waveform.

peak voltage—The maximum value present in a varying or alternating voltage. This value may be either positive or negative.

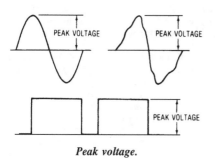

Peak voltage.

peak voltmeter—A voltmeter that reads peak values of an alternating voltage.

peak wavelength—The wavelength at the peak of the radiated spectrum of an emitter.

pea lamp—An incandescent lamp with a bulb about the size of a pea. Its small size makes it ideal for use by doctors, on instrument panels, and in small flashlights.

pedal clavier—In an organ, the pedal keyboard that supplies the bass accompaniment for the other manuals.

pedal keyboard—*See* pedal clavier.

pedestal—1. A substantially flat-topped pulse that elevates the base level for another wave. 2. The base of a radar antenna.

pedestal level—*See* blanking level.

pedestal pulse—A square-wave pulse or gate on which a video signal or sweep voltage may be superimposed.

peek-a-boo—In a computer, a method of determining the presence or absence of holes in identical locations on punched cards by placing one card on top of another. *See also* Batten system.

peel strength (peel test)—A measure of adhesion between a conductor and the substrate. The test is performed by pulling or peeling the conductor off the substrate and observing the force required.

peel-strength adhesion—*See* bond strength.

peg-count meters—In telephone practice, meters or registers used to indicate the number of trunks tested, circuits passed busy, test failures, and repeated tests completed.

pel—Contraction of the phrase picture element; synonymous with pixel.

pellicle—A thin membrane that has the capability of splitting beams, polarizing light, and reflecting images with few or no optical side effects.

Peltier coefficient—The quotient of the rate of Peltier heat absorption by the junction of two dissimilar conductors divided by the current through the junction. The Peltier coefficient of a couple is the algebraic difference between either the relative or absolute Peltier coefficients of the two conductors making up the couple.

Peltier effect—The production or absorption of heat at the junction of two metals when current is passed through the junction. Reversing the direction of the current changes a production of heat to an absorption, and vice versa.

Peltier electromotive force—1. The component of voltage produced by a thermocouple after being heated by the Peltier effect at the junction of the different metals. It adds to the Thomson electromotive force to produce the total voltage of the thermocouple. 2. The boundary emfs produced across the junctions of two different metals. Associated with the heating and cooling effects of the two junctions.

Peltier heat—The thermal energy absorbed or produced as a result of the Peltier effect.

pen centering—An electrical or mechanical adjustment by which an oscillator pen is positioned to channel center.

pencil beam—A radar beam in which the energy is confined to a narrow center.

pencil-beam antenna—A unidirectional antenna in which those cross sections of the major lobe perpendicular to the maximum radiation are approximately circular.

pencil tube—A small vacuum tube designed for operation in the ultrahigh-frequency band and used as an oscillator or rf amplifier.

pendant—The type of plug and/or receptacle that is not mounted in a fixed position or attached to a panel or side of equipment.

pendant station—A push-button station suspended from overhead and connected by means of flexible cord or conduit, but supported by a separate cable.

pendulous accelerometer—A device that measures linear accelerations by means of a restrained unbalanced mass. Two pivots and jewels support the unbalanced gimbal, and the torsion bar functions as the spring.

penetrating frequency—*See* critical frequency.

penetration depth—1. In induction heating, the effective depth of the induced current. The skin effect causes this to be nearer the surface with high frequencies than with low frequencies. 2. The extent to which an external magnetic field penetrates a superconductor.

Penning discharge—A type of discharge in which electrons are forced to oscillate between two opposed cathodes and are prevented from going to the surrounding anode by the presence of a magnetic field. It is sometimes referred to as a pig discharge because the device producing it was first used as an ionization gage called the Penning Ionization Gage.

Penning Ionization Gage—*See* Penning discharge.

pen plotter—A recorder specifically designed for computer graphics that may be used online, remote batch, or in a time-sharing environment. The plotter will yield reproducible records of graphs, charts, and drawings on plain paper annotated with alphanumerics.

pen position—An electrical or mechanical adjustment by which an oscillograph pen is positioned to any desired amplitude grid mark on the chart to represent zero signal.

pent—Abbreviation for pentode.

Pentaconta switching system—Trade-marked designation of an ITT common-controlled electromechanical telephone switching system that uses relays and crossbar switches.

pentagrid converter—A pentagrid vacuum tube used as a combination oscillator and mixer in a superheterodyne receiver.

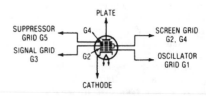

Pentagrid converter.

pentagrid mixer — A pentagrid vacuum tube used to mix the rf and local-oscillator signals in a superheterodyne receiver.

pentagrid tube — An electron tube having five grids, plus an anode and a cathode.

pentatron — A five-electrode vacuum tube that provides push-pull amplification with a single tube. It has one cathode, two grids, and two anodes. In effect, it is two tubes in one.

pentode — A five-electrode vacuum tube containing an anode, a cathode, a control electrode, and two grids.

pentode field-effect transistor — A five-lead transistor with three gates. It can be connected like a pentode if each of the gates is supplied from an independent bias source.

pentode transistor — A point-contact transistor in which there are four point-contact electrodes; the body serves as a base, and there are three emitters and one collector.

PEP — Abbreviation for peak envelope power.

perceived noise level — An empirical measure that includes allowance for the subjective reaction of people to noise in the various frequency ranges. It is expressed in decibels (PNdB).

percentage differential relay — A differential relay that functions when the difference between two quantities of the same nature exceeds a fixed percentage of the smaller quantity. This term includes relays formerly known as ratio-balance relays, biased relays, and ratio-differential relays.

percentage modulation — See percent of modulation.

percentage of meter accuracy — The ratio of the actual meter reading to the true reading, expressed as a percent.

percentage ripple — See percent of ripple voltage.

percentage supervision — A method of line supervision in which the current in, or resistance of, a supervised line is monitored for changes. When the change exceeds a selected percentage of the normal operating current or resistance in the line, an alarm signal is produced.

percentage sync — Ratio of the amplitude of the synchronizing signal to the peak-to-peak amplitude of the picture signal between blanking and reference white level, expressed in percent.

percentage timer — A repeat-cycle non-reset timer with fixed cycle length, having a dial adjustment of the percentage of the cycle time for which the contacts are operated.

percent break — The period of time, expressed as a percentage, that a dial circuit stands open compared to the total time of the dial signals.

percent conductivity — Conductivity of a material expressed as a percentage of that of copper.

percent make — 1. In pulse testing, the length of time a circuit is closed compared to the duration of the test signal. 2. The portion (in percent) of a pulse period during which telephone-dial pulse springs make contact.

percent-modulation meter — An instrument that indicates the modulation percentage of an amplitude-modulated signal, either on a meter or a cathode-ray tube.

percent of deafness — See percent of hearing loss.

percent of harmonic distortion — A measure of the harmonic distortion in a system or component. It is equal to 100 times the ratio of the square root of the sum of the squares of the root-mean-square harmonic voltages (or currents) to the root-mean-square voltage (or current) of the fundamental.

percent of hearing — At a given frequency, 100 minus the percent of hearing loss at that frequency.

percent of hearing loss — Also called percent of deafness. At a given frequency, 100 times the ratio of the hearing loss in decibels to the number of decibels between the normal threshold levels of audibility and feeling.

percent of modulation — 1. In AM, the ratio of half the difference between the maximum and minimum amplitudes, expressed in percentage. 2. In FM and TV audio transmission, the ratio of the actual frequency swing to the frequency swing defined as 100-percent modulation, expressed in percentage. For FM broadcast stations, a frequency swing of ±75 kHz is defined as 100-percent modulation. For television, it is ±25 kHz.

percent of ripple voltage — Ratio of the effective (root-mean-square) value of the ripple voltage to the average value of the total voltage, expressed in percent.

percent of syllabic articulation — See syllable articulation.

percent ripple (rms) — The ratio of the effective (rms) value of the ripple voltage, expressed in percent. The new term is PARD.

perceptron — A system capable — either in theory or in practice — of performing knowledgeable functions such as recognition, classification, and learning. These functions may exist as mathematical analyses, computer programs, or hardware.

percussion — Musical sounds characterized by sudden or sharp transients. Organ percussion is achieved by causing the tone to start to decay the instant it is played rather than waiting until the key is released.

percussive arc welding — A process in which the surfaces to be welded are held at a fixed gap while rf energy is applied. This ionizes the air gap between the two surfaces, causing the air gap to become a conductor. When the air gap is conductive, a capacitor bank dumps a controlled amount of energy into the system for a controlled time period. This results in an electric arc that scarifies the surfaces to be welded and heats them to welding temperature. As the pulse from the capacitor bank decays, a mechanical system drives the two hot surfaces together, consummating the weld.

perfect dielectric — Also called ideal dielectric. A dielectric in which all the energy required to establish the electric field in it is returned to the electric system when the field is removed. A perfect dielectric has zero conductivity and exhibits no absorption phenomena. A vacuum is the only known perfect dielectric.

perforated tape — See punched tape.

perforator — In telegraphy, a device that punches code signals into paper tape for application to a tape transmitter.

performance — Degree of effectiveness of operation.

performance characteristic — A characteristic measurable in terms of some useful denominator — e.g., gain, power output, etc.

perimeter alarm system — An alarm system that provides perimeter protection.

perimeter protection — Protection of access to the outer limits of a protected area, by means of physical barriers, sensors on physical barriers, or exterior sensors not associated with a physical barrier.

period — Sometimes called periodic time. 1. The time required for one complete cycle of a regular, repeating signal, function, or series of events. 2. The time between two consecutive transients of the pointer or indicating means of an electrical indicating instrument in the same direction through the rest position. 3. The time elapsing between two consecutive passages of a satellite through a characteristic point on its orbit. 4. Of an underdamped instrument, the time required, following an abrupt change in the measurand, for the pointer or other indicating means to make two consecutive transits in the same direction through the rest position.

periodic — Repeating itself regularly in time and form.

periodic antenna — An antenna in which the input impedance varies as the frequency does (e.g., open-end wires and resonant antennas).

periodic current — Oscillating current, the values of which recur at equal time intervals.

periodic damping — Also called underdamping. Damping in which the pointer of an instrument oscillates about the final position before coming to rest. The point of change between periodic and aperiodic damping is called critical damping.

periodic duty — Intermittent duty in which the load conditions recur at regular intervals.

periodic electromagnetic wave — A wave in which the electric field vector is repeated in detail — either at a fixed point, after a lapse of time known as the period; or at a fixed time, after the addition of a distance known as the wavelength.

periodic electromotive force — An oscillating electromotive force that repeats its sequence of values over equal intervals of time.

periodicity — The variations in the insulation diameter of a transmission cable that result in reflections of a signal when its wavelength or a multiple thereof is equal to the distance between two diameter variations.

periodic line — A line consisting of identical, similarly oriented sections, each section having nonuniform electrical properties.

periodic PARD — Pertains to that portion of the total PARD in an electronic power supply whose frequency is identical or harmonically related to the input frequency and/or intentionally internally generated signal frequencies. This phenomenon is frequently referred to as ripple.

periodic pulse train — A pulse train made up of identical groups of pulses repeated at regular intervals.

periodic quantity — An oscillating quantity in which any value it attains is repeated at equal time intervals.

periodic rating — The load that can be carried for the alternate periods of load and rest specified in the rating without exceeding the specified heating limits.

periodic resonance — Also called natural resonance. Resonance in which the applied agency maintaining the oscillation has the same frequency as the natural period of oscillation in a system.

periodic time — See period.

periodic vibration — 1. A vibration having a regularly recurring waveform, e.g., sinusoidal vibration. 2. An oscillator motion whose amplitude pattern repeats after fixed increments of time.

periodic wave — 1. Wave in which the displacement has a periodic variation with time, distance, or both. 2. A wave that repeats itself at regular intervals of time.

peripheral — 1. Having to do with a device by means of which a computer communicates to the outside world. Auxiliary memories, such as tape, disk, and drum, may also be considered to be peripheral devices. 2. Any external device used in a computer system. 3. A device used to extend the operation of a computer, either with regards to functionality or capacity. Peripherals are connected to the central processor of a computer system by appropriate data paths. 4. Any human interface device connected to a computer. 5. A noncomputing input or output device, external to the CPU and main memory of a computer but connected by the appropriate electrical connections, thus allowing the computer to perform various external functions.

peripheral bus — For a computer, a dedicated bus for high-speed transfer of blocks of data between a system and its peripherals, including tape and disk drives.

peripheral control unit — An intermediary control device that links a peripheral unit to the central processor, or, in the case of offline operation, to another peripheral unit.

peripheral device — 1. Any instrument or machine that enables a computer to communicate with the outside world or that otherwise aids the operation of the computer, but does not form part of the basic installation. 2. A general term designating various kinds of machines that operate in combination or conjunction with a computer but are not physically part of the computer. Peripheral devices typically display computer data, store data from the computer and return the data to the computer on demand, prepare data for human use, or acquire data from a source and convert it to a form usable by a computer. Peripheral devices include printers, keyboards, graphic display terminals, paper tape reader/punches, analog-to-digital converters, disks, and tape drives. 3. Any device distinct from the computer that can provide input to and/or accept output from the computer.

peripheral electron — Also called a valence electron. One of the outer electrons of an atom. Theoretically, it is responsible for visible light, thermal radiation, and chemical combination.

peripheral equipment — 1. In a data processing system, any unit of equipment, distinct from the central processing unit, that may provide the system with outside communication. 2. Equipment that is external to and not a part of the central processing instrumentation. Includes such equipment as tape punches and readers, magnetic tape or disk storage units, digital printers, graphic recorders, and typewriters. 3. Equipment used in connection with or attached to electronic computers. This includes typewriters, calculators and accounting machines, cash registers, tape readers, and similar devices. 4. Units that may communicate with the programmable controller, but are not part of the programmable controller, e.g., Teletype, cassette recorder, CRT terminal, tape reader, etc. 5. Equipment that works in conjunction with a computer or a communication system but is not a part of it. 6. Machines such as card readers, magnetic tape units, fast driver scanners, and other equipment bearing a similar working relationship to the computing device. In general they can be considered a sink and source of information external to the information process.

peripheral processor — A general term for a laser computer associated with a large machine. Among the functions may be multiplexing, data formatting, concentrating, polling, and the handling of simple routines to increase the capacity of a communications channel or relieve the main (often called host) computer.

peripherals — Accessory parts of a computer system not considered essential to its operation, such as printers and modems.

peripheral transfer — The transmission of data between two peripheral units.

peripheral units — Equipment that works in conjunction with a data terminal or computer but is not a part of that unit, such as cards, paper-tape readers, punches, or keyboards.

periphonic system — See ambisonic reproduction.

Permalloy — A high-permeability magnetic alloy composed mainly of iron and nickel.

permanent circuit — An alarm circuit that is capable of transmitting an alarm signal whether the alarm control is in access mode or secure mode. Used, for example, on foil fixed windows, tamper switches, and supervisory lines. *See also* permanent protection; supervised lines; supervisory alarm system; supervisory circuit.

permanent echo — Signal received and displayed by a radar, indicating reflections from fixed objects.

permanent-field synchronous motor — A type of synchronous motor in which the member carrying the secondary laminations and windings also carries permanent-magnet field poles that are shielded from the alternating magnetic flux by the laminations. It behaves as an induction motor when starting but runs at synchronous speed.

permanent magnet — Abbreviated pm. A piece of hardened steel or other magnetic material that has been so strongly magnetized that it retains the magnetism indefinitely.

permanent-magnet centering — Vertical or horizontal shifting of a television picture by means of magnetic fields from permanent magnets mounted around the neck of the picture tube.

permanent-magnet focusing — 1. Focusing of the electron beam in a television picture tube by means of one or more permanent magnets located around the neck. 2. The focusing of an electron beam by a magnetic field that permanently retains the majority of its magnetic properties.

permanent-magnet material — Ferromagnetic material that, once having been magnetized, resists external demagnetizing forces (i.e., requires a high coercive force to remove the magnetism).

permanent-magnet motor — A direct-current motor that has a permanent magnet field (stationary member) and a wound armature (rotating member). Its speed can be changed by varying the armature voltage.

permanent-magnet moving-coil instrument — Also called D'Arsonval instrument. An instrument in which a reading is produced by the reaction between the current in a movable coil or coils and the field of a fixed permanent magnet.

Permanent-magnet moving-coil instrument.

permanent-magnet moving-iron instrument — Also called polarized-vane instrument. An instrument in which a reading is produced by an iron vane as it aligns itself in the magnetic field produced by a permanent magnet and by the current in an adjacent coil of the instrument.

permanent-magnet speaker — A moving-conductor speaker in which the steady magnetic field is produced by a permanent magnet.

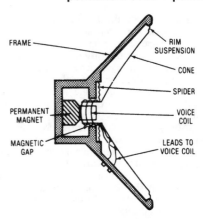

Permanent-magnet speaker.

permanent-magnet stepping motor — A type of motor in which a permanent magnet serves as the rotor. Current is switched sequentially through different stator coils, and the rotor aligns itself with the energized stator poles.

permanent magnistor — A saturable reactor that has the properties of memory and the ability to handle appreciable power.

permanent memory — A type of storage device that retains data intact when the computer has been shut down.

permanent-memory computer — A computer in which the stored information remains intact even after the power has been turned off.

permanent protection — A system of alarm devices, such as foil, burglar alarm pads, or lacings, connected in a permanent circuit to provide protection whether the control unit is in the access mode or secure mode.

permanent set — The deformation that remains in a specimen after it has been stressed in tension for a definite time interval and released for a definite time interval.

permanent storage — A computer storage device that retains the stored data indefinitely.

permatron — A thermionic gas diode, the discharge of which is controlled by an external magnetic field. It is used mainly as a controlled rectifier and functions like a thyratron.

permeability — Symbolized by the Greek letter mu (μ). The measure of how much better a given material is than air as a path for magnetic lines of force. (Air is assumed to have a permeability of 1.) It is equal to the magnetic induction (B) in gausses divided by the magnetizing force (H) in oersteds.

permeability tuning — A method of tuning a circuit by moving a magnetic core into or out of a coil to vary its inductance.

permeameter — An apparatus for determining the magnetizing force and flux density in a test specimen. From these values, the normal induction curves or hysteresis loops can then be plotted and the magnetic permeability computed.

permeance — The ratio of the flux through any cross section of a tabular portion of a magnetic circuit bounded by the lines of force and by two equal potential surfaces to the magnetic potential of the difference between the surfaces, taken within the portion under consideration. Permeance is the reciprocal of reluctance and can be considered analogous to electrical conductance. In the cgs system it is equal to the magnetic flux (in maxwells) divided by the magnetomotive force (in gilberts).

permeance coefficient — Also called demagnetizing coefficient. Describes operating conditions of a magnet and is the slope of the magnetic load line.

permissive control device — Generally a two-position, manually operated switch that in one position permits the closing of a circuit breaker or the placing of an equipment into operation, and in the other position prevents the circuit breaker or the equipment from being operated.

permittivity — 1. *See* dielectric constant. 2. That property of a dielectric which determines the electrostatic energy stored per unit volume for a unit potential gradient.

permutation modulation — Proposed method of transmitting digital information by means of additive white Gaussian noise. Pulse-code modulation and pulse-position modulation are considered simple special cases of permutation modulation.

permutation table — In computers, a table for use in the systematic construction of code groups. It may also be used in the correction of garbles in groups of code text.

peroxide of lead — A lead compound that forms the principal part of the positive plate in a charged lead-acid cell.

perpendicular magnetization — In magnetic recording, magnetization that is perpendicular to the line of travel and parallel to the smallest cross-sectional dimension of the medium. Either single- or double-pole-piece magnetic heads may be used.

persistence — 1. The length of time a phosphor dot glows on the screen of a cathode-ray tube before going out; i.e., the length of time it takes to decay from initial brightness (reached during fluorescence) until it can no longer be seen. 2. In a cathode-ray tube, the period that a phosphor continues to glow after excitation is removed.

persistence characteristic (of a luminescent screen) — Also called the decay characteristic. The relationship (usually shown by a graph) between the time a luminescent screen is excited and the time it emits radiant power.

persistence of vision — The physiological phenomenon whereby the eye retains a perception of an image for a short time after the field of vision has disappeared.

persistent current — A current that is magnetically induced and flows undiminished in a superconducting material or circuit.

persistor — A bimetallic circuit used for storage or readout in a computer. It is operated near absolute zero, and changes from a resistive to a superconductive state at a critical current value.

persistron — A device in which electroluminescence and photoconductivity are combined into a single panel capable of producing a steady or persistent display with pulsed signal input.

personal computer — Abbreviated PC. 1. A microcomputer that is designed to be used by a single user. 2. Self-contained, relatively small computer for individual users that contains all the hardware and software necessary to perform the required task(s).

persuader — In a storage tube, an element that directs secondary emission toward the electron-multiplier dynodes.

PERT — Acronym for program evaluation and review technique. A management tool for comparing actual with scheduled progress of a program.

perveance — The space-charge-limited cathode current divided by the three-halves power of the anode voltage in a diode.

petallized dish — A parabolic satellite signal receiving dish that is shipped in sections, or petals, and assembled at the installation site.

petticoat insulator — An insulator having an outward flaring lower part that is hollow to increase the length of the surface leakage path and keep part of the path dry at all times.

pF — Letter symbol for picofarad.

pg — Abbreviation for power gain.

p-gate thyristor — A thyristor in which the gate terminal is connected to the p-region nearest the cathode and which is normally switched to the on state by applying a positive signal between the gate and cathode terminals.

pH — A measure of the degree of acidity or alkalinity of a solution. In a neutral solution the pH value is 7. In acid solutions it ranges from 0 to 7, and in alkaline solutions it ranges from 7 to 14.

phanastron — An electronic circuit of the multivibrator type that is normally used in the monostable form. It is a stable trigger generator in this connection and is used in radar systems for gating functions and sweep-delay functions.

phanotron — A term used primarily in industrial electronis to mean a hot-cathode gas diode.

phantom — A signal derived from two sources in such a way as to appear located from a third source. Stereo signals "appearing" between speakers are said to be phantomed.

phantom channel — In a stereo system, an electrical combination of the left and right channels fed to a third, centrally located speaker.

phantom circuit — A superimposed circuit derived from two suitably arranged pairs of wires called side circuits. Each pair of wires is a circuit itself, and at the same time acts as one conductor of the phantom circuit.

phantom-circuit loading coil — A loading coil that introduces the desired amount of inductance into a phantom circuit and a minimum amount into the constituent side circuits.

phantom-circuit repeating coil — A repeating coil used at a terminal of a phantom circuit, in the terminal circuit extending from the midpoints of the associated side-circuit repeating coils.

phantom coil — A coil originally used in a phantom circuit for impedance matching. Most generally, any coil, side or phantom, in a phantom circuit. When the term is used, the meaning should be made clear.

phantom group — 1. A group of four open-wire conductors suitable for the derivation of a phantom circuit. 2. Three circuits that are derived from simplexing two physical circuits to form a phantom circuit.

phantom OR and AND — *See* wired OR.

phantom repeating coil — A side-circuit repeating coil or a phantom-circuit repeating coil, when discrimination between these two types is not necessary.

phantom signals — Signals appearing on the screen of a cathode-ray-tube indicator; their cause cannot readily be determined, and they may be due to circuit fault, interference, propagation anomalies, jamming, etc.

phantom target — *See* echo box.

phase — 1. The angular relationship between current and voltage in alternating-current circuits. 2. The number of separate voltage waves in a commercial alternating-current supply (e.g., single-phase, three-phase, etc.). Symbolized by the Greek letter *phi* (ϕ). 3. In a periodic function or wave, the fraction of the period that has elapsed, measured from some fixed origin. If the time for one period is represented as 360° along a time axis, the phase position is called phase angle. 4. The relative timing of a signal in relation to another signal; if both signals occur at the same instant, they are in phase; if they occur at different instants, they are out of phase. 5. That part of a period through which the independent variable has advanced, as measured from an arbitrary origin. 6. A particular stage, or point of advancement, in an electrical cycle. The fractional part of the period through which the time has advanced, measured from some arbitrary point, usually expressed in electrical degrees, where 360° represents one cycle. 7. The difference between the zero crossing or starting reference point between a standard waveform and the measured waveform. Phase is usually measured in degrees.

phase advancer — A phase modifier that supplies leading reactive volt-amperes to the system to which it is connected. Phase advancers may be either synchronous or nonsynchronous.

phase angle — 1. Of a periodic function, the angle obtained by multiplying the phase by 2π if the angle is to be expressed in radians, or 360 for degrees. 2. The angle between the vectors representing two periodic functions that have the same frequency. 3. The phase difference, in degrees, between corresponding stages of progress of two cyclic operations. 4. The angle between two vectors that represent two simple periodic quantities that vary sinusoidally and that have the same frequency. 5. A notation for phase position when the period is denoted as 360.

phase-angle correction factor — That factor by which the reading of a wattmeter or watthour meter operated from the secondary of a current or potential transformer, or both, must be multiplied to correct for the effective phase displacement of current and voltage due to the measuring apparatus.

phase-angle measuring relay — Also called out-of-step protective relay. A device that functions at a predetermined phase angle between two voltages or currents, or between voltage and current.

phase-angle meter — *See* phase meter.

phase angle of a current transformer — The angle between the primary current vector and the secondary current vector reversed. This angle is conveniently considered as positive when the reversed secondary current vector leads the primary current vector.

phase angle of a potential (voltage) transformer — The angle between the primary voltage vector and the secondary voltage vector reversed. This angle is conveniently considered as positive when the reversed secondary voltage vector leads the primary voltage vector.

phase anomaly — A sudden irregularity in the phase of a low-frequency or very low frequency signal.

phase balance — In a chopper, the phase-angle difference between positive and negative halves of the square wave; the difference in degrees between 180° and the measured angle between square-wave midpoints.

phase-balance current relay — *See* reverse-phase current relay.

phase-balance relay — A relay that functions by reason of a difference between two quantities associated with different phases of a polyphase circuit.

phase center (center of radiation) — Pertaining only to antenna types that have radiation characteristics such that while they are radiating energy, one can observe the antenna from a distance of many wavelengths and see the energy radiating from a point within the antenna array. The position of a point-source radiator that would replace the antenna and produce the same far-field phase contour.

phase characteristic — A graph of phase shift versus frequency, assuming sinusoidal input and output.

phase comparator — *See* phase detector.

phase-comparison tracking system — A system that provides target-trajectory information by the use of cw phase-comparison techniques.

phase-compensation network — A network used to provide closed-loop stability in an operational amplifier. No greater than 12-dB/octave rolloff of the open-loop gain is allowed.

phase conductors — Those conductors other than the neutral conductor of a polyphase circuit.

phase constant — 1. The imaginary component of the propagation constant. For a traveling plane wave at a given frequency, the rate in radians per unit length at which the phase lag of a field component (for the voltage or current) increases linearly in the direction of propagation. 2. With respect to a traveling plane wave at a known frequency, the space rate of decrease of phase of a field component in the direction of propagation, measured in radians per unit length.

phase control — Also called horizontal parabola control. 1. One of three controls for adjusting the phase of a voltage or current in a color television receiver employing the magnetic-convergence principle. Each control varies the phases of the sinusoidal voltages applied, at the horizontal-scanning frequency, to the coils of the magnetic-convergence assembly. 2. A technique for proportional control of an output signal by conduction only during certain parts of the cycle of the ac line voltage. 3. A method of regulating a supply of alternating current by use of a switching device such as a thyristor, by varying the point in each ac cycle or half-cycle at which the device is switched on.

phase-controlled rectifier — A rectifier circuit in which the rectifying element is a thyratron having a variable-phase sine-wave grid bias.

phase correction — The process of keeping synchronous telegraph mechanisms in substantially correct phase relationship.

phase corrector — 1. A network designed to correct for phase distortion. 2. *See* pc, 2.

phased array — 1. A group of simple radiating elements arranged over an area called an aperture. A beam (or beams) can be formed by superposition of the radiation from all the elements, and the direction of the beam can be adjusted by varying the relative phase of the signal applied to each element or by varying the frequency of the main oscillator. 2. An antenna consisting of a plurality of individual antennas, called elements, that are arrayed in a grid and interconnected so that a specific phase relationship exists between them, forming a narrow beam pattern for the reception of electromagnetic signals. 3. A technique of improving the gain of an antenna system by combining the outputs of several similar VHF/UHF/FM antennas in an array in such a way that the output signals from each one are exactly in phase with one another.

phase delay — 1. In the transfer of a single frequency wave from one point to another in a system, the delay of part of the wave identifying its phase. 2. The insertion phase shift (in cycles) divided by the frequency (in cycles per second, or hertz). *See also* delay distortion, 2.

phase-delay distortion — The difference between the phase delay at one frequency and the phase delay at a reference frequency.

phase detector — Also called phase discriminator or phase comparator. 1. A TV circuit in which a dc correction voltage is derived to maintain a receiver oscillator in sync with some characteristic of the transmitted signal. 2. A circuit that detects both the magnitude and the sign of the phase angle between two sine-wave voltages or currents. 3. A circuit that creates an output level which is a function of the phase angle between two ac input signals. Most phase detectors are also amplitude sensitive. 4. A circuit that compares the relative phase between two inputs and produces an error voltage dependent on the difference. This error voltage corrects the VCO frequency during tracking. Sometimes called a phase comparator or mixer. 5. In both digital and analog phase-locked loops, a circuit that compares the input with the signal from either a VCO (analog) or a divide-by-*n* counter (digital) and generates an error voltage based on the phase difference. The error signal corrects the VCO frequency (analog) or controls the direction of an up/down counter (digital).

phase-detector gain factor — In analog devices, the conversion factor relating the phase detector output voltage and the phase difference between the input and voltage-controlled oscillator (VCO) signals. In digital phase-locked loops, the output of a divide-by-*n* counter is compared with the input signal, since digital chips do not use VCOs.

phase deviation — In phase modulation, the peak difference between the instantaneous phase angle of the modulated wave and the phase angle of the sine-wave carrier, both expressed in radians or degrees.

phase difference — The time in electrical degrees by which one wave leads or lags another.

phase discriminator — *See* phase detector.

phase distortion — The alteration of a complex waveform produced as it passes through a network or transducer whose phase shift is a function of frequency. *See also* phase-frequency distortion.

phase-distortion coefficient — In a transmission system, the difference between the maximum and minimum transit times for frequencies within a specified band.

phase encoded — Abbreviated PE. A method of magnetic recording in which the phase of a flux reversal determines the binary content.

phase equalizer — A circuit employed to neutralize the effect of phase-frequency distortion in a particular range of frequencies.

phase-frequency distortion — Also called phase distortion. Distortion that occurs when the phase shift is not directly proportionate to the frequency over the range required for transmission, or the effect of such departure on a transmitted signal.

phase hits — Abrupt shifts in the phase of a transmitted carrier. Excessive phase hits can cause errors in high-speed phase-modulated modems. Phase hits generally originate in radio carrier systems.

phase inversion — The condition whereby the output of a circuit produces a wave of the same shape and frequency but 180° out of phase with the input.

phase inverter — 1. A stage that functions chiefly to change the phase of a signal by 180°, usually for feeding one side of a following push-pull amplifier. 2. *See* vented baffle. 3. A network or device such as a paraphrase amplifier, which produces two output signals that differ in phase by half a cycle.

phase jitter — 1. An analog-line impairment caused by power and communication equipment along the line that shifts the signal phase relationship back and forth. 2. Peak-to-peak phase deviation (expressed in degrees) of a transmitted carrier signal. An excessive phase jitter causes errors in high-speed phase-modulated modems. Phase jitter generally originates in frequency-division multiplexers in carrier systems.

phase localizer — An airfield runway localizer in which lateral guidance is obtained by comparing the phases of two signals.

phase lock — The technique of making the phase of an oscillator signal follow exactly the phase of a reference signal by comparing the phases between the two signals to adjust the frequency of the reference oscillator.

phase-locked loop — Abbreviated pll or PLL. 1. A communications circuit in which a local oscillator is synchronized in phase and frequency with a received signal. 2. A closed-loop electronic servomechanism whose output locks onto and tracks a reference signal. Phase lock is accomplished by comparing the phases of the output signal (or a multiple of it) and the reference signal. Any phase difference between these signals is converted into a correction voltage that causes the phase of the output signal to change so that it tracks the reference. 3. A closed-loop system, consisting of a phase detector, a filter, and a voltage-controlled oscillator. The phase detector provides an error signal that locks the voltage-controlled oscillator to the frequency of an incoming signal. 4. A circuit whose output locks onto and tracks a reference signal. Phase locking is accomplished by comparing the phases of the output signal and the reference signal, and then converting any difference into a correction voltage that changes the phase of the output so it matches that of the reference or input signal. 5. An electronic network consisting of a voltage-controlled oscillator, a phase comparator, low-pass filter, and an amplifier. The PLL can be used as an FM detector of extreme linearity, as a tunable filter, and as an extremely stable oscillator. When combined with an external reference oscillator it will function as a frequency synthesizer, yielding stable outputs at various frequencies. 6. A voltage-controlled oscillator and a comparator feeding back a control signal forcing the oscillator to lock onto an input signal. 7. A circuit containing a voltage-controlled oscillator whose output phase or frequency can be steered to keep it in sync with a reference source. A PLL circuit is generally used to lock onto and "up-convert" the frequency of a stable source. 8. A circuit for synchronizing a variable local oscillator with the phase of a transmitted signal.

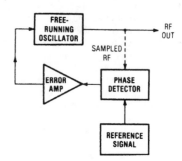

Phase-locked loop, 2.

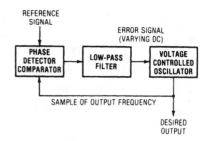

Phase-locked loop, 3.

phase magnet — Also called trip magnet. A magnetically operated latch used to phase a facsimile transmitter or recorder.

phase margin — 1. A safety factor in phase shift. When the loop gain is 1.0 or more and the phase shifts total 180°, instability will occur. The amount that the total phase shift is less than 180° is called the phase margin.

2. The additional amount of phase shift of the output signals in an operational amplifier at the open-loop unity-gain crossover frequency that would produce instability.

phase meter — Also called phase-angle meter. An instrument for measuring the difference in phase between two alternating quantities of the same frequency.

phase modifier — A device that supplies leading or lagging voltamperes to the system to which it is connected.

phase-modulated transmitter — A transmitter whose output is a phase-modulated wave.

phase-modulated wave — A wave whose phase angle has been caused to deviate from its original (no-signal) angle by an amount proportional to the modulating signal amplitude.

phase modulation — Abbreviated pm. 1. Modulation in which the angle of a sine-wave carrier deviates from the original (no-signal) angle by an amount proportional to the instantaneous value of the modulating wave. Phase and frequency modulation in combination are commonly referred to as frequency modulation. 2. Method of modulation in which the amplitude of the modulated wave remains constant, while varying in phase with the amplitude of the modulating signal. A phase-modulated wave is electrically identical to a modified frequency-modulated wave and vice versa. 3. Variation of the phase of the audio-modulating signal in accordance with the superimposed intelligence. Unlike amplitude modulation, in both phase modulation and frequency modulation the average energies of the modulated signals are the same. 4. Variation of an analog signal's phase in direct relationship to digital input information.

phase modulator — A circuit that modulates the phase of a carrier signal.

phase multiplexing — The process of encoding two (or more) information channels on a single tone.

phase multiplier — A device that multiplies the frequency of signals used for phase comparison so that phase differences may be measured to a higher degree of resolution.

phase noise — A measure of the random phase instability of a signal.

phase offset — The difference between voltage and current in an ac power line with a lagging or leading power factor.

phase-propagation ratio — In wave propagation, the propagation ratio divided by its magnitude. Expressed as a unit vector of the same angle as the propagation ratio.

phaser — 1. A device for adjusting facsimile equipment so that the recorded area bears the same relationship to the record sheet as the corresponding transmitted area of the scanning line. 2. A microwave ferrite phase shifter that employs a longitudinal magnetic field along a rod or rods of ferrite in a waveguide.

phase-recovery time (tr and pre-tr tubes) — The time required for a fired tube to deionize to such a level that a specified phase shift is produced in the low-level radio-frequency signal transmitted through the tube.

phase resonance — Also called velocity resonance. Resonance in which the angular phase difference between the fundamental components of the oscillation or vibration and the applied agency is 90°.

phase-response characteristic — The phase displacement versus frequency properties of a network or system.

phase reversal — A 180° change in phase (or one half cycle) such as a wave might undergo upon reflection under certain conditions.

phase-reversal protection — In a polyphase circuit, the interruption of power whenever the phase sequence of the circuit is reversed.

phase-reversal switch — A switch used on a stereo amplifier or in a speaker system to shift the phase 180° on one channel.

phase-rotation relay — See phase-sequence relay.

phase-sensitive amplifier — A servoamplifier whose output signal polarity or phase is dependent upon the polarity or phase relationship between an error (input) voltage and a reference voltage.

phase-sensitive detector — A system that produces a dc output signal in response to an ac input signal of a defined frequency equal to the frequency of ac reference signal. The dc output is proportional to both the amplitude of the ac input signal and the cosine of its phase angle relative to that of the reference signal. Used as synchronous rectifiers in chopper dc amplifiers and for the accurate measurement of small ac signals obscured by noise.

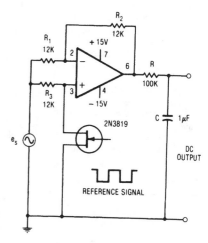

Phase-sensitive detector.

phase-sequence indicator — A device that indicates the sequence in which the fundamental components of a polyphase set of potential differences or currents successively reach some particular value (e.g., their maximum positive value).

phase-sequence relay — Also called phase-rotation relay. A relay that functions according to the order in which the phase voltages successively reach their maximum positive values.

phase-sequence voltage relay — A device that functions upon a predetermined value of polyphase voltage in the desired phase sequence.

phase shift — 1. The difference between corresponding points on input and output signal waveshapes (not affected by a magnitude) expressed as degrees lead or lag. 2. A change in the phase of a periodic quantity. 3. The changing of phase of a signal as it passes through a filter. A delay in time of the signal is referred to as phase lag; in normal networks, phase lag increases with frequency, producing a positive envelope delay. It is possible for an output signal to experience a time shift ahead of the input signal, which is called phase lead. The phase shift is always dependent on frequency.

phase-shift circuit — A network that shifts the phase of one voltage with respect to another voltage of the same frequency.

phase-shift discriminator — A circuit that produces an output proportional to the phase difference between two input signals. When used as an FM demodulator, the input is a set of push-pull signals and a reference voltage 90° displaced from each of them. These are taken across tuned circuits in such a way that the phase difference between the push-pull signals and the reference is very nearly proportional to the difference between the input frequency and the resonant frequency of the tuned circuits.

phase shifter — A device in which the output voltage (or current) may be adjusted to have some desired phase relationship with the input voltage (or current).

phase-shifting transformer — Also called a phasing transformer. A transformer connected across the phases of a polyphase circuit to provide voltages of the proper phase for energizing varmeters, varhour meters, or other instruments. *See also* rotatable phase-adjusting transformer.

phase-shift keying — 1. A form of phase modulation in which the modulating function shifts the instantaneous phase of the modulation wave between predetermined discrete values. 2. The encoding of digital information as different phases of signal elements having constant amplitude and frequency.

phase-shift microphone — A microphone whose directional properties are provided by phase-shift networks.

phase-shift oscillator — An oscillator in which a network having a phase shift of an odd multiple of 180° (per stage) at the oscillation frequency is connected between the output and input of an amplifier. When the phase shift is obtained by resistance-capacitance elements, the circuit is called an *RC* phase-shift oscillator.

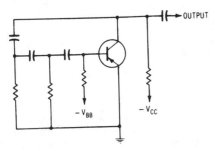

Phase-shift oscillator.

phase simulator — A precision test instrument that generates reference and data signals on the same frequency but precisely separated in phase. It is normally used to check out the precision phase meters.

phase splitter — 1. A device that produces, from a single input wave, two or more output waves that differ in phase from one another. 2. In color television, the stage that takes I and Q signals from demodulators, produces four signals — positive and negative I and Q — and feeds them to the matrix. 3. A circuit that generates out of an ac input signal two equal-amplitude outputs, one of which is 180° out of phase with the other; i.e., one is the other inverted. The dc levels may not be identical.

phase-tuned tube (tr tubes) — A fixed tuned broadband tr tube in which the phase angle through it and the reflection it introduces are kept within limits.

phase undervoltage relay — A relay that is tripped by the reduction of one phase voltage in a polyphase circuit.

phase velocity — 1. The velocity at which a point of constant phase is propagated in a progressive sinusoidal wave. 2. The velocity with which a point where there exists an electromagnetic wave of a certain fixed phase moves through space in the direction of propagation of the wave.

phase-versus-frequency response characteristic — A graph or other tabulation of the phase shift occurring, in an electrical transducer, at several frequencies within a band.

phasing — 1. Causing two systems or circuits to operate in phase or at some desired difference from the in-phase condition. 2. Adjusting a facsimile-picture position along the scanning line. 3. In stereo application, the establishment of the correct relative polarity in the connection between amplifier output and speakers so that one speaker tends to reinforce rather than cancel the output of the other (particularly evident at low frequencies).

phasing capacitor — A capacitor used in a crystal-filter circuit for neutralizing the capacitance of the crystal holder.

phasing line — In facsimile, the portion of the scanning line set aside for the phasing signal.

phasing pulse — A short pulse or signal employed for phasing the recorder with the transmitter in a television or facsimile system.

phasing signal — In facsimile, a signal used for adjusting the position of the picture along the scanning line.

phasing transformer — *See* phase-shifting transformer.

phasitron — A tube designed to produce a frequency-modulated audio signal, which is induced by a varying field from a magnet placed around the glass envelope of the tube.

phasmajector — *See* monoscope.

phasor — An entity that includes the concepts of magnitude and direction in a reference plane.

pH electrode — Transducer sensitive to hydrogen ion concentration. The sensor comprises a thin-walled glass membrane (glass electrode) or spongy platinum exposed to gaseous hydrogen (hydrogen electrode) or platinum exposed to quinhydron (quinhydrone electrode), all of which develop an electric force proportional to the hydrogen-ion concentration of a solution when immersed in the solution.

phenolic material — Any one of several thermosetting plastic materials available that may be compounded with fillers and reinforcing agents to provide a broad range of physical, electrical, chemical, and molding properties.

phi polarization — In an electromagnetic wave, the state in which the E vector of the wave is tangential to the lines of latitude of some given spherical frame of reference.

pH meter — An instrument used with a probe to determine the alkalinity or acidity of a solution.

phon — The subjective unit for measuring the apparent loudness level of a sound. Numerically equal to the sound-pressure level, in decibels relative to 0.0002 microbar, of a 1000-hertz tone that is considered by listeners to be equivalent in loudness to the sound under consideration.

phone — 1. A telephone. 2. An earphone. 3. A headphone.

phone jack — Also called telephone jack. 1. A jack designed for use with phone plugs. 2. Receptacle having two or more through circuits. May also have shunt circuits and/or isolated switching circuits. Used for extending circuits through mating plugs. Phone jacks are short or long types, depending on physical dimensions.

phonemes — 1. The minimal set of shortest segments of speech that, if substituted one for another, convert one

Phone jack.

word to another. Phonemes are the elements on which computer speech are based. 2. A class of speech sounds that are alike, except as they are modified by the sound of adjacent letters. 3. A set of abstract units that can be used for writing a language in a systematic and unambiguous way. English has about 40 phonemes: 16 vowel phonemes and 24 consonent phonemes.

phone phreak — 1. A person involved in exploring a telephone or computer network by routing calls manually. 2. One who makes telephone calls without paying for them.

phone plug — Also called telephone plug. A plug used with headphones, microphones, and other audio equipment. It is a male connecting device (almost always connected to a cable) that connects with a phone jack. Consists usually of finger and handle that comprise the through circuit, terminals, insulators, and handle. A cable clamp may or may not be part of a phone plug design.

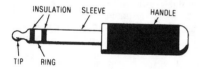

Phone plug.

phonetic alphabet — A list of standard words, one for each letter in the alphabet. It is used for distinguishing the letters in a spoken radio or telephone message. The list is as follows:

Alfa	November
Bravo	Oscar
Charlie	Papa
Delta	Quebec
Echo	Romeo
Foxtrot	Sierra
Golf	Tango
Hotel	Uniform
India	Victor
Juliet	Whiskey
Kilo	X-ray
Lima	Yankee
Mike	Zulu

phonocardiogram — A graphic recording of the sounds produced by the heart and its associated parts (e.g., its mitral or aortic valves).

phonocardiograph — An instrument for recording sounds of the heart on a strip chart.

phonocardiography — The recording and interpretation of the sounds of the heart. A typical instrument for this purpose consists of a microphone, an amplifier, a cathode-ray tube or strip-chart recorder, and sometimes a speaker or headset.

phono cartridge — The means by which stylus movements are converted into an electrical signal. Various versions of magnetic (moving iron, magnet, or coil), ceramic-crystal, capacitive (electret), and strain-gauge devices are in use.

phonocatheter — A catheter-microphone combination that is inserted through the artery into the heart. It picks up inner cardiac sounds.

phonoelectrocardioscope — A dual-beam oscilloscope that displays both ECG signals and heart-sound signals.

phonograph — An instrument for reproducing sound. It consists of a turntable on which the grooved medium containing the impressed sound is placed, a needle that rides in the groove, and an electrical (formerly mechanical) amplifying system for taking the minute vibrations of the needle and converting them into electrical (formerly mechanical) impulses that drive a speaker.

phonograph oscillator — An rf oscillator circuit whose output is modulated by a phonograph pickup and sent through space to a receiver. Thus, no wires to the receiver are needed.

phonograph pickup — Also called mechanical reproducer, pickup, or phono pickup. A mechanoelectric transducer that is actuated by modulations present in the groove of a recording medium and that transforms this mechanical input into an electrical output.

phono jack — A jack designed to accept a phono plug. Receptacle having two through circuits (coaxial oriented) primarily intended for connecting audio signals between phonograph and amplifiers. Now widely used for many other types of signal, including occasionally rf.

phonon — 1. A lattice vibration with which a discrete amount (quantum) of energy is associated. Some thermal and electrical properties of the lattice are theoretically treated in terms of electron-phonon interactions. 2. Quantum of thermal energy used to help calculate the thermal vibration of a crystal lattice. 3. Sharply tuned radiation of superhigh-frequency sound waves.

phonons — Packets of sound energy vibrating in a solid at ultrahigh frequencies — so high that the energy is commonly thought of as heat.

phono pickup — *See* phonograph pickup.

phono plug — A plug used at the end of a shielded conductor for feeding audio-frequency signals to a mating phono jack on an audio preamplifier or amplifier.

Phono plug.

phonoselectroscope — A stethoscopic device that suppresses low frequencies (characteristic of the normal heart function) to permit detection of higher-frequency sounds.

phosphor — 1. A layer of luminescent material applied to the inner face of a cathode-ray tube. During bombardment by electrons it fluoresces, and after the bombardment, it phosphoresces. 2. A material that emits light when excited (energized) by radiant energy. 3. Generic name for the class of substances deposited on the inner surface of television tubes and other cathode-ray tubes that exhibit luminescence when stimulated by an electron beam. 4. The compound, typically polycrystalline, that produces photons upon stimulation — for instance, by high-energy electrons or high electrostatic fields.

phosphor bronze — A frequently used and easily formed connector contact material, with good corrosion resistance and fair conductivity.

phosphor-dot faceplate — 1. The glass viewing screen on which the trios of color phophor dots are mounted in a three-gun picture tube. 2. A glass plate in a tricolor picture tube. May be the front face of the tube or a separate internal plate. In either case, its rear surface is covered with an orderly array of tricolor lines or tricolor phosphor dots. When excited by electron beams in proper sequence, the phosphors glow in red, green, and blue to produce a full-color picture.

phosphor dots — Minute particles of phosphor on the viewing screen of a picture tube. On a tricolor picture tube, the red, green, and blue phosphor dots are placed on the viewing screen in a pattern of dot triads — a phosphor dot of each color forming one-third of the triad.

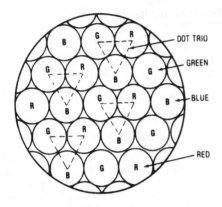

Phosphor dots.

phosphorescence — 1. The emission of light from a substance for a long period after excitation has been removed. Essentially, it is time of persistence that distinguishes phosphorescence from fluorescence. For example, the hands of clocks coated in radium salts glow after exposure to ordinary light. *See also* afterglow. 2. The emission of light from a source that is delayed by over 10^{-8} seconds following excitation. 3. Luminescence of a material that occurs while, and for some time after, the material has been stimulated by radiation energy. *See also* luminescence.

phosphorescence spectroscopy — The spectroscopic study of the radiation emitted in the lifetime of phosphorescence.

phosphors — Chemical substances that exhibit fluorescence when excited by ultraviolet radiation, X-rays, or an electron beam. The amount of visible light is proportional to the amount of excitation energy. If fluorescence decays slowly after the exciting source is removed, the substance is said to be phosphorescent.

phosphor trio — In the phosphor screen of a tricolor kinescope, closely spaced triangular groups of three phosphor dots accurately deposited in interlaced positions.

phot — Abbreviated pt. The unit of illumination when the centimeter is taken as the unit of length; it is equal to 1 lumen per square centimeter.

photoacoustic effect — Generation of an acoustical signal by a sample exposed to modulated light.

photobiology — The study of the effects on living matter (or substances derived therefrom) of electromagnetic radiation extending from the ultraviolet through the visible light spectrum into the infrared. The conversion of electromagnetic energy into chemical energy. Photosynthesis is an important branch of photobiological investigation.

photocathode — 1. An electrode that releases electrons when exposed to light or other suitable radiation. Used in phototubes, television camera tubes, and other light-sensitive devices. 2. An electrode used to release photoelectric emission when irradiated, making it then the irradiated negative electrode of a phototube. 3. An electrode used for obtaining photoelectric emission.

photocathode luminous sensitivity — The responsivity of a photocathode to luminous energy; equal to the ratio of the photoelectric emission to the incident luminous flux. The measurement is made under specified conditions of illumination, usually with radiation from a tungsten-filament lamp operated at a color temperature of 2870 K. The cathode is usually illuminated by a collimated beam at normal incidence.

photocathode radiant sensitivity — The ratio of the photoelectric emission current from the photocathode to the incident radiant flux. It is usually measured at a given wavelength under specified conditions of irradiation with a collimated beam at normal incidence.

photocathode tube response — The photoemission current resulting from a specified luminous flux from a tungsten lamp filament at a color temperature of 2870 K when the flux is filtered by a specified blue filter.

photocell — *See* photoelectric cell.

photochemical radiation — Energy in the ultraviolet, visible, and infrared regions used to produce chemical changes in materials.

photochromic — Pertaining to a single-crystal inorganic material used as a display and storage element. The material can be from one of several families of materials, such as fluorides or titanates. The display color and storage time are determined by the amount and kind of doping of the material.

photoconduction — An increase in the electrical conduction capability resulting from the absorption of electromagnetic radiation by the material.

photoconductive cell — Also called photoresistive cell and photoresistor. 1. A photoelectric cell whose electrical resistance varies inversely with the intensity of light that strikes its active material. 2. A device for detecting or measuring electromagnetic radiation intensity by variation of the conductivity of a substance caused by absorption of the radiation. 3. Photosensor normally made of cadmium sulfide (CdS) or cadmium selenide (CdSe). Unlike the junction types, it has no junction. The entire layer of material changes in resistance

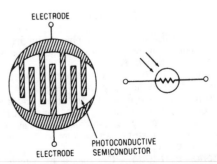

Photoconductive cell.

when it is illuminated (analogous to a thermistor except the heat is replaced by light). 4. *See* photoconductive detector.

photoconductive detector — A detector used in the spectral region from the optical to the far infrared. It is based on photoconductive effects in semiconductors that are sensitive to photons from light to lower energies of microwave quanta. *See* intrinsic photoconductivity; extrinsic photoconductivity.

photoconductive device — A device that makes use of photoconductivity, such as a photoconductive cell.

photoconductive effect — 1. The change of electrical conductivity of a material when exposed to varying amounts of radiation. 2. The phenomenon in which some nonmetallic materials exhibit a marked increase in electrical conductivity on absorption of photon energy. Photoconductive materials include gases (ionization) as well as crystals. They are used in conjunction with semiconductor materials that are ordinarily poor conductors but become distinctly conducting when subjected to photon absorption. The photons excite electrons into the conduction band, where they move freely, resulting in good electrical conductivity. The conductivity increase is due to the additional free carriers generated when photon energies are absorbed in energy transitions. The rate at which free carriers are generated and the length of time they persist in conducting states (their lifetime) determine the amount of conductivity change. 3. The alteration of electric conductivity produced by the absorption of varying amounts of radiation composed of photons.

photoconductive film — A film of material whose electrical current-carrying ability is enhanced when illuminated by electromagnetic radiation, particularly in the visible region of the frequency spectrum.

photoconductive gain factor — The ratio of the number of electrons per second flowing through a circuit containing a cube of semiconducting material, whose sides are of unit length, to the number of photons per second of incident electromagnetic radiation absorbed in this volume.

photoconductive material — Material having a high resistance in the dark, and a low resistance when exposed to the light.

photoconductive meter — An exposure meter in which a battery supplies power through a photoconductive cell to an electrical current-measuring device, such as a millimeter, to measure the intensity of radiation, such as light intensity, incident on its active surface.

photoconductive photodetector — A photodetector that makes use of the phenomenon of photoconductivity in its operation. Thus, it detects the presence of electromagnetic radiation, particularly in the visible region of the frequency spectrum, by changing its electrical resistance in accordance with the intensity of the incident radiation, thus controlling the current from an applied bias voltage power source. Usually a source of voltage is needed to drive a current that will vary according to the variation in conductivity resulting from the variation in incident electromagnetic radiation.

photoconductivity — 1. The greater electrical conductivity shown by some solids when illuminated. The incoming radiation transfers energy to an electron, which then takes on a new energy level (in the conduction band) and contributes to the electrical conductivity. 2. The ability of a material to hold a charge of electricity (insulator) in the absence of light yet act as a conductor of electricity when exposed to light. 3. The increase in electrical conductivity displayed by many materials, particularly

nonmetallic solids, when they absorb electromagnetic radiation. 4. Characteristic of a material that produces changes in the electrical conductivity of the material as a result of absorption of photons.

photoconductor — 1. A passive, high-impedance device composed of thin single-crystal or polycrystalline films of compound semiconductor materials. When the sensitive surface is illuminated, its resistance decreases and, hence, its conductivity increases. 2. A light-sensitive resistor whose resistance decreases with increases in light intensity, when illuminated. Consists of a thin single crystal or polycrystalline films of compound semiconductor substances. 3. A device whose electrical resistance varies in relationship with exposure to light. 4. A material, usually a nonmetallic solid, whose conductivity increases when it is exposed to electromagnetic radiation. 5. A metallic substance, such as selenium or cadmium sulfide, that is capable of conducting and retaining electrical charges. If any portion of a photoconductor is exposed to light, that part will lose its charge. Photoconductive materials are employed in the electrostatic process to retain a latent image (charge) of a document, which is subsequently imbued with toner particles to create an image.

photocoupled solid-state relay — A relay in which the control signal activates the load circuit via a light source and photosensitive semiconductors; electrical isolation between input and output is complete.

photocurrent — 1. The difference between light current and dark current in a photodetector. 2. *See* photoelectric current.

photo-Darlington — A light-sensitive Darlington-connected transistor pair. The photo-Darlington has a very high sensitivity to illumination or radiation.

Photo-Darlington symbol.

photodetector — 1. A device that senses incident illumination. 2. Any device that utilizes the photoelectric effect to detect the presence of light. 3. A device that provides an electrical output signal when subjected to radiation in the visible, infrared, or ultraviolet regions of the electromagnetic spectrum. Photodetectors are square-law detectors that respond to the intensity of light averaged over some time period. Photodiodes, phototransistors, and photo-Darlingtons are the most common types of photodetectors. 4. A device capable of extracting the information from an optical carrier, i.e., a thermal detector or a photon detector, the latter being used for communications more than the former. *See* photoconductive photodetector; photoelectromagnetic photodetector; photoemissive photodetector; photovoltaic photodetector.

photodetector responsivity — The ratio of the rms value of the output current or voltage of a photodetector to the rms value of the incident optical power input. Note: In most cases, detectors are linear in the sense that the responsivity is independent of the intensity of the incident radiation. Thus, the detector response in amperes or volts is proportional to incident optical power, watts. Differential responsivity applies to small variations in optical power.

photodielectric effect — The effect, present in particular phosphors, that is defined as a transformation in the dielectric constant and of a material when illuminated. The effect is observed only in phosphors that show photoconductivity during luminescence.

photodiffusion effect — 1. *See* Dember effect. 2. The potential difference between two areas of a semiconductor when one is exposed to light.

photodiode — 1. A pn semiconductor diode designed so that light falling on it greatly increases the reverse leakage current, so that the device can switch and regulate electric current in response to varying intensity of light. 2. A fiber optics receiver that changes light signals back to electrical signals. 3. A two-electrode radiation-sensitive junction formed in a semiconductor material in which the reverse current varies with illumination. 4. A junction diode that is responsive to radiant energy. Photodiodes have a high degree of linearity between the input radiation and the output current. The resistance across the semiconductor junction changes as a function of the light falling on it. Photodiodes are very fast in response, but limited in sensitivity due to the small area of the junction. They have faster switching speeds than phototransistors.

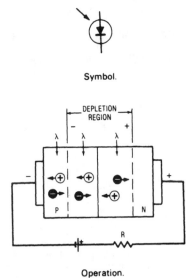

Symbol.

Operation.

Photodiode.

photodiode sensor array scanner — A type of scanner that employs a stationary configuration of tiny photodiodes arranged in a matrix that equals in width one lateral scan line of a document. An original is roller fed from a flatbed tray and passes by the sensor array one line at a time. Light is reflected from each line and focused through a lens onto the face of the photodiode array. Each diode acts as an independent photosensor in converting a small picture element into part of the total electrical signal for a scan line.

photodischarge spectroscopy — Abbreviated PDS. A spectroscopic process that detects and analyzes the discharge from an extrinsic surface with less than band-gap light. This method may determine the extrinsic surface-state energy levels of a semiconductor surface.

photoelasticity — Changes in the optical properties of transparent isotropic dielectrics subject to stress.

photoelectric — Pertaining to the electrical effects of light or other radiation — i.e., emission of electrons, generation of a voltage, or a change in electrical resistance upon exposure to light.

photoelectric absorption — Conversion of radiant energy into photoelectric emission.

photoelectric alarm system — An alarm system that employs a light beam and photoelectric sensor to provide a line of protection. Any interruption of the beam by an intruder is sensed by the sensor. Mirrors may be used to change the direction of the beam. The maximum beam length is limited by many factors, some of which are the light source intensity, number of mirror reflections, detector sensitivity, beam divergence, fog, and haze.

photoelectric beam-type smoke detector — A smoke detector that has a light source which projects a light beam across the area to be protected onto a photoelectric cell. Smoke between the light source and the receiving cell reduces the light reaching the cell, causing actuation.

photoelectric cathode — A cathode whose primary function is photoelectric emission.

photoelectric cell — Also called photocell. 1. A cell, such as a photovoltaic or photoconductive cell, whose electrical properties are affected by illumination. The term should not be used for a phototube, which is a vacuum tube and not a cell. 2. A resistive, bulk-effect type of photosensor. Used when it is desirable to wire several photoreceivers in series or in parallel.

photoelectric colorimeter — A colorimeter that uses a photoelectric cell and a set of color filters to determine, by the output current for each filter, the chromaticity coordinates of light of a given sample.

photoelectric conductivity — The increased conductivity exhibited by certain crystals when struck by light (e.g., a selenium cell).

photoelectric constant — A quantity that, when multiplied by the frequency of the radiation causing the emission of photoelectrons, gives (in centimeter-gram-second units) the voltage absorbed by the escaping photoelectron. The constant is equal to h/e, where h is Planck's constant and *see* is the electron's charge.

photoelectric control — 1. The control of a circuit or piece of equipment in response to a change in incident light impinging on a photosensitive device. 2. The control of an instrument or electrical circuit by the current produced by varying radiation incident to a photoelectric cell.

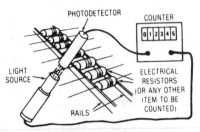

Photoelectric control.

photoelectric counter — A device that registers a count whenever an object breaks the light beam shining on its phototube or photocell. An amplifier then boosts the minute energy to register on a mechanical or other type of counter. Slow counting rates can be recorded by a mechanical counter, but high rates require an electronic counter.

photoelectric current — The stream of electrons emitted by a phototube when the cathode is exposed to light.

photoelectric cutoff control — A photorelay circuit used in machines for cutting long strips of paper, cloth, metal, or other material accurately into predetermined lengths or at predetermined positions.

photoelectric effect — 1. The transfer of energy from incident radiation to electrons in a substance. This phenomenon includes photoelectric emission of electrons from the surface of a metal, the photovoltaic effect, and photoconductivity. 2. Interaction between radiation and matter resulting in the absorption of photons and the consequent generation of mobile charge carriers. 3. The changes in material electrical characteristics due to photon absorption. 4. The emission of electrons as the result of the absorption of photons in a material. This effect is the means by which photons may be detected. (The photons can be of any energy, and the electrons can be released into a vacuum or into a second material. The material itself may be solid, liquid, or gas. Thus, photoelectromagnetic effects are photoelectric effects.)

photoelectric electron-multiplier tube — A vacuum phototube that employs secondary emissions to amplify the electron stream emitted from the illuminated photocathode.

photoelectric emission — 1. Electron emission due directly to the incidence of radiant energy on the emitter. 2. The phenomenon of emission of electrons by certain materials on exposure to radiation in and near the visible region of the spectrum.

photoelectric exposure meter — A device consisting of a microammeter, a photovoltaic cell, and a battery. Used for the measurement of scene brightness and the determination of correct exposure for photographic processes.

photoelectric flame-failure detector — An industrial electronic control employing a phototube and amplifier to actuate an electromagnetic or other valve that cuts off the fuel flow when the fuel-consuming flame is extinguished and light no longer falls on the phototube.

photoelectric inspection — Quality control of a product by means of a phototube, light-beam system, and associated electronic equipment.

photoelectric intrusion detector — A burglar-alarm system in which interruption of a light beam by an intruder reduces the illumination on a phototube, phototransistor, or photocell and thereby closes an alarm circuit.

photoelectric liquid-level indicator — A level indicator in which the rising liquid interrupts the beam of light in a photoelectric control system.

photoelectric material — Any material that will emit electrons when illuminated in a vacuum (e.g., barium, cesium, lithium, potassium, rubidium, sodium, and strontium).

photoelectric multiplier — A phototube in which the primary photoemission current, prior to being extracted at the anode, is multiplied many times.

photoelectric phonograph pickup — A phonograph reproducing device consisting essentially of a light source, a jewel stylus to which a very thin mirror is attached, and a selenium cell that picks up light reflected from the mirror. Sidewise movements of the stylus in the

record groove cause the amount of reflected light to vary, and accordingly the resistance of the selenium cell. The light source is fed by a radio-frequency oscillator rather than from the power line, to eliminate hum caused by a 60-hertz flicker from the light beam.

photoelectric photometer — Also known as electronic photometer. A photometer with a photocell, phototransistor, or phototube for measuring the intensity of light. *See also* electronic photometer.

photoelectric photometry — The use of photoelectric sensors to detect and measure the intensity of a light source. This application, as compared with human observation, results in higher speeds of operation, a greater consistency of results, and superior accuracy in certain photometric situations.

photoelectric pickup — A transducer that transforms a change in light into an electric signal.

photoelectric pyrometer — 1. An instrument for measuring high temperatures from the intensity of the light given off by the heated object. 2. An instrument used to measure the temperature of a source through the use of photoelectric cells to detect and measure the intensity of the light emitted by the source.

photoelectric reader — A device that reads information stored in the form of holes punched in paper tape or cards, by sensing light passed through the holes.

photoelectric receiver — An instrument that uses a photocell to detect and measure the intensity of incident light.

photoelectric recorder — An optical recording instrument employing a light source and phototube for the basic measuring element.

photoelectric reflectometer — A photoelectric photometer used to measure the reflectance of a surface.

photoelectric register control — A photoelectric device used for controlling the position of a strip of paper, cloth, metal, etc., with respect to the machine through which it is being passed.

photoelectric relay — Also called light relay. 1. A relay combined with a phototube (and amplifier, if necessary), so arranged that changes in incident light on the phototube cause the relay contacts to open or close. 2. A relay that opens or closes an electrical circuit depending on the intensity of the light incident to a photoelectric device connected to the relay.

photoelectric scanner — A light source, lens system, and one or more phototubes in a single, compact housing. It is mounted a few inches above a moving surface, where it actuates control equipment when the amount of light reflected from the surface changes.

photoelectric sensitivity — Also called photoelectric yield. 1. The rate at which electrons are emitted from a metal per unit radiant flux at a given frequency. 2. That property of a material that determines its ability to release electrons when absorbing photons.

photoelectric sensor — A device that detects a visible or invisible beam of light and responds to its complete or nearly complete interruption. *See also* modulated photoelectric alarm system; photoelectric alarm system.

photoelectric smoke detector — A photoelectric instrument used to measure the density of smoke and to sound an alarm when a predetermined smoke density is exceeded.

photoelectric sorter — An industrial-electronic control employing a light beam, phototube, and amplifier to sort objects according to color, size, shape, or other characteristics.

photoelectric spectrophotometer — A system consisting of a spectrophotometer with a photoelectric detector for measurement of radiant energy.

photoelectric spot-type smoke detector — A smoke detector that contains a chamber with covers that prevent the entrance of light but allow the entrance of smoke. The chamber contains a light source and a photosensitive cell so placed that light is blocked from it. When smoke enters, the smoke particles scatter and reflect the light into the photosensitive cell, causing an alarm.

photoelectric threshold — The quantum energy just sufficient to release photoelectrons from a given surface. The corresponding frequency is the critical, or threshold, frequency.

photoelectric timer — An electronic instrument that automatically turns off an X-ray machine or other photographic device when the film reaches the correct exposure.

photoelectric transducer — A transducer that converts changes in light energy into electrical changes.

photoelectric tube — *See* phototube.

photoelectric work function — The energy required to transfer electrons from a given metal to a vacuum or other adjacent medium during photoelectric emission. It is sometimes expressed as energy in ergs or joules per unit of emitted charge, and sometimes as energy per electron in electron volts.

photoelectric yield — *See* photoelectric sensitivity.

photoelectromagnetic effect — Also called photomagnetoelectric effect. 1. The production of a potential difference by virtue of the interaction of a magnetic field with a photoconductive material that is subjected to incident radiation. (The incident radiation creates hole-electron pairs that diffuse into the material. The magnetic field causes the pair components to separate, resulting in a potential difference across the material. In most applications the light is made to fall on a flat surface of an intermetallic semiconductor located in a magnetic field that is parallel to the surface, excess hole-electron pairs are created, and these carriers diffuse in the direction of the light but are deflected by the magnetic field to give a current through the semiconductor that is at right angles to both the light rays and the magnetic field. This is due to transverse forces acting on electrons and holes diffusing into the semiconductor from the surface.) 2. Interaction of a magnetic field with a photoconductive substance exposed to light waves to create a potential difference. 3. *See* Dember effect.

photoelectromagnetic photodetector — A photodetector that makes use of the photoelectromagnetic effect; namely, it uses an applied magnetic field.

photoelectromotive force — 1. Electromotive force caused by photovoltaic action. 2. The force that stimulates the emission of an electric current when photovoltaic action creates a potential difference between two points.

photoelectrons — 1. The electrons emitted from a metal in the photoelectric effect. 2. The electrons released in photoelectric activity.

photoemission — *See* photoelectric emission.

photoemissive — Capable of emitting electrons when under the influence of light or other radiant energy.

photoemissive cell — 1. A device that detects or measures radiant energy by measurement of the resulting emission of electrons from a surface that has or displays a photoemissive effect. 2. *See also* photoemissive detector.

photoemissive detector — 1. An electron tube in which the anode current varies with the intensity of light incident on the cathode. 2. A photosensor that measures light by the emission in a vacuum of one electron per photon impinging on a metal photocathode.

photoemissive effect — The emission of electrons as a result of incident radiation.

photoemissive photodetector — A photodetector that makes use of the photoemissive effect. Usually an applied electric field is necessary to attract or collect the emitted electrons.

photoemissive tube photometer — A photometer that uses a tube made of a photoemissive material. It is highly accurate, requires electronic amplification, and is used mainly in laboratories.

photoemissivity — The property of a substance that causes it to emit electrons when electromagnetic radiation in the visible region of the frequency spectrum is incident on it. Normally an electric field is applied to collect the emitted electrons.

photofabrication — The production of precise shapes in metal or other substances by recording a precise photographic image on the surface of the substance and etching away the unprotected areas by chemical or electrical means.

photoflash — A means of firing expendable flashbulbs with an instantaneous surge of current supplied by two or more 1.5-volt single-cell batteries, or by the discharge of a capacitor that has been charged to full capacity by a medium-voltage battery.

photoflash bulb — 1. An oxygen-filled glass bulb containing a metal foil or wire. A surge of current heats the metal to incandescence, and a brilliant flash of light is produced when the wire burns in the oxygen. 2. A glass bulb packed with a highly inflammable mixture of zirconium, wool, and glass. The mixture is ignited and produces an instantaneous flash when a small powder-filled primer cap in the base of the bulb is set off by a small hammer.

photoflash tube — *See* flash tube.

photoflood lamp — An incandescent lamp employing excess voltage to give brilliant illumination. Used in television and photography, it has a life of only a few hours.

photogalvanic cell — A cell that generates an electromotive force when light falls on either of the electrodes immersed in an electrolyte.

photogalvanometer — An instrument for recording the electric variations produced by emotional stresses.

photogenerator — A semiconductor-junction device that emits light when pulsed.

photoglow tube — A gas-filled phototube used as a relay. This is done by making the operating voltage so high that ionization and a glow discharge occur, accompanied by considerable current, when a certain illumination is reached.

photographic writing speed — A figure of merit used to describe the ability of a particular combination of camera, film, oscilloscope, and phosphor to record a high-speed trace. It expresses the maximum single-event spot velocity (usually in centimeters per microsecond) that can be recorded on film as an image just discernible to the eye.

photoionization — Ionization occurring in a gas as a result of visible light or ultraviolet radiation.

photoisland grid — A photosensitive surface in the storage-type Farnsworth dissector tube used with television cameras. It comprises a thin, finely perforated (about 400 holes per square inch or 62 per square centimeter) sheet of metal. (No longer used in current equipment.)

photojunction battery — A nuclear-type battery in which the radioactive material, promethium 147, irradiates a phosphor that converts nuclear energy into light. The light is then converted to electrical energy by a small silicon junction.

photolithography — 1. The process whereby patterns are etched into an oxide or similarly passive layer coating a semiconductor crystal wafer, using a photoresist process followed by an etchant. The remaining oxide material thus forms a precisely located miniature mask, used to control impurity doping and contact metallization. 2. The process used to print the masks on the wafer. It involves a photosensitive emulsion and selective etching analogous to commercial lithography.

photoluminescence — Luminescence stimulated by visible light or ultraviolet radiation. A special case of electroluminescence that involves a light-to-light conversion, the emitted light differing from the stimulating radiation in frequency.

photomagnetic effect — The direct effect of light on the magnetic susceptibility of certain substances.

photomagnetoelectric effect — 1. The production in a semiconductor of an electromotive force normal to both an applied magnetic field and to a photon flux of proper wavelength. 2. *See* photoelectromagnetic effect.

photomask — 1. A photographic template through which images are transferred by light onto a photoresist coating. 2. A transparent plate slightly larger than a silicon slice, containing numerous tiny opaque spots, used in the planar diffusion process as a shadow mask over a slice coated with photoresist to expose the surface of the slice to acid in desired spots in a later step. 3. A square, flat, glass substrate, coated with a photographic emulsion or a very thin layer of metal, on which appear several hundred circuit patterns (each containing thousands of images). The patterns are exposed onto semiconductor wafers. 4. A square-shaped glass substrate, coated with either an extremely thin metal layer or a photographic emulsion, on which hundreds of complex circuit patterns have been precisely reproduced. Each circuit pattern contains literally thousands of geometric images, some as small as 50 millionths of an inch (1.27 µm). In much the same way as a photograph is made from a conventional negative in a darkroom, the patterns on the mask are exposed onto a photosensitive wafer surface during each step of the semiconductor wafer processing operation. In projection printing of wafers, the photomasks used are called master plates. In contact printing copies of the master plates, called working plates, are used. As many as 7 to 10 different photomasks typically are required to manufacture a single type of multilayered semiconductor device.

photometer — 1. An instrument for measuring the intensity of a light source or the amount of illumination, usually by comparison with a standard light source. 2. A device used to compare the luminous intensities of two sources by comparing the illuminance they produce. 3. *See* photoemissive tube photometer.

photometric — Related to measurements of light.

photometric equipment — Photocells of various kinds used to measure photometric quantities, i.e., intensity, luminance, and illuminance. Meter readings are used to express illuminance and, by calibration, to measure intensity and luminance. Other photometric quantities requiring measurement include transmission and reflection densities in photography, the reflectance of a mirror or a screen, and the transmittance of an optical instrument.

photometry — 1. The techniques for measuring luminous flux and related quantities (e.g., luminous intensity, illuminance, luminance, luminosity). 2. The science of the measurement of light intensity, where *light* refers to the total integrated range of radiation to which the eye is sensitive. It is distinguished from radiometry, in which each separate wavelength in the electromagnetic spectrum is detected and measured, including the ultraviolet and infrared. 3. The science devoted to the measurement of the effects of electromagnetic radiation on the eye. Photometry is an outgrowth of psychophysical aspects and involves the determination of visual effectiveness by considering radiated power and the sensitivity of the eye to the frequency in question.

photomultiplier pulse-height resolution — A measure of the smallest change in the number of electrons emitted during a pulse from the photocathode that can be discerned as a change in output-pulse height.

photomultiplier tube — *See* multiplier phototube.

photon — 1. A quantum of electromagnetic energy. The equation is $E = h\nu$, where E is the energy, h is Planck's constant and ν is the frequency associated with the photon. The momentum of the photon in the direction of propagation is $h\nu/c$, where c is the velocity of light. 2. A quantum of light; used to help describe characteristics of light in conjunction with the wave theory of light. 3. A quantum of electromagnetic energy carried in a small amount and moving with the speed of light. Optical photons have energies corresponding to wavelengths between 120 and 1800 nanometers. 4. The smallest unit of radiant energy. 5. Particle carrying the energy associated with electromagnetic radiation. 6. An elementary particle of light.

photon-coupled isolator — A circuit-coupling device consisting of an infrared emitter diode coupled to a photon detector over a short shielded light path, which provides extremely high circuit isolation.

photon coupling — Coupling between circuits by a beam of light.

photon detector — A device that responds to incident photons.

photonegative — Having a negative photoconductivity — hence, decreasing in conductivity (increasing in resistance) under the action of light. Selenium sometimes exhibits this property.

photon energy — Planck's constant times photon frequency.

photon flux — Incident photons per second. The product of photon energy and flux is radiant power.

photonics — The technology of generating and harnessing light and other forms of radiant energy, whose unit is the photon. The range of applications of photonics extends from energy generation to detection to communications and information processing.

photo-optic memory — A memory that uses an optical medium for storage. For example, a laser might be used to record on photographic film.

photo-optics — The combination of an input light source and a photoreceiver producing an output signal, assembled either separately or in a single package.

photoparametric diode — A pill-sized device for simultaneously detecting and amplifying optical energy modulated at microwave frequencies.

photophone — A device for converting variations in light intensity into sound.

photoplotter — Device used to generate artwork photographically for printed circuit boards.

photopolymer hologram — A holographic plate coated by photopolymeric mixtures that are composed of one or more monomers and a photoredox catalyst sensitized to visible light. The material becomes a plastic solid when exposed to ultraviolet light and, thus, produces a nonerasable hologram.

photopolymer material — A photosensitive plastic that can be formed optically by photographic methods into three-dimensional configurations to reproduce a planar image precisely and in depth.

photopositive — Having a positive photoconductivity — hence, increasing in conductivity (decreasing in resistance) under the action of light. Selenium ordinarily has this property.

photoradiometer — Instrument for measuring the intensity and penetrating power of radiation.

photoreceiver — A unit consisting of photosensor, focusing lens, and protective enclosure.

photorelay circuit — A form of on-off control actuated by a change of illumination.

photoresist — 1. A solution that when exposed to ultraviolet light becomes extremely hard and resistant to etching solutions that dissolve materials such as silicon dioxide. 2. A liquid plastic that hardens into a tough acid-resistant solid when exposed to ultraviolet light. 3. A chemical substance rendered insoluble by exposure to light. By means of a photoresist, a selected pattern can be imaged on a metal. The unexposed areas are washed away and are then ready for etching by acid or doping to make a microcircuit. 4. A material that selectively allows etching of a wafer when photographically exposed. With a negative resist, the resist film under the clear area of a photomask undergoes physical and chemical changes that render it insoluble in a developing solution. In a positive resist, the same areas after exposure are soluble in the developing solution, so they disappear, permitting development of the exposed pattern underneath. 5. A light-sensitive liquid that is spread as a uniform thin film on a wafer or substrate. After baking to solidify the liquid, exposure of specific patterns is performed using a photomask. Material remaining after development shields regions of the wafer from subsequent etch or implant operations.

photoresistive cell — *See* photoconductive cell.

photoresistive or **photoconductive transduction** — Conversion of the measurand into a change in the resistance of a semiconductor material (by changing the illumination incident on the material).

photoresistor — 1. A semiconductor resistor that, when illuminated, drops in resistance. 2. *See* photoconductive cell.

photosensitive — Capable of emitting electrons when struck by light rays.

photosensitive field-effect transistor — A special unipolar field-effect transistor (FET) structure that is positioned on a header to receive illumination transmitted through a lens in the top of the header can. It combines the circuit and device characteristics of a photodiode and a high-impedance low-noise amplifier.

photosensitive recording — Recording by the exposure of a photosensitive surface to a signal-controlled light beam or spot.

photosensitive semiconductor — A semiconductor material in which light energy controls the current-carrier movement.

photosensitivity — That property of materials determining that they react when exposed to light energy.

photosensor — The light-sensitive device in a photoelectric control that converts a light signal into an electric signal.

photosphere — The outermost luminous layer of the gaseous body of the sun.

photoswitch — A solid-state device that functions as a high-speed power switch activated by incident radiation. *See also* LASCR; LASCS.

phototelegraphy — In facsimile, the process of sending photographs over a wire.

photothyristor — A thyristor whose switching action is controlled by light applied to the thyristor gate.

phototransistor — 1. A junction transistor with its base exposed to light through a lens in the housing. The collector current increases as the light intensity increases, because of the amplification of the base current by the transistor structure. The device may have only collector and emitter leads, or it may also have a base lead. 2. A light-sensitive transistor that delivers an electrical output proportional to the light intensity at its input. The low-level photocurrent that is generated is amplified by the current gain of the transistor. The base region (or gate in a FET) may or may not be brought out as an external terminal. 3. Transistor having a base-collector acting as a photodiode. Light generates a base current that turns on the working current through the transistor. This gives a much larger current than a simple photodiode. 4. A transistor whose electrical output current is proportional to the intensity and wavelength of a beam of electromagnetic radiation applied to its input. 5. A solid-state device similar to an ordinary transistor except that incident light on the pn junctions regulates the response of this device. It has built-in gain and a greater sensitivity than that of photodiodes. 6. A type of photosensor, typically used where speed of response is critical or ambient temperature variations are great. 7. A transistor utilizing the photoelectric effect.

With base connection.

Without base connection.

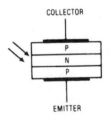

Layering.

Phototransistor.

phototronic cell — 1. A type of photovoltaic cell in which a voltage is generated in a layer of selenium during exposure to visible or other radiation. 2. *See* photovoltaic cell.

phototronic photocell — *See* photovoltaic cell.

phototube — Also called photoelectric tube. An electron tube containing a photocathode. Its output depends on the total photoelectric emission from the irradiated area of the photocathode.

phototube bridge circuit — A circuit in which a phototube is one arm of a bridge circuit. With such a circuit, a balanced condition (no signal output) can be reached under either a black-signal or white-signal condition, depending on the impedance adjustments in the other arms.

phototube relay—An electrical relay in which the action of a beam of light on a phototube operates mechanical devices such as counters and safety controls.

phototypesetting—The setting of type via electronic or electromechanical optical systems onto photographic paper or film. Input to such devices can be by direct keyboard entry, paper tape, or magnetic media. Fonts are contained on reduced character matrices through which light is flashed and focused to expose the photographic medium.

photovaristor—A varistor in which the current-voltage relation may be modified by illumination. Cadmium sulfide and lead telluride exhibit such properties.

photovoltaic—Capable of generating a voltage when exposed to visible or other light radiation.

photovoltaic cell—Also called barrier-layer photocell, boundary-layer photocell, self-generating cell, solar cell, and phototronic photocell. 1. A semiconductor device that converts light into electrical energy. 2. A photosensor that generates a voltage across a pn junction as a function of the photons impinging on it. Usually made of selenium or silicon. Photosensors require no external power supply. 3. A photosensor that converts radiant flux directly to electrical current. The self-generating solar cell is very similar to an electrical battery. Barrier-layer cells are manufactured in large quantities for use in photographic exposure meters. The voltage generated by each cell is small, but series connection of a large number of cells in an array is possible. 4. A device that detects or measures radiant energy by the production of a source

of voltage proportional to the incident radiation intensity. It is possible to operate a photovoltaic cell without an additional source of voltage, since it develops a voltage. The cell detects or measures electromagnetic radiation by generating a potential at a junction (barrier layer) between two types of material on absorption of radiant energy.

photovoltaic converter—A device for converting light to electric energy by means of the photovoltaic effect.

photovoltaic effect—1. The generation of a voltage (or an electric field) in a material that is illuminated with radiation of a suitable wavelength. 2. The production of a voltage across a semiconductor pn junction due to the absorption of photon energy. The potential is caused by the diffusion of hole-electron pairs (which the incident photons cause to shift or increase) across the junction potential barrier, leading to direct conversion of a part of the absorbed energy into usable electromotive force (voltage). Usually the photovoltaic effect involves the production of a voltage in a nonhomogeneous semiconductor, such as silicon, or at a junction between two types of material. 3. The generation of a difference in electric potential between two electrodes when radiation is incident to one of them.

photovoltaic meter—An exposure cell in which a photovoltaic cell produces a current proportional to the light intensity (or area exposed) falling on the cell, and this current is measured by a sensitive current-measuring device, such as a microammeter.

photovoltaic photodetector—A photodetector that makes use of the photovoltaic effect. A source of voltage is usually not needed for the photovoltaic photodetector, since it is its own source of voltage.

photovoltaic transduction—Conversion of the measurand into a change in the voltage generated when a junction of dissimilar material is illuminated.

photox cell—A type of photovoltaic cell in which a voltage is generated between a copper base and a film of cuprous oxide during exposure to visible or other radiation.

photran—A triode pnpn-type switch of the reverse-blocking pnpn-type switch class. The photran provides optical triggering in addition to standard gate-terminal triggering.

physical configuration—The arrangement of materials, components, and wires on a structural assembly for physical strength, conservation of size, and minimal electrical interaction.

physical device—An actual peripheral hardware device, such as a line printer, terminal, card reader, or paper-tape punch, that is attached to a computer system.

physiological patient monitor—A device for automatically measuring and/or recording one or several physiological variables and responses of a patient, including heart potential, blood flow, blood pressure, pulse rate, respiration rate, temperature, etc.

pi—The Greek letter π. It designates the value of the ratio of the circumference of a circle to its diameter, approximately 3.14159.

PIC—*See* position-independent code.

picket fencing—Rapid flutter of the signal from a mobile station. Caused by reflections of the signal from trees and telephone poles.

pickoff—A device that produces a signal output, generally a voltage, as a function of the angle between two gimbals or between a gimbal and a base.

pickup—1. A device that converts a sound, scene, or other form of intelligence into corresponding electric

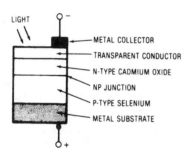

Selenium.

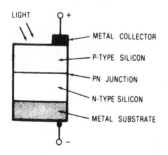

Silicon.

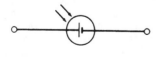

Symbol.

Photovoltaic cell.

signals (e.g., a microphone, television camera, or phono-graph pickup). 2. The minimum current, voltage, power, or other value that will trip a relay. 3. Interference from a nearby circuit or system.

pickup arm — *See* tonearm.

pickup cartridge — The removable portion of a pickup arm. It contains the electromechanical transducing elements and the reproducing stylus.

Pickup cartridge.

pickup current — Also called pull-in current. That current at which a magnetically operated device starts to operate.

pickup head — The end of the pickup arm containing the cartridge. It is often removable from the arm and is usually called the head shell.

pickup spectral characteristic — In television, the spectral response of the camera tube including the optical parts. It converts radiation into electric signals, which are measured at the output terminals of the pickup tube.

pickup tube — A television camera image pickup tube.

pickup value (voltage, current, or power) — The minimum value that will energize the contacts of a relay. *See also* dropout value; hold current.

pickup voltage — That voltage at which a magnetically operated device starts to operate.

pico- — Prefix meaning 10^{-12}. (Formerly micromicro.) Letter symbol: p.

picoammeter — A sensitive ammeter that indicates current values in picoamperes.

picoampere — One-millionth of a microampere (10^{-12} ampere). Letter symbol: pA.

picocoulomb — One-millionth of a microcoulomb (10^{-12} coulomb). Letter symbol: pC.

picofarad — One-millionth of a microfarad (10^{-12} farad); formerly micromicrofarad. Called puff in England. Letter symbol: pF.

picosecond — A micromicrosecond (10^{-12} second); one-thousandth of a nanosecond. Letter symbol: ps.

picowatt — One-millionth of a microwatt (10^{-12} watt). Letter symbol: pW. Formerly called micromicrowatt.

pictorial wiring diagram — A wiring diagram containing actual sketches of components and clearly showing all connections between them.

picture black — Also called black signal. In facsimile, the signal produced at any point by the scanning of a selected area of subject copy having maximum density.

picture brightness — A measure of the brightness of the highlights in a television picture, usually measured in foot-lamberts.

picture carrier — Also called luminance carrier. The carrier frequency 1.25 MHz above the lower frequency limit of a standard NTSC television signal. In color television, this carrier is used for transmitting luminance information; the chrominance subcarrier, which is 3.579545 MHz higher in frequency, transmits the color information.

picture element — Also called elemental area, critical area, scanning spot, or recording spot. 1. In facsimile, that portion of the subject copy that is seen by the scanner at any instant. It can be considered a square area having dimensions equal to the width of the scanning line. 2. In television, any segment of a scanning line whose dimension along the line is exactly equal to the nominal line width. The area being explored at any instant in the scanning process. 3. *See* pixel.

picture freeze — A TV mode that makes it possible to "freeze" an image to study specific details at ease.

picture frequency — The number of complete pictures scanned per second in a television system.

Picture-in-Picture — Abbreviated PIP.

picture line standard — The total number of horizontal lines in a complete television image. The standard in the United States is 525 lines (NTSC standard).

picture monitor — A cathode-ray tube and its associated circuits arranged for viewing a television picture.

picture signal — In television, the signal resulting from the scanning process.

picture-signal amplitude — The difference between the white peak and the blanking level of a video signal.

picture-signal polarity — The polarity of the signal voltage that represents a dark area of a scene given with respect to the signal voltage representing a light area. Expressed as black negative or black positive.

picture size — The usable viewing area on the screen of a television receiver, measured in square inches.

picture-synchronizing pulse — *See* vertical-synchronizing pulse.

picture transmission — Electric transmission of a shaded (halftone) picture.

picture transmitter — *See* visual transmitter.

picture tube — 1. The cathode-ray tube in a TV monitor or receiver on which the picture is produced by variation of the beam intensity as the beam scans the raster. 2. *See* kinescope, 1.

picture-tube brightener — An accessory added to an aging picture tube to increase the image brightness and thereby extend its useful life. The brightener raises the filament voltage and thereby increases the electron emission from the cathode.

PIDD — Acronym for *p*ositive *id*entification and *d*etection. A passive terminal homing device for identifying ships by radar signatures.

Pierce oscillator — Basically a Colpitts oscillator in which a piezoelectric crystal is connected between the plate and grid (or collector and base, or drain and gate). Voltage division is provided by the interelectrode capacitances of the circuit.

pie winding — A winding constructed from individual washer-shaped coils called pies.

piezodielectric — Pertaining to a change in dielectric constant under mechanical stress.

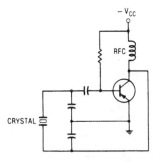

Pierce oscillator.

piezoelectric — 1. The property of certain crystals that produce a voltage when subjected to a mechanical stress, or undergo mechanical stress when subjected to a voltage. (The word is derived from the Greek *piezein*, meaning to squeeze or press.) 2. A speaker drive principle using a ceramic element that expands or bends under the application of a signal voltage. This deflection generates a sound output. Used in some tweeter designs. 3. The conversion of mechanical energy, such as mechanical stress, to electric energy.

piezoelectric accelerometer — 1. Basically a crystalline material that, when force is applied, generates a charge. Through the incorporation of a mass in direct contact with the crystal, an acceleration transducer is produced. 2. A transducer that converts dynamic force (vibration) into electrical energy, i.e., it is a charge generator.

piezoelectric axis — With respect to a crystal, one of the paths or axes that will exhibit a piezoelectric charge when subject to tension or compression.

piezoelectric crystal — 1. A piece of natural quartz or other crystalline material capable of demonstrating the piezoelectric effect. A quartz crystal, when ground to certain dimensions, will vibrate at a desired radio frequency when placed in an appropriate electric circuit. 2. A substance that has the ability to become electrically polarized and has strong piezoelectric properties in it, so cut as to emphasize the coupling to some distinct mechanical mode of the crystal. Used as an electromechanical transducer.

piezoelectric crystal cut — The orientation of a piezoelectric crystal plate with respect to the axes of the crystal. It is usually designated by symbols; e.g., GT, AT, BT, CT, and DT identify certain quartz-crystal cuts having very low temperature coefficients.

piezoelectric crystal element — A piece of piezoelectric material cut and finished to a specified shape and orientation with respect to the crystallographic axes of the material.

piezoelectric crystal plate — A piece of piezoelectric material cut and finished to specified dimensions and orientation with respect to the crystallographic axes of the material and having two essentially parallel surfaces.

piezoelectric crystal unit — A complete assembly comprising a piezoelectric crystal element mounted, housed, and adjusted to the desired frequency, with means for connecting it into an electric circuit. Such a device is commonly employed for frequency control or measurement, electric wave filtering, or interconversion of electric and elastic waves.

piezoelectric device — A substance that generates an electric voltage when bent, squeezed, or twisted. Conversely, when a voltage is applied, it will twist, bend, expand, or contract.

piezoelectric effect — 1. The mechanical deformity of certain natural and synthetic crystals under the influence of an electric field. This effect is used in high-precision oscillators and certain high-frequency filters. 2. The property of certain natural and synthetic crystals to produce a voltage when subjected to mechanical stress (compression, expansion, twisting, etc.). 3. The interaction between electrical and mechanical stress-strain factors in a material. In particular natural and synthetic crystals, i.e., those that have no center of symmetry, it is a property, and the effect is linear in a field of strength. Thus, when a quartz crystal is compressed, an electrostatic voltage is generated across it, or when an electric field is applied, the crystal may expand or contract in particular directions. 4. The property of some crystals such as quartz, tourmaline, and Rochelle salt, when compressed in certain directions, of showing electric charges of opposite polarity on pairs of the crystal face. The charges are proportional to the pressure, and disappear when the pressure is removed.

piezoelectricity — The phenomenon whereby certain crystalline substances, such as barium titanate, generate electrical charges when subjected to mechanical deformation. The effect was first noticed by the Curie brothers in the 1880s as a result of extensive study of the symmetry of crystalline materials. The reverse effect also occurs.

piezoelectric loudspeaker — *See* crystal loudspeaker.

piezoelectric material — A material that generates an electrical output when subjected to a mechanical stress.

piezoelectric microphone — *See* crystal microphone.

piezoelectric oscillator — A crystal-oscillator circuit in which the frequency is controlled by a quartz crystal. *See also* Pierce oscillator.

piezoelectric pickup — 1. A type of cartridge whose generating element is a ceramic, crystal, or electret that generates electricity when bent, twisted, or stressed. The output of such cartridges can be fairly high. It is also proportional to the amplitude of the stylus motion, rather than stylus velocity, and so requires no equalization. Both these factors allow the use of simpler input circuits, one reason why piezoelectric (chiefly ceramic) cartridges are used in low-cost equipment. 2. *See* crystal pickup.

piezoelectric pressure gage — An apparatus for measuring or recording very high pressures. The pressure is applied to quartz discs or other piezoelectric crystals. The resultant voltage, after amplification, is then measured or is recorded with an oscillograph.

piezoelectric speaker — *See* crystal speaker.

piezoelectric transducer — Also called ceramic or crystal transducer. A transducer that depends for its operation on the interaction between the electric charge and the deformation of certain asymmetric crystals having piezoelectric properties.

piezoelectric transduction — The conversion of the measurand into a change in the electrostatic charge or voltage generated by mechanically stressed crystals.

piezoid — The finished crystal blank. It may also include the electrodes for making contact with the crystal.

piezo-optical transducer — A structure consisting of a thin film of liquid crystal sandwiched between light-polarizing filters that have received a surface lubricant. Depending on the motion, the transducer acts as a highly sensitive detector by virtue of the sensitivity of the liquid crystal that is made to float by the lubricant between the plates. It produces a readily visible and colorful optical pattern that is dependent on the amount of disturbance.

piezoresistance — Resistance that changes with pressure.

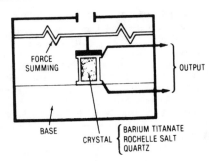

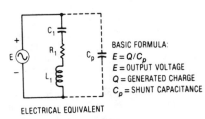

BASIC FORMULA:
$E = Q/C_p$
E = OUTPUT VOLTAGE
Q = GENERATED CHARGE
C_p = SHUNT CAPACITANCE

ELECTRICAL EQUIVALENT

Piezoelectric transducer.

piezoresistive transduction—A change in the resistance of a conductor or semiconductor caused by a change in the mechanical stress applied to it.

piggyback—*See* voltage corrector.

piggyback control—*See* cascade control.

piggyback twistor—An electrically alterable, non-destructive-readout information-storage device that consists of a thin, narrow tape of magnetic material wound spirally around a fine copper conductor. A second similar tape is wrapped on top of the first to sense the stored information. A binary digit is stored at the intersection of a copper strap and a pair of these twistor wires.

pigtail—1. Either a wire attached for terminating purposes to a shield, or a conductor extending from a small component. 2. The termination of a capacitor winding to its lead. 3. The disc-shaped head, at the end of a lead, that is attached to a capacitor winding. 4. A short wire extending from an electric or electronic device to serve as a jumper or ground connection.

pigtail splice—A splice made by tightly twisting the bared ends of parallel conductors together.

pigtail wire—1. Fine-stranded, extra-flexible, rope-lay lead wire. 2. A short piece of wire attached to a shield for terminating purposes, or the conductor extending from a small component.

pile—1. A nuclear reactor. So called because early reactors were piles of graphite blocks and uranium slugs. 2. (Voltaic) Invented by Volta in 1800, it was the first primary battery known to the modern world. The pile consisted of an arrangement of pairs of discs of copper and zinc, each pair separated by a disc of moistened pasteboard. Later Volta arranged a series of cups filled with brine, each containing a zinc and copper plate, and by connecting these together obtained an electric current.

pileup—Also called stack. 1. On a relay, a set of contact arms, assemblies, or springs fastened one on top of the other with layers of insulation separating them. 2. A departure from the base line because of a rapid accumulation of pulses.

pill—A microwave stripline termination.

pillbox antenna—A cylindrical parabolic reflector enclosed by two plates perpendicular to the cylinder and spaced to permit propagation of only one mode in the

desired direction of polarization. It is fed on the focal line.

pilot—1. In a transmission system, a signal wave, usually a single frequency, transmitted over the system to indicate or control its characteristics. 2. In a tape relay, instructions appearing in a routing line, relative to the transmission or handling of that message.

pilot cell—The storage-battery cell selected because its temperature, voltage, and specific gravity are assumed to be those of the entire battery.

pilot channel—A very narrow-band channel over which a single frequency, the pilot frequency, is transmitted to operate trouble alarms or automatic level regulators, or both.

pilot contacts—Contacts whose opening, closing, or transfer govern an operation of relays or similarly controlled devices.

pilot holes—Also called manufacturing holes or fabrication holes. Holes on a printed circuit board used as guides during manufacturing operations. Mounting holes in a part are sometimes used as pilot holes.

pilot lamp—A light that indicates whether a circuit is energized.

pilot light—Also known as a monitor light. 1. A light that, by means of position or color, indicates whether a control is functioning. 2. A light that indicates which of a number of normal conditions of a system or device exists. It is unlike an alarm light, which indicates an abnormal condition.

pilot regulator—A device for maintaining the receiving level of a carrier-derived circuit constant under varying attenuation conditions of the transmission line.

pilot spark—A low-power preliminary spark used in a gas-discharge tube to produce an ionized path for the large main spark discharge.

pilot subcarrier—A subcarrier used as a control signal in the reception of compatible FM stereophonic broadcasts.

pilot tone—1. A single frequency sent over a narrow channel to cause an alarm or automatic control to operate. 2. The signal included with the stereo information and applied in a subchannel on the FM signal, which is used to reform the suppressed subcarrier and synchronize the switching of the detectors in the stereo decoder. Pilot-tone frequency is 19 kHz (broadcast at the equivalent of 10 percent FM modulation), and doubling this (38 kHz) gives the subcarrier frequency. The pilot tone also activates the stereo indicator light.

pilot-type device—A device used in a circuit for control apparatus that carries electrical signals for directing the performance but does not carry the main power current.

pilot wire—An auxiliary conductor used with remote measuring devices or for operating apparatus at a distance.

pilot-wire regulator—An automatic device for controlling adjustable gains or losses associated with transmission circuits to compensate for transmission changes caused by temperature variations, the control usually depending on the resistance of a conductor or pilot wire having substantially the same temperature conditions as the conductors of the circuit being regulated.

pi mode—A mode of magnetron operation in which the fields of successive anode openings facing the interaction space differ in phase by π radians.

pin—Also called a prong or base pin. A terminal on a connector, plug, or tube base.

pinboard—A perforated board into which pins may be inserted manually to control the operation of equipment.

pinch effect — 1. The result of an electromechanical force that constricts, and sometimes momentarily ruptures, a molten conductor that is carrying a high-density current. 2. In the reproduction of lateral recordings, the pinching of the reproducing stylus tip twice each cycle due to a decrease in the groove angle cut by the recording stylus during the swing from a negative to a positive peak. 3. The self-contraction of a plasma column carrying a large current due to the interaction of this current with the magnetic field it produces. The current required for such an effect is on the order of 105 amperes. If the current is in the form of a short pulse, a radically imploding shock wave is generated.

pinch-off — 1. In a field-effect transistor, a condition in which the gate bias causes the depletion region to extend completely across the channel, with a resulting cessation of drain current. 2. The voltage required to stop majority current flow in an FET.

pinch-off voltage — The gate voltage of a field-effect transistor that blocks the current for all source-drain voltages below the junction breakdown value. Pinch-off occurs when the depletion zone completely fills the area of the device.

pinch resistor — A monolithic silicon resistor derived from a p-base diffusion resistor by superimposing an n+ emitter diffusion over the resistor. As a result, the surface of the p resistor reverts to n-type material, and a narrow "pinched" resistor is left under the n+ diffusion. A pinch resistor features 10,000 ohms per square, but it is limited to low voltages, and it has a high temperature coefficient.

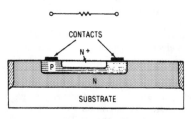

Pinch resistor.

pinch roller — 1. An idling roller used to press a tape against the capstan to cause advance. A friction clamp is used with the roller, as a brake to stop perforated tape. 2. A rubber or neoprene wheel that presses the tape against the capstan during recording or play.

pinch wheel — 1. *See* pressure roller. 2. *See* idler.

pin connection — Connections made to the base of pins in a vacuum tube. They are identified by the following abbreviations: NC, no connection; IS, internal shield; IC, internal connection (not an electrode connection); P, plate; G, grid; SG, screen grid; SU, suppressor; K, cathode; H, heater; F, filament; RC, ray-control electrode; and TA, target.

pin contact — A male-type contact, usually designed to mate with a socket of female contact. It is normally connected to the "dead" side of a circuit.

pincushion distortion — 1. Distortion that results in a monotonic increase in radial magnification in the reproduced image away from the axis of symmetry of the electron optical system. The four sides of the raster are curved inward, leaving the corners extending outward. 2. Distortion in a television picture that makes all sides appear to bulge inward. 3. A form of geometrical distortion in a CRT image that results in the picture resembling a pincushion rather than a well-defined rectangle. Some monitors include controls to correct this effect.

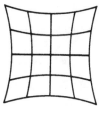

*Pincushion
distortion.*

pincushioning — A video-display defect in which the image appears to bend inward toward the middle of the screen.

pin diode or **PIN diode** — 1. A diode made by diffusing the semiconductor with p dopant from one side and n dopant from the opposite side, with the process so controlled that a thin intrinsic (i) region separates the p and n regions. The storage time of the pin diode is long enough that it cannot rectify at microwave frequencies. Instead, it behaves as a variable resistor with its value controlled by a dc bias current. Therefore, this type of diode is well suited for use as a variable microwave attenuator. 2. A pn silicon junction with a layer of intrinsic or high-sensitivity silicon between the p and n regions. When forward biased, the pin diode behaves as a resistance from dc to microwave frequencies. The value of the resistance depends on the forward current. Under reverse bias, the resistance is very high. The pin diode may be used as an electrically variable resistance or attenuator. Its uses at microwave frequencies are for switching and modulating microwave signals and as an electrically variable attenuator. It can perform switching in a few nanoseconds and handle peak powers of hundreds of watts. 3. A current-controlled resistor at radio and microwave frequencies. A silicon semiconductor diode in which a high-resistivity intrinsic i-region is sandwiched between a p-type and n-type region. When the pin diode is forward biased, holes and electrons are injected into the i-region. These charges do not immediately annihilate each other; instead, they stay alive for an average time called the carrier lifetime. This results in an average stored charge, Q, that lowers the effective resistance of the i-region. When the pin diode is at zero or reverse bias, there is no stored charge in the i-region and the diode appears as a capacitor, shunted by a parallel resistance. 4. A diode that has an undoped semiconductor region, referred to as an intrinsic layer, between the n and p doped layers. The intrinsic layer provides low junction capacitance that changes very little as reverse bias is changed. Used mostly as a microwave switch or attenuator.

pin-diode attenuator — A two-port network consisting of two or more pin diodes controlled by a driver circuit. At microwave frequencies, the diodes act as a small value of capacitance shunted by a resistance that can be varied over a range of about 2 to 10,000 ohms through control of the bias current by the driver circuit.

pine-tree array — An array of dipole antennas aligned vertically (termed the radiating curtain), behind which and approximately a quarter wavelength away is a parallel array of dipole antennas forming the reflecting curtain.

pi network — 1. A network composed of three branches, all connected in series with each other to form a mesh. The three junction points form an input terminal, an output terminal, and a common input and output terminal. 2. A network consisting of three impedance elements, one bridged across the input, one bridged across the output, and the third in series between one input terminal and one output terminal. 3. A tuned circuit at the output stage of a transmitter to match it to the antenna or feedline.

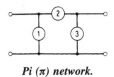

Pi (π) network.

pinfeed platen — In a computer, a cylindrical platen having integral rings of pins that engage perforated holes in the paper, thus permitting feeding of the paper.

ping — A sonic or ultrasonic pulse of predetermined width.

ping-pong — A programming technique in which two magnetic tape units are used for multiple-reel files. Automatic switching between the two units is carried out until the entire file is processed.

ping-pong-ball effect — The bouncing of sound back and forth between the two sides of a stereophonic reproducing system.

pinhole — 1. A small opening, occurring as an imperfection, extending through the thickness of a substance. 2. A minute hole through a layer or pattern. 3. Small hole occurring as an imperfection that penetrates entirely through an applied material to the substrate (e.g., holes in semiconductor insulating oxides, screened resistors, thin-film elements).

pinhole detector — A photoelectric device that detects extremely small holes and other defects in moving sheets of material, and often actuates sorting equipment that automatically rejects defective sheets.

pinion — Of two gears that mesh, the one with the fewer teeth.

pin jack — A single-conductor jack having a small opening into which a plug tipped with a metal pin can be inserted.

pink noise — 1. Noise whose amplitude is inversely proportional to frequency over a specified range. Equal energy distribution occurs in any octave bandwidth within that range. Pink noise is very pleasing to the human ear. Many people feel relaxed listening to the patter of rain (a close approximation of pink noise). Other examples include the sound of surf and a shower stream. 2. A complete mixture of all frequencies in one signal characterized by equal energy per octave. This means that there is an equal amount of energy between 20 and 40 Hz, 80 and 160 Hz, or 620 and 1240 Hz. 3. Random noise modified to have an equal amount of energy in each octave. The octave from 10,000 to 20,000 Hz occupies a 10,000-Hz bandwidth, whereas the octave from 5000 Hz to 10,000 Hz occupies only 5000 Hz bandwidth, but due to a 6-dB/octave roll-off, both octaves contain equal levels of pink noise energy. 4. Random electrical noise that has constant energy per octave. Pink noise can be produced from white noise by putting it through a filter that has a slope of −3 dB per octave.

PINO — Acronym for positive input, negative output.

pinout — 1. A list or diagram that shows how the individual wires in a cable or connector are used. 2. A diagram showing connections to an integrated circuit or transistor.

pinouts — The external wires or pins on a module (generally having a circuit function).

pins — The leads emerging from an IC package.

pin sensing — A process in which a device using a punched card generates digital data by sensing the opening and closing of switches.

pip — *See* blip, 1.

PIP — Abbreviation for Picture-in-Picture. A feature available in some TV receivers that makes it possible to watch one picture on the screen, while at the same time keeping track of another program in the PIP "window." The two images are readily switched back and forth.

piped program — A program transmitted over telephone wires, usually from one studio to another.

pipeline — A processor design approach whereby instruction execution takes place in a series of units arranged so that several units can be simultaneously processing the appropriate parts of several instructions.

pipeline computers — Computers that execute serial programs only.

pipelining — 1. A hardware arrangement that permits different sections of a bit-slice processor to work simultaneously instead of sequentially, thus speeding up processing. 2. Beginning one instruction sequence before another has been completed. Once a technique used on supercomputers, pipelining is now used to speed execution on machines of all sizes. 3. A computer design technique in which execution of a new task is started before the preceding task is completed.

pip-matching display — A navigational display in which the received signal appears as a pair of blips. The desired quantity is measured by comparing the characteristics.

pi point — The frequency at which the insertion phase shift of an electric structure is 180° or an integral multiple of 180°.

piracy — Illegal copying of diskettes or other copyright material that is then offered for sale.

Pirani gage — A bolometric vacuum gage for measuring pressure. Its operation depends on the thermal conduction of the gas present. The pressure being measured is a function of the resistance of a heated filament, ordinarily over a range of 10^{-1} to 10^{-4} mm Hg.

pirate — An unlicensed, unauthorized, and illegal broadcasting station.

piston — Also called a plunger. In high-frequency communications, a conducting plate that can be moved along the inside of an enclosed transmission path to short out high-frequency currents.

piston action — The movement of a speaker cone or diaphragm when driven at the bass audio frequencies.

piston attenuator — An attenuator, generally used at microwave frequencies, whose amount of attenuation can be varied by moving an output coupling device along its longitudinal axis.

pistonphone — A small chamber equipped with a reciprocating piston of measurable displacement. In this way, a known sound pressure can be established in the chamber.

PIT — Abbreviation for programmable interval timer. An IC chip with a separate clock and several registers, used to count time independently of the MPU, for real-time applications. At the end of a time period, it sets a flag or generates an interrupt, or merely stores the time elapsed.

pitch — 1. That attribute of auditory sensation by which sounds may be ordered on a scale extending from low to high (e.g., a musical scale). 2. The distance between two adjacent corresponding threads of a screw measured parallel to the axis. 3. The distance between the peaks of two successive grooves of a disc recording. 4. A term applied to a musical tone that is used as a standard for tuning, singing, etc. Standard U.S. and European pitch is based on A = 440 Hz. When the pitch is raised one octave, the frequency is twice the original.

pitch control — 1. A circuit that permits the speed of a tape transport's motor to be varied slightly to raise and lower the musical pitch of the recording or to slightly lengthen or shorten playing time. 2. A circuit that permits a turntable's speed to be varied slightly to raise and lower the musical pitch of the recording being played (hence, the name) or to slightly lengthen or shorten playing time.

pits — 1. Small holes occurring as imperfections that do not penetrate entirely through the printed element. 2. Depressions produced in metal or ceramic surfaces by nonuniform deposition.

pitted contact — An electrical contact that has numerous discrete hollows in its surface.

PIV — Abbreviation for peak inverse voltage.

pivot — A low-friction bearing in the support of a tonearm that allows it freedom of movement in vertical and horizontal planes. In lower-priced tonearms, it may be a simple point-in-cup pivot. More expensive tonearms usually have precision ball bearings or knife-edge pivots.

pivot and jewel — A method of suspending the moving coil or moving iron vane of a meter in a magnetic field. A glass jewel and steel pivot.

PIV rating — *See* PRV rating.

pixel — Contraction of picture element. 1. A spatial resolution element. It is the smallest distinguishable and resolvable area in an image, as, for example, displayed on a CRT monitor. It can also describe the smallest distinguishable variation over time in a signal sequence. (The term pixel is not, strictly speaking, applicable to an analog image, but it is sometimes equated to limiting resolution. In general, however, actual pixel resolution is less than limiting resolution.) 2. The smallest controllable picture element in a digital video image. The CRT screen is divided into a rectangular grid of pixels of the same size and shape. In the vertical direction, the highest picture resolution occurs when each pixel is one scan line high. Horizontal resolution is typically limited by the speed at which the CRT electron gun can switch on and off during a horizontal scan. For square pixels, which are normally more desirable than horizontally elongated rectangles, the smallest possible pixel size is usually determined by this horizontal resolution limit. Normally, a square pixel is two or more scan lines high. (In North America, the standard television picture is normally 525 lines high, though partly cropped.) 3. Picture element or picture cell. A term used to describe the information contained in one unit of display surface. (For example, a horizontal resolution of 1700 pixels per line.) 4. The smallest picture element, made up of tricolor phosphor cells. In raster-scan systems the computer divides the screen into such points, whose number depends on the resolution selected. 5. A small element of a scene, or picture element, in which an average brightness value is determined and used to represent that portion of the scene. Pixels are arranged in a rectangular array to form a complete image of the scene. 6. The smallest unit of a video display, usually too small to be detected by the naked eye. Pixels, illuminated dots of glowing phosphor, can vary in size and shape depending on the monitor and the graphics mode. Display quality is often measured by pixel resolution — the number of dots on the screen measured in width-by-height fashion. 7. Smallest independently addressable display area; one full-color pixel requires three phosphor dots, one for each primary color.

PLA — Abbreviation for programmable logic array.

place — *See* column.

plaintext — A message in ordinary language, such as English.

planar — 1. Lying essentially in a single plane. 2. Constructed in layers or planes. 3. A semiconductor fabrication technique in which the semiconductor device chips are protected by an oxide passivation layer throughout the various stages of fabrication. The planar process thus represents a synthesis of the separate oxide-layer functions involved in the photolithographic etching of diffusion masking and in chip passivation.

planar ceramic tube — An electron tube constructed with parallel planar electrodes and a ceramic envelope.

planar devices — *See* planar process.

planar diffusion — Technique used to manufacture semiconductors having diffused pn junctions. All the junctions emerge at the top surface of the wafer.

planar diode — A diode containing planar electrodes lying in parallel planes.

planar display — A display in which the light-emitting segments or elements are all mounted in a single plane.

planar mask — A shadow or aperture mask that has no curvature; one that is perfectly flat.

planar module — A packaged module wherein the individual components are positioned and terminated flat or parallel with the plane of the substrate.

planar network — A network in which no branches cross when drawn on the same plane.

planar process — 1. The technology used in fabricating semiconductor devices wherein all pn junctions terminate in the same geometric plane. An oxide is formed at the surface for the purpose of stabilizing the parameters (passivating). 2. Semiconductor fabrication technology that uses silicon dioxide as a masking agent and produces components on a single plane.

planar silicon photoswitch — Abbreviated PSPS. Essentially a complementary silicon-controlled rectifier. Like the LASCR, it can be triggered by light. In addition, a negative signal (with reference to the anode) at the anode gate terminal can trigger the device.

planar soldering — A soldering method in which the printed circuit assembly is held loosely in a carrier. This freedom of movement allows the printed circuit assembly to float on the still surface of the solder bath, equalizing the thermogradient throughout the entire assembly.

planar technique — The formation of p-type and/or n-type regions in a semiconductor crystal by diffusing impurity atoms into the crystal through holes in an oxide mask, which is on the surface. The latter is left to protect the junctions so formed against surface contamination.

planar transistor — 1. A diffused transistor in which the emitter, base, and collector regions come to the same plane surface. Their junctions are protected by a material such as silicon oxide. The manufacturing process consists of an oxide-masking technique in which the

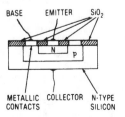

Planar transistor.

silicon oxide is formed by adding oxygen or water vapor to the atmosphere of a diffusion furnace. The thickness of the oxide layer is a function of time, temperature, and the amount of oxidizing agent. 2. A junction transistor manufactured by a process in which the surface of a chip is passivated with a thin film of oxide, dopants being introduced by successive etching and diffusion.

planchet — A small metal container or sample holder for radioactive materials undergoing radiation measurements in a proportional counter or scintillation detector.

Planckian locus — A line drawn on a chromaticity diagram to represent light radiation from a reference blackbody at 2000 to 10,000 kelvins (K).

Planck's constant — Symbolized by h. The constant representing the ratio of the energy of any radiation quantum to its frequency. It has the dimension of action (energy $\times$ *time*) and a numerical value of 6.547×10^{-27} erg-second. Its significance was first recognized by the German physicist Max Planck in 1900.

Planck's distribution — An equation that describes the entire distribution of energy radiated from a blackbody as a function of wavelength, based on quantum mechanics.

Planck's radiation law — An expression representing the spectral radiance of a blackbody as a function of the wavelength and temperature.

plane — 1. A screen of magnetic cores. Planes are combined to form stacks. 2. A surface such that a straight line that joins any two of its points lies entirely in that surface.

plane earth — Earth that is considered to be a plane surface. Used in ground-wave calculations.

plane-earth attenuation — Attenuation of an electromagnetic wave over an imperfectly conducting plane earth in excess of that over a perfectly conducting plane.

plane-earth factor — Ratio of the electric field strength that would result from propagation over an imperfectly conducting plane earth to that over a perfectly conducting plane.

plane of a loop — An infinite imaginary plane that passes through the center of a loop and is parallel to its wires.

plane of polarization — For a plane-polarized wave, the plane containing the electric field vector and the direction of propagation.

plane-polarized wave — At any point in a homogeneous isotropic medium, an electromagnetic wave with an electric field vector that at all times lies in a fixed plane containing the direction of propagation.

planetary electron — One of the electrons moving in an orbit or shell around the nucleus of an atom.

plane wave — A wave in which the wavefronts are everywhere parallel planes normal to the direction of propagation.

plan-position indicator — Abbreviated PPI. Also called P-display. A type of presentation on a radar indicator. The signal appears as a bright spot, with range indicated by the distance of the spot from the center of the screen, and the bearing indicated by the radial angle of the spot.

plasma — 1. A wholly or partially ionized gas in which the positive ions and negative electrons are roughly equal in number. Hence, the space charge is essentially zero. 2. The region in which gaseous conduction takes place between the cathode and anode of an electric arc. 3. An electrically conductive gas comprised of neutral particles, ionized particles, and free electrons but which, taken as a whole, is electrically neutral. 4. A gas at an extremely high (20,000 kelvins) temperature and completely ionized. It is therefore conductive and affected

by magnetic fields. Plasma is sometimes referred to as the fourth state of matter.

plasma-cathode electron gun — An electron-beam gun in which plasma that is generated within a low-voltage hollow-cathode discharge serves as the source of electrons.

plasma deposition — The spraying of highly excited atomic particles onto a surface. Heat, commonly obtained by electric arc or hydrogen flame, is the source of excitation. Almost any elementary material can be applied to almost any surface by this method. Disadvantages are grossness of the spray process and high temperatures.

plasma diode — A thermodynamic engine with electrons as the working fluid, the potential energy of which is converted to a useful output.

plasma display — A display in which the emitted light is produced by ionized gas.

plasma engine — A reaction engine using electrically accelerated plasma as the propellant.

plasma etching — An etching process using a cloud of ionized gas as the etchant.

plasma frequency — A natural frequency for coherent electron motion in a plasma.

plasma jet — A high-temperature stream of electrons and positive ions produced by the magnetohydrodynamic effect of a strong electrical discharge.

plasma laser — A laser that operates with light collectively emitted by free electrons in the plasma state. This plasma is a dense beam of electrons traveling at speeds approaching the speed of light.

plasma length — *See* Debye length.

plasma oscillation — Electrostatic or space-charge oscillations in a plasma that are closely related to the plasma frequency. There is usually enough damping due to electron collisions to prevent self-generation of the oscillations. They can be excited, however, by such techniques as shooting a modulated electron beam through the plasma.

plasma physics — The study of highly ionized gases. Many phenomena not exhibited by uncharged gases are associated with plasma physics.

plasma sheath — An envelope of ionized gas that surrounds an object moving through an atmosphere at a hypersonc velocity. The plasma sheath affects radio-wave transmission, reception, and diffraction.

plasma thermocouple — An electronic device in which the heat from nuclear fission is converted directly into electric power.

plasmatron — A helium-filled current amplifier that combines the grid-control characteristics and linearity of a vacuum triode with the extremely low internal impedance of a thyratron.

plastic-clad silica fiber — A fiber composed of a silica glass core with a transparent plastic cladding.

plasticizer — A substance added to a plastic to produce softness and adhesiveness in the finished product.

PLAT — Acronym for pilot landing aid television. A system in which television cameras cover aircraft landings on a carrier from several angles, allowing the landing personnel to talk the pilot down with increased precision. Recordings can be made for future reference.

plate — Preferably called the anode. 1. The principal electrode to which the electron stream is attracted in an electron tube. 2. One of the conductive electrodes in a capacitor. 3. One of the electrodes in a storage battery. 4. *See* printed circuit board.

plateau — In the counting-rate-versus-voltage characteristic of a radiation counter tube, that portion in which the counting rate is substantially independent of the applied voltage.

plateau length — The applied-voltage range over which the plateau of a radiation counter tube extends.

plate-bypass capacitor — A capacitor connected between the plate and cathode of a vacuum tube to bypass high-frequency currents and, thus, keep them out of the load. *See also* anode-bypass capacitor.

plate characteristic — A graph showing how changes in plate voltage affect the plate current of a vacuum tube.

plate circuit — The complete external electrical circuit between the plate and cathode of an electron tube.

plate-circuit detector — A detector that functions by virtue of its nonlinear plate-current characteristic.

plate conductance — The in-phase component of the alternating plate current divided by the alternating plate voltage, all other electrode voltages being maintained constant.

plate current — Electron flow from the cathode to the plate inside an electron tube.

plate-current saturation — An occurrence in a vacuum tube when the plate overcomes any other charge in the tube.

plate detection — The operation of a vacuum-tube detector at or near plate-current cutoff so that the input signal is rectified in the plate circuit.

plate dissipation — The amount of power lost as heat in the plate of a vacuum tube.

plated resist — A material electroplated on conductive areas to make them impervious to etching.

plated-through hole — 1. A connection between upper and lower conductive patterns formed when metal is deposited on the walls of a hole in a double-sided or multilayer printed-circuit board. Through-board continuity is thus established. 2. A hole formed by deposition of metal on the sides of the hole and on both sides of a printed circuit board to provide electrical connection from the conductive pattern on one side to that on the opposite side of the printed circuit board.

plated-wire memory — A memory consisting of wires that are coated with a magnetic material. The magnetic material may be magnetized in either of two directions to represent 1s and 0s.

plate efficiency — Also called the anode efficiency. Ratio of load-circuit power (alternating current) to plate power input (direct current).

plate impedance — Also called plate-load impedance.

plate-input power — In the last stage of a transmitter, the direct plate voltage applied to the tubes times the total direct current flowing to their plates, measured without modulation.

plate keying — Keying done by interrupting the plate supply circuit.

plate-load impedance — Also called the anode-load impedance and plate impedance. The total impedance between the anode and cathode of a vacuum tube, exclusive of the electron stream.

plate modulation — Also called anode modulation. Modulation produced by applying the modulating voltage to the plate of any tube in which the carrier is present.

platen — 1. A backing structure (usually cylindrical) against which a printing mechanism strikes in producing an impression. 2. In a printer, a metal plate or roller that forms a surface for the striking mechanism.

plate neutralization — Also called anode neutralization. Neutralizing an amplifier by shifting a portion of the plate-to-cathode ac voltage 180° and applying it to the grid-to-cathode circuit through a capacitor.

plate power input — Also called the anode power input. The dc power (mean anode voltage times current) delivered to the plate (anode) of a vacuum tube.

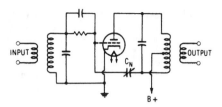
Plate neutralization.

plate power supply — *See* anode power supply.

plate pulse modulation — Also called anode pulse modulation. Modulation produced in an amplifier or oscillator by applying externally generated pulses to the plate circuit.

plate resistance — The plate-voltage change divided by the resultant plate-current change in a vacuum tube, all other conditions being fixed.

plate saturation — Also called anode or voltage saturation. The point at which the plate current of a vacuum tube no longer increases as the plate voltage does.

plate supply — *See* anode supply.

plate-to-plate impedance — The load impedance between the two plates in a push-pull amplifier stage.

plate voltage — The dc voltage between the plate and cathode of a vacuum tube.

plate winding — A transformer winding connected to the plate circuit of a vacuum tube.

plating — 1. The deposition of a metal layer on a substrate surface by electrolytical or certain chemical means. The materials include gold, copper, solder, etc. The functions of the metal plate vary, including corrosion protection, solderability enhancement, etch resist, bonding for lead frames, and electrical connection, among others. 2. The overlaying of a thin coating of metal or metallic components to improve conductivity, provide for easy soldering, or prevent rusting or corrosion. 3. The application of one metal over another by electrolysis.

plating anode — Usually, a pure form of the metal being plated. The workpiece being plated is the cathode.

plating resist — Any material that when deposited on conductive areas prevents plating of the areas it covers.

plating up — The process consisting of the electrochemical deposition of a conductive material on the base material (surface, holes, etc.) after the base material has been made conductive.

plating void — The area of absence of a particular metal from a specific cross-sectional area.

platinotron — A cross-field vacuum tube used to generate and amplify microwave energy. It resembles the magnetron, except that it has no resonant circuit and has two external rf connections instead of only one.

platinum — A heavy, almost white metal that resists practically all acids and is capable of withstanding high temperatures.

platinum contacts — Used where currents must be broken frequently (e.g., in induction coils and electric bells). Sparking does not damage platinum as much as it does other metals. Hence, a cleaner contact is assured with minimum attention.

platter — 1. A popular term for phonograph records and transcriptions. 2. The flat disc that supports the record and is turned by a motor at a constant speed. Usually machined from a nonferrous alloy but is sometimes a lightweight stamped or pressed disc.

playback — The reproduction of a tape recording or disc through an amplifier and speaker or phones.

playback head—1. The magnetic assembly on a tape recorder that responds to the recorded pattern on the tape and develops a signal representing that pattern to feed to the preamplifier. In some tape machines, the playback and recording head are the same device; in others they are separate units. 2. Magnetic head used to pick up a signal from a tape.

playback loss—*See* translation loss.

player—A software application that permits viewing or playback of content such as audio, video, or movie files in primarily a linear fashion, although random indexing or control of playback may be permitted. As opposed to a browser. *See also* viewer.

playing weight—1. Downward force of a pickup on a record. Sometimes called stylus pressure. 2. The downward force required on the pickup stylus to keep it in the groove and to counter the mechanical reactions of replay.

PLCC—Abbreviation for plastic leaded chip carrier. A leaded quad package—a replacement for the plastic DIP (dual in-line package) in surface-mount applications. External connections consist of leads around all four sides of the package.

PLD—Abbreviation for programmable logic device. A semiconductor device containing transistors that can be interconnected electronically by users to perform various logic functions. *See also* FPLA.

plethysmograph—An instrument for detecting variations of blood volume in the tissues during the cardiac cycle. *See also* electrical-impedance cephalography; finger plethysmograph.

pliers—1. A small pair of pincers. 2. An instrument having two short handles extended into pivoted jaws suitable for grasping or cutting.

pliodynatron—A four-element vacuum tube with an additional grid, which is maintained at a higher voltage than the plate to obtain negative-resistance characteristics.

pliotron—An industrial-electronic term for a hot-cathode vacuum tube having one or more grids.

PLL or pll—Abbreviation for phase-locked loop. 1. A circuit containing a voltage-controlled oscillator whose output phase or frequency can be steered to keep it in sync with a reference source. A PLL circuit is generally used to lock onto and up-convert the frequency of a stable source. 2. An electronic circuit that consists of a phase detector, low-pass filter, and voltage-controlled oscillator. A PLL can be used as an FSK demodulator or to synchronize a terminal's internal clock to the received bit stream.

plot—*See* print.

plotter—1. A device that produces an inscribed visual display of the variation of a dependent variable as a function of one or more other variables. 2. A device for presenting computer output in graphical form instead of a printed listing. 3. A visual display or board in which a dependent variable is graphed by an automatically controlled pen or pencil as a function of one or more variables. 4. A device used to make a permanent copy of a display image. 5. An output device that provides data in pictorial form. A pen controlled by two motors moves in the *x* and *y* directions, drawing a picture that is defined in terms of *x* and *y* coordinates.

plotting—The practice of mechanically converting *x*, *y* positional information into a visual pattern, such as artwork.

plotting board—A device that plots one or more variables against one or more other variables.

ploy effect—In surface-channel charge-coupled devices (CCDs), the tendency for charges to be captured by surface effects, thus resulting in a loss of signal. By continuously introducing a charge into all CCD channels through a diffusion at the beginning of the channel, the areas that trap charges are filled by the induced charges rather than the signal charges, thus increasing transfer efficiency.

PLT—Abbreviation for power-line transient—one kind of conducted noise, generally caused by switching inductive loads measured on the power line.

plug—1. The part of the two mating halves of a connector that is free to move when not fastened to the mating half. The plug is usually thought of as the male portion of the connector. This is not always the case. The plug may have female contacts if it is the free-to-move member. *See also* connector. 2. Also called plugging and plug reverse. A method of braking a motor by applying partial or full rated voltage in reverse in an attempt to quickly bring the motor to zero speed.

plugboard—In a computer, a removable board having many electric terminals into which connecting cords may be plugged in patterns varying for different programs. To change the program, one wired plugboard is replaced by another.

plugboard computer—A computer that has a punch-card input and output, and to which program instructions are delivered by means of interconnecting patch cords on a removable plugboard.

plug braking—A method of braking an electric vehicle in which the kinetic energy of the vehicle is dissipated as heat, either in a traction motor or a special resistor.

plug-compatible—A term used to indicate when devices may be effectively interchanged without any modifications.

plug connector—An electrical fitting containing male, female, or male and female contacts and constructed so that it can be affixed to the end of a cable, conduit, coaxial line, cord, or wire for convenience in joining with another electrical connector or connectors. It is not designed for mounting on a bulkhead, chassis, or panel.

plug fuse—A fuse of small rating (5 to 30 amperes) with a screw thread like that on an electric lamp base; used in a standard screw receptacle.

pluggable unit—A chassis that can be removed from or inserted into the rest of the equipment by merely plugging in or pulling out a plug.

plugging—*See* plug, 2.

plug-in—1. Any device to which connections can be completed through pins, plugs, jacks, sockets, receptacles, or other ready connectors. 2. A small software program that plugs into a larger application to provide added functionality.

plug-in coil—A coil that can be easily interchanged and used for varying the tuning range of a receiver or transmitter. It is wound around a form often resembling an elongated tube base, with the coil leads connected to pins on the base.

plug-in device—A component or group of components and their circuitry that can be easily installed or removed from the equipment. Electrical connections are made by mating contacts.

plug-in unit—A standard subassembly of components that can be readily plugged into or pulled out of a circuit as a unit.

plug reverse—*See* plug, 2.

Plumbicon—A vidicon with a lead-oxide target; its major advantage is its lack of image retention. It is a tube with the simplicity of a vidicon, and the sensitivity and lag of a glass target image orthicon. The tube is used for live black-and-white and color broadcasting. Trademark of N. V. Philips of Holland.

plumbing—Coaxial lines or waveguides and accessory equipment for transmission of radio-frequency energy.

plume— The hot gaseous material ejected briefly from a highly absorbent material after bombardment by an intense laser pulse. The plume emits broadband white light and is the most prominent feature in most irradiation experiments.

plunger— See piston.

plunger relay— A relay consisting of a movable core or plunger surrounded by a coil. Solenoid action causes the plunger or core to move and, thus, energize the relay whenever current flows through the coil.

plunger-type instrument— A moving-iron instrument for measuring current. It consists of a pointer attached to a plunger inside a coil. The current being measured flows through the coil and pulls the plunger down. How far it goes into the coil depends on the magnitude of the current.

plutonium— A heavy element that undergoes fission when bombarded by neutrons. It is a useful fuel in nuclear reactors. Its symbol is Pu; its atomic number, 94.

pm— Abbreviation for phase modulation or permanent magnet.

pm erasing head— A head that uses the fields of one or more permanent magnets for erasing.

PMOS— 1. P-channel metal-oxide semiconductor (MOS) having p-type source and drain regions diffused into an n-type substrate to create a p channel for conduction. 2. MOS devices made on an n-type silicon substrate in which the active carriers are holes (p) flowing between p-type source and drain controls. 3. An MOS (unipolar) transistor in which the working current consists of positive (p) electrical charges. 4. Pertaining to MOS devices made on n-type substrates in which the active carriers flow between p-type source and drain contacts. The n-type channel inverts to p-type at the surface with the application of the proper voltage to the gate terminal. 5. A type of metal-oxide silicon field-effect transistor using holes to conduct current in the semiconductor channel. The channel has a predominantly positive charge.

pm speaker— Abbreviation for permanent-magnet speaker.

pn boundary— The surface where the donor and acceptor concentrations are equal in the transition region between p- and n-type materials.

PNdB— Perceived noise level expressed in decibels. See perceived noise level.

pn diode— A diode that has no intrinsic region and a short storage time. It functions as a normal diode rectifier into the high microwave frequencies. If the diode is given a dc bias that is large compared to the rf signal, it ceases to be a rectifier; thus, it can be used as a reflective microwave switch. It also can be employed as a variable reflective attenuator, except in that operating region for which the bias and rf voltages are comparable and rectification occurs.

pneumatic bellows— A gas-filled bellows sometimes used to provide delay time in plunger-type relays.

pneumatic robot— Programmable machine that usually employs vane motors, and often combines them with cylinders. Both the motors and the cylinders may be standard components that can be serviced on the robot as on any other machine. Each such component represents a degree of freedom, i.e., a way in which the robot can move, either by extending or retracting a cylinder or by rotating a motor-driven joint.

pneumatic speaker— A speaker in which the acoustic output is produced by controlled variation of an air stream.

pn hook transistor— Also called hook transistor or hook collector transistor. A junction transistor that secures increased current amplification by means of an extra pn junction.

pnin— A transistor in which an intrinsic region is between two n-regions.

pnip transistor— A pnp transistor in which a layer of high-purity germanium has been placed between the base and collector to extend the frequency range. When the same process is applied to an npn transistor, the resulting device is called an npin transistor.

pn junction— 1. The region of transition between p-type and n-type material in a single semiconductor crystal. 2. The boundary surface between p-type and n-type materials. 3. A relatively abrupt transition between p-type and n-type semiconductor regions within a crystal lattice. Such a junction possesses unique electrical properties, including the ability to conduct substantially in only one direction. Single and multiple pn junctions form the basis for most semiconductor devices. These junctions are fundamental to the performance of switching, rectification, and amplification functions in electronic devices and circuits.

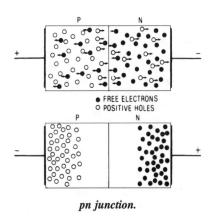

pn junction.

pn-junction laser— See injection laser.

pn-junction luminescence— Discharge that results when a doped semiconductor crystal with a pn junction is charged with a low-voltage direct current. The usual process depends on excitation caused by electrical energy absorption and recombination where release of the absorbed energy occurs.

pnpn diode— Also called four-layer diode. A semiconductor device that may be regarded as a two-transistor structure with two separate emitters feeding a common collector. This combination constitutes a feedback loop that is unstable for loop gains greater than unity. The instability results in a current that increases until ohmic circuit resistances limit the maximum value. This gives rise to a negative-resistance region that may be utilized for switching or for waveform generation.

pnpn transistor— A hook transistor with an n-type base, p-type emitter, and a hook collector. The electrodes are connected to the four end layers of the n- and p-type semiconductor materials.

pnpn-type switch— A bistable semiconductor device made up of three or more junctions, at least one of which is able to switch between reverse and forward voltage polarity within a single quadrant of the anode-to-cathode voltage-current characteristic.

pnp transistor— A transistor consisting of two p-type regions separated by an n-type region. When a small forward bias is applied to the first junction and a large reverse bias to the second junction, the system behaves much like a vacuum-tube triode.

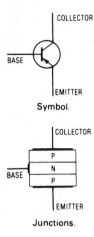

Symbol.

Junctions.

pnp transistor.

Pockel's effect — The alternation in the refractive properties of a transparent piezoelectric crystal by the application of an electric field. *See also* modulator crystal.

Pockel's-effect modulation — A phenomenon that occurs when a transparent dielectric is a piezoelectric crystal. The crystal tends to strain whenever an electric field is applied, rotating the plane of polarization of the incident wave. Some 7500 V/m causes a 90° rotation of light.

poid — The curve traced by the center of a sphere when it rolls or slides over a surface having a sinusoidal profile.

point — Called the binary point in binary notation, and the decimal point in decimal notation. In positional notation, the character or location of an implied symbol that separates the integral part of a number from its fractional part.

point availability — The percent of time an equipment is available for use when an operator requires it.

point-based linearity — Nonlinearity expressed as the deviation from a straight line that passes through a given point or points.

point contact — A pressure contact between a semiconductor body and a metallic point.

point-contact crystal diode — A crystal diode whose rectifying activity is determined by the touching of the crystal to a finely pointed wire surrounded by a material of opposite type.

point-contact diode — 1. A diode that consists of a semiconductor against which the end of a fine wire (cat whisker) is pressed. Such a diode has a very low reactance and can be used as a detector or mixer over most of the microwave range. It has a square-law response at low power levels. 2. Device consisting of a metal whisker making pressure contact with the semiconductor chip, normally tungsten for silicon and phosphorus bronze for germanium and gallium arsenide. Point-contact diodes are generally encapsulated in axial lead glass, axial prong ceramic, cartridge-type ceramic, or metal coaxial enclosures. The electrical characteristics of the device are determined by the size, shape, and pressure of the whisker and the thickness and resistivity of the epitaxial layer.

point-contact transistor — A transistor having a base electrode and two or more point-contact electrodes.

point defect — An imperfection caused by the presence of an extra atom or the absence of an atom from its proper place in the crystal.

point effect — The phenomenon whereby a discharge will occur more readily at sharp points than elsewhere on an object or electrode.

pointer — Also called a needle. 1. A slender rod that moves over the scale of a meter. 2. Registers in a CPU that contain memory addresses. *See also* data pointer; program counter.

pointer address — The address of a core-memory location that contains the actual effective address.

pointer register — A register that contains the absolute address of an item of data in its memory. Data can be accessed at this address or relative to it via the pointer register. The value of the pointer register can be updated to access a different block of data, where the data can be one or several bytes.

point impedance — Ratio of the maximum E-field to the maximum H-field observed at a given point in a waveguide or transmission line.

pointing — A method of allowing a nontypist operator to enter data items. A menu of items is displayed on the screen; the operator chooses one by pointing at it with a system device, such as a lightpen, stylus, or even the terminal's cursor.

pointing and flying — The method of navigating through virtual reality when wearing a virtual reality glove by pointing and then "flying" in that indicated direction.

point-junction transistor — A transistor having a base electrode and both point-contact and junction electrodes.

point-of-sale terminal — Abbreviated POS terminal. 1. An intelligent input/output device that is used to capture data in retail stores, i.e., supermarkets or department stores. POS is a term used to indicate that data regarding a sale is entered directly into the computerized system without having to be converted to another form first. 2. Electronic terminal that can serve as a conventional cash register but has the capacity to capture sales data and store or transmit it to a computer.

point-plane rectifier — *See* glow-tube rectifier.

point source — 1. A radiation source whose dimensions are small compared with the distance from which it is observed. 2. Radiation source whose maximum dimension is less than 1/10 the distance between source and receiver.

point-to-point — 1. Describing communication between two fixed stations. 2. A limited network configuration with communication between two terminal points only, as opposed to multipoint and multidrop.

point-to-point network — A communications network consisting of a single communications link that connects two terminals and is not shared by other terminals.

point-to-point radio communication — Radio communication between two fixed stations.

point-to-point transmission — Direct transmission of data between two points without using an intermediate terminal or computer.

point-to-point wiring — 1. A method of forming circuit paths by connecting the various devices, components, modules, etc., with individual pieces of wire or ribbon. May be soldered, welded, or attached by other means. 2. Wiring done in a direct path from one point to another without dressing wiring in parallel runs. Crosstalk is thus reduced. Used for high-speed logic panels.

point transposition — Transposition, usually in an open-wire line, that is executed within a distance comparable to the wire separation, without material distortion of the normal wire configuration outside this distance.

Poisson distribution — A statistical distribution similar to the normal distribution except that the standard deviation is equivalent to the square root of the mean.

Poisson's ratio—The ratio of the lateral strain to the longitudinal strain in a specimen subjected to a longitudinal stress.

polar—Pertaining to, measured from, or having a pole (e.g., the poles of the earth or a magnet).

polar capacitor—A capacitor intended for use with a direct voltage connected according to the polarity indicated on the terminations.

polar circuit—A teletypewriter circuit in which current flows in one direction on a marking impulse and in the opposite direction during a spacing impulse.

polar coordinates—A system of coordinates in which a point is located by its distance and direction (angle) from a fixed point on a reference line (called the polar axis).

polar crystals—Crystals having a lattice composed of alternate positive and negative ions.

polar diagram—A diagram in which the magnitude of quantity is shown by polar coordinates.

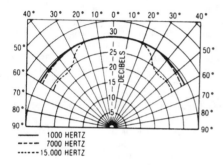

Polar diagram.

polar grid—A type of circular grid on which range and azimuth are represented from a central reference point.

polarimetry—The measurement of the rotation of plane of polarization of radiant energy.

polarity—1. A condition by which the direction of current can be determined in an electrical circuit (usually batteries and other direct-voltage sources). 2. Having two opposite charges, one positive and one negative. 3. Having two opposite magnetic poles, one north and the other south. 4. The condition of positiveness or negativeness in an electrical circuit. 5. The positive and negative orientation of a signal or power source. 6. Any condition in which there are two opposing voltage levels or charges, such as positive and negative.

polarity of picture signal—Stated as black negative or black positive. The particular potential state of a portion of the signal representing a dark area of a scene relative to the potential representing a light area.

polarization—1. The process of making light or other radiation vibrate perpendicular to the ray. The vibrations are straight lines, circles, or ellipses — giving plane, circular, or elliptical polarization, respectively. 2. The increased resistance of an electrolytic cell as the potential of an electrode changes during electrolysis. In dry cells, this shortens their useful life. 3. The slight displacement of the positive charge in each atom whenever a dielectric is placed into an electric field. 4. The magnetic orientation of molecules in a piece of iron or other magnetizable material placed in a magnetic field, whereby the tiny internal magnets tend to line up with the magnetic lines of force. 5. The direction of the electric vector in a linearly

polarized wave radiated from an antenna. 6. A mechanical arrangement of inserts and shell configuration (referred to as clocking in some instances) that prohibits the mating of mismatched plugs and receptacles. 7. A technique of eliminating symmetry within a plane so that parts can be engaged in only one way in order to minimize the possibility of electrical and mechanical damage or malfunction. 8. A property of a radiated electromagnetic wave that describes the direction of its electric field.

polarization diversity—A term that designates a method of transmission and reception used to minimize the effects of selective fading of the horizontal and vertical components of a radio signal. It is usually accomplished through the use of separate vertically and horizontally polarized receiving antennas.

polarization-diversity antenna—An antenna in which any of a number of types of polarization can be readily selected. The polarization can be horizontal, vertical, right-hand circular, left-hand circular, or any combination of these four.

polarization-diversity reception—Diversity reception that uses separate vertically and horizontally polarized receiving antennas.

polarization error—In navigation, the error arising from the transmission or reception of radiation having other than the intended polarization for the system.

polarization fading—Fading as the result of changes in the direction of polarization in one or more of the propagation paths of waves arriving at a receiving point. *See also* Faraday effect, 1.

polarization in a dielectric—The slight displacement of the positive charge in each atom whenever a dielectric is placed into an electric field.

polarization index—A practical measure of dielectric absorption expressed numerically as the ratio of the insulation after 10 minutes to the insulation resistance after 1 minute of voltage application.

polarization modulation—A technique in which modulation is produced by changing the direction of polarization of circularly polarized layer energy.

polarization receiving factor—Ratio of the power received by an antenna from a given plane wave of arbitrary polarization to the power received by the same antenna from a plane wave of the same power density and direction of propagation whose state of polarization has been adjusted for the maximum received power.

polarization unit vector (for a field vector)—A complex field vector at one point, divided by the magnitude of the vector.

polarize—1. To cause to be polarized. 2. To arrange mating connectors so that they can be joined in only one way.

polarized capacitor—An electrolytic capacitor with the dielectric film formed adjacent to only one metal electrode. The opposition to the passage of current is then greater in one direction than in the other. The polarity is established for minimum current; operation with reversed polarity can result in damage to the capacitor if excessive current occurs.

polarized double-biased relay—Also called magnetic-latch relay. A relay whose operation depends on the polarity of the energizing current and which is magnetically biased, or latched, in either of two positions. Its coil symbol is usually marked + and DB.

polarized light—Light that has the electric-field vector of all the energy vibrating in the same plane. Looking into the end of a beam of polarized light, one would see the electric-field vectors as parallel or coincident lines.

polarized no-bias relay — A three-position or a center-stable polarized relay. Its coil symbol is usually marked + and NB.

polarized plug — 1. A plug so constructed that it may be inserted in its receptacle only in a predetermined position. 2. A multiconductor plug that can be inserted into a jack or receptacle in only one position.

polarized receptacle — A receptacle into which a polarized plug can be inserted only in a predetermined position.

polarized relay — Also called a polar relay. A relay in which the armature movement depends on the direction of the current. Its coil symbol is sometimes marked +.

polarized-vane instrument — *See* permanent-magnet moving-iron instrument.

polarizer — A substance that, when added to an electrolyte, increases the polarization.

polarizing pin — A pin located on one half of a two-piece connector in such a position that, by mating with an appropriate hole on the other half during assembly of the connector, it will ensure that only related connector halves can be assembled.

polarizing slots — Also called indexing slots. One or more slots placed in the edge of a printed circuit board to accommodate and align certain types of connectors.

polar keying — A form of telegraph signal in which circuit current in one direction is used for marking, and current in the other direction is used for spacing.

polar modulation — A form of amplitude modulation in which the positive excursions of the carrier are modulated by one signal and the negative excursions by another.

polar mount — 1. A common mount used with satellite dishes. One axis is aligned with the true north pole so that the satellites in the Clarke belt can be scanned with the movement of only one axis. 2. An antenna mounting and aiming system in which one pivot is positioned only one time and the other (hour axis) is positioned to sweep the satellite arc. Some fine adjustments may be required on the first pivot (declination axis), but this mount is much easier to aim than the azimuth/elevation mount.

polar orbiting satellites — A satellite that orbits over the north and south poles. Since the earth rotates beneath it, the satellite sees a different view on each rotation.

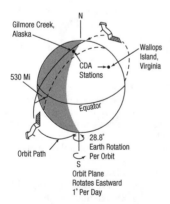

Polar orbiting satellites.

polarotor — A small motor mounted on a dish that rotates the microwave probe (LNA) to receive signals of either horizontal or vertical polarity.

polar radiation pattern — A diagram that shows the relative strength of the radiation from a source in all directions in a given plane.

polar relay — 1. A relay containing a permanent magnet that centers the armature. The direction of movement of the armature is governed by the direction of the current. *See also* polarized relay. 2. A permanent-magnet-core relay that is designed to operate only when current flows in a specified direction. 3. A springless relay built by winding a magnetic core with two equal but opposite windings. The armature stands theoretically in the middle of the two windings. This kind of relay permits the operation of an rtty circuit that has current in one direction for marks and in the opposite direction for spaces.

polar response — 1. Polar diagram or circular graph that shows the sensitivity of an antenna or microphone or the output from speakers in an angular mode through 360°. 2. The variation of output, at any given frequency, at different angles to the forward axis of symmetry of the speaker. In general, it will be different in horizontal and vertical planes, as well as with frequency.

polar signal — 1. A signal whose information is transmitted by means of directional currents. 2. A signal in which the current in the transmission line is reversed in polarity in changing from marking to spacing.

pole — 1. One end of a magnet. 2. One electrode of a battery. 3. An output terminal on a switch. 4. An item that controls one path of the circuit. 5. A combination of no and/or nc mating contacts.

pole face — In a relay, the end of the magnetic core nearest the armature.

pole piece — One or more pieces of ferromagnetic material forming one end of a magnet and so shaped that the distribution of the magnetic flux in the adjacent medium is appreciably controlled.

pole shoe — The portion of a field pole facing the armature of the machine. It may be separable from the body of the pole.

poles of a network function — Those real or complex values of p for which the network function is infinite.

pole-type transformer — A transformer suitable for mounting on a pole or similar structure.

police calls — Broadcasts (usually orders) issued by police radio stations.

police connection — The direct link by which an alarm system is connected to an annunciator installed in a police station. Examples of a police connection are an alarm line and a radiocommunications channel.

poling — 1. The adjustment of polarity. Specifically, in wire-line practice, it signifies the use of transpositions between transposition sections of open wire or between lengths of cable to cause the residual crosstalk couplings in individual sections or lengths to oppose one another. 2. A step in the production of ceramic piezoelectric bodies that orients the axes of the crystallites in the preferred direction. In general, a process similar to magnetizing ferromagnetic materials.

polishing — 1. A mechanical finishing operation conducted upon solid-state substrates to achieve smoothness and desired surface qualities. *See also* lapping. 2. Act of smoothing ends of fibers to an optically smooth finish, generally using abrasives. Optically smooth surfaces allow maximum transmission of light between fibers at connections and minimum coupling loss.

Polish notation — A system for writing and evaluating logical and arithmetic expressions without the use of parentheses. So called because it was originated by the Polish logician J. Lukasiewicz.

polling — 1. A communications technique that determines when a terminal is ready to send data. The computer continually interrogates all of its attached terminals in a round-robin sequence. A terminal acknowledges the poll when it has data to send. 2. A means of controlling devices on a multipoint line. 3. Controlling communication lines by designating one station as the master. This station then gives control of the line to each of the other stations, in turn, for a predetermined amount of time. 4. A control message sent from a master site to a slave site that serves as an invitation to transmit data to the master site. 5. Scheduling technique for I/O devices, whereby the program interrogates the status of each peripheral in turn and gives service when required. The other essential techniques are interrupts and direct memory access. 6. A process in which a number of peripheral devices, remote stations, or nodes in a computer network are interrogated one at a time to determine if service is required. 7. In data communications, the action of the central system periodically requesting input from multiple terminals on a line by sending a predetermined message (known as a poll sequence) to the terminals.

poll response — *See* train time.

polycarbonate — An amorphous thermoplastic used in the connector industry and offering high impact strength over a broad temperature range. Polycarbonates are excellent electrical insulators over a wide range of humidity and temperature. They are used as dielectrics in film capacitors. They have a high resistance to creep.

polychromatic radiation — Electromagnetic radiation consisting of two or more frequencies or wavelengths.

polycrystalline cell — A photovoltaic cell made of crystalline semiconductor compounds (two or more different atoms).

polycrystalline ceramic — A ceramic material, such as barium titanate, with a crystalline structure in which all molecules are similarly oriented and regularly arranged. (It may be made piezoelectric by pretreatment with a polarizing electric field.)

polycrystalline material — Material, typically an element like silicon or germanium, made up of many single crystals having a random orientation. The term may be applied to a twin crystal as well as to a heterogeneous growth of many crystals.

polycrystalline structure — The granular structure of crystals that have nonuniform shapes and arrangements.

polyergic — A type of emission in which the groups of energies or velocities are produced simultaneously (e.g., simulated micrometeoroids in varying charge states separated by velocity where accelerated by the same potential).

polyester — Polyethylene glycol terephthalate, the material most often used as a base film for precision magnetic tape. The chief advantages of this material compared to other materials are its stability with respect to humidity and time, its resistance to solvents, and its mechanical strength. It is used as a dielectric in film capacitors.

polyester backing — A plastic-film backing added to magnetic tape to make it stronger and more resistant to changes in humidity.

polyester base — A plastic-film backing for magnetic tape used for special purposes where strength and resistance to temperature and humidity change are important. (Mylar is a Du Pont trade name for their brand of polyester.)

polyester films — A broad category of films that differ in chemical composition, properties, and processibility, but which exhibit very good electrical properties.

polyesters — A class of thermosetting synthetic resins having great strength and good resistance to moisture and chemicals.

polyethylene — Short for polymerized ethylene, a tough white plastic insulator with low moisture absorption. It is often used as a dielectric.

polygon — A closed figure with straight edges; often used as the underlying 3D data structure for shaded 3D systems.

polygraph — Also called a lie detector. A recorder of several signals simultaneously, such as blood pressure, respiratory motion, galvanic skin resistance, etc., commonly used for study of emotional reactions involving deception (lie detection).

polyimide film — A plastic film exhibiting excellent physical and electrical properties over a wide temperature range. Produced from pyromellitic dianhydride and an aromatic diamine, it is used as a printed circuit substrate.

polymer — A compound formed by polymerization that results in the chemical union of monomers or the continued reaction between lower molecular weight polymers.

polynomial — An algebraic expression that contains two or more terms and in which the dependent variable is represented by a linear combination of powers of the independent variable, with the degree of the polynomial determined by the highest power in the expression.

polyphase — Describing an electrical circuit or electrical equipment that uses two or more phases. Polyphase circuits that have two, three, and six phases are common. Thus, a polyphase motor operates from a power line having several phases of alternating current.

polyphase circuit — A group of alternating-current circuits (usually interconnected) that enter (or leave) a delimited region at more than two points of entry. They are intended to be so energized that, in the steady state, the alternating potential differences between them all have exactly equal periods but have differences in phase, and may have differences in waveform.

polyphase motor — An induction motor wound for operation on two- or three-phase alternating current.

polyphase synchronous generator — A generator with its ac circuits so arranged that two or more symmetrical alternating electromotive forces with definite phase relationships to each other are produced at its terminals.

polyphase transformer — A transformer designed for use in polyphase circuits.

polyphase voltages — In an ac electrical system, voltages having a definite phase relationship to each other.

polyplexer — Radar equipment that combines the functions of duplexing and lobe switching.

polypropylene — A thermoplastic with good electrical characteristics, high tensile strength, and resistance to heat. It is used as a dielectric in film capacitors.

polyrod antenna — An end-fire dielectric microwave antenna made of tapered polystyrene rods.

polysilicon — A multicrystalline form of silicon used in silicon-gate MOS technology that is electrically conductive and optically transparent. It is commonly used to form electrodes of solid-state imaging devices.

polystyrene — A clear thermoplastic material having excellent dielectric properties, especially at ultrahigh frequencies.

polystyrene capacitor — A low-loss precision capacitor with a polystyrene dielectric.

polysulfones — Plastics that are transparent and have high dimensional stability and high heat-deflection temperature. They are tough and strong and have excellent dimensional stability and electrical properties.

polyvinyl chloride — Abbreviated PVC. A general-purpose thermoplastic used for insulations and jackets on components, wire, and cable.

pony circuit — A local, on-base circuit that does not have direct entry into a relay network.

pool cathode — A cathode at which the principal source of electron emission is a cathode spot on a metallic-pool electrode.

pool tube — A gas tube with a pool cathode.

popcorn noise — Also called burst noise. 1. So named for the audible characteristic, popcorn noise is randomly occuring random-amplitude noise, lasting from a few microseconds to several seconds. A type of noise generally associated with operational amplifiers. When popcorn noise is present, it does not occur in all devices made by a given process or even in all devices on the same wafer. It becomes worse at low temperatures, and it disappears above some threshold temperature. 2. An undesired source of interference associated with semiconductor devices that usually exceeds the background noise by at least a factor of 2 and occurs randomly at very long intervals. Popcorn noise is characterized by pulses of current in a semiconductor device operating in an electrical circuit. It can be observed by placing a high-value resistor in the base circuit of a transistor and observing the amplified noise signal present at the collector.

Pope cell — *See* sulfonated polystyrene sensor.

pop filter — 1. A cloth, foam, or similar shield placed over a microphone to prevent popping sounds resulting from sudden bursts of breath. 2. A filter that attenuates low frequencies where the popping sounds exist. Typically, a high-pass filter with its cutoff at approximately 70 Hz to 100 Hz. 3. *See* blast filter.

POPI — Post Office Position Indicator. A British long-distance navigational system for providing bearing information. It is a continuous-wave, low-frequency system in which the phase difference between sequential transmissions on a single frequency is measured.

pops — A form of record noise that usually results from disc imperfections and the ever-present electrostatic charge (with its attendant crackling sounds) on the disc.

population — Sometimes called universe. The entire group of items being studied, from which samples are drawn..

population inversion — 1. A nonequilibrium condition that exists when there are more atoms in the excited state than in the ground state. Atoms return to the lower-energy level, and release energy and emit photons. 2. A condition in a stimulated material, such as a semiconductor, in which the upper energy level of two possible electronic energy levels in a given atom, distribution of atoms, molecule, or distribution of molecules has a higher probability (usually only slightly higher but nevertheless higher) of being occupied by an electron. When population inversion occurs, the probability of downward energy transitions giving rise to radiation is greater than the probability of upward energy transitions. This thus brings about stimulated emission, i.e., laser action. 3. The condition in which there are more atomic systems in the upper of two energy levels than in the lower, so stimulated emission will predominate over stimulated absorption. 4. In the context of semiconductors, any situation in which the normal majority/minority carrier ratio of an impurity semiconductor region is disturbed, to a degree that the nominal minority carriers are actually present in larger numbers than the nominal majority carriers. 5. A redistribution of energy levels in a population of elements such

that instead of having more atoms with lower energy level electrons there are fewer atoms with higher energy level electrons, i.e., an increase in the total number of electrons in the higher excited states occurs at the expense of the energy in the electrons in the ground or lower state and at the expense of the resonant energy source, i.e., the pump. This is not an equilibrium condition. The generation of population inversion is caused by pumping.

porcelain — A glazed ceramic insulating material made from clay, quartz, and feldspar.

porcelain capacitor — A fixed electrostatic capacitor whose dielectric is a high grade of porcelain molecularly fused to alternate layers of fine silver electrodes so as to form a monolithic capacitor.

porcelainize — To coat and fire a metal with glass material, forming a hybrid circuit substrate.

port — 1. A point of access into a computer (such as the serial and parallel ports on the back of most PCs), a network, or other electronic system; the physical or electrical interface through which one gains access; the interface between a process and a communications or transmission facility. 2. The hardware that permits data to enter or exit a computer, network node, or communications device. *See also* communications port. 3. A place of access to a system or circuit. Through it, energy can be selectively supplied or withdrawn or measurements can be made. Examples are the port in a waveguide or in a base-reflex speaker enclosure. 4. A fluid connection to the servovalve (e.g., supply port, return port, control port). 5. An external opening of an internal passage. 6. The physical circuit that interfaces a computer to a peripheral, and the numerical address the computer uses in communicating through it. 7. An opening in a speaker enclosure, permitting the bass radiation from the back of the woofer cone to be combined with its forward radiation to enhance the total response. 8. An input/output channel, including the physical connector and control logic, to interface a peripheral device to a mainframe. 9. A set of lines for internal and external communication with a computer. 10. An outlet in a processor where a peripheral plugs in, in equipment in which the peripheral is compatible with the processor. 11. The gateway that connects the computer to its outside world. 12. To convert software to run in a different computer environment.

portable data medium — A data medium intended to be transportable easily and independently of the mechanism used in interpreting it.

portable duress sensor — A device carried on a person that may be activated in an emergency to send an alarm signal to a monitoring station.

portable intrusion sensor — A sensor that can be installed quickly and that does not require the installation of dedicated wiring for the transmission of its alarm signal.

portable operation — Radiocommunication conducted from a specific geographical location other than that shown on the station license.

portable recorder — A sound and/or video recorder designed for easy mobility, but which may require connection to a 120-volt ac supply for operation. Also applied to battery-powered recorders that do not require external power for operation.

portable standard meter — A portable meter used principally as a standard for testing other meters.

portable transmitter — A transmitter that can be readily carried on one's person and operated while in motion (e.g., walkie-talkies, Handie-Talkies, and similar personal transmitters). *See also* transportable transmitter.

portamento — The continuous change of a tone from one pitch to another. *See also* glissando.

port selector — A switching device that extends the capability of a computer to handle more data traffic without additional ports. It eliminates dedicated line-to-port interfacings, so fewer ports can handle more data lines.

pos — Abbreviation for positive.

posistor — A thermally sensitive resistor that has a positive temperature characteristic of resistance.

position — 1. The location of an object with respect to a specific reference point or points. 2. In a string, a location that can be occupied by a character or bit and that can be identified by a serial number.

positional crosstalk — In a multibeam cathode-ray tube, the deviation of an electron beam from its path under the influence of another electron beam within the tube.

positional notation — One of the schemes for representing numbers. It is characterized by the arrangement of digits in sequence, with successive digits forming coefficients of successive powers of an integer called the base of the number system.

position-changing mechanism — The mechanism used to move a removable circuit-breaker unit to and from the connected, disconnected, and test positions.

position control system — A discrete or point-to-point control in which the controlled motion is used as a means of arriving at a given end point without path control during the movement between end points.

position dialing — Dialing over the regular position cord circuits by means of a relay circuit controlled by a dial of the regular cord circuits.

position feedback — A feedback signal that is proportional to the position or deflection of some object.

position-independent code — Abbreviated PIC. Machine-coded programs using only relative addressing, permitting the program to reside in any portion of system memory.

position of effective short — The distance between a specified reference plane and the apparent location of the short circuit of a fixed switching tube in its mount.

position sensor — A device that measures position and converts the measurement into a form convenient for transmission as a feedback signal.

position-type telemeter — See ratio-type telemeter.

positive — Abbreviated pos 1. Any point to which electrons are attracted — as opposed to negative, from where they come. 2. An artwork, artwork master, or production master in which the intended conductive pattern is opaque to light, and the areas intended to be free from conductive material are transparent.

positive bias — The condition in which the control grid of a vacuum tube is more positive than the cathode.

positive charge — An electrical charge with fewer electrons than normal.

positive column — The luminous glow, often striated, between the Faraday dark space and the anode in a glow-discharge, cold-cathode tube.

positive electricity — The electricity that predominates in a glass body after it has been electrified by rubbing with silk. See positive charge.

positive electrode — The conductor that is connected to the positive terminal of a primary cell and serves as the anode when the cell is discharging. Electrons flow to it through the external circuit.

positive electron — See positron.

positive feedback — 1. The process by which the amplification is increased by having part of the power in the output circuit returned to the input circuit in order to reinforce the input power. 2. Recycling of a signal that is in phase with the input to increase amplification. Used in digital circuits to standardize the waveforms in spite of any anomalies in the input. See also regeneration, 1.

positive ghost — A television ghost-signal display with the same tonal variations as those of the image.

positive-going — Increasing toward a positive direction (e.g., a current or waveform).

positive grid — A grid with a more positive potential than the cathode in a vacuum tube.

positive-grid multivibrator — A multivibrator that has one or more grids connected to the plate-voltage supply, usually through a large resistance.

positive-grid oscillator — See retarding-field oscillator.

positive-grid oscillator tube — Also called a Barkhausen tube. An oscillating triode in which the grid has a more positive quiescent voltage than either of the other electrodes.

positive ground — The positive battery terminal of a vehicle is connected to the body and frame.

positive-intrinsic-negative photodiode coupler — A coupling device that enables the coupling of light energy from an optical fiber or cable onto the photosensitive surface of a positive-intrinsic-negative (pin) diode of a photon detector (photodetector) at the receiving end of an optical-fiber data link. The coupler may be only a fiber pigtail epoxied to the photodiode.

positive ion — An atom that has lost one or more electrons and, thus, has an excess of protons, giving it a positive charge.

positive-ion emission — Thermionic emission of positive particles from the cathode of a vacuum tube. They either are made up of ions from the metal in the cathode or are due to some impurity in it.

positive-ion sheath — A collection of positive ions on the control grid of a gas-filled triode tube. If too high a negative bias is applied to the grid, this positive sheath will block the plate current.

positive light modulation — Modulation in which the transmitted power increases as the light intensity does, and vice versa.

positive logic — A form of logic in which the more positive logic level represents 1 and the more negative level represents 0.

positive magnetostriction — Magnetostriction in which a material expands whenever a magnetic field is applied.

positive modulation — Also called positive picture modulation. In an AM television system, modulation in which the brightness increases as the transmitted power does, and vice versa.

positive phase-sequence relay — A relay that is energized by the positive phase-sequence component of the current, voltage, or power of a circuit.

positive picture modulation — See positive modulation.

positive picture phase — 1. The condition in which the picture-signal voltage goes positive above the zero level whenever a positive scene or picture increases in brilliance. 2. Positioning of the composite video signal so that the maximum point of the sync pulses is at zero voltage. The brightest illumination is caused by the most positive voltages.

positive plate — 1. A hollow lead grid filled with active material and connected to the positive terminal of a storage battery. When the battery is discharging, electrons flow toward it through the external circuit. 2. In a charged capacitor, the plate that has fewer electrons.

positive ray — See canal ray.

positive temperature coefficient — The condition whereby the resistance, capacitance, length, or other characteristic of a substance increases as the temperature does.

positive terminal — In a battery or other voltage source, the terminal toward which electrons flow through the external circuit from the negative terminal.

positive transmission — Transmission of television signals in such a way that the transmitted power increases whenever the initial light intensity does.

positive-true logic — A logic system in which the voltage representing a logical 1 has a higher or more positive value than that representing a logical 0.

positron — Fundamental particle equal in mass and energy to an electron, but having a positive charge. It has a very short life, being usually lost in the formation of a photon by combination with an electron. It occurs in the radiations from a few radioactive isotopes.

post — 1. In a computer, to place a unit of information in a record. 2. To submit a message in a newsgroup or other online forum.

postaccelerating electrode — *See* intensifier electrode.

postacceleration — Also called postdeflection acceleration. Acceleration of the beam electrons in a tube after they have been deflected.

postconversion bandwidth — In a telemetry receiver, the bandwidth presented to the detector.

postcuring — Heat aging of a film circuit after firing to stabilize the resistor values through stress relieving.

postdeflection accelerating electrode — *See* intensifier electrode.

postdeflection acceleration — *See* postacceleration.

postedit — In a computer, to edit output data resulting from a previous computation.

postemphasis — *See* deemphasis.

postequalization — *See* deemphasis.

POS terminal — *See* point-of-sale terminal.

postfiring — Refiring of a film circuit after having gone through the firing cycle. Sometimes used to change the values of the already fired resistors.

postmortem — A diagnostic computer routine for locating a malfunction in the computer or an error in coding a problem. Should a problem tape come to a standstill, the computer will print out — either automatically or when called for — any information concerning the contents of all or part of the registers in the computer.

post processor — A software subsystem that converts stimulus and response information obtained from a simulator into the machine language of a particular automatic test equipment.

postregulator — A circuit that performs the functions of reference, comparison, and control in power supplies. So called because it follows the transformer, rectifier, and (usually) the ripple filter.

pot — 1. Short for potentiometer. 2. A solder pot. 3. To embed a component in a liquid resin within a casing that becomes part of the product.

potassium — An alkali metal having photosensitive characteristics, especially to blue light. It is used on the cathodes of phototubes whenever maximum response to blue light is desired.

pot core — A magnetic structure consisting of a rod and a sleeve arranged so that the rod fits inside a coil and the sleeve fits around the coil. The sleeve and rod are connected at one end by a plate. The open end (opposite the plate) is usually, but not necessarily, ground so that two pot cores or a pot core and a separate plate can be put together around a suitable coil to form a low-reluctance magnetic path and/or shield for the coil.

potential — 1. The difference in voltage between two points of a circuit. Frequently one point is assumed to be ground, which has zero potential. 2. In general, the electrical voltage difference between two bodies. When bodies of different potentials are brought into communication, a current is set up between them.

potential barrier — A semiconductor region through which electric charges attempting to pass will encounter opposition and may be turned back.

potential coil — The shunt coil in a measuring instrument or other device having series and shunt coils — i.e., the coil connected across the circuit and affected by changes in voltage.

potential difference — 1. A voltage existing between two points (e.g., the voltage drop across an impedance, from one end to another). 2. The voltage that can be measured between any two points in a circuit. 3. The algebraic difference between the voltages at two points in an electrical circuit.

potential divider — *See* voltage divider.

potential drop — The difference in potential between the two ends of a resistance with a current through it.

potential energy — Energy due to the position of one body with respect to another or to the relative parts of the same body.

potential galvanometer — A galvanometer with such a high resistance that it takes practically no current. It has been replaced by the vacuum-tube voltmeter.

potential gradient — 1. The rate of change of potential with distance. Units such as volts per meter or kilovolts per centimeter may be used. 2. Voltage gradient due to the diffusion of holes and electrons across the space charge region.

potential transformer — Also called a voltage transformer. An instrument transformer whose primary winding is connected in parallel with the circuit whose voltage is to be measured or controlled.

potential wells — A voltage placed on MIS capacitor electrodes causes a voltage gradient zone to be formed under the electrode so as to collect minority carriers.

potentiometer — 1. A resistor provided with a tap that can be moved along it in such a way as to put the tap effectively at the junction of two resistors whose sum is the total resistance, the ratio of the two effective resistors being a function of the position of the tap. 2. A measuring instrument in which a potentiometer is used as a voltage divider in order to provide a known voltage that can be balanced against an unknown voltage. 3. A variable voltage divider. A resistor that has a variable contact arm so that a portion of the potential applied between its ends may be selected. 4. A variable voltage divider used for measuring an unknown electromotive force or potential difference by balancing it, in whole or in part, by a known potential difference. 5. An instrument used to measure or compare voltages.

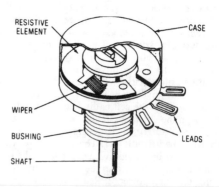

Potentiometer.

potentiometer circuit—A network arranged so that when two or more electromotive forces (or potential differences) are present in as many branches, the response of a suitable detecting device in any branch can be made zero by adjusting the electrical constants of the network.

potentiometer recorder—A null-balance type of recorder using a servo-operated voltage-balancing device; the sliding contact of a precision measuring potentiometer is adjusted automatically by a servomechanism so that the difference in voltage of the circuit becomes zero. Main feature is high sensitivity.

potentiometric transducer—A transducer in which displacement of a force-summing member is transmitted to the slider in a potentiometer, thus changing the ratio of output resistance to total resistance. Transduction is accomplished by changing the ratios of a voltage divider.

potentiometric transduction—The conversion of the measurand into a change in the position of a contact on a resistance element across which excitation is applied, the output usually being given as a voltage ratio.

pothead—An insulator for making a sealed joint between an underground cable and overhead line.

Potier diagram—A vector diagram showing the voltage and current relationships in an ac generator.

pot life—Also called working life. 1. The period after the addition of a catalyst to a potting compound during which the potting operation must be completed. 2. The time period during which a compound remains suitable for the intended use after compounding ingredients, such as solvents or catalysts, have been added.

potted circuit—A circuit that has been encapsulated in a nonconductive material.

potted line—A pulse-forming network immersed in oil and enclosed in a metal container.

Potter oscillator—A cathode-coupled multivibrator.

potting—1. An embedding process for parts that are assembled in a container or can into which the insulating material is poured, with the container remaining an integral part as the outer surface of the finished unit. 2. Sealing of a component (e.g., the cable end of a multiple-contact connector) with a plastic compound or material to exclude moisture, prevent short circuits, and provide strain relief. 3. A process of embedding a part or assembly by complete immersion in the potting compound. A container, can, or shell used as a mold is retained as an integral part of the finished product. Usually refers to protective encasements of greater than 100 mils (2.54 mm) thickness.

potting compound—A sealing material used to fill the case or enclosure in which a component is contained.

powdered-iron core—A core consisting of fine particles of magnetic material mixed with a suitable bonding material and pressed into shape.

power—1. The energy dissipated in an electrical or electronic circuit or component that is conducting either ac or dc. 2. Electrical energy developed to do work, such as the voltage from an amplifier used to drive a speaker. Also acoustical energy or sound pressure developed in a room by a speaker. 3. Rate of doing work. Some units of power are the foot-pound per second, or (in the cgs system) 1 erg/sec, the watt (joule/second), and the kilowatt.

power amplification—See power gain, 1.

power amplifier—1. An amplifier intended for driving one or more speakers or other transducers. 2. The final stage in a multistage amplifier circuit, designed to give power to the load, rather than to be used mainly as a voltage amplifier. 3. A fluidic device that causes a change in output power following a change of sufficient

magnitude in control power. 4. An amplifier driven by a relatively low voltage, low-power signal (of the order of 0.1 milliwatt or less) that delivers a substantial power output to low-impedance speaker loads. Typical power outputs may range from a few watts to several hundred watts, into impedances in the range of 2 to 16 ohms. The term power amplifier is commonly used to distinguish an amplifier that does not handle source signals directly and does not have volume or tone control functions. 5. A device that causes a change in output power following a change in control power. 6. An amplifier that can deliver a relatively large current and that can often operate from fairly high voltages. Power amplifiers are widely used in audio, but can also be used to drive servomotors and the like.

power amplifier input sensitivity—The input that will drive an amplifier to its maximum rated output power. Usually measured in dBm, dBV, or volts.

power-amplifier stage—1. An audio-frequency amplifier stage capable of handling considerable audio-frequency power without distortion. 2. A radio-frequency amplifier stage used in a transmitter primarily to increase the power of the carrier signal.

power attenuation—See power loss, 1.

power bandwidth—The range of audio frequencies over which an amplifier can produce half its rated power without exceeding its rated distortion. It is determined by using a measurement procedure standardized by the Institute of High Fidelity (IHF). This specification indicates how much power is available at the critical high and low frequencies. The wider the power bandwidth, the better the amplifier.

power (temperature-rise) coefficient—The maximum rise in hot-spot temperature of a resistor above ambient, per watt of dissipation, assuming free-air convection and negligible loss of heat through the leads, after thermal equilibrium has been reached, usually expressed in degrees Celsius per watt.

power conditioning—The process of maintaining uniform voltage on a power line.

power connection—British term used for the constant horsepower connection in a multispeed motor.

power consumption—The maximum wattage used by a device within its operating range during steady-state signal conditions.

power cord—Flexible, insulated cable used in such applications as supplying line power to power tools and electronic equipment.

power density—1. The radiated field strength set up by a radiating source, expressed in microvolts per meter or in dB above 1 μV/m. 2. Power in a band per hertz, or the total power in a band of frequencies divided by the bandwidth in hertz. 3. The power output of a battery, expressed in watts, per unit weight of the battery.

power derating—Use of computed curves to determine the correct power rating of a device or component to be used above its reference ambient temperature.

power detection—Detection in which the power output of the detector is used for supplying a substantial amount of power directly to a device such as a speaker or recorder.

power detector—A vacuum tube detector operating with such a high plate voltage that strong input signals can be handled without appreciable distortion.

power dissipation—1. The dispersion of the heat generated within a device or component when a current flows through it. This is accomplished by convection to the air, radiation to the surroundings, or conduction. 2. The supply power consumed by a logic circuit operating with a 50-percent duty cycle (equal times in the logical 0 and logical 1 states).

power dissipation rating — The maximum average power that can be continuously dissipated under stated temperature conditions.

power divider — 1. A device that provides a desired distribution of power at a branch point in a waveguide system. 2. A passive resistive network that equally divides power applied to the input port between any particular number of output ports without substantially affecting the phase relationship or causing distortion.

power driver — An amplifier with two outputs that can be used to drive a pair of complementary power transistors. The output power available from these transistors can be very high. An ordinary operational amplifier is not very suitable for use as a power driver, since only one output is available and this cannot be used easily to drive the power transistors, which require a bias difference between their input for low crossover distortion.

power dump — Also called dump. The removal of all power either accidentally or intentionally.

power factor — 1. Ratio of the actual power of an alternating or pulsating current, as measured by a wattmeter, to the apparent power, as indicated by an ammeter and voltmeter. 2. Ratio of resistance to impedance; therefore, a measure of the loss in an inductor, capacitor, or insulator. 3. The cosine of the phase angle between the voltage applied to a load and the current passing through it. (Sometimes the cosine is multiplied by 100 and expressed as a percentage.) 4. The ratio of actual power being used in a circuit, expressed in watts or kilowatts, to the power that is apparently being drawn from the line, expressed in voltamperes or kilovoltamperes. Actual power is working, or real or true, power used to produce heat or work. Apparent power is the product of volts times amperes, and may or may not be more than the actual power. When the two values are equal, their ratio is 1 : 1, or 1.0, or 100 percent. This is the highest power factor (unity) that can be obtained.

power-factor correction — Adding capacitors to an inductive circuit in order to increase the power factor by making the total current more nearly in phase with the applied voltage.

power-factor meter — A direct-reading instrument for measuring power factor. Its scale is graduated directly in power factor.

power-factor regulator — A regulator that maintains the power factor of a line or apparatus at a predetermined value, or varies it according to a predetermined plan.

power-factor relay — A device that operates when the power factor in an ac circuit becomes above or below a predetermined value.

power-fail circuit — A logic circuit that protects an operating program if primary power fails. A typical power-fail circuit informs the computer when power failure is imminent, initiating a routine that saves all volatile data. After power has been restored, the circuit initiates a routine that restores the data and restarts computer operation.

power foldback — Reduction of the power input to a power supply under fault conditions to less than the full rate input power under normal conditions.

power frequency — The frequency at which electric power is generated and distributed. Throughout most of the United States, this frequency is 60 Hz.

powerful — Usually refers either to a computer with a lot of memory or a lot of processing speed (a personal computer with 256 K RAM is powerful) or to a program with an unusual versatility (a spreadsheet is a powerful business tool).

power gain — Abbreviated pg. Also called power amplification. 1. The ratio of the signal power developed at the output(s) of a device to the signal power applied at the input(s). 2. Of an antenna in a given direction, 4π times the ratio of the radiation intensity to the total power delivered to the antenna. (The term is also applied to receiving antennas.)

power ground — 1. The ground between units that is part of the circuit for the main source of power to or from these units. 2. The potential of the terminal or circuit point to which the output of a power supply and often an amplifier output load is returned (i.e., power-supply "zero").

power-handling capability — 1. A measure of the maximum power input a speaker can absorb without damage or unreasonable distortion. In speaker systems the power-handling capacity will vary depending on the frequency and length of time the signal is applied. 2. The maximum power rating of a component, which determines how much current can be passed safely without adverse effects.

power IC — A monolithic integrated circuit that combines, on the same chip, signal-level digital logic and analog functions, for signal processing or interface, with one or more output power devices. The power IC is essentially an extension of the prevalent small-signal integrated circuit concept. A power IC is more specifically defined as a single-chip circuit that operates at 2 or more amperes and/or 2 or more watts.

power keys — A computer interface in which certain simple key-stroke combinations perform the same function as opening a pull-down menu and then selecting an entry.

power level — At any point in a transmission system, the difference between the measure of the steady-state power at that point and the measure of an arbitrarily specified amount of power chosen as a reference.

power-level indicator — An ac voltmeter calibrated to read audio power level.

power line — Two or more wires conducting electric power from one location to another.

power-line transient — See PLT.

power loss — Also called power attenuation. 1. Ratio of the power absorbed by the input circuit of a transducer to the power delivered to a specified load under specified operating conditions. 2. Also called watt loss. In the circuit of an instrument for measuring current or voltage, the active power at its terminals for nominal full-scale indication. For other instruments, for example, wattmeters, the power loss is expressed at a stated value of current or voltage.

power modulation factor — Ratio of the maximum positive departure of the envelope of an amplitude-modulation wave from its average value to its average value. This rating is used when the modulating signal wave has unequal positive and negative peaks.

power output — 1. The power in watts delivered by a power amplifier to a load such as a speaker. 2. Amplifier power measured with all channels operating, after a standard preconditioning period that brings amplifier components to their maximum working temperature.

power output (continuous watts) — In an amplifier, the power output at a total maximum harmonic distortion of 0.5 percent with a pure-tone (sine-wave) input.

power pack — A unit for converting power from an alternating- or direct-current supply into alternating- or direct-current power at voltages suitable for supplying the proper operating power to an electronic device.

power programmer — A device for controlling the output power of a radar automatically as a function of the target range.

power rating — The maximum power that can be dissipated in a component or device for a specified period.

power ratio — Ratio of the power output to the power input of a device. Usually expressed as the number of decibels loss or gain.

power relay — 1. A relay that functions at a predetermined value of power. It may be an overpower relay, an underpower relay, or a combination of both. 2. General-purpose relay with high ratings, generally 10 to 20 amperes or greater. Some relays in this category may be open frame in that they are not protected with individual dust covers; they are covered with a common dust cover in the end-use equipment.

power response — The frequency-response capabilities of an amplifier running at or near its full rated power.

power semiconductor device — Solid-state device capable of handling 1 watt of power or more at room temperature. Included are rectifiers, transistors, and thyristors.

power spectral density function — A measure of the power distribution of a signal with respect to frequency.

power-speed product — The product of a semiconductor device's propagation delay and its power dissipation.

power supply — 1. A unit that supplies electrical power to another unit. Generally, a circuit that accepts alternating current and converts it into direct current that is regulated precisely enough to drive electronic circuits and that maintains a constant voltage output within limits. For most electronics, the source of power is line voltage (117 V to 220 V, 60 Hz). Most electronic circuits today require low-voltage dc, typically 12 V or less. At present there are three different design approaches to providing this regulated dc: the series or linear regulated supply, the ferroresonant supply, and the switching regulated supply. 2. Energy source that provides power for operating electronic apparatus.

power-supply rejection ratio — The ratio of the change in input offset voltage of an operational amplifier to the change in power-supply voltage that causes it.

power switch — Often called an on-off switch. The switch that connects or disconnects a radio receiver, transmitter, or other equipment from its power line.

power switchboard — Part of a switch gear consisting of a panel or panels on which the switching-control, measuring, protective, and regulatory equipment is mounted. The panel or panel supports also may carry the main switching and interrupting devices and their connections.

power transformer — A transformer used for raising or lowering the supply voltage to the various values required by vacuum-tube plate, heater, and bias circuits.

power transistor — 1. A transistor designed to handle large currents and safely dissipate large amounts of power. 2. A transistor that can dissipate more than

1 watt of power. General-purpose types are used for low-frequency service (below 3 MHz) as amplifiers, switches, or current regulators. Rf types are used to amplify high-frequency signals (above 3 MHz) that reach up to VHF, UHF, and microwave regions. 3. A transistor that handles power levels of about 0.25 watt and above. Units handling about 0.25 to 10 watts are called medium-power transistors, whereas high-power transistors are those handling above 10 watts.

power tube — An electron tube designed to handle more current and power than a voltage-amplifier tube.

power winding — A saturable-reactor winding to which the power to be controlled is supplied. Commonly, the output and power are furnished by the same winding, then termed the output winding.

Poynting's law — The transfer of energy can be expressed as the product of the values of the magnetic field and of the components of the electric field that are perpendicular to the magnetic field, and the flow of energy at any point is perpendicular to both fields.

Poynting's theorem — The rate of flow of electromagnetic energy into or out of a closed region is at any instant proportional to the surface integral of the vector product of the electric and magnetic intensities.

Poynting's vector — 1. The vector product of the electric and magnetic intensities at one point and at a given instant in a wave. 2. In remote sensing technology, represents the intensity of energy flow in the direction of wave propagation.

PPI — Abbreviation for plan-position indicator.

PPI repeater — Also called remote plan-position indicator. A unit that repeats a plan-position indicator at a place remote from the radar console.

PPI scope — A cathode-ray oscilloscope arranged to present a PPI display.

pp junction — A region of transition between two regions having different properties in a p-type semiconducting material.

p+ region — The region created by diffusing into a silicon crystal a group III element, which creates a deficiency of electrons or an excess of holes.

p+ semiconductor — A p-type semiconductor with an extremely large excess mobile hole concentration.

p+-type material — Heavily doped p-type material, formed by introducing acceptor impurities into a silicon substrate. Conduction takes place by the movement of the holes.

PPM — Abbreviation for pulse-position modulation.

ppm — Abbreviation for parts per million.

PPM/AM — Amplitude modulation of a carrier by pulses that are position modulated by data.

pps — Abbreviation for pulses per second.

practical system of electrical units — A system in which the units are multiples or submultiples of the units of the centimeter-gram-second electromagnetic system.

praetersonic — The higher region of the sonic spectrum.

praetersonics — The propagation and signal processing of acoustic waves in solids at frequencies that extend into the microwave region.

preamplifier — 1. An amplifier that primarily raises the output of a low-level source so that the signal may be further processed without appreciable degradation in the signal-to-noise ratio. A preamplifier may also include provision for equalizing and/or mixing. 2. Also known as control amplifier or control center. A switching, amplification, and equalization component designed to select input signals, amplify them by amounts from 0 to 60 dB, and deliver an output voltage compatible with the input requirements of a power amplifier.

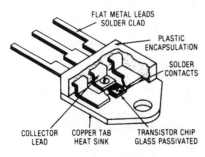

FLAT METAL LEADS
SOLDER CLAD

PLASTIC
ENCAPSULATION

SOLDER
CONTACTS

COLLECTOR   COPPER TAB   TRANSISTOR CHIP
LEAD   HEAT SINK   GLASS PASSIVATED

Power transistor.

preburning — Stabilizing vacuum tubes by operating their heaters continuously for a given number of hours. Cathode current may be drawn and the tubes vibrated at the same time.

precession — The effect resulting when a torque is applied to a rotating body, such as a gyroscope, causing it to wobble. The wobbling frequency is determined by the gravitational field strength and the mass of the body.

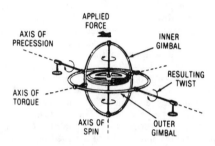

Precession of a gyroscope.

precious metal — One of the relatively scarce and valuable metals: gold, silver, and the platinum group metals.

precipitation attenuation — A reduction in radio energy as it passes through a volume of the atmosphere that contains precipitation. Part of the energy loss results from scattering, and part results from absorption.

precipitation noise — Noise generated in an antenna circuit, generally in the form of a relaxation oscillation, caused by the periodic discharge of the antenna or conductors in the vicinity of the antenna into the atmosphere.

precipitation static — A type of interference experienced in a receiver during snow, rain, and dust storms. Often caused by the impact of dust particles against the antenna or the creation of induction fields by nearby corona discharges.

precipitator — Sometimes called a precipitron. An apparatus for removing small particles of smoke, dust, oil, mist, etc. from the air by electrostatic precipitation.

precipitron — *See* precipitator

precise (or precision) measurement — Careful measurement under controlled conditions that can be repeated again and again with similar results. It also means that small differences can be detected and measured with confidence. Precision is related, then, to confidence in successive measurements with the same equipment and operating conditions.

precision — 1. The quality of being sharply or exactly defined — i.e., the number of distinguishable alternatives from which a representation was selected. This is sometimes indicated by the number of significant digits the representation contains. *See also* accuracy. 2. Also called reproducibility or repeatability. The degree with which repeated measurements of a given quantity agree when obtained by the same method and under the same conditions.

precision approach radar — Abbreviated PAR. 1. A rapid-scanning airport radar system so located that aircraft on approach to the runway are presented on the display in terms of linear deviation from a desired glide path and in terms of distance from the point of touchdown on the runway. 2. Radar used by traffic controllers in a ground-controlled approach to talk down a pilot on final approach.

precision-balanced hybrid circuit — A circuit for interconnection of a four-wire telephone circuit with a particular two-wire circuit, in which the impedance of the balancing network is adjusted so that a relatively high degree of balance is obtained.

precision comparator — A high-gain amplifier circuit whose output changes decisively between two definite levels whenever the sum of the input voltages change sign.

precision device — A device that operates within prescribed limits and will consistently repeat operations within those limits.

precision gate — A circuit that may be switched from closed to open circuit or vice versa without error (time, bias, impedance) in response to a command signal (voltage or current).

precision limit switches — Usually snap-acting switches that require a very small known movement to actuate. Many machines depend on the accuracy and repeatability of the precision limit switch. (Travel of 0.001 inch, or 25.4 µm, is possible.) The force needed to close a limit switch is varied from fractions of an ounce (a few grams) to pounds.

precision net — In a four-wire terminating set or similar device employing a hybrid coil, an artificial line designed and adjusted to provide an accurate balance for the loop and subscriber's set or line impedance.

precision potentiometer — A mechanical electrical transducer dependent upon the relative position of a moving contact (wiper) and a resistance element for its operation. It delivers a voltage output that is some specified function of the applied voltage and shaft position, to a high degree of accuracy.

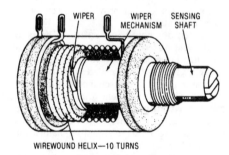

Precision potentiometer.

precision snap-acting switch — An electromechanical switch having predetermined and accurately controlled characteristics and having a spring-loaded quick make-and-break contact action.

precision sweep — A delayed expanded radar sweep for high resolution and range accuracy.

preconduction current — The low value of plate current in a thyratron or other grid-controlled gas tube prior to conduction.

precursor — Also called undershoot. The initial transient response to a unidirectional change in input. It precedes the main transition and is opposite in sense.

predefined process — A named process that consists of one or more operation or program steps that are specified in another part of a routine.

predetection combining — A method for producing an optimum signal from multiple receivers in a diversity reception.

predetection recording — The recording of telemetry receiver intermediate-frequency signals.

predetermined counter — A device that automatically stops an instrument to which it is attached when a preset limit is reached.

predictive control — A type of computer control that allows a digital computer to include a dynamic control loop for repetitive comparison of pertinent factors.

predissociation — The dissociation that occurs in a molecule that has absorbed energy before it has had an opportunity to lose energy by radiation.

predistortion — *See* preemphasis.

pre-Dolbyed tape — A prerecorded tape with Dolby compression added for low-noise playback in the home via a Dolby B stretcher.

preemphasis — Also called preequilization or predistortion. 1. In a system, a process designed to emphasize the magnitude of some of the frequency components. Preemphasis is applied at the transmitting end (with deemphasis at the receiving end) in order to improve the signal-to-noise ratio. 2. In recording, an arbitrary change in the frequency response from its basic response (e.g., constant velocity or amplitude) in order to improve the signal-to-noise ratio or to reduce distortion. *See also* accentuation. 3. A scheme sometimes adopted during recording to nullify the effect of tracing distortion on replay. The distortion deliberately introduced is the reciprocal of that produced on replay.

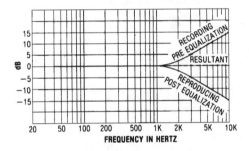

Preemphasis/postemphasis curves.

preemphasis network — A network inserted into a system to emphasize one range of frequencies.

preempted process — A process that, because of scheduling policy, must relinquish the processor to another process.

preemptor — A feature of some automated networks by which a high-precedence message, call, or transmission preempts a line from a use of lower precedence if all other lines are busy.

preequalization — *See* preemphasis.

preferred tube types — Tube types recommended to designers of electronic equipment to minimize the number of tube types that must be stocked by the manufacturer or by service agencies.

preferred values — A series of resistor and capacitor values adopted by the EIA and military. In this system, the increase between any two steps is the same percentage as between all other steps. Increases may be in steps of 20 percent, 10 percent, or 5 percent each.

prefired — Conductors fired in advance of the screening of resistors on a substrate.

prefix multipliers — Prefixes that designate a greater or smaller unit than the original, by the factor indicated. These prefixes are as follows:

Prefix	Symbol	Factor
yotta-	Y	10^{24}
zeta-	Z	10^{21}
exa-	E	10^{18}
peta-	P	10^{15}
tera-	T	10^{12}
giga-	G	10^{9}
mega-	M	10^{6}
kilo-	k	10^{3}
hecto-	h	10^{2}
deka-	da	10
deci-	d	10^{-1}
centi-	c	10^{-2}
milli-	m	10^{-3}
micro-	μ	10^{-6}
nano-	n	10^{-9}
pico-	p	10^{-12}
femto-	f	10^{-15}
atto-	a	10^{-18}
zepto-	z	10^{-21}
yocto-	y	10^{-24}

preform — Also called a biscuit. 1. In disc recording, the small slab of record material used in the presses. 2. Small circle or square of the solder or epoxy punched out of thin sheets. Preforms are placed on the spot to be soldered or bonded, prior to the placing of the object to be attached, to aid in soldering or adhesion.

prelasing condition (or state) — The condition of an injection laser corresponding to the emission of predominantly incoherent or spontaneous radiation.

preliminary contacts — In a relay, contacts that open or close before other contacts when the relay is actuated.

premium service — Extra television programming service for which a subscriber must pay an additional fee above and beyond the basic subscription fee.

preohmic alignment — In a semiconductor, the positioning of the oxide opening into which the metallization is placed.

preohmic window — The opening etched through the oxide in a semiconductor for metallization contact to the emitter and base regions.

preprocessor — 1. A device for placing source records into a format that facilitates system processing in a computer. 2. A method of converting data into computer-usable form for processing and output. 3. Software program or procedure that interprets graphical data and formats it into data readable by a numerical-control machine or by other computer programs.

prerecorded tape — *See* recorded tape.

prescaler — A circuit that generates an output signal related to the input signal by a fractional scale factor such as $1/2$, $1/8$, $1/10$, etc. An example of a digital prescaler is a decade frequency divider, which has an output frequency one-tenth of the input frequency.

preseal visual — The process of visual inspection (of a completed hybrid circuit assembly) for defects prior to sealing the package.

preselection — 1. The use of a preselector. 2. In buffered computers, a time-saving technique in which a block of information is read into the computer memory ahead of time from whichever input tape will next be called on. 3. In digital computers, a technique whereby data from the next input tape is stored while the computer is still processing other data.

preselector — 1. A device placed ahead of a frequency converter or other device to pass signals of desired frequencies but reduce all others. 2. In automatic switching, a device that makes its selection before seizing an idle trunk.

preselector stage — A radio-frequency amplifier stage in the input of a superheterodyne receiver.

presence — The quality of naturalness in sound reproduction. When the presence of a system is good, the illusion is that the sounds are being produced intimately at the speaker.

presence control — A potentiometer used in a three-way speaker system for controlling the volume of the middle-range speaker.

presentation — The form that the radar echo signals take on the screen, depending on the nature of the sweep circuit.

presenting — Displaying data in a form that human intelligence can comprehend and use.

preset — 1. To establish an initial condition, for example the control values of a loop. 2. An asynchronous input of a flip-flop by which the Q output is set to a logical 1 and the $\overline{Q}$ output is set to a logical 0.

preset guidance system — A guidance system in which the flight path is determined before the missile is launched and cannot be altered after launch.

preset parameter — In a computer, a parameter that is fixed for each problem at a value established by the programmer.

preshoot — The initial transient response to a unidirectional change in input that precedes the main transmission and may be of the same or opposite polarity.

press-button switch — British term for push-button switch.

pressed alumina — Aluminum-oxide ceramic formed by applying pressure to the ceramic powder and a binder prior to firing in a kiln.

pressed stem — An obsolete method of vacuum-tube construction in which all support wires are formed into a flattened piece of glass tubing (actually a relic from the lampmaker's era). *See also* button stem.

press-fit contact — An electrical contact that can be pressed into a hole in an insulator, printed board (with or without plated-through holes), or a metal plate.

press-fit pin — A connector pin that is forced into a wiring board or substrate hole, forming a gas-tight connection point without solder or welds.

pressing — A disc recording produced in a record-molding press from a master or stamper.

press-to-talk switch — Also called a push-to-talk switch. A spring-loaded switch that must be held down as long as the operator talks. Releasing the switch deactivates the microphone. It is used on transmitter and dictating-machine microphones.

pressure — Force per unit area. Measured in pounds per square inch (psi) or by the height (in feet, inches, or centimeters) of a column of water or mercury that the force will support. Absolute pressure is measured with respect to zero pressure. Gage pressure is measured with respect to atmospheric pressure.

pressure alarm system — An alarm system that protects a vault or other enclosed space by maintaining and monitoring a predetermined air-pressure differential between the inside and outside of the space. Equalization of pressure resulting from opening the vault or cutting through the enclosure will be sensed and will initiate an alarm signal.

pressure amplifier — A fluidic device that causes a change in output pressure following a change of sufficient magnitude in control pressure.

pressure amplitude — For a sinusoidal sound wave, the maximum absolute value of the instantaneous sound pressure at a point during any given cycle. The unit is the dyne per square centimeter.

pressure connector — A conductor terminal applied under pressure to make the connection mechanically and electrically more secure. *See also* solderless connector.

pressure-gradient hydrophone — A type of hydrophone in which the electric output is essentially determined by a component of the gradient (space derivative) of the sound pressure.

pressure-gradient microphone — A microphone that reproduces only the difference in acoustic pressure between the front and back of its element. Such microphones are designed to create various pickup patterns through specific acoustic ducting in the microphone case.

pressure hydrophone — A type of hydrophone in which the electric output is essentially determined by the instantaneous sound pressure of the impressed sound wave.

pressure microphone — A microphone in which the electric output corresponds substantially to the instantaneous sound pressure of the impressed sound waves. It is a gradient microphone of zero order, and is nondirectional when its dimensions are smaller than a wavelength.

pressure pad — 1. In single-motor tape recorders, a device that forces the tape into intimate contact with the head gap, usually by direct pressure at the head assembly. Felt or similar material, occasionally protected with self-lubricating plastic, is used to apply pressure uniformly and with a minimum of drag to the backing side of the tape. 2. A small piece of cloth or felt that holds the recording tape against the tape heads.

pressure potentiometer — A pressure transducer in which the electrical output is derived by moving a contact arm along a resistance element.

pressure roller — Also called pinch roller, puck, or capstan idler. A spring-loaded rubber-tired roller that holds the magnetic tape tightly against the capstan, permitting the latter to draw the tape off the stock reel and past the heads at a constant speed.

pressure-sensing element — In a pressure transducer, the part that converts the measured pressure into mechanical motion.

pressure spectrum level — The effective sound-pressure level for the sound energy contained within a band 1 hertz wide and centered at a specified frequency. Ordinarily this level is not significant except for sound having a continuous distribution of energy within the frequency range under consideration.

pressure switch — A switch actuated by a change in the pressure of a gas or liquid.

pressure transducer — An instrument that converts a static or dynamic pressure input into the proportionate electrical output.

pressure-type capacitor — A fixed or variable capacitor used chiefly in transmitters. It is mounted inside a metal tank filled with nitrogen at a pressure that may be as great as 300 pounds per square inch (21 kg/cm^2). The high pressure permits a voltage rating several times that of air.

pressure unit — A moving-coil speaker drive unit that usually has as its diaphragm a small dome of plastic or metal. It is designed for use in the throat of a horn.

pressurization — The process by which the critical parts of equipment designed for high-altitude operation are surrounded with dry air or an inert gas under pressure (about 5 pounds per square inch, or 0.35 kg/cm^2, at sea level). Thus, breakdowns from the impaired insulating properties of air at reduced pressure are prevented.

prestore — To store a quantity in an available or convenient location in a computer before it is required in a routine.

pretinned — Refers to solder applied to an electrical component or printed circuit board prior to soldering.

pre-tr — In a radar set, an additional tr box that provides additional attenuation of transmitted pulse to prevent damage to the crystal mixer.

pretravel — The distance or angle through which the actuator moves from the actuator free position to the actuator operating position.

pretrigger — 1. A timed pulse used to start a sequence of operations prior to the main trigger. 2. In random-sampling oscilloscope technique, a trigger that occurs or arrives prior to a related signal event. 3. The amount of information that will be stored in a digital oscilloscope memory preceding the trigger event. Typically, up to one sweep interval of pretrigger information can be placed in memory.

pre-tr tube — A gas-filled radio-frequency switching tube used to protect the tr tube from excessive power and the receiver from frequencies other than the fundamental.

preventive maintenance — Precautionary measures taken on a system to forestall failures rather than to eliminate them after they have occurred.

preview — Also called page preview. In word processing, displaying a formatted document on the screen to see exactly what the printed page will look like.

previous-element coding — A method of signal coding for digital television transmission in which each transmitted picture element depends on the similarity of the preceding element.

prewound core — A motor core (stator laminations) that can be removed and replaced by a factory-wound (prewound) stator core.

PRF — Abbreviation for pulse-repetition frequency.

pri — Abbreviation for primary.

primaries — See primary colors.

primary — Abbreviated pri and symbolized by P. Also called a primary winding. 1. A transformer winding that carries current and normally sets up a current in one or more secondary windings. 2. Pertaining to the high-voltage conductors of a power-distribution system. 3. Any one of three lights in terms of which a color is specified by giving the amounts required to duplicate it by additive combination. 4. A transformer winding that receives energy from a supply source and uses it to create a magnetic flux in the transformer core.

primary area — See primary service area.

primary battery — A battery consisting of primary cells.

primary breakdown — Also called avalanche breakdown. The sustaining mode of a transistor — unlike second breakdown, not a failure mode. The transistor collector-to-emitter voltage is relatively constant for different collector supply voltages.

primary calibration — Calibration in which the transducer output is observed or recorded while direct stimulus is applied under controlled conditions.

primary-carrier flow — Also called primary flow. The current flow responsible for the major properties of a semiconductor device.

primary cell — A cell that converts chemical energy into electrical energy by irreversible chemical reactions. It cannot be recharged (like a secondary cell) by passing an electric current through it.

primary circuit — The first, in electrical order, of two or more coupled circuits, wherein a change in current will induce a voltage in the other, or secondary, circuit.

primary colors — Also called primaries. 1. A set of colors from which all other colors are derived; hence, any set of stimuli from which all colors may be produced by mixture. 2. The three colors used in color TV, no two of which can be combined to produce the third: red, green, and blue.

primary current — The current flowing through the primary winding of a transformer. Changes in this current cause a voltage to be induced in the secondary winding of the transformer.

primary detector — Also called a sensing, primary, or initial element. The first system element or group of elements that responds quantitatively to the measurand and performs the initial measurement operation. A primary detector performs the initial conversion or control of measurement energy. It does not include those transformers, amplifiers, shunts, resistors, etc., used as auxiliary means.

primary distribution voltage — Voltage at the primary (high) side of a transformer at a distribution substation.

primary electron — 1. After a collision between two electrons, the one with the greater energy. The other is called the secondary electron. 2. The electron produced in a detector or counter tube after ionization.

primary element — See primary detector.

primary emission — Emission of electrons due to primary causes (e.g., heating of a cathode) rather than secondary effects (e.g., electron bombardment).

primary failure — A failure occurring under normal environmental conditions and having no significant relationship to a previous failure but whose occurrence imposes abnormal stress on some other part or parts, which may then undergo a secondary failure.

primary fault — In an electric circuit, initial breakdown of the insulation of a conductor, usually followed by a flow of power current.

primary flow — See primary-carrier flow.

primary frequency — The frequency assigned for normal use on a particular circuit or communications channel.

primary grid emission — See thermionic grid emission.

primary instrument concept — A theory in which single (telephone) line households and businesses would have at least one company-provided telephone; additional ones could be bought elsewhere.

primary insulation — The layer of material that is designed to do the electrical insulating, usually the first layer of material applied over the conductor.

primary ionizing event — See initial ionizing event.

primary power cable — Power service cables connecting the outside power source to the main-office switch and metering equipment.

primary radar — See radar.

primary radiation — Radiation direct from the source without interaction.

primary radiator — The antenna element from which the radiated energy leaves the transmission system.

primary relay — A relay that produces the initial action in a sequence of operations.

primary service area — Also called the primary area. The area within which radio or TV reception is not normally subject to objectionable interference or fading.

primary skip zone — The area beyond the ground-wave range around a transmitter, but within the skip distance. Radio reception is possible in this zone by sporadic and zigzag reflections.

primary standard — 1. A unit directly defined and established by some authority, and against which all secondary standards are calibrated. 2. A standard of voltage, current, frequency, etc., precisely defined by the National Bureau of Standards. Portable secondary standards are calibrated against this primary standard.

primary station — The principal amateur radio station at a specific land location shown on the station license.

primary storage — In a computer, the main internal storage.

primary voltage — 1. The voltage applied to the terminals of the primary winding in a transformer. 2. The voltage produced by a primary cell.

primary winding — *See* primary, 1 and 4.

priming illumination — A small, steady illumination applied to a phototube or photoelectric cell to make it more sensitive to variations in the illumination being measured.

primitive period — The smallest increment of time during which a quantity repeats itself.

principal axis — A reference direction for angular coordinates, used in describing the directional characteristics of a transducer employed for sound emission or reception. It is usually an axis of structural symmetry or the direction of maximum response. If these two do not coincide, however, the reference direction must be described explicitly.

principal current — The current through the main terminals of a bidirectional thyristor or through the anode and cathode terminals of a reverse-blocking or a reverse-conducting thyristor.

principal *E*-plane — The plane containing the direction of maximum radiation and in which the electric vector lies.

principal focus — For a lens or spherical mirror, the point of convergence of light coming from a source at an infinite distance.

principal *H*-plane — A plane containing the direction of maximum radiation; the electric vector is everywhere normal to the plane, and the magnetic vector lies in it.

principal mode — *See* dominant mode.

principal voltage — The voltage between the main terminals of a bidirectional thyristor or between the anode and cathode terminals of a reverse-blocking or a reverse-conducting thyristor.

print — The possible output formats of a teletypewriter or electric typewriter terminal. Print refers to tabulated output data; plot refers to a graphical arrangement of the output data performed by the typewriter with symbols such as "*x*" or "***" used to indicate data points — the plot is discontinuous. Print-plot refers to the availability of both formats at the same terminal.

print and fire — The process wherein the ink is printed on a substrate and is subsequently fired.

printed — Reproduced on a surface by some process (e.g., letterpress, lithography, silk screen, etching).

printed board — The general term for completely processed printed circuit or printed wiring configurations. It includes single, double, and multilayer boards, both rigid and flexible.

printed-board assembly — A printed circuit board to which separable components have been attached. Also an assembly of one or more printed circuit boards that may include several components.

printed cable — A cable having a thin film of copper laid onto insulation and deriving its strength from the insulation rather than from the conductor.

printed circuit — Abbreviated pc. 1. A circuit in which the interconnecting wires have been replaced by conductive strips printed, etched, etc., onto an insulating board. It may also include similarly formed components on the baseboard. 2. A substrate on which a predetermined pattern of printed wiring and printed elements has been formed. 3. A board on which a predetermined pattern of printed connections has been formed.

printed-circuit assembly — 1. A printed circuit board to which separate components have been attached.

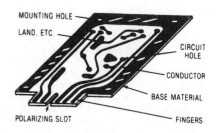

Printed circuit.

2. An assembly of one or more printed circuit boards that may include several components.

printed circuit board — Abbreviated pcb. Also called a card, chassis, or plate. An insulating board (usually Fiberglas® or plastic) onto which a circuit has been printed. *See also* printed circuit.

printed-circuit chemicals — All the cleaning solutions, resists, etchants, plating solutions, and similar materials that are specifically applied to the manufacturing of printed circuit boards.

printed-circuit motor — A motor that has a flat plastic rotor with photoetched conductors. The brushes bear directly on these conductors, and a separate commutator is not required. The rotor diameter is relatively large, resulting in wide pole faces. A large magnet is needed to create the necessary magnetic field. In addition, the winding is limited by the number of effective turns that can be printed, limiting the voltage that the motor can be designed to meet. The printed-circuit motor has some distinct advantages. Low armature inertia and low winding resistance give this motor the highest acceleration available. Since there is no iron in the rotor, the motor does not cog.

printed-circuit switch — A special rotary switch that can be connected directly to a mating printed circuit board without wires.

printed component — A type of printed circuit intended primarily for electrical and/or magnetic functions other than point-to-point connections or shielding (e.g., printed inductor, resistor, capacitor, transmission line, etc.).

printed contact — The portion of a printed circuit that connects the circuit to a plug-in receptacle and performs the function of a plug pin.

printed element — An element, such as a resistor, capacitor, or transmission line, that is formed on a circuit board by deposition, etching, etc.

printed wiring — 1. A pattern of conductors printed (screened) onto the surface of an insulating base to provide interconnection of active and passive devices to make an electronic circuit. 2. A printed circuit, or a portion thereof, intended primarily to provide point-to-point electrical connections. 3. A conductive pattern within or bonded to the surface of a base material intended for point-to-point connection of separate components and not containing printed components. 4. Conductive patterns etched or bonded to the surface of the base material to establish point-to-point connection of components.

printed wiring substrate — A conductive pattern printed on a substrate.

printer — Also called a teleprinter and teletypewriter. 1. A telegraph instrument with a signal-actuated mechanism for automatically typing received messages. It may have a keyboard similar to that of a typewriter for sending messages. The term *receiving-only* is applied to a printer with no keyboard. 2. A device that prints the output from a computer. Generally categorized as impact or nonimpact types and serial printers or line printers. 3. A device to provide hard-copy output. Unlike a terminal, there is virtually no communication from printer to central processor. Printers usually output copy more rapidly than hard-copy terminals. 4. An output device that converts electronic signals from the computer into human-readable form, or hard copy. The two major printer types are impact printers and nonimpact printers.

printer telegraph code — A five- or seven-unit code used for operation of a teleprinter, teletypewriter, or similar telegraph printer.

printing — The reproduction of a pattern on a surface by any of various processes, such as vapor deposition, photo etching, embossing, or diffusion.

printing demand meter — An integrated demand meter that prints on a paper tape the demand for each interval and indicates the time it occurred.

printing recorder — An electromechanical device used at a monitoring station that accepts coded signals from alarm lines and converts them to an alphanumeric printed record of the signal received.

printing telegraphy — Telegraph operation in which the received signals are automatically recorded as printed characters.

printout — 1. The output of a computer program as recorded by a line printer. 2. *See* display. 3. Computer output printed on paper.

print-plot — *See* print.

print server — A printing processor that provides an interface between compatible peripheral devices on a local area network.

print-through — 1. The transfer of the magnetic field from one layer to the next when recorded tape is stored on a reel. (It tends to be more of a problem with thin tapes [less than 1 mil, or 25 μm] and at very high recording levels.) Print-through causes faint "echoes" preceding and following loudly recorded passages, and is aggravated by recording at excessively high levels or by exposing recorded tapes to alternating magnetic fields, as from nearby power transformers.

print wheel — In a wheel printer, the single element providing the character set at one printing position.

priority — 1. A property that designates a process' relative urgency. 2. The order in which a system satisfies simultaneous service requests. 3. In a computer program, the number assigned to an event or device that determines the order in which it will receive service. By convention, 0 is the highest priority (usually assigned to power-failure detection). 4. The sequence in which various entries and tasks are processed or peripheral devices are serviced. Priorities are based on analyses of codes associated with an entry or task, or the positional assignment of a peripheral device within a group of devices. Order of importance.

priority indicator — A character group that indicates the relative urgency, and, thus, the order of transmission, of a message.

priority interrupt — *See* interrupt.

priority paging — Pager address that has been designated as a "priority address." This designation overrides the unit's silent mode of operation.

privacy system — In radio transmission, a system designed to make unauthorized reception difficult.

private-aircraft station — A mobile radio station on board an aircraft not operated as an air carrier.

private automatic branch exchange — *See* PABX.

private automatic exchange — *See* PAX.

private branch exchange — *See* PBX.

private exchange — A telephone exchange that serves a single organization and has no means of connection to a public telephone system.

private line — 1 A telephone line that does not go through the central office and is reserved for exclusive use of a single customer. 2. Line of communications dedicated to one customer who leases it for exclusive use. Businesses often have such lines to their offices around the country. 3. *See* dedicated line. 4. A telephone line that serves only one party. An individual line. 5. A communication line that is used only for communication between two points and does not connect with a public telephone system.

private radio carrier — A radio carrier owned and controlled by the central station organization.

privileged instruction — A computer instruction that is available only to the operating system or supervisory programs, and not to the general user.

probability — Mathematically, a number between 0 and 1 that estimates the fraction of experiments (if the same experiment were being repeated many times) in which a particular result would occur. This number can either be a subjective guess or it can be based on the empirical results of some experimentation. It can also be derived for a process so as to give the probable outcome of experimentation.

probability distribution — 1. A mathematical model showing a representation of the probabilities for all possible values of a given random variable. 2. A table of numbers or a mathematical expression that indicates the frequency with which each of all possible results of an experiment should occur.

probability of success — The likelihood that an article will function satisfactorily for a stated period of time when subjected to a specified environment.

probability theory — The study of the likelihood of occurrence of chance events. Used to predict behavior of a group, not of a single item in the group.

probable error — The amount of error that, according to the laws of probability, is most likely to occur during a measurement.

probe — 1. A resonant conductor that can be placed into a waveguide or cavity resonator to insert or withdraw electromagnetic energy. 2. A test lead that contains an active or passive network and is used with certain types of test equipment. 3. A rod placed into the slotted section of a transmission line to measure the standing-wave ratio or to inject or extract a signal. 4. The method of making a temporary electrical connection to a die so that its electrical properties can be determined. 5. An electrical device used for making contact with a circuit test point for test or debug purposes. 6. The driven element in a microwave dish antenna system. It is located in the feed and converts rf energy in the waveguide to a signal on a transmission line. 7. A pointed conductor used in making electrical contact to a circuit board pad for testing.

Probe, 2 (high-voltage).

probing — 1. The determination of radio interference by obtaining the relative interference level in the immediate area of a source by the use of a small insensitive antenna in conjunction with a receiving device. 2. Electrical testing of a semiconductor chip before it is broken out of the wafer. Electrical contact is made to the chip bonding pads so that defective circuits can be marked to eliminate them from further processing. Only low-current dc tests can be carried out by probing. 3. A testing technique that uses finely tipped probes to make electrical connections to a sample chip.

problem check — A test or tests used to aid in obtaining the correct machine solution to a problem.

problem description — In information processing, a statement of a problem. The statement may include a description of the method of solution.

problem language — The language a computer programmer uses in stating the definition of a problem.

problem-oriented language — In a computer, a source language suited to the description of a specific class of problems.

problem-solving language — A language that can be used to specify a complete solution to a problem.

procedure — Also called an algorithm. 1. In a computer, the course of action taken in solving a problem. 2. A precise step-by-step method for effecting a solution to a problem.

procedure-oriented language — 1. A programming language in which the operations to be performed are all executable and their sequence is specified by the user. This term applies to most familiar programming languages. 2. A programming language designed for the convenient expression of procedures used in the solution of a wide class of problems, e.g., FORTRAN, COBOL, APL, and C.

process — 1. Any operation or sequence of operations involving a change of energy state, composition, dimension, or other property that may be defined with respect to a datum. The term *process* is used in this standard to apply to all variables other than instrument signals. 2. The basic unit of computation within an operating system. Also termed a software process to distinguish it from an abstract process, which is the task the software process implements.

process control — 1. Automatic control of continuous operations, contrasted with numerical control, which provides automatic control of discrete operations. 2. The regulation or manipulation of variables influencing the conduct of a process in such a way as to obtain a product of desired quality and quantity in an efficient manner.

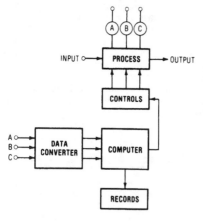

Process-control system.

process-control block — The data structure that defines a software process and its status.

processing — Additional handling, manipulation, consolidation, compositing, etc., of information to change it from one format to another or to convert it to a manageable and/or intelligible form.

processing section — The portion of a computer that does the actual changing of input into output. This includes the arithmetic and logic sections.

processor — 1. In hardware, a data processor. 2. In software, a computer program that includes the compiling, assembling, translating, and related functions for a particular programming language, including logic, memory, arithmetic, and control. 3. A unit in the programmable controller that scans all the inputs and outputs in a predetermined order. The processor monitors the status of the inputs and outputs in response to the user-programmed instructions in memory, and it energizes or deenergizes outputs as a result of the logical comparisons made through these instructions. 4. A computer or part of a computer capable of receiving data, manipulating it, and supplying results.

processor status word — Abbreviated PSW. A special-purpose CPU register that contains the status of the most recent instruction execution result, trap bit, and interrupt priority.

producer's reliability risk — The risk faced by the producer (usually set at 10 percent) that a product will be rejected by a reliability-acceptance test even though the product is actually equal to or better than a specified value of reliability.

product detector — A demodulator whose output is the product of the input signal voltage and the signal voltage of a local oscillator operating at the input frequency.

production lot — A group of (electronic) parts manufactured during the same period from the same basic raw materials, processed under the same specifications and procedures, produced with the same equipment, and identified by the documentation defined in the manufacturer's reliability assurance program through all significant manufacturing operations, including final assembly operations. Final assembly operation is considered the last major assembly operation, such as casing, hermetic sealing, or lead attachment, rather than painting or marking.

production sampling tests — Those tests normally made by either the vendor or the purchaser on a portion of a production lot for the purpose of determining the general performance level.

production tests — Those tests normally made on 100 percent of the items in a production lot by the vendor and normally on a sampling basis by the purchaser.

product modulator — A modulator whose output is substantially equal to the carrier times the modulating wave.

professional channel — Subcarrier channel in FM broadcasting. Professional channels are usually 6.5 times the frequency of the pilot carrier, or they may be interspersed between the stereo position and 102 kHz, if there is no SAP (second audio program) conflict.

professional engineer — An engineer whose education and experience qualify him or her to be responsible for important engineering work, and who is registered as a professional engineer by a state authority.

profile chart — A vertical cross-sectional drawing of the microwave path between two stations. Terrain, obstructions, antenna-height requirements, etc., are indicated on the drawing.

program — 1. A sequence of instructions that tells a computer how to receive, store, process, and deliver information. 2. A plan for solving a problem, including instructions that cause the computer to perform the desired operations and such necessary information as data description and tables. 3. A prepared list of instructions, written in a special language or code, to be carried out in sequence by a computer or other programmable device. 4. To design, write, and test such a set of coded instructions. 5. A series of actions proposed in order to achieve a certain result. 6. In a calculator, a sequence of detailed instructions for the operations necessary to solve a problem. Programmable electronic calculators can "learn" the steps of a problem so that, after the first sequence of entries, only the variable numbers need be entered on the keyboard without manual activation of control keys. 7. The statement of an algorithm in some well-defined language. Thus, a computer program represents an algorithm, although the algorithm itself is a mental concept that exists independently of any representation. 8. A set of coded instructions that direct a computer to perform some specific function, yield the solution to some specific problem, or control a machine or process. 9. A sequence of instructions that will execute a predetermined sequence of operations. 10. An organized set of instructions used to control operations of an electronic switching system. 11. A sequence of user-specified instructions that result in the execution of an algorithm. Programs are essentially written at three levels: (*a*) binary (can be directly executed by the MPU), (*b*) assembly language (symbolic representation of the binary), and (*c*) high-level language (such as BASIC; requires a compiler or interpreter). 12. A meaningful assembly of encoded instructions and data formats and data values internal to the program. 13. A sequence of audio signals alone, or audio and video signals, transmitted for entertainment or information.

program amplifier — *See* line amplifier.

program assembly — Also called translation. A process that translates a symbolic program into a machine-language program before the working program is executed. Several sections or different programs can also be integrated during this process. *See also* assembler.

program break — The length of a program; the first location not used by a program (before relocation); the relocation constant for the following program (after relocation).

program circuit — A telephone circuit that has been equalized to handle a wider range of frequencies than ordinary speech signals require. In this way, music can be transmitted over telephone wires.

program control — A control system that automatically holds or changes its target value on the basis of time, to follow a prescribed program for the process.

program counter — A CPU register that specifies the address of the next instruction to be fetched and executed. Normally it is incremented automatically each time an instruction is fetched. *See also* program register.

program-distribution amplifiers — A group of amplifiers fed by a bridging bus from a single source. Each amplifier then feeds a separate line or other service.

program element — The part of a central computer system that performs the sequence of instructions scheduled by the programmer.

program failure alarm — In broadcasting stations, a relay circuit that gives a visual and aural alarm when a program fails. A delay prevents the relay from giving a false alarm during the silence before and after station identification or other short breaks.

program flowchart — A flowchart that describes the control flow — the order in which the various program steps are executed — within any computer program or module.

program generator — A program that enables a computer to write other programs automatically.

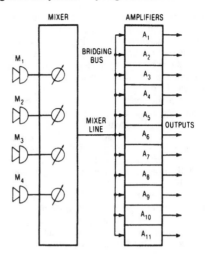

Program-distribution amplifiers.

program instruction — A group of letters, symbols, or numbers that direct a computer to perform an operation. (The instruction may also include one or more addresses.)

program level — The measure of the program signal in an audio system. It is expressed in volume units (VU).

program library — A collection of available computer programs and routines.

program linkage — In a computer, efficient use of all registers and development of subroutines so that there is smooth, economical transition from one program segment to another, and memory capacity is conserved.

program loop — A series of computer instructions that are repeated until a terminal condition is achieved.

programmable — That characteristic of a device that makes it capable of accepting data to alter the state of its internal circuitry to perform a specific task(s).

programmable calculator — 1. A calculator whose operation is controlled by programs stored in its memory. 2. Electronic calculator capable of performing preset sequences of computations. 3. One that can learn a repetitive series of operations; that is, can be programmed by various means to handle a series of steps so that only variable information need be entered into the calculator.

programmable communications processor — A digital computer that has been specifically programmed to perform one or more control and/or processing functions in a data communications network. As a self-contained system, it may or may not include communications line multiplexers, line adapters, a computer system interface, and online peripherals. It always includes a specific set of user-modifiable software for the communications function.

programmable controller — Abbreviated PC. 1. A control machine based on solid-state digital logic and built of computer subsystems, and primarily intended to take the place of electromechanical relay panels in applications in which rewiring would be made necessary by periodic changes in sequence. This type of controller is particularly useful in the control of processes, materials handling, and certain machine functions. 2. A controller whose operation is determined by codes or instructions programmed into it by the user. 3. A solid-state control system that has a user-programmable memory for storage of instructions to implement specific functions, such as I/O control logic, timing, counting, arithmetic, and data manipulation.

A PC consists of central processor, input/output interface, memory, and programming device, which typically uses relay-equivalent symbols. It is purposely designed as an industrial control system that can perform functions equivalent to a relay panel or a wired solid-state logic control system. 4. A device that provides control of a machine or process on the basis of input conditions received from the machine or process, or other inputs. It is comprised of a central processing unit, memory, and input/output. The central processor accepts inputs, and on the basis of these inputs and the instructions programmed into the unit, determines the appropriate output. The memory portion of the PC stores the instructions. 5. A digitally operating electronic apparatus that uses a programmable memory for the internal storage of instructions for implementing specific functions, such as logic, sequencing, timing, counting, and arithmetic to control, through digital or analog input/output modules, various types of machines or processes. A digital computer that is used to perform the functions of a programmable controller is considered to be within this scope. Excluded are drum and similar mechanical-type sequencing controllers. 6. A device or transmission control unit in which hardware functions have been replaced with software or microcode, so that capabilities can be added, changed, or tailored to the user's needs.

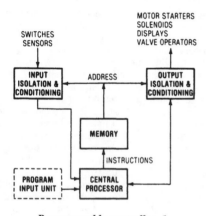

Programmable controller, 1.

programmable counter — Also called modulo-n counter. A device capable of being programmed so that it counts to any number from zero to its maximum possible modulus.

programmable logic array — Abbreviated PLA. 1. A general-purpose logic structure consisting of an array of logic circuits. The way in which these circuits are programmed determines how input signals to the PLA are processed. Programming is done on a custom basis at the factory and permanently establishes the functional operation of the PLA. 2. An LSI chip that can implement a combinatorial logic circuit involving usually over ten inputs and eight outputs. The logic is determined by the internal masking of an AND matrix and an OR matrix. 3. An array of logic gates (ANDs, ORs, NANDs, NORs, etc.) all formed on a single IC chip. The gates can be joined together to form any combinatorial logic function desired. Given a certain digital input, the collection of gates will deliver a particular digital output. 4. An arrangement of two device arrays, each with a different type of gate. The PLA's logic functions are determined by the interconnections made between the two

component arrays. PLAs are not space efficient and are more expensive than dedicated circuits, but are easier to design and can be programmed by the system integrator.

programmable logic level — Digital data with high and low levels capable of being varied under computer control.

programmable operational amplifier — An operational amplifier in which some of the parameters (such as input bias current, slew rate, power consumption, noise, etc.) can be set by means of an external resistor.

programmable read-only memory — Abbreviated PROM. 1. An integrated circuit memory array that is manufactured with a pattern of either all logical 0s or 1s and has a specific pattern written into it by the user by a special hardware programmer. (Some PROMs, called EAROMs, can be erased and reprogrammed.) 2. A field-programmable read-only memory that can have the data content of each memory cell altered only once. PROMs are generally bipolar devices programmed by blowing a fusible link.

programmable robot — A servorobot directed by a programmable controller that memorizes a sequence of arm and gripper movements; this routine can then be repeated perpetually. The robot is reprogrammed by leading its gripper through the new task.

programmable unijunction transistor — Abbreviated PUT. A four-layer device similar to a silicon-controlled retifier except that the anode gate rather than the cathode gate is brought out. It is used in conventional unijunction transistor (UJT) circuits. The characteristics of both devices are similar, but the triggering voltage of the PUT is programmable and can be set by an external resistive voltage divider network. The PUT is faster and more sensitive than the UJT. It finds limited application as a phase-control element and is most often used in long-duration timer circuits. In general, the PUT is a more versatile and more economical device than the UJT. *See also* PUT.

programmatics — The branch of learning that has to do with the study of programming methods and languages.

programmed check — A means of testing for the correctness of a computer program and machine functioning, either by running a similarly programmed sample problem with a known answer (including mathematical or logical checks) or by building a checking system into the actual program being run.

programmed logic array — Abbreviated PLA. An orderly arrangement of logical AND and logical OR functions. Its application is very much like a glorified ROM. It is primarily a combinational logic device.

programmed marginal check — A computer program that provides voltage variation to check a tube or other computer equipment during a preventive-maintenance check.

programmed operators — Computer instructions that make it possible for subroutines to be accessed with a single programmed instruction.

programmed wiring — Method by which conductors are attached to a multicontact termination panel by a programmable machine. Applicable to highly dense wiring and high production quantities. Wire attachment is by automatically wrapping the wire around a solid, square, or rectangular terminal.

programmer — 1. A person who prepares the sequences of instructions for a computer or other data-handling system. He or she may or may not convert them into detailed codes. 2. A device for timed switching of several interrelated functions or set of functions. 3. A machine or interface that will allow the programming of PROMs.

programming — 1. Definition of a computer problem resulting in a flow diagram. 2. Preparing a list of instructions for the computer to use in the solution of a problem. 3. Selecting various circuit patterns by interconnecting or "jumping" the appropriate contacts on one side of a connector plug. 4. The control of a power-supply parameter, such as output voltage, by means of a remotely located or internally located control element (usually resistance) or signal (voltage). 5. The process of translating a problem from its physical environment to a language that a computer can understand and obey. 6. The process of planning the procedure for solving a problem. This may involve, among other things, an analysis of the problem, establishing input/output formats, establishing testing and checkout procedures, allocation of storage, preparation of documentation, and supervision of the running of the program on a computer. 7. Selecting a circuit pattern from the variety of circuit options offered by a given device. Programmable devices include pin boards, programming systems, and various switches. In addition, programming is sometimes achieved by interconnecting appropriate contacts on one side of a panel or connector plug.

programming device — A device by which a series of mechanical or electrical operations or events may be preset to be performed automatically in a predetermined sequence and at specified time intervals.

programming module — A set of instructions that is discrete and identifiable and usually is handled as a unit by an assembler, compiler, linkage editor, loading routine, or other routine or subroutine.

programming station — A system configured to perform programming tasks only.

programming system — Any method of programming problems, consisting of a language (other than machine language) and its associated processor(s).

program panel — A device for inserting, monitoring, and editing a program in a programmable calculator.

program parameter — In a subroutine of a computing or other data-handling system, an adjustable parameter that can be given different values on several occasions when the subroutine is used.

program register — Also called program counter or control register. The computer control-unit register into which is stored the program instruction being executed, hence controlling the computer operation during the cycles required to execute that instruction.

program scan — The time required for a programmable calculator processor to execute all instructions in the program once. The program scan repeats continuously. The program monitors inputs and controls outputs through the input and output image tables.

program selector — Control that switches an eight-track recorder from one set of tracks to another.

program-sensitive error — In a computer, an error arising from unforeseen behavior of some circuits, discovered when a comparatively unusual combination of program steps occurs.

program signal — In audio systems and components, the complex electric wave — corresponding to speech, music, and associated sounds — destined for audible reproduction.

program step — An increment, usually one instruction, of a computer program.

program storage — A portion of the internal computer storage reserved for programs, routines, and subroutines, as contrasted with temporary storage. In many systems, protective devices are used so that the contents of the program storage cannot be altered inadvertently.

program stubs — Code sections added to a subprogram to make it executable; stubs are usually substitutes for parts of the missing main program.

program tape — In a computer, a magnetic or punched paper tape that contains the sequence of instructions for solving a problem.

program time — The phase of computer operation during which an instruction is being interpreted so that the required action can be performed.

program timer — 1. A loosely used term sometimes referring to a complex multicircuit timing device in which the program is readily changed, such as tape-controlled timers, cam timers, time switches, or to any other type of timer. 2. A multiple-circuit repeat-cycle timer that repeats a preset program continuously as long as power is applied.

progressive scanning — A rectilinear process in which adjacent lines are scanned in succession. In television, the scanning process in which the distance from center to center of successively scanned lines is equal to the nominal line width.

projected peak point — The point on the characteristic of a tunnel diode where the current is equal to the peak-point current, but where the voltage is greater than the valley-point voltage.

projected peak-point voltage — The voltage value at the projected peak point.

project engineer — Engineer in charge of a project; may be the designer of the system, and even in charge of purchasing for project.

projection cathode-ray tube — A cathode-ray tube that produces an intense but relatively small image, which can be projected onto a large viewing screen by an optical system consisting of lenses or a combination of lenses and mirrors.

projection PPI — A unit in which the image of a 4-inch (10.1-cm) dark-trace cathode-ray tube is projected on a 24-inch (61-cm) horizontal plotting surface. The echoes appear as magenta-colored arcs on a white background. *See also* dark-trace tube.

projection television — A combination of lenses and mirrors for projecting an enlarged television picture onto a screen.

projection welding — Resistance welding in which heat is localized through embossed or coined projections on the fastener. During the welding process, the projections coalesce with the part surface to form the weld.

projector — 1. A device used in an underwater sound system to radiate sound pulses through the water from the bottom of a ship. 2. A horn designed to direct sound chiefly in one direction from a speaker. 3. A lighting unit that, by means of mirrors and lenses, concentrates the light to a limited solid angle so as to obtain a high value of luminous intensity.

PROM — Abbreviation for programmable read-only memory (programmable ROM). 1. A read-only memory that can be written to only once. Programmed after manufacture by external equipment. Typically, PROMs utilize fusible links that are burned open to set a specific memory location to a specific logic level. After a PROM is programmed, it effectively becomes a ROM. 2. Similar to the conventional ROM (read-only memory). A write-once memory. When an instruction is written via a memory-write cycle into the programmable ROM, certain kinds of fusing take place and the data is written permanently into the memory. 3. A digital storage device that can be written into only once but then can be continually read. 4. An integrated-circuit memory array manufactured with a pattern of either all logical 0s or a specific pattern written into it by the user using a special hardware programmer. Some PROMs, called EAROMs (electrically alterable read-only memory), can be erased and reprogrammed.

promethium cell — A low-power cell containing a radioactive isotope called promethium 147, which emits beta particles that strike a phosphor. Two photocells then convert the light output from the phosphor into electrical energy.

prompt — 1. A symbol that alerts the user that a device is online and connected to a channel for transmission. 2. Information sent to a PC from the remote computer requesting more input. 3. An on-screen hint to the user about what to do next.

prong — *See* pin.

proof pressure — 1. The maximum pressure that may be applied to the sensing element of a transducer without changing the transducer performance beyond specified tolerances. 2. The maximum pressure that a diaphragm, capsule, or element can withstand without permanent deformation. Expressed in terms of input pressure.

propagated error — An error that is carried through succeeding computer operations.

propagation — Also called wave propagation. 1. The travel of electromagnetic waves or sound waves through a medium. Propagation does not refer to the flow of current in the ordinary sense. 2. The means by which radio signals are carried from one location to another.

propagation anomaly — An irregularity introduced into an electromagnetic or other sensing device by discontinuities in the propagation medium.

propagation constant — 1. The transmission characteristic that indicates the effect of a line on a wave being transmitted along the line. It is a complex quantity having a real term called the attenuation constant and an imaginary term called the phase constant. 2. Also called transfer factor. The natural logarithm of the ratio of the current into an electric transducer to the current out of the transducer, with the transducer terminated in its iterative impedance.

propagation delay — 1. A measure of the time required for a logic signal to travel through a device or a series of devices forming a logic string. It occurs as the result of four types of circuit delays — storage, rise, fall, and turn-on delay — and is the time between when the input signal crosses the threshold-voltage point and when the responding voltage at the output crosses the same voltage point. 2. Time delay occurring between the application of an input to a digital logic circuit and the change of state at the output. 3. The time necessary for a signal to travel from one point in a circuit to another. 4. The time required for a pulse or a level transition to propagate through a device. 5. *See* insertion delay.

propagation error — For ranging systems, the algebraic sum of propagation-velocity error and curved-path error. Except at long ranges and low angles, the curved-path component of propagation error is generally negligible.

propagation factor — *See* propagation ratio.

propagation loss — The loss of energy suffered by a signal while passing between two points.

propagation ratio — Also called the propagation factor. For a wave that has been propagated from one point to another, the ratio of the complex electric-field strength at the second point to that at the first point.

propagation time — 1. The time it takes for a signal to travel from point to point. In a communications channel, the velocity of signal propagation is less than that of radio. A signal delay of 20 ms per thousand miles or 1600 kilometers is a reasonable maximum. 2. The time required for transmission of a unit of binary information (high or low voltage) from one physical point in a system or subsystem to another, such as from the input to the output of a device.

propagation time delay — The time required for a wave to travel from one point to another along a transmission line. It varies according to the type of line.

propagation velocity — *See* velocity of propagation.

property sort — In a computer, a technique for the selection of records meeting a certain criterion from a file.

proportional — 1. A linear (straight line) relationship between two variables. 2. Method of printing characters in which each character takes up only the horizontal space it needs, rather than a fixed amount of space, e.g., the character "i" takes up less space than "m."

proportional band — The range of the controlled variable corresponding to the full range of operation of the final control element.

proportional control — 1. A method of control in which the intensity of action varies linearly as the condition being regulated deviates from the conditions prescribed. 2. Also called galloping ghost. An advanced type of radio-control system in which the rudder (and sometimes the elevator) can move as much (or as little) as the operator wishes. 3. A control system in which corrective action is always proportionate to any variation of the controlled process from its desired value. For example, instead of snapping directly open-closed in the manner of two-position control, a proportional valve will be always positioned at some point between open and closed, depending on the flow requirement of the system at any given moment.

proportional counter tube — A sealed tube containing an inert gas such as argon, krypton, xenon, methyl bromide, etc. It is used like a Geiger-Mueller counter and operated at about 100 volts in the proportional region.

proportional linearity — A manner of expressing nonlinearity as the deviation from a straight line in terms of a given percentage of the transducer output at the stimulus point under consideration (i.e., as a percentage of the reading).

proportional region — In a radiation counter tube, the applied-voltage range in which the gas amplification is greater than unity and is independent of the charge liberated by the initial ionizing event.

proportional temperature control — A method of stabilizing (an oscillator) by providing heater power that is directly proportional to the difference between the desired operating temperature and the ambient temperature.

proprietary alarm system — An alarm system that is similar to a central station alarm system except that the annunciator is located in a constantly manned guard room maintained by the owner for his or her own internal security operations. The guards monitor the system and respond to all alarm signals or alert local law enforcement agencies, or both.

protected area — An area monitored by an alarm system or guards, or enclosed by a suitable barrier.

protected location — A computer storage location reserved for special purposes, in which data cannot be stored without being subjected to a screening procedure to establish suitability for storage at that location.

protected memory — Storage (memory) locations reserved for special purposes, in which data cannot be entered directly by the user.

protected port — A point of entry, such as a door, window, or corridor, that is monitored by sensors connected to an alarm system.

protected wireline distribution system — Also known as approved circuit. A communications system to which electromagnetic and physical safeguards have been applied to permit secure electrical transmission of unencrypted, classified information, and that has been approved by the cognizant department or agency. The associated facilities include all equipment and wirelines so safeguarded. Major components are wirelines, subscriber sets, and terminal equipment.

protection — An operating system service that prevents a task from interfering with the execution of another task.

protection device — 1. A sensor such as a grid, foil, contact, or photoelectric sensor connected into an intrusion alarm system. 2. A barrier that inhibits intrusion, such as a grill, lock, fence, or wall.

protective cable — Small-gage quadded cable used in toll cables to serve as fuses, usually at building entrances.

protective device — Any device for keeping an undesirably large current, voltage, or power out of a given part of an electric circuit.

protective gap — A spark gap provided between a conductor and the earth by suitable electrodes. High-voltage surges due to lightning are thus permitted to pass harmlessly to earth through the gap.

protective relay — A relay whose principal function is to protect services from interruption or to prevent or limit damage to apparatus.

protective resistance — A resistance placed in series with a device (e.g., a gas tube) to limit the current to a safe value.

protective signaling — The initiation, transmission, and reception of signals involved in the detection and prevention of property loss due to fire, burglary, or other destructive conditions. Also, the electronic supervision of persons and equipment concerned with this detection and prevention. *See also* line supervision; supervisory alarm system.

protector — 1. A device to protect equipment or personnel from high voltage or current. 2. A protective device used on communication systems to limit the magnitude of extraneous overvoltages. The discharge device within a protector may consist of closely spaced carbon electrodes discharging in air, or metallic electrodes discharging in a hermetically sealed gaseous atmosphere at reduced pressure. A protector does not contain an element to prevent holdover, as in the case of an arrester.

protector block — A rectangular piece of carbon with an insulated metal insert, or porcelain with a carbon insert, that makes one element of a protector. It forms the gap that will break down and provide a path to ground for voltages over 350 volts.

protector tube — A glow-discharge cold-cathode tube in which a low-voltage breakdown is employed between two or more electrodes to protect the circuit against overvoltage.

protocol — 1. A set of conventions or rules governing the format and timing of message exchanges to control data movements and correct errors. It is important to ensure that the protocol is valid, makes sense, works, and is adhered to by all users of the network in question. 2. The set of multiprocessor system rules that define response sequences (handshaking) and maintain the servicing priorities of the system. 3. A defined means of establishing criteria for receiving and transmitting data through communication channels. 4. A formal set of conventions governing the format and control of inputs and outputs between two communicating processes, including handshaking and line discipline. 5. A set of conventions for the transfer of information between devices. The simplest protocols define only the hardware configuration. More complex protocols define timings, data formats, error detection and correction techniques, and software structures. The most powerful protocols describe each level of the transfer process as a layer separate from the rest, so that certain layers such as the interconnecting

hardware can be changed without affecting the whole. 6. The convention by which data is transmitted over a line. Most word proccessors employ either an asynchronous or synchronous protocol. With an asynchronous protocol, data is transmitted a character at a time, with a start and stop bit transmitted before and after each character to ensure correct receipt. Synchronous protocol is a form of transmission that uses no redundant information (such as the start and stop bits in asynchronous transmission) to identify the beginning and end of each character.

proton — One of the three basic subatomic particles, with a positive charge equivalent to the negative charge of the electron, but with approximately 1845 times the mass. The proton has a positive electric charge of 1.6×10^{-19} coulomb and a mass of 1 amu. The proton is the positive nucleus of the hydrogen atom.

prototype — 1. Original design or first operating model. 2. A development or first production model of a circuit, device, or system.

prototype model — A working model, usually hand-assembled, that is suitable for complete evaluation of mechanical and electrical form, design, and performance. Approved parts are employed throughout, so that it will be completely representative of the final, mass-produced equipment.

prototyping kit — A hardware system used to bread-board a microprocessor-based product. Contains CPU, memory, basic I/O, power supply, switches and lamps, provisions for custom I/O controllers, memory expansion, and, often, a utility program in fixed memory (ROM).

proximity alarm system — *See* capacitance alarm system.

proximity detector — A sensing device that gives an indication when approaching or being approached by another object.

proximity effect — The redistribution of current brought about in a conductor by the presence of another current-carrying conductor.

proximity fuse — A fuse designed to detonate a projectile, bomb, mine, or charge when activated by an external influence in the vicinity of a target. The variable time fuse is one type of proximity fuse.

proximity switch — A device that reacts to the proximity of an actuating means without physical contact or connection.

PRR — Abbreviation for pulse-repetition rate.

PRV (or PIV) rating — Abbreviation for peak reverse voltage or peak inverse voltage. The maximum instantaneous value of reversing voltage allowable across a diode or similar device.

ps — Letter symbol for picosecond (10^{-12} second).

PSD — Abbreviation for power spectral density. The function of the variance of a random process with frequency, used to describe distribution of signal power in the frequency domain.

pseudocode — 1. An instruction that is not meant to be followed directly by a computer. Instead, it initiates the linking of a subroutine into the main program. 2. A series of natural-language statements arranged in a manner resembling a program. There is no firm connection with any actual programming language. Used for instructional purposes or as a preliminary rough sketch of a program.

pseudodifferential input — Amplifier input configuration in which the reference point is other than ground.

pseudoinstruction — A user-defined instruction (such as a macro) that does not belong to the basic instruction set of an MPU.

pseudo-op — 1. An operation that is not part of the computer's operation repertoire as realized by hardware; hence, an extension of the set of machine operations.

2. A directive used in assembly language to control the operation of the assembler. The pseudo-op does not appear in the machine-language program produced by the assembler. A typical pseudo-op is the word "END" written at the end of a program to tell the assembler its work is done.

pseudoprogram — A program that is written in a pseudocode and may include short coded logical routines.

pseudorandom — 1. Having the property of being produced by a definite calculation process while simultaneously satisfying one or more of the standard tests for statistical randomness. 2. An apparently random output that has been created by an algorithm.

pseudorandom binary sequence — A two-level signal that has a repetitive sequence, but a random pattern within the sequence. Such a signal finds use as a test signal since it has the basic characteristics of noise, but in terms of parameters that are easily controlled.

pseudorandom number sequence — A sequence of numbers, determined by some defined arithmetic process, that is satisfactorily random for a given purpose, such as by satisfying one or more of the standard statistical tests for randomness. Such a sequence may approximate any one of several statistical distributions, such as the uniform distribution or normal Gaussian distribution.

pseudorandom patterns — 1. A sequence of digital patterns, determined by some defined process that is satisfactorily random to perform its intended function. 2. A repeatable sequence of input test patterns that appears statistically random. 3. A repeatable sequence of digital patterns that has the appearance of being random.

pseudoshot noise — *See* partition noise.

pseudostereo — Devices and techniques for obtaining stereo qualities from one channel.

psophometric emf — The electromotive force (or voltage) generated by a source having an internal resistance of 600 ohms and no internal reactance, which, when connected across a standard receiver having 600 ohms of resistance and no reactance, produces the same sinusoidal current as an 800-hertz generator having the same impedance. *See also* psophometric voltage.

psophometric voltage — The voltage that would appear across a 600-ohm resistance connected between any two points in a telephone circuit. (This value is one-half the psophometric emf, since the latter is essentially the open-circuit potential necessary from a source to produce the psophometric voltage if the source has a 600-ohm internal resistance.)

PSOS — A term used by some companies to denote p-channel silicon gate devices.

PSPS — Abbreviation for planar silicon photoswitch.

PSW — Abbreviation for processor status word.

psychoacoustics — A relatively new branch of audio that concerns itself with personal and subjective factors in hearing and in evaluating the performance of high-fidelity equipment.

PTM — Abbreviation for pulse-time modulation.

PTM/PPM/AM — A system in which a number of pulse-position or pulse-time modulated subcarriers are used to amplitude modulate the carrier.

PTT — 1. Abbreviation for postal telephone and telegraph. A generic term for European telephone companies, which are generally operated by the country's postal service. 2. Abbreviation for press-to-talk or push-to-talk, generally with reference to a push-button switch on a microphone that is operated to turn the transmitter on and is released to enable the receiver.

p-type — Pertaining to semiconductor material that has been doped with an excess of acceptor impurity atoms so that free holes are produced in the material.

p-type conductivity — The conductivity associated with the holes in a semiconductor.

p-type conductor — A positive-type conductor, one with electron holes as the principal carriers. This implies the presence of acceptors.

p-type crystal rectifier — A crystal rectifier in which forward current flows whenever the semiconductor is more positive than the metal.

p-type material — 1. A pure crystal of semiconductor material to which an impurity has been added (electron acceptor such as boron or gallium) to give it a deficiency of electrons and alter its electrical characteristics. 2. Semiconductor material having holes as the majority charge carriers; formed by doping with acceptor atoms. 3. Refers to an excess of positive electrical charges in a semiconductor material. Natural silicon is made to be p-type by the addition of an acceptor impurity.

p-type region — Portion of semiconductor material containing a small number of dopant atoms that have an electron deficiency (an empty space) in their outer orbit. The deficiency is called a hole and behaves like a positively charged particle. The p-type region is a source of mobile positive charges.

p-type semiconductor — 1. An extrinsic semiconductor in which the hole density exceeds the conduction-electron density. By implication, the net ionized impurity concentration is an acceptor type. 2. A semiconductor material doped so that it has a net deficiency of free electrons. It therefore conducts electricity through movement of holes. 3. Impurity semiconductor material containing a predominance of acceptor dopants, and in which valence-band holes normally form the principal current carriers.

public-address system — Also called a PA system. One or more microphones, an audio-frequency system, and one or more speakers used for picking up and amplifying sounds to a large audience, either indoors or out.

publication language — A well-defined form of a programming language suitable for printing. A publication language is necessary because some languages use special characters that are not available in normal type.

public communications service — Telephone or telegraph service provided for the transmission of unofficial communications for the public.

public correspondence — Any telecommunication that the offices and stations, by reason of their being at the disposal of the public, must accept for transmission.

public radiocommunication services — Land mobile or fixed radio services, the stations of which are open to public correspondence.

public-safety radio service — Any radio-communication service essential either to the discharge of non-federal governmental functions relating to public-safety responsibilities or to the alleviation of an emergency endangering life or property. The radio transmitting facilities in this service may be fixed, land, or mobile stations.

public switched network — Any switching system that provides circuit switching to many customers. In the United States there are four such networks: Telex, TWX, telephone, and Broadband Exchange.

puck — 1. Manually operated directional control device used to input coordinate data. 2. *See* pressure roller.

pucker pocket — A small, angular vacuum column that isolates the magnetic tape and air mass of a vacuum-column drive's main (large) columns from the large accelerations of the tape in the head area.

puff — British abbreviation for picofarad.

pull — 1. To cause an oscillator to depart from its designed frequency of operation. 2. To depart from the designed frequency of operation, as an oscillator. 3. Retrieving a byte of data from the data stack.

pull curves — The characteristics relating force to displacement in the actuating system of a relay.

pull-down resistor — 1. A resistor connected across the output of a device or circuit to hold the output equal to or less than the zero input level of the following digital device. Also used to lower the output impedance of a device. 2. A resistor connected to a negative voltage or to ground.

pull-in current (or voltage) — The maximum current (or voltage) required to operate a relay. *See also* pickup current.

pulling — 1. In an oscillator, the undesired change from the desired frequency. It is caused either by coupling from another source of frequency or by the influence of the load impedance. 2. In television, partial loss of synchronization.

pulling figure — The difference between the maximum and minimum frequencies of an oscillator whenever the phase angle of the load-impedance reflection coefficient varied through 360°. The absolute value of this coefficient is constant and equal to 0.20.

pull-in rate — The maximum stepping rate at which a stepper motor can start its load without missing a step.

pull-in torque — 1. Torque that a synchronous motor can exert to bring its driven load into synchronous speed. There is no corresponding term for induction motors. 2. A measure of the maximum torque that can be applied to the shaft of a stepper motor without causing it to miss a step when starting.

pull-out force — The tensile force required to separate a conductor from a contact or terminal, or to separate a contact from a connector.

pull-out rate — The maximum stepping rate at which a stepper motor can move its load without losing synchronism with the field.

pull-out torque — Also called breakdown torque, or maximum torque. 1. The maximum torque a motor can deliver without stalling. 2. *See* running torque. 3. The maximum torque that a synchronous motor develops at synchronous speed at rated frequency and normal excitation. 4. A measure of the maximum torque that can be applied to the shaft of a stepper motor running at a constant speed within its pull-out ratings before the motor loses synchronism with the field.

pull strength — The values of the pressure achieved in a test in which a pulling stress is applied to determine breaking strength of a lead or bond.

pull test — A test for bond strength of a lead, interconnecting wire, or a conductor.

pull the plug — CB radio term for shut off the radio.

pull-up — 1. The placing of the output voltage of a logic circuit at the high level by means of an internal current sink or source. 2. A dc voltage imposed on the input of an amplifier to move the amplifier's operating point out of the offset range. Pull-up is usually accomplished by means of a voltage divider network.

pull-up resistor — 1. A resistor connected to the positive supply voltage of a transistor circuit, as from the collector supply to the output collector. 2. A resistor connected across the output of a device or circuit to hold the output voltage equal to or greater than the input transition level of a digital device. It is usually connected to a positive voltage or to the plus supply.

pull-up torque — 1. The minimum torque developed by an alternating-current motor during the period of acceleration from rest to the speed at which breakdown torque occurs. For motors that do not have a definite breakdown torque, the pull-up torque is the minimum torque developed up to rated speed. 2. Lowest value of torque produced by a motor between zero speed and full-load speed.

pulsating current — Current that varies in amplitude but does not change polarity.

pulsating direct current — A direct current that changes its value at regular or irregular intervals but flows in the same direction at all times.

pulsating electromotive force — A direct electromotive force and an alternating electromotive force combined.

pulsating quantity — A periodic quantity that can be considered the sum of a continuous component and an alternating component in the quantity.

pulsation welding — A form of resistance welding in which the power is alternately applied and removed.

pulse — 1. A variation of a quantity whose value is normally constant; this variation is characterized by a rise and a decay and has finite amplitude and duration. 2. An abrupt change in voltage, either positive or negative, that conveys information to a circuit. *See also* impulse. 3. A brief excursion of a quantity from normal. 4. Signal characterized by the rise and decay in time of a quantity whose value is normally constant. 5. Voltage level, typically 5 volts of very short duration, used in computers to represent a bit. 6. Single impulse of a telephone dial. Generally transmitted in groups of one to ten to represent dialed digits or unique tones to represent digits. 7. A sudden and abrupt jump in an electrical quantity from its usual level to a higher or lower value, quickly followed by an equally abrupt return. 8. A voltage or current that lasts for a short period and is square or Gaussian in shape.

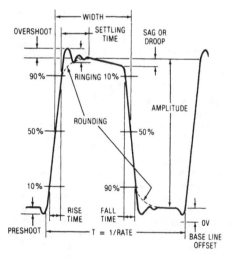

Pulse, 1.

pulse amplification — The compression and intensification of a laser pulse of a specific width into a smaller pulse width. A spherical cavity, in conjunction with a beam compressor, is efficient for pulse amplification. Cones and flats are highly effective when used in conjunction with swept-line foci.

pulse amplifier — A wideband amplifier used to amplify square waves without appreciably changing their shape.

pulse amplitude — A general term for the magnitude of a pulse. For more specific designation, adjectives such as average, instantaneous, peak, rms (effective), etc., should also be used.

pulse-amplitude modulation — Abbreviated PAM. Modulation in which the modulating wave is caused to amplitude modulate a pulse carrier.

pulse-amplitude modulation/frequency modulation — A system in which a carrier is frequency modulated by pulse-amplitude-modulated subcarriers.

pulse analyzer — 1. Equipment for analyzing pulses to determine their time, amplitude, duration, shape, etc. 2. The instrument used to analyze a pulsed electromagnetic wave to determine its time, amplitude, duration, and shape, and to display this information in some appropriate form, either visually or digitally.

pulse arc welding — A type of welding in which the material to be welded is positioned together, forming one electrode. The other electrode is positioned to form a gap with one of the workpieces. An arc is struck and the current heats the work-pieces to the melting point at their interface. *See also* arc percussive welding.

pulse average time — The duration of a pulse, measured between two points at 50 percent of the maximum amplitude on the leading and trailing edges.

pulse bandwidth — The smallest continuous frequency interval outside of which the amplitude of the spectrum does not exceed a prescribed fraction of the amplitude at a specified frequency.

pulse capacitor — A capacitor for use with pulses of current or voltage.

pulse carrier — A carrier consisting of a series of pulses. Usually employed as a subcarrier.

pulse code — 1. A pulse or series of pulses that, by means of waveform, pulse width, pulse time, pulse numbers, or pulse sequences, may be used to convey information. 2. Loosely, a code consisting of pulses; e.g., Morse, Baudot, binary.

pulse code modulation — Abbreviated PCM. 1. Pulsed modulation in which the signal is sampled periodically and each sample is quantized and transmitted as a digital binary code. 2. A digital technique by which information may be carried from one point to another. The signal is carried as a series of separate pulses or digits. No distortion is introduced and no information is lost from a signal unless a complete pulse disappears or unless a spurious noise pulse is formed that is large enough to be accepted by the equipment as a genuine pulse. Thus, many channels of communication can be made available along a single connecting line. 3. A method of quantizing audio-range analog signals into a digital form for transmission in digital communications systems, or for processing in DSP. Effectively the same as analog-to-digital conversion. 4. A time-division modulation technique in which analog signals are stamped and quantized at periodic intervals into digital signals. The values observed are typically represented by a coded arrangement of 8 bits, of which one may be for parity.

pulse-code modulation/frequency modulation — A system in which pulse-code-modulated subcarriers are used to frequency modulate a second carrier. Binary digits are formed by the absence or presence of a pulse in an assigned position.

pulse coder — A circuit that sets up pulses in an identifiable pattern.

pulse coding and correlation — A general technique concerning a variety of methods used to change the transmitted waveform and then decode upon its reception. Pulse compression is a special form of pulse coding and correlation.

pulse compression — A matched filter technique used to discriminate against signals that do not correspond to the transmitted signal.

pulse counter — A device that gives an indication or record of the total number of pulses that it has received during a given time interval.

pulse counter detector — A device designed to detect frequency-modulated signals by forming a unidirectional pulse from each sine wave. The direct current of the pulse is proportional to the frequency of the frequency-modulation signal.

pulsed Doppler system — A pulsed radar system that utilizes the Doppler effect to obtain information about the target (not including simple resolution from fixed targets).

pulse decay time — The amount of time required for the trailing edge of a pulse to decay from 90 percent to 10 percent of the peak pulse amplitude.

pulse delay time — The time interval between the leading edges of the input and output pulses, measured at 10 percent of their maximum amplitude.

pulse demoder — Also called a constant delay discriminator. A circuit that responds only to pulse signals with a specified spacing between them.

pulse dialing — Older form of phone dialing, utilizing breaks in dc current to indicate the number being dialed.

pulse digit — A code element comprising the immediately associated train of pulses.

pulse-digit spacing — The time interval between the end of one pulse digit and the start of the next.

pulse discriminator — A device that responds only to pulses having a particular characteristic (e.g., duration, amplitude, period). One that responds to period is also called a time discriminator.

pulse dispersion — In fiber optics, separation or spreading of input optical signals along the length of the optical fiber. Pulse dispersion is expressed in time and distance as nanoseconds and is sometimes called pulse spreading.

pulsed laser — A laser that emits energy in a wave of short bursts or pulses and remains inactive between each burst or pulse. The frequency of the pulses is termed the pulse repetition frequency.

pulsed light — A beam of visible radiant energy of finite duration. The beam rises to a finite amplitude and decays to the same value from which it rose.

pulsed oscillator — 1. An oscillator in which oscillations are sustained by either self-generated or external pulses. 2. An oscillator that generates a carrier-frequency pulse or a train of pulses.

pulsed-oscillator starting time — The interval between the leading-edge times of the pulse at the oscillator control terminals and the related output pulse.

pulsed power supply — A supply that delivers power in pulses, rather than continuously.

pulsed radiance — The integral of the radiance over exposure time.

pulse droop — Distortion characterized by a slanting of the top of an otherwise essentially flat-topped rectangular pulse.

pulsed ruby laser — A laser that uses ruby as the active material. The extremely high pumping power required is obtained by discharging a bank of energy storage capacitors through a special high-intensity flash tube. *See* figure on page 599.

pulse duration — Also called pulse length or pulse width. 1. The time interval between the points at which the instantaneous value on the leading and trailing edges bears a specified relationship to the peak pulse amplitude. 2. The lifetime of a (laser) pulse, generally defined as the time interval between the half-power points on the leading and trailing edges of the pulse.

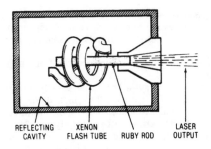

Pulsed ruby laser.

REFLECTING CAVITY   XENON FLASH TUBE   RUBY ROD   LASER OUTPUT

pulse-duration discriminator — A circuit in which the sense and magnitude of the output is a function of the deviation of the pulse length from a reference.

pulse-duration modulation — Abbreviated PDM. Also called by the less preferred terms pulse-width modulation and pulse-length modulation. 1. Pulse-time modulation in which the duration of a pulse is varied. 2. Pulse-time modulation in which the value of each instantaneous sample of the modulating wave is caused to modulate the duration of a pulse.

pulse-duration modulation/frequency modulation — Also called pulse-width modulation/frequency modulation. A system in which pulse-duration-modulated subcarriers are used to frequency modulate a second carrier.

pulse duty factor — Ratio of the average pulse duration to the average pulse spacing. Equivalent to the average pulse duration times the pulse-repetition rate.

pulse emission — Emission drawn for short periods; it may or may not follow a regular repetition rate.

pulse emitter load — The load seen by the collector of an inverter that drives the pulse input to a flip-flop, pulse amplifier, or delay.

pulse equalizer — A circuit that produces output pulses of uniform size and shape when driven by input pulses that vary in size and shape.

pulse fall time — That time during which the trailing edge of a pulse is decreasing from 90 percent to 10 percent of its maximum amplitude.

pulse-forming line — A combination of circuit components used to produce a square pulse of controlled duration.

pulse-forming network — A network that converts either an ac or dc charging source into an approximately rectangular wave output. By means of a high-speed switch, it alternately stores energy from the charging source and releases the energy through the load to which it is connected. This network supplies the accurately shaped pulse required by the magnetron or klystron oscillator of a radar modulator.

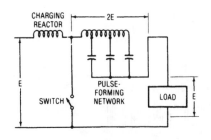

CHARGING REACTOR   2E
E   SWITCH   PULSE-FORMING NETWORK   LOAD   E

Pulse-forming network.

pulse-frequency modulation — Abbreviated PFM. More precisely called pulse-repetition-rate modulation. 1. Pulse-time modulation in which the pulse repetition rate is varied. 2. A form of modulation in which the pulse repetition rate is varied as a function of the instantaneous value of the modulating wave.

pulse-frequency spectrum — *See* pulse spectrum.

pulse generator — 1. A device for generating a controlled series of electrical pulses. 2. A device that produces a single pulse or a train of repetitive pulses.

pulse group — *See* pulse train.

pulse-height analyzer — Also called kicksorter. An instrument that indicates the number or rate of occurrence of pulses within each of one or more specified amplitude ranges.

pulse-height discriminator — Also called a pulse-height selector. A circuit that selects and passes only those pulses that exceed a certain minimum amplitude.

pulse-height selector — *See* pulse-height discriminator.

pulse-improvement threshold — In a constant-amplitude pulse-modulation system, the condition existing when the peak pulse voltage is at least twice the peak noise voltage after selection and before any nonlinear process such as amplitude clipping and limiting. The ratio of peak to rms noise voltage is ordinarily assumed to be 4 to 1.

pulse-interference eliminator — A device that removes pulse signals that are not precisely on the radar operating frequency.

pulse-interference separator and blanker — A circuit that automatically blanks all video signals that are not synchronous with the radar pulse-repetition frequency.

pulse interleaving — A process in which pulses from two or more time-division multiplexers are systematically combined in time division for transmission over a common path.

pulse interrogation — 1. The triggering of a transponder by a pulse or pulse mode. Interrogations by the latter may be employed to trigger one or more transponders. 2. Periodic electrical activation and observation, either synchronous or asynchronous, of the shaft position of an encoder.

pulse interval — *See* pulse spacing.

pulse-interval jitter — The time or displacement band within which lie all of the transitions in the same direction of a signal through a specified amplitude. This is a dynamic value to be determined at one or more specified speeds. Pulse-interval jitter is expressed as a percentage of one pulse interval at a specified speed.

pulse-interval modulation — Pulse-time modulation in which the pulse spacing is varied. *See also* pulse-spacing modulation.

pulse jitter — A relatively slight variation of the pulse spacing in a pulse train. It may be random or systematic, depending on its origin, and is generally not coherent with any imposed pulse modulation.

pulse length — *See* pulse duration.

pulse-length modulation — *See* pulse-duration modulation.

pulse-link repeater — In a telephone signaling system, equipment that receives pulses from one E and M signaling circuit and transmits corresponding pulses into another E and M signaling circuit.

pulse load — The load presented to a pulse source.

pulse mode — 1. A finite sequence of pulses in a prearranged pattern, used for selecting and isolating a communication channel. 2. The prearranged pattern in (1) above.

pulse-mode multiplex — A process or device for selecting channels by means of pulse modes. In this way, two or more channels can use the same carrier frequency.

pulse moder — A device for producing a pulse mode. *See also* pulse demoder.

pulse-modulated jamming — The use of jamming pulses that have various widths and repetition rates.

pulse-modulated radar — Radar in which the radiation consists of a series of discrete pulses.

pulse-modulated waves — Recurrent wave trains used extensively in radar. In general, their duration is shorter than the interval between them.

pulse modulation — 1. The use of a series of pulses modulated to convey information. Modulation may involve changes of pulse amplitude (PAM), position (PPM), or duration (PDM). 2. Modulation of a carrier by a train of pulses: the generation of carrier-frequency pulses. 3. Modulation of a characteristic or characteristics of a pulse carrier: transmission of information on a pulse carrier.

pulse modulator — A device that applies pulses to the element being modulated.

pulse numbers modulation — A type of modulation in which the pulse density per unit time of a carrier is varied in accordance with a modulating wave; omissions of pulses are made systematically without changing the phase or amplitude of the transmitted pulses. For example, the omission of every other pulse could represent zero modulation; the reinsertion of pulses would then represent positive modulation, and the omission of more pulses would represent negative modulation.

pulse operation — The method whereby the energy is delivered in pulses. Usually described in terms of the shape and the frequency of the pulses.

pulse oscillator — An oscillator in which the oscillations are sustained by self-generated or external pulses.

pulse packet — In radar, the volume of space occupied by the pulse energy.

pulse period — Also called impulse period. In telephony, the time required for the dial pulse springs to open and close one time.

pulse-position modulation — Abbreviated PPM. 1. Pulse-time modulation in which the value of each instantaneous sample of the wave modulates the position in time of a pulse. 2. Modulation in which a pulse is delayed from its normal position in time as a function of the modulating wave.

pulse-position modulator — A device that converts analog information to variations in pulse position.

pulser — A generator that produces extremely short, high-voltage pulses at definite recurrence rates for use in radar transmitters and similar pulsed systems.

pulse rate — 1. *See* pulse-repetition rate. 2. The rate at which pulses are fed to a drive of a stepper motor. In most cases, the pulse rate equals the stepping or running rate.

pulse ratio — Ratio of the length of any pulse to its total period.

pulse recovery — The time, usually in microseconds, required for electron flow in a diode to start or stop when voltage is suddenly applied or removed.

pulse-recurrence counting-type frequency meter — A device for measuring frequency. It uses a direct-current ammeter calibrated in pulses per second.

pulse-recurrence time — The time elapsing between the start of one transmitted pulse and the next pulse. The reciprocal of the pulse-recurrence frequency.

pulse regeneration — The restoring of a series of pulses to their original timing, form, and relative magnitude.

pulse repeater — Also called a transponder. A device that receives pulses from one circuit and transmits corresponding pulses at another frequency, waveshape, etc., into another circuit.

pulse-repetition frequency — Abbreviated PRF. The rate (usually given in hertz or pulses per second) at which pulses or pulse groups are transmitted from a radar set.

pulse-repetition period — The reciprocal of the pulse-repetition frequency.

pulse-repetition rate — Abbreviated PRR. Also called pulse rate. The average number of pulses per unit of time.

pulse-repetition-rate modulation — *See* pulse-frequency modulation.

pulse reply — The transmission of a pulse or pulse mode by a transponder as the result of an interrogation.

pulse resolution — The minimum time separation, usually in microseconds or milliseconds, between input pulses that permits proper circuit or component response.

pulse rise time — The interval of time required for the leading edge of a pulse to rise from 10 percent to 90 percent of its peak amplitude, unless some other percentage is stated.

pulse sample-and-hold circuit — A circuit that holds the final amplitude of an integrated pulse until the final amplitude of the succeeding integrated pulse is reached. A less desirable sample-and-hold circuit resets after each hold period to a fixed level before integrating a succeeding pulse.

pulse scaler — A device capable of producing an output signal whenever a prescribed number of input pulses has been received. It frequently includes indicating devices that facilitate interpolation.

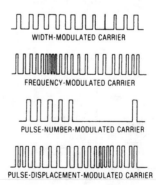

INTELLIGENCE

UNMODULATED PULSE-CHAIN CARRIER

AMPLITUDE-MODULATED CARRIER

WIDTH-MODULATED CARRIER

FREQUENCY-MODULATED CARRIER

PULSE-NUMBER-MODULATED CARRIER

PULSE-DISPLACEMENT-MODULATED CARRIER

Pulse modulation, 3.

pulse selector — A circuit or device that selects the proper pulse from a sequence of (telemetering) pulses.

pulse separation — The interval between the trailing-edge pulse time of a pulse and the leading-edge pulse time of the succeeding pulse.

pulse shaper — Any transducer (including pulse regenerators) used for changing one or more characteristics of a pulse.

pulse shaping — Intentionally changing the shape of a pulse.

pulse soldering — Soldering a connection by melting the solder in the joint area by pulsing current through a high-resistance point applied to the joint area and the solder.

pulse spacing — The time interval from one pulse to the next — i.e., between the corresponding times of two consecutive pulses. (The term *pulse interval* for this concept is ambiguous — it may be taken to mean the duration of a pulse instead of the space or interval between pulses.)

pulse-spacing modulation — Formerly called pulse-interval modulation. A form of pulse-time modulation in which the pulse spacing is varied.

pulse spectrum — Also called pulse-frequency spectrum. The frequency distribution, in relative amplitude and phase, of the sinusoidal components of a pulse.

pulse spike — A relatively short-duration pulse superimposed on the main pulse.

pulse-spike amplitude — The peak amplitude of a pulse spike.

pulse spreading — 1. The cumulative effect of material and modal dispersions in an optical fiber. 2. *See* dispersion, 7.

pulse stepper — A stepper motor that responds directly to a pulse of specified length and amplitude. The positioning of the motor shaft is directly proportionate to the number of pulses applied. Rotational direction is controlled by electrical shading.

pulse-storage time — The time interval from a point at 90 percent of the maximum amplitude on the trailing edge of the input pulse to the same 90-percent point on the trailing edge of the output pulse.

pulse stretcher — 1. A circuit designed to extend the duration of a pulse — primarily so that its pulse modulation will be more readily discernible in an audio presentation. 2. In a computer, a circuit that generates a long pulse when triggered by a short pulse. The width of the output pulse is determined by the value of the coupling capacitor. The maximum width of the output pulse cannot exceed 50 percent of the clock rate.

pulse tilt — A distortion characterized in an otherwise essentially flat-topped rectangular pulse by either a decline or a rise of the pulse top.

pulse time — The time interval from a point at 90 percent of the maximum amplitude on the leading edge of a pulse to the 90-percent point on the trailing edge.

pulse-time-modulated radiosonde — Also called time-interval radiosonde. A radiosonde that transmits the indications of the meteorological sensing elements in the form of pulses spaced in time. The meteorological data is evaluated from the intervals between the pulses.

pulse-time modulation — Abbreviated PTM. Modulation (e.g., pulse-duration and pulse-position) in which the values of instantaneous samples of the modulating wave are made to modulate the occurrence time of some characteristic of a pulse carrier.

pulse timer — A timer whose time cycle is started with a continuous input voltage and application or removal of an additional positive input voltage pulse. At the end of the time-delay period, the contacts transfer.

pulse train — Also called pulse group or impulse train. 1. A group or sequence of pulses of similar characteristics. 2. A succession of pulses, usually at equal intervals.

pulse-train frequency spectrum — *See* pulse-train spectrum.

pulse-train spectrum — Also called pulse-train frequency spectrum. The frequency distribution, in amplitude and in phase angle, of the sinusoidal components of the pulse train.

pulse transformer — 1. A special type of transformer designed to pass pulse waveforms as distinguished from sine waves. The major features of a pulse transformer are high-voltage insulation between windings and to ground, low capacitance between windings, and low reactance in the windings. 2. A transformer capable of passing a wide band of frequencies; thus, it can pass a pulse with minimal distortion.

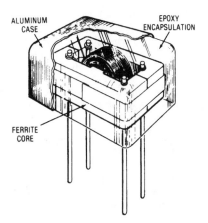

Pulse transformer.

pulse transmitter — 1. A pulse-modulated transmitter in which the peak power-output capabilities are usually larger than the average power-output rating. 2. A transmitter used to generate and transmit pulses over a telemetering or pilot-wire circuit to the remote indicating or receiving device.

pulse-triggered binary — A flip-flop in which a change of state results from application of a pulse or waveform of short duration to the input.

pulse valley — *See* pulse duration.

pulse width — In a pulse, the portion between two specified maxima.

pulse-width discriminator — A device that measures the pulse length of video signals and passes only those whose time duration falls into some predetermined design tolerance.

pulse-width modulation — *See* pulse-duration modulation.

pulse-width modulation/frequency modulation — *See* pulse-duration modulation/frequency modulation.

pulse-width modulator amplifier — *See* class D amplifier.

pulsing key — 1. A method of transmitting voice-frequency pulses over a line under control of a key at the original office. It is used with E and M supervision in intertoll dialing. 2. A system of signaling in which numbered keys are used instead of a dial.

pulsing transformer — A transformer designed to supply pulses of voltage or current.

pump — 1. An external source used to increase the electron population of excited energy states. 2. Of a parametric device, the source of alternating-current power that causes the nonlinear reactor to behave as a time-varying reactance. 3. To supply high-frequency energy to a maser, laser, parametric amplifier, etc.

pumped tube — An electron tube (chiefly a pool cathode) that is continuously connected to evacuating equipment during operation.

pumping — Of a laser, the application of radiation of appropriate frequency to invert the distribution of systems of electrons of the laser media so that the levels of higher-energy states are more populated.

pumping band — A group of energy states to which ions in the ground state are excited at first when pumping radiation is applied to a laser medium. The pumping band usually is higher in energy than the levels that are to be inverted.

pumping frequency — The frequency at which pumping is provided in a maser, quadrupole amplifier, or other amplifier requiring high-frequency excitation.

pumping radiation — Light applied to the sides or end of a laser crystal for excitation of the ions to the pumping band.

pump oscillator — An alternating-current generator that supplies pumping energy for maser and parametric amplifiers. Operates at twice or some higher multiple of the signal frequency.

punch-card machine — See key punch.

punched card — A heavy, stiff paper of constant size and shape, suitable for punching in a pattern that has meaning and for being handled mechanically. The punched holes are sensed electrically by wire brushes, mechanically by metal fingers, or photoelectrically by photocells. The standard card measures $3\frac{1}{4} \times 7\frac{3}{8}$ inches (8.25×18.73 cm) and contains 80 columns and 12 rows in which information may be punched. 2. A piece of lightweight cardboard on which information is represented by holes punched in specific positions.

punched tape — Also called tape, perforated tape, or punched paper tape. Paper tape punched in a coded pattern of holes, which convey information.

punched-tape recorder — A recorder that records data in the form of holes punched in a tape strip.

punch-through — 1. A permanently destructive short circuit through the thin dielectric insulator located between the gate and substrate in a metal-oxide semiconductor device. 2. A disruptive discharge through a dielectric layer in a semiconductor caused by the application of an excessive electric field.

punch-through voltage — 1. That voltage at which two adjacent diffused transistor beds become shorted together, causing a sharp rise in current. 2. The value of the collector-base voltage of a transistor, above which the open-circuit emitter-base voltage increases almost linearly with increasing collector-base voltage. (Reach-through voltage is a term also used in the United States.) 3. A form of transistor failure in which an internal short develops between emitter and collector across the base, usually as a result of excessive voltages.

puncture — 1. A disruptive discharge of current through insulation, which breaks down under electrostatic stress and permits the flow of a sudden large current through the opening. See also breakdown. 2. Breakdown of a solid dielectric or insulation, frequently resulting in a hole.

puncture voltage — The voltage at which insulation fails by disruptive discharge through the insulation sample. It is assumed that the sample area is large enough to prevent flashover.

Pupin coil — An iron-core loading coil inserted into telephone lines at regular intervals to balance out the effect of capacitance between the lines.

pup jack — See tip jack.

pure code — Another term for reentrant code. Code that is never modified in the process of execution. Hence, it is possible to let many users share the same copy of a program.

pure tone — See simple tone, 2.

purity — Physically complete saturation of a hue — i.e., uncontaminated by white and other colors. See also excitation purity.

purity coil — 1. A coil consisting of two current-carrying windings. In a color television receiver, they produce a magnetic field that directs the three electron beams so that each one will strike only the proper set of phosphor dots. 2. An electromagnetic device placed about the neck of a three-gun tricolor picture tube. Its function is to control the angle at which all three beams approach the aperture mask. Its correct adjustment produces pure colors of red, green, and blue on the phosphor-dot faceplate.

purity control — A variable resistor that controls the current through the purity coil mounted around the neck of a color picture tube.

purity magnet — Two adjustable magnetic rings used in place of a purity coil.

Purkinje effect — As ambient lighting is dimmed, the response of the eye shifts away from the red region and toward the blue region of the color spectrum. Maximum response then tends toward blue-green, rather than the yellow-green maxima of bright daylight.

purple boundary — The straight line drawn between the ends of the spectrum locus on a chromaticity diagram.

purple plague — 1. A compound that forms as a result of intimate contact between gold and aluminum, and appears on silicon planar devices and integrated circuits in which gold leads are bonded to aluminum thin-film contacts and interconnections. It causes serious degradation of the reliability of semiconductor devices. 2. An intermetallic gold-aluminum compound formed when gold wires are bonded to aluminum bonding pads. As the compound forms, aluminum migrates from the bonding pad until depleted, and the bond ultimately fails. The name stems from the deep purple color associated with the compound. 3. The tendency of a gold-to-aluminum wire junction to develop a small electrical potential, especially in the presence of moisture. It thus becomes a small battery, which encourages a migration of aluminum molecules toward the gold. The subsequent dark growth on the gold side of the bond — AlAu, "the purple plague" — is believed to raise the contact resistance and weaken the strength of the gold-aluminum junction. 4. One of the several gold-aluminum compounds formed when bonding gold to aluminum; activated by reexposure to moisture and high temperature (>340°C). Purple plague is purplish in color and is very brittle, potentially leading to time-based failure of the bonds. Its growth is highly enhanced by the presence of silicon to form ternary compounds.

PUSH — 1. A computer instruction used to deposit a word on top of the stack. 2. (Lowercase) Storing a byte of data on the data stack.

pushback hookup wire — Tinned copper wire covered with loosely braided insulation, which can be pushed back with the fingers to expose enough bare wire for making a connection.

push button — Device mounted on a plunger (or actuator) that interfaces the operator's fingertip with the internal mechanism of the switch.

push-button control — Control of equipment by means of push buttons, which in turn operate relays, etc.

push-button dialing — The use of keys or push buttons to generate a sequence of digits to establish a telephone circuit connection.

push-button dialing pad — A twelve-key device for originating tone keying signals. Usually, it is attached to a rotary-dial telephone for origination of data signals.

push-button switch — A switch in which a button must be depressed each time the contacts are to be opened or closed.

push-button tuner — A series of push-button switches that connect into a circuit, with the correct tuning frequency corresponding to the depressed button.

pushdown dialing — Also called tone dialing and touch call. The use of keys or push buttons instead of a rotary dial to generate a signal, usually in the form of tones, representing a sequence of digits and used to set up a circuit connection.

pushdown list — 1. A list that is made up and maintained in such a way that the next item to be retrieved is the item most recently stored (last in, first out). 2. An alternative name for a stack.

pushdown stack — Also called P-stack. 1. A circuit that operates in the reverse of a shift register. Whereas, a shift register is a first-in first-out (FIFO) circuit, pushdown stacks are last-in, first-out (LIFO) memories. When data is requested, the stack will read the last data stored, and all other data will move one step closer to the output. Unless memory is emptied, the first data in will never be retrieved. 2. A register that receives information from the program counter and stores the address locations of the instructions that have been pushed down during an interrupt. This stack can be used for subroutining. Its size determines the level of subroutine nesting (one less than its size, or 15 levels of subroutine nesting in a 16-word register). When instructions are returned they are popped back on a last-in, first-out (LIFO) basis. 3. Dedicated consecutive temporary storage registers in a computer, sometimes part of system memory, structured so that the data items retrieved are the most recent items stored on the stack. See LIFO. 4. Essentially a last-in, first-out buffer. As data is added, the stack moves down with the last item, added taking the top position. Stack height varies with the number of stored items, increasing or decreasing with the entering or retrieving of data. The words push (move down) and pop (retrieve the most recently stoked item) are used to describe its operation. In actual practice, a hardware-implemented pushdown stack is a collection of registers with a counter that serves as a pointer to indicate the most recently loaded register. Registers are unloaded in the reverse of the sequence in which they were loaded.

pushdown stack architecture — An organization of memory technique for overcoming certain minicomputer limitations. It provides the high storage efficiency of assembly-language programs while sacrificing little of the simplicity and lower software cost of using higher-level languages. Pushdown stack architecture greatly simplifies the work of the high-level language complier. See pushdown stack.

pushing figure — The change in oscillator frequency due to a specified change in plate current (excluding thermal effects).

push-off strength — The amount of force required to dislodge a chip device from its mounting pad by application of the force to one side of the device, parallel to the mounting surface.

push operation — Refers to the storing of operand(s) from a general register(s) into the most current top location in a pushdown memory stack. See also stack.

push-pull amplifier — See balanced amplifier.

push-pull circuit — A circuit containing two like elements that operate in 180° phase relationship to produce additive output components of the desired wave and cancellation of certain unwanted products. Push-pull amplifiers and oscillators use such a circuit.

push-pull configuration — A fundamental oscillator design in which each half of the circuit operates during a portion of the rf cycle. Primary advantages are increased power output over a single or parallel pair of transistors or tubes and reduction of second harmonic content in the output.

push-pull currents — Balanced currents.

push-pull doubler — An amplifier used for frequency doubling. It consists of two transistors or vacuum tubes; the latter have their grids (input) connected in push-pull configuration and their plates (output) in push-pull or parallel configuration.

push-pull microphone — A microphone comprising two like elements actuated by the same sound waves and operated 180° out of phase.

push-pull oscillator — A balanced oscillator employing two similar tubes or transistors in phase opposition.

push-pull transformer — An audio-frequency transformer that has a center-tapped winding and is used in a push-pull amplifier circuit.

push-pull voltages — Balanced voltages.

push-push circuit — A circuit usually used as a frequency multiplier to emphasize even-order harmonics. Two similar transistors are employed, or two tubes with their grids connected in phase opposition and their plates in parallel to a common load.

push-push configuration — A harmonic oscillator design in which the signals from each output transistor or tube operating at f_o are combined to produce an output signal at $2f_o$. The main advantage of this configuration is the extension of transistor operating frequency limits without the use of an extra frequency-doubler circuit.

push-push currents — Currents that are equal in magnitude and that flow in the same direction at every point in the two conductors of a balanced line.

push-push voltages — Voltages that are equal in magnitude and have the same polarity (relative to ground) at every point on the two conductors of a balanced line.

push rod — A shaft that connects a servo or other actuator with a part of the controlled device.

push-to-talk switch — See press-to-talk switch.

pushup list — A list that is made up and maintained in such a way that the next item to be retrieved and removed is the oldest item remaining in the list (first in, first out).

PUT — Abbreviation for programmable unijunction transistor.

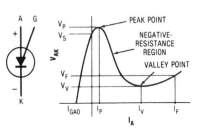

PUT.

put — To insert a single data record into an output file.

PVC — Abbreviation for polyvinyl chloride.

pW — Letter symbol for picowatt.

PWM — Abbreviation for pulse-width modulator. 1. A digital logic circuit that can be programmed to produce pulses having any desired period or duty cycle. A means of controlling variable-speed motors. 2. A form of analog control in which the duration of digital pulses is varied analogously with the signal of interest.

PWM/FM — A system in which a number of pulse-width-modulated subcarriers are used to frequency modulate the carrier.

pylon antenna — A vertical antenna constructed of one or more sheet-metal cylinders with a lengthwise slot. The gain depends on the number of sections.

pyramidal horn — An electromagnetic horn whose sides form a pyramid. The electromagnetic field in such a horn would be expressed basically in a family of spherical coordinates.

pyramid wave — A triangular wave whose sides are approximately equal in length.

pyroelectric effect — Also called pyroelectricity. The redistribution of the charge in a crystal that has been heated. The crystal is left with a net electric dipole moment — i.e., the centers of the positive and negative charges are separated.

pyroelectric infrared detector — A current source with an output proportional to the rate of change of its temperature. Widely used in radiometric systems, from industrial temperature-measuring systems to environmental satellite instruments to the analysis of infrared lasers.

pyroelectricity — *See* pyroelectric effect.

pyroelectric material — A material that produces electrical output when subjected to a change in temperature.

pyroelectric pulse detector — A current-source thermal detector used to detect and study the pulses obtained from particular lasers.

pyroheliometer — A device for the measurement of infrared radiation.

pyrolysis — The thermal decomposition of a volatile chemical compound into nonvolatile and volatile byproducts. It is generally carried out in an inert carrier gas, and this fact distinguishes it from vapor plating.

pyromagnetic — Pertaining to the effect of heat and magnetism on each other.

pyrometer — 1. An instrument used to measure elevated temperatures (beyond the range of mercury thermometers) by electric means. These include immersion optical, radiation, resistance, and thermoelectric pyrometers. 2 A temperature-measuring instrument incorporating a sensor and a readout device.

pyrone detector — A crystal detector in which rectification occurs between iron pyrites and copper (or other metallic points).

pyrotechnic — Pertaining to explosive-actuated devices, especially those that burn rather than producing a shattering effect.

Pythagorean scale — A musical scale in which the frequency intervals are represented by the ratios of integral powers of 2 and 3.

Q

Q—1. Also called quality factor or *Q* factor. A measure of the relationship between stored energy and rate of dissipation in certain electric elements, structures, or materials. In an inductor or capacitor, the ratio of its reactance to its effective series resistance at a given frequency. 2. A measure of the sharpness of resonance or frequency selectivity of a mechanical or electrical system. 3. Figure of merit for energy-storing device, tuned circuit, or resonant system. 4. Half-width of power spectrum of bandpass filter response in hertz divided by the center frequency in hertz. 5. The figure of merit of a resonator, defined as twice the average energy stored in the resonator divided by the energy dissipated per cycle. The higher the reflectivity of the surfaces of an optical resonator, the higher the *Q* and the less energy loss from the desired mode. 6. Symbol for quantity of electric charge.

QAM—Abbreviation for quadrature amplitude modulation. A high-speed modem modulation technique employing both differential phase modulation and amplitude modulation.

Q antenna—A dipole matched to its transmission line by stub matching.

Q band—A band of frequencies extending from 36 to 46 GHz, corresponding to wavelengths of 0.834 to 0.652 cm.

Q (Q-bar) output—The second output of a flip-flop; its logic level is always opposite to that of the Q output.

Q channel—The 0.5-MHz-wide band used in the American NTSC color television system to transmit green-magenta color information.

QCW—A 3.58-MHz continuous-wave signal having Q phase. The term is generally limited to refer to the color-television receiver local oscillator and associated circuits.

QCW signal—*See* quadrature-phase subcarrier signal.

Q demodulator—A demodulator circuit whose inputs are the chrominance signal and the signal from the local 3.58-MHz oscillator after it has been shifted 90°. This phase shift is necessary so that the local signal will be an accurate representation of the Q subcarrier that was suppressed at the transmitter. The output of the Q demodulator is a color video signal representing colors in the televised scene.

q8—A quadraphonic eight-track tape cartridge.

Q-8—Formerly known as Quad-8; term applied to tape cartridges when they contain four-channel programming.

Q factor—1. Of a tuned circuit, the ratio of the inductive reactance of the circuit at the resonant frequency to its radio-frequency resistance. It is a measure of the increase in voltage that is developed across the tuned circuit at resonant frequency, and so the term "magnification factor" is sometimes used for *Q*. If the *Q* factor of a tuned circuit is high, the voltages developed across it are high and its selectivity is good. 2. *See Q*, 1.

Q-meter—Also called a quality-factor meter. An instrument for measuring the *Q*, or quality factor, of a circuit or circuit element.

Q-multiplier—A special filter that has a sharply peaked response curve or a deep rejection notch at a particular frequency.

Q output—The reference output of a flip-flop. That is, the flip-flop is said to be in the 1 state when this output is 1, and it is said to be in the 0 state when this output is 0.

Q-phase—Also called quadrature carrier. A color-television signal carrier having a phase difference of 147° from the color subcarrier.

QPL—Abbreviation for qualified products list. Military qualified products list for high-reliability applications.

QRM—An obsolete term for any type of man-made interference.

QRS complex—That portion of the waveform in an electrocardiogram extending from point Q to point S; it includes the maximum amplitude shown in the trace.

QS—A matrix system developed by Sansui Electronics.

Q signal—1. In color television, the signal formed by the combination of R–Y and B–Y color-difference signals having positive polarities of 0.48 and 0.41, respectively. It is one of the two signals used to modulate the chrominance subcarrier, the other being the I signal. *See also* coarse-chrominance primary. 2. In a color TV system, the color sidebands produced by modulating the color subcarrier at a phase 147° removed from the burst reference phase (sometimes known as the quadrature signal). This signal is capable of reproducing the range of colors from purple to yellow-green. 3. One of a special group of abbreviations used in radiocommunications.

QSL—1. Written acknowledgement of two-way radiocommunication. 2. A card or letter from a station used to confirm listener's reception reports. A QSL, or verification, confirms that a listener's report was correct. Often a colorful, well-designed postcard. Also exchanged by radio amateurs to confirm contacts with each other.

QSL card—A card exchanged by radio amateurs to confirm radiocommunications with each other.

Q-switch—A device used to rapidly change the *Q* of an optical resonator. Used in the optical resonator of a laser to prevent lasing action until the high level of inversion (optical gain and energy storage) is achieved in the laser rod. A giant pulse is generated when the switch rapidly increases the *Q* of the cavity.

Q-switched laser—A laser in which the *Q* of the resonant cavity is spoiled, using rotating mirrors or saturable absorbers, until large quantities of energy are accumulated in the active medium. Return of the *Q* causes a high-energy short pulse.

Q-switched pulse—The output of a laser when the *Q* of the cavity resonator initially is kept very low so that the population inversion achieved is much larger than that which normally characterizes laser operation. On restoration of the Q to its normal high value, a high-power, short-duration pulse of coherent radiation (called a giant pulse) is emitted. Used most often in conjunction with pulsed pump radiation.

quad — 1. A structural unit employed in cables. A quad consists of four separately insulated conductors twisted together. These conductors may take the form of two twisted pairs. 2. A combination of four elements, either electronic components or complete circuits, in series-parallel or parallel-series arrangement. 3. A (series-parallel) combination of four transistors. 4. *See* quadraphonic.

quadded cable — A cable in which some or all of the conductors are in the form of quads.

quadding — Connecting transistors in a series-parallel configuration to achieve greater reliability.

Quad-8 — *See* Q-8.

quad latch — A group of four flip-flops, each of which has the capability of storing a true or false logic level, and all of which normally are enabled by a single control line. When the flip-flops are all enabled new information may be stored in each of them.

quadradisc — Another name for CD-4 disc.

quadrant — 1. A sector, arc, or angle of 90°. 2. An instrument for measuring or setting vertical angles.

quadrantal error — The error in magnetic-compass readings by the magnetic field of the steel hull of a ship, or by metal structures near the loop antenna of radio direction finders aboard a vessel or aircraft.

quadrant electrometer — An electrometer for measuring voltages and charges by means of electrostatic forces. A metal plate or needle is suspended horizontally inside a vertical metal cylinder that is divided into four insulated parts, each connected electrically to the one opposite it. The two parts of quadrants are connected to the two terminals between which the potential difference is to be measured. The resultant electrostatic forces displace the suspended indicator a certain amount, depending on the voltage.

quadraphonic — Also spelled quadriphonic, quadrasonic; sometimes contracted to quad. A term used to describe four-channel sound systems and equipment. Sounds recorded and reproduced from four different directions to produce a field of sounds coming from an apparent 360° around the listener. Generally, any system of sound reproduction using more than the two usual stereo signals to recreate an impression of sounds coming from the rear of the listener as well as from the front.

quadraphony — A scheme of extended stereo whereby ambient and dimensional information is fed directly or via a matrix to a set of four speaker systems suitably oriented in the listening room. Various modulation or matrix systems are sometimes used so that four channels can be obtained by using some two-channel (stereo) equipment. The signals are then decoded so that four channels of sound can be reproduced through four speakers.

quadrasonic — *See* quadraphonic.

quadratic programming — In operations research, a particular case of nonlinear programming in which the function to be maximized or minimized and the constraints are quadratic functions of the controllable variables.

quadrature — The state or condition of two related periodic functions or two related points separated by a quarter of a cycle, or 90 electrical degrees.

quadrature amplifier — A stage used to supply two signals of the same frequency but with phase angles that differ by 90 electrical degrees.

quadrature amplitude modulation — *See* QAM.

quadrature carrier — *See* Q-phase.

quadrature component — 1. The reactive current or voltage component due to inductive or capacitive reactance in a circuit. 2. A vector representing an alternating quantity that is in quadrature (at 90°) with some reference vector.

quadrature modulation — The modulation of two carrier components 90° apart in phase by separate modulating functions.

quadrature phase detector — A phase detector operated in quadrature (90° out of phase) with the loop detector.

quadrature-phase subcarrier signal — Abbreviated QCW signal. That portion of the chrominance signal that leads or lags the in-phase portion by 90°.

quadrature portion — In the chrominance signal, the portion with the same or opposite phase from that of the subcarrier modulated by the Q signal. This portion of the chrominance signal may lead or lag the in-phase portion by 90 electrical degrees.

quadrature sensitivity — Also called side sensitivity, lateral sensitivity, or crosstalk sensitivity. The sensitivity of a transducer to motion normal to the principal axis. Commonly expressed in percent of the sensitivity in the principal axis.

quadriphonic — *See* quadraphonic.

quadripole network — *See* two-terminal-pair network.

quadruple diversity — The operation of combining four identical signals received over diverse paths to obtain an improvement of up to 6 dB in signal-to-noise ratio.

quadruple-diversity system — A receiving system in which space-diversity and frequency-diversity techniques are employed simultaneously.

quadruple play — Magnetic recording tape that is thinner than standard-play tape and consequently makes possible recordings four times longer than the standard-play tape.

quadruplex circuit — A telegraph circuit designed for carrying two messages in each direction simultaneously.

quadrupole — A combination of two dipoles that produces a force varying in inverse proportion to the fourth power of the distance from the generating charge.

quadrupole network — *See* two-terminal-pair network.

qualification — The entire procedure by which electronic parts are examined and tested to obtain and maintain approval at specified failure rate levels, and then identified on the qualified products lists.

qualified products list — Abbreviated QPL. A listing of manufacturers qualified by test and performance

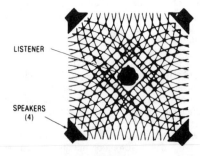

Quadraphony.

verification to produce items listed in the military specifications.

qualifying activity — The military activity or its agent delegated to administer the qualification program.

qualitative analysis — A study that reveals how a system works.

quality — 1. The extent of conformance to specifications of a device, or the proportion of satisfactory devices in a lot. 2. The extent to which a device meets the required specifications when it is shipped to the user.

quality assurance — 1. A planned, systematic pattern of actions necessary to provide suitable confidence that a system or component will perform satisfactorily in actual operation. 2. A systematic pattern of actions throughout design and production, to ensure confidence in a product's conformance with specifications.

quality control — The control of variation of workmanship, processes, and materials in order to produce a consistent, uniform product.

quality engineering — An engineering program whose purposes are to establish suitable quality tests and quality acceptance criteria and to interpret quality data.

quality factor — See Q, 1.

quality-factor meter — See Q-meter.

quantitative analysis — A study that determines how well a system performs.

quantity — 1. Any positive or negative number. It may be a whole number, a fraction, or a whole number and a fraction. 2. A constant, variable, function name, or expression. 3. In computers, a positive or negative real number; the term *quantity* is preferred to the term *number*.

quantity of electricity — See electrical charge.

quantization — 1. The process whereby the range of values of a wave is divided into a finite number of subranges, each represented by an assigned (quantized) value. 2. The process of converting a continuous analog input into a set of discrete output levels.

quantization distortion — Also called quantization noise. Inherent distortion introduced during quantization.

quantization error — 1. The difference between the actual values of data and corresponding discrete values resulting from quantization. 2. Error generated because an analog-to-digital converter can be no more accurate than $\pm 1/2$ lsb because of its resolution.

quantization level — 1. A particular subrange in quantization. 2. The symbol designating the subrange of (1).

quantization noise — See quantization distortion.

quantize — 1. To convert a continuous variable, such as a waveform, into a series of levels or steps. There are no in-between values in such a quantized waveform. All values of signal are represented by the nearest standard of value or code position. 2. To sample the amplitudes of a complex wave periodically, and to represent each amplitude thus sampled by one of a finite number of values. 3. To restrict a variable to a discrete number of possible values.

quantized pulse modulation — Pulse modulation that involves quantization (e.g., pulse numbers modulation or pulse code modulation).

quantizer — 1. A device that partitions a continuum of analog values into discrete ranges to be represented by a digital code. An analog-to-digital converter. 2. A device with a restricted quantity of possible output values that will assimilate any range of input values into its limited range.

quantizing — Expressing an analog value as the nearest one of a discrete set of prechosen values.

quantizing error — 1. The basic uncertainty associated with digitizing an analog signal, due to the finite resolution of an a/d converter. An ideal converter has a maximum quantizing error of $\pm 1/2$ lsb. In an a/d converter, an infinite number of possible input voltages can occur, but only 2^n output codes (n = number of bits) can exist. Because of this quantizing effect, there will always be an error as great as $1/2$ lsb, and the greatest error will occur in the value of the transition voltage at the point where the output changes state. 2. A measure of the ability of a digital meter to discriminate between an incremental value and values slightly above or below this value. 3. An error caused by conversion of a variable having a continuous range of values to a quantized form having only discrete values, as in analog-to-digital conversion. The error is the difference between the original (analog) value and its quantized (digital) representation.

quantum — 1. A discrete portion of energy of a definite amount. It was first associated with intra-atomic or intermolecular processes involving changes among electrons, and the corresponding radiation. 2. If the magnitude of a quantity is always an integral multiple of a definite unit, then that unit is called the quantum of the quantity. 3. The angular increment of input-shaft rotation of an encoder, subtended by one code position. 4. The unit carrier of energy: the photon for light, and the electron for electricity. 5. The smallest amount into which the energy of a wave can be divided. The quantum is proportional to the frequency of the wave. See photon.

quantum efficiency — 1. In a phototube, the average number of electrons photoelectrically emitted from the photocathode per incident photon of a given wavelength. 2. The fraction of those ions or atoms excited to a higher energy level by pumping radiation that decays with light emission in a particular desired range of frequencies. The remaining ions decay by various radiationless mechanisms (such as phonon emission) or by undesired radiative transitions. 3. The ratio of the number of carriers generated to the number of photons incident upon the active region. 4. The ratio of the number of quanta radiant energy (photons) emitted per second to the number of electrons flowing per second, e.g., photons/electrons. 5. Ratio of generated current carriers to the number of absorbed photons.

quantum-limited operation — In the operation of a photodetector, the inability of the detector to measure incident radiation levels below a threshold level because of fluctuations in the output current that are not due to incident radiation, i.e., not due to incident photons.

quantum mechanics — 1. The study of atomic structure and other related problems in terms of quantities that can actually be measured. 2. The science of all complex elements of atomic and molecular spectra, and the interaction of radiation and matter.

quantum noise — 1. A random variation or noise signal due to fluctuations in the average rate of incidence of quanta on a detector. The basic electromagnetic quantum of noise power is just 1 photon per electromagnetic mode. 2. Noise generated within an optical communications system link that has both internal (dark current) and external (background noise, noise in signal) components.

quantum number — Any set of numbers assigned to the particular values of a quantized quantity in its discrete range. The state of a particle or system may be described by a set of compatible quantum numbers.

quantum theory — The theory that an atom or molecule does not emit or absorb energy continuously. Rather, it does so in a series of steps, each step being the emission or absorption of an amount of energy called the quantum. The energy in each quantum is directly proportionate to the frequency.

quantum transition — A transition between two quantum states.

quarter phase — *See* two-phase.

quarter-squares multiplier — An analog multiplier unit that makes use of the relationship $xy = \frac{1}{4}[(x + y)^2 - (x - y)^2]$.

quarter track — *See* four-track.

quarter-track tape — *See* four-track tape.

quarter wave — One-quarter cycle of a wave.

quarter-wave antenna — 1. An antenna whose electrical length is one-quarter the wavelength of the transmitted or received signal. 2. A dipole antenna with an electrical length equal to one-quarter of its working wavelength. Its physical length is somewhat shorter than one-quarter wavelength.

quarter-wave attenuator — Two energy-absorbing grids or other structures placed in a transmission line and separated by an odd number of quarter wavelengths. As a result, the wave reflected from the first grid annuls the wave reflected from the second grid.

quarter-wavelength — That distance which corresponds to an electrical length of a quarter of a wavelength of the frequency under consideration.

quarter-wave line — *See* quarter-wave stub.

quarter-wave plate — A mica or other double-refracting crystal plate of such thickness that a phase difference of one-quarter cycle is introduced between the ordinary and extraordinary components of light passing through.

quarter-wave resonance — In a quarter-wave antenna, the condition in which its resonant frequency is equal to the frequency at which it is to be used.

quarter-wave stub — Also called a quarter-wave line or quarter-wave transmission line. A section of transmission line equal to one-quarter of a wavelength at the fundamental frequency. It is commonly used to suppress even harmonics. This is done by shorting the far end so that the open end presents a high impedance to the fundamental frequency and all odd harmonics, but not to the even-order harmonics.

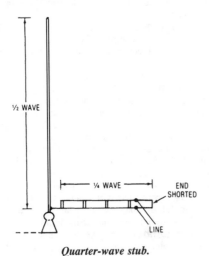

Quarter-wave stub.

quarter-wave support — A quarter-wave metallic stub used in place of dielectric insulators between the inner and outer conductors of a coaxial transmission line.

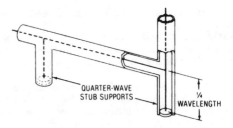

Quarter-wave supports.

quarter-wave termination — A waveguide termination consisting of a metal plate and a wire grating (or semiconducting film) spaced one-quarter wavelength apart. The plate is the terminating element. The wave reflected by the grating (or film) is cancelled by the wave reflected by the plate.

quarter-wave transformer — A one-quarter-wavelength section of transmission line used for impedance matching.

quarter-wave transmission line — *See* quarter-wave stub.

quartz — A mineral (silicon dioxide) occurring in hexagonal crystals in nature and having piezoelectric properties that are highly useful in radio and carrier communication. The crystals from which slabs are cut for oscillators are transparent and almost colorless. When excited electrically, they vibrate and maintain extremely accurate and stable frequencies.

quartz crystal — Also called a crystal. 1. A thin slab cut from quartz and ground to the thickness at which it will vibrate at the desired frequency when supplied with energy. It is used to accurately control the frequency of an oscillator. 2. The naturally occurring crystalline form of silicon dioxide. It is slightly birefringent and exhibits rotary dispersion of light rays transmitted along the crystal axis, both right-hand and left-hand forms being known. Quartz transmits light from about 0.18 to 4.5 micrometers in the infrared. It is very hard and takes a high polish, and has a low thermal expansion coefficient. 3. A piezoelectrical crystal cut from natural quartz (silicon dioxide, SO_2).

quartz delay line — A delay line in which fused quartz is the medium for delaying sound transmission or a train of waves.

quartz lamp — A mercury-vapor lamp having a transparent envelope made from quartz instead of glass. Quartz resists heat (permitting a higher current) and passes ultraviolet rays, which glass will absorb.

quartz light source — A lamp with a quartz bulb that transmits radiation generally rich in ultraviolet.

quartz plate — A crystalline-quartz section completely finished to specifications, with its two major faces essentially parallel.

quartz resonator — A piezoelectric resonator with a quartz plate.

quasi-complementary symmetry circuit — A push-pull power amplifier that utilizes npn output transistors driven by pnp and npn driver transistors.

quasi-linear feedback-control system — A feedback-control system in which the relationships between the pertinent measures of the system input and output signals are substantially linear despite the existence of nonlinear elements.

quasi-monochromatic light — Light radiation that behaves like ideal monochromatic radiation. The frequencies of quasi-monochromatic radiation are strongly peaked about a certain frequency.

quasi-optical — Having properties similar to those of light waves. The propagation of waves in the television spectrum is said to be quasi-optical (i.e., cut off by the horizon).

quasi-random code generator — A high-speed PCM information source that provides a means of closed-loop testing for use in designing and evaluating wideband communications links.

quasi-rectangular wave — A wave nearly, but not quite, rectangular in shape.

quasi-single-sideband transmission — Simulated single-sideband transmission done by transmitting parts of both sidebands.

quasi-steady-state vibration — A nearly periodic vibration in which the amplitude and phase relationships of the component sinusoids vary slowly with time.

quaternary — A coding scheme that uses four different voltage levels to represent information, used over the local loop with basic ISDN.

quaternary signaling — The communications mode in which information is passed by the presence and absence, or plus and minus variations, of four discrete levels of one parameter of the signaling medium.

quench — To stop an oscillation abruptly.

quenched spark — A spark consisting of only a few sharply defined oscillations because the gap is deionized almost immediately after the initial spark has passed.

quenched spark gap — A spark gap with provision for producing a quenched spark. One form consists of many small gaps between electrodes that have a relatively large mass and, thus, are good radiators of heat. As a result, they cool the gaps rapidly and thereby stop conduction.

quenched spark-gap converter — A spark-gap generator or other power source in which the oscillatory discharge of a capacitor through an inductor and a spark gap provides the radio-frequency power. The spark gap comprises one or more closely spaced gaps in series.

quench frequency — 1. An ac voltage applied to an electrode of a tube used as a superregenerative detector to alternately vary its sensitivity and thereby prevent sustained oscillations. The quench frequency is usually lower than the signal frequency to be received. 2. The number of times per second a circuit goes in and out of oscillation.

quenching — 1. The process of terminating a discharge in a radiation-counter tube by inhibiting the reignition. 2. A process of rapid cooling from an elevated temperature, in contact with liquids, gases, or solids. 3. The inhibition or elimination of one process by another process. The stimulated emission of a laser oscillator can be quenched by a pulse of radiation of the same frequency traversing the oscillator in a different direction. This pulse induces the excited ions to emit radiation in a direction apart from the oscillating mode and, hence, the oscillation is decreased.

quenching circuit — A circuit that inhibits multiple discharges from an ionizing event by suppressing or reversing the voltage applied to a counter tube.

quenching frequency — That frequency at which oscillations in a superregenerative receiver are suppressed (quenched).

quench oscillator — A superregenerative receiver circuit that produces the quench-frequency signal.

queue — 1. A line of items waiting for service in a system, such as messages to be transmitted in a message-switching system. 2. To arrange in or form a queue. 3. A waiting line for execution of computer or peripheral operations. 4. A multi-element data structure in which the first element in is the first element out.

This data structure works in the same manner as a supermarket checkout line — items are added at one end and removed at the other. Compare with a stack, in which items are added and removed only from one end. 5. An area in the temporary call store memory used to record a writing list for some particular function. For example, the writing list or queue for customer dial pulse receiver circuits. 6. A list of processes to be executed in sequential order, information blocks to be processed in sequential order, or a mixture of the two. 7. Orderly access to a system; generally, "first in, first out" prioritization.

queue control block — A control block for regulating the sequential use of a programmer-defined facility for a number of tasks.

queued access method — An access method that provides automatic synchronization of data transfer between programs using the access method and input/output devices. Delays for input/output operations are thereby eliminated.

queuing theory — A research technique having to do with the correct order of moving units, such as sequence assignments for bits of information, whole messages, assembly-line products, or automobiles in traffic.

quick-break — A characteristic of a switch or circuit breaker, whereby it has a fast contact-opening speed that is independent of the operator.

quick-break fuse — A fuse that draws out the arc and rapidly breaks the circuit when its wire melts. Usually a spring or weight is used to quickly separate the broken ends.

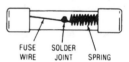

FUSE   SOLDER   SPRING
WIRE   JOINT

Quick-break fuse.

quick-break switch — A switch that minimizes arcing by breaking a circuit rapidly, independent of the rate at which the switch handle is moved.

quick-connect terminal — 1. A plug-in type of terminal designed to make possible rapid wiring. 2. A type of connection, similar to a lamp plug and wall socket, used where connections are frequently removed, then connected again. Quick-connect terminals are found on home appliances, hi-fi equipment, TV sets, and in other applications.

quick disconnect — A type of connector designed to facilitate rapid locking and unlocking of two contacts or connector halves.

quickening — A characteristic of a display with compressed time scale (such as that employed in time-lapse photography). Used to exaggerate trends.

quick-flashing light — A rhythmic light that shows very quick, regular, alternate displays of light and darkness.

quick-make — A characteristic of a switch or circuit breaker, whereby it has a fast contact-closing speed that is independent of the operator.

quick-make switch — A switch or circuit breaker that has a high contact-closing speed independent of the operator.

quick-stop control — On some tape recorders, and on all recorders used for dictation, a control with which the operator can stop the tape without taking the machine out of the play or record position.

quiescence — 1. The state of a transistor amplifier with no signal applied. 2. The operating condition that exists in a circuit when no input signal is applied to the circuit.

quiescent — At rest; specifically, the condition of a circuit when no input signal is being applied to it.

quiescent-carrier modulation — A modulation system in which the carrier is suppressed during intervals when there is no modulation.

quiescent-carrier telephony — Telephony in which the carrier is suppressed whenever no modulating signals are to be transmitted.

quiescent current — *See* idling current.

quiescent dissipation — The power dissipated by a component or circuit in the absence of dynamic activating signals applied to the input or inputs.

quiescent input voltage — The dc voltage at the input of an amplifier that has one input terminal when that terminal is not connected to any source.

quiescent operating point — 1. The state or condition that exists under any specified external condition when the signal is zero. 2. *See* quiescent point.

quiescent output voltage — The dc voltage at the output terminals of an amplifier when the input is grounded for ac through a resistance equal to the resistance of the signal source.

quiescent period — The resting period; e.g., the period of no activity between pulses in pulse transmissions.

quiescent point — On the characteristic curve of an amplifier, the point representing the conditions existing when there is no input signal. *See also* operating point.

quiescent push-pull — In a radio receiver, a push-pull output stage in which practically no current flows when no signal is being received. Thus, there is no noise while the radio is being tuned between stations.

quiescent state — The time during which a tube or other circuit element is not performing its active function in the circuit.

quiescent value — The voltage or current value of a vacuum-tube electrode when no signals are present.

quiescing — The process of stopping a multiprogrammed system by rejection of new jobs.

quiet automatic volume control — *See* delayed automatic volume control.

quiet AVC — *See* delayed automatic volume control.

quiet battery — Also called talking battery. A source of energy of special design, or with added filters, that is sufficiently quiet and free from interference that it may be used for speech transmission.

quieting — 1. The decrease in noise voltage at the output of an FM receiver in the presence of an unmodulated carrier. 2. A measure of the usable sensitivity of an FM tuner, expressed as the least rf signal level (100 percent modulated with a 400-Hz tone) that reduces the receiver internal noise and distortion to 30 dB below the output level obtained with the modulated tone present $[(s + n)/n = 30 \text{ dB}]$. A null filter tuned to 400 Hz is used to remove the tone. For an AM receiver, the carrier is modulated by 30 percent and the field strength (microvolts per meter) is measured that is necessary to provide a 20 dB $(s + n)/n$ ratio.

quieting sensitivity — In an FM receiver, the minimum input signal that will give a specified output signal-to-noise ratio.

quiet tuning — In a radio receiver, a form of tuning in which the output is silenced except when the receiver is tuned to the precise frequency of the incoming carrier wave.

QWERTY keyboard — The layout on a standard typewriter-style keyboard; the first six letters on the top alphabetic line.

R

R— 1. Symbol for resistor, resistance, or reluctance. 2. Letter symbol for roentgen.

race— 1. The condition that exists when a signal is propagated through two or more memory elements during the same clock period. 2. The condition that occurs when changing the state of a system requires a change in two or more state variables. If the final state is affected by which variable changes first, the condition is a critical race. 3. An improper condition in which data that is supposed to move in steps, as in a shift register, goes through a whole string of stages at one step. Usually caused by incorrect timing pulses.

raceway— Any channel designed and used solely for holding wires, cables, or bus bars.

rack and panel connector— 1. A connector that is attached to a panel or side of equipment so that when these members are brought together, the connector is engaged. 2. A connector that connects the inside back end of the cabinet (rack) with the drawer containing the equipment when it is fully inserted. The drawer permits convenient removal of portions of the equipment for repair or examination. Special design and rugged construction of the connector allows for variations in rack-to-panel alignment.

rack and pinion— A toothed bar (rack) that engages a gear (pinion) to convert the back-and-forth motion of the rack into rotary motion, or the rotary motion of the pinion into back-and-forth motion.

rackmount— Designed to be installed in a cabinet that is usually 19-in. wide.

racon— *See* radar beacon.

rad— Abbreviation for radiation absorbed dose. 1. The amount of radiation that delivers 100 ergs of energy to 1 gram of a substance. It is approximately equivalent to 1 roentgen in the case of body tissue. 2. The amount of energy transferred to a material by ionizing radiation. One rad is equal to the energy of 100 ergs per gram of material. The material must be specified, because the energy differs with each material.

radar— Acronym for radio detection and ranging. A system that measures distance (and usually the direction) to an object by determining the amount of time required by electromagnetic energy to travel to and return from an object. Called primary radar when the signals are returned by reflection. Called secondary radar when the incident signal triggers a responder beacon and causes it to transmit a second signal. Predicted in the early part of the 20th century, the first important system was built in England in 1938. Basic building blocks of a radar are the transmitter, the antenna (normally used for both transmission and for reception), the receiver, and the data-handling equipment.

radar alarm system— An alarm system that employs radio-frequency motion detectors.

radar altitude— Also called radio altitude. Absolute altitude measured by a radar altimeter.

radar and television aid to navigation— A device that converts a circular-scan radar presentation to a horizontally scanned television presentation. It provides a continuous bright display with target trails for course and speed indications of moving targets.

radar antenna— Any of the many types of antennas used in radar.

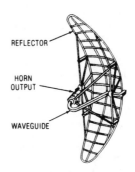

Radar antenna.

radar attenuation— Ratio of the transmitted power to the reflected (received) power— specifically, the ratio of the power that the transmitter delivers to the transmission line connected to the transmitting antenna to the power reflected from the target and delivered to the transmission line connected to the receiving antenna.

radar beacon— Also called a racon. An automatic transmitter-receiver that receives signals from a radar transmitter and retransmits coded signals that enable the radar operator to determine his own position.

radar beam— The space where a target can be effectively detected and/or tracked in front of a radar. Its boundary is defined as the locus of points measured radially from the beam center at which the power has decreased to one-half.

radar calibration— Taking measurements on various parts of electronic equipment (e.g., radar, IFF, communications) to determine its performance level.

radar camouflage— The use of coverings or surfaces on an object to considerably reduce the reflected radio energy and, thus, conceal the object from the radar beam.

radar cell— The volume enclosed by dimensions of one radar-pulse length by one radar beamwidth.

radar clutter— The image produced on a radar indicator screen by sea or ground return. If not of particular interest, it tends to obscure the target indication.

radar-confusion reflectors — Metallic devices (e.g., chaff, corner reflections) employed to return false signals in order to confuse enemy radar receivers. The use of radar-confusion reflectors is termed reflective jamming.

radar control area — The designated space within which aircraft approach, holding, stacking, and similar operations are performed under guidance of a surveillance radar system.

radar controller — An air traffic controller or other responsible person proficient in the use and interpretation of radar and capable of performing one or more of the following functions: surveillance controller, traffic director, and final controller.

radar countermeasures — Interception, jamming, deception, and evasion of enemy radar signals to obtain information about the enemy from his radar and to prevent him from obtaining accurate, usable information from his or her radar.

radar-coverage indicator — A device showing how far a radar station should track a given aircraft. It also provides a reference (detection) range for quality control. Aircraft size and altitude, screening angle, site elevation, type of radar, antenna radiation pattern, and antenna tilt are taken into account.

radar cross section — That portion of the backscattering cross section of a target associated with a specified polarization component of the scattered wave.

radar data filtering — A quality-analysis process in which the computer rejects certain radar data and alerts personnel at the mapping and surveillance consoles to the rejection.

radar deception — Radiation or reradiation of radar signals in order to confuse or mislead an enemy operator when he or she interprets the data shown on his or her scope.

radar decoy — An object that has the same reflective characteristics as a target and is used in radar deception.

radar display — The spontaneous visual presentation of radar information by electronic traces on a CRT.

radar distribution switchboard — A switching panel for connecting video, trigger, and bearing from any one of five systems to any or all of 20 repeaters. Also contains order lights, bearing cutouts, alarms, test equipment, etc.

radar dome — A weatherproof cover for protection of a primary radiation element of a radar or radio device; the cover is transparent to radio-frequency energy and permits operation of the radiating element, including rotation or other physical movement. *See also* radome.

radar echo — 1. The radio-frequency energy received after it has been reflected from an object. 2. The deflection or change of intensity that a radar echo produces in the display of a cathode-ray tube.

radar equation — A mathematical expression relating the transmitted and received powers and antenna gains of a primary-radar system to the echo area and distance of the target.

radar fence — A network of radar warning stations that maintains constant watch against surprise attack (e.g., the DEW line).

radar field gradient — Abbreviated RFG. The rate at which the strength of the field from a primary radar decreases as the distance from the transmitting antenna increases.

radar frequency bands — The frequency bands for radar are shown in the table.

radargrammetry — The analysis of the photographs taken from the radar display of a survey aircraft and used when recording terrain that is obscured by clouds.

radar homing — Missile guidance for which the intelligence is provided by a radar aboard the missile.

Radar bands

Band	Frequency range
HF	3–30 MHz
VHF	30–300 MHz
UHF	300–1000 MHz
L-band	1.0–2.0 GHz
S-band	2.0–4.0 GHz
C-band	4.0–8.0 GHz
X-band	8.0–12.0 GHz
K_u-band	12.0–18.0 GHz
K-band	18.0–27.0 GHz
K_a-band	27.0–40.0 GHz
millimeter	40–300 GHz

radar horizon — The most distant point (from the radar antenna along a given azimuth) on the earth's surface illuminated by the radar on purely geometric conditions. The conditions are that the illumination occurs along a straight-line path, where the path is taken over an effective earth's radius of 4/3 its true radius and where the illuminating power of the radar is considered unlimited.

radar illumination — The subjection of an object (target) to electromagnetic radiation from a radar.

radar indicator — A cathode-ray tube with its associated equipment that provides a visual indication of the echo signals picked up by the radar set.

radar marker — A fixed facility that continuously emits a radar signal so that a bearing indication appears on a radar display.

radar nautical mile — The time interval of approximately 12.67 microseconds required for radio-frequency energy to travel one nautical mile (1.853 km) and return, a total of two nautical miles.

radar paint — A radar-energy-absorbent material that can be applied to an object to reduce the possibility of detection.

radar-performance figure — Ratio of the pulse power of the radar transmitter to the power of the minimum signal detectable by the receiver.

radar picket — Early-warning aircraft that flies at a distance from a ship or other force being protected, to increase the radar detection range.

radar pulse modulator — A modulator that turns an rf energy source off and on in a precise, known manner. In essence, the modulator supplies the energy that causes an rf source to oscillate or amplify, thus creating a burst, or pulse, of rf energy.

radar range — The maximum range at which a radar can ordinarily detect objects.

radar receiver — The receiver that amplifies the returned radar signal and demodulates the rf carrier before further amplifying the desired signal and delivering it, in a form suitable for presentation, to the indicator. Unlike a radio receiver, it is more sensitive, has a lower noise level, and is designed to pass a pulse signal.

radar-reflection interval — The length of time required for a radar pulse to reach a target and return.

radar reflectivity — A measure of the ability of a radar target to intercept and return a radar signal.

radar relay — 1. Equipment for relaying radar video and appropriate synchronizing signals to a remote location. 2. Process or system by which radar echoes and synchronization data are transmitted from a search radar installation to a receiver at a remote point.

radar repeaters — Remote indicators used to reproduce radar data from a primary source.

radar resolution — The ability of a radar to distinguish between the desired target and its surroundings.

radarscope — A cathode-ray tube serving as an oscilloscope, the face of which is the radar viewing screen.

radar selector switch — A manual or motor-driven switch that transfers a plan-position-indicator repeater from one system to another, switching video, trigger, and bearing data.

radar shadow — An area shielded from radar illumination by an intervening reflecting or absorbing medium. This region appears as an area that is void of targets on a radar display.

radar silence — 1. A period of time during which radar operations are stopped. 2. An imposed discipline under which the transmission by radar of electromagnetic signals on some or all frequencies is prohibited.

radar speed detector — A broadband, high-sensitivity, rf-level detector. It has a tuned microwave horn antenna that is designed to pass signals in the X- and K- (several thousand megahertz) bands into a diode. The radio-frequency energy is then rectified into a voltage that upsets a delicately balanced alarm circuit. Law enforcement agencies can outwit the detectors. They may use short bursts of radar to monitor specific vehicles momentarily — hardly enough time for a radar detector to resolve whether the incoming signal is a random bit of rf interference or is actually a transmitted radar signal.

radar target — Any reflecting object of particular interest in the path of the radar beam (usually, but not necessarily, the object being tracked).

radar trace — The pattern produced on the screen of the cathode-ray tube in a radar unit.

radar transmitter — The transmitter portion of a radar system. The unit of the radar system in which the rf power is generated and keyed.

radiac — Acronym for radioactive detection, identification, and computation. A descriptive term referring to the detection, identification, and measurement of nuclear radiation.

radiac instrument — See radiac set.

radiac meter — See radiac set.

radiac set — Also called radiac meter or radiac instrument. Equipment for detecting, identifying, and measuring the intensity of nuclear radiation.

radiac test equipment — Equipment for testing radiac sets.

radial — 1. Pertaining to or placed like a radius (i.e., extending or moving outward from a central point, like the spokes of a wagon wheel). 2. One of a number of lines of position defined by an azimuthal navigation facility and identified in terms of its bearing (usually magnetic) from that facility.

radial-beam tube — A vacuum tube producing a flat, radial electron beam that can be rotated about the axis of the tube by an external magnetic field.

radial component — A component that acts along (parallel to) a radius; as contrasted to a tangential component, which acts at right angles (perpendicular) to a radius.

radial component of the electric field — The component of an electric field in the direction of the slant-range vector or the radius vector at which an antenna pattern is measured. The radial component of the electric field is relatively small in the far field and is normally neglected.

radial field — A field of force that is directed toward or away from a point in space.

radial field cathode-ray tube — A CRT in which a fine-grained, curved, high-transmission metallic mesh is placed on the exit side of the deflection area. The mesh establishes a ground plane for the postaccelerating field so that the resulting equipotential surfaces are truly spherical, creating a radial electrostatic field.

radial grating — A conformal grating consisting of wires arranged radially in a circular frame, like the spokes of a wagon wheel. The radial grating is placed inside a circular waveguide to obstruct E waves of zero order, but not the corresponding H waves.

radial lead — A lead extending out the side of a component, rather than from the end. (The latter is called an axial lead.) Some resistors and capacitors have radial leads.

radial tonearm — Sometimes called straight-line tracking arm. A tonearm that moves along a track parallel to the record radius, maintaining perfect tangency to the groove.

radial transmission line — A pair of parallel conducting planes used for propagating uniform cylindrical waves whose axes are normal to the planes.

radian — In a circle, the angle included within an arc equal to the radius of the circle. Numerically it is equal to $57°$, $17'$, $44.8''$. A complete circle contains 2π radians. One radian equals $57.3°$, and $1°$ equals 0.01745 radian.

radiance — Also called emittance. 1. The apparent radiation of a surface. It is the same as luminance except that radiance applies to all kinds of radiation instead of only light flux. 2. Surface radiation. Solid angle times radiant flux per area. 3. The radiant intensity of electromagnetic radiation per unit projected area of a source or other area, i.e., it is the radiant power of electromagnetic radiation per unit solid angle and per unit surface area normal to the direction considered. The surface may be that of a source detector, or it may be any other real or virtual surface intersecting the flux. The unit of measure is watts/steradian-square meter. The concept is usually applicable to the visible or near visible region of the electromagnetic frequency spectrum.

radian frequency — See angular velocity.

radian length — The distance, in a sinusoidal wave, between phases differing by an angle of one radian. It is equal to the wavelength divided by 2π.

radian per second — A unit of angular velocity.

radiant — Pertaining to electromagnetic radiation, with the contributions at all wavelengths of interest weighted equally.

radiant efficiency — The ratio of the radiant flux emitted by a source to the power supply.

radiant emittance — The light flux radiated per unit area of a source.

radiant energy — Energy transmitted in the form of electromagnetic radiation (e.g., radio, heat, or light waves). It is measured in units of energy such as kilowatt-hours, ergs, joules, or calories. There is no associated transfer of matter per se under this concept.

radiant-energy detecting device — A device employing radiant energy to detect flaws in the surface and/or volume of solids.

radiant exitance — The radiant power emitted into a full sphere (4π steradians) by a unit area of source.

radiant flux — 1. Time rate of flow of radiant energy, expressed in watts or in ergs per second. 2. The radiant energy crossing or striking a surface per unit time, usually measured in watts.

radiant heat — Infrared radiation from a body not hot enough to emit visible radiation.

radiant heater — An electric heating appliance with an exposed incandescent heating element.

radiant intensity — 1. The energy emitted within a certain length of time per unit solid angle about the direction considered. 2. The radiant power per unit solid angle in the direction considered, i.e., the time rate of transfer of radiant energy per unit solid angle, or the flux radiated per unit solid angle about the specified direction. The unit of measure is watts per steradian or joules per steradian-second.

radiant power — 1. The rate of transfer of radiant energy. 2. The time rate of flow of electromagnetic energy. The unit is watts or joules per second.

radiant reflectance — The ratio of reflected radiant power to incident radiant power.

radiant sensitivity — The output current of a phototube or camera tube divided by the incident radiant flux of a given wavelength at constant electrode voltages. The term "output current" as here used does not include the dark current.

radiant transmittance — The ratio of transmitted radiant power to incident radiant power.

radiate — To emit rays from a center source; e.g., electromagnetic waves emanating from an antenna.

radiated — Energy transfer by propagation of electromagnetic fields.

radiated interference — 1. Interference that is transmitted through the atmosphere according to the laws of electromagnetic wave propagation. (The term radiated interference is generally considered to include the transfer of interfering energy by inductive or capacitive coupling.) 2. Any unwanted electrical signal that is radiated from the equipment under test or from any lines connected to that equipment under test.

radiated output noise — Electrical interference signals radiated by switching the ac/dc converters and by the transformer in line-operated power supplies. These signals can cause malfunctioning of analog and digital circuits within their fields.

radiated power — The total energy, in the form of Hertzian waves, radiated from an antenna.

radiated spurious transmitter output — A spurious output radiated from a radio transmitter. (The associated antenna and transmission lines are not considered part of the transmitter.)

radiating curtain — An array of dipoles in a vertical plane, positioned to reinforce each other. Usually placed one-quarter wavelength ahead of a reflecting curtain of corresponding half-wave reflecting antennas.

radiating element — Also called radiator. A basic subdivision of an antenna. It, by itself, is capable of radiating or receiving radio-frequency energy.

radiating guide — A waveguide designed to radiate energy into free space. The waves may emerge through slots or gaps in the guide, or through horns inserted into its wall.

radiation — 1. The emission and propagation of energy through space or a material medium. The energy may be in the form of electromagnetic waves (Y-rays and X-rays) or particles (beta particles, alpha particles, and neutrons). 2. A general term for energy emitted from a substance and traveling across space in straight lines. (Originally used only for electromagnetic waves, it now includes streams of particles, such as alpha particles, beta particles, etc.) 3. The transfer of heat from a hot body to a cooler body through space, without heating any medium that may be in the space. 4. The emission of electromagnetic energy into space regardless of the frequency. Thus, radiation includes thermal, radio, visual, and X-ray energy. The propagation of energy by radiation does not require a medium to support transmission. 5. The electromagnetic waves or photons emitted from a source.

6. Act of giving off electromagnetic energy, particularly by an antenna when excited by a transmitter.

radiation angle — The vertical angle at which maximum energy is radiated by an antenna with respect to the earth. Low-angle radiation delivers more energy in the local area. High-angle radiation augments skip transmission and is wasteful of power.

radiation belt — See Van Allen radiation belts.

radiation characteristic — An identifying feature, such as frequency or pulse width, of a radiated signal.

radiation counter — 1. A device for counting radiation particles (alpha, beta, gamma, neutrons, etc.) or photons of energy (X-rays, etc.), usually using either scintillation or ionization resulting from the presence of the particle or photon to be measured. 2. An instrument used to recognize and identify incident radiation by the ionizing or stimulating properties of the radiation.

radiation counter tube — See counter tube.

radiation damage — A loss in certain physical properties of organic substances, such as elastomers, caused principally by ionization of the long-chain molecule. This ionization process (i.e., loss of electrons) is believed to result in redundant cross-linking and possible scission of the molecule. The effect is cumulative.

radiation detection (fluorescence) — Radiation detection using fluorescence produced in an efficient screen.

radiation detector — Any of the many devices used to detect the presence of radiation from a specific region of the electromagnetic spectrum.

radiation-detector tube — A tube in which current passes between its electrodes whenever the tube is exposed to penetrating radiation. The amount of this current corresponds to the intensity of radiation.

radiation dosage — The total radiation energy absorbed by a substance, usually expressed in ergs per gram. See also rad; roentgen; roentgen equivalent man; roentgen equivalent physical.

radiation dosimetry — The detection and measurement of the presence of nuclear and X-ray radiation.

radiation efficiency — In an antenna, the ratio of the radiated power to the total power supplied to the antenna at a given frequency.

radiation field — The electromagnetic field that breaks away from a transmitting antenna and radiates outward into space as electromagnetic waves.

radiation filter — A transparent body that transmits only selected wavelengths.

radiation flux density (irradiance) — The total incident radiation energy measured in power per unit area (e.g., milliwatts per square centimeter).

radiation hard — The characteristic of a material that is insensitive to nuclear or X-ray radiation.

radiation-hardened circuit — Integrated circuit manufactured with special devices and isolation techniques such that when it is exposed to heavy radiation it remains operational. Conventional circuits will short out under such conditions because the radiation generates electrical currents inside the semiconductor material.

radiation hardening — 1. Manufacturing techniques applied to a device so that its performance is not degraded significantly by exposure to high gamma and neutron radiation environments. Examples are the use of dielectric isolation techniques and nichrome thin-film resistors. 2. A process of preparing components or circuits so they will be better able to withstand radiation and yet will operate. 3. A preconditioning technique used in manufacturing various electronic components. The process is employed to be sure the product's performance in the equipment is not degraded excessively by exposure to the high levels of neutron and gamma rays that are encountered in space. The principle involved in applying this hardening to the device is to change the characteristics of the component by the preapplication of gamma or neutron rays, so as to permanently fix its electrical characteristics. Entering the hostile space environment, the preconditioned or hardened components will no longer be affected by additional gamma and neutron ray exposure.

radiation hazard — 1. The health hazard caused by exposure to ionizing radiation. 2. The possible harmful effect of powerful electromagnetic radiation on the human body or on electrical components.

radiation intensity — In a given direction, the power radiated from an antenna per unit solid angle in that direction.

radiation lobe — *See* lobe.

radiation loss — In a transmission system, the portion of the transmission loss due to radiation of the radio-frequency power.

radiation monitor — A device for determining amount of exposure to radioactivity. May be periodic or continuous, may monitor an area or an individual's breath, clothing, etc.

radiation pattern — 1. *See* directional pattern. 2. For a fiber or bundle, a curve of the output radiation intensity plotted against the exit angle. 3. For an optical fiber or fiber bundle, the curve of the output radiation intensity plotted as a function of the angle between the optical axis of the fiber or bundle and a normal to the surface on which the radiation intensity is being measured, i.e., the output radiation versus direction of measurement relative to the optical axis.

radiation potential — The voltage required to excite an atom or molecule and cause the emission of one of its characteristic radiation frequencies.

radiation pyrometer — Also called a radiation thermometer. 1. A pyrometer that uses the radiant power from the object or source whose temperature is being measured. Within wide- or narrow-wavelength bands filling a definite solid angle, the radiant power impinges on a suitable detector — usually a thermocouple, thermopile, or a bolometer responsive to the heating effect of the radiant power, or a photosensitive device connected to a sensitive electric instrument. 2. A temperature-measuring device that uses an optical system to focus radiant energy from an object onto a detector. The detector converts this energy into an electrical signal that varies with the temperature of the object.

radiation report — A formal report of radiation measurements made by an engineer skilled in interference control techniques. Usually required by the FCC prior to certification of industrial heating equipment.

radiation resistance — 1. The power radiated by an antenna, divided by the square of the effective antenna current referred to a specified point. 2. The resistance that, if inserted in place of the antenna, would consume the same amount of power radiated by the antenna. 3. The characteristic of a material that enables it to retain useful properties during or after exposure to nuclear radiation.

radiation sensitivity — The ratio of photoinduced current to incident radiant energy, the latter measured at the plane of the lens of a photodevice.

radiation sickness — An illness resulting from exposure to radiation.

radiation survey meter — An instrument that measures instantaneous radiation.

radiation temperature — 1. The temperature to which an ideal blackbody must be heated so it will have the same emissive power as a given source of thermal radiation. 2. The temperature of a complete radiator that has a total radiant emittance identical with that of an unknown source.

radiation thermocouple — A thermocouple that is used in infrared spectroscopy to detect a sample's infrared emittance. *See* thermocouple.

radiation thermometer — *See* radiation pyrometer.

radiation transfer index — Abbreviated RTI. A parameter that describes the transmission performance of optical fiber cables. It measures cable performance and includes both coupling and propagation losses.

radiation trapping — That process whereby radiation spontaneously emitted by a volume of optical material is resonantly reabsorbed within the same volume before it escapes. This effect is manifested in a reduction in the observed rate of spontaneous emission from the material relative to the rate for single atoms or ions.

radiative equilibrium — The constant-temperature condition that exists in a material when the radiant energies absorbed and emitted are equal.

radiative recombination — In an electroluminescent diode in which electrons and holes are injected into the p-type and n-type regions by application of a forward bias, the recombining of injected minority carriers with the majority carriers in such a manner that the energy released on recombination results in the emission of photons of energy $h\nu$, which is approximately equal to the bandgap energy. Radiative recombination produces the light in a LED, which can be modulated for signaling purposes using optical fibers for transmission or integrated optical circuits for switching.

radiator — 1. Any device that emits radiation. *See also* radiating element. 2. Any of the parts of an antenna that radiate electromagnetic waves, either directly into space or against a reflector.

radio — 1. Communication by electromagnetic waves transmitted through space. 2. A general term, principally an adjective, applied to the use of electromagnetic waves between 10 kHz and 3000 GHz. 3. Electronic equipment for the wireless transmission or reception, or both, of electromagnetic waves, especially when used to transmit and receive sounds, activate a remote-control mechanism, etc.;

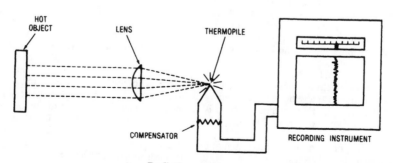

Radiation pyrometer.

a radio set. 4. The science of communicating over a distance by converting sounds or signals to electromagnetic waves and radiating these through space.

radioacoustic position finding—A method of determining distance through water. This is done by closing a circuit at the same instant a charge is exploded under water. The distance to the observing station can then be calculated from the difference in arrival times between the radio signal and the sound of the explosion.

radioacoustics—A study of the production, transmission, and reproduction of sounds carried from one place to another by radiotelephony.

radioactive—Pertaining to or exhibiting radioactivity.

radioactive isotope—*See* radioisotope.

radioactive series—A succession of radioactive elements, each derived from the disintegration of the preceding element in the series. The final element, known as the end product, is not radioactive.

radioactivity—A property exhibited by certain elements whose atomic nuclei spontaneously disintegrate and gradually transmute the original element into stable isotopes of that element or into another element with different chemical properties. The process is accompanied by the emission of alpha particles, beta particles, gamma rays, positrons, or similar radiations.

radioactivity detector—An instrument used to detect radioactive materials: alpha particles, or helium nuclei; beta particles, or free electrons; and gamma rays, which are X-rays of very short wavelength. They may be detected by their chemical effects, by ionization produced in gases at low pressure, and by their tracks formed in a cloud chamber.

radio altitude—*See* radar altitude.

radio approach aids—Equipment making use of radio to determine the position of an aircraft with considerable accuracy from the time it is in the vicinity of an airfield or carrier until it reaches a position from which a landing can be carried out.

radioastronomy—The branch of astronomy in which the radio waves emitted by certain celestial bodies are used for obtaining data about them.

radio attenuation—For one-way propagation, the ratio of the power delivered by the transmitter to the transmission line connecting it with the transmitting antenna, to the power delivered to the receiver by the transmission line connecting it with the receiving antenna.

radio beacon—Also called a radiophone or, in air operations, an aerophare. A radio transmitter, usually nondirectional, that emits identifiable signals for direction finding.

radio-beacon station—In the radionavigation service, a station whose emissions are intended to enable a mobile station to determine its bearing or direction in relation to the radio-beacon station.

radio beam—1. A radio wave in which most of the energy is confined within a relatively small angle. 2. A low-frequency radio transmitter used in direction finding for determining fixes and homing—a process of navigation whereby the pilot directs the aircraft toward the station to which it is tuned.

radio bearing—The angle between the apparent direction of a source of electromagnetic waves and a reference direction determined at a radio direction-finding station. In a true radio bearing, this reference direction is true north. Likewise, in a magnetic radio bearing, it is magnetic north.

radiobiology—The study of the effects on living matter (or substances derived therefrom) of high-energy radiation extending from X-rays to gamma rays, including

high energy beams of neutrons and charged particles, e.g., alpha particles, electrons, protons, deuterons.

radio breakthrough—The breakthrough of modulated radio signals into the channels of an audio amplifier due to the presence of high-level radio signal fields. The effect is that the base/emitter junction of the low-level input transistor rectifies the signals picked up by the wiring or circuit components, and the resulting audio is then handled by the amplifier in the ordinary way so that the radio program appears as a disconcerting background on the wanted source signal.

radio broadcast—A program of music, voice, and/or other sounds broadcast from a radio transmitter for reception by the general public.

radio broadcasting—*See* radio broadcast.

radio channel—A band of frequencies wide enough to be used for radiocommunication. The width of a channel depends on the type of transmission and on the tolerance for the frequency of emission.

radio circuit—1. A means for carrying out one radiocommunication at a time in either direction between two points. 2. A communication circuit between two points via radio. One circuit may be comprised of many channels, which may be used for teletypewriter, voice, or data communication.

radiocommunication—An overall term for transmission by radio of writing, signs, signals, pictures, and sounds of all kinds.

radiocommunication circuit—A radio system for carrying out one communication at a time in either direction between two points.

radiocommunication guard—A communication station designated to listen for and record transmission and to handle traffic on a designated frequency for a certain unit or units.

radio compass—*See* direction finder.

radio control—Remote control of apparatus by radio waves (e.g., model airplanes, boats).

radio deception—Sending false dispatches, using deceptive headings or enemy call signs, etc., by radio to deceive the enemy.

radio detection—Also called radio warning. Determining the presence of an object by radiolocation, but not its precise position.

radio detection and location—Use of an electronic system to detect, locate, and predict future positions of an earth satellite.

radio detection and ranging—Abbreviated radar. 1. Any of certain methods or systems of using beamed and reflected electromagnetic energy for detecting and locating objects; for measuring distance, velocity, or altitude; or for other purposes such as navigating, homing, bombing, missile tracking, mapping, etc. 2. In Federal Communications Commission regulations, a radiodetermination system based on the comparison of reference signals with radio signals reflected or retransmitted from the position to be determined. *See also* radar.

radio direction finder—A radio receiver that pinpoints the line of travel of the received waves.

radio direction finding—Abbreviated RDF. Radiolocation in which only the direction, not the precise location, of a source of radio emission is determined by means of a directive receiving antenna.

radio direction-finding station—A radiolocation station that determines only the direction of other stations, not their location, by monitoring their transmission.

radio Doppler—A device for determining the radial component of the relative velocity of objects by observing the frequency change due to such velocity.

radioelectrocardiogram—A broadcast electrocardiograph signal from the subject to a remote receiver. It makes an ECG practical while the subject is exercising or during natural work or home activities.

radioelectroencephalograph — An electroencephalograph in which a radio link is used so that the patient may move about while the electroencephalogram is being recorded.

radio engineering — The branch of engineering concerned with the generation, transmission, and reception of radio (and television) waves, and with the design, manufacture, and testing of associated equipment.

radio fadeout — Also called the Dellinger effect. The partial or complete absorption of substantially all radio waves normally reflected by the ionospheric layers in or above the E region.

radio field intensity — Also called radio field strength. The maximum (unless otherwise stated) electric or magnetic field intensity at a given location associated with the passage of radio waves. It is commonly expressed as the electric field intensity in microvolts, millivolts, or volts per meter. For a sinusoidal wave, its root-mean-square value is commonly stated instead.

radio field strength — *See* radio field intensity.

radio field-to-noise ratio — The ratio of the field intensity of the desired wave to the noise field intensity at a given location.

radio fix — 1. A method by which the position source of radio signals can be determined. Two or more radio direction finders monitor the transmission and obtain cross bearings. The position can then be pinpointed by triangulation. 2. The method by which a ship, aircraft, etc., equipped with direction finding equipment can determine its own position. This it does by obtaining radio bearings from two or more transmitting stations of known location. The position can then be pinpointed by triangulation as in (1) above.

radio-fixing aids — Equipment making use of radio to assist a user in determining his or her geographical position.

radio frequency — Abbreviated rf. 1. Any frequency at which coherent electromagnetic radiation of energy is possible. 2. A term describing incoming radio signals to a receiver or outgoing signals from a radio transmitter. There are no finite limits in the rf range, but it is usually considered to denote frequency above 150 kHz and extending up to the infrared range.

radio-frequency alternator — A rotating generator that produces radio-frequency power.

radio-frequency amplification — Amplification of a signal by a receiver before reflection, or by a transmitter before radiation.

radio-frequency choke — An inductor used to impede the flow of radio-frequency currents. Its core is generally air or pulverized iron.

radio-frequency component — In a signal or wave, the portion consisting of the rf alternations only — not its audio rate of change in amplitude or frequency.

radio-frequency converter — A power source for producing electrical power at frequencies of 10 kHz and above.

radio-frequency generator — In industrial and dielectric heaters, a power source comprising an electron-tube oscillator, an amplifier (if used), a power supply, and associated control equipment.

radio-frequency heating — 1. The process of heating a substance by subjecting it to a high-frequency energy field. *See also* dielectric heating. 2. A heating and drying process utilizing radio-frequency energy to generate heat in a dielectric material (nonmetallic) by molecular friction.

radio-frequency interference — Abbreviated RFI. Any electrical signal capable of being propagated into and interfering with the proper operation of electrical or electronic equipment. The frequency range of such interference may be taken to include the entire electromagnetic spectrum.

radio-frequency motion detector — A sensor that detects the motion of an intruder through the use of a radiated radio-frequency electromagnetic field. The device operates by sensing a disturbance in the generated rf field caused by intruder motion, typically a modulation of the field referred to as a Doppler effect, which is used to initiate an alarm signal. Most radio-frequency motion detectors are certified by the FCC for operation as "field disturbance sensors" at one of the following frequencies: 0.915 GHz (L-band), 2.45 GHz (S-band), 5.8 GHz (X-band), 10.525 GHz (X-band), or 22.125 GHz (K-band). Units operating in the microwave frequency range are usually called microwave motion detectors.

radio-frequency oscillator — An oscillator that generates alternating current at radio frequencies.

radio-frequency pacemaker — A pacemaker that consists of an implanted circuit designed to receive pacing signals from an extracorporeal transmitter.

radio-frequency preheating — A method of preheating used in the molding of materials so that the molding operation may be facilitated or the molding cycle reduced. The frequencies used most often are between 10 and 100 MHz.

radio-frequency pulse — A radio-frequency carrier that is amplitude modulated by a pulse. Between pulses, the modulated carrier has zero amplitude. (The coherence of the carrier with itself is not implied.)

radio-frequency resistance — *See* skin effect.

radio-frequency signal generator — *See* rf signal generator.

radio-frequency suppressor — A device that absorbs radiated energy that might interfere with radio reception.

radio-frequency transformer — A transformer used with radio-frequency currents.

radio-frequency welding — Also called high-frequency welding. A method of welding thermoplastics using a radio-frequency field to apply the necessary heat.

radiogoniometer — In a radio direction finder, the part that determines the phase difference between the two received signals. The Bellini-Tose system has two loop antennas, both at right angles to each other and connected to two field coils in the radiogoniometer. Bearings are obtained by rotating a search coil inductively coupled to the field coils.

radiogram — A message sent via radio telegraphy.

radiograph — 1. An X-ray film image that shows internal structural features of the body. 2. An X-ray or radium photograph illustrating the nonuniformity or density of the structure the rays penetrate. *See* radiophoto.

radiography — Any nondestructive method of internal examination in which metal objects are exposed to a beam of X-ray or gamma radiation. Differences in thickness, density, or absorption caused by internal defects or inclusions are apparent in the shadow image, either on a fluorescent screen or on photographic film placed behind the object.

radio guard — A ship, aircraft, or radio station designated to listen for and record transmissions and to handle traffic on a designated frequency for a certain unit or units.

radio guidance system — A system that makes use of radio signals in guiding a missile or vehicle in flight. The system includes both the flightborne equipment and the guidance-station equipment on the ground.

radio homing aids — Radio equipment used to assist in the location of an area with sufficient accuracy that an approach may be effected.

radio horizon — The boundary line beyond which direct rays of the radio waves cannot be propagated over the earth's surface. This distance is not a constant; rather, it is affected by atmospheric refraction of the waves.

radio inertial-guidance system — A command type of guidance system consisting essentially of the following: (*a*) A radar tracking unit, comprising radar equipment on the ground, one or more transponders in the missile, and necessary communications links to the guidance station. (*b*) A computer that accepts missile position and velocity information from the tracking system and furnishes appropriate signals to the command link to steer the missile. (*c*) The command link, which consists of a transmitter on the ground and an antenna and receiver on the missile; actually, the command link is built into the tracking unit. (*d*) An inertial system for partial guidance in case of radio guidance failure.

radio influence — Radio-frequency interference that originates on and from power lines.

radio intelligence — Interception and interpretation of enemy radio transmissions.

radio intercept — 1. An act or instance of interception of a radio message. 2. An intercepted radio message. 3. A service or agency that intercepts radio messages.

radio interference — Undesired conducted or radiated electrical disturbances, including transients, which can interfere with the operation of electrical or electronic equipment. These disturbances fall between 14 kHz and 10 GHz.

radioisotope — Also called radioactive isotope. 1. The isotope produced when an element is placed into a nuclear reactor and bombarded with neutrons. Radioisotopes are used as tracers in many areas of science and industry. Like all isotopes, they decay spontaneously with the emission of their radiation at a definite rate measured by their half-lives. 2. Tracer form of an element having similar chemical behavior but "tagged" with a radioactive substance (chromium, iodine, phosphor, etc.) so that it emits gamma rays that can be counted with a scintillation counter.

radio jamming — Blocking communications by sending overpowering interference signals.

radio knife — A form of surgical knife that uses a high-frequency electric arc at its tip to cut tissue and, at the same time, also sterilizes the edges of the wound.

radio landing aids — Radio equipment used in assisting an aircraft in making its actual landing.

radio landing beam — A distribution of vertical directional radio waves used for guiding aircraft into a landing.

radio link — A radio system used to provide communication between two specific points.

radiolocation — 1. Use of the constant-velocity or rectilinear propagation characteristics of radio waves to detect an object or to determine its direction, position, or motion. 2. With respect to Federal Communications Commission regulations, radiodetermination used for purposes other than those of radionavigation.

radiolocation service — A radio service in which radiolocation is used.

radiolocation station — A radio station in the radiolocation service.

radio log — A record of all messages sent and received, transmitter tests made, and other important information pertaining to the operation of a particular station.

radiologist — A specialist in the field of radiology.

radiology — The branch of medical science that deals with the use of radiant energy in the diagnosis and treatment of disease. While radiology originally involved only the use of radiant energy in visualizing tissues, this science now covers a wide spectrum of diagnostic and therapeutic applications, such as diagnosis using roentgen rays, diagnosis using radioisotope scanning techniques, genetic radioisotopes in research, roentgen cinematography, angiography, spectroscopy, and treatment of tumors using roentgen rays, implanted radioactive pellets, cyclotrons, or lasers.

radioluminescence — Luminescence produced by radiant energy (e.g., by X-rays, radioactive emissions, alpha particles, or electrons).

radiomagnetic indicator — Abbreviated RMI. A navigational instrument used by land vehicles. It presents a display combining the heading and the relative and magnetic bearings of the vehicle with the relative bearing of a radio station whose location is known.

radio marker beacon — In the aeronautical radionavigation service, a land station that provides a signal to designate a small area above the station.

radio marker station — Station marking a definite location on the ground as an aid to air navigation.

radiometallography — X-ray examination of the crystalline structure and other characteristics of metals and alloys.

radiometeorograph — A meteorograph which, when carried into the stratosphere by an unmanned gas-filled rubber balloon, automatically reports atmospheric conditions by radio as it ascends into the stratosphere. The ultrahigh-frequency signals are transmitted so that they can be recorded and interpreted in terms of pressure, temperature, and humidity. *See also* radiosonde.

radiometer — 1. A device for measuring radiant flux density. Generally, a blackened thermocouple or bolometer is employed, but the simplest type employs a rotating vane. 2. An instrument for measuring intensity of radiant energy, including X-ray. A Nicols or Crookes radiometer detects the energy by the torsional twist of suspended

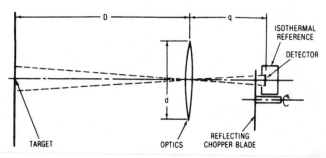

Radiometer.

vanes that are blackened on one side and exposed to radiant energy.

radiometric— Pertaining to the measurement of radiation.

radiometry— 1. The measurement of the spectral emission characteristics of sources of electromagnetic radiation. 2. The science of radiation measurement concerned with the detection and measurement of radiant energy either at separate wavelengths or integrated over a broad wavelength band, and the interaction of radiation with matter, such as absorption, reflectance, and emission. 3. The measurement of radiation in the infrared, visible, and ultraviolet portion of the spectrum.

radiomicrometer— *See* microradiometer.

radionavigation— Navigational use of radiolocation for determining position or direction, or for providing a warning of obstructions. Radionavigation includes use of radio direction finding, radio ranges, radio compasses, radio homing beacons, etc.

radionavigation land station— A fixed station in the radionavigation service—i.e., one not intended for mobile operation.

radionavigation mobile station— A radionavigation station operated from a vehicle.

radionavigation service— A radiolocation service used for radionavigation.

radio net— A system of radio stations that communicate with each other. A military net usually is made up of a radio station of a superior unit and stations of all subordinate or supporting units.

radio-noise field intensity— A measure of the field intensity of interfering electromagnetic waves at some point (e.g., a radio receiving station). In practice, the field intensity itself is not measured, but some proportionate quantity.

radiopaque— Not penetrable by X-rays or other radiation. (The opposite of radioparent.)

radioparent— Penetrable by X-rays or other radiation. (The opposite of radiopaque.)

radiophone— *See* radio beacon; radiotelephone.

radiophoto— Also called radiophotography, radiophotograph, radiophotogram, facsimile, or radiograph. The transmission of photographs and other illustrations by radio.

radiophotogram— *See* radiophoto.

radiophotograph— A photograph transmitted by radio waves.

radiophotography— The transmission of photographic images or pictures by radio waves.

radiophotoluminescence— The property whereby the previous exposure of certain materials to nuclear radiation enables them to give off visible light when irradiated with ultraviolet light.

radio position finding— Determining the location of a radio station by using two or more direction finders and a process of triangulation.

radiopositioning land station— A station, other than a radionavigation station, in the radiolocation service not intended to be operated while in motion.

radiopositioning mobile station— A station, other than a radionavigation station, in the radiolocation service intended to be operated while in motion or while stopped at unspecified points.

radio prospecting— The use of radio equipment to locate mineral or oil deposits.

radio proximity fuse— A radio device that detonates a missile by electromagnetic interaction within a predetermined distance from the target.

radio pulse— An intense, split-second burst of electromagnetic energy.

radio range— A radionavigational facility whose emissions provide radial lines of position by having special characteristics that are recognizable as bearing information and useful in lateral guidance of aircraft.

radio range beacon— A radionavigation land station in the aeronautical radionavigation service providing radio equisignal zones.

radio range finding— Determination of range by means of radio waves.

radio range leg— The space within which an aircraft will receive an on-course signal from a radio range station.

radio range monitor— An instrument that automatically monitors the signal from a radio range beacon and warns attending personnel when the transmitter deviates from its specified current bearings. It also transmits a distinctive warning to approaching planes whenever trouble exists at the beacon.

radio range station— A land station that operates in the aeronautical radionavigation service and provides radial equisignal zones.

radio receiver— A device for converting radio waves into signals perceptible to humans.

radio reception— Reception of radioed messages, programs, or other intelligence.

radio recognition— The use of radio means to determine the friendly or hostile character or the individuality of another.

radio relay— *See* radio-relay system.

radio-relay system— Also called radio relay. A radio transmission system in which the signals are received and retransmitted from point to point by intermediate radio stations.

radio set— A radio transmitter, radio receiver, or a combination of the two.

radio shielding— A metallic covering over all electric wiring and ignition apparatus that is grounded at frequent intervals for the purpose of eliminating electric interference with radiocommunications.

radio signal— A signal that is transmitted by a radio.

radio silence— A period during which all or certain radio equipment capable of radiation is kept inoperative.

radiosonde— A balloonborne instrument for the simultaneous measurement and transmission of meteorological data. The instrument consists of transducers for the measurement of pressure, temperature, and humidity; a modulator for conversion of the output of the transducers to a quantity that controls a property of the radio-frequency signal; a selector switch that determines the sequence in which the parameters are to be transmitted; and a transmitter that generates the radio-frequency carrier.

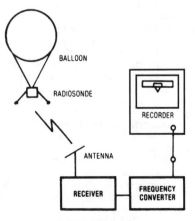

Radiosonde system.

Radio spectrum

Band number	Frequency range*	Metric subdivision	Band designation
4	3–30 kHz	Myriametric waves	VLF
5	30–300 kHz	Kilometric waves	LF
6	300–3000 kHz	Hectometric waves	MF
7	3–30 MHz	Decametric waves	HF
8	30–300 MHz	Metric waves	VHF
9	300–3000 MHz	Decimetric waves	UHF
10	3–30 GHz	Centimetric waves	SHF
11	30–300 GHz	Millimetric waves	EHF
12	300–3000 GHz or 3 THz	Decimillimetric waves	—

*Lower limit is exclusive, upper limit inclusive.

radiosonde recorder — An instrument that is located at the surface observing station and is used to record the data presented by a radiosonde aloft. The mechanism of the recorder depends upon the type of radiosonde system used.

radiosonde transmitter — The component of the radiosonde that includes the modulating blocking oscillator and the radio-frequency carrier oscillator.

radiosonobuoy — See sonobuoy.

radio spectrum — The range of frequencies of electromagnetic radiation usable for radiocommunication. The radio spectrum may range from 3 kHz to over 300 GHz. Corresponding wavelengths are 100 km to 1 mm.

radio station — An assemblage of equipment for radio transmission, reception, or both.

radio-station interference — Interference caused by reception of radio waves from other than the desired station.

Radio Technical Commission for Aeronautics — A cooperative, nonprofit association of all telecommunications agencies of the U.S. government and industry. Its purpose is to advance the art and science of aereonautics through investigation of all available or potential applications of the telecommunications art, coordination of these applications with allied arts, and the adaptation of them to recognized operational requirements.

radiotelegraph transmitter — A radio transmitter capable of handling code signals.

radiotelegraphy — Radiocommunication by means of the International Morse code or some other, similarly coded, signal.

radiotelephone — Also called radiophone. The complete radio transmitter, receiver, and associated equipment required at one station for radiotelephony.

radiotelephone distress signal — The spoken word MAYDAY (phonetic spelling of the French expression *m'aidez*, or "help me"). It corresponds to SOS in radiotelegraphy and is used by aircraft, ships, etc., needing help.

radiotelephone transmitter — A radio transmitter capable of handling audio-frequency modulation (e.g., voice and music).

radiotelephony — Two-way transmission and reception of sounds by radio.

radio telescope — 1. A very sensitive radio receiver that is used in conjunction with a large and highly directional antenna to receive signals from an object or region in space. 2. An instrument designed to collect naturally formed extraterrestrial electromagnetic radiation within the radio-frequency range of the spectrum in amounts sufficient to be measured.

radioteletype — See RTTY.

radiotherapy — The treatment of disease with radiation, especially ultraviolet, infrared, X-rays, and gamma rays.

radiothermics — The application of heat generated by radio waves (e.g., in diathermy or electronic heating).

radio transmission — Transmission of signals by electromagnetic waves other than light or heat waves.

radio transmitter — A device capable of producing radio-frequency power and used for radio transmission.

radiotransparent — Permitting the passage of X-rays or similar radiation.

radio tube — A general term for any type of electron tube used in electronic equipment.

radiovision — An early name for television.

radio warning — See radio detection.

radio watch — Also called watch. The vigil maintained by an operator when on duty in the radio room of a vessel and listening for signals — especially on the international distress frequencies.

radio wave — 1. Electromagnetic wave at a frequency lower than 3000 GHz and propagated through space without an artificial guide. 2. Combination of electric and magnetic fields varying at a radio-frequency rate and capable of travelling through space at the speed of light. It is produced by feeding the output of a radio transmitter into a transmitting antenna.

radio wavefront distortion — A change in the direction of advance of a radio wave.

radio-wave propagation — The transfer of energy by electromagnetic radiation at frequencies below approximately 3×10^{12} hertz.

radist — The radionavigation system in which the position of a vehicle is determined by comparing the arrival times of transmitted pulses at three or more ground stations.

radix — Also called the base. 1. The total number of distinct marks or symbols used in a numbering system. For example, since the decimal numbering system uses ten symbols (0, 1, 2, 3, 4, 5, 6, 7, 8, 9), the radix is 10. In the binary numbering system the radix is 2, because there are only two marks or symbols (0, 1). 2. In positional representation, that integer, if it exists, by which the significance of the digit place must be multiplied to give the significance of the next higher digit place. For example, in decimal notation, the radix of each place is 10. 3. Total number of characters available to each position of a digital numeric system.

radix complement — Also called true complement. A number obtained by subtracting each digit of the given number from one less than the radix, then adding to the least significant digit, executing all required carries. Examples are the tens complement in decimal notation and the twos complement in binary notation.

radix-minus-one complement — Also called diminished-radix complement. A number obtained by subtracting each digit of the given number from one less than the radix. Examples are the nines complement in decimal notation and the ones complement in binary notation.

radix notation — A positional representation in which any two adjacent digit positions have significances in an integral ratio called the radix of the least significant of the two positions; the digit in any position may have a value from zero to one less than the radix of that position.

radix point — Also called base point, and binary point, decimal point, etc., depending on the numbering system. The index that separates the integral and fractional digits of the numbering system in which the quantity is represented.

radome — Also called radar dome. The housing that protects a radar antenna from the elements, but does not block radio frequencies.

radux — A long-distance, low-frequency navigational system that provides hyperbolic lines of position. It is of the continuous-wave, phase-comparison type.

railing — Radar pulse jamming at high recurrence rates (50 to 150 kHz). On a radar indicator, it results in an image resembling fence railing.

railings — The pattern produced on an A-scope by continuous waves modulated with a high-frequency signal. It appears as a series of vertical lines resembling target echoes along the baseline.

rain-barrel effect — The characteristic sound noted on an equalized line that is overcompensated.

rainbow — The technique that applies pulse-to-pulse frequency changing to identify and discriminate against decoys and chaff.

rainbow generator — A signal generator with which the entire color spectrum can be produced on the screen of a color television receiver. The colors merge together as in a rainbow. *See also* keyed rainbow generator.

rake angle — The angle that a stylus makes with a phonograph record when viewed from the side.

RAM — Abbreviation for random-access memory.

ram — The moving part of the head of a crimping tool.

ramark — Also called a radar marker. A fixed radar transmitter that emits a continuous signal that is used as a bearing indication on a radar display.

ramp — A voltage or current that varies at a constant rate; for example, that portion of the output waveform of a time/linear sweep generator used as a time base for scope display.

ramping — Control method used to vary the pulse rate from one value to another, to produce either acceleration or deceleration of a stepper motor.

ramp input — A change, in an input signal, that varies at a constant rate.

Ramsauer effect — The absorption of slow-moving electrons by intervening matter.

random — Irregular; having no set pattern.

random access — 1. Access to a computer storage under conditions whereby there is no rule for predetermining the position from where the next item of information is to be obtained 2. The process of obtaining information from the computer storage with the access time independent of the location of the information. 3. Access method in a computer in which each word can be retrieved directly by its address. 4. The ability to gain access to any location of a memory unit in a time that is independent of the location itself. 5. A term used to describe computer files that do not have to be searched sequentially to find a particular record but can be addressed directly. 6. The ability of a computer to directly access random bits of memory without going through a predefined sequence of locations.

random-access device — Also called direct-access device. A data-storage device in which the access time is effectively independent of the location of the stored data.

random-access memory — Abbreviated RAM. 1. A storage arrangement from which information can be retrieved with a speed that is independent of the location of the information in the storage. For example, a core memory is a random-access memory, but a magnetic tape memory is not. 2. A device that permits individual interrogation of any memory cell in a completely random sequence. Any point in the total memory system can be accessed without looking at any other bit. 3. A memory that may be written to, or read from, any address location in any order. May refer specifically to the integrated circuit method of implementation. 4. A read/write memory, usually a permanent memory, that stores information in such a way that each bit of information may be retrieved within the same amount of time as any other bit. As opposed to serial memory. 5. A type of memory that offers access to storage location within it by means of X and Y coordinates. 6. The main memory of a computer. Instructions can be written into and read out of RAM, and its contents can be changed at any time. But its contents are wiped clean when power is shut off. *See* ROM. 7. A type of memory that is used to store and retrieve digital data that is constantly being changed, updated, or modified in some way. 8. A memory in which the central processor can access all locations with equal facility. Sometimes called read/write memory because data can be added to or removed from such memory as an ordinary procedure. RAM is used for both storage of a computer operation program and for data. 9. A memory that permits access to any of its address locations in any desired sequence with similar access time to each location. The term has come to mean only a read/write memory, not a ROM or EPROM, which, strictly speaking, are also random-access memories.

random-access programming — Programming a problem for a computer without regard to the access time to the information in the registers called for in the program.

random-access storage — A form of storage in which information can be recovered immediately, regardless of where it was stored. For example, magnetic-core memory devices will usually yield any bit of information with almost no time penalty regardless of where it is located, while magnetic tape must be run until the required information can be found.

random bundle — A fiber-optics bundle that scrambles an image but can convey light from a source to a relatively inaccessible point, and then again from that location to a photocell.

random error — The inherent imprecision of a given process of measurement; the unpredictable component of repeated independent measurements on the same object under sensible uniform conditions, usually of an approximately normal (Gaussian) frequency distribution, other than systematic or erratic errors and mistakes, sometimes called short-period errors.

random experiment — An experiment that can be repeated a large number of times but may yield different results each time, even when performed under similar circumstances.

random failure — Also called chance failure. 1. Any chance failure whose occurrence at a given time is unpredictable. 2. Any failure whose cause and/or mechanism make its time of occurrence unpredictable, but which is predictable only in a probablistic or statistical sense.

random-function generator — A device that generates nonrepetitive signals that are distributed over a broad frequency range.

random-impulse generator — A generator of electrical impulses that occur at random rather than at specific intervals.

random interlace — A technique for scanning in which there is no fixed relationship between adjacent lines in successive field; often used in closed-circuit television systems. It offers somewhat reduced precision to that employed in commercial broadcast service.

randomized jitter — Jitter produced by means of noise modulation.

random logic — A collection of various gates interconnected to perform a variety of functions. Random-logic circuits form no repetitive pattern and are more difficult to fabricate by conventional LSI/CMOS techniques than is repetitive logic. However, new high-density CMOS fabricating techniques overcome the old restrictions and allow large-scale integration of random logic.

randomness — 1. A condition of equal chance for the occurrence of any of the possible outcomes. 2. The occurrence of an event in accordance with the laws of chance.

random noise — Also called fluctuation noise. 1. A signal whose instantaneous amplitude is determined at random and therefore is unpredictable. It contains no periodic frequency components and its spectrum is continuous. *See also* broadband electrical noise. 2. Noise generated in a circuit by random movement of electrons caused by thermal agitation. In tape recording it can be caused by uneven distribution of magnetized particles and is reproduced as a background hiss. 3. Transient perturbations occurring at random with a Gaussian spectral distribution equivalent to thermal noise. Thermal and shot noise are forms of random noise that may be present in all portions of a telecommunications system, being more significant, of course, at points of lower signal-to-noise ratio. 4. Essentially, noise that cannot be predicted. Therefore, even if the magnitude of sound or oscillation in a system is known at a given moment, random noise can change in a short time. 5. Noise composed of many uncoordinated and overlapping transient disturbances occurring at random.

random-noise generator — A generator of a succession of random signals that are distributed over a wide frequency spectrum.

random-number generator — A special machine routine or hardware that produces a random number or series of random numbers in accordance with specified limitations.

random numbers — 1. A set of digits such that each successive digit is equally likely to be any of n digits to the base n of the number. 2. Numbers composed of digits selected from an orderless sequence of digits. 3. Array of independent digits having no logical interrelationship, so that the occurrence of any particular one is totally unpredictable. 4. A sequence of integers or group of numbers (often in the form of a table) that show absolutely no relationship to each other anywhere in the sequence. At any point, all integers have an equal chance of occurring, and they occur in an unpredictable fashion.

random PARD — Pertains to that portion of the total PARD in an electric power supply that is not periodic. This phenomenon is frequently referred to as noise.

random processing — The treatment of information without respect to where it is located in external storage and in an arbitrary sequence determined by the input against which it is to be processed.

random pulsing — Varying the repetition rate of pulses by noise modulation or continuous frequency change.

random sample — A sample in which every item in the lot is equally likely to be selected in the sample.

random sampling — 1. A sampling process in which there is a significant time uncertainty between the signal being sampled and the taking of samples. 2. A selection of observations taken from all of the observations of a phenomenon in such a way that each chosen observation has the same possibility of selection.

random-sampling oscilloscope — An oscilloscope that functions by constructing a coherent display from samples taken at random.

random sequential memory — A memory in which one reference can be found immediately; the other reference is found in a fixed sequence.

random signals — Waveforms having at least one parameter (usually amplitude) that is a random function of time (e.g., thermal noise or shot noise).

random variable — Also called variate or stochastic variable. 1. The result of a random experiment. 2. A discrete or continuous variable that may assume any one of a number of values, each having the same probability of occurrence. 3. Any signal whose amplitude or phase cannot be predicted by a study of previous values of the signal.

random variation — A fluctuation in data that is due to uncertain or random occurrences.

random velocity — The instantaneous velocity of a particle without regard to direction. It may be characterized by its distribution function or by its average, root-mean-square, or most probable value.

random vibration — A vibration generally composed of a broad, continuous spectrum of frequencies, the instantaneous magnitude of which cannot be specified to any given moment of time. Instantaneous amplitude can only be defined statistically by a probability distribution function that gives the fraction of the total time that the amplitude lies within specified amplitude intervals. (If random vibration has instantaneous magnitudes distributed according to the Gaussian distribution, it is called Gaussian random vibration.) *See also* white noise.

random winding — A coil winding in which the turns and layers are not regularly positioned or spaced but are positioned haphazardly.

random wound — Describing a coil wound without care to ensure that the wire is in layers. Random-wound coils have fewer turns for a given volume.

range — 1. The maximum useful distance of a radar or radio transmitter. 2. The difference between the maximum and the minimum value of a variable. 3. The set of values that may be assumed by a quantity or function. 4. *See* receiving margin.

range-amplitude display — A radar display in which a time base provides the range scale from which echoes appear as deflections normal to the base.

range calibration — Adjustment of radar-range indications by use of known range targets or delayed signals so that, when on target, the radar set will indicate the correct range.

range coding — A method of coding a beacon response so that the response appears as a series of pulses on a radarscope. The coding provides identification.

range finder — 1. A movable, calibrated unit of the receiving mechanism of a teletypewriter that can be used to move the selecting interval relative to the start signal. 2. An optical distance finder that depends on triangulation of two convergent beams on an object from disparate viewpoints. A pair of unaided human eyes coupled with a computerlike brain can estimate distance, at least for nearby objects, with some accuracy. 3. A device that depends on the measurement of time of wave travel from an object to a point, as in radar and sonar.

range gate — A gate voltage used to select radar echoes from a very short range interval.

range-height indicator — Abbreviated RHI. A radar display on which an echo appears as a bright spot on a rectangular field. The slant range is indicated along the x-axis, and the height above the horizontal plane (on a magnified scale) along the y-axis. A cursor shows the height above the earth.

range mark — *See* distance mark.

range marker — A variable or movable discontinuity in the range time base of a radar display (in the case of a PPI, a ring). It is used for measuring the range of an echo or calibrating the range scale.

range of an instrument — *See* total range of an instrument.

range of gain — The minimum and maximum gain to which the amplifier can be set.

range resolution — The minimum difference in range between two radar targets along the same line of bearing for which an operator can distinguish between targets.

range ring — An accurate, adjustable ranging mark on a plan-position indicator corresponding to a range step on a type-M indicator.

range step — The vertical displacement on an M-indicator sweep to measure range.

range surveillance — Surveillance of a missile range by means of electronic and other equipment.

range unit — A radar-system component used for control and indication (usually counters) of range measurements.

range zero — Alignment of the start of a sweep trace with zero range.

ranging oscillator — An oscillator circuit containing an *LC* resonant combination in the cathode circuit, usually used in radar equipment to provide range marks.

rank — To arrange in a series in ascending or descending order of importance.

rapid memory — *See* rapid storage.

rapid storage — Also called rapid memory, fast-access storage, and high-speed storage. Computer storage in which the access time is very short; rapid access usually is gained by limiting the storage capacity.

rare gas — *See* noble gas.

raser — Acronym for radio amplification by stimulated emission of radiation, a chemical pumping process that is accomplished without external radiation.

raster — 1. On the screen of a cathode-ray tube, a predetermined pattern of scanning lines that provide substantially uniform coverage of an area. 2. The illuminated area produced by the scanning lines on a television picture tube when no signal is being received. 3. Rectangular line pattern of light produced on the screen of a cathode-ray tube with no signal present. It is formed by deflecting the electron beam rapidly from left to right and relatively slowly from top to bottom. 4. The pattern of lines traced by rectilinear scanning in display systems. 5. The visible part of the picture displayed by a monitor or TV set.

raster burn — In camera tubes, a change in the characteristics of the area that has been scanned. As a result, a spurious signal corresponding to that area will be produced when a larger or tilted raster is scanned.

raster display — 1. A display in which the entire display surface is scanned at a constant refresh rate. 2. A refresh graphics system in which the electron beam sweeps horizontally across the face of the CRT from left to right, drawing the picture as a series of scan lines. At the end of each line, the beam is turned off and repositioned down and to the left (the horizontal retrace) for the start of the next line. After the last line has been drawn, the beam is repositioned at the upper left corner (the vertical retrace) and the scan is repeated. Interlaced displays require two such scans to complete each picture; the first scan draws the odd-numbered lines, and the second scan draws the even-numbered lines. Horizontal vectors tend to flicker at 30 Hz on interlaced displays since they fall on a single scan line. Noninterlaced displays draw the entire picture every scan. The display scan rate always matches the ac line frequency, resulting in a 60-Hz refresh rate for noninterlaced displays and 30-Hz for interlaced displays.

raster scan — 1. The most basic method of sweeping a CRT one line at a time to generate and display images. This method is used in commercial television in the United States. 2. Line-by-line sweep across the entire display surface to generate elements of a display image.

raster scanning — 1. Radar antenna scanning similar to electron-beam scanning in a television picture tube; a horizontal sector scan that is changed in elevation. 2. A process whereby displays are generated through repetitive video scan lines. For best appearance, 10 raster scans per 5×7 character and 12 scans per 7×9 character are required with underlined cursors.

raster-scan technology — A technique similar to that used in consumer color TVs, whereby an electron beam scans across and down the screen, turning on and off at selected points or addresses. Because the screen of a color CRT uses three phosphor colors, it provides an unlimited palette of colors and gray-scale values.

ratchetjaw — CB radio term for one who often talks on radio.

ratchet relay — A stepping relay actuated by an armature-driven ratchet.

rate action — Also called derivative action. Corrective action whose rate is determined by how fast the error being corrected is increasing.

rate center — A specified geographic location used by telephone companies to determine mileage measurements for the application of interexchange mileage rates.

rated capacitance — The capacitance value for which a capacitor has been designed; usually indicated on the capacitor.

rated contact current — The current that contacts are designed to handle for their rated life.

rated current (for ac capacitors) — The rms value of the alternating current at rated frequency and rated temperature for which the capacitor is designed.

rated operational voltage — The voltage on which a specific application rating of a device is based.

rated output — 1. The output power, voltage, current, etc., at which a machine, device, or apparatus is designed to operate under normal conditions. 2. The power, voltage, or current that a device (or machine) can put out for long periods without becoming overheated under specified conditions of ambient temperature and ventilation.

rated power output — Also called continuous power output or rms power output. 1. The normal radio-frequency power-output capability (peak or average) of a transmitter under optimum adjustment and operation conditions. 2. The maximum power that an amplifier will deliver continuously without exceeding its specified distortion rating.

rated range—The nominal range within which a device can be operated and still maintain the level of performance specified for it by the manufacturer.

rated ripple current—The maximum rms alternating current of specified frequency that may be applied continuously to a capacitor at a specified temperature. This is a specific term used for electrolytic capacitors.

rated ripple voltage—The rms value of the maximum alternating voltage of a specified frequency superimposed on the direct voltage that may be applied continuously to a capacitor at a specified temperature.

rated-safe plate dissipation—The number of watts that can be used to heat the plate of a vacuum tube without damaging the tube.

rated temperature—The maximum temperature at which an electric component can operate for extended periods without loss of its basic properties.

rated thermal current—In a contactor, that current at which the permissible temperature rise is reached.

rated voltage—1. The voltage at which a device or component is designed to operate under normal conditions. 2. The maximum voltage at which an electric component can operate for extended periods without undue degradation or safety hazard.

rate effect—The anode-voltage transients that cause pnpn devices to switch into high conduction.

rate generator—A proportional element that converts angular speed into a constant-frequency output voltage. *See also* angular velocity.

rate-grown junction—A grown junction, in a semiconductor, produced by varying the rate of crystal growth periodically so that n-type impurities alternately predominate.

rate-grown transistor—Also called graded-junction transistor. A variation of the double-doped transistor, in which n- and p-type impurities are added to the melt.

rate gyro—A particular kind of gyroscope used for measuring angular rates. A system of three rate gyros, each oriented to one of three mutually perpendicular axes—roll, pitch, and yaw—can control a missile or aircraft by detecting angular rates and then generating proportional corrective signals.

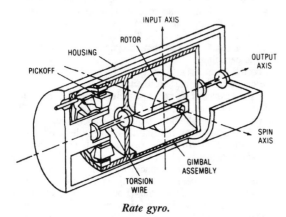

Rate gyro.

rate limit—*See* slew rate, 2.

rate limiting—Nonlinear behavior in an amplifier due to its limited ability to produce large, rapid changes in output voltage (slewing), restricting it to rates of change of voltage lower than might be predicted by observing the small signal frequency response.

ratemeter—1. An instrument for measuring the rates at which counts are received, usually in counts per minute. 2. A type of radiation detector whose output is proportional to instantaneous radiation intensity (rate of radioactivity emission).

rate of decay—The rate at which the sound-pressure level (velocity level, or sound-energy density level) is decreasing at a given point and at a given time. The practical unit is the decibel per second.

rate-of-rise relay—*See* instantaneous overcurrent relay.

rate-of-rise timer—A percentage timer as applied in a temperature control system for controlling the rate of temperature rise. Usually a standard percentage timer with a dial marked in various scales of degrees per hour rise.

rate of transmission—*See* speed of transmission.

rate receiver—A device for receiving a signal giving the rate of speed of a launched missile.

rate signal—A signal proportional to the time derivative of a specified variable.

rate test—A test to verify that the time constants of the integrators in an analog computer are correct.

rate transmitter—A guidance antenna used to signal the desired speed for a missile in flight.

rating—A value that establishes either a limiting capability or a limiting condition for an electronic device. It is determined for specified values of environment and operation, and may be stated in any suitable terms. (Limiting conditions may be either maxima or minima.)

rating system—The set of principles on which electronic device ratings are established and which determines their interpretation. (The rating system indicates the division of responsibility between the device manufacturer and the circuit designer, with the object of ensuring that the working conditions do not exceed the ratings.)

ratio—The value obtained by dividing one number by another. This value indicates their relationship to each other.

ratio arms—Two adjacent arms of a Wheatstone bridge, both having an adjustable resistance and so arranged that they can be set to have any of several fixed ratios to each other.

ratio calibration—1. The calibration of a dimensionless quantity that represents the ratio of one of its values to another. 2. A method by which potentiometric transducers may be calibrated, in which the value of the measurand is expressed in terms of decimal fractions representing the ratio of output resistance to total resistance.

ratio detector—An FM detector that inherently discriminates against amplitude modulation. Two diodes are connected such that the audio output is proportional to the ratio of the FM voltages applied to them.

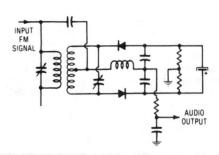

Ratio detector.

ratio meter — 1. An instrument that measures electrically the quotient of two quantities. It generally has no mechanical control means such as springs. Instead, it operates by the balancing of electromagnetic forces, which are a function of the position of the moving element. 2. Instrument for measuring the ratio of transformation of a transformer by means of a resistance bridge arrangement. 3. Moving-coil type of instrument in which the deflection is proportional to the ratio of the current sent through two coils.

ratio of transformation — The ratio of the secondary voltage of a transformer to the primary voltage under no-load conditions, or the corresponding ratio of currents in a current transformer.

ratio squelch — *See* squelch circuit.

ratio-type telemeter — Also called a position-type telemeter. A telemeter in which the relative phase position between, or magnitude relation of, two or more electrical quantities is the translating means.

rat race — A magic-T modification for the acceptance of higher power. A circular loop of coaxial line closed upon itself and having four branching connections. *See also* hybrid ring.

raw data — 1. Data that has not been processed. 2. Data as received, which has not been processed by a machine and which may not yet be in the form that a machine can accept.

raw tape — Also called virgin tape or blank tape. A term sometimes used to describe tape that has not been recorded. *See also* blank tape.

ray — 1. A line of propagation of any form of radiant energy. 2. In fiber optics, a straight line, representing light, perpendicular to the light wavefront and traveling in the same direction. At a boundary surface or interface, such as the surface between a fiber core and cladding, the ray may change direction suddenly, but it remains a straight line. 3. A line of light extending between two points.

ray angle — The angle between a light ray and a reference line or plane.

ray-control electrode — An electrode that controls the position of the electron beam on the screen of a cathode-ray tuning-indicator tube.

Raydist — A system using cw transmission to provide hyperbolic lines of position through rf phase comparison techniques. The system is used for surveying or ship positioning in a two-dimensional array. The frequency band is 1.7 to 2.5 MHz. Similar to LORAC in principle.

Rayleigh disc — A special acoustic radiometer used for the fundamental measurement of particle velocity.

Rayleigh distribution — Frequency distribution for an infinite number of quantities of the same magnitude, but of random phase relationships. Sky-wave field intensities follow the Rayleigh distribution for intervals of one minute or less.

Rayleigh distribution (fading) — A statistical distribution describing the magnitude of a phasor composed of the sum of many component phasors randomly distributed in amplitude and phase. Fading of signals caused by cancellation and reinforcement of contributions received over separate paths often exhibits a Rayleigh distribution.

Rayleigh line — In scattered radiation, a spectrum line that has the same frequency as the corresponding incident radiation.

Rayleigh reciprocity theorem — The reciprocal relationship for an antenna when transmitting or receiving. The effective heights and the radiation resistance and pattern are alike, whether the antenna is transmitting or receiving.

Rayleigh scattering — Scattering of radiation by minute particles suspended in air (e.g., by dust).

Rayleigh wave — A surface wave associated with the free boundary of a solid. The wave is of maximum intensity at the surface, but diminishes quite rapidly as it proceeds into the solid.

ray path — An imaginary line, perpendicular to the wavefront, that describes the path along which the energy associated with a point on a wavefront moves.

Raysistor — Raytheon trade name for a device that contains a photosensitive semiconductor element and a light source that can be used to control the conductivity of the semiconductor.

RBOC — Abbreviation for Regional Bell Operating Company. Also called Bell Operating Companies (BOC). Local telephone operating companies that were split off from AT&T and which provide most local and intrastate telephone service in the United States.

RC — Symbol for resistance-capacitance, resistance-coupled, or ray-control electrode.

RC amplifier — Abbreviation for resistance-capacitance coupled amplifier.

RC circuit — A time-determining network of resistors and capacitors in which the time constant is defined as resistance times capacitance.

RC constant — The time constant of a resistor-capacitor circuit, equal in seconds to the resistance value in ohms multiplied by the capacitance value in farads.

RC coupling — *See* resistance-capacitance coupling.

RC filter — *See* resistance-capacitance filter.

RC network — A circuit containing resistances and capacitances arranged in a particular manner to perform a specific function.

RC oscillator — An oscillator in which the frequency is determined by resistance-capacitance elements. *See also* resistance-capacitance oscillator.

RCTL — Abbreviation for resistor-capacitor-transistor logic.

rd — Letter symbol for rutherford.

RDF — Abbreviation for radio direction finder (or finding).

R-display — Essentially an expanded radar A-display in which the trace of an echo can be expanded for more detailed examination.

reach-through voltage — That value of reverse voltage for which the depletion layer in a reverse-biased pn junction spreads sufficiently to make electrical contact with another junction. *See also* punch-through voltage.

reacquisition time — The time required for a tracking radar to relock on the target after the radar's automatic tracking mechanism has been disengaged.

reactance — Symbolized by X. 1. Opposition to the flow of alternating current. Capacitive reactance (X_C) is the opposition offered by capacitors, and inductive reactance (X_L) is the opposition offered by a coil or other inductance. Both reactances are measured in ohms. 2. The opposition offered an alternating flow by a capacitance or inductance related to the frequency of the alternating current. The reactance of a capacitor decreases with increasing frequency, but the reactance of nn inductance increases with frequency.

reactance drop — The voltage drop in quadrature with the current.

reactance factor — In a conductor, the ratio of its ac resistance to its ohmic resistance.

reactance frequency multiplier — A frequency multiplier whose essential element is a nonlinear reactor; the nonlinearity of the reactor is used to generate harmonics of a sinusoidal source.

reactance grounded — Grounded through a reactance.

reactance modulator — A modulator whose reactance can be varied in accordance with the instantaneous amplitude of the modulating electromotive force applied. It is normally an electron-tube circuit that usually modulates the phase or frequency.

reactance relay — A form of impedance relay whose operation is a function of the reactance of the circuit.

reactance tube (transistor) — A vacuum tube or transistor connected so as to appear as a reactance to the rest of the circuit. This is accomplished by deriving the grid (base) excitation from the plate (collector) voltage through an RC network that changes its phase by approximately 90°. Thus, since the plate (collector) current is in phase with the grid (base) excitation, the tube (transistor) draws an apparently reactive current. The magnitude of the apparent reactance is controlled by variation of the gain of the tube (transistor).

reactivation — Application of an above-normal voltage to a thoriated filament for a few seconds to bring a fresh layer of thorium atoms to the filament surface and thereby improve electron emission.

reactive — Pertaining to either inductive or capacitive reactance. A reactive circuit has a higher reactance than resistance.

reactive attenuator — An attenuator that dissipates almost no energy.

reactive balance — 1. The capacitive or inductive balance that is often required to null the output of certain transducers or systems when the excitation and/or the output is given in terms of alternating current. 2. The condition of an ac circuit when the phase angle between the voltage and current is zero. 3. The amount of corrective capacitance or inductance required to null the output of certain transducers or systems having ac excitation.

reactive factor — The ratio of reactive power to total power in a circuit.

reactive-factor meter — An instrument for measuring reactive factor.

reactive load — A load having reactance (i.e., a capacitive or inductive load), as opposed to a resistive load.

reactive near-field region — That region of the field of an antenna immediately surrounding the antenna wherein the reactive field predominates.

reactive power — Also called wattless power. The reactive voltage times the current, or the voltage times the reactive current, in an ac circuit. Unit of measurement is the var.

reactive power flow — The component of energy lost over transmission lines due to reactive power.

reactive sputtering — A sputtering technique that involves the introduction of reactive gases such as oxygen, nitrogen, and hydrogen into the glow discharge so that oxides, nitrides, and hydrides of the evaporant are deposited on the substrate.

reactive voltampere — See voltampere reactive.

reactive voltamperehour meter — See varhour meter.

reactive voltampere meter — See varmeter.

reactor — A physical device used primarily to introduce reactance or susceptance into a branch.

reactor-start motor — A form of split-phase motor designed for starting with a reactor in series with the main winding. The reactor is short-circuited (or otherwise made ineffective) and the auxiliary circuit is opened as soon as the motor attains a predetermined speed.

read — In a computer: 1. To copy, usually from one form of storage to another. 2. To sense the meaning of an arrangement of hardware representing information. 3. To exact information. 4. To transmit data from an input device to a computer. 5. To sense the presence of information in some type of storage, which includes RAM, magnetic tape, punched tape, etc.

readability — The ability to be understood — specifically, the understandability of signals sent by any means of telecommunications.

read-around — See read-around ratio.

read-around ratio — Also called read-around. In electrostatic storage tubes, the number of successive times information can be recorded as an electrostatic charge on a single spot in the array without producing the necessity for restoring the charge on surrounding spots.

reader — 1. In a computer, a device that converts information in one form of storage to information in another form of storage. 2. Any device capable of transcribing data from an input medium.

read head — In a computer, a device that converts digital information stored on a magnetic tape or on a drum into electrical signals usable by computer arithmetic.

read in — To sense information in a source and transmit it to an internal storage.

reading access time — In a computer, the time before a word may be used during the reading cycle.

reading rate — 1. In a computer, the number of characters, cards, etc., that can be sensed by an input unit in a given time. 2. In a digital meter, the number of successive readings (including measurement and display) performed per second, on a continuous basis.

read-in program — A program that itself can be put into a computer in simple binary form but that makes it possible for other programs to be read into the computer in more complex forms.

read-mainly memory — See RMM.

read-mostly memory — An integrated array of amorphous and crystalline semiconductor devices that is capable of being programmed, read, and reprogrammed repeatedly. Once reprogrammed, this type of memory retains data unless it is altered intentionally.

read-only memory — Acronym: ROM. 1. A digital memory that, after its initial programming by the manufacturer, can only be read and no longer written on; a permanent memory. 2. A memory that cannot be altered in normal use of a computer. Usually a small memory that contains often-used instructions such as microprograms or system software as firmware. Peripheral equipment uses ROM for character generation, code translation, and for designing peripheral processors. 3. A memory in which information is stored permanently, e.g., a math function or a microprogram. A ROM is programmed according to the user's requirements during memory fabrication and cannot be reprogrammed. A ROM is analogous to the dictionary, in which a certain address results in predetermined information output. 4. A digital building block usually classified as a memory type that contains information permanently written into it during manufacture that can be read at the outputs but not changed. 5. A random-access storage in which the data pattern is unchangeable after manufacture. 6. A computer memory that can be written into during manufacture or installation, but is only read out after that. The read-only memory is usually a nonvolatile form of memory. 7. A storage arrangement primarily for information-retrieval applications. The information may be wired in when the storage device is made, or it may be written in at a speed much less than the retrieval speed. 8. A digital storage device specified for a single function. Data is loaded permanently into the ROM when it is manufactured. This data is available whenever the ROM address lines are scanned. 9. A memory in which the contents are not intended to be altered during normal operation. The term read-only memory implies that the content is determined by its structure and is unalterable (e.g., mask programmable ROM). Most ROMs are $n \times 8$ (words $\times$ bits/word) to work with popular microprocessors. There are also special-purpose ROMs, such as character generators, with a 7-bit-wide output word and addressing structure to output one-hundred twenty-eight 9×7 characters.

readout — 1. The manner in which a computer displays processed information — e.g., digital visual display, punched tape, automatic typewriter, etc. *See also* display. 2. The information extracted from a memory device, such as a program store or call store. 3. The visual display of the output of a measuring instrument, or of a memory, or of a computer.

readout device — In a computer, a device, consisting usually of physical equipment, that records the computer output either as a curve or as a set of printed numbers or letters.

readout equipment — The electronic apparatus that provides indications and/or recordings of transducer output.

readout station — A recording or receiving radio station at which information is received as it is read out by the transmitter in a missile, probe, satellite, or other spacecraft. (The same station may serve also as a tracking station.)

read pulse — 1. A pulse applied to one or more binary cells to determine whether a bit of information is stored there. 2. A pulse that causes information to be read out of a memory cell.

readthrough — The continuous recovery in an audio channel of the target modulation, making possible rapid evaluation of the effectiveness of a jamming effort.

read time — More commonly called access time. With respect to a memory, the interval between the time the read control and the address or location are present and the time the data output changes state.

read/write check indicator — A device incorporated in some computers to indicate, on interrogation, whether there was an error in reading or writing. The machine can be made to stop, attempt the operation again, or follow a special subroutine, depending on the result of the interrogation.

read/write cycle — The sequence of operations required to read and write (restore) memory data.

read/write cycle time — *See* cycle time.

read/write head — 1. The device that reads and writes information on tape, drum, or disk storage devices. 2. The mechanism that writes data to or reads data from a magnetic recording medium. 3. An electromagnet capable of producing a switchable magnetic field to read and record bit streams.

read/write memory — 1. A memory whose contents can be continuously changed quickly and easily during system operation. It differs from a read-only memory (ROM), whose contents are fixed and not subject to change, and a reprogrammable ROM, whose contents can be changed but only periodically. 2. A memory in which each cell may be selected by applying appropriate electrical input signals, and the stored data may be either (*a*) sensed at appropriate output terminals, or (*b*) changed in response to other similar electrical input signals.

ready-to-receive signal — In a facsimile system, a signal returned to the transmitter to indicate that the receiver is ready to accept a transmission.

real estate — Slang for the area on a printed circuit board or the surface of a wafer on which circuits can be built.

real power — The component of apparent power that represents true work in an ac circuit. It is expressed in watts and is equal to the apparent power times the power factor.

real time — 1. Having to do with the actual time during which physical events take place. 2. The performance of a computation during the actual time that the related physical process transpires in order that results of the computations are useful in guiding the physical process. 3. Refers to a type of operating system that supports online equipment having critical time constraints. Events must be handled promptly (within set timing limits). Most process control and military command/control systems are real-time systems. 4. A computer process executed with sufficient speed so that the results of a process being monitored appear to be presented instantaneously. The computer generally is able to present the results with sufficient speed to permit control changes to be made. *See* closed loop; open loop. 5. A computation or process by a computer using inputs derived from time-initiated events; the output resulting from the computation or processing can have an effect on and/or predict trends concerning those events. 6. Responding instantaneously, rather than delayed by transmission format. 7. The operation of a program in the same time frame as a human being would.

real-time clock — 1. A clock that indicates the passage of actual time, such as elapsed time in the flight of a missile, as opposed to some fictitious time established by a computer program. 2. A system clock that indicates actual elapsed time from some reference time (e.g., noon). 3. A timing device used by a computer to derive elapsed time between events and to control processing of time-initiated event data. 4. A device that measures time at a rate consistent to the tasks being performed. Sometimes used for pacing the occurrence of events within a system.

real-time data processing — The processing of transactions as they occur, rather than batching them.

real-time executive — 1. Supervisory software that allocates system resources among several tasks to allow them to perform necessary calculations in real time. The executive has responsibility for all priority scheduling, interrupt handling, timer service, physical control, interprogram communication, and queue maintenance required by real-time applications. 2. An operating system that runs the system in a real-time mode, typically required by online data communications or process-control systems.

real-time input — Input information inserted into a system at the time it is generated by another system.

real-time operating system — Operating system capable of real-time task management. Includes event scheduling, interrupt management, and real-time event counters.

real-time operation — 1. Operations performed on a computer in time with a physical process so that the answers obtained are useful in controlling that process. 2. The use of a computer to control a process as it is actually occurring, necessitating, in general, relatively rapid operation on the part of the computer. 3. Data-processing technique in which information is utilized as events occur and the information is generated, as opposed to batch processing at a time unrelated to the time the information was generated.

real-time output — Output information removed from a system at the time it is needed by another system.

real-time reaction — A computer function that immediately responds to, or causes, a physical application action.

real-time spectrum analyzer — A device in which analysis of the spectrum of the incoming signal is performed continuously, with the time sequence of events preserved between input and output.

real-time system — 1. A computer system in which data processing is performed so that the results are available in time to influence the controlled or monitored process. 2. An information system whose input or output rate is not controlled by the system, but depends on external factors. 3. A system in which transactions are processed as they occur.

rear projection — A projection television system in which the picture is projected on a ground-glass screen to be viewed from the opposite side of the screen.

rear suspension — In moving-coil speakers, a pliable support situated near the apex of the cone. Assists in keeping the coil in a concentric position in the air gap between the magnet poles.

rebatron — A relativistic electron-bunching accelerator that produces a very tightly bunched beam with little velocity modulation, but high harmonic content. The beam can be used to excite structures that are large compared to a wavelength.

Rebecca — An airborne interrogator-responsor of the British Rebecca-Eureka navigation system. It can also be used with a special ground beacon known as Babs to provide low-approach facilities.

Rebecca-Eureka system — A British radar navigational system employing an airborne interrogator (Rebecca) and a ground transponder beacon (Eureka). It provides homing to an airfield from distances of up to 90 miles (145 km).

rebond — A second bonding attempt after a bond has been removed or failed to bond on the first attempt.

rebonding-over bond — A second bond made on top of a removed or damaged bond or a second bond immediately adjacent to the first bond.

reboot — 1. To restart a computer by reloading the operating system. 2. To restart a computer after it has been operating for some time, usually in an attempt to clear an error condition.

rebroadcast — The reception and the simultaneous or subsequent retransmission of a radio or television program by a broadcast station.

recalescent point — The temperature at which heat is suddenly liberated as the temperature of a heated metal drops.

recall — In a calculator, to retrieve from a register a previously entered number, for checking or use in further calculations.

receive current — The amount of current drawn by a transceiver when receiving radio signals.

received power — The power of a returned target signal received at the radar antenna.

received signal level — The strength of an intercepted radio signal at the antenna terminals of the receiver, expressed in microvolts or dBm.

receive only — Abbreviated RO. A teletypewriter-type terminal having no keyboard or tape reader.

receive-only typing reperforator — Also called rotor. A teletypewriter receiver whose output is a perforated tape that has characters along the edge of the tape.

receiver — 1. The portion of a communications system that converts electric waves into a visible or audible form. 2. An electromechanical device for converting electrical energy into sound waves. *See also* earphone. 3. A device for the reception and, if necessary, demodulation of electronic signals. 4. The electromagnetic unit in a telephone handset used to convert electrical energy to sound energy.

receiver bandwidth — The spread in frequency between the half-power points on the response curve of a receiver.

receiver gating — Application of operating voltages to one or more stages of a receiver only during the part of a cycle when reception is desired.

receiver images — Undesired signals in a receiver caused by the heterodyning process.

receiver incremental tuning — A control feature to permit receiver tuning (of a transceiver) up to 3 kHz to either side of the transmitter frequency.

receiver lockout system — In mobile communications, an arrangement of control circuits whereby only one receiver can feed the system at one time, to avoid distortion.

receiver noise figure — The ratio of noise voltage in a given receiver to that of a theoretically perfect receiver.

receiver noise threshold — The level that must be exceeded by the minimum discernible signal. External noise reaching the front end of a receiver and the noise added by the receiver itself determine the noise threshold.

receiver primaries — Constant-chromaticity, variable-luminance colors that are produced by a television receiver and that when mixed in proper proportions produce other colors. Usually three primaries — red, green, and blue — are used.

receiver radiation — Radiation of interfering electromagnetic signals by any oscillator of a receiver.

receiver sensitivity — The lower limit of useful signal input to the receiver. It is set by the signal-to-noise ratio at the output.

receiving amplifier — The amplifier used at the receiving end of a system to raise the level of the signal.

receiving antenna — A device for converting received space-propagated electromagnetic energy into electrical energy.

receiving circuit — An apparatus and connections used exclusively for the reception of messages at a radiotelephone or radiotelegraph station.

receiving equipment — The equipment (amplifiers, filters, oscillator, demodulator, etc.) associated with incoming signals.

receiving-loop loss — That part of the repetition equivalent assignable to the station set, subscriber line, and battery-supply circuit on the receiving end of a telephone line.

receiving margin — Also called range or operating range. In telegraphy, the usable range of adjustment of the range finder; for a machine that is adjusted properly, approximately 75 points on a 120-point scale.

receiving perforator — In printing telegraph systems, an apparatus that punches a paper strip automatically in accordance with the arriving signals. When the paper strip is later passed through a printing telegraph machine, the signals will be reproduced as printed messages, ready for delivery to the customer.

receptacle — 1. Usually the fixed or stationary half of a two-piece multiple-contact connector. Also, the connector half usually mounted on a panel and containing socket contacts. 2. A contact device installed at the outlet. Allows the connection of external electric cords from lamps or appliances.

receptacle connector — An electrical connector intended to be mounted or installed onto a fixed structure, such as a panel, electrical case, or chassis, and which couples or mates to a plug connector.

reception — Listening to, copying, recording, or viewing any form of emission.

rechargeable — Capable of being recharged. Usually used in reference to secondary cells or batteries.

rechargeable primary cell — A cell that is ordinarily used in one-shot service, but which is capable of a limited number of charge/discharge cycles.

reciprocal — The number 1 (unity) divided by a quantity; e.g., the reciprocal of 2 is $1/2$; of 4, $1/4$, etc.

reciprocal-energy theorem — If an electromotive force E_1 in one branch of a circuit produces a current I_2 in any other branch, and if an electromotive force E_2 inserted into this other branch produces a current I_1 in the first branch, then $I_1 E_1 = I_2 E_2$.

reciprocal ferrite switch — A ferrite switch that can be placed in a waveguide to route an input signal to either of two output waveguides. Switching is accomplished by a Faraday rotator under the influence of an external magnetic field.

reciprocal impedance — Two impedances Z_1 and Z_2 are said to be reciprocal impedances with respect to an impedance Z (invariably a resistance) if they are so related as to satisfy the equation $Z_1 Z_2 = Z^2$.

reciprocal transducer — A transducer that satisfies the principles of reciprocity — i.e., if the roles of excitation and response are interchanged, the ratio of excitation to response will remain the same.

reciprocation — The process of deriving a reciprocal impedance from a given impedance or finding a reciprocal network for a given network.

reciprocity theorem — In any system composed of linear bilateral impedances, if an electromotive force E is applied between any two terminals and the current I is measured in any branch, their ratio (called the transfer impedance) will be equal to the ratio obtained if the positions E and I are interchanged.

reclosing relay — Any voltage, current, power, etc., relay that recloses a circuit automatically.

recognition device — A device that can identify any number of a set of distinguishable entities.

recognition differential — For a specified listening system, the amount by which the signal level exceeds the noise level that is presented to the ear when a 50 percent probability of detection of the signal exists.

recombination — 1. A collision within a semiconductor crystal lattice of a conduction-band electron and a valence-band hole. The ability of each to function as a current carrier is lost, due to mutual cancellation, so that a recombination effectively destroys the hole-electron carrier pair. 2. A process in which current carriers of opposite signs combine and form stable, neutral entities, resulting in zero net charge.

recombination coefficient — The quantity that results from dividing the time rate of recombination of ions in an ionized gas by the product of the positive-ion density and the negative-ion density.

recombination radiation — The radiation produced in a semiconductor when electrons in the conduction band recombine with holes in the valence band. If an actual population inversion between portions of the valence and conduction bands (or between adjacent localized states of acceptors or donors near these bands) is achieved, stimulated emission and laser amplification or oscillation can take place. This is the radiation process of importance in injection lasers.

recombination velocity — On a semiconductor surface, the normal component of the electron (or hole) current density at the surface divided by the excess electron (or hole) charge density at the surface.

reconditioned-carrier reception — Also called exalted-carrier reception. Reception in which the carrier is separated from the sidebands in order to eliminate amplitude variations and noise, and then is increased and added to the sidebands in order to provide a relatively undistorted output. This method is frequently employed with a reduced-carrier signal-sideband transmitter.

record — 1. A character or characters that are grouped together in the flow of data in a system; for example, one line of type of the contents of a punched card. A record may be of fixed length, as with punched cards, or of variable length, as with a line of type. 2. A group of related facts or fields of information handled as a unit; thus a listing of information, usually printed or in printable form. 3. The process of putting data into a computer storage device. 4. To preserve for later reproduction. 5. Relating to data that is treated as a unit of logical information. The delineation of a record may be arbitrary and determined by the designer of the information format. (A record may be recorded on all or part of a block or more than one block.) 6. A collection of related items of data (fields) treated as a unit.

record changer — 1. A device that will automatically play a number of phonograph records in succession. 2. A type of automatic turntable capable of playing a number of records (usually 6 to 10) in sequence.

record code — A special control code used to mark the separation between adjacent records.

record compensator — Also called a record equalizer. An electrical network that compensates for different frequency-response curves in various recording techniques.

recorded tape — Also called a prerecorded tape. 1. A tape that contains music, dialogue, etc., and is sold to audiophiles and others for their listening pleasure. 2. A commercially available recorded tape.

recorded value — The value recorded by the marking device on a chart with reference to the division lines marked on the chart.

recorded wavelength — In a phonograph record, the length of groove required for a signal of given frequency to complete one cycle. At any particular distance from the record center, i.e., at a particular groove velocity, the recorded wavelength decreases with increasing frequency. Similarly for a given frequency, the recorded wavelength decreases with progress toward the record center (i.e., as groove velocity decreases).

record equalizer — See record compensator.

recorder — Also called recording instrument. 1. An instrument that makes a permanent record of varying electrical impulses — e.g., a code recorder, which punches code messages into a paper tape; a sound recorder, which preserves music and voices on disc, film, tape, or wire; a facsimile recorder, which reproduces pictures and text on paper; and a video recorder, which records television pictures on film or tape. 2. A device that makes a record of changes in varying electrical quantities or signals. 3. An instrument that makes a graphic record of the value of one or more quantities as a function of another variable (usually time).

record gap — In a computer, a space between records on a tape. It is usually produced by acceleration or deceleration of the tape during the write operation.

recording ammeter — An ammeter that provides a permanent recording of the value of either an alternating or a direct current.

recording blank — See recording disc.

recording channel — One of several independent recorders in a recording system, or independent recording tracks on a recording medium.

recording curve — See equalization.

recording demand meter — Also called demand recorder. An instrument that records the average value of the load in a circuit during successive short periods.

recording density — The number of bits recorded per unit of length in a single linear track in a recording medium.

recording disc — Also called a recording blank. A (unrecorded) disc made for recording purposes.

recording head — A magnetic head that transforms electrical variations into magnetic variations for storage on magnetic media. *See also* cutter.

recording instrument — Also called a recorder or graphic instrument.

recording lamp — A light source used in the variable-density system of sound recording on movie film. Its intensity varies in step with the variations of the audio-frequency signal sent through it.

recording level — The amplifier output required to provide a satisfactory recording.

recording-level meter — An indicator on a tape or disc recorder that provides some idea of the signal levels being applied to the recording medium from moment to moment. It is intended as an aid in setting the recording levels to ensure that the tape or disc is neither overloaded with excessive levels or under-recorded with too little signal, allowing hiss and other noise to intrude. Recording-level meters come in a variety of types, including meters that register the approximate average value of the signal (of which the professional vu meter is an example), those designed to show the instantaneous peak levels of the signal, and some not readily classifiable into any specific group.

recording loss — In mechanical recording, the loss that occurs in the recorded level because the amplitude executed by the recording stylus differs from the amplitude of the wave in the recording medium.

recording noise — Noise induced by the amplifier and other components of a recorder.

recording preamplifier — *See* preamplifier.

recording-reproducing head — A dual-purpose head used in magnetic recording.

recording spot — An instantaneous area acted on by the registering system of a facsimile recorder.

recording storage tube — A type of cathode-ray tube in which the equivalent of an image can be stored in the form of a pattern of electrostatic charge on a storage surface. There is no visual display, but the stored information can be read out at any later time in the form of an electric output signal.

recording stylus — The tool that inscribes a groove into the recording medium.

recording trunk — A trunk that extends between a local central office or private branch exchange and a toll office and that is used only for communication with toll operators and not for the completion of toll connections.

recording voltmeter — A voltmeter that provides a permanent record of the value of either alternating or direct voltage.

record layout — The arrangement and structure of information in a record, including the sequence and size of the components. By extension, the description of such an arrangement.

record length — A measure of the size of a record, usually expressed in terms of such units as words or characters.

record mark — A means of marking the separation between adjacent records: on magnetic tape, a record gap; on paper tape, a record code; in data transmission, a record pause. Often, the record code is used along with a gap or a pause to allow for different devices throughout a data-transmission system, each of which can recognize only one of the three types of record marks.

record medium — In a facsimile recorder, the physical medium onto which the image of the subject copy is formed.

record player — A motor-driven turntable, pickup arm, and stylus for converting the signals impressed onto a phonograph record into a corresponding audio-frequency voltage. This voltage is then applied to an amplifier (usually contained within the record player cabinet) for amplification and conversion to sound waves.

record/play head — A single magnetic-tape head that is used for recording and playback of a tape.

record separator — A character intended as an identifier of a logical boundary between records.

record sheet — In a facsimile recorder, a sheet or medium upon which the image of the subject copy is recorded.

recover — To restore the computer to a previous state.

recovered audio — The value of the audio voltage measured at the detector output under the specified circuit conditions.

recovered charge — The total charge recovered from the diode after switching from a specified forward current condition to a specified reverse condition. This charge includes components due to both carrier storage and depletion-layer capacitance.

recovered voice quality — Quality of the voice signal after the process of decoding or scrambling.

recovery — 1. In an electronic device, the time required to enable the device to react to new signals. 2. In fluidic devices, a generally percentile representation of output capture as related to supply, such as output pressure versus input pressure. 3. The actions required to bring a system to a predefined level of operation after a degradation or failure.

recovery time — 1. The time required for a fired atr tube in its mount to deionize to the level at which its normalized conductance and susceptance are within the specified ranges. 2. The time required for the control electrode in a gas tube to regain control after the anode current has been interrupted. 3. In Geiger-Mueller counters, the minimum time from the start of a counted pulse to the instant a succeeding pulse can attain a specified percentage of the maximum amplitude of the counted pulse. 4. The time required for a fired tr or pre-tr tube to deionize to the level at which the attenuation of a low-level radio-frequency signal transmitted through the tube drops to the specified value. 5. In a radar or its component, the time required — after the end of the transmitted pulse — for recovery to a specified relation between receiving sensitivity or received signal and the normal value. 6. Also called response time. The interval required, after a sudden decrease in input-signal amplitude to a system or component, for a specified percentage (usually 63 percent) of the ultimate change in amplification or attenuation to be attained. 7. In thermal time-delay relay, the cooling time required from heater deenergization to reenergization such that the new time delay is 85 percent of that exhibited from a cold start. 8. Also called response time. In a power supply, the time required for recovery of the load voltage from a step change in load current or line voltage. 9. The length of time that a condition remains in effect after the control signal that initiated the effect has disappeared.

rectangular array — 1. Multiple copies of an entity at intervals of vertical and horizontal spacing, e.g., a matrix. 2. *See* matrix.

rectangular-loop material — A material that has a rectangular hysteresis loop, i.e., the difference between the coercive force and the magnetizing force at saturation induction is small.

rectangular scanning — A two-dimensional sector scan in which a slow sector scan in one direction is superimposed perpendicularly onto a rapid sector scan.

rectangular wave — A periodic wave that alternately assumes one of two fixed values, the time of transition being negligible in comparison with the duration of each fixed value.

rectangular waveguide — A hollow enclosure of rectangular cross section, normally with the dimensions of the sides in a ratio of $2:1$. With dimensions so proportioned, the dominant mode has a free-space wavelength range between one and two times the longer side dimension. Rectangular waveguides normally can be used only over less than octave ranges.

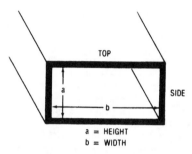

Rectangular waveguide.

Rectenna — A device that converts microwave power into dc power. It consists of a number of small dipoles, each of which has an associated diode rectifier network connected to a dc bus.

rectification — Conversion of alternating current into unidirectional or direct current by means of a rectifier.

rectification efficiency — Ratio of the direct-current power output to the alternating-current power input of a rectifier.

rectification factor — The change in average current of an electrode divided by the change in amplitude of the alternating sinusoidal voltage applied to the same electrode, the direct voltages of all electrodes being maintained constant.

rectified value — The average of all the positive or negative values of an alternating quantity over a whole number of periods.

rectifier — 1. Device having an asymmetrical conduction characteristic employed in a way to convert alternating current into unidirectional current. In amplitude-modulation detection, recovery of original signals is frequently accomplished by a rectifier. 2. Device that converts alternating current into unidirectional current by permitting appreciable current in one direction only. 3. A two-element tube or a solid-state device that is used to convert alternating current to direct current. Usually rated above one-half ampere.

rectifier diode — A diode that exhibits an asymmetrical voltage-current characteristic and is used for current and voltage rectification.

rectifier instrument — The combination of an instrument sensitive to direct current and a rectifying device whereby alternating currents or voltages can be measured.

rectifier meter — *See* rectifier instrument.

rectifier stack — A dry disc rectifier made up of layers of individual rectifier discs (e.g., a selenium or copper-oxide rectifier).

rectifier transformer — A transformer whose primary operates at the fundamental frequency of the ac system and whose secondary has one or more windings conductively connected to the main electrodes of the rectifier.

rectifying element — A circuit element that conducts current in one direction only.

rectigon — A hot-cathode gas-filled diode that operates at a high pressure. Used most frequently in battery-charging circuits.

rectilineal compliance — A mechanical element that opposes a change in the applied force (e.g., the springiness that opposes a force on the diaphragm of a speaker or microphone).

rectilinear — In a straight line; specifically, moving, forming, or bounded by a straight line.

rectilinear scanning — The scanning of an area in a predetermined sequence of narrow, straight, parallel strips.

rectilinear writing recorder — An oscillograph that records in rectilinear coordinates.

recuperability — The ability to continue operating after partial or complete loss of the primary communications facility.

recurrence rate — *See* repetition rate, 1.

recursion — The continuous repeating of the same operation or group of operations.

recursive — Capable of being repeated.

recursive filter — *See* digital filter.

recyclability — The capability of a battery system to be recharged after it has been discharged.

recycling modulo-*n* counter — A counter that has *n* distinct states and that counts to a maximum number and returns, when the next input pulse is applied, to its minimum number.

red gun — In a three-gun color television picture tube, the electron gun whose beam strikes only the phosphor dots that emit the red primary.

redistribution — In a charge-storage tube or television camera tube, the alteration of charges on one area of a storage surface by secondary electrons from any other area of the same surface.

redox cell — A cell designed to convert the energy of the reactants to electrical energy. An intermediate reductant, in the form of a liquid electrolyte, reacts at the anode in a conventional manner; it is then regenerated by reaction with a primary fuel.

red-tape operation — In a computer, operations that do not directly contribute to the results, i.e., those internal operations that are necessary to process data, but do not in themselves contribute to any final answer.

reduced generator efficiency — The ratio of a given thermoelectric-generator efficiency to the corresponding Carnot efficiency.

reduced instruction set computer — Abbreviated RISC. A CPU in which processing capabilities have been reduced to increase speed.

reduced telemetry — Telemetry data transformed from raw form into a usable form.

reduction technique — A technique for simplifying or restructuring a Boolean expression for easier, lower-cost implementation in circuitry.

redundancy — 1. The employment of several devices, each performing the same function, in order to improve the reliability of a particular function. 2. Added or repeated information employed to reduce ambiguity or error in a transmission of information. As the signal-to-noise ratio decreases, redundancy may be employed to prevent an increase in transmission error. 3. In information transmission, the fraction of the total information content of a message that can be eliminated without losing essential information. 4. The property of having functionally equivalent systems, subsystems, or components in place to back up one another. 5. Also called backup control. A technique that provides a secondary method of control for certain critical functions. 6. The use of additional equipment and facilities to provide for continuity of service during trouble situations. 7. A method of ensuring higher reliability or performance in which more than one of the same item is used. 8. A configuration of multiple elements that can still produce a correct output when one or more elements are not functioning correctly.

redundancy check — In a computer, an automatic or programmed check that makes use of components or characters inserted especially for checking purposes.

redundant code — A code that uses more signal elements than are necessary to represent the intrinsic information.

redundant data — 1. Data that is not necessary for the information content of a transmission but is usually added to transmitted information to aid in the detection of communications errors. 2. A data sample so similar to the preceding sample from the same source that it is of no interest in connection with subsequent analysis of the experiment or test, except for the fact of the similarity.

red video voltage — The signal voltage that controls the grid of the red gun in a three-gun picture tube. This signal is a reproduction of the output from the red camera at the transmitter.

reed — A thin bar located in a narrow gap and made to vibrate electrically, magnetically, or mechanically by forcing air through the gap.

reed frequency meter — See vibrating-reed meter.

reed relay — 1. A relay in which two flat magnetic strips mounted inside a coil are attracted to each other when the coil is energized. The relay contacts are mounted on the strips. 2. A device that uses two (sometimes three) strips of magnetizable metal, enclosed in glass, as the contacts. The control member is a coil surrounding the glass capsule. 3. One or more reed switches operated by a single coil. 4. An assembly that combines a reed switch with an electromagnetic operating coil. 5. An electromechanical relay based on the use of a reed switch, an assembly of partially overlapping magnetic metal strips or reeds hermetically sealed in a glass tube. In the influence of a magnetic field, the contact ends of the reeds may be caused to open or close. This permits extremely fast making or breaking of an external circuit. It is important to distinguish between a reed switch and a reed relay. Reed switches, as in keyboards, normally depend on the motion of an external permanent magnet moved with respect to the glassed reed capsule. By contrast, reed relays are assemblies of the glassed reed capsule within a coil or solenoid.

reed switch — A type of magnetic switch consisting of contacts formed by two thin, movable, magnetically actuated metal vanes or reeds, held in a normally open position within a sealed glass envelope.

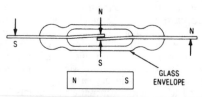

Reed switch.

reel-to-reel — See open-reel.

reentrancy — 1. A type of feedback employed in microwave oscillators. In most magnetrons, a circuit is used that can be described as a slow-wave structure that feeds back into itself. In beam reentrancy, the beam may be circulated repeatedly through the interaction space. 2. That characteristic of a computer subroutine that permits a second task to enter the subroutine before completion of its execution by the first task.

reentrant cavity — A resonant cavity in which one or more sections are directed inward with the result that the electric field is confined to a small area or volume.

reentrant code — 1. The instructions forming a single copy of a program or subroutine that is shared by two or more programs, as opposed to the conventional method of embedding a copy of a subroutine within each program. Characteristically, reentrant routines are composed completely of instructions and constants that are not subject to modification during execution. 2. Code that may be executed simultaneously by more than one task. Thus, the code cannot be self-modifying, and each task must maintain its own data area. This does not imply that tasks will actually execute the code simultaneously, although this could occur in a multiprocessor system. 3. A program task or routine that can be executed simultaneously by more than one process.

reentrant program — A subroutine written in such a way that it can be used by many users concurrently. (The alternative is to let each user have his or her own copy of the program, which uses up a lot of memory space.) A reentrant program cannot be allowed to modify its own instructions.

reentrant subroutines — Computer subroutines that can be executed from any of several application programs operating on different levels of priority.

reentrant winding — An armature winding that returns to its starting point, thus forming a closed circuit.

reference acoustic pressure — Also called reference sound level. That magnitude of a complex sound that produces a sound-level meter reading equal to the reading that results from a sound pressure of 0.0002 dyne per square centimeter at 1000 hertz.

reference address — An address used in digital-computer programming as a reference for a group of relative addresses.

reference angle — The angle formed between the center line of a radar beam as it strikes a reflecting surface and the perpendicular drawn to that reflecting surface.

reference black level — 1. The picture-signal level corresponding to a specified maximum limit for black peaks. 2. The maximum negative-polarity amplitude of the video signal.

reference boresight — A direction defined by an optical, mechanical, or electrical axis of an antenna, established as a reference for the purpose of beam direction or tracking axis alignment.

reference burst — See color burst.

reference dipole — A half-wave straight dipole tuned and matched for a given frequency and used as a unit of comparison in antenna measurement work.

reference electrode — In pH measurements, an electrode, usually hydrogen-filled, used to provide a reference potential. See also glass electrode.

reference frequency — A frequency coinciding with, or having a fixed and specified relation to, the assigned frequency. It does not necessarily correspond to any frequency in an emission.

reference language — The definitive description of a programming language, completely specifying all its concepts and the relations between them. It may be in

English, or the language itself may be the reference language.

reference level — The starting point for designating the value of an alternating quantity or a change in it by means of decibel units. For sound loudness, the reference level is usually the threshold of hearing. For communications receivers, 60 microwatts is normally used. A common reference in electronics is 1 milliwatt, and power is stated as so many decibels above or below this figure.

reference line — A line from which angular measurements are made.

reference monitor — A receiver (or other similar device of known performance capabilities) used for judging the transmission quality.

reference noise — The magnitude of circuit noise that will produce a noise-meter reading equal to that produced by 10^{-12} watt of electric power at 1000 hertz.

reference oscillator — The high stability, usually crystal and temperature controlled, rf signal source used as a phase reference in phase-locked oscillators.

reference phase — The phase of the color burst transmitted with color-television carriers. It is used in synchronizing the receiver reference oscillator with the transmitted color signals.

reference point — 1. A terminal that is common to both the input and the output circuits. 2. The point of a chromaticity diagram that represents the chromaticity of a reference stimulus.

reference record — In digital-computer programming, a compiler output that lists the operations and their positions in the final specific routine, plus information describing the segmentation and storage allocation of the routine.

reference recording — A recording of a radio program for future reference or checking.

reference signal — A stable signal used as a standard against which other variable signals may be compared and adjusted.

reference sound level — *See* reference acoustic pressure.

reference time — In a computer, an instant chosen near the beginning of switching as an origin for time measurements. It is taken as the first instant at which either the instantaneous value of the drive pulse, the voltage response of the magnetic cell, or the integrated voltage response reaches a specified fraction of its peak pulse amplitude.

reference tone — A stable tone of known frequency continuously recorded on one track of multitrack signal recordings and intermittently recorded on signal-track recordings by the collection-equipment operators for subsequent use by the data analysts as a frequency reference.

reference voltage — Alternating-current voltage in a synchro servosystem used to determine the in-phase or 180° out-of-phase condition to provide directional sense.

reference volume — The magnitude of a complex electric wave, such as that corresponding to speech or music, that gives a reading of 0 vu on a standard volume indicator. The sensitivity of the volume indicator is adjusted so that the reference volume of 0 vu is read when the instrument is connected across a 600-ohm resistance to which there is delivered a power of 1 milliwatt at 1000 hertz.

reference white — The light from a nonselective diffuse reflector as a result of the normal illumination of the scene to be televised.

reference white level — 1. In television, the picture-signal level corresponding to a specified maximum limit for white peaks. 2. The maximum positive-polarity amplitude of the video signal.

refiring — Recycling a thick-film resistor through the firing cycle to change the resistor value.

reflectance — 1. Ratio of reflected flux to incident flux. Unless otherwise specified, the total reflectance is meant; it is sometimes convenient to divide this into the sum of the specular and the diffuse reflectance. 2. *See* reflection factor.

reflected impedance — 1. The apparent impedance across the primary of a transformer when current flows in the secondary. 2. The impedance at the input terminals of a transducer as a result of the impedance characteristics at the output terminals. 3. The impedance seen at the input of a network when its output is terminated in an impedance of a specified value. (Most often, the network referred to in this connection is a transformer.)

reflected power or signal — The power flowing back to the generator from the load.

reflected resistance — The apparent resistance across the primary of a transformer when a resistive load is across the secondary.

reflected wave — The wave that has been reflected from a surface, a junction of two different media, or a discontinuity in the medium in which it is traveling (e.g., the echo from a target in radar, the sky wave in radio, the wave traveling toward the source from the termination of a transmission line).

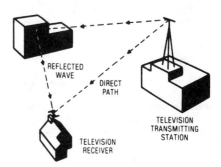

Reflected and direct waves.

reflecting curtain — A vertical array of half-wave reflecting antennas placed one-quarter wavelength behind a radiating curtain of dipoles to form a pine-tree array.

reflecting electrode — A tubular outer electrode or the repeller plate in a microwave oscillator tube corresponding in construction but not function to the plate of an ordinary triode. It is capable of generating extremely high frequencies.

reflecting galvanometer — A galvanometer with a small mirror attached to the moving element. The mirror reflects a beam of light onto a scale.

reflecting grating — An arrangement of wires placed in a waveguide to reflect the desired wave while freely passing one or more other waves.

reflection — 1. The phenomenon in which a wave that strikes a medium of different characteristics is returned to the original medium with the angles of incidence and reflection equal and lying in the same plane. 2. The change in direction imparted to light rays that impinge on a surface but do not penetrate therein. (Those rays which penetrate are either absorbed as heat energy or refracted within the material.) Equally applicable to other radiation. 3. Return of radiation by a surface, without change in wavelength. The reflection may be specular, from a smooth surface; diffused, from a rough surface or from within the specimen; or mixed, a combination of the two.

reflection altimeter — An aircraft altimeter that determines altitude by the reflection of sound, supersonic, or radio waves from the earth.

reflection coefficient — 1. At the junction of a uniform transmission line and a mismatched terminating impedance, the vector ratios between the electric fields associated with the reflected and the incident waves. 2. At any specified plane in a uniform transmission medium, the vector ratios between the electric fields associated with the reflected and the incident waves. 3. At any specified plane in a uniform transmission line between a source and an absorber of power, the vector ratio between the electric fields associated with the reflected and the incident waves. It is given by the formula

$$\frac{Z_2 - Z_1}{Z_2 + Z_1}$$

where the impedances Z_1 and Z_2 are the impedances of the source and load, respectively. 4. A measure of the difference between the driving-source impedance and the input impedance of a filter. 5. Parametric measurement for elliptical fiber and cable expressed as a ratio of the two-directional flow of power through the cable at any chosen point.

reflection color tube — A color picture tube that produces an image electron reflection in the screen region.

reflection Doppler — A system utilizing the Doppler frequency shift to measure the position and/or velocity of an object not carrying a transponder.

reflection error — In navigation, the error due to the wave energy that reaches the receiver as a result of undesired reflections.

reflection factor — 1. Also called mismatch factor, reflectance reflectivity, or transition factor. The ratio of the current delivered to a load whose impedance is not matched to the source to the current that would be delivered to a load of matched impedance. Expressed as a formula:

$$\frac{\sqrt{4Z_1 Z_2}}{Z_1 + Z_2}$$

where Z_1 and Z_2 are the unmatched and the matched impedances, respectively. 2. Also called reflectance and reflectivity. A measure of the effectiveness of a surface in reflecting light; the ratio of reflected lumens to the incident lumens.

reflection grating — A wire grating that is placed inside a waveguide so as to reflect a desired wave while at the same time allowing one or more other waves to pass freely.

reflection hologram — A hologram that is illuminated by a source from the viewer's side.

reflection law — For any reflected object, the angle of incidence is equal to the angle of reflection.

reflection loss — 1. That part of transmission loss due to the reflection of power at the discontinuity. 2. The ratio (in decibels) of the power incident upon the discontinuity to the difference between the powers incident to and reflected from the discontinuity.

reflection sounding — Echo depth sounding in which the depth is measured by reflecting sound or supersonic waves off the bottom of the ocean.

reflective code — A code that appears to be the mirror image of a normal counting code. The most useful property of reflective codes is that only one digit changes at a time in increasing or decreasing by 1. The reflective binary code is called a Gray code.

reflective jamming — See radar-confusion reflectors.

reflective optics — A system of mirrors and lenses used in projection television.

reflectivity — See reflection factor, 2.

reflectometer — A microwave system arranged to measure the incidental and reflected voltages and indicate their ratio.

reflector — Also called a reflector element. 1. One or more conductors or conducting surfaces for reflecting radiant energy — specifically, a parasitic antenna element located in other than the general direction of the major lobe of radiation. 2. See repeller. 3. An element in a VHF antenna that is situated to the rear of the main dipole. Reflects radio waves onto the dipole. 4. The dish often employed to reflect quiet or distant sounds in an open air setting onto a microphone. 5. A metal plate, metal screen, or any of a group of spaced tuned rods (elements) placed back of the active element of an antenna to make it directive. 6. A type of conducting surface or material used to reflect radiant energy.

reflector element — See reflector.

reflector module — A device containing a light source and a photodetector that detects any object that reflects light produced by the source back to the detector.

reflector satellite — A satellite so designed that radio or other waves bounce off its surface.

reflector voltage — The voltage between the reflector electrode and the cathode in a reflex klystron.

reflex baffle — A speaker baffle in which a portion of the radiation from the rear of the diaphragm is propagated forward after a controlled phase shift or other modification. This is done to increase the overall radiation in some portion of the frequency spectrum.

reflex bunching — In a microwave tube, a type of bunching that is brought about when the velocity-modulated electron stream is made to reverse its direction by means of an opposing dc field.

reflex cabinet — A type of speaker enclosure fitted with a vent or port through which out-of-phase signals from the rear of the cone are "reflexed" by allowing the enclosed air in the cabinet to be tuned for a coupled resonance effect with the cone of the drive unit. The signals are then brought into phase with the front radiation from the cone of the speaker so as to reduce the "boomy" effect of resonance.

reflex circuit — 1. A circuit through which the signal passes for amplification both before and after detection. 2. A single stage of amplification that operates on two signals in widely separated frequency ranges.

reflex klystron — A klystron with a reflector (repeller) electrode in place of a second resonant cavity, to redirect the velocity-modulated electrons through the resonant cavity that produced the modulation. Such klystrons are well suited for use as oscillators, because the frequency is easily controlled by repositioning the reflector.

reflowing — The melting of an electrodeposit followed by solidification. The surface has the appearance and physical characteristics of being hot-dipped (especially tin or tin alloy plates).

reflow soldering — A method of soldering involving application of solder, in the form of preforms, paste, or as a solder plate, prior to the actual joining. To solder, the parts are joined and heated, causing the solder to remelt, or reflow. The joint can be made by resistance soldering, hot gas soldering, forced hot air oven, radiant heating, liquid immersion, or condensation soldering.

refracted wave — Also called the transmitted wave. In an incident wave, the portion that travels from one medium into a second medium.

refraction — 1. The change in direction of propagation of a wavefront due to its passing obliquely from one medium into another in which its speed is different. Refraction may also occur in a single medium of varying characteristics. 2. The deflection from a straight path undergone by a light ray passing from one medium into another having a different index of refraction. 3. Light rays that impinge on a surface and thereafter continue through the material or substance generally have their direction of travel altered. This bending constitutes refraction. Example: light entering water from the air. 4. The bending of sound, radio, or light waves as they pass obliquely from a medium of one density to a medium of another density in which their speed is different. 5. The bending of light at the boundary of two surfaces.

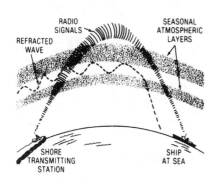

Refraction (of radio waves).

refraction error — In navigation, the error due to the bending of one or more wave paths by undesired refraction.

refraction loss — That part of the transmission loss due to refraction resulting from nonuniformity of the medium.

refractive index — Also called index of refraction. 1. Of a wave-transmission medium, the ratio between the phase velocity in free space and in the medium. 2. The ratio of the velocity of light in a vacuum to its velocity in a material, such as a fiber. Also, the ratio of the sine of the angle of incidence of light on the material to the angle of refraction of the light. The refractive index of any material varies with the wavelength of light. In a fiber, core refractive index must be greater than that of the cladding.

refractive modulus — Also called the excess modified index of refraction. The excess over unity of the modified index of refraction. It is expressed in millionths and is given by the equation

$$M = (n + h/a - 1)10^6$$

where

M = the refractive modulus,
n = the index of refraction at a height H above sea level, and
a = the radius of the earth.

refractivity — Ratio of phase velocity in free space to that in the medium, minus 1.

refractometer — An instrument for measuring the refractive index of a liquid or solid, usually from the critical angle at which total reflection occurs.

refractor metal — A metal that has an extremely high melting point; in the broad sense, a metal that has a melting point above those of iron, nickel, and cobalt.

refractory metal-oxide semiconductor — *See* RMOS.

refrangible — Capable of being refracted.

refresh — 1. The periodic renewing of data, or data-carrying electrical charges, in a semiconductor memory. The data in various semiconductor memories would quickly be lost if this action were not taken periodically. 2. CRT display technology requiring continuous restroking of the display image. 3. Repeated writing of display frames of a CRT display surface. Since the image is retained by nonstoring phosphor for only short intervals, it must be continually rewritten (refreshed) in order to remain visible.

refresh cycle — The periodic recharging of the capacitors in a dynamic RAM (DRAM) array.

refresh display — A CRT device that requires the refresh of its screen presentation at a high rate in order that the image will not fade or flicker. The refresh rate is proportional to the decay rate of the phosphor. Refresh displays require continuous interaction with the host computer and display controller.

refresh rate — The number of times per second that a complete image is formed on the screen by a monitor or TV. Refresh rate is proportional to the decay rate of a monitor's phosphor compound. A slow-decay phosphor produces bright images, but often a ghosting effect as well. A fast-decay phosphor's images leave no such trailing effect, but are not as bright and may flicker. Refresh rates less than 70 Hz produce an annoying flicker, which can contribute to eye strain.

refresh time interval (refresh period) — The time between the successive refresh operations that are required to restore the charge in a dynamic memory cell.

regeneration — Also called regenerative feedback and positive feedback. 1. The gain in power obtained by coupling from a high-level point back to a lower-level point in an amplifier or in a system that encloses devices having a power-level gain. 2. In a computer storage device whose information storing state may deteriorate, the process of restoring the device to its latest undeteriorated state. *See also* rewrite. 3. The replacement of a charge in a charge-storage tube to overcome decay effects, including a loss of charge by reading.

regeneration control — A variable capacitor, variable inductor, potentiometer, or rheostat used in a regenerative receiver to control the amount of feedback and thereby keep regeneration within useful limits.

regenerative amplification — Amplification in which increased gain and selectivity are given by a feedback arrangement similar to that in a regenerative detector. However, the operation is always kept just below the point of oscillation.

regenerative braking — 1. Dynamic braking in which the momentum, as the equipment is being braked, causes the traction motors to act as generators. A retarding force is then exerted by the return energy to the power-supply system. 2. A method of braking of an electric vehicle in which the vehicle's kinetic energy is converted to electric current that, in turn, is used to recharge the battery rather than being lost as heat.

regenerative detector — A detector circuit in which regeneration is produced by positive feedback from the output to the input circuit. In this way the amplification and sensitivity of the circuit are greatly increased.

regenerative divider — Also called a regenerative modulator. A frequency divider in which the output wave is produced by modulation, amplification, and selective feedback.

regenerative feedback — *See* regeneration, 1.

regenerative modulator — *See* regenerative divider.

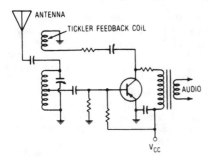

Regenerative detector.

regenerative receiver — A receiver in which controlled regeneration is used to increase the amplification provided by the detector stage.

regenerative repeater — 1. A repeater that regenerates pulses to restore the original shape. Used in teletypewriter and other code circuits; each code element is replaced by a new code element with specified timing, waveform, and magnitude. 2. Normally, a repeater utilized in telegraph applications. Its function is to retime and retransmit the received signal impulses restored to their original strength. These repeaters are speed and code sensitive and are intended for use with standard telegraph speeds and codes. 3. A circuit that samples incoming signal pulses and retransmits them with perfect timing and no distortion.

Regional Bell Operating Company (RBOC) — Also called Bell Operating Companies (BOC). Local telephone operating companies that were split off from AT&T and which provide most local and intrastate telephone service in the United States.

regional channel — A standard broadcast channel within which several stations may operate at 5 kilowatts or less. However, interference may limit the primary service area of such stations to a given field-intensity contour.

regional interconnections — *See* interconnection, 1.

region of limited proportionality — The range of applied voltage, below the Geiger-Mueller threshold, where the gas amplification depends on the charge liberated by the initial ionizing event.

register — 1. A short-term, fast-access circuit used to store bits or words in a CPU; its capacity usually is one computer word. Variations may include provisions for shifting, calculating, etc. Registers play a key role in CPU operations. In most applications the efficiency of programs is related to the number of registers. *See also* static shift register; dynamic shift register. 2. The relative position of all or part of the conductive pattern with respect to a mechanical feature of the board or to another pattern on the obverse side of the printed-circuit board (e.g., pattern-to-hole register or pattern-front-to-pattern-back register). 3. Also called registration. The accurate matching of two or more patterns such as the three images in color television. 4. A range of notes used for playing a particular piece or part of it (e.g., melody or harmony), particularly the range covered by a clavier or manual. 5. In an automatic-switching telephone system, the part of the system that receives and stores the dialing pulses that control the additional operations necessary to establish a telephone connection. 6. A device that can store information, usually that contained in a small subset or word of the total within a digital computer system. 7. Logic elements (gates, flip-flops, shift registers) that, taken together, store 4-, 8-, or 16-bit numbers. They are essentially for temporary storage, in that the contents usually change from one instruction cycle to the next. In fact, much of a microprocessor's operation can be learned by studying the registers, which take part in nearly all operations. 8. An electromechanical device that marks a paper tape in response to signal impulses received from transmitting circuits. A register may be driven by a prewound spring mechanism, an electric motor, or a combination of these. 9. One word of memory. Usually implemented in fast flip-flops, directly accessible to a processor. Most MPUs include a set of internal registers, which can be accessed much faster than the main memory. 10. A special section of primary storage in a computer where data is held while it is being worked on.

register control — Any device that provides automatic register. In photoelectric register control, a light source and phototube from a scanning head. Whenever a special mark or a part of the design printed on a continuous web of paper arrives at the scanning head, the amount of light reaching the phototube changes. If necessary, the web is then moved slightly to bring it back into register.

register file — 1. A small area of memory in which several data elements, or registers, can be accessed simultaneously, rather than one by one. 2. A bank of multiple-bit registers that can be used as temporary storage locations for data or instructions (sometimes referred to as a stack).

register length — The number of digits, characters, or bits that a computer register can store.

register mark — In printed circuits, a mark used to establish the relative position of one or more printed-wiring patterns or portions of patterns with respect to their desired locations on the base.

register of a meter — In a meter, the part that registers the revolutions of the rotor, or the number of impulses received from or transmitted to the meter, and gives the answer in units of electric energy or other quantity measured.

registration — 1. The accuracy of relative position or concentricity of all functional patterns on any mask with the corresponding patterns of any other mask of a given device series of masks when the masks are properly superimposed. 2. The degree of proper alignment of a circuit pattern on the substrate. 3. The degree of accuracy of pattern position with respect to patterns on other layers of double-sided or multilayer boards.

registration marks — The marks used for aligning successive processing masks.

registration of a meter — The apparent amount of electric energy (or other quantity being measured) that has passed through the meter, as shown by the register reading. It is equal to the register reading times the register constant. During a given period, it is equal to the register constant times the difference between the register readings at the beginning and end of the period.

registry — The superposition of one image onto another (e.g., in the formation of an interlaced scanning raster).

regular — Pertaining to reflection, refraction, or transmission in a definite direction rather than in a diffused or scattered manner.

regulated power supply — A unit that maintains a constant output voltage of current for changes in line voltage, output load, ambient temperature, or time.

regulating device — A device that functions to regulate a quantity or quantities such as voltage, current, power, speed, frequency temperature, and load, at a certain value or between certain limits for machines, tie lines, or other apparatus.

regulating transformer — A transformer for adjusting the voltage or the phase relation (or both) in steps, usually without interrupting the load. It comprises one or more windings excited from the system circuit or a separate source, and one or more windings connected in series with the system circuit.

regulating winding — A supplementary transformer winding connected in series with one of the main windings and used for changing the ratio of transformation or the phase relationship, or both, between circuits.

regulation — 1. The difference between the maximum and minimum voltage drops within a specified anode-current range in a gas tube. 2. The holding constant of some condition (e.g., voltage, current, power, or position). 3. In a power supply, the ability to maintain a constant load voltage or load despite changes in line voltage or load impedance. 4. The change in value of dc output voltage of a power supply resulting from a change in ac input voltage over the specified range from low (105 V ac) to high (125 V ac) or from high line to low line. Normally specified as the plus or minus change around the nominal ac input voltage.

regulation of a constant-current transformer — The maximum departure of the secondary current from its rated value, expressed in percent of the rated secondary current, with the rated primary voltage and frequency applied and at the rated secondary power factor and with the current variation taken between the limits of a short circuit and rated load.

regulation of a constant-potential transformer — The change in secondary voltage, expressed in percent of rated secondary voltage, that occurs when the rated kVA output at a specified power factor is reduced to zero, with the primary impressed terminal voltage maintained constant. In the case of a multiwinding transformer, the loads on all windings or specified power factors are to be reduced from the rated kVA to zero simultaneously.

regulator — 1. A device whose function is to maintain a designated characteristic at a predetermined value or to vary it according to a predetermined plan. 2. A device used to maintain a desired output voltage or current constant regardless of normal changes to the input or to the output load.

regulator tube — A two-electrode glow-discharge gas tube that has an essentially constant voltage drop. When series-connected with a resistance across a dc source, the tube will maintain a constant dc voltage across its terminal, with wide variations in the dc source voltage.

reignition — The generation of multiple counts within a radiation-counter tube by the atoms or molecules excited or ionized in the discharge accompanying a tube count.

reignition voltage — Also called restriking voltage. That voltage that is just sufficient to reestablish conduction of a gas tube if applied during the deionization period. It varies inversely with time during the deionization period.

Reike diagram — A polar-coordinate load diagram for microwave oscillators, particularly for klystrons and magnetrons.

Reinartz crystal oscillator — A crystal-controlled vacuum-tube oscillator in which the crystal current is kept low by placing in the cathode lead a resonant circuit tuned to half the crystal frequency. The resultant regeneration at the crystal frequency improves the efficiency, but without the problem of uncontrollable oscillations at other frequencies.

reinforced insulation — An insulation providing protection against electrical shock hazard; equivalent to double insulation.

reinsertion of a carrier — Combining a locally generated carrier signal in a receiver with an incoming suppressed-carrier signal.

reinstalling — Installing a program onto a computer for a second time in order to overwrite the old software and potentially fix a problem.

rejection band — The frequency range below the cutoff frequency of a uniconductor waveguide.

rejector — Filter or part of a circuit that rejects a particular frequency or band of frequencies.

rejector circuit — A circuit that suppresses or eliminates signals of the frequency to which it is tuned.

rejuvenator — A device or instrument for restoring the emissivity of a thermionic cathode by running it at an elevated temperature for a short period of time.

rel — A unit of reluctance, equal to one ampere-turn per magnetic line of force.

relative accuracy — 1. The possible deviation among the standards in a group. 2. The input-to-output error as a fraction of full scale with gain and offset errors adjusted to zero. Relative accuracy is a function of linearity.

relative address — 1. A designation used to identify the position of a memory location in a computer routine or subroutine. 2. A label used to identify a word in a routine or subroutine with respect to its position in that routine or subroutine. 3. An address of a machine instruction that is referred to an origin address. For example, consider the relative address 14, which is translated into the absolute address Origin R + 14, where R is, typically, the contents of the program counter. Relative addressing allows the generation of position-independent code.

relative addressing — A method of memory addressing in which the desired information is located by adding a distance from a pointer to the pointer. The location in relation to the base address or pointer is known.

relative bearing — The bearing in which the direction of the reference line is the heading of the vehicle.

relative binary — The primary form in which information is generated by a link editor and as such has all internal links resolved and is capable of holding address references such that they are relative, usually to one single base address.

relative coding — In a computer, coding in which all addresses refer to an arbitrarily selected position, or in which all addresses are represented symbolically.

relative damping of an instrument — Also called specific damping. Ratio of the actual damping torque at a given angular velocity of a moving element to the damping torque that would produce critical damping at this same angular velocity.

relative detector response — A plot showing how the response (ability to detect a signal) varies with wavelength.

relative dielectric constant — Ratio of the dielectric constant of a material to that of a vacuum. The latter is arbitrarily given a value of 1.

relative gain of an antenna — The gain of an antenna in a given direction when the reference antenna is a half-wave loss-free dipole isolated in space whose equatorial plane contains the given direction.

relative humidity — 1. Ratio of the quantity of water vapor in the atmosphere to the quantity that would saturate at the existing temperature. Usually expressed as a percentage. 2. Ratio of the pressure of water vapor to that of saturated water vapor at the same temperature.

relative interference effect — With respect to a single-frequency electric wave in an electroacoustic system, the ratio (usually expressed in decibels) of the amplitude of a wave of specified reference frequency to that of the given wave when the two waves produce equal interference effects.

relative luminosity — Ratio of the actual luminosity at a particular wavelength to the maximum luminosity at the same wavelength.

relative Peltier coefficient — The Peltier coefficient of a couple made up of the given material as the first-named conductor and a specified standard conductor, commonly platinum, lead, or copper.

relative permeability — 1. Ratio of the magnetic permeability of one material to that of another, or of the same material under different conditions. 2. The permeability of a body relative to that of a vacuum.

relative plateau slope — The average percentage change in the counting rate of a radiation-counter tube near the midpoint of the plateau, per increment of applied voltage. It is usually expressed as the percentage change in counting rate per 100-volt change in applied voltage.

relative power — A power level referred to another power level.

relative power gain — Of one transmitting or receiving antenna over another, the measured ratio of the signal power one antenna produces at the receiver input terminals to the signal power produced by the other, the transmitting power level remaining fixed.

relative refractive index — Ratio of the refractive indices of two media.

relative response — The ratio, usually expressed in decibels, of the response under some particular conditions to the response under reference conditions (which should be stated explicitly).

relative Seebeck coefficient — The Seebeck coefficient of a couple made up of the given material as the first-named conductor and a specified standard conductor such as platinum, lead, or copper.

relative spectral response — The output or response of a device as a function of wavelength normalized to the maximum value.

relaxation — An action requiring an observable length of time for initiation in response to a sudden change in conditions.

relaxation circuit — A circuit arrangement, usually of vacuum tubes, reactances, and resistances, that has two states or conditions, one, both, or neither of which may be stable. The transient voltage produced by passing from one to the other, or the voltage in the state of rest, can be used in other circuits.

relaxation inverter — An inverter that uses a relaxation-oscillator circuit to convert dc power to ac power.

relaxation oscillator — 1. An oscillator that generates a nonsinusoidal wave by gradually charging and quickly discharging a capacitor or an inductor through a resistor. The frequency of a relaxation oscillator may be self-determined or determined by a synchronizing voltage derived from an external source. 2. An oscillator characterized by two semistable states such that when changed to either state the system will, after a time, recover the other state without external excitation. 3. An oscillator whose frequency is determined by the time required to charge a capacitor through a resistor. It produces a sawtooth output.

relaxation time — 1. The time an exponentially decaying quantity takes to decrease in amplitude by a factor of 0.3679. 2. The average time between collisions of an electron with the lattice.

relay — 1. An electromechanical device in which contacts are opened and/or closed by variations in the conditions of one electric circuit and thereby effect the operation of other devices in the same or other electric circuits. 2. A transmission forwarded by way of an intermediate action. 3. An intermediate station on a multilink radio system. 4. An electromechanical device that opens or closes electrical contacts when energized by an isolated electrical coil circuit. Basically an electromagnetic solenoid arranged to open or close a spring-loaded armature. The construction and insulation of this device permits it to switch high-current or high-voltage power when an electromagnet is actuated by a low-power coil. Relays are also used for the remote switching of low signal levels in communications and telecommunications and may have many different contact combinations. 5. A device used to connect (make) or interrupt (break) an electrical current path. Unlike a switch, which is usually actuated by hand or by external mechanical action, a relay has its own built-in actuating mechanism that is responsive to a controlling voltage or current.

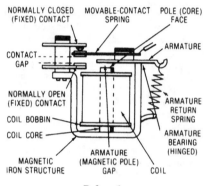

Relay, 1.

relay bias — Bias produced by a spring on an electromagnet. By acting on the relay armature, the spring tends to hold it in a given position.

relay broadcast station — A station licensed to retransmit, from points where wire facilities are not available, the programs from one or more broadcast stations.

relay center — A central point at which switching of messages takes place.

relay contacts — Contacts that are closed or opened by the movement of a relay armature.

relay driver — A circuit with the high-voltage and high-current switching capability necessary for actuation of electromechanical relays.

relay drop — A relay activated by an incoming ringing current to call an operator's attention to a telephone subscriber's line.

relay flutter — Erratic rather than positive operation and release of a relay.

relay function — Control of power in one circuit by means of a low-power, isolated signal in another circuit. *See* figure on page 639.

relay magnet — A coil and iron core forming an electromagnet that, when energized, attracts the armature of a relay and thereby opens or closes the relay contacts.

relay receiver — A specific assembly of apparatus that accepts a sound or television relay signal at its input terminals and delivers the amplified signal at its output terminals.

relay selector — A relay circuit associated with a selector, consisting of a magnetic impulse counter, for registering digits and holding a circuit.

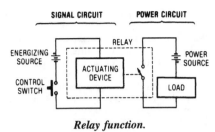

Relay function.

relay station — *See* relay transmitter.

relay-station satellite — An artificial earth satellite intended to receive radio signals from the earth and retransmit them on command to other receiving stations.

relay system — Dial switching equipment made up principally of relays instead of mechanical switches.

relay transmitter — Also called a repeater or relay station (but only if the signal is reduced to a composite picture signal at a standard impedance level and polarity between the receiver and transmitter). The specific assembly of apparatus that accepts a sound or television relay input signal from the relay receiver and rebroadcasts it to another station outside the range of the operating station.

release — 1. An electromagnetic device that opens a circuit breaker automatically or allows a motor starter to return to its off position when tripped by hand, by an interruption of power-supply operation, or by an excessive current. 2. The condition attained by a relay when it has been deenergized, all contacts have functioned, and the armature (if applicable) has attained a fully opened position. 3. *See* disconnect, 3. 4. A system service that informs the operating system that a task no longer requires use of a resource.

release current — 1. The maximum current needed to fully release a relay after it has been fully closed. 2. That value of current at which a relay will just release after having been operated for an appreciable time.

release force — The value to which the force on the actuator of a momentary-contact switch must be reduced to allow the contacts to snap from the operated contact position to the normal contact position.

release time — 1. The time interval from coil deenergization of a reed relay to the functioning time of the last contact to function. Where not otherwise stated, the functioning time of the contact in question is taken as its initial functioning time (that is, it does not include contact bounce time). 2. Also called hold time. The time interval during which data must not change after a flipflop has been clocked. 3. The time for a gain-changing device (such as a compressor/limiter or an expander) to recover from changes in signal level.

releasing position — That position of the actuator of a momentary-contact switch at which the contacts snap from the operated contact position to the normal contact position.

reliability — 1. The probability that a device will perform adequately for the length of time intended and under the operating environment encountered. 2. The characteristic of an item expressed by the probability that it will perform a required function under stated conditions for a stated period. 3. The probability of failure-free performance of a device system or component for a given period under certain stress conditions. Probability of success. 4. A measure of a good device's ability to meet the required specifications for a given period. 5. The ability of an item to perform a required function under stated conditions for a specified period of time.

reliability assurance — The management and technical integration of the reliability activities essential in maintaining reliability achievements, including design, production, and product assurance.

reliability control — The scientific coordination and direction of technical reliability activities from a system viewpoint.

reliability data — 1. Data related to the frequency of failure of an item, equipment, or system. This data may be expressed in terms of failure rate, mean time between failures (MTBF), or probability of success. 2. Data contained in comprehensive documents that provide a detailed history of the reliability evaluation of component parts, component assemblies, etc., or the entire program during the design, development, production, and major product improvement phases of an equipment, weapon, or weapon system in which engineering studies have been performed to select the most reliable product for the intended application.

reliability engineering — 1. The establishment, during design, of an inherently high reliability in a product. 2. The science of including those factors in the basic design that will ensure the required degree of reliability.

reliability index — A quantitive measure of the reliability of equipment or systems.

reliability test — Test and analyses carried out in addition to other type tests and designed to evaluate the level of reliability in a product, etc., as well as the dependability, or stability, of this level relative to time and use under various environmental conditions.

reliable control — Equipment that performs control functions similar to those performed by conventional relay panels and electromechanical devices. Commands are stored in the PC's built-in program. Control actions (outputs) are initiated by signals from sensing devices (inputs) located in the machine or process being controlled. Input can be supplied by limit switches, photoelectric cells, timers, temperature and pressure sensors, or similar devices.

relieving anode — In a pool-cathode tube, an auxiliary anode that provides an alternative conducting path to reduce the current to another electrode.

relocatability — That characteristic of a compiled or assembled computer program segment that makes possible loading (locating) it into any region of memory.

relocatable — 1. Computer program or routine that does not contain fixed addresses, and can therefore be easily relocated elsewhere in the memory. 2. Object programs that can reside in any part of system memory. The actual starting address is established at load time by adding a relocation offset to the starting address. Relocatable code is typically composed of positionindependent code.

relocatable assembler — A program that translates object code from an assembly-language source program, using memory locations specified as displacements from a relative origin or as external references. This facilitates the running of programs in any memory area.

relocatable binary — The form in which information is generated by a compiler or assembler and which is the primary input to a link editor.

relocatable macro assembler — A program that permits instruction groups to be combined for execution, using symbolic addresses.

relocatable program — A software program so written that it can be moved to and executed from many different areas of memory.

relocate — In computer programming, to move a routine from one part of storage to another and adjust the necessary address references to permit execution of the routine in its new location.

relocation dictionary — In a computer, the part of an object or load module that provides identification of all relocatable address constants in the module.

reluctance — 1. The resistance of a magnetic path to the flow of magnetic lines of force through it. It is the reciprocal of permeance and is equal to the magnetomotive force divided by the magnetic flux. 2. Property of a magnetic circuit that determines the total magnetic flux in the circuit when a given magnetomotive force is applied.

reluctance pickup — A pickup that depends for its operation on variations in the reluctance of a magnetic circuit due to the movements of an iron stylus assembly that is a part of the magnetic circuit. The reluctance variations alternately increase and decrease the flux through two series-connected coils, inducing in them the desired ac output voltage.

reluctance-type synchronous motor — An ac motor that runs at synchronous speed without excitation because of the salient pole rotor punchings of the laminations. The rotor has a squirrel cage type winding. Made in single- and three-phase types.

reluctive transduction — The conversion of the measurand into a change in ac voltage by changing the reluctance path between two or more coils when ac excitation is applied.

reluctivity — The ability of a magnetic material to conduct magnetic flux. It is the reciprocal of permeability.

rem — A measure of physiological damage to the human body caused by radiation, calculated as the absorbed dose in rads multiplied by various factors that qualify the type of radiation and the way in which it is absorbed. *See* roentgen equivalent man.

remanence — 1. The extent to which a body remains magnetized after removal of a magnetizing field that has brought the body to its saturation (maximum) magnetization. A substance with remanence is called ferromagnetic. 2. The magnetic induction that remains in a magnetic circuit after the removal of an applied magnetomotive force. (If there is an air gap in the magnetic circuit, the remanence will be less than the residual induction.)

remanent magnetization — The magnetization retained by a substance after the magnetizing force has been removed.

reminder chirp — Optional feature of a pager that emits a "chirp" every two minutes when there are unread messages.

remodulator — A device for converting amplitude modulation to audio-frequency-shift modulation for transmission over voice radio-frequency channels.

remote — *See* field pickup.

remote access — 1. Having to do with communication with a data-processing facility by stations at a distance from the facility. 2. Pertaining to communication with a computer processor by terminal stations that are distant from that processor.

remote alarm — An alarm signal that is transmitted to a remote monitoring station. *See also* local alarm.

remote batch — A method of entering jobs into a computer from a remote terminal in a conversational mode, for processing later in a batch processing mode. In this mode a plant or office geographically distant from the central computer can load in a batch of transactions, transmit them to the computer, and receive the results by mail or via direct transmission to a printer or other output device at the remote site.

remote control — 1. Any system of control performed from a distance. The control signal may be conveyed by intervening wires, sound (ultrasonics), light, or radio. 2. Manual control, with the control operator monitoring the operation on duty at a control point located elsewhere than at the station transmitter, such that the

associated operating adjustments are accessible through a control link. 3. A method whereby a device is programmable via its electrical interface connection in order to enable the device to perform different tasks.

remote-cutoff tube — Also called a variable-mu or extended-cutoff tube. An electron tube used mainly in rf amplifiers. The control-grid wires are farther apart at the center than at the ends. Therefore, the amplification of the tube does not vary in direct proportion to the bias. Also, some plate current will flow regardless of the negative bias on the grid.

remote error sensing — A means by which the regulator circuit of a related power supply senses the voltage directly at the load. This connection is used to compensate for a voltage drop in the connecting wires.

remote indicator — 1. A radar indicator that is connected in parallel with a primary indicator. 2. An indicator located some distance away from the data-gathering element.

remote job entry — The process of entering data-processing jobs or tasks for execution from an input device, such as a terminal that is remote from the processing computer and connected to the computer by a communication line.

remote line — A program transmission line between a remote-pickup point and the studio or transmitter site.

remote/local — Refers to device connection to a given computer with remote devices attached directly over communications lines and local devices attached directly to a computer channel; in a network environment the computer itself may be a remote device to the CPU controlling the network.

remotely adjustable timer — A time delay having external leads to which a variable or fixed resistor can be attached.

remote metering — *See* telemetering.

remote pickup — Transmitting a program that originates away from the studio to the studio or transmitter over telephone lines or a radio link.

remote plan-position indicator — *See* PPI repeater.

remote programming — A power supply feature whereby the controlled output parameter, voltage, or current may be controlled through the application of external resistance, voltage signals, or current signals to designated external terminals.

remote sensing — 1. In a power supply, terminations that allow the regulator to sense and regulate the output voltage at a remote location, usually the load. This connection is used to compensate for voltage drops in the power leads. The maximum voltage drop in both power leads must be specified. 2. Technique that utilizes electromagnetic energy to detect and quantify information about an object that is not in contact with the sensing apparatus.

remote station — Data terminal equipment for communication with a data-processing system that is distant electrically or in terms of time or space.

remote-station alarm system — An alarm system that employs remote alarm stations, usually located in building hallways or on city streets.

remote subscriber — A network subscriber without direct access to the switching center, but with access to the circuit through a facility such as a base message center.

remote terminal — 1. A terminal that is a substantial distance from the central computer. Usually it accesses the computer through a telephone line or other type of communication link. 2. An input-output control unit and one or more input/output devices at some distance from the central processing unit.

removable contact — A contact that can be mechanically joined to or removed from an insert.

Usually, special tools are required to lock the contact in place or remove it for repair or replacement.

render — A computer graphics process that removes hidden surfaces and jagged edges from computer-generated 3D models by adding shades, texture, depth, and other realistic attributes.

renewable fuse — A fuse that may be readily restored to operation by replacing the fused link.

reoperate time — The release time of a thermal relay.

rep — Abbreviation for roentgen equivalent physical.

repaint — To redraw a display image on a CRT to reflect its updated status.

repair — To restore a failed module or system to a specified operable condition.

repairing — The act of restoring the functional capability of a defective part without necessarily restoring appearance, interchangeability, and uniformity.

repeatability — 1. The ability of a device or instrument to come back to the same reading after a certain length of time. 2. The ability of an instrument to repeat its readings taken when deflecting the pointer up-scale, compared to the readings taken when deflecting the pointer down-scale, expressed as a percentage of the full-scale deflection. 3. *See* precision, 2. 4. The ability of a transducer to reproduce output readings when the same measurand value is applied to it consecutively, in the same direction. (Short time duration.) 5. The precision with which a sensor will give the same output or value among repeated temperature measurements.

repeatability error — The inability to reproduce the same output readings when a given level of measurand is applied repeatedly in the same direction.

repeat-cycle timer — A timer that has any number of load contacts and that continues to repeat its time program as long as power is applied.

repeater — 1. An FM, TV, facsimile, or similar station that receives a signal on some input frequency and automatically transmits the received signal on some output frequency. The purpose of a repeater is to extend the communication range between a group of stations. The repeater generally consists of a receiver with its antenna, a control unit, and a transmitter with its antenna. 2. Switch by which originating central-office, calling-telephone dialed pulses are repeated to switches at a distant office. 3. Relay circuit in dial signaling, which amplifies and repeats dial pulses received from one circuit into another. 4. A device used to amplify and/or reshape signals. 5. A sensitive receiver coupled to a high-power transmitter located on top of a tall building or high hill. 6. A device that serves to amplify signals that have become too weak. 7. A device that converts a received optical signal to its electrical equivalent, reconstructs the source signal format, amplifies it, and reconverts it to an optical output signal. 8. A device that serves as an interface between two circuits, receiving signals from one circuit and transmitting them to the other circuit.

repeater control operator — A licensed amateur designated by a repeater trustee who offers assistance with autopatch and listens for inappropriate use of the repeater. (This is different from the FCC's definition of a control operator, which is anyone in control of an amateur transmitter.)

repeater facility — Radio equipment needed to relay radio signals between central stations, satellite stations, and/or protected premises.

repeater jammer — Equipment intended to confuse or deceive the enemy by causing its equipment to present false information regarding azimuth, range, number of targets, etc. This result is achieved by a system that intercepts and reradiates a signal on the frequency of the enemy equipment.

repeater station — 1. Station licensed to automatically retransmit the radio signals of other (amateur) radio stations. 2. *See* relay transmitter.

repeating coil — 1. An audio-frequency transformer, usually with a 1:1 ratio, for connecting two sections of telephone line inductively to permit the formation of simplex and phantom circuits. 2. Telephone industry term for a voice-frequency transformer.

repeating-coil bridge cord — A way of connecting the common office battery to the midpoints of a repeating coil, which is bridged across the cord circuit.

repeating flash tube — A flash tube that, by producing rapid, brilliant flashes, permits night aerial photographs to be taken from as high as two miles.

repeating timer — A timer that repeats each operating cycle automatically until excitation is removed.

repeller — Sometimes called a reflector. 1. An electrode whose primary function is to reverse the direction of an electron stream. 2. The element in a reflex klystron tube that reflects the electrons back toward the grid.

reperforator — 1. A device that converts teletypewriter signals into perforations on tape instead of the usual typed copy on a roll of paper or ticker tape. 2. A machine that reads one punched paper tape or card and punches the same information into another paper tape or card.

reperforator/transmitter — An integrated unit for temporarily storing traffic for retransmission. It consists of a paper-tape punch and a paper-tape reader.

repertory dialer — A device that automatically places a telephone call to any one of the phone numbers stored therein.

repertory dialing — The process of dialing a complete telephone number, or area code plus number, by pressing a single button key. The telephone set (or accessory dialer) has a number of buttons that can be programmed to produce an individual telephone number.

repertory instruction — 1. A set of instructions that a computing or data-processing system can perform. 2. A set of instructions assembled by an automatic coding system.

repetition equivalent — A measure of the quality of transmission experienced by the subscribers using a complete telephone connection. It represents a combination of the effects of volume, distortion, noise, and all other subscriber reactions and usage.

repetition frequency — The number of repetitions of an event per unit time.

repetition instruction — A computer instruction that calls for one or more instructions to be executed an indicated number of times.

repetition rate — 1. The number of repetitions of an event per unit time. For example, in radar, the rate (usually expressed in pulses per second) at which pulses are transmitted (also called pulse repetition frequency, recurrence rate, or repetition frequency). 2. The number of repetitions per unit time requested by users of a telephone connection.

repetitive logic — A large-scale repetition of logic functions (or cell structures), such as in a memory or shift register. A semiconductor memory, for example, is simply a repetition of identical structures of transistors and their interconnects. As a result, repetitive-logic circuits on a chip generally have a checkerboard pattern and, thus, are relatively easy to fabricate.

repetitively pulsed laser — A pulsed laser that emits a recurring pulsed output. Frequency of the pulses emitted is known as pulsed recurrence frequency, PRF. When the PRF or duty cycle is very high, repetitively pulsed lasers illustrate properties like those of the cw laser.

repetitive peak inverse voltage — The maximum allowable instantaneous value of reverse (negative) voltage that may be repeatedly applied to the anode of an SCR with the gate open. This value of peak inverse voltage does not represent a breakdown voltage, but it should never be exceeded (except by the transient rating if the device has such a rating).

repetitive peak off-state voltage — The highest instantaneous value of the off-state voltage that occurs across a thyristor, including all repetitive transient voltages but excluding all nonrepetitive transient voltages.

repetitive peak on-state current — The peak value of the on-state current, including all repetitive transient currents.

repetitive unit — A type of circuit that appears more than once in a computer.

replacement theory — The mathematics of deterioration and failure, used in estimating replacement costs and selecting optimum replacement policies.

reply — In transponder operation, the radio-frequency signal or signals transmitted as a result of an interrogation.

repolarization — The process by which the normal resting potential of a biological cell is restored after the cell has fired, or depolarized.

report — 1. The output document produced by a data-processing system. 2. An application data display or print-out containing information in a user-designed format. Reports include operator messages, part records, production lists, etc. Initially entered as messages, reports are stored in a memory area separate from the user's program.

report generation — 1. In a computer, production of complete output reports from only a specification of the desired content and arrangement and from specifications regarding the input file. 2. The printing or displaying of user-formatted application data by means of a data terminal. Report generation can be initiated by means of either the user's program or a data terminal keyboard. *See* message; report.

report generator — A special computer routine designed to prepare an object routine that, when later run on the computer, produces the desired report.

representative calculating time — A measure of the performance speed of a computer; it is the time required for performance of a specified operation or series of operations.

reproduce — In a computer, to prepare a duplicate of stored information.

reproduce head — An electromagnetic transducer that converts the remanent flux pattern in a magnetic tape into electric signals during the reproduce (playback) process.

reproducer — A device used to translate electrical signals into sound waves.

reproducibility — 1. The exactness with which measurement of a given value can be duplicated. 2. *See* precision, 2.

reproducing stylus — A mechanical element that hollows the modulations of a record groove and transmits the mechanical motion thus derived to the pickup mechanism.

reproduction speed — In facsimile, the area of copy recorded per time.

reprogrammable ROM — A ROM that can be programmed any number of times. Generally, however, the information stored in a reprogrammable ROM is changed very seldom.

repulsion — The mechanical force that tends to push apart adjacent conductors that carry currents in opposite directions, or tends to separate bodies having like electrostatic charges or magnetic poles of like polarity.

repulsion-induction motor — A constant- or variable-speed repulsion motor with a squirrel-cage winding in the rotor in addition to the regular winding.

repulsion motor — A single-phase motor in which the stator winding is connected to the source of power, and the rotor winding to the commutator. Brushes on the commutator are short-circuited and are placed so that the magnetic axis of the brush winding is inclined to that of the stator winding. This type of motor has a varying speed characteristic.

repulsion-start, induction-run motor — A single-phase motor that has the same windings as a repulsion motor but operates at a constant speed. The rotor winding is short-circuited (or otherwise connected) to give the equivalent of a squirrel-cage winding. It starts as a repulsion motor, but operates as an induction motor with constant-speed characteristics.

request repeat system — A system that uses an error-detecting code and is arranged so that when a signal is detected as being in error, a request for retransmission of the signal is initiated automatically.

request-response time — The interval between an operator's request for a display and the moment it appears on the screen.

required voltage (engine or vehicle) — The voltage required to fire the spark plugs. Requirement data is obtained using the same instrumentation as for available voltage runs, but with all spark plug leads connected. Values plotted are the maximum voltage observed, viewing all cylinders simultaneously. Requirements are usually determined with the engine operating at full load (wide-open throttle) and under a variety of part-load conditions to ascertain the maximum required voltages.

reradiation — 1. Scattering of incident radiation. 2. Radiation from a radio receiver resulting from insufficient isolation between the antenna circuit and the local oscillator and causing undesirable interference in other receivers.

rerecording — The process of making a recording by reproducing a recorded sound source and recording this reproduction. *See also* dubbing.

rerecording system — An association of reproducers, mixers, amplifiers, and recorders capable of being used for combining or modifying various sound recordings to provide a final sound record. Recording of speech, music, and sound effects may be so combined.

rerun — Also called rollback. 1. To run a computer program (or a portion of it) over again. 2. To repeat an entire transmission. 3. In a computer, a system that will restart the running program after a system failure. Snapshots of data and programs are stored at periodic intervals and the system rolls back to restart at the last recorded snapshot.

rerun point — In a computer program, one of a set of preselected points located in a computer program such that if an error is detected between two such points, the problem may be rerun by returning to the last such point instead of returning to the start of the problem.

rerun routine — A computer routine designed to be used, in the event of a malfunction or mistake, to reconstitute a routine from the previous rerun point.

reset — 1. To restore a storage device to a prescribed state. 2. To place a binary cell in the initial, or 0, state. *See also* clear. 3. An input to a binary counter or register that causes all binary elements to assume the 0 logical state or the minimum binary state. 4. The technique of returning an electronic device to its starting point. Some equipment offers automatic reset; other apparatus offer manual or automatic and manual reset. 5. To restore a device to its original (normal) condition after an alarm or trouble signal.

reset action — A type of control in which correction is proportional to both the length of time and the amount that a controlled process has deviated from the desired value, and provision is made to ensure that the process is returned to its set point.

reset button — A button that restarts a computer without turning power off and on.

reset pulse — A drive pulse that tends to reset a magnetic cell.

reset rate — The number of corrections made per unit of time by a control system; it usually is expressed in terms of the number of repeats per minute.

reset terminal — Also called clear terminal or zero-input terminal. In a flip-flop, the input terminal used to trigger the circuit from its second state back to its original state.

reset time — 1. A timer that can be reset by electrical means. May be either an on-delay or off-delay type. 2. A timer with one or more circuits that spring-reset to zero when the clutch is disengaged. 3. The period required from the time of the reset command until a timer is fully returned to the before-start conditions ready for the next cycle.

resident assembler — An assembler that runs on the machine for which it generates code. Eliminates the need for another computer system or time-sharing service, as required by a cross assembler.

residual — 1. The difference between any value and some estimate of the mean of such values; e.g., residuals from a curve of regression. (Residual unbalance.) 2. The complement of a set is called a residual set. 3. Pertaining to a measure of the output of a transducer under static conditions and with no stimulus applied.

residual charge — 1. The charge remaining on the plates after an initial discharge of a capacitor. 2. The small charge that remains in the dielectric of a capacitor after a quick discharge.

residual current — 1. The vector sum of the currents in the several wires of an electric supply circuit. 2. Current through a thermionic diode in the absence of anode voltage; it results from the velocity of the electrons emitted by the heated cathode.

residual deviation — Apparent modulation due to noise and/or distortion in the transmitter.

residual discharge — A discharge of the residual charge of a capacitor remaining after the initial discharge.

residual error — The direction-finding errors remaining after errors due to site and antenna effects have been minimized.

residual field — The magnetic field left in an iron field structure after excitation has been removed.

residual flux — 1. The value of magnetic induction that remains in a magnetic circuit when the magnetomotive force is reduced to zero. 2. In a uniformly magnetized sample of magnetic material, the product of the residual flux density and the cross-sectional area. Residual flux is indicative of the output that can be expected from a tape at long wavelengths.

residual flux density — The magnetic flux density that exists at zero magnetizing field strength when a sample of magnetic material has undergone symmetrically cyclical magnetization.

residual FM — *See* incidental FM.

residual gases — The small amounts of gases remaining in a vacuum tube despite the best possible exhaustion by vacuum pumps.

residual induction — 1. The magnetic induction that remains in a magnetized material when the effective magnetizing force has been reduced to zero. When the

material is in a symmetrically cyclic magnetic condition, the residual induction is termed the "normal residual induction." Letter symbol: B_r. 2. The magnetic induction corresponding to zero magnetizing force in a magnetic material that is in a symmetrically cyclically magnetized condition. Note: The European definitions of residual induction and remanence are interchanged. Thus, the meaning of these two words can vary, depending on the author.

residual ionization — Ionization of air or other gas not accounted for in a closed chamber by recognizable neighboring agencies.

residual losses — In a magnetic core, the difference between the total losses and the sum of the eddy-current and hysteresis losses.

residual magnetic induction — Magnetic induction remaining in a ferromagnetic object after the magnetizing force has been removed. The amount depends on the material, shape, and previous magnetic history.

residual magnetism — The magnetism that remains in the core of an electromagnet after the operating circuit has been opened.

residual modulation — *See* carrier noise level.

residual screw — A brass screw in the center of a relay armature. It is used to adjust the residual air gap between the armature and the coil core to prevent residual magnetism from holding the armature operated after the relay operating circuit has opened.

residual torque — Nonenergized detent torque arising from permanent-magnet effects and bearing friction of a stepper motor.

residual voltage — 1. The vector sum of the voltages to ground of the several phase wires in an electric supply circuit. 2. The voltage caused by dielectric absorption that appears in the capacitor terminations after the capacitor is discharged and then left with the terminations disconnected.

resin — 1. One of a class of solid or semisolid, natural or synthetic organic products, generally with a high molecular weight and no definite melting point. Most resins are polymers. 2. An inorganic substance of natural or synthetic origin that is polymeric in structure and predominantly amorphous.

resist — 1. A material placed on the surface of a copper-clad base material to prevent the removal by etching of the conductive layer from the area covered. 2. A material deposited on conductive areas to prevent plating of the areas covered. 3. Coating that masks areas of a board from the effects of an etchant used during manufacture.

resistance — 1. A property of conductors that — depending on their dimensions, material, and temperature — determines the current produced by a given difference of potential; that property of a substance which impedes current and results in the dissipation of power in the form of heat. The practical unit of resistance is the ohm. It is defined as the resistance through which a difference of potential of 1 volt will produce a current of 1 ampere. 2. A circuit element designed to offer a predetermined resistance to current. 3. Ratio of the applied electromotive force to the resulting current in a circuit. It is a measure of the resistance of the circuit to the passage of an electric current and is measured in ohms. Its value is determined by $R = E/I$, where E is voltage in volts, and I is current in amperes.

resistance balance — The amount of resistance that is required to null the output of certain transducers or input systems.

resistance box — An assembly of resistors and the necessary switching or other means for changing the resistance connected across its output terminals by known, fixed amounts. *See also* Wheatstone bridge.

resistance brazing — Brazing by resistance heating, the joint being part of the electrical circuit.

resistance bridge — A common form of Wheatstone bridge employing resistances in three arms. *See also* Wheatstone bridge.

resistance-bridge pressure pickup — A pressure transducer in which the electrical output is derived from the unbalance of a resistance bridge, which is varied according to the applied pressure.

resistance-bridge smoke detector — A smoke detector that responds to the particles and moisture present in smoke. These substances reduce the resistance of an electrical bridge grid and cause the detector to respond.

resistance-capacitance-coupled amplifier — An amplifier whose stages are connected by a suitable arrangement of resistors and capacitors.

resistance-capacitance coupling — Also called RC coupling. Coupling between two or more circuits, usually amplifier stages, by a combination of resistive and capacitive elements.

resistance-capacitance filter — Abbreviated RC filter. A filter made up only of resistive and capacitive elements.

resistance-capacitance oscillator — Abbreviated RC oscillator. An oscillator whose output frequency is determined by resistance and capacitance elements.

resistance coupling — Also called resistive coupling. The association of circuits with one another by means of the mutual resistance between circuits.

resistance drop — Also known as an *IR* drop. The voltage drop occurring across two points on a conductor when current flows through the resistance between those points. Multiplying the resistance in ohms by the current in amperes gives the voltage drop in volts.

resistance furnace — An electric furnace in which the heat is developed by the passage of current through a suitable resistor, which may be the charge itself or a resistor imbedded in or surrounding the charge.

resistance grounded — Grounded through a resistance, so as to limit the current that will flow in the event of a ground fault.

resistance lamp — An electric lamp used as a resistance to limit the amount of current in a circuit.

resistance loss — The power lost when current flows through a resistance. Its value in watts is equal to the resistance in ohms multiplied by the square of the current in amperes ($W = R \times I^2$).

resistance magnetometer — A magnetometer that depends for its operation on the variation in the electrical resistance of a material immersed in the field to be measured.

resistance material — A material having sufficiently high resistance per unit length or volume to permit its use in the construction of resistors.

resistance noise — *See* thermal noise, 1.

resistance pad — A network employing only resistances. It is used to provide a fixed amount of attenuation without altering the frequency response.

resistance ratio — A thermistor specification for determining resistance at a specific temperature. It is defined as the ratio of zero power resistance R_o measured at two specified reference temperatures, usually 0°C and 25°C.

resistance soldering — A method of soldering in which a current is passed through and heats the soldering area by contact with one or more electrodes.

resistance standard — *See* standard resistor.

resistance-start motor — 1. A form of split-phase motor having a resistance connected in series with the auxiliary winding. The auxiliary circuit opens whenever the motor attains a predetermined speed. 2. A split-phase induction motor employing a resistive auxiliary starting winding (copper alloy or iron wire) to shift the phase angle approximately 90°. Used almost exclusively for low-starting-power requirements, this winding is switched out of the circuit when the armature approaches near-synchronous speed.

resistance strain gage — A strain gage consisting of a small strip of resistance material cemented to the part under test. Its resistance changes when the strip is compressed or stretched.

resistance temperature coefficient — 1. The ratio of the resistance change of an element between two temperatures to the product of the temperature change and the original resistance. A positive value of the coefficient indicates an increase of resistance with an increase in temperature; a negative value indicates a decrease in resistance with an increase in temperature; a zero value indicates no resistance change with temperature. 2. The magnitude of change in resistance due to temperature, usually expressed in percent per degree Celsius or parts per million per degree Celsius (ppm/°C). If the changes are linear over the operating temperature range, the parameter is known as temperature coefficient.

resistance temperature detector — Also called resistance-thermometer detector. A resistor made of some material for which the electrical resistivity is a known function of the temperature. It is intended for use with a resistance thermometer and is usually in such a form that it can be placed in the region where the temperature is to be determined.

resistance temperature meter — *See* resistance thermometer.

resistance thermometer — Also called resistance temperature meter. 1. An electric thermometer that has a temperature-responsive element called a resistance temperature detector. Since the resistance is a known function of the temperature, the latter can be readily determined by measuring the electrical resistance of the resistor. 2. Thermometer using variation of resistance with temperature of some material, usually platinum, considered the standard for temperature measurement over its range.

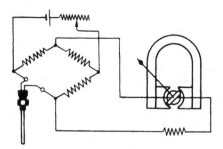

Resistance thermometer.

resistance-thermometer detector — *See* resistance temperature detector.

resistance-thermometer resistor — *See* resistance temperature detector.

resistance thermometry — A temperature-measuring technique that utilizes the temperature coefficient of a wirewound resistor. Known as a resistance thermometer, this resistor consists of a spiral of nickel or platinum wire.

Since the ohmic value of the wire varies with temperature, the resistance of the spiral is thus an indication of temperature.

resistance tolerance — The permissible deviation of the manufactured resistance value (expressed in percent) from the specified nominal resistance value at standard (or stated) environmental conditions.

resistance weld — The junction produced by heat obtained from the resistance of the work to the flow of electric current in a circuit of which the work is a part, and by the application of pressure before and during the flow of current. The term includes all types of bonds produced by the process, which may or may not be classified metallurgically as welds.

resistance welding — Welding in which the metals to be joined are heated to melting temperature at their points of contact by a localized electric current while pressure is applied.

resistance wire — A wire made from a metal or alloy having a high resistance per unit length (e.g., Nichrome). It is used in wirewound resistors, heating elements, and other high-resistance circuits.

resist etchant — Any material deposited onto a copper-clad base material to prevent the conductive area underneath from being etched away.

resistive conductor — A conductor used primarily because of its high electrical resistance.

resistive coupling — See resistance coupling.

resistive cutoff frequency — The frequency at which the real part of the tunnel-diode admittance at its terminals is zero at the specified bias point.

resistive load — A load in which the voltage is in phase with the current.

resistive transduction — The conversion of the measurand into a change in resistance.

resistive unbalance — 1. Unequal resistance in the two wires of a transmission line. 2. The difference in resistances of two or more conductors in a cable. It is expressed as a percentage of the resistance of some single conductor.

resistivity — 1. A measure of the resistance of a material to electric current either through its volume or on a surface. The unit of volume resistivity is the ohm-centimeter; the unit of surface resistivity is the ohm. 2. The ability to resist current; the reciprocal of conductivity. Resistivity is normally defined in terms of the resistance in ohms between opposite faces of a cube of the material measuring 1 centimeter on each side.

resistor — 1. A component made of a material (like carbon) that has a specified resistance, or opposition, to the flow of electrical current. Resistors are used to control (or limit) the amount of current in a circuit or to provide a voltage drop. The majority of resistors consist of a resistive element to which axial or radial leads are attached and to which a protective coating or molding is applied. Multiple-resistor networks, consisting of resistive inks screen printed on a flat ceramic base, reduce assembly time in printed-circuit applications. 2. A device having electrical resistance and utilized in an electric current for purposes of protection, operation, or control of current.

resistor-capacitor-transistor logic — Abbreviated RCTL. A logic circuit design that employs a resistor and a speedup capacitor in parallel for each input of the gate. A transistor's base is connected to one end of the RC network. A positive voltage on the RC input will energize the transistor and turn it on, so that the output voltage is nearly zero volts. This circuit is a positive NOR or negative NAND when npn transistors are used in the circuit.

resistor color code — A code adopted by the Electronic Industries Association to indicate the values of resistance on resistors in a readily recognizable manner.

RESISTOR COLOR CODES			
Color	Digit	Multiplier	Tolerance
BLACK	0	1	± 20%
BROWN	1	10	+ 1%
RED	2	100	± 2%
ORANGE	3	1000	± 3%
YELLOW	4	10000	GMV
GREEN	5	100000	± 5% (EIA ALTERNATE)
BLUE	6	1000000	± 6%
VIOLET	7	10000000	± 12 ½ %
GRAY	8	0.01 (EIA ALTERNATE)	± 30%
WHITE	9	0.1 (EIA ALTERNATE)	± 10% (EIA ALTERNATE)
GOLD		0.1 (JAN AND EIA PREFERRED)	± 5% (JAN AND EIA PREFERRED)
SILVER		0.01 (JAN AND EIA PREFERRED)	± 10% (JAN AND EIA PREFERRED)
NO COLOR			± 20%

GMV = Guaranteed minimum value, or − 0 + 100% tolerance.
± 3, 6, 12 ½, and 30% are ASA 40, 20, 10 and 5 step tolerances.

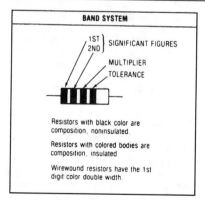

BAND SYSTEM

1ST } SIGNIFICANT FIGURES
2ND }

MULTIPLIER
TOLERANCE

Resistors with black color are composition, noninsulated.

Resistors with colored bodies are composition, insulated.

Wirewound resistors have the 1st digit color double width.

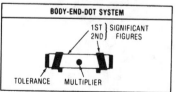

BODY-END-DOT SYSTEM

1ST } SIGNIFICANT
2ND } FIGURES

TOLERANCE   MULTIPLIER

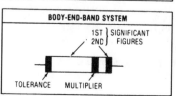

BODY-END-BAND SYSTEM

1ST } SIGNIFICANT
2ND } FIGURES

TOLERANCE   MULTIPLIER

Resistor color code.

The first color represents the first significant figure of the resistor value, the second color the second significant figure, and the third color represents the number of zeros following the first two figures. A fourth color is sometimes added to indicate the tolerance of the resistor.

resistor core — An insulating support around which a resistor element is wound or otherwise placed.

resistor element — 1. That portion of a resistor which possesses the property of electric resistance. 2. That portion of a potentiometer which provides the change in resistance as the shaft is rotated.

resistor geometry — The film resistor outline.

resistor housing — The enclosure around the resistance element and the core of a resistor.

resistor network — 1. A combination of several resistors contained in a single package, the resistors being accessible for individual measurement and use or configured to produce a specific circuit function in which some resistors may be inaccessible. 2. A group of fixed resistors manufactured on a single substrate using thick- and/or thin-film technology. Most resistor networks are produced with thick film. More expensive thin-film networks are used for precise applications such as instrumentation.

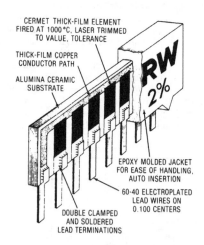

CERMET THICK-FILM ELEMENT FIRED AT 1000°C, LASER TRIMMED TO VALUE, TOLERANCE

THICK-FILM COPPER CONDUCTOR PATH

ALUMINA CERAMIC SUBSTRATE

EPOXY MOLDED JACKET FOR EASE OF HANDLING, AUTO INSERTION

60-40 ELECTROPLATED LEAD WIRES ON 0.100 CENTERS

DOUBLE CLAMPED AND SOLDERED LEAD TERMINATIONS

SIP.

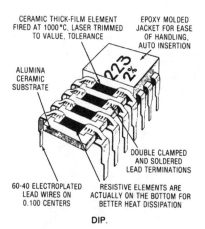

CERAMIC THICK-FILM ELEMENT FIRED AT 1000°C, LASER TRIMMED TO VALUE, TOLERANCE

EPOXY MOLDED JACKET FOR EASE OF HANDLING, AUTO INSERTION

ALUMINA CERAMIC SUBSTRATE

DOUBLE CLAMPED AND SOLDERED LEAD TERMINATIONS

60-40 ELECTROPLATED LEAD WIRES ON 0.100 CENTERS

RESISTIVE ELEMENTS ARE ACTUALLY ON THE BOTTOM FOR BETTER HEAT DISSIPATION

DIP.

Resistor networks.

resistor spark plug — A spark plug containing a resistor, designed to shorten both the capacitive and inductive phases of the spark. This will suppress radio interference and lengthen electrode life.

resistor starting (or starter) — A motor starter using resistance to limit inrush current. The resistors are shorted by a paralleling contactor on the final step. A nontransition type of starting.

resistor termination — The contact area between a film resistor and a film conductor.

resistor-transistor logic — Abbreviated RTL. 1. A form of logic that has a resistor as the input component that is coupled to the base of an npn transistor. As in RCTL, the transistor is an inverting element that produces the positive NOR gate or the negative NAND gate function. 2. One of the earliest forms of semiconductor logic, in which the basic logic element is a resistor-transistor network. RTL is now little used.

resist plating — Any material that, when deposited on a conductive area, prevents the areas underneath from being plated.

resnatron — A high-power cavity-resonator tetrode for high-efficiency operation in the very high frequency and ultrahigh frequency bands. It is water cooled, and the cavities form an integral part of the tube.

resolution — 1. The deriving of a series of discrete elements from a sound, scene, or other form of intelligence so that the original may subsequently be synthesized. 2. The degree to which nearly equal values of a quantity can be discriminated. 3. The degree to which a system or a device distinguishes fineness of detail in a spatial pattern. 4. In facsimile, a measure of the narrowest line width that may be transmitted and reproduced. 5. A measure of the smallest possible increment of change in the variable output of a device. 6. In a potentiometer, the smallest possible incremental resistance change. 7. The reciprocal number of steps per revolution of a motor shaft, expressed in degrees per step. 8. The degree to which the distance separating different states of magnetization recorded along a tape can be reduced and still permit these states to be distinguished usefully on reproduction. 9. In radar, the minimum angular or distance separation between two targets that permits them to be distinguished on the radar screen. 10. In television, the maximum number of lines discernible on the screen in a distance equal to the tube height. 11. The degree to which significant signals can be extracted from comparatively random signals, as with a radio telescope. 12. A measure of ability to delineate picture detail; also, the smallest discernible or measurable detail in a visual presentation. Resolution may be stated in terms of modulation transfer function, spot diameter, line width, or raster lines. 13. A measure of the smallest possible increment of change in the variable output of a device. 14. Of a DMM (digital multimeter) the ratio of 1 to the maximum display. For example, a three-digit DMM with 100-percent overranging can display from 000 to 1999. Its resolution is then 1 part in 2000. 15. The magnitude of output step changes (expressed in percent of full scale output) as the measurement is continuously varied over the range. 16. The ability of a meter to discriminate between two adjacent values of the quantity being measured. 17. The degree of setability. 18. The number of bits on the input or output of an a/d or d/a converter. The number of discrete steps or states is equal to 2^n, where n is the resolution of the converter. (Note that n bits of resolution do not guarantee n bits of accuracy.) 19. Describes the smallest standard incremental change in output voltage of a DAC or the amount of input voltage change required to increment the output of an ADC between one code change and the next adjacent code change. A converter with n switches can resolve 1 part in $2n$. The least significant increment is then

2^{-n}, or one least significant bit. In contrast, the most significant bit carries a weight of 2^{-1}. Resolution applies to DACs and ADCs, and may be expressed in percent of full scale or in binary bits. 20. The smallest change in measured value to which an instrument will respond. 21. The smallest spacing between points on a graphic device at which the points can be detected as distinct. 22. The measurement of image sharpness and clarity, usually in the number of pixels per square inch; for example, standard VGA has a resolution of 640×480. The higher the resolution, the better the picture. 23. The smallest detectable increment of measurement. Resolution is usually limited by the number of bits used to quantize the input signal. For example, a 12-bit ADC can resolve to one part in 4096 (2 to the 12 power equals 4096).

resolution chart — 1. A pattern of black and white lines used to determine the resolution capabilities of equipment. 2. A chart used to examine the definition, linearity, and contrast of television systems.

resolution noise — The noise due to the stepped character of the resistance element in wirewound potentiometric transducers.

resolution wedge — A narrow-angled, wedge-shaped pattern calibrated for the measurement of resolution. It is composed of alternate contrasting strips that gradually converge and taper individually to preserve equal widths along a line drawn perpendicular to the axis of the wedge.

resolver — 1. A means for resolving a vector into two mutually perpendicular components. 2. A transformer whose coupling between primary and secondary can be varied. 3. A small section with a faster access than the remainder of the magnetic-drum memory in a computer. 4. A device that separates or breaks up a quantity into constituent parts or elements. 5. An electromechanical transducing device that develops an output voltage proportional to the product of an input voltage and the sine of the shaft angle.

resolving cell — In radar, a volume in space whose diameter is the product of slant range and beam width and whose length is the pulse length.

resolving power — 1. The reciprocal beam width in a unidirectional antenna, measured in degrees. It may differ from the resolution of a directional radio system, since the latter is affected by other factors as well. 2. The ability of an optical instrument to distinguish closely spaced points in an optical image, small angles between light beams, or components of light beams with small wavelength differences.

resolving time — 1. The minimum time interval by which two events must be separated to be distinguishable. 2. In computers, the shortest time interval between trigger pulses for which reliable operations of a binary cell can be obtained.

resonance — 1. A circuit condition whereby the inductive and capacitive reactance (or impedance) components of a circuit have been balanced. In usual circuits, resonance can be obtained for only a comparatively narrow frequency band or range. 2. In a mechanical system, the frequency at which the maximum displacement occurs. 3. A condition that exists between an oscillating system coordinate and its maintaining periodic agency when a small amplitude of the periodic agency produces relatively large amplitudes of oscillation in the system. 4. Condition existing in a body when the frequency of an applied vibration equals the body's natural frequency. 5. The tendency of an electrical or mechanical system to vibrate or oscillate at a certain frequency.

resonance bridge — A four-arm alternating-current bridge normally used for measuring inductance, capacitance, or frequency. An inductor and a capacitor are both present in one arm, the other three arms being (usually) nonreactive resistors. The adjustment for balance includes the establishment of resonance for the applied frequency. Two general types — series or parallel — can be distinguished, depending on how the inductor and capacitor are connected.

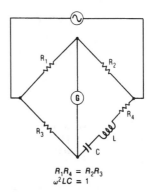

$$R_1 R_4 = R_2 R_3$$
$$\omega^2 LC = 1$$

Resonance bridge (series).

resonance characteristics — *See* resonance curve.

resonance curve — Also called a resonant curve or resonance characteristic. A graphical representation of how a tuned circuit responds to the various frequencies at and near resonance.

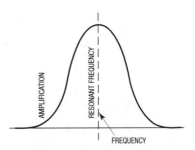

Resonance curve.

resonance indicator — A meter, neon lamp, headphone, etc., that indicates when a circuit is at resonance.

resonance radiation — Radiation from a gas or vapor due to excitation and having the same frequency as the exiting source (e.g., sodium vapor irradiated with sodium light).

resonant capacitor — A tubular capacitor that is purposely wound so as to have inductance in series with its capacitance to resonate at a predetermined (IF) frequency. Used as a bypass capacitor in amplifiers for more effective bypassing.

resonant cavity — 1. A form of resonant circuit in which the current is distributed on the inner surface of an enclosed chamber. By making the chamber of the proper dimensions, it is possible to give the circuit a high Q at microwave frequencies. The resonant frequency can be changed by adjusting screws that protrude into the cavity, or by changing the shape of the cavity. 2. A metal box, cylindrical or rectangular, of such dimensions that it will support electromagnetic oscillations if excited at a microwave frequency.

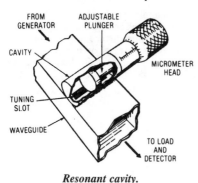

Resonant cavity.

resonant-chamber switch — A waveguide switch in which a tuned cavity is placed in each waveguide branch; detuning of a cavity prevents the flow of energy in the associated branch.

resonant charging choke — A modulator inductor that sets up an oscillation of a given charging frequency with the effective capacitance of a pulse-forming network in order to charge a line to a high voltage.

resonant circuit — A circuit that contains both inductance and capacitance and is therefore tuned to resonance at a certain frequency. The resonant frequency can be raised or lowered by changing the inductance and/or capacitance values.

resonant-circuit-type frequency indicator — A frequency-indicating device that depends for its operation on the frequency-versus-reactance characteristics of two series-resonant circuits. The circuit is arranged so that the deflecting torque is independent of the amplitude of the signal to be measured.

resonant current step-up — The ability of a parallel-resonant circuit to circulate a much higher current through its inductor and capacitor than the current fed into the circuit.

resonant curve — *See* resonance curve.

resonant diaphragm — In waveguide technique, a diaphragm so proportioned that it does not introduce reactive impedance at the design frequency.

resonant frequency — 1. The frequency at which a given system or object will respond with maximum amplitude when driven by an external sinusoidal force of constant amplitude. For an LC circuit, the resonant frequency is determined by the formula

$$f = \frac{1}{2\pi\sqrt{LC}}$$

where f is in hertz, L is in henrys, and C is in farads. 2. The frequency of a crystal unit for a particular mode of vibration to which, discounting dissipation, the effective impedance of the crystal unit is zero. 3. The frequency at which the total inductive and capacitive reactances of components in a circuit are equal. This results in the circuit impedance being equivalent to a pure resistance.

resonant gap — The small region where the electric field is concentrated in the resonant structure inside a tr tube.

resonant gate transistor — A surface field-effect transistor incorporating a cantilevered beam that resonates at a specific frequency to provide high-Q frequency discrimination.

resonant line — A transmission line in which the distributed inductance and capacitance are such that the line is resonant at the frequency it is handling.

resonant-line oscillator — An oscillator in which one or more sections of transmission line are employed as tanks.

resonant mode — In the response of a linear device, a component characterized by a certain field pattern and, when not coupled to other modes, representable as a single-tuned circuit. When modes are coupled together, the combined behavior is similar to that of single-tuned circuits that have been correspondingly coupled.

resonant-reed relay — 1. A relay with multiple contacts, each actuated by an ac voltage of the frequency at which the reeds resonate. 2. A relay that operates in response to signals of the proper frequency, power level, and time duration. Usually used for multicircuit control.

resonant resistance — The resistance value to which a resonant circuit is equivalent.

resonant voltage step-up — The ability of an inductor and a capacitor in a series-resonant circuit to deliver a voltage several times greater than the input voltage.

resonant window — A parallel combination of inductive and capacitive diaphragms used in a waveguide structure so that transmission occurs at the resonant frequency and reflection occurs at other frequencies.

resonate — To bring to resonance — i.e., to maximize or minimize the amplitude or other characteristic of a steady-state quantity.

resonating cavity — A waveguide that is adjustable in length and terminates in a metal piston, diaphragm, or other wave-reflecting device at either or both ends. It is used as a filter, a means of coupling between guides of different sizes, or an impedance network.

resonator — An apparatus or system in which some physical quantity can be made to oscillate by oscillations in another system.

resonator cavity — A section of coaxial line or waveguide completely enclosed by conductive lines.

resonator grid — A grid attached to a cavity resonator in a velocity-modulated tube to couple the resonator and the electron beam.

resonator mode — A condition of operation corresponding to a particular field configuration for which the electron stream introduces negative conductance into the coupled circuit.

resonator wavemeter — A resonant circuit for determining wavelength (e.g., a cavity-resonator wavemeter).

resource — 1. Assets of a computer system that the operating system can use and/or allocate to tasks for their use. Assets such as memory, disk storage space, printers, and terminals, as well as processors in multiprocessing systems, are typical system resources. 2. Any device or item used by a computer, including special areas of memory such as buffers.

resource-sharing — The sharing of one central processor by both several users and several peripheral devices. Principally used in connection with the sharing of time and memory.

responder — The part of a transponder that automatically transmits a reply to the interrogator-responser. By contrast, the responser is the receiver that accepts and interprets the signals from the transponder.

response — 1. A quantitative expression of the output of a device or system as a function of the input, under conditions that must be explicitly stated. The response characteristic, often presented graphically, gives the response as a function of some independent variable such as frequency or direction. 2. The output signal of a filter, referenced to the input, or excitation, signal. It is used as a measure of filter performance. Usually a particular type of response is of interest, such as impulse response, forced (steady-state) response, or transient response. 3. The speed at which a device can be cycled. 4. The fidelity with which the output waveform of a device corresponds to the input waveform.

response curve — 1. A plot of output versus frequency for a specific device. 2. A plot of stimulus versus output. 3. A graphical representation of frequency response. Usually measured in decibels, with reference to a given level on a vertical scale. When the response curve of an amplifier, pickup, microphone, etc., is accurately plotted it represents the relative levels of amplitude at all frequencies within a specified bandwidth.

responser — *See* responsor.

response range — The range over which a stepper motor can start, stop, or reverse without missing a step.

response rate — The maximum stepping rate of a stepping motor at which an unloaded stepper can run from a standing start without missing a step.

response speed — The time for a control action to start after a temperature change has occurred at the sensor.

response time — 1. The time (usually expressed in cycles of the power frequency) required for the output voltage of a magnetic amplifier to reach 63 percent of its final average value in response to a step-function change of signal voltage. 2. The time required for the pointer of an instrument to come to apparent rest in its new position after the measured quantity abruptly changes to a new, constant value. 3. *See* recovery time, 8. 4. The time required to reach a specified percentage (e.g., 90, 98, 99 percent) of the final output value. 5. *See* transient recovery time. 6. The elapsed time between generation of an inquiry at a computer terminal and receipt of a response at the same terminal. This includes the time for transmission to the computer, processing at the computer, and transmission back to the terminal. 7. The time it takes for a device to respond to an input signal. 8. The elapsed time from the entry of a command until its execution is complete. This term is usually used to represent the time it takes for a time-sharing system to respond to a user command line. 9. The interval between a stated step change in the line voltage or load current and the restoration of the output voltage of a power supply to its normal operating range. 10. The time a system requires to respond to an operator command in supplying stored data or completing a processing cycle.

responsitivity — 1. In a photosensor, the ratio of the change in photocurrent to the change in incident radiant flux density. 2. A measure of how much output current can be obtained from a photodetector for a given optical energy input.

responsor — Also spelled responser. 1. The receiver used to receive and interpret the signals from a transponder. 2. An electronic device used to receive an electronic challenge and to display a reply thereto.

restart — 1. To reestablish performance of a computer routine, using the information recorded at a checkpoint. 2. Resuming execution of an interrupted program at the point of interruption, commonly done by taking status checkpoints during execution and then specifying during rerun the particular checkpoint at which to resume.

resting frequency — *See* center frequency, 1.

resting potential — The voltage (typically about 80 millivolts) between the inside and outside of a nerve cell or muscle cell.

restore — 1. To do periodic charge regeneration of a volatile computer storage system. 2. In computers, to regenerate. 3. In a computer, to return a cycle index or variable address to its initial value. 4. To store again.

restorer — *See* dc restorer.

restorer pulses — In a computer, pairs of complement pulses applied to restore the charge of the coupling capacitor in an ac flip-flop.

restoring spring — Also called return spring. A spring that moves the armature to the normal position and holds it there when the relay is deenergized.

restoring torque — *See* holding torque.

rest potential — The residual potential difference remaining between an electrode and an electrolyte after the electrode has become polarized.

restrictor — Equipment for insertion in an outgoing telephone line that counts the digits dialed and restricts calls to forbidden codes.

restriking voltage — *See* reignition voltage.

resultant — The effect produced by two or more forces or vectors.

retained image — Also called image burn. A change that is produced on the target of a television camera tube by a stationary light image and that results in the production of a spurious electrical signal corresponding to the light image for a large number of frames after the image is removed.

retard — In ignition timing, to set the ignition timing so that a spark occurs later or less degrees before top dead center.

retardation coil — 1. A high-inductance coil used in telephone circuits to permit passage of dc or low-frequency current while blocking audio-frequency currents. *See also* inductor. 2. An audio-frequency choke coil used to limit voice-frequency currents in a telephone circuit.

retarding-field oscillator — Also called a positive-grid oscillator. An oscillator tube in which the electrons move back and forth through a grid that is more positive than the cathode and plate. The frequency depends on the electron-transit time and sometimes on the associated circuit parameters. The field around the grid retards the electrons and draws them back as they pass through it in either direction. Barkhausen-Kurz and Gill-Morrell oscillators are examples of a retarding-field oscillator.

retarding magnet — Also called a braking magnet or drag magnet. A magnet used for limiting the speed of the rotor in a motor-type meter.

retard transmitter — A transmitter in which a delay is introduced between the time it is actuated and the time transmission begins.

retention time — The maximum time after writing into a storage tube that an acceptable output can be obtained by reading.

retentivity — That property of a material measured by the normal residual induction remaining after the removal of an applied magnetizing force corresponding to the saturation induction for the material.

retentivity of vision — The image retained momentarily by the mind after the view has left the field of vision. *See also* persistence of vision.

reticle — 1. A glass-emulsion or chrome plate having an enlarged image of a single IC pattern. The reticle in usually stepped and repeated across a chrome plate to form the master mask. 2. A pattern of intersecting lines, wires, filaments, or the like placed in the focus of the objective element of an optical system. This pattern is used for sighting and alignment of the system.

RETMA — Abbreviation for Radio-Electronics-Television Manufacturers Association, now the Electronic Industries Association (EIA).

retrace — *See* flyback, 2.

retrace blanking — The blanking of a television picture tube during vertical-retrace intervals to prevent the retrace lines from being visible on the screen.

retrace interval — *See* return interval.

retrace line — Also called the return line. The line traced by the electron beam in a cathode-ray tube as it travels from the end of one line or field to the start of the next line or field.

retrace time — *See* return interval.

retractile cord — A cord having specially treated insulation or jacket so that it will retract like a spring. Rectractability may be added to all or part of a cord's length.

retractile spring — The spring that tends to open the armature of an electromagnetic device and that holds the armature open when its force is not overcome by magnetic attraction.

retransmission unit — A control unit used at an intermediate station for automatically feeding one radio receiver-transmitter unit from another for two-way communications.

retrieve — In a computer, to obtain specific information from memory or a storage device.

retrodirective reflector — A reflector that redirects incident flux back toward the point of origin of the flux.

retrofit — 1. To fit an earlier system so it can be compatible with later technology. 2. The modification of existing equipment to incorporate recent developmental changes in design and use; derives from the term retroactive refitting.

retroreflective scan — The reflective scan technique that uses a special reflector (retroreflector) to return light along the same path it is sent.

retro zoom — A lens assembly designed to reduce the focal length range of a zoom lens.

retry — Repetition of search or read/write operations to recover from "soft" (correctable) errors.

return — 1. A received radar signal. 2. To go back to a planned point in a computer program and run part of the program again (usually because an error has been detected). 3. Instruction used to terminate a computer subroutine. It forces "return" from the subroutine. The next instruction to be executed is the one following the subroutine call.

return code — In a computer, a code used to influence the carrying out of following programs.

return-code register — In a computer, a register used for storage of a return code.

return interval — Also called retrace interval, retrace time, or return time. The interval corresponding to the direction of sweep not used for delineation.

return line — *See* retrace line.

return lines — Conductors connecting the loads to the lowest-potential power-supply terminal. Lowest potential means most nearly zero with respect to the ground point, regardless of polarity. The term ground is reserved for a single point; return lines include all means of connecting the low-potential terminal of the power supply to the load, such as ground buses and chassis.

return loss — 1. At a discontinuity in a transmission system, the difference between the power incident on, and the power reflected from, the discontinuity. 2. The ratio in decibels of (1) above.

return spring — *See* restoring spring.

return time — *See* return interval.

return trace — The path of the scanning spot in a cathode-ray tube during the return interval.

return transfer function — In a feedback control device, the transfer function that relates a loop return signal to the corresponding loop input signal.

return wire — The ground, common, or negative wire of a direct-current circuit.

reusable routine — A computer routine that can be used by more than one task.

reverb — Contraction for reverberation. An electronic sound effect similar to echo, used to create a fuller sound or to duplicate the ambience of a room.

reverberation — 1. The persistence of sound due to the repeated reflections from walls, ceiling, floor, furniture, and occupants in a room or auditorium. 2. A slight, tapering prolongation of sounds due to multiple reflections in a large auditorium. As distinguished from echo, which is (acoustically) a sudden return of sound rather than a smooth decay. 3. The act of sound or pressure waves being reflected by the surfaces of an enclosure.

reverberation chamber — An enclosure in which all surfaces have been made as sound-reflective as possible. It is used for certain acoustic measurements.

reverberation period — The time required for the sound in an enclosure to die down to one millionth (60 dB) of its original intensity.

reverberation strength — The difference between (*a*) the level of a plane wave that produces in a nondirectional transducer a response equal to that produced by the reverberation corresponding to a 1-yard (0.91-m) range from the effective center of the transducer and (*b*) the index level of the pulse transmitted by the same transducer on any bearing.

reverberation time — The time required for the average sound energy in an enclosure to decay by 60 dB, i.e., fall to one-millionth of its initial intensity. This time is directly proportional to the volume of the enclosure and inversely proportional to the total absorption in the enclosure. It should be less than 1 second for speech, and 2 to 3.5 seconds for a large concert orchestra.

reverberation-time meter — An instrument for measuring the reverberation time of an enclosure.

reverberation unit — A circuit or device that adds an artificial echo to a sound being reproduced or transmitted.

reverberator — Any of various electromechanical devices that process an audio signal in such a way as to simulate the effects of reverberation.

reversal — A change in the direction of transmission or polarity.

reverse — As distinguished from high-speed rewind, a distinct function that allows a tape recorder to change head configuration and tape direction at the end of a tape so as to continue playing (or recording) on the reverse tracks without switching reels or turning over the cassette.

reverse bias — Also called back bias. 1. An external voltage applied to a semiconductor pn junction to reduce the current across the junction and thereby widen the depletion region. It is the opposite of forward bias. 2. The polarity of external voltage applied to a semiconductor junction that tends to reinforce the internal potential barrier set up in equilibrium, resulting in either a marked reduction or complete extinction of the diffusion currents.

reverse-blocking diode thyristor — A two-terminal thyristor that for negative anode-to-cathode voltage does not switch, but exhibits a reverse-blocking state.

reverse-blocking pnpn-type switch — A pnpn-type switch that exhibits a reverse-blocking state when its anode-to-cathode voltage is negative and does not switch in the normal manner of a pnpn-type switch.

reverse-blocking state — The condition of a reverse-blocking thyristor corresponding to the portion of the anode characteristic for reverse currents of lower magnitude than the current at the reverse breakdown voltage.

reverse-blocking thyristor — A thyristor that, for negative anode voltage, cannot be made conductive, i.e., it exhibits a reverse-blocking state.

reverse-blocking triode thyristor — A three-terminal thyristor that for negative anode-to-cathode voltage does not switch, but exhibits a reverse-blocking state. This is the device that is often known in the power field as an SCR (semiconductor-controlled rectifier).

reverse-breakdown voltage — The voltage that produces a sharp increase in reverse current in a semiconductor, without a significant increase in voltage.

reverse channel — A technique for providing means of simultaneous communication from the receiver to the transmitter on two-wire transmission facilities. It is an optional feature of modems, intended to facilitate certain kinds of error control.

reverse-conducting diode thyristor — 1. A two-terminal thyristor that for negative anode-to-cathode voltage does not switch, but conducts large currents at voltages comparable in magnitude to the on-state voltage. 2. A diode thyristor that is inherently conductive when the anode voltage is negative.

reverse-conducting triode thyristor — 1. A three-terminal thyristor that for negative anode-to-cathode voltage does not switch, but conducts large currents at voltages comparable in magnitude to the on-state voltage. 2. A triode thyristor that is inherently conductive when the anode voltage is negative.

reverse coupler — A directional coupler used for sampling reflected power.

reverse current — *See* back current.

reverse-current relay — A relay that operates whenever current flows in the reverse direction.

reversed feedback — *See* negative feedback.

reversed feedback amplifier — An amplifier in which inverse feedback is employed to reduce harmonic distortion and otherwise improve fidelity.

reverse direction — Also called inverse direction. The direction of greater resistance to current through a diode or rectifier.

reverse-direction flow — In flowcharting, a flow in a direction other than from left to right or top to bottom.

reverse emission — *See* back emission.

reverse gate-to-source breakdown voltage — The breakdown voltage between the gate and source terminals of an insulated gate field-effect transistor with a reverse gate-to-source voltage applied, and all other terminals short-circuited to the source terminal.

reverse gate voltage — The positive gate-to-anode voltage for n-gate thyristors. The negative gate-to-cathode voltage for p-gate thyristors.

reverse key — A key used in a circuit to reverse the polarity of that circuit.

reverse leakage current — The current through a device when a voltage of polarity opposite to that normally specified is impressed across the device. The term is commonly used with electrolytic capacitors.

reverse open-circuit voltage amplification factor — In a transistor, the ratio of incremental values of input voltage to output voltage measured with the input ac open-circuited.

reverse-phase current relay — Also called phase-balance current relay. A device that functions when the polyphase currents are of reverse-phase sequence, or when the polyphase currents are unbalanced or contain negative phase-sequence components above a given amount.

reverse recovery time — In a semiconductor diode, the time required for the current or voltage to reach a specified state after being switched instantaneously from a specified reversed bias condition.

reverse resistance — The resistance measured at a specified reverse voltage or current in a diode or rectifier.

reverse saturation current — The reverse current that flows in a semiconductor because of a specified reverse voltage.

reverse voltage — 1. The voltage applied in the reverse direction to a diode or rectifier. 2. In the case of two opposing voltages, the voltage with the polarity that results in the smaller current. 3. The voltage applied to a semiconductor diode or rectifier diode that causes the respective current in the reverse direction.

reversible booster — A booster capable of adding to or subtracting from the voltage of a circuit.

reversible capacitance — For a capacitor, the limit, as the amplitude of the applied sinusoidal voltage approaches zero, of the ratio of the amplitude of the in-phase, fundamental-frequency component of transferred charge to the amplitude of the applied voltage, with a specified constant bias voltage superimposed on the sinusoidal voltage.

reversible capacitance characteristic — The function giving the relation of reversible capacitance to bias voltage.

reversible counter — *See* up/down counter.

reversible motor — A motor in which the rotation can be reversed by a switch that changes the motor connections.

reversible permeability — Also called swingback permeability. 1. The limit approached by the incremental permeability as the alternating field strength approaches zero. 2. The slope of the hysteresis loop at the residual induction. For a permanent magnet when the induction is increased, the operating point (B_d, H_d) does not return along the demagnetization curve but moves along a line having the slope.

reversible transducer — *See* bilateral transducer.

reversing rate — The maximum stepping rate at which a stepper motor can reverse direction while maintaining synchronism with the field.

reversing switch — A switch used for changing the direction of any form of motion — specifically, the direction of motor rotation or the polarity of circuit connections.

reverting call — In telephony, a call made by one party on a line to another party on the same line.

revolving-lens fiber-optic scanner — A sequential scanning device, utilizing a revolving lens, in which the CRT image is transformed into a circle of fibers. The rotating lens focuses each fiber successively on a multiplier phototube.

rewind — To return the tape to its starting point in a magnetic recorder.

rewind control — A button or lever for rapidly rewinding magnetic recording tape from the take-up reel to the feed reel.

rework — To repeat one or more manufacturing operations for the purpose of improving the yield of acceptable parts.

rewrite — Also called regeneration. 1. In a storage device in which the information is destroyed by being read, the restoring of information into the storage. 2. To restore a binary cell prior to being read.

rf — Abbreviation for radio frequency. A frequency at which coherent electromagnetic radiation of energy is useful for communication purposes. Also, the entire range of such frequencies.

rf amplifier — An amplifier capable of operation in the radio-frequency portion of the spectrum.

rf bandwidth — The band of frequencies comprising 99 percent of the total radiated power extended to include any discrete frequency on which the power is at least 0.25 percent of the total radiated power.

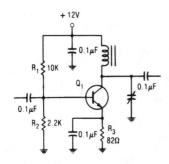

Bipolar transistor.

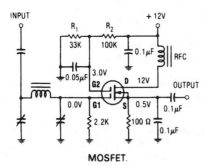

MOSFET.

Rf amplifiers.

rfc — Abbreviation for radio-frequency choke.

rf cavity preselector — An ultrahigh-frequency circuit component that is similar in function to a tuned resonant circuit. A tunable cavity.

rf choke — A coil designed to have a high inductive reactance at radio frequencies and used to prevent currents at these frequencies from passing from one circuit to another.

rf component — The portion of a signal or wave which consists only of the radio-frequency alternations, and not including its audio rate of change in amplitude or frequency.

rf connector — Connector used for connecting or terminating coaxial cable.

rf current — Alternating current having a frequency higher than 10,000 hertz.

rf energy — Alternating-current energy generated at radio frequencies.

RFG — Abbreviation for radar field gradient.

rf gain control — A manual control that sets the gain of a receiver. It is included on some receivers to supplement the AGC circuit. Some signals are simply too strong for the AGC to handle and will overload the receiver unless gain can be further reduced by this control.

rf generator — A generator that produces sufficient rf energy at its assigned frequency for induction or dielectric heating.

rf head — A unit consisting of a radar transmitter and part of a radar receiver, the two contained in a package for ready removal and installation.

RFI — Abbreviation for radio-frequency interference. Usually unintentionally radiated electromagnetic energy that may interfere with the operation of, or even damage, electronic equipment.

rf indicator — A device that shows the presence of rf energy. It may consist of a tuned circuit or parallel line connected to an incandescent lamp or other indicator.

rf interference shield ground — The grounding technique for all shields that are used to suppress the radiation of interference from leads.

rf intermodulation distortion — Intermodulation distortion that has its origin in the rf stages of a receiver.

RFI suppression — Radio-frequency-interference suppression. Generally consists of a frequency-discriminating element or circuit in the control. The purpose of such a circuit is to keep undesirable high-frequency energy waves generated by thyristor switching and arcing at the motor brushes from being conducted back into the supply conductors. If unattenuated, these waves will cause radio reception interference to units connected to the same power lines.

rf line — 1. A system of metallic tubes (waveguides and/or coaxial lines) that conduct radio-frequency energy from one point to another. 2. A metallic conductor used to transmit radio-frequency energy from one point to another.

rf mixer — A circuit that changes a high rf (radio frequency) to a lower IF (intermediate frequency) by mixing it with a local oscillator (LO) frequency.

rf oscillator — *See* radio-frequency oscillator.

rf pattern — A term used to describe a fine herringbone pattern in a picture that is caused by a high-frequency interference. This pattern may also cause a slight horizontal displacement of scanning lines, which results in a rough or ragged vertical edge on the picture.

rf plumbing — Radio-frequency transmission lines and associated equipment in the form of waveguides.

rf power supply — A high-voltage power supply consisting of an rf oscillator whose output voltage is stepped up and then rectified. Used in television receivers or other equipment to supply the high dc voltage required by the second anode of cathode-ray tubes.

rf preheating — *See* radio-frequency preheating.

rf preselectors — Bandpass filters that improve the selectivity by rejecting unwanted frequencies at the radio-frequency input state.

rf probe — 1. A resonant conductor that is placed in a waveguide or cavity resonator for the purpose of inserting or withdrawing electromagnetic energy. 2. A detecting device used with a voltmeter to measure rf voltages.

rf pulse — A radio-frequency carrier amplitude modulated by a pulse. The carrier amplitude is zero before and after the pulse. Coherence of the carrier with itself is not implied.

rf resistance — *See* high-frequency resistance.

rf shift — *See* frequency shift.

rf signal generator — Also called service oscillator. A test instrument that generates several bands of radio frequencies necessary for the alignment and servicing of radios, television, and other electronic equipment.

rf sputtering — A deposition process wherein a high-frequency potential is applied directly to a metal electrode behind the target. The target, which is an insulator, is bonded to the metal electrode, forming a capacitor. The insulator surface in contact with the plasma is alternately bombarded by electrons and positive ions during each rf cycle. When the surface is positive it attracts electrons; when it is negative it attracts ions. Since the electrons in the plasma have a higher mobility than the ions, the electron current to the target is initially much greater than the ion current. The cathode acts as a diode and charges the coupling capacitor to the peak value of the rf input voltage, then attains a negative bias.

rf tolerance — The amount of rf energy the human body can receive without injury.

rf transformer — *See* radio-frequency transformer.

RGB — Abbreviation for red, green, blue. 1. A video standard in which the color signals for red, green, and blue are carried on separate lines, then combined to form a color video picture. Horizontal and vertical sync are imposed on one of the colors, usually green. 2. Separate red, green, and blue video signals. When combined, they make up a complete color video image. The quality of the final image depends upon the size of the signals (4-bit, 8-bit, 16-bit or 24-bit). At 24-bit resolution, the image is considered "true-color" and, when displayed on a high resolution display, is photo-quality.

RG/U — Abbreviation for radio guide/universal. In MIL-C-17, RG is the military designation for coaxial cable, and U stands for general utility.

R/h — Abbreviation for roentgens per hour.

rheo — Abbreviation for rheostat.

rheoencephalography — *See* electrical-impedance cephalography.

rheostat — 1. A variable resistor that has one fixed terminal and a movable contact (often erroneously referred to as a two-terminal potentiometer). Potentiometers may be used as rheostats, but a rheostat cannot be used as a potentiometer because connections cannot be made to both ends of the resistance element. 2. An adjustable resistor so constructed that its resistance may be changed without opening the circuit in which it is connected. A liquid rheostat employs a conduction liquid, rather than a metal.

RHI — Abbreviation for range-height indicator. A radar display in which the abscissa represents the range to the target, and the ordinate indicates height.

rhombic antenna — An antenna composed of long wire radiators comprising the sides of a rhombus. The antenna usually is terminated in an impedance. The sides of the rhombus and the angle between them, the elevation, and the termination are proportioned to give the desired directivity.

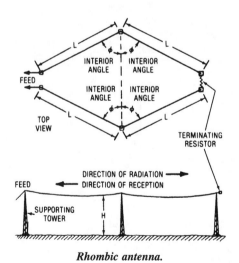

Rhombic antenna.

rho-theta system — 1. Any electronic navigation system in which position is defined in terms of distance (ρ) and bearing (θ) relative to a transmitting station. 2. A polar-coordinate navigational system providing sufficiently accurate data so that a computer can be used to provide arbitrary course lines anywhere within the coverage area of the system.

rhumbatron — A resonant cavity consisting of lumped inductance and capacitance. It is used, instead of circuits, to act as an oscillator capable of giving an output of several kilowatts at frequencies of several thousand megahertz.

rhythm bar — In some organs, a bar used to permit rhythmic playing of a chord without interrupting the pressure on the keys or chord button.

RIAA — Abbreviation for Recording Industry Association of America. The official association of the disc recording field.

RIAA curve — 1. A standard recording characteristic curve approved for long-playing records by the Recording Industry Association of America. 2. The equalization curve for playback of records recorded as in (1).

ribbon — A common intraconnecting material, usually nickel, of rectangular cross section, that is used to connect electronic component parts or modules together into the form of a functioning circuit.

ribbon cable — 1. Cable made of more than one conductor, laid parallel. 2. A flat cable of individually insulated conductors lying parallel and held together by means of adhesive or woven textile yarn. Structure is usually characterized by individual colors of insulation for each conductor, although a single color may be used for all conductors. 3. Round hookup cable with stranded conductors. Laid side by side, the cables are held together either by fusing or bonding of the insulation.

ribbon-cable connectors — Designed to terminate ribbon cable with round connectors. The term ribbon cable can refer to etched flexible-circuit flat cable, round-conductor laminated cable, round-conductor extruded cable, or bonded round-conductor cable. The word cable is usually omitted.

ribbon contact connectors — A rectangular connector with a self-wiping contact. Not to be confused with ribbon-cable connector.

ribbon interconnect — A flat, narrow ribbon of metal such as nickel, aluminum, or gold used to interconnect circuit elements or to connect the element to the output pins.

ribbon microphone — 1. A microphone in which the moving conductor is in the form of a ribbon driven directly by the sound waves. 2. A microphone that uses a narrow corrugated aluminum alloy strip suspended in a magnetic field. Sound makes the strip vibrate in a direction perpendicular to the magnetic field, resulting in an ac current being induced in a coil. The natural response of a ribbon unit is bidirectional, and a ribbon microphone has certain areas of minimum sensitivity. Only a very small unit of response is produced by quite loud noises from the sides. The polar diagram for this type is known as "figure of eight." Some ribbon microphones have cardioid and hypercardioid responses. Widely used in studios.

ribbon tweeter — 1. A high-frequency speaker, usually horn loaded, in which a stretched, straight flat ribbon is used instead of a conventional voice coil. The magnetic gap is a straight slit that can be made quite narrow so that a maximum amount of flux is concentrated in it. The ribbon serves both as an extremely light driven element and as a diaphragm. 2. A form of high-frequency driver using a light ribbon suspended in a magnetic field to generate sound when current is passed through it. In its basic form, a very high quality but fragile high-frequency driver.

Rice neutralizing circuit — A radio-frequency amplifier circuit that neutralizes the grid-to-plate capacitance of the amplifier tube.

Richardson effect — *See* Edison effect.

Richardson equation — An expression for the density of the thermionic emission at saturation current, in terms of the absolute temperature of the filament.

ride gain — To continually adjust the volume level of a program while observing a volume indicator so that the resulting audio-frequency signal will have the necessary magnitude for proper operation of the transmission equipment.

ridge waveguide — A circular or rectangular waveguide with one or more longitudinal ridges projecting inwardly from one or both sides. The ridges increase the transmission bandwidth by lowering the cutoff frequency.

Rieke diagram — A special polar-coordinate chart whereby load conditions can be determined for oscillators such as klystrons and magnetrons. Information from the chart indicates where optimum operation is located, as well as the limitations oscillators have with various loads.

rig — 1. A system of components. 2. An amateur station consisting of receiver, transmitter, and all the necessary accessory equipment.

Righi-Leduc effect — The phenomenon whereby when a metal strip is placed with its plane perpendicular to a magnetic field and heat flows through the strip, a temperature difference is developed across the strip.

right-hand rule — See Fleming's rule.

right-hand polarized wave — Also called clockwise polarized wave. An elliptically polarized transverse electromagnetic wave in which the electric intensity vector rotates clockwise as an observer looks in the direction of propagation.

right-hand taper — The characteristic whereby a potentiometer or rheostat has a higher resistance in the clockwise half of its rotational range then in its counterclockwise half (looking at the shaft end).

rigid disk — Disk storage wherein the medium is a magnetic alloy mounted on a thick metallic substrate. Rigid disks may take the form of nonremovable disks, which have the medium in a sealed container, and disk packs or disk cartridges, wherein the medium may be removed from the drive mechanism. Rigid disks have a capacity range from 5 M to greater than 200 M bytes.

rigid metal conduit — A raceway specially constructed for the purpose of the pulling in or the withdrawing of wires or cables after the conduit is in place. It is made of metal pipes of standard weight and thickness, permitting the cutting of standard threads.

rigid microdisk drive — See micro-Winchester drive.

rim drive — The method of driving a phonograph or sound-recorder turntable by means of a small, rubber-covered wheel that contacts the shaft of an electric motor and the rim of the turntable.

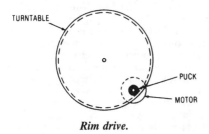

Rim drive.

rim magnet — See field-neutralizing magnet.

ring — 1. A ring-shaped contacting part of a plug, usually placed in back of but insulated from the tip. 2. An audible alerting signal on a telephone line. 3. Also called a circularly linked list or a cycle. In a computer, a special kind of linked list, in which the last item of a string of items points back to the first item.

ring-around — In a secondary radar: 1. The undesired triggering of a transponder by its own transmitter. 2. The triggering of a transponder at all bearings, causing a ring presentation on a PPI.

ring circuit — 1. In waveguide practice, a hybrid-T having the physical configuration of a ring with radial branches. 2. A communication network in the form of a ring. Opening any part of the ring will not interrupt communications to any node on the ring.

ring connection — Connection of a group of components or circuit elements in series, with the output of the last connected to the input of the first to form a closed ring.

ring counter — 1. A loop of interconnected bistable elements arranged so that only one is in a specified state at any given time. As input signals are counted, the specified state moves in an ordered sequence around the loop. 2. A device that can store several bits of information. A ring counter accepts shift instructions that cause all the information to shift one position at a time. If information is being shifted left in a register, the value of the leftmost bit shifts into the position of the rightmost bit of the register. Similarly, if the shift is to be to the right, the value of the rightmost bit shifts into the leftmost bit position in the register. In a ring counter, the information recycles every n shift pulses, where n is the number of bits in the ring counter.

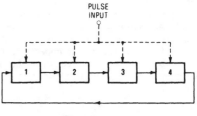

Block diagram.

PULSES	STAGE			
	1	2	3	4
0	ON			
1		ON		
2			ON	
3				ON
4	ON			
5		ON		
6			ON	
7				ON
8	ON			

Sequence.

Ring counter.

ringdown — A method of signaling subscribers and operators using either a 20-hertz ac signal, a 135-hertz ac signal, or a 1000-hertz signal interrupted 20 times per second.

ringer — The bell unit in a subscriber's telephone set.

ring head — A magnetic head in which the magnetic material forms an enclosure with one or more air gaps. The magnetic-recording medium bridges one of these gaps and contacts or is close to the pole pieces on one side only.

ringing — 1. The production of an audible or visible signal at a station or switchboard by means of an alternating or pulsating current. 2. A damped oscillation in the output signal of a system as a result of a sudden change in the input signal. 3. High-frequency damped oscillations caused by shock excitation of high-frequency resonances. 4. Transient decaying oscillation about high or low limits induced by unmatched impedance reflections. 5. Causing a telephone bell to ring, by application of ringing current to the line. 6. A damped oscillatory transient at or near the frequency of cutoff, caused by a sudden change in signal level. 7. In receivers, an oscillatory transient occurring in the output of a system as a result of a sudden change in input.

ringing code — A sequence of long and short rings that is used to signal a particular party on a party telephone line.

ringing current — 1. An alternating current that may or may not be superimposed onto a direct current for telephone ringing. 2. A 75- to 105-volt, 20-Hz ac voltage supplied by the central office to ring a telephone subscriber's bell.

ringing key — A key that, when operated, sends a ringing current over its circuit.

ringing out — The process of locating or identifying specific conductive paths by means of passing current through selected conductors.

ringing signal — Any ac or dc signal transmitted over a line or trunk for the purpose of alerting a party at the distant end of an incoming call. The signal may operate a visual or aural device.

ring magnet — A ceramic permanent magnet in which the axial length is no greater than the wall thickness, and the wall thickness is no less than 15 percent of the outside diameter.

ring modulator — 1. A modulator used as a balanced modulator, demodulator, or phase detector that has four diodes connected in series to form a ring around which current can easily flow in one direction. Input and output connections are made at the four nodal points of the ring. 2. A modulator consisting of four diodes connected in a closed square, or ring.

ring network — A network topology that connects its terminals in a loop or ring.

ring oscillator — A circuit configuration in which two or more pairs of tubes are operated as push-pull oscillators in a ringlike arrangement. Usually alternate successive pairs of plates and grids are connected to tank circuits, and the load is coupled to the plate circuits.

ring retard — Also called slug retard. A heavy conductor surrounding the iron flux path in a relay to retard the establishment or the decay of flux in the path.

ring time — In radar, the time during which the output of an echo box remains above a specified level. It is used in measuring the performance of radar equipment.

riometer — Acronym for relative ionospheric opacity meter. An instrument for recording the level of extraterrestrial cosmic noise at selected frequencies in the HF and VHF regions.

ripple — 1. That portion of the output voltage of a power supply harmonically related in frequency to the input power and to any internally generated switching frequency. Normally ripple is expressed as an rms percentage, but it can also be expressed as peak to peak. Ripple has been replaced by a new term, PARD, which includes hum and noise as well as spikes in the output. 2. The wavelike variations in the amplitude response of a filter. Ideally, Tchebychev and elliptic-function filters have characteristics such that the differences in peaks and valleys of the amplitude response in the passband are always the same. Butterworth, Gaussian, and Bessel functions do not have ripple. Ripple usually is measured in decibels. 3. The serial transmission of data; a serial reaction that may be compared to a bucket brigade or a row of falling dominoes. 4. The alternating superimposed

component of a unidirectional voltage applied to a capacitor. 5. Amplitude variations in the output voltage of a power supply caused by insufficient filtering.

ripple adder — A binary adding system in which the column of lowest order is added, the resulting carry is added to the column of the next highest order, and so on for all columns. It is necessary to wait for propagation of the signal even though all columns are present at the same time (parallel).

ripple counter — An asynchronously controlled counter; the clock is derived from a previous-stage output.

ripple current — The alternating component of a substantially steady current.

ripple-current rating — The rms value of the ac component of the current through a capacitor.

ripple filter — Also called smoothing circuit and smoothing filter. A low-pass filter designed to reduce the ripple current while freely passing the direct current from a rectifier or generator.

ripple frequency — The frequency of the ripple current. In a full-wave rectifier it is twice the supply frequency. In a generator it is a function of the speed and the number of poles.

ripple quantity — The alternating component of a pulsating quantity when this component is small relative to the continuous component.

ripple regulation — An efficient power supply regulating method in which the output ripple is sampled to maintain energy balance, hence voltage regulation, in an *LC* circuit.

ripple-through counter — *See* serial counter.

ripple voltage — The alternating component of a unidirectional voltage (this component is small relative to the continuous component).

RISC — Abbreviation for reduced instruction set computer. 1. A type of computer architecture that has a small, or reduced, set of instructions that execute very quickly. 2. A microprocessor that processes information faster by concentrating on the functions most often performed.

rise cable — In communication practice: 1. The vertical portion of a house cable extending from one floor to another. 2. Sometimes, any other vertical sections of cable.

risers — In a multilayer substrate, the conductive paths that vertically connect various levels.

rise time — 1. The time required for the leading edge of a pulse to rise from 10 percent to 90 percent of its final value. It is proportionate to the time constant and is a measure of the steepness of the wavefront. 2. The measured length of time required for an output voltage of a digital circuit to change from a low voltage level (0) to a high voltage level (1) after the change has started. 3. The time required for the pointer of an electrical indicating instrument to attain 90 percent (within a specified tolerance) of end-scale deflection following sudden application of constant electric power from a source with sufficiently high impedance so as not to influence damping (100 times the impedance of the instrument). 4. For a switching transistor, the time interval between the instants at which the magnitude of the pulse at the output terminals reaches specified lower and upper limits, respectively, when the transistor is being switched from its nonconducting to its conducting state. The lower and upper limits are usually 10 percent and 90 percent, respectively, of the amplitude of the output pulse. 5. The time taken for the radiant flux to increase from 10 percent to 90 percent of its peak value when a laser is subjected to a step function current pulse of specified amplitude. 6. The rate at which a signal changes from a logical 0 to a logical 1 (or from 1 to 0), usually expressed in volts per nanosecond.

rising-sun magnetron — A multicavity vane-type magnetron in which resonators of two different resonant frequencies are arranged alternately for the purpose of mode separation.

rising-sun resonator — A magnetron anode structure in which large and small cavities alternate around the perimeter of the structure.

risk — The probability of making the wrong decision based on pessimistic data or analysis.

RMA — Abbreviation for Radio Manufacturers Association, now the Electronic Industries Association (EIA).

RMA color codes — A term formerly used to designate the EIA color codes.

RMI — Abbreviation for radio-magnetic indicator.

R/min — Abbreviation for roentgens per minute.

R–Y signal — In color television, the red-minus-luminance color-difference signal. When combined with the luminance (Y) signal, it produces the red primary signal.

RMM — Abbreviation for read-mainly memory. A nonvolatile memory used much as a ROM or PROM except that the data contained therein may be altered through the use of special techniques (often involving external action) that are much too slow for read/write use.

RMOS — Abbreviation for refractory metal-oxide semiconductor. An MOS device that uses refractory metals like molybdenum instead of aluminum or silicon as the gate metal.

rms — Abbreviation for root-mean-square.

rms amplitude — Root-mean-square amplitude, also called effective amplitude. The value assigned to an alternating current or voltage that results in the same power dissipation in a given resistance as dc current or voltage of the same numerical value. The rms value of a periodic quantity is equal to the square root of the average of the squares of the instantaneous values of the quantity taken throughout one period. If the quantity is a sine wave, its rms amplitude is 0.707 of its peak amplitude.

rms PARD — The value of the output waveform of a regulated power supply, omitting the dc component value.

rms power — *See* continuous power.

rms pulse amplitude — Also called effective pulse amplitude. The square root of the average of the squares of the instantaneous amplitudes taken over the duration of the pulse.

rms value — The root-mean-square value of ac voltage, current, or power. Calculated as 0.707 of peak amplitude of a sine wave at a given frequency.

rms voltage — The effective value of a varying or alternating voltage. That value which would produce the same power loss as if a continuous voltage were applied to a pure resistance. In sine-wave voltages, the rms voltage is equal to 0.707 times the peak voltage.

RO — *See* receive only.

roadmap — A printed pattern of nonconductive material by which the circuitry and components are delineated on a board to aid in service and repair of the board.

roaming — Using a cellular phone in a location other than the one in which the phone is registered.

Roberts rumble — The nickname given to a phenomenon whereby certain radio disturbances appear to be connected with the passage of satellites through the earth's ionosphere. Investigations have been made to determine the feasibility of exploiting this effect for the tracking of satellites.

Robinson antenna — A microwave scanning antenna consisting of an astigmatic reflector and a feed system in which a parallel-plate region is fed by a waveguide. The parallel-plate region is made so that the feed waveguide end is circular, to permit rotation of the feed guide, and is in approximately the same plane as the output end, or larger aperture of the parallel-plate region.

robot — 1. A programmable, multifunction manipulator designed to move materials, parts, tools, or specialized devices through variable programmed motions for the performance of a variety of tasks. 2. Mobile, manipulative machine controlled remotely by a human operator. 3. A mechanism, fixed or mobile, possessing the ability to manipulate objects external to itself under the constant control of a human being, a computer, or some other external intelligence. 4. A machine devised to function in place of a living agent. In Gothic the word *robot* is akin to a word meaning inheritance; in German, to work. An old Slavic word that is equivalent is *rabota*; and in Czech and Polish *robota* means servitude or forced labor. 5. A "mechanical man," originally from Karel Capek's *R.U.R.* Often interchangeable with *android* (from whence came *Star Wars'* lovable droids) and *humanoid*, though the later creations generally are closer mimics of the human form (using artificial skin, for example) than the more obviously mechanical robots.

robot device — An instrumented mechanism used in science or industry to take the place of a human being. It may or may not physically resemble a human or perform its tasks in a human way, and the line separating robot devices from merely automated machinery is not always easy to define. In general, the more sophisticated and individualized the machine is, the more likely it is to be classed as a robot device.

robotics — A term that describes the discipline that designs and creates robot device structures and subassemblies.

robot pilot — *See* autopilot.

robot vision — The use of a vision system to provide visual feedback to an industrial robot. Based on the vision system's interpretation of a scene, the robot may be commanded to move in a certain way.

robustness — The ability of a computer program to withstand stresses (input quantity and quality) beyond the range for which it was designed.

Rochelle-salt crystal — A crystal made of sodium potassium tartrate. Because of its pronounced piezoelectric effect, it is used extensively in crystal microphones and phonograph pickups. Perfect Rochelle-salt crystals up to 4 inches (10.1 cm) and even more in length can be grown artificially.

rock — To move a control back and forth (or make other adjustments as necessary) in order to obtain the best alignment or on-station tuning.

rocking — Rotating the tuning control in a superheterodyne receiver back and forth while adjusting the oscillator padder near the low-frequency end of the tuning dial to obtain more accurate alignment.

rocky-point effect — Transient but violent discharges between electrodes in high-voltage transmitting tubes.

rod gap — A spark gap in which the electrodes are two coaxial rods, with ends between which the discharge takes place, cut perpendicularly to the axis.

roentgen — 1. A unit of exposure to radiation defined as the amount of gamma or X-rays that produce ions carrying one electrostatic unit of charge in 1 cubic centimeter of air that is surrounded by an infinite mass of air at standard temperature and pressure conditions; 1 roentgen equals 2.58×10^{-4} coulomb per kilogram. 2. A unit of exposure dose that is a measure of the ability of X-rays or gamma rays to produce ionization in air. Ionization is the creation of ion pairs — positively and negatively charged parts of atoms — by the impact of radiation on those atoms. One roentgen of radiation has the ability to produce an amount of ionization that represents the absorption of approximately 83 ergs of energy from radiation per gram of air.

roentgen densitometer — A device for recording changes in concentration of a radiopaque indicator injected into circulation for evaluating circulatory function.

roentgen equivalent man — Abbreviated rem. 1. A radiation-exposure dose that produces the same effects on human tissue as one roentgen of X-ray radiation. 2. The product of the absorbed dose at the point of interest in a tissue, multiplied by modifying factors. Indicates the effect on a given organ of a radiation dose absorbed in that organ. 3. May be thought of as an abbreviation for radiation effect, man. Its meaning has undergone considerable change since first conceived. It now represents the absorbed dose of any radiation that has the same biological effect as a rad of "standard" X-rays. (Since various radiations such as alpha, beta, and gamma rays and neutrons have different biological effects per rad of absorbed energy, they are ascribed a relative biological effectiveness, or rbe.) Using this concept, the number of rem equals rads × rbe. 4. A measure of physiological damage to the human body, calculated as the absorbed dose in rads multiplied by various factors that qualify the type of radiation and the way in which it is absorbed.

roentgen equivalent physical — Abbreviated rep. An amount of ionizing radiation that results in an absorption of energy of approximately 83 to 93 ergs per gram of tissue.

roentgen meter — Also called a roentgenometer. An instrument for measuring the quantity or intensity of roentgen rays (X-rays or gamma rays).

roentgenogram — Also called an X-ray photograph or an X-ray. A photograph taken by showering an object or the human body with X-rays (roentgen rays). Depending on the transparency of the object or body, the interior can thus be seen and recorded.

roentgenology — That branch of science related to the application of roentgen rays (X-rays) for diagnostic or therapeutic purposes.

roentgenometer — *See* roentgen meter.

roentgen rays — *See* X-rays.

roger — A code word used in communications to mean: 1. Your message has been received and is understood. 2. OK — an expression of agreement.

Roget spiral — A helix of wire that contracts in length when a current is sent through, as a result of the mutual attraction between adjacent turns.

roll — 1. Also called flip-flop, especially when intermittent. The upward or downward movement of a television picture due to lack of vertical synchronization. 2. The process in which alphanumeric text moves across a CRT screen to the left or right. As a character disappears on one end of the screen, a new character appears at the other end. Roll can also be extended to include nontextual graphical constructions, although this is more properly called translation.

rollback — *See* rerun, 1.

roll bonding — *See* yield-strength-controlled bonding.

roll in — In a computer, to restore in main storage information previously transferred from main to auxiliary storage.

rolling transposition — The method by which two or more conductors of an open-wire circuit are spiral-wound. With two wires, a complete transposition can be executed utilizing two consecutive suspension points.

roll mode — A digital scope display feature whereby new information acquired by the scope constantly updates the screen display. The effect is similar to that of a strip chart recorder.

rolloff — 1. A gradual increase in attenuation over a range of frequencies; sometimes called slope. 2. An attenuation that varies with frequency, generally increasing at a constant rate beyond the corners of the amplitude-frequency characteristic of a system. 3. The rate of attenuation of a filter at the bass end (high-pass) or treble end (loss-pass). The crossover frequency is that frequency where the response is — 3 dB of the midfrequency response. The rolloff of an amplifier is similarly defined but in terms of decrease in amplification. 4. A gradual increase in attenuation of a signal voltage.

roll out — To read out of a computer storage by simultaneously increasing by 1 the value of the digit in each column, repeating this r times (where r is the radix) and, at the instant the representation changes from $(r - 1)$ to 0, either generating a particular signal, terminating a sequence of signals, or originating a sequence of signals.

rollover indexing — In a calculator, allows depression of a second key before releasing first key.

ROM — Abbreviation for read-only memory. 1. A memory in which the binary information located at each address is fixed and cannot be changed subsequently. Permanently stores information repeatedly used, such as tables of data, characters for electronic displays, etc. In its virgin state, the ROM consists of a mosaic of undifferentiated cells. One type of ROM is programmed by mask pattern as part of the last fabrication stage. Another popular type, known as PROM, is programmable in the field with the aid of programmer equipment. Programmed data stored in ROMs are often called firmware. 2. A source of permanent data that is not erasable or changeable, and is not lost when power is removed from the system. The use of ROM signifies a program that is mask-programmed right at the semiconductor manufacturer. Data is permanently stored when the device is manufactured and cannot be altered. Newer memories, called PROMs, employ techniques that allow the device to be manufactured without programmed data, and allow the user access to the PROM. The user can thus convert the original blank into a ROM programmed to the custom requirements of the user. Once the program patterns have been burned into the memory, they become permanent and cannot be changed. 3. Nonvolatile memory that will not be lost if power is removed. It may be factory programmable (ROM), field programmable (PROM), field programmable and erasable with ultraviolet exposure (EROM), or field programmable with electrical erase (EEROM or EAROM). It is most often randomly accessible. 4. A memory assembly wherein the contents have been preprogrammed in a manner that precludes modification. Used where a specific routine is to be used but never altered.

Romex cable — A moisture-resistant, flame-resistant, flexible cable that contains one or more wires.

roof filter — A low-pass filter used in a carrier telephone system to limit the frequency response to that needed for normal transmission, thereby blocking unwanted high-frequency interference induced in the circuit by external sources. Use of such a filter gives improved runaround crosstalk suppression and minimized high-frequency singing.

room acoustics — The quality of a room that affects how sounds will be heard in it. Room acoustics are a function of the room's size, geometry, structural materials, and furnishings. A live room is one in which sounds are fairly reverberant; a dead room is one in which sounds are fairly absorbed.

room noise — *See* ambient noise.

root mean square — Abbreviated rms. The square root of the average of the squares of the values of a periodic quantity taken throughout one complete period. It is the effective value of a periodic quantity.

root-mean-square amplitude — *See* rms amplitude.

root-mean-square current — The alternating value that corresponds to the direct current value that will produce the same heating effect.

root-mean-square power — *See* continuous power.

root-mean-square value — 1. Of alternating currents and voltages, the effective current or voltage applied. It is that value of alternating current or voltage that produces the same heating effect as would be produced by an equal value of direct current or voltage. For a sine wave, it is equal to 0.707 times the peak value. 2. The dc voltage or current that will generate, in a resistive circuit, the same amount of energy (heat) as the ac waveform; its amplitude is the square root of the sum of the squares of all the instantaneous amplitudes.

root segment — 1. In a computer, the segment of an overlay program that remains in main storage throughout the execution of the overlay program. 2. The first segment in an overlay program.

root-sum square — The square root of the sum of the squares. A common expression of the total harmonic distortion.

rope — Similar to chaff, but longer. Electromagnetic-wave reflectors used to confuse enemy radar. They consist of long strips of metal foil, to which small parachutes may be attached to reduce their rate of fall.

rope-lay conductor or cable — A cable consisting of one or more layers of helically laid groups of wires surrounding a central core.

rope-lay strand — A conductor made of multiple groups of filaments. A 7×19 rope lay strand has 19 wires laid into a group and then 7 such groups laid cabled into a conductor.

rosin connection — Also called a rosin joint. A defective connection of a conductor to a piece of equipment or to another conductor. Supposedly the joint is tightly soldered, but actually it is held together only by unburnt rosin flux.

rosin-core solder — 1. Self-fluxing solder consisting of a hollow center filled with rosin. 2. Hollow-wire form of tin-lead solder whose core is filled with pockets of rosin flux, so that the required flux is always supplied with the molten solder.

rosin flux — The mildest and least effective of solder fluxes. To increase rosin flux efficiency, small amounts of organic activating agents are added. Type RA, fully activated rosin flux, is the flux most commonly used for electrical connections.

rosin joint — *See* rosin connection.

rotary-beam antenna — A highly directional short-wave antenna system. It is mounted on a mast and can be rotated manually or by an electric-motor drive to any desired position.

rotary converter — *See* dynamotor.

rotary coupler — *See* rotating coupler.

rotary dial — 1. In a switched telephone system, the conventional dialing method that creates a series of pulses to identify the called station. 2. A rotary mechanism with a ten-hole finger wheel that, when wound up and released, causes pulsing contacts to interrupt the line current and operate central telephone office selecting equipment in accordance with the digit (1 through 0) dialed.

rotary generator (induction-heating usage) — An alternating-current generator adapted to be rotated by a motor or other prime mover.

rotary joint — *See* rotating coupler.

rotary phase converter — A machine that converts power from an alternating-current system of one or more phases to an alternating-current system of a different number of phases, but of the same frequency.

rotary plunger relay — A relay in which the linear motion of the plunger is converted mechanically into rotary motion.

rotary relay — 1. A relay in which the armature rotates to close the gap between two or more pole faces (usually with a balanced armature). 2. A term sometimes used for stepping relay.

rotary-solenoid relay — A relay in which the linear motion of the plunger is converted into rotary motion by mechanical means.

rotary spark gap — A device used to produce periodic spark discharges. It consists of several electrodes that are mounted on a wheel and rotate past a fixed electrode.

rotary stepping relay — *See* stepping relay.

rotary stepping switch — *See* stepping relay.

rotary switch — An electromechanical device that is capable of selecting, making, or breaking an electrical circuit. It is actuated by a rotational torque applied to its shaft.

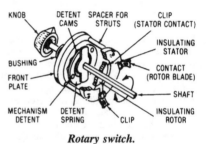

Rotary switch.

rotary transformer — A term sometimes applied to a rotating machine used to transform direct-current power from one voltage to another

rotary-vane attenuator — A device designed to introduce attenuation into a waveguide circuit through variation of the angular position of a resistive material placed in the guide.

rotary variable capacitor — A capacitor with a rotatable electrode that enables the capacitance to be varied continuously over its complete range.

rotary voltmeter — *See* generating voltmeter.

rotatable phase-adjusting transformer — A transformer in which the secondary voltage may be adjusted to have any desired phase relation with the primary voltage by mechanically orienting the secondary winding with respect to the primary. The latter winding consists usually of a distributed symmetrical polyphase winding and is energized from a polyphase circuit. *See also* phase-shifting transformer.

rotate — A computer instruction that causes the bits in a word to be shifted a certain number of places left or right. The bits that get "pushed off the end" reappear at the other end of the word. For instance, 1011011 rotated two places to the right is 1110110.

rotating-anode tube — An X-ray tube in which the anode rotates continually to bring a fresh area of its surface into the beam of electrons. This procedure allows a greater output without melting the target.

rotating coupler — Also called rotary coupler and rotary joint. A joint that permits one section of a waveguide to rotate while passing rf energy.

rotating cylinder scanner — Type of facsimile transmitter in which the original is mounted around a cylindrical drum and scanned by an optics/photocell assembly (the scan head) parallel to the length of the cylinder. The drum rotates and the scan head moves across the document, scanning one line width per revolution.

rotating disc — *See* drum memory.

rotating element of a meter — *See* rotor of a meter.

rotating field — The magnetic field in the stator of induction motors. Because of excitation from a polyphase source, the field appears to rotate around the stator from pole to pole.

rotating helical aperture scanner — A type of facsimile transmitter in which the original is roller-fed over a flatbed copy platen and illuminated by an area lamp. Lens and mirror optics reflect and focus one scan line of the moving document at a time first through a fixed horizontal-slit aperture, and then through a rotating helical aperture. The rotating of the helical aperture produces one lateral scan of the original per revolution, with the passing light being focused a final time onto a photocell for conversion to an electrical signal.

rotating joint — A device that permits one section of a transmission line to rotate continuously with respect to the other while still maintaining radio-frequency continuity.

rotating radio beacon — A radio transmitter that rotates a concentrated beam horizontally at a constant speed. Different signals are transmitted in each direction so that ships and aircraft without directional receiving equipment can determine their bearings.

rotational life — The ability of a potentiometer to be operational after a defined number of wiper sweeps across the element.

rotational wave — *See* shear wave, 1.

rotation spectrum — An X-ray spectrum of the diffraction pattern obtained when X-rays are sent through a rotating crystal.

rotator — 1. A motor-driven assembly that turns an antenna so that it can be aimed in the direction of best reception. 2. In waveguides, a means of rotating the plane of polarization. In rectangular waveguides it is done by simply twisting the guide itself.

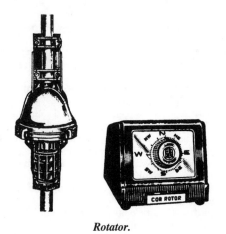

Rotator.

rotoflector — In radar, an elliptically shaped rotating reflector used to divert a vertically directed radar beam at right angles so that it radiates horizontally.

rotor — 1. The rotating member of an electric machine. In a motor, it is connected to and turns the drive shaft. In a generator, the rotor is turned to produce electricity by cutting magnetic lines of force. 2. The movable plates of a variable capacitor. 3. *See* receive-only typing reperforator.

rotor contact resistance — The resistance between the rotor contact terminal of a variable capacitor and the rotor shaft.

rotor of a meter — Also called the rotating element of a meter. The portion driven directly by electromagnetic action.

rotor plates — The movable plates of a variable capacitor.

round — To adjust the least significant digits retained in truncation to partially reflect the dropped portion. For example, when rounded to three digits the decimal number 2.7561 becomes 2.76.

round chart recorder — Data recorder that provides a record in the form of a graph on a circular piece of paper. Typical are recorders that give a record of temperature and/or humidity over a 24-hour period. Single-pen types measure a single variable; multipen models record more than one variable on a single chart.

round conductor — A solid or stranded conductor with a substantially circular cross section.

rounding — A lack of a sharp corner of a waveform, or a smooth transition from the leading or trailing edge to the limiting final value.

rounding error — The error that results when the less significant digits of a number are dropped and the most significant digits are then adjusted.

round off — Also known as truncation. To delete less significant digits from a number and possibly apply some rule of correction to the part retained. For example, if the discarded part is 5, 6, 7, 8, or 9 (or 50. . ., etc.), the new final digit is raised by 1 (e.g., 30.7 would be rounded off to 31, and 519.2 to 519). This is done for ease of calculation, where an estimate will suffice.

round-off error — *See* rounding error.

round-trip echoes — Multiple-reflection echoes produced when the radar pulse is reflected from a target strongly enough that the echo is reflected back to the target, where it produces a second echo.

round up — In a calculator, the last digit displayed in an answer is increased by 1 if the following digit would have been a 1 or greater.

route — A path from a signaling point to a destination.

routed wiring — Wiring done in channels or patterns characterized by long, parallel runs. Used when it is necessary to keep wiring close to the ground plane. Some automated wiring systems perform routed wiring. *See also* point-to-point wiring.

router — 1. A special-purpose computer (or software package) that handles the connection between two or more networks. Routers look at the destination addresses of the packets passing through them and select the appropriate routes on which to send the packets. 2. A program that automatically determines the routing path for the component connections on a printed circuit board. 3. Network node, which, alone or in tandem with other routers, ensures the delivery of a message from a station on one network to a station on another network. The sending station message includes the address of the receiving station and the address of the local router. The local router determines the best route to the receiving station, which may or may not involve intermediate routers. 4. Program that automatically determines the routing path for the component connections on a printed circuit board.

routine—A set of computer instructions arranged in a correct sequence and used to direct a computer in performing one or more desired operations.

routine library—An ordered set of standard and proven computer routines that may be used to solve problems or parts of problems.

routing—1. The assignment of the communications path by which information is carried to its destination. 2. A sequence of passing packets through various store-and-forward packet switches in a network to the desired destination. 3. Placement of interconnections on a printed circuit board. 4. In production, the sequence of steps to be performed in the production of a part or assembly. 5. The process of selecting the correct circuit path for a message.

routing indicator—An address, or group of characters, used in the header of a message to specify the final circuit or terminal to which the message is to be delivered.

routing tables—Customer-defined tables that describe the sequence of lines a telephone call may select to reach its destination. Routing tables also establish the points at which a call may hold or overflow to higher-cost lines.

row—1. A horizontal arrangement of a number of characters or other expressions. 2. In computers, the characters or corresponding bits of binary-coded characters that make up a word. 3. A path, perpendicular to the edge of a tape, along which storage of information may be accomplished by means of the presence or absence of holes or magnetized areas. 4. A predetermined number of consecutive functional patterns lying along a line parallel to the X axis of a photomask.

row binary—Having to do with the binary representation of data on cards by a method in which adjacent positions in a row correspond to adjacent bits of data; for example, the representation of 80 consecutive bits of two 40-bit words may be contained in each row of an 80-column card.

row pitch—The distance between corresponding points in adjacent rows.

row scanning—Decoding technique that determines which key of a keyboard was pressed. Each row is scanned in turn by outputting a 1. The output on the columns is examined, resulting in identification of the key.

RPG—Abbreviation for report program generator. 1. A computer language that can be used on several types of computers. The language stresses complex output reports based on information that describes the input files, operations, and format. 2. A language designed with built-in logic to produce report-writing programs given input and output descriptions.

rpm—Abbreviation for revolutions per minute.

rps—Abbreviation for revolutions per second.

R-S flip-flop—A flip-flop having two inputs, designated R and S. At the application of a clock pulse, a 1 on the S input will set the flip-flop to the 1 or on state, and 1 on the R input will reset it to the 0 or off state. It is assumed that 1s will never appear simultaneously at both inputs.

RSSI—Abbreviation for received signal strength indicator. A dc signal from the IF amplifier of a receiver. Its magnitude represents the level of the rf input signal.

R-S-T flip-flop—A flip-flop having three inputs: R, S, and T. The R and S inputs produce states as described for the R-S flip-flop; the T input causes the flip-flop to change states.

RS-232C—1. A widely used interface standard for computers, printers, modems, and test equipment. Switching occurs via levels, with zero defined as $+3$ to $+9$ volts or more and 1 as -3 to -9 volts or less. The teletype current-loop standard, on the other hand, opens and closes a 20-mA (or 60-mA) circuit to generate code. 2. A serial interface standard for communicating with byte-oriented units, such as teletypewriters. 3. A de facto standard, originally introduced by the Bell System, for the transmission of data over a twisted-wire pair less than 50 feet (15.24 m) in length. It defines pin assignments, signal levels, etc., for receiving and transmitting devices. Other RS-standards cover the transmission of data over distances in excess of 50 feet. 4. A standard form for serial computer interfaces. 5. The industry standard for a 25-pin interface that connects computers and various forms of peripheral equipment, e.g., modems and printers. 6. A standard developed by the Electronics Industry Associations (EIA) specifying what signals and voltages will be used to transmit data from a computer to a modem. The full standard covers some 25 pins on the RS-232C plug interface found on a serial card, but most personal computers make use of only a handful of these (actually about seven fingers worth). (The "C" is frequently dropped when using this term.)

RS-232 port—A standardized serial port for connecting a computer to peripheral equipment such as a printer, mouse, scanner, or modem.

RTL—Abbreviation for resistor-transistor logic.

RTMA—Abbreviation for Radio-Television Manufacturers Association, now Electronic Industries Association (EIA).

RTS—Request to send. An RS-232C control signal between a modem and user's digital equipment that initiates the data-transmission sequence on a communication line.

RTS/CTS delay—*See* train time.

rtty—Abbreviation for radioteletype. A wireless method of communication whose end result is printed messages. The originator types his or her message on a typewriterlike device. The message is then sent over the air via his or her transmitting system, whereupon it is automatically typed in final readable form on the recipient's typewriterlike device.

rubber—A material that is capable of recovering from large deformations quickly and forcibly and can be, or already is, modified to a state in which it is essentially insoluble (but can swell) in boiling solvents, such as benzene, MEK, etc.

rubber banding—A technique for displaying a straight line that has one end fixed and the other end following a stylus or some input device.

rubber-covered wire—A wire with rubber insulation.

rubber ducky—Common term for the flexible rubber-covered antenna generally supplied with handheld radios.

ruby—A type of aluminum-oxide crystal used to produce one form of solid-state laser.

ruby laser—1. An optically pumped solid-state laser in which a ruby crystal produces an extremely narrow and intense beam of coherent red light. Used for localized heating and for light-beam communication. 2. The optically pumped, solid-state laser that uses sapphire as the host lattice and chromium as the active ion. The emission of the laser takes place in the red portion of the spectrum. *See* figure on page 661.

rubylith—Ulano Company's trade name for a laminate consisting of a thin, red "light-safe" stripping film with a heavier clear polyester backing, used to produce master artwork for thick-film patterns by scribing and peeling away portions of the red layer.

ruby maser—A maser that has a ruby crystal in the cavity resonator.

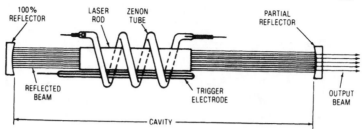

Ruby laser.

ruggedization—The redesign of a piece of equipment or its components to make them able to withstand prolonged vibration and mechanical shock.

Ruhmkorff coil—An induction coil having a magnetic interrupter. It is used to produce a spark discharge across an air gap.

rumble—1. Also called turntable rumble. A descriptive term for a low-frequency vibration that is mechanically transmitted to the recording or reproducing turntable and superimposed onto the reproduction. 2. Low-frequency noise caused by a tape transport. 3. The audible effect of low-frequency vibration transmitted from the motor or other moving parts to the record or the tonearm. Heard (as a hum or rumbling sound) only when the pickup stylus is on a rotating record. Rumble is measured in decibels below a specified signal level; the farther below the level (i.e., the larger the number), the less audible the rumble.

run—A single, continuous execution of a program by a computer.

runaround crosstalk—Crosstalk resulting from the coupling of the high-level end of one repeater to the low-level end of another repeater. Often a third repeater or line is the means of coupling; therefore, runaround crosstalk may be a form of interaction crosstalk.

runaway—1. Any additive condition to which continued exposure will eventually destroy a device. 2. A condition in which one of the dynamic variables of a system makes an unintended increase to a level beyond the design limits, often with destructive consequences.

run book—All the material needed to document a run of a program on a computer.

rung—A grouping of programmable controller instructions that controls one output. This is represented as one section of a logic ladder diagram.

run-in—The start of a groove at the beginning of a side of a phonograph record that runs in to the recorded section.

run motor—In facsimile equipment, a motor that supplies the power to drive the scanning or recording mechanisms. A synchronous motor is used to limit the speed.

running circuit breaker—A device whose principal function is to connect a machine to its source

of running voltage after having been brought up to the desired speed on the starting connection.

running open—In telegraph applications, a condition in which a machine is connected to an open line or a line without battery (constant space condition). Under this condition, the telegraph receiver appears to be running because the machine continually decodes the open line as the Baudot character "blank" or the U.S. ASCII character "null," and the type hammer continually strikes the type box but does not move across the page.

running torque—1. The turning power of a motor when running at its rated speed. 2. Force movement produced by a stepper motor after it has been accelerated to a running rate (sometimes also called slew or pull-out torque).

run time—1. The time required to complete a single continuous execution of an object program. 2. The time period during which a program is running.

runway localizing beacon—A small radio-range beacon that provides accurate directional guidance along the runway of an airport and for some distance beyond it.

rupture—The ability of contacts to break apart or rupture the electrical flow without welding under excessive currents.

rupture (or interrupting) capacity—The maximum current that a protective device will interrupt. Specified as the number of interruptions in amperes (adjusted circuit) without a change in calibration or a failure of dielectric strength.

rural line—A telephone subscriber's line in rural territory designed to serve from five to ten parties per line.

rush-box—A superregenerative receiver.

rutherford—Abbreviated rd. A quantity of radioactive material that produces one million disintegrations per second.

R value—A value of thermal resistance assigned to any given material of a given thickness or density, which is indicative of the ability of that material to resist or retard the flow of heat.

RX meter—An impedance meter that measures resistance and equivalent capacitance.

ryotron—A thin-film inductive superconductive device. An inductive switch capable of inductance variation of better than three orders of magnitude.

S

s — Letter symbol for second.

S — Symbol for secondary and source electrode.

SA — Abbreviation for signature analysis.

sabin — Also known as square-foot unit of absorption. A measure of the sound absorption of a surface. It is equivalent to 1 square foot of a perfectly absorptive surface. *See also* equivalent absorption.

saddle — Insulation placed under a splice in a coil lead.

safe operating area — Abbreviated SOA. The limit range in which a semiconductor device can be operated without damage. Boundaries of this area are maximum voltage, maximum current, and the secondary breakdown region.

safety — The conservation of human life and its effectiveness, and the prevention of damage to items, consistent with mission requirements.

safety factor — The amount by which the normal operating rating of a device can be exceeded without causing failure of the device.

safety service — A radiocommunication service used permanently or temporarily for the safeguarding of human life and property.

SAGE system — *See* Semiautomatic Ground Environment.

SAG MOS — Abbreviation for self-aligning-gate MOS.

St. Elmo's fire — A visible electric discharge sometimes seen at the tips of aircraft propellers or wings, the mast of a ship, or any other metal point where there is considerable atmospheric difference of potential due to concentration of the electric field at the points of the conductor.

sal-ammoniac cell — A cell in which the electrolyte consists primarily of a solution of ammonium chloride.

salient pole — 1. A pole consisting of a separate radial projection having its own iron pole piece and its own field coil, used in the field system of a generator or motor. 2. In an electric motor or generator, a magnetic field pole projecting toward the armature.

Salisbury darkbox — An isolating chamber used for test work in connection with radar equipment. The walls of the chamber are specially constructed to absorb all impinging microwave energy at a certain frequency.

Sallen-Key filter — A design technique that utilizes both positive and negative feedback to achieve single-amplifier realizations of multiple low-pass, high-pass, and bandpass responses.

SAM — Acronym for sequential access memory.

sample — 1. One or more units of product drawn from a lot, the units being selected at random without regard to their quality. 2. An instantaneous value of a variable obtained at regular intervals. 3. To obtain sample values of a complex wave at periodic intervals.

sample and hold — 1. A circuit used in an analog-to-digital converter whenever it is desirable to make a measurement of a signal and to know precisely when the input signal corresponds to the results of the measurement. It is also used to increase the duration of the signal. 2. A system in which a sample of an analog input signal is frozen in time (is stored in a capacitor) and held while it is converted to a digital representation or otherwise processed. 3. A circuit that holds, or freezes, a changing analog input signal voltage. Usually, the voltage thus frozen is then converted into another form, either by a voltage-controlled oscillator, an analog-to-digital converter, or some other device. 4. To capture and retain a signal so it may be converted by an analog-to-digital converter.

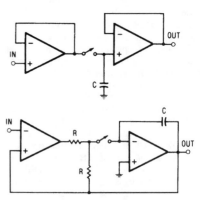

Sample-and-hold circuits.

sample-and-hold amplifier — 1. An analog circuit element that is the analog equivalent of the digital latch. Used to sample an analog signal and then hold it steady at a particular point so that a voltage of interest may be measured or used elsewhere in a system. The most common use of sample-and-hold devices is to sample and hold an analog signal at a particular point while it is measured with an analog-to-digital converter. 2. Analog amplifier, usually with a gain of 1, that has two modes of operation. In the sample mode the output follows the input; in the hold mode the output retains the input value present when the amplifier mode was switched from sample to hold by control logic.

sampled date — Data in which the information content is determined only at discrete time intervals. Sampled data may be either analog or digital in form.

sample pulse — *See* strobe pulse, 1.

sampler — 1. A directional coupler that has a detector attached to the auxiliary arm so that a video output sample proportional to the input power level is obtained. When the sampler is used to monitor power or drive a closed-loop source leveling system, the directional coupler must have a flat coupling coefficient. 2. *See* sampling circuit.

sample rate — 1. The rate at which the analog sample is measured and/or displayed per second. 2. The rate at which samples are taken in the analog-to-digital conversion process.

sample-to-hold offset error — In a sample-and-hold circuit, the difference in output voltage between the time the switch starts to open and the time when the output has settled completely. It is caused by charge being transferred to the hold capacitor from the switch as it opens.

sampling — 1. To obtain values of a function that correspond to discrete, regularly or irregularly spaced values of the independent variable. 2. In pulse-code modulation, selecting samples of an analog wave at recurring intervals so that the original wave can later be reconstructed with reasonable fidelity from the samples. 3. Measuring an input value at intervals.

sampling circuit — Also called a sampler. A circuit whose output is a series of discrete values representative of the values of the input at a series of points in time.

sampling distribution — In random-sampling oscilloscope technique, a function that describes the manner in which the density of a large number of randomly placed samples varies across the signal period.

sampling gate — A device that must be activated by a selector pulse before it will extract information from the input waveform.

sampling oscilloscope — An oscilloscope technique very similar in principle to the use of stroboscopic light to study fast mechanical motion or other very high frequency occurrences. Progressive samples of adjacent portions of successive waveforms are taken; then they are "stretched" in time, amplified by relatively low-bandwidth amplifiers, and finally shown, one sample at a time, on the screen of a cathode-ray tube. The graph produced is a replica of the sampled waveforms. The principal difference in appearance between displays made by sampling techniques and conventional displays is that those made by sampling comprise separate segments or dots. This technique is limited to depicting repetitive signals, since no more than one sample is taken and displayed each time the signal occurs, and provides a means for examining fast-changing signals of low amplitude that cannot be examined in any other way.

sampling plan — A program for the acceptance or rejection of a lot based on tests or inspections indicating the quality of predetermined sample sizes.

sampling rate — The number of times that a particular data channel is sampled by a commutator in one second.

sampling theorem — A theorem (developed by Nyquist in 1928) which states that two samples per cycle will completely characterize a band-limited signal; that is, the sampling rate must be twice the highest-frequency component. (In practice, the sampling rate is ordinarily from five to ten times the highest frequency.)

sand load — An attenuator used as a terminating section on a transmission line to dissipate power. The space between inner and outer conductors is filled with a sand and carbon mixture that acts as the dissipative element.

sandwich — A packaging method in which components are placed between boards or layers.

sanitary motor — A type of motor used in the food industry. It usually has a frame that is so shaped that

deposits of material cannot collect to contaminate nearby food, and can be easily kept clean.

SAP — An acronym for second audio program. Transmitted in FM at a bandwidth of only 10 kHz, this is generally used for dual-language purposes or other audio.

sapphire — A gem used on the tip of quality phonograph needles and also for bearings in precision instruments.

sapphire substrates — Materials that provide a uniform dielectric constant, controlled orientation, thermal conductivity, and the single crystal surface desired for SOS, hybrid IC, and other microcircuit systems. The material may be grown directly in ribbons, tubes, filaments, and sheets.

SARAH — Acronym for search and rescue and homing. A radio homing device, originally designed for personnel rescue, used in operations for the recovery of spacecraft at sea.

SATCOM — The name of satellites built by RCA Astro-Electronics and operated by RCA Americom. They distribute programming to cable TV systems and provide network radio transmissions, as well as commercial and government voice, data, and video services.

SATCOM F1 — Also referred to as just F1. American TV satellite (operated by RCA) to supply most of cable TV programming on 24 transponders (12 are vertically and 12 are horizontally polarized). It is located at 135° west longitude.

SATCOM F2 — Also referred to as F2. American TV satellite (operated by RCA) to supply assorted video and data programming to Alaska and other points in the United States. Like its sister, F1, it has 24 transponders. It is located at 119° west longitude.

satellite — 1. Orbiting system in space that receives communications radio signals from ground bases on earth and then retransmits them to distant locations. 2. A TV station licensed to rebroadcast the programming of a parent station. It differs from a translator in that satellite power limits are much higher, and satellites may also originate some programming. *See* figure on p. 664.

satellite communication systems — A remote communications technique using a satellite in orbit to receive signals from one location and then retransmit them to another location.

satellite receiver — 1. The electronic component of an earth station used indoors that downconverts, processes, and prepares satellite signals for viewing or listening. 2. A component used for tuning in a selected satellite transponder. It may contain one or two downconverters, or none. The receiver recovers the original baseband signals and delivers them to a remodulator. The receiver can also supply the dc operating voltages for an external LNA and downconverter.

SATO — Abbreviation for self-aligned thick oxide.

saturable-core magnetometer — A magnetometer in which the change in permeability of a ferromagnetic core provides a measure of the field.

saturable-core oscillator — A relaxation oscillator in which the occurrence of saturation in a magnetic core initiates a change in the conductive state of amplifying or switching elements.

saturable-core reactor — *See* saturable reactor.

saturable ferrite-core switch — A keyboard switch. In the off position, the magnets straddling the ferrite core saturate it and inhibit a drive signal from being picked up by the sense wire through transformer action. When the key is depressed, the magnetic field is removed, allowing the analog drive signal to be coupled to the sense wire, and the switch turns on.

saturable reactor — Also called a saturable-core reactor. A magnetic-core reactor whose reactance is

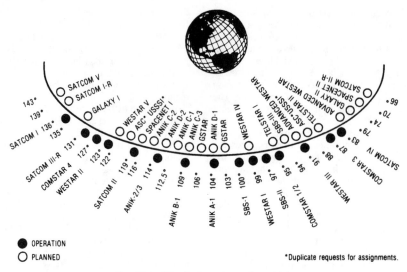

Satellites (North American domestic).

● OPERATION
○ PLANNED

*Duplicate requests for assignments.

controlled by changing the saturation of the core by varying a superimposed unidirectional flux.

saturable transformer — A saturable reactor with an additional winding to provide voltage transformation or isolation from the ac supply.

saturated — That operating state of a transistor in which there is no further increase in collector current when the base current increases; in this state, the collector-emitter voltage is low, typically less than 1 volt.

saturated color — A pure color — i.e., one not contaminated by white.

saturated logic — A type of logic in which one output state is the saturation voltage of a transistor. Examples are resistor-transistor logic (RTL), diode-transistor logic (DTL), and transistor-transistor logic (TTL). *See also* unsaturated logic.

saturated recovery time — The recovery time of a thermal relay measured when the relay is deenergized after temperature saturation (equilibrium) has been reached.

saturated reoperate time — Reoperate time of a thermal relay when temperature saturation (equilibrium) is reached before the relay is deenergized.

saturating reactor — A magnetic-core reactor capable of operating in the region of saturation without independent control means.

saturating signal — In radar, a signal of greater amplitude than the dynamic range of the receiving system.

saturation — 1. A term applied to a color that is "pure" to the extent that it is not mixed with white light. The less white light, the more saturated the color is said to be. This is suggestive of signal-to-noise ratio, where the "noise" in this case is white light. 2. The operating condition of a transistor when an increase in base current produces no further increase in collector current. 3. The state of magnetism beyond which a metal or alloy is incapable of further magnetization — i.e., the point beyond which the *B/H* curve is a straight line. 4. A circuit condition whereby an increase in the driving or input signal no longer produces a change in the output. 5. The condition in which a transistor is driven so hard that it becomes biased in the forward direction. In a switching application, the charge stored in the base region prevents the transistor from turning off quickly under saturation conditions. 6. The condition in an electron tube when maximum current is passing through the cathode circuit. 7. An effect that occurs when a tape is fully magnetized, and further increase of signal input level does not produce a corresponding increase in recorded level. Saturation can also occur in the magnetic structure of the heads. 8. In color TV, the degree to which a color is undiluted with white light or is pure. Saturation is directly related to the amplitude of the chrominance signal. 9. Generally, that state in which a semiconductor device is conducting most heavily for a given applied voltage. In many devices it is also a state in which the normal amplification mechanisms have become swamped and inoperative.

saturation absorption — The condition obtainable between pairs of electron energy levels whereby, under very high radiation intensity, the absorption coefficient gradually decreases to zero because of the absence of empty levels to which transition can occur.

saturation control — In a color television receiver, a control that regulates the amplitude of the chrominance signal. The latter, in turn, determines the color saturation.

saturation current — 1. The current in the plate circuit of a vacuum tube when all electrons emitted by the cathode pass on to the plate. 2. The current between the base and collector of a transistor when an increase in the emitter-to-base voltage causes no further increase in the collector current. 3. The maximum current obtainable as the applied voltage is increased. 4. Current passed by a reverse-biased semiconductor pn junction, which is composed of minority carriers drifting across the potential barrier of the depletion layer. The term saturation is used because in material that is even moderately doped the number of minority carriers present in the material is almost solely determined by the lattice excitation, so that once all the minority carriers available at a given excitation level are involved in the reverse conduction, further increases in reverse bias produce virtually no increase in current.

saturation curve — A magnetization curve for a ferromagnetic material.

saturation flux density — *See* saturation induction.

saturation induction — Sometimes loosely referred to as saturation flux density. The maximum intrinsic induction possible in a material.

saturation limiting — Limiting the minimum output voltage of a vacuum-tube circuit by operating the tube in the region of plate-current saturation.

saturation magnetization — The magnetic condition of a body when an increase in the magnetizing force produces practically no change in the intensity of magnetization.

saturation moment — The greatest magnetic moment possible in a sample of magnetic material.

saturation noise — The noise arising when a uniformly saturated tape is reproduced. This is often some 15 dB higher than the bulk-erased noise and is associated with imperfect particle dispersion.

saturation point — The point beyond which an increase in one of two quantities produces no increase in the other.

saturation recording — Direct digital recording. So called because the magnetic coating on the recording tape is fully saturated by the recording process. Normal audio recording uses only a small portion of the tape's "magnetic energy" to reduce harmonic distortion to acceptable levels. By magnetically saturating the tape, however, variations in tape sensitivity are masked and the higher-level playback is better able to overcome noise.

saturation resistance — 1. Ratio of voltage to current in a saturated semiconductor. 2. The resistance between collector and emitter terminals under specified conditions of base current and collector current, when the collector current is limited by the external circuit.

saturation value — 1. The highest value that can be obtained under given conditions. 2. The value of magnetic-flux density beyond which increases in the magnetizing force have no appreciable effect on the flux density in a particular sample of magnetic material.

saturation voltage — 1. The voltage drop appearing across a switching transistor (collector-emitter) that is fully turned on. 2. Generally, the voltage excursion at which a circuit self-limits (i.e., is unable to respond to excitation in a proportional manner). In operational amplifiers, the output-voltage saturation limits may be imposed by any stage, from the input to the output, depending in part on the external loading and feedback parameters. 3. The voltage between base and emitter required to cause collector current saturation. 4. The residual voltage between collector and emitter terminals under specified conditions of base current and collector current, the collector current being limited by the external circuit.

SAW — Abbreviation for surface acoustic wave.

SAW device — Surface acoustic-wave device. A technology for broad-bandwidth signal delay, custom-designed filters, and complex generation and correlation at IF frequencies. SAW devices make use of a single-crystal, planar substrate with aluminum or gold electrode patterns fabricated by photolithography of a substrate of piezoelectric material. The electrode patterns are used to excite and detect minute acoustic waves that travel over the surface of the substrate much like earthquake waves travel over the crust of the earth.

SAW (surface acoustic wave) filter — An electronic device that allows a sharp transition between regions of allowed and attenuated frequencies.

sawtooth — A waveform increasing approximately linearly as a function of time for a fixed interval, returning to its original state sharply, and repeating the process periodically.

sawtooth current — A current that has a sawtooth waveform.

sawtooth generator — An oscillator providing an alternating voltage with a sawtooth waveform.

sawtooth-modulated jamming — An electronic-countermeasure technique in which a high-level jamming signal is transmitted so that large AGC voltages are developed at the radar receiver and the target pip and receiver noise are caused to disappear completely.

sawtooth voltage — A voltage that varies between two values in such a manner that the waveshape resembles the teeth of a saw.

sawtooth wave — A periodic wave whose amplitude varies linearly between two values. A longer interval is required for one direction of progress than for the other.

Sawtooth wave.

sawtooth waveform — A waveform that has a slow or sloping rise time and a sharp or sudden fallback to the starting point, resembling the teeth of a saw.

saxophone — A linear-array antenna with a cosecant-squared radiation pattern.

SBA — Abbreviation for standard beam approach.

S-band — Microwave band in which the wavelengths are at or near 10 cm.

SBT — Abbreviation for surface-barrier transistor.

SCA — Abbreviation for Secondary (or Subsidiary) Communications Authorization. Permission granted by the FCC for an FM broadcaster to send out, on the same carrier frequency and simultaneously with the regularly heard program, a program in addition to the one heard with ordinary receivers. The purpose is to provide special programming to a limited audience without the cost of an entire new transmitter. SCA is primarily for the transmission of programs that are of a broadcast nature but which are of interest primarily to limited segments of the public wishing to subscribe. This includes background music, storecasting, detailed weather forecasting, special time signals, and other material of a broadcast nature expressly designed and intended for business, professional, educational, religious, trade, labor, agriculture, or other groups engaged in any lawful activity. SCA facilities may also be used in transmission of signals directly related to the operation of FM broadcast stations.

SCA channel — Abbreviation for subsidiary carrier authorization channel. A channel that carries program material modulated onto a radio frequency subcarrier on a CATV system or an FM broadcast signal.

SCA demodulator — Abbreviation for subsidiary carrier authorization demodulator. A (low-frequency) receiver tuned to a subcarrier signal broadcast by an FM radio station. It is connected to the multiplex output in an FM receiver and demodulates the subcarrier program carrying subscription music, stock reports, etc.

scalar — 1. A quantity that has magnitude but no direction (e.g., real numbers). 2. A circuit with two stable states, which can be triggered to the opposite state by appropriate means (a bistable circuit).

scalar feed — A type of horn antenna feed that uses a series of concentric rings to capture signals that have been reflected toward the focal point of a parabolic antenna.

scalar function — A function that has magnitude only. Thus, the scalar product of two vectors is a scalar function, as is a real function of a real variable.

scalar operations—Mathematical operations performed on random data elements rather than on sequential data elements.

scalar processing—Calculations performed one at a time.

scalar quantity—Any quantity that has magnitude only—e.g., time, temperature, quantity of electricity.

scale—1. A series of musical notes, symbols, sensations, or stimuli arranged from low to high by a specified scheme of intervals suitable for musical purposes. 2. The theoretical basis of a numerical system. 3. A series of markings used for measurement or computation. 4. A defined set of values, in terms of which different quantities of the same nature can be measured. 5. In a computer, to change the units of a variable so that the problem is within its capacity. 6. To change a quantity by a factor in order to bring its range within prescribed limits.

scale division—The space between two adjacent markings on a scale.

scale error (full-scale error)—The departure from design output voltage of a DAC for a given input code, usually full-scale code. In an ADC, it is the departure of actual input voltage from design input voltage for a full-scale output code. Scale errors can be caused by errors in reference voltage, ladder resistor values or amplifier gain, etc.

scale factor—1. In analog computing, a proportionality factor that relates the magnitude of a variable to its representation within a computer. 2. In digital computing, the arbitrary factor that may be associated with numbers in a computer to adjust the position of the radix point so that the significant digits occupy specified columns. 3. The factor by which the number of scale divisions indicated or recorded by an instrument must be multiplied to compute the value of the measurand. 4. A value used to convert a quantity from one notation to another. 5. The amount by which a measured quantity must change in order to produce unit deflection of a recording pen.

scale length (of an indicating instrument)— The length of the path described by the tip of the pointer (or other indicating means) in moving from one end of the scale to the other. Pointers that extend beyond the scale division marks are considered to end at the outer end of the shortest division marks; in multiscale instruments, the longest scale is used in determining the scale length.

scale-of-ten circuit—See decade scaler.

scale-of-two counter—A flip-flop circuit in which successive similar pulses are applied at a common point, causing the circuit to alternate between its two conditions of permanent stability.

scaler—Also called a scaling circuit. A circuit that produces an output after a predetermined number of input pulses have been received.

scaler frequency meter—A frequency meter in which electronic circuits are used for counting and gating electrical signals to indicate their number and/or rate.

scale span—The algebraic difference between the values of the actuating electrical quantity corresponding to the two ends of the scale of an instrument.

scaling—1. An electronic method of counting electrical pulses occurring too fast to be handled by mechanical recorders. 2. The changing of a quantity from one notation to another. 3. Adjusting the coefficient of a circuit to each of its one or more input-signal terminals. The relative scaling of one input to another is called weighting. 4. In computing, relating problem variables to machine variables.

scaling circuit—See scaler.

scaling factor—Also called the scaling ratio. The number of input pulses per output pulse required by a scaler.

scaling ratio—See scaling factor.

scalloping distortion—In videotape recording, a series of small, vertical curves in the recorded image. Caused by unequal stretching across the width of the tape.

SCA modulator—Abbreviation for subsidiary carrier authorization modulator. A device that takes an audio input and modulates a supersonic frequency that can be combined with the regular audio input to an FM transmitter, which will thus broadcast two programs simultaneously. The SCA program for which the subscriber pays will not be heard unless the FM receiver is equipped with an SCA demodulator.

scan—1. In facsimile, to analyze the density of successive elemental areas of the subject copy in a predetermined pattern at the transmitter, or to record these areas at the recorder. 2. To examine point by point—e.g., in converting a televised scene or image into a methodical sequence of elemental areas. 3. One sweep of the mosaic in a camera tube or of the screen in a picture tube. 4. To sample each of a number of inputs intermittently. A scanning device may provide additional functions such as record or alarm.

scan-coded tracking system—See monopulse tracking.

scan converter—1. A device that converts computer video to a TV video format. 2. Equipment that samples radar images at a 3-kHz to 10-kHz rate that can be sent over telephone lines or narrow bandwidth radio circuits and converted into a slow-scan image by a similar converter. See also slowed-down video.

scan-converter tube—A device consisting of a cathode-ray tube and a vidicon imaging tube assembled face to face in the same envelope.

scanistor—An integrated semiconductor optical-scanning device that converts images into electrical signals. The output analog signal represents both the amount and the position of the light shining on its surface.

scan line—One of the many horizontal lines that make up the picture on a video screen.

scan moiré—See moiré effect.

scanner—1. A device that converts images or text on paper into data that can be manipulated by a computer. 2. An instrument that automatically samples or interrogates the state of various processes, conditions, or physical states and initiates action in accordance with the information obtained. 3. In a facsimile transmitter, the part that systematically translates the densities of the subject copy into the signal waveform. 4. The moving parts of an antenna that cause the beam to scan. 5. A switching device that sequentially samples a number of points.

scanner amplifier—An amplifier in a facsimile transmitter used to amplify the output-signal voltage of the scanner.

scanner radio—A receiver that can automatically check the airwaves for signals from preselected stations that are broadcasting at the time. When it picks up such signals, the scanner pauses at the station setting for as long as the signals last—or for as long as you care to listen. When the signals stop for a few seconds, the scanner automatically tunes itself to the next preselected channel that may be broadcasting.

scanning—1. In television, facsimile, or picture transmission, the successive analyzing or synthesizing, according to a predetermined method, of the light values or equivalent characteristics of elements constituting a picture area. 2. In radar, the directing of a beam of radio-frequency energy successively over the elements of a given region, or the corresponding process in reception. 3. The comparison of input variables with some reference to determine a particular action. 4. The successive exposure of small portions of an object to a sensing device of some type. Television, radioactive scanning, facsimile transmission, and photoelectric scanning are all examples of this technique. 5. Moving the electron beam of an image pickup tube or a picture tube diagonally across the target or screen area of tube. 6. The process of translating images into digital form that can be recognized by a computer.

scanning-antenna mount—An antenna support that provides a mechanical means for scanning or tracking with the antenna, and a means for taking off information and using it for indication and control.

scanning beam—A beam of light, a radar beam, or an electron beam that is used in scanning.

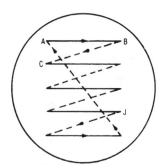

Scanning beam.

scanning circuit—A circuit that produces a linear, circular, or other movement of the beam in a cathode-ray tube at regular intervals.

scanning disc—1. A rotating tricolor wheel used between the camera lens and subject, or picture tube and viewer, in field-sequential color television. 2. A Nipkow disc.

scanning frequency—*See* stroke speed.

scanning head—A light source and a phototube combined as a single unit for scanning a moving strip of paper, cloth, or metal in photoelectric side-register control systems.

scanning line—A single, narrow, continuous strip containing highlights, shadows, and halftones of the picture area, as determined by the scanning process.

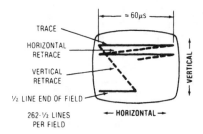

Scanning lines.

scanning linearity—In television, the uniformity of the scanning speed during the trace interval.

scanning-line frequency—The number of scanning lines per second. *See also* stroke speed.

scanning loss—In a radar system employing a scanning antenna, the reduced sensitivity that occurs in scanning across a target compared with the sensitivity when a constant beam is directed at the target. This loss is expressed in decibels.

scanning sonar—An echo-ranging system in which the sound pulse is transmitted simultaneously through the entire angle to be searched and a rapidly rotating transducer having a narrow beam angle scans for the returning echoes.

scanning speed—The number of inches per second explored by the spot of light or other source of energy in television, facsimile, radar, etc.

scanning spot—1. The immediate area being explored at any instant by a spot of light or other energy source in television, facsimile, radar, etc. *See also* picture element. 2. The spot illuminated on a cathode-ray tube by the initial impact of the scanning ray and the screen.

scanning yoke—A yoke-shaped iron core that supports the electromagnetic deflecting coils around the neck of some cathode-ray tubes.

scannogram—The recording made on paper by a scanner.

scan rate—The rate at which a control computer periodically checks a controlled quantity.

scan time—1. The time necessary to completely execute the entire programmable controller program one time. 2. The time necessary to completely execute an entire program one time. 3. The time required to examine the state of all inputs.

SCA rejection—The ratio of the 67-kHz SCA signal at the output to the desired output with the standard FCC signal input.

SCART—A 20-pin rectangular connector used on TV receivers in PAL areas. The connector provides composite video inputs and outputs, stereo audio inputs and outputs and RGB input. It is also known as Euro-connector.

scatter—1. A disordered change in the direction of propagation when radio waves encounter matter. 2. Spurious radar echoes due to reflections from layers of the ionosphere.

scatterband—In pulse systems, the total bandwidth occupied by the frequency spread of numerous interrogations operating on the same nominal radio frequency.

scattered reflections—Reflections from portions of the ionosphere at different virtual heights. These reflections interfere with each other and cause rapid fading of the signal.

scattering—1. The change in direction, frequency, or polarization of radio waves when they encounter matter. 2. In a narrower sense, a disordered change in the incident energy of (1) above. 3. The change of direction of particles or systems. 4. The diffusion of a sound or light beam due to discontinuities in the transmitting medium. 5. Change of the spatial distribution of a beam of radiation when it interacts with a surface or a heterogeneous medium, in which process there is no change of wavelength of the radiation.

scattering loss—That part of the transmission lost because of scattering within the medium or the roughness of the reflecting surface.

scatter loading—In a computer, the form of fetch that may result in placement of the control sections of a load module in nonadjoining main-storage positions.

scatterometer—A wide-sweep radar for terrain mapping.

scatter propagation—Also called beyond-the-horizon propagation, beyond-the-horizon transmission, or over-the-horizon transmission. Transmission of high-power radio waves beyond line-of-sight distances by reflecting them from the troposphere or ionosphere.

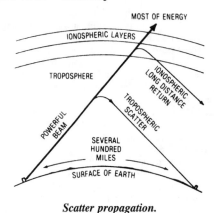

Scatter propagation.

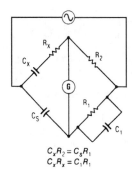

$$C_x R_2 = C_s R_1$$
$$C_x R_x = C_1 R_1$$

Schering bridge.

scatter read — The ability of a computer to distribute data into several memory areas as it is being entered into the system from magnetic tape.

SCC — Abbreviation for spark control computer. An analog electronic system (Chrysler Corp.) that accurately controls spark timing in response to various inputs, including engine temperature, manifold pressure, air temperature, and air pressure. Used with EGR, it provides for cleaner exhaust and better fuel economy.

scc wire — Abbreviation for single-cotton-covered wire.

SCEPTRON — Acronym for spectral comparative pattern recognizer. A device that automatically classifies complex signals derived from any type of information that can be changed into an electrical signal.

scheduled maintenance — Maintenance performed according to an established plan.

scheduling — Determining the order in which job programs will use the available computer facilities.

schematic — 1. A diagram that shows, by means of graphic symbols, the electrical connections and functions of a specific circuit arrangement. 2. A representation of the components of an electrical circuit and their interconnections by symbols and lines.

schematic circuit diagram — *See* schematic diagram.

schematic diagram — Also called a schematic circuit diagram, diagram, or schematic. 1. A diagram of the electrical scheme of a circuit, with components represented by graphical symbols. 2. A drawing that shows by means of graphic symbols the electrical connections, components, and functions of a specific circuit arrangement. 3. A functional diagram of an electrical circuit in which the components are represented by conventional symbols, and wires interconnecting them by lines.

Schering bridge — A four-arm alternating-current bridge used for measuring capacitance and dissipation factor. The unknown capacitor and a standard loss-free capacitor form two adjacent arms, the arm adjacent to the standard capacitor consists of a resistor and capacitor in parallel, and the fourth arm is a nonreactive resistor.

schlieren — An optical system that produces images in which the illumination or hue at a given point is related to the angular deflection a light ray undergoes in passing through the corresponding point in the object. The object is back-illuminated, and a straightedge, circular aperture, or graded density or multicolored filter is employed in the system to discriminate between deflected and undeflected rays.

Schmidt antenna — A microwave scanning antenna similar in principle to the optical Schmidt camera. The spherical reflector has a spheric microwave lens at the center of curvature and a scanner is located approximately halfway between these elements.

Schmidt optical system — An optical system for magnifying and projecting a small, brilliant image from a projection-type cathode-ray tube onto a screen.

Schmitt limiter — *See* Schmitt trigger.

Schmitt trigger — Also called Schmitt limiter. 1. A bistable pulse generator in which an output pulse of constant amplitude exists only as long as the input voltage exceeds a certain dc value. The circuit can convert a slowly changing input waveform to an output waveform with sharp transitions. Normally, there is hysteresis between an upper and a lower triggering level. 2. A regenerative circuit that changes state abruptly when the input signal crosses specified dc triggering levels. 3. A bistable device that utilizes the effect of hysteresis. In the Schmitt trigger, hysteresis is a form of nonlinear operation that forces the output to be dependent not only on the absolute value of the input, but also on the most recent prior value of the input. This hysteresis is characterized by two different switching theshold levels: one for positive-going input transitions and the other for negative-going input transitions.

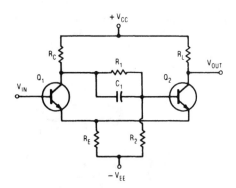

Schmitt trigger.

schmoo plot — An x, y plot giving the pass/fail region for a specific test while varying the parameters in x and y coordinates.

schooping — Spraying metal onto the ends of the roll of a metallized capacitor. This provides a metal surface on the ends of the roll to which the leads are then soldered.

Schottky — A bipolar technology that is faster than standard TTL but often uses more power.

Schottky barrier—1. A simple metal-to-semiconductor interface that exhibits a nonlinear impedance. 2. A metal and a semiconductor. A metal-insulator-semiconductor (MIS) barrier contains a very thin film of oxide between the metal and semiconductor layers; in a semiconductor-insulator-semiconductor (SIS) barrier, a high-conductivity, transparent, large-bandgap semiconductor layer replaces the metal of the MIS cell. 3. A junction diode with the junction formed between the semiconductor and a metal contact rather than between dissimilar semiconductor materials, as in the case of an ordinary pn diode.

Schottky-barrier detector—A detector based on a Schottky barrier—a current-rectifying contact at a junction between a semiconductor and a metal.

Schottky barrier diode—Also called Schottky diode and hot carrier diode. 1. A junction diode with the junction formed between the semiconductor and a metal contact rather than between dissimilar semiconductor materials, as in the case of an ordinary pn diode. 2. A special diode characterized by nanosecond switching speed, but relatively low voltage (45 V max) and limited temperature range (125°C to 150°C).

Schottky diode—*See* Shottky barrier diode.

Schottky rectifier—A high-speed rectifier that makes use of the rectification effect of a metal-to-silicon barrier. Low forward-voltage characteristics provide high rectification efficiency, while majority carrier forward conduction enhances switching speed.

Schottky transistor logic—Abbreviated STL. An improved version of integrated injection logic that has a power-delay product that is three times lower than that for I²L.

Schottky TTL—1. A TTL circuit that incorporates Schottky diodes to greatly speed up TTL circuit operation. 2. A very high speed TTL circuit using metal/semiconductor diodes (Schottky diodes) to prevent transistors from saturating. 3. Integrated circuits modified by having a special diode built into each transistor. These diodes enable the transistor to be turned off much faster than would otherwise be possible.

scientific notation—In a calculator, the number entered or a result displayed in terms of a power of 10.

Schematic.

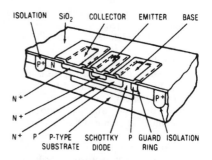

Cross section.

Schottky barrier diode.

For example, the number 1234 is entered as 1.234×10^3, and the number 0.001234 would appear as 1.234×10^{-3}.

scintillate—To emit flashes of light.

scintillation—1. In radio propagation, a random and usually relatively small fluctuation of the received field about its mean value. 2. Also called target glint or wander. On a radar display, a rapid apparent displacement of the target from its mean position. 3. The flash of light produced by an ionic action. 4. A momentary breakdown of a tantalum oxide film in a capacitor, accompanied by rapid heating of the dielectric. Such events are caused by capacitor overvoltages or improper techniques of capacitor manufacture. 5. The flash of light produced by certain crystalline materials when a charged particle is passed through them. 6. Rapid fluctuation in parameters such as the amplitude or the phase of a wave passing though a medium with small-scale irregularities that cause irregular changes in the transmission path with time; akin to twinkling.

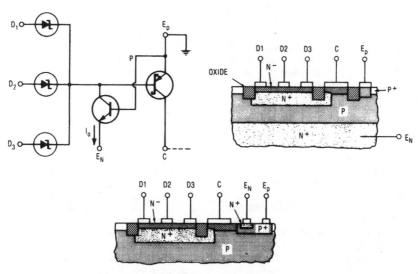

Schottky transistor logic.

scintillation conversion efficiency — In a scintillator, the ratio of the optical photon energy emitted to the energy of the incident particle or photon of ionizing radiation.

scintillation counter — A device that indirectly detects charged particles and gamma rays and neutrons by using a photomultiplier tube to convert the short flashes of light produced as the particle passes through a transparent scintillating material into electric signals that can be recorded. One advantage of scintillation counters is that they are very fast; by this is meant that they have a very small resolving time.

scintillation-counter cesium resolution — The scintillation-counter energy resolution for the gamma ray or conversion electron emitted from cesium-137.

scintillation-counter energy resolution — In a scintillation counter, a measure of the smallest discernible difference in energy between two particles or photons of ionizing radiation.

scintillation-counter energy-resolution constant — The product of the square of the scintillation-counter energy resolution times the specified energy.

scintillation-counter head — The combination of scintillators and photosensitive devices that produces electrical signals in response to ionizing radiation.

scintillation-counter time discrimination — In a scintillation counter, a measure of the smallest time interval between two successive individually discernible events. Quantitatively, the standard deviation of the time-interval curve.

scintillation crystals — Special crystals that emit flashes of light when struck by alpha particles.

scintillation decay time — The time required for the decrease of the rate of emission of optical photons in a scintillation from 90 percent to 10 percent of the maximum value.

scintillation duration — The interval from the time of emission of the first optical photon of a scintillation to the time when 90 percent of the optical photons of the scintillation have been emitted.

scintillation rise time — The time interval occupied by the increase of the rate of emission of optical photons of a scintillation from 10 percent to 90 percent of the maximum value.

scintillator — The combination of the body of scintillator material and its container.

scintillator material — A material that exhibits the property of emitting optical photons in response to ionizing radiation.

scintillator-material total-conversion efficiency — In a scintillator material, the ratio of the produced optical photon energy to the energy of a particle or photon of ionizing radiation that is entirely absorbed in the scintillator material.

scissor — To apportion a drawing into segments that can be viewed on a CRT screen.

scissoring — The ability of the vector generator to blank the beam whenever it is moved outside of the screen (where the image becomes distorted).

scope — Slang for a cathode-ray oscilloscope.

scophony television system — A mechanical television projection system developed in England. In it, ingenious optical and mechanical methods provide large, bright images suitable for theater installation as well as home television receivers. The apparent screen brightness is multiplied several hundred times because several hundred picture elements are projected simultaneously.

scored substrate — A substrate that has been scribed with a thin cut at the break lines. *See* snap-strate.

scoring system — In motion-picture production, a system for recording music in time with the action on the film.

Scott connection — A method of connecting transformers to convert two-phase power to three-phase or vice versa.

Scott's breakdown theory — Breakdown is due to the attainment of a critical avalanche size that leads to a conducting path.

SCR — Abbreviation for silicon controlled rectifier. The formal name is reverse-blocking triode thyristor. 1. A thyristor that can be triggered into conduction in only one direction. Terminals are called anode, cathode, and gate. 2. A semiconductor device that functions as an electrically controlled switch for dc loads. The SCR is one type of thyristor. 3. A power switching device in which a pulse at the gate (input) initiates a switching action in the output circuit. It conducts only during the positive half-cycle of ac. 4. A type of thyristor that is designed for forward bias, unidirectional power switching, and control. *See also* thyristor.

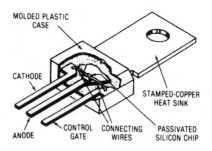

In molded plastic case.

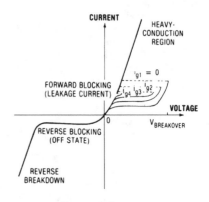

Characteristic curve.

SCR.

scramble — 1. To transpose and/or invert bands of frequencies, or otherwise modify the form of the intelligence at the transmitting end, according to a prearranged scheme in order to obtain secrecy. 2. To mix, in cryptography, in a random or quasi-random fashion.

scrambled speech — Also called inverted speech. Speech that has been made unintelligible (e.g., for secret transmission) by inverting its frequency. At the receiving end, it can then be converted back into intelligible speech by reinverting the frequency.

scrambler — 1. A device used to electronically alter a signal so that it can only be viewed or heard on a receiver equipped with a special decoder. 2. An electrical device that uses one or more methods to convert speech or video into a secret, unintelligible form. 3. A device that changes data so it appears to be in a random pattern. A descrambling device can change (unscramble) this data back to its original pattern.

scrambler circuit — Also called a speech scrambler. A circuit in which essential speech frequencies are divided into several ranges by filters and then inverted to produce scrambled speech. *See also* speech inverter.

scrambling — A method of altering the identity of a video or audio signal in order to prevent its reception by persons not having authorized decoders.

scratch filter — A low-pass filter, often an integral part of an amplifier circuit, that attenuates the higher frequency noise derived from disk recordings. The scratch filter is also suitable for the suppression of background noise produced by tape background hiss.

scratch pad — 1. Information that the processing unit of a computer stores or holds temporarily. It is a memory containing subtotals for various unknowns that are needed for final results. 2. An area of RAM used for short-term storage of data during a process. 3. Group of general-purpose registers without specific function, providing a high-speed workspace. Usually, an internal RAM in a computer.

scratch-pad memory — 1. A high-speed, limited-capacity computer information store that interfaces directly with the central processor. It is used to supply the central processor with the data for the immediate computation, thus avoiding the delays that would be encountered by interfacing with the main memory. (The function of the scratch-pad memory is analogous to that of a pad of paper used for jotting down notes.) 2. A high-speed memory used to temporarily store small amounts of data that may be needed often and without delay. 3. Any memory space used for the temporary storage of data. Typically, scratch-pad memories are high-speed integrated circuits that are addressed as internal registers.

screen — 1. The surface on which the visible pattern is produced in a cathode-ray tube. 2. A metal partition that isolates a device from external electric or magnetic fields. 3. *See* screen grid. 4. Surface on which the master artwork of the circuit pattern is projected. Screen fabrics include polyester, stainless steel, nylon, and silk. 5. The surface of a medium on which a visible image, pattern, or picture is produced, generally within a camera or cathode-ray tube.

screen angle — A vertical angle bounded by a straight line from the radar antenna to the horizon and the horizontal at the antenna, assuming a 4/3 earth's radius.

screen deposition — The laydown of a circuit pattern on a substrate using the silk screening technique.

screen dissipation — The power that the screen grid dissipates as heat after bombardment by the electron stream.

screen grid — Also called a screen. A grid placed between a control grid and an anode in a vacuum tube and usually maintained at a fixed positive potential. By reducing the electrostatic influence of the anode, it prevents the electrons from bunching in the space between the screen grid and the cathode.

screen-grid modulation — Modulation produced by introducing the signal into the screen-grid circuit of any multigrid tube where the carrier is present.

screen-grid tube — A vacuum tube in which a grid is placed between the control grid and the anode to prevent the latter from reacting with the control grid.

screen-grid voltage — The direct-voltage value applied between the screen grid and the cathode of a vacuum tube.

screening — 1. The process whereby the desired film circuit patterns and configurations are transferred to the surface of a substrate during manufacture by forcing a material through the open areas of the screen, using the wiping action of a soft squeegee. 2. The display of raw or processed data for operator verification. 3. The process of performing 100-percent inspection on product lots and removing defective units.

screening test — A test or combination of tests intended to remove unsatisfactory items or those likely to exhibit early failure.

screen printing (thick film) — 1. The art of depositing conductive, resistive, and insulating materials on a dielectric base. This deposition is made through selected open areas in screens with inks or pastes forced through the open areas of the screen by squeegee motion onto the substrate base. In some cases, masks instead of conventional mesh screens may be used. 2. The basic thick-film deposition process in which the paste is squeegeed through a fine-mesh stencil screen to produce a prescribed pattern on a substrate.

screen savers — Software that automatically blanks a monitor screen or displays a moving pattern if there has been no user interaction with the computer for a specified time. Screen savers are used to preventing a fixed pattern from being burned into the screen phosphor. As soon as the operator touches any key or moves the mouse, the screen saver disappears and the original display returns to the screen.

scribe and break — The procedure used to separate a processed semiconductor wafer into individual ICs. Narrow channels between individual ICs are mechanically weakened by scratching with a diamond tip (scribe), sawing with a diamond blade, or burning with a laser. The wafer is mechanically stressed and broken apart along the channels (called scribe lines), thereby separating the individual ICs (dice).

scribe projection — A method of automatic information presentation in which information is placed on a small metallic-coated glass slide by using a movable, servocontrolled, fine-pointed scribe to remove the coating. Light passed through the scribed area is projected onto a screen.

scribing — 1. A process, similar to glass cutting, in which a slice of semiconductor devices is scored in rows and columns so that it may be separated easily into individual devices. The process is performed in a machine called a scriber by repeated movement of a weighted diamond stylus across the slice to form the scored pattern. 2. Scratching a tooled line or laser path on a brittle substrate to allow a wafer to be cleft or broken along the line, producing IC, transistor, or diode chips when all breaks are completed.

scribing machines and tools — Equipment used to separate wafers into individual devices, chips, or dice. This has been done by crude techniques similar to glass cutting, but is now accomplished by more efficient methods using truncated pyramid diamond scribers, automated machines, conical tools, or lasers.

script — Much like a macro, a set of program commands used to automate routine computing tasks.

scroll — 1. To move a video display up or down, line by line, or side to side, character by character. 2. To move all or part of the screen material up or down, or left or right, to allow new information to appear.

scrolling — 1. Moving the contents of the screen of a CRT up or down by one line at a time. 2. A multiple-rung display function that allows all displayed rungs to be moved up or down, adding the next or preceding rung at the bottom (or top) of the display. (As determined by the user, the display may be changed either one rung at a time or continuously.) 3. The vertical movement of information on a CRT screen, caused by the dropping of one line of displayed information for each new line added; the movement appears as an upward rolling if the new line is added at the bottom of the screen, and vice versa.

scrubbing action — Rubbing of a chip device around on a bonding operation to break up the oxide layer and improve wetability of the eutectic alloy used in forming the bond.

SCS — Abbreviation for silicon-controlled switch.

SCU — Abbrevitaion for subscriber channel unit. A telephone interface circuit.

S-curve — An S-shaped frequency-response curve showing how the output of a frequency-modulation detector or circuit varies with frequency.

SDLC — Abbreviation for Synchronous Data Link Control. 1. A communications line discipline that initiates, controls, checks, and terminates information exchanges or communications lines. SDLC is designed for full-duplex operation (simultaneously sending and receiving data). 2. A protocol specifying a layered approach to serial data communications.

sea clutter — See sea return.

seal — Any device used to prevent gases or liquids from passing through.

sealed contacts — A contact assembly enclosed in a sealed compartment separate from the other parts of the relay.

sealed-gage pressure transducer — A pressure transducer that has the sensing element sealed in its case at room ambient pressure. The sealing method holds the original internal pressure for long periods of time.

sealed meter — A meter constructed so that moisture or vapor cannot enter the meter under specified test conditions.

sealed relay — A relay that has both coil and contacts enclosed in a relatively airtight cover.

sealed tube — A hermetically sealed electron tube used chiefly for pool-cathode tubes.

sealing compound — A type of wax or pitch compound used in dry batteries, capacitor blocks, transformers, or circuit units to keep out air and moisture.

sealing off — The final closing of the bulb of a vacuum tube or lamp after evacuation.

seam welding — A resistance welding process in which overlapping spot welds are made progressively along a joint by means of circular electrodes. The circular seam-welding wheels roll along the overlapping edges to be welded, and the control circuit is arranged to pass current at sufficiently close intervals to produce the desired degree of overlapping of the spot welds. The primary purpose of a seam-welding joint is to produce liquid or airtight containers from comparatively thin sheet metal.

search — 1. In a radar operation, the directing of the lobe (beam of radiated energy) in order to cover a large area. A broad-beam antenna may be used, or a rotating or scanning antenna. 2. A systematic examination of the available information in a specific field of interest. 3. To scan available stored information. 4. The process of applying a sweeping tuning signal to the free-running oscillator portion of a phase-locked oscillator, causing the oscillator output frequency to pass within the capture bandwidth of the feedback network and ensuring a locked condition in response to a new reference frequency. 5. The process of finding a particular item in a file. Search techniques include the sequential or linear search, the binary or logarithmic search, and direct lookup.

search coil — See magnetic test coil.

search engine — 1. A computer program that helps users find information in most databases. 2. A database or index that can be queried to help find information on the World Wide Web.

search gate — A gate pulse that is made to search back and forth over a certain range.

searchlighting — In radar, the opposite of scanning. Instead, the beam is projected continuously at an object.

searchlight-type sonar — An echo-ranging system employing the same narrow beam pattern for both transmission and reception.

search radar — A radar intended primarily for displaying targets as soon as possible after their entrance into the coverage area.

search receiver — See intercept receiver.

search time — The time required for location of a particular data field in a computer storage device. The process involves comparison of each field with a predetermined standard until an identity is obtained. Contrasted with access time.

sea return — Also called sea clutter. In radar, the aggregate received echoes reflected from the sea.

seasonal factors — Factors that are used to adjust sky-wave absorption data for seasonal variations. Those variations are due primarily to seasonal fluctuations in the heights of the ionospheric layers.

seasoning — Overcoming a temporary unsteadiness of a component that may appear when the component is first installed.

seating time — The elapsed time after the coil of a relay has been energized until the armature of the relay is seated.

sec — Abbreviation for secondary winding of a transformer.

SEC — Abbreviation for secondary-electron conduction. The transport of charge by secondary electrons moving through the interparticle spaces of a porous material under the influence of an externally applied electric field.

SECAM — Abbreviation for Sequential Color and Memory System (Séquential Couleur avec Mémoire). A color TV system with 625 lines per frame and 50 fields per second, used primarily in France, Russia, parts of Africa, and the former states of the Soviet Union. SECAM uses alternating lines of U, V chroma information to modulate the frequency. It is quite different from NTSC and PAL standards.

secondary — 1. The transformer output winding in which the current is due to inductive coupling with another coil called the primary. 2. Low-voltage conductors of a power-distributing system.

secondary area — See secondary service area.

secondary breakdown — A condition that occurs in bipolar transistors brought on by hot spots occurring within the device structure. This is a type of thermal runaway and in most cases causes permanent device damage.

secondary calibration — Also called sense step. Calibration of accessory equipment in which a transducer is deliberately unbalanced electrically to change the output, voltage, current, or impedance. Generally performed by means of a calibration resistor that is placed across one leg of the bridge.

secondary cell — A voltaic cell that, after being discharged, may be restored to a charged position by an electric current sent through the cell in a direction opposite that of the discharge current. See also storage cell, 1.

secondary circuit — The high-voltage part of an ignition system.

secondary color — A color produced by combing any two primary colors in equal proportions. In the light-additive process, the three secondary colors are cyan, magenta, and yellow.

Secondary Communications Authorization — See SCA.

secondary electron — An electron emitted from a material as a result of bombardment by electrons or of the collision of a charged particle with a surface.

secondary-electron conduction tube — A sensitive TV tube that uses a two-step process to convert the invisible image to a charge image. In the image intensifier stage, light ejects electrons from a photoemitter. After imaging and amplification in the middle or imaging section, the primary photoelectrons fall on the thin-film face of the secondary target. This charge image modulates the scanning beam current from the reading section. The electric field applied across the film causes a majority of the secondary electrons to be transported through a potassium chloride low-density film layer to produce a secondary conduction current.

secondary-electron multiplier — An amplifier tube in which the electron stream is focused onto a succession of targets, each of which adds its secondary electrons to the stream. In this way, considerable amplification is provided.

secondary emission — The liberation of electrons from an element other than the cathode as a result of being struck by other high-velocity electrons. In a vacuum tube there are usually more secondary than primary electrons — a desirable phenomenon in electron-multiplier or dynatron-oscillator tubes. However, pentodes have a suppressor grid to nullify the undesirable effect of secondary emission.

secondary-emission ratio — The average number of secondary electrons emitted from a surface per incident primary electron.

secondary-emission tube — A tube that makes use of secondary emission to achieve a useful end. The photomultiplier is an example.

secondary failure — 1. A failure occurring as a direct result of the abnormal stress on a component brought about by the failure of another part or parts. 2. Any failure that is the direct or indirect result of a primary failure.

secondary grid emission — Emission from the grid of a tube as a result of high-velocity electrons being driven against it and knocking off additional electrons. The effect is the same as for primary grid emission.

secondary insulation — A nonconductive material whose prime functions are to protect the conductor against abrasion and provide a second electrical barrier. Placed over the primary insulation.

secondary line — The conductors connected between the secondaries of distribution transformers and the consumer service entrances.

secondary radar — See radar.

secondary radiation — Random reradiation of electromagnetic waves.

secondary service area — Also called the secondary area. The service area of a radio or television broadcast station within which satisfactory reception can be obtained only under favorable conditions.

secondary standard — A unit (e.g., length, capacitance, weight) used as a standard of comparison in individual countries or localities, but checked against the one primary standard in existence somewhere in the world.

secondary station — Station licensed for a land location other than the primary station location, i.e., for use at a subordinate location such as an office, vacation home, etc.

secondary storage — Also called auxiliary storage. 1. Storage that is not an integral part of a computer, but which is directly linked to and controlled by it. 2. Devices that are used to store large quantities of data and programs. To be processed, these data and programs must first be loaded into primary storage.

secondary voltage — The voltage across the secondary winding of a transformer.

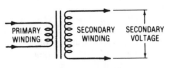

Secondary voltage.

secondary winding — Abbreviated sec. 1. The winding on the output side of a transformer. 2. A transformer winding that receives its energy by electromagnetic induction from a primary winding.

secondary X-rays — X-rays given off by an object irradiated with X-rays. Their frequency depends on the material in the object.

second breakdown — 1. A condition in which the output impedance of a transistor changes almost instantaneously from a large value to a small limiting value. It may be distinguished from normal transistor operation by the fact that once it occurs, the base no longer controls normal collector characteristics. Second breakdown is associated with imperfections in the device structure, usually being more severe in multiple-diffused, high-speed devices. 2. Lateral current instability through a transistor when operating at relatively high voltages and current. It has its greatest effect under dc conditions, but falls off with increasing temperature and frequencies; the breakdown caused is usually permanent. 3. A potentially destructive phenomenon that occurs in all bipolar transistors. This phenomenon may occur when the transistor operates in the active region with a forward-biased emitter-to-base junction, or with the application of reverse bias during the cutoff mode of transistor operation.

second-channel attenuation — Alternate-channel attenuation. See also selectance, 2.

second-channel interference — Also called alternate-channel interference. Interference in which the extraneous power originates from an assigned (authorized) signal two channels away from the desired channel.

second detector — Also called a demodulator. In a superheterodyne receiver, the portion that separates the audio component from the modulated intermediate frequency.

second generation — In reference to computers, the period during which transistors took the place of vacuum tubes; the period began in the mid-1950s.

second-generation computer — A computer in which solid-state components are used.

second source — 1. The manufacturer of a device, other than the original manufacturer. 2. A reference to the manufacture by a competitive company of a part that is electrically and mechanically identical to the original part.

second-time-around echo — An echo received after a time interval greater than the pulse interval.

section — 1. A four-terminal network that cannot be divided into a cascade of two simpler four-terminal networks. 2. One individual span of a radio relay system;

the number of sections in a system is one more than the number of repeaters.

sectional center — A toll switching point to which are connected a number of primary outlets.

sectionalized vertical antenna — A vertical antenna separated into parts by insulators at one or more points along its length. When suitable reactances or driving voltages are placed across the insulated points, the current distribution is modified to give a more desirable radiation pattern in the vertical plane.

sectionalizing — Breaking up of a distribution feeder into smaller sections to minimize the extent of power interruption to customers.

sector — 1. The smallest recordable unit on a CD. A disc can contain [(75 sectors per second) × (60 seconds per minute) × (number of minutes on disc)] sectors. The amount of data contained in the sector depends on the physical format and mode in which it is recorded; for regular CD-ROM (Mode 1) data, 2048 bytes (2 kilobytes) of data can fit. 2. A segment of a magnetic disk's track, typically occupied by one block of data. 3. The smallest contiguous storage area on a magnetic secondary storage medium. In microprocessor systems with flexible diskette drives as the secondary storage medium, sector size is typically 128 bytes.

sectoral horn — A horn with two parallel and two diverging sides.

Sectoral horn.

sector cable — A multiconductor cable in which the cross section of each conductor is essentially a sector of a circle, an ellipse, or some figure intermediate between them. Sector cables are used in order to make possible the use of larger conductors in a cable of given diameter.

sector conductor — A stranded conductor whose cross section is approximately the shape of a sector of a circle. A multiconductor insulated cable with sector conductors has a smaller diameter than the corresponding cable with round conductors.

sector display — A range-amplitude display used with a radar set. The antenna system rotates continuously, and the screen (of the long-persistence type) is excited only while the beam is within a narrow sector centered on the object.

sector scan — A scan in which the antenna oscillates through a selected angle.

sector scanning — Modified circular scanning in which only a portion of the plane or flat cone is generated.

secular variation — A slow variation in the strength of the earth's magnetic field.

secure mode — The condition of an alarm system in which all sensors and control units are ready to respond to an intrusion.

secure voice — Voice message that is scrambled or coded, therefore not transmitted in the clear.

SED — Abbreviation for spectral energy distribution.

Seebeck coefficient — The ratio of the open-circuit voltage to the temperature difference between the hot and cold junctions of a circuit exhibiting the Seebeck effect.

Seebeck coefficient of a couple — For homogeneous conductors, the limit, as the difference in temperature approaches zero, of the quotient of the Seebeck emf divided by the temperature difference between the junctions. By convention, the Seebeck coefficient of a couple is considered positive if, at the cold junction, the first named conductor has a positive potential with respect to the second. It is the algebraic difference between either the relative or absolute Seebeck coefficients of the two conductors.

Seebeck effect — 1. The production of an electromotive force (emf) in a circuit composed of two dissimilar metals when their two junctions are at different temperatures. The emf is considered to be the resultant of the Peltier and Thomson emfs around the circuit. 2. Characteristic of dissimilar metals in thermoelectric solar cells whereby separate junctions exhibiting distinct temperatures transform incident voltage into a current.

Seebeck emf — Also called thermal emf. The emf produced by the Seebeck effect.

seed — A special single crystal from which large single crystals are grown by the Czochralski technique.

seed crystal — A crystal used to start the growing of a large semiconductor ingot.

seek — 1. With reference to a computer, to look for data according to information given with respect to that data. 2. Moving a set of read/write heads so that one of them is over the desired track.

seek time — The time needed to position the head of a disk drive over the specified track.

segment — 1. In a routine, the part short enough to be stored entirely in the internal storage of a computer, yet containing all the coding necessary to call in and jump automatically to other segments. 2. To divide a program into an integral number of parts, each of which performs a part of the total program and is short enough to be completely stored in internal memory.

segmental conductor — A round, stranded conductor composed of three or four sectors slightly insulated from one another. This construction has the advantage of lower ac resistance (less skin effect).

segmentation — A technique for managing variable-sliced areas of memory, termed segments, that contain logical program parts.

segmented thermoelectric arm — A thermoelectric arm made up of two or more materials that have different compositions.

segmenting — *See* partitioning.

segment table — A table that describes all segments of a task and is used by the operating system for memory allocation, relocation, and paging.

seismic mass — The force-summing member for applying acceleration and/or gravitational force in an accelerometer.

seismic sensor — A sensor, generally buried under the surface of the ground for perimeter protection, that responds to minute vibrations of the earth generated as an intruder walks or drives within its detection range.

seismograph — An instrument for recording the time, direction, and intensity of earthquakes or of earth shocks produced by explosions.

seize — To access or connect to a communications circuit while at the same time making it busy to prevent intrusion.

selcal — Acronym for selective calling system. This system allows a teleprinter to be remotely controlled by a unique code.

selectance — 1. A measure of the drop in response as a resonant device loses its resonance. It is the ratio of the amplitude of response at the resonant frequency to the response at some other specified frequency. 2. Often expressed as adjacent-channel attenuation (ACA) or second-channel attenuation (SCA). The reciprocal of the ratio of the sensitivity of a receiver tuned to a specified channel to its sensitivity at another channel a specified number of channels away.

selected mode — A mode of operation for an encoder selector circuit in which one set of brushes is selected to be read and another inhibited from being read; also a mode of operation for a system controlling several encoder outputs in which the encoder is selected to be read and all others inhibited from being read.

selection check — A verification of a computer instruction, usually automatic, to ensure that the correct register or device has been chosen.

selection ratio — The ratio of the least magnetomotive force used to select a cell or core to the maximum magnetomotive force used that is not intended to select a cell or core.

selection sort — A simple sorting algorithm for putting all the elements in a file in order.

selective — The characteristic of responding to a desired frequency to a greater degree than to other frequencies.

selective absorption — Absorption of rays of a certain group of frequencies only.

selective calling — 1. A means of calling in which code signals are transmitted for the purpose of activating the automatic attention device at the station being called. 2. A type of operation in which the transmitting station can specify which of several stations on a line is to receive a message.

selective diffusion — The process in which specified isolated regions in a semiconductor material are doped. The components in a silicon integrated circuit are formed in this way.

selective dump — A dump of a selected area of internal computer storage.

selective fading — Fading in which the received signal does not have the same variation in strength for all frequencies in the band. Selective fading usually occurs during multipath transmission.

selective interference — Interference whose energy is concentrated within narrow frequency bands.

selective ringing — An arrangement used on telephone party lines so that only the bell of the called subscriber rings.

selective squelch — *See* squelch circuit.

selectivity — 1. The characteristic that determines the extent to which the desired signal can be differentiated from disturbances of other frequencies. 2. A tuner's ability to discriminate between a wanted signal and an interfering signal on adjacent frequency settings of the tuning dial. It is the ratio (in decibels) of the signal strength that produces a standard output on the desired channel to the strength of a signal on a nearby channel needed to produce an output 30 dB below the standard level. Selectivity measurements usually refer to signals on the alternate channel (400 kHz from the desired channels in FM and 20 kHz in AM); selectivity measured on the adjacent channel (200 kHz for FM, 10 kHz for AM)

is usually lower. The higher the selectivity, the less interference there will be from signals on nearby channels, which is most important in crowded metropolitan areas. 3. The characteristic that determines the extent to which the desired frequency can be differentiated from other frequencies.

selectivity control — The control for making a receiver more selective.

select lines — In a core memory array, the wires that pass through magnetic cores and carry the selecting coincident currents.

selector pulse — A pulse used to identify one event of a series.

selector relay — A relay capable of automatically selecting one or more circuits.

selector switch — A multiposition switch that permits one or more conductors to be connected to any of several other conductors.

selectron — A computer-memory tube capable of storing 256 binary digits and permitting very rapid selection and access.

selenium — A chemical element with marked photosensitive properties and a resistance that varies inversely with illumination. It is used as a rectifier layer in metallic rectifiers.

selenium cell — A photoconductive cell consisting of a layer of selenium on a substrate whose electrical resistance varies with the illumination falling on the cell. (Selenium cells have been largely replaced by photocells of one kind or another.)

selenium rectifier — A metallic rectifier in which a thin layer of selenium is deposited on one side of an aluminum plate and a highly conductive metal is coated over it. Electrons flow more freely from the coating to the selenium than in the opposite direction, thereby providing rectification.

Selenium rectifier.

self-adapting — Pertaining to the ability of a system to change its performance characteristics in response to its environment.

self-adaptive system — A system that can exhibit the qualities of reorganization and/or learning.

self-adjusting communication — *See* adaptive communication.

self-aligned thick oxide — A term used to describe a proprietary low-voltage, self-aligned gate process.

self-aligning-gate MOS — Abbreviated SAG MOS. 1. An MOS device in which a polycrystalline silicon layer is substituted for the usual aluminum metal gate. The key feature is a different processing technology in which the gate is automatically aligned. 2. A process in which materials like polycrystalline silicon or refractory metals are used in place of aluminum at the gate. These materials act as a mask and result in the gate being automatically aligned between source and drain regions.

self-balancing recorder — A recording device operating on the servomechanism principle.

self-bias — Also called automatic bias. The voltage developed by the flow of vacuum-tube current through a resistor in a grid or cathode lead.

self-capacitance — *See* distributed capacitance.

self-checking code — In computers, a code in which errors produce forbidden combinations. A single self-checking code produces a forbidden combination if a digit gains or loses a single bit. A double self-checking code produces a forbidden combination if a digit gains or loses either one or two bits, and so forth.

self-cleaning contact — *See* wiping contact.

self-complementing code — A machine language in which the code of the complement of a digit is the complement of the code of the digit.

self-contained instrument — An instrument that has all the necessary equipment built into the case or made a corporate part thereof.

self-demagnetization — The process by which a magnetized sample of magnetic material tends to demagnetize itself by virtue of the opposing fields created within it by its own magnetization. Self-demagnetization inhibits the successful recording of signal components having short wavelengths or sharp transitions.

self-diagnostic — The hardware and firmware within a controller that allows it to continuously monitor its own status and indicate any fault that could occur within it.

self-discharge — The loss of useful capacity of a cell or battery in storage due to internal chemical reactions; for example, chemicals evaporating, or electrolyte slowly reacting with the anode even on open circuit.

self-energizing — A type of electrostatic phone that uses the stepped-up signal voltage to supply the dc polarizing voltage required for operation.

self-erasure — The tendency for strongly magnetized areas of the tape coating to erase adjacent areas of opposite-polarity magnetization. This is a major cause of loss of high frequencies at reduced tape speeds.

self-excitation — The supplying of required exciting voltages by a device itself rather than from an external source.

self-excited — A type of generator that provides the current for its own field coils.

self-excited oscillator — An oscillator that operates without external excitation and solely by the direct voltages applied to the electrodes. It depends on its resonant circuits for frequency determination (i.e., not crystal controlled).

self-extinguishing — Material that ignites and burns when exposed to flame or elevated temperature, but which stops burning when the flame or high temperature is removed.

self-focused picture tube — A television picture tube with an automatic electrostatic focus designed into the electron gun.

self-generating transducer — A transducer that requires no external electrical excitation to provide an output.

self-healing — 1. The characteristic of metallized capacitors by which faults or shorts occurring during operation are removed, or healed, by an internal clearing action, and the part continues to function. *See also* clearing, 1. 2. The process by which the electrical properties of the capacitor, after local breakdown of the dielectric, are instantaneously and essentially restored to the values before the breakdown.

self-healing capacitor — A capacitor that restores itself to operation after a breakdown caused by excessive voltage. The electrode layer is so thin that a small part of the energy released in a breakdown condition is quite sufficient to evaporate the metal layer around the breakdown point and, thus, to terminate the breakdown in the shortest possible time. Only a small healed area is produced in the dielectric, and the capacitor voltage only drops to a minimal extent during the breakdown.

self-heated thermistor — A thermistor whose body temperature is significantly higher than the temperature of its ambient medium as a result of the power being dissipated in it.

self-heating coefficient of resistivity — The maximum change in resistance due to temperature change caused by power dissipation, at constant ambient temperature. Usually expressed in percent or per-unit (ppm) change in nominal resistance per watt of dissipation. This parameter is actually the product of the power coefficient and the resistor temperature coefficient.

self-impedance — At any pair of terminals of a network, the ratio of an applied potential difference to the resultant current at these terminals (all other terminals open).

self-inductance — 1. The property that determines the amount of electromotive force induced in a circuit whenever the current changes in the circuit. 2. At any pair of terminals of a network, the ratio of an applied potential difference to the resultant current at these terminals, all other terminals being open.

self-induction — The property that causes a counter-electromotive force to be produced in a conductor when the magnetic field produced by the conductor collapses or expands with a change in current.

self-inductor — An inductor used for changing the self-inductance of a circuit.

self-instructed carry — A system of executing the carry process in a computer by allowing information to propagate to succeeding places as soon as it is generated, without receipt of a specific signal.

self-latching relay — A relay in which the armature remains mechanically locked in the energized position until deliberately reset.

self-optimizing communication — *See* adaptive communication.

self-organizing — Having to do with the ability of a system to arrange its own internal structure.

self-organizing machines — Machines that can recognize, or learn to recognize, such stimuli as patterns, characters, and sound, and which can then adapt to a changing environment.

self-passivating glaze — The glassy material in a thick-film resistor that comes to the surface and seals the surface against moisture.

self-powered — Equipment containing its own power supply. It may be either a combination of wet and dry cells, or dry cells in conjunction with a spring-driven motor.

self-pulse modulation — Modulation accomplished by using an internally generated pulse. *See also* blocking oscillator, 1.

self-pulsing — A special type of grid-pulsing circuit that automatically stops and starts the oscillations at the pulsing rate.

self-quenched counter tube — A radiation-counter tube in which reignition of the discharge is inhibited.

self-quenched detector — A superregenerative detector in which the grid-leak grid-capacitor time constant is sufficiently large to cause intermittent oscillation above audio frequencies. As a result, normal regeneration is stopped just before it spills over into a squealing condition.

self-quenching oscillator — An intermittent self-oscillator producing a series of short trains of rf oscillations separated by intervals of quiescence. The quiescence is caused by rectified oscillatory currents, which build up to the point at which they cut off the oscillations.

self-rectifying X-ray tube — An X-ray tube operating with an alternating anode potential.

self-refresh — In a dynamic RAM, a method of asynchronously refreshing during a battery backup mode of operation with only one control pin held active.

self-regulation — The tendency of a component or system to resist change in its condition or state of operation.

self-repeating timer — A form of time-delay circuit in which relay contacts are used to restart the time delay.

self-reset — Automatically returning to the original position when normal conditions are resumed (applied chiefly to relays and circuit breakers).

self-saturating rectifier — A half-wave rectifying-circuit element connected in series with the output windings of a saturable reactor in a self-saturating magnetic-amplifier circuit.

self-saturation — The saturation obtained in a magnetic amplifier by rectifying the output current of a saturable reactor.

self-screening range — The range at which a target can be detected by a radar in the midst of its jamming mask, with a certain specified probability.

self-selecting V scan — The V-scan method of reading a polystrophic code (primarily the binary code), in which diode logic circuits are used internally in the encoder to perform the necessary bit-to-bit selection to prevent ambiguity in the encoder output data.

self-starting synchronous motor — A synchronous motor provided with the equivalent of a squirrel-cage winding so it can be started like an induction motor.

self-stopping modulo-n counter — A counter having n distinct states that stops when it reaches a predetermined maximum number; it then does not accept count pulses until it is reset to a number less than the maximum number.

self-sustained oscillations — Oscillations maintained by the energy fed back from the output to the input circuit.

self-testing — The ability of a piece of equipment to automatically verify the proper operation of its components or subsystems.

self-threading reel — A device for spooling and storing tape; in particular, one that does not require external aid to affix or start the first turn of tape on its winding surface or hub. Flanges on such reels can be continuous and free of windows; winding surfaces are continuous and free of distortion-producing threading slots.

self-wiping contact — *See* wiping contact.

selsyn — Electrical remote-indicating instrument operating on direct current, in which the angular position of the transmitter shaft carrying a contact arm moving on a resistance strip controls the pointer on the indicator dial.

SEM — 1. Abbreviation for standard electronic module. A subassembly configuration format that meets a particular U.S. Navy set of specifications. 2. Abbreviation for scanning electron microscope.

semantics — The relationships between symbols and their meanings.

semaphore — *See* lock byte.

semiactive homing guidance — A system of homing guidance in which radiations used by the receiver in the missile are reflected from a target being illuminated by an outside source.

semiactive repeater — A communications satellite that uses a minimum of onboard electronics to take a modulated signal beamed at it from a ground station and

transfer its information (modulation) into an unmodulated beam (on a different frequency) set up by the receiving station. For this transfer it uses Van Atta or other directive arrays and nonlinear elements.

Semiautomatic Ground Environment — Abbreviated SAGE. An air-defense system in which data from air surveillance is processed for transmission to computers at direction centers.

semiautomatic keying circuits — Mechanization that provides torn-tape switching systems in teleprinter links. Incoming and outgoing messages are placed on tapes that are inserted manually into a distributor that provides automatic mechanical keying of the circuit.

semiautomatic message switching center — A center at which messages are routed by an operator on the basis of information contained in them.

semiautomatic phonograph — A phonograph having automatic arm return and motor shutoff at the end of a record, but no automatic start and tonearm setdown at the beginning of play.

semiautomatic starter — A starter in which some of the operations are not automatic, but selected portions are automatic.

semiautomatic tape relay — A method of communication whereby messages are received and retransmitted in teletypewriter tape form involving manual intervention in the transfer of the tape from the receiving reperforator to the automatic transmitter.

semiautomatic telephone system — A telephone system in which operators receive orders from the calling parties verbally, but use automatic apparatus in making connections.

semiconducting material — A solid or liquid having a resistivity midway between that of an insulator and a metal.

semiconductor — 1. A class of materials, such as silicon and germanium, whose electrical properties lie between those of conductors (such as copper and aluminum) and insulators (such as glass and rubber), in which the electrical charge carrier concentration increases with increasing temperature over some temperature range. Over most of the practical temperature range, the resistance has a negative temperature coefficient. Certain semiconductors possess two types of carriers: negative electrons and positive holes. The charge carriers are usually electrons, but there may be also some ionic conductivity. 2. An electronic device whose main functioning parts are made from semiconductor materials. Examples include germanium, lead sulfide, lead telluride, selenium, silicon, and silicon carbide. Used in diodes, photocells, thermistors, and transistors. 3. A device (or material) with an electrical conductivity that lies between metal conductors and insulator devices. 4. A material whose resistivity is between that of insulators and conductors. The resistivity is often changed by light, heat, and electric field, or a magnetic field. Current is often achieved by transfer of positive holes as well as by movement of electrons. 5. A material that exhibits relatively high resistance in a pure state and much lower resistance when it contains small amounts of certain impurities. The term is also used to denote electronic devices made from semiconductor materials.

semiconductor carrier — A permanent protective structure that provides for mounting and for electrical continuity in application of a semiconductor chip to a major substrate.

semiconductor chip — A single piece of semiconductor material of any dimension.

semiconductor device — 1. A device in which the characteristic distinguishing electron conduction takes place within a semiconductor material, ranging from the single-unit transistor to multiple-unit devices such as the semiconductor rectifier. Other devices are diodes, photocells, thermistors, and thyristors. 2. A device in which n-type and p-type materials are used in combination to obtain specific characteristics for controlling the flow of current. 3. A device, including its encapsulation and terminals, whose essential characteristics are governed by the flow of charge carriers within a semiconductor material.

semiconductor diode — 1. A device consisting of n-type and p-type semiconductor material joined together to form a pn junction, which passes current in the forward direction (from anode to cathode) and blocks current in the reverse direction. High reverse voltages, such as transients greater than a specified limit, can destroy the junction due to excessive reverse leakage currents. *See also* crystal diode. 2. A light-emitting diode that emits coherent light by suitably arranged geometry. Gallium arsenide is used for lasers of this type. 3. A two-electrode semiconductor device that conducts current more easily in one direction.

semiconductor-diode parametric amplifier — A parametric amplifier using one or more varactors.

semiconductor ignition system — An ignition system for internal combustion engines that employs solid-state semiconductors for switching purposes.

semiconductor integrated circuit — Abbreviated SIC. 1. Complex circuits fabricated by suitable and selectively modifying areas on and within a wafer of semiconductor material to yield patterns of interconnected passive as well as active elements. The circuit may be assembled from several chips and use thin-film elements or even discrete components to achieve a specified performance when the necessary device parameters cannot be achieved by materials modification. 2. The physical realization of a number of electric elements inseparably associated on or within a continuous body of semiconductor material to perform the function of a circuit.

semiconductor intrinsic properties — Properties of a semiconductor that are characteristic of the ideal crystal.

semiconductor junction — The region of transition between semiconducting regions of different electrical properties, usually between p-type and n-type materials.

semiconductor laser — 1. A device in which laser action takes place through stimulated recombination of free electrons in the conduction band with holes in the valence band of a direct-gap semiconductor such as gallium arsenide. 2. A light-emitting diode that uses stimulated emission to produce a coherent-light output. *See also* diode laser.

semiconductor lead wire — Fine wire used to connect semiconductor chips to substrate patterns, packages, other chips, etc. Usually made from an aluminum alloy or gold.

semiconductor material — 1. A material in which the conductivity ranges between that of a conductor and an insulator. The electrical characteristics of semiconductor materials such as silicon are dependent on the small amounts of added impurities, or dopants. 2. A chemical element, such as silicon or germanium, that has a crystal lattice whose atomic bonds are such that the crystal can be made to conduct an electric current by means of free electrons or holes.

semiconductor memory — 1. A memory whose storage medium is a semiconductor circuit. Often used for high-speed buffer memories and for read-only memories. 2. A memory in which semiconductors are used as the storage elements; characterized by low-to-moderate cost storage and a wide range of memory operating speed,

from very fast to relatively slow. Almost all semiconductor memories are volatile. 3. A memory with storage elements formed by integrated semiconductor devices, as opposed to a memory composed of ferrite cores. Semiconductor read/write memories are characterized by low cost, wide speed ranges, and data volatility. Semiconductor read-only memories are nonvolatile. 4. A computer memory that uses silicon integrated-circuit chips.

semiconductor photodiode — A semiconductor diode utilizing the photoelectric effect.

semiconductor rectifier diode — A semiconductor diode designed for rectification and including its associated mounting and cooling attachments if integral with it.

semicustom IC — An LSI circuit that incorporates either linear or digital components. Semicustom ICs are designed to serve as replacements for small- to medium-scale ICs and are based on the concept of integrating extremely complex functions onto a single IC to fulfill a particular custom function. One semicustom IC may replace anywhere from several to more than 100 individual ICs. The semicustom IC, initially processed up to, but not including, the interconnect level, is adapted to specific requirements with relative ease through patterning of the metal interconnect layer. By connecting the transistors (which may number in the thousands) on a single IC in different ways, various functions, such as flip-flops, gates, adders, and parity control, are created. These blocks are then interconnected to form an entire system function on the chip.

semicustom large-scale integrated circuit — *See* gate array, 3.

semicustom logic — Chips whose logic functions can be determined in the final stages of manufacturing, such as PLAs and gate arrays.

semidirectional microphone — A microphone whose field response is determined by the angle of incidence in part of the frequency range but is substantially independent of the angle of incidence in the remaining part.

semiduplex — In a communications circuit, a method of operation in which one end is duplex and one end simplex. This type of operation is sometimes used in mobile systems, with the base station duplex and the mobile station or stations simplex. A semiduplex system requires two operating frequencies.

semimagnetic controller — An electrical controller whose basic functions are not all performed by electromagnets.

semimetals — Materials, such as bismuth, antimony, and arsenic, having characteristics that class them between semiconductors and metals.

semiremote control — Radio-transmitter control performed near the transmitter by devices connected to but not an integral part of the transmitter.

semiselective ringing — An arrangement in which the bells of two stations on a telephone party line are rung simultaneously; differentiation is made by the number of rings.

semitone — Also called half step. The interval between two sounds. Its basic frequency ratio is equal to approximately the twelfth root of 2.

semitransparent photocathode — A photocathode in which radiant flux incident on one side produces photoelectric emission from the opposite side.

sender — That part of an automatic-switching telephone system that receives pulses originated by a dial or other source and, in accordance with the pulses received, controls the further operations necessary to establish the connection.

sending-end impedance — Also called the driving-point impedance. The ratio of an applied potential difference of a transmission line to the resultant current at the point where the potential difference is applied.

sending filter — A filter used at the transmitting terminal to restrict the transmitted frequency band.

sensation level — *See* level above threshold.

sense — 1. In navigation, the relationship between the change in indication of a radionavigational facility and the change in the navigational parameter being indicated. 2. In some navigational equipment, the property of permitting the resolution of 180° ambiguities. 3. To examine or determine the status of some system components. 4. To read holes in punched tape or cards.

sense amplifier — 1. A circuit used to sense low-level voltages such as those produced by magnetic or plated-wire memories and to amplify these signals to the logic voltage levels of the system. 2. A circuit used in communications-electronics equipment to determine a change of phase or voltage and to provide an automatic control function.

sense finder — In a direction finder, that portion which permits determination of direction without 180° ambiguity.

sense-reversing reflectivity — The characteristic of a reflector that reverses the sense of an incident ray. (For example, a perfect corner reflector is invisible to a circularly polarized radar because it reverses the sense.)

sense step — *See* secondary calibration.

sense switch — One of a series of switches on the console of the digital computer that permits the operator to control some parts of a program externally.

sense wire — A wire threaded through the core of a magnetic memory to detect whether a logical 1 or 0 is stored in the core when the core is interrogated by a read pulse. This technology is no longer in use.

sensing — 1. The process of determining the sense of an indication. 2. A technique used in a power supply regulator for monitoring the output voltage or current. In local sensing, the monitor points are the output terminals. In remote sensing, the monitor points are located at appropriate locations in the circuit being powered, connected by wire to sensing input terminals on the supply.

sensing element — *See* primary detector.

sensing field — The zone in which an object can be sensed by a proximity switch.

sensistor — A silicon resistor whose resistance varies with temperature, power, and time.

sensitive relay — 1. A relay requiring only a small current. It is used extensively in photoelectric circuits. 2. Any of a number of different types of relays requiring very low pickup power. Generally considered to be one requiring less than 100 milliwatts of pickup power.

sensitive volume — In a radiation-counter tube, the portion responding to a specific radiation.

sensitivity — 1. The minimum input signal required in a radio receiver or similar device to produce a specified output signal having a specified signal-to-noise ratio. This signal input may be expressed as power or voltage at a stipulated input network impedance. 2. Ratio of the response of a measuring device to the magnitude of the measured quantity. It may be expressed directly in divisions per volt, milliradians per microampere, etc., or indirectly by stating a property from which sensitivity can be computed (e.g., ohms per volt for a stated deflection). 3. The signal current developed in a camera tube per unit incident radiation density (i.e., per watt per unit area). Unless otherwise specified, the radiation is understood to be that of unfiltered incandescent source of 2854 K, and its density, which is generally measured in watts per unit area, may then be expressed in lumens per foot. 4. The degree of response of an instrument or control unit to a change in the incoming signal. 5. In tape recording, the relative intensity of the magnetic signal recorded by a magnetizing field of a given intensity. 6. A measurement of the electrical output of a microphone for a given sound pressure level at its diaphragm. 7. The smallest input change that a DMM is able to display. It is equal to the least significant digit on the lowest measurement range. For example, a three-digit DMM with a 100-mV range has 100 μV sensitivity. 8. Generally expressed in dBm at a specified impedance (usually 600 ohms), sensitivity is a measure of the lowest DTMF signal level that a receiver can detect. It represents an absolute threshold below which detection of a single frequency is not generated. 9. Measure of the ability of a device or circuit to react to a change in some input. 10. In television, a factor expressing the incident illumination on a specified scene required to produce a specified picture signal at the output terminals of a television camera. 11. A measure of relative output for a given input of a tape, microphone, etc. 12. Characteristic of a receiver that determines the minimum input signal strength required for a given signal output. Sensitivity is usually measured in microvolts (μV).

sensitivity adjustment — Also called span adjustment. The control of the ratio of output signal to excitation voltage per unit measurand. Generally accomplished in a system by changing the gain of one or more amplifiers. The practice of placing excitation control components (such as potentiometers or rheostats) in series with the excitation to a transducer is a sensitivity adjustment for the system. However, in the latter case no significant change is introduced in the output-to-input ratio of the transducer.

sensitivity control — The control that adjusts the amplification of the radio-frequency amplifier stages and thereby makes the receiver more sensitive.

sensitivity-time control — Also called gain-time control or time gain. The portion of a system that varies the amplification of a radio receiver in a predetermined manner.

sensitizing (electrostatography) — The establishing of an electrostatic surface charge of uniform density on an insulating medium.

sensitometer — An instrument used to measure the sensitivity of light-sensitive materials.

sensitometry — Measurement of the light-response characteristics of photographic film.

sensor — 1. In a navigational system, the portion that perceives deviations from a reference and converts them into signals. 2. A component that converts mechanical energy into an electrical signal, either by generating the signal or by controlling an external electrical source. 3. *See* primary detector. 4. An information-pickup device. 5. A transducer designed to produce an electrical output proportional to some time-varying quantity, as temperature, illumination, pressure, etc. 6. The component of an instrument that converts an input signal into a quantity that is measured by another part of the instrument. 7. Any device that can detect the presence of, or a change in the level of, light, sound, capacitance, magnetic field, etc. 8. A device or component that reacts to a change; the reaction is then used to cause a control or instrument to function. For example, a thermistor changes resistance as temperature changes, and the resistance changes can be used in an electric circuit to vary current. 9. A transducer that converts a parameter at a test point to a form suitable for measurement by the test equipment. 10. A sensing element. The basic element of a transducer that usually changes some physical parameter to an electrical signal. 11. A device that is designed to produce a signal or offer indication in response to an event or stimulus within its detection zone. 12. A component that provides an electrical signal in response to a specific physical or chemical stimulus such as heat, pressure, magnetic field, or a particular chemical vapor. Microsensors are fabricated using processes similar to those for manufacturing ICs, or extensions of such processes. Integrated microsensors incorporate an integrated circuit on the same die as that used for the sensor element.

sensory robot — A computerized robot with one or more artificial senses, usually sight or touch.

sentinel — 1. A symbol marking the beginning or end of some piece of information in digital-computer programming. 2. *See* tag.

separate excitation — Excitation in which generator field current is provided by an independent source, or motor field current is provided from a source other than the one connected across the armature.

separately instructed carry — Executing the carry process in a computer by allowing carry information to propagate to succeeding places only when a specific signal is received.

separation — The degree to which two stereo signals are kept apart. Stereo realism is dependent on the successful prevention of their mixture before reaching the output terminals of the power amplifier. Tape systems have a separation capability inherently far superior to that of the disc systems.

separation circuit — A circuit that separates signals according to their amplitude, frequency, or some other selected characteristic.

separation filter — A combination of filters used to separate one band of frequencies from another — often, to separate carrier and voice frequencies for transmission over individual paths.

separation loss — The loss that occurs in output when the surface coating of a tape fails to make perfect contact with the surfaces of either the record or reproduce head.

separator — 1. An insulating sheet or other device employed in a storage battery to prevent metallic contact between plates of opposite polarity within a cell. 2. An insulator used in the construction of convolutely wound capacitors. 3. *See* delimiter.

septate coaxial cavity — A coaxial cavity with a vane or septum added between the inner and outer conductors. The result is a cavity that acts as if it had a rectangular cross section bent transversely.

septate waveguide — A waveguide with one or more septa placed across it to control microwave power transmission.

septum — A thin metal vane that has been perforated with an appropriate wave pattern. It is inserted into a waveguide to reflect the wave. Plural: septa.

sequence — 1. The order in which objects or items are arranged. 2. To place in order. 3. A succession of terms so related that each may be derived from one or more of the preceding terms in accordance with some fixed law.

sequence checking routine — A checking routine that examines every instruction executed and prints certain data concerning this check.

sequence control — Automatic control of a series of operations in a predetermined order.

sequencer — 1. The component of a processor that controls the program flow by implementing branches for subroutine processing and handling interrupts. 2. A device or computer program that records, edits, and plays back MIDI data much like a word processor for music, such that you can fix wrong notes or lengthen or shorten notes.

3. A mechanical or electronic device that may be set to initiate a series of events and to make the events follow in sequence. 4. A circuit that pulls information from the control store memory, based on external conditions. 5. In a bit-slice system, the module in charge of providing the next microprogram address to the microprogram memory. Essentially a complex multiplexer, but may include stack facilities and a loop counter.

sequence relay — A relay that controls two or more sets of contacts in a predetermined sequence.

sequencer register — In a computer, a counter that is pulsed or reset following the execution of an instruction to form the new memory address that locates the next instruction.

sequence timer — A succession of time-delay circuits arranged so that completion of the delay in one circuit initiates the delay in the following circuit.

sequencing equipment — A special selecting device by means of which messages received from several teletypewriter circuits may be subsequently selected and retransmitted over a smaller number of trunks or circuit.

sequency of operation — A detailed written description of the order in which electrical devices and other parts of the equipment should function.

sequential access — 1. An access mode in which records are retrieved in the same order in which they were written. Each successive access to a file refers to the next record in the file. 2. A term used to describe files such as magnetic tape that must be searched sequentially to find any desired record. 3. Computer access method in which a word is accessed by scanning sequential blocks or records. For example: a tape. 4. Data on storage, such as magnetic tape, that much be searched serially from the beginning to find any desired record.

sequential-access file — A type of file structure in which data may only be accessed sequentially, one record at a time. Data stored on magnetic tape is an example of a sequential file.

sequential-access memory — Abbreviated SAM. 1. A serial-type memory in which words are selected in a fixed order. The addressing circuit steps from word to word in a predetermined order, with the result that the access time for the stored information (words) is variable. 2. A method of information retrieval in which the complete memory is scanned and each word is, in its turn, read out, worked on, then rewritten.

sequential circuit — A digital circuit that changes state according to an input signal (normally under clock control); it must be tested with a sequence of signals.

sequential color television — A color television system in which the three primary colors are transmitted in succession and reproduced on the receiver screen in the same manner.

sequential color transmission — The transmission of television signals that originate from variously colored parts of an image in a particular sequential order.

sequential computer — A computer in which events occur in time sequence with little or no simultaneous occurrence or overlap of events.

sequential control — Digital-computer operation in which the instructions are set up in sequence and fed to the computer consecutively during the solution of a problem.

sequential element — A device having at least one output channel and one or more input channels, all characterized by discrete states, such that the state of each output channel is determined by the previous states of the input channels.

sequential interlace — A method of interlacing in which the lines of one field are placed directly under the corresponding lines of the preceding field.

sequential lobing — A direction-determining technique utilizing the signals of overlapping lobes existing at the same time.

sequential logic — 1. A circuit arrangement in a computer in which the output state is determined by the previous state of the input. *See also* combinatorial logic. 2. Part of a circuit in which the output values are a function of the inputs and data stored within the circuit.

sequential logic element — A device that has one or more output channels and one or more input channels, all of which have discrete states, such that the state of each output channel depends on the previous states of the input channel.

sequential operating connector — A form of connector that has two or more groups of contacts that open and close in a predetermined sequence. For example, a connector that is designated for use with ground connections, power distribution, and signal circuits. Operates in such a way that when the connector is closed, the ground contacts close first, power contacts second, and signal contacts last. This sequence is reversed when the connector is opened.

sequential operation — The carrying out of operations one after the other.

sequential relay — A relay that controls two or more sets of contacts in a predetermined sequence.

sequential sampling — Sampling inspection in which the decision to accept, reject, or inspect another unit is made following the inspection of each unit.

sequential scan — A system of TV scanning in which each line of the raster is scanned sequentially.

sequential scanning — In television, rectilinear scanning in which the distance from center to center of successively scanned lines is equal to the nominal line width.

sequential switcher — A device that automatically permits the viewing of pictures from a number of CCTV cameras on one CCTV monitor in a selected sequence.

sequential timer — A timer in which each interval is initiated by the completion of the preceding interval. All intervals may be independently adjusted.

sequential with memory — *See* SECAM.

serial — 1. Pertaining to time-sequential transmission of, storage of, or logical operations on the parts of a word in a computer — the same facilities being used for successive parts. 2. The technique for handling a binary data word that has more than one bit. The bits are acted upon one at a time, analogous to a parade passing a review point. 3. Typically refers to a port on a computer for transmitting one bit at a time. Modems and mice typically connect to a serial port.

serial access — 1. Pertaining to transmission of data to or from storage in a sequential or consecutive manner. 2. Pertaining to the process in which information is obtained from or placed into storage with the time required for such operations dependent on the location of the information most recently obtained or placed in storage. *See also* random access.

serial adder — A device in which additions are performed in a series of steps: the least significant addition is performed first, and progressively more significant additions are performed in order until the sum of the two numbers is obtained.

serial arithmetic unit — In a computer, a unit in which the digits are operated on sequentially. *See also* parallel arithmetic unit.

serial bit — Pertaining to computer storage in which the individual bits making up a word appear in time sequence.

serial computer — A computer having a single arithmetic and logic unit.

serial counter — Also called ripple-through counter. A counter in which each flip-flop cannot change state until after the preceding flip-flop has changed state; relatively long delays after an input pulse is applied to the counter can occur before all flip-flops reach their final states.

serial data — Data transmitted sequentially, one bit at a time.

serial digital computer — A computer in which the digits are handled serially. Mixed serial and parallel machines are frequently called serial or parallel, according to the way the arithmetic processes are performed. An example of a serial digital computer is one that handles decimal digits serially, although the bits that comprise a digit might be handled either serially or in parallel. *See also* parallel digital computer.

serial interface — 1. A data channel that transfers digital data (1s and 0s) in a serial fashion, one bit after another. Serial interfaces save space by requiring fewer lines compared with parallel interfaces, but at the sacrifice of data transfer speeds. 2. A port that sends or receives the eight bits in each byte one by one, much like beads on a string. Printers located far from a computer usually require a serial interface.

serial I/O — A method of data transfer between a computer and a peripheral device in which data is transmitted for input to the computer (or output to the device) bit by bit over a single circuit.

serialize — To convert from parallel-by-bit to serial-by-bit.

serially reusable routine — A computer routine in main storage that can be used by another task following conclusion of the current use.

serial memory — 1. A memory in which information is stored in series and reading or writing of information is done in time sequence, as with a shift register. Compared with a RAM, a serial memory has slow to medium speed and lower cost. *See* sequential-access memory. 2. A memory whose contained data is accessible only in a fixed order, beginning at some prescribed reference point. Data in any particular location is not available until all data ahead of that location has been read. Such a memory is inherently slow compared with a random-access memory.

serial mode — A type of computer operation that is performed bit by bit, generally with the least significant bit handled first. Read-in and readout are accomplished bit after bit by shifting the binary data through the register.

serial operation — 1. In a digital computer, information transfer such that the bits are handled sequentially, rather than simultaneously as they are in parallel operation. Serial operation is slower than parallel operation, but it is accomplished with less complex circuitry. 2. Type of information transfer within a programmable controller whereby the bits are handled sequentially rather than simultaneously, as they are in parallel operation. Serial operation is slower than parallel operation for equivalent clock rate. However, only one channel is required for serial operation.

serial-parallel — Having the property of being partially serial and partially parallel.

serial port — A method of data communication in which bits of information are sent consecutively through one wire.

serial printer — A device that can print characters one at a time across a page.

serial processing — The sequential or consecutive execution of more than one process into a single device such as a channel or processing unit. Opposed to parallel processing.

serial programming — Programming of a digital computer in such a manner that only one arithmetical or logical operation can be executed at one time.

serial storage — 1. In a computer, storage in which time is one of the coordinates used in the location of any given bit, character, or word. 2. A storage media organization in which data or text is serially recorded one character or text block after another. Text access points are retrieved by serially searching through the medium (usually a magnetic-tape cassette or cartridge).

serial transfer — Data transfer in which the characters of an element of information are transferred in sequence over a single path.

serial transmission — 1. Information transmission in which the characters of a word are transmitted in sequence over a single line. 2. The transmission of a character's bits one at a time (implies a single transmission pathway). 3. A method of transferring information in which the code elements or pulses are sent sequentially, one after another. 4. Moving data in sequence one at a time, as opposed to parallel transmission.

series — 1. The connecting of components end to end in a circuit, to provide a single path for the current. 2. An indicated sum of a set of terms in a mathematical expression (e.g., in an alternating or arithmetic series).

series circuit — 1. A circuit in which resistances or other components are connected end to end so that the same current flows throughout the circuit. 2. A circuit in which the current has only one path to follow. 3. An electric circuit in which all the receptive devices are arranged in succession, as distinguished from a parallel circuit. The same current flows through each part of the circuit.

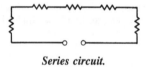

Series circuit.

series connection — A way of making connections so as to form a series circuit.

series excitation — The field excitation obtained in a motor or generator by allowing the armature current to flow through the field winding.

series-fed vertical antenna — A vertical antenna that is insulated from ground and energized at the base.

series feed — The method by which the dc voltage to the plate or grid of a vacuum tube is applied through the same impedance in which the alternating current flows.

series field — In a machine, the part of the total magnetic flux due to the series winding.

series-gate noise limiter — *See* noise limiter.

series loading — Loading in which reactances are inserted in series with the conductors of a transmission circuit.

series modulation — Modulation in which the plate circuits of a modulating tube and a modulated amplifier tube are in series with the same plate-voltage supply.

series motor — Also called series-wound motor. A motor in which the field and armature circuits are connected in series. In small motors with laminated field frames, the performance will be similar when operated on direct current or alternating current. For this reason, the series motor is frequently called a universal motor. A series motor has a high starting torque, but its speed varies with the load.

series operation — The connection of two or more power supplies together to obtain an output voltage of the combination equal to the sum of the individual supplies. A common current passes through all the supplies.

series-parallel network — Any network that contains only resistors, inductors, and capacitors and in which successive branches are connected in series and/or in parallel.

series-parallel switch — A switch that changes the connections of lamps or other devices from series to parallel or vice versa.

series peaking network — *See* peaking network.

series regulator — A device that is placed in series with a source of power and is able to automatically vary its series resistance, thereby controlling the voltage or current output.

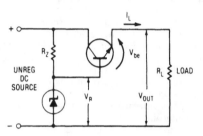

Series-regulated power supply.

series resistance — Any sum of resistances installed in sequential order within one circuit.

series resistor — A resistor generally used for adapting an instrument so that it will operate on some designated voltage or voltages. It forms an essential part of the voltage circuit and may be either internal or external to the instrument.

series resonance — The condition existing in a circuit when the source of electromotive force is in series with an inductance and capacitance whose reactances cancel each other at the applied frequency, thereby reducing the impedance to minimum.

series-resonant circuit — A circuit in which an inductor and capacitor are connected in series and have values such that the inductive reactance of the inductor will be equal to the capacitive reactance of the capacitor at the desired resonant frequency. At resonance, the current through a series-resonant circuit is at maximum.

series-shunt network — *See* ladder network.

series T-junction — *See* E-plane T-junction.

series winding — In a motor or generator, a field winding that carries the same current as the armature; i.e., this winding is in series with the armature rather than in parallel with it. Series-wound motors are used in fractional-horsepower ac-dc applications, such as fans and electric mixers. Their other chief use is in heavy-duty dc traction equipment, such as electric locomotives, because of their extremely high starting torque.

series wound — Characteristic of a generator or motor whose armature and field windings are connected in series.

series-wound motor — A commutator motor with field and armature circuits in series. *See also* Series motor.

serpentine cut — A trim cut in a film resistor that follows a serpentine or wiggly pattern to effectively increase the resistor length and increase resistance.

serrated pulse — A vertical synchronizing pulse divided into a number of small pulses, each acting for the duration of half a line in a television system.

serrated rotor plate — Also called a slotted or split rotor plate. A rotor plate with radial slots that permit different sections of the plate to be bent inward or outward so that the total capacitance of a variable-capacitor section can be adjusted during alignment.

serration — 1. The sawtooth appearance of vertical and near-vertical lines in a television picture. This is caused by their starting at different points during the horizontal scan. 2. A designed irregular surface used as a reservoir to retain excess infiltrating material and/or multiple points to obtain high-current-density resistance welding or resistance brazing.

serrodyne — A frequency translator or frequency converter based on linear sawtooth modulation of phase shift or time delay. One convenient modulable device for serrodyne use is a traveling-wave tube, which provides gain as well as frequency translation.

serve — 1. With reference to cable construction, a type of separator applied directly over the conductor or conductors. The serve may consist of one or more materials such as paper, cotton, silk, nylon, or rayon. These materials may be applied spirally or laterally. 2. A filament or group of filaments, such as fibers or wires, wound around a central core.

server — 1. A computer on a network that serves as a central repository for data and programs and which can be accessed over the network by other computers, called clients. 2. A computer or processor that holds applications, files, or memory shared by users on a network. 3. A computer, or a software package, that provides a specific kind of service to allow client software to run on other computers. The term can refer to a particular piece of software, such as a WWW server, or to the machine on which the software is running. A single server could have several different server software packages running on it, thus providing many different servers to clients on the network.

server PC — Microcomputer used by a network as a source of disk drives and information.

service — 1. A function offered by some part of an open system to communicating application processes. 2. The conductors and equipment for delivering energy from the electricity supply system to the wiring system of the premises served.

serviceability — Those properties of an equipment design that facilitate service and repair in operation.

service area — 1. The area within which a navigational aid is of use. 2. The area surrounding a broadcasting station where the signal is strong enough for satisfactory reception at all times (i.e., not subject to objectionable interference or fading).

service band — The band of frequencies allocated to a class of radio service.

service channel — A band of frequencies, usually including a voice channel, utilized for maintenance and fault indication on a communication system.

service circuit — An interconnecting circuit in a switching network that may be connected to lines or trunks as required to perform various functions, such as dial pulse detection and audible ringing.

service conductors — The supply conductors that extend from the street main or from the transformers to the service equipment of the premises served.

service drop — The overhead service conductors from the last pole or other aerial support to and including the splices or taps, if any, connecting to the service-entrance conductors at the premises served.

service entrance — The conductors and equipment used for delivering energy from the utility pole to the premises.

service life — 1. The period of time during which a device is expected to perform in a satisfactory manner. 2. The length of time a primary cell or battery needs to

reach a specified final electrical condition on a service test that duplicates normal usage.

service oscillator — *See* rf signal generator.

service provider — A company that provides Internet access. For end users, this service can be as simple as providing Internet e-mail accounts, access to Usenet news groups, ftp, and web browser access via the service provider's servers. A service provider can also provide a way for connecting an enterprise's LAN to the Internet, allowing an enterprise to place their own servers on the Internet.

service rating — The maximum voltage or current that a component is designed to carry continuously.

service request — 1. The appeal by a process or task for access to a system resource. 2. A notification to a system that one element wants to access a resource.

service routine — 1. In digital computer programming, a routine designed to assist in the actual operation of the computer. 2. A set of instructions to perform a programmed operation, typically in response to an interrupt.

service switch — A switch, usually in a box, for disconnecting the line voltage from the circuits it services.

service unit — In a microwave system, the equipment or facilities used for maintenance communications and transmission of fault indications.

serving — 1. Of a cable, wrapping applied around the core before a cable is leaded, or over the lead if the cable is armored. Some common materials are jute, cotton, or duct tape. 2. A wrapping of thread or yarn over a relay coil to protect it from damage.

servo — Short for servomotor. A device that contains and delivers power to move a control or controls.

servoamplifier — A servo unit in which information from a synchro is amplified to control the speed and direction of the servomotor output.

servo control — A technique by which the speed or position of a moving device is forced into conformity with a desired or standard speed or position. For example, the speed of a servo-controlled turntable is established by a precision voltage or frequency standard to which it is compared and automatically adjusted to reduce the difference to a minimum. In a servo-controlled tonearm, a small departure of the cartridge from tangency to the groove is sensed and used to activate a motor drive that moves the tonearm to minimize the error.

servo loop — In a servoamplifier, the entire closed loop formed by feedback from output to input. In a position servo, the output position is compared to a command signal at the input.

servomechanism — 1. An automatic feedback-control system in which one or more of its signals represent mechanical motion. 2. A system in which output is compared to input to control error according to desired relationship, or feedback. 3. A self-contained system (except for inputs) in which the feedback signal is subtracted from a desired value so that the difference is reduced to zero. 4. A control system that provides the following: a command instrument to control or program the final process; amplification to strengthen and modify the command signal; work instrumentation to manipulate the controlled process; and feedback provision to initiate corrective action when needed. Since feedback signals go from the controlled process back to the original command station, a servo system is said to operate closed-loop. 5. A device in a closed-cycle system that controls a process based on direct feedback from the process.

servomotor — A motor used in a servo system. Its rotation or speed (or both) are controlled by a corrective electric signal that has been amplified and fed into the motor circuit.

servo noise — The hunting of the tracking servomechanism of a radar as a result of backlash and compliance in the gears, shafts, and structures of the mount.

servo oscillation — An unstable condition in which the load tends to hunt back and forth about the ordered position.

servo system — An automatic control system for maintaining a condition at or near a predetermined value by activation of an element such as a control rod. It compares the required condition (desired value) with the actual condition and adjusts the control element in accordance with the difference (and sometimes the rate of change of the difference).

servo techniques — Methods of studying the performance of servomechanisms or other control systems.

servovalve — Electrohydraulic flow control. An electrical-input, fluid-control valve capable of continuous control.

sesquisideband transmission — A system in which the carrier, one full sideband, and half of the other sideband are transmitted.

set — 1. To place a storage device in a prescribed state that is opposite to the reset state. 2. To place a binary cell in the 1 state. 3. A permanent change, attributable to any cause, in a given parameter. 4. See equipment. 5. Pertaining to a flip-flop input used to affect the Q output. Through this input, signals can be entered to change the Q output from 0 to 1. It cannot be used to change Q to 0. Opposite of clear. 6. An input to a binary counter or register that forces all binaries to assume the maximum binary state. 7. To place a binary device into a given logic state, usually the 1 state.

set analyzer — A test instrument designed to permit convenient measurement of voltages and currents.

set breakpoint — A user debug command that causes the setting of a breakpoint in a specified memory location. At program execution, encountering this breakpoint causes temporary program suspension and a transfer of control to the system debug routine. See breakpoint.

set composite — A signaling circuit in which two signaling or telegraph legs may be superimposed on a two-wire interoffice trunk by means of one of the balanced pairs of high-impedance coils connected to each side of the line with an associated capacitor network.

set input — An asynchronous input to a flip-flop used to force the Q output to its high state.

set noise — Inherent random noise caused in a receiver by thermal currents in resistors and by variations in the emission currents of vacuum tubes.

set point — In a feedback control loop, the point that determines the desired value of the quantity being controlled.

set pulse — A drive pulse that tends to set a magnetic cell.

set-reset flip-flop — A standard flip-flop except that if both the set and reset inputs are a 1 at the same time, the flip-flop will assume a prescribed state.

set terminal — The flip-flop input terminal that triggers the circuit from its first state to its second state.

setting accuracy — The ability to set a knob, switch, or other adjustment to the desired time delay, speed, light, sound, or other parameter. Normally specified in percent of maximum or at set point.

settling time — 1. The time interval, following the initiation of a specified stimulus to a system, required for a specified variable to enter and remain within a specified narrow band centered on the final value of the variable. 2. In an operational amplifier, the interval between the time of application of an ideal step input and the time at which the closed-loop amplifier output enters and remains within a specified band of error, usually symmetrical about the final value. Settling time includes a propagation delay and the time needed for the output to slew to the vicinity of the final value, recover from the overload condition associated with slewing, and settle within the specified error range. 3. In a feedback control system, the time required for an error to be reduced to a specified fraction, usually 2 percent or 5 percent, of its original magnitude. 4. The time required for the output frequency of a voltage- or current-tuned oscillator to change from the initial value to within a specified window around the final value in response to a voltage or current stop on the tuning input port. 5. The period required for a digital multimeter's input circuits to reach a steady-state condition before analog-to-digital conversion is started. There are actually three settling times to evaluate. They are the settling time required with an input step change when operating in a fixed range, the settling time required with a range change, and the settling time required with a function change, such as a switch in measurement from ac to dc volts. 6. Time necessary for a multiplexer's output to be within a certain error percentage of the input signal once the channel is selected, or turned on. It may be specified as either the semiconductor switch's switching time plus analog output settling time, or an analog output settling time alone. 7. The time delay between a change of input-signal value and the resultant change in the output signal. Usually expressed in terms of how long it takes the output to arrive at, and remain within, a certain error band around the final value. Often given for several different magnitudes of input-step change. 8. The time required for an amplifier to approach within a percentage of its final steady state value after the application of a step input. It can be specified as 1 percent, 0.1 percent, or 0.01 percent. 9. The elapsed time after a code transition for DAC output to reach a final value within specified limits, usually $\pm \frac{1}{2}$ lsb. 10. The time elapsing between the start of a measurement and the instant that the indicator reaches, and remains within, a certain percentage of the final measured value (typically ± 1 percent).

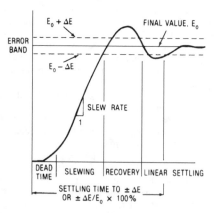

Settling time.

set-top unit — Abbreviated STU. Typically refers to the unit/device that sits on top of a television for current cable TV reception. Also refers to digital set-top units, which accept digital media and sometimes support interactivity.

setup — 1. The ratio of the difference between black level and blanking level to the difference between reference white level and blanking level, usually expressed in percent. 2. An arrangement of data or devices for the solution of a particular problem. 3. The difference between the reference black level and the blanking level of a composite video signal.

setup diagram — A diagram that specifies a given computer setup.

setup time — 1. The time, measured from the point of 10-percent input change, required for a capacitor-diode gate to open or close after the occurrence of a change of input level. 2. The length of time that data must be present and unchanging before a flip-flop is clocked. 3. Time required before a signal can be changed from its prior state.

seven-segment display — A display format consisting of seven bars so arranged that each digit from 0 to 9 can be displayed by energizing two or more bars. LED, LCD, and gas-discharge displays all use seven-segment display formats.

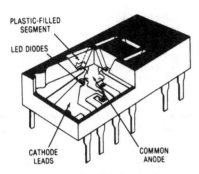

Seven-segment LED display.

seven-unit teleprinter code — Frequently called Teletype code. A code that represents the letters of the alphabet, the numerals, the punctuation marks, and the various control functions necessary on a teleprinter (such as line feed, carriage return, upshift, and downshift) by five-unit combinations of mark and space conditions. In addition to the five units that indicate the letter or other data, the code contains a start unit, which is always a space, and a stop unit, which is always a mark, to indicate the beginning and ending of each character.

sexadecimal — Pertaining to the number system that has a radix of 16.

sexadecimal notation — Also called dekahexadecimal notation. A scale of notation for numbers with base 16.

sexless connector — Also called hermaphroditic connector. An interconnecting device in which the mating parts are exactly alike at their mating surfaces.

sferics — 1. Contraction of the term *atmospherics*, meaning interference. 2. Radio interference from atmospherics, or static.

sferics receiver — Also called lightning recorder. A type of radio direction finder that electronically measures the direction of arrival, intensity, and rate of occurrence of atmospherics. In its simplest form the instrument consists of two orthogonally crossed antennas. Their output signals are connected to an oscillograph so that one loop measures the east-west components. These are combined vertically to give the azimuth.

sg — Abbreviation for screen-grid electrode (of a vacuum tube).

shaded-pole motor — A single-phase induction motor provided with one or more auxiliary short-circuited

stator windings that are displaced magnetically from the main winding.

shading — 1. A brightness gradient in the reproduced picture not present in the original scene, but caused by the camera tube. 2. Compensating for the spurious signal generated in a camera tube during the trace intervals. 3. Controlling the directivity pattern of a transducer through the distribution of phase and amplitude of the transducer action over the active face.

shading coil — *See* shading ring.

shading compensation — Dynamic sensitivity control of the picture signal to reduce the amount of video level change from center to edge of the picture.

shading ring — 1. A heavy copper ring sometimes placed around the central pole piece of an electrodynamic loudspeaker to act as a shorted turn for cancellation of the hum voltage of the field coil. 2. A copper ring set into part of the pole piece of a small alternating-current motor to produce the lagging component of a rotating magnetic field for starting purposes.

shading signal — A signal that increases the gain of the amplifier in a television camera while the electron beam is scanning a dark portion.

shadow attenuation — Attenuation of radio waves over a sphere in excess of that over a plane when the distance over the surface and other factors are the same.

shadow factor — The ratio of the electric field strength that would result from propagation over a sphere to that which would result from propagation over a plane (other factors being the same).

shadowgraph — A device arranged in such a manner as to enable photography and/or visual observation of the silhouette of back-illuminated objects placed within the object field of the device and of gradations in luminous intensity resulting from variations in the opacity or in the index of refraction of media contained within the object field.

shadow mask — *See* aperture mask.

shadow mask tube — A type of color-generating cathode-ray tube that utilizes a shadow mask, a thin perforated electrode located close to the display screen. Each hole in the mask coincides with a triad of three phosphor dots, each representing red, blue, or green. A specific electron beam from the gun energizes phosphor dots of single color.

shadow region — A region in which, under normal propagation conditions, an obstruction reduces the field strength from a transmitter to the point at which radio reception or radar detection is ineffective or virtually so.

shadow tuning indicator — A vacuum tube in which a moving shadow shows how accurately a radio receiver is tuned.

shaft — The axial member to which torque is applied to cause rotation of an adjustable component.

shaft angle encoder — An electromechanical device that has a means for counting equally spaced radii that represent angular increments around the periphery of a disk. Usually, the measurement is in degrees, minutes, and seconds, since 2π radians is not a prime number. Usually the disc is divided into an even number of equal increments.

shaft position encoder — Also called converter or coder. An analog-to-digital converter that transduces a mechanical analog shaft rotation to an electrical digital representation.

shakedown test — An equipment test carried out during the installation work.

shaker — An electromagnetic device capable of imparting known and/or controlled vibratory acceleration to a given object.

shake table — A laboratory tester in which an instrument or component is subjected to vibration that stimulates operating conditions.

shake-table test — A laboratory test in which a device or component is placed in a vibrator to determine the reliability of the device or component when subjected to vibration.

shank — 1. The part of a phonograph needle that is clamped into position in the pickup or cutting head. 2. The cylindrical or rodlike portion of a connector or contact. 3. That part of a fastener lying between the head and the extreme opposite end. 4. *See* cantilever.

Shannon limit — The maximum signal-to-noise ratio improvement that can be achieved by the best modulation technique, as implied by Shannon's theorem relating channel capacity to signal-to-noise ratio.

Shannon's theorem — *See* sampling theorem.

shaped-beam antenna — An antenna whose directional pattern over a certain angular range is designed to a specific shape for some particular use.

shaped-beam display tube — A cathode-ray tube in which the beam is first deflected through a matrix, then repositioned along the axis of the tube, and finally deflected into the desired position on the faceplate. A typical tube is the Charactron.

shape factor — 1. For a filter, the ratio (usually maximum) comparing a high-attenuation-level bandwidth and a low-attenuation-level bandwidth. 2. The ratio of the 60-dB bandwidth to the 6-dB bandwidth. Defines the selectivity of an amplifier stage. 3. A factor used to take the shape of a coil into account when its inductance is computed.

shaping — Adjustment of a plan-position-indicator pattern set up by a rotating magnetic field.

shaping network — 1. An electrical network designed to be inserted into a circuit to improve its transmission or impedance properties, or both. *See also* corrective network. 2. A network inserted in a telegraph circuit for improving the waveshape of the signals.

shared data — Data in memory or on a secondary storage device that is used by more than one task.

shared file — A direct-access device that two systems may use at the same time; a shared file may link two systems.

shareware — Software that is freely shared on public networks and BBSs. Users are frequently asked, on the honor system, to remit a small amount of money to the software developer.

sharp-cutoff tube — The opposite of a remote-cutoff tube. A tube in which the control-grid spirals are uniformly and closely spaced. As the grid voltage is made more and more negative, the plate current decreases steadily to cutoff.

sharp tuning — Response to a limited range of frequencies.

shaving — In mechanical recording, the removal of material from the surface of a recording medium for the purpose of obtaining a new surface.

shear wave — 1. Also called a rotational wave. A wave, usually in an elastic solid, that causes an element of the solid to change its shape but not its volume. 2. A wave in which particle displacement is at right angles to the direction of propagation.

sheath — 1. The external conducting surface of a shield transmission line. 2. A metal wall of a waveguide. 3. Part of a discharge in a rarefied gas, in which there is a space charge due to an accumulation of electrons or ions. 4. The outer covering or jacket of a cable.

sheath-reshaping converter — A converter in which the pattern of the wave is changed by gradual reshaping of the waveguide sheath and the metal sheets mounted longitudinally in the guide.

sheet grating — A three-dimensional grating consisting of thin metal sheets extending along the inside of a waveguide for about one wavelength. It is used to stop all but the predetermined wave, which passes unimpeded.

sheet-metal contact — A contact made by stamping and bending sheet metal rather than by the machining of metal stock. It is available in a wide variety of configurations and is usually less expensive than machined contacts.

sheet resistance — The electrical resistance of a thin sheet of material with uniform thickness as measured across opposite sides of a unit square pattern. Expressed in ohms per square.

sheet resistivity — Also called ohms per square. 1. The electrical resistance measured across the opposite sides of a square pattern of deposited film material. 2. The resistance of a unit area of printed and fired thick-film material. Expressed in ohms per square per unit of film thickness.

Sheffer-stroke function — The Boolean operator that gives a truth-table value of true only when both of the variables that the operator connects are not true.

shelf aging — The change with time of the properties of a stored component or material.

shelf corrosion — Consumption of the negative electrode of a dry cell as a result of local action.

shelf life — 1. The length of time under specified conditions that a material or component retains its usability. 2. The time period that an item can be stored, under specified conditions, and still meet specifications. 3. The length of time a battery is expected to retain charge when stored.

shell — 1. A group of electrons having a common energy level that forms part of the outer structure of an atom. 2. The outer section of a plug or receptacle that mechanically supports the assembly and in some cases provides coupling and locking. 3. That part of a phonograph pickup which carries the cartridge. The head shell can often be detached from the pickup arm.

shell-type transformer — A transformer in which the magnetic circuit completely surrounds the windings.

shelving equalizer — An equalizer whose boost or cut characteristic response curve resembles a shelf. Maximum boost or cut occurs at the indicated frequency and remains constant at all frequencies beyond.

SHF — Abbreviation for superhigh frequency.

shield — 1. A device designed to protect a circuit, transmission line, etc., from stray voltages or currents induced by electric or magnetic fields, consisting, in the case of an electric field, of a grounded conductor surrounding the protected object. At high frequencies this will provide magnetic shielding as well. At low frequencies (through the audio range) magnetic shielding is accomplished by surrounding the object with a material of high magnetic permeability. *See* braid, 2. 2. Any barrier to the passage of interference-causing electrostatic or electromagnetic fields. An electrostatic shield is formed by a conductive layer (usually foil) surrounding a cable core. An electromagnetic shield is a ferrous metal cabinet or wire-way. 3. In cables, a metallic layer placed around a conductor or group of conductors to prevent electrostatic or electromagnetic interference between the enclosed wires and external fields. 4. A metallic covering, usually copper or aluminum, placed around or between electric circuits, around cables or their components, to suppress the effects of undesired signals that may originate from adjacent or external sources. The shield may be braided wires, foil wrap, foil backed tape, a metallic tube, or conductive vinyl or rubber.

shield coverage — *See* shield percentage.

shielded building — In modern practice, a building in which shielding was incorporated in the basic architectural design. Shielded buildings often employ structural steel members as an integral part of RFI shielding.

shielded cable — 1. A single- or multiple-conductor cable surrounded by a separate conductor (the shield) intended to minimize the effects of adjacent electrical circuits. 2. A cable in which each insulated conductor is enclosed in a conducting envelope so constructed that substantially every point of the surface of the insulation is at ground potential with respect to ground under normal operating conditions.

shielded-conductor cable — A cable in which the insulated conductor or conductors are enclosed in a conducting envelope or envelopes, almost every point on the surface of which is at ground potential with respect to ground.

shielded enclosure — A room, hangar, or box that is shielded or screened so as to provide a controlled electromagnetic environment.

shielded joint — A cable joint having its insulation so enveloped by a conducting shield that substantially every point on the surface of the insulation is at ground potential, or at some predetermined potential with respect to the ground.

shielded line — A transmission line whose elements confine propagated radio waves to an essentially finite space inside a tubular conducting surface called the sheath, thus preventing the line from radiating radio waves.

shielded pair — A two-wire transmission line surrounded by a metallic sheath to isolate the pair from electrical interference.

shielded room — An enclosed area made free from electrical interference that would affect the sensitivity of electrical equipment.

shielded transmission line — A transmission line whose elements confine the propagated electrical energy inside a conducting sheath.

shielded wire — An insulated wire covered with a metal shield — usually of tinned, braided copper wire.

shielded X-ray tube — An X-ray tube enclosed in a grounded metal container, except for a small opening through which the X-rays emerge.

shield effectiveness — The relative ability of a shield to screen out undesirable radiation.

shield factor — Ratio between the noise (or induced current or voltage) in a telephone circuit when a source of shielding is present and when it is not.

shield grid — In a glass tube, a structure that shields the control electrode from the anode or cathode, or both. It prevents the radiation of heat from them and the depositing of thermionic activating material on them. It also reduces the electrostatic influence of the anode, and may be used as a control electrode in some applications.

shield-grid thyratron — A thyratron that contains a shield grid, usually operated at the same potential as the cathode.

shielding — 1. The practice of confining the dielectric field of an electric cable to the inside of the cable insulation or insulated conductor assembly by surrounding the insulation or assembly with a grounded conducting medium called a shield. 2. Metal covering used on a cable; also a metal can, case partition, or plates enclosing an electronic circuit or component. Shielding is used to prevent undesirable radiation, pickup of signals, magnetic induction, stray current, ac hum, or radiation of an electrical signal. 3. A physical barrier, usually electrically conductive, designed to reduce the interaction of electric or magnetic fields on devices, circuits, or portions of circuits.

shielding effectiveness — The relative reduction of radiated electromagnetic energy levels occasioned by the use of an enclosure either to contain or exclude the energy.

shield percentage — Also called shield coverage. The physical area of a circuit or cable actually covered by shielding material, expressed in percent.

shield wire — A wire employed for reducing the effects of extraneous electromagnetic fields on electric supply or communication circuits.

shift — 1. Displacement of an ordered set of computer characters one or more places to the left or right. If the characters are the digits of a numerical expression, a shift is equivalent to multiplying a power of the base. 2. The process of moving information from one place to another in a computer; generally, a number of bits are moved at once. A word can be shifted sequentially (generally referred to as shifting left or right), or all bits of a word can be shifted at the same time (called parallel load or parallel shift). 3. In a computer, an operation whereby a number is moved one or more places to the left or right. The number 110, for instance, becomes 1100 if shifted one place left or 11 if shifted one place right. The operation is of considerable use in digital computer operations. 4. The difference between the mark and space frequencies. For example, if the mark frequency is 2125 Hz and the space frequency is 2295 Hz, the difference of 170 Hz is referred to as the shift. The 170-Hz and 850-Hz shifts have become two widely used standards.

shift counter — 1. A shift register in which the first stage, through logic feedback, produces a pattern of 1s or 0s as a function of the state of other stages in the register. The pattern of 1s and 0s so produced is termed a ring code. 2. See ring counter, 2.

shift-frequency modulation — A form of frequency modulation in which the modulating wave shifts the output frequency between predetermined values and the output wave is coherent, with no phase discontinuity.

shift-in character — In a computer, a code extension character that can be used by itself to bring about a return to the character set in effect before substitution of another set was caused by a shift-out character.

shift out — In a computer, to move information within a register toward one end so that, as the information leaves this one end, 0s are entered into the other end.

shift-out character — In a computer, a code extension character that can be used by itself to cause substitution of some other character set for the standard set, usually to access additional graphic characters.

shift pulse — A drive pulse that initiates the shifting of characters in a register.

shift register — 1. A digital storage circuit in which information is shifted from one flip-flop of a chain to the adjacent flip-flop on application of each clock pulse. Data may be shifted several places to the right or left, depending on additional gating and the number of clock pulses applied to the register. Depending on the number of positions shifted, the rightmost characters are lost in a right shift, and the leftmost characters are lost in a left shift. See dynamic shift register; static shift register. 2. A program, entered by the user into the memory of a programmable controller, in which the information data (usually single bits) is shifted one or more positions on a continual basis. There are two types of shift registers: asynchronous and synchronous. 3. A register in which binary data bits are moved as a contiguous group a prescribed number of positions to the right or to the left. 4. A memory in which data words are entered serially and shifted to successive storage locations. The data word can be read when it has been sequentially shifted to the output. 5. A digital circuit consisting of flip-flops, which is used to convert parallel data (where several binary digits arrive at once) to serial (where the same digits travel one after another), or vice versa.

ship error — A radio direction-finder error that occurs when radio waves are reradiated by the metal structure of a ship.

ship-heading marker — On a PPI scope, an electronic radial sweep line indicating the heading of the ship on which the equipment is installed.

ship's emergency transmitter — A ship's transmitter to be used exclusively on a distress frequency for distress, urgency, or safety purposes.

ship station — A radio station operated in the maritime mobile service and located on board a vessel that is not permanently moored.

ship-to-shore communication — Communication by radio between a ship at sea and a shore station.

shock — 1. An abrupt impact applied to a stationary object. It is usually expressed in gravities (g). 2. An acceleration transient of short duration and nonrepetitive occurrence. 3. Sudden stimulation of the nerves and convulsive contraction of the muscles caused by a discharge of electricity though the body.

shock absorber — A device for dissipating vibratory energy, to modify the response of a mechanical system to an applied shock.

shock excitation — See impulse excitation.

shock-excited oscillations — See free oscillations.

shock hazard — A hazardous condition that exists at a part of a ground-fault circuit interrupter if (a) there would be current of 5 milliamperes or more in a resistance of 500 ohms connected between the part in question and the grounded supply conductor, and (b) the device would not operate to open the circuit to the 500-ohm resistor within a specified time.

shock isolator — Also called a shock mount. A resilient support that tends to isolate a system from applied mechanical shock.

Shockley diode — A four-layer controlled semiconductor rectifier diode without a base connection, used as a trigger or switching diode.

shock motion — In a mechanical system, transient motion characterized by suddenness and by significant relative displacements.

shock mount — See shock isolator.

shockover capacitor — A capacitor connected between the grid and cathode of a thyratron to prevent premature firing.

shock pulse — Usually a single disturbance that is characterized by an increase and a decrease of acceleration in a relatively short period.

shock test (mechanical) — A test to determine the ability of a device to withstand suddenly applied forces of specified magnitude and duration.

shodop — Acronym for short-range Doppler.

shoran — Acronym for short-range air navigation. A precision distance-measuring system employing the pulse timing principle to measure the distance from an aircraft to one or more fixed responder ground stations. Fundamentally, the system consists of a mobile transmitter-receiver-indicator unit and a fixed receiver-transmitter unit (transponder). Pulses are sent from the mobile transmitter and returned to the originating point by the transmitter. The indicator measures the time interval required for the travel of a pulse between stations and converts this information into distance to the nearest thousandth of a mile.

shore effect — The bending of radio waves toward the shore line when traveling over water, due presumably to the slightly greater velocity of radio waves over water than over land. This effect causes errors in radio direction-finder indications.

shore-to-ship communication — Communication by radio between a shore station and a ship at sea.

short — See short circuit.

short base-line system — A system that uses continuous waves and that has a base-line length that is short in comparison to the target distance.

short check — See flash test.

short circuit — Also called a short. 1. An abnormal connection of relatively low resistance between two points of a circuit. The result is excess (often damaging) current between these points. 2. An accidental or intentional near-zero resistance connection between two sides of a circuit. 3. An abnormal connection of relatively low impedance, whether made accidentally or intentionally, between two points of different potential. 4. A low-resistance path in a system that allows the current to flow outside of the proper circuit in an appliance, junction box, or in the wires. Due to a failure in insulation or improper hookup, the current takes the short low-resistance path rather than flowing through the entire circuit. 5. An abnormal condition that occurs when there is an unwanted electrical connection between two wires. It results in a flow of excess current.

short-circuit current — 1. The current a power supply delivers when its output terminals are short-circuited. 2. An electronic regulator's output current with the output terminal shorted to the negative supply. Short-circuited current and maximum regulator output current depend on the pass circuitry. The maximum output current is also limited by the maximum allowable package power dissipation and, hence, by quiescent current.

short-circuit driving-point admittance — The driving-point admittance between the j terminal of an n-terminal network and the reference terminal when all other terminals have zero alternating components of voltage with respect to the reference point.

short-circuit feedback admittance (of an electron-device transducer) — The short-circuit transfer admittance from the output terminals to the input terminals of a specified socket, the associated filters, and the electron device.

short-circuit forward admittance (of an electron-device transducer) — The short-circuit transfer admittance from the input terminals to the output terminals of a specified socket, the associated filters, and the electron device.

short-circuit impedance — 1. The driving-point impedance of a line or four-terminal network when its far end is short-circuited. 2. The input impedance to a line or four-terminal network when the far end or output terminals are shorted.

short-circuiting or grounding device — A power or stored-energy operated device that functions to short-circuit or ground a circuit in response to automatic or manual means.

short-circuit output admittance (of an electron-device transducer) — The short-circuit driving-point admittance at the output terminals of a specified socket, the associated filters, and the electron device.

short-circuit output capacitance (of an n-terminal electron device) — The effective capacitance determined from the short-circuit output admittance.

short-circuit parameters — In an equivalent circuit of a transistor, the resultant parameters when independent variables are selected for the input and output voltages.

short-circuit protection (automatic) — A current-limiting system that enables a power supply to continue operating without damage into any output overload, including short circuits. The output voltage must be restored to normal when the overload is removed, as distinguished from a fuse or circuit-breaker system, which opens at overload and must be reclosed to restore power.

short-circuit transfer admittance — The transfer admittance from terminal j to terminal l of an n-terminal network when all terminals except j have zero complex alternating components of voltage with respect to the reference point.

short-circuit transfer capacitance (of an electron device) — The effective capacitance determined from the short-circuit transfer admittance.

short code — A system of instructions that causes an automaton to behave as if it were another, specified automaton.

short-contact switch — A selector switch in which the movable contact is wider than the distance between its clips, so that the new circuit is made before the old one is broken.

short-distance navigational aid — An equipment or system that provides navigational assistance to a range not exceeding 200 miles (322 km).

shorted — Prevented from operating by a short circuit.

shorted out — Made inactive by connecting a heavy wire or other low-resistance path around a device or portion of a circuit.

short-flash light source — An electronic flash tube in which the flash recurs at a frequency extending to many thousands per second. A stroboscopic light source is a short-flash light source.

short-gate gain — Video gain on a short-range gate.

shorting noise — A noise that occurs in wirewound potentiometric transducers, even when no current is drawn from the device. It is due to the shorting out of adjacent turns of the wire as the slider traverses the winding. The portion of the interturn current that flows through the slider appears as noise.

shorting switch — A switch type in which contact is made for a new position before breaking contact with the previous position. Classified as a make-before-break switch.

short-range navigation — A precision position-fixing system using a pulse transmitter and receiver in connection with two transponder beacons at fixed points.

short-range navigation aid — A navigation aid that is usable only at distances within radio line of sight.

short-slot coupler — A 3-dB coupler.

short-time duty — A service requirement that demands operation at a substantially constant load for a short, specified time.

short time limits — Values of minimum and maximum trip time measured at various percentages of overload.

short-time rating — The rating that defines the load that a machine, apparatus, or device can carry at approximately the room temperature for a short, specified time.

shortwave — Abbreviated sw. Radio frequencies from 1.6 to 30 MHz, which fall above the commercial broadcasting band and are used for sky-wave communication over long distances.

shortwave converter — An electronic unit designed to be connected between a receiver and its antenna system to permit reception of frequencies higher than those the receiver ordinarily handles.

shortwave transmitter — A radio transmitter that radiates shortwaves, which ordinarily are shorter than 200 meters.

shot effect — Noise voltages developed by the random travel of electrons within a tube. The effect is characterized by a steady hiss from a radio, and by snow or grass in a television picture.

shotgun mike — Also called a hyperdirectional or long-reach mike. An extremely unidirectional microphone, used for spotting a speaker or soloist from a considerable distance.

shot noise — 1. Noise generated due to the random passage of discrete current carriers across a barrier or discontinuity, e.g., a semiconductor junction, as well as by thermal agitation in a base resistor. Shot noise is characteristic of all transistors and diodes and is directly proportional to the square root of the applied current. 2. Noise voltages developed in a thermionic tube as a result of random variations in the number and velocity of electrons emitted by the cathode. The effect is characterized by the presence of a steady hiss in audio reproduction and of snow or grass in video reproduction. The shot noise (German, *schotteffekt*) concept stems from the random manner in which electrons in a vacuum tube collide with the plate (anode) — not unlike the sound effect produced by casting a handful of BB-shot against a wall, with their slightly differing and overlapping impact times. Additionally, the electrons from a vacuum-tube cathode are not emitted uniformly, producing fluctuation noise. 3. Noise that is inherent in an electric current because it consists of a stream of finite particles, i.e., electrons. 4. The noise generated by a charge crossing a potential barrier. For medium and high frequencies it is the dominant noise mechanism in bipolar devices. Shot noise has a constant spectral density.

shrinkable tubing — 1. A nonmetallic tubing that is fabricated to allow a nonreversible decrease in its diameter on the application of heat. It is used to provide insulation or mechanical protection to conductors, cables, splices, and terminations. 2. Plastic or elastomeric tubing used to protect wires, cables, and splices from mechanical damage. The tubing shrinks to a predetermined size on application of heat or solvent evaporation.

shunt — 1. A precision low-value resistor placed across the terminals of an ammeter to increase its range. The shunt may be either internal or external to the instrument. 2. Any part connected, or the act of connecting any part, in parallel with some other part. 3. In an electric circuit, a branch whose winding is in parallel with the external or line circuit. 4. An alternate path in parallel with a part of a circuit. 5. To place a circuit element in parallel with another. 6. To bypass a portion of a circuit. 7. A deliberate shorting-out of a portion of an electric circuit. 8. A key-operated switch that removes some portion of an alarm system from operation, allowing entry into a protected area without initiating an alarm signal. A type of authorized access switch. 9. A calibrated low resistance connected in parallel with the input terminals of an ammeter, to enable measurement of a wider range of currents.

shunt calibration — A procedure in which a parallel resistance is placed across a (similar) element to obtain a known and deliberate electrical change.

shunt-fed vertical antenna — A vertical antenna connected to the ground at the base and energized at a suitable point above the grounding point.

shunt feed — *See* parallel feed.

shunt field — Part of the magnetic flux produced in a machine by the shunt winding connected across the voltage source.

shunt-field relay — A special polarized relay with two coils on opposite sides of a closed magnetic circuit. The relay operates only when the currents in its two windings flow in the same direction.

shunting effect—A reduction in signal amplitude caused by the load that an amplifier or measuring instrument imposes on the signal source. For dc signals the shunting effect is directly proportional to the output impedance of the signal source and inversely proportional to the input impedance of the amplifier.

shunting or discharge switch—A switch that serves to open or to close a shunting circuit around any piece of apparatus (except a resistor), such as a machine field, a machine armature, a capacitor, or a reactor.

shunt leads—Those leads that connect the circuit of an instrument to an external shunt. The resistance of these leads must be taken into account when the instrument is adjusted.

shunt loading—Loading in which reactances are applied in parallel across the conductors of a transmission circuit.

shunt neutralization—See inductive neutralization.

shunt peaking network—See peaking network.

shunt regulator—A device placed across the output of a regulated power supply to control the current through a series-dropping resistance in order to maintain a constant output voltage or current.

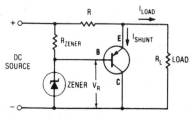

Shunt regulator.

shunt T-junction—See H-plane T-junction.

shunt wire—A conductor joining two parts of an electric circuit to divert part of the current.

shunt-wound generator—A direct-current generator in which the field coils and armature are connected in parallel.

shunt-wound motor—1. A direct-current motor that has its field (stationary member) and armature (rotating member) circuit connected in parallel. Its speed can be regulated by varying either the applied armature or field voltage. 2. A motor whose armature and field windings are connected in parallel. It has a fairly constant speed, but a low starting torque.

shutoff—A provision whereby a recorder will automatically go into the stop mode at the end of a tape. In some recorders, the automatic shutoff can be made to turn off the entire unit as well as any other components powered by it.

shutter—A movable cover that prevents light from reaching the film or other light-sensitive surface in a still, movie, or television camera except during the exposure time.

shuttle—A high-speed tape-running mode that permits fast cuing or rewinding of the tape.

sibilance—The strong emphasis in pronunciation of the letters "s" and "sh" in speech. It can be exaggerated by microphones having peaks in their high-frequency response.

SIC—Abbreviation for semiconductor integrated circuit.

side armature—An armature that rotates about an axis parallel to that of the core, with the pole face on a side surface of the core of a relay.

sideband attenuation—Attenuation in which the relative transmitted amplitude of one or more components of a modulated signal (excluding the carrier) is smaller than the amplitude produced by modulation.

sideband power—The power contained in the sidebands. This is the power to which a receiver responds when receiving a modulated wave, not the carrier power.

sidebands—1. The frequency bands on both sides of the carrier frequency. The frequencies of the wave produced by modulation fall within these bands. 2. The wave components lying within such bands. During amplitude modulation with a sine-wave carrier, the upper sideband includes the sum (carrier plus modulating) frequencies, and the lower sideband includes the difference (carrier minus modulating) frequencies. 3. The modulation bands of frequencies that are both above and below the carrier frequency during modulation. 4. A band of frequencies on each side of the carrier frequency of an amplitude-modulated wave. Each sideband contains all of the information that was in the modulating wave. (The upper sideband contains frequencies that are the sums of the carrier and modulation frequencies, and the lower sideband contains the difference frequencies.)

sideband splatter—1. Those portions of the modulation sidebands that lie beyond the limits of the assigned channel. 2. In radio communications, interference on other channels caused by spurious sidebands resulting from overmodulation.

side circuit—A circuit arrangement for deriving a phantom circuit. In four-wire circuits, the two wires associated with the "go" channel form one side circuit, and those associated with the return channel form another. See also phantom circuit.

side-circuit loading coil—A loading coil for introducing a desired amount of inductance into a side circuit while introducing a minimum amount of inductance into the associated phantom circuit.

side-circuit repeat coil—See side-circuit repeating coil.

side-circuit repeating coil—Also called side-circuit repeat coil. A device that functions as a transformer at a terminal of a side circuit, and acts simultaneously as a device for superposing one side of a phantom circuit on the side circuit.

side echo—An echo due to a side lobe of an antenna.

side frequency—One of the frequencies of a sideband.

side lobe—A portion of the beam from an antenna, other than the main lobe. It is usually much smaller than the main lobe. See figure on page 691.

side-lobe blanking—A technique that compares relative signal strengths between an omnidirectional antenna and the radar antenna.

side-lobe cancellation—A technique designed to exclude or greatly attenuate jamming signals introduced through the side or back lobes of a receiving antenna.

side-looking airborne radar—A high-resolution airborne radar system in which the beam from the antenna is directed at right angles to the direction of flight.

siderial day—The time it takes for the earth to rotate exactly 360° about its axis with respect to the "fixed" stars. The siderial day contains 1436.07 minutes. *Compare* solar day.

sideswiper—A telegraph key that operates from side to side rather than up and down.

side thrust—1. In disc recording, the radial component of force on a pickup arm caused by the stylus drag. 2. The tendency of a stylus to skate toward the center of a

Side lobes.

record, causing increased wear on the inner groove wall. With low tracking weight, side thrust can cause the stylus to jump the record's groove.

sidetone — 1. The reproduction, in a telephone receiver, of sounds received by the transmitter of the same telephone set (e.g., hearing one's own voice in the receiver of a telephone set when speaking into the mouthpiece). 2. That portion of a speaker's voice that is fed back to his or her receiver.

sidetone telephone set — A telephone set with no balancing network for reducing sidetone.

siemens — 1. International standard unit of conductance that replaces and is identical with the term *mho*. The reciprocal of resistance in ohms. 2. The unit of electric conductance of a conductor in which a current of 1 ampere is produced by an electric potential difference of 1 volt.

sight check — To verify the sorting or punching of punched cards by looking through the pattern of punched holes.

sign — 1. A symbol that distinguishes negative from positive quantities. 2. A binary indicator of the position of the magnitude of a number relative to zero.

signal — 1. A visible, audible, or other conveyor of information. 2. The intelligence, message, or effect to be conveyed over a communication system. 3. The physical embodiment of a message or of information. 4. An electrical wave used to convey information. 5. An alerting signal; that is, an acoustic device (such as a bell) or a visual device (such as a lamp) that calls the attention. 6. To transmit an information signal or alerting signal. 7. The event, phenomenon, or electrical quantity that conveys information from one point to another. 8. A current used to convey information, either digital, analog, audio, or video. 9. An electrical impulse of a predetermined voltage, current, polarity, and pulse width. 10. Any electronic visual, audible, or other indication used to convey information. In semiconductors, an electrical quantity (typically voltage, current, or light level) corresponding to some physical quantity. Signals are coded in frequency or amplitude to separate them from unwanted noise.

signal attenuation — The reduction in the strength of electrical signals.

signal averaging — A technique for extracting a signal waveform (generally a time-varying voltage) from a background of unwanted noise. Simple frequency-domain filtering with passive or active circuit elements is the most widely used method for accomplishing this result. But this type of filtering is effective only when the frequency spectrum of the signal and the frequency

spectrum of the noise do not overlap. A signal averager is a special kind of filter, sometimes referred to as a comb filter. It can be used effectively only if the desired signal, with its contaminating noise, can be repeated a number of times. In addition, a synchronization of that pulse must have a fixed time relationship to the desired signal, preferably, but not necessarily, ahead of the signal.

signal-averaging computer — An electronic averager that filters out signals of interest from background noise.

signal bias — A form of teletypewriter signal distortion brought about by the lengthening or shortening of pulses during transmission. When marking pulses are all lengthened, a marking signal bias results; when marking pulses are all shortened, a spacing signal bias results.

signal-carrier FM recording — A method of recording in which the input signal is frequency modulated onto a carrier, and the carrier is recorded on a single track at saturation and without bias.

signal conditioner — A device placed between a signal source and a readout device for the purpose of conditioning the signal. Some examples are damping networks, attenuator networks, preamplifiers, excitation and demodulation circuits, converters for changing one electrical quantity into another (such as voltage to current), instrument transformers, equalizing or matching networks, and filters.

signal conditioning — 1. To process the form or mode of a signal so as to make it intelligible to, or compatible with, a given device, including such manipulation as pulse shaping, pulse clipping, digitizing, and linearizing. 2. Any operation that prepares a transducer signal for subsequent display or control functions. Depending on the application, a transducer signal might require any one or a combination of several conditioning operations such as filtering, amplification, isolation, integration, differentiation, and rectification. For extracting low-level signals from electrical noise, an instrumentation amplifier is often the best choice.

signal conductor — An individual conductor used to transmit an impressed signal.

signal converter — A circuit that reduces, filters, and (if necessary) rectifies incoming signals to logic system levels.

signal delay — The transmission time of a signal through a network. The time is always finite, and it may be undesired or purposely introduced.

signal diode — 1. A semiconductor diode used for the purpose of extracting or processing information contained in an electrical signal that varies with time and may be either analog or digital in nature. 2. A diode that exhibits an asymmetrical voltage-current characteristic and is used for signal detection.

signal-distortion generator — An instrument furnished and designed to apply distortion on a signal for the purpose of ranging and adjusting teletypewriter equipment or for furnishing a clear signal.

signal dropout — The loss of signal that occurs when the signal becomes too weak to be usable.

signal electrode — The electrode from which the signal output of a camera tube is taken.

signal element — Also called a unit interval. That part of a signal which occupies the shortest interval of the signaling code. It is considered to be of unit duration in building up signal combinations.

signal encoding device — A system component located at the protected premises that will initiate the transmission of an alarm signal, supervisory signal, trouble signal, or other signals the central station is prepared to receive and interpret.

signal enhancement — Ensemble averaging of time-domain signals, whereby a set of time domain samples are digitized and then averaged. In order to enhance the signal due to averaging, the time function must be repetitive, and the start of the ensemble average must have a known relationship to some repetitive event (trigger). Such a repetitive signal is the vibration from one rotation of an engine (where the firing of spark plug 1 serves as the trigger).

signal filtering — The shaping of amplitude or phase characteristics with respect to frequency, for the purpose of meeting an operational requirement. This usually is accomplished by analog methods.

signal-frequency shift — In a frequency-shift facsimile system, the numerical difference between the frequencies corresponding to the white and black signals at any point in the system.

signal generator — 1. Also called a standard voltage generator. A device that supplies a standard voltage of known amplitude, frequency, and waveform for measuring purposes. 2. An instrument that provides
(*a*) Calibrated and variable frequency over a broad range.
(*b*) Calibrated and variable output level over a wide dynamic range.
(*c*) One or more forms of calibrated modulation.
Not all frequency sources or synthesizers are signal generators. Sweepers, test oscillators, and traditional frequency synthesizers cannot be classified as signal generators because they usually lack a calibrated output or some form of calibrated modulation. 3. A portable test oscillator that can be adjusted to provide a test signal at some desired frequency, voltage, modulation, or waveform.

signal ground — The ground return for low-level signals, such as inputs to audio amplifiers or other circuits, that are susceptible to coupling through ground-loop currents.

signal highlighting — Identifying the connection points of a net in a printed circuit board.

signaling — 1. The process by which a caller on the transmitting end of a line informs a particular party at the receiving end that a message is to be communicated. Signaling is also that supervisory information that lets the caller know that the called party is ready to talk, that his or her line is busy, or that he or she has hung up. Signaling also holds the voice path together while a conversation goes on. 2. Indicating to the receiving end of a communication circuit that intelligence is to be transmitted. 3. In a circuit-switched telecommunications network, the exchange of information that is concerned with the establishment, control, and management of a telephone connection.

signaling channel — A tone channel used for signaling purposes.

signaling key — A key used in wire or radiotelegraphy to control the sequence of current impulses that form the code signals.

signal injector — A test instrument, usually small, that contains an audio-frequency pulse oscillator. A signal injected at points in the circuitry aids in troubleshooting.

signal intelligence — A generic term that includes both communications intelligence and electronics intelligence.

signal interpolation — *See* interpolation.

signal lamp — A lamp that indicates, when lit or out, the existence of certain conditions in a circuit (e.g., signal lamps on switchboards, or pilot lamps in radio sets).

signal leakage — Interference in a given playback channel that has its origin in the recording system. Such interference occurs during simultaneous record/reproduce

and has a leading time displacement with reference to the signal on the tape.

signal level — 1. The difference between the measure of the signal at any point in a transmission system and the measure of an arbitrary reference signal. (Audio signals are often stated in decibels — thus, their difference can be conveniently expressed as a ratio.) 2. The magnitude of signal compared with an arbitrary reference magnitude.

signal line — One of a set of signal conductors in an interface system used to transfer messages among interconnected devices.

signal matching — Inserting buffers near the target system when the target microprocessor is replaced for emulation, so that the signals at the target microprocessor are reproduced exactly.

signal-muting switch — A switch used on a record changer to ground (mute) the signal from the pickup during a change cycle.

signal-noise ratio — *See* signal-to-noise ratio.

signal parameter — That parameter of an electrical quantity whose values or sequence of values conveys information.

signal plane — A conductor layer intended to carry a signal, rather than serve as a ground or other fixed-voltage function.

signal plate — A metal plate that backs up the mica sheet containing the mosaic in one type of cathode-ray television camera tube. The electron beam acts on the capacitance between this plate and each globule of the mosaic to produce the television signal.

signal plus noise and distortion — A radio-receiver sensitivity measurement based on the signal input required to produce 50 percent of the rated output at a 12-decibel ratio (4 : 1 voltage ratio) of signal plus noise and distortion to noise and distortion alone.

signal-point stereo microphone — A housing containing two, usually directional, microphones angled so that each picks up sound from one side of the stereo field, with both picking up sounds from the middle.

signal processing — A broad class of electronic functions that enhance the representations of physical or electrical phenomena. Temperature, pressure, vibration, acceleration, and flow are examples of physical properties that rely on signal processing enhancements. The detection and conversion of rf, X-ray, or ultrasonic energy into images and sound is another form of signal processing.

signal processing equipment — Any equipment or circuit used to intentionally change the characteristics (but not the overall level) of a signal. Includes such devices as equalizers, limiters, phasers, flangers, and delay lines.

signal reconditioning — The act of partially or completely restoring the original form of a distorted signal.

signal regeneration — The process of demodulating a received signal to recover its baseband data (thus removing received noise but creating bit errors) and remodulating the baseband signal onto a carrier for retransmission.

signal report — A report given in numerical values of signal strength and quality.

signal-separation filter — A bandpass filter that selects the desired signal or channel from a composite signal.

signal-shaping network — An electric network inserted into a telegraph circuit, usually at the receiving end, to improve the waveshape of the signals.

signal-shield ground — A ground technique for all shields used for the protection from stray pickup of leads carrying low-level, low-frequency signals.

signal shifter — A variable-frequency oscillator for shifting amateur transmitters to a less crowded frequency within a given band.

signal splitter — A passive device that enables two or more TV sets to divide a TV signal between them with proper balancing and isolation. Available in either 75- or 300-ohm impedances.

signal strength — 1. The strength of the signal produced by a transmitter at a particular location. Usually it is expressed as so many millivolts per meter of the effective receiving-antenna length. 2. The intensity of the television signal measured in volts, millivolts, microvolts, or decibels, using 0 dB as a reference. Equal to 1000 microvolts in rf systems; generally 1 volt in video systems.

signal-strength meter — Also called an S meter. A meter connected in the AVC circuit of a receiver and calibrated in dB or arbitrary "S" units to read the strength of a received signal.

signal-to-distortion ratio — The ratio of desired to undesired signal in a transmitted single-sideband signal.

signal-to-noise ratio — Abbreviated SNR or s/n ratio. Also called signal-noise ratio. 1. Ratio of the magnitude of the signal to that of the noise (often expressed in decibels). 2. In television transmission, the ratio in decibels of the maximum peak-to-peak voltage of the video television signal (including the synchronizing pulse) to the rms voltage of the noise at any point. 3. The ratio of the amplitude of a signal after detection to the amplitude of the noise accompanying the signal. It may also be considered as the ratio, at any specific point of a circuit, of signal power to total circuit-noise power. 4. Ratio of the root-mean-square facsimile signal level to the root-mean-square noise level. 5. The difference, measured in decibels, between a specified signal reference level and the level of unwanted noise. The higher the ratio, the better the equipment. 6. The span, measured in decibels, of signal intensity between a device's overload point at the upper limit and its background noise at the lower limit. (In tape recording, the s/n ratio usually lies between the permissible limit of saturation distortion and the tape's background hiss.) 7. The ratio, in decibels, between a reference power output (usually an amplifier's rated power) and the hum and noise power in the output of the amplifier. The higher this ratio, the better. 8. The ratio, usually in decibels, between the level of the loudest undistorted tone that can be recorded and the noise that is generated and recorded when no signal is present. 9. The ratio of the power in a desired signal to the undesirable noise present in the absence of a signal. 10. The difference between the nominal or maximum operating level and the noise floor; usually specified in decibels. 11. The ratio of the peak value of the video signal to value of the noise. Usually expressed in decibels.

signal tracer — A test instrument used for tracing a signal through a circuit in order to find faulty wiring or components.

signal tracing — The process of locating a fault in a circuit by injecting a test signal at the input and checking each stage, usually from the output backwards.

signal voltage — The effective (root-mean-square) voltage value of a signal.

signal wave — A wave with characteristics that permit it to carry intelligence.

signal-wave envelope — The contour of a signal wave that is composed of a series of wave cycles.

signal winding — Also called an input winding. In a saturable reactor, the control winding to which the independent variable (signal wave) is applied.

signature (target) — The characteristic pattern of the target displayed by detection and classification equipment.

signature analysis — Abbreviated SA. 1. A synchronous process, whereby activity at an electrical node (referenced to a clock signal) is monitored for a particular stimulus over a given time. The analysis that follows the nodal monitoring is based on data compression. The long, complex pattern of a data stream is reduced to a 16-bit, 4-digit "signature." Correct signatures for a particular circuit are determined empirically from tests on a known-good product. 2. A means of isolating digital logic faults at the component level applicable to all digital systems. The technique involves the tracing of signals and the conversion of lengthy bit streams into 4-digit hexadecimal "signatures." Using logic diagrams and schematics specially annotated with correct signatures at each trace node, and guided by troubleshooting trees, a circuit can be traced back to a point in the circuit that has a correct input signature and incorrect output signature. Signatures are traced under the direction of a test PROM. 3. A patented troubleshooting technique (Hewlett-Packard Co.) based on the principle that a good digital circuit in a known (initialized) state will produce the same output when stimulated repeatedly by the same input. If the repeated output of a device is not the one it has been designed to produce, it has failed. 4. A specific digital-circuit-testing troubleshooting technique that makes use of coded representations of serial bit streams. Using a known input signal, a signature-analysis system generates such a coded representation at each point on a known-good pc board. The signature at each point on a board under test should then be the same as the signature at the corresponding point on the known-good board. 5. A technique for compressing large amounts of digital data into a relatively short data word, or signature. If the signature matches a reference signature it can be assumed, to an extremely high probability, that the data stream is valid. Signature analysis, therefore, can be used to verify the integrity of data streams.

signature analyzer — An instrument that compares stored patterns (signatures) against actual received patterns.

signature testing — Comparison of the actual output digital signatures, such as transition counts, with the expected correct signatures recorded from a known-good device.

sign bit — 1. In complementary arithmetic, the leftmost bit of a number. If the sign bit is 1, the number is negative; if it is 0, the number is positive. 2. The leftmost bit of a computer word, which is sometimes used to indicate whether the number it contains is positive or negative. A 0 usually means a positive number, and a 1 a negative number.

sign-control flip-flop — In computers, a flip-flop in the arithmetic unit used for storing the sign of the result of an operation.

sign digit — A character (+ or −) used to designate the algebraic sign of a number.

significance — Weight. In positional representation, the factor by which a digit must be multiplied to obtain its additive contribution to the value of a number; the factor is determined by the digit position.

significant digits (of a number) — 1. A set of digits from consecutive columns, beginning with the most significant digit other than zero, and ending with the least significant digit whose value is known or assumed to be relevant. The digits of a number can be ordered according to their significance, which is greater when occupying a column corresponding to a higher power of the radix. 2. A digit that contributes to the precision of a number. Significant digits are counted from the first digit on the left that is not zero and continue to the last accurate digit on the right. (A right-hand zero may be counted if it is an accurate part of the numeral.) For example, 2500.0 has five significant digits, 2500 probably has only two (it is not known that the last two digits are accurate) but 2501 has four, and 0.0025 has two.

sign position — A position, normally at one end of a number, that contains an indicator of the algebraic sign of the number.

silent alarm — A remote alarm without an obvious local indication that an alarm has been transmitted.

silent-alarm system — An alarm system that signals a remote station by means of a silent alarm.

silent discharge — The gradual and nondisruptive discharge of electricity from a conductor into the atmosphere. It is sometimes accompanied by the production of ozone.

silent period — An hourly period during which ship and shore radio stations must remain silent and listen for distress calls.

silica gel — A moisture-absorbent chemical used for dehydrating waveguides, coaxial lines, pressurized components, shipping containers, etc.

silicon — 1. A metallic element often mixed with iron or steel during smelting to provide desirable magnetic properties for transformer-core materials. In its pure state, it is used as a semiconductor. 2. A brittle, gray, crystalline chemical element that, in its pure state, serves as a semiconductor substrate in microelectronics. It is naturally found in compounds, such as silicon dioxide.

silicon bilateral switch — A device that has characteristics similar to those of the silicon unilateral switch, but exhibits the same characteristics in both directions.

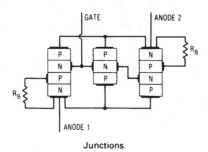

Junctions.

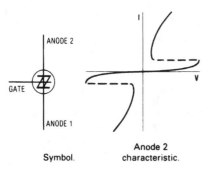

Symbol.

Anode 2 characteristic.

Silicon bilateral switch.

silicon capacitor — *See* varactor.

silicon cell — A solid-state device composed of silicon that is used to convert radiation into electrical energy.

silicon compiler — A software package that takes over chip creation from design to mask production.

silicon controlled rectifier — Abbreviated SCR. Also called reverse-blocking triode thyristor. 1. A four-layer pnpn semiconductor device that, when in its normal

state, blocks a voltage applied in either direction. The device is enabled to conduct in the forward direction when an appropriate signal is applied to the gate electrode. When such conduction is established, it continues even with the control signal removed until the anode supply is removed, reduced, or reversed. The SCR is the solid-state equivalent of the thyratron tube. 2. A semiconductor device that functions as an electrically controlled switch for dc loads. The SCR is one type of thyristor. 3. A reverse-blocking triode thyristor that can be triggered into conduction in only one direction. Terminals are anode, cathode, and gate. 4. A three-junction semiconductor device that is normally an open circuit until an appropriate gate signal is applied to the gate terminal, at which time it rapidly switches to the conducting stage. Its operation is similar to that of a gas thyratron, which conducts current in one direction only.

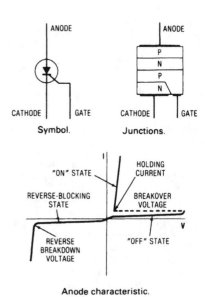

Anode characteristic.

Silicon controlled rectifier.

silicon controlled switch — Abbreviated SCS. A four-terminal pnpn semiconductor switching device; it can be triggered into conduction by the application of either a positive or negative pulse.

silicon detector — *See* silicon diode.

silicon diffused epitaxial mesa transistor — A silicon transistor that has high voltage and power ratings and low storage time and saturation voltage.

silicon diode — Also called a silicon detector. A crystal detector used for rectifying or detecting UHF and SHF signals, It consists of a metal contact held against a piece of silicon in a particular crystalline state.

silicon diode array tube — A highly sensitive vidicon-type tube used in CCTV cameras designed for low light level applications.

silicon dioxide — 1. A compound that results from oxidizing silicon quartz. Selective etching of silicon dioxide makes possible selective doping for the generation of components in monolithic integrated circuits. 2. An abundant material found in the form of quartz and agate and as one of the major constituents of sand. The silicates of sodium, calcium, and other metals can be readily fused, and on cooling do not crystallize but instead form transparent material glass.

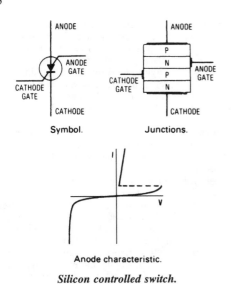

Symbol.   Junctions.

Anode characteristic.

Silicon controlled switch.

The thickness of SOS films is comparable with diffusion depths commonly used in MOS/LSI fabrication. Consequently, doping impurities penetrate completely through the silicon, so that the only component of the pn junction is that normal to the surface. Since the principal area contributions to a pn junction come from the underside and side walls of a diffusion well, the SOS vertical junction area — and, hence, capacitance — is reduced considerably. 4. A technology whereby MOSFETs are deposited on a sapphire substrate to increase the transistor switching speed. Silicon is grown on a passive insulating base (sapphire) and then selectively etched away to form a solid-state device.

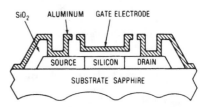

Silicon-on-sapphire transistor.

silicon double-base diode — *See* unijunction transistor.

silicone — A member of the family of polymeric materials characterized by a recurring chemical group that contains silicon and oxygen atoms as links in the main chain. These compounds are presently derived from silica (sand) and methyl chloride. One of their important properties is resistance to heat.

silicon foundry — A facility that fabricates an integrated circuit from a design supplied by an independent party.

silicon gate — 1. An MOS process that uses silicon rather than metal as one of the transistor elements. This permits the use of lower operating voltages and increases the dynamic response of the device. 2. MOS technology that uses silicon as the metal for the gate of the transistor. An alternative is aluminum gate. 3. One of several methods for fabricating metal-oxide semiconductor circuits.

silicon-gate MOS — A process using polycrystalline silicon to replace the metal layer as the gate electrode. It offers high speed and low threshold.

silicon monoxide — A dielectric material often used in the fabrication of a microelectronic device to form an insulator, substrate, or a thin-film capacitor dielectric.

silicon nitride — A compound that is deposited on the surface of a silicon monolithic integrated circuit to improve the stability of the integrated circuit. Silicon nitride is relatively impervious to some ions that penetrate silicon dioxide; best stability is obtained through the use of a combination of silicon nitride and silicon dioxide. Charge storage at the interface between layers of silicon nitride and silicon dioxide has resulted in memory devices in which the retention times are extremely long.

silicon on sapphire — Abbreviated SOS. Also called spinel. 1. A semiconductor manufacturing process in which a silicon layer is grown on a sapphire substrate. The silicon layer is divided into a number of electrically isolated islands for individual transistors. Silicon on sapphire is used for high-performance LSI circuits. 2. A semiconductor manufacturing process that uses an insulated material (sapphire) instead of silicon as a substrate on which the epitaxial layer is grown. With the process, MOS or bipolar performance can be significantly improved over that of conventional devices. 3. A fabrication technique in which thin crystalline films of silicon are deposited on a single crystal alumina (sapphire) substrate.

silicon oxide — 1. A dielectric material commonly used in the surface passivation of microelectronic circuits. Silicon oxide contains various combinations of silicon monoxide and silicon dioxide. 2. Silicon monoxide or dioxide or a mixture, the latter of which can be deposited on a silicon IC as insulation between metalization layers.

silicon photodiode — Semiconductor pn or pin junction that utilizes absorbed photon energy in the range of 1.06 to 1.03 electronvolts to excite carriers from one energy level to a higher state. The resultant change in the charge across the junction is monitored as a current in the external photodiode circuit.

silicon planar transistor — A silicon transistor produced by the planar process and consisting of a series of etchings and diffusions to produce a transistor with a thin oxide layer within the planes of a silicon substrate.

silicon rectifier — 1. One or more silicon rectifying cells or cell assemblies. 2. Semiconductor diode that converts alternating current to direct current and which can be designed to withstand large currents and high voltages.

silicon rectifying cell — An elementary two-terminal silicon device that consists of a positive and a negative electrode and conducts current effectively in only one direction.

silicon solar cell — A photovoltaic cell designed to convert light energy into power for electronic and communication equipment. It consists essentially of a thin wafer of specially processed silicon.

silicon steel — Steel containing 3 percent to 5 percent silicon. Its magnetic qualities make it desirable for use in the iron cores of transformers and in other ac devices.

silicon surface-barrier detectors — Silicon radiation detectors based on a rectifying contact at the silicon's surface.

silicon symmetrical switch — A thyristor modified by the addition of a semiconductor layer to make the device into a bidirectional switch. It is used as an ac phase control for synchronous switching and control of motor speed.

silicon target — A high-sensitivity TV image pickup tube of the direct readout type utilizing a silicon diode array photoconductive target. Suitable for low-light applications. High sensitivity extends through the visible range,

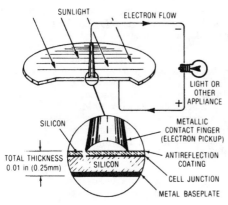

Silicon solar cell.

with extended sensitivity to the near-infrared region. Has low lag and high burn resistance.

silicon transistor — A transistor in which silicon is used as the semiconducting material.

silicon unijunction transistor — *See* unijunction transistor.

silicon unilateral switch — Abbreviated SUS. A device similar to the silicon-controlled switch, except that a Zener junction is added to the anode gate so that the silicon unilateral switch is triggered into conduction at approximately 8 volts. The SUS can also be triggered by application of a negative pulse to the gate.

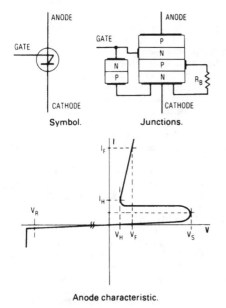

Silicon unilateral switch.

Silicon Valley — Also called Silicon Gulch. The area around Sunnyvale, California, where most of the American semiconductor manufacturers are located.

silk-covered wire — A wire covered with one or more layers of fine floss silk. It is a better insulator than cotton. Also, it is more moisture-resistant and permits more turns of wire within a given space.

silk screen — A screen of a closely woven silk mesh stretched over a frame and used to hold an emulsion outlining a circuit pattern and used in screen printing of film circuits. Used generally to describe any screen (stainless steel or nylon) used for screen printing.

silver — A precious metal that is more conductive than copper. Because it does not readily corrode, it is used for contact points of relays and switches. Its chemical symbol is Ag.

silvered-mica capacitor — A mica capacitor that has a coating of silver deposited directly on the mica sheets instead of using conducting metal foil.

silvering — *See* silver spraying.

silver migration — A process by which silver in contact with an insulating surface, under conditions of high humidity and with an electrical potential applied, is removed ionically from one location and redeposited as a metal in another location. This transfer results in reduced insulation resistance and dielectric failure.

silver oxide cell — A small dry cell giving a constant voltage of 1.5 volts. Used in low-current applications, such as hearing aids, calculators, and electric watches. It consists of a pure zinc anode, a depolarizing silver oxide cathode, and a potassium hydroxide or sodium hydroxide electrolyte.

silver solder — A solder that is composed of copper, silver, and zinc. It has a melting point lower than that of silver, but higher than that of lead-tin solder.

silver soldering — Brazing with a silver-based filler metal.

silver spraying — Also called silvering. Metallizing the surface of an original master disc recording by using a dual spray nozzle in which ammoniated silver nitrate and a reducer are combined in an atomized spray to precipitate the metallic silver.

silverstat — An arrangement of closely spaced contactors. Sometimes used as a step-by-step device to unbalance the arms of a resistance bridge.

silverstat regulator — A multitapped resistor whose taps are connected to single-leaf silver contacts. Variation of voltage causes a solenoid to open or close these contacts, shorting out more or less of the resistance in the exciter circuit as a means of regulating the output voltage to the desired value.

simple buffering — A technique for buffer control such that the buffers are assigned to a single data control block and remain assigned to it until it is closed.

simple-gate IC — An integrated circuit that consists of one or more gate circuits formed on a single chip. The input and output of each gate are brought out to separate pins on the integrated-circuit package.

simple harmonic current — Also called sinusoidal current. A symmetrical alternating current whose instantaneous value is equal to the product of a constant and the sine or cosine of an angle having a value varying linearly with time.

simple harmonic electromotive force — A symmetrical alternating electromotive force that is equal to the product of a constant and the cosine or sine of an angle that varies linearly with time.

simple harmonic motion — A periodic motion whose displacement varies as a sinusoidal function of time.

simple quad — *See* s-quad.

simple scanning — Scanning of only one scanning spot at a time.

simple sound source — A source that radiates sound uniformly in all directions under free-field conditions.

simple steady-state vibration — A periodic motion made up of a single sinusoid.

simple target — In radar, a target whose reflecting surface does not cause the amplitude of the reflected signal to vary with the aspect of the target (e.g., a metal sphere).

simple tone — 1. A sound wave whose instantaneous sound pressure is a simple sinusoidal function of time. 2. Also called a pure tone. A sound sensation characterized by its singleness of pitch.

simplex — 1. Transmission in one direction only. 2. A transmission facility in which the transmission is restricted to only one direction. 3. A form of communications satellite operation that involves a communication in only one direction at a time (mainly for facsimile, television, and some data).

simplex channel — A path for electrical transmission of information in one direction between two or more terminals.

simplex coil — A repeating coil used on a pair of wires to derive a commercial simplex circuit.

simplexed circuit — A two-wire metallic circuit from which a simplex circuit is derived, the metallic and simplex circuits being capable of simultaneous use.

simplex mode — Operation of a communication channel in one direction only, with no capability for reversing.

simplex modem — A two-wire modem that can transmit in only one direction.

simplex modem with backward channel — Two-wire modem that can transmit simultaneously in both directions, with the primary direction being reasonably high speed and the secondary (or backward) direction being rather low speed.

simplex operation — Communication that takes place in only one direction at a time between two stations. Included in this classification are ordinary transmit-receive or press-to-talk operation, voice-operated carrier, and other forms of manual or automatic switching from transmit to receive.

simplex software — One-way transmission of data. A program that can be in the form of ROM, floppy disk data, cassette data, or hard-copy (firmware), or in the form of a machine code or high-level language in RAM.

simplex transmission — Data transmission in one direction only.

simulate — 1. To use the behavior of another system to represent certain behavioral features of a physical or abstract system. 2. To represent the functioning of a device, system, or computer program by another; e.g., to represent one computer by another, to represent the behavior of a physical system by the execution of a computer program, or to represent a biological system by a mathematical model. To represent, by imitation, the functions of one system or process by means of another. *See* emulate.

simulation — Also called digital simulation. 1. A type of problem in which a physical model and the conditions to which the model may be subjected are all represented by mathematical formulas. 2. The substitution of instrumentation (often a computer) for actual operational conditions, so that valid data can be obtained. 3. Modeling of the operation of a logic circuit by a computer program containing device models and topology information about their interconnections. 4. The technique of utilizing representative or artificial data to reproduce in a model various conditions that are likely to occur in the actual performance of a system. Frequently used to test the behavior of a system under operating policies. 5. Representation of either an abstract or a physical system's features by computer operations. Often the operating environment of a program must be simulated during

the software testing. 6. Modeling a target microprocessor with a software interpreter so that object code can be checked as if it were actually executing in the target microprocessor. Simulation usually can't duplicate timing problems, glitches, or microprocessor idiosyncrasies. Input/output devices are often simulated so microcomputer development can proceed before the actual devices are available. 7. Representation of physical systems by computers.

simulator — 1. A device that represents a system or phenomenon and that reflects the effects of changes in the original so that it may be studied, analyzed, and understood from the behavior of that device. 2. A cross-computer program that allows the user to test the object program by simulating the action of the microcomputer when the actual circuitry is unavailable. Simulators often provide certain kinds of diagnostic information unavailable with a debugger program running on the actual microcomputer: warning of the overflow of a processor stack or of an attempt by the program to write into a location in the ROM, for example. They usually allow manipulation and display of the simulated microcomputer memory and CPU registers; setting of breakpoints, whereby processing can be stopped at a certain program address or when the program reads or writes into a specified memory location; and tracing, in which each instruction in a certain address range is printed out as it is executed. Often they provide timing information, such as the number of instructions or machine cycles executed from program start to stop. 3. Program that helps to evaluate a microprocessor by duplicating all logic operations within the software of a large computer. Software simulators are sometimes used in the debug process to simulate the execution of machine-language programs using another computer (often a time-sharing system). These simulators are especially useful if the actual computer is not available. They may facilitate the debugging by providing access to internal registers of the CPU that are not brought out to external pins in the hardware. 4. A device or a computer program that simulates the operation of another device or computer.

simulator program — A program that causes one computer to imitate the logical operation of another computer for purposes of measurement and evaluation. Primarily used to exercise program logic independent of hardware environment. Extremely useful for debugging logic prior to committing it to ROM.

simulcast — 1. To broadcast a program simultaneously over more than one type of broadcast station, e.g., to broadcast a stereophonic program over an AM and FM station. 2. A program so broadcast.

simulcasting — Broadcasting a stereo program over an AM and FM station. An AM and FM tuner are required for stereo reception.

simultaneous — Pertaining to the occurrence of events at the same instant of time.

simultaneous access — *See* parallel access.

simultaneous computer — A computer in which there is a separate unit to perform each portion of the complete computation concurrently, the units being interconnected in a manner that depends on the computation. At different times during a run, a given interconnection carries signals that represent different values of the same variable. For example, the simultaneous computer is a differential analyzer.

simultaneous lobing — In radar, a direction-determining technique utilizing the received energy of two concurrent and partially overlapped signal lobes. The relative phase of power of the two signals received from a target is a measure of the angular displacement of the target from the equiphase or equisignal direction.

simultaneous transmission — Transmission of control characters or data in one direction at the same time that information is being received in the opposite direction.

SINAD or sinad — Abbreviation for signal-to-noise and distortion. 1. A measurement of the signal-to-noise ratio of a receiver system in which the signal level measurement includes the system noise and distortion: (s + n + d)/n.

sine — The sine of an angle of a right triangle is equal to the side opposite that angle divided by the hypotenuse (the long side opposite the right angle).

sine galvanometer — An instrument resembling a tangent galvanometer except that its coil is in the plane of the deflecting needle. The sine of the angle of deflection will then be proportionate to the current.

sine law — The law which states that the intensity of radiation in any direction from a linear source varies in proportion to the sine of the angle between a given direction and the axis of the source.

sine potentiometer — A dc voltage divider (potentiometer) whose output is proportionate to the sine of the shaft-angle position.

sine wave — 1. A wave that can be expressed as the sine of a linear function of time, space, or both. 2. A waveform (often viewed on an oscilloscope) of a pure alternating current or voltage. It is drawn on a graph of amplitude versus time or radial degrees and follows the rules of sine and cosine values in relation to angular rotation of an alternator. It can be simulated by means of an electronic oscillator. 3. The only waveform that cannot be considered to be a pulse. All other waveforms consist of more than one frequency, all harmonically related. A sine wave has a single frequency and therefore occupies a very small bandwidth. It passes through circuitry of any bandwidth with no change in waveform, but it may be changed in amplitude.

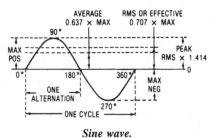

Sine wave.

sine-wave modulated jamming — A jamming signal consisting of a cw signal modulated by one or more sine waves.

singing — 1. An undesired self-sustained oscillation at a frequency in or above the passband of a system or component. 2. An unwanted self-sustained audio-frequency oscillation in an audio system or device.

singing margin — Also called gain margin. The excess of loss over gain around a possible singing path at any frequency, or the minimum value of such excess over a range of frequencies.

singing point — 1. The condition of a circuit or transmission path where the sum of the gains exceeds the sum of the losses. When expressed in decibels, it is the gain that can be added to the circuit equivalent before singing will begin. 2. The amount of total gain in the transmission system (most commonly used in connection with two-wire repeaters) that causes the system to begin to

lose efficiency of performance because the self-oscillating point is too closely approached. 3. The singing point of a circuit that is coupled back to itself is the point at which the gain is just sufficient to make the circuit break into oscillation.

singing-stovepipe effect — Reception and reproduction of radio-signal modulation by ordinary pieces of metal, such as sections of stovepipe, in contact with each other. It is caused by mechanically poor connections, such as rusty bolts or faulty welds, that act as nonlinear diodes and produce intermodulation distortion when subjected to strong radiated fields near transmitters.

single-address code — An instruction that contains the location of the data as well as the operation or sequence of operations to be performed on this data.

single amplitude — With reference to vibratory conditions, the peak displacement of an oscillating structure from its average or mean position.

single-anode tank — See single-anode tube.

single-anode tube — Also called a single-anode tank. An electron tube with one anode (used chiefly for pool-cathode tubes).

single-assignment language — A programming language that allows only one value to be assigned to a variable in a single expression.

single-axis gyro — A type of gyro in which the spinning rotor is mounted in a gimbal arranged so as to tilt about only one axis relative to the stable element.

single-board microcomputer — Also called monoboard microcomputer. A single printed circuit board containing, as a minimum, processor, memory (ROM and/or RAM), and input/output — usually a combination of serial and parallel ports. May also include a counter/timer function and bus interconnection scheme. A single-board microcomputer family may also include other functional system elements (such as memory and I/O functions) on circuit boards of the same format as the microcomputer board.

single-button carbon microphone — A microphone having a carbon-filled buttonlike container on one side of its flexible diaphragm. As the sound waves move the diaphragm, the resistance of the carbon changes, and the microphone current constitutes the desired audio-frequency signal.

single-carrier FM recording — The method of recording in which the input signal is frequency modulated onto a carrier and the carrier is recorded on a single track at saturation and without bias.

single-channel — A carrier-only or single-tone modulated radio control transmitter and matching receiver installation.

single-channel monopulse tracking system — See monopulse tracking.

single-channel simplex — Nonsimultaneous communication between stations over the same frequency channel.

single circuit — A telegraph circuit capable of non-simultaneous two-way communication.

single-circuit system — An alarm circuit that routes only one side of the circuit through each sensor. The return may be through either ground or a separate wire.

single-conversion receiver — A receiver employing a superheterodyne circuit in which the input signal is downconverted once.

single crystal — A piece of material in which the crystallographic orientation of all the basic groups of atoms is the same.

single-degree-of-freedom system — A system for which only one coordinate is required to define the configuration of the system.

single-dial control — Control of a number of different devices or circuits by means of a single adjustment (e.g., in tuning all variable-capacitor sections of a radio receiver).

single-ended — Unbalanced, such as grounding one side of a circuit or transmission line.

single-ended amplifier — An amplifier in which only one tube or transistor normally is employed in each stage — or if more than one is used, they are connected in parallel so that operation is asymmetric with respect to ground.

single-ended input — Amplifier input configuration in which all analog inputs are referenced to system ground.

single-ended input impedance — The impedance between one amplifier input terminal and ground (with the other input terminal, if any, grounded for ac) when the amplifier is balanced.

single-ended input voltage — The signal voltage applied to one amplifier input terminal with the other input terminal at signal ground.

single-ended output voltage — The signal voltage between one amplifier output terminal and ground.

single-ended push-pull amplifier circuit — An amplifier circuit having two transmission paths designed to operate in a complementary manner and connected to provide a single unbalanced output. (No transformer is used.)

single-ended signal — As opposed to a difference-mode signal, a signal that is at ground potential when it is at zero level.

single-ended tube — A metal tube in which all electrodes — including the control grid — are connected to base pins and there is no top connection. The letter *S* after the first numerals in a receiving-tube designation (e.g., 6SN7) indicates a single-ended tube.

single-ended voltage gain — Within the linear range of an amplifier, the ratio of a change in output voltage to the corresponding change in single-ended input voltage.

single-frequency duplex — A method that provides communications in opposite direction over a single-frequency carrier channel, but not at the same time. The change between transmitting and receiving conditions is controlled automatically by the voices of the communicating parties.

single-frequency simplex — A system of single-frequency carrier communications in which the change from transmission to reception is accomplished by manual rather than automatic means.

single-grip terminal — A solderless terminal designed to permit a crimp to the wire only.

single-groove stereo — *See* monogroove stereo.

single-gun color tube — A color picture tube with a single electron gun that produces only one beam, which is sequentially deflected across the phosphor dots.

single harmonic distortion — The ratio of the power at the fundamental frequency measured at the output of the transmission system considered to the power of any single harmonic observed at the output of the system because of its nonlinearity, when a single-frequency signal of specified power is applied to the input of the system. It is expressed in decibels.

single-hop propagation — Transmission in which the radio waves are reflected only once in the ionosphere.

single in-line package — *See* SIP.

single-junction photosensitive semiconductor — Two layers of semiconductor materials with an electrode connection to each material. Light energy controls the amount of current.

single-line diagram — Also called single-line drawing. A form of schematic diagram in which single lines are used to show component interconnections even though two or more conductors are required in the actual circuit.

single-line telephone — A telephone that provides access to one telephone line.

single-loop feedback — A loop in which feedback may occur only through one electrical path.

single-mode fiber — 1. A fiber waveguide that supports only one mode of propagation. 2. An optical glass fiber that consists of a step core of very small diameter, approximately 6 μm, and a cladding approximately 20 times the thickness of the core. Tremendous information rates (great bandwidth) are possible with single-mode fibers. The primary disadvantages of this type of fiber are cost of manufacture, difficulty in launching signals into the fiber, and difficulty in splicing and general handling in the field.

single-operand instruction — An instruction containing a reference to one register, memory location, or device.

single-phase circuit — Either an alternating-current circuit with only two points of entry, or one with more than two points of entry but energized in such a way that the potential differences between all pairs of points of entry are either in phase or 180° out of phase. A single-phase circuit with only two points of entry is called a single-phase two-wire circuit.

single-phase synchronous generator — A generator that produces a single alternating electromotive force at its terminals.

single phasing — The tendency of the rotor (of a motor tach generator) to continue to rotate when one winding is opened and the other winding remains excited.

single-point ground — *See* uniground.

single-point grounding — A grounding system that attempts to confine all return currents to a network that serves as the circuit reference. It does not imply that the grounding system is limited to one earth connection. To be effective, no appreciable current is allowed to flow in the circuit reference; i.e., the sum of the above return currents is zero.

single-polarity pulse — A pulse that departs from normal in one direction only.

single pole — A contact arrangement in which all contacts in the arrangement connect, in one position or another, to a common contact.

single-pole, double-throw — Abbreviated SPDT. A three-terminal switch or relay contact for connecting one terminal to either of two other terminals.

single-pole-piece magnetic head — A magnetic head with only one pole piece on one side of the recording medium.

single-pole, single-throw — Abbreviated SPST. 1. A two-terminal switch or relay contact that either opens or closes one circuit. 2. A switch with only one moving and one stationary contact. Available either normally open (no) or normally closed (nc).

single rail — The method of data transfer in a computer on only one line or wire. The device at the destination must be able to handle the data in either the high-level or low-level value. The return path is by way of common or ground.

single-rank binary — A flip-flop that requires no more than one full clock pulse from a single clock system to transfer the logic from a synchronous input to the output of the binary. It contains only one memory stage.

single sampling plan — The plan that consists of a single sample size with associated acceptance and rejection criteria.

single-shield solid enclosure — An all-metal enclosure providing higher attenuation than cell-type units. It is usually a rigid, free-standing enclosure.

single-shot — *See* monostable.

single-shot blocking oscillator — A blocking oscillator modified to operate as a single-shot trigger circuit.

single-shot multivibrator — Also called a single-trip multivibrator. 1. A multivibrator modified to operate as a single-shot trigger circuit. *See also* monostable multivibrator. 2. A monostable multivibrator that, after being triggered to the quasi-stable state, will "flop" back by itself to the stable state after a certain period of time.

single-shot trigger circuit — Also called a single-trip trigger circuit. A trigger circuit in which the pulse initiates one complete cycle of conditions, ending with a stable condition.

single sideband — Abbreviated SSB. An AM radio transmitter technique in which only one sideband is transmitted. The other sideband and the carrier are suppressed. This gives SSB a 6:1 efficiency advantage over AM, and, thus, greater range per watt of output power. SSB occupies half of a conventional AM (double-sideband) channel.

single-sideband filter — A bandpass filter in which the slope on one side of the response curve is greater than on the other side. So called because it is used in systems to suppress a carrier frequency and transmit one or both sidebands.

single-sideband modulation — 1. Modulation whereby the spectrum of the modulating wave is translated in frequency by a specified amount, either with or without inversion. 2. A form of amplitude modulation in which only one of the two sidebands is transmitted. Either of the two sidebands may be transmitted, and the carrier may be transmitted, reduced, or suppressed.

single-sideband suppressed carrier — Modulation resulting from the partial or complete elimination of the carrier and all components of one sideband from an amplitude-modulated wave.

single-sideband system — A type of radiotelephone service in which one set of sidebands (either the upper or lower) is completely suppressed and the transmitted carrier is partially suppressed.

single-sideband transmission — Transmission of only one sideband, the other sideband being suppressed. The carrier wave may be transmitted or suppressed.

single-sideband transmitter — A transmitter in which only one sideband is transmitted.

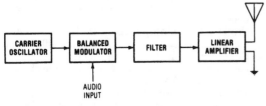

Single-sideband transmitter.

single-signal receiver — A superheterodyne receiver equipped for single-signal reception. A highly selective filter is placed in the intermediate-frequency amplifier, and provision is included for varying the selectivity of the receiver to suit the requirements of the band condition.

single-signal reception — Use of a piezoelectric quartz crystal and associated coupling circuits as a crystal filter to provide the high degree of selectivity required for reception in a crowded band.

single step — Pertaining to a method of computer operation in which each step is carried out in response to a single manual operation.

single-stroke bell — A bell that is struck once each time its mechanism is activated.

single-stub transformer — A shorted section of coaxial line connected to a main coaxial line near a discontinuity so that impedance matching at the discontinuity is achieved.

single-stub tuner — A section of transmission line that is terminated by a movable short-circuitry plunger or bar and that is attached to a main transmission line to provide impedance matching.

single sweep — The operating mode for a triggering-sweep oscilloscope in which the sweep must be reset for each operation, thus preventing unwanted multiple display; it is particularly useful for trace photography. In the interval after the sweep is reset and before it is triggered, the oscilloscope is said to be armed.

single throw — A contact arrangement in which each contact form included is a single contact pair.

single-throw circuit breaker — A circuit breaker in which only one set of contacts need be moved to open or close the circuit.

single-throw switch — A switch in which only one set of contacts need be moved to open or close the circuit.

single-tone keying — Keying in which the carrier is modulated with a single tone for one condition, either marking or spacing, but is unmodulated for the other condition.

single-track magnetic system — A magnetic-recording system whose medium has only one track.

single-track recorder — A tape recorder that records or plays only one track at a time on or from the tape. *See also* monaural recorder.

single-trip multivibrator — *See* single-shot multivibrator.

single-trip trigger circuit — *See* single-shot trigger circuit.

single-tuned amplifier — An amplifier characterized by resonance at a single frequency

single-tuned circuit — A circuit that may be represented by a single inductance and capacitance, together with associated resistances.

single-turn potentiometer — A potentiometer in which the slider travels the complete length of the resistive element with only one revolution of the shaft.

single-unit semiconductor device — A semiconductor device having one set of electrodes associated with a single carrier stream.

single-wire line — A transmission line that uses the ground as one side of the circuit.

single-wound resistor — A resistor in which only one layer of resistance wire or ribbon is wound around the base or core.

sink — In communication practice: 1. A device that drains off energy from a system. 2. A place where energy from several sources is collected or drained away. 3. Anything into which power of some kind is dissipated. 4. The component or network into which energy (usually current) flows. 5. A device that switches ground or minus to a load. The current flows from the load into the sensor. *See* interface, 2.

sinker — An n+ region that extends down from the collector contact area on an integrated transistor to the n+ island under the collector for the purpose of reducing the collector resistance.

sink load — A load with a current whose direction is away from its input. A sink load must be driven by a current sink.

sins — Acronym for ship's inertial navigational system; especially applicable to submarine use.

sinter — A ceramic material or mixture fired so that it is not completely fused but is a coherent mass.

sintered plate — A powder that holds the active plate material used for both the anode and cathode in secondary cells. This provides a large surface area for the active material, allowing better cycle life, higher discharge rates, and better efficiency than the pocket-type plate design.

sintering — 1. The process in which metal or other powders are bonded by cold-pressing them into the desired shape and then heating them so that a strong, cohesive body is formed. 2. The welding together of powdered particles at temperatures below the melting or fusion point. Particles are fused together to form a mass, but the mass, as a whole, does not melt.

sinusoid — A curve having ordinates proportional to the sine of the abscissa.

sinusoidal — Varying in proportion to the sine of an angle or time function (e.g., ordinary alternating current).

sinusoidal current — *See* simple harmonic current.

sinusoidal electromagnetic wave — In a homogeneous medium, a wave with an electric field strength proportionate to the sine (or cosine) of an angle that is a linear function of time, distance, or both.

sinusoidal field — A field in which the magnitude of the quantity at any point varies as the sine or cosine of an independent variable such as time, displacement, or temperature.

sinusoidal quantity — A quantity that varies in the manner of a sinusoid.

sinusoidal vibration — A cyclical motion in which the object moves linearly. The instantaneous position is a sinusoidal function of time.

sinusoidal wave — A wave whose displacement varies as the sine (or cosine) of an angle that is proportional to time, distance, or both.

SIP — Abbreviation for single in-line package. 1. A package having a single row of external leads, usually mounted vertically with leads through the printed circuit board, but can be surface mounted with leads bent in gullwing fashion. 2. A package for electronic components that is suited for automated assembly into printed circuit boards. The SIP is characterized by a single row of external connecting terminals, or pins, which are inserted into the holes of the printed circuit board.

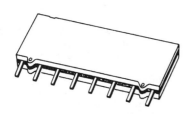

Single in-line package.

SI system — International System of Units. Includes mks and mksa units.

site error — In navigation, the error that occurs when the radiated field is distorted by objects near navigational equipment.

situation-display tube — A large cathode-ray tube used for displaying tubular and vector information having to do with the various functions of an air-defense mission.

six-phase circuit — A combination of circuits energized by alternating electromotive forces that differ in phase by one-sixth of a cycle (60°).

size control — On a television receiver, a control that varies the size of the picture either horizontally or vertically.

skating — The tendency of a pivoted tonearm to be pulled to the center spindle. It is caused by friction between the stylus and the record surface.

skating force — A frictional force between a pickup stylus and the record material, tending to move the pickup toward the center of the record. It is present only when the cartridge is offset at an angle to reduce tracking error, and is a function of tracking force, offset angle, stylus size and shape, record material, and recorded amplitude. In a stereo recording, the effect of skating force is to increase the stylus-to-groove contact force on the left channel and decrease it on the right channel.

skein winding — A method of winding single-phase motors in which each pole is a long skein of wire, formed by winding around two headless nails or bolts (smooth) set some distance apart on a piece of wood. The entire pole is wound in place by twisting the coil to form a concentric chain. No internal connection is made between coils of the same pole. The winding is measured with a single turn of wire; then, using this endless wire, the location of the winding pins can be found.

skeletal coding — Sets of computer instructions in which some addresses and other parts are undetermined. These items usually are determined by routines designed to modify them according to given parameters.

skew — 1. In facsimile, the nonrectangular received frame due to asychronism between the scanner and recorder. Numerically it is the tangent of the angle of this deviation. 2. The motion characterized on a magnetic tape by an angular velocity between the gap center line and a line perpendicular to the tape center line. 3. In magnetic thin film, the deviation of the easy axis during fabrication. 4. The angular displacement of a printed character, character group, or other data from the intended or ideal placement. 5. A measure (expressed in percent) of the departure of each individually received DTMF signal frequency from its nominal value. A function of component tolerances, aging, environmental conditions, and certain types of transmission-multiplexing equipment, skew is measured at the DTMF receiver. 6. The time difference between the logic-state changes on different input pins within a particular test pattern. 7. The time difference of corresponding digital information on separate lines, measured at the rising or falling edges of the digital data.

skewed distribution — A frequency distribution of any natural phenomenon in which zero or infinity is one of its limits.

skewing — The time delay or offset between two signals with respect to each other.

skewness — A statistical measure of the asymmetry existing in a distribution.

skew ray — A ray that is skewed to the axis of an optical fiber. If the fiber is straight, a skew ray travels along a helical path around, but not crossing, the fiber's axis.

skiatron — 1. A dark-trace oscilloscope tube. *See also* dark-trace tube. 2. A display employing an optical system with a dark-trace tube.

skin antenna — A flush-mounted aircraft antenna made by isolating a portion of the metal skin of the aircraft with insulating materials.

skin depth — Also called depth of penetration. In a current-carrying conductor, the depth below the surface at which the current density has decreased one neper below the current density at the surface; that is, the field has decreased to $1/\varepsilon$ (36.8 percent) of its surface value.

skin effect — Also called radio-frequency resistance. 1. The tendency of rf currents to flow near the surface of a conductor. Thus, they are restricted to a small part of the total sectional area, which has the effect of increasing the resistance. 2. The phenomenon that occurs when an alternating current forces the ac current to flow mostly in the outer parts of a conductor. 3. The phenomenon in which the depth of penetration of electric currents into a conductor decreases as the frequency increases. 4. A characteristic of current in a conductor whereby as the frequency increases more and more current flows near the conductor surface and less at the center.

skinner — A wire brought out at the end of a cable prepared for soldering to a terminal.

skinning — Peeling the insulation from a wire.

skin tracking — Radar tracking of an object without the aid of a beacon or other signal device on board the object.

skip — 1. A digital-computer instruction to proceed to the next instruction. 2. In a computer, a "blank" instruction. 3. To ignore one or more of the instructions in a sequence. 4. Term referring to propagation of radio signals over considerable distances due to reflection back to earth from the ionosphere.

skip distance — The distance separating two points on the earth between which radio waves are transmitted by reflection from the ionized layers of the ionosphere.

skip fading — Fading due to fluctuations of ionization density at the place in the ionosphere where the wave is reflected, which causes the skip distance to increase or decrease.

skip-if-set instructions — In computers, a class of instructions in which provision is made for examining particular logic conditions. Usually they are used in conjunction with a jump (branch) instruction. For example, a skip-if-word-register-ready instruction would allow the program to check for a ready condition of the word register and then permit the program to continue along one of two different paths, depending on the condition of the word register.

skip keying — The reduction of the radar pulse-repetition frequency to a submultiple of that normally used, to reduce the mutual interference between radars or to increase the length of the radar time base.

skip zone — Also called zone of silence. A ring-shaped space or region within the transmission range wherein signals from a transmitter are not received. It is the distance between the farthest point reached by the ground wave and the nearest point at which the refracted sky waves come back to earth.

skirt selectivity — A measure of the resolution capability of a spectrum analyzer when displaying signals of unequal amplitude. A unit of measure would be the bandwidth at some level below the 6-dB down points.

SKU — Stockkeeping unit. Abbreviation used in many computer reports to define an individual stock item.

sky error — See ionospheric error.

sky hook — Amateur term for antenna.

sky noise — 1. Noise produced by radio energy from stars. 2. Background microwave radiation coming from deep space. It can be a noise source for dish antennas and sets a lower boundary for the possible noise temperature of any dish antenna of approximately 16 to 20 K.

sky wave — See indirect wave; ionospheric wave.

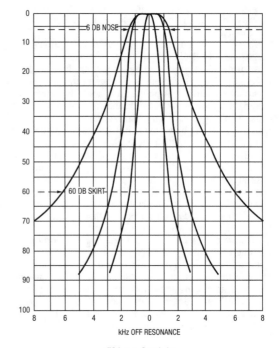

Skirt selectivity.

sky-wave correction — In navigation, a correction for sky-wave propagation errors applied to measured positional data. The amount of the correction is established on the basis of an assumed position and on the height of the ionosphere.

sky-wave station error — In sky-wave-synchronized loran, the station-synchronization error due to the effect of the ionosphere on the synchronizing signal transmitted from one station to the other.

sky-wave-synchronized loran — A loran system in which the range is extended by using ionosphere-reflected signals for synchronizing the two ground stations.

sky-wave transmission delay — The longer time taken by a transmitted pulse when carried by sky waves reflected once from the E layer compared with the same pulse carried by ground waves.

slab — A relatively thick crystal from which blanks are cut.

slab line — A double-slotted coaxial line whose outer shield has been unwrapped and extended to infinity in both directions so that the resulting configuration is a cylindrical conductor between two parallel conductors.

slab wafer — A slice of semiconductor material that has straight edges, as opposed to a conventional rounded wafer that has 21 percent less area than a square with comparable dimensions.

slant range — 1. In radar, the line-of-sight distance from the measuring point to the target, particularly an aerial target. 2. Line-of-sight distance between two points not at the same elevation.

slap-back — An echo effect wherein the original signal reappears as distinct echoes that decay in level each time they appear. One way of creating slap-back is to feed the output signal from the play head back into the record head, at a slightly lower level. For example, if the sound "la" is originally fed to the recorder, then "la-la-la ... la" will be heard, with each "la" slightly lower in level until the signal fades away.

slave — 1. A component in a system that does not act independently, but only under the control of another similar component. 2. A device that follows an order given by a master remote control.

slave antenna — A directional antenna that is positioned in azimuth and elevation by a servo system. The information controlling the servo system is supplied by a tracking or positioning system.

slave drive — *See* follower drive.

slaved tracking — A method of interconnecting two or more regulated power supplies so that the master supply operates to control other power supplies called slaves.

slave operation — A method of interconnecting two or more stabilized power supplies so that coordinated control of the assembly by controlling the master supply alone is achieved, and essentially proportional outputs are obtained from all units.

slave relay — *See* auxiliary relay, 2.

slave station — 1. A radionavigational station whose emissions are controlled by a master station. 2. A device on a communication facility that is prevented from initiating transmission in order to organize and control communications between two or more stations.

slave sweep — A time base that is synchronized or triggered by a waveform from an external source. It is used in navigational systems for displaying or utilizing the same information at different locations, or in displaying or utilizing different information with a common or related time base.

slaving — The use of a torque to maintain the orientation of the spin axis of a gyro relative to an external reference such as a pendulum or magnetic compass.

sleeping sickness — In transistors, the gradual appearance of leakage.

sleeve — 1. A cylindrical contacting part, usually placed in back of the tip or ring of a plug and insulated from it. 2. An iron core (usually a thin-walled cylinder) used as an electromagnetic shield around an inductor. 3. A lead tube placed over cable conductors that have been spliced. 4. A tube of woven cotton pushed over a twisted wire joint in a cable. 5. A brass or copper tube or paired tubes for fastening line or drop wires together by twisting, crimping, or rolling. 6. A tube of copper or iron placed over a relay winding to make the relay slow acting.

sleeve antenna — A vertical half-wave antenna whose lower half is a metallic sleeve through which the concentric feed line runs. The upper radiating portion, which is one-quarter wavelength, is connected to the center of the line.

sleeve-dipole antenna — A dipole antenna with a coaxial sleeve around the center.

sleeve-stub antenna — An antenna consisting of half of a sleeve-dipole antenna projecting from an extended conducting surface.

sleeve wire — 1. A third conductor when associated with a pair. 2. Wire that connects to the sleeve of a plug or jack. By extension, it is common practice to designate by this term the conductors having similar functions or arrangements in circuits where plugs or jacks may not be involved.

sleeving — Any preformed tubular insulation, of braided or extruded construction, that is intended for placement over portions of cables or conductors during their installation to insulate them or hold them together.

slewing — 1. Rapid change of a mechanism associated with either end of a data-transmission system when it stops following one target and takes up another. 2. In random-sampling-oscilloscope technique, the process of incrementally delaying successive samples or a set of samples with respect to the signal under examination.

slewing rate — *See* slew rate.

slew range — 1. The high-speed range in which a motor can run continuously but cannot stop, start, or reverse without losing step count. 2. The maximum stepping rate at which a stepper motor will run without losing synchronism with the field.

slew rate — Also called rate limit or voltage velocity limit. 1. The maximum rate of change of the output voltage of an amplifier operated within its linear region. 2. The maximum rate of change of the output voltage of a closed-loop amplifier under large-signal conditions (the conditions that exist when an ac input voltage causes saturation of an amplifier stage, resulting in current limiting of that stage). 3. The maximum rate at which an amplifier output can be driven between limits in response to a step change in input voltage while accurately reproducing the input waveform. Usually measured in volts per second or volts per microsecond. For example, a 0.5-V/μs slew rate means that the output rises or falls no faster than 0.5 V every microsecond. Slew rate is caused by current limiting and saturation of an op-amp internal stage. That limited current is the maximum current available to charge the compensation capacitance network. 4. The rate at which the output can be driven from limit to limit under overdrive conditions. Generally specified in volts per microsecond. 5. The rate at which an amplifier follows a fast-rising waveform. Usually measured in volts per microsecond. 6. In a sample-and-hold circuit, the fastest rate at which the sample hold output can change. Specified in volts per microsecond.

slew torque — *See* running torque.

SLIC — Acronym for Subscriber Line Interface Circuit. 1. In digital transmission of voice, the circuit that performs some or all of the interface functions at the central office. *See* BORSCHT. 2. An analog device incorporated in a telephone transmission network and in the interface of a telephone with a central office. It also performs conversions from the two-wire subscriber loop to the four-wire central-office switch, and vice versa.

slice — 1. A single wafer cut from a silicon ingot and forming a thin substrate on which have been fabricated all the active and passive elements for multiple integrated circuits. A completed slice usually contains hundreds of individual circuits. *See also* chip, 2. 2. Those parts of a waveform between two given amplitude limits on the same side of the zero axis. 3. A type of chip architecture that permits the cascading or stacking of devices to increase word bit size.

slicer — Also called an amplitude gate or a clipper-limiter. A transducer that transmits only portions of an input wave lying between two amplitude boundaries.

slicked switch — An alacritized mercury switch in which the rolling surface has been treated with an oily material.

slideback — The technique of applying a dc voltage to one input of a differential amplifier in order to change the vertical position on the CRT screen of the signal applied to the other input.

slideback voltmeter — A vacuum-tube voltmeter that measures effective voltage values indirectly by measuring the change in grid bias voltage required to restore the plate current of the vacuum tube to the value it had before the unknown voltage was applied to the grid circuit.

slider — A sliding contact.

slide-rule dial — A tuning dial in which a pointer moves in a straight line over a straight scale. So called because it resembles a slide rule.

slide switch — A switch that is actuated by sliding a control lever from one position to another.

Slide switch.

slide wire—A bare resistance wire and a slider that can be set anywhere along the wire to provide a continuously variable resistance.

slide-wire bridge—A simplified Wheatstone bridge in which the resistance ratio is determined by the position of a slider on a resistance wire.

slide-wire rheostat—A long single-layer coil of a resistance wire with a sliding contact. The resistance is varied by moving the slider.

sliding contact—*See* wiping contact.

sliding load—A length of transmission line containing a matched electrical load at a distance from the connector end that can be varied.

sliding short—A length of transmission line containing an electrical short at a distance from the connector end that can be varied.

slip—1. The difference between the synchronous speed of a motor and the speed at which it operates. Slip may be expressed as a percentage or decimal fraction of synchronous speed or directly in revolutions per minute. 2. Distortion produced in a recorded facsimile image as a result of slippage in the mechanical drive system. 3. A method of interconnecting multiple wiring between switching units so that trunk 1 becomes the first choice for the first switch, trunk 2 becomes the first choice for the second switch, and so on.

slip process—*See* wet process.

slip ring—A device for making electrical connections between stationary and rotating contacts. *See also* collector rings.

slip-ring motor—Term usually applied to an induction motor with a wound secondary. (The correct term is a wound rotor motor.)

slip speed—The speed difference between speed at any load and the synchronous speed.

slope—1. The essentially linear portion of the grid-voltage, plate-current characteristic curve of a vacuum tube. This is where the operating point is chosen when linear amplification is desired. 2. *See* roll-off, 1. 3. The rate of attenuation of frequencies beyond the passband of a crossover network. Usually either 6 dB or 12 dB per octave of frequency.

slope-based linearity—A manner of expressing nonlinearity as the deviation from a straight line for which only the slope is specified.

slope detection—A discriminator operation on one of the slopes of the response curve for a tuned circuit. It is rarely used in FM receivers because the linear portion of the response curve is too narrow for large-signal operation.

slope detector—A detector in which slope detection is employed.

slot—1. One of the grooves formed in the iron core of a motor or generator armature for the conductors forming the armature winding. 2. A unit of time in a time-division multiplex (TDM) frame during which a subchannel bit or character is carried to the other end of the circuit and extracted by the receiving TDM unit. 3. That longitudinal position in the geosynchronous orbit into which a communications satellite is parked. Above the United States, communications satellites are typically

positioned in slots that are based at two- to three-degree intervals.

slot antenna—A radiating element formed by a slot in a conducting surface.

slot armor—An insulator in the slot of a magnetic core of a machine; it may be on the coil or separate from it.

slot cell—A formed sheet of insulation that is separate from the coil and placed in the slot of a magnetic core.

slot coupling—A method of transferring energy between a coaxial cable and a waveguide by means of two coincident narrow slots, one in the sheath of the coaxial cable. *E*- or *H*-waves are launched into the guide, depending on whether the cable and guide are parallel or perpendicular to each other.

slot-discharge resistance—*See* corona resistance.

slot effect—The minimum voltage of rated frequency applied to the control-voltage winding of a motor tach generator necessary to start the rotor turning at no-load conditions with rated voltage and frequency on the fixed-voltage winding.

slot insulation—Flexible sheet-type insulation inserted into the slots of armatures and stators to insulate the windings from the core.

slot radiator—A primary radiating element in the form of a slot cut in the walls of a metal waveguide or cavity resonator or in a metal plate.

slotted line—*See* slotted section.

slotted rotor plate—*See* serrated rotor plate.

slotted section—Also called a slotted line or slotted waveguide. A section of a waveguide or shielded transmission line, the shield of which is slotted to permit examination of the standing waves with a traveling probe.

slotted SWR measuring equipment—A device in which standing and/or reflected waves are measured with a slotted line and a detecting probe.

slotted waveguide—*See* slotted section.

slow-acting relay—*See* slow-operating relay.

slow-action relay—*See* time-delay relay.

slow death—The gradual change of transistor characteristics with time. This change is attributed to ions that collect on the surface of the transistor.

slowed-down video—A technique of transmitting radar data over narrow-bandwidth circuits. The radar video is stored over the time required for the antenna to move through one beam width, and is subsequently sampled at such a rate that all range intervals of interest are sampled at least once each beam width or once per azimuth quantum. The radar-return information is quantized at the gap-filler radar site.

slow memory—*See* slow storage.

slow-operate, fast-release relay—A relay designed specifically for a long make and short release time.

slow-operate, slow-release relay—A slow-speed relay designed specifically for both a long make and a long release time.

slow-operating relay—Also called slow-acting relay. A relay that is slow to attract its armature after its winding is energized. A copper slug, or collar, at the armature end of the core delays the operation momentarily after the operating circuit is completed. Such a relay is often marked SO on circuit diagrams.

slow-release relay—*See* slow-releasing relay.

slow-releasing relay—Also called a slow-release relay. A slow-acting relay in which a copper slug, or collar, at the heelpiece end of the core delays the restoration momentarily after the operating circuit is opened. Such a relay is often marked SR on circuit diagrams.

slow-scan television — Abbreviated SSTV.

slow-speed relay — A relay designed specifically for long operate or release time, or both.

slow storage — Also called slow memory. Computer storage in which the access time is relatively long. *See also* secondary storage.

slow-wave circuit — A microwave circuit in which the phase velocity of the waves is considerably below the speed of light. Such waves are used in traveling-wave tubes.

slow-wave structure — A circuit composed of selected inductance and capacitance that causes a wave to be propagated at a speed slower than the speed of light.

SLSI — Abbreviation for super large-scale integration (100,000 transistors per chip).

slug — 1. A heavy metal ring or short-circuited winding used on a relay core to delay operation of the relay. 2. A metallic core that can be moved along the axis of a coil for tuning purposes.

slug tuner — A waveguide tuner containing one or more longitudinally adjustable pieces of metal or dielectric.

slug tuning — Varying the frequency of a resonant circuit by introducing a slug of material into the electric or magnetic fields, or both.

slumber switch — A circuit arrangement whereby a radio or a recorder automatic shutoff provision can be made to turn off the apparatus itself as well as any other equipment plugged into its ac outlet.

SMAC — Abbreviation for Scene Matching Area Correlator. An optical terminal homing system that matches images with stored maps.

small-business computer — 1. A system with a price tag low enough to fit the budget of a small business, simple enough for clerks and typists to operate, yet sophisticated enough to perform a variety of complex transactions. 2. A computer that is affordable by a small business, rather than a relatively low-performance system for business.

small-scale integration — *See* SSI.

small-outline transistor — *See* SOT.

small signal — That value of an ac voltage or current which, when halved or doubled, will not affect the characteristic being measured beyond the normal accuracy of the measurement of that characteristic.

small-signal analysis — Consideration of only small excursions from the no-signal bias, so that a vacuum tube or transistor can be represented by a linear equivalent circuit.

small-signal characteristics — The characteristics of an amplifier operating in the linear amplification region.

small-signal current gain (current-transfer ratio) — The output current of a transistor with the output circuit shorted, divided by the input current. The current components are understood to be small enough that linear relationships hold between them.

small-signal drain-to-source on-state resistance — The small-signal resistance between the drain and source terminals of a field-effect transistor with a specified gate-to-source voltage applied to bias the device to the on state. For a depletion-type device, this gate-to-source voltage may be zero.

small-signal gain — The gain characteristics of an amplifier operating in the linear amplification region. Small-signal gain is typically measured at least 10 dB below the input power level that creates 1 dB gain compression.

small-signal open-circuit forward-transfer impedance — In a transistor, the ratio of the ac output voltage to the ac input current when the ac output current is zero.

small-signal open-circuit input impedance — In a transistor, the ratio of the ac input voltage to the ac input current when the ac output current is zero.

small-signal open-circuit output admittance — In a transistor, the ratio of the ac output current to the ac voltage applied to the output terminals when the ac input current is zero.

small-signal open-circuit output impedance — In a transistor, the ratio of the ac voltage applied to the output terminals to the ac output current when the ac input current is zero.

small-signal open-circuit reverse-transfer impedance — In a transistor, the ratio of the ac input voltage to the ac output current when the ac input current is zero.

small-signal open-circuit reverse-voltage transfer ratio — In a transistor, the ratio of the ac input voltage to the ac output voltage when the ac input current is zero.

small-signal power gain — In a transistor, the ratio of the ac output power to the ac input power under specified small-signal conditions. Usually expressed in dB.

small-signal short-circuit forward-current transfer ratio — In a transistor, the ratio of the ac output current to the ac input current when the ac output voltage is zero.

small-signal short-circuit forward-transfer admittance — In a transistor, the ratio of the ac output current to the ac input voltage when the ac output voltage is zero.

small-signal short-circuit input admittance — In a transistor, the ratio of the ac input current to the ac input voltage when the ac output voltage is zero.

small-signal short-circuit input impedance — In a transistor, the ratio of the ac input voltage to the ac input current when the ac output voltage is zero.

small-signal short-circuit output admittance — In a transistor, the ratio of the ac output current to the ac output voltage when the ac input voltage is zero.

small-signal short-circuit reverse-transfer admittance — In a transistor, the ratio of the ac input current to the ac output voltage when the ac input voltage is zero.

small-signal transconductance — In a transistor, the ratio of the ac output current to the ac input voltage when the ac output voltage is zero.

smart terminal — 1. *See* intelligent terminal. 2. A computer terminal that can do more than simply send data to a computer and display data from a computer. Depending on how many chips are built into it, a smart terminal can edit and temporarily store up to several thousand words, produce graphics, and even change the color of the background on its monitor.

smear — 1. Television-picture distortion in which objects appear stretched out horizontally and are blurred. 2. Small frequency and time distortion introduced into a radio signal by a dispersive reflector such as the moon.

smectic phase — A parallel arrangement of liquid crystal molecules arranged in layers. The physical appearance of the smectic state is that of a highly viscous, turbic fluid.

S-meter — A meter provided in some communications receivers to give an indication of the relative strength of the received signal in terms of arbitrary units. It is calibrated in "S" units and decibels. Nominally, each S unit equals 6 dB. Above S9, most meters are calibrated in 10-dB increments. Sometimes a manufacturer will specify what input level (usually between 50 and 100 µV) is required for a reading of S9. S-meters are intended to be relative, rather than absolute, indicators. They are useful to an extent in comparing the strength of two stations, or the performance of two antennas at one location. *See also* signal-strength meter.

smileys — Also called emoticons. Little symbols in a text message meant to be viewed sideways; used to express emotion. Smileys are often used in e-mail messages and newsgroup postings. Examples include :) (happy), :-((sad), :-< (mad), :-o (wow!), :-@ (yell), and ;-) (wink).

Smith chart — A special polar diagram used in the solution of transmission-line and waveguide problems. It consists of constant-resistance circles, constant-reactance circles, circles of constant standing-wave ratio, and radius lines that represent constant line-angle loci.

smoke detector — A device that detects visible or invisible products of combustion. *See also* ionization smoke detector; photoelectric beam-type smoke detector; photoelectric spot-type smoke detector; resistance-bridge smoke detector.

smooth — To apply procedures that bring about a decrease in or the elimination of rapid fluctuations in data.

smoothing — Averaging by a mathematical process or by curve fitting, such as the method of least squares or exponential smoothing.

smoothing choke — An iron-core choke coil that filters out fluctuations in the output current of a vacuum-tube rectifier or direct-current generator.

smoothing circuit — Also called ripple filter. A combination of inductance and capacitance employed as a filter circuit to remove fluctuations in the output current of a vacuum-tube or semiconductor rectifier or direct-current generator.

smoothing factor — The factor expressing the effectiveness of a filter in smoothing out ripple voltages.

smoothing filter — 1. Also called ripple filter. A filter used to remove fluctuations in the output current of a vacuum-tube or semiconductor rectifier or direct-current generator. 2. A low-pass filter in the vertical-deflection amplifier of a spectrum analyzer. It is used to smooth amplitude fluctuations in order to display spectral density and the average level of random signals as single lines. (The time constant of a smoothing filter is generally variable.)

SMPTE — Abbreviation for the Society of Motion Picture and Television Engineers. (Formerly it was the SMPE, Society of Motion Picture Engineers.)

SMT — Abbreviation for surface-mount technology. The mounting of components on the surface of a printed circuit board; as contrasted with through-hole mounting, in which component leads extend through the board.

snake — A tempered steel wire, usually of rectangular cross section. The snake is pushed through a run of conduit or through an inaccessible space such as a partition and used for drawing in wires.

snap-acting switch — A switch in which there is a rapid motion of the contacts from one position to another position, or their return. This action is relatively independent of the rate of travel of the actuator.

snap-action — 1. In a mercury switch, the rapid motion of the mercury pool from one position to another. 2. A rapid motion of the contacts from one position to another position, or their return (differential storing of energy). This action is relatively independent of the rate of travel of the actuator.

snap-action contacts — A contact assembly such that the contacts remain in one of two positions of equilibrium with substantially constant contact pressure

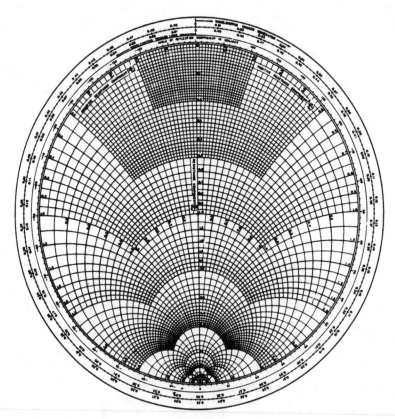

Smith chart.

during the initial motion of the actuating member until a point is reached at which stored energy causes the contacts to move abruptly to a new position of equilibrium.

snap magnet — A permanent magnet used in thermostatic, pressure, and other control instruments to provide quick make-and-break action at the contact and thereby minimize sparking. The magnet pulls the armature in suddenly against the spring to close the contacts and hold them closed until the spring is compressed enough to make them fly apart.

snap-off diode — A planar epitaxial passivated silicon diode that is processed in such a way that a charge is stored close to the junction when the diode conducts. Upon application of a reverse voltage, the stored charge forces the diode to switch quickly to its blocking state, or snap off.

snapshot — 1. In a computer, a dynamic printout of selected data in storage that occurs at breakpoints and checkpoints during the computing operations, as opposed to a static printout. 2. A printout of intermediate results part way through a program run.

snapshot dump — A selective dynamic dump carried out at various points in a machine run.

snapshot routine — Special type of debugging routine that includes provisions for dynamic printout of selected data at various checkpoints in a computing operation.

snap-strate — A scribed substrate that can be processed by gang deposition in multiples of circuits and snapped apart afterward.

snap switch — A switch (e.g., a light switch) in which the contacts are separated or brought together suddenly as the operating knob or lever compresses or releases a spring.

snap varactor — *See* step-recovery diode.

sneak circuit — That part of a complete electrical circuit which carries an unintentional (sneak) current. Sneak currents may prevent proper operation of interconnected equipment.

sneak current — A leakage current that enters telephone circuits from other circuits; it is not strong enough to cause immediate damage, but it can produce harmful heating effects if permitted to continue.

sneak path — In computers, an unwanted circuit through a series-parallel configuration.

Snell's law — The sine of the angle of incidence divided by the sine of the angle of refraction equals a constant called the index of refraction when one of the media is air.

sniperscope — Also called snooperscope. 1. An instrument for seeing in the dark by means of infrared radiation. A high-aperture lens forms an image of distant objects on the photocathode of an infrared-sensitive image tube, the visible final image being viewed through a magnifier (eyepiece). The scene must be illuminated with infrared if such a tube is to be usable. These instruments find extensive use in warfare and also for operations in rooms that have to be kept dark because film emulsions are exposed. 2. A snooperscope attached to a rifle and used to aid in the sighting and shooting of targets in low-light-level conditions.

snivet — A straight, jagged, or broken vertical black line that appears near the right edge of the screen of a television receiver, caused by discontinuity in the plate-current characteristic of the horizontal amplifier tube under conditions of zero bias.

snivitz — A small pulse of noise. *See also* glitch.

snooperscope — 1. A night viewing device to permit a user to see objects in total darkness. It consists of an infrared source, an infrared image converter, and a battery-operated high-voltage dc-to-dc converter. Infrared radiation sent out is reflected back to the snooperscope, where it is then converted into a visible image on the fluorescent screen of the image-converter tube. 2. An instrument used for viewing in low levels of illumination by means of infrared radiation. A high-aperture lens forms an image of distant objects on the photocathode of an infrared-sensitive image tube, the visible final image being viewed through a magnifier. The scene must be illuminated with infrared energy. 3. *See also* sniperscope.

snow — Also known as sparklies. 1. A speckled background caused by random noise on an intensity-modulated display, such as alternate dark and light dots appearing randomly in a television picture (usually indicative of a weak signal). 2. Heavy random noise in a video picture.

snowflake transistor — A high-speed, medium-power switching transistor for use as a thin-film or core driver. It employs a six-pointed emitter geometry that permits the optimum ratio of emitter periphery to emitter area.

snow static — Precipitation static caused by falling snow.

SNR — Abbreviation for signal-to-noise ratio. 1. Expressed in decibels, it relates how much louder a signal is than the background noise. It is measured at the speaker, and is used in sensitivity ratings. Often, $(s + n)/n$ is used, as it is easier to measure; it is the ratio of the signal and background noise to the background noise, and will yield an apparently higher sensitivity than the s/n ratio. 2. Usable information in a signal (optical or electrical) compared to the noise that tends to interfere with the transmission of the information. Also abbreviated s/n ratio.

s/n ratio — *See* SNR.

snubber capacitor — A capacitor incorporated in a rapidly switched *LC* circuit to reduce EMI by lowering the circuit resonant frequency and characteristic impedance.

snubber circuit — 1. A form of suppression network that consists basically of a series-connected resistor and capacitor connected in shunt with an SCR. The snubber circuit combined with the effective circuit series inductance controls the maximum rate of change of voltage and the peak voltage across the device when a stepped forward voltage is applied to it. 2. A circuit designed to reduce the sensitivity of a solid-state relay to spikes and other high-dV/dt transients in the load circuit voltage. Snubbers prevent false firing of solid-state relay switching elements. One such circuit is often referred to as an RC circuit.

SO — Abbreviation for small outline (package). Similar to a miniature plastic flat pack, but with gull-wing lead forms primarily or wholly constructed for surface mounting. Typical lead spacing is 0.05 inch.

SOA — Abbreviation for safe operating area.

soak — 1. In an electromagnetic relay, the condition that exists when the core is approximately saturated. 2. To increase the current through a relay winding until the core is saturated with flux, and then hold the magnetization at that level for a certain time.

soakage — The inability of a capacitor to come up to voltage instantaneously without voltage lag or creep during or after charging. The lower the soakage, the lower the lag and creep.

soak time — 1. The period of time required following activation for the electrolyte in a cell or battery to be sufficiently absorbed into the active materials. 2. The length of time a ceramic material (such as a substrate or thick-film composition) is held at the peak temperature of the firing cycle.

soak timer — A reset timer, usually dial-adjustable, as applied in a temperature-control system for controlling the length of time the temperature is held at a predetermined level.

soak value — The voltage, current, or power applied to the coil of the relay coil to ensure that a condition approximating magnetic saturation exists.

socket — 1. An opening that supports and electrically connects to vacuum tubes, bulbs, or other devices or components when they are inserted into it. 2. Also known as the female contact. A mechanical electrical connector.

socket adapter — A device placed between a tube and its socket so that the tube can be used in a socket designed for some other base, or so that current or voltage can be measured at the electrodes while the tube is in use.

socket connector — A connector that contains socket contacts and that receives a plug connector containing male contacts.

socket contact — A hollow female contact designed to mate with a male contact. It is normally connected to the live side of a circuit.

sodar — Acronym for sound detecting and ranging. A device that detects large changes in temperature overhead by the amount of sound returned as echoes (the colder the atmosphere, the louder the echoes). The sound, which is within the range of human hearing, is launched upward, and the echoes are changed into oscilloscope patterns.

sodium amalgam-oxygen cell — A fuel-cell system in which materials functioning in the dual capacity of fuel and anode are consumed continuously. Low operating temperatures and high power-to-weight ratios are significant characteristics of the system.

sodium-vapor lamp — A gas-discharge lamp containing sodium vapor. It is used chiefly for highway illumination.

sofar — Acronym for sound fixing and ranging. An underwater sound system with which air and ship survivors can be located within a square mile and as far as 2000 miles (3200 km) away. Survivors drop a TNT charge into the water. The charge, which is timed to explode at 3000 to 4000 feet (914 to 1220 m), sets up underwater sound waves that can be picked up by hydrophones at shore stations.

soft copy — Alphanumeric or graphical data, or both, presented in nonpermanent form, such as on a video screen.

soft error — 1. An occasional random loss of data in a memory. The reason for the error is normally not identifiable, nor is it repeatable on a device tester. 2. A dynamic error normally caused by some transient condition. Retrying the failed operation will often result in successful completion. 3. Alteration of the information in a memory cell resulting from an alpha particle striking the cell.

soft key — A key on the terminal keyboard that may be labeled on the lower part of the terminal screen and whose function can be programmed.

soft magnetic material — Also called low-energy material. Ferromagnetic material that, once having been magnetized, is very easily demagnetized (i.e., requires only a slight coercive force to remove the resultant magnetism).

soft phototube — A gas phototube.

soft radiation — Term applied to radiation composed of particles or photons that will not easily penetrate a material because of their low energy levels.

soft soldering — Process of joining two metals with a fusible alloy or solder that melts below 800°F (427°C). *See* hard soldering.

soft start — A method of increasing the duty cycle of the switching element in a power supply from zero to its normal operating point during system startup. Using soft start eliminates output voltage overshoot and magnetizing

current imbalances in the transformer. *See also* warm boot.

soft tube — 1. A high-vacuum tube that has become defective because of the entry of a small amount of gas. 2. An electronic tube into which a small amount of gas has purposely been put to obtain the desired characteristics.

software — 1. Programs, routines, codes, and other written information for use with digital computers, as distinguished from the equipment itself, which is referred to as hardware. 2. A set of computer programs, procedures, rules, and associated documentation concerned with the operation of a data-processing system, e.g., compilers, monitors, editors, utility programs. 3. A computer program or set of programs held in some storage medium and loaded into read/write memory (RAM) for execution. 4. The set or package of programs that instructs a computer to perform certain predefined functions. 5. The user program that controls the operation of a programmable controller. 6. Totality of programs and routines used to extend the capabilities of computers, such as compilers, assemblers, narrators, routines, and subroutines. 7. The computer programs, procedures, and documentation concerned with the operation of a computer system, e.g., assemblers, compilers, operating systems, diagnostic routines, program loaders, manuals, library routines, and circuit diagrams. Software is the name given to the programs that cause a computer to carry out particular operations. Contrasted with hardware. 8. The modifiable (to some extent) binary bit patterns in the memory of a computer that control the operation of the processing portion of the computer. Software programs are usually written in one of three general classes of language: machine, assembler, or higher level. 9. A computer's programs. If a particular bit of data manipulation is done through a program rather than by special circuitry, it is said to be "in software." Doing things in software is cheap and flexible, since a program can be easily changed. 10. An expression for programs and also tapes or disks with recorded data. There are various kinds of software, including applications and executive. The executive is the "heart" of the system and is responsible for scheduling and controlling activity in the system. Applications or task programs perform specific functions whose activity is controlled by the executive. 11. Programs that control the operation of computer hardware. Operating systems, executives, monitors, compilers, editors, utility routines, and user programs are considered software.

software buffer — A location or set of locations in memory given by name by the resident program and used to hold information until it can be used.

software compatible — Describes an MPU that executes the same software as another one.

software documentation — Program listings and/or technical manuals describing the operation and use of programs.

software house — A company that offers software programs and support service to users. Such support can range from simply supplying manuals and other information to a complete counseling and computer part-time programming service.

software interrupt — The interruption of a user-level program in response to the acknowledgment of a hardware interrupt by the operating system. In high-level-language programs, software interrupts can safely occur only at the end of a program line.

software list — A document that defines all the next lower level modules that are contained in the module being documented.

software maintenance — 1. Improvements and changes made in software. 2. The task of keeping software up to date and working properly.

software programmable — A system whose functions are defined by a program, generally supplied by the manufacturer, that may be redefined or updated by changing or replacing the program.

software tool — A computer program, rules, and associated documentation that assist a data-processing technologist in designing, developing, maintaining, and managing data and software.

softwire or computer numerical control — A numerical control system wherein a dedicated stored-program computer is used to perform some or all of the basic numerical control functions. The control program can be read in and stored from data on tape, cards, manual switches, etc. Changes in the response, sequence, and/or functions can be made by reading in a different control program.

soft X-rays — X-rays with comparatively long wavelengths and, hence, poor penetrating power.

solar absorber — A surface that has the property of converting solar radiation into thermal energy.

solar absorption index — A quantity that relates the angle of the sun at different latitudes and local times with ionospheric absorption.

solar battery — A series of solar cells arranged to collect solar radiation and to generate a given amount of electrical energy.

solar cell — 1. A device capable of converting light or other radiant energy directly into electrical energy. 2. Silicon photovoltaic cell that can be used to generate electricity from direct sunlight. Such cells are especially useful in space vehicles, for which no other source of electricity is available. 3. A photovoltaic cell designed to respond to wavelengths in solar radiation. Basically a semiconductor diode that absorbs light and converts it to electrical power. This ability depends on the material of the semiconductor and on the photovoltaic effect. In this effect a photon of solar radiation is absorbed by the semiconductor material by giving up its total energy to produce an excess charge carrier (electron or hole). The absorption can happen only if the photon energy is greater than the energy of the bandgap of the material — the difference between the conduction band and the valence band (the excited state and the ground state).

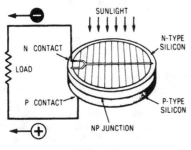

Solar cell.

solar concentrator — A device that increases the intensity of solar energy by optical means.

solar constant — The standard intensity of solar radiation impinging on the earth's atmosphere at the mean distance between earth and sun. It is the energy equivalent of 1.353 kilowatts of electricity per square meter. Under clear skies some 80 to 90 percent of this radiation reaches the ground; under heavy cloud cover, 10 percent or less. The average energy rate at the earth's surface is about 1 kW per square meter.

solar day — A day that contains exactly 24 hours (1440 minutes). During a solar day the sun rotates slightly more than 360° about its axis with respect to "fixed" stars. *Compare* siderial day.

solar-energy conversion — The process of changing solar radiation into electrical or mechanical power, either directly or by using a heat engine.

solar furnace — An optical system that is designed to produce a high temperature in a specified area by the optical direction and concentration of solar radiation on that area. The system usually consists of a collecting system that may have reflective optics concentrators, or a combination of both, and a tracking system that keeps the collecting and directing system properly oriented to the sun.

solar noise — Electromagnetic radiation from the sun at radio frequencies.

solar radiation — Radiation from the sun that is made up of a very wide range of wavelengths, from the long infrared to the short ultraviolet, with its greatest intensity in the visible green at about 5000 angstroms (500 nm). The solar radiation received on the earth's surface is restricted to the visible and near infrared, since the air strongly absorbs the wavelengths located at either end of the spectrum.

solar simulation — The simulation of solar radiation in the infrared and ultraviolet spectrum for the analysis of extraterrestrial sunlight and its effects on spacecraft, materials, and components.

solder — 1. A readily meltable metal or alloy that produces a bond at a junction of two metal surfaces. True solder must have a lower melting point than the metals being joined and must also be capable of uniting with the metals to be joined. 2. An alloy that melts at relatively low temperatures and is used to join or seal metals with higher melting points. Solder alloys melt over a range of temperatures; the temperature at which a solder begins to melt is the solidus, and the temperature at which it is completely molten is the liquidus.

solderability — The ability of a conductor to be "wetted" by hot solder and to form a strong low-resistance bond with the solder.

solder bumps — The round solder balls bonded to a transistor contact area and used to make connection to a conductor by face-down bonding techniques.

solder contact — A contact having a cup, hollow cylinder, eyelet, or hook to accept a wire for a conventional soldered termination.

solder cup — The end of a terminal or similar device into which a contact is inserted before being soldered.

solder dam — A dielectric composition screened across a conductor to limit molten solder from spreading farther onto solderable conductors.

solder eye — A solder-type terminal provided with a hole at its end through which a wire can be inserted prior to being soldered. A ring-shaped contact termination of a printed-circuit connector for the same purpose.

solder eyelet — An eyelet, or hole, in a contact through which a wire may be mechanically connected to the contact prior to soldering.

solder flux — A substance that transforms a passive, contaminated metal surface into an active, clean, solderable surface. Generally, all such fluxes should prevent oxidation during heating, lower interfacial surface tensions, be thermally stable, be easily displaced by molten solder, be noninjurious to components, remove easily if desired, remove oxide, and penetrate films.

solder ground — A conducting path to ground due to dripping or overhanging solder.

soldering — 1. The joining of metallic surfaces (e.g., electrical contacts) by melting a metal or an alloy (usually tin and lead) over them. 2. A process of joining metallic surfaces with solder, without the melting of the base metals. A soldered connection has metallic continuity and therefore excellent long-term reliability. *See* soft soldering; hard soldering.

soldering dross — Usually a combination of tin oxide and lead oxide in a ratio closely resembling the parent metal. In practice this floating mass of tarnish also contains other metal reaction products — sulfides and organic residues such as burned flux, etc. Dross contains the metallic impurities picked up during soldering. This occurs only when the impurity concentrations exceed the solubility limits and when intermetallics are formed that float to the surface.

soldering fluid — A liquid used with wave solder systems that can be intermixed with solder to reduce the surface tension of solder, promote wetting, and eliminate the formation of dross. *See also* soldering oils.

soldering gun — 1. A pistol-shaped soldering tool having a trigger switch to turn it on. Operates from 117 volts ac and has an integral stepdown transformer with a single-turn secondary that quickly heats the copper soldering loop. 2. Category of soldering tool. Pistol-grip guns have trigger action control of low or high tip temperatures for general use, with output ranges commonly rated at 100/140, 145/210, or 240/325 watts; temperature-controlled guns designed for solid-state electronics work with interchangeable power heads for varying fixed-tip temperatures.

soldering iron — 1. A soldering tool consisting of a heating element to heat the tip and melt the solder, plus a heat-insulated handle. 2. An electrically heated copper-tipped tool that heats the work and melts solder to make a soldered joint.

soldering-iron tip — A high-purity copper-substrate form, iron plated 0.006 to 0.030 inch (0.254 to 1.127 mm) thick, hot tin dipped in the working area, with the remaining surface immunized by nickel-chromium plating. The working area of the tip is usually fabricated for access and maximum heat transfer to the work point.

soldering oils — Liquid compounds formulated for use as the oil in oil intermix wave-soldering equipment and as pot coverings on still solder pots.

solder joint — The point of bonding between solder and component surfaces.

solderless connection — The joining of two metallic parts by pressure only, without soldering, brazing, or using any method that requires heat.

solderless connector — A device for clamping two wires firmly together to provide a good connection without solder. A common form is a cap with tapered internal threads that are twisted over the exposed ends of the wires.

solderless contact — *See* crimp contact.

solderless lug — A terminal lug that holds the conductor it terminates by compressing it under a screw.

solderless terminals — Small metal parts used for joining a wire to another wire or to a stud by the method of crimping.

solderless wrap — Also called wire wrap. A method of connection in which a solid wire is tightly wrapped around a rectangular, square, or V-shaped terminal by means of a special tool.

solderless wrapped connection — Also called wire-wrapped connection or wrapped connection. A connection made by wrapping wire that is under tension around a square or rectangular terminal.

solder lug — Device to which wire is secured by soldering. Solder lugs are attached to a printed circuit board, termination strip, chassis, or electrical component.

solder mask — A printed-circuit-board technique in which everything is coated with a plastic except the contacts to be soldered.

solder short — A defect that occurs when solder forms a short-circuit path between two or more conductors.

sole — In a magnetron or a backward-wave oscillator, an electrode used to carry a current that produces a magnetic field in the desired direction.

solenoid — 1. An electric conductor wound as a spiral with a small pitch, or as two or more coaxial spirals. 2. An electromagnet having an energized coil approximately cylindrical in form and an armature whose motion is reciprocating within and along the axis of the coil. 3. A coil of wire surrounding a movable iron bar that is located in such a way that when the coil is energized the core is drawn into it. 4. An electromagnet with a movable steel core (plunger), used to translate electrical energy into linear mechanical motion. The wire coil is wound on a cylindrical or rectangular tube, and the similarly shaped plunger (which may be solid or laminated steel) extends about half way into the tube, in its deenergized position. When the coil is energized by direct current, the plunger is drawn into the tube by magnetic attraction, and the external mechanism connected to the plunger is activated accordingly. Solenoids designed for operation on alternating current are constructed like the dc types, except that the plunger is equipped with the "shading" coil that maintains the magnetic field between alternations of the current, to reduce hum and eliminate chatter. 5. An electromagnet with a movable core, or plunger, which, when it is energized, can move a small mechanical part a short distance. 6. A device that converts electrical energy into mechanical work by providing force only during a linear closing stroke.

Solenoid.

solenoid valve — 1. A combination of an electromagnet plunger and an orifice to which a disc or plug can be positioned to either restrict or completely shut off a flow. (Orifice closure or restriction occurs when the electromagnet actuates a magnet plunger.) 2. A combination of two basic functional units: (*a*) a solenoid (electromagnet) with its plunger (or core), and (*b*) a valve containing an orifice in which a disc or plug is positioned to stop or allow flow. The valve is opened or closed by movement of the magnetic plunger (or core), which is drawn into the solenoid when the coil is energized.

solid — A state of matter in which the motion of the molecules is restricted. They tend to remain in one position, giving rise to a crystal structure. Unlike a liquid or gas, a solid has a definite shape and volume.

solid circuit — A semiconductor network fabricated in one piece of material by alloying, diffusing, doping, etching, and cutting, and using jumper wires as necessary.

solid conductor — An electrical conductor consisting of a single wire.

solid electrolyte — A solid semiconductor in direct contact with a thin nonconductive oxide coating.

solid-electrolyte fuel cell — A self-contained fuel cell in which oxygen is the oxidant and hydrogen is the fuel. The oxidant and fuel are kept separated by a solid electrolyte that has a crystalline structure and a low conductivity.

solid-electrolyte tantalum capacitor — Also called solid tantalum capacitor. A tantalum capacitor with a solid semiconductor electrolyte instead of a liquid. A wire anode is used for low capacitance values, and a sintered pellet for higher values.

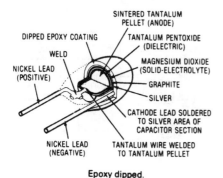

Hermetically sealed.

Epoxy dipped.

Solid-electrolyte tantalum capacitor.

solid logic technology — The use in computers of miniaturized modules that make possible faster circuitry because of the reduced distances current must travel.

solidly grounded — Also called directly grounded. Grounded through an adequate grounded connection in which no impedance has been inserted intentionally.

solid metal mask — A thin sheet of metal with an etched pattern used in contact printing of film circuits.

solid silicon circuit — Semiconductor circuit that employs a single piece of silicon material in which the various circuit elements (transistors, diodes, resistors, and capacitors) are formed by diffusion in the planar configuration. By combining oxide masking, diffusion, metal deposition, and alloying, a complex network with active and passive components is made completely within a die that is part of a single semiconductor wafer. Often, thin-film devices are applied to the surface of the silicon wafer to provide passive circuit elements beyond the range of solid silicon technology. External

connections are made through small wires soldered, welded, or thermocompression-bonded to selected points of the surface.

solid state — 1. Pertaining to circuits and components using semiconductors. 2. The physics of materials in their solid form. Examples of solid-state materials are transistors, diodes, solid-state lasers, metals, and alloys, etc. 3. Technology utilizing solid semiconductors in place of vacuum tubes for amplification, rectification, and switching. 4. A silicon or germanium semiconductor device, such as a diode, transistor, or integrated circuit; or circuits, equipment, or systems made from such devices. 5. A circuit or system that does not rely on vacuum or gas-filled tubes to control or modify voltages and currents. 6. The electronic components that convey or control electrons within solid materials, e.g., transistors, germanium diodes, and integrated circuits. 7. Refers to the electronic properties of crystalline materials, generally semiconductors. As opposed to vacuum and gas-filled tubes that function by flow of electrons through space, or by flow through ionized gases, solid-state devices involve the interaction of light, heat, magnetic field, and electric currents in crystalline materials. Compared with earlier vacuum-tube devices, solid-state components are smaller, less expensive, more reliable, use less power, and generate less heat.

solid-state atomic battery — A device in which a radioactive material and a solar cell are combined. The radioactive material emits particles that enter the solar cell, which in turn produces electrical energy.

solid-state bonding — The process of forming a metallurgical joint between similar or dissimilar metals by causing adjoining atoms at the joint interface to combine by interatomic attraction in the solid state. (This process is different from diffusion bonding in that no atomic diffusion is required.) The adjoining surfaces to be bonded must be atomically clean and must be brought within atomic distances before such a bond can become established.

solid-state circuit — A complete circuit formed from a single block of semiconductor material. *See also* monolithic integrated circuit.

solid-state component — A component whose operation depends on the control of electric or magnetic phenomena in solids (for example, a transistor, crystal diode, or ferrite).

solid-state computer — A computer built primarily from solid-state electronic circuit elements.

solid-state device — 1. Any element that can control current without moving parts, heated filaments, or vacuum gaps. All semiconductors are solid-state devices, although not all solid-state devices (for example, transformers) are semiconductors. 2. An electronic device that operates by virtue of the movement of electrons within a solid piece of semiconductor material. 3. Electronic component that controls electron flow through a solid material such as a crystal; for example, a transistor, diode, or integrated circuit.

solid-state imaging system — An imaging system that uses a mosaic of tiny light-sensitive semiconductors (phototransistors) to produce individual outputs that are then converted into a coherent video signal.

solid-state integrated circuits — The class of integrated components in which only solid-state materials are used.

solid-state lamp — 1. A pn junction that emits light when forward biased. Made from a complex compound of gallium, arsenic, and phosphorus called gallium arsenide phosphide. Its light output is typically at 670 nanometers. It characteristically looks like a forward-biased diode with a breakdown voltage in the region of 1.6 volts. 2. An electroluminescent semiconductor that emits low-intensity radiation in the green or red regions. Used as an indicator lamp.

solid-state laser — A laser using a transparent substance (crystalline or glass) as the active medium, doped to provide the energy states necessary for lasing. The pumping mechanism is the radiation from a powerful light source, such as a flash tube. Ruby lasers are solid-state lasers.

solid-state physics — The branch of physics that deals with the structure and properties of solids, including semiconductors (i.e., a material whose electric resistivity is between that of insulators and conductors). Generally used semiconductors are silicon and germanium.

solid-state relay — Abbreviated SSR. 1. A relay that employs solid-state semiconductor devices as components. 2. A factory-built and packaged product used for switching an ac load. It has an isolated input/output construction that permits the current to be switched electronically without moving parts. Optical and transformer coupling are among the methods used to achieve input/output isolation and permit coupling from the control circuit to the trigger circuit. SSRs are used in place of electromechanical relays when there is a need for compatibility with digital logic drive circuits, where an electromagnetic coil and contacts would cause interference problems. 3. An on/off control device in which the load current is conducted by one or more semiconductors, e.g., a power transistor, SCR, or triac.

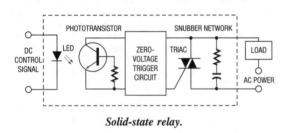

Solid-state relay.

solid-state switch — A no-contact switch that completes a circuit by means of solid-state components.

solid-state watch — A timepiece that uses a quartz crystal or other precise frequency resonator in conjunction with low-power MOS integrated circuits. Employs liquid crystals or light-emitting diodes to indicate hours, minutes, seconds, data, months, days of the week, etc., in a digital and/or alphanumeric format.

solid tantalum capacitor — *See* solid-electrolyte tantalum capacitor.

solidus — The highest temperature at which a metal or alloy is completely solid.

solid wire — Wire that consists of a single conductor, not of multiple strands.

solion — Contraction of solution ion. An electrochemical sensing and control device in which ions in solution carry electric charges to give amplification corresponding to that of vacuum tubes and transistors.

solion integrator — A precision electrochemical cell housed in glass and containing four small platinum electrodes in a solution of potassium iodide and iodine. The integrator anode and cathode make up the covers of a small cylindrical volume (less than 0.00025 in.^{3}, or 0.0041 cm^3) for storing electrical information in the form of ions. The integrator cathode contains a fixed amount of hydraulic porosity for completing the internal-solution path to the other two electrodes.

solo manual — *See* swell manual.

Sommerfeld formula — An approximate wave-propagation relationship that may be used when distances are short enough that the curvature of the earth may be neglected in the computations.

Sommerfeld's equation — Equation for ground-wave propagation that relates field strength at the surface of the earth at any distance from a transmitting antenna to the field strength at unit distance for given ground losses.

Sonalert — A solid-state tone-emitting device (P. R. Mallory & Co., Inc.).

sonar — Acronym for sound navigation and ranging. Also called active sonar if it radiates underwater acoustic energy, or passive (listening) sonar if it merely receives the energy generated from a distant source. 1. Apparatus or technique of obtaining information regarding underwater objects or events through the transmission and reception of acoustic energy. Two well-known uses are to detect submarines and fish. 2. A system that uses underwater sound, at sonic or ultrasonic frequencies, to detect and locate objects in the sea. Sonar signals can also be used as a communication medium. The various systems generally may be divided into three basic classifications: passive (listening only), echo-ranging (active), and communication.

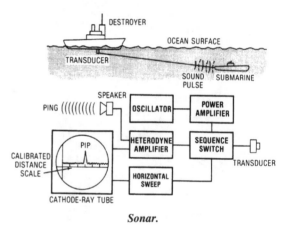

Sonar.

sonar background noise — In sonar, the total noise presented to the final receiving element that interferes with the reception of the desired signal.

sone — 1. A unit of loudness. A simple tone of frequency 1000 hertz, 40 dB above a listener's threshold, produces a loudness of 1 sone. The loudness of any sound that is judged by the listener to be *n* times that of the 1-sone tone is *n* sones. 2. A value for loudness. May be used for overall evaluation of a sound or of a frequency band. The sone scale is linear (in contrast to decibels, which are logarithmic).

sones per bark — Loudness density defined as a function of subjective pitch. When loudness density is integrated over subjective pitch, the result is total loudness.

sonic — 1. Pertaining to the speed of sound. 2. Utilizing sound waves.

sonic altimeter — An altimeter that determines the height of an aircraft above the earth by measuring the time the sound waves take to travel from the aircraft to the ground and back, based on the fact that the velocity of sound at sea level is 1080 feet per second (329 m/s) through dry air at 0°C (32°F).

sonic applicator — A self-contained electromechanical transducer for local application of sound for therapeutic purposes.

sonic cleaning — The cleaning of contaminated materials by the action of intense sound waves produced in the liquid into which the material is immersed.

sonic delay line — A device in which electroacoustic transducers and the propagation of an elastic wave through a medium are used to produce the delay of an electrical signal.

sonic drilling — The cutting or shaping of materials with an abrasive slurry driven by a reciprocating tool attached to an electromechanical transducer.

sonic frequencies — Vibrations that can be heard by the human ear (from about 15 hertz to approximately 20,000 hertz).

sonic motion detector — A sensor that detects the motion of an intruder by his or her disturbance of an audible sound pattern generated within the protected area.

sonic soldering — The method of joining metals by the use of mechanical vibration to break up the surface oxides.

sonic speed — *See* speed of sound.

sonic thermocouple — A thermocouple so designed that gas moves past the junction with a velocity of mach 1 or greater, resulting in maximum heat transfer to the junction.

sonne — Also called consol. A radionavigational aid that provides a number of rotating characteristic signal zones. A bearing may be determined by observation (and interpolation) of the instant when transition occurs from one zone to the following zone.

sonobuoy — Also called a radiosonobuoy. 1. A device used to locate a submerged target (e.g., a submarine). By means of a hydrophone system in the water, a sonobuoy detects the noises and converts them into radio signals, which are transmitted to a receiver in an airplane. Each sonobuoy transmits on one of several possible frequencies, and the receiver in the airplane has a channel selector so the operator can switch from one to another. 2. A passive sonar device (to distinguish it from an active sonar system, which transmits a signal and listens for an echo) that uses hydrophones to convert acoustic signals from natural phenomena, such as wave motion, or from ships or submarines into electrical signals. These signals are filtered and sent to the surface for transmission to a surface vessel or aircraft for further processing.

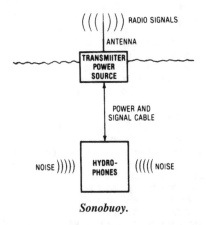

Sonobuoy.

sonoluminescence — 1. The creation of light in liquids by sonically induced cavitation. 2. The luminescence

of a substance resulting from its exposure to ultrasonic waves.

sonometer — A frequency meter whose operation depends on its mechanical resonance with the vibrations of a variable length of stretched wire.

sonoptography — The use of sound waves to obtain a 3D image of an object. No lenses are required. In essence, a two-stage process in which the diffraction pattern of an object irradiated by sound waves is biased by a coherent sound wave and recorded. The resultant pattern, like the hologram, then is interrogated with a suitable coherent light source to obtain a three-dimensional image.

sonoradiography — The diagnostic procedure that uses ultrasonic energy to probe the body and, with the help of laser beams, a reflecting membrane to produce a three-dimensional picture.

sophisticated — A piece of equipment, system, etc., that is complex and intricate, or requires special skills to operate.

sophisticated vocabulary — An advanced and elaborate set of computer instructions, enabling the computer to perform such intricate operations as linearizing, extracting square roots, selecting the highest number, etc.

sorption — The combination of absorptive and adsorptive processes in the same material.

sort — 1. To arrange items of information according to rules that depend on a key or field contained by the items. 2. A function performed by a program, usually part of a utility package; items in a data file are arranged or rearranged in a logical sequence designated by a keyword or field in each item in the file. 3. A procedure designed to arrange a group of elements in random order into some kind of a sequence. Examples include the bubble sort, the selection sort, and the partition sort. 4. The process of arranging data in a specific order.

sorter — A machine that sorts cards according to the position of coded holes.

SOS — 1. A distress signal used in radiotelegraphy. 2. Abbreviation for silicon on sapphire. A CMOS technology in which a layer of silicon is epitaxially grown on a sapphire wafer, with specific regions subsequently etched away between individual transistors. Each device is thus totally isolated from other devices. Since sapphire is an insulator, SOS is a subset of SOI (silicon-on-insulator) technology. Both SOI and SOS technologies provide for high levels of radiation hardness.

SOT — Abbreviation for small outline transistor. A small plastic package with gull wing leads for mounting discrete semiconductor devices on printed circuit boards.

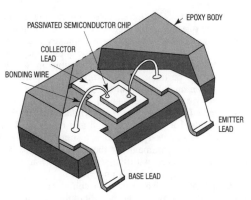

SOT — small outline transistor (SOT 23 size).

sound— 1. Also called a sound wave. An alteration in pressure, stress, particle displacement or velocity, etc., propagated in an elastic material, or the superposition of such propagated alterations. 2. Also called a sound sensation. The auditory sensation usually evoked by the alterations described in (1) above. 3. An undulatory motion of air or other elastic medium, which can produce the sensation of hearing when incident on the ear. (Sound requires a medium for propagation, for unlike electromagnetic waves, sound cannot travel through a vacuum.)

sound absorption— The conversion of sound energy into some other form (usually heat) in passing through a medium or on striking a surface.

sound-absorption coefficient— The incident sound energy absorbed by a surface or medium, expressed in the form of a fraction.

sound analyzer— A device for measuring the amplitude and frequency of the components of a complex sound. It usually consists of a microphone, an amplifier, and a wave analyzer.

sound articulation— The percent of articulation obtained when the speech units are fundamental sounds (usually combined into meaningless syllables).

sound bars— Alternate dark and light horizontal bars caused in a television picture by audio-frequency voltage reaching the video-input circuit of the picture tube.

sound board— A hardware adapter that dramatically adds digital sound reproductions capabilities to IBM-compatible computers.

sound carrier— The frequency-modulated carrier that transmits the sound portion of television programs.

sound concentrator— A parabolic reflector used with a microphone at its focus to obtain a highly directive pickup response.

sound-effect filter— A filter, usually adjustable, designed to reduce the passband of a system at low and/or high frequencies in order to produce special effects.

sound energy— The total energy in a given part of a medium, minus the energy that would exist there if no sound waves were present.

sound-energy density— At a point in a sound field, the sound energy contained in a given infinitesimal part of the medium, divided by the volume there. The commonly used unit is the erg per cubic centimeter.

sound-energy flux— The average rate at which sound energy flows through any specified area for a given period. The commonly used unit is the erg per second.

sound-energy flux density— *See* sound intensity.

sounder— *See* telegraph sounder.

sound field— A region in any medium containing sound waves.

sound film— Motion-picture film having a sound track along one side of the picture frames for simultaneous reproduction of the sounds that accompany the film. A beam of light is projected through the sound track and is modulated at an audio rate by the variations in the width or density of the track. A phototube and amplifier then convert these modulations into sound.

sound gate— A mechanical device through which film is passed in a projector to convert the sound track into audio signals that can be amplified and reproduced. In a television camera used for pickups, a sound gate provides the sound accompaniment for the motion picture being televised. Associated with the sound gate are an exciter lamp, a lens assembly, and a phototube.

sound head— The part of a sound motion-picture projector that converts the photographic or magnetic sound track on the film into audible sound signals.

sounding— Determination of the depth of water or the altitude above the earth.

sound intensity— Also called specific sound-energy flux or sound-energy flux density. The average rate of sound energy transmitted in a specified direction through a unit area normal to this direction at the point considered. The common unit is the erg per second per square centimeter, although sound intensity expressed in watts per square centimeter may occasionally be used.

sound intensity level— The amount of sound power passing through a unit area.

sound level— A measure of the overall loudness of sounds on the basis of approximations of equal loudness of pure tones. It is expressed in decibels with respect to 0.0002 microbar.

sound-level meter— An instrument— including a microphone, amplifier, output meter, and frequency-weighting networks— for the measurement of noise and sound levels. The measurements approximate the loudness level obtained for pure tones by the more elaborate ear-balance method.

sound-on-sound recording— 1. A method by which material previously recorded on one track of a tape may be recorded on another track while simultaneously adding new material to it. 2. A method of tape recording in which an original sound track may be impressed with an added sound track for special effects, such as one performer appearing to play two instruments, etc.

sound-powered telephone set— 1. A telephone set in which the transmitter and receiver are passive transducers; operating power is obtained from the speech input only. The voice sound waves operate a moving coil or variable-reluctance transmitter to produce the current waves transmitted to the telephone line. No external power is used, but the output level is lower than that of a battery-powered phone. 2. A telephone set whose transmitter is a dynamic microphone. It requires no power other than that of the voice of the user.

sound-power level— 1. The ratio, expressed in decibels, of the sound power emitted by a source to a standard reference power of 10^{-13} watt. 2. The number of watts of acoustic power radiated by a noise source.

sound power of a source— The total sound energy radiated by the source per unit of time. The common unit is the erg per second, but the power may also be expressed in watts.

sound pressure— The instantaneous pressure minus the static pressure at some point in a medium when a sound wave is present.

sound-pressure level— Abbreviated SPL. 1. In decibels, 20 times the logarithm of the ratio of the pressure of a sound to the reference pressure, which must be explicitly stated (usually, either 20 µPa, 2 × 10^{-4} microbar, or 1 microbar). 2. The pressure of an acoustic wave stated in terms of newtons/square meter, dynes/square centimeter, or microbars. (One microbar is approximately equal to one millionth of the standard atmospheric pressure.)

sound probe— A small microphone (or tube added to a conventional microphone) for exploring a sound field without significantly disturbing it.

sound recordings— Records, tapes, or other sonic components upon which audio intelligence is inscribed or recorded or can be reproduced.

sound-recording system— A combination of transducing devices and associated equipment for storing sound in a reproducible form.

sound-reflection coefficient— Also called acoustical reflectivity. Ratio at which the sound energy reflected from a surface flows on the side of incidence to the incident rate of flow.

sound-reproducing system— A combination of transducers and associated equipment for reproducing prerecorded sound.

sound sensation— *See* sound, 2.

sound-sensing detection system— An alarm system that detects the audible sound caused by an attempted forcible entry into a protected structure. The system consists of microphones and a control unit containing an amplifier, accumulator, and a power supply. The unit's sensitivity is adjustable so that ambient noises or normal sounds will not initiate an alarm signal. However, noises above this preset level or a sufficient accumulation of impulses will initiate an alarm.

sound sensor— A sensor that responds to sound; a microphone.

sound spectrum— The frequency components included within the range of audible sound.

sound stage— The area between the two or more speakers of a stereo or quad setup where subjective sound images or imaginary speakers seem to be, providing a wide area of apparent sound source.

sound takeoff— The connection or coupling at which the 4.5-MHz frequency-modulated sound signal in a television receiver is obtained.

sound track— The narrow band that carries the sound in a movie film. It is usually along the margin of the film, and more than one band may be used (e.g., for stereophonic sound).

sound-transmission coefficient (of an interface or septum)— Also called acoustical transmittivity. The ratio of the transmitted to the incident sound energy. Its value is a function of the angle of incidence of the sound.

sound wave— *See* sound, 1.

sound-with-sound— 1. A special provision in some recorders that allows the record head for one channel to be used for listening to that track while adding new material in exact synchronism on the adjacent track. Playing and mixing both simultaneously produces a composite sound without the degradation of one sound otherwise caused by a dubbing step in sound-on-sound mixing. 2. A process by which a program is recorded on one track, then monitored as a second program is recorded on another track.

source— 1. The device that supplies signal power to a transducer. 2. In a field-effect transistor, the electrode that corresponds to the cathode of a vacuum tube. 3. Supply of energy, or device upstream from a sink. *See also* sink. 4. Terminal that usually sources carriers. In MOS devices, which are usually symmetrical, it can be interchanged with the drain terminal in a circuit. 5. The working-current terminal (at one end of the channel in a FET) that is the source of holes (p-channel) or free electrons (n-channel) flowing in the channel. Corresponds to the emitter in a bipolar transistor. 6. The device that switches positive dc to a load. The current flows from the sensor into the load. 7. The origin of radiant energy, such as a light-emitting diode. 8. User-written instruction statements prior to translation by a computer into machine-executable form. 9. One of the three regions that form a field-effect transistor. Majority-carriers (electrons in an n-channel FET or holes in a p-channel FET) originate at the source and flow across the channel to the drain as a result of the electric field applied between source and drain.

source address— In computer systems having a source-destination architecture, the source address is the address of the device address or memory location from which data is being transferred.

source code— 1. A nonexecutable computer program written in a high-level language. A compiler or assembler must translate the source code into object code (machine language) that the computer can understand and process. 2. Virtually any computer language, from assembly to high-level, that doesn't fit the definition of object code.

source connector— One of three classifications of fiber-optic connectors that interconnect a light source (typically a LED) to a fiber-optic cable. A metal connector shell that is mounted on a circuit board or attached to a panel provides heat sinking and EMI protection. *See* bundle connector.

source-cutoff current— The current into the source terminal of a depletion-type transistor with a specified gate-to-drain voltage applied to bias the device to the off state.

source data automation— The methods of recording information in coded forms on paper tapes, punched cards, or tags that can be used repeatedly to produce many other records without rewriting.

source document— A paper containing information that is to be read into the computer.

source electrode— One of the electrodes in a field-effect transistor. It is analogous to the emitter in a transistor or the cathode in a vacuum tube. Represented by the symbol S.

source impedance— 1. The impedance that a source of energy presents to the input terminals of a device. 2. The impedance that a meter or other instrument sees; i.e., the impedance of the driving circuit when measured from the input terminals of the meter.

sourcing— Redesign or modification of existing equipment to eliminate a source of electromagnetic interference. When sourcing is not feasible, engineers are forced to resort to suppression, filtering, or shielding.

source language— 1. The language used to prepare a problem as the input for a computer operation. 2. In a computer, the language from which a statement is translated. 3. The original language used by the programmer, on which a translator program operates. 4. In general, any language that is to be translated into another (target) language; usually, however, it refers to the language used by a programmer to program a system.

source load— A load with a current whose direction is toward its input. A source load must be driven by a current source.

source machine— The computer used to translate the source program into the object program.

source module— In a computer, a series of statements expressed in the symbolic language of an assembler or compiler and constituting the entire input to a single execution of the assembler or compiler.

source program— 1. A computer program written in a language designed for ease of expression of a class of problems or procedures by humans. A generator, assembler, translator, or compiler routine translates the source program into an object program in machine language. 2. The original program, as written by the programmer, from which a working program system is derived. 3. A program, in either hard-copy or stored form, written in a high-level language (source language) other than machine language, which requires translation by the assembler, compiler, or interpreter program before it can be run on the computer. 4. A program written by a human, either in assembly or in a high-level language.

source recording— The recording of information in machine-readable form, such as punched cards or tape, magnetic tape, etc. Once in this form, the information may be transmitted, processed, or reused without a need for manual processing.

source statement — A computer program written in other than machine language, usually in three-letter mnemonic symbols that suggest the definition of the instruction. There are two kinds of source statements: executive instructions, which translate into operating machine code (op code), and assembly directives, which are useful in documenting the source program but generate no code.

source/tape switch — A control found on control amplifiers with tape monitor jacks, and on recorders with monitor heads; allows comparison of the signal being fed to the tape (source) with the signal just recorded.

source terminal — The terminal electrically connected to the region from which majority carriers flow into the channel of a field-effect transistor.

south pole — In a magnet, the pole into which magnetic lines of force are assumed to enter after emerging from the north pole.

space — 1. An impulse that, in a neutral circuit, causes the loop to open or causes the absence of a signal, whereas in a polar circuit it causes the loop current to flow in a direction opposite to that for a mark impulse. A space impulse is equivalent to a binary 0. 2. In some codes, a character that causes a printer to leave a character width with no printed symbol. 3. The signaling state used to represent a binary 0. *See* mark. 4. One of two states describing the condition of a teleprinter's loop. The space is characterized by an open-loop condition. The mark state indicates a closed loop.

space attenuation — 1. The loss of energy, expressed in decibels, of a signal in free air caused by such factors as absorption, reflection, scattering, and dispersion. 2. Reduction of a TV satellite signal strength due to the fact that the beam spreads out after leaving the antenna. It is a major factor in path loss.

space charge — 1. The negative charge caused by the cloud of electrons that forms in the space between the cathode and plate of a vacuum tube because the cathode emits more electrons than are attracted immediately to the plate. 2. An electrical charge distributed throughout a volume or space.

space-charge debunching — In a microwave tube, a process in which the bunched electrons are dispersed due to the mutual interactions between electrons in the stream.

space-charge effect — Repulsion of electrons emitted from the cathode of a thermionic vacuum tube by electrons accumulated in the space charge near the cathode.

space-charge field — The electric field that occurs inside a plasma due to the net space charge in the volume of the plasma.

space-charge grid — A grid, usually positive, that controls the position, area, and magnitude of a potential minimum or of a virtual cathode adjacent to the grid.

space-charge region — The region around a pn junction in which holes and electrons recombine, leaving no mobile charge carriers and a net charge density different from zero. *See also* depletion layer.

space-charge tube — A tube in which the space charge is used to greatly increase the transconductance. A positively charged grid is placed next to the cathode, in front of the control grid. This enlarges the space charge, moving it out to where the control grid can have a greater effect on it and hence on the plate current.

space coordinates — A three-dimensional system of rectangular coordinates. The x and y coordinates lie in a reference plane tangent to the earth, and the z coordinate is perpendicular.

space current — The total current between the cathode and all other electrodes in a vacuum tube.

spaced antenna — An antenna system used for minimizing local effects of fading at shortwave receiving stations. So called because it consists of several antennas spaced a considerable distance apart.

spaced-antenna direction finder — A direction finder comprising two or more similar but separate antennas coupled to a common receiver.

space detection and tracking system — A system that can detect and track space vehicles from the earth and report the orbital characteristics of such vehicles to a central control facility.

space diversity — *See* space diversity reception.

space diversity gain — The improvement in radio reception, expressed in decibels, obtained by combining the signals from two receiving antennas physically separated by not less than 5 wavelengths.

space diversity reception — Also called space diversity. 1. Diversity reception from receiving antennas placed in different locations. 2. A method of diversity radio reception in which the receiving antennas are physically separated, vertically or horizontally, by 5 wavelengths or more.

spaced-loop direction finder — A spaced-antenna direction finder in which the individual antennas are loops.

space factor — Ratio of the effective area utilized to the total area in a winding section.

space harmonics — Harmonics in the distribution of flux in the air gap of a resolver. They may be determined as a percent of fundamental by a Fourier analysis of the flux distribution antenna. Space harmonics cause angular inaccuracy.

space-hold — In a computer, the normal no-traffic line condition by which a steady space is transmitted.

space pattern — On a test chart, a pattern designed for the measurement of geometric distortion. The EIA ball chart is an example.

space permeability — The factor that expresses the ratio of magnetic induction to magnetizing force in a vacuum. In the cgs electromagnetic system of units, the permeability of a vacuum is arbitrarily taken as unity.

space phase — Reaching corresponding peak values at the same point in space.

space quadrature — The difference in the position of corresponding points of a wave in space, the points being separated by one-quarter of the wavelength in question.

space radio station — An amateur radio station located on an object that is beyond, is intended to go beyond, or has been beyond the major portion of the earth's atmosphere.

spacer cable — A means of primary power distribution that consists of three partially insulated or covered phase wires and a high-strength messenger-ground wire, all mounted in plastic or ceramic insulating spacers.

space-to-mark transition — The transition, or switching, from a spacing impulse to a marking impulse. (Teletypewriter term.)

space wave — The radiated energy consisting of the direct and ground waves.

spacing — The distance between stereo microphones or speakers.

spacing end distortion — End distortion that lengthens the spacing impulse by advancing the mark-to-space transition. (Teletypewriter term.)

spacing interval — The interval between successive telegraph signal pulses. During this interval, either no current flows or the current has the opposite polarity from that of the signal pulses.

spacing pulse — In teletypewriter operation, the signal interval during which the selector unit does not operate.

spacing wave — Also called back wave. In telegraphic communication, the emission that takes place between the active portion of the code characters or while no code characters are being transmitted.

spacistor — 1. A semiconductor device consisting of one pn junction and four electrode connections. It is characterized by a low transient time for carriers to flow from the input to the output. 2. Multiple-terminal solid-state device, similar to a transistor, that achieves high-frequency operation up to about 10,000 MHz by injecting electrons or holes into a space-charge layer that rapidly forces these carriers to a collecting electrode.

spade connector — A terminal with a slotted tongue and nearly square sides.

spade contact — A contact with fork-shaped female members designed to dovetail with spade-shaped male members. Alignment in this type of connection is very critical if good conductivity is to be achieved.

spade lug — A solder lug that has an open end so that it can be slipped under the head of a binding screw.

spade tips — Notched, flat metal strips connected to the end of a cord or wire so that it can be fastened under a binding screw.

spade-tongue terminal — A slotted-tongue terminal designed to be slipped around a screw or stud without removal of the nut.

spaghetti — 1. Heavily varnished cloth tubing sometimes used to provide insulation for circuit wiring. 2. A form of tubular insulation that can be slipped over wires before they are connected to terminals.

span — 1. The part or space between two consecutive points of support in a conductor, cable, suspension strand, or pole line. 2. The reach or spread between two established limits, such as the difference between high and low values in a given range of physical measurements.

span adjustment — See sensitivity adjustment.

s-parameters — Abbreviation for scattering parameters. A group of measurements taken at different frequencies that represent the forward and reverse gain and the input and output reflection coefficients of a microwave component when the input and output ports of the component are terminated in specified impedances — usually 50 ohms.

spark — 1. The abrupt, brilliant phenomenon that characterizes a disruptive discharge. 2. A single, short electrical discharge between two electrodes. 3. The brief flash of light from the arc. 4. The bridging or jumping of a gap between two electrodes by a current of electricity.

spark capacitor — A capacitor connected across a pair of contact points, or across the inductance that causes the spark, for the purpose of diminishing sparking at these points.

spark coil — An induction coil used to produce spark discharges.

spark duration — 1. The time between the moment when the electrons first jump a gap and the moment when the current ceases to flow across it. 2. The length of time a spark is established across a spark gap or the length of time current flows in a spark gap.

spark electrodes — The conductive element on each side of a spark gap through which current flows to and from the gap.

spark energy — The amount of energy dissipated between the electrodes of a spark gap. This is normally expressed as a steady-state wattage as though dissipated for a full second. Normally expressed in milliwatt seconds or millijoules.

spark frequency — The total number of sparks occurring per second in a spark transmitter (not the frequency of the individual waves).

spark gap — The arrangement of two electrodes between which a disruptive discharge of electricity may occur, and such that the insulation is self-restoring after the passage of a discharge.

spark-gap modulation — Modulation in which a controlled spark-gap breakdown produces one or more pulses of energy for application to the element in which the modulation is to take place.

spark-gap modulator — A modulator employed in certain radar transmitters. A pulse-forming line is discharged across either a stationary or a rotary spark gap.

spark-gap oscillator — A type of oscillator consisting essentially of an interrupted high-voltage discharge and a resonant circuit.

sparking — Intentional or accidental spark discharge, as between the brushes and commutator of a rotating machine, between the contacts of a relay or switch in a solid tantalum capacitor, or at any other point, in which a circuit is broken.

sparking voltage — The minimum voltage at which a spark discharge occurs between electrodes of a given shape at a given distance apart under given conditions.

spark killer — Also called spark suppressor. An electric network, usually a capacitor and resistor in series, connected across a pair of contact points (or across the inductance that causes the spark) to diminish sparking at these points.

spark lag — The interval between attainment of the sparking voltage and passage of the spark.

sparklies — Also known as snow. 1. Weak-signal noise that appears as dot or streak interference in a satellite TV picture. Loss of lock in an FM video demodulator causes this. In extreme cases, tearing or loss of the picture may result. 2. Small black and/or white blips or dots in a television picture, indicating an insufficient signal-to-noise ration.

sparkover — Breakdown of the air between two electrical conductors, permitting the passage of a spark.

spark plate — In an automobile radio, a metal plate insulated from the chassis by a thin sheet of mica. It bypasses the noise signals picked up by the wiring under the hood.

spark plug — An assembly that includes a pair of electrodes and insulator. It provides a spark gap in an engine cylinder.

spark-quenching device — See spark suppressor.

spark recorder — A recorder in which the recording paper passes through a spark gap formed by a metal plate underneath and a moving metal pointer above the paper. Sparks from an induction coil pass through the paper, periodically burning small holes that form the record trace.

spark source — A device used to produce a short-circuit pulse of luminous energy by an electrical discharge between two closely spaced electrodes either in air or in a controlled atmosphere at a pressure usually greater than half an atmosphere.

spark spectrum — The spectrum produced in a substance when the light from a spark passes between terminals made of that substance or through an atmosphere of that substance.

spark suppression — The use of a capacitor or resistor across contacts that break currents in inductive circuits, to prevent excessive sparking when the contacts break. A capacitor and resistor in series may also be used.

spark suppressor — Also called a spark-quenching device, spark killer, or an arc suppressor. An electric network, such as a capacitance and resistance in series or a diode connected across a pair of contacts, to diminish sparking (arcing) at these contacts.

spark test — A test performed on wire and cable to determine the amount of detrimental porosity or defects in the insulation.

spark transmitter — A radio transmitter in which the source of radio-frequency power is the oscillatory discharge of a capacitor through an inductor and a spark gap.

spatial coherence — The phase relationship of two wave trains in space.

spatial distribution — The directional properties of a speaker, transmitting antenna, or other radiator.

spatial frequency — The number of black and white line pairs displayed on a screen per degree of visual angle. Spatial frequency is expressed as cycles per degree or as lines per inch for a given viewing distance.

SPDT — Abbreviation for single-pole, double-throw. A three-contact switching arrangement that connects a circuit to one of two alternate connections.

speaker — Abbreviated spkr. Also called a loudspeaker. An electroacoustic transducer that radiates acoustic power into the air with essentially the same waveform as that of the electrical input.

Speaker.

speaker efficiency — Ratio of the total useful sound radiated from a speaker at any frequency to the electrical power applied to the voice coil.

speaker impedance — The rated impedance of the voice coil of a speaker.

speaker-reversal switch — A switch for connecting the left channel to the right speaker and vice versa on a stereo amplifier. It is a means of correcting for improper left-right orientation in the program source.

speaker system — A combination of one or more speakers and all associated baffles, horns, and dividing networks used to couple the driving electric circuit and the acoustic medium together.

speaker voice coil — In a moving-coil speaker, the part that is moved back and forth by electric impulses and is fastened to the cone in order to produce sound waves.

speaking arc — A dc arc on which audio-frequency currents have been superimposed. As a result, the arc reproduces sounds in a manner similar to a speaker, and its light output will vary at the audio rate required for sound-film recording.

special-effects generator — An apparatus used in the production of videotapes, this unit makes possible smooth switching of camera inputs and provides a wide variety of screen techniques, such as split and wipes.

special-event station — Station licensed at a specific land location for operation related to the celebration of an event, past or present, that is unique, distinct, and of general interest to either the public or to amateur radio operators, for the purpose of bringing public notice to the amateur radio service.

special-function key — A key on a keyboard that often has no predefined purpose but that can be programmed to send one or more useful commands.

special-purpose computer — A computer designed to solve a restricted class of problems, as contrasted with a general-purpose computer.

special-purpose language — A programming language that is designed to satisfy a single specific objective.

special-purpose logic — Proprietary features of a programmable controller that allow it to perform logic not normally found in relay ladder logic.

special-purpose motor — A motor possessing special operating characteristics and/or special mechanical construction designed for a particular application and not included in the definition of a general-purpose motor.

special-purpose relay — A relay whose application requires special features not characteristic of general-purpose or definite-purpose relays.

specific acoustic impedance — Also called unit-area acoustic impedance. The complex ratio of sound pressure to particle velocity at a point in a medium.

specific acoustic reactance — The imaginary component of the specific acoustic impedance.

specific acoustic resistance — The real component of the specific acoustic impedance.

specifications — 1. A published set of instructions for doing work in a uniform, standard manner. 2. Detailed description of the characteristics of a component, device, or system. 3. A clear, complete, and accurate statement of the technical requirements descriptive of a material, an item, or a service, and of the procedure to be followed to determine if the requirements are met.

specific coding — Digital-computer coding in which all addresses refer to specific registers and locations.

specific conductance — *See* electrolytic conduction.

specific conductivity — The conducting ability of a material in mhos per cubic centimeter. It is the reciprocal of resistivity.

specific damping of an instrument — *See* relative damping of an instrument.

specific dielectric strength — The dielectric strength per millimeter of thickness of an insulating material.

specific gravity — The weight of a substance compared with the weight of the same volume of water at the same temperature.

specific heat — 1. The capacity of a material to be heated at a given temperature (expressed as calories per degree Celsius per gram) compared with water, which has a specific heat of 1. 2. The amount of heat required to raise a specified mass by one unit of a specified temperature.

specific inductive capacity — *See* dielectric constant.

specific magnetic moment — The saturation moment of a magnetic material per unit weight. It is expressed in terms of emu/gram.

specific program — Digital-computer programming for solving a specific problem.

specific repetition rate — In loran, one of a set of closely spaced repetition rates derived from the basic rate and associated with a specific set of synchronized stations.

specific resistance — The resistance of a conductor. It is expressed in ohms per unit length per unit area, usually circular mil feet. *See also* resistivity, 1.

specific routine — A digital-computer routine expressed in specific computer coding and used to solve a specific mathematical, logical, or data-handling problem.

specific sound-energy flux — *See* sound intensity.

spectral — 1. Being a function of wavelengths. 2. Pertaining to or as a function of wavelength. Spectral quantities are evaluated at a single wavelength.

spectral bandwidth — The difference between the wavelengths of single-peak devices at which the radiant intensity is 50 percent (unless otherwise stated) of the maximum value.

spectral characteristic — 1. The relationship between the radiant sensitivity of a phototube and the wavelength of the incident radiant flux. It is usually shown by a graph. 2. The relation, usually shown by a graph, between the emitted radiant power per wavelength interval and the wavelength.

spectral coherence — A measure of the extent to which the output of a photodetector is restricted to a single wavelength or band of wavelengths; color response.

spectral contour plotter — A spectrum analyzer that presents a three-dimensional contour plot of analog signals, heart sounds, brain waves, etc.

spectral density — A value of a function whose integral over a frequency interval represents the contributions of the signal components within that frequency interval.

spectral energy distribution — Abbreviated SED. A plot of energy as a function of wavelength for a given light.

spectral intensity — A function that precisely defines the spectrum and has the units of voltage squared per unit frequency.

spectral output (of a light-emitting diode) — A description of the radiant-energy or light-emission characteristic versus wavelength. This information is usually given by stating the wavelength at peak emission and the bandwidth between half-power points or by means of a curve.

spectral purity — A measure of how much the spectrum of the output of a signal generator deviates from that of a pure sine wave.

spectral radiant flux — The radiant flux per unit wavelength interval, usually in watts per nanometer.

spectral radiant reflectance — That fraction of the power in a light beam that is reflected from a surface. In less precise usage, called reflectivity.

spectral response — Also called spectral sensitivity characteristic. 1. The relative amount of visual sensation produced by one unit of radiant flux of any one wavelength. The human eye or a photocell exhibits greatest spectral response to the wavelengths producing yellow-green light. 2. The variation of responsivity of the detector with the wavelength of the impinging radiation.

spectral sensitivity — The color response of a photosensitive device.

spectral sensitivity characteristic — *See* spectral response

spectral voltage density — The rms voltage corresponding to the energy contained in a frequency band having a width of 1 hertz. For the spectral voltage density at a given frequency, the band is centered on the given frequency.

spectrogram — A machine-made graphic representation of sounds in terms of their component frequencies. Time is shown on the horizontal axis, frequency on the vertical axis, and intensity by the darkness of the mark.

spectrograph — An instrument with an entrance slit and dispersing device that uses photography to obtain a record of the spectral range. The radiant power passing through the optical system is integrated over time, and the quantity recorded is a function of the radiant energy.

spectrometer — A test instrument that determines the frequency distribution of the energy generated by any source and displays all components simultaneously.

spectrophotoelectric — 1. Pertaining to the dependence of photoelectric phenomena on the wavelength of the incident radiation. 2. Characteristic of the relationship between photoelectric activity and the wavelength of incident radiation.

spectrophotometer — An instrument for measuring spectral transmittance or reflectance.

spectrophotometric analysis — The detection and measurement, relative to wavelength, of spectral reflectance, spectral transmittance, or spectral emittance.

spectrophotometry — 1. A process of making comparisons between parts of light spectra by means of a photometric device in combination with a spectrometer. 2. Study of the reflection or transmission properties of specimens as a function of wavelength.

spectroradiometer — An instrument for measuring the radiant energy from a source at each wavelength through the spectrum. Spectral regions are separated either by calibrated filters or by a calibrated monochromator. The detector is usually an energy receiver, such as a thermocouple.

spectroscope — An instrument used to disperse radiation into its component wavelengths and to observe or measure the resultant spectrum.

spectroscopy — 1. The branch of optics that deals with radiations in the infrared, visible, and ultraviolet regions of the spectrum. 2. The measurement and interpretation of the electromagnetic radiation absorbed or emitted when molecules, atoms, or ions change from one internal energy level to another. Such a change could be caused by incident radiation of the kind supplied by a laser. In the case of molecules, the wavelengths corresponding to the energy transitions that occur are mainly established by mechanical motions like rotation and vibration. These motions show up in the infrared region of the spectrum — from beyond the red end of the visible region at 0.8 µm to about 100 µm, where the microwave region begins. The molecules' electrical properties determine the intensities of the energy transitions. In order to determine whether a given type of molecule is present, it is only necessary to pass a range of IR wavelengths through the substance and note the position and intensity of the peaks and valleys of the detected beam. Use of a spectrum "dictionary" then shows what materials are present. 3. The branch of science dealing with the theory and interpretation of spectra.

spectrum — 1. A continuous range of electromagnetic radiations, from the longest known radio waves to the shortest known cosmic rays. Light, which is the visible portion of the spectrum, lies about midway between these two extremes. 2. The frequency components that make up a complex waveform. The band of frequencies necessary for transmission of a given type of intelligence. 3. The range of frequencies considered in a system. 4. The distribution of the amplitude of the components of a time-domain signal as a function of frequency. 5. The wavelengths or frequencies associated with any system one may wish to describe, such as the visible spectrum, the infrared spectrum, the entire electromagnetic spectrum, etc. 6. The entire range of electromagnetic energy used in transmission of voice, data, and television. When applied to light, and/or lasers, it includes the frequencies from shortwave ultraviolet through infrared energy.

spectrum analysis — 1. The study of energy distribution across the frequency spectrum for a given electrical signal. 2. The process of determining the magnitude of frequency components of a signal, i.e., magnitude of the Fourier transform.

spectrum analyzer — 1. A scanning receiver that automatically tunes through a selected frequency spectrum and displays on a CRT or a chart a plot of amplitude versus frequency of the signals present at its input. A spectrum analyzer is, in effect, an automatic Fourier analysis plotter. 2. A test instrument that shows the frequency distribution of the energy emitted by a pulse magnetron. It also is used in measuring the Q of resonant cavities and lines, and in measuring the cold impedance of a magnetron. 3. Test equipment that displays the energy present at all input signal frequencies over a finite period of time. Spectrum analyzers can be synchronized with a sweeping oscillator to display the frequency response of audio equipment or sound systems. 4. An instrument that displays the power or voltage of a time-domain signal as a function of frequency.

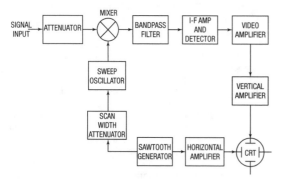

Spectrum analyzer.

spectrum intervals — Frequency bands represented as intervals on a frequency scale.

spectrum level — For a specified signal at a particular frequency, the level of that part contained within a band 1 hertz wide, centered at the particular frequency.

spectrum locus — The locus of a point representing the colors of the visible spectrum in a chromaticity diagram.

spectrum-selectivity characteristic — A measure of the increase in the minimum input-signal power over the minimum detectable signal required to produce an indication on a radar indicator, if the received signal has a spectrum different from that of the normally received signal.

spectrum signature analysis — The evaluation of electromagnetic interference from transmitting and receiving equipment in order to determine operational and environmental compatibility.

spectrum utilization characteristics — The compiled data of either transmitters or receivers that describes operational parameters such as bandwidth, sensitivity, stability, antenna pattern, power output, etc. Also describes the capabilities of the equipment to either reject or suppress unwanted electromagnetic energy.

specular reflection — Reflection of light, sound, or radio waves from a surface so smooth that its inequalities are small in comparison with the wavelength of the incident rays. As a result, each incident ray produces a reflected ray in the same plane.

SPEDAC — Acronym for solid-state, parallel, expandable, differential-analyzer computer. A high-speed digital differential analyzer using parallel logic and arithmetic, solid-state circuitry, and modular construction and capable of being expanded in computing capacity, precision, and operating speed.

speech amplifier — A voltage amplifier made specifically for a microphone.

speech audiometer — An audiometer for measuring either live or recorded speech signals.

speech clipper — 1. A speech-amplitude-limiting circuit that permits the average modulation percentage of an amplitude-modulated transmitter to be increased. 2. Circuit using one or more biased diodes to limit the wave crests of speech frequency signals. Used in speech amplifiers of transmitters to maintain a high average modulation percentage.

speech compression — 1. A modulation technique that makes use of certain properties of the speech signal in transmitting adequate information regarding quality, characteristics, and the sequential pattern of a speaker's voice over a narrower frequency band than otherwise would be necessary. 2. A means of boosting the average level of voice signals to provide increased "talk power" of higher average modulation levels.

speech frequency — *See* voice frequency.

speech-interference level — A value for rating the effect of background noise on the intelligibility of speech.

speech interpolation — The method of obtaining more than one voice channel per voice circuit by giving each subscriber a speech path in the proper direction only at the times when his or her speech requires it.

speech inverter — An apparatus that interchanges high and low speech frequencies by removing the carrier wave and transmission of only one sideband in a radiotelephone. This renders the speech unintelligible unless picked up by apparatus capable of replacing the carrier wave in the correct manner. *See also* scrambler circuit.

speech level — The energy of speech (or music), measured in volume units on a volume indicator.

speech scrambler — 1. A device to provide privacy to voice communications via wire or radio. (The simplest is an inverter that changes low frequencies to high, and vice versa at the receiving end.) 2. *See* scrambler circuit.

speech synthesizer — A system used in research to generate speech from electrical signals in order to study human vocal patterns.

speed calling — A feature that allows numbers to be dialed with abbreviated codes.

speed control — The act of precisely increasing or decreasing the speed of a driven gear or shaft by an electrical or mechanical means.

speed limit — A control function that prevents the controlled speed from exceeding prescribed limits.

speed of light — The speed at which light travels: 186,284 miles per second, or 3×10^8 m/s.

speed of sound — Also called sonic speed. The speed at which sound waves travel through a medium (in air and at standard sea-level conditions, about 750 miles per hour or 1080 feet per second, or 329 m/s).

speed of transmission — Also called rate of transmission. The instantaneous rate of processing information by a transmission facility. Usually measured in characters or bits per unit time.

speed-ratio control — A control function that maintains a preset ratio of the speeds of two drives.

speed regulation — A figure of merit indicating the change in motor speed from no load to full load expressed as a percentage of full-load speed. Speed regulation is generally established at rated speed.

speed regulator — A regulator that maintains or varies the speed of a motor at a predetermined rate.

speedup capacitor — 1. A capacitor used in RCTL to permit faster turn-on of the transistor in response to a change in input; it also helps overcome the storage delay of the transistor itself. 2. *See* commutating capacitor.

spelling checker — A program that checks the words in a computer file against a previously recorded list, querying or correcting words not on that list.

sphere gap — A spark gap with spherical electrodes. It is used as an excess-voltage protective device.

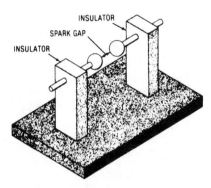

INSULATOR
SPARK GAP
INSULATOR

Sphere gap.

sphere-gap voltmeter — An instrument for measuring high voltages. It consists of a sphere gap, and the electrodes are moved together until the spark will just barely pass. The voltage can be calculated from gap spacing and the electrode diameter, or read directly from a calibrated scale.

spherical aberration — 1. Image defects (e.g., blurring) due to the spherical form of a lens or mirror. These defects cause a blurred image because the lens or mirror brings the central and marginal rays to different focuses. Common types of spherical aberration are astigmatism and curvature of the field. 2. The optical error introduced by the fact that incident rays at different distances from the optical axis are focused at different points along the axis by reflection from spherical mirror surfaces or refraction by spherical lenses.

spherical candlepower — In a lamp, the average candlepower in all directions in space. It is equal to the total luminous flux of the lamp, measured in lumens, divided by 4π.

spherical coordinates — A system of polar coordinates that originate in the center of a sphere. All points lie on the surface of the sphere, and the polar axis cuts the sphere at its two poles.

spherical-earth attenuation — Attenuation of radio waves over an imperfectly conducting spherical earth in excess of the attenuation that would occur over a perfectly conducting plane.

spherical earth factor — Ratio between the electrical field strengths that would result from propagation over an imperfectly conducting spherical earth and a perfectly conducting plane.

spherical stylus — A cartridge stylus whose shape is conical, with the downward-facing point of the cone rounded to a specified radius of curvature, usually 0.5 or 0.7 mil (12.7 or 17.8 μm).

spherical wave — A wave in which the wavefronts are concentric spheres.

sphygmocardiograph — A device for simultaneous recording of heartbeat and pulse.

sphygmogram — A graphic recording of the movements, forms, and forces of an arterial pulse.

sphygmomanometer — An instrument for measuring blood pressure, especially that in the arteries.

spider — Highly flexible ring, washer, or punched flat member used in a dynamic speaker to center the voice coil on the pole piece without appreciably hindering the in-and-out motion of the voice coil and its attached diaphragm.

spider bonding — A method used for connecting an integrated-circuit chip to its package leads or to a film-type substrate for hybrid constructions. Instead of running individual wires from each bonding pad on the chip to the corresponding package lead, a preformed lead frame is placed over the chip, and all connections are made by a single operation of a bonding machine.

spider-web antenna — An all-wave receiving antenna having several lengths of doublets connected somewhat like the web of a spider to give favorable pickup characteristics over a wide range of frequencies.

spider-web coil — A flat coil having an open weave somewhat like the bottom of a woven basket. It was used in older radio receivers.

spike — Also called a transient. 1. An abrupt transient that comprises part of a pulse but exceeds its average amplitude considerably. 2. A pulse of short duration and of greater amplitude than an average pulse. 3. A distortion in the form of a pulse waveform of relatively short duration superimposed on an otherwise regular or desired pulse waveform. 4. A burst of extra voltage in a power line, lasting only a fraction of a second.

spike discriminator — A circuit used in a transponder to discriminate against pulses of extremely short duration, such as might be caused by ignition noise.

spike-leakage energy — The radio-frequency energy per pulse transmitted through tr and pre-tr tubes before and during the establishment of the steady-state radio-frequency discharge.

spike noise — In a chopper, the static field noise caused by insulating material; can be observed if the chopper is followed by a wideband amplifier.

spiking — Short, multiple, irregular bursts of laser-output radiation. Spiking is characteristic of pulse lasers, especially flash-pumped solid-dielectric types (e.g., ruby, neodymium in glass). The spike duration is typically 0.2 to 2 microseconds.

spill — The redistribution and hence loss of information from a storage element of a charge-storage tube.

spillover positions — Storage positions in which backlogged traffic that accumulates when a send channel is inoperative or unusually busy is held for transmission immediately on the availibility of a channel.

spin block — A loop created when a process keeps checking the state of a flag or status bit while waiting for an event to occur.

spindle — The upward-projecting shaft used on a phonograph turntable for positioning and centering the record.

spinel — *See* silicon on sapphire.

spinner — 1. An automatically rotatable radar antenna, together with its associated equipment. 2. The part of a mechanical scanner that is rotated about an axis; generally, use of the term is restricted to cases of relatively high-speed rotation.

spinning electron — An electron that spins with an angular momentum.

spinthariscope — An instrument for viewing the scintillations of alpha particles on a luminescent screen.

spin wave — A moving magnetic disturbance in a ferrite. The presence of an alternating magnetic field with its frequency near that of the natural precessional frequency of a spin tends to increase the precession angle. This increase in precession angle is passed from atom to atom by dipole and exchange interactions.

spin-wave amplitude — The difference between the precession angles of two spins.

spiral distortion — In camera tubes or image tubes using magnetic focusing, a form of distortion in which image rotation varies with distance from the axis of symmetry of the electron optical system.

spiral four — A quad in which the four conductors are twisted about a common axis, the two sets of opposite conductors being used as pairs.

spiral scanning — Scanning in which the maximum radiation describes a portion of a spiral, with the rotation always in one direction.

spkr — Abbreviation for speaker.

SPL — Abbreviation for sound-pressure level.

splashproof — A device or machine so constructed and protected that external splashing will not interfere with its operation.

splashproof motor — An open motor in which the ventilating openings are so constructed that drops of liquid or solid particles falling on it, or coming toward it in a straight line at any angle not greater than 100° from the vertical, cannot enter either directly or by striking and running along a surface of the motor.

splatter — Adjacent-channel interference due to over-modulation of a transmitter by abrupt peak audio signals. It is particularly noticeable for sounds containing high-frequency harmonics.

splice — 1. A device used for joining two or more conductors. 2. A joint connecting conductors with good mechanical strength and good conductivity; a terminal that permanently joins two or more wires. 3. Nonseparable junction joining optical conductor to optical conductor. 4. An in-line connection between two wires that results in a single conductor. 5. A physical join between pieces of tape 6. A connection of two cables in which each pair in one cable is connected to the corresponding pair in the other.

splice insulation — Insulation used over a splice.

splicer — Any device that holds magnetic recording tape ends in place for properly aligned splices. Many splicers are automatic to some extent, emitting the tape and trimming the edges of the splice by means of built-in cutting edges.

splicing block — 1. A nonautomatic recording tape splicer consisting of an elongated block of metal or plastic with a shallow groove to hold the tape and a narrow slot, usually diagonal, across the middle of the block to guide the cutting blade. Splices are made with a special splicing tape that generally needs no edge trimming. 2. A tool for tape editing consisting of a nonmagnetic metal block with a guide channel that holds magnetic recording tape in precise alignment. Additional straight and diagonal grooves provide a path for a razor blade to follow for cutting the tape so it can be spliced effectively.

splicing loss — See coupling loss, 1.

splicing tape — A pressure-sensitive, nonmagnetic tape designed for joining two pieces of magnetic tape; characterized by high flexibility and an adhesive that will not flow out from under the splice and, thus, cause adhesion between adjacent layers on the reel.

spline — A piecewise curve adhering to certain continuity conditions at the node points.

split — 1. Of radar tracks, the separation of radar data from a single track to such a degree that one or more additional tracks can be initiated manually or automatically; a similar condition exists when a raid is separated to such an extent that it can be represented on a situation display as two or more raids. 2. Initiation of split tracks, raids, or groups by a direction center track monitor or by programming means.

split-anode magnetron — A magnetron with an anode divided into two segments, usually by parallel slots.

split-conductor cable — A cable in which each conductor is composed of two or more insulated conductors normally connected in parallel.

split fitting — A conduit fitting, bend, elbow, or tee split longitudinally so that it can be positioned after the wires have been drawn into the conduit. The two parts are held together usually by screws.

split frequency — The frequencies between the 10-kHz channels assigned by the FCC for use by U.S. broadcasters.

split gear — A type of gear designed to minimize backlash. The method consists of splitting one gear of a meshing pair and so connecting a spring between the two halves that pressure is exerted on both sides of the teeth of the other gear.

split hydrophone — A direction hydrophone in which the electroacoustic transducers are divided and arranged so that each division can induce a separate electromotive force between its own terminals.

split image — Two or more scenes appearing on a television screen as a result of trick photography at the studio.

split-phase — Characteristic of any device (such as an induction motor) that derives a second phase from a single-phase alternating current source by using a capacitive or inductive reactor.

split-phase motor — 1. A single-phase induction motor having an auxiliary winding connected in parallel with the main winding, but displaced in magnetic position from the main winding so as to produce the required rotating magnetic field for starting. The auxiliary circuit is generally opened when the motor has reached a predetermined speed. 2. A single-phase induction motor that is momentarily converted to two-phase (polyphase, self-starting) motor by inclusion of an additional primary power winding displaced 90° from the main windings to provide required starting torque. Occasionally the auxiliary windings are designed to operate full time rather than momentarily.

split projector — A directional projector in which electroacoustic transducing elements are divided and arranged so that each division can be energized separately through its own terminals.

split rotor plate — See serrated rotor plate.

split-sound system — An early television receiver IF system in which the audio and video IF signals were separated right after the mixer stage and amplified in separate IF stages. Replaced by the more current intercarrier sound system.

split-stator capacitor — A capacitor comprising two isolated stators that function in series with a common rotor.

split-stator variable capacitor — A variable capacitor with a rotor section common to two separate stator sections. Used for balancing in the grid and plate tank circuits of transmitters.

split-streaming — See bandsplitting.

splitter — 1. A device with one input that provides two or more outputs to allow multiple receiver hookups to one antenna. Can be passive (an antenna coupler) or active, providing gain. 2. A passive device similar to an antenna coupler but designed to match a 75-ohm impedance. 3. A device that takes a signal and splits it into two or more identical but lower power signals.

split transducer — A directional transducer in which electroacoustic transducing elements are divided and arranged so that each division is electrically separate.

split winding — An equal division of a winding that will allow series or parallel external connection of the divided winding (four external leads) of a servomotor.

spoiler — A rod grating mounted on a parabolic reflector to cause the radiation pattern to change from a pencil beam to a cosecant-squared pattern. When the reflector and grating are rotated through 90° with respect to the feed antenna, one pattern changes to the other.

spontaneous emission — 1. Emission occurring without stimulation or quenching after excitation. 2. Radiation produced when a quantum mechanical system falls spontaneously from an excited state to a lower state. Emission of this radiation occurs in accordance with the laws of probability and without regard for the presence of similar radiation at the same time.

spoofing — The defeat or compromise of an alarm system by tricking or fooling its detection devices, such as by short circuiting part or all of a series circuit, cutting wires in a parallel circuit, reducing the sensitivity of a sensor, or entering false signals into the system. Spoofing contrasts with circumvention.

spool — A flanged form serving as the foundation on which a coil is wound.

spooler — A system program that permits I/O transfers to be queued for an I/O device, thereby permitting the requesting task to continue executing even when it cannot immediately use the I/O device. Spoolers are commonly used with very slow sequential-output-only devices, such as printers.

spooling — Abbreviation for simultaneous peripheral operations online. Process of allowing programs using slow output devices to complete execution rapidly. Data is temporarily stored in buffers or queues for later low-speed transmission concurrent with normal system operation.

sporadic E ionization — Ionization that appears in the atmosphere at E-layer heights, is more noticeable at higher latitudes, and occurs at all times of the day. It may be caused by particle radiation from the sun.

sporadic E layer — A portion that sometimes breaks away from the normal E layer in the ionosphere and exhibits unusual erratic characteristics.

sporadic propagation — Abnormal and unpredictable radio transmission that only occurs occasionally. Caused by unusual intense ionization in some part of the E layer of the ionosphere.

sporadic reflections — Also called abnormal reflections. Sharply defined, intense reflections from the sporadic E layer. Their frequencies are higher than the critical frequency of the layer, and they occur anytime, anywhere, and at any frequency.

spot — 1. The area instantaneously affected by the impact of an electron beam of a cathode-ray tube. 2. See land, 2. 3. The smallest luminescent area of the screen surface instantaneously excited by the impact of an electron beam.

spot beam — A satellite beam that covers a relatively small geographical area.

spot bonding — See yield-strength-controlled bonding.

spot distortion — Undesirable asymmetry or defect in the spot shape.

spot jamming — Jamming of a specific frequency or channel.

spot noise factor — See spot noise figure.

spot noise figure — Also called spot noise factor. Ratio of the output noise of a transducer to the portion attributable to the thermal noise in the input termination when the termination has a standard noise temperature (290°K). The spot noise figure is a point function of input frequency.

spot projection — In facsimile: 1. An optical method in which the scanning or recording spot is delineated by an aperture between the light source and the subject copy or record sheet. 2. The optical system in which the scanning or recording spot is the size of the area being scanned or reproduced.

spot protection — Protection of objects such as safes, art objects, or anything of value that could be damaged or removed from the premises.

spot size — The smallest area of light that can be produced by a CRT.

spot speed — 1. In facsimile, the length of the scanning line times the number of lines per second. 2. In television, the product of the length (in units of elemental area, i.e., in spots) of the scanning line and the number of scanning lines per second.

spottiness — Bright spots scattered irregularly over the reproduced image in a television receiver due to human-made or static interference entering the television system at some point.

spot welding — 1. A resistance welding process whereby welds are made between two or more overlapping sheets of metal by pressing them together between two electrodes arranged to conduct current to the outer surfaces of the overlapped sheets. The tips of one or both of the electrodes are restricted in area to approximately the diameter of the spot weld desired. 2. Resistance welding in which the current is directed through the entire area under the electrode tip. Welding is usually performed by a rocker-arm type spot welder.

spot wobble — An externally produced oscillating movement of an electron beam and its resultant spot. Spot wobble is used to eliminate the horizontal lines across the screen and thus make the picture more pleasing.

spray fluxing — A specialized wave-solder fluxing technique in which a fine stainless-steel screen drum is rotated in liquid flux. The amount of flux transferred is controlled by the rotational speed of the drum and air pressure. The drum contains air jets. See also brush fluxing; foam fluxing; wave fluxing.

spray-gun soldering — Soldering fluxed and heated parts by blowing molten solder on them from a gas or electrically heated gun.

spreader — 1. An insulating crossarm used to hold the wires of a transmission line apart. 2. The crossarm separating the parallel wire elements of an antenna.

spread groove — A groove cut between recordings. The groove, which has an abnormally high pitch, separates the recorded material but still enables the stylus to travel from one to the next.

spreading anomaly — That part of the propagation anomaly that is identifiable with the geometry of the ray pattern.

spreading loss — The transmission loss suffered by radiant energy. The effect of spreading, or divergence, is measured by this loss.

spreadsheet — 1. A computer program that sets up an electronic spreadsheet in which the lines and columns are automatically calculated according to formulas chosen by the user. When one number is changed, the program will automatically change all the sums and multiples that are affected. 2. A program that provides automatic calculations on any numbers and formulas that are input.

spread spectrum transmission — 1. A communications technique that involves transmitting information that has been multiplied by a pseudo-random noise (PN) sequence that essentially spreads it over a relatively wide frequency bandwidth. The receiver detects and uses the same PN sequence to "despread" the frequency bandwidth and decode the transmitted information. This communications technique allows greater signal density within a given transmission bandwidth and also provides a high degree of signal encryption and security in the process. 2. A communications technique in which many different signal waveforms are transmitted in a wide band. Power is spread thinly over the band so narrow-band radios can operate within the wide band without interference.

spring — A resilient, flat piece of metal forming or supporting a contact member in a jack or a key.

spring-actuated stepping relay — A stepping relay in which cocking is done electrically and operation is produced by spring action.

spring contact — 1. A relay or switch contact, usually of phosphor bronze and mounted on a flat spring. 2. A device employing a current-carrying cantilever spring that monitors the position of a door or window.

spring contact probe — Also called spring pin or spring probe. The element that provides the electrical connection between a particular node on the product to be tested and the verifier electronics. Usually spring loaded to allow some thickness, lead length, and other product variations.

spring curve — A plot of the spring force on the armature of a relay versus armature travel.

spring-finger action — The design of a contact, as used in a printed circuit connector or a socket contact, permitting easy stress-free spring action to provide contact pressure and/or retention.

spring pile-up — An assembly of all contact springs operated by one armature lever.

spring-return switch — A switch that returns to its normal position when the operating pressure is released.

spring stop — In a relay, the member used to control the position of a pretensioned spring.

spring stud — In a relay, an insulating member that transmits the armature motion from one movable contact to another in the same pileup.

sprocket holes — Holes punched on each line of a perforated tape used as a timing reference and for driving certain transports.

sprocket pulse — 1. A pulse generated by one of the magnetized spots that accompany every character recorded on magnetic tape. During read operations, sprocket pulses permit regulation of the timing of the read circuits and provide a count of the number of characters read from the tape. 2. A pulse generated by a sprocket or driving hole in a paper tape; this pulse serves as the timing pulse for reading or punching the tape.

SPST — Abbreviation for single-pole, single-throw. A two-contact switching arrangement that opens or closes one circuit; the circuit may be normally open or normally closed.

spurious counts — *See* spurious tube counts.

spurious emanations — Unintentional and undesired emissions from a transmitting circuit.

spurious emission — *See* spurious radiation.

spurious modulation — Undesired modulation of an oscillator; for example, frequency modulation resulting from mechanical vibration.

spurious pulse — In a scintillation counter, a pulse not purposely generated or directly due to ionizing radiation.

spurious pulse mode — An unwanted pulse mode that is formed by the chance combination of two or more pulse modes and is indistinguishable from a pulse interrogation or reply.

spurious radiation — Also called spurious emission. Emissions from a radio transmitter at frequencies outside its assigned or intended emission frequency. Spurious emissions include harmonic emissions, parasitic emissions, and intermodulation products, but exclude emissions in the immediate vicinity of the necessary band, which are a result of the modulation process for the transmission of information.

spurious response — 1. Any undesired response from an electric transducer or similar device. 2. The sensitivity of a circuit to signals of frequencies other than the frequency to which the circuit is tuned. 3. In electronic warfare, undesirable signal images in the intercept receiver as a result of mixing of the intercepted signal with harmonics of the receiver local oscillators. 4. A characteristic of a spectrum analyzer wherein displays appear that do not conform to the calibration of the radio-frequency dial. 5. Response of a frequency-selective system to an undesired frequency.

spurious-response attenuation — The ability of a receiver to discriminate between a desired signal to which it is resonant and an undesired signal at any other frequency to which it is simultaneously responsive.

spurious-response ratio — Ratio of the field strength at the frequency that produces a spurious response to the field strength at the desired frequency, each field being applied in turn to produce equal outputs. Image ratio and intermediate-frequency response ratio are special forms of spurious-response ratio.

spurious-response rejection — The ability of an FM tuner to reject spurious signals falling outside the tuned frequency, or the immunity of the tuner itself to the production of spurious signals as the result of intermodulation, etc.

spurious signal — 1. An unwanted signal generated either in the equipment itself or externally and heard (or seen) as noise. 2. Undesired signals appearing external to an equipment or circuit. They may be harmonics of existing desired signals, high-frequency components of complex waveshapes, or signals produced by incidental oscillatory circuits. 3. An undesired false signal.

spurious transmitter output — Any component of the radio-frequency output that is not implied by the type of modulation and the specified bandwidth.

spurious transmitter output, conducted — A spurious output of a radio transmitter that is conducted over a tangible transmission path such as a power line, control circuit, radio-frequency transmission line, waveguide, etc.

spurious transmitter output, extraband — A spurious transmitter output that lies outside the specified band of transmission.

spurious transmitter output, inband — A spurious transmitter output that lies within the specified band of transmission.

spurious transmitter output, radiated — A spurious output radiated from a radio transmitter. (The associated antenna and transmission lines are not considered part of the transmitter.)

spurious tube counts — Also called spurious counts. The counts in radiation-counter tubes other than background counts and those caused directly by the radiation to be measured. They are caused by electrical leakage, failure of the quenching process, etc.

spurt tone — A short-duration audio-frequency tone used for signaling or dialing selection.

sputtering — Also called cathode sputtering. 1. A process sometimes used in the production of a metal master disc. In this process the original is coated with an electric conducting layer by means of an electric discharge in a vacuum. 2. A thin-film technique in which material for the film is ejected from the surface of the bulk source when the source is subjected to ion bombardment. 3. Dislocation of surface atoms of a material bombarded by high-energy atomic particles. 4. A method of depositing a thin film or material onto a substrate. The substrate is placed in a large demountable vacuum chamber having a cathode made of the metal or ceramic to be sputtered. The chamber is then operated so as to bombard the cathode with positive ions. As a result, small particles of the material fall uniformly on the substrate. 5. A deposition process wherein a surface, or target, is immersed in an inert-gas plasma and is bombarded by ionized molecules that eject surface atoms. The process is based on the disintegration of the target material under ion bombardment. Atoms broken away from the target material by gas ions deposit on the part (substrate), forming a thin film.

SQ — A matrix system developed by CBS, Inc.

s-quad — Also called simple quad. An arrangement of two parallel paths, each of which contains two elements in series.

square — A unit area, i.e., the ratio of length/width = 1.

square-law demodulator — See square-law detector.

square-law detection — Detection in which the output voltage is substantially proportional to the square of the input voltage over the useful range of the detector.

square-law detector — Also called square-law demodulator. A detector in which the output signal current is proportional to the square of the radio-frequency input voltage. Operation of this circuit depends on nonlinearity of the detector characteristic, rather than on rectification.

square-law scale meter — A meter in which the deflection is proportional to the square of the applied energies.

square-loop ferrite — A ferrite with a rectangular hysteresis loop.

square-loop material — Ferromagnetic material having a squareness ratio approaching 1. Compare rectangular loop.

squareness — The ratio of the residual induction to the maximum induction obtained at the maximum magnetizing field being used. It can be considered as a figure of merit of the flatness of the decaying magnetizing curve.

squareness ratio — 1. For a magnetic material in a symmetrically cyclically magnetized condition, the ratio of the flux density at zero magnetizing force to the maximum flux density. 2. The ratio of the flux density to the maximum flux density when the magnetizing force has changed halfway from zero toward its negative limiting value.

square wave — 1. A square- or rectangular-shaped periodic wave that alternately assumes two fixed values for equal lengths of time, the transition time being negligible in comparison with the duration of each fixed value. 2. An ac periodic waveform in which voltage alternates rapidly from a positive peak value to the negative peak value, and vice versa, after a delay.

square-wave amplifier — A resistance-coupled amplifier (in effect, a wideband video amplifier) that amplifies a square wave with a minimum of distortion.

square-wave generator — A signal generator for producing square or rectangular waves. It is useful for testing the frequency response of wideband devices. See also square-wave testing.

square-wave response — In camera tubes, the ratio of the peak-to-peak signal amplitude given by a test pattern consisting of alternate black and white bars of equal widths to the difference in signal between large areas of black and white having the same illuminations

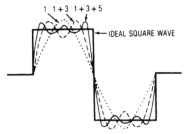

1 = FUNDAMENTAL
1 + 3 = FUNDAMENTAL PLUS THIRD HARMONIC
1 + 3 + 5 = FUNDAMENTAL PLUS THIRD PLUS FIFTH HARMONIC

Square wave showing harmonic composition.

as the bars. Horizontal square-wave response is measured if the bars are perpendicular to the horizontal scan, and vertical square-wave response is measured if they are parallel.

square-wave testing — The use of a square wave signal to test the frequency response of a wideband device. The output wave seen on an oscilloscope will show how much the square wave was distorted in passing through the device.

squaring circuit — 1. A circuit that changes a sine wave or other wave into a square wave. 2. A circuit that contains nonlinear elements and that produces an output voltage proportional to the square of the input voltage.

squawker — The midrange speaker of a three-way system.

squeal — 1. Audible tape vibrations, primarily in the longitudinal node, caused by a frictional excitation at the heads and guides. 2. In a radio receiver, a high-pitched tone heard together with the wanted signal.

squealing — The high-pitched noise heard along with the desired intelligence in a radio receiver. It is due to interference between stations or to oscillation in one of the receiver circuits.

squeegee — The part of a screen printer used in thick-film circuit manufacturing that pushes the composition across the screen and through the mesh onto the substrate.

squeezable waveguide — In radar, a variable-width waveguide for shifting the phase of the radio-frequency wave traveling through it.

squeeze section — A length of waveguide whose critical dimension can be altered to correspond to changes in the electrical length.

squeeze track — A variable-density sound track in which variable width with greater signal-to-noise ratio is obtained by means of adjusting masking of the recording light beam and simultaneously increasing the electric signal applied to the light modulator.

squegger — A self-quenching oscillator in which the suppression occurs in the grid circuit.

squegging — A self-blocking condition in an oscillator circuit.

squegging oscillator — See blocking oscillator, 1.

squelch — To automatically quiet a receiver by reducing its gain in response to a specified characteristic of the input.

squelch circuit — 1. A circuit for preventing a radio receiver from producing an audio-frequency output in the absence of a signal having predetermined characteristics. A squelch circuit may be operated by signal energy in the receiver passband, by noise quieting, or by a combination of the two (ratio squelch). It may also be operated by a signal having modulation characteristics (selective squelch). 2. A circuit that silences a receiver in the absence of signals above a certain level of signal strength. This squelch threshold is usually adjustable and will stop background noise from reaching the speaker but will activate the receiver when an intelligible signal is detected.

squib — Contained powder charge and initiating device that, when energized, produces heat and pressure.

squint — In radar, an ambiguous term meaning either the angle between the two major-lobe axes in a lobe-switching antenna or the angular difference between the axis of antenna radiation and a selected geometric axis such as the axis of the reflector.

squint angle — The angle between the physical axis of the antenna center and the axis of the radiated beam.

squirrel-cage induction motor — An induction motor in which the secondary circuit, usually the rotor, consists of a squirrel-cage winding (two discs connected along their circumference with copper bars) arranged in slots in the iron core.

squirrel-cage motor — A rugged electric motor that basically consists of two components: a wound stator and the rotor assembly, which most typically consists of a laminated cylindrical iron core with slots for conductors. The unit rotates when a moving magnet field induces a current in the shorted rotor conductors.

squirrel-cage winding — A permanently short-circuited winding that is usually uninsulated and has its conductors uniformly distributed around the periphery of the machine, and is joined by continuous end rings.

squitter — In radar, random firing (intentional or otherwise) of the transponder transmitter in the absence of interrogation.

SRAM — Abbreviation for static random-access memory. A read/write memory in which the data are latched and retained. SRAMs do not lose their contents as long as power is on. This memory does not need to be refreshed as does a DRAM.

S/rf meter — An indicator on some CB transceivers to indicate relative strength of an intercepted signal when receiving, and the relative rf power output when transmitting.

SSB — *See* single sideband.

SSFM — A system of multiplex in which single-sideband subcarriers are used to frequency modulate a second carrier.

SSI — Abbreviation for small-scale integration. 1. A classification of ICs by size, applicable to chips that contain less than 12 gates or circuits of equivalent complexity. 2. A circuit of under 10 gates, generally involving one metallization level implementing one circuit function in monolithic silicon. 3. The earliest form of integrated-circuit technology. A typical SSI circuit contains from one to four logic circuits.

SS loran — Sky-wave-synchronized loran. Loran in which the slave station is controlled by the sky wave from the master station rather than the ground wave. This method is used with unusually long base lines.

SSPM — A system of multiplex in which single-sideband subcarriers are used to phase-modulate a second carrier.

SSR — Abbreviation for solid-state relay.

SSTV — Abbreviation for Slow Scan Television. 1. Sending still images (usually black and white) by means of audio tones over telephone lines or on the MF/HF bands. Transmission times vary from a few seconds to several minutes. 2. A television system that employs a slow rate of horizontal scanning suitable for transmission of printed matter, photographs, and illustrations.

stability — 1. The ability of a component or device to maintain its nominal operating characteristics after being subjected to changes in temperature, environment, current, and time. It is usually expressed in either percent or parts per million for a given period of time. 2. The ability of a power supply to maintain a constant output voltage (or current) over a period of time under fixed conditions of input, load, and temperature. Usually expressed in terms of a voltage (or current) change over a fixed length of time. 3. The ability to maintain effectiveness within reasonable bounds in spite of large changes in environment. 4. For a feedback-control system or element, the property such that its output will ultimately attain a steady-state dc level within the linear range and without continuing external stimuli. 5. The overall ability of a resistor to maintain its initial resistance value over extended periods when subjected to any combination of environmental conditions and electrical stresses. 6. The ability of a motor-control system to operate at or near a constant speed over a wide load range without oscillations in speed (hunting). 7. A measure of deviations from a rated output over long periods. 8. Fluctuations in accuracy, or drift, over a given period.

stability factor — The measure of the bias stability of a transistor amplifier. It is defined as the change in collector current, I_C, per change in cutoff current, I_C.

stabilivolt — A gas-filled tube containing a number of concentric, coated iron electrodes. It is used as a source of practically constant voltage for apparatus drawing only small currents.

stabilization — 1. The introducing of stability into a circuit. 2. A process by means of which the output of electromechanical transducers is optimized by adjusting magnetic and mechanical parameters for its maximum magnetic permanency, to maintain its output stable under changing environmental and external conditions. 3. The reduction of variations in voltage or current not due to prescribed conditions. 4. The treatment of a permanent-magnet material designed to increase the permeance of its magnetic properties. This process may include such conditions as heat, shock, or demagnetizing fields so that the magnet will produce a constant magnetic field. Stabilization generally refers only to magnetic stability wherein if the disturbing influence were removed and the magnet remagnetized, any magnetic changes can be completely restored. Flux changes caused by internal structural changes are permanent in character and cannot be restored simply by remagnetization.

stabilization bake — A preconditioning performed on finished material to ensure reliability in commercial end use. In semiconductors this procedure accelerates any inherent metallurgical and/or chemical degradation within the device and takes care of over 95 percent of the failures due to infant mortality. In addition, it acts to stabilize such semiconductor parameters as current gain (beta), breakdown voltage, and diode leakage.

stabilization network — A network used to prevent oscillation in an amplifier with negative feedback.

stabilized feedback — *See* negative feedback.

stabilized flight — A type of flight in which control information is obtained from inertia-stabilized references such as gyroscopes.

stabilized local oscillator — An extremely stable radio-frequency oscillator used as a local oscillator in the superheterodyne radar receiver in a moving-target indicator system.

stabilized master oscillator — The master oscillator in complex microwave systems that acts as the frequency reference for all rf signals within the system. Usually crystal and temperature controlled for maximum frequency stability.

stabilized shunt-wound motor — A shunt-wound motor to which a light series winding has been added to prevent a rise in speed, or to reduce the speed when the load increases.

stabilized winding — Also called tertiary winding. An auxiliary winding used particularly in star-connected transformers to (*a*) stabilize the neutral point of the fundamental frequency voltage, (*b*) protect the transformer and the system from excessive third-harmonic voltages, or (*c*) prevent telephone interference caused by third-harmonic currents and voltages in the lines and earth.

stabistor — A voltage-limiting semiconductor. A diode designed to break over and conduct at a certain voltage. This is the normal forward conduction of a diode and is also characteristic of Zener diodes, which avalanche into conduction when breakdown (backward) voltage is exceeded.

stable element — In navigation, an instrument or device that maintains a desired orientation independently of the vehicle motion.

stable oscillation — A response that does not increase indefinitely with time; the opposite of an unstable oscillation.

stable platform — Also called a gyrostabilized platform. A gyro instrument that provides accurate azimuth, pitch, and roll attitude information. In addition to serving as reference elements, they are used for stabilizing accelerometers, star trackers, and similar devices in space.

stable strobe — A series of strobes that behaves as if caused by a single jammer.

stack — 1. That portion of a computer memory and/or registers used to temporarily hold information, usually the contents of the internal registers within a microprocessor chip. 2. *See* pileup, 1. 3. A block of successive memory locations that is accessible from one end on a last-in first-out basis (LIFO). The stack is coordinated with the stack pointer, which keeps track of storage and retrieval of each byte of information in the stack. A stack may be any block of successive information locations in the read/write memory. 4. A series of extra data registers, found especially in calculators using reverse Polish notation. The stack is used as a "first-in, last-out"

type of memory. Data is shifted into the stack by pressing an operational key, and is shifted down by pressing an operational key. The lowest registers and any subsequent registers on a nonrandomly accessible basis allow storage of intermediate results prior to their reuse with a later completing operation. Thus, access to parentheses (brackets) is automatic on pressing the ENTER key. A four-level stack (X, Y, Z, t) has the capability of three levels of parentheses (three sets of brackets). 5. A data area allocated to a process from which individual stack frames are allocated. 6. A region of memory that works by special rules. Each time the computer stores a word there, it goes on top of the stack and all the previously stored words move down one level. When a computer takes a word off the top of the stack, everything moves up one level, until the stack is empty. The computer has access only to the top of the stack. Piling a word on the stack is called a push, and taking a word off is called a pull or a pop. 7. A sequence of memory locations used in LIFO (last-in, first-out) fashion that stores, or stacks, computer words when the computer receives an interrupt request. A stack pointer specifies the last-in entry (or where the next-in entry will go). 8. A dynamic, sequential data list, usually contained in system memory, having special provisions for program access from one end or the other. Storage and retrieval of data from the stack is generally automatically performed by the processor. 9. A last-in, first-out structure that preserves the chronological ordering of information. Stacks are necessary for subroutines and interrupt management. A stack is manipulated by two basic instructions: push and pop.

stacked array — 1. An antenna system consisting of two or more antennas connected together and placed with respect to each other to increase the gain in a specific direction or directions. 2. A group of several identical microwave antennas placed one above the other to increase the gain of an antenna. They are connected in proper phase relationship, so that their signals are additive.

stacked-beam radar — A three-dimensional radar system in which elevation information is derived by

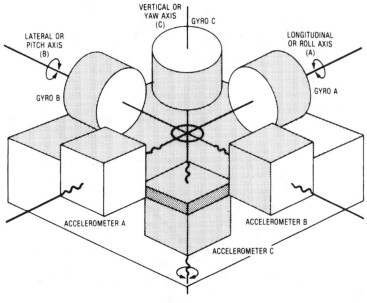

Stable platform.

emitting narrow beams placed one above the other to cover a vertical segment, azimuth information is obtained by horizontal scanning of the beam, and range information is obtained from the echo-return time.

stacked-diode laser — A type of laser used when a great amount of power is required. Avoiding the bulk of large numbers of optical lenses, this instrument offers high output intensity and a small emitting region at reasonable drive currents.

stacked dipole antenna — Antenna in which the antenna directivity is increased by providing a number of identical dipole elements, excited either directly or parasitically. The resultant radiation pattern will depend on the number of dipole elements used, the spacing and phase difference between the elements, and the relative magnitudes of the currents.

stacked heads — Also called inline heads. An arrangement of magnetic recording heads used for stereophonic sound. The two heads are directly in line, one above the other.

stack frame — A contiguous data area allocated for every activation of a routine; holds parameter values, local variables, temporary variables, and return-linkage information.

stack pointer — 1. A register that comes into use when the microprocessor must service an interrupt — a high-priority call from an external device for the central processing unit to suspend temporarily its current operations and divert its attention to the interrupting task. A CPU must store the contents of its registers before it can move on to the interrupt operation. It does this in a stack, so named because information is added to its top, with the information already there being pushed further down. The stack thus is a last-in, first-out type of memory. The stack-pointer register contains the address of the next unused location in the stack. 2. The counter or register used to address a stack in the memory.

stage — 1. A term usually applied to an amplifier to mean one step, especially if part of a multistep process; or the apparatus employed in such a step. 2. A hydraulic amplifier used in a servovalve. Servovalves may be single-stage, two-stage, three-stage, etc. 3. A single section of a multisection circuit or device.

stage-by-stage elimination — A method of locating trouble in electronic equipment by using a signal generator to introduce a test signal into each stage, one at a time, until the defective stage is found.

stage efficiency — Ratio of useful power (alternating current) delivered to the load to the power at the input (direct current).

stagger — Periodic positional error of the recorded spot along a recorded facsimile line.

staggered heads — An infrequently used arrangement of magnetic recording heads for stereophonic sound. The heads are $1\,7/32$ inch (30.95 mm) apart. Stereo tapes

recorded with staggered heads cannot be played on recorders using stacked heads, and vice versa.

staggered tuning — A means of producing a wide bandwidth in a multistage IF amplifier by tuning to different frequencies by a specified amount.

staggering — The offsetting of two channels of different carrier systems from exact sideband-frequency coincidence in order to avoid mutual interference.

staggering advantage — A reduction in intelligible crosstalk between identical channels of adjacent carrier systems as a result of using slightly different frequency allocations for the different systems.

stagger time — The interval between the times of actuation of any two contact sets.

stagger-tuned amplifier — An amplifier consisting of two or more stages, each tuned to a different frequency.

stagnation thermocouple — A type of thermocouple in which a high recovery factor is achieved by stagnating the flow in a space surrounding the junction. This results in a high response time as compared with an exposed junction.

staircase — A video test signal containing several steps at increasing luminance levels. The staircase signal is usually amplitude modulated by the subcarrier frequency and is useful for checking amplitude and phase linearities in video systems.

staircase generator — A special-purpose signal generator that produces an output that increases in steps; thus, its output waveform has the appearance of a staircase.

staircase signal — A waveform consisting of a series of discrete steps resembling a staircase.

stall torque — 1. The torque that the rotor of an energized motor produces when restrained from motion. 2. The torque developed by a servomotor at speed in excess of 1 rpm but less than 0.5 percent of the synchronous speed with a rated voltage and frequency of the proper phase relationship applied to both windings. 3. *See* holding torque.

stalled-torque control — A control function used to control the drive torque at zero speed.

stalo — Acronym for stabilized local oscillator. A highly stable oscillator, usually stabilized by feedback from a very high-Q LC circuit such as a high-Q cavity. Used as part of a moving-target indication device in conjunction with a radar.

stamped printed wiring — Wiring that is produced by die stamping and that is bonded to an insulating base.

stamper — A negative (generally made of metal by electroforming) from which finished records are molded.

stand-alone — 1. Pertaining to a device that requires no other piece of equipment along with it to complete its own operation or function. 2. A system or piece of equipment that is capable of doing its job without being

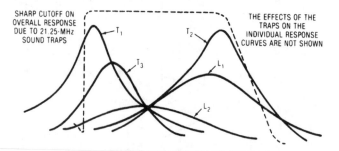

Staggered tuning.

connected to anything else. 3. Self-contained, not requiring any special add-ons or supports, e.g., a stand-alone word processor. 4. Word-processing equipment that contains all the necessary components within itself, allowing it to operate entirely by itself. This and shared logic are the two major types of word-processing equipment.

stand-alone system — A microcomputer software development system that runs on a microcomputer without connection to another computer or a time-sharing system. This system includes an assembler, editor, and debugging aids. It may include some of the features of a prototyping kit.

stand-alone terminal — *See* intelligent terminal.

standard — 1. An exact value, or a concept established by authority, custom, or agreement, that serves as a model or rule in the measurement of a quantity or in the establishment of a procedure. 2. A device used to maintain continuity of value in the units of measurement by periodic comparison with higher-echelon or national standards. 3. An agreement on a definition. By referencing standards, which may be either test methods or physical or electrical descriptions, a component or system having desired properties may be obtained. 4. A specification for data communication that is widely accepted and implemented by communications vendors. Standards may be formal (published by a recognized standards organization) or de facto (accepted without formal publication).

standard antenna — An open single-wire antenna, including the lead-in wire, having an effective height of four meters.

standard beam approach — Abbreviated SBA. A VHF, 40-MHz continuous-wave low-approach system using a localizer and markers. The two main-signal lobes are tone-modulated with the Morse-code letters E and T (● and —). These modulations form a continuous tone when the aircraft is on its course. The airborne equipment is usually instrumented for visual reference, but may be used aurally in some applications.

standard broadcast band — The band of frequencies extending from 530 to 1710 kilohertz.

standard broadcast channel — The band of frequencies occupied by the carrier and two sidebands of a broadcast signal. The carrier frequency is at the center, with the sidebands extending 5 kHz on either side.

standard broadcast station — A radio station operated on a frequency between 530 and 1710 kHz for the purpose of transmitting programs intended for reception by the general public.

standard candle — A unit of candlepower equal to a specified fraction of the visible light radiated by a group of 45 carbon-filament lamps preserved at the National Bureau of Standards, the lamps being operated at a specified voltage. The standard candle was originally the amount of light radiated by a tallow candle of specified composition and shape.

standard capacitor — A capacitor whose capacitance is not likely to vary. It is used chiefly in capacitance bridges.

standard cell — 1. A primary cell that serves as a standard of voltage. 2. Predefined circuit elements that may be selected and arranged to create a custom or semicustom integrated circuit more easily than through original (custom) design. Cell libraries provide the building blocks from which designers create ASICs (application-specific integrated circuits).

standard component — A component that is regularly produced by some manufacturer and is carried in stock by one or more distributors.

standard deflection — 1. In a galvanometer having an attached scale, one scale division. 2. In a galvanometer

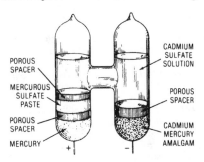

Standard cell.

without an attached scale, 1 millimeter when the scale distance is 1 meter.

standard deviation — 1. A measure of the variation of data from the average. It is equal to the root mean square of the individual deviations from the average. 2. A measure of dispersion of data or of a variable. The standard deviation is computed by finding the difference between the average and actual observations, squaring each difference, summing the squared differences, finding the average squared difference (called the variance), and then taking the square root of the variance. 3. A statistical measure used most often when analyzing experimental data. The standard deviation of a set of data is the measure of the dispersion of data values about the mean.

standard electrode potential — An equilibrium electrode potential for an electrode in contact with an electrolyte in which all of the components of a specified chemical reaction are in their standard states. The standard state for an ionic constituent is unit ion activity.

standard error — Applied to statistics such as the mean to provide a distribution within which samples of the statistics are expected to fall.

standard eye — An observer that has red and infrared luminosity functions.

standard facility — In programming actions, a basic communications electronics functional entity that is engineered to satisfy a specific communications electronics operational requirement. An associated standard facility equipment list describes the facility functionally and indicates the material required for the standard facility.

standard-frequency service — A radiocommunication service that transmits for general reception specified standard frequencies of known high accuracy.

standard-frequency signal — One of the highly accurate signals broadcast by the National Bureau of Standards radio station WWV on 2.5, 5, 10, 15, 20, and 25 MHz.

standard-gain horn — A waveguide device that has essentially flared out its waveguide dimensions to specific lengths that match (with certain gain) the incoming energy to the atmosphere. The applications include using these horns for reflectors and lenses, pickup horns for sampling power, and receiving and transmitting antennas.

standardization — The process of establishing by common agreement engineering criteria, terms, principles, practices, materials, items, processes, equipment, parts, subassemblies, and assemblies to achieve the greatest practicable uniformity of items of supply and engineering practices, to ensure the minimum feasible variety of such items and practices, and to effect optimum interchangeability of equipment parts and components.

standard luminosity curve — An empirically derived function that describes the response of the eye to radiation of different wavelengths. (The terms luminous and illumination indicate that this function is taken into account.)

standard microphone — A microphone whose response is known for the condition under which it is to be used.

standard noise temperature — A standard reference temperature (T) for noise measurements, taken as 290 K.

standard observer — A hypothetical observer who requires standard amounts of primaries in a color mixture to match every color.

standard pitch — The tone A at 440 hertz.

standard play — An arbitrary description given to identify a spool of "thick" recording tape with a specified playing time according to reel size. For example, a 7-inch (17.8-cm) spool will hold 1200 ft (366 m) of standard-play tape; a 5-inch (12.7-cm) spool will hold 600 ft (182 m). The actual playing time is then dependent on tape speed.

standard propagation — Propagation of radio waves over a smooth, spherical earth of uniform dielectric constant and conductivity, under standard atmospheric refraction.

standard reference temperature — In a thermistor, the body temperature for which the nominal zero-power resistance is specified.

standard refraction — The refraction that would occur in an idealized atmosphere; the index of refraction decreases uniformly with height at the rate of 39×10^{-6} per Kilometer.

standard register of a motor meter — Also called a dial register. A four- or five-dial register, each dial being divided into ten equal parts numbered from 0 to 9. The dial pointers are geared so that adjacent ones move in opposite directions at a 10-to-1 ratio.

standard resistor — Also called a resistance standard. A resistor that is adjusted to a specified value, is only slightly affected by variations in temperature, and is substantially constant over long periods of time.

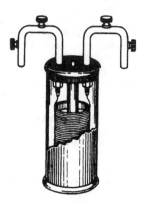

Standard resistor.

standard rod gap — A gap between the ends of two half-inch square rods. Each rod is cut off squarely and mounted on supports so that it overhangs the inner edge of each support by a length equal to or greater than half the gap spacing. It is used for approximate measurements of crest voltages.

standard seawater conditions — Seawater with a static pressure of 1 atmosphere, a temperature of 15°C, and a salinity such that the velocity of sound propagation is exactly 1500 meters per second.

standard source — In fiber optics, the reference optical power source to which emitting and detecting devices are compared for calibration purposes.

standard sphere gap — A gap between two metal spheres of standard dimensions. It is used for measuring the crest value of a voltage by observing the maximum gap spacing at which sparkover occurs when a voltage is applied under known atmospheric conditions.

standard subroutine — In a computer, a subroutine that is applicable to a class of problems.

standard television signal — A signal that conforms to accepted specification.

standard test conditions — The environmental conditions under which measurements should be made when disagreement of data obtained by various observers at different times and places may result from making measurements under other conditions

standard test-tone power — One milliwatt (0 dBm) at 1000 hertz.

standard voltage generator — *See* signal generator, 1.

standard volume indicator — A volume indicator with the characteristics prescribed by the American Standards Association.

standby — 1. The condition of equipment that will permit complete resumption of stable operation within a short period of time. 2. A duplicate set of equipment to be used if the primary unit becomes unusable because of malfunction.

standby battery — A storage battery held in reserve to serve as an emergency power source in event the regular power facilities at a radio station, hospital, etc., fail.

standby power generator — An alternating-current generator, held in reserve and used to supply the necessary ac power requirements when there is failure of commercially supplied ac power.

standby power supply — Equipment that supplies power to a system in the event the primary power is lost. It may consist of batteries, charging circuits, auxiliary motor generators, or a combination of these devices.

standby redundancy — That redundancy wherein the alternative means of performing the function is inoperative until needed and is switched upon failure of the primary means of performing the function.

standby register — A register in which accepted or verified information can be stored to be available for a rerun in the event of a mistake in the program or a malfunction in the computer.

standby transmitter — A transmitter installed and maintained ready for use whenever the main transmitter is out of service.

standing current — The current present in a circuit in the absence of signal.

standing-on-nines carry — A system of executing the carry process in a computer. If a carry into a given place produces a carry from there, the incoming carry information is routed around that place.

standing wave — Also called stationary wave. 1. The distribution of current and voltage on a transmission line formed by two sets of waves traveling in opposite directions, and characterized by the presence of a number of points of successive maxima and minima in the distribution curves. At points where the voltages of the two waves add, there will be a voltage antinode; at points where the voltages subtract, there will be a voltage node. The nodes and antinodes do not move, and the resultant wave is called a standing wave. 2. The combination of two waves having the same frequency and amplitude and traveling in opposite directions. Standing waves are indicated by a stationary set of nodes spaced one-half wavelength apart along the propagation direction of the waves. The modes of optical resonators are standing waves produced as the radiation travels back and forth between highly reflecting walls.

standing-wave detector — *See* standing-wave meter.

standing-wave indicator — *See* standing-wave meter.

standing-wave loss factor — Ratio of the transmission loss in a waveguide when it is unmatched compared with the loss when it is matched.

standing-wave meter — Also called a standing-wave indicator or detector. An instrument for measuring the standing-wave ratio in a transmission line. It may also include means for finding the location of maximum and minimum amplitudes.

standing-wave ratio — Abbreviated SWR. The ratio of current (or voltage) at a loop (maximum) in the transmission line to the value at a (minimum) node. It is equal to the ratio of the characteristic impedance of the line to the impedance of the load connected to the output end of the line.

standing-wave-ratio bridge — Abbreviated SWR bridge. A bridge for measuring the standing-wave ratio on a transmission line to check the impedance match.

standing waves — The behavior of air pressure waves in an enclosed room or box, giving rise to resonances that occur. They are created by the effects of multiple sound reflections between opposite walls, and cycle at frequencies determined by the distance between them. In effect, the room acts as a resonator.

standoff insulator — 1. An insulator used to hold a wire or other radio component away from the structure on which it is mounted. 2. An insulator used to support a high-voltage lead, such as an antenna feeder, at some distance from the mounting surface.

star chain — A group of navigational radio transmitting stations comprising a master station about which three or more slave stations are symmetrically located.

star-connected circuit — A polyphase circuit in which all current paths within the region that limits the circuit extend from each of the points of entry of the phase conductors to a common conductor (which may be the neutral conductor).

star connection — *See* wye connection.

Stark effect — The splitting or shifting of spectral lines or energy levels due to an applied electric field.

starlight — A scene illumination of 1/10,000 of a footcandle.

star network — A set of three or more branches with one terminal of each connected at a common node.

star-quad cable — Four wires laid together and twisted as a group.

start bit — In asynchronous transmission, the first bit used to indicate the beginning of a character; normally, a space condition that serves to prepare the receiving equipment for the reception and registration of the character.

start dialing signal — A signal transmitted from the incoming end of a circuit, after receipt of a seizing signal, to indicate that the circuit conditions necessary for receiving the numerical routine information have been established.

start element — In certain serial transmissions, the initial element of a character, used for the purposes of synchronization. In Baudot teletypewriter operation, the start element is one space bit.

starter — 1. An auxiliary electrode used to initiate conduction in a glow-discharge cold-cathode tube. 2. Sometimes referred to as a trigger electrode. A control electrode whose principal function is to establish sufficient ionization to reduce the anode breakdown voltage

in a gas tube. 3. An electric controller for accelerating a motor from rest to normal speed.

starter breakdown voltage — The voltage required to initiate conduction across the starter gap of a glow-discharge cold-cathode tube, all other tube elements being held at cathode potential before breakdown.

starter gap — The conduction path between a starter and the other electrode to which the starting voltage is applied in a glow-discharge cold-cathode tube.

starter voltage drop — The voltage drop across the starter gap after conduction is established there in a glow-discharge cold-cathode tube.

starting anode — The anode that establishes the initial arc in a mercury-arc rectifier tube.

starting circuit breaker — A device whose principal function is to connect a machine to its source of starting voltage.

starting current of an oscillator — The value of oscillator current at which self-sustaining oscillations will start under specified loading.

starting electrode — The electrode that establishes the cathode spot in a pool-cathode tube.

starting reactor — A reactor for decreasing the starting current of a machine or device.

starting torque — Also called pull-in torque. 1. The maximum load torque with which motors can start and come to synchronous speed. 2. The torque necessary to initiate motion of a system.

starting voltage — The voltage necessary for a gaseous voltage regulator to become ionized or to start conducting. As soon as this happens, the voltage drops to the operating value.

start lead — Also called inside lead. The inner termination of a winding.

startover — A program function that causes an inactive computer to become active.

start-record signal — In facsimile transmission, the signal that starts the converting of the electrical signal to an image on the record sheet.

start signal — The signal that converts facsimile-transmission equipment from standby to active.

start-stop multivibrator — *See* monostable multivibrator; flip-flop multivibrator.

start-stop system — A system in which each group of code elements that represents an alphabetical signal is preceded by a start signal and followed by a stop signal. The start signal prepares the receiving mechanism to receive and register the character. The stop signal causes the receiving mechanism to come to rest in preparation for reception of the next character.

start-stop transmission — The method of transmission used in a start-stop system.

start-up — The time between equipment installation and the full operation of the system.

start-without-error rate — Stepping rate at which a stepper motor with no external load inertia can start and stop without losing a step.

start-without error torque — Also called pull-in torque. Force moment produced by a stepper motor when started at a fixed frequency or step rate.

starved amplifier — An amplifier employing pentode tubes in which the screen voltage is set 10 percent below the plate voltage and the plate-load resistance is increased to 10 times the normal value. Thus, the amplification factor is greatly increased — often a stage gain of 2000 is achieved.

stat- — A prefix used to identify electrostatic units in the cgs system. *See also* statampere; statcoulomb; statfarad; stathenry; statmho; statohm; statvolt.

statampere — The cgs electrostatic unit of electric current, equal to 3.3356×10^{-10} ampere (absolute).

statcoulomb — The cgs electrostatic unit of charge, equal to 3.3356×10^{-10} coulomb (absolute).

state — 1. The condition of a circuit, system, etc. 2. The condition at the output of a circuit that represents logical 0 or logical 1. 3. A condition or set of conditions considered together, especially one of the two normal sets of operating conditions of a gate or flip-flop. 4. The logical 0 or 1 condition in PC memory or at a circuit's input or output.

state code — A coded indication of what state the CPU is in — responding to an interrupt, servicing a DMA request, executing an I/O instruction, etc.

state diagram — A representation of the stable states of a process, and the vectored paths by which the process passes to and from these states.

state machine — A sequential-logic system whose outputs depend on previous and present inputs (for example, a counter), as opposed to processes that are functions of present inputs alone.

statement — In computer programming, a meaningful expression or generalized instruction written in a source language.

state of charge — The condition of a storage cell or battery in terms of the remaining capacity.

state table — Also called state transition table. A list of the outputs of a logic circuit based on the inputs and previous outputs. Such a circuit has memory and cannot be described by a simple truth table.

state transition table — See state table.

statfarad — The cgs electrostatic unit of capacitance, equal to 1.11263×10^{-12} farad (absolute).

stathenry — The cgs electrostatic unit of inductance, equal to 8.98766×10^{-11} henrys (absolute).

static — 1. See atmospherics. 2. A form of information storage in shift registers and memories whereby information will be retained as long as power is applied. 3. Capable of maintaining the same state indefinitely (with power applied) without any change of condition. Not requiring a continuous refreshing. 4. A state in which a quantity does not change appreciably within an arbitrarily long time interval. 5. In burn-in, the quality of a test wherein the device is subject to either forward or reverse bias applied to appropriate terminals; voltages are unvarying throughout test.

statically balanced arm — A type of tonearm whose masses are first balanced about the pivot, then deliberately unbalanced by approximately 1 gram in order to provide the required tracking force.

static analysis — Examination of a program (usually via computer) for errors and inconsistencies, without actual execution.

static behavior — The behavior of a control system or an individual unit under fixed conditions (as contrasted to dynamic behavior, under changing conditions).

static breakdown voltage — The voltage at which a transient suppressor begins to conduct when subjected to slow-rising dc. Does not account for transient rise-time rates.

static burn-in — High-temperature test with device subjected to unvarying voltage rather than to operating conditions; either forward or reverse bias.

static cell — A memory cell; basically a cross-coupled flip-flop. Some power is consumed at all times through the two loads. Some newer designs reduce the power required to maintain the state of the flip-flop by using high-resistance polysilicon load resistors.

static characteristic — The relationship between a pair of variables such as electrode voltage and electrode current, all other voltages being maintained constant. This relationship is usually represented by a graph.

static charge — 1. The accumulated electric charge on an object. 2. An electric charge held on the surface of an object.

static check — Of a computer, consists of one or more tests of computing elements, their interconnections, or both, performed under static conditions.

static control — A control system in which control functions are performed by solid-state devices.

static convergence — Convergence of the three electron beams at the center of the aperture mask in a color picture tube. The term static applies to the theoretical paths the beams would follow if no scanning forces were present.

static decay — In a storage tube, decay that is a function only of storage surface properties, such as lateral or transverse leakage.

static detector — A device used to detect presence of static charges of electricity, which could cause explosions in hazardous atmospheres.

static device — As associated with electronic and other control or information-handling circuits, the term static refers to devices with switching functions that have no moving parts.

static dump — In a computer, a dump performed at a particular time with respect to a machine run, often at the end of the run.

static electricity — Stationary electricity — i.e., in the form of a charge in equilibrium, or considered independently of the effects of its motion.

static electrode potential — The electrode potential measured when there is no net current between the electrode and electrolyte.

static eliminator — 1. A device for reducing atmospheric static interference in a radio receiver. 2. One of a broad range of devices that neutralize nonconductive materials by producing a region of ionized air through which the charged material can pass. Induction static bars consist of a row of grounded metallic points or tufts placed as close as possible to the moving material without touching it. Radioactive static bars employ a coating of radioactive substance facing toward the material to be discharged at a distance of about 1 inch. Electrical static bars have a series of points maintained at high voltage, most frequently capacitively coupled to the voltage source so that they are shock-free even when touched. A small power unit energizes the electrical bars, which are about 1 inch (2.54 cm) from the moving material. Ionizing air guns and blowers also neutralize static. See antistatic sprays.

static error — An error that does not depend on the time-varying nature of a variable.

static field — A field that is present between the poles of either a permanent magnet or an electromagnet that has a direct current passing through its coils.

static focus — The focus attained when the electron beam is theoretically at rest or is at the position it would occupy if scanning energy were not applied.

static forward-current transfer ratio — In a transistor, the ratio, under specified test conditions, of the dc output current to the dc input current.

static input resistance — In a transistor, the ratio of the dc input voltage to the dc input current.

staticize — 1. In a computer, to perform a conversion of serial or time-dependent parallel data into a static form. 2. Occasionally, to retrieve an instruction and its operands from storage prior to executing the instruction.

staticizer — A storage device that is able to take information sequentially in time and put it out in parallel.

static line regulation—The output voltage variation of a power supply as the line voltage is varied slowly from rated minimum to rated maximum, with the load current held at the nominal value.

static load regulation—The output voltage variation of a power supply as load current is varied slowly from 0 to 100 percent of rating, with the input line voltage held at the nominal value.

static machine—A machine for generating an electric charge, usually by induction.

static measurement—A measurement taken under conditions in which neither the stimulus nor the environmental conditions fluctuate.

static memory—1. A type of semiconductor memory in which the basic storage element can be set to either of two states, in which it will remain so long as the power stays on. *See also* dynamic memory. 2. A type of semiconductor read/write random-access memory that does not require periodic refresh cycles. 3. An MOS memory that uses a flip-flop as a storage element. It does not need to be refreshed, does not require a clock, and does not lose its contents as long as power is applied.

static MOS array—A circuit made up of MOS devices that does not require a clock signal.

static noise—Noise resulting from lightning, aurora, or other atmospheric discharges.

static power conversion equipment—Any equipment that converts electrical power from one form to another without the use of moving parts such as rotors or vibrators. *Static* implies the use of semiconductors.

static pressure—Also called hydrostatic pressure. The pressure that would exist at a certain point in a medium with no sound waves present. In acoustics, the commonly used unit is the microbar.

static printout—In a computer, a printout of data that is not one of the sequential operations and occurs after conclusion of the machine run.

static RAM—Memory that does not require refreshing because of the cell circuit design although it is still volatile, losing data on removal of power. SRAM is more expensive than DRAM, but is faster and requires less control circuitry.

static register—A computer register that retains its information in static form.

static regulator—A transmission regulator in which the adjusting mechanism is in self-equilibrium at any setting, and control power must be applied to change the setting.

static sensitivity—In phototubes, the direct anode current divided by the incident radiant flux of constant value.

static shift register—A shift register in which logic flip-flops are used for storage. This technique, in integrated form, results in greater storage-cell size and consequently in shorter shift-register lengths. Its primary advantage is that information is retained as long as power is supplied to the device. A minimum clock rate is not required, and, in fact, the device can be unclocked.

static skew—1. A measure of the distance that the output from one track is ahead or behind (i.e., leading or lagging) the output of another track as a tape is transported over the read head. 2. The long-term or average misalignment of a drive's read head as referenced to a master skew tape. Results from gap scatter and misadjustment.

static storage—In computers, storage in which the information does not change position (e.g., electrostatic storage, flip-flop storage, binary magnetic-core storage, etc.). The opposite of dynamic storage.

static subroutine—A digital-computer subroutine involving no parameters other than the address of the operands.

static switch—A semiconductor switching device in which there are no moving parts.

static torque—*See* locked-rotor torque.

static transconductance—In a transistor, the ratio of the dc output current to the dc input voltage.

station—1. One or more transmitters, receivers, and accessory equipment required to carry on a definite radiocommunication service. The station assumes the classification of the service in which it operates. 2. An input or output point in a communications system, such as the telephone set in a telephone system or the point at which a business machine interfaces the channel on a leased private line.

stationarity—The absence of variations with time in the special intensity and amplitude distribution of random noise.

stationary appliance—An appliance that is not easily moved from one place to another in normal use.

stationary battery—A storage battery designed for service in a permanent location.

stationary contacts—Those members of contact pairs that are not moved directly by the actuating system.

stationary field—A constant field—i.e., one in which the scalar (or vector) at any point does not change during the time interval under consideration.

stationary wave—*See* standing wave.

station authentication—Security measure designed to establish the authenticity of a transmitting or receiving station.

station battery—The electrical power source for signaling in telegraphy.

station break—1. A cue given by the station originating a program, to notify network stations that they may identify themselves to their audiences, broadcast local items, etc. 2. The actual time taken in (1) above.

station license—The instrument of authorization for a radio station in the amateur radio service.

station-to-station call—A telephone call in which the calling party does not specify that he or she wishes to reach a particular person at the called point.

statistical multiplexer—*See* ITDM.

statistical multiplexing—Multiplexing by providing bandwidth on the multiplexed data line only for those channels that have data available for transmission. No bandwidth is wasted on terminals that are not sending data.

statmho—The cgs electrostatic unit of conductance, equal to 1.11263×10^{-12} mho (absolute).

statohm—The cgs electrostatic unit of resistance, equal to 8.98766×10^{-11} ohms (absolute).

stator—1. The nonrotating part of the magnetic structure in an induction motor. It usually contains the primary winding. 2. The stationary plates of a variable capacitor. 3. The conducting surfaces of a switch. Similar to the commutator in electrical rotating mechanisms.

stator of an induction watthour meter—A voltage circuit, one or more current circuits, and a magnetic circuit combined so that the reaction with currents induced in an individual, or a common, conducting disc exerts a driving torque on the rotor.

stator plates—The fixed plates of a variable capacitor.

status—The present condition of a device. Usually indicated by flag flip-flops or special registers. *See* flag.

status line—A simple method of representing some state of a device in an interconnection scheme.

status register — Register used in a computer to hold status information inside a functional unit, such as an MPU, a PIC, a DMAC, or an FDC. A typical MPU's status register provides carry, overflow, sign ("negative"), zero, and interrupt. It could also include parity, enable (interrupts), or mask.

status word register — A group of binary numbers that informs the user of the present condition of the microprocessor.

statvolt — The cgs electrostatic unit of voltage, equal to 299.796 volts (absolute).

stave — One of the number of individual longitudinal elements that comprise a sonar transducer.

stay cord — A component of a cable, usually a high-tensile textile, used to anchor the cable ends at their points of termination to keep any pull on the cable from being transferred to the electrical connection.

steady state — 1. A condition in which circuit values remain essentially constant, occurring after all initial transients or fluctuating conditions have settled down. 2. A term used to specify the current through a load or electric circuit after the inrush current is complete. A stable run condition. 3. A characteristic of a condition exhibiting only minor change over an arbitrarily long period.

steady-state deviation — The difference between the final value assumed by a specified variable after the expiration of transients and its ideal value.

steady-state oscillation — Also called steady-state vibration. Oscillation in which the motion at each point is a periodic quantity.

steady-state regulation — Slow changes in the output voltage of a power supply following an input-voltage and/or load-impedance variation; it is usually expressed in volts (ΔV) or as a percentage of the nominal output.

steady-state value — The value of a current or voltage after all transients have decayed to negligible values.

steady-state vibration — *See* steady-state oscillation.

steatite — A ceramic consisting chiefly of a silicate of magnesium. Because of its excellent insulating properties — even at high frequencies — it is used extensively in insulators and as a circuit substrate.

steer — To adjust by electrical means the polar response pattern of an antenna or the direction of current in a circuit.

steerable antenna — 1. An antenna whose major lobe can be readily shifted in direction. 2. A multielement antenna in which the phase relationship between the elements is electronically adjustable. Thus, the antenna beam can be steered in direction and adjusted for beam width.

Stefan-Boltzmann constant — 1.38×10^{-23} J·K^{-1}. The constant of proportionality between blackbody radiated power and temperature.

Stefan-Boltzmann law — The total emitted radiant energy per unit of a blackbody is proportionate to the fourth power of its absolute temperature.

Steinmetz coefficent — A factor by which the 1.6 power of the magnetic flux density must be multiplied to give the approximate hysteresis loss of an iron or steel sample in ergs per cubic centimeter per cycle when that sample is undergoing successive magnetization cycles having the same maximum flux density.

stenode circuit — A superheterodyne receiving circuit in which a piezoelectric unit is used in the intermediate-frequency amplifier to balance out all frequencies except signals at the crystal frequency, thereby giving very high selectivity.

step — To use the step-and-repeat method.

step and repeat — 1. A method of dimensionally positioning multiples of the same or intermixed functional patterns on a given area of a photoplate or a film by repetitions, or contact or projection printing of a single original pattern of each type. 2. A method of positioning multiples of the same pattern on a mask or wafer. 3. A process wherein the conductor or resistor pattern is repeated many times in evenly spaced rows onto a single film or substrate.

step-and-repeat camera — A type of camera that has scales or other arrangements by which successive exposures can be lined up and equally spaced on a sheet of film. It is used in manufacturing microcircuits.

step-and-repeat fix technique — A mechanical technique that provides for linear indexing of a movable platform carrying a wafer or photographic plate. Applications of this technique are in the testing of a device wafer, and in the masking operation that is part of the process of fabricating microelectronic devices.

step-and-repeat printer — A projection printer that is capable of reproducing a multiplicity of images from a master transparency on a single support coated with a photosensitive layer by indexing the receiving material from position to position. Such accessories or attachments find use in microcircuit production.

step angle — Also called increment. The nominal angle through which the rotor shaft of a stepper motor moves between adjacent step positions. The angle is specified in degrees. For a 200-step per revolution motor, the step angle is 1.8°.

step-by-step automatic telephone system — A switching system that employs successive step-by-step selector switches that are actuated by current impulses from a telephone dial.

step-by-step switch — A bank-and-wiper switch in which the wipers are moved by individual electromagnetic ratchets.

step calibration — Also called interval calibration. Often confused with sense step, that is, the application of a calibration resistor to produce a deliberate electrical unbalance.

step counter — In a computer, a counter used in the arithmetical unit to count the steps in multiplication, division, and shift operations.

step down — To decrease the value of some electrical quantity, such as a voltage.

step-down transformer — A transformer in which the voltage is reduced as the energy is transferred from its primary to its secondary winding.

step fiber — A glass optical fiber that has a core of uniform optical density. That is, there is a step function change between the optical density of the cladding and the optical density of the core. When attenuation of signal was the primary practical problem, step fiber was acceptable, and still is acceptable, for limited information rate application.

step function — 1. A signal characterized by instantaneous changes between amplitude levels. The term usually refers to a rectangular-front waveform used for making tests of transient response. 2. An essentially nonperiodic waveform that has a transistion from one voltage level to another, the time of which is negligible compared with the total duration of the waveform.

step-function response — *See* transient response, 2.

step generator — A device for testing the linearity of an amplifier. A step wave is applied to the amplifier input and the step waveform observed, on an oscilloscope, at the output.

step input — A sudden but sustained change in an input signal.

step load change — An instantaneous change in the magnitude of the load current.

stepped index fiber — 1. A fiber composed of a core glass of one index of refraction and a different cladding material of a lower index of refraction. The fiber is characterized by a sharp change in index of refraction at the interface of the two materials. 2. An optical fiber that confines most photons to the fiber's core because they ricochet off the surface, back toward the center. Since the photons skip from side to side at different angles while other photons are traveling more or less unimpeded straight down the center of the fiber, the photons constituting a piece of a message gradually become more and more scattered and eventually deteriorate into meaningless noise unless picked up, amplified, and retransmitted by repeaters every few hundred yards or so.

stepped oxide — A technique of forming the SiO_2 layer of each electrode in two thicknesses so that a two-level potential well can be formed with one voltage. (MIS technology term.)

stepper — Photo equipment used to transfer a reticle pattern onto a wafer. Because of its limited field of view, low throughput, and high cost, such equipment is usually used only for feature size smaller than 1.5 microns, where resolution and line-width control are critical.

stepper motor — 1. A motor whose normal operation consists of discrete angular motions of essentially uniform magnitude, rather than continuous rotation. 2. A digital device that converts electrical pulses into proportionate mechanical movement. Each revolution of the motor shaft is made in a series of discrete identical steps. The design of the motor usually provides for clockwise and/or counterclockwise rotation. (Thus, the stepper is ideally suited for many positional and control applications.) 3. Electromagnetical prime mover that rotates through fixed angles in response to applied pulses. The motor accordingly permits use of digital signals to control mechanical motion or position. In addition, the high holding torque associated with each step permits a stepping motor to replace devices such as brakes and clutches, with a gain in system reliability. 4. A device that converts pulsating direct current into rotary mechanical motion. Each dc pulse rotates the stepper a certain fraction of one revolution. The rotor is magnetically held at its last position. 5. A bidirectional permanent-magnet motor that turns through one angular increment for each pulse applied to it. 6. An electric motor that moves incrementally. 7. A device that translates electrical pulses into precise mechanical movement. The output shaft may deliver rotary or linear motion.

steppers — Specially designed electric motors that revolve from pole to pole under control of sequentially energized field windings. Two principle designs are currently available: variable reluctance and permanent magnet.

stepping — See zoning, 1.

stepping rate — A measure of stepper motor speed — the number of steps through which the motor rotates in a specified time, usually 1 second.

stepping relay — Also called rotary stepping switch (or relay) or stepping switch. 1. A multiposition relay in which moving wiper contacts mate with successive sets of fixed contacts in a series of steps, moving from one step to the next in successive operations of the relay. 2. A switch that electromechanically steps its wipers across a bank of contacts.

stepping switch — See stepping relay.

step-recovery diode — Also called snap varactor. 1. A varactor in which forward voltage injects carriers

Stepping relay.

across the junction, but before the carriers can combine, the voltage reverses and carriers return to their origin in a group. The result is the abrupt cessation of reverse current and a harmonic-rich waveform. 2. A special form of pn junction in which the charge storage and switching characteristics are optimized for use in microwave-frequency multipliers and comb generators. The device stores charge while in forward conduction, and a large reverse bias current can be obtained until all the charge is removed. The device impedance then goes from a low value to a very high one in transition times as low as 50 picoseconds.

step response — Motor-shaft rotational response of a stepper motor to a step command related to time.

step-servo motor — A device that, when properly energized by dc voltage, indexes in definite angular increments.

steps per revolution — The total number of steps required for the output shaft of a dc stepping motor to rotate 360°, or one complete revolution. Steps per revolution is calculated by dividing the step angle into 360°.

steps per second — The number of angular movements accomplished by the motor of a dc stepping motor in 1 second. This figure replaces the rpm figure of a standard drive motor.

step stress test — A test consisting of several stress levels applied sequentially for periods of equal duration to a sample; during each period a stated stress level is applied, which is increased from one step to the next.

step-strobe marker — A form of strobe marker in which the discontinuity is in the form of a step in the time base.

step up — To increase the value of an electrical quantity, such as a voltage.

step-up transformer — A transformer in which the voltage is increased as the energy is transferred from the primary to the secondary winding.

step voltage — The potential difference between two points on the earth's surface separated by a distance of one pace, or about 3 feet or 91.4 cm, in the direction of maximum potential gradient.

step-voltage regulator — A device consisting of a regulating transformer and a means for adjusting the voltage or the phase relation of the system circuit in steps, usually without interrupting the load.

steradian — 1. The unit of solid angular measure, being the subtended surface area of a sphere divided by the square of the sphere's radius. There are 4π steradians in a sphere. The steradian is the unit of solid angular measurement often used in problems of illumination. 2. The solid angle described by a unit area on a sphere from a point source at unity distance and located at the center of the sphere, e.g., 1 sq ft illuminated at a distance of 1 ft.

sterance — Describes the intensity per unit area of a source.

Sterba antenna — A series-fed array of adjacent, broadside-firing, transposed square loops that have half-wave sides and are spaced a distance of approximately one-half wavelength.

Sterba curtain — A stacked dipole antenna array that consists of one or more phased half-wave sections and a quarter-wave section at each end. The array can be oriented for either vertical or horizontal radiation and can be fed at either the center or the end.

stereo- — A prefix denoting having or dealing with three dimensions.

stereo — See stereophonic.

stereo adapter — Also called a stereo control unit. A device used with two sets of monophonic equipment to make them act as a single stereo system.

stereo amplifier — An audio-frequency amplifier with two or more channels, for a stereo sound system.

stereo broadcasting — See stereocasting.

stereo cartridge — A phonograph pickup for reproduction of stereophonic recordings. Its high-compliance needle is coupled to two independent voltage-producing elements.

stereocasting — Also called stereo broadcasting. Broadcasting over two sound channels to provide stereo reproduction. This may be done by simulcasting, multicasting, or multiplexing.

stereocephaloid microphone — Two or more microphones arranged to simulate the acoustical patterns of human hearing.

stereo control unit — See stereo adapter.

stereo microphones — Two or more microphones spaced as required for stereo recording.

stereophonic — 1. Designating a sound reproduction system in which sound is delivered to the listener through at least two channels, creating the illusion of depth and of locality of source. 2. A two-channel recording and reproduction system more popularly referred to as *stereo*. At the recording studio, separate microphones are used for each recorded channel. The correct reproduction of stereo signals in the home gives to the listener a sense of direction of sound and, thus, realism. 3. A multiple-channel sound system or recording in which each channel carries a unique version of the total original performance. When the channels are blended acoustically, they recreate the breadth and depth of the original, adding a new dimension to reproduced sound. At least two channels are required for playback, although more than two may be used in recording. 4. Using two or more channels to create a spatial effect.

stereophonic reception — Reception involving the use of two receivers having a phase difference in their reproduced sounds. The sense of depth given to the received program is analogous to the listener's being in the same room as the orchestra or other medium.

stereophonic separation — The ratio of the electrical signal caused in the right (or left) stereophonic channel to the electrical signal caused in the left (or right) stereophonic channel by the transmission of only a right (or left) signal.

stereophonic sound system — A sound system with two or more microphones, transmission channels, and speakers arranged to give depth to the reproduced sound.

stereophonic subcarrier — A subcarrier employed in FM stereophonic broadcasting that has a frequency that is the second harmonic of the pilot subcarrier frequency.

stereophonic subchannel — The band of frequencies from 23 to 53 kilohertz containing the stereophonic subcarrier and its associated sidebands.

stereophonic system — A sound-reproducing system in which a plurality of microphones, transmission channels, and speakers (or earphones) are arranged to afford a listener a sense of the spatial distribution of the sound sources.

stereo pickup — A phonograph pickup used with stereo records, wherein a single stylus actuates two transducer elements, one of which reproduces the left channel and the other the right channel.

stereo recording — The impressing of signals from two channels onto a tape or disc in such a way that the channels are heard separately on playback. The result is a directional, three-dimensional effect.

stereoscope — A small instrument containing a picture support and a pair of magnifying lenses, so arranged that the left eye sees only the picture taken with the left lens of the stereo camera, and the right eye only the right picture. If the camera lenses are separated by the interocular distance, and if the focal lengths of camera and stereoscope lenses are equal, then a true-to-scale, or orthoscopic, reproduction of the scene will be obtained. Stereoscopes are used extensively in photogrammetry for plotting the contour lines in a terrain recorded by aerial photography.

stereoscopic — Imagery giving the illusion of being three-dimensional.

stereoscopic television — A system of television broadcasting in which the images appear to be three-dimensional.

stereo separation — A stereo receiver's ability to keep the audio information in each channel separate from the other.

stereosonic system — A recording technique using two closely spaced directional microphones with their maximum directions of reception 45° from each other. In this way, one picks up sound largely from the right and the other from the left, similar to middle-side recording.

stick — CB radio term for any nondirectional antenna.

stick circuit — A circuit used to maintain energization of a relay or similar unit through its own contacts.

stickiness — The condition of physical interference with the operation of the moving part of an electrical indicating instrument.

sticking — In computers, the tendency of a flip-flop to remain in, or to spontaneously switch to, one of its two stable states.

stiff — A voltage source whose value is largely independent of the current drawn having a relatively low impedance.

stiffness factor — The angular lag between the input and output of a servosystem.

stilb — Abbreviated sb. The unit of luminance (photometric brightness) equal to 1 candela per square centimeter.

still — Photographic or other stationary illustrative material used in a television broadcast.

stimularity — An arbitrary measure of sensitivity to stimulation. It is proportional to the quantum efficiency relative to incident radiation.

stimulated emission — 1. The emission of radiation by a system going from an excited electron energy level to a lower energy level under the influence of a radiation field. The emitted radiation is in phase with the stimulating radiation and produces a negative absorption condition. 2. Radiation similar in origin to spontaneous emission, but determined by the presence of other radiation having the same frequency. Since the phase and amplitude of the stimulated wave depend on the stimulating wave, this radiation is coherent with the stimulating wave. The rate of stimulated emission is proportional to the intensity of the stimulating wave. 3. In a laser, the emission of light caused by a signal applied to the laser such that the response is directly proportional to, and in phase coherence with, the electromagnetic field.

stimulus — 1. *See* excitation, 1. 2. An input parameter to a unit under test, e.g., voltage or current. *See also* measurand.

stirring effect — The circulation in a molten conductive charge due to the combined motor and pinch effects.

stitch bond — 1. A wire bond made by laying the wire on the bonding pad and using thermocompression or ultrasonic scrubbing to form the joint. Note that in thermocompression wire bonding, the first termination is a ball bond, while subsequent terminations of the same wire are stitch bonds. 2. A bond made with a capillary-type bonding tool when the wire is not formed into a ball prior to bonding.

stitch bonding — A bonding technique in which wire is fed through a capillary tube. A bent section of the wire is bonded to the contact area by the capillary. The capillary is removed and a cutter severs the wire, forming a new bend for the next bonding operation.

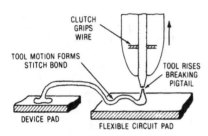

Stitch bonding.

stitching — The process of welding thermoplastic materials by successive applications of two small electrodes that are connected to the output of a radio-frequency generator; a mechanism similar to that of a conventional sewing machine is used.

stitch wire — A semiautomatic system of point-to-point interconnections in which gold-plated steel pins are pressed into holes in conventional printed-circuit boards. Teflon-insulated 30 AWG nickel wire is bonded to the pins. Electronic components are then soldered to terminal projections on the opposite side of the board.

STL — Abbreviation for Schottky transistor logic.

stochastic — The characteristic of events changing the probabilities of various responses.

stochastic process — A random process.

stock reel — Also called supply reel or storage reel. On a tape recorder, the reel from which unrecorded or unplayed tape is taken as the machine records on or plays it.

stoichiometric impurity — A crystalline imperfection caused in a semiconductor by a deviation from the stoichiometric composition.

stopband — That part of the frequency spectrum that is subjected to specified attenuation of signal strength by a filter. (The part of the spectrum between the passband and the stopband is called the transition region.)

stop bit — 1. A signal following a character or block that prepares the receiving device to receive the next character or block. 2. In asynchronous transmission, the last bit, used to indicate the end of a character (normally a mark condition) that serves to return the line to its idle or rest state.

stop-copy — *See* copyguard.

stop-cycle timer — A timer that runs through a single cycle and then stops until the starting signal is reinitiated.

stop element — Also called stop signal. In certain types of serial transmission, the last element of a character, used to ensure that the next start element will be recognized.

stop instruction — A machine operation or routine that requires some manual action other than operation of the start key to continue processing.

stop opening — In a camera, the size of the aperture that controls the amount of light passing through the lens.

stopping potential — The voltage required to stop the outward movement of electrons emitted by photoelectric or thermionic action.

stop-record signal — A facsimile signal used for stopping the conversion of the electric signal into an image on the record sheet.

stop signal — The signal that transfers facsimile equipment from active to standby.

storage — 1. The act of storing information. *See also* store, 1 and 2. 2. A computer section used primarily for storing information in electrostatic, ferroelectric, magnetic, acoustic, optical, chemical, electronic, electrical, mechanical, etc., form. Such a section is sometimes called a memory, or a store, in British terminology. 3. In an oscilloscope, the ability to retain the image of an electrical event on the cathode-ray tube (CRT) for further analysis after that event ceases to exist. This image retention may be for only a few seconds with variable persistence storage, or it may be for hours with bistable storage. 4. The retention of data so that the data can be obtained at a later time. 5. A computer-oriented medium in which data is retained. Primary storage is the internal storage area where the data and program instructions are retained for active use in the system — normally core storage. Auxiliary or external storage is for less active data. These may include magnetic tape, disk, or drum. 6. Synonymous with memory.

storage access time — In a computer, the time required to transfer information from a storage location to the local storage register or other location, where the information then becomes available for processing.

storage allocation — The assignment of specific sections of computer memory to blocks of data or instructions.

storage battery — Two or more storage cells connected in series and used as a unit.

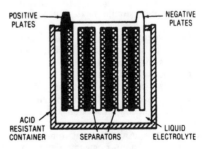

Storage battery.

storage capacity — 1. The amount of information that a storage (memory) device can retain. It is often expressed as the number of words the device can retain (given the number of digits and the base of the standard word). In the case of comparisons among devices that use different bases and word lengths, the capacity is customarily expressed in bits. For example, a 1 K semiconductor memory can store 1024 bits, a 2 K semiconductor memory can store 2048 bits. Fixed memories usually contain instructions and therefore their capacity is sometimes expressed as the number of words of a certain length that they can hold. For example, 256×4 means that the memory can store 256 four-bit words, which makes it a 1 K memory (1024 bits). The same 1 K memory could be a 128×8 memory. 2. The number of items of data that a storage device is capable of containing. Frequently defined in terms of computer words, or by a specific number of bytes or characters.

storage cell — 1. Also called a secondary cell. A cell that, after being discharged, can be recharged by sending an electric current through it in the opposite direction from the discharging current. 2. An elementary unit of storage (e.g., binary cell, decimal cell). 3. A cell that converts chemical energy into electrical energy by a reversible chemical reaction and that may be recharged by passing a current through it in the direction opposite to that of its discharge.

storage counter — A counter in which a series of current pulses charge a capacitor, with each pulse raising the voltage to a higher level. A comparator circuit determines when the capacitor voltage reaches a predetermined level. Special techniques are frequently used to linearize the charging curve of the capacitor.

storage CRT — A CRT that can retain a visual image for some length of time so that it is not necessary to refresh to avoid flicker. Thus, the picture can be written at a slower rate. The absence of refresh eliminates the refresh memory and reduces display-deflection and video-bandwidth requirements. The resultant system is available at a price less than that of most other systems. However, it does have deficiencies, such as low luminance and contrast and the need to rewrite an entire picture if any element is changed.

storage cycle — 1. The periodic sequence of events that occur when information is transferred to or from a computer storage device. 2. Storing, sensing, and regeneration from parts of the storage sequence.

storage device — 1. A device into which data can be inserted, in which it can be retained, and from which it can be retrieved 2. A device used for storing data within a computer system, e.g., core storage, magnetic disk unit, magnetic tape unit, etc.

storage element — 1. An area that retains information distinguishable from that of adjacent areas on the storage area of charge-storage tubes. 2. The smallest part of a digital-computer storage facility, used for storage of a single bit.

storage integrator — In an analog computer, an integrator used to store a voltage in the hold condition for future use while the rest of the computer assumes another computer control state.

storage key — In a computer, an indicator that is associated with a storage block or blocks and that requires tasks to have a matching protection key in order to use the blocks.

storage laser — Any laser that stores unusually high energy prior to discharge. For example, a storage diode laser is a laser in which some carriers are electrically excited for a time longer than the lasing period.

storage life — Also called shelf life. 1. The minimum length of time over which a device, system, or transducer can be exposed to specified environmental storage conditions without changing the performance beyond a specified tolerance. 2. The period during which a battery can be stored and remain suitable for use. 3. The length

of time an item can be stored under specified conditions and still meet specified requirements.

storage location — 1. A computer storage position that holds one machine word and usually has a specific address. 2. A position in storage where a character, byte, or word may be stored.

storage medium — Any recording device or medium into which data can be copied and held until some later date, and from which the entire original data can be obtained.

storage node — Also called a storage electrode. The structure in a DRAM cell on which the cell's charge is stored.

storage oscilloscope — 1. An instrument that has the ability to store a CRT display in order that it may be observed for any required time. This stored display may be instantly erased to make way for storage of a later event. 2. An oscilloscope used as the output device of analog and digital computers or in the analysis of nonperiodic events and the monitoring of slow signals.

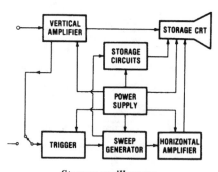

Storage oscilloscope.

storage print — In computers, a utility program that causes the requested core image, core memory, or drum locations to be recorded in either absolute or symbolic form on either the line printer or the delayed printer tape.

storage protection — Also called memory protection. In a computer, an arrangement by which access to storage is prevented for reading, writing, or both.

storage register — A collection of electronic circuits that allows data (usually one or more computer words) to be stored until needed.

storage surface — The part of an electrostatic storage tube on which storage of information takes place.

storage temperature — The temperature of the medium immediately adjacent to the device at which the device, without power applied, may be stored indefinitely without deterioration.

storage time — 1. The time during which the output current or voltage of a pulse is falling from maximum to zero after the input current or voltage has been removed. 2. An increase in the time needed to turn off a transistor that has been driven into saturation. It results from the fact that a transistor in heavy conduction has many excess charge carriers moving in the collector region. When the base signal is changed to the cutoff level, collector current continues until all the excess charge carriers have been removed from the collector region.

storage tube — 1. An electron tube into which information can be stored and later read out, usually a cathode ray tube with a storage screen that will retain charges impressed on it and that can control an electron beam in some way, allowing changes to be read out. 2. A CRT that stores images on a separate storage screen behind the viewing screen in the tube. Images then remain on the viewing screen until the storage screen is erased. Since a storage tube does not have to be refreshed, it can display an extremely large amount of data without flicker. 3. A cathode-ray tube combined with an electrostatic storage unit that is used to introduce, store, and retrieve information translated into electric charge form. 4. A CRT that retains an image for a considerable period without redrawing.

store — 1. To retain information in a device from which the information can later be withdrawn. 2. To introduce information into the device in (1) above. 3. A British synonym for storage. *See* storage, 2.

store and forward — 1. A data communications technique that accepts packets, stores them until they are validated and complete, and then forwards them to the next node on the packet path. 2. Process of message handling used in a message-switching system. 3. Communications system in which messages received at intermediate routing points are recorded for later retransmission to a further routing point or the ultimate recipient.

stored energy — The amount of energy stored in the primary of an electonic ignition system. In an inductive system the stored energy is

$$W_p = 1/2LI^2$$

where

W_p = energy stored in the primary field, in joules
L = primary inductance, in henrys
I = current in the primary winding, in amperes

In a capacitor discharge system,

$$W_p = 1/2CE^2$$

where

W_p = energy stored in the primary capacitor, in joules
C = primary capacitance, in farads
E = peak primary voltage, in volts

stored-energy welding — A method of welding in which electric energy is accumulated (stored) electrostatically, electromagnetically, or electrochemically at a relatively slow rate and is then released at the required rate for welding.

stored program — A set of instructions in the computer memory specifying the operations to be performed and the location of the data on which these operations are to be performed.

stored-program computer — Also called general-purpose computer. 1. A computer in which the instructions specifying the program to be performed are stored in the memory section along with the data to be operated on. 2. A digital computer that, under control of its own instructions, can synthesize, and sometimes alter, stored instructions as though they were data and can subsequently execute these new instructions.

stored program logic — A program stored in a memory unit containing logical commands to the remainder of the memory so that the same processes are performed on all problems.

stored-response testing — Comparison of the actual output responses of the device under test with the expected correct output responses stored within the tester. The expected correct responses can be recorded from a known-good device or determined by manual analysis or software simulation. Stored-response testing often implies storage of the actual logic states, although such digital signatures as transition counts could be the stored responses.

stored routine — In computers, a series of stored instructions for directing the step-by-step operation of the machine.

stored writing rate (of a storage oscilloscope) — The highest rate of spot movement that will leave behind a stored image on the face of the cathode-ray tube. Faster spot movement will not leave an image, as in step-response displays with no vertical edges or sine-wave displays with the zero-crossing edges missing.

store transmission bridge — A transmission bridge that consists of four identical impedance coils (the two windings of the back-bridge relay and the live relay of a connector, respectively) separated by two capacitors. It couples the calling and called telephones together electrostatically for the transmission of voice-frequency (alternating) currents, but separates the two lines for the transmission of direct current for talking purposes (talking current).

storm loading — The mechanical loading imposed on the components of a pole line by wind, ice, etc., and by the weight of the components themselves.

straight dipole — A half-wave antenna consisting of one conductor, usually centerfed.

straightforward circuit — A circuit in which signaling is performed automatically and in one direction.

straightforward trunking — In a manual telephone switchboard system, that method of operation in which one operator gives the order to another operator over the trunk that later carries the conversation.

straight-line capacitance — The variable-capacitor characteristic obtained when the rotor plates are shaped so that the capacitance varies directly with the angle of rotation.

straight-line code — The repetition of a sequence of instructions, with or without address modification, by explicitly writing the instructions for each repetition. Generally straight-line coding will require less execution time and more space than equivalent loop coding. If the number of repetitions is large, this type of coding is tedious unless a generator is used. The feasibility of straight-line coding is limited by the space required as well as the difficulty of coding a variable number of repetitions.

straight-line frequency — The variable-capacitor characteristic obtained when the rotor plates are shaped so that the resonant frequency of the tuned circuit containing the capacitor varies directly with the angle of rotation.

straight-line tracking arm — *See* radial tonearm.

straight-line wavelength — The variable-capacitor characteristic obtained when the rotor plates are shaped so that the wavelength of resonance in the tuned circuit containing the capacitor varies directly with the angle of rotation.

strain — The physical deformation, deflection, or change in length resulting from stress (force per unit area). The magnitude of strain is normally expressed in microinches per inch.

strain anisotropy — A force that directs the magnetization of a particle along a preferred direction relative to the strain.

strain gage — 1. A resistive transducer whose electrical output is proportional to the amount it is deformed under strain. 2. A measuring element for converting force, pressure, tension, etc., into an electrical signal. 3. A device for measuring the expansion or contraction of an object under stress, comprising wires that change resistance with expansion or contraction. *See also* load cell. 4. A sensor that produces a voltage or resistance change when a mechanical force is applied.

strain-gage alarm system — An alarm system that detects the stress caused by the weight of an intruder as

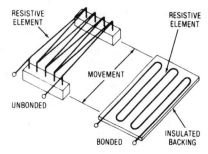

Strain gage, 1.

he or she moves about a building. Typical uses include placement of the strain-gage sensor under a floor joist or under a stairway tread.

strain-gage based — An instrument or transducer with a sensing element composed of bonded or unbonded strain gages.

strain-gage sensor — A sensor that, when attached to an object, will provide an electrical response to an applied stress on the object, such as a bending, stretching, or compressive force.

strain-gage transduction — The conversion of the measurand into a change in resistance caused by strain in four or, more rarely, two arms of a Wheatstone bridge.

strain insulator — A single insulator, an insulator string, or two or more strings in parallel designed to transmit the entire pull of the conductor to, and insulate the conductor from, the tower or other support.

strain pickup — A phonograph pickup cartridge using the principle of the strain gage.

strain-sensitive cable — An electrical cable that is designed to produce a signal whenever the cable is strained by a change in applied force. Typical uses including mounting it in a wall to detect an attempted forced entry through the wall, fastening it to a fence to detect climbing on the fence, or burying it around a perimeter to detect walking or driving across the perimeter.

strain wire — Wire having a composition such that it exhibits favorable strain-gage performance. Also the strain-sensitive filments that constitute the transfer electrical elements in certain transducers.

strand — 1. One of the wires or groups of wires of any stranded conductor. 2. A bundle of wires that are twisted together to form a flexible cable capable of withstanding large tensile stress.

stranded conductor — 1. A conductor composed of a group of noninsulated wires, usually twisted. 2. *See* stranded wire.

stranded wire — Also called stranded conductor. A conductor composed of a group of wires or any combination of groups of wires. The wires in a stranded conductor are usually twisted or braided together.

stranding — The twisting together of small wires to form a single larger conductor. Used to provide long flexing life, limpness, ease of handling, or vibration resistance.

stranding effect — The property of a stranded conductor to exhibit a higher dc resistance than does a solid conductor of the same material and cross-sectional area. It is due to the relatively longer distance that current must travel when following a stranded conductor's helically configured wires.

strand lay — The distance of advance of one strand of a spirally stranded conductor in one turn, measured axially.

strap — 1. A wire or strip connected between the ends of the segments in the anode of a cavity magnetron to promote operation in the desired mode. 2. A bare or insulated conductor run from terminal to terminal of the same or adjacent components. 3. A connecting link or wire between two terminals. 4. To interconnect two or more terminals with strapping wire.

strapping — 1. In a multicavity magnetron, the connecting together of resonator segments that have the same polarity, so that undesired modes of oscillation are suppressed. 2. An interconnecting strap or straps.

stratification — The separation of nonvolatile components of a thick film into horizontal layers during firing, due to large differences in density of the component. It is more likely to occur with a glass-containing conductor paste, and under prolonged or repeated firing.

stratosphere — A calm region of the upper atmosphere characterized by little or no temperature change throughout. It is separated from the lower atmosphere (troposphere) by a region called the tropopause.

stray capacitance — 1. The capacitance introduced into a circuit by the leads and wires connecting the circuit components. 2. Any of the small unintentional capacitances that exist between wires or components in a circuit, particularly capacitance to ground. 3. The capacitance introduced into a circuit by the relative proximity of cables, wires, components, etc.

stray current — A portion of the total current that flows over paths other than the intended circuit.

stray-current corrosion — Corrosion that results when a direct current from a battery or their external source causes a metal in contact with an electrolyte to become anodic with respect to another metal in contact with the same electrolyte. Accelerated corrosion will occur at the electrode where the current direction is from the metal to the electrolyte and will generally be in proportion to the total current.

stray field — The leakage magnetic flux that spreads outward from an inductor and does no useful work.

stray magnetic field — Stray magnetic flux from nearby transformers or inductors that can link with conductors or other inductors to produce undesired and interfering noise voltages.

strays — *See* atmospherics.

streaking — Distortion in which televised objects appear stretched horizontally beyond their normal boundaries. It is most apparent at the vertical edges, where there is a large transition from black to white or white to black, and is usually expressed as short, medium, or long streaking. Long streaking may extend as far as the right edge of the picture and, in extreme low-frequency distortion, even over a whole line interval.

stream deflection amplifier — A fluidic device that utilizes one or more control streams to deflect power stream, altering the output.

streamer breakdown — Breakdown caused by an increase in the field due to the accumulation of positive ions produced during electron avalanches.

streaming — 1. The production of a unidirectional flow of currents in a medium where sound waves are present. 2. A nonstop save-and-restore technique wherein data is transmitted between a tape backup system and Winchester disk in a continuous stream without starting and stopping between data blocks. When all of the data has been transferred, the streaming drive stops. This technology allows for low-cost drives, since the requirement for search and overwrite of a single record is eliminated. A streaming drive does not need high-performance start/stopability, and the expensive servo electronics inherent in that function are also eliminated. 3. A modem's condition when it is sending a carrier signal on a multidrop communication line and hasn't been polled.

strength of a simple sound source — The rms magnitude of the total air flow at the surface of a simple source in cubic meters per second (or cubic centimeters per second), where a simple source is taken to be a spherical source whose radius is small compared with one-sixth wavelength.

strength of a sound source — The maximum instantaneous rate of volume displacement produced by the source when emitting a sinusoidal wave.

stress — The force producing strain in a solid.

stretched display — A PPI display having the polar plot expanded in one rectangular dimension. The equal-range circles of the normal PPI display become ellipses.

stria — A defect in optical materials, such as glass, plastic, or crystals, consisting of a more or less sharply defined streak of material having a slightly different index of refraction than the main body of the material.

striation technique — Rendering sound waves visible by using their individual ability to refract light waves.

striking an arc — Starting an electric arc by touching two electrodes together momentarily.

striking distance — The effective separation of two conductors having an insulating fluid between them.

striking potential — 1. The voltage required to start an electric arc. 2. The lowest grid-to-cathode potential at which plate current begins flowing in a gas-filled triode.

striking voltage — The value of voltage required to start current in a gas tube.

string — 1. In a list of items, a group of items that are already in sequence according to a rule. 2. A connected sequence of entities, such as characters, in a command string. 3. A contiguous set of memory addresses each of which contains a single alphanumeric character. Strings are often enclosed in quotes when written in program statements. 4. A line of symbols of indefinite length treated as a single unit. 5. The way a computer sees text. 6. A sequence of characters.

string electrometer — An electrostatic voltage-measuring instrument consisting of a conducting fiber stretched midway between and parallel to two conducting plates. The electrostatic field between the plates displaces the fiber laterally in proportion to the voltage between the plates.

string-shadow instrument — An instrument in which the indicating means is the shadow (projected or viewed through an optical system) of a filamentary conductor whose position in a magnetic or an electric field depends on the measured quantity.

strip — To remove insulation from a wire or cable.

strip chart recorder — 1. An instrument for recording variations in the measurement of a quantity over time, using a moving pen on a long strip of paper. 2. Data recorder that provides a record in the form of a graph on a strip of paper. Used to measure parameters whose values do not change rapidly with time (one typical device has a specified response time of 0.25 second, full scale). Strip chart recorders are available in multipen models.

strip contacts — Formed contacts in a continuous length, or strip, for use in an automatic installation machine.

striping — In flowcharting, the use of a line across the upper portion of a symbol to indicate the presence of a detailed representation elsewhere in the same set of flowcharts.

stripline — 1. Also called microstrip. Layout and interconnection method used for circuits that operate at very high frequencies, usually above 1 gigahertz. 2. Strip transmission line. 3. A type of microwave transmission line used in integrated circuits, consisting of a narrow,

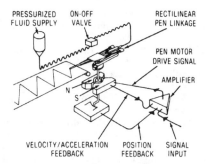

Strip chart recorder.

flat conductor sandwiched between, and insulated from, wider flat grounded conductors. 4. A type of transmission line configuration that consists of a single narrow conductor parallel and equidistant to two parallel ground planes. 5. A microwave conductor on a substrate.

strip-loaded diffused optical waveguide — A three-dimensional optical waveguide constructed from a two-dimensional diffused optical waveguide on whose surface has been deposited a dielectric strip of a lower refractive index material, thus confining the electromagnetic fields of the propagating mode to the vicinity of the strip, hence achieving a three-dimensional guide.

stripper — A hand-operated or motor-driven tool for removing insulation from wires.

stripping — 1. A process using either acids or plasma to remove the resist coating of a wafer after the exposure, development, and etching steps. 2. Removing the insulation from a wire. 3. Removing the outer sheath or covering of a cable or wire, thus exposing the insulated wires.

strip printer — A peripheral device used with a programmable controller to provide a hard copy of process number, status, and functions.

strip transmission line — A microwave transmission line in the form of a thin, narrow rectangular strip adjacent to a wide ground-plane conductor or between two wide ground-plane conductors. Separation of the conductors usually is achieved by using a low-loss dielectric material on which the conductors are formed by etching.

strobe — 1. An intensified spot in the sweep of a deflection-type indicator, used as a reference mark for ranging or expanding the presentation. 2. An intensified sweep on a plan-position indicator or B-scope. Such a strobe may result from certain types of interference or it may be purposely applied as a bearing or heading marker. 3. On a console oscilloscope, a line representing the azimuth data generated by a jammed radar site. 4. *See* electronic flash. 5. An input to a counter or register that permits the asynchronous entry of parallel data. 6. A selection signal that is active when data is correct on a bus. 7. A control signal used to effect information transfers at the hardware level.

strobe hold time — The time necessary to strobe parallel data into a counter or register completely.

strobe marker — A small bright spot, a short gap, or other discontinuity produced on the line trace of a radar display to indicate the part of the time base that is receiving attention.

strobe pulse — Also called sample pulse. 1. A pulse used to gate the output of a core-memory sense amplifier into a trigger in a register. 2. A pulse of duration less than the time period of a recurrent phenomenon, used for making a close investigation of that phenomenon. The frequency of the strobe pulse bears a simple relation to that of the phenomenon, and the relative timing is usually adjustable. 3. A pulse that enables a system for a fixed period only.

strobe release time — The time that must elapse after the strobe input is disabled before a clocking transition will be recognized by a counter.

Strobolume — A trade name used to describe a high-intensity electronic stroboscope manufactured by General Radio Company.

stroboscope — 1. A light that flashes at a frequency that can be adjusted to coincide with any repeating motion. When the flashing rate is synchronized with the motion, the moving device appears to be stopped. Since the flashing rate is known, the speed of the device can be determined without physically contacting the device. A flashing rate slightly different from the synchronized rate makes the device appear to operate in slow motion. With the stroboscope, such characteristics of motion as whip, vibration, rotation, or chatter can be observed readily. 2. Device that administers brief flashes of light for the purpose of observing the behavior of an object during a short interval of time. One of the most effective means for accomplishing this purpose is a gaseous tube energized by the discharge of a capacitor. Light flashes as short as 1 microsecond have been produced in this fashion. By illuminating moving machinery with intermittent flashes of light at the correct frequency, the machine can be made to appear stationary; by reducing or increasing the flash rate slightly, the machine will appear to move slowly in the forward or reverse direction. Many turntable platters carry a band of dots or lines around their rims, or on their under surfaces, lit by a neon lamp. When the platter speed is adjusted to exactly $33\frac{1}{3}$ or 45 rpm, the dots or lines appear to stand still.

stroboscopic disc — A printed disc having several rings, each with a different number of dark segments. The pattern is placed on a rotating phonograph turntable and illuminated at a known frequency by a flashing discharge tube. The speed can then readily be determined by noting which pattern appears to stand still or rotate the slowest.

Stroboscopic disc.

stroboscopic light source — An electronic flash tube capable of repeated operation at hundreds or thousands of flashes per second for long periods.

stroboscopic tachometer — A stroboscope with a scale calibrated in flashes or in revolutions per minute. The stroboscopic lamp is directed onto the rotating device being measured, and the flashing rate is adjusted until the device appears to be standing still. The speed can then be read directly from the scale.

strobotron — 1. A type of glow lamp that produces intense flashes of light when fed with accurately timed voltage pulses. It is used in electronic stroboscopes for visual inspection of high-speed moving parts. 2. A specified cold-cathode gas tube used to apply a short-duration, high-power arc for a stroboscope.

stroke — In character recognition, a straight line segment or arc that forms a part of a graphic character.

stroke centerline — In character recognition, a line equidistant from the two stroke edges.

stroke edge — In character recognition, the region of discontinuity between a side of a stroke and the background, determined by averaging, over the length of the stroke, the irregularities produced by the printing and detection processes.

stroke pattern — The pattern, formed by a character-generation CRT system, in which the characters are composed of a sequence of line segments (strokes) generated by the electron beam motion with time intervals between successive coordinates. The characters are formed within a rectangle by different combinations of vertical, horizontal, and diagonal strokes, coupled with intensity control signals.

stroke speed — Also called scanning-line frequency or scanning frequency. The number of times per minute that a fixed line, perpendicular to the direction of scanning, is crossed in one direction by a scanning or recording spot in a facsimile system.

stroke width — In character recognition, the distance, measured in the direction perpendicular to the stroke centerline, between the two edges of a stroke.

structured logic design — A method of describing chip design through a hierarchical structure by which each level of component functions can be described by a lower level of functions.

structured programming — Also called top-down programming. 1. A systematic procedure for writing programs in modular form with a clear logical structure. Such programs are easy to understand and modify. 2. A set of conventions and rules that, when followed, yields programs that are easy to write, test, modify, and read. 3. Techniques concerned with improving the programming process through better organization of programs and better programming notation to facilitate correct and clear description of data and control structures.

stub — 1. A short length of transmission line or cable joined as a branch to another transmission line or cable. 2. A short section of transmission line, open or shorted at the far end, connected in parallel with a transmission line to match the impedance of the line to that of an antenna or transmitter.

stub angle — A right-angle elbow for a coaxial rf transmission line, the inner conductor being supported by a quarter-wave stub.

stub cable — A short branch from a principal cable. The end is often sealed until used. Pairs in the stub are referred to as stubbed-out pairs.

stub matching — Using a stub to match a transmission line to an antenna or load.

stub-supported coaxial — A coaxial cable whose inner conductor is supported by short-circuited coaxial stubs.

stub tuner — A stub terminated by movable short-circuiting means and used for matching impedance in the line to which it is joined as a branch.

stuck-at-1, stuck-at-0 — A particular fault model in which a faulty node remains at a logical 1 or 0 state regardless of the inputs applied.

stud — Threaded or serrated insert or post used for connecting wires or terminals.

stud welding — A process in which the heat of an electric arc drawn between the fastener and the work melts a quantity of metal, after which the two heated parts are brought together under pressure.

stunt box — 1. A device for controlling the nonprinting functions of a Teletype terminal. 2. A mechanical unit that allows a teletype operator to add control features to the teleprinter; for example, an automatic nonoverline control that prevents printing over a message already typed. Controlling a remote device is another use for the stunt box.

stutter — In facsimile, a series of undesired black and white lines sometimes produced when the signal amplitude changes sharply.

STV — *See* subscription TV.

stylus — Also called a needle. 1. The needlelike object used in a sound recorder to cut or emboss the record grooves. Generally it is made of sapphire, stellite, or steel. The plural is styli. 2. The pointed element that contacts the record sheet in a facsimile recorder. 3. A small piece of industrial-grade diamond or artificial sapphire, conically shaped, that tracks the groove in a phonograph record. Stylus motion is transmitted through the supporting cantilever to the generating elements in the cartridge. Styli come in several shapes, such as elliptical or spherical. 4. A hand-held object that provides coordinate input to the display device.

stylus alignment — The position of the stylus with respect to the record. The correct position is perpendicular.

stylus drag — Also called needle drag. The friction between the reproducing stylus and the surface of the recording medium.

stylus force — Also called vertical stylus force or tracking force, and formerly called needle pressure or stylus pressure. The downward force, in grams or ounces, exerted on the disc by the reproducing stylus.

stylus oscillograph — An instrument in which a pen or stylus records, on paper or another suitable medium, the value of an electrical quantity as a function of time.

stylus pressure — *See* stylus force.

stylus radius — The radius in mils of the spherical tip that contacts the groove wall of a phonograph record. In an elliptical stylus there are two radii. The smaller radius applies to the sides of the stylus looking down on it, and the larger radius applies to the curvature that contacts the V groove, looking along the groove.

subassembly — 1. Parts and components combined into a unit for convenience in assembling or servicing. A subassembly is only part of an operating unit; it is not complete in itself. 2. A portion of a complete equipment that has one particular function in a single unit for ease of replacement. 3. A part of a complete assembly. Several subassemblies are usually required to complete an assembly. A subassembly is not normally capable of delivering an output by itself.

subatomic — Smaller than atoms — i.e., electrons or protons.

subatomic particles — The particles that make up the atom — i.e., protons, electrons, and neutrons.

subcarrier — 1. A carrier used to generate a modulated wave that is applied, in turn, as a modulating wave to modulate another carrier. 2. The carrier used in stereo broadcasting to accommodate the subchannel stereo components. Frequency is 38 kHz and is suppressed at the transmitter, leaving only the sidebands, but is reformed at the receiver for detection of the stereo components by doubling the synchronized 19-kHz pilot tone. 3. A second signal "piggybacked" onto a main signal to carry additional information. In satellite television transmission, the video picture is transmitted over the main carrier. The corresponding audio is sent via an FM subcarrier. Some satellite transponders carry as many as four special audio or data subcarriers whose signals may or may not be related to the main programming. 4. A carrier wave that modulates another higher-frequency carrier. In satellite transmissions, a 6.8 MHz audio subcarrier is often used to modulate the C-band carrier. In television, a 3.58 MHz subcarrier modulates the video carrier on each channel.

subcarrier band — A band associated with a given subcarrier and specified in terms of maximum subcarrier deviation.

subcarrier channel — The channel required to convey the telemetric information of a subcarrier band.

subcarrier discriminator tuning unit — A device that tunes the discriminator to a particular subcarrier.

subcarrier frequency shift — The use of an audio-frequency shift signal to modulate a radio transmitter.

subcarrier oscillator — 1. In a telemetry system, the oscillator that is directly modulated by the measurand or its equivalent in terms of changes in the transfer elements of a transducer. 2. In a color television receiver, the crystal oscillator operating at the chrominance subcarrier frequency of 3.579545 MHz.

subcarrier substrate — A small substrate of a film circuit that is mounted in turn onto a larger substrate.

subcarrier transmission — A subdivision of carrier transmission designed to increase the capacity of the telephone system. The term id used in other circumstances in a similar fashion to indicate a transmission modulated upon a carrier or subcarrier of a communication channel.

subchannel — In a telemetry system, the route for conveying the magnitude of one subcommutated measurand.

subchassis — The chassis on which closely associated components, such as those of an amplifier or power supply, are mounted. A subchassis is a building block, easily changed and usable in a variety of systems.

subcircuit — A group of physically realizable components that performs a specific function, typically treated as a black box.

subclutter visibility — The characteristic that relates to how well a radar equipped with a moving-target indicator can see through clutter.

subcommutation — In a computer, the act of connecting one data source to a sampled data system less frequently than other data sources.

subcommutation frame — In PCM systems, a recurring integral number of subcommutator words that include the single subcommutation frame synchronization word. The number of subcommutator words is equal to an integral number of primary commutator frames. The length of a subcommutation frame is equal to the total number of words or bits generated as a direct output of the subcommutator.

subcycle generator — A frequency-reducing device used in telephone equipment to furnish ringing power at a submultiple of the power-supply frequency.

subdivided capacitor — A capacitor in which several capacitors, known as sections, are mounted so that they may be used individually or in combination.

subelement — A distinguishable portion of a circuit element (e.g., the emitter, collector, and base are subelements of an integrated bipolar transistor).

subfigure — User-defined symbols that are used repetitively in a drawing.

subframe — A complete sequence of frames during which all subchannels of a specific channel are sampled once.

subharmonic — 1. A sinusoidal quantity whose frequency is an integral submultiple of the fundamental frequency of its related periodic quantity. A wave with half the frequency of the fundamental of another wave is called the second subharmonic of that wave; one with a third of the fundamental frequency is called a third subharmonic, etc. 2. A harmonic signal whose frequency is less than that of the fundamental, being the fundamental frequency divided by an integral number. 3. A fractional multiple of the fundamental frequency.

subjamming visibility — The characteristic that relates to the ability of a particular radar antijam technique to see through jamming signals.

subject copy — Also called copy. In facsimile, the material in graphic form to be transmitted for reproduction by the recorder.

submarine cable — A cable designed for service under water; usually a lead-covered cable with a steel armor applied between layers of jute.

submerged-resistor induction furnace — A furnace for melting a metal. It comprises a melting hearth, a descending melting channel closed through the hearth, a primary induction winding, and a magnetic core that links the melting channel and primary winding.

submersible transformer — A transformer so constructed that it will operate when submerged in water under predetermined conditions of pressure and time — e.g., a subway transformer.

subminiature tube — A small electron tube used generally in miniaturized equipment.

subminiaturization — The technique of packaging discrete miniaturized parts, using assembly techniques that result in increased volumetric efficiency (e.g., a hybrid circuit).

submultiple resonance — Resonance at a frequency that is a submultiple of the frequency of the exciting impulses.

subnanosecond — Less than a nanosecond.

subpanel — An assembly of electrical devices connected together that forms a simple functional unit in itself.

subprogram — An independently compilable part of a larger computer program.

subrefraction — Atmospheric refraction that is less than standard refraction.

subroutine — 1. In computer technology, the portion of a routine that causes a computer to carry out a well-defined mathematical or logical operation. 2. Usually called a closed subroutine. One to which control may be transferred from a master routine, and returned to the master routine at the conclusion of the subroutine. 3. A set of instructions and constants necessary to direct a computer to carry out a well-defined mathematical, special, or logical operation. A subroutine usually resides in mass storage and can be called by the computer whenever that specific function must be performed. It may be used over and over in the same program and in different programs. 4. A subprogram (group of instructions) reached from more than one place in a main program. The process of passing control from the main program to a subroutine is a subroutine call, and the mechanism is a subroutine linkage. Often data or data addresses are made available by the main program to the subroutine. The process of returning control from the subroutine to the main program is subroutine return. The linkage automatically returns control to the original position in the main program or to another subroutine. *See also* nesting. 5. A short program segment that performs a specific function and is available for general use by other programs and routines. 6. Computer program segment identified by name and bracketed by a "subroutine" and a "return" statement. Execution is transferred to a subroutine when a subroutine call occurs. Subroutines save memory space at the expense of execution speed.

subroutine call — *See* subroutine, 4.

subroutine linkage — *See* subroutine, 4.

subroutine reentry — In a computer, initiation of a subroutine by a program before the subroutine has completed its response to another program that called for it. This can occur when a control program is subjected to a priority interrupt.

subroutine return — *See* subroutine, 4.

subscriber line — 1. A line that connects a central office and a telephone station, private branch exchange, or other end equipment. 2. A telephone circuit from the central office to a subscriber's telephone.

subscriber loop — *See* local loop, 1.

subscriber multiple — A bank of manual-switchboard jacks that provides outgoing access to subscriber lines and usually has more than one appearance across the switchboard.

subscriber's drop — The line from a telephone cable to a subscriber's building.

subscriber's equipment — That portion of a central station alarm system installed in the protected premises.

subscriber set — Also called a customer set. An assembly of apparatus for originating or receiving calls on the premises of a subscriber to a communication or signaling service.

subscript — A notation for use in a computer to specify one member of an array in which each member is referenced only in terms of the array name.

subscripted variable — 1. A variable with one or more following subscripts enclosed in parentheses. 2. A variable that has one or more numbers attached to it, indicating its place in a series or array.

subscription television — Abbreviated STV. 1. *See* pay television. 2. An over-the-air or broadcast pay TV service in which scrambled signals are sent to decoder boxes on TV sets for which viewers pay a fee to receive movies and sports programming.

subscription television broadcast program — A television broadcast program intended to be received in intelligible form by members of the public only for a fee.

subset — 1. In a telephone system, the handset or deskset at the station location. 2. Also known as a modem, data set, or subscriber set. A modulation/demodulation device designed to make business-machine signals compatible with communications facilities. 3. The subscriber's telephone instrument.

Subsidiary Communications Authorization — *See* SCA.

subsonic frequency — *See* infrasonic frequency.

subsonic speed — A speed less than the speed of sound.

substation — 1. Any building or outdoor location at which electric energy in a power system is transformed, converted, or controlled. 2. A junction point where several transmission and distribution lines are tied together.

substep — A part of a computer step.

substitute — In a computer, to replace one element of information by another.

substitute character — An accuracy-control character, intended to be used in place of a character determined to be invalid, in error, or not representable on a particular device.

substitution emulation — *See* emulation.

substitution interference measurement — A measurement in which the noise level of the source being measured is compared to a known level from a calibrated source; for example, an impulse generator for broadband noise or a sine-wave generator for cw.

substitution method — A three-step method of measuring an unknown quantity in a circuit. First, some circuit effect dependent on the unknown quantity is measured or observed. Then a similar but measurable quantity is substituted in the circuit. Finally, the latter quantity is adjusted to produce a like effect. The unknown value is then assumed to be equal to the adjusted known value.

substrate — Also called base material. 1. The supporting material on or in which the parts of an integrated circuit are attached or made. The substrate may be passive (thin film, hybrid) or active (monolithic compatible). 2. A material on the surface of which an adhesive substance is spread for bonding or coating; any material that provides a supporting surface for other materials, especially materials used to support printed-circuit patterns. 3. The physical material upon which an electronic circuit is fabricated. Used primarily for mechanical support but may serve a useful thermal or electrical function. Also, a material on whose surface an adhesive substance is spread for bonding or coating, or any material that provides a supporting surface for other materials. 4. The base or support layer of a transistor or monolithic chip, which usually constitutes a major proportion of the total volume. When composed of ceramic, glass, or sapphire, the substrate functions mainly as a support during the operations of fabrication and encapsulation. However, when composed of heavily doped semiconductor material it normally performs the additional function of a distributed low-resistance connection to the physically lowest region of the device. 5. That part of an integrated circuit that acts as a support. 6. A slab of insulating material used for structural support of thick-film depositions and assembly components, usually high-purity (96 to 99 percent) alumina. *See* alumina. 7. The material on which the chips and other components are mounted, comparable to a printed circuit board. Substrate materials in common use include glass, sapphire, silicon, alumina, beryllia, and porcelainized steel. 8. The underlying material on which a microelectronic device is built.

substrate base material — The supporting material on which the elements of a thick-film circuit are deposited or attached.

subsurface wave — An electromagnetic wave propagated through water or land. Operating frequencies for communications may be limited to approximately 35 kHz due to attenuation of high frequencies.

subsynchronous — Having a frequency that is a submultiple of the driving frequency.

subsynchronous reluctance motor — A form of reluctance motor with more salient poles in the primary winding. As a result, the motor operates at a constant average speed that is a submultiple of its apparent synchronous speed.

subsystem — 1. A major, essential, functional part of a system. The subsystem usually consists of several components. 2. A part or division of a system that in itself has the properties of a system. 3. An organization of computer components (e.g., a tape drive and controller) that comprises a functional unit that is part of a larger system.

subtractive filter — An optical filter that is of a certain color and eliminates that color when placed in the path of white light.

subtractive process — A printed circuit manufacturing process in which a conductive pattern is formed by the removal of portions of the surface of a metal-clad insulator by chemical means (etching).

subtractor — An operational amplifier circuit in which the output is proportional to the difference between its two input voltages or between the net sums of its positive and negative inputs.

subvoice-grade channel — A channel whose bandwidth is less than that of a voice-grade channel. Such a channel usually is a subchannel of a voice-grade line.

(According to common usage, a telegraph channel is excluded from this definition.)

subwoofer — A speaker that is specifically made to reproduce the lowest of audio frequencies, between approximately 20 Hz and 100 Hz.

success ratio — The ratio of the number of successful attempts to the total number of trials. It is frequently used as a reliability index.

suckout — A hole in the response pattern of a tuned circuit due to the self-resonance of components at certain frequencies.

sudden commencement — Magnetic storms that start suddenly (within a few seconds) and simultaneously all over the earth.

sudden ionospheric disturbances — The sudden increase in ionization density in lower parts of the ionosphere, caused by a bright solar chromospheric eruption. It gives rise to a sudden increase of absorption in radio waves propagated through the low parts of the ionosphere, and sometimes to simultaneous disturbances of terrestrial magnetism and earth current. The change takes place within one or a few minutes, and conditions usually return to normal within one or a few hours.

Suhl effect — When a strong transverse magnetic field is applied to an n-type semiconducting filament, the holes injected into the filament are deflected to the surface. Here they may recombine rapidly with electrons and, thus, have a much shorter life, or they may be withdrawn by a probe as though the conductance had increased.

suicide control — A control function that uses negative feedback to reduce and automatically maintain the generator voltage at approximately zero.

sulfating — The accumulation of lead sulfate on the plates of a lead-acid storage battery. This reduces the energy-storing ability of the battery and causes it to fail prematurely.

sulfation — The lead sulfate that forms on battery plates as a result of the battery action that produces electric current.

sulfonated polystyrene sensor — Also called Pope cell. An ion-exchange device with good response, accuracy, and long-term stability whose resistance changes exponentially with humidity and temperature.

sum — The combination of two electrical signals of the same electrical polarity. The total electrical energy produced by combining the two different signals of a stereo program.

sum channel — A combination of left and right stereo channels identical to the program, which may be recorded or transmitted monophonically.

summary punch — A punch-card machine that may be attached to another machine in such a way that it will punch information produced, calculated, or summarized by the other machine.

summary recorder — In computers, output equipment that records a summary of the information handled.

summation check — A redundant computer check in which groups of digits are summed, usually without regard to overflow. The sum is then checked against a previously computed sum to verify the accuracy of the computation.

summation frequency — A frequency that is the sum of two other frequencies that are produced simultaneously.

summation tone — A combination tone, heard under certain circumstances, whose pitch corresponds to a frequency equal to the sum of the frequencies of the two components.

summing junction — The input terminal of an operational amplifier that is inverted and has both input and feedback connected to it.

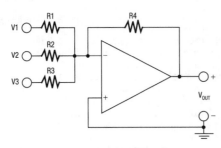

Summing amplifier.

summing point — 1. A mixing point whose output is obtained by adding its inputs (with the prescribed signs). 2. The input terminal to which the feedback loop from the output of an amplifier is returned.

sum of products — A general form of Boolean-algebraic expression that can be implemented readily through the use of electronic gate circuits.

sum operand — In a calculator, automatically adds first factors of any sequence of multiplication or division problems. Used when obtaining average unit price and standard deviation.

superband — The frequency band from 216 to 300 MHz, used for fixed and mobile radios and additional television channels on a cable system.

supercardioid — Similar to cardioid but with a narrower response lobe.

supercardioid microphone — A microphone having a cardioid "polar pattern" with an unusually high discrimination between sounds from the front and rear.

supercommutation — 1. Commutation at a higher rate by connection of a single data input source to equally spaced contacts of the commutator (cross patching). Corresponding cross patching is required at the decommutator. 2. The connection of one data source of a computer to a sampled-data system more frequently than other data sources are connected.

superconducting — Exhibiting superconductivity.

superconductivity — 1. The ability of certain alloys and metals, such as lead, tin, and vanadium, to conduct electricity with zero resistance. In general, electrical resistance falls as the temperature of a conductor decreases. For certain metals and alloys, resistance abruptly drops to nil when the temperature reaches a specific point near absolute zero. In recent years, superconductivity has been achieved at temperatures as high as −140°C. 2. The decrease in resistance of certain materials (lead, tin, thallium, etc.) as their temperature is reduced to nearly absolute zero. When the critical (transition) temperature is reached, the resistance will be almost zero.

superconductors — Materials that exhibit superconductivity. These are materials in which the resistance drops to almost zero at a temperature near absolute zero.

Superconductivity is exhibited by many of the metallic elements, their alloys, and intermetallic compounds.

super-emitron — *See* image iconoscope.

supergroup — In carrier telephony, five groups (60 voice channels) multiplexed together and handled as a single unit. A basic supergroup occupies the 312 to 552 kHz band.

superhet — Slang for a superheterodyne receiver.

superheterodyne receiver — 1. A receiver in which the incoming modulated rf signals are usually amplified in a preamplifier and then fed into the mixer for conversion into a fixed, lower carrier frequency (called the intermediate frequency). The modulated IF signals undergo very high amplification in the IF-amplifier stages and are then fed into the detector for demodulation. The resultant audio or video signals are usually further amplified before being sent to the output. 2. A receiver that converts the incoming signal to an intermediate frequency and amplifies the signal at an intermediate frequency before detection.

superheterodyne reception — A method of receiving radio waves in which heterodyne reception converts the voltage of the intermediate, but usually superaudible, frequency, which is then detected.

superhigh frequency — Abbreviated SHF. The frequency band extending from 3 to 30 GHz (100 to 10 mm).

superimpose — In a tape recorder, to record one or more signals over another without erasure, so that when a tape is played back, all recordings can be heard simultaneously. (Particularly useful if one wishes to have a spoken commentary with a musical background.)

Supermalloy — Trade name of Arnold Engineering Company for a magnetic alloy with a maximum permeability greater than 1,000,000.

supermode laser — A frequency-modulated laser whose output is passed through a second phase modulator driven 180° out of phase and with the same modulation index as the first modulator. All of the energy of the previously existing laser modes is compressed into a single frequency, with nearly the full power of the laser concentrated in that signal.

superposed circuit — An additional channel obtained in such a manner from one or more circuits normally provided for other channels that all channels can be used simultaneously, without mutual interference.

superposed ringing — Also called superimposed ringing. Party-line telephone ringing accomplished by using a combination of alternating and direct currents; direct current of both polarities is used to provide selective ringing.

superposition — The process of adding two or more signals together and having each signal retain its unique identity.

superposition theorem — When a number of voltages (distributed in any manner throughout a linear network) are applied to the network simultaneously, the current that flows is the sum of the component currents

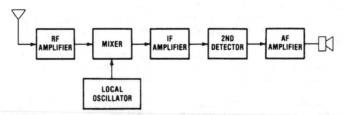

Superheterodyne receiver.

that would flow if the same voltages had acted individually. Likewise, the potential difference that exists between any two points is the component potential difference that would exist there under the same conditions.

superpower — A comparatively large power (sometimes over 1,000,000 watts) used by a broadcasting station in its antenna.

superradiance — In lasers, a rapid increase in intensity of fluorescent line emission with increasing excitation (pump) power. This intensity increase and associated line narrowing are attributed to coherent reinforcement of spontaneously emitted photons during a single pass through the active region.

superrefraction — Abnormally large refraction of radio waves in the lower layers of the atmosphere, leading to abnormal ranges of operation.

superregeneration — 1. A form of regenerative amplification frequency used in radio-receiver detecting circuits. Oscillations are alternately allowed to build up and are quenched at a superaudible rate. 2. Method used to produce greater regeneration than otherwise possible without the harmful effects of oscillation. *See also* quench frequency.

superregenerative detector — A detector that functions on superregeneration to achieve extremely high sensitivity with a minimum of amplifier stages.

superregenerative receiver — A receiver in which the regeneration is varied in such a manner that the circuit is periodically rendered oscillatory and nonoscillatory.

supersensitive relay — A relay that operates on extremely small currents (usually less than 250 microamperes).

supersonic — Faster than the speed of sound (approximately 750 mph or 1200 km/h). These speeds are usually referred to by the term *mach* or *mach number*. Mach 1 equals the speed of sound; mach 2, twice the speed of sound, etc.

supersonic communication — Communication through water by manually keying the sound output of echo-ranging equipment used on ships.

supersonic frequency — *See* ultrasonic frequency.

supersonics — 1. The general term covering phenomena associated with speeds higher than that of sound (e.g., aircraft and projectiles that travel faster than sound). 2. The general term covering the use of frequencies above the range of normal hearing.

supersonic sounding — A system of determining ocean depths by measuring the time interval between the production of a supersonic wave just below the surface of the water and the arrival of the echo reflected from the bottom. The sounds are transmitted and received by either magnetostriction or piezoelectric units, and electronic equipment is employed to provide a continuous indication of depth (sometimes with a permanent recording).

supersync signal — A combination horizontal- and vertical-sync signal transmitted at the end of each scanning line in commercial television.

superturnstile antenna — A stacked antenna array in which each element is a batwing antenna.

supertweeter — A tweeter used usually in four-way or five-way systems only for extremely high frequencies.

Super Video Graphics Array — See SVGA.

supervised lines — Interconnecting lines in an alarm system that are electrically supervised against tampering. *See also* line supervision.

supervisor — 1. A routine or routines carried out in response to a requirement for changing or interrupting the flow of operation through a central processing unit, or for performance of input-output operations; therefore, the medium for coordinating the use of resources and maintaining the flow of operations through the central processing unit. Hence, a control routine executed in supervisor state. 2. A program that helps manage the operation of a computer system.

supervisory alarm system — An alarm system that monitors conditions or persons or both and signals any deviation from an established norm or schedule. Examples are the monitoring of signals from guard patrol stations for irregularities in the progression along a prescribed patrol route, and the monitoring of production or safety conditions such as sprinkler water pressure, temperature, or liquid level.

supervisory circuit — An electrical circuit or radio path that sends information on the status of a sensor or guard patrol to an annunciator. For intrusion alarm systems, this circuit provides line supervision and monitors devices. *See also* supervisory alarm system.

supervisory control — 1. A system by which selective control and automatic indication of remote units is provided by electrical means over a relatively small number of common transmission lines. (Carrier-current channels on power lines can be used for this purpose.) 2. An arrangement for selectively controlling remotely located equipment by electrical means.

supervisory control signaling — Characters or signals that actuate equipment or indicators at a remote terminal automatically.

supervisory signal — A signal for attracting the attention of an attendant in connection with switching apparatus, etc.

supervoltage — A voltage applied to X-ray tubes operating between 500 and 2000 kilovolts.

supplementary group — In wire communications, a group of trunks that provides direct connection of local or trunk switching centers by way of other than a fundamental route.

supplementary insulation — An independent insulation provided in addition to the functional insulation to ensure protection against electrical shock hazard in the event that functional insulation should fail.

supply port — In a fluidic device, the port at which power is provided to an active device.

supply reel — 1. In a tape recorder, the feed reel from which the tape unwinds while playing or recording. 2. On a tape recorder, the reel generally on the left side of the recording head stack. It has a full reel of tape prior to beginning of the recording.

supply voltage — The voltage obtained from a power supply to operate a circuit.

support chips — 1. Integrated circuits, exclusive of the microprocessor unit (the central processing unit of the microcomputer), that are required to complete a system. These chips must be compatible with the selected microprocessor and are generally part of the same family as the microprocessor unit. The degree of the support chip's flexibility and availability is what will determine the total capability of a microcomputer system and, thus, becomes a significant factor in the selection of one system over another. 2. Semiconductor chips required to make the MPU functional, such as a clock generator. Includes chips that directly interface with the MPU to enhance its function, such as a DMA or a dedicated MCU supplied by a manufacturer as a standard part. Also includes chips that are not connected directly to the MPU but which enhance its function, such as a modem. A support chip may or may not contain memory. A memory with I/O ports would be considered a support chip, whereas a chip containing only memory would be classified as a memory device.

support software — A library of software tools used selectively or in total to produce software. This library may consist of compiler, link, editors, source update programs, etc.

suppressed carrier — 1. That type of system which results in the suppression of the carrier frequency from the transmission medium. The intelligence of a carrier wave after modulation is contained in either sideband, and normally only one sideband is transmitted; the other sideband and carrier frequency are suppressed. The intelligence is recovered at the receiving end by inserting a carrier frequency from a local source that, when combined with the incoming signal, produces the original frequencies with which the transmitting carrier was modulated. 2. A method of transmission in which one or more sidebands are transmitted but the carrier is not transmitted.

suppressed-carrier operation — *See* suppressed-carrier transmission.

suppressed-carrier transmission — Transmission in which the carrier frequency is either partially or totally suppressed. One or both sidebands may be transmitted.

suppressed time delay — Deliberate displacement of the zero of the time scale with respect to the time of emission of a pulse, in order to simulate electrically a geographical displacement of the true position of a transponder.

suppressed-zero instrument — An indicating or recording instrument in which the zero position is below the end of the scale markings.

suppression — 1. Elimination of any component of an emission — e.g., a particular frequency or group of frequencies in an audio- or radio-frequency signal. 2. Reduction or elimination of noise pulses generated by a motor or motor generator. 3. Elimination of unwanted signals or interference by means of shielding, filtering, grounding, component relocation, or sometimes redesign. 4. An optical function in online or offline printing devices by which they can ignore certain characters or character groups transmitted through them. 5. The reduction to an acceptable level of a certain frequency or frequencies.

suppression pulse — The pulse generated in an airborne transponder by coincidence of the first interrogation pulse and the control pulse. This pulse, also known as a killer pulse, is used to suppress unwanted interrogations from the side lobes.

suppressor — 1. A resistor used in an electron-tube circuit to reduce or prevent oscillation or the generation of unwanted rf signals. 2. A resistor in the high-tension lead of the ignition system in a gasoline engine.

suppressor grid — A grid interposed between two positive electrodes (usually the screen grid and the plate) primarily to reduce the flow of secondary electrons from one to the other.

suppressor pulse — The pulse used to disable an ionized flow field or beacon transponder during intervals when interference would be encountered.

surface acoustic wave — A sound or acoustic wave that travels on the surface of the optically polished surface of a piezoelectric material. This wave travels at the speed of sound but can pass frequencies as high as several gigahertz. Its amplitude decays exponentially with substrate depth.

surface analyzer — An instrument that measures or records irregularities in a surface. As a crystal-pickup stylus or similar device moves over the surface, the resulting voltage is amplified and fed to an indicator or recorder that magnifies the surface irregularities as high as 50,000 times.

surface asperities — Small projecting imperfections on the surface coating of a tape that limit and cause variations in head-to-tape contact. A term useful in discussions of friction and modulation noise.

surface barrier — A barrier formed automatically at a surface by the electrons trapped there.

surface-barrier transistor — Abbreviated SBT. A wafer of semiconductor material into which depressions have been etched electrochemically on opposite sides. The emitter- and collector-base junctions or metal-to-semiconductor contacts are then formed by electroplating a suitable metal onto the semiconductor in the etched depressions. The original wafer constitutes the base region. No longer manufactured.

surface conductance — Conductance of electrons along the outer surface of a conductor.

surface-controlled avalanche transistor — A transistor in which the avalanche breakdown voltage is controlled by an external field applied through surface insulating layers, and which permits operation at frequencies up to the 10-GHz range.

surface diffusion — The high-temperature injection of atoms into the surface layer of a semiconductor material to form the junctions. Usually a gaseous diffusion process.

surface duct — An atmospheric duct for which the lower boundary is the surface of the earth.

surface electromagnetic waves — Waves that propagate along the interface between two different media without radiation, with exponentially decaying evanescent fields on both sides of the interface.

surface insulation — Also called oxalizing or insulazing. A coating applied to magnetic-core laminations to retard the passage of current from one lamination to another.

surface leakage — The passage of current over the surface of an insulator rather than through it. Surface leakage in new components is very low, but when a component is installed in equipment and exposed to dust, dirt, moisture, and other degrading environments, leakage current can increase and cause problems.

surface mounting — 1. Connecting discrete components directly to foil patterns without using holes. 2. A packaging technique that attaches components directly to a circuit board rather than using leads, thus saving space.

surface noise — Also called needle scratch. 1. In mechanical recording, the noise caused in the electrical output of a pickup by irregular contact surfaces in the groove. 2. Noise generated by contact of a phonograph stylus with minute particles of dust or other irregularities in a record groove. Can also be caused by excessive wear of a disc or by poor-quality coating on recording tape.

surface of position — Any surface defined by a constant value of some navigational coordinate.

surface recombination rate — The rate at which free electrons and holes recombine at the surface of a semiconductor.

surface recording — Storage of information on a coating of magnetic material such as that which is used on magnetic tape, magnetic drums, etc.

surface reflection — Also called Fresnel loss. The part of the incident radiation that is reflected from the surface of a refractive material. It is directly proportional to the refractive index of the material and is reduced for a given wavelength by application of an appropriate surface coating.

surface resistance — The ratio of the direct voltage applied to an insulation system to the current that passes across the surface of the system. In this case, the surface consists of the geometric surface and the material immediately in contact with it. The thin layer of moisture at the interface between a gas and a solid usually has the greatest effect on surface resistance.

surface resistivity — 1. The resistance between opposite edges of a surface film 1 cm square. Since the length and width of the path are the same, the centimeter terms cancel. Thus, units of surface resistivity are actually ohms. To avoid confusion with usual resistance values, surface resistivity is normally given in ohms per square. It is measured by determining the resistance between two straight conductors 1 cm apart, pressed upon the surface of a slab of the material. Water-absorbent materials usually show a lower resistivity than nonabsorbent ones. 2. The resistance of the surface of an insulating material to the dc current.

surface states — Discontinuities and contaminants at the surface of a semiconductor device, which tend to change the surface characteristics and promote device parameter instability.

surface-temperature resistor — A platinum resistance thermometer designed for installation directly on the surface whose temperature is being measured.

surface wave — 1. A subclassification of the ground wave. So called because it travels along the surface of the earth. 2. A wave that travels along the interface between two media without radiation, such as an electric wave on the surface of a wire surrounded by air. 3. A radio wave that reaches the reception point from the transmitter by traveling along the surface of the earth rather than by reflection from the ionosphere.

surface-wave filter — A filter whose operation is based on the use of the interlaced, deposited electrode pairs and acoustic surface waves on a single crystal substrate. The device operates on traveling-wave principles, not a lumped-element concept.

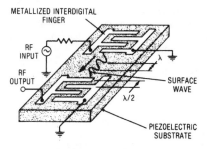

Surface-wave device.

surface-wave transmission line — Abbreviated SWTL. Ideally, a nonradiating broadband transmission line that functions by guiding electromagnetic energy in the surrounding air. A surface line allows coupling to the field anywhere along the guide (in contrast to an ordinary waveguide, in which the internal fields are completely screened from external space by metal walls). A noncontacting device can be used to couple some or all of the energy to or from the line.

surfing — 1. Casually roaming around the Internet. 2. Exploring the World Wide Web. Commonly called surfing the Net.

surfing the Net — Navigating the Internet; usually, random Web browsing.

surge — 1. Sudden current or voltage changes in a circuit. 2. A transient variation in the current and/or potential at a point in the circuit. 3. An oversupply of voltage from the power company, lasting as long as several seconds. A strong surge can damage electronic equipment.

surge admittance — Reciprocal of surge impedance.

surge-crest ammeter — A special magnetometer used with magnetizable links to measure the crest value of transient electric currents.

surge current rating — The maximum current pulse that can be carried by a semiconductor diode for the specified length of time, repetition frequency, waveform, and temperature.

surge generator — *See* impulse generator, 1.

surge impedance — *See* characteristic impedance.

surge-limiting capacitor — A capacitor intended to limit the maximum voltage across the terminations by acting as a low impedance to surges.

surge suppressor — Also called voltage clipper or thyrector. A two-terminal device (pnp) that will conduct in either direction above a specified voltage and polarity, but otherwise acts as a blocking device to current. It essentially is a back-to-back diode with avalanche characteristics for protecting circuitry from high alternating voltage peaks or transients.

surge voltage (or current) — 1. A large, sudden change of voltage (or current), usually caused by the collapse of a magnetic field or by a shorted or open circuit element. 2. The maximum voltage (or current) to which the capacitor should be subjected under any conditions. This includes transients and peak ripple at the highest line voltage.

surge-voltage recorder — *See* Lichtenberg figure camera.

surround — The part of a speaker cone by which its outside edge is anchored, usually corrugated.

surround sound — 1. A system of audio reproduction that uses four or more speakers to simulate the full three-dimensional effect of a live musical performance or cinematic environment. 2. An audio system that includes front and rear channels and sometimes a front and center channel to stimulate the full three-dimensional effect of a live performance.

surveillance — 1. Systemic observation of air, surface, or subsurface areas by visual, electronic, photographic, or other means. 2. Control of premises for security purposes through alarm systems, closed-circuit television (CCTV), or other monitoring methods. 3. Supervision or inspection of industrial processes by monitoring those conditions that could cause damage if not corrected. *See also* supervisory alarm system.

surveillance radar — In air traffic control systems, a radar set or system used in a ground-controlled approach system to detect aircraft within a certain radius of an airport and to present continuously to the radar operator information as to the position, in distance and azimuth, of these aircraft.

surveillance radar station — In the aeronautical radionavigation service, a land station employing radar to detect the presence of aircraft.

survivability — The measure of the degree to which an item will withstand hostile human-made environments and not suffer abortive impairment of its ability to accomplish its designated mission.

SUS — *See* silicon unilateral switch.

susceptance — 1. The reciprocal of reactance, and the imaginary part of admittance. It is measured in siemens. 2. The component of (sinusoidal) current in quadrature with the terminal voltage of a circuit, divided by that voltage.

susceptance standard — A standard with which small, calibrated values of shunt capacitance are introduced into 50-ohm coaxial transmission arrays.

susceptibility — 1. Ratio of the induced magnetization to the inducing magnetic force. 2. The undesired response of an equipment to emissions, interference, or transients, or to signals other than those to which the equipment is intended to be responsive.

susceptibility meter — A device for measuring low values of magnetic susceptibility.

susceptiveness — The tendency of a telephone system to pick up noise and low-frequency induction from a power system. It is determined by telephone-circuit balance, transpositions, wire spacing, and isolation from ground.

suspension — A wire that supports the moving coil of a galvanometer or similar instrument.

suspension galvanometer — An early type of moving-coil instrument in which a coil of wire was suspended in a magnetic field and would rotate when it carried an electric current. A mirror attached to the coil deflecting a beam of light, causing a spot of light to travel on a scale some distance from the instrument. The effect was a pointer of greater length but no mass.

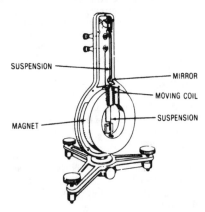

Suspension galvanometer.

sustain — In an organ, the effect produced by circuitry that causes a note to diminish gradually after the key controlling the note has been released.

sustained oscillation — 1. Oscillation in which forces outside the system but controlled by it maintain a periodic oscillation at a period or frequency that is nearly the natural period of the system. 2. Continued oscillation due to insufficient attenuation in the feedback path.

sustained start — An electrical signal for starting a timer that is of any duration longer than the timer setting.

sustaining current — The current required to maintain ionization across a spark gap.

SVGA — Abbreviation for Super Video Graphics Array. An enhancement from VGA monitors. SVGA monitors can display a resolution up to 1028×768 and up to 16.7 million colors.

sw — Abbreviation for shortwave.

swamping resistor — In transistor circuits, a resistor placed in the emitter lead to mask (minimize the effects of) variations caused in the emitter-base junction resistance by temperature variations.

swamp resistance — A small amount of resistance, provided by a resistor with a negative or small positive resistance-temperature coefficient, placed in series with the coil of an electrical indicating instrument to reduce the overall temperature coefficients of the instrument. The resistor also may be used for adjustment of the terminal resistance of the meter.

swap — Refers to the process of moving part of a program from the main memory to external storage (tape or disk) so that another piece of program can be moved from external storage to main memory. Swapping reduces computer efficiency, but it allows the system to get along with less memory.

swapping — 1. A time-sharing term referring to the transfer of a currently operating program from system memory to an external storage device, the replacement of that program by a program of higher priority, and the restoration of the temporarily stored program following execution of the higher-priority program. 2. A feature of an operating system that permits suspended tasks to be moved to secondary storage in order to generate enough memory space so that the next task of the ready list can be loaded into memory (if it is not already resident) and executed. The task will be swapped back into system memory when it is scheduled to resume execution. This movement to and from secondary storage may occur many times to a given task before its executing is complete.

sweating — Soldering by assembling and heating parts previously coated with solder.

SWECS — Abbreviation for small wind energy conversion systems. A system with a generating capacity less than 100,000 kW.

sweep — The crossing of a range of values of a quantity for the purpose of delineating, sampling, or controlling another quantity. Examples of swept quantities are the displacement of a scanning spot on the screen of a cathode-ray tube, and the frequency of a wave.

sweepable equalizer — An equalizer whose center frequency is continuously variable over a given frequency range, but the Q is not variable.

sweep accuracy — The accuracy of the trace horizontal displacement in an oscilloscope compared with the reference independent variable, usually expressed in terms of average rate error as a percent of full scale.

sweep amplifier — An amplifier stage designed to increase the amplitude of the sweep voltage.

sweep circuit — A circuit that produces, at regular intervals, an approximately linear, circular, or other movement of the beam in a cathode-ray tube.

sweep delay — The time between the application of a pulse to the sweep-trigger input of an oscilloscope and the start of the sweep.

sweep-delay accuracy — The accuracy of an indicated sweep delay in an oscilloscope, usually specified in error terms.

sweep expander — *See* sweep magnifier.

sweep-frequency generator — 1. A signal source capable of changing frequency automatically and in synchronism with a display device. The frequency sweep can be obtained by either mechanical or electronic means. 2. An oscillator that generates an audio or radio frequency that is repetitively swept, low to high or high or low, over a preset band of frequencies to provide a test signal for wideband devices.

sweep-frequency record — A test record on which a series of constant-amplitude frequencies have been recorded. Each frequency is typically repeated 20 times per second, starting at 50 Hz and continuing up to 10 kHz or higher.

sweep generator — Also called timing-axis oscillator. A circuit that applies voltages or currents to the deflection elements in a cathode-ray tube in such a way that the deflection of the electron beam is a known function of time against which other periodic electrical phenomena may be examined, compared, and measured.

sweeping receivers — Automatically and continuously tuned receivers designed to stop and lock on when a signal is found or to continually plot band occupancy.

sweep jammer — An electric jammer that sweeps a narrow band of electronic energy over a broad bandwidth.

sweep linearity — The maximum displacement error of the independent variable between specified points on the display area in an oscilloscope.

sweep lockout — A means for preventing multiple sweeps when operating an oscilloscope in a single-sweep mode.

sweep magnifier — Also called sweep expander. A circuit or control for expanding part of the sweep display of an oscilloscope.

sweep oscillator — An oscillator used to develop a sawtooth voltage that can be amplified to deflect the electron beam of a cathode-ray tube. *See also* sweep generator.

sweep switching — The alternate display of two or more time bases or other sweeps using a single-beam CRT. Comparable to dual- or multiple-trace operation of a deflection amplifier.

sweep test — Pertaining to cable, checking the frequency response by generating an rf voltage whose frequency is varied back and forth through a given frequency range at a rapid constant rate while observing the results on an oscilloscope.

sweep-through — A jamming transmitter that sweeps through a radio-frequency band and jams each frequency briefly, producing a sound like that of an aircraft engine.

sweep voltage — The voltage used for deflecting an electron beam. It may be applied to either the magnetic deflecting coils or the electrostatic plates.

swell manual — Also called solo manual. In an organ, the upper manual normally used to play the melody. *See also* manual, 2.

swept resistance — The portion of the total resistance of a potentiometric transducer over which the slider travels when the device is operated through its total range.

swim — The phenomenon in which the constructs on a CRT screen appear to move about their normal position. It can be observed when the refresh rate is slow and is not some multiple or submultiple of line frequency. In some cases, swim is a result of instability in the digital-to-analog converters in the display controller

swimming — Lateral shifting of a thick-film conductor pattern on molten glass crossover patterns.

swing — 1. The variation in frequency or amplitude of an electrical quantity. 2. The total variation of voltage, current, or frequency. 3. The arc traversed by the needle of a meter.

swingback permeability — *See* reversible permeability.

swinger — 1. A swinging short. 2. *See* swing short.

swinging — 1. Momentary variations in frequency of a received wave. 2. Existing only for short periods.

swinging arm — A type of mounting and feed used to move the cutting head at a uniform rate across the recording disc in some recorders. All phonograph pickups are of the swinging-arm type.

swinging choke — 1. A filter inductor designed with an air gap in its magnetic circuit so its inductance decreases as the current through it increases. When used in a power-supply filter, a swinging choke can maintain approximately critical inductance over wide variation in load current 2. An audio-frequency choke whose core is operated saturated with flux. It is used at the input of a power-supply filter for improved voltage regulation. Its inductance is at a maximum for small currents, and charges (swings) to a minimum for large currents.

swing short — Also called swinger. A come-and-go (intermittent) short produced by a pair of wires swinging together in the wind.

swiss-cheese packaging — Also called imitation 2D. A high-density packaging technique in which passive and active components are inserted into holes punched in printed circuit board substrates and attached by soldering or thermocompression bonding or by means of conductive epoxy adhesive.

switch — 1. A mechanical or electrical device that completes or breaks the path of the current or sends it over a different path. 2. In a computer, a device or programming technique by means of which selections are made. 3. A device that connects, disconnects, or transfers one or more circuits and is not designated as a controller, relay, or control valve. The term is also applied to the functions performed by switches. 4. A mechanical component for opening or closing (interrupting or completing) one or more electrical circuits. In electronics, as opposed to the electrical industry, switches tend to be low-voltage, low-current units scaled to the size of the equipment in which they function. Switches suitable for opening and closing 120- and 240-volt ac line current and various dc and signal-level voltages under 100 volts dc predominate. 5. A mechanical or electronic device designed for conveniently interrupting, completing, or changing connections in electrical circuits whenever desired or necessary. Mechanical types may control more than one circuit by incorporating multiple-contact elements that are controlled by the same actuator. Electronic switches ordinarily control only a single electrical circuit.

switchboard — 1. A manually operated apparatus at a telephone exchange. The various circuits from subscribers and other exchanges terminate here, so that operators can establish communications between two subscribers on the same exchange or on different exchanges. 2. A single large panel or an assembly of panels on which are mounted the switches, circuit breakers, meters, fuses, and terminals essential to the operation of electrical equipment. 3. An attended console where telephone subscribers' lines appear for answering and calling. An operator interconnects lines and trunks and supervises the connections.

switch detector — A detector that extracts information from the input waveform only at instants determined by a selector pulse.

switched capacitor — A technique commonly used in analog signal processing to create filtering and signal conditioning circuits.

switched line — Also called dial-up line. A communications link for which the physical path may vary with each usage, such as the public telephone network.

switched network — Also called public switched network and switched message network. 1. The network by which switched telephone service is provided to the public. 2. A multipoint network with circuit switching capabilities. The telephone network is a switched network, as are Telex and TWX.

switcher — 1. A catchall term for a power source that employs switching techniques to achieve higher-efficiency regulation. Can include line switchers and conventional transformer/rectifier ac-operated power sources employing switching regulator techniques. 2. A device that allows the pictures from a number of cameras to be viewed on one monitor. 3. *See* switching power supply.

switch-fader — A control that permits each of two or more cameras to be selectively fed into the distribution system. The fader permits gradual transition from one camera to another.

switch gear — A general term covering switching, interrupting, control, metering, protective, and regulating devices; also assemblies of these devices and associated interconnections, accessories, and supporting structures, used primarily in connection with the generation, transmission, and distribution of electric power.

switch hook — A switch associated with the structure on a telephone set that supports the receiver or handset. The switch is operated when the receiver or handset is removed from or replaced on the support.

switching — 1. Making, breaking, or changing the connections in an electrical circuit. 2. The action of turning a device on and off.

switching amplifier — 1. One whose output stage rapidly switches output power between transistor saturation and cutoff. Average output current is varied by controlling switching frequency, pulse amplitude, duty cycle, or any combination of these factors. Usually, however, switching amplifiers control power by modulating either pulse width or pulse frequency. Switching rates can be varied up to about 100 kHz. Since a switching amplifier controls output power by varying on-off time ratio of the output pulses, amplifier heat dissipation varies little from a nominal rating and is much less than that of amplifiers that consume unused power. However, switching amplifiers generate transients that can upset the operation of nearby electronic circuits. 2. See class D amplifier.

switching center — 1. A location at which data from an incoming circuit is routed to the proper outgoing circuit. 2. A group of equipment within a relay station for automatically or semiautomatically relaying communications traffic. 3. A location where an incoming call/message is automatically or manually directed to one or more outgoing circuits. 4. See switching office.

switching characteristics — An indication of how a device responds to an input pulse under specified driving conditions.

switching circuit — A circuit that performs a switching function. In computers, this is performed automatically by the presence of a certain signal (usually a pulse signal). When combined, switching circuits can perform a logical operation.

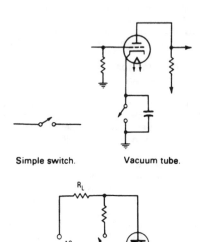

Simple switch.   Vacuum tube.

Triac.

Switching circuits.

switching coefficient — The derivative of applied magnetizing force with respect to the reciprocal of the resultant switching time. It is usually determined as the reciprocal of the slope of a curve of reciprocals of switching times versus the values of applied magnetizing forces, which are applied as step functions.

switching control — An installation in a wire system where telephone or teletypewriter switchboards are installed to interconnect circuits.

switching current — The current through a device at the switching voltage point.

switching device — Any device or mechanism, either electrical or mechanical, that can place another device or circuit in an operating or nonoperating state.

switching differential — The difference between the operate and release points of a switch, caused by hysteresis. It can be in units of amperes, volts, inches, gausses, etc. See hysteresis.

switching diode — A diode that has a high resistance (corresponding to an open switch) below a specified applied voltage but changes suddenly to a low resistance (closed switch) above that voltage.

switching equipment — Equipment located in the telephone company offices that makes the interconnection between the station equipment of two or more subscribers.

switching hysteresis — The principle associated with sensors, such that the operate point is not at the same level as the release point. In solid-state sensors it is accomplished with negative-resistance devices, and in mechanical switches it results from the storing of potential energy before the transition occurs.

switching mode — A way of utilizing a vacuum tube or transistor so that (except for negligibly small transition times) it is either in cutoff or saturation. A transistor operated in this mode can switch large currents with little power dissipation.

switching office — Also called switching center. A location where either toll or local telephone traffic is switched or connected from one line or circuit to another.

switching pad — A transmission-loss pad automatically inserted into or removed from a toll circuit for different desired operating conditions.

switching power supply — Also called switcher. A power supply (usually dc output) that achieves its output regulation by means of one or more active power-handling devices that are alternately placed in the off and on states. Distinguished from linear or dissipative power supplies, in which regulation is achieved by power-handling devices whose conduction is varied continuously over a wide range that seldom (if ever) includes the full off or full on condition. See figure on page 753.

switching regulator — A power supply design that achieves efficient regulation by commuting the input voltage into a filter circuit.

switching time — 1. The interval between the reference time and the last instant at which the instantaneous-voltage response of a magnetic cell reaches a stated fraction of its peak value. 2. The interval between the reference time and the first instant at which the instantaneous integrated-voltage response reaches a stated fraction of its peak value. 3. The time for a multiplexer to change from one channel to the next with the new output signal remaining within a specified percentage of its final value. Expressed for a maximum voltage transition.

switching transients — Transient voltage spikes that appear at a multiplexer's output when the multiplexer is switched from one channel to another and one of the switches is turned off. Such spikes may cause inaccurate measurements if output is sampled, digitized, or integrated during this time.

switching transistor — A three-terminal device with one terminal controlling the electrical impedance

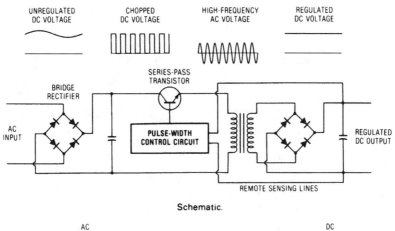

UNREGULATED DC VOLTAGE   CHOPPED DC VOLTAGE   HIGH-FREQUENCY AC VOLTAGE   REGULATED DC VOLTAGE

Schematic.

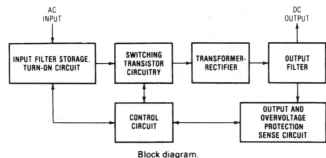

Block diagram.

Switching power supply.

between the other two. Typical transistor switching circuits include inverters, converters, switching voltage regulators, and relay and solenoid drivers.

switching trunk — A trunk that runs between a long-distance office and a local exchange office and is used for completing a long-distance call.

switching voltage — The maximum forward voltage a device can sustain without breaking over into full conduction.

switchplate — A small plate attached to a wall to cover a push-button or other type of switch.

switch register — A physical register made up of a number of manual switches, typically equal to the number of bits in the computer, and generally located on the computer control panel. The switch register is used to manually enter addresses and data into the computer's memory and to manually intervene in program execution. The function performed by a physical switch register can also be implemented by software, with switches being set through a terminal device or a memory location.

switch room — That part of a telephone central office building that houses switching mechanism and associated apparatus.

switchtail ring counter — A type of ring counter in which the output of one stage is inverted before being applied as an input to the next stage. An even number of states equal to $2n$ (where n is the number of flip-flops) normally is produced. For example, a modulo-10 counter can be made from five flip-flops. Each flip-flop changes states on every fifth count. Decoding of all 10 states is accomplished conveniently with 10 two-input gates. A switchtail ring counter will contain the complement of the information it contained initially after n clock pulses, and will contain the initial information again after $2n$ clock pulses.

switch train — A sequence of switches through which connection must be made when a circuit between a calling telephone and a called telephone is established.

SWL — Shortwave listener; one who tunes the short-wave bands as a hobby.

SWR — Abbreviation for standing-wave ratio.

SWR bridge — *See* standing-wave-ratio bridge.

SWR meter — An external or built-in circuit that measures the standing-wave ratio at the transceiver end of the antenna transmission line.

SWTL — Abbreviation for surface-wave transmission line.

syllabic companding — Companding in which the effective gain variations are made at speeds allowing response to the syllables of speech but not to individual cycles of the signal wave.

syllable articulation — Also called percent of syllabic articulation. The percent of articulation obtained when the speech units considered are syllables (usually meaningless and usually of the consonant-vowel-consonant type).

symbol — 1. A simplified design representing a part in a schematic circuit diagram. 2. A letter representing a particular quantity in formulas.

symbolic — 1. Having to do with the representation of something by a conventional sign. 2. Represented by the usual alphanumeric symbols.

symbolic address — Also called a floating address. In digital-computer programming, a label chosen in a routine to identify a particular word, function, or other information independent of the location of the information within the routine.

symbolic code — Also called pseudocode. A code by which programs are expressed in source language; that is, storage locations and machine operations are referred to by symbolic names and addresses that do not depend on their hardware-determined names and addresses. Contrasted with computer code.

symbolic coding — In digital computer programming, any coding system using symbolic rather than actual computer addresses.

symbolic debugger — A system software interactive debugging utility in which the debugging software has access to program symbol tables, and a programmer can refer to memory location names rather than absolute addresses. This is a valuable facility for use with relocatable code or paged systems in which task code may not be loaded at the same memory address each time it executes.

symbolic deck — A deck of cards punched in programmer coding language rather than binary language.

symbolic execution — Reconstructing the logic and actions along a program path via symbolic rather than actual data.

symbolic language — Human-oriented programming language. Any programming language prepared in coding other than a specific machine language, and which thus must be translated by compiling, assembly, etc.

symbolic-language programming — Also called assembly-language programming. The writing of program instructions in a language that facilitates the translation of programs into the binary code through the use of mnemonic convention.

symbolic layout editors — Software design tools that allow the designer to place symbols representing circuit elements (for example, transistors, logic gates, or even registers) instead of having to redesign them each time they are used.

symbolic logic — A special computer or control system language composed of symbols that the instrumentation can accept and handle. Combinations of these symbols can be fed in to represent many complex operations.

symbolic programming — A program using symbols instead of numbers for the operations and locations in a computer. Although the writing of a program is easier and faster, an assembly program must be used to decode the symbol into machine language and assign instruction locations.

symbol table — A table constructed by an assembler or compiler to bind symbolic labels to their actual addresses.

symmetrical — Balanced; i.e., having equal characteristics on each side of a central line, position, or value.

symmetrical alternating quantity — An alternating quantity for which all values separated by a half period have the same magnitude but opposite sign.

symmetrical avalanche rectifier — An avalanche rectifier that can be triggered in either direction. After triggering, it presents a low impedance in the triggered direction.

symmetrically cyclically magnetized condition — The condition of a cyclically magnetized material when the limits of the applied magnetizing forces are equal and of opposite sign.

symmetrical transducer (with respect to specified terminations) — A transducer in which all possible pairs of specified terminations can be interchanged without affecting the transmission.

symmetrical transistor — A transistor in which the collector and emitter are made identical, so either can be used interchangeably.

sync — 1. Short for synchronous, synchronization, synchronizing, etc. 2. The maintenance of correct time relationships between events, i.e., synchronization.

syncable — Capable of being synchronized.

sync compression — 1. The reduction in gain applied to the sync signal over any part of its amplitude range with respect to the gain at a specified reference level. 2. The reduction in the amplitude of the sync signal, with respect to the picture signal, occurring between two points of a circuit.

sync generator — 1. An electronic device that supplies pulses to synchronize a television system. 2. A device for generating a synchronizing signal.

synchro — 1. A small motorlike device containing a stator and a rotor and capable of transforming an angular-position input into an electrical output, or an electrical input into an angular output. When several synchros are correctly connected together, all rotors will line up at the same angle of rotation. 2. A range of ac electromechanical devices that are used in data transmission and computing systems. A synchro provides mechanical indication of its shaft position as the result of an electrical input, or an electrical output that represents some function of the angular displacement of its shaft. Such components are basically variable transformers. As the rotor of a synchro rotates it causes a change in synchro voltage outputs. Major types or classes of synchros include torque synchros, control synchros, resolvers, and induction potentiometers (linear synchro transmitters).

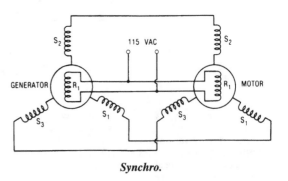

Synchro.

synchro control differential generator — A rotary component for modifying the synchro control generator output signal to correspond to the addition or subtraction from the generator shaft angle. Usually used with a synchro control generator and synchro control transducer.

synchro control generator — A rotary component for transforming the shaft angle to a corresponding set of electrical signals for ultimate retransformation to the shaft position in a remote location.

synchro control transformer — A rotary component that accepts signals from a generator or differential generator for reconversion to the shaft angle with the aid of a servomechanism. Often used by itself as an angle-to-signal transducer, but usually with a synchro control generator and synchro control differential generator.

synchro differential generator — A synchro unit that receives an order from a synchro generator at its primary terminals, modifies this order mechanically by any desired amount according to the angular position of the rotor, and transmits the modified order from its secondary terminals to other synchro units.

synchro differential motor — A motor that is electrically similar to the synchro differential generator except that a damping device is added to prevent oscillation. Its rotor and stator are both connected to synchro generators, and its function is to indicate the sum of or difference between the two signals transmitted by the generators.

synchro generator—A synchro that has an electrical output proportional to the angular position of its rotor.

Synchroguide—A type of control circuit for horizontal scanning in which the sync signal, oscillator voltage pulse, and scanning voltage are compared and kept in synchronism.

synchro motor—A synchro in which the rotor-shaft position is dependent on the electrical input.

synchronism—1. The phase relationship between two or more quantities of the same period when their phase difference is zero. 2. Applied to the synchronous motor, the condition under which the motor runs at a speed that is directly related to the frequency of the power applied to the motor and is not dependent upon other variables.

synchronization—1. The precise matching of two waves or functions. 2. The process of keeping the electron beam on the television screen in the same position as the scanning beam at the transmitter. 3. In a carrier, that degree of matching in frequency between the carrier used for modulation and the carrier used for demodulation that is sufficiently accurate to permit efficient functioning of the system. 4. The maintenance of correct time relationships between events. Examples in recording include synchronization of sound and film for motion-picture use, synchronization of a slide-changing projector with a tape by means of signals recorded on the tape, and selective synchronization, used to synchronize several tracks when they are recorded one at a time.

synchronization error—In navigation, the error due to imperfect timing of two operations (may or may not include the signal transmission time).

synchronization mode—The level at which data on a communications link is synchronized. Start-stop mode denotes character-by-character synchronization; synchronous mode denotes frame-by-frame (or block level) synchronization.

synchronization pulses—Pulses originated by the transmitting equipment and introduced into the receiving equipment to keep the equipment at both locations operating in step.

synchronize—1. To adjust the periodicity of an electrical system so that it bears an integral relationship to the frequency of the periodic phenomenon under investigation. 2. To lock one element of a system into step with another. The term usually refers to locking a receiver to a transmitter, but it can refer to locking the data terminal equipment bit rate to the data set frequency. 3. In television, to maintain two or more scanning processes in phase.

synchronized sweep—A sweep that would free-run in the absence of an applied signal, but is synchronized by the presence of the signal.

synchronized waveforms—A condition whereby the timing of one waveform is related to the timing of another waveform.

synchronizer—1. The component of a radar set that generates the timing voltage for the complete set. See also timer, 3. 2. A computer storage device used to compensate for a difference in a rate of flow of information or time of occurrence of events when information is being transmitted from one device to another.

synchronizing-pulse selector—A circuit used to separate synchronizing pulses from commutated pulse trains.

synchronizing reactor—A current-limiting reactor that is connected momentarily across the open contacts of a circuit-interrupting device for synchronizing purposes.

synchronizing relay—A relay that functions when two alternating-current sources are in agreement within predetermined limits of phase angle and frequency.

synchronizing separator—See amplitude separator.

synchronizing signal—See sync signal.

synchronous—1. In step or in phase, as applied to two devices or machines. 2. A term applied to a computer in which the performance of a sequence of operations is controlled by equally spaced clock signals or pulses. 3. Having a constant time interval between successive bits, characters, or events. The term implies that all equipment in the system is in step. 4. Operation of a switching network by a clock pulse generator. More critical than asynchronous timing but requires fewer and simpler circuits. 5. Describes modems, terminals, and transmissions in which all bits are of equal duration. Synchronous transmissions must always be associated with a clocking signal to keep the receiver and transmitter in step. 6. A communications protocol in which the data characters and bits are transmitted at a fixed rate with the transmitter and receiver synchronized. This eliminates the need for start-stop elements and provides greater transmission efficiency. 7. Term used to describe a device or system in which all events occur in a predetermined timed sequence.

synchronous booster converter—A synchronous converter connected in series with an ac generator and mounted on the same shaft. It is used for adjusting the voltage at the commutator of the converter.

synchronous capacitor—A rotating machine running without mechanical load and designed so that its field excitation can be varied in order to draw a leading current (like a capacitor) and thereby modify the power factor of the ac system or influence the load voltage through such change in power factor.

synchronous clock—An electric clock driven by a synchronous motor, for operation on an ac power system in which the frequency is accurately controlled.

synchronous communication—A method of transferring serial binary data between computer systems or between a computer system and a peripheral device; binary data is transmitted at a fixed rate, with the transmitter and receiver synchronized. Synchronization characters are located at the beginning of each message or block of data to synchronize the flow.

synchronous communications satellite—A communications satellite whose orbital speed is adjusted so that the satellite remains above a particular point on the surface of the earth.

synchronous computer—A digital computer in which all ordinary operations are controlled by clock pulses from a master clock.

synchronous converter—A synchronous machine that converts alternating current to direct current and vice versa. The armature winding is connected to the collector rings and commutator.

synchronous data communications—A serial I/O hardware protocol in which the transmitter and receiver are synchronized to a common clock signal.

synchronous demodulation—In a color television receiver, the process of separately detecting the I and Q sidebands of the color subcarrier system.

synchronous demodulator—Also called a synchronous detector. 1. A demodulator in which the reference signal has the same frequency as the carrier or subcarrier to be demodulated. It is used in color television receivers to recover either the I or the Q signals from the chrominance sidebands. 2. A double-balanced, full-wave demodulator. The basic operation is similar to that of the chroma demodulation system except that the

45.75-MHz reference carrier is generated inside of the IC rather than in a separate crystal oscillator referenced to burst signal as is done with the chroma. The only signals that are detected are the ones that are in phase with the demodulation signal produced by the carrier generator. Signals that are not synchronized with the IF carrier, such as the adjacent-channel video and sound carriers and random noise, are not detected. This requires less extensive trapping of interfering signals in earlier IF stages. 3. A detector sensitive to signals close to or at a particular frequency that is the same as the frequency of a control signal, applied independently. The synchronous detector is also phase-sensitive and is used in bridge and other null circuits as an antinoise device.

synchronous detector — *See* synchronous demodulator.

synchronous device — A device that transfers information at its own rate and not at the convenience of any interconnected device.

synchronous gate — A time gate in which the output intervals are synchronized with the incoming signal.

synchronous generator — A circuit designed to synchronize an externally generated signal with a train of clock pulses. The generator produces precisely one output pulse for each cycle of the input signal. The output pulse thus has a width equal to that of the period of the clock pulse train.

synchronous idle character — A communication control character used to provide a signal for synchronization of the equipment at the data terminals.

synchronous induction motor — A motor with the rotor laminations cut away, exposing the rotor and definite poles. Also, a motor that has definite poles, but the poles are permanent magnets. This term is also applied to a wound-rotor (slip-ring) motor that is started as an induction motor and, when near synchronism, has dc applied to two of the rings to operate as a true synchronous motor.

synchronous inputs — 1. Those inputs of a flip-flop that do not control the output directly, as do those of a gate, but only when the clock permits and commands. Called J and K inputs or ac set and reset inputs. 2. Inputs to a flip-flop that affect the output only at the time the clock signal changes from a particular level to the other level.

synchronous inverter — *See* dynamotor.

synchronous logic — The type of digital logic used in a system in which logical operations take place in synchronism with clock pulses.

synchronous machine — 1. A machine that has an average speed exactly proportionate to the frequency of the system to which it is connected. 2. A system whose operation is tied to data-independent timing or a system clock.

synchronous modem — A line-termination unit that uses a derived clocking signal to perform bit synchronization with incoming data.

synchronous motor — 1. An induction motor that runs at synchronous speed. Its stator windings have the same arrangement as in nonsynchronous induction motors, but the rotor does not slip behind the rotating magnetic stator field. 2. Type of ac electric motor in which rotor speed is related directly to frequency of power supply. 3. A motor that runs at a speed in step with the power-station frequency (fundamental or harmonic of the alternator speed). Often started as an induction motor and converted to synchronous operation as it approaches synchronous speed. 4. An alternating-current motor that operates at a speed determined solely by the frequency of the supply power and does not slow down as its load increases.

synchronous multiplexer — A multiplexer that can time-interleave two data streams into one higher-speed stream. In a system using these types of multiplexers, all peripheral equipment in the system must be under the control of a master synchronizing device or clock.

synchronous operation — Operation of a system under the control of clock pulses.

synchronous rectifier — A rectifier in which contacts are opened and closed at the correct instant by either a synchronous vibrator or a commutator driven by a synchronous motor.

synchronous shift register — Shift register that uses a clock for timing of a system operation and in which only one state change per clock pulse occurs.

synchronous speed — A speed value related to the frequency of an ac power line and the number of poles in the rotating equipment. Synchronous speed in revolutions per minute is equal to the frequency in hertz divided by the number of poles, with the result multiplied by 120.

synchronous system — 1. A system in which the sending and receiving instruments are operating continuously at substantially the same frequency and are maintained in a desired phase relationship. 2. A system in which all events are synchronized by a common clock pulse.

synchronous torque — 1. The maximum load torque with which a motor can be loaded after it comes to synchronous speed. These torques are usually higher than starting torques. 2. Torque that a synchronous motor develops after dc field excitation is applied; the total steady-state torque available to drive the load.

synchronous transfer — An I/O transfer that takes place in a certain amount of time without regard to feedback from the receiving device.

synchronous transmission — 1. Data transmission in which characters and bits are transmitted at a fixed rate, with the transmitter and receiver synchronized by a clock source. This eliminates the need for individual start bits and stop bits surrounding each byte, thus providing greater efficiency. Compare with asynchronous transmission. 2. Transmission in which the sending and receiving instruments operate continuously at the same frequency and are held in a desired phase relationship by correction devices.

synchronous vibrator — An electromagnetic vibrator that simultaneously converts a low dc voltage to a low alternating voltage and applies it to a power transformer, from which a high alternating voltage is obtained and rectified. In power packs, it eliminates the need for a rectifier tube.

synchroscope — 1. An instrument used to determine the phase difference or degree of synchronism of two alternating-current generators or quantities. 2. An oscilloscope on which recurrent pulses or waveforms may be observed, and which incorporates a sweep generator that produces one sweep for each pulse.

synchro system — A system for obtaining remote indication or control by means of self-synchronizing motors such as selsyns and equivalent types.

synchro-torque receiver — A relatively low-impedance positioning device that generates its own torque when driven by a suitable synchro-torque transmitter.

synchro-torque transmitter — A positioning device that generates electrical information of sufficient power to drive a suitable torque receiver.

synchrotron — A device for accelerating charged particles (e.g., electrons) in a vacuum. The particles are guided by a changing magnetic field while being accelerated many times in a closed path by a radio-frequency electric field.

synchrotron noise — Radio noise caused by the acceleration of charged particles to high speeds.

synchrotron radiation — Also called magnetic bremsstrahlung. The radiation produced by relativistic electrons as they travel in a region of space containing magnetic fields.

sync level — The level of the peaks of the synchronizing signal.

sync limiter — A circuit used in television circuits to prevent sync pulses from exceeding a predetermined amplitude.

sync pulse — Part of the sync signal in a television system.

sync section — A color TV circuit comprising a keyer, burst amplifier, phase detector, reactance tube, subcarrier oscillator, and quadrature amplifier.

sync separator — The circuit that separates the picture signals from the control pulses in a television system.

sync signal — Also called a synchronizing signal. The signal employed for synchronizing the scanning. In television it is composed of pulses at rates related to the line and field frequencies.

sync-signal generator — A synchronizing signal generator for a television receiver or transmitter.

syntax — 1. The rules that govern the structure of expressions in a language. 2. The grammar of a programming language, that is, rules about how commands may be used and how they fit together. 3. Set of grammatical rules defining valid constructs of a language. 4. Structure of expressions in a language and the rules governing the structure of a language. 5. The way in which words are put together to form valid computer commands.

synthesis — The combination of parts to form a whole.

synthesizer — 1. A device that can generate a number of crystal-controlled frequencies for multichannel communications equipment. 2. A system for generating a precise and stable frequency whose accuracy is determined by quartz crystal oscillators, instead of inductance/capacitance tuned circuits. As compared with the latter, a synthesizer circuit can result in a tuner or transmitter whose frequency setting is known with great accuracy and that is free from drift or other tuning errors. True digital tuners (as opposed to those which tune conventionally but have digital frequency displays) use synthesizers in order to advance in discrete steps from one exact channel frequency to another without passing through the unwanted frequencies in between.

synthesizer frequency meter — A device for measuring frequency by utilizing a synthesized crystal-based signal for the internally generated signal.

synthetic display generation — Logical and numerical processing to display collected or calculated data in symbolic form.

synthetic speech — Artificially reproduced acoustic signals that are recognizable as human speech.

syntony — The condition in which two oscillating circuits have the same resonant frequency.

system — 1. An assembly of component parts linked together by some form of regulated interaction into an organized whole. 2. A collection of consecutive operations and procedures required to accomplish a specific objective. 3. A collection of units combined to work a larger integrated unit having the capabilities of all the separate units. 4. The complete computer assembly, with CPU, memory, I/O, plus any required devices or peripherals for the application intended. 5. A set of interconnected elements constituted to achieve a given objective by performing a specific function.

system analysis — The examination of an activity, procedure, method, technique, or business to determine what must be accomplished and how the necessary operations may best be accomplished by using electronic data-processing equipment.

systematic distortion — Distortion of a periodic or constant nature, such as bias or characteristic distortion; the opposite of fortuitous distortion.

systematic error — 1. The magnitude and direction of the tendency of a measuring process to measure some quantity other than the one intended. 2. An error of the type that has an orderly character that can be corrected by calibration.

systematic inaccuracies — Those inaccuracies due to inherent limitations in the equipment.

system bus — A general-purpose backbone used to connect processors, memory, and peripherals to form a computer system.

system deviation — The instantaneous difference between the value of a specified system variable and the ideal value of the same system variable.

system effectiveness — A measure of the degree to which an item can be expected to fulfill a set of specified mission requirements, which may be expressed as a function of availability, dependability, and capability.

system element — One or more basic elements, together with other components necessary to form all or a significant part of one of the general functional groups into which a measurement system can be classified.

system engineering — A method of engineering analysis whereby all the elements in a system, including the process itself, are considered.

system failure rate — The number of occasions during a given time period on which a given number of identical systems do not function properly.

system ground — One common point to which the grounds for various pieces of equipment in a system are connected. The system ground is generally the best point to connect to earth ground.

system input unit — A device defined as a source of an input job stream.

system integration — The process of matching physical, electrical, and logical characteristics of different components so that they work together.

system layout — In a microwave system, a chart or diagram showing the number, type, and termination of circuits used in the system.

system library — The assemblage of all cataloged data sets at an installation.

system macroinstruction — A predefined macroinstruction that makes available access to operating system facilities.

system master tapes — Magnetic tapes that contain programmed instructions necessary for preparation of a computer before programs are run.

system network diagram — A diagram showing each station and its relationship to the other stations in a network of stations and to the control point(s).

system noise — The output of a system when it is operating with zero input signal.

system of beams — The three electron beams emitted by the triple electron-gun assembly in a color tube. They occupy positions equidistant from a common axis and are spaced 120° apart around the axis.

system of units — An assemblage of units for expressing the magnitudes of physical quantities.

system output unit — An output device that is shared by all jobs and onto which specified output information is transcribed.

system overshoot — The largest value of system deviation following the dynamic crossing of the ideal value as a result of a specified stimulus.

system reliability — The probability that a system will perform its specified task properly under stated conditions of environment.

system residence volume — The volume in which the nucleus of an operating system and the highest-level index of the catalog are located.

system resonance — The fundamental resonance of the woofer/enclosure combination. Related to the low-frequency performance of the systems, but not without ambiguity, especially when comparing different types of enclosures. Not to be confused with free-air resonance of the woofer.

systems analyst — A person skilled in solving problems with a digital computer. He or she analyzes and develops information systems.

systems-implementation languages — Essentially compilers with assemblerlike features. These allow the programmer to take advantage of compiler-level languages, but revert to assembler languages when necessary.

system software — 1. Software that is intimately associated with the operating system, e.g., kernel routines, system services, and system support software. 2. All the programs that tie together and coordinate the devices that make up the computer system. It includes such programs as loaders, compilers, interpreters, and input/output routines. 3. That collection of programs which controls the computer and helps people use the computer. Generally, supervisory and support modules, as opposed to application programs. This includes assemblers, editors, debuggers, operating systems, compilers, I/O drivers, loaders, and other utility programs. 4. A program that acts to assist the applications program and is usually not visible to the user. It takes the form of controlling and utilizing the internal resources of the computer. It keeps track of time, sequences, and events, and performs the input and output functions. This is the program that is used for general applications and, when made part of a memory medium such as disk, tape, semiconductor, or bubble memory, is referred to as the system firmware.

systems programs — Computer programs provided by a computer manufacturer. Examples are operating systems, assemblers, compilers, debugging aids, and input/output programs.

system support — Functions such as language translators, debugging tools, diagnostics, and libraries that enable a system user or programmer to write and test tasks in an efficient manner.

T

T— 1. Symbol for transformer or absolute temperature. 2. Abbreviation for prefix *tera-* (10^{12}).

tab— 1. *See* land, 2. 2. A nonprinting spacing action on a typewriter or tape preparation device, the code of which is necessary to the tab sequential format method of programming. 3. Term used to describe construction of wound capacitors whereby the two electrode foils, separated by dielectric, are positioned one above the other with a margin of dielectric completely surrounding the edges of both foils. Foil tabs, placed in the wound capacitor during winding, span the margin for connection to the leads.

table— 1. A collection of data, each item of which is uniquely identified by a label, its position with respect to the other items, or some other means. 2. A set of data values.

table lockup— 1. A computer technique that stores a table of data in a computer so that the data can be used during the running of the program. 2. In a computer, a method of controlling the location to which a jump or transfer is made. It is used especially when there is a large number of alternatives, as in function evaluation in scientific computations.

tablet— An input device that digitizes coordinate data indicated by stylus position.

tab sequential format— A means for identification of a computer word by the number of tab characters in the block preceding the word. The initial character in each word is a tab character. Words must be presented in a certain order, but all characters in a word except the tab character may be omitted when the command that word represents is not desired.

tabulate— 1. To arrange data into a table. 2. To print totals.

tabulated cylinder— Translation of a curve along a direction line with upper and lower limits on the distance of translation.

tabulator— A machine (e.g., a punch-card machine) that reads information from one medium and produces lists, totals, or tabulations on separate forms or continuous paper strips.

tachometer— 1. An instrument used to measure the frequency of mechanical systems by the determination of angular velocity. 2. A transducer that gives an electric output signal proportional to the rotational speed of a shaft. 3. A device for measuring rate by counting the number of pulses that occur in a given period. 4. An instrument designed to measure the rate of rotation of components such as shafts.

tachometer generator— A small generator attached to a rotating shaft for the purpose of generating a voltage proportional to the shaft speed. In speed-control circuits, a tachometer may be coupled to the shaft of the motor whose speed is to be controlled. A change of motor speed will produce a change of tachometer output voltage. This

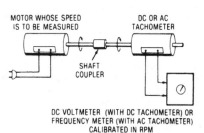

Tachometer.

change of voltage can be used as an error signal to restore the speed to the desired value.

tacky state— The condition of a material when it exhibits an adhesive bond to another surface.

TACS— Abbreviation for Total Access Communication System. An analog cellular radio protocol used in the United Kingdom.

tactical— 1. Of or relating to combat tactics. 2. Of or occurring at the battlefront. 3. Using or being weapons of force employed at the battlefront.

tactical air navigation— A short-range UHF air-navigation system that presents accurate information to a pilot in two dimensions, distance and bearing from a selected ground station.

tactical call sign— Call sign that identifies a tactical command or tactical communication facility.

tactical communications system— Systems that provide internal communications within tactical air elements, composed of transportable and mobile equipment assigned as unit equipment to the supporting tactical unit.

tactical frequency— Radio frequency assigned to a military unit to be used in the accomplishment of a tactical mission.

tag— 1. A label attached to a piece of data in a data-flow computer that says where the data is to be used in a program. 2. Also called a sentinel. In digital-computer programming, a unit of information whose composition differs from that of other members of the set so that it can be used as a marker or label.

Tagged Image File Format— Abbreviated TIFF. A high-resolution bit-mapped graphics format for storing scanned images on IBM PCs and compatibles and Macintosh computers.

tail— 1. A small pulse following the main pulse and in the same direction, or the slow decay following the main body of the pulse. 2. The free end of wire extending beyond the bond impression of a wire bond from the heel.

tail clipping— A method of sharpening the trailing edge of a pulse.

tailing— *See* hangover, 1.

tailor-made — Also called custom. Referring to a program or programs that are specially written for one particular task, for one set of people. Tailor-made software is usually commissioned by an individual customer, and not sold to anyone else.

tail pull — The act of removing the excess wire left when a wedge or ultrasonic bond is made.

tail pulse — A pulse in which the decay time is much longer than the rise time.

tails-out — 1. Storage of a nonreversing tape on the takeup reel rather than on the supply reel, to avoid the tape-distorting interval winding stresses and uneven wind of high-speed rewinding. 2. In the case of reversing tapes (i.e., cassettes or four track open-reel tapes), storing the tape wound on whichever reel (or hub) it ends up on after having been played or recorded.

tail-warning radar set — A radar set placed in the tail of an aircraft to warn of aircraft approaching from the rear.

takeup reel — 1. The reel that accumulates the tape as it is recorded or played on a tape recorder. 2. The reel on a tape recorder that is on the right side of the head stack and is empty prior to beginning the recording.

talbot — A unit of luminous energy in the mksa system equal to 1 lumen-second.

talk-back — A voice intercommunicator; an intercom.

talker echo — An echo that reaches the ear of the person who originated the sound.

talking battery — The dc voltage supplied by the central office to the subscriber's loop to operate the carbon transmitter in the handset.

talking path — The transmission path of a telephone circuit, making up the tip and ring conductors and the equipments connected to them.

talk-listen switch — A switch on an intercommunication unit to switch the speaker as required to function either as a reproducer or as a microphone.

talk-off — The tendency of a dual-tone multifrequency (DTMF) system to respond falsely to other-than-valid DTMF signals. Talk-off criteria are generally specified very subjectively — sometimes in such broad terms as "good" or "poor."

tally light — Signal lights installed at the front and back of television cameras to inform performers and crew members when a particular camera is on the air.

tamper device — 1. Any device, usually a switch, that is used to detect an attempt to gain access to intrusion alarm circuitry, such as by removing a switch cover. 2. A monitor circuit to detect any attempt to modify the alarm circuitry, such as by cutting a wire.

tamper switch — A switch that is installed in such a way as to detect attempts to remove the enclosure of some alarm system components, such as control box doors, switch covers, junction box covers, or bell housings. The alarm component is then often described as being "tampered."

tandem — *See* cascade.

tandem office — In a telephone system, an office that interconnects the local end offices over tandem trunks in a densely settled exchange area where it is not economical to provide direct interconnection between all end offices. The tandem office completes all calls between the end offices but is not connected directly to subscriber's stations.

tandem transistor — Two transistors in one package and internally connected together.

tangent — A straight line that touches the circumference of a circle at one point.

tangent galvanometer — A galvanometer consisting of a small compass mounted horizontally in the center of a large vertical coil of wire. The current through the coil is proportional to the tangent of the angle at which the compass needle is deflected.

tangential component — A component acting at right angles to a radius.

tangential pickup arm — A pickup arm that maintains the longitudinal axis of the stylus tangent to the record grooves throughout the entire movement of the arm across the record.

tangential sensitivity — A term generally applied as an indication of quality in a receiving system. This term can be used to define the minimum signal level that can be detected above the background noise. However, it is usually expressed as that signal power level that causes a 3-dB rise above the noise-level reading.

tangential sensitivity on look-through — The strength of the target signal, measured at the receiver terminals, required to produce a signal pulse having twice the apparent height or the noise.

tangential wave path — In radio-wave propagation of a direct wave over the earth, a path that is tangential to the surface of the earth. The tangential wave path is curved by atmospheric refraction.

tangent of loss angle — The ratio of the equivalent series resistance and the capacitive reactance of a capacitor obtained at a sinusoidal alternating voltage of specified frequency.

tangent sensitivity — The slope of the line tangent to the response curve at the point being measured.

tank — 1. A unit of acoustically operating delay-line storage containing a set of channels. Each channel forms a separate recirculation path. 2. *See* tank circuit, 2.

tank circuit — 1. A circuit capable of storing electrical energy over a band of frequencies continuously distributed about a single frequency at which the circuit is said to be resonant or tuned. The selectivity of the circuit is proportionate to the ratio between the energy stored in the circuit and the energy dissipated. This ratio is often called the Q of the circuit. 2. Also called a tank. A parallel-resonant circuit connected in the plate circuit of an electron-tube generator.

tantalum (electrolytic) capacitor — An electrolytic capacitor with a tantalum foil or sintered-slug anode.

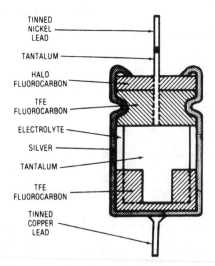

Tantalum capacitor.

tantalum-foil electrolytic capacitor — A capacitor that consists of two tantalum foil electrodes with an oxide on the anode, separated by layers of absorbent paper saturated with an operating electrolyte.

tantalum oxide — A dielectric material used in capacitors; it is formed electrochemically in a thin film on surfaces of tantalum metal.

T-antenna — Any antenna consisting of one or more horizontal wires, with the lead-in connected approximately in the center.

tap — 1. A fixed electrical connection to a specified position on the element of a potentiometer, transformer, etc. 2. A branch. Applied to conductors, such as a battery tap, and to miscellaneous general use. 3. A connection of the terminal end of one conductor to another conductor at some point along its run; the process of making such a connection. Also, a connection or connection point brought out of a winding, as in a transformer, at some point between the ends of the winding, for controlling output ratios. 4. Connection on the phone line that lets a third party listen to what is being said in a telephone conversation.

tap center — A connection to the electrical midpoint of a coil, resistor, or transformer winding.

tap conductor — A conductor, usually short but longer than a pigtail, that serves to connect utilization equipment to the serving circuit.

tap crystal — A compound semiconductor that stores current when stimulated by light and then gives up energy in the form of flashes of light when subjected to mechanical tapping.

tape — 1. Plastic ribbons with one side coated metallically to receive impressions from a recording head, or, if already recorded, to induce signals in a playback head. Tape is regularly wound on reels or packaged in magazines or cartridge form. 2. *See* punched tape. 3. A ribbon of flexible material — e.g., friction, magnetic, punched, etc. 4. A strip of material, which may be punched, coated, or impregnated with magnetic or optically sensitive substances, used for data input, storage, or output. The data is stored serially in several channels across the tape transversely to the reading or writing motion. Tape has a far higher storage capacity than disk storage of similar volume, but it takes much longer to write or recover data to tape than to disk.

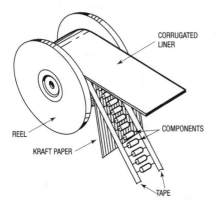

Tape-and-reel packaging.

tape cable — Also called flat flexible cable. A form of flexible multiple conductor in which parallel strips of metal are imbedded in an insulating material.

tape cartridge — A magazine or holder for a length of magnetic tape that by its design avoids the necessity for manual threading or handling. Usually compatible only with one specific type of machine. *See also* tape magazine.

tape character — Information consisting of bits stored across the several longitudinal channels of a tape.

tape-controlled carriage — A paper-feeding device automatically controlled by a punched paper tape.

tape copy — A message received in tape form as the result of a transmission.

tape deck — 1. The basic assembly of a tape recorder, consisting of the tape-moving mechanism (the tape transport) and a head assembly. Some decks also include recording and playback preamplifiers; these properly are called tape recorders. Some have playback-only preamplifiers; these have no standard name but are often called tape players. 2. A tape recorder that does not include power amplifiers or speakers.

tape distributor — A machine that reads prerecorded paper tape.

tape drive — 1. A mechanism for moving tape past a head; tape transport. 2. The motor and associated mechanism that pulls the tape past the playing or recording heads.

tape feed — A mechanism that feeds the tape to be read or sensed by a computer or other data-handling system.

tape guides — Grooved pins of nonmagnetic material mounted at either side of the recording-head assembly. Their function is to position the magnetic tape on the head as it is being recorded or played.

tape head — The transducer on a tape recorder past which the tape runs during record or replay. It applies a magnetic field to the tape during the recording process and provides electrical output during replay.

tape hiss — Sibilant background noise heard when a tape is played. Although some of this noise is directly attributable to irregularities of the oxide coating, some also is contributed by the recording circuitry.

tape lifters — A system of movable guides that automatically divert tape from contact with the recorder heads during the fast forward or rewinding mode of operation, thus preventing head wear.

tape limited — Pertaining to a computer operation in which the time required for the reading and writing of tapes is greater than the time required for computation.

tape loop — A length of magnetic tape with the ends joined together to form an endless loop. Used either on a standard recorder, a special message-repeater unit, or in conjunction with a cartridge device, it enables a recorded message to be played back repetitively; there is no need to rewind the tape.

tape magazine — Also called a tape cartridge. A container holding a reel of magnetic recording tape, which can be played without being threaded manually.

tape mark — A special record indicating end of file.

tape monitor — 1. A circuit that permits the checking of recordings by taking the signal directly from the tape a moment after the recording is made. This is only possible on three-head recorders. 2. An interruption in the signal path of a preamplifier, from which the selected input signal is supplied to an external tape recorder, and to which the playback output of the recorder is returned for further amplification and processing. Tape monitor circuits allow one to listen to a tape as it is recorded, ensuring it is being recorded properly. They also allow the use of external signal processing devices such as equalizers, noise reducers, and expanders, and are sometimes known as external processor loops for that reason.

tape-on surface-temperature resistor — A surface-temperature resistor installed by adhering the sensing element to the surface with a piece of pressure-sensitive tape.

tape parity — Parity error that occurs when information is transferred to or from magnetic tape.

tape-path center line — The locus or path traced by an imaginary point, located on the recording tape midway between its edges, as it travels from reel-to-reel through guides, past heads, and between capstan and pressure rollers. For correct tracking and minimum tape distortion, the entire path should lie in a single plane located above the motor board at a height compatible with head-gap locations.

tape perforator — A device that records a teletype-writer or data message by punching holes in paper tape, under control of either a local keyboard or an incoming signal.

tape phonograph — See tape player.

tape player — Sometimes called a tape phonograph or a tape reproducer. A unit for playing recorded tapes. It has no facilities for recording.

tape punch — A peripheral device that generates (punches) holes in a paper tape to produce a hard copy of PC memory contents.

taper — In communication practice, a continuous or gradual change in electrical properties with length — e.g., as obtained by a continuous change of cross section of a wavelength or by the distribution and change of resistance of a potentiometer or rheostat.

tape reader — 1. A peripheral device for converting information stored on punched paper tape to electrical signals for entry into PC memory. 2. A unit that is capable of sensing data from punched tape.

tape recorder — A mechanical-electronic device for recording voice, music, and other audio-frequency material. Sound is converted to electrical energy, which in turn sets up a corresponding magnetic pattern on iron oxide particles suspended on paper or plastic tape. During playback, this magnetic pattern is reconverted into electrical energy and then changed back to sound through the medium of headphones or a speaker. The recorded material may be converted to a visual display by the use of an oscilloscope, other visual indicator, or a graphic recorder. See also videotape recording.

tapered potentiometer — A continuously adjustable potentiometer whose resistance varies nonuniformly along the element, being greater or less for equal slider movement at various points along the resistance element.

tapered transmission line — See tapered waveguide.

tapered waveguide — Also called a tapered transmission line. A waveguide in which a physical or electrical characteristic changes continuously with distance along the axis of the guide.

tape relay — A method in which perforated tape is used as the intermediate storage in the process of relaying messages between transmitting and receiving stations.

tape-relay station — A component of a communications center that carries out the function of receiving and forwarding messages by means of tape relay.

tape reproducer — See tape player.

tape reservoir — That part of a magnetic tape system used to isolate the tape storage inertia (i.e., tape reels, etc.) from the drive system.

taper pin — A pin-type contact having a tapered end designed to be inserted into a tapered hole.

tape skew — The deviation of a tape from following a linear path when transported across the heads, causing a time displacement between signals recorded on different tracks and amplitude differences between the outputs from individual tracks owing to variations in azimuth alignment. The adjectives *static* and *dynamic* are used to distinguish between the steady and fluctuating components of the tape skew.

tape speed — The speed at which tape moves past the head in the recording or playback mode. The standard tape speed for home use is $7\frac{1}{2}$ ips (19.05 cm/s) or half this speed ($3\frac{3}{4}$ ips). One-fourth or even one-eighth this speed is also used, but usually only for music when special high-quality tape is utilized. The professional recording speed for music mastering is usually 15 ips (38.1 cm/s).

tape-speed errors — Any variation in tape speed from the normal speed over the record or reproduce head, regardless of cause.

tape-speed variations — See flutter, 1.

tape splicer — A device for splicing magnetic tape automatically or semiautomatically.

tape station — See tape unit.

tape threader — A device that makes easier the threading of magnetic recording tape onto the reel.

tape-to-card — Pertaining to equipment or methods used to transfer data from magnetic or punched tape to punched cards. (No longer in use.)

tape-to-head speed — The relative speed of the tape and head during normal recording or replay. (The tape-to-head speed coincides with the tape speed in conventional longitudinal recording, but it is considerably greater than the tape speed in systems where the heads are scanned across or along the tape.)

tape-to-tape converter — A device for changing from one form of input/output medium or code to another, i.e., magnetic tape to paper tape (or vice versa) or eight-channel code to five-channel code.

tape transmitter — 1. A machine actuated by previously punched paper tape and used for high-speed code transmission. 2. A facsimile transmitter designed for transmission of subject copy printed on narrow tape.

tape transport — See transport.

tape-transport mechanism — See motor board.

tape unit — Also called tape station. A device that contains a tape drive and the associated heads and controls.

tape-wound core — Also known as bimag. A magnetic core consisting of a plastic or ceramic toroid around which is wound a strip of thin magnetic tape possessing a square hysteresis-loop characteristic. A tape-wound core is used principally as a shift-register element.

tap lead — The lead connected to a tap on a coil winding.

tapped control — A rheostat or potentiometer having a fixed tap at some point along the resistance element, usually to provide fixed grid bias or automatic tone compensation.

tapped line — A delay line in which more than two terminal pairs are associated with a single sonic-delay channel.

tapped resistor — A wirewound fixed resistor having one or more additional terminals along its length, generally for voltage-divider applications.

tapped winding — A coil winding with connections brought out from turns at various points. See figure on page 763.

tapper bell — A single-stroke bell designed to produce sound of low intensity and relatively high pitch.

tap switch — A multicontact switch used chiefly for connecting a load to any one of a number of taps on a resistor or coil.

target — 1. In a camera tube, a structure employing a storage surface that is scanned by an electron beam to generate an output-signal current corresponding to the charge-density pattern stored thereon. 2. Also called an

Tapped winding.

anticathode. In an X-ray tube, an electrode or part of an electrode on which a beam of electrons is focused and from which X-rays are emitted. 3. In radar, a specific object of radar search or surveillance. 4. Any object that reflects energy back to the radar receiver. 5. In image pickup tubes, a structure employing a storage surface that is scanned by an electron beam to generate a signal output current corresponding to a charge-density pattern stored thereon. The structure may include the storage surface that is scanned by an electron beam, the backplate, and the intervening dielectric.

target acquisition — In radar operation, the first appearance of a recognizable and useful signal returned from a new target.

target capacitance — In camera tubes, the capacitance between the scanned area of the target and backplate.

target cutoff voltage — In camera tubes, the lowest target voltage at which any detectable electrical signal, corresponding to a light image on the sensitive surface of the tube, can be obtained.

target discrimination — The characteristic of a guidance system that permits it to distinguish between two or more targets in close proximity.

target fade — The loss or decrease of signal from the target due to interference or other phenomena.

target glint — *See* scintillation, 2.

target identification — A visual procedure by which a radar target is positively identified as either hostile or friendly.

target integration — A system of increasing the sensitivity of a TV camera when viewing a static scene by cutting off the beam for a predetermined number of frames and reading out the information in the first frame after beam turn-on.

target language — The language into which some other language is to be properly translated. In computers, any language (such as object code) into which another language (such as source code) is translated.

target noise — Reflections of a transmitted radar signal from a target that has a number of reflecting elements randomly oriented in space.

target reflectivity — The degree to which a target reflects electromagnetic energy.

target scintillation — The apparent random movement of the center of reflectivity of a target observed during the course of an operation.

target seeker — In a missile homing system, the element that senses some feature of the target so that the resulting information can be used to direct appropriate maneuvers to maintain a collision course.

target signature — The characteristic pattern of a given target when displayed by detection and classification equipment.

target system — 1. The computer that is being emulated on another system. 2. The microcomputer under development: the prototype.

target voltage — In a camera tube with low-velocity scanning, the potential difference between the thermionic cathode and the backplate.

tariff — The published rate for utilization of a specific unit of equipment, facility, or type of service provided by a communications common carrier. Also the vehicle by which the regulating agency approves or disapproves such facilities or services. Thus, the tariff becomes a contract between customer and common carrier.

task — 1. A unit of work for the central processing unit as determined by the control program; therefore, the basic multiprogramming unit under the control program. 2. A software module in which code is executed in a sequential manner.

task control block — The consolidation of the control information that has to do with a task.

task dispatcher — The control-program function that selects a task from the task queue and gives control of the central processing unit to that task.

tasking — The capability of a computer to process more than one set of instructions, or task, at a time.

task list — A system data structure containing a list of tasks within the system.

task management — Those functions of the control program by which the use of the central processing unit and other resources by tasks is regulated.

task queue — A queue that contains control information for all tasks in a system at any given time.

task state — The status of a task; it may be undefined, ready to execute, executing, or suspended awaiting some event.

taut-band galvanometer — A galvanometer whose moving coil is suspended between two taut ribbons.

taut-band suspension — In an indicating instrument, a mechanical arrangement in which the moving element is suspended by means of a thin, flat conducting ribbon at each end. The ribbons normally are in tension sufficient to maintain the lateral motion of the moving element within limits that permit the freedom of useful motion for any mounting position of the instrument. A restoring torque is produced within the ribbons when the moving element rotates.

TBC — Abbreviation for time base corrector. A device that corrects time base stability errors (errors in the rate at which a signal is occurring) during tape playback.

TC — 1. Abbreviation for teleconferencing. The use of telecommunications to link physically separated individuals or groups for purposes of holding a meeting. The technology can range from a simple speakerphone to elaborate custom-designed teleconference rooms equipped with two-way audio, graphics, and video. *See also* teleconference. 2. Abbreviation for temperature coefficient.

T carrier — A time-division-multiplexed service, normally supplied by the telephone company, that usually operates a digital transmission facility at an aggregate data rate of 1.544 Mbps and above.

TC capacitor — Abbreviation for temperature-compensating capacitor.

Tchebychev filters — Filter networks that are designed to exhibit a predetermined ripple in the passband (ripple amplitudes from 0.01 dB to 3 dB are common) in exchange for which they provide a more rapid attenuation above cutoff — which, unlike their passband response, is monotonic.

Tchebychev function — A mathematical function whose curve ripples within certain bounds (*See* ripple, 2). This produces an amplitude response more square than that of the Butterworth function, but with less desirable phase and time-delay characteristics. There is an entire family of Tchebychev functions (0.1 ripple, 0.5 ripple, etc.).

T-circulator — A circulator consisting of three identical rectangular waveguides joined asymmetrically to form a T-shaped structure, with a ferrite post or wedge at the center of the structure. Power that enters any waveguide emerges from only one adjacent waveguide.

TCR — Abbreviation for temperature coefficient of resistance.

TDD/TTY — Acronym for tone dialing for the deaf/teletype. A method by which the hearing impaired can type messages over normal phone lines using special equipment.

TDM — *See* time-division multiplexing.

TDMA — Abbreviation for time division multiple access. A method of sharing a transmission medium by dividing the time into "slots" and allocating those slots to users on a network.

TDR — Abbreviation for time-delay relay.

teach box — A hand-held control with which a robot can be programmed.

tearing — Distortion observed on the television screen when the horizontal synchronization is unstable. Groups of horizontal lines are displaced in an irregular manner, creating the appearance of parts of the image having been torn away.

teaser transformer — A transformer of two T-connected, single-phase units for three-phase to two-phase or two-phase to three-phase operation; it is connected between the midpoint of the main transformer and the third wire of the three-phase system.

teasing — In the life testing of switches, the slow movement of rotor contacts making and breaking with stator contacts.

technical control board — In a switch center or relay station, a testing position at which there are provisions for making tests on switches and associated access lines and trunks.

technical load — The portion of the operational power load of a facility that is required for communications-electronics, tactical-operations, and ancillary equipment. It includes power for lighting, air conditioning, or ventilation necessary for full continuity of communications-electronics operation.

technician — A person who works directly with scientists, engineers, and other professionals in every field of science and technology. Technicians' duties vary greatly, depending on their field of specialization. But in general, the scientist or engineer does the theoretical work, and the technician translates theory into action.

technician license — A class of amateur radio license issued in the United States by the FCC for the primary purpose of operation and experimentation on frequencies above 50 MHz.

technology — 1. To organize the means for satisfying needs and desires. 2. Know-how applied to get a job done.

tecnetron — A high-power multichannel field-effect transistor similar to a triode tube in that it has anode and cathode connections on opposite ends of a small germanium rod (and a grid connection between them).

TED — Abbreviation for transferred electron device. A gate-controlled, Gunn-effect unit. The triode TED consists of an anode, cathode, and Schottky barrier gate. The device resembles a Gunn diode. It has a threshold and a negative-resistance region that produces oscillation at the transit-time frequency, which depends on the anode-cathode spacing.

tee junction — Also spelled T-junction. A junction of waveguides in which the longitudinal guide axes form a T. The guide that continues through the junction is called the main guide; the one that terminates at a junction, the branch guide.

telautograph — Also called a telewriter. A writing telegraph instrument in which the movement of a pen in the transmitting apparatus varies the current and thereby causes the corresponding movement of a pen at the remote receiving instrument.

tele- — A prefix, from Greek, that means distant or at a distance.

telecamera — Abbreviation for a television camera.

telecardiophone — An amplifying stethoscope that permits heart sounds to be heard at a distance.

telecast — Abbreviation for television broadcasting — specifically, a television program, or the act of broadcasting a television program.

telecasting — The broadcasting of a television program.

telecommand — One-way transmission to modify, initiate, or terminate functions of a device at a distance.

telecommunication — 1. All types of systems in which electric or electromagnetic signals are used to transmit information between or among points. Transmission media may be radio, light, or waves in other portions of the electromagnetic spectrum; wire; cable; or any other medium. 2. Data transmission between a computing system and remotely located devices via a unit that performs the necessary format conversion and controls the rate of transmission. 3. Pertaining to the art and science of telecommunication. 4. Any transmission, emission, or reception of signs, signals, writing, images, or sounds, or intelligence of any nature by wire, radio, visual, or other electromagnetic systems. 5. Generally refers to the reception and/or transmission of information — whether in digital or analog form — by telephone, telegraph, and the like. 6. The transmission/reception between terminals, or between terminals and computers, of (digitized) information or messages over telephone lines or by wireless transmission. 7. Communication by electrical transmission, such as television, radio, telephone and telegraph, that is carried over a distance. 8. The conversion and transmission of information from a computer to another device at a distance. 9. The process of moving data from one point to another. 10. Synonym for data communication. The transmission of information from one point to another.

telecommunication lines — Telephone and other communication lines that are used to transmit messages from one location to another.

teleconference — 1. A conference between persons who are remote from one another but linked together by a telecommunications system. 2. Electronically linked meeting conducted among groups of people in geographically separated locations. Three forms of teleconference currently exist. A typical video mode includes an audio system, a facsimile device, and a storage unit for the display of prepared material. An audio teleconference lacks live-video capability but nonetheless usually utilizes graphic-display equipment. In a computer teleconference, data is transmitted by keyboard and then printed out. 3. A meeting conducted via telephone lines between two or more groups of individuals who are physically separated. It provides a low-cost technique of bringing people together via telephone channels and also overcomes problems imposed by distance and time.

telegenic — The suitability of a subject or model for televising.

telegram — Any written matter that is transmitted by telegraphy.

telegraph — 1. A system that employs interruptions or polarity changes of direct current for the transmission of signals. 2. A system of communication using coded signals.

telegraph channel — 1. The transmission media and intervening apparatus involved in the transmission of telegraph signals in a given direction between two intermediate telegraph installations. A means of one-way transmission of telegraph signals. 2. A communication path that is suitable for transmitting telegraph signals between two points. The required bandwidth depends on the signaling speed.

telegraph circuit — A complete circuit over which signal currents flow between transmitting and receiving apparatus in a telegraph system. It sometimes consists of an overhead wire or cable and a return path through the ground.

telegraph concentrator — A switching arrangement by means of which a number of branch or subscriber lines or station sets may be connected to a lesser number of trunk lines, operating positions, or instruments through the medium of manual or automatic switching devices to obtain more efficient use of facilities.

telegraph distributor — A device that effectively associates one direct current or carrier telegraph channel in rapid succession with the elements of one or more signal-sending or signal-receiving devices.

telegraph-grade circuit — A circuit suitable for transmission by teletypewriter equipment. Normally the circuit is considered to employ dc signaling at a maximum speed of 75 bauds.

telegraph key — A single-pole, single-throw switch that can be rapidly operated by hand so as to form the dots and dashes of telegraph code signals by opening and closing contacts to modulate current.

telegraph-modulated waves — Continuous waves whose amplitude or frequency is varied by means of telegraphic keying.

telegraph repeater — Apparatus that receives telegraph signals from one line and retransmits corresponding signals on another line.

telegraph selector — A device that performs a switching operation in response to a definite signal or group of successive signals received over a controlling circuit.

telegraph signal distortion — The time displacement of transitions between conditions, such as marking and spacing, with respect to their proper relative positions in perfectly timed signals. The total distortion is the algebraic sum of the bias and the characteristic and fortuitous distortions.

telegraph sounder — A telegraph receiving instrument in which an electromagnet attracts an armature each time a pulse arrives. This armature makes an audible sound as it hits against its stops at the beginning and end of each current impulse, and the intervals between these sounds are translated from code into the received message by the operator.

telegraph transmission speed — The rate at which signals are transmitted. This may be measured by the equivalent number of dot-cycles per second or by the average number of letters or words transmitted and received per minute.

telegraph transmitter — A device for controlling a source of electric power in order to form telegraph signals.

telegraph wave — A completely random two-level signal.

telegraphy — 1. A system of telecommunication for the transmission of graphic symbols, usually letters or numerals, by the use of a signal code. It is used primarily for record communication. 2. Any system of telecommunication for the transmission of graphic symbols or images for reception in record form, usually without gradation of shade values. 3. The science of communicating at a distance by means of coded current pulses sent over wire circuits.

telelectrocardiograph — A device for transmission and remote reception of electrocardiograph signals.

telematics — Also called telematique. 1. A capability in telecommunications, computers, and semiconductors.

2. The integration of computer-processing applications with telecommunications capabilities.

telemeter — 1. To transmit analog or digital reports of measurements and observations over a distance (e.g., by radio transmission from a guided missile to a control or recording station on the ground). 2. A complete measuring, transmitting, and receiving apparatus for indicating, recording, or integrating the value of a quantity at a distance by electric translating means.

telemetering — A measurement accomplished with the aid of intermediate means that allow perception, recording, or interpretation of data at a distance from a primary sensor. The most widely employed interpretation of telemetering restricts its significance to data transmitted by means of electromagnetic propagation.

telemeter service — Metered telegraph transmission between paired telegraph instruments over an intervening circuit adapted to serve a number of such pairs on a shared-time basis.

telemetry — 1. The science of sensing and measuring information at some remote location and transmitting the data to a convenient location to be read and recorded. 2. The transmission of measurements obtained by automatic sensors and the like over communications channels. 3. The practice of transmitting and receiving the measurement of a variable for readout or other uses. The term is most commonly applied to electric signal systems. 4. Transmission and collection of data obtained by sensing conditions in a real-time environment. 5. Transmission of coded analog data, often real-time parameters, from a remote site.

telemetry beacon — A system whereby two or more reply pulses are transmitted by the beacon for the transmission of data from the test vehicle to the ground station.

telemetry cable — Cable used for the transmission of information from instruments to the peripheral recording equipment.

telemetry frame — In PCM systems, one complete sampling of words or channels of information at a given rate; in time-division multiplexing, one complete commutator revolution.

telemetry frame rate — The frequency derived from the period of one frame.

teleoperator — A mobile robot controlled by a human operator. Teleoperators are most often used in areas that would be hazardous to human beings.

telephone — Combination of apparatus for converting speech energy into electrical waves, transmitting the electrical energy to a distant point, and there reconverting the electrical energy to audible sounds.

telephone address — The complete 10-digit number that specifies the location of a particular telephone. Consists of a 3-digit area code, a 3-digit central office code, plus a 4-digit station number.

telephone answering machine — A device that answers a telephone in the subscriber's absence, plays a recorded message, and then records messages from callers before disconnecting. When the subscriber returns he or she can rewind and play back all messages recorded in his or her absence, or the subscriber can interrogate the telephone answering machine remotely by sending appropriate tones from another telephone and then listening to the recorded messages.

telephone capacitor — A fixed capacitor connected in parallel with a telephone receiver to bypass rf and higher audio frequencies and thereby reduce noise.

telephone carrier current — A carrier current used for telephone communication so that more than one channel can be obtained on a single pair of wires.

telephone central office— A switching unit, installed in a telephone system that provides service to the general public, that has the necessary facilities for terminating and interconnecting lines and trunks.

telephone channel— 1. A channel suitable for the transmission of telephone signals. 2. A communication path suitable for carrying voice message traffic between two points.

telephone circuit— A complete circuit over which audio and signaling currents travel between two telephone subscribers communicating with each other in a telephone system.

telephone company— Any common carrier providing public telephone system service. There are about 2500 telephone companies in the United States.

telephone current— An electric current produced or controlled by the operation of a telephone transmitter.

telephone instrument— A subscriber's complete telephone set.

telephone jack— See phone jack.

telephone network— The arrangement of trunks, subscriber lines, centralized switching locations, etc., used to switch calls and make connections between various customers.

telephone number— In the United States, a seven-digit number assigned to a telephone. It contains a three-digit central office code.

telephone pickup— Any of several devices used to monitor telephone conversations, usually without direct connection to the telephone line and operating on the principle of mutual magnetic coupling.

telephone plug— See phone plug.

telephone receiver— The portion of the telephone instrument held closely to the human ear that converts the analog electrical speech signal into sound waves.

telephone relay— 1. A relay used in telephone system switching equipment. 2. Any high-density, multipole relay intended for signal level switching in communications and data processing. These relays may have self-wiping contacts, that is, bifurcated pole blades that scrape over mating surfaces to remove minor oxide buildup.

telephone repeater— An assemblage of amplifiers and other equipment employed at points along the line to rebuild the signal strength in a telephone circuit.

telephone ringer— An electric bell that operates on low-frequency alternating or pulsating current and is used for indicating a telephone call to a station being alerted.

telephone system— A group of telephones plus the lines, trunks, switching mechanisms, and all other accessories required to interconnect the telephones.

telephone transmitter— 1. A microphone used in a telephone system. See also microphone. 2. A transducer that uses voice sound pressure on a diaphragm to translate the acoustic message into an analog electrical signal.

telephone wire— A general term referring to many different types of communication wire. It refers to a class of wires and cables, rather than a specific type.

telephony— 1. The transmission of speech current over wires, enabling two persons to converse over almost any distance. 2. A telecommunications system for transmitting speech or other sounds.

telephoto— Also called telephotography. A photoelectrical transmission system for point-to-point or air-to-ground transmission of high-definition pictorial information.

telephotography— See telephoto.

telephoto lens— A lens system that is physically shorter than its rated focal length. It is used in still, movie, and television cameras to enlarge images of objects photographed at comparatively great distances.

telepoint— A cordless telephone system in which a subscriber can make but not receive phone calls in public areas that have been equipped with telepoint base stations. The system is not mobile — the user must remain essentially in a fixed location throughout the duration of the call. Both service and equipment are less expensive than cellular.

teleprinter— 1. See printer, 1. 2. Trade name used by Western Union for its telegraph terminal equipment. 3. See teletypewriter.

teleprocessing— 1. A form of information handling in which the data-processing system operates in conjunction with communication facilities (originally a trademark of International Business Machines Corp.). 2. The processing of data that is received from or sent to remote locations by way of telecommunication lines. Such systems are essential to hook up remote terminals or connect geographically separated computers.

teleprompter— A device mounted on the front of a camera that projects copy on a semireflective mirror placed directly in front of the lens so the performer can read lines while appearing to look at the camera.

teleran— A navigational system in which radar and television transmitting equipment are employed on the ground, with television receiving equipment in the aircraft, to televise the image of the ground radar PPI scope to the aircraft along with map and weather data.

telering— A frequency-selector device for the production of ringing power.

telerobotics— Remote operation systems that mimic human movements through machinery.

telesis— Progress intelligently planned and directed; the attainment of desired ends by the application of intelligent human effort to the means.

telesynd— Telemeter or remote-control equipment that is synchronous in both speed and position.

Teletype— 1. A trademark of Teletype Corporation for a series of teleprinter equipment such as tape punches, reperforators, page printers, etc., used in communications systems. Now often incorrectly used as a generic term to indicate any similar piece of equipment. 2. A peripheral electromechanical device for inserting or recording a program into or from a PC memory in either a punched paper tape or printed ladder diagram format.

teletypewriter— Also known as a teleprinter. 1. See printer, 1. 2. A generic term referring to the basic equipment made by Teletype Corporation and to teleprinter equipment. The teletypewriter uses eletromechanical functions to generate codes (Baudot) in response to a human input to a manual keyboard. 3. A keyboard machine that can transmit and receive alphabetical, numerical, and certain control (nonprinting) characters as a train of pulses on two wires. Attachments can be fitted for punching paper tape and printing on a roll of paper at the same time, also for reading tape and printing the message that is read. 4. A keyboard printing unit that is often used to enter information into a computer and to accept output from a computer. The widest usage of the teletypewriter is as an input/output device in minicomputer systems, a remote terminal in a time-sharing system, and an operator console for computer systems.

teletypewriter code— A special code in which each code group is made up of five units, or elements, of equal length that are known as marking or spacing impulses. The five-unit start-stop code consists of five signal impulses preceded by a start impulse and followed by a stop impulse. Each impulse except the stop impulse is 22 milliseconds in length; the stop impulse is 32 milliseconds (based on 60-word-per-minute operation).

teletypewriter exchange service — 1. A commercial service that provides teletypewriter communication on the same basis as telephone service, through central switchboards to stations in the same city or other cities. 2. Abbreviated TWX. A public teletypewriter exchange (switched) service in the United States and Canada formerly owned by AT&T but now belonging to Western Union. Baudot and ASCII-coded machines are used.

teletypewriter signal distortion — With respect to a stop-start teletypewriter signal, a shift of the transition points of the signal pulses from their proper positions in relation to the beginning of the start pulse. The magnitude of the distortion is expressed as a percentage of a perfect unit pulse length.

teletypewriter switching system — A total message switching system whose terminals are teletypewriter equipment.

teletypewriter test tape — A tape perforated so that it contains the identification of the transmitting station followed by repetitions of the letters RY and a test that consists of letters and figures.

televise — The act of converting a scene or image field into a television signal for transmission.

television — Abbreviated TV. 1. A telecommunication system for transmission of transient images of fixed or moving objects. 2. A system for converting visual images into electrical signals that can be transmitted (by radio or wire) to distant receivers, where the signals are reconverted to the original visual images and displayed on a CRT.

television and radar navigation — A navigational system that (a) employs ground-based search radar equipment along an airway to locate aircraft flying near that airway; (b) transmits, by television means, information pertaining to these aircraft and other information to the pilots of properly equipped aircraft; and (c) provides information to the pilots appropriate for use in the landing approach.

television bandwidth — The range of frequencies within which a television broadcast transmission must fall, generally assigned to be from 54 to 890 MHz.

television broadcast band — The frequencies assignable to television broadcast stations in the band extending from 54 to 806 MHz. These frequencies are grouped into channels, as follows: channels 2 through 4, 54 to 72 MHz; channels 5 and 6, 76 to 88 MHz; channels 7 through 13, 174 to 216 MHz; and channels 14 through 69, 470 to 806 MHz.

television broadcast station — A radio station for transmitting visual signals, and usually simultaneous aural signals, for general reception.

television camera — Abbreviated TV camera. A camera that contains an electronic image tube in place of a photographic film. The image formed on the tube face by a lens is scanned rapidly by a moving electron beam. The beam current varies with the local brightness of the image, which is transmitted to the viewer's set, where it controls the brightness of the scanning spot in a cathode ray tube. The scanning spots at the camera and the viewing tube must be accurately synchronized.

television channel — A band of frequencies 6 megahertz wide in the television broadcast band, assigned to television broadcasting stations. The channel for associated sound signals may or may not be considered part

Television channels

Channel number	Band (megahertz)	Channel number	Band (megahertz)	Channel number	Band (megahertz)
2	54–60	29	560–566	57	728–734
3	60–66	30	566–572	58	734–740
4	66–72	31	572–578	59	740–746
5	76–82	32	578–584	60	746–752
6	82–88	33	584–590	61	752–758
7	174–180	34	590–596	62	758–764
8	180–186	35	596–602	63	764–770
9	186–192	36	602–608	64	770–776
10	192–198	37	608–614	65	776–782
11	198–204	38	614–620	66	782–788
12	204–210	39	620–626	67	788–794
13	210–216	40	626–632	68	794–800
14	470–476	41	632–638	69	800–806
15	476–482	42	638–644	70*	806–812
16	482–488	43	644–650	71*	812–818
17	488–494	44	650–656	72*	818–824
18	494–500	45	656–662	73*	824–830
19	500–506	46	662–668	74*	830–836
20	506–512	47	668–674	75*	836–842
21	512–518	48	674–680	76*	842–848
22	518–524	49	680–686	77*	848–854
23	524–530	50	686–692	78*	854–860
24	530–536	51	692–698	79*	860–866
25	536–542	52	698–704	80*	866–872
26	542–548	53	704–710	81*	872–878
27	548–554	54	710–716	82*	878–884
28	554–560	55	716–722	83*	884–890
		56	722–728		

*The frequencies between 806 and 890 MHz, formerly allocated to television broadcasting, are now allocated to the land mobile services. Operation, on a secondary basis, of some television translators may continue on these frequencies.

of the television channel. There are a total of 12 VHF and 56 UHF television channels, beginning with channel 2 and ending with channel 69. Former UHF channels 70 through 83 are now either used by repeaters or are directly assigned to the land mobile service. Frequency ranges begin at 54 MHz and extend to 806 MHz.

television engineering — *See* radio engineering.

television interference — Abbreviated TVI. Interference in the reception of the sound and/or video portion of a television program by a transmitter or another device.

television monitor — A television receiver that receives the signals generated by a television camera directly, or remotely through a radio link. It may be used to check image and sound reception continuously, as a closed-circuit television system for security and surveillance, or in editing.

television pickup station — A land mobile station used for the transmission of television program material and related communications from the scenes of events occurring at points removed from television broadcast station studios to the television broadcast stations.

television picture monitor — A special-purpose television set for displaying picture signals in broadcast or closed-circuit television systems. Applications are in studio master control, for tape monitoring, for control of picture quality in studios and intercity network relays, and for the display of pictures for audiences.

television radar air navigation — A system in which aircraft positions are determined by ground radar, and the resulting PPI display, superimposed on a map, is transmitted to the aircraft by television. By this means, each pilot can observe the position of his aircraft in relation to others.

television receive only — Abbreviated TVRO. Small satellite earth stations designed to receive satellite-relayed television programming, but having no provision for transmitting to the satellite. Typically small home systems.

television receiver — A radio receiver for converting incoming electric signals into television pictures and the associated sound.

television reconnaissance — Air reconnaissance by optical or electronic means to supplement photographic and visual reconnaissance.

television relay system — A system of two or more stations for transmitting television relay signals from point to point, using radio waves in free space as a medium. Such transmission is not intended for direct reception by the public.

television repeater — A repeater used in a television circuit.

television screen — In a television receiver, the fluorescent screen of the picture tube.

television signal — The audio signal and video signal that are broadcast simultaneously to produce the sound and picture portions of a televised scene.

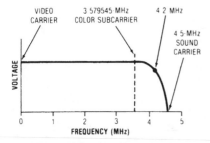

Television signal.

television transmitter — The aggregate radio-frequency and modulating equipment necessary to supply to an antenna system the modulated radio-frequency power by which all component parts of a complete television signal (including audio, video, and synchronizing signals) are concurrently transmitted.

televoltmeter — A telemeter that measures voltage.

telewattmeter — A telemeter that measures power.

telewriter — *See* telautograph.

telex — 1. An audio-frequency teleprinter system used in Great Britain to provide teletypewriter service over telephone lines. 2. An automatic teleprinter exchange service available worldwide through various common carriers. Western Union is the carrier in the United States. It is similar to TWX. Only Baudot equipment is provided, but business machines may be used also. 3. A dial-up telegraph service enabling its subscribers to communicate directly and temporarily among themselves by means of start-stop apparatus and circuits of the public telegraph network. The service operates worldwide. Computers can be connected to the telex network.

telluric current — *See* earth current, 1.

telpak — A communications-carrier service for the leasing of wideband channels between points.

Telstar — A low-altitude active communications satellite used for microwave communication and satellite tracking.

tempco — Abbreviation for temperature coefficient.

temperature coefficient — Abbreviated TC or tempco. 1. A factor used to calculate the change in the characteristics of a substance, device, or circuit element with changes in its temperature. 2. The percentage change in the output voltage (or current) of a regulated power supply due to a variation of ambient temperature. The values are usually expressed as a percentage per degree Celsius and restricted to the specified ambient range of the unit. 3. A measure of a sensor's sensitivity, specifically the rate at which a parameter (such as resistance or voltage) changes with temperature. It is usually expressed as a percent per degree (either Fahrenheit or Celsius), and may be positive or negative.

temperature coefficient of capacitance — Often referred to as the TC characteristic. The amount of capacitance change of a capacitor per degree change in temperature. For capacitors that exhibit a relatively linear capacitance change with temperature, the temperature coefficient is commonly expressed in parts per million per degree Celsius (ppm/°C). Capacitors with a nonlinear capacitance change are more often described in terms of percent capacitance change for a given temperature.

temperature coefficient of frequency — The rate at which the frequency changes with temperature, generally expressed in hertz per megahertz per degree Celsius at a given temperature.

temperature coefficient of permeability — A coefficient expressing the change in permeability as the temperature rises or falls. It is expressed as the rate of change in permeability per degree.

temperature coefficient of resistance — Abbreviated TCR. 1. The ratio of the change in resistance (or resistivity) to the original value for a unit change in temperature. The temperature coefficient over the temperature range from t to t_1, referred to the resistance R_t at temperature t, is determined by the following ratio:

$$\frac{R_{t1} - R_t}{R_t(t_1 - t)}$$

where t is the temperature, preferably in degrees Celsius. The value will be positive unless otherwise indicated by a negative sign. 2. The maximum change in resistance per unit change in temperature, usually referred to in parts per million (ppm) per degree Celsius and specified over a temperature range. The temperature is that of the resistor itself, not ambient temperature.

temperature coefficient of voltage drop — The change in the voltage drop of a glow-discharge tube, divided by the change in ambient temperature or in the temperature of the envelope.

temperature coefficient value — The expected percentage change per degree of temperature difference from a specified temperature.

temperature-compensated zener diode — A positive-temperature-coefficient reversed-bias zener diode (pn junction) connected in series with one or more negative-temperature forward-biased diodes within a single package.

temperature-compensating capacitor — Abbreviated TC capacitor. A capacitor whose capacitance varies with temperature in a known and predictable manner. Normally this characteristic is specified with a P or N (positive or negative: to indicate the direction of change) followed by a number that indicates the change in parts per million per degree Celsius (centigrade). Such capacitors are used extensively in oscillator circuits to compensate for changes due to temperature variations in the values of other components.

temperature compensation — The process whereby the effects of an increase or decrease in ambient temperature are canceled (e.g., as in the case of an oscillator that is required to maintain a stable output frequency regardless of ambient temperature changes).

temperature control — 1. A switch actuated by a thermostat responsive to changes in temperature; used to maintain temperature within certain limits. 2. A control device responsive to temperature.

temperature cycling — A type of accelerated test in which systems or devices are subjected alternately to high and low temperatures to simulate diurnal temperature fluctuations.

temperature derated voltage — The maximum voltage that may be applied continuously to the terminals of a capacitor at a stated temperature between the rated and the maximum temperature.

temperature derating — Lowering the voltage, current, or power rating of a device or component when it is used at elevated temperatures.

temperature detector — An instrument used to measure the temperature of a body. Any physical property that is dependent on temperature may be employed, such as the differential expansion of two bodies, thermoelectromotive force at the junction of two metals, change of resistance of a metal, or the radiation from a hot body.

temperature-limited — The condition of a cathode when all the electrons emitted from it are drawn away by a strong positive field. The only way to increase the flow of electrons is to raise the cathode temperature.

temperature relay — A relay that functions at a predetermined temperature.

temperature rise — Also called T rise. 1. The difference between the initial and final temperature of a component or device. Temperature rise is expressed in degrees Celsius or Fahrenheit, usually referred to an ambient temperature, and equals the hot-spot temperature minus the ambient temperature. 2. Temperature change of a terminal from a no-load condition to a full-current load.

temperature saturation — See filament saturation.

temperature sensor — See thermistor; thermocouple, 3.

temperature shock — A rapid change from one temperature extreme to another.

temperature-wattage characteristic — In a thermistor, the relationship, for a specific ambient temperature, between the temperature of the thermistor and the applied steady-state power.

temporary magnet — A magnetized material having a high permeability and low retentivity.

temporary memory — A read and write memory containing information that can be changed by the internal circuitry of the electronic switching system.

temporary storage — 1. Internal storage locations in a computer reserved for intermediate and partial results. 2. Memory locations or registers reserved for immediate and partial results obtained during the execution of a program.

TEM wave — Abbreviation for transverse electromagnetic wave.

tens complement — 1. An arithmetic process employed in a computer to perform decimal subtractions through the use of addition techniques. The tens-complement negative of a number is obtained by individually subtracting each digit in the number from 9 and adding 1 to the result. 2. The radix complement in decimal notation.

tensilized tape — A variety of polyester backing that has been "prestretched" to prevent further severe elongation when subjected to excessive tension. A tape that does not stretch before breaking can be spliced back together without loss of program material.

tensiometer — A device for determining the tautness of a supporting wire or cable.

tension — 1. Mechanical — the condition of strain that tends to stretch. 2. Electrical — the potential or electrostatic voltage.

tension arm — An arm, or feeler, over which a recording tape rides as it enters or leaves the heads. It is lightly spring loaded to take up any tape slack and maintain a uniform tension in order to reduce flutter. Should the tape end or break, the arm causes the transport to shut off.

tenth-power width — In a plane containing the direction of the maximum of a lobe, the full angle between the two directions in that plane, about the maximum, in which the radiation intensity is one-tenth the maximum value of the lobe.

tera- — Abbreviated T. Prefix for the numerical quantity of 10^{12}.

terabyte — A unit of memory equal to approximately 1 trillion bytes.

terahertz — One million megahertz, or 10^{12} hertz. Letter symbol: THz.

teraohm — One million megohms, or 10^{12} ohms.

teraohmmeter — An instrument used to measure extremely high resistance.

terminal — 1. A point of connection for two or more conductors in an electrical circuit. 2. A device attached to a conductor to facilitate connection with another conductor. 3. A point in a system or communication network at which data can be either inserted or removed. 4. A device that permits access to a central computer, usually including a typewriterlike keyboard and either CRT or typing device (teletypewriter) on which the operator's input and the computer's output can be displayed. 5. Any device capable of sending and/or receiving information over a communications channel. 6. The means by which data is entered into a computer system and by which the decisions of the system are communicated to the environment it affects. A wide variety of terminal devices have been built, including teleprinters, special keyboards, light displays, cathode-ray tubes, thermocouples, pressure gauges and other instrumentation, radar units, and telephones. 7.

A device designed to terminate a conductor that is to be affixed to a post, stud, chassis or another conductor, or the like, to establish an electrical connection. Types of terminals include ring, tongue, spade, flag, hook, blade, quickconnect, offset, and flanged.

terminal area – 1. A portion of a microelectronic circuit used for making electrical connections to the conductor pattern – e.g., an enlarged pad (area) on a semiconductor die. 2. A portion of a printed circuit used for making electrical connections to the conductive pattern, such as the enlarged portion of conductor material surrounding a component mounting hole.

terminal block – An insulating base or slab equipped with one or more terminal connectors for the purpose of making electrical connections thereto.

terminal board – Also called a terminal strip. 1. An insulating base or slab equipped with terminals for connecting wiring. 2. Board fabricated from an insulating material containing a single or multiple row or arrangement of termination points for the purpose of making connections. 3. An insulated mounting for terminal connections. Terminal strips are available with threaded holes to accept threaded screws or with threaded studs to accept fastening washers and nuts. If terminal areas are separated by an insulating barrier, the terminal strips are called barrier blocks.

Terminal boards.

terminal box – A housing in which cable pairs are brought out to terminations for connections.

terminal brush – A brush with long bristles for cleaning fuses and terminals in a terminal box.

terminal connector – A connector that joins a conductor to a lead, terminal pad (solid or laminated block), or round terminal stud of electrical apparatus.

terminal cutout pairs – Numbered, designated pairs brought out of a cable at a terminal.

terminal emulation – 1. A situation in which special software makes a computer behave as though it were a terminal connected to another computer. 2. The process by which an asynchronous terminal, such as a PC, can communicate with a synchronous computer system, such as a mainframe or minicomputer. This is done by imitating (emulating) the operations of the mainframe's or mini's synchronous terminals.

terminal equipment – 1. At the end of a communications channel, the equipment essential for controlling the transmission and/or reception of messages. 2. Telephone and teletypewriter switchboards and other centrally located equipment to which wire circuits are terminated. 3. Assemblage of communications-type equipment required to transmit and/or receive a signal on a channel or circuit, whether it be delivery or relay. 4. In radio relay systems, usually refers to equipment used at points where intelligence is inserted or derived, as distinct from equipment used to relay a reconstituted signal. 5. Any equipment at the end of communications lines that sends and/or receives certain signals for specific services.

For example, telephones and teletypewriters are terminal devices; so are many datasets.

terminal guidance – 1. Guidance applied to a guided missile between midcourse guidance and arrival at the target. 2. Electronic, mechanical, visual, or other assistance given an aircraft pilot to facilitate arrival at, landing upon, or departure from an air landing or air-drop facility.

terminal hole – Also called component hole. A hole used for the attachment and electrical connection of component terminations, including pins and wires, to a printed board.

terminal impedance – 1. The complex impedance seen at the unloaded output or input terminals of transmission equipment or a line in otherwise normal operating condition. 2. See terminal resistance.

terminal leg – See terminal stub.

terminal lug – 1. A threaded lug to which a wire may be fastened in a terminal box. 2. A cylindrical piece of metal, either solid or hollow and of two or more diameters, that can be stacked, flared, swaged, or pressed into a hole for the purpose of connecting leads or external wires to the conductive pattern.

terminal mode – Condition of a PC when it is connected to a host computer.

terminal pad – An alternate term for terminal area.

terminal pair – An associated pair of accessible terminals (e.g., the input or output terminals of a device or network.)

terminal repeater – 1. An assemblage of equipment designed specifically for use at the end of a communication circuit – as contrasted with the repeater, which is designed for an intermediate point. 2. Two microwave terminals arranged to provide for the interconnection of separate systems, or separate sections of a system.

terminal resistance – Also called terminal impedance. The total resistance measured between the input terminals of a meter. For an ac meter, it is the effective dc resistance measured by the voltage-doubling or substitution technique with rated end-scale input of the appropriate frequency applied.

terminal resistor – A resistor used as a terminating device.

terminal room – In telephone practice, a room associated with a central office, private branch exchange, or private exchange that contains distributing frames, relays, and similar apparatus.

terminal station – The microwave equipment and associated multiplex equipment employed at the ends of a microwave system.

terminal strip – See terminal board.

terminal stub – Also called terminal leg. A piece of cable that comes with a cable terminal for splicing into the main cable.

terminal unit – Abbreviated TU. 1. An apparatus that terminates the considered interface system and by means of which a connection (and translation, if required) is made between the considered interface system and another external interface system. 2. The RTTY equivalent of a modem. It contains a modulator, demodulator, and loop power supply. 3. Equipment usable on a communication channel for either input or output.

terminal VHF omnirange – Very high frequency omnirange, normally low powered, complete with a local monitoring device that will automatically shut down the facility if it is not operating properly.

terminated line – A transmission line terminated in a resistance equal to the characteristic impedance of the line, so that there is no reflection or standing waves.

terminating – The closing of the circuit at either end of a line or transducer by connection to some device.

Terminating does not imply any special condition, such as the elimination of reflection.

terminating capacitor — A capacitor sometimes used as a terminating device for a capacitance sensor antenna. The capacitor allows the supervision of the sensor antenna, especially if a long wire is used as the sensor.

terminating device — A device that is used to terminate an electrically supervised circuit. It makes the electrical circuit continuous and provides a fixed impedance reference (end of line resistor) against which changes are measured to detect an alarm condition. The impedance changes may be caused by a sensor, tampering, or circuit trouble.

termination — 1. A load connected to a transmission line or other device. To avoid wave reflections, it must match the characteristic impedance of the line or device. 2. A waveguide technique; the point at which energy flowing along a waveguide continues in a nonwaveguide mode of propagation. 3. The terminals at an antenna to which the transmission line is connected (screw terminals, solder connections, coaxial connector, etc.).

termination block — A nonconductive material on which are provided several termination points.

ternary — 1. A numerical system of notation using the base 3 and employing the characters 0, 1, and 2. 2. Able to assume three distinct states.

ternary code — A code in which each element may be any one of three distinct kinds or values.

ternary gates — Ternary circuits that operate on three logic states at a time — that is, in base 3 arithmetic instead of base 2.

ternary incremental representation — A type of incremental representation in which the value of an increment is rounded to one of three values plus or minus one quantum or zero.

ternary pulse-code modulation — A form of pulse-code modulation in which each element of information is represented by one of three distinct values, e.g., positive pulses, negative pulses, and spaces.

terrain-avoidance radar — Airborne radar that provides a display of terrain ahead of a low-flying airplane to permit horizontal avoidance of obstacles.

terrain-clearance indicator — A device for measuring the distance from an aircraft to the surface of the sea or earth.

terrain error — In navigation, the error resulting from distortion of the radiated field by the nonhomogeneous characteristics of the terrain over which the radiation in question has been propagated.

terrain-following radar — Airborne radar that provides a display of terrain ahead of a low-flying aircraft to permit manual control, or signals for automatic control, to maintain constant altitude above the ground.

terrestrial interference — Abbreviated TI. Interference of earth-based microwave communications with reception of satellite broadcasts.

terrestrial-reference flight — Stabilized flight in which control information is obtained from terrestrial phenomena (e.g., flight in which basic information derived from the magnetic field of the earth, atmospheric pressure, and the like is fed into a conventional automatic pilot).

tertiary coil — A third coil used in the output transformer of an audio amplifier to supply a feedback voltage.

tertiary winding — 1. A winding added to a transformer, in addition to the conventional primary and secondary windings, to suppress third harmonics or to make connections to a power-factor-correcting device. 2. See stabilized winding.

tesla — The SI unit of magnetic flux density, equal to 1 weber per square meter. Letter symbol: T.

Tesla coil — An air-core transformer used for developing high-voltage discharge at a very high frequency. It has a few turns of heavy wire as the primary and many turns of fine wire as the secondary.

test — A procedure or sequence of operations for determining the manner in which equipment is functioning or the existence, type, and location of any trouble.

test bed — A test site that either contains or simulates all hardware and software interfaces.

test bench — Equipment designed specifically for making overall bench tests on equipment in a particular test setup under controlled conditions.

test board — A switchboard equipped with testing apparatus arranged so that connections can be made from it to telephone lines or central office equipment for testing purposes.

test clip — A spring clip fastened to the end of an insulated wire to enable quick temporary connections when circuits or devices are being tested.

test driver — Tool providing the facilities needed to execute a program, e.g., inputs or files, and commands. May also evaluate outputs and produce reports.

testing level — The value of power used for reference, represented by 0.001 watt working into 600 ohms.

test jack — 1. A jack that makes a circuit or circuit element available for testing purposes. 2. In recent practice, a jack that is multiplied with the operating jack on the switchboard.

test language — A procedure or programming language designed or adapted for the development of test specifications and routines.

test lead — 1. A flexible, insulated lead wire that usually has a test prod on one end. It is ordinarily used for making temporary electrical connections. The insulation normally is rubber; the standard colors are red and black. 2. A flexible insulated lead used chiefly for connecting meters and test instruments to a circuit under test.

test loop — A cycle of tests that can be repeated over and over, e.g., to locate intermittent faults.

test oscillator — A test instrument that can be set to generate an unmodulated or tone-modulated radio-frequency signal at any frequency needed for aligning or servicing receivers and/or amplifiers.

test pattern — A geometric pattern containing a group of lines and circles, used for testing the performance of a television receiver or transmitter by revealing the following video-signal characteristics: horizontal linearity, vertical linearity, contrast, aspect ratio, interlace, streaking, ringing, vertical resolution, and horizontal resolution. The camera is focused on the chart, and the pattern is viewed for fidelity at the monitor.

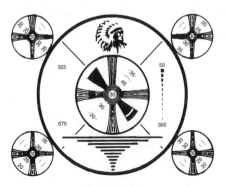

Test pattern.

test point — 1. A connection to which no instrument is permanently connected, but which is intended for temporary, intermittent, or future connection of an instrument. 2. A convenient, safe access to a circuit or system.

test prod — A sharp metal point used for making a touch connection to a circuit terminal. It has an insulated handle and a means for electrically connecting the point to a test lead.

test program — 1. A software component that implements a test procedure. 2. A particular group of test sequences or test patterns.

test record — A phonograph disc designed to test the quality and characteristics of turntables, pickups, amplifiers, etc.

test routine — In a computer: 1. A synonym for check routine. 2. Generally both the check and the diagnostic routines.

test sequence — A group of test steps or test patterns.

test set — 1. One or more instruments required for servicing of a particular type of equipment. 2. Also called analyzer, circuit verifier, and tester. An assembly of instruments that electrically simulate and monitor electronic equipment. Some test sets are used to service equipment.

test to failure — The practice of inducing increased electrical and mechanical stresses in order to determine the maximum capability of a device so that conservative use in subsequent applications will thereby increase its life through the derating determined by these tests.

test tone — A tone used in circuit identification for purposes of locating trouble or making adjustments.

tetrad — A group of four, especially a group of four pulses used to express a digit in the scale of 10 or 16.

tetrode — A four-electrode electron tube containing an anode, a cathode, a control electrode (grid), and one additional electrode that is ordinarily a screen grid.

tetrode junction transistor — See double-base junction transistor.

tetrode transistor — A junction transistor with two electrode connections to the base to reduce interelement capacitance (in addition to the normal emitter and collector elements, each having one connection).

Te value — The temperature at which the resistance of a centimeter cube is 1 megohm.

TE wave — Abbreviation for transverse electric wave.

text — 1. In U.S. ASCII and communications, a sequence of characters treated as an entity if it is preceded by one STX communication control character and terminated by one EXT communication control character. 2. The control sections of an object or load module, considered together. 3. The part of a message or transaction between the control information of the header and that of the trace section or tail that constitutes the information to be processed or delivered to the addressed location. 4. The main part of the message that is sent from a data source (host) to a data link (terminal). It is usually preceded by a header and followed by an "End of Text" signal.

text editing — A general term that encompasses any rearrangement or change performed on textual material, such as deleting, adding, or reformatting.

text fonts — A complete set of one-character type.

T flip-flop — 1. Also called binary. A type of flip-flop whose outputs change state each time the input-signal voltage falls from 1 to 0 and remain unchanged when the input-signal voltage rises from 0 to 1. Thus, there is one change in output state for every two changes in input signal, and the frequency of the output is half the frequency of the input. 2. A flip-flop with only one input. When a pulse appears on the input, the flip-flop changes

states. (Used in ripple counters.) 3. Also called toggle or trigger flip-flop. A circuit that performs the divide-by-two function; frequently used in counting applications.

thallofide cell — A photoconducting cell that has thallium oxysulfide as the light-sensitive agent.

THD — Abbreviation for total harmonic distortion.

theoretical acceleration at stall — A figure of merit derived from the stall-torque to rotor-inertia ratio, which indicates how rapidly a motor will accelerate from stall.

theoretical cutoff — See theoretical cutoff frequency.

theoretical cutoff frequency — Also called theoretical cutoff. The frequency at which, disregarding the effects of dissipation, the attenuation constant of an electric structure changes from zero to a positive value or vice versa.

theoretical electrical travel — The shaft travel over which the theoretical function characteristic of a precision potentiometer extends, as determined from the index point.

theremin — An electronic musical instrument consisting of two radio-frequency oscillators that beat against each other to produce an audio-frequency tone, in a manner similar to a beat-frequency audio oscillator. The pitch and volume are varied by hand capacitance.

thermal — A general term for all forms of thermoelectric thermometers, including a series of couples, thermopiles, and single thermocouples.

thermal agitation — 1. Movement of the free electrons in a material. In a conductor they produce minute pulses of current. When these pulses occur at the input of a high-gain amplifier in the conductors of a resonant circuit, the fluctuations are amplified together with the signal currents and heard as noise. 2. Also called thermal effect. Minute voltages arising from random electron motion, which is a function of absolute temperature expressed in kelvins. 3. In a semiconductor, the random movement of holes and electrons within a crystal due to the thermal (heat) energy.

thermal-agitation voltage — The potential difference produced in circuits by thermal agitation of the electrons in the conductor.

thermal alloying — The act of uniting two different metals to make one common metal by the use of heat.

thermal ammeter — See hot-wire ammeter.

thermal breakdown — 1. A form of breakdown in which decomposition or melting occurs due to the temperature rise resulting from the applied electric stress. 2. A runaway condition in a dielectric, the loss factor of which increases with temperature. Dielectric loss heats the material, producing an increase in temperature. Therefore, the dielectric loss increases still more, producing a further increase in temperature, and so on.

thermal circuit breaker — A circuit breaker whose operation depends on temperature expansion due to electrical heating.

thermal coefficient of resistance — The change in the resistivity of a substance due to the effects of temperature only. Usually expressed in ohms per ohm per degree change in temperature.

thermal compensation — A method employed to reduce or eliminate the thermal effects on one or more of the performance parameters of a transducer.

thermal compression bonding — Diffusion bonding whereby two carefully prepared surfaces are brought into intimate contact under carefully controlled conditions of temperature, time, and clamping pressure. Plastic deformation is induced by the combined effects of pressure and temperature, which in turn results in atom movement causing the development of a crystal lattice bridging the gap between the facing surfaces and results in bonding. (Time is a critical factor in controlling the ambient temperature at the area to be bonded and the size of the bond that is formed.) Generally, the process is performed under a protective atmosphere of inert gas to keep the surfaces to be bonded clean while they are being heated.

thermal conduction — 1. The transfer of thermal energy by processes having no net movement of mass and having rates proportional to the temperature gradient. 2. The rate of flow of heat through a material by thermal conduction.

thermal conductivity — A measure of the ability of a substance to conduct heat. Expressed in terms of calories of heat conducted per second per square centimeter per centimeter of thickness per degree Celsius difference in temperature from one surface to the other.

thermal conductor — A material that readily transmits heat by conduction.

thermal contraction — The shrinkage exhibited by most metals when cooled.

thermal converter — Also called thermocouple converter, thermoelectric converter, thermoelectric generator, or thermoelement. One or more thermojunctions in thermal contact with, or an integral part of, an electric heater, so that the electromotive force developed by thermoelectric action at the output terminals gives a measure of the input current in the heater.

thermal cutout — 1. An overcurrent protective device that contains a heater element that affects a fusible member and thereby opens the circuit. 2. A heat-sensitive switch that automatically opens the circuit of an electrical device when the operating temperature of the device exceeds a predetermined value.

thermal derating factor — The factor by which the power dissipation rating must be reduced with an increase of ambient or case temperature.

thermal detector — *See* bolometer.

thermal drift — A change in the output of a regulated power supply over a period of time, due to changes in internal ambient temperatures not normally related to environmental changes. Thermal drift is usually associated with changes in line voltage and/or load changes.

thermal effect — *See* thermal agitation, 2.

thermal emf — The electromotive force generated when the junction of two dissimilar metals is heated. *See also* Seebeck emf.

thermal endurance — An indication of the relative life expectancy of a product when exposed to operating temperatures much higher than normal room temperature.

thermal equilibrium — The condition that exists when a system and its surroundings are at the same temperature.

thermal expansion — 1. Physical expansion resulting from an increase in temperature; it may be linear and volumetric. 2. The expansion of a material when subjected to heat.

thermal flasher — An electric device that automatically opens and closes a circuit at regular intervals, owing to alternate heating and cooling of a bimetallic strip heated by a resistance element in series with the circuit being controlled.

thermal generation — The creation of a hole and a free electron by freeing a bound electron through the addition of heat energy.

thermal instrument — An instrument that depends on the heating effect of an electric current for its operation (e.g., thermocouple and hot-wire instruments).

thermal ionization — Ionization due to high temperature (e.g., in the electrically conducting gases of a flame).

thermal junction — *See* thermocouple.

thermal lag — The time expended in raising the entire mass of a cathode structure to the temperature of the heater.

thermal life — The operating life of a device under varying ambient temperatures.

thermal microphone — A microphone depending for its action on the variation in the resistance of an electrically heated conductor that is being alternately increased and decreased in temperature by sound waves.

thermal noise — Also called resistance, Johnson, and white noise. 1. Random circuit noise associated with the thermodynamic interchange of energy necessary to maintain thermal equilibrium between the circuit and its surroundings. *See also* Johnson noise. 2. Noise generated by the random thermal motion of charged particles. 3. A voltage, produced by the random motion of free electrons in a conductor, whose value is instantaneously fortuitous with time and whose spectral components uniformly embrace the electromagnetic gamut — albeit the bandwidth of many practical circuits limit both the noise spectrum and magnitude as well. Thermal noise is characterized by a normal distribution of levels. 4. A type of electromagnetic noise produced in conductors or in electronic circuitry that is proportional to temperature. 5. Noise that occurs in all transmission media and communications equipment as a result of random electron motion (which is a function of temperature). Thermal noise sets the lower limit for the sensitivity of a receiving system.

thermal noise level — The equivalent rms voltage value, over a stated bandwidth, of all energy components generated by a resistor at a stated resistor temperature with no externally supplied current through the resistor.

thermal oxidation — The formation of a self-oxide on the exposed surfaces of a semiconductor.

thermal printer — 1. A nonimpact printing device that utilizes a special heat-sensitive paper. The paper passes over a matrix of dot heating elements. As data is fed to the printer, dot elements relating to specific characters are heated, which changes the color of the paper at that point to reveal individual characters. 2. A printer that forms characters in a dot matrix by selectively heating printhead elements as they move across the paper.

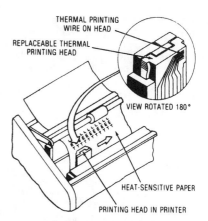

Thermal (matrix) printer.

thermal protector — A current- and temperature-responsive device used to protect another device against overheating due to overload.

thermal radiation — Commonly known as heat. 1. Radiation produced by the action of heat on molecules or atoms. Its frequency extends between the extremes of infrared and ultraviolet. 2. The process of electromagnetic emission in which the radiated energy is extracted from the thermal excitation of atoms or molecules.

thermal rating — 1. A statement of the permissible temperature rating, beyond which unsatisfactory performance occurs. 2. The maximum or minimum temperature at which a material or component will perform its function without undue degradation.

thermal regenerative cell — A fuel-cell system in which there is continuous regeneration of the reactants from the products formed during the cell reaction.

thermal relay — A relay that responds to the heating effect of an energizing current, rather than to the electromagnetic effect. Delay between the start of the energizing current and the switching response is generally predictable and sometimes adjustable.

thermal resistance — 1. Ratio of the temperature rise to the rate at which heat is generated within a device under steady-state conditions. 2. The resistance of a substance to the conductivity of heat. 3. That change in the electrical resistance of a material when subjected to heat. 4. Of a semiconductor device, the quotient of the temperature difference between two specified points or regions and the heat flow between these two points or regions under conditions of thermal equilibrium.

thermal resistivity — Thermal resistance of a unit cube of material.

thermal resistor — An electronic device that makes use of the change in resistivity of a semiconductor with changes in temperature. *See* thermistor.

thermal response time — The time from the occurrence of a step change in power dissipation until the junction temperature reaches 90 percent of the final value of junction-temperature change, when the device-case or ambient temperature is held constant.

thermal runaway — 1. A regenerative condition in a transistor, whereby heating at the collector junction causes collector current to increase, which in turn causes more heating, etc. The temperature can rapidly approach levels that are destructive to the transistor. 2. An unstable condition common in bipolar transistors occurring because collector current and gain both increase with device temperature. Permanent damage can result if circuit precautions are not taken.

thermal sensitivity set — A permanent change in sensitivity due to temperature effects only. Usually expressed as the difference in sensitivity at room temperature before and after a temperature cycle over the operating temperature range of the transducer.

thermal shock — 1. A sudden, marked change in the temperature of the medium in which a component or device operates. 2. The effect of heat or cold applied at such a rate that nonuniform thermal expansion or contraction occurs within a given material or combination of materials. In connectors, the effect can cause inserts and other insulation materials to pull away from metal parts.

thermal telephone receiver — Also called thermophone. An electroacoustic transducer, such as a telephone receiver, in which the temperature of a conductor is caused to vary in response to the current input, thereby producing sound waves as a result of the expansion and contraction of the adjacent air.

thermal time constant — 1. The time from the occurrence of a step change in power dissipation until the junction temperature reaches 63.2 percent of the final value of junction-temperature change, when the device-case or ambient temperature remains constant. 2. In a thermistor, the time required for 63.2 percent of the change from initial to final body temperature after the application of a step change in temperature under zero-power conditions.

thermal time-delay relay — A type of relay in which the time interval between energization and actuation is determined by the thermal storage capacity of the actuator critical operating temperature, power input, and thermal insulation.

thermal time-delay switch — 1. An overcurrent protective device containing a heater element and thermal delay. 2. A switch whose contacts control a load circuit and are delayed from operating for a predetermined time interval. Operation of the contacts is by the effect of heat generated by current through a heater.

thermal tuning — Adjusting the frequency of a cavity resonator by using thermal expansion to vary its shape.

thermic — Pertaining to heat.

thermion — An ion, either positive or negative, that has been emitted from a heated body. Negative thermions are electrons (thermoelectrons).

thermionic — 1. Pertaining to the emission of electrons by heat. 2. Pertaining to the emission of electrons or ions from an incandescent body.

thermionic cathode — *See* hot cathode.

thermionic converter — Also called thermionic generator or thermoelectron engine. A device that produces electrical power directly from heat. One type contains a heated cathode to emit electrons and a cold anode to collect them, thereby causing a current. Both electrodes are enclosed in a vacuum or gas-filled envelope.

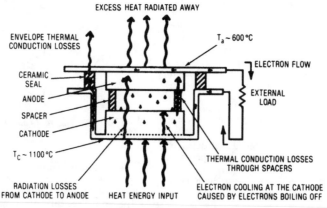

Thermionic converter.

thermionic current — Current due to directed movements of thermions (e.g., the flow of electrons from the cathode to the plate in a thermionic vacuum tube).

thermionic detector — A detector circuit in which a thermionic vacuum tube delivers an audio-frequency signal when fed with a modulated radio-frequency signal.

thermionic diode — A diode electron tube that has a heated cathode.

thermionic emission — 1. Emission of electrons from a solid body as a result of elevated temperature. *See also* Edison effect. 2. That portion of the emission of electrons from a hot cathode in a vacuum tube which is due solely to the elevated temperature of the cathode.

thermionic energy conversion — The direct production of electricity by means of the electron emission from a heated substance.

thermionic generator — *See* thermionic converter.

thermionic grid emission — Also called primary grid emission. The current produced by the electrons thermionically emitted from a grid. Generally it is due to excessive grid temperatures or to contamination of the grid wires by cathode coating material.

thermionic rectifier — A rectifier utilizing a thermionic vacuum tube to convert alternating current into unidirectional current.

thermionic tube — *See* hot-cathode tube.

thermionic work function — The energy required to transfer electrons from a given metal to a vacuum or some other adjacent medium during thermionic emission.

thermistor — Also called thermal resistor. 1. A thermally sensitive solid-state semiconducting device made by sintering mixtures of the oxide powders of various metals. (Made in many shapes, such as beads, disks, flakes, washers, and rods, to which contact wires are attached.) As its temperature is increased, the electrical resistance (typically) decreases. The associated temperature coefficient of resistance is extremely high, nonlinear, and most frequently negative. The large temperature coefficients and the nonlinear resistance temperature characteristics of thermistors enable them to perform many unique regulatory functions. 2. A passive semiconductor device whose electrical resistance varies with temperature. Its temperature coefficient of resistance is high, nonlinear, and usually negative. 3. A solid-state semiconducting structure (basically one of the bolometers) that changes electrical resistance with temperature. Materially, some kind of ceramic composition is used. It is of much higher electrical resistance than metallic bolometers and, hence, requires much higher voltages to become useful. 4. A resistor with a high temperature coefficient of resistance, exhibiting a definite, reliable, repeatable response to temperature change. Negative temperature coefficient (TC) types show an exponential decrease in resistance as temperature increases. Positive TC types show an increase in resistance with increasing temperature. 5. A resistor that is sensitive to temperature changes. It will change in ohmic value as a function of changing temperature.

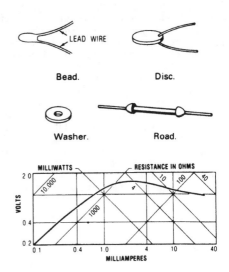

Typical characteristic.

Thermistors.

thermoammeter — Also called a thermocouple ammeter. An ammeter that is actuated by the voltage generated in a thermocouple through which the current to be measured is sent. It is used chiefly for measuring radio-frequency currents.

thermocompensator — In pH meters, a temperature-sensitive device sometimes used to make electronic adjustments in the circuit that are required due to changes in the temperature of the solution.

thermocompression bond — A bond formed by two elements through the simultaneous application of heat and pressure. No additional materials are used to assist the fusing. Common types of thermocompression bonds are wedge, ball (nailhead), and stitch. *See* ball bond; stitch bond.

thermocompression bonding — 1. A method of interconnecting ICs in a circuit by bonding thin gold wires between conducting patterns of a circuit and to the IC chip's metal preforms by means of heat and pressure. 2. The joining of two materials by the combined effects of heat and pressure.

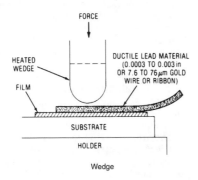

Wedge

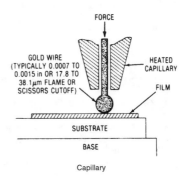

Capillary

Thermocompression bonding.

thermocouple—Also called thermal junction. 1. Temperature transducer comprising a closed circuit made of two different metals. If the two junctions are at different temperatures, an electromotive force is developed that is proportional to the temperature difference between the junctions. This is called the Seebeck effect. 2. Dissimilar metals that, when welded together, develop a small voltage dependent on the relative temperature between the hotter and the colder junction. Banks of thermocouples connected together in series of parallel make up a thermopile. Either may be thought of as a weak battery that converts radiant to electrical energy. 3. A device for measuring temperature in which two electrical conductors of dissimilar metals are joined at the point of heat application and a resulting voltage difference, directly proportional to the temperature, is developed across the free ends and is measured potentiometrically. 4. A pair of dissimilar-metal wires, joined so that when their junction is heated, the thermoelectric effect causes a voltage to be generated that is proportional to the temperature.

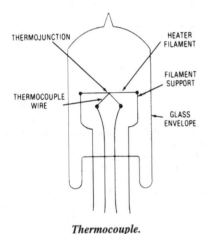

Thermocouple.

thermocouple ammeter—*See* thermoammeter.

thermocouple contact—A contact of special material used in connectors employed in thermocouple applications. Materials often used are iron, constantan, copper, chromel, alumel, and others.

thermocouple converter—*See* thermal converter.

thermocouple instrument—An electrothermic instrument in which one or more thermojunctions are heated by an electric current, causing a direct current to flow through the coil of a suitable direct-current mechanism such as one of the permanent-magnet, moving-coil type.

thermocouple junction—A pair of electrically dissimilar metals that make intimate contact and generate a voltage related to temperature.

thermocouple lead wire—An insulated pair of wires used from the thermocouple to a junction box or to the recording instrument.

thermocouple thermometer—*See* thermoelectric thermometer.

thermocouple vacuum gage—A vacuum gage that depends for its operation on the thermal conduction of the gas present. The pressure being measured is a function of the electromotive force of a thermocouple whose measuring junction is in thermal contact with a heater carrying a constant current. Thermocouple vacuum

gages ordinarily are used over a pressure range of 10^{-1} to 10^{-3} mm Hg.

thermocouple wire—A wire drawn from special metals or alloys and calibrated to established specifications for use as a thermocouple pair (e.g., iron, constantan, alumel, etc.).

thermodynamics—1. The study of the relationship between heat and other forms of energy. 2. Examination of the processes whereby heat energy is converted into other forms of energy.

thermoelectric arm—Also called thermoelectric leg. The portion of a thermoelectric device having the electric current density and the temperature gradient approximately parallel or antiparallel and having electrical connections made at its extremities to a part in which the opposite relation between the direction of the temperature gradient and the electric current density exists.

thermoelectric converter—A device capable of converting heat energy directly into electrical energy. *See also* thermal converter.

thermoelectric cooler—1. A device utilizing the Peltier phenomenon to provide a silent, nonmoving cooler having a controllable cooling rate. 2. A solid-state device that cools as current passes through it.

thermoelectric couple—A thermoelectric device in which there are two arms of unlike composition.

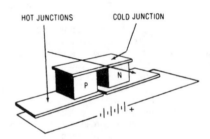

Thermoelectric couple.

thermoelectric device—A general term for thermoelectric heat pumps and generators.

thermoelectric effect—The electromotive force produced by the difference in temperature between two junctions of dissimilar metals in the same circuit.

thermoelectric engine—*See* thermoelectric converter.

thermoelectric generator—*See* thermal converter.

thermoelectric heating device—A thermoelectric heat pump used for adding thermal energy to a body.

thermoelectric heat pump—A device in which the direct interaction of an electrical current and heat flow is used to transfer thermal energy between bodies.

thermoelectricity—1. The direct conversion of heat into electricity. 2. The reciprocal use of electricity to create heat or cold. *See also* Peltier effect; Seebeck effect; Thomson effect. 3. Electricity produced by the agency of heat alone. *See also* thermocouple.

thermoelectric junction—1. A thermojunction, as in a thermocouple. 2. A junction between dissimilar wires in a thermocouple or thermopile. A potential difference is created at that junction by the application of heat.

thermoelectric leg—*See* thermoelectric arm.

thermoelectric manometer—A manometer (pressure-measuring instrument) that depends on the variation of thermoelectromotive force (voltage due to heat) with pressure.

thermoelectric series— A series of metals arranged in the order of their thermoelectric powers.

thermoelectric solar cell— A solar cell that uses a thermoelectric converter, consisting of two sheets of metal with a semiconductor sandwiched between them, to generate electricity.

thermoelectric thermometer— Also called a thermocouple thermometer. A thermometer employing one or more thermocouples, of which one set of measuring junctions is in thermal contact with the body whose temperature is to be measured, while the temperature of the reference junctions is either known or otherwise taken into account.

thermoelectromotive force— 1. The voltage developed due to the differences in temperature between parts of a circuit containing two or more different metals. 2. The algebraic sum of the Peltier emf at a thermocouple junction and the Thomson emf in the thermocouple metals.

thermoelectron— The electron emitted from a heated body.

thermoelectron engine— See thermionic converter.

thermoelement— A device consisting of a thermocouple and a heating element arranged for measuring small currents. See also thermal converter.

thermogalvanometer— An instrument for measuring small high-frequency currents from their heating effect. Generally it consists of a dc galvanometer connected to a thermocouple that is heated by a filament carrying the current to be measured.

thermogram— High-resolution images resulting from a series of thermal scans using a bolometer.

thermograph— Radiation chart that plots heat emitted by the body.

thermography— 1. The process of recording the distribution of temperature over the surface of an object by detecting the heat radiation from it. The detection may be by the effect of radiation on an infrared-sensitive phosphor or by the use of cholesteric liquid crystals that exhibit brilliant colors with temperature change. 2. The recording of a scanned pattern on a photographic medium, utilizing the infrared radiation naturally emitted by the object, as well as infrared receptors, such as photoelectric cells. 3. The technique of producing pictures from heat.

thermojunction— One of the contact surfaces between the two conductors of a thermocouple. The thermojunction in thermal contact with the body under measurement is called the measuring junction, and the other thermojunction is called the reference junction.

thermojunction battery— A nuclear-type battery that converts heat into electrical energy directly by the thermoelectric or the Seebeck effect.

thermoluminescence— 1. The production of light in a material by moderate heat. 2. An alternative term for incandescence.

thermomagnetic— 1. Pertaining to the effect of temperature on the magnetic properties of a substance. 2. Pertaining to the effect of a magnetic field on the temperature distribution in a conductor.

thermometer— An instrument for measuring temperature. Electrical versions depend on the change in resistance of a material with temperature, the voltage produced in a thermocouple, or various other effects of temperature.

thermophone— An electroacoustic transducer in which sound waves of calculable magnitude are produced by the expansion and contraction of the air adjacent to a conductor whose temperature varies in response to a current input. See also thermal telephone receiver.

thermopile— A group of thermocouples connected in series. Specifically: 1. A device used to measure radiant power or energy. 2. A source of electric energy. 3. Battery of thermocouples, consisting of alternate rods of antimony and bismuth suitably joined and connected to a galvanometer. Thermopiles are used for the measurement of heat where thermometers cannot be employed.

thermoplastic— 1. A term describing plastic materials that can be made to flow repeatedly by applying heat. Lowering the temperature causes hardening. See thermosetting. 2. A type of plastic that can be remelted a number of times without any important change in properties. For example, nylon, lexan, and PVC are thermoplastic plastics. Such plastics are resilient after molding. 3. A classification of resin that can be readily softened and resoftened by repeated heating.

thermoplastic flow test— For an insulating material, a measure of the resistance to deformation when subjected to heat and pressure.

thermoplastic material— A plastic material that can be softened by heat and rehardened into a solid state by cooling. This remelting and remolding can be done many times.

thermoplastic polyesters— Family of plastics with excellent dimensional stability, electrical properties, toughness, and chemical resistance.

thermoplastic recording— A recording process in which information is placed onto plastic tape electronically. A special electron gun, fed by a digital or scanner input, writes a charge in narrow bands on a moving film coated with a plastic that has a low melting point. The film is heated by an rf heater that melts the plastic coating, permitting it to be deformed by electrostatic and surface-tension forces in proportion to the charge laid down by the beam of the electron gun. The ridges cool quickly and form a diffraction grating that can then be viewed, projected by suitable optics, or read out by a flying-spot scanner. The system operates in a high vacuum.

thermosetting— 1. A term describing plastic materials that are capable of becoming essentially insoluble or infusible when cured. Once cured, these materials do not soften or flow. See thermoplastic. 2. A term used to describe materials that are processed using steam or radiation. After processing the materials cannot be made to flow under the application of heat. 3. A classification of resin that cures by chemical reaction when heated and, when cured, cannot be resoftened by heating.

thermosetting material— Plastic that hardens when heat and pressure are applied. Unlike a thermoplastic, it cannot be remelted or remolded.

thermosetting plastic— A type of plastic in which an irreversible chemical reaction takes place while it is being molded under heat and pressure. This type of plastic cannot be reheated or softened without adversely affecting its properties. Melamine and diallyl-phthalate are thermosetting plastics.

thermostat— A mechanism that can be set to operate at definite temperatures and can convert the expansion of heated metal or fluid into sufficient movement and power to operate small devices, control electric circuits or small valves, etc.

thermostatic switch— 1. A temperature-operated switch that receives its operating energy by thermal conduction or convection from the device being controlled or operated. 2. A switch whose function is controlled by variations of temperature and whose contacts make or break a load circuit automatically when the temperature of the ambient space in which its sensing element is placed or the temperature of the surface on which it is fixed reaches a predetermined value.

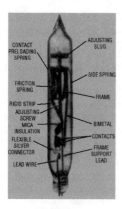

Thermostat.

thermostat wire — Single or multiple-conductor wire, bare soft solid-copper conductor, usually PVC insulated. May be twisted and/or jacketed. May have enameled or nylon covered conductors and may have a metal armor covering. May also have asbestos insulations. It is used to transmit electrical signals between the thermostat and the heating or cooling unit.

theta — Brain wave signals whose frequency is approximately 3.5 to 7.5 Hz. The associated mental state is fuzzy, unreal, uncertain, daydreamlike, or ambiguous.

theta polarization — The state of the wave whereby the E vector is tangential to the meridian lines of some given spherical frame of reference.

Thevenin's theorem — The current that will flow through an impedance Z_1 when connected to any two terminals of a linear network between which an open-circuit voltage E and impedance Z previously existed is equal to the voltage E divided by the sum of Z and Z_1.

thick film — 1. Pertaining to a film pattern usually made by applying conductive and insulating materials to a ceramic substrate by a silk-screen process. Thick films can be used to form conductors, resistors, and capacitors. 2. Successive layering of resistive, dielectric, and conductive inks on a substrate by a type of screening process. 3. Technology using pastes to form conductor, resistor, and insulator patterns; usually screened onto the substrate and cured by firing at temperatures of 800°C to 900°C. While less expensive than thin films, thick-film resistors exhibit somewhat poorer matching and temperature coefficients. 4. A layer of resistive, dielectric, or conductive paste that is deposited on a substrate by screen printing, then fired at an elevated temperature to drive off the binder and sinter the solids. By definition, the thickness is greater than 10^4 angstroms (0.0004 inch),

although characteristically it is 0.0004 to 0.001 inch (6.3 to 25 μm).

thick-film assembly — Network deposited on a ceramic substrate, in combination with discrete active elements, forming a complete hybrid functional circuit.

thick-film circuit — 1. A microelectronic assembly in which the passive circuit elements and their interconnections are defined on a ceramic substrate by using the silk-screen process. The active elements are added as discrete chips or packaged devices. 2. Conductive, resistive, and/or capacitive network deposited on a substrate using a metallic or resistive film that is more than 5 micrometers in thickness. 3. A completed assembly of a thick-film substrate and possible add-on components, with its appropriate terminations and packaging. 4. A microcircuit whose passive components consist of a ceramic-metal combination deposited on a given substrate by screening and firing processes.

thick-film hybrid integrated circuits — The physical realization of a hybrid integrated circuit fabricated on a thick-film network.

thick-film hybrids — 1. A hybrid circuit in which conductive and resistive pastes are deposited on a substrate and heated. Heating hardens the paste and fuses it to the substrate. Once hardened, the resistive pastes become thick-film resistors, named because the thickness is about 1 mil (25 μm). Capacitors, discrete semiconductors, and ICs are then added to the circuit. 2. A thick-film circuit with add-on components other than encapsulated chips.

thick-film network — A thick-film circuit that does not have add-on components.

thick-film process — 1. A method in which electronic circuit elements — resistors, conductors, capacitors, etc. — are produced by applying specially formulated pastes to a ceramic substrate in a defined pattern and sequence. The substrate containing the "green pattern" is fired at a relatively high temperature to mature the circuit elements and bond them integrally to the substrate. The term thick film distinguishes this technology from thin-film practice, in which circuit elements are made by evaporation or sputtering in a high-vacuum environment. Circuits produced by this technology are identified by several descriptive phrases: thick-film passive circuits or thick-film hybrid circuits when active devices are included. 2. A method of manufacturing hybrid circuits by screen deposition of conductive, resistive, or insulating films thicker than 0.005 inch (127 μm). Thick-film structures generally contain only conductors, resistors, and capacitors, with any other required components added as discrete devices.

thick-film resistor — A fixed resistor whose resistance element is a film considerably more than 0.001 inch (25 μm) thick.

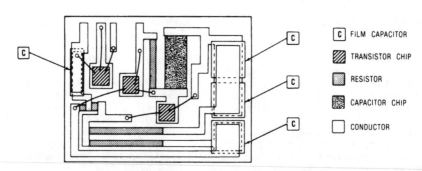

Thick-film hybrid integrated circuit.

thick-film resistor, conductor, and dielectric compostions — The principal materials for making thick-film circuits, available in paste form and consisting of mixtures of metal, oxide, and glass powders.

thick-film resistor network — A combination of several resistors manufactured by incorporating the use of precious metal inks, screen printed and fired onto alumina substrates.

thick films — Layers of resistive, dielectric, and conductive inks that are deposited on a substrate. The deposition process, similar to graphic silk screening, employs a fine-mesh screen to hold the pattern for the components that are to be deposited. The pattern is produced by photographic means, and wherever the inks are not to be deposited, the holes in the mesh are blocked by an emulsion.

thick-film substrate — A supporting material (commonly alumina) on which circuit elements such as conductors, resistors, inductors, and capacitors have been realized by a printing and firing technique.

thickness vibration — Vibration of a piezoelectric crystal in the direction of its thickness.

thin film — 1. A film of conductive or insulating material, usually deposited by sputtering or evaporation, that may be made in a pattern to form electronic components and conductors on a substrate or used as insulation between successive layers of components. 2. The deposition of thin layers of a substance on an insulating base in a vacuum by a microelectronic process. 3. A conductive, resistive, and/or capacitive passive network deposited on a substrate using a metallic or resistive film that is less than 5 micrometers in thickness. 4. Metal film used to fabricate conductors and resistors, usually a few hundred angstroms thick. Commonly used resistor materials are nickel-chromium, tantalum nitride, and silicon-chromium. These films are usually deposited using sputtering, vacuum evaporation, or electron-beam techniques. 5. A method of manufacturing hybrid circuits in which evaporation or sputtering techniques are used to deposit very thin films of material onto a substrate. These film structures generally contain only conductors, resistors, and capacitors, with any other required components added as discrete devices.

thin-film assembly — Passive or active network deposited on a suitable substrate using vapor or vacuum deposition techniques. The components may be produced individually, they may be totally integrated, or they may be combined as hybrids with the advantages of extremely small size and weight plus a low cost for a high-volume production. Thin-film assembly requires considerable engineering effort and sophisticated production equipment.

thin-film capacitor — A capacitor utilizing a metal oxide as the dielectric or insulating material. Both the electrodes and the dielectric are deposited in layers on a substrate. This device is usually associated with microelectronics and integrated and thin-film circuits.

thin-film circuit — 1. A microcircuit in which the component element and interconnections are fabricated from thin deposited films of metal, semiconductor, and dielectric material, generally on an insulating substrate such as ceramic or sapphire. The term *thin* is usually taken to imply films having a thickness on the order of 1 micrometer. 2. A circuit whose passive components are deposited on a given substrate by sputtering or vacuum processes.

thin-film deposition (chemical vapor type) — A technique that involves a decomposition and reaction between gases on the surface of a heated substrate such that a solid layer is nucleated and grown. Metals are generally derived from the decomposition of the metal halides. Insulators may be formed by reacting metal halides with oxygen (oxides), ammonia (nitrides), diborane (borides), etc.

thin-film deposition (evaporation type) — Popular technique for depositing thin film in vacuum, accomplished by heating the source material in a low-pressure chamber so that it vaporizes and then condenses onto all cooler surfaces in line of sight from the source.

thin-film deposition (sputtering type) — Evaporation produced by ion bombardment of the source material, known as cathode sputtering.

thin-film deposition materials (conductors and resistors) — Metals such as aluminum, gold, chromium, nickel, platinum, tungsten, alloys, and cermets deposited as electrical conductors and resistors on silicon or other substrates.

thin-film deposition materials (inorganic dielectrics) — Film compounds produced by various vacuum evaporation processes and deposited on substrates to perform electrical functions. Examples include silicon monoxide, ZnS, CaF, SiO_2, Al_2O_3, Si_3N_4, and other chemical compounds.

thin-film deposition materials (organic dielectrics) — Insulating film compounds produced when organic vapors are heated under conditions in which polymerization and deposition occur. Examples are parylent, butadene, acrolein, and divinylbenzene.

thin-film deposition materials (semiconductors) — Polycrystalline films deposited by vacuum or flash evaporation to produce high-purity single-crystal silicon or other semiconductor substances.

thin-film formation — A process that is either additive, by pattern formation through masks, or subtractive, by selective etching of predeposited films from a substrate.

thin-film hybrid integrated circuit — The physical realization of a hybrid integrated circuit fabricated on a thin-film network.

thin-film hybrids — Hybrid circuits that are made by placing the substrate in a vacuum chamber and depositing films of conductive and resistive material on the entire substrate. The material is photoetched in selected patterns to form conductive interconnections and thin-film resistors. These resistors are approximately 0.1 mil (25 μm) thick. Unencapsulated components are then added either by soldering or chemical bonding.

thin-film integrated circuit — 1. An integrated circuit consisting of a passive substrate on which the various passive elements (resistors and capacitors) are deposited in the form of thin-patterned films of conductive or nonconductive material. Active components (transistors and diodes) are attached separately as individually packaged devices or in unpackaged (chip) form, or may be formed integrally by thin-film techniques. *See also* hybrid thin-film circuit. 2. A device consisting of a number of electrical elements entirely in the form of thin films deposited in a pattern on a supporting material. 3. The physical realization of a number of electric elements entirely in the form of thin films deposited in a patterned relationship on a structural supporting material.

thin-film memory — In a computer, a storage device made of thin disks of magnetic material deposited on a nonmagnetic base. Its operation is similar to the core memory. *See also* core memory, 1; storage, 2.

thin-film microelectronics — Circuits made up of two-dimensional passive and essentially two-dimensional active elements mounted or deposited on thin wafers of an insulating substrate material.

thin-film optical modulator — A device made of mulitlayered films of material of different optical characteristics capable of modulating transmitted light by using electroacoustic, electro-optic, or magneto-optic effects to obtain signal modulation. Thin-film optical modulators are used as component parts of integrated optical circuits.

thin-film optical multiplexer — A multiplexer consisting of layered optical materials that make use of electroacoustic, electro-optic, or magneto-optic effects to accomplish the multiplexing. Thin-film optical multiplexers may be component parts of integrated optical circuits.

thin-film optical switch — A switching device for performing logic operations using light waves in thin films, usually supporting only one propagation mode, making use of electroacoustic, electro-optic, or magneto-optic effects to perform switching functions, such as are performed by semiconductor gates (AND, OR, NOT). Thin-film optical switches may be component parts of integrated optical circuits.

thin-film optical waveguide — An optical waveguide consisting of thin layers of differing refractive indices, the lower indexed material on the outside or as a substrate, for supporting usually a single electromagnetic wave propagation mode with laser sources. The thin-film waveguide lasers, switches, modulators, filters, directional couplers, and related components need to be coupled from their integrated optical circuits to the optical waveguide transmission media, such as optical fibers and slab dielectric waveguides.

thin-film resistor network — A combination of several resistors manufactured by deposition of appropriate alloys and metals onto glass silicon and alumina substrates using vacuum techniques.

thin films — Resistive materials of submicrometer thickness made of nichrome or tantalum (or other exotic metal). They are deposited onto a substrate by vacuum deposition or by rf sputtering under a vacuum. Thin-film resistors provide excellent tolerances, are very stable with temperature changes, but are costly to apply, necessitating the use of expensive vacuum equipment.

thin-film semiconductor — A semiconductor produced by the deposition of an appropriate single-crystal layer on a suitable insulator.

thin-film solar cell — A solar cell that is lightweight and flexible due to its construction by vacuum deposition of a semiconductor material (e.g., gallium arsenide or cadmium sulfide) onto a thin plastic or metal substrate. It is used as a power source in spacecraft because of its light weight.

think time — The time the user of a time-sharing system spends sitting at his or her terminal but not using the computer.

thin-walled conduit — Metallic tubing used to enclose insulated wires in an electrical circuit.

thin-wall ring magnet — A type of ceramic permanent magnet in which the axial length is greater than the wall thickness or the wall thickness is less than 15 percent of the outside diameter.

third harmonic — A sine-wave component having three times the fundamental frequency of a complex wave.

third-harmonic distortion — The rms third-harmonic voltage divided by the rms fundamental voltage. This value often is used as a measure of distortion in an essentially symmetrical system, such as ac-biased recording.

third-party traffic — Amateur radiocommunication by or under the supervision of the control operator at an amateur radio station on behalf of anyone other than the control operator.

thixotropic — 1. A property of a paste or liquid that describes its ability to flow more readily when agitated or sheared. Thixotropy of a paste is a necessary condition for

screen ability. 2. Description of materials that are gel-like at rest but fluid when agitated.

thixotropy — The property of a coating compound that provides low viscosity during agitation, but will thicken to reduce runoff when allowed to stand for a short period.

Thomson bridge — *See* Kelvin bridge.

Thomson coefficient — The ratio of the voltage between two points on a metallic conductor to the difference in temperature between the same points.

Thomson effect — The production or absorption of heat (in addition to the I^2R loss) by a current between two points within a temperature gradient in a homogeneous conductor. Whether the heat is given off or absorbed depends on the direction of the current.

Thomson electromotive force — The voltage that exists between two points that are of different temperatures in a conductor.

Thomson heat — The thermal energy absorbed or produced due to the Thomson effect.

thoriated filament — A tungsten vacuum-tube filament to which a small amount of thorium has been added to improve emission. The thorium comes to the surface and is primarily responsible for the electron emission.

thread — *See* chip, 1.

three-address — Pertaining to an instruction format that contains three address parts.

three-address code — Also called instruction code. In computers, a multiple-address code that includes three addresses, usually two addresses from which data is taken and one address where the result is entered. Location of the next instruction is not specified, and instructions are taken from storage in preassigned order.

three-address instruction — 1. In computers, an instruction that includes an operation and specifies the location of three registers. *See also* three-address code. 2. An instruction that can separately specify two sources and a destination.

three-channel stereo — A stereo recording or reproduction system that uses three spaced microphones for recording and three sound reproducers for playback.

three-conductor jack — Receptacle having three through circuits: tip, ring, and sleeve.

three-layer diode — Also called diac. A two-terminal voltage-controlled device exhibiting a bilateral negative resistance characteristic. The device has symmetrical switching voltages ranging from 20 to 40 volts and is specifically designed for use as a trigger in ac power-control circuits such as those using triacs.

three-level maser (or laser) — A maser or laser system that involves the ground state and two other energy levels. Laser action usually takes place between the intermediate and ground states. The pump populates the intermediate state by way of the highest state. When applied to a maser, this pump brings the population of the highest state and the ground state into equilibrium. Maser action may take place either between the upper and intermediate levels or between the intermediate and ground levels, depending on the relaxation times of the different transitions.

three-phase circuit — A combination of circuits energized by alternating electromotive forces that differ in phase by one-third of a cycle, or 120 electrical degrees. In practice, the phases may vary several degrees from the specified angle.

three-phase current — A current delivered through three wires — each wire serving as the return for the other two, and the three current components differing in phase successively by one-third of a cycle, or 120 electrical degrees.

three-phase, four-wire system — An ac supply system comprising four conductors: three connected as in a three-phase, three-wire system and the fourth to the neutral point of the supply, which may be grounded.

three-phase motor — 1. An ac motor operated from a three-phase circuit. 2. A true polyphase induction motor with three or more power windings designed specifically to operate from a three-phase power source. It is self-starting.

three-phase, seven-wire system — A system of ac supply from groups of three single-phase transformers connected in a Y. Thus, a three-phase, four-wire, grounded-neutral system of a higher voltage for power is obtained, the neutral wire being common to both systems.

three-phase, three-wire system — An ac supply system comprising three conductors, between successive pairs of which are maintained alternating differences of potential successively displaced in phase by one-third of a cycle.

three-plus-one instruction — In digital computer programming, a four-address instruction in which one of the addresses always specifies the location of the next instruction to be performed.

three-pole switch — An arrangement of three single-pole switches coupled together to operate three contacts simultaneously.

three-position relay — Sometimes called a center-stable polar relay. A relay that may be operated to three distinct positions.

three-pulse cascade canceler — A moving-target indicator technique in which two 2-pulse cancelers are cascaded together. This improves the velocity response.

three-quarter bridge — A bridge connection in which one of the diode rectifiers has been replaced by a resistor.

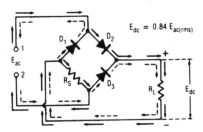

Three-quarter bridge.

three-state — *See* tristate.

three-state logic — A logic family that may be in one of three states rather than the usual two — high, low, or high impedance. In the high-impedance state the output voltage is unaltered. Logic in a high-impedance state may be easily and permanently connected to a bus.

three-way speaker system — A sound-reproducing system using three separate speakers, each designed for a specific portion of the audio spectrum (high, low, and middle frequencies). The high- and low-frequency speakers are known as the tweeter and woofer, respectively.

three-way switch — A switch that can connect one conductor to any one of two other conductors.

three-wire system — A system of electric supply comprising three conductors, one of which (known as the neutral wire) is maintained at a potential midway between the potential of the other two (referred to as the outer conductors). Part of the load may be connected directly between the outer conductors, the remainder being divided as evenly as possible into two parts, each of which is connected between the neutral and one outer conductor. There are thus two distinct supply voltages, one being twice the voltage of the other.

threshold — 1. The point at which an effect is first produced, observed, or otherwise indicated. 2. In a modulation system, the smallest value of carrier-to-noise ratio at the input of the demodulator for all values above which a small percentage change in the input carrier-to-noise ratio produces a substantially equal or smaller percentage change in the output signal-to-noise ratio. 3. That point at which an indication exceeds the background or ambient. 4. A minimal signal-to-noise input required to allow a video receiver to deliver an acceptable picture.

threshold current — 1. The minimum current at which a gas discharge becomes self-sustaining. 2. The minimum forward current for which the laser is in a lasing state at a specified temperature.

threshold decoding — A decoding procedure so arranged that the decision on the symbol that was transmitted is based on a majority count of the parity-check equations involving that symbol.

threshold element — A device that performs the logic threshold operation, but in which the contribution to the output determination by the truth of each input statement has a weight associated with that statement.

threshold field — The least magnetizing force in a direction that tends to decrease the remanence, which, when applied either as a steady field of long duration or as a pulsed field appearing many times, will cause a stated fractional change of remanence.

threshold frequency — The frequency at which the quantum energy is just sufficient to release photoelectrons from a given surface.

threshold of audibility — Also called threshold of detectability or threshold of hearing. For a specified signal, the minimum effective sound pressure of a signal capable of evoking an auditory sensation in a specified fraction of the trials. The characteristics of the signal, the manner in which it is presented to the listener, and the point at which the sound pressure is measured must all be specified. This threshold is usually expressed in decibels relative to 0.0002 microbar.

threshold of detectability — *See* threshold of audibility.

threshold of discomfort — Also called threshold of feeling. For a specified signal, the minimum effective sound-pressure level that, in a specified fraction of the trials by a battery of listeners, will stimulate the ear to the point at which the sensation of feeling becomes uncomfortable. This threshold is customarily expressed in decibels relative to 0.0002 microbar.

threshold of feeling — *See* threshold of discomfort.

threshold of hearing — *See* threshold of audibility.

threshold of luminescence — *See* luminescence threshold.

threshold of sensitivity — The smallest stimulus or signal that will result in a detectable output. This phrase is frequently used to describe the voltage point at which an operations monitor or event marker will trigger.

threshold signal — In navigation, the smallest signal capable of producing a recognizable change in the positional information.

threshold-triggered flip-flop — A flip-flop whose state changes when the actuating signal passes through a certain voltage level, regardless of the rate at which the voltage changes.

threshold value — 1. The minimum input that produces a corrective action in an automatic control system. 2. The minimum level for which there is a measurable output.

threshold voltage — 1. The level of input voltage at which a binary logic circuit changes from one logic state to the other. 2. The voltage at which a pn junction begins to conduct current. 3. In a solid-state lamp, the voltage at which emission of light begins. 4. The minimum gate voltage needed to turn on an MOS enhancement-mode device.

throat — 1. Part of the flare or tapered parallel-plate guide immediately adjacent to and connected to the main run of a waveguide. 2. The smaller cross-sectional area of a horn.

throat microphone — A microphone worn around the throat and actuated by vibrations of the larynx as the user talks. It is used in jet airplanes, tanks, and other places where background noise would drown out the conversation.

through connection — 1. Electrical continuity between patterns on double-sided or multilayer boards established by means of plated-through holes or jumper wires. 2. An electrical connection between conductive patterns on opposite sides of an insulating base, e.g., plated-through hole or clinched jumper wire.

through path — The transmission path from the loop input signal to the loop output signal in a feedback control loop.

throughput — 1. A measure of the efficiency of a system; the rate at which the system can handle work. 2. The speed with which problems or segments of problems are performed in a computer. Throughput varies from application to application and is meaningful only in terms of a specific application. 3. The total useful information processed or communicated over a given period; expressed in bits per second, packets per second, or some similar measurement. This quantity is frequently used for system comparisons. For example, a communication system with a throughput of 50 messages per second is superior to a system with a throughput of 30 messages per second. 4. The total amount of data that can be manipulated per second. Throughput takes into consideration such machine characteristics as basic architecture, execution speed, logic capability, and software sophistication, but is more directly related to the size of a computer's real addressable memory

throughput rate — 1. The highest rate at which a multiplier can switch from channel to channel at its specified accuracy. This rate is determined by the settling time. 2. An a/d converter or a data-acquisition system can convert a finite number of points in any given time. Throughput rate is an expression of that quantity and depends both on the time required to make a conversion and the time required to set up the unit to make the next conversion. In a data-acquisition system, throughput rate includes the effects of the composite delay arising from switching and settling times of the MUX, settling time of the amplifier, and acquisition time of the sample-and-hold.

through repeater — A microwave repeater that is not equipped to provide for connections to any local facilities other than the service channel.

through transfer function — The transfer function of the through path in a feedback control loop.

throw — 1. In an electric motor or generator, the number of core slots spanned between the bottom leg of a coil and the top leg of the same coil. 2. Movement of a contact from one stationary point to another. A single-throw switch has a normally open or a normally closed circuit per pole. A double-throw switch has a normally open and a normally closed circuit per pole.

throwing power — Referring to the ability of an anode used in an impressed-current cathodic-protection system to distribute its current over a large surface. It is principally dependent upon the anode voltage, current,

surface area, position of the anode with respect to the cathode, and the salinity and flow velocity of the water.

throw-out spiral — *See* lead-out groove.

thru-hole connection — Also called feed-thru connection and plated-through hole. A conductive material used to make electrical and mechanical connections between the conductive patterns on opposite sides of a printed circuit board.

thru repeater — In a microwave system, a repeater station that is not equipped to be connected to any local facilities other than the service channel.

thumbwheel switch — A rotating numeric switch used to input numeric information to a controller.

thump — 1. A low-frequency transient disturbance in a system or component. 2. The noise caused in a receiver by telegraph currents when the receiver is connected to a telephone circuit on which a direct-current telegraph channel is superimposed. 3. A brief, unwanted low-frequency noise (transient) that may occur when system power is turned on or off and is due to a lack of control in certain amplifier stages at that time.

thyratron — 1. A hot-cathode gas tube in which one or more control electrodes initiate the anode current, but do not limit it except under certain operating conditions. 2. A gas-filled triode in which a sufficiently large positive pulse applied to the control grid ionizes the gas and initiates conduction. Thereafter, the grid has no further effect, but conduction can be halted by reducing the plate voltage to zero (or less), or by reducing plate current to a value too small to maintain ionization.

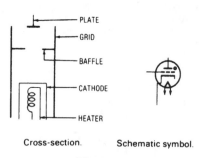

Cross-section.   Schematic symbol.

Thyratron.

thyratron gate — In computers, an AND gate consisting of a multielement gas-filled tube in which conduction is initiated by the coincident application of two or more signals. Conduction may continue after one or more of the initiating signals have been removed.

thyratron inverter — An inverter circuit in which thyratron tubes convert the dc power to ac power.

thyrector — A silicon diode that acts as an insulator until its rated voltage is reached, and as a conductor above that voltage. It is used for ac surge-voltage protection.

thyristor — 1. A bistable device comprising three or more junctions. At least one of the junctions can switch between reverse and forward voltage polarity within a single quadrant of the anode-to-cathode voltage-current characteristics. Used in a generic sense to include silicon controlled rectifiers and gate-control switches, as well as multilayer two-terminal devices. (Many countries use the term thyristor in place of the term pnpn-type switch.) 2. Member of a family of semiconcuctor switching devices of which the silicon controlled rectifier (SCR) and the triac are the most commonly used. Thyristors are fabricated from four alternate layers of positive (p) and negative (n) semiconductor material. 3. An SCR or triac

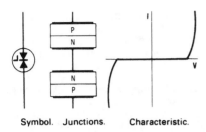

Symbol.   Junctions.   Characteristic.

Thyrector.

(which is a form of SCR) that can be used for switching both positive and negative half cycles of ac. 4. A member of a class of semiconductor devices (made of at least four alternate p-n-p-n layers) that snap to a completely on state for working current when a momentary pulse of control current is received and can (typically) be turned off only by interrupting the working current. Thyristors are mostly high-power devices. 5. A bistable semiconductor device comprising three or more junctions that can be made conductive at any instant when the anode (or principal) voltage is positive. The term thyristor may be used either as a generic term or as an abbreviation for reverse-blocking triode thyristor. 6. A device with turn-on characteristics that can be controlled by externally applied voltage or current. These devices generally consist of several interconnected layers of p- and n-type semiconductor material.

thyrite — A silicon-carbide ceramic material with nonlinear resistance characteristics. Above a critical voltage, the resistance falls considerably.

THz — Letter symbol for terahertz (10^{12} hertz).

tickler — A small coil connected in series with the anode circuit of an electron tube and inductively coupled to a grid-circuit coil. The tickler coil is used chiefly in regenerative detector circuits, to establish feedback or regeneration.

tie cable — 1. A cable between two distributing frames or distributing points. 2. A cable between private branch exchanges. 3. A cable between a private branch exchange switchboard and the main office. 4. A cable connecting two other cables.

tie line — 1. *See* interconnection, 1. 2. A private-line communication channel of the type communications common carriers provide for joining two or more points.

tie point — 1. An insulated distributing point (other than an active terminal connection) where junctions of component leads are made in circuit wiring. 2. An insulated terminal to which two or more wires are connected.

tier array — An array of antenna elements, one above the other.

ties — 1. Electrical connections or straps. 2. Tie wires.

tie trunk — A telephone line or channel directly connecting two private branch exchanges.

tie wire — Wire that connects a number of terminals together.

tie wires — Short pieces of wire used to tie open-line wires to insulators.

TIFF — *See* tagged image file format.

.TIF files — Tagged image file format files. Bit-mapped graphic images popular among desktop publishers.

TIG — Abbreviation for tungsten inert-gas welding. A technique using a tungsten electrode, generally without a filler material. Popular for light-metal gages and high-precision work.

tight coupling — *See* close coupling.

tilt — 1. In radar, the angle between the axis of radiation in the vertical plane and a reference axis, which is normally the horizontal. 2. *See* wave tilt. 3. The angle an antenna makes with the horizontal. 4. A deviation from the ideal low-frequency response; unsatisfactory low-frequency response.

tilt angle — In radar, the angle between the vertical axis of radiation and a reference axis (normally the horizontal).

tilt controls — In a color television receiver employing the magnetic-convergence principle, the three controls used to tilt the vertical center rows in the three colored patterns produced by a dot-generator signal.

tilt error — 1. In navigation, the ionospheric error component due to nonuniform height. 2. The difference between the true tilt and the mechanical tilt. The sign is such that when it is algebraically added to the mechanical tilt, the result is the true tilt. *See also* antenna tilt error.

tilting — 1. Forward inclination of the wavefront of radio waves traveling along the ground. The amount of tilt depends on the electrical constants of the ground. 2. Changing the angle of a television camera to follow a moving object being televised. 3. Changing the vertical angle of a directional antenna.

TIM — Abbreviation for transient intermodulation distortion.

timbre — Also called tone color or musical quality. 1. The character of a musical tone that distinguishes one musical instrument from another playing the same note. The difference between two steady tones having the same pitch and degree of volume is called the difference in timbre. Timbre depends mostly on the relative intensity of the different harmonics and the frequencies of the most prominent harmonics. 2. The tonal quality of sound based on the pitch and the relative mix of fundamental and harmonic frequencies.

time — The measure of the duration of an event. The fundamental unit of time is the second.

time assignment speech interpolation — Telephone switching equipment whereby a person is connected to idle circuits when he starts talking, and is disconnected when he stops talking.

time base — A voltage generated by the sweep circuit of a cathode-ray-tube indicator. Its waveshape is such that the trace is either linear with respect to time or, if nonlinear, is still at a known timing.

time clock — A loosely used term sometimes referring to time switches, sometimes to interval timers, or to any type of timer.

time-code generator — A timer with an absolute time reference having digital outputs suitable for machine interpretation. Dial or numeric display may be included.

time-compression multiplexing — A method of providing the appearance of full-duplex communication over a single twisted-pair half-duplex copper loop. Data are buffered at each end and sent across the line at double the subscriber data rate, with the two ends taking turns.

time constant — 1. The time required for an exponential quantity to change by an amount equal to 0.632 times the total change that will occur. Specifically: (*a*) In a capacitor-resistor circuit, the number of seconds required for the capacitor to reach 63.2 percent of its full charge after a voltage is applied. The time constant of a capacitor having a capacitance C in farads in series with a resistance R in ohms is equal to $R \times C$. (*b*) In an inductor-resistor circuit, the number of seconds required for the current to reach 63.2 percent of its final value. The time constant of an inductor having an inductance L in henrys and resistance R in ohms is equal to L/R. 2. The time required for a motor to accelerate from 0 to 63.2 percent of its final no-load speed when the rated voltage, with the proper phase relationship, is applied. 3. A measure of the rapidity with which a transducer responds to a change in temperature. It is the time required for the sensor to complete 63.2 percent of its exponential response to a step change in temperature and is highly dependent on the thermal conductivity of the medium measured. Time constant must match the speed of which it is a part. In general, small sensors respond more rapidly than large ones.

time constant of a capacitor — The product of the insulation resistance and the capacitance of a capacitor.

time constant of fall — The time required for a pulse to fall from 70.7 percent to 26.0 percent of its maximum amplitude, excluding spikes.

time-current characteristics of a fuse — The relation between the root-mean-square alternating current or direct current and the time for the fuse to perform the whole or some specified part of its interrupting function. Usually shown as a curve.

timed acceleration — A control function that automatically controls the speed increase of a drive as a function of time.

time date generator — A device used in surveillance that electronically produces a single-row display of day, time, and date information on any standard TV screen (or raster).

timed deceleration — A control function that automatically controls the speed decrease of a drive as a function of time.

time delay — 1. The time required for a signal to travel between two points in a circuit. 2. The time required for a wave to travel between two points in space. 3. The total elapsed time or lag required for a given command to be effected after the command is given. 4. Also called envelope delay. The slope of the phase-versus-frequency curve at a specified frequency. In a loose sense, this is the time it takes a designated point in a wave to pass through a filter.

time-delay circuit — A circuit that delays the transmission of an impulse signal, or the performance of a transducer, for a definite desired length of time.

time-delay closing relay — See time-delay starting relay.

time-delay generator — A device that accepts an input signal and provides a delay in time before the initiation of an output signal.

time-delay relay — Also called slow-action relay. 1. A relay in which there is an appreciable interval of time between the energizing or deenergizing of the coil and the movement of the armature (e.g., slow-operating relays and slow-release relays). 2. A device in which a defined delay occurs between the application or removal of the control current and the switching of the load current. 3. An electronic device with either relay or solid-state output that performs a retarding function on receipt of instruction. 4. A reset timer with delayed contacts. 5. A relay with electromagnetic action that provides a specified time interval between the energizing of the coil and the actual contact closure. The timing interval can be preset and the delay may be achieved either by mechanical clockwork or electronic means.

time-delay spectrometry — An acoustic measurement technique utilizing a real-time spectrum analyzer.

time-delay starting relay — Also called time-delay closing relay. A device that functions to give a desired amount of time delay before or after any point or operation in a switching sequence or protective relay system.

time-delay stopping or opening relay — A time-delay device that serves in conjunction with the device that initiates the shut-down, stopping, or opening operation in an automatic sequence.

time-derived channel — Any of the channels obtained by time-division multiplexing of a channel.

time discriminator — A circuit in which the sense and magnitude of the output is a function of the time difference of, and relative time sequence between, two pulses.

time-distribution analyzer — Also called time sorter. An instrument that indicates the number or rate of occurrence of time intervals falling within one or more specified ranges. The time interval is defined by the separation between members of a pulse pair.

time-division data link — Radiocommunications that use time-division techniques for channel separation.

time-division multiplex channel — A path established for a segment of time during which the central station equipment obtains a signal indicating the status of a signal-encoding device.

time-division multiplexer — A device that samples all data input from different low-speed devices, and retransmits all the samples in an equal amount of time.

time-division multiplexing — Abbreviated TDM. 1. Multiplexing by assigning each low-speed channel its own time slot during which it transmits data. 2. A process by which two or more channels of information are transmitted over the same link by allocating a different time interval for the transmission of each channel. 3. A signaling method characterized by the sequential and noninterfering transmission of more than one signal in a communication channel. Signals from all terminal locations are distinguished from one another by each signal occupying a different position in time with reference to synchronizing signals. 4. A system of multiplexing in which channels are established by connecting terminals one at a time at regular intervals by means of an automatic distribution. 5. The multiplexing technique that provides for the independent transmission of several pieces of information on a time-sharing basis by sampling, at frequent intervals, the data to be transmitted. 6. A data-communication technique for combining several lower-speed channels into one facility or transmission path at a higher speed; each low-speed channel is allotted a specific position in the signal stream based on time. Thus, the information on the low-speed input channels is interleaved at higher speed on the multiplexed channel. At the receiver, the signals are separated to reconstruct the individual low-speed channels.

time domain — 1. An analysis of a waveform in terms of how the parameter of interest, voltage for example, varies with time. (The sine wave can help to visualize this concept. At any instant in time the voltage has only a single unique value. But as time progresses, the voltage rise and fall describes the familiar smooth curve. A square wave, on the other hand, is a picture in the time domain of a voltage that repeatedly changes abruptly from one value to another.) 2. A measurement technique where the results are plotted or shown against a scale of time, in contrast to frequency domain where the results are plotted on a scale of frequency.

timed-release circuit — A circuit designed to automatically release other connected circuits after a preset interval.

time-edit — To prevent the time code in a computer from being recorded or transmitted when no data samples are being passed, or, conversely, to record or transmit the time code with each data sample passed.

time flutter — A variation in the synchronization of components of a radar system, leading to variations in the position of the observed pulse along the time base and reducing the accuracy with which the time of arrival of a pulse may be determined.

time frame — In telemetry, the time period containing all elements between corresponding points of two successive reference markers.

time gain — *See* sensitivity-time control.

time gate — A transducer that has an output during chosen time intervals only.

time harmonics — High-frequency components generated by nonlinearity in a resolver magnetic circuit. In ac resolvers with no dc components present, time harmonics are odd multiples of the fundamental frequency. Time harmonics appear as a component of null voltage.

time-interval selector — A circuit that functions to produce a specified output pulse when and only when the time interval between two input pulses is between set limits.

time lag — 1. The interval between application of any force and full attainment of the resultant effect. 2. The interval between two phenomena.

time-lapse VTR — A videotape recorder that provides a continuous record of events over a period of from 12 to 48 hours on one reel of videotape.

time-mark generator — A circuit that produces accurately spaced pulses for display on the screen of an oscilloscope.

time modulation — Modulation in which the time of occurrence of a definite part of a waveform is made to vary in accordance with a modulating signal.

time of persistence — The time that elapses between the instant of removal of the excitation and the instant at which the luminance or radiance has dropped to a stated fraction of its initial value, usually 10 percent.

time-out — The interval of time allotted for certain operations to occur (for example, response to polling or addressing) before operation of the system is interrupted and must be started again.

time pattern — A picture-tube presentation of horizontal and vertical lines or rows of dots generated by two stable frequency sources operating at multiples of the line and field frequencies.

time phase — Reaching corresponding peak values at the same instants of time, though not necessarily at the same points in space.

time pulse distributor — A device or circuit for allocating timing pulses or clock pulses to one or more conducting paths or control lines in a specified sequence.

time quadrature — Differing by a time interval corresponding to one-fourth the time of one cycle of the frequency in question.

timer — 1. A special clock mechanism or motor-operated device used to perform switching operations at predetermined time intervals. 2. An assembly of electric circuits and associated equipment that provides the following: trigger pulses, sweep circuits, intensifier pulses, gate voltage, blanking voltages, and power supplies. 3. The part of a radar set that initiates pulse transmission and synchronizes this with the beginning of indicator sweeps, timing of gates, range markers, etc. *See also* synchronizer, 1. 4. An instrument that measures the interval between events. The term *event* may encompass anything from persons moving through a turnstile to oscillations in the highest-frequency radar bands.

time resolution — The smallest interval of time that can be measured with a given system.

time response — An output, expressed as a function of time, that results from a specified input applied under specified operating conditions.

timer motor — A synchronous clock motor having a definite output speed determined by the number of poles in the stator and the reduction of the associated gear train.

time sequencing — In a computer, switching signals generated by a program purely as a function of accurately measured elapsed time.

time series analysis — Analysis of any variable classified by time, in which the values of the variable are functions of the time periods.

time share — To use a device for two or more purposes on a time-sharing basis.

time sharing — 1. A method of operation in which a computer facility is shared by several users for different purposes at (apparently) the same time. Although the computer actually services each user in sequence, the high speed of the computer makes it appear as if the users were all handled simultaneously. 2. A means of making more efficient use of a facility by allowing more than one using activity access to the facility on a sequential basis. 3. A computer system in which CPU time and system resources are shared simultaneously by several operators engaged in unrelated tasks and located at terminals remote from the central processing unit. The speed of the computer gives the appearance of simultaneously performing multiple jobs. Programs performing the individual jobs are swapped at high speeds under direction of the scheduling formulas or plan for the time-sharing system, thus assuring each user a system response within a few seconds (almost the exact opposite of a batch system). 4. The use of the same computer memory for two or more simultaneous tasks.

time-sharing system — *See* TSS.

time signals — Time-controlled radio signals broadcast by government-operated radio station WWV at regular intervals each day on several frequencies.

time signal service — Radiocommunication service for the transmission of time signals of stated high precision, intended for general reception.

time slicing — 1. The sharing of CPU time among several software processes by giving each process a defined interval (slice) of the CPU's time. 2. A type of operation in which a computer works on one program for a short time, then goes to another program and works on that for a short time, and so forth. 3. A technique that shares microprocessor time among several processes. A quantum of time allocated to a process is termed a time slice.

time sorter — *See* time-distribution analyzer.

time stability — The degree to which a component value is maintained to a stated degree of certainty (probability) under stated conditions of use over a stated period of time. It is usually expressed in $\pm$ percent or $\pm$ per unit (ppm) change per 1000 hours of continuous use.

time switch — A clock-controlled switch used to open or close a circuit at one or more predetermined times. Usually refers to repeat cycle timers with a dial graduated with time of day (in 24-hour type) or with time and day (in 7-day type). Has means of adjusting the time when circuits are turned on or off, usually with pins or clips. Used for turning lights on at night, ringing bells in offices, schools, etc.

time-to-digital conversion — The process of converting an interval of time into a digital number.

time window — In random-sampling-oscilloscope technique, the part of the signal period that is displayed.

timing — The procedure of adjusting the ignition distributor so the sparks will occur at the spark plugs at the correct time in relation to the position of the piston on the compression stroke.

timing axis oscillator — *See* sweep generator.

timing lag — The number of engine degrees retard in timing caused by electrical lag in the system. Generally stated as number of engine degrees per 1000 rpm engine speed.

timing light — A stroboscope unit that is connected to the secondary ignition circuit of an internal-combustion engine to produce flashes of light in unison with the firing of a specific spark plug. By directing these flashes of light on the whirling timing marks, the marks appear to stand still. By adjusting the distributor, the timing marks may be properly aligned, thus setting the timing.

timing-pulse distributor — Also called waveform generator. A computer circuit driven by pulses from the master clock. It operates in conjunction with the operation decoder to generate timed pulses needed by other machine circuits to perform the various operations.

timing relay — A form of auxiliary relay used to introduce a definite time delay in the performance of a function.

timing signal — Any signal recorded simultaneously with data to provide a time index.

timing tape — A variety of leader tape having printed indications of intervals of $7\frac{1}{2}$ inches (19.05 cm) along its length to facilitate measuring off a desired number of seconds' worth of leader.

tinkertoy — An attempt at modularization whereby wafers, with one or more component parts printed or mounted on them, are stacked vertically, with interconnecting wiring stiff enough to provide support running through holes around the periphery of the wafer.

tin-lead — An alloy used for the majority of soldering operations in the electronics industry. Usually, an alloy close to the eutectic composition (62 percent tin, 38 percent lead) is chosen to permit usage of the lowest possible soldering temperature, thereby reducing risk of damage to temperature-sensitive components.

tinned — Covered with metallic tin to permit easy soldering.

tinned wire — A copper wire that has been coated with a layer of tin or solder to prevent corrosion and to simplify soldering.

tinning — A process of coating a conductor surface with solder.

tinsel — Flat ribbons of bronze, silver, or copper alloy spiraled around a textile core of cotton, nylon, etc. Used in telephone and electronics applications as conductors in line cords, microphone cords, and retractile cords.

tinsel conductor — A type of electrical conductor comprising a number of tiny threads, each thread having a fine, flat ribbon of copper or cadmium bronze closely spiraled about it. Used for small cables requiring limpness and extra-long flex life (e.g., used with headsets or handsets).

tinsel wire — 1. A very flexible conductor made by wrapping one or more thin ribbon conductors over a nylon string core. Used to make telephone cords. 2. A low-voltage stranded wire in which each strand is a very thin conductor ribbon spirally wrapped around a textile yarn. Insulation is generally a textile braid. Intended usage is for severe flexing.

tint — A mixture of a color with white light.

tip — 1. The contacting part at the end of a plug or probe. 2. Also called tip side. The end of the plug used to make circuit connections in a manual switchboard. The tip is the connector attached to the positive side of the common battery that powers the station equipment. By extension, it is the positive battery side of a communications line.

tip and ring — Traditional telephone terminology for *positive* and *negative*. In old-style telephone switchboards, the tip (positive) wire was the one that connected to the tip of the plug; the ring (negative) wire was connected to a slip ring in the jack.

tip jack — Also called a pup jack. A small single-hole jack for a single-pin contact plug.

tip mass — The effective mass at the tip of the stylus of a pickup cartridge. (Modern techniques have reduced this mass toward 1 milligram or below.)

tipoff — The last portion of a vacuum-tube bulb to be melted and sealed after evacuation of the bulb.

tip-ring-sleeve (TRS) phone plug — A three-conductor phone plug commonly used for balanced audio connections.

tip side — Also called a tip wire. The conductor of a circuit associated with the tip of a plug or the tip spring of a jack. *See also* tip, 2.

tip-sleeve (TS) phone plug — A two-conductor phone plug (or jack) commonly used for unbalanced audio connections.

tip wire — *See* tip side.

T-junction — *See* tee junction.

TM wave — Abbreviation for transverse magnetic wave.

Tm$_{m,n}$ wave — In a rectangular waveguide, the transverse magnetic wave for which m and n are the number of half-period variations of the magnetic field along the longer and shorter transverse dimensions, respectively.

TNC — A threaded connector for miniature coax; TNC is said to be short for threaded-Neill-Concelman. (Contrast with BNC.)

T-network — A network composed of three branches. One end of each branch is connected to a common junction point. The three remaining ends are connected to an input terminal, an output terminal, and a common input and output terminal.

T-network.

TO can — Abbreviation for transistor-outline metal-can package.

TO-5 — A small cylindrical IC package about 0.375 inch (9.5 mm) in diameter and made of metal; it can have up to 12 pins protruding from the bottom of the can.

to-from indicator — An instrument that forms part of the omnirange facilities and is used for resolving the 180° ambiguity.

toggle — 1. A flip-flop. The term implies that the flip-flop will change state upon receipt of a clock pulse, and is used mainly with reference to flip-flops connected as a counter, in which the process of changing state is called toggling. 2. To use switches to enter data into the memory of a computer. 3. To change states abruptly in such a way that the process goes to completion once a critical point is reached, even if the actuating cause is removed; in logic circuits, a change of state of a flip-flop or Schmitt trigger. 4. *See* T flip-flop. 5. The ability, through the use of a windowing program, to allow a user to switch back and forth between different PC applications programs, as well as between PC and mainframe or mini applications.

toggle frequency — In a digital circuit, the number of times per second that the circuit changes state.

toggle rate — Twice the frequency at which a flip-flop completes a full cycle encompassing both states. Usually used to denote the maximum input frequency that a flip-flop can follow.

toggle switch — 1. A switch with a projecting lever whose movement through a small arc opens or closes one or more electric circuits. 2. A panel-mounted switch with an extended lever; normally used for on/off switching. 3. A switch having a lever (toggle), the movement of which results either directly or indirectly in the connection or disconnection of the switch terminations in a specified manner. Any indirect action through an actuating mechanism should be such that the speed of connection and/or disconnection is independent of the speed of lever movement.

tokamak — A doughnut-shaped plasma confinement device for research into the problems associated with nuclear fusion.

token — 1. A group of bits used in some bus networks to signal network access by a particular station. 2. The information package that contains a piece of data and a description of its location in a computer program.

token passing — A protocol that gives a terminal permission to transmit on a token ring LAN. A unique bit pattern, called a token, circulates around the ring from terminal to terminal. The terminal that possesses the token has permission to transmit.

token ring network — A local area network (LAN) in a circle configuration. Messages are sent when one computer attaches a message to a special bit pattern (called a token) as it travels around the circle.

tolerance — 1. A permissible deviation from a specified value. A frequency tolerance is expressed in hertz or as a percentage of the nominal frequency; an orientation tolerance, in minutes of arc; a temperature tolerance, in degrees Celsius; and a dimensional tolerance, in decimals or fractions. 2. A permissible deviation from specified capacitance (resistance, inductance, etc.) values, characteristics, and physical dimensions. 3. A specified allowance for error from a desired or measured quantity. 4. Maximum error or variation from the standard permissible in a measuring instrument. 5. Maximum electrical or mechanical variation from specifications that can be tolerated without impairing the operation of a device.

toll — 1. In public switched systems, a charge for a connection beyond an exchange boundary, based on time and distance. 2. Any part of telephone plant, circuits, or services for which toll charges are made.

toll call — 1. Telephone call to points beyond the area within which telephone calls are covered by a flat monthly rate or are charged for on a message unit basis. 2. A telephone call, subject to charge, for a destination outside of the local service area of the calling station.

toll center — Also called toll office and toll point. 1. The basic toll switching entity; a central office where channels and toll-message circuits terminate. While this is usually one particular central office in a city, larger cities may have several central offices where toll-message circuits terminate. 2. A major telephone distribution center that distributes calls from one major metropolitan area to another.

toll-free number — *See* enterprise number.

toll office — *See* toll center.

toll point — *See* toll center.

toll restriction — The action of preventing restricted telephones from making calls to toll or other restricted points. A restrictor counts the first three digits dialed and diverts calls to forbidden codes either to a busy tone, a recorded announcement, or an operator.

toll-terminal loss — On a toll connection, that part of the overall transmission loss attributable to the facilities from the toll center through the tributary office to and including the subscriber's equipment.

tone — 1. A sound wave capable of exciting an auditory sensation having pitch. 2. A sound sensation having pitch.

tonearm — 1. The pivoted arm of a record player that extends over the record and holds the pickup cartridge. Wires from the cartridge run through the arm to the preamplifier, usually located on the underside of the turntable mounting board. 2. The portion of a record player that supports the phono cartridge and maintains it in the correct relationship to the record surface and the spiral groove. On conventional pivoted tonearms, the cartridge is mounted at an offset angle and with a slight overhang beyond the turntable center to reduce the tracking error. The mass of the cartridge and the forward portion of the arm is balanced by an adjustable counterweight, and the desired vertical tracking force is supplied by a slight mass unbalance or a spring.

tone burst — A single sine-wave frequency, 50 to 500 microseconds long, having a rectangular envelope, used for testing the transient response of speakers.

tone channel — An intelligence or signaling circuit in which on-off or frequency-shift modulation of a frequency (usually an audio frequency) is used as a means of transmission.

tone control — 1. A control, usually part of a resistance-capacitance network, used to alter the frequency response of an amplifier so that the listener can obtain the most pleasing sound. In effect, a tone control accentuates or attenuates the bass or treble portion of the audio-frequency spectrum. 2. A circuit designed to increase or decrease the amplification in a specific frequency range, with little or no effect at other frequencies. Bass tone controls usually affect frequencies below a turnover frequency that may vary between 100 and 1000 Hz. Treble tone controls are typically hinged to affect frequencies above 1500 Hz. The range of a tone control (the maximum amount by which it can vary the amplification within its operating range) is typically about ±15 dB, but may be as low as ±7 dB or as great as ±20 dB. (Equalizers can provide tone control over five or more frequency bands.)

tone dialing — *See* pushdown dialing.

tone generator — A device for providing an audio-frequency current suitable for testing audio-frequency equipment or for signaling.

tone keyer — An instrument device that converts direct-current impulses to audio tones, for line transmission or for keying a transmitter.

tone localizer — *See* equisignal localizer.

tone-modulated waves — Waves obtained from continuous waves by amplitude-modulating them at an audio frequency in a substantially periodic manner.

tone modulation — A type of code-signal transmission obtained by causing the radio-frequency carrier amplitude to vary at a fixed audio frequency.

tone pad — An array of 12 or 16 numbered keys that generate the standard telephone dual-tone multifrequency (DTMF) dialing signals.

toner — 1. Charged carbon particles, dry or suspended in a liquid solvent, used to produce a dark image on a light medium. 2. Minute, dry particles of resin and carbon black that are used to create images. In the electrostatic copying process, toner is capable of accepting an electrical charge. It is carried to the photoconductor by a developer medium and transferred to the surface of a copy sheet by a series of successively greater electrical charges.

tone reversal — Distortion of the recorder copy in facsimile. It causes the various shades of black and white not to be in the proper order.

tone signaling — The transmission of supervisory, address, and alerting signals over a telephone circuit by means of voice frequency tones. Also used in "touch tone" dialing.

tongue — The portion of a solderless terminal that projects from the barrel.

Tonotron — A multimode, selective-erasure storage tube.

tool function — In automatic control of machine tools, a command that identifies a specific tool and calls for the use of that tool.

TO package — Can-type IC chip configuration, an outgrowth of the original TO transistor package. Most common are the TO-5, TO-18, and TO-47. The IC chip is mounted within the package, interconnected to terminals on the can, and then hermetically sealed. TO stands for transistor outline.

top cap — A terminal in the form of a metal cap at the top of some vacuum tubes and connected to one of the electrodes.

top-down programming — *See* structured programming.

top-hat resistors — Film resistors having a projection on one side allowing a notch to be cut into the center of the projection to effectively form a serpentine resistance and thereby increase the resistivity.

top-loaded vertical antenna — A vertical antenna that is larger at the top, resulting in a modified current distribution that gives a more desirable radiation pattern vertically. A series reactor may be connected between the enlarged portion of the antenna and the remaining structure.

topology — 1. The physical arrangement and relationship of interconnected nodes and lines in a network. Linear bus, mesh, star, token ring, and point-to-point are the major LAN topologies. 2. The surface layout of the elements comprising an IC.

tornadotron — A millimeter wave device that generates radio-frequency power from an enclosed, orbiting electron cloud excited by a radio-frequency field when subjected to a strong, pulsed magnetic field.

torn-tape relay — A method of receiving messages in tape form, breaking the tape, and retransmitting the message in tape form.

torn-tape switching center — A location at which operators tear off the incoming printed and punched paper tape and transfer it manually to a machine for transmission over the proper outgoing circuit.

toroid — 1. A surface, or its closed solid, generated by any closed plane rotating about a straight line in its own plane — the resulting configuration being doughnut shaped. 2. A highly efficient type of coil wound on a ring or doughnut type of core. The toroid provides for high concentrated magnetic field within itself, and has a minimum magnetic flux leakage (external field).

toroidal coil — A coil wound in the form of a toroidal helix.

toroidal core — A ring-shaped core.

toroidal permeability — Under stated conditions, the relative permeability of a toroidal body of the given material. The permeability is determined from measurements of a coil wound on the toroid such that stray fields are minimized or can be neglected.

torque — A force that tends to produce rotation or twisting.

torque amplifier — A device with input and output shafts and supplying work to rotate the output shaft so that its position corresponds to that of the input shaft but does not impose any significant torque on the latter.

torque-coil magnetometer — A magnetometer that depends for its operation on the torque developed by a known current in a coil capable of turning in the field to be measured.

torque gradient — The torque required in inch-ounces to pull a specific energized synchro 1° away from its normal position.

torque motors — A motor that is designed to provide its maximum torque under the condition of stall or locked-rotor. A second criteria of torque motors is that they must be capable of remaining in a stalled condition for prolonged periods.

torque of an instrument — Also called deflecting torque. The turning moment produced on the moving element by the quantity to be measured or by some quantity dependent thereon acting through the mechanism.

torquer — A device that produces torque about an axis of freedom in response to a signal input.

torque-to-inertia ratio — *See* acceleration at stall.

torr — The unit of pressure used in the measurement of a vacuum. It is equal to $^1/_{760}$ of a standard atmosphere, and for practical purposes may be considered equivalent to one millimeter of mercury (mm Hg).

torsiometer — An instrument for measuring the amount of power that a rotating shaft is transmitting.

torsion galvanometer — A galvanometer in which the force between the fixed and moving systems is measured by the angle through which the supporting head of the moving system must be rotated to return the moving system to zero.

torsion-string galvanometer — A sensitive galvanometer in which the moving system is suspended by two parallel fibers that tend to twist around each other.

total capacitance — The capacitance between a given conductor and all other conductors in a system when all other conductors are connected together.

total combined regulation — The change in output of a regulated power supply arising from simultaneous changes in all of the specified operating conditions, when the direction of such changes is such as to make their effects additive. It may be stated as a percentage of the specified output and/or as an absolute value.

total connected load — The total current drawn in amperes, or power consumption in watts, of all of the utilization equipment connected to all or any given segment of a wiring system.

total distortion — The sum total of all forms of signal distortions.

total emission — The magnitude of the current produced when electrons are emitted from a cathode under the influence of a voltage such that all the electrons emitted are drawn away from the cathode.

total emissivity — The ratio of radiation emitted by a surface to the radiation emitted by the surface of a blackbody under identical conditions. Important conditions that affect emissivity of a material are surface finish, color, temperature, and wavelength of radiation. Emissivity may be expressed for radiation of a single wavelength (monochromatic emissivity), for total radiation of a specified range of wavelengths (total spectral emissivity), or for total radiation of all wavelengths (total emissivity).

total excursion — The application of a stimulus, in a controlled manner, over the span of an instrument.

total flux — The luminous flux emitted by a light source in all directions.

total harmonic distortion — Abbreviated THD. 1. The ratio of the power at the fundamental frequency, measured at the output of the transmission system considered, to the power of all harmonics observed at the output of the system because of its nonlinearity, when a single-frequency signal of specified power is applied to the input of the system. It is expressed in decibels. 2. The square root of the sum of the squares of the root-mean-square harmonic voltages divided by the root-mean-square fundamental voltage. 3. The ratio of the sum of the amplitudes of all signals harmonically related to the fundamental and the amplitude of the fundamental signal. 4. A measure of the distortion produced in an audio system as a result of the production, by the system, of harmonics or multiples of the original audio signal.

total internal reflection — When light passes from one medium to another that is optically less dense (e.g., from glass to air), the ray is bent away from the normal. If the incident ray meets the surface at such an angle that the refracted ray must be bent away at an angle of more than 90°, the light cannot emerge at all, and is totally internally reflected.

totalizing — To register a precise total count from mechanical, photoelectric, electromagnetic, or electronic inputs or detectors.

total losses of a ferromagnetic part — Under stated conditions, the power absorbed and then dissipated as heat when a body of ferromagnetic material is placed in a time-varying magnetic field.

total losses of a transformer — The losses represented by the sum of the no-load and load loses.

total luminous flux — The total light emitted in all directions by a light source.

totally enclosed motor — A motor so enclosed as to prevent the free exchange of air between the inside and the outside of the case, but not sufficiently enclosed to be termed airtight.

totally unbalanced currents — See push-push currents.

total range of an instrument — Also called the range of an instrument. The region between the limits within which the quantity measured is to be indicated or recorded.

total regulation — The arithmetic sum of changes in output of a regulated power supply arising from changes in each of the specified operating conditions (current, voltage, or power) when such changes are applied individually and in a manner to make their effects additive. It may be stated as a percentage of the specified output and/or as an absolute value.

total resistance — The dc resistance of a (precision) potentiometer between the input terminals with the shaft positioned so as to give a maximum resistance value.

total spectral emissivity — See total emissivity.

total telegraph distortion — Telegraph transmission impairment expressed in terms of time displacement of mark-space and space-mark transitions from their proper positions and given in percent of the shortest perfect pulse, called the unit pulse.

total transition time — In a circuit, the time interval between the point of 10 percent input change and the point of 90 percent output change. It is equal to the sum of the delay time and rise (or fall) time.

totem pole amplifier — A push-pull amplifier circuit that provides a single-ended output signal without the use of a transformer. Used mainly in the output stages of transistor audio amplifiers.

TO-3 — A large metal IC housing, approximately 1.5 × 11 inch (3.8 × 2.5 cm) having up to eight pins. When properly heat-sinked, this case is used where high-power dissipation takes place.

touch call — See pushdown dialing.

touch control — A control circuit that actuates a circuit when two metal areas or a preselected area are bridged by one's finger or hand.

touchscreen — A touch-sensitive display screen, whereby users can interact with a computer and select options by simply pointing with a finger instead of using a pointing device such as a mouse. Typically used in a kiosk.

touch sensitivity — The sensitivity of a capacitance sensor at which the alarm device will be activated only if an intruder touches or comes in very close proximity (about 1 cm or 0.39 in) to the protected object.

Touch-Tone — A service mark of the American Telephone and Telegraph Company that identifies its push-button (DMTF) dialing service. The signal form is multiple tones in the audio-frequency range.

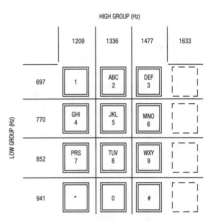

Touch-Tone frequencies.

Touch-Tone dialing — Push-button telephone dialing in which the direct-current pulsing technique of the rotating dial is replaced by a combination of tones to provide proper automatic switching.

Touch-Tone telephone set — Designation for a telephone set using a push-button multifrequency dial. Each digit consists of combining one lower frequency with one higher frequency, e.g., the digit 8 consists of combining 852 Hz and 1336 Hz.

touch voltage — The potential difference between a grounded metallic structure and a point on the earth's surface equal to the normal maximum horizontal reach (approximately 3 feet or 0.91 m).

tourmaline — A strongly piezoelectric natural or synthetic crystal.

tower — A structure usually used when an antenna must be mounted higher than 50 feet (15.24 m).

tower loading — The load placed on a tower by its own weight, the weight of the wires and insulators with or without ice covering, the wind pressure acting on both the tower and the wires, and the tension in the wires.

tower radiator — A metal tower structure used as a transmitting antenna.

Townsend criterion — The relationship expressing the minimum requirement for breakdown in terms of the ionization coefficients.

Townsend discharge — An electrical discharge in a gas at moderate pressure (above about 0.1 millimeter of mercury). It corresponds to corona, and is free from space charges.

Townsend ionization coefficient — The average number of ionizing collisions made by an electron as it drifts a unit distance in the direction of an applied electric force.

T-pad — A pad made up of resistance elements arranged in a T-network (two resistors inserted in one line, with a third between their junction and the other line).

TP tape — See triple-play tape.

tr— 1. Abbreviation for transient response. 2. Abbreviation for transmit-receive.

trace— 1. The pattern on the screen of a cathode-ray tube. 2. Software used for extremely detailed testing of the validity of an application program or other software. Data is stored in RAM as a program executes; when the program halts, a predetermined number of steps are available for examination.

traceability— 1. Symbol identification that permits symbols to be traced throughout a software system. Also, continuity between program versions. *See also* auditing. 2. The ability to track a manufactured item through the steps of its manufacture.

trace interval— The interval corresponding to the direction of sweep used for delineation.

trace line— The visible or recordable path traced on the screen of a CRT by a moving spot.

tracer— 1. A radioisotope that is mixed with a stable material to trace the material as it undergoes chemical and physical charges. 2. A thread of contrasting color woven into the insulation of a wire for identification purposes. 3. In automatic machine control, a sensing element that is made to follow the outline of a contoured template and produce the desired control signals. 4. *See* signal tracer.

trace routine— Special type of debugging routine that develops a sequential record of specified events in the execution of a program.

tracer program— Analysis tool that searches programs for unreachable (dead) code.

trace width— The distance between two points on opposite sides of a trace on an oscilloscope at which luminance is 50 percent of maximum. If the trace departs from a well-behaved (approximately Gaussian) form, it should be smoothed for the purpose of measurement.

tracing— The process of determining from what location in a telephone or computer network a call has originated.

tracing distortion— The nonlinear distortion introduced in the reproduction of a mechanical recording when the curve traced by the reproducing stylus is not an exact replica of the modulated groove.

tracing routine— *See* trace routine.

track— 1. A path that contains reproducible information left on a medium by recording means energized from a single channel. 2. In electronic computers, that portion of a moving-type storage medium accessible to a given reading station (e.g., film, drum, tapes, disks). *See also* band, 3; channel, 3. 3. To follow a point of radiation to obtain guidance. 4. The path on the magnetic tape along which a single channel of sound is recorded. 5. One of the ring-shaped sequence of sectors defined on the magnetic surface of a disk or drum. A floppy disk has 77 tracks, numbered from 0 to 76. Tracks are arranged concentrically. 6. The actual line of movement of an aircraft or a rocket over the surface of the earth; a projection of the history of the flight path on the surface. 7. The trace of a moving target on a plan-position-indicator radar screen or an equivalent plot. *See also* tracking.

trackability— How well the pickup system will track high amplitude and velocity modulation on a phonograph record at a given tracking weight.

trackball— A data-entry device consisting of a sphere mounted in a small open-topped box. This sphere is hand operated by rotating it about either its x or y axis. The rotation is translated into digital data and stored in special input registers. The registers are then read by the host computer. The most common use of a trackball is in the control of cursor and tracking-symbol movements. It can also be used wherever it is desired to continuously vary display or process parameters.

track-command guidance— A missile guidance method in which both target and missile are tracked by separate radars and commands are sent to the missile to correct its course.

track configuration— The number of separate recorded tracks that a machine can play or record. Thus, a two-track machine makes or plays two parallel tracks, each occupying slightly less than half the width of the tape. A four-track machine makes or plays four parallel tracks, two in the same playing direction on a two-channel machine, or four in the same playing direction on a quadraphonic machine. Monophonic machines can be full-track, half-track, or quarter-track.

track-hold memory— A circuit that, in its track mode, develops an output that follows (ideally) the input exactly, or is proportional to it; and then, in its hold mode, maintains the output constant (ideally) at the value it had at the instant the circuit was commanded to change from track to hold.

track homing— The process of following a line of position known to pass through an object.

tracking— 1. The process of keeping a radio beam, or the cross hairs of an optical system, set on a target — usually while determining the range of the target. 2. The maintenance of proper frequency relationships in circuits designed to be simultaneously varied by ganged operations. 3. The accuracy with which the stylus of a phonograph pickup follows a prescribed path over the surface of the record. 4. The ability of a multigun tube to simultaneously superimpose information from each gun. Tracking error is the maximum allowable distance between the displays of any two guns. 5. The ability of an instrument to indicate at the division line being checked when energized by corresponding proportional values of actual end-scale excitation. 6. The interconnecting of power supplies in such a manner that one unit serves to control all units in series operation. In this manner, the output voltage of the slave units follows the variations of the master. 7. The difference at any shaft position between the output ratios of any two commonly actuated similar electrical elements expressed as a percentage of the single total voltage applied to them. 8. The process of following the movement of a satellite or rocket by radar, radio, and photographic observations. 9. The moving of a tracking symbol on the face of a CRT with a light-pen. The meaning can also be extended to include the movement of any graphical image on the screen, using any of the data entry devices. 10. The ability of a power supply to maintain a fixed ratio between multiple outputs regardless of variations in line, load, or ambient temperature. In supplies with symmetrical plus-and-minus outputs, tracking maintains equal amplitudes of opposite polarity, referred to the common terminal. 11. The accuracy of indication of an analog meter having a nonlinear scale.

tracking accuracy— The ability to indicate at a division line being checked when a meter is energized by corresponding proportional values of actual end-scale excitation, expressed as a percentage of actual end-scale value. Tracking accuracy is usually within one and one-half times the rated full-scale accuracy of the meter.

tracking antenna— A directional antenna system that automatically changes in position or characteristics to follow the motion of a moving signal source.

tracking error— 1. In lateral mechanical recording, the angle between the vibration axis of the mechanical system of the pickup and a plane that contains the tangent to the unmodulated record groove and is perpendicular to the recording surface at the point of needle contact. 2. In horizontal recording, the angle between the body of a phonograph cartridge and the tangent of a groove at the point of a stylus contact. Ideally the angle should be zero, but can be maintained at less than 0.5° per inch (0.19° per centimeter) of playing radius in a well-designed tonearm. Excessive error can cause increased distortion on heavily recorded passages, especially near the inner grooves of the record.

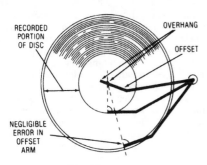

Tracking error.

tracking filter — An electronic device that attenuates unwanted signals and passes the desired signal by the use of phase-lock techniques that reduce the effective bandwidth of the circuit and eliminate amplitude variations.

tracking force — The downward force applied to a stylus by means of an adjustment on the tonearm. Sometimes specified as a recommended range or a single force, in grams. It is usually advisable to track near the upper rather than the lower suggested value, since distortional record damage from mistracking at too low a force is more objectionable than the slightly increased wear from the higher force. A large stylus has a large groove contact area, and thus can tolerate a higher tracking force. *See also* stylus force.

tracking jitter — *See* jitter, 2.

tracking range — Also called lock-in range. The range of input frequency about f_0 under which a phase-locked loop, once locked, will remain locked. This is a function of loop gain, since the phase detector output is bounded and can drive the oscillator only over a certain frequency range. The tracking range may be controlled, if desired, by limiting the input signal to the oscillator.

tracking resistance — *See* arc resistance.

tracking spacing — The distance between the center line of adjacent tracks on a recording tape. In the typical case of longitudinal tracks 50 mils (1.27 mm) wide, the tracking spacing is 70 mils (1.78 mm).

tracking spot — A moving spot used for target indication on a radar.

tracking weight — The downward force of a pickup stylus that ensures optimum reproduction of recorded groove modulation with minimum wear of groove wall and stylus.

track-while scan — A radar system utilizing electronic-computer techniques whereby raw data is used to track an assigned target, compute target velocity, and predict its future position without interfering with the scanning rate.

track width — The width of the track on a recording tape, corresponding to a given record gap. The most common track widths encountered in longitudinal recording are 48 and 50 mils (1.22 and 1.27 mm), several such tracks being accommodated on a half-inch-wide (1.27 cm) tape.

TRACON — Abbreviation for Terminal Radar Approach Control. An air traffic control facility that is responsible for areas designated as terminal control areas, the exact size of which differs depending on the facility, but typically extends to about 30 miles (48 kilometers) from the airport tower.

tractive force — The force that a permanent magnet exerts on a ferromagnetic object.

traffic — 1. Messages handled by communication or amateur stations. 2. The messages sent and received over a communication channel.

traffic diagram — A chart or drawing that shows the movement and control of traffic over a communication system.

traffic distribution — The routing by which communications traffic passes through a terminal to a switchboard or dialing center.

traffic-flow security — Protection of the contents of messages by transmitting an uninterrupted flow of random text over a wire or radio link, with no indication that an interceptor may be used to determine which portions of this steady stream constitute encrypted messages and which portions are nearly random filler.

traffic forecast — Traffic level prediction on which communication system management decisions and engineering effort are based.

trailer — 1. Tough, nonmagnetic tape spliced at the end of the recorded material on a tape that is expected to receive rough or frequent handling. Usually has one matte-finished surface for writing, and is often available in a variety of colors for coding purposes. *See also* leader.

trailer record — A record that follows one or more records and contains information related to those records.

trailers — Bright streaks at the right of large dark areas, or dark areas or streaks at the right of bright areas, in a televised picture. The usual cause is insufficient gain at low video frequencies.

trailing blacks — *See* following blacks.

trailing edge — The transition of a pulse that occurs last, such as the high-to-low transition of a high clock pulse.

trailing-edge pulse time — The time at which the instantaneous amplitude last reaches a stated fraction of the peak pulse amplitude.

trailing reversal — *See* following blacks; following whites.

trailing whites — *See* following whites.

train — A sequence of units of apparatus linked together to forward or complete a telephone call.

trainer — 1. The representation of an operating system by computers and its associated equipment and personnel. 2. Electronic equipment used for training operators of radar or sonar apparatus by simulating signals received under operating conditions in the field.

train time — The period required for synchronous modems used on two- or four-wire leased lines to equalize the line and recover timing from the received data. This time, which applies primarily to full-duplex initialization, is especially important for multipoint operations. For either two- or four-wire circuits, actual transmission and reception cannot occur until both receiving and transmitting modems are synchronized. Train time is often referred to as RTS/CTS delay or poll response. It varies from 8.5 to about 40 ms. Faster modems use longer equalization times, and have longer train times.

tramlines — A pattern on an A-scope appearing as a number of horizontal lines above the baseline. The effect is produced by a continuous wave modulated with a low-frequency signal.

trans — Abbreviation for transmitter. Also abbreviated xmtr or xmitter.

transaction file — In a computer, a file containing information relating to current activities or transactions.

transaction processing — A processing method in which transactions are executed immediately after they are received by the computer. The largest single application for fault-tolerant computers, which involves repetitive, typically financial, tasks in banking, stock trading, and the like.

transadmittance — From one electrode to another, the alternating components of the current of the second electrode divided by the alternating component of the voltage of the first electrode, all other electrode voltages being maintained constant. As most precisely used, the term refers to infinitesimal amplitudes.

transceiver — The combination of radio transmitting and receiving equipment in a common housing, usually for portable or mobile use and employing some common circuit components for both transmitting and receiving.

transceiving data link — Integrated data processing by means of punched cards, using transceivers as terminal equipment. The transmission path can be wire or radio.

transconductance — Symbolized g_m. Also called mutual conductance. An electron-tube rating equal to the change in anode current divided by the change in grid voltage causing the anode-current change. The unit of transconductance is the siemens; a more commonly used unit is the microsiemens (10^{-6} siemens).

transconductance amplifier — An amplifier that supplies an output current proportional to its input voltage. From its output terminals the amplifier appears to be a current source with a high output impedance Z_o. Its input impedance $Z_{in} \gg Z_g$, and the output impedance $Z_o \gg Z_L$, where Z_g and Z_L are the source and load impedances.

transconductance meter — Also called a mutual-conductance meter. An instrument for indicating the transconductance of a grid-controlled electron tube.

transconductance tube tester — Also called a mutual-conductance or dynamic mutual-conductance tube tester. A tube tester with circuits set up in such a manner that the test is made by applying an ac signal of known voltage to the control grid, and the tube amplification factor or mutual conductance is measured under dynamic operating conditions.

transconductor — An active or passive network whose short-circuit output current is a specific, accurately known, linear or nonlinear function of the input voltage, thereby establishing a predetermined relationship between input voltage and output current.

transcribe — 1. To copy, with or without translating, from one external storage medium of a computer to another. 2. To copy from a computer into a storage medium or vice versa.

transcriber — Equipment associated with a computing machine for the purpose of transferring input or output data from a record of information in a given language to the medium and language used by a digital computing machine, or from a computing machine to a record of information.

transcription — 1. An electrical recording (e.g., a high-fidelity, $33\frac{1}{3}$-rpm record) containing part or all of a (radio) program. It may be either an instantaneous recording disc or a pressing. 2. A phonograph record, especially the $33\frac{1}{3}$-rpm 16-inch (40.6-cm) record, used to record broadcast programs.

transdiode — A transistor so connected that the base and collector are actively maintained at equal potentials, although not connected together. The logarithmic transfer relationship between collector current and base-emitter voltage very closely approximates that of an ideal diode.

transducer — 1. A device that when actuated by signals from one or more systems or media can supply related signals to one or more other systems or media.

2. A device, component, machine, system, or combination of these that converts energy from one form to another. The energy may be in any form, such as electrical, mechanical, acoustical, etc. The term transducer is often restricted to a device in which the magnitude of an applied stimulus is converted into an electrical signal proportional to the quantity of the stimulus. 3. A device used to convert physical parameters, such as temperature, pressure, or weight, into electrical signals. 4. A device that converts information from one physical form to another. Examples include the phono cartridge (mechanical to electrical), speaker (electrical to acoustical), and microphone (acoustical to electrical). 5. A force-measuring device. It has the characteristics of providing an output, usually electrical, that serves as the measurement of load, force, compression, pressure, etc., when placed along the sensitive axis of the force cell. 6. An electromechanical device for translating a physical quantity such as pressure, temperature, displacement, acceleration, etc., into a proportional electrical quantity, such as voltage or current. 7. A device that converts something measurable into another form; often, a physical property such as temperature or flow into an electrical signal. Frequently used interchangeably with *sensor*.

transducer-coupling system efficiency — The power output at the point of application, divided by the electrical power input into the transducer.

transducer efficiency — Ratio of the power output to the electrical power input at the rated power.

transducer equivalent noise pressure (of an electroacoustic transducer or system used for sound reception) — Also called equivalent noise pressure. For a sinusoidal plane-progressive wave, the rms sound pressure that, if propagated parallel to the principal axis of the transducer, would produce an open-circuit signal voltage equal to the rms of the inherent open-circuit noise voltage of the transducer in a transmission band having a bandwidth of 1 hertz and centered on the frequency of the plane sound wave.

transducer gain — Ratio of the power that the transducer delivers to the load under specified operating conditions to the available power of the source.

transducer insertion loss — *See* insertion loss.

transducer loss — Ratio of the available power of the source to the power that the transducer delivers to the load under specified operating conditions.

transducer pulse delay — The interval of time between a specified point on the input pulse and on its related output pulse.

transfer — 1. To transmit, or copy, information from one device to another. 2. To jump. 3. The act of transferring. 4. In electrostatography, the act of moving a developed image, or a portion thereof, from one surface to another without altering its geometrical configuration (e.g., by electrostatic forces or by contact with an adhesive-coated surface). 5. In a computer, to terminate one sequence of instructions and start another.

transfer accuracy — The input-to-output error as a percentage of the input. Transfer accuracy depends on the source impedance, switch resistance, load impedance if the multiplexer is not buffered, and the signal frequency.

transfer admittance — 1. The complex ratio of the current at the second pair of terminals of an electrical transducer to the electromotive force applied between the first pair, all pairs of terminals being terminated in any specified manner. 2. The reciprocal of transfer impedance.

transfer characteristic — 1. The relationship, usually shown by a graph, between the voltage of one electrode and the current to another electrode, all other electrode voltages being maintained constant. 2. The characteristics pertaining to the degree of output activity for an input stimulus, or vice versa. 3. Function that when multiplied by an input magnitude will give a resulting output magnitude. 4. Relation between the illumination on a camera tube and the corresponding output-signal current, under specified conditions of illumination.

transfer charge — The net electric charge moved from one terminal of a capacitor to another through an external circuit.

transfer check — In a computer, verification of transmitted data by temporarily storing, retransmitting, and comparing. Also a check to see if the transfer or jump instruction was properly performed.

transfer circuit — A circuit that connects communication centers of two or more separate networks in order to transfer the traffic between the networks.

transfer constant — Also called the image-transfer constant. One-half the natural logarithm of the complex ratio of the voltage times current entering a transducer to that leaving the transducer when it is terminated in its image impedance.

transfer contact — *See* break-make contact.

transfer control — *See* jump.

transfer current — The starter-gap current required to cause conduction across the main gap of a glow-discharge cold-cathode tube.

transfer efficiency — The percentage of total charge that is transferred in a charge-coupled device from one position to another during readout.

transfer factor — Number relating the input and output currents of a transducer. *See* propagation constant, 2.

transfer function — 1. A characterization of a system (for example, mechanical structure of filter) by determining how the system responds to input energy at different frequencies. It may be measured by one of three methods: (*a*) applying a single sine wave and measuring the response to that sine wave, then varying the frequency of the sine wave and noting the change in the system's response, (*b*) applying specially shaped random noise (many frequencies simultaneously) and observing the system response, or (*c*) applying an impulse (also containing a broad range of frequencies). The transfer function can be derived by dividing the cross-spectral density by the power spectral density of the input or driven signal. 2. A mathematical expression that describes the relationship between the values of a set of conditions at two different times, typically at the beginning and end of a process. 3. The mathematical expression that relates the output and input of any characteristics of a filter as a function of frequency. The function usually is complex and therefore usually contains components corresponding to both attenuation and phase. 4. The mathematical expression of the relationship between a simple scale factor, representing the effects of any linear transmitting or recording medium when a waveform is evaluated as frequency components, and the frequency.

transfer-function analyzer — An instrument that measures the amplitude and phase response of the output of a circuit relative to the input over a range of frequencies.

transfer impedance — Ratio of the potential difference applied at one pair of terminals of a network to the resultant current at another pair of terminals (all terminals being terminated in any specified manner).

transfer instruction — An instruction, signal, or operation that conditionally or unconditionally specifies the location of the next instruction and directs a computer to that instruction.

transfer loss — *See* coupling loss.

transfer operation — 1. An operation that moves data from one storage location or one storage medium

to another. Transfer is sometimes taken to refer specifically to movement between different media; storage to movement within the same medium. 2. In a computer, transferring information from an accumulator to memory, or from memory to an accumulator; pushing into or popping out of the stack; or moving information to the program counter, stack pointer, etc.

transfer rate — The rate at which a transfer of data between the computer registers and storage, input, or output devices may be performed. It is usually expressed as a number of characters per second.

transfer ratio — From one point to another in a transducer at a specified frequency, the complex ratio of the generalized force or velocity at the second point to that applied at the first point.

transferred charge — The net electric charge moved from one terminal of a capacitor to another through an external circuit.

transferred charge characteristic — The mathematical function that relates transferred charge to capacitor voltage.

transfer switch — A form of air switch arranged so that a conductor connection can be transferred from one circuit to another without interrupting the current.

transfer time — 1. The time required for a transfer to be made in a digital computer. 2. In a relay, the total time — after contact bounce has ceased — between the breaking of one set of contacts and the making of another.

Transfluxor — A trade name (RCA) for a binary magnetic core having two or more openings. Control of the transfer of magnetic flux between the three or more legs of the magnetic circuits provides ways to store information, gate electrical signals, etc.

transform — 1. To convert a current or voltage from one magnitude to another, or from one type to another. 2. In digital-computer programming, to change information in structure or composition without significantly altering the meaning or value. 3. To pass through a transformer. 4. To change data or equations from rectangular to polar coordinates, or vice versa.

transformation point — The temperature at which an alloy or metal changes from one crystal state to another as it heats or cools.

transformation ratio — The ratio between electrical output and input under certain specified conditions.

transformer — 1. An electrical device that, by electromagnetic induction, transforms electric energy from one or more circuits to one or more other circuits at the same frequency, but usually at a different voltage and current value. 2. An electrical device that changes voltage in direct proportion to the ratio of the number of turns of its primary and secondary windings.

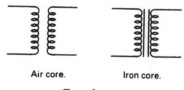

Air core.　　　　Iron core.

Transformer.

transformer build — The amount of window area used in constructing a transformer.

transformer-coupled amplifier — An amplifier whose stages are coupled together by transformers.

transformer-coupled solid-state relay — A relay in which the control is applied to the primary of a low-power transformer, and the resulting secondary voltage triggers the thyristor switch to the load.

transformer coupling — Use of a transformer between stages of an amplifier, i.e., to connect the anode circuit of one stage to the grid circuit of the following stage.

transformerless receiver — A receiver in which the power-line voltage is applied directly to series-connected tube heaters or filaments and to a rectifier circuit instead of first being stepped up or down in voltage by a power transformer.

transformer load loss — Losses in a transformer that are incident to the carrying of the load. Load losses include I^2R loss in the windings due to load current, stray loss due to stray fluxes in the winding core, clamps, etc., and to circulating current, if any, in parallel windings.

transformer loss — Expressed in decibels, the ratio of the signal power that an ideal transformer of the same impedance ratio would deliver to the load impedance to the signal power delivered by the actual transformer.

transformer oil — A high-quality insulating oil in which windings of large power transformers are sometimes immersed to provide high dielectric strength, insulation resistance, and flash point, plus freedom from moisture and oxidation.

transformer read-only store — In computers, a type of read-only store in which the presence or absence of mutual inductance between two circuits is the condition that determines whether a binary 1 or 0 is stored.

transformer vault — An isolated enclosure, either above or below ground, with fire-resistant walls, ceiling, and floor for unattended transformers and their auxiliaries.

transformer voltage ratio — The ratio of the root-mean-square primary terminal voltage to the root-mean-square secondary terminal voltage under specified conditions of load.

transforming section — A length of waveguide or transmission line of modified cross section, or with a metallic or dielectric insert, used for impedance transformation.

transhybrid loss — 1. The transmission loss at a given frequency measured across a hybrid circuit when connected to a given two-wire termination and balancing network. 2. The loss (isolation) between the transmit and receive branches of a four-wire line. Transhybrid loss is directly dependent on the degree of balance between the two-wire line and its balancing network.

transient — 1. A phenomenon caused in a system by a sudden change in conditions, and which persists for a relatively short time after the change. 2. A distinct line or series of lines perpendicular to the direction of scanning produced in the recorded copy immediately following a sudden change in density. 3. A momentary surge on a signal or power line. It may produce false signals or triggering impulses and cause insulation or component breakdowns and failures. 4. Signal component of fast-rising leading side and usually of short duration. 5. A pulse, damped oscillation, or other temporary phenomenon occurring in a circuit or system. 6. Intermediate unstable state. 7. A change in a variable that disappears during transition from one steady-state operating condition to another. 8. Signals that exist for a brief period prior to the attainment of a steady-state condition. May include overshoots, damped sinusoidal waves, etc.

transient analyzer — An electronic device for repeatedly producing a succession of equal electric surges of small amplitude and of adjustable waveform in a test circuit and presenting this waveform on the screen of an oscilloscope.

transient behavior — In a power supply, the general attitude of response in terms of amplitude and time.

transient distortion — 1. Distortion due to the inability of a system to reproduce or amplify transients linearly. 2. Distortion of a transient signal component such as produced by resonance and lack of damping, the effect being overshoot or ringing following the fast-rising leading side of the signal component.

transient intermodulation distortion — Abbreviated TIM 1. Distortion that occurs in an amplifier having a large amount of negative feedback in its main feedback loop and a certain amount of time or phase delay between the input and output signals. If a very fast transient musical signal or pulse is fed to such amplifiers, the feedback needed to reduce the amplitude of that signal at the input (and later) stages arrives too late, and an overload or momentary clipping occurs. At the instant of such clipping, other program-signal elements are also distorted or even obliterated. 2. Intermodulation distortion that occurs only momentarily during brief transients or signal peaks.

transient magnistor — A high-speed saturable reactor in which an alternating electric current in the form of a sine-wave carrier or pulses is passed through a signal winding and modulated by variations of current passing through a control coil.

transient motion — Any motion that has not reached, or has ceased to be, a steady state.

transient oscillation — A momentary oscillation that occurs in a circuit during switching.

transient overshoot — The largest transient deviation of the measured quantity following the dynamic crossing of the final value as a result of the application of a specified stimulus.

transient peak-inverse voltage — Under specified conditions, the maximum allowable instantaneous value of nonrecurrent reverse (negative) voltage that may be applied to the anode of an SCR with the gate open.

transient phenomena — Rapidly changing actions occurring in a circuit during the interval between closing of a switch and settling to steady-state conditions, or any other temporary actions occurring after some change in a circuit or its constants.

transient recovery time — Also called recovery time, transient response time, or response time. 1. The interval between the time a transient deviated from a specified amplitude range and the time it returns and remains within the specified amplitude range. The amplitude range is centered about the average of the steady-state values that exist immediately before and after the transient. 2. The time required for the output voltage of a power supply to come back to within a level approximating the normal dc output following a sudden change in load current.

transient regulation — The degree of freedom of a system from variations in voltage and frequency caused by sudden changes, operational switching, emergency system faults, power transfers, etc.

transient response — 1. The transient of a dependent variable resulting from an abrupt change of an independent variable with all other variables constant. 2. Also called step-function response. Commonly, the characteristic response of a system to a unit sweep or unit impulse. The elements of transient response most commonly specified are rise time, fall time, overshoot, undershoot, preshoot, and ringing. 3. (Input line changes) The transient of an output variable (voltage, current, or power) of an electronic power supply resulting from an abrupt specified change of input line voltage, with all other conditions held constant. 4. (Load changes) The transient of an output variable (voltage, current, or power) of an electronic power supply resulting from an abrupt specified change of load, with all conditions held constant. 5. Ability to respond to percussive signals cleanly and instantly. 6. The ability of an amplifier or transducer to handle and faithfully reproduce rapid changes in signal amplitudes. A very short rise time is a measure of this characteristic. 7. The closed-loop step function response of an amplifier under small-signal conditions. 8. The ability of an amplifier, microphone, or speaker to follow sudden changes in signal levels.

transient response time — *See* transient recovery time.

transient state — The condition in which a variable temporarily behaves in an erratic manner instead of changing smoothly and predictably.

transistance — 1. The characteristic of an electrical element that makes possible the control of voltages, currents, or flux so as to produce gain or switching action in a circuit. Examples of the physical realization of transistance occur in transistors, diodes, and saturable reactors. 2. An electronic characteristic exhibited in the form of voltage or current gain or in the ability to control voltages or currents in a precise and nonlinear manner. (Examples of parts showing transistance: transistors, diodes, vacuum tubes.) 3. A function of transistors. An electrical property that causes applied voltage to create amplification or accomplish switching.

transistor — 1. An active semiconductor device having three or more electrodes and capable of performing almost all the functions of tubes, including rectification and amplification. Germanium and silicon are the main materials used, with impurities introduced to determine the conductivity type (n-type has an excess of free electrons; p-type, a deficiency). Conduction is by means of electrons (elementary particles having the smallest negative electrical charge that can exist) and holes (mobile electron vacancies equivalent to a positive charge). 2. A tiny chip of crystalline material, usually silicon, that amplifies or switches electric current. It is a three-terminal semiconductor device. A small current (base current) applied to one terminal can control a larger current (collector current) between the other two terminals. For elementary purposes, a transistor can be looked at as a resistor whose value can be changed from a high to a low value by applying a current to the base lead. 3. A three-terminal active semiconductor device that provides current amplification. A bipolar transistor is comprised of base, emitter, and collector and is a current-controlled device with a low input impedance. A field-effect transistor has gate, source, and drain electrodes and is a high-impedance, voltage-controlled device. The first transistor was invented at Bell Laboratories in 1947 by Nobel-Prize physicists John Bardeen, William Shockley, and Walter Bratain.

transistor action — The physical mechanism of amplification in a junction transistor.

transistor amplifier — An amplifier in which the required amplification is produced by one or more transistors.

transistor base — The region between an emitter and a collector of a transistor, into which minority carriers are injected.

transistor chip — An unencapsulated transistor element of very small size used in microcircuits.

transistor dissipation — The power dissipated in the form of heat by the collector. The difference between the power supplied to the collector and the power delivered by the transistor to the load.

transistorized — Pertaining to equipment or a design in which transistors instead of vacuum tubes are used.

transistor oscillator — An oscillator that uses a transistor in place of an electron tube.

transistor-outline metal-can package — Abbreviated TO can. A type of package that resembles a transistor can, but generally is larger and has more leads. The pins are arranged in a circular pattern in the base. This type of package is often used for MSI, LSI, and MOS ICs.

transistor parameters — The performance characteristics of a transistor or class of transistors.

transistor pentode — A transistor designed for mixing, modulating, or switching and containing the equivalent of three emitters, a base, and a collector.

transistor radio — A radio receiver that uses transistors in place of electron tubes.

transistor region — The region around a pn junction in which the majority carriers from each side of the junction diffuse across it to recombine with their respective counterparts.

transistor-resistor logic — Abbreviated TRL. An early form of logic-circuit design that was employed prior to the advent of monolithic circuits. The input elements are resistors; however, unlike RTL there is only one active output transistor. In discrete circuit designs, this logic form was the least expensive since it required a minimum number (one) of active devices.

transistor seconds — Also called fallouts. Those transistors that remain after the firsts (units meeting rigid specifications for a specific application) have been removed from the production line.

transistor symbols — Symbols used to represent transistors on schematic diagrams.

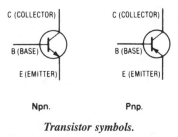

Npn.   Pnp.

Transistor symbols.

transistor testers — Equipment and instruments that detect or measure leakage current, breakdown voltage, gain, or saturation voltage. Some testers are computer operated.

transistor-transistor logic — Abbreviated TTL or T^{2}L. Also called multiemitter transistor logic. 1. A logic-circuit design similar to DTL, with the diode inputs

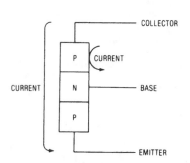

Transistor (typical).

replaced by a multiple-emitter transistor. In a four-input DTL gate, there are four diodes at the input. A four-input TTL gate will have four emitters of a single transistor as the input element. TTL gates using npn transistors are positive-level NAND gates or negative-level NOR gates. 2. A relatively high-speed and inexpensive bipolar logic circuit most commonly used in today's logic designs. 3. Any logic gate circuit that uses a multiemitter npn transistor to perform the positive AND function, followed by one or more transistors to add power to (and possibly invert) the output. Operating speed is generally faster than that of DTL. 4. A widely used form of semiconductor logic. Its basic logic element is a multiple-emitter transistor. TTL is characterized by fairly high speed and medium power dissipation.

transit angle — The product of angular frequency and the time taken for an electron to cross a given path.

transition — 1. The process in which a quantum-mechanical system makes a change between energy levels. During this process, energy is either emitted or absorbed, usually in the form of photons, phonons, or kinetic energy of particles. Transitions that involve photons alone are called direct radiative transitions, whereas those that require a combination of a photon and a phonon are called indirect transitions. 2. The instance of changing from one state (such as a positive voltage) to a second state (such as a negative voltage) in a serial transmission.

transitional function — A transfer function that represents a compromise made between the classical types of functions (Tchebychev, Butterworth, Gaussian, etc.) as a result of the fact that each classical function has certain advantages and disadvantages. In general, however, better functions than the transistional functions are available if the transfer function is optimized with respect to its own particular requirement.

transition band — The range of frequencies that bound a passband/stopband interface.

transition card — A card that signals the computer that the reading-in of a program has ended and that the carrying out of the program has started.

transition counts — A particular digital signature used in logic-board testing.

transition element — An element used for coupling different types of transmission systems, e.g., for coupling a coaxial line to a waveguide.

transition error — In an encoder, the difference between the shaft angle at which a code-position change should occur and the angle at which it actually does occur.

transition factor — *See* reflection factor, 1.

transition frequency — Also called crossover frequency. 1. In a disc-recording system, the frequency corresponding to the point of intersection of the asymptotes to the constant-amplitude and the constant-velocity portions of its frequency-response curve. This curve is plotted with output-voltage ratio in decibels as the ordinate, and the logarithm of the frequency as the abscissa. 2. In a transistor, the product of the magnitude of the small-signal, common-emitter, forward-current transfer ratio times the frequency of measurement when this frequency is high enough that the magnitude is decreasing with a slope of approximately 6 dB per octave.

transition layer — *See* transition region, 1.

transition-layer capacitance — *See* depletion-layer capacitance.

transition loss — At any point in a transmission system, the ratio of the available power from that part of the system ahead of the point under consideration to the power delivered to that part of the system beyond point under consideration.

transition oscillator — A negative-transconductance oscillator employing a pentode tube with a capacitor connected between the screen and suppressor grids. The suppressor grid periodically divides the current between the screen grid and anode, thereby producing oscillations.

transition point — A point at which the circuit constants change in such a way that a wave being propagated along the circuit is reflected.

transition region — 1. Also called transition layer. The region, between two homogeneous semiconductor regions, in which the impurity concentration changes. 2. The area, between the passband and the stopband, in which the attenuation of a filter is neither great nor small. The narrower the transition region, the more difficult is the design of the filter. 3. The region around a pn junction in which the majority carriers of each side diffuse across the junction to recombine with their respective counterparts.

transition temperature — The temperature below which the electrical resistance of a material becomes too small to be measured.

transition time — The time required for a voltage change from one logic level to the opposite level, as measured between specified points on the transition waveform. The transition times from high to low level and from low to high level generally are different.

transitron — A thermionic tube circuit whose action depends on the negative transconductance of the suppressor grid of a pentode with respect to the screen grid.

transit time — 1. The time taken for a charge carrier to cross a given path. 2. The average time a minority carrier takes to diffuse from emitter to collector in a junction transistor. 3. The time an electron takes to cross the distance between the cathode and anode.

transit-time mode — A condition of oscillator operation, corresponding to a limited range of drift-space transit angle, for which the electron stream introduces a negative conductance into the coupled circuit.

translate — 1. To change computer information from one language to another without significantly affecting the meaning. 2. To change one binary word to another, or to change to a different pair of binary signal levels, for the purpose of achieving compatibility between parts of a system. 3. To transform information in a way that preserves meaning, i.e., from one code to another or from one language to another.

translation — The conversion of a message to a different set of symbols. For example, in instrumentation, the changing of conventional algebraic expressions into machine language for computer programming.

translational morphology — The structural characterization of an electronic component in which it is possible to identify the areas of patterns of resistive, conductive, dielectric, and active materials in or on the surface of the structure with specific corresponding devices assembled to perform an equivalent function.

translation frequency — The 2220-MHz (2.2-GHz) frequency difference between an uplink and a downlink signal of a satellite TV system.

translation loss — Also called playback loss. The loss in the reproduction of a mechanical recording, whereby the amplitude of motion of the reproducing stylus differs from the recorded amplitude in the medium.

translator — 1. A device that transforms signals from the form in which they were generated into a useful form for the purpose at hand. 2. A television receiver and low-power transmitter that receives television signals on one channel and retransmits them on another channel (usually a UHF channel) to valleys and like areas that cannot receive the direct signals. 3. In telephone equipment, the device that converts dialed digits into call-routing information. 4. In electric computers, a network or system that has a number of inputs and outputs and is connected so that input signals representing information expressed in a certain code result in output signals that represent the input information in a different code. 5. A generic term for any system that converts one language into another.

translator package — A computer program that allows a user program (in binary) to be converted into a usable form for computer manipulation. *See also* program assembly.

transliterate — To change the characters of one alphabet to the corresponding characters of another alphabet.

translucent — Partially transparent. In optics the notion of imperfect, or diffused, light transmission is generally implied. Thus, a translucent material may conduct light but not clear images.

transmissibility — Ratio of the response amplitude of the system in steady-state forced vibration to the excitation amplitude. The ratio may be between forces, displacements, velocities, or accelerations.

transmission — 1. Conveying electrical energy from point to point along a path. 2. The transfer of a signal, message, or other form of intelligence from one place to another by electrical means. 3. The dispatching of a signal, message, or other form of intelligence by wire, radio, telegraphy, telephony, facsimile, or other means; the signaling of data over communications channels. 4. A series of characters, messages, or blocks, including control information and user data, sent over a communication channel.

transmission anomaly — The difference, in decibels, between the total loss in intensity and the reduction in intensity that would be due to an inverse-square divergence.

transmission band — The range of frequencies above the cutoff frequency of a waveguide, or the comparable useful frequency range for any other type of transmission device.

transmission cable — 1. Two or more transmission lines. If the structure is flat, it is sometimes called flat transmission cable to differentiate it from a round structure, such as a jacketed group of coaxial cables. *See* transmission line. 2. A signal-carrying circuit composed of conductors and dielectric material with controlled electrical characteristics, used for the transmission of high-frequency or narrow-pulse type signals.

transmission coefficient — For a transition or discontinuity between two transmission media at a given frequency, the ratio of some quantity associated with the transmitted wave at a specified point in the second medium to the same quantity associated with the incident wave at a specified point in the first medium.

transmission facility — 1. The transmission medium and all the associated equipment required to transmit a message. 2. Cable and/or radio facility constituting inter-switch trunks and subscriber access lines.

transmission factor — Ratio of flux transmitted to flux incident. Inverse of opacity.

transmission function — The ratio of the output voltage to the input voltage of a filter expressed in terms of magnitude and phase (or delay).

transmission gain — 1. *See* gain. 2. The increase in the power of an electrical signal from one point in the circuit to another.

transmission gate — The solid-state equivalent of a relay, having terminals that are connected to each other or not depending on the application of a separate control voltage. (In CMOS logic it is bidirectional.)

transmission level — The level of signal power at any point in a transmission system. It is equal to the ratio of the power at that point to the power at some point in the system chosen as a reference point. This ratio is usually expressed in decibels.

transmission line — 1. A material structure forming a continuous path from one place to another, and used for directing the transmission of electromagnetic energy along this path. 2. A conductor or series of conductors used for conveying electrical energy from a source to a load. 3. One or more insulated conductors arranged to transmit electrical energy signals from one locality to another. 4. A signal-carrying circuit with controlled electrical characteristics used to transmit high-frequency or narrow-pulse signals.

transmission-line coupler — A coupler that allows the passage of electric energy in either direction between balanced and unbalanced transmission lines.

transmission-line loss — 1. The difference in the amount of energy delivered at the output of a transmitter and the amount of energy absorbed by the antenna. For example, if a transmitter delivers 3 watts and the transmission line (coaxial cable) loss is 1 dB, the antenna will absorb approximately 2.4 watts since 20 percent of the available power is lost in the transmission line. 2. The power lost when a radio signal is fed from the antenna to the receiver through coaxial cable; expressed in decibels.

transmission-line-tuned frequency meter — A frequency meter using a tuned length of wire or a coaxial cavity as the frequency-determining element.

transmission link — The path over which information flows from sender to receiver.

transmission loss — The decrease or loss in power during transmission of energy from one point to another. Usually expressed in decibels.

transmission measuring set — A measuring instrument comprising a signal source and a signal receiver having known impedances, for measuring the insertion loss or gain of a network or transmission path connected between those impedances.

transmission mode — A form of waveguide propagation along a transmission line characterized by the presence of transverse magnetic or transverse electromagnetic waves. Waveguide transmission modes are designated by integers (modal numbers) associated with the orthogonal functions used to describe the waveform. These integers are known as waveguide-mode subscripts. They may be assigned from observations of the tranverse field components of the wave and without reference to mathematics.

transmission modes — Simplex transmission goes only one way; simplex with back-channel, one way with limited talk-back handshaking. Half-duplex transmission is bidirectional but in one direction at a time; full-duplex is simultaneously bidirectional.

transmission modulation — Amplitude modulation of the current in the reading beam of a charge-storage tube as the beam passes through aperatures in the storage surface. The degree of modulation depends on the stored charge pattern.

transmission primaries — The set of three primaries, either physical or nonphysical, each chosen to correspond in amount to one of the three independent signals contained in the color-picture signal. The I, Q, and Y signals.

transmission regulator — A device that maintains the transmission substantially constant over a system.

transmission response — The conversion of electrical energy traveling at the speed of light to sonic energy traveling at the speed of sound.

transmission security — That part of communications security resulting from all measures intended for protection of transmissions from unauthorized interception, traffic analysis, and imitative deception.

transmission selective — The transmission of electromagnetic energy at wavelengths other than those that are reflected or absorbed.

transmission speed — The number of information elements sent per unit time, usually expressed as bits, characters, word groups, or records per second or per minute.

transmission system — 1. An assembly of elements capable of functioning together to transmit signal waves. 2. One of two broad groups of communication means, distinguished by their information-carrying bandwidth and referred to as either wideband or narrow-band transmission systems. The wideband system includes tropospheric scatter radio relay and submarine cables, and has an information bandwidth of more than four voice-frequency (3 to 4 kHz bandwidth) channels. The narrow-band system includes high-frequency radio (single sideband, amplitude modulated, ionospheric scatter) and landlines, and has an information-carrying bandwidth of four voice channels or less.

transmission time — The absolute time interval between transmission and reception of a signal.

transmission-type frequency meter — A frequency meter in which a tuned electrical circuit or a cavity is used to transmit the energy from the signal source under test to a detecting load.

transmission wavemeter — A device that makes use of a cavity to transmit maximum power at resonance so that maximum deflection is obtained on a readout meter at the frequency of resonance.

transmit — To send a program, message, or other information from one location to another.

transmit current — The current drawn by a transceiver when in the transmit mode.

transmit-on-alarm — A security device that activates only when triggered. Most security devices are of this nature; however, in multiplexing systems, it is usually impossible to determine whether or not transmit-on-alarm devices are actuated or whether the transmitter has failed or the communication channel is inoperative.

transmit-on-interrogation — The response of a security device that stores an alarm until it is interrogated in an appropriate manner. Used in multiplexed systems to avoid the problem of two simultaneous transmissions of alarm by two or more different stations.

transmit-receive switch — Also called a tr switch, tr box, tr tube, or duplexing assembly. An automatic device employed in a radar to prevent the transmitted energy from reaching the receiver, but allowing the received energy to do so without appreciable loss.

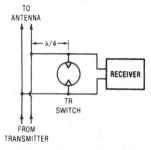

Transmit-receive switch.

transmit-receive tube — *See* tr tube.

transmittal mode — The method by which the contents of an input buffer are made available to the program and the method by which records are made available by a program for output.

transmittance — 1. Of a material, the ratio of the radiant power transmitted through the material to the incident radiant power. 2. As a transfer function, a response function for which the variables are measured at different ports (terminal pairs).

transmitted-carrier operation — Amplitude-modulated carrier transmission in which the carrier wave is transmitted.

transmitted wave — *See* refracted wave.

transmitter — 1. Equipment used to generate and amplify an rf carrier signal, modulate this carrier with intelligence, and radiate the modulated rf carrier into space. 2. In telephony, the microphone that converts sound waves into electrical signals at an audio-frequency rate. 3. Apparatus for converting electrical energy received from a source into radio-frequency electromagnetic energy capable of being radiated. 4. The device in a telephone handset used to convert speech to electrical energy. 5. A device that translates the low-level output of a sensor or transducer to a higher-level signal suitable for transmission to a site where it can be further processed.

transmitter distributor — In teletypewriter operations, a motor-driven device that translates teletypewriter code combinations from perforated tape into electrical impulses and transmits these impulses to one or more receiving stations.

transmitter frequency tolerance — The extent to which the carrier frequency of a transmitter may legally depart from the assigned frequency.

transmitter start code — Usually a two-letter call that is sent to an outlying machine to automatically turn on its tape transmitter.

transmitting antenna — A device for converting electrical energy into electromagnetic radiation capable of being propagated through space.

transmitting current response — Of an electroacoustic transducer used for sound emission, the ratio of the sound pressure apparent at a distance of 1 meter in a specified direction from the effective acoustic center of the transducer to the current flowing at the electric input terminals. Usually expressed in decibels above a reference-current response of 1 microbar per ampere.

transmitting efficiency (projector efficiency) — Ratio of the total acoustic power output of an electroacoustic transducer to the electric power input.

transmitting element — In fiber optics, the light-radiating side of the terminus in an optical conductor.

transmitting loop loss — That part of the repetition equivalent assignable to the station set, subscriber line, and battery supply circuit that are on the transmitting end.

transmitting power response (projector power response) — Of an electroacoustic transducer used for sound emission, the ratio of the effective sound pressure apparent at a distance of 1 meter in a specified direction from the effective acoustic center of the transducer to the electric power input. Usually expressed in decibels above a reference response of 1 microbar square per watt of electric input.

transmitting station — A location at which the transmitter of a radio system and its antenna and associated equipment are grouped.

transmitting voltage response — Of an electroacoustic transducer used for sound emission, the ratio of the sound pressure apparent at a distance of 1 meter in a specified direction from the effective acoustic center of the transducer to the signal voltage applied at the electric input terminals. Usually expressed in decibels above a reference-voltage response of one microbar per volt.

transolver — A servo unit with a three-phase stator and a two-phase rotor that can be used either as a control transmitter (with one rotor phase excited and the other shorted) or as a control transformer with the signal obtained from one phase of the rotor, making sure the other phase is equally loaded to prevent unbalances.

transonic speed — A speed of 600 to 900 miles per hour (965 to 1448 km/h), corresponding to about Mach 0.8 to 1.2.

transparent — 1. Implying clear and unimpeded transit of light. Window glass is a common example. 2. A computer process or function that is invisible to the user.

transparent computer voice — Digitized and reconstructed voice that 80 percent of uncued listeners would not recognize as having come from a computer.

transparent device — A device on a communications network that functions without making its presence known to the end terminals.

transponder — 1. A combination receiver, frequency converter, and transmitter package designed to receive a signal and to retransmit it on another frequency. Broadcast and telecommunication satellites are equipped with transponders, which have a typical output of 5 to 10 watts and operate over a frequency band with a 36 to 72 megahertz bandwidth in the L, C, Ku, and sometimes Ka bands. Communication satellites typically have between 12 and 24 onboard transponders, although the INTELSAT VI has 50. 2. A component designed to receive a signal and to retransmit it on another frequency. 3. A radio transmitter-receiver that transmits identifiable signals automatically when the proper interrogation is received. *See also* pulse repeater. 4. A device (usually in the form of a radio transmitter) which sends a confirming or identifying message or releases a stored message after being triggered by an interrogator. Used on communication satellites. 5. A microwave repeater that receives, amplifies, downconverts, and retransmits signals from a communication satellite.

transponder dead time — The time interval between the start of a pulse and the earliest instant at which a new pulse can be received or produced by a transponder.

transponder efficiency — A ratio expressed as a percentage of the number of replies to the number of interrogations from a transponder.

transponder suppressed time delay — The overall fixed delay between reception of an interrogation and the transmission of a reply.

transport — Also called tape transport. 1. Platform or deck of a tape recorder on which the motor (or motors), reels, heads and controls are mounted. It includes those parts of the recorder other than the amplifier, preamplifier, speaker, and case. 2. That portion of a tape recorder which holds the reels of magnetic tape and draws the tape at a constant speed past the recording and playback heads.

transportable transmitter — Sometimes called a portable transmitter. A transmitter designed to be readily carried from place to place, but normally not operated while in motion.

transport time — In an automatic system, the time required to move an object, element, or information between two predetermined positions.

transposition — The interchanging of the relative positions of the conductors in an open-wire line to reduce noise, interference, and crosstalk.

transposition blocks — Spreaders used to space and reverse the relative positions of two conductors at fixed intervals.

transposition section — A length of open-wire line to which a fundamental transposition design or pattern is applied as a unit.

transradar — A bandwidth-compression system for use in long-range narrow-band transmission of radio signals from a radar receiver to a remote location.

transrectification — The rectification that occurs in one circuit when an alternating voltage is applied to another circuit.

transrectification characteristic — The graph obtained by plotting the direct voltage values for one electrode of a vacuum tube as abscissas against the average current values in the circuit of that electrode as ordinates, for various values of alternating voltage applied to another electrode as a parameter. The alternating voltage is held constant for each curve, and the voltages on other electrodes are maintained constant.

transrectification factor — The change in average current of an electrode, divided by the change in the amplitude of the alternating sinusoidal voltage applied to another electrode (the direct voltages of this and other electrodes being maintained constant).

transrectifier — A device, ordinarily a vacuum tube, in which rectification occurs in one electrode circuit when an alternating voltage is applied to another electrode.

transresistance amplifier — An amplifier that supplies an output voltage proportional to an input current. The transfer function of the amplifier is $e_o/i_{in} = R_m$, where R_m is the transresistance.

transverse-beam traveling-wave tube — A traveling-wave tube in which the electron beam intersects the signal wave rather than moving in the same direction.

transverse crosstalk coupling — Between a disturbing and a disturbed circuit in any given section, the vector summation of the direct couplings between adjacent short lengths of the two circuits, without dependence on intermediate flow in nearby circuits.

transverse electric wave — Abbreviated TE wave. In a homogeneous isotropic medium, an electromagnetic wave in which the electric field vector is everywhere perpendicular to the direction of propagation. The dominant mode in a rectangular waveguide is TE_{10}.

transverse electromagnetic wave — Abbreviated TEM wave. In a homogeneous isotropic medium, an electromagnetic wave in which the electric and magnetic field vectors both are everywhere perpendicular to the direction of propagation. This is the normal mode of propagation in coaxial line, open-wire line, and stripline.

transverse-field traveling-wave tube — A traveling-wave tube in which the traveling electric fields that interact with the electrons are essentially transverse to the average motion of the electrons.

transverse interference — Interference occurring across terminals or between signal leads.

transverse magnetic E-mode — A type of mode in which the longitudinal component of the magnetic field is zero and the longitudinal component of the electric field is not zero.

transverse magnetic wave — Abbreviated TM wave. In a homogeneous isotropic medium, an electromagnetic wave in which the magnetic field vector is everywhere perpendicular to the direction of propagation.

transverse magnetization — Magnetization of a recording medium in a direction perpendicular to the line of travel and parallel to the greater cross-sectional dimension.

transverse recording — The technique of rotating heads that are oriented perpendicular to the edge and surface of the tape.

transverse wave — A wave in which the direction of displacement is perpendicular to the direction of propagation at each point of the medium. When the direction of displacement forms an acute angle with the direction of propagation, the wave is considered to have both a longitudinal and a transverse component.

trap — 1. A selective circuit that attenuates undesired signals but does not affect the desired ones. *See also* wave trap. 2. A crystal imperfection that can trap carriers. 3. An unprogrammed conditional jump to a known location, the jump being activated automatically by hardware; the location from which the jump occurs is recorded. 4. A device, usually a switch, installed within a protected area, which serves as secondary protection in the event a perimeter alarm system is successfully penetrated. Examples are a trip-wire switch placed across a likely path for an intruder, a mat switch hidden under a rug, or a magnetic switch mounted on an inner door. 5. A volumetric sensor installed so as to detect an intruder in a likely traveled corridor or pathway within a security area. 6. A CPU-initiated interrupt that is automatically generated when a predetermined condition, such as an illegal instruction, a breakpoint, a specified error, or power failure, is detected. Two vector locations are dedicated for each trap type. The vector locations contain the pc and ps for the trap service routine. 7. A method of catching computer program error when illegal instructions are executed or illegal memory locations are accessed. 8. A frequency-sensitive device that is used to attenuate specific signals that cause interference.

TRAPATT diode — Abbreviation for trapped plasma avalanche transit time diode. 1. A microwave avalanche diode that has either an n+pp+ or a p+nn+ structure. It may be manufactured from either silicon or germanium. When the diode is biased into breakdown, an electron-hole plasma fills the entire p or n region, and the voltage across the diode drops. A large current, induced by the low residual electric field, extracts the plasma from the diode. After the plasma is removed, the current drops and voltage rises. Energy stored in the resonant circuits of an oscillator raises the voltage above breakdown, and the cycle repeats. 2. A semiconductor microwave diode that, when its junction is biased into avalanche, exhibits a negative resistance at frequencies below the transit-time frequency range of the diode due to generation and dissipation of trapped electron-hole plasma resulting from the intimate interaction between the diode and a multiresonant microwave cavity.

TRAPATT oscillator — *See* trapped avalanche triggered transit (TRAPATT) oscillator.

trapezoidal distortion — Distortion in which a televised picture has the shape of a trapezoid (wide at top or bottom) instead of a rectangle. It is due to the interaction between the vertical- and horizontal-deflection coils (or plates) of the cathode-ray tube.

trapezoidal generator — An electronic circuit that produces a trapezoidal voltage wave.

trapezoidal pattern — An oscilloscope pattern that indicates the percentage of modulation in an amplitude-modulated system.

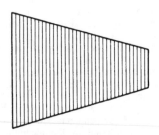

Trapezoidal pattern.

trapezoidal wave — 1. A trapezoidal-shaped waveform. 2. A square wave onto which a sawtooth has been superimposed. It is a voltage wave necessary to give a linear deflection current through the coils of a magnetically deflected cathode-ray tube.

trapped avalanche triggered transit (TRAPATT) oscillator — Oscillator device composed of a semiconducting diode in a coaxial resonating cavity. When the biasing current is applied to the diode, high-frequency waves are emitted into the cavity, where they are reflected back and forth and eventually produce a radio-frequency output.

trapped flux — In a material in the superconducting state, magnetic flux linked with a closed superconducting loop.

trapped plasma avalanche transit time diode — *See* TRAPATT diode.

trapping — 1. The holding of electrons or holes by any of several mechanisms in a crystal, thereby preventing them from moving. 2. In a computer, instructions that cause initiation by the central processing unit of an internal interrupt that transfers control to a subroutine which activates the desired operation of the instruction. Also, the subroutine can be changed, thereby causing the operation on the instruction to be changed.

trap wire — A low-voltage wire used at hinge points where severe flexing occurs, usually in burglar alarm systems. It is made with tinsel conductor.

traveling detector — A probe mounted on a slider and free to move along a longitudinal slot cut into a waveguide or coaxial transmission line. The traveling detector is connected to auxiliary measuring apparatus and used for examining the relative magnitude of any standing-wave system.

traveling plane wave — A plane wave in which each frequency component has an exponential variation of amplitude and a linear variation of phase in the direction of propagation.

traveling wave — The resulting wave when the electric variation in a circuit takes the form of translation of energy along a conductor, such energy being always equally divided between current and potential forms.

traveling-wave amplifiers — A two-port amplifier in which the input signal excites a space-charge wave at one end of a slab of negative-differential-resistivity bulk material. The wave travels through the material between ohmic contacts that serve as a transmission line. The charge builds as it moves, so the output is a larger version of the input.

traveling-wave magnetron — A traveling-wave tube in which the electrons move in crossed static electric and magnetic fields that are substantially normal to the direction of wave propagation.

traveling-wave magnetron oscillations — Oscillations sustained by the interaction between the space-charge cloud of a magnetron and a traveling electromagnetic field with approximately the same phase velocity as the mean velocity of the cloud.

traveling-wave parametric amplifier — A parametric amplifier that has a continuous iterated structure incorporating nonlinear reactors and in which the signal, pump, and difference-frequency waves are propagated along the structure.

traveling-wave phototube — A traveling-wave tube that has a photocathode and a window that admits a modulated laser beam, which causes the emission of a current-modulated photoelectron beam. This beam is then accelerated by an electron gun and directed into the helical slow-wave structure of the tube.

traveling-wave tube — Abbreviated TWT. A tube in which a stream of electrons interacts continuously or

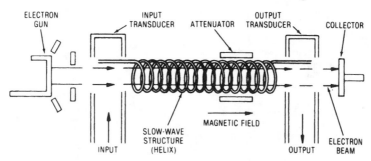

Traveling-wave tube.

repeatedly with a guided electromagnetic wave moving substantially in synchronism with it, and in such a way that there is a net transfer of energy from the stream to the wave.

traveling-wave-tube amplifier — A power-amplifying piece of equipment yielding about 30 dB of gain over broad bandwidths. The units are built around certain TWTs in the most commonly used frequency bands, and they have external modulation circuitry incorporated.

traveling-wave-tube interaction circuit — An extended electrode arrangement used in a traveling-wave tube to propagate an electromagnetic wave in such a manner that the traveling electromagnetic fields are retarded to the point at which they extend into the space occupied by the electron stream.

tr box — *See* transmit-receive switch.

tr cavity — The resonant portion of a tr switch.

treble — 1. The higher part in harmonic music or voice; of high or acute pitch. In music, the frequencies from middle C (261.63 hertz) upward. 2. The upper range of audio frequencies.

treble boost — Deliberate adjustment of the amplitude-frequency response of a system or component to accentuate the higher audio frequencies.

tree — A set of connected branches without meshes.

treeing — A progressive type of insulation failure in which branching hollow channels slowly penetrate the insulation at rather low applied voltages.

tree network — *See* hierarchical network.

tremolo — 1. The amplitude modulation of an audio tone, widely used in musical-instrument amplifiers to create effects that add to the capability of the instrument itself. It is a musical effect characterized by a subaudio modulation of the musical tone. When amplitude modulation is used, it is called tremolo; a frequency modulation effect is termed vibrato. 2. A warbling or fluctuating effect, approximately seven times per second, in the tone of an instrument characterized by a variation in intensity rather than pitch. *See also* vibrato. 3. A regular variation in the amplitude of a sound, generally at a frequency between 0.5 and 20 Hz.

TRF — Abbreviation for tuned radio frequency.

tri — Abbreviation for triode.

triac — Also called bidirectional triode thyristor. 1. A bidirectional rectifier (essentially two SCRs in parallel) that functions as an electrically controlled switch for ac loads and having an npnpn structure that can be triggered into either forward or reverse conduction by a pulse applied to its gate electrode. A triac will pass an alternating current. 2. A thyristor that can be triggered into conduction in either direction. Terminals are called main terminal 1 and gate. 3. A semiconductor device that functions as an electrically controlled switch for ac loads. The triac is one type of thyristor.

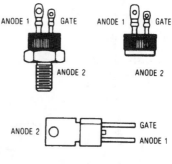

Construction and connections.

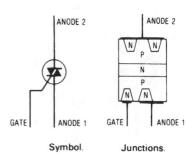

Symbol.   **Junctions.**

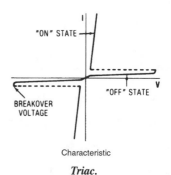

Characteristic

Triac.

triad — 1. Three radio stations operated as a group for determining the position of aircraft or ships. 2. A group of three dots, one of each color-emitting phosphor, on the screen of a color picture tube. 3. A group of three bits or three pulses, usually in sequence on one wire or simultaneously on three wires. 4. Also called a triplex. A group of three insulated conductors twisted together without (or with) a sheath overall. Usually color-coded for identification. 5. In a color CRT, one set of red, green, and blue phosphors.

triangulation — A method of finding the location of a third point by taking bearings from two fixed points a known distance apart. The third point will be at the intersection of the two bearing lines.

triax — A type of shielded conductor that employs a shield and jacket over the primary insulation, plus a second shield and jacket over all. Aside from applications requiring maximum attenuation of radiated signals or minimum pickup of external interference, this cable can also be used to carry two separate signals.

triaxial cable — 1. A special form of coaxial cable containing three conductors. 2. A three-conductor cable with one conductor in the center, a second circular conductor shield concentric with the first, and a third circular conductor shield insulated from and concentric with the first and second, and a braid or impervious sheath overall.

Triaxial® speaker — Trade name applied to the three-way speakers manufactured by Jensen Sound Laboratories, Division of Pemcor, Inc., consisting of three electrically independent speaker elements mounted on a common (single) speaker frame.

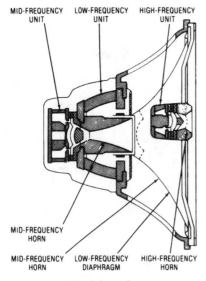

MID-FREQUENCY UNIT   LOW-FREQUENCY UNIT   HIGH-FREQUENCY UNIT

MID-FREQUENCY HORN

MID-FREQUENCY HORN   LOW-FREQUENCY DIAPHRAGM   HIGH-FREQUENCY HORN

Triaxial speaker.

tribo- — A prefix meaning due to or pertaining to friction.

triboelectric — Pertaining to electricity generated by friction.

triboelectricity — Electrostatic charges generated due to friction between different materials.

triboelectric series — A list of substances arranged so that any of them can become positively electrified when rubbed with one farther down the list, or negatively charged when rubbed with one farther up the list.

triboluminescence — Luminescence that arises from friction. Usually occurs in crystalline materials.

tributary circuit — A circuit that connects an individual drop, or drops, to a switching center.

tributary station — A communications terminal consisting of equipment compatible for the introduction of messages into, or reception from, its associated relay station.

Triboelectric Series

Positive

Asbestos
Rabbit fur
Glass
Mica
Nylon
Wool
Cat fur
Silk
Paper
Cotton
Wood
Lucite
Sealing wax
Amber
Polystyrene
Polyethylene
Rubber balloon
Sulfur
Celluloid
Hard rubber
Vinylite
Saran Wrap
Negative

trickle charge — 1. A continuous charge of a storage battery at a slow rate approximately equal to the internal losses and suitable to maintain the battery in a fully charged condition. 2. A continuous direct current, usually very low, that is applied to a battery to maintain it at peak charge or to recharge it after it has been partially or completely discharged. Usually applied to nickel-cadmium (nicad) or wet-cell batteries.

trickle charger — A device for charging a storage battery at a low rate continuously, or for several hours at one time.

tricolor camera — A television camera designed to separate reflected light into three frequency groups, each corresponding to the light energies of the three primary colors. The camera transforms the intensity variations of each primary into amplitude variations of an electrical signal.

tricolor picture tube — A picture tube that reproduces a scene in terms of the three light primaries.

tricon — A radionavigational system in which an airborne receiver accepts pulses from a triplet (chain of three stations) in a variable time sequence so that the pulses arrive at the same time even though traveling over paths of various lengths.

trigatron — An electronic switch in which breakdown of an auxiliary gap initiates conduction.

trigger — 1. To cause, by means of one circuit, action to start in another circuit, which then functions for a certain length of time under its own control. 2. A pulse that starts an action or function; for example, a triggered sweep or delay ramp. 3. A timing pulse used to initiate the transmission of signals through the appropriate circuit signal paths.

trigger action — The instantaneous initiation of main current flow by a weak controlling impulse in a device.

trigger circuit — 1. A circuit with two conditions of stability, with means for passing spontaneously or through application of an external stimulus from one to the other when certain conditions are satisfied. *See also* flip-flop, 1. 2. A circuit in which an electron tube performs the function of a relay. Impulses applied to the input of the

803

tube produce corresponding impulses in the output circuit, starting a chain of events.

trigger countdown — A process that reduces the repetition rate of a triggering signal.

trigger diode — A symmetrical three-layer avalanche diode used to control SCRs and triacs. It has a symmetrical switching mode and therefore fires whenever the breakover voltage is exceeded in either polarity of applied voltage.

triggered blocking oscillator — A blocking oscillator that can be reset to its starting condition by the application of a trigger voltage.

triggered spark gap — A fixed spark gap in which the discharge passes between two electrodes and is struck (started) by an auxiliary electrode called the trigger, to which low-power pulses are applied.

triggered sweep — In a cathode-ray oscilloscope, a sweep initiated by a signal pulse.

trigger electrode — See starter, 2.

trigger flip-flop — See T flip-flop.

trigger gap — A gas-discharge device with three electrodes that can be triggered upon command. Two electrodes provide the main conduction path, and the third electrode serves as the trigger.

triggering — The starting of circuit action, which then continues for a predetermined time under its own control.

triggering signal — The signal from which a trigger is derived.

trigger level — 1. In a transponder, the minimum receiver input capable of causing the transmitter to emit a reply. 2. The instantaneous level of a triggering signal at which a trigger is to be generated. Also the name of the control that selects the level.

trigger point — The amplitude point on the input pulse at which triggering of the sweep of a cathode-ray oscilloscope occurs.

trigger pulse — A pulse used for triggering.

trigger-pulse steering — In transistors, the routing or directing of trigger signals (usually pulses) through diodes or transistors (called steering diodes or transistors) so that the signals affect only one of several associated circuits.

trigger recognition — In random-sampling-oscilloscope technique, the process of making a response to a suitably applied trigger, such as the time reference for the time window.

trigistor — A bistable pnpn semiconductor component with characteristics comparable to those of a flip-flop or bistable multivibrator.

tri-gun color picture tube — See tri-color picture tube.

trim — To make a fine adjustment in a circuit or a circuit element.

trim control — On some regulated power supplies, a control used to make minor adjustments of output voltage.

trimmer capacitor — Also called a trimmer. 1. A small variable capacitor associated with another capacitor and used for fine adjustment of the total capacitance of the combination. 2. A small capacitor, usually adjustable by screwdriver, that is placed in parallel with a larger capacitor and used to tune the high-frequency end of the circuit more precisely. 3. A small adjustable circuit element connected in series or parallel with a circuit element of the same kind such that its adjustment sets the combination of the two to a desired value.

trimmer potentiometer — A lead-screw-actuated potentiometer.

trimmer resistor — A small rheostat used in place of a fixed resistor to permit adjustment of resistance values in a circuit during initial calibration of an equipment or when recalibration is required.

Trimmer capacitor.

trimming — 1. The fine adjustment of capacitance, resistance, or inductance in a circuit. 2. The process of using abrasive or laser technique to accurately remove film material from a substrate to change (increase) the resistance value. Thick films are trimmed using either abrasives or lasers; thin-film resistors are laser trimmed. Automated laser trimming is possible using computer-controlled systems. 3. The process of cutting away a portion of a thick-film resistor in order to raise the nominal resistance value. Resistors can often be trimmed to within 1 percent of a desired value. Cutting is done with an air abrasive tool or with a laser beam. The width of the cut is called the kerf.

trimming potentiometer — An electrical mechanical device with three terminals. Two terminals are connected to the ends of a resistive element, and one terminal is connected to a movable conductive contact that slides over the element, thus allowing the input voltage to be divided as a function of the mechanical input. It can function as either a voltage divider or rheostat.

trim notch — The notch made in a resistor by trimming to obtain the design value. See kerf.

trinistor — A three-terminal silicon semiconductor device with characteristics similar to those of a thyratron; used for controlling large amounts of power.

trinoscope — 1. Any assembly of three kinescopes producing the red, green, and blue images required for tricolor television optical projection (e.g., for theater TV). 2. A color television viewing system with three kinescopes, three lenses, and three deflection yokes used to form the red, green, and blue images required for a tricolor television projection.

triode — A three-electrode electron tube containing an anode, a cathode, and a control electrode (grid).

triode amplifier — An amplifier in which only triode tubes are used.

triode field-effect transistor — A field-effect transistor having a gate region, a source region, and a drain region. Where no confusion is possible, the term may be abbreviated to field-effect triode.

triode-heptode converter — A superheterodyne converter circuit that uses a triode local oscillator and a heptode converter, both contained in one envelope.

triode-hexode converter — A superheterodyne converter circuit that uses a triode local oscillator and a hexode converter, both contained in one envelope.

triode laser — A gas laser whose light output may be modulated by applying signal voltages to an integral grid.

triode-pentode — A dual-purpose vacuum tube containing a triode and a pentode in the same envelope.

triode pnpn-type switch — A pnpn-type switch having an anode, cathode, and gate terminal.

triode thyristor — A three-electrode thyristor, of which one electrode is the gate. Commonly it has three terminals.

trip coil — An electromagnet in which a moving armature trips a circuit breaker or other protective device and thereby opens a circuit under abnormal conditions.

trip computer — Digital and push-button dashboard electronic system (Chrysler Corp.) that displays readout of speed, miles, fuel, time, and performs computations involving any of these parameters.

trip-free relay — *See* tripping relay.

triple-conversion receiver — A communications receiver in which three different intermediate frequencies are employed to give better adjacent-channel selectivity and greater image-frequency suppression.

triple detection — *See* double superheterodyne reception.

triple diffused — Pertaining to transistors fabricated within a monolithic substrate by three diffusion steps.

triple-play tape — Abbreviated TP tape. Very thin recording tape of which it is possible to wind 3600 ft (1097 m) onto a 7-inch (17.8-cm) spool. This would give 180 minutes of playing time at 3 ips (9.525 cm/s).

triple-stub transformer — A microwave transformer consisting of three stubs placed a quarter-wavelength apart on a coaxial line. The stubs are adjusted in length to compensate for impedance mismatch.

triplet — 1. Three radionavigational stations operated as a group for the determination of position. 2. The waveform of the output voltage of a delay line when the input pulse has a width approximately equal to the resolution of the delay line.

triplex — *See* triad, 4.

triplex cable — A cable made up of three insulated single-conductor cables twisted together with or without a common insulating covering.

triplexer — A dual duplexer that permits the use of two receivers simultaneously and independently in a radar system by disconnecting the receivers during the transmitted pulse.

triplex system — A system for simultaneously sending two messages in one direction and one message in the other direction over a single telegraph circuit.

tripod — A three-legged camera support.

tripping device — A mechanical or electromagnetic device used for opening (turning off) a circuit breaker or starter, either when certain abnormal electrical conditions occur or when a catch is actuated manually.

tripping relay — Also called trip-free relay. A device that functions to trip a circuit breaker, contactor, or equipment, or to permit immediate tripping by other devices, or to prevent immediate reclosure of a circuit interrupter in case it should open automatically even though its closing circuit is maintained closed.

trip protection circuit — A protective circuit that electrically interrupts the output when an overload occurs.

trip voltage — Also called firing voltage. The voltage at which ionization occurs under any circumstances.

trip-wire switch — A switch that is actuated by breaking or moving a wire or cord installed across a floor space.

trisistor — A fast-switching semiconductor consisting of an alloyed junction pnp device in which the collector is capable of electron injection into the base. Its characteristics resemble those of a thyratron electron tube, and the switching time is in the nanosecond range.

tristate — Also called three-state. 1. Logic systems utilizing three conditions on one line: a definite applied high voltage (logical 1); a definite low voltage (logical 0); and an open circuit or undefined state, permitting another part of the circuit to determine whether the line will be high or low. Usually refers to device outputs or systems using outputs of the tristate type. Useful in a bus-organized system. 2. An output configuration found in several logic families that is capable of assuming three output states: high, low, and high impedance. The feature is useful for interconnecting large numbers of devices on the same wires while allowing only one to control the levels of the lines at a given time.

tristate device — A device that has three states: on, off, and electrically disconnected.

tristimulus values — The amounts of each of the three primary colors that must be combined to match a sample.

tri-tet oscillator — A crystal-controlled, electron-coupled vacuum-tube oscillator that is isolated from the output circuit through use of the screen-grid electrode as the oscillator anode. Used for multiband operation because it generates strong harmonics of the crystal frequency.

tritium — A radioactive isotope of hydrogen with an atomic number of 3.

TRL — Abbreviation for transistor-resistor logic.

Trojan horse — 1. Software that, while performing a useful function, has hidden in it code for a program that performs security-breaching or fraudulent functions. 2. A program or data that seems innocuous when loaded into a system or network but later facilitates an attack by a hacker or a virus.

trombone — An adjustable U-shaped coaxial line matching assembly.

tropicalization — A chemical treatment developed to combat the fungi that ruin electronic equipment in hot, humid jungle regions.

tropo — *See* tropospheric scatter communication.

troposphere — The lower layer of the earth's atmosphere, extending to about 60,000 feet (18,300 m) at the equator and 30,000 feet (9150 m) at the poles. In this area the temperature generally decreases with altitude, clouds form, and convection is active.

tropospheric scatter — The propagation of radio waves by scattering as a result of irregularities or discontinuities in the physical properties of the troposphere.

tropospheric scatter communication — Also called tropo. A method or system of transmitting, within the troposphere, microwaves in the UHF or SHF bands to effect radiocommunication between two points on the earth's surface separated by moderate distances of from 70 to 600 miles (112 to 965 km). Such a span or hop may be augmented by other spans in tandem to permit end-to-end or through circuits up to many thousands of miles. More specifically, this method of communication is now generally understood to embrace a radio system that permits communication over the distances indicated, with excellent reliability and good information capacity, using relatively high transmitted power, frequency modulation, and highly sensitive receiving apparatus.

tropospheric superrefraction — The phenomenon occurring in the troposphere whereby radio waves are bent sufficiently to be returned to the earth.

tropospheric wave — A radio wave that is propagated by reflection from a place of abrupt change in the dielectric constant of its gradient in the troposphere. In some cases, the ground wave may be so altered that new components appear to arise from reflections in regions of rapidly changing dielectric constants; when these components are distinguishable from the other components, they are called tropospheric waves.

trouble — Failure of a circuit or element to perform in a standard manner.

trouble-location problem — A computer test problem whose incorrect solution supplies information on the location of faulty equipment. It is used after a check problem has shown that a fault exists.

troubleshoot — *See* troubleshooting.

troubleshooting — Also troubleshoot. 1. Locating and diagnosing malfunctions or breakdowns in equipment by means of systematic checking or analysis. 2. The work necessary to discover the cause of trouble; also implies the correction of the trouble by elimination of the cause.

trouble unit — A weighting figure applied to indicate the expected performance of a telephone circuit or circuits in a given period of time.

tr switch — *See* transmit-receive switch.

tr tube — Transmit-receive tube. A gas-filled rf switching tube that permits use of the same antenna for both transmitting and receiving by preventing the transmitted power from damaging the receiver. (Usually, a tr unit consists of a cavity containing a discharge gap that connects the transmitter to the antenna, and a coupling circuit that connects the antenna to the receiver when the discharge gap is not fired, indicating that the transmitting tube is quiescent.)

true bearing — A bearing given in relation to geographic north, as opposed to a magnetic bearing.

true complement — *See* radix complement.

true course — A course in which the direction of the reference line is true rather than magnetic north.

true credit balance — In a calculator, when the answer is negative, the minus sign automatically appears in the display.

true homing — The following of a course in such a way that the true bearing of an aircraft or other vehicle is held constant.

true north — Geographic north.

true ohm — The actual value of the practical unit of resistance. It is equal to 10^9 absolute electromagnetic units of resistance.

true power — The average power consumed by a circuit during one complete cycle of alternating current.

true radio bearing — *See* radio bearing.

true random noise — A noise characterized by a normal or Gaussian distribution of amplitudes.

true value — The value of a physical quantity that would be attributed to an object or physical system if that value could be determined with no error.

truncate — 1. To drop digits of a number of terms in a series, thereby lessening precision. For example, the value 3.14159265 (π) when truncated to five figures is 3.1415, whereas it could be rounded off to 3.1416. 2. To conclude a computational process according to some rule; for example, to stop the evaluation of a power series at a specified term. 3. The dropping of digits or characters from one end of a data item, causing loss of precision or information.

truncated paraboloid — A paraboloid reflector in which a portion of the top and bottom have been cut away to broaden the main radiated lobe in the vertical plane.

truncation — The process of dropping one or more digits at the left or right of a number without changing any of the remaining digits. For example, in most operations the number 3847.39 would become 3847.3 when truncated one place at the right, whereas the same number would become 3847.4 when rounded correspondingly. *See also* round off.

truncation error — The error resulting from the use of only a finite number of terms of an infinite series, or from the approximation of operations in the infinitesimal calculus by operations in the calculus of finite differences.

trunk — 1. A single message circuit between two points, both of which are switching centers and/or individual message distribution points. 2. A communications channel between two different offices, or between groups of equipment within the same office. 3. A telephone line or channel between two central offices or switching devices, which is used in providing telephone connections between subscribers. 4. A major link in a communication system, usually between two switching centers.

trunk circuit — A circuit that connects two switching centers.

trunk facility — That part of a communication channel connecting two or more leg facilities to a central or satellite station.

trunk group — Those trunks connecting two points, both of which are switching centers and/or individual message distribution points and which make use of the same multiplex terminal equipments.

trunk hunting — A method by which an incoming call is switched to the next consecutive number if the first called number is busy.

trunk loss — That part of the repetition equivalent assignable to the trunk used in the telephone connection.

truth table — 1. A tabulation that shows the relation of all output logic levels of a digital circuit to all possible combinations of input logic levels in such a way as to characterize the circuit functions completely. 2. A matrix that describes a logic function by listing all possible combinations of inputs and by indicating the outputs for each combination.

TSC — Abbreviation for transmitter start code.

TSS — Abbreviation for time-sharing system. A computer operating system in which the processor's time is shared among simultaneous users, resulting in an interactive facility.

TTL or **T²L** — Abbreviation for transistor-transistor logic. A bipolar technology used for producing logic gates. Positioned in the evolution of logic families after RTL (resistor-transistor logic) and DTL (diode-transistor logic) and before ECL (emitter-coupled logic) and CMOS. *See* gate.

TTL or **T²L** **compatible** — The ability of a device or circuit to be connected directly to the input or output of TTL logic devices. Such compatibility eliminates the need for interfacing circuitry.

TTL-compatible circuits — Circuits that meet the input/output interface standards of transistor-transistor logic (TTL) circuits. These can be TTL circuits themselves, or circuits made with other technologies that conform to TTL standards.

TTY — An abbreviation of Teletype or teletypewriter unit.

TU — *See* terminal unit.

tuba — A powerful land-based radar jamming transmitter operated between 480 and 500 MHz. It was developed during World War II for use against night fighter planes.

tube — A hermetically sealed glass or metal envelope in which conduction of electrons takes place through a vacuum or gas.

tube bridge — An instrument used in the precise measurement of vacuum-tube characteristics. It contains one or more bridge-type measuring circuits, plus power supplies and signal sources for all possible electrode combinations.

tube coefficients — Constants that describe the characteristics of a thermionic vacuum tube (e.g., amplification factor, mutual conductance, ac plate resistance).

tube complement — The number and types of electron tubes required in an electronic equipment.

tube count — A terminated discharge produced by an ionizing event in a radiation-counter tube.

tube drop — The voltage measured across a tube, from plate to cathode, when the tube is conducting at its normal current rating.

tube electrometer — A thermionic vacuum tube adapted for use as an electrometer to measure potential difference.

tube heating time — The time required for the coolest portion of a mercury-vapor tube to attain its operating temperature.

tubing — Extruded nonsupported plastic materials designed as protection and electrical insulation for exposed components of electrical and electronic assemblies; as opposed to a coated, braided, or woven tube, termed sleeving.

tube noise — Noise originating in a vacuum tube (e.g., from shot effect, thermal agitation, etc.).

tube shield — A metallic enclosure placed over a vacuum tube to prevent external fields from interfering with the function of the tube.

tube socket — A receptacle that provides mechanical support and electrical connection for a vacuum tube.

tube tester — A test instrument for indicating the condition of vacuum tubes used in electronic equipment.

tube voltage drop — The anode voltage in an electron tube during conduction.

tubular capacitor — A paper, ceramic, or electrolytic capacitor shaped like a cylinder; leads or lugs project from one or both ends.

tunable-cavity filter — A microwave filter in which tuning can be accomplished by adjustment of one or more screws that project into the cavity or by adjustment of the positions of one or more rectangular or circular irises in the cavity or waveguide.

tunable echo box — An echo box consisting of an adjustable cavity operating in a single mode. When the echo box is calibrated, the setting of the plunger at resonance will indicate the wavelength.

tunable magnetron — A magnetron that can be tuned mechanically or electronically over a limited band of frequencies.

tuned — Adjusted to resonate or operate at a specified frequency.

tuned amplifier — An amplifier in which the load is a tuned circuit. Thus, the load impedance and amplifier gain vary with the frequency.

tuned-anode oscillator — *See* tuned-plate oscillator.

tuned antenna — An antenna designed, by means of its own inductance and capacitance, to provide resonance at the desired operating frequency.

tuned-base oscillator — A transistor oscillator comparable to a tuned-grid electron-tube oscillator. The frequency-determining device (resonant circuit) is located in the base circuit.

tuned circuit — 1. A circuit consisting of inductance and capacitance that can be adjusted for resonance at the desired frequency. 2. A circuit that is adjusted to be resonant at a particular frequency.

tuned-collector oscillator — A transistor oscillator comparable to the tuned-plate electron-tube oscillator. The frequency-determining device is located in the collector circus.

tuned dipole — A dipole antenna that resonates at its operating frequency.

tuned filter — A resonant arrangement of electronic components that either attenuates signals at a particular frequency and passes signals at other frequencies, or vice versa.

tuned-filter oscillator — An oscillator in which a tuned filter is used.

tuned-grid oscillator — An oscillator whose frequency is determined by a parallel-tuned tank in the grid circuit. The tank is coupled to the plate to provide the required feedback.

tuned-grid, tuned-anode oscillator — *See* tuned-grid, tuned-plate oscillator.

tuned-grid, tuned-plate oscillator — Also called tuned-grid, tuned-anode oscillator. An oscillator having parallel-resonant circuits in both the plate and the grid circuits. The necessary feedback is provided by the plate-to-grid interelectrode capacitance.

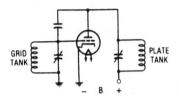

Tuned-grid, tuned-plate oscillator.

tuned-plate oscillator — Also called tuned-anode oscillator. An oscillator whose frequency is determined by a parallel-tuned tank in the plate circuit. The tank is coupled to the grid to provide the required feedback.

tuned radio-frequency amplifier — Abbreviated TRF amplifier. A tuned amplifier using resonant-circuit coupling and designed to operate at radio frequencies.

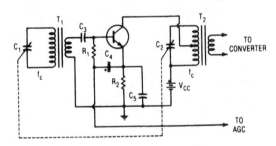

Tuned radio-frequency amplifier.

tuned radio-frequency receiver — Abbreviated TRF receiver. A radio receiver consisting of several amplifier stages that are tuned to resonate at the carrier frequency of the desired signal by a ganged variable-tuning capacitor. The amplified signals at the original carrier frequency are fed directly into the detector for demodulation. The resultant audio-frequency signals are again amplified, and are then reproduced by a speaker.

tuned radio-frequency transformer — Abbreviated TRF transformer. A transformer used for selective coupling in radio-frequency stages.

tuned-reed frequency meter — A vibrating-reed instrument for measuring the frequency of an alternating current.

tuned relay — A relay having mechanical or other resonating arrangements that limit the response to currents at one particular frequency.

tuned resonating cavity — A resonating cavity half a wavelength long, or some multiple of a half wavelength, used in connection with a waveguide to produce a resultant wave with the amplitude in the cavity greatly exceeding that of the wave in the guide. For reception of waves, a detecting grating can be placed at the point of maximum amplitude in the cavity to convert the energy to a form suitable for amplification in a telephone or television circuit. A tuned cavity is a nonreflecting termination for a guide.

tuned rope — Long lengths of chaff cut to the various lengths necessary for tuning to different wavelengths.

tuned transformer — A transformer whose associated circuit elements are adjusted as a whole to be resonant at the frequency of the alternating current supplied to the primary, thereby causing the secondary voltage to build up to higher values than would otherwise be obtained.

tuner — In the broad sense, a device for tuning. Specifically, in radio-receiver practice: 1. A packaged unit capable of producing only the first portion of the functions of a receiver and delivering either rf, IF, or demodulated information to some other equipment. 2. That portion of a receiver that contains the circuits that are tuned to resonance at the received-signal frequency and those which are tuned to the local-oscillator frequency. 3. A radio or TV receiving circuit; a high-fidelity component containing such circuits.

tungar rectifier — A gaseous rectifier containing argon gas. It is employed in battery chargers and low-voltage power supplies.

tungar tube — A phanotron (hot-cathode gas-filled rectifier tube) having a heated filament serving as the cathode and a graphite disc as the anode in a bulb filled with low-pressure argon. Used chiefly in battery chargers.

tungsten — A metal used in the manufacture of filaments for vacuum tubes and in making contact points for switches and other parts where sparking may occur. After the tungsten is made ductile by rolling, swaging, and hammering, it is very tough.

tungsten filament — A filament used in incandescent lamps, and in thermionic vacuum tubes and other tubes requiring an incandescent cathode. Smaller tungsten filaments are operated in a vacuum, whereas those for larger lamps are used in an inert gas at about ordinary atmospheric pressure.

tungsten lamp — An evacuated bulb containing a tungsten filament that is heated by passing an electric current through it.

tuning — 1. The adjustment relating to frequency of a circuit or system to secure optimum performance. Commonly, the adjustment of a circuit or circuits to resonance. 2. Making changes in a program to improve its performance without altering its results.

tuning capacitor — 1. A variable capacitor for adjusting the natural frequency of an oscillatory or resonant circuit. 2. A variable capacitor that is intended to be mechanically actuated frequently throughout its life.

tuning circuit — A circuit containing inductance and capacitance, either or both of which may be adjusted to make the circuit responsive to a particular frequency.

tuning coil — A variable inductance for adjusting the natural frequency of an oscillatory or resonant circuit.

tuning control — A control knob that adjusts all tuned circuits simultaneously.

tuning core — Normally a molded iron core for permeability tuning, into which an adjusting screw has been cemented or molded.

tuning diode — A varactor diode used for rf tuning. This includes functions such as automatic frequency control and automatic fine tuning.

tuning eye — Slang for a cathode-ray tuning indicator.

tuning fork — A two-pronged hard-steel device that vibrates at a definite natural frequency when struck or when set in motion by electromagnetic means. Used in some electronic equipment as an accurately controllable source of signals, because its vibrations can be formed

readily into audio-frequency signals by means of pickup coils.

tuning-fork contact — A U-shaped female contact that resembles a tuning fork. It may be stamped or formed.

tuning-fork drive — Control of an oscillator by continuous vibrations of a tuning fork. A high harmonic of the oscillating signal obtained from the fork is selected by filter circuits and is strongly amplified to determine the main oscillator frequency in a transmitter or other equipment.

tuning in — Adjusting the tuning controls of a receiver to obtain maximum response to the signals of the station it is desired to receive.

tuning indicator — A device that indicates whether or not a receiver is tuned accurately. It is connected to some circuit in which current or voltage is maximum or minimum when the receiver is accurately tuned to give the strongest output signal.

tuning meter — A direct-current meter connected to a receiver circuit and used for determining whether the receiver is accurately tuned to a station.

tuning probe — An essentially lossless probe of adjustable penetration extending through the wall of a waveguide or cavity resonator.

tuning range — The frequency range over which a tuned circuit can be adjusted.

tuning screw — A screw or probe inserted into a transmission line (parallel to the E field) to produce susceptance of magnitude and sign that depend on the depth of penetration of the screw.

tuning stub — 1. A short length of transmission line, usually with the free end shorted, connected to a transmission line to provide impedance matching. 2. A type of inductor element, usually adjustable, connected to a transmission line at intervals to improve the voltage distribution.

tuning susceptance — The normalized susceptance of an atr tube in its mount due to its deviation from the desired resonance frequency.

tuning wand — *See* neutralizing tool.

tunnel action (in a pn junction) — A process whereby conduction occurs through the potential barrier due to the tunnel effect and in which electrons pass in either direction between the conduction band in the n-region and the valence band in the p-region. (Tunnel action, unlike the diffusion of charge carriers, involves electrons only, and for all practical purposes the transit time is negligible.)

tunnel cathode — A metal-insulator-metal sandwich. Electrons tunnel from the metal substrate and appear in the metal film as hot electrons. Some of the hot electrons have sufficient energy to pass over the cathode surface barrier into the vacuum.

tunnel diode — Also called Esaki diode. 1. A pn diode to which has been added a large amount of impurity. The tunnel diode has high-speed charge movement and a negative-resistance region above a minimum level of applied voltage. With the addition of suitable external circuits, it can be used as an oscillator or amplifier. 2. A two-layer device similar to the rectifier diode. As a small voltage is applied, current starts to flow. Increase the voltage a little more and current drops to zero. Add still a little more voltage and the current increases again. At still higher voltages it responds like an ordinary diode. That first surge of current is called tunneling. 3. A heavily doped junction diode based on tunneling effect: the piercing of a potential barrier by a low-energy electron. Has a forward negative resistance, and is an excellent low-noise microwave amplifier. It has no upper-frequency limitation. 4. A diode utilizing the tunnel effect. It gives rise to negative differential conductance in a

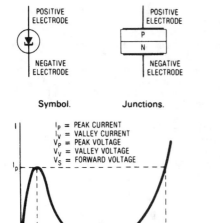

Symbol.   Junctions.

I_P = PEAK CURRENT
I_V = VALLEY CURRENT
V_P = PEAK VOLTAGE
V_V = VALLEY VOLTAGE
V_S = FORWARD VOLTAGE

Characteristic.

Tunnel diode.

certain range of the forward direction of the current-voltage characteristic.

tunnel effect — 1. The piercing of a potential hill by a carrier, which would be impossible according to classical mechanics, but the probability of which is not zero according to wave mechanics, if the width of the hill is small enough. The wave associated with the carrier is almost totally reflected on the first slope, but a small fraction crosses the hill. 2. In a pn transition region, a process whereby conduction occurs through a potential barrier and in which electrons pass in either direction between the conduction band in the n-region and the valence band in the p-region. The tunnel effect, unlike the diffusion of charge carriers, involves electrons only and is for all purposes an instantaneous process.

tunneling — An observed effect of the ability of certain atomic particles to pass through a barrier that they cannot pass over because of the required energy level. It is based on a law of quantum mechanics that predicts that the particles have a finite probability for tunneling according to their quantum-mechanical nature.

tunnel rectifier — A tunnel diode that has a relatively low peak-current rating in comparison with other tunnel diodes employed in memory-circuit applications.

tunnel resistor — A resistor containing a thin layer of metal plated across a tunneling junction, so that the characteristics of a tunnel diode and an ordinary resistor are combined.

tunnel triode — A transistorlike device in which the emitter-base junction is a tunnel diode and the collector-base junction is a conventional diode.

tunneluminescence — The emission of light from a phosphor film deposited on the surface of a thin-film metal-oxide-metal sandwich.

turbidimeter — *See* opacimeter.

turbulence amplifier — 1. Fluidic digital element using laminar-to-turbulent flow transition to create the control effect. 2. A fluidic device in which the power jet is at a pressure such that it is in the transition region of laminar stability and can be caused to become turbulent by a secondary jet or by sound.

turn — One complete loop of wire.

turnaround time — 1. The interval between the time at which a job is submitted to a batch processing system

and the time at which the results are returned. 2. The actual time required to reverse the direction of transmission in a half-duplex circuit. For most communications facilities, time is required for line machine reaction. (A typical time is 200 milliseconds on a half-duplex telephone connection.) 3. The period required for half-duplex modems to reverse the direction of transmission on dial-up or two-wire leased lines. This includes the time for reversing the echo suppressors used on the two-wire simplex paths between telephone central offices. These suppressors are placed between central offices to prevent signals from one transmission path from crossing over into another. Also, for synchronous modems, turnaround time covers line equalization and bit synchronization. A full-duplex modem has no turnaround time when used in a point-to-point link, since it can transmit and receive information simultaneously.

turn factor — Under stated conditions, the number of turns that a coil of given shape and dimensions placed on a core in a given position must have for a coefficient of self-inductance of 1 henry to be obtained for the core.

turnkey — 1. A computer console containing a single control, usually a power switch, that can be turned on and off only with a key. 2. A design and/or installation in which the user receives a complete running system 3. Pertaining to a computer system sold in a ready-to-use state.

turn-off delay time — The time interval between occurrence of the trailing edge of a fast input pulse and the occurrence of the 90-percent point of the negative-going output waveform.

turn-off reversal — A polarity reversal of the output of the electronic power supply occurring from the turning off of the power of a regulated power supply.

turn-off thyristor — A thyristor that can be turned from the on state to the off state, and vice versa, with appreciable gain by applying control signals of appropriate polarities to the gate terminal.

turn-off time — The time that a switching circuit (gate) takes to stop the current in the circuit it is controlling.

turn-on delay time — The time interval from the occurrence of the leading edge of a fast input pulse to the occurrence of the 10-percent point of the positive-going output waveform, assuming that the rise time of the incoming pulses is 0.1 of the rise time of the element to be measured under loaded conditions.

turn-on overshoot — Overshoots occurring from the turning on of power of a regulated power supply.

turn-on reversal — A polarity reversal of the output of the power supply occurring from the turning on of the power supply of a regulated power supply.

turnover cartridge — A phonograph cartridge adapted, by the use of two styli, to play both large- and fine-groove records.

turnover frequency — 1. In disc recording, the frequency below which constant-amplitude recording is used and above which constant-velocity recording is employed. 2. The knee of the tone control or filter frequency response curve. Though normally thought of as the frequency at which the control begins to have its effect, the turnover point is actually the frequency at which response has already been altered by 3 dB relative to the unfiltered signal. Some tone controls and filters offer a choice of turnover frequencies that are usually switch-selected.

turnover pickup — Also called dual pickup. A pickup designed for playing both standard and micro-groove records, using a single magnetic structure. *See also* turnover cartridge.

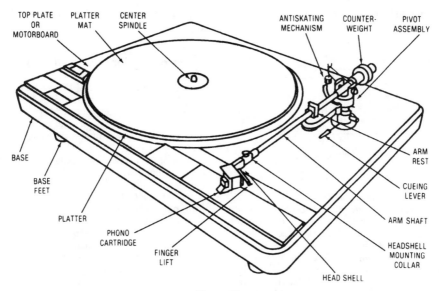

Turntable.

TOP PLATE OR MOTORBOARD • PLATTER MAT • CENTER SPINDLE • ANTISKATING MECHANISM • COUNTER-WEIGHT • PIVOT ASSEMBLY • BASE • BASE FEET • PLATTER • PHONO CARTRIDGE • FINGER LIFT • HEAD SHELL • ARM REST • CUEING LEVER • ARM SHAFT • HEADSHELL MOUNTING COLLAR

turns ratio — The ratio of the number of turns in the primary winding to the number in the secondary winding of a transformer.

turnstile antenna — An antenna composed of two dipole antennas normal to each other and with their axes intersecting at their midpoints. Usually the currents are equal and in-phase quadrature.

turntable — 1. The round platter on which a phonograph record rests during cutting or playback. Also refers to the platter and its driving motor and associated parts, as a high-fidelity component. 2. In tape recording, the rotating flat disc on which the reel (or, in some professional machines, the tape "pie") lies slightly raised above the recorder's front or top panel. The turntable is usually fitted with a center spindle and gripping keys of some kind to keep the reel centered and prevent it from slipping against the rotation of the turntable.

turntable rumble — *See* rumble, 1.

turret — A revolving plate mounted at the front of some television cameras and carrying two or more lenses of different types to permit rapid interchange of lenses.

turret tuner — A television-receiver tuner containing a separate set of resonating circuit elements for each channel. Each set is mounted on an insulating strip or strips placed on a drum rotated from the channel position for the desired channel.

TV — Abbreviation for television.

TV camera — An optical device, consisting of lens, electron beam tube, and preamplifier, that converts an optical image into an electrical signal.

TV channel — *See* television channel.

TVI — Abbreviation for television interference.

TV recording — A permanent record of video signals recorded photographically, electronically, or by other means, and which may be displayed through a television system or projected as a motion-picture film.

TVRO — *See* television receive only.

TVT — *See* TV terminal.

TV terminal — Abbreviated TVT. A keyboard and electromagnetic deflection-type display unit that utilizes an ordinary (or modified) TV set as a display.

tweaking — 1. The process of adjusting an electronic receiver circuit to optimize its performance. 2. The fine tuning of a circuit with overly critical adjustments,

repeated several times until no further improvement in performance is obtained. This usually requires considerable skill and is not recommended as a requirement in circuits for production or construction by a hobbyist with limited technical resources.

tweeter — Also called a high-frequency unit. 1. A speaker intended to reproduce the very high frequencies, usually those above 3000 Hz, in a high-fidelity audio system. Units may be ionic, ribbon, electrostatic, or dynamic types. 2. A high-frequency speaker (driver) specializing in treble reproduction.

twelve punch — A punch in the top row of a card.

twenty-one type repeater — A two-wire telephone repeater in which one amplifier serves to amplify the telephone currents in both directions. The circuit is arranged so that the input and output terminals of the amplifier are in one pair of conjugate branches, while the lines in the two directions are in another pair.

twenty-two type repeater — A two-wire telephone repeater with two amplifiers. One amplifies the telephone currents being transmitted in one direction, and the other the telephone currents being transmitted in the other direction.

twilight — A scene illumination of approximately one footcandle.

twin axial cable — A single shielded twisted-pair cable that has low-loss signal transmission and high noise immunity.

twin cable — 1. A cable composed of two parallel insulated stranded conductors having a common covering. 2. A pair of insulated conductors twisted and/or sheathed or held together mechanically and not identifiable from each other in a common covering.

twin check — A continuous check of computer operations accomplished by duplication of equipment and automatic comparison of results.

twin coaxial cable — A configuration containing two separate, complete coaxial cables laid parallel or twisted around each other in one complex.

twinflex — British word for twisted pair of wires or twin lead.

twin lead — Also called twin line. A type of transmission line covered by a solid insulation and comprising two parallel conductors, the impedance of which is determined by their diameter and spacing. The three most common impedance values are 75, 150, and 300 ohms.

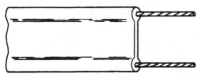

Twin lead.

twin line — *See* twin lead.

twinning — 1. The intergrowth of two crystal regions having opposite oriented axes. Two types of twinning may occur: electrical and optical. In electrical twinning, the electrical senses of the crystal axes are reversed and the twinned regions will interfere with one another piezoelectrically. It is this effect that limits the high-temperature utility of quartz as a piezoelectric material. 2. A defect of natural quartz crystals, in which both the right and left quartz are in the same crystal.

twin-T network — *See* parallel-T network.

twin triode — Two triode vacuum tubes in a single envelope.

twin wire — A cable composed of two small, parallel, insulated conductors having a common covering.

twist — 1. The progressive rotation of the cross section of a waveguide about the longitudinal axis. 2. The deviation from a plane surface measured from one corner to the corner diagonally opposite. 3. The difference (in decibels) between DTMF high-group and low-group signal levels, mathematically defined as 10 log (high-group power)/(low-group power). Measured at the DTMF receiver, it's a function of both the level difference generated by the signal source and the gain-frequency characteristic of the transmission facility.

twisted joint — A union of two conductors wound tightly around each other. A sleeve may be used, and it and the conductors twisted.

twisted pair — 1. A cable composed of two small insulated conductors twisted together without a common covering. The two conductors of a twisted pair are usually substantially insulated, so that the combination is a special case of a cord. 2. Two insulated wires (signal and return) that are twisted around each other in a spiral pattern mainly to cancel the effects of electrical noise. Since both wires have nearly equal exposure to any electrostatic or electromagnetic interference, the differential noise is slight.

twister — A piezoelectric crystal that generates a voltage when twisted.

twistor — A computer memory element containing inclined helical windings of magnet wire on a nonmagnetic wire, with another winding over the helix. Information is stored in the form of polarized helical magnetization.

two-address — In a computer, having the property that each complete instruction includes an operation and specifies the location of two registers, usually one containing an operand and the other the result of the operation.

two-address code — A computer code that uses two address instructions.

two-address instruction — A computer instruction that includes an operation and specifies the location of two registers.

two-conductor jack — Receptacle having two circuits, tip and sleeve.

two-cored screened cable — British term for two-conductor shielded cable.

two-dimensional circuitry — *See* thin-film integrated circuit.

two-fluid cell — A cell having unlike electrolytes at the positive and negative electrodes.

two-hole directional coupler — A directional coupler that consists of two parallel coaxial lines in contact, with holes or slots through their contacting walls at two points one-quarter wavelength apart. With this device, a portion of the rf energy traveling in one direction through the main line may be extracted, while energy traveling in the opposite direction is rejected. It is necessary that one end of the secondary line be terminated in its characteristic impedance.

two-level system — A laser that uses only two electron energy levels. Electrons in the ground state (level 1) are pumped to the excited state (level 2). The electrons then surrender their energy by stimulated emission and return to the ground state.

two-out-of-five code — A type of positional notation in which each decimal digit is represented by five binary digits; two of the five are of one kind (for example, 1s) and three are of the other kind (for example, 0s).

two-part code — A randomized code with an encoding section and a decoding section. In the encoding section, the plaintext groups are arranged in alphabetical or other significant order, accompanied by their code groups in nonalphabetical or random order. In the decoding section, the code groups are arranged in alphabetical or numerical order and accompanied by their meanings given in the encoding section.

two-phase — Also called quarter phase. Having a phase difference of 90 electrical degrees, or one quarter-cycle.

two-phase current — Two currents delivered through two pairs of wires at a phase difference of one quarter-cycle (90°) between them.

two-phase dynamic — Pertaining to a dynamic logic circuit that uses two clock signals to control the processing of information through the circuit or logic system.

two-phase, five-wire system — An alternating-current supply in which four of its conductors are connected as in a four-wire, two-phase system and the fifth is connected to the neutral points of each phase and usually grounded. Despite its name, it is strictly a four-phase, five-wire system.

two-phase, four-wire system — A system of alternating-current supply comprising two pairs of conductors, between one pair of which is maintained an alternating difference of potential displaced in phase by one-quarter of a period from an alternating difference of potential of the same frequency maintained between the other pair.

two-phase, three-wire system — An alternating-current supply consisting of three conductors. Between one conductor (known as the common return) and each of the other two, alternating differences of potential that are 90° out of phase with each other are maintained.

two-piece contact — A contact made of two or more separate parts joined by swedging, brazing, or other means of fastening to form a single contact. This type provides the mechanical advantages of two metals but also has the inherent electrical disadvantage of difference in conductivity.

two-pilot regulation — The use of two pilot frequencies within a transmitted band so that the change in attenuation due to twist can be detected and compensated for by a regulator.

two-plus-one address — Pertaining to an instruction that contains two operand addresses and one control address.

two-port network — A network with two ports.

two-pulse canceler — A moving-target indicator canceler that compares the phase variation of two successive pulses received from a target. It discriminates against signals with radial velocities that produce a Doppler frequency equal to a multiple of the pulse-repetition frequency.

two-quadrant multiplier — Of an analog computer, a multiplier in which operation is restricted to a single sign of one input variable only.

twos complement — Pertaining to a form of binary arithmetic used in a computer to perform binary subtractions with addition techniques. The twos-complement negative of a binary number is formed by complementing each bit in the number and adding 1 to the result.

twos-complement binary — An alternate and more widely used code to represent negative values. With this code, zero and positive values are represented as in natural binary, and all negative values are represented in a twos-complement form. That is, the twos complement of a number represents a negative value so that interface to a computer or microprocessor is simplified.

two-source frequency keying — Keying in which the modulating wave shifts the output frequency between predetermined values derived from independent sources.

two-state device — A mechanical or electronic device that, except during the time it is changing between states, is intended to be operated in either of two states or conditions.

two-terminal network — A network that is connected by only two terminals to an external system.

two-terminal-pair network — Also called a four-pole, quadripole, or quadrupole network. A network with four accessible terminals grounded in pairs. One terminal of each pair may coincide with a network node.

two-tone keying — Keying in which the modulating wave causes the carrier to be modulated with one frequency for the marking condition and a different frequency for the spacing condition.

two-tone modulating — A method of modulating in which two different carrier frequencies are used for the two signaling conditions.

two track (half track) — A tape format in which the width of the tape is recorded in two parallel magnetic tracks, separated by an unrecorded guard band. As compared with four-track recording, the two-track system gives improved dynamic range and can be edited without loss of program, since the tape is passed in a single direction only.

two-track recorder — *See* dual-track recorder.

two-track recording — On quarter-inch-wide tape, the arrangement by which only two channels of sound may be recorded, either as a stereo pair in one direction or as separate monophonic tracks (usually in opposite directions).

two-value capacitor motor — A capacitor motor that uses different values of effective capacitance for starting and running.

two-wattmeter method — A method of measuring total power in a balanced or unbalanced three-phase system by adding the readings of two wattmeters, each with its current coil in one phase and its voltage coil connected between it and the third phase.

two-way amplifier — An amplifier in which the right and left channels of a stereo system are both amplified simultaneously by the same tubes, using push-pull circuitry but feeding one signal to the input grids in parallel instead of push-pull. The parallel and push-pull signals are then separated by two output transformers in a matrixing circuit.

two-way communication — Communication between radio stations, each having both transmitting and receiving equipment.

two-way radio — Radiotelephone communications between fixed points (base stations) and portable units.

two-way repeater — *See* repeater.

two-way switch — A switch used for controlling electrical or electronic equipment, components, or circuits from either of two positions.

two-way system — A speaker in which the low and the high frequencies are reproduced separately by two electrically independent speaker elements, each of which is provided with a suitable sound-radiating system.

two-way, three-way, etc. — Refers to the number of frequency bands into which a speaker's output is divided. A two-way system divides the spectrum into two such bands, one of which is handled by a woofer or woofers, the other by a tweeter or tweeters. A three-way system would have one or more woofers, midrange speakers, and tweeters. Systems up to five-way are available.

two-wire channel — A two-way circuit for transmission in either direction.

two-wire circuit — 1. A metallic circuit formed by two conductors insulated from each other. It is possible to use the two conductors as either a one-way transmission path, a half-duplex path, or a duplex path. Also used in contrast with a four-wire circuit to indicate a circuit using one line or channel for transmission of electric waves in both directions. 2. A communications circuit that uses a single pair of wires for both transmitted and received information.

two-wire repeater — A repeater that can be used for transmission in both directions over a two-wire circuit. In carrier operation, it usually makes use of the principle of frequency separation for the two directions of transmission.

two-wire system — 1. A system of electric supply comprising two conductors, with the load connected between them. 2. A system in which all communication takes place over a two-wire circuit or the equivalent.

TWT — Abbreviation for traveling-wave tube.

TWX — *See* teletypewriter exchange service, 2.

Twystron — A very high power, hybrid microwave tube combining the input section of a high-power klystron with the output section of a traveling-wave tube. It is characterized by high operating efficiency and wide bandwidths.

.TXT files — Text files. These are usually just plain text that can be read by most programs.

type acceptance — Equipment authorization granted by the FCC to ensure that equipment will function properly in the service for which it has been accepted.

type A facsimile — Facsimile communication in which the images are built up of lines of constant-intensity dots.

type A waves — Continuous waves.

type A₁ waves — Unmodulated, keyed, continuous waves.

type A₂ waves — Modulated, keyed, continuous waves.

type A₃ waves — Continuous waves modulated by music, speech, or other sounds.

type A$_4$ waves — Superaudio frequency-modulated continuous waves, as used in a facsimile system.

type A$_5$ waves — Superaudio frequency-modulated continuous waves, as used in television.

type A$_9$ waves — Composite transmissions and cases not covered by type A through type A$_5$ waves.

typebar — A linear type element that contains all printable symbols.

type B facsimile — Facsimile communication in which the images are built up of lines of dots having a varying intensity (e.g., in telephotography and photoradio).

type B waves — Keyed, damped waves.

type-printed telegraphy — Telegraphy in which the message is automatically printed at the receiving station.

U

UART—*See* universal asynchronous receiver/transmitter.

ubitron—An amplifier or oscillator in which an undulating electron beam interacts with an rf wave. The kinetic energy of the beam is converted into rf energy (0-type interaction). The undulation of the beam is produced by a periodic magnetic field. This field gives the beam a transverse velocity component that interacts with the rf wave.

U-bolt—A U-shaped bolt threaded on both ends, for fastening antennas to masts.

UG—The two-letter designation that precedes the number on connectors for coaxial cable. It means Universal Government.

UHF—1. Abbreviation for ultrahigh frequency. 2. In television, a term used to designate channels 14 through 69.

UJT—Abbreviation for unijunction transistor.

UL—Abbreviation for Underwriters Laboratories, Inc.

UL certificated—For certain types of products that have met UL requirements and for which it is impractical to apply the UL Listing Mark or Classification Marking to the individual product, a certificate is provided that the manufacturer may use to identify quantities of material for specific job sites or to identify field-installed systems.

UL listed—Signifies that production samples of the product have been found to comply with established Underwriters Laboratories' requirements and that the manufacturer is authorized to use the Laboratories' Listing Marks on the listed products that comply with the requirements, contingent on the follow-up services as a check of compliance.

ultimately controlled variable—The quantity whose control is the end purpose of an automatic control system.

ultimate sensitivity or threshold—One-half the dead band in a graphic recorder. When the instrument is balanced at the center of the dead band, it denotes the minimum change in measured quantity required to initiate pen response.

ultimate trip current—The smallest value of current that will cause tripping of a circuit breaker under a given set of ambient conditions.

ultimate trip limits—The values of overload of a circuit breaker at which the minimum and maximum limits of the time-current curve become asymptotic; i.e., the limits of current that will trip or not trip the breaker "ultimately." Minimum and maximum limits of ultimate trip are often called calibration.

ultor—An adjective used to identify the picture-tube anode or element farthest from the cathode, the anode to which the highest voltage is applied, or the voltage itself (e.g., the ultor anode is the second anode of the picture tube, and the ultor voltage is the voltage applied to it).

ultor element—The element that receives the highest dc voltage in a cathode-ray tube.

ultra-audible frequency—*See* ultrasonic frequency.

ultra-audion—Any of several special vacuum-tube circuits employing regeneration.

ultra-audion circuit—A regenerative detector circuit in which a parallel-resonant circuit is connected between the grid and the plate of a vacuum tube, and a variable capacitor is connected between the plate and cathode to control the amount of regeneration.

ultra-audion oscillator—A variation of the Colpitts oscillator in which the resonant circuit employs a transmission-line section.

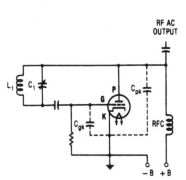

Ultra-audion oscillator.

Ultrafax—A trade name of RCA for a system in which printed information is transmitted by radio, facsimile, and television at high speeds.

ultrahigh frequency—Abbreviated UHF. Frequency band: 300 to 3000 MHz. Wavelength: 100 to 10 centimeters.

ultrahigh-frequency converter—A circuit used to convert UHF television signals to VHF in order to permit UHF television reception on a VHF receiver.

ultrahigh-frequency generator—Any device for generating ultrahigh-frequency alternating currents (e.g., a conventional negative-grid generator; a positive-grid, or Barkhausen, generator; a magnetron; and a velocity-modulation, or electron-beam, generator such as the klystron).

ultrahigh-frequency loop—Generally a single-loop antenna used in ultrahigh-frequency work to secure a nondirectional radiation pattern in the plane of the loop. The doughnut-shaped pattern is perpendicular to the loop.

ultrahigh-frequency translator—A television-broadcast translator station that transmits on a UHF TV-broadcast channel.

ultralinear amplifier—A class AB or B audio amplifier using pentodes or high-power-output beam-power tubes whose screen voltages are taken from taps on a specially wound output transformer rather than from a fixed dc source. This form of operation results in a considerable decrease in distortion in high-fidelity systems.

ultramicrometer—An instrument for measuring very small displacements by electrical means (e.g., by the variation in capacitance produced by the movement being measured).

ultramicrowave—Having wavelengths of about 10^{-1} to 10^{-4} cm.

ultrashort waves—Radio waves shorter than 10 meters in wavelength (about 30 MHz in frequency).

ultrasonic—1. Having a frequency above that of audible sound—i.e., between sonic and hypersonic. 2. Sound waves that vibrate at frequencies beyond the hearing power of human beings (above 16,000 Hz). Commercial and military applications include ultrasonic cleaning, gauging, cutting, detection instruments, and welding.

ultrasonic bond—A contact area where two materials are joined by means of ultrasonic energy and pressure.

ultrasonic bonding—1. A process for joining metal parts by the scrubbing action and energy transfer of a tool vibrating at an ultrasonic rate. This method is used to attach leads to pads on silicon devices. 2. A process involving the use of ultrasonic energy and pressure to join two materials. 3. A joining technique for wire and lead attachment that employs pressure plus an ultrasonically introduced scrubbing action to form a molecular bond.

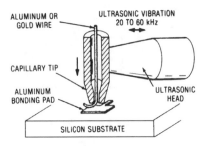

Ultrasonic bonding.

ultrasonic brazing—*See* ultrasonic soldering.

ultrasonic cleaner—A device using ultrasonic pressure waves to clean objects.

ultrasonic cleaning—A method of cleaning that uses cavitation in fluids caused by applying ultrasonic vibrations to the fluid.

ultrasonic cleaning equipment—Ultrasound used in the cleaning of metal and optical parts by virtue of its vibration rates. Large acoustic forces break off particles and contaminants from surfaces.

ultrasonic cleaning tank—A heavy-gage, polished stainless-steel tank with transducers mounted on the bottom or sides.

ultrasonic coagulation—The bonding together of small particles by the action of ultrasonic waves.

ultrasonic cross grating—Also called grating. The two- or three-dimensional space grating produced

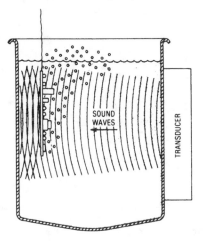

Ultrasonic cleaning tank.

when ultrasonic beams having different directions of propagation intersect.

ultrasonic delay line—Also called an ultrasonic storage cell. A contained medium (usually a liquid such as mercury) in which the signal is delayed because of the longer propagation time of the sound waves in the medium.

ultrasonic densitometer—An instrument device for determining the thickness or density of an object or material based on the time required for an ultrasonic signal to penetrate to a receiver and/or echo back to a receiver adjacent to another transmitter.

ultrasonic detector—A device—either mechanical, electrical, thermal, or optical—for detecting and measuring ultrasonic waves.

ultrasonic diagnosis—A method of obtaining information from within the body in a visual presentation without employing ionizing radiation. It differs from X-ray in that the form of energy used is high-frequency sound, or ultrasound, which is inaudible. The sound is transmitted in very brief pulses followed by relatively lengthy "silent" intervals. Also, in contrast to X-ray techniques, in which the film is placed behind the tissue being examined, ultrasonic information is picked up at the original point of transmission in the form of echoes from internal structures. These returning echoes are converted to electrical energy and displayed in either a static or dynamic pattern, as desired, on a cathode-ray-tube screen.

ultrasonic disintegrator—An apparatus for using the pressure wave produced by an ultrasonic generator to tear cells apart.

ultrasonic drill—A special type of drill that has a magnetostrictive transducer attached to a tapered cone that serves as a velocity transformer. With an appropriate tool, practically any shape of hole can be drilled in brittle and hard material.

ultrasonic flaw detector—Equipment comprising an ultrasonic generator, transducer, detector, and display; used to detect flaws or cracks in solids from the reflection pattern of ultrasonic signals observed on a cathode-ray tube.

ultrasonic frequency—Also called an ultra-audible frequency. 1. Any frequency above the audio range, but commonly applied to elastic waves propagated in gases, liquids, or solids. 2. Sound frequencies that are above the range of human hearing; approximately 20,000 Hz and higher.

ultrasonic generator—A device for producing mechanical vibrations at frequencies above the range of human hearing. Typically, such a device consists of an rf oscillator whose output is applied to a piezoelectric crystal.

ultrasonic grating constant—The distance between diffracting centers of the sound wave producing particular light-diffraction spectra.

ultrasonic immersion—Cleaning technique depending on cavitation (rapid formation of tiny bubbles in a cleaning liquid). Cavitation is created by ultrasonic, high-intensity sound waves. The agitation of imploding bubbles scrubs the immersed part.

ultrasonic inspection—A nondestructive testing method of locating internal defects in a part by sending ultrasonic impulses (inaudible high-frequency sound waves of 0.5 to 11 megahertz) into the part and measuring the time required for these impulses to penetrate the material, be reflected from the opposite side or from the defect, and return to the sending point.

ultrasonic level detector—A level detector consisting of an ultrasonic receiver and transmitter located in one wall of a container or vessel. With nothing to obstruct the beam, it is reflected from the opposite wall. When the level of the liquid or other material in the container reaches the beam, the liquid or material acts as a reflector, thus reducing the reflection time and indicating that a given level has been reached.

ultrasonic light diffraction—The formation of optical diffraction spectra when a beam of light is passed through a longitudinal sound-wave field. The diffraction results from the periodic variation of the light refraction in the sound field.

ultrasonic light modulator—A device containing a fluid that, by action of ultrasonic waves passing through the fluid, modulates a beam of light passed transversely through the fluid.

ultrasonic material dispersion—The production of suspensions or emulsions of one material in another by the action of high-intensity ultrasonic waves.

ultrasonic motion detector—A sensor that detects the motion of an intruder through the use of ultrasonic generating and receiving equipment. The device operates by filling a space with a pattern of ultrasonic waves; the modulation of these waves by a moving object is detected and initiates an alarm system.

ultrasonic plating—The chemical or electrochemical deposition and bonding of one or more solid materials to the surface of another material by the use of vibrational wave energy.

ultrasonic probe—A rod for directing ultrasonic force, used in a disintegration or foreign-body location application.

ultrasonic rejection—The level of rejection of the 19-kHz pilot tone and 38-kHz voltage-controlled oscillator frequency in a stereo FM receiver. The intrinsic rejection of a stereo decoder is the logarithmic ratio of the level of 19-kHz and 38-kHz reference tones with only the standard deemphasis filter at the decoder outputs.

ultrasonics—The general subject of sound in the frequency range above 15 kilohertz.

ultrasonic sealing—A film sealing method based on the application of vibratory mechanical pressure at ultrasonic frequencies (20 to 40 kHz). Electrical energy is converted into ultrasonic vibrations by a magnetostrictive or piezoelectric transducer. The vibratory pressures at the film interface in the sealing area produce localized heat losses that cause the plastic surfaces to melt, thereby forming the seal.

ultrasonic soldering—1. A method of forming a nonporous, continuously metallic connection between metal or alloy parts without necessarily employing chemicals or mechanical abrasives. Instead, vibrational wave energy, heat, and a separate alloy or metal having a melting point below 800°F (427°C) and also below that of the metals or alloys being joined are used. 2. Soldering in which the surface oxide coating of the base metal is removed by ultrasonic vibration.

ultrasonic space grating—Also called grating. A periodic spatial variation in the index of refraction caused by the presence of acoustic waves within the medium.

ultrasonic storage cell—*See* ultrasonic delay line.

ultrasonic stroboscope—A light interrupter in which the light beam is modulated by an ultrasonic field.

ultrasonic therapy—The use of ultrasonic vibrations for therapeutic purposes.

ultrasonic thickness gage—A thickness gage in which the propagation time of an ultrasonic beam through a sheet of material is translated into a measure of the thickness of the material.

ultrasonic transducer—A device that takes the electrical oscillations produced by the ultrasonic generators and transforms them into mechanical oscillations. Typical transducer materials are piezoelectric (e.g., quartz or barium titanate) or magnetostrictive (e.g., nickel).

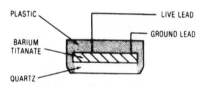

Ultrasonic transducer.

ultrasonic waves—Waves having a frequency in the ultrasonic range.

ultrasonic welding—1. A process that joins two pieces of metal by a form of diffusion bonding. The metals to be welded are clamped between a rigid anvil and a probe, which is vibrated at ultrasonic frequency. The vibration removes any surface oxide film by a simple mechanical scrubbing action, thus exposing the metal surfaces. Then plastic deformation caused by the imposed mechanical clamping load causes the atom movement necessary to join the two crystal lattices, creating a strong, bonded, monolithic structure at the joint line. Some heat is generated by friction caused by the rubbing of the one surface upon the other. This heat undoubtedly aids the diffusion mechanism that occurs but is obviously insufficient to cause welding by itself. 2. Method of fusing two plastic parts by ultrasonic vibrations that are produced by mechanical motion of a converter, expanding and contracting some 20,000 times per second. The vibratory energy is channeled through a horn and applied to thermoplastic materials. This creates the frictional heat to produce a molecular interaction and weld materials.

ultrasonic wire bonder—Equipment unit that fastens fine wire onto a substrate by use of ultrasonic energy.

ultrasonography—A medical diagnostic technique in which pulses of ultrasonic energy are directed into the body, and returning echoes are detected.

ultraviolet—1. Pertaining to electromagnetic radiations at wavelengths beyond the violet end of the spectrum of visible radiation. Because of the shorter wavelengths (200 to 4000 angstrom units), the photons of ultraviolet light have enough energy to initiate most chemical reactions and to degrade some plastics. 2. The invisible region of the spectrum immediately beyond the violet end and between the wavelengths of approximately 1000 to 3800 angstroms.

ultraviolet erasable PROM — *See* UV erasable PROM.

ultraviolet lamp — 1. A lamp providing a high proportion of ultraviolet radiation (e.g., arc lamps, mercury-vapor lamps, or incandescent lamps in bulbs of a special glass that is transparent to ultraviolet rays). 2. A type of lamp that emits a great quantity of ultraviolet radiation. This may be an arc lamp encased in a bulb of a glass that is transparent to ultraviolet rays.

ultraviolet rays — Radiation in the ultraviolet region.

umbilical cable — A lifeline cable used for the main power supply to a missile in order to launch it. It is attached by means of a connector, which detaches as the missile becomes airborne. (Usually seen as a cable waving like a snake alongside the missile as it moves off the launching pad.)

umbilical connector — A device for connecting cables to a rocket or missile prior to launch. It is removed (unmated) from the missile at the time of launching.

umbrella antenna — An antenna in which the wires are guyed downward in all directions from a central pole or tower to the ground, somewhat like the ribs of an open umbrella.

unamplified back bias — A degenerative voltage developed across a fast time-constant circuit within an amplifier stage itself.

unbalanced — 1. Lacking the conditions for balance. 2. Frequently, a circuit having one side grounded. 3. Differential mutual impedance or mutual admittance between two circuits that ideally would have no coupling.

unbalanced circuit — A circuit whose two sides are electrically unlike.

unbalanced line — A transmission line in which the voltages on the two conductors are not equal with respect to ground (e.g., a coaxial line).

unbalanced output — An output in which one of the two output terminals is substantially at ground potential.

unbalanced wire circuit — A circuit whose two sides are electrically unlike.

unblanking — The turning on of the CRT beam.

unblanking generator — A circuit for producing pulses that turn on the beam of a cathode-ray tube.

unblanking pulse — A pulse that turns on the beam of a cathode-ray tube.

unblocked record — A record contained in a file in which each block contains only one record or record segment.

unbonded strain gage — A pressure-sensing element made up of resistance strain-gage wire elements arranged in a Wheatstone bridge. It can have two or four active arms, which respond to a pressure applied to the transducer. The unbonded strain-gage wires are suspended in air and are activated by a mechanism attached to a diaphragm or other pressure-responding element.

unbundling — Pricing certain types of software and services separately from the hardware.

uncertainty — A number or numbers assigned to a measurement as an assessment of all the errors associated with the process producing the measurement. *See also* accuracy.

uncharged — Having a normal number of electrons and, hence, no electrical charge.

unconditional — In a computer, not subject to conditions external to the specific instruction.

unconditional jump — A computer instruction that interrupts the normal process of obtaining the instructions

in an ordered sequence and specifies the address from which the next instruction must be taken.

unconditional transfer of control — In a digital computer that obtains its instructions serially from an ordered sequence of addresses, an instruction that causes the following instruction to be taken from an address that becomes the first of a new sequence.

uncontrolled terminal — A user terminal that is online all the time and does not contain line-control logic for polling and calling.

undamped natural frequency — The frequency at which a system with a single degree of freedom will oscillate, in the absence of damping, upon momentary displacement from the rest position by a transient force.

undamped oscillations — Oscillations that have a constant amplitude for their duration.

undamped wave — A wave whose amplitude does not change.

undefined record — A record contained in a file in which the records have not been defined as being fixed-length records or variable-length records.

underbunching — The condition whereby the buncher voltage of a velocity-modulation tube is lower than the value required for optimum bunching of the electrons.

undercompounded — A generator in which the output voltage drops as the load is increased.

undercurrent relay — A relay that functions when its coil current falls below a predetermined value.

undercut — In a printed circuit board, the reduction of the cross section of a metal-foil conductor due to the removal of metal from beneath the edge of the resist by the etchant.

undercutting — A cutting with too shallow a groove or with insufficient lateral movement of the stylus during sound disc recordings.

underdamped — A degree of damping that is not sufficient to prevent oscillation in the output of a system following application of an abrupt stimulus.

underdamping — 1. In a system, the condition whereby the amount of damping is so small that the system executes one or more oscillations when subjected to a single disturbance (either constant or instantaneous). 2. Oscillation of the transducer output about a final steady value in response to a step change in the measurand. After an initial overshoot, the oscillation amplitude decreases. 3. *See* periodic damping.

underflow — 1. In a computer, the generation of a quantity smaller than the accepted minimum (e.g., floating-point underflow). 2. Pertaining to the condition that arises when a machine computation yields a nonzero result that is smaller than the smallest nonzero quantity that the intended unit of storage is capable of storing. (Contrast with overflow.) 3. When a calculator's capacity is exceeded, some of the least significant digits are discarded and the resulting display is sometimes zero.

underglaze — A glass or ceramic glaze applied to a substrate prior to the screening and firing of a resistor.

underground cable — A cable installed below the surface of the earth.

under insulation — The insulation under wire that is brought from the center of a coil over the top or bottom wall.

underlap — Recorded elemental areas that are smaller than normal — specifically, the space between the recorded elemental area in one recording line of a facsimile system and the adjacent elemental area in the next recording line, or the elemental areas in the direction of the recording line.

underload relay — A relay that operates when the load in a circuit drops below a certain value.

undermodulation—Insufficient modulation of a transmitter due to misadjustment or to insufficient modulation signal.

underpass—A semiconductor component that permits two conductors to cross each other without a short circuit between them. Generally, it is in the form of a low-value resistor covered by a silicon dioxide layer that isolates the top conductor; the resistor is part of the bottom conductor.

underpower relay—A relay that functions when the power decreases below a predetermined value.

underscan—Reducing the height and width of the video picture so that the edges, and, thus, portions of blanking, can be observed.

undershoot—1. The initial transient response to a unidirectional change in input, which precedes the main transition and is opposite in sense. *See also* precursor. 2. The crossing of the base line in the direction opposite to that of the principal pulse, but with insufficient amplitude to be considered a bipolar pulse.

underthrow distortion—Distortion resulting when the maximum amplitude of the signal wavefront is less than the steady-state amplitude that would be attained by a prolonged signal wave.

undervoltage protection—Also called low-voltage protection. The effect of a device to cause and maintain the interruption of power to the main circuit upon the reduction or failure of voltage.

undervoltage relay—A relay that operates when its coil voltage falls below a predetermined value.

underwater sound projector—An electroacoustic transducer designed to convert electric waves into sound waves, which are radiated in water for reception at a distance.

Underwriters Laboratories, Inc.—Abbreviated UL. 1. An independent, nonprofit product-safety testing and certification organization that operates laboratories for the examination and testing of devices, systems, and materials to ensure that they meet standards for safety. Acceptance is usually indicated by tags or by labels on devices showing the words "UL Approved" or "UL Listed." 2. A corporation supported by some underwriters for the purpose of establishing safety standards on types of equipment and components.

undistorted wave—A periodic wave in which both the attenuation and the velocity of propagation are the same for all sinusoidal components, and in which the same sinusoidal component is present at all points.

undisturbed-one output—A 1 output of a magnetic cell to which no partial-read pulses have been applied since that cell was last selected for writing.

undisturbed-zero output—A 0 output of a magnetic cell to which no partial-write pulses have been applied since that cell was last selected for reading.

unfired tube—The condition of tr, atr, and pre-tr tubes when there is no radio-frequency glow discharge at either the resonant gap or the resonant window.

unfurlable antenna—A device that can be unfolded to form a larger antenna.

ungrounded—Not intentionally connected to ground except through high-impedance devices.

ungrounded system—A system in which no point is directly connected to earth except through potential or ground-detecting transformers or other very high impedance devices.

uniaxial magnetic anisotropy—A property of magnetic thin film in which the direction of magnetization is always parallel to the easy axis unless an external force acts upon it.

uniconductor waveguide—A waveguide consisting of a rectangular or cylindrical metallic surface surrounding a uniform dielectric medium.

unidirectional—Flowing in only one direction (e.g., direct current).

unidirectional antenna—An antenna with a single, well-defined direction of maximum gain.

unidirectional bus—A bus used by any individual device for one-way transmission of messages only, that is, either input only or output only.

unidirectional coupler—A directional coupler that samples only one direction of transmission.

unidirectional current—A direct current; i.e., one that is always positive or always negative—never alternating.

unidirectional log-periodic antenna—A broadband antenna in which the cut-out portions of a log-periodic antenna are placed at an angle to each other to produce a unidirectional radiation pattern whose major lobe is in the backward direction, off the apex of the antenna. The impedance and the radiation pattern are essentially constant for all frequencies.

unidirectional microphone—A microphone that is most sensitive to sounds arriving at it from one direction.

unidirectional pulses—Single-polarity pulses that all rise in the same direction.

unidirectional pulse train—A pulse train in which all pulses rise in the same direction.

unidirectional transducer—A transducer that responds to stimuli in only one direction from a reference zero or rest position.

uniform corrosion—A form of corrosion that results from shifting anodic and cathodic areas evening out metallic thinning. In more complex cases, erosion constantly deprives the metal of its protective coating (e.g., oxides), exposing it to continuing corrosion.

uniform field—A field in which the scalar (or vector) has the same value at every point in the region under consideration at that instant.

uniformity—In terms of magnetic tape properties, a figure of merit relating to the ability of the tape to deliver a steady and consistent output level upon being recorded with a constant input. Usually expressed in decibel variation from average at a midrange frequency.

uniform line—A line with substantially identical electrical properties throughout its length.

uniform plane wave—A plane wave with constant-amplitude electric and magnetic field vectors over the equiphase surfaces. Such a wave can only be found in free space, at an infinite distance from the source.

uniform precession—The condition in which the magnetic moments of all atoms in a sample are parallel and precess in phase about the magnetic field. Uniform precession occurs in the regions where the magnetic field is uniform. A spin wave is a phase distortion of this condition.

uniform waveguide—A waveguide whose physical and electrical characteristics do not change with distance along its axis.

uniground—Also called single-point ground. A single point in an electrical system connected to ground to eliminate noise currents.

unijunction transistor—Abbreviated UJT. 1. Formerly called a double-base diode. A three-terminal semiconductor device that exhibits a stable negative-resistance characteristic between two of its terminals. It is this negative-resistance feature that makes the UJT suitable for the applications with which it is associated—thyristor trigger circuits, oscillator circuits, timing circuits, bistable

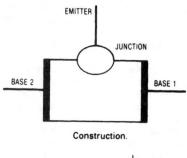

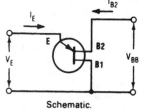

Unijunction transistor.

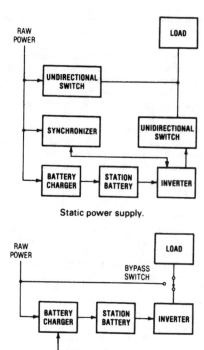

Static power supply.

Buffered power supply.

Uninterruptible power system.

circuits, etc. 2. A type of transistor that has easily controllable on-off characteristics and excellent voltage-sensing characteristics. 3. A three-terminal semiconductor having only one pn junction and exhibiting a stable open-circuit negative-resistance property.

unilateral area track — A sound track in which only one edge of the opaque area is modulated in accordance with the recorded signal. However, there may be a second edge modulated by a noise-reduction device.

unilateral bearing — A bearing obtained with a radio direction finder having unilateral response, eliminating the chance of a 180° error.

unilateral conductivity — Conductivity in only one direction (e.g., in a perfect rectifier).

unilateral element — A two-terminal element with a zero voltage-to-current characteristic (or the equivalent) on one side of the origin.

unilateralization — A special case of neutralization in which the feedback parameters are completely balanced out. In transistors, these feedback parameters include a resistive in addition to a capacitive component. Unilateralization changes a network from bilateral to unilateral.

unilateral network — 1. A network in which any driving force applied at one pair of terminals produces a response at a second pair, but yields no response when the driving force is applied in the other direction. 2. A network that does not pass currents and signals equally well in both directions. For example, a network containing a rectifying element.

unilateral switch — A semiconductor device similar to a miniature SCR. It switches at a fixed voltage that depends on its internal construction.

unilateral transducer — *See* unidirectional transducer.

uninterruptible power system — A solid-state power conversion system to provide regulated ac power to critical loads. Such a system provides uninterrupted power even during brownouts and blackouts.

unipolar — 1. Having but one pole, polarity, or direction. With respect to amplifier power supplies, having an output that varies in only one polarity from zero and, therefore, always contains a dc component. 2. *See* neutral transmission. 3. Refers to transistors in which the working current flows through only one type of semiconductor

material, either n-type or p-type. In unipolar transistors, the working current consists of either positive or negative electrical charges, but never both. All MOS IC transistors are unipolar. Unipolar (MOS) IC transistors operate slower than bipolar IC transistors, but take up much less space on a chip and are much more economical to manufacture.

unipolar field-effect transistor — A field-controlled majority-carrier device wherein the conductance of a semiconductor channel is modulated by a transverse electric field. The field is controlled by the combination of gate bias V_{gs} and the net voltage V_{ds} between channel drain and source.

unipolar pulse — A pulse that has appreciable amplitude in only one direction.

unipolar transistor — 1. A transistor in which charge carriers are of only one polarity. *See also* field-effect transistor. 2. Transistor formed from a single type of semiconductor material, either n-channel or p-channel, as employed in field-effect transistors.

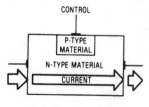

Unipolar transistor.

unipole — 1. An all-pass filter section with one pole and one zero. 2. A hypothetical antenna that radiates and receives equally in all directions. *See also* isotropic antenna.

unipotential cathode — *See* indirectly heated cathode.

unipotential electrostatic lens — A simple electrostatic lens having a focus that is controlled by a single potential difference.

unit — 1. A computer portion or subassembly that constitutes the means of accomplishing some inclusive operation or function (e.g., an arithmetic unit). 2. The specific magnitude of a quantity set apart by appropriate definition and serving as a basis for the comparison or measurement of like quantities; the lowest standard quantity in any system of measurement. The unit of electrical energy, for example, is the kilowatthour. 3. One of the transceivers covered by a CB station license when more than one transceiver is used. 4. An assembly or device capable of independent operation.

unit-area acoustic impedance — *See* specific acoustic impedance.

unitary code — A code having only one digit, the number of times it is repeated determining the quantity it represents.

unit charge — The electrical charge that will repel a force of 1 dyne on an equal and like charge 1 centimeter away in a vacuum, assuming each charge is concentrated at a point.

United States of America Standards Institute — *See* American National Standards Institute.

unit interval — *See* signal element.

unit length — The basic element of time for determining code speeds in message transmission.

unit load — The electrical load imposed on a driver output by the receiver inputs. The actual values for a unit load are specified by the device manufacturer. The unit load is helpful to a logic system designer in determining bus drive requirements.

unit magnetic pole — A pole with a strength such that when it is placed 1 centimeter away from a like pole, the force between the two is 1 dyne.

unitor — In computers, a device or circuit that performs a function corresponding to the Boolean operation of union. *See also* OR gate.

unit pulse — *See* baud.

unit record equipment — Equipment using punched cards as input data, such as collators, tabulating machines, etc.

unit sequence starting relay — A device that functions to start the next available unit in a multiple-unit equipment on the failure or on the nonavailability of the normally preceding unit.

unit sequence switch — A switch used to change the sequence in which units may be placed in and out of service in multiple-unit equipment.

unit step current (or voltage) — A current (or voltage) that undergoes an instantaneous change in magnitude from one constant level to another.

unit substation transformer — A transformer that is mechanically and electrically connected to and coordinated in design with one or more switch-gear or motor-controlled assemblies or combinations thereof.

unit torque gradient — The torque gradient of a synchro, measured when the synchro is electrically connected to another synchro of the same size.

unit under test — Abbreviated UUT. Any system, set, subsystem, assembly, or subassembly undergoing testing.

unitunnel diode — 1. A diode similar to a tunnel diode, but specially treated to give peak reverse currents in the microampere region while providing high forward conductance at low voltage levels. 2. A tunnel diode whose peak and valley point currents are approximately equal.

unity coupling — Perfect magnetic coupling between two coils, so that all the magnetic flux produced by the primary winding passes through the entire secondary winding.

unity gain — An amplifier or active circuit in which the output level is the same as the input level has unity gain.

unity-gain bandwidth — The frequency at which the open-loop gain reaches unity, based on a 6-dB-per-octave crossing. It is a measure of the gain-frequency product of an amplifier.

unity-gain crossover frequency — The frequency at which the curve of open-loop voltage gain of an amplifier crosses through unity gain, or zero decibels.

unity power factor — A power factor of 1.0. It is obtained only when current and voltage are in phase (e.g., in a circuit containing only resistance, or in a reactive circuit at resonance).

universal asynchronous receiver/transmitter — Abbreviated UART. 1. A device that will interface a word-parallel controller or data terminal to a bit-serial communication network. 2. A serial-to-parallel and parallel-to-serial converter. 3. An integrated circuit designed to handle serial/parallel/serial conversion and transmission of data. 4. Commonly used LSI circuit that serves as a universal, single-package, TTL-compatible, full-duplex, serial communication line controller and data interface. 5. A sophisticated integrated circuit that accepts serial data and retransmits it as parallel data and vice versa. 6. A two-way, serial-to-parallel (and reverse) converter chip. UARTs are either dedicated or programmable; the latter permits selection of variables such as type of parity and word length. UART clock frequency is generally 16 times the baud rate. 7. A logic circuit that converts parallel information to an asynchronous serial format, and serial information to a parallel format. Useful for connecting processors having parallel data buses to serial I/O lines. 8. A logic circuit that can connect a parallel I/O bus to either an asynchronous or a synchronous serial I/O line.

universal motor — 1. A series-wound motor designed to operate at approximately the same speed and output on direct current or on a single-phase alternating current of not more than 60 hertz and approximately the same rms voltage. 2. A motor with wound field and armature, with a commutator to make dc operation possible. When all windings are connected in series, it will operate on ac as well as dc, although the higher the ac frequency the more carefully the iron must be laminated to prevent excess eddy-current heating loss.

universal output transformer — An output transformer having a number of taps on its winding. By proper choice of connections, it can be used between the audio-frequency output stage and the speaker of practically any radio receiver or audio amplifier.

universal pattern — Circuit board pattern accommodating standard package configurations, such as DIPs.

universal product code — *See* UPC.

universal receiver — Also called an ac/dc receiver. A receiver with no power transformer and, thus, capable of operating from either ac or dc power lines without changes in its internal connections.

universal shunt — *See* Ayrton shunt.

universal time — Abbreviated UT. Also called Greenwich mean time and Greenwich civil time. A standard based on the rotation of the earth on its axis, with reference to the position of the sun.

universe — *See* population.

UNIX — A complex and powerful multiuser computer operating system written in the C language originally developed, marketed, and trademarked by AT&T. It needs a computer with a large amount of RAM (random-access memory or storage capacity). UNIX allows a computer to handle multiple users and programs simultaneously and has TCP/IP built-in. It is the most common operating system for servers on the Internet. It also allows software to be moved (known as porting) to computers of different sizes or types. UNIX is available in several related versions.

unload — In a computer: 1. To remove the tape from the columns of a recorder by raising or lowering the recording head. 2. To remove a portion of the address part of an instruction. 3. *See also* dump.

unloaded antenna — An antenna with no added inductance or capacitance.

unloaded applicator impedance (dielectric heaters) — The complex impedance measured at the point of application and at a specified frequency without the load material in position.

unloaded line — A line with no loading coils.

unloaded Q (switching tubes) — Also called the intrinsic Q. The Q of a tube unloaded by either the generator or termination.

unloading amplifier — An amplifier capable of reproducing or amplifying a given voltage signal while drawing negligible current from the voltage source.

unloading circuit — In an analog computer, a computing element or combination of computing elements capable of reproducing or amplifying a given voltage signal while drawing negligible current from the voltage source, thus decreasing the loading errors.

unmodulated — Having no modulation; e.g., a carrier that is transmitted during moments of silence in radio programs, or a silent groove in a disc recording.

unmodulated groove — Also called a blank groove. In mechanical recording, the groove made in the medium with no signal applied to the emitter.

unoriented — A structure in which the crystallographic axes of the grains of a metal are not aligned to give directional magnetic properties.

unpack — In a computer, to separate combined items of information, each into a separate machine word.

unsaturated logic — A form of logic containing transistors operated outside the region of saturation; for example, current-mode logic (CML) and emitter-coupled logic (ECL).

unserved energy — The amount of energy not delivered as a result of an equipment outage.

untuned — Not resonant at any of the frequencies being handled.

unusable samples — In random-sampling-oscilloscope technique, those samples not falling within the time window.

unweighted noise — The measured noise level in electronic equipment, with a measuring device that is sensitive to a wide range of frequencies that extend beyond the audible spectrum.

unwind — In a computer, to code all the operations of a cycle, at length and in full, for the express purpose of eliminating all red-tape operations.

UPC — Abbreviation for universal product code. A product identification system designed to assign a unique number to every product in distribution. A 10-digit bar code, with the first 5 digits identifying the manufacturer, the second 5 identifying the item. Each digit is represented by the ratio of the widths of adjacent stripes and white areas. Used with optical checkout scanning devices that retrieve item price from a computer.

upconverter — 1. A device that increases the frequency of a transmitted signal. 2. A type of parametric amplifier that is characterized by the frequency of the output signal being greater than the frequency of the input signal.

update — 1. To search a file (such as a particular record in a computer tape) and select one entry, then perform some operation to bring the entry up-to-date. 2. In a computer, to modify an instruction so that the address numbers in it are increased by a specified amount each time the instruction is executed. 3. Generally applied to computer files in which records are added, deleted, or amended to ensure that the latest information is contained in the file.

update-response time — The interval between the entry of new data into a system and the display of that data.

updating — The act of bringing information up to the current value.

up/down counter — Also called reversible counter. A counter with the capability of counting in an ascending or descending order, depending on the logic present at the up/down inputs.

uplink — 1. An rf link from a site on the earth or from an aircraft to a satellite. 2. The earth-to-geosynchronous satellite microwave link and related components, such as earth station transmitting equipment. The satellite contains an uplink receiver; uplink components in the earth station are involved with the processing and transmission of signals to the satellite. 3. The communications path from the earth to the satellite. 4. The earth station electronics and antenna that transmit information to a communication satellite for relay back to the ground.

upload — 1. The process of transferring communications instructions or data from terminals, including PCs, into a mainframe or host computer system. 2. To send a file from one computer to another via modem or other telecommunication method. *See also* download.

upper operating temperature — The maximum temperature to which a material can be subjected and still maintain specified operating characteristics within limits.

upper sideband — 1. The higher frequency or group of frequencies produced by an amplitude-modulation process. 2. In carrier transmission, the band of frequencies that is higher than the carrier frequency. It is the sum of the instantaneous values of the carrier frequency and the modulating frequency.

upset-duplex system — A direct-current telegraph system in which a station between any two pieces of duplex equipment may transmit signals by opening and closing the line circuit and thereby upsetting the duplex balance.

upset welding — A resistance-welding process wherein the weld is made simultaneously over the entire area of abutting surfaces or progressively along the joint with the aid of rolls or clamps that force the abutting surfaces together. The pressure is applied before heating starts and is maintained throughout the heating period.

up time — 1. The time during which an equipment is either operating or available for operation, as opposed to down time, when no productive work can be accomplished. 2. That element of active time during which an item is either alert, reacting, or performing a mission.

up-time ratio — The quotient of up time divided by up time plus down time.

urea plastic material — A thermosetting plastic material, with good dielectric qualities, used for radio-receiver cabinets, instrument housing, etc.

urgency — The degree to which a process requires attention; determined by the process's priority.

URL — Abbreviation for Uniform Resource Locator. An HTTP address used by the World Wide Web to specify a certain site.

usable samples — In random-sampling-oscilloscope technique, those samples falling within the time window.

USART — Acronym for universal synchronous/asynchronous receiver/transmitter. *See* universal asynchronous receiver/transmitter.

USASCII — Abbreviation for USA Standard Code for Information Interchange. The standard code, using a coded character set consisting of 7-bit coded characters (bits including parity check), for information interchange among data-processing communication systems and associated equipment. The USASCII set consists of control characters and graphic characters. Synonymous with ASCII.

USASCSOCR — The United States of America Standard Character Set for Optical Characters.

USASI — Abbreviation for United States of America Standards Institute, the successor to ASA (American Standards Association).

USB-compatible — Universal Serial Bus Compatible. A new technology that allows connection of many different kinds of peripherals to a PC using "hubs" to allow as many as 127 devices to be connected to a single attachment point. It can handle traffic up to 12 Mb/sec.

U scan — A parallel reading method to prevent ambiguity in the readout of polystrophic codes at the code-position transitions by reading one to two sets of brushes, depending on the state of a control or selector bit.

useful life — The total time a device operates between debugging and wearout.

Usenet — A worldwide system of discussion groups, with comments passed among hundreds of thousands of machines. Usenet is completely decentralized, with over 10,000 discussion areas, called newsgroups.

user-defined key — A key whose function or program can be changed, so that a command or sequence of commands can be executed with a single keystroke. Same as programmable key and soft-function key. Unlike a special-function key, a user-defined key may have a predefined purpose.

user-friendly — Term used to describe computer hardware or software that is easy to use, by virtue of its design and the facilities that are offered to the user.

user interface — The collection of screen formats, editing tools, commands, and software tools by which a user interacts with a computer.

user-to-user service — A switching method that permits a direct user-to-user connection that does not include provision for message store-and-forward service.

utilities — 1. Programs used to perform a routine task. 2. Standard routines of often-used functions, usually supplied as part of system software.

utility — A (usually small) application or computer subroutine designed with a very particular task in mind, like converting between two formats.

utility program — 1. A program providing basic conveniences, such as capability for loading and saving programs, for observing and changing values in a computer, and for initiating program execution. The utility program eliminates the need to rewrite a program every time a designer wants to perform a common function. 2. A computer program made available by the operating system to save programmers the bother of writing their own programs to do often-needed tasks.

utility routine — A standard routine, usually part of a larger software package, that performs a service and/or program maintenance function, such as file maintenance, file storage and retrieval, media conversions, and production of memory and file printouts.

utilization factor — In electrical power distribution, the ratio of the maximum demand of a system (or part of a system) to the rated capacity of the system (or part) under consideration.

Utilogic — A line of digital ICs built around a basic AND and a basic NOR circuit. The AND has multiple emitter inputs; the NOR has emitter-follower inputs. The output for the AND is an emitter follower, and for the NOR, a totem-pole arrangement. A J-K binary element is also included in the line.

UUT — Abbreviation for unit under test.

UV erasable PROM — Abbreviation for ultraviolet erasable PROM. A programmable read-only memory that can be cleared (set to 0) by exposure to intense ultraviolet light. After being cleared, it may be reprogrammed.

UVROM — An ultraviolet light erasable read-only memory. The memory is totally erased when exposed to ultraviolet light for a minimum of 20 minutes.

V

V—1. Letter symbol for volt. 2. Symbol for voltmeter. 3. Schematic symbol for vacuum tube.

V_{CC}—Symbol for the supply voltage to an integrated circuit with respect to ground.

V_{DD}, V_{SS}, V_{CC}, V_{EE}—In an MOS circuit, the designation of the power-supply terminal serving the drain, source, collector, or emitter. The double subscript refers to the power-supply terminal, while a single subscript references the parameter at the element of a device. For example, V_C is the voltage measured on the collector itself, while V_{CC} is the (constant) voltage supplied to the collector circuit. Note: In CMOS, the term V_{DD} has been adopted as a convention referring to the positive power-supply terminal, although it is actually applied to the source of a p-channel transistor.

VA—Letter symbol for voltampere.

vac—Abbreviation for vacuum.

vaccine—A program to detect the presence of a computer virus.

vacuum—Abbreviated vac. Theoretically, an enclosed space from which all air and gases have been removed. However, since such a perfect vacuum is never attained, the term is taken to mean a condition whereby sufficient air has been removed so that any remaining gas will not affect the characteristics beyond an allowable amount.

vacuum capacitor—A capacitor consisting usually of two concentric cylinders enclosed in a vacuum to raise the breakdown voltage.

vacuum deposition—A process in which a substance is heated in a vacuum enclosure until the substance vaporizes and condenses (deposits) on the surface of another material in the enclosure. This process is used in the manufacture of resistors, capacitors, microcircuits, and semiconductor devices. The deposited material is called a thin film.

vacuum envelope—The airtight envelope that contains the electrodes of an electron tube.

vacuum evaporation—A process in which a material is vaporized and the vapor deposits itself, through openings in a mask, onto a substrate to form a thin film.

vacuum gage—A device that indicates the absolute gas pressure in a vacuum system (e.g., in the evacuated parts of a mercury-arc rectifier).

vacuum impregnation—Filling the spaces between electric parts or turns of a coil with an insulating compound while the coil or parts are in a vacuum.

vacuum level—The degree of a vacuum, as determined by the pressure: rough vacuum (760 torr to 1 torr), medium vacuum (1 torr to 10^{-3} torr), high vacuum (10^{-3} torr to 10^{-6} torr), very high (hard) vacuum (10^{-6} torr to 10^{-9} torr), ultrahigh (ultrahard) vacuum (below 10^{-9} torr).

vacuum metalizing—A process in which surfaces are given a thin coating of metal by exposing them to metallic vapor produced by evaporation under vacuum (one millionth of normal atmospheric pressure).

vacuum phototube—1. A phototube that is evacuated to such a degree that its electrical characteristics are essentially unaffected by gaseous ionization. 2. A phototube that functions within a vacuum and therefore eliminates the effect of gaseous ionization on its electrical properties.

vacuum pickup—A handling instrument with a small vacuum cup on one end, used to pick up chip devices.

vacuum range—For a communications system, the maximum range computed for an atmospheric attenuation of zero.

vacuum seal—An airtight junction between component parts of an evacuated system.

vacuum switch—A switch in which the contacts are enclosed in an evacuated bulb, usually to minimize sparking.

vacuum tank—An airtight metal chamber that contains the electrodes and in which the rectifying action takes place in a mercury-arc rectifier.

vacuum tight—*See* hermetic.

vacuum tube—An electron tube evacuated to such a degree that its electrical characteristics are essentially unaffected by the presence of residual gas or vapor.

vacuum-tube amplifier—An amplifier in which electron tubes are used to control the power from the local source.

vacuum-tube characteristics—Data that shows how a vacuum tube will operate under various electrical conditions.

vacuum-tube keying—A code-transmitter keying system in which a vacuum tube is connected in series with the plate-supply lead going to the winding in the plate circuit of the final stage. The grid of the tube is connected to its filament through the transmitting key so that when the key is open, the tube is blocked, interrupting the plate supply to the output stage. Closing the key allows plate current once more through the keying tube and the output tubes.

vacuum-tube modulator—A modulator in which a vacuum tube is the modulating element.

vacuum-tube oscillator—A circuit in which a vacuum tube is used to convert dc power into ac power at the desired frequency.

vacuum-tube rectifier—A tube that changes an alternating current to an unidirectional pulsating direct current.

vacuum-tube transmitter—A radio transmitter in which electron tubes are utilized to convert the applied electric power into radio-frequency power.

vacuum-tube voltmeter — Abbreviated VTVM. *See* electronic voltmeter.

valence — A number representing the proportion in which an atom is able to combine with other atoms. It generally depends on the number and arrangement of electrons in the outermost shell of each type of atom.

valence band — 1. In the spectrum of a solid crystal, the range of energy states containing the energies of the valence electrons that bind the crystal together. In a semiconductor material, it is just below the conduction band, separated from it by the forbidden gap. 2. The band of atomic energy levels containing the valence electrons, i.e., those electrons in the outer shell of an atom. In an insulating or semiconductor material, the valence band energy level is below the conduction band. In a conducting material — for example, copper, aluminum, silver, gold, and lead — the valence-band energy level is above the conduction-band energy, thus allowing the electrons to be more free to move as an electric current.

valence bond — Also called a bond. The bond formed between the electrons of two or more atoms.

valence electrons — 1. The electrons of an atom in the outer shell that determine the chemical valency of the atom. 2. Electrons in the valence band of a semiconductor, where they are free to move under the influence of an electric field.

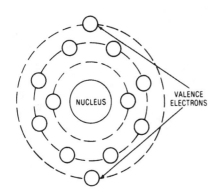

Valence electrons.

valence shell — The electrons that form the outermost shell of an atom.

validate — To ensure correctness of data that has been (previously) or is being entered by any of a number of means, including check digit, batch total, numeric only field, and verification.

validation — The process of verifying that execution of a system in its specified environment causes no operational problems. Includes prevention, diagnosis, recovery, and correction of errors.

validity — Correctness; specifically, how closely repeated approximations approach the desired (i.e., correct) result.

validity check — 1. A check to determine that a code group actually represents a character in the particular code being used. 2. A computer input-data check based on known limits for variables in given fields.

valley — A dip between two peaks in a curve.

valley current — In a tunnel diode, the current measured at the positive voltage for which the current has a minimum value from which it will increase if the voltage is further increased.

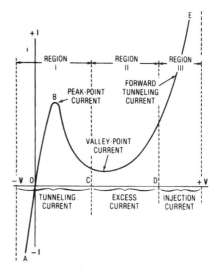

Valley current.

valley point — The point on the characteristic of a tunnel diode corresponding to the lowest voltage greater than the peak point voltage for which the differential conductance is zero.

valley-point current — The current value at the valley point.

valley-point emitter current — The current flowing in the emitter of a unijunction transistor when the device is biased to the valley point.

valley-point voltage — The voltage value at the valley point.

valley voltage — In a tunnel diode, the voltage corresponding to the valley current.

value — 1. The magnitude of a physical quantity. 2. The quantitative measure of a signal or variable.

value analysis — The systematic use of techniques that serve to identify a required function, establish a value for that function, and finally to provide that function at the lowest overall cost. This approach focuses on the functions of an item rather than the methods of producing the present product design.

value theory — The assignment of numerical significance to the worths of alternative choices.

valve — 1. A British term for a vacuum tube. 2. A device permitting current flow in one direction only (e.g., a rectifier). 3. A device or system that is capable of flow diversion, cutoff, or modulation.

valve tube — *See* kenotron, 1.

Van Allen radiation belts — Two doughnut-shaped belts of high-energy particles that surround the earth and are trapped in its magnetic field. They were first discovered by Dr. James A. Van Allen of Iowa State University.

Van Atta antenna — A retrodirective array of antenna elements so interconnected that an incident wave produces a radiated beam from the array in the direction of the incident wave reversed over a significant range of angles of incidence, as in the case of an optical autocollimator. Versions of the array may contain amplifiers, circulators, and mixers, and may direct the transmitted beam, with or without modifications, in a direction or directions other than that of the incident wave reversed.

Van Atta array — An antenna array designed so that the received signal is reflected back toward its source in a narrow beam to provide signal enhancement without amplification. It consists of pairs of corner reflectors or other elements equidistant from the center of the array and connected together by means of low-loss transmission line.

Van de Graaff accelerator — An electrostatic-generator type of particle accelerator from which the voltage is obtained by picking up static electricity at one end of the machine (on a rubber belt) and carrying it to the other end, where it is stored.

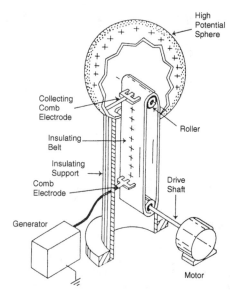

High Potential Sphere

Collecting Comb Electrode

Roller

Insulating Belt

Insulating Support

Comb Electrode

Drive Shaft

Generator

Motor

Van de Graaff generator.

vane-anode magnetron — A cavity magnetron in which the walls between adjacent cavities have parallel plane surfaces.

vane attenuator — A waveguide device designed to preset attenuation in a circuit by sliding a resistive element from the side wall of the waveguide to the center for maximum attenuation. This method of attenuation is used in precision calibrated attenuation readings, and resetting must be made. Some of its countless applications are calibrations of other attenuators, directional couplers, filters, and other lossy components, in antenna-pattern measurements, noise-level measurements, and for setting power levels to desired values.

vane-type instrument — A measuring instrument in which the pointer is moved by the force of repulsion between fixed and movable magnetized iron vanes, or by the force between a coil and a pivoted vane-shaped piece of soft iron.

vane-type magnetron — A cavity magnetron in which the walls between adjacent cavities have plane surfaces.

V-antenna — A V-shaped arrangement of conductors, the two branches being fed equally in opposite phase at the apex.

vapor pressure — The pressure of the vapor accumulated above a confined liquid (e.g., in a mercury-vapor rectifier tube).

vaporware — A jocular term for products or components that have been announced by a vendor that do not yet exist and may never exist.

var — Letter symbol for voltampere reactive. The unit of reactive power, as opposed to real power in watts. One var is equal to one reactive voltampere.

VAR — Abbreviation for visual-aural range.

varactor — Also called varactor diode, silicon capacitor, voltage-controlled capacitor, and voltage-variable capacitor. 1. A two-terminal solid-state device that utilizes the voltage-variable capacitance of a pn junction. In the normal semiconductor diode, efforts are made to minimize inherent capacitance, while in the varactor, this capacitance is emphasized. Since the capacitance varies with the applied voltage, it is possible to amplify, multiply, and switch with this device. 2. Semiconductor diode that exhibits a change in capacitance with a change in applied voltage when operated in a reverse-biased condition. Varactors are used as voltage-variable capacitors in tuned circuits.

varactor diode — A two-terminal semiconductor device in which use is made of the fact that its capacitance varies with the applied voltage. *See also* varactor.

varactor-tuned oscillator — An oscillator in which a varactor diode is used in the frequency-determining networks that encompass the circuit's active device(s).

varhour meter — Also called a reactive voltampere-hour meter. An electricity meter that measures and registers the integral (usually in kilovarhours) of the reactive power of the circuit into which the meter is connected.

variable — 1. Any factor or condition that can be measured, altered, or controlled (e.g., temperature, pressure, flow, liquid level, humidity, weight, chemical composition, color). 2. A quantity that can take on any of a given set of values. 3. In a computer, a symbol whose numeric value changes from one iteration of a program to the next or within each iteration of a program.

variable-area track — A sound track divided laterally into opaque and transparent areas. A sharp line of demarcation between these areas forms an oscillographic trace of the waveshape of the recorded signal.

variable-capacitance diode — Abbreviated VCD. A semiconductor diode in which the junction capacitance present in all semiconductor diodes has been accentuated. An appreciable change in the thickness of the junction-depletion layer and a corresponding change in the capacitance occur when the dc voltage applied to the diode is changed.

variable-capacitance transducer — A transducer that measures a parameter or a change in a parameter by means of a change in capacitance.

variable capacitor — 1. A capacitor that can be changed in capacitance by varying the useful area of its plates, as in a rotary capacitor, or by altering the distance between them, as in some trimmer capacitors. 2. A capacitor whose capacitance can be varied by varying the separation between a pair of plates, or by varying the depth of insertion of interleaved plates. (Most widely used for tuning radio-frequency circuits.)

variable-carrier modulation — *See* controlled-carrier modulation.

variable-compression capacitor — A capacitor in which the capacitance can be varied by compressing a stack of electrode and dielectric layers.

variable concentric capacitor — An air dielectric capacitor in which the capacitance can be varied by the axial movement of a rotor in a stator.

variable connector — 1. A flowchart symbol representing a sequential connection that is not fixed, but which can be varied by the flowcharted procedure itself. 2. The device that inserts instructions in a program corresponding to the selection of paths appearing in a flowchart. 3. The computer instructions that cause a logical chain to take one of several alternative paths.

variable coupling—Inductive coupling that can be varied by moving the windings.

variable-cycle operation—Computer operation in which any cycle is started at the completion of the previous cycle, instead of at specified clock times.

variable-density track—A sound track of constant width and usually, but not necessarily, of uniform light transmission on any instantaneous transverse axis. The average light transmission varies along the longitudinal axis in proportion to some characteristic of the applied signal.

variable-depth sonar—A sensor that can be lowered by cable from a ship or helicopter, through the thermal layers, to detect submarines operating at deeper levels.

variable disc capacitor—A solid-dielectric capacitor in which the capacitance can be varied by rotating a metal or metallized disc.

variable-erase recording—The method of recording on magnetic tape by selective erasure of a prerecorded signal.

variable field—A field in which the scalar (or vector) at any point changes during the time under consideration.

variable-frequency oscillator—Abbreviated VFO. A stable oscillator whose frequency can be adjusted over a given range.

variable-frequency synthesizer—An instrument that translates the stability of a single frequency, usually obtained from a frequency standard, to any one of many other possible frequencies. In common usage, such instruments are called simply frequency synthesizers.

variable inductance—Also called variable inductor. A coil whose inductance can be varied.

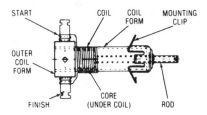

Variable inductance.

variable-inductance pickup—A phonograph pickup in which the movement of a stylus causes the inductance to vary accordingly.

variable-inductance transducer—A transducer in which the output voltage is a function of the change in a variable-inductance element.

variable-length record—In a computer, pertaining to a file in which there is no constraint on the record length. (Opposite of fixed-length record.)

variable monoergic—A type of emission in which the magnitude of the homogeneous particle or radiation energy is continuously variable over broad limits; e.g., the proton energy from some types of accelerators can be controlled by varying the high voltage.

variable-mu pressure transducer—A device that converts mechanical input pressure to a proportional electrical output based on the change of the mu of its magnetic circuit due to the applied pressure (Villari effect).

variable-mu tube—*See* remote-cutoff tube.

variable point—Pertaining to a system of numeration in which the position of the radix point is indicated by a special character at that position.

variable radio-frequency radiosonde—A radiosonde whose carrier frequency is modulated by the magnitude of the meteorological variables being sensed.

variable reluctance—Principle employed in certain phonograph pickups. Deflections of the stylus when playing a record make an armature vibrate between the poles of an electromagnet. The reluctance is the ratio of magnetic force to magnetic flux in a magnetic field. Variations due to stylus movement create variations in the current through the electromagnet.

variable-reluctance microphone—Also called a magnetic microphone. A microphone that depends for its operation on the variations in reluctance of a magnetic circuit.

variable-reluctance pickup—1. A phonograph cartridge that derives its electrical output signal from change effected in a magnetic circuit by means of some mechanical device such as a moving coil or magnet. 2. A type of cartridge that generates its signal from the relative motions of a magnetic field and a coil or coils (either the field or the coils may move, depending on cartridge design). The output is proportional to the velocity of the stylus motion. This requires an equalization circuit in the preamplifier to restore proper frequency balance, since records are cut with more nearly constant-amplitude than constant-velocity characteristics. 3. A phonograph pickup that depends for its operation on the variation of a resistance.

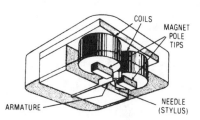

Variable-reluctance pickup.

variable-reluctance stepping motor—A motor with a soft-iron rotor that is made to step by sequential excitation of stator coils. It requires complementary equipment to furnish the sequencing pulses.

variable-reluctance transducer—Also called a magnetic transducer. A transducer that depends for its operation on the variations in reluctance of a magnetic circuit.

variable-resistance transducer—A transducer in which the signal output depends on the change in a resistance element.

variable resistor—A wirewound or composition resistor whose resistance may be changed. *See also* potentiometer, 1; rheostat.

Variable resistors.

variable-speed motor— A motor whose speed can be adjusted within certain limitations, regardless of load.

variable-speed scanning— A scanning method whereby the optical density of the film being scanned determines the speed at which the scanning beam in the cathode-ray tube of a television camera is deflected.

variable transformer— An iron-core transformer with provision for varying its output voltage over a limited range, or continuously from zero to maximum — generally by the movement of a contact arm along exposed turns of the secondary winding.

variable tubular capacitor— A solid-dielectric capacitor in which the capacitance can be varied by the axial movement of an electrode within a tube.

variable vane capacitor— A capacitor in which the capacitance can be varied by rotating the rotor vanes between the stator vanes.

Variac— An autotransformer that contains a toroidal winding and a rotating carbon brush so that the output voltage is continuously adjustable from zero to line voltage plus 17 percent. Trade name of General Radio Company.

variance— A measure of the fluctuations of data around the mean.

variants— 1. Two or more cipher or code groups that have the same plain-language equivalent. 2. Two or more plaintext meanings that are represented by a single code group.

variate— *See* random variable, 1.

variation— The angular difference between a true and a magnetic bearing or heading.

varicap— *See* varactor.

varindor— An inductor whose inductance varies markedly with the current in the winding.

variocoupler— A radio transformer with windings that have an essentially constant self-impedance, but the mutual impedance between them is adjustable.

variolosser— A device whose loss can be controlled by a voltage or current.

variometer— A variable inductor consisting of a pair of series- or parallel-connected coils whose axes can be varied one with respect to the other. The change in mutual inductance causes a change in the total inductance.

varioplex— A telegraph switching system that establishes connections on a circuit-sharing basis between a number of transmitters in one locality and corresponding receivers in another locality over one or more intervening channels. Maximum use of channel capacity is obtained by employing momentary storage of signals and allocating circuit time in rotation among the transmitters that have information in storage.

varistor— From "variable resistor." 1. A two-electrode semiconductor device with a voltage-dependent nonlinear resistance that drops markedly as the applied voltage is increased. 2. A passive resistorlike circuit element whose resistance is a function of the current through it. The current through it is a nonlinear function of the voltage across its terminals; hence, a self-varying resistance. *See also* voltage-dependent resistor. 3. A voltage-dependent symmetrical resistor with a high degree of nonlinearity (the resistance does not change linearly as a function of the applied voltage). The value of the resistance is very high at voltages lower than nominal and very quickly changes to an extremely low value of resistance as the applied voltage is increased above the nominal voltage. The device is used as a voltage-transient or voltage-surge suppressor to improve the reliability of a voltage-sensitive circuit. It provides instantaneous response to high voltage, with good temperature stability and excellent clamping characteristics. The device is normally connected across its input terminals or across the ac power source, providing protection with an effective low resistance somewhat above the normal input or line voltage values. One of the varistor's main uses is to protect electronic equipment against lightning. It is also used to protect against sudden or transient surges of voltage in an ac power line. *See also* MOV.

Varley loop— A type of Wheatstone-bridge circuit that gives, in one measurement, the difference in resistance between two wires of a loop.

varmeter— Also called a reactive voltampere meter. An instrument for measuring reactive power in either vars, kilovars, or megavars. If the scale is graduated in kilovars or megavars, the instrument is sometimes designated a kilovarmeter or megavarmeter.

varnished cambric— A linen or cotton fabric that has been impregnated with varnish or insulating oil and baked. It is used as insulation in coils and other radio parts.

varying duty— A requirement of service that demands operation at loads and for intervals of time, both of which may be subject to wide variation.

varying-speed motor— A motor that slows down as the load increases (e.g., a series motor, or an induction motor with a large amount of slip).

varying voltage control— A form of armature-voltage control obtained by impressing, on the armature, a voltage that varies considerably with a change in load and consequently changes the speed of the motor (e.g., by using a differentially compound-wound generator or a resistance in the armature circuit).

V-beam system— A radar system for measuring elevation. The antenna emits two fan-shaped beams, one vertical and the other inclined, which intersect at ground level. Each beam rotates continuously about a vertical axis, and the time elapsing between the two echoes from the target provides a measure of its elevation.

VCD— Abbreviation for variable-capacitance diode.

v-chip— A semiconductor that blocks the display of violent and/or pornographic material on a TV. Transmitters have to incorporate a signal for the system to function.

VCO— Abbreviation for voltage-controlled oscillator.

VCR— 1. Abbreviation for voltage-controlled resistor. 2. Abbreviation for videocassette recorder (or player).

V-cut— A type of oscillator-crystal cut in which the major plane surfaces are not parallel to the X, Y, or Z planes.

VDR— *See* voltage-dependent resistor.

VDT— Abbreviation for video display terminal. A CRT or gas-plasma tube display screen terminal or keyboard console that allows keyed or stored text to be viewed for manipulation or editing.

VDU— 1. *See* video display unit. 2. Abbreviation for visual display unit. A display of data for use by human operators. Data is usually displayed as several rows of alphanumeric characters.

vector— A quantity that has both magnitude and direction. Vectors commonly are represented by a line segment with a length that represents the magnitude and an orientation in space that represents the direction. *See* vector quantity.

vector admittance— The ratio for a single sinusoidal current and potential difference in a portion of a circuit of the corresponding complex harmonic current to the corresponding complex potential difference.

vector-ampere— The unit of measurement of vector power.

vector cardiograph— An instrument that measures both the magnitude and the direction of heart signals by displaying cardiograph signals on any desired set of axes, usually x, y, and z.

vector diagram — An arrangement of vectors showing the relationships between alternating quantities having the same frequency.

vectored interrupt — 1. An interrupt in a computer that carries its identity number or the address of its handler. 2. An interrupt scheme in which each interrupting device causes the operating system to branch to a different interrupt routine. This scheme is useful for a very fast interrupt response.

vector field — In a given region of space, the total value of some vector quantity that has a definite value at each point of the region (e.g., the distribution of magnetic intensity in a region surrounding a current-carrying conductor).

vector function — A function that has both magnitude and direction (e.g., the magnetic intensity at a point near an electric circuit is a vector function of the current in that circuit).

vector generator — That part of the display controller which draws vectors on the screen. Control codes within the display list specify whether the vector generators will move the beam in a blanked or unblanked mode. If the beam is unblanked, the vector will be drawn in a specified texture.

vector impedance — The ratio for a simple sinusoidal current and potential difference in a portion of a circuit of the corresponding complex harmonic potential difference to the corresponding complex current.

vector interrupt — A term used to describe a microprocessor system in which each interrupt, both internal and external, has its own uniquely recognizable address. This enables the microprocessor to perform a set of specified operations that are preprogrammed by the user to handle each interrupt in a distinctively different manner.

vector power — A vector quantity equal to the square root of the sum of the squares of the active and reactive powers. The unit is the vector-ampere.

vector power factor — Ratio of the active power to the vector power. In sinusoidal quantities, it is the same as power factor.

vector processing — A method for carrying out many repetitive mathematical operations with a single computer instruction.

vector quantity — A quantity that has both magnitude and direction. Examples of quantities that are vectors are displacement, velocity, force, and magnetic intensity.

vector refresh display — *See* calligraphic display.

vector scan — Technique of displaying images on a screen, particularly suitable for precision drawings and animation. It is available only on special visual display units. Images are created by moving an electron beam in the CRT to any place on the screen, just as a pen is moved in any direction on paper. The positions are specified as x and y coordinates.

vectorscope — 1. An oscilloscope with a circular time base of extreme stability (determined by the frequency of the color subcarrier). The instrument can be used to check the time delay between two signals because the phase difference at a particular frequency can be related to time difference. 2. Special synchroscope used in color TV camera and color encoder calibration. The vectorscope will graphically indicate on a CRT the absolute angles the different color signals describe in respect to a reference and to each other. These angles as read on the vectorscope represent the phase differences of the signals.

vector stroke display — *See* calligraphic display.

velocimeter — A continuous-wave reflection Doppler system used to measure the radial velocity of an object.

velocity — A vector quantity that includes both magnitude (e.g., speed) and direction in relation to a given frame of reference.

velocity error — The amount of angular displacement existing between the input and output shafts of a servomechanism when both are turning at the same speed.

velocity filter — A storage-tube device that blanks all targets that do not move more than one resolution cell in less than a predetermined number of antenna scans.

velocity hydrophone — A type of hydrophone in which the electric output is substantially proportional to the instantaneous particle velocity in the incident sound wave.

velocity-lag error — A lag between the input and output of a device, proportional to the rate of variation of the input.

velocity level — In decibels of a sound, 20 times the logarithm to the base 10 of the ratio of the particle velocity of the sound to the reference particle velocity. The latter must be stated explicitly.

velocity microphone — A microphone in which the electric output corresponds substantially to the instantaneous particle velocity in the impressed sound wave. It is a gradient microphone of order one, and is inherently bidirectional.

velocity-modulated amplifier — Also called a velocity-variation amplifier. An amplifier in which velocity modulation is employed for amplifying radio frequencies.

velocity-modulated oscillator — Also called a velocity-variation oscillator. An electron-tube structure in which the velocity of an electron stream is varied (velocity-modulated) in passing through a resonant cavity called a buncher. Energy is extracted from the bunched electron stream at a higher level in passing through a second cavity resonator called the catcher. Oscillations are sustained by coupling energy from the catcher cavity back to the buncher cavity.

velocity modulation — Also called velocity variation. 1. Modification of the velocity of an electron stream by the alternate acceleration and deceleration of the electrons with a period comparable to that of the transit time in the space concerned. 2. Modulation of an electron stream by alternately accelerating and decelerating the electrons, thus grouping them into bunches.

velocity of light — A physical constant equal to 2.99796×10^{10} centimeters per second. (More conveniently expressed as 186,280 statute miles per second, or 161,750 nautical miles per second, or 328 yards per microsecond.)

velocity of propagation — 1. The speed at which a disturbance (e.g., sound, radio, light waves) is radiated through a medium. 2. The ratio of the speed of the flow of an electric current in an insulated cable to the speed of light, expressed in percentage. In the case of coaxial cables, this ratio is 65 to 66 percent, where the insulation is polyethylene. 3. The speed of transmission of electrical energy through a cable compared with its speed through air, expressed as a percentage of the speed in free space. Radio-frequency signals are transmitted in free space at a speed of 186,280 statute miles per second (2.99796×10^8 m/s). Velocity of propagation is determined by the ratio of the dielectric constant of air to the square root of the dielectric constant of the insulation and is expressed in percent. 4. The speed with which a signal wave travels through a particular transmission.

velocity pickup — A magnetic pickup whose output increase is a function of recorded velocity. (This differs from the ceramic pickup, whose output is a function of amplitude or deflection of the stylus.)

velocity resonance — *See* phase resonance.

velocity sorting — The selecting of electrons according to their velocity.

velocity spectrograph — An apparatus for separating an emission of electrically charged particles into distinct streams in accordance with their speed, by means of magnetic or electric deflection.

velocity transducer — A transducer that generates an output proportionate to the imparted velocities.

velocity variation — *See* velocity modulation.

velocity-variation amplifier — *See* velocity-modulated amplifier.

velocity-variation oscillator — *See* velocity-modulated oscillator.

Venn diagrams — Diagrams in which circles or ellipses are used to give a graphic representation of basic logic relations. Logic relations between classes, operations on classes, and the terms of the propositions are illustrated and defined by the inclusion, exclusion, or intersection of these figures. Shading indicates empty areas, crosses indicate areas that are not empty, and blank spaces indicate areas that may be either. Named for English logician John Venn, who devised them.

vent — A controlled weakness somewhere in the enclosure of an aluminum electrolytic capacitor to permit pressure relief in case of failure due to shorting or improper installation of the capacitor.

vented baffle — An enclosure designed to properly couple a speaker to the air.

ventilated transformer — A dry-type transformer that is so constructed that the ambient air may circulate through its enclosure to cool the transformer core and windings.

venturi tube — A short tube with flaring ends and a constricted throat. It is used for measuring flow velocity by measurement of the throat pressure, which decreases as the velocity increases.

verification — The process of checking the results of one data transcription against those of another, both transcriptions usually involving manual operations. *See also* check.

verifier — 1. Of computers, a device on which a record can be compared or tested for identity character by character with a retranscription or copy as it is being prepared. 2. A machine that is used to check the correctness of manually recorded data.

verify — 1. To check, usually with an automatic machine, one recording of data against another in order to minimize the number of human errors in the data transcription. 2. To make certain that the information being prepared for a computer is correct.

vernier — 1. An auxiliary scale comprising subdivisions of the main measuring scale and, thus, permitting more accurate measurements than are possible from the main scale alone. 2. An auxiliary device used for obtaining fine adjustments.

vernier capacitor — A variable capacitor placed in parallel with a larger tuning capacitor and used to provide a finer adjustment after the larger one has been set to the approximate desired position.

vernier dial — A type of tuning dial used chiefly for radio equipment. Each complete rotation of its control knob moves the main shaft only a fraction of a revolution and thereby permits fine adjustment.

vernitel — A precision device that makes possible the transmission of data with high accuracy over standard frequency-modulated telemetering systems.

vertex — *See* node, 1.

vertex plate — A matching plate placed at the vertex of a reflector.

vertical — In 45/45 recording, the signal produced by a sound arriving at the two microphones simultaneously and 180° out of phase, causing the cutting stylus to move vertically.

vertical amplification — Signal gain in the circuits of an oscilloscope that produces vertical deflection on the screen.

vertical-amplitude controls — *See* parabola controls.

vertical antenna — A vertical metal tower, rod, or suspended wire that is used as a receiving and/or transmitting antenna.

vertical blanking — Blanking of a television picture tube during the vertical retrace.

vertical-blanking interval — The brief time between television fields required for the scanning electron gun to retrace from the bottom of the image to the top to begin scanning the next field. The interval occupies 16,667 milliseconds and occurs at the end of each 262.5 field scan lines. There are 21 usuable lines in this interval, which are devoted to 6 vertical synch pulses and 12 equalizing pulses, as well as additional horizontal pulse intervals.

vertical-blanking pulse — In television, a pulse transmitted at the end of each field to cut off the cathode-ray beam while it returns to the start of the next field.

vertical-centering control — A control provided in a television receiver or cathode-ray oscilloscope to shift the entire image up or down on the screen.

vertical compliance — The ability of a reproducing stylus to move vertically while in the reproducing position on a record.

vertical-deflection electrodes — The pair of electrodes that move the electron beam up and down on the screen of a cathode-ray tube employing electrostatic deflection.

vertical dynamic convergence — Convergence of the three electron beams at the aperture mask of a color picture tube during the scanning of each point along a vertical line at the center of the tube.

vertical-field-blanking interval — *See* vertical-blanking interval

vertical field-strength diagram — A representation of the field strength at a constant distance from, and in a vertical plane passing through, an antenna.

vertical frame transfer — A CCD configuration in which all charges accumulated during an integration period are rapidly moved out of the optically active area and to an identical optically shielded CCD area where the accumulation is read out at a slower pace during the next integration period.

vertical-frequency response — In an oscilloscope, the band of frequencies passed, with amplification between specified limits, by the amplifiers that produce vertical deflection on the screen.

vertical-hold control — *See* hold control.

vertical-incidence transmission — The transmission of a radio wave vertically to the ionosphere and back. The transmission remains practically the same for a slight departure from the vertical (e.g., when the transmitter and receiver are a few kilometers apart).

vertical-lateral recording — A technique of making stereo phonograph discs by recording one signal laterally, as in monophonic records, and the other vertically, as in hill-and-dale-transcriptions.

vertical linearity control — A control that permits adjustment of the spacing of the horizontal lines on the upper portion of the picture to effect linear vertical reproduction of a television scene.

vertically polarized wave — 1. An electromagnetic wave with a vertical electric vector. 2. A linearly polarized wave with a horizontal field vector.

vertical MOS—*See* VMOS.

vertical polarization—1. Transmission in which the transmitting and receiving antennas are placed in a vertical plane, so that the electrostatic field also varies in a vertical plane. 2. Transmission of radio waves whose undulations vary vertically with respect to the earth.

vertical quarter-wave stub—An antenna with a vertical portion that is electrically one quarter-wavelength long. It is used generally with a ground plane at the base of the stub.

vertical radiator—A transmitting antenna perpendicular to the earth's surface.

vertical recording—Also called hill-and-dale recording. Mechanical recording in which the groove modulation is perpendicular to the surface of the recording medium.

vertical redundance—In a computer, an error condition that exists when a character fails a parity check, i.e., has an even number of bits in an odd-parity system or vice versa.

vertical resolution—1. On a television test pattern, the number of horizontal wedge lines that can be clearly discerned by the eye before they merge together. 2. The number of horizontal lines that can be seen in the reproduced image of a television pattern.

vertical retrace—1. The return of the electron beam from the bottom of the image to the top after each vertical sweep. 2. The return of the electron beam to the top of the picture tube screen or the pickup tube target at the completion of the field scan.

vertical scanning—Scanning that proceeds in a series of vertical lines.

vertical scrolling—The ability of a word processor to move vertically, a line at a time, up and down through a display page or more of text. This allows text that will not fit on a video display screen to be accessed for review or editing. Used in systems that have a display buffer (memory area) larger than the display screen capacity.

vertical speed transducer—An instrument that furnishes an electrical output that is proportionate to the vertical speed of the aircraft or missile in which it is installed.

vertical structured metal-oxide silicon power field-effect transistor—*See* VMOS power FET.

vertical stylus force—*See* stylus force.

vertical sweep—The downward movement of the scanning beam from top to bottom of the televised picture.

vertical sync—The synchronizing pulses used to define the end of one television field and the start of the next, occurring at a rate of approximately 59.94 Hz (color) and 60 Hz (black and white).

vertical-synchronizing pulse—Also called picture-synchronizing pulse. One of the six pulses transmitted at the end of each field in a television system. It maintains the receiver in field-by-field synchronism with the transmitter.

vertical tracking—The phonograph pickup stylus motion path that is near vertical. Applies to hill-and-dale recordings on stereo discs.

vertical tracking force—The minimum downward force at which a phonograph cartridge works without mistracking, i.e., a momentary hop by the stylus (making a raspy sound) or a jump into the next groove.

very high frequency—Abbreviated VHF. Frequency band: 30 to 300 MHz. Wavelength: 10 to 1 meters.

very large-scale integration—*See* VLSI.

very long range—A classification of ground radar sets by slant range, applied to those with a maximum range exceeding 250 miles (402 km).

very low frequency—Abbreviated VLF. Frequency band: below 30 kHz. Wavelength: above 10,000 meters.

very short range—Classification of ground radar sets by slant range, applied to those with a maximum range of less than 25 miles (40.2 km).

vestigial—Pertaining to a remnant or remaining part.

vestigial sideband—Amplitude-modulated transmission in which a portion of one sideband has been largely suppressed by a transducer having a gradual cutoff in the neighborhood of the carrier frequency.

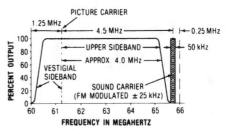

Vestigial-sideband television channel.

vestigial-sideband filter—A filter that is inserted between an AM transmitter and its transmitting antenna to suppress part of one of the sidebands.

vestigial-sideband transmission—Also called asymmetric-sideband transmission. Signal transmission in which one normal sideband and the corresponding vestigial sideband are utilized.

vestigial-sideband transmitter—A transmitter in which one sideband and only a portion of the other are transmitted.

v/f converter—*See* voltage-to-frequency converter.

VFO—Abbreviation for variable-frequency oscillator.

VGA—Abbreviation for Video Graphics Array. A VGA monitor has a 480 × 640 display resolution and can display up to 256 colors simultaneously. A video standard for IBM PC and compatible computers.

V-groove metal-oxide silicon—*See* VMOS.

VHF—1. Abbreviation for very high frequency. 2. In television, a term used to designate channels 2 through 13.

VHF omnirange—Abbreviated VOR. A specific type of range operating at VHF and providing radial lines of position in a direction determined by the bearing selection within the receiving equipment. A nondirectional reference modulation is emitted, along with a rotation pattern that develops a variable modulation of the same frequency as the reference modulation. Lines of position are determined by comparing the phase of the variable with that of the reference.

VHLL—Abbreviation for very high-level language. Usually a problem or requirements description language, ranging in form from the highly abstract to plain English.

VHSIC—Abbreviation for very high-speed integrated circuits.

via—1. A vertical conductor or conductive path forming the interconnection between multilayer hybrid circuit layers. 2. Means of passing from one layer or side of a printed circuit board to the other.

via hole—A plated-through hole that establishes electrical continuity but which is not intended for a component lead. *See* plated-through hole.

vibrating bell—A bell having a mechanism designed to strike repeatedly and as long as it is actuated.

vibrating-reed meter—Also called reed frequency meter. A frequency meter consisting of a row of steel reeds, each having a different natural frequency. All are excited by an electromagnet fed with the alternating current whose frequency is to be measured. The reed whose frequency corresponds most nearly with that of the current vibrates, and the frequency is read on a scale beside the row of reeds.

vibrating-reed relay — 1. A type of relay in which an alternating or a self-interrupted voltage is applied to the driving coil so as to produce an alternating or pulsating magnetic field that causes a reed to vibrate. 2. A type of relay that is actuated by sound frequency. Can be triggered by an electrical resonant circuit, or simply by a mechanically induced sound vibration.

vibrating-wire tranducer — A transducer that utilizes a thin wire suspended in a magnetic field; the change in tension of the wire reflects a frequency-modulating output.

vibration — 1. A continuously reversing change in the magnitude of a given force. 2. A mechanical oscillation or motion about a reference point of equilibrium.

vibration analyzer — A device used to analyze mechanical vibrations.

vibration-detection system — An alarm system that employs one or more contact microphones or vibration sensors that are fastened to the surfaces of the area or object being protected to detect excessive levels of vibration. The contact microphone system consists of microphones, a control unit containing an amplifier and an accumulator, and a power supply. The unit's sensitivity is adjustable so that ambient noises or normal vibrations will not initiate an alarm signal. In the vibration sensor system, the sensor responds to excessive vibration by opening a switch in a closed-circuit system.

vibration galvanometer — An ac galvanometer in which a reading is obtained by making the natural oscillation frequency of the moving element equal to the frequency of the current being measured.

vibration isolator — A resilient support that tends to isolate a system from steady-state excitation or vibration.

vibration meter — Also called a vibrometer. An apparatus comprising a vibration pickup, calibrated amplifier, and output meter, for the measurement of displacement, velocity, and acceleration of a vibrating body.

vibration pickup — A microphone that responds to mechanical vibrations rather than to sound waves. In one type, a piezoelectric unit is employed; the twisting or bending of a Rochelle-salt crystal generates a voltage that varies with the vibration being analyzed.

vibration sensitivity — The peak instantaneous change in output at a given sinusoidal vibration level for any one stimulus value within the range of an instrument or equipment. It is usually expressed in percentage of full-scale output per vibratory "g" over a given frequency range. It may also be specified as a total error in percentage of full-scale output for a given vibratory acceleration level.

vibration sensor — A sensor that responds to vibrations of the surface on which it is mounted. It has a normally closed switch that will momentarily open when it is subjected to a vibration with sufficiently large amplitude. Its sensitivity is adjustable to allow for the different levels of normal vibration, to which the sensor should not respond, at different locations. *See also* vibration-detection system.

vibration survey — A method of determining the natural frequency of a transducer by observation of the output waveform upon the application of a shock or tapping of sufficient magnitude to initiate oscillation of the instrument.

vibration test — A test to determine the ability of a device to withstand physical oscillations of specified frequency, duration, and magnitude.

vibration welding — Method of fusing two plastic parts by vibrating (rubbing) the mating surfaces together at relatively low frequencies, 90 to 120 Hz.

vibrato — 1. A musical embellishment that depends primarily on periodic variations of frequency, often accompanied by variations in amplitude and waveform. The quantitative description of vibrato is usually in terms of the corresponding modulation of frequency (typically 5 to 7 Hz), amplitude, or waveform, or all three. 2. Regular variation in the frequency of a sound, generally at a frequency between 2 and 15 times per second. This frequency modulation by a low-frequency oscillator is commonly used with guitars, organs, synthesizers, and the human voice.

vibrator — 1. A vibrating reed that is driven like a buzzer and has contacts arranged to interrupt direct current to the winding(s) of a transformer, resulting in an alternating current being supplied from another winding to the load. 2. Electromagnetic device that is used to change a continuous steady current into a pulsating current. 3. An electromagnetic device for converting a direct voltage into an alternating voltage.

vibrator power supply — A power supply incorporating a vibrator, step-up transformer, rectifier, and filters for changing a low dc voltage to a high dc voltage.

vibratron — A triode with an anode that can be moved or vibrated by an external force. Thus, the anode current will vary in proportion to the amplitude and frequency of the applied force.

vibrometer — *See* vibration meter.

video — 1. Pertaining to the bandwidth and spectrum position of the signal resulting from radar or television scanning. In current usage, video means a bandwidth on the order of several megahertz and a spectrum position that goes with a dc carrier. 2. A prefix to the name of television parts or circuits that carry picture signals. 3. Radar or television signals that actuate a cathode-ray tube. 4. Composite video contains color and luminance (brightness information) as well as horizontal and vertical synch pulses during the horizontal and vertical blanking intervals. One line of video, for instance, amounts to approximately 52 microseconds in visible horizontal scan time, and 63.5 microseconds for a complete line, including the color burst and horizontal synch.

video adapter — An expansion board in a computer that provides enhanced display capabilities.

video amplifier — 1. An amplifier that provides wideband operation in the frequency range of approximately 15 hertz to approximately 5 megahertz. 2. An amplifier designed for linear amplification over a wide range of frequencies from dc (zero) to over about 10 MHz. 3. A wideband amplifier used for passing picture signals.

video band — The frequency band utilized to transmit a composite video signal.

video carrier — 1. The television signal whose modulation sidebands contain the picture, sync, and blanking signals. 2. A frequency that is 1.25 MHz above the lower edge of the assigned 6-MHz frequency band of a TV channel. At channel 2, for instance, with a bandwidth of 54 to 60 MHz, the video carrier rests at 55.25 MHz.

videocassette — A plastic shell containing two reels and a given length of videotape.

videocassette recorder — Abbreviated VCR. A device for recording and playing video images on magnetic tapes that are contained in plastic cases.

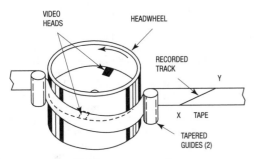

Videocassette recorder.

videocast — 1. To broadcast a program by means of television. 2. A program so broadcast.

VideoCipher I, II — A trademark for a video/audio encryption system developed by M/A-COM LINKABIT, Inc. VCI uses digital video and audio encryption; VCII uses analog video and digital audio encryption.

video compression — 1. The reduction of the number of bits needed to describe a video image. 2. A digital technique to compress several video channels into the bandwidth normally required by one channel.

video correlator — A radar circuit that enhances the capability for automatic target detection; supplies data for digital target plotting; and provides improved immunity to noise, interference, and jamming.

video data digital processing — Digital processing of video signals for pictures transmitted by way of a television link. A computer is used to compare each scanned line with adjacent lines so that extreme changes resulting from electromagnetic interference can be eliminated.

video detector — The demodulator circuit that extracts the picture information from the amplitude-modulated intermediate frequency in a television receiver.

video dialtone — An FCC term for the new generation of home-distribution video systems that deliver compressed digital signals over switched networks to an interactive set-top box in the user premises.

video digitizing — The process of capturing, converting, and storing video images for use by a computer.

video disc — *See* optical video disc.

video discrimination — A radar circuit that reduces the frequency band of the video-amplifier stage in which it is used.

video display unit — Abbreviated VDU. A device for visual presentation of information. A CRT is a typical video display unit.

video freeze — A TV mode which makes it possible to "freeze" an image to study specific details at ease, e.g., to make notes of certain program information like phone numbers.

video frequency — 1. The frequency of the signal voltage containing the picture information that arises from the television scanning process. In the present United States television system, these frequencies are limited from approximately 30 Hz to 4 MHz. 2. A band of frequencies extending from less than 100 hertz to several megahertz.

video-frequency amplifier — A device capable of amplifying those signals that comprise the periodic visual presentation.

video-gain control — A control for adjusting the amplitude of a video signal. Two such controls are provided in the matrix section of some color television

receivers so that the proper ratios between the amplitudes of the three color signals can be obtained.

video impedance — Of a detector diode, the output impedance measured at the video frequency when the diode is operating under specified bias conditions.

video integration — A method of improving the output signal-to-noise ratio by utilizing the redundancy of repetitive signals to sum the successive video signals.

video integrator — A device that uses the redundance of repetitive signals to improve the output signal-to-noise ratio by summing the successive video signals.

video inversion — A type of encoding or scrambling in which the transmitted downlink video signals are inverted.

video mapping — The procedure whereby a chart of an area is electronically superimposed on a radar display.

video masking — A method for the removal of chaff echoes and other extended clutter from radar displays.

video mixer — A circuit or device used to combine the signals from two or more television cameras.

video moiré — *See* moiré effect.

video monitor — A high-quality television set (without rf circuits) that accepts video baseband inputs directly from a TV camera, VTR, or satellite TV receiver with no rf modulator required.

video pair cable — A transmission cable containing low-loss pairs with an impedance of 125 ohms. Used for TV pickups, closed circuit TV, telephone carrier circuits, etc.

video recording (magnetic tape) — The methods of recording data having a bandwidth in excess of 100 kHz on a single track.

video signal — 1. The picture signal in a television system — generally applied to the signal itself and the required synchronizing and equalizing pulses. 2. In television, the signal that conveys all of the intelligence present in the image, together with the necessary synchronizing and equalizing pulses. 3. That portion of the composite video signal that varies in gray-scale levels between reference white and reference black. Also referred to as the picture signal, this is the portion that can be seen. 4. The picture signal. A signal containing visual information and horizontal and vertical blanking. *See also* composite video signal. 5. The output from a video graphics adapter incorporating the red (R), green (G), and blue (B) signals and the luminance signal, or combinations of these signals, that pass to the video input of a monitor.

video stretching — In navigation, a procedure whereby the duration of a video pulse is increased.

video synthesizer — A video analog computer that accepts standard video signals from a camera, film chain, videotape or graphics generator, which it then processes and applies a combination of effects to in order to reshape and add motions (animation) to fixed graphics or live scenes.

videotape — A wide magnetic tape designed for recording and playing back a composite black and white or color television signal.

videotape recorder — Abbreviated VTR. 1. A device that permits audio and video signals to be recorded on magnetic tape and then played back without any processing, as with films, on a CCTV monitor. 2. A device that translates the electrical signals of a VTR television camera into corresponding magnetic variations, which are recorded on magnetic tape; the original image can be displayed on a television receiver at a later time by reconversion of the magnetic variations, stored on the tape, into appropriate electrical signals.

videotape recording — Abbreviated VTR. A method of recording television picture and sound signals on tape for reproduction at some later time.

videotex — 1. A system that links TV screens to mass databases through television or telephone-based communication channels. 2. A generic term for interactive services delivered to personal computers or adapted television sets in the home or office. 3. Technology that connects computerized text and graphics to a television screen via telephone lines. 4. A terminal-oriented communication network that links users' equipment to a computer that maintains a database of information services. Videotex services include shop-at-home catalogs, listings of community events, and message forwarding.

videotext system — A system for the widespread dissemination of textual and graphic information by wholly electronic means for display on low-cost terminals (often suitably equipped television receivers) under the selective control of the recipient using control procedures easily understood by untrained users.

vidicon — 1. A camera tube in which a charge-density pattern is formed by photoconduction and stored on that surface of the photoconductor which is scanned by an electron beam, usually of low-velocity electrons. 2. A vacuum tube capable of changing light images into electrical voltage variations corresponding to the brightness of those images; a particular type of cathode-ray pickup tube used in some video cameras.

viewer — A software application that permits visual display of content. *See also* browser; player.

viewfinder — An auxiliary optical or electronic device attached to a television camera so the operator can see the scene as the camera sees it.

viewfinder monitor — *See* electronic viewfinder.

viewing angle — The angle through which an LCD display has acceptable contrast.

viewing area — The area of the CRT face that can be directly seen by the user. Its size is determined by the size of the CRT and the area covered by the face mask.

viewing mirror — A mirror used in some television receivers to reflect the image formed on the screen of the picture tube at an angle convenient to the viewer.

viewing screen — The face of a cathode-ray tube on which the image is produced.

viewing time — The time during which a storage tube presents a visible output that corresponds to the stored information.

Villari effect — A phenomenon in which a change in magnetic induction occurs when a mechanical stress is applied along a specified direction to a magnetic material having magnetostrictive properties.

vinyl resin — A soft plastic used for making phonograph records.

virgin tape — *See* raw tape.

virtual address — In a computer, an immediate, or real-time, address.

virtual cathode — An electron cloud that forms around the outer grid in a thermionic vacuum tube when the inner grid is maintained slightly more positive than the cathode.

virtual circuit — In packet switching, a network facility that gives the appearance to the user of an actual end-to-end circuit; a dynamically variable network connection in which sequential data packets may be routed differently during the course of a virtual connection. Virtual circuits enable transmission facilities to be shared by many users simultaneously.

virtual connection — A packet-switched data path between two terminals that performs as if the two devices were linked by a switched circuit.

virtual height — 1. The height of the equivalent reflection point that will cause a wave to travel to the ionosphere and back in the same time required for an actual reflection. In determining the virtual height, the wave is assumed to travel at uniform speed and the height is determined by the time required to go to the ionosphere and back at the assumed velocity of light. 2. The apparent height of a layer of the ionosphere. It is determined from the time interval between the transmitted signals and the ionospheric echo at vertical incidence (the radio wave penetrating the ionosphere perpendicular to it).

virtual image — The optical counterpart of an object, formed at imaginary focuses by prolongations of light rays (e.g., the image that appears to be behind an ordinary mirror).

virtual memory — 1. The use of techniques by which the computer programmer may use the memory as though the main memory and mass memory were available simultaneously. 2. A technique that permits the user to treat secondary (disk) storage as an extension of core memory, thus giving the virtual appearance of a larger core memory to the programmer. 3. A technique for managing a limited amount of high-speed memory and a (generally) much larger amount of lower-speed memory in such a way that the distinction is largely transparent to a computer user. The technique entails some means of swapping segments of program and data from the lower-speed memory (which would commonly be a drum or disk) into the high-speed memory, where it would be interpreted as instructions or operated on as data. The unit of program or data swapped back and forth is called a page. The high-speed memory from which instructions are executed is real memory, while the lower-speed memory (drums or disks) is called virtual memory. 4. Any of several schemes in which the user-visible address space is divided into several subspaces, so that each of the subspaces in use can be placed anywhere in main memory. In some schemes, parts of the virtual memory may be on a mass-storage device.

virtual PPI reflectoscope — A device for superimposing a virtual image of a chart onto the PPI pattern. The chart is usually prepared with white lines on a black background to the scale of the PPI range scale.

virtual reality — Abbreviated VR. 1. A technology that is computer generated and allows the user to interact with data that gives the appearance of a three-dimensional environment. The user can enter and navigate the 3D world portrayed as graphic images, and change viewpoints and interact with objects in that world as if inside that world. A virtual reality environment can be experienced using a headset and electronic gloves, or simply viewed on a monitor. 2. Sophisticated, multidimensional imaging systems and high-speed processing capabilities that create environments that users can interact with and manipulate directly.

virtual storage — Storage space that may be viewed as addressable main storage, but is actually auxiliary storage (usually peripheral mass storage) mapped into read addresses; the amount of virtual storage is limited by the addressing scheme of the computer.

virus — 1. A software program that attaches itself to another program in computer memory or on a disk, and spreads from one program to another. Viruses may damage data, cause the computer to crash, display messages, or lie dormant. 2. Any destructive self-replicating program. 3. A computer program written to secretly reproduce itself across many computer systems. Viruses can cause serious software damage.

viscometer — Also called a viscosimeter. A device for measuring the degree to which a liquid resists a change in shape.

viscosimeter — *See* viscometer.

viscosity — 1. The frictional resistance offered by one part or layer of a liquid as it moves past an adjacent part or layer of the same liquid. 2. The property of a liquid that resists internal flow; measured in units of poise or centipoise. Low-viscosity materials are usually thin, while high-viscosity materials are thick.

viscous and magnetic damping — Damping by virtue of the viscosity of a fluid around the sensing element or of a magnetic field.

viscous-dampened arm — A phonograph pickup arm mounted on a liquid cushion of oil, which provides high damping to eliminate arm resonances. It also protects the record groove and stylus; the arm does not fall on the record when dropped, but floats down gently.

visibility factor — Also called display loss. Ratio of the minimum input-signal power detectable by ideal instruments connected to the output of a receiver to the minimum signal power detectable by a human operator through a display connected to the same receiver. The visibility factor may include the scanning loss.

visible emission — Also called visible light. Radiation that is characterized by wavelengths of about 0.38 to 0.77 micrometer.

visible light — Electromagnetic wavelengths that can be seen by the human eye, ranging from 380 to 770 nanometers.

visible radiation — Radiation with wavelengths ranging from about 0.4 and 0.7 micrometer, corresponding to the visible spectrum of light.

visible spectrum — That region of the electromagnetic spectrum to which the retina is sensitive and by which the eye sees. It extends from about 400 to about 750 nanometers in wavelengths of the radiation.

vision — The series of processes in which luminous energy incident on the eye is perceived and evaluated.

visual-aural range — Abbreviated VAR. A special type of VHF range providing a pair of radial lines of position that are reciprocal in bearing and are displayed to the pilot on a zero-center, left-right indicator. This facility also provides a pair of reciprocal radial lines of position located 90° from the above visually indicated lines. These are presented to the pilot as aural A-N radio-range signals, which provide a means for differentiating between the two visually indicated lines (and vice versa).

visual carrier frequency — The frequency of the television carrier that is modulated by the picture information.

visual communication — Communication by optical signs such as flags and lights.

visualization — Graphic representation of abstract data usually relayed in text and numbers.

visual radio range — Abbreviated VRR. A radio range whose course is followed by means of visual instruments.

visual scanner — 1. A device that generates an analog or digital signal by optically scanning printed or written data. 2. *See* scanner, 2.

visual signal device — A pilot light, annunciator, or other device that provides a visual indication of the condition of the circuit or system being supervised.

visual storage tube — An electron tube that stores and visually displays information by means of a cathode-ray beam scanning and charge storage mechanism.

visual telephony — The transmission of picture information (television) by means of telephone lines.

visual transmitter — Also called a picture transmitter. In television, the radio equipment for transmission of the picture signals only.

visual transmitter power — The peak power output during transmission of a standard television signal.

vitreous — 1. Having the nature of glass. 2. A term used in ceramic technology to indicate fired characteristics approaching being glassy, but not necessarily totally glassy.

vitreous binder — A glassy material used in compounds to bind other particles together. This takes place after melting the glass and cooling.

vitrification — 1. The progressive reduction in porosity of a ceramic material as a result of heat treatment or some other process. 2. The reduction of porosity in a ceramic product through the formation of a glassy bond.

VLF — Abbreviation for very low frequency.

VLSI — Abbreviation for very large-scale integration. Generally considered to be an IC with more than 1000 gates.

VLSI circuit — Abbreviation for very large-scale integrated circuit. *See* gate array, 3.

VMOS — 1. Abbreviation for V-groove metal-oxide silicon — a variation of MOS technology. A manufacturing process (called isotropically etched double-diffusion MOS process) based on the technique of using V-shaped grooves in the silicon wafer to provide a third-dimensional surface. This allows for a smaller chip geometry than the (older) process using a two-dimensional area technique of manufacture. With smaller geometry, the same diameter wafer can, theoretically, yield more chips (roughly 50 percent more) and thereby reduce the unit cost per chip considerably. The cost of processing a wafer generally remains constant no matter how many chips it contains. 2. Abbreviation for vertical MOS. A semiconductor device in which current travels vertically in the semiconductor chip. A cross section shows a four-layer diffusion consisting of n+ at the top (source), p as the channel, n− as the drift region, and n+ as the drain. The layers' dimensions are precisely controlled by diffusion processes. Unlike conventional bipolar transistors, VMOS devices exhibit high density because of their short channel spacing: 1 μm compared with bipolar devices typical 5 μm.

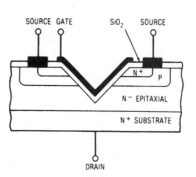

Structure.

Symbol.

VMOS transistor.

VMOS power FET — Abbreviation for vertical structured metal-oxide silicon power field-effect transistor. In these semiconductor devices, current flows vertically from source to drain. This structure results in high current densities, low saturation resistance, excellent heat dissipation and power handling capabilities, low chip capacitance, and excellent wideband performance.

vocabulary — A list of operating codes or instructions available for writing the program for a given problem and for a specific computer.

vocoder — Abbreviation for voice-operated coder. A device used to compress the frequency bandwidth requirement of voice communications. It consists of an electronic speech analyzer, which converts the speech waveform to several simultaneous analog signals, and an electronic speech synthesizer, which produces artificial sounds in accordance with analog control voltages.

vodas — Abbreviation for voice-operated device, antising. A system for preventing the overall voice-frequency singing of a two-way telephone circuit by disabling one direction of transmission at all times.

voder — Abbreviation for voice-operation demonstrator. An electronic device capable of artificially producing voice sounds. It uses active devices in connection with electrical filters controlled through a keyboard.

vogad — Abbreviation for voice-operated gain-adjusting device. A voice-operated device used to give a substantially constant volume output for a wide range of inputs.

voice activated (sound activated) — Refers to a bug that radiates radio frequencies or a recorder that is activated only when sound is present in target area.

voice analyzer — An electronic instrument for printing out waveforms corresponding to vocal characteristics; an aid in identifying speech problems as well as speakers.

voice channel — A transmission path suitable for carrying analog voice signals (covering a frequency band of 250 to 3400 Hz) between two points.

voice coder — A device that converts a speech signal into digital form prior to encipherment for secure transmission, and converts the digital signals back into speech at the receiving point.

voice coil — Also called a speaker voice coil. 1. A coil attached to the diaphragm of a dynamic speaker and moved through the air gap between the pole pieces. 2. The moving coil in a dynamic speaker. It is suspended in the field of a permanent magnet and is fixed to the speaker diaphragm.

voice/data system — An integrated communications system for transmission of both voice and digital data signals.

voice filter — A parallel-resonant circuit connected in series with a line feeding several speakers. Its purpose is to remove the tubbiness of the male voice. The frequency of resonance is adjusted somewhere between 125 and 300 Hz.

voice frequency — Also called speech frequency. 1. Any of the frequencies within the band of 32 to 16,000 Hz that are audible to the human ear. 2. Any of the frequencies within the band of 300 to 3500 Hz that are normally used for telephone communication.

voice-frequency carrier telegraphy — Carrier telegraphy in which the carrier currents have frequencies such that the modulated currents may be transmitted over a voice-frequency telephone channel.

voice-frequency dialing — A method of dialing by which the direct-current pulses from the dial are transformed into voice-frequency alternating-current pulses.

voice-frequency telegraph system — A telegraph system by which many channels can be carried on a single circuit. A different audio frequency is used for each channel and is keyed in the conventional manner. Each frequency is generated by an oscillator. At the receiving end, the various audio frequencies are separated by filter circuits and are fed to their respective receiving circuits.

voice-frequency telephony — Telephony in which the frequencies of the components of the transmitted electric waves are substantially the same as the frequencies of corresponding components of the actuating acoustical waves.

voice grade — A telephone circuit suitable for transmitting a bandpass from 300 to approximately 2700 Hz, or greater, with certain standards of noise and interference such that intelligible speech can be transmitted.

voice-grade channel — 1. A channel suitable for the transmission of speech, digital or analog data, or facsimile, generally with a frequency range of about 300 to 3000 hertz. 2. A telephone circuit normally used for speech communication, accommodating frequencies from 300 to 3000 Hz.

voice-grade circuit — A switched (dial-up) or leased (dedicated) telephone circuit suitable for the transmission of speech, digital or analog data, or facsimile, generally with a frequency range of about 300 to 3000 hertz.

voice-grade line — A local telephone loop, or trunk, having a bandpass of approximately 300 to 3000 Hz.

voice mail — Sophisticated telephone voice messages that are recorded and translated into digital bits for storage and manipulation. Voice mail systems use specialized hardware and software and can be incorporated into a PBX or used as stand-alone systems.

voice-operated coder — See vocoder.

voice-operated device — A device that permits the presence of voice or sound signals to affect a desired control.

voice-operated device, antising — A system for preventing the overall voice-frequency singing of a two-way telephone circuit by disabling one direction of transmission at all times. Acronym: vodas.

voice-operated gain-adjusting device — A voice-operated device used to give a substantially constant-volume output for a wide range of inputs.

voice-operated loss control and suppressor — A voice-operated device that switches the loss out of the transmitting branch and inserts the loss into the receiving branch under control of the subscriber's speech.

voiceprint — A speech spectrograph sufficiently sensitive and detailed to identify individual human voices.

voice recognition — The conversion of spoken words into computer text. Speech is first digitized and then matched against a dictionary of coded waveforms. The matches are then converted into text as if the words were typed on the keyboard.

voice-recognition equipment vocabulary size — The number of utterances (words or short phrases) that can be distinctly recognized and digitally encoded. The larger the vocabulary, the more complex the source data that can be handled without special software.

voice spectrum — The total fundamental frequency range of the human voice.

voice synthesis — Technology that produces sound or voices by processing compressed digital signals and storing them in a memory in the same manner a human voice generates sound processed by the vocal tract and other organs.

void — The absence of substance in a localized area.

volatile — 1. A computer storage medium in which information cannot be retained without continuous power dissipation. 2. Capable of evaporating.

volatile memory — 1. In computers, any memory that can return information only as long as energizing power is applied. The opposite of nonvolatile memory. 2. A read/write memory whose content is irretrievably lost when operating power is removed. Virtually all types of read/write semiconductor memories are volatile. *See also* nonvolatile memory.

volatile storage — 1. A computer storage device in which the stored information is lost when the power is removed. 2. A storage medium in which data cannot be retained without continuous power dissipation (e.g., acoustic delay lines, electrostatics, capacitors).

volatile store — A storage device in which stored data is lost when the applied power is removed (e.g., an acoustic delay line).

volatility — With respect to memory, an inability to retain stored data in the absence of external power.

Voldicon — The trade name of Adage, Inc., for a family of high-speed, all-semiconductor, current-balancing devices that use digital logic and readout for high-speed precision measurement of analog signals.

volt — Letter symbol: V. 1. The unit of measurement of electromotive force. It is the difference of potential required to make a current of 1 ampere flow through a resistance of 1 ohm. 2. The difference of electric potential between two points of a conductor carrying a constant current of 1 ampere when the power dissipated between these points is equal to 1 watt.

Volta effect — *See* contact potential.

voltage — 1. Electrical pressure; i.e., the force that causes current through an electrical conductor. 2. Symbolized by E. The greatest effective difference of potential between any two conductors of a circuit. 3. The term most often used in place of electromotive force, potential, potential difference, or voltage drop, to designate electric pressure that exists between two points and is capable of producing a flow of current when a closed circuit is connected between the two points. 4. Standard unit of magnitude of an electrical signal, named after Count A. Volta, inventor of the battery about 1800.

voltage amplification — Also called voltage gain. 1. Ratio of the voltage across a specified load impedance connected to a transducer to the voltage across the input of the transducer. 2. The ratio of the voltage at the input of a device to the voltage at the output from the device, expressed in decibels.

voltage amplifier — An amplifier used specifically to increase a voltage. It is usually capable of delivering only a small current.

voltage-amplifier tube — A tube that is designed primarily as a voltage amplifier. It has high gain, but delivers very little output power.

voltage and power directional relay — A device that permits or causes the connection of two circuits when the voltage difference between them exceeds a given value in a predetermined direction, and causes these two circuits to be disconnected from each other when the power flowing between them exceeds a given value in the opposite direction.

voltage attenuation — Ratio of the voltage across the input of a transducer to the voltage delivered to a specified load impedance connected to the transducer.

voltage balance relay — A device that operates on a given difference in voltage between two circuits.

voltage breakdown — 1. The voltage necessary to cause insulation failure. 2. A rapid increase of current flow, from a relatively low value to a relatively high value, upon the application of a voltage to a pn junction or dielectric.

voltage-breakdown test — A test whereby a specified voltage is applied between given points in a device to ascertain that no breakdown will occur at that specified voltage.

voltage calibrator — Test equipment that supplies accurate ac voltages for comparison on a scope screen with other waveforms to determine their voltage level.

voltage coefficient of capacitance — Also called voltage sensitivity. The quotient of the derivative with respect to voltage of a capacitance characteristic at a point divided by the capacitance at that point.

voltage coefficient of resistivity — The maximum change in nominal resistance value due to the application of a voltage across a resistor, after correcting for self-heating effects; usually expressed in percent or per-unit (ppm) change in nominal resistance per volt applied.

voltage comparator — 1. An amplifying device with a differential input that will provide an output polarity reversal when one input signal exceeds the other. When operating with open loop and without phase compensation, operational amplifiers make fast and accurate voltage comparators. 2. A circuit that compares two analog voltages and develops a logic output when the voltages being compared are equal or one is greater or less than the reference level.

voltage control — A method of varying the magnitude of voltage in a circuit by means of amplitude control, phase control, or both.

voltage-controlled capacitor — *See* varactor.

voltage-controlled crystal oscillator — A crystal oscillator whose operating frequency can be changed by applying a controlling voltage to introduce a phase shift in the oscillator circuit.

voltage-controlled oscillator — Abbreviated VCO. 1. Any oscillator for which a change in tuning voltage results in a predetermined change in output frequency. Frequency tuning is accomplished by either changing the bias voltage on a varactor diode in the frequency-determining resonant network or the bias voltage to the active device. The former approach, although more complex than the latter method of tuning, is capable of multioctave bandwidth. 2. A circuit that creates an ac output signal whose frequency is a function of the dc input voltage. 3. An oscillator whose frequency can be changed by means of an external control voltage. It is commonly found in synthesizers.

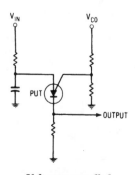

Voltage-controlled oscillator.

voltage-controlled resistor — Abbreviated VCR. A three-terminal variable resistor in which the resistance value between two of the terminals is controlled by a voltage potential applied to the third.

voltage corrector — An active source of regulated power placed in series with the output of an unregulated supply. The voltage corrector senses changes in the output voltage (or current) and corrects for these changes automatically by varying its own output in the opposite direction so as to maintain the total output voltage constant.

voltage/current crossover — The characteristic of a power supply that automatically converts the mode of operation of a power supply from voltage regulation to current regulation (or vice versa) as required by preset limits. The constant-current and constant-voltage settings are independently adjustable over specified limits. The region near the intersection of the constant-voltage and constant-current curves is designated by the term "crossover characteristics."

voltage-dependent resistor — Abbreviated VDR. Also called varistor or metal oxide varistor (MOV). A special type of resistor whose resistance changes appreciably, nonlinearly, and consistently in response to voltage across its terminals.

voltage-directional relay — 1. A relay that functions in conformance with the direction of an applied voltage. 2. A device that operates when the voltage across an open circuit breaker or contactor exceeds a given value in a given direction.

voltage divider — Also called a potential divider. 1. A resistor or reactor connected across a voltage and tapped to make a fixed or variable fraction of the applied voltage available. *See also* potentiometer, 1; rheostat. 2. High-voltage resistance string, tapped resistor, potentiometer, adjustable resistors, or a series arrangement of two or more fixed resistors connected across a voltage source. Of the total voltage, a desired fraction is obtained from the intermediate tap, movable contact, or resistor junction.

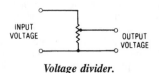

Voltage divider.

voltage doubler — A voltage multiplier that rectifies each half cycle of the applied alternating voltage separately, and then adds the two rectified voltages to produce a direct voltage having approximately twice the peak amplitude of the applied alternating voltage.

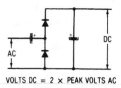

VOLTS DC = 2 × PEAK VOLTS AC

Voltage doubler.

voltage drop — 1. The difference in voltage between two points due to the loss of electrical pressure as a current flows through an impedance. 2. The voltage developed across the component or conductor by current through the resistance or impedance of that component or conductor. 3. The voltage developed between the terminals of a circuit component by current through the resistance or impedance of that part. *See IR* drop. 4. The decrease in voltage as a current traverses a resistance. 5. The voltage measured across a resistance through which a current is flowing. 6. The voltage existing across each element of a series circuit. (Also true of a contact in series with its load where a voltage drop also exists.) The voltage present across the load will be the line voltage less the voltage drops across each element in series with the load.

voltage endurance — *See* corona resistance.

voltage feed — Excitation of a transmitting antenna by applying voltage at a point of maximum potential (at a voltage loop or antinode).

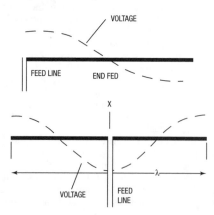

Voltage feed.

voltage feedback — A form of amplifier feedback in which the voltage drop across part of the load impedance is put in series with the input-signal voltage.

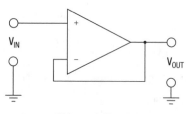

Voltage follower.

voltage frequency converter — A circuit that produces an output frequency that varies with the voltage applied to its input.

voltage gain — *See* voltage amplification.

voltage generator — A two-terminal circuit element with a terminal voltage independent of the current through the element.

voltage gradient — The voltage per unit length along a resistor or other conductive path.

voltage inverter—A circuit having a response (output) proportional to a constant (the gain) times the input signal, but opposite in sign to it. In a unity-gain inverter, the output is −1 times the input.

voltage jump—An abrupt change or discontinuity in the tube voltage drop during operation of glow-discharge tubes.

voltage level—Ratio of the voltage at any point in a transmission system to an arbitrary value of voltage used as a reference. In television and other systems where waveshapes are not sinusoidal or symmetrical about a zero axis and where the sum of the maximum positive and negative excursions of the wave is important in system performance, the two voltages are given as peak-to-peak values. This ratio is usually expressed in dBV, signifying decibels referred to 1 volt peak-to-peak.

voltage limit—A control function that maintains a voltage between predetermined values.

voltage loop—A point of maximum voltage in a stationary-wave system. A voltage loop exists at the ends of a half-wave antenna.

voltage loss—The voltage between the terminals of a current-measuring instrument when the applied current has a magnitude corresponding to nominal end-scale deflection. In other instruments, the voltage loss is the voltage between the terminals at rated current.

voltage-measuring equipment—Equipment for measuring the magnitude of an alternating or direct voltage.

voltage multiplier—1. A rectifying circuit that produces a direct voltage approximately equal to an integral multiple of the peak amplitude of the applied alternating voltage. 2. A series arrangement of capacitors charged by rapidly rotating brushes in sequence, giving a high direct voltage equal to the source voltage multiplied by the number of capacitors in series. 3. A precision resistor used in series with a voltmeter to extend its measuring range.

Voltage multiplier.

voltage node—1. A point having zero voltage in a stationary-wave system (e.g., at the center of a half-wave antenna). 2. In a transmission system having standing waves, a point at which the voltage is a minimum.

voltage offset—The amount of dc voltage present at an (instrumentation) amplifier's output with a 0-volt input. Initial offset may be adjusted to zero; however, offset-voltage shifts during operation can cause errors. Systems containing microcomputers can often correct offset with an auto-zero cycle, or the offset can be removed by a control adjustment.

voltage plane—A conductor or portion of a conductor layer on or in a printed board that is maintained at other than ground potential. It can also be used as a common voltage source for heat sinking or for shielding.

voltage quadrupler—A rectifier circuit in which four diodes are employed to produce a dc voltage of four times the peak value of the ac input voltage.

voltage-range multiplier—Also called an instrument multiplier. 1. A series resistor installed external to

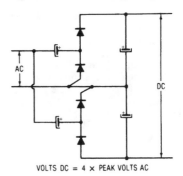

VOLTS DC = 4 × PEAK VOLTS AC

Voltage quadrupler.

the measurement device to extend its voltage range. 2. A precision resistor placed in series with a voltmeter to enable measurement of a high voltage with a voltmeter having a lower voltage range.

voltage rating—Also called the working voltage. 1. The maximum voltage that an electrical device or component can sustain without breaking down. 2. The maximum sustained voltage that can be safely applied to a capacitor up to a specified temperature without risking capacitor failure. 3. The maximum voltage at which a given device may be safely maintained during continuous use in a normal manner.

voltage ratio (of a transformer)—Ratio of the rms primary terminal voltage to the rms secondary terminal voltage under specified load conditions.

voltage-ratio box—*See* measurement voltage divider.

voltage reference—A highly regulated voltage source used as a standard to which the output voltage of a power supply is continuously compared for purposes of regulation.

voltage-reference diode—A diode that develops across its terminals a reference voltage of specified accuracy when biased to operate within a specified current range.

voltage-reference tube—A gas tube in which the voltage drop is essentially constant over the operating range of current and is relatively stable at fixed values of current and temperature.

voltage reflection coefficient—The ratio of the complex electric field strength or voltage of a reflected wave to that of the incident wave.

voltage-regulating transformer—A saturated-core type of transformer that holds the output voltage to within a few percent, with input variations up to ±20 percent. Considerable harmonic distortion results unless extensive filters are employed.

voltage regulation—A measure of the degree to which a power source maintains its output voltage stability under varying load conditions.

voltage regulator—1. A circuit that holds an output voltage at a predetermined value or causes it to vary according to a predetermined plan, regardless of normal input-voltage change or changes in the load impedance. 2. A gas-filled electronic tube that has the property of maintaining a nearly constant voltage across its terminals over a considerable range of current through the tube. It is used in an electronic voltage regulator. 3. An electronic circuit used for controlling and maintaining a voltage at a constant level.

voltage-regulator diode—1. A diode that develops across its terminals an essentially constant voltage throughout a specified current range. 2. A diode that is normally biased to operate in the breakdown region of its voltage-current characteristic and that develops across its terminals an essentially constant voltage throughout a specified current range.

voltage-regulator tube — Also called a VR tube. A glow-discharge cold-cathode tube in which the voltage drop is essentially constant over the operating range of current, and which is designed to provide a regulated direct-voltage output.

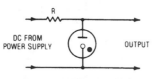

Voltage-regulator tube.

voltage relay — A relay that functions at a predetermined value of voltage.

voltage saturation — *See* plate saturation.

voltage-sensitive resistor — A resistor (e.g., a varistor) whose resistance varies with the applied voltage.

voltage sensitivity — 1. The voltage that produces standard deflection of a galvanometer when impressed on a circuit made up of the galvanometer coil and the external critical-damping resistance. The voltage sensitivity is equal to the product of the current sensitivity and the total circuit resistance. 2. *See* voltage coefficient of capacitance.

voltage spectrum — A function that is the square root of the spectral intensity; it is expressed in terms of voltage in a unit frequency band.

voltage-stabilizing tube — A gas-filled glow-discharge tube normally working in that part of its characteristic where the voltage is practically independent of current drop within a given range.

voltage standard — An accurately known voltage source (e.g., a standard cell) used for comparison with or calibration of other voltages.

voltage standing-wave ratio — Abbreviated VSWR. In a stationary-wave system (such as in a waveguide or coaxial cable), the ratio of the amplitude of the electric field or voltage at a voltage maximum to that at an adjacent voltage minimum.

voltage stress — That stress found within a material when subjected to an electrical charge.

voltage-to-frequency converter — Abbreviated v/f converter. An electronic circuit that converts an input voltage into a train of digital output pulses at a rate that is directly proportional to the input.

voltage to ground — The voltage between any live conductor of a circuit and earth (or common reference plane).

voltage transformer — *See* potential transformer.

voltage transients — Unpredictable and usually unavoidable spikes and surges of electrical power.

voltage tripler — A rectifier circuit in which three diodes are employed to produce dc voltage equal to approximately three times the peak ac input voltage.

voltage-tunable magnetron — A high-frequency continuous-wave oscillator operating in the microwave region. Power outputs begin in the milliwatts range and extend through hundreds of watts.

voltage-tunable tube — An oscillator tube whose operating frequency can be changed by varying one or more of its electrode voltages (e.g., a backward-wave magnetron).

voltage-tuned cavity oscillator — Primarily a cavity oscillator with the addition of a varactor diode to slightly modify the cavity's resonant frequency. These oscillators are capable of FM and/or phase-locking.

voltage-tuned crystal oscillator — A crystal-controlled oscillator with a varactor diode in the frequency-determining network that is used to slightly vary the crystal frequency.

voltage-type telemeter — A telemeter in which the translating means is the magnitude of a single voltage.

voltage-variable capacitance diode — Another name for varactor diode.

voltage-variable capacitor — *See* varactor.

voltage velocity limit — *See* slew rate, 2.

voltaic cell — An electric cell having two electrodes of unlike metals immersed in a solution that chemically affects one or both of them, thus producing an electromotive force. The name is derived from Volta, a physicist who discovered this effect.

voltaic couple — Two dissimilar metals in contact, resulting in a contact potential difference.

voltaic pile — A voltage source consisting of alternate pairs of dissimilar metal discs separated by moistened pads, forming a number of elementary primary cells in series.

volt-ammeter — An instrument calibrated to read both voltage and current.

voltampere — Letter symbol: VA. A unit of apparent power in an ac circuit containing reactance. It is equal to the potential in volts multiplied by the current in amperes, without taking phase into consideration.

voltamperehour meter — An electricity meter that measures the integral, usually in kilovoltamperehours, of the apparent power in the circuit where the meter is connected.

voltampere loss — *See* apparent power loss.

voltampere meter — An instrument for measuring the apparent power in an alternating-current circuit. Its scale is graduated in voltamperes or kilovoltamperes.

voltampere reactive — Also called wattless power. Component of the apparent power in an alternating-current circuit that is delivered to the circuit during part of a cycle, but is returned to the source during another part of the cycle. The practical unit of reactive power is the var, equal to 1 reactive voltampere.

Volta's law — When two dissimilar conductors are placed in contact, the same contact potential is developed between them, whether the contact is direct or through one or more intermediate conductors.

volt box — *See* measurement voltage divider.

volt-electron — An obsolete expression for electron-volt.

voltmeter — 1. An instrument for measuring potential difference between two points. Its scale is usually graduated in volts. If graduated in millivolts or kilovolts, the instrument is usually designated as a millivoltmeter or a kilovoltmeter. 2. An instrument used for the measurement of electric voltage. The instrument may be of the electrostatic or tube type, but usually consists of a moving-coil ammeter connected in series with a high resistance. The resistance of the meter being fixed, the current passing through it will be directly proportional to the voltage at the points where it is connected; thus, the instrument can be calibrated in volts.

voltmeter-ammeter — A voltmeter and an ammeter combined into a single case, but with separate circuits.

voltmeter sensitivity — The ratio, expressed in ohms per volt, of the total resistance of a voltmeter to its full-scale reading.

volt-ohm-milliammeter — A test instrument with several ranges, for measuring voltage, current, and resistance.

volume — Also called power level. 1. The magnitude (measured on a standard volume indicator) of a complex audio-frequency wave, expressed in volume units. In addition, the term *volume* is used loosely to signify either the intensity of a sound or the magnitude of an audio-frequency wave. 2. The amount or a measure of energy in an electrical or acoustical train of waves. 3. A physical unit of storage media; for example, a reel of magnetic tape. A volume may contain part of a file, a complete file, or more than one file. Sections of one or more files may be contained in a volume, but not multiple sections of the same file. 4. The level of an audio signal or the intensity of a sound. 5. A unit of secondary storage media, such as a magnetic tape, disk pack, or flexible diskette.

volume bar — When activated, a TV feature that displays a volume setting bar on-screen, showing the adjustment being made whenever volume levels are adjusted.

volume compression — Also called automatic volume compression. The limiting of the volume range to about 30 to 40 decibels at the transmitter to permit a higher average percentage modulation without overmodulation. Also used in recording to raise the signal-to-noise ratio.

volume compressor — Audio-frequency control circuit that limits the volume range of a (radio) program at the transmitter to permit using a higher average percent modulation without risk of overmodulation.

volume conductivity — *See* conductivity, 1.

volume control — A variable resistor for adjusting the loudness of a radio receiver or amplifying device.

volume equivalent — A measure of the loudness of speech reproduced over a complete telephone connection. It is expressed numerically in terms of the trunk loss of a working reference system that has been adjusted to give equal loudness.

volume expander — A circuit that provides volume expansion.

volume expansion — *See* automatic volume expansion.

volume indicator — An instrument for indicating the volume of a complex electric wave such as that corresponding to speech.

volume lifetime — The average time interval between the generation and recombination of minority carriers in a homogeneous semiconductor.

volume limiter — An amplifier whose gain is automatically reduced when the average input volume to the amplifier exceeds a predetermined value, so that the output volume is maintained substantially constant. The normal gain is restored whenever the input volume drops below the predetermined limit.

volume-limiting amplifier — An amplifier that reduces the gain whenever the input volume exceeds a predetermined value, so that the output volume is maintained substantially constant. The normal gain is restored whenever the input volume drops below the predetermined limit.

volume magnetostriction — The relative volume change of a body of ferromagnetic material when the magnetization of the body is increased from zero to a specified value (usually saturation) under specified conditions.

volume range — 1. Of a transmission system, the difference, expressed in dB, between the maximum and minimum volumes that the system can satisfactorily handle. 2. Of a complex audio-frequency signal, the difference, expressed in dB, between the maximum and minimum volumes occurring over a specified period.

volume recombination rate — The rate at which free electrons and holes recombine within the volume of a semiconductor.

volume resistance — Ratio of the dc voltage applied to two electrodes in contact with or embedded in a specimen to the portion of the current between them distributed through the specimen.

volume resistivity — 1. Ratio of the potential gradient parallel to the current in a material to the current density. *See also* resistivity, 1. 2. Also called specific insulation resistance. The electrical resistance between opposite faces of a 1-cm cube of insulating material, commonly expressed in ohm-centimeters. 3. The resistance of a material to dc current as a measure of volume. Expressed in ohm-centimeters.

volumetric displacement — The change in volume required to displace the diaphragm of a pressure transducer from its rest position to a position corresponding to the application of a stimulus equal to the rated range of the transducer.

volumetric efficiency — Also called packing factor. The ratio of parts volume to total equipment volume, expressed in percent. In modules, it is usually taken as the volume of component bodies only (not including leads, other interconnecting media, insulators, heat sinks, and so on) and based on nominal sizes of components and nominal outside module dimensions.

volumetric radar — A radar capable of producing three-dimensional position data on several targets.

volumetric sensor — A sensor with a detection zone that extends over a volume such as an entire room, part of a room or a passageway. Ultrasonic motion detectors and sonic motion detectors are examples of volumetric sensors.

volume unit — Abbreviated vu. 1. The unit of transmission measurement for measuring the level of non-steady-state currents. Zero level is the steady-state reference power of 1 milliwatt in a circuit of 600 ohms characteristic impedance. 2. A measure of the power level of the voice wave. Zero volume unit is equivalent to +4 dBm for simple electrical waves (single frequencies). 3. An arbitrary sound-level standard related to the decibel and used for calibrating recording levels.

volume-unit indicator — Also called a vu indicator or volume-unit meter. An instrument calibrated to read audio-frequency power levels directly in volume units.

volume-unit meter — *See* vu meter.

volume velocity — The rate at which a medium flows through a specified area due to a sound wave.

Von Hippel breakdown theory — Also called low-energy criterion. Breakdown occurs at fields for which the rate of recombination of electrons and positive holes is less than the rate of collisional ionization. Assumes no distribution in electron energies.

VOR — Abbreviation for VHF omnirange.

vortex amplifier — 1. A fluidic device in which the angular rate of a vortex is controlled to alter the output. 2. A fluidic amplifier using momentum interaction to produce stream rotation and consequent output modification.

vowel articulation — The percent of articulation obtained when the speech units considered are vowels, usually combined with consonants into meaningless syllables.

VR — *See* virtual reality.

V-ring — In commutator construction, a specially shaped insulating structure having one or more V-shaped sections.

VRR — Abbreviation for visual radio range.

VR tube — Abbreviation for voltage-regulator tube.

VSWR — Abbreviation for voltage standing-wave ratio.

VTF — Abbreviation for vertical tracking force.

VTL — Abbreviation for variable threshold logic.

VTM — *See* voltage-tunable magnetron.

VTR — Abbreviation for videotape recording or video-tape recorder.

VTVM — Abbreviation for vacuum-tube voltmeter.

vu — Abbreviation for volume unit.

vu indicator — Abbreviation for volume-unit indicator.

vulcanized fiber — A dense, homogeneous, cellulosic material that has been partially gelantinized by swelling with a zinc chloride solution. It may be made in the form of sheets, rods, coils, or tubes, and it may be used for electrical insulation as well as in mechanical applications.

vu meter — Abbreviation for volume-unit meter. 1. A volume indicator with a decibel scale and specified dynamic and other characteristics. It is used to obtain correlated readings of speech power necessitated by the rapid fluctuations in a level of voice currents. 2. Strictly, a recording-level meter whose indicator needle's motion is damped according to a specified standard to allow it to respond at a certain speed to sudden impulses without overshooting the mark by more than a certain amount. The term vu meter is loosely applied to practically any recording-level indicator that uses an indicator needle. 3. A type of recording-level indicator that shows average signal levels in decibels relative to a fixed 0-dB reference level or in percent of maximum recommended modulation. The term is frequently used for any level meter using this scale, but it applies mostly to meters having a specified, standard degree of damping.

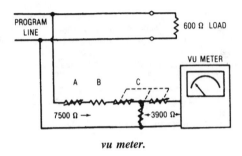

vu meter.

W

W — 1. Letter symbol for watt. 2. Symbol for energy or work.

wafer — 1. A thin semiconductor slice of silicon or germanium with parallel faces on which matrices of microcircuits or individual semiconductors can be formed. After processing, the wafer is separated into dice or chips containing individual circuits. 2. A single section of a wafer switch. 3. Slice of semiconductor crystal material used as substrate for monolithic ICs, diodes, and transistors. 4. The thin slice of semiconductor crystal from which many hundreds or thousands of single monolithic device chips are ultimately obtained. Normally all techniques, such as epitaxy, photolithography, diffusion, and passivation, are carried out on the wafer before it is scribed and broken into individual dice. 5. A thin slice, typically 5 to 30 mils thick, sawed from a cylindrical ingot (boule) of bulk semiconductor material (silicon or gallium arsenide), 4 to 8 or more inches in diameter. Arrays of ICs or discrete devices are fabricated in the wafers during the manufacturing process. The wafer is then tested, scribed, and broken apart to produce semiconductor chips.

wafer and die sorters — Equipment that automates the testing and sorting of semiconductor devices from wafer form.

wafer fabrication — The process of epitaxial growth, impurity diffusion, oxide deposition, and metallization that forms the semiconductor devices on the wafer.

wafer-handling equipment — Equipment used for processing silicon wafers using methods that include batch processing in a common carrier, air-bearing single-wafer processing, and a combination of batch and single-wafer processing.

wafer probing — An electrical test of devices on the wafer, utilizing tiny probes to make contact with the metallized pads on the die.

wafer socket — A vacuum-tube socket that consists of two punched sheets or wafers of an insulating material, separated by spring-metal clips that grip the terminal pins of the inserted tube.

wafer switch — A rotary multiposition switch with fixed terminals on ceramic or Bakelite wafers. The rotor arm, in the center, is positioned by a shaft. Several decks may be stacked onto one switch and rotated by a common shaft.

Wagner ground — A bridge with an additional pair of ratio arms, onto which the ground connection to the bridge is moved in order to effect a perfect balance, free from error.

WAIS — *See* wide area information service.

wait — A computer system service that causes a task to be suspended for a specified time or pending the occurrence of an event.

waiting time — 1. In certain tubes (e.g., thyratrons), the time that must elapse between the turning on of their heaters and the application of plate voltage. 2. *See* access time, 1.

waldo — Mechanical hands that function as an extension of human hands, such as those used to handle radioactive materials. Now a scientific reality, the idea was first used in Robert A. Heinlein's classic story "Waldo," from whence came the name.

walkie-lookie — A compact, portable television camera used for remote broadcasts. The resultant electrical pulses are transmitted by microwave radio to a local control point for retransmission over a standard television station.

walkie-talkie — 1. A two-way radio communication set designed to be carried by one person, usually strapped to the back, and capable of being operated while in motion. 2. A hand-held transceiver.

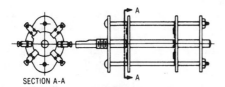

Wafer switch.

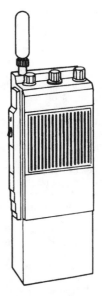

Walkie-talkie, 2.

walk test light — A light on motion detectors that comes on when the detector senses motion in the area. It is used while setting the sensitivity of the detector and during routine checking and maintenance.

walkthrough — Programming by giving a robot instructions one by one, with the robot executing each before receiving the next. The speed of the robot is increased when programming is satisfactory. A teach box is usually used.

wall-attachment amplifier — 1. A family of fluidic elements that make use of flow-created low-pressure regions, causing fluid to adhere to an amplifier wall in a controlled manner. 2. A fluidic device in which the control of the attachment of a stream to a wall(s) (Coanda effect) alters the output.

wall box — A metal box placed in the wall and containing switches, fuses, etc.

wall outlet — A spring-contact device to which a portable lamp or appliance is connected by means of a plug attached to a flexible cord. The wall outlet is installed in a box and connected permanently to the power-line wiring of a home or building.

walls — The sides of the groove in a disc record.

Walmsley antenna — An array of vertical rectangular loops with a height of one wavelength and a spacing of one-half wavelength. The loops are arranged in parallel planes and are series fed at the centers of the longer, vertical sides, with appropriate transpositions of the feed lines.

Walsh radiator — A type of speaker driver invented by Lincoln Walsh, in which a gently sloping cone, moving up and down, so displaces the air with its sloped sides as to radiate a cylindrical wavefront in a 360° horizontal circle.

wamoscope — Abbreviation for a wave-modulated oscilloscope. A cathode-ray tube that includes detection, amplification, and display of a microwave signal in a single envelope, thus eliminating the local oscillator, mixer, IF amplifier, detector, video amplifier, and associated circuitry in a conventional radar receiver. Tubes are available for a range of 2 to 4 GHz.

WAN — Abbreviation for wide area network. A network that serves an area of hundreds or thousands of miles, using common-carrier-provided lines. Contrast with LAN.

wander — *See* scintillation, 2.

warble-tone generator — An oscillator whose frequency is varied cyclically at a subaudio rate over a fixed range. It is usually used with an integrating detector to obtain an averaged transmission or crosstalk measurement.

warm boot — Also called soft start or warm start. 1. Partial restarting of a computer under operating system control without turning it off and on again. 2. Resetting the computer while power is on. *See also* soft start.

warm-up time — 1. In an indirectly heated tube, the time that elapses, after the heater is turned on, before the cathode reaches its optimum operating temperature. 2. The time, following power application to a device, required for the output to stabilize within specifications.

warning net — Communications system established for the purpose of disseminating warning information of enemy movement or action to all interested commands.

washout emitter process — A process used in the manufacture of semiconductor devices. It involves use of the same mask for both the emitter diffusion and the deposition of the ohmic contact. The aluminum metallization for the contact is permitted to form in the same hole that was used for the diffusion.

waste instruction — *See* no-operation instruction.

watch — *See* radio watch.

watchdog — In control systems, a combination of hardware and software that acts as an interlock scheme, disconnecting the system's output from the process in event of system malfunction.

water-activated battery — A primary battery that contains the electrolyte but requires the addition of (or immersion in) water before it becomes usable.

water-cooled tube — A vacuum tube having an anode structure projecting through the glass envelope and constructed to permit circulation of water around the anode for cooling purposes during operation.

water load — A matched waveguide termination in which the electromagnetic energy is absorbed in water. The output power is calculated from the difference in temperature between the water at the input and output.

waterproof motor — A totally enclosed motor so constructed that it will exclude water applied in the form of a stream from a hose. Leakage may occur around the shaft, provided it is prevented from entering the oil reservoir and provision is made for automatically draining the motor. The means for automatic draining may be a check valve, or a tapped hole at the lowest part of the frame that will serve for application of a drain pipe.

WATS — Abbreviation for Wide Area Telephone Service. Also called 800 or 888 service. 1. A service provided by telephone companies in the United States that permits a customer to make calls to or from telephones in a specific nonlocal zone for a flat monthly charge. 2. A service that, among other things, allows people almost anywhere in the United States to call — without charge to them — a certain number. A familiar arrangement has a business paying monthly fees for a so-called 800 number that can be dialed by the public. In-WATS allows a central customer to call out.

watt — Letter symbol: W. 1. A unit of the electric power required to do work at the rate of 1 joule per second. It is the power expended when 1 ampere of direct current flows through a resistance of 1 ohm. In an alternating-current circuit, the true power in watts is effective voltamperes multiplied by the circuit power factor. (There are 746 watts in 1 horsepower.) 2. A measure of electrical or acoustical power. The electrical wattage rating of an amplifier describes the power it can develop to drive a loudspeaker. Acoustical wattage describes the actual sound produced by a loudspeaker in the given environment. (The two figures, in any given amplifier-speaker system, are necessarily very widely divergent inasmuch as the low efficiency of speakers necessitates their receiving relatively large amounts of amplifier power in order to produce satisfactory sound levels.)

wattage rating — 1. The maximum power that a device can safely handle. 2. The maximum power that the resistor can dissipate, assuming a specific life, a standard ambient temperature, and a stated long-term drift from its no-load value. Increasing the ambient temperature or reducing the allowable deviation from the initial value (more-stable resistance value) requires derating the allowable dissipation. With few exceptions, resistors are derated linearly from full wattage at rated temperature to zero wattage at the maximum temperature.

watthour — A unit of electrical work indicating the expenditure of 1 watt of electrical power for 1 hour. Equal to 3600 joules.

watthour capacity — The number of watthours delivered by a storage battery at a specified temperature, rate of discharge, and final voltage.

watthour constant of a meter — The registration, expressed in watthours, corresponding to one revolution of the rotor.

watthour-demand meter — A combined watthour meter and demand meter.

watthour meter — 1. An electricity meter that measures and registers the integral, usually in kilowatthours, of the active power of the circuit into which the meter is connected. This power integral is the energy delivered to the circuit during the integration interval. 2. A totalizing meter that registers the total electrical energy used, usually in kilowatthours.

wattless component — A reactive component.

wattless power — *See* reactive power.

watt loss — *See* power loss, 2.

wattmeter — An instrument for measuring the magnitude of the active power in an electric circuit. Its scale is usually graduated in watts. If graduated in kilowatts or megawatts, the instrument is usually designated as a kilowattmeter or megawattmeter.

wattsecond — The amount of energy corresponding to 1 watt acting for 1 second. It is equal to 1 joule.

wattsecond constant of a meter — The registration, in wattseconds, corresponding to one revolution of the rotor.

wave — 1. A physical activity that rises and falls, or advances and retreats, periodically as it travels through a medium. 2. Propagated disturbance, usually periodic, such as a radio wave or sound wave. If the periodic motion is regular and recurring, it is said to be aperiodic or damped.

wave amplitude — The maximum change from zero of the characteristic of a wave.

wave analyzer — An electric instrument for measuring the amplitude and frequency of the various components of a complex current or voltage wave.

wave angle — The angle at which a wave is propagated from one point to another.

wave antenna — Also called a Beverage antenna. A directional antenna composed of a parallel horizontal conductor one-half to several wavelengths long, and terminated to ground in its characteristic impedance at the far end.

wave band — 1. A band of frequencies, such as that assigned to a particular type of communication service. 2. The band of frequencies comprising an electrical wave.

wave-band switch — A multiposition switch for changing the frequency band tuned by a receiver or transmitter.

wave clutter — Clutter caused on a radar screen by echoes from sea waves.

wave converter — A device for changing a wave from one pattern to another (e.g., baffle-plate, grating, and sheath-reshaping converters for waveguides).

wave duct — 1. A tubular waveguide capable of concentrating the propagation of waves within it. 2. A natural duct formed in air by atmospheric conditions. Waves of certain frequencies travel through it with more than average efficiency.

wave equation — An equation that gives a mathematical specification of a wave process or describes the performance of a medium through which a wave is passing.

wave filter — A transducer for separating waves on the basis of their frequency. It introduces a relatively small insertion loss to waves in one or more frequency bands, and a relatively large insertion loss to waves of other frequencies. *See also* filter, 1.

wave fluxing — A wave solder fluxing method. Flux is applied using the liquid wave principle to form a double-sided parabolic wave. Washing action of the wave promotes flux coverage of the underside surfaces, while capillary forces promote thru-hole penetration. *See also* brush fluxing; foam fluxing; spray fluxing.

waveform — 1. The shape of an electromagnetic wave. 2. A graphical representation of the relationship

between voltage, current, or power against time. It also provides a picture of the behavior of signals at given frequencies. 3. A geometrical shape as obtained by displaying a characteristic voltage or current as a function of time. 4. The variations in magnitude and polarity of a current or voltage with respect to time, plotted in graphic form.

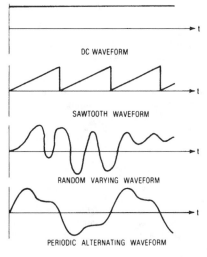

Waveforms, 2.

waveform-amplitude distortion — Sometimes called amplitude distortion. Nonlinear waveform distortion caused by unequal attenuation or amplification between the input and output of a device.

waveform analyzer — 1. An instrument that measures the amplitude and frequency of the components in a complex waveform. 2. A frequency-selective voltmeter that can be used to determine the frequency and amplitude of each sine-wave component of a complex wave. 3. An instrument that measures the amplitude of a waveform over a specific period, with resolution usually in the low nanosecond range.

waveform digitization — The earliest approach taken for speech synthesis. The technique relies on sampling of the waveform in the time domain at twice the highest frequency of interest (this is known as the Nyquist rate). Critical to the use of this technique is data compression; otherwise, memory requirements are prohibitive.

waveform encoding — A synthesis technique that reproduces speech by reconstructing the original speech waveform.

waveform error — A change in measurement accuracy of an average-reading ac meter, occurring when the waveform of the measured voltage or current departs from a pure sine-wave form.

waveform generator — Also called arbitrary waveform generator. 1. *See* timing-pulse distributor. 2. An instrumentation that does not necessarily employ mathematically derived relationships between the output voltage and the time base to produce its output waveshape. Most generators of this type develop an output that follows a waveshape drawn out or copied manually from some existing natural phenomenon (e.g., an EKG trace or the stress versus elongation curve from a pull-to-fracture test of a metal sample) or specified via keyboard entries. The waveform may also follow tracings made manually on an accompanying digitizer tablet.

waveform influence — The change in meter indication of the applied current and/or voltage caused solely by a change in waveform from a specified waveform.

waveform monitor — 1. An oscilloscope designed for viewing signal waveforms. 2. An oscilloscope designed especially for viewing the waveform of a video signal.

waveform recorder — A general-purpose, waveform-capturing instrument with high-speed, high-resolution analog-to-digital converters and digital memory.

waveform synthesizer — Equipment for generating a signal of a desired waveform.

wavefront — 1. With respect to a wave in space, a continuous surface at every point of which the displacement from zero in the positive or negative direction is the same at any instant. In a periodic wave, the displacements of the points on a wavefront are in phase. For a surface wave, the wavefront is a continuous line whose points have the same properties as do the points in the wavefront of a wave in space. 2. That part of a signal-wave envelope between the initial point of the envelope and the point at which the envelope reaches its crest. 3. For a field of electromagnetic energy emanating from a source, the wavefront is a surface connecting all field points that are equidistant from the source. 4. A surface at right angles to rays that proceed from the wave source. The surface passes through those parts of the wave that are in the same phase and travel in the same direction. For parallel rays, the wavefront is a plane; for rays that radiate from a point, the wavefront is spherical.

wave function — In a wave equation, a point function that specifies the amplitude of a wave.

waveguide — 1. A system of material designed to confine direct electromagnetic waves in a direction determined by its physical boundaries. 2. A transmission line comprising a hollow conducting tube (rectangular or tubular) within which electromagnetic waves are propagated on a solid dielectric or dielectric-filled conductor.

waveguide attenuator — A waveguide device for producing attenuation by some means (e.g., by absorption and reflection).

waveguide connector — Also called a waveguide coupling. A mechanical device for electrically joining parts of a waveguide system together.

waveguide coupling — *See* waveguide connector.

waveguide critical dimension — The dimension of waveguide cross section that determines the cutoff frequency.

waveguide cutoff frequency — Also called the critical frequency. The frequency limit of propagation, along a waveguide, for waves of a given field configuration.

waveguide dummy load — Sections of waveguide for dissipating all the power entering the input flange.

waveguide elbow — A bend in a waveguide.

Waveguide elbows.

waveguide flange — *See* flange.

waveguide lens — A microwave device in which the required phase changes are produced by refraction through suitable waveguide elements acting as lenses.

waveguide mode suppressor — A filter used to suppress undesired modes of propagation in a waveguide.

waveguide phase shifter — A device for adjusting the phase of the output current or voltage relative to the phase at the input of a device.

waveguide plunger — 1. In a waveguide, a plunger used for reflecting the incident energy. 2. A movable shorting plate used to adjust the length of a resonant waveguide section.

waveguide post — In a waveguide, a rod placed across the waveguide and behaving substantially like a shunt susceptance.

waveguide propagation — Long-range communications in the 10-kHz to 35-kHz frequency range by the waveguide characteristics of the atmospheric duct formed by the ionospheric D layer and the surface of the earth. *See also* atmospheric duct.

waveguide resonator — A waveguide device intended primarily for storing oscillating electromagnetic energy.

waveguide shim — A thin metal sheet inserted between waveguide components to ensure electrical contact.

waveguide shutter — A vane within a waveguide, used to protect the receiver system from adjacent radar power by establishing an electrical short across the waveguide when a companion transmitter is operating.

waveguide stub — An auxiliary section of waveguide that has an essentially nondissipative termination and is joined to the main section of the waveguide.

waveguide switch — A transmission-line switch for connecting a transmitter or receiver from one antenna to another or to a dummy load.

waveguide taper — A section of tapered waveguide.

waveguide tee — A junction for connecting a branch section of waveguide in series or parallel with the main transmission line.

waveguide transformer — A device, usually fixed, added to a waveguide for the purpose of impedance transformation.

waveguide tuner — An adjustable device added to a waveguide for the purpose of impedance transformation.

waveguide twist — A waveguide section in which the cross section rotates about the longitudinal axis.

waveguide wavelength — Also called a guide wavelength. For a traveling plane wave at a given frequency, the distance along the waveguide between the points at which a field component (or the voltage or current) differs in phase by 2π radians.

wave heating — The heating of a material by energy absorption from a traveling electromagnetic wave.

wave impedance (of a transmission line) — At every point in a specified plane, the complex ratio between the transverse components of the electric and magnetic fields. (Incident and reflected waves may both be present.)

wave interference — The phenomenon that results when waves of the same or nearly same type and frequency are superimposed. It is characterized by variations in the wave amplitude that differ from that of the individual superimposed waves.

wavelength — 1. In a periodic wave, the distance between points of corresponding phase of two consecutive cycles. The wavelength (λ) is related to the phase velocity (v) and frequency (f) by the formula $\lambda = v/f$. 2. The physical distance between cycles. Wavelength also may be stated as being equal to the distance traveled by a wave in the time required for one cycle. Wavelength equals the speed divided by the frequency. (For most purposes, we assume that all electromagnetic waves travel at the speed of light in a vacuum, approximately 186,000 miles per second or 3×10^8 m/s.) 3. The distance between the beginning and the end of a complete cycle of any spatial periodic phenomenon. In acoustics, it is the distance occupied by one cycle of a repetitive sound traveling through the air at a velocity of about 1100 feet per second or 335 m/s. (An 1100-Hz tone has a wavelength of 1 foot or 30.48 cm.) In magnetic recording, it refers to the length of tape occupied by a full cycle of recorded signal (at $7\frac{1}{2}$ ips tape speed, a recorded frequency of 1000 Hz has a wavelength on the tape of 0.0075 inch). 4. The basic characteristic of electromagnetic waves that describes the distance that a wave travels during one complete cycle. Units most frequently used in light measurements are the angstrom, Å (10^{-10} meter), and the nanometer, nm (10^{-9} meter). 5. The distance between three consecutive nodes of a wave, equal to 360 electrical degrees. It is equal to the velocity of propagation divided by the frequency, when both are in the same units. 6. In tape recording, the shortest distance between two peaks of the same magnetic polarity; also, the ratio of the tape speed to recorded frequency.

wavelength constant — The imaginary part of the propagation constant — i.e., the part that refers to the retardation in phase of an alternating current passing through a length of transmission line.

wavelength of peak radiant intensity — The wavelength at which the spectral distribution of radiant intensity is a maximum.

wavelength shifter — A photofluorescent compound employed with a scintillator material. Its purpose is to absorb photons and emit related photons of a longer wavelength, thus permitting more efficient use of the photons by the phototube or photocell.

wave mechanics — A general physical theory whereby wave characteristics are assigned to the components of atomic structure, and all physical phenomena are interpreted in terms of hypothetical waveforms.

wavemeter — An instrument for measuring the wavelength of a radio-frequency wave. Resonant-cavity, resonant-circuit, and standing-wave meters are representative types.

wave normal — A unit vector normal to an equiphase surface, with its positive direction taken on the same side of the surface as the direction of propagation. In isotropic media, the wave normal is in the direction of propagation.

wave number — 1. The number of waves per unit length. The usual unit of wave number is the reciprocal centimeter, cm^{-1}. In terms of this unit, the wave number is the reciprocal of the wavelength when the latter is in centimeters in a vacuum. 2. Reciprocal of wavelength. The number of waves per centimeter. Wave number is directly proportional to energy.

wave oils — Liquid compounds formulated for use as the oil in oil-intermix wave-soldering equipment and as pot coverings on still solder pots.

wave packet — A short pulse of waves (e.g., spin waves).

wave propagation — *See* propagation.

waveshape — A graph of a wave as a function of time or distance.

wave soldering — 1. A manufacturing process for connecting components to a printed circuit board. After the components are inserted and the copper side of the board is fluxed, the board moves on a conveyor across a tank that is so adjusted that the peak of a standing wave of molten solder just makes contact with the copper side of the board and, thus, solders all components to the pc board in the pan. 2. A process wherein printed boards are brought in contact with a gently overflowing

wave of liquid solder that is circulated by a pump in an appropriately designed solder pot reservoir. The prime functions of the molten wave are to serve as a heat source and heat transfer medium and to supply solder to the joint area.

wave-soldering equipment — Systems that achieve wave soldering and that consist of stations.

wave tail — That part of a signal-wave envelope between the steady-state value (or crest) of an envelope and the end.

wave tilt — The forward inclination of a radio wave due to its proximity to ground.

wave train — A limited series of wave cycles caused by periodic short-duration disturbances.

wave trap — Also called a trap. 1. A device used to exclude unwanted signals or interference from a receiver. Wave traps are usually tunable to enable the interfering signal to be rejected or the true frequency of a received signal to be determined. 2. A tuned parallel resonant circuit for the purpose of eliminating one undesired frequency from the circuit to which it is connected.

wave velocity — A quantity that specifies the speed and direction at which a wave travels through a medium.

wax — In mechanical recording, a blend of waxes with metallic soaps.

wax master — *See* wax original.

wax original — Also called a wax master. An original recording on a wax surface, from which the master is made.

way-operated circuit — A circuit shared by three or more stations on a party-line basis. One of the stations may be a switching center. May be a single or duplex circuit.

way point — A selected point having some significance on a radionavigational course line.

way station — In telegraphy, one of the stations on a multipoint network.

weak coupling — Loose coupling in a radio-frequency transformer. *See also* loose coupling.

wearout — 1. The point at which the continued operation and repair of an item becomes uneconomical because of the increased frequency of failure. The end of the useful life of the item. 2. The process of attrition that results in an increase of hazard rate with increasing age (cycles, time, miles, events, and so on as applicable for the item).

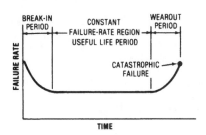

Wearout.

wearout failure — A failure that is predictable on the basis of known wearout characteristics. This type of failure is due to deterioration processes or mechanical wear, the probability of occurrence of which increases with time.

weather-protected motor — An open motor whose ventilating passages are so designed as to minimize the entrance of rain, snow, and airborne particles to the electric parts.

web — Web, used as a proper noun (capitalized), is short for the World Wide Web (WWW). As an adjective, it denotes relationship to activities pertaining to the World Wide Web.

web browser — 1. A software application that permits browsing, retrieval, and viewing of content from the World Wide Web. A web browser is a client for the HTTP, FTP, and gopher protocols, as well as others. 2. A type of software that allows one to navigate the Web.

WebCrawler — A tool for searching WWW sites for any documents containing the user-supplied search string.

weber — The practical unit of magnetic flux, equal to the magnetic flux that, linking a circuit of one turn, produces in it an electromotive force of 1 volt as it is reduced to zero at a uniform rate in one second. One weber equals 10^8 maxwells. Symbolized Wb.

webmaster — The person in charge of a web site.

web page — A hypertext page in the WWW system. *See* home page; HTML; hypertext.

web server — A networked host computer that contains HTML pages and possibly other forms of content served to clients via HTTP.

web site — 1. A site (location) on the WWW with a web page. 2. The primary web server or collection of web servers on the Internet that represents an entity such as a company, university, organization, or other institution. The term does not usually refer to a web server on an internal LAN. Many sites on the Internet have adopted the naming convention of using the hostname or hostname alias to denote the primary web server in the domain name; e.g., www.northcountryradio.com.

WECS — Large wind-energy conversion systems, with a generating capacity of more than 100,000 kilowatts.

wedge — 1. The fan-shaped pattern of equidistant black and white converging lines in a television test pattern. 2. A waveguide termination consisting of a tapered length of dissipative material, such as carbon, which is introduced into the guide.

wedge bond — 1. Metal to metal lead bond formed by a wedge-shaped tool. It may be a cold weld, ultrasonic, or thermal compression bond. 2. A bond between a gold wire and a gold metallized substrate using a wedge-shaped tool. Thermocompression bonding combines temperature and pressure to make a wedge bond. Ultrasonic bonding combines ultrasonic energy with the pressure of the tool.

wedge bonding — 1. A type of thermocompression bonding used in integrated-circuit manufacturing whereby a wedge-shaped tool is used to press a small section of the lead wire onto a bonding pad. 2. A bond formed when a heated wedge is brought down on a wire prepositioned on a heated contact. The wedge's heat and the pressure of the wedge in combination with the heat applied to the mounting contact form the bond.

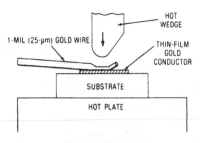

Wedge bonding.

wedge-type connector — A connector in which the contact between the conductor and connector is made by pressure exerted by a wedge.

Wehnelt cathode — A hot cathode that consists of a metallic core coated with alkaline-earth oxides. It is widely used in vacuum tubes. *See also* oxide-coated cathode.

weight — 1. The force with which a body is attracted toward the earth. 2. *See* significance.

weight coefficient of a thermoelectric generator — The quotient of the electrical power output of the thermoelectrical generator divided by the weight of the generator.

weight coefficient of a thermoelectric generator couple — The quotient of the electrical power output of the thermoelectric couple divided by the weight of the couple.

weighted average — An averaging technique in which the data to be averaged are nonuniformly weighted. The weights must always sum to 1.00 or 100 percent.

weighted distortion factor — The weighting of harmonics in proportion to their harmonic relationship.

weighted noise — 1. The measured noise level in electronic equipment or in an acoustic environment with a measuring device that uses any of several standard filters that restrict response to the audio-frequency spectrum or a selected portion thereof. 2. The noise measured within the audio-frequency band using a measuring instrument that has a frequency-selective characteristic adjusted to correspond to that of the average human hearing response.

weighted noise level — The noise level weighted in accordance with the 70-decibel equal-loudness contour of the human ear and expressed in dBm.

weighted value — The numerical value assigned to any single bit as a function of its position in a code word.

weighting — 1. The artificial adjustment of measurements in order to account for factors that, during normal use of a device, would otherwise differ from the conditions during measurement. For example, background-noise measurements may be weighted by applying factors or introducing networks to reduce the measured values in inverse ratio to their interference. 2. Any correction factor added to a measurement to make it correlate more accurately with subjective perceptions. A noise measurement may be weighted at various parts of the audio spectrum to reflect the ear's acute sensitivity around 3000 Hz and relative lack of sensitivity at 60 Hz.

weighting network — A network whose loss varies with frequency in a prescribed standard manner.

weightlessness — 1. A condition in which no acceleration, whether of gravity or other force, can be detected by an observer within the system in question. 2. A condition in which gravitational and other external forces acting on a body produce no stress, either internal or external, in the body. Weightlessness occurs when gravity forces are exactly balanced by other forms of acceleration (zero *g*).

weld — The consolidation of two metals, usually by application of heat to the proposed joint.

welded circuit — A circuit of electronic parts in which leads are interconnected by welding techniques.

weldgate pulse — A waveform used in controlling the flow of welding current.

welding — 1. A process in which the oxide layer covering the surface of a metal is eliminated, and the irregular surfaces are brought into intimate contact over the entire surface of the bond. Welding may be accomplished with the application of heat, mechanical force, or heat and mechanical force. 2. Joining thermoplastic pieces by one of several heat-softening processes.

welding transformer — A power transformer with a secondary winding consisting of only a few turns of very heavy wire. It is used to produce high-value alternating currents at low voltages for welding purposes.

weld junction — A junction formed by heat or metallurgical fusion of conductors. It provides a strong electrical connection with good conductivity. It is widely used in microelectronic packaging. Wires, ribbons, or films as small as 0.0005-inch (12.7-μm) thick can be joined by resistance and electron-beam welding methods.

weld-on surface-temperature resistor — A surface-temperature resistor installed by welding the sensing element to the surface being measured.

weld polarity — Certain material combinations have a different resistance to a weld current, depending on the direction of the current. In dc welding, a suitable weld may be possible in only one direction of current. A weld schedule must define the proper polarity for such cases.

weld time — The interval during which current is allowed to flow through the work during the performance of one weld. In pulsation welding, the weld period includes the "cool" time intervals.

well — A region of silicon formed by introducing impurities of opposite polarity, usually in the substrate; used to separate MOS transistors there.

Wertheim effect — When a wire placed in a longitudinal magnetic field is twisted, there will be a transient voltage difference between the ends of the wire.

WESTAR I — American TV satellite operated by Western Union. It has 12 transponders and is located at 99° west longitude.

WESTAR II — American TV satellite (in the WESTAR series) located at 123.5° west longitude. It has 12 transponders.

WESTAR III — American TV satellite in the WESTAR series. It has 12 transponders and is located at 91° west longitude.

Western Union joint — A strong, highly conductive splice made by crossing the cleaned ends of two wires, twisting them together, and soldering.

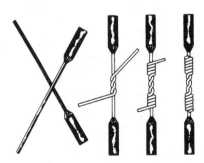

Western Union joint.

Weston normal cell — A standard cell of the saturated cadmium type, in which the positive electrode is cadmium and the electrolyte is a cadmium-sulfate solution.

Westrex system — *See* forty-five/forty-five.

wet — Term describing the condition in which the liquid electrolyte in a cell is free-flowing.

wet cell — A cell whose electrolyte is in liquid form and free to flow and move.

wet-charged stand — The period of time that a charged wet secondary cell can stand before losing a specified, small percentage of its capacity.

wet circuit — 1. A circuit that carries direct current. 2. Circuit having current flow to melt (microscopically) contact material at point of contact, thereby dissolving and evaporating away contaminants.

wet contact — A contact through which direct current flows.

wet dross — Metallic droplets of molten solder suspended in oxides, tarnish, and charred organic compounds. Wet dross normally is formed as a result of some mechanical agitation, as in the case of the wave when droplets of molten solder are retained in the dross without being able to coalesce with adjacent droplets and/or through gravity rejoin the parent metal underneath. Wet dross often appears as a spongy metallic mass floating on the surface.

wet electrolytic capacitor — An electolytic capacitor that has a liquid electrolyte.

wet flashover voltage — The voltage at which the air surrounding a clean, wet insulator shell breaks down completely between electrodes. This voltage will depend on the conditions under which the test is made.

wet process — Also called slip process. A method of preparing a ceramic body in which the constituents are blended in a form that is sufficiently liquid to produce a suspension for use as is or in subsequent processing.

wet-reed relay — A reed-type relay containing mercury at the relay contacts to reduce arcing and contact bounce.

wet shelf life — The period that a wet secondary cell can remain discharged without deteriorating to a point at which it cannot be recharged.

wet tantalum capacitor — A polar electrolytic capacitor whose cathode is a liquid electrolyte (a highly ionized acid or salt solution). Characteristics: highest capacity per unit volume, low impedance, lowest dc leakage, excellent shelf life. These are polar devices. They are generally used in voltage applications up to 125 volts in the temperature range −55°C to +200°C.

wetted surface — A surface on which solder flows uniformly to make a smooth, continuous, adherent layer.

wetting — 1. The formation of a uniform, smooth, unbroken, and adherent film of solder to a base metal. 2. A phenomenon involving a solid and a liquid in such intimate contact that the adhesive force between the two phases is greater than the cohesive force within the liquid. Thus, a solid that is wet, on being removed from the liquid bath, will have a thin continuous layer of liquid adhering to it. Foreign substances such as grease may prevent wetting. (Other agents, such as detergents, may induce wetting by lowering the surface tension of the liquid.)

Wheatstone bridge — Also called resistance bridge. A null-type resistance-measuring circuit in which resistance is measured by direct comparison with a standard resistance. *See* figure on page 848.

wheeling — A technique by which a utility, located between two others, transfers power from its neighbor on one side to its neighbor on the other side.

wheel static — Auto-radio interference due to a static charge buildup between the brake drum and the wheel spindle.

whiffletree switch — In computers, a multiposition electronic switch composed of gate tubes and flip-flops. It is so named because its circuit diagram resembles a whiffletree.

whip antenna — 1. A simple vertical antenna consisting of a slender whiplike conductor supported on a

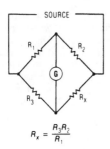

$$R_x = \frac{R_3 R_2}{R_1}$$

Wheatstone bridge.

base insulator; used mainly on motor vehicles. 2. The flexible metal pole used as an antenna on automobiles.

whisker — 1. *See* catwhisker. 2. A very small, hair-like metallic growth (a micron-size single crystal with a tensile strength of the order of one million psi) on a metallic circuit component. 3. Ultrapure elongated metal and ceramic filaments of extremely high tensile strengths. 4. Single-crystal growth resembling fine wire that appears most frequently on components that have been electroplated with tin. Whisker growth requires no voltage. 5. A slender acicular (needle-shaped) metallic growth on a printed board.

whistler mode propagation — 1. The transmission of radio waves between conjugate points with respect to the geomagnetic equator (i.e., points of opposite geomagnetic latitude and equal geomagnetic longitude) by the apparent ducting of waves along the flux lines of the geomagnetic field. 2. The transmission of radio signals along the flux lines of the earth's magnetic field from the northern hemisphere to the southern hemisphere.

whistlers — 1. High-frequency atmospherics that decrease in pitch and then tend to rise again. 2. Audio-frequency waves from lightning-stroke radiation that have penetrated the ionosphere.

white — 1. For color TV, white is a mixture of red, green, and blue in the picture this is produced by exciting all three dots in each phospor trio. Since the eye cannot distinguish the individual dots, the mixture appears white. 2. The facsimile signal produced when an area of subject copy having minimum density is scanned.

white balance — The adjustment of the red, green, and blue channels in a color camera to produce the correct balance (and, thus, white) when shooting a flat white field.

white circuit — A cathode follower with another tube replacing the cathode resistor. By driving this additional tube with a signal that is out of phase with the original signal, low-impedance broadband characteristics are obtained.

white compression — 1. Also called white saturation. In facsimile or television, a reduction in gain (relative to the gain at the level for a midrange light value) at signal levels corresponding to light areas of the picture. The overall effect of white compression is reduced contrast in the highlights of the picture. 2. Amplitude compression of the signals corresponding to the white regions of the picture; results in differential gain.

white-dot pattern — *See* dot pattern.

white level — 1. The carrier-signal level that corresponds to maximum picture brightness in television and facsimile. 2. The picture signal level corresponding to a specified maximum limit for white peaks.

white light — 1. Radiation that has a spectral energy distribution that produces the same color stimulus to the unaided eye as that of noon sunlight. 2. Light perceived as achromatic, that is, without hue.

white noise — Also called Gaussian noise. 1. Random noise (e.g., shot and thermal noise) whose constant energy per unit bandwidth is independent of central frequency at the band. The name is taken from the analogous definition of white light. 2. The random motion of electrons in a conductor, which, when reproduced through a loudspeaker or phones, sounds like noise and covers a wide frequency range. It is used to test loudspeakers and phones for resonance and sensitivity. 3. Noise whose amplitude (strength) is a random (Gaussian) variable but which has equal energy distribution over all frequencies of interest, regardless of the center frequency of the frequency range being considered. 4. A complete mixture of all frequencies in one signal characterized by equal energy per bandwidth. This means that there is an equal amount of energy between 100 and 200 Hz as between 200 and 300 Hz, or 1000 to 1100 Hz, and so on. 5. Random noise that contains equal energy for any equal bandwidths. (For example, the 5000-Hz bandwidth between 5000 Hz and 10,000 Hz or between 10,000 Hz and 15,000 Hz has equal white noise energy.) The total energy per octave increases 6 dB/octave with increasing frequency. 6. Random noise having a special density that is substantially independent of frequency over a specified range. It is widely used in the random vibration testing of devices. 7. Random variations in voltage, current, or data, therefore often thermal in origin. Audible manifestation is a hissing sound. It usually exists for the entire duration of a connection. At low levels, it may not impair operation at all. At high levels, it may render both voice and data communication impossible. Electrical noise, (specifically of thermal origin), is also known as Johnson noise.

white object — An object that reflects all wavelengths of light with substantially equal high efficiencies and considerable diffusion.

white peak — A peak excursion of the picture signal in the white direction.

white peak clipping — Limiting the amplitude of the picture signal to a preselected maximum white level.

white plume — The result of target lag in an image pickup tube that causes a moving highlight to trail a highlight and resemble a white plume.

white raster — *See* chroma-clear raster.

white recording — 1. In an amplitude-modulation system, that form of recording in which the maximum received power corresponds to the minimum density of the record medium. 2. In a frequency-modulation system, that form of recording in which the lowest received frequency corresponds to the minimum density of the record medium.

white room — An area in which the atmosphere is controlled to eliminate dust, moisture, and bacteria. It is used in the production and assembly of components and systems whose reliability or functions might be adversely affected by the presence of foreign matter.

white saturation — *See* white compression.

white signal — The facsimile signal produced when a minimum-density area of the subject copy is scanned.

white-to-black amplitude range — 1. In a facsimile system employing positive amplitude modulation, the ratio of signal voltage (or current) for picture white to that for picture black at any point in the system. 2. In a facsimile system employing negative amplitude modulation, the ratio of the signal voltage (or current) for picture black to that for picture white.

white-to-black frequency swing — In a facsimile system employing frequency modulation, the numerical difference between the signal frequencies corresponding to picture white and picture black at any point in the system.

white transmission — 1. In an amplitude-modulation system, that form of transmission in which the maximum transmitted power corresponds to the minimum density of the subject copy. 2. In a frequency-modulation system, that form of transmission in which the lowest transmitted frequency corresponds to the minimum density of the subject copy.

whole step — *See* whole tone.

whole tone — Also called a whole step. The interval between two sounds with a basic frequency ratio approximately equal to the sixth root of 2.

wicking — 1. The flow of solder up under the insulation on covered wire. 2. The act of drawing moisture through a fabric or thread, like the action of a wick in an oil lamp when it draws oil up to the flame. 3. Capillary absorption of liquid (including water) along the fibers of the base material. 4. The tendency for flux and solder to run in under the insulation of a wire when its end is being soldered to a terminal. 5. Desoldering method utilizing prefluxed braid or stranded wire, or braid used with flux. The wick material is placed on the solder joint and a heated iron tip is applied to the wick. Capillary action draws the solder up into the wick material. 6. The longitudinal flow of a liquid in a wire or cable due to capillary action.

wide-angle lens — A lens that picks up a wide area of a television stage setting at a short distance.

wide area data service — Automatic wide teletypewriter data exchange service by way of leased commercial lines.

wide area information service — Abbreviated WAIS. A Net-wide system for looking up specific information in Internet databases.

wide area telephone service — *See* WATS.

wideband — 1. Capable of passing a broad range of frequencies (said of a tuner or amplifier). Especially vital to good multiplex reception and for faithful audio reproduction. 2. Having a bandwidth greater than a voice band. 3. A communications channel having a bandwidth characterized by data transmission speeds of 10,000 to 500,000 bits per second. 4. Implies data speeds requiring the equivalent of more than one voice-frequency channel for operation; broadband. 5. *See* broadband.

wideband amplifier — An amplifier capable of passing a wide range of frequencies with equal gain.

wideband axis — In phasor representation of the chrominance signal, the direction of the phasor representing the fine-chrominance primary.

wideband circuit — A transmission facility having a bandwidth greater than that of a voice-grade line.

wideband communications system — A communications system that provides numerous channels of communications on a highly reliable and secure basis; the channels are relatively invulnerable to interruption by natural phenomena or countermeasures. Included are multichannel telephone cable, tropospheric scatter, and multichannel line-of-sight radio systems such as microwave.

wideband improvement — Ratio of the signal-to-noise ratio of the system in question to the signal-to-noise ratio of a reference system.

wideband ratio — Ratio of the occupied-frequency bandwidth to the intelligence bandwidth.

wideband repeater — An airborne system that receives an rf signal, and conditions it, translates it in frequency, and amplifies it for transmission. Such a repeater is used in reconnaissance missions when low-altitude aircraft require an airborne relay platform for transmission of data to a readout station beyond the line of sight.

wide open — Refers to the untuned characteristic or lack of frequency selectivity.

wide-open receiver — A receiver that has essentially no tuned circuits and is designed to receive all frequencies simultaneously in the band of coverage.

width — 1. The distance between two specified points of a pulse. 2. The horizontal dimension of a television or facsimile display.

width coding — Modifying the duration of the pulses emitted from the transponder according to a prearranged code for recognition in the display.

width control — A television-receiver or an oscilloscope control that varies the amplitude of the horizontal sweep and, hence, the width of the picture.

Wiedemann effect — The direct Wiedemann effect is the twist produced in a wire placed in a longitudinal magnetic field when a current flows through the wire; the twist is due to the helical resultant of the impressed longitudinal field and the circular field of the wire. The magnetic material expands (or contracts) parallel to the helical lines of force, hence the twist. The inverse Wiedemann effect is the axial magnetization of a current-carrying wire when twisted.

Wiedemann-Franz law — A theoretical result that states that the ratio of thermal conductivity to electrical conductivity is the same for all metals at the same temperature.

Wiegand wire — A single strand of ferromagnetic material fabricated to produce a permanently work-hardened outer layer called the shell and a relatively soft inner layer called the core. Under the influence of a strong magnetic field, the shell and core have the same magnetic polarity. Wire-core polarity is reversed by placing the wire in a weak external magnetic field of the opposite polarity that leaves the original shell polarity intact. Increasing the external field strength then reverses the shell polarity. A small wire coil placed near the Wiegand wire can sense these polarity changes and produce a corresponding voltage pulse. This phenomenon can be used in a variety of self-powered sensors that produce voltage pulses without any other supporting circuitry.

Wien bridge — An alternating-current bridge used to measure inductance or capacitance in terms of resistance and frequency. *See also* Wien capacitance bridge; Wien inductance bridge.

Wien-bridge oscillator — 1. An oscillator whose frequency is controlled by a Wien bridge. 2. A type of phase-shift oscillator that uses resistance and capacitance in a bridge circuit to control the frequency.

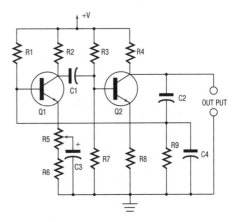

Wien-bridge oscillator.

Wien capacitance bridge—A four-arm alternating-current capacitance bridge used for measuring capacitance in terms of resistance and frequency. Two adjacent arms contain capacitors—one in series and the other in parallel with a resistor—while the other two are normally nonreactive resistors. The balance depends on the frequency, but the capacitance of either or both capacitors can be computed from the resistances of all four arms and from the frequency.

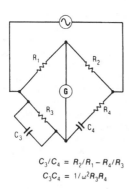

$$C_3/C_4 = R_2/R_1 - R_4/R_3$$
$$C_3 C_4 = 1/\omega^2 R_3 R_4$$

Wien capacitance bridge.

Wien displacement law—The relationship between the temperature of a black body and the wavelength for its emission maximum. The wavelength of maximum emission may be found from the expression

$$\lambda_{max} = \frac{2898 \ \mu m - degrees}{T}$$

where T is in kelvins.

Wien inductance bridge—A four-arm alternating-current inductance bridge used for measuring inductance in terms of resistance and frequency. Two adjacent arms contain inductors—one in series and the other in parallel with a resistor—while the other two are normally nonreactive resistors. The balance depends on the frequency, but the inductances of either or both inductors can be computed from the resistors of the four arms and from the frequency.

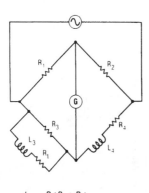

$$\frac{L_3}{L_4} = \frac{R_1(R_L + R_3)}{R_2 R_3 - R_1 R_4}$$

$$\omega^2 L_3 L_4 = R_4(R_L + R_3) - R_L R_3(R_2/R_1)$$

Wien inductance bridge.

Wien radiation law—An expression representing approximately the spectral radiance of a black body as a function of its wavelength and temperature.

Wien's law—The wavelength of maximum radiation intensity is inversely proportional to the absolute temperature of a black body, and the intensity of radiation at this maximum wavelength varies as the fifth power of the absolute temperature. *See also* Wien displacement law.

wild card—Characters that represent one or more other characters in a computer. For example, in DOS and OS/2, an asterisk is a wild card that stands for any combination of letters; a question mark stands for any single character.

Williamson amplifier—A high-fidelity push-pull audio-frequency amplifier using triode-connected tetrodes. The circuit was developed by D. T. N. Williamson.

Williams-tube storage—A type of electrostatic storage using a cathode-ray tube.

Wimshurst machine—A common static machine or electrostatic generator consisting of two coaxial insulating discs rotating in opposite directions. Sectors of tinfoil are arranged, with respect to a connecting rod and collecting combs, so that static electricity is produced for charging Leyden jars or discharging across a gap.

Wimshurst static machine.

Winchester disk—Originally an IBM code name for a small, hard disk. Now a generic name for any (permanently sealed) hard-disk system for computer use. Winchester disks are available in 14-inch (35.6-cm), 8-inch (20.3-cm), or $5^1/_4$-inch (13.3-cm) diameters.

wind—The way in which recording tape is wound onto a reel. An A wind is one in which the tape is wound so that the coated surface faces toward the hub; a B wind is one in which the coated surface faces away from the hub. A uniform, as opposed to an uneven, wind is one giving a flat-side tape pack free from laterally displaced, protruding layers.

wind charger—A wind-driven dc generator for charging batteries (e.g., 32-volt batteries formerly used on many farms).

winding—1. A conductive path, usually wire, inductively coupled to a magnetic core or cell. Windings may be designated according to function—e.g., sense, bias, drive, etc. 2. One or more turns of wire forming a continuous coil for a transformer, relay, rotating machine, or other electromagnetic device.

winding arc—In an electrical machine, the length of a winding stated in terms of degrees.

winding factor—The ratio of the total area of wire in the center hole of a toroid to the window area of a toroid or transformer core.

wind loading—1. The force exerted by the wind on the surface of a dish antenna. It can cause misalignment or damage to the system. 2. The maximum wind an antenna is rated to withstand without being damaged. Expressed in miles per hour.

Windom antenna—A horizontal half-wave dipole located above ground and fed by a vertical or nearly vertical single wire connected at a point approximately one-twelfth wavelength from the center of the dipole.

window—1. In digital filter design, a type of weighting function. 2. In graphical interface terminology, any area of a computer display temporarily dedicated to running a particular software-controlled task. 3. One of several possibly overlapping areas of a terminal screen that communicates with an independent process or program. 4. Strips of metal foil, wire, or bars dropped from aircraft or fired from shells or rockets as a radar countermeasure. 5. The small area through which beta rays enter a Geiger-Mueller tube. 6. Aperture in a photresist coating produced by exposure and development. 7. In computer graphics, a defined area in the system not bounded by any limits; unlimited "space" in graphics.

window area—The opening in the laminations of a transformer.

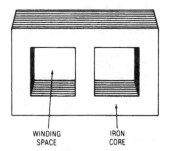

Window area.

window corridor—Also called the infected area or lane. An area in which window has been sown.

windowing—The division of a CRT display into sections (by means of software), allowing the display of data from several different sources.

window jamming—Reradiation of electromagnetic energy by reflecting it from a window to jam enemy electronic devices.

Windows—A graphical user interface developed by Microsoft for DOS (Disk Operating System), sometimes called MS-DOS, the standard operating system for IBM PCs. The operating system is the software that controls the computer hardware, manages program operations, and handles the flow of data to and from storage devices and peripherals.

windshield—In radar, a streamlined cover placed in front of airborne paraboloid antennas to minimize wind resistance. The cover material is such as to present no appreciable attenuation to the radiation of the radar energy.

wing spot generator—An electronic circuit that grows wings on the video target signal of a type-G indicator. These wings are inversely proportional in size to the range.

wipe—A transition from one scene to another wherein the new scene is revealed by a moving line or pattern.

wiped joint—A joint heated by wiping molten solder on the area to be joined.

wiper—1. The moving contact that makes contact with a terminal in a stepping relay or switch. 2. In a potentiometer, the contact that moves along the element, dividing the resistance according to its mechanical position.

wiper arm—In a pressure potentiometer, the movable electrical contact that is driven by the sensing element and moves along the coil.

wiping action—The action that occurs when contacts are mated with a sliding action. Wiping has the effect of removing small amounts of contamination from the contact surfaces, thus establishing better conductivity.

wiping contact—Also called self-cleaning contact, sliding contact, and self-wiping contact. 1. A switch or relay contact designed to move laterally with a wiping motion when engaging with or disengaging from a mating contact. 2. Contact that has sliding motion during opening and closing.

wire—1. A solid or stranded group of solid cylindrical conductors having a low resistance to current flow, together with any associated insulation. 2. A single metallic conductor of round, square, or rectangular section, either bare or insulated. 3. A slender rod or filament of drawn metal. The term is a generally used one, which may refer to any single conductor. If larger than 9 AWG or having multiple conductors, it is usually referred to as cable.

wire barrel—*See* barrel.

wire bond—1. The method by which very fine wires are attached to semiconductor components for interconnection of those components with each other or with package leads. *See also* beam leads. 2. The fastened union point between a conductor or terminal and the semiconductor die. 3. Includes all the constituent components of a wire electrical connection, such as between the terminal and the semiconductor die. These components are the wire, metal bonding surfaces, the adjacent underlying insulating layer (if present), and substrate.

wire bonding—1. A lead-covered tie used for connecting two cable sheaths until a splice is closed and covered permanently. 2. The method used for connecting chips to substrate conductor patterns, package pins, or to other chips. Commonly used techniques include thermocompression ball and wedge types, and ultrasonic bond. The wires are typically made of either aluminum or gold

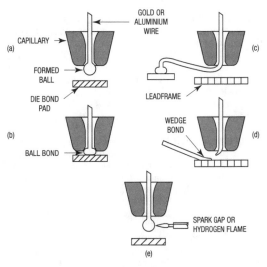

Wire bonding.

and are usually 1.25 to 1.50 mils in diameter for ICs, but can be as large as 15 mils in diameter for power devices. 3. The method used to attach very fine wire to semiconductor components to interconnect these components with each other or with package leads.

wire communication — Transmission of signs, signals, pictures, and sounds of all kinds over wire, cable, or other similar connections.

wired AND — Also called dot AND or implied AND. The external connection of separate circuits or functions in such a way that the combination of their outputs results in an AND function. The logic level at the point at which the separate circuits are wired together is 1 if all circuits feed is into this point.

wired OR — Also called dot OR or implied OR. The external connection of separate circuits or functions in such a way that the combination of their outputs results in an OR function. The logic level at the point at which the separate circuits are wired together is 1 if any of the circuits feeds a 1 into this point.

wired OR and AND — The connection of two or more (open-collector or tristate) logic outputs to a common bus so that any 1 can pull the bus down to 0 level. Depending on the logic convention used (positive or negative), it will be the logical OR or the logical AND function.

wired-program computer — A computer in which nearly all instructions are determined by the placement of interconnecting wires held in a removable plugboard. This arrangement allows for changes of operations by simply changing plugboards. If the wires are held in permanently soldered connections, the computer is called a fixed-program type.

wired radio — Communication whereby the radio waves travel over conductors.

wire drawing — The pulling of wire through dies made of tungsten carbide or diamond, with a resultant reduction in the diameter of the wire.

wire dress — Arranging of wires or corductors in preparation for a mechanical hookup.

wire gage — Also called American wire gage (AWG), and formerly Brown and Sharpe (B&S) gage. A system of numerical designations of wire sizes, starting with 0000 as the largest size and going to 000, 00, 0, 1, 2, and beyond for the smaller sizes.

wire grating — An arrangement of wires set into a waveguide to pass one or more waves while obstructing all others.

wire-guided — In missile terminology, guided by electrical impulses sent over a closed wire circuit between the guidance point and the missile.

wire-lead termination — The method by which wire leads are fastened at a circuit termination; for example, soldering, wire wrapping, or crimping.

wireless — 1. A British term for radio. 2. Used in the United States, in the sense of (1) above, when the word *radio* might be misinterpreted (e.g., wireless record player).

wireless device — Any apparatus (e.g., a wireless record player) that generates a radio-frequency electromagnetic field for operating associated apparatus not physically connected and at a distance in feet not greater than 157,000 divided by the frequency in kilohertz. Legally, the total electromagnetic field produced at the maximum operating distance cannot exceed 15 microvolts per meter.

wireless microphone — A microphone connected to a small (frequently hidden) radio transmitter that sends a signal to a suitable receiver located a short distance away.

wireless networks — A network that does not require twisted-pair hubs and the wire. It is not totally wireless, however, because it requires a connection from transmitter/receiver units to PCs or workstations.

wireless record player — *See* wireless device.

wire-link telemetry — Also called hard-wire telemetry. Telemetry in which a hardwire link is used as the transmission path, no radio link being used.

wire mile — The unit in which the length of two-conductor wire between two points is expressed. The number of wire miles is obtained by multiplying the length of the route by the number of circuits. This figure does not include the slack for ties, overheads, etc.; for computer purposes, the slack accounts for an additional 50 percent per wire mile.

wire net — Subset of electrical connections in a logical net having the same characteristics and common identifiers. No physical order of connection is implied.

wire nut — A conically wound brass spring, usually with a plastic insulating cover, which can be screwed onto a pair of power conductors to connect them without twisting, soldering, or wrapping with tape.

wire ORed — A logic technique in which the inverted OR function is produced by connecting separate logic function outputs to a common point.

wirephoto — Transmission of a photograph or other single image over a telegraph system. The image is scanned into elemental areas in orderly sequence, and each area is converted into proportional electric signals that are transmitted in sequence and reassembled in correct order at the receiver.

wirephoto facsimile — A facsimile photograph.

wireprinter — A high-speed printer that prints characterlike configurations of dots through the proper selection of wire ends from a matrix of wire ends, rather than conventional characters through the selection of type faces.

wire recorder — A magnetic recorder in which the recording medium is a round stainless-steel wire about 0.004 inch (100 μm) in diameter.

wire recording — A recording method in which the medium is a thin stainless-steel wire (instead of a tape or disc).

wire solder — Solder in the form of a long, slender rod, usually wound on a spool.

wiresonde — An atmospheric sounding instrument that is supported by a captive balloon and used to obtain temperature and humidity data from ground level to a height of a few thousand feet. Height is determined by means of a sensitive altimeter, or from the amount of cable released and the angle that the cable makes with the ground. The information is telemetered to the ground station through a wire cable.

wire splice — An electrically sound and mechanically strong junction of two or more conductors.

wire stripper — A tool used to remove the insulation from a wire.

wiretapping — The act of connecting to a telephone circuit for the purpose of surreptitiously obtaining information or evidence from the intelligence it carries.

wireways — Sheet-metal troughs with hinged covers for housing and protecting electrical conductors and cable, and in which conductors are laid in place after the wireway has been installed as a complete system.

wirewound resistor — 1. A resistor in which the resistance element is a length of high-resistance wire or ribbon wound onto an insulating form. 2. Resistor made from lengths of resistance wire wound on a central ceramic core and used where high-temperature stability, power-handling capability, or low-resistance values are factors. Most frequently used in the 10-ohm to 1-megohm region. Resistance tolerances are generally ±2 percent or better and TCRs are generally ±10 ppm/°C or better. Power wirewounds capable of handling more than 50 watts are of coarse-grade wire that is uninsulated at the time of winding to afford better heat dissipation after the completed unit is given an overall protective jacket.

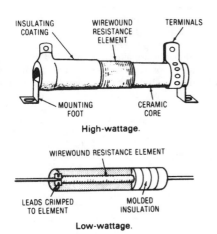

Wirewound resistors.

wirewound trimming potentiometer — A trimming potentiometer characterized by a resistance element made up of turns of wire on which the wiper contacts only a small portion of each turn.

wire wrap — 1. Method of connecting a solid wire to a square, rectangular, or V-shaped terminal by tightly wrapping or winding it with a special automatic or hand-operated tool. Trademark of Gardner-Denver. *See also* wrap post. 2. A way of making electrical connections without soldering. A special tool wraps the wire tightly around a square post. The sharp edges bite into the wire, producing a reliable connection.

wire-wrapped connection — Also called solder-less wrapped connection, wrapped connection, or wrap-post connection. A solderless connection made by wrapping bare wire around a square or rectangular terminal with a power or hand tool.

wire wrapping — 1. A technique for terminating conductors. 2. A type of electrical connection made by tightly coiling wire around a square terminal having sharp corners.

wire-wrapping tools — Portable electric tools and automatic stationary machines used to make solderless wrapped connections of wires to terminals. The operation is fast and reliable. Insulation is removed from the wire and the bare wire is inserted into a wire slot of the wrapping bit, butting the insulation of the wire against the bit face. With the wire positioned in the notch of the wrapping sleeve, the tool is positioned over the terminal to be wrapped and the connection made.

wiring connector — A device for joining one or more wires together.

wiring diagram — A drawing that shows electrical equipment and/or components, together with all interconnecting wiring.

wiring harness — A group of coded insulated wires, cut to length, bent to shape, and laced together. Installed as a unit to form the back-of-panel wiring for a unit of equipment.

wobble — *See* flicker, 3.

wobble bond — A thermocompression, multicontact bond produced by rocking (wobbling) a bonding head on the beam leads. *See also* beam lead.

wobble stick — A rod extending from a pendant station to operate the "stop" contacts; it will function when pushed in any direction.

wobbulator — More commonly called a sweep generator. 1. A signal generator whose frequency is varied automatically and periodically over a definite range. It is used, together with a cathode-ray tube, for testing frequency response. One form consists of a motor-driven variable capacitor, which is used to vary the output frequency of a signal generator periodically between two limits. 2. A test oscillator that continually and periodically varies its output frequency between two limits, so as to give an indication of response over a band of frequencies.

woofer — 1. A speaker designed primarily for reproduction of the lower audio frequencies. Woofers may operate up to several thousand hertz, but their output becomes quite directional at these frequencies. Woofers are characterized by large, heavy diaphragms and large voice coils that overhang the magnetic gap. 2. A low-frequency or bass speaker for reproducing musical notes in the approximate range of 25 to 2500 Hz. Employed with a tweeter and crossover network to reproduce a range of frequencies for audio reproduction. 3. A low-frequency driver. 4. A speaker or driver in a two-way (or more complex) speaker system that reproduces only the bass or lower part of the audible spectrum.

word — 1. An ordered set of characters or data that occupies one storage location and is treated by the computer circuits as a unit. Ordinarily a word is treated by the control unit as a quantity. Word lengths may be fixed or variable, depending on the particular computer. 2. In telegraphy, six operations or characters (five characters and one space). Also called group. 3. The number of bits needed to represent a computer instruction, or the number of bits needed to represent the largest data element normally processed by a computer. 4. A number of consecutive characters. 5. A set of binary bits handled by the computer as the primary unit of information. The length of a computer word is determined by the hardware design. Typically, each system memory location contains one word. 6. A term used in digital systems to indicate the number of bits treated as a single entity by the system. The length of a word may be fixed in some computers or variable in others.

word address format — In a computer, the addressing of each word in a block of information by a character or characters that identify the meaning of the word.

word code — A word that, by prearrangement, conveys a meaning other than its conventional one.

word format — The way in which characters are arranged in a word, with each position or group of positions in the word containing certain specified information.

word generator — An instrument that generates a data stream of 1s and 0s with bit position, bit frequency, etc., completely under the control of the operator. It may be considered to be a read-only memory, a substitute for a paper-tape reader, a computer simulator, a tester for a data-transmission line, a programming device, or a programmable pulse generator.

word length — 1. The number of bits in a sequence that is handled as a unit and that normally can be stored in one location in a memory. A greater word length implies high precision and more intricate instructions. 2. The size of a field. 3. The number of bits in a computer word. The longer the word length, the greater the precision (number of significant digits). 4. The number of bits, bytes, or characters in a word.

word pattern — The smallest meaningful language unit recognized by a machine. It is usually composed of a group of syllables and/or words.

word processing — Abbreviated WP. 1. A system for converting ideas and concepts into hard copy. 2. A combination of electronics and communications to facilitate dictation and the production of typewritten pages with automatic text-editing capabilities. 3. The transformation of ideas and information into a readable form. 4. Using a microcomputer to write, edit, and print text. Most word processing software requires at least 64 K bytes of RAM. Although some machines can function with 48 K, memory that small is usually limiting. 5. The creation and manipulation of text documents.

word processing system — The hardware, software, and peripheral devices employed to perform word processing tasks.

word processor — 1. A text editor system for writing and formatting letters, reports, and books. 2. An automated, computerized system incorporating variously an electronic typewriter, CRT terminals, memory, printer, and the like. It is used to prepare, edit, store, transmit, or duplicate letters, reports, records, etc., as for a business. Some programs have spelling and syllabification verifiers.

word rate — The frequency derived from the elapsed period between the beginning of the transmission of one word and the beginning of transmission of the next word.

word size — In computer terminology, the number of decimal or binary bits comprising a word.

word time — 1. The time required to move one word from one storage device to a second storage device. 2. In a storage device providing serial access to storage positions, the interval of time between the appearance of corresponding parts of successive words.

work — 1. The magnitude of a force times the distance through which that force is applied. 2. *See* load, 6.

work area — A portion of computer storage in which an item of data may be processed or temporarily stored. Often, the term work area is used to refer to a place in storage used to retain intermediate results of a calculation, particularly those results that will not appear directly as output from the program.

work coil — *See* load coil.

work function — 1. The minimum energy (commonly expressed in electronvolts) required to remove an electron from the Fermi level of a material and send it into field-free space. 2. A general term applied to the energy required to transfer electrons or other particles from the interior of one medium, across a boundary, and into an adjacent medium.

working memory — *See* working storage.

working plates — *See* photomask, 4.

working Q — *See* loaded *Q*, 1.

working storage — Also called the working memory. In a computer storage (internal), a portion reserved by the program for the data upon which the operations are being performed.

working voltage — 1. *See* voltage rating. 2. The recommended maximum voltage of operation for an insulated conductor. It is usually set at approximately one-third of the breakdown voltage. 3. The voltage rating of a fixed capacitor. It is the recommended maximum voltage at which the unit should be operated. 4. The rated voltage that can be applied to a device or conductor continuously without danger of breakdown. It is usually well below the test or dielectric withstanding voltage so as to provide a safety factor for transient voltages.

workload — The mix of different types of program typically run at a given worksite; major characteristics include I/O requirements, amount and kinds of computation, and degree of vectorization.

workstation — 1. A personal computer or terminal device, usually but not necessarily operating within a local area network, which is used by someone to perform the greater part of his or her everyday work. 2. Originally, a terminal and keyboard remotely connected to a mainframe, but now refers to the combination of a powerful computer, graphics terminal, and keyboard at one location. Replaces the term minicomputer, which is larger than a microcomputer (or PC) but smaller than a mainframe.

World Wide Web — Abbreviated WWW. Often referred to as simply "the Web." An easy-to-use but powerful global information system based on a combination of information retrieval and hypertext, which allows users to graphically browse through documents on sites throughout the Internet and follow pointers (called links or hyperlinks) to other documents that can be anywhere. These documents can contain text, graphics, sounds, and even movies. The original idea was developed at CERN (the European Laboratory for Particle Physics) between 1989 and 1992.

worm — A self-replicating program that consumes processor time but cannot destroy data, software, or other system resources.

WORM — Abbreviation for write once, read many. A common type of cartridge-based optical storage drive allowing the storage of hundreds of megabytes of data. The disk can be written to only once; after that, the data is permanently stored there.

worst-case circuit analysis — A type of circuit analysis used to determine the worst possible effect on the output parameters due to changes in the values of circuit elements. The circuit elements are set to the values within their anticipated ranges that produce the maximum detrimental changes in the output.

worst-case design — An extremely conservative design approach in which the circuit is designed to function normally even though all component values have simultaneously assumed the worst possible condition that can be caused by initial tolerance, aging, etc. Worst-case techniques are also applied to obtain conservative derating of transient and speed specifications.

worst-case noise pattern — Sometimes called checkerboard or double-checkerboard pattern. Maximum noise appearing when half of the selected cores are in a 1 state and the other half are in a 0 state.

wound capacitor — A capacitor made by winding foils and dielectric material on a mandrel.

wound-rotor induction motor — An induction motor in which the secondary circuit consists of a polyphase winding or coils with either short-circuited terminals or ones closed through suitable circuits.

wound-rotor motor — *See* slip-ring motor.

woven-screen storage — A digital storage plane woven from wires coated with thin films of magnetic material. When currents are passed through a selected pair of wires that lie at right angles in the screen, storage and readout occur at the intersection of those wires.

wow — 1. Distortion caused in sound reproduction by variations in speed of the turntable or tape. *See also* flutter, 1. 2. The audible effect of a low-frequency flutter, occurring at a rate of 0.5 to 10 Hz. Most audible and objectionable on sustained tones. 3. A slow, periodic change in the pitch of frequency of a sound during recording or playing, usually produced by mechanical deviations in the tape transports.

wow and flutter — Audible periodic variations in the pitch of a sound from an audio system. The low-frequency variations (up to about 10 Hz) are wow, while the higher-frequency variations are flutter.

wow meter — An instrument that indicates the instantaneous speed variation of a turntable or similar equipment.

WP — Abbreviation for word processing.

wpm — Abbreviation for words per minute, a measure of speed in telegraph systems and printers.

wrap — 1. One winding of ferromagnetic tape. 2. The length of the path of a magnetic recording tape along which the tape and head are in intimate physical contact. It is sometimes measured as the angle of arrival and departure of the tape with respect to the head. 3. A measure of the length of recording tape that is in intimate contact with the surface of the record or play head. The better the tape-to-head contact over the head gap, the better the high-frequency response; the better the contact over the rest of the area of the head, the better the response at middle and low frequencies.

wrap and fill — A method of capacitor encasement in which the capacitor element is wrapped with plastic tape and sealed on the ends with an epoxy resin.

wrap-around — The amount of curvature exhibited by the magnetic tape or film in passing over the pole pieces of the magnetic heads.

wrapped termination — A gastight, separable connection formed by helically wrapping insulated copper wire around sharp-edged rectangular posts (typically 0.025 in^2, or 0.16 cm^2), either manually, semiautomatically, or automatically by means of numerically controlled machines.

wrapper — An insulating barrier applied to a coil by wrapping a sheet of insulating material around the coil periphery so as to form an integral part of the coil.

wrapping — A method of applying insulation to wire by wrapping insulating tapes around the conductor.

wrap post — A square, rectangular, or V-shaped terminal to which a bare solid wire is tightly wrapped or wound around it to establish electrical continuity by means of a special automatic or hand-operated tool. *See also* wire wrap.

wratten filter — An optical filter used for filtering a given band of light. It is used in film-recorder optical systems when recording directly on color film (direct positive). It is also extensively used in photography.

wrinkle finish — An exterior paint that dries to a wrinkled surface when applied to cabinets or panels.

write — 1. In a computer, to copy, usually from internal to external storage. 2. In a computer, to transfer elements of information to an output medium. 3. In a computer, to record information in a register, location, or other storage device or medium. 4. The process of loading information into memory. 5. In a charge-storage tube, to establish a charge pattern corresponding to the input.

write after read — Writing (restoring) previously read data into a core memory following completion of the read cycle.

write head — A device that stores digital information by placing coded pulses on a magnetic drum or tape.

write pulse — 1. In a computer, a pulse that is used to enter information into one or more magnetic cells for storage purposes. 2. A pulse that causes information to be stored in a memory cell.

write time — *See* access time, 2.

writing rate — The maximum speed at which the spot on a cathode-ray tube can move and still produce a satisfactory image.

writing speed — 1. The rate of writing on successive storage elements in a charge-storage tube. 2. In a cathode-ray tube, the maximum linear speed at which the electron beam can produce a visible trace.

Wullenweber antenna — An antenna array that consists of two concentric circles of masts so connected as to permit electronic steering.

wvdc — Abbreviation for working voltage, direct current. This is the maximum safe dc operating voltage that can be applied across the terminals of a capacitor at its maximum operating temperature.

WWV — Call letters of the radio station of the National Bureau of Standards at Ft. Collins, Colorado. WWV provides radio-broadcast technical services, including time signals, standard radio and audio frequencies, and radio-propagation disturbance warnings at carrier frequencies of 2.5, 5, 10, 15, 20, and 25 megahertz.

WWVH — Call letters of the National Bureau of Standards radio station at Maui, Hawaii. It broadcasts on 2.5, 5, 10, and 15 MHz for many locations not served by WWV.

WWW — *See* World Wide Web.

wye — A network consisting of three branches meeting at a common node; an alternate form of tee network.

wye connection — Also called a star connection. A Y-shaped (Y = "wye") winding connection.

wye junction — A Y-shaped junction of waveguides.

WYSIWYG — Acronym for what you see is what you get. A description of computer software whose screen display is nearly identical to its printed output.

X

X— Symbol for reactance.

X~C~ — Symbol for capacitive reactance.

X~L~ — Symbol for inductive reactance.

X and Z demodulation — A system of color TV demodulation in which the two reinserted 3.58-MHz subcarrier signals differ by approximately 60° rather than the usual 90°. The R — Y, B — Y, and G — Y voltages are derived from the demodulated signals, and these voltages control the three guns of the picture tube. An important advantage of this system is that receiver circuitry is simpler than that required with I and Q demodulation.

X-axis — 1. The reference axis in a quartz crystal. 2. The horizontal axis in a system of rectangular coordinates. 3. The horizontal or left-to-right direction in a two-dimensional system of coordinates. X-X signifies one direction followed in a step-and-repeat method.

X-band — A radio-frequency band of 5200 to 11,000 MHz, with wavelengths of 5.77 to 2.75 cm.

X-bar — A rectangular crystal bar, usually cut from a Z-section, elongated parallel to X and with its edges parallel to X, Y, and Z.

X-capacitor — A radio interference suppression capacitor intended for applications in which failure of the capacitor would not lead to danger of electric shock.

X-cut crystal — A crystal cut so that its major surfaces are perpendicular to an electrical (X) axis of the original quartz crystal.

xenon — A rare gas used in some thyratron and other gas tubes.

xenon flash tube — A high-intensity source of incoherent white light; it operates by discharging a capacitor through a tube of xenon gas. Such a device is used frequently as a source of pumping radiation for various optically excited lasers.

xerographic printer — A device for printing an optical image on paper; light and dark areas are represented by electrostatically charged and uncharged areas on the paper. Powdered ink, dusted on the paper, adheres to the charged areas and is subsequently melted into the paper by the application of heat.

xerographic recording — A recording produced by xerography.

xerography — 1. That branch of electrostatic electrophotography in which images are formed onto a photoconductive insulating medium by infrared, visible, or ultraviolet radiation. The medium is then dusted with a powder, which adheres only to the electrostatically charged image. Heat is then applied in order to fuse the powder into a permanent image. 2. A printing process of electrostatic electrophotography that uses a photoconductive insulating medium, in conjunction with infrared, visible, or ultraviolet radiation, to produce latent electrostatic-charge patterns for achieving an observable record.

xeroprinting — That branch of electrostatic electrophotography in which a pattern of insulating material on a conductive medium is employed to form electrostatic-charge patterns for use in duplicating.

xeroradiography — A printing process of electrostatic electrophotography that uses a photoconductive insulating medium, in conjunction with X-rays or gamma rays, to produce latent electrostatic-charge patterns for achieving an observable pattern.

xeroradiography equipment — Equipment employing principles of electrostatics and photoconductivity to record X-ray images on a sensitized plate in a short time after exposure.

xfmr — Abbreviation for transformer.

xistor — Abbreviation for transistor.

XLR connector — A shielded three-conductor microphone plug or socket with finger-release lock to prevent accidental removal. The standard connector for professional microphone users.

xmitter — Abbreviation for transmitter. Also abbreviated trans or xmtr.

xmsn — Abbreviation for transmission.

xmtr — Abbreviation for transmitter. Also abbreviated trans or xmitter.

X-off — Transmitter off.

X-on — Transmitter on.

X-particle — A particle having the same negative charge as an electron, but a mass between that of an electron and a proton. It is produced by cosmic radiation impinging on gas molecules or actually forming a part of cosmic rays.

X-ray apparatus — An X-ray tube and its accessories, including the X-ray machine.

X-ray crystallography — 1. Use of X-rays in studying the arrangement of the atoms in a crystal. 2. Study of the structure of crystalline materials, which makes use of the interaction of X-rays and the crystal's electron density (diffractions).

X-ray detecting device — A device that detects surface and volume discontinuities in solids by means of X-rays.

X-ray diffraction camera — A camera that directs a beam of X-rays into a sample of unknown material and allows the resultant diffracted rays to act on a strip of film.

X-ray diffraction pattern — The pattern produced on film exposed in an X-ray diffraction camera. It is made up of portions of circles having various spacings, depending on the material being examined.

X-ray goniometer — An instrument that determines the position of the electrical axes of a quartz crystal by reflecting X-rays from the atomic planes of the crystal.

X-rays — Also called roentgen rays. Penetrating radiation similar to light, but having much shorter wavelengths

$(10^{-7}$ to 10^{-10} cm). They are usually generated by bombarding a metal target with a stream of high-speed electrons.

X-ray spectrograph — An instrument that is used to chart X-ray diffraction patterns, such as an X-ray spectrometer having photographic or other recording implements.

X-ray spectrometer — 1. An instrument for producing an X-ray spectrum and measuring the wavelengths of its components. 2. An instrument designed to produce an X-ray spectrum of a material as an aid in identifying it. This technique is particularly useful when the material cannot be physically broken down.

X-ray spectrum — An arrangement of a beam of X-rays in order of wavelength.

X-ray thickness gage — A contactless thickness gage used to measure and indicate the thickness of moving cold-rolled sheet steel during the rolling process. An X-ray beam directed through the sheet is absorbed in proportion to the thickness of the material and its atomic number, and measurement of the amount of absorption gives a continuous indication of sheet thickness.

X-ray tube — A vacuum tube in which X-rays are produced by bombarding a target with high-velocity electrons accelerated by an electrostatic field.

X-ray tube target — Also known as an anticathode. An electrode or electrode section that is focused on by an electron beam and that emits X-rays.

xso — Abbreviation for crystal-stabilized oscillator.

xtal — Abbreviation for crystal.

X-wave — One of the two components into which the magnetic field of the earth divides a radio wave in the ionosphere. The other component is the ordinary, or O-, wave.

XY-cut crystal — A crystal cut so that its characteristics fall between those of an X- and a Y-cut crystal.

XY plotter — 1. A device used in conjunction with a computer to plot coordinate points in the form of a graph. 2. A computer output device that responds to digital signals of prerecorded and/or processed data by printing arrangements of line segments. This data, which can include alphanumerics, charts, tables, or drawings, is fed from computer memory at speeds slow enough for a plotter to handle. An XY plotter cannot be used to directly record analog signals without having digitizers.

XY recorder — 1. A recorder that traces, on a chart, the relationship between two variables, neither of which is time. Sometimes the chart moves and one of the variables is controlled so that the relationship does increase in proportion to time. 2. A recorder in which two signals are recorded simultaneously by one pen, which is driven in one direction (X-axis) by one signal, and in the other direction (Y-axis) by the second signal. 3. A data recorder that is used to record the variation of one parameter with respect to another. For example, the change of pressure with temperature. For these recorders, a wide range of transducers is available to convert physical parameters into electrical signals usable to the recorder. Pressure transducers, thermocouples, strain gages, and accelerometers are a few examples. 4. A type of recorder that responds to incoming analog signals as they occur. The signals print on a predetermined chart size that can cover test periods from a few seconds to as much as a year. An XY recorder records via continuous lines. In addition, the instrument's speed of response is important to the fidelity of the record.

XY switch — A remote-controlled bank-and-wiper switch arranged so that the wipers move back and forth horizontally.

Y

y—Abbreviation for the prefix yocto-(10^{-24}).
Y—Abbreviation for the prefix yotta-(10^{24}).
Y—Symbol for admittance.
Yagi antenna—Also called Yagi-Uda antenna. 1. An end-fire antenna that consists of a driven dipole (usually a folded dipole), a parasitic dipole reflector, and one or more parasitic dipole directors. All the elements usually lie in the same plane; however, the parasitic elements need not be coplanar, but can be distributed on both sides of the plane of symmetry. 2. Antenna system exhibiting directional properties, in which several elements may be mounted perpendicular to a common boom. One element is connected to the transmission line and the rest are parasitic radiators.

Yagi antenna.

YAG laser—A solid-state laser using yttrium aluminum garnet as the matrix material. YAG works most efficiently in small or moderately sized lasers.
Yahoo—A popular World Wide Web search engine. It can be accessed at http://www.yahoo.com.
Y antenna—*See* delta matched antenna.
Y-axis—1. A line perpendicular to two parallel faces of a quartz crystal. 2. The vertical direction, perpendicular to the X-axis, in a two-dimensional system of coordinates. Y-Y signifies one direction followed in a step-and-repeat method. 3. The vertical axis on a graph or cathode-ray-tube screen. 4. One of the three mutually perpendicular axes of a crystal.
Y-bar—A crystal bar cut in Z-sections, with its long direction parallel to Y.
Y-capacitor—A radio interference suppression capacitor intended for applications in which failure of the capacitor could lead to danger of electric shock.
Y circulator—A circulator consisting of three identical rectangular waveguides joined in a symmetrical Y-shaped configuration with a ferrite post or wedge at the center. Power that enters any waveguide emerges from only one adjacent waveguide.
Y-connected circuit—A star-connected, three-phase circuit.
Y-connection—*See* Y-network.
Y connector—A connector that joints two branch conductors to the main conductor at an angle. The three conductors are in the same plane.
Y-cut crystal—A crystal cut in such a way that its major flat surfaces are perpendicular to the Y-axis of the original quartz crystal.
yield—1. In a production process, the quantity or percentage of finished parts that conform to specifications, relative to either the quantity started into production or to time. 2. The ratio of usable chips to the total number available on a single wafer of semiconductor material. The greater the yield, the more efficient the manufacturing process and the greater its profitability. 3. The number of usable IC dice coming off a production line divided by the total number of dice going in. Yield tends to be reduced at every step in the manufacturing process by wafer breakage, contamination, mask defects, and processing variations. 4. The percentage of operational chips in a batch or on a wafer.
yield map—A microcircuit or semiconductor wafer on which dots indicate those devices that failed the test criteria.
yield strength—1. *See* yield value. 2. The minimum stress at which a material will start to physically deform without further increase in load.
yield-strength-controlled bonding—The method of diffusion bonding based on the use of pressures that exceed the yield stress of the metal at the bonding temperature. The process is characterized by the use of high unit loads for brief time cycles, ranging from fractions of a second to a few minutes. Typical examples are spot bonding and roll bonding.
yield value—Also called yield strength. The lowest stress at which a material undergoes plastic deformation. Below this stress, the material is elastic; above it, viscous.
YIG—Abbreviation for yttrium iron garnet. 1. Crystalline material that resonates at microwave frequencies when immersed in a magnetic field. Small spheres of YIG material are mounted in resonant structures for tuning applications. 2. A synthetic crystalline ferrite containing yttrium and iron $(Y_3Fe_6O_{16})$. If a single crystal sphere of YIG is immersed in a magnetic field, and rf energy is coupled into it via a magnetic loop, the crystal will resonate at a frequency linearly proportional to the magnetic field strength. In practical YIG-tuned oscillators and filters, the magnetic field is derived from an electromagnet and the resonant frequency of the YIG sphere is proportional to the current flowing through the magnetic coil.
YIG devices—Small solid-state filters, discriminators, and multiplexers that contain yttrium-iron-garnet

crystals used in combination with a variable magnetic field to accomplish wideband tuning in microwave circuits.

YIG filter — A filter that consists of a yttrium-iron-garnet crystal positioned in the field of a permanent magnet and a solenoid. Tuning is accomplished by controlling the direct current through the solenoid. The bias magnet tunes the filter to the center of the band and thus minimizes the solenoid power required to tune over wide bandwidths.

YIG-tuned filter — A microwave filter using a YIG sphere as the resonant element.

YIG-tuned oscillator — A microwave tunable oscillator using the YIG sphere as the frequency-determining element. YIG-tuned oscillators can be made with Gunn-diode technology or an added internal buffer amplifier to minimize frequency pulling and produce additional output power capability. YIG-tuned oscillators are fundamental oscillators; they do not contain frequency multiplication circuitry.

YIG-tuned parametric amplifier — A parametric amplifier in which tuning is accomplished by controlling the direct current through the solenoid of a YIG filter.

YIG-tuned tunnel-diode oscillator — A microwave oscillator in which precisely controlled wideband tuning is accomplished through control of the current through a tuning solenoid that acts on a YIG filter in the circuit of a tunnel-diode oscillator.

Y-junction — A junction of waveguides in which their longitudinal axes form a Y.

Y match — Also called a delta match. A method of connecting to an unbroken dipole. The transmission line is fanned out and connected to the dipole at the points where the impedance is the same as that of the line.

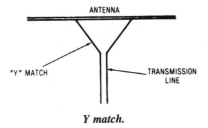

Y match.

Y-network — Also called a Y-connection. A star network of three branches.

yocto — Abbreviated y. Prefix for the numerical quantity of 10^{-24}.

yoke — 1. A set of coils placed over the neck of a magnetically deflected cathode-ray tube to deflect the electron beam horizontally and vertically when suitable currents are passed through them. 2. A piece of ferromagnetic material that does not have windings and that is used to connect two or more magnet cores permanently.

Yoke.

yotta — Abbreviated Y. Prefix for the numerical quantity of 10^{24}.

Young's modulus — A constant that expresses the ratio of unit stress to unit deformation for all values within the proportional limit of the material.

Y-punch — On a Hollerith punched card, a punch in the top row, two rows above the zero row.

Y signal — 1. A luminance transmission primary that is 1.5 to 4.2 MHz wide and equivalent to a monochrome signal. *See also* luminance signal. 2. A signal transmitted in color television containing brightness information. This signal produces a black and white picture on a standard monochrome receiver. In a color picture it supplies fine detail and brightness information. It is made up of 0.30 red, 0.59 green, and 0.11 blue.

yttrium iron garnet — *See* YIG.

Z

z — Abbreviation for the prefix zepto (10^{-21}).

Z — Abbreviation for prefix zetta (10^{21}).

Z — Symbol for impedance.

Z_o — Symbol for characteristic impedance; i.e., the ratio of the voltage to the current at every point along a transmission line on which there are no standing waves.

Zamboni pile — A primary electrochemical system capable of supplying very high electrical potentials in comparatively little space. The anode material in the pile is aluminum, and the cathode is manganese dioxide and carbon black. The electrolyte in the chemical system is aluminum chloride.

Z-angle meter — An electronic instrument for measuring impedance in ohms and phase angle in electrical degrees.

zap — In a computer, to wipe out, delete.

Z-axis modulation — Also called beam modulation or intensity modulation. 1. Varying the intensity of the electron stream of a cathode-ray tube by applying a pulse or square wave to the control grid or cathode. 2. The intensity regulation of a cathode-ray tube by alteration of the grid-cathode voltage.

Z-bar — A rectangular crystal bar usually cut from X sections and elongated parallel to Z.

Zebra time — An alphabetic expression denoting Greenwich mean time.

Zeeman effect — 1. If an electric discharge tube or other light source emitting a bright-line spectrum is placed between the poles of a powerful electromagnet, a very powerful spectroscope will show that the action of the magnetic field has split each spectrum line into three or more closely spaced but separate lines, the amount of splitting or the separation of the lines being directly proportional to the strength of the magnetic field. 2. The splitting of energy levels of an atom, ion, or molecule due to a magnetic field.

Zener breakdown — 1. A breakdown caused in a semiconductor device by the field emission of charge carrier in the depletion layer. 2. One of the mechanisms responsible for voltage breakdown of semiconductor junctions and devices. When breakdown occurs, the electric field intensity in the material has become so great that electrons are effectively ripped from the valency bonding system. Another name for this mechanism is field emission. Provided the current increase that tends to occur is limited externally, Zener breakdown causes no permanent damage.

Zener diode — 1. A two-layer device that, above a certain reverse voltage (the Zener value), has a sudden rise in current. If forward-biased, the diode is an ordinary rectifier; when reverse-biased, the diode exhibits a typical knee, or sharp break, in its current-voltage graph. The voltage across the device remains essentially constant for any further increase of reverse current, up to the allowable

dissipation rating. The Zener diode is a good voltage regulator, overvoltage protector, voltage reference, level shifter, etc. True Zener breakdown occurs at less than 6 volts. 2. A pn-junction, two-terminal, single-junction semiconductor device reverse biased into the breakdown region and providing high impedance under less than breakdown voltage but conduction with no impedance above breakdown voltage level. 3. A general term used to describe any semiconductor diode intended to be operated in the reverse-biased breakdown condition. Low-voltage devices of this type do in fact exploit the Zener breakdown mechanism, but with most devices having a breakdown voltage about 6 volts, breakdown is due to the avalanche mechanism. 4. A semiconductor pn-junction diode that has a controlled reverse-bias breakdown voltage and is used to supply (clamp) a specific voltage for other protected components (for example, in an IC). The Zener effect describes a tunnel breakdown phenomenon that is restricted to less than 5V. However, Zener diodes are traditionally used to describe any reverse-bias pn-junction device used to supply a specific voltage, even those of several hundred volts.

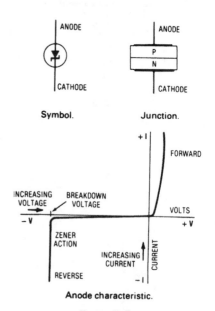

Zener diode.

Zener effect — A reverse-current breakdown due to the presence of a high electric field at the junction of a semiconductor or insulator.

Zener impedance — *See* abreakdown impedance.

Zener knee current — The reverse current that flows through a Zener diode at the breakdown point or Zener knee. Typically, knee currents range from 0.25 to 5 milliamperes.

Zener voltage — *See* breakdown voltage, 2.

Zener zapping — An IC trimming technique approach used for several years by Precision Monolithics that involves depositing Zener diodes in parallel with the collector resistors in the op amp's input stage. Initially they do not affect the circuit, but when each diode experiences a high applied current pulse, it shorts out the resistor it parallels.

zenith — A point directly overhead. Zenith angles are angles measured from this point.

zenith adjustment — A mechanical adjustment of a magnetic tape head to obtain uniform contact with the top and bottom of the magnetic tape. (Zenith refers to the forward-backward tilting of the head.)

zeppelin antenna — A horizontal antenna that is a multiple of a half-wavelength long. One end is fed by one lead of a two-wire transmission line that is also a multiple of a half-wavelength long.

zepto — Abbreviated z. Prefix for the numerical quantity of 10^{-21}.

zero (in a computer) — Positive binary 0 is indicated by the absence of a digit or pulse in a word. In a coded-decimal computer, decimal 0 and binary 0 may not have the same configuration. In most computers, there are distinct representations for plus and minus zero conditions.

zero-access storage — Computer storage for which the waiting time is negligible (e.g., flip-flop, trigger, or indicator storage).

zero-address instruction — A digital-computer instruction specifying an operation in which the locations of the operands are defined by the computer code; no explicit address is required.

zero adjuster — A device for adjusting a meter so that the pointer will rest exactly on zero when the electrical quantity is zero.

zero adjustment — 1. The act of nulling out the output from a system or device. 2. The circuit or other means by which a no-output condition is obtained from an instrument when properly energized.

zero-axis symmetry — A type of symmetry in which a waveform is symmetrical about an axis and does not exhibit a net dc component.

zero beat — The condition whereby two frequencies being mixed are exactly the same and therefore produce no beat note.

zero-beat reception — *See* homodyne reception.

zero bias — 1. The absence of a potential difference between the control grid and the cathode. 2. When the received teletypewriter signal is equal to the transmitted signal (neither long nor shorter), the circuit is said to have zero bias.

zero-bias tube — A vacuum tube designed to be operated as a class B amplifier with no negative bias applied to its control grid.

zero compensation — A method by which, in certain transducers, the effects of temperature on the output at zero measurand may be minimized and maintained within known limits.

zero compression — In computers, any of several techniques used to eliminate the storage of nonsignificant leading 0s.

zero-current turnout — A characteristic of thyristors whereby turnoff is delayed until the next zero-current crossing of the ac line. This "soft turnoff" eliminates arcing associated with inductive loads.

zero-cut crystal — A quartz crystal cut in such a direction that its temperature coefficient with respect to the frequency is essentially zero.

zero drift — *See* zero shift.

zero elimination — In a computer, the editing or deleting of nonsignificant 0s appearing to the left of the integral part of a quantity.

zero error — The delay time occurring within the transmitter and receiver circuits of a radar system. For accurate range data, this delay time must be compensated for when the range unit is calibrated.

zero-field emission — Thermionic emission from a hot conductor that is surrounded by a region of uniform electric potential.

zero-gravity switch — Also called weightlessness switch. A switch that closes when weightlessness or zero gravity is approached.

zero-input terminal — Also called a reset terminal. The terminal that, when triggered, will put a flip-flop in the zero (starting) condition — unless the flip-flop is already in a zero condition, in which event it will not change.

zero insertion force connector — Abbreviated ZIF connector. 1. A connector whose contacts in the plug and receptacle do not initially touch each other while the connector halves are being engaged. Instead, the halves are physically positioned together and then a turn of an actuating cam arrangement mates all the contacts at once. This also locks the connector halves together. Used in applications in which there are an unusually large number of contacts, in which the contacts are extremely delicate, and/or in which it is desirable to mate all the contacts at one time. 2. A connector in which the contact surfaces do not mechanically touch until it is completely mated, thus requiring no insertion force. After mating, the contacts are actuated in some fashion to make intimate electrical contact. 3. An integrated-circuit socket in which the user moves a lever that "opens" a socket to insert or remove the chip, instead of pressing and prying the chip manually. This reduces the chances of damaging the integrated circuit's pins.

zero level — 1. A reference level for comparing sound or signal intensities. In audio-frequency work, it is usually a power of 0.006 watt; in sound, the threshold of hearing. 2. The transmission power at a reference point in a circuit, to which all other power measurements in the circuit are compared. 3. Zero decibels above 1 milliwatt, or 1 milliwatt of power.

zero line stability — An absence of drift in an indicating instrument when it is registering zero.

zero method — *See* null method.

zero-modulation noise — The noise arising when reproducing an erased tape with the erase and record heads energized as they would be in normal operation, but with zero input signal (usually 3 to 4 dB higher than the bulk-erased noise).

zeros of a network function — Those values of p (real or complex) for which the network function is zero.

zero output — The voltage response obtained by a reading or resetting process from a magnetic cell that is in a zero state.

zero-output terminal — The terminal that produces an output (of the correct polarity to trigger a following circuit) when a flip-flop is in the zero condition.

zero phase-sequence relay — A relay that functions in conformance with the zero phase-sequence component of the current, voltage, or power of the circuit.

zero pole — 1. A reference point for an open-wire pole line. 2. The dead-end pole at the origin of the line. 3. The lowest-numbered pole.

zero potential — The potential of the earth, taken as a convenient reference for comparison.

zero-power resistance — In a thermistor, the resistance at a specified temperature when the electrical power dissipation is zero (usual reference temperature is 25°C).

zero-power resistance-temperature characteristic — In a thermistor, the function relating the zero-power resistance and body temperature.

zero-power temperature coefficient of resistance — In a thermistor, the ratio, at a specified temperature, of the rate of change with temperature of zero-power resistance to the zero-power resistance.

zero set — 1. A permanent change in the output of a device at zero measurand due to any cause. 2. A control on a vacuum-tube voltmeter to set the pointer to zero. 3. A control for adjusting a range counter to give the correct range.

zero shift — Also called zero drift. 1. The amount by which the zero or minimum reading of an instrument deviates from the calibrated point as a result of aging or the application of an external condition to the instrument. 2. The maximum deviation from zero reading of a meter having zero input, over a given temperature range. Expressed as a percent of full-scale reading per degree Celsius.

zero-shift error — In an electrical indicating instrument, error manifested as a difference in deflection between an initial position of the pointer, such as zero, and the deflection after the instrument has remained deflected up-scale for an extended length of time. The error is expressed as a percentage of the end-scale deflection.

zero stability — The ability of an instrument to withstand effects that might cause zero shift. Usually expressed as a percentage of full scale.

zero state — 1. In a magnetic cell, the state wherein the magnetic flux through a specified cross-sectional area has a negative value, from an arbitrarily specified direction. *See also* one state. 2. The condition of a binary memory cell when a logical 0 is stored.

zero-subcarrier chromaticity — The chromaticity normally displayed when the subcarrier amplitude is zero in a color television system.

zero suppression — 1. In a recording system, the injection of a controllable voltage to balance out the steady-state component of the input signal. 2. In a computer, the elimination of 0s to the left of the significant integral part of a quantity. 3. Internal circuits in a calculator that prevent the nonsignificant 0s that precede whole numbers from being displayed; thus, an uncluttered number is shown. 4. In a calibrated zero-suppression system, the magnitude of the component that is being bucked out is indicated by the setting of the zero suppression control.

zero time reference — In a radar, the time reference of the schedule of events during one cycle of operation.

zero transmission-level reference point — An arbitrarily chosen point in a circuit, whose level is used as a reference for all relative transmission levels.

zero-voltage switch — A circuit designed to switch on at the instant the ac supply voltage passes through zero, thereby minimizing the radio-frequency interference generated at switch closure.

zero-voltage turn-on — A delay in application of the control signal to the thyristor until the ac power-line voltage next passes through zero.

zetta — Abbreviated Z. Prefix for the numerical quantity of 10^{21}.

Z factor — In thermoelectricity, an accepted figure of merit that denotes the quality of the material.

ZIF connector — *See* zero insertion force connector.

zigzag reflections — From a layer of the ionosphere, high-order multiple reflections that may be of abnormal intensity. They occur in waves that travel by multihop ionospheric reflections and finally turn back toward their starting point by repeated reflections from a slightly curved or sloping portion of an ionized layer.

zinc — A bluish-white metal that, in its pure form, is used in dry cells.

Z-marker — Also called a zone marker. A marker beacon that radiates vertically and is used for defining a zone above a radio range station.

zone — 1. Any of the three top positions (12, 11, or 10) on a punch card. 2. A part of internal computer storage allocated for a particular purpose.

zone bits — 1. The two leftmost binary digits in a digital computer in which six binary digits are used for characters and the four rightmost are used for decimal digits. 2. The bits in a group of bit positions that are used to indicate a specific class of items (e.g., numbers, letters, special signs, and commands).

zone blanking — A method of turning off the cathode-ray tube during part of the sweep of the antenna.

zoned circuit — A circuit that provides continual protection for parts or zones of the protected area while normally used doors and windows or zones may be released for access.

zone leveling — In semiconductor processing, the passage of one or more molten zones along a semiconductor body, for the purpose of uniformly distributing impurities throughout the material.

zone marker — *See* Z-marker.

zone of silence — 1. An area — between the points at which the ground wave becomes too weak to be detected and the sky wave first returns to earth — where normal radio signals cannot be heard. *See also* skip zone. 2. A region within the skip distance in which the signals of a particular radio station cannot be heard.

zone-position indicator — An auxiliary radar set for indicating the general position of an object to another radar set with a narrower field.

zone punch — *See* overpunch.

zone purification — In semiconductor processing, the passage of one or more molten zones along a semiconductor to reduce the impurity concentration of part of the ingot.

zone refining — 1. A purification process in which an rf heating coil is used to melt a zone, or portion, of a silicon billet. The molten section is moved the length of the billet, and the impurities are deposited at the end. 2. A technique used to reduce the impurity content of raw semiconductor materials to an extremely low level, relying on the phenomenon of segregation. A zone of molten material is swept repeatedly through the ingot in the same direction, collecting the impurities.

zones — Smaller subdivisions into which large areas are divided to permit selective access to some zones while maintaining other zones secure and to permit pinpointing the specific location from which an alarm signal is transmitted.

zoning — 1. Also called stepping. Displacement of the various portions of the lens or surface of a microwave reflector so that the resulting phase front in the near field remains unchanged. 2. Purifying a metal by passing it through an induction coil; the impurities are swept ahead of the heating effect. Specifically used in purifying semiconductor crystals.

zoom — 1. To enlarge or reduce on a continuously variable basis the size of a televised image. Zooming may be done electronically or optically. 2. To control, by magnifying or reducing, the size of an image. 3. To enlarge proportionately or decrease the size of the displayed entities by resealing.

zooming — In computer graphics, causing an object to appear smaller or larger by moving the window and specifying various window sizes.

zoom lens — 1. An optical lens with some elements made movable so that the focal length or angle of view can be adjusted continuously without losing the focus. 2. An optical system of continuously variable focal length with the focal plane remaining in a fixed position.

Numbers

1/*f* noise — *See* current noise.

10-code — Abbreviations used by CBers and other radio communications users to minimize use of air time.

2-D system — A computer graphics system whose data represents only a plane.

2¹⁄₂-D system — A computer graphics system whose objects are flat, parallel surfaces that can cover each other.

3-D system — A computer graphics system whose data objects are in three dimensions.

73 — Abbreviation for "best regards" in radiocommunications.

International System of Units (SI)

The SI is constructed from seven base units for independent quantities plus supplementary units for plane solid angles as given in Table 1.

Units for all other quantities are derived from these nine units. Table 2 lists seventeen SI derived units with special names. These units are expressed as products and ratios of the nine base and supplementary units without numerical factors.

Table 1. SI base and supplementary units

Quantity	Name	Symbol
Base Units		
length	meter	m
mass	kilogram	kg
time	second	s
electric current	ampere	A
thermodynamic temperature	kelvin	K
amount of substance	mole	mol
luminous intensity	candela	cd
Supplementary Units		
plane angle	radian	rad
solid angle	steradian	sr

Table 2. SI Derived units with special names

Quantity	SI unit Name	SI unit Symbol	Expression in terms of other units
frequency	hertz	Hz	s^{-1}
force	newton	N	$kg \cdot m/s^2$
pressure, stress	pascal	Pa	N/m^2
energy, work, quantity of heat	joule	J	$N \cdot m$
power, radiant flux	watt	W	J/s
quantity of electricity, electric charge	coulomb	C	$A \cdot s$
electric potential, potential difference electromotive force	volt	V	W/A
capacitance	farad	F	C/V
electric resistance	ohm	Ω	V/A
conductance	siemens	S	A/V
magnetic flux	weber	Wb	$V \cdot s$
magnetic flux density	tesla	T	Wb/m^2
inductance	henry	H	Wb/A
luminous flux	lumen	lm	$cd \cdot sr$
illuminance	lux	lx	lm/m^2
activity of ionizing radiation source	becquerel	Bq	s^{-1}
absorbed dose	gray	Gy	J/kg

All other SI derived units, such as those in Tables 3 and 4, are similarly derived in a coherent manner from the 26 base, supplementary, and special-name SI units.

Table 3. Examples of SI derived units expressed in terms of base units

Quantity	SI unit	Unit symbol
area	square meter	m^2
volume	cubic meter	m^3
speed, velocity	meter per second	m/s
acceleration	meter per second squared	m/s^2
wave number	1 per meter	m^{-1}
density, mass density	kilogram per cubic meter	kg/m^3
current density	ampere per square meter	A/m^2
magnetic field strength	ampere per meter	A/m
concentration (of amount of substance)	mole per cubic meter	mol/m^3
specific volume	cubic meter per kilogram	m^3/kg
luminance	candela per square meter	cd/m^2

Table 4. Examples of SI derived units expressed by means of special names

Quantity	Name	Unit symbol
dynamic viscosity	pascal second	Pa·s
moment of force	newton meter	N·m
surface tension	newton per meter	N/m
heat flux density, irradiance	watt per square meter	W/m^2
heat capacity, entropy	joule per kelvin	J/K
special heat capacity, specific entropy	joule per kilogram kelvin	J/(kg·K)
specific energy	joule per kilogram	J/kg
thermal conductivity	watt per meter kelvin	W/(m·K)
energy density	joule per cubic meter	J/m^3
electric field strength	volt per meter	V/m
electric charge density	coulomb per cubic meter	C/m^3
electric flux density	coulomb per square meter	C/m^2
permittivity	fared per meter	F/m
permeability	henry per meter	H/m
molar energy	joule per meter	J/mol
molar entropy, molar heat capacity	joule per mole kelvin	J/(mol·K)

For use with the SI units there is a set of 16 prefixes (see Table 5) to form multiples and submultiples of these units. It is important to note that the kilogram is the only SI unit with a prefix. Because double prefixes are not to be used, the prefixes of Table 5, in the case of mass, are to be used with gram (symbol g) and not with kilogram (symbol kg).

Table 5. SI prefixes

Factor	Prefix	Symbol	Factor	Prefix	Symbol
10^{24}	yotta	Y	10^{-1}	deci	d
10^{21}	zeta	Z	10^{-2}	centi	c
10^{18}	exa	E	10^{-3}	milli	m
10^{15}	peta	P	10^{-6}	micro	µ
10^{12}	tera	T	10^{-9}	nano	n
10^{9}	giga	G	10^{-12}	pico	p
10^{6}	mega	M	10^{-15}	femto	f
10^{3}	kilo	k	10^{-18}	atto	a
10^{2}	hecto	h	10^{-21}	zepto	z
10^{1}	deka	da	10^{-24}	yocto	y

Schematic Symbols

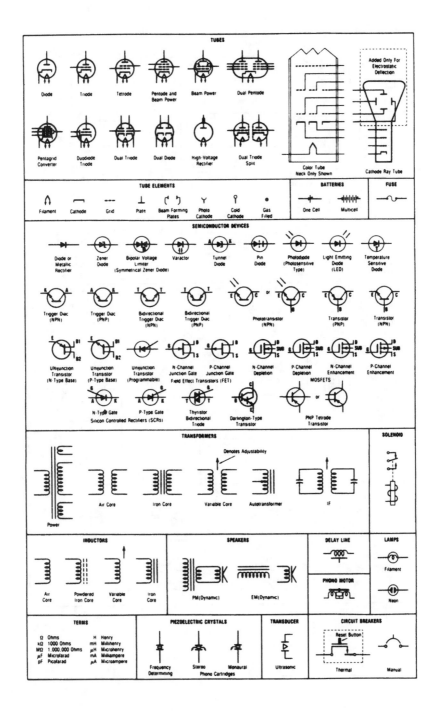

Schematic Symbols

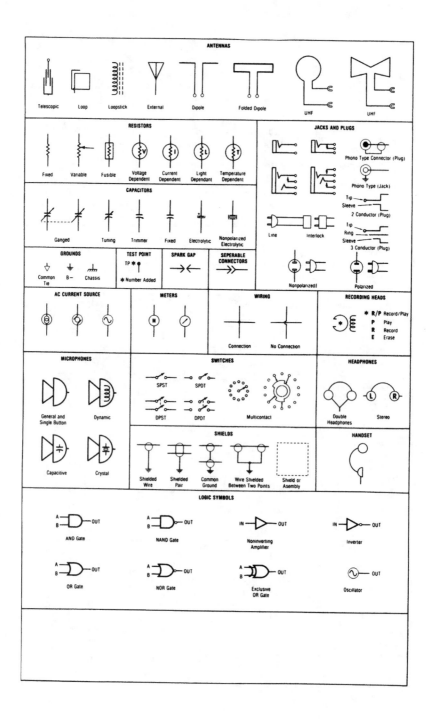

Greek Alphabet

| Letter | | Name | Designates |
Small	Capital		
α	A	Alpha	Angles, coefficients, attenuation constant, absorption factor, area.
β	B	Beta	Angles, coefficients, phase constant.
γ	Γ	Gamma	Specific quantity, angles, electrical conductivity, propagation constant, complex propagation constant (cap).
δ	Δ	Delta	Density, angles, increment or decrement (cap or small), determinant (cap), permittivity (cap).
ε	E	Epsilon	Dielectric constant, permittivity, base of natural (Napierian) logarithms, electric intensity.
ζ	Z	Zeta	Cooridnate, coefficients.
η	H	Eta	Intrinsic impedance, efficiency, surface charge density, hysteresis, coordinates.
θ	Θ	Theta	Angular phase displacement, time constant, reluctance, angles.
ι	I	Iota	Unit vector
κ	K	Kappa	Susceptibility, coupling coefficient.
λ	Λ	Lambda	Wavelength, attenuation constant, permeance (cap).
μ	M	Mu	Prefix *micro-*, permeability, amplification factor.
ν	N	Nu	Reluctivity, frequency.
ξ	Ξ	Xi	Coordinates
o	O	Omicron	—
π	Π	Pi	3.1416 (circumference divided by diameter).
ρ	P	Rho	Resistivity, volume charge density, coordinates.
σ	Σ	Sigma	Surface charge density, complex propagation constant, electrical conductivity, leakage coefficient, sign of summation (cap).
τ	T	Tau	Time constant, volume resistivity, time-phase displacement, transimission factor, density.
υ	Υ	Upsilon	—
ϕ	Φ	Phi	Magnetic flux, angles, scalar potential (cap).
χ	X	Chi	Electric susceptibility, angles.
ψ	Ψ	Psi	Dielectric flux, phase difference, coordinates, angles.
ω	Ω	Omega	Angular velocity ($2\pi f$), resistance in ohms (cap), solid angles (cap).

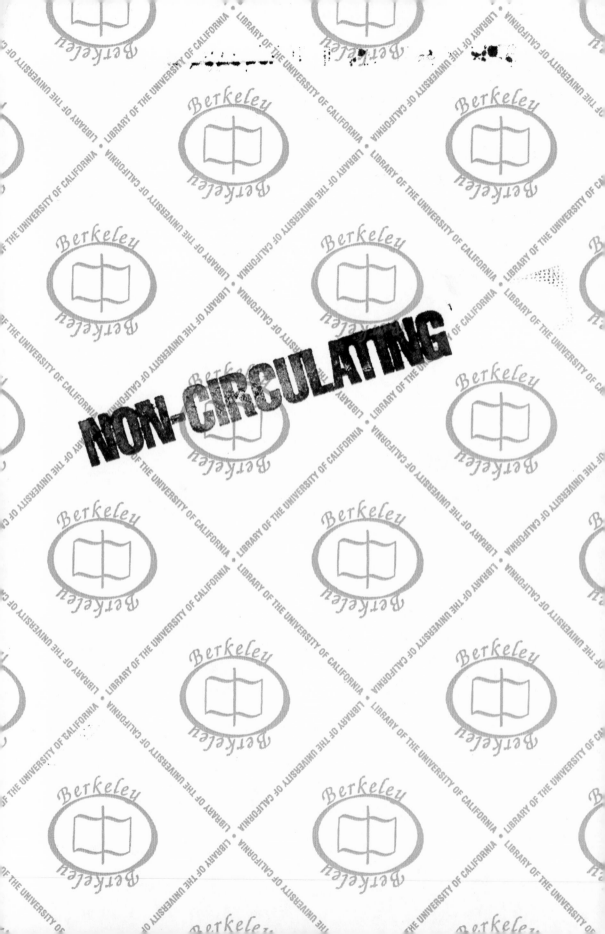